MICRO BIOLOGY

THE HUMAN EXPERIENCE

Second Edition

John W. Foster
University of South Alabama

Zarrintaj Aliabadi
University of South Alabama

Joan L. Slonczewski
Kenyon College

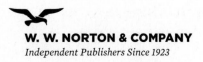

W. W. NORTON & COMPANY
Independent Publishers Since 1923

W. W. Norton & Company has been independent since its founding in 1923, when William Warder Norton and Mary D. Herter Norton first published lectures delivered at the People's Institute, the adult education division of New York City's Cooper Union. The firm soon expanded its program beyond the Institute, publishing books by celebrated academics from America and abroad. By midcentury, the two major pillars of Norton's publishing program—trade books and college texts— were firmly established. In the 1950s, the Norton family transferred control of the company to its employees, and today—with a staff of five hundred and hundreds of trade, college, and professional titles published each year—W. W. Norton & Company stands as the largest and oldest publishing house owned wholly by its employees.

Editor: Betsy Twitchell
Project Editor: Jennifer Barnhardt
Developmental Editor: Lisa Bess Kramer
Copyeditors: Stephanie Hiebert and Heather Cammarn
Assistant Editor: Danny Vargo
Managing Editor, College: Marian Johnson
Associate Director of Production, College: Benjamin Reynolds
Media Editor: Kate Brayton
Associate Media Editor: Jasmine Ribeaux
Media Project Editor: Jesse Newkirk
Media Editorial Assistant: Kara Zaborowsky
Content Development Specialist: Todd Pearson
Managing Editor, College Digital Media: Kim Yi
Ebook Production Manager: Kate Barnes
Marketing Manager, Biology: Ruth Bolster
Design Director: Lissi Sigillo
Design, Chapter Openers, and Life-Cycle Figures: DeMarinis Design, LLC
Director of College Permissions: Megan Schindel
Photo Editor: Catherine Abelman
Photo Researcher: Dena Digilio Betz
Composition: MPS North America LLC / MPS Project Manager: Jackie Strohl
Illustrations: Dragonfly Media Group
Manufacturing: Transcontinental—Beauceville, QC

Permission to use copyrighted material is included on page C-1.

Library of Congress Cataloging-in-Publication Data

Names: Foster, John W. (John Wade), author. | Aliabadi, Zarrintaj, author.
   | Slonczewski, Joan, author.
Title: Microbiology: the human experience / John W. Foster, University of
   South Alabama, Zarrintaj Aliabadi, University of South Alabama, Joan L.
   Slonczewski, Kenyon College.
Description: Second edition. | New York, NY: W.W. Norton & Company, [2021]
   | Includes bibliographical references and index.
Identifiers: LCCN 2020039928 | ISBN 9780393533248 (paperback) | ISBN 9780393680805 (epub)
Subjects: LCSH: Microbiology.
Classification: LCC QR41.2 .F69 2021 | DDC 579–dc23
LC record available at https://lccn.loc.gov/2020039928

W. W. Norton & Company, Inc., 500 Fifth Avenue, New York, NY 10110
wwnorton.com
W. W. Norton & Company Ltd., 15 Carlisle Street, London W1D 3BS

1 2 3 4 5 6 7 8 9 0

We dedicate this edition to the many thousands of health care workers around the world who are risking their lives fighting the COVID-19 pandemic of 2019–2020. One young, exhausted ICU nurse from St. Louis said this about their plight:

"We know we are going to get through this. . . . At the end of the day none of my co-workers regret being a nurse, being a respiratory therapist, a doctor. Pandemic or not, we are here . . . and we will always be here."

She won't admit it, but the rest of us know, that's exactly what heroes say.

Contents

6
Bacterial Growth, Nutrition, and Differentiation 150

Metabolic Fingerprints of a Brain Pathogen 151

Part II Essential Biology and Control of Infectious Agents

7
Bacterial Metabolism 184

A Spring Break Mishap 185

Part III The Immune System

15

The Immune System: Inflammation and Innate Immunity 438

A Tale of Two Tourists 439

16

The Immune System: Adaptive Immunity 468

Born in a Bubble 469

17

Vaccines, Diagnostics, and Immune Disorders 498

Kindergarten Contagion 499

Part IV Infectious Diseases

18

Microbial Pathogenesis 540

Panic, September 2001 541

19
Infections of the Skin and Eye 578

Rash Ruins Vacation 579

20
Infections of the Respiratory Tract 614

The Ominous Cough 615

21
Systemic Infections 650

Too Tired for Soccer 651

22
Infections of the Digestive System 690

Dehydration: A Toddler's Plight 691

23
Infections of the Urinary and Reproductive Tracts 738

Frequent Urination 739

24
Infections of the Central Nervous System 776

Diagnosis Comes Too Late 777

Part V Epidemiology and Biotechnology

25
Diagnostic Clinical Microbiology 812

Dazed and Confused in New Jersey 813

26
Epidemiology: Tracking Infectious Diseases 848

Mystery of the Microcephalic Newborns 849

27
Environmental and Food Microbiology 878
Explosive Toxic Soil 879

About the Authors

John W. Foster became intrigued with microbiology at 12 years old, after reading *Microbe Hunters* by Paul de Kruif, a book that chronicles the exploits of early microbiologists. Eventually, he received his BS from the University of the Sciences in Philadelphia and his PhD from Hahnemann University (now Drexel University School of Medicine), also in Philadelphia, where he worked with physiologist Albert G. Moat. After postdoctoral work at Georgetown University, he joined the Marshall University School of Medicine in West Virginia and then moved south to the Department of Microbiology and Immunology at the University of South Alabama College of Medicine in Mobile, Alabama. Dr. Foster has taught microbiology and infectious disease for 40 years and was recently awarded professor emeritus status for his efforts. His work includes over 100 journal articles revealing the physiology and genetics of microbial stress responses, and coauthorship of the textbook *Microbial Physiology*. He has chaired the Microbial Physiology and Metabolism division of the American Society for Microbiology and was a member of the editorial advisory board of the journal *Molecular Microbiology* (2008–2009, 2013).

Zarrintaj (Zari) Aliabadi is a physician assistant and a microbiologist. She received her PharmD from the University of Tehran College of Pharmacy, in Iran, and her PhD in biomedical sciences from Marshall University in Huntington, West Virginia, where she worked with John W. Foster. After her postdoctoral work at the University of South Alabama (USA), in Mobile, Alabama, she joined the Department of Biochemistry at the USA College of Medicine, where she taught biochemistry and conducted research on sickle cell disease. Dr. Aliabadi then earned a master's in health sciences from the USA Physician Assistant Studies Program, practiced medicine as a PA in endocrinology, served as director of the USA Diabetic Foot Clinic, and became chair of the USA Department of Physician Assistant Studies. She was named professor emeritus for her contributions. Dr. Aliabadi has taught extensively on infectious disease, pathophysiology, and clinical medicine to undergraduate pre–health profession students, graduate physician assistant students, and medical students. Her publications span the realms of microbiology and medicine.

Joan L. Slonczewski received her BA from Bryn Mawr College and her PhD in molecular biophysics and biochemistry from Yale University, where she studied bacterial motility with Robert M. Macnab. Since completing postdoctoral work at the University of Pennsylvania, she has taught undergraduate microbiology in the Department of Biology at Kenyon College, where she earned a Silver Medal in the National Professor of the Year program of the Council for Advancement and Support of Education. She has published numerous research articles with undergraduate coauthors on microbial pH stress and antibiotic resistance. She has also published six science fiction novels, including *A Door into Ocean* and *The Highest Frontier*, both of which earned the John W. Campbell Memorial Award. She served as at-large member representing divisions on the Council Policy Committee of the American Society for Microbiology and as a member of the editorial board of the journal *Applied and Environmental Microbiology*.

[Preface]

When we started writing the First Edition of **Microbiology: The Human Experience** in 2010, we decided that human experiences should drive all discussions, from basic microbiological concepts to infectious diseases. That approach was well received by faculty and students alike, and in 2018 we began work on this, the Second Edition.

Then, December 2019 happened.

As we were ending work on the Second Edition, the world was suddenly brought to its knees by a tiny virus named SARS-CoV-2. This virus, composed of a mere wisp of RNA wrapped in protein and a thin membrane, has killed hundreds of thousands of people, crippled the world's economy, and forced people everywhere to wear face masks and socially distance themselves for months, hiding from one another for fear of being the virus's next victim. During this time, the closest many people have been to human contact is on a computer screen, digitally trapped within tiny rectangles (thank you, Zoom).

Nothing in the past 100 years has so powerfully revealed the importance of microbiology and its impact on humans as the COVID-19 pandemic you've lived through. The last time everyone on Earth was threatened by a single infectious disease was in 1918, when a powerful strain of influenza killed about 20 million people worldwide in one year and over 50 million over two years. In 2020, SARS-CoV-2, the viral cause of COVID-19, killed around 1.2 million people in its first 10 months, with no end in sight. A scary number to be sure, but it's far fewer deaths than the 1918 flu claimed in a similar time frame. Why hasn't SARS-CoV-2 claimed as many lives as the 1918 flu virus? There are several reasons, but three stand out. First, people around the world listened to health professionals and wore face masks to prevent transmitting the virus to others. Second, we have more effective treatments for severe lung inflammation than we had back in 1918, enabling more critically ill patients to survive. Third, and perhaps most importantly, COVID-19 patients were cared for by courageous, highly-skilled health care professionals who, as part of their training, studied microbiology.

Your own personal experiences with COVID-19 might be motivating you to enter the field of medicine as a nurse, a respiratory therapist, a physician, a physician assistant, or other health care professional. We hope the human stories we tell in this book inspire you to achieve your goals. After all, COVID-19 will not be the last infectious disease to threaten humanity.

We are honored to have so many early fans of our book. But more important, we are thankful for the feedback offered by many adopters—feedback we were able to use in preparing this edition.

John W. Foster
Zarrintaj (Zari) Aliabadi
Joan L. Slonczewski
Second Edition (October 2020)

Major Features and What's New

WHY FOCUS ON THE HUMAN EXPERIENCE? Learning the basic concepts of microbiology from a human health perspective is something we find allied health students really enjoy. After all, microbiology as a science was forged largely from a desire to ease human suffering, so why not embrace that theme while teaching microbiology to students who share that desire? From anthrax to smallpox, measles to mumps, and influenza to COVID-19, infectious diseases have been a driving force in our biological and intellectual evolution for millions of years. Even today, infectious diseases kill two-thirds of the nearly 9 million children who die each year. Understanding how microbes live, grow, and die is a good way to gain some control over our own mortality.

We also find that students interested in pursuing health careers become energized when aspects of medicine are injected into the discussion early and often. In Chapter 2, for instance, we quickly introduce the basics of infectious disease and then employ multiple case histories to illustrate those concepts. This strategy, pairing the explanation of core microbiology concepts with applications to medicine and human health, is maintained throughout the book. More than 100 clinical case histories propel our coverage of basic microbiological concepts such as microbial growth, metabolism, genetics, differentiation, ecology, and immunology. The chapters on immunology, for instance, use patient case histories to highlight what happens when one part of the immune system fails and then build discussions of the immune system around those failures. Cases in every chapter provide a window through which students can unravel the mystery of infections: how they happen, how they are diagnosed, and how they are treated. Sections of several chapters also reveal important connections between the environment, climate change, and emerging infections—a growing concern among national and world health organizations.

A FLEXIBILITY TO MATCH TEACHING PARADIGMS Many features of our book lend themselves to new as well as traditional teaching paradigms. The writing style is engaging and conversational, and includes a touch of humor to drive learning. Clearly written explanations and a lavish art program support the "flipped" classroom approach, in which students learn independently through reading before attending interactive classroom sessions. The case histories themselves can be used to design innovative, team-based, active-learning exercises. We also know that health career students learn microbiology most easily when it relates to their own infectious disease experiences. Toward that end, frequent thought exercises embedded in each chapter revolve around infections that students may have had themselves or may wonder about. These exercises link concepts from earlier chapters while developing a student's critical thinking skills. For these and many other reasons, we think you will find the pedagogical tools included in this book superior to those of other undergraduate allied health textbooks.

OUR REVISION STRATEGY Revisions in this Second Edition were driven by advances in the field and by thoughtful critiques submitted by many reviewers. Every chapter was updated, and several chapters were reorganized to maximize information flow. However, we preserved the unique features of the First Edition that set our work apart from other microbiology textbooks. These features include the following:

CHAPTER-OPENING CASE HISTORIES Each chapter opens with an elegantly illustrated case history that sets the theme. For example, the chapter on biochemistry (Chapter 4) begins with the case of a small boy stricken with cholera; the bacterial metabolism chapter (Chapter 7) begins with the case of a college student vacationing in Mexico who contracts shigellosis; and the chapter on vaccines, diagnostics, and immune disorders (Chapter 17) opens with the case of a kindergarten teacher on chemotherapy who catches measles pneumonia from one of her students. Later, we explain how these cases connect to the basic concepts presented in the chapter. In addition to updating the content of these cases, we have written new chapter-opening cases for several chapters.

CHAPTER AND SECTION OBJECTIVES Every chapter still begins with a set of objectives outlining the major concepts that will be covered as the chapter unfolds. The chapter objectives are followed by individual section objectives and section summaries that alert students about what they should be able to explain, discuss, and compare after reading each section.

CONCEPTUAL LEARNING THROUGH EMBEDDED CASE HISTORIES Besides the chapter-opening cases, more than 100 additional patient case histories are integrated into this textbook; all have been updated, and several are new. The cases have been taken from the literature or told to us by colleagues. Some are embellished, but all of them convey the human toll of infectious disease and why we as a species chose to explore the hidden world of microbes in the first place. Each story provides a focal point for discussing:

- Basic concepts of microbiology (Chapters 1–13), such as microbe structure, genetics, biochemistry, biotechnology, antibiotics, and disinfectants.

- The human microbiome and our immune responses to infection (Chapters 14–17). Topics include innate immunity, adaptive immunity, immunological diseases, and immunological tools.

- Microbial pathogenesis and infectious diseases by organ system (Chapters 18–24), including skin, respiratory, systemic, gastrointestinal, urinary, reproductive, and central nervous systems.

- Clinical microbiology and epidemiology (Chapters 25 and 26).
- Environmental and food microbiology (Chapter 27).

The cases not only provide a framework for concept building but also repeatedly outline how diseases are diagnosed and how they are tracked and treated. Some cases explore the intimate connections between evolution and emerging diseases, between climate change and epidemiology, and between the immune system and the severity of disease.

END-OF-CHAPTER ASSESSMENT QUESTIONS Questions at the end of each chapter are tied to learning outcomes for each section and help students quickly evaluate their mastery of the material. Additional online questions supplement and expand the end-of-chapter assessments.

THOUGHT QUESTIONS Opportunities for critical thinking are central to student learning. Scattered throughout every chapter are eight to ten Thought Questions that help students think about the concepts they just read and connect them to concepts in other chapters. Answers to these questions are sometimes difficult but always insightful, and if students get stumped, the answers can be found at the back of the book.

CLINICAL CORRELATION QUESTIONS Because this textbook was designed in large part for students interested in health science careers, each chapter ends with Clinical Correlation Questions. These questions are designed around clinical scenarios that students must evaluate by explaining how they would diagnose, treat, or track a pathogen.

IMPACT FEATURES Each chapter also includes a highlighted feature called IMPACT that delves deeper into a basic concept and reveals its current or future societal impact. For example, the IMPACT feature for antibiotics (Chapter 13) explains how temporary antibiotic tolerance (persistence) can produce drug treatment failures; the IMPACT feature for virology (Chapter 12) reveals how scientists can trace the origins of pathogenic viruses and foretell the emergence of never-before-seen pandemic viruses; and the IMPACT feature for gastrointestinal infections (Chapter 22) illuminates how something called fecal microbiome transplantation (also known as fecal transplants) has become a lifesaving procedure for patients suffering with a chronic intestinal infection caused by *Clostridioides difficile*. Each IMPACT feature ends with a classroom discussion question such as this one from Chapter 13:

How would you go about identifying new antimicrobials that specifically target persister cells?

THE ORGAN SYSTEM APPROACH TOWARD TEACHING INFECTIOUS DISEASE Infectious disease is taught in two basic ways: the taxonomic approach, in which all the diseases a single microbe can cause are presented together; and the organ system approach, which focuses on an individual organ system and the array of pathogens that infect it. From the health care provider's standpoint, the organ system approach is the most useful way to compile information about infections because it reflects how a clinician interacts with a patient. Sick patients visiting a clinician will describe their symptoms ("I have a fever, cough, and chest pain"). The patient's description tells the clinician which organ system is principally involved (the respiratory system in this case). Then the clinician, knowing which pathogens can infect that organ system, will prescribe a broad antimicrobial treatment directed toward all of the most likely pathogens. The laboratory, after collecting relevant clinical specimens, will identify the precise organism involved and steer the clinician to a more targeted antibiotic therapy. For these reasons we use the organ system approach coupled with real-world medical examples to hook students on learning microbiology, and do it in a way that does not require the instructor to learn medicine.

Art Program

The vast majority of concepts introduced in our book take place at a scale not visible to the naked eye. Yet, one of the most important skills that introductory students can master is the ability to visualize key microbial processes and structures. As a result, we focused considerable effort on making sure that the images selected and developed for our book achieve the highest level of visual engagement and pedagogical value.

- Figures depicting processes and structures are colorful and engaging and include helpful bubble captions that provide essential information students need to learn from the figure.
- Meticulously designed life-cycle figures combine drawn art, photographs, and a system of helpful schematic arrows that convey the key steps in the progression of important diseases.
- Carefully chosen micrographs include scale bars with size information and an acronym indicating the kind of microscope used to capture the image.
- Ample images of the physical manifestations of disease on the human body draw students in and help future clinicians become comfortable recognizing the outward signs of important diseases.

Case Histories are drawn from diverse sources and include both cases that clinicians will encounter every day and cases that focus on global health issues.

There are two types of **Case Histories**. Each chapter begins with a chapter-opening case history that establishes a theme for the chapter. For example, the Chapter 26 (Epidemiology) opener explains how epidemiologists discovered that Zika virus was the cause of a spike in babies born with microcephaly.

(26)

Epidemiology: Tracking Infectious Diseases

26.1 Principles and Tools of Epidemiology

26.2 Tracking Health Care–Associated Infections

26.3 Tracking Global Disease Trends

26.4 Detecting Emerging Microbial Diseases

26.5 Bioterrorism and Biodefense

Mystery of the Microcephalic Newborns

SCENARIO One morning in 2015, a young woman from Recife, Brazil, was in the fourth month of her pregnancy when she awoke with a fever, a fiery red rash, and pain in her joints. Her symptoms went away within a few days, so she went about her life, oblivious to what was to come. Five months later, in September, she delivered her baby and immediately knew from the look on the obstetrician's face that something was terribly wrong. Her son had an abnormally small head and incomplete brain development—something called microcephaly. The young mother was devastated.

A child born with microcephaly in Brazil
Stunted brain development results in an unusually small cranium at birth.

Computed tomography image of brain calcifications (arrow) caused by congenital Zika virus infection

INITIAL CLUES In August 2015 a neurologist in Recife, started not of newborns with microcephaly. N without her seeing a single case. N any given day. Serological tests rul measles and toxoplasma, but imagi calcification in affected brains—a p Then Vanessa's mother, Ana, called. in another hospital. Ana had just seen seven microcephalic babies in <u>one day</u>. Many of the mothers reported having a rash early during pregnancy. Was a new infectious agent involved?

EPIDEMIOLOGY Vanessa alerted the state health secretary about the spike in microcephaly cases. This was the first step in the long epidemiological journey linking Zika virus, a positive-sense RNA virus that normally causes benign, mosquito-transmitted disease in adults, to the development of microcephaly in Zika virus–infected fetal brains. Between 2015 and 2018, at least 4,000 babies in Brazil (116 in the USA) had been born with microcephaly and other birth defects linked to Zika virus. Estimates are that about 1 in 7 babies exposed to Zika virus in the womb will develop at least one developmental defect within a year.

Zika virus

TEM 100 nm

Zika Virus
This enveloped, single-stranded RNA virus (Flavivirus family) is usually transmitted by mosquito bite and typically causes a mild rash with fever. However, the virus can cross the placental barrier and cause devastating fetal disease.

STATUS Fortunately, the number of Zika infections overall has dropped significantly because of mosquito control measures and personal protection strategies such as using mosquito repellent. The World Health Organization (WHO) and the US Centers for Disease Control and Prevention (CDC) continue to carefully monitor the spread of Zika virus throughout the Americas by intensively screening mosquitoes.

CASE HISTORY 26.3

It Began in Wuhan (but Didn't Stay There)

Peng Yinhua was a 29-year-old respiratory and critical illness doctor at First People's Hospital in the Jiangxia District of the city of Wuhan, China. Near the end of December 2019, Peng found himself heading to work, daydreaming about his upcoming wedding; a long-awaited ceremony that would take place in just a matter of days. But he couldn't know what was about to happen. Later that morning the patients started arriving. There were only a few at first, but the few grew to nearly 20 that day. Each of them had a high fever and a persistent cough and struggled to breathe. Radiographs showed that most had bilateral lung infiltrates characteristic of pneumonia. The next day more patients arrived, and even more the day after that. What was happening? Peng found himself on the front lines of what would be called the COVID-19 pandemic, caused by a new emerging coronavirus eventually called SARS-CoV-2 (see thumbnail image). Peng knew he would have to postpone his wedding because he was needed here. Sadly, despite taking medical precautions, Peng himself was infected while treating his patients. By January 25, 2020, nearly 700 patients with COVID-19 disease had been admitted to the hospitals of Wuhan. One of those admitted was Peng. Five days later, this selfless doctor died from respiratory failure. His wedding invitations, never mailed, were still in his desk.

Many other case histories are embedded in each chapter. Case History 26.3 describes the very beginning of the COVID-19 pandemic in China. The case kicks off a discussion of emerging diseases: how they are discovered, how they are dealt with, and how humans react to the prospect of a seemingly unstoppable (and invisible) infectious agent.

LEARNING OBJECTIVES IN THE TEXT AND MEDIA HELP YOU PICK RESOURCES THAT MATCH YOUR TEACHING

CHAPTER OBJECTIVES

After reading this chapter, you will be able to:

- Describe the relationships between a host, its microbiome, and pathogens.
- Apply basic concepts of pathogenesis to specific infections.
- Discuss how infectious diseases impact communities and how communities shape emerging pathogens.

Each chapter opens with a set of **big-picture objectives**, outlining the most important concepts.

To further ensure student comprehension, a set of objectives is provided for each section as well. These objectives correspond to the learning objectives in Smartwork, the Interactive Instructor's Guide, and the Test Bank, helping you pick the perfect resource for class.

SECTION OBJECTIVES

- Describe differences between "normal" microbiota and pathogens.
- Discuss the relationship between infection and disease and between virulence and pathogenicity.
- Differentiate between infectious dose and lethal dose.
- Discuss the fundamental attributes of a successful pathogen.

SECTION SUMMARY

- **Diseases** are recognized by their signs and symptoms.
- **Symptoms** are caused by bacterial products and by the host immune response (immunopathology).
- **Stages of an infectious** disease include the following:
  - **Incubation period**, when the organism begins to grow but symptoms have not developed
  - **Prodromal phase**, which can be unapparent or show vague symptoms
  - **Illness phase**, when signs and symptoms are apparent and the immune system is fighting the disease
  - **Decline phase**, when the numbers of pathogens decrease and symptoms abate
  - **Convalescence**, when symptoms are gone and the patient recovers
- **Morbidity** is a measure of how many people are sick from an infectious disease. Mortality is a measure of how many died.

An extensive **Section Summary** at the end of each section reiterates key concepts.

End-of-chapter review questions are explicitly tied to each section's learning objectives in a visual road map of the chapter. Clinical Correlation Questions on the following page help students effectively apply what they have learned to real-world situations.

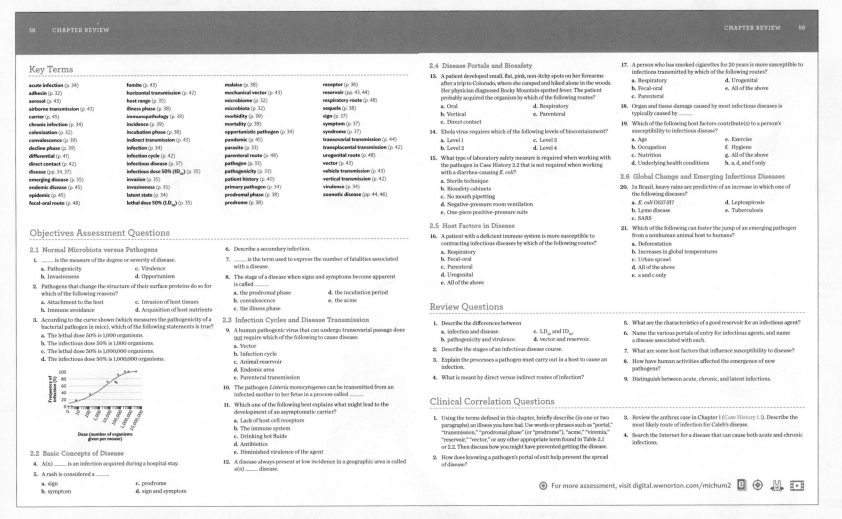

Key Terms

acute infection (p. 34)
adhesin (p. 32)
aerosol (p. 43)
airborne transmission (p. 43)
carrier (p. 45)
chronic infection (p. 34)
colonization (p. 32)
convalescence (p. 39)
decline phase (p. 39)
differential (p. 41)
direct contact (p. 42)
disease (pp. 34, 37)
emerging disease (p. 55)
endemic disease (p. 45)
epidemic (p. 45)
fecal-oral route (p. 48)

fomite (p. 43)
horizontal transmission (p. 42)
host range (p. 35)
illness phase (p. 38)
immunopathology (p. 38)
incidence (p. 39)
incubation phase (p. 38)
indirect transmission (p. 43)
infection (p. 34)
infection cycle (p. 42)
infectious disease (p. 37)
infectious dose 50% (ID₅₀) (p. 35)
invasion (p. 35)
invasiveness (p. 35)
latent state (p. 34)
lethal dose 50% (LD₅₀) (p. 35)

malaise (p. 38)
mechanical vector (p. 43)
microbiome (p. 32)
microbiota (p. 32)
morbidity (p. 39)
mortality (p. 39)
opportunistic pathogen (p. 34)
pandemic (p. 45)
parasite (p. 33)
parenteral route (p. 48)
pathogen (p. 33)
pathogenicity (p. 33)
patient history (p. 40)
primary pathogen (p. 34)
prodromal phase (p. 38)
prodrome (p. 38)

receptor (p. 36)
reservoir (p. 43, 44)
respiratory route (p. 48)
sequela (p. 38)
sign (p. 37)
symptom (p. 37)
syndrome (p. 37)
transovarial transmission (p. 44)
transplacental transmission (p. 42)
urogenital route (p. 48)
vector (p. 43)
vehicle transmission (p. 43)
vertical transmission (p. 42)
virulence (p. 34)
zoonotic disease (pp. 44, 46)

Objectives Assessment Questions

2.1 Normal Microbiota versus Pathogens

1. _____ is the measure of the degree or severity of disease.
   a. Pathogenicity
   b. Invasiveness
   c. Virulence
   d. Opportunism

2. Pathogens that change the structure of their surface proteins do so for which of the following reasons?
   a. Attachment to the host
   b. Immune avoidance
   c. Invasion of host tissues
   d. Acquisition of host nutrients

3. According to the curve shown (which measures the pathogenicity of a bacterial pathogen in mice), which of the following statements is true?
   a. The lethal dose 50% is 1,000 organisms.
   b. The infectious dose 50% is 1,000 organisms.
   c. The lethal dose 50% is 1,000,000 organisms.
   d. The infectious dose 50% is 1,000,000 organisms.

2.2 Basic Concepts of Disease

4. A(n) _____ is an infection acquired during a hospital stay.

5. A rash is considered a _____.
   a. sign
   b. symptom
   c. prodrome
   d. sign and symptom

6. Describe a secondary infection.

7. _____ is the term used to express the number of fatalities associated with a disease.

8. The stage of a disease when signs and symptoms become apparent is called _____.
   a. the prodromal phase
   b. convalescence
   c. the illness phase
   d. the incubation period
   e. the acme

2.3 Infection Cycles and Disease Transmission

9. A human pathogenic virus that can undergo transovarial passage does not require which of the following to cause disease:
   a. Vector
   b. Infection cycle
   c. Animal reservoir
   d. Endemic area
   e. Parenteral transmission

10. The pathogen *Listeria monocytogenes* can be transmitted from an infected mother to her fetus in a process called _____.

11. Which one of the following best explains what might lead to the development of an asymptomatic carrier?
    a. Lack of host cell receptors
    b. The immune system
    c. Drinking hot fluids
    d. Antibiotics
    e. Diminished virulence of the agent

12. A disease always present at low incidence in a geographic area is called a(n) _____ disease.

2.4 Disease Portals and Biosafety

13. A patient developed small, flat, pink, non-itchy spots on her forearms after a trip to Colorado, where she camped and hiked alone in the woods. Her physician diagnosed Rocky Mountain spotted fever. The patient probably acquired the organism by which of the following routes?
    a. Oral
    b. Vertical
    c. Direct contact
    d. Respiratory
    e. Parenteral

14. Ebola virus requires which of the following levels of biocontainment?
    a. Level 1
    b. Level 2
    c. Level 3
    d. Level 4

15. What type of laboratory safety measure is required when working with the pathogen in Case History 2.2 that is not required when working with a diarrhea-causing *E. coli*?
    a. Sterile technique
    b. Biosafety cabinets
    c. No mouth pipetting
    d. Negative-pressure room ventilation
    e. One-piece positive-pressure suits

2.5 Host Factors in Disease

16. A patient with a deficient immune system is more susceptible to contracting infectious diseases by which of the following routes?
    a. Respiratory
    b. Fecal-oral
    c. Parenteral
    d. Urogenital
    e. All of the above

17. A person who has smoked cigarettes for 20 years is more susceptible to infections transmitted by which of the following routes?
    a. Respiratory
    b. Fecal-oral
    c. Parenteral
    d. Urogenital
    e. All of the above

18. Organ and tissue damage caused by most infectious diseases is typically caused by _____.

19. Which of the following host factors contribute(s) to a person's susceptibility to infectious disease?
    a. Age
    b. Occupation
    c. Nutrition
    d. Underlying health conditions
    e. Exercise
    f. Hygiene
    g. All of the above
    h. a, d, and f only

2.6 Global Change and Emerging Infectious Diseases

20. In Brazil, heavy rains are predictive of an increase in which one of the following diseases?
    a. *E. coli* O157:H7
    b. Lyme disease
    c. SARS
    d. Leptospirosis
    e. Tuberculosis

21. Which of the following can foster the jump of an emerging pathogen from a nonhuman animal host to humans?
    a. Deforestation
    b. Increases in global temperatures
    c. Urban sprawl
    d. All of the above
    e. a and c only

Review Questions

1. Describe the differences between
   a. infection and disease.
   b. pathogenicity and virulence.
   c. LD₅₀ and ID₅₀.
   d. vector and reservoir.

2. Describe the stages of an infectious disease course.

3. Explain the processes a pathogen must carry out in a host to cause an infection.

4. What is meant by direct versus indirect routes of infection?

5. What are the characteristics of a good reservoir for an infectious agent?

6. Name the various portals of entry for infectious agents, and name a disease associated with each.

7. What are some host factors that influence susceptibility to disease?

8. How have human activities affected the emergence of new pathogens?

9. Distinguish between acute, chronic, and latent infections.

Clinical Correlation Questions

1. Using the terms defined in this chapter, briefly describe (in one or two paragraphs) an illness you have had. Use words or phrases such as "portal," "transmission," "prodromal phase" (or "prodrome"), "acme," "viremia," "reservoir," "vector," or any other appropriate term found in Table 2.1 or 2.2. Then discuss how you might have prevented getting the disease.

2. How does knowing a pathogen's portal of exit help prevent the spread of disease?

3. Review the anthrax case in Chapter 1 (Case History 1.1). Describe the most likely route of infection for Caleb's disease.

4. Search the Internet for a disease that can cause both acute and chronic infections.

⊙ For more assessment, visit digital.wwnorton.com/michum2

smartwork

Smartwork is an online assessment platform that improves student problem-solving skills and makes it easy to assign the case histories and animations that accompany the text. Students receive answer-specific feedback and hints that help them understand key concepts.

InQUIZITIVE

Norton's formative adaptive quizzing program, InQuizitive, reinforces core concepts in the text through a variety of interactive questions. When assigned before class, InQuizitive helps students build their knowledge and come to lecture better prepared.

Figure 21.26 Life Cycle of *Toxoplasma gondii*

The only known definitive hosts for *T. gondii* are domestic cats and their relatives. Infected cats can release up to 100 million gametes (unsporulated oocytes) every day for 7–21 days. The oocytes will sporulate in the environment and become infective, which will complete their growth cycle in the intermediate host.

Bradyzoite

Tachyzoite

Cyst

LM   20 μm

4
Cats (definitive hosts) become infected by eating material containing toxoplasma tissue cysts.

DEFINITIVE HOST (cat)

3
The ingested sporulated oocytes transform into tachyzoites and move into brain and muscle tissue, where they become cyst bradyzoites.

The oocyte contains two sporocysts

1
Infected cats shed unsporulated oocytes.

2
Oocytes sporulate, become infective, and are ingested by the intermediate hosts.

Cat feces

Fecal oocyte

LM   20 μm

INTERMEDIATE HOSTS

Tissue cyst

5
Fecal oocytes contaminate kitty litter, food, and water.

6
Humans can become infected from handling contaminated materials via hand-to-mouth transfer, . . .

ingesting fecal-contaminated fruits, vegetables, or water, . . .

7
ingesting meat from infected animals, . . .

8
by blood transfusion, . . .

9
or by transplacental transmission to a fetus.

VISUALLY DYNAMIC ART BRINGS MICROBIAL PROCESSES AND STRUCTURES TO LIFE

4

Cats (definitive hosts) become infected by eating material containing toxoplasma tissue cysts.

Bubble captions help students interpret what they see without having to toggle between the figure and the caption.

Photos emphasize the real-life physical manifestations of human disease.

Meticulously developed and visually stunning life-cycle drawings highlight the stages of important diseases.

Micrographs improve student understanding of and appreciation for the microbes underlying human diseases.

Chapter features help students get the most out of the text

◉ **Note:** Bacterial names include designations for genus (*Salmonella*) and species (*enterica*). Both are written in italics, with the genus capitalized and species lowercase. Serovar subclass names of a species are not italicized but are capitalized (for instance, Typhimurium).

Notes provide tips that students may need to develop a fuller understanding of a concept or to navigate the text more effectively.

Thought Questions designed for discussion both inside and outside the classroom encourage students to consider the big picture.

Thought Question 2.1 Is a microbe with an LD_{50} of 5×10^4 more or less virulent than a microbe with an LD_{50} of 5×10^7? Why?

Thought Question 2.2 The bacterium *Salmonella enterica* serovar Typhimurium (*S.* Typhimurium) is usually transmitted to humans by contaminated food and causes a diarrhea that typically resolves without medical intervention. During infection, the organism remains localized to the intestine. A different serovar of *Salmonella*, *S.* Typhi, is also transmitted via contaminated food but can penetrate the intestinal wall and enter the lymphatic system and bloodstream, causing high fever and, in some cases, death. Clearly, *S.* Typhi is more virulent than *S.* Typhimurium to humans. However, when tested for lethal dose (LD_{50}) in mice, *S.* Typhi fails to kill any mice, whereas *S.* Typhimurium has a very low lethal dose (it takes very few organisms to kill mice). How might you broadly explain this apparent contradiction?

RESOURCES FOR STUDENTS AND INSTRUCTORS

smartwork

digital.wwnorton.com/michum2

Smartwork is an interactive assessment tool that improves student problem-solving skills. Assign visual, animation, critical thinking, case history questions and more, all with rich answer-specific feedback that builds knowledge and skills. The Second Edition course includes new tutorial lessons for some of the most challenging course topics that prompt students to build their microbiology problem-solving skills and foundational knowledge through a set of practice questions based on the main topic's core concepts. All Smartwork questions are book-specific, matching the terminology and conventions that students see in the textbook; include ebook links; and have been developed to address the learning objectives in the book. Prebuilt activities and intuitive settings make getting started easy, and instructors can customize and create assignments to best fit their classrooms. Smartwork integrates with the most popular campus learning-management systems, and intuitive performance reports for both individual students and entire classes help instructors gauge student comprehension and adjust their teaching accordingly.

INQUIZITIVE

digital.wwnorton.com/michum2

Norton's formative adaptive quizzing platform, easily accessible via any mobile device, preserves valuable lecture and lab time by helping students practice and master core concepts. A variety of question types test students' knowledge in different ways across the sections in each chapter. The Second Edition course includes a variety of interactive question types, including image-based questions and animation-based questions to promote visual understanding of microbiology concepts. Students are motivated by the engaging, gamelike elements that allow them to set their confidence level on each question to reflect their knowledge, track their own progress, earn bonus points for high performance, and review section content that they might not have mastered. Instructors can assign InQuizitive activities out of the box or use simple tools to customize the section content they want students to work on and to specify how much time they want students to spend on each module.

Resources for Your LMS

Easily add high-quality Norton digital resources to your online, hybrid, or lecture courses. Get started building your course with our easy-to-use integrated resources; all activities can be accessed within your existing LMS (learning management system). The ebook, Smartwork assignments, InQuizitive activities, animations, PowerPoint lecture outlines, and Case Mystery PowerPoints are all available, organized by chapter. Instructors can also add customizable multiple-choice and short-answer questions to their LMS by using Norton Testmaker. Explore a bank of over 400 questions to build quizzes and tests that can be easily exported as Common Cartridge files for import into your LMS.

Ebook

The Norton ebook for *Microbiology: The Human Experience*, Second Edition, includes dynamic features that engage students, including process animations and drop-down answers to the Thought Questions. Instructors can focus student reading by sharing notes with their classes, including embedded images and video. Art expands, pop-up key terms are linked to definitions, and students can search, highlight, and take notes with ease. The ebook can be viewed on any device and will sync across devices.

Ⓖ Interactive Instructor's Guide

Searchable by chapter, phrase, topic, or ASM (American Society for Microbiology) guideline, the Interactive Instructor's Guide compiles the many valuable teaching resources available with *Microbiology: The Human Experience*. For every chapter of the book, instructors will find two multimedia suggestions (videos or podcasts) with suggested classroom uses and discussion questions; two active-learning activities with premade, student-facing handouts; animations with discussion questions; as well as Case Mystery PowerPoints, lecture PowerPoints, and image files. Each resource is tagged to learning objectives in the text, making it easy to find the best resource for class. This repository of lecture and teaching materials functions both as a course-prep tool and as a means of tracking the latest ideas in teaching the allied health microbiology course.

Case Mystery PowerPoints

For every chapter in the text, a new case study is introduced in PowerPoint form, including the symptoms of a fictional patient, the various tests that would be administered by a doctor, the "results" of these tests, and finally, the diagnosis and treatment. Integrated clicker and discussion questions throughout the PowerPoint help instructors gauge student comprehension of key concepts and encourage students to develop the clinical reasoning skills needed to predict patient outcomes. Instructor notes throughout the file answer discussion questions and elaborate on key topics.

Lecture PowerPoints

Visually dynamic, chapter-specific lecture PowerPoints include outlines, notes, and clicker questions, and can be used in the classroom or for student self-study.

Art Slides

All the art in *Microbiology: The Human Experience*, Second Edition, is available for instructor use and presentation, in both labeled and unlabeled versions, as JPEGs and in PowerPoint.

Squarecap Classroom Response Questions

Squarecap is a next-generation classroom response system that promotes a high level of student engagement in the classroom through quality assessment in a variety of question types. Students can answer questions directly from their smartphones; the Squarecap system offers instructors in-depth analytics on student performance in real time. *Microbiology: The Human Experience*, Second Edition, can be packaged with review questions specific to each chapter.

Test Bank

Every chapter of the test bank includes at least 50 multiple-choice, fill-in-the-blank, and short-answer questions, all designed to evenly cover text learning objectives. Up to ten questions per chapter are based on modified art from the text, and three questions tie in case histories from the book. Every test question is classified by Bloom's taxonomy, text section, learning objective, and difficulty level, making it easy to construct meaningful and diagnostic tests and quizzes.

Infectious Disease Flashcards

digital.wwnorton.com/michum2

These online interactive flashcards contain information on 50 diseases that affect different body systems, now including Zika fever and COVID-19. Each card includes photographs and information on a specific disease. The cards can be used for individual study and review. Printed cards can be packaged as a supplement but are available at no additional cost as an online resource.

Animations

In response to feedback from reviewers regarding the topics that students find most challenging, 37 high-quality, ADA-compliant animations bring the course's most complex topics to life. Questions based on the animations are embedded in Smartwork and InQuizitive, and the animations are available on the digital landing page and linked to directly from the ebook. Animation topics include:

FUNDAMENTALS

Microscopy and Staining
Oxidation-Reduction Reactions
Osmosis and Water Balance
Endocytosis in a Eukaryotic Cell
Active Transport
Bacterial Cell Division
Dilution Streaking Technique
Biofilm Formation
Endospore Formation

ESSENTIAL BIOLOGY

Glycolysis and the TCA (Krebs) Cycle
The Electron Transport System
Photosynthesis
DNA Replication
Mutations and Base Excision Repair
Polymerase Chain Reaction
Operons
Transcription
Translation
Conjugation
Transduction
Influenza Virus Replication
HIV Replicative Cycle

IMMUNOLOGY

Phagocytosis
Inflammation
Clonal Selection Theory
Antigen Processing and Presentation
Humoral Immunity/Antibody-Mediated Immunity
Cell-Mediated Immunity
Allergic Response: Type I Hypersensitivity
Type II Hypersensitivity
Type III Hypersensitivity
Type IV Hypersensitivity
Enzyme-Linked Immunosorbent Assay (ELISA)

INFECTIOUS DISEASES

Intracellular Pathogens
Varicella-Zoster Virus (VZV)
Neisseria meningitidis

FOOD MICROBIOLOGY

Food Fermentation

PACKAGE OPTIONS

Contact your Norton representative for information on pricing, packaging, and customizing.

Microbiology: The Laboratory Experience, by Steve Keating (Pennsylvania State University)

Students get more out of their microbiology lab experience because the manual has thorough introductions that emphasize important concepts and applications—written in a uniquely engaging authorial voice—and is accompanied by an unparalleled visual program.

ISBN: 978-0-393-92364-3

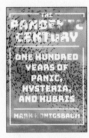

The Pandemic Century: One Hundred Years of Panic, Hysteria, and Hubris, by Mark Honigsbaum (City University of London)

How can we understand the COVID-19 pandemic? Ever since the 1918 Spanish influenza pandemic, scientists have dreamed of preventing such catastrophic outbreaks of infectious disease. Yet despite a century of medical progress, viral and bacterial disasters continue to take us by surprise, inciting panic and dominating news cycles. In *The Pandemic Century*, medical historian Mark Honigsbaum combines reportage with the history of science and medical sociology to artfully reconstruct epidemiological mysteries and the ecology of infectious diseases. Now updated with a new chapter and epilogue on coronavirus.

ISBN: 978-0-393-54131-1

Zika: The Emerging Epidemic, by Donald G. McNeil Jr. (*New York Times*)

As early as August 2015, doctors in northeast Brazil began to notice a trend: many mothers who had recently experienced symptoms of the Zika virus were giving birth to babies with microcephaly, a serious disorder characterized by unusually small heads and brain damage. The following year, scientists confirmed that Zika, which can be contracted through mosquito bites or sexual transmission, was the cause of microcephaly. The virus has been known to be transmitted by *Aedes aegypti* (the yellow fever mosquito), but now *Aedes albopictus* (the Asian tiger mosquito) has been found to carry it as well, which means it might affect regions as far north as New England and the Great Lakes. In *Zika: The Emerging Epidemic*, Donald G. McNeil Jr. sets the facts straight in a fascinating exploration of Zika's origins, how it spreads, the race for a cure, and what we can do to protect ourselves now.

ISBN: 978-0-393-35396-9

Infectious Disease Flashcards, by Julie Harless (Lone Star College)

This deck of cards includes information on 50 diseases that affect different body systems. One side of each card includes two photographs: one showing the disease agent itself and one showing the effect of the disease on the human body. The other side includes an organized summary of information about the disease. The cards can be used for individual study and review or for several students to play with together. An online version of these flashcards is included with the purchase of the book.

ISBN: 978-0-393-26323-7

[Acknowledgments]

We thank the dozens of people whose dedication and nurturing brought this project to fruition. Special thanks go to our marvelous editor, Betsy Twitchell, whose talents and personality kept everyone focused and who skillfully guided both the First and now Second Editions from their inception to the printed page; and to Jen Barnhardt, our tireless project editor, who kept the countless moving parts of this project from seizing and whose attention to detail was, well, detailed. We again thank Kathy Gillen for her expert biochemistry content incorporated into Chapter 4. Our developmental editor, Lisa Bess Kramer, used impressive skills and vision to polish our writing and clarify our logic. In addition, we are immensely grateful to our copyeditors, Stephanie Hiebert and Heather Cammarn, whose uncanny (and a little scary) skills uncovered numerous examples of grammatical malfeasance and small but critically important inconsistencies in the text. Once again, a huge debt of gratitude goes to Dena Betz, who miraculously tracked down elusive images from sources all around the world (fortunately, her "travels" did not require donning a COVID-19 mask), and to photo editor Cat Abelman for being with her every step of the way. We are lucky, too, to have worked with an incredibly talented artist, Anne DeMarinis, whose captivating art and design grace the cover of this book and every chapter within it. We must also thank the unsung heroes of the project, Katie Callahan and Danny Vargo, who relentlessly tracked the countless emails, chapter drafts, and Excel files that constantly zipped back and forth between editors, authors, and reviewers, and whose electronic savvy helped guide us to and through the Norton website. Production manager Ben Reynolds adeptly managed the process of translating our raw material into the polished final product; for that he has our deepest thanks. The amazing folks at Precision Graphics (a Lachina company) deserve medals for the excellent work they did. Thank you to Eric Bramer, Megan Calderwood, Terri Hamer, and Rebecca Marshall.

We have an absolutely tireless team at Norton creating the print and digital supplementary resources for our book. Media editor Kate Brayton, associate editor Jasmine Ribeaux, and media editorial assistant Kara Zaborowsky worked on every element of the package as a team, and the content meets our very high standards as a result. The Second Edition's resources have also greatly benefited from Todd Pearson's teaching background and scientific expertise. Marketing manager Ruth Bolster's expertise in direct marketing has helped ensure this important book makes it into the hands of as many instructors and students as possible. Thank you also to Kim Yi's media project editorial group for the invaluable work they do shepherding content through many stages of development. We thank everyone involved in Norton's sales and marketing team for their unflagging support of our book, as well as Erik Fahlgren, Roby Harrington, Marian Johnson, Drake McFeely, Julia Reidhead, and Mike Wright, for believing in us all these years.

We also thank the many colleagues who encouraged us and the dozens of reviewers across the country who lent their keen eyes and deep experience to critique early drafts of our work. We are grateful, too, to the countless scientists who graciously contributed the micrographs that appear in this textbook; and to our students, who remain for us a constant source of wonder and unbridled energy. Finally, we thank our families—the parents, spouses, children, grandchildren, and yes, even great-grandchildren, who fill our lives with joy and just the right amount of chaos.

Reviewers

SECOND EDITION

Philips Akinwole, DePauw University
Mary Allen, Hartwick College
Tammy Atchinson, Pitt Community College
Sankhiros Babapoor, Sacred Heart University
Dave Bachoon, Georgia College
John Battista, Louisiana State University
Lynn Bedard, DePauw University
Samantha Beebe, Penn State
Ralph Benedetto, Wayne Community College
Jacqueline Biscardi, Virginia Western Community College
Constantine Bitsaktsis, Seton Hall University
Gilles Bolduc, Massasoit Community College
Jennifer Brigati, Maryville College
Sandra Buerger, Boston University
Kristin Burkholder, University of New England
Robert Cattolica, Chabot College
Emma Chaput, Central Oregon Community College
Sujata Chiplunkar, Cypress College
Elizabeth Co, Boston University
Francie Coblentz, Brunswick Community College
Joseph Colosi, DeSales University
Tanya Crider, Holmes Community College
Candice Damiani, University of Pittsburgh
Charles Daniel, Bay State College
Deborah Dardis, Southeastern Louisiana University
Debra Davis, Wingate University

Cathy Dobbs, Joliet Junior College

Lisa Draughon, Johnston Community College

Melissa Elliott, Butler Community College

Ashley Fink, College of Saint Benedict and Saint John's University

Diane Gibson, Hazard Community & Technical College

Zaida Gomez-Kramer, University of Central Arkansas

Neil Greene, University of Rhode Island

Paul Himes, University of Louisville

Jacob Hornby, Lewis-Clark State College

Jeff Hucko, Glen Oaks Community College

Eric Hugo, Northern Kentucky University

James Hutcherson, Southeastern Community College

Bob Iwan, Inver Hills Community College

Eddie Johnson, Central Oregon Community College

Cheng Kao, Indiana University

Judy Kaufman, Monroe Community College

Carrie Ketel-Opheim, Normandale Community College

Susan King, Gadsden State Community College

Heather Klenovich, Community College of Allegheny County

Ashley Konal, Coastal Carolina Community College

Michael Kovach, Baldwin Wallace University

Rachel Larsen, University of Southern Maine

Scott Layton, Cowley College

Bill McCleary, Brigham Young University

Danielle McGrath, San Jacinto College

Emily Messina, Husson University

Kathy Monroe, Blue Ridge Community and Technical College

Jeanelle Morgan, University of North Georgia

Jonathan Morris, Manchester Community College

Archana Mudbidri, Quinsigamond Community College

Holly Pacquette, College of Western Idaho

Daniela Payne, Aiken Technical College

Krista Peppers, University of Central Arkansas

Jocelyn Ramos, Ivy Tech Community College

Mark Randa, Cumberland County College

Juan Rojo, University of New Hampshire

Cleo Rolle, Capitol Community College

Rachel Romain, Columbus State Community College

Ben Rowley, University of Central Arkansas

Katie Ruggieri, Bristol Community College

Deborah Scheiwe, Tarrant County College

Melissa Schreiber, Valencia College

Sheila Schreiner, Salem State University

Polly Schulz, Clackamas Community College

Heather Seitz, Johnson County Community College

Nilesh Sharma, Western Kentucky University

Sandra Specht, Sinclair Community College

Monica Strzempko, Springfield Technical Community College

Dustin Stutts, Roane State Community College

Sundeep Talwar, Harry S. Truman College

Sheila Taylor, Ozarks Community College

Cindy Trussell, Kodiak College

Mitchell Walkowicz, University of Massachusetts Amherst

Ginny Webb, University of South Carolina Upstate

Amy White, Virginia Western Community College

FIRST EDITION

Mari Aanenson, Western Illinois University

Lawrence R. Aaronson, Utica College

Sherrice V. Allen, Fayetteville State University

Cindy B. Anderson, Mt. San Antonio College

Daniel Aruscavage, Kutztown University of Pennsylvania

Dave Bachoon, Georgia College

Glenn Barnett, Central College

Jennifer Bess, Hillsborough Community College

R. Clark Billinghurst, St. Lawrence College

Will Blackburn, Tarrant County College

Lisa Ann Blankinship, University of North Alabama

Lanh Bloodworth, Florida State College at Jacksonville

Shaun Bowman, Clarke University

Jacqueline Brown, Angelo State University

Brad A. Bryan, Worcester State College

Sybil K. Burgess, Brunswick Community College

Evan Burkala, Oklahoma State University—Oklahoma City

Kristin Burkholder, University of New England

Kari Cargill, Montana State University

Bradley W. Christian, McLennan Community College

Tin-Chun Chu, Seton Hall University

Jenny H. Clark, Saddleback College

Bela Dadhich, Delaware County Community College

H. Kathleen Dannelly, Indiana State University

Deborah Dardis, Southeastern Louisiana University

Margaret Das, Southern Connecticut State University

Joyce Davis, Carroll Community College

Sondra Dubowsky, McLennan Community College

Elizabeth A. B. Emmert, Salisbury University

Tracey Emmons, Sandhills Community College

Clifton Franklund, Ferris State University

Ashley D. Frazier, Walters State Community College

Monica Friedrich, Saddleback College

David E. Fulford, Edinboro University of Pennsylvania

Ellen F. Fynan, Worcester State University

Julie Galvin, Halifax Community College

Ronald Girmus, New Mexico State University—Carlsbad

Carla J. Guthridge, Cameron University

Chris Guyer, Henderson State University

Kimberly Hale, Quincy University

Georgia A. Hammond, Radford University

Leanne Hanson, Clarke University

Steven Harris, University of Nebraska—Lincoln

Anne Hemsley, Antelope Valley College

Diane Hilker, Mercer County Community College

Dale Holen, Penn State Scranton

Sara Reed Houser, Jefferson College of Health Sciences

Jane E. Huffman, East Stroudsburg University

Karen Huffman-Kelly, Genesee Community College

Julie Huggins, Arkansas State University

Abdallah M. Isa, Tennessee State University

Sayna A. Jahangiri, Folsom Lake College

V. Karunakaran, St. Francis Xavier University

D. Sue Katz, Rogers State University

Judy Kaufman, Monroe Community College

Christine A. Kirvan, California State University, Sacramento

Laurieann Klockow, Marquette University

Richard Knapp, University of Houston

Malda Kocache, George Mason University

Hari Kotturi, University of Central Oklahoma

Anne Kruchten, Linfield College

Ashwini Kucknoor, Lamar University

Melissa Lail-Trecker, Western New England University

Maia Larios-Sanz, University of St. Thomas

Carol Lauzon, California State University, East Bay

Leo G. Leduc, Laurentian University

Kate LePore, Monroe Community College

Heather Lofton-Garcia, Troy University

Suzanne Long, Monroe Community College

Aaron Lynne, Sam Houston State University

Shannon R. Mackey, St. Ambrose University

Bernard MacLennan, Cape Breton University

Barry Margulies, Towson University

Nancy Marthakis, Purdue University Northwest

Carolyn F. Mathur, York College of Pennsylvania

Meghan May, Towson University

Sherry L. Meeks, University of Central Oklahoma

Karin Melkonian, LIU Post

Brian J. Merkel, University of Wisconsin—Green Bay

Stephanie Leah Molloy, California State University, East Bay

Ellyn Mulcahy, Johnson County Community College

Jacqueline Nesbit, University of New Orleans

Tanya Noel, University of Windsor

Lourdes Norman, Florida State College at Jacksonville

Maura Pavao, Worcester State University

Nancy A. Perigo, William Paterson University of New Jersey

Wendy L. Picking, Oklahoma State University

Laraine Powers, East Tennessee State University

Madhura Pradhan, The Ohio State University

Gregory Pryor, Francis Marion University

Mark Randa, Cumberland County College

Syed Raziuddin, Richard J. Daley College

Sarah M. Richart, Azusa Pacific University

Lori Rink, Penn State Abington

Frances Sailer, University of North Dakota

Tony Schountz, University of Northern Colorado

Melissa Schreiber, Valencia College

Larry J. Scott, Central Virginia Community College

Heather Seitz, Johnson County Community College

Debbie Sesselmann, Fox Valley Technical College

Josh Sharp, Northern Michigan University

Michael E. Shea, Hudson Valley Community College

Shana Shields, Curry College

Jean M. Shingle, Immaculata University

Uma Singh, Valencia College

Jennifer Staiger, Mount St. Mary's University

Paula Steiert, Southwest Baptist University

David Straus, Virginia Commonwealth University

Anne O. Summers, University of Georgia

Sonia Suri, Valencia College

Sundeep Talwar, Richard J. Daley College

Renato V. Tameta, Schenectady County Community College

Steven J. Thurlow, Jackson College

Michael Troyan, Penn State

Jorge Vasquez-Kool, Wake Technical Community College

Charles Vo, University of Arkansas—Fort Smith

Jeremiah Wagner, Mid Michigan College

Michael H. Walter, University of Northern Iowa

Mark Watson, University of Charleston

John E. Whitlock, Hillsborough Community College

Michael L. Womack, Gordon State College

Accuracy Reviewers

SECOND EDITION

Charles Daniel, Bay State College

Debra Davis, Wingate University

Carrie Ketel-Opheim, Normandale Community College

Rachel Larsen, University of Southern Maine

Emily Messina, Husson University

Juan Rojo, University of New Hampshire

Dustin Stutts, Roane State Community College

FIRST EDITION

Bradley W. Christian, McLennan Community College

Elizabeth A. B. Emmert, Salisbury University

Sherry L. Meeks, University of Central Oklahoma

Madhura Pradhan, The Ohio State University

Focus Group Participants

Dale Amos, University of Arkansas—Fort Smith

Lance Bowen, Truckee Meadows Community College

Donald Breakwell, Brigham Young University

Amy Warenda Czura, Suffolk County Community College

Sharon Gusky, Northwestern Connecticut Community College

Michael Ibba, The Ohio State University

Nastassia N. Jones, Philander Smith College

D. Sue Katz, Rogers State University

Suzanne Long, Monroe Community College

Aaron Lynne, Sam Houston State University

Stephanie Leah Molloy, California State University, East Bay

Francine Norflus, Clayton State University

Kathleen A. Page, Southern Oregon University

Christopher Parker, Texas Wesleyan University

Todd P. Primm, Sam Houston State University

Sarah Richart, Azusa Pacific University

Melissa Schreiber, Valencia College

Heather Seitz, Johnson County Community College

Eric Yager, Albany College of Pharmacy and Health Sciences

Sonja B. Yung, Sam Houston State University

Contributors to the Supplementary Resources

SECOND EDITION

Philips Akinwole, DePauw University

Becky Buckley, Metropolitan State University

Bradley Christian, McClennan University

Charles Daniel, Bay State College

Ashley Day, Isothermal Community College

Jenny Gernhart, Iowa Central

Julie Grainy, University of Georgia

Cecily Haley, San Jacinto College

Heather Klenovich, Community College of Allegheny County

Erin Lentz, Galen College of Nursing

Karla Lightfield, University of Kentucky

Danielle McGrath, San Jacinto College

Jeanelle Morgan, University of North Georgia

Jordan Parker, University of California Los Angeles

Veronica Riha, Madonna University

Juan Rojo, University of New Hampshire

Matthew Schmidt, SUNY Empire State College

Debra Scheiwe, Tarrant County Community College

Heather Seitz, Johnson County Community College

Amy Stone, Touro University

Shyra Tedesco, Ivy Tech Community College

Jarrett Whelan, Lenoir Community College

FIRST EDITION

Holly Ahern, SUNY Adirondack

Alexandra Armstrong, University of Arizona

Jennifer Bess, Hillsborough Community College

Carrie Bottoms, Collin College

Nancy Boury, Iowa State University

Bradley Christian, McLennan Community College

Elizabeth Collins, Iowa Central Community College

Jamie Cunningham, Johnson County Community College

Bela Dadhich, Delaware County Community College

Jenny Gernhart, Iowa Central Community College

Kathy Gillen, Kenyon College

Julie Harless, Lone Star College—Montgomery

Amy Helms, Collin College

Meg Howard, University of Alaska, Anchorage

Erin Lentz, Galen College of Nursing

Jacqueline Nesbit, University of New Orleans

Tiffany Randall, John Tyler Community College

Deborah Scheiwe, Tarrant County College

Heather Seitz, Johnson County Community College

Michael Shea, Hudson Valley Community College

Shana Shields, Curry College

Jennifer Smith, Tarrant County College

Suzanne Wakim, Butte College

Eric Warrick, State College of Florida

Preliminary Edition Survey Respondents

Lynn Bedard, DePauw University

Kristen Butela, Seton Hill University

Shauni Calhoun, Mt. San Jacinto College

Kathleen Dannelly, Indiana State University

Larry Feinstein, University of Maine at Presque Isle

Tom Firak, Oakton Community College, Des Plaines

David Fulford, Edinboro University of Pennsylvania

Eileen Gregory, Rollins College

Tray Hamil, University of South Alabama

Katina Harris-Carter, Paul D. Camp Community College

Amber Huber, Madison Area Technical College

Cristi Hunnes, Rocky Mountain College—Montana

Dave Johnson, Genesee Community College

George Keller, Samford University

Rachel Larsen, University of Southern Maine

Antje Lauer, California State University, Bakersfield

Shannon Mackey, St. Ambrose University

Matt Ruddy, Onondaga Community College

Zenda Rushing, Brunswick Community College

Josh Sharp, Northern Michigan University

John L. Sloyer Jr., Cape Cod Community College

Mary Ann Smith, Marywood University

Larry Weiskirch, Onondaga Community College

Matthew Wood, Lake Sumter State College

Ruth Wrightsman, Flathead Valley Community College

Daniel Zamzow, University of Wisconsin—Rock County

MICRO BIOLOGY

THE HUMAN EXPERIENCE

Second Edition

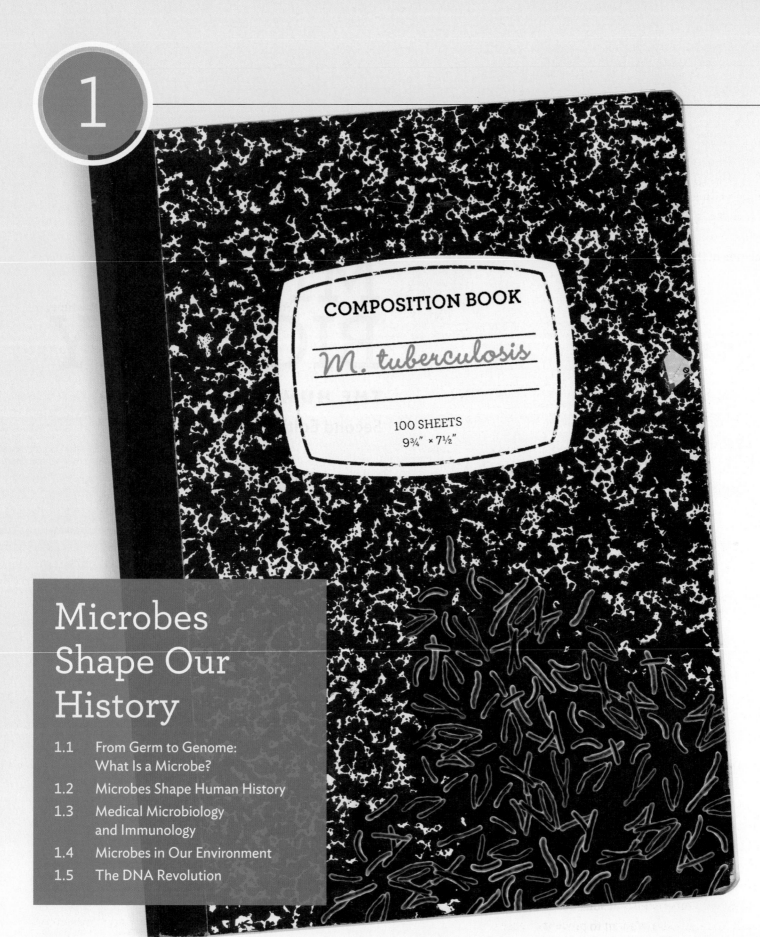

1

Microbes Shape Our History

Half a Lung Is Better than None

SCENARIO Debi was an ordinary teenager attending an affluent American public high school when she contracted tuberculosis (TB). She did not know the person who infected her. Infection requires inhalation of the causative bacteria.

SIGNS AND SYMPTOMS
For several weeks, Debi coughed all the time, felt tired, and was losing weight. Her coughing brought up blood. An X-ray revealed the signs of infection in her lung, including a large hole eaten away by the bacteria.

X-ray of Lung
The X-ray reveals infiltration into the lungs (white wisps) caused by *Mycobacterium tuberculosis,* and a hole where dead tissue was semiliquified.

DIAGNOSIS From Debi's sputum sample, a DNA sequence was amplified by PCR (polymerase chain reaction). The DNA sequence revealed *Mycobacterium tuberculosis*, the cause of TB. Doctors prescribed isoniazid and rifampin, antibiotics that kill most strains of *M. tuberculosis*. But Debi's TB strain proved resistant to nearly all known drugs (MDR-TB, multidrug-resistant TB).

TREATMENT Because drugs failed to eliminate the MDR strain, surgeons removed nearly half of Debi's right lung to help the antibiotics overcome the infection. Debi recovered and returned to high school. She would have to continue taking antibiotics for 20 months afterward. All the teachers and students in Debi's school were screened, and over 200 were found to have been infected by a student with tuberculosis misdiagnosed for 2 years. All required treatment to prevent disease.

Mycobacterium tuberculosis
Bacteria revealed by scanning electron microscopy.

The disease that threatened Debi, a healthy American teenager, was tuberculosis, an ancient killer. The cause is a **bacterium** (plural, **bacteria**), a single-celled microorganism lacking a nucleus, called *Mycobacterium tuberculosis. M. tuberculosis* can be found anywhere; for instance, in 2019, cases were reported in an Arizona prep school and at a university in Texas. Today, we fight this ancient disease by using modern **antibiotics**, chemicals that kill or slow growth of bacterial pathogens. But *M. tuberculosis* evolves to resist every antibiotic we use (discussed in Section 20.3). The emergence of multidrug-resistant (MDR) strains of disease-causing microbes is just one of the microbial challenges faced by health professionals. Despite all the advances of modern medicine and public health, tuberculosis and other microbial infections remain the world's leading cause of death in humans.

Yet bacteria that cause disease are surprisingly related to other kinds of bacteria that inhabit our environment without causing us harm—and actually enable us to live. All the plants in natural environments require nitrogen fixed by bacteria. Much of the oxygen we breathe is released by cyanobacteria and algae. And soil bacteria such as streptomycetes produce most of the antibiotics we use to kill disease-causing pathogens.

In this chapter we will introduce the concept of a microbe. We will then describe how humans discovered the roles of microbes in disease and in our environment. Finally, we will see how advances in microbial science, especially the discovery of the structure of DNA, have transformed medicine and society.

1.1 From Germ to Genome: What Is a Microbe?

Where did life come from? Life on Earth began early in our planet's history with microscopic organisms, or microbes. Microbial life has since shaped our atmosphere, our geology, and the energy cycles of all ecosystems. Some early microbes eventually evolved into multicellular plants and animals, including ourselves. Today, microbes generate the very air we breathe, including nitrogen gas and much of the oxygen and carbon dioxide. They fix (or combine) nitrogen into forms used by plants, and they make essential vitamins that we consume, such as vitamin B_{12}. Microbes are the primary producers of food webs, particularly in the oceans; when we eat fish, we indirectly consume tons of algae at the base of the food chain.

The human body contains as many microbes as it does human cells, including numerous bacteria on the skin and in the digestive tract. Throughout history, humans have had a hidden partnership with microbes ranging from food production and preservation to mining for precious minerals. Today, microbes serve as tools for biotechnology in fields from medicine to microscopic robots. Nevertheless, a small but critical proportion of all microbes are **pathogens**, the causative agents of disease. Diseases caused by pathogens, commonly called "germs," remain the principal cause of human mortality.

A Microbe Is a Microscopic Organism

A **microbe** is commonly defined as a living organism that requires a microscope to be seen. Microbial cells range in size from millimeters (mm) down to 0.2 micrometer (μm), and viruses may be tenfold smaller (Figure 1.1). Some microbes consist of a single **cell**. A cell is the smallest unit of life, composed of a membrane-enclosed compartment of water solution containing molecules that carry out metabolism.

Each microbe contains in its genome the capacity to reproduce its own kind. Microbes are found throughout our biosphere, from the superheated black smoker vents at the depths of the ocean floor to the subzero ice fields of Antarctica. Bacteria such as *Escherichia coli* live in our intestinal tract, whereas algae and cyanobacteria turn ponds green.

Our simple definition of a microbe, however, leaves us with contradictions. Most single-celled organisms require a microscope to render them visible, and thus they fit the definition of "microbe." Nevertheless, some protists and algae are large enough to be seen with the naked eye. An example is the ameba *Pelomyxa*, which can span several millimeters (Figure 1.1A). Some amebas can cause meningitis; others can harbor thousands of *Legionella* bacteria, the cause of legionellosis, a severe form of pneumonia.

Other kinds of microbes form complex multicellular assemblages, such as mycelia (multicellular filaments) and biofilms.

Figure 1.1 Representative Microbes

A. *Pelomyxa* sp., a large ameba.

B. *Escherichia coli* bacteria colonizing a stomate cell in a lettuce leaf (colorized scanning electron micrograph).

C. *Methanocaldococcus jannaschii*, an archaeon that produces methane.

D. Human papillomavirus, the cause of genital warts and cervical or penile cancer (model based on electron microscopy).

In a biofilm, cells are differentiated into distinct types that complement one another's function, as in multicellular organisms. On the other hand, some complex multicellular animals, such as mites and roundworms, require a microscope for us to see, but they are not considered microbes. Allied health courses may cover parasitic invertebrates under "microbiology" because these infectious agents are transmitted in a manner similar to that of disease-causing microbes. For example, hookworms infect about 700 million people worldwide, including some in US communities with poor sanitation.

Different Kinds of Microbes

Microbes, like other organisms, are classified as members of a **species** according to a shared set of genes and traits. The scientific name of the species, such *Staphylococcus epidermidis* for a common skin bacterium, consists of a capitalized genus name (*Staphylococcus*) and a lowercase species name (*epidermidis*), both italicized. In addition, members of a genus are often referred to informally by a romanized vernacular term, such as "staphylococci." The names of some microbial species are sometimes changed to reflect our new understanding of genetic relationships. For example, the causative agent of bubonic plague was formerly called *Bacillus pestis* (1900) and *Pasteurella pestis* (1923), but it is now called *Yersinia pestis* (since 1944). The older names, however, still appear in the literature.

Microbes are classified according to their genetic relatedness. The more closely related two organisms are, the more recently they diverged from a common ancestor. Relatedness is important for understanding how microbes respond to treatment. For example, an antibiotic used against an intestinal pathogen will also kill many beneficial bacteria that normally live in the intestine; consequently, the antibiotic may cause digestive problems. The degree of genetic relatedness between microbes is calculated by comparing DNA sequences in the **genome**—the total DNA sequence content of each organism. Genome comparison is now the fundamental basis for classifying all life forms.

A major trait distinguishing microbes is possession or lack of a membrane-enclosed nucleus. Microbes that lack a nuclear membrane are called **prokaryotes**, which include bacteria and archaea. Microbial **eukaryotes** (cells with a nucleus) include fungi, protozoa, and algae.

Bacteria are prokaryotic cells, usually 0.2–20 μm in size—or, one-tenth to one-hundredth the size of a sentence period. Different species may grow as single cells, as filaments (chains), or as communities with simple differentiated forms. An example is *Escherichia coli* (Figure 1.1B), a bacterial species that grows in the human intestine. Most strains of *E. coli* are harmless members of the microbial community that aid human digestion, but some strains cause acute gastroenteritis that may lead to kidney failure. Bacteria are found in every habitat of our biosphere, even several kilometers underground.

Archaea (singular, **archaeon**) are a genetically distinct group of prokaryotes that evolved by diverging from bacteria and eukaryotes more than 3 billion years ago. Some archaea are "extremophiles" that live in seemingly hostile environments, such as the boiling sulfur springs of Yellowstone. Other archaea are methanogens, whose metabolism releases methane (natural gas; Figure 1.1C). Methanogens are common in the gut of humans and animals, the source of the "gas" passed by one's intestinal tract. Their metabolism increases the efficiency of digestion. A remarkable feature of archaea is that none cause disease. The absence of pathogenesis (disease causation) in archaea is of great interest to medical researchers studying the cause and prevention of disease.

Eukaryotic microbes include **protozoa** (singular, **protozoan**), which are motile heterotrophs (consuming organic food), usually single-celled. A protozoan such as an ameba (see Figure 1.1A) may be free-living or parasitic. **Algae** (singular, **alga**) are eukaryotic microbes containing chloroplasts that conduct photosynthesis. Algae form an essential base of the food web, although overgrowth causes "algal blooms" that poison fish. Protozoa and algae together are classified as **protists**. Distinct from protists are **fungi** (singular, **fungus**), heterotrophic organisms that are usually nonmotile and grow by absorbing nutrients from their surroundings. Fungi may grow as single cells (yeast) or as filaments (bread mold), or they may form complex structures such as mushrooms. Some fungi cause infections, especially in people with a depressed immune system.

Eukaryotic microbial pathogens are also called **parasites**. Parasites are organisms that live at the expense of a host they inhabit, debilitating the host. By convention, the word "parasite" is used for both single-celled and multicellular eukaryotes. Multicellular parasites include, for example, worms and mites (presented in Chapter 11).

Viruses are noncellular microbes. A virus particle contains genetic material (DNA or RNA) that takes over the metabolism of a cell to generate more virus particles. Some viruses, such as papillomaviruses (Figure 1.1D), consist of only a few molecular parts. Other viruses, such as herpesviruses, show complexity approaching that of a cell, although none are fully functional cells. Engineered viruses are used as tools for gene therapy. For example, in 2017 the Food and Drug Administration (FDA) approved a nonpathogenic derivative of the human immunodeficiency virus (HIV) for gene delivery to the white blood cells of a child, enabling the child's immune system to overcome leukemia.

SECTION SUMMARY

- **Microbes are microscopic**; that is, they are organisms too small to be seen without a microscope. Different species of microbes grow as single cells, in filaments, in biofilms, or in simple differentiated structures.

- **Bacteria** are cells lacking a nucleus (prokaryotes). Bacteria grow in all habitats. Most human-associated species are harmless, but some cause disease.

- **Archaea** are nonbacterial cells that lack a nucleus (prokaryotes) and are distantly related to other microbes. Methanogens live in the human intestine (among other places), where their metabolism releases methane. No archaea cause disease.

- **Eukaryotic microbes** include protists (protozoa and algae) and fungi. Parasitic protozoa and fungi may infect humans.

- **Viruses** are noncellular microbes that must infect a host cell.

1.2 Microbes Shape Human History

SECTION OBJECTIVES

- Explain how microbial diseases have changed human history.
- Describe how microbes participate in human cultural practices such as production of food and drink.

Have microbes changed the course of human history? Until microscopes were developed in the seventeenth century, we humans were unaware of the unseen living organisms that surround us, that float in the air we breathe and the water we drink, and that inhabit our own bodies. Yet microbes have molded human culture since our earliest civilizations. Yeasts and bacteria made foods such as bread and cheese, as well as alcoholic beverages (Figure 1.2A). "Rock-eating" bacteria known as **lithotrophs** leached copper and other metals from ores exposed by mining, enabling ancient human miners to obtain these metals. Today, bacterial leaching produces about 20% of the world's copper. Unfortunately, microbial acid consumes the stone of ancient monuments (Figure 1.2B), a process intensified by airborne acidic pollution.

As we humans became aware of microbes, our relationship with the microbial world changed in important ways (Table 1.1). Early scientists in the seventeenth and eighteenth centuries used microscopes to discover the existence of microbes and revealed their means of reproduction and death. In the nineteenth century, the "golden age" of microbiology, key principles of disease pathology and microbial ecology were established that scientists still use today. This period laid the foundation for modern science, in which genetics and molecular biology provide powerful tools for scientists to manipulate microorganisms for medicine and industry.

Microbial Diseases Devastate Human Populations

Historians traditionally emphasize the role of warfare in shaping human history. Yet throughout history, more soldiers have died of microbial infections than of wounds in battle. Microbes have often determined the fate of human societies. For example, smallpox, carried by European invaders, exterminated much of the native population of North America.

Throughout history, microbial diseases such as tuberculosis and leprosy have profoundly affected human demographics and cultural practices (Figure 1.3). The bubonic plague, which wiped out a third of Europe's population in the fourteenth century, was caused by *Yersinia pestis*, a bacterium spread by rat fleas.

Figure 1.2   Production and Destruction by Microbes

A. Roquefort cheeses ripening in France.

B. Statue at the Cathedral of Cologne, in Germany, undergoing decay from the action of lithotrophic microbes. Acid rain accelerates the process.

Table 1.1
Microbes and Human History

Date	Microbial Discovery	Discoverer(s)
Microbes Impact Human Culture without Detection		
10,000 BCE	Food and drink are produced by microbial fermentation.	Egyptians, Chinese, and others
1500 BCE	Tuberculosis, polio, leprosy, and smallpox are evident in mummies and tomb art.	Egyptians
50 BCE	Copper is recovered from mine water acidified by sulfur-oxidizing bacteria.	Roman metal workers under Julius Caesar
1000 CE	Smallpox immunization is accomplished by transfer of secreted material.	Chinese and Africans
1025 CE	Diseases are observed to be contagious. The basis of hygiene and quarantine is proposed.	Avicenna, or Ibn Sina (Persia)
1300–1400 CE	The Black Death (bubonic plague) kills 17 million people in Europe and Asia.	Catherine of Siena nursed plague victims; canonized as patron saint of nurses
1546 CE	Syphilis and other diseases are observed to be contagious.	Girolamo Fracastoro (Padua)
Early Microscopy and the Origin of Microbes		
1676	Microbes are observed under a microscope.	Antonie van Leeuwenhoek (Netherlands)
1717	Smallpox is prevented by inoculation of pox material, a rudimentary form of immunization.	Africa and Asia; Turkish women taught Lady Montagu, who brought the practice to England
1765	Microbes fail to grow after boiling in a sealed flask; evidence against spontaneous generation.	Lazzaro Spallanzani (Padua)
1798	Cowpox vaccination prevents smallpox.	Edward Jenner (England)
1835	Fungus causes disease in silkworms (first pathogen to be demonstrated in animals).	Agostino Bassi de Lodi (Italy)
1847	Chlorine as antiseptic wash for doctors' hands decreases pathogens.	Ignaz Semmelweis (Hungary)
1881	Bacterial spores survive boiling but are killed by cyclic boiling and cooling.	John Tyndall (Ireland)
"Golden Age" of Microbiology		
1855	Poor sanitation leads to mortality (Crimean War).	Florence Nightingale (England)
1866	Microbes are defined as a class distinct from animals and plants.	Ernst Haeckel (Germany)
1867	Antisepsis during surgery prevents patient death.	Joseph Lister (England)
1857–1881	Microbial fermentation produces lactic acid or alcohol. Microbes fail to appear spontaneously, even in the presence of oxygen. The first artificial vaccine is developed (against anthrax).	Louis Pasteur (France)
1877–1884	Bacteria are a causative agent in developing anthrax. First pure culture of colonies on solid medium, *Mycobacterium tuberculosis*. Koch's postulates demonstrate the microbial cause of a disease (anthrax and tuberculosis).	Robert Koch (Germany)
1884	Gram stain is devised to distinguish bacteria from human cells.	Hans Christian Gram (Denmark)
1886	Intestinal bacteria include *Escherichia coli*, the future model organism.	Theodor Escherich (Austria)
1889	Bacteria oxidize iron and sulfur (lithotrophy).	Sergei Winogradsky (Russia)
1889–1899	Bacteria isolated from root nodules fix nitrogen. The concept of a virus is proposed to explain tobacco mosaic disease.	Martinus Beijerinck (Netherlands)

(continued on next page)

Table 1.1
Microbes and Human History (*continued*)

Date	Microbial Discovery	Discoverer(s)
Biochemistry, Genetics, and Medicine		
1900	Yellow fever is shown to be transmitted by mosquitoes.	Walter Reed (USA) and Carlos Finlay (Cuba)
1908	Antibiotic chemicals are synthesized and identified (chemotherapy).	Paul Ehrlich (USA)
1911	Cancer in chickens can be caused by a virus.	Francis Peyton Rous (USA)
1917	Bacteriophages are recognized as viruses that infect bacteria.	Frederick Twort (England) and Félix d'Herelle (France)
1918	Influenza A pandemic kills 50 million people worldwide.	
1928	*Streptococcus pneumoniae* bacteria are transformed by a genetic material from dead cells.	Frederick Griffith (England)
1929	Penicillin, the first widely successful antibiotic, is made by a fungus. The molecule is isolated in 1941.	Alexander Fleming (Scotland), Howard Florey (Australia), and Ernst Chain (Germany)
1937	The tricarboxylic acid cycle is discovered.	Hans Krebs (England)
1938	The microbial "kingdom" is subdivided into prokaryotes and eukaryotes.	Herbert Copeland (USA)
1938	*Bacillus thuringiensis* spray is produced as the first bacterial insecticide.	Insecticide manufacturers (France)
1941	One gene encodes one enzyme in *Neurospora*.	George Beadle and Edward Tatum (USA)
1941	Poliovirus is grown in human tissue culture.	John Enders, Thomas Weller, and Frederick Robbins (USA)
1944	DNA is the genetic material that transforms *S. pneumoniae*.	Oswald Avery, Colin MacLeod, and Maclyn McCarty (USA)
1946	Bacteria transfer DNA by conjugation.	Edward Tatum and Joshua Lederberg (USA)
1951	Transposable elements are discovered in maize and later shown in bacteria, where they play key roles in evolution.	Barbara McClintock (USA)
1952	DNA is injected into a cell by a bacteriophage.	Martha Chase and Alfred Hershey (USA)
1946–1956	X-ray diffraction crystal structures are obtained for the first complex biological molecules: penicillin and vitamin B_{12}.	Dorothy Hodgkin, John Bernal, and co-workers (England)
Molecular Biology and Medicine		
1953	The overall structure of DNA is a double helix, based on X-ray crystallography.	Rosalind Franklin and Maurice Wilkins (England)
1953	Double-helical DNA consists of antiparallel chains connected by the hydrogen bonding of AT and GC base pairs.	James Watson (USA) and Francis Crick (England)
1959	Expression of the messenger RNA for the *E. coli lac* operon is regulated by a repressor protein.	Arthur Pardee (England); François Jacob and Jacques Monod (France)
1960	Radioimmunoassay for detection of biomolecules is developed.	Rosalyn Yalow and Solomon Berson (USA)
1961	The chemiosmotic theory, which states that biochemical energy is stored in a transmembrane proton gradient, is proposed and tested.	Peter Mitchell and Jennifer Moyle (England)
1966	The genetic code by which DNA information specifies protein sequence is deciphered.	Marshall Nirenberg, H. Gobind Khorana, and others (USA)
1967	Bacteria can grow at temperatures above 80°C (176°F) in hot springs at Yellowstone National Park.	Thomas Brock (USA)

(continued on next page)

Table 1.1
Microbes and Human History (*continued*)

Date	Microbial Discovery	Discoverer(s)
1968	Serial endosymbiosis explains the evolution of mitochondria and chloroplasts.	Lynn Margulis (USA)
1969	Retroviruses contain reverse transcriptase, which copies RNA to make DNA.	Howard Temin, David Baltimore, and Renato Dulbecco (USA)
1953–1971	Oral rehydration therapy (ORT) is developed for dehydration due to diarrhea and cholera, saving millions of lives in developing countries.	Hemendra Chatterjee and Dilip Mahalanabis (India); Robert Phillips (USA)
1972	Artemisinin, the antimalarial agent, is discovered from an ancient Chinese herb.	Tu Youyou and co-workers (China)
1973	A recombinant DNA molecule is made in vitro (in a test tube).	Annie Chang, Stanley Cohen, Robert Helling, and Herbert Boyer (USA)
1974	The bacterial flagellum is driven by a rotary motor.	Howard Berg, Michael Silverman, and Melvin Simon (USA)
1975	The dangers of recombinant DNA are assessed at the Asilomar Conference.	Paul Berg, Maxine Singer, and others (USA)
1977	A DNA sequencing method is invented and used to sequence the first genome of a virus.	Frederick Sanger, Walter Gilbert, and Allan Maxam (USA)
1977	Archaea are a third domain of life, the others being eukaryotes and bacteria.	Carl Woese (USA)
1978	Biofilms are a major form of existence of microbes.	J. William Costerton and others (Canada)
1979	Smallpox is declared eliminated—the culmination of worldwide efforts of immunology, molecular biology, and public health.	World Health Organization
Genomics and Medicine		
1981	The polymerase chain reaction (PCR) makes available large quantities of DNA.	Kary Mullis (USA)
1982	Uncultured bacteria contribute to ecology and pathology.	Rita Colwell, Norman Pace, and others (USA)
1982	Prions, infectious agents consisting solely of protein, are characterized.	Stanley Prusiner (USA)
1981–1983	AIDS pandemic begins (continues into present). Human immunodeficiency virus (HIV) is discovered as the cause of AIDS.	Françoise Barré-Sinoussi and Luc Montagnier (France), Robert Gallo (USA), and others
1984	*Helicobacter pylori*, the cause of gastritis, is discovered in the stomach.	Barry Marshall and Robin Warren (Australia)
1988	Molecular Koch's postulates are devised to show that a gene found in a pathogenic microorganism contributes to the disease caused by the pathogen.	Stanley Falkow (USA)
1993	Gene therapy using a vector derived from HIV succeeds in treating severe combined immunodeficiency disorder (SCID).	Donald Kohn and others (USA)
1995	The first bacterial genome is sequenced: *Haemophilus influenzae*.	Craig Venter, Hamilton Smith, Claire Fraser, and others (USA)
1988–2013	Fecal microbiota transplant cures intestinal infections of drug-resistant *Clostridioides difficile* (CDI).	Thomas Borody and others (Australia and USA)
2014	Human Microbiome Project releases first compilation of microbes associated with healthy human bodies.	National Institutes of Health (USA)
2019	CRISPR gene editing is used to treat human patients for multiple myeloma, a cancer of white blood cells.	University of Pennsylvania
2019–2020	COVID-19 coronavirus is discovered; causes pandemic respiratory illness.	Li Wenliang (China)

Figure 1.3 Microbial Disease in History and Culture

A. Medieval church procession to ward off the Black Death (bubonic plague).

B. The AIDS Memorial Quilt spread before the Washington Monument. Each panel of the quilt memorializes an individual who died of AIDS.

Ironically, the plague-induced population decline enabled the social transformation that led to the Renaissance, a period of unprecedented cultural advancement. In the nineteenth century, the bacterium *Mycobacterium tuberculosis* stalked overcrowded cities, and tuberculosis became so common that the pallid appearance of tubercular patients became a symbol of tragic youth in European literature. Today, societies throughout the world have been profoundly shaped by the pandemic of acquired immunodeficiency syndrome (AIDS), caused by the human immunodeficiency virus (HIV); in 2019, more than 38 million people worldwide were living with HIV.

Before the nineteenth century, the role of microbes as infectious agents was unknown. But people had a sense that those suffering from diseases such as plague and leprosy were to be avoided. The rare individuals who chose to nurse such people were considered spiritual heroes. A prominent example is Catherine of Siena (1347–1380), who served God by nursing the sick (**Figure 1.4**). When a wave of bubonic plague swept the Italian city of Siena, Catherine and her followers stayed to care for the ill and bury the dead. After her death, Catherine of Siena was canonized and became known as the patron saint of nurses. In modern times, nursing continues to require personal courage—for instance, during epidemics of influenza and AIDS, when ill individuals have often been abandoned by their communities.

Figure 1.4 Catherine of Siena

Catherine of Siena nursed victims of plague and leprosy. She became known as the patron saint of nurses.

Microscopes Reveal the Microbial World

The seventeenth century was a time of growing inquiry and excitement about the "natural magic" of science and patterns of our world, such as the laws of physics and chemistry formulated by Isaac Newton (1642–1727) and Robert Boyle (1627–1691). Physicians attempted new treatments for disease involving the application of "stone and minerals" (that is, chemicals); today, we would call this chemotherapy. Minds were open to consider the astounding possibility that our surroundings, indeed our very bodies, were inhabited by tiny living beings.

ROBERT HOOKE OBSERVES THE MICROSCOPIC WORLD The first microscopist to publish a systematic study of the world as seen under a microscope was Robert Hooke (1635–1703). As curator of experiments for the Royal Society of London, Hooke built a compound microscope, a magnifying instrument containing two or more lenses that multiply their magnification in series. With his microscope, Hooke observed biological materials such as nematodes ("vinegar eels"), mites, and mold filaments. He then published his illustrations of these materials in *Micrographia* (1665), the first publication that illustrated objects observed under a microscope (**Figure 1.5**).

Hooke was the first to observe distinct units of living material, which he called "cells." Hooke first named the units cells because the shape of hollow cell walls in a slice of cork reminded him of the shape of monks' cells in a monastery. But his crude lenses achieved at best 30-fold power (30×), so he never observed single-celled microbes such as bacteria.

ANTONIE VAN LEEUWENHOEK OBSERVES BACTERIA WITH A SINGLE LENS Robert Hooke's *Micrographia* inspired other microscopists, including Antonie van Leeuwenhoek (1632–1723; **Figure 1.6A**). Leeuwenhoek became the first individual to observe single-celled microbes. As a young man, Leeuwenhoek lived in the

Figure 1.5
Robert Hooke's *Micrographia*

An illustration of mold sporangia, drawn by Hooke in 1665, from his observations of objects with a compound microscope.

Dutch city of Delft, where he worked as a cloth draper, a profession that introduced him to magnifying glasses. (The magnifying glasses were used to inspect the quality of the cloth, enabling the worker to count threads.) Later in life, Leeuwenhoek took up the hobby of grinding ever-stronger lenses to see into the world of the unseen.

He ground lenses stronger than Hooke's, which he used to build single-lens magnifiers, complete with sample holder and focus adjustment (Figure 1.6B). Leeuwenhoek observed first insects, including lice and fleas; then the relatively large single cells of protists and algae; and finally bacteria. One day he used his microscope to observe matter extracted from between his teeth. He wrote, "To my great surprise [I] perceived that the aforesaid matter contained very many small living Animals, which moved themselves very extravagantly."

Leeuwenhoek recorded page after page on the movement of microbes, reporting their size and shape so accurately that we can often determine the species he observed (Figure 1.6C). He performed experiments, comparing, for example, the appearance of "small animals" from his teeth before and after drinking hot coffee. The disappearance of microbes from his teeth after drinking a hot beverage suggested that heat kills microbes—a profoundly important principle for the study and control of microbes ever since.

Historians have often wondered why it took so many centuries for Leeuwenhoek and his successors to determine the link between microbes and disease. The very ubiquity of microbes—most of them harmless—may have obscured the deadlier roles of pathogens (disease-causing microbes). In addition, it was hard for the microscopist to distinguish between microbes and the single-celled components of the human body, such as blood cells and sperm. It was not until the nineteenth century that human tissues could be distinguished from microbial cells by the application of differential chemical stains. Microscopy is discussed further in Chapter 3.

Spontaneous Generation: Do Microbes Have Parents?

The observation of microscopic organisms led priests and philosophers to wonder where these organisms came from. In the eighteenth century, scientists and church leaders intensely debated the question of **spontaneous generation**, the theory that living microbes can arise spontaneously, without parental organisms. Some chemists supported spontaneous generation, arguing that microbes appear the same way chemicals precipitate from a solution. Christian church leaders, however, argued the biblical view that all organisms have "parents" going back to the first week of creation.

The Italian priest Lazzaro Spallanzani (1729–1799) sought to disprove the spontaneous generation of microbes. Spallanzani showed that a sealed flask of meat broth sterilized by boiling failed to grow microbes. The priest also noticed that microbes often appeared in pairs. Were these two parental microbes coupling to produce offspring, or did one microbe become two? Through long observation, Spallanzani watched a single microbe grow until it split in two. Thus he demonstrated cell fission, the process by which cells arise by the splitting of preexisting cells.

Figure 1.6 Antonie van Leeuwenhoek

A. Portrait of Leeuwenhoek, the first person to observe individual microbes.

Leeuwenhoek microscope (circa late 1600s)

Lens
Sample holder
Focus knob
Sample mover

B. "Microscope" (magnifying glass) used by Leeuwenhoek.

C. Spiral bacteria viewed through a replica of Leeuwenhoek's instrument.

Spallanzani's experiments, however, did not put the matter to rest. Supporters of spontaneous generation argued that the microbes in the priest's flask lacked access to oxygen and therefore could not grow. The pursuit of this question was left to future microbiologists, including the famous French microbiologist Louis Pasteur (1822–1895; **Figure 1.7A**). In addressing spontaneous generation and related questions, Pasteur and his contemporaries laid the foundations for modern microbiology.

Pasteur began his career as a chemist. He was asked to help with a chemical problem encountered by French manufacturers of wine and beer. Alcoholic beverages are made by **fermentation**, a process by which microbes gain energy by converting sugars to alcohol. In the time of Pasteur, however, the role of microbes was unknown. The conversion of grapes or grain to alcohol was believed to be a spontaneous chemical process. No one could explain why some fermentation mixtures produced vinegar (acetic acid) instead of alcohol. Pasteur discovered that fermentation is actually caused by living yeast, a single-celled fungus. In the absence of oxygen, yeast produces alcohol as a terminal waste product. But when the yeast culture is contaminated with bacteria, the bacteria outgrow the yeast and produce acetic acid instead of alcohol.

Pasteur's work on fermentation led him to test a key claim made by proponents of spontaneous generation. The proponents claimed that Spallanzani's failure to find spontaneous appearance of microbes was due to a lack of oxygen. From his studies of yeast fermentation, Pasteur knew that some microbial species do not require oxygen for growth. So, he devised an unsealed flask with a long, bent "swan neck" that admitted air but kept the boiled contents free of microbes (**Figure 1.7B**). The famous swan-necked flasks remained free of microbial growth for many years. Yet, when a flask was tilted to enable contact of broth with microbe-containing dust, microbes immediately grew. As a result, Pasteur disproved

that lack of oxygen was the reason for the failure of spontaneous generation in Spallanzani's flasks.

But even Pasteur's work did not prove that microbial growth requires preexisting microbes. The Irish scientist John Tyndall (1820–1893) attempted the same experiment as Pasteur but sometimes found the opposite result. Tyndall found that the broth sometimes gave rise to microbes, no matter how long it was sterilized by boiling. The microbes appeared because some kinds of organic matter, particularly hay infusion (broth of dried grass, boiled), are contaminated with a heat-resistant form of bacteria called endospores (or spores). The spore form can be eliminated only by repeated cycles of boiling and resting, in which the spores germinate to the growing, vegetative form that is killed at 100°C (212°F).

It was later discovered that endospores could be killed by boiling under pressure, as in a pressure cooker, which generates higher temperatures than can be obtained at atmospheric pressure. The use of a steam pressure device called the **autoclave** became a standard way to sterilize materials required for the controlled study of microbes and for surgical procedures. (Microbial control and antisepsis are discussed further in Chapter 13.)

The Origin of Life

If all life on Earth shares descent from a microbial ancestor, how did the first microbe arise? Although spontaneous generation has been discredited as a continual source of microbes, at some point in the past the first living organisms must have originated from nonliving materials. The earliest fossil evidence of cells in the geological record appears in sedimentary rock that formed over 2 billion years ago.

The components of the first living cells may have formed from spontaneous reactions sparked by ultraviolet absorption or electrical discharge. Such "early-Earth" conditions were simulated in 1953 during famous experiments by chemist Stanley Miller (1930–2007). Miller combined hydrogen gas, methane, and ammonia and applied an electrical discharge (comparable to a lightning strike), which generated simple amino acids such as glycine and alanine. Similar experiments conducted in 1961 by Spanish-American researcher Juan Oró (1923–2004) combined hydrogen cyanide and ammonia under electrical discharge to obtain adenine, a fundamental component of DNA and of the energy carrier adenosine triphosphate (ATP). These small organic molecules are also found in meteorites. Thus, chemical reactions both on Earth and in outer space could have generated the fundamental components of life—but how they assembled into living cells remains a mystery.

Figure 1.7 Louis Pasteur, Founder of Medical Microbiology and Immunology

A. Pasteur's contributions to the science of microbiology and immunology earned him lasting fame.

Open to air

S curve excludes dust and microbes

Growth medium

B. Pasteur's swan-necked flask. After boiling, the contents in the flask remain free of microbial growth, despite access to air.

SECTION SUMMARY

- **Microbes affected human civilization** for centuries, long before humans guessed at their existence, through their contributions to our environment, food and drink production, and infectious diseases.

- **Robert Hooke and Antonie van Leeuwenhoek** were the first to record observations of microbes through simple microscopes.

- **Spontaneous generation** is the idea that microbes arise spontaneously, without parental organisms.

- **Lazzaro Spallanzani** showed that microbes arise from preexisting microbes and demonstrated that heat sterilization can prevent microbial growth.
- **Louis Pasteur** discovered the microbial basis of fermentation. He also showed that supplying oxygen does not enable spontaneous generation.
- **John Tyndall** showed that repeated cycles of heat are necessary to eliminate spores formed by certain kinds of bacteria.
- **All life evolved from microbial cells.** Simple organic molecules can form out of inorganic chemicals, but how they assembled into the first cells is unknown.

1.3 Medical Microbiology and Immunology

SECTION OBJECTIVES

- Define the germ theory of disease.
- Explain how Florence Nightingale first drew a statistical correlation between infectious disease and human mortality.
- Explain how Koch's postulates can show that a specific kind of microbe causes a disease. Explain the problems in interpreting Koch's postulates in practice.

How did medical workers first figure out the connection between microbes and disease? As early as the eleventh century, the Persian physician and philosopher Avicenna (Ibn Sina) discovered the role of contagion in sexually transmitted diseases and established the principle of quarantine. Over the centuries, thoughtful observers, such as the Venetian physician Girolamo Fracastoro in the sixteenth century, noted a connection between disease and some kind of transmissible entity. Early physicians, however, could never see the actual agent of transmission. In the eighteenth century, researchers combined the tools of microscopy, microbial culture, and statistical analysis to develop the **germ theory of disease**. The germ theory holds that specific diseases are caused by specific kinds of microbes.

Linking Infectious Disease with Mortality

The significance of disease in warfare and other conditions of overcrowding, such as cities, was first recognized by the British nurse and statistician Florence Nightingale (1820–1910; **Figure 1.8A**). Born into a wealthy British family, Nightingale rejected her family's expectations for her future and chose a career as a nurse. At the time, nursing was not well codified as a professional discipline.

In 1859, Nightingale published *Notes on Nursing*, the first major textbook for nursing instruction. She raised funds to establish the Nightingale Home and Training School for Nurses at St. Thomas' Hospital, which is now called the Florence Nightingale Faculty of Nursing, Midwifery, and Palliative Care and is part of King's College London.

Nightingale also founded the science of medical statistics. Before Nightingale, health professionals lacked the analytical tools to calculate the proportions of populations that succumbed to infectious diseases. Nightingale used methods invented by French statisticians to quantify the role of disease in population mortality. She first applied her methods to demonstrate the high mortality rate due to disease among British soldiers during the Crimean War. In 1854, Nightingale arrived at the war hospital with a group of nurses seeking to improve the soldiers' conditions. Initially, their main aim was to improve nutrition and decrease overwork of the soldiers.

To Nightingale's surprise, however, the death rates continued to rise. To represent the deaths of soldiers due to various causes, Nightingale devised the "polar area chart" (**Figure 1.8B**). The area of each wedge, measured from the center, represents the proportion of deaths due to a particular cause: blue wedges represent deaths due to infectious diseases such as typhus and cholera; red wedges represent deaths due to wounds; and black wedges represent all other causes of death. After Nightingale compiled the data and made such charts, she discovered that infectious disease, rather than poor nutrition, accounted for more than half of all mortality. The death rates due to disease were highest in the summer months, when the pathogens multiply fastest. Her statistics persuaded the British government to improve army

Figure 1.8 Florence Nightingale, Founder of Medical Statistics

A. Florence Nightingale was the first to use medical statistics to demonstrate the significance of mortality due to disease.

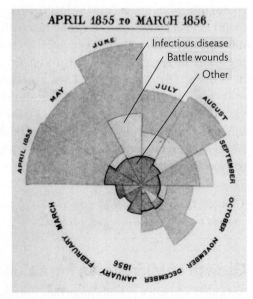

B. Nightingale's polar area chart of mortality data during the Crimean War.

Figure 1.9 Epidemiological Chart of the H1N1 Influenza Epidemic

The number of people in the United States hospitalized for complications related to H1N1 influenza, and deaths reported, from August 30, 2009, to April 3, 2010.

hygiene and to upgrade the standards of army hospitals. So, for the first time, a statistical analysis led to improved health policies.

In epidemiology today, statistical analysis continues to serve as a crucial tool in determining the causes of disease. Assessing the role of infectious diseases in the health of large populations is now a major field of service known as **public health**. Public health is monitored and managed by government agencies, most prominently the Centers for Disease Control and Prevention (CDC), based in Atlanta. For example, the CDC tracks epidemics of seasonal influenza and tries to predict which influenza strains will require vaccination in a given year. **Figure 1.9** plots the course of the H1N1 influenza epidemic in 2009. During the epidemic, these data were tracked daily across the country and were used to allocate supplies of vaccine and antiviral drugs.

Thought Question 1.1   Why do you think it took so long for humans to connect microbes with infectious disease?

Growth of Microbes in Pure Culture

Although statistics reveal important correlations, diagnosis for a patient requires more direct evidence that a given microbe causes a given disease.

CASE HISTORY 1.1

Sickened by Dead Cattle

In 2000, on a farm in North Dakota, 67-year-old Caleb helped bury five cows that had died of anthrax. Wearing heavy leather gloves, Caleb placed chains around the heads and hooves of the carcasses and moved them to the burial site. Four days later, he noticed a small lump on his neck. Over 2 days the lump enlarged, and a lesion opened. Caleb then sought medical attention. The physician reported a firm, superficial nodule surrounded by a purple ring, with an overlying black eschar (piece of dead tissue sloughed from the skin; **Figure 1.10A**). The eschar is a hallmark symptom of cutaneous (affecting skin) anthrax. The physician prescribed ciprofloxacin, the standard antibiotic for cutaneous anthrax. Testing the patient's serum with a bacterial antigen revealed the presence of antibodies, confirming the diagnosis of anthrax. The ciprofloxacin was continued, and the patient slowly improved over several weeks.

The patient in this case was infected by the bacterium *Bacillus anthracis* (**Figure 1.10B**), a well-known veterinary pathogen that is common in soil and causes anthrax outbreaks among livestock. Anthrax bacteria produce tough, drought-resistant forms called

Figure 1.10 Anthrax Infection

B. *B. anthracis* (scanning electron micrograph, colorized). Yellow indicates growing cells; pink indicates a developing endospore.

A. Cutaneous anthrax; skin nodule infected by *Bacillus anthracis*.

endospores that can remain viable for decades in soil. Human infection is rare, but it does occur, as Caleb found out. Caleb was fortunate that the infection in his case remained limited to the skin (cutaneous anthrax) and was not a respiratory infection, which becomes systemic and progresses rapidly with high mortality. Respiratory anthrax is of concern to the government for use in bioterrorism because specially treated endospores can survive dry shipment through the mail. In 2001, mailed anthrax spores contaminated post offices throughout the northeastern United States, as well as an office building of the US Senate, causing several deaths.

In the nineteenth century, without antibiotics, anthrax was a scary business for farmers. The disease seemed to appear out of nowhere when the spores came up in plowed soil; the animal deaths threatened the livelihood of communities, and farmers became infected and died. Thanks to the German physician Robert Koch (1843–1910; **Figure 1.11**), anthrax became the first infectious disease for which the scientific method established the microbial cause.

As a college student, Koch had conducted biochemical experiments on his own digestive system. Koch's curiosity about the natural world led him to develop principles and techniques crucial to modern microbial investigation, including the pure-culture technique and the famous Koch's postulates for identifying the causative agent of a disease. He applied his methods to many lethal diseases around the world, including anthrax and tuberculosis in Europe, malaria in Africa and the East Indies, and bubonic plague in India.

Unlike Pasteur, who was a university professor, Koch took up a medical practice in a small Polish-German town. To make space in his home for a laboratory to study anthrax and other deadly diseases, his wife curtained off part of his patients' examining room. Anthrax interested Koch because its epidemics in sheep and cattle caused economic hardship.

To investigate whether anthrax was a transmissible disease, Koch used blood from an anthrax-infected carcass to inoculate a rabbit. When the rabbit died, he used the rabbit's blood to inoculate a second rabbit, which then died in turn. The blood of the unfortunate animal had turned black with long, rod-shaped cells of the bacterium *Bacillus anthracis*. Upon introduction of these bacteria into healthy animals, the animals became ill with anthrax. Accordingly, Koch demonstrated an important principle of epidemiology—the **chain of infection**, or transmission, of a disease. In retrospect, his choice of anthrax was fortunate, for the microbes that cause it generate disease very quickly, multiply in the blood to an extraordinary concentration, and remain infective outside the body for long periods.

Koch and his co-workers then applied their experimental logic and culture methods to a more challenging disease: tuberculosis. In Koch's day, tuberculosis caused one-seventh of reported deaths in Europe; today, tuberculosis bacteria continue to infect millions of people worldwide (see **IMPACT**). Koch's approach to anthrax, however, was less applicable to tuberculosis, a disease that develops

Figure 1.11 Robert Koch, Founder of the Scientific Method of Microbiology

A. Robert Koch as a university student.

B. Koch's sketch of anthrax bacilli in mouse blood.

C. Koch (center, standing) investigated malaria in New Guinea in 1899.

impact

Tuberculosis: From Mummies to Multidrug Resistance

Tuberculosis is an ancient human disease. Traces of the causative bacterium, *Mycobacterium tuberculosis*, are found in Egyptian mummies 4,000 years old. The mummified lungs show lesions characteristic of the disease. During the seventeenth and eighteenth centuries, tuberculosis caused up to 30% of the deaths in Europe. Famous individuals who died of tuberculosis include Anton Chekhov and George Orwell.

As nutrition and housing standards improved during the twentieth century, people had healthier bodies to resist infection, and the rate of tuberculosis began to decline. In 1944, the first antibiotic against mycobacteria, streptomycin, was used to cure a tuberculosis patient who was near death. Antibiotics largely eliminated tuberculosis from developed countries. But in poorer countries throughout Asia and Africa, tuberculosis intensified with growing populations. New strains of the bacteria developed resistance to antibiotics. And the rise of AIDS accelerated the spread of tuberculosis through patients with impaired immune systems. Today, all countries are at risk for the epidemic spread of tuberculosis—including strains resistant to all known antibiotics.

How do bacteria infect a person and cause tuberculosis? The *M. tuberculosis* bacteria have one of the thickest and chemically complex cell envelopes of any bacterium, enabling their survival for days in air or water. By talking or sneezing, a person with tuberculosis spreads droplets of mucus and saliva into the air. Another person can inhale these droplets, even days later. Fewer than ten bacteria may be enough to establish infection. The inhaled bacteria reach the alveoli, tiny sacs of

Figure 1 Tuberculosis

A. *Mycobacterium tuberculosis* bacteria are engulfed by a human macrophage, a white blood cell (scanning electron micrograph, colorized).

SEM 1 μm

B. Patient being treated for tuberculosis.

slowly after many years of dormancy. Furthermore, the causative bacterium, *Mycobacterium tuberculosis*, is small and difficult to detect in human tissue or distinguish from different bacteria of similar appearance associated with the human body. How could Koch prove that a particular bacterium causes a particular disease?

What he needed was to isolate a **pure culture** of microorganisms—a culture grown from a single "parental" cell. Previous researchers had achieved pure cultures by the laborious process of serially diluting suspended bacteria. In serial dilution, each tube is diluted tenfold into sterile medium; then the diluted culture is diluted another tenfold into the next tube, and so on until a culture tube contains only a single cell. Alternatively, inoculation of a solid surface such as a sliced potato could produce isolated **colonies** (sing., **colony**). Colonies are distinct populations of bacteria, each grown from a single cell. For *M. tuberculosis,* Koch inoculated serum, which then formed a solid gel after heating. Later he refined

the solid-substrate technique by adding gelatin to a defined liquid medium, which could then be chilled to form a solid medium in a glass dish. A colleague, Julius Richard Petri (1852–1921), invented a covered version of this dish called a **Petri dish** (or Petri plate). The Petri dish consists of a round dish with vertical walls covered by an inverted dish of slightly larger diameter. Today, the Petri dish, generally made of disposable plastic, remains an indispensable part of the microbiological laboratory.

Another improvement in solid-substrate culture was the replacement of gelatin with materials that remain solid at higher temperatures, such as the gelling agent **agar** (a sugar polymer). The use of agar was recommended by Angelina Hesse (1850–1934), a microscopist and illustrator, to her husband, Walther Hesse (1846–1911), a young medical colleague of Koch (**Figure 1.12**). Agar comes from red algae (seaweed), which East Indian birds use to build nests; it is the main ingredient in the delicacy "bird's nest tissue where oxygen and carbon dioxide are exchanged. There, macrophages, white blood cells that protect the body (Figure 1A), engulf the bacteria.

But the bacterium's thick envelope helps it resist degradation by enzymes within the white blood cell's lysosome, which is a dedicated compartment that enables digestion of pathogenic microbes. Instead of being digested by the macrophage, some of the bacteria grow and multiply within the white blood cell's cytoplasm. The macrophage dies, releasing bacteria that can infect other macrophages. The body's immune response may then seal the bacteria and infected macrophages within a lump called a "tubercle," where they can remain dormant for many years. As the immune system weakens, however, the bacteria may break out and multiply, causing an active infection. An active case of tuberculosis can then spread bacteria to other people.

How is tuberculosis treated? A standard course of treatment for tuberculosis requires 6 months of antibiotic therapy (Figure 1B). A combination of antibiotics must be taken because certain drugs (such as isoniazid and pyrazinamide) are most effective against the rapidly multiplying bacteria, whereas other drugs (such as ethambutol) are more effective against long-term persistent bacteria. Simultaneous use of multiple drugs also decreases the chance of survival of mutant strains resistant to any one drug. Unfortunately, patients often stop taking the drugs once they feel well, failing to complete the full course of treatment. The result is that bacteria resume growing after a period of selection for drug resistance.

To avoid the spread of drug-resistant strains, many countries have adopted the program of directly observed therapy, short course (DOTS). The DOTS program requires a nurse to visit the patient and observe the patient swallowing every pill of the medication. DOTS requires a substantial commitment to public health, but it is an effective way to curb the spread of tuberculosis. Today, the World Health Organization (WHO) recommends DOTS for all countries. The United States requires all immigrants to be tested for tuberculosis in their country of origin; if they test positive, they must undertake DOTS before entering the United States.

DISCUSSION QUESTION

Suppose a patient who tests positive for tuberculosis refuses treatment, claiming to feel no symptoms of disease. What should be done?

Figure 1.12 Angelina and Walther Hesse

A. Portrait of the Hesses, who first used agar to make solid plate media for bacterial growth.

B. Colonies, each containing millions of bacteria, growing on an agar plate.

soup." Dutch colonists used agar to make jellies and preserves, and a Dutch colonist from Java introduced it to Angelina Hesse. The Hesses used agar to develop the first effective growth medium for tuberculosis bacteria. (Pure culture is discussed further in Chapter 6.)

Some kinds of microbes cannot be grown in pure culture without other organisms. Bacteria such as *Chlamydia* species, which cause trachoma and sexually transmitted diseases, grow only within living, eukaryotic host cells. All viruses must be cultured with their host cells. Viruses are discussed in Chapter 12.

Koch's Postulates Link a Pathogen with a Disease

For his successful discovery of the bacterium that causes tuberculosis, *Mycobacterium tuberculosis*, Koch was awarded the Nobel Prize in Physiology or Medicine in 1905. Koch formulated his famous set of criteria for establishing a causative link between an infectious agent and a disease (**Figure 1.13**). These four criteria are known as **Koch's postulates**:

1. The microbe is found in all cases of the disease.

2. The microbe is isolated from the diseased host and grown in pure culture.

3. When the microbe is introduced into a healthy, susceptible host (or animal model), the same disease occurs.

4. The same strain of microbe is obtained from the newly diseased host. When cultured, the strain shows the same characteristics as before.

Koch's postulates continue to be used to determine whether a given strain of microbe causes a disease. A modern example is Lyme disease, a tick-borne infection that has become widespread in New England and the mid-Atlantic states. Another example is rodent-borne hantavirus, the cause of hantaviral pneumonia, an emerging disease particularly prevalent in the southwestern United States. Nevertheless, the postulates remain only a guide; individual diseases and pathogens may confound one or more of the criteria.

A disease for which Koch's postulates were difficult to establish is AIDS. The first patients identified with AIDS had blood concentrations of HIV (the causative virus) so low that initially no virus could be detected. It took the invention of the PCR, a method of specifically amplifying the number of copies of any DNA or RNA sequence, to detect the presence of HIV. Today, PCR test kits are used routinely to identify pathogens in patients with many kinds of illnesses, such as COVID-19, the pandemic caused by a coronavirus in 2020. (**Figure 1.14**).

A difficulty with studying certain human diseases is the absence of an animal host that exhibits the same disease. For AIDS, even chimpanzees, our closest relatives, are not susceptible. Chimpanzees do, however, exhibit a similar disease from a related pathogen, simian immunodeficiency virus (SIV). Testing of retroviruses on humans is prohibited because they cannot be cured, but in rare instances researchers have voluntarily exposed themselves to a proposed pathogen. For example, Australian researcher Barry Marshall ingested *Helicobacter pylori* to convince skeptical colleagues that this organism could colonize the extremely acidic stomach. *H. pylori* turned out to be the causative agent of gastritis and stomach ulcers, conditions that had long been thought to be caused by stress rather than infection. For the discovery of *H. pylori*, Marshall and colleague Robin Warren won the 2005 Nobel Prize in Physiology or Medicine.

Immunization Prevents Disease

Identifying the cause of a disease is only the first step toward developing an effective therapy and preventing further transmission. Early microbiologists achieved some remarkable insights on how to control pathogens (again, see Table 1.1).

The first clue as to how to protect an individual from a deadly disease came from the dreaded smallpox, which we now know is caused by a virus. Smallpox was an ancient disease that originated in Africa about 10,000 BCE and spread throughout Europe and Asia. By 1000 BCE, in India and China it was known that individuals

Figure 1.13 Koch's Postulates Defining the Causative Agent of a Disease

1 The microbe is found in all cases of the disease but is absent from healthy individuals.

No microbe

2 The microbe is isolated from the diseased host and grown in pure culture.

3 When the microbe is introduced into a healthy, susceptible host, the same disease occurs.

4 The same strain of microbe is obtained from the newly diseased host.

Figure 1.14 COVID-19 Patient

A patient undergoes evaluation for COVID-19, a respiratory disease caused by a coronavirus.

could be made immune to the disease if secretions from a diseased individual were transferred to healthy individuals. Presumably, if some time elapsed between removal of the secretions and transfer to the next patient, the virus became naturally "attenuated"— meaning that its form deteriorated, rendering it less infectious but still capable of inducing an immune response.

In the eighteenth century, smallpox infected a large fraction of the European population, killing or disfiguring many people. In Turkey, however, the incidence of smallpox was decreased by the deliberate inoculation of children with material from smallpox pustules. Inoculated children usually developed a mild case of the disease and were protected from smallpox thereafter. The practice of smallpox inoculation was introduced from Turkey to other parts of Europe in 1717 by Lady Mary Montagu, a smallpox survivor (**Figure 1.15A**). Stationed in Turkey with her husband, the British ambassador, Lady Montagu learned that many elderly women there had perfected the art of inoculation: "The old woman comes with a nut-shell full of the matter of the best sort of small-pox, and asks what vein you please to have opened." Lady Montagu arranged for the procedure on her own son and then brought the practice back to England, where it became widespread.

A similar practice of smallpox inoculation was introduced to the American colonies by an enslaved man, Onesimus, from the Coromantee people of Africa. Onesimus persuaded his enslaver, Reverend Dr. Cotton Mather, to promote smallpox inoculation while an epidemic was devastating Boston.

Preventive inoculation with smallpox was dangerous, however, as some infected individuals still contracted serious disease and were contagious. So, doctors continued to seek a better method of prevention. In England, milkmaids claimed they were protected from smallpox after they contracted cowpox, a related but much milder disease. This claim was confirmed by English physician Edward Jenner (1749–1823), who deliberately infected patients with matter from cowpox lesions (**Figure 1.15B**). The practice of cowpox inoculation was called "vaccination," after the Latin word *vacca* for "cow." To this day, cowpox, or vaccinia virus, remains the basis of the modern smallpox vaccine.

Pasteur was aware of vaccination as he studied the course of various diseases in experimental animals. In the spring of 1879, he was studying fowl cholera, a transmissible disease of chickens with a high death rate. He had isolated and cultured the bacterium that killed the chickens, but he left his work during the summer for a long vacation. No refrigeration was available to preserve cultures, and when he returned to work, the aged bacteria failed to cause disease in his chickens. Pasteur then obtained fresh bacteria from an outbreak of disease elsewhere. But the fresh bacteria failed to make the original chickens sick (those that had been exposed to the aged bacteria). When he tested new chickens, exposed only to the fresh bacteria, the new chickens contracted the disease. Grasping the clue from his mistake, Pasteur had the insight to recognize that an attenuated ("weakened") strain of microbe, altered somehow to eliminate its potency to cause disease, could still confer immunity to the virulent disease-causing form.

Pasteur was the first to recognize the significance of attenuation and extend the principle to other pathogens. We now know that the molecular components of pathogens generate **immunity**, the resistance to a specific disease, by stimulating the **immune system**, an organism's exceedingly complex cellular mechanisms of defense (discussed in Chapters 15 and 16). An understanding of the immune system awaited the techniques of molecular biology a century later, but nineteenth-century physicians developed several effective examples of **immunization**, the stimulation of an immune response by inoculation with an attenuated pathogen.

The way to attenuate a strain for vaccination varies greatly among pathogens. Heat treatment or aging for various periods often

Figure 1.15 Smallpox Vaccination

A. Lady Mary Wortley Montagu, displayed in Turkish dress. The artist avoided showing Montagu's facial disfigurement from smallpox.

B. Dr. Edward Jenner, depicted vaccinating 8-year-old James Phipps with cowpox matter from the hand of milkmaid Sarah Nelmes, who had caught the disease from a cow.

C. Newspaper cartoon depicting public reaction to cowpox vaccination.

Figure 1.16 Pasteur Cures Rabies

This cartoon from a French newspaper depicts Pasteur protecting children from rabid dogs.

turned out to be the most effective approach. A far more elaborate treatment was required for the most famous disease for which Pasteur devised a vaccine: rabies. The rabid dog with the foaming mouth loomed large in folklore, and the disease was dreaded for its particularly horrible and inevitable course of death. Pasteur's vaccine for rabies required a highly complex series of heat treatments and repeated inoculations. Its success led to his instant fame (**Figure 1.16**). Grateful survivors of rabies founded the Pasteur Institute, one of the world's greatest medical research institutions, whose scientists in the twentieth century discovered HIV, the virus that causes AIDS.

Today, researchers continually develop new vaccines to control and eliminate emerging pathogens. For example, in 2019 an experimental vaccine was introduced by Merck to combat Ebola virus. Endemic to sub-Saharan Africa, Ebola virus causes a deadly disease that includes fever and hemorrhages. The Ebola vaccine proved remarkably effective at protecting individuals exposed to this terrifying pathogen. (Viruses are presented in Chapter 12.)

Antiseptics and Antibiotics Control Pathogens

Before the work of Koch and Pasteur, many patients died of infections transmitted unwittingly by their own doctors. In 1847, Hungarian physician Ignaz Semmelweis (1818–1865) noticed that the death rate of women in childbirth due to puerperal fever (uterine infection following childbirth) was much higher in his own hospital than in a birthing center run by midwives. He hypothesized that the doctors in his hospital were transmitting pathogens from cadavers they had dissected. So, he ordered the doctors to wash their hands in chlorine, an **antiseptic** agent (a chemical that kills microbes on living tissues). The mortality rate fell, but this revelation displeased other doctors, who refused to accept Semmelweis's findings. The doctors felt insulted by the handwashing order, and they argued that Semmelweis had no direct proof of how the contagion was transmitted.

In 1865, the British surgeon Joseph Lister (1827–1912) noted that half his amputee patients died of sepsis. Lister knew from Pasteur that microbial contamination might be the cause. As a result, he began experiments to develop the use of antiseptic agents, most successfully carbolic acid, to treat wounds and surgical instruments. After initial resistance, Lister's work—with the support of Pasteur and Koch—drew widespread recognition, as did the contribution of Semmelweis. In the twentieth century, surgeons

developed fully **aseptic** environments for surgery—that is, environments completely free of microbes. Nevertheless, even in hospitals today, ensuring that physicians and other health care workers wash their hands between patient visits remains a challenge.

Although the use of antiseptic chemicals was a major advance, most of these substances could not be taken internally, since they would kill the human patient. Researchers sought a "magic bullet," an **antibiotic** molecule that would kill only the bacteria, leaving their host unharmed.

An important step in the search for antibiotics was the realization that microbes themselves produce antibiotic compounds with highly selective effects. This finding followed from the famous accidental discovery of penicillin by the English medical researcher Alexander Fleming (1881–1955). In 1929, Fleming was culturing *Staphylococcus*, a bacterium that can infect wounds. He found that one of his plates of *Staphylococcus* was contaminated with a mold, *Penicillium notatum*. This mold was surrounded by a clear region free of *Staphylococcus* colonies (**Figure 1.17**). Following up on this observation, Fleming showed that the mold produced a substance that killed bacteria. We now know this substance as penicillin.

In 1941, British biochemists Howard Florey (1898–1968) and Ernst Chain (1906–1979) purified the penicillin molecule, which we now know inhibits formation of the bacterial cell wall. Penicillin saved the lives of many Allied troops during World War II—the first war in which an antibiotic became available to soldiers.

The second half of the twentieth century saw the discovery of many powerful antibiotics. Most of the new antibiotics, however, were produced by newly discovered bacteria and fungi from dwindling ecosystems—a circumstance that focused attention on wilderness preservation worldwide. Furthermore, the widespread and often indiscriminate use of antibiotics has selected for antibiotic-resistant pathogens. As a result, antibiotics have lost their effectiveness against

Figure 1.17 Alexander Fleming, Discoverer of Penicillin

A. Alexander Fleming in his laboratory.

B. Fleming's original plate of bacteria with *Penicillium* mold inhibiting the growth of bacterial colonies.

Figure 1.18 Viruses

A. Tobacco mosaic virus (TMV). The RNA chromosome is surrounded by protein subunits.

B. Herpesvirus structure. 3D structure of herpes simplex virus type 1.

certain strains of major pathogens. For example, multidrug-resistant *Mycobacterium tuberculosis* is now a serious threat to public health.

Fortunately, biotechnology provides new approaches to antibiotic development, including genetic engineering of microbial producers and artificial design of antimicrobial chemicals. This industry has become ever more critical because the indiscriminate use of antibiotics has led to a "molecular arms race" in which our only hope is to succeed faster than the pathogens develop resistance. (Antibiotics are described in Chapter 13, and their medical uses are discussed in Chapters 19–24.)

The Discovery of Viruses

The discovery of ever-smaller and more elusive microorganisms continues to this day. From the nineteenth century on, researchers were puzzled to find contagious diseases whose agent of transmission could pass through a filter with tiny pores that blocked known microbial cells. In 1892, the Russian botanist Dmitri Ivanovsky (1864–1920) studied tobacco mosaic disease, a condition in which the tobacco leaves become mottled and the crop yield is decreased or destroyed altogether. Ivanovsky concluded that because the agent of disease passed through a filter that retained bacteria, it could not be a bacterial cell. The Dutch plant microbiologist Martinus Beijerinck (1851–1931) replicated Ivanovsky's work and called the new infectious agent a "virus."

The filterable agent was ultimately purified by the American scientist Wendell Stanley (1904–1971), who processed 4,000 kilograms (kg) of infected tobacco leaves and crystallized the infective particle. What he had crystallized was the tobacco mosaic virus (TMV), the causative agent of tobacco mosaic disease. TMV infects many kinds of plants, such as lettuce and tomatoes; the virus is so infectious that plants can be infected by people smoking TMV-contaminated

cigarettes. The crystallization of this virus particle earned Stanley the 1946 Nobel Prize in Chemistry. The fact that an entity capable of biological reproduction could be inert enough to be crystallized amazed scientists and ultimately led to a new, more mechanical view of living organisms.

The individual TMV particle consists of a helical tube of protein subunits containing its genetic material coiled within (Figure 1.18A). Stanley thought the virus was a catalytic protein, but colleagues later determined that it contained RNA as its genetic material. The structure of the coiled RNA was solved through X-ray diffraction crystallography by the British scientist Rosalind Franklin (1920–1958), who had previously obtained the X-ray data that revealed the structure of DNA. Since then, X-ray analysis has solved the structure of other viruses, including herpesvirus (Figure 1.18B). The herpesvirus contains a DNA chromosome coiled within an icosahedral (20-sided) protein capsid.

Today, RNA viruses are some of our most challenging pathogens, including HIV (human immunodeficiency virus, the cause of AIDS), influenza virus (the cause of the famous 1918 pandemic), and the SARS-CoV-2 coronavirus. In 2019, this new coronavirus swept the globe, causing COVID-19 (coronavirus disease 2019). The new coronavirus leapt from an animal host into the human population, where it caused a respiratory disease that killed nearly one million people (as of October 2020) and sickened many millions more (see Figure 1.14). To slow its spread, schools and factories closed, and NCAA basketball games were played in empty stadiums.

Naming a Pathogen

What's in a name? A lot is at stake—especially if it's named after your own home town, such as Lyme disease, first described in the town of Lyme, Connecticut. Today, the World Health Organization recommends standards for avoiding naming any new pathogen in a way that stigmatizes a place or group of persons. The name for an emerging pathogen (or disease it causes) must avoid geographic locations, people's names, the names of animals, and any cultural or occupational references. The disease name must also be short and easy to pronounce and relevant to the public awareness of the disease; otherwise its usage will not catch on, despite the intentions of health professionals.

SECTION SUMMARY

- **Florence Nightingale** statistically quantified the impact of infectious disease on human populations.
- **Robert Koch** devised techniques of pure culture to study a single species of microbe in isolation. A key technique is to culture the microbe on a solid

medium by using agar, as developed by Angelina and Walther Hesse, in a double-dish container devised by Julius Petri.

- **Koch's postulates** provide a set of criteria to establish a causative link between an infectious agent and a disease.
- **Edward Jenner** established the practice of vaccination—inoculation of cowpox to prevent smallpox. Jenner's discovery was based on earlier observations by Lady Mary Montagu and others that a mild case of smallpox could prevent future infection.
- **Louis Pasteur** developed the first vaccines based on attenuated strains, including the famous rabies vaccine.
- **Ignaz Semmelweis and Joseph Lister** showed that antiseptics could prevent the transmission of pathogens from doctor to patient.
- **Alexander Fleming** discovered that the *Penicillium* mold generates a substance that kills bacteria. **Howard Florey and Ernst Chain** purified the substance, penicillin, the first commercial antibiotic to save human lives.
- **Dmitri Ivanovsky and Martinus Beijerinck** discovered that tobacco mosaic disease could be transmitted by a virus, an infectious agent capable of passing through a filter that retains bacteria.

Thought Question 1.2 How could you use Koch's postulates to demonstrate the causative agent of influenza? What problems would you need to overcome that were not encountered with anthrax?

Thought Question 1.3 Why do you think some pathogens stimulate immunity readily, whereas others evade the immune system?

1.4 Microbes in Our Environment

SECTION OBJECTIVES
- Describe examples of how microbes contribute to natural ecosystems.
- Explain how mitochondria and chloroplasts evolved by endosymbiosis.

How do microbes shape Earth's environment? Koch's growth of microbes in pure culture was a major technological advance that led to amazing revelations in microbial physiology and biochemistry. In hindsight, this discovery eclipsed the equally important study of microbial ecology. Microbes cycle the many minerals essential for all life. Yet barely 0.1% of all microbial species can be cultured in the laboratory—and the rest make up most of Earth's entire biosphere. Only the outer skin of Earth supports complex multicellular organisms. The depths of Earth's crust, to at least 2 miles down—as well as the atmosphere 10 miles out into the stratosphere—remain the domain of microbes. So, for the most part, Earth's ecology is microbial ecology.

Microbes Support Natural Ecosystems

The first microbiologists to culture microbes in the laboratory selected the kinds of nutrients that feed humans, such as beef broth or potatoes. Some of Koch's contemporaries, however, suspected that other kinds of microbes living in soil or wetlands existed on more exotic fare. Soil samples were known to oxidize hydrogen gas,

Figure 1.19 Wetland Habitat for Microbes

The wetlands of the Everglades act as natural microbial filters for the aquifers that supply drinking water for southern Florida. Microbes are part of an aquatic food web that includes birds and alligators.

and this activity was eliminated by treatment with heat or acid, suggesting microbial origin. Ammonia in sewage was oxidized to nitrate, and this process was eliminated by antibacterial treatment. These findings suggested the existence of microbes that "ate" hydrogen gas or ammonia instead of beef or potatoes, but no one could isolate these microbes in culture.

Among the first to study microbes in natural habitats was the Russian scientist Sergei Winogradsky (1856–1953). Winogradsky waded through marshes (wetlands) to discover new forms of microbes. In wetlands Winogradsky discovered microbes whose metabolism is very different from human digestion. For example, species of the bacterium *Beggiatoa* oxidize hydrogen sulfide (H_2S) to sulfuric acid (H_2SO_4). *Beggiatoa* fixes carbon dioxide into biomass without consuming any organic food. Organisms that feed solely on inorganic materials such as iron or ammonia are known as lithotrophs. Today, wetland microbes are known for their critical roles in environmental quality (**Figure 1.19**). Wetland microbes support the food web of animals and plants, and they filter the groundwater that we eventually drink.

The lithotrophs that Winogradsky studied could not be grown on Koch's plate media containing agar or gelatin. The bacteria that Winogradsky isolated could grow only on inorganic minerals. For example, nitrifiers convert ammonia to nitrate, forming a crucial part of the nitrogen cycle in natural ecosystems. Winogradsky cultured nitrifiers on a completely inorganic solution containing ammonia and silica gel, which supported no other kind of organism. This experiment was an early example of **enrichment culture**, the use of selective growth media that support certain classes of microbial metabolism while excluding others (discussed in Chapter 6). Enrichment culture is important in clinical microbiology labs to help

Figure 1.20 Biofilm of Digestive Bacteria in the Human Intestine

SEM   10 µm

Bacteria growing on the surface of residual food particles aid our digestion.

Figure 1.21 Lynn Margulis and the Serial Endosymbiosis Theory

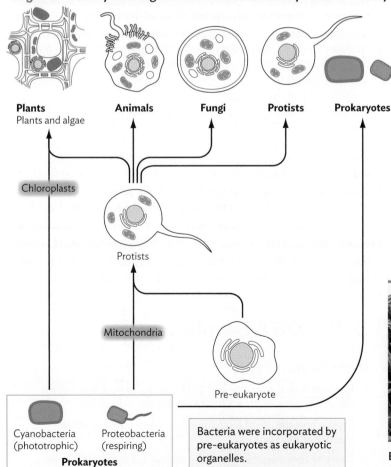

Plants
Plants and algae

Animals

Fungi

Protists

Prokaryotes

Chloroplasts

Protists

Mitochondria

Pre-eukaryote

Cyanobacteria (phototrophic)   Proteobacteria (respiring)
Prokaryotes

Bacteria were incorporated by pre-eukaryotes as eukaryotic organelles.

A. Five-kingdom classification scheme, modified by the endosymbiosis theory.

B. Lynn Margulis proposed that organelles evolve through endosymbiosis.

identify the specific microbes that cause disease. A kind of enrichment culture used today in hospitals is MacConkey agar, a formulation that permits growth of Gram-negative bacteria such as *E. coli* but excludes Gram-positive soil bacteria such as *Bacillus* species.

Winogradsky and later microbial ecologists showed that bacteria perform unique roles in the global interconversion of inorganic and organic forms of nitrogen, sulfur, phosphorus, and other minerals. Without these essential conversions (nutrient cycles), no plants or animals could live. Bacteria and archaea fix nitrogen (N_2) by reducing it to ammonia (NH_3), the form of nitrogen assimilated by plants. This process is called **nitrogen fixation**. Other bacterial species oxidize ammonium ions (NH_4^+) in several stages back to nitrogen gas. (Global cycles of nutrients are presented in Chapter 27.)

Within plant cells, certain bacteria fix nitrogen as partners of a host organism, a relationship called **mutualism**. Mutualism is a form of **symbiosis**, growth of two species in intimate association; in mutualism, the partners receive essential benefits from each other. Bacteria that fix nitrogen in a mutualism within a plant cell are also called **endosymbionts** (meaning "inside symbionts"), organisms living inside a larger organism. Endosymbiotic bacteria provide nutrients to their host cell, which in turn provides protection and nutrients to the endosymbionts. Endosymbiotic bacteria known as rhizobia induce the roots of legumes (beans and lentils) to form special nodules to facilitate bacterial nitrogen fixation. Rhizobial endosymbiosis was first observed by Martinus Beijerinck.

Communities of bacteria serve animals as digestive endosymbionts (**Figure 1.20**). Such a community is known as a digestive **microbiome**. Animals such as cattle and termites require a digestive microbiome to break down cellulose and other plant polymers. Even we humans obtain as much as 15% of our nutrition from bacteria growing in our colon. Increasingly, the human digestive microbiome is considered an organ of the human body, with a role in medical conditions ranging from obesity to depression. Amazingly, bacteria from feces can actually cure a life-threatening disease. The transfer of fecal bacteria from one person to another, known as fecal microbiota transplant, was approved in 2013 by the FDA to treat infections of drug-resistant *Clostridioides difficile*.

Animals and Plants Evolved through Endosymbiosis

Microbial endosymbiosis, in many diverse forms, is widespread in all ecosystems. Many interesting cases involve animal or human hosts. The endosymbiotic origin of eukaryotic cells was proposed by Lynn Margulis (1938–2011), of the University of Massachusetts, Amherst (**Figure 1.21**). Margulis tried

Figure 1.22 Archaea, a Newly Discovered Form of Life

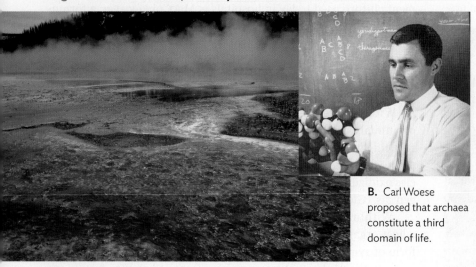

B. Carl Woese proposed that archaea constitute a third domain of life.

A. Yellowstone National Park hot springs contain archaea growing above 80°C (176°F) in water containing sulfuric acid.

New Microbes Continue to Emerge

We continue to discover surprising new kinds of microbes deep underground and in places previously thought uninhabitable, such as the hot springs of Yellowstone National Park (**Figure 1.22A**). Microbes shape our biosphere and provide new tools that impact human society. For example, a bacterial DNA polymerase (a DNA-replicating enzyme) from a Yellowstone hot spring is used for the PCR technology that identifies pathogens in ill patients.

In 1977, Carl Woese (1928–2012), of the University of Illinois (**Figure 1.22B**), discovered that some of the microbes from Yellowstone hot springs have genomes very different from those of all other known life forms. The genomes of these microbes had diverged so far from that of any known bacteria that the newly discovered prokaryotes were seen as a distinct form of life—the **archaea** (discussed in Section 1.1). Archaea living in extreme environments produce exceptionally sturdy enzymes that can be used for biotechnology.

SECTION SUMMARY

- **Sergei Winogradsky** first developed a system of enrichment culture to grow microbes from natural environments.
- **Lithotrophs** metabolize inorganic minerals, such as ammonia, instead of the organic nutrients used by the microbes isolated by Koch.
- **Endosymbionts** are microbes that live within a larger organism. Some endosymbionts are **mutualists** that provide essential functions for their host, such as nitrogen fixation for a legume plant.
- **Martinus Beijerinck** first demonstrated that nitrogen-fixing rhizobia grow as endosymbionts within leguminous plants.
- **The human body has a microbiome,** a community of associated bacteria that play important roles in human health.
- **Lynn Margulis** proposed that eukaryotic organelles such as mitochondria and chloroplasts evolved by endosymbiosis from prokaryotic cells engulfed by pre-eukaryotes.
- **Carl Woese** discovered a domain of life, Archaea, whose genetic sequences diverge equally from those of bacteria and those of eukaryotes.

Thought Question 1.4 Suppose a microbiologist isolates microbes from 2 miles below Earth's surface. Would these microbes be likely to cause disease in humans?

----

to explain how it is that eukaryotic cells contain mitochondria and chloroplasts, membranous organelles that possess their own chromosomes. She proposed that eukaryotes evolved by merging with bacteria to form composite cells by intracellular endosymbiosis, in which one cell internalizes another that grows within it. The endosymbiosis may ultimately generate a single organism whose formerly independent members are now incapable of independent existence.

Margulis proposed that, early in the history of life, respiring bacteria similar to *E. coli* were engulfed by pre-eukaryotic cells, where they evolved into mitochondria, the eukaryote's respiratory (energy-generating) organelle. Similarly, a phototroph related to cyanobacteria was taken up by a eukaryote, giving rise to the chloroplasts of phototrophic algae and plants. Ultimately, DNA sequence analysis produced compelling evidence of the bacterial origin of mitochondria and chloroplasts. Both these classes of organelles contain circular molecules of DNA, whose sequences show unmistakable homology (similarity) to those of modern bacteria. DNA sequences and other evidence established the common ancestry between mitochondria and respiring bacteria and between chloroplasts and cyanobacteria.

The endosymbiotic origin of mitochondria has medical importance because mitochondria still share key properties with bacteria. Such properties include the structure and function of the respiratory complex. Mitochondrial defects cause human diseases, including a form of epilepsy associated with malformed muscle fibers (myoclonic epilepsy with ragged-red fibers, or MERRF). The similarity between mitochondria and bacteria also limits the use of certain antibiotics, such as those that kill bacteria by inhibiting respiratory proteins embedded in the membrane. Bacteria have the respiratory proteins in their cell membrane, whereas human mitochondria have similar proteins in the mitochondrial inner membrane.

1.5 The DNA Revolution

SECTION OBJECTIVES

- Describe how the structure of DNA was discovered, and explain the significance of DNA for determining the traits of life.
- Explain how the use of DNA information has transformed the practice of medicine.
- Describe how microbial molecular biology led to editing human DNA for therapy.

How did science change medicine during the twentieth century?
Amid world wars and societal transformations, the field of microbiology exploded with new knowledge (see Table 1.1). More than 99% of what we know about microbes today was discovered since 1900 by scientists too numerous to cite in this book. Advances in biochemistry and microscopy revealed the fundamental structure and function of cell membranes and proteins. Scientists learned to read the cell's central information molecule, deoxyribonucleic acid, better known as DNA (**Figure 1.23**). Beyond microbiology, these advances produced the technology of "recombinant DNA," or genetic engineering—the construction of molecules that combine DNA sequences from unrelated species. These microbial tools offered unprecedented applications to human medicine and industry.

The Discovery of the Structure of DNA

The famous double helix of DNA was determined by X-ray crystallography, a method developed by British physicists in the early 1900s. The field of X-ray analysis included an unusual number of women. Dorothy Hodgkin (1910–1994), for example, won a Nobel Prize for determining the structures of penicillin and vitamin B_{12}. In 1953, crystallographer Rosalind Franklin (1920–1958) joined a laboratory at King's College to study the structure of DNA (**Figure 1.24A**). As a woman and as a Jew who supported relief work in Palestine, Franklin felt socially isolated at the male-dominated Protestant university. Nevertheless, the exceptional quality of her X-ray micrographs enabled her to propose that the DNA molecule is a double helix. The X-shaped pattern in her micrograph (**Figure 1.24B**) arises from the double-helical form of the DNA molecule.

Without Franklin's knowledge, her colleague Maurice Wilkins (1916–2004) showed her data to a competitor, James Watson (born 1928). The X-shaped pattern led Watson and Francis Crick (1916–2004) to guess that the four bases of the DNA "alphabet" are

Figure 1.23
DNA: The Central Information Molecule
The sequence of base pairs in DNA encodes all the genetic information of an organism.

paired in the interior of Franklin's double helix (**Figure 1.24C**). They published their model in the journal *Nature*, without acknowledging their use of Franklin's data.

The discovery of the double helix earned Watson, Crick, and Wilkins the 1962 Nobel Prize in Physiology or Medicine. Franklin died of ovarian cancer before the Nobel Prize was awarded. Before her death, however, she turned her efforts to the structure of ribonucleic acid (RNA). She determined the form of the RNA chromosome within tobacco mosaic virus, the first viral RNA to be characterized.

The structure of DNA base pairs led to the development of techniques for **DNA sequencing**, the reading of the sequence of DNA base pairs (the DNA sequencing technique is outlined in Section 8.6). **Figure 1.25** shows an example of DNA sequence data, in which each color represents one

Figure 1.24 Discovering the DNA Double Helix

A. Rosalind Franklin discovered that DNA forms a double helix.

B. X-ray crystallography pattern of DNA obtained by Rosalind Franklin

C. James Watson and Francis Crick proposed a structure for the complementary pairing between bases of DNA.

Figure 1.25 DNA Sequence Data

C T C T T A G T G G C G C A G C T A A C G C A T T A A G T T G A C C G

300   310   320   330

Sequence obtained from genomic DNA of a microbe isolated from a discarded beverage container. Each colored trace represents the concentration of one of the four bases terminating an interrupted chain of DNA.

of the four bases and each peak represents a DNA fragment terminating in that particular base. The order of fragment lengths yields the sequence of bases in one strand. Reading the sequence enabled microbiologists to determine the beginning and endpoint of genes and ultimately entire genomes.

A method of DNA sequencing was devised by Frederick Sanger (**Figure 1.26A**) and was used to reveal the first genome of a virus in 1977. This method uses a bacterial enzyme, DNA polymerase. Sanger shared the 1980 Nobel Prize in Chemistry with Walter Gilbert and Paul Berg. The first genome sequence of a cellular microbe was obtained in 1995 for *Haemophilus influenzae*, a bacterium that causes ear infections and meningitis in children. The *H. influenzae* genome has nearly 2 million base pairs that specify about 1,700 genes. It was sequenced by a team of scientists led by Hamilton Smith and Craig Venter, who devised a special computational strategy for assembling large amounts of sequence data. This strategy was later applied to sequencing the human genome. Another group of scientists, led by Claire Fraser (**Figure 1.26B**), sequenced the genomes of many microbes, including *Bacillus anthracis*, the cause of anthrax, and *Treponema pallidum*, the cause of syphilis. Today, entire communities of microbial genomes are sequenced at once. The sequence of DNA from a microbial community is called a metagenome (discussed in Chapter 9).

Microbial Discoveries Transform Medicine and Industry

Twentieth-century microbiology transformed the practice of medicine and generated entire industries of biotechnology and bioremediation (the use of living organisms to decontaminate pollutants). After the discovery of penicillin, Americans poured millions of dollars of private and public funds into medical research. The March of Dimes campaign for private donations to prevent polio led to a vaccine that has nearly eliminated the disease. With the end of World War II, research on microbes and other aspects of biology drew increasing financial support from US government agencies—such as the National Institutes of Health and the National Science Foundation—and from governments of other countries, particularly the European nations and Japan. Further support has come from private foundations, such as the Pasteur Institute, the Wellcome Trust, and the Howard Hughes Medical Institute.

Today, microbial molecules continue to enable advances in medicine that until now could only have been imagined in science fiction. Human genome sequences are being used to determine the gene variants of given patients and thus optimize their treatment according to their genes. And tools from microbial molecular biology provide surprising ways to edit human DNA. Most important are these innovations:

- The AIDS virus HIV was engineered to make DNA carrier molecules that can insert genes into cells of the human body. These DNA carriers, called lentivectors, are the basis of CAR-T (chimeric antigen receptor T-cell) therapy. In 2017, CAR-T therapy was first approved to treat pediatric patients with acute lymphoblastic leukemia (see Chapter 12). Now CAR-T is being developed rapidly for other cancers.

Figure 1.26 Microbial Genome Sequencers

A. Frederick Sanger helped devise the method of DNA sequence analysis that is the basis of modern genome sequencing.

B. Claire Fraser led a team that completed the sequences of many microbial genomes.

Table 1.2
Careers Involving Microbiology

Profession	Primary Activity
Nursing	Providing comprehensive care for patients with illness
Physician assistant	Practicing medicine under supervision by a licensed physician
Medical laboratory science	Diagnosing, treating, and developing therapies for human diseases
Veterinary laboratory science	Diagnosing, treating, and developing therapies for animal diseases
Public health	Monitoring, assessing, and promoting policies influencing the health of populations
Food and industrial microbiology	Developing microbial and industrial food products; addressing food contamination and bioremediation
Agricultural microbiology	Managing plant pathogens and plant-associated microbes
Environmental microbiology	Environmental remediation through microbial processes
Forensic microbiology	Analyzing microbial strains as evidence in criminal investigations
Experimental microbiology	Investigating fundamental questions about microbial form and function, genetics, and ecology

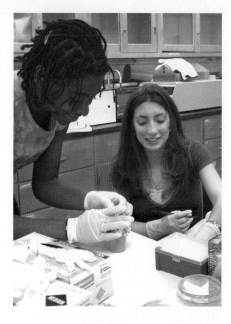

Figure 1.27
Microbiology Students at Kenyon College Prepare for Careers in Nursing and Medical Research

SECTION SUMMARY

- **The structure and function of DNA** were revealed by a series of experiments in the twentieth century.
- **Genome sequence determination** has shaped the study of biology in the twenty-first century.
- **Microbial discoveries transformed medicine and industry.** Biotechnology enables the production of new kinds of pharmaceuticals and industrial products.
- **Microbiology today offers many different careers** in the health field and in industry.

- CRISPR (clustered regularly interspaced short palindromic repeats) is a DNA editing tool that was engineered from the bacterial protein-RNA complex that defends against infection by bacterial viruses (bacteriophages). In 2019, the first therapeutic use of CRISPR was approved for cancer therapy.

Careers in diverse fields recruit microbiologists (**Table 1.2**; **Figure 1.27**). Nurses, physician assistants, pharmacists, and other health care professionals use microbiological science to keep patients healthy and diagnose their illnesses. Public health workers apply statistical models to manage the health of communities and nations. Forensic technologists use DNA and microbial analysis to solve crimes. In environmental science, newly discovered microbes provide innovative ways to bioremediate wastes and control insect pests. On a global level, the challenges of pollution and climate change increasingly involve contributions of environmental microbiology to human health.

Perspective

Microbiology: The Human Experience is your guide to microbes and the enormous impact they have on human life. Our focus is medical by design, but we also explore intriguing connections between microbes, Earth's ecology, and human existence. Part I of the book lays the foundation for studying microorganisms and for understanding infectious disease. Chapter 1 provides the human time line of discovery. Next, Chapter 2 introduces the basic concepts of microbial disease, and then Chapter 3 describes the tools of microscopy that enable us to see cells and their inner workings. Chapter 4 explains the fundamental biochemistry needed to understand microbial life (and infectious disease), while Chapter 5 reveals the intricate structures of the microbes themselves. Chapter 6 introduces microbial nutrition and growth within the human body and human-related environments. Throughout, we invite readers to share with us the excitement of discovery in microbiology and the potential applications of that knowledge to improving human health.

Key Terms

agar (p. 16)
alga (p. 5)
antibiotic (pp. 4, 20)
antiseptic (p. 20)
archaeon (p. 5)
aseptic (p. 20)
autoclave (p. 12)
bacterium (p. 4)
cell (p. 4)
chain of infection (p. 15)

colony (p. 16)
DNA sequencing (p. 25)
endosymbiont (p. 23)
enrichment culture (p. 22)
eukaryote (p. 5)
fermentation (p. 12)
fungus (p. 5)
genome (p. 5)
germ theory of disease (p. 13)
immune system (p. 19)

immunity (p. 19)
immunization (p. 19)
Koch's postulates (p. 18)
lithotroph (p. 6)
microbe (p. 4)
microbiome (p. 23)
mutualism (p. 23)
nitrogen fixation (p. 23)
parasite (p. 6)
pathogen (p. 4)

Petri dish (p. 16)
prokaryote (p. 5)
protist (p. 5)
protozoan (p. 5)
public health (p. 14)
pure culture (p. 16)
species (p. 5)
spontaneous generation (p. 11)
symbiosis (p. 23)
virus (p. 6)

Objectives Assessment Questions

1.1 From Germ to Genome: What Is a Microbe?

1. Which trait is absent from cellular microbes such as bacteria?
   a. A nucleus with a nuclear membrane
   b. A cell membrane (plasma membrane)
   c. One or more chromosomes of DNA
   d. Enzymes that conduct metabolism

2. Which of the following applies only to Archaea?
   a. They have no nucleus or nuclear membrane.
   b. They possess mitochondria.
   c. They are not true cells.
   d. They have never been shown to cause disease.

3. Which trait of cells do viruses share?
   a. A genome
   b. Ribosomes that synthesize protein
   c. Active metabolism
   d. Motility

1.2 Microbes Shape Human History

4. Which person was admired for tending victims of a deadly disease?
   a. Robert Hooke
   b. Antonie van Leeuwenhoek
   c. Catherine of Siena
   d. Lazzaro Spallanzani

5. Which person first demonstrated that the failure of spontaneous generation of microbes was not due to lack of oxygen?
   a. Catherine of Siena
   b. Lazzaro Spallanzani
   c. Robert Hooke
   d. Louis Pasteur

6. Which epidemic disease inspired a large demonstration at the Washington Monument?
   a. Bubonic plague
   b. Influenza
   c. AIDS
   d. Smallpox

1.3 Medical Microbiology and Immunology

7. Florence Nightingale's polar area chart (see Figure 1.8) and the CDC chart of influenza (see Figure 1.9) have what in common?
   a. Both charts graphically represent patterns of public nutrition.
   b. Both charts represent the rise and fall of an infectious disease.
   c. Both charts depict the role of disease in an army campaign.
   d. Both charts include data from culturing the causative agent of disease.

8. Which of Koch's postulates requires modification for ethical study of a human disease?
   a. The microbe is found in all cases of the disease.
   b. The microbe is isolated from the diseased host and grown in pure culture.
   c. When the microbe is introduced into a healthy, susceptible host, the same disease occurs.
   d. The same strain of microbe is obtained from the newly diseased host. When cultured, the strain shows the same characteristics as before.

9. Which statement about antibiotics is not correct?
   a. Antibiotics reproduce themselves to form more antibiotics.
   b. Antibiotics may be chemicals produced by specific microbes.
   c. A specific antibiotic can kill certain classes of microbes.
   d. Fewer antimicrobial agents are known for viruses than for bacterial pathogens.

1.4 Microbes in Our Environment

10. Microbes of which class gain energy from oxidizing inorganic minerals?
    a. Enteric bacteria
    b. Lithotrophs
    c. Rhizobia
    d. Algae

11. Microbes of which class most closely resemble the ancestor of chloroplasts?
    a. Enteric bacteria
    b. Mycobacteria
    c. Cyanobacteria
    d. Archaea

12. Which important cellular process requires bacteria or archaea?

    a. Fermentation

    b. Nitrogen fixation

    c. DNA replication

    d. Membrane formation

1.5 The DNA Revolution

13. The person whose X-ray crystallography data first revealed the fundamental structures of DNA and RNA was _____.

    a. James Watson

    b. Francis Crick

    c. Dorothy Hodgkin

    d. Rosalind Franklin

14. Who led the team of scientists that sequenced the genomes of numerous microbes, including *Bacillus anthracis* and *Treponema pallidum*?

    a. Frederick Sanger

    b. Walter Gilbert

    c. Claire Fraser

    d. Hamilton Smith

15. Which molecular technique is now used to identify a very specific strain of a disease-causing microbe?

    a. Pathogen inoculation for immunization

    b. Plate culture

    c. Polymerase chain reaction (PCR) amplification of DNA

    d. Filtration with microscopic pore size

Review Questions

1. Explain the apparent contradictions in defining microbiology as the study of microscopic organisms or the study of single-celled organisms.

2. What is the genome of an organism? How do genomes of viruses differ from those of cellular microbes?

3. Under what conditions might microbial life have originated? What evidence supports current views of microbial origin?

4. List the ways in which microbes have affected human life throughout history.

5. Summarize the key experiments and insights that shaped the controversy over spontaneous generation. What key questions were raised, and how were they answered?

6. Explain how microbes are cultured on solid media.

7. Compare the culture methods of Koch and Winogradsky. How did their different approaches to microbial culture address different questions in microbiology?

8. Explain how a series of observations of disease transmission led to the development of immunization to prevent disease.

Clinical Correlation Questions

1. A patient comes to the emergency ward of a hospital complaining of shortness of breath. The patient also has fever, coughing, and shaking chills. The patient demands antibiotics, but after testing, the physician declines to supply them. Instead, the hospital provides corticosteroids, fluids, and oxygen. Why does the physician not prescribe antibiotics?

2. A patient visits the doctor with symptoms of gastrointestinal illness, including diarrhea and vomiting. A stool culture is performed, revealing *Escherichia coli* bacteria. Does this test prove that *E. coli* caused the illness? Which of Koch's postulates might not be fulfilled?

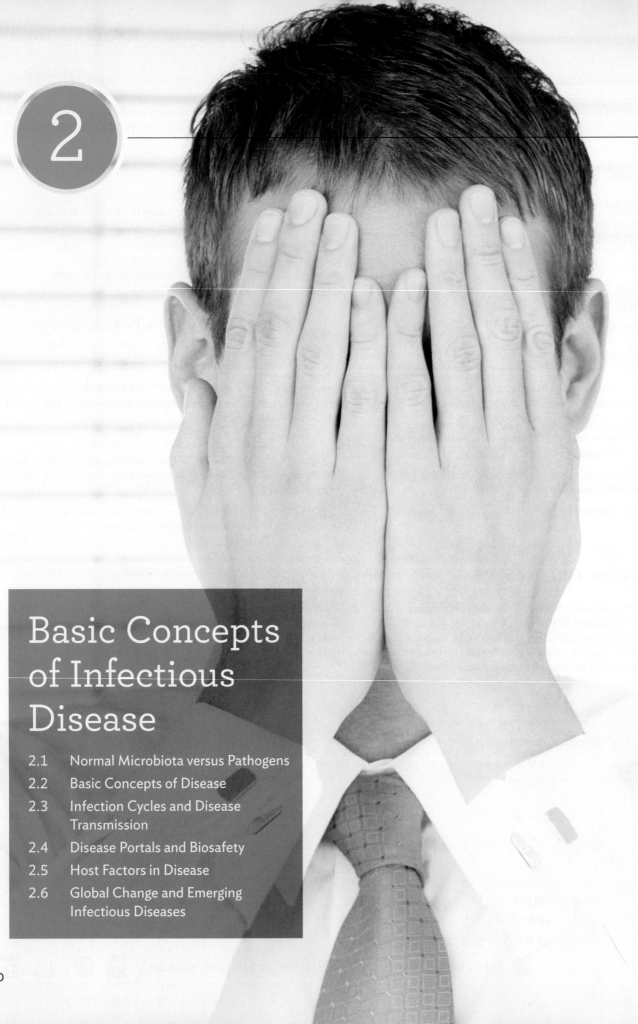

2

Basic Concepts of Infectious Disease

[The Broker's Secret]

SCENARIO Brandon, a 30-year-old stockbroker living in Chicago, visited his physician's office. When the nurse asked Brandon why he was there, he blushed and said he wanted to talk only to the physician about his problem.

SIGNS AND SYMPTOMS Once the doctor entered the room, Brandon explained that he had a small, round lesion on his penis. When asked about his sexual partners, Brandon initially said he was dating only one woman, but when pressed, he admitted he had been intimate with two women and one man over the previous month. Upon examination, the lesion appeared to cause no pain but exuded a clear fluid.

Chancre of Syphilis
Note single ulcer on the foreskin, usually painless and indurated (hardened) with a well-defined margin and a clear base.

TESTING The physician quickly sent a sample of the fluid to the clinical laboratory. There, the sample was found to contain fast-moving, corkscrew-shaped bacteria.

DIAGNOSIS The diagnosis was syphilis, caused by the bacterium *Treponema pallidum*. Left untreated, the disease could eventually cause horrible disfiguration and death.

TREATMENT Confident that he knew the cause, the physician gave Brandon a shot of long-acting penicillin. The broker recovered and returned to the Chicago Stock Exchange; he was somewhat embarrassed, but considerably wiser.

SEM   2 μm

Treponema pallidum
A sexually transmitted, spiral-shaped bacterium that can grow only in a host.

CHAPTER OBJECTIVES

After reading this chapter, you will be able to:

• Describe the relationships between a host, its microbiome, and pathogens.

• Apply basic concepts of pathogenesis to specific infections.

• Discuss how infectious diseases impact communities and how communities shape emerging pathogens.

In 1493, a horrifying new disease emerged in Spain shortly after Columbus and his crew returned from the Americas. Victims developed disfiguring rashes, ulcers, violent fevers, and painful bone aches. Internal organs were irreparably damaged, and patients frequently died. After starting in Spain, the disease afflicted Italy, and by 1500 it had spread throughout Europe. Everyone believed it was a new disease (it was) and realized it was transmitted by venereal (sexual) contact. This was almost 200 years before Antonie van Leeuwenhoek even discovered bacteria. Meanwhile, the French blamed the Italians, and the Italians said it must be the French; the English blamed both the French and Spanish or simply referred to it as the pox. By the 1520s, it was called syphilis, the same disease Brandon the stockbroker had in the opening case of this chapter.

We tell this tale to emphasize that you do not need to know much about infectious agents to understand important concepts of infectious disease, such as disease transmission. Brandon's case illustrates that you can easily grasp the basics of an infectious disease and host-pathogen interactions even without knowing much about bacterial structure, genetics, or metabolism. However, once you understand the fundamentals of infection, you will better appreciate the elegance of bacterial form and function. You will then see how a pathogen's biology guides the way it causes disease and exposes ways to treat it.

Perhaps the following question, more than any other, captures the mystique of microbiology: How can an organism too small to see with the naked eye kill a human 1 million times larger than itself? In this introductory chapter we will discuss the lingo (or terminology) of infectious disease, introduce some basic types of pathogens, and describe how they manage to colonize parts of our bodies. We'll outline the stages of an infectious disease and discuss where microbes that cause disease reside when not causing disease, and how these infectious agents are transmitted to humans. Some familiar and not-so-familiar microbial diseases will be mentioned along the way, including examples of new, emerging infectious diseases. Remember, however, that even though we focus on pathogens in most of the chapters, the vast majority of microbes are not pathogens, and many are beneficial to humans and their endeavors.

2.1 Normal Microbiota versus Pathogens

SECTION OBJECTIVES

• Describe differences between "normal" microbiota and pathogens.

• Discuss the relationship between infection and disease and between virulence and pathogenicity.

• Differentiate between infectious dose and lethal dose.

• Discuss the fundamental attributes of a successful pathogen.

What is a pathogen? You may not want to think about it, but we humans harbor at least two to three times more microbial hitchhikers than we have human cells in our body. The collection of bacteria, archaea, and eukaryotic microbes—our normal **microbiota** or **microbiome**—that call us home will be discussed in more depth in Chapter 14. Although they are commonly called "commensal organisms" (organisms that take, but do not give, sustenance from their host), most members of our microbiota both derive benefit from and give benefit to their hosts; this give-and-take defines mutualism, or a mutualistic relationship.

To stay with their host, normal microbiota have proteins called **adhesins** on their surface that enable them to attach to and colonize epithelial cells lining mucous membranes (intestine, urinary tract, mouth, nose). **Colonization** refers to the ability of a microbe to stay affixed to a body surface and replicate (**Figure 2.1**). Think about your own intestine. Without an ability to attach, some very useful bacteria would too easily be swept out of your body with your feces. However, membership in an individual's microbiome can fluctuate. Different bacteria and strains of microbial species can enter or leave our bodies, depending on what we eat, how we exercise, and whom we meet. A changing microbiota can alter our susceptibility to pathogens, strengthen or weaken our immune system, and—believe it or not—influence our weight and mental health.

Defining which species are members of our normal microbiota can be tricky. Take, for example, the bacterium *Escherichia coli*. *E. coli* is a part of our normal intestinal microbiome. We all harbor

Figure 2.1 Bacterial Attachment to Human Cells

Escherichia coli

SEM | 1 μm

3D micrograph showing a small cluster of bacteria (*Escherichia coli*) adhering tightly to human cells. Note that these bacteria are rod-shaped. Some are shown dividing.

this organism. But news outlets carry alarming stories about *E. coli* causing death and disease. So, does *E. coli* cause disease or doesn't it? The answer is yes—to both questions. Every species of bacteria, including *E. coli*, is composed of different strains that display unique characteristics. Some strains of *E. coli* contribute to our health as residents of the colon and are a natural part of our microbiome. Other strains, however, can cause diarrhea, septicemia (blood infection), meningitis, and urinary tract infections. Those strains of *E. coli* are pathogens. The difference lies in each strain's genes.

Defining a **pathogen** is easy; it is any bacterium, virus, fungus, protozoan, or worm (helminth) that causes disease. Defining what actually makes a pathogen a pathogen is not so easy. The **pathogenicity** of an organism—that is, its ability to cause disease—depends on a combination of the organism's genetic makeup, its location in or on the host's body, and the effectiveness of the host's immune response. For example, an *E. coli* strain that is innocuous in you could possibly kill your immunocompromised neighbor.

The term "parasite" is often used in casual conversation but requires some definition and context. **Parasites**, in the broadest sense, include any organism that colonizes and harms its host. In medical usage, the word "parasite" is reserved for disease-causing protozoa (single-celled eukaryotic organisms) and worms. Insects such as fleas or lice

A. Athlete's foot is caused by the fungus *Trichophyton rubrum*.

that live on the body surface are also called parasites—ectoparasites. The term "pathogen" is typically used by health professionals to refer to bacterial, viral, and fungal agents of disease.

Pathogens and parasites infect their animal and plant hosts in different ways and enter into different host-pathogen relationships, depending on the site of colonization and the capabilities of the organism. For instance, some organisms can live only on the surface of a host; an example is the fungus *Trichophyton rubrum*, one cause of athlete's foot, also called tinea pedis (**Figure 2.2A**). *T. rubrum*, which grows on solid media as filaments called hyphae (**Figure 2.2B**; Section 11.2), does not have the molecular tools needed to grow inside a host. In contrast, *Wuchereria bancrofti*, the parasitic worm that causes elephantiasis, cannot grow on the skin but causes disease by living inside the body (**Figure 2.3**). Parasites such as *W. bancrofti* that live inside the body are called endoparasites. Although not a problem in the United States, *W. bancrofti* is found throughout middle Africa, Asia, and New Zealand.

- -

◉ **Note:** The various types of microscopy used to see the microbes in this chapter will be described in Chapter 3. For now, let's focus on appreciating the appearance and size of these organisms rather than on the techniques used to capture their images. To appreciate size, visually compare the dimensions of the magnified microbe with the length of the scale bar shown in the inset photo in Figure 2.3. The scale bar is labeled with an actual, unmagnified length.

- -

Figure 2.2 A Skin Fungus

B. *T. rubrum* colony shape (approx. 3 inches, or 75 mm, in diameter). Inset is a magnification of *T. rubrum* showing the branching filaments (called hyphae) with small spores (called microconidia) that form the colony.

Figure 2.3 An Endoparasite

The disease filariasis, commonly known as "elephantiasis" for obvious reasons, is caused by the worm *Wuchereria bancrofti* (inset), which enters the lymphatics (vessels connecting lymph nodes) and blocks lymphatic circulation. Adult worms are threadlike and measure 4–10 cm (1.5–4 inches) in length. The young microfilaria (shown) are approximately 0.5 mm in length.

The Lingo of Pathogenesis

Before we talk about how microbes cause disease, it is helpful to know the terminology of pathogenesis. An **infection** occurs when a pathogen or parasite enters or begins to grow on a host. The term "infection," however, does not necessarily imply overt **disease**, in which a patient develops symptoms. Any potential pathogen growing in or on a host is said to cause an infection, but that infection may be temporary and undetected if immune defenses kill the pathogen before symptoms develop. Indeed, most infections go unnoticed. For example, every time you have your teeth cleaned by a dentist, your gums bleed and your oral microbiota transiently enter the bloodstream, but these brief infections very rarely cause disease.

Diseases in which symptoms develop rapidly and resolve quickly (within a few days or weeks) are called **acute infections**. The common cold, for example, is an acute infection often caused by rhinoviruses. A **chronic infection**, however, is one in which symptoms develop gradually (over weeks or months) and are slow to resolve. Tuberculosis, the lung disease described in the Chapter 1 opening case, is typically a chronic infection that can last for many months or even years if untreated.

How are pathogens classified? **Primary pathogens** are disease-causing microbes that can breach the defenses of a healthy host. For example, the rod-shaped (bacillus) *Shigella flexneri*, the cause of bacillary dysentery, is a primary pathogen. When ingested, it can survive the natural defensive barrier of an acidic stomach, enter the intestine, and begin to replicate. The organism in the opening case of this chapter, *Treponema pallidum*, is an example of a primary pathogen because it can survive initial attack by a healthy immune system and spread through the body in the bloodstream. **Opportunistic pathogens**, on the other hand, cause disease only in a compromised host. The fungus *Pneumocystis jirovecii* (previously known as *Pneumocystis carinii*; **Figure 2.4A**) easily infects people but rarely causes disease. It is, however, an opportunistic pathogen that causes life-threatening infections in AIDS patients, whose immune systems have been eroded by HIV.

Some microbes even enter into a **latent state** during infection, where the organism cannot be found by culture. Herpesvirus, for instance, can enter the peripheral nerves and remain dormant for years, then suddenly emerge to cause cold sores (**Figure 2.4B**). The bacterium *Rickettsia prowazekii* causes epidemic typhus, but it can also enter a mysterious latent phase and emerge months or years later to cause a disease relapse called recrudescent typhus. Where in the body does this bacterium hide? With all our technology, evidence of a latent form of *R. prowazekii* has never been found—yet it must be there.

Pathogenicity is defined in terms of how easily an organism causes disease (infectivity) and how severe that disease is (virulence). Pathogenicity, overall, is shaped by the genetic makeup of the pathogen. In other words, an organism is more—or less—pathogenic depending on the tools at its disposal (such as toxins) and their effectiveness.

Virulence is a measure of the degree or severity of disease. For instance, Ebola virus (**Figure 2.5**) and the closely related Marburg virus have case fatality rates of about 50%. This means that they are highly virulent. On the other hand, rhinovirus, the cause of the common cold, is very effective at causing disease but never kills its

Figure 2.4 Opportunistic and Latent Infections

A. *Pneumocystis jirovecii* cysts in bronchoalveolar material. Notice how the fungi look like crushed Ping-Pong balls.

B. Cold sore produced by a reactivated herpesvirus hiding latent in nerve cells.

Figure 2.5 Highly Virulent Ebola Virus

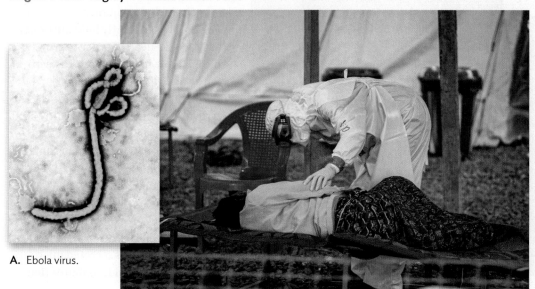

A. Ebola virus.

B. A health care worker tends to an Ebola patient in Kenema, Sierra Leone. Ebola causes hemorrhagic infections in which patients bleed from the mouth, nose, eyes, and other orifices. The mortality rate for this disease is approximately 50%.

victims. So, it is highly infective but has a low virulence. Both organisms are pathogenic, but with one you live and with the other you probably die.

One way to measure virulence is to calculate the **lethal dose 50% (LD$_{50}$)**. The LD$_{50}$ is the number of bacteria or virus particles (virions) required to kill 50% of an experimental group of animal hosts (usually mice or guinea pigs). An organism with a low LD$_{50}$, in which very few organisms are required to kill 50% of the hosts, is more virulent than one with a high LD$_{50}$ (**Figure 2.6**). For organisms that colonize but do not kill the hosts, the infectious dose needed to colonize 50% of the experimental hosts—the **infectious dose 50% (ID$_{50}$)**—can be measured. The ID$_{50}$ is measured by determining how many microbes are required to cause disease symptoms in half of an experimental group of hosts.

Two other aspects of pathogenesis are invasion and invasiveness. These terms sound similar but describe very different things. **Invasion** is the entry of a pathogen into a living human or nonhuman animal cell, where it then lives. Some pathogens live inside small, membrane-enclosed, spherical structures within the host cell called vacuoles (**Figure 2.7A**). Examples are *Salmonella*, a cause of diarrhea; *Coxiella*, the cause of Q fever; and *Legionella*, the cause of Legionnaires' disease. Other bacteria prefer to live directly in the cytoplasm of the host cell, not in a vacuole (**Figure 2.7B**). Examples of bacteria that grow in cytoplasm are *Shigella*, a cause of bloody diarrhea, and *Listeria*, which can cause diarrhea and neonatal meningitis.

Note: The disease processes and pathogenic mechanisms of the many microbes mentioned in this chapter will be explained more completely in later chapters. The basic concepts described here, however, can be applied to all the case histories that anchor the chapters of this book.

Invasiveness, in contrast to invasion, refers to the ability of a bacterial pathogen to rapidly spread through tissue. For instance, some strains of *Streptococcus pyogenes* are said to be highly invasive because they secrete enzymes that degrade host tissues. These strains are called "flesh-eating bacteria" because of how quickly they destroy tissue as the organism spreads.

It is also important to remember that different microbes have different **host ranges** in terms of the animals they can infect and produce disease in. Pathogens can have a very narrow host range; for example, *Salmonella enterica* serovar Typhi, the bacterium that causes typhoid fever, can infect only humans. (Serovar is a subclass within a species.) But a close relative, *Salmonella enterica* serovar Typhimurium, which produces diarrhea in humans, can infect many different animals. *S.* Typhimurium, then, has a broad host range. *T. pallidum*, in the chapter-opening syphilis case, has a very narrow host range. It infects only humans.

Note: Bacterial names include designations for genus (*Salmonella*) and species (*enterica*). Both are written in italics, with the genus capitalized and species lowercase. Serovar subclass names of a species are not italicized but are capitalized (for instance, Typhimurium).

Figure 2.6 Measurement of Virulence

Each LD$_{50}$ measurement requires infecting small groups of animals with increasing numbers of infectious agent and observing how many animals die. The number of microbes that kill half the animals is called the LD$_{50}$. In this example, agent 1, which requires fewer organisms to kill half the animals than does agent 2, is more virulent than agent 2.

Figure 2.7 Bacterial Life in a Host Cell

TEM   1 μm

Actin tail

Listeria

FM   7 μm

A. Invasion of intestinal cells (enterocytes) by *Salmonella enterica* serovar Typhimurium. Intracellular bacteria are located within membrane-enclosed vacuoles (arrows). Bacteria are first captured by cytoplasmic projections coming from the enterocyte surface (arrowhead).

B. The bacterium *Listeria* similarly invades eukaryotic host cells by being engulfed in a vacuole, but the pathogen then escapes the vacuole to live in the cytoplasm. *Listeria* (green) can propel itself in the cytoplasm by linking together (polymerizing) many copies of a host protein (actin) at one end of the bacterium. The structure is called an actin tail (red). Regions where the red and green fluorescence overlap appear yellow. The host cell is invisible in this image but encompasses the bacteria.

When Pathogen Meets Host

All pathogens—whether viral, bacterial, or eukaryotic—must interact with a host in several ways to be successful. The pathogen must attach to a host tissue, avoid the host's immune system, and steal nutrients from the host. We will delve more deeply into these interactions later (especially in Chapters 18–24), but let's discuss the basic concepts here.

Attachment. Specific proteins (adhesins) on the surface of microbes help them adhere to host cells, like thumbtacks to corkboard. Microbial adhesins bind to host cell surface structures called **receptors** (proteins, lipids, or polysaccharides) that have roles critical to normal host physiology. Because different host cells in different tissues have different cell-surface structures (in other words, different receptors), a pathogen will have a preference (or predilection) to infect tissues whose cells are "decorated" with the right surface receptors. Pathogens that fail to attach to the host can be expelled.

Immune avoidance. Different pathogens possess unique tools to avoid the immune system. Some, such as *Salmonella*, periodically change the molecular shape of their surfaces to confuse the immune system. Others, such as *Staphylococcus aureus*, secrete molecules that "tell" immune system cells, "All is well; no infection here." Still others (*Shigella*, for instance) secrete proteins that convince immune system cells to kill themselves in a process known as apoptosis (pronounced "ay-pop-TOH-sis"). Several of the ingenious strategies that microbes use to evade the immune system will be described in Chapters 15 and 16.

Stealing nutrients from the host. Animals, including humans, have evolved ways to tightly hold nutrients that bacteria need in order to grow. One example involves iron. To keep iron away from invading bacteria, the human body makes compounds (lactoferrin and transferrin) that tightly bind this metal. Consequently, most bacteria that enter the bloodstream cannot access the iron in the blood and cannot grow. Pathogens, however, have ways of stealing the iron from host tissues to use for their own purposes. One iron-stealing mechanism, described in Section 5.2, involves secreted microbial compounds called siderophores that tightly bind iron.

SECTION SUMMARY

- **Infection** with a microbe does not always lead to disease.
- **Primary pathogens** have mechanisms that help the organism circumvent host defenses.
- **Opportunistic pathogens** cause disease only in a compromised host.
- **Pathogenicity** refers to the mechanisms a pathogen uses to produce disease and how efficiently the organism causes disease, whereas **virulence** is a measure of disease severity.
- **Invasion** refers to a pathogen's entry into and then living inside of host cells. **Invasiveness** describes the ability of an organism to spread through tissues.
- **Host range** refers to how readily a pathogen can infect different animals.
- **Pathogens** must be able to attach to host cells, evade the immune system (at least temporarily), and obtain nutrients from the host.

Thought Question 2.1 Is a microbe with an LD_{50} of 5×10^4 more or less virulent than a microbe with an LD_{50} of 5×10^7? Why?

Thought Question 2.2 The bacterium *Salmonella enterica* serovar Typhimurium (*S.* Typhimurium) is usually transmitted to humans by contaminated food and causes a diarrhea that typically resolves without medical intervention. During infection, the organism remains localized to the intestine. A different serovar of *Salmonella*, *S.* Typhi, is also transmitted via contaminated food but can penetrate the intestinal wall and enter the lymphatic system and bloodstream, causing high fever and, in some cases, death. Clearly, *S.* Typhi is more virulent than *S.* Typhimurium to humans. However, when tested for lethal dose (LD_{50}) in mice, *S.* Typhi fails to kill any mice, whereas *S.* Typhimurium has a very low lethal dose (it takes very few organisms to kill mice). How might you broadly explain this apparent contradiction?

2.2 Basic Concepts of Disease

How do we define disease? A **disease** is a disruption of the normal structure or function of any body part, organ, or system that can be recognized by a characteristic set of symptoms and signs (described shortly). An **infectious disease**, then, is a disease caused by a microorganism (bacterial, viral, or parasitic) that can be transferred from one host to another. **Table 2.1** presents the terms commonly used to describe the features of infectious disease. It is critically important for you to learn these terms because they will be used repeatedly throughout this book and, very likely, in your future career.

Signs, Symptoms, and Sequelae

Most diseases are recognized by their signs and symptoms. A **sign** is something that can be observed by a person examining the patient; a runny nose, a rash, sweating, fever, and blood pressure are examples of signs. Brandon's lesion in the chapter-opening case was an obvious disease sign. A **symptom** is something that can be felt only by the patient, such as pain or general discomfort (malaise). Some signs can also be called symptoms if the patient alone witnessed it. For example, a rash that went away before a clinician saw it would be a symptom that is reported to the clinician. A **syndrome** is a collection of signs and symptoms that occur together and signify a particular disease.

Table 2.1
Terms Used to Describe Infections and Infectious Diseases

Term	Definition	Example(s)
Acute infection	Infection in which symptoms develop rapidly; its course can be rapid or protracted.	Strep throat (*Streptococcus pyogenes*)
Chronic infection	Infection in which symptoms develop gradually, over weeks or months, and are slow to resolve (heal), taking 3 months or more.	Tuberculosis (*Mycobacterium tuberculosis*)
Subacute disease	Infection in which symptoms take longer to develop than in an acute (rapid) infection but arise more quickly than for a chronic infection.	Subacute bacterial endocarditis (*Enterococcus faecalis*)
Latent infection	Type of infection that may occur after an acute episode; the organism is present but symptoms are not; after time, the disease can reappear.	Cold sores due to herpesvirus
Focal infection	Initial site of infection, from which organisms can travel via the bloodstream to another area of the body.	Boils (*Staphylococcus aureus*)
Disseminated infection	Infection caused by organisms traveling from a focal infection; when affecting several organ systems, it is called a "systemic" infection.	Tularemia (*Francisella tularensis*)
Metastatic infection	Site of infection resulting from dissemination.	Blastomycosis, fungal infection of the lung; can disseminate to form abscesses in the extremities (arms/legs)
Bacteremia	Presence of bacteria in blood; usually transient, little, or no replication.	May occur during dental procedures (*Streptococcus mutans*)
Septicemia	Presence and replication of bacteria in the blood (blood infection).	Plague (*Yersinia pestis*)
Viremia	Presence of viruses in the blood.	HIV
Toxemia	Presence of toxin in the blood.	Diphtheria, toxic shock syndrome
Primary infection	Infection in a previously healthy individual.	Syphilis (*Treponema pallidum*)
Secondary infection	Infection that follows a primary infection; damaged tissue (e.g., lung) is more susceptible to infection by a different organism.	Bacterial infection following viral influenza (*Haemophilus influenzae*)
Mixed infection	Infection caused by two or more pathogens.	Appendicitis (*Bacteroides fragilis* and *Escherichia coli*)
Iatrogenic infection	Infection transmitted from a health care worker to a patient.	Some septicemias (*Staphylococcus aureus*)
Nosocomial infection	Infection acquired during a hospital stay (hospital-acquired infection).	MRSA (methicillin-resistant *Staphylococcus aureus*)
Community-acquired infection	Infection acquired in the community, not in a hospital.	Some MRSA strains, influenza, HIV

It is important to realize that many of the signs and symptoms of an infectious disease are actually caused by the host's response to the infection (called **immunopathology**). Cells given the task of killing the microbe can also damage nearby host tissue, causing signs and symptoms such as runny nose, rash, and headaches. Even after an infectious disease has resolved, pathological consequences called **sequelae** (singular, **sequela**) may develop. For instance, the immune response to strep throat (caused by *Streptococcus pyogenes*) can produce heart damage (rheumatic fever) or kidney disease (acute glomerulonephritis) weeks <u>after</u> the infection has resolved.

Stages of an Infectious Disease

The stages (or phases) of a disease reflect how the contest between an infectious agent and a host progresses (**Figure 2.8**). At first, the microbe has the "upper hand" and the patient displays only minor, if any, signs or symptoms. After the immune system recognizes the threat, it begins to attack the invader, which can initially cause symptoms to <u>worsen</u>. As the immune system kills the pathogen, the symptoms decline in severity, and the host recovers. Most infections go through these stages, even when treated with antibiotics (although symptoms are much less severe).

The **incubation phase** is the time after a microbe first infects a person until the first signs of disease appear. Depending on the disease, patients may be contagious—that is, able to transmit the organism to someone else—during the incubation period, even though they themselves appear healthy. For instance, a child with chickenpox can start to shed virus 5 days before the rash appears. During the incubation period, the microbe is trying to replicate to larger and larger numbers before the immune system knows it's under attack. Actually, most infections never progress to active disease, because the invading microbe succumbs to early host defenses. This explains why people who have never had symptoms of Lyme disease, for example, can have antibodies against the bacterial agent, *Borrelia burgdorferi*, in their blood. How long the incubation period lasts before signs of disease appear depends on many factors, including the infectious dose, the host's health status, and the tools of pathogenicity available to the microbe.

The **prodromal phase** (or **prodrome**), which develops after the incubation phase, is short and may not even be apparent. It involves vague symptoms, such as headache or a general feeling of **malaise** (fatigue and mild body discomfort), that warn of more serious symptoms to come. For example, a child infected with measles virus will experience a prodromal phase marked by fever and a runny nose before the characteristic red, spotty rash develops. As with the incubation period, people in the prodromal phase can spread the microbe to others.

The **illness phase** begins when typical symptoms and signs of the disease appear. The pathogen reaches peak numbers during this phase. Symptoms start to develop once the immune system begins attacking the organism or virus. The point at which disease symptoms are most severe is called the acme. The battle between microbe and host is at its peak during this time. Fever is a sign or symptom often seen during the illness phase.

Fever, though uncomfortable, is actually a protective mechanism to fight infection. High body temperatures can help our immune system function better and can inhibit microbial growth. Children can generally tolerate a fever of 41°C (106°F), but for adults, 39.4°C (103°F) is about the safe limit. Above that temperature, brain damage can occur. How fever is triggered and why it is important is discussed further in Section 15.6, where inflammation and the immune response are covered.

Figure 2.8 Stages of an Infectious Disease

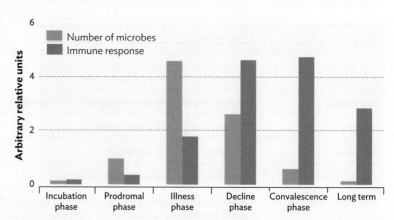

After initial infection, during the incubation phase, the number of organisms (or virus particles) rises in the body until the immune system recognizes that an invader is present; this is the prodromal phase. Symptoms start to develop at the same time the immune system begins attacking the organism or virus. During the illness phase, the pathogen reaches peak numbers. Once the immune system becomes more efficient at killing the pathogen, its numbers decrease (the decline phase). The immune system remains at the ready during convalescence, but its ability to attack the same invader will wane over the long term (years).

Once the immune system becomes efficient at killing the pathogen, the invader's numbers decrease (the decline phase). The immune system remains at the ready as the patient recovers (the convalescence phase), but its ability to attack the same invader will wane over the years.

The **decline phase** of a disease begins as the symptoms subside. Host defenses have won. As the infection recedes, fewer pyrogens (the compounds that trigger fever) are made, and the body's thermostat is reset to the lower temperature. To restore body temperature, blood vessels will dilate to lose heat (vasodilation), and the patient will start to sweat. These are the signs that a fever is "breaking."

Convalescence is the period after the decline phase, once symptoms have completely disappeared. The patient begins to recover strength and normal health.

When discussing infectious diseases, we often use the terms "morbidity" and "mortality." **Morbidity** refers to the existence of a disease state and the rate of incidence of the disease. **Incidence** refers to new cases of the disease that develop during the course of a year (see Chapter 26). You do not have to die to be included in a morbidity statistic; you only need to be sick. **Mortality**, on the other hand, is a measure of how many patients have died from a disease. The Centers for Disease Control and Prevention (CDC) publishes a report called the *Morbidity and Mortality Weekly Report (MMWR)* that discusses current outbreaks, statistics, and other health topics. Table 2.1 lists several other terms often used by medical professionals that you should master before proceeding.

SECTION SUMMARY

- **Diseases** are recognized by their signs and symptoms.
- **Symptoms** are caused by bacterial products and by the host immune response (immunopathology).
- **Stages of an infectious** disease include the following:
  - **Incubation period**, when the organism begins to grow but symptoms have not developed
  - **Prodromal phase**, which can be unapparent or show vague symptoms
  - **Illness phase**, when signs and symptoms are apparent and the immune system is fighting the disease
  - **Decline phase**, when the numbers of pathogens decrease and symptoms abate
  - **Convalescence**, when symptoms are gone and the patient recovers
- **Morbidity** is a measure of how many people are sick from an infectious disease. Mortality is a measure of how many died.

Thought Question 2.3 Why would quickly killing a host be a bad strategy for a pathogen?

Thought Question 2.4 Chapter 1 explained Koch's postulates and how they are used to determine the etiology of a disease (see Figure 1.13). The postulates say that you must find the same organism in each case of a disease in order to say that the organism causes the disease. Suppose three individuals with septicemia (a blood infection) have been admitted to the hospital. The clinical laboratory identifies *E. coli* in the blood of one, *Enterococcus* in the blood of another, and *Yersinia pestis* in the blood of the third. Do these findings defy Koch's postulates? After all, the same organism was not identified in each case.

2.3 Infection Cycles and Disease Transmission

SECTION OBJECTIVES

- Describe complex versus simple infection cycles.
- Differentiate between endemic, epidemic, and pandemic disease.
- Explain animal reservoirs and incubators.

How are diseases spread? Somehow, pathogens must pass from one person, or nonhuman animal, to another, if a disease is to spread. The route of transmission an organism takes to infect additional hosts is called the infection cycle. These routes can be direct or roundabout, as seen in the case history presented next.

CASE HISTORY 2.1
A Hike, a Tick, and a Telltale Rash

Emma Katherine, a 21-year-old college student in Massachusetts, suddenly developed fever, nausea, headache, and muscle pain. A few days later, a rash appeared on her forearms—small, flat, pink, non-itchy spots (**Figure 2.9**). She went to her physician, who took a patient history. The clinician asked, among other things, what she had been doing over the past month. (On pp. 40–41 we discuss the role of patient histories in diagnosing infectious diseases.) She told him about a trip she had taken to North Carolina 3 weeks earlier, where she hiked through the

Figure 2.9 Rash Associated with Rocky Mountain Spotted Fever

The rash is caused by the immune system's reaction to the presence of the organisms. Rocky Mountain spotted fever will be described in more detail in Chapter 21.

The Crucial Art of Diagnosing Infectious Disease

How Do Health Care Providers Determine What Ails You?

S uddenly you feel very ill. You don't think it's serious, but you go to the clinic, trusting that the clinician there will diagnose your illness and prescribe an effective treatment. How does the clinician make the right diagnosis?

This book describes many infections that afflict humans and the pathogens that cause those diseases. At the book's core are numerous case histories woven into every chapter, not just into the infectious disease chapters (Chapters 19–24). Here we describe a skill that is critical for diagnosing disease and integral to every case history: the taking of a patient history.

Patient Histories: A "Gold Mine" of Diagnostic Clues

When you go to the clinic, what is the first thing that happens when the physician, physician assistant, or nurse practitioner walks into the room? You are asked questions such as "What brings you here today?" or "How long have you had these symptoms?" and perhaps "Where have you traveled recently?" This is not idle conversation; the clinician is taking a patient history (Figure 1).

But what is a patient history, and why is it so important? Many infectious diseases display similar symptoms, making diagnosis of a specific disease difficult. However, the patient history provides critical clues that can guide the clinician to the possible culprit. For instance, *Vibrio cholerae* and enterotoxigenic *E. coli* both produce diarrheal diseases characterized by cramps, lethargy, and liters of watery stool each day. So, when a patient presents these symptoms, how does the clinician intuit which microbe is the cause (even before seeing lab results)? Although cholera is not commonly seen in the United States, a clinician might suspect cholera if the patient recently traveled to or emigrated from an endemic area where the disease is regularly observed (India, for example). Of course, this travel information is not gained by examining the patient but comes from talking with the patient and taking a patient history.

A **patient history**, or medical history, is the written summary of a carefully choreographed dialogue between clinician and patient. It is composed of several parts:

- **Chief complaint.** A statement of what prompted the patient's visit

- **History of present illness.** A chronological summary of the current illness from its inception to the present (often prompted by the question, "When did your symptoms start?")

- **Medical history.** A list of prior illnesses

- **Social history.** Marital status, occupation, travel history, hobbies, stresses, diet, sexual history, or alcohol or drug use (legal or otherwise)

Figure 1 **Clinician Taking a Patient History**

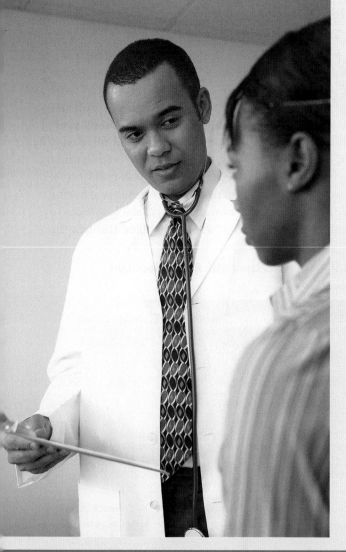

- **Family history.** Present health or cause of death of parents and any siblings (very useful for determining genetic diseases)

Many questions asked while taking a patient's social history can seem irrelevant or intrusive to the patient, who only wants relief from the symptoms. For instance:

- Do you have any hobbies?
- What is your occupation?
- What foods have you eaten recently?
- Does your child attend day care?
- How long have you been in your current sexual relationship?

Although these questions may sometimes raise eyebrows, the answers can provide important diagnostic clues to the astute clinician. For example, learning that a man suffering from enlarged glands, fever, and headaches also likes to hunt and recently killed rabbits can be an important clue (**Figure 2**). The patient may have been exposed to the Gram-negative rod-shaped bacterium *Francisella tularensis*, a facultative intracellular pathogen that infects various wild animals and is the cause of tularemia, an illness also known as "rabbit fever." The patient could have accidentally infected a cut while cleaning his "kill."

Or consider the case of a woman with acute pneumonia. The clinician discovers from taking a patient history that the woman is a sheep farmer. Knowing this, the clinician considers *Coxiella burnetii* as one of the <u>possible</u> causes of the infection (the list of possible causes is called the **differential**). *C. burnetii* is a pathogenic bacterial species that infects sheep and is shed in large quantities in the animal's amniotic fluid or placenta. Dried soil contaminated with *C. burnetii* can become aerosolized (airborne) whenever the dirt is disturbed, and any human (such as a farmer) who inhales the dried particles can develop the lung infection Q fever. *C. burnetii* would be lower on the differential if the woman were an accountant.

Keep the idea of patient history in mind as you read the numerous case histories to come. Take note of how often a patient history can tilt a diagnosis away from one suspected pathogen and toward another. In the end, you will have a greater understanding of infectious diseases and will be better able to appreciate your clinician's diagnostic skills. Perhaps you can even see yourself as a health care provider or a clinical laboratory scientist.

Figure 2 Patient with Tularemia

Francisella tularensis is a Gram-negative bacillus that causes tularemia, a zoonotic bacterial disease typically transmitted by direct contact with rodents such as rabbits. The man in this photo had a swollen lymph node in his neck (called cervical lymphadenitis), a fever of 38.8°C (101°F), fatigue, and weakness.

SEM 1 μm

Francisella tularensis
A murine macrophage infected with *Francisella tularensis* strain LVS (arrow).

woods for a couple of days. One day she found an ugly tick on her back, which she quickly removed. She didn't encounter any other people or animals during her trek. About a week later, she developed the symptoms that drove her to the doctor's office. Emma's signs, symptoms, and travel history led the doctor to suspect that Emma might have Rocky Mountain spotted fever. How could Emma have contracted this disease if she never encountered anyone who might be infected?

Rocky Mountain spotted fever is caused by the bacterium *Rickettsia rickettsii*. Although initially recognized in the Rocky Mountain states, most cases of the disease occur on the East Coast. The organism lives in infected ticks without harming them. However, the bite of an infected tick can transmit the organism (present in tick salivary secretions) to unsuspecting humans and cause serious, life-threatening disease. This case illustrates several concepts of disease transmission and infection cycles.

⊙ **Note:** This chapter is an introductory chapter. In-depth discussions of the infection cycles of specific organisms are presented in Chapters 19–24. Chapter 26 includes thorough coverage of epidemiology, including hospital- and community-acquired infections.

Modes of Transmission

For an infectious disease to spread through a community, the pathogen must pass, or be transmitted, from one person or animal to another. The route of transmission is called an **infection cycle** (**Figure 2.10**). Terms used to describe infection cycles are provided in **Table 2.2**. The two major types of transmission are **horizontal transmission**, in which the infectious agent is transferred from one person or animal to the next (Figure 2.10A and B), and **vertical transmission**, whereby an infectious agent is transferred directly from parent to offspring (Figure 2.10C). One example of vertical transmission is provided by the cause of syphilis (*Treponema pallidum*), which can travel through the placenta to the fetus (**transplacental transmission**). In another example, the organism that causes gonorrhea (*Neisseria gonorrhoeae*) doesn't pass through the placenta but can contaminate a newborn during birth (parturition; see Figure 2.10C). Both of these pathogens can be horizontally transmitted as well.

There are at least five forms of horizontal transmission. Pathogens that spread horizontally from person to person, such as rhinovirus, are often transferred by **direct contact** via handshaking or even sexual contact (see Figure 2.10A). Handshaking is a very efficient way to horizontally transmit pathogens. An infected

Figure 2.10 Horizontal and Vertical Infection Cycles

A. Direct horizontal transmission happens between people (sneezing or touch) or between a person and an animal reservoir.

B. Indirect transmission is carried out by vehicles such as inanimate objects (fomites), air (sneeze), food, or water.

C. Vertical transmission includes transplacental transmission to the fetus and contamination of the newborn during birth.

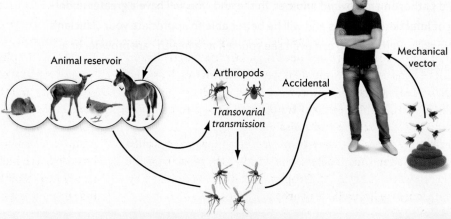

D. Arthropod vectors (mosquitoes or ticks) feed on animal reservoir hosts and infect human hosts when humans unintentionally encounter an animal host-insect vector cycle. Some insects can infect their own eggs during their formation via transovarial transmission. The transport of a pathogen to a susceptible host when an insect vector (fly) lands on fecal matter is an example of mechanical transmission.

Table 2.2

Terms Used to Describe Disease Transmission and Frequency

Term	Definition	Examples
Direct contact	Intimate interaction between two people.	Touching, kissing, sexual intercourse (mononucleosis or gonorrhea)
Indirect contact or transmission	Transmission of an infectious agent from one person to another by an insect vector or an inanimate object.	Mosquito (malaria); sharing a spoon or fork (strep infection)
Vehicle transmission (fomites)	Form of indirect transmission whereby an infectious agent is transferred to an inanimate object (fomite) by one person touching it and then transferred to another person touching the same object, or by ingesting contaminated food or water, or by inhaling the agent in air.	Doorknobs, shared utensils (influenza, strep infection)
Vehicle	Means of pathogen transmission, as by air, food, or liquid.	Air (*anthrax*), food (*Salmonella*), water (*Leptospira*)
Vector	Living carrier of an infectious organism.	Mosquitoes (West Nile), fleas (*Yersinia pestis*), body lice (*Rickettsia*)
Aerosol	Organisms in air suspension.	Sneezing (rhinovirus), air-conditioning or heating (*Legionella*), water from a shower (*Mycobacterium avium*)
Reservoir	Nonhuman animal, plant, human, or environment that can harbor the organism; a reservoir may or may not exhibit disease.	Cattle (*E. coli* O157:H7), horses (West Nile virus), alfalfa sprouts (*Salmonella*)
Fecal-oral route	Pathogen exits the body in feces, which contaminates food, water, or fomite; pathogen is introduced into a new host by ingestion.	*Shigella, E. coli, Salmonella*, rotavirus
Respiratory route	Pathogen enters the body through breathing.	Influenza, *Streptococcus pneumoniae*
Urogenital route	Pathogen enters the body through urethra or vagina.	Urinary tract infections (nonsexual transmission), syphilis (sexual contact)
Parenteral route	Pathogen enters the body through insect bite or needle injection.	Malaria, HIV (AIDS)
Endemic	Disease (humans) is present in a specific geographic location; pathogen is usually harbored in an animal or human reservoir.	Lyme disease, common cold
Epidemic	Number of disease cases exceeds the endemic level.	Influenza, plague
Pandemic	Worldwide epidemic.	Influenza, plague, acquired immunodeficiency syndrome (AIDS), COVID-19 (Coronavirus)

person whose hand is contaminated with a viral or bacterial pathogen can pass the microbes to several people just by shaking their hands. Newly contaminated people can then infect themselves by touching their eyes or mouth.

Rhinovirus and other pathogens can also be passed person-to-person <u>indirectly</u> through air by sneezing (**airborne transmission**) <u>or</u> through the sharing of inanimate objects (**fomites**) such as contaminated utensils (fork or pen), towels, cloth handkerchiefs, and doorknobs (Figure 2.10B). **Indirect transmission** via fomites, food, or water is also called **vehicle transmission**.

Sneezing produces an **aerosol** of secretion particles, called droplet nuclei, laden with bacterial or viral pathogens from the respiratory tract (**Figure 2.11**). Droplet nuclei can be inhaled by, <u>and</u> can infect, a new host who is nearby.

Many gastrointestinal infectious agents can be indirectly transmitted between people or animals when food or water becomes contaminated with fecal material containing the pathogen. This is why handwashing after using the bathroom is an important control measure.

More complex horizontal transmission cycles involve **vectors**, usually insects or ticks (arthropods), and **reservoirs**, animals or environments that harbor an infectious agent (discussed more in the next section). The housefly is a simple **mechanical vector** that can transmit disease by landing on contaminated material (for example, feces) and then carrying any pathogens present to a living host such as a human or nonhuman animal, or to food (for example, an oatmeal cookie). In contrast, mosquito and tick vectors transmit infectious agents by biting an infected reservoir host. They ingest

Figure 2.11 **Spreading Respiratory Pathogens by Sneezing or Coughing**

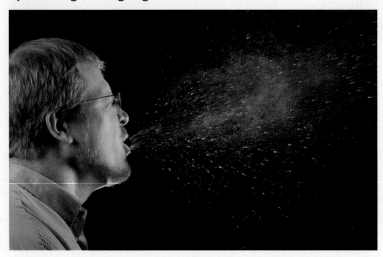

Sneezing or coughing will aerosolize respiratory pathogens, such as *Mycobacterium tuberculosis* (the cause of tuberculosis) and rhinoviruses (which cause common colds). Droplets from a sneeze can travel far in some circumstances (when carried by the wind, for example), but generally you can limit your exposure to aerosols by standing 10–15 feet away from someone who is coughing or sneezing. Contaminated aerosolized droplets tend to arc toward the ground over that distance.

the pathogen, then pass it to a new host during a subsequent blood meal (Figure 2.10D). The tick in Case History 2.1 served as the vector that passed *Rickettsia rickettsii* from infected rodents in the forest to Emma Katherine (Case History 2.1). An infectious disease that is primarily seen in animals but can be transmitted to humans, either by vector or by other means, is called a **zoonotic disease** (examples include West Nile virus and Lyme disease).

Figure 2.12 **Insect Vector and Yellow Fever**

A. Walter Reed, a member of the University of Virginia Medical School class of 1869, proved in 1901 that the mosquito *Aedes aegypti* is the vector of transmission for yellow fever, a disease named for the jaundice produced by liver damage.

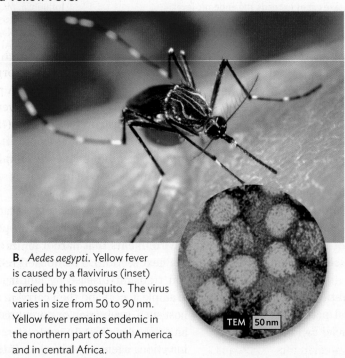

B. *Aedes aegypti*. Yellow fever is caused by a flavivirus (inset) carried by this mosquito. The virus varies in size from 50 to 90 nm. Yellow fever remains endemic in the northern part of South America and in central Africa.

TEM 50 nm

Sometimes a vector can also serve as the reservoir for an infectious disease. The mosquito vector *Aedes* can horizontally transfer many viral pathogens, including the yellow fever virus, from infected to uninfected individuals (**Figure 2.12**). A major outbreak of yellow fever is currently (2019) plaguing areas of Africa as well as Central and South America. An infected *Aedes* mosquito can also bequeath the yellow fever virus to its offspring via infected eggs. This is a form of vertical transmission called **transovarial transmission** (again, see Figure 2.10D).

◉ Note: Only female mosquitoes and ticks take blood meals. They need the blood to develop their eggs. Male mosquitoes do not bite but feed on plant nectar. Adult male ticks will attach to a host but do not feed and cannot transmit disease. Once a male tick mates with a female, the male dies.

Although yellow fever is not a problem in the United States today, a closely related virus called West Nile virus (WNV) claims several victims each year. WNV is also transmitted to humans by a mosquito vector (*Culex*) that transovarially passes the virus though its eggs. Because insects and ticks can transmit many pathogens, strategies to repel or kill arthropods, or prevent them from laying eggs, are effective ways to halt the spread of disease. Interventions include spraying insecticide in a community during egg-hatching season or using other microbes as assassins "trained" to kill the vector. For example, the bacterium *Bacillus thuringiensis* will kill many types of insects that carry infectious agents. More recently, an insect virus called baculovirus has been developed that kills the *Culex* mosquito vectors carrying West Nile virus. The advantage of these vector-targeting microbes is that, unlike many chemical insecticides, they do not kill other insects or animals.

Reservoirs of Infection

As mentioned earlier, another critical factor in an infection cycle is the reservoir of infection. A **reservoir** is an animal (such as a bird, rat, horse, insect, or arthropod) or an environment (such as soil or water) that normally harbors the pathogen. For yellow fever, monkeys serve as the animal reservoir. However, the mosquito vector is also a reservoir for this virus because the insect can pass the virus to future generations of mosquitoes through vertical transmission. Likewise, the tick in the Rocky Mountain spotted fever case history served as both reservoir and vector.

In contrast to the yellow fever virus, the virus causing eastern equine encephalitis (EEE, a potentially lethal brain infection seen in some parts of the United States) uses birds as a reservoir. The microbe is normally a bird pathogen transmitted from bird to bird via a mosquito vector. The virus does not persist in the insect, but transmission to new avian hosts keeps the virus alive. Humans or horses entering geographic areas harboring the disease (called endemic areas) can also be bitten by the mosquito. When this happens, they become accidental hosts and contract the disease. The EEE virus does not replicate to large numbers (titers) in mammals, so horses and humans are poor reservoirs for the virus. In avian hosts, however, the virus does replicate to large numbers. Reservoirs are crucial for the survival of a pathogen and serve as a source of infection. If the EEE virus had to rely on humans to survive, it would cease to exist because of limited replication potential and limited access to mosquitoes. Remember: the reservoir of a given pathogen might not exhibit disease.

A special type of reservoir is the asymptomatic **carrier**—a person who harbors a potential disease agent but does not have the disease. This is how *Neisseria meningitidis*, an important cause of meningitis, remains in a population. The bacterium has no animal reservoir other than humans, so where does it go between outbreaks? It colonizes the nasopharynx (the area behind the nose down to the throat) of unsuspecting hosts whose immune systems keep the bacterium out of the bloodstream. A simple sneeze from the carrier, however, can aerosolize the pathogen and transfer it to a susceptible host, who then contracts the disease and passes the organism to others.

Endemic, Epidemic, or Pandemic?

CASE HISTORY 2.2
The Third Pandemic

In 1884, Yong Ding was a 38-year-old male who worked as a cook in Canton, located in southern China. Life was hard, but he still managed to support his family. As he walked to his restaurant each day, he barely noticed the small bands of rats scurrying through the streets. Disease was rampant that year; victims of the *Shuyi* (rat epidemic) were stacked like firewood in the streets, five bodies high in places, waiting to be taken for burial (Figure 2.13). Yong Ding knew that the disease started as a swollen gland in the armpit that often had a black appearance (now called a bubo). He checked himself daily for these swellings and, seeing none, always felt relief. Then, one morning, he found one. Within days, Yong Ding began coughing blood as the agent (a mystery at the time) spread though his bloodstream to his lungs. Once that happened, he knew death was not far behind. Within weeks, more than 60,000 people died this way in Canton—victims of what would be called the Third Pandemic of bubonic plague. Yong Ding's body was one of many lining the streets that year.

This case is about bubonic plague, caused by the bacterium *Yersinia pestis*. *Y. pestis* is transmitted from infected rats to humans

Figure 2.13 **Chinese Plague Victims, 1910**

To stem the plague pandemic, the bodies of plague victims were piled into "plague pits" and burned. The homes of these people were also burned to the ground.

by the rat flea, whose bite delivers bacteria-laden saliva. Similar to the earlier case of rickettsiosis (Case History 2.1), this case involves vectors, reservoirs, and infection cycles, but it also introduces the concept of pandemics. Several devastating pandemics of plague have occurred over the centuries, including the Black Death that killed more than a third of Europe's population in the fourteenth century. Plague is still with us today. It is endemic in Madagascar, where 200–700 cases are seen every year, and even the United States sees a few cases most years (see Chapter 21).

We have all heard the terms "endemic," "epidemic," and "pandemic" in reference to disease, but what do they mean? An **endemic disease** is one that is always present in a community and seen at a low rate. Often an animal reservoir harbors the disease-causing organism. An **epidemic** occurs when many cases develop in a community over a short time. An organism need not be endemic to an area to cause an epidemic. A **pandemic** is essentially a worldwide epidemic, although not every country needs to be affected. The Third Pandemic of plague (1855–1959) that started in China spread to many other areas of the globe, carried by people (and rats) as they traveled. In India alone, an estimated 12 million people died of plague over 30 years spanning the late nineteenth and early twentieth centuries.

Although many microbes can cause epidemics, surprisingly few have proved capable of causing pandemics. The major organisms known to have caused pandemics are the bacteria *Y. pestis* (bubonic plague) and *Vibrio cholerae* (cholera); and the viruses influenza, HIV, and, most recently, the coronavirus SARS-CoV-2 (COVID-19 disease) (Table 2.3). You might wonder why more organisms do not cause pandemics. There are several reasons. The organism must be easily transmitted from person to person. Such transmission is most readily accomplished if the microbe causes a respiratory

Table 2.3

Examples of Deadly Pandemics

Disease	Organism	Years (Number of Deaths)	Link to Disease Coverage
Influenza	Influenza virus	1918–1919 (40 million–50 million dead) 1957 (2 million dead) 1968 (1 million dead) 2009 (150,000–500,000 dead)	Chapter 20
Bubonic plague	*Yersinia pestis*	541–542 CE (Plague of Justinian[a]; the number of dead is unclear but is in the millions) 1347–1351 (Black Death; 25 million dead) 1855–1959 (Third Pandemic; 137 million dead)	Chapter 21
Cholera	*Vibrio cholerae*	Seven pandemics: the first, in 1816, began in India (over 15 million dead); the third, in 1852, began in Russia (over 1 million dead); the seventh, in 1961, began in Indonesia (ongoing)	Chapter 22
AIDS	Human immunodeficiency virus	1981– (over 25 million dead)	Chapter 23
COVID-19	Coronavirus SARS-CoV-2	2019– (ongoing)	Chapter 26

[a]Justinian I (483–565) was an emperor of the Byzantine Empire.

disease and becomes airborne as a result of coughing. The organism must also cause disease at a relatively low infectious dose.

HIV is not an airborne infectious agent, so why is AIDS a pandemic? HIV does not rapidly kill its host, and because the virus tampers with the host immune system, the virus is not readily killed. Thus, over a person's life, ample opportunity exists to transmit the virus to a partner, especially considering that the symptoms of AIDS can take years to appear. During this extended prodromal period, the virus is present in an assortment of body fluids and is easily transmitted by various sexual practices. Having multiple sexual partners, straight or gay, combined with frequent air travel and an inability to be cured has led infected people to rapidly spread this virus around the globe.

Note that in some parts of the world, such as sub-Saharan Africa, HIV has actually become endemic. An infection like HIV that relies on person-to-person transmission becomes endemic when each person who becomes infected with the disease, on average, passes it to one other person. In this way, the infection doesn't die out, nor does the number of infected people increase exponentially; the infection is now said to be in an endemic steady state. An infection that starts as an epidemic will eventually either die out or reach the endemic steady state, depending on disease virulence and mode of transmission.

You might now ask, if AIDS is considered a pandemic, why isn't the common cold considered a pandemic? Every country around the globe deals with seasonal epidemics of the common cold. Why not call it a pandemic? The reason here is technical. Pandemics are caused by a single strain or closely related strains of a single microbe—for example, H1N1 influenza. The common cold, however, is caused by many different types of viruses (rhinoviruses, coronaviruses, adenoviruses) and many strains of viruses (160 for rhinoviruses alone). Plus, common-cold symptoms are not severe,

and people don't typically die. Viral replication, pathogenesis, and the diseases of influenza and the common cold will be described in more detail in Chapters 12, 18, and 20.

Note: Two potentially lethal strains of coronavirus, MERS-CoV and SARS-CoV, have emerged over the past two decades in regions of the Middle East and Asia, respectively. However, the outbreaks involved small numbers of people and never became a pandemic. A new respiratory coronavirus, SARS-CoV-2, has caused a pandemic disease called COVID-19. How COVID-19 became a pandemic is described in Chapter 26.

Zoonotic Diseases

As noted earlier, infections that normally afflict animals but can be transmitted to humans are called **zoonotic diseases**. Humans typically contract a zoonotic disease after accidentally encountering the animal reservoir. Bubonic plague is a case in point. The bacterium usually infects rodents but can be transferred to humans by a flea bite. Other examples include Lyme disease (*Borrelia burgdorferi*), eastern equine encephalitis (EEE virus), and Rocky Mountain spotted fever (*Rickettsia rickettsii*), whose agents typically infect white-footed mice, birds, and ticks, respectively, but can infect humans if they get close enough.

Animal and insect reservoirs can also function as "incubators" for new infectious diseases yet to emerge in humans. For example, two different strains of influenza virus can infect the same animal (usually pigs) at the same time. When both virus strains infect the same cell, the viruses can exchange discrete chunks of their RNA genomes, and a new virus—more infectious and deadly than either of the original forms—can emerge. Here are some other examples of how infectious agents are thought to have crossed between species.

- We gave lions ulcers. It is thought that a large cat dining on the entrails of one of our ancestors became infected with the bacterium *Helicobacter pylori*, the major cause of human stomach ulcers. Today, large cats suffer from ulcers.

- Cats gave us *Toxoplasma gondii*, a parasite that infects the brains of more than half the human population. It is thought that this organism may even contribute to schizophrenia.

- HIV, the virus that causes acquired immunodeficiency syndrome (AIDS), may have originated from chimpanzees and crossed to humans a century ago.

- Rabies virus, which causes a deadly and painful neurological disease, is transmitted from dogs to humans by dog bite. (However, there is no report of a human transferring the virus back to dogs by human bite.)

Clearly, the animal kingdom serves as an important incubator of new infectious diseases that can cross species.

Figure 2.14 Portals of Entry/Modes of Transmission

Entry via the eye. The conjuctiva (a membrane lining the eye) is subject to infection by some organisms carried there by fingers or air.

Respiratory route. Airborne pathogens can be inhaled.

Fecal-oral route. Pathogens can be ingested (e.g., through contamination of food or water).

Parenteral route. Pathogens can be injected into the bloodstream (e.g., by insect vector or contaminated needle).

Entry through skin. Skin can be colonized by some pathogens and penetrated by others. Wounds, too, offer pathogens access to the body's interior.

Urogenital route. Pathogens can enter via direct contact with genital mucosa or sexual transmission.

SECTION SUMMARY

- **Infectious agents can be spread** by aerosols, inanimate objects (fomites), direct contact, or vectors.

- **Insect vectors can horizontally transmit disease** from animals to humans (zoonotic disease) or vertically from insect parent to offspring (vertical transmission).

- **Transplacental transmission** is the transfer of a pathogen across the placenta from human mother to fetus.

- **A disease reservoir** is a site where an infectious agent survives and thus serves as a source of infection. Animals (including humans), plants, soil, and water can serve as reservoirs of infectious agents.

- **Endemic disease** is always present in an area, usually in small numbers, compared with **epidemic disease**, in which the number of victims suddenly rises above endemic levels. **Pandemics** are worldwide epidemics.

Thought Question 2.5 Yellow fever is a viral disease found primarily in South America and Africa. The symptoms include a high fever, headache, and muscle aches, and the disease can lead to kidney and liver failure. Liver damage produces a yellowing of skin (jaundice), giving the disease its name. The virus can be spread directly from one person to another by way of a mosquito whose bite can transmit the virus from an infected person to an uninfected person. However, HIV, another virus found in blood, cannot be transmitted by a mosquito bite. Why might that be?

2.4 Disease Portals and Biosafety

SECTION OBJECTIVES

- Describe the various portals of entry and exit for microbial pathogens.

- Discuss concepts of biosafety and biocontainment.

How do infectious agents gain access to the body? Each organism is adapted to enter the body in different ways (**Figure 2.14**). Food-borne pathogens (for example, *Salmonella, E. coli, Shigella,* and rotavirus) are ingested by mouth and ultimately colonize the

Figure 2.15 Guarding a Portal of Entry

Seattle policemen wearing protective gauze face masks during the influenza pandemic of 1918, which claimed millions of lives worldwide.

intestine. They have an oral portal of entry (**fecal-oral route** of infection). Airborne organisms, in contrast, infect through the respiratory tract (**respiratory route**). Some microbes enter through the conjunctiva of the eye; others, through the mucosal surfaces of the genital and urinary tracts (**urogenital route**). Agents that are transmitted by mosquitoes or other insects enter their human hosts via the **parenteral route**, meaning injection into the bloodstream. Wounds and needle punctures can also serve as portals of entry for many microbes. Shared needle use between drug addicts has been a major factor in the spread of HIV and hepatitis C.

Denying a pathogen access to its portal of entry is an effective way to halt the spread of disease. For example, condoms greatly limit the spread of sexually transmitted diseases. And face masks were used at the height of the deadly 1918 influenza pandemic in an attempt to cloak the respiratory tract (**Figure 2.15**). Simple behaviors like covering a sneeze and handwashing can limit the spread of respiratory diseases.

"Picking" a Portal

Why does a pathogen use one portal of entry and not another? Several factors impact why a pathogen "prefers" one portal over another. One factor is the reservoir for the organism. Is the agent found in contaminated food (as for *Salmonella* bacteria) or can it be aerosolized by a sneeze (like influenza virus)? The first example favors ingestion; the second, inhalation. Is its natural reservoir an animal (as with *Francisella*, the subject of Case History 2.3)? If so, then organisms can be transmitted to hunters who cut themselves while cleaning a dead animal (the parenteral route) or by insect vectors that can feed on the animal and then on humans.

Portal-of-entry preference is also related to the pathogen's attachment capability. Some pathogens have access to many different entry points but cannot establish infection unless they enter by the appropriate route. Different body tissues have

different receptor structures on their surface. The bacterium or virus must have a compatible receptor in order to bind. If it can't bind, the disease-causing organism can be expelled by the host. HIV, for instance, is not transmitted through respiration or by ingestion because the virus must enter the bloodstream to infect susceptible T lymphocytes, a type of white blood cell (Chapter 16).

Once a pathogen gets in the body, how does it get out? Hosts have a limited life span. When the host dies, the pathogen can die with it, owing to the lack of food. Even in a live host, microbial growth can be limited by the space in which the pathogen can grow. So, to continue propagating, most pathogens need to leave and find new hosts. You will learn more about microbial growth in Section 6.2.

Do pathogens leave the same way they came in? Not always. Obvious examples are diarrhea-causing bacteria and viruses, which enter by ingestion but leave by defecation, where they can again be ingested when food, water, or hands become contaminated (this is the **fecal-oral route** of transmission). Some organisms gain entrance through one route but disseminate to other organs, through which they then exit. A good example is *Yersinia pestis*, the agent in Case History 2.2. The bacterium enters the body by the parenteral route (via a flea bite) but spreads to the lung and is expelled by coughing (the respiratory route). On the other hand, some microbes do use the same portal to enter and exit the body. Influenza virus both comes and goes via the respiratory tract. Other microbes can enter and exit through multiple portals. For instance, HIV can enter and leave the body by the urogenital route or parenterally through injection and shared needles.

Note: In Chapters 19–24, which describe infectious diseases in greater detail, you will learn more about the portals of entry and exit for specific pathogens. Each of these chapters concludes with a summary figure that diagrams the portals of entry and exit of the key pathogens discussed in the chapter.

Biosafety and Universal Precautions

CASE HISTORY 2.3
Risky Business in the Lab

In 2004, three people in Boston came down with a virulent form of pneumonia. An investigation by public health officials revealed that all three worked at the same laboratory studying *Francisella tularensis*, a highly infectious bacterium (although not usually spread by person-to-person contact). Under specific conditions, the organism can be aerosolized, inhaled, and cause deadly pneumonia—making it a possible bioterrorism agent. Its handling is highly restricted by the US Department of Homeland Security. Scientists studying this bacterium must use extreme precautionary measures to prevent infecting themselves and to ensure that it cannot escape the laboratory. The researchers in this case, however, thought they were working with a safer vaccine strain of *F. tularensis* that required only moderate-level safety conditions. The investigation determined that they had indeed contracted tularemia and that the samples were somehow contaminated with highly virulent *F. tularensis*. Because the researchers thought the material was safe, they handled the organisms in several instances without wearing or using proper protective gear—for example, agar plates containing the organism were examined outside of a biosafety containment hood. The type of disease and the laboratory procedures performed suggested that the victims, who all fully recovered, had inhaled the organism while working with it.

Medical and laboratory personnel are exposed to extremely dangerous pathogens on a daily basis. The microbiologists in this case history did not protect themselves properly because they thought they were working with a nonvirulent strain of *F. tularensis*. As a result, they ended up with a laboratory-acquired infection leading to tularemia. The CDC has published a series of regulations designed to protect laboratory workers and medical personnel who are at risk of infection by human pathogens. These are called "Universal Precautions" and "Standard Precautions," respectively. Universal Precautions are based on the idea that all blood and body fluids should be treated as infectious because patients with blood-borne infections can be asymptomatic. Standard Precautions are similar but apply to all patient care, regardless of suspected or confirmed infection status. The eight Standard Precautions are as follows:

1. **Perform proper hand hygiene.** This is the most important measure to prevent the spread of infections. Handwashing with soap and water or with a waterless hand sanitizer is the most effective method.

2. **Use personal protective equipment (PPE).** Gloves, masks, and/or eyewear should be used when exposure to infectious material is possible (**Figure 2.16**).

3. **Maintain respiratory and coughing hygiene.** Cover your mouth and nose when coughing or sneezing; wear a surgical mask when examining a patient with a respiratory infection, especially if fever is present. (In the laboratory, use a biohazard hood when handling potentially hazardous materials.)

4. **Use safe injection practices.** Maintain aseptic technique during injection procedures (wear gloves, disinfect injection area).

5. **Handle sharps safely** (syringes, needles, others). Medical and laboratory personnel who accidentally stick themselves with a syringe needle used on a patient can become infected by any blood-borne pathogen (known or unknown) that may be present. For example, if you accidentally stick

Figure 2.16 Clinician Examining a Patient Hospitalized with Suspected Flesh-Eating Disease (Necrotizing Fasciitis)

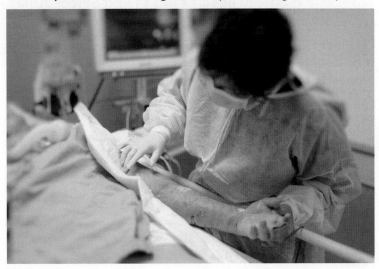

Notice the disposable gown, gloves, and mask.

yourself with a needle used on a hepatitis B patient and have not received the hepatitis B vaccine, you have a one in five chance of contracting HepB. Precautions include never removing a needle from a syringe by hand and carefully discarding sharps in a tamper-resistant, red plastic container marked "sharps."

6. **Practice patient isolation precautions** to prevent transmission of infectious agents. Steps include sanitizing hands before entering and after leaving the infected patient's room; donning protective clothing, masks, and eye protection (as needed) before entering the room; and removing PPEs and disposing of them in marked containers before leaving the room.

7. **Sterilize instruments and devices.**

8. **Clean and disinfect environmental surfaces.**

We will discuss sterilization and disinfection techniques more fully in Chapter 13.

In the laboratory, it is important to prevent the transmission of an infectious agent to laboratory personnel and the accidental release of agents into the environment. Infectious agents are ranked by the severity of disease and ease of transmission. On the basis of this ranking, four levels of agent containment are employed, as indicated in Table 2.4.

Biosafety level 1 organisms have little to no pathogenic potential and require the lowest level of containment. Standard sterile techniques and laboratory practices are sufficient.

Biosafety level 2 agents have greater pathogenic potential, but vaccines and/or therapeutic treatments (for example, antibiotics) are readily available. Biosafety level 2 agents require somewhat rigorous containment procedures, such as limiting laboratory

Table 2.4
Biological Safety Levels (BSLs) and Select Agents[a]

	Biosafety Level			
	BSL-1	**BSL-2**	**BSL-3**	**BSL-4**
Class of Disease Agent	Agents not known to cause disease.	Agents of moderate potential hazard; also required if personnel may have contact with human blood or tissues.	Agents may cause disease by inhalation route.	Dangerous and exotic pathogens with high risk of aerosol transmission; only 13 labs in the United States handle these.
Recommended Safety Measures	Basic sterile technique; no mouth pipetting.	Level 1 procedures plus limited access to lab; biohazard safety cabinets used; hepatitis vaccination recommended.	Level 2 procedures plus ventilation providing directional airflow into room, exhaust air directed outdoors; restricted access to lab (no unauthorized people); respiratory protection and wraparound gowns.	Level 3 procedures plus one-piece positive-pressure suits; lab is completely isolated from other areas present in the same building or is in a separate building.
Representative Organisms in Class	*Bacillus subtilis* *E. coli* K-12 *Saccharomyces* spp.	*Bordetella pertussis* *Campylobacter jejuni* *Chlamydia* spp. *Clostridium* spp. *Corynebacterium diphtheriae* *Cryptococcus neoformans* *Cryptosporidium parvum* Dengue virus Diarrheagenic *E. coli* *Entamoeba histolytica* *Giardia lamblia* *Haemophilus influenzae* *Helicobacter pylori* Hepatitis virus *Legionella pneumophila* *Listeria monocytogenes* *Mycoplasma pneumoniae* *Neisseria* spp. *Salmonella* spp. *Shigella* spp. *Staphylococcus aureus* *Toxoplasma gondii* Pathogenic *Vibrio* spp. *Yersinia enterocolitica*	*Bacillus anthracis* (anthrax) *Brucella* spp. (brucellosis) *Burkholderia mallei* (glanders) California encephalitis virus *Coxiella burnetii* (Q fever) EEE (eastern equine encephalitis) virus *Francisella tularensis* (tularemia) Japanese encephalitis virus La Crosse encephalitis virus LCM (lymphocytic choriomeningitis) virus *Mycobacterium tuberculosis* Rabies virus *Rickettsia prowazekii* (typhus fever) Rift Valley fever virus SARS (severe acute respiratory syndrome) virus Variola major (smallpox) and other poxviruses VEE (Venezuelan equine encephalitis) virus West Nile virus (WNV) Yellow fever virus *Yersinia pestis*	Ebola virus Guanarito virus Hantavirus Junin virus Kyasanur Forest disease virus Lassa fever virus Machupo virus Marburg virus Tick-borne encephalitis viruses

[a]Organisms in blue are on the list of CDC select agents that are considered possible agents of bioterrorism.

access when experiments are in progress and using biological laminar flow cabinets if aerosolization is possible (**Figure 2.17**).

Biosafety level 3 pathogens produce a serious or lethal human disease, but vaccines or therapeutic agents may be available. For safe handling of these organisms, level 2 procedures are supplemented with a lab design that ensures ventilation air flows only into the room (negative pressure) and exhaust air vents directly to the outside. Negative pressure will keep any organism that may aerosolize from escaping into hallways. In addition, access to

the lab is strictly regulated and includes double-door air locks at the entrance. This was the containment level needed to prevent accidental infection by *F. tularensis* in Case History 2.3.

By law, extremely dangerous pathogens such as the Ebola virus, for which no treatment or vaccine is available, are biosafety level 4 pathogens and may be studied only at a biosafety level 4 containment facility. Practices here dictate that lab personnel wear positive-pressure lab suits connected to a separate air supply (**Figure 2.18**). The positive pressure ensures that if the suit is

Figure 2.17 Biological Safety Cabinet

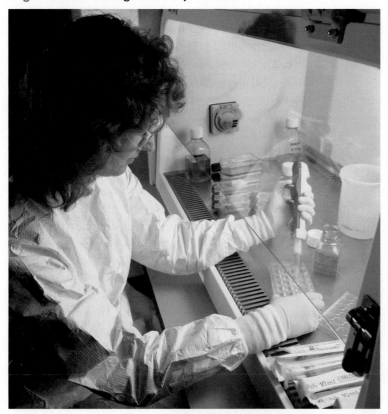

A. A scientist examines a sample under the hood.

Side view

⟶ Blue = room air
➡ Red = contaminated air
➡ Green = filtered air

C. Microbes trapped in a HEPA filter.

B. Schematic of the safety cabinet. Air from the room enters the cabinet through the cabinet opening (1) or is pumped in (2) through a HEPA filter (3). Air then passes behind the negative-pressure exhaust plenum (4) and is passed out of the cabinet through another HEPA filter (5).

Figure 2.18 Biosafety Level 4 Containment

Dr. Kevin Karem, at the CDC, performs viral plaque assays to determine the neutralization potential of serum from smallpox vaccination trials. Working in a biosafety level 4 laboratory, he is protected by a positive-pressure suit. The airflow into his suit is so loud that he must wear earplugs to protect his hearing.

penetrated, organisms will be blown <u>away</u> from the breach and not sucked into the suit. As of this writing, 13 biosafety level 4 laboratories are operating in the United States.

Although these regulations may seem reasonable or even obvious, they were not always in effect. Before 1970, some scientists had an almost cavalier approach toward handling pathogens. For instance, culture material was routinely transferred from one vessel to another by mouth pipetting (essentially using a glass or plastic pipette as a straw). This practice is now forbidden for obvious reasons and has been replaced with mechanical pipettors that have disposable tips.

SECTION SUMMARY

- **Diseases can spread** by direct or indirect contact between infected and uninfected people/animals or by insect vectors.
- **Pathogens use portals of entry** best suited to their mechanisms of pathogenesis.
- **Various levels of protective measures** are used for handling potentially infectious biological materials.
- **Four BSL levels** correlate the hazard potential of an infectious agent with the level of containment and precaution necessary to work safely with each.

2.5 Host Factors in Disease

SECTION OBJECTIVES
- Define the biological features of human hosts that influence the course of an infection.
- Explain how host behavior can impact susceptibility to disease.

How does health or lifestyle factor into contracting an infectious disease? Until now we have focused on the pathogen as the main culprit in infectious disease. The host, however, is not blameless in the process. Many host factors can influence a person's susceptibility to a disease, and more often than not, host factors actually contribute to the disease process itself. Which host factors influence susceptibility?

Age. As a general rule, the people most susceptible to an infection are either very young (under 3 years old) or beyond middle age (over 60 years old). Those younger than 3 or older than 60 are more susceptible, largely because of the natural development of our immune system. The immune system is still developing in newborns and babies and is waning in the elderly. As a result, these two populations have a harder time fending off infections than do healthy children or adults under 60.

Host genetic makeup (genotype). Genetics not only determines whether we have blue or brown eyes but also influences our susceptibility to infectious diseases. The host receptors to which bacteria and viruses bind are all encoded by host genes. Losing or altering a host receptor will affect susceptibility to a given pathogen. As one example, people with blood group O, rather than A, B, or AB, are at higher risk of cholera caused by the bacterium *Vibrio cholerae*, although the reason is not clear. This relationship was evident during the cholera epidemic striking Peru in 1991, where the indigenous people were virtually all blood type O and suffered with severe cholera. The good news, however, is that people with blood type O are more resistant to malaria. Malarial parasites release a protein that can make red blood cells sticky, causing them to clump and form rosettes that clog blood vessels leading to the brain. Malaria-infected type O red blood cells, however, are less likely to form rosettes. Other examples of genetics influencing disease susceptibility will be discussed in the chapters on immunity (Chapters 15 and 16).

Hygiene and behavior. Hygiene plays an important role in preventing disease. For instance, among a population of people exposed to influenza virus, those who frequently wash their hands or use hand sanitizers are less likely to contract the flu. The simple act of shaking a flu victim's hand and then scratching your nose can transmit the virus, but using a hand sanitizer <u>after</u> the handshake and before scratching your nose can block that transfer. This measure also works if flu victims sanitize their hands before shaking. Handwashing or sanitation blocks the transfer of many respiratory as well as gastrointestinal diseases. Washing hands after a bowel movement will prevent you from transmitting diseases, such as hepatitis A or shigellosis, to someone else. Cooking food properly can also interrupt the fecal-oral cycle of many food-borne pathogens. In addition to hygiene, behavior obviously plays a big role in the transfer of sexually transmitted diseases such as syphilis and gonorrhea.

Nutrition and exercise. Several reports suggest that good nutrition and moderate exercise can enhance a person's immune system. In an extreme example, people who are starving produce less acid in their stomach (a condition called achlorhydria) and, as a result, become more susceptible to intestinal pathogens. The reason is this: an acidic stomach kills most bacteria, making it an important first line of defense against food-borne pathogens such as *Salmonella* and *Vibrio cholerae*. The more acidic a stomach is, the more organisms must be ingested to ensure that enough bacteria survive to reach the intestine and cause disease. Consequently, when a stomach becomes <u>less</u> acidic, fewer organisms are needed to cause disease. Beyond nutrition, recent studies also suggest that moderate exercise can boost the immune system and decrease the number and severity of colds. The mechanism by which exercise boosts immunity is unknown.

Underlying noninfectious diseases or conditions. Genetic defects in the immune system, chronic infections (such as HIV), and cancer, as well as drugs designed to prevent transplant rejection will compromise a person's immune system, making them extremely susceptible to infectious diseases. Substance abuse can also place an individual at higher risk of infections. For example, alcoholics are very susceptible to infections because, along with decreasing a person's libido, alcohol use can depress facets of the immune system. Diabetics with elevated blood glucose (hyperglycemia) are at increased risk for many infections because of hyperglycemia-related impairment of immune responses, vascular insufficiency (decreased blood flow), increased skin and mucosal colonization, and other factors. Serious infections on the soles of the feet are common in diabetic patients because sensory nerve damage (neuropathy) caused by hyperglycemia prevents them from feeling the beginning of an infection. Smoking increases not only the risk of lung cancer but also of respiratory infections because smoke paralyzes the lung mechanisms that sweep bacteria up and out of the lung. Even sleep deprivation can dampen immune responses.

Occupation. The job, or hobby, a person pursues can also influence the exposure and thus the susceptibility to certain infectious agents. Health care workers are an obvious example. Not only are these individuals in almost constant contact with sick people, but they also place themselves at risk of exposure to blood-borne pathogens (for example, HIV) every time they use a hypodermic needle to administer a drug or draw a blood sample. Other, lesser-known examples of at-risk endeavors also exist. Agricultural workers may be exposed to anthrax spores—nearly ubiquitous in soil, as well as the agent of Q fever (*Coxiella burnetii*)—present in the placenta of infected pregnant cattle. Hunters can contract zoonotic diseases such as tularemia and plague. People who work with animal hides, which may carry anthrax spores, can occasionally contract anthrax. And, of course, sex workers (male or female) expose themselves to a host of sexually transmitted agents.

Immune status and immunopathogenesis. Do pathogens <u>directly</u> cause the signs and symptoms of a disease? Sometimes, but most often the pathogen is not the direct cause of symptoms. Usually, damage results from the immune response trying to kill the invader. Our immune system reacts to the presence of any pathogen by sending specific types of white blood cells to the infected area. Some of these white blood cells become quite aggressive when attacking the infection and cause a considerable amount of collateral damage to healthy host tissues. The result is called immunopathology. The paradox is that the better a host's immune system is at ridding infection, the more damage can be inflicted on the host itself.

Ebola virus is an extreme example of a pathogen that can disable some features of the immune system and overstimulate others. Ebola virus causes a hemorrhagic fever in which the patient's blood pressure drops and bleeding occurs from multiple orifices (**Figure 2.19**). The overall syndrome is called sepsis, described in Chapter 15 IMPACT. The immunopathology of Ebola begins as the virus infects the white blood cells sent to fight it. The infected white cells will carry the virus to more body sites and release several potent signaling molecules. Some of these molecules activate more white blood cells, others trigger a high fever, and still others increase the permeability of blood vessels throughout the body. The more permeable blood vessels leak blood, leading to internal and external hemorrhaging. The consequence is that 50%–70% of those infected will die. Fortunately, few infectious agents trigger that degree of immunopathology. Immunopathogenesis and the strategies that pathogens use to avoid the immune response are further described in Chapter 18.

Figure 2.19 Ebola Hemorrhagic Fever

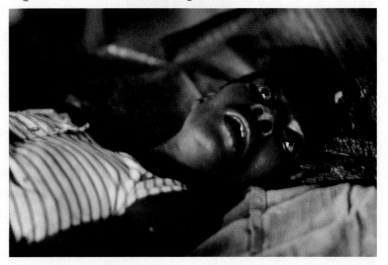

Ebola patient with red eyes.

SECTION SUMMARY

- **The people most susceptible to infectious disease** are the very old, the very young, and the immunocompromised.
- **An individual's genetic makeup** can influence the person's susceptibility to pathogens.
- **Hygiene** (washing hands, covering your mouth when sneezing, properly cooking and storing foods) is an important factor in preventing the spread of disease.
- **Proper nutrition and exercise can boost the immune system** and lessen susceptibility to infections. In contrast, smoking, diabetes, and alcohol consumption can increase susceptibility by weakening the immune response.
- **Exposure to infectious agents is higher for certain occupations** (health care workers) and hobbies (hunting).
- **Tissue damage caused by an infection is due mostly to the immune response** trying to combat the infection.

Thought Question 2.6 Unbeknownst to you, you have contracted the flu (influenza), a viral disease whose prodromal period is followed by fever, muscle aches, coughing, and lethargy. However, before you develop symptoms you attend a party with 100 guests. During the evening you greet, hug, and speak to each person for an equal amount of time. A week later, you find yourself quite ill from flu and are worried that you transmitted the disease to all the party guests. After 10 more days, you are feeling better and begin to ask about the health of the other guests. Much to your surprise, only 20 of them became ill. No one had been vaccinated with this year's flu shot. How could 80% of the guests have escaped disease?

impact

Unintended Consequences: Rebirth of Tuberculosis Follows Death of the Aral Sea

A coughing woman stands outside a tuberculosis sanatorium in Nukus, Uzbekistan, near the Aral Sea in central Asia. She is one of many waiting for treatment. The gray hospital building, now supervised by Doctors Without Borders (an international medical humanitarian organization), is a remnant of the Soviet era. Tuberculosis, a lung disease caused by the bacterium *Mycobacterium tuberculosis* (Figure 1), was once uncommon in this area, but its incidence has risen to alarming levels in recent decades as a result of a preventable environmental disaster.

The Aral Sea was once the world's fourth-largest inland body of water. Now it is barely a lake. The people who lived near the Aral Sea of old now live in the desert, many of them dying of tuberculosis (Figure 2). The World Health Organization defines the lower limit for a tuberculosis epidemic at 50–70 infected people per 100,000. In cities near the shrinking Aral Sea, the number is as high as 265 per 100,000. What led to this environmental and infectious disease disaster?

In the middle of the twentieth century, the Soviet Union built an elaborate series of irrigation channels taking water from the Amu Darya, a river that once fed the Aral Sea with freshwater. Soon after the project was completed, the river no longer reached the sea. Unreplenished with freshwater, the sea began to dry up. As the sea began to shrink, fishing communities were left miles away from the water that traditionally provided a livelihood—some as far as 30 miles (Figure 3). The collapse of

Figure 1 Tuberculosis

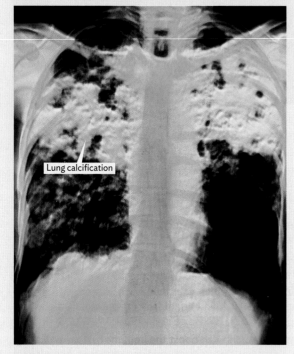

Lung tissue calcification, resulting from pulmonary tuberculosis, appears as artificially colored yellow patches within the chest area of this human X-ray. When airborne sputum (expectorated phlegm) contaminated with the bacillus *Mycobacterium tuberculosis* is inhaled, nodular lesions, called tubercles, form in the lungs and spread through the nearest lymph node. About a quarter of the world's population (not US population) is infected with the bacillus that causes tuberculosis, but most of these people show no signs of the disease as long as their immune systems can keep it in check.

Figure 2 The Aral Sea in Uzbekistan

July–September 1989   March 26, 2010

Soviet Union irrigation projects starting in the 1960s have reduced what was the fourth-largest body of inland water in the world to 10% of its former size.

Figure 3 The Aral Sea Ship Cemetery

Old, rusting ships lie in the sand where the Aral sea used to be.

the fishing industry in the early 1980s was the first in a chain of ecological reactions that included a change in the climate. Once noted for its mild temperatures, the area now has shorter, hotter, and rainless summers, while winters tend to be longer, colder, and without much snow. Arid conditions combined with wind produce windblown sand that damages lungs and makes them more susceptible to infection.

Deeply affected by the closing of the fishing industry, traditional communities have been slow to adapt, leaving economies once reliant on the sea in deep recession. Massive unemployment has led to extreme poverty. Most families have five or six children on average, and nutritional standards have fallen sharply. As a result of the arid conditions and deepening poverty, many infections and diseases traditionally found in desert or semidesert areas and among malnourished people have entered the area. They include tuberculosis.

Adding to the nightmare, many of the new strains of *M. tuberculosis* isolated from patients in this region of Uzbekistan are resistant to almost every anti-tuberculosis antibiotic available. And treatment itself is arduous. Patients must stay in the sanatorium for 3 months—until they are no longer contagious—and then return 3 days a week for 6 months to receive medication under direct observation by hospital staff.

Worldwide, there are many examples of climate change caused by ill-advised human endeavors. Several of these actions have led to new emerging infectious diseases or the reemergence of ancient ones, such as tuberculosis. As for the fate of the Aral Sea, efforts to reconstruct canals and other waterworks since 2006 have led to an approximate 7-foot rise in water levels in the sea's northernmost part. The situation for the South Aral Sea remains dire.

DISCUSSION QUESTION

Why would medical staff have to directly observe patients taking their medications?

2.6 Global Change and Emerging Infectious Diseases

SECTION OBJECTIVES

- Explain how civilization has impacted emerging infectious diseases.
- Describe how climate change can alter infectious disease patterns.

Are all infectious diseases known to us? With all that we currently know about infectious diseases, we will never understand them all because new diseases are continually emerging or reemerging. An **emerging disease** is one that has recently appeared in a population. A reemerging disease, in contrast, is a known disease that was controlled but whose incidence or geographic range is increasing or is threatening to increase in the near future. Since 1976, the World Health Organization (WHO) has recorded more than 40 emerging and reemerging infectious diseases. Lyme disease, *E. coli* O157:H7 (one cause of bloody diarrhea), drug-resistant tuberculosis, and hantavirus (the cause of a lethal respiratory infection) are but a handful of diseases and disease-causing microbes that have emerged (or reemerged) over the past several decades. Where do these new diseases come from?

Global increases in death and disease are often connected to human activities that bring humans closer to disease reservoirs and vectors. Deforestation, human settlement sprawl, industrial development, road construction, large water control projects (such as dams, canals, irrigation systems, and reservoirs), and climate change alter the environment in ways that can expose humans to agents of infectious disease (see **IMPACT**). Towns built on the margins of woodlands that harbor deer populations, for example, make it easier for deer ticks to transmit the Lyme disease agent from deer to humans. Here are other recent examples:

- Malaria has spread in South America because deforestation in the Amazon altered mosquito breeding patterns. A similar situation on the island of Borneo in the South China Sea led to a human outbreak of monkey malaria.

- In Malaysia, pig farms built where fruit bats live led to the transfer of the Nipah virus from bats to pigs and then to humans. Nipah viruses can cause severe respiratory and brain infections. Fruit bats are also natural reservoirs for both Australian bat rhabdovirus (similar to rabies) and Hendra virus.

- The hunting of small civet cats, an Asian delicacy, is thought to have introduced the virus causing severe acute respiratory syndrome (SARS) to humans, leading to the SARS outbreak in Asia in 2005. A similar scenario is suspected for the COVID-19 pandemic of 2019–2020.

Figure 2.20 A Link between Deforestation and Malaria

Colored areas in panels A and B represent 795 municipalities encompassing the Brazilian Amazon. Gray indicates the Brazil boundary. Green outlined areas mark the interior of the Brazilian Amazon, an area where malaria and forest cover are the highest.

Deforestation
Kilometers lost, 2008

- 0–17
- 17–47
- 47–99
- 99–206
- 206–545
- 545–1,287
- Brazil boundary

A. The amount of deforestation for each municipality in 2008.

P. falciparum
Incidence, 2008

- 0.0
- 0.0–7.3
- 7.3–15.3
- 15.3–27.2
- 27.2–51.5
- 51.5–92.6
- Brazil boundary

B. The incidence of malaria caused by *Plasmodium falciparum* in each municipality in 2008.

- Leptospirosis, a zoonotic bacterial disease transmitted by water contaminated with animal urine, is a globally reemerging infectious disease now spreading from rural areas to cause urban epidemics in industrialized and developing nations. In urban Brazil, for instance, outbreaks of leptospirosis can be predicted by heavy rain and flooding, which brings increased risk of human contact with waterborne *Leptospira*.

The global rate of tropical deforestation, one factor contributing to increased malaria infections, continues at staggering levels, with nearly 2%–3% of forest habitats lost per year. Some of the loss is the result of fire, but most is caused by agriculture and urban sprawl—which is also on the rise, especially in developing countries. For example, between 1940 and 1960 the number of malaria cases in the Amazon basin had dropped from 6 million to 50,000, thanks to malaria control efforts. By the year 2000, however, that number had risen to 600,000—an increase that paralleled large-scale deforestation for land use. A recent study has correlated the increase in malaria incidence (the number of new cases of malaria per year per 1,000 population) to the amount of deforestation that took place in the Amazon from 2003 to 2015. Data for 2008 are shown in **Figure 2.20A and B**. Increases in deforestation strongly correlated to the increased incidence of malaria, as shown in **Figure 2.20C**. The study concluded that a 10% rise in deforestation led to a 3.3% average increase in malaria transmission.

C. Correlation between malaria incidence and the amount of deforestation. Each dot represents a different municipality over different years (2003–2015). The incidence of malaria was shown to increase proportionally to the increase in deforestation.

Note: The authors also found that deforestation had the greatest effect on malaria incidence in the interior regions (green outline in Figures 2.20A and B), where forest cover remains high. The relationship was not significant in the outer Amazonian states, where most of the forest has already been cleared.

Deforestation practices will continue to create breeding areas for mosquitoes and bring humans closer to disease reservoirs and vectors. We can also expect global warming to change worldwide disease patterns. For instance, as a result of rising global temperatures, insect vectors now limited to tropical areas of the world will be able to migrate to higher (historically colder) latitudes, bringing infectious microbes with them. All this stirring up of the global environment not only fosters the spread of existing diseases but also builds breeding grounds in which new pathogens will evolve.

SECTION SUMMARY

- **Emerging infectious diseases** are diseases that have not previously been recognized and are the result of either newly evolved organisms or organisms that have made the "leap" from animal to human host.

- **Reemerging diseases** are diseases that have existed but are exhibiting a recent increase in incidence or geographic range.

- **Deforestation and urban sprawl** bring insect vectors and animal reservoirs of disease into closer proximity to humans.

- **Temperature changes and/or drought or excessive rain** affect the geographic distribution of insect and other animal vectors.

Perspective

The interaction between host and pathogen can be likened to a war, with each side evolving new weapons to outwit the opponent—a biological arms race. Of course, pathogens do not engage in this warfare out of ill intent, but out of the need to survive. Even though infectious agents continue to evolve, the stages of an infectious disease remain the same: entry, attachment, immune avoidance, recognition by the host, immunopathology, exit from the host, and resolution of disease (recovery or death). To understand the strategies that different pathogens use to invade a host and cause disease, you must first appreciate the physical tools and rich diversity of their biochemical and genetic capabilities. The next several chapters will describe the basic structure, biochemistry, and growth of microbes, and illustrate how these features impact infectious disease.

Key Terms

acute infection (p. 34)
adhesin (p. 32)
aerosol (p. 43)
airborne transmission (p. 43)
carrier (p. 45)
chronic infection (p. 34)
colonization (p. 32)
convalescence (p. 39)
decline phase (p. 39)
differential (p. 41)
direct contact (p. 42)
disease (pp. 34, 37)
emerging disease (p. 55)
endemic disease (p. 45)
epidemic (p. 45)
fecal-oral route (p. 48)

fomite (p. 43)
horizontal transmission (p. 42)
host range (p. 35)
illness phase (p. 38)
immunopathology (p. 38)
incidence (p. 39)
incubation phase (p. 38)
indirect transmission (p. 43)
infection (p. 34)
infection cycle (p. 42)
infectious disease (p. 37)
infectious dose 50% (ID_{50}) (p. 35)
invasion (p. 35)
invasiveness (p. 35)
latent state (p. 34)
lethal dose 50% (LD_{50}) (p. 35)

malaise (p. 38)
mechanical vector (p. 43)
microbiome (p. 32)
microbiota (p. 32)
morbidity (p. 39)
mortality (p. 39)
opportunistic pathogen (p. 34)
pandemic (p. 45)
parasite (p. 33)
parenteral route (p. 48)
pathogen (p. 33)
pathogenicity (p. 33)
patient history (p. 40)
primary pathogen (p. 34)
prodromal phase (p. 38)
prodrome (p. 38)

receptor (p. 36)
reservoir (pp. 43, 44)
respiratory route (p. 48)
sequela (p. 38)
sign (p. 37)
symptom (p. 37)
syndrome (p. 37)
transovarial transmission (p. 44)
transplacental transmission (p. 42)
urogenital route (p. 48)
vector (p. 43)
vehicle transmission (p. 43)
vertical transmission (p. 42)
virulence (p. 34)
zoonotic disease (pp. 44, 46)

Objectives Assessment Questions

2.1 Normal Microbiota versus Pathogens

1. _____ is the measure of the degree or severity of disease.
   a. Pathogenicity
   b. Invasiveness
   c. Virulence
   d. Opportunism

2. Pathogens that change the structure of their surface proteins do so for which of the following reasons?
   a. Attachment to the host
   b. Immune avoidance
   c. Invasion of host tissues
   d. Acquisition of host nutrients

3. According to the curve shown (which measures the pathogenicity of a bacterial pathogen in mice), which of the following statements is true?
   a. The lethal dose 50% is 1,000 organisms.
   b. The infectious dose 50% is 1,000 organisms.
   c. The lethal dose 50% is 1,000,000 organisms.
   d. The infectious dose 50% is 1,000,000 organisms.

2.2 Basic Concepts of Disease

4. A(n) _____ is an infection acquired during a hospital stay.

5. A rash observed by a clinician is considered a _____.
   a. sign
   b. symptom
   c. prodrome
   d. sign and symptom

6. Describe a secondary infection.

7. _____ is the term used to express the number of fatalities associated with a disease.

8. The stage of a disease when signs and symptoms become apparent is called _____.
   a. the prodromal phase
   b. convalescence
   c. the illness phase
   d. the incubation period
   e. the acme

2.3 Infection Cycles and Disease Transmission

9. A human pathogenic virus that can undergo transovarial passage does not require which of the following to cause disease:
   a. Vector
   b. Infection cycle
   c. Animal reservoir
   d. Endemic area
   e. Parenteral transmission

10. The pathogen *Listeria monocytogenes* can be transmitted from an infected mother to her fetus in a process called _____.

11. Which one of the following best explains what might lead to the development of an asymptomatic carrier?
    a. Lack of host cell receptors
    b. The immune system
    c. Drinking hot fluids
    d. Antibiotics
    e. Diminished virulence of the agent

12. A disease always present at low incidence in a geographic area is called a(n) _____ disease.

2.4 Disease Portals and Biosafety

13. A patient developed small, flat, pink, non-itchy spots on her forearms after a trip to Colorado, where she camped and hiked alone in the woods. Her physician diagnosed Rocky Mountain spotted fever. The patient probably acquired the organism by which of the following routes?

a. Oral
b. Vertical
c. Direct contact
d. Respiratory
e. Parenteral

14. Ebola virus requires which of the following levels of biocontainment?

a. Level 1
b. Level 2
c. Level 3
d. Level 4

15. What type of laboratory safety measure is required when working with the pathogen in Case History 2.2 that is not required when working with a diarrhea-causing *E. coli*?

a. Sterile technique
b. Biosafety cabinets
c. No mouth pipetting
d. Negative-pressure room ventilation
e. One-piece positive-pressure suits

2.5 Host Factors in Disease

16. A patient with a deficient immune system is more susceptible to contracting infectious diseases by which of the following routes?

a. Respiratory
b. Fecal-oral
c. Parenteral
d. Urogenital
e. All of the above

17. A person who has smoked cigarettes for 20 years is more susceptible to infections transmitted by which of the following routes?

a. Respiratory
b. Fecal-oral
c. Parenteral
d. Urogenital
e. All of the above

18. Organ and tissue damage caused by most infectious diseases is typically caused by _____.

19. Which of the following host factors contribute(s) to a person's susceptibility to infectious disease?

a. Age
b. Occupation
c. Nutrition
d. Underlying health conditions
e. Exercise
f. Hygiene
g. All of the above
h. a, d, and f only

2.6 Global Change and Emerging Infectious Diseases

20. In Brazil, heavy rains are predictive of an increase in which one of the following diseases?

a. *E. coli* O157:H7
b. Lyme disease
c. SARS
d. Leptospirosis
e. Tuberculosis

21. Which of the following can foster the jump of an emerging pathogen from a nonhuman animal host to humans?

a. Deforestation
b. Increases in global temperatures
c. Urban sprawl
d. All of the above
e. a and c only

Review Questions

1. Describe the differences between
   a. infection and disease.
   b. pathogenicity and virulence.
   c. LD_{50} and ID_{50}.
   d. vector and reservoir.

2. Describe the stages of an infectious disease course.

3. Explain the processes a pathogen must carry out in a host to cause an infection.

4. What is meant by direct versus indirect routes of infection?

5. What are the characteristics of a good reservoir for an infectious agent?

6. Name the various portals of entry for infectious agents, and name a disease associated with each.

7. What are some host factors that influence susceptibility to disease?

8. How have human activities affected the emergence of new pathogens?

9. Distinguish between acute, chronic, and latent infections.

Clinical Correlation Questions

1. Using the terms defined in this chapter, briefly describe (in one or two paragraphs) an illness you have had. Use words or phrases such as "portal," "transmission," "prodromal phase" (or "prodrome"), "acme," "viremia," "reservoir," "vector," or any other appropriate term found in Table 2.1 or 2.2. Then discuss how you might have prevented getting the disease.

2. How does knowing a pathogen's portal of exit help prevent the spread of disease?

3. Review the anthrax case in Chapter 1 (Case History 1.1). Describe the most likely route of infection for Caleb's disease.

4. Search the Internet for a disease that can cause both acute and chronic infections.

For more assessment, visit digital.wwnorton.com/michum2

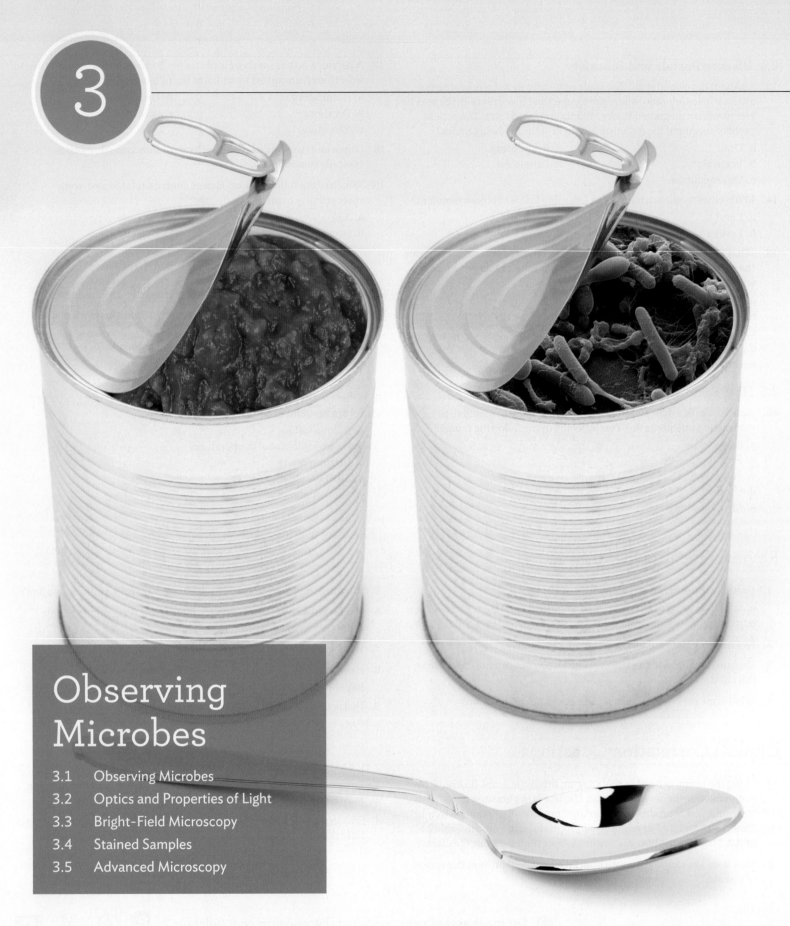

3

Observing Microbes

[Pathogen in a Can]

SCENARIO Marianna and Julio, two children in Oklahoma aged 8 and 11, fell ill with food poisoning after consuming commercially canned hot dog chili sauce. The children complained of double vision and an inability to move their facial muscles.

SIGNS AND SYMPTOMS
Examination showed normal vital signs, symmetrical weakness (equal on right and left sides), and fixed pupils due to cranial nerve palsy (paralysis). Upper body paralysis gradually spread downward, and breathing became labored.

TESTING Botulinum toxin was identified in their blood by fluorescent antibodies.

TREATMENT The children were placed on a mechanical ventilator (a machine that assists breathing) and treated with botulinum antitoxin. After several days, the children were removed from mechanical ventilation. They underwent physical rehabilitation for a full recovery.

FOLLOW-UP The chili sauce was traced to a canning facility where six swollen cans tested positive for botulinum toxin A. From the cans, inspectors cultured a bacterium that grew only without oxygen, as in a closed can. A microscope showed that the bacterium had a distinctive club-shaped appearance. The bacterium was *Clostridium botulinum*, which produces botulinum toxin, the cause of botulism. One end of the club-shaped cell contains a developing endospore, an inert form of the cell that can germinate and grow in a closed container of food. Growing cells produce botulinum toxin, leading to botulism, a life-threatening form of paralysis. Microscopy of the pathogen's unique form confirms the identification.

LM 2 µm

Clostridium botulinum
The distinctive club-shaped cells (purple) each show the endospore (pink bulge) near one end. Bright-field microscopy with Gram stain.

CHAPTER OBJECTIVES

After reading this chapter, you will be able to:

- Explain how light microscopy reveals information about pathogenic microbes isolated from patients.
- Explore a variety of microbial cells under a microscope.
- Identify the type of microscopy used to obtain images of microbes.
- Explain the information provided by images from bright-field microscopy, fluorescence microscopy, and electron microscopy.

A can of food that looks harmless can hide deadly life-forms, too small for you to see. But a microscope reveals the food-borne contaminant *Clostridium botulinum*, the cause of botulism. Food-borne botulism has low incidence in the United States (about 24 cases per year), but its onset is rapid and it can be fatal without a prompt diagnosis.

The bacterium *C. botulinum* has a distinctive club-shaped cell. The swollen end of the "club" contains an endospore, a dormant particle in which the cell's DNA is condensed with a few enzymes protected by a thick coat. Endospores of certain types of bacteria can survive many years until they encounter the right conditions for growth—for this species, access to food and moisture in the absence of oxygen. The club-shaped form of the sporulating cell is unique to *C. botulinum*. You can see *C. botulinum* by using bright-field microscopy with the simple technique of the Gram stain, which we will discuss later in the chapter.

Microscopy reveals the vast realm of bacteria, fungi, and protozoa invisible to our unaided eyes. The microbial world spans a wide range of sizes over several orders of magnitude. For different size ranges, we use different instruments, from the simple bright-field microscope to the electron microscope. The microscope enables us to count the microbes in the human bloodstream or in natural environments like the ocean. It detects emerging pathogens, such as the *Salmonella* transmitted from backyard poultry in 2019, an epidemic across the United States. The microscope can actually follow pathogenic bacteria as they attach, invade, and move through a human host cell.

In this chapter, you will learn how to use the microscope to observe and distinguish different microbes. Microscopes are used in many settings, from hospitals and veterinary clinics to industrial plants and wastewater treatment facilities. You will also learn about the capabilities of more advanced techniques, such as fluorescence and electron microscopy.

3.1 Observing Microbes

SECTION OBJECTIVES

- Explain how magnification improves resolution of a microscopic image.
- Explain what can be learned from different kinds of microscopy.

CASE HISTORY 3.1
Bacteria Invade the Brain

Eleanor, a 28-year-old woman, was a buyer for an upscale clothing store in New York City. While at work, she developed a sudden headache, and light bothered her eyes. She was concerned about missing work, but her co-workers urged her to get checked.

Eleanor presented in the urgent-care facility with headache, fever, and neck stiffness. She told the clinician that light hurt her eyes, and she had difficulty bending her head forward. On examination, her temperature was 38°C (100.8°F) and her blood pressure was low, making her feel dizzy when she stood up. Eleanor underwent a lumbar puncture, or spinal tap, a procedure in which a spinal needle is inserted between the lumbar vertebrae in order to withdraw a sample of cerebrospinal fluid (CSF). This fluid bathes the meningeal lining of the brain. Normal CSF is clear and sterile, free of cells, but Eleanor's CSF sample was cloudy. The cloudy fluid was indicative of bacterial meningitis.

Bacterial meningitis develops rapidly, and some pathogens confer a 40% percent fatality rate. Thus, rapid diagnosis is critical. Fortunately, Eleanor received a prompt diagnosis and antibiotic therapy. Her partner also received antibiotic therapy for prophylaxis (prevention). Eleanor asked about her co-workers. The clinician assured her that friends and co-workers who do not share intimate contact are not at risk for transmission.

In Case History 3.1, what was in the cloudy CSF? (Figure 3.1A) Cerebrospinal fluid normally is clear and sterile, but Eleanor's sample was full of white blood cells (WBCs). The cells scattered light, resulting in a cloudy appearance. The WBCs were responding to a bacterial infection (to be presented in Chapter 15). Microscopy revealed the individual WBCs, along with the much smaller cells of bacteria (Figure 3.1B). The bacteria have a shape called diplococcus (paired spherical cells), consistent with the species *Neisseria meningitidis*. Both the WBC nuclei and the *Neisseria* bacteria appear pink when Gram-stained, discussed later in this chapter.

Even before being viewed in a microscope, the WBCs and bacteria in Eleanor's cerebrospinal fluid were <u>detected</u> in the tube as

Figure 3.1 Cloudy Cerebrospinal Fluid

Nucleus of white blood cell

Pair of diplococci

LM 10 μm

A. Cerebrospinal fluid was obtained through a lumbar puncture (spinal tap). The cloudy appearance is caused by suspended white blood cells infiltrating the cerebrospinal fluid in response to bacterial infection.

B. Light micrograph of cerebrospinal fluid containing white blood cell nuclei and *Neisseria meningitidis* bacteria. Both appear pink under Gram staining.

a distinct entity separate from other cells. The cells cannot be resolved individually because they are too small.

In Case History 3.1, microscopy was needed to resolve the size and shape of the diplococcus bacteria. The specific shape of the bacteria confirms the diagnosis of *N. meningitidis*. Most microbial cells are "microscopic," requiring the use of a **microscope** to be seen. In a microscope, **magnification** increases the apparent size of an image. "Useful magnification" enables resolution of smaller separations between objects and thus increases the information obtained by our eyes. Why do we need magnification to resolve individual microbes? What makes a cell too small for unaided eyes to see? Surprisingly, the answer depends on the observer. Our definition of "microscopic" is based not on inherent properties of the organism under study but on the properties of the human eye. What is "microscopic" actually lies in the eye of the beholder.

The size at which objects become visible depends on the resolution of the observer's eye. **Resolution** is the smallest distance by which two objects can be separated and still be distinguished. In the eyes of humans and other animals, resolution is achieved by focusing an image on a retina packed with light-absorbing photoreceptor cells (rods and cones; **Figure 3.2**). A group of photoreceptors with their linked neurons forms one unit of detection, comparable to a pixel on a computer screen. The distance between two retinal "pixels" limits resolution.

The resolution of the human retina (that is, the length of the smallest object most human eyes can see) is about 150 μm, or one-seventh of a millimeter. In the retinas of eagles, photoreceptors are more closely packed, so an eagle can resolve objects eight times as small or eight times as far away as a human can—hence the phrase "eagle-eyed," meaning sharp-sighted.

suspended particles causing a cloudy appearance. Detection means to see that an object exists, even if you cannot see its details. For example, our eyes can detect a large group of microbes such as a spot of mold on a piece of bread (about a million cells).

Within the tube or the colony, the presence of cells may be detected, yet no individual cell can be <u>resolved</u>—that is, observed as

Figure 3.2 Defining the Microscopic

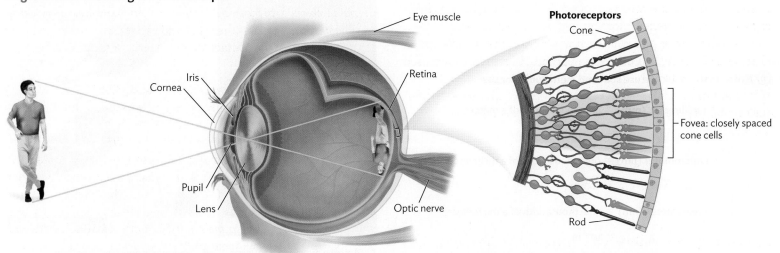

Eye muscle

Photoreceptors
Cone

Iris
Cornea

Retina

Fovea: closely spaced cone cells

Pupil

Lens

Optic nerve

Rod

We define what is visible and what is microscopic in terms of the human eye. Within the human eye, the lens focuses an image on the retina. Light receptors are close enough to resolve the image of a human being, even at a distance where its apparent height shrinks to a millimeter.

Figure 3.3 Relative Sizes of Different Microbes

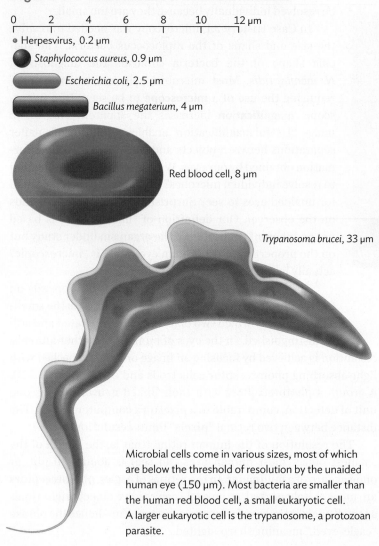

0   2   4   6   8   10   12 µm

- Herpesvirus, 0.2 µm
- *Staphylococcus aureus*, 0.9 µm
- *Escherichia coli*, 2.5 µm
- *Bacillus megaterium*, 4 µm

Red blood cell, 8 µm

Trypanosoma brucei, 33 µm

Microbial cells come in various sizes, most of which are below the threshold of resolution by the unaided human eye (150 µm). Most bacteria are smaller than the human red blood cell, a small eukaryotic cell. A larger eukaryotic cell is the trypanosome, a protozoan parasite.

◉ **Note:** In this book, we use standard metric units for size:

1 millimeter (mm) = one-thousandth of a meter

= **10^{-3} m**

1 micrometer (µm) = one-thousandth of a millimeter

= **10^{-6} m**

1 nanometer (nm) = one-thousandth of a micrometer

= **10^{-9} m**

1 picometer (pm) = one-thousandth of a nanometer

= **10^{-12} m**

Microbial Size and Shape

What is the actual size of a microbe? Different kinds of microbes differ in size, over a range of several orders of magnitude, or powers of 10 (**Figure 3.3**). Eukaryotic microbes are often large enough that we can resolve their compartmentalized structure under a light microscope. An example is *Trypanosoma brucei* (33 µm), which causes sleeping sickness. This trypanosome has a nucleus and a whiplike flagellum. Prokaryotic microbes (bacteria and archaea) are generally smaller (0.4–10 µm); as a result, their overall shape can be seen under a light microscope, but their internal structures are too small to be resolved. Some eukaryotic microbes are as small as bacteria—for example, parasitic microsporidia are 1–40 µm—so the actual size range of eukaryotic microbes overlaps that of prokaryotes.

EUKARYOTIC MICROBIAL CELLS Many eukaryotes, such as protozoa, can be resolved with **light microscopy (LM)** to reveal complex shapes and internal structures such as the nucleus. For example, the ameba in **Figure 3.4A**, from an aquatic ecosystem, shows a large nucleus and a pseudopod that can engulf prey. Pseudopods can be seen to move by the streaming of their cytoplasm. A very different eukaryotic microbe is *Leishmania donovani*, a parasite of humans and animals, transmitted by sand flies (**Figure 3.4B**). Here we see the parasite multiplying amid human red blood cells (RBCs). Parasite infections are discussed in Chapter 11.

The appearance of a microbe in an image depends on the type of microscopy. **Table 3.1** shows the protozoan *Giardia lamblia* imaged by common types of microscopy. *Giardia* is an intestinal parasite that often contaminates water supplies and spreads in child-care centers. The cell has two nuclei and multiple flagella that generate a whiplike movement. But different kinds of microscopy show different aspects of the cell.

Figure 3.4 Eukaryotic Microbial Cells, under Light Microscopy

A. *Amoeba proteus*, a free-living ameba.

B. *Leishmania donovani,* a parasite of humans and animals, transmitted by sand flies. Purple color and contrast are provided by Giemsa stain.

Table 3.1
Giardia lamblia Visualized by Different Forms of Microscopy

Type of Microscopy	Characteristics of Image	Typical Image
Light microscopy (LM) with stain	Dark object (partly transparent) appears against bright background. Stain (methylene blue) darkens some parts more than others.	 Nuclei Flagella LM 10 μm
Phase-contrast microscopy (PCM)	Object looks transparent. Organelles appear as light-dark patterns caused by variation in refractive index. This type of imaging gives an illusion of three-dimensionality. Background may be dark or light.	 PCM 10 μm
Fluorescence microscopy (FM)	Specific fluorescent molecules label parts of the cell—in this case, the cell surface and flagella. Fluorophore-labeled parts of the object show bright color against a dark background.	 FM 10 μm
Scanning electron microsopy (SEM)	A high resolution of detail is present, much higher than with light microscopy. Shadowing of the object's surface approximates three-dimensionality.	 SEM 10 μm
Transmission electron microscopy (TEM)	A thin section of the specimen appears. Only some of the cell's parts appear, those that happen to fall within the section. High resolution of detail reveals organelles within the cell.	 Nucleus Base of flagella TEM 5 μm

- **Light microscopy (LM) with stain.** The stained specimen absorbs light and appears dark against a bright background. The specimen is fixed (dead and attached to the slide), so no movement can be seen. Light microscopy with stain reveals the two nuclei and multiple flagella of *Giardia*.

- **Phase-contrast microscopy (PCM).** Different parts of the specimen with different refractive index appear darker or lighter. The light-dark pattern gives an illusion of three dimensions. Live specimens may be observed.

- **Fluorescence microscopy (FM).** The background and specimen appear dark, except for the fluorophores. Fluorophores are molecules that fluoresce (give off light at a specific wavelength). Only fluorescent-labeled structures are visible; in the example shown, the unlabeled nuclei cannot be seen.

- **Scanning electron microscopy (SEM).** Beams of electrons scatter from the stained surface of a fixed (dead) specimen. The image shows much greater detail than with light microscopy, but only the specimen's external surface is enhanced. The appearance of three-dimensionality is approximately correct.

- **Transmission electron microscopy (TEM).** Electrons pass through a fixed, stained section. Very fine details may appear, such as individual ribosomes, but we see only a small fraction of the whole cell.

Note: Each micrograph in this book has a label for the type of microscopy (black) with a **scale bar** (yellow). The length of the scale bar corresponds to the actual size magnified, such as 5 μm. Like a ruler, the scale bar applies to length measurements throughout the micrograph.

BACTERIAL CELLS How do bacterial cells compare with cells of eukaryotes? Bacterial cell structures are generally simpler than those of eukaryotes (discussed further in Chapter 5). **Figure 3.5** shows representative members of several common cell types as visualized by either light microscopy or scanning electron microscopy (see Section 3.5). With light microscopy, the cell shape is barely discernible under the highest power. With scanning electron microscopy, the shapes appear clearer, but we still see no subcellular structures. Subcellular structures are best visualized with transmission electron microscopy and fluorescence microscopy (again, see Section 3.5).

Certain shapes of bacteria are common to other taxonomic groups (Figure 3.5). For example, both bacteria and archaea form cells of similar shapes: **bacillus** (plural, **bacilli**, rods) and **coccus** (plural, **cocci**, spheres). Thus, rods and spherical shapes have evolved independently within different taxa (groups of related organisms).

Figure 3.5 Common Shapes of Bacteria

Light microscopy (LM) with stain

A. Filamentous rods (bacilli): *Lactobacillus lactis*, Gram-positive bacteria.

C. Spirochetes: *Borrelia burgdorferi*, cause of Lyme disease, among human blood cells.

E. Cocci in chains (filaments): *Streptococcus pneumoniae*, a cause of pneumonia, stained with methylene blue.

Scanning electron microscopy (SEM, colorized)

B. Rods (bacilli): *Lactobacillus acidophilus*, Gram-positive bacteria.

D. Spirochetes: *Leptospira interrogans*, cause of leptospirosis in animals and humans.

F. Cocci in chains (filaments): *Anabaena* spp., cyanobacteria, producers for the marine food chain.

On the other hand, a unique spiral shape is seen in the bacterial **spirochetes**, which cause diseases such as syphilis and Lyme disease. Spirochetes possess an elaborate spiral structure with internal flagella as well as an outer sheath. Spiral axial filaments are found only within the spirochete group of closely related species.

⊙ **Note:** The genus name *Bacillus* refers to a specific taxonomic group of bacteria, but the term "bacillus" (plural, bacilli) refers to any rod-shaped bacterium or archaeon.

Microscopy for Different Size Scales

What kind of microscope do we need to observe microbes? The answer depends on the size of the object. **Figure 3.6** shows how different techniques resolve microbes and structures of various sizes. For example, a single paramecium can be resolved under a light microscope, but an individual ribosome requires electron microscopy.

- **Light microscopy (LM)** resolves images of individual bacteria based on their absorption of light. The specimen is commonly viewed as a dark object against a light-filled field or background; this is called **bright-field microscopy** (seen in Figure 3.6A and B). Advanced techniques, based on special properties of light, include dark-field, phase-contrast, and fluorescence microscopy.

- **Electron microscopy (EM)** uses beams of electrons to resolve details several orders of magnitude smaller than those seen with light microscopy. For example, EM can resolve the shape of ribosomes and viruses. In **scanning electron microscopy (SEM)**, the electron beam is scattered from the metal-coated surface of an object, generating an appearance of three-dimensional depth. In **transmission electron microscopy (TEM)** (Figure 3.6C and D), the electron beam travels through the object, where the electrons are absorbed by an electron-dense metal stain.

- **X-ray crystallography (X-ray)** detects the interference pattern of X-rays, the pattern of light and dark as reflected wavefronts intersect. The X-ray interference occurs as wavefronts enter the array of molecules in a crystal. From the interference pattern, researchers build a computational model of the individual molecules that are stacked in the crystal. This molecule could be a protein or a nucleic acid, or even a molecular complex such as a ribosome (Figure 3.6E).

Of all these methods, scanning electron microscopy (SEM) visualizes an exceptionally wide range of sizes (**Figure 3.7**). For example, SEM reveals the 1-to-2-μm-sized cells of *Shigella* species, a cause of shigellosis diarrhea (Figure 3.7A). The bacteria become entwined by filopodia (cytoplasmic extensions) of the gut epithelial cells, which internalize the *Shigella* cells. This enables the invasive growth of *Shigella* in the gut. The green color is added by the photo

Figure 3.6 Microscopy and X-ray Crystallography, Range of Resolution

A. Internal structures of *Paramecium* (a eukaryote) can be seen. Magnification 100×.

B. Rod-shaped *Pseudomonas* bacteria are barely resolved. Magnification 1,200×.

C. Internal structures of the bacterium *Escherichia coli* are revealed by transmission electron microscopy. Magnification 32,000×.

D. Individual ribosomes attached to the messenger RNA molecules that are translated to make peptides can be distinguished. Magnification 150,000×.

E. A ribosome (diameter 21 nm) can be modeled by X-ray crystallography. Ribosome models reveal targets for new antibiotics.

10^{-3} m	10^{-4} m	10^{-5} m	10^{-6} m	10^{-7} m	10^{-8} m	10^{-9} m
1 mm	100 μm	10 μm	1 μm	0.1 μm = 100 nm	10 nm	1 nm = 10 Å

Human eye

Light microscopy

Scanning electron microscopy

Transmission electron microscopy

X-ray crystallography

Figure 3.7 Scanning Electron Microscopy: Visualization over a Wide Range of Sizes

SEM   1 µm

A. *Shigella* bacteria, cause of dysentery, internalized by intestinal cell filopodia.

SEM   25 µm

B. Scolex (head) of a tapeworm from a cat intestine.

artist (that is, the image is "colorized") to emphasize the bacteria. At a much larger scale, SEM depicts the multicellular scolex (head) of a tapeworm (Figure 3.7B, colorized). The tapeworm's hooks and suckers enable it to attach to the interior lining of the gut. When interpreting a micrograph, always check the scale bar to be aware of what size you are seeing.

When you use a microscope, how does it make a tiny object appear large? Light microscopy extends the lens system of your own eyes. Light is a form of energy that is propagated as waves (**Figure 3.8**). Regions of the spectrum are defined by **wavelength**, the distance between one peak of a wave and the next peak. The wavelength of visible light is about 400–750 nm, about 100 times less than the width of a human hair. Radiation of longer wavelengths includes infrared and radio waves, whereas shorter wavelengths include ultraviolet rays and X-rays.

Light Carries Information

All forms of radiating energy that interact with an object carry information about the object. The information carried by radiation can be used to detect objects; for example, radar uses radio waves to detect a speeding car. For energy radiation to resolve an object (distinguish it) from

SECTION SUMMARY

- **Detection** is the ability to determine the presence of an object.
- **Magnification** means an increase in the apparent size of an image so as to resolve smaller separations between objects.
- **Resolution** is the smallest distance by which two objects can be separated and still be distinguished.
- **Eukaryotic microbes may be large enough for subcellular structures to be resolved** under a light microscope. Other eukaryotes are as small as bacteria.
- **Bacteria are usually too small for subcellular resolution.** Their shapes include characteristic forms, such as rods and spheres.
- **Different kinds of microscopy** are required to resolve cells and subcellular structures of different sizes.

3.2 Optics and Properties of Light

SECTION OBJECTIVES

- Explain how light interacts with an object.
- Describe how refraction enables magnification, and explain the limits to magnification and resolution.

Figure 3.8 The Spectrum of Electromagnetic Radiation

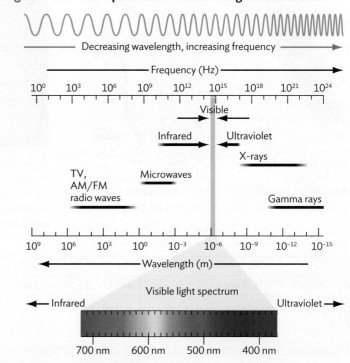

Decreasing wavelength, increasing frequency →

Frequency (Hz)

10^0 10^3 10^6 10^9 10^{12} 10^{15} 10^{18} 10^{21} 10^{24}

Visible

Infrared   Ultraviolet

X-rays

TV, AM/FM radio waves   Microwaves

Gamma rays

10^9 10^6 10^3 10^0 10^{-3} 10^{-6} 10^{-9} 10^{-12} 10^{-15}

← Wavelength (m) →

Visible light spectrum

← Infrared   Ultraviolet →

700 nm   600 nm   500 nm   400 nm

Different forms of energy radiate at different wavelengths. The spectrum of radiation includes the visible range (400–750 nm) used in light microscopy.

neighboring objects or from its surrounding medium, certain conditions must exist:

- **Contrast must exist between the object and its surroundings. Contrast** is the difference in light and dark. If an object and its surroundings absorb or reflect radiation equally, then the object will be undetectable. Consequently, observing a cell of transparent cytoplasm floating in water is hard because the aqueous cytoplasm and the extracellular water tend to transmit light similarly, producing little contrast.

- **Wavelength must be smaller than the object.** For most kinds of microscopy, the wavelength of the radiation must be smaller than the size of the object you wish to detect. If the wavelength of the radiation is larger than the object, then most of the wave's energy will simply pass through it, like an ocean wave passing around a dock post. This is why radar, with a wavelength of 1–100 centimeters (cm), cannot resolve microbes, although it easily resolves cars and people.

- **A detector must resolve the given wavelength.** Your eye has a retina with photoreceptors that absorb radiation within a narrow range of wavelengths, 400–750 nm (0.40–0.75 μm), which we define as "visible light." But the distance between your retinal photoreceptors is 150 μm, about 500 times the wavelength of light. As a result, your eyes cannot access all the information contained in the light that enters. For you to see objects smaller than the distance between your photoreceptors, the light rays coming from the object must be spread apart before they reach your retina. The spreading of the light rays results in magnification of the image.

Light Interacts with an Object

How does light interact with an object? The physical behavior of light in some ways resembles a beam of particles and in other ways resembles a waveform. The particles of light are called photons. Each photon has an associated wavelength that determines how the photon will interact with a given object. The combined properties of particle and wave enable light to interact with an object in several different ways: absorption, reflection, refraction, and scattering.

- **Absorption** occurs when the photon's energy is acquired (absorbed) by the object (**Figure 3.9A**). The energy is converted to a different form, usually heat. When a microbial specimen absorbs light, it can be observed as a dark spot against a bright field, as in bright-field microscopy.

- **Reflection** occurs when the wavefront "bounces off" the surface of an object at an angle equal to its incident angle (**Figure 3.9B**). Reflection of light waves is analogous to the reflection of water waves. Reflection from a silvered mirror or a glass surface is used in the optics (light path) of microscopy.

- **Refraction** is the bending of light as it enters a substance that slows its speed (**Figure 3.9C**). Such a substance is said to be

Figure 3.9 Interaction of Light with Matter

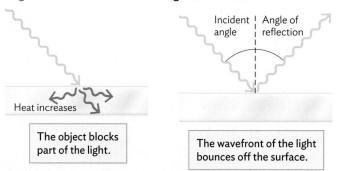

A. Absorption: The object blocks transmission of incident light and acquires some of its energy.

B. Reflection: The wavefront leaves an object at an angle mirror-reversed from its incident angle.

C. Refraction: The light slows as it enters an object, and the wavefront bends in toward the object.

D. Scattering: Most of the wavefront passes through a small object, while a small portion of the light energy propagates away from the object in all directions.

"refractive" and, by definition, has a higher **refractive index** than air. Refraction is the key property that enables a lens to magnify an image.

- **Scattering** occurs when a portion of the wave energy passing through an object is converted to rays that depart from the object in all directions (**Figure 3.9D**). The light rays are said to be scattered. If many small objects simultaneously scatter light, we observe a haze—for example, the haze of bacteria suspended in a culture tube.

Refraction Enables Magnification

Magnification requires refraction of light through a medium of high refractive index, such as glass. As a wavefront of light enters a refractive material, the region of the wave that first reaches the material is slowed, like the wheel of a car that enters a bank at the side of the road. Meanwhile, the rest of the wave continues at its original speed; thus the wavefront bends until it passes into the refractive material (**Figure 3.10**). As the entire wavefront continues

Figure 3.10 Refraction of Light Waves

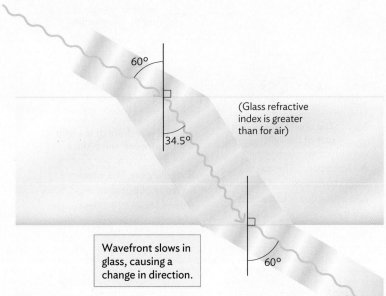

60°

(Glass refractive index is greater than for air)

34.5°

Wavefront slows in glass, causing a change in direction.

60°

As one side of a wavefront enters the glass, which has a higher refractive index than air, the wave's speed decreases. But the rest of the wavefront continues at the original speed until it, too, enters the glass. The result is that the wavefronts of light shift direction as they enter the glass. As the light waves exit the glass at the same angle, they bend in reverse, traveling parallel to the original direction.

through the refractive material, its path remains at an angle from its original direction. As the wave reaches the opposite face of the refractive material, the wavefront bends in the opposite direction and resumes its original speed.

A refractive material with a curved surface can serve as a lens to focus light. Light rays entering a lens are bent at an angle such that all rays meet at a common point. Parallel light rays entering the lens emerge at angles, intersecting each other at the **focal point** (Figure 3.11). From the focal point, the light rays continue

Figure 3.11 Focal Point

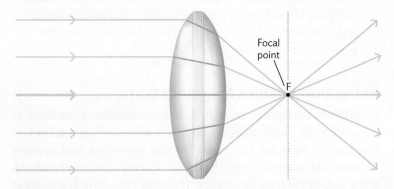

Focal point

F

Light rays pass through a lens and converge at the focal point.

onward in an expanding wavefront. This expansion magnifies the image carried by the wave. The distance from the lens to the focal point (called the focal distance) is determined by the degree of parabolic curvature of the lens and by the refractive index of its material.

In **Figure 3.12**, the object under observation is placed just outside the **focal plane** (a plane containing the focal point of the lens). The lens bends all light rays from the object. Parallel rays converge at the opposite focal point. At the focal point, the light rays continue through until they converge with light rays refracted by the lens. The inverted image expands, magnified by the spreading out of refracted rays. The distance between parts of the magnified image is enlarged, enabling our eyes to resolve finer details.

Magnification and Resolution

How does magnification enable our eyes to see increased resolution? Magnification increases the distance between peaks of light intensity, so that they are detected by separate photoreceptors on your retina (see Figure 3.2). But the resolution of detail in microscopy is limited by the degree to which details "expand" with magnification. The spreading of light rays does not necessarily increase resolution. For example, an image composed of dots does not gain detail when enlarged on a photocopier, nor does an image composed of pixels gain detail when enlarged on a computer screen. In such cases, resolution fails to increase because the individual details of the image expand in proportion to the expansion of the overall image. Magnification without an increase in resolution is called **empty magnification**.

- -

Thought Question 3.1   Explain what happens to a refracted light wave as it emerges from a piece of glass of even thickness. How do its new speed and direction compare with its original (incident) speed and direction?

- -

SECTION SUMMARY

- **Electromagnetic radiation** interacts with an object and acquires information we can use to detect the object.
- **Contrast** between an object and its background makes it possible to detect the object and resolve its parts.
- **The wavelength of the radiation** must be equal to or smaller than the size of the object if we are to resolve the object's shape.
- **Absorption** means that the energy from light (or other electromagnetic radiation) is acquired by the object.
- **Reflection** means that the wavefront bounces off the surface of a particle at an angle equal to its incident angle.
- **Refraction** is the bending of light as it enters a substance that slows its speed.

Figure 3.12 Generating an Image with a Lens

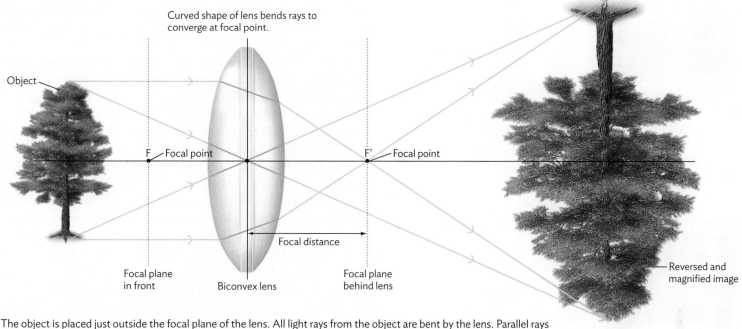

Curved shape of lens bends rays to converge at focal point.

Object

F Focal point

F' Focal point

Focal distance

Focal plane in front

Biconvex lens

Focal plane behind lens

Reversed and magnified image

The object is placed just outside the focal plane of the lens. All light rays from the object are bent by the lens. Parallel rays converge at the opposite focal point. The light rays continue through the focal point, generating an image of the object that is reversed and magnified.

3.3 Bright-Field Microscopy

SECTION OBJECTIVES

- Explain the use of a bright-field microscope.
- Describe how to observe an object in focus by using a compound microscope.

What do bacteria look like under your microscope? In **bright-field microscopy**, an object such as a bacterial cell is perceived as a dark silhouette blocking the passage of light. Details of the dark object are defined by the points of light surrounding its edge. The optics of a bright-field microscope are designed to maximize detail under magnification by a lens.

To increase resolution, we must consider these factors:

- **Wavelength.** The increase in resolution with magnification is limited by the wavelength of light (to about 1,000× magnification, for a typical student microscope). Any greater magnification expands only the width of the intensity peak; the image becomes larger but with no greater resolution (empty magnification).

- **Light and contrast.** For any given lens system, there is an optimal amount of light that yields the highest contrast between the dark specimen and the light background. High contrast is needed to achieve the maximum resolution at high magnification.

- **Lens quality.** All lenses possess inherent aberrations—shape defects that detract from perfect parabolic curvature. Instead of trying to grind a "perfect" single lens, modern manufacturers construct microscopes with a series of lenses that multiply each other's magnification and correct for aberrations.

Achieving the highest magnification requires a lens of high curvature and high refractive index. **Figure 3.13** shows an **objective lens**, a lens situated directly above an object or specimen that we wish to observe at high resolution. An object at the focal point of the objective lens sits at the tip of an inverted light cone formed by rays of light through the lens converging at the object. The angle of the light cone is determined by the curvature and refractive index of the lens. The lens fills an aperture, or hole, for the passage of light; and for a given lens, the light cone is defined by an angle projecting from the midline. As the light cone angle increases, the horizontal width of the light cone increases, and a wider cone of light passes through the specimen. As the width of the light cone increases, the resolution improves. Therefore, the greater the angle of the light cone from a given lens, the better the resolution.

Resolution also depends on the refractive index of the medium containing the light cone, which is usually air. The refractive index (n) is the ratio of the speed of light in a vacuum to its speed in a refractive medium. For air, the refractive index is defined as 1, whereas lens material has a refractive index greater than 1. The lens bends the light, spreading it at an angle (θ, the Greek letter theta).

Figure 3.13 Resolution and Numerical Aperture

Low-power magnification (10×)

Objective

$\theta = 15°$

Specimen

Light intensity

Poorly resolved

$$\begin{aligned} NA &= n \sin \theta \\ &= 1.0 \sin 15° \\ &= 0.25 \end{aligned}$$

High-power magnification (100×)

$\theta = 72°$

Objective

Specimen

Resolved

$$\begin{aligned} NA &= n \sin \theta \\ &= 1.0 \sin 72° \\ &= 0.95 \end{aligned}$$

NA = Numerical aperture
n = Refractive index
= 1.00 (air)
θ = Angle of light cone

The resolution depends on numerical aperture (NA), which equals the refractive index (n) of the medium containing the light cone multiplied by the sine of the angle of the light cone (θ). Higher NA yields higher resolution.

The product of the refractive index of the medium (n) and the sine of that angle ($\sin \theta$) is the **numerical aperture (NA)**:

$$NA = n \sin \theta$$

In Figure 3.13, we see the calculation of NA for an objective lens magnification of 10× and for a lens magnification of 100×. As NA increases, the magnification power of the lens increases.

As the lens power increases, the light cone widens, and the lens must come closer to the object. Defects in lens curvature become more of a problem, and focusing becomes more challenging. As the angle θ becomes very wide, too much of the light from the object is lost owing to refraction at the glass-to-air interface. The highest-power objective lens, generally 100×, poses a problem in that the interface between lens and air would refract at such a wide angle that too much light would be lost (**Figure 3.14**). To minimize loss of light, we maintain a zone of uniform refractive index by inserting **immersion oil** between the lens and the slide. Immersion oil has a refractive index comparable to that of glass ($n = 1.5$). Using immersion oil minimizes loss of light rays at the widest angles and makes it possible to reach 100× magnification.

The Compound Microscope

The manufacture of higher-power lenses is difficult because of increased aberration (as you recall, aberrations are shape defects that undermine perfect parabolic curvature). For this reason, we rarely observe through a single lens. Instead we use a **compound microscope**, a system of multiple lenses designed to correct or compensate for aberration. A typical arrangement of a compound microscope is shown in **Figure 3.15**. In this arrangement, the light source is placed at the bottom, shining upward through a series of lenses, including the condenser, objective, and ocular lenses.

The **condenser** lens focuses light rays from the light source onto a small area of the slide. Between the light source and the condenser sits a **diaphragm**, a device to vary the diameter of the light column. Lower-power lenses require operation at lower light levels because the excess light makes it impossible to observe the darkening effect of absorbance by the specimen (contrast). Higher-power lenses require more light and accordingly, an open diaphragm.

The nosepiece of a compound microscope typically holds three or four objective lenses of different magnifying power, such as 4×, 10×, 40×, and 100× (requiring immersion oil). These lenses are arranged so as to rotate in turn into the optical column. A high-quality instrument will have the lenses set at different heights from the slide so as to be **parfocal**. In a parfocal system, when an object is focused using one lens, it remains in focus, or nearly so, when another lens is rotated to replace the first.

The image from the objective lens is amplified by a secondary focusing step through the **ocular lens**. The ocular lens sits directly beneath the observer's eye. In magnification, each light ray traces a path toward a position opposite its point of origin, thus creating a mirror-reversed image (reversal of right and left). We need to keep this reversal in mind when exploring a field of cells.

Figure 3.14 Use of Immersion Oil in Microscopy

Objective lens

Immersion oil

Glass slide

Microscope stage

Light

Refracted light

No oil

Light

If the light rays exit the glass into air (which has a lower refractive index than glass), their direction bends away from the objective lens (red arrow), as explained in Figure 3.9. Immersion oil, with a refractive index comparable to that of glass ($n = 1.5$), maintains a straight path of light rays (blue arrow). With immersion oil, more light is collected, NA increases, and resolution improves.

Figure 3.15 The Anatomy of a Compound Microscope

Ocular lens (eyepiece)

Image of object magnified only by objective lens

Objective lens

Object or specimen

Virtual image seen by eye, magnified by objective and ocular lenses

A. Light path through a compound microscope.

Eyepiece

Focusing knob

Objectives on revolving nosepiece

Specimen

Stage

Condenser

Diaphragm

Light source

B. Cutaway view of a compound microscope.

The magnification factor of the ocular lens is multiplied by the magnification factor of the objective lens to generate the **total magnification** (power). For example, a 10× ocular times a 100× objective generates 1,000× total magnification.

Observing a specimen under a compound microscope requires several steps:

- **Position the specimen centrally in the optical column.** Only a small area of a slide can be visualized within the field of view of a given lens. The higher the magnification, the smaller the field of view that will be seen.

- **Optimize the amount of light.** At lower power, too much light will wash out the light absorption of the specimen without contributing to magnification. At higher power, more light needs to be collected by the condenser. To optimize light, the condenser must be set at the correct vertical position to focus on the specimen, and the diaphragm must be adjusted to transmit the amount of light that produces the best contrast.

- **Focus the objective lens.** The focusing knob permits adjustment of the focal distance between the objective lens and the specimen on the slide so as to bring the specimen into the focal plane. Typically, we focus first by

Wet mount of *Oenococcus oeni*.

using a low-power objective, which generates a greater **depth of field**—that is, a region along the optical column over which the object appears in reasonable focus. After focusing under low power, we can rotate a higher-power lens into view and then perform a smaller adjustment of focus.

Is the Object in Focus?

An object appears in focus (that is, is situated within the focal plane of the lens) when its edge appears sharp and distinct from the background. The shape of the dark object is defined by the points of light surrounding its edge. The partial resolution of these points of light generates extra rings of light surrounding an object whose dimensions are close to the resolution limit.

Figure 3.16 presents microscopic images of *Oenococcus oeni*, a species of bacteria that ferment grapes during winemaking. The bacteria are observed in a **wet mount**, a drop of water on a slide with a coverslip. The advantage of the wet mount is that the organism is viewed in as natural a state as possible, without artifacts resulting from chemical treatment.

Figure 3.16 *Oenococcus oeni* Observed at Different Levels of Focus: Wet Mount of Live Bacteria Swimming

Too close to lens; image is blurred.

A. When the bacteria drift too close to the lens, the image blurs.

In focus; image appears sharp.

B. When the bacteria lie within the focal plane, the image appears sharp.

Too far beyond focal point; image appears "empty," with interference rings.

C. When the bacteria lie too far from the lens, the image appears "empty" or "hollow," surrounded by rings of interference.

Object extends through focal planes.

D. When the chain of bacteria extends through the focal plane, different parts show different focal effects (in focus, too near, or too far).

The disadvantage is that most living cells are transparent and therefore show little contrast with the external medium.

At highest magnification, the focal plane is so narrow that only part of each chain of bacteria can be seen in focus. As bacterial cells drift in and out of the focal plane, their appearance changes through optical effects. When the bacteria move out of the focal plane too close to the lens, resolution declines and the image blurs (Figure 3.16A). When the bacteria move within the focal plane, the image appears sharp, with a bright line along the edge of the cells. Each chain of cells shows well-focused portions along with hazier portions that are too close to the lens (Figure 3.16B). When the cells move too far past the focal plane, the shape may appear bright or "hollow" (Figure 3.16C). In fact, the bacteria are not hollow at all; only the magnified image has changed.

At high magnification, the depth of field narrows such that the bacterial cells extends through the focal plane. As a result, different parts of the cells appear out of focus (either too near or too far from the lens; Figure 3.16D). When the end of a rod or chain points toward the observer, it appears round and dark. Light travels through the length of the cell before reaching the observer, so the cell absorbs more light and appears darker.

SECTION SUMMARY

- **In bright-field microscopy,** resolution depends on:
  - —The wavelength of light, which limits resolution to about 0.2 μm.
  - —The magnifying power of a lens, which depends on its numerical aperture ($n \sin \theta$).
  - —The position of the focal plane, which is the location where the specimen is "in focus" (that is, where the sharpest image is obtained).
- **A compound microscope** achieves magnification and resolution through the objective and ocular lenses.
- **A wet mount** specimen is the way to observe living microbes.

- -

Thought Question 3.2 In theory, which angle of the light cone would produce the highest resolution? What practical problem would you have in designing a lens to generate this light cone?

Thought Question 3.3 Under starvation conditions, bacteria such as *Bacillus thuringiensis*, the biological insecticide, repackage their cytoplasm into endospores, leaving behind an empty cell wall. Suppose, under a microscope, you observe what appears to be a hollow cell. How can you tell whether the cell is indeed hollow or whether it is simply out of focus?

- -

3.4 Stained Samples

SECTION OBJECTIVES

- Explain how a stain reveals additional information about a microscopic object.
- Describe how the Gram stain distinguishes two classes of bacteria.

CASE HISTORY 3.2
400,000 Sickened by Parasite

Purna, a 6-year-old boy in India's state of Assam, fell ill with fever. At the time, prolonged rainfall had left vast pools of water, breeding mosquitoes. First the boy felt intense chills, and his temperature rose to 40°C (104°F). He experienced nausea and intense headaches and vomited repeatedly. The fever subsided, only to return 3 days later. After several weeks of cycling fevers, Purna was at last brought to a clinic for examination. He showed extreme anemia, with enlarged spleen and liver. A blood film slide revealed *Plasmodium falciparum*, a parasite that causes malaria. In the Assam outbreak of malaria, 400,000 people were infected. Some travelers who returned to the United States experienced malarial symptoms months afterward.

The diagnosis of malaria offers a clinical application of microscopy using a blood sample that is fixed and treated with the Giemsa stain (**Figure 3.17**). Stained blood samples play a crucial role in malaria diagnosis, which is often a challenge to obtain in remote rural regions where malaria is prevalent. New technology such as the CellScope cell phone microscope aims to make such tests available in developing countries (see **IMPACT**).

The Giemsa stain contains eosin, a negatively charged dye that binds red blood cells (RBCs), and methylene blue, a positively

Figure 3.17 Blood Film Showing Red Blood Cells Infected with Malaria Parasite

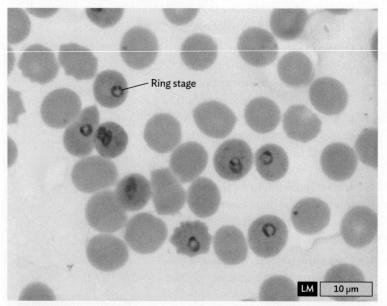

Ring stage

LM   10 μm

Plasmodium falciparum grows within red blood cells (ring stage). Emerging parasites develop into gametocytes (not shown here). Bright-field microscopy, Giemsa stain.

charged dye that binds the negatively charged phosphate groups of DNA. The methylene blue specifically stains the DNA-containing nucleus of the malarial parasite. Giemsa stain reveals both the "ring form" stage of the parasite, growing within a RBC, and the "gameto-cyte" stage that develops outside the host cell, ready for transmission to the next host.

Different Kinds of Stains

Fixation and **staining** are procedures that enhance detection and resolution of microbial cells. These procedures kill the cells and stabilize the sample on a slide. In fixation, cells are made to adhere to a slide in a fixed position. Cells may be fixed with methanol or by heat treatment to denature (unfold) the cell's proteins, whose exposed side chains then adhere to the glass. A stain absorbs much of the incident light, usually over a wavelength range that results in a distinctive color. The use of chemical stains was first developed in the nineteenth century, when German chemists used organic synthesis to invent new coloring agents for clothing. Clothing was made of natural fibers such as cotton or wool, so a substance that dyed clothing was likely to react with biological specimens.

How do stains work? Most stain molecules contain double-bond systems that absorb visible light (**Figure 3.18**) and one or more positive charges that bind to negative charges on the bacterial cell envelope (discussed in Section 5.3). Different stains vary with respect to their strength of binding and the degree of binding to different parts of the cell.

A simple stain adds dark color specifically to cells but not to the external medium or surrounding tissue. The most commonly used simple stain is methylene blue, originally used by Robert Koch to stain anthrax bacteria. A typical procedure for fixation and staining is shown in **Figure 3.19**. First, a drop of culture is fixed on a slide by treatment with methanol or by heat on a slide warmer. Either treatment denatures cell proteins, exposing side chains that bind to the glass. The slide is then flooded with methylene blue solution.

Figure 3.18 Chemical Structure of Stains

Methylene blue   **Crystal violet**

Methylene blue and crystal violet are cationic (positively charged) dyes. The positively charged groups react with the bacterial cell envelope, which carries mainly negative charge.

Figure 3.19 Procedure for Simple Staining with Methylene Blue

1 Place a loopful of broth culture on a clean slide.

2 Spread in a thin film over the slide.

3 Air-dry.

4 Fix cells to slide by adding drop of methanol; air-dry.

5 Stain with methylene blue, 1 min.

6 Wash off stain with water.

7 Blot off excess water.

8 View under microscope.

impact

The CellScope: A Cell Phone Microscope

Microscopy plays a key role in diagnosing many diseases, including malaria and tuberculosis. But microscopy is often unavailable in the developing world, particularly in remote rural areas where few medical facilities exist. A solution to this problem was devised by Daniel Fletcher, professor of bioengineering at UC Berkeley. Fletcher and his students designed an inexpensive optical system that can be attached to a cell phone camera (Figure 1). The phone is thus converted to a handheld microscope, called the CellScope. The CellScope obtains remarkably high-quality images of blood films, showing malaria infection and sickle-shaped cells. The optics can even include fluorescence filters allowing fluorescent antibody detection of tuberculosis bacteria in a sputum sample. **Fluorescence** requires illumination with light at one wavelength, followed by detection of light emitted by the sample at a different wavelength (explained in Section 3.5).

The optical attachment includes an eyepiece, an objective lens, and a sample holder. Illumination is provided by a low-energy light-emitting diode (LED). For fluorescence, the attachment includes special filters for detecting the excitation and emission wavelengths. Even with inexpensive lenses, the CellScope has achieved resolution of 1.2 μm, enabling detection of malaria parasites and *Mycobacterium tuberculosis*.

The cell phone's computer elements can manipulate images obtained through the CellScope. The CellScope start-up company, in partnership with UC Berkeley, develops applications of "telemicroscopy" for remote communities around the globe. Users can transmit images through the mobile wireless network, which now reaches many remote regions. Distant medical professionals can interpret the transmitted images. So far, workers have field-tested the device in several countries, and clinical comparison trials are under way. For example, the CellScope has been used to detect tuberculosis at Hanoi Lung Hospital, Vietnam; *Schistosoma* parasites at the University of Yaoundé, Cameroon; and cytomegalovirus retinitis at Chiang Mai University, Thailand.

DISCUSSION QUESTION

What other kind of infections do you think you could diagnose using a CellScope?

Figure 1 The CellScope: A Microscope Made from a Cell Phone

Cell phone

Slide holder

CellScope

3D-printed stand

The positively charged molecule binds to the negatively charged cell envelope. After excess stain is washed off and the slide has been dried, it is observed under high-power magnification using immersion oil.

A **differential stain** stains two different kinds of cells in a distinguishable manner. The most famous differential stain is the **Gram stain**, devised in 1884 by the Danish physician Hans Christian Gram (1853–1938). Gram first used the Gram stain to distinguish pneumococcal bacteria (*Streptococcus pneumoniae*) from human lung tissue. A similar use of the Gram stain is seen in **Figure 3.20A**, where *S. pneumoniae* bacteria appear dark purple against the pink background of human epithelial cells.

Figure 3.20 Gram Staining of Bacteria (a Type of Differential Staining)

A. Gram stain of a sputum specimen from a patient with pneumonia, containing Gram-positive *Streptococcus pneumoniae* (purple diplococci) and Gram-negative epithelial cells of the patient (pink nuclei).

B. Gram-negative *Pseudomonas aeruginosa* (pink rods).

▶ Figure 3.21 The Gram Stain Procedure

Gram-positive   **Gram-negative**

1 Add methanol to fix cells to surface; air-dry.

Cells are fixed to slide surface   Cells are fixed to slide surface

2 Add crystal violet stain (1 min).

Crystal violet stains cells reversibly   Crystal violet stains cells reversibly

3 Add iodine, which binds stain to cells (1 min).

Iodine complex with crystal violet increases stain retention   Stain remains reversible

4 Wash with ethanol to decolorize (20 s).

No change   Stain is removed from Gram-negative cells

5 Add safranin counterstain (1 min).

Gram-positive cells remain dark purple   Safranin counterstains Gram-negative cells

The Gram stain distinguishes between Gram-positive cells, which have thick cell walls and retain the crystal violet stain; and Gram-negative cells, which have thinner cell walls and lose the crystal violet stain but are counterstained by safranin.

However, not all species of bacteria retain the purple stain. An example of a bacterium that fails to retain the stain (Gram-negative) is *Pseudomonas aeruginosa*, which as recently as 2019 caused a series of infections of surgical wounds in the United States and Mexico (**Figure 3.20B**). Different bacterial species are classified as Gram-positive or Gram-negative, depending on whether or not they retain the purple stain.

Gram Staining Distinguishes Two Classes of Bacteria

In the Gram stain procedure (**Figure 3.21**), a dye such as crystal violet binds to the bacteria; it also binds to the surface of human cells, but less strongly. After the excess stain is washed off, a mordant, or binding agent, is applied. The mordant used is iodine solution, which contains iodide ions (I^-). The iodide ions form complexes with the positively charged crystal violet molecules trapped inside the cells. The crystal violet–iodide complex is now held more strongly within the cell wall. The thicker the cell wall, the more crystal violet–iodide complexes are retained.

Next, a decolorizer—here, ethanol—is added for a precise interval (typically 20 seconds). The decolorizer removes loosely bound crystal violet–iodide complexes, but Gram-positive cells retain the stain tightly. The **Gram-positive** cells that retain the stain appear dark purple, whereas the **Gram-negative** cells are colorless. The timing of the decolorizer step is critical because if it lasts too long, the Gram-positive cells, too, will release their crystal violet stain. In the final step, a **counterstain**, safranin, is applied. This process allows the visualization of Gram-negative material, which the safranin stains pale pink.

The Gram stain procedure was originally devised to distinguish bacteria (Gram-positive) from human cells (Gram-negative). Microscopists soon discovered, however, that many important

Figure 3.22 Differential Stains

A. *Mycobacterium tuberculosis* colonizing mouse lung, visualized by Ziehl–Neelsen acid-fast stain. Chains of bacteria are stained purple; lung cell nuclei are stained blue.

B. *Clostridium tetani*, visualized by malachite green endospore stain. Endospores are stained blue-green, 2 μm long; vegetative cells are stained pink.

Gram-variable) and are thus not distinguished by the Gram stain. Bacterial diversity and identification are explored further in Chapter 10.

Other Differential Stains

Other differential stains applied to various prokaryotes are illustrated in **Figure 3.22**. These include **acid-fast stains**, **endospore stains**, **negative stains**, and **antibody stains**:

- **Acid-fast stain.** Carbolfuchsin with heat treatment specifically stains mycolic acids of *Mycobacterium tuberculosis* and *Mycobacterium leprae*, the causative agents of tuberculosis and leprosy, respectively (Figure 3.22A).

- **Endospore stain.** When samples are boiled with malachite green, the stain binds specifically to the endospore coat (Figure 3.22B). This method detects endospores of *Bacillus* species such as *B. thuringiensis*, an insecticide, and *B. anthracis*, the causative agent of anthrax. It also detects endospores of *Clostridium botulinum*, the source of botulinum toxin, and *Clostridium tetani*, the causative agent of tetanus.

- **Negative stain.** Some bacteria synthesize a capsule of extracellular polysaccharide filaments, which protects the cell from predation or from engulfment by WBCs. The capsule is transparent and invisible in suspended cells. However, adding a suspension of opaque particles such as India ink or Congo red can darken the surrounding medium.

- **Antibody stains.** Specialized stains use antibodies to identify precise strains of bacteria or even specific molecular components of cells. The antibody (which binds a specific cell protein) is linked to a reactive enzyme for detection or to a fluorophore (fluorescent molecule) for fluorescence microscopy (to be discussed shortly, in Section 3.5).

species, such as the intestinal bacterium *Escherichia coli* and the nitrogen-fixing symbiont *Sinorhizobium meliloti*, do not retain the Gram stain. It turns out that Gram-negative species of bacteria possess a thinner and more porous cell wall than Gram-positive species. A Gram-negative cell wall has only a single layer of peptidoglycan (sugar chains cross-linked by peptides), which more readily releases the crystal violet–iodide complexes. By contrast, a Gram-positive cell such as *Staphylococcus aureus*, a cause of toxic shock syndrome and wound infections, has multiple layers of peptidoglycan. The multiple layers retain enough stain to make the cell appear purple.

Among bacteria, the Gram stain distinguishes many types of bacteria, including two groups with distinctive cell wall structures: Proteobacteria (such as *E. coli* and *Proteus*), with a thin cell wall plus an outer membrane (Gram-negative), and Firmicutes (such as *Staphylococcus* and *Streptococcus*), with a multiple-layered cell wall but no outer membrane (Gram-positive). The outer membrane of Proteobacteria (see Chapter 5) often possesses important pathogenic factors, such as the lipopolysaccharide endotoxins that cause toxic shock syndrome. During the Gram stain procedure, however, the decolorizing wash with ethanol disrupts the outer membrane, allowing most of the crystal violet–iodide complex to leak out.

The Gram stain emerged as a key tool for the biochemical identification of species, and it remains essential in the clinical laboratory. However, the Gram stain is reliable only for freshly cultured bacteria. Some Gram-positive cells, such as *Bacillus anthracis* (the cause of anthrax), fail to retain stain once the cells run out of nutrients. Still other groups of bacteria, including *Lactobacillus* species common in the oral cavity and gastro-intestinal tract, may stain either positive or negative (known as

SECTION SUMMARY

- **Fixation and staining** kill the specimen but improve contrast and resolution.
- **Differential stains** distinguish between bacteria with different structural features.
- **The Gram stain** differentiates between two major bacterial taxa, Proteobacteria (Gram-negative) and Firmicutes (Gram-positive). Other bacteria and archaea vary in their Gram stain appearance. Eukaryotes stain negative.

3.5 Advanced Microscopy

SECTION OBJECTIVES

- Describe the different kinds of information obtained by dark-field, phase-contrast, and fluorescence microscopy.

- Describe how electron microscopy enables visualization of objects below the resolution limit of light, such as viruses and parts of cells.

Advanced optical techniques enable us to see microbes that are difficult or impossible to detect under a bright-field microscope. Detecting these microbes is a challenge, either because their size is below the limit of resolution of light or because their cytoplasm is transparent. These advanced techniques take advantage of special properties of light waves, including scattering and interference patterns.

Dark-Field Observation Detects Unresolved Objects

In **dark-field microscopy**, the condenser contains an opaque central disk that blocks transmitted light, so the field appears dark. The only light rays we detect are those scattered by the object, as shown earlier in Figure 3.9D. For dark-field optics, an intense light source allows a microbe smaller than the wavelength of light to scatter enough light for detection. Detection does not allow resolution; in effect, the microbes are visualized as halos of bright light against a dark field, just as stars are observed against the night sky.

One use of dark-field microscopy is to detect *Treponema pallidum*, the bacterium that causes syphilis. A *T. pallidum* cell is called a "spirochete," a spiral so narrow (0.1 μm) that its shape cannot be fully resolved by light microscopy. Nevertheless, the spiral form of *T. pallidum* can be detected by dark-field microscopy (**Figure 3.23**). In the dark-field image, the length of the cell is accurate, but the width of the cell body appears "overexposed," or wider than the actual cell, because the scattered object lacks resolution.

A limitation of dark-field microscopy is that any tiny particle, including specks of dust, can scatter light and interfere with visualization of the specimen. Unless the medium is extremely clear, it can be difficult to distinguish microbes of interest from particulates. Other methods of contrast enhancement, such as phase contrast and fluorescence, avoid this problem.

Phase-Contrast Microscopy

Phase-contrast microscopy exploits small differences in refractive index between the cytoplasm and the surrounding medium or between different organelles. This technique is particularly useful for eukaryotic cells, such as parasites, which contain many intracellular compartments. For example, **Figure 3.24** shows a phase-contrast image of *Paramecium*, a freshwater protozoan that is part of the food web feeding invertebrates and fish (see Chapter 27). The paramecium

Figure 3.23 Dark-Field Observation of Bacteria

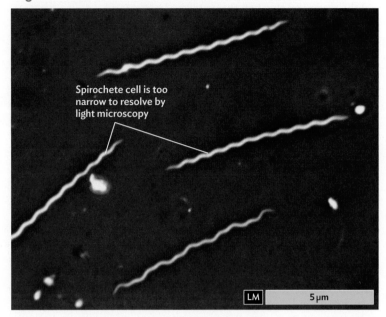

Spirochete cell is too narrow to resolve by light microscopy

LM   5 μm

Treponema pallidum specimen from a patient with syphilis.

Figure 3.24 Phase-Contrast Microscopy

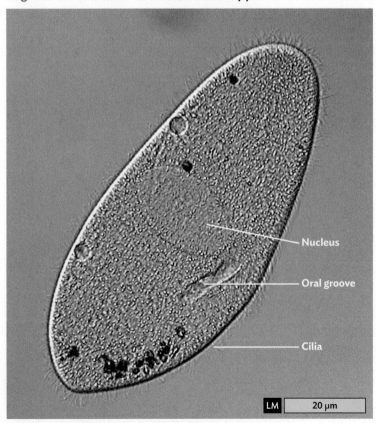

Nucleus

Oral groove

Cilia

LM   20 μm

Phase-contrast micrograph of *Paramecium*, a freshwater protozoan. The abrupt changes in refractive index at the cell's edge makes it appear three-dimensional. Subcellular organelles show similar effects.

is covered with cilia (visible in Figure 3.24) that beat in waves to propel the organism through the water and move food.

Phase-contrast microscopy makes use of the fact that living cells have relatively high contrast due to their high concentration of solutes. Given the size and refractive index of commonly observed cells, light is slowed by approximately one-quarter of a wavelength when it passes through the cell. The optical system slows the refracted light by an additional one-quarter of a wavelength, so that the light refracted through the cell is slowed by a total of half a wavelength compared with the light transmitted through the medium. When two light rays are out of phase by half a wavelength, they cancel each other's amplitude, and that portion of the image appears dark. **Figure 3.25** illustrates the addition of wave amplitudes in phase and the subtraction of wave amplitudes out of phase.

Phase-contrast optics generates alternating regions of light and darkness in the image. Even small differences in refractive index can produce dramatic differences in contrast between the offset phases of light. In Figure 3.24, at the edge of the cell's image, the abrupt change in refractive index makes it appear three-dimensional. Similar effects reveal subcellular organelles such as the nucleus and the oral groove.

Interference Microscopy

Other kinds of optical systems have been devised that use light interference to enhance cytoplasmic contrast. Interference microscopy enhances contrast by superimposing the specimen's image on a second beam of light that generates an interference pattern. The interference pattern produces an illusion of three-dimensional shadowing across the specimen. For example, **Figure 3.26** shows the shape of the dust-borne bacterium *Bacillus subtilis* illuminated

Figure 3.26 Interference Micrograph of *Bacillus subtilis*

LM   2 μm

The apparent 3D effect is illusory. One bacterium appears as if crossed through another.

by interference contrast. The shape of the cells is more clearly defined than it would be in conventional bright-field microscopy.

Fluorescence Microscopy

CASE HISTORY 3.3

From Hot Tub to Hospital

Jared was a 48-year-old healthy man who ran a business servicing whirlpool baths and hot tub spas in Columbus, Ohio. He began to experience dyspnea (shortness of breath) and watery diarrhea. Jared's symptoms increased, including a dry cough, myalgia (muscle pain), nausea, and vomiting. After 6 days, he was admitted to a hospital in Columbus. The clinician noticed that Jared appeared flushed and had dry mucous membranes. His temperature was 39°C (102.2°F), and his pulse rate was 103 beats per minute. Chest radiographs showed pneumonia with consolidation (formation of a firm mass) of the left lower lobe. His C-reactive protein (CRP, a blood protein produced by the liver) was greater than 220 milligrams per liter (mg/l), a high level that indicates inflammation.

Before onset of symptoms, Jared recalled having cleaned the filter of a hot tub spa at an elite downtown Columbus hotel. The water in hot tubs is filtered and sanitized by chlorination, but it can become contaminated by Gram-negative pathogens. Culture from Jared's sputum and from the spa filter revealed Gram-negative bacilli consistent with *Legionella pneumophila*. The diagnosis was confirmed by fluorescence microscopy using direct fluorescent antibody (DFA) stain. The patient received intravenous amoxicillin and clarithromycin antibiotics for 8 days before his condition returned to normal.

Figure 3.25 Constructive (Additive) and Destructive (Subtractive) Interference of Light Waves

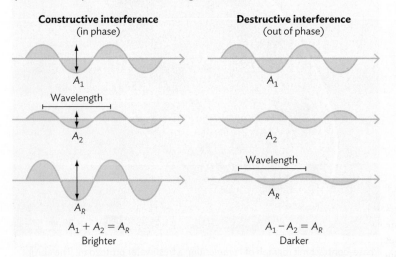

Constructive interference
(in phase)

A_1

Wavelength

A_2

A_R

$A_1 + A_2 = A_R$
Brighter

Destructive interference
(out of phase)

A_1

A_2

Wavelength

A_R

$A_1 - A_2 = A_R$
Darker

In constructive (additive) interference, the peaks of the two light waves are in phase and rise together; their amplitudes combine to create a wave of greater amplitude. In destructive (subtractive) interference, the peaks of the waves are out of phase; their amplitudes cancel to create a wave of lesser amplitude.

The bacteria from Jared's sputum in this case history were shaped like small rods—just like many other species. How can we distinguish specific types of bacteria that look alike? **Fluorescence microscopy (FM)** is a tool that enables us to identify a single species, such as the Gram-negative, rod-shaped *L. pneumophila* (Figure 3.27). *L. pneumophila* is the cause of legionellosis, a form of pneumonia that can be rapidly fatal if not diagnosed in time. The bacteria grow intracellularly within amebas that contaminate water sources such as air-conditioning units or hot tub spas (as in the case history). During human infection, *L. pneumophila* grows within lung macrophages.

Fluorescence microscopy offers a powerful way to detect specific microbes and subcellular structures while avoiding signals from dust and other nonspecific materials. Bacteria such as *L. pneumophila* and *Mycobacterium tuberculosis* can be identified by fluorescence microscopy of samples stained with a fluorescent antibody specific to the bacterial species.

In fluorescence microscopy, incident light is absorbed by the specimen and then emitted at a lower energy, and thus at a longer wavelength. Fluorescence occurs when a molecule absorbs light of one color, then emits light at another color of longer wavelength (lower energy). Light absorbed at a specific wavelength (the excitation wavelength), which excites an electron to a higher energy level. The electron then falls to its original level by emitting a photon of lower energy and longer wavelength (the emission wavelength). The emitted photon has a longer wavelength (less energy) because part of the electron's energy of absorption was lost as heat.

The wavelengths of excitation and emission are determined by the **fluorophore**, the fluorescent molecule used to stain the specimen. For detecting *L. pneumophila*, the fluorophore used is fluorescein, conjugated (attached) to an antibody that specifically binds the cell surface of *L. pneumophila*. In Figure 3.27, the fluorescence is brightest around the rim of each cell, where the maximal "depth" of the envelope fluoresces.

Fluorescent molecules offer amazing kinds of information about what happens inside cells and within microbial environments. In **Figure 3.28A**, *Escherichia coli* K-12 cells express tetracycline resistance from a gene encoded on a plasmid, a small circle of DNA. The plasmid also encodes a protein that fluoresces red. As the plasmid is donated to tetracycline-sensitive cells (green), the sensitive cells become resistant to tetracycline (red). This experiment shows

Figure 3.27 *Legionella pneumophila* **Visualized by Direct Fluorescent Antibody Stain**

Fluorescent labeled cells appear brightly colored against a dark background.

us how pathogens may acquire drug resistance in a hospital. In **Figure 3.28B**, an advanced type of fluorescence microscopy provides 3D images of pathogens within the body. Individual cells of *Vibrio cholerae*, the cause of cholera, are attached to the mouse intestinal epithelium.

Figure 3.28 **Fluorescence Microscopy Can Tell Us What Happens inside Cells**

A. *E. coli* K-12 cells express tetracycline resistance from a gene encoded on a plasmid, which also encodes a protein that fluoresces red. As the plasmid is donated to tetracycline-sensitive cells (green), the sensitive cells become resistant to tetracycline (red).

B. Three-dimensional visualization of mouse intestinal tissue, with cell nuclei (blue fluorescence) secreting mucin (red) into the gut lumen. *Vibrio cholerae* cells fluoresce green. Source: William DePas and Dianne Newman, Caltech.

Electron Microscopy

In an electron microscope, beams of electrons generate images at resolution levels up to a thousandfold greater than that possible for light microscopy. Electron microscopy reveals microbes in remarkable detail. In research, we use the electron microscope to observe virus infection and propagation and to study subcellular structures such as ribosomes and toxin-secreting organelles.

An important use of electron microscopy is observation of the life cycle of human immunodeficiency virus (HIV), the cause of acquired immunodeficiency syndrome (AIDS). **Figure 3.29** shows an image generated by **transmission electron microscopy (TEM)**. To generate a TEM image, the electron beams penetrate a thin section of tissue stained with an electron-dense, heavy-atom salt. The image shows HIV particles budding out of an infected lymphocyte. The dark spot within each virus particle is its ribonucleic acid (RNA) genome, compacted by special proteins. But note that, in TEM, only a slice through the virion is seen. The TEM resolution is high enough to reveal individual ribosomes (small dark spots) within the cytoplasm of the infected cell.

How does an electron microscope work? The apparatus requires an evacuated sample chamber and a series of magnetic lenses. The transmission electron microscope parallels the design of a bright-field microscope (**Figure 3.30**). The image is generated by a beam of electrons, instead of light. Electrons are ejected from a metal subjected to a voltage potential. The electrons travel in a straight line, like photons. And like photons, electrons interact with matter and exhibit the properties of waves. Beams of electrons are focused by means of a magnetic field. The shape of the magnet is designed to focus the beam of electrons in a manner analogous to the focusing of photons by a refractive lens. Overall, the electron beam passes

Figure 3.29 Transmission Electron Micrograph (TEM) of HIV Particles Emerging from an Infected Lymphocyte

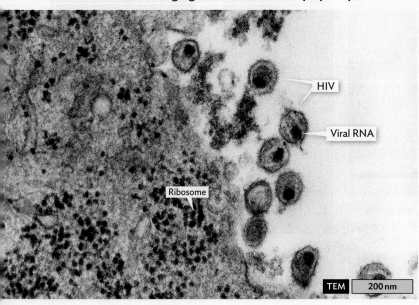

Figure 3.30 Transmission Electron Microscopy (TEM) Compared with Light Microscopy

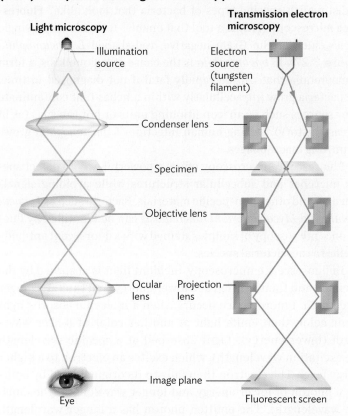

In a transmission electron microscope (right), the light source is replaced by an electron source consisting of a high-voltage current applied to a tungsten filament, which gives off electrons when heated. The electron beam is focused by magnetic lenses equivalent to the condenser, objective, and ocular lenses of a light microscope, and the image is projected onto a fluorescent screen by a projection lens.

through a magnetic condenser lens, a specimen, and a magnetic objective lens. A projection lens projects the image onto a fluorescent screen, and the final images are obtained by a digital camera.

Another type of electron microscopy is **scanning electron microscopy (SEM)**. In SEM, the electron beams scan the specimen and are reflected by stain molecules to reveal the contours of the specimen's 3D surface. SEM can be performed for a wide range of microbe sizes. SEM was used to define the shape and habitat of *Helicobacter pylori*, the major cause of gastritis. *H. pylori* inhabits the human stomach, a region long believed too acidic to harbor life. In the 1980s, however, Australian scientist Barry Marshall reported finding a new species of bacterium in the stomach. The bacterium, *Helicobacter pylori*, proved difficult to isolate and culture, and for over a decade, medical researchers refused to believe Marshall's report. Ultimately, electron microscopy (EM) confirmed the existence of *H. pylori* in the stomach and helped document its role in gastritis and stomach ulcers.

The scanning electron micrograph in **Figure 3.31** shows *H. pylori* growing upon the gastric crypt cells. *H. pylori* bacteria are helical rod-shaped cells. The cell contours are well resolved by

Figure 3.31 *Helicobacter pylori* **within the Crypt Cells of the Stomach Lining**

H. pylori bacteria grow on the lining of the human stomach, a location previously believed too acidic to permit microbial growth. Colorized to indicate bacteria (green).

the scanning beam of electrons, an achievement far beyond a light microscope's capacity. In Figure 3.31, note also the use of **colorization**, or "false color," to emphasize particular aspects of the image. Colorization is a process by which the observer adds color to aid interpretation. The color added does not indicate any natural coloration, which cannot be observed through electron microscopy.

An important limitation of traditional electron microscopy, whether TEM or SEM, is that usually it can be applied only to fixed, stained specimens. The fixatives and heavy-atom stains can introduce **artifacts** (false structures not present in the original organism), especially at finer details of resolution. Different preparation procedures have sometimes led to substantially different interpretations of structure.

The development of exceptionally high-strength electron beams now permits low-temperature cryo-electron microscopy (cryo-EM). In cryo-EM, the specimen is flash-frozen—that is, suspended in water and frozen rapidly in a refrigerant of high heat capacity (ability to absorb heat). The rapid freezing avoids ice crystallization, leaving the water solvent in a glass-like phase. The specimen retains water content and thus closely resembles its living form, although it is still ultimately destroyed by electron bombardment. The sample does not require staining because the

high-intensity electron beams increase the signal intensity compared to that of earlier instruments. Another innovation in cryo-EM is tomography, in which images are projected from different angles of a transparent specimen. Tomography avoids the need to physically slice the sample for thin-section TEM. The images from EM tomography are combined digitally to visualize the entire object in three dimensions.

SECTION SUMMARY

- **Dark-field microscopy** uses scattered light to detect objects too small to be resolved by light rays.
- **Phase-contrast microscopy (PCM)** superimposes refracted light and transmitted light shifted out of phase so as to show small differences in refractive index as distinct patterns of light and dark. This method is used to observe organelles that are transparent, with low contrast.
- **Interference microscopy** superimposes interference bands on an image, accentuating small differences in refractive index.
- **Fluorescence microscopy (FM)** detects specific cells or cell parts on the basis of fluorescence by a fluorophore. Cell parts can be labeled by a fluorophore attached to an antibody stain.
- **Transmission electron microscopy (TEM)** is based on the focusing of electron beams that penetrate a thin sample, usually stained with a heavy-metal salt that blocks electrons. TEM obtains much higher resolution than light microscopy.
- **Scanning electron microscopy (SEM)** involves scanning a 3D surface with an electron beam.

Thought Question 3.4   Compare fluorescence microscopy with dark-field microscopy. What similar advantage do they provide, and how do they differ?

Thought Question 3.5   You have discovered a new kind of microbe, never observed before. What kind of questions about this microbe might be answered by light microscopy? What questions would be better addressed by electron microscopy?

Perspective

You can use the tools of microscopy described in this chapter to discover microbes in our environment and to reveal pathogens in human samples. Identifying pathogens such as the malarial parasite or tuberculosis bacterium is a crucial aspect of diagnosis. In research, microscopy has shaped our understanding of microbial cells—how they grow and divide, organize their DNA and cytoplasm, and respond to antibiotics. Advanced microscopy reveals how pathogens infect their host cells: do they remain attached to the outside of the cell, or do they invade the cytoplasm and consume the host from within?

Next, in Chapter 4, you will learn the molecular building blocks of the microbes you observe by microscopy. You will see how small molecules, such as phospholipids, form flexible membranes for cell compartments and how large macromolecules, such as DNA and proteins, store information. These molecules all function together within a cell, as we explore further in Chapter 5.

Key Terms

absorption (p. 69)
acid-fast stain (p. 78)
antibody stain (p. 78)
artifact (p. 83)
bacillus (p. 66)
bright-field microscopy (p. 67, 71)
coccus (p. 66)
colorization (p. 83)
compound microscope (p. 72)
condenser (p. 72)
contrast (p. 69)
counterstain (p. 77)
dark-field microscopy (p. 79)
depth of field (p. 73)

diaphragm (p. 72)
differential stain (p. 76)
electron microscopy (EM) (p. 67)
empty magnification (p. 70)
endospore stain (p. 78)
fixation (p. 75)
fluorescence (p. 76)
fluorescence microscopy (FM) (p. 81)
fluorophore (p. 81)
focal plane (p. 70)
focal point (p. 70)
Gram-negative (p. 77)
Gram-positive (p. 77)

Gram stain (p. 76)
immersion oil (p. 72)
light microscopy (LM) (pp. 64, 67)
magnification (p. 63)
microscope (p. 63)
negative stain (p. 78)
numerical aperture (NA) (p. 72)
objective lens (p. 71)
ocular lens (p. 72)
parfocal (p. 72)
phase-contrast microscopy (p. 79)
reflection (p. 69)
refraction (p. 69)

refractive index (p. 69)
resolution (p. 63)
scale bar (p. 66)
scanning electron microscopy (SEM) (pp. 67, 82)
scattering (p. 69)
spirochete (p. 67)
staining (p. 75)
total magnification (p. 73)
transmission electron microscopy (TEM) (pp. 67, 82)
wavelength (p. 68)
wet mount (p. 73)
X-ray crystallography (X-ray) (p. 67)

Objectives Assessment Questions

3.1 Observing Microbes

1. Which property of bacteria determines that they are "microscopic"?
   a. Bacteria can consist of isolated single cells.
   b. Bacteria can grow as chains or biofilms of cells.
   c. Bacteria are less than a centimeter long.
   d. Bacteria are smaller than the resolution distance of the human retina.

2. Which kind of microscopy reveals microbes as dark objects against a field of light?
   a. Bright-field light microscopy
   b. Transmission electron microscopy
   c. Scanning electron microscopy
   d. Fluorescence microscopy

3.2 Optics and Properties of Light

3. Which condition is <u>not</u> required for an object to be seen under a light microscope?
   a. The dimensions of the object must be greater than the wavelength of light.
   b. The object must be fixed to a slide.
   c. The object must be magnified to a size resolvable by the human retina.
   d. The object must show contrast with its surroundings.

4. Which property of light allows magnification by a lens?
   a. Photons of light are absorbed by an opaque substance.
   b. When a light wave encounters a small particle, some of the light energy is scattered in the form of a spherical wave.
   c. Light bends, or "refracts," as it enters a substance that slows its speed.
   d. Light rays travel through a vacuum at a speed of about 3×10^8 meters per second (m/s).

3.3 Bright-Field Microscopy

5. The part of the microscope that concentrates light onto the field of view in a slide is _____.
   a. the condenser
   b. the objective lens
   c. the ocular lens
   d. the optical column

6. The part of the microscope that magnifies the image, placed nearest the object in a slide is _____.
   a. the condenser
   b. the objective lens
   c. the ocular lens
   d. the optical column

3.4 Stained Samples

7. Giemsa stain can reveal _____.
   a. the Gram-positive cell envelope of *Bacillus anthracis*
   b. the ring form and gametocyte stage of the malaria parasite *Plasmodium falciparum*
   c. the endospores of *Clostridium botulinum*
   d. the extracellular polysaccharide capsule of *Flavobacterium capsulatum*

8. Which of these statements about the Gram stain procedure is not correct?
   a. It requires fixation of the sample to a slide.
   b. It requires a dye such as crystal violet, followed by a mordant to strengthen the binding of the dye.
   c. It requires a counterstain such as safranin for completion.
   d. It requires an antibody stain.

3.5 Advanced Microscopy

9. In a patient's sample, the intracellular pathogen *Legionella pneumophila* can best be identified using _____.
   a. fluorescence microscopy, with a fluorophore attached to a cell-surface–specific antibody
   b. dark-field microscopy, using a dust-free medium
   c. phase-contrast microscopy
   d. electron microscopy

10. To visualize individual particles of human immunodeficiency virus (HIV) requires _____.

   a. fluorescence microscopy
   b. dark-field microscopy
   c. phase-contrast microscopy
   d. electron microscopy

11. State the type of microscopy used to generate each of the images shown.

   a. *Mycobacterium avium* subspecies *paratuberculosis*, associated with Johne's disease of cattle.

   b. Motile *E. coli* cells with flagella.

   c. *Streptococcus mutans*, bacteria found in dental plaque.

20 µm

   d. *Bacillus anthracis*, cause of anthrax.

Glycoprotein

Cell wall

250 nm

Review Questions

1. What principle defines an object as "microscopic"?

2. Explain the difference between detection and resolution.

3. How do eukaryotic and prokaryotic cells differ in appearance under the light microscope?

4. Define how light interacts with an object through absorption, reflection, refraction, and scattering.

5. Explain how refraction enables magnification of an image.

6. Explain how magnification increases resolution and why "empty magnification" fails to increase resolution.

7. Explain how angle of aperture and resolution change with increasing lens magnification.

8. Summarize the optical arrangement of a compound microscope.

9. Explain how to focus an object and how to tell when the object is in or out of focus.

10. Explain how the Gram stain distinguishes between two major categories of bacteria.

11. Explain the relative advantages and limitations of wet mount and stained preparations for observing microbes.

12. Explain the significance (and limitations) of the Gram stain for bacterial taxonomy.

13. Explain the basis of dark-field, phase-contrast, and fluorescence microscopy. Give examples of applications of these advanced techniques.

14. Compare the resolving power of a light microscope versus an electron microscope.

15. Explain the difference between transmission and scanning electron microscopy, and name the different applications of each.

Clinical Correlation Questions

1. After eating peanut butter, many people get sick with vomiting and diarrhea. Samples of the peanut butter show a rod-shaped bacterium that stains Gram-negative. Unstained bacteria (live mount) swim with multiple flagella. Which pathogen do these results suggest?

2. A patient with HIV infection has failed to keep up with medications. The patient has a recurring cough with shortness of breath, and his sputum shows rod-shaped bacteria that stain acid-fast. The patient should be tested to confirm possible infection by which pathogen?

For more assessment, visit digital.wwnorton.com/michum2

4

Living Chemistry: From Atoms to Cells

Lethal Chemistry from a Waterborne Pathogen

SCENARIO Water is the most abundant molecule of life—but too much water can be deadly. In the summer of 2010, heavy rains flooded Pakistan, submerging one-fifth of the country. Millions of people were displaced and exposed to water contaminated by microbial pathogens. Diarrheal disease threatened many, especially children.

PATIENT A 10-year-old boy named Akal developed profuse diarrhea. He showed extreme thirst, but clean drinking water was unavailable. His diarrhea became watery and cloudy with white flecks of intestinal mucus (known as rice-water stool). Akal became listless, his eyes sunken and his skin wrinkled from dehydration.

SIGNS AND SYMPTOMS A nurse observed a sample of Akal's diarrheal fluid under dark-field microscopy. Numerous bacteria swam rapidly, appearing like "twinkling stars." The bacteria, comma-shaped bacilli, stained Gram-negative.

DIAGNOSIS Later testing confirmed *Vibrio cholerae*, the cause of cholera. These bacteria produce cholera toxin, a protein complex that causes intestinal epithelial cells to secrete large amounts of chloride ion (Cl^-). The high chloride concentration draws out the positively charged sodium ions (Na^+). The increased extracellular salt (Na^+ and Cl^-, or NaCl) causes water to diffuse out of the epithelial cells, leading to dehydration and diarrhea. Without treatment, Akal's chance of death was 50%.

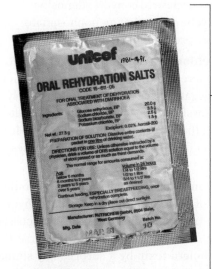

Oral rehydration therapy
Replaces the water and electrolytes lost during severe diarrhea.

TREATMENT Akal was treated with oral rehydration therapy (ORT), a method invented by Bangladeshi physicians. ORT involves gradual feeding of a solution containing glucose and NaCl to replace ions and water lost through the diarrhea. Within hours, Akal's flesh had rehydrated, and within a week he made a full recovery.

SEM   1 μm

Vibrio cholerae
A comma-shaped, Gram-negative bacterium whose enterotoxin drives water out of the intestine.

CHAPTER OBJECTIVES

After reading this chapter, you will be able to:

• Explain that cells consist of molecules and describe the properties of these molecules.

• Describe an example of how a pathogen alters host cell biochemistry.

• Explain how chemical reactions interconvert biological molecules, including the role of energy.

• Describe how membranes mediate the transport of molecules and particles.

• Describe examples of how pathogens take advantage of transport processes.

Cholera threatens impoverished people living in crowded conditions such as refugee camps, where bacteria flushed out of an infected person by diarrhea spread rapidly to new hosts. The causative bacteria, *Vibrio cholerae* (shown on the preceding page), are tiny curved rods that stain negative in the classic Gram stain, as we learned in Chapter 3. Remarkably, the cause of mortality is a single type of bacterial toxin, cholera toxin (**Figure 4.1A**). Cholera toxin, a protein complex, kills the patient by extreme water loss (dehydration) and sodium loss (hyponatremia). The cellular loss of these simple chemicals leads to death.

How does one protein complex trick the body into losing its essential water and salt? The mechanism of cholera toxin demonstrates several key principles of cell biochemistry. **Figure 4.1B** shows how cholera toxin AB5 acts within a cell of the intestinal lining:

1. **The cholera toxin AB5 binds a ganglioside.** The AB5 toxin consists of one protein subunit A with five copies of subunit B. AB5 binds to a ganglioside, a special sugar-lipid molecule found on the host cell surface. (Sugars and lipids are discussed in Section 4.2.)

2. **AB5 toxin enters the cell by endocytosis.** In endocytosis, the membrane pinches in, forming a vesicle called an endosome (discussed in Sections 4.5 and 4.6). The endosome carries AB5 to the endoplasmic reticulum (ER).

3. **AB5 toxin releases part of its A subunit, called A1.** The A1 returns to the cytoplasm.

4. **A1 binds host G protein.** The A1-bound G protein overstimulates adenylate cyclase to convert adenosine triphosphate (ATP) to cyclic adenosine monophosphate (cAMP). (ATP reactions are presented in Section 4.4.)

5. **cAMP stimulates the chloride (Cl⁻) transporter CFTR.** Chloride (Cl⁻) buildup increases the negative charge outside the cell. The increased negative charge draws out positively charged ions such as potassium (K⁺) and sodium (Na⁺). The high concentration of all these ions outside causes osmotic imbalance, drawing water out of the cell. Water loss causes dehydration and death.

Under the dire circumstances of Pakistan's flood, how could a health worker save the lives of cholera patients suffering dehydration? By understanding ion transport and osmolarity (discussed in Section 6.4), we will see how doctors learned to reverse the life-threatening effects of cholera toxin by a surprisingly simple treatment of glucose and salt solution. The treatment is called oral rehydration therapy (discussed in this chapter's IMPACT).

This chapter introduces the biochemistry that a health professional needs to know. Section 4.1 reviews the chemistry of small biological molecules, with an emphasis on the special properties of water. We then discuss macromolecules, including the large information-bearing molecules that define living organisms, such as proteins and nucleic acids. Finally, we explore important chemical reactions that keep cells alive and the membrane compartmentalization of these reactions inside a cell. These molecules and reactions are important both for microbes and for the human body.

Figure 4.1 Cholera Toxin Kills Cells

1 AB5 toxin binds ganglioside.

A subunit

B subunit

A subunit

B5 subunits

Adenylate cyclase

Ganglioside

G protein

ATP

Cl⁻ K⁺ H₂O

CFTR Na⁺

cAMP

GM1   Intestinal cell surface

A. Cholera toxin is composed of protein A and five subunits of protein B. The AB5 complex binds a host surface sugar, ganglioside.

2 AB5 toxin enters cell by endocytosis and travels to ER.

4 A1 modifies G protein; adenylate cyclase catalyzes ATP ⟶ cAMP

5 cAMP activates Cl⁻ efflux; K⁺ and water rush out. Diarrhea results.

Nucleus

3 AB5 releases A1, which returns to the cytoplasm.

Endoplasmic reticulum (ER)

B. AB5 toxin binds ganglioside (step 1) and enters the cell by endocytosis (step 2). At the endoplasmic reticulum, an enzyme cleaves the A subunit, releasing A1 peptide (step 3). The A1 peptide returns to the cytoplasm and binds G protein (step 4), thus activating adenylate cyclase. Adenylate cyclase converts ATP to cAMP, which stimulates the chloride transporter CFTR (step 5). Chloride efflux increases negative charge outside the cell, drawing out positively charged ions such as potassium (K^+) and sodium (Na^+). High salt concentration outside draws water out of the cell, causing diarrhea.

4.1 Elements, Bonding, and Water

SECTION OBJECTIVES

- Describe how atoms form the molecules of life.
- Describe the types of bonds atoms can form to make a molecule.
- Explain the special properties of the water molecule, and distinguish between hydrophilic and hydrophobic molecules.

What are living cells made of? Living cells are remarkably complex machines, able to integrate and respond to multiple stimuli, to catalyze (speed up) reactions, and to replicate themselves. Yet despite all the various tasks that cells perform, 98% of the mass of living organisms consists of just six elements: hydrogen (H), oxygen (O), nitrogen (N), phosphorus (P), sulfur (S), and carbon (C). For example, all the proteins and membranes involved in the cholera toxin cycle are composed of these elements.

The most abundant compound in cells is water (H_2O). The rest of the cell consists, for the most part, of just four different kinds of organic (carbon-based) macromolecules: lipids, proteins, carbohydrates (such as sugars), and nucleic acids (such as DNA and RNA). Cells are compartments that organize chemical reactions between these organic molecules. To understand life, we must understand the properties of water and organic molecules and of their building blocks, the chemical elements.

Cells consist mostly of water, inorganic ions (such as K^+ and Cl^-), and organic molecules. These cellular components are composed of atoms, the fundamental unit of matter. An **atom** consists of a positively charged nucleus that contains protons and neutrons, surrounded by negatively charged electrons. The hydrogen nucleus consists of a single proton. Protons and neutrons have a mass approximately 2,000-fold greater than that of electrons. Yet the charge on a proton (+1) is of the same magnitude as the negative charge on the electron (−1).

Different types of atoms contain different numbers of protons and electrons. The number of protons determines the type of atom, called an **element** (Table 4.1). The elements can be organized into a periodic table (see Appendix 1). The periodic table indicates each element's **atomic number** (the number of protons), which is the property that defines an element. For example, all carbon atoms have six protons in their nucleus, so the atomic number of carbon is 6. To maintain neutrality, atoms have negatively charged electrons equal in number to the positively charged protons.

The periodic table also shows each element's **mass number**. The mass number of an element is the mass of the nucleus—the sum of the number of protons and neutrons. The weighted average mass number of all known isotopes of an atom (with different numbers of neutrons) gives the standard **atomic weight**.

Isotopes are forms of an element that differ in their number of neutrons. For example, the most abundant isotope of carbon is carbon-12 (six protons plus six neutrons), which makes up 99% of all carbon in natural sources. But naturally occurring isotopes of carbon-13 (seven neutrons) and carbon-14 (eight neutrons) also exist. Because carbon-13 and carbon-14 are rare, the average atomic mass is close to, but not exactly, 12.

Although some isotopes are stable, others decay at a known rate. The decay of an isotope is known as radioactivity. Carbon-14 has a **half-life** (the amount of time it takes for half the sample to decay) of 5,700 years. The decay of carbon-14 is used in radiocarbon dating to determine the age of fossil bones and of ancient materials such as the Dead Sea Scrolls. Scientists use carbon-14 and shorter-lived isotopes such as tritium (hydrogen with a mass of 3—one proton and two neutrons) as tracers to follow specific atoms in metabolic pathways.

Chemical Bonds Form Molecules

How do atoms form molecules? The three-dimensional shape, or structure, of a molecule such as a protein determines how it functions in a cell. The structure depends on which atoms are present and how they are bonded together by interactions of their electrons. Pairs of electrons occupy regions of space called **orbitals**. The geometry of orbitals determines the shape of a molecule.

Atoms that undergo chemical reactions have unfilled "places" for electrons in their orbitals (see Table 4.1). Usually, the unfilled place occurs where an orbital for two electrons contains only one. Hydrogen has one place available to fill its orbital. Larger atoms, such as carbon, nitrogen, and oxygen, possess four orbitals in a shell

Figure 4.2 Covalent Bonding of Hydrogen, Carbon, and Oxygen

Methane (CH₄)   Water (H₂O)

Electron sharing in the four single bonds of methane (CH_4) and in the two single bonds of water (H_2O). In the structural formulas, each line represents a covalent bond, and in the diagrams, dots represent electrons.

that could hold eight electrons—the "octet rule." For example, carbon has four orbitals, each containing only one electron; as a result, carbon has four places available for electrons to complete a shell of eight. To attain a complete outer shell of electrons, atoms combine in one of two ways: (1) by sharing electrons (covalent bond) or (2) by gaining or losing electrons (becoming ions).

COVALENT BONDS Atoms that share electrons form a **molecule**. The sharing of electrons between two atoms is called a **covalent bond**. For example, an atom of carbon combines with four hydrogen atoms to form a molecule of methane (CH_4), the main component of natural gas (**Figure 4.2**). The carbon now has a full outer shell of eight electrons, and each hydrogen has a full shell of two electrons. Methane thus has four covalent bonds. Much larger molecules, such as the cholera toxin presented earlier, have thousands of atoms linked by covalent bonds. The bonds of a protein link together chains of carbon atoms as well as atoms of hydrogen, oxygen, nitrogen, and sulfur.

The symbols CH_4 and H_2O are examples of **molecular formulas**. A molecular formula is a shorthand notation indicating the number and type of atoms present in a molecule. Atoms can also share more than one pair of electrons with another atom, forming double or triple bonds. In water (H_2O), each hydrogen shares a pair of electrons with oxygen. The bonding in molecules can be represented by a **structural formula**, in which covalent bonds are shown as a solid line between two atoms (shown in Figure 4.2).

A structural formula represents the sequence of bonds in a molecule but not its actual three-dimensional shape. The bonding orbitals in the outer shell of carbon are arranged so that they point to the vertices of a tetrahedron; methane, the simplest hydrocarbon, actually has the shape of a tetrahedron, with the hydrogen atoms at the vertices (**Figure 4.3A**). This space-filling model shows the volume filled by the outer shells of the atoms. **Figure 4.3B** shows a different kind of model used to depict the three-dimensional shape of molecules: the "ball-and-stick" model. Each stick shows the length and orientation of bonds between the nuclei of each pair of atoms.

Table 4.1

Major Elements of Living Cells

Element	Symbol	Atomic Number	Mass Number[a]	Outer Electron Shell[b]	Number of Bonds Shared in Molecules	Common Ionic State	Example(s) of Compound
Hydrogen	H	1	1		1	+1	H_2O (water); HCl (hydrochloric acid)
Carbon	C	6	12		4		CH_3–CH_2–OH (ethanol)
Nitrogen	N	7	14		3		NH_2–CH_2–COOH (glycine, an amino acid)
Oxygen	O	8	16		2		$C_6H_{12}O_6$ (six-carbon sugar such as glucose)
Sulfur	S	16	32		2		H_2S (hydrogen sulfide); SO_4^{2-} (sulfate ion)
Phosphorus	P	15	31		5		PO_4^{3-} (phosphate ion)
Sodium	Na	11	23			+1	NaCl (Na^+ and Cl^-) (sodium chloride)
Chlorine	Cl	17	35			−1	$MgCl_2$ (Mg^{2+} and $2Cl^-$) (magnesium chloride)
Potassium	K	19	39			+1	K_3PO_4 ($3K^+$ and PO_4^{3-}) (potassium phosphate)
Calcium	Ca	20	40			+2	$CaCl_2$ (Ca^{2+} and $2Cl^-$) (calcium chloride)
Magnesium	Mg	12	24			+2	Mg-ATP (magnesium-ATP)
Iron[c]	Fe	26	56			+2 or +3	$Fe(OH)_3$ (Fe^{3+} and $3OH^-$) [iron(III) hydroxide]

[a]Mass number shown is that of the most common isotope of an element.

[b]Electrons in an outer, or "valence," shell of an atom can participate in forming chemical bonds with other atoms. For some atoms, an unpaired electron can share a covalent bond. Other atoms can lose an electron (ionize) to form a cation.

[c]Iron can lose two electrons from its outer shell, plus a third electron from an inner shell (not shown), making a net charge of +3.

Figure 4.3 **Molecular Models of Methane**

A. A space-filling model of methane.   **B.** A ball-and-stick model of methane.

IONIC BONDS The molecules we have just described fill their orbitals by sharing electrons. A different way that atoms can obtain full outer shells is by gaining or losing electrons (**Figure 4.4A**). For example, Na loses its one outer-shell electron to become positively charged (Na$^+$) with its inner shell full. By contrast, Cl gains an electron to fill its outermost shell (Cl$^-$). Both Cl$^-$ and Na$^+$ are charged atoms, called **ions**. For an ion, the numbers of electrons and protons are unequal. **Anions** are negatively charged ions, and **cations** are positively charged ions. These ions are also known as **electrolytes** because their solutions conduct electricity. The appropriate concentrations of ions such as Cl$^-$ and Na$^+$ are essential for microbial life and for the human body; for example, the excessive efflux of Cl$^-$ causes the pathology of cholera. In the bloodstream, the appropriate Na$^+$ concentration is vital for blood pressure.

As water evaporates, anions and cations can form ionic crystals held together by **ionic bonds**—charge attractions between anions and cations. For example, sodium ion (Na$^+$) and chlorine ion, or chloride (Cl$^-$), interact to form crystals of table salt, NaCl (**Figure 4.4B**). Some kinds of crystal formation in the body can cause problems such as gout (uric acid crystals) and kidney stones (calcium oxalate or calcium phosphate).

Figure 4.4 **Formation of Ions and Ionic Crystals**

A. The loss of an electron from an atom of sodium forms the cation Na$^+$, and the gain of an electron by a chlorine atom forms the chloride anion Cl$^-$.

B. Oppositely charged anions and cations—in this case, Cl$^-$ and Na$^+$—are attracted to one another and form crystals of table salt (sodium chloride).

As water content increases and the ion concentrations decline, ionic crystals are destabilized and begin to dissolve. To understand why, we need to understand the structure of water.

Water Is the Solvent of Life

Living organisms consist mostly of water; that is why the loss of water (dehydration) in cholera has fatal results. Water has many unique properties that make it especially suitable for sustaining life. For example, water molecules have uneven charge distribution, and their partial charges attract each other.

In some molecules, such as H$_2$ and O$_2$, no charge distribution exists because both atoms have equal electronegativity (tendency to hold electrons). Therefore, the electrons in the covalent bond are shared equally and form **nonpolar covalent bonds**. In water, however, the shared electrons of H$_2$O spend more of their time around the highly electronegative oxygen than near the less electronegative hydrogen (**Figure 4.5A**). A bond with unequal electron sharing is a **polar covalent bond**, so called because the molecule has partial positive and negative poles. In a water molecule, the polar covalent bonds point at an angle from each other, so their charge difference forms a partial charge separation within the molecule.

The charge separation within each water molecule causes water molecules to attract each other. This partial charge attraction, between the hydrogen of one water molecule and the oxygen of another molecule, is known as a **hydrogen bond** (**Figure 4.5B**). Hydrogen bonds occur not just in water but also in other molecules. For example, in proteins this occurs between a hydrogen bound to an oxygen or nitrogen and a second oxygen or nitrogen, in either the same or a different molecule. Hydrogen bonds are short-lived and constantly break and re-form in liquid water. Figure 4.5B shows the polar covalent bonds and hydrogen bonds present in water.

Figure 4.5 **Polar Covalent Bonds and Hydrogen Bonds in Water**

A. Polar covalent bonds in an individual water molecule. The displacement of electrons toward oxygen causes oxygen to have a partial negative charge and the hydrogen atoms to have a partial positive charge. Hence, water is polar.

B. A hydrogen bond between two water molecules.

Hydrogen bonds are weaker than covalent bonds and are therefore represented by dotted lines.

The hydrogen bonds in water contribute to its unique properties that support life. Water is a liquid over a large temperature range because hydrogen bonds cause water molecules to attract each other, favoring the liquid state over the gaseous state. The polar nature of water also defines its properties as a solvent. Compounds that are ionic or polar themselves tend to dissolve in water and are termed **hydrophilic** ("water loving"). For example, the ionic bonds in NaCl are very strong in the absence of water but are easily dissolved in the presence of water. This is because the polar water can surround and interact with the sodium and chloride ions and shield them from each other (**Figure 4.6A**). Water also dissolves polar compounds. In this case, water does not break the polar covalent bonds; rather, individual, intact polar molecules form hydrogen bonds with the surrounding water.

By contrast, compounds that are mostly nonpolar do not dissolve in water and are termed **hydrophobic** ("water fearing"). Examples of hydrophobic molecules include long-chain hydrocarbons such as vegetable oil and gasoline. Nonpolar molecules have no partial charges to attract water. Because water forms hydrogen bonds with other water molecules, it tends to exclude nonpolar compounds, forcing them together (**Figure 4.6B**). Nonpolar compounds are said to attract each other by "hydrophobic" forces.

Hydrophobic attraction is weaker than hydrogen bonding, but it contributes to the stability of important cell features. For example, the components of cell membranes are held together by hydrophobic attraction (see Section 4.5).

SECTION SUMMARY

- **The major atoms of life are the elements carbon, hydrogen, oxygen, nitrogen, phosphorus, and sulfur.**
- **Atoms form bonds.** Bonds can be covalent or ionic. Weaker bonds include hydrogen bonds and hydrophobic interactions.
- **The molecular formula indicates atomic composition.** In a covalent molecule, the structural formula indicates the order of covalent bonds.
- **Water is the solvent of life.** Water is the most prevalent molecule within most living organisms. Water molecules form hydrogen bonds with each other and with dissolved polar molecules or ions.
- **Molecules can be polar or nonpolar.** Polar molecules are hydrophilic (soluble in water), whereas nonpolar molecules tend to be hydrophobic—more soluble in a nonpolar medium such as a membrane.

Thought Question 4.1 Use the properties of molecules described in this section to explain why table salt (NaCl) dissolves readily in water but vegetable oil (a hydrocarbon) does not.

Figure 4.6 Interactions between Water and Solutes

A. Water surrounds and interacts with ions or polar molecules, causing them to dissolve.

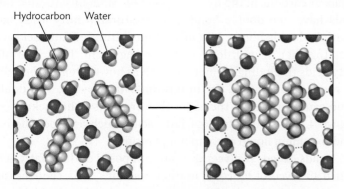

B. Nonpolar (hydrophobic) molecules do not dissolve in water. Instead, to minimize the disruption of hydrogen bonding among water molecules, nonpolar molecules in water associate with each other.

4.2 Lipids and Sugars

SECTION OBJECTIVES

- Describe organic molecules, and explain the role of functional groups and the formation of macromolecules.
- Describe the structure of lipids and their key functions within a cell.
- Describe the structure of sugars and polysaccharides.

What kinds of molecules form the structure of cells? Cells are largely made of **organic molecules**, defined as molecules that contain a carbon-carbon bond. Some organic molecules are small; for example, ethanol (CH_3–CH_2–OH), the active component of alcoholic beverages, has a single carbon-carbon bond. Other organic molecules, such as proteins and DNA, consist of thousands of atoms with many carbon-carbon bonds. Such large organic molecules are called **macromolecules.** Each type of macromolecule is a polymer composed of smaller units called monomers. Major macromolecules of cells are lipids, polysaccharides, nucleic acids, and proteins.

An important feature of any organic molecule is its functional groups. A **functional group** is a small group of atoms with characteristic bonding, shape, and reactivity (**Table 4.2**). Functional groups enable a molecule to undergo important chemical reactions. For example, in molecules of DNA, a chain-terminal hydroxyl group reacts with an additional nucleotide to add a "letter" onto the DNA chain. Information about the functional groups present in a molecule is often indicated by a molecule's name; for example,

Table 4.2

Common Functional Groups of Organic Molecules[a]

Functional Group	General Structure	Example
Alkane	R—C—H (with H above and below C)	H—C—C—H (Ethane)
Amino	R—N (with H above and below)	H—C—C—NH₂ (Ethylamine)
Carboxyl	R—C—O—H (with O double bonded)	H—C—C—OH (Acetic acid)
Ester	R—C—O—R (with O double bonded)	Methyl formate
Hydroxyl	R—O—H	H—C—C—OH (Ethanol)
Carbonyl	R—C—R (with O double bonded)	CH₂OH / C=O / H—C—OH / H (Dihydroxyacetone)

[a]R stands for an organic group.

amino acids (the building blocks of proteins) each contain an amino group (–NH₂) and a carboxylic acid group (–COOH).

Lipids

Lipids are organic molecules that serve as structural components of membranes and other parts of cells. Lipids store lots of energy (as in "calories") because their large number of C–H bonds release energy when metabolized with the oxygen we breathe. All lipids contain a substantial number of nonpolar C–H and C–C covalent bonds. Because lipids are nonpolar, they are hydrophobic and do not dissolve in water. Many toxic pollutants such as dioxins are also hydrophobic and therefore concentrate in the lipids of fat tissues.

Lipid building blocks include the fatty acids (**Figure 4.7**). Each fatty acid has a hydrophobic portion (the hydrocarbon tail) with a

Figure 4.7 Fatty Acids

The carboxyl group on the end of a fatty acid can act as an acid by releasing a proton (H⁺).

Palmitic acid

Linoleic acid

A. Palmitic acid is a saturated fatty acid. Only single bonds connect the carbon atoms.

B. Linoleic acid is an unsaturated fatty acid. Note that the double bonds (highlighted in yellow) form kinks in the hydrocarbon tail.

long chain of CH₂ units, like the hydrocarbon chains of gasoline. But the hydrocarbon tail of a fatty acid ends with a hydrophilic component, the carboxylic acid, which ionizes (releases a hydrogen ion, H⁺). The remaining ionized molecule, called a carboxylate, now bears a negative charge (R–COO⁻), where R stands for the rest of the molecule. The hydrocarbon tails avoid interacting with water, but the carboxylic acid end is highly water-soluble. Molecules with a charged head and a hydrophobic tail can interact with each other to form soap-like films or bubbles (called micelles).

Cells contain many fatty acids, which differ in the number of carbons (usually an even number). Some fatty acids are **saturated**; that is, the carbon-carbon bonds are all single bonds, and the carbons are bonded to the maximum number of hydrogen atoms ("saturated" with hydrogen; see Figure 4.7A). **Unsaturated** fatty acids, on the other hand, contain one or more double bonds between adjacent carbons in the hydrocarbon tail. Monounsaturated fatty acids have one double bond; polyunsaturated fatty acids have multiple double bonds. In a membrane, saturated fatty acids can pack together in an orderly, stable arrangement. Thus, saturated fatty acids (including animal fats such as butter) tend to be solids at room temperature. In unsaturated fatty acids, the double bond causes a kink in the chain that prevents tight packing (see Figure 4.7B), so unsaturated fatty acids, such as vegetable oils, tend to remain liquid at room temperatures.

Lipids called "triglycerides" are composed of three fatty acids joined to glycerol by condensation reactions. In a **condensation reaction**, two molecules join to form a covalent bond while losing atoms that form H₂O. The three hydroxyl groups on glycerol can each undergo condensation with the carboxylic acid portion of a

Figure 4.8 Formation of a Triglyceride

Condensation reactions form covalent ester linkages between the carboxyl groups of three fatty acids and the three hydroxyl groups of glycerol, with loss of three molecules of H_2O.

Figure 4.9 Structures of Two Hexoses: Glucose and Fructose

D-Glucose   **D-Fructose**

D designation is determined by orientation at this carbon.

The numbering of the carbons starts at the carbon of the terminal carbonyl group (highlighted) or at the terminal carbon closest to the carbonyl (as in fructose). The D designation is determined from the mirror form of the hydroxyl group on the carbon (shown by the leader line) that is farthest from the carbonyl group (highlighted).

fatty acid to form a triglyceride (**Figure 4.8**). Triglycerides are a compact energy source for cells; they are highly concentrated in fat tissue. Phospholipids, which are key components of membranes, contain fatty acids attached to two of the hydroxyl groups of glycerol and a phosphate attached to the third hydroxyl group. Their structure determines how drug molecules get into cells. Phospholipid membranes are discussed further in Section 4.5.

Carbohydrates

Carbohydrates (sugars and polysaccharides) store energy, and they serve as part of cell surfaces. Sugar structures convey specific information; for example, the ganglioside mentioned in the opening case is a sugar chain on the surface of intestinal epithelial cells that unfortunately binds cholera toxin (see Figure 4.1). Carbohydrates are an energy source for cells and also serve structural roles, as in cell walls. Bacteria often coat their cells with long sugar chains that prevent phagocytosis (engulfment) by white blood cells.

A **monosaccharide**, or "simple sugar," has several linked carbons with multiple hydroxyl groups (R–OH) and typically one carbonyl group (R–C=O), as shown in **Figure 4.9**. Sugars differ from each other in the number of carbons present, in the position of the carbonyl group, and in the orientation of the hydroxyl groups. In a sugar structure, the carbons are numbered starting with the carbon of the terminal carbonyl group or the terminal carbon closest to the carbonyl group. The

six-carbon sugars (hexoses) are particularly important for energy storage and cell wall formation, whereas five-carbon sugars (pentoses) such as ribose form the backbone of nucleic acids (discussed in the next section). Hexose sugars include, for example, glucose and fructose (Figure 4.9). Hexoses all have the molecular formula $C_6H_{12}O_6$, but the atoms in various hexoses are linked differently. The contrast between glucose and fructose lies in the position of the carbonyl, either at C-1 (glucose) or at C-2 (fructose).

A sugar such as glucose may take one of two "mirror" forms—that is, forms that are mirror opposites of each other. The two mirror forms, D and L, depend on the order of bonds around a carbon with four different connections. The D-form or L-form is extremely important because an enzyme can catalyze reactions for only one mirror form. Most enzymes are completely inactive for the opposite mirror form of their substrate, like a shoe that can "fit" only the right foot, not the left.

Sugars with five or more carbons may also form a ring structure in which one of the hydroxyl groups reacts with the carbonyl group (**Figure 4.10**). Depending on how the bond between C-1 and C-2 was oriented during the formation of

Figure 4.10 Straight-Chain and Cyclic Forms of Glucose

The straight-chain form of glucose can cyclize to form two different ring structures, alpha-glucose and beta-glucose. During cyclization, the carbonyl group (C=O) is rearranged to carbon-hydroxyl (C–OH).

Straight-chain form   **Intermediate form**   **α-Glucose**   **β-Glucose**

Figure 4.11 Formation of the Disaccharide Maltose

Maltose forms by a condensation reaction, in which two molecules of glucose are joined and H_2O is lost. Two molecules of glucose condense by linking the C-1 of one molecule of glucose via an oxygen atom to the C-4 of a second molecule of glucose. The hydroxyl groups of the second glucose can be either alpha or beta (here the beta form is shown).

the ring, the cyclic form of glucose will be one of two types—alpha-glucose or beta-glucose. Alpha-and beta-linked sugars are found on cell-surface proteins that bind to viruses. For example, deadly influenza viruses bind alpha-linked sugar proteins (glycoproteins) on respiratory cells.

Two monosaccharides can link and form a **disaccharide** by a condensation reaction (**Figure 4.11**). In condensation, two molecules form a new covalent bond through loss of water (H_2O). The bond typically forms between the C-1 carbon of one monosaccharide and the C-4 carbon of another monosaccharide. Such a reaction requires an enzyme that recognizes the specific sugars (discussed shortly). The structure of the disaccharide varies, depending on which monosaccharides are involved and whether the C-1 carbon participating in the bond is in the alpha or beta form.

Several sugars can condense to form short chains. Longer chains are called **polysaccharides** or **glycans**. An example of a polysaccharide or glycan is starch, found in cereals and potatoes. The starch polymer consists of a long chain of condensed glucose sugars. Starch is a linear glycan, but other glycans are branched. For example, glycogen, the liver's main compound for storing energy, is a branched polymer of glucose molecules. The interconversion of glycogen and glucose is a key process for athletic performance and for blood sugar diseases such as diabetes.

SECTION SUMMARY

- **Organic compounds contain one or more carbon-carbon bonds.** Organic molecules make up much of a living organism.

- **Large biological organic molecules are called macromolecules.** Macromolecules contain functional groups that participate in various chemical reactions.

- **Lipids are hydrophobic molecules with mainly C–C and C–H bonds.** Triglycerides, a storage form of fat, consist of glycerol condensed with three fatty acids. Phospholipids, a component of cellular membranes, contain glycerol condensed with two fatty acids and a phosphoryl group.

- **Sugars have several linked carbons with multiple hydroxyl groups (R–OH) and one carbonyl group (R–C=O).** Sugars may form a ring structure in which one hydroxyl group reacts with the carbonyl group. Chains of sugars form polysaccharides or glycans.

4.3 Nucleic Acids and Proteins

SECTION OBJECTIVES

- Describe the structure of DNA, and explain how DNA stores information in its sequence of base pairs.

- Describe the structure and diverse functions of RNA.

- Describe the fundamental structure of proteins.

How do living cells write information? Some macromolecules, particularly nucleic acids and proteins, contain complex information encoded in the pattern of their repeated units. The most famous kind of nucleic acid in cells is deoxyribonucleic acid (DNA). DNA is a cell's hereditary material, forming the chromosomes and containing the entire **genome** of a cell. The information content of DNA includes genes that encode proteins. In some cases, the DNA of a human genome can tell a physician which diseases a patient may get. And bacterial genomic DNA determines which antibiotics will work.

A related macromolecule, equally important for life, is ribonucleic acid (RNA). Cells make RNA copies of DNA genes for immediate use, like making photocopies from a book. Some RNA consists of messenger RNA (mRNA), which specifies the synthesis of proteins (discussed in Chapter 8). But some RNA molecules serve as a cell's building blocks, such as the rRNA subunits of the ribosome—the cell's protein-building factory. And viruses such as influenza and HIV actually have genomes of RNA. RNA viral genomes mutate faster than those of DNA, and these new strains can cause pandemic outbreaks such as that of COVID-19 in 2020 (discussed in Chapter 12).

Figure 4.12 Nucleotide Structure

Nucleobase

Phosphate group

5-Carbon sugar

Ribose

Deoxyribose

Purines

NH₂

Adenine

Guanine

Pyrimidines

Cytosine
(both DNA and RNA)

Thymine
(DNA only)

Uracil
(RNA only)

A nucleotide consists of a five-carbon sugar (ribose or deoxyribose), a phosphate group, and a nucleobase (also called nitrogenous base). The circled nitrogen (N) of the nucleobase forms a covalent bond to the sugar.

Nucleic Acids

Nucleic acids are long chains composed of nucleotides. A **nucleotide** has three components—a five-carbon sugar (pentose), a **nucleobase** (nitrogenous base, often just called a base), and a phosphoryl group (phosphate) (**Figure 4.12**). For RNA, the sugar is ribose, and for DNA, the sugar is 2-deoxyribose—that is, ribose with oxygen removed from the 2′ (two prime) carbon. The prime indicates that the numbering refers to the sugar portion of the nucleotide. In both DNA and RNA, the 1′ (one prime) carbon of the sugar forms a covalent bond to a nitrogen of the base. Bases come in two structural classes (Figure 4.12): purines (adenine, A, and guanine, G) and pyrimidines (cytosine, C; thymine, T, found only in DNA; and uracil, U, found only in RNA). The 5′ (five prime) carbon usually bonds to a phosphoryl group, whereas the 3′ (three prime) carbon usually has a free hydroxyl group (–OH).

A nucleotide has a phosphoryl group with a covalent ester bond (C–O–P–O–C) to the 5′ carbon of the sugar. Nucleotides can have one, two, or three phosphoryl groups and are called monophosphate (MP), diphosphate (DP), or triphosphate (TP), respectively. Adenine nucleotides, for instance, are abbreviated AMP, ADP, or ATP (**Figure 4.13**). Nucleotide structure serves as a model for many antiviral drugs—such as azidothymidine, or AZT, the famous first effective antiviral agent discovered for HIV/AIDS. AZT acts by mimicking the adenine nucleotide structure and thus blocking the action of the enzyme that copies the HIV genome.

Ribose nucleotides are the precursors for nucleic acid synthesis, but they also have other functions in the cell. For example, ATP is an energy carrier (see Figure 4.13). Energy is released by the reaction of removing one or two phosphoryl groups from ATP. Other forms of ribose nucleotides, such as guanosine triphosphate (GTP) and cyclic adenosine monophosphate (cAMP), act as signals. For example, cAMP activates a transporter that exports chloride ion from the intestine. As we discussed earlier, cholera toxin acts by inappropriately turning on an enzyme that forms cAMP.

Adenosine triphosphate (ATP), shown in Figure 4.13, is a ribonucleotide capable of forming a unit of RNA. The formation of DNA requires deoxyribonucleotides, in which the 2′ carbon bonds to a hydrogen atom, not to OH. Deoxyribonucleotides are designated by a lowercase "d" in front of the nucleotide name. For example, cytosine triphosphate is CTP, and deoxycytosine triphosphate is dCTP. Nucleotide monomers with three phosphoryl groups (collectively designated NTPs or dNTPs) form nucleic acids through reactions catalyzed by a polymerase enzyme (discussed in Chapter 8.) The enzyme that assembles dNTPs into DNA is now used in the famous polymerase chain reaction (PCR) to amplify DNA for forensic and paternity analysis.

To extend a DNA chain, the alpha phosphoryl group at the C-5 position of a dNTP condenses with the chain's 3′ OH (C–O–P–O–C). This reaction releases pyrophosphate (double phosphoryl group)

Figure 4.13 ATP, an Important Energy-Carrying Molecule

NH₂

Adenine

Phosphate groups

Gamma   Beta   Alpha

Ribose

OH OH

Adenosine

AMP (adenosine monophosphate)

ADP (adenosine diphosphate)

ATP (adenosine triphosphate)

ATP is a nucleotide containing the nucleobase adenine. A nucleotide may be a monophosphate, diphosphate, or triphosphate. The phosphate attached to the sugar is designated alpha, the beta phosphate is in the middle, and the gamma phosphate is farthest from the sugar.

Figure 4.14 DNA Structure

A. The orientation of a DNA strand is determined by the presence of a free phosphoryl group on the 5′ end and a free hydroxyl on the 3′ end. The sequence of a nucleic acid is read from the 5′ end to the 3′ end, so for this portion of the DNA strand, the primary structure is GCTA.

B. Hydrogen bonding (represented by red dots) between complementary base pairs. Two hydrogen bonds form between A and T, whereas three hydrogen bonds form between G and C. The DNA strands are antiparallel (point in opposite directions).

from the dNTP. Successive sugar-phosphoryl bonds form the "backbone" of the molecule. The information of DNA is contained in the sequence of bases attached to the backbone (**Figure 4.14A**). The sequence end with a free 5′ phosphate is called the 5′ (pronounced "5-prime") end, and the end with the free 3′ hydroxyl is called the 3′ ("3-prime") end. This linear sequence of nucleotides, from the 5′ end to the 3′ end, is called the primary structure of the nucleic acid.

The DNA Double Helix

The chromosomes of all cells consist of DNA that is double-stranded, or duplex (**Figure 4.14B**). The two strands of DNA are held together by hydrogen bonds between a purine and a pyrimidine on opposite strands. Two hydrogen bonds form between each adenine and thymine pair, and three hydrogen bonds form between each cytosine and guanine pair. The strands twist around each other to form a helix—the famous double helix of DNA (**Figure 4.15A**). The double-helical form was first demonstrated in 1953 by Rosalind Franklin and Maurice Wilkins, whereas the nucleotide base pairing was proposed and reported by James Watson and Francis Crick. The sugar-phosphate backbone of DNA is on the outside of the helix (**Figure 4.15B**), its negative charges interacting with the partial positive charges on water molecules. Within the helix, the hydrophobic bases "avoid" water by stacking together like steps of a ladder.

Although the base pairs are mostly buried within the helix, portions of them remain accessible to proteins that regulate transcription (discussed in Chapter 8). The accessible portions take the form of two grooves in the helix—a wide major groove and a narrower minor groove (see Figure 4.15B). Regulatory proteins play many roles in pathogenesis. For example, *Vibrio cholerae* proteins that detect environmental signals such as temperature and pH can tell when the pathogen has entered a host. These regulatory proteins then bind the pathogen's DNA to start the disease process.

Figure 4.15 The DNA Double Helix

A. The DNA double helix, showing the purine-pyrimidine base pairs.

B. A space-filling model of DNA. Note that the base pairs stack on top of each other like the rungs of a ladder.

Figure 4.16 RNA Primary and Secondary Structure

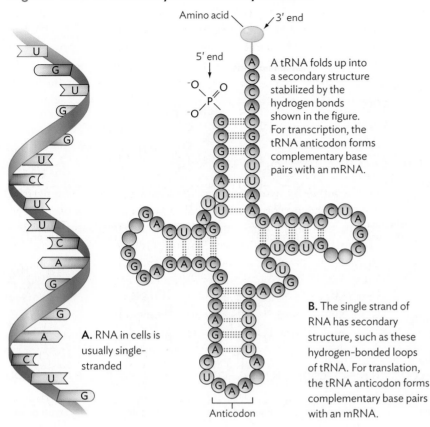

A. RNA in cells is usually single-stranded

A tRNA folds up into a secondary structure stabilized by the hydrogen bonds shown in the figure. For transcription, the tRNA anticodon forms complementary base pairs with an mRNA.

B. The single strand of RNA has secondary structure, such as these hydrogen-bonded loops of tRNA. For translation, the tRNA anticodon forms complementary base pairs with an mRNA.

RNA Takes Diverse Forms

RNA can form base pairs like those of DNA, except that in RNA, the nucleotide uracil replaces thymine. Some RNA molecules form a double helix like DNA; an example is the RNA of rotaviruses, the most common cause of diarrhea and "stomach flu" in young children. But in cells, the most common structure of RNA molecules is single-stranded (**Figure 4.16A**). Messenger RNA (mRNA) encodes the information from a gene to specify a protein (again, see Chapter 8). Single-stranded RNA can "double back" and form <u>intra</u>molecular base pairs, such as those in a tRNA molecule (**Figure 4.16B**). These base-paired forms (called secondary structure) enable RNA to assume a globular protein–like form with enzymatic functions, such as protein synthesis within a ribosome. Base pairing between RNA and DNA is used to make mRNA from a DNA template.

Proteins

A cell makes thousands of different proteins and may contain more than 2 million protein molecules. Proteins perform many functions, from catalyzing biochemical reactions to serving as receptors and transporters to providing structure and aiding movement. Proteins fold into three-dimensional structures that may fit together in a **complex**. The cholera toxin is an example of a protein complex composed of two types of protein chain (one A chain and five copies of B chain).

AMINO ACIDS All proteins, or polypeptides, are composed of the same building blocks: amino acids. All 20 of the amino acids used in proteins have the same general structure. Each **amino acid** contains a central carbon atom (the alpha carbon) covalently bound to four different parts—a hydrogen atom, an amino group, a carboxyl group, and a side chain (R) that is unique for each amino acid (**Figure 4.17A**). The alpha carbon is an example of a chiral carbon—a carbon with four different groups attached to it. Chiral carbons exist in two different forms called **isomers** that are mirror images of each other, like your left and right hands; the two forms cannot be superimposed (**Figure 4.17B**). Each amino acid isomer (or mirror form) is designated L or D, based on the configuration at the alpha carbon. Only L-amino acids are used by ribosomes to make proteins. But bacteria use D-amino acids (the opposite mirror form) to build their cell walls. D-Amino acids are found in peptide antibiotics, such as D-alanine in vancomycin, a powerful antibiotic used to treat flesh-eating staphylococcus infections.

The side chains of amino acids can be grouped according to their hydrophobicity (how hydrophobic), charge, and/or presence of specific functional groups. **Figure 4.18** shows the 20 amino acids used to make protein, along with their three-letter abbreviations and single-letter codes. The amino acid side chains vary greatly in their chemistry. They may be acidic (such as aspartic acid, whose

Figure 4.17 Amino Acid Structure and Chiral Carbons

A. All amino acids contain a central carbon (the alpha carbon) bonded to an amino group, a carboxyl group, a hydrogen, and a variable group or side chain, designated R. At cellular pH, the amino and carboxyl groups are ionized.

Molecule fits template.

Fit to template is impossible for isomer.

Asymmetrical carbon atoms

Molecule   Mirror image

B. The alpha carbon is bound to four different molecules and thus has two mirror forms. Only one mirror form can interact with a given enzyme.

Figure 4.18 The 20 Amino Acids for Protein Synthesis

The categories of amino acids are based on their side chains, highlighted in yellow.

Charged amino acids

Basic (alkaline) amino acids

Lysine (Lys or K)   Arginine (Arg or R)   Histidine (His or H)

Acidic amino acids

Aspartate (Asp or D)   Glutamate (Glu or E)

Polar amino acids

Asparagine (Asn or N)   Glutamine (Gln or Q)   Serine (Ser or S)   Threonine (Thr or T)

Hydrophobic amino acids

Alanine (Ala or A)   Valine (Val or V)   Isoleucine (Ile or I)   Leucine (Leu or L)   Methionine (Met or M)   Phenylalanine (Phe or F)   Tyrosine (Tyr or Y)   Tryptophan (Trp or W)

Special amino acids

Cysteine (Cys or C)   Glycine (Gly or G)   Proline (Pro or P)

carboxyl group deprotonates at neutral pH) or basic (such as lysine, whose amino group, $R-NH_2$, becomes protonated to $R-NH_3^+$). Other R groups are polar (serine) or hydrophobic (phenylalanine). Proline has a ring structure that interrupts the regular geometry of the polypeptide chain. The thiol (–SH) side chain of cysteine can form both intermolecular and intramolecular disulfide (–S–S–) bonds. For example, in antibodies, disulfide bonds hold together the different combinations of heavy chains and light chains that generate millions of different antibodies (discussed in Chapter 16).

PROTEIN ORGANIZATION The first level of organization, the **primary structure**, is simply the linear sequence of amino acids, specified by the DNA code. During protein synthesis, amino acids are connected by a condensation reaction to form a covalent peptide bond. The peptide bond forms between the carboxyl group of one amino acid and the amino group of a second amino acid (**Figure 4.19A**). The portion of the amino acid incorporated into

Figure 4.19 The Peptide Bond and Primary Structure of a Protein

A. Formation of a peptide bond. Amino acids join by a condensation reaction between the carboxyl group (–COOH) of one amino acid and the amino group (–NH₂) of another to form a covalent linkage called a peptide bond.

B. The primary structure of a short peptide, a chain of five amino acid residues.

Figure 4.20 Secondary Structure of a Protein

The alpha helix and beta sheet are each held together by hydrogen bonds.

α Helix   β Sheet

Normal conformation   Abnormal conformation

Figure 4.21 Prion Conformation Change

A normal prion consists of mainly alpha helices (red coils). Under pathological conditions, the prion transitions to the abnormal, disease-causing conformation, consisting of beta sheets (cyan arrows).

causing an alpha-to-beta transition in the secondary structure of proteins within infected cells (**Figure 4.21**). The alpha helices are coiled structures formed by a series of hydrogen bonds between the oxygen in a carboxyl group and a hydrogen in an amino group, four amino acids farther along the chain. The beta sheets form as a result of extensive hydrogen bonding between regions of the protein lying next to each other.

The **tertiary structure** of a protein is its unique three-dimensional shape (**Figure 4.22**). At the tertiary level of organization, regions distant in the primary structure may be brought close together. The nature and order of the side chains (the

the peptide after condensation is called an amino acid residue. The peptide chain has an amino terminus (N terminus) and a carboxyl terminus (C terminus); numbering of the amino acid residues starts at the amino terminus (**Figure 4.19B**). A chain of five residues is shown in Figure 4.19B, but actual proteins may include hundreds of residues.

The **secondary structure** of a protein is the regular pattern of amino acid residues that form over short regions of the polypeptide chain. **Figure 4.20** depicts two common secondary structures, the **alpha helix** and the **beta sheet**. These secondary structures are extremely important for how antibodies recognize antigens and for how pathogens cause disease. An example is the prion, an infectious protein that causes a brain disease known as "mad cow disease" (bovine spongiform encephalopathy). The prion acts by

Figure 4.22 Tertiary Structure of a Protein

Various types of bonds maintain protein tertiary structure.

primary structure) determine how a protein folds into its final tertiary structure. The large number of amino acid side chains is responsible for the diversity of protein structures and hence functions. As proteins emerge from the ribosome, they are bound by chaperone proteins that help them fold into their functional shape, known as the **native conformation**. In soluble proteins, amino acids with hydrophobic side chains tend to cluster inside the protein to minimize reactions with water, whereas polar amino acids are exposed on the surface. Tertiary structure is stabilized by hydrogen bonding between side chains. Ionic interactions between acidic (negatively charged) and basic (positively charged) side chains also contribute to tertiary structure. The one covalent interaction found stabilizing tertiary structure is that of disulfide bonds. Disulfide bonds (R–S–S–R) form between two cysteine residues (R–SH).

Some proteins form stable, functional complexes with other proteins, such as in the cholera toxin AB5 (see Figure 4.1). A multisubunit protein complex shows the fourth level of protein organization, quaternary structure. The forces that hold interacting polypeptide chains together are the same noncovalent interactions and disulfide bonds that maintain tertiary structure within a folded chain. An example of quaternary structure is shown in the space-filling model of an antibody, composed of two heavy chains and two light chains (**Figure 4.23**).

Proteins are dynamic structures with parts that move and bend. The weak bonds, such as hydrogen bonds and ionic bonds, contribute to protein structure but can come apart and re-form. Nevertheless, because these weak bonds do not break all at once,

the large number of such bonds act collectively to maintain surprisingly strong protein integrity.

SECTION SUMMARY

- **A nucleotide is made of a five-carbon sugar, a nucleobase, and a phosphoryl group.** In RNA, the sugar is ribose; in DNA, the sugar is 2-deoxyribose (one fewer oxygen). The nucleobases include adenine, guanine, cytosine, and thymine (for DNA) or uracil (for RNA).

- **Nucleic acids are chains of nucleotides** linked by phosphoryl groups. Nucleic acids such as DNA and RNA carry information in the sequence of their bases.

- **DNA in cells forms a double-stranded helix.** Adenine pairs with thymine, and cytosine pairs with guanine.

- **RNA is single-stranded in cells** but can form intramolecular base pairs. RNA molecules include mRNA (messenger RNA), tRNA (transfer RNA), and rRNA (ribosomal RNA).

- **Proteins are composed of amino acids.** In a protein, the 20 types of L-form amino acids are condensed through peptide bonds, generating a unique order of amino acid residues.

- **The peptide chain (primary structure) is folded into secondary and tertiary structure.** A complex of multiple subunits has quaternary structure.

Thought Question 4.2   What kinds of amino acid residues would you find on the exterior surface of a protein embedded among the hydrocarbon chains of the membrane? What kinds of amino acid residues would you find buried within the protein?

4.4 Biochemical Reactions

SECTION OBJECTIVES

- Explain how energy and entropy determine the direction of reactions.

- Describe the oxidation-reduction changes that occur during reactions.

- Explain how enzymes catalyze reactions and how various factors influence the rate of reaction.

How do cells change shape and grow? A cell grows by breaking down molecules (catabolism) and building more complex ones (anabolism). Cellular catabolism and anabolism are described in Chapter 7.

Each change of a molecule occurs through chemical reactions. Chemical reactions are always catalyzed by the cell's enzymes, proteins that catalyze reactions; remember that, even when the enzyme is not shown. **Catalysis** means to greatly increase the rate of a reaction. The practical effect of catalysis is to turn on a reaction that otherwise would be turned off. A cell's reactions all require enzymes to ensure that reactions occur only as needed by the cell.

An example of a biochemical reaction catalyzed by an enzyme is the formation of cAMP, the reaction that leads to excess Cl^- efflux in

Figure 4.23 Antibody Complex

Antigen-binding region

Light chains

Heavy chain

Heavy chain

An example of protein quaternary structure: The antibody complex comprises two light chains and two heavy chains.

cholera (seen in Figure 4.1). cAMP forms by the conversion of ATP, catalyzed by adenylate cyclase:

ATP

Pyrophosphate (diphosphate)

cAMP

The enzyme removes pyrophosphate, and the remaining adenosine monophosphate (AMP) is then cyclized at the 3′ OH, forming cAMP. cAMP mediates many important cell signals, such as that of the hormone adrenaline.

What determines whether a reaction goes forward? In the preceding reaction, the removed pyrophosphate is a high-energy compound that is rapidly hydrolyzed—that is, cleaved with input of H_2O. (Hydrolysis in biochemical reactions is discussed in Chapter 7.) Overall, the reaction has a favorable **free energy change (ΔG)**.

Energy Change Powers Life

Where do living cells get the fuel to drive their reactions? Like the engine of a car, reactions in organisms, such as the synthesis of proteins from amino acids by condensation, must spend energy. An endless cycle connects reactions that yield energy for cells (such as those that oxidize sugars or lipids) with reactions that spend energy (such as synthesis of DNA and proteins). When the energy chain fails, as when your heart stops pumping blood, the rest of your reactions fail, and your body undergoes rigor mortis—in other words, death.

All systems—including living cells, organisms, and ecosystems—must conserve energy. **Energy** is the ability to do work. Changes in energy must obey the laws of thermodynamics (**Figure 4.24**). The **first law of thermodynamics** states that matter and energy are neither created nor destroyed. In a closed system (a system that is not exchanging energy with its environment), the total amount of energy remains constant. While the amount of energy remains constant,

Figure 4.24 Energy Can Be Changed into Different Forms but Can Never Be Created or Destroyed

Energy transformation

Usable energy

Usable energy (free energy)

Unusable energy (entropy)

The first law of thermodynamics states that, in an energy transformation, the total amount of energy remains constant. Both sides have the same amount of energy (the balance reads zero). The second law of thermodynamics states that, in an energy transformation, the amount of usable energy, or free energy, decreases and the amount of unusable energy, or entropy, increases.

portions of energy can be converted from one form to another. For example, some of the energy released by oxidizing food can be captured by enzymes building proteins and DNA for a cell to grow.

The **second law of thermodynamics** states that, in all energy transformations, some energy becomes unavailable to do work and is lost as **entropy**, or disorder. Overall, entropy tends to increase. For example, when you dilute a concentrated drug to the correct dosage for therapy, the diluted solution has increased entropy. Entropy is increased because the drug will never spontaneously become concentrated. The only way to increase the concentration again would be to add energy—for instance, by adding heat to evaporate the solvent.

Reactions that increase entropy may go forward. For example, the breakdown of food molecules into a larger number of smaller molecules increases entropy. This reaction also yields energy. A living cell can couple energy-yielding processes to energy-spending processes such as building proteins. While energy is spent to build proteins, the overall process of food breakdown plus protein building releases net energy and increases entropy. So, although life itself is a highly ordered process, it generates even greater disorder (entropy) outside the living organism.

For a reaction to go forward, there needs to be a change in "free energy" (G), a term that incorporates changes in both energy and entropy. The free-energy change is designated ΔG. Reactions are favored when the free energy of the products is less than the free energy of the reactants; that is, when ΔG is negative. Reactions with a negative ΔG are termed **exergonic**, or "spontaneous," because they are energetically favorable and can occur without an input of energy. Reactions with a positive ΔG are **endergonic**. They are not energetically favorable, and energy needs to be added for them to proceed. Reactions with a ΔG of zero are at equilibrium, and neither products nor reactants are favored. An example of an exergonic reaction that yields energy is the oxidation, or "burning," of methane (CH_4) to release carbon dioxide and water:

$$CH_4 + 2O_2 \rightarrow CO_2 + 2H_2O + \text{energy}$$

The bonds in methane and oxygen have more energy than do the bonds in CO_2 and H_2O. Because the products CO_2 and H_2O have less energy than the reactants, the ΔG is negative and the reaction goes forward. As it goes forward, the reaction releases heat.

Every living organism must behave as a chemical reaction. That is, we must oxidize enough food to yield energy for building our bodies. Plants and photosynthetic microbes must absorb enough sunlight to yield energy. When an organism can no longer transform energy, or lacks a supply of energy, it dies.

Free Energy Change and Chemical Equilibrium

The ΔG (net energy yield) of a reaction depends on the intrinsic energy and entropy changes of the molecules (just discussed) as well as the concentrations of reactants and products. Concentration refers to how much of something is present in a given volume. As reactant concentration increases, the reaction is more likely to proceed and yield energy. For example, a high concentration of sugar in grapes enables yeast to ferment the sugar, converting it to alcohol. The alcohol-producing reaction yields energy for the yeast to grow.

Concentration is reported in units of **molarity**. Molarity is defined as the number of moles of substance per liter of solution (usually water). One **mole** is composed of 6.022×10^{23} molecules (known as Avogadro's number). The weight of a mole of a substance (in grams) equals the average molecular mass in atomic mass units. For example, NaCl has a mass of 58.5 atomic mass units, so a mole of NaCl weighs 58.5 grams. To make a 1-molar solution of NaCl, we take 58.5 grams of NaCl and add water up to a liter.

A free energy change can be related to an equilibrium constant, K_{eq}, that indicates whether products or reactants will be favored at equilibrium. Reactions between biomolecules can be written as chemical equations. For example, in a given reaction, reactants A and B form products C and D. The value of K_{eq} at which the reaction is at equilibrium is given by

$$K_{eq} = [C][D]/[A][B]$$

The value of K_{eq} tells us how the concentrations of reactants and products affect the ability of a reaction to go forward. A K_{eq} of 1 means that neither products nor reactants are favored. A K_{eq} greater than 1 indicates that products are favored at equilibrium and corresponds to a negative free energy change that yields energy. A K_{eq} less than 1 means that reactants predominate and corresponds to a positive free energy change in which energy must be spent for the reaction to go forward.

A reaction can be made spontaneous and driven forward by keeping the concentration of reactants high or the concentration of products low. The cell uses this strategy in metabolism, where the product of one reaction is constantly removed by a subsequent reaction (see Chapter 7). For example, the synthesis of cAMP is driven forward by the rapid hydrolysis of the product pyrophosphate to inorganic phosphate; as a result, the concentration of the immediate product pyrophosphate is kept low, and cAMP formation is driven forward.

Activation Energy and the Rate of Reaction

Although a reaction with a large negative free energy change is spontaneous, it may be slow. The size of the energy change says nothing about the rate of a reaction. An everyday example of a spontaneous but slow reaction is how iron rusts by gradual oxidation. A spontaneous reaction will be slow if it must pass through a high-energy transition state on the way to forming products. The **activation energy** (E_a) is the energy needed to reach this transition state (**Figure 4.25**). For reactions with low activation energies, random collisions between reactants may provide enough energy to boost them over the E_a. For reactions with high activation energies, collisions between molecules may rarely provide enough energy for them to reach the transition state. Even though such a reaction goes forward, it is so slow that, on the timescale of a cell, the reaction does not occur.

Figure 4.25 Enzymes Decrease Activation Energy and Increase Reaction Rate

The negative value of ΔG makes the reaction spontaneous, but the high activation energy makes the reaction extremely slow. The enzyme makes the reaction proceed fast enough to be useful.

Enzymes

A living cell can increase the rate of a specific reaction by using a specific enzyme. For example, the DNA polymerase enzyme of *E. coli* bacteria attaches nucleotides at the rate of 1,000 nucleotides per second. Most enzymes consist of protein, although some are RNA, such as the RNA subunits of the ribosome. An **enzyme** is a biological **catalyst**; that is, the enzyme participates in the reaction while increasing the reaction rate but is not itself consumed.

An enzyme catalyzes the conversion of substrates (reactants) to products (see Figure 4.25). Each enzyme is highly specific, able to bind only one type of substrate. Like a puzzle piece, the substrate fits into the "active site" of the enzyme, where functional groups of the enzyme stabilize the transition state. By stabilizing the transition state, holding the molecules in a specifically shaped binding pocket, the enzyme lowers the activation energy. The activation energy is the amount of energy that must be put in before the reaction goes forward, releasing more energy than the amount put in. Although enzymes increase reaction rates, they do not change the net ΔG of a reaction (see Figure 4.25).

The rate of reactions catalyzed by enzymes is affected by many factors, including temperature and pH.

TEMPERATURE   Higher temperature increases the rate of collision of reactants and thus the rate of reaction. Most bacteria grow faster as temperature increases; this is why warm milk spoils faster than milk in the refrigerator. At too high a temperature, however, the bacterium's enzymes become **denatured** (lose their structure), and growth slows or stops. So, the human body generates a fever to fight pathogenic bacteria.

pH (ACID OR BASE)   Another factor affecting the rate of reaction is the hydrogen ion concentration, represented by pH, the negative logarithm of H^+ concentration. Most enzymes work only within a narrow range of pH. The pH of human blood is regulated very closely by our rate of breathing. When we hold our breath, carbon dioxide builds up in the blood, where it forms acid (decreasing pH). Decreased pH, or acidosis, may impair the brain and lead to coma. When we breathe too fast, or hyperventilate, carbon dioxide is lost too fast; acid is lost, and blood pH rises (alkalosis). Alkalosis may cause fainting and seizures.

Hydrogen ions (H$^+$) are also referred to as protons because they have lost their single electron and consist of only a proton. Hydrogen ion concentration is reported using a pH (power of hydrogen) scale, where pH is the negative logarithm of the hydrogen ion concentration: pH = $-\log$ [H$^+$]. The designation "log" refers to logarithm, or power of 10. In pure water, [H$^+$] is 1×10^{-7} molar (M), a pH of 7, or neutral pH (**Figure 4.26**). A pH value below pH 7 (greater H$^+$ concentration) is acidic; an example of an acidic substance is lemon juice (pH 2). A pH value above pH 7 (lower H$^+$ concentration) is alkaline; an example of an alkaline substance is baking soda (pH 9).

Because the pH scale is logarithmic, every pH unit corresponds to a tenfold change in hydrogen ion concentration. Thus, a solution with a pH of 6 has a hydrogen ion concentration of 1×10^{-6} M—ten times the hydrogen ion concentration at pH 7 (1×10^{-7} M). Remember that 10^{-6} is a larger number than 10^{-7} whereas 10^{+7} is a larger number than 10^{+6}.

Figure 4.26 pH Values of Common Substances

Concentration of hydrogen ions compared with distilled water		Example of solutions at this pH
10,000,000	pH = 0	Battery acid (strong), hydrofluoric acid
1,000,000	pH = 1	Hydrochloric acid secreted by stomach lining
100,000	pH = 2	Lemon juice, gastric acid, vinegar
10,000	pH = 3	Grapefruit, orange juice, soda
1,000	pH = 4	Acid rain, tomato juice
100	pH = 5	Soft drinking water, black coffee
10	pH = 6	Urine, saliva
1	pH = 7	"Pure" water
1/10	pH = 8	Seawater
1/100	pH = 9	Baking soda solution
1/1,000	pH = 10	Great Salt Lake, milk of magnesia
1/10,000	pH = 11	Ammonia solution
1/100,000	pH = 12	Soapy water
1/1,000,000	pH = 13	Bleaches, oven cleaner
1/10,000,000	pH = 14	Liquid drain cleaner

The pH value equals the negative logarithm of the hydrogen ion concentration.

Figure 4.27 Organic Acids and Bases

The carboxyl group of a carboxylic acid dissociates, releasing H$^+$.

Acid   Base   Proton

The amino group of an organic base binds a proton from water, leaving OH$^-$ in solution.

Base   Acid

In living cells, where do hydrogen ions come from? The largest source of hydrogen ions is water (H$_2$O), which always exists in equilibrium with hydrogen ions (H$^+$) and hydroxide ions (OH$^-$). In water solution, the hydrogen ion concentration [H$^+$] multiplied by the hydroxide ion concentration [OH$^-$] always equals 1×10^{-14}. Hence, the concentrations of H$^+$ and OH$^-$ are reciprocally related (if one goes up, the other goes down). If the pH is less than 7, a solution is acidic (more H$^+$, less OH$^-$). If the pH is greater than 7, the solution is basic, or alkaline (less H$^+$, more OH$^-$).

Acids (for example, carboxyl groups) donate protons, and bases (for example, amino groups) accept protons (**Figure 4.27**). At intracellular pH (near 7), the carboxyl group and the amino group of each amino acid are both ionized (charged), with the carboxyl group carrying a negative charge and the amino group carrying a positive charge (see Figure 4.18). The ionization of these groups can change if the pH changes. At lower pH (more acidic solution), a proton moves back onto the ionized carboxylic acid. For example, the side chain of an acidic amino acid such as glutamate (see Figure 4.18) is ionized (with a negative charge) at normal cell pH but may regain a proton and become glutamic acid at a lower pH. At higher pH values (lower proton concentrations), protonated amino groups (with a positive charge) lose their extra protons to become uncharged. Changes in the ionization state of carboxyl or amino groups on the side chains of amino acid residues can disrupt ionic bonds between these groups and lead to protein denaturation. This effect of pH on protein tertiary structure is one reason why cells can tolerate only a narrow range of intracellular pH and why acidosis or alkalosis in our own tissues is harmful.

Thought Question 4.3 Bacteria that normally inhabit your intestinal tract, such as *Bacteroides* species, have large numbers of genes that encode many different types of digestive enzymes. Why would intestinal bacteria have more different types of digestive enzymes than a pathogen infecting your blood?

Oxidation and Reduction

An important class of reactions catalyzed by enzymes is redox reactions (reactions involving reduction and oxidation). Redox reactions involve the transfer of electrons from one molecule to

Figure 4.28 Oxidation-Reduction Reactions

Reduced compound A

Oxidized compound B

Oxidized compound A

Reduced compound B

A. A donates electrons to reduce compound B. Because A loses electrons, it becomes oxidized itself in the process.

$$\text{NAD}^+ + \text{H}^+ + 2\,e^- \rightleftharpoons \text{NADH}$$

NAD⁺   **NADH**

B. NADH is a common biological reducing agent.

another or from one atom to another. Redox reactions yield the majority of the energy we obtain from oxidizing the food we eat (again, see Chapter 7). Many cancer drugs and antiparasitic drugs target the cell's redox balance.

In a redox reaction, **reduction** refers to the gain of electrons, and **oxidation** refers to loss of electrons. The molecule that gains electrons becomes reduced, and the molecule that loses electrons becomes oxidized (**Figure 4.28A**). (A useful mnemonic device to remember this is "LEO the lion says GER." LEO: lose electrons, oxidation; GER: gain electrons, reduction.) An example from metabolism is the reduction of nicotinamide adenine dinucleotide (NAD⁺) by a pair of hydrogen atoms removed from a food molecule such as glucose (**Figure 4.28B**). Each hydrogen atom transfers one electron to NAD⁺. The reduction of NAD⁺ (that is, addition of electrons) to form NADH enables NADH to reduce a protein of the electron transport chain, an important process by which cells acquire energy from food (explored further in Chapter 7).

Oxidation and reduction reactions always occur together. That is, wherever one atom loses electrons, another atom must gain electrons. The oxygen gas that we breathe has a very strong tendency to gain electrons from the food molecules we eat. Oxygen usually gains electrons and becomes reduced in redox reactions. Molecules that become reduced are called **oxidizing agents** because they cause something else to become oxidized. In contrast, **reducing agents** can donate electrons, reduce other molecules, and become oxidized themselves in the process.

SECTION SUMMARY

- **Life requires energy.** If the organism lacks a source of energy, or cannot transform energy, it dies.
- **Biochemical reactions are driven by negative values of free energy change, ΔG.** The reactions of a living organism must build order within the organism but decrease order (increase entropy) in the universe as a whole.
- **Concentrations of substrates and products determine whether a reaction will go forward.** Increasing concentration of substrates

or decreasing concentration of products favors the reaction going forward.

- **Most biochemical reactions require catalysis by enzymes.** An enzyme catalyzes a specific reaction. Enzymes increase the rate of reactions by lowering the activation energy.
- **Reaction rates are affected by temperature and pH.**
- **Some reactions involve reduction and oxidation—the gain and loss of electrons.**

Thought Question 4.4   Saline is a salt solution used for intravenous (IV) procedures; it contains approximately the same concentration of NaCl as in the blood. The concentration of NaCl in blood is approximately 0.154 mole per liter. How much salt would you weigh out to make 1 liter of saline? (*Hint:* See earlier definition of the atomic weight of atoms.)

Thought Question 4.5   Your muscle cells contain ATP at concentrations of 1–10 mM (millimolar; thousandths of molar). The reaction of ATP hydrolysis (forming ADP and phosphate) has a large negative value of free energy change, ΔG. So why doesn't all the ATP hydrolyze?

4.5 Membranes and Transport

SECTION OBJECTIVES

- Describe the structure of cellular membranes.
- Explain the concept of selective permeability of a membrane.
- Explain how membranes compartmentalize cells and control the movement of a cell's molecules.
- Describe the effect of osmosis on membrane compartments.

How do cells organize their molecules and their molecular reactions? How do they keep the substrates or reactants they need within the cell and exclude waste products or harmful toxins? All cells are enclosed by a **cell membrane** (also called the plasma membrane or cytoplasmic membrane) that creates an internal environment distinct from the external environment. The aqueous fluid inside the membrane is called **cytoplasm** (or cytosol). The cell membrane proteins regulate uptake of nutrient molecules and pump out toxins and waste products. Some membrane proteins receive signals from outside the cell, while other membrane proteins conduct processes that yield energy. Some antibiotics act by disrupting bacterial membranes; an example is polymyxin B, found in the topical ointment Neosporin.

Membranes Consist of Lipids and Proteins

As shown in **Figure 4.29A**, membranes consist mainly of lipids and proteins. Many components have attached sugar chains; there are glycolipids (sugars joined to lipids) and glycoproteins (sugars joined to proteins). For example, we saw in the chapter-opening case how a membrane glycolipid ganglioside is used by cholera toxin to attach to the intestinal epithelial cell (see Figure 4.1). Membranes contain a greater number of lipid molecules than protein molecules, but the

Figure 4.29 The Cell Membrane

A. A cutaway view of the cell membrane.

Extracellular fluid

Carbohydrate attached to protein
Glycoprotein
Glycolipid

Peripheral protein associates with the membrane but does not span it.

Transmembrane protein spans the membrane.

Cytoskeletal proteins

Phospholipid bilayer

Cytoplasm

Approximate width of the membrane

B. Ribbon diagram from X-ray crystallographic data of a multidrug transporter protein. Red portions are alpha helices buried in the membrane, and yellow portions are loops within the watery cytoplasm or exterior.

proteins are larger and contribute about half the membrane mass. Proteins confer specific properties on membranes; for example, cell-surface proteins enable important communication with other cells. But viruses evolve to bind these surface proteins and initiate infection. For example, the human immunodeficiency virus (HIV) binds CD4, a T-cell surface protein required to activate the immune response.

In bacteria and eukaryotes, the predominant lipids in membranes are **phospholipids**. Phospholipids consist of a core of glycerol, to which two fatty acids and a modified phosphate group are attached via ester linkages (**Figure 4.30A**). Archaea, unlike eukaryotes and bacteria, have very different kinds of lipids that enable survival in extreme environments such as high temperature or acidity (discussed in Chapter 10). Phospholipids are "amphipathic"; the fatty acid hydrocarbon tails are hydrophobic, and the phosphate head group is hydrophilic. Amphipathic lipids are most stable in water when the hydrophilic portions interact with water and the hydrophobic portions cluster together away from water.

One way phospholipids can achieve stability is by forming a bilayer. Indeed, the cell membrane is a **phospholipid bilayer**, two

layers of phospholipids whose hydrocarbon fatty acid tails face the interior of the bilayer and whose charged phospholipid head groups face the aqueous cytoplasm and extracellular environment (**Figure 4.30B**). The phospholipid layer in contact with the cytoplasm is called the inner leaflet (facing into the cell), and the layer in contact with the environment is called the outer leaflet (facing outside the cell), although additional layers of envelope may cover it. Cells contain several phospholipids that differ in how the phosphate head group is modified. The fatty acids composing phospholipids differ in the length of their fatty acid chains and the number of carbon-carbon double bonds (unsaturation).

Membranes also contain molecules that act as stiffening agents, providing mechanical stability to the phospholipid bilayer.

Figure 4.30 Phospholipids and the Lipid Bilayer

A. A phospholipid composed of glycerol, phosphate, and two saturated fatty acids.

Phosphate

Glycerol

Fatty acids

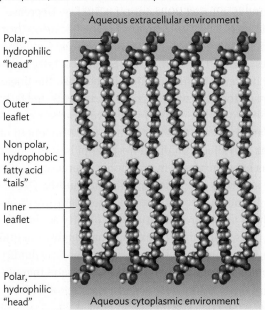

Aqueous extracellular environment

Polar, hydrophilic "head"

Outer leaflet

Non polar, hydrophobic fatty acid "tails"

Inner leaflet

Polar, hydrophilic "head"

Aqueous cytoplasmic environment

B. Orientation of phospholipids in the bilayer.

In eukaryotic membranes, the main stiffening agents are cholesterols; bacteria do not have cholesterol but have "hopanoids"—molecules of analogous structure that stabilize membranes.

Membrane proteins can be classified on the basis of how they interact with membranes (Figure 4.29A). **Transmembrane proteins** (integral proteins) span the bilayer. Transmembrane proteins are amphipathic, meaning they have both hydrophobic and hydrophilic portions; the hydrophilic portions face the cytoplasm and the extracellular environment, and the hydrophobic components span the membrane. The transmembrane domains are often alpha helices containing 15–20 amino acid residues (**Figure 4.29B**). **Peripheral membrane proteins** are associated with the cell membrane but are not directly inserted into the bilayer. Some peripheral membrane proteins are bound to membranes through noncovalent interactions with transmembrane proteins, others by noncovalent interactions with the lipids of the membrane.

Membranes are dynamic structures; their components, such as lipids and proteins, can move rapidly, even rotating like a wheel. Membrane components are free to diffuse (move around) in the plane of the membrane. Certain membrane proteins may be restricted to specific regions of the membrane by interactions with other proteins. Many phospholipids and membrane proteins can move laterally within a leaflet. However, they rarely "flip-flop" from one leaflet of the bilayer to the other, because the charged head groups would have to move through the membrane's hydrophobic interior. So, pathogenic bacteria with an outer membrane, such as *Salmonella* species, have special membrane lipids that extend only from the outer membrane's outer leaflet, whereas completely different lipids face the interior.

Membrane fluidity refers to the movement of membrane phospholipids within the plane of the membrane; this fluidity is important for proper membrane function. For example, protein transport across the membrane is affected by membrane fluidity. Decreased fluidity is associated with decreased transport rates. Because a drop in temperature decreases fluidity, cold temperatures may slow transport processes across the membrane. Saturated fatty acids (such as in butter) decrease membrane fluidity because the linear hydrocarbon tails pack together well. In contrast, unsaturated fatty acids (such as in olive oil) have kinks in the hydrocarbon chains that limit packing and increase fluidity (Figure 4.7). The length of the fatty acid chains also affects fluidity. Phospholipids with longer hydrocarbon chains have increased hydrophobic interactions with neighboring lipids and thus decreased membrane fluidity. Organisms can alter membrane fluidity in response to temperature stress by changing the length and degree of saturation of fatty acids present in membrane phospholipids. For example, as environmental temperatures drop, some kinds of cells maintain membrane fluidity by replacing long-chain fatty acids with shorter chains and increasing the percentage of unsaturated fatty acids in their membranes.

Figure 4.31 Selective Permeability of Cell Membranes

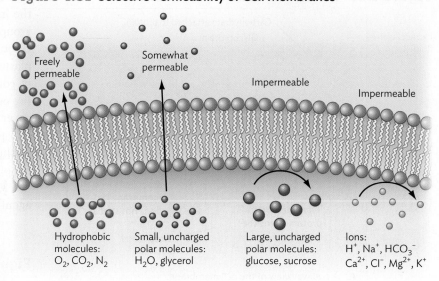

Hydrophobic molecules: O_2, CO_2, N_2

Small, uncharged polar molecules: H_2O, glycerol

Large, uncharged polar molecules: glucose, sucrose

Ions: H^+, Na^+, HCO_3^- Ca^{2+}, Cl^-, Mg^{2+}, K^+

Membranes Are Selectively Permeable

How do cells allow some molecules to enter while excluding others? Membranes are selectively permeable; that is, they are permeable to some substances but not to others. In general, the cell membrane is permeable to hydrophobic molecules and impermeable to charged molecules (**Figure 4.31**). Diffusion across the membrane also depends on the size of the molecule. Small nonpolar molecules such as O_2 freely pass through the membrane, but larger nonpolar molecules diffuse across more slowly. Molecules that are polar but small (like ethanol and water) can also diffuse across the membrane. The membrane is impermeable to large polar molecules, such as glucose, and to charged molecules, regardless of their size. So membranes help a cell keep its nutrients and enzymes, while efflux proteins export wastes and toxins.

An interesting question in medicine is, how do drug molecules administered through the blood (a watery fluid) cross the hydrophobic membranes of cells to affect them? If a drug is soluble in water, how can it cross a hydrophobic membrane? Many kinds of drugs are **membrane-permeant weak acids** or **membrane-permeant weak bases** (see earlier Figure 4.27). Weak acids and weak bases exist in equilibrium between charged and uncharged forms (**Figure 4.32**):

Weak acid: $HA \rightleftharpoons H^+ + A^-$

Weak base: $B + H_2O \rightleftharpoons BH^+ + OH^-$

Weak acids and weak bases cross the membrane in their uncharged forms: HA (weak acid) and B (weak base). On the other side, upon entering the aqueous cytoplasm, they dissociate (HA to A^- and H^+) or reassociate with H^+ (B to BH^+). As a result, the membrane-soluble HA conducts acid (H^+), or B conducts base (OH^-), across the membrane. The same effect occurs when fermenting bacteria release acids such as lactic acid. The lactic acid crosses the membrane in the hydrophobic protonated form (HA) and then releases H^+ (strong acid) on the other side. Lactic acid acts as a

Figure 4.32 Weak Acids and Weak Bases Cross the Cell Membrane

A weak acid crosses the membrane in the protonated form (HA). A weak base crosses the membrane in the deprotonated form (B).

natural food preservative. For example, during yogurt production, bacterial lactic acid builds up and limits further growth, preserving food value. Other membrane-permeant weak acids are commercial food preservatives such as benzoic acid (benzoate) and propionic acid (propionate).

Many key substances in cellular metabolism are membrane-permeant weak acids, such as acetic acid, and membrane-permeant weak bases, such as the nucleobases of DNA. Drugs that deprotonate within a cell (weak acids, acquiring a negative charge, A^-) include aspirin (acetylsalicylic acid) and penicillin. Drugs that protonate within a cell (weak bases, acquiring a positive charge, BH^+) include fluoxetine (Prozac) and tetracycline.

Transport across Cell Membranes

If large nutrient molecules such as glucose cannot diffuse directly across the cell membrane, how do cells acquire them? In both eukaryotes and prokaryotes, movement of molecules across the cell membrane occurs through specific transmembrane proteins such as channels and transporters. Many types of transporters exist, each differing in its energy requirements and in the types of molecules that it transfers across the membrane. For example, in the intestinal epithelium, a transporter protein cotransports glucose with sodium ions. Glucose-sodium transport is key to the success of oral rehydration therapy in treating cholera (see **IMPACT**). Transport proteins are discussed further in Chapter 5.

DIFFUSION   If a molecule (such as water) can penetrate the membrane or if a transport protein is available (such as for chloride ion), the rate of transport depends on the concentrations of the substance inside and outside the cell or cell compartment. **Diffusion** is the net movement of molecules from an area of high concentration to one of low concentration. Diffusion increases entropy and releases energy. Diffusion occurs by the random, thermal movement of molecules. Factors that influence the diffusion rate include:

- **Temperature.** Increased temperatures mean faster motion. The faster the molecules are moving, the faster they will arrive at the membrane and cross it.

- **Solubility of the molecules in the membrane.** To cross the membrane, the molecules must penetrate it. Hydrophobic molecules will dissolve in the membrane and cross it; charged molecules will not.

- **Size and shape of the molecule.** Friction between a molecule and its medium is a source of resistance that slows down motion. Larger molecules with more mass experience more resistance and cross the membrane more slowly.

- **Surface area of the membrane.** To cross the membrane, molecules must first encounter it. The chances of this happening are increased with greater membrane surface area.

- **Thickness of the membrane.** The thinner the membrane, the faster the molecules can get across.

- **Concentration gradient of the dissolved molecules.** A larger concentration difference between the two sides of the membrane increases the rate of diffusion. The more molecules there are in a given volume, the more molecules will encounter the membrane and cross it.

OSMOSIS   **Osmosis** is the diffusion of water across a selectively permeable membrane from regions of high water concentration (low solute) to regions of low water concentration (high solute). Osmosis across the cell membrane must be tightly controlled to preserve the form of the cell. So, cells must maintain osmotic balance with the surrounding environment. The direction of water movement depends on the concentration of dissolved solutes in the cell in relation to the cell's environment (**Figure 4.33**). When a cell is in an **isotonic** or **isosmotic** environment (equal concentrations of dissolved solutes inside the cell and out), osmotic balance exists and water will enter and exit the cell at equal rates (that is, no net movement of water occurs). In a **hypertonic** or **hyperosmotic** environment (higher concentration of solutes outside the cell), osmosis causes net loss of water from a cell. The cell shrinks, and the concentration of cell solutes (inorganic ions as well as organic molecules) increases. This is what happens in cholera: the increased chloride concentration in the intestinal lumen draws water out of the epithelial cells and the blood circulation, ultimately from the body as a whole.

In a **hypotonic** or **hypoosmotic** environment (lower concentration of solutes outside the cell), osmosis causes net uptake of water. The cell swells with water, and the cell components are diluted (see Figure 4.33). If enough water enters, the cell is destroyed by **lysis**—a rupturing of the cell membrane and dispersal of cell contents. The sudden increase of osmotic pressure is called **osmotic shock**. That is why intravenous fluids must always include saline, not distilled water—to prevent blood cell lysis by osmotic shock.

Many free-living cells, such as microbes, live in environments that are hypotonic (lower concentration of solutes outside the cell than inside). To survive a dilute environment, most bacteria and many eukaryotes have a rigid cell wall external to the cell membrane (discussed in Section 5.3). As water enters by osmosis and pushes

impact

Oral Rehydration Therapy: A Miracle Cure

Atherapy so simple that schoolchildren are taught to perform it became one of the major lifesaving medical advances of the twentieth century. This therapy, called oral rehydration therapy (ORT), is now saving hundreds of thousands of lives in countries such as Bangladesh that have pioneered widespread education about and access to ORT.

Diarrheal diseases cause 20% of all childhood deaths worldwide, particularly in communities lacking access to clean water (**Figure 1**). Causes of life-threatening diarrhea include cholera and pathogenic *E. coli*, viruses such as Norwalk and rotaviruses, and parasites such as *Giardia*. In the developing world, especially in remote communities lacking hospitals, dehydration from diarrhea is often a death sentence. Before 1950, it was thought that no oral treatment could reverse the dehydration and sodium loss (hyponatremia). Simply giving a saline solution (NaCl) by mouth has no benefit because the normal route of absorption of Na^+ is impaired in the presence of excess Cl^-, and if the Na^+ is not absorbed, then the water is not absorbed either. Thus, industrialized nations with advanced medical care developed intravenous fluids as the standard treatment for dehydration; this treatment is simply unavailable in parts of the world most vulnerable to diarrheal disease.

In the late 1950s, Dr. Hemendra Nath Chatterjee, in India, discovered a modified oral solution that promotes water uptake. Chatterjee added glucose to the saline solution. The glucose molecules are absorbed by a membrane protein that cotransports glucose with 2 Na^+ (**Figure 2**). Normally, the transporter is used to take up glucose driven by a gradient of Na^+, but in diarrhea, the glucose drives sodium uptake. The increase in solute uptake then draws water back into the cells, reversing dehydration. So, a simple solution of salt and glucose can reverse the water loss of diarrhea while allowing the patient's immune system to eliminate the pathogenic bacteria.

In 1971, during the Bangladesh Liberation War, ORT was used to treat cholera patients. As a result, the death rate from cholera was less than 4%, compared with the 30% expected. Thereafter, Bangladesh and other countries with communities prone to waterborne disease adopted public education programs encouraging mothers and schoolchildren to learn to administer ORT. **Figure 3** shows a mother in Bangladesh administering ORT to a severely dehydrated child. Within hours, the child recovers. In 2001, the Gates Award for Global Health was conferred on the Centre for Health and Population Research, in Dhaka, Bangladesh, for its role in development of ORT.

Figure 1 Dehydrated by Diarrheal Illness

Pinched skin retains the shape.

Figure 2 Cotransport of Sodium Ions and Glucose

The sodium-glucose cotransporter protein carries glucose together with two sodium ions (2 Na^+) across the plasma membrane. A gradient of glucose drives uptake of Na^+, and vice versa.

Glucose   2 Na^+

Cotransport

Figure 3 Oral Rehydration Therapy (ORT)

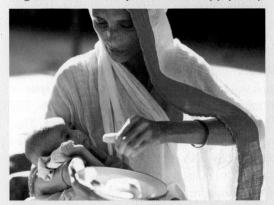

A mother provides ORT to an infant dehydrated by cholera, in Dhaka, Bangladesh.

against the cell wall (turgor pressure), the wall resists the tension and pushes back with an equal but opposite force known as wall pressure. The wall pressure is an inward pressure exerted by the cell wall against the cell membrane. When turgor pressure and wall pressure are equal in magnitude, the cell is at equilibrium with respect to water movement. Organisms that lack a cell wall employ other strategies to deal with hypotonic environments. For example, freshwater protists that lack a cell wall, such as amebas and paramecia, expel excess water through a contractile vacuole (an organelle shown in Figure 5.31).

What happens to marine microbes that live in high-salt conditions? Ocean-dwelling microbes such as *Vibrio cholerae* face the problem of water loss. These organisms accumulate solutes known as osmolytes, for example potassium ion, to ensure that they are isotonic to the external environment, thus preventing water loss.

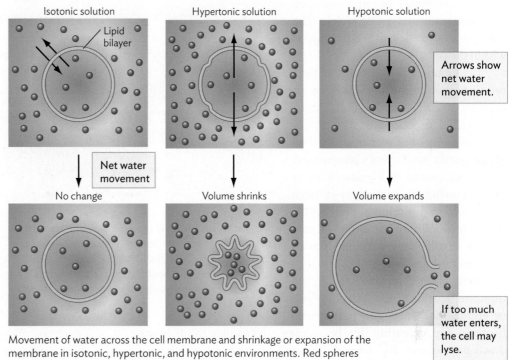

ANIMATION **Figure 4.33 Osmosis and Water Balance**

Movement of water across the cell membrane and shrinkage or expansion of the membrane in isotonic, hypertonic, and hypotonic environments. Red spheres represent solutes. Arrows indicate net water movement.

SECTION SUMMARY

- **Membranes consist of lipids, proteins, and small molecules.** The phospholipid bilayer includes inward- and outward-facing leaflets. The membrane is a fluid in which components can move in two directions but rarely flip sides.

- **Transmembrane proteins span the bilayer.** Peripheral membrane proteins are associated with one side of the bilayer.

- **Membranes are selectively permeable.** Small hydrophobic molecules can diffuse through a membrane, whereas polar or charged molecules cannot. Membrane-permeant weak acids and bases can cross the membrane in their hydrophobic forms, but their ionized forms are soluble only in water.

- **Transport proteins can move molecules across the membrane.**

- **Osmosis is the diffusion of water across a membrane from a region of low solute concentration to a region of high solute concentration.** Cells that live in a hypotonic (hypoosmotic) environment, where water is diffusing in, need structures to exclude water.

4.6 Endocytosis and Phagocytosis

SECTION OBJECTIVES

- Describe how endocytosis and phagocytosis transport material into a eukaryotic cell.
- Explain how pathogens take advantage of endocytosis and phagocytosis.

What happens in different cell compartments? The phospholipid bilayer of a cell membrane encloses a compartment of water solution (the cytoplasm) whose properties differ from those outside the membrane. Many bacteria have just the one compartment, the cytoplasm; some have a second compartment, the periplasm of Gram-negative bacteria (presented in Chapter 5). Photosynthetic bacteria such as cyanobacteria possess intracellular membrane compartments for photosynthesis.

Eukaryotic cells may hold volumes a thousandfold larger than those of typical bacteria. These larger eukaryotic cells face extra challenges of transport: how to get nutrients efficiently to all parts of the cell and how to export wastes and toxins. In a eukaryotic cell, extensive structures of intracellular membranes called **organelles** manage intracellular transport. Remarkably, these distinctive and intricate organelles are all composed of phospholipid membranes. The function of the organelles is presented in Chapter 5. Here in Chapter 4, we consider two eukaryotic cell processes that form the compartments.

The cells of animals and eukaryotic microbes (such as amebas) need to take up objects that are too large for protein transporters, such as food particles or bacteria. The uptake mechanism requires a portion of the membrane phospholipids to rearrange in the bilayer, which bends and folds to form a pocket that contains the object (**Figure 4.34A**). This pocket formation is called "invagination." Typically, the pocket contains a region of membrane in which cell-surface proteins have bound an object or particles useful for the cell. Ultimately, the pocket closes in as a small hollow sphere called a **vesicle**, which separates from the original membrane while now contained within the cytoplasm. For smaller particles or prey bacteria, uptake by membrane invagination is called **endocytosis**

(**Figure 4.34B**). The particles taken up by endocytosis become enclosed in an **endosome**, a type of vesicle. The incorporated particles are now trapped within the cell.

Taking up larger particles or prey requires membrane invagination on a larger scale, often by cell extensions called **pseudopods**. This large-scale engulfment of an object is called **phagocytosis**, generating a **phagosome**. The giant ameba in **Figure 4.35** is phagocytosing a paramecium—a large protozoan that similarly phagocytoses bacteria. Within the ameba, a previously engulfed paramecium is seen undergoing digestion.

Human macrophages (cells of the immune system) use phagocytosis or endocytosis to engulf pathogenic bacteria for digestion by **lysosomes** (organelles that break down pathogens). But some pathogens can subvert the uptake process—as we see next.

 Figure 4.34 **Endocytosis in a Eukaryotic Cell**

A. The phospholipids of the cell membrane rearrange to allow the membrane to bend in. The bent portion of the cell membrane pinches in, forming a pocket that contains extracellular material bound to cell surface proteins. The membrane pocket closes off as a vesicle within the cell.

Endocytosis forms endocytic vesicle.

B. An endocytic vesicle forms within a rat capillary epithelial cell.

CASE HISTORY 4.1

Chicken Surprise

In October 2013, Kelsey was a 28-year-old computer programmer who worked for a prominent technology firm in Silicon Valley. In the middle of the night, Kelsey awoke with abdominal cramps and diarrhea. Her temperature spiked as high as 40°C (104°F). She had a bad headache and could not retain sufficient fluids. After 3 days, her fever subsided, but the diarrhea continued (about 20 times a day) and began showing blood. Kelsey was admitted to the hospital, where she was given intravenous fluids to control her dehydration. The physician asked her about her dietary history preceding the onset of symptoms. Kelsey recalled that, the night before onset, she had eaten rotisserie chicken cooked at the supermarket. An antibody test confirmed the diagnosis of *Salmonella enterica* serovar Heidelberg. Because of the severity of her symptoms, Kelsey was treated with antibiotics, although sometimes antibiotics may prolong the persistence of *Salmonella*. When amoxicillin failed to control the infection because of bacterial resistance, ciprofloxacin was used. Ultimately, Kelsey's illness was one of more than 400 cases in 23 states traced to contaminated chicken from one California farm. Kelsey recovered, but 3 months later, *S.* Heidelberg could still be detected in her feces.

How could enteric bacteria make someone as healthy as Kelsey so sick and persist in the body for months despite treatment with antibiotics? Unlike *Vibrio cholerae* (the culprit in our chapter-opening case history), *Salmonella* bacteria do more than alter the intestinal fluid balance. *Salmonella* actually invades the host cells of the intestinal lining, including enterocytes (absorptive epithelial cells) and macrophages (defensive white blood cells). Within the host cells, the bacteria grow and multiply in protected compartments called *Salmonella*-containing vacuoles.

Figure 4.35 **Phagocytosis**

Ameba

Paramecia

LM   0.25 mm

Ameba engulfing paramecia.

Figure 4.36 *Salmonella*-containing vesicles

Salmonella
bacterium

TEM 1 μm

Macrophage infected by *Salmonella* bacteria (dark rods) growing within individual vesicles.

To invade host cells, the bacteria take advantage of the host's own processes of uptake and transport mediated by membrane vesicles. These pathogens (called facultative intracellular bacteria) can stabilize the endosome to allow bacterial growth (**Figure 4.36**). The stabilized endosomes multiply as the bacteria grow, forming a protective vacuole surrounding each bacterium. Because the bacteria multiply within host cells, protected from antibiotics, they may persist for long periods. Other examples of pathogens that enter cells by phagocytosis include *Listeria* bacteria, a food-borne pathogen, and influenza virus.

Even when a pathogen remains outside the host cell, endocytosis can bring in toxins such as cholera toxin. As shown at the beginning of the chapter (see Figure 4.1), the pathogen *V. cholerae* remains extracellular, but its toxin must gain entry to the intestinal cells. The cholera toxin first binds to the intestinal cell-surface molecule ganglioside; then it enters the cell by endocytosis, and the endocytic vesicle merges into the endoplasmic reticulum (ER; transport membranes within the cell). In the ER membrane, a transport protein is co-opted to release a portion of the toxin (A1) into the cytoplasm, where it gains access to the protein complex that leads to Cl^- efflux. ER is discussed further in Chapter 5.

How can a cell disable endocytosed pathogens and toxins? Normally, the endosome or phagosome must fuse with a lysosome. **Lysosomes** help eukaryotic cells obtain nourishment from macromolecular nutrients, and they help destroy pathogenic bacteria and viruses. For digestion, lysosomes contain many hydrolytic enzymes (for example, proteases, nucleases, and lipases) and have an acidic pH of around 5. Ultimately, the digested contents of the lysosome are delivered to the plasma membrane and expelled outside the cell (discussed in Chapter 5).

SECTION SUMMARY

- **Endocytosis takes up small particles by membrane invagination.** Invaginating membranes form vesicles called endosomes that are enclosed within the cytoplasm.
- **Phagocytosis is the uptake of larger particles and bacteria.** Phagocytosis may include the formation and fusion of pseudopods.
- **Lysosomes fuse with endosomes and acidify their contents, adding factors that kill pathogenic bacteria.** Lysosomes deliver their contents to the endomembrane system.

Thought Question 4.6 A pathogenic bacterium such as *Mycobacterium tuberculosis* gets taken up by a cell through phagocytosis. What do you think the bacterium has to do to avoid getting destroyed and instead reproduce within the host cell?

Perspective

The human body is made of molecules and chemical reactions. A simple change in salt concentration can dramatically impact health, causing the water loss and dehydration seen in cholera. A small change in blood pH, either too acidic or too alkaline, can lead to convulsions or fainting. Overall, we see how the various kinds of biological molecules, from water and inorganic ions to macromolecular complexes, interact with one another to compose a living cell. Enzymes catalyze the condensation of small organic molecules such as sugars to form polysaccharides. Other molecules build cell walls and other components of the cell, including a wide range of proteins with various specialized functions. Phospholipids interact by hydrophobic forces to form membranes that define cell compartments.

In Chapter 5, you will learn how the structures and processes discussed in Chapter 4 fit together within living cells. For example, you will see that sugars and amino acids (presented in Sections 4.2 and 4.3) are assembled to form the cagelike cell wall that encloses a bacterial cell. You will also learn how the membranes of a cell form its organelles and how organelles mediate the cell's interactions with pathogens.

Key Terms

activation energy (p. 104)
alpha helix (p. 101)
amino acid (p. 99)
anion (p. 92)
atom (p. 90)
atomic number (p. 90)
atomic weight (p. 90)
beta sheet (p. 101)
carbohydrate (p. 95)
catalysis (p. 102)
catalyst (p. 104)
cation (p. 92)
cell membrane (p. 106)
complex (p. 99)
condensation reaction (p. 94)
covalent bond (p. 90)
cytoplasm (p. 106)
denatured (p. 104)
diffusion (p. 109)
disaccharide (p. 96)
electrolyte (p. 92)
element (p. 90)
endergonic (p. 103)
endocytosis (p. 111)

endosome (p. 112)
energy (p. 103)
entropy (p. 103)
enzyme (p. 104)
exergonic (p. 103)
first law of thermodynamics (p. 103)
free energy change (ΔG) (p. 103)
functional group (p. 93)
genome (p. 96)
glycan (p. 96)
half-life (p. 90)
hydrogen bond (p. 92)
hydrophilic (p. 93)
hydrophobic (p. 93)
hyperosmotic (p. 109)
hypertonic (p. 109)
hypoosmotic (p. 109)
hypotonic (p. 109)
ion (p. 92)
ionic bond (p. 92)
isomer (p. 99)
isosmotic (p. 109)
isotonic (p. 109)
isotope (p. 90)

lipid (p. 94)
lysis (p. 109)
lysosome (p. 112, 113)
macromolecule (p. 93)
mass number (p. 90)
membrane-permeant weak acid (p. 108)
membrane-permeant weak base (p. 108)
molarity (p. 104)
mole (p. 104)
molecular formula (p. 90)
molecule (p. 90)
monosaccharide (p. 95)
native conformation (p. 102)
nonpolar covalent bond (p. 92)
nucleobase (p. 97)
nucleotide (p. 97)
orbital (p. 90)
organelle (p. 111)
organic molecule (p. 93)
osmosis (p. 109)
osmotic shock (p. 109)
oxidation (p. 106)

oxidizing agent (p. 106)
peripheral membrane protein (p. 108)
phagocytosis (p. 112)
phagosome (p. 112)
phospholipid (p. 107)
phospholipid bilayer (p. 107)
polar covalent bond (p. 92)
polysaccharide (p. 96)
primary structure (p. 100)
pseudopod (p. 112)
reducing agent (p. 106)
reduction (p. 106)
saturated (p. 94)
secondary structure (p. 101)
second law of thermodynamics (p. 103)
structural formula (p. 90)
tertiary structure (p. 101)
transmembrane protein (p. 108)
unsaturated (p. 94)
vesicle (p. 111)

Objectives Assessment Questions

4.1 Elements, Bonding, and Water

1. Which atom is <u>not</u> an essential part of most living cells?
   a. Carbon
   b. Hydrogen
   c. Oxygen
   d. Nitrogen
   e. Helium

2. Which compound would dissolve less well in water than in lipids?
   a. Butane ($CH_3-CH_2-CH_2-CH_3$)
   b. NaCl
   c. Glycine
   d. CO_2

3. Which compound would dissolve in water by forming ions?
   a. NaCl
   b. O_2
   c. Glycine
   d. CO_2

4.2 Lipids and Sugars

4. Which part of a lipid is the most hydrophobic?
   a. Glycerol
   b. Hydrocarbon chain
   c. Carboxyl group
   d. Ester linkage of a triglyceride

5. A simple sugar contains all of these types of chemical structures, except for _____.
   a. a carbonyl group
   b. a hydroxyl group
   c. an amino group
   d. a carbon-carbon bond

6. Which molecule contains a sugar component?
   a. Fatty acid
   b. Methane
   c. Glycine
   d. Glycoprotein

4.3 Nucleic Acids and Proteins

7. A nucleotide contains all of these chemical structures, except for _____.
   a. a thiol (–SH)
   b. a phosphoryl group
   c. a nucleobase
   d. a pentose sugar

8. The alpha helix is a form of protein structure at which level?
   a. Primary structure
   b. Secondary structure
   c. Tertiary structure
   d. Quaternary structure

9. Which nucleobase is usually a component of RNA but not DNA?
   a. Uracil
   c. Guanine
   b. Cytosine
   d. Adenine

4.4 Biochemical Reactions

10. Which process does not yield energy for a cell?
    a. DNA synthesis
    b. Oxidation of sugar
    c. Oxidation of lipids
    d. Breakdown of carbohydrates into smaller molecules

11. For an intestinal bacterium, which factor is most likely to increase the rate of a reaction catalyzed by an enzyme?
    a. Decreasing concentration of the substrate
    b. pH increase from pH 8 to pH 9
    c. Salt saturation
    d. Increase in temperature from 25°C to 35°C

12. Which reaction involves increasing entropy?
    a. Building a protein
    b. Replicating DNA
    c. Breaking down a protein
    d. Coiling DNA into a chromosome

4.5 Membranes and Transport

13. Which part of a phospholipid bilayer faces inward, toward the middle of the bilayer?
    a. Phosphoryl group
    c. Glycerol
    b. Hydrocarbon chain
    d. Carboxyl group

14. Which molecule requires a transport protein in order to cross a membrane?
    a. H_2O
    c. Alanine
    b. O_2
    d. Ethanol

15. Which process involves osmosis?
    a. A transport protein carries a nutrient across a membrane.
    b. CO_2 crosses a membrane.
    c. A cell placed in distilled water takes up water and expands.
    d. A protein is synthesized in the cytoplasm.

4.6 Endocytosis and Phagocytosis

16. Which pathogen does not enter a host cell by endocytosis?
    a. *Salmonella enterica*
    b. *Listeria monocytogenes*
    c. Influenza virus
    d. *Vibrio cholerae*

17. An endocytic vesicle must be composed of which kind of molecules?
    a. Nucleic acids
    c. Sodium ions
    b. Phospholipids
    d. Sugars

18. Which cell can conduct phagocytosis?
    a. Plant cell
    b. *Escherichia coli*
    c. Nerve cell
    d. Ameba

Review Questions

1. Which are the major elements of life? Which of life's molecules contain atoms of these elements?

2. Explain the different kinds of molecular bonding. Which types are stronger, and which are weaker?

3. Describe the composition of carbohydrates and lipids. What types of monomers must condense to form longer chains?

4. What is the composition of a nucleotide?

5. Explain how DNA and RNA are related and how they differ. Explain the difference between intramolecular and intermolecular base pairing.

6. For a protein, describe the primary, secondary, tertiary, and quaternary levels of structure.

7. Explain the role of energy and entropy in driving a chemical reaction. What is necessary for a reaction to power life?

8. Explain the role of enzymes in biochemical reactions. Why do all life's reactions need enzymes?

9. Describe the formation and composition of a membrane. How does a membrane participate in endocytosis?

Clinical Correlation Questions

1. A malnourished cholera patient is given clean water to drink but fails to improve. The nurse considers what additional components may be needed, such as sugar for energy, antibiotics to kill the bacteria, or vitamins to reverse malnutrition. Which additional component of oral therapy is most needed, and why?

2. A patient is found in whom *Salmonella enterica* bacteria persist for several months in the digestive tract. By contrast, in cholera patients, *Vibrio cholerae* bacteria do not persist. Why not?

3. Researchers discover a new drug molecule that mimics the structure of an RNA nucleotide. How does this molecule shorten the course of influenza in a patient?

 For more assessment, visit digital.wwnorton.com/michum2

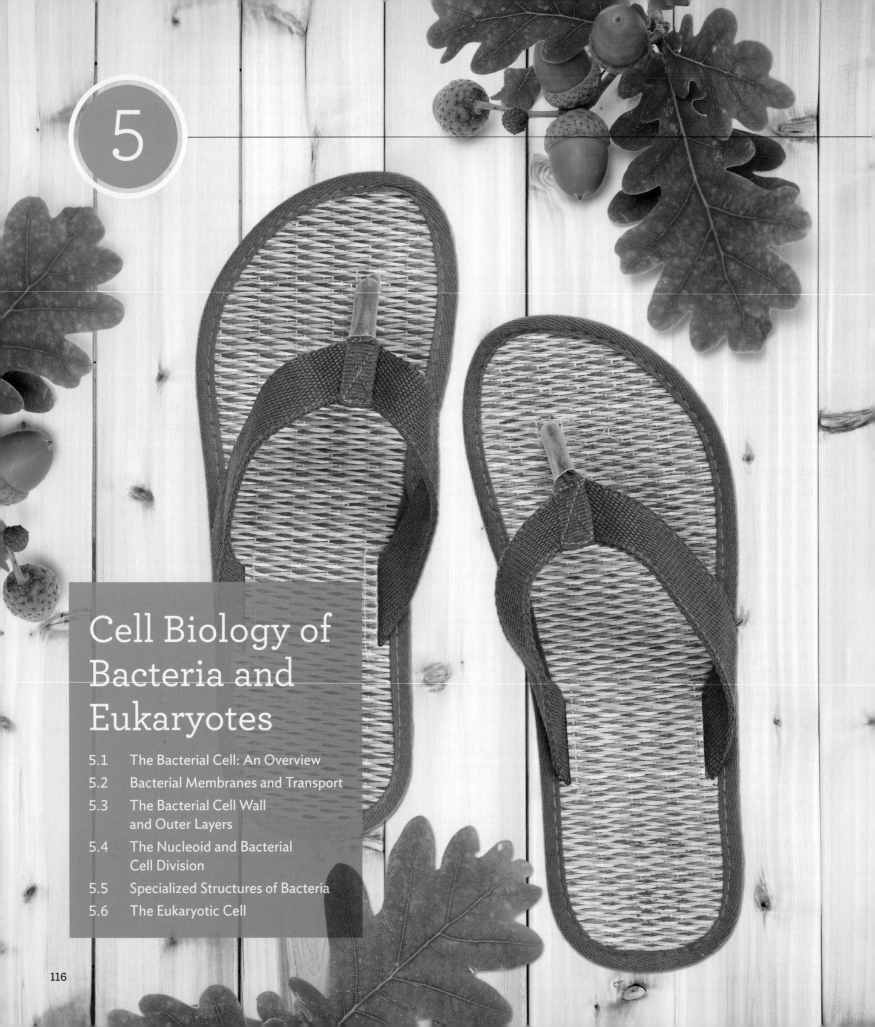

5

Cell Biology of Bacteria and Eukaryotes

SCENARIO Sharon, the CEO of a start-up company, lives in Westchester County, a wooded community north of New York City. She spends her summer weekends emailing her managers from the outdoor deck of her home, shaded by tall oak trees. The acorns attract mice and deer, and the leaf litter is full of ticks (*Ixodes scapularis*).

SIGNS AND SYMPTOMS One evening, Sharon's husband noticed a red rash on the back of her arm. The rash consisted of a ring shape several centimeters across, surrounding another red spot in the middle. Sharon recalled seeing this "bull's-eye" type of rash on the Internet. The rash was described as the hallmark of Lyme disease (Lyme borreliosis), which in the United States is caused by the tick-borne bacterium *Borrelia burgdorferi* (in Europe, by the closely related species *Borrelia afzelii*). The distinctive rash, erythema migrans ("migrating redness"), begins at the site of a tick bite and expands concentrically as the bacteria migrate outward. Many infected individuals, however, never see the rash. Sharon recalled that a neighbor's child had suffered crippling arthritis caused by an undetected case of Lyme disease. Another neighbor who contracted Lyme disease had suffered from meningitis (inflammation of the brain lining) and neurological abnormalities, including facial paralysis, ultimately losing his job and his million-dollar home.

DIAGNOSIS AND TREATMENT The next day, Sharon went to her doctor and was treated with the antibiotic doxycycline, a tetracycline derivative that targets the bacterial ribosome. Doxycycline is the antibiotic of choice for *B. burgdorferi*. Sharon was fortunate to make a full recovery before serious symptoms appeared.

Ixodes scapularis (nymphs; actual size)

Erythema Migrans
The migrating red rash of Lyme disease, caused by the spiral-shaped bacterium *Borrelia burgdorferi*.

Ixodes scapularis (adult females; actual size)

SEM   1 μm

Borrelia burgdorferi
The helical shape of *Borrelia burgdorferi* is typical of spirochetes.

CHAPTER OBJECTIVES

After reading this chapter, you will be able to:

- Describe the key parts of a bacterial cell.
- Explain the function of a bacterial cell's parts and their importance for antibiotic therapy.
- Describe the key parts of a eukaryotic cell, and explain how pathogens take advantage of these to infect a eukaryotic host.

Lyme disease (Lyme borreliosis) was first named for the town of Lyme, Connecticut, where an outbreak began in 1975; it is now the most common reported tick-borne disease in the Northern Hemisphere, with the heaviest concentration in New England. The causative agent, *Borrelia burgdorferi*, is transmitted to humans by the bite of an infected tick of the genus *Ixodes*. The *Ixodes* ticks are adapted to mice and deer, which maintain a permanent reservoir for the bacteria infecting humans—especially in locales such as Westchester, where hunting is restricted and deer have no natural predators. In Sharon's case, the rash was observed early and the pathogen was eliminated by antibiotics. In cases where the rash is not observed, the clinical laboratory can reveal the bacteria by detecting serum antibodies (antibodies produced by the patient in response to bacteria) or by PCR amplification of bacterial DNA.

The causative bacterium, *Borrelia burgdorferi*, was first identified in 1982 by Swiss entomologist Willy Burgdorfer (1925–2014). *B. burgdorferi* shows a flexible spiral shape, like a coiled audio cable. The complex cell form is called a "spirochete" (**Figure 5.1**). Bacterial flagella (rotary tails) propel the spirochete through body fluids. The spirochete's flagella do not extend from the cell as they do for other bacteria. Instead, they coil along the spiral cell body, where they are plugged into rotary motors embedded in the cell membrane. The membrane and flagella are surrounded by a

thick outer sheath (colorized brown in Figure 5.1). As the motors rotate, the helical flagella cause the entire cell body to flex in a corkscrew motion. The motion enables the spirochete to twist its way through host structures such as the capillary vesicles. This twisting motility helps *Borrelia* bacteria migrate outward from a tick bite, causing the red bull's-eye rash called erythema migrans.

In this chapter we describe the structural differences that distinguish prokaryotic cells (bacteria and archaea) from eukaryotic cells (such as plants and animals). The major difference between these two groups of organisms is the presence of a defined nucleus. Prokaryotic cells have no nucleus; bacterial DNA extends in loops throughout the cytoplasm. In contrast, the DNA of eukaryotic cells is packaged by a nuclear membrane. Some distinctive structures of bacteria, such as the peptidoglycan cell wall, enable us to use antibiotics without harming the human host.

5.1 The Bacterial Cell: An Overview

SECTION OBJECTIVES

- Describe the structure and function of the bacterial cell and how it differs from the cell of a eukaryote.
- Explain the biochemical composition of a bacterial cell.

How does a microbial cell survive assault by host defenses or the stress of extreme conditions in its natural environment? Bacterial cells build structures that protect the cell from what's outside. The bacterial cell wall acts like a molecular cage, protecting the cell membrane from osmotic shock. For this reason, the cell wall is a vital drug target; antibiotics such as penicillins and vancomycin block its synthesis. Structures in the bacterial cytoplasm also provides drug targets; for example, tetracyclines, aminoglycosides, and macrolides target the bacterial ribosome.

We distinguish two major classes of cells: **prokaryotes** and **eukaryotes**. Prokaryotic cells are cells of bacteria and archaea. Eukaryotic cells have their DNA in linear chromosomes enclosed by a nucleus. Eukaryotic cells include those of plants and animals (including humans), as well as fungi and protozoa. This chapter focuses on the fundamental traits of bacteria and eukaryotes

Figure 5.1 A Cell of *Borrelia* Bacteria

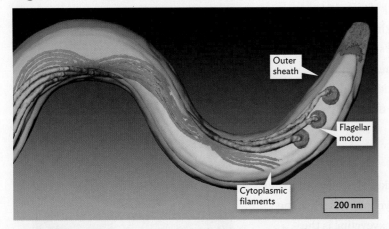

The spiral cell body is surrounded by a thick outer sheath (colorized brown). The cell membrane (colorized yellow), internal flagella (violet), cytoplasmic filaments (orange), and outer sheath (brown) are visualized by cryo-electron tomography.

Table 5.1
Bacteria and Eukaryotes: Cell Traits

Bacteria	Eukaryotes
Small cell size (usually 0.2–10 μm)	Wide range of cell size, from very small to very large (usually 0.2 μm–1 mm)
DNA organized in nucleoid throughout the cytoplasm	DNA contained in nucleus, enclosed by nuclear membrane
Small genome (0.5–15 million base pairs)	Wide range of genome size, including very large (0.5 million–20 billion base pairs)
Circular chromosome (usually), although may have multiple circular and linear chromosomes	Linear chromosomes (in nucleus); mitochondria (derived from bacteria) have one circular chromosome
Chromosomes replicate and segregate during cell growth	Chromosomes segregate by mitosis and meiosis after replication during interphase
Few intracellular membranes (such as thylakoids of photosynthetic bacteria)	Many types of intracellular membranous organelles (such as endoplasmic reticulum, Golgi, and lysosomes)
No intracellular endosymbiotic organelles	Mitochondria and chloroplasts, organelles that evolved from endosymbiotic bacteria
Cell wall composed of peptidoglycan	Cell walls of plants and fungi composed of various carbohydrates (such as cellulose or chitin), but never peptidoglycan
Rotary flagella for motility, driven by proton motive force	Whiplike flagella for motility, with microtubule contraction driven by ATP

(Table 5.1). Both kinds of cells show a wide range of form in different species; their diversity is explored in Chapters 10 and 11.

Most bacteria share the following traits:

- **Complex cell envelope.** The envelope (cell membrane, cell wall, and outer layers, if present) protects the cell from environmental stress and predators. The cell membrane allows a bacterium to secrete toxins as well as signaling molecules that communicate with fellow pathogens.

- **Small genome.** Prokaryotic genomes are compact, with relatively little noncoding DNA. Small genomes help a bacterium make as many offspring as possible from the limited resources of its host.

- **Tightly coordinated cell parts.** The cell's parts work together in a highly coordinated mechanism. Coordinated action helps a pathogen reproduce quickly, outracing the host's defensive response.

In the early twentieth century, the cell was envisioned as a bag of "soup" full of floating ribosomes and enzymes. Today, however, microscopy shows that the cell's parts fit together in a structure that is ordered, though flexible.

A Model of the Bacterial Cell

The model of the bacterial cell in **Figure 5.2** shows how the major components of a bacterial cell fit together. This model represents *Escherichia coli*, but its general features apply to many kinds of bacteria. Remember that we cannot actually "see" the molecules within a cell, but microscopy and subcellular analysis generate a remarkably detailed model.

Within the bacterial cell, the cytoplasm consists of a gel-like network of proteins and other macromolecules. The cytoplasm is surrounded by a **cell membrane** (also called a **plasma membrane** or **cytoplasmic membrane**). As discussed in Chapter 4, the membrane is composed of phospholipids, hydrophobic proteins, and small organic molecules. The cell membrane keeps cytoplasmic proteins from leaking out, allows nutrients to enter, and maintains concentration gradients of ions and nutrients. The cell membrane is also where the process of oxidative phosphorylation (respiration) pulls electrons from food molecules and uses that energy to pump hydrogen ions (H$^+$) across the membrane (discussed in Section 7.4). The resulting H$^+$ gradient then drives ATP synthesis by the ATP synthase complex (the knob-like structures shown in Figure 5.5).

Outside the bacterial membrane, the cell body is enclosed by a **cell wall**—a rigid, mesh-like structure that surrounds the cell like a cage (**Figure 5.3**). The molecule has been isolated from *E. coli* and visualized by TEM. In the image shown in Figure 5.3A, the cell wall appears flattened, like a deflated balloon. The wall consists of **peptidoglycan**, a material consisting of sugar chains linked covalently by peptides (as discussed in Section 5.3). In Gram-negative bacteria such as *E. coli*, the entire cell wall is a single layer of peptidoglycan. Gram-positive bacteria have multiple layers of peptidoglycan.

Outside the cell wall, Gram-negative bacteria have an **outer membrane**. Between the outer membrane and cell membrane (or "inner membrane") lies a region of water solution called the **periplasm**, containing proteins such as sugar transporters. The outer membrane consists of phospholipids and **lipopolysaccharides (LPS)**,

Figure 5.2   Model of a Bacterial Cell (*Escherichia coli*)

Envelope   Nucleoid   Ribosomes

TEM   0.25 μm

Ribosome

mRNA

50S

30S   Polypeptide

Flagellum

Flagellar motor

DNA-binding protein

RNA polymerase

DNA

RNA

Nucleoid

Envelope: The cell membrane contains embedded proteins for structure and transport. The cell membrane is supported by the cell wall. In this Gram-negative cell, the cell wall is coated by the outer membrane, whose sugar chain extensions protect the cell from attack by the immune system or by predators. The rotary motor of a flagellum is plugged into the envelope.
Cytoplasm: Molecules of messenger RNA (mRNA) extend out of the nucleoid to the region of the cytoplasm rich in ribosomes. Ribosomes translate the mRNA to make proteins, which are folded into their active form by chaperones (proteins that assist folding).
Nucleoid: The chromosomal DNA is wrapped around binding proteins. Replication by DNA polymerase and transcription by RNA polymerase occur at the same time within the nucleoid.

a class of lipids attached to long polysaccharides (sugar chains). When a bacterium is killed by antibiotics, its LPS may be released as an **endotoxin**. Endotoxin is dangerous to the human host, causing toxic shock and possible death.

For the intact bacterium, LPS and other polysaccharides can generate a thick **capsule** surrounding the cell. The capsule polysaccharides form a slippery mucus layer that inhibits phagocytosis by white blood cells (WBCs). Pathogens stripped of their capsule can become vulnerable to our body's defenses.

The cell membrane, cell wall, and outer membrane (for Gram-negative species) constitute the **cell envelope**. The envelope includes cell-surface proteins that enable the bacteria to interact with the surfaces of specific host organisms. For example, some *E. coli* cell-surface proteins enable colonization of the human intestinal epithelium, whereas *Sinorhizobium* cell-surface proteins enable colonization of legume plant roots. *Sinorhizobium* bacteria then capture nitrogen from the air and provide it to the plant in a process called nitrogen fixation.

ATP synthase

Lipopolysaccharide

Outer membrane

Cell wall

Periplasm

Inner membrane (cell membrane)

Envelope

Ribosome

Peptide

RNA

RNA polymerase

DNA

DNA-bridging protein

DNA-binding protein

Cytoplasm

Nucleoid

40 nm

Figure 5.3 The Peptidoglycan Cell Wall

TEM   1 μm

A. Entire cell wall isolated from *Escherichia coli*.

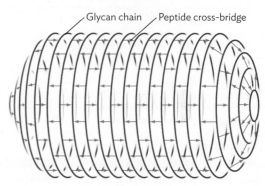

Glycan chain   Peptide cross-bridge

B. The cell wall consists of glycan chains (parallel rings) linked by peptides (arrows). The spaces between links are open, porous to large molecules.

⊙ **Note:** In some texts, the term "envelope" refers only to structures outside the cell membrane. This book considers the cell membrane (or Gram-negative inner membrane) to be part of the bacterial envelope.

Within the cell cytoplasm, the cell envelope provides an attachment point for one or more chromosomes. The chromosome is organized within the cytoplasm as a system of looped coils called the **nucleoid**. Unlike the round, compact nucleus of eukaryotic cells, the bacterial nucleoid is not enclosed by a membrane, so the coils of DNA can extend throughout the cytoplasm. Loops of DNA from the nucleoid are transcribed by the enzyme RNA polymerase to form messenger RNA (mRNA). As the bacterial mRNA transcripts grow—and before they are complete—they already bind ribosomes to start synthesizing polypeptide chains (proteins). This bacterial cell organization differs greatly from that of a eukaryotic cell, in which most mRNA molecules are completed and leave the nucleus before translation (see Section 5.6).

Another common external structure is the **flagellum** (plural, **flagella**), a helical protein filament whose rotary motor (seen in Figure 5.2) propels the cell in search of a more favorable environment. For *E. coli* and other bacteria, the flagella extend freely outside the cell. For spirochete bacteria such as *Borrelia burgdorferi* (the cause of Lyme disease) or *Treponema pallidum* (the cause of syphilis), the flagella wrap around the cell body, enclosed by the outer sheath (see Figure 5.1).

Table 5.2
Molecular Composition of a Bacterial Cell, *Escherichia coli*, during Balanced Exponential Growth[a]

Component	Percentage of Total Weight[b]	Approximate Number of Molecules per Cell	Number of Different Kinds
Water	70%	20,000,000,000	1
Proteins	16%	2,400,000	2,000[c]
RNA			
rRNA, tRNA, and other small regulatory RNAs (sRNAs)	6%	250,000	200
mRNA	0.7%	4,000	2,000[c]
Lipids			
Phospholipids (membrane)	3%	25,000,000	50
Lipopolysaccharide (outer membrane)	1%	1,400,000	1
DNA	1%	2[d]	1
Metabolites and biosynthetic precursors	1.3%	50,000,000	1,000
Peptidoglycan (murein sacculus)	0.8%	1	1
Inorganic ions	0.1%	250,000,000	20

[a]Values shown are for a hypothetical "average" cell cultured with aeration in glucose medium with minimal salts at 37°C (98.6°F).
[b]The total weight of the cell (including water) is about 10^{-12} gram (g), or 1 picogram (pg).
[c]The number of kinds of mRNA and proteins is difficult to estimate because some genes are transcribed at extremely low levels and because some RNA and proteins are rapidly degraded.
[d]In rapidly growing cells, cell fission typically lags approximately one generation behind DNA replication; hence there are two identical DNA copies per cell.

Biochemical Composition of Bacteria

Our model of the bacterial cell shows shape and size but tells little about chemical composition. All cells share the common chemical components introduced in Chapter 4:

- Water, the fundamental solvent of life
- Inorganic ions, such as potassium, magnesium, and chloride ions
- Small organic molecules, such as lipids and sugars, that are incorporated into cell structures and that provide nutrition by chemical reactions
- Macromolecules, such as nucleic acids and proteins, which contain information needed to make cells, catalyze reactions, and mediate transport, among many other functions

Cell composition varies with species, growth phase, and environmental conditions. Table 5.2 lists the chemical components of a cell for the model bacterium *Escherichia coli* during exponential growth.

SMALL MOLECULES The *E. coli* cell consists of about 70% water. Inorganic ions such as potassium (K^+) store energy in the form of a transmembrane gradient (concentration difference across the membrane). Other ions serve essential roles in enzymes;

for example, RNA polymerase requires a magnesium ion (Mg^{2+}) at its active site to help extend an RNA chain.

The cell also contains many kinds of small charged organic molecules, such as phospholipids and enzyme cofactors. Organic cations include **polyamines**, molecules with multiple amine groups that are positively charged. Polyamines balance the negative charges of the cell's DNA and stabilize ribosomes during translation. Some bacteria digest protein and release polyamines with names such as "cadaverine" and "putrescine"; these polyamines cause the odors of decaying flesh or meat.

MACROMOLECULES A specific type of cell requires specific kinds of macromolecules (long-chain molecules), especially its proteins. Proteins vary among different species, and even a single species will make very different proteins when its environment changes. For example, a pathogen such as *Yersinia pestis* (the cause of bubonic plague) will make different proteins as it moves from the cold flea host to the warm human host and when it encounters an increase or decrease in nutrients such as iron. Individual proteins, encoded by specific genes, are found in very different amounts, from 10 per cell to 10,000 per cell. The total proteins encoded by a genome and capable of expression in the cell are known collectively as the **proteome**. The collection of

Figure 5.4 A Toxin Efflux Complex in *E. coli*

Model of the TolC efflux protein complex, based on X-ray crystallography of isolated components. AcrA, AcrB, and TolC are protein subunits of the complex.

proteins is unique to each bacterial strain and culture conditions, such as the presence of an antibiotic. Proteome analysis is being developed to diagnose the drug resistance of pathogens infecting a patient, such as methicillin-resistant *Staphylococcus aureus* (MRSA).

Electron microscopy largely defines how we "see" the cell's interior as a whole (discussed in Chapter 3). A closer look at any of these components requires breaking the cell into parts. The cell parts include protein complexes we can study by tools such as X-ray crystallography. For example, X-ray crystallography was used to reconstruct a model of the *E. coli* TolC protein complex embedded in the bacterial envelope (**Figure 5.4**). A TolC complex forms a tube that exports some toxins and antibiotics out of the bacterial cytoplasm and across the cell membrane (inner membrane), periplasm, and outer membrane (discussed in the next section). The proteins show alpha helix and beta sheet structures, as introduced in Chapter 4.

SECTION SUMMARY

- **Bacterial cells** are protected by a thick cell envelope. The envelope can protect pathogens from phagocytosis by macrophages.
- **The cell envelope includes the cell membrane and cell wall.** In addition, the Gram-negative cell envelope includes an outer membrane outside the cell wall.

- **The nucleoid** has loops of DNA that extend throughout the bacterial cell. In the growing bacterial cell, DNA replication, RNA transcription, and protein synthesis occur together.
- **Proteins in the cell vary**, depending on the species and environmental conditions.
- **The biochemical composition of bacteria** includes nucleic acids, proteins, phospholipids, and other molecules. The chemistry of these molecules determines a pathogen's ability to cause disease.

Thought Question 5.1   Which molecules occur in the greatest number in a prokaryotic cell? The smallest number? Why does a prokaryotic cell contain 100 times as many lipid molecules as strands of RNA?

5.2 Bacterial Membranes and Transport

SECTION OBJECTIVES

- Describe the functions of a cell membrane.
- Explain how nutrients are transported and how energy is spent to drive transport.
- Describe how pathogens use ion gradients and ATP to obtain nutrients from the host.

Which structure defines the existence of a cell? The defining structure is the cell membrane (**Figure 5.5**). As explained in Chapter 4, the membrane separates cytoplasm from the external

Figure 5.5 Bacterial Cell Membrane and ATP Synthase

The cell membrane consists of a phospholipid bilayer, with hydrophobic fatty acid chains directed inward, away from water. Embedded proteins can transport nutrients. The proton motive force drives H^+ across the membrane through protein complexes such as ATP synthase, which makes ATP. P_i = inorganic phosphate.

environment and mediates transport between the two. The cell membrane is a phospholipid bilayer containing proteins. Membrane proteins transport molecules in and out of the cell. Other membrane proteins detect information from outside the cell, such as the level of iron in the blood of a host.

Bacterial Membrane Proteins

The cell membrane is a two-dimensional fluid that contains many floating proteins. The portion of the protein contained within the membrane must be hydrophobic (soluble in lipid), whereas portions extending into the cytoplasm are hydrophilic (water-soluble). Membrane proteins determine the capabilities of the cell, such as:

- **Structural support.** Some membrane proteins anchor together different layers of the cell envelope. Other proteins form the base of structures that extend out from the cell, such as pili and flagella (discussed in Section 5.5), which enable adherence and motility.

- **Detecting environmental signals.** In *Vibrio cholerae*, the causative agent of cholera, ToxR is a transmembrane protein whose amino-terminal part reaches into the cytoplasm. When ToxR detects acidity and elevated temperature—signs of the host digestive tract—its amino-terminal domain binds to a specific sequence of DNA. The binding of ToxR to the control sequence activates expression of cholera toxin and other virulence factors (molecules involved in disease).

- **Secreting virulence factors and communication signals.** Some membrane proteins form secretion complexes to export toxins. Other membrane proteins enable cell signaling across the envelope. For example, symbiotic nitrogen-fixing bacteria use membrane proteins to signal their host plant roots to form nodules. The nodules incorporate the nitrogen-fixing bacteria into the plant cells, where they receive nutrients and provide the plant with fixed nitrogen.

- **Selective transport across the cell membrane.** Membrane proteins determine which substances move between the cytoplasm and the outside. The proteins take up scarce nutrients and exclude harmful toxins.

- **Energy storage and transfer.** The membrane allows the cell to maintain different concentrations of molecules inside and outside the cell (a concentration gradient). The cell stores energy in these gradients, and membrane proteins harness that energy. For example, the TolC-AcrAB complex (see Figure 5.4) uses energy from a H^+ gradient to pump antibiotics out of the cell.

Transport of Nutrients

Whether a microbe is propelled by flagella toward a favorable habitat or, lacking motility, drifts through its environment, it must find nutrients and move them across the membrane into the cytoplasm. A few compounds, such as oxygen and carbon dioxide, can passively diffuse across the membrane, but most cannot. Nutrients may cross a membrane by binding a transporter protein, called a **permease**, that deposits the nutrient on the other side of the membrane. Alternatively, nutrients such as peptides and sugars may cross a membrane through a protein channel.

Microbes must also overcome the problem of low nutrient concentrations in the natural environment (for example, in lakes or streams). If the intracellular concentration of nutrients were no greater than the extracellular concentration, the cell would remain starved of most nutrients. To solve this dilemma, most organisms have evolved efficient transport systems that concentrate nutrients inside the cell relative to outside. However, to move molecules against a concentration gradient requires energy.

Some habitats have plenty of nutrients, but those nutrients may be locked in a form that cannot be transported into the cell. Starch is an example of a large, complex nutrient that cannot be transported across a membrane. To obtain complex nutrients, microbes secrete digestive enzymes that break them down into smaller compounds that are easier to transport.

PASSIVE AND ACTIVE TRANSPORT   A few nutrients, such as O_2, can be taken across a membrane by passive transport. **Passive transport** uses the concentration gradient of a compound to drive that compound across the membrane (from higher to lower concentration). The rate of passive transport can be increased by a membrane protein that carries the compound; this process is called **facilitated diffusion**. For most nutrients, however, the cell has to take up molecules present at low concentration outside the cell and concentrate them inside. Transport against a concentration gradient (that is, from lower to higher concentration) must spend energy. Transport that spends energy is called **active transport**. The energy spent can take the form of an energy carrier molecule, such as ATP, or an energy potential across the membrane, such as the **proton motive force** (the force driving H^+ across the membrane, discussed shortly). In either case, energy is spent to bring compounds into the cell.

The simplest way to use energy to move molecules across a membrane is to exchange the energy of one chemical gradient for that of another. The most common chemical gradients used are those of ions, particularly the positively charged ions Na^+ and K^+. These ions are kept at different concentrations on either side of the cell membrane. When an ion moves down its concentration gradient (from high to low), energy is released. Some transport proteins harness that released energy and use it to drive the transport of a second molecule up, or against, its concentration gradient. The use of energy from one gradient to drive transport up another gradient is called **coupled transport**.

Two types of coupled transport systems are **symport**, in which the two molecules travel in the same direction, and **antiport**, in which the actively transported molecule and the driving ion move

Figure 5.6 Coupled Transport

Symport

1 Energy is released as one substrate (red) moves down concentration gradient.

2 This energy moves a second substrate (blue) against its gradient and into the cell.

Outside

Inside

Antiport

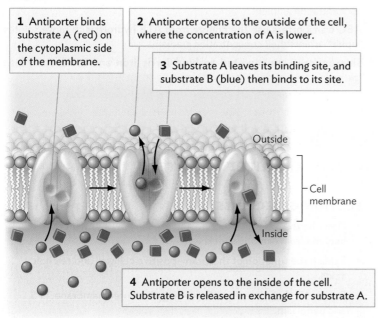

1 Antiporter binds substrate A (red) on the cytoplasmic side of the membrane.

2 Antiporter opens to the outside of the cell, where the concentration of A is lower.

3 Substrate A leaves its binding site, and substrate B (blue) then binds to its site.

Outside

Cell membrane

Inside

4 Antiporter opens to the inside of the cell. Substrate B is released in exchange for substrate A.

The energy released by substrate A (red) traveling down its gradient drives uptake of substrate B (blue) against its gradient, with the two substrates moving either in the same direction (symport) or in opposite directions (antiport).

in opposite directions (**Figure 5.6**). An example of a symporter is the lactate/H⁺ symporter of the food pathogen *Campylobacter jejuni*, commonly acquired from poultry. This symporter brings the food molecule lactate into the cell, coupled to H^+ moving inward down its gradient (symport). The lactate/H^+ symporter helps the pathogen rapidly colonize its host.

TRANSPORT POWERED BY ATP The largest family of energy-driven transport systems is the ATP-binding cassette superfamily, also known as **ABC transporters**. Some ABC transporters pick up nutrients and take them into the cell. For example, *Mycobacterium tuberculosis* needs an ABC transporter to pick up iron during growth in the lungs. Others, called **efflux transporters**, expel hazardous wastes. The efflux ABC transporters are generally used as multidrug efflux pumps that help microbes expel antibiotics. *Lactococcus*, for example, can use one pump, LmrP, to export many types of antibiotics, such as tetracyclines, quinolones, macrolides, and aminoglycosides.

An ABC transporter (**Figure 5.7**) consists of two proteins that form a membrane channel and two cytoplasmic proteins that contain a conserved amino acid motif that binds ATP (called an ATP-binding cassette). An additional membrane protein, called a **substrate-binding protein**, initially binds the substrate (also called a solute). The substrate-binding protein then delivers its substrate

Figure 5.7 ABC (ATP-Binding Cassette) Transporters

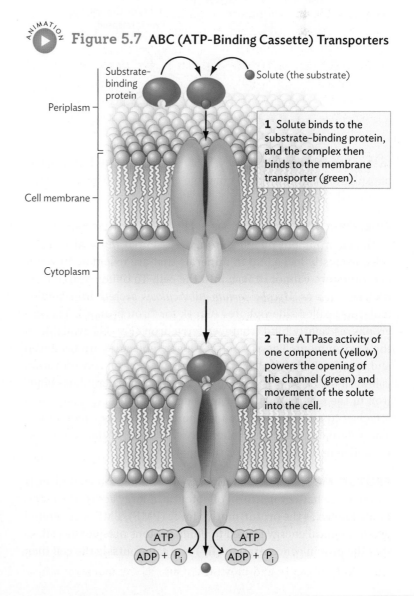

Substrate-binding protein

Solute (the substrate)

Periplasm

Cell membrane

Cytoplasm

1 Solute binds to the substrate-binding protein, and the complex then binds to the membrane transporter (green).

2 The ATPase activity of one component (yellow) powers the opening of the channel (green) and movement of the solute into the cell.

ATP   ATP

ADP + P$_i$ ADP + P$_i$

Figure 5.8 Iron Transport by a Siderophore and an ABC Transport Complex

1 Bacterium secretes siderophore that binds Fe^{3+}.

Siderophore

2 ABC transporter brings Fe^{3+} across the membrane.

Cell membrane

ATPase

ATP   ATP

ADP + P_i ADP + P_i

3 Inside the cell, the iron is released.

Siderophore

to the channel protein. Hydrolysis of ATP to ADP plus P_i (inorganic phosphate) yields energy to open the channel, allowing the solute to enter the cell.

SIDEROPHORES Minerals needed by cells are often inaccessible in the environment outside. Iron, an essential nutrient of most cells, is mostly locked up in minerals that are insoluble in water and therefore cannot diffuse into the cell. To obtain scarce iron, bacteria such as *Mycobacterium tuberculosis* secrete iron-binding molecules called siderophores (Greek for "iron bearer"). The siderophores tightly bind whatever ferric iron (Fe^{3+}) is available in the environment. These iron scavenger molecules are produced and sent forth by cells when the intracellular iron concentration is low. In the example shown in **Figure 5.8**, the siderophore binds iron in the environment, and the siderophore-iron complex then attaches to specific receptors in the bacterial outer membrane. The siderophore-Fe^{3+} complex may then be transported across the cell membrane by an ABC transporter.

PROTON MOTIVE FORCE Energy from a cell's metabolism is stored like a battery in a form of potential energy across the membrane known as the proton motive force (PMF). PMF is generated when chemical energy is used to pump protons outside the cell, so that the proton concentration (H^+) is greater outside the cell than

inside. Because protons are positively charged, proton movement across the cell membrane produces an electrical gradient, making the inside of the cell more negatively charged than the outside. The energy stored in the proton motive force (the H^+ concentration difference plus the charge difference) can be used to move nutrients into the cell, to drive motors that rotate flagella, and to drive synthesis of ATP (see Figure 5.5). The proton-powered ATP synthase is a target for antibiotics; for example, the diarylquinoline antibiotics target the ATP synthase of *Mycobacterium tuberculosis*. Proton motive force and energy use are discussed further in Chapter 7.

SECTION SUMMARY

- **The bacterial cell membrane** contains proteins for functions such as transport and cell communication.
- **Active transport requires input of energy** from a chemical reaction or from an ion gradient across the membrane. Ion gradients generated by membrane pumps store energy for cell functions.
- **Antiporters and symporters are coupled transport systems** in which energy released by moving an ion from a region of high concentration to one of low concentration is used to move a desired solute against its concentration gradient.
- **ABC transporters use the energy from ATP hydrolysis** to move solutes "uphill" against their concentration gradients.
- **Siderophores are secreted to bind ferric iron** and transport it into the cell.

5.3 The Bacterial Cell Wall and Outer Layers

SECTION OBJECTIVES

- Describe the cell wall structure and explain how it protects bacteria from osmotic shock.
- Explain the function of the Gram-positive cell wall and teichoic acids.
- Explain the function of the Gram-negative outer membrane, LPS, and periplasm.

How do bacteria protect their cell membranes from falling apart under pressure? Most bacteria have at least one structural supporting layer outside the cell membrane: the cell wall. A few classes of bacteria lack cell walls—most notably the mycoplasmas that cause pneumonia, and the chlamydias that cause the blinding eye disease trachoma and sexually transmitted infections. Mycoplasmas and chlamydias lost their cell walls through reductive, or degenerative, evolution. Bacteria that do possess cell walls often have additional coverings outside, such as the outer membrane of Gram-negative bacteria (as seen in Figure 5.1).

The Cell Wall Is a Single Molecule

The cell wall makes a bacterial cell rigid and helps it withstand the intracellular turgor pressure that can build up as a result of osmotic shock, a sudden increase in osmotic pressure (see Section 4.5). The cell wall is the target of penicillin, the first antibiotic discovered. The cell wall remains one of the most important antibiotic targets.

The bacterial cell wall consists of a single interlinked molecule that encloses the entire cell (Figure 5.9). Unlike the cell membrane, the cell wall is highly porous to ions and small organic molecules. Most bacterial cell walls are composed of peptidoglycan, also called murein. Peptidoglycan consists of parallel polymers of disaccharides called **glycan** chains. The glycan chains are cross-linked with short peptides. Peptidoglycan is unique to bacteria, never found in human cells. For this reason, antibiotics that target peptidoglycan synthesis typically have minimal side effects for a patient.

The long glycan chains of peptidoglycan consist of repeating units of two amino sugars: *N*-acetylglucosamine (NAG) and

Figure 5.9 The Cell Wall Consists of Peptidoglycan

A disaccharide unit of glycan has an attached peptide of four to six amino acids. The peptide bonds with *N*-acetylmuramic acid. On the peptide, the extra amino group of *m*-diaminopimelic acid can cross-link to the carboxyl terminus of a neighboring peptide. The addition of D-alanine to the peptide is blocked by vancomycin, and the cross-bridge formation by transpeptidase is blocked by penicillin.

impact

Vancomycin: The Gift of Soil Bacteria

Vancomycin, which targets the bacterial cell wall, is the antibiotic of last resort for deadly infections such as pseudomembranous enterocolitis, a common hospital-acquired infection caused by the Gram-positive bacterium *Clostridioides difficile*, and abscesses caused by methicillin-resistant *Staphylococcus aureus* (MRSA). Deadly MRSA infections can arise from hospital complications or from a seemingly innocuous sports-related injury, such as a scrape on the gym floor. Both *C. difficile* and MRSA are resistant to other drugs and therefore require treatment with last-resort drugs such as vancomycin. Where did vancomycin originate—and where will other new drugs be found?

It may come as a surprise that most of the antibiotics we use against bacteria today were discovered out in the environment, in soil bacteria. A handful of soil is more than dirt; it contains one of the most complex natural environments (Figure 1A). Soil combines minerals from powdered bedrock with decaying leaves (detritus) and animal bodies, plus the waste products of many living organisms. These materials provide a habitat for countless species of microbes, including fungi, protists, and bacteria, as well as viruses that infect them. Any single sample of soil contains species of microbes unknown to science. And unknown microbes may produce unknown antibiotics.

Why do microbes produce antibiotic molecules? The limited nutrients of soil lead to a fierce competition. And some bacteria play a particularly devious trick. The actinomycetes (discussed in Chapter 10) are a group of bacteria that grow in the soil as long, fungus-like filaments of cells (Figure 1B). Actinomycetes have large genomes that encode many enzymes for digesting many kinds of organic molecules—including the complex constituents of dead fellow bacteria. As an actinomycete filament grows new cells at one end, the older cells at the other end age and die, lysing to release their contents. The lysed cell contents attract unsuspecting soil bacteria to come and feed. But then the newer actinomycete cells release antibiotics that kill the feeding bacteria—so instead, the actinomycete feeds on them.

Figure 1
Soil Bacteria

A. Each handful of soil contains many species of microbes unknown to science.

B. Filaments of the actinomycete *Amycolatopsis*, which produces vancomycin.

SEM   0.5 μm

Of course, the other soil bacteria evolve resistance to the actinomycete antibiotics; so, actinomycete bacteria must continually evolve new ones. We can discover these antibiotics by isolating and culturing new species of actinomycetes. In 1953, a missionary in Borneo sent a sample of jungle soil to Eli Lilly and Company, a pharmaceutical firm. At Lilly, microbiologist Edmund Kornfeld (1919–2012) isolated a species, now called *Amycolatopsis orientalis*, that produced a new antibiotic molecule (**Figure 2**). The molecule was called vancomycin because it vanquishes bacteria resistant to older antibiotics such as penicillins and metronidazole. Vancomycin binds to the terminal alanine dipeptide of a growing cell wall, preventing cross-bridge formation (see Figure 5.9). The structure of vancomycin is so complex that no organic chemist would have invented it; yet bacteria in nature continually evolve by natural selection to produce such new molecules.

But just as previous strains of *S. aureus* mutated and evolved resistance to methicillin, the previous wonder drug, today some MRSA strains have already grown resistant to vancomycin. So we must continue to discover new antibiotics.

DISCUSSION QUESTION

How would you go about discovering a new antibiotic? What kinds of natural environments, besides soil, might contain microbes that produce undiscovered antibiotics?

Figure 2 Vancomycin

N-acetylmuramic acid (NAM). Parallel glycan strands are linked by short peptide cross-bridges containing four to six amino acid residues. The sequence of the cross-bridge-forming peptide is L-alanine, D-glutamic acid, *m*-diaminopimelic acid, and D-alanine (sometimes two of these). Note that this peptide contains two amino acids in the unusual D mirror form (discussed in chapter 4): D-glutamic acid and D-alanine. This means the peptide cannot be formed by a ribosome but requires special enzymes for synthesis. These wall-building enzymes are targets for antibiotics.

The L-alanine of a peptide cross-bridge (Figure 5.9) is attached to NAG on one glycan strand, whereas the second amino group of *m*-diaminopimelic acid attaches to D-alanine from a neighboring peptide on a parallel glycan chain (attached to NAM). Formation of a cross-link removes the second D-alanine (fifth in the chain). The cross-linked peptides of neighboring glycan strands form the cage of the peptidoglycan.

The details of peptidoglycan structure vary among bacterial species. Some Gram-positive species, such as *Staphylococcus aureus* (a cause of toxic shock syndrome), have peptides linked by bridges of pentaglycine instead of the D-alanine link to *m*-diaminopimelic acid. In Gram-negative species, the diaminopimelic acid is linked to the outer membrane, as discussed shortly.

Does the cell wall alone determine the cell shape? In addition to the cell wall, bacteria, like eukaryotic cells, possess a **cytoskeleton**. The cytoskeleton consists of cytoplasmic proteins that mold the cell into a form that is straight or curved, even helical. The bacterial proteins show intriguing homology (genetic similarity) to proteins of the better-known eukaryotic cytoskeleton (discussed in Section 5.6).

Peptidoglycan Synthesis as a Target for Antibiotics

Peptidoglycan requires many different enzymes to build the special sugars and peptides and to seal the cross-bridges. These biosynthetic enzymes make great targets for antibiotics (see Chapter 13). For example, the transpeptidase that cross-links the peptides is the target of penicillin (Figure 5.9). Vancomycin, a major defense against *Clostridioides difficile* and drug-resistant staphylococci, prevents cross-bridge formation by binding the terminal D-Ala-D-Ala dipeptide, thus preventing release of the terminal D-alanine.

Unfortunately, the widespread use of such antibiotics selects for evolution of resistant strains. One of the most common agents of resistance is the enzyme beta-lactamase, which cleaves penicillin, preventing it from inhibiting transpeptidase. Strains resistant to vancomycin contain an altered enzyme that adds lactic acid to the end of the branch peptides in place of the terminal D-alanine. The cross-bridge formation is no longer blocked by vancomycin. Vancomycin is now the drug of last resort for MRSA.

Most of our antibiotics, including vancomycin, were actually discovered in harmless environmental bacteria, particularly actinomycete bacteria growing in soil (see **IMPACT**). As new forms of drug resistance emerge, researchers continue to seek new antibiotics that target cell wall formation.

Gram-Positive and Gram-Negative Outer Layers

Outside the cell membrane, most bacteria have additional layers of cell envelope—materials that provide structural support and protection from predators and host defenses. Outer structures are attached to the cell wall and cell membrane, in some cases interpenetrating them. These structures define two major categories of bacteria distinguished by the Gram stain (discussed in Section 3.4):

- **Gram-positive** bacteria have a thick cell wall with multiple layers of peptidoglycan, threaded by teichoic acids (**Figure 5.10A**). Gram-positive species include species in the phylum Firmicutes. The Firmicutes include many pathogens, such as *Bacillus anthracis*, the cause of anthrax, and *Streptococcus pyogenes*, the cause of "strep throat."

- **Gram-negative** bacteria have a thin cell wall (one or two layers of peptidoglycan) enclosed by an outer membrane (**Figure 5.10B**).

Figure 5.10   Gram-Positive and Gram-Negative Cell Envelopes

A. The Gram-positive cell has a thick cell wall with multiple layers of peptidoglycan, threaded by teichoic acids. The cell wall may be covered by an S-layer (surface array of proteins). In some Gram-positive species, carbohydrate filaments form a capsule. The micrograph shows the Gram-positive envelope of *Bacillus subtilis*.

B. The Gram-negative cell has a single layer of peptidoglycan covered by an outer membrane, which includes lipopolysaccharide (LPS). The cell membrane of Gram-negative species is called the inner membrane. The micrograph shows the Gram-negative envelope of *Pseudomonas aeruginosa*. Some Gram-negative cells have an S-layer attached outside the outer membrane, as shown.

Gram-negative species include species in the phylum Proteobacteria, such as *Escherichia coli* and nitrogen-fixing rhizobia. Among the many Gram-negative pathogens is the species *Pseudomonas aeruginosa*, a cause of wound infections.

Besides these two groups, however, many kinds of bacteria do not fit the Gram stain models. For example, the mycobacteria (discussed shortly) have exceptionally complex cell walls containing waxy outer layers that exclude antibiotics.

THE GRAM-POSITIVE CELL ENVELOPE A section of a Gram-positive cell envelope is shown in Figure 5.10A. The Gram-positive cell wall consists of multiple layers of peptidoglycan, up to 40 in some species. The peptidoglycan is reinforced by **teichoic acids** threaded through its multiple layers. Teichoic acids are chains of phosphodiester-linked sugar alcohols such as glycerol or ribitol. The phosphodiester links are deprotonated (ionized) and thus have a negative charge. The negatively charged cross-threads of teichoic acids, as well as the overall thickness of the Gram-positive cell wall, help retain the positively charged crystal violet of the Gram stain. Teichoic acids act as "bacterial signature" molecules that alert the immune system of an impending infection.

Outside the cell wall, Gram-positive cells may have a slippery capsule consisting of loosely bound polysaccharides (not shown in the diagram). The capsule can be visualized under the light microscope by using a negative stain consisting of suspended particles of India ink, which the capsule excludes.

An additional protective layer commonly found in many bacterial cells is the surface layer, or **S-layer**. The S-layer is an important virulence factor for pathogens, such as *Bacillus anthracis*, the cause of anthrax. Pathogens may need S-layer proteins to adhere to host tissues. The S-layer is a crystalline layer of thick subunits consisting of protein or glycoprotein (proteins with attached sugars). Each subunit of the S-layer contains a pore large enough to admit a wide range of molecules. The subunits form a smooth layer on the cell wall or outer membrane (see Figure 5.10). The proteins are arranged in a highly ordered array that can exclude predators and parasites. The S-layer is permeable, allowing substances to pass through it in either direction.

Many bacterial species have lost parts of their ancestral outer layers through evolution. A trait can be lost if natural selection for the genes encoding it is absent. This evolutionary loss is called **reductive evolution** or degenerative evolution (see Chapter 9). For example, the mycoplasmas that cause pneumonia have permanently lost their cell walls. Mycoplasmas are close relatives of Gram-positive bacteria, but they have no need for cell walls because they are parasites living in host environments, such as the human lung, where they are protected from osmotic shock. Pathogens that have an S-layer, such as *Bacillus anthracis*, may lose the genes to synthesize S-layer proteins when cultured in the laboratory.

THE GRAM-NEGATIVE OUTER MEMBRANE A Gram-negative cell envelope is seen in Figure 5.10B. The thin cage of peptidoglycan

consists of only one or two layers, much thinner than in Gram-positive bacteria. But a Gram-negative bacterium has an extra layer, the outer membrane. The outer membrane confers defensive abilities and toxigenic properties on many pathogens, such as *Salmonella* species and enterohemorrhagic *E. coli* (strains that cause hemorrhaging of the colon). Between the outer and inner (cell) membranes is the periplasm. The periplasm contains distinct proteins not found in the cytoplasm, such as proteins that protect the cell from dangerous oxidants (molecules that strip electrons) produced by WBCs.

The inward-facing leaflet of the outer membrane has a phospholipid composition similar to that of the cell membrane. (In Gram-negative species, the cell membrane is called the **inner membrane** or inner cell membrane.) The outer membrane's inward-facing leaflet includes lipoproteins that connect the outer membrane to the peptide bridges of the cell wall. The major lipoprotein is called **murein lipoprotein** (Figure 5.11).

Figure 5.11 Lipoprotein and Lipopolysaccharide in the Gram-Negative Envelope

Murein lipoprotein consists of a protein embedded in the inward-facing leaflet of the outer membrane. The protein's C-terminal lysine bonds with the peptidoglycan (murein) cell wall. GlcNAc = N-acetylglucosamine; MurNAc = N-acetylmuramic acid; m-A$_2$pm = m-diaminopimelic acid.

Murein lipoprotein consists of a protein inserted in the inward-facing leaflet of the outer membrane that forms a peptide bond with the *m*-diaminopimelic acid of peptidoglycan (murein).

The outward-facing leaflet, however, consists of a special kind of membrane lipid called **lipopolysaccharide**, or **LPS** (Figure 5.11). LPS is of crucial medical importance because its lipid A component acts as an **endotoxin**. An endotoxin is a cell component that is harmless as long as the pathogen remains intact. When released by a lysed cell, however, the endotoxin (lipid A) overstimulates host defenses, inducing potentially lethal endotoxic shock (dangerously low blood pressure and high fever). So, antibiotic treatment of an LPS-containing pathogen can kill the cells but can also cause a reaction that kills the patient.

The LPS has three parts: the **lipid A** anchored in the outer membrane, a core polysaccharide, and the O antigen (O polysaccharide). The lipid A in LPS has shorter fatty acid chains than those of the inner cell membrane, and some are branched. Lipid A is attached to a core polysaccharide, a sugar chain that extends outside the cell. The core polysaccharide consists of about five sugars with side chains. It extends to the O antigen, a polysaccharide chain of as many as 200 sugars. The O antigen polysaccharide may extend longer than the cell itself. These polysaccharide chains form a layer that helps a pathogen resist phagocytosis by white blood cells.

The outer membrane also contains unique proteins not found in the inner membrane. Outer membranes contain a class of transporters called **porins** that permit entry of nutrients such as sugars and peptides. Cells express different outer membrane porins under different environmental conditions. In a dilute environment, cells express porins of large pore size, maximizing the uptake of nutrients. In a rich environment—for example, within a host—cells down-regulate the expression of large porins and express porins of smaller pore size, selecting only smaller nutrients and avoiding the uptake of toxins. For example, the porin regulation system of Gram-negative bacteria enables them to grow in the intestinal region containing bile salts—a hostile environment for Gram-positive bacteria.

Some Gram-negative species may also have an S-layer of proteins attached to the outer membrane. However, during lab culture the energy cost of synthesizing an S-layer leads to its rapid loss by reductive evolution. We do not know which selective pressures maintain the S-layer of bacteria in natural environments.

The Mycobacterial Envelope

Mycobacteria include important pathogens, such as *Mycobacterium tuberculosis* (the cause of tuberculosis) and *Mycobacterium leprae* (the cause of the disfiguring skin disease leprosy). These bacteria are considered Gram-positive, but the mycobacterial envelope is exceptionally thick and complex, including extra layers not found in other Gram-positive cells.

CASE HISTORY 5.1

A Rash Reveals an Ancient Disease

In the Indian state of Bihar, a school nurse noticed that a 13-year-old boy, Naranjan, had swellings on his nose and lips. When questioned, Naranjan complained of a pruritic (itchy) rash on his arms and legs. Upon examination, the nurse found pale patches of skin on his back. The pale patches lacked sensation, and there was partial loss of sensation in Naranjan's wrists and forearms. Naranjan was diagnosed with leprosy, caused by the bacterium *Mycobacterium leprae*. The bacteria stain acid-fast. They are spread primarily by nasal secretions. For unknown reasons, only 5% of people exposed to the bacteria are susceptible to infection. In a susceptible host, the bacteria can grow slowly for many years without symptoms. The bacteria infect the peripheral nerves, causing loss of sensation, and the body's immune reaction generates skin lesions.

To halt the disease, Naranjan was started on a 2-year course of multidrug therapy combining dapsone, clofazimine, and rifampin. To monitor hepatotoxicity, liver function tests were ordered. Naranjan's family and neighbors were screened; one member, an uncle, showed similar skin lesions. The uncle received treatment along with Naranjan. Leprosy is difficult to eradicate because of the long incubation period and because people hide their symptoms, fearing stigma as "lepers."

Figure 5.12A shows typical facial lesions of leprosy. The bacteria need to grow at a temperature lower than that controlled by the

Figure 5.12 Leprosy Is Caused by *Mycobacterium leprae*

A. The characteristic skin lesions of leprosy take many years to develop.

B. Thermal image of heat radiating from a face shows cooler areas of the nose and cheeks (blue) where leprosy lesions most commonly form.

body, so lesions typically occur in superficial parts of the body with lower temperature, as shown by the infrared scan in **Figure 5.12B**. However, leprosy can develop in various forms in different parts of the body, often mistaken for other conditions. In most cases, lesions develop over many years because the bacteria grow so slowly; in laboratory culture, cell doubling may take as long as 6 weeks. Transmission is rare because only 5% of the human population is susceptible; thus, most industrial countries have eliminated leprosy simply by treating all known cases. But in societies where leprosy is well established, and where people hide their symptoms, the disease is hard to eradicate.

Why do *M. leprae* and *M. tuberculosis* grow so slowly? The mycobacterial envelope is unusually thick, with several extra layers (**Figure 5.13**). The envelope enables identification by the acid-fast stain (see Chapter 3). The outermost layer of capsule is a "waxy" coating (with benzene-like aromatic rings) that is extremely hydrophobic and retards phagocytosis by lymphocytes. The waxy coating excludes many antibiotics and offers exceptional protection from host defenses, enabling the pathogens of tuberculosis and leprosy to colonize their hosts over long periods. However, the thick, waxy layer also retards nutrient uptake. As a result, *M. tuberculosis* and *M. leprae* grow extremely slowly. Complex cell walls are also found in the related group of actinomycetes, soil bacteria that produce many important antibiotics, such as streptomycin and vancomycin.

Mycobacterial cell envelopes include unusual lipids called **mycolic acids** (see Figure 5.13). Mycolic acids contain a backbone with two hydrocarbon chains—one comparable in length to typical membrane lipids (about 20 carbons), the other about threefold longer. Mycolic acids provide the basis for acid-fast staining, an important diagnostic test for mycobacteria and actinomycetes (described in Section 3.4). In *M. tuberculosis* and *M. leprae*, the mycolic acids form a kind of bilayer interleaved with "waxy" phenolic glycolipids. The mycolic acids and peptidoglycan are linked to chains of arabinogalactan, composed of galactose and the five-carbon sugar arabinose. Ethambutol, a major drug against tuberculosis, can inhibit arabinogalactan biosynthesis.

Other drugs effective against mycobacteria include dapsone (which inhibits synthesis of nitrogenous bases for DNA), clofazimine (which binds to guanine and blocks DNA synthesis), and rifampin (which inhibits synthesis of messenger RNA). Development of such drugs requires understanding the cellular processes they target, as introduced in this chapter.

Figure 5.13 Mycobacterial Envelope Structure

A complex cell wall includes a peptidoglycan layer linked to a chain of galactose polymer (galactan) and arabinose polymer (arabinan). Arabinan forms ester links to mycolic acids, which form an outer bilayer with phenolic glycolipids. Outside the outer bilayer is a capsule of loosely associated phospholipids and phenolic glycolipids.

SECTION SUMMARY

- **The cell wall maintains turgor pressure.** The cell wall is porous, but its rigid network of covalent bonds protects the cell from osmotic shock.

- **The Gram-positive cell envelope** has multiple layers of peptidoglycan, interpenetrated by teichoic acids.

- **The S-layer**, composed of proteins, is highly porous but can prevent phagocytosis.

- **The capsule**, composed of polysaccharide and glycoprotein filaments, protects cells from phagocytosis. Both Gram-positive and Gram-negative cells may possess a capsule.

- **The Gram-negative outer membrane** regulates nutrient uptake and excludes toxins. The envelope layers include protein pores, transporters, and lipopolysaccharides (LPS). Lysed cells release LPS as endotoxin.

- **Mycobacteria have an exceptionally thick and complex cell envelope, including mycolic acids.** The mycobacterial envelope excludes most antibiotics and prevents phagocytosis, but it also slows nutrient entry, leading to slow growth.

Thought Question 5.2   Why do antibiotics differ in their effectiveness against Gram-positive and Gram-negative bacteria?

Thought Question 5.3   A patient with Gram-negative bacteremia (bacterial infection of the blood) was treated with an antibiotic and experienced septic shock. Why did this happen?

5.4 The Nucleoid and Bacterial Cell Division

SECTION OBJECTIVES

- Describe how DNA is organized within the bacterial cell.
- Explain how DNA replication is coordinated with cell growth and division.

How does a bacterium protect its genomic DNA? An important function of the cell envelope is to contain and protect the cell's genome. **Figure 5.14** shows the organization of chromosomal material in enteropathogenic *E. coli* (EPEC) cells that are adherent and invading a HeLa cell in human tumor tissue culture. Thin-section transmission electron microscopy (TEM) was performed as discussed in Chapter 3. For the image in

Figure 5.14 Bacterial Nucleoid and Eukaryotic Nucleus

Enteropathogenic *Escherichia coli* (EPEC) bacteria, infecting HeLa cells (cultured human cancer cells). The bacterial DNA is loosely arranged in the nucleoid region.

Figure 5.14, each bacterium was sliced through, revealing a filamentous region of DNA that extends throughout the cytoplasm. This DNA-containing region is called the **nucleoid**. In the image shown, each bacterium either adheres to the host human cell or is engulfed by it. Only a fraction of the much larger human cell is visible.

What is the physiological and reproductive state of a bacterial cell? A bacterium may be growing and reproducing new cells, or it may be nongrowing. A nongrowing cell may exist in a "stationary" state, metabolically active without expanding its cytoplasm; alternatively, its metabolism may be suspended completely, as in the case of a *Bacillus anthracis* spore that can persist for decades in the soil. But for bacteria, a growing cell is always a dividing cell. Unlike the cells of eukaryotes—which undergo mitotic division at a time separate from cell expansion—bacterial cells grow and divide as a continuous process. In a growing bacterial cell, DNA replicates in coordination with expansion of the cell wall and, ultimately, separation of the cell into two daughter cells.

Here we describe the organization of genetic material within a bacterial cell. The genetic processes involving DNA, RNA, and protein are discussed further in Chapter 8.

Bacterial DNA Is Organized in a Nucleoid

All living cells on Earth possess chromosomes consisting of DNA. In bacteria and archaea, the DNA genome usually consists of a single circular, ringlike chromosome. However, some bacterial species have a linear chromosome or multiple chromosomes. In this chapter we focus on the simple case of a single circular chromosome.

The bacterial nucleoid has no membrane (as for a eukaryotic nucleus), but it does have structural organization (**Figure 5.15**). The nucleoid contains loops of DNA held together by DNA-binding proteins. All the loops connect back to a central point called the **origin of replication (*ori*)**, which is attached to the cell envelope at a point on the cell's "equator," halfway between the two poles.

The chromosome includes several extra twists, called supercoils or superhelical turns, beyond those inherent in the structure of the DNA duplex (double helix). The supercoiling causes portions of DNA to double back and twist upon itself, like a twisted garden hose. The extra supercoils compact the DNA. Supercoiling is generated by enzymes such as DNA gyrase and maintained by DNA-binding proteins (shown in green in Figure 5.15B). The enzyme gyrase is a major target for antibiotics such as quinolones, which are used, for example, to treat bacterial pneumonia and urinary tract infections.

The information encoded in DNA is "read" by the processes of transcription and translation to yield gene products: RNA and protein (see Chapter 8). As bacteria grow, both transcription and translation occur at top speed while the DNA itself is being replicated. This remarkable coordination of replication, transcription, and translation is a hallmark of the bacterial cell; it explains why some bacteria can double their population size in as little as 10 minutes.

Bacterial Cell Division

Most bacteria have a circular chromosome that begins to replicate at its origin, proceeding in both directions (bidirectionally) all around the circle. At the origin, the DNA double helix begins to unzip, forming two replication forks (see Chapter 8). At each replication fork, DNA polymerase synthesizes DNA (**Figure 5.16**). Growing bacteria replicate their DNA continuously, with no resting phase. Note that the cell must expand its cytoplasm at the same time that it synthesizes DNA. In a rod-shaped cell, the cell envelope and cell wall must elongate as well to maintain progeny of even girth and length.

Once DNA has been replicated completely to form two daughter genomes, the cell may divide to form two daughter cells. Replication of the DNA termination site triggers growth of the dividing partition of the envelope, called the **septum** (plural, **septa**). The septum grows inward from the sides of the cell, at last constricting and sealing off the two daughter cells in the process of **septation**. Envelope extension and septation require rapid biosynthesis of all envelope components, including membranes and cell wall. The pattern of septation differs among bacterial species, causing differences in the appearance of cells that provide clues to diagnosis.

Figure 5.15 **The Nucleoid**

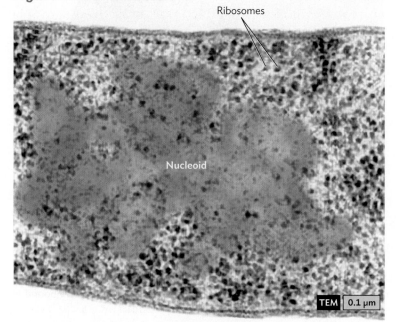

A. The *E. coli* nucleoid (colorized violet) appears as clear regions that exclude the dark-staining ribosomes and contain DNA strands.

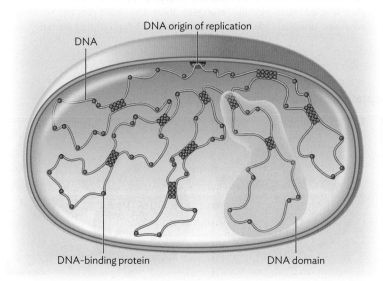

B. The nucleoid forms approximately 50 loops of chromosome called domains (shaded yellow), which radiate from the center. Within each domain, the DNA is supercoiled and partly compacted by DNA-binding proteins (shown in green).

Figure 5.16 **Replisome Movement within a Dividing Cell**

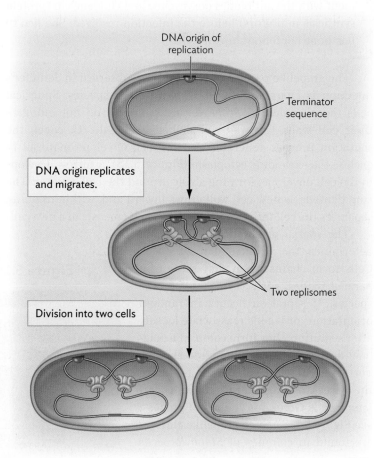

The DNA origin-of-replication sites (green) move apart in the expanding cell as the replisomes (DNA polymerase complexes) proceed in opposite directions around the circular chromosome until they meet at the terminator sequence. Replication of the termination sequence triggers septation.

CASE HISTORY 5.2
Deadly Infection Acquired from a Hospital

 Jennifer, age 23, had just graduated from college with a Fulbright fellowship to study sociology in Africa. Before leaving the United States, she underwent a routine physical exam at a large city hospital. Upon returning home from the hospital, she developed a swelling in her leg. When she went back to the hospital, she was told that the swelling indicated an allergic reaction and was given anti-inflammatory agents. The swelling grew, and within a day the skin ruptured with a bloody discharge (Figure 5.17A). This time when she returned to the emergency room, Jennifer at last received the correct diagnosis: methicillin-resistant staphylococcus infection (MRSA; Figure 5.17B). A nurse commented that hospital visitors can acquire MRSA; the infectious agent is endemic at many hospitals in the United States and is very difficult to eradicate. She showed Jennifer a micrograph of *Staphylococcus aureus* bacteria, which septate (divide) in alternating division planes, forming clusters of cells. Jennifer required many weeks of treatment with the antibiotics doxycycline and clindamycin (protein synthesis inhibitors) before the infection resolved. She had to postpone her fellowship for a year.

The grapelike pattern of cells that was observed in Jennifer's infection is characteristic of *Staphylococcus* species. Spherical cells (cocci), such as *Staphylococcus aureus*, do not elongate their cell walls the way rod-shaped bacilli do. In cocci, the septation process generates the new cell envelope material to enclose the expanding cytoplasm (Figure 5.18). A furrow develops in the cell envelope, forming a ring around the cell equator. The ring tightens as new cell wall grows inward and, within minutes, separates the cell into two halves. Each half consists of a new cell.

The spatial orientation of septation determines the arrangement of cocci. Cells form chains when they septate in parallel planes, such as in *Streptococcus* species. If septation occurs in random orientations or if cells reassociate loosely after septation, they form compact hexagonal arrays similar to the grape clusters portrayed in classical paintings—hence the Greek-derived term **staphylococci** (*staphyle* is Greek for "bunch of grapes"; see Figure 5.17B). Such grapelike clusters are found in colonies of *Staphylococcus aureus*. In other species, where septation occurs at precise right angles to the previous division, the cells form tetrads (groups of four) and even cubical groups

Figure 5.17 Methicillin-Resistant *Staphylococcus aureus* (MRSA) Infection

A. Abscess from patient infected with MRSA.

B. MRSA cells. Note the clustering of cells that divide in different planar directions.

of eight. *Micrococcus* bacteria, commonly found in dust and in the air we breathe, form tetrads.

SECTION SUMMARY

- **The nucleoid region contains loops of DNA**, supercoiled and bound to DNA-binding proteins.
- **Bacterial DNA is transcribed to RNA in the cytoplasm**, often simultaneously with DNA replication.
- **DNA is replicated bidirectionally by DNA polymerase.**
- **Cell expansion and septation are coordinated with DNA replication.** Rod-shaped cells (bacilli) elongate their cell walls before septation, whereas spherical cells (cocci) expand their cell walls during septation.

 ### Figure 5.18 Septation in *Staphylococcus aureus*

A. Furrows appear in the cell envelope all around the cell equator as a new cell wall grows inward.

B. Two new envelope partitions are complete.

C. The two daughter cells peel apart. The facing halves of each cell contain entirely new cell wall.

5.5 Specialized Structures of Bacteria

SECTION OBJECTIVES

- Describe how pili and stalks enable bacteria to adhere to a substrate where conditions are favorable.
- Explain how flagellar motility and chemotaxis enable bacteria to respond to environmental change.
- Describe the functions of thylakoids, storage granules, and magnetosomes.

Do bacteria have special structures for special tasks? The cell envelope, the nucleoid, and the gene expression complexes are needed for all bacterial cells. But other structures are needed to colonize different habitats, from the human colon to the hot springs of Yellowstone. In addition to the fundamental structures shared by all bacteria, different species have evolved specialized devices adapted to specific metabolic strategies and environments. Particularly important are structures for attachment and motility.

Pili and Stalks Enable Attachment

In a favorable habitat, such as a running stream full of fresh nutrients or the epithelial surface of a host, it is advantageous for a cell to adhere to a substrate such as a rock or an intestinal epithelial cell. **Adherence** is the ability to attach to a substrate, using specific structures such as pili (protein filaments). As bacteria grow and proliferate, however, they face the dilemma of whether to stay where nutrients are declining or leave their present habitat and seek a better one. In rapidly changing environments, cell survival requires **motility**, the ability to move and relocate. Motility requires structures such as rotary flagella. Some bacteria, such as the sewage bacterium *Caulobacter crescentus*, have it both ways by producing two kinds of progeny: one adherent, the other motile.

The most common structures that bacteria use to attach to a substrate are **pili** (singular, **pilus**); short attachment pili are called **fimbriae** (singular, **fimbria**). Pili are straight filaments of protein monomers called **pilin**. For example, pili attach the oral pathogen *Porphyromonas gingivalis* to gum epithelium, where they are associated with periodontal disease (**Figure 5.19**). A different kind of pili, the **sex pili**, attach a "male" donor cell to a "female" recipient cell before transferring DNA in a process called conjugation.

A different kind of bacterial attachment organelle is a membrane-enclosed extension of the cytoplasm called a **stalk**. The tip of the stalk secretes adhesion factors called **holdfasts**, which firmly attach the bacterium in an environment that has proved favorable. The mechanism of stalk and holdfast attachment has been

Figure 5.19 Pili: Protein Filaments for Attachment

TEM   200 nm

Porphyromonas gingivalis, a causative agent of gum disease, or gingivitis. The *P. gingivalis* cells show fimbriae (pili), along with vesicles budding from the cell's outer membrane.

extensively studied in iron-oxidizing bacteria such as *Gallionella ferruginea*, which clog iron mines with massive biofilms of adherent cells.

Rotary Flagella Enable Motility and Chemotaxis

Bacteria that are motile generally swim using rotary **flagella** (singular, **flagellum**; **Figure 5.20A**). Flagella are helical propellers that drive the cell forward, like the motor of a boat. Different bacterial species have different numbers and arrangements of flagella. Flagella enhance virulence; for example, *Proteus* species use flagella to swim up the urethra and infect the bladder and kidneys.

Each flagellum is a spiral filament of protein monomers called flagellin. The filament is actually rotated by means of a rotary motor (**Figure 5.20B**). The flagellar motor is embedded in the layers of the cell envelope and is powered by the cell's proton motive force (PMF). Flagellar motility benefits the cell by dispersing progeny and thus decreasing competition. In addition, most flagellated cells have an elaborate sensory system, known as **chemotaxis**, that enables them to swim toward favorable environments (attractant signals, such as nutrients) and away from inferior environments (repellent signals, such as waste products).

Figure 5.20 Flagella on *Salmonella* Bacteria

A. *Salmonella enterica* has multiple flagella (colorized TEM).

B. Micrograph and diagram of the bacterial flagellar motor, including major protein components.

Figure 5.21 Chemotaxis: Swimming toward Attractants

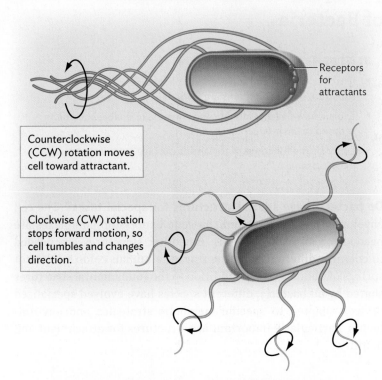

Counterclockwise (CCW) rotation moves cell toward attractant.

Clockwise (CW) rotation stops forward motion, so cell tumbles and changes direction.

Receptors for attractants

A. Multiple flagella are oriented in a bundle extending behind one pole, whereas their chemotactic receptors are concentrated at the opposite pole. When the cell veers away from the attractant, the receptors send a signal that enables one or more flagella to switch rotation from counterclockwise (CCW) to clockwise (CW). This switched rotation disrupts the bundle of flagella, causing the cell to tumble briefly before it swims off in a new direction.

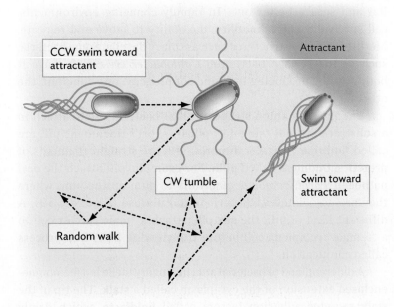

CCW swim toward attractant

Attractant

CW tumble

Swim toward attractant

Random walk

B. The pattern of movement resulting from alternating swimming and tumbling is a "biased random walk" in which the cell moves randomly but overall tends to migrate toward the attractant.

Chemotaxis requires a way for the cell to move toward attractants (positive chemotaxis) and away from repellents (negative chemotaxis). Counterclockwise rotation of the flagella propels the cell forward; clockwise rotation propels it backward (**Figure 5.21**). When a cell is swimming toward an attractant chemical, the flagella rotate counterclockwise (CCW) and form a rotating helical bundle. The cell swims smoothly for a long stretch. When the cell veers away from the attractant, receptors send a signal that enables one or more flagella to switch rotation clockwise (CW), against the twist of the helix. This switch in the direction of rotation disrupts the bundle of flagella, causing the cell to tumble briefly and end up pointed in a random direction. The cell then swims off in the new direction. The resulting pattern of movement generates a "biased random walk" in which the cell tends to migrate toward the attractant.

Note that bacterial flagella differ completely from the whiplike flagella and cilia of eukaryotes, to be discussed in Section 5.6.

Figure 5.22 Structures for Different Habitats

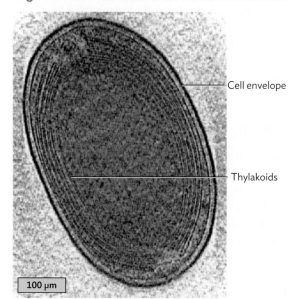

A. Photosynthetic thylakoid membranes of *Prochlorococcus*, a genus of the smallest and most abundant marine phototrophic bacteria.

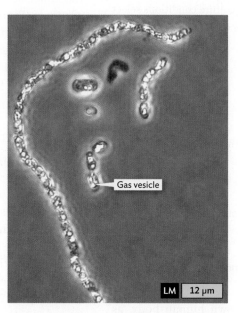

B. Gas vesicles enable *Halochromatium* bacteria to stay at the surface of salt water, where they absorb light.

C. Magnetosomes (crystals of magnetite, colored red) enable a magnetotactic anaerobic bacterium to swim downward in the pond, along magnetic field lines.

Bacterial Structures for Different Habitats

Other structures that perform specialized functions for bacteria include photosynthetic membranes for light absorption and even magnetic structures for orientation of motility in a magnetic field (Figure 5.22). Here are some examples:

- **Light-harvesting membranes that conduct photosynthesis.** Photosynthetic bacteria (phototrophs) need to collect as much light as possible to drive photosynthesis. To do this, they grow extensively folded intracellular membranes called thylakoids. The thylakoids contain chlorophyll and other proteins that absorb light energy and convert it into the chemical energy needed by the cell.

- **Gas vesicles.** Aquatic phototrophs possess gas vesicles to increase buoyancy and keep themselves high in the water column, near the sunlight. The vesicles trap and collect gases such as hydrogen and carbon dioxide produced by the cell's metabolism.

- **Storage granules.** Some bacteria store carbon and energy in storage granules composed of glycogen or other polymers called polyhydroxyalkanoates (PHAs). PHA polymers are of commercial interest as a biodegradable plastic, which bacteria are engineered to produce industrially.

- **Sulfur granules.** Some sulfur-metabolizing bacteria deposit granules of solid sulfur within the cytoplasm or as "globules" attached outside the cell. The presence of toxic sulfur granules may help these cells avoid predation.

- **Magnetosomes that direct motility.** Some anaerobic pond-dwelling bacteria can swim along a magnetic field in a process called magnetotaxis. Magnetotactic species of bacteria possess tiny magnets called **magnetosomes**. Magnetosomes are microscopic membrane-enclosed crystals of the magnetic mineral magnetite. Because Earth's magnetic field lines point downward in the northern latitudes, magnetotactic bacteria swim "downward" toward magnetic north. Thus, magnetosome-equipped anaerobes swim toward the bottom of the pond, where oxygen concentration is low.

- **Endospores.** Some bacteria, mainly *Bacillus* and *Clostridium* species, form heat- and chemical-resistant dormant structures called endospores. These will be described in Section 6.6.

SECTION SUMMARY

- **Adherence structures** such as pili, or fimbriae (protein filaments), and the stalk (a cell extension) enable prokaryotes to remain in a favorable environment.

- **Flagellar motility** of bacteria involves rotary motion of helical flagella.

- **Chemotaxis** involves a biased random walk up a gradient of an attractant substance or down a gradient of a repellent.

- **Phototrophs** possess thylakoid membranes packed with photosynthetic apparatus.

- **Storage granules** store polymers for energy. Sulfur granules store solid sulfur.

- **Magnetosomes** orient the swimming of magnetotactic anaerobic bacteria.

Thought Question 5.4   Why would laboratory culture conditions select for the evolution of cells lacking an S-layer? Or for cells lacking flagella?

5.6 The Eukaryotic Cell

SECTION OBJECTIVES

- Explain the structure and interconnection of membranous organelles in the endomembrane system.
- Describe the functions of the Golgi complex, endoplasmic reticulum, and nuclear membrane.
- Explain how the evolutionary process of endosymbiosis led to mitochondria and chloroplasts.
- Describe the eukaryotic cytoskeleton.
- Describe the motility of eukaryotic flagella and cilia.

Eukaryotic cells include the cells of our own human body, as well as the cells of all other animals, plants, and fungi. They also include single-celled parasites that cause disease (such as malaria and giardiasis). How do eukaryotic cells compare with the cells of bacteria? Most well-known eukaryotic cells are much larger than bacteria; for instance, some amebas (alternative spelling, "amoebas") are visible to the unaided eye. The eukaryotic cells of animal bodies can contain a millionfold larger volume than that of a typical bacterium. A mammalian neuron (**Figure 5.23A**) can possess dendrites that extend for several centimeters and an axon that extends over meters.

Yet DNA-based surveys of natural environments, such as the ocean and soil, reveal other microbial eukaryotes as small as the smallest bacteria. An example is the marine alga *Ostreococcus tauri* (**Figure 5.23B**). *O. tauri* is barely longer than an *E. coli* cell, yet it manages to squeeze within its tiny volume all the major eukaryotic organelles: nucleus, mitochondria, endoplasmic reticulum (ER), and Golgi. So, eukaryotic cells have sizes over the range encompassed by bacteria and archaea, but they also reach much larger sizes.

Organelles of the Eukaryotic Cell

Compared with prokaryotic cells, the larger eukaryotic cells have much more extensive intracellular membrane systems. These membranes form various organelles consisting of vesicles and folded pockets of membrane (**Figure 5.24**). Membranous organelles provide many advantages to the eukaryotic cell. Membrane folds increase the membrane surface area without increasing the cell volume, thus expanding the area for membrane-embedded complexes such as the electron transport system for respiration (see Chapter 7). Organelles composed of membranes enable different reactions to occur simultaneously under very different conditions. For example, proteins can be assembled by ribosomes in the cytoplasm at the same time that proteins are being broken down within the acidic lysosomes. Compartmentalization also protects cytoplasmic components from harmful substances. For instance, organelles called peroxisomes generate hydrogen peroxide (H_2O_2) to detoxify harmful substances. Localizing the reaction within the peroxisome keeps the toxic hydrogen peroxide away from other cell components, such as proteins and DNA, that are sensitive to oxidative stress.

DYNAMIC TRANSPORT BETWEEN MEMBRANOUS ORGANELLES   As we saw in Section 4.6, a eukaryotic cell can take up larger objects such as prey bacteria by pinching in their cell membrane to enclose the object. The process of membrane invagination to form vesicles is called endocytosis (for small particles) or phagocytosis (for large particles or bacteria). Endocytosis is a controlled, energy-requiring process that relies on many proteins, including cytoskeletal proteins. Human neutrophils and macrophages (white blood cells of the immune system) use phagocytosis to engulf pathogenic bacteria for digestion by lysosomes (see Section 4.6). However, pathogenic bacteria such as *Mycobacterium tuberculosis* or viruses such as influenza can subvert phagocytosis or endocytosis to gain entry and multiply within a cell.

Figure 5.23 Eukaryotic Cells

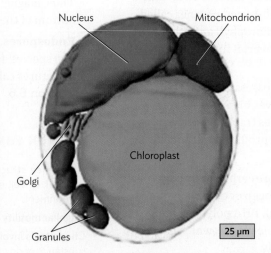

A. Rat spinal neuron, stained green with fluorescent anti-neurofilament antibody. Blue stain shows DNA; orange stain shows axons of surrounding neurons. The cell body is approximately 100 μm across; dendrites may extend several millimeters.

B. *Ostreococcus tauri*, a marine microbe barely bigger than *E. coli*, visualized by cryo-EM tomography.

What happens to endocytic vesicles (endosomes) or phagosomes within the cell? Endosomes normally fuse with a **lysosome**, a membranous organelle containing enzymes that kill bacteria and process organic material for nutrition (**Figure 5.25**). Lysosomal vesicles ultimately fuse with the plasma membrane for **exocytosis**. Exocytosis is a process by which vesicle fusion forms an opening to release the vesicle contents outside the cell.

The major system of intracellular transport is the **endoplasmic reticulum (ER)**. The ER is continuous with the outer nuclear membrane, and the **lumen** (interior) of the ER is continuous with the space between the two nuclear membranes (see **Figure 5.26**). This means that the contents of membrane compartments can mix by fusion of vesicles. Thus, material in the ER does not need to cross a membrane to enter these other compartments. For example, fusion between the membranes of ER and the Golgi complex enables material contained within the ER to mix with the contents of the Golgi complex (Figure 5.24).

The ER is used to sequester substances that must be held at low concentrations in the cytoplasm (for example, calcium ions for nerve signals). These substances can be transferred by vesicles to the Golgi. Golgi form vesicles for exocytosis. Exocytosis can mediate chemical communication; releasing neurotransmitters across a synapse to a receiving neuron is one example.

THE ENDOMEMBRANE SYSTEM The ER forms a transport network among several kinds of membrane compartments that are collectively termed the **endomembrane system** (Figure 5.26). Organelles of the endomembrane system include the endoplasmic reticulum (ER), the Golgi complex (also called Golgi apparatus), lysosomes, and peroxisomes, as well as the outer membrane of the nucleus. Different organelles contain unique subsets of proteins that contribute to their function. Their membranes are distinct from one another and from the cell membrane, although vesicles may pinch off one organelle and dissolve into the membrane of another. The function of the ER and the Golgi complex is to direct proteins to their proper cellular location, such as lysosomes, the cell membrane, or vesicles for secretion from the cell.

Figure 5.24 Membranous Organelles of a Eukaryotic Cell

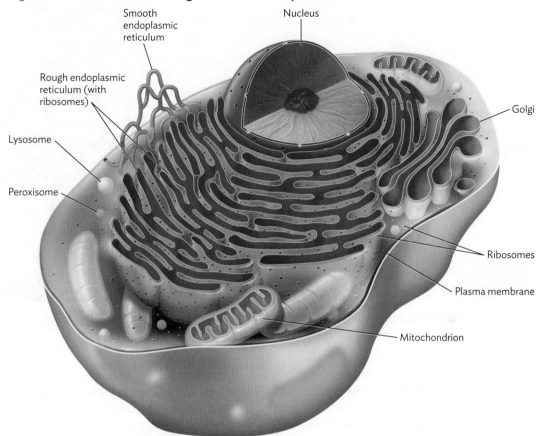

Organelles of the endomembrane system include the smooth endoplasmic reticulum (smooth ER), rough ER (ER with attached ribosomes), lysosomes, and Golgi complex. Other membranous organelles include the nucleus and the mitochondria.

Figure 5.25 Cellular Trafficking

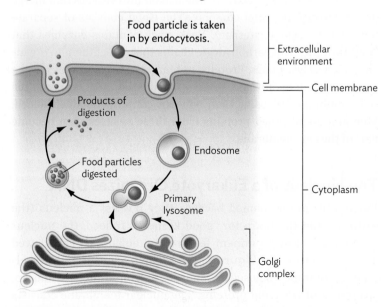

Lysosomes fuse with endosomes, providing enzymes to digest material brought into the cell by endocytosis.

Figure 5.26 The Endomembrane System

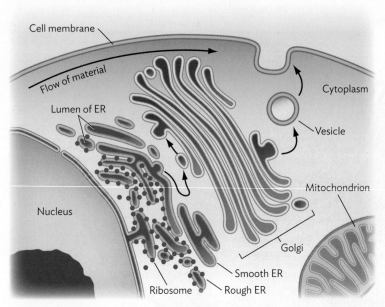

A. The relationship of the ER to other cellular membranes, and the flow of material through vesicles from the rough ER to the cell membrane.

B. Cell section showing the chromosomal DNA, nuclear pores, ER, and Golgi.

A special category of endoplasmic reticulum is the "rough ER," so called because of the "rough" appearance of the ribosome-studded membrane surface. The rough ER is where ribosomes translate RNA sequence into membrane-embedded proteins and secreted proteins. Secreted proteins and proteins destined for the lumen of an organelle are threaded completely through the ER membrane and into the lumen of the ER. Membrane-embedded (transmembrane) proteins have peptide threaded back and forth across the membrane, while portions usually extend outside.

Proteins to be secreted are pinched off into vesicles that merge into the **Golgi**, or Golgi complex. The Golgi consists of separate membrane stacks (cisternae). Each stack contains enzymes that "tag" proteins with carbohydrate chains that direct them to organelles such as lysosomes. Proteins not targeted to lysosomes may be sent to the cell membrane. Vesicles leaving the Golgi fuse with the cell membrane, releasing their aqueous contents outside the cell. Membrane-embedded proteins from these vesicles then become part of the cell membrane.

The Nucleus of a Eukaryote Organizes DNA

Eukaryotes are so named because they possess a nucleus (the word is from the Latin for "good kernel"). Indeed, the nucleus is often the most prominent feature of a eukaryotic cell viewed under a microscope (**Figure 5.27A**). The nucleus is an intracellular membrane-enclosed compartment containing chromatin, a complex of DNA and proteins. The nuclear membrane consists of two concentric phospholipid membranes. The outer nuclear

membrane is continuous with the endomembrane ER, and the space between the two nuclear membranes is continuous with the lumen (inside) of the ER folds (**Figure 5.27B**). Nuclear function is important for infection by viruses, such as herpesvirus and human immunodeficiency virus (HIV), which must manipulate nuclear processes to replicate their own genomes.

The nuclear membrane contains **nuclear pore complexes** that allow for transport of material into and out of the nucleus. Metabolites and small proteins can diffuse through the nuclear pores, but larger proteins and organelles require active transport. These selectively imported proteins contain a nuclear localization signal, a sequence of amino acids that acts like a zip code to direct them through a nuclear pore into the nucleus. Besides importing proteins, nuclear pores export mRNAs to the cytoplasm. But viruses such as herpes can "hijack" a nuclear pore to inject the viral genome into the nucleus.

Within the nucleus, a structure called the **nucleolus** (plural, **nucleoli**) assembles ribosomes. At the nucleolus, multiple genes encoding ribosomal RNA (rRNA) are transcribed, and the resulting rRNA combines with ribosomal proteins imported into the nucleus from the cytoplasm to form the ribosomal subunits. Unlike bacteria, in eukaryotic cells most mRNA molecules are completed and transported outside the nucleus for translation by ribosomes. Eukaryotic ribosomes are found in the cytoplasm or on the surface of ER membranes ("rough" ER).

In eukaryotes, unlike bacteria, cells divide at a different time from cell growth and gene expression. Eukaryotic transcription and translation occur only during interphase, an interval of the

Figure 5.27 **The Nucleus**

TEM   500 nm

A. Electron micrograph of a yeast cell, showing the nucleus (colorized blue) and mitochondria (colorized orange).

The nuclear envelope is a double membrane.

Outer membrane
Inner membrane
Nucleolus

The nuclear outer membrane is continuous with the endoplasmic reticulum.

Chromatin
Nuclear envelope

Endoplasmic reticulum (ER)

Nuclear pore complexes

B. Diagram of a nucleus.

cell cycle during which the cell is not undergoing division. The cell grows during interphase, replicating its DNA. Then the daughter cells come apart through mitosis. Sexual reproduction requires meiosis, a modified form of mitosis that generates haploid ($1n$) gametes (Section 11.1). Gamete fertilization restores the $2n$ diploid chromosome number.

Mitochondria and Chloroplasts Yield Energy

Mitochondria and chloroplasts are organelles that obtain energy for a eukaryotic cell. Nearly all eukaryotes have mitochondria, which conduct oxidative phosphorylation and make ATP (see Chapter 7). Chloroplasts are found only in photosynthetic eukaryotes, such as plants and algae. Chloroplasts perform photosynthesis (also making ATP).

Mitochondria and chloroplasts are <u>not</u> considered part of the endomembrane system because their membranes undergo little traffic with the endomembrane organelles. Mitochondria and chloroplasts evolved early in the history of life through a form of evolution called **endosymbiosis**. In endosymbiosis, one cell is engulfed by a larger cell but not digested (**Figure 5.28A**). A mutualistic relationship evolves. In the evolution of mitochondria and chloroplasts, the larger eukaryotic cell provided protection to an intracellular bacterium, and the bacterium provided energy to the eukaryote.

MITOCHONDRIA Mitochondria have two membranes: an outer membrane and an inner membrane (**Figure 5.28B**). The inner membrane has many infoldings, called cristae, that increase its surface area. The mitochondrial inner membrane is derived from the cell membrane of the ancestral bacterium; it contains the protein complexes for oxidative phosphorylation, including the ATP synthase. The mitochondrial outer membrane is derived from the surrounding cell. Thus, the inner membrane has phospholipids similar to those of a bacterial membrane, whereas the outer membrane composition resembles that of the host cell.

Mitochondria contain two distinct compartments: the intermembrane space (between the two membranes) and the matrix enclosed by the inner membrane. Different stages of oxidative respiration occur in specific compartments. The arrangement of these processes is similar in bacteria and mitochondria. For example, in bacteria the tricarboxylic acid cycle (TCA, or Krebs cycle) occurs in the cytoplasm, and in mitochondria it takes place inside the matrix, the metabolic equivalent of the prokaryotic cytoplasm. Mitochondria retain a small circular chromosome containing bacteria-like genes, as well as their own ribosomes. During evolution, however, most of the endosymbiont's original genes were transferred to the nuclear chromosomes of the host cell. Consequently, inherited disorders of human mitochondria, such as Leigh syndrome (failure of oxidative metabolism), may arise from mutations either in mitochondrial genes or in nuclear genes.

CHLOROPLASTS Chloroplasts convert light energy from the Sun to ATP and reduced NADPH in a process known as the "light reactions" or light-dependent reactions. In the subsequent light-<u>in</u>dependent reactions, the ATP and NADPH are used to reduce CO_2 to sugar. The endosymbiotic precursor of chloroplasts was a common ancestor of modern cyanobacteria. Like eukaryotic chloroplasts, cyanobacteria perform oxygenic photosynthesis by using thylakoids, internal membrane pockets packed with chlorophyll to absorb light.

Chloroplasts have three membranes whose topology can be understood in light of endosymbiosis (**Figure 5.28C**). The outer membrane appears to be derived from the host eukaryotic cell, the inner membrane is equivalent to the bacterial cell membrane,

Figure 5.28 Endosymbiosis Led to Mitochondria and Chloroplasts

Endosymbiosis

Membrane of larger cell

Membrane of smaller cell

Double membrane

A. Origin of organelles through endosymbiosis.

Mitochondrion

Outer membrane
Inner membrane
Intermembrane space

TEM 300 nm

Matrix
Cristae

B. Structures in a mitochondrion.

Chloroplast

Outer membrane
Inner membrane

TEM 1 μm

Granum
Thylakoids
Stroma

C. Chloroplast structure.

and the thylakoid membrane is derived from the bacterial thylakoid membrane. The region enclosed by the inner membrane is called the stroma and is equivalent to the bacterial cytoplasm. ATP and NADPH are produced in the stroma and used there in the light-independent reactions of CO_2 fixation. As a result of endosymbiosis, chloroplasts, like mitochondria, contain their own circular DNA and their own ribosomes.

The Cytoskeleton Maintains Shape

Eukaryotic cells contain a network of proteins that determine the shape of the cell, collectively termed the cytoskeleton. Cytoskeletal proteins are multifunctional, also involved in whole-cell movements and the movement of substances within the cell. Three major classes of cytoskeletal proteins are the microfilaments, intermediate filaments, and microtubules.

Microfilaments, also known as actin filaments, have a diameter of 7 nm (**Figure 5.29A**). They are formed when individual actin monomers (globular actin, or G-actin) polymerize—in a process powered by ATP hydrolysis—to form chains of filamentous actin (F-actin). Two F-actin chains twist around each other, forming a microfilament with a plus end and a minus end.

Microfilaments are dynamic structures, growing and shrinking in a controlled manner by polymerization and depolymerization of actin. For example, an ameba's pseudopod starts to move by polymerizing actin at the leading edge of growth. Actin polymerization can be manipulated by invading bacteria—for example, by *Listeria monocytogenes*, the cause of food poisoning (listeriosis)

acquired from melons and other produce. The *Listeria* bacterium alters host actin polymerization in order to enter a host cell, and once inside, it induces host actin to polymerize behind one end of the bacterium. The actin tails then propel bacteria through the host and beyond, into neighboring host cells.

Intermediate filaments consist of various fibrous proteins that have a diameter of about 10 nm (**Figure 5.29B**). These filaments form a meshwork under the cell membrane, where they help maintain cell shape and strengthen the cell by resisting tension placed on the cell membrane.

Microtubules have a larger diameter (25 nm) than microfilaments and intermediate filaments (**Figure 5.29C**). The hollow microtubule structure consists of 13 dimers of tubulin; one alpha-tubulin protein plus one beta-tubulin protein forms one tubulin dimer. Microtubules are dynamic structures that can polymerize and depolymerize. They segregate (pull apart) the duplicated chromosomes during mitosis. Other microtubules form parts of cilia and flagella (discussed shortly).

Microtubules help transport substances within the cell. Protein traffic through the endomembrane system relies on the controlled movement of vesicles from one cell compartment to the next. Microtubules provide tracks that can move vesicles from one organelle to the next. Unfortunately, they also provide tracks for invading viruses. For example, particles of herpesvirus (such as herpes simplex, the cause of cold sores and genital herpes) can enter the cytoplasm and travel down a track of microtubules to the nucleus. The virus then injects its DNA through a nuclear pore complex for replication within the nucleus.

Figure 5.29 Cytoskeletal Proteins

Microfilament

7 nm

Actin monomer

A. Microfilaments consist of two strands of actin polymers twisted together.

Intermediate filament

8–12 nm

Fibrous subunit

B. Intermediate filaments are ropelike assemblages of various proteins.

Microtubule

25 nm

β α

α-Tubulin monomer

Tubulin dimer

β-Tubulin monomer

C. Microtubules are polymers of tubulin dimers.

Specialized Structures

As we saw for bacteria, different kinds of eukaryotic cells have different kinds of specialized structures, depending on their environment. Many eukaryotic microbes have motility organelles such as flagella or cilia, which help the microbe find and ingest food. Free-living eukaryotes, such as *Paramecium*, that lack a cell wall require structures to counteract osmotic pressure; for example, contractile vacuoles pump out water. And parasites form invasive structures to penetrate host cells, such as the apical complex of the malaria-causing *Plasmodium*. Diverse eukaryotic parasites are discussed in Chapter 11.

EUKARYOTIC FLAGELLA AND CILIA   Flagella and cilia consist of bundles of microtubules within thin extensions of the cell membrane (**Figure 5.30A**). Each flagellum, or each cilium, moves in a whiplike fashion, driven by interactions between microtubules and the motor protein dynein. Flagella of eukaryotes are relatively long, and a cell usually has only one or two of them; see, for example, the green alga *Chlamydomonas* (**Figure 5.30B**). Cilia (singular, cilium)

Figure 5.30 Eukaryotic Flagella and Cilia

Basal body   Flagellum   Dynein   Microtubules

Cell membrane

Microtubules

A. Structure of flagella and cilia.

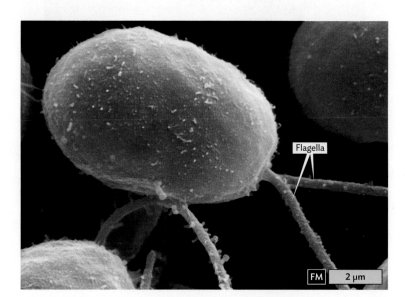

Flagella

FM   2 μm

B. The alga *Chlamydomonas* has two long flagella.

SEM   10 μm

C. The protozoan *Tetrahymena* has many cilia.

are shorter and more numerous, as seen in the ciliated protozoan *Tetrahymena* (**Figure 5.30C**). Both flagella and cilia can move an entire cell. Cilia may also be used to capture food by sweeping extracellular fluid into the gullet of a protozoan such as a paramecium or a vorticella. Eukaryotic cilia and flagella use ATP hydrolysis by dynein to generate the whiplike motion along the length of the cilium or flagellum.

⊙ **Note:** Distinguish eukaryotic flagella from bacterial flagella, which have a completely different structure. Eukaryotic flagella are large, containing multiple microtubules enclosed by a membrane. They move with a whiplike motion, powered by ATP hydrolysis all along the flagellum. By contrast, bacterial flagella consist of a single coiled tube of protein, whose rotary motion is driven by the flagellar motor embedded in the cell envelope.

PELLICLE   Like bacteria, microbial eukaryotes typically possess a thick outer covering. Fungi possess cell walls of the polysaccharide chitin, whereas green algae have cell walls of cellulose. Parasitic protozoa such as *Entamoeba histolytica* (a pathogenic ameba) and *Balantidium coli* (a pathogenic ciliate) lack a rigid cell wall. Instead, these protozoa have a thick, flexible pellicle, consisting of membranous layers reinforced by protein microtubules. The pellicle allows great flexibility of shape; it enables the uptake of large particles by endocytosis and of larger objects, even entire cells, by phagocytosis. So, protozoa have the nutritional option of engulfing prey—an option unavailable to prokaryotes and to eukaryotes with cell walls. The protozoan pellicle often contains additional structures, such as extrusomes, which secrete toxins or digestive enzymes. Other kinds of eukaryotic cell surfaces are reinforced by inorganic materials, such as the silica shells of diatoms.

CONTRACTILE VACUOLE   Free-living microbial eukaryotes that lack a cell wall need special organelles to prevent osmotic shock (see Section 4.5). An organelle that removes excess water from the cytoplasm is the **contractile vacuole** (**Figure 5.31**). As water enters the cell from a low-solute environment, the contractile vacuole takes up water through an elaborate network of intracellular channels extending throughout the cytoplasm. When the vacuole fills, it then contracts and expels the water through a pore, preventing osmotic shock and cell lysis.

Figure 5.31 Contractile Vacuole

Contractile vacuoles

PCM 20 µm

Vacuole filled with water

Vacuole after water has been expelled

A. A paramecium, showing two contractile vacuoles. The magnified images show radiating channels that collect water and expel it from the cytoplasm.

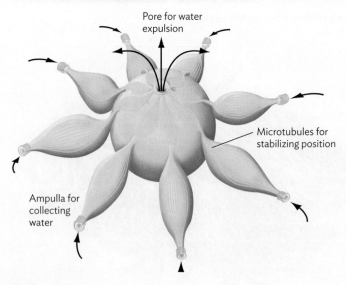

Pore for water expulsion

Microtubules for stabilizing position

Ampulla for collecting water

B. Diagram of a contractile vacuole.

SECTION SUMMARY

- **Membranous organelles** such as the endoplasmic reticulum, lysosomes, and Golgi complex compartmentalize the functions of a eukaryotic cell.
- **Endocytosis and phagocytosis** mediate transport between the cell and the external environment and between organelles.
- **The nuclear membrane** encloses eukaryotic chromosomes. Nuclear pore complexes mediate transport in and out of the nucleus.
- **The nucleolus** assembles ribosomes, which are sent to the cytoplasm.
- **Mitochondria and chloroplasts** are organelles that evolved from endosymbiotic bacteria. Mitochondria conduct oxidative phosphorylation, and chloroplasts conduct photosynthesis (for plant cells).
- **The cytoskeleton of eukaryotic cells** is composed of microfilaments, intermediate filaments, and microtubules.
- **Flagella and cilia** of eukaryotes are whiplike organelles powered by ATP hydrolysis.

Thought Question 5.5   What are the advantages and disadvantages of a contractile vacuole, compared with a bacterial cell wall?

Thought Question 5.6   What kinds of human cells possess cilia or flagella, and what are their functions within the human body?

Perspective

Every new cell structure and process we discover offers new clues to the mystery of life—and new targets for lifesaving drugs. The structures of bacteria even reveal surprising insights into human cells, since our own mitochondria evolved from bacteria.

Bacterial cells are adapted to grow in all kinds of hostile environments, including the immune system defenses of human and animal hosts. Bacteria that thrive in natural environments, such as the soil or ocean, also provide exciting new capabilities that scientists try to harness for lifesaving molecular therapies.

In the next chapter we focus on growth and nutrition: how do bacteria obtain energy and materials from their environment, including host environments such as the human body? We also learn how bacteria use their energy to build their intricate cells. The chapters that follow reveal how genetics shapes microbial cells and populations and how evolution generates the bewildering diversity of microbial species.

Key Terms

ABC transporter (p. 125)
active transport (p. 124)
adherence (p. 137)
antiport (p. 124)
capsule (p. 120)
cell envelope (p. 120)
cell membrane (p. 119)
cell wall (p. 119)
chemotaxis (p. 137)
contractile vacuole (p. 146)
coupled transport (p. 124)
cytoplasmic membrane (p. 119)
cytoskeleton (p. 129)
efflux transporter (p. 125)
endomembrane system (p. 141)
endoplasmic reticulum (ER) (p. 141)
endosymbiosis (p. 143)

endotoxin (pp. 120, 132)
eukaryote (p. 118)
exocytosis (p. 141)
facilitated diffusion (p. 124)
fimbria (p. 137)
flagellum (pp. 121, 137)
glycan (p. 127)
Golgi (p. 142)
Gram-negative (p. 130)
Gram-positive (p. 130)
holdfast (p. 137)
inner membrane (p. 131)
intermediate filament (p. 144)
lipid A (p. 132)
lipopolysaccharide (LPS)
(pp. 119, 132)
lumen (p. 141)

lysosome (p. 141)
magnetosome (p. 139)
microfilament (p. 144)
microtubule (p. 144)
motility (p. 137)
murein lipoprotein (p. 131)
mycolic acid (p. 133)
nuclear pore complex (p. 142)
nucleoid (pp. 121, 134)
nucleolus (p. 142)
origin of replication (*ori*) (p. 134)
outer membrane (p. 119)
passive transport (p. 124)
peptidoglycan (p. 119)
periplasm (p. 119)
permease (p. 124)
pilin (p. 137)

pilus (p. 137)
plasma membrane (p. 119)
polyamine (p. 122)
porin (p. 132)
prokaryote (p. 118)
proteome (p. 122)
proton motive force (PMF) (p. 124)
reductive evolution (p. 131)
septation (p. 135)
septum (p. 135)
sex pilus (p. 137)
S-layer (p. 131)
stalk (p. 137)
staphylococcus (p. 136)
substrate-binding protein (p. 125)
symport (p. 124)
teichoic acid (p. 131)

Objectives Assessment Questions

5.1 The Bacterial Cell: An Overview

1. Which cell component(s) extend far outside a Gram-negative cell and help prevent phagocytosis by blood cells?
   a. Cell wall
   b. Periplasmic proteins
   c. Polyamines
   d. Lipopolysaccharide (LPS)

2. Which intracellular molecule(s) are present in the greatest number?
   a. Proteins
   b. Messenger RNA
   c. Water molecules
   d. Potassium ions

3. Which part of the Gram-negative cell holds the ATP synthase?
   a. Cell wall
   b. Cell membrane (inner membrane)
   c. Periplasm
   d. Capsule

5.2 Bacterial Membranes and Transport

4. Which of the following is <u>not</u> a function of the cell membrane or membrane-embedded proteins?
   a. Protein synthesis in the cytoplasm
   b. Detecting changes in temperature and pH that tell a pathogen it is within a host
   c. Secreting toxins to counteract host defenses
   d. Transporting nutrients from the environment into a cell

5. Which transport process would <u>not</u> require expenditure of energy?
   a. Uptake of sugar by an ABC transporter complex
   b. Facilitated diffusion of oxygen across a membrane
   c. Transport of an amino acid from a lower concentration to a higher concentration
   d. Symport of lactose coupled to the movement of H^+ inward down its gradient

6. Which structure can export antibiotics?
   a. TolC efflux complex
   b. Ribosome
   c. Amino acid transporter
   d. Lactose transporter

5.3 The Bacterial Cell Wall and Outer Layers

7. Which cell component consists of a single macromolecule that surrounds the cell and protects it from osmotic shock?
   a. Cell wall
   b. Cell envelope
   c. Cell membrane or plasma membrane
   d. Lipopolysaccharide (LPS)

8. Which compound(s) are found in the outer cell layers of *Mycobacterium tuberculosis*, but not in most other kinds of bacteria?
   a. Teichoic acids
   b. S-layer proteins
   c. Murein lipoprotein
   d. Mycolic acids

9. Which structure of a Gram-negative cell contains porins?
   a. Cell wall
   b. Inner membrane
   c. Periplasm
   d. Outer membrane

5.4 The Nucleoid and Bacterial Cell Division

10. Which aspect of DNA organization does <u>not</u> occur in bacteria?
    a. DNA is double-stranded.
    b. DNA has an origin of replication attached to the cell envelope.
    c. DNA is contained within a nuclear membrane.
    d. Most species have a genome that is circular.

11. Which process can occur <u>only</u> after the bacterial DNA terminates replication?
    a. Completion of the septum to separate two daughter cells
    b. Transcription of genes to make messenger RNA
    c. Translation of RNA to make proteins
    d. Insertion of proteins into the cell membrane

12. How is DNA most commonly arranged inside a bacterial cell?
    a. Multiple pieces in the cytoplasm
    b. A circular molecule tangled throughout the cytoplasm
    c. A circular molecule with loops bound by DNA-binding proteins
    d. A circular molecule within a nuclear membrane

5.5 Specialized Structures of Bacteria

13. Which extracellular structure helps a bacterial cell attach to a substrate?
    a. Magnetosome
    b. Stalk
    c. Gas vesicle
    d. Thylakoid

14. Which statement does <u>not</u> describe bacterial motility?
    a. Bacterial motility enables pathogens to colonize human organs.
    b. Helical flagella rotate, using a rotary motor embedded in the cell membrane.
    c. Flagella generate a whiplike motion powered by ATP.
    d. Cells alternate swimming and tumbling to migrate up the gradient of an attractant.

15. What is the process by which cells migrate toward nutrients?
    a. Chemotaxis
    b. Contraction of contractile vacuole
    c. Photosynthesis
    d. Active transport

5.6 The Eukaryotic Cell

16. Which organelle is <u>not</u> a part of the endomembrane system?
    a. Cell membrane
    b. Lysosome
    c. Golgi complex
    d. Mitochondrion

17. Which structure(s) do <u>not</u> contribute to the eukaryotic cytoskeleton?
    a. Microtubules
    b. Peptidoglycan cell wall
    c. Microfilaments
    d. Intermediate filaments

18. Which organelle evolved from a bacterium?
    a. Golgi
    b. Chloroplast
    c. Nucleus
    d. Endoplasmic reticulum (ER)

Review Questions

1. What are the major features of a bacterial cell, and how do they fit together for cell function as a whole?

2. Which fundamental traits do most bacteria have in common with eukaryotic microbes? Which traits are different?

3. Compare and contrast the structure of Gram-positive and Gram-negative cell envelopes. Explain the strengths and weaknesses of each kind of envelope.

4. Explain how DNA replication is coordinated with cell wall septation.

5. Compare and contrast bacterial flagella with eukaryotic flagella.

6. Describe the organelles of the endomembrane system. Explain the process of transport that transfers membranes between them.

7. Explain the function of mitochondria, chloroplasts, and the cytoskeleton.

Clinical Correlation Questions

1. A skin biopsy sample reveals bacteria with extremely narrow, flexible, spiral cells. The cells have a thick sheath, beneath which rotary flagella embedded in the cell membrane cause the cells to twist in a corkscrew motion. The bacteria are associated with a tick bite. What disease may the patient have?

2. A skin biopsy sample reveals rod-shaped bacteria with a thick cell wall and capsule; the bacteria stain acid-fast. The thick outer layers of the capsule prevent phagocytosis but may also decrease nutrient uptake. The bacteria can be cultured but require several weeks to grow colonies. The patient shows skin lesions with loss of sensation. What disease may the patient have?

 For more assessment, visit digital.wwnorton.com/michum2

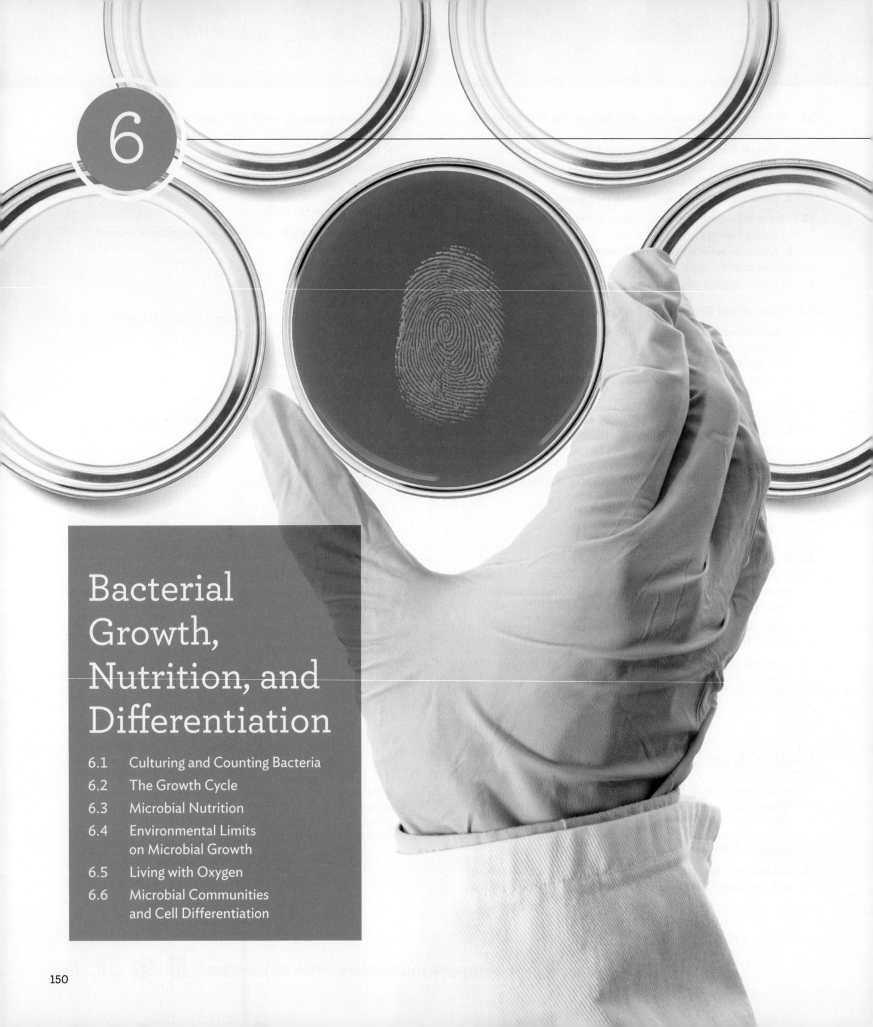

6

Bacterial Growth, Nutrition, and Differentiation

[Metabolic Fingerprints of a Brain Pathogen]

SCENARIO Deeqa was a 16-year-old girl when she and several of her Somali classmates were captured by the terrorist group Al Shabaab. During her 2 years of captivity, Deeqa was forced to marry one of her kidnappers. One night she managed to escape and started a long, arduous journey to the refugee camp in Kenya. There, aid workers helped her apply for asylum in the United States, which was granted. Eventually she relocated to St. Cloud, Minnesota, where she married another Somali refugee. One morning their 18-month-old daughter Mirembe woke with a very high fever and was crying in pain. Rushed to an emergency room, the infant cried uncontrollably, especially when the physician tried to move her head. Meningitis (an infection of the spinal fluid and membrane surrounding the brain) was suspected, and a spinal tap was performed to collect cerebrospinal fluid (CSF). The fluid, which appeared cloudy, was sent to the laboratory. While waiting for the results, the doctor learned that Deeqa's daughter had never been vaccinated.

TESTING CSF should be free of microbes, but a microscopic examination revealed Gram-negative bacilli. The technician cultured the sample on various solid media. The organism failed to grow on blood agar plates containing intact red blood cells (RBCs) but did grow on a medium containing lysed RBCs (called chocolate agar because of its brown color). The organism recovered from the CSF also grew on a relatively simple medium containing growth factors known as X and V.

DIAGNOSIS The diagnosis was meningitis caused by the bacterium *Haemophilus influenzae*. Mirembe received antibiotics and made a full recovery. She is now in preschool.

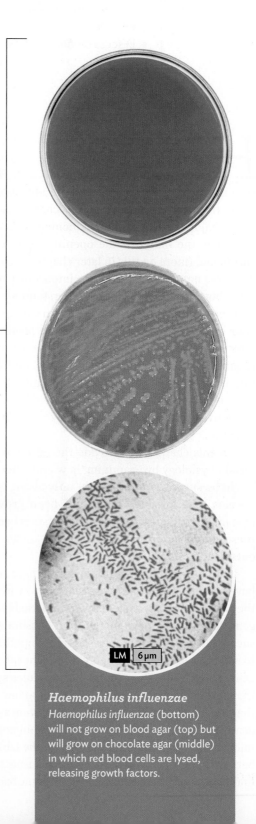

Haemophilus influenzae
Haemophilus influenzae (bottom) will not grow on blood agar (top) but will grow on chocolate agar (middle) in which red blood cells are lysed, releasing growth factors.

LM 6 μm

CHAPTER OBJECTIVES

After reading this chapter, you will be able to:

- Explain how nutrition and the environment can impact microbial growth and differentiation.
- Discuss microbial classifications based on nutritional needs and environmental limits.
- Describe how understanding microbial growth helps identify disease-causing pathogens.
- Discuss biofilms and their importance to infectious diseases.

Haemophilus influenzae is a pathogenic bacterium that spreads from person to person by direct contact or by aerosolization from someone with *H. influenzae* in their respiratory tract. Before 1988 and the advent of an effective vaccine made from the organism's polysaccharide capsule, *H. influenzae* was a major cause of childhood meningitis in developed countries. Mirembe, the infant in the opening scenario, however, was not vaccinated (more on this in later chapters), so she was susceptible. This case not only points out the importance of childhood vaccinations but also highlights why understanding bacterial physiology is crucial to people entering the health care field. The adage "know your enemy" applies to all health care professionals dealing with infectious diseases. Had the technician not known that *H. influenzae* requires two factors to grow—X (hemin) and V (nicotinamide adenine dinucleotide, NAD)—the cause of this infection might not have been identified. The child could have died. Clinical microbiology saves lives, and bacterial physiology is the foundation of clinical microbiology.

In this chapter, we provide the basis for understanding bacterial physiology by examining how and where bacteria grow. You might be surprised to learn that **only about 0.1% of the bacterial species that inhabit Earth have been grown in a laboratory**. One important quest for microbiologists is to discover how to grow these microbes on artificial medium under artificial conditions outside their native environment.

Although most of Earth's microbes have not been cultured, we have learned much from those that do grow in the lab; in particular, they capture carbon and energy in remarkably diverse ways. Pathogens, for example, gain energy from consuming organic molecules such as glucose. In stark contrast, some nonpathogenic bacteria native to soil or water get their energy from light (photosynthesis). Others gain energy by oxidizing inorganic molecules such as sulfur. Indeed, there are so many microbes consuming so many different nutrients that you might wonder how Earth's biosystem avoids "eating" itself to death! As we explore these facets of microbiology, you will learn how microbes are experimentally grown (in the lab) and how their numbers are measured. We will also spotlight aspects of growth critical to finding, growing, and identifying pathogenic bacteria.

6.1 Culturing and Counting Bacteria

SECTION OBJECTIVES

- Explain how pure cultures are obtained and why they are important in medicine.
- Distinguish between synthetic, complex, selective, and differential media and their uses in clinical microbiology.
- Describe the ways bacterial growth is measured, and explain the advantages and disadvantages of each method.

How do we grow, culture, and separate microbes in the laboratory? Microbes in nature usually exist in complex, multispecies communities. For detailed laboratory studies, however, they must be grown separately in pure culture—that is, as a single species. But can all microbes be grown in the lab? It is nothing short of amazing—and humbling—that after nearly 140 years of trying to grow microbes in the lab, we have succeeded in culturing only 0.1% of the microorganisms around us. Since the time of Koch (see Chapter 1) in the late nineteenth century, microbiologists have used the same fundamental techniques to culture bacteria. Improvements have certainly been made, but almost all of the microbial world has yet to be tamed.

Growing Bacteria in Culture Media

For those organisms that can be cultured, a variety of techniques are used, each for a different purpose. Bacterial culture medium may be either liquid or solid. In a liquid (broth) medium, organisms can move about freely. This approach is useful when studying the growth of a single, genetically homogeneous strain of a single species (that is, a **pure culture**). Liquid medium is also convenient for examining growth rates and changes in microbial biochemistry under different conditions. Solid medium, usually prepared in petri dishes, is useful when trying to separate mixtures of different organisms found in the natural environment or in clinical specimens.

Obtaining Pure Cultures

Solid medium is simply made by adding a solidifying agent to a liquid medium. The most versatile and widely used solidifying agent is agar (the development of agar medium is covered in Section 1.3). Derived from seaweed, agar forms an unusual gel that liquefies at 100°C (212°F) but does not solidify until cooled to about 40°C (104°F). Liquefied agar medium is poured into shallow, covered petri dishes. As the liquid cools, it hardens to form a large, flat surface onto which a drop of microorganisms can be streaked. If the medium contains the proper nutrients and growth factors, a single cell will multiply, producing many millions of offspring that together form a visible mound of cells called a **colony** (Figure 6.1). As shown in Figure 6.2A, a drop of liquid culture is collected with an inoculating loop and streaked across the agar plate surface in a pattern called **isolation streaking** (also known as **dilution streaking**). Organisms fall off the loop as it moves along the agar surface. Toward the end of the streak, few bacteria remain on the loop; at this point, individual cells will land and stick to different places on the agar surface (Figure 6.2B). A pure culture of the species can then be obtained by touching a single colony with a sterile inoculating loop and inserting that loop into fresh liquid medium.

Another way to isolate pure colonies is the **spread plate** technique. Starting from a liquid culture of bacteria, a series of tenfold dilutions is made. This process entails transferring 1 milliliter (ml) of a broth containing bacteria into 9 ml of sterile broth and repeating the dilution into subsequent tubes. Next, a small, measured amount of each dilution is placed directly on the surface of separate agar plates (Figure 6.3). The sample is spread over the surface of the plate by using a heat-sterilized (but cooled) bent glass rod (cell spreader). The early dilutions, containing the most bacteria, will produce **confluent** (continuous) growth that covers the entire agar surface. The later dilutions, containing fewer and fewer organisms, yield individual colonies. Each colony on an agar plate represents one viable organism present in the diluted sample. A **viable** organism is one that successfully replicates to form a colony. Spread plates, therefore, enable us to isolate pure cultures and can be used to back-calculate the number of viable bacteria in the original, undiluted sample. The technique is invaluable for public health by determining the level of bacterial contamination in food and water.

Figure 6.1 Separation and Growth of Microbes on an Agar Surface

This agar plate culture was made from swabbing patient elevator buttons and escalator rails at a university hospital. Note the various colony shapes and colors made by different microbial species. Courtesy Wikimedia Commons
https://commons.wikimedia.org/wiki/File:VCU_agar_plate_colonies.jpg
https://creativecommons.org/licenses/by-sa/3.0/legalcode

Figure 6.2 Isolation (Dilution) Streaking Technique

A. A liquid culture is sampled with a sterile inoculating (holding) loop, which is streaked across the plate. After the first section is streaked, the loop is passed through a flame to resterilize it and then cooled. The resterilized loop is passed three to five times through the original section and into an adjacent section, as demonstrated, and this process is repeated several times. Dragging the loop across the agar to each new section diminishes the number of organisms clinging to the loop. Eventually, only single cells are deposited at a given location.

B. *Salmonella enterica* culture obtained by isolation streaking.

Figure 6.3 Tenfold Dilutions, Plating, and Viable Counts

A. A culture containing an unknown concentration of cells is serially diluted. One milliliter (ml) of culture is added to 9.0 ml of clear broth (diluent) and mixed, and then 1 ml of this dilution is added to another 9.0 ml of diluent (10^{-2} dilution). These steps are repeated for further dilution, each of which lowers the cell number tenfold. After dilution, 0.1 ml of each dilution is spread onto an agar plate.

B. Plates prepared as in (A) are incubated at 37°C (98.6°F) to grow colonies. Plates containing between 30 and 300 colonies are countable. Numbers outside that range are not statistically reliable. Multiplying the number of countable colonies (107 colonies per 0.1 ml on the 10^{-5} plate) by 10 gives the number of colonies in 1.0 ml of the 10^{-5} dilution. Multiplying that number by the reciprocal of the dilution factor ($1/10^{-5}$) gives the number of cells (colony-forming units, or CFUs) per milliliter in the original broth tube ($107 \times 10 \times 10^5 = 1.07 \times 10^8$ CFUs/ml). TNTC = too numerous to count.

Note, however, that the "one cell equals one colony" paradigm does not hold for all bacteria. Some organisms, such as *Streptococcus* or *Staphylococcus*, exist in chains or clusters rather than as single cells. Thus, a cluster of ten *Staphyloccocus* cells will remain together and form only one colony on an agar medium. For this reason, the cell or cells that form a colony are called a **colony-forming unit (CFU)**.

Growth in Complex and Synthetic Media

Bacteria can be grown in a precisely defined **synthetic medium** that contains only known components. A **minimal medium** is a synthetic medium in which the components are limited to only those nutrients the organism needs in order to grow. A synthetic medium starts with water, and then various salts, carbon, nitrogen, and energy sources are added in precise amounts. For self-reliant organisms such as *E. coli* or *Bacillus subtilis*, those few chemicals are all that is needed. Other organisms, such as *Shigella* species or mutant strains of *E. coli* or *B. subtilis*, require additional ingredients to satisfy requirements imposed by the absence of specific metabolic pathways.

Bacteria can also be grown in a nutrient-rich but less defined **complex medium**, also called **rich medium**. Recipes for complex media usually contain one or more poorly defined ingredients, such as yeast extract or beef extract, whose exact composition is not known. These additives include a rich variety of amino acids, peptides, nucleosides, vitamins, and some sugars. Some organisms are particularly fussy, or fastidious, about their nutritional requirements. *H. influenzae* and *Neisseria meningitidis*, for example, require that certain components of blood (or blood itself) be added to a basic complex medium, which would then be called an **enriched medium**.

Complex, or rich, media provide many of the chemical building blocks that a cell would otherwise have to synthesize on its own. For example, instead of making proteins that synthesize the amino acid tryptophan, all the cell needs is a membrane transport system to harvest tryptophan already present in complex medium. Likewise, fastidious organisms that require blood in their media will claim the heme released from red blood cells as their own. The captured heme is then incorporated into certain enzymes as a prosthetic group critical to enzyme function (for example, the heme group in cytochromes). All of this saves the scavenging cell a tremendous amount of energy, and as a result, bacteria tend to grow fastest in complex media. More about nutrients will be discussed in Section 6.3.

Selective and Differential Media

In a clinical diagnostic laboratory, it is important to quickly determine whether a pathogen is present in a patient's specimen and, if so, to identify it. This process may be easy if the specimen is expected to be sterile, as in the case of cerebrospinal fluid, but challenging if the specimen contains normal microbiota, as in fecal specimens.

CASE HISTORY 6.1

Differential Medium: Finding the Infectious Needle in a Haystack

 Fifteen-year-old Olivia arrived at the emergency room complaining of severe abdominal cramping and diarrhea that had lasted for 2 days. A fecal sample from the patient was taken to the laboratory and streaked onto MacConkey lactose agar medium. MacConkey medium selectively prevents the growth of Gram-positive bacteria but will differentiate Gram-negative bacteria that can ferment lactose from those that cannot. The next day, several different types of bacterial colonies were visible on the agar plate; some were red, others were white. The presence of the white colonies led the clinical microbiologist to suspect the patient was infected with the Gram-negative bacillus *Salmonella enterica* serovar Typhimurium. So, what might the red colonies represent?

Salmonella enterica is an important and widespread cause of diarrhea. But our intestines are crammed with many other bacteria that are considered normal microbiota. How, then, does the laboratory find *Salmonella* among all those other bacterial species in a sample of feces? Once again, the answer lies in understanding bacterial physiology.

Microorganisms are remarkably diverse with respect to their metabolic capabilities and resistance to certain toxic agents. These differences are exploited in clinical microbiology through the use of both selective and differential media. A **selective medium** favors the growth of one organism over another, while a **differential medium** exposes biochemical differences between two species that grow equally well.

Several media used in clinical microbiology are both selective and differential. **MacConkey medium**, for example, is selective because it contains bile salts and crystal violet that inhibit the growth of Gram-positive bacteria. However, the distinctive outer membranes of Gram-negative bacteria keep the detergent (bile salts) away from their critical cytoplasmic membrane, allowing these bacteria to grow. MacConkey medium also differentiates bacteria that ferment lactose from those that do not (nonfermenters). The medium contains lactose as a carbon source, the dye neutral red, and a mix of peptides (peptones) as an alternative source of carbon. Lactose fermenters such as *E. coli* degrade lactose and secrete acidic products that lower the pH around the colony. The acidic pH allows neutral red dye to enter cells and produce a red colony. So, the red colonies in Case History 6.1 were probably *E. coli.*

In stark contrast, the pathogen *S. enterica* does not ferment lactose and grows only on the nonfermentable peptides. No acid is produced, neutral red does not enter cells, and the colonies are white. As a result, a culture grown on MacConkey medium will consist of Gram-negative organisms that can be identified as lactose fermenters (red colonies) or nonfermenters (white colonies;

Figure 6.4 MacConkey Medium, a Culture Medium both Selective and Differential

Lactose-fermenting (Lac$^+$) bacteria

Nonfermenting (Lac$^-$) bacteria

Isolation (dilution) streak of a mixture of lactose-fermenting (Lac$^+$) and non-lactose-fermenting (Lac$^-$) bacteria. Only Gram-negative bacteria grow on lactose MacConkey medium (selective). Only a species that can ferment lactose produces red colonies (differential).

Figure 6.4). This medium is of particular benefit for diagnosing the etiology (cause) of diarrheal disease. Normal intestinal bacteria that grow on MacConkey agar are lactose fermenters, whereas two important pathogens, *Salmonella* and *Shigella*, are lactose nonfermenters. You will learn more about the enriched, selective, and differential media used in clinical microbiology in Chapter 25.

CASE HISTORY 6.2

Urine Cultures (Where Numbers Count)

 Melissa, a 25-year-old woman, visited her doctor's office complaining of a terrible burning sensation while urinating. The physician assistant (PA) suspected the patient was suffering from a urinary tract infection (UTI) in her bladder and sent a sample of urine to the laboratory. The lab technician took a very small but exact volume of urine (1 µl) and spread it onto agar medium. After 24 hours of incubation, the clinical microbiologist observed many *Escherichia coli* colonies (a Gram-negative bacillus), equivalent to 100,000 bacteria per milliliter of the patient's urine. Because urine samples from healthy patients usually have almost no bacteria, the PA determined that the woman did indeed have a UTI and prescribed an antibiotic appropriate for the etiologic (causative) agent, uropathogenic *E. coli.*

This case history illustrates how bacterial numbers can be determined in clinical samples as well as in laboratory culture. Counting (or quantifying) organisms that are invisible to the naked eye is surprisingly difficult because each of the available techniques measures a different physical or biochemical aspect of growth.

Figure 6.5 The Petroff-Hausser Chamber for Direct Microscopic Counts

1 Slide with shallow wells and inscribed grid (~400 squares, 0.0025 mm² each).

2 Coverslip is placed over slide.

3 Bacterial suspension is added to wells and seeps under the coverslip to fill the shallow space of known volume over the grid.

4 Bacterial cells in each square are counted under a microscope.

0.2 mm apart

Special slides have a precision grid etched on the surface. The organisms in several squares are counted, and their numbers are averaged. Knowing the dimensions of the grid and the height of the coverslip over the slide makes it possible to calculate the number of organisms in a milliliter of culture.

For this reason, a cell density value (given as cells per milliliter) derived from one technique will not necessarily agree with the value obtained by a different method. In Chapter 23, you will learn more about how a clinician makes a presumptive diagnosis of urinary tract infection and how lab results help confirm that diagnosis.

Distinguishing the Living from the Dead

Though it is not the technique used in Case History 6.2, microorganisms can be counted directly by use of a microscope. A dilution of a bacterial culture is placed on a special microscope slide called a hemocytometer (or, more specifically for bacteria, a Petroff-Hausser counting chamber; **Figure 6.5**). A grid of precise dimensions is etched on the surface of the slide, and placing a coverslip over the grid creates a space of precise volume. The number of organisms within that specific volume is used to calculate the concentration of cells in the original culture.

However, "seeing" an organism under the microscope does not mean the organism is alive. Living cells may be distinguished from dead cells by fluorescence microscopy with fluorescent chemical dyes (discussed in Chapter 3). For example, a red dye called propidium iodide inserts between DNA bases but cannot freely penetrate the energized membranes of living cells. So, only <u>dead</u> cells appear red under a fluorescence scope. Another dye, Syto-9, enters both living and dead cells, staining them both green. By combining Syto-9 with propidium iodide, microbiologists can distinguish living cells from dead cells: living cells will appear green, whereas dead cells will appear red or yellow-orange because the combination of dyes, Syto-9 (green) plus propidium (red), appears yellow-orange (**Figure 6.6**).

Direct counting of cells (without microscopy) can also be accomplished by using sophisticated electronic instruments called a **flow cytometer** or a **fluorescence-activated cell sorter (FACS)**. A flow cytometer counts cells, but a FACS instrument both counts and separates bacterial cells that have different properties. For instance, FACS analysis can count and separate bacterial cells that

synthesize a fluorescent protein from cells that do not. The mixture of cells flow single-file past a laser (**Figure 6.7**), which counts the cells that fluoresce versus those that do not and measures the fluorescence intensity. The FACS instrument also places different charges on the two types of cells, enabling them to be separated and collected for further analysis.

Figure 6.6 LIVE/DEAD Stain

FM   10 μm

LIVE/DEAD *Bac*Light Bacterial Viability Kit. Dead bacterial cells fluoresce red or yellow-orange because propidium iodide (red) can enter the cells and slip between the base pairs of DNA. Live cells fluoresce green because Syto-9 (green) can enter the cells and propidium iodide cannot.

Figure 6.7 **Fluorescence-Activated Cell Sorter**

1 Laser allows detection and counting of fluorescent cells.

2 Fluorescent cells are charged and their paths are deflected in an electric field.

3 Deflected cells are collected.

A. Schematic diagram of a FACS apparatus (bidirectional sorting).

B. Separation of green fluorescent protein (GFP)–producing *E. coli* from non-GFP-producing *E. coli*. The low-level fluorescence in the cells on the left is baseline fluorescence (autofluorescence). The scatterplot displays the same FACS data, showing the size distribution of cells (*x*-axis) with respect to level of fluorescence (*y*-axis). The larger cells may be ones that are about to divide. FL1 stands for the instrument channel used to measure fluorescence.

"Proof of Life" (Viable Counts)

Viable cells, as noted previously, are those that can replicate and form colonies on an agar plate. To obtain a viable cell count, dilutions of a liquid culture are plated directly on an agar surface. After colonies form, they are counted, and the original cell number is calculated (again, see Figure 6.3). This procedure was used in diagnosing Melissa's urinary tract infection in Case History 6.2. For example, if 100 colonies are observed on an agar plate that was spread with 100 microliters (µl) of a 10^{-3} dilution of a culture, then 1 ml (1000 µl) of that dilution would contain 1000 cells:

$$100 \text{ cells}/100 \text{ µl} = X \text{ cells}/1000 \text{ µl}; X = 1000 \text{ cells/ml}$$

Because the tube was a 10^{-3} dilution of the original culture, the original culture contained 10^6 organisms per milliliter:

$$1000 \text{ cells/ml} = Y \text{ cells}/10^{-3} \text{ dilution}; Y = 1 \times 10^6 \text{ cells per ml}$$

Although viable counts are widely used in research, using this method to measure cell number is not perfect. One issue is that viable counts usually underestimate the number of living cells in a culture. For example, metabolically active but damaged cells will not form colonies on agar plates, so they will not be counted as alive. Comparing a viable count with a direct count obtained from a live/dead stain can expose the presence of damaged cells. As mentioned earlier, organisms such as streptococci that grow in chains will cause actual cell numbers to be underestimated, which is why the results of this counting technique are reported as colony-forming units (CFUs) rather than cells.

See the Light (Optical Density)

As you will learn, replicating bacteria change their metabolism as cell number increases (crowding) or when they encounter environmental stress (heat, starvation, osmolarity). Using plate counts or direct counting to monitor cell number takes too long to carry out when you want to immediately study cells in a specific growth phase. In such instances, population size can be calculated by other methods, such as measuring the **optical density** of a cell culture—the degree to which the liquid medium has become cloudy because of the presence of organisms within it.

Much as creek water turns cloudy after a rainstorm because of suspended soil particles, the optical density of liquid medium will increase as bacteria increase in number. Optical density can be measured in real time by an instrument called a **spectrophotometer**, which passes a beam of light through a sample of the culture. The more particles, or bacteria, there are in the liquid, the more the light beam gets scattered and the less light reaches the detector. So, as cell numbers increase in a culture, so does optical density. Observing the change in optical density as bacteria grow in culture is very useful for estimating population size, especially if you have previously developed a standard curve that plots number of cells (viable count) versus optical density.

SECTION SUMMARY

- **A pure culture** contains only a single bacterial species that is usually genetically homogeneous. A pure culture is needed for detailed molecular and biochemical studies of that species.
- **Bacteria can be cultured** on solid or liquid medium.
- **Minimal defined medium** contains only those nutrients essential for growth.
- **Complex, or rich, medium** contains many nutrients. Other media exploit specific physiological differences between organisms and can be defined as selective or differential.
- **Bacteria in culture may be counted directly** under a microscope, with or without staining, or by using a fluorescence-activated cell sorter.
- **Bacteria can be counted indirectly** through viable counts or optical density measurements.
- **A viable bacterial organism** is defined as a cell with the ability to replicate and form a colony on the surface of a solid medium.

Thought Question 6.1 The addition of sheep blood to agar produces a very rich medium called blood agar. Do you think blood agar can be considered a selective medium? A differential medium? *Hint:* Some bacteria can lyse red blood cells.

6.2 The Growth Cycle

SECTION OBJECTIVES

- Understand the phases of a typical bacterial growth curve.
- Explain how bacterial growth correlates to disease.
- Describe the purpose of continuous culture and how it correlates to the human digestive tract.

How do microbes grow, and what determines their rate of growth? In nature, the answers to these questions are extremely complex. Most microbes in nature do not exist alone but in complex communities of microbes or multicellular organisms that influence one another's growth. Nevertheless, all species at one time or another exhibit both rapid growth and nongrowth, as well as many phases in between. For clarity, we present the principles of rapid growth, but bear in mind that ever-changing conditions in nature can broadly impact an organism's growth.

The ultimate goal of any species is to make more of its own kind. This fundamental law ensures survival of the species. A typical bacterium (at least the typical bacterium that can be cultured in the laboratory) grows by increasing in length and mass. This facilitates expansion of the bacterium's nucleoid as its DNA replicates (see Section 5.4). As DNA replication nears completion, the cell—in response to complex genetic signals—begins to synthesize an equatorial septum (cell wall) that ultimately separates the two daughter cells. In this overall process, called **binary fission**, one parental cell splits into two equal daughter cells (Figure 6.8A).

Although most culturable bacteria divide symmetrically into halves, some species divide asymmetrically. For example, the bacterium *Caulobacter* forms a stalked cell that remains fixed to a solid surface but reproduces by budding from one end to produce small, unstalked motile cells that then break off from the parental cell. The marine organism *Hyphomicrobium* also replicates asymmetrically by budding, releasing a smaller cell from a stalked parent (Figure 6.8B).

Eukaryotic microbes divide by a special form of cell fission involving mitosis (described in Chapter 11). Some eukaryotes also undergo more complex life cycles involving budding and diverse morphological forms. No matter how reproduction is accomplished, similar mathematical patterns of population growth are seen in all organismal populations, both prokaryotic and eukaryotic—including our own.

Exponential Growth

How does replication affect the growth of a population? Growth of any population obeys a simple law: as the cell number doubles, whatever parameter you use to measure growth also doubles (for example, optical density). Growth in which population size

Figure 6.8 Symmetrical and Asymmetrical Cell Division

A. Symmetrical cell division, or binary fission, in *Lactobacillus* species.

B. Asymmetrical budding of the marine bacterium *Hyphomicrobium*.

doubles at a fixed rate (say, every 20 minutes) is called **exponential growth** because the number of new cells produced after each cell division is larger each time the population divides. For example, two cells divide to become four cells (an increase of two cells), four cells divide to become eight cells (an increase of four cells), and eight cells divide to become 16 cells (an increase of eight cells). Plotting the increase in the number of cells over time generates an exponential curve, a curve whose slope increases continually. Exponential growth explains how ingesting only ten cells of a pathogenic strain of *E. coli* can lead to bloody diarrhea within only 2 days.

In an environment with few bacteria but plenty of resources, bacteria will divide at a constant interval called the **generation time**. The generation time for any organism varies with respect to many parameters, including the bacterial species, type of medium, temperature, and pH. The generation time for cells in culture is also known as the **doubling time** because the population of cells doubles over one generation. For example, one cell of *E. coli* placed into a complex medium will divide every 20 minutes. After 1 hour of growth (three generations), that one cell will have become eight (one cell to two, two cells to four, four cells to eight). Because cell number (N) doubles with each division, the increase in cell number over time is exponential, not linear. (Imagine if your bank account started at \$1 but doubled every day. In 20 days, you would have over \$1 million.) A linear increase would occur if cell number rose by a fixed amount after every generation (for example, one cell to two, two cells to three, three cells to four). In the bank analogy, you would have only \$21 after 20 days.

Starting with any number of organisms at time zero (N_0), the number of organisms after n generations will be $N_0 \times 2^n$. For example, a single cell after three generations ($n = 3$) will produce

$$1 \text{ cell} \times 2^3 = 8 \text{ cells}$$

The mathematics of exponential growth is relatively straightforward. However, remember that microbes grow differently in pure culture (which is very rare in nature) than they do in mixed communities, where neighboring cells produce all kinds of substances that may feed or poison other microbes. In mixed communities, the microbes may grow individually floating in liquid, as in the open ocean, or as a biofilm on solid matter suspended in that ocean (biofilms are described in Section 6.6). In each instance, the mathematics of exponential growth applies, at least until the

community reaches a density at which different species begin to compete.

Phases of Growth

Exponential growth never lasts indefinitely. Consumption of nutrients and accumulation of toxic by-products eventually slow the generation time until growth halts altogether. The simplest way to culture bacteria is in flasks or test tubes containing liquid medium, a technique called **batch culture**. In batch culture, nutrient concentrations decline and waste products accumulate as bacteria grow.

Batch culture illustrates the remarkable ability of bacteria to adapt to their environment. As conditions in the medium deteriorate, cells will change their membrane composition, cell size, and metabolism, all of which impact generation time. The goal for the cells is to slow growth before they lose viability. You can see the effect of changing conditions on generation time by plotting growth as the logarithm of the number of cells versus incubation time. Such a plot reveals several phases of growth (**Figure 6.9**).

Lag phase. Cells transferred from spent culture medium to fresh medium need time to detect the new environment, express new genes, and synthesize new components needed to continue growth. As a result, bacteria inoculated into fresh

Figure 6.9 Bacterial Growth Curves

1 Lag phase. Bacteria are preparing their cell machinery for growth.

2 Log phase. Growth approximates an exponential curve (straight line, on a logarithmic scale).

3 Stationary phase. Cells stop growing and shut down their growth machinery while turning on stress responses to help retain viability.

4 Death phase. Cells begin to die at an exponential rate.

Phases of bacterial growth that occur in a typical batch culture are represented by different shadings under the curve. Note: in this graph, the y-axis is a logarithmic scale that reflects order of magnitude differences between cell numbers, not simple increases in cell number. One order of magnitude difference equals a tenfold increase in cell number. For example, growing from 2×10^6 cells (bottom arrow) to 4×10^6 cells (middle arrow) is a 100% increase in cell number. This increase in percentage is greater than the 50% change when growing from 4×10^6 cells to 6×10^6 cells (top arrow). In both cases, however, the increase in the number of cells is the same, 2×10^6. A logarithmic scale (the y-axis) presents these order of magnitude differences as distances between tick marks. In our example, the distance between the tick marks indicating 2×10^6 and 4×10^6 is greater than the distance between the tick marks indicating 4×10^6 and 6×10^6. However, when cell numbers double (from 2×10^6 to 4×10^6 cells, or from 4×10^6 to 8×10^6 cells), the distances between those tick marks are equal. When we plot order of magnitude differences in this type of graph, exponential growth appears as a linear function.

medium typically experience a lag period, or **lag phase**, during which cells do not divide. For example, cells taken from an aged culture may be damaged and require time to repair themselves before they can replicate. Changes in carbon, nitrogen, or energy source in the new medium will force the organism to retool its metabolism before growth can resume. Transferring cells from spent to <u>fresh</u> complex medium requires only a very short lag phase. However, when cells are switched from a rich, complex medium to a nutritionally poor, minimal medium, they experience a long lag phase. During the lag, cells must activate new systems to synthesize all the amino acids, nucleotides, and other metabolites originally supplied by the complex medium.

Log, or exponential, phase. Once cells have retooled their physiology to accommodate the new environment, they begin to grow exponentially and enter what is called **exponential phase**, or **logarithmic (log) phase**. Exponential growth is balanced growth, where all cell components are synthesized at constant rates in relation to each other. At this stage, cells grow and divide at the maximum rate possible based on the medium and growth conditions provided (such as temperature, pH, and osmolarity; see Section 6.4 for more on osmolarity). Cells are largest at this stage of growth.

As the cell density (number of cells per milliliter) rises during log phase, the rate of doubling eventually slows (called **late log phase**), and a new set of growth phase–dependent genes are expressed. At this point, some bacteria can also detect the presence of other members of their own species or even completely different species. Bacteria do this by sending and receiving chemical signals in a process known as quorum sensing (think of it as Twitter for microbes; discussed in Section 6.6 and Chapter 9).

Stationary phase. Eventually, viable cell numbers stop rising, owing to lack of a key nutrient or buildup of waste products. This phase occurs for bacteria grown in a complex medium when cell density rises above 10^9 cells/ml, but it can occur at lower cell densities if nutrients are limiting. At this point, the growth curve levels off and the culture enters what is called **stationary phase**. In stationary phase, the growth of individual cells slows, and the number of cells dividing approximately equals the number of cells dying. As bacterial pathogens enter stationary phase, many of them develop some resistance to antibiotics and host defenses.

If microbes did not change their physiology when entering stationary phase, they would be vulnerable to deteriorating conditions in the medium. Damage from highly reactive by-products of oxygen metabolism (oxygen radicals such as superoxide; see Section 6.5) and other toxic metabolic by-products would readily kill them. As an avoidance strategy, some bacteria develop resistant spores (also called endospores) in response to nutrient depletion (see Section 6.6),

whereas other bacteria undergo less dramatic but still effective molecular reprogramming. The bacterium *E. coli*, for example, adjusts to stationary phase by decreasing its size. One benefit of these changes is that fewer nutrients are required to sustain the smaller cell. New stress resistance enzymes are also synthesized to detoxify toxic oxygen radicals, protect DNA and proteins, and increase cell wall strength through increased peptidoglycan cross-linking. As a result, *E. coli* cells in stationary phase become more resistant to heat, osmotic pressure, pH changes, and other stresses they might encounter while waiting for a new supply of nutrients.

Death phase. Without reprieve in the way of new nutrients, cells in stationary phase will eventually succumb to toxic chemicals present in the environment. Like the growth rate, the **death rate** (the rate at which cells die) is exponential. Although death curves are basically exponential, exact death rates are difficult to define because mutations arise that promote survival, and some cells grow by cannibalizing others. Consequently, the **death phase** is extremely prolonged. A portion of the cells will often survive for months, years, or even decades.

We have just described how bacteria live, grow, and die in batch culture. But how do these stages apply to growth in a human host? All growth stages do occur in a colonized or infected host, but their measurements are confounded by microbial competition, collaboration with resident microbiota, and the host's immune system. In vivo growth (growth in a live host) will be discussed further in Chapter 14.

Continuous Culture

In the classic growth curve that develops in a closed system (such as a test tube), the exponential phase spans only a few generations. In an open system, fresh medium is continually added to a culture and an equal amount of culture is constantly siphoned off (**Figure 6.10A**). This process is called **continuous culture**. Continuous culture can keep a bacterial population in exponential phase at a constant cell mass for long periods of time. All cells in the population achieve a steady state, which permits detailed analysis of microbial physiology at different growth rates. The **chemostat** is a continuous culture system in which the diluting medium contains a growth-limiting amount of an essential nutrient (**Figure 6.10B**). Increasing the flow rate increases the amount of nutrients available to the microbe. The more nutrients available, the faster a cell's mass will increase. The faster the cell's mass increases, the more quickly it will divide. It follows, then, that the growth rate in a chemostat is directly related to the dilution rate, or flow rate (milliliters per hour divided by the vessel volume). The more nutrient a culture receives as a result of increasing flow rate, the faster those cells can replicate (that is, the shorter the generation time). Your intestine, which is normally populated by billions of bacteria, functions much like a chemostat (see Figure 6.10A).

Figure 6.10 Chemostats and Continuous Culture

A. The basic chemostat ensures exponential growth by constantly adding and removing equal amounts of culture medium. The human gastrointestinal tract is engineered much like a chemostat in that new nutrients in food enter from the mouth while corresponding amounts of bacterial culture exit in fecal waste, although the process is not continuous.

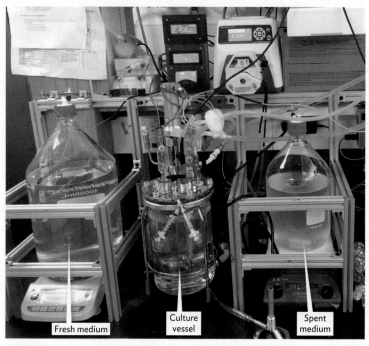

B. A modern chemostat.

Continuous cultures are used in the laboratory to study large numbers of cells at constant growth rate and cell mass for both research and industrial applications. Most bacteria in nature grow at very slow rates, a situation that can be mimicked in a chemostat. The physiology of these slow-growing cells is quite different from what is typically observed in batch culture.

SECTION SUMMARY

- **The growth cycle** of organisms grown in liquid batch culture consists of lag phase, log phase, stationary phase, and death phase.
- **The physiology of a bacterial population** changes throughout the phases of growth.
- **Continuous culture** can be used to sustain a population of bacteria at a specified growth rate and cell density.

Thought Question 6.2 If bacteria divide every 20 minutes, why is the exponential growth curve (measured as viable count) smooth and not shaped like a series of steps?

Thought Question 6.3 Suppose 1,000 bacterial cells are inoculated into a tube of minimal salts medium, where they double once an hour, and 10 bacteria are inoculated into a separate tube of rich medium, where they double every 20 minutes. Which tube will contain more bacteria after 2 hours? After 4 hours?

6.3 Microbial Nutrition

SECTION OBJECTIVES

- Compare heterotrophy, autotrophy, phototrophy, and chemotrophy.
- Describe the importance of the nitrogen and carbon cycles and the role of microbes in the maintenance of each.
- Discuss biofilms and their relevance to infectious diseases.

Are there any limits to bacterial growth? Bacterial cells, for all their apparent simplicity, are remarkably complex and efficient replication machines. One cell of *E. coli*, for example, can divide into two cells every 20–30 minutes. At a rate of 30 minutes per division, one cell could potentially multiply to over 1×10^{14} cells in 24 hours—that is, 100 trillion organisms! Although 100 trillion cells would weigh only about 1 gram, after another 24 hours (a total of 2 days) of continuous growth, the mass of cells would explode to 10^{14} grams, or 10^7 tons, a weight equivalent to 4 million dump trucks piled on top of one another! So why aren't we buried under mountains of *E. coli*? The answer is limiting nutrients and ecological balance.

Nutrients and Environmental Niches

Do all cells build themselves the same way? And if not, is there a core set of nutrients that all cells must use to grow? **Essential nutrients** are those compounds a microbe cannot make itself but must gather from its immediate environment if the cell is to grow and divide. Some essential nutrients, such as carbon (C), nitrogen (N), phosphorus (P), hydrogen (H), oxygen (O), and sulfur (S), are needed in large quantities because they make up the carbohydrates, lipids, nucleic acids, and proteins of the cell; these are known as macronutrients. The enormous number of microbial species that exist today evolved to use different environmental sources of elements. As a result, all cells do not build themselves the same way. However, all cells indeed use the same essential nutrients.

In addition to essential nutrients, a variety of cations are needed for growth. Cations such as magnesium (Mg^{2+}), ferrous iron (Fe^{2+}), and potassium (K^+) are needed in large amounts because they serve as enzyme cofactors (Mg^{2+}, Fe^{2+}) or help balance cell osmolarity (K^+). In addition, all cells require small amounts of cobalt, copper, manganese, molybdenum, nickel, and zinc—known as the trace elements (micronutrients)—as essential components of some enzymes. Cobalt, for example, is part of the cofactor vitamin B_{12}, which aids the catalytic process of specific enzymes.

In addition to the simple blend of elements mentioned above, some organisms, such as the intestinal bacterium *E. coli*, need only a carbon source (like glucose) and an energy source (also glucose) to make all their proteins, nucleic acids, and cell wall and membrane components. These organisms can be grown in a **chemically defined minimal medium** that contains only those compounds needed for an organism to propagate (Table 6.1).

Many microbes, however, cannot synthesize all the compounds they need for growth. For instance, all organisms require glutamic acid to grow. Many bacteria make their own glutamic acid because the amino acid is not always available in their surroundings. But if glutamic acid became <u>permanently</u> available in the organism's natural environment, the organism would no longer need to make that amino acid. Consequently, any cell that became defective in making glutamic acid through random mutation would survive and might even flourish. But that mutant and its offspring will now <u>require</u> an external supply of glutamic acid for growth. Any such compound is called a **growth factor** for that microbe, and it must be included in laboratory medium if the microbe is to grow. Table 6.2 presents the growth factors needed by some common microbes associated with disease.

Consider *Haemophilus influenzae*, the causative agent of Mirembe's meningitis in our chapter-opening case history. This pathogen cannot synthesize NAD (V factor, a cofactor needed to harvest energy from nutrients) or heme (derived from hemin, X factor, needed to make cytochromes for energy production). Why should *H. influenzae* make these compounds if they are readily available in its normal environment, the human respiratory tract? *H. influenzae* cannot grow on typical laboratory media unless those specific growth factors are added. Knowing what growth factors are required by certain pathogenic bacteria can help clinical laboratory technicians identify the agent causing an infection.

Although we can grow many bacteria in or on laboratory media, some bacteria have such complex nutrient requirements in their

Table 6.1
Compositions of Some Commonly Used Media

Medium	Ingredients per Liter	Organisms Cultured
Lysogeny broth; also called Luria-Bertani broth (complex)	Bacto tryptone[a], 10 g Bacto yeast extract, 5 g NaCl, 10 g pH 7	Many Gram-negative and Gram-positive organisms
M9 medium (defined)	Glucose, 2.0 g Na_2HPO_4, 6.0 g (42 mM) KH_2PO_4, 3.0 g (22 mM) NH_4Cl, 1.0 g (19 mM) NaCl, 0.5 g (9 mM) $MgCl_2$, 2.0 mM $CaCl_2$, 0.1 mM pH 7	Gram-negative organisms such as *E. coli*
Azotobacter medium (defined)	Mannitol, 20g $(NH_4)_2SO_4$, 0.2g K_2HPO_4, 0.8 g $MgSO_4$, 0.1g $CaSO_4$, 0.1g $MoO_4 \cdot 2H_2O$, 0.002g $FeSO_4$, 0.1 mg	*Azotobacter*
Sulfur oxidizers medium (defined)	NH_4Cl, 0.52 g KH_2PO_4, 0.28 g $MgSO_4 \cdot 7H_2O$, 0.25 g $CaCl_2$, 0.07 g Elemental sulfur, 1.56 g CO_2, 5% pH 3	*Acidithiobacillus thiooxidans*

[a]Bacto tryptone is a pancreatic digest of casein (bovine milk protein).

natural habitat that we still cannot grow them using artificial media. For instance, *Rickettsia* species that cause Rocky Mountain spotted fever and typhus grow only within the cytoplasm of eukaryotic cells (Figure 6.11). This obligate intracellular bacterium lost key pathways needed for independent growth because the host cell supplies them. We have yet to discover what those factors are.

Next, we provide an introductory look at the ways different bacterial life-forms obtain the carbon, energy, and nitrogen needed for their growth while maintaining ecological balance. Deeper discussion of metabolic pathways used to harvest carbon and nitrogen and to generate energy will take place in Chapter 7.

Obtaining Carbon: Autotrophy and Heterotrophy

How do microbes obtain carbon to build biomass? All Earth's life-forms are based on carbon. But if all life-forms used carbon in the same way, the carbon would become trapped in an unusable form. The carbon cycle involves two counterbalancing metabolic groups of organisms—heterotrophs and autotrophs—that use different metabolic pathways to obtain carbon. Each group converts carbon to a form the other group can use.

Table 6.2
Growth Factors and Natural Habitats of Bacterial Pathogens

Organism[a]	Disease	Natural Habitats	Growth Factors
Shigella	Bloody diarrhea	Human intestine	Nicotinamide (from NAD)[b]
Haemophilus influenzae	Meningitis	Humans and other animal species, upper respiratory tract	Hemin, NAD
Staphylococcus aureus	Boils, osteomyelitis	Nasal passages; widespread but short-lived in nature	Multiple requirements
Abiotrophia	Osteomyelitis	Humans and other animal species	Vitamin K, cysteine
Legionella pneumophila	Legionnaires' disease	Soil, refrigeration cooling towers	Cysteine
Bordetella pertussis	Whooping cough	Humans and other animal species	Glutamic acid, proline, cysteine
Francisella tularensis	Tularemia	Wild deer, rabbits	Multiple requirements, notably cysteine
Mycobacterium tuberculosis *Mycobacterium leprae*	Tuberculosis Leprosy	Humans	Nicotinic acid (from NAD),[b] alanine *M. leprae* has not yet been cultured
Streptococcus pyogenes	Pharyngitis, rheumatic fever	Humans	Glutamic acid, alanine

[a]These are just examples. More complete coverage of the diseases caused by these organisms is found in Chapters 19–24.

[b]Both nicotinamide and nicotinic acid are derived from NAD, nicotinamide adenine dinucleotide.

Heterotrophs (like *E. coli*) rely on other organisms to make the organic compounds (such as glucose) they use as carbon sources. All bacterial pathogens are heterotrophs. During heterotrophic metabolism, organic carbon sources are disassembled to generate energy and then reassembled to make cell components such as proteins and carbohydrates. This process converts a large amount of the organic carbon source to carbon dioxide (CO_2), which is then released to the atmosphere. Consequently, if left alone, heterotrophs would deplete the world of organic carbon sources (converting them to unusable CO_2), and they would starve to death. For life to continue, CO_2 must be recycled.

Autotrophs use the CO_2 discarded by heterotrophs to make organic compounds such as carbohydrates and amino acids, and then they use those compounds to build complex cell components. To complete the carbon cycle, the organic compounds synthesized by autotrophs can, after death, be used as carbon sources by heterotrophs.

The interdependence of autotrophs and heterotrophs is an integral part of the carbon cycle (**Figure 6.12**). Keep in mind that the metabolism of any one group of organisms depends on the metabolism of other groups of organisms. Even in the microbial world, "it takes a village."

Figure 6.11 *Rickettsia prowazekii* Growing within Eukaryotic Cells

A. *R. prowazekii* growing within the cytoplasm of a chicken embryo fibroblast.

B. Fluorescent stain of *R. prowazekii* (yellow-green, 0.5 μm long) growing within a cultured human cell (outline marked by dotted line). Host nucleus is shown in blue; mitochondria are red.

Figure 6.12 The Carbon Cycle

The carbon cycle requires both autotrophs and heterotrophs.

A. Heterotrophs use organotrophy to gain energy by degrading complex organic compounds, such as polysaccharides, to smaller compounds, such as glucose and pyruvate. The carbon from pyruvate moves through the tricarboxylic acid (TCA) cycle, or Krebs cycle, and is released as CO_2. Without a TCA cycle, the carbon can end up as fermentation products, such as ethanol or acetic acid.

B. Autotrophs use light energy or energy derived from the oxidation of minerals to capture CO_2 and convert it to complex organic molecules.

C. *Chloroflexus aggregans*, originally isolated from hot springs in Japan, possesses extraordinary metabolic versatility. It grows anaerobically (without oxygen) as a photoheterotroph (able to capture light for energy and use organic compounds for carbon) and aerobically (with oxygen) as a chemoheterotroph (able to use organic compounds for sources of energy and carbon).

Capturing Energy: Phototrophy and Chemotrophy

How do microbes obtain energy? The cellular synthesis of proteins, nucleic acids, and other structures from carbon requires an energy source. Organisms capture energy in two basic ways. Some microbes extract energy from chemical reactions triggered by the absorption of light (called **phototrophy**, or photosynthesis). Other microbes use oxidation-reduction reactions to remove electrons from high-energy compounds (like glucose) to produce lower-energy compounds. This process is called **chemotrophy** and it comes in two forms. One form (**lithotrophy**) uses inorganic chemicals for energy. These might include hydrogen (H_2), hydrogen sulfide (H_2S), ammonium (NH_4), nitrite (NO_2^-), and ferrous iron (Fe_2^+). The other form (**organotrophy**) oxidizes organic compounds like sugars to obtain energy. Most pathogenic bacteria use carbohydrates as energy *and* carbon sources, but some can also use inorganic compounds (for example, hydrogen gas) as energy sources.

In chemotrophy, a reduced compound, such as glucose, can donate one or more of its electrons to a less reduced (that is, more oxidized) compound, such as nicotinamide adenine dinucleotide (NAD). During this process, energy is released in the form of donated electrons. NAD is a cell molecule critical to energy metabolism and is discussed, along with oxidation-reduction reactions, in Chapter 7. An analogy that explains oxidation-reduction is a parent giving you money to buy a car. Your parent, who has money (making them reduced in this analogy), donates money (electrons)

to you. Your parent becomes a little poorer (becoming slightly "oxidized"), but you become a little richer ("reduced") as a result.

Different species of bacteria and archaea can carry out these varied forms of carbon and energy acquisition. Remember, however, that only bacteria—not archaea—cause disease, and among the bacteria, only certain heterotrophic organotrophs are pathogenic.

⊙ **Note:** In biology, the suffix "-trophy" refers to the acquisition of nutrients. The following prefixes for "-trophy" terms help distinguish different forms of biomass building (carbon source) and energy-yielding (energy source) metabolism.

Carbon Source for Biomass	
Auto-	CO_2 is fixed and assembled into organic molecules.
Hetero-	Preformed organic molecules are acquired from outside the cell and broken down for carbon, and the carbon is reassembled to make biomass.
Energy Source	
Photo-	Light absorption captures energy.
Chemo-	Preformed organic or inorganic electron donors are oxidized.
Electron Source	
Litho-	Inorganic molecules donate electrons.
Organo-	Organic molecules donate electrons.

Figure 6.13 The Nitrogen Cycle Simplified

Dinitrogen gas (N_2) is fixed in ammonium ions (NH_4^+) by species of bacteria (nitrogen fixers) that possess the enzyme nitrogenase. Other bacteria (nitrifiers) oxidize NH_4^+ to generate energy. Still others (denitrifiers) use oxidized forms of nitrogen, such as NO_3^-, as an alternative electron acceptor in place of O_2.

Obtaining Nitrogen

How do different microbes obtain nitrogen, another essential nutrient? Nitrogen is a critical component of proteins, nucleic acids, and other cellular constituents and is required in large amounts by living organisms. Nitrogen gas (N_2) makes up nearly 79% of Earth's atmosphere. However, the nitrogen in N_2 is unavailable for use by most organisms because the triple bond between the two nitrogen atoms requires considerable energy to break. Nitrogen from N_2, to be used for growth, must first be "fixed," meaning it must be converted to ammonium ions (NH_4^+). As with the carbon cycle, various groups of organisms collaborate to convert nitrogen gas to ammonium, then to nitrate (NO_3^-), and back again to nitrogen gas in what is called the **nitrogen cycle** (Figure 6.13).

One type of bacteria "fixes" atmospheric nitrogen to make NH_4^+ (**nitrogen fixation**), while other groups of bacteria do the opposite, transforming ammonia to nitrate (**nitrification**) and then converting nitrate to N_2 (**denitrification**).

Nitrogen-fixing bacteria may be free-living in soil or water, or they may form symbiotic associations with plants or other organisms. A **symbiont** is an organism that lives in intimate association with a second organism. For example, *Rhizobium*, *Sinorhizobium*, and *Bradyrhizobium* species are nitrogen-fixing symbionts present in leguminous plants such as soybeans, chickpeas, and clover (Figure 6.14). These plant symbionts convert atmospheric nitrogen to

ammonium, which the plant needs to form proteins and other essential compounds. Although symbionts are the most widely known nitrogen-fixing bacteria, free-living bacteria and archaea fix the majority of nitrogen in soil and marine environments. Even some bacterial pathogens can carry out nitrification, denitrification, or nitrogen fixation.

This overview—the "big picture" of how bacterial cells of all types gather carbon, generate energy, and recycle nitrogen—is preparation for the deeper exploration of bacterial metabolism to come in Chapter 7. For now, appreciate that all these life-forms coexist and collaborate to carry out the continuous nutrient cycling that sustains Earth. Without these life-forms, even humans would not exist.

SECTION SUMMARY

- **Essential nutrients fuel the growth** of all microorganisms.
- **Microbial genomes evolve in response to nutrient availability.** As a result, some disease-causing bacteria have developed requirements for growth factors supplied by their hosts.
- **Autotrophs use CO_2 as a carbon source.**
- **Microbes obtain energy through** phototrophy, lithotrophy, or organotrophy.
- **Heterotrophs consume organic compounds** made by autotrophs; the global carbon cycle is made possible by the interactions between heterotrophs and autotrophs.
- **Nitrogen fixers, nitrifiers, and denitrifiers** contribute to the nitrogen cycle.

- -

Thought Question 6.4 Red blood cells contain NAD and hemin. Why, then, in our chapter-opening case history, did *H. influenzae* not grow on blood agar medium?

Thought Question 6.5 You have designed a mixed ecosystem of autotrophs and heterotrophs. The ecosystem contains an amount of organic substrate that will support growth of the heterotroph for just a short

Figure 6.14 *Rhizobium* and a Legume

A. Symbiotic *Rhizobium* cells forming a nodule on a pea plant (SEM). The rhizobia (about 0.9 mm × 3 mm) invade the root to begin a symbiotic partnership that will benefit both organisms.

B. After the rhizobia invade the plant root, symbiosis between plant and microbe produces nodules (indicated by the arrow).

period of time. The system also contains very little CO_2 for the autotroph to use. The system is supplied with oxygen but no extra CO_2. What happens to the autotroph and heterotroph in this ecosystem?

6.4 Environmental Limits on Microbial Growth

> ## SECTION OBJECTIVES
> - Describe how environmental changes can alter the shape of a growth curve.
> - List different classes of microbes on the basis of their preferred environmental niches (pH, temperature, salt).
> - Identify the biological properties that allow different classes of microbes to grow in extreme environments.

What are the environmental limits of growth? We have explored bacterial growth and nutrition, but another important factor is the impact of environment on growth. Within our human frame of reference, we tend to think of "normal" growth conditions as those that exist at sea level: for example, a temperature between 20°C (68°F) and 40°C (104°F), a near-neutral pH, a salt concentration of 0.9%, and ample nutrients. Any ecological niche outside this window is called <u>extreme</u>. **Extremophiles** are bacteria, archaea, and some eukaryotic microbes that can grow in extreme environments. For instance, one group of organisms can grow at temperatures above boiling, while another group's growth relies on a pH 2 acidic environment. According to our definition of <u>normal</u> conditions, conditions on Earth when life began were certainly extreme. Consequently, the earliest microbes probably grew in these extreme

environments. Organisms that grow under conditions that seem "normal" to humans probably evolved from an ancient extremophile that gradually adapted as the environment evolved to that of our present-day Earth.

Be aware that multiple extremes in an environment can be encountered simultaneously. For instance, in Yellowstone National Park, we can find a very hot acid pool next to a very hot alkali pool. Thus, extremophiles typically evolve to survive multiple extreme environments. At Yellowstone, this means that some extremophiles can grow in highly alkaline pH at high temperature while others grow in highly acidic pH at high temperature. No extremophile, however, can grow in both highly acidic and highly alkaline environments.

Why do differences in physical conditions (temperature, pH, and osmolarity) favor the growth of specific groups of organisms? Remember, the folded shape of an enzyme is critical to its ability to convert substrates (reactants) to products. However, the folding process is profoundly affected by temperature, pH, osmolarity, and the reactive by-products of oxygen consumption (see Section 6.5). So, each enzyme has an <u>optimal</u> condition at which it folds into proper shape, or conformation. Deviations from these conditions will cause a protein to unfold (denature) or fold incorrectly, lessening its ability to catalyze reactions. Different enzymes within a cell can have different physical optima (temperature, pH, and salt concentration) for folding, but these optima must be similar and matched to the organism's environment if the organism is to grow. Note that many disinfection procedures actually work by denaturing bacterial proteins (see Chapter 13).

Given what we have learned so far in this chapter, it's no surprise that microbes can be classified by their environmental niche. **Table 6.3** lists various classes of microorganisms and the environmental criteria that define them.

Table 6.3
Basic Environmental Classification of Microorganisms

Environmental Parameter	Classification[a] (optimal growth condition)			
Temperature	Hyperthermophile (above 80°C [176°F])	Thermophile (between 50°C [122°F] and 80°C [176°F])	Mesophile (between 15°C [59°F] and 45°C [113°F])	Psychrophile (below 15°C [59°F])
pH	Alkaliphile (above pH 9)	Neutralophile (between pH 5 and pH 8)	Acidophile (below pH 3)	
Osmolarity	Halophile (high salt, greater than 2 M NaCl)			
Oxygen	Strict aerobe (only in oxygen)	Facultative microbe (with or without oxygen)	Microaerophile (only in small amounts of oxygen)	Strict anaerobe (only without oxygen)
Pressure	Barophile (at high pressure, greater than 380 atm)		Barotolerant (between 10 and 500 atm)	

[a]Classes shown in red are considered extremophiles.

Variations in Temperature

Unlike humans (and mammals in general), microbes cannot control their temperature, so bacterial cell temperature will match that of the environment. But because temperature affects the rate of molecular motion, any change in temperature will impact every aspect of microbial physiology. This includes membrane fluidity, nutrient transport, DNA and RNA stability, and enzyme structure and function. Every organism, therefore, has an <u>optimum</u> temperature at which it grows most quickly, as well as minimum and maximum temperatures that define the limits of its growth. The fastest (or optimal) growth rate for a species occurs at a temperature that allows all the cell's proteins to work most efficiently as a group. Growth stops when rising temperatures cause critical enzymes or cell structures (like the cell membrane) to fail. At cold temperatures, growth ceases because enzymatic processes become too sluggish and the cell membrane less fluid. The membrane needs to remain fluid so it can expand as cells grow larger and so proteins needed for solute transport can be inserted into the membrane.

Great biological diversity exists among microbes, as different groups have evolved to grow within different thermal niches. A species grows within a specific thermal range because its proteins have evolved to tolerate that range. Outside that range, however, proteins will denature or function too slowly for growth. The upper limit for protists (single-celled eukaryotic microbes) is around 50°C (122°F), whereas some fungi can grow at temperatures as high as 60°C (140°F). Examples of bacteria and archaea, however, have been found to grow at temperatures ranging from below 0°C (32°F) to above 100°C (212°F). Temperatures over 100°C are usually found around superheated thermal vents deep in the ocean, where water temperature can rise to 350°C (662°F) but the pressure is sufficient to keep water liquid.

By their ranges of growth temperature, microorganisms can be classified as liking it hot (thermophiles), cold (psychrophiles), or in between (mesophiles), as shown in **Figure 6.15**.

Thermophiles have adapted to growth at high temperature, typically 55°C (131°F) and higher. **Hyperthermophiles** grow at temperatures as high as 121°C (250°F), which occur under extreme pressure (for example, at the ocean floor). These organisms flourish in hot environments such as composts, near thermal vents that penetrate Earth's crust on land (for example, hot springs, as shown in **Figure 6.16A**), or on the ocean floor (**Figure 6.16C**). Most thermophiles have membranes and protein sequences specially adapted to life at high temperatures. For example, proteins in thermophiles fold more tightly than those in mesophiles, so they do not unfold (denature) at high temperatures.

One thermophile that has greatly benefited humans is *Thermus aquaticus,* discovered in a hot spring at Yellowstone National Park. A high-temperature thermophilic polymerase purified from this organism was critical to developing the polymerase chain reaction (PCR) used to amplify DNA in a test tube (discussed in Chapter 8). PCR has revolutionized molecular biology.

Figure 6.15 Relationship between Temperature and Growth Rate

Each curve represents the temperature growth range for a different group of bacteria. Growth rate increases linearly with temperature until an optimum is reached, and then it decreases. These curves are only an approximation. Growth rates and optimal ranges for individual species within a group will vary.

Figure 6.16 Thermophilic Environments and Thermophiles

A. Yellowstone National Park hot spring.

B. *Thermus aquaticus,* a hyperthermophile first isolated at Yellowstone National Park by Thomas Brock. Cell length varies from 3 to 10 μm.

C. A "smoker," a hydrothermal vent 2 miles deep in the Pacific Ocean.

D. *Methanocaldococcus jannaschii,* isolated in 1983 near this vent, grown at 78°C (172.4°F) and 30 psi.

Figure 6.17 Psychrophilic Environment and a Psychrophile

A. An iceberg, in which psychrophilic organisms can be found.

B. Psychrotolerant *Flavobacterium* (grows between 0°C and 22°C) from a South Pole lake made from glacial meltwater in summer (high temperature = 0.9°C). Novel compounds made by members of the polar microbiome are screened for anticancer and antimicrobial potential.

SEM   5 µm

Psychrophiles are microbes that grow at temperatures as low as –10°C (14°F), but their optimum growth temperature is usually around 15°C (59°F). Because they are adapted to growth at cold temperatures, psychrophiles are prominent flora beneath icebergs in the Arctic and Antarctic (**Figure 6.17**). Psychrophilic microbes can grow in the cold because their proteins are more flexible than those of mesophiles and require less energy (heat) to function. The downside to increased flexibility is that it does not take much heat to denature psychrophilic proteins. As a result, psychrophiles grow poorly, if at all, when temperatures rise above 20°C (68°F). Another reason psychrophiles can grow in cold is that their membranes are more fluid than mesophilic membranes at low temperature. At higher temperatures, the membranes of psychrophiles are too flexible and fail to maintain cell integrity.

Mesophiles include the typical "lab rat" microbes, such as *E. coli* and *Bacillus subtilis*. Their growth optima range between 20°C (68°F) and 40°C (104°F), with a minimum of 15°C (59°F) and a maximum of 45°C (113°F). Because they are easy to grow and because human pathogens are also mesophiles, much of what we know about protein, membrane, and DNA structure came from studying this group of organisms. Though they prefer growth at moderate temperatures, cold-resistant mesophiles of clinical importance do exist. For instance, the modern practice of refrigeration has selected for cold-resistant mesophilic pathogens that can grow in refrigerated foods. One example is *Listeria monocytogenes* (a cause of food poisoning and septic abortions; see Chapter 22).

Variations in Pressure

Creatures living at Earth's surface (sea level) are subjected to a pressure of 1 atmosphere (atm), which is equal to 0.101 megapascal (MPa) or 14 pounds per square inch (psi). These include bacteria that can destroy various foods. Because high pressure can kill these bacteria, many food processes are carried out at high pressure to minimize bacterial contamination. High-pressure processing has been used on cheeses, yogurt, luncheon meats, and oysters to kill contaminating bacteria without destroying the flavor or texture of the food.

At the bottom of the ocean, however, thousands of meters deep, hydrostatic pressure averages a crushing 400 atm and can go as high as 1,000 atm (101 MPa, or 14,600 psi) in ocean trenches (**Figure 6.18**). Microorganisms at locations deep within the ocean that have adapted to grow at oppressively high pressures are called **barophiles** or **piezophiles**. Barophiles actually require elevated pressure to grow, whereas barotolerant organisms grow well over the range of 10–500 atm (1–50 MPa), but their growth falls off thereafter. Many barophiles are also psychrophilic because the average temperature at the ocean's floor is 2°C (35.6°F).

Water Availability and Salt Concentration

Water is critical to life, but environments differ in terms of how much water is actually available to growing organisms; microbes can use only available water that is not bound to ions or other solutes in solution. Water availability is also known as water potential. Water availability is measured as water activity (a_w), a quantity approximated by concentration. Because interactions with solutes (such as NaCl) lower water activity, the more solutes there are in a solution (for example, the higher the salt concentration), the less water becomes available for microbes to use for growth. **Osmolarity** is a measure of the number of solute molecules in a solution and is inversely related to a_w. The more particles that are in a solution, the greater the osmolarity and the lower the water activity. The food industry manipulates water availability in some foods as a way to prevent bacterial growth and limit spoilage.

Oceans, freshwater lakes, and estuarine environments (where ocean and freshwater mix) exhibit different concentrations of salt and water activity. Thus, the bacteria and archaea that inhabit these niches exhibit different salt tolerances. Several bacteria from these environments can be pathogenic. For example, *Vibrio vulnificus*, a Gram-negative species that can cause fatal septicemia, and *Vibrio cholerae*, a cause of extreme diarrhea, are marine microorganisms that prefer growth at an elevated salt concentration

Figure 6.18 Barophilic Environment and a Barophile

A. Ocean depths. The deepest part of the ocean is at the bottom of the Mariana Trench, a depression in the floor of the western Pacific Ocean, just east of the Mariana Islands. The Mariana Trench is 2,500 km (1,554 miles) long and 70 km (44 miles) wide. Near its southwestern extremity, about 340 km (210 miles) southwest of Guam, lies the deepest point on Earth. This point, referred to as "Challenger Deep," plunges to a depth of nearly 11,035 meters (nearly 7 miles). The pressure there (110 MPa) is over 1,000 times higher than what we experience on land (0.1 MPa).

0.296 atm

0.98 atm

Average height of the land, 870 m

Mount Everest, 8,848 m

365 atm

Average depth of the ocean, 3,730 m

1,085 atm

B. Barophile *Shewanella.*

Mariana Trench, 11,035 m

TEM   1 µm

Microbial Responses to Changes in pH

As with salt and temperature, the concentration of hydrogen ions (H$^+$) also has a direct effect on the cell's macromolecular structures. Actually, hydrogen ions associate with water making their proper name hydronium ions (H$_3$O$^+$). Extreme concentrations of either hydronium or hydroxide (OH$^-$) ions in a solution will limit growth. In other words, too much acid (hydronium) or base (hydroxide) is harmful to cells. Despite this sensitivity to pH extremes, living cells can <u>tolerate</u> a greater range of H$^+$ concentration than of virtually any other chemical substance. *E. coli*, for example, tolerates an extracellular pH range of 2–9, a 10,000,000-fold difference!

How does pH influence growth? The intramolecular bonds that shape proteins into active forms are highly sensitive to proton concentrations. Low pH (that is, excess protons) will transform negatively charged carboxyl groups (R-COO$^-$) on an amino acid to their uncharged form (R-COOH). Likewise, low pH can change uncharged amino groups (R-NH$_2$) to a positively charged form (R-NH$_3^+$). The reverse processes happen at high pH, when fewer protons

Figure 6.19 Halophilic Environment and a Halophile

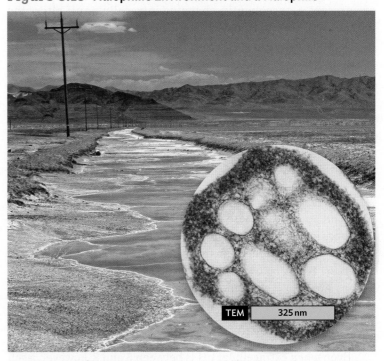

TEM   325 nm

(up to 1 M NaCl). Other diarrhea-causing organisms, such as certain strains of *E. coli* and *Salmonella*, prefer growth at much lower salt concentrations (approximately 0.2 M NaCl). These pathogens inhabit freshwater contaminated by fecal material from humans and other animals.

Some species of archaea have evolved to require high concentrations of salt (NaCl); these are called **halophiles** (Figure 6.19). In striking contrast to most bacteria, which prefer salt concentrations in the range of 0.05–1 M (0.2%–5%) NaCl, the moderately halophilic microbes grow optimally at 0.85–3.4 M (5%–20%) NaCl. For comparison, seawater is about 3.5% NaCl. Extremely halophilic archaea require 3.4–5.1 M (20%–30%) NaCl to grow. Halotolerant organisms can grow in conditions of high or low salt (0%–30% NaCl).

All cells, even halophiles, prefer to keep a relatively low intracellular concentration of sodium ions (Na$^+$). To achieve a low internal Na$^+$ concentration, halophilic microbes use special ion pumps to excrete sodium and replace it with other cations, such as potassium. Potassium is a compatible solute—one that can accumulate to high levels intracellularly without affecting the functions of enzymes.

A. The halophilic salt flats along Highway 50, east of Fallon, Nevada, are colored pinkish red by astronomical numbers of halophilic bacteria.

B. *Halobacterium* species showing an intracellular cluster of gas vesicles; cross-sectional cell width 0.5–0.8 µm.

Figure 6.20 Classification of Organisms Grouped by Optimum Growth pH

Note that pH and pOH are the reciprocal values of hydronium (H_3O^+) and hydroxide (OH^-) concentrations, respectively. Although the range of intracellular pH compatible with life is between pH 6 and pH 8, acidophiles grow well below that pH range and alkaliphiles grow well above it.

are present. Consequently, changing intracellular pH alters the ionic charges on proteins, which in turn alters protein shape. A misshapen protein, if it is an enzyme, will exhibit lower—or even no—activity.

The result is that all enzyme activities exhibit optima, minima, and maxima with regard to pH, much as they do for temperature. As we saw with temperature, groups of microbes have evolved to inhabit environments with diverse pH conditions that can range from pH 0 to pH 11.5 (**Figure 6.20**). However, the optimum pH at which a particular species grows is <u>not</u> dictated by the pH limits of critical cell proteins.

In general, most enzymes, regardless of the range of growth pH, tend to operate best between pH 5 and pH 8.5. Yet many microbes grow in environmental pH conditions that are even more acidic or alkaline than this range. How can that be? Growth above pH 8.5 or below pH 5 is possible because biological membranes are relatively impermeable to protons. This fact allows the cell to maintain an internal pH compatible with protein function even when growing in extremely acidic or alkaline environments.

Even when extracellular pH is between 5 and 8, however, a cell can struggle to maintain an intracellular pH suitable for growth. The presence of membrane-permeant organic acids such as citric acid or propionic acid (also called weak acids) can accelerate the

leakage of H^+ into cells (discussed in Section 4.5). The food industry has taken advantage of this phenomenon by preemptively adding citric acid or sorbic acid to certain foods, such as jams, canned fruits, candy, and carbonated beverages. This allows manufacturers to stop microbial growth under food pH conditions that do not destroy the flavor or quality of the food. (Food microbiology is discussed more fully in Chapter 27.)

Neutralophiles, Acidophiles, and Alkaliphiles

How do microbes grow at pH extremes, and how are they classified with regard to their growth pH? Cells have evolved to live under different pH conditions not by drastically changing the pH optima of their enzymes but by using novel pH homeostasis strategies. Such strategies maintain intracellular pH above pH 5 and below pH 8, even when the cell is immersed in pH environments well above or below that range (refer to Figure 6.20). Microbes are divided into three categories based on their success in maintaining intracellular pH under different extracellular pH extremes.

Neutralophiles are bacteria that generally grow between pH 5 and pH 8; this group includes most human pathogens. Many neutralophiles, including *E. coli* and *Salmonella enterica*, adjust their metabolism to maintain an internal pH slightly above neutrality, which is where their enzymes work best. Other species of neutralophiles allow their internal pH to fluctuate with external pH but usually maintain a pH difference (ΔpH) of about 0.5 pH unit across the membrane at the upper and lower limits of growth pH. The value of ΔpH is an important factor in energy production, which is discussed in Chapter 7.

Acidophiles are bacteria and archaea that live in acidic environments. They are often **chemoautotrophs** (lithotrophs) that oxidize reduced metals and generate strong acids such as sulfuric acid. Consequently, they grow between pH 0 and pH 5. The ability to grow at this pH is due partly to altered membrane lipid profiles that decrease proton permeability and partly to ill-defined mechanisms that move protons out of the cell.

Alkaliphiles occupy the opposite end of the pH spectrum, growing best at values ranging from pH 9 to pH 11. They are commonly found in saline soda lakes, which have high salt concentrations and pH values as high as pH 11. Soda lakes, such as Lake Magadi in Africa (**Figure 6.21A**), are steeped in carbonates, which explains their extraordinarily alkaline pH.

The cyanobacterium *Spirulina* is an alkaliphile that grows in soda lakes (**Figure 6.21B**). Its high concentration of carotene gives the organism a distinctive pink color (note the color of the lake in Figure 6.21A). *Spirulina* is also a major food for the famous pink flamingos indigenous to these African lakes and is, in fact, the reason pink flamingos are pink. After the birds ingest these organisms, digestive processes release the carotene pigment to the circulation, which then deposits it in the birds' feathers, turning them pink (**Figure 6.21C**).

Figure 6.21 A Soda Lake Ecosystem

A. Lake Magadi, in Kenya. Its pink color is due to *Spirulina*.

C. Pink flamingo colored by ingestion of *Spirulina*.

LM | 10 μm

B. *Spirulina* is a coiled, filamentous cyanobacterium that grows in the alkaline environment of Lake Magadi.

Note: Humans who consume *Spirulina* as a health food supplement do not turn pink because the cyanobacteria are only a small component of their diet.

The key to the survival of alkaliphiles is the cell membrane barrier that sequesters fragile cytoplasmic enzymes away from harsh extracellular pH. In contrast to proteins <u>within</u> the cytoplasm, the enzymes <u>secreted</u> from alkaliphiles work in very alkaline environments. Similar base-resistant enzymes such as lipases and cellulases are included in laundry detergents to whiten our whites and brighten our brights.

pH Homeostasis Mechanisms

When cells are placed in pH conditions below their optimum growth pH, the excessive external protons can "force" their way into the cell and lower internal pH to lethal levels. Microbes can prevent the unwanted influx of protons in several ways. *E. coli*, for example, can "throw" protons out of the cell by linking their expulsion to the import of a variety of cations such as K^+ or Na^+. Internal charge will be the same, but the number of protons will be less. Under extremely alkaline conditions, microbes can use Na^+/H^+ antiporters (see **Section 5.2**) to recruit protons into

the cell in exchange for expelling Na^+. Some organisms can also change the pH of the medium by using various amino acid decarboxylases and deaminases. Decarboxylases remove a carboxyl group from an organic molecule to produce an alkaline product. Deaminases, however, remove an amine group from organic molecules to yield acidic products. For example, *Helicobacter pylori*, the causative agent of gastric ulcers, employs an exquisitely potent urease (an enzyme that catalyzes the hydrolysis of urea) to generate massive amounts of ammonia. The ammonia neutralizes acid pH and allows the organism to survive and grow along the stomach lining.

Many, if not all, microbes also have the ability to sense a change in environmental pH and alter their cell physiology in response. The levels of many proteins increase, while the levels of others decrease. The result includes modifications in membrane lipid composition, enhanced pH homeostasis, and many other changes. Some pathogens, such as *Salmonella*, sense a change in external pH as part of the signal indicating the bacterium has entered a host cell environment (see **IMPACT**).

SECTION SUMMARY

- **Extremophiles** inhabit "fringe" environmental conditions that do not support human life.
- **The environmental niche** (such as conditions of high salt or acidic pH) inhabited by a particular species is defined by how tolerant that organism's proteins and other macromolecular structures are to the physical conditions within that niche.
- **Optimum growth values** of temperature, pH, and osmolarity vary among microbial species. Mesophiles, psychrophiles, and thermophiles are groups of organisms that grow at moderate, low, and high temperatures, respectively.
- **Barophiles (also known as piezophiles)** can grow at pressures up to 1,000 atm but fail to grow at low pressures.
- **Water activity (a_w)** is a measure of how much water in a solution is available for a microbe to use.
- **Osmolarity** is a measure of the number of solute molecules in a solution and is inversely related to a_w.
- **Halophilic organisms** grow best at high salt concentration.
- **Hydrogen ion concentration affects protein structure and function.** Accordingly, enzymes have pH optima, minima, and maxima.
- **Microbes use pH homeostasis mechanisms** to keep their internal pH near neutral when in acidic or alkaline environments.
- **The addition of weak acids** to certain foods undermines bacterial pH homeostasis mechanisms, preventing food spoilage and killing potential pathogens.
- **Neutralophiles, acidophiles, and alkaliphiles** prefer growth under neutral, low, and high pH conditions, respectively.

Thought Question 6.6 How might the concept of water availability be used by the food industry to control spoilage?

---

impact

Can Pathogens Sense a Host?

This chapter has discussed how microbes cope with environmental conditions in nature, outside a human body. An intriguing question posed by infectious disease scientists is: What conditions do pathogens sense when they enter a suitable host? Most, if not all, pathogens express certain genes only <u>after</u> infection, so they must sense that something in their environment has changed. A change in environment that causes increased virulence is called a virulence signal. One of these virulence signals is pH. On their way to causing disease, many pathogens travel through acidic host compartments. The most obvious are the stomach, which is extremely acidic (pH 1–3), and infected host cell phagosomes and phagolysosomes, which are mildly acidic (pH 4.5–6). *Salmonella enterica* serovar Typhimurium, a major cause of diarrheal disease, provides an intriguing example of virulence that is triggered by low pH and other environmental cues. After gaining entrance to the intestine via ingestion and subsequently penetrating the intestinal epithelia, this pathogen tricks certain white blood cells, called macrophages, into making elegant membrane ruffles that engulf the bacterium (**Figure 1A**) and place it within a phagosome in the cytoplasm (**Figure 1B**). Recall from Section 4.6 that a phagosome is an intracellular membrane vesicle that holds material captured by phagocytosis; in this case the material is *Salmonella*.

Ordinarily, a bacteria-containing phagosome fuses with a lysosome vesicle that contains antimicrobial chemicals and enzymes. The result is a harsh phagolysosome in which the microbe is killed and digested. But *Salmonella* has other plans. Once engulfed into a phagosome, *Salmonella* senses the changes that occur in the vesicle: acidification from pH 6 to 4.5 and a decrease in iron and magnesium concentrations. *Salmonella* responds by synthesizing virulence proteins and then injects those proteins through the phagosome membrane into the host cytoplasm. Some virulence proteins recruit some lysosome membrane proteins to the phagosome but prevent complete fusion between the two vesicles. The result is a *Salmonella*-containing vacuole (SCV)—a safe space in which the pathogen can freely replicate.

How do we know that pH is an important signal for triggering *Salmonella* pathogenesis? Drugs that prevent phagosome acidification (bafilomycin, for example) render *Salmonella* helpless to prevent phagosome-lysosome fusion. *Salmonella* "knows" it is in a macrophage phagosome in part because the pH drops. Many studies have mapped the complex bacterial regulatory circuits that sense acid, iron, and magnesium and then translate those signals into virulence.

Figure 1 *Salmonella* Invading a Macrophage

A. *Salmonella* Typhimurium (red) enters macrophages by inducing patches of ruffling on the cell surface. SEM. Size bar = 3 mm

B. After it is engulfed into phagosomes, S. Typhimurium engineers and resides in *Salmonella*-containing vacuoles. TEM. Size bar = 1 mm.

DISCUSSION QUESTION

You are thinking about using bafilomycin as a treatment for *Salmonella* intestinal infections. What facets about the drug or the host should you consider or evaluate before trying this?

6.5 Living with Oxygen

SECTION OBJECTIVES

- Differentiate anaerobes from aerobes and describe how each group is cultured.
- Explain how both aerobes and anaerobes can cause disease.
- Discuss the basic differences between respiration and fermentation and how this impacts where an organism grows.

Imagine having the ability to swim underwater without breathing oxygen. This sounds like science fiction for humans, but many microorganisms can grow <u>only</u> in the absence of molecular oxygen (O_2). Environments that lack oxygen, such as sediments at the bottom of the ocean, hot springs, and vertebrate intestines, are described as anaerobic. Microbes that grow in those environments are called anaerobes. One class of anaerobes, designated as facultative anaerobes, can grow with or without molecular oxygen. Aerobes, however, grow only in the presence of oxygen. In the next section we describe the aerobes' need for oxygen, why anaerobes cannot grow in oxygen, and how facultative anaerobes grow with or without oxygen. You will learn more about the metabolism of aerobes and anaerobes in Chapter 7.

CASE HISTORY 6.3
Attack of the Anaerobe

Robert, a 21-year-old software salesman from Dallas, Texas, presented to the Parkland Hospital emergency room with 3 days of cramping pain in the midgastric and lower abdominal areas. He had no nausea, vomiting, or diarrhea. Upon physical examination, the patient screamed in severe pain when the clinician pressed on his lower right abdomen. MRI imaging revealed an inflamed appendix. Blood samples were immediately obtained and sent for bacteriological culture. While laboratory scientists processed the blood, surgeons performed an exploratory laparoscopy on Robert to examine, and possibly remove, his appendix. A laparoscopy is when surgery is performed through a small incision in the skin to minimize scarring and speed patient recovery. The procedure revealed a grossly inflamed appendix that released a liter of pus after it was removed. Back in the clinical lab, technicians knew that blood should not contain any microbes, and indeed, after 24 hours of incubation, the blood culture bottle incubated in air appeared sterile. However, the blood culture bottle incubated anaerobically grew thick with bacteria. Additional tests identified the organism as *Bacteroides fragilis*—a common intestinal anaerobe. The final diagnosis was appendicitis caused by *Bacteroides fragilis*. After surgery, the patient was placed on antibiotics to kill any remaining bacteria and made a full recovery.

Not all bacteria can use or even tolerate oxygen. *Bacteroides fragilis*, for example, is killed when it is exposed to air. The bacterium is a beneficial part of our normal intestinal microbiome, but it can cause serious disease if it escapes to other tissues. *B. fragilis* grows in the intestine because the intestinal lumen lacks oxygen. In Case History 6.3, knowing that the cultured bacterium did not grow in the presence of oxygen enabled the laboratory technician to identify the cause of Robert's disease.

So how do bacteria use oxygen, and why are some, like *B. fragilis*, killed by it? Electrons pulled from various energy sources (for example, glucose) possess intrinsic, or potential, energy—an energy that cytoplasmic membrane–associated electron transport proteins (called cytochromes) can harness. A group of cytochromes cluster as an **electron transport chain** (discussed further in Chapter 7) that extracts the energy in incremental stages and uses this energy to move protons out of the cell. The result is proton motive force (introduced in Section 5.2). Once the cell has drained as much energy as possible from an electron, that electron must be passed to a diffusible, final electron acceptor molecule floating in the medium. This transfer clears the way for another electron to be passed down the electron transport chain. One such terminal electron acceptor is dissolved oxygen. The bad news is that certain by-products of oxygen metabolism, such as superoxide radical ($\cdot O_2^-$), hydroxyl radicals ($\cdot OH$), and hydrogen peroxide (H_2O_2), can be dangerously reactive—a serious problem for all cells. As a result, different microorganisms have evolved to either tolerate or avoid oxygen altogether. Table 6.4 gives examples of microbes that grow at different levels of oxygen.

Aerobes versus Anaerobes

We can illustrate the varied relationships between bacteria and oxygen by using a standing test tube filled with growth medium (Figure 6.22). The top of the tube, closest to air, is oxygenated; the bottom of the tube has almost no oxygen. Some microbes grow only at the top of the tube because they require oxygen to respire and make energy. Others grow only at the bottom, where oxygen by-products won't kill them.

A **strict aerobe** is an organism that not only exists in oxygen but also uses oxygen as a terminal electron acceptor. A strict aerobe grows <u>only</u> with oxygen present and consumes oxygen during metabolism (a process known as **aerobic respiration**). An aerobe will grow at the top of the tube shown in Figure 6.22. In contrast, a **strict anaerobe** dies if the least bit of oxygen is present (greater than 5 μM dissolved oxygen). The organism in Case History 6.3 was an anaerobe. Strict anaerobes do not use oxygen as an electron acceptor, but this is not why they die in air. Anaerobes die in air because they are vulnerable to the highly toxic, chemically reactive oxygen species produced by their own metabolism when exposed to oxygen. Reactive oxygen species (ROS) are essentially oxygen molecules or ions with one too few or one too many electrons. As a result, anaerobes will grow only at the bottom of the tube shown in Figure 6.22.

Aerobes, but not anaerobes, destroy ROS with enzymes such as superoxide dismutase (to remove superoxide) and peroxidase

Table 6.4
Examples of Aerobic and Anaerobic Microbes

Aerobic	Anaerobic	Facultative	Microaerophilic
Neisseria spp. (causative agents of meningitis, gonorrhea)	*Azoarcus tolulyticus* (degrades toluene)	*Escherichia coli* (normal gastrointestinal flora and additional pathogenic strains)	*Helicobacter pylori* (causative agent of gastric ulcers)
Pseudomonas fluorescens (found in soil and water; degrades pollutants such as TNT and aromatic hydrocarbons)	*Bacteroides* spp. (normal gastrointestinal flora)	*Saccharomyces cerevisiae* (yeast; used in baking)	*Lactobacillus* spp. (used to ferment milk to form yogurt)
Mycobacterium leprae (causative agent of leprosy)	*Clostridium* spp. (found in soil; causative agents of tetanus, botulism, and gas gangrene)	*Bacillus anthracis* (causative agent of anthrax)	*Campylobacter* spp. (causative agents of gastroenteritis)
Azotobacter spp. (found in soil; fix atmospheric nitrogen)	*Actinomyces israelii* (found in soil; used to synthesize antibiotics)	*Vibrio cholerae* (causative agent of cholera)	*Treponema pallidum* (causative agent of syphilis)
Rhizobium spp. (found in soil; function as plant symbionts)	*Desulfovibrio* spp. (reduce sulfate)	*Staphylococcus* spp. (found on skin; causative agents of boils, impetigo, toxic shock syndrome, others)	

and catalase (to remove hydrogen peroxide). Aerobes also have resourceful enzyme systems that detect and repair macromolecules damaged by oxidation.

Most microbes in the world are anaerobic. They grow buried in the soil, within our anaerobic digestive tract, or in the biofilms that coat our teeth. Anaerobic microbes fall into several categories. Some anaerobes respire via electron transport systems, but instead of using oxygen, they rely on other terminal electron acceptors, such as nitrate, to conduct **anaerobic respiration** and produce energy. Other anaerobes do not possess electron transport chains and cannot respire at all. Instead, these anaerobes must rely on

carbohydrate **fermentation** for energy (that is, they conduct fermentative metabolism). In fermentation, ATP energy is produced through substrate-level phosphorylation in a process that does not involve cytochromes (as discussed in Section 7.3).

In addition to aerobes and anaerobes, several other categories of microbes are also distinguished by their oxygen tolerance. **Facultative** organisms can live with or without oxygen because they come equipped with enzymes that destroy toxic oxygen by-products (enzymes such as superoxide dismutase, catalase, and peroxidase). Facultative organisms will grow throughout the tube shown in Figure 6.22. They include the **facultative anaerobes** (such as *E. coli*). In addition to having enzymes that destroy toxic oxygen radicals, facultative anaerobes also possess the machinery for both fermentation and aerobic respiration. Whether a member of this group uses aerobic respiration, anaerobic respiration, or fermentation depends on the availability of oxygen and the amount of carbohydrate present. For example, the more carbohydrate present in a medium, the more likely a facultative anaerobe will be to ferment that carbohydrate, even in the presence of oxygen (see MacConkey medium in Figure 6.4). When carbohydrates become scarce, however, the facultative anaerobe prefers oxygen (aerobic respiration) over fermentation.

Aerotolerant bacteria that use fermentation (not aerobic respiration) to provide energy represent another subclass of facultative organisms. They also have superoxide dismutase and catalase (or peroxidase) to protect them from ROS. Consequently, aerotolerant anaerobes will also grow throughout the tube in Figure 6.22. *Streptococcus pyogenes*, a common human pathogen, is an aerotolerant anaerobe. The final category of microbes classified by their oxygen tolerance is said to be **microaerophilic**, meaning they will grow only at low oxygen concentrations. Microaerophilic

Figure 6.22 Oxygen-Related Growth Zones in a Standing Test Tube Containing Culture Medium

Growth zones		Dissolved oxygen	
	Aerobic	High oxygen	300 µM O$_2$
Facultative or aerotolerant anaerobes	Microaerophilic	Low oxygen	50 µM O$_2$
	Anaerobic	"No" oxygen	5 µM O$_2$

organisms possess <u>decreased</u> levels of superoxide dismutase and/ or catalase and grow toward the middle of the tube in Figure 6.22.

Culturing Anaerobes

The discovery of anaerobic bacteria by pioneering microbiologists such as Louis Pasteur hinged on finding innovative ways to remove most if not all oxygen from culture environments. An ability to generate anaerobic conditions quickly led to the discovery that anaerobes are important human pathogens. Many anaerobic bacteria cause horrific human diseases, such as tetanus, botulism, and gangrene. Some of these organisms or their secreted toxins are even potential weapons of terror (for example, *Clostridium botulinum*).

Several techniques for culturing anaerobes are used today. One popular way to culture anaerobes, especially on agar plates, employs an anaerobe jar (**Figure 6.23A**). Agar plates streaked with the organism are placed into a sealed jar along with a foil packet that releases H_2 and CO_2 gases. Pellets of the element palladium, suspended under the jar's lid, catalyze a reaction between the H_2 released from the packet and the O_2 trapped in the jar. This reaction forms H_2O, which effectively removes O_2 from the chamber. The CO_2 released from the packet is required by the anaerobe to produce key metabolic intermediates.

For a strict anaerobe exquisitely sensitive to oxygen, a special anaerobic chamber with glove ports is required to establish an oxygen-free environment. The atmosphere inside the chamber is removed by vacuum and replaced with a precise mixture of N_2 and CO_2 gases (**Figure 6.23B**).

In Case History 6.3, a sample of Robert's blood (3–5 ml) was injected into two sealed blood culture bottles. One bottle contained oxygen, while the other contained an anaerobic mix of CO_2 and N_2. Both media contained resins, to bind any antibiotics in the blood that could prevent growth, and a chemical sensor that fluoresces when microbial growth produces CO_2. The inoculated bottles were placed in an automated incubator that can hold 100–200 bottles. The incubator electronically monitors all bottles simultaneously, and it sounds an alarm when it detects growth in a bottle. In Robert's case, the alarm helped save his life.

SECTION SUMMARY

- **Strict aerobes** require oxygen for growth, using it as a terminal electron acceptor to extract energy from nutrients. Aerobes cannot ferment or conduct anaerobic respiration, but they do have enzymes capable of destroying reactive oxygen molecules.
- **Oxygen is toxic to strict anaerobes**, which do not have enzymes capable of efficiently destroying reactive oxygen species.
- **Anaerobic metabolism can be either fermentative or respiratory.** Anaerobic respiration requires the organism to possess electron transport chains that can use compounds other than oxygen as terminal electron acceptors.
- **Facultative organisms** can grow in the presence or absence of oxygen because they possess enzymes that destroy reactive oxygen species.
- **Facultative anaerobes** can grow with or without oxygen, have enzymes that destroy reactive oxygen molecules, and possess the ability for both

Figure 6.23 Anaerobic Growth Technology

Catalyst in lid mediates reaction: $H_2 + \frac{1}{2}O_2 \rightarrow H_2O$

GasPak envelope generates H_2 and CO_2.

A. An anaerobe jar. Inoculated plates are placed in the jar. Water is added to the GasPak foil packet, and the top is sealed shut. Hydrogen and carbon dioxide emerge from the foil packet, and the hydrogen reacts with oxygen in the chamber to form water, effectively removing free oxygen from the chamber.

B. Student using a mini–anaerobic chamber with glove ports. Materials such as plates and tubes are passed through the air lock and into the chamber, whose environment (fed from gas tanks) is mainly nitrogen with some hydrogen and carbon dioxide. Glove ports allow the investigator to manipulate the materials in the box without breaking a seal. An incubator is normally included in the box.

fermentation and respiration (anaerobic and aerobic). They can use oxygen as a terminal electron acceptor.

- **Aerotolerant organisms** can grow with or without oxygen, but they use fermentation as their primary, if not only, means of gathering energy. These microbes also have enzymes that destroy reactive oxygen species, allowing them to grow in oxygen.

Thought Question 6.7   If anaerobes cannot live in oxygen, how do they incorporate oxygen into their cellular components?

Thought Question 6.8   How can anaerobes grow in the human mouth when so much oxygen is there?

6.6 Microbial Communities and Cell Differentiation

SECTION OBJECTIVES

- Discuss how biofilms develop and the role of quorum sensing in the process.
- Explain the importance of biofilms to infection.
- Describe the process of sporulation, and explain how spores impact certain infections.

Can bacteria cooperate with each other? Bacteria were once thought of as unicellular organisms incapable of differentiation or communication. We now realize that many, if not most, bacteria form specialized, surface-attached communities called **biofilms**. Indeed, within aquatic environments, bacteria are found mainly associated with surfaces.

However, bacterial specialization is not limited to biofilms. Many bacteria undergo complex differentiation programs when faced with environmental stress. Some species, such as *E. coli*, undergo relatively simple changes in cell structure involving the formation of smaller cells or thicker cell surfaces that help them survive. Other species initiate elaborate cell differentiation processes. For example, endospore formers such as *Bacillus anthracis* (the cause of anthrax) generate heat-resistant spores that can remain dormant for thousands of years. Another group of bacteria, the actinomycetes, are rich sources of many antibiotics used today and form complex multicellular structures analogous to those of eukaryotic fungi. The production of these antibiotics is tied to the differentiation these organisms undergo as they adapt to growth on agar.

We begin this section with the landmark clinical case (Case History 6.4) that exposed how microbial biofilm communities play critically important roles in infectious disease.

CASE HISTORY 6.4

Death by Biofilm

Cystic fibrosis (CF) is an inherited disease in which chloride ion (Cl⁻) transport is compromised in many organ systems. The lungs are especially affected by the production of thick, sticky mucous buildup. In 1951, two women with CF developed acute respiratory distress. From their chest X-rays, they were diagnosed as having infectious bronchopneumonia.

Treatment at the time consisted of intramuscular administration of the antibiotics streptomycin and penicillin. Repeated bronchial lavage (washing) was attempted in a futile bid to clear airway secretions. Tragically, the two women did not respond to therapy and died within hours. *Pseudomonas aeruginosa*, a Gram-negative bacillus common in soil and water, was subsequently isolated from both victims. This was the first report of *Pseudomonas* infection in CF patients.

Lung infection with the Gram-negative organism *Pseudomonas aeruginosa* is a frequent cause of death in cystic fibrosis (CF) patients. Lung damage caused by CF provides a setting in which mucoid strains of *P. aeruginosa* can attach to injured tissue and form tenacious biofilm communities consisting of billions of bacteria. A *Pseudomonas* lung biofilm makes breathing more difficult for the CF patient and increases inflammation, which further damages the lung. Treating any biofilm infection is challenging because a biofilm, like a fortress, is nearly impenetrable to antibiotics, as you will see. Biofilms, in fact, are a major factor in many bacterial infections. All bacteria can make them. Consequently, understanding how biofilms form is a critical first step in devising treatments for many types of bacterial infections.

Biofilms: Multicellular Microbes?

A **biofilm** is a mass of bacteria that stick to and multiply on a solid surface, whether a stone in a lake or a lung in a cystic fibrosis patient. Biofilms can be constructed by a single species or by multiple, collaborating species and can form on a range of organic or inorganic surfaces. **Figure 6.24** shows a common mixed-species biofilm—called plaque—that forms on teeth. More serious examples include the single-species biofilm formed by *P. aeruginosa* on the lungs of cystic fibrosis patients (as in Case History 6.4) or by staphylococci on medical implants (for instance, knee replacements).

Why would a microbe "want" to make a biofilm? The goal of biofilms in nature is to stay where food is plentiful. Why should a

Figure 6.24 Dental Biofilm

A. The biofilm that forms on teeth is called plaque.

B. Magnified dental plaque.

microbe travel off to hunt for food when food is already present? Once nutrients become scarce, however, individuals detach from the community to forage for new sources of nutrient.

Biofilms typically develop in the manner illustrated in **Figure 6.25**. When conditions are right, unattached (planktonic) single cells start to attach to nearby inanimate surfaces by means of flagella, pili, lipopolysaccharides, or other cell-surface structure (step 1). These cells multiply to form microcolonies. Cells of some species (like *Pseudomonas*) move across the surface by a process called twitching motility (step 2), which involves the extension and retraction of a specific type of pilus. Ultimately, these cells stop moving and firmly attach to the surface.

Next, the cells form a thick extracellular matrix of polysaccharide and other polymers collectively called **extracellular polymeric substances (EPS)** that can entrap miscellaneous organic and inorganic materials (step 3). Exopolysaccharides such as alginate, produced by *P. aeruginosa,* and colanic acid, produced by *E. coli*, increase antibiotic resistance by limiting antibiotic access to the center of the biofilm. The chemical signaling molecules in the biofilm also stimulate production of some antibiotic resistance mechanisms. (See Chapter 13 for a deeper discussion of biofilms and antibiotic resistance.)

As more and more cells bind to the surface, they begin to communicate with one another by sending and receiving chemical signals in a process called **quorum sensing** (step 4). A quorum in any organization is the minimum number of people necessary to hold a meeting and vote on an issue. Microbes "vote" by means of chemical signals. Individual cells continually produce the chemical signaling molecules, but once the population attains a certain number, the chemical signal, called an **autoinducer**, reaches a concentration that the cells can sense. Reception of this signal triggers genetically regulated changes that cause bacteria to bind doggedly to the surface substrate and to one another (Section 8.4).

As the biofilm matures, the amalgam, or combination of adherent bacteria and matrix, forms columns and streamers, creating channels through which nutrients flow. Finally, when a mature biofilm (or a part of it) begins to starve or experiences oxygen depletion, some cells start making enzymes that dissolve the matrix. The biofilm then sends out "scouts" called dispersal cells that will initiate new biofilms (step 5).

A hallmark of all infectious diseases in which biofilms play a prominent role is their chronic nature. A biofilm infection may linger for months, years, or even a lifetime. Although it may never kill its victim, the infection can compromise quality of life. Biofilm infections can smolder indefinitely, even in people with healthy immune systems.

Endospores: Time Capsules for Bacteria

Certain Gram-positive genera, including important pathogens such as *Clostridium tetani* (tetanus), *Clostridium botulinum* (botulism), and *Bacillus anthracis* (anthrax), have the remarkable ability to develop dormant heat- and desiccation-resistant spores. Resistance to heat and desiccation help make the spores of *B. anthracis* a potential bioweapon.

Most of our knowledge of bacterial sporulation comes from the nonpathogenic, Gram-positive soil bacterium *Bacillus subtilis*. When growing in rich medium, this microbe undergoes normal vegetative growth and can replicate every 30–60 minutes. However, starvation initiates an elaborate 8-hour genetic program that directs an asymmetrical cell division process and ultimately yields a spore.

As shown in **Figure 6.26**, sporulation begins when the chromosome replicates and stretches into a long axial filament spanning the length of the cell. The cell makes a molecular "decision" to divide at one of the two cell poles (stage I) instead

Figure 6.25 Biofilm Development

Swimming planktonic cells

Biofilm towers

Twitching motility

| **1** Cellular flagella attach to the monolayer. | **2** Microcolonies form. Some cells undergo twitching motility. | **3** Cells secrete a matrix of extracellular polymeric substances (EPS). | **4** The biofilm matures. Cells communicate by quorum sensing. | **5** The biofilm dissolves and cells disperse. |

A. A mucoid environmental strain of *P. aeruginosa* produces uneven, lumpy biofilms in a continuous-flow cell. Cells in the biofilm were stained green with LIVE/DEAD viability stain (3D confocal laser scanning microscopy).

B. The stages of biofilm development in *Pseudomonas* generally apply to the formation of many kinds of biofilms.

Figure 6.26 The Seven Stages of Endospore Formation

Origin

Vegetative growth

Sporulation

Germination

Spore

Spore coat

Exosporangium

Cortex

Axial filament DNA

Mother cell

Forespore

Engulfment

Stage I. Septum forms near one pole. DNA replicates and extends into an axial filament.

Stage VII. Mother cell releases spore.

Stage VI. Dipicolinic acid is synthesized, and calcium is incorporated into the spore core.

Stage II. Septum separates forespore from mother cell. DNA is pumped through septum until each compartment gets a chromosome.

Stage III. Mother cell engulfs the forespore, surrounding it with a second membrane.

Stage V. Cortex is completed and coat proteins are deposited on outer membrane. Mother cell chromosome begins to break down.

Stage IV. Forespore starts to develop a cortex layer of peptidoglycan between the original forespore membrane and the membrane of the mother cell.

of in the middle, as normally occurs during growth. Septation divides the cell into two unequal compartments (stage II): the forespore, which will ultimately become the spore, and the larger mother cell, from which it is derived. Each compartment contains a chromosome.

The mother cell membrane engulfs the forespore to form a double membrane (stage III). Next, a thick peptidoglycan layer (cortex) is placed between the two membranes surrounding the forespore (stage IV). Layers of coat proteins are then deposited on the outer membrane and the mother cell chromosome begins to break down (stage V). Stage VI completes the development of spore resistance to heat and chemical insults. This process includes the synthesis of dipicolinic acid (which aids in heat resistance) and the uptake of calcium (which stabilizes DNA) into

the core of the spore. The mother cell, now called a sporangium, dies and releases the mature spore (stage VII).

A mature endospore can exist in soil for at least 50–100 years. Some last thousands and even millions of years. Once proper nutrient conditions arise, another genetic program, called **germination**, is triggered that wakes the dormant cell, dissolves the spore coat, and releases a viable vegetative cell.

Why do we think endospores can last millions of years? Because bacterial endospores that were extracted from the stomachs of ancient bees trapped and preserved in amber millions of years ago have successfully germinated. The results of that study, however, were controversial. To definitively analyze endospore survival, scientists from the United Kingdom and Germany launched a 500-year microbiology experiment in 2014. The goals

Figure 6.27 Start of the 500-Year Spore Survival Experiment

A. Glass vials containing 100 μL of *Bacillus subtilis* spore stock solution (10^6 CFU/ml) are dried down on silicon bead desiccators (red spheres) before being sealed.

B. Phase-contrast microscopy of spores present at hour 0 of experiment. Arrow points to spores.

C. Same spores after 2.5 hours of germination. Nongerminating spores appear more gray (arrows).

of the experiment are to determine how long *B. subtilis* endospores remain alive during desiccation as compared to desiccated cells of a hardy cyanobacterium, and whether long-term storage will affect the kinetics of spore germination. Preparation of the desiccated spores is shown in **Figure 6.27**. Who will be around in 2514 to complete the experiment is unclear.

SECTION SUMMARY

- **Biofilms** are complex multicellular surface-attached microbial communities.
- **Chemical signals** enable bacteria to communicate (via quorum sensing) and in some cases to form biofilms.
- **Biofilm development** involves adherence of cells to a substrate, formation of microcolonies, and ultimately formation of complex channeled communities that generate new planktonic cells.
- **Endospore development** in *Bacillus* and *Clostridium* involves production of dormant, stress-resistant endospores.

Thought Question 6.9 *Clostridium botulinum* is an anaerobic, Gram-positive, spore-forming bacillus, normally found in soil, that secretes a poisonous toxin. Why would this organism be a potential problem when Nana and Papa decide to can vegetables such as green beans?

Perspective

Knowing how microbes grow and develop is an important part of understanding infectious diseases. Many facets of microbial physiology impact a pathogen's ability to sidestep a host's immune system, establish itself at certain body sites, survive at those sites, and collect nutrients—sometimes killing the host, sometimes not. As we proceed through future chapters, you will learn how bacterial pathogens evolve and new ones suddenly emerge, why different organisms focus on specific body sites, and how bacteria "know" when to eject or inject specific proteins (toxins) that hijack host cells. You will also appreciate that our ability to isolate and grow pathogens on (or in) laboratory media is critical to identify a disease-causing pathogen and determine how to stop its rampage through a host.

Key Terms

acidophile (p. 170)

aerobic respiration (p. 173)

aerotolerant (p. 174)

alkaliphile (p. 170)

anaerobic respiration (p. 174)

autoinducer (p. 177)

autotroph (p. 163)

barophile (p. 168)

batch culture (p. 159)

binary fission (p. 158)

biofilm (p. 176)

chemically defined minimal
medium (p. 162)

chemoautotroph (p. 170)

chemostat (p. 160)

chemotrophy (p. 164)

colony (p. 153)

colony-forming unit (CFU)
(p. 154)

complex medium (p. 154)

confluent (p. 153)

continuous culture (p. 160)

death phase (p. 160)

death rate (p. 160)

denitrification (p. 165)

differential medium (p. 155)

dilution streaking (p. 153)

doubling time (p. 159)

electron transport chain (p. 173)

enriched medium (p. 154)

essential nutrient (p. 161)

exponential growth (p. 159)

exponential phase (p. 160)

extracellular polymeric
substances (EPS) (p. 177)

extremophile (p. 166)

facultative (p. 174)

facultative anaerobe (p. 174)

fermentation (p. 174)

flow cytometer (p. 156)

fluorescence-activated cell
sorter (FACS) (p. 156)

generation time (p. 159)

germination (p. 178)

growth factor (p. 162)

halophile (p. 169)

heterotroph (p. 163)

hyperthermophile (p. 167)

isolation streaking (p. 153)

lag phase (p. 160)

late log phase (p. 160)

lithotrophy (p. 164)

logarithmic (log) phase (p. 160)

MacConkey medium (p. 155)

mesophile (p. 168)

microaerophilic (p. 174)

minimal medium (p. 154)

neutralophile (p. 170)

nitrification (p. 165)

nitrogen cycle (p. 165)

nitrogen fixation (p. 165)

optical density (p. 157)

organotrophy (p. 164)

osmolarity (p. 168)

phototrophy (p. 164)

piezophile (p. 168)

psychrophile (p. 168)

pure culture (p. 152)

quorum sensing (p. 177)

rich medium (p. 154)

selective medium (p. 155)

spectrophotometer (p. 157)

spread plate (p. 153)

stationary phase (p. 160)

strict aerobe (p. 173)

strict anaerobe (p. 173)

symbiont (p. 165)

synthetic medium (p. 154)

thermophile (p. 167)

viable (p. 153)

Objectives Assessment Questions

6.1 Culturing and Counting Bacteria

1. From which of the following agar plate(s) can you directly obtain a
   pure culture?
   a. Plate A
   b. Plate B
   c. Plate C
   d. Plates A and B
   e. Plates B and C

A.   B.   C.

2. In question 1, image C best illustrates which of the following concepts?
   a. Selective medium
   b. Viable count
   c. Differential medium
   d. Enriched medium

3. A 21-year-old woman exhibits painful urination and a pain in her side.
   You are asked to determine how many bacteria are present in 1.0 ml
   of this patient's urine to help establish whether she has a urinary
   tract infection. You take 1 μl of her collected urine and streak it over
   the surface of a blood agar plate. You see 35 similar-looking colonies.
   How many bacteria are in 1.0 ml of this woman's urine?
   a. 3.5×10^2
   b. 3.5×10^3
   c. 3.5×10^4
   d. 3.5×10^5
   e. 3.5×10^6

4. Which of the following counting methods will yield the lowest number
   of bacteria per milliliter in a stationary-phase culture?
   a. Direct count
   b. Viable count
   c. Propidium iodide and Syto-9 staining
   d. Fluorescence-activated cell sorting
   e. They will all give approximately the same number.

6.2 The Growth Cycle

5. With regard to the figure below, which phase exhibits balanced growth?
   a. 1
   b. 2
   c. 3
   d. 4

6. The pathogen *Salmonella enterica* serovar Typhi has a generation time of 20 minutes in rich medium. If you start with 10 bacteria/ml, <u>approximately</u> how many bacteria would be present in 1 ml after 12 hours (assuming unlimited nutrients)? (You may go online and use an anti-log calculator to get the answer.)
   a. 1×10^3
   b. 1×10^5
   c. 7×10^6
   d. 7×10^8
   e. 7×10^{10}
   f. 7×10^{11}

7. Decreasing the flow rate of medium in a chemostat will have which of the following effects?
   a. Decrease growth rate
   b. Increase growth rate
   c. Shorten lag phase
   d. Increase exponential (log) phase
   e. Increase cell size

6.3 Microbial Nutrition

8. A medium that contains only those compounds needed for an organism to grow is called a(n) _____.

9. A bacterial species that can make its own organic compounds from CO_2 is called a(n) _____.
   a. organotroph
   b. heterotroph
   c. autotroph
   d. phototroph
   e. chemotroph

10. Label the following figure with these terms: nitrifiers, denitrifiers, nitrogen fixers, NH_4^+, NO_3^-, N_2.

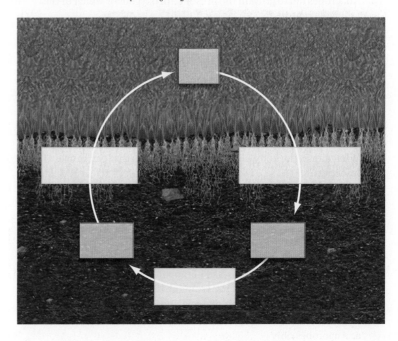

6.4 Environmental Limits on Microbial Growth

11. Microorganisms that grow at very high pressures are called _____.

12. With respect to their growth temperature requirements, pathogenic bacteria are considered _____.
   a. extremophiles
   b. psychrophiles
   c. mesophiles
   d. thermophiles
   e. halophiles

13. Which of the following will take place when growth temperature for a bacterium is set below the organism's growth optimum?
   a. The culture will grow to a lower maximum cell density.
   b. The growth rate in exponential phase will be faster.
   c. Stationary phase will be shorter.
   d. The growth rate in exponential phase will be slower.
   e. Lag phase will be shorter.

14. Psychrophiles grow poorly at high temperatures for which of the following reasons?
   a. Ribosomes synthesize proteins too quickly for balanced growth.
   b. Proteins become less flexible.
   c. DNA replication becomes more error prone.
   d. Water activity becomes too low.
   e. Membranes become too flexible.

6.5 Living with Oxygen

15. Anaerobes fail to grow in the presence of oxygen for which of the following reasons?

   a. Anaerobes cannot use oxygen to generate proton motive force.
   b. Oxygen cannot be incorporated into an anaerobe's proteins.
   c. Anaerobes lack enzymes that detoxify oxygen by-products.
   d. Anaerobes lack enzymes that detoxify oxygen.

16. Anaerobes cannot conduct respiration. True or false?

17. In order to grow, anaerobes must carry out fermentation. True or false?

18. What site in the body has the most anaerobes that can cause infection?

   a. Skin
   b. Lung
   c. Vagina
   d. Intestine
   e. Oral cavity

19. A patient with appendicitis had his appendix removed. The surgeon removed almost 500 ml of pus from the infected appendix. Bacteriological examination of this material most likely found which of the following?

   a. Anaerobes only
   b. Strict aerobes only
   c. Facultative organisms only
   d. Anaerobes and strict aerobes
   e. Anaerobes and facultative organisms

20. While working at a construction site, you cut your hand on an old nail buried in dirt. To avoid infection, you apply H_2O_2 to the cut. This chemical will be MOST effective against which of the following classes of bacteria?

   a. Aerobe
   b. Heterotroph
   c. Facultative anaerobe
   d. Strict anaerobe
   e. Organotroph

6.6 Microbial Communities and Cell Differentiation

21. The transmission and sensing of chemical signaling molecules between bacteria is called _____.

22. The secreted polymers that trap microbes within a biofilm are called _____.

23. Biofilms are typically associated with which of the following types of infections?

   a. Acute
   b. Chronic
   c. Septicemia
   d. Latent
   e. b and c only

24. Spores are important to the infectious cycle of certain pathogens for which of the following reasons?

   a. They resist heat.
   b. They resist disinfectants.
   c. They remain viable for long periods.
   d. They inhibit the host's immune response.
   e. All of the above
   f. a, b, and c only

Review Questions

1. What classes of nutrients do microbes need in order to grow?

2. Explain the differences between autotrophy, heterotrophy, phototrophy, and chemotrophy.

3. Briefly describe the carbon and nitrogen cycles.

4. Why is it important to grow bacteria in pure culture?

5. What factors define the growth phases of bacteria grown in batch culture?

6. Describe the important features of biofilms.

7. What are extremophiles, and why are these organisms important?

8. What parameters define any growth environment?

9. List and define the classifications used to describe microbes that grow in different physical growth conditions.

10. Why is water activity important to microbial growth? What changes water activity?

11. Why do changes in H^+ concentration affect cell growth?

12. How do acidophiles and alkaliphiles manage to grow at the extremes of pH?

13. If an organism can live in an oxygenated environment, does that mean the organism uses oxygen to grow? Why or why not?

14. If an organism can live in an anaerobic environment, does that mean it cannot use oxygen as an electron acceptor? Why or why not?

15. What happens when a cell exhausts its available nutrients?

16. Why are biofilms important to infections?

17. Why are endospores important in medicine? Outline the stages of sporulation.

18. Review previous Case Histories 3.1 and 5.2. Classify the pathogens as to their nutritional categories (phototroph or chemotroph, heterotroph or autotroph, and so forth).

Clinical Correlation Questions

1. Brianna, a 4-year-old girl, is taken to the emergency room with a chief complaint of diarrhea. Bacteriological culture of her stool reveals the Gram-negative pathogen *Salmonella enterica*. The organism is known to contaminate poultry. When her mother is asked, she says that Brianna ate some homemade chicken salad at a party the day before but that her twin brother Brian had the same chicken salad 2 days earlier when it was first made, and he was fine. Can you explain why Brian might not have gotten ill but Brianna did after eating from the same preparation of chicken salad?

2. Two people happen to drink equal amounts of water while swimming in the Ganges River, in India. Unbeknownst to them, the water is contaminated with *Vibrio cholerae*, which causes the diarrheal disease cholera. Gagandeep, a 53-year-old man, has been taking aspirin to prevent clots from forming in his blood; Padman, another 53-year-old man, has been taking a proton pump inhibitor for chronic indigestion. Three days later, Padman develops severe diarrhea but Gagandeep does not. Propose a plausible explanation for why one man became ill but the other did not.

3. *Helicobacter pylori*, a bacterial cause of stomach ulcers, can grow in medium with a pH above pH 5. It can survive but not grow in pH 2 medium as long as the medium contains urea. You inoculate a pH 2 broth with 1×10^7 *H. pylori* organisms, and for 8 hours you confirm that the organism does not grow. The next morning, however, you find that the medium has become quite turbid, suggesting that the organism has grown. Explain how this could happen.

4. A 67-year-old woman had a knee replacement 1 year ago. About 6 months ago, she developed pain, swelling, and reddening of the knee. Her surgeon suspected an infection of her prosthetic knee and prescribed antibiotics. The antibiotics seemed to work and her knee recovered. However, her knee became infected twice more. Each time the infection resolved with antibiotic treatment. Explain why the antibiotics never fully cured the infections, which continue to return.

7

Bacterial
Metabolism

[A Spring Break Mishap]

SCENARIO Shane, a 21-year-old college student, spent spring break at the beach in Cancún, where he frequently ordered drinks from the beachfront stands. Upon his return to classes, Shane fell ill with a fever of 38.9°C (102°F), severe abdominal cramps, and watery diarrhea.

SIGNS AND SYMPTOMS By the second day, Shane had blood in his stools—a condition known as dysentery. Upon admission to the hospital, Shane showed dehydration, and his rectal exam was very painful, with bleeding.

Hektoen Agar, Growing *Shigella*
Shigella produces pale green colonies (no sugar fermentation and no thiosulfate reduction), whereas *Salmonella* colonies would appear black (production of H$_2$S from thiosulfate reduction).

TESTING The clinician ordered a fecal culture on Hektoen agar, a selective medium for Gram-negative bacteria; bile salts exclude Gram-positives. Hektoen agar also differentiates among the enteric Gram-negative pathogens *Salmonella*, *Shigella*, and *Escherichia*. Most *Salmonella* species show black colonies because of H$_2$S formation from thiosulfate, whereas *Escherichia* species form colonies that are orange from lactose fermentation. Shane's culture produced translucent colonies, indicating *Shigella* species, which neither ferment lactose nor reduce thiosulfate to H$_2$S.

DIAGNOSIS AND TREATMENT Serotyping (determining which antibodies react with the pathogen) confirmed the species *Shigella flexneri*, a common cause of "traveler's diarrhea" from drinking contaminated water. Shane received intravenous rehydration and was given the antibiotic quinolone. The choice of quinolone was based on lab tests of the pathogen's antibiotic sensitivity. Afterward, Shane made a full recovery.

SEM 1 µm

Shigella flexneri
Shigella flexneri causes shigellosis, a form of dysentery.

Shigellosis is a common illness in day-care centers and in developing countries with contaminated water supplies. *Shigella* species are Gram-negative pathogens of the Enterobacteriaceae (see photo accompanying the chapter-opening case history). The pathogen has an unusually low infective dose, requiring as few as ten organisms to cause infection. Because they are found in feces, among many normal microbiota, the culture and identification of *Shigella* require selective and differential media (see Section 6.1). An example of selective media for Gram-negative enteric organisms is Hektoen agar (see the case history photo). Hektoen agar includes indicators that use bacterial metabolism. Different bacteria conduct different kinds of metabolism to gain energy from food molecules. In the Hektoen formula, one such food molecule is lactose, a sugar fermented by *Escherichia coli* but not by most species of *Salmonella* and *Shigella*.

Fermentation is a process that bacteria use to obtain energy from sugars by breaking the molecules into smaller ones—a process that increases entropy (disorder) and therefore yields energy (discussed in Section 7.1). Often, the products of fermentation include acids, such as lactic acid, which decrease the extracellular pH. The pH decrease causes a change in the indicator dye of Hektoen agar—hence the appearance of orange colonies of *E. coli*.

A different Gram-negative organism, *Salmonella*, fails to ferment lactose; instead, *Salmonella* species grow on peptides in the media. Metabolism of peptides as carbon and energy sources does not produce acid, so the pH indicator in Hektoen does not change. However, *Salmonella* can combine food molecules with the thiosulfate present in Hektoen; the food becomes oxidized, releasing H_2S. The oxidation of food by an inorganic molecule is called respiration, a different way for growing cells to gain energy (see Section 7.4). The H_2S produced by *Salmonella* respiration reacts with iron in Hektoen agar to produce a black FeS precipitate seen in the colonies. Thus, black colonies on Hektoen agar would indicate *Salmonella* as the pathogen tested.

Both fermentation and respiration are chemical reactions that yield energy to "do the work" of growing cells. The energy released by a reaction is transferred by enzymes to other reactions that build simple molecules into a complex cell. In this chapter we present the many kinds of reactions that enable bacteria to grow—and that reveal their identity.

7.1 Energy for Life

Have you ever run out of money before your next paycheck? As money is to your budget, energy is to life. If money stops coming in, the household goes bankrupt; if energy stops flowing through an organism, it dies. On a broader scale, if money falls short in an economy, the economy collapses. Similarly, if energy fails in an ecosystem, a community of organisms collapses. Thus, energy is key to the survival of all life in our environment.

This chapter presents the biochemical reactions by which different kinds of microbes obtain energy, along with the reactions they use to build biomass. In the chapter-opening case history, a lab technician used knowledge of these reactions to identify a pathogen on the basis of its metabolic reactions.

Energy and Entropy for Living Cells

How do living organisms assemble themselves out of small, nonliving molecules? In everyday life we know that simple objects do not spontaneously assemble themselves into something complex. As you learned in Chapter 4, the universe overall becomes more random, or less ordered; this disorder, or randomness, is called **entropy**. And yet, every living organism, from soil microbe to human, must assemble simple molecules such as H_2O and CO_2 into complex cells (**Figure 7.1**). To build something complex requires spending **energy**. In Chapter 4 we introduced energy as the property that enables a chemical reaction to go forward. For a living organism, energy enables the cell to perform all its functions, such as swimming or growing.

As a cell grows and builds complex molecules, entropy, or disorder, decreases. But the decrease in entropy is local to the cell. The cell releases heat and waste products to its surroundings, where disorder actually increases. So, the local, temporary spending of energy enables a cell to build order and grow while increasing entropy (disorder) outside. Continued growth requires continual spending of energy and dissipation of heat.

Entropy increases beyond the cell's environment, throughout the biosphere. To overcome entropy, life on Earth captures energy from the Sun (**Figure 7.2**). Chapter 6 introduced the nutritional strategies by which microbes obtain the Sun's energy, either directly (as producers) or indirectly (by consuming the producers). Photosynthetic microbes and plants generate biomass, which is consumed by heterotrophs and decomposers. The consumers—including humans—store a small fraction of their energy in biomass. At each successive level of consumers, most of the energy is lost, radiated from Earth as heat. But a key fraction of energy is stored in the bodies of living organisms.

Figure 7.1 Living Organisms Assemble Simple Molecules into Complex Cells
To build complex cells, living organisms must spend energy.

Metabolic Reactions That Yield Energy

Different kinds of metabolism are like different occupations: each offers microbes a way to "make a living." The major categories of metabolism are summarized in **Table 7.1**. All these kinds of

Figure 7.2 Energy Flows through the Biosphere
Solar radiation reaches Earth, where a small fraction is captured by photosynthetic microbes and plants. The microbial and plant biomass enters heterotrophs and decomposers, which convert a small fraction to biomass at each successive level while losing energy as heat.

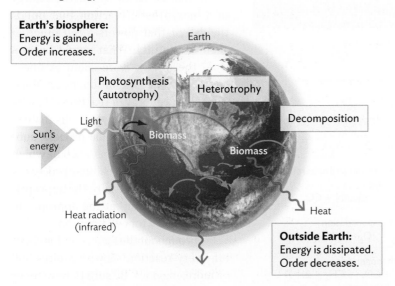

metabolism play important roles in the environment—and have consequences for human health.

- **Phototrophy** refers to biochemical reactions that capture energy from light. Light energy captured by chlorophyll is used to excite electrons from a stable molecule such as water (H_2O). In oxygenic photosynthesis, the excited electrons are used to reduce CO_2 to form sugar ($C_6H_{12}O_6$) while releasing oxygen gas (O_2). Oxygenic photosynthesis produces the food we eat and the oxygen we breathe. Most organisms conducting oxygenic photosynthesis are **autotrophs**, capable of forming all their own organic molecules from CO_2 plus minerals.

- **Chemotrophy** refers to reactions that yield energy stored in the chemical bonds of molecules, such as the biomass of a consumed prey organism. The transfer of an electron from one molecule to another (or to another part of the same molecule) releases a combination of products more stable than the original; this reaction yields energy. Classes of chemotrophy include organotrophy and lithotrophy, which are defined by the nature of their electron donors and acceptors.

- **Organotrophy** refers to chemotrophic metabolism in which the food molecules (electron donors) are organic (linked carbon molecules). Most organisms that conduct organotrophy are **heterotrophs**, requiring preformed organic compounds.

  — Aerobic organotrophy consists of reactions in which an organic molecule donates electrons to O_2, forming the stable molecule H_2O. Most forms of aerobic organotrophy are also classified as **aerobic respiration**. An example is the oxidative breakdown of sugar releasing H_2O and CO_2. Human aerobic respiration requires us to breathe oxygen continuously—and to release CO_2, which serves as the essential pH buffer for our blood.

— Anaerobic organotrophy involves reactions <u>without</u> oxygen. In anaerobic respiration, oxygen is replaced by an alternative electron acceptor, such as nitrate. In fermentation, partial breakdown of food molecules releases small molecules such as ethanol or lactic acid. Microbial fermentation plays a key role in the production of foods such as cheese and alcoholic beverages.

• **Lithotrophy** refers to chemotrophic metabolism in which the food molecules (electron donors) are inorganic. Organisms that grow solely by lithotrophy are autotrophs, forming their own organic molecules from minerals.

— Aerobic lithotrophy involves reactions with oxygen. An example of aerobic lithotrophy is ammonia oxidation by O_2 in which electrons from ammonia are donated to oxygen. Ammonia oxidation commonly occurs in the soil, especially where farmers spread nitrogen-rich fertilizers such as manure. A medical consequence is the formation of harmful levels of nitrites and nitrates that enter groundwater.

— Anaerobic lithotrophy is lithotrophic metabolism <u>without</u> O_2. For example, methanogenesis is a reaction by which certain archaea convert hydrogen and CO_2 to methane and water. Methanogens are common in soil and in our own digestive tract.

To yield energy, a biochemical reaction must go forward from reactants to products. The direction of a reaction, whether it goes forward or in reverse, can be predicted by the free energy change, $\Delta G = \Delta H - T\Delta S$. If the value of free energy change (ΔG) is negative, a given metabolic reaction yields energy for the cell.

The free energy change, ΔG, includes two components: the change in chemical energy (ΔH) and the change in entropy multiplied by temperature ($-T\Delta S$). For example, the oxidation of glucose by O_2 (aerobic respiration) generates products with increased chemical stability (value of ΔH). Aerobic respiration also involves the breakdown of one molecule to a greater number of small molecules, thus increasing entropy (value of $-T\Delta S$). Other kinds of metabolism tend to rely on one component of free energy more than the other. For example, when baker's yeast in bread dough ferments sugar, the sugar molecule breaks down to two molecules of ethanol and two molecules of CO_2; the gas forms bubbles, causing the dough to rise. The bread-making reaction is driven more by the increase in entropy (S), and its efficiency increases with temperature (T).

Enzymes Control Reactions

When a chemical reaction releases energy (such as by oxidizing a food molecule), the energy must be captured by another reaction that builds order (such as protein synthesis). Enzymes manage this energy transfer from one reaction to another (discussed in Chapter 4). An enzyme allows a specific reaction to occur only as needed, in the right amount at the right time. Enzymes act by lowering the activation energy—the input energy needed to generate the high-energy transition state on the way to products (Figure 4.25). Enzymes determine the precise rearrangement of chemical bonds and the rate at which biochemical reactions occur in the cell.

For an organism to grow, the chemical reactions of the food must be controlled so as to spend the energy in a biosynthetic reaction that builds the cell or that powers a cell function such as motility. What would happen if all the energy were spent at once? In uncontrolled oxidation, or combustion, all the energy of reaction is lost by the release of heat. Living cells instead oxidize food in a gradual process, subdivided into many steps, each of which transfers energy to a cellular process. A cell's enzymes make sure the appropriate reaction occurs at the appropriate time and place for cell function.

Throughout the figures in this chapter, every reaction shown requires one or more enzymes. Be sure to remember this, even when no enzyme is shown.

Table 7.1
Energy-Yielding Metabolism

Class of Metabolism	Source of Energy	Example (Simplified)
Phototrophy	Light energy is absorbed to make high-energy molecule that donates electrons to acceptor	Photosynthesis Light, $H_2O + CO_2 \rightarrow$ sugar $+ O_2$
Chemotrophy	High-energy food molecule donates electrons to acceptor	
Organotrophy, aerobic	Organic molecule donates electrons to O_2	Aerobic respiration Sugar $+ O_2 \rightarrow H_2O + CO_2$
Organotrophy, <u>anaerobic</u>	Organic molecule donates electrons to itself or other molecule, <u>not</u> O_2	Fermentation and anaerobic respiration Sugar $\rightarrow$ ethanol $+ CO_2$
Lithotrophy, aerobic	Inorganic molecule donates electrons to O_2	Ammonia oxidation $NH_3 + 2O_2 \rightarrow HNO_3 + H_2O$
Lithotrophy, <u>anaerobic</u>	Inorganic molecule donates electrons to other molecule, <u>not</u> O_2	Methanogenesis $4H_2 + CO_2 \rightarrow CH_4 + 2H_2O$

- **Different kinds of metabolism yield energy for different microbes.** Important classes of metabolism include phototrophy, organotrophy (aerobic and anaerobic), and lithotrophy (aerobic and anaerobic).

- **Energy-yielding reactions must show a negative value of free energy change (ΔG).** The free energy change includes changes in chemical energy and in entropy.

- **Enzymes catalyze all biochemical reactions.** Enzymes increase the rate of reaction and make sure the reaction occurs at the appropriate time and place for cell function.

7.2 Catabolism: The Microbial Buffet

SECTION OBJECTIVES

- Describe how catabolism converts many complex food molecules into a few kinds of catabolites.

- Explain how catabolism yields energy and how the energy is stored for use.

- Describe how the energy carriers ATP and NADH transfer energy between energy-yielding and energy-spending reactions.

What kind of metabolism yields energy for the human body— and for most of our intestinal bacteria? We and our bacteria break down complex organic food molecules—a process called **catabolism**. In principle, catabolism can break down any organic substance found in the environment. A given species, however, can catabolize only those substances for which its DNA encodes the right enzymes. Consequently, humans express enzymes enabling catabolism of starches, lipids, and proteins, but not cellulose or lignin (a structural substance in woody plants). Many soil bacteria, however, have enzymes to catabolize cellulose and lignin. Other bacteria are pathogens that catabolize components of the human body. For example, the bacterium that causes acne, *Cutibacterium acnes* (formerly called *Propionibacterium acnes*), catabolizes skin cell molecules such as sialic acids, matrix proteins, and lipids.

Certain kinds of food molecules are widely catabolized by many species. Carbohydrates, lipids, and proteins are common substrates for catabolism. In general, food molecules such as starch (a polymer of glucose units) are broken down by specific enzymes to smaller molecules, called **catabolites**, that feed into shared central pathways (**Figure 7.3**). For example, the catabolite glucose from starch breakdown enters **glycolysis**, a pathway of step-by-step catabolism in which energy is transferred to the energy carriers ATP and NADH (discussed shortly). Glucose also arises from the breakdown of cellulose, a glucose polymer found in plants. Cellulose is broken down by the enzyme cellulase; the glucose produced enters glycolysis, as does the glucose from starch. Other materials, such as pectin, a sugar acid polymer in fruits, are broken down to sugar acids. Sugar acids are further catabolized by a modified pathway of sugar breakdown called the Entner-Doudoroff pathway.

Lipids and proteins are catabolized to products that enter the central pathways of glycolysis and the **tricarboxylic acid cycle (TCA cycle)**, also known as the **Krebs cycle** or the **citric acid cycle**. Lipids are broken down by **hydrolysis** to glycerol and fatty acids (compounds introduced in Chapter 4). Hydrolysis is the cleavage of a molecule while incorporating H_2O. Glycerol is broken down to pyruvic acid (pyruvate). Fatty acids are broken down to two-carbon units of acetic acid, which is tagged with coenzyme A (acetyl-CoA) to enter the TCA cycle (see Section 7.3). Proteins are hydrolyzed to peptides and ultimately to amino acids. The amino acids are broken down and converted to intermediates such as acetyl-CoA and TCA cycle intermediates such as 2-oxoglutarate (presented in Figure 7.10).

Note: Many intermediates of metabolism are carboxylic acids (R—COOH) such as acetic acid, pyruvic acid, and succinic acid. Each acid exists in the cell as interchangeable forms that may be considered equivalent in our diagrams. The forms are shown here for acetic acid:

Acetic acid   **Acetate** (deprotonated; accompanied by H^+)   **Acetyl-CoA** (coenzyme A reversibly replaces –OH)

Although many food molecules are digested by humans and heterotrophic microbes, other substances, such as complex plant materials and petroleum compounds, are digested only by bacteria or fungi. Within our digestive tract, as much as 15% of our caloric intake is provided by bacteria such as *Bacteroides* species. These bacteria catabolize the complex plant fibers we ingest. In the oceans, petroleum seeping from the ocean floor or from damaged wellheads is broken down by oil-eating bacteria. Most of the oil breakdown products eventually are converted to acetyl-CoA (acetyl group attached to coenzyme A).

Energy Carriers: ATP and NADH

As food molecules break down and are oxidized, these reactions yield energy. The energy-yielding reactions are regulated by enzymes that transfer the energy into biosynthesis. Let's imagine a system in which each enzyme catalyzes an energy-yielding reaction coupled directly to a biosynthetic reaction, such as building an amino acid. In this manner, energy yield could be coupled directly to energy use, and thus little or no energy would be lost. How many enzymes would such a system require?

Suppose an organism could use one of 100 possible substrates to yield energy and needed to build 100 different compounds that make up its cell. For each substrate yielding energy, 100 different enzymes would be needed to couple that energy from any substrate to build each of the 100 different products. In all, the organism would need $100 \times 100 = 10{,}000$ different enzymes, each encoded by a specific gene. But a bacterium often has fewer than 5,000 genes in its entire genome. So, how could all the needed enzymes be made?

Figure 7.3 Many Carbon Sources Enter Central Pathways of Catabolism

All reactions are catalyzed by specific enzymes. Polysaccharides are broken down into disaccharides and then to monosaccharides (six-carbon sugars) such as glucose. Glucose and sugar acids are converted to pyruvate, which releases acetyl groups. Acetyl groups and acetate are also the breakdown products of fatty acids, amino acids, and complex aromatic plant materials such as lignin.

Now suppose instead that all of the 100 substrates would be converted by enzymes to a common energy carrier, such as ATP. Then this same type of energy carrier, holding the energy from any of the 100 substrates, could be used by an enzyme to build one of 100 products. Now, this system requires only 100 + 100 = 200 different enzymes. So, the use of common energy carriers enables a smaller genome to direct many kinds of metabolism.

Another important function of energy carriers is to gain and release energy in small amounts. If all the energy of food molecules were released at once (as in combustion), it would dissipate as heat without building biomass. In living cells, food molecules never break down in one step. Instead, the energy yield is divided among a large number of stepwise reactions with smaller energy changes.

Thus the cell can be thought of as "making change" by converting a large energy source to numerous smaller sources that can be "spent" conveniently for cell function and biosynthesis.

All organisms use common energy carriers such as **adenosine triphosphate (ATP)** and reduced **nicotinamide adenine dinucleotide (NADH)**. Molecules of ATP and NADH are formed by many reactions of the catabolic pathways shown in Figure 7.3.

As described in Chapter 4, the energy carrier ATP is composed of a base (adenine), a sugar (ribose), and three phosphoryl groups. The adenine-ribose-phosphate molecule (adenosine nucleotide) is equivalent to a ribonucleotide, the monomer (unit) of RNA. The base adenine is a fundamental molecule of life—one that forms spontaneously from methane and ammonia in experiments simulating

Figure 7.4 Energy Carriers: ATP and NADH

ADP is phosphorylated to ATP

A. The chemical reaction of ADP (adenosine <u>di</u>phosphate) with inorganic phosphate makes ATP (adenosine <u>tri</u>phosphate). The reaction requires energy input because the negatively charged oxygens of the phosphates are forced near each other. In the reversed reaction, removal of each phosphate by hydrolysis (inserting H_2O) yields energy.

NAD^+ is reduced to NADH + H^+

B. In the reduction of NAD^+, the ring (shaded pink) loses a double bond as two electrons are gained from an electron donor; the resulting NADH carries energy. The reaction consumes two hydrogen atoms; one bonds to NAD^+, forming NADH, while the other ionizes in solution (H^+).

the origin of life on early Earth. ATP is an ancient component of cells, used by all living organisms. Presumably, ATP was used by the ancestral cell from which all life evolved, including humans.

During catabolism (see Figure 7.3), many of the enzymes couple an energy-yielding reaction to the formation of ATP (**Figure 7.4A**). ATP is formed by phosphorylation of adenosine diphosphate (ADP); the ADP molecule condenses with inorganic phosphate (P_i) to form ATP. Forming ATP is like getting cash from a bank teller, ready to use.

ATP formation requires spending energy (positive ΔG) because condensing ADP with phosphate brings negatively charged oxygen atoms together, where they repel each other. Because ATP formation requires spending energy, its breakdown yields energy. ATP breaks down by hydrolysis to form ADP and inorganic phosphate (P_i).

One molecule of NADH carries about three times as much energy as one molecule of ATP. Besides carrying energy, NADH carries two electrons ($2e^-$) from a food molecule, which participate in the oxidation-reduction reactions of metabolism. Oxidation-reduction reactions, or redox reactions, are those that involve a transfer of electrons between two molecules. As discussed in Section 4.4:

- **Oxidation is <u>loss</u> of an electron.** A molecule that loses an electron is said to be **oxidized** because oxygen has a strong ability to "take" electrons, but the electron can be lost to other atoms besides oxygen.

- **Reduction is <u>gain</u> of an electron.** A molecule that gains an electron is said to be **reduced** because the charge on the atom becomes slightly more negative.

For example, as the sugar glucose is broken down to carbon dioxide, it loses its hydrogen atoms, which consist of electrons (e^-) and protons (H^+). Sugar breakdown includes reactions in which a pair of electrons ($2e^-$) from a food molecule are transferred to the oxidized form of the carrier, NAD^+. The electrons are accompanied by two hydrogen ions, also called protons ($2H^+$). Typically, the oxidized form NAD^+ receives two electrons ($2e^-$) plus a proton (H^+) from glucose. Meanwhile, a second H^+ from glucose enters water solution. Overall, reduction of NAD^+ consumes two hydrogen atoms to make NADH (**Figure 7.4B**):

$$NAD^+ + 2H^+ + 2e^- \rightarrow NADH + H^+ \qquad \Delta G = 62\,kJ/mol$$

For the reduction of NAD^+, the value of ΔG is positive; therefore, NADH formation requires an input of energy from the catabolized food molecule. The reduced energy carrier NADH can then donate two electrons ($2e^-$) to another molecule, restoring the oxidized form NAD^+. NADH is central to life because it provides a way of carrying electrons to the **electron transport system (ETS)**, also known as the electron transport chain (introduced in Section 6.5; further discussed in Section 7.4). The ETS is a series of membrane-embedded proteins that store the energy from NADH in a H^+ ion gradient across the membrane. The energy from NADH is harvested by the transfer of electrons via the ETS to molecular oxygen, which is combined with $2H^+$ to form water.

The overall series of reactions from glucose to water is called aerobic respiration (presented in Chapter 6). Aerobic respiration takes place in the cell membrane of aerobic bacteria and also in the inner membrane of our own mitochondria. Aerobic respiration is the reason we need to breathe oxygen.

In general, for food breakdown remember that an atom of hydrogen removed from a C–H bond consists of a proton (H^+) plus an electron (e^-). In a metabolic reaction, the proton and electron may be transferred together to a molecule, or the proton may be released as a hydrogen ion (as in NADH + H^+). The electron transfer defines redox reactions; the loss of H^+ is reversible and does not affect the redox change.

SECTION SUMMARY

- **Catabolic pathways organize the breakdown of large molecules** in a series of sequential steps coupled to reactions that store energy in small energy carriers such as ATP and NADH.

- **ATP stores energy in the bonds** formed by condensation of ADP with inorganic phosphate (P_i). Hydrolysis of these bonds yields energy.

- **NADH stores energy associated with a pair of electrons** that can reduce a substrate, such as a protein of the electron transport system (ETS). Oxidizing NADH to NAD^+ yields energy and transfers two electrons ($2e^-$) to a substrate molecule.

7.3 Glucose Catabolism, Fermentation, and the TCA Cycle

SECTION OBJECTIVES

- Describe how sugar is catabolized to pyruvate via glycolysis, and explain how these catabolic reactions generate ATP and NADH.

- Explain how pyruvate is further catabolized by fermentation or by the TCA cycle (when oxygen is available).

- Explain how a clinical laboratory uses bacterial catabolic reactions in tests to identify a pathogen.

We now summarize the central pathways of catabolism that generate ATP and NADH from glucose. Glucose is converted to two molecules of pyruvic acid (during glycolysis), which are decarboxylated to acetic acid (acetate). Acetate releases two molecules of CO_2 via the TCA cycle. These pathways of glycolysis and the TCA cycle are essential for heterotrophic bacteria (including pathogens) and for human cells. In human cells, the TCA cycle and respiration are conducted within mitochondria. Mitochondria are organelles that evolved from Gram-negative bacteria related to *Escherichia coli*, and their respiration mechanism is similar to that of bacteria. In external environments,

Figure 7.5 **Energy Flow in Glycolysis: The Embden-Meyerhof-Parnas (EMP) Pathway** Glucose is activated by 2 ATP that phosphorylate the sugar. The breakdown of glucose to two molecules of pyruvate is coupled to production of 2 NADH and 4 ATP. Since 2 ATP were spent originally, the net yield is $4 - 2 = 2$ ATP. Inorganic phosphate is shown as P_i. G3P = Glyceraldehyde 3-phosphate; 3C = three-carbon molecule; 6C = six-carbon molecule.

bacteria obtain glucose for glycolysis by breaking down complex substrates (as shown in Figure 7.3). Blood-borne pathogens, however, such as the malaria parasite, take glucose directly from the plasma.

Glycolysis: From Glucose to Pyruvate

For many bacteria, the main pathway of glucose catabolism is glycolysis, also known as the Embden-Meyerhof-Parnas (EMP) pathway. In glycolysis, one molecule of D-glucose (the D mirror form) breaks down into two molecules of pyruvic acid (**Figure 7.5**). The reaction yields a net gain of energy carriers: 2 ATP and 2 NADH.

Note: Pyruvic acid exists in equilibrium with pyruvate, its anion (with negative charge), plus a hydrogen ion (H^+). For simplicity, pyruvic acid and other acids may be labeled as the anion ("-ate" suffix), although only some of the compound is ionized.

Glucose breaks down in two stages (**Figure 7.6**).

- **Stage 1 of glycolysis actually <u>consumes</u> energy (2 ATP).** The added energy prepares the glucose for a controlled breakdown that will release more energy (in stage 2) than was added. The six-carbon (6C) glucose molecule is primed for breakdown by two steps of phosphorylation by 2 ATP, yielding fructose 1,6-bisphosphate. Each ATP phosphotransfer step consumes energy. The sugar with two phosphoryl groups (fructose 1,6-bisphosphate) then splits. The two halves of the molecule

Figure 7.6 Glycolysis (EMP Pathway)

Glucose is phosphorylated by 2 ATP, which input energy to prime the pathway. The reactions include rearrangement to form fructose 1,6-bisphosphate. Fructose 1,6-bisphosphate is cleaved into two three-carbon molecules of glyceraldehyde 3-phosphate—a reaction that yields energy. Each molecule of glyceraldehyde 3-phosphate loses two hydrogen atoms by transferring $2e^-$ to NAD^+, forming NADH. Each three-carbon molecule now incorporates inorganic phosphate (P_i) to form 1,3-bisphosphoglycerate. Subsequent reactions transfer both phosphoryl groups to 2 ADP to form 2 ATP. (More details of glycolysis are presented in Appendix 2, Figure A2.1.)

consist of three-carbon (3C) molecules of glyceraldehyde 3-phosphate (G3P).

- **Stage 2 of glycolysis <u>yields</u> energy (4 ATP).** Each of the two G3P molecules loses $2e^-$ (from two hydrogen atoms) and converts NAD^+ to NADH (in all, 2 NADH). Subsequent steps from

Figure 7.7 From Glucose to Pyruvate

The Embden-Meyerhof-Parnas (EMP) pathway of glycolysis, the Entner-Doudoroff (ED) pathway, and the pentose phosphate pathway (PPP).

each three-carbon molecule lead to pyruvate, yielding enough energy to form 2 ATP from each pyruvate (in all, 4 ATP from the original glucose).

Since 2 ATP were spent originally to phosphorylate glucose, **the net gain of energy carriers is two molecules of ATP (2 ATP) plus two molecules of NADH (2 NADH).**

Key intermediates of glucose metabolism are presented in Figure 7.6 (full details are presented in Appendix 2, Figure A2.1). Most of the conversion steps have only a small change in energy—so small that the sign of ΔG depends on the concentrations of substrates or products; thus, some individual steps are reversible. In the cytoplasm, however, as intermediate products form they are quickly consumed by the next step, so the pathway flows in only one direction.

Alternatives for Sugar Catabolism

Besides glycolysis (EMP pathway), there are other pathways microbes can use to break down sugar (**Figure 7.7**). Different species use different pathways. And different pathways are used for different environmental needs. For example, within our intestines *E. coli* uses the Entner-Doudoroff pathway to consume mucus secreted by the intestinal epithelium. Intestinal mucus contains sugar acids—sugars that possess one carboxylic acid group:

Figure 7.8 Glucose Fermentation Pathways

Alternative pathways from glucose catabolism to end products, many of which we use for food or industry.
Different species conduct different portions of the pathways shown. CoA = coenzyme A.

An example is gluconic acid (deprotonated as gluconate). Gluconate can be broken down in a series of steps leading to glyceraldehyde 3-phosphate and ultimately pyruvate. The Entner-Doudoroff pathway enables *E. coli* and other bacteria to "farm" intestinal mucus.

Another pathway, the **pentose phosphate pathway (PPP)**, allows bacteria to divert sugars into biosynthesis. In the pentose phosphate pathway, the early steps of glucose breakdown release one carbon as CO_2, forming a five-carbon sugar, ribulose 5-phosphate. The reaction also leads to formation of two molecules of NADPH (an NADH molecule with an extra phosphoryl group). The extra phosphate "tags" NADPH for use in biosynthesis reactions rather than for the electron transport system (Section 7.6). The NADPH can be used to build amino acids and cell walls, whereas the ribulose 5-phosphate can be used to construct nucleotide bases. Remaining steps of the pentose phosphate pathway lead back to G3P and pyruvate, but sugar intermediates of various sizes can be shunted off into biosynthesis (see Section 7.6).

Fermentation Completes Catabolism

So far, all our catabolic pathways have ended up with the product pyruvic acid (or its ionized form, pyruvate). Are these pathways complete? Actually, no, because along with pyruvate remain a large number of NADH molecules containing electrons removed from the sugars. For metabolism to be complete, the electrons, as well as the waste carbon from pyruvate, need to be released from the cell. As introduced in Chapter 6, catabolism can be completed in different ways, depending on the availability of oxygen (O_2):

- **Fermentation** occurs when the electrons of NADH are put back onto pyruvate, forming waste products such as lactic acid. No oxygen is consumed.

- **Respiration** occurs when the NADH donates electrons to the electron transport system (ETS), which ultimately transfers them to O_2 or to an alternative inorganic molecule serving as a terminal electron acceptor (see Section 7.4).

Without O_2 or other inorganic electron acceptors, cells must recycle the electrons from NADH back onto the pyruvate in the process of fermentation. Fermentation forms partly oxidized products that have the same redox level (balance of O and H) as the original glucose (**Figure 7.8**). For example, glucose may be fermented to two molecules of lactic acid (lactate fermentation) or two molecules of ethanol plus two molecules of CO_2 (ethanolic fermentation) or to one molecule of lactic acid, one molecule of ethanol, and one molecule of CO_2 (heterolactic fermentation). These fermentation products are then excreted from the cell. This excretion of organic products seems wasteful, since products such as ethanol and lactic acid still contain a lot of food value—that is, potential for catabolism, yielding energy. But without oxygen or other electron acceptors, the cell has no choice.

Fermentation has a long history in human civilization. Because fermentation is incomplete digestion, it offers us a way for limited microbial treatment of a food source to make a stable product that retains food value. The microbial "wastes," such as ethanol or acids, retard further growth of microbes, including pathogens, so they preserve foods such as wine and cheese for extended storage by humans. Different species of microbes convert pyruvate to different combinations of products, yielding different properties and flavors. For example, *Propionibacterium* species generate propionic acid, the source of the flavor of Swiss cheese.

Because digestion is incomplete, fermenting microbes must catabolize a large amount of sugar to grow, producing a large amount of waste products. The waste products most commonly made by human-associated pathogens are acids such as lactic acid and acetic acid.

Fermentation acids help the clinician identify pathogens isolated from patients. The appearance of acids in diagnostic media changes the color of pH indicators such as the phenol red dye of phenol red broth (**Figure 7.9A**). Phenol red dye appears red above pH 7.4 but turns orange and then yellow in media acidified by fermentation acids (below pH 6.8). A culture of *Escherichia coli* turns phenol red broth yellow, showing fermentation to acid. In addition, the gas produced indicates production of CO_2 and H_2. A different enteric pathogen, *Alcaligenes faecalis*, ferments sugar poorly but converts peptides in the broth to alkaline amines; this culture turns deep red.

Other kinds of media show a color change indicating a strain's ability to catabolize a specific sugar (**Figure 7.9B**). For example, MacConkey agar (see Section 6.1) containing sorbitol allows *E. coli* K-12 to grow as red colonies (ferment sorbitol), whereas *E. coli* O157:H7 (the cause of hemorrhagic disease) forms white colonies (fail to ferment sorbitol).

Figure 7.9 Clinical Tests Based on Fermentation

Uninoculated control

Gas

E. coli fermentation generates acid; phenol red turns yellow

Alcaligenes faecalis does not ferment

A. Organisms are cultured in a tube of broth containing phenol red. The dye turns yellow below pH 7. Phenol red is orange-red at neutral pH, yellow at lower pH (acid), and red at higher pH (base). A small inverted tube (Durham tube) collects gaseous fermentation products (CO_2 and H_2).

B. In the sorbitol fermentation test for the pathogen *E. coli* O157:H7, red colonies indicate fermentation by nonpathogenic *E. coli* whereas white colonies (strain O157:H7) fail to ferment sorbitol.

Tricarboxylic Acid Cycle: Transferring All Electrons to NADH

Unlike fermentation, the complete breakdown of glucose requires the breakdown of two molecules of pyruvate to six molecules of CO_2 (**Figure 7.10**). The hydrogen atoms from glucose must ultimately reduce an electron acceptor, such as oxygen gas (O_2). Again, this breakdown process takes several stages, each yielding useful amounts of energy.

- **Pyruvate is broken down to form acetyl-CoA and CO_2.** The breakdown of pyruvate also removes two hydrogens, transferring $2e^-$ to form NADH.

   Pyruvate breakdown is catalyzed by the pyruvate dehydrogenase complex (PDC). The PDC is a key component of metabolism in bacteria and in mitochondria because it is the first enzyme to direct sugar catabolism into respiration instead of fermentation or biosynthetic pathways. In human mitochondria, the PDC is especially important for organs with a high metabolic rate, such as the heart and brain. Defects in mitochondrial PDC cause heart failure and neurodegeneration.

   Acetyl-CoA consists of acetic acid in which the OH group is replaced (reversibly) with coenzyme A. Coenzyme A acts as a signal for enzymes. Like a tag for baggage at an airport, coenzyme A tells enzymes the correct pathway to send the acetyl group. In this case, enzymes must direct the acetyl group into the TCA cycle (Figure 7.10) instead of other reactions, such as biosynthesis of fatty acids. Recall that acetyl-CoA can also arise from many nonsugar foods (Figure 7.3), such as fatty acids from lipids, and from complex aromatic derivatives, such as lignin and petroleum (see IMPACT). Whatever its source, acetyl-CoA is directed by its "tag" into the TCA cycle.

   In the TCA cycle (see Figure 7.10), the two carbons of acetyl-CoA are ultimately converted to $2CO_2$. One turn of the cycle yields enough energy to form ATP. In addition, the eight electrons from acetate's four hydrogen atoms are transferred to 3 NAD^+ (forming 3 NADH) and 1 FAD (forming $FADH_2$, an energy carrier similar to NADH). These carriers transfer the electrons to the electron transport system and ultimately to oxygen.

- **The TCA cycle condenses acetyl-CoA with oxaloacetate, releasing CoA (Figure 7.10).** The two-carbon acetyl plus four-carbon oxaloacetate form citrate. Citrate, or citric acid, is a common energy storage molecule; it confers part of the flavor of citrus fruits such as oranges and lemons.

- **Citrate releases another CO_2 and transfers $2H^+ + 2e^-$ to make NADH.** Citrate breaks down by a series of reactions that form 2-oxoglutarate. The 2-oxoglutarate loses CO_2 and transfers $2e^-$ to form NADH. The remaining four-carbon molecule is succinate.

Figure 7.10 Pyruvate Is Converted to Acetyl-CoA and Feeds into the TCA (Krebs) Cycle

The TCA cycle starts with acetyl-CoA from the breakdown of pyruvate. The acetyl group is combined with oxaloacetate, forming citrate. The electrons from acetyl-CoA are then used to make NADH and $FADH_2$, while the two carbons are oxidized to CO_2. (Full details of the TCA cycle are presented in Appendix 2, Figure A2.4.)

- **Succinate transfers $2e^-$ to FAD (an energy carrier similar to NAD^+), forming $FADH_2$.** The subsequent reactions incorporate H_2O and transfer the final $2e^-$ to form NADH. Each $2e^-$ transferred to NAD^+ to form NADH comes from the loss of two hydrogen atoms from the acetyl group.

- **Oxaloacetate is regenerated.** The four-carbon oxaloacetate is ready to condense with another acetyl group and start a new round of the TCA cycle.

   Overall, in each round of the TCA cycle, oxaloacetate condenses with acetate (as acetyl-CoA) to form citrate; the citrate then loses two carbons as CO_2, regenerating oxaloacetate (**Figure 7.11**). The series of reactions stores increments of energy in 3 NADH, $FADH_2$, and ATP. Each reaction step couples energy-yielding events (such as CO_2 release) to energy-storing events (such as NADH formation).

   While a full turn of the TCA cycle releases all the remaining carbons from the original pyruvate molecule, the metabolic pathway is still not complete. The NADH and $FADH_2$ produced from the TCA cycle must be recycled (to NAD^+ and FAD, respectively) if more glucose or other carbohydrate is to be consumed. The energy carriers are recycled by donating their electrons to the electron transport system (ETS). Electron transport through the proteins yields a large amount of energy.

The electrons ultimately reach a terminal electron acceptor of the ETS. For aerobic respiration, the terminal electron acceptor is oxygen. Anaerobic microbes rely on alternative terminal electron acceptors, such as nitrate. The energy from electron transfer through the ETS is used to pump H^+ ions outside, forming a H^+ gradient (presented in next section.) The H^+ ions then flow back through the membrane-embedded ATPase, making ATP.

The electron transfer reactions through the ETS produce a much larger number of ATP molecules than does fermentation. So, glucose respiration yields far greater energy than fermentation, which essentially "gives back" the potential energy of NADH and captures only the energy from 2 ATP.

Other Functions of Citrate and Oxaloacetate

Some of the TCA intermediates serve multiple functions in the cell, such as substrates for biosynthesis (Figure 7.11). For instance, oxaloacetate can be converted to the amino acid aspartic acid (aspartate). The reaction requires only a single enzyme-catalyzed step—the incorporation of an amine group replacing the carboxylate (double-bonded oxygen atom). Similarly, one amine incorporation converts 2-oxoglutarate to glutamic acid (glutamate). Further details of the TCA cycle, including all intermediate molecules, are shown in Appendix 2, Figure A2.4.

Suppose a cell were to obtain any of the TCA intermediates from a source other than glucose, instead of incorporating acetyl-CoA to form citrate. Could the molecules be processed by the TCA enzymes to yield energy? They could if the cell had transporters to take up such molecules from the environment. In fact, the ability of bacteria to catabolize TCA intermediates such as citrate is used in diagnostic media, along with sugar fermentations, to identify pathogens.

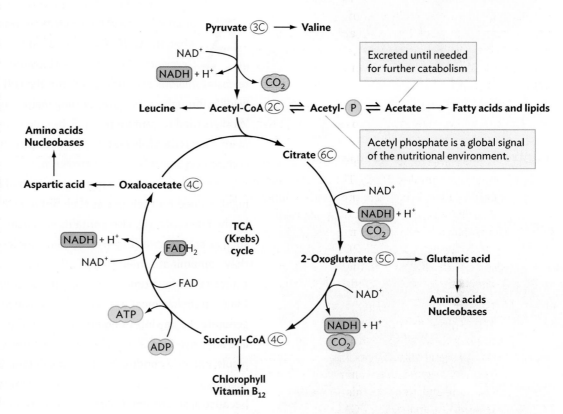

Figure 7.11 **The TCA Cycle Intermediates Act as Substrates for Biosynthesis** TCA cycle intermediates are used to build amino acids, fatty acids, and nucleobases. (For other TCA intermediates, see Appendix 2, Figure A2.4.)

normal, but within an hour he developed respiratory distress due to prematurity; he required intubation and mechanical ventilation (assisted breathing). A blood culture showed no bacterial growth, but ampicillin and gentamicin were begun in case of infection. By day 5, Mark showed metabolic acidosis (blood pH less than 7.35), thrombocytopenia (low platelet count), and the presence of leukocytes and erythrocytes in the cerebrospinal fluid (CSF). The CSF was cultured and grew a Gram-negative enteric rod; the same organism was now isolated from blood. To identify the isolate, the clinician performed diagnostic tests. Simmons citrate agar tested positive (color change from raised pH due to acid consumption). Phenol red tests for fermentation acid from dulcitol and melibiose were positive; sucrose was negative. Tests for arginine dihydrolase and ornithine decarboxylase were positive, but the organism did not possess lysine decarboxylase. These and other tests confirmed identification of *Citrobacter sedlakii*, a rare cause of septicemia and meningitis (inflammation of the brain lining). Antibiotic susceptibility was tested, and the ampicillin was replaced with cefotaxime. After 10 weeks of antibiotic therapy, Mark made a full recovery.

In this case history an infant was born prematurely, as a result of the mother's young age and lack of prenatal care. Premature infants requiring tracheal intubation are susceptible to respiratory

CASE HISTORY 7.1

Fermentation Products Identify a Meningeal Pathogen

In a Chicago hospital, a male infant, Mark, was born prematurely at 30 weeks (weight, 1.7 kg, or 3.75 pounds) to a 15-year-old woman. The infant's mother had received no prenatal care. The infant's physical scores at birth were

impact

Bacteria Remediate Oil Spills

A dramatic case of bacterial catabolism on a massive scale occurred in 2010, when the Gulf of Mexico faced an unprecedented spill of oil from the *Deepwater Horizon* oil well blowout. The offshore oil platform exploded, releasing millions of barrels of oil into the gulf over 3 months. The leaked oil killed wildlife throughout the gulf, causing major environmental damage (**Figure 1A**). Workers tried to contain the spill, but ultimately, most of the oil was consumed by marine bacteria. Only certain microbes possess the enzymes needed to degrade the complex organic mixture of petroleum.

Petroleum contains a high proportion of long-chain hydrocarbons, which are highly valued for their use as fuel. But unrefined petroleum, such as that released by the *Deepwater* rig, also contains significant amounts of aromatic molecules such as anthracene (**Figure 1B**). Such molecules are poisonous and carcinogenic; they contaminate shellfish and other organisms in the food chain. Aromatic molecules take a long time to degrade in the environment; eventually, bacteria and fungi catabolize most of them. Catabolism involves several enzyme pathways that generate the key intermediate catechol. Catechol is then catabolized to acetyl-CoA and succinyl-CoA, both of which can enter the TCA cycle (recall Figure 7.10). As a result, bacteria conducting aromatic catabolism can convert petroleum pollutants such as anthracene to CO_2, removing them from contaminated water. However, because aromatic catabolism is slower than sugar catabolism, microbial remediation of spilled oil takes many years. And the release of CO_2, although necessary, adds to global warming.

A surprising discovery is that many of the marine bacteria involved in degrading spilled oil are related to pathogens. Some oil from the *Deepwater Horizon* spill was degraded by bacteria of the genus *Vibrio*, such as *Vibrio parahaemolyticus* and *Vibrio vulnificus*, which cause serious human infections. So, even a bacterium that has negative consequences for human health may have a positive role in our environment.

DISCUSSION QUESTION

Suppose we wish to engineer a strain of bacteria that can help clean up an oil spill. How could we do this?

Figure 1 Bacteria Catabolize Spilled Oil

A. Marine bird contaminated by petroleum from an offshore wellhead.

Anthracene and related aromatic compounds are found in petroleum.

Catechol is an intermediate of aromatic catabolism.

Acetyl-CoA and succinyl-CoA enter the TCA cycle.

B. Petroleum contains aromatic components, such as anthracene, that can be digested by microbes through pathways generating the key intermediate catechol. Catechol is then catabolized to acetyl-CoA and succinyl-CoA, which enter the TCA cycle for conversion to CO_2.

pathogens. In this case the infectious agent was a rare opportunistic pathogen that normally inhabits the intestinal tract without causing problems (**Figure 7.12A**). Because the organism was so rare, its identification required multiple diagnostic tests. Diagnostic tests often make use of bacterial catabolic enzymes. Simmons citrate agar tests the pathogen's ability to respire by using a TCA intermediate (citric acid) as the sole carbon source. As the acidic carbon source is consumed, the pH of the agar rises, causing the color to change from green to blue (**Figure 7.12B**). In phenol red broth, the sugars dulcitol and melibiose showed positive reactions (yellow), as seen earlier, in Figure 7.9A.

Another test, sucrose fermentation, addresses the organism's ability to break down the disaccharide sucrose into glucose and then ferment to acids (lactic acid and/or acetic acid). In Case History 7.1, the result for sucrose fermentation was negative (red color, neutral pH). The decarboxylation of certain amino acids (arginine and ornithine) produced alkaline amines that raised pH for a color change. A similar test for lysine decarboxylase was negative. Combining the results of many such tests revealed a pattern highly consistent with a particular pathogen. Today, as genetic data are accumulated, species can also be identified by DNA tests, which we will discuss in Chapter 8.

Although we focus on sugar catabolism, recall that in nature, microbes catabolize many different kinds of organic molecules. Within our intestines, bacteria such as *Bacteroides* species digest complex plant material into acetyl groups and other small molecules that our intestines can absorb. And in soil and water, bacteria consume toxic pollutants such as petroleum spilled by an oil well blowout (**IMPACT**).

Figure 7.12 Diagnostic Testing of a Respiratory Pathogen

A. Premature infants requiring tracheal intubation are susceptible to respiratory pathogens.

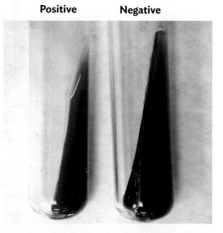

Positive   Negative

B. Simmons citrate agar test of a Gram-negative pathogen. Growth on citrate (citric acid) as a sole carbon source consumes acid, thereby raising the pH, with a color change to blue.

SECTION SUMMARY

- **Glycolysis (EMP pathway) breaks down glucose to two molecules of pyruvic acid, storing energy in 2 ATP and 2 NADH.** In stage 1, glucose receives two phosphoryl groups from ATP and then is cleaved to two three-carbon sugars. In stage 2, both three-carbon sugars eventually are converted to pyruvate. The process generates 4 ATP; subtracting the 2 ATP spent priming glucose leaves a net yield of 2 ATP.

- **The pentose phosphate pathway generates 2 NADPH for biosynthesis plus a five-carbon sugar** that may be used to build nucleic acids. Further reactions yield other intermediate substrates for biosynthesis of amino acids and other cell components.

- **Fermentation completes catabolism** by donating electrons from NADH back to pyruvate to form waste products that are excreted from the cell.

- **Pyruvate dehydrogenase converts pyruvate to acetyl-CoA.** The coenzyme A (CoA) directs the acetyl group into the TCA cycle.

- **The TCA cycle breaks down the acetyl group to CO_2.** Each acetyl group condenses with oxaloacetate to form citrate, which is successively broken down by enzymes until regenerating oxaloacetate. Energy is stored in the form of NADH, $FADH_2$, and ATP.

- **Sugar fermentation and organic acid respiration are used as tools in diagnostic media to identify pathogens.**

Thought Question 7.1 Why are glucose catabolism pathways ubiquitous, even in bacterial habitats where glucose is scarce? Give several reasons.

Thought Question 7.2 How is sugar catabolism used as a diagnostic indicator? Why are so many different sugars used as convenient indicators?

7.4 Respiration and Lithotrophy

SECTION OBJECTIVES

- Describe how NADH transfers electrons to electron transport proteins of the ETS and ultimately to the terminal electron acceptor, such as O_2.
- Explain how the transfer of electrons from one ETS protein to the next yields energy and pumps H^+ across the membrane.
- Explain how lithotrophy yields energy by oxidizing mineral electron donors instead of organic molecules. State an example of how environmental lithotrophy affects human health.

The overall process of catabolism from substrate breakdown to reduction of a terminal electron acceptor (such as O_2) is called **respiration** (**Figure 7.13**). Human respiration is conducted by our mitochondria. The overall equation for the respiration of glucose with oxygen is:

$$C_6H_{12}O_6 + 6H_2O + 6O_2 \rightarrow 12H_2O + 6CO_2$$

Respiration completes catabolism by donating electrons from NADH to the ETS in a process that stores energy by pumping

ANIMATION ▶ **Figure 7.13 Respiration of Glucose**

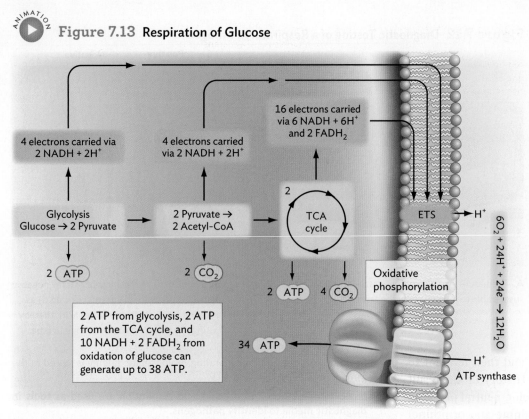

Bacterial glucose catabolism generates ATP by ADP phosphorylation from glycolysis and by the ETS pumping H^+ across the cell membrane to drive the ATP synthase. The complete oxidative breakdown of glucose to CO_2 and H_2O could theoretically generate up to 38 ATP; the actual number of ATP molecules observed experimentally is smaller (approximately 30).

hydrogen ions (H^+) across the membrane, up to $3H^+$ per NADH. The hydrogen ions return to the cell via ATP synthase, driving ATP formation. The process of ATP formation by respiration is called **oxidative phosphorylation**. The "phosphorylation" part refers to phosphorylation of ADP to ATP by the ATP synthase complex driven by a proton gradient generated through the ETS.

Glucose respiration with oxygen can generate up to 38 molecules of ATP per molecule of glucose—far more than in fermentation. In bacteria, the actual number of ATP molecules formed varies widely with the availability of a carbon source and oxygen. For example, as oxygen concentration decreases, the ability to oxidize NADH decreases, so the cell may make only 1 or 2 ATP per NADH.

Electron Transport System and Proton Motive Force

The redox energy carriers NADH and $FADH_2$ transfer their electrons to a membrane-embedded enzyme (electron transport protein) of the ETS. Each ETS enzyme is called an **oxidoreductase** because it receives electrons from a stronger electron <u>donor</u> (becoming reduced) and then transfers them to a stronger electron <u>acceptor</u> (becoming oxidized). These transfer reactions yield energy. The final oxidoreductase of the chain transfers $2e^-$ onto oxygen (the terminal electron acceptor). Two rounds of electron transfer ($4e^-$ in all) from 2 NADH allow incorporation of four

hydrogen ions ($4H^+$) to form two molecules of H_2O:

$$O_2 + 4e^- + 4H^+ \rightarrow 2H_2O$$

The reduction of O_2 to water by all the electrons from glucose hydrogen atoms finally completes the aerobic catabolism of glucose.

Along the way, each redox step of an ETS yields energy. How is this energy stored? Some of the ETS oxidoreductases use the energy of electron transfer to pump hydrogen ions (protons) across the bacterial cell membrane. **Figure 7.14** shows an example of an ETS in the membrane of *E. coli*. In this ETS, NADH donates its pair of electrons ($2e^-$) to an oxidoreductase called NADH dehydrogenase 1 (NDH-1). NDH-1 then transfers the $2e^-$ onto a quinone (Q), generating quinol (QH_2). The redox step from NADH through NDH-1 to quinol (QH_2) yields enough energy to pump an extra $4H^+$ across the membrane.

Quinones and quinols are small aromatic molecules soluble in the membrane. Quinols diffuse laterally within the membrane but do not cross the membrane between cytoplasm and exterior. Each quinol reduced by NDH-1 can transfer its $2e^-$ onto the next ETS complex, cytochrome *bo* oxidoreductase. (A cytochrome is a protein containing an iron heme group that readily transfers electrons.)

Cytochrome *bo* oxidoreductase pumps $2H^+$ across the membrane and releases the $2H^+$ received from the quinol. Then, as terminal electron acceptor, the cytochrome *bo* oxidoreductase puts $2e^-$ onto molecular oxygen, incorporating one oxygen atom into water. Overall, for each NADH oxidized, up to $8H^+$ may be pumped across the membrane.

The ETS generates a gradient of hydrogen ions across the membrane—that is, a larger concentration of H^+ outside than inside. This hydrogen ion gradient, or **proton motive force (PMF)**, then drives the ATP synthase "machine" to phosphorylate ADP to ATP (**Figure 7.15**). Bacterial ATP synthase is the target of important new antibiotics that combat tuberculosis bacteria and methicillin-resistant *Staphylococcus aureus* (MRSA).

The flow of protons (H^+) through the ATP synthase and back into the cell is driven both by the charge difference (more positive outside than inside) and by the H^+ concentration (greater outside; see **Section 5.2**). The sum of these two forces gives the proton motive force (PMF). The PMF generates a current of protons, analogous to an electrical current of electrons. The ATP synthase complex actually rotates with proton flow, like a water wheel powered by a river (**Figure 7.16**). The "wheel" part is the F_o complex, a

Figure 7.14 A Bacterial ETS for Aerobic NADH Oxidation

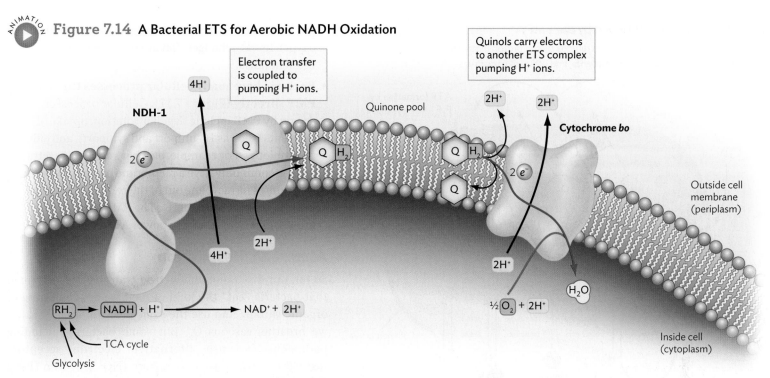

In *E. coli*, a pair of electrons ($2e^-$) from the oxidoreductase NDH-1 are transferred onto a quinone (Q), generating quinol (QH_2). This redox step yields enough energy to pump $4H^+$ across the membrane. Each quinol can transfer its $2e^-$ onto the next oxidoreductase (cytochrome *bo*), which pumps $2H^+$ across the membrane in addition to $2H^+$ from the quinol. For each NADH oxidized, up to $8H^+$ may be pumped across the membrane.

Figure 7.15 The ETS Generates the Proton Motive Force (PMF)

Some of the ETS oxidoreductases pump protons out of the cell. The resulting electrochemical gradient of protons (proton motive force) drives conversion of ADP to ATP through the ATP synthase.

Figure 7.16 Bacterial Membrane-Embedded ATP Synthase

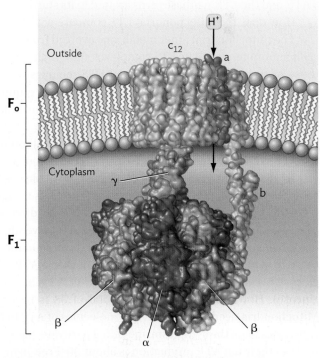

The F_1F_o complex is plugged into the *E. coli* plasma membrane. The three pairs of alpha (α) and beta (β) subunits rotate around the gamma (γ) axle, catalyzing formation and hydrolysis of ATP. The ring of 12 c subunits rotates while translocating three protons.

Figure 7.17 Processes Driven by the PMF

Besides ATP synthesis, processes driven directly by the proton motive force include flagellar rotation, uptake of nutrients, and efflux of toxic drugs.

membrane-embedded circle of 12 c subunits. For every three protons ($3H^+$) that flow through the F_o, the F_1 rotates one-third of a turn and catalyzes ADP conversion to ATP. So, the flow of $9H^+$ generates 3 ATP, and the ATP synthase turns around to its starting position.

How does the mitochondrial ETS compare with that of bacteria? The mitochondrial inner membrane evolved from the bacterial cell membrane, and all its ETS complexes show high similarity with those of bacteria. The mitochondrial ETS includes four oxidoreductases: complex I transfers $2e^-$ from NADH to a quinol; complex II receives $2e^-$ from $FADH_2$; complex III receives $2e^-$ from a quinol; and complex IV transfers $2e^-$ to oxygen (see the diagram in Appendix 2, Figure A2.6). The final mitochondrial complex (V) is the ATP synthase, highly similar to its bacterial homolog.

In all, how many molecules of ATP can be made from glucose catabolism? In theory, under optimal conditions (see Figure 7.13), 1 NADH yields enough energy for the PMF to drive synthesis of 3 ATP; and 1 $FADH_2$ yields energy for 2 ATP. Overall, the PMF made from one molecule of glucose drives the membrane ATP synthase to synthesize as many as 34 ATP. Another 4 ATP come from glucose breakdown via glycolysis and the TCA cycle (38 total per glucose). However, this theoretical total of 38 is never reached. Mitochondria need to spend several molecules of ATP to transport substrates from the cytoplasm, including pyruvate, phosphate, and ADP, so their net ATP production is about 30. Free-living bacteria make less ATP than do mitochondria, for several reasons:

- **Bacteria sacrifice efficiency for flexibility.** Bacteria spend some of their PMF energy to maintain stable ion

gradients during extreme changes in external pH and oxygen levels—changes that are not experienced by mitochondria.

- **Bacteria have many cellular processes that use PMF directly (Figure 7.17).** Cellular processes driven by PMF include flagellar rotation, uptake of nutrients (such as H^+ uniport and symport), and efflux of antibiotics. For example, bacteria resistant to tetracycline commonly have a membrane protein complex that pumps tetracycline out of the cell, powered by an inwardly directed flow of H^+. The PMF spent on such processes is unavailable to power ATP synthase.

Anaerobic Respiration

So far, our discussion of electron transport has assumed that bacteria "eat" the kind of food we do, consisting of organic carbon, and use the same electron acceptor that we breathe, gaseous O_2. But some bacteria—including pathogens—can use alternative electron acceptors instead of oxygen, such as nitrate (NO^{3-}) or iron (Fe^{3+}; Table 7.2). For example, *Pseudomonas aeruginosa* is the most common pathogen infecting the lungs of cystic fibrosis patients. In cystic fibrosis, the lung sputum becomes anaerobic, thus requiring *P. aeruginosa* to donate electrons to nitrate. Soil bacteria use even more exotic electron acceptors. For example, *Geobacter metallireducens* transfers electrons to iron, uranium, and other metals (Figure 7.18A). *Geobacter* anaerobic respiration can remove pollutant metals from the environment by converting them to soluble form. These bacteria can actually donate electrons to a metal electrode, generating electricity in a fuel cell.

Table 7.2
Anaerobic Terminal Electron Acceptors

Reaction	Species and Environment
$Fe^{3+} + e^- \rightarrow Fe^{2+}$	*Geobacter metallireducens*; transfers electrons from catabolism to iron, uranium, and other metals; can remediate pollutants; can generate electricity in a fuel cell
$NO^{3-} + 2e^- \rightarrow NO^{2-}$ $N_2O + 2e^- \rightarrow N_2$	*Pseudomonas aeruginosa*, a soil bacterium and opportunistic pathogen; colonizes the lungs of cystic fibrosis patients
$SO_4^{2-} + 2e^- \rightarrow SO_3^{2-}$	*Desulfovibrio* spp., found in soil and in marine sediments; some strains can degrade pollutants
$S^0 + 2e^- \rightarrow H_2S$	*Pyrococcus furiosus*, a hyperthermophilic archaeon; grows in marine thermal vents at temperatures above 100°C (212°F)
$S_4O_6^{2-} + 2e^- \rightarrow 2S_2O_3^{2-}$	*Salmonella enterica*, an enteric pathogen; induces white blood cells to produce tetrathionate ($S_4O_6^{2-}$)

Figure 7.18 Anaerobic Respirers and Aerobic Lithotrophs

A. *Geobacter* conducts anaerobic respiration. *Geobacter metallireducens* transfers electrons to a graphite electrode.

B. *Nitrosomonas* species oxidize ammonia to nitrite.

and archaea that can break down the rock. This rock breakdown can lead to "leaching" of metals into solution, where they can be recovered by miners. Another form of lithotrophy involves oxidation of ammonia to nitrite and nitrate. An example of an ammonia-oxidizing bacterium in the soil is *Nitrosomonas* (**Figure 7.18B**). Ammonia oxidizers thrive in soil that is artificially fertilized with ammonia-rich material.

In our environment, bacterial lithotrophy and anaerobic respiration make many positive contributions. But certain kinds of bacterial metabolism can have unfortunate medical consequences, as the following case history illustrates.

Note that each electron acceptor (oxidant) must be reduced by a different terminal oxidoreductase. Oxidants Fe^{3+} and NO^{3-} are particularly useful electron acceptors in anaerobic soil and wetlands. In marine environments, especially near the ocean floor, sulfate (SO_4^{2-}) is commonly used as an electron acceptor because its concentration in seawater is high.

A surprising medical example of anaerobic respiration is that of the enteric Gram-negative pathogen *Salmonella enterica* serovar Typhimurium. *Salmonella* bacteria can respire using tetrathionate ($S_4O_6^{2-}$) as an electron acceptor. Tetrathionate is not normally available in the host mucosa, and most bacteria cannot use it. But *Salmonella* bacteria induce an inflammatory response in white blood cells, which in turn produce reactive tetrathionate to kill pathogens. With tetrathionate available as an electron acceptor, *Salmonella* can grow in the intestinal lumen, whereas other mucosal bacteria cannot. As a result, *Salmonella* has a growth advantage during pathogenesis.

Lithotrophs Eat Rock

Suppose that bacteria have O_2 available but lack organic food such as sugars. Could they transfer electrons to their ETS from inorganic electron donors such as minerals or even hydrogen gas? Yes. This kind of "rock eating" metabolism is called lithotrophy (see Table 7.1). In lithotrophy, an inorganic reduced molecule such as ferrous ion (Fe^{2+}), ammonium ion (NH_4^+), or hydrogen gas (H_2) donates electrons to the first oxidoreductase of the ETS (Table 7.3). Compared with catabolism of an organic compound, lithotrophy involves fewer intermediate reactions preceding the ETS. That is because an inorganic compound such as Fe^{2+} or H_2 can donate electrons directly to the ETS without requiring catabolic breakdown of a complex molecule.

Lithotrophy involving metals such as iron is important for mining, where reduced metals from underground are uncovered. Exposed to oxygen, the reduced metals now feed lithotrophic bacteria

CASE HISTORY 7.2

Nitrite Poisons an Infant

Brianna was a 2-month-old, formula-fed infant in a rural community in Nebraska. Brianna's mother noticed signs of blueness around the baby's mouth and fingers. One night after feeding, Brianna had trouble breathing, and her face had a peculiar lavender color. The infant became lethargic, salivated excessively, and had diarrhea and vomiting. An EMT came and took blood samples, which appeared brown and failed to turn pink when exposed to air. The EMT diagnosed methemoglobinemia, a condition in which toxic levels of nitrite oxidize hemoglobin and prevent it from receiving oxygen. The nitrite-oxidized hemoglobin is called methemoglobin; elevated levels can cause asphyxiation and death. A solution of methylene blue, a reducing agent, was given by IV to reduce (add electrons to) the hemoglobin in Brianna's blood, which can now carry oxygen again.

Nitrite forms in the digestive tract by bacterial reduction of nitrate during anaerobic respiration. The well water from which Brianna's formula was prepared was tested and found to contain high levels of nitrate, enough to cause methemoglobinemia. Brianna's parents were surprised because they had felt no ill effect from the water. But an infant's stomach pH is less acidic than that of an adult, allowing bacteria to reduce nitrate to nitrite, which may then oxidize hemoglobin.

The nitrite formed in Brianna's blood because her intestinal bacteria were using nitr<u>ate</u> (NO_3^-) as a terminal electron acceptor for anaerobic respiration, with nitr<u>ite</u> (NO_2^-) as the product. But where did so much nitrate come from in the well water? Nebraska and other farm states have a problem with the runoff of fertilizer spread on fields, from which ammonia drains into the water supply.

Table 7.3
Lithotrophic Reactions

Reaction	Species and Environment
$4Fe^{2+} + O_2 + 4H^+ \rightarrow 4Fe^{3+} + 2H_2O$	*Gallionella ferruginea*, a bacterium; oxidizes reduced iron in mine drainage and in water pipes; causes brown deposits in home plumbing
$FeS_2 + 14Fe^{3+} + 8H_2O \rightarrow 15Fe^{2+} + 2SO_4^{2-} + 16H^+$	*Ferroplasma acidarmanus*, an archaeon; oxidizes pyrite (iron-sulfur ore) in iron mines, generating strong acid
$O_2 + 2H_2 \rightarrow 2H_2O$	*Hydrogenomonas eutropha*, a soil bacterium; grows by oxidizing H_2 or an organic substrate
$2NH_3 + 3O_2 \rightarrow 2NO_2^- + 2H^+ + 2H_2O$	*Nitrosomonas europaea*, a bacterium; oxidizes ammonia to nitrite, causing nitrite pollution in lake water contaminated by ammonia-rich fertilizer
$CO_2 + 4H_2 \rightarrow 2H_2O + CH_4$	*Methanobrevibacter smithii*, an archaeon; grows in the human colon, in association with mixed-acid-fermenting bacteria

Much of the ammonia gets oxidized to nitrate by lithotrophic bacteria such as *Nitrosomonas* (Figure 17.18B), which oxidize ammonia by using O_2 (discussed earlier). The nitrate can permanently contaminate the drinking water, as in the case of the well water used for Brianna's formula.

A very different form of lithotrophy associated with normal human microbiota is **methanogenesis**, an anaerobic reaction of CO_2 and H_2 to form methane (CH_4; see Table 7.3). Methanogenesis is performed solely by methanogenic archaea, or methanogens. Species of methanogens are highly diverse, ranging from deep-sea-vent thermophiles to Antarctic psychrophiles. They are highly active in the anaerobic soil of wetlands and rice paddies. Surprisingly, some methanogens grow in the gut of humans and other animals, especially cattle. *Methanobrevibacter smithii* grows in the human colon by using the CO_2 and H_2 released by enteric bacteria through mixed-acid fermentation. The methanogens produce methane, which does not harm the host but does act as a greenhouse gas in global warming. Some methanogens also are isolated from dental plaque, but their contribution to disease is unknown.

SECTION SUMMARY

- **An electron transport system (ETS) transfers electrons from one membrane-embedded oxidoreductase to the next.** The final oxidoreductase transfers the electron to the terminal electron acceptor. Some of the oxidoreductases pump H^+ across the membrane, generating proton motive force.

- **Respiration is catabolism with transfer of electrons through an ETS**, whose terminal electron acceptor is O_2 or another inorganic oxidant. Anaerobic respiration involves terminal electron acceptors other than O_2.

- **Oxidative phosphorylation is the use of the proton motive force to drive ATP synthase.** Three protons crossing through ATP synthase power synthesis of one molecule of ATP.

- **Lithotrophy is the oxidation of an inorganic electron donor through an ETS.** Electron donors may be reduced metals, H_2, or NH_3.

Thought Question 7.3 The opportunistic pathogen *Pseudomonas aeruginosa* colonizes wounds and the lungs of cystic fibrosis patients. *P. aeruginosa* also grows in soil. The microbe can respire aerobically or else anaerobically by using nitrate. In which habitat (the lung or the soil) would *P. aeruginosa* respire aerobically or with nitrate? What part of its ETS would need to change?

7.5 Photosynthesis and Carbon Fixation

SECTION OBJECTIVES

- Describe how photosynthesis harvests light to generate ATP and NADPH while releasing O_2.
- Describe how a photosynthetic microbe fixes CO_2 into biomass.
- Explain the importance of photosynthesis for human health.

How long can you hold your breath? All the oxygen we breathe comes from **photosynthesis**—much of it by microbes. Photosynthesis fundamentally supports all the ecosystems of our biosphere. Every year, photosynthesis converts more than 10% of atmospheric carbon dioxide to biomass, most of which then feeds microbial and animal heterotrophs. As you're now aware, most of Earth's carbon fixation, especially in the oceans, comes from phototrophic microbes such as cyanobacteria (**Figure 7.19**).

All organisms use electron transfer (redox reactions) to yield energy for growth and function. But the process must start with a high-energy molecule. A phototrophic organism obtains energy from light to excite (add energy to) electrons of molecules such as chlorophyll. The energy gathered from photons exciting chlorophyll is transferred to an ETS that is remarkably similar to the ETS used to transfer electrons from organic food or from minerals. The process of light absorption by chlorophyll and energy transfer

Figure 7.19 **Cyanobacteria Conduct Photosynthesis**

Green colonies of the cyanobacterium *Merismopedia*.

LM | 100 μm

to build biomass is called photosynthesis. Some microbes, such as cyanobacteria, gain all their energy from photosynthesis; they are "photoautotrophs," existing entirely on water and inorganic ions. Other microbes only supplement their energy gained by catabolism with light absorption; they are "photoheterotrophs."

Some bacterial photosynthesis can use the energy of light absorption to split electrons from a sulfur compound such as hydrogen sulfide (H_2S). But the most advanced form of photosynthesis, **oxygenic photosynthesis**, splits a water molecule (H_2O; **Figure 7.20A**). The **photolysis** of water releases oxygen atoms, which form oxygen gas:

$$2H_2O \text{ (light absorbed)} \rightarrow O_2 + 4H^+ + 4e^-$$

Water-splitting photosynthesis is performed only by cyanobacteria and by plants and algae, whose chloroplasts evolved from an ancient cyanobacterium. The water-splitting reaction of oxygenic photosynthesis consumes large amounts of energy—a highly positive ΔG (energy change). But the energy consumed from light is then stored in biomass: four electrons are used to form NADPH, which then donates the electrons to fix CO_2. And besides fixing carbon into organic biomolecules, water-based photosynthesis produces oxygen gas, O_2.

Oxygen gas is a highly reactive substance that threatens to destroy photosynthetic cells unless it is released or is detoxified by antioxidant proteins. The oxygen released by cyanobacteria and by plants is used by heterotrophic organisms for aerobic respiration. Virtually all the oxygen we breathe—all the O_2 available on our planet—was released as a "toxic waste" from photosynthetic bacteria and from algae and plant chloroplasts that evolved from bacteria.

Photolysis: Light Absorption and Electron Transfer

What is the mechanism of photolysis? Scientists eagerly seek to understand photolysis, looking for clues to build better light-harvesting devices, mechanisms that harvest light energy by splitting electrons from a molecule such as water (**Figure 7.20B**). Oxygenic photolysis involves chlorophyll photoexcitation in a reaction center (**photosystem II**), electron transfer through an ETS, a second photoexcitation and reaction center (**photosystem I**), and generation of energy carriers ATP and NADPH. The

Figure 7.20 **Photosynthesis in Cyanobacteria**

A. Light is absorbed by chlorophyll molecules in thylakoid membrane pockets. The absorbed energy splits water, releasing oxygen that forms O_2 gas. Light energy converts ADP to ATP, and the electrons from water reduce NAD^+ to NADPH. The ATP and NADPH provide energy and reducing power for the Calvin cycle (see text for details) to fix CO_2. The CO_2 is reduced and condensed to form sugars ($C_6H_{12}O_6$), amino acids, and other biomass for the cell.

B. The Z pathway of photosynthesis shows how two photosystems with two ETS complexes obtain energy from light to make ATP and NADPH.

overall reaction—absorption of 4 photons splitting $2H_2O$ to form O_2—drives the formation of 3 ATP and 2 NADPH:

$$2H_2O + 2\,NADP^+ + 3\,[ADP + P_i] \rightarrow$$
$$O_2 + 2\,[NADPH + H^+] + 3\,ATP + 3H_2O$$

The stages of photolysis include:

- **Chlorophyll photoexcitation.** Light is absorbed by pigments called **chlorophylls**. Because the cell has no way to "reach out" to grab light rays, it needs to pack as many chlorophyll molecules as possible into its surface area in the hope of "capturing" photons. Most cyanobacteria have folded membranes called lamellae that multiply the surface area for light capture. The lamellae are packed with antenna complexes of proteins and chlorophyll that (1) absorb light and (2) transfer the absorption energy to the reaction center of the ETS.

- **Photosystem II (PS II).** In oxygenic photosynthesis, the photon energy absorbed by PS II is used to split $4H^+$ plus $4e^-$ from $2H_2O$. Each H^+ is released outside the cell, increasing the proton motive force (H^+ concentration and charge difference across the membrane). Each electron from H_2O reduces a quinol and enters an ETS in the bacterial cell membrane. Energy yield through the ETS is converted to pump $2H^+$ per electron across the membrane, for a total of $12H^+$ added outside. This proton motive force drives ATP synthase, making about 3 ATP.

- **Photosystem I (PS I).** The electrons from water that released energy through the ETS now get another energy boost from photoexcitation of chlorophyll in PS I. This transfer of electrons from PS II to PS I gives this mechanism the name "Z pathway." The newly energized electrons now enter another ETS, which transfers $4e^-$ onto $2\,NADP^+$ to form $2\,NADPH + 2H^+$.

The energy carriers ATP and NADPH are now available to reduce and energize CO_2 (CO_2 fixation), building sugars, amino acids, and other biomass for the cell.

Carbon Dioxide Fixation

The incorporation of CO_2 into glucose and other metabolites, such as amino acids, is called **carbon fixation** or **carbon dioxide (CO_2) fixation**. The fixation of CO_2 completes oxygenic photosynthesis, as shown in Figure 7.20. The overall reaction to make glucose is

$$6CO_2 + 12\,[NADPH + H^+] + 18\,ATP + 18H_2O \rightarrow$$
$$C_6H_{12}O_6 + 6H_2O + 12\,NADP^+ + 18\,[ADP + P_i]$$

Note that nowhere in this equation is light absorbed. In fact, CO_2 can be fixed without light absorption and photolysis if another source of energy (ATP and NADPH) is available; that is why CO_2 fixation is sometimes called the "dark reaction." CO_2 is fixed by lithotrophs (see Section 7.4) using ATP and NADPH derived from oxidation of minerals. Various lithotrophs fix CO_2 by different pathways. Some lithotrophs spend energy to reverse the TCA cycle and glycolysis, "running backward" to build acetate, pyruvate, sugars, and amino acids. Other lithotrophs fix CO_2 through the same

pathway used by phototrophs: the **Calvin cycle** (see Figure 7.20A), named for Melvin Calvin (1911–1997), who was awarded the Nobel Prize in Chemistry in 1961.

The importance of the Calvin cycle for our global ecosystem can scarcely be overstated. Most of the biomass on Earth—the body parts of all living things—is built from carbon fixed by bacteria, algae, and plants using the Calvin cycle. Its key enzyme, ribulose 1,5-bisphosphate carboxylase oxygenase, or **Rubisco**, may be the most abundant enzyme on Earth. The Calvin cycle plays the leading role in removing atmospheric CO_2; it's what climate scientists estimate when they model carbon flux and global warming.

Like the TCA cycle, the Calvin cycle starts with a key intermediate molecule to "recycle"—in this case the five-carbon sugar ribulose 1,5-bisphosphate. The five-carbon sugar condenses with CO_2 and H_2O, catalyzed by the enzyme Rubisco. Because this sugar—like the six-carbon sugar in glycolysis—has two phosphoryl "tags," the newly formed six-carbon sugar can now split into two three-carbon molecules—each tagged with a phosphoryl group and each equivalent to the other in the rest of the cycle. The three-carbon molecules now become reduced to the sugar glyceraldehyde 3-phosphate (G3P), with energy and electrons from ATP and NADPH.

In six turns of the "cycle," six molecules each of CO_2 and H_2O are fixed and reduced by 6 NADPH (with energy from 6 ATP) to form 6 G3P. A complex series of sugar exchange reactions regenerates the five-carbon intermediate ribulose 1,5-bisphosphate and assembles G3P into sugars, amino acids, or other biomolecules needed by the cell. Overall, photosynthetic microbes and plants form the basis of all ecosystems.

Photosynthetic microbes are rarely pathogens. Some marine algae, however, can generate toxins that cause human disease, as seen in the following case history.

CASE HISTORY 7.3
Food Poisoned by Red Tide

Tonya, age 32, was spending a warm June weekend at a Massachusetts beach. The water near shore looked clear, although farther offshore the water had a reddish tint. Tonya dug up a pailful of clams and cooked them thoroughly. After dinner, her lips grew numb and her fingers tingled. She felt nauseated and had trouble walking across the room. Her speech became slurred, but she managed to dial 911. The ambulance got her to the hospital in time for her to have her stomach pumped and receive help with breathing. The physician explained to Tonya that she had contracted paralytic shellfish poisoning from eating clams contaminated by an algal toxin. The toxin, called saxitoxin, accumulates in the clams as they feed on the algae; cooking does not affect the toxin. At that time, Massachusetts had a ban on shellfish because of a high population density of saxitoxin-producing algae, *Alexandrium tamarense*, a cause of "red tide."

Figure 7.21 Red Tide Caused by Dinoflagellates

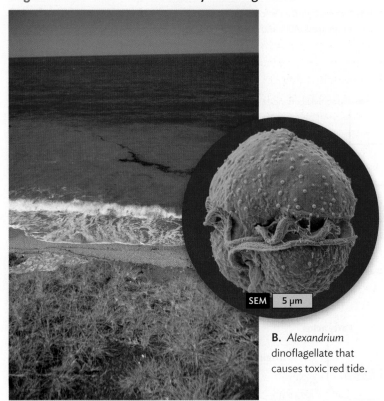

B. *Alexandrium* dinoflagellate that causes toxic red tide.

SEM | 5 µm

A. "Red tide," a bloom of dinoflagellate algae.

The cause of Tonya's poisoning was saxitoxin, a potent neurotoxin that inhibits the sodium channels of human and avian neurons, causing paralysis. Saxitoxin is produced by dinoflagellate algae (see Chapter 11), photoheterotrophs whose red color causes the appearance of a red tide (**Figure 7.21A**). Not all red tides cause poisoning, but off the New England coast, blooms of *Alexandrium tamarense* (**Figure 7.21B**) produce the toxin. The toxin concentrates in durable cysts of the dinoflagellate, which are consumed by filter feeders such as clams and other shellfish. The toxin concentrates in the clam's tissues but does not harm the clam; it poisons only the consumer of the clam's flesh.

Dinoflagellates are algae with an intriguing lifestyle that combines heterotrophy (flagellar motility and predation) with photosynthesis. They evolved from heterotrophic protozoa that engulfed green algae, whose own chloroplasts had evolved by a similar engulfing of a cyanobacterium. Such events of "endosymbiosis" contribute widely to the evolution of different eukaryotic microbes (discussed in Chapter 11). The oxygenic photosynthesis of these mixed-metabolism microbes (called **mixotrophs**) contributes greatly to the O_2 production and CO_2 consumption of marine ecosystems.

SECTION SUMMARY

- **Photosynthesis is the absorption of light to excite an electron** and transfer the absorbed energy to an ETS. The energy is converted to make NADPH and ATP, which then reduce CO_2 to build biomass.

- **Photolysis is the use of energy absorbed from light to split a molecule** such as H_2O or H_2S.

- **CO_2 fixation is the reduction and incorporation of inorganic carbon** into the carbon skeleton of a biological molecule. Energy and reducing power come from ATP and NADPH, which are produced either by photolysis or by lithotrophy.

7.6 Biosynthesis and Nitrogen Fixation

SECTION OBJECTIVES

- Describe how intermediates of metabolism are reduced and assembled into biomolecules.
- Explain how bacteria and archaea fix nitrogen gas into ammonium ion and incorporate nitrogen into biomolecules.

How do microbes use their energy to grow? Cells connect their energy-yielding metabolism to reactions that use the energy. Building cell parts is known as **biosynthesis**. Biosynthesis generates new cells, as well as secreted products such as toxins and antibiotics. Engineered microbes produce most of the antibiotics we use in medicine.

Biosynthesis is also called **anabolism**, the reverse of catabolism, which degrades complex molecules. For biosynthesis, some catabolic reactions, such as those of the TCA cycle and glycolysis, can run in reverse (with energy input) to build sugars and polysaccharides. Other anabolic reactions, such as those building amino acids, are distinct from those of catabolism.

What other kinds of biosynthesis do we find in pathogens? Some pathogens, such as *Mycobacterium tuberculosis*, have many extra genes to make virulence factors, such as unusual envelope lipids that defeat host defenses. Biosynthetic pathways offer targets for antibiotics; for instance, the tuberculosis antibiotic isoniazid interrupts the biosynthesis of mycolic acids in the envelope of *M. tuberculosis*. Other pathogens, such as the intracellular pathogen *Chlamydia trachomatis* (the cause of sexually transmitted infections), have lost many genes for biosynthesis. *C. trachomatis* thus depends on the host for various fundamental substrates.

Requirements for Biosynthesis

To build new parts, a cell must do the following:

- **Assemble carbon skeletons.** Microbes obtain carbon either by fixing CO_2 (autotrophy) or by consuming organic molecules made by other organisms (heterotrophy). Autotrophs may reverse catabolic pathways such as the TCA cycle and glycolysis, whereas heterotrophs assemble products of catabolism such as acetyl-CoA and TCA cycle intermediates (**Figure 7.22**). Glycolysis forms glycerol and acetyl-CoA, which are assembled into fatty acids. Both glycolysis and the TCA cycle provide precursors for amino acids.

Figure 7.22 Biosynthesis: An Overview

Microbes obtain carbon skeletons through CO_2 fixation or through catabolism of compounds formed by other microbes. Nitrogen is obtained by N_2 fixation or by assimilation of organic amines. Biosynthesis spends energy from ATP and reduction energy from NADPH.

- **Fix nitrogen, sulfur, and phosphorus.** Besides carbohydrate, biosynthesis requires nitrogen, sulfur, phosphorus, and other essential elements. The availability of such elements often limits growth.

- **Spend energy.** Building complex structures requires energy. Biosynthetic enzymes spend energy by coupling their reactions to the hydrolysis of ATP and the oxidation of NADPH. NADPH is produced from catabolism via the pentose phosphate pathway (Section 7.3).

- **Reduce carbon skeletons (add electrons).** Cell parts such as lipids and amino acids are more reduced than the substrates available—for example, sugars. So, the biosynthesis of lipids and amino acids includes reduction of sugar-derived carbon skeletons by a reducing agent such as NADPH.

Nitrogen Fixation and Ammonia Incorporation

An important limit for biosynthesis is the incorporation of nitrogen. Many biological molecules, such as DNA and amino acids, contain nitrogen as well as carbon. This hunger for nitrogen has substantial impacts on human society and environmental quality. Human agriculture is so intensive that most crops require the spreading of nitrogenous fertilizers. The production of such fertilizers incurs great costs, including the consumption of petroleum to reduce atmospheric nitrogen. And fertilizer runoff pollutes water supplies with ammonia and nitrate (see Case History 7.2).

But a few crops, such as soybeans and other legumes, can grow without fertilizer because their nitrogen is fixed by symbiotic bacteria. These bacterial symbionts, called **rhizobia** (singular, **rhizobium**), convert atmospheric nitrogen to ammonium ion, which gets assimilated into plant proteins and nucleic acids. Exciting research focuses on expansion of the host range of rhizobia so that someday they may grow in symbiosis with other crop plants, such as wheat and corn.

Nitrogen gas (N_2) constitutes 78% of Earth's atmosphere; yet this form of nitrogen is surprisingly unusable by organisms, except for certain species of bacteria and archaea. As a result, all life on Earth depends on bacterial and archaeal nitrogen fixation. The process of nitrogen fixation requires a high energy input because N_2 has a highly stable triple bond and because the process requires a large amount of reduction:

$$N_2 + 8H^+ + 8e^- + 16\,ATP \rightarrow 2NH_3 + H_2 + 16\,ADP + 16\,P_i$$

The reaction needs a total of eight electrons from 4 NADPH or from some other reducing factor, such as the protein ferredoxin. Of the eight electrons, six are needed to hydrogenate the nitrogens to ammonia and two are needed to form H_2 in order to prime the reaction. The reaction occurs in four rounds of hydrogenation by the enzyme hydrogenase. Because each NADPH carries the energy equivalent of 3 ATP, the total amount of energy needed is $16 + (4 \times 3) = 28$ ATP equivalents.

A problem with nitrogen fixation is that its final product, ammonia, is toxic to cells. So, although cells need to incorporate ammonia (or the ionized form, ammonium ion, NH_4^+), they need to keep the concentration of free ammonia very low. Cells do this by combining ammonia immediately with one of the TCA cycle intermediates to form an amino acid. Most often, the TCA intermediate 2-oxoglutarate incorporates ammonium ion to form glutamic acid (**Figure 7.23**). In this reaction, NADPH donates two electrons to replace the ketone oxygen (R=O) with an amine group from ammonium ion. Once glutamate is formed, this amino acid can be maintained safely at a much higher concentration than ammonia. Furthermore, glutamate can transfer its amine group to other molecules to form other amino acids—a process known as "transamination." Consequently, glutamate serves as a safe storage form for nitrogen.

SECTION SUMMARY

- **Biosynthesis, or anabolism, requires assembly of carbon skeletons** from intermediates of metabolism. All carbon skeletons derive from CO_2 fixed by autotrophs.
- **Biosynthesis requires energy from ATP and reduction by energy carriers such as NADPH.**
- **Nitrogen gas is fixed into ammonia**, which is assimilated into amino acids by condensation with a TCA intermediate.

Thought Question 7.4 Disease-causing bacteria vary widely in their ability to synthesize amino acids. What classes of pathogens are likely to make their own amino acids, and what kinds are not?

Thought Question 7.5 *Mycoplasma genitalium*, a cell wall–less bacterium that grows in human serum, lacks the ability to synthesize fatty acids. How do you think *M. genitalium* makes its cell membrane?

Perspective

In this chapter we have seen how the energy-yielding reactions introduced in Chapter 4 enable living cells to build order out of disorder. Living organisms acquire energy by catalyzing reactions that form energy carriers and then spend the energy from these carriers to build cells. Within communities of organisms, some microbes absorb light energy for photosynthesis, whereas others consume biomass via catabolism. Much of human civilization is built on harnessing microbial catabolism for waste treatment and food production. In the diagnostic laboratory, clinicians use microbial metabolism for indicators to identify pathogens.

All forms of metabolism involve chemical exchange of electrons through oxidation and reduction. The transfer of electrons is fundamental to respiration, the oxidative catabolism of organic molecules to CO_2 and water. It also underlies the autotrophic means of building biomass: the gain of energy from photosynthesis and from lithotrophy. In living cells, the pathways of energy yield and of biosynthesis occur together, sharing common substrates and products. For example, acetyl-CoA produced by glycolysis is then used by pathways that synthesize fatty acids, amino acids, carbohydrates, and other biomolecules. Together, all of this energy yield and biosynthesis enables cells to grow and reproduce.

Figure 7.23 Assimilation of NH_4^+ into Glutamate

The key TCA cycle intermediate 2-oxoglutarate (alpha-ketoglutarate) incorporates one molecule of NH_4^+ at the ketone (C=O) to form glutamate.

Key Terms

adenosine triphosphate (ATP) (p. 190)
aerobic respiration (p. 188)
anabolism (p. 207)
autotroph (p. 187)
biosynthesis (p. 207)
Calvin cycle (p. 206)
carbon dioxide (CO₂) fixation (p. 206)
carbon fixation (p. 206)
catabolism (p. 189)
catabolite (p. 189)

chemotrophy (p. 187)
chlorophyll (p. 206)
citric acid cycle (p. 189)
electron transport system (ETS) (p. 191)
energy (p. 186)
entropy (p. 186)
fermentation (p. 195)
glycolysis (p. 189)
heterotroph (p. 187)
hydrolysis (p. 189)
Krebs cycle (p. 189)

lithotrophy (p. 188)
methanogenesis (p. 204)
mixotroph (p. 207)
nicotinamide adenine dinucleotide (NADH) (p. 190)
organotrophy (p. 187)
oxidative phosphorylation (p. 200)
oxidized (p. 191)
oxidoreductase (p. 200)
oxygenic photosynthesis (p. 205)
pentose phosphate pathway (PPP) (p. 195)

photolysis (p. 205)
photosynthesis (p. 204)
photosystem I (p. 205)
photosystem II (p. 205)
phototrophy (p. 187)
proton motive force (PMF) (p. 200)
reduced (p. 191)
respiration (pp. 195, 199)
rhizobium (p. 208)
Rubisco (p. 206)
tricarboxylic acid cycle (TCA cycle) (p. 189)

Objectives Assessment Questions

7.1 Energy for Life

1. The class of metabolism most useful for food production because it yields important microbial "waste products" is _____.
   a. respiration
   b. fermentation
   c. lithotrophy
   d. photosynthesis

2. These compounds all require energy to build, except _____.
   a. DNA
   b. peptidoglycan
   c. fatty acid
   d. carbon dioxide

3. This chemical reaction does not require an enzyme.
   a. Combustion
   b. ATP formation
   c. Photosynthesis
   d. Sugar catabolism

7.2 Catabolism: The Microbial Buffet

4. Which of these food substrates must first be broken down to hexose sugars in order to enter glycolysis?
   a. Cellulose
   b. Lignin
   c. Lipids
   d. Proteins

5. Which of these statements is true of NADH?
   a. It is a common food substrate for catabolism.
   b. It is an electron acceptor for catabolite electrons.
   c. It is an electron donor formed during catabolism.
   d. It is a waste product of fermentation.

6. Which catabolic pathway generates NADPH (but no NADH) for biosynthesis?
   a. The Embden-Meyerhof-Parnas pathway (glycolysis)
   b. The Entner-Doudoroff pathway

   c. The pentose phosphate pathway
   d. The TCA cycle

7.3 Glucose Catabolism, Fermentation, and the TCA Cycle

7. Which of the following statements does not describe glycolysis?
   a. Glycolysis spends 2 ATP to phosphorylate glucose during the breakdown to pyruvate, with reactions forming 4 ATP.
   b. The pathway of glucose breakdown to pyruvate is incomplete because it generates two NADH molecules that need to transfer electrons to pyruvate or to a terminal electron acceptor.
   c. Glycolysis has a cycle of intermediates that incorporate a carbon skeleton and regenerate the original intermediate.
   d. Glycolysis occurs with an increase of entropy because one large molecule is broken down to two molecules of pyruvate.

8. Which molecule from glucose catabolism directly enters the TCA cycle?
   a. Acetyl-CoA
   b. Fructose 1,6-bisphosphate
   c. Glyceraldehyde 3-phosphate
   d. Glucose

7.4 Respiration and Lithotrophy

9. Which of the following statements does not describe an electron transport system (ETS)?
   a. NADH transfers two electrons to an oxidoreductase embedded in the bacterial cell membrane.
   b. The oxidoreductase pumps H^+ out of the cytoplasm, generating an H^+ gradient that stores energy.
   c. The oxidoreductase directly phosphorylates ADP to make ATP.
   d. The oxidoreductase transfers electrons to a quinol (Q), forming quinone (QH_2), which carries the electrons to a terminal oxidoreductase.

10. Which of the following environmental reactions is not an example of lithotrophy?
    a. In a landfill, methanogens obtain energy by combining hydrogen gas plus CO_2 (products of bacterial fermentation) to form methane.

b. *Geobacter* catabolizes small organic molecules and donates electrons to an electrode in a fuel cell.

c. In soil spread with nitrogen-rich fertilizer, bacteria use O_2 to oxidize ammonia, forming nitrite (NO_2^-).

d. In Yellowstone National Park, archaea use O_2 to oxidize hydrogen sulfide (H_2S), forming sulfuric acid.

11. Which metabolism can cause methemoglobinemia?

a. Fermentation

b. Respiration using oxygen

c. Lithotrophy oxidizing iron

d. Anaerobic respiration reducing nitrate to nitrite

7.5 Photosynthesis and Carbon Fixation

12. Which major class of bacteria conducts oxygenic photosynthesis, releasing much of the oxygen we breathe?

a. Cyanobacteria

b. Spirochetes

c. Proteobacteria

d. Bacteroidetes

13. Which portion of photosynthesis can occur in the dark and also occurs in many lithotrophic bacteria?

a. Chlorophyll electron excitation

b. ATP production from the H^+ gradient generated by the ETS

c. NADPH production by electrons from the ETS

d. CO_2 fixation via the Calvin cycle

14. Which kind of phototrophs cause food poisoning?

a. Dinoflagellates in red tide

b. Algal symbionts of coral

c. Marine photosynthetic bacteria producing oxygen

d. Purple sulfur-reducing bacteria

7.6 Biosynthesis and Nitrogen Fixation

15. Bacteria synthesize amino acids by obtaining carbon skeletons from _____.

a. glycolysis

b. the pentose phosphate pathway

c. the TCA cycle

d. all of the above

16. Bacteria incorporate ammonia into a TCA cycle intermediate to form the amino acid _____.

a. cysteine

b. glutamate

c. tryptophan

d. histidine

Review Questions

1. Why must the biosphere continually take up energy from outside? Why can't all the energy be recycled among organisms, like the fundamental elements of matter?

2. How do organisms determine which of their catabolic pathways to use? How does catabolism depend on environmental factors?

3. Beer is produced by yeast fermentation of grain to ethanol. Why must the process of beer production be anaerobic? Why are such large quantities of ethanol produced, with relatively small production of yeast biomass?

4. Explain how glycolysis breaks down glucose to pyruvate. Why is it necessary to start by spending two molecules of ATP?

5. Explain the source of electrons and the sink for electrons (terminal electron acceptor) in respiration, lithotrophy, and photolysis.

6. What are the sources of substrates for biosynthesis? From what kinds of pathways do they arise?

7. How do different organisms obtain nitrogen? How is nitrogen assimilated into biomass?

Clinical Correlation Questions

1. A stool culture was obtained for a patient suffering from diarrhea and blood in his stool (dysentery). The cultured bacterium was used to inoculate Hektoen agar. The colonies on Hektoen agar were orange. Which bacterial pathogen was identified?

2. A family at the beach is having a seafood picnic. Two family members experience instantaneous numbness of the mouth and paralysis, followed by nausea and respiratory distress. What possible cause should be tested for immediately?

8

Bacterial Genetics and Biotechnology

[Dental Biofilm Inflames the Gums]

PATIENT HISTORY Tina, a 33-year-old store clerk from Peoria, Illinois, had her first dental exam in 5 years. She told the hygienist that her gums hurt and that stains appeared on her pillow where her jaw had rested. She also told the hygienist that she did not smoke or drink.

EXAMINATION The hygienist found that Tina's gums were swollen and bled upon probing. The gums had receded from Tina's teeth, forming pockets about 5 mm deep, and X-rays revealed some loss of bone.

DIAGNOSIS Tina then saw the dentist, who told her she had periodontitis, inflammatory disease of the gums and bone supporting the teeth. Periodontitis is caused by dental plaque, a biofilm of mixed bacterial species that grow on the teeth. Without regular oral hygiene, the biofilm grows beneath the gum and eventually causes loss of teeth.

Tina expressed surprise, as she thought that only elderly people suffered gum disease. The dentist asked her again whether she smoked, perhaps two packs a day? Tina denied smoking that much, but she admitted to one pack a day. The dentist told her that smoking is a common factor in early gum disease, as are diabetes and genetic susceptibility. Under predisposing conditions, many different kinds of bacteria can cause gum disease.

TESTING To determine the bacterial species causing Tina's periodontitis, the dentist ordered a DNA test. The test works by the polymerase chain reaction (PCR), a technique in which a short piece of DNA is amplified (copied many times), making it possible to read the sequence of DNA base pairs. The DNA sequence reveals the bacterial species. Tina's DNA test revealed *Porphyromonas gingivalis* and *Aggregatibacter actinomycetemcomitans*, two Gram-negative anaerobes. These bacteria are sensitive to metronidazole and amoxicillin, respectively.

TREATMENT The DNA test enabled the dentist to target the most effective antibiotics for Tina's condition. The dentist also enrolled Tina in a smoking cessation program.

Gums Showing Periodontitis
Note the reddened gums and the receding gumline (arrows).

SEM 2 μm

Periodontal Biofilm
The diverse forms of bacteria can be identified based on their DNA sequence.

CHAPTER OBJECTIVES

After reading this chapter, you will be able to:

- Describe the structure and replication of bacterial genomes and plasmids.
- Describe techniques of biotechnology that aid medical diagnosis and treatment, including restriction enzymes, gel electrophoresis, recombinant DNA, and polymerase chain reaction (PCR).
- Explain how genes are transcribed to RNA and translated to protein.
- Explain how gene expression is regulated.

Some diseases, such as tuberculosis, are caused by a single pathogen. A single pathogen can be identified by a battery of tests based on colony appearance and biochemical reactions, as shown in Chapter 7. But periodontitis involves a mixture of different species, which may be impossible to separate or culture. Each species has different virulence factors (factors associated with disease, explored in Chapter 2). And each organism is sensitive to different antibiotics. For a given patient, how do we know which bacteria are present and which antibiotics to use?

A mixed-species sample can be sorted by its DNA. Each species of the mixture has a unique DNA sequence in its **genome**, the genetic material that defines the organism (see Figure 8.2). The genomic DNA determines all the organism's inherited traits. The genome also provides a unique signature of that organism. As we saw in the case history that opened this chapter, a key gene sequence from the genome can be amplified (copied many times) by a technique called the polymerase chain reaction (PCR). Using PCR, we can obtain enough DNA to "read" its sequence. The DNA sequence is then matched with a known database of bacterial sequences, and the species of the pathogen is identified.

For the case of periodontitis, we can identify several different species in a mixed sample. The most prominent pathogens can be targeted by the right antibiotics. Thus, PCR helps identify multiple organisms in a condition that involves more than one infectious agent.

Knowing how DNA works is an important part of understanding the interactions between microbes and humans. Pathogens entering a host, for instance, use their DNA information to make virulence factors and secrete them into the host tissues. Outside a host, environmental bacteria use different genes to survive in different habitats. And over generations, microbial DNA sequence gradually changes by mutation, leading to evolution. Organisms evolve into new forms with new traits, including new kinds of pathogens (discussed in Chapter 9).

In this chapter we explain how bacteria replicate their DNA and express their genes, and we describe the consequences for human health and the environment. We also show how bacterial and viral components, such as enzymes and plasmids, provide tools for biotechnology—an exciting field that gives us powerful tools for diagnosis and therapy.

8.1 Bacterial Genomes

SECTION OBJECTIVES

- Describe the structure of a bacterial genome, and explain how it differs from a eukaryotic genome.
- Explain what a plasmid is, and describe the role of plasmids in the spread of antibiotic resistance genes.

How does DNA define a unique species of organism? As discussed in Chapter 4, all living cells maintain their "blueprint" genetic information in **chromosomes** that consist of double-stranded DNA. Each strand of DNA carries a sequence of **nucleobases** (bases) of four types: adenine (A), thymine (T), cytosine (C), and guanine (G). Each base is attached to a sugar (deoxyribose) and a phosphoryl group, forming a **nucleotide**, the fundamental unit of DNA (**Figure 8.1**).

The nucleotides assemble into long strands, joined by sugar-phosphoryl links called a phosphodiester bond. (The process of DNA synthesis is described in Section 8.2). Two strands of DNA are connected by complementary base pairs, in which A pairs with T and G pairs with C (**Figure 8.2**; also review Figure 4.12). The order of the base pairs determines the order of amino acids that form the protein specified by a gene. A double-stranded DNA helix is formed when two strands of complementary single-stranded DNA molecules base-pair with each other.

Most bacteria have a genome in the form of a circular chromosome containing several million base pairs. The entire genome of

Figure 8.1 Structure of a Nucleotide, a Unit of DNA

Figure 8.2
Chemical Structure of the DNA Molecule

The 2-deoxy position that distinguishes DNA from RNA

2-Deoxyribose

Phosphodiester bond

Thymine (T)

Adenine (A)

3' end

Cytosine (C)

Guanine (G)

White, hydrogen; gray, carbon; red, oxygen; blue, nitrogen; gold, phophorus.

3' end

5' end

bacterium itself. How can this enormous molecule fit into a single cell? And how can enzymes duplicate that much DNA without the whole DNA molecule getting tangled up? These questions are addressed in Section 8.2.

Genome of a Species

Each species has a genome with a set of genes unique to the species. A **gene** is a unit of information encoded by DNA that can be copied (expressed) to form an RNA product (**Figure 8.4**). In most cases, the RNA product of a gene directs the synthesis of a unique protein. The use of gene information to make RNA and protein is discussed in Section 8.3. The information determining <u>when</u> a gene is expressed is contained in a DNA control sequence called a **promoter** (colored red in Figure 8.4). In the simplest case, a single gene can stand alone in the DNA; in other cases, multiple genes in tandem are controlled by one promoter. The RNA produced from a stand-alone gene encodes a single protein.

Bacterial genes and genomes are organized differently from those of eukaryotic organisms (such as humans). Figure 8.4 compares the organization of a genome section of the bacterium *Escherichia coli* with a section of the human genome. The *E. coli* genome is packed with genes that encode proteins (yellow in Figure 8.4), with very little unused sequence between them (gray). By contrast, the human genome contains vast stretches of noncoding sequence, much of which evolved

the bacterium *Escherichia coli* consists of a circular chromosome of 4,600,000 base pairs. In a remarkable experiment, a cell of *E. coli* was lysed by osmotic shock, releasing its chromosome to visualize by electron microscopy (**Figure 8.3**). What spewed out of this single cell was a chromosome of DNA 800 times longer than the

Figure 8.3 DNA Spewing out of a Bacterial Cell Lysed by Osmotic Shock

DNA

TEM 1 μm

Figure 8.4 Gene Structure in Bacteria and in Humans

Coding region

5'

Bacterial gene

A. A bacterial gene contains a coding sequence (yellow). The control sequence is red. The gene is flanked by short, noncoding sequences (gray).

Coding regions (exons)

Noncoding regions (introns)

Eukaryotic gene

B. A section of the human genome. Human coding genes are interrupted by noncoding sequences called introns, and their coding portions (called exons) are spread out amid even longer stretches of noncoding DNA.

from infecting viruses. The human coding genes may be interrupted by noncoding stretches called introns; the separated coding regions are then called exons. Bacterial genes lack introns; they consist of uninterrupted coding regions.

Besides the main genomic DNA, bacterial cells may contain additional smaller pieces of DNA called **plasmids**. A plasmid is a small circle of DNA that is replicated separately from the main genome. A plasmid may enter or leave a bacterial cell without changing the species. Plasmids typically contain a small number of genes that are useful in a given environment or virulence genes that confer disease properties on the bacterium. A particularly serious problem is that of plasmid-borne genes that confer resistance to antibiotics.

CASE HISTORY 8.1
An ESKAPE Pathogen Escapes Antibiotics

Maryanne, a 76-year-old retired teacher, was in a rehab center recovering from a procedure for kidney stones. She developed endocarditis (heart infection) and was transferred to a Chicago hospital for treatment. Meanwhile, her kidneys still drained through a nephrostomy tube (a catheter that had been inserted previously, during treatment for kidney stones). Five days after arriving at the hospital, Maryanne experienced septic shock, a condition in which infecting pathogens reach the bloodstream, leading to inflammation, loss of blood pressure, and respiratory distress. The patient required intubation (assisted breathing) and administration of vasopressors to restore blood pressure.

The physician ordered cultures of Maryanne's blood and urine. Blood cultures came back negative, but the urine tested positive for *Klebsiella pneumoniae*, presumably acquired through the nephrostomy tube. The finding of *K. pneumoniae* alarmed the physician, who knew that extensively drug-resistant *K. pneumoniae* (XDR-KP) was an emerging pathogen appearing at other hospitals. The physician ordered testing of the patient's isolate for susceptibility to all available antibiotics. The MIC (minimum inhibitory concentration) results showed that the pathogen was resistant to all cephalosporins (advanced forms of penicillin), aminoglycosides (protein synthesis inhibitors such as streptomycin), gentamicin (an inhibitor of a different step of protein synthesis), carbapenems (cell wall synthesis inhibitors that usually overcome penicillin resistance), and colistin (an outer membrane disruptor). The lab also reported borderline susceptibility to tigecycline, an advanced derivative of tetracycline (protein synthesis inhibitor) called a glycylcycline.

The physician administered tigecycline intravenously—a treatment that initially cleared the infection. On the fifteenth day of hospitalization, however, the patient again experienced septic shock and respiratory failure, leading to death. *K. pneumoniae* was cultured from the patient's bronchial fluid, urine, and blood.

In this case history, an all-too-common occurrence, Maryanne's weakened condition and her nephrostomy tube left her vulnerable to opportunistic infection by an extensively drug-resistant (XDR) strain of *Klebsiella pneumoniae* (XDR-KP). *K. pneumoniae* is one of six **ESKAPE pathogens** designated by US medical authorities as the leading causes of life-threatening hospital-acquired infections. The other five species of the ESKAPE acronym are *Enterococcus faecium*, *Staphylococcus aureus*, *Acinetobacter baumannii*, *Pseudomonas aeruginosa*, and *Enterobacter* species. Typically, these ESKAPE pathogens start out possessing numerous genes for resistance to specific drugs, then acquire still further resistance during treatment. How do bacteria acquire so many resistance genes so fast?

To analyze the resistance of XDR-KP, doctors obtained the sequence of its entire genome, including the main chromosome plus five plasmids (**Figure 8.5**), using methods described later, in Section 8.6. The sequence of the XDR-KP genome reveals four genes on the main chromosome that confer resistance to different classes of antibiotics. But an additional four genes on plasmid 1—a relatively small circle of DNA—confer resistance to four other drug classes. Plasmid 2 carries ten genes conferring resistance to a combination of four classes. Finally, the XDR-KP genome includes three more plasmids, whose properties remain unknown.

Such drug resistance plasmids are typical of ESKAPE pathogens, as well as other emerging XDR organisms. Bacteria containing these plasmids circulate in hospitals, carrying genes for resistance to five or more antibiotics at once. Such spread is a problem because when a patient is treated with one antibiotic, the antibiotic

Figure 8.5 Genome of an ESKAPE Pathogen, *Klebsiella pneumoniae* **(XDR-KP)**

Resistance to various classes of antibiotics is conferred by genes carried on the main chromosome and on different plasmids. The genes are revealed by whole-genome sequencing (WGS).

enhances the survival of pathogens that have previously acquired a plasmid conferring resistance to several other antibiotics.

Plasmids may carry genes from one species to a completely unrelated species. When a hospital patient receives antibiotic therapy over the long term, the patient's normal biota (normally occurring, harmless bacteria) will become antibiotic resistant by natural selection for strains that happen to have picked up such plasmids. A plasmid of a normally occurring organism may then acquire the antibiotic resistance genes and transfer them to disease-causing bacteria. Transfer of genes from one organism to another is discussed in Chapter 9.

Genome Structural Diversity

Not all bacteria have simple circular genomes. Some species have complex genomes that include multiple circular and linear pieces of DNA. In these species, the smaller DNA components are called plasmids, although the distinction between plasmid and genome is not always clear. We do not know, for example, why the genome of *Borrelia burgdorferi* (the cause of Lyme disease) is organized into 22 separate segments or how the replication of this multipart genome is coordinated.

In eukaryotes, by contrast, the genome always consists of linear chromosomes organized within the nucleus. This is true of microbial eukaryotic pathogens, such as the protist *Giardia* and the yeast *Candida*, as well as of their human hosts. Eukaryotic chromosome segregation during meiosis is discussed in Chapter 11.

SECTION SUMMARY

- **"Genome" refers to the total DNA content of a cell.** A bacterial genome includes its main chromosome plus, in some cases, shorter DNA molecules.
- **The DNA of bacterial genomes is typically circular**, but it can be linear or a combination of circular and linear DNA chromosomes.
- **Plasmids are small DNA circles** containing supplementary genes that may encode pathways that degrade materials from the environment or confer drug resistance.

- **Plasmids are often transferable between bacteria of different species.** Plasmid transfer is one way that pathogens evolve different capabilities, such as resistance to multiple antibiotics.
- **ESKAPE pathogens are considered the most dangerous XDR (extensively drug-resistant) pathogens acquired in hospitals.** Genome sequence analysis shows that plasmids play a key role in the evolution of ESKAPE pathogens.

8.2 DNA Replication

SECTION OBJECTIVES

- Explain how bidirectional semiconservative DNA replication copies the circular chromosome of a bacterium during cell division.
- Explain the steps of replication by DNA polymerase, including initiation, elongation, and termination.

We've all heard about community meals, such as a holiday barbecue, after which everyone gets sick with food poisoning. The culprit is typically *Salmonella* bacteria or a norovirus. How can a few bites of food make us sick within a day? The answer lies in DNA replication. Enteric pathogens have evolved to reproduce themselves and replicate their DNA as fast as possible to outcompete other gut microbiota. Replication efficiency explains why bacterial pathogens can cause disease so quickly after ingestion (usually within 24–48 hours). Viral replication can be faster yet, sacrificing accuracy for speed, as discussed in Chapter 12.

The process of bacterial replication involves an amazing number of proteins and genes working together. The molecular details are important because they provide targets for new antibiotics and anti-tumor agents (Table 8.1), as well as tools for biotechnology, such as the polymerase chain reaction (PCR). In addition, the study of bacterial DNA replication and repair gives clues to DNA function in humans. This is because the enzymes that replicate DNA are encoded by homologous genes, or homologs. **Homologs** are genes with shared ancestry and related function—in this case, shared

Table 8.1
Antibiotics and Anti-tumor Agents That Target DNA Synthesis

Enzyme and/or Process of DNA Synthesis	Antimicrobial Molecule	Source of Molecule	Organisms Affected
DNA polymerase, beta subunit: Adds nucleotides to growing DNA chain	Bleomycin	*Streptomyces verticillus*	Humans (tumors)
DNA topoisomerase I	Topostins	*Flexibacter topostinus*	Humans (cancer)
DNA stability	Metronidazole (redox reactions cause DNA breakage)	Artificial design	Anaerobic bacteria and protozoa
Gyrase: Adds superturns to newly completed DNA helix	Quinolones	Artificial design	Bacteria
Helicase: Unwinds helix turns and superturns	Heliquinomycin	*Streptomyces* sp. MJ929-SF2	Humans (tumors)

from the common ancestor of bacteria and humans. The human homologs for bacterial DNA replication enzymes protect our own DNA from damage that causes cancer.

Semiconservative Replication

To replicate a molecule containing millions of base pairs poses formidable challenges. How does DNA replication generate two identical copies? How are the copies checked for accuracy?

DNA replication is catalyzed by an enzyme complex called **DNA polymerase**. Replication of a cell's DNA is semiconservative, meaning that each new daughter helix receives one original strand (called the parental strand) and one newly synthesized strand. To form the new strand, each new nucleotide must pair with a complementary base on the parental strand (see Figure 8.2): adenine pairs with thymine (AT), and guanine pairs with cytosine (GC). How is correct pairing achieved? The active site of DNA polymerase constrains the bases to fit in a planar position (**Figure 8.6A**). Incorrect bases may enter the site but fail to fit in a pair (**Figure 8.6B**). The incoming nucleotide is then rejected by the enzyme. The DNA polymerase then waits for a correct nucleotide to enter the active site.

To replicate DNA, the double helix must unwind, exposing the template strand to form new pairs with a growing strand (**Figure 8.7**). The original DNA strands are called **parental strands**; the growing strands, complementary to each parental strand, are called **daughter strands**. The unwinding requires initiation proteins and a "helicase" enzyme that spends ATP to unwind helical DNA. The partial unwinding of the helix generates a **replication fork** at each end of the replication "bubble." At each replication fork, the advancing DNA synthesis machine separates the parental strands while extending the new, growing strands. Each daughter duplex will be checked for accuracy against its parental strand—first by the DNA polymerase and then later by proofreading enzymes.

Figure 8.7 **Bidirectional Semiconservative Replication**

DNA replication with two replication forks. Parental strands are gray, and newly synthesized daughter strands are purple. Replication is called "semiconservative" because one parental strand is conserved and inherited by each daughter cell genome. It is called "bidirectional" because it begins at a fixed origin and progresses in both directions from there.

The process of DNA synthesis is divided into three phases:

- **Initiation**, which involves unwinding of the helix, priming, and loading of the DNA polymerase enzyme complex
- **Elongation**, the sequential extension of DNA by adding deoxyribonucleoside triphosphates (dNTPs) with release of pyrophosphate, followed by proofreading
- **Termination**, in which the DNA duplication is complete and replication stops

We now examine each phase of replication in turn. Note that all of the steps shown for DNA synthesis require catalysis by enzymes and DNA-binding proteins. Most of these proteins have homologs in the DNA synthesis of our own cells. Their function is so important and ancient that they date back to the common ancestor of bacteria and of human cells.

Figure 8.6 **Pairing of a New Base: Correct versus Misaligned**

The models shown are based on X-ray crystallography of exceptionally stable enzymes obtained from *Sulfolobus solfataricus*, an archaeon that grows at volcanic vents in near-boiling sulfuric acid.

A. The DNA polymerase active site contains a template strand T (yellow) pairing with a new deoxy ATP (dATP; red) to be added to the growing complementary chain (orange).

B. Deoxy GTP (dGTP) can enter the binding site, but it's misaligned with the template. The DNA polymerase will reject the misaligned base.

Initiation of DNA Synthesis

In most bacteria, replication begins at a single defined DNA sequence called the **origin**, *oriC* (see Figure 8.7). Enzymes that synthesize DNA or RNA can connect nucleotides only in a 5′-to-3′ direction. That is, every newly made strand begins with a 5′ triphosphate and ends with a 3′ hydroxyl group. A **DNA polymerase** (a chain-lengthening enzyme complex) joins the 5′ phosphate of an incoming nucleoside triphosphate (NTP) to the 3′ OH end of the growing daughter strand (**Figure 8.8**). The new bond of the chain (C–O–P–O–C) is called a phosphodiester bond; that is, an acid with an ester link to two sugars. To form this bond, the reaction releases a pair of phosphoryl groups (called pyrophosphate) from the NTP. Loss of pyrophosphate releases a lot of energy and thus drives the reaction forward.

Figure 8.8 Nucleotide Condensation with 3′ OH End of DNA

Pyrophosphate is released.

Nucleoside triphosphate

The nucleotide releases its terminal and middle phosphoryl groups as pyrophosphate. The phosphoryl group nearest the sugar condenses with the 3′ OH of the growing DNA chain.

If every new NTP must add on to a 3′ OH group, then how is the very first nucleotide added? A short **RNA primer** molecule with a 3′ OH end base-pairs with the DNA (**Figure 8.9**). The RNA primer is synthesized by a special enzyme, called primase, that does not need a 3′ OH end. The use of RNA for a primer instead of DNA may seem surprising, since eventually the RNA will have to be replaced by DNA nucleotides. The universal use of RNA primers by all cellular chromosomes suggests that the process is a vestige of replication in the ancient ancestral cells from which all life evolved. The chromosomes of those ancient cells consisted entirely of RNA.

Once replication has begun, the cell is committed to completing a full round of DNA synthesis. Therefore, the decision of when to start is critical. If replication starts too soon, the cell accumulates unneeded copies of its DNA. If it starts too late, the dividing cell's septum (partition across the middle) cuts through the chromosome, killing both daughter cells.

Elongation of a DNA Strand

After initiation, a circular bacterial chromosome replicates bidirectionally. As illustrated in Figure 8.9, in bidirectional replication the helix unwinds and chain elongation proceeds in both directions away from the origin (**Figure 8.10**, step 1). The chromosome region of active DNA replication is called a "replication bubble." At each end of the replication bubble, the two replication forks (sites of helix unwinding and new-strand synthesis) move in opposite directions around the chromosome (step 2). Although not all enzymes are shown in Figure 8.10, remember that all the steps of DNA synthesis require catalysis by enzymes.

The 5′-to-3′-directed elongation of DNA leads to an interesting puzzle: if polymerases can synthesize DNA only in a 5′-to-3′ direction, and the two phosphodiester backbones of the double helix are

Figure 8.9 RNA Primers Provide the First 3′ OH for Each Strand

Each leading strand (pointing toward the replication fork) needs only one primer, but each lagging strand (pointing back from the fork) needs multiple primers. The RNA primers are made one by one and used to prime DNA synthesis before the next primer is made.

antiparallel (opposite directions), how are both strands of a moving replication fork synthesized simultaneously?

At each fork, one strand is synthesized directly in a 5′-to-3′ direction toward the fork (Figure 8.10, step 3). But synthesizing the other new strand in a 5′-to-3′ direction requires that it move away from the fork. Thus, at each replication fork, one of the two daughter strands needs to elongate by "jogging back" toward the fork at regular intervals. The smoothly elongating strand is called the **leading strand**, whereas the jogging-back strand is called the **lagging strand**.

The fragments of DNA that form as the lagging strand is synthesized are called Okazaki fragments (Figure 8.10, step 4), named for Reiji and Tsuneko Okazaki, the scientists who discovered the fragments of the DNA lagging strand in 1968.

Both leading and lagging strands are extended by a DNA polymerase. The main polymerase for DNA synthesis is called DNA

Figure 8.10 Replication of a Bacterial Genome

1 Replication begins at origin.

Origin, *oriC*

Terminus, *ter*

2 Replication bubble forms. Replication forks progress in opposite directions.

Replication bubble

Leading strand

3 Leading strand at each fork is synthesized continuously 5′ to 3′.

Fork movement

4 Lagging strand at each fork is synthesized discontinuously in Okazaki fragments 5′ to 3′.

Lagging strand
Okazaki fragment

Fork movement

Terminus

5 Replication ends at terminus.

Ligase seals nicks

Terminus, *ter*

polymerase III; the number reflects the fact that this polymerase happened to be the third one purified from cells by microbiologists early in the twentieth century. The first discovered, DNA polymerase I, actually catalyzes a later step of DNA replication; it is also known as the Kornberg enzyme, for Arthur and Sylvy Kornberg, who discovered it in 1956. DNA polymerases are of enormous importance for all biology and biotechnology; their applications have transformed the practice of medicine and forensic science.

Figure 8.10 shows an overview of initiation at the origin and formation of the replication bubble with two replication forks. Note that the two replication forks each generate a leading strand and a lagging strand, oriented opposite to the other pair. Late in DNA replication, when most of the complementary daughter strand has been elongated, the RNA primers need to be replaced with DNA. The RNA is removed by an enzyme that specifically recognizes RNA-DNA hybrid molecules and cleaves away only the RNA strand. After the removal of RNA, the enzyme DNA polymerase I "fills in" the complementary DNA nucleotides. Completion of DNA synthesis requires replication at the defined termination (*ter*) site located on the opposite side of the molecule from the origin.

ACCURACY OF DNA SYNTHESIS An important problem for DNA synthesis is accuracy. The enzyme DNA polymerase III (abbreviated DNA Pol III) places the correct complementary base only about 99.9% of the time. This rate of success may sound good, but one error in a thousand would lead to thousands of errors in the genome as a whole and many mutant genes. To avoid mutations, the newly made DNA must be proofread. Proofreading is done by several kinds of enzymes. For example, when DNA polymerase I runs into the 5′ back end of the next Okazaki fragment, it cleaves the 5′ base off. The base is then replaced by a new base condensed with the 3′ OH of the DNA "behind" the enzyme. In this process of base replacement, any incorrectly paired bases get replaced by correct ones. Other repair enzymes conduct repair by recognizing a "bump" in the DNA where the incorrect base fails to make a smooth pair. The enzyme cleaves the backbone at either side of the incorrect base and removes a stretch of DNA that contains it, to be filled in correctly by polymerase I. Error-correcting enzymes are important for repairing mutations, as discussed in Section 8.5.

Given that all DNA molecules look very similar, how do repair enzymes "know" which strand of an incorrect pair is the "new" DNA to be fixed? During DNA replication, each "old" parental DNA strand contains a methyl group on some of its cytosine bases. Repair enzymes recognize methylated DNA as "old" DNA, containing the original sequence. Thus, at the conclusion of replication, all the newly synthesized DNA needs to be methylated by methylase enzymes—a process akin to putting a seal of approval on the completed DNA. If the error is not repaired before methylation, the cell doesn't know which is the wrong base and may "fix" the correct one, resulting in a mutation.

Another requirement as DNA synthesis nears completion is to join the backbone of all the Okazaki fragments to make a continuous, unbroken strand. The "nicks" (positions of unsealed backbone between two adjacent nucleotides) are joined by **DNA ligase**, an enzyme that condenses the 3′ hydroxyl "back end" of each fragment with the 5′ phosphoryl "front end" of the adjacent fragment.

Termination of DNA Synthesis

In bacteria with circular chromosomes, the two replication forks eventually circle around to approach each other. The two forks meet at a DNA sequence called the terminus, or *ter* site (Figure 8.10, step 5). The terminus requires special proteins to complete the growing DNA strands while disengaging the DNA polymerases.

As DNA replication terminates, the cell forms a partition called the septum across the cell equator. The septum splits the cell in half. Once the septum is complete, the two daughter cells, with their new chromosomes, can separate.

Another important process for completion of replication is that the DNA must regain its negative **supercoils**. Supercoils, or superturns, are extra turns in the DNA beyond those generated by the shape of the base pairs. DNA needs to be "unwound" slightly, that is, negatively supercoiled, for genes to be transcribed to RNA (see Section 8.3). Negative supercoils are generated by the enzyme **gyrase**—a particularly important drug target (Figure 8.11). Bacterial gyrase is the target of quinolones, a group of broad-spectrum antibiotics used to treat anthrax and pneumonia.

How does gyrase introduce a supercoil into a circular DNA molecule, lacking ends? The gyrase complex cleaves the backbones of both strands of DNA at the same position, causing a double-strand break. Holding on to both ends, the enzyme passes another double-stranded portion of the DNA through the break and then reseals the two backbones—a process that spends ATP. The net result is one turn "against" the twist of the helix—that is, a negative supercoil. Bacteria have several other kinds of enzymes that generate positive or negative supercoils for various functions; the general class of enzymes that change supercoiling is called topoisomerases.

So, how long does it take for the bacteria in your contaminated chicken salad to double their DNA and undergo cell division? At optimum temperature and resources, the entire process of cell division in an enteric bacterium (digestive tract) may take as little as 20 minutes (see Figure 5.16).

SECTION SUMMARY

- **DNA is synthesized semiconservatively.** Each daughter helix contains one "old" parental strand (the template) and one newly formed complementary strand.

- **Initiation of DNA synthesis** occurs at the replication origin and proceeds bidirectionally around the cell until both replication forks meet at the terminus.

- **Enzymes catalyze all steps of replication.** Helicase unwinds the helix, primase forms RNA primers with a 3′ OH end, DNA polymerase III extends the growing DNA strand, DNA polymerase I removes the primer RNA,

Figure 8.11 Gyrase Adds Negative Supercoils to DNA
The enzyme gyrase is the target of quinolone antibiotics.

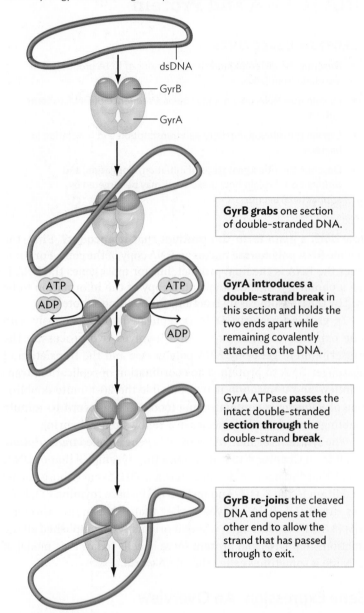

dsDNA
GyrB
GyrA

GyrB grabs one section of double-stranded DNA.

ATP
ADP
ATP
ADP

GyrA introduces a double-strand break in this section and holds the two ends apart while remaining covalently attached to the DNA.

GyrA ATPase **passes** the intact double-stranded **section through** the double-strand **break**.

GyrB re-joins the cleaved DNA and opens at the other end to allow the strand that has passed through to exit.

proofreading enzymes correct mismatched bases, and DNA ligase seals the nicks between segments of each growing strand.

- **Leading and lagging strands are elongated at each growing fork.** The lagging strand requires that DNA polymerase backtrack at regular intervals to elongate from a 3′ OH end.

- **Completed DNA requires methylation and restoration of negative supercoils.** Gyrase, an important antibiotic target, forms the supercoils.

Thought Question 8.1 Suppose that a rapidly growing pathogen can be killed by an antibiotic that blocks DNA replication, but then the bacteria form a biofilm on a catheter in a patient. Why might the bacteria have become resistant to the antibiotic?

8.3 Gene Expression: From DNA to RNA and Protein

How does a gene form the product that it encodes? First, the enzyme **RNA polymerase** makes an RNA copy of the gene. For some genes the RNA is the final product, but for most genes the RNA is like a photocopy of the blueprint, a copy of the information to be used by the ribosome as instructions for building a protein.

In growing bacteria, DNA is replicated simultaneously with gene expression. Gene expression involves two processes: the transcription of genes by RNA polymerase and the translation of messenger RNA to protein. The coordination of replication, transcription, and translation makes possible the 20-minute doubling time of *Salmonella* contaminating food and the record 10-minute doubling time of *Bacillus* species in a Yellowstone hot spring.

The synthesis of an RNA "copy" of a gene is called **transcription**. Recall from Chapter 4 that RNA structure resembles that of DNA, except that the nucleotide sugar has a 2′ OH group (ribose sugar) in addition to the 3′ OH. And instead of the base thymine (T), RNA uses uracil (U). RNA evolved earlier than DNA in our ancestral cells (3.8 billion years ago), where it originally accomplished all the functions that DNA and protein serve in cells today. This model of early life is sometimes called the "RNA world."

Gene Expression: An Overview

How are genes organized on a chromosome, and what products are expressed? Many genes have as their final product an RNA molecule—for example, the ribosomal RNA molecules that form the main structural components of the small and large subunits of ribosomes (to be described shortly). Other "functional" or "small, noncoding" RNA molecules found in bacteria include transfer RNAs (tRNAs) and regulatory small RNAs (sRNAs). Eukaryotes have a much larger number of noncoding RNAs, including several thousand that serve as enzyme subunits.

The majority of genes, however, are transcribed to **messenger RNA (mRNA)**, which encodes one or more peptides (proteins) to be translated by a ribosome. For example, the *mdtABCD* operon of *E. coli* encodes proteins of a membrane-embedded complex that pumps out toxic molecules such as the antibiotic novobiocin (**Figure 8.12A**). The operon has six genes in tandem controlled by

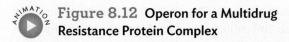

Figure 8.12 Operon for a Multidrug Resistance Protein Complex

A. Genes *mdtA*, *mdtB*, *mdtC*, and *mdtD* encode proteins of an efflux pump that exports antibiotics such as novobiocin. Gene *baeS* encodes an environmental sensor; *baeR* encodes a regulator that binds the promoter.

B. Alignment of the drug pump–encoding gene *mdtA* with its mRNA transcript and start of peptide (protein).

one promoter. The right-angled red arrow in the figure indicates a **promoter** sequence. The promoter plus the adjacent genes it controls make an operon. Since all the protein parts work together and they have no function apart from one another, it makes sense to express the genes all together at the same rate, from one common promoter.

Within the operon, each structural gene (gene encoding a protein) is designated by an italic three-letter name, which may have a fourth letter in upper case (for example, *mdtA*, *mdtB*, *mdtC*, *mdtD*). The final two structural genes (*baeS*, *baeR*) form a two-component regulator that activates the operon—a type of regulation discussed in Section 8.4. All the genes of the operon shown are transcribed together, forming one mRNA. The genes are transcribed starting from the promoter, a short DNA sequence "upstream" of the gene *mdtA*.

How does the DNA sequence relate to the corresponding mRNA? The alignment between the DNA sequence of a structural gene and the resulting mRNA transcript is shown in **Figure 8.12B**. The double-helical DNA includes the **coding strand** (or **sense strand**), which contains the base sequence equivalent to that of the mRNA, substituting U's for T's; and its complement, the **template strand**. The template strand is the one that actually enters the RNA polymerase to base-pair with the growing chain of mRNA. The mRNA contains triplet codons that specify amino acids of the protein to make, starting with AUG, the start codon specifying methionine (discussed later in this section).

RNA Synthesis

Like DNA synthesis, transcription to RNA includes stages of initiation, elongation, and termination. Each stage requires specific molecules to interact with the RNA and RNA polymerase. These steps are targets for antibiotics, such as rifamycin, which blocks RNA polymerase; rifamycin is an important drug for tuberculosis.

RNA polymerase is the enzyme complex that adds nucleotides onto the growing RNA chain. In bacteria, the enzyme complex is composed of four subunits, as seen in **Figure 8.13** (step 1). For initiation, the RNA polymerase complex includes a fifth subunit, called sigma (the Greek letter σ). The **sigma factor** can be one of several different versions that regulate genes responding to major environmental conditions, such as heat shock (fever, within a host), oxidation stress (phagocytosis by a macrophage), or starvation (competition with other microbes in the colon). Genes whose products are required under most typical conditions are regulated by a "housekeeping" sigma factor.

When RNA polymerase complexes with a sigma factor, the resulting "holoenzyme" RNA polymerase binds loosely to any sequence of DNA. Thus, within the cell the RNA polymerase binds and comes off DNA rapidly, as if scanning the sequence. The RNA polymerase stops when it finds a promoter sequence and binds tightly (Figure 8.13, step 2). The promoter sequence is commonly found upstream (left) of the +1 start site for RNA synthesis (transcription). The "+1" label in Figure 8.13 marks the DNA base where the mRNA transcript starts. The promoter sequences where RNA polymerase first binds are approximately −10 and −35 bases upstream of the +1 start. The exact sequence of the promoter can influence the amount of mRNA synthesized.

To start transcribing a gene, the RNA polymerase–DNA complex must open through the unwinding of one helical turn of DNA, which causes DNA to become unpaired (Figure 8.13, step 3). Now the RNA polymerase can fit a ribonucleotide to base-pair with the DNA. To elongate the chain, the polymerase catalyzes the addition of new ribonucleoside triphosphates (rNTPs). Just as for DNA, the 5′ phosphoryl end of each incoming rNTP condenses with a 3′ OH and releases pyrophosphate—the step that yields energy. Chain elongation is the step that can be blocked by rifamycin antibiotics.

The RNA polymerase holoenzyme (containing sigma factor) can add only a few ribonucleotides. Remember, the whole complex is still bound tightly to the promoter. So, what happens? The sigma factor falls off the complex (Figure 8.13, step 4), enabling the core RNA polymerase to continue to move along the template, synthesizing RNA at approximately 45 bases per second. The unwinding of DNA ahead of the moving complex forms a 17-base-pair "transcription bubble" in which RNA is being synthesized, while behind

 Figure 8.13 Transcription Initiation and Elongation
The RNA polymerase holoenzyme scans the DNA sequence until it binds to a promoter. The promoter DNA helix unzips, exposing bases of the template to base-pair with incoming nucleotides. The polymerase catalyzes the addition of nucleotides. The sigma factor must fall off before elongation continues.

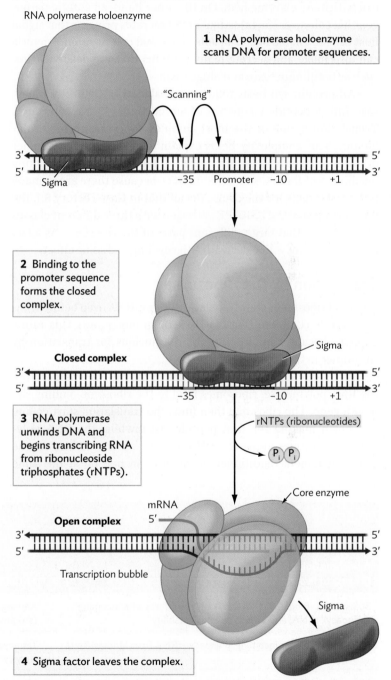

it the two DNA strands "snap" back together. Because of the DNA unwinding, positive supercoils are formed ahead of the advancing bubble. These positive supercoils are removed by topoisomerase enzymes that return the DNA to its normal negative-supercoiled state. Finally, transcription must finish at the end of the gene where a transcription "stop" sequence in the mRNA causes the mRNA and RNA polymerase to fall off.

Protein Synthesis

Proteins determine most of the distinguishing traits of a cell. For instance, a thermophilic archaeon that grows at a deep-sea ocean vent at temperatures above 100°C (212°F) makes proteins that can withstand extreme heat. On the other hand, an enteric pathogen that colonizes the intestinal lining secretes proteins that signal host cells to enable bacterial adherence and colonization. Bacteria can also make and secrete many proteins that are deadly toxins, such as botulinum toxin and Shiga toxin.

All protein synthesis requires **translation** of the mRNA message into a peptide (sequence of amino acids) by the ribosome. Translation is one of the most complex processes in molecular biology. This complexity helps explain why such a large number of antibiotics have been discovered to block the protein-making machinery of the ribosome (Table 8.2), because there are so many potential targets for the drugs. Recall that in Case History 8.1, the physician tested the ESKAPE pathogen with three different classes of antibiotics that target different parts of the ribosome. As a last resort, the tetracycline derivative tigecycline had some effect.

Translation by the Ribosome

Bacterial ribosomes start translating their mRNA even before transcription is complete (Figure 8.14A). In eukaryotes, this rarely happens because the mRNA exits the nucleus for translation by ribosomes in the cytoplasm.

As the bacterial 5′ leader sequence of the mRNA emerges from the RNA polymerase, ribosomes bind to the ribosome-binding site near a gene. The ribosome then finds the translation start codon, and begin making the first peptide. Meanwhile, the RNA polymerase continues along the DNA, elongating the mRNA. As the mRNA lengthens, additional ribosomes can hop on at the same

Figure 8.14 Coupled Transcription and Translation in Prokaryotes

A. Ribosomes attach at mRNA ribosome-binding sites and start synthesizing protein before transcription of the gene is complete.

B. Electron micrograph of a polysome (ribosome is 21 nm across). Several ribosomes may translate a single mRNA molecule at the same time. Each ribosome starts at the beginning (5′ end) of the mRNA and travels toward the 3′ end. Note that the synthesized protein molecule grows longer and longer as the ribosome nears the 3′ end of the mRNA. The protein molecule is seen most clearly at the 3′ end of the mRNA (right).

Table 8.2

Antibiotics That Target Protein Synthesis

Ribosome Subunit (30S or 50S): Process of Protein Synthesis	Antimicrobial Molecules	Source of Molecule	Organisms Affected
30S subunit: Binds 16S rRNA at A site; interferes with incoming aminoacyl-tRNA; also affects envelope stability	Aminoglycosides (streptomycin)	*Streptomyces griseus*	Aerobic bacteria (Gram-positive and Gram-negative)
30S subunit: Blocks incoming aminoacyl-tRNA from binding A site	Tetracyclines and glycylcyclines (tigecycline)	*Streptomyces* spp.	Bacteria
50S subunit: Binds peptide release tunnel and blocks translocation from A site to P site	Macrolides (erythromycin)	*Saccharopolyspora erythraea*	Bacteria
50S subunit: Inhibits peptidyltransferase activity	Lincosamides	*Streptomyces lincolnensis*	Bacteria
50S subunit: Inhibits complex formation	Everninomicins	*Micromonospora carbonacea*	Gram-positive bacteria
50S subunit: Binds 23S rRNA near P site; inhibits translation	Oxazolidinones (linezolid)	Chemical synthesis	Gram-positive bacteria
50S subunit: Binds peptidyltransferase domain and blocks elongation	Streptogramins	*Streptomyces* spp.	Bacteria

Figure 8.15 The Genetic Code

Codons within a single box encode the same amino acid. Color-highlighted amino acids are encoded by codons in two boxes. Stop codons are highlighted pink. Note: Before transcription, the DNA gene sequence had thymine (T) to specify each uracil (U) in the RNA.

Figure 8.16 Aminoacyl-tRNA Base-Pairs with mRNA

The tRNA anticodon consists of three nucleotides at the base of the anticodon loop. The anticodon hydrogen-bonds with the mRNA codon in an antiparallel direction. This tRNA is "charged" with an amino acid covalently attached to the 3' end.

ribosome binding site. The electron micrograph in **Figure 8.14B** shows a "polysome," an mRNA with a series of ribosomes chugging along, synthesizing peptide. Notice how the ribosomes farthest from the mRNA upstream leader (those that have traveled longest) show the longest growing peptide chains.

The mRNA of protein-encoding genes specifies amino acid sequences according to the genetic code (**Figure 8.15**). In the diagram, each possible triplet group of RNA bases forms a codon. The codons will each specify an amino acid to add onto a protein. The codon AUG that specifies the amino acid methionine also specifies the start site of translation; it's called a "start codon." In bacteria, the initial methionine has a special formyl group attached (fMet). In eukaryotes, there is no formyl group but other mechanisms specify initiation.

In addition, three codons—UGA, UAA, and UAG—specify no amino acid. These are called "stop codons." In an mRNA transcript containing multiple genes, each gene has its own translation start and stop codons.

How are the three-letter codons of the genetic code "decoded" to add amino acids into a peptide? The decoder for the genetic code is **transfer RNA** (**tRNA**; **Figure 8.16**). Each tRNA molecule has its own aminoacyl-tRNA synthetase, which attaches the correct

amino acid to the correct tRNA. A tRNA attached to an amino acid is called a "charged" tRNA.

Each tRNA also possesses a loop, called the anticodon loop, that exposes three unpaired bases. The anticodon loop can base-pair with a codon in the mRNA—the codon that specifies the amino acid attached to the tRNA. Each anticodon of a charged tRNA pairs with the mRNA codon at a special site deep within the ribosome. As each tRNA anticodon pairs with the mRNA codon, an amino acid is added onto the growing peptide chain.

The ribosome is composed of two complex subunits, each of which includes rRNA and protein components (**Figure 8.17**). In bacteria, the subunits are named 30S and 50S for their "size" in Svedberg units. (A Svedberg unit reflects the rate at which a molecule sediments under the centrifugal force of a centrifuge.) Each subunit is an enormous complex consisting of an rRNA (16S rRNA for the 30S subunit; 23S and 5S rRNAs for the 50S subunit), plus a large number of proteins that stabilize the rRNA. Within the living cell, the two ribosomal subunits exist separately but come together on mRNA to complete the ribosome.

INITIATION The key event of initiation is binding of the 30S subunit to the translation initiation sequence on the mRNA

Figure 8.17 Initiation of mRNA Translation to Synthesize a Peptide

As we saw for DNA and RNA synthesis, protein synthesis occurs in three phases: initiation, elongation, and termination.

1 Formyl-Met tRNA binds AUG codon of mRNA on 30S subunit of ribosome.

fMet
GTP
U A C
A U G
IF1
5′ mRNA 3′
30S subunit

GDP + P_i

50S subunit

2 50S subunit binds 30S, completing ribosome.

fMet
E P A
50S
U A C
A U G
5′ mRNA 3′

3 GTP is hydrolyzed.

(see Figure 8.17). Completing the 30S initiation complex requires spending energy in the form of GTP. The initiation complex includes proteins called initiation factors (not shown in the figure). In bacteria, initiation also involves hybridization (base pairing) between the mRNA ribosome-binding site and the 3′ terminal sequence of the 16S rRNA. In eukaryotes, the mechanisms of initiation are different; that is one reason why so many antibiotics can disrupt function of the bacterial ribosome without harming the human host.

When the 30S subunit binds the mRNA, the shape of the complex forms a "pocket" to fit the initiator tRNA. The initiator tRNA specifies methionine, while its anticodon base-pairs with the AUG codon of the mRNA (see Figure 8.17, step 1). Thus AUG is termed the start codon.

The initial methionine attached to the initiator tRNA is a special methionine with an extra formyl (R-CHO) group. Peptides with formylmethionine (fMet) are made only by bacteria and by our mitochondria, which evolved from bacteria. Interestingly, this difference is used by our white blood cells to detect the presence of invading bacteria and of mitochondria leaking from damaged tissue. White blood cells called leukocytes can detect a concentration of formylmethionyl peptides as low as 10^{-12} M and migrate toward the source.

Next comes the large (50S) subunit of the ribosome, suspended in the cytoplasm. The 50S subunit joins the complex, fitting around

the 30S subunit like a pot fits a lid (Figure 8.17, step 2). The 50S subunit encloses the initiator tRNA and completes formation of the three tRNA sites: the acceptor (A) site, also called "aminoacyl" site; the peptide (P) site; and the exit (E) site. To drive the reaction, GTP is hydrolyzed and initiation factors are released (step 3). Now the ribosome is ready to add amino acids, elongating the chain.

ELONGATION To add an amino acid to the chain, an aminoacyl-tRNA binds to the A site (**Figure 8.18**). The aminoacyl-tRNA is assisted by elongation factor Tu (EF-Tu), an abundant protein in the cell. EF-Tu carries a GTP with energy to "spend" for elongation (Figure 8.18, step 1). EF-Tu can bring any tRNA to the A site. However, the tRNA will bind only if its anticodon base-pairs

Figure 8.18 Translation: Elongation and Termination

EF-Tu–GTP
GTP
Amino acid
tRNA
Anticodon

1 EF-Tu brings aminoacyl-tRNA to the A site.

E P A
Direction of ribosome movement
5′ mRNA 3′

Peptidyltransferase

2 Peptide bond forms between the amino acid or peptide in the P site and the amino acid in the A site.

E P A
5′ mRNA 3′

EF-G–GTP

3 The EF-G–GTP complex binds to the ribosome, GTP is hydrolyzed, and the ribosome advances one codon. The tRNA in the A site moves into the P site.

E P GTP
5′ mRNA 3′

4 The A site is now empty and ready to receive a new charged tRNA.

E P A
5′ mRNA 3′

correctly with the mRNA codon within the site; otherwise, the tRNA will slip out again. The anticodon allows some ambiguity at its third position; for instance, either U or C in the codon may be acceptable. Ambiguity at the third position means that fewer than 64 different tRNAs are actually needed to serve the entire code. Thus, the code has evolved to be more efficient.

During peptide bond formation, the chain of amino acids is transferred from the tRNA in the P site to the tRNA in the A site (Figure 8.18, step 2). The 50S subunit catalyzes formation of a **peptide bond** (amide bond, Chapter 4) between the amino end of the new amino acid and the carboxyl of the formylmethionine sitting in the P site. To drive peptide bond formation, the GTP brought by EF-Tu is hydrolyzed, and the resulting EF-Tu-GDP is expelled. Finally, the mRNA must "move over" one position, so that each tRNA enters a new "pocket" in the ribosome (step 3). An elongation factor G (EF-G) with GTP inserts into the A site. The GTP is hydrolyzed to GDP, spending energy to shift the mRNA forward relative to the 50S subunit. Now the A site can receive the next aminoacyl-tRNA. This move is a crucial step that can be blocked by antibiotics such as streptomycin. Now the peptidyl-tRNA again resides in the P site, ready to engage the next incoming aminoacyl-tRNA (step 4).

TERMINATION The peptide chain terminates when a stop codon in the mRNA enters the A site. A stop codon is one of three codons (UGA, UAA, UAG) for which no tRNA anticodon exists. Instead, a protein called "release factor" enters the A site. The release factor triggers hydrolysis of the peptide-tRNA ester link, thus releasing the peptide. Finally, the 30S and 50S subunits fall off the mRNA and separate in the cytoplasm, available once again to form a new translation complex at some other mRNA.

As the peptide chain (protein) emerges from the ribosome, its various amino acid side chains interact very differently with the cytoplasmic water. Some peptides fold in the cytoplasm, whereas others must be synthesized directly at the cell membrane, where they will reside as membrane-embedded proteins. Finally, some important peptides are **secreted**; that is, they are transported out of the cell in order to interact with the cell's environment. Secreted proteins include signaling molecules and toxins, which can help a pathogen cause disease.

SECTION SUMMARY

- **The first gene of an operon is preceded (upstream) by a noncoding sequence called a promoter.** RNA polymerase with a sigma factor (holoenzyme) scans DNA (binds loosely and comes off) until it binds a promoter, and then transcription begins.

- **After the first few RNA bases are added, sigma factor falls off, allowing RNA polymerase (core enzyme) to slide along the DNA and continue elongation.** Elongation terminates at a pause site, where the RNA polymerase falls off.

- **Transfer RNA (tRNA) decodes the genetic code.** The tRNA is charged by an aminoacyl-tRNA synthetase enzyme with a specific amino acid that corresponds to the anticodon of the tRNA.

- **Initiation of translation occurs when the 30S subunit binds mRNA.** The formylmethionyl-tRNA binds to an AUG codon of the first gene in the mRNA.

- **Elongation includes the successive entry of aminoacyl-tRNA molecules, followed by addition of each new amino acid forming a peptide bond.** Codon–anticodon pairings guide which new aminoacyl-tRNA binds the A site. If the pairing is correct, the 50S subunit catalyzes peptide bond formation. The mRNA translocates one position, so that now the "new" tRNA (holding on to the growing peptide) occupies the P site.

- **Termination occurs when a stop codon in the mRNA enters the A site.** The completed peptide is released, and the 30S and 50S ribosomal subunits fall off, to commence translation elsewhere.

- **The newly formed peptide must fold into its final conformation.** Some peptides fold in the cytoplasm, whereas others must fold in the membrane, for membrane function or for secretion outside the cell.

Thought Question 8.2 In pathogenic bacteria, some genes contain multiple promoters that are used under different cell conditions, such as particular temperatures or nutrient availability. Why might this ability to adapt to a variety of conditions be advantageous to the pathogen?

8.4 Regulation of Gene Expression

SECTION OBJECTIVES

- Describe how and why a cell regulates expression of a virulence gene.
- Outline the levels of gene regulation. Explain why it is useful to regulate gene expression at different levels.
- Explain how regulator proteins interact with chemical inducers and signaling molecules to alter the rate of gene expression.

A bacterial genome encodes thousands of different proteins needed to handle many different environments, including those of a host. How does the bacterium decide which proteins to make at what time? Some proteins, such as RNA polymerase, are always required for growth. Most proteins, however, are needed only under a limited set of conditions. For example, proteins that degrade the sugar lactose are useful only when lactose is present. Likewise, *Corynebacterium diphtheriae* needs to make diphtheria toxin only when growing inside the human body.

Protein synthesis is expensive. Each amino acid added to the chain requires 1 ATP to charge the tRNA and 2 GTP to admit the tRNA and form the peptide bond. To compete successfully with others, microbes cannot waste energy making unneeded proteins. Energy waste is minimized by elegant genetic controls that selectively increase or decrease transcription of a gene (that is, determine whether RNA is made) or control mRNA translation to protein, mRNA degradation, or degradation of a regulatory protein. But how does a cell "know" when to alter a gene's expression? How does *C. diphtheriae* "know" whether it needs to make diphtheria toxin?

Microbes use numerous mechanisms to sense their internal and external environments and then translate that information into action. The cell's surface, for example, contains an array of sensing proteins that monitor osmolarity, pH, temperature, and the chemical content of the surroundings or that detect the presence of a host or competitor. Sharing of chemical signaling molecules makes it possible for members of microbial communities to communicate and cooperate in a process known as quorum sensing (presented in Section 6.6). Even virulence genes have tapped into these sensing systems, using them to determine whether the microbe has entered a susceptible host.

Levels of Gene Regulation

Hundreds of intracellular and extracellular sensing mechanisms monitor the overall health of the cell and its environment. But how do these systems control gene expression? Expression of genes and operons and their mRNA products can be controlled at various levels. Different levels of control offer different advantages to the cell. Gene expression can be regulated by DNA sequence controls, transcriptional controls, translational controls, and posttranslational modifications. Here are some examples:

- **Changing the DNA sequence.** Some microbes can change the DNA sequence to activate or disable a particular gene. An example of regulation by DNA change is phase variation. In phase variation, a DNA segment containing a promoter can flip so that the promoter is pointing toward or away from a downstream gene. Because the immune system targets bacterial surface proteins, turning off the genes that encode those proteins is one way the organism stays alive in a host. However, reversing a DNA change is a relatively rare event, so the rate of control is slow.

- **Control of transcription.** Many types of gene regulation occur at the level of transcription (discussed shortly). Transcription can be regulated by protein repressors, activators, and alternative sigma factors (the class of proteins that helps RNA polymerase recognize promoters).

- **Translational control.** Translation by ribosomes can be regulated in the mRNA by translation initiation sequences, which recognize specific translational repressor proteins.

- **Posttranslational control.** Once proteins are made, their activity can be controlled by modifying their structure—for example, by protein cleavage, phosphorylation (adding a phosphate), or methylation (adding a methyl group, $-CH_3$). These modifications can activate or deactivate a protein or even lead to its destruction.

Transcriptional Control

Gene expression can be turned on (**induction**, or up-regulation) or off (**repression**, or down-regulation). Some genes respond to changes inside the cell, such as the presence of a food molecule,

Figure 8.19 Transcriptional Regulation by Regulator Proteins

The regulatory protein will bind a DNA regulatory sequence near the promoter of the target gene. The binding of regulator to DNA controls whether the target gene gets transcribed.

whereas others respond to outside signals, such as a dangerous antibiotic. How do these responses work?

SENSING THE INTRACELLULAR ENVIRONMENT Sensing conditions within the cell requires that regulatory proteins bind to specific small-molecular-weight compounds (ligands). **Figure 8.19** shows a general model of a regulatory protein that can sense an intracellular metabolite (not shown). The regulatory protein binds to a control sequence and thus determines whether the gene gets transcribed. Depending on the system, the target gene could be induced (transcribed) or repressed (not transcribed).

Different regulatory proteins bind to different metabolite signaling molecules (ligands). Take, for example, the expression of *dtx*, the diphtheria toxin gene. In diphtheria pathogenesis, expression of virulence genes is controlled by the **repressor** protein DtxR. DtxR senses the concentration of a ligand, iron—a scarce metal ion (Fe^{2+}) that *Corynebacterium diphtheriae* needs to grow. Host cells killed by diphtheria toxin release iron that the pathogen can now use to grow. When iron is plentiful, as can happen in the environment, no toxin is made. DtxR binds iron, attaches to the regulatory region of the diphtheria toxin gene, and then represses the gene's transcription (**Figure 8.20A**). However, when the bacterium grows within a human host, much of the host's iron is tied up by the iron-binding proteins transferrin and lactoferrin, and very little iron is left free for *C. diphtheriae* to use. Because iron is scarce, DtxR cannot bind iron, and therefore cannot repress the *dtx* gene. Without repression, diphtheria toxin will be made to obtain host iron (**Figure 8.20B**).

In contrast to repressor proteins that stop a gene's expression, other regulatory proteins called transcriptional **activators** can <u>stimulate</u> gene expression. For example, a transcriptional activator protein might bind to a carbohydrate that has entered the cytoplasm. The activator protein then activates the transcription of genes whose products metabolize that carbohydrate.

Figure 8.20 Transcriptional Regulation of the Diphtheria Toxin Gene (*dtx*) by the DtxR Repressor

High iron levels

DtxR repressor

Fe²⁺

DtxR-Fe2+   No expression

DNA   *dtxR*   *dtx*

Regulatory region   Toxin gene

Iron binds *dtxR*. DtxR binds regulatory region. No toxin expressed.

A. Outside a human host, *Corynebacterium diphtheriae* will usually encounter sufficient iron for growth. It does not need to produce diphtheria toxin.

Low iron levels: Human throat

RNA polymerase

Toxin gene

DNA   *dtxR*   *dtx*

Regulatory region   mRNA

Diphtheria toxin

Kill host cells to release iron

Without iron, DtxR comes off DNA. Toxin is expressed.

B. In the human throat, iron is sequestered in host cells. Diphtheria toxin will kill those cells and release iron that the bacteria can use.

Figure 8.21 Induction of the Lactose (*lac*) Operon

P_{lacI} P_{lacZYA}

lacI   *lacZ*   *lacY*   *lacA*   DNA

lacO_I   *lacO*

mRNA

β-Galactosidase (LacZ)   Lactose permease (LacY)   Thiogalactoside transacetylase (LacA)

LacI repressor

lacO_I   *lacO*

lacZ   *lacY*   *lacA*   DNA

Repressor LacI binds to *lacO*. The bound protein overlaps the *lacZYA* promoter (P) and prevents transcription.

lacI   *lacZ*   *lacY*   *lacA*   DNA

mRNA

Inducer (allolactose)

Inducer (lactose converted to allolactose) binds LacI repressor, which falls off the DNA. The *lac* operon is transcribed.

A. Organization of the *lac* operon. Bent arrows mark promoters. Green regions indicate LacI protein-binding sites on DNA. The products of the operon have different roles: LacY transports lactose into the cell; LacZ degrades lactose to glucose and galactose; LacA's role in lactose metabolism is unknown.

Regulators can act upon an operon containing multiple genes, such as the multidrug resistance *mdtABCD* operon (see Figure 8.12). Gene expression for an operon can be turned on when an inducer molecule <u>relieves repression</u> imposed by a repressor protein. A classic example is the lactose operon (**Figure 8.21A**), whose gene products degrade lactose for use as a carbon source. Lactose catabolism gives us an important indicator of the presence of fecal bacteria in water supplies. Water bacteria samples are collected on a filter and cultured on a medium containing lactose. Induced by lactose, the sugar breakdown produces acid, which turns the pH indicator dark red (**Figure 8.21B**), indicating coliforms (*E. coli*–related fecal bacteria).

When lactose is provided in media, a form of lactose (called allolactose) acts as the inducer. The inducer binds to the Lac repressor

Dark red colony: Fecal coliforms catabolize lactose

B. Coliform count showing high contamination of drain water with fecal bacteria. The culture medium (mEndo agar) contains lactose, which *E. coli* and related fecal bacteria metabolize, forming acid. A pH indicator changes color, leading to dark red colonies.

protein (called LacI or LacR), causing the repressor to release from the *lac* operon. The *lac* operon, now free to be transcribed, will produce the enzymes needed to degrade lactose. When the lactose inducer declines in concentration, it is no longer available to bind repressor. Now the repressor binds the promoter sequence of DNA and halts further transcription. So, the cell avoids wasting energy when lactose-catabolizing enzymes have no useful task.

Some regulatory proteins control many different genes and operons throughout a genome. Proteins that affect the expression of many different genes are called **global regulators**. One example is the cyclic AMP receptor protein (CRP) of *Escherichia coli* and related species. Cyclic adenosine monophosphate (cyclic AMP, or cAMP) is a chemical signaling molecule synthesized from ATP. Its synthesis is, in part, a signal to the genome that the cell needs more carbon and energy. When present at high levels, cAMP binds to CRP, which in turn binds to many different regulatory DNA sequences throughout the genome. For example, CRP controls certain virulence genes in pathogens such as *Yersinia pestis*, the cause of bubonic plague. Global regulators, in general, are important for balancing the cellular response to changes in the environment, such as the temperature increase when a bacterium enters the human body.

SENSING THE EXTRACELLULAR ENVIRONMENT Sensing what goes on outside the cell is challenging because intracellular regulatory proteins cannot reach through the membrane and touch what is outside. A common mechanism bacteria use to sense the outside of the cell and transmit that information inside relies on a series of two-component regulators. Each two-component protein pair activates a different set of genes. For example, the VanS-VanR two-component regulator (**Figure 8.22**) is used by the ESKAPE pathogen *Enterococcus faecium* to turn on genes expressing proteins that resist vancomycin. Drug resistance systems consume energy and often decrease the growth rate of pathogens in the absence of the antibiotic. Without regulation, the vancomycin operon would quickly be lost from the population by competition with faster-growing strains.

The first protein, VanS, is a sensor kinase (protein that adds a phosphoryl group) that spans the membrane. One end of this sensor protein reaches outside the cell membrane, while the other end protrudes into the cytoplasm. A sensor protein recognizes a specific signaling molecule or condition; the VanS sensor binds vancomycin (Figure 8.22, step 1). Binding of vancomycin triggers phosphorylation of the sensor kinase (step 2). The phosphorylated sensor kinase then transfers the phosphate to a response regulator protein, VanR, in the cytoplasm. The VanR response regulator binds to target promoter DNA sequences and stimulates expression of vancomycin resistance proteins (step 4). To down-regulate (repress) the system, an enzyme (phosphatase) removes phosphate from VanR (step 5).

Note that in most two-component systems, only "information" is transmitted from outside the cell to inside; no molecule

Figure 8.22 Vancomycin Resistance Is Activated by a Two-Component Regulator

1 VanS sensor kinase detects vancomycin.

Vancomycin

Sensor kinase

VanS

Outside cell

2 Signal triggers autophosphorylation.

Cell membrane

ATP

ADP

P

Response regulator

3 Phosphate is transferred to VanR response regulator.

VanR — P

5 Phosphate is removed from VanR, which then comes off the DNA.

P

RNA polymerase

P

Inside cell

P

Phosphatase

Promoter   Operator   Gene

Vancomycin resistance gene

4 Genes are expressed to make mRNA encoding vancomycin resistance proteins.

actually crosses the membrane. By contrast, inducer systems such as the *lac* operon require uptake of an inducer molecule (such as lactose) into the cell.

SECTION SUMMARY

- **Microbes sense their environment and regulate gene expression** to optimize growth and survival.

- **Transcriptional regulation** is carried out by specialized DNA-binding proteins (called activators and repressors) that stimulate (activate) or inhibit (repress) the ease with which RNA polymerase initiates transcription at a given promoter.

- **Induction of a gene's transcription occurs** when a signaling molecule binds to a repressor DNA-binding protein and pulls it off the operator, thus enabling transcription of that gene.

- **Two-component signal transduction** involves a **sensor kinase** protein that monitors one aspect of the external or internal cell environment and transmits a signal via transphosphorylation to a response regulator protein that then acts as an activator or repressor.

- **Additional mechanisms of control** include DNA rearrangements, translation regulation, posttranslational modifications, and degradation.

Thought Question 8.3 What would happen to the expression of the diphtheria toxin gene if the *dtxR* gene were deleted from the genome?

8.5 Mutation and Repair

SECTION OBJECTIVES

- Explain the different kinds of mutations and how they occur.
- Describe different ways that a cell repairs DNA damage, preventing or reversing mutations.

How do organisms evolve into so many kinds, with unexpected new traits? A species is maintained by producing offspring that have the DNA sequence inherited from their parents. But DNA replication is never perfect. For humans, each of us carries approximately 150 differences in DNA sequence compared with our parents—or even compared with an "identical" twin.

The error rate of DNA replication in enteric bacteria such as *E. coli* is estimated at 1 change per 10 million base pairs (bp). Thus, in a colony of bacteria whose genome is 5 million base pairs, about half the bacteria should contain a **mutation**—that is, a base pair different from that of the preexisting parental cell (discussed shortly). The "mutant" base pair now exists in the genome, just like other base pairs in the rest of the DNA sequence. Any permanent, heritable alteration in a DNA sequence, whether harmful, beneficial, or neutral, is called a mutation. Over many generations, such mutations add up, generating variant strains of the organism that are highly similar but may differ in crucial ways, such as their degree of virulence and their susceptibility to antibiotics.

How do mutations occur? Producing a heritable mutation has two basic requirements: the base sequence must change, and the cell must fail to repair the change before the next round of replication. Different kinds of mutations have vastly different effects on the organism, as we will see.

CASE HISTORY 8.2

Sepsis by a Rare Strain

In San Francisco, a 3-year-old boy named Luke was admitted to the hospital after an abrupt onset of fever and a generalized seizure. The boy's cerebrospinal fluid was normal, and he had a normal chest radiograph, but his white blood cell count was 20,000/μl (normal is 4,000–10,000), with 85% polymorphonuclear leukocytes (PMNs). Luke's blood culture revealed bacteremia (bacteria in the blood). The bacteria were identified as "nontypeable" (lacking a capsule) *Haemophilus influenzae* bacteria. The nontypeable *H. influenzae* strain was beta-lactamase-positive (expressed an enzyme to inactivate penicillin G, ampicillin, and amoxicillin) but proved sensitive to cefixime, a third-generation cephalosporin.

The diagnosis surprised the boy's two fathers. They showed the doctor Luke's full record of immunizations and booster shots, which included the standard *Haemophilus influenzae*

b (Hib). Why was Luke not protected by the Hib vaccine? The doctor explained that Hib vaccine protects only from *H. influenzae* type b, the most common virulent strain. Some rarer virulent strains arose by mutation and evolution from a common ancestral strain. Six strains, or serotypes, designated a–f, are "typed" based on the type of polysaccharide capsule surrounding the cell envelope. The Hib vaccine targets capsule type b. Nontypeable strains of *H. influenzae* (those lacking a capsule) can be identified by PCR amplification of a DNA sequence, but they are unaffected by the vaccine. Fortunately, after 7 days of intravenous antibiotic, Luke recovered fully.

Immunization against *Haemophilus influenzae* is very important—but it protects children from only the most common strains. Now that immunization is widespread (and protects most children most of the time), we recognize the existence of rarer strains that the immunization misses. How do these other strains arise?

The bacterium *H. influenzae* was originally named for the mistaken idea that it causes influenza. In fact, *H. influenzae* causes a different kind of respiratory infection, as well as meningitis and bacteremia. The main victims are young children who have not yet developed protective antibodies. The first virulent strains to be characterized had a capsule with distinctive polysaccharides, first observed by Margaret Pittman in 1931. These strains were "typed" according to antibody reactions to their polysaccharides. Other strains that lack a capsule were called "nontypeable" because they could not be immunologically typed according to polysaccharide. Nontypeable, or unencapsulated, strains are unaffected by the body's immune response to the Hib vaccine, which targets capsule polysaccharide. Because most children in the United States now receive the Hib vaccine, a growing proportion of the *H. influenzae* infections seen in hospitals are caused by nontypeable strains. Vaccines and immunization will be discussed more deeply in Section 17.1.

At first, only a few strains of *H. influenzae* were known to cause disease, particularly type b (the strain that caused meningitis in the Chapter 6 opening case). Today, genome sequencing reveals many strains with various disease properties that all arose from mutations in a common ancestral strain (**Figure 8.23**). Building a "family tree," or phylogeny of related strains, is explained in Chapter 9. The colored bars in the figure mark strains "typeable" by six different antibody reactions (types a–f). But many other virulent strains, such as those shown in Figure 8.23, cannot be typed. Their genomes are highly similar—but they turn out to differ in key respects, such as virulence genes that confer aspects of disease.

Differences in gene sequence arise by mutation. An example of such a mutation is shown in **Figure 8.24**. A nontypeable strain of *H. influenzae* isolated from a child's nasopharynx has a virulence gene, *hsdM3*. This gene confers virulence (disease) by an unknown mechanism. But a closely related strain, isolated from bronchial fluids, shows a deletion of five bases (GACGA) from this gene. The number of bases deleted was not divisible by three, so this deletion

Figure 8.23 Relatedness of *Haemophilus influenzae* Strains

A phylogenetic tree was constructed from a partial DNA sequence of 400 isolates of *H. influenzae* from children in Finland with otitis media (ear infection). The sequences were compared by computer analysis. The length of horizontal gray bars represents the sequence difference between each pair of strains, in base pairs per generation (cell division). The longer the bar, the more base differences separate the strains. Colored bars indicate letter types for strains that are encapsulated (typeable by antibody reactions to capsule).

Figure 8.24 Mutation of a *Haemophilus influenzae* Virulence Gene

The initial sequence is shown for the *H. influenzae* virulence gene *hsdM3*. Certain strains of the pathogen isolated from pediatric lung infection show a deletion of five nucleotides that causes a frameshift. The shifted reading frame generates a stop codon, which prematurely terminates synthesis of the peptide.

H. influenzae gene *hsdM3* sequence with deletion mutation

shifts the "reading frame" of downstream codons. The ribosome reads the wrong triplets as codons, one of which looks like a stop codon. The peptide downstream of the frameshift includes the wrong amino acids and terminates prematurely at the inadvertent stop codon. Thus, no functional protein is made. Nonetheless, the *H. influenzae* strain retains virulence, presumably conferred by other genes.

Classes of Mutations

Mutations come in several different physical and structural forms, resulting from different kinds of "mistakes" made by the replication enzymes. The changed sequence of bases changes the information content of the DNA.

How the mutation affects information content depends on how the codons of that particular sequence will be "read" to make a protein. Review the diagram of the genetic code (Figure 8.15). Mutations can change the sequence of DNA bases in several ways, which have various impacts on the sequence of codons:

- In a **point mutation**, or **base substitution**, a single nucleotide is changed in a DNA sequence (**Figure 8.25A and B**). For example, thymine may be replaced by adenine. In the complementary strand, the paired adenine is replaced by thymine.

- **Insertions** and **deletions** involve, respectively, the addition or subtraction of one or more nucleotides (**Figure 8.25C and D**), making the sequence either longer or shorter than it was originally.

- An **inversion** occurs when a fragment of DNA is flipped in orientation relative to DNA on either side (**Figure 8.25E**).

Changes in DNA sequence can result in different amino acids being incorporated into the translated protein. The effects of a given codon change on protein function can be described as follows:

- A **silent mutation** occurs when a base-pair substitution changes a codon, but the new codon specifies the same amino acid as the original codon. For example, all the codons for leucine (Leu in Figure 8.25) start with the same first and second bases (see all possible codons in Figure 8.15). So, any point mutation changing the third letter of a codon for leucine will still encode leucine. The mutation is "silent" because no difference occurs in the protein.

- A **missense mutation** changes the codon and leads to the expression of an altered protein (see Figure 8.25A). For example, in the bacterium *Mycobacterium tuberculosis*, the *rpoD* gene for RNA polymerase can have a mutation replacing C with T, which changes a codon CAC (histidine) to TAC (tyrosine). The *rpoD* gene then expresses a protein containing tyrosine in place of histidine. The result is a rifampicin-resistant bacterium.

- A **nonsense mutation** replaces an amino acid–coding codon with a codon that specifies no amino acid and terminates translation (see Figure 8.25B). The result is a truncated protein, usually nonfunctional.

- A **frameshift mutation** (as in the *Haemophilus influenzae* virulence gene) occurs when insertion or deletion of one or more base pairs shifts the **reading frame** (see Figure 8.25C and D). The reading frame determines which base pairs are read in the first position of the codons. A shift of the reading frame leads to a sequence of completely different amino acids; the drastically altered protein is usually nonfunctional.

How does a mutation (change in **genotype**) affect the **phenotype**, an observable trait of the organism? The effect of a mutation is hard to predict. For example, a small mutation, such as a single base substitution causing a missense codon, can eliminate an entire gene and consequently have a large effect on phenotype. Conversely, a large mutation—such as moving a gene from a plasmid onto a main chromosome—may have little or no effect. For mutations, it's not just the size that counts; it's how the mutation affects gene expression.

Mechanisms of Mutation

What causes mutation? Some mutations happen spontaneously, the result of a "mistake" by the DNA polymerase that fails to be repaired by a proofreading enzyme. Other mutations are caused by physical or chemical agents that damage DNA. Physical agents that damage DNA include cosmic rays, X-rays, and ultraviolet (UV) radiation. X-rays cause large mutations by breaking the DNA backbone and deleting a major part of a chromosome. Such major mutations can destroy tissues. For example, in 2010 the US Nuclear Regulatory Commission reported that a major hospital had accidentally irradiated cancer patients at levels so high that they suffered burns. This incident illustrates why it is so important for an X-ray technician to correctly set the level of an X-ray machine to minimize the dose received by the patient.

UV radiation, such as that received from sunlight or from tanning salons, causes a different kind of DNA damage. UV absorption by an adjacent pair of pyrimidines (T or C) causes electrons to rearrange and form a covalent bond between the two bases. The fused pair of adjacent bases is called a **pyrimidine dimer** (**Figure 8.26**). The pyrimidine dimer causes a "bump" in the helix that fails to be read properly by the DNA polymerase, leading to a gap or an insertion of several incorrect bases in the complementary daughter strand. Such damage may appear subtle, but it can lead to deadly cancers such as melanoma. Melanoma resulting from tanning salons has been seen in people as young as age 18.

Chemical agents that can damage DNA include reactive oxygen molecules such as the hydrogen peroxide (H_2O_2) and superoxide

Figure 8.25 Changes in a DNA Sequence That Result in Different Mutations

A. Missense point mutation: T-to-A is a point mutation that converts the phenylalanine codon TTT to the leucine codon TTA.

B. Nonsense point mutation: C-to-G converts the TCA codon, encoding serine, to the TGA stop codon.

C. Insertion frameshift: The addition of AT into the middle of the TCC serine codon causes a shift in the reading frame that changes all the downstream amino acids.

D. Deletion frameshift: Removing two nucleotides from the arginine codon shifts the frame down.

E. Inversion: Flipping a DNA sequence to face in the opposite direction changes the amino acids produced.

Figure 8.26 **Pyrimidine Dimer Caused by UV Absorption**

The energy from UV irradiation can be absorbed by pyrimidines (thymine and/or cytosine). Two adjacent pyrimidines react to form a covalent connection.

Figure 8.27 **Colonies of Mutator Bacteria**

In an *E. coli* mutator strain, cells are lactose-negative mutants that do not ferment lactose and do not turn the indicator blue. But because the strain has a mutation in one of the mutator genes, a high rate of reversion to the lactose fermenter strain occurs. The mutation is revealed by blue papillae (tiny spots) on the white colony.

radicals ($\cdot O_2^-$) that are produced by white blood cells to kill pathogens. Other agents, such as acridine orange, are "intercalating agents" that slip between two base pairs. An intercalating agent can be "mistaken" for a base pair by the DNA polymerase. The polymerase may then insert an extra base in the daughter strand.

Biological processes can also increase the rate of mutation. Certain strains of a microbe are called **mutators** because their DNA replication process allows a high frequency of mutations. Mutators can be observed when they produce "sectored colonies," in which some cells have inherited a recent mutation, causing a color change. Mutator strains of *Escherichia coli* are shown in **Figure 8.27**. A surprisingly high percentage of mutators arise in pathogens such as *Salmonella*. Mutators have the disadvantage of incurring lots of genetic defects. But when infecting a host, the mutator may express altered surface proteins that evade the host's immune system.

The most highly mutable pathogens known are the RNA viruses, such as influenza virus and human immunodeficiency virus (HIV), discussed in Chapter 12. The high mutation rates of these viruses (caused by error-prone polymerases) prevent their eradication with vaccines.

Repair of Mutations

Mutations are more likely to decrease fitness than to enhance fitness. Some pathogens allow high mutation rates to generate proteins that evade the immune system, but pathogens with the most extreme mutation rates produce surprisingly low proportions of

infectious progeny. For example, less than 1% of HIV virions may be able to infect the next cell.

By contrast, most cellular microbes have repair enzymes that balance the rate of mutation with the need for accuracy. Such enzymes can repair DNA damage before the damage is inherited as a mutation. Many microbial repair enzymes share homologs in the human body, descended from our common ancestral microbe. Inherited defects in human DNA repair enzymes cause the disease xeroderma pigmentosum, in which the skin and eyes are completely sensitive to UV radiation, developing cancer at the slightest exposure.

The type of repair that bacteria use (and when they use it) depends on two things: the type of mutation needing repair and the extent of damage involved. Some repair enzymes excise a whole fragment of DNA that contains damaged bases, whereas others precisely excise just the damaged bases or directly reverse the damage. After extensive damage, special "emergency" DNA polymerases sacrifice accuracy to rescue the broken chromosome. Some examples of DNA repair include:

- **Base excision repair.** In **base excision repair**, an excision repair enzyme recognizes a specific damaged base and removes it from the DNA backbone (**Figure 8.28**). The base excision leaves a phosphoryl sugar (with no base) that a different repair enzyme, an endonuclease, can recognize and cleave. To complete the repair, DNA polymerase I synthesizes a replacement strand containing the proper base, complementary to the template DNA strand. Then DNA ligase seals the strand.

- **Methyl mismatch repair.** The process of **methyl mismatch repair** requires recognition of the methylation pattern in DNA bases. When DNA polymerase misincorporates a base during replication, a mismatch forms. This mismatch occurs between the incorrect base in the newly synthesized but unmethylated strand and the correct base residing in the parental, methylated strand. Methyl-directed mismatch repair enzymes bind to the mismatch and excise the stretch of DNA <u>opposite</u> the methylated strand. DNA polymerase I then fills in the correctly matched nucleotides, and ligase seals the gaps.

- **SOS ("Save Our Ship") repair.** When DNA damage is extensive, the repair strategies that excise pieces of DNA cannot work. The cell must take more drastic measures, known as the SOS response. The SOS response is a coordinated cellular response to DNA damage that, in order to save the cell, can introduce mutations into severely damaged DNA.

Figure 8.28 Base Excision Repair

Specialized enzymes can recognize specific damaged bases and remove them from DNA without breaking the phosphodiester backbone. The result is an abasic site (a phosphoryl sugar with no base) that a specific endonuclease can recognize and cleave. After excision and removal, DNA polymerase I synthesizes a replacement strand containing the proper base.

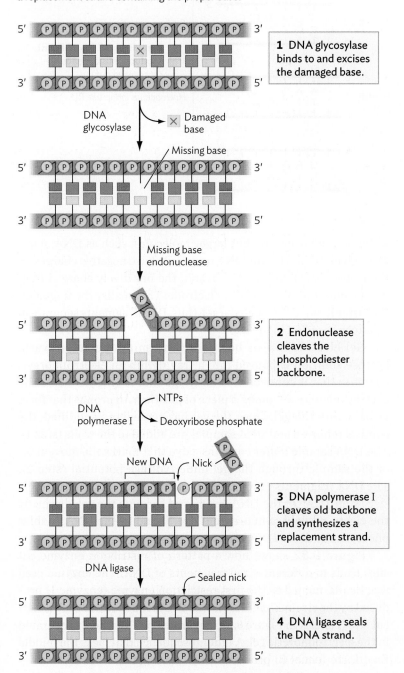

1 DNA glycosylase binds to and excises the damaged base.

2 Endonuclease cleaves the phosphodiester backbone.

3 DNA polymerase I cleaves old backbone and synthesizes a replacement strand.

4 DNA ligase seals the DNA strand.

This error-prone repair does not correct mutations but does maintain the circular chromosome.

Another important repair process involves DNA homologous recombination, the process of crossing over and exchange of two DNA helices that are homologs with nearly identical sequence. (Do not confuse "natural" DNA recombination within cells with "artificial" recombination of DNA through biotechnology—to be discussed in Section 8.6.) The mechanism of DNA recombination is discussed in Chapter 9.

SECTION SUMMARY

- **Mistakes in DNA replication that fail to be corrected result in mutations.** Different kinds of mutations include base substitutions, insertions and deletions, and inversions.
- **Pathogens may have high mutation rates, which help them evade the immune system.**
- **The effect of a mutation on the gene product depends on the genetic code.** A base change in a protein-coding gene may cause a silent mutation, a missense mutation, or a nonsense mutation. A base insertion may cause a frameshift.
- **Mutagens are chemical or physical agents that damage DNA and increase the rate of mutation.** Chemical agents include oxygen radicals that react with DNA. Physical agents include X-rays and UV radiation.
- **Repair systems include base excision repair and methyl mismatch repair.** Error-prone repair such as the SOS system allows base substitutions but preserves the DNA backbone to keep the chromosome intact.

Thought Question 8.4 When bacteria mutate and become resistant to an antibiotic, several different mechanisms of resistance are possible. How many different ways can you think of for a bacterial mutation to cause resistance?

Thought Question 8.5 Explain how it is possible for a very large mutation, involving a large part of a chromosome with many genes, to have no effect on the phenotype of the bacterium.

8.6 Biotechnology

SECTION OBJECTIVES

- Describe the use of plasmids, restriction enzymes, and gel electrophoresis to analyze DNA.
- Explain the formation of a recombinant DNA molecule, and describe applications in medical technology.
- Explain the use of DNA hybridization and DNA sequencing technology.

How can we use DNA to make discoveries in the laboratory?
Isolated DNA can tell us many things, such as the identity of a participant in a crime scene or the presence of a pathogen in food. Analysis of DNA, together with construction of artificial DNA

molecules for an applied purpose, is called **biotechnology**. The tools of biotechnology include physical and chemical methods, such as electrophoresis (discussed shortly), as well as enzymes isolated from bacteria, such as DNA polymerases. Here we describe key tools of DNA analysis: plasmid analysis, DNA hybridization, polymerase chain reaction (PCR) amplification, and DNA sequence analysis.

Plasmids in Biotechnology

Biotechnology makes use of plasmids, such as that shown in **Figure 8.29**. Ironically, our most useful plasmids carry drug resistance genes like those circulating in hospitals. In the lab, a drug resistance gene enables us to culture engineered bacteria with useful properties while preventing the growth of nonengineered strains lacking the plasmid.

A plasmid may have DNA sequences called **restriction enzyme sites**. A restriction enzyme site is a short sequence of base pairs that can be cut only by a specific type of enzyme called a **restriction endonuclease** (**Figure 8.30**). Usually, the restriction sequence is a "palindrome," in which the sequence of base pairs reads the same forward and back.

In nature, restriction enzymes are made by various kinds of bacteria to cleave foreign invading DNA that could be from a virus or an intracellular parasite. In biotechnology, we culture recombinant bacteria engineered to make the enzymes, then purify the enzymes and use them to cut DNA at specific sequences. A cut piece of DNA, such as a plasmid, can be joined to DNA extracted from another organism, forming "recombinant DNA" (discussed shortly).

Gel Electrophoresis

Restriction endonucleases generate DNA fragments that can be separated by a technique called **gel electrophoresis** to yield a predictable pattern (**Figure 8.31**). The gel for DNA separation is formed from agarose, a long-chain polysaccharide that is soluble in water above 50°C (122°F) but forms a solid gel at room temperature. The agarose gel matrix has molecular holes through which a

Figure 8.29 Map of Plasmid pBR322

This plasmid contains an origin of DNA replication and genes encoding resistance to ampicillin (*amp*) and tetracycline (*tet*). p = promoter for each gene.

Figure 8.30 Bacterial DNA Restriction Endonucleases

Target sequences of sample restriction enzymes. The names of these enzymes reflect the genus and species of the source organism. For example, EcoRI comes from *Escherichia coli*.

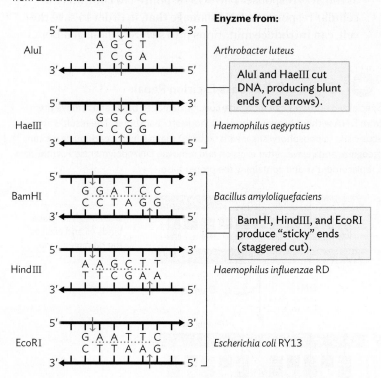

water solution can flow, but larger molecules, such as DNA, move through more slowly. The DNA fragments have negative charges on their phosphoryl groups. As a result, the negatively charged fragments move toward a positive electrode. The smaller the fragment, the faster it moves through the gel; thus, DNA fragments separate in a pattern based on their size.

Gel electrophoresis begins with forming the gel. A warm agarose solution is poured into a tray within an electrophoresis chamber that is connected to a voltage source (Figure 8.31). The gel tray includes a "comb," a piece of plastic with prongs that form wells within the gel. Once the gel has cooled and solidified, the comb is removed and DNA samples are added to the wells (step 1). The DNA sample buffer contains a dye that marks the movement of the sample through the gel under voltage potential (step 2). The DNA fragments move different distances, depending on their size (step 3). After the DNA has moved through the gel, driven by the electrical field, a fluorescent dye bound to the DNA enables photography of the DNA bands.

Figure 8.32 shows how a pattern of restriction enzyme cut sites leads to different-sized fragments of DNA. The enzyme used was EcoRI, named as the first restriction enzyme (enzyme I) produced by the strain *E. coli* R. A linear DNA sequence with one EcoRI cut site (panel A in Figure 8.32) generates two fragments separated in the electrophoretic gel. A mutant sequence that lacks the middle EcoRI site (panel B) produces only one fragment, equal in size to

Figure 8.31 Agarose Gel Electrophoresis Separation of DNA Fragments

DNA samples are loaded into wells at one end of an agarose gel (step 1). When voltage is applied (step 2), the DNA fragments (which have negative charge) are attracted toward the positive electrode (step 3). The fragments move according to size; the smaller fragments move fastest. When the fragments have run far enough to separate, they can be visualized by fluorescence when exposed to UV light (step 4). In the lane at the far left, marker DNA provides fragments of defined length to calibrate the size of the fragments cut in the experiment.

1 DNA sample loading

DNA sample in gel loading buffer

Cathode

Plastic tray

Electrical field (E)

Agarose gel

Anode

2 Application of electrical field (DC)

$t = 0$ min

Power supply (80–120 V)

3 DNA separation

$t = 30$–45 min

Separated DNA fragments

Marker DNA   DNA samples

4 UV transillumination and documentation

UV

Figure 8.32 Restriction Enzyme Cut Sites

A. A DNA sequence with one cut site for the EcoRI restriction enzyme generates two fragments.

EcoRI

7 kb | 3 kb

DNA sequence with one EcoRI site

C. A mixture of both DNA types shows fragments of all three sizes.

EcoRI

7 kb | 3 kb

10 kb

Mix of DNA (A) and (B)

B. A mutant sequence that lacks the middle EcoRI site produces only one large fragment.

10 kb

DNA sequence with no EcoRI site

A   C   B

10 kb
7 kb
3 kb

the sum of the two smaller fragments in panel A. A mixture of both DNA types shows fragments of all three sizes (panel C).

Recombinant DNA

Another use of restriction enzymes is to construct **recombinant DNA**, a DNA molecule containing sequences from different source organisms, combined "in a tube" outside any living cell. The idea of recombining DNA from different organisms leads to fanciful ideas of reshaping life (**Figure 8.33**).

In fact, recombinant microbes are a major industry because they are engineered to produce commercial products ranging from insulin to vaccines. For example, the vaccine against human papillomavirus (HPV), now recommended for all boys and girls before puberty, consists of virus coat proteins expressed by viral genes in recombinant yeast (Gardasil, by Merck) or in recombinant baculovirus in insect cell culture (Cervarix, by GlaxoSmith-Kline). These vaccines are much safer than traditional vaccines using live or inactivated viruses because their manufacture uses no viruses, but only coat proteins made from recombinant microbes that cannot infect humans.

Recombinant DNA generally requires a **vector** molecule to carry the DNA sequence of interest (**Figure 8.34**). The vector may

Figure 8.33 **Results of Recombinant DNA?**

The imagined results of using recombinant DNA to modify animal genomes.

Figure 8.34 **Recombinant DNA**

A plasmid vector is cut with a restriction enzyme. The same restriction enzyme is used to cut foreign DNA. The "sticky ends" of the restriction-cut foreign DNA hybridize to the plasmid ends. The enzyme DNA ligase seals the backbone of the DNA molecule, completing the recombinant plasmid.

Figure 8.35 **DNA Hybridization**

Hybridization of a labeled DNA probe (short DNA sequence) to its complementary sequence in a target DNA.

be a plasmid, such as the plasmid pBR322, shown in Figure 8.29, or alternatively, the genome of a virus. The vector DNA can be cut with a restriction enzyme that has only one cut site (unique sequence) in the plasmid DNA sequence. The plasmid is now linearized, with "sticky ends" that can **hybridize**, or base-pair—that is, bind to a complementary sequence of bases from a different helix (**Figure 8.35**). The ends hybridize to the single-stranded ends of other DNA cut with the same enzyme.

A piece of foreign DNA—from any species of organism—can be cut with the same restriction enzyme, or alternatively, nucleotides can be added onto the ends of a DNA fragment to generate the same sticky ends. Now the piece of foreign DNA is combined in the same test tube with the cut vector plasmid. Some of the cut plasmids will hybridize to both ends of a piece of foreign DNA, re-forming a circle. The DNA circle is now completed by the enzyme DNA ligase, which seals the "nicks" in the backbone where the sticky ends hybridized. The recombinant plasmid can then be put back into a microbial cell by a process called **transformation**. Transformation generates a cell that can now express the protein product of a gene in the recombinant plasmid.

A somewhat different form of recombinant DNA is called "complementary DNA," abbreviated cDNA. Complementary DNA is reverse-transcribed from a messenger RNA molecule and then inserted or cloned into a vector. The cDNA technology requires use of a reverse transcriptase enzyme that makes a DNA copy from RNA.

DNA Hybridization

How can we use DNA to identify a pathogen? In the clinical lab, a quick test for certain pathogens can be performed by DNA hybridization with a known "probe" of a pathogen gene. For example, food can be tested for DNA from enteric pathogens such as *E. coli* O157:H7 or other strains having a gene that encodes Shiga toxin.

To analyze a DNA sample, researchers can hybridize the sample (called the "target") to a short sequence (the probe), which has an attached fluorophore (fluorescing molecule; Figure 8.35). For example, the target DNA could be a sample from food, whereas the probe DNA could be a gene encoding Shiga toxin. The target DNA is first denatured (the helix melted) so that single strands come apart, exposing their bases. The DNA is then bathed in a solution containing the labeled probe DNA, which can base-pair specifically (hybridize) to a complementary sequence in the target DNA. The hybridization reaction is then detected by fluorescence. Alternative labels can be a radioisotope or an enzyme that reacts with a detectable substrate.

Hybridization techniques are rapid and inexpensive and can detect microbial pathogens in a large sample. But in clinical isolates, especially from infected patients, the pathogenic cells are few, with vanishingly small amounts of DNA. A more sensitive detection method is needed.

Polymerase Chain Reaction Amplification

Imagine how easily you could diagnose illness if you could detect one single copy of DNA in a patient's tissue sample. You might be able to do this by first making a million copies in just 2 hours. A million copies of DNA are easy to detect by fluorescent dyes. This "amplification" of DNA offers a way to detect a pathogen.

CASE HISTORY 8.3

Spotting Surprise

In Wichita, Kansas, a 20-year-old college student named Lillian visited the college health practitioner for a routine gynecological exam. Lillian's periods had been regular, but she had recently noted some spotting between periods. Her most recent menstrual period had started 4 weeks earlier. The practitioner asked Lillian whether she was sexually active. Lillian reported two male partners within the previous 6 months. She denied having any vaginal discharge, dyspareunia (painful intercourse), genital lesions, or sores. Her breast, thyroid, and abdominal exam were within normal limits, as were her vital signs: blood pressure 118/68, pulse 74, respiration 18, temperature 37.1°C (98.8°F).

The genital exam revealed healthy vulva and vagina. The practitioner found no cervical motion pain and no uterine or adnexal tenderness (areas that include and surround the fallopian tubes and ovaries). But Lillian's cervix appeared inflamed and bled easily, with a purulent discharge coming from the cervical os (opening).

The practitioner performed nucleic acid amplification tests (NAATs) for *Chlamydia trachomatis* and *Neisseria gonorrhoeae*. Within a day, the NAAT results were positive for *Chlamydia* and negative for *N. gonorrhoeae*. Wet mount microscopy revealed no pathogens but many white blood cells (WBCs). The diagnosis was confirmed as chlamydia infection, an infection reportable to the Centers for Disease Control and Prevention (CDC).

Doxycycline was prescribed to clear up the infection. To curtail transmission, the practitioner asked Lillian to identify all her sexual partners for partner notification. Lillian recalled a third partner 7 months previously. The practitioner also recommended use of condoms to decrease further infections.

DNA amplification tests are now routine in thousands of laboratories around the world—such as the laboratory available for Lillian's testing in this case history. The DNA amplification in the NAAT did not require culture or even microscopic detection of the pathogen, an obligate intracellular bacterium. The test enabled Lillian's practitioner to diagnose her condition within a day of Lillian's visit to the clinic. Detection of the correct pathogen is important because if Lillian had been treated with an inappropriate antibiotic, her chlamydia infection could have persisted, ultimately causing scarring of her fallopian tubes that could lead to infertility. Undiagnosed chlamydia infection is a major cause of infertility in the United States (see Chapter 23).

NAAT is based on a method of DNA amplification called the **polymerase chain reaction (PCR)**. The DNA amplification technique makes use of heat-stable DNA polymerase isolated from thermophiles, bacteria, and archaea adapted to live in extreme heat. PCR was used in the opening case history of this chapter to detect the various kinds of bacteria present in the patient's diseased gums. PCR has revolutionized biological research, medicine, and forensics; it commonly identifies or exonerates criminal suspects.

In PCR, specific DNA primers (usually 20–30 bases) are hybridized to known DNA sequences flanking the target gene (**Figure 8.36**). A heat-stable DNA polymerase is used, such as the DNA polymerase I of *Thermus aquaticus*, a thermophilic bacterium (Taq polymerase). Heat stability enables Taq polymerase to survive the first step of the reaction cycle: heating to 95°C (203°F), a temperature at which the DNA helix melts and the strands separate so that their bases are exposed to pair with primers. Next comes step 2, cooling to 55°C (131°F) to allow primer hybridization, followed by step 3, heating to 72°C (162°F), the optimum reaction temperature for the polymerase. The PCR reaction is then repeated for 25–30 cycles, with precisely regulated temperatures, in a machine called a thermal cycler. Each cycle doubles the number of copies of the DNA sequence positioned between the forward and reverse primer ends. Thus, a single DNA template can generate tens of millions of DNA copies—enough for us to see a band on an electrophoretic gel.

Figure 8.36 Polymerase Chain Reaction (PCR)
The cyclic PCR reaction makes a large number of copies of a small piece of DNA.

2 55°C, 30 seconds: Primers hybridized.

Segment to amplify

Primers

72°C, Taq polymerase

30 cycles

1 95°C, 30 seconds: Denaturation separates strands.

3 72°C, 60 seconds: Taq polymerase replicates sequence.

95°C, denaturation
55°C, hybridize primers

95°C, denaturation;
55°C, primer hybridization;
72°C, extension by Taq polymerase

PCR amplification offers an exciting way to detect pathogens present in minute amounts in samples taken from patients who do not yet show symptoms, such as patients with tuberculosis or with HIV infection. The technique can even determine whether the pathogens possess a gene encoding antibiotic resistance. See **IMPACT** for an example of an advanced PCR

machine developed to screen for tuberculosis in developing countries.

DNA Sequencing

How can we "read" the actual sequence of a piece of DNA? DNA sequencing can now tell us the entire genome of a pathogen, revealing unknown genes encoding novel toxins, or it can reveal antibiotic production genes in soil microbes such as *Streptomyces* species.

To read a DNA sequence, we start with a template DNA in a reaction similar to that of PCR. The reaction contains DNA polymerase and dNTPs (deoxy nucleotide triphosphate). But in addition to the dNTPs, a small amount of "dideoxy" nucleotide is included for one of the four dNTPs, such as dATP. "Dideoxy" means that the 3′ OH is replaced by H. Thus, after the dideoxynucleotide is incorporated into DNA, no further 3′ OH is available to add the next nucleotide. Four reactions are conducted in the same tube, with a different fluorescent label for each of the four dideoxynucleotides (ddATP, ddTTP, ddCTP, or ddGTP).

The DNA polymerase then synthesizes DNA on the basis of the template, such as a PCR-amplified DNA sample from a bacterium. But each sequencing reaction contains just enough dideoxy ATP (or CTP, TTP, or GTP) to cause DNA polymerase occasionally to substitute the dead-end base for the natural one, at which point the chain stops growing. The result is a population of DNA strands of various size, each one cut short at a different adenine position.

Now imagine using four different dideoxynucleotides, corresponding to A, T, C, and G, each one tagged with a different-colored fluorescent dye (**Figure 8.37**). The result would be a series of different-sized strands tagged at their 3′ end with different colors, depending on the base incorporated. Electrophoresis for each of the four reactions will separate the various fragments according to size. A laser and detector positioned at the bottom of the electrophoretic gel read the individual fragments as they pass by. The computer

Figure 8.37 DNA Sequencing

The DNA molecule is used as a template to synthesize complementary DNA in a reaction containing fluorescently tagged nucleotides of one of the four types (A, T, C, or G). The tagged nucleotide terminates a DNA sequence as a short piece. The short pieces of DNA are separated by size by electrophoresis, appearing as bands in the gel. A laser detects the different fluorescently tagged DNA pieces, using their four different wavelengths of light emission.

Single-stranded DNA to be sequenced
5′ C T G A C T T C G A C A A 3′

Add DNA polymerase, dATP, dGTP, dCTP, dTTP; plus limiting amounts of fluorescently labeled **ddATP, ddGTP, ddCTP, ddTTP**

Chain elongation stops with incorporation of dideoxynucleotide.

Electrophoresis is performed. Bands pass through a laser to activate the fluorescent dideoxynucleotide and then through a detector to distinguish the colors.

Larger fragments → Smaller fragments

3′ G A C T G A A G C T G T T 5′

So the sequence of the template strand is

5′ C T G A C T T C G A C A A 3′

then prints a series of colored peaks whose order corresponds to the template DNA sequence.

The method described here to read DNA sequences is called Sanger sequencing. Today, even faster technologies, such as Illumina sequencing, allow rapid sequencing of whole genomes.

SECTION SUMMARY

- **A restriction endonuclease is an enzyme that cleaves DNA at a specific sequence,** which is usually a palindrome.

- **Gel electrophoresis separates DNA fragments,** enabling identification of patterns of restriction enzyme cut sites in a DNA sequence.

- **Plasmids are used as vectors.** For biotechnology, a vector carries a short piece of DNA from one organism into another.

- **Recombinant DNA is made by inserting a piece of DNA into a plasmid or viral DNA cut with the same restriction enzyme.** The DNA backbone is sealed with DNA ligase. The inserted piece of DNA has a gene that can be expressed by a cell containing the recombinant plasmid or infected by a recombinant virus.

- **DNA hybridization can identify a source of DNA by base-pairing between a target DNA and a labeled probe DNA.** The probe DNA must be complementary to a sequence within the target.

- **The polymerase chain reaction (PCR) amplifies (makes many copies of) a short DNA sequence** between two primer sequences directed toward each other. A heat-resistant polymerase catalyzes the synthesis of DNA from each primer, and thermal cycling allows exponential amplification.

- **DNA sequencing** is a technique to read the sequence of base pairs in a DNA molecule. The sequence of a template DNA can be read by using reactions of DNA synthesis by a DNA polymerase.

Perspective

In this chapter you have learned how microbial DNA is maintained through reproduction. The DNA genes are then expressed as RNA to make protein products. Gene expression is regulated by the environment and by host factors for disease.

Despite accuracy and proofreading, DNA replication makes occasional errors. Mutation provides one of many mechanisms that pathogens use to colonize a host and evade the immune system. Protein and RNA synthesis machines critical for bacterial growth are targets for many antibiotics—a subject discussed more in Chapter 13.

The genes and enzymes of bacterial and viral genetics provide tools for biotechnology. Restriction endonucleases enable us to define pieces of DNA, whereas PCR may generate a million copies of a gene. Genes that express valuable proteins can be cloned into plasmid vectors for production of the protein for therapy. In Chapter 9 you will see how bacteria transfer genes to other cells and how bacterial gene transfer mechanisms provide additional tools to manipulate DNA.

impact

Finding Tuberculosis DNA in a Drop of Spit

Tuberculosis is a global plague; in India alone, 2 million people fall ill with the disease every year. But how do we detect this insidious disease, identify the victims, and treat them? The pathogen, *Mycobacterium tuberculosis*, grows slowly and can be present in extremely low numbers. The most common test, a sputum slide assay, misses half the cases of disease. Even when a case is identified, how do we provide the right antibiotics? It takes 2 weeks to culture *M. tuberculosis* to know which antibiotics will be effective. If the wrong antibiotic is prescribed, it will select only for the spread of a resistant strain.

An international consortium of scientists, led by the Foundation for Innovative New Diagnostics, in Geneva, Switzerland, has developed a new instrument to identify minute quantities of *M. tuberculosis* DNA. The instrument, called the Xpert MTB/RIF (Figure 1), uses advanced PCR technology to amplify tiny amounts of DNA in a drop of sputum (deep lung material that can be coughed up). As few as five bacterial genomes can provide enough DNA for amplification (Figure 2).

Figure 1 Detecting Tuberculosis Bacteria with or without Rifampicin Resistance

1 Sputum liquefaction and inactivation with 2:1 sample reagent.

5 Ultrasonic lysis of filter-captured organisms to release DNA.

4 Sample automatically filtered and washed.

6 DNA molecules mixed with dry PCR reagents.

7 Real-time PCR amplification and detection in integrated reaction tube.

2 Transfer of 2 ml of material into test cartridge.

3 Cartridge inserted into MTB/RIF test platform (end of hands-on work).

8 Printable test result.

Time to result: 1 hour, 45 minutes

Figure 2 Real-Time PCR Detects Tuberculosis Bacteria

Mycobacterium tuberculosis genomic DNA samples were amplified.

In the Xpert instrument, PCR primers are used that contain sequences flanking the gene *rpoD*, which encodes RNA polymerase. As discussed in Section 8.3, RNA polymerase is a central enzyme of gene expression, transcribing all bacterial genes to make RNA copies; messenger RNA copies are then decoded by ribosomes to make proteins.

The unique *rpoD* sequence is amplified by an advanced technology called real-time PCR. "Real-time" means that each round of amplification can be measured as it occurs because the fluorescent label detection is extremely sensitive. Thus, we observe the DNA concentration as it rises. The earlier the DNA signal rises (the lower the cycle number), the greater the concentration of DNA in the original sample.

Once the *rpoD* DNA is detected, its sequence is checked for a key mutation that confers resistance to the drug rifampicin. If the resistance sequence appears, then physicians know immediately to exclude rifampicin from the drug combination that is prescribed. The entire process of the Xpert instrument takes under 2 hours; all the steps are automated, requiring no advanced knowledge to run. In a study reported in the *New England Journal of Medicine*, the instrument detected tuberculosis in 98% of a patient group in Uganda and correctly identified all rifampicin-resistant cases. It can be run in remote villages by individuals with modest training. Could this process be made available widely? Providers such as the Bill & Melinda Gates Foundation are trying to do just that.

DISCUSSION QUESTION

Once a physician receives the test result for rifampicin resistance, what next steps will she need to take to select the best therapy for the tuberculosis patient?

Key Terms

activator (p. 228)
base excision repair (p. 234)
base substitution (p. 232)
biotechnology (p. 236)
chromosome (p. 214)
coding strand (p. 222)
daughter strand (p. 218)
deletion (p. 232)
DNA ligase (p. 221)
DNA polymerase (p. 218)
elongation (p. 218)
ESKAPE pathogen (p. 216)
frameshift mutation (p. 233)
gel electrophoresis (p. 236)
gene (p. 215)
genome (p. 214)
genotype (p. 233)

global regulator (p. 230)
gyrase (p. 221)
homolog (p. 217)
hybridize (p. 238)
induction (p. 228)
initiation (p. 218)
insertion (p. 232)
inversion (p. 232)
lagging strand (p. 219)
leading strand (p. 219)
messenger RNA (mRNA) (p. 222)
methyl mismatch repair (p. 234)
missense mutation (p. 232)
mutation (p. 231)
mutator (p. 234)
nonsense mutation (p. 233)
nucleobase (p. 214)

nucleotide (p. 214)
origin (oriC) (p. 218)
parental strand (p. 218)
peptide bond (p. 227)
phenotype (p. 233)
plasmid (p. 216)
point mutation (p. 232)
polymerase chain reaction (PCR) (p. 239)
promoter (pp. 215, 222)
pyrimidine dimer (p. 233)
reading frame (p. 233)
recombinant DNA (p. 237)
replication fork (p. 218)
repression (p. 228)
repressor (p. 228)
restriction endonuclease (p. 236)

restriction enzyme site (p. 236)
RNA polymerase (p. 222)
RNA primer (p. 219)
secreted (p. 227)
sense strand (p. 222)
sigma factor (p. 223)
silent mutation (p. 232)
supercoil (p. 221)
template strand (p. 222)
termination (p. 218)
transcription (p. 222)
transfer RNA (p. 225)
transformation (p. 238)
translation (p. 224)
vector (p. 237)

Objectives Assessment Questions

8.1 Bacterial Genomes

1. Which aspect of bacterial genome structure is the <u>same</u> in eukaryotes?
   a. The genome is composed of double-helical DNA.
   b. Genes are often arranged in tandem, within an operon.
   c. The genome is compact, with genes interspersed with relatively short noncoding sequences.
   d. Most genomes are circular.

2. Which statement about plasmids is <u>not</u> correct?
   a. Plasmids often carry genes encoding enzymes that confer resistance to an antibiotic.
   b. Plasmids are small circular pieces of DNA.
   c. When a plasmid is transferred into a new cell, it usually changes the species of the host organism.
   d. Plasmids can be used as vectors for biotechnology.

3. Which kind of sequence interrupts genes of eukaryotes but not bacteria?
   a. Plasmid
   b. Intron
   c. Primer
   d. Promoter

8.2 DNA Replication

4. Which step of DNA replication occurs <u>before</u> DNA polymerase extends the chain?
   a. An RNA primer is synthesized.
   b. A lagging strand is synthesized.
   c. Proofreading enzymes check the DNA.
   d. The RNA primers are replaced with DNA sequence.

5. The function of gyrase is _____.
   a. to prime DNA synthesis
   b. to connect Okazaki fragments
   c. to generate negative supercoils in DNA
   d. to initiate RNA synthesis

6. The function of ligase is _____.
   a. to prime DNA synthesis
   b. to connect Okazaki fragments
   c. to generate negative supercoils in DNA
   d. to initiate RNA synthesis

8.3 Gene Expression: From DNA to RNA and Protein

7. Which kind of RNA product of a gene will be translated by a ribosome to make protein?
   a. Transfer RNA (tRNA)
   b. Ribosomal RNA (rRNA)
   c. Messenger RNA (mRNA)
   d. Regulatory small RNA (sRNA)

8. Which step of translation is blocked by tetracycline?
   a. The ribosome 30S subunit binds to mRNA.
   b. The incoming aminoacyl-tRNA binds the A site of the ribosome.
   c. The peptide bond forms, catalyzed by the ribosome.
   d. The terminated peptide chain falls off the ribosome.

9. Which step of translation requires only the 30S subunit, not yet the entire ribosome?
   a. The first formylmethionyl-tRNA binds to the mRNA.
   b. The A site forms.
   c. A peptide bond forms.
   d. The peptide chain moves out of the P site.

8.4 Regulation of Gene Expression

10. Which kind of gene regulation does <u>not</u> change the rate of gene transcription to RNA?

    a. A repressor binds a promoter sequence.

    b. An activator binds a promoter sequence.

    c. A two-component regulator binds a signaling molecule.

    d. A translation initiation sequence binds a translational repressor.

11. Which kind of regulation has a long-term effect on a pathogen's genome and helps it evade the immune system?

    a. An activator binds a promoter.

    b. DNA phase variation occurs.

    c. A two-component regulator detects iron.

    d. A virulence protein is modified.

12. Which kind of gene regulation must involve the transfer of protein phosphoryl groups?

    a. A corepressor binds a repressor to halt transcription.

    b. An inducer removes a repressor to enable transcription.

    c. An activator protein enhances transcription.

    d. A two-component regulator transmits information from an external signal to the DNA inside the cell.

8.5 Mutation and Repair

13. Which kind of mutation does not change the protein encoded by a gene?

    a. Missense mutation

    b. Silent mutation

    c. Deletion mutation

    d. Frameshift mutation

14. Which kind of bacterial DNA repair leads to multiple mutations?

    a. Base excision repair

    b. SOS repair

    c. Methyl mismatch repair

    d. Proofreading of DNA during replication

15. One type of mutation that can cause a frameshift is _____.

    a. deletion of a base pair

    b. substitution of a base pair

    c. silent mutation

    d. missense mutation

8.6 Biotechnology

16. Which technique is commonly used to reveal the size of a DNA fragment?

    a. Gel electrophoresis

    b. Recombinant DNA formation using a vector

    c. DNA hybridization

    d. DNA sequencing

17. Which technique allows same-day identification of a clinical isolate from a patient?

    a. Recombinant DNA formation

    b. Culturing the isolate

    c. Gel electrophoresis of a genome fragmented by restriction enzymes

    d. PCR amplification of a bacterial gene

18. For DNA sequencing, the nucleotides may be labeled with _____.

    a. Gram stain

    b. methylene blue stain

    c. electron-dense metal stain

    d. fluorophores (fluorescent molecules)

Review Questions

1. What is a genome? What is the structure of a genome in bacteria and in eukaryotes?

2. What are plasmids? How do they function in cells and in microbial communities? What kinds of traits are commonly encoded on plasmids?

3. What is a restriction endonuclease? Explain how restriction endonucleases can be used to define specific fragments of DNA.

4. Explain the polymerase chain reaction (PCR) and how it is used to identify a pathogen.

5. For DNA synthesis, explain the steps of initiation, elongation, and termination.

6. Explain how different kinds of mutation occur.

7. How do mutations change the information contained in DNA?

8. How are mutations repaired?

9. Explain how RNA is synthesized on the basis of the DNA gene. What is required to initiate, elongate, and terminate RNA synthesis?

10. How does the ribosome translate a message to make protein? Explain the steps of initiation, elongation, and termination for this process.

11. How are proteins such as toxins made by a bacterial pathogen?

Clinical Correlation Questions

1. A patient presents with bacteremia caused by *Enterococcus faecium*. Why is this pathogen extremely concerning for hospitals? For treatment, what kind of tests must be performed?

2. A patient reports spotting between periods, and she undergoes a genital exam. Her cervix is inflamed, with a purulent discharge. Microscopy reveals Gram-negative paired cocci (diplococci). What is the likely pathogen? Which test will confirm it?

9

Bacterial Genomes and Evolution

Diphtheria, Toxins, and Natural Disasters

SCENARIO When 15-year-old Oriel arrived at the clinic in Port-au-Prince, the doctors were not expecting diphtheria, an ancient disease almost never seen today. But this was Haiti. A major earthquake in 2010 had destroyed much of the country's infrastructure and many thousands of homes. The city lacked adequate sanitation, and tens of thousands of people, including Oriel, were living in tent cities—in conditions ripe for spreading infectious diseases. Oriel had started feeling ill 6 days earlier with a sore throat and fever. Everyone thought it was a cold, but Oriel kept getting worse.

Diptheria
A symptom of diphtheria is an inflammatory response in the throat that produces a grayish membrane. This pseudomembrane can obstruct airflow into the trachea.

SIGNS AND SYMPTOMS Details are sketchy, but when he arrived at the clinic, Oriel had a fever of 38.9°C (102°F) and a "bull neck" from enlarged lymph nodes. He felt very weak and had a thick, gray membrane (pseudomembrane) forming at the back of his throat, making breathing difficult.

DIAGNOSIS Oriel had contracted diphtheria, a disease easily prevented by vaccination. But only 50–60 out of every 100 Haitian children were vaccinated before the quake. The disease is caused by the Gram-positive rod *Corynebacterium diphtheriae*. This pathogen grows in the throat but secretes a powerful toxin that circulates and destroys the victim's cellular protein synthesis machinery. The gene encoding diphtheria toxin (*dtx*) is actually carried by a virus called beta phage, whose DNA inserts itself into the chromosome of *Corynebacterium*.

TREATMENT As Oriel's condition worsened, a frantic search began for lifesaving antitoxin. This search, through relief efforts led by actor Sean Penn, fell short. Oriel died on May 7, 2010. Thankfully, as of 2017, the diphtheria vaccination rate had risen to about 70%.

SEM 1 μm

Corynebacterium diphtheriae
The causative agent of diphtheria, *Corynebacterium diphtheriae*, harbors DNA from a bacterial virus that carries the gene for diphtheria toxin.

CHAPTER OBJECTIVES

After reading this chapter, you will be able to:

• Describe how genes move between genomes by the processes of transformation, transduction, conjugation, and transposition.

• Explain the evidence for how microbial life originated and diversified.

• Describe how bioinformatics uses genomic information to predict traits of related organisms.

• Explain microbial taxonomy.

The unfortunate case of Oriel's diphtheria infection illustrates two major topics we will discuss in this chapter: the horizontal transfer of genes and traits between bacteria, and concepts of pathogen evolution. **Horizontal gene transfer** is the transfer of genes from one organism to another coexisting organism. Oriel succumbed to infection by *Corynebacterium*, a species that would be harmless without the "hitchhiking" beta phage DNA that had horizontally transferred the *dtx* toxin gene. The *dtx* gene is an example of a virulence gene—a gene whose product helps a pathogen cause disease. Horizontal transfer of virulence genes is a common way that pathogens evolve from otherwise harmless species.

In this chapter we will discuss how bacterial genes, such as *dtx*, are moved between bacteria by bacteriophages (bacterial viruses, also called "phages"). You will also learn how we identify genes that were transferred many thousands of years ago. In addition, we will consider how the horizontal transfer of genes is involved in the evolution of microbes, including modern-day pathogens and essential host microbiota.

9.1 Bacterial Gene Transfers

SECTION OBJECTIVES

• Explain how genetic information moves between bacteria by transformation, phage-mediated transduction, and conjugation.

• Describe how DNA molecules recombine by homologous recombination and site-specific recombination.

• Describe how gene mobility and DNA recombination lead to the emergence of new pathogens.

How does nature play Frankenstein, making genomes change as microbes evolve? Dr. Frankenstein in Mary Shelley's nineteenth-century novel created his monster by stitching together body parts from the graves of townsfolk. Nature does something in a similar, but less sinister, way. As we learned in Chapter 8, genomes change and evolve gradually through spontaneous mutations, such as errors in DNA replication or proofreading. Such mutations alter the DNA sequence that cells inherit from their parents.

We discover genome change by comparing DNA sequences—a field of study called bioinformatics. Bioinformatic analysis has shown that microbial genomes evolve via small spontaneous mutations, but also by gaining or losing entire genes, and by gene rearrangements and duplications. Much like Shelley's conjuring of Frankenstein's monster, genetic material from various species of bacteria can be acquired randomly and stitched by recombination into a living cell's genome to produce a new species. These rapid, large-scale changes are usually mediated by genetic elements such as viruses and transposons, which you will learn about in this chapter.

Large-scale gene traffic between species is known as horizontal gene transfer. Horizontal gene transfer, recombination, mutation, and mistakes during DNA repair are all processes that accelerate natural selection (described later in the chapter). In this section we will describe the mechanisms used to move genes between microbes and then explain how mobile genes contribute to microbial evolution.

Transformation

In 1928, a perceptive English medical officer, Fred Griffith (1879–1941), found that he could kill mice by injecting them with <u>dead</u> cells of a virulent pneumococcus (*Streptococcus pneumoniae*), a cause of pneumonia, but only if he also injected <u>live</u> cells of an avirulent (nonvirulent) mutant (**Figure 9.1**). Even more extraordinary was the discovery that the bacteria recovered from the dead mice were of the live, <u>virulent</u> type. Were the dead bacteria brought back to life? Unfortunately, Griffith was killed by a German bomb during an air raid on London in 1941 and never learned the answer.

In a landmark series of experiments published in 1944, physician and medical researcher Oswald Avery (1877–1955), and geneticists Colin MacLeod (1909–1972) and Maclyn McCarty (1911–2005) proved that Griffith's experiment was not a case of reviving dead cells. Instead, DNA released from the dead, virulent bacteria was being transferred into the harmless, but living, strain of *S. pneumoniae*—an event that transformed the live strain into a killer. To prove this, they added an enzyme that degraded DNA (DNase) to extracts of virulent *S. pneumoniae* and then added the treated extract to live, avirulent pneumococci. Extracts that were <u>not</u> treated with DNase transformed avirulent pneumococci into virulent forms; however, DNase-treated extracts could not

Figure 9.1 The Griffith Experiment

Virulent bacteria that possess a virulence capsule and avirulent mutant bacteria that do not were injected into mice. Control experiments 1–3 produced the expected results; live virulent streptococci killed mice (2), but heat-killed virulent bacteria (1) and live avirulent bacteria (3) did not. Experiment 4, however, revealed that mixing dead virulent bacteria with live avirulent bacteria enabled genetic material to be transferred from the dead virulent bacteria to the live avirulent bacteria. The material somehow transformed the live avirulent bacteria into a capsule-producing, virulent form, which was recovered from affected mice.

achieve this transformation. The results proved that the genetic material was DNA, not protein, RNA, lipids, or polysaccharides. The process of importing free DNA into bacterial cells is now known as **transformation**. Transformation provided the first clue that gene transfer can occur in microorganisms.

Transformation can happen in two ways: naturally or forced. Many bacteria naturally import DNA fragments and plasmids released from nearby dead cells via transformation. These species use specific protein complexes in their membranes to take up DNA from the surrounding environment. Examples include Gram-positive bacteria such as *Streptococcus* and *Bacillus,* as well as Gram-negative organisms such as *Haemophilus* and *Neisseria* species.

Other bacteria, however, such as *E. coli* and *Salmonella,* do not import DNA efficiently. These bacteria require artificial manipulations in the lab to take up DNA into the cell. These laboratory techniques include disturbing the membrane by chemical and electrical methods. The chemical mechanism involves using calcium

chloride ($CaCl_2$) and heat shock, which alters the cell membrane, allowing DNA to pass. **Electroporation**, on the other hand, uses a brief electrical pulse to "shoot" DNA across the membrane. These methods are used in the laboratory to move recombinant DNA plasmids (described in Section 8.1) between bacteria.

Why do some species, such as *Neisseria,* undergo natural transformation? Organisms that indiscriminately import DNA may consume the transformed DNA as food. More finicky species transform only donor DNA molecules with sequences very similar (**homologous**) to their own. These microbes use DNA released from dead companions to repair their own damaged genomes. Yet other bacteria may use transformation to obtain new genes from different species. These new genes may help the transformed bacteria survive in a changing environment by providing a new function that confers a growth advantage. Pathogens such as *Neisseria gonorrhoeae,* the cause of gonorrhea, have used transformation to acquire new genes whose protein products help the organism evade the host immune system. Horizontal gene transfer is different from **vertical gene transfer**, the inherited passing of genes from parent to offspring, as in cell division.

Gram-positive organisms transform DNA by using a protein complex, called a transformasome, that spans the membrane (**Figure 9.2**). The transformasome complex assembles only after a certain number of cells (called a quorum) have gathered. But how do they know? It's much like our sense of smell. A certain number of odor molecules from a pizza must interact with our olfactory sensors before we know to look for the pizza shop. As for transformation, all bacterial cells in a population synthesize a peptide, called **competence factor**, that progressively accumulates in the medium Figure 9.1, step 1). Above a certain concentration threshold, competence factor binds to the sensory protein of a two-component signal transduction system (Section 8.4) that then stimulates expression of the sigma factor sigH and the transformasome genes (steps 2 and 3). The transformasome is made, and the cells are now competent, meaning they are ready to import DNA. Note that the more cells per volume in a culture, the more quickly competence factor will accumulate and trigger transformasome synthesis.

The accumulation of competence factor is yet another example of the phenomenon of quorum sensing (see Section 6.6), in which a small molecule made by the cell accumulates outside the cell before it affects transcription inside the cell. It is called quorum sensing because a certain number of cells (a quorum) must be reached before enough signal factor is made to evoke a change in gene expression. Many pathogens use quorum sensing to activate the transcription of virulence genes.

Once made, the transformasome captures extracellular DNA floating in the environment. A nuclease (DNA-cutting enzyme) degrades one strand of the incoming double-stranded DNA molecule while pulling the other strand intact into the cell. The energy for taking up DNA is supplied by hydrolysis of ATP. Once inside, the strand can be incorporated into the chromosome by recombination, a process we will discuss shortly.

Figure 9.2 **Transformation in *Streptococcus***

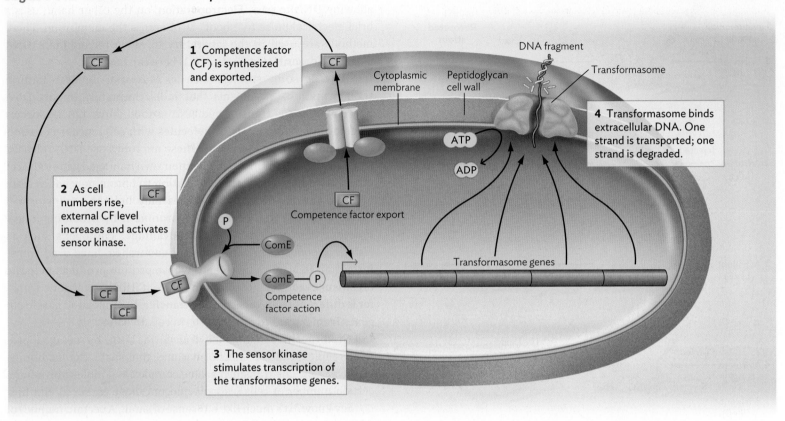

1 Competence factor (CF) is synthesized and exported.

Cytoplasmic membrane

Peptidoglycan cell wall

DNA fragment

Transformasome

4 Transformasome binds extracellular DNA. One strand is transported; one strand is degraded.

ATP

ADP

2 As cell numbers rise, external CF level increases and activates sensor kinase.

Competence factor export

ComE

ComE

Competence factor action

Transformasome genes

3 The sensor kinase stimulates transcription of the transformasome genes.

The process of transformation in this organism begins with the synthesis of a signaling molecule (competence factor) and concludes with the import of a single-stranded piece of DNA through a transformasome-protein complex.

Some bacteria, such as Gram-negative species, transform without competence factors. Either they are always competent, like *Neisseria*, or they become competent when starved (as do *Haemophilus* species, a cause of pediatric sepsis and meningitis, presented in Case History 8.2).

Conjugation

Another way DNA can be transferred from one bacterium to another is by **conjugation**. DNA transfer by conjugation requires intimate cell-to-cell contact between a donor and a recipient. DNA from the donor passes into the recipient through an assemblage of proteins on the donor's surface. It sounds like sex, but it's not. Conjugation occurs in many species of bacteria and archaea. In nature, mixed-species biofilms may enable cross-species DNA transfer, albeit at a low frequency. Conjugation among members of mixed-species biofilms is one way antibiotic resistance genes move between species. For instance, subinhibitory concentrations of antibiotics such as gentamicin or ciprofloxacin have been reported to stimulate conjugational transfer of antibiotic resistance between the common gut microbe *Escherichia coli* and the opportunistic pathogen *Pseudomonas aeruginosa*.

Bacterial conjugation in Gram-negative bacteria requires the presence of a special plasmid that carries all the genes needed for

the process. The transferable *E. coli* plasmid called **fertility factor (F factor)** is a well-studied model for conjugation. Cells that carry the F-factor plasmid are called **F⁺ cells**. The conjugation process begins when a specialized pilus on an F⁺ donor cell contacts a plasmidless recipient cell called the **F⁻ cell** (**Figure 9.3** and **Figure 9.4**, step 1).

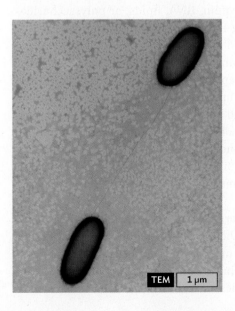

Figure 9.3
"Sex" Pilus Connecting Two *E. coli* Cells

TEM   1 µm

Figure 9.4 The Conjugation Process

Some plasmids, such as F factor in *E. coli*, can mediate cell-to-cell transfer of DNA. The plasmid is nicked at the nick site, *oriT* (step 3), and the 5′ end is transferred to the other cell. Purple arrowheads depict the 3′ end of replicating DNA. The 5′ end of the nick (black dot) will move through the pore and remain attached to the membrane while the rest of the single-stranded DNA passes into the recipient.

Mating-pair formation

1 Sex pilus from the F⁺ plasmid donor (left) attaches to receptors on the recipient cell (right).

Chromosome   F factor

Sex pilus

Donor (F⁺ cell)   Recipient (F⁻ cell)

Bridge

2 Contraction of the pilus draws the two cells together and forms a protein bridge.

Sex pilus contracts

Nick at *oriT*

3 The F factor is nicked at *oriT*, and the 5′ end begins transfer through the bridge.

5′
3′

DNA transfer

4 The strand remaining in the donor is replicated.

5 Once in the recipient, the transferred strand circularizes and replicates.

3′ 5′

6 The recipient has been converted to a donor.

F⁺ cell   F⁺ cell

Once attached, the pilus contracts to draw the two cells together (step 2). Their membranes fuse, and the F⁺ cell forms a protein bridge to connect to the F⁻ cell.

Bridge formation and membrane fusion trigger the nicking (cutting) of one strand of the plasmid DNA. The nick is made at the DNA transfer origin *oriT* (Figure 9.4, step 3). One end (the 5′ end) of the nicked strand begins to transfer through the protein bridge into the recipient cell. DNA replication in the donor simultaneously replaces the strand transferred (step 4). Meanwhile, DNA replication in the recipient cell makes a strand complementary to the one it received (step 5). The F-factor DNA that was transferred into the recipient cell then circularizes to form a new F-factor plasmid. Ultimately, the protein bridge joining donor and recipient cells spontaneously breaks, and the two cells separate. The F⁻ recipient cell has now been converted to a new F⁺ donor cell (step 6).

Within the recipient cell, the F factor can remain as a plasmid or integrate into the chromosome (**Figure 9.5**, steps 1 and 2). When the chromosome contains an integrated F factor, the entire chromosome can potentially be transferred to a new cell. However, only host genes closest to one end of the inserted F factor are transferred efficiently. The reason is that the conjugation bridge usually breaks long before the 100 minutes needed to transfer the entire chromosome. Cells carrying integrated F factors are called **Hfr cells** because of the high likelihood (compared with F⁺ cells) that genes from the Hfr cell's chromosome will be transferred into the recipient and replace homologous (similar) genes in the recipient's chromosome (Hfr = <u>h</u>igh-<u>f</u>requency <u>r</u>ecombination of chromosomal genes). The ability to move chromosomal genes by conjugation helped scientists understand distance between genes on a chromosome, chromosome structure (are they linear or circular?), and mutations.

An integrated F factor can also be excised (dis-integrated) from an Hfr chromosome and restored to its original F⁺ form (Figure 9.5, step 3). Occasionally, however, the plasmid excision is imprecise, taking some adjacent host chromosome with it. An F plasmid that contains host DNA is called an **F-prime (F′) plasmid** or F-prime (F′) factor (step 4). It, too, can be transferred through conjugation.

Note: Bacterial conjugation is completely different from the process called conjugation in eukaryotes. Microbial eukaryotes actually form a cytoplasmic bridge in which two genomes undergo <u>reciprocal</u> exchange (discussed in Chapter 11).

Transduction

Bacterial viruses (**bacteriophages**, or **phages**, to be discussed in Chapter 12) are composed of nucleic acid enclosed within a symmetrical protein structure called a capsid. A DNA phage injects its DNA into a live bacterial cell, usually of the same, or related, species of the bacterium from which the phage came. The phage DNA commandeers host protein synthesis machinery to make more capsid protein and uses bits of the host's DNA replication machinery

Figure 9.5 Integration and Excision of F Factor

Chromosome   F factor

1 F-factor plasmid integrates into host chromosome.

Integration   Excision

Integration   Excision

2 Cell becomes an Hfr capable of transferring genes.

3 F factor can be excised, re-forming a plasmid.

Hfr

4 F-factor plasmid can include host DNA.

Imprecise excision

5 Plasmid carrying host DNA is called F-prime factor.

F-prime (F′)

Figure 9.6 Generalized Transduction

Generalized transduction by a phage can move any segment of donor chromosome to a recipient cell. The number of genes transferred in any one phage capsid is limited, however, to what can fit in the phage head.

Virion genome   **Donor host cell**
Capsid   Host DNA

1 P22 phage DNA infects a host cell and makes subunits for more phage.

Phage DNA

Tails

2 DNA is packaged into capsid heads. Some capsids package host DNA by mistake.

3 New phage assembly is completed.

4 Cell lyses; phages are released.

5 Transducing phage particle injects host DNA into new cell, where it may recombine into the chromosome.

Recombination crossover events exchange host DNA for donor DNA.

Recipient host cell

to make more phage DNA. Newly made phage DNA is slipped into a capsid, which acts as a container. The completed phage particles are then released into the environment, often by lysing the host cell.

The process of forming a phage particle is not perfect. Some capsids may accidentally package a small piece of host bacterial DNA. The process by which a bacteriophage carries a payload of host DNA from one cell to another is known as **transduction**.

In **generalized transduction**, the phage capsid mistakenly fills with a fragment of bacterial DNA instead of phage DNA (**Figure 9.6**, steps 1, 2, and 3). Any host genome segment can be randomly packaged into a transducing phage particle and moved to a recipient cell (steps 4 and 5). The transducing phage particles then randomly attach to other bacterial cells and inject genes from the "donor" cell into the recipient cell. It's as though Amazon.com blindly packaged different items (genes) from the warehouse (donor cell) into boxes (phages) and haphazardly delivered each one to random addresses (recipients). DNA recombination mechanisms can then replace the homologous DNA segment of the recipient host genome with the donor DNA.

In another process, called **specialized transduction**, the phage genome initially integrates into a specific DNA sequence in the host genome (**Figure 9.7**). The integrated phage genome is called a **prophage**. In the chapter-opening case, the beta phage that carried the diphtheria toxin gene formed a prophage in the *Corynebacterium* cell. Prophage DNA replicates along with the host DNA as the host cell multiplies. But at some point, the phage DNA may excise from the host chromosome and start making more phage particles (step 1). Occasionally, the excision process mistakenly includes a small piece of host DNA that flanks the phage DNA insertion site (step 2). The flanking host DNA becomes packaged into a capsid along with the phage genome (step 3) and transferred to a new host cell (steps 4, 5, and 6).

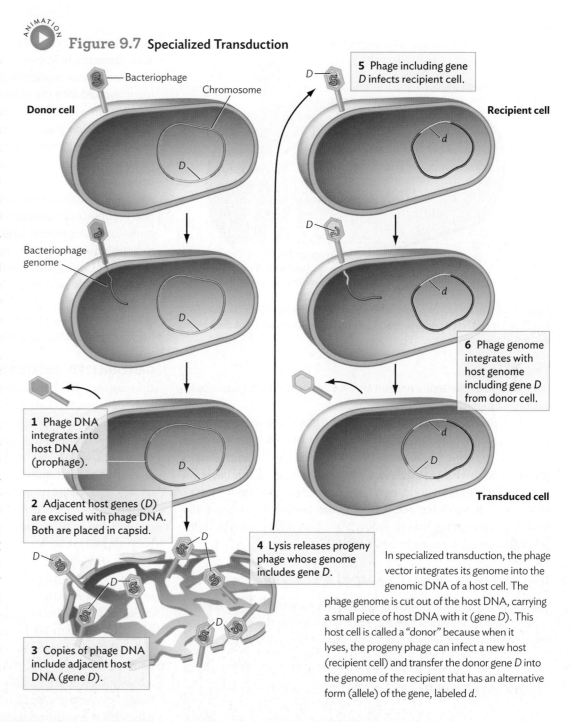

Figure 9.7 Specialized Transduction

Bacteriophage
Chromosome
Donor cell

5 Phage including gene *D* infects recipient cell.

Recipient cell

Bacteriophage genome

1 Phage DNA integrates into host DNA (prophage).

2 Adjacent host genes (*D*) are excised with phage DNA. Both are placed in capsid.

3 Copies of phage DNA include adjacent host DNA (gene *D*).

4 Lysis releases progeny phage whose genome includes gene *D*.

6 Phage genome integrates with host genome including gene *D* from donor cell.

Transduced cell

In specialized transduction, the phage vector integrates its genome into the genomic DNA of a host cell. The phage genome is cut out of the host DNA, carrying a small piece of host DNA with it (gene *D*). This host cell is called a "donor" because when it lyses, the progeny phage can infect a new host (recipient cell) and transfer the donor gene *D* into the genome of the recipient that has an alternative form (allele) of the gene, labeled *d*.

DNA Restriction and Modification

As you might imagine, certain dangers await cells that are invaded by foreign DNA. The most obvious risk involves bacteriophages whose replication will kill the target cells. A bacterium that can digest invading phage DNA while protecting its own chromosome has a far better chance of surviving in nature than do cells unable to make this distinction.

Phages are but one danger. The unrestricted incorporation of foreign DNA can also be an energetic drain on the cell. Genes encoding competing or useless products would squander resources (for instance, ATP) and lower the overall fitness of the cell. As a result, bacteria have developed a kind of "safe sex" approach to gene

exchange. It is an imperfect approach, however, that leaves room for beneficial genetic exchanges.

This protection system, called restriction and modification, is an immune system for bacteria that attacks foreign DNA without attacking its own genome. The system involves enzymatic cleavage (restriction) of foreign DNA and protective methylation (a chemical modification) of self DNA (**Figure 9.8**). DNA restriction endonucleases (introduced in **Section 8.6**) recognize short but specific DNA sequences known as restriction sites. The enzymes cleave, and so destroy, invading DNA at or near those sequences (**Figure 9.9**). In the laboratory, restriction enzymes

Figure 9.8 Restriction of Invading Phage DNA

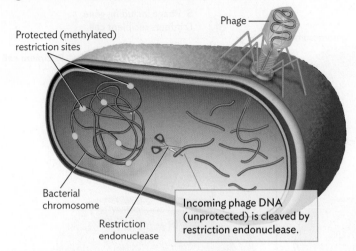

Phage DNA is injected into a host, where restriction endonucleases can digest it. Host DNA is protected because specific methylations of its own DNA prevent the enzymes from cutting it.

from different bacteria are used as DNA "scissors" in various applications of genetic engineering.

How do bacteria protect their own DNA from these endonucleases? During host DNA replication, DNA modification enzymes attach methyl groups to the host DNA sequences. Methylation makes the sequence invisible to the restriction enzyme. But, if this bacterial immune system is so good, how does horizontal gene transfer ever take place? It turns out the system is not perfect. Sometimes, the modification enzyme finds a methylation site on an invading DNA <u>before</u> the restriction enzyme can cut it.

Recombination between DNA Helices

Once a new piece of DNA has entered a cell and escaped restriction, what happens to it? The answer depends on the nature of the acquired DNA. If the new DNA is in the form of a plasmid capable

Figure 9.9 EcoRI Restriction Site

EcoRI is an example of a restriction-modification system. Shown here are cleavage (top) and methyl modification (bottom). The DNA sequence shown is specifically recognized by the endonuclease and methylase enzymes.

of autonomous replication, the plasmid will coexist in the cell separate from the host chromosome. If the foreign DNA is not capable of autonomous replication, it is either degraded by nucleases or incorporated into the chromosome through recombination.

Within a cell, two separate DNA molecules can recombine. **Recombination** is an enzyme-mediated process in which two DNA molecules exchange portions by breaking and re-forming their sugar-phosphate backbones. Mechanisms of recombination can be either "generalized" or "site-specific." **Generalized recombination**, also called **homologous recombination**, requires that the two recombining molecules have long stretches of DNA sequence homology. In contrast to generalized recombination, **site-specific recombination** requires very little sequence homology between the recombining DNA molecules. However, site-specific recombination does require a short sequence, 10–20 base pairs (bp), recognized by a special recombination enzyme. Site-specific recombination is what phages such as the beta phage of *Corynebacterium diphtheriae* use to integrate into the host genome.

GENERALIZED RECOMBINATION In generalized recombination, foreign DNA that enters a cell (by transformation, conjugation, or transduction) may recombine with the recipient genome if the two sequences are homologous. A protein called RecA finds and aligns homologous stretches of DNA sequence in donor and recipient DNA molecules (**Figure 9.10**). The enzyme then catalyzes an exchange of strands (similar to replacing a section of pipe in your home). RecA can recombine donor DNA that is single-stranded (entering by transformation) or double-stranded (entering by conjugation or transduction). Homologs of RecA are found in many bacteria and archaea.

So, why should health care professionals care about all this? Imagine your intestine, with its millions of harmless *E. coli*—normal in a healthy intestine. What if your resident *E. coli* also contained a mutant DNA gyrase gene whose product (needed for DNA replication) is resistant to the antibiotic ciprofloxacin? (Ciprofloxacin is used to treat serious bouts of diarrhea caused by pathogenic strains of *E. coli*.) You are still healthy, but not if you eat a hamburger tainted with a pathogenic *E. coli* strain. In your intestinal microbiome's mixed community, your *E. coli* Hfr strains conjugate with the pathogenic newcomers and transfer the ciprofloxacin resistance gene. The pathogen grows, you develop bloody diarrhea, and the clinician prescribes ciprofloxacin; thanks to conjugation, the pathogenic strains are now resistant to that antibiotic. The disease persists and you end up hospitalized. Gene transfer and recombination mechanisms have human consequences.

When trying to understand recombination (and gene transfer overall), it is important to remember that the processes are generally random. The lucky bacterium that gets the right piece of DNA and recombines the right segment of that DNA will benefit. This is why a genetic experiment usually requires screening hundreds of millions of cells to find a few recombinants that grow into visible colonies.

Figure 9.10 Generalized Recombination

In both single-strand and double-strand recombination, sections of the recipient strand are replaced by the donor DNA. The recipient DNA removed from the genome is degraded in both cases. A single-stranded molecule introduced by transformation undergoes single-strand crossovers that replace only one strand of the homologous region of the donor chromosome. The fate of the resulting hybrid molecule depends on whether it is replicated or repaired. If it is replicated, one daughter cell chromosome will look like the original before transformation. The other daughter cell, however, will have replicated the donor strand and inherit any traits carried on it. In double-strand recombination, double-strand crossovers replace both strands of the recipient DNA.

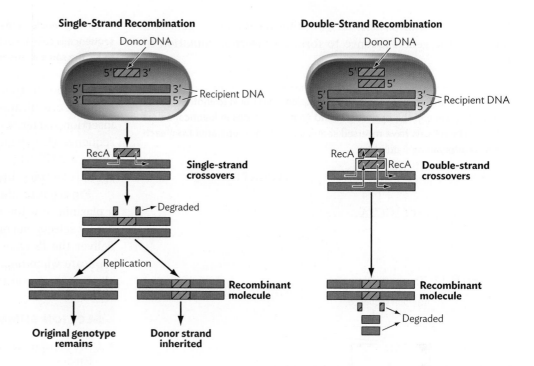

Why is recombination advantageous to a microbe? Generalized recombination serves three functions in the microbial cell:

- Recombination probably first evolved as a method of DNA repair, useful to fix mutations or restart stalled replication forks. This role does not involve foreign DNA.

- Cells with damaged chromosomes use DNA donated by others of the same species to repair their damaged genes.

- Recombination is also part of a "self-improvement" program to obtain genes from other organisms that might enhance the competitive fitness of the cell. Thus, recombination makes major contributions to evolution.

SITE-SPECIFIC RECOMBINATION In contrast to generalized recombination mechanisms that require RecA protein and can recombine any region of the chromosome, site-specific recombination does not use RecA and moves only a limited number of genes. This form of recombination involves short regions of homology between donor and target DNA molecules. Special recombinase enzymes recognize those sequences and catalyze a crossover between them to produce a cointegrate molecule. The integration of beta phage into *Corynebacterium diphtheriae*, the organism in the opening case of this chapter, is one example of site-specific recombination. A short DNA sequence called the *att* site is present on both the beta phage DNA and the *Corynebacterium* chromosome. A phage-encoded protein called integrase engages the two *att* sites and, through a strand breakage and re-joining mechanism, combines the two molecules. The process is similar to what happens when the F-factor plasmid integrates (see Figure 9.5).

Mobile Genetic Elements

Antibiotic resistance is a confounding problem in infectious disease treatment. Besides the mechanisms of genetic exchange that can move antibiotic resistance genes between bacterial cells, scientists discovered that some antibiotic resistance genes could "jump" from one place in a genome to another, or to a plasmid residing in the same cell. Once a resistance gene has moved to a plasmid, resistance can be transferred to another cell. These types of genes, referred to as **transposable elements**, or **transposons**, exist in virtually all life forms and can jump by a process called **transposition**. In addition to antibiotic resistance, transposable elements can include, and therefore transfer, genes other than antibiotic resistance genes between chromosomes and plasmids.

Unlike plasmids, transposable elements cannot replicate outside a larger DNA molecule. They exist only as hitchhikers integrated into some other DNA molecule (**Figure 9.11**). All transposable elements include a transposase gene whose enzyme product (transposase) moves the element from one location to another within the same DNA molecule or from one DNA molecule into a different DNA molecule in the same cell. The simplest transposable elements are short DNA sequences (700–1,500 bp) called **insertion sequences (IS)**. An insertion sequence has a transposase gene flanked by short inverted DNA repeat sequences (see Figure 9.11). The transposase, once made from the transposase gene, binds to these inverted repeats and to a more or less random DNA target site. The transposon then replicates and inserts itself into the target DNA. The insertion sequence has now moved, or "hopped," from one segment of DNA to another.

Note that any transposable element that inserts into a gene will interrupt the gene sequence to form an insertion mutation in that gene.

⊙**Note:** An inverted repeat is a DNA sequence identical to another downstream sequence in the opposite strand (note the bases in identical colors below). The repeats have reversed sequences and are separated from each other by intervening sequences:

5′-AATCGAT............ATCGATT-3′
||||||| |||||||
3′-TTAGCTA............TAGCTAA-5′

Figure 9.11 Transposition

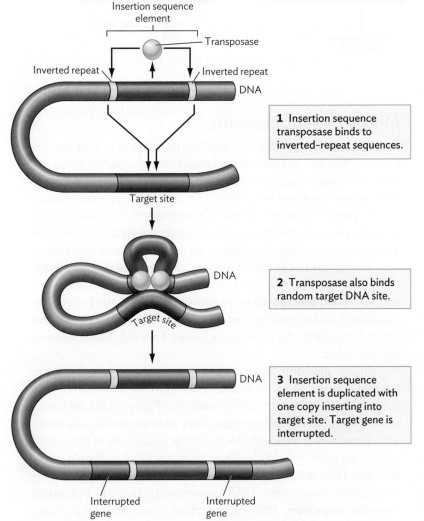

1 Insertion sequence transposase binds to inverted-repeat sequences.

2 Transposase also binds random target DNA site.

3 Insertion sequence element is duplicated with one copy inserting into target site. Target gene is interrupted.

How a simple transposable element (an insertion sequence element) hops from one piece of DNA to another. Transposases (enzymes that catalyze transposition) bind to inverted repeats at the ends of an established IS element and interact with a random target site. The IS element duplicates and inserts a copy into the target sequence. The new insertion, if it occurs inside a gene, interrupts (mutates) the gene sequence.

When no nucleotides intervene between the inverted sequences, the sequence is called a palindrome. The inverted bottom strand reads, from 5′ to 3′, the same as the top strand.

Composite transposons have, as the name suggests, a more complex organization; they include gene sequences in addition to insertion sequences. Tn*10*, for instance, consists of two insertion sequence (IS) elements that flank a tetracycline resistance gene (Figure 9.12). This transposon is important because it mediates the spread of tetracycline resistance through populations of bacteria.

Figure 9.13 illustrates the transposition of a transposon from a plasmid into the chromosome. Note that the transposase randomly selects one of many possible target sequences where it will deliver the IS element. In this example, the transposon did not replicate when hopping. Today, we know that these mobile DNA elements have contributed to the evolution of all species.

SECTION SUMMARY

- **DNA transfer mechanisms** enable microbes to obtain genes from other microbes.
- **Horizontal gene transfers** between species occur by conjugation, transformation, and transduction.
- **Transformation** is the uptake by living cells of free-floating DNA from dead, lysed cells.
- **Conjugation** is a DNA transfer process mediated by a transferable plasmid that requires cell-to-cell contact and formation of a protein complex between mating cells.
- **Transduction** is the process whereby bacteriophages transfer fragments of bacterial DNA from one bacterium to another. In generalized transduction, a phage preparation can move any gene in a bacterial genome to another bacterium. In specialized transduction, a phage can move only a limited number of bacterial genes.
- **Restriction endonucleases** protect bacteria from invasion by foreign DNA.
- **Recombination** is the process by which DNA molecules are cleaved and reattached to other DNA. Recombination may be generalized (affecting a long stretch of DNA homology) or site-specific (requiring only short homology between ends of two DNA molecules).
- **Transposons** ("jumping genes") move randomly from one DNA molecule to another. Composite transposons can carry a variety of genes, including antibiotic resistance genes.

Figure 9.12 Example of a Transposon That Includes an Antibiotic Resistance Gene

IS element *10* | Tetracycline resistance | IS element *10*
Tn*10*

Transposon Tn*10* is a composite transposon. Composite transposons contain genes (in this case, a gene for resistance to the antibiotic tetracycline) flanked by two insertion elements similar to the one shown in Figure 9.11. The transposase acts only on the outermost inverted repeats (red arrows) to transpose the entire transposon, including the antibiotic resistance gene, to a new target site.

Figure 9.13 A Transposon "Hops" from a Plasmid to a Host Chromosome

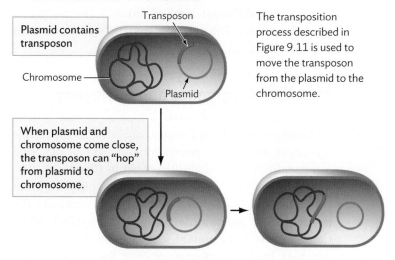

Plasmid contains transposon

Transposon

Chromosome

Plasmid

The transposition process described in Figure 9.11 is used to move the transposon from the plasmid to the chromosome.

When plasmid and chromosome come close, the transposon can "hop" from plasmid to chromosome.

Thought Question 9.1   Can a *recA* mutant of *Streptococcus pneumoniae* experience horizontal gene transfer?

Thought Question 9.2   Transfer of an F factor from an F+ cell to an F- cell converts the recipient to F+. How might the timing of DNA transfer explain why an Hfr cell does not do the same?

Thought Question 9.3   How do you think phage DNA that contains a restriction site recognized by its host evades degradation by the host's restriction enzyme?

Thought Question 9.4   How might a transposon containing a gene for tetracycline resistance make its way from the chromosome of one cell to the chromosome of a second cell?

9.2 Origins and Evolution

SECTION OBJECTIVES

- Describe how all life originated from microbes, how life forms evolved over time, and how species continue to evolve.
- Explain how microbial genomes change via processes of random mutation and natural selection.
- Explain how a molecular clock works to measure phylogeny, and describe the key attributes of a useful molecular clock gene.
- Describe the three domains of life.

Has modern technology influenced the way we evaluate evolutionary processes? We now have the ability to sequence the whole genome of any organism, including humans, in less than a day (DNA sequencing is discussed in Chapter 8). Scientists can use desktop computers and the Internet to quickly compare DNA sequences from thousands of sequenced genomes. This analysis has led to startling discoveries about evolution and the interconnectedness

of life. Archaea, for example, possess many traits in common with eukaryotes and other traits more like those of bacteria. In fact, most proteins found in the thermophilic archaeon *Methanocaldococcus jannaschii* are also found in Eukarya and/or Bacteria. As we just learned in Section 9.1, microbial genomes not only inherit parental genes but also may acquire genes from other species. The genome of *E. coli*, an organism under scrutiny for more than 100 years, turns out to be a mosaic of DNA segments acquired from other species. For example, *E. coli* O157:H7, a strain that causes bloody diarrhea, acquired several virulence operons from other species, including the genes for Shiga toxin (from *Shigella*), which causes kidney failure.

How do we know where genes come from? And how do we know that all organisms share a common ancestry? Our knowledge of the history of life is founded on a massive collection of evidence from geology, physiology, and DNA sequences.

Origins of Life

Earth formed approximately 4.5 billion years ago as the Sun and other planets condensed out of interstellar matter. The major elements of Earth—including the atoms of life, such as carbon, nitrogen, and oxygen—could not have originated in our Sun, which is too young to form elements heavier than helium. Heavier elements formed through nuclear reactions much earlier, from stars that were born and died billions of years before. Thus, when poets say "we are made of stardust" . . . we actually are.

How did inert atoms give rise to living organisms? We do not know, but there are signs of life's processes in the geological record. Our earliest evidence for microbial life comes from carbon isotope ratios—that is, the ratio of carbon isotopes (atoms) of atomic mass 12 (^{12}C) versus 13 (^{13}C), which differ by one neutron. Photosynthesis (discussed in Chapter 7) preferentially fixes carbon-12 into organic material rather than carbon-13. So, a carbon deposit depleted of ^{13}C in relation to ^{12}C probably came from living cells that fixed CO_2 into biomass. The oldest rock on Earth showing depletion of carbon-13, the footprint of ancient life, is 3.7 billion years old (3.7 gigayears, Gyr).

Some early rock formations called **microfossils** show organic patterns typical of living creatures (**Figure 9.14**). For example, ancient microfossils resemble microbial biofilms and stromatolites, towerlike layers of cyanobacteria that grow in shallow, salty seas such as Shark Bay, in Australia. Well-preserved microfossils date back at least 2 billion years. This time frame is determined from the age of the geological strata in which the microfossil was found. Geological strata are dated by radioisotope decay. More recent fossils, 500 million years old, record the emergence of multicellular animals such as trilobites (three-lobed arthropods, now extinct) that evolved from microbes. Ultimately, microbial ancestors gave rise to all multicellular plants and animals, including ourselves.

How did all the different species of life diverge from common ancestors? And how do we know that all modern life

Figure 9.14 Geological Evidence for Early Life

Early in Earth's history, the geological record shows evidence of microbial life. Photos on the right include microfossils from the Proterozoic eon that show the presence of ancient microbes.

Bil'yakh, Siberia, 1.5 Gyr

Doushantuo, China, 0.6 Gyr

A trilobite from Morocco, 340–440 Myr

forms possess a common microbial ancestor some billions of years ago? The answers to these questions come from a very different kind of "fossil": the remnants of millennia-old information that lie within modern sequences of DNA.

Microbial Divergence and Phylogeny

A unifying doctrine of modern biology is that genetic relatedness arises from common ancestry. As organisms reproduce over many generations, individual offspring can acquire mutations (as discussed in Chapter 8). If a mutation provides a survival advantage, offspring from the individual mutant cell will begin to dominate the population—a phenomenon known as natural selection. As a result, the mutant sequences become "fixed" in genomes, leading to **divergence** from the original population. Divergent organisms show less and less similarity the longer their lineage has diverged. Besides mutation, another source of divergence is gene mobility, discussed earlier. The gain or loss of a large number of genes can lead to a more sudden change, even to "instant speciation," in which a new strain of organism shows traits very different from those of the parental strain. An example of this is a bacterium that might integrate DNA from a bacterial virus and, in the process, acquire a gene conferring virulence, generating a pathogen. This is how *Corynebacterium diphtheriae* gained its toxin gene.

Several mechanisms of evolution determine how populations of organisms diverge, including:

- **Random mutation.** DNA sequence can change through rare replication mistakes (in bacteria and archaea, this typically happens to one of a million replicated base pairs). Replication errors result in mutation. Most mutations have no effect on gene function and are called neutral mutations.

- **Natural selection.** In a given environment, natural selection favors the organism that produces more offspring in that environment (competitive advantage). Therefore, genes encoding traits that improve the survival of offspring are said to be under "selection pressure."

- **Reductive evolution (degenerative evolution).** A trait no longer needed for growth or survival is no longer under selection pressure. As a result, genes encoding the trait can accumulate mutations without affecting the organism's reproductive success. For example, genes for Heme and NAD biosynthesis are absent in *Haemophilus influenzae* (Table 6.2). Because mutations that decrease function are more common than mutations that improve function, a

Figure 9.15 The Molecular Clock

As DNA replicates, mutations accumulate at random. The number of mutations accumulated is proportional to the number of generations and thus the time since divergence. In each sequence designation (for example, 2A), the numeral indicates the generation and the letter identifies a specific organism's DNA. Red dots mark base differences between organisms 3A and 3B.

gene that is no longer needed for survival will accumulate mutations that ultimately cause the loss of that trait.

Molecular Clocks

How do we tell time on an evolutionary scale (in the range of millions of years)? The answer, again, is found in DNA. Random mutations with neutral effects are not subject to selection and thus accumulate at a steady rate over generations. The result of random mutations is **genetic drift**. In populations that do not interbreed, genetic drift causes genome sequences to diverge over time. If the mutation rate is <u>constant</u>, then genetic drift in a DNA sequence measures the time it took two species to diverge. A gene in which random mutations accumulate at a constant rate is called a **molecular clock**.

Molecular clocks have revolutionized our understanding of evolution, including that of human beings. The most important molecular clocks are genes that exist in all life forms, such as ribosomal RNA (rRNA) genes. Ribosome structure and function are highly similar across all organisms. Carl Woese and others used rRNA genes to reveal the divergence of three domains of life: Archaea, Bacteria, and Eukarya (see Section 1.4). Other gene sequences are now used to measure divergence at different classification levels.

A molecular clock measures the time of divergence based on the accumulation of new random mutations during each round of DNA replication. For example, in **Figure 9.15** each offspring in generation 2 acquires two new mutations; their sequences now differ from each other in 4 out of 16 bases, or 25%. In the next generation, each individual propagates the earlier mutations while acquiring two more random mutations. Strain 3A now differs in 8 out of 16 bases, or 50%, from strains 3B (marked by red dots) and 3C, which differ by 25% from each other. (Actual mutation frequencies, of course, are much lower, about one base per million per generation.) The key assumption is that each generation of offspring acquires a consistent number of random mutations from the parents, so the number of sequence differences between two species should be proportional to the time of divergence between them.

A gene used for a molecular clock must be one that is inherited vertically, from parent to offspring, and does not tend to be transmitted horizontally (by transformation or transduction, as previously discussed). Genes horizontally transferred from an unknown species to another known species would give inaccurate information about the divergence of the genome they entered.

In real genomes, the molecular clock works best if:

- **The molecular clock gene has the same function across all species being compared (no selection pressure).**

- **The generation time is the same for all species compared.** By necessity, cells with shorter generation times replicate their DNA more often than do cells with longer generation times. Consequently, species that replicate DNA more often than another species will accumulate more mutations over a set length of time. The difference would lead to overestimates of how far back in time two species diverged.

- **The average mutation rate remains constant among species and across generations.** If different species mutate at different rates, species with more rapid rates of mutation will appear to have diverged over a longer time than is actually the case.

Using a molecular clock to tell evolutionary time requires aligning the sequences of similar genes from divergent species. The number of sequence differences can be used to propose a tree of divergence (family tree) that describes the historical relationship, or **phylogeny**, of related organisms. Those relationships are depicted in diagrams called **phylogenetic trees** (**Figure 9.16**).

Figure 9.16 DNA Phylogenetic Tree

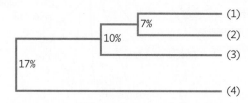

This phylogenetic tree was derived by aligning sequences of the same gene from four different species. The length of each tree branch is proportional to the number of differences (divergence) between two sequences. For instance, homologous sequences showing 10% difference diverged further back in time than homologous sequences showing a 7% difference. The start of a fork represents the ancestor (which may no longer exist) of the actual species compared.

Figure 9.17 Phylogeny of Selected Gammaproteobacteria

The phylogenetic tree was derived from sequences of highly conserved "housekeeping" proteins. The length of the scale bar corresponds to 5% amino acid sequence divergence, which takes 500 Myr to change. *Shewanella* is the beginning of an "outgroup" that originated from the same ancestral root organism that the group containing both *Vibrio* and *Escherichia* did.

Figure 9.18 The Three Domains of Life: Bacteria, Eukarya, and Archaea

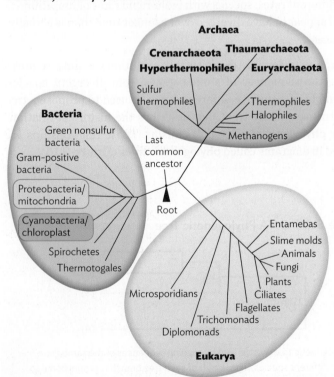

Mitochondria were derived from proteobacteria that became endosymbionts of an ancestral eukaryotic cell and so are placed with proteobacteria in this phylogenetic tree. Cyanobacteria and chloroplasts are similarly related.

Phylogenetic Trees

A phylogenetic tree can be used to compare any set of organisms. We can compare organisms from a single habitat, such as the human digestive tract or a Yellowstone hot spring. In **IMPACT**, you will see how phylogeny was used to discover surprising pathogens in soap scum from a shower curtain.

A set of related organisms may also be compared from different habitats. **Figure 9.17** shows the divergence of selected gammaproteobacteria, a group (clade) of related Gram-negative bacteria (presented in Chapter 10). The tree is based on a set of **housekeeping genes**—that is, genes encoding functions essential for all cells and usually transmitted vertically (parent to offspring). *Escherichia, Shigella,* and *Salmonella* are closely related genera of intestinal bacteria. Their clade includes pathogenic species, as well as commensal species normally present in the human body. The genus *Klebsiella* includes species that cause pneumonia. *Photorhabdus luminescens* is a bacterium that infects nematodes (roundworms) and helps them parasitize insects. *Yersinia pestis* causes bubonic plague. *Photobacterium profundum* is a marine barophile (high pressure) growing in a deep-sea trench.

Diverse as they are, all these species diverged at some point from a common ancestor (which no longer exists). This common ancestor, in turn, diverged from an even more ancient "root" organism, which also spawned an "outgroup" (most divergent sequence). The outgroup in Figure 9.17 is *Shewanella*, a marine bacterium associated with fish spoilage. Accordingly, species having diverged from a recent common ancestor can show surprisingly diverse traits and effects on humans.

The Three Domains of Life

The analysis of rRNA gene sequences shows that all life forms on Earth, including humans and other animals, plants, and microbes, share a common ancestral cell (**Figure 9.18**). An amazing fact of life is that all living cells on Earth share profound similarities. From a molecular point of view, all cells consist of membrane-enclosed compartments that shelter DNA, RNA, and protein synthesis machines. The fundamental components of these machines evolved <u>before</u> the three domains diverged from their last universal common ancestor. From a molecular standpoint, all cells on Earth appear more similar than they are different.

As you learned in Chapter 1, early in life's history, organisms diverged into three major groups called **domains**: Bacteria, Archaea, and Eukarya (including humans). Most data indicate that bacteria diverged from the ancient root organism before archaea and eukaryotes diverged from each other. Even though the domains arose from an ancient common ancestor, each domain shows unique traits that distinguish it from the other two domains, such as:

- **Eukaryotes possess many complex membranous organelles.** Organelles include the nuclear membrane and the endomembrane complex, as well as mitochondria and (in plants) chloroplasts. Mitochondria and chloroplasts are descended from endosymbiotic bacteria (discussed in Chapter 5). By contrast, bacteria and archaea have relatively simple structures in their cytoplasm. In most bacteria and archaea, the nucleoid extends throughout the cytoplasm, accessible to DNA-binding proteins floating in cytoplasm and embedded in the membrane.

- **Bacteria have a distinctive gene expression apparatus.** The DNA and RNA polymerases of bacteria are streamlined compared with those of the eukaryotic and archaeal domains, perhaps because the bacterial ancestor diverged first from the ancestor of the other two domains. Only bacteria use the modified amino acid formylmethionine (instead of plain methionine) to fill the start codon in protein synthesis (discussed in Chapter 8).

- **All bacterial cell walls are made of peptidoglycan,** a molecular cage consisting of repeating disaccharide units. Although some archaea and eukaryotes have cell walls, they do not contain peptidoglycan.

- **Archaea have distinctive membrane lipids.** The membrane lipids of archaea are completely different from those of bacteria and eukaryotes. For instance, archaea have more durable ether links (–O–) instead of the ester links (–COO–) of bacteria and eukaryotes presented in Chapter 3. The greater durability of archaeal lipid structures enables some archaea to grow in extreme environments, such as deep-sea thermal vents.

- **There are no known archaeal pathogens.** Why no archaea have evolved into pathogens remains a mystery. Many archaea, such as methanogens, live within the human body but never seem to cause disease. The lack of known archaeal pathogens seems strange, given that numerous bacteria and eukaryotes originating from many different habitats and lifestyles evolved as pathogens.

Overall, the three-domain phylogeny based on ribosomal genes divides life usefully into three distinctive clades (a clade is a series of related organisms that originate from a common ancestor in a phylogenetic tree). But when we look at other genes, we find remarkable evidence of horizontal gene flow even between domains. A most dramatic example of horizontal gene transfer is seen in the eukaryotes, which contain mitochondria derived from entire assimilated bacteria (see Section 1.4). While some bacterial genes persist within the mitochondria in mitochondrial DNA, thousands of other bacterial genes "migrated" to nuclear chromosomes through mobile gene transfer mechanisms like those described earlier. Between bacteria and archaea, many examples of gene transfer are known, especially between those organisms sharing a common environment, such as high-temperature habitats.

SECTION SUMMARY

- **Life originated early in the history of Earth.** Evidence for early life includes geochemical signs, as well as microfossils more than 2 billion years old.
- **All life on Earth shares a common ancestral cell.** Early life diverged into three domains: Bacteria, Archaea, and Eukarya. Each domain shows traits distinctive from those of the other two.
- **Evolution occurs by steady accumulation of random mutations.** Even without natural selection, gene sequences gradually diverge.
- **A molecular clock is a gene whose sequence can be used to measure divergence time.** Molecular clock genes must have a constant mutation rate and must share a common function in the organisms compared.
- **Phylogenetic trees show the relatedness between different kinds of organisms.** Phylogeny can be determined by comparing the DNA sequences of molecular clocks.
- **Evolution of organisms includes vertical descent along a phylogenetic clade for some genes and horizontal transfer for others.**

9.3 Natural Selection

SECTION OBJECTIVES
- Explain how a trait's effect on fitness depends on the environment and how environment-dependent fitness leads to the evolution of antibiotic resistance in pathogens.
- Describe the evolution of endosymbiosis between a host cell and an internalized partner organism.

How do new species arise? In Section 9.2 we described how the gradual accumulation of random mutations causes genomes to diverge and evolve. The rate of nonselected (neutral) sequence changes enables us to measure evolutionary time. But for organisms, meaningful evolution occurs when a changed trait affects survival, which is the basis of **natural selection**. In natural selection, the variant strains that result from random mutation will differ in their chance of survival. Variants that survive and leave more offspring than other strains undergo "selection"; that is, their particular traits are overrepresented in the next generation. Over many generations, the descendants develop traits that are best suited for their environment and very different from those of their ancestors. At some point, traits can change enough for the descendant to be considered a new species.

Natural selection is driven by environmental factors such as nutrient availability, temperature, and, in the medical realm, the presence of an antibiotic. Even over short periods, such as a few weeks, the natural selection of bacteria exposed to antibiotics can yield antibiotic-resistant descendants that will threaten a patient's life.

impact

What's That Lurking in My Shower?

This chapter describes the ways bacteria diverge and evolve to fill available nutritional niches. Some of those niches are places you may not suspect but interact with often, and the creatures you find there could be making you sick!

Suppose you finally get around to changing that disgusting shower curtain—and your allergy symptoms disappear. What was causing your allergic reaction? It just might have been a microbe clinging to the soap film that coated the shower curtain—perhaps even a species never seen before. It seems that microbial biofilm communities lurking in your shower can harbor potentially dangerous pathogens.

Such organisms were discovered by Norman Pace and colleagues at the University of Colorado. Pace, renowned for his studies of extremophile bacteria and archaea plucked from the ocean floor, has explored an even nastier aquatic habitat: the "soap scum" biofilm that coats our shower curtains and doors (Figure 1). Domestic water systems such as cooling towers and whirlpool spas have long been known to harbor pathogens such as *Legionella pneumophila*, the cause of a deadly form of pneumonia (legionellosis). Lesser-known reservoirs for microbes are the biofilms that collect on household surfaces such as shower curtains, an item often neglected during cleaning. Viable cells of a shower curtain biofilm are revealed by fluorescence microscopy using a stain for DNA (see Figure 1B). Two of Pace's students, Scott Kelley and Ulrike Thiesen, scraped biofilms from several used shower curtains and extracted microbial DNA from them. PCR (see Section 8.6) was used to amplify the 16S rRNA genes from the extracts, and the amplified genes were sequenced. By comparing the gene sequences recovered from the biofilm with those of known species, Kelley and Thiesen exposed the shower curtain inhabitants.

The results revealed 117 unique 16S rRNA sequences, implying that the shower curtain biofilm harbored 117 different types of organisms. The sequences were aligned, and their similarity was measured. Then the data were used to propose a phylogenetic tree. The tree shows what types of microbes can colonize shower curtains and offers clues as to how those microbes might interact with humans. The most abundant groups of species include two genera of Gram-negative bacteria: *Sphingomonas* and *Methylobacterium*. The phylogeny of the shower curtain sphingomonads is shown in Figure 2.

Sphingomonas species grow in a variety of soil and water habitats. The organisms break down (catabolize) complex chemicals and are of interest for bioremediation of environmental pollutants. Some species, however—particularly *S. paucimobilis* (highlighted in Figure 2)—cause opportunistic infections, including bacteremia, peritonitis, and abscesses, in immunocompromised individuals.

Species of the next-most-abundant genus, *Methylobacterium*, are versatile heterotrophs known for their ability to metabolize chemicals such as methanol and methylamine. These species grow in soil and water, as well as within plant tissues, but they are also isolated from automobile air-conditioning systems, paper-printing machines (printers), and dental-unit water lines. Their pink color may be the source

Figure 1 Bacterial Diversity on a Shower Curtain

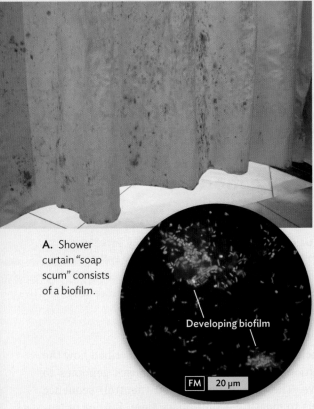

A. Shower curtain "soap scum" consists of a biofilm.

Developing biofilm

FM 20 µm

B. Biofilm from a shower curtain is visualized by fluorescence microscopy using a DNA stain (DAPI).

Figure 2 *Sphingomonas* Phylogeny from a Shower Curtain Biofilm

A. Phylogeny of sphingomonad bacteria was based on rRNA gene sequence comparison. Samples were taken from four different shower curtains, designated sc1–4. Numbers that follow the "sc" designation represent unknown species.

B. Norman Pace characterized the first sequence phylogeny of thermophiles from high-temperature environments. His laboratory has since characterized the microbial diversity of other environments, such as shower curtain biofilms.

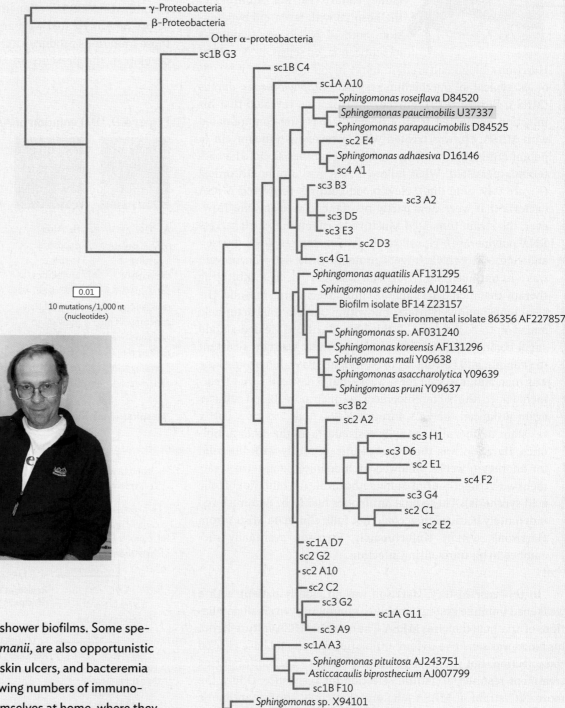

of the pinkish hue commonly seen in shower biofilms. Some species, such as *M. extorquens* and *M. zatmanii*, are also opportunistic pathogens able to cause pneumonia, skin ulcers, and bacteremia in immunocompromised people. Growing numbers of immunocompromised individuals care for themselves at home, where they need to control their exposure to possible pathogens. Therefore, finding opportunistic pathogens in home settings is worrisome. Pace and colleagues conclude with a reminder that "exposure can be minimized by regular cleaning or by changing shower curtains." So, what's hanging out in <u>your</u> shower?

DISCUSSION QUESTION

Why might the composition of the biofilm on your shower curtain differ from that on the curtain next door, where your best friend lives?

CASE HISTORY 9.1

The Bacteria Kept Coming Back: Persistent Pathogen

Harrison was a 73-year-old man with kidney failure. He was admitted to the hospital with fever and hypotension, both of which are symptoms of septicemia, or sepsis. A sample of Harrison's blood cultured on horse blood agar grew large colonies typical of methicillin-resistant *Staphylococcus aureus* (MRSA; Figure 9.19). Further investigation revealed that his infection originated from a home dialysis line contaminated with MRSA. He was treated with intravenous vancomycin (a potent antibiotic that targets cell wall synthesis), but the bacteremia persisted. What followed next was a 6-month ordeal (Figure 9.19 time line). New blood cultures recovered MRSA bacteria that were now partly resistant to vancomycin. However, the strain remained sensitive to rifampin (which targets RNA polymerase), ciprofloxacin (which targets DNA gyrase), and linezolid (which targets protein synthesis). Vancomycin was discontinued, and Harrison was started on a multidrug therapy that included rifampin, ciprofloxacin, and linezolid. His symptoms improved briefly, but within 3 weeks the bacteremia and symptoms returned. A third blood culture yielded an odd small-colony variant (SCV) of *S. aureus* that was now resistant to rifampin and ciprofloxacin (see Figure 9.19). Linezolid alone was then administered for 6 weeks, but 5 days after that treatment ended, the bacteremia recurred once more. Blood cultures again produced the SCV form of MRSA that was now partly resistant to linezolid and fully resistant to all the other antibiotics. Harrison was then treated intravenously with linezolid for another 6 weeks, followed by indefinite, long-term treatment with trimethoprim-sulfamethoxazole (an inhibitor of folic acid synthesis). These final antibiotics had to be administered indefinitely because they could not fully eliminate MRSA from Harrison's system. Unfortunately, Harrison eventually succumbed to his unrelenting infection.

In this case history, Harrison was an elderly patient with a weakened immune system, unable to eliminate even small populations of a pathogen such as MRSA. The original MRSA cultured from his blood was sensitive to four antibiotics that should have cleared the infection. But prolonged exposure to these antibiotics during treatment resulted in natural selection for resistance. Oddly, the more "fit" strain of MRSA—the one that could survive antibiotic therapy—grew very slowly, even in the absence of antibiotic. Isn't this counterintuitive?

Fitness Depends on Environment

What determines the "fitness" of a trait? In Case History 9.1, the drug-resistant strain made slow-growing small-colony variants when growing on laboratory media. So, how could the "fitness" of this strain have been greater than that of its fast-growing parent? Natural selection yielded a slow-growing population of cells that was the "fittest" under a particular environmental condition—the presence of linezolid in an immunocompromised host. Random mutations in MRSA developed while the bacteria grew in the patient before he was treated with antibiotics. In this case, one or two of those mutations produced a few slow-growing cells that could still develop in the presence of the antibiotics (once administered),

Figure 9.19   Evolution of Antibiotic Resistance in MRSA

A. The initial MRSA isolate infecting the patient showed partial resistance to vancomycin but was sensitive to rifampin, ciprofloxacin, and linezolid.

B. After exposure to these antibiotics, a new strain was isolated that grew more slowly (the SCV) but was resistant to all the antibiotics.

Hospital admission

Culture → MRSA
Sensitive to vanc, rif, cipro, line

Vanc (2 weeks)
Treatment failure

Culture → MRSA
Vanc resistant

Rif, cipro, lzd (3 weeks)
Treatment failure

Culture → SCV MRSA
Resistant to vanc, rif, cipro

lzd (6 weeks)
Treatment failure

Culture → SCV MRSA
Resistant to vanc, rif, cipro, and partially to lzd

Line IV (6 weeks), then trimethoprim-sulfamethoxazole indefinitely

↓

Death

The time line of Harrison's hospital stay and course of therapy. Images to the left are plate cultures of the original MRSA isolate and the antibiotic-resistant small-colony variant (SCV). cipro = ciprofloxacin; lzd = linezolid; rif = rifampin; vanc = vancomycin.

while the faster-growing parental strain succumbed. However, in the absence of antibiotics the original strain would outcompete the small-colony variant. Thus, natural selection for the "fittest" depends entirely on the environment in which selection occurs. Different environments lead to selection for different traits that are "fittest" for that particular environment.

How did the multidrug resistance come about? Researchers compared the DNA sequences of Harrison's original MRSA genome with that of the later small-colony variant. Just three point mutations in the small-colony variant's genome accounted for the drug resistance, as well as its slow growth. One of these mutations causes accumulation of the bacterial signaling molecule ppGpp (guanosine tetraphosphate). Accumulation of ppGpp alters the expression of many bacterial genes. Genes involved in ribosome synthesis are expressed less, whereas genes that enable a cell to survive in the presence of antibiotics are expressed more. But the cost to the cell of maintaining this unbalanced gene expression is a slower growth rate; like a community continually under "terror alert," the cell's normal, everyday processes are slowed by the demands of the stress response.

Evolution of Endosymbionts: The Guests That Never Leave

We have seen how a pathogen can evolve multidrug resistance at the expense of growth rate in a short period of time. Over a longer period, pathogens evolve specific traits that enhance their ability to colonize a host, such as the production of toxins that disrupt the host's immune system. In addition, a pathogen can evolve by losing cell functions that are supplied by the host cell. Ultimately, the pathogen may become completely dependent on its host. For example, *Treponema pallidum*, the cause of syphilis, is an ancient pathogen of humans. The genome of *T. pallidum* shows loss of numerous genes found in related bacteria, including genes required for energy production and amino acid biosynthesis. The pathogen obtains carbon sources and amino acids from the bloodstream of its host. So, there is no selection pressure for this pathogen to maintain these traits. Loss of traits in the absence of selection pressure is called **reductive evolution** or **degenerative evolution**.

An even more remarkable kind of evolution occurs between parasites and their intracellular partner species, called **endosymbionts**. An intriguing medically important class of endosymbionts is the *Wolbachia* bacteria that live inside the cells of invertebrate hosts, such as the nematode worms that cause filariasis, also known as elephantiasis (see Figure 2.3). The filarial nematode *Brugia malayi* harbors *Wolbachia* endosymbionts. These *Wolbachia* strains may have originally entered the nematode as a pathogen and then persisted because of its metabolic contributions to the host. A mutualistic relationship eventually evolved in which *Wolbachia* is required for the nematode's embryonic development. These endosymbiotic bacteria grow

Figure 9.20 The Filarial Endosymbiont *Wolbachia*

Cross section of the nematode *Brugia malayi*, showing endosymbiotic *Wolbachia* bacteria (stained pink) within the dermis and the uterine tubes.

within tissues below the nematode's skin and within the uterine tubes of females, where they enter the developing offspring (Figure 9.20).

Over time, the *Wolbachia* bacteria gained genetic traits that enabled growth within nematode hosts. But the *Wolbachia* genome also lost many genes through reductive evolution. With barely a million base pairs, the endosymbiont's genome retained genes for glycolysis and the TCA cycle but jettisoned the pathways needed to make amino acids and most vitamins. It nonetheless retains pathways to make purines, pyrimidines, and the coenzymes riboflavin and FAD—essential pathways not provided by its host nematode. Overall, *Wolbachia* appears to be evolving into an organelle of its host, just as the ancestors of mitochondria and chloroplasts did.

Filarial disease is very difficult to treat with antiparasitic agents. But because the nematodes require *Wolbachia* to live, human patients infected by the nematodes can be treated with antibiotics such as tetracycline that target the *Wolbachia*. The *Wolbachia* disappear from worm tissues, and the worm burden gradually decreases. Antibacterial agents eliminate the worms sooner and better than does treatment with anti-nematode agents.

SECTION SUMMARY

- **Natural selection favors traits that increase survival and reproduction.** Traits that increase the number of offspring inheriting the trait increase in the population.

- **Natural selection depends on the environment.** A trait favored under one set of conditions may be disfavored under other conditions.

- **Use of an antibiotic can lead to rapid selection of microbial strains resistant to the drug being used.** Carefully limited use of antibiotics prolongs their period of effectiveness.
- **Pathogens and parasites evolve specific traits enabling them to take advantage of a host and cause disease.**
- **Pathogens and parasites undergo reductive (degenerative) evolution.** Reductive evolution involves loss of traits that the parasite no longer needs because the functions are carried out by its host.

Thought Question 9.5   Parasitic microbes, such as *Giardia* (the cause of the diarrheal disease giardiasis) and *Plasmodium falciparum* (which causes malaria), undergo reductive evolution, as well as selection for complex traits. Explain why both kinds of natural selection occur.

9.4 Bioinformatics

SECTION OBJECTIVES

- Explain how bioinformatic analysis helps reveal the pathogenesis and physiology of an emerging pathogen.
- Describe how gene duplication leads to different kinds of homologous genes, including orthologs and paralogs.

If we cannot culture most species of bacteria that reside in, on, or around us, how do we even know they exist? As described in Chapter 2, most bacteria that call humans "home" cannot be cultured in the laboratory. Nevertheless, through various DNA techniques, we know that they exist and can even sequence their genomes, or at least parts of them. Our knowledge about how microbes evolve can now be used to "interrogate" these DNA sequences and predict the traits of our hidden microbiota. The analysis of DNA involves comparing thousands of sequences through the use of computer algorithms—a field of study called **bioinformatics**. This section describes how bioinformatics can lead to startling discoveries about the nature of deadly pathogens and how they evolve.

CASE HISTORY 9.2

Tracking a "Patchwork Quilt" Pathogen

Nine-month-old Eva had 12 clinic visits over a 7-month period because of a runny nose and/or ear infections. During each episode, nasopharyngeal swabs were obtained for bacterial culture, and *Streptococcus pneumoniae* was revealed. Although all the infections were resolved with antibiotics, scientists gained insight into pathogen evolution by seeing how the isolates compared. Each isolate was first analyzed for seven genes that can indicate strain variability—a process called **multilocus sequence typing (MLST)**. The lab found two MLST types, indicating that the patient was infected with two different strains of *S. pneumoniae*. However, when the complete genomes of several of the isolates were sequenced and compared, <u>four</u> different genotypic strains were identified—all of which evolved as a result of 16 distinct recombination events between strains (**Figure 9.21**). Curiously, the technicians noticed that some recombined sequences did not seem

---

Figure 9.21   Horizontal Gene Transfer and Evolution of *Streptococcus pneumoniae*

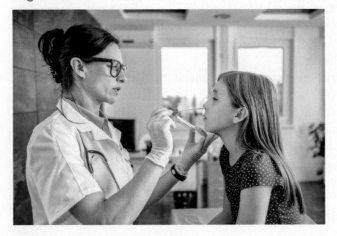

A. Nasopharyngeal sampling of a child.

B. Genome of the patient's isolate, *S. pneumoniae* ST13, and the relative positions of recombination events that incorporated horizontally transferred DNA (purple boxes).

C. Representation of horizontal gene transfers that may have led to the evolution of the last infecting strain (ST13, strain 5) isolated from the patient. Strains that donated DNA for this evolution are shown in red. Donor strains were no longer found in the patient. Note the large size of the horizontally transferred DNA. Almost 10% of the whole-genome sequence—about 2.1 million bases (2.1 Mb)—was replaced.

to come from either of the two primary coinfecting strains. Instead, it appeared that this new DNA came from uncultured and undetected strains of *S. pneumoniae* that had colonized Eva's throat at one time or another. Horizontal gene transfers (transformations) between the undetected and detected isolates appear to have produced, like a patchwork quilt, the new recombined strains of *S. pneumoniae*. The genome sequence data clearly support the idea that *S. pneumoniae* evolution is characterized by high rates of horizontal gene transfer and recombination.

The scientists in this case history propose that *S. pneumoniae* and other related bacterial species use horizontal gene transfers to defeat the adaptive immunity of individual hosts, possibly explaining Eva's recurring infections. These types of bioinformatic comparisons suggest that horizontal gene transfers produce most of the genetic variation that pathogens develop while persisting within a host. As for Eva, her immune system matured as she grew older and successfully fended off any further infections by *S. pneumoniae*.

Annotating the Genome Sequence

How do we compare genomes, such as those of the *S. pneumoniae* strains that colonized Eva's throat? Knowing the sequence is just a first step; meaningful information comes from **annotation** of the DNA sequence—that is, understanding what the sequence means. Annotation is analogous to identifying separate sentences and words in an unknown language. For most languages, you would look for punctuation marks like periods or exclamation points to identify sentences and then scan for blank spaces between letters to signify the beginning and end of a word. You may not initially know what the sentence says, but defining what makes a sentence is the first step.

Annotating a DNA sequence includes finding the translational start codons (usually ATG) and stop codons (TAA, TAG, and TGA) of genes, as well as predicting the function of the gene product. Computers use established rules to mark potential genes—called **open reading frames (ORFs)**. ORF finder programs use the universal genetic code to deduce all possible protein sequences of a double-stranded DNA sequence, examining both strands in the 5′-to-3′direction. Because three bases make a codon, each strand has three reading frames. The goal is to find an ORF, which is defined as a DNA sequence that can potentially encode a peptide of minimum length—say, 50 amino acids or more. Once an ORF is identified, its function is computationally predicted by comparing the deduced amino acid sequences of the ORF with the sequences of proteins having known functions. As discussed in Section 9.1, sequences common to two different protein or DNA molecules are called homologous sequences. Genes for transfer RNAs and ribosomal RNAs do not encode proteins but can be identified because

their sequences are highly conserved across vast phylogenetic distances. The results of annotation can be represented in many different forms.

Sequence alignments can be done either with DNA sequences or with protein sequences deduced from the DNA sequences. Because of the degeneracy of the code, however, it is usually easier to pick up evolutionary relationships by comparing deduced protein products rather than DNA sequences.

Exposing Evolutionary Connections

Finding sequence homologies between genes or proteins suggest an evolutionary relatedness. Genes or proteins that are homologous probably evolved from a common ancestral gene. Homologous genes or proteins can be classified as either orthologs or paralogs (**Figure 9.22**), two designations that further define evolutionary relationships. Genes that are **orthologs** of each other encode proteins that have similar functions but occur in two or more species. For example, the gene for glutamine synthase (*gltB*) in *E. coli* is orthologous to *gltB* in *Vibrio cholerae*. Two species having orthologous genes probably diverged from a common ancestor.

Genes that are **paralogs** arise when one gene, or set of genes, in a single species (or its ancestor) duplicates, and one gene copy evolves to carry out a function different from that of the other copy (Figure 9.22, step 2). If the original gene is essential for survival, only one of the duplicates can evolve to carry out a new function. For example, the intestinal pathogen *Salmonella enterica* uses flagella for motility. However, an unknown ancestor of *S. enterica* long ago experienced a duplication of the flagellar genes, one set of which genetically evolved into a type of protein secretion system called the type III secretion system. This particular system injects virulence proteins directly into eukaryotic cells. These virulence proteins commandeer host cell signaling pathways, forcing the host cell to engulf the pathogen and accommodate its needs. Because the flagellar and secretion genes are genetically related but have different functions, they are called paralogous.

However, the ancestor of *Salmonella* that contained the paralogous genes eventually diverged into two new species: *Salmonella* and *Yersinia*. Each species still has paralogous flagellar and secretion genes, but because mutations accumulated in the genes after each species diverged, the flagellar genes in *Salmonella* are also considered orthologous to the flagellar genes in *Yersinia*; the same is true for the secretion genes.

Many computer programs and resources used to analyze DNA and protein sequences are freely available on the Internet. For example:

- **BLAST** (Basic Local Alignment Search Tool) compares a sequence of interest with all other DNA or protein sequences deposited in sequence databases.

Figure 9.22 Paralogous versus Orthologous Genes

Ancestral flagellar genes

1 Gene A is duplicated.

2 Divergence. Copies of A mutate and diverge to have different functions. A1 gene (flagella) is paralogous to new gene A2 (secretion).

Species w

Speciation

3 Divergence of other genes leads to two spieces, X (Salmonella) and Y (Yersinia) with different versions of A1 and A2.

Paralogous — Orthologous — Paralogous — Orthologous

Salmonella   **Yersinia**

4 Gene A1 in species X is **orthologous** to gene A1 in species Y. Gene A2 in species X is **orthologous** to gene A2 in species Y.

An ancestral gene can undergo a duplication to evolve an orthologous (same function, different species) or paralogous (same species, different function) gene. Subsequent divergence of gene functions can produce different species. The example begins with an ancestral species that has the ancestral flagellar gene marked A. (1) During evolution the A gene is duplicated, forming two copies. (2) Mutations in the duplicate gene, over millennia, change its function, in this case to a paralogous secretion system (A2). (3) Additional mutations, additions, and horizontal gene transfers occur as the cells proliferate, eventually yielding two genera: X (Salmonella) and Y (Yersinia). (4) Each contains flagellar genes that are orthologous to each other, and secretion system genes that are also orthologous. However, the secretion genes are paralogous to the flagellar genes.

- **National Microbial Pathogen Data Resource, or** <u>Patho</u>-systems <u>R</u>esource <u>I</u>ntegration <u>C</u>enter **(PATRIC),** includes annotations of microbial pathogen genomes.

A word of caution: It is enticing to make definitive proclamations about gene or protein function on the basis of computer analysis of a DNA sequence. As good as a prediction may seem, it is only a prediction, a well-educated guess. Biochemical confirmation of function must be made, if possible. Nevertheless, the predictions we can make are powerful. As one example, a metabolic model was constructed by genome annotation for *Helicobacter pylori*, the causative agent of gastritis (ulcers). Scientists noted that genes for some amino acid biosynthesis pathways were missing from the *H. pylori* genome and could predict which amino acids or other compounds (for example, purines) the organism would require for growth (**Figure 9.23**). The metabolic requirements of *H. pylori* suggest possible target enzymes for new antibiotics. Predicting possible gene function(s) from bioinformatic data is an example of **functional genomics**.

Bioinformatics provides new insight into pathogenesis, physiology, and evolution. For example, the DNA sequence of the bacterium *Rickettsia prowazekii* (the cause of epidemic typhus) revealed many similarities to genes within eukaryotic mitochondrial DNA. This finding led to the conclusion that eukaryotic mitochondria evolved from a rickettsial predecessor that became an endosymbiont (an intracellular symbiotic organism), rather than food, for a eukaryotic cell.

How does bioinformatics expose horizontal gene transfer? Case History 9.2 revealed one example of scientists tracking horizontal gene transfer between multiple strains of a single species, *Streptococcus pneumoniae*, that infected a single person. But how do we find evidence of horizontal gene transfer between the <u>ancestors</u> of today's species? One approach is to compare the proportion of GC versus AT base pairs (also known as GC content) along a chromosome. A horizontally acquired gene often has a GC content very different from that of the rest of the chromosome. Within a chromosome, a **genomic island** is a section of DNA sequence that appears to have originated by horizontal transfer but no longer shows mobility. A genomic island generally contains many genes serving a common function. For example, a **pathogenicity island** is a genomic island that encodes virulence factors (discussed in Section 18.1). These islands are found in bacteria pathogenic for plants and animals but not in related nonpathogenic strains. *Salmonella enterica* serovar Typhimurium (previously called *Salmonella typhimurium*) contains three major pathogenicity islands, each with a role specific to a different stage of disease (discussed in Chapter 18).

Figure 9.23 *Helicobacter pylori* Metabolism Predicted from DNA Sequence Information

Notice the vast amount of information you can learn about a cell just from sequencing the genome—a process involving DNA sequencing, sequence assembly, and gene annotation that can now be completed in just a few days. The large rectangle represents a cell. Many transport systems predicted from genome annotations are shown in the cell's membrane. The main components of *H. pylori*'s central metabolism are presented within the rectangle. This scheme is based on using glucose as the sole carbohydrate source. A question mark is attached to pathways that could not be completely elucidated. Red arrows represent pathways or steps for which no enzymes were predicted from the sequence (for example, phenylalanine, methionine, and purine synthesis). Pathways for macromolecular biosynthesis (RNA, DNA, and fatty acids) were found but are not shown in this model.

Legend:
- Amino acids, peptides, and amines
- Anions
- Carbohydrates, organic alcohols, and acids
- Cations
- Nucleosides, purines, and pyrimidines
- ATP-binding cassette (ABC) transporters

SECTION SUMMARY

- **Annotation** requires computers that look for patterns in DNA sequence. Annotation predicts regulators and genes that specify protein and RNA products. Similarities in protein sequence (deduced from the DNA sequence) are used to predict protein structure and function.

- **An open reading frame (ORF)** is a sequence of DNA predicted by various sequence cues to encode an actual protein.

- **DNA alignments** of similar genes or proteins can reveal gene functions and evolutionary relationships. Homology between two genes in the same organism whose products have <u>different</u> functions (**paralogous genes**) indicates an ancestral gene duplication. Homology between two genes with the same function in different organisms (**orthologous genes**) suggests that the two organisms are of different species that evolutionarily diverged.

- A DNA sequence with a **GC content** different from that of flanking chromosomal DNA is one sign of horizontal gene transfer.

- **Pathogenicity islands** are the result of horizontal gene transfers that improve the pathogenicity of a recipient.

Thought Question 9.6   An ORF 1,200 base pairs long can encode a protein of what size and molecular weight? (*Hint*: Find the molecular weight of an average amino acid.)

9.5 Microbial Taxonomy

SECTION OBJECTIVES

- Describe how we define distinct species of bacteria, and explain why the definition of species is problematic.
- Explain how a microbe is identified, and describe the relative advantages and limitations of a dichotomous key, compared with a probabilistic indicator.

What is a microbial species? Among eukaryotes, a species is defined by the principle that members of different species do not normally breed with each other. Bacteria and archaea, however, reproduce asexually, so interbreeding is not a basis for classification. And even without interbreeding in the traditional sense, microbes manage to transfer genes horizontally between distantly related clades.

Classifying Microbial Species

As genome sequence data became available, scientists thought that quantitative measures of divergence could provide a consistent basis for defining species of microbes. But in some bacteria, such as *Helicobacter pylori*, the genomes of different strains that cause the same disease (gastritis) differ by as much as 7%. On the other hand, strains of *Bacillus* with nearly identical genomes cause completely different diseases, such as anthrax (*B. anthracis*) and caterpillar infection (the biological pesticide *B. thuringiensis*). Because of the rather fluid, if not rapid, nature of DNA exchanges among bacteria and archaea, some researchers argue that the species concept holds little meaning for microbes.

Despite the arguments, microbiologists generally agree on the importance of two criteria: phylogeny (based on DNA relatedness) and ecology (based on shared traits and ecological niche). Phylogeny is used to define a species as a group of individuals with similar sequences of key "housekeeping genes," such as ribosomal and transcriptional genes. Such similarity indicates that they were inherited vertically from parent to offspring. In addition, a species should include individuals that share common traits and a common ecological niche, or "ecotype." Shared traits generally include cell shape and nutritional requirements. A shared niche includes a common habitat and life history (for example, causing the same disease). By these criteria, highly divergent strains of *H. pylori* causing gastritis make up one species, whereas *B. anthracis* (anthrax) and *B. thuringiensis* (caterpillar infection) are different species, despite their highly similar genomes.

Many microbiologists accept the following criteria for a working definition of a microbial species:

- **Small-subunit (16S) rRNA gene sequence similarity ≥95%.** Two bacteria with 95% or greater similarity in 16S rRNA sequence are considered the same genus.

- **Average nucleotide identity of orthologs ≥95%.** If whole genomes are available and annotated, we can identify all the orthologous genes (genes of the same function) shared by a pair of strains. If two strains share 95% or greater sequence identity for these genes, they are considered the same species.

- **Shared ecological niche (ecotype).** If two organisms with 95% or greater identity share the same environment (or cause the same disease), have a similar cell shape, and have a common metabolism, they are considered the same species.

Defining a species is part of the task of **taxonomy**, the description of distinct life forms and organizing them into different categories with shared traits. Taxonomy includes <u>classification</u>, the recognition of different classes of life; <u>nomenclature</u>, the naming of different classes; and <u>identification</u>, the recognition of the class of a given microbe isolated in pure culture. Taxonomy is critical for every microbiological pursuit, from diagnosing a patient's illness to understanding microbial ecosystems.

Classification generates a hierarchy of taxa (groups of related organisms; singular, taxon) based on successively narrow criteria. The basis of modern taxonomy is DNA sequence similarity, but historically, taxa were defined and named according to genetic and phenotypic traits. Levels of taxonomic hierarchy, or rank, above genus were defined as phylum, class, order, and family (Table 9.1).

Table 9.1

Classification: Hierarchy of Taxa for an Antibiotic-Producing Bacterium

Taxon Rank	A Long-Studied Taxon
Domain	Bacteria
Division (phylum)	Actinobacteria Filamentous Gram-positive
Class	Actinobacteria High GC; Gram-positive
Subclass	Actinobacteridae
Order	Actinomycetales Filamentous; acid-fast stain
Family	Streptomycetaceae Filamentous; hyphae produce spores
Genus	*Streptomyces*
Species (date first described)	*Streptomyces coelicolor* (1908)

Some levels of rank are designated by certain suffixes—for example, "-ales" (order) and "-aceae" (family). The ultimate designation of a type of organism is that of species. A species name includes the capitalized genus name (group of closely related species) followed by the lowercase species name—for example, *Streptomyces coelicolor*, for a soil bacterium that produces streptomycin and other antibiotics.

The nomenclature of microbes remains surprisingly fluid. As new traits are identified and genomes are sequenced, species are all too often renamed. For example, the species *Escherichia coli*, a normal member of our gut microbiota, was called "*Bacterium coli commune*" in 1885. It took more than 70 years to be renamed in 1958 for its discoverer, Theodor Escherich. In 2016, the notorious cause of pseudomembranous enterocolitis, *Clostridium difficile*, was renamed *Clostridioides difficile* because 16S rRNA and ribosomal protein sequences indicated that the organism had actually diverged from the *Clostridium* genus. Physicians, much to their relief, can still call it *C. diff*.

⊙ **Note:** Taxonomic categories generally have two forms: formal and informal. The formal term is capitalized, with a latinized suffix: Actinomycetales, *Pseudomonas*, *Micrococcus*. The informal term is lowercase, sometimes with an anglicized ending, and informal references to genera are not italicized: actinomycetes, pseudomonads, micrococci.

Identifying a Microbe

Once a species has been described and classified, we need a way to identify it from natural and host environments. Identification poses special difficulties with microbes, which by definition are invisible to the unaided eye. Even under the electron microscope, thousands of different species may show similar shape and form.

The most conclusive way to identify an isolate is to sequence part or all of its genome. In clinical practice, DNA-based methods are used more and more. Nevertheless, in the clinical lab—as well as in field and environmental microbiology—it's convenient to narrow down the "suspects" by using various easily determined traits. Such traits include cell shape, staining properties, and metabolic reactions. Thus, practical identification is based on a combination of phylogeny (relatedness based on DNA sequence divergence) and phenotypic traits.

A common strategy of practical identification is the **dichotomous key**, in which a series of yes/no decisions successively narrows down the possible categories of species. **Figure 9.24** shows a dichotomous key used to identify filamentous bacteria found during wastewater treatment. Filamentous bacteria are important in wastewater because their entangled filaments interfere with the settling of bacteria out of the treated water. In the key shown, most traits are phenotypic, such as cell size and motility. Most branch decisions have two choices, although one juncture (cell motility and shape) has four. The key identifies some organisms down to the species level (for example, *Sphaerotilus natans*), others to the genus level (such as *Flexibacter*), and still others only to numbered samples whose characterization remains incomplete (for instance, filamentous bacteria types 0914 and 021N).

A disadvantage of the dichotomous key is that it requires time-consuming steps. In the clinical setting, time is critical in identifying a potential pathogen and prescribing appropriate treatment. Another means of identification is the **probabilistic indicator**—a battery of biochemical tests performed simultaneously on an isolated strain. The indicator requires a predefined database of known bacteria from a well-studied habitat, such as Gram-negative bacteria from the human intestinal tract. The database can be used to identify a newly isolated specimen from a patient. When the test results are obtained, the probabilities for all results, negative or positive, for a given species are multiplied to generate a probability score for the species. For each species in the database, a score is computed, and the scores are sorted. The top-scoring species is then taken as the most probable identification. **Figure 9.25** shows an example of a Biolog test battery that identifies an isolate as *Stenotrophomonas maltophilia*, an antibiotic-resistant organism that often infects tracheostomy tubes and urinary catheters. In contrast, identifying this organism by a dichotomous key would have required several steps over

Figure 9.24 Dichotomous Key for Wastewater Bacteria

A. A section of a dichotomous key for identifying Gram-negative bacteria from wastewater. The designations 0961, 0914, and 021N refer to unidentified species.

B. *Thiothrix* species, a filamentous wastewater bacterium. Rossetti et al., *Int J Syst Evol Microbiol*. 2003;**53**(Pt 5):1271–1276.

C. Filament 021N, an unclassified organism that forms starburst filaments.

several days. The rapid identification of such pathogens is critical in order to save the patient.

Another advantage of the probabilistic indicator is its inherent redundancy; if an isolate gives one odd result for its species, it may still show the highest probability score in the database. Note, however, that both the dichotomous key and the probabilistic indicator assume a known set of organisms. To explore and characterize a completely unknown organism—or an isolate with a new property, such as antibiotic resistance—requires sequencing the organism's DNA.

SECTION SUMMARY

- **Microbial species are defined by sequence similarity of vertically transmitted genes.**
- **The species definition should be consistent with the ecological niche or pathogenicity.**
- **Taxonomy is the description and organization of life forms into classes (taxa).** Taxonomy includes classification, nomenclature, and identification.
- **Classification is traditionally based on a hierarchy of ranks.** Groups of organisms long studied tend to have many ranks, whereas recent isolates have few.

Figure 9.25 **Probabilistic Indicator: Biolog Test**

ID = *Stenotrophomonas maltophilia*

The Biolog GEN III plate performs multiple enzyme assays simultaneously. Each positive result involves a redox reaction that converts the dye to purple. The probabilities of each result are multiplied, and a computer program yields the most probable identification.

- **Practical identification includes DNA-based tests and phenotypic traits.** Methods of identification include the dichotomous key and the probabilistic test battery. Both methods assume a predefined set of organisms.

- **Defining a new strain of a microbe requires sequencing its DNA.**

Thought Question 9.7 According to Figure 9.24, how would you identify a straight, nonsheathed, Gram-negative bacterium that has sulfur granules, is motile, and is 1.0 μm wide? What would happen if you assigned the bacterium a width of 0.9 μm?

Perspective

The evolutionary history of microbes is fascinating. With today's whole-genome DNA sequencing technologies and analytical capabilities, scientists can unravel how species diverge to inhabit new ecological niches, including the human body. We now understand, to a large extent, how pathogens evolved to secrete toxins, manipulate their hosts, and develop antibiotic resistance. The evolutionary history of every pathogenic and nonpathogenic microorganism includes random mutations that sculpt new genes from a duplication of an old gene. This history also includes the addition of new genes from other members of their habitat. Both vertical transmission of mutant sequences and horizontal acquisition of mobile genes contribute to the constant evolution of microbes. It's hard to say which is more astonishing: the idea that all living cells evolved from a common ancestor or knowing that modern life forms have evolutionarily adapted to exploit every possible niche of our planet. In Chapter 10 we will explore some of the amazingly diverse adaptations of different groups of bacteria, including the most important pathogens known to medicine.

Key Terms

annotation (p. 267)
bacteriophage (p. 251)
bioinformatics (p. 266)
competence factor (p. 249)
conjugation (p. 250)
degenerative evolution (p. 265)
dichotomous key (p. 271)
divergence (p. 258)
domain (p. 260)
electroporation (p. 249)
endosymbiont (p. 265)
F⁺ cell (p. 250)
F⁻ cell (p. 250)

fertility factor (F factor) (p. 250)
F-prime (F′) plasmid (p. 251)
functional genomics (p. 268)
generalized recombination (p. 254)
generalized transduction (p. 252)
genetic drift (p. 259)
genomic island (p. 268)
Hfr cell (p. 251)
homologous (p. 249)
homologous recombination (p. 254)
horizontal gene transfer (p. 248)
housekeeping gene (p. 260)
insertion sequence (IS) (p. 255)

microfossil (p. 257)
molecular clock (p. 259)
multilocus sequence
typing (MLST) (p. 266)
natural selection (p. 261)
open reading frame (ORF) (p. 267)
ortholog (p. 267)
paralog (p. 267)
pathogenicity island (p. 268)
phage (p. 251)
phylogenetic tree (p. 259)
phylogeny (p. 259)
probabilistic indicator (p. 271)

prophage (p. 253)
recombination (p. 254)
reductive evolution (p. 265)
site-specific recombination (p. 254)
specialized transduction (p. 253)
taxonomy (p. 270)
transduction (p. 252)
transformation (p. 249)
transposable element (p. 255)
transposition (p. 255)
transposon (p. 255)
vertical gene transfer (p. 249)

Objectives Assessment Questions

9.1 Bacterial Gene Transfers

1. A process of DNA transfer into a recipient cell that requires bacteriophage packaging is called _____.
   a. transformation
   b. transduction
   c. conjugation
   d. transposition

2. All of the following processes require a step of DNA recombination to integrate the incoming DNA into the host chromosome except _____.
   a. transfer of a multidrug resistance plasmid
   b. generalized transduction
   c. specialized transduction
   d. transposition of a mobile genetic element

3. The reason bacteria make restriction endonucleases is _____.
   a. to degrade excess genomic DNA
   b. to prevent replication of an invading bacteriophage
   c. to obtain nucleotides for DNA synthesis
   d. to synthesize DNA

9.2 Origins and Evolution

4. Which statement does not describe a process by which life forms actually change?
   a. Mutations accumulate at random, even in populations under the same selection pressure.
   b. Natural selection favors the prevalence of traits with a fitness advantage in a given environment.
   c. When a pathogen associates with a host, the pathogen undergoes reductive evolution, losing the genes that specify the functions provided by the host.
   d. Genetically distinct organisms emerge from environments, and they prevail over preexisting organisms.

5. Which property is required for a molecular clock gene?
   a. The gene must encode different functions in different organisms.
   b. The gene must acquire mutations at a steady rate over time in all organisms compared.
   c. The gene must encode functions under strong environmental selection, such as resistance to antibiotics.
   d. The gene must transfer readily between distantly related organisms.

6. Refer to Figure 9.17. Of the given genera, which genus is the most distantly related to Escherichia coli?
   a. Salmonella
   b. Klebsiella
   c. Shigella
   d. Vibrio

9.3 Natural Selection

7. Suppose a pathogenic bacterium acquires a plasmid expressing an enzyme that modifies the target of an antibiotic. What is most likely to happen when the antibiotic disappears?
   a. Because the plasmid has a useful trait, the bacterium will increase copy number of the plasmid.
   b. The bacterium will outcompete other bacteria, even when the antibiotic disappears.
   c. The number of bacteria containing the resistance plasmid will decline because replicating the plasmid costs energy.
   d. In the absence of antibiotic, the plasmid will mutate to provide resistance to a different antibiotic.

8. Which statement does not describe an example of reductive (degenerative) evolution?
   a. Treponema pallidum, the causative agent of syphilis, has lost many of the genes that encode enzymes for amino acid biosynthesis.
   b. T. pallidum has evolved virulence genes that enable it to take advantage of host cells.
   c. Wolbachia bacteria that live within host insect cells lack genes for biosynthesis of vitamins.
   d. Chlamydia bacteria are obligate intracellular pathogens that fail to synthesize a peptidoglycan cell wall.

9. When drug-resistant bacteria are isolated from a hospital patient, they may make smaller colonies than the original strain made. The reason is that _____.
   a. drug-resistant bacteria always grow more slowly
   b. smaller colony size provides an advantage in resisting antibiotics
   c. the random mutation that conferred resistance also conferred deleterious effects on growth
   d. the drug-resistant bacteria require different kinds of nutrients

9.4 Bioinformatics

10. Which statement best describes an example of bioinformatic analysis to reveal the capabilities of a pathogen?

   a. *Streptococcus pneumoniae* was cultured on various agar media containing different combinations of nutrients in order to analyze the bacterium's metabolism.

   b. *S. pneumoniae* was cultured with bacteriophage in order to transfer genes between bacteria.

   c. *S. pneumoniae* isolates were selected for genome analysis, which revealed the gene functions that were shared and the ones that were distinct to different species.

   d. Novel isolates of *S. pneumoniae* were constructed by genetic engineering.

11. Which statement best describes a result of gene duplication within a genome?

   a. Two genes within an organism's genome show very similar DNA sequence, but they encode enzymes that act on different substrates.

   b. Two organisms each have a gene that encodes the same function.

   c. An organism has two genes with low homology that have the same function.

   d. The presence of an antibiotic can trigger gene amplification that results in the presence of several copies of an antibiotic resistance gene.

9.5 Microbial Taxonomy

12. Which statement best describes evidence of highly divergent or distantly related organisms?

   a. The genomes of two organisms include distinctly different plasmids.

   b. Between two genomes, the gene sequences that are transcribed to form 16S rRNA show greater than 97% similarity.

   c. Orthologous genes of two genomes share greater than 95% of the same sequence.

   d. Two genomes show very different proportions of GC versus AT content.

13. Pathogen identification in a clinical laboratory is most commonly based on _____.

   a. a combination of DNA-based tests and phenotypic traits

   b. sequencing the entire genome of an isolate directly from a patient

   c. microscopy and staining only because these procedures are quick and inexpensive

   d. immunological tests only because these tests are highly strain-specific

Review Questions

1. What are the basic ways microorganisms exchange DNA?

2. Discuss horizontal versus vertical gene transfer.

3. Describe competence and how it comes about in a population.

4. What is an F factor, and how does it (and other factors like it) contribute to gene exchange?

5. Compare specialized versus generalized transduction.

6. Discuss how bacteria protect themselves from invading bacteriophages.

7. Explain the basic process of transposition. Why are insertion sequences always flanked by direct repeats of host DNA? How are transposons different from plasmids?

8. What is a pathogenicity island?

9. What is the common ancestor of all life, including humans?

10. What are the three domains of life? Which traits are unique to each domain, distinguishing it from the other two?

11. Explain the connection between phylogeny and evolution.

12. What is random mutation? How are the results of random mutation modified by natural selection?

13. Explain how a molecular clock works. What properties are required of a gene to be used for a molecular clock?

14. What is meant by homologous genes? Explain the difference between orthologs and paralogs.

15. What is endosymbiosis? How has endosymbiosis contributed to the evolution of eukaryotic organisms?

16. What is reductive (degenerative) evolution? Why does it occur?

Clinical Correlation Questions

1. Why might a pathogen strain that is defective in its ability to repair mutations (mutator phenotype; see Chapter 8) be a more successful pathogen than a related strain that is highly efficient in mutation repair?

2. *Salmonella enterica* is a Gram-negative species with more than 2,000 serovars (variant strains differing in surface proteins and carbohydrate antigens) that can cause gastrointestinal or bloodstream infections. Most serovars of *S. enterica* can infect human and nonhuman animals alike. However, one serovar, *S. enterica* serovar Typhi (the cause of a blood infection called typhoid fever), infects only humans. Propose an evolutionary model for how *S.* Typhi may have developed this narrow host range.

3. Joshua, a 14-year-old boy, fell seriously ill during the summer of 2013. Fluid had accumulated in his brain, and he was experiencing seizures. Joshua had encephalitis. Sick for 6 weeks, he fell critically ill and was placed into a medically induced coma. Standard tests failed to identify any bacteria, viruses, or fungi in his spinal fluid as the cause. However, there was evidence in his spinal fluid; the clinicians just had to find it. Which of the techniques you've learned about in Chapters 8 and 9 would you use to discover the cause? This is a real case. Once the medical team came up with the strategy, it took only 48 hours to find the cause and determine effective treatment.

 For more assessment, visit digital.wwnorton.com/michum2

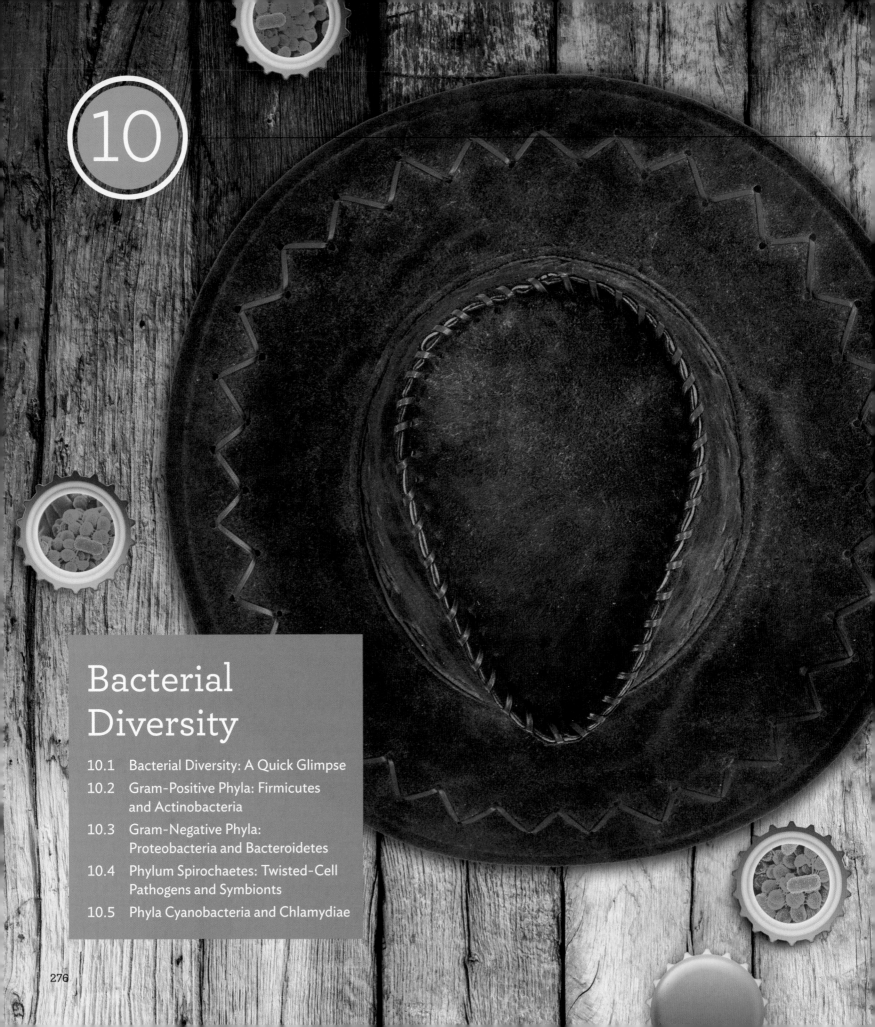

10

Bacterial Diversity

"You Should See the Other Guy!"

SCENARIO On a Friday night, Cody, a 32-year-old Fort Worth, Texas, rancher, drove to Malone's Pub to shoot pool. After drinking several beers, he got into a brawl with Ty, who was also drunk. A quick right cross to the face made short work of Ty, but Cody received a deep bite wound to his fist in the process. He washed and wrapped the wound, but by Wednesday, the wound was extremely painful and warm, and had pus oozing from it. Reluctantly, he drove to the local hospital for treatment.

A clenched-fist human bite wound
The infection is polymicrobial, involving multiple species from diverse phyla colonizing the human mouth.

TREATMENT Even before knowing the culture results, the physician debrided the wound (removed dead tissue) and cleaned the area by irrigating it with dilute hydrogen peroxide solution. Surgical consultation indicated the wound should not be sutured but instead wrapped in sterile gauze. The antibiotic amoxicillin was prescribed, and because Cody was unsure of his tetanus immune status, he was given a tetanus vaccination. Within 2 weeks, the wound healed and Cody returned to the Pub. Ty, who was also there, walked up to Cody, apologized, and bought him a beer.

SIGNS AND SYMPTOMS At the emergency department, Cody's hand had a deep, inch-long wound above one knuckle. The wound was red, swollen, warm to the touch, and so painful that Cody couldn't move his fingers. He had a fever of 37.9°C (100.2°F), but his lymph nodes were not swollen. The examining physician also noted crepitus, a bubbling feel or sound when the wound area was pressed. Crepitus can indicate the presence of gas from an anaerobic infection.

TESTING An X-ray ruled out broken bones. Meanwhile, swabs of the wound material were sent to the lab for culture. If Cody's fever continued to rise, the physician would have to order a complete blood count and blood cultures to test for sepsis, a serious blood infection. Culture results after 48 hours yielded a polymicrobial infection that included mouth and skin bacteria identified as *Eikenella corrodens*, *Streptococcus pyogenes*, *Fusobacterium nucleatum*, and *Porphyromonas gingivalis*.

DIAGNOSIS Cody had cellulitis, an infection of the soft tissues, excluding muscle.

TEM   0.5 μm

Eikenella corrodens
A Gram-negative member of the phylum Proteobacteria. It is a common participant in clenched-fist human bite infections and a sign that the bite came from a human, not a dog.

Cody's bite wound is a common result of many ill-advised, bare-knuckle fistfights. The skin on the hand of a person throwing a punch is easily broken by the teeth of the person being punched. On impact, a variety of microbes from the skin and the oral cavity enter the wound, causing a **polymicrobial** infection—that is, an infection involving diverse kinds of bacteria.

Cody's polymicrobial infection included members of several widely dissimilar bacterial phyla. *Eikenella corrodens* belongs to the Gram-negative, highly diverse phylum Proteobacteria. *E. corrodens* is the most common organism isolated from human bite wound infections. *Streptococcus pyogenes* is a tiny, spherical, Gram-positive member of the phylum Firmicutes that causes a wide variety of infections. Firmicutes are overwhelmingly Gram-positive bacteria. Finally, *Porphyromonas* and *Fusobacterium* are anaerobic, Gram-negative, rod-shaped organisms belonging to the Bacteroidetes and Fusobacteria phyla, respectively. Their production of gas through fermentation contributed to the crepitus felt by the physician.

So, one infection was caused by bacteria from three widely diverse phyla, all coming from one source, a human mouth. But it could have been worse. Bacteria from many other phyla residing on the skin and in the mouth can cause similar clenched-fist wound infections. In fact, a person's teeth can contain members of nearly all the major branches of the bacterial family tree. People who fail to brush their teeth regularly and lack regular dental care exhibit the most diversity—and the most cavities.

Why is understanding bacterial diversity important? Aside from the ways diverse microbes maintain the health of Earth (introduced in Chapter 6), you will soon learn that they also nurture, and sometimes threaten, human health. Almost every bacterial phylum is represented in the human microbiome, which is the collection of bacteria that naturally colonize our bodies. These diverse bacterial hitchhikers, with their varied proteins, lipids, enzymes, and so on, hone our immune system, help digest our food, and communicate with our organ systems, including the brain. Without a microbiome, we don't exist. But the ability of these diverse microbes to colonize the human body also comes with some danger. Human microbiota from many different phyla can sometimes cause disease. That's what happened to Cody.

The bewildering diversity of bacteria in, on, and around us poses a challenge to the physician who has to diagnose disease—and to the environmental microbiologist who studies how microbes contribute to ecosystems. Deadly pathogens often turn out to be close relatives of bacteria that enhance our environment and improve human health. For example, the Shiga toxin–producing strains of diarrhea-causing *Escherichia coli* are closely related to *E. coli* strains that help digest our food. Similarly, *Bacillus anthracis*, a soil microbe that causes anthrax, is very closely related to *Bacillus thuringiensis*, a natural insecticide (Bt) that doesn't infect humans but kills inchworms. Understanding how different phyla produced today's pathogens may help scientists predict future emerging pathogens. In this chapter we will explore the diversity of bacteria that evolved by the processes outlined in Chapter 9.

10.1 Bacterial Diversity: A Quick Glimpse

What is out there? The genetic diversity of the bacterial domain is overwhelming. Many bacteria differ more from each other than any animal differs from any plant! Yet there exists another entire domain of microbes, the archaea, that look superficially like bacteria but diverged from both eukaryotes and bacteria more than 3 billion years ago (Figure 10.1). The archaea include microbes living in extreme habitats, such as hyperthermophiles in boiling-hot springs and acidophiles in mine drainage acidic enough to dissolve a shovel. Some archaea, the methanogens, even live within our intestines, converting bacterial fermentation products into the methane we release by passing gas (explored in Chapter 14). Archaea play interesting roles in our environment, as you'll learn in Chapter 27.

Even among the bacteria, some types, called "deeply branching" forms, diverged soon after bacteria diverged from archaea.

Figure 10.1 Bacterial Phylogeny

Bacterial phylogeny (relatedness) determined on the basis of 16S rRNA sequence comparisons. Red font marks phyla and genera that include human pathogens. Blue font indicates phyla without known human pathogens. Alpha, beta, gamma, and delta/epsilon are classes of the Proteobacteria phylum.

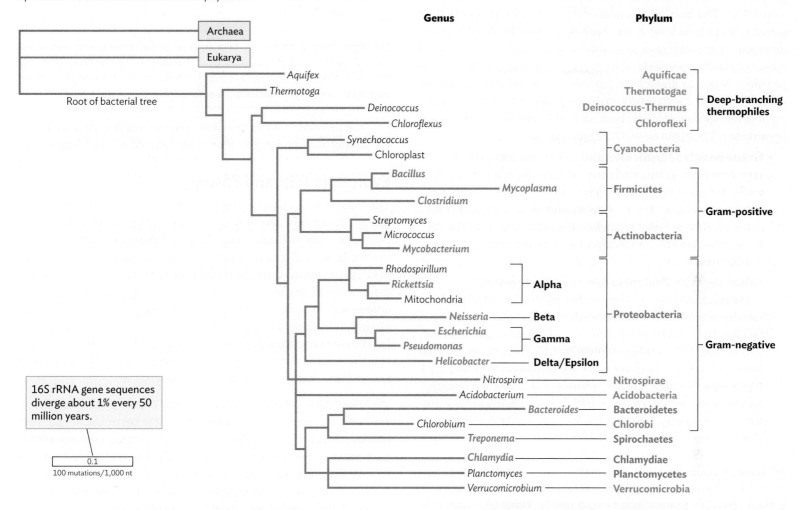

Deeply branching bacterial phyla include *Aquifex* species, which grow in the hot springs of Yellowstone National Park. Another example is the *Deinococcus* species, bacteria that resist extremely high doses of ionizing radiation. These superbacteria are being used to remediate (detoxify) waste material contaminated with radioactivity released from nuclear power plants.

The Diversity of Life

How much diversity exists in the microbial world? There are 30 phyla of bacteria from which species can be cultured in the laboratory. DNA sequencing data suggest the existence of 20 or so other "candidate" phyla that have yet to be cultured. But most amazingly, microbiologists estimate there may really be over 1,000 more phyla we have yet to see.

Why don't we accurately know how much diversity exists? The reason is that we can find only those organisms whose genes can be amplified by PCR (discussed in Chapter 8). But the DNA primer sequences we use to amplify those genes are based on genome sequences of organisms we already know. So, any new bacterial genes we amplify from environmental samples will, by definition, be related to organisms we've previously found. And of those we do find, less than 1% can be cultured to study their metabolism and physiology. Nevertheless, that 1% includes a vast range of species, from psychrophiles growing in your refrigerator to thermophiles growing in your compost pile.

Note: Genome sequences that guide the choice of PCR primers come from microorganisms that have been "isolated" in pure culture from clinical or environmental samples. The isolated organism is called an **isolate**. Isolates whose rRNA sequences are 97% similar usually share a common genus. Two or more isolates of the same species that display genetic differences are called **strains**.

Evolutionary connections for major phyla that impact human health are shown in Figure 10.1. As described in Chapter 9, evolutionary relatedness is measured by comparing the sequences of genes encoding 16S rRNA, the highly conserved small-subunit ribosomal RNA. The number of sequence differences in rRNA-encoding genes reflects how long it has been since two kinds of bacteria diverged from a common ancestor. Here we introduce the seven major phyla or superphyla of bacteria that you should know. You probably hoped to find all the pathogens lumped into one group. As you will see, nearly every major phylum of bacteria includes pathogenic species that evolved alongside species of ecological importance. The phyla we will highlight include:

- **Gram-positive Firmicutes and Actinobacteria.** Most members of these two phyla are heterotrophs with thick cell walls that resist drying. Some **Firmicutes**, such as *Clostridium*, form dormant endospores. **Actinobacteria**, also called actinomycetes, grow as radial filaments (long chains of cells) similar to fungi and include *Streptomyces*, soil bacteria that produce antibiotics.

- **Gram-negative Proteobacteria and Bacteroidetes.** **Proteobacteria** is a very diverse phylum that includes species capable of a wide range of metabolism, from heterotrophs (such as the *E. coli* that inhabits animal digestive tracts) to soil and wetland lithotrophs and photosynthesizers. Many are used in biotechnology. Members of the **Bacteroidetes** phylum are all Gram-negative rods; most are obligate anaerobes. *Bacteroides* species, for example, are gut microbiota that help us digest our food. After surgery, however, these same species may cause abscesses.

⊙ **Note:** Proteobacteria is a large phylum that is subdivided into nine classes. The classes, and some representative pathogens, are Alphaproteobacteria (*Brucella*), Betaproteobacteria (*Bordetella*, *Neisseria*), Gammaproteobacteria (*Escherichia*, *Salmonella*, *Yersinia*), and so on. Recent genome sequence data have led taxonomists to propose redefining the classes as phyla and elevating Proteobacteria to superphylum status.

- **Spirochaetes.** Also called spirochetes, **Spirochaetes** share a distinctive form: tightly coiled cells enclosed by a sheath, with periplasmic flagella that run underneath the sheath along the cell body. Some spirochetes live in soil or water or as digestive symbionts of termites. Other spirochetes are pathogens, such as *Borrelia burgdorferi*, the cause of Lyme disease.

- **Cyanobacteria.** These phototrophs fix carbon dioxide into biomass. **Cyanobacteria** include tiny marine bacteria, as well as massive filamentous species. Although there are no human pathogens in this phylum, cyanobacteria are the only bacteria that produce oxygen—and we need oxygen to breathe.

- **Chlamydiae.** The bacteria of the phylum **Chlamydiae** are obligate intracellular pathogens that grow as inclusion bodies within host cells and cause pneumonia, the eye disease trachoma, and the sexually transmitted infection called chlamydia.

⊙ **Note:** Formal names for taxa such as phyla, orders, and genera are capitalized (for example, phylum Cyanobacteria and genus *Streptococcus*). Informal names are lowercase roman (cyanobacteria, streptococci), as are adjectival forms (cyanobacterial, streptococcal).

The sections that follow include examples of how pathogens and colonizing bacteria can emerge out of the vast bacterial family.

Evolution: Past and Future

The evolution of today's pathogens took place over millennia, and just as typewriters gave way to computers, the ancestors of today's pathogens may no longer exist. Chapter 9 described how we traced the evolution of today's species by uncovering roles for mutations, gene duplications, horizontal gene transfers, and reductive evolution—all in the name of one species improving its fitness to compete in heavily populated niches or to expand into new, previously uninhabited ones. These studies have produced intriguing hypotheses about how pathogens evolved and continue to emerge. One important line of inquiry asks, Where do pathogens go to evolve? If we know that, maybe we can stop them.

INCUBATORS OF PATHOGEN EVOLUTION Pathogens today continue to evolve, and occasionally a new version emerges that causes human or animal disease. Examples include the organisms that cause Lyme disease, cat scratch disease, and some forms of bloody diarrhea. You might wonder where these microbes came from and how they managed to "adopt" genes from other phyla. What are the "incubators" of pathogen evolution?

New pathogens brew in three basic evolutionary incubators: zoonotic hosts, shelter species, and the environment (**Figure 10.2**). As described in Chapter 2, some organisms cause disease primarily in animals and can occasionally infect humans. These animal diseases are called **zoonoses** (singular, **zoonosis**). Within the infected animal, the pathogen will interact with all kinds of bacterial species that make up natural animal microbiota. The genes of these microbes can be sampled by the pathogen. Incorporating the right gene, or combination of genes, could help the pathogen evolve into a new species able to infect humans.

In one scary example, scientists recently discovered a new strain of *Enterococcus faecalis*, a common gut microbe, which contains a gene whose product is a botulism-like exotoxin. Botulism toxin, normally made by *Clostridium botulinum*, interferes with neuromuscular nerve transmissions to render its victim unable to move or breathe. So far, this toxigenic strain of *Enterococcus* is found only in cows, not humans. However,

Figure 10.2 "Incubators" of Pathogen Evolution

Pathogens evolve from nonpathogens by horizontal gene transfers of DNA from microorganisms. Horizontal gene transfers into an ancestor organism (probably now-extinct) happen in zoonotic situations, inside of shelter organisms (worm), or in the natural environment (soil). The new pathogen, now capable of infecting humans, can be directly transferred to humans (zoonotic or natural) or indirectly transferred from shelter organisms by intermediate passage through another animal.

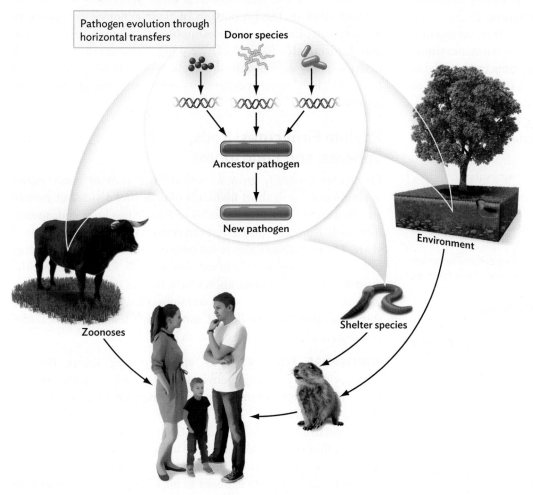

Of course, the natural environment is an obvious genetic melting pot where genes are sampled from thousands of species. The natural environment is where some genes causing antibiotic resistance originated, where plasmids and transposons first evolved, and where phages became phages. These niches and opportunities all shape the evolution of bacteria, including pathogens, and contribute to their diversity.

MANIPULATING EVOLUTION If the genomes of today's life forms reveal their past descent, what can they tell us about the future? Could this information help us predict, or even direct, future evolution? And what about engineering completely new organisms? Is this even possible?

Bioinformatic insights into evolution explained in Chapter 9 have expanded the limits for manipulating evolution in the lab. Imagine a chemical industry that contaminates soil with pollutants, such as chlorinated aromatics, that no current species of soil bacteria can eat. Future scientists might be able to accelerate evolution by predicting which gene from which species could best evolve by duplication and mutational divergence into a new paralogous gene whose enzyme product could break down the chlorinated pollutant (revisit Section 9.4 for more on the evolution of paralogs). The scientists could accelerate natural selection by duplicating that gene in the soil microbe, introducing targeted mutations into it, and selecting for a new strain (species?) that could use the pollutant as a carbon or energy source. Such designer bacteria would make valuable agents of bioremediation, the process of using bacteria to remove chemical contamination (discussed in Chapter 27). If you think what we've described is pure science fiction, google "Frances Arnold" and "directed evolution."

E. faecalis is also a member of the human gut microbiome. Fortunately, the botulism-like toxin doesn't currently work on human cells (or on cows). However, it is frightening to think we might someday become colonized by a microbe whose toxin could evolve into one that could paralyze us.

Some hosts, called **shelter species**, are organisms that do not interact directly with humans but do provide shelter for bacteria to evolve. Shelter species include nematodes, earthworms, insects, protozoa, and even plants and larger animals. Mixed communities of bacteria and phages within a worm or other shelter species can lead to genetic exchange and evolution of the bacterium. Over time, some combination of genes could enable a bacterium to infect across a species barrier—to an animal, plant, or human.

What about pathogens? Certainly, we must ban all genetic tinkering that would make a new, or more deadly, pathogen. But might we use evolutionary strategies to render a pathogen harmless or even extinct? Could we one day molecularly "vaccinate" someone with transducing bacteriophages able to infect a specific pathogen and delete a critical virulence gene—a toxin, perhaps? In fact, the genetic targeting apparatus of the new gene excision technology CRISPR (see Chapter 1) can be packaged into a phage capsid. Lacking its virulence gene, a targeted pathogen would be crippled, unable to cause disease.

A final question to ponder is: What about synthetic organisms? Scientists have successfully engineered new microbes that have unique synthetic bases incorporated into their DNA. Soon researchers will be able to create organisms with wholly unnatural attributes and traits not found elsewhere in nature. These bacteria could produce new kinds of proteins that might be turned into drugs designed to treat a range of diseases. To which phylum would they belong? And can we prevent these organisms from evolving on their own? These are not questions for the future; we need to start addressing them now.

Now that we've discussed why taxonomy is important, in the next section let's examine the major bacterial phyla that most directly impact human existence.

SECTION SUMMARY

- **Bacteria and archaea** are two groups of prokaryotes that diverged very early in the history of Earth's biosphere.
- **Major phylogenetic groups of bacteria** include Gram-positive Firmicutes and Actinobacteria, Gram-negative Proteobacteria and Bacteroidetes, Spirochaetes, Cyanobacteria, and Chlamydiae.
- **New species and strains continue to emerge** through evolution.
- **Pathogens evolve by sampling genes** spread by zoonoses, within shelter species, and from the environment. Modern technologies can even direct evolution in the lab.

Thought Question 10.1   What impact might global climate change have on bacterial diversity and pathogen evolution?

10.2 Gram-Positive Phyla: Firmicutes and Actinobacteria

SECTION OBJECTIVES

- Describe the traits that enable Firmicutes to survive dryness and persist for long periods.
- Describe how diverse Actinobacteria have both positive and negative influences on human health.

Which phyla include bacteria that make spores able to persist for thousands of years? A large group of related bacteria are called the "Gram-positive bacteria" because most members stain Gram-positive (Table 10.1). The Gram-positive bacteria include two major subgroups: the phyla Firmicutes and Actinobacteria. The name "Firmicutes" derives from Latin meaning "tough skin." Indeed, these bacteria have thick, tough cell walls, with multiple layers of peptidoglycan threaded by supporting molecules such as teichoic acids, a type of sugar-phosphate chain (see Figure 5.10).

The thick, reinforced cell wall is what retains the Gram stain, especially in Firmicutes. Some Actinobacteria (*Mycobacterium* species) also possess a thick, waxy coat that actually excludes the Gram stain, despite their genetic affiliation with Gram-positive bacteria.

Firmicutes and Actinobacteria differ genetically in their GC content—that is, the proportion of their genomes consisting of guanine-cytosine base pairs (as opposed to adenine-thymine base pairs). Firmicutes have a relatively <u>low</u> GC content (45%–60%), whereas many actinobacteria have a relatively <u>high</u> GC content (60%–70%). GC content differences commonly indicate evolutionary divergence.

Phylum Firmicutes: Rods, Spheres, and Endospores

Firmicutes generally grow as well-defined rods or cocci, either as single cells or in short filaments of cells that divide but remain attached end to end. Some species of Firmicutes form **endospores**, inert heat-resistant spores that can remain viable for thousands of years. The endospores also resist desiccation (drying), high-energy radiation, freezing, and chemical disinfectants. Endospores are the most durable type of spore known to be formed by bacteria. By comparison, other bacteria, such as actinomycetes, form spores that are starvation resistant but do not survive high temperatures. Endospore-forming bacteria are common in soil and air.

ENDOSPORE FORMERS   The spore-forming genus *Bacillus* includes species of major environmental and economic importance. For example, caterpillars can destroy certain crops. *Bacillus thuringiensis* (Figure 10.3A) kills caterpillar pests by infecting their digestive tract; the high pH of the insect gut interior activates the crystalline toxin released by the bacteria. *Bacillus* species grow best by aerobic respiration; they form large, cream-colored colonies on a nutrient agar plate exposed to air (Figure 10.3B). Scientists

Figure 10.3 *Bacillus* Species

A. Sporulating cell of *Bacillus thuringiensis*, showing crystalline inclusion of insecticidal toxin (colorized TEM).

B. Colonies of *Bacillus subtilis* cultured on broth agar. Note the flat, irregular form of each colony.

Table 10.1
Selected Gram-Positive Firmicutes and Actinobacteria

Firmicutes	Endospore Formers	Characteristics
	Bacillus anthracis	Rod-shaped soil bacterium; causes anthrax
	Bacillus subtilis	Rod-shaped soil bacterium; major model system for research
	Bacillus thuringiensis	Rod; insecticide used against moth caterpillars
	Clostridioides difficile	Rod with bulging spores; causes intestinal disease in patients whose normal biota are diminished by antibiotics
	Clostridium acetobutylicum	Rod; used industrially to produce butanol
	Clostridium botulinum	Rod with bulging spores; causes botulism; produces Botox
	Clostridium tetani	Rod with bulging spores; causes tetanus
	Non–Spore Formers	**Characteristics**
	Enterococcus spp.	Cocci; enteric microbiota
	Lactobacillus spp.	Rods; dairy culture
	Lactococcus spp.	Cocci; dairy culture
	Listeria spp.	Rods; intracellular pathogens; grow at refrigerator temperature
	Ruminococcus spp.	Cocci; cellulose-digesting symbionts of cattle
	Staphylococcus aureus	Coccus; infects skin; causes toxic shock syndrome and MRSA (methicillin-resistant *S. aureus*)
	Staphylococcus epidermidis	Coccus; skin microbiota
	Mollicutes (Lack Cell Wall)	**Characteristics**
	Mycoplasma genitalium	Ameboid shape; genital microbiota
	Mycoplasma pneumoniae	Ameboid shape; causes pneumonia
Actinobacteria	**Actinomycetes**	**Characteristics**
	Actinomyces israelii	Forms branched mycelial filaments; causes actinomycosis
	Streptomyces coelicolor	Forms branched mycelial filaments; produces many antibiotics
	Other Actinobacteria	**Characteristics**
	Corynebacterium diphtheriae	Irregular rod; causes diphtheria
	Cutibacterium acnes (formerly *Propionibacterium acnes*)	Rod; causes acne
	Micrococcus luteus	Coccus; common airborne bacterium; forms yellow colonies
	Mycobacterium leprae	Short rod; acid-fast; causes leprosy
	Mycobacterium tuberculosis	Short rod; acid-fast; extremely thick cell wall with mycolic acids; causes tuberculosis
	Propionibacterium freudenreichii	Rod; makes Swiss cheese

can isolate *Bacillus* species from soil or food by suspending a sample in water and heating it at 80°C (176°F) for half an hour. The elevated temperature kills vegetative cells (cells undergoing binary fission). But endospores remain until the temperature cools, at which point they can germinate and grow.

If the plate of *B. subtilis* shown in Figure 10.3B sits for several days at room temperature, the cells will starve for nutrients; many of them will sporulate, developing endospores (discussed in

Figure 10.4 Sporulation and Germination Visualized in *Bacillus subtilis*

A. Unequal cell division leads to septation near one pole, forming the forespore, which develops into the endospore.

Chapter 6). **Figure 10.4A** shows some of the stages of sporulation of *B. subtilis*. To produce endospores, the cell divides near one of its poles instead of at the middle. The polar compartment becomes the forespore, directed by unique regulatory proteins such as the one that is fluorescently labeled green in **Figure 10.4B**. The larger compartment, called the mother cell, provides DNA and nutrients to the growing forespore. Once the mature endospore is released, the mother cell compartment disintegrates. The free endospore may endure for days, years, even centuries, until it encounters favorable moisture and nutrients that allow it to germinate and restart vegetative growth (**Figure 10.4C and D**).

Unlike *Bacillus*, most species of the genus *Clostridium* are obligate anaerobes. Their cells sporulate in a form distinct from that of *Bacillus* species. The growing endospore swells the end of the cell, giving the cell a club shape (**Figure 10.5A**). *Clostridium* species include the causative agents of two well-known diseases: tetanus (*C. tetani*) and botulism (*C. botulinum*). The botulism toxin, **botulinum**, or "Botox," is famous for its therapeutic use to relax muscle spasms and smooth wrinkles in the skin (**Figure 10.5B**). Other species, such as *C. acetobutylicum*, have economic importance as producers of industrial solvents such as butanol and acetone.

Clostridium spores are found in soil and water, ready to germinate and grow when the environment becomes anaerobic. Many harmless species grow in the human colon, particularly in infants.

C. Malachite green stain reveals endospores (green).

Spore shell   Vegetative cell

D. Cell emerging from a germinating spore.

B. Fluorophores label the membranes (red) and the chromosomal DNA (blue). The small green patches indicate the position of a protein essential for forespore engulfment to form the endospore. Progression of the engulfment is shown from left to right. DNA translocase moves chromosomal DNA from the mother cell into the forespore during polar septation.

Figure 10.5 *Clostridium* Species: Spore-Forming Anaerobes

A. As *Clostridium* cells sporulate, the developing endospore causes cells to swell into a characteristic club shape.

Before   After

B. The deadly botulism toxin (Botox) from *C. botulinum* is used at low concentration to relax muscle spasms and smooth wrinkles caused by muscle movement. This woman had wrinkles in her forehead because of muscle movement (left frame)—a condition that was improved by the injection of Botox (right frame).

But pathogenic *C. botulinum* can grow within the colon of very young infants, and cause **infant botulism** (also known as floppy baby syndrome). Another species, *Clostridioides difficile* (formerly called *Clostridium difficile*), is emerging as a life-threatening intestinal pathogen, resistant to most antibiotics. There are nearly 500,000 cases per year, with 29,000 deaths, in the United States. Normal human gut biota outcompete *C. difficile*, but when patients are treated with antibiotics in a hospital or as outpatients, the antibiotics will diminish the microbiome, thereby providing an opportunity for this antibiotic-resistant pathogen to emerge and overwhelm the host (as in the disease pseudomembranous enterocolitis; discussed further in Chapter 14).

NON-SPORE FORMERS The Firmicutes include many species that do not form spores; they probably lost the spore-forming ability

through reductive evolution (see Section 9.3). For example, the lactic acid bacteria, especially *Lactococcus* and *Lactobacillus* species, ferment milk to make yogurt and cheese. Most lactic acid bacteria are obligate fermenters; that is, they generate ATP by glycolysis and substrate-level phosphorylation. They can grow in the presence or absence of oxygen but cannot use oxygen to respire (see Section 7.3). Lactic acid bacteria ferment by converting sugars to lactic acid. As the lactic acid builds up, the pH of the medium decreases until it halts bacterial growth. Thus, the carbon source in yogurt or cheese retains much of its food value for human consumption.

Other non-spore-forming Firmicutes are pathogens such as *Listeria*. *Listeria* species are facultative anaerobic rods, named for the British surgeon Joseph Lister (1827–1912), who was the first to promote antisepsis during surgery. They include enteric pathogens such as *Listeria monocytogenes*, the cause of a gastrointestinal illness (listeriosis) that can spread to the central nervous system. Initial infection with *L. monocytogenes* generally begins after eating contaminated soft cheeses, lunch meats, or sauerkraut. Unlike most other food-associated organisms, *Listeria* species grow at temperatures as low as 4°C (39.2°F). Prior to the invention of refrigeration, food contaminated by many species of bacteria would spoil long before *L. monocytogenes* could grow to infectious levels. But the rise of refrigeration prevented spoilage and extended the food's shelf life. Unfortunately, extended shelf life in the cold enabled the psychrophilic *L. monocytogenes* to grow to numbers high enough to cause disease.

Listeria grows in soil but can also invade and grow inside mammalian cells, where they can evade the host immune system. Bacteria are taken up by macrophages into phagocytic vesicles (phagosomes), but escape the vesicles to grow in the cytoplasm (**Figure 10.6**). The bacteria travel through the host cytoplasm by generating "tails" of polymerized host actin. The actin tails eventually propel *Listeria* directly into a neighboring host cell. This cell-to-cell projection strategy is surprisingly common among pathogens; for example, the Gram-negative bacterium *Shigella* and HIV (AIDS virus) disseminate by similar mechanisms (see Sections 12.6 and 18.3).

Other non-spore-forming Firmicutes are spherical cocci, such as the paired cocci of *Enterococcus*. Normally found in the intestinal tract, *Enterococcus* species can become invasive in immunocompromised hospital patients, causing urinary tract infections and bacteremia. Enterococcal infections are commonly treated with the antibiotic vancomycin, but hospitals have now seen the rise of vancomycin-resistant enterococci (VRE).

Another genus of non-spore-forming Firmicutes is *Staphylococcus*. Staphylococci are facultative anaerobic cocci that grow in clusters, often hexagonal in arrangement (**Figure 10.7A and B**). They include common skin microbiota such as *Staphylococcus epidermidis*. The staphylococci are generally salt tolerant, and their fermentation generates short-chain fatty acids that inhibit the growth of skin pathogens. Diverse Gram-positive cocci favor different parts of the skin, as discussed in Chapter 19. For example,

Figure 10.6 *Listeria monocytogenes*: An Intracellular Pathogen That Travels on Tails of Actin

A. Fluorescent antibodies mark the tails of polymerized actin (green) behind the *Listeria* cell bodies (red–yellow) traveling within an infected macrophage.

B. Invading bacteria polymerize host actin at one cell pole to form an actin tail. Actin tails propel them through the host cytoplasm and out through the cell membrane to invade a neighboring cell.

certain genera of Gram-positive anaerobes are prevalent in the folds of the umbilicus, or belly button (see **IMPACT**).

Some Gram-positive cocci, however, are serious pathogens. *Staphylococcus aureus* causes many different types of infections. These include impetigo and toxic shock syndrome, as well as pneumonia, mastitis (inflammation of the mammary gland), and osteomyelitis (bone infection). Disease involves several toxins, such as the toxic shock syndrome toxin (TSST). *S. aureus* is a major cause of hospital-acquired infections, especially infections of surgical wounds. The most dangerous strains are termed MRSA (methicillin-resistant *S. aureus*; Section 19.3).

Members of another genus, *Streptococcus*, generally form chains instead of clusters because their cells divide in a single plane (**Figure 10.7C and D**). They are aerotolerant (grow in the presence of oxygen) but metabolize by fermentation (they lack electron transport chains needed for aerobic or anaerobic respiration). Many live on oral or dental surfaces, where they cause

Figure 10.7 Staphylococci and Streptococci

A. *Staphylococcus* species.

B. *Staphylococcus* species.

C. *Streptococcus* species (Gram stain).

D. *Streptococcus* species.

Figure 10.8 Hemolysis Reactions

A. *Streptococcus pneumoniae* (alpha hemolysis is the green zone around colonies; see arrow).

B. *Staphylococcus aureus* (beta hemolysis is the clear yellow zone around colonies; see arrow).

caries (tooth decay), but some also grow on skin and in the intestine. Their fermentation of sugars in the oral cavity produces such high concentrations of lactic acid that the pH at the tooth surface can fall to pH 4, which demineralizes teeth. *Streptococcus* species cause many serious diseases, including pneumonia (*S. pneumoniae*), strep throat, erysipelas, and scarlet fever (*S. pyogenes* or group A strep). Different species of *Streptococcus* can be identified, or grouped, by antibody reactions to their cell-surface glycoproteins (bacterial antigens). These reactions are codified as the Lancefield grouping, named for the physician Rebecca Lancefield (1895–1981).

Different strains of staphylococci and streptococci may be distinguished by their abilities to hemolyze (lyse) red blood cells—a process called **hemolysis**. Hemolysis can be tested on sheep blood agar. Colonies that partly hemolyze red blood cells, via hydrogen peroxide reaction, generate a greenish halo (alpha hemolysis; **Figure 10.8A**). Colonies that fully hemolyze red blood cells, via a hemolysin enzyme, generate a clear halo (beta hemolysis; **Figure 10.8B**).

BACTERIA LACKING A CELL WALL Members of one group formerly belonging to the phylum Firmicutes—the mycoplasmas (class Mollicutes)—entirely lack a cell wall. Therefore, the class Mollicutes has been moved into a new phylum, Tenericutes, of cell wall–deficient bacteria. Mycoplasmas have completely lost their cell wall through reductive evolution, retaining only their cell membrane (**Figure 10.9**). Because they lack a cell wall, mycoplasmas cannot stain Gram-positive. In nature, mycoplasmas grow only within tissues of a host organism. The loss of the energy-expensive cell wall may have enhanced the cell's reproductive rate in a host environment protected from osmotic shock. On agar plates, some mycoplasmas form colonies with a characteristic fried-egg appearance (Figure 10.9B). Every known class of

Figure 10.9 Mycoplasmas: Bacteria without Cell Walls

A. *Mycoplasma mobile*, a parasite of fish.

B. Colonies of mycoplasmas cultured on agar have a fried-egg shape.

multicellular organism is parasitized by mycoplasmas, including vertebrates, insects, and vascular plants; in humans, mycoplasmas cause pneumonia and meningitis.

Phylum Actinobacteria: Branching Bacteria

Actinobacteria, the "high-GC Gram-positives," include major pathogens and antibiotic producers, as well as essential decomposers in natural environments. Most actinobacteria can be identified under the microscope by the acid-fast stain, a procedure in which cells are penetrated with a dye that is retained in the cell even under treatment with acid alcohol (see Section 3.4). The acid-fast property is associated with unusual cell wall lipids, such as the mycolic acids of *Mycobacterium tuberculosis* (Section 5.3).

The actinomycetes, a major group of Actinobacteria, undergo complex life cycles forming filamentous mycelia (branching filaments). **Figure 10.10A** shows colonies of *Streptomyces coelicolor*;

Figure 10.10 *Streptomyces* Bacteria

A. Colonies of *S. coelicolor* show sky-blue mycelia. The dark blue droplets on the colonies contain secreted antibiotics.

B. *Streptomyces* cells form coiled filaments.

impact

Contemplating Your Navel

This chapter explores the diversity of bacteria that exist throughout Earth's biosphere. The opening case history pointed out that vastly different bacterial phyla can populate our skin and oral cavity. But what about the belly button?

Everyone is aware of that mysterious nub on the body known as the belly button, umbilicus, or navel. The belly button is actually an indented scar left over from when your umbilical cord was removed at birth. In 90% of people, the belly button dips inward (Figure 1A), forming a moist, protective pocket inhabited by unique species of microbes—the umbilical microbiome. In very rare cases, the umbilicus can become infected. Most of the time, though, the microbial inhabitants simply maintain their own little ecosystem.

So, what kinds of microbes live down there in the navel? And what might they be doing? Do different people have different umbilical microbiomes? To address these questions, scientists founded the Belly Button Biodiversity project. A team of scientists led by Robert Dunn, at North Carolina State University, recruited more than 500 volunteers to provide samples of their umbilical microbiomes. Each volunteer sampled the umbilical indentation by using a sterile cotton swab soaked in phosphate saline solution. The suspended microbes were collected and processed to isolate their genomic DNA. Another part of the sample was streaked on an agar plate to isolate colonies of growing strains (Figure 1B). The many different microbes formed colonies that varied in size, color, and texture—differences attributable to different DNA sequences in the microbial genomes.

Figure 1 The Umbilicus, or Belly Button: Host of Diverse Microbes

A. The indented scar of the umbilicus.

B. Plate culture of bacteria and fungi from an umbilical swab—sample 982 from the Belly Button Biodiversity project. The many colonies of different species vary in size, texture, and color.

---

the dark blue droplets on the colonies contain secreted antibiotics. *Streptomyces* bacteria grow as filaments (Figure 10.10B) that develop specialized cells at their tips called arthrospores, which are released and disperse in the air to form new colonies. The mycelia of *Streptomyces* and other actinomycetes produce molecules we use as antibiotics, antitumor agents, and immunosuppressants. Pharmaceutical companies test soil samples all over the world for new actinomycete species that may produce the next new wonder drugs.

Other filamentous actinomycetes are pathogens. For example, *Nocardia* species are opportunists that cause pneumonia in hospital patients.

Actinobacteria that are <u>not</u> actinomycetes lack the mycelial life cycle. Some have thick, partially acid-fast cell walls. They grow as isolated rods and cocci, such as the Gram stain–variable coccobacillus *Gardnerella vaginalis*, which causes bacterial vaginosis. They also grow as cells of variable shape, such as the Gram-positive *Corynebacterium diphtheriae*, the cause of diphtheria. Another important actinobacterial genus is *Propionibacterium*, whose members are rod-shaped bacteria named for their fermentation pathway, which releases propionic acid. Propionic acid is the component that gives Swiss cheese its distinctive taste. *Propionibacterium freudenreichii* ferments

In the project, scientists identified bacteria by the sequences of genes encoding small-subunit ribosomal RNA (16S rRNA), which are highly conserved among all life forms. In all, more than 2,000 genetically unique strains were isolated, most appearing in only a few samples. But six isolates were found in more than 80% of human navels sampled. The most frequent isolates were typed to genera *Staphylococcus* (Firmicutes), *Corynebacterium* (Actinobacteria), and *Bacillus* (Firmicutes). Also common were actinobacteria such as *Micrococcus* and the *Clostridium*-related bacteria *Anaerococcus*, *Finegoldia*, and *Peptidophilus* (all Firmicutes). The genus *Anaerococcus* (Figure 2) is an anaerobic Gram-positive coccus commonly found as normal biota within creased regions of skin. The related anaerobic species *Finegoldia magna* is found as normal intestinal biota but can cause infections of the skin and bone. The presence of umbilical anaerobes suggests that the inner folds of the belly button consistently avoid exposure to air. The phylum Proteobacteria was also well represented in all navels examined, but the genera found varied greatly between individuals.

Why these bacteria? The skin is a very unforgiving environment. It is acidic and salty, possesses antimicrobial compounds, and constantly sheds. Certain members of the Firmicutes, Actinobacteria, and Proteobacteria are best suited to thrive under these conditions. The next important question is whether, and how, the umbilical microbes interact with their host. The details remain unclear,

Figure 2 *Anaerococcus* **Species**

SEM | 1 μm

These anaerobic Gram-positive cocci are often found within creased skin (colorized SEM).

but the scientists hypothesize that, as seen elsewhere in the skin, the dominant organisms maintain a stable community by helping to exclude potential pathogens from the moist, protected habitat of the belly button.

DISCUSSION QUESTION

What personal habits could change the bacterial diversity in a person's skin microbiome?

---

milk during cheese production, but the close relative *Cutibacterium acnes* (formerly *Propionibacterium acnes*) causes acne (skin infection).

Actinobacteria also include the causative agents of tuberculosis (*Mycobacterium tuberculosis*) and leprosy (*Mycobacterium leprae*). Cells of *M. tuberculosis* can be detected by the acid-fast stain as tiny rods associated with sloughed cells in sputum (Figure 10.11A). The bacteria form crinkled colonies after 2 weeks of growth on agar-based media (Figure 10.11B). *M. tuberculosis* has exceptionally thick, complex membranes that include some of the longest-chain acids known, up to 90 carbons.

The mycolic acids are linked to arabinogalactan, a polymer of the sugars arabinose and galactose built onto the peptidoglycan. This mycolyl-arabinogalactan-peptidoglycan complex forms a waxy coat that impedes the entry of nutrients through porins. As a result, this complex slows the growth rate, but like a suit of armor, the coat also protects the bacterium from host defenses and antibiotics. For this reason, the cure for tuberculosis requires an exceptionally long course of antibiotic therapy.

Tuberculosis is one of the oldest documented diseases of humankind. How did *Mycobacterium tuberculosis* evolve? And are there clues to help us predict future emerging strains?

Figure 10.11 *Mycobacterium tuberculosis*

A. Acid-fast stain of tissue sample containing *M. tuberculosis*.

B. Crinkled appearance of *M. tuberculosis* colonies

Evolution of *Mycobacterium tuberculosis*

CASE HISTORY 10.1

Tuberculosis: "I'll Be Back"

 Michael is a 45-year-old cook at a fast-food restaurant in Rome, Georgia. He went to see his physician because he was feeling tired, couldn't sleep, couldn't eat, and had been coughing for several weeks. A chest X-ray revealed scattered, small (1–5 mm) granulomas (nodules of inflammation) throughout both lobes of his lung (**Figure 10.12**). Material he coughed up from deep in his lung (called sputum) contained acid-fast bacilli (see Figure 10.11). The physician diagnosed Michael with secondary tuberculosis. Secondary tuberculosis is caused by bacteria reactivated from old lesions in which the bacteria have persisted in a dormant state. The causative bacterium is *Mycobacterium tuberculosis*, a facultative intracellular pathogen that infects only humans and is transmitted though respiratory droplets. Michael was placed in isolation and started on a long-term, four-drug treatment regimen consisting of isoniazid (fatty acid synthesis inhibitor), rifampin (transcription inhibitor), pyrazinamide (unclear mode of action), and streptomycin (protein synthesis inhibitor). The four-drug therapy continued for 2 months, followed by 4 months with isoniazid and rifampin only. Michael responded well to treatment. Luckily, he was not infected with one of the highly dangerous multidrug-resistant strains (MDR-TB).

Tuberculosis is one of the most devastating and ancient of human diseases. The Greek physician Hippocrates (460–370 BCE) describes a disease causing small, hard, knobby lumps called

Figure 10.12 Lung X-Ray Showing Small Lesions Filled with *Mycobacterium tuberculosis* Cells

"tubercles" in the lungs, and abscesses in the lumbar region of the spine. Modern molecular techniques such as PCR have shown clear evidence of *Mycobacterium tuberculosis* DNA in ancient mummies and in bones dating back to 3000 BCE. Because this organism has caused human disease for millennia, the evolutionary rise of *M. tuberculosis* is particularly interesting.

How did *M. tuberculosis* evolve from a strictly environmental bacterium into a pathogen whose only reservoir, and victim, is humans? Evolution into a pathogen required several steps. In natural environments, the ancestral mycobacteria were engulfed by predators such as amebas, single-celled protozoa that use finger-like projections of protoplasm to move and catch food. By evolving to survive intracellularly, this mycobacterial ancestor essentially "tamed" its predator into being its host. The next step was to acquire the capacity to transfer between hosts. Once an intracellular niche was established, *M. tuberculosis* was able to colonize new hosts and maintain a persistent chain of infection.

Genome comparisons between *M. tuberculosis* and nontuberculosis mycobacteria reveal a multistep process that led to *M. tuberculosis* (**Figure 10.13**). Its evolution started from an ancestral organism that also spawned actinomycete genera such as *Corynebacterium* and *Streptomyces*. The ancestral mycobacterium

Figure 10.13 Evolution of "Slow-Growing Mycobacteria," Including *Mycobacterium tuberculosis*

Slow-growing mycobacteria

This unrooted tree shows lineages of the phylum Actinobacteria. The red arrow shows the *M. tuberculosis* main lineage; the black arrows show where the lineage gained genes by horizontal transfer from unknown sources. Gene loss also occurred but is not shown. Note that *M. hominissuis* and *M. paratuberculosis* are actually subspecies of *M. avium*.

further diverged into species such as *M. avium*, a pathogen of birds; *M. marinum*, a marine organism causing human skin ulcers; *M. kansasii*, a group of environmental organisms causing tuberculosis in mammals but rarely in humans; and *M. tuberculosis* and *M. bovis*, which cause tuberculosis in humans. Along the way, these strains acquired new genetic material via horizontal gene transfers from unknown sources (marked by arrows in Figure 10.13). Later, *M. tuberculosis* lost genes it no longer needed to live in its human host. Consequently, the ancestral mycobacterial strains were environmental microbes, one of which adapted to inhabit an intracellular host environment and become *M. tuberculosis*.

Interestingly, *M. bovis*, an organism that produces a tuberculosis-like disease in cattle, is not the progenitor of *M. tuberculosis*, as previously thought. Instead, it is more recently derived from a common, but unknown, predecessor. An attenuated strain of *M. bovis* called Bacillus Calmette-Guérin is used in some countries as a vaccine (the BCG vaccine) to protect against *M. tuberculosis*. In the United States, the Centers for Disease Control and Prevention (CDC) does not recommend BCG because the vaccine's protection rate varies (from 0% to 60% in various trials) and because the vaccine stimulates antibody production that interferes with the tuberculin test used to identify people exposed to the pathogen.

M. tuberculosis continues to evolve today. This ongoing evolution is especially evident from the emergence of new forms of antibiotic resistance and from genomic studies comparing strains from around the world.

⊙ **Note:** Certain bacterial species have names that are coincidentally similar. Distinguish between:

- *Strepto**coccus***, the firmicute cocci, and *Strepto**myces***, the actinomycetes that produce antibiotics.
- *Myco**plasma***, the Firmicutes that lack a cell wall, and *Myco**bacterium***, the Actinobacteria with an extremely complex cell wall.

SECTION SUMMARY

- **Firmicutes** are Gram-positive bacteria that include endospore-forming genera such as *Bacillus* and *Clostridium*.
- **Non-spore-forming firmicutes** include pathogenic rods such as *Enterococcus* and *Listeria*, as well as food-producing bacteria such as *Lactobacillus* and *Lactococcus*.
- ***Staphylococcus*** and ***Streptococcus*** are Gram-positive cocci that include innocuous members of the human microbiome, as well as serious pathogens causing toxic shock syndrome, pneumonia, and scarlet fever.
- **Mycoplasmas**, previously classified as a firmicute, is now in the phylum Tenericutes. They lack a cell wall and show ameboid motility. Species cause diseases such as meningitis and pneumonia.
- **Actinobacteria** include actinomycetes, mycelial spore-forming soil bacteria such as *Streptomyces*; and *Nocardia*, an opportunistic pathogen. Other actinobacteria have irregularly shaped cells; this group includes the *Corynebacterium* species that causes diphtheria.
- **Mycobacteria** are acid-fast-staining actinomycete rods whose cell envelope contains a diverse assemblage of complex mycolic acids. Mycobacterial species cause tuberculosis and leprosy.

Thought Question 10.2   How do you think *Mycobacterium tuberculosis* manages to grow even though its thick envelope screens out most nutrients?

10.3 Gram-Negative Phyla: Proteobacteria and Bacteroidetes

SECTION OBJECTIVES

- Describe the diverse environmental and pathogenic roles of Proteobacteria and Bacteroidetes.
- Identify the diseases caused by diverse Gram-negative bacteria.

Which phylum of bacteria has the widest range of shapes and can "eat" nearly anything? And which phylum enables humans to digest foods that would otherwise be indigestible? The phylum Proteobacteria is so diverse and large that taxonomists have proposed calling it a superphylum. Its members are all Gram-negative. They include rod-shaped bacilli and spherical cocci with wide-ranging physiologies. Some eat rocks, others live in our intestines, and still others carry out photosynthesis. Members of the phylum Bacteroidetes are all Gram-negative rods, most are anaerobes, and those that live in the intestine broaden the menu of what humans can eat. Proteobacteria and Bacterioidetes both harbor human pathogens.

Phylum Proteobacteria: A Diverse Superphylum

The Proteobacteria show an amazing diversity of form and metabolism—as varied as cocci and spiral cells and including heterotrophs, lithotrophs, and photosynthesizers (Table 10.2). Yet all the Proteobacteria share a common form of cell envelope, which stains Gram-negative. As discussed in Chapter 5, the Gram-negative envelope includes an outer membrane, a cell wall of peptidoglycan permeated by the periplasm, and an inner membrane (comparable to the plasma membrane of other bacteria). The outer membrane is packed with receptor proteins and porins, making up two-thirds of the membrane's mass. The outer membrane lipids contain long sugar polymer extensions (lipopolysaccharide, or LPS). In pathogens, LPS has toxic effects on the host.

Although our focus will be pathogenic Proteobacteria, one intriguing nonpathogenic genus, *Geobacter* (family Geobacteraceae), has evolved to transfer electrons to iron or magnesium instead of oxygen. *Geobacter* species can be used to build a fuel cell in which the bacteria transfer electrons to an electrode, thus generating electricity. Such bacterial circuits can drive a clock or a research instrument.

The Enterobacteriaceae

The best-studied family of Proteobacteria is the Enterobacteriaceae, a family composed of a large number of nonpathogenic species but also some important pathogens. The most famous species is *Escherichia coli*. Some strains of *E. coli* grow normally in the human intestine, feeding on our mucous secretions and producing vitamins, such as vitamin K. But other strains, such as *E. coli* O157:H7, cause serious illness. A large proportion of the world's children die of *E. coli*–related intestinal illness before the age of 5. Other genera of Enterobacteriaceae that cause disease include *Salmonella* species, the cause of enteritis and typhoid fever; *Klebsiella*, a leading cause of hospital-acquired pneumonia; and *Yersinia pestis*, the cause of bubonic plague.

Many of the Enterobacteriaceae appear as rods, very similar under the microscope, but they can be distinguished by a variety of biochemical tests (Table 10.3). These tests, for example, can include pH indicator chemicals that change color when the medium pH decreases or increases during fermentation of particular food molecules (such as sugars and amino acids). Fermentation pathways distinguish between closely related species, such as *Escherichia coli*, which carries out mixed acid fermentation (tested using methyl red as a pH indicator), and the environmental organism *Enterobacter* (which tests positive for acetoin; Voges-Proskauer test).

All Enterobacteriaceae (sometimes called enterobacteria or enterics) are facultative anaerobic rods that may grow in human or animal digestive tracts or in lakes and streams. They grow singly, in chains, or in biofilms. Many species are motile, with many

Table 10.2
Selected Gram-Negative Proteobacteria and Bacteroidetes

Proteobacteria	Characteristics
Agrobacterium tumefaciens	Rod; induces plant tumors
Bdellovibrio bacteriovorus	Curved rod; invades periplasm and consumes prey bacteria
Brucella melitensis	Rod; causes brucellosis disease of cattle and sheep
Coxiella burnetii	Rod; causes Q fever
Escherichia coli	Rod; main vehicle for biotechnology; strains cause intestinal and bladder infections
Geobacter metallireducens	Rod with electron-conducting nanowires; generates electricity in fuel cells
Helicobacter pylori	Wide spiral cell; survives in stomach acid; causes gastritis
Legionella pneumophila	Rod; grows within amebas or macrophages; causes lung infection legionellosis
Myxococcus xanthus	Rod; soil bacterium that aggregates to form fruiting bodies
Neisseria gonorrhoeae	Diplococcus (pair of cocci); causes gonorrhea
Neisseria meningitidis	Diplococcus; causes meningitis
Nitrobacter winogradskyi	Rod; nitrite-oxidizing lithotroph active in soil
Rhodopseudomonas palustris	Rod; photoheterotroph in pond water
Rickettsia rickettsii	Rod; intracellular pathogen; causes Rocky Mountain spotted fever
Salmonella enterica	Rod; enteric pathogen; may grow intracellularly
Sinorhizobium meliloti	Rod; forms bacteroids that fix nitrogen for legumes
Vibrio cholerae	Comma-shaped rod; causes cholera
Yersinia pestis	Rod; causes bubonic plague

Bacteroidetes	Characteristics
Bacteroides fragilis	Rod; obligate anaerobe; found in gut biota, where it modulates innate immune responses
Bacteroides thetaiotaomicron	Rod; obligate anaerobe; major component of gut biota; digests complex plant fibers
Porphyromonas gingivalis	Rod; obligate anaerobe; causes periodontal disease

Table 10.3
Enteric Gram-Negative Bacteria: Examples of Biochemical Reactions[a]

	Indole Production	Methyl Red	Voges-Proskauer	Simmons Citrate	Hydrogen Sulfide	Urea Hydrolysis	Phenylalanine Deaminase	Lysine Decarboxylase	Arginine Dihydrolase	Ornithine Decarboxylase	Motility (36°C)	Gelatin Hydrolysis (22°C)	D-Glucose, Acid	D-Glucose, Gas	Lactose Fermentation	Sucrose Fermentation	D-Mannitol Fermentation	Dulcitol Fermentation	Adonitol Fermentation	D-Sorbitol Fermentation	L-Arabinose Fermentation
Enterobacter aerogenes	0	5	98	95	0	2	0	98	0	98	97	0	100	100	95	100	100	5	98	100	100
Escherichia coli	98	99	0	1	1	1	0	90	17	65	95	0	100	95	95	50	98	60	5	94	99
Klebsiella oxytoca	99	20	95	95	0	90	1	99	0	0	0	0	100	97	100	100	99	55	99	99	98
Klebsiella pneumoniae	0	10	98	98	0	95	0	98	0	0	0	0	100	97	98	99	99	30	90	99	99
Proteus mirabilis	2	97	50	65	98	98	98	0	0	99	95	90	100	96	2	15	0	0	0	0	0
Salmonella, most serotypes	1	100	0	95	95	1	0	98	70	97	95	0	100	96	1	1	100	96	0	95	99
Salmonella Typhi	0	100	0	0	97	0	0	98	3	0	97	0	100	0	1	0	100	0	0	99	2
Shigella flexneri	50	100	0	0	0	0	0	0	5	1	0	0	100	2	0	0	93	2	0	30	60
Shigella sonnei	0	100	0	0	0	0	0	0	2	98	0	0	100	0	2	1	99	0	0	2	95

[a]For each species, numbers indicate the percentage of clinical isolates showing a positive reaction for a given test.

flagella. Most strains grow well with or without oxygen, by either respiration (aerobic or anaerobic) or fermentation. They ferment rapidly on carbohydrates, generating fermentation acids, ethanol, and gases (CO_2 and H_2) in various proportions, depending on the species. Their presence in the intestine supports the growth of organisms that use these gases, including methanogens (discussed in Chapters 1 and 7). Many strains form biofilms. Biofilm formation explains the persistence of drug-resistant infections, such as those associated with urinary catheters in long-term hospital patients.

One interesting uropathogen is *Proteus mirabilis*, which causes serious bladder and kidney infections in humans, particularly as a complication of urethral catheterizations performed in hospitals. *Proteus* species are heavily flagellated and display a remarkable swarming behavior (**Figure 10.14**) that enables migration along

Figure 10.14 **The Enteric Rod *Proteus mirabilis*:
Isolated Swimmer or Cooperative Swarmer**

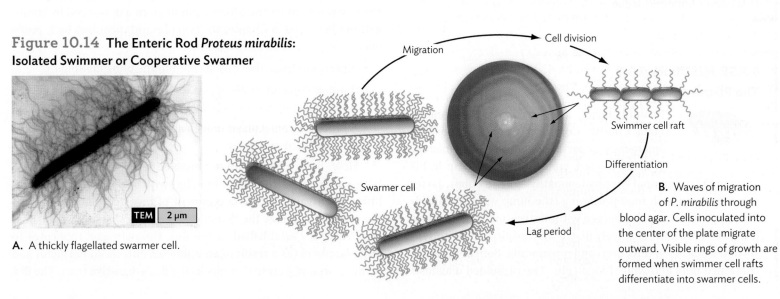

A. A thickly flagellated swarmer cell.

B. Waves of migration of *P. mirabilis* through blood agar. Cells inoculated into the center of the plate migrate outward. Visible rings of growth are formed when swimmer cell rafts differentiate into swarmer cells.

a catheter. Upon encountering certain amino acids in urine, the flagellated rods grow into long-chain swarmer cells. The swarmers gather, forming "rafts" that swim together. Once inside a host, the bacteria form a complex biofilm that resists antibiotics and can be hard to eradicate from a patient.

Some Enterobacteriaceae are pathogens with complex life cycles involving more than one host. A famous example is *Yersinia pestis*, the cause of bubonic plague, which remains prevalent among prairie animals in the western United States. *Y. pestis* evolved among a group of animal pathogens, members of which persist today. The following section describes how *Y. pestis* came to be.

Yersinia pestis and the Evolution of Flea-Borne Transmission

TALES OF TWO DISEASES I

CASE HISTORY 10.2

Pseudoappendicitis

Kimiko, a 6-year-old girl from Shodo Island, in Japan, had a 4-day history of fever, severe abdominal pain, and loose stools. She was diagnosed with appendicitis, but an exploratory laparotomy revealed a healthy appendix. The doctors then suspected that she had enteritis (intestinal inflammation), even though early lab results failed to grow any pathogens from her stool. They treated her with the antibiotic tetracycline, hoping they were right. Luckily, Kimiko responded well to the antibiotic, and eventually her stool culture grew *Yersinia pseudotuberculosis*, an occasional cause of enteritis. The organism, which is transmitted when contaminated food or water is ingested, was detected in the family's well water and in the stool of animals found on their property. Kimiko's parents were startled to learn that *Y. pseudotuberculosis* is related to the organism that causes bubonic plague.

CASE HISTORY 10.3

The Plague

In July, Javier, a 28-year-old man from La Plata County, Colorado, went to a hospital emergency department with a 3-day history of fever, nausea, vomiting, and a lump in his groin area (an inguinal lymphadenopathy called a bubo). Javier was diagnosed with food poisoning (the lump was not considered important) and discharged without treatment. Three days later, Javier returned, gravely ill. He was hospitalized with sepsis (a serious blood infection) and pneumonia. Both of his lungs contained fluid and white blood cells. The presumed diagnosis

was now plague. Once the organism causing plague is in the lung, it is easily transmitted to others by coughing, so the patient was placed in respiratory isolation. Javier was treated with the antibiotic gentamicin (a protein synthesis inhibitor) and fortunately recovered. Cultures of blood and of the aspirate from the swollen lymph node in his groin grew *Yersinia pestis*, confirming the diagnosis. Medical investigators then discovered *Y. pestis* in fleas collected near the patient's home. Javier had contracted plague from the bite of a flea. Fleas serve as vectors that transmit the organism from infected wild rodents (such as prairie dogs) to pets and humans. It so happened that Javier lived in Colorado, where the disease is endemic.

Yersinia pseudotuberculosis and *Yersinia pestis* are both Gram-negative bacilli and members of the family Enterobacteriaceae. *Y. pseudotuberculosis* causes gastrointestinal disease that can mimic appendicitis and is transmitted from animals to humans by contaminated food or water. *Y. pestis* causes a deadly disease (plague) that can be transmitted from animals to humans by an infected flea or, when the disease progresses to the pneumonic (lung) stage, from one human to another by respiratory droplets produced from coughing. Despite large differences in pathogenesis, the two bacterial species are actually closely related. Which organism evolved first?

Comparative genome analyses indicate that *Y. pestis* is a variant of *Y. pseudotuberculosis* that diverged only within the last 1,500–20,000 years. Just a few discrete genetic changes separate the two species. But those changes were sufficient for the *Y. pseudotuberculosis* ancestor to evolve from food-borne and waterborne transmission (and the cause of mild infection) to flea-borne transmission and the agent (*Y. pestis*) of an aggressive, life-threatening infection (**Figure 10.15**). What small genetic changes could account for such a dramatically altered pathogenesis?

Part of the answer is horizontally transferred plasmids. Plasmids are short, circular pieces of DNA that can be acquired from the environment by transformation or from a donor cell by conjugation (discussed in Chapter 9). Two plasmids acquired by *Y. pestis* encode important virulence factors:

- A phospholipase that digests membrane phospholipids
- A unique capsule made of peptide rather than the polysaccharides that compose most bacterial capsules
- A protease, called plasminogen activator, that dissolves clots in the human bloodstream

How does phospholipase enable flea-borne transmission? When a flea bites an animal infected with *Y. pestis*, the phospholipase inactivates a toxic component of ingested blood and helps the microbe colonize the flea midgut—a critical step for transmission. After it's established in the flea, the pathogen replicates to large numbers (as a result of an unknown chromosomal gene) and forms a large aggregate that blocks the flea's digestive tract. The flea

Figure 10.15 Transmission of *Yersinia pestis* by a Flea Bite

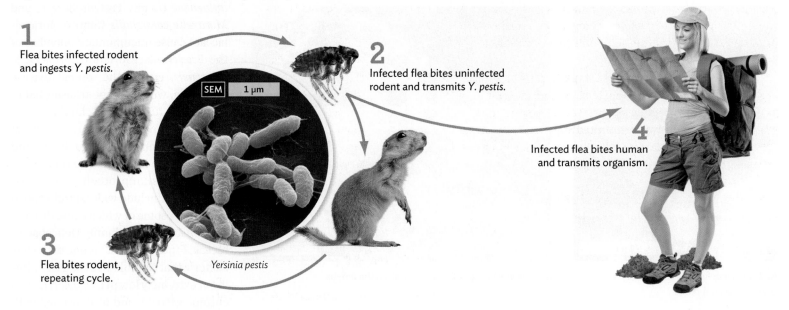

1 Flea bites infected rodent and ingests *Y. pestis*.

SEM 1 μm

2 Infected flea bites uninfected rodent and transmits *Y. pestis*.

4 Infected flea bites human and transmits organism.

3 Flea bites rodent, repeating cycle.

Yersinia pestis

constantly feels hungry and tries to take a blood meal on a new host but regurgitates the bacteria into the bite. If the host is human, the human is infected.

Once transmitted to the host by the flea bite, *Y. pestis* becomes trapped at the site by blood clots. The bacterial protease (plasminogen activator) is secreted and, in part, activates host plasmin. Activated host plasmin will dissolve the clots and allow *Y. pestis* to disseminate in the human host's bloodstream. Once within the bloodstream, the horizontally acquired peptide capsule of *Y. pestis* protects the bacterium from the human immune system.

From what we know of the two species of *Yersinia*, an evolutionary sequence has been proposed for the switch from food-borne/waterborne transmission to vector-borne transmission. We know that *Y. pseudotuberculosis* can cause a blood infection in rodents and will be taken up by fleas during a blood meal. But because the organism cannot colonize the flea, this is a dead-end transfer. Consequently, the first step in the switch must have been the horizontal transfer of the phospholipase gene enabling flea colonization. The second step in the process would have been acquisition of the unknown gene required for the flea-gut biofilm formation that blocks flea digestion. The third important step in the evolution of flea-borne transmission occurred when the progenitor cell captured the small plasmid containing the plasminogen activator gene. This new species, *Y. pestis*, could dissolve clots, move from the flea bite to the bloodstream of its human host, and disseminate to the lung.

Other Pathogenic Proteobacteria

Beyond Enterobacteriaceae, many other Gram-negative families of Proteobacteria include important pathogens. Closely related to Enterobacteriaceae are several genera of rod-shaped bacteria that are obligate respirers, such as *Pseudomonas* (family Pseudomonadaceae). An obligate respirer requires a terminal electron acceptor like oxygen to make energy. Many pseudomonads require oxygen for growth, although some can use alternative electron acceptors, such as nitrate. These genera metabolize an extraordinary range of natural compounds, including aromatic derivatives of lignin, an organic compound found in plant cell walls. Because of this metabolic diversity, pseudomonads have important roles in natural recycling and soil turnover. *Pseudomonas aeruginosa* commonly grows in soil as a decomposer. But in humans, *P. aeruginosa* can infect surgical wounds, and it forms biofilms on the pulmonary lining of cystic fibrosis patients.

An important intracellular pathogen related to the pseudomonads is *Legionella pneumophila* (family Legionellaceae). Incapable of growth on sugars, *L. pneumophila* requires oxygen to grow on amino acids. The organism exhibits an unusual dual lifestyle, alternating between intracellular growth within human macrophages and intracellular growth within freshwater amebas (Figure 10.16). The amebas help maintain the pathogen in fresh water ponds and in the cooling water of large-scale air-conditioning units perched atop large buildings. Bacteria released from the ameba into the cooling tower water are then dispersed from room vents as aerosols. A human who unknowingly inhales the aerosol can develop a serious lung disease called legionellosis.

Vibrio cholerae (family Vibrionaceae) is a comma-shaped organism with a single flagellum (Figure 10.17A). It causes the diarrheal disease cholera, a major problem for countries where sanitation is limited. For example, the epidemic of cholera in Haiti that followed the 2010 earthquake continues to hamper the island's recovery. A related organism, *Vibrio parahaemolyticus*, causes gastrointestinal illness through seafood contamination.

Figure 10.16 *Legionella pneumophila* Colonizing an Ameba

A. *L. pneumophila* cell is caught by an ameba's pseudopod.

B. *L. pneumophila* cells have colonized the ameba.

The Gram-negative rod *Bordetella pertussis* (family Alcaligenaceae) causes pertussis, or whooping cough, an illness that is prevented by immunization but is a growing problem as our current vaccines lose effectiveness with age. *Neisseria gonorrhoeae* (family Neisseriaceae), which occurs as paired cocci (**Figure 10.17B**), causes the sexually transmitted infection gonorrhea; and *Neisseria meningitidis* causes meningitis, infection of the brain and spinal cord lining. *Haemophilus influenzae* (family Pasteurellaceae) and *Moraxella catarrhalis* (family Moraxellaceae) cause pneumonia. A member of the Francisellaceae family, *Francisella tularensis*, causes the animal disease tularemia, which can be transmitted to humans. *F. tularensis* is classified as a potential biowarfare agent because it easily disseminates through air and causes severe illness and death.

The Epsilonproteobacteria (see Figure 10.1) include wide, spiral-shaped organisms, some of which cause disease. *Helicobacter pylori* (family Helicobacteraceae; **Figure 10.17C**) is one of the few bacteria that can colonize the gastric lining at extremely low pH. *H. pylori* causes chronic gastritis and is associated with stomach cancer. The related organism *Campylobacter jejuni* colonizes the intestinal tract and is a major cause of food poisoning.

Lithotrophic Proteobacteria Eat Rock and Metals

Many kinds of nonpathogenic Proteobacteria don't have to eat organic food; instead, they oxidize minerals such as iron, sulfur, or nitrogen. Mineral oxidation for energy is called **lithotrophy**.

Figure 10.17 Diverse Gram-Negative Pathogens

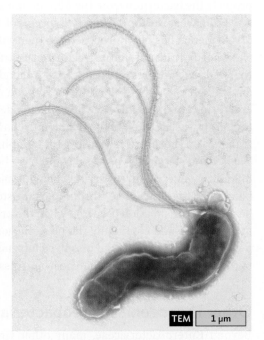

A. *Vibrio cholerae*, a flagellated, comma-shaped bacterium that causes cholera, attaching to a villus in the small intestine (colorized SEM).

B. *Neisseria gonorrhoeae*, paired cocci with attachment pili, the cause of gonorrhea.

C. *Helicobacter pylori*, a corkscrew-shaped bacterium that causes gastritis.

Some lithotrophs, such as ammonia-oxidizing bacteria and methanogenic archaea, actually grow within human and animal digestive tracts. Lithotrophs in natural ecosystems have surprising uses in human technology. In wastewater treatment, for example, nitrogen lithotrophs (nitrifiers) oxidize ammonia (NH_3) to nitrite (NO_2^-), or nitrite to nitrate (NO_3^-; Figure 10.18A). Nitrifiers such as *Nitospira* are of enormous economic and practical importance for wastewater treatment because they decrease the reduced nitrogen (NH_3) content of sewage. Special systems have been developed to retain nitrifier bacteria behind filters as one stage of water treatment. In the system shown in Figure 10.18B, nitrifying bacteria are encapsulated in pellets to retain them within the bioreactor while the treated water flows through a filter.

Acidithiobacillus is a genus of Proteobacteria whose species oxidize iron and sulfur and thus produce extreme acids such as sulfuric acid. *Acidithiobacillus* metabolism can be used to solubilize metal ores during mining for iron or copper; this process is called "leaching" the metals. Metal leaching was performed by miners as early as ancient Roman times, although they did not understand the role of bacteria in their technology.

Intracellular Symbionts and Predators

Proteobacteria have evolved many intimate relationships with other organisms, both mutualistic and parasitic. Some of these relationships led to the bacteria evolving into organelles of eukaryotes. The chloroplasts in plant cells, for example, evolved from the lineage of cyanobacteria. The mitochondria of our own cells evolved from the lineage of the intracellular pathogen *Rickettsia* (family Rickettsiaceae). Today, species of *Rickettsia* are carried by tick vectors, causing Rocky Mountain spotted fever (see Case History 2.1). A similar tick-borne intracellular pathogen, *Ehrlichia chaffeensis* (family Anaplasmataceae), causes ehrlichiosis. An obligate intracellular pathogen with a different transmission route (inhalation) is *Coxiella burnetii* (family Coxiellaceae), the cause of Q fever.

Other pathogens may grow harbored inside a host cell or free-living in extracellular fluids; these bacteria are called facultative intracellular pathogens. Examples of facultative intracellular pathogens include *Brucella* species (family Brucellaceae), the cause of the livestock disease brucellosis, and *Salmonella enterica* (family Enterobacteriaceae), which can invade white blood cells in humans. Still other bacteria are capable of a free-living existence but also engage in mutualism of vital importance to ecosystems and human agriculture. The most important form of mutualism may be nitrogen fixation. **Nitrogen fixation** is a process by which the nitrogen atoms in atmospheric nitrogen are made available for other uses.

Nitrogen is often a limiting nutrient for plants, which lack the enzymes needed to incorporate nitrogen gas from the air into amino acids and proteins. Nitrogen-fixing endosymbionts of plants include bacterial genera such as *Rhizobium*, *Bradyrhizobium*, and *Sinorhizobium* (collectively known as rhizobia). Rhizobia can live freely in the soil, but they also colonize plants—usually legumes such as peas or alfalfa—by forming distinctive nodule structures (Figure 10.19A). In the nodules, the plant cells provide the bacteria with nutrients, as well as the pink-colored "leghemoglobin," an oxygen-binding protein that maintains the anaerobic conditions needed to reduce (or "fix") N_2 to NH_4^+, which the plant can use. Each species of bacteria colonizes and infects a particular range of plants. Within the host cell, the bacteria lose their own cell walls and become elongated "bacteroids," specialized for nitrogen fixation (Figure 10.19B).

Other bacteria have evolved as voracious predators that prey even on other bacteria. Bdellovibrios (phylum Proteobacteria) are bacteria that attack other proteobacterial host cells, such as *E. coli* (Figure 10.20). The structure of the "attack cell" is a small comma-shaped rod with a single flagellum. The

Figure 10.18 Ammonia-Oxidizing Bacteria

A. *Nitrosomonas europaea* oxidizes ammonia to nitrite. Internal membranes contain the electron transport complexes.

B. Wastewater treatment uses nitrifiers (ammonia-oxidizing bacteria) to remove ammonia.

Figure 10.19 Rhizobia: Legume Endosymbionts

A. Legume nodules cut open to show pink regions where the plant cells produce leghemoglobin to maintain anaerobic conditions for bacteroid nitrogen fixation.

B. Within the host cells, *Sinorhizobium meliloti* cells form bacteroids that fix nitrogen while receiving nutrients from the plant.

attack cell attaches to the envelope of its host and then penetrates into the periplasm, where it uses host resources to grow. Within the periplasm, the invading bdellovibrio elongates as a spiral filament while replicating its DNA. When most nutrients have been exhausted, the filament divides into multiple short cells. The cells develop flagella, and the periplasmic compartment bursts, releasing the newly formed attack cells. Bdellovibrios can be isolated from human sewage—a sign that vicious predatory behavior is going on in our intestines.

Phylum Bacteroidetes: Obligate Anaerobes

An important group of rod-shaped bacteria that stain Gram-negative but diverged earlier from the phylum Proteobacteria is the phylum Bacteroidetes. Most genera are anaerobic, but some, like species of *Flavobacterium*, are aerobic. The most clinically important genus of this group is *Bacteroides*. *Bacteroides* species, such as *B. fragilis* and *B. thetaiotaomicron*, are the major inhabitants of the human colon. These bacteria actually communicate with our immune system, telling our immune cells which responses to make while avoiding immune defenses that could harm them.

Bacteroides species are obligate anaerobes, incapable of growing in the presence of more than a tiny trace of oxygen. Their main

Figure 10.20 Parasites of Bacteria: *Bdellovibrio* Species

Bdellovibrio bacteriovorus

A. *Escherichia coli* under attack by *Bdellovibrio bacteriovorus* (note predator cell within *E. coli* periplasm).

Bdellovibrio

E. coli periplasm   *E. coli* cytoplasm

AFM   1 μm

**B.
LIFE CYCLE
OF
*Bdellovibrio***

1 *Bdellovibrio* finds host by chemotaxis.

2 *Bdellovibrio* binds to host receptors.

3 *Bdellovibrio* invades periplasm.

4 Spiral chain grows.

5 Chain fragments into flagellated cells.

6 Host lysis releases *Bdellovibrio* cells.

source of energy is fermentation of a wide range of sugar derivatives from plant material—compounds indigestible by humans and potentially toxic to us. *Bacteroides* species convert these substances to simple sugars and fermentation acids, some of which are absorbed by the intestinal epithelium. Consequently, *Bacteroides* species serve two important functions for their host: they break down potential toxins in plant food, and their fermentation products make up as much as 15% of the caloric value we obtain from food. Yet another benefit of *Bacteroides* species is their ability to remove side chains from bile acids, enabling the return of bile acids to the hepatic circulation. In effect, our *Bacteroides* species and associated intestinal bacteria constitute a functional organ of the human body.

Bacteroides species cause trouble, however, when they reach parts of the body not designed to host them. During abdominal surgery that involves the gastrointestinal tract or in which the colon is accidentally nicked (even C-sections), bacteria can escape the colon and invade the surrounding tissues. The displaced bacteria can form an abscess, a localized mass of bacteria and pus contained in a cavity of dead tissue. The interior of the abscess is anaerobic and often impenetrable to antibiotics.

SECTION SUMMARY

- **Members of the phylum Proteobacteria stain Gram-negative and have a thin cell wall and an LPS-containing outer membrane.** The varied families of Proteobacteria demonstrate wide diversity of form and metabolism, including phototrophy, lithotrophy, and heterotrophy on diverse organic substrates.

- **The family Enterobacteriaceae evolved to include a number of important pathogens.** One example, *Yersinia pestis* (the cause of plague), evolved from *Yersinia pseudotuberculosis* within the past 1,500–20,000 years. Flea-borne transmission of *Y. pestis* evolved following the horizontal transfer of virulence genes. Acquisition of these genes enabled *Y. pestis* to colonize the flea, block its intestine, and, once transmitted by flea bite, disseminate in the new host.

- **Other families of Proteobacteria also include pathogens**, such as *Vibrio cholerae*, the cause of cholera; *Neisseria gonorrhoeae*, the cause of gonorrhea; and *Helicobacter pylori*, the cause of gastritis.

- **Proteobacteria that are mutualistic endosymbionts** include nitrogen-fixing bacteria such as rhizobia, which form bacteroids within a legume root host cell.

- **Members of the phylum Bacteroidetes are Gram-negative bacilli; most are anaerobes.** Species of *Bacteroides* ferment complex plant materials in the human colon but can accidentally enter body tissues through wounds and cause abscesses.

Thought Question 10.3 The Gram-negative alphaproteobacterium *Rickettsia prowazekii* is an obligate intracellular pathogen that causes the disease typhus. It is transmitted by the body louse. What impact would an obligate intracellular lifestyle have on opportunities for horizontal gene transfer (transfer of genes between bacteria)?

10.4 Phylum Spirochaetes: Twisted-Cell Pathogens and Symbionts

SECTION OBJECTIVES

- Describe the unique coiled cells of the spirochetes.
- Identify spirochetes that cause important human diseases.
- Identify spirochetes that grow as environmental symbionts.

Which spiral bacteria cause some of the world's most famous diseases? Spirochetes are a distinctive clade of bacteria with a cell structure consisting of a long, tight spiral that is flexible like a coiled telephone cord (**Figure 10.21**). In many species the spiral is so thin that its width cannot be resolved by bright-field microscopy. The best-known species are *Treponema pallidum*, the cause of syphilis (Chapter 23); *Borrelia burgdorferi*, the cause of Lyme disease (Chapter 21); and *Borrelia recurrentis*, which causes relapsing fever. Most spirochetes are slow-growing heterotrophs and difficult to culture. Besides the pathogens, however, many spirochetes are free-living organisms in soil or water. Many nonpathogenic spirochetes are digestive symbionts of hosts as diverse as termites and cattle, assisting their breakdown of plant material.

The spirochete cell is surrounded by a thick outer sheath of lipopolysaccharides and proteins. The spirochete sheath is similar to a proteobacterial outer membrane, except that the periplasmic space, containing a very thin peptidoglycan cell wall, completely separates the sheath from the plasma membrane. At each end of the cell, one or more flagella extend and double back around the cell body within the periplasmic space (Figure 10.21A). The periplasmic flagella rotate on proton-driven motors, as do extracellular flagella (see Section 5.5). But because the spirochete's flagella twine back around the cell body, their rotation forces the entire cell to twist around, corkscrewing through the medium. This corkscrew motion

Figure 10.21 Spirochete Structure

Protoplasmic cylinder   Periplasm   Outer membrane   Flagella

A. Spirochete cell structure, showing the arrangement of periplasmic flagella (one red, one yellow).

Outer envelope

Flagellum

TEM   0.2 μm

B. Cross section through a spirochete, showing outer envelope and flagella (axial fibrils). The unidentified spirochete was obtained from a human lesion of acute necrotizing ulcerative gingivitis.

of the cell body turns out to have a physical advantage in highly viscous environments, such as human mucous secretions.

The bacterial domain contains many examples in which random gene alterations and horizontal gene transfers yield new pathogens. The earlier example of *Yersinia pestis* and the plague is one. Syphilis is another.

TALES OF TWO DISEASES II

CASE HISTORY 10.4
A Vaginal Ulcer That Seemed to Heal

Diane, a 20-year-old accountant from Cedar City, Utah, came to the Valley View Medical Center urgent-care clinic with a low-grade fever, malaise, and headache. She was sent home with a diagnosis of influenza but again sought treatment 7 days later with a macular rash (small, flat, red spots) on her trunk, arms, hands, and feet (**Figure 10.22A**). Further questioning revealed that Diane had noticed a painless ulcer on her vagina a month earlier that eventually healed, even without treatment. Serology results indicated that Diane had contracted syphilis, not influenza. Syphilis is caused by *Treponema pallidum* subspecies *pallidum*. Because this disease is sexually transmitted, Diane provided a list of her sexual partners over the previous several months so that the health department could contact them for testing and, if necessary, treatment.

CASE HISTORY 10.5
An Arm Wound That Seemed to Heal

Fabrice, an 11-year-old boy from the village of Ngbongbo Tokonzi, in the Democratic Republic of Congo (Africa), was seen at a nearby clinic. He had a cluster of raised, papular lesions on his arm (**Figure 10.22B**). When the doctor asked Fabrice whether he had ever seen lesions like these before, Fabrice remembered having a single lesion about 2 months earlier, which had spontaneously healed. Serology revealed that the boy had secondary yaws disease caused by the spirochete *Treponema pallidum* subspecies *pertenue*, a spirochete closely related to the strain that causes syphilis. However, *T. pallidum pertenue* is transmitted by direct contact with a cut or wound.

As Diane found out in case 10.4, a strain of the spirochete *Treponema pallidum* (**Figure 10.22C**) causes the sexually transmitted infection syphilis (discussed further in Chapter 23). Syphilis arises only from strains of the subspecies *pallidum* (*T. pallidum pallidum*). Strains of a different *T. pallidum* subspecies, subspecies *pertenue* (*T. pallidum pertenue*), produce a different disease, called yaws, which is transmitted by direct skin contact. Yaws, like syphilis, is a multistage disease; thus, Fabrice in Case History 10.5 thought his infection had gone away, but new lesions appeared 2 months later. Whereas syphilis is sexually transmitted and affects people worldwide, yaws is transmitted by direct skin contact predominantly in developing countries with a tropical climate. Besides their different modes of transmission and geographic distribution, these strains of *T. pallidum* differ dramatically in their abilities to cross the placenta and infect a developing fetus. Strains that cause syphilis can infect a fetus, but yaws strains cannot.

So, what distinguishes the two subspecies? Are they perhaps a single species infecting hosts under different conditions? The *T. pallidum pallidum* and *T. pallidum pertenue* treponemes (spirochetes of the species *T. pallidum*) cannot be distinguished by morphology or physiology. Consequently, it had been proposed that syphilis and yaws were caused by the same organism and that climate or cultural differences resulted in the different diseases. However, a single strain causing both

Figure 10.22 Lesions of Secondary Syphilis and Yaws

A. The macular lesions of secondary syphilis. Syphilis is a sexually transmitted infection caused by *Treponema pallidum* subspecies *pallidum*.

C. *Treponema pallidum*.

B. The very different papular lesions of secondary (daughter) yaws caused by *Treponema pallidum* subspecies *pertenue*. This subspecies infects cuts and wounds but is not sexually transmitted.

diseases seemed unlikely because syphilis and yaws have distinctive lesions and patterns of pathogenesis. Furthermore, an immune response to one will not fully cross-protect against the other in experimental animal infections.

What do we find in the subspecies genomes? The genomes of *T. pallidum pallidum* and *T. pallidum pertenue* are similar in both genome size and structure, but they differ in six regions, amounting to about 0.4% of the genome (about 4,500 of 1,138,006 bp). The alterations involve small deletions and some repeated sequences. No plasmid differences or major gene differences are evident. It is likely that these minor differences cause the distinct infection patterns observed in syphilis and yaws. As of this writing, none of the *T. pallidum* subspecies have been cultured continuously in the laboratory, so we cannot yet mutate the genes to determine their potential roles in pathogenesis. Remarkably, all of the pathogenic differences that distinguish *T. pallidum pallidum* from *T. pallidum pertenue* arise from a few very small and poorly understood differences in the genome sequences of these subspecies.

In the course of evolution, which came first: *T. pallidum pertenue* or *T. pallidum pallidum*? Phylogenetic analysis examining the nature and frequency of differences in the two strains suggests that *T. pallidum pertenue* evolved first, followed by the syphilis subspecies. This finding, which suggests that syphilis evolved from the organism that causes yaws in tropical areas, has strengthened the theory that, upon discovering the Americas, members of Columbus's crew contracted syphilis from sexual encounters with the native people and transported *T. pallidum pallidum* back to Europe more than 500 years ago.

◉ **Note:** Distinguish **spirochetes** (long, tightly coiled spirals with unique intracellular flagella) from **spirilla** (singular, **spirillum**; wide, helical Gram-negative cells such as *Helicobacter* and *Rhodospirillum*, both members of the phylum Proteobacteria).

SECTION SUMMARY

- **The spirochete cell** is a tight coil surrounded by a sheath and periplasmic space containing periplasmic flagella.
- **Spirochetes move by a flexing motion** caused by rotation of the periplasmic flagella, propagated along the length of the coil.
- **Spirochetes include free-living fermenters** in water or soil. Others are endosymbionts of animal digestive tracts, such as the termite gut.
- **Spirochetes include important pathogens.** *Treponema pallidum* subspecies *pallidum* causes syphilis, and *Treponema pallidum* subspecies *pertenue* causes yaws.

Thought Question 10.4   For motility, what are the relative advantages of external flagella versus the flexible spiral cells of spirochetes?

10.5 Phyla Cyanobacteria and Chlamydiae

SECTION OBJECTIVES

- Explain the key environmental contributions of cyanobacteria.
- Describe different forms of marine cyanobacteria.
- Describe the cell cycle of chlamydias.
- Describe the diseases caused by chlamydias.

Is it true that one group of bacteria makes most of the oxygen we breathe? And do any bacteria grow <u>only inside</u> of eukaryotic host cells? The answer to both questions is yes! Members of the Cyanobacteria phylum, sometimes called blue-green algae, are bacteria that make vast amounts of oxygen through photosynthesis. They are widespread around the world but do not cause human infection. Members of the Chlamydiae phylum, on the other hand, have evolved to grow <u>only</u> inside eukaryotic host cells, never in the natural environment. Several members of the Chlamydiae can cause serious respiratory diseases and sexually acquired infections.

Phylum Cyanobacteria: CO_2 Fixers and O_2 Producers

Cyanobacteria are the only bacteria that produce oxygen gas, a by-product of oxygenic photosynthesis. Cyanobacterial photosynthesis also fixes much of the carbon dioxide for consumers in ecosystems, particularly marine ecosystems supporting fish. These same cyanobacteria also fix much of the biosphere's nitrogen.

Cyanobacteria commonly appear green because of the predominant blue and red absorption by chlorophylls *a* and *b*. Essentially the same chlorophylls are found in plant chloroplasts, which evolved from internalized cyanobacteria that became endosymbionts. Oxygenic phototrophy is so successful that it drives all members of the phylum Cyanobacteria, which account for about a quarter of Earth's entire biomass production. Despite their common metabolism of photosynthesis, cyanobacteria vary widely in cell form and behavior, as well as in genetic diversity. Vast blooms of cyanobacteria occur in the ocean (**Figure 10.23A**). Cyanobacteria include species that are edible for humans and are components of spirulina salad and nostoc soup (also known as "bird's nest soup"). No cyanobacterial pathogens are known, although some species produce substances that endanger aquatic consumers. For example, blooms of *Microcystis* produce microcystins, peptide toxins that damage the liver.

Cyanobacteria have evolved incredibly diverse forms. The simplest single-celled forms include *Synechococcus* and *Prochlorococcus*, the most abundant phototrophs in the oceans. Other genera, such as *Oscillatoria*, grow as long filaments of hundreds or even thousands of cells that can associate in thick biofilms. *Oscillatoria* cells are stacked like plates. Filamentous genera such as *Nostoc* develop special cells

Figure 10.23 Cyanobacteria

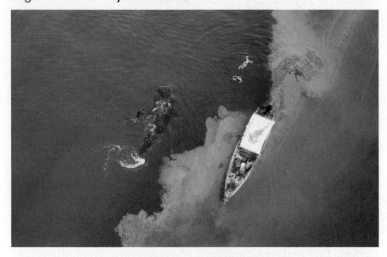

A. A ship plows through an ocean bloom of *Trichodesmium*.

B. Filamentous *Nostoc* makes heterocysts (round cells).

C. *Merismopedia* forms extended quartets, octets, and so on.

Figure 10.24 Lichens with Cyanobacteria

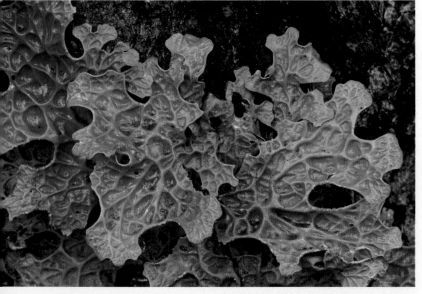

Lichens with symbiotic cyanobacteria provide a food source for forest animals.

called heterocysts that fix nitrogen (**Figure 10.23B**). Other species, such as *Merismopedia,* form regular sheets of cells (**Figure 10.23C**), while still others form globular aggregates.

Some cyanobacteria form a unique mutualism with fungi, called a **lichen** (**Figure 10.24**). In a lichen, the cyanobacterium fixes carbon dioxide and nitrogen into nutrients while the fungus provides physical protection. Lichens form important portions of ground cover and essential food for animals, particularly in dry northern forests. Overall, cyanobacteria serve the ecosystem in several fundamental ways: by fixing carbon and nitrogen into biomass and by producing oxygen that we can breathe.

Phylum Chlamydiae: Intracellular Pathogens

Chlamydiae is a large group of bacteria with cell walls that have very little peptidoglycan. Once the chlamydias were thought to completely lack peptidoglycan. Pathogenic species of Chlamydiae lack the classic cagelike form of peptidoglycan shown in **Figure 5.8B**, but they do have a narrow ring of peptidoglycan at midcell to guide cell division. Chlamydias evolved independently of the mycoplasmas (cell wall–less Firmicutes), and their alternative cell forms show very different environmental adaptations. Chlamydias are obligate parasites or pathogens. *Chlamydia trachomatis* (**Figure 10.25A**) is the causative agent of a major sexually transmitted infection in the United States (Chapter 23). It is also a cause of trachoma, an eye disease dating back to ancient Egypt. The related species *Chlamydophila pneumoniae* causes pneumonia and has been implicated in cardiovascular disease. Chlamydias infect a wide range of host cell types, from respiratory epithelium to macrophages.

Chlamydias show a complex developmental life cycle of parasitizing host cells (**Figure 10.25B**). Chlamydias alternate between two developmental stages with different functions: elementary bodies and reticulate bodies. The form of a chlamydia transmitted outside host cells is called an elementary body. Elementary bodies resemble endospores in that they are metabolically inert, with a compacted chromosome. Though they lack a classic cell wall, chlamydias possess an outer membrane whose proteins are cross-linked by disulfide bonds, making a tough coat that provides osmotic stability.

To reproduce, the elementary body must adhere to a host cell surface and be endocytosed into the host cell. Once in an endosome, the elementary body must transform itself into a larger reticulate body, named for the netlike appearance of its uncondensed DNA. The reticulate body has active metabolism and divides rapidly, but it is incapable of infection and is vulnerable to osmotic shock. Therefore, to complete the infection cycle, the reticulate bodies must divide to develop into new elementary bodies before exiting the host. When the host cell lyses, the elementary bodies are released to infect new cells.

Without treatment, chlamydial infections may persist indefinitely. In the female reproductive tract, chlamydial infection may lead to inflammation of the fallopian tubes and ovaries (**Figure 10.25C**), a condition known as pelvic inflammatory disease

Figure 10.25 *Chlamydia* Life Cycle within a Cell of Urogenital Epithelium

Chlamydia trachomatis

A. *Chlamydia trachomatis* multiplying within a human cell (colorized TEM). Infected cell, equivalent to step 5 in part B, contains reticulate bodies (yellow) growing and dividing, as well as newly formed elementary bodies (red).

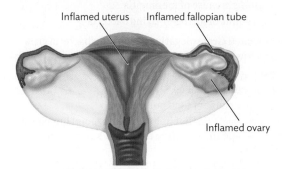

C. Pelvic inflammatory disease (PID). Chlamydial infection may cause permanent damage to the fallopian tubes and ovaries, resulting in infertility.

B. *Chlamydia* species persist outside the host as a spore-like elementary body that is metabolically inactive and contains compacted DNA. Upon endocytosis, the elementary body avoids lysosomal fusion and develops into a reticulate body in which the DNA, now uncondensed, has a reticular (netlike) appearance. The reticulate body replicates within the host cytoplasm and then develops into new elementary bodies that are released when the cell lyses.

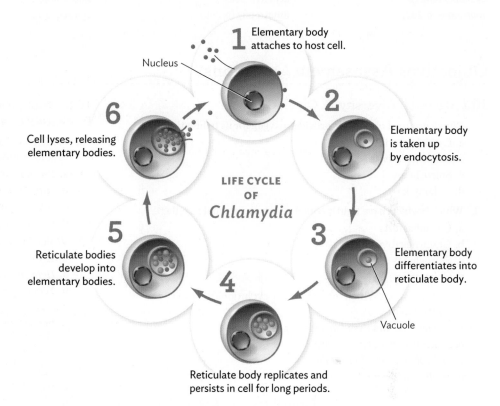

(PID; see Section 23.5). PID can leave permanent adhesions (fibrous tissue across the organs), a common cause of infertility.

SECTION SUMMARY

- **Cyanobacteria produce** much of the biosphere's **oxygen gas**.
- **Cyanobacteria fix** much of the biosphere's **carbon and nitrogen**.
- **Lichens** may include cyanobacteria in mutualism with fungi.
- **Chlamydias are obligate intracellular parasites** that undergo a complex developmental progression, culminating in a spore-like form, called an elementary body, that can be transmitted outside the host cell. Chlamydial cell walls have very little peptidoglycan.
- *Chlamydia trachomatis* **causes a sexually transmitted infection**, as well as the eye disease trachoma.
- *Chlamydophila pneumoniae* **causes pneumonia**.

Perspective

Several themes emerge in the diversity of the domain Bacteria. Some phyla, such as Proteobacteria and Actinobacteria, have evolved highly diverse metabolism and cell form. Other phyla show remarkable uniformity in cell structure (Spirochaetes) or metabolism (Cyanobacteria). Many bacterial phyla include human pathogens, some of which are closely related to organisms that enhance human health or contribute to environmental ecosystems. New species continually emerge through evolution, displaying innovative capabilities for environmental change or pathogenesis. In Chapter 11 we will explore the diversity of eukaryotic microbes, again focusing on human pathogens, such as those that cause malaria (*Plasmodium vivax*), amebic dysentery (*Entamoeba histolytica*), and African sleeping sickness (*Trypanosoma brucei*).

Key Terms

Actinobacteria (p. 280)

Bacteroidetes (p. 280)

botulinum (p. 284)

Chlamydiae (p. 280)

Cyanobacteria (p. 280)

endospore (p. 282)

Firmicutes (p. 280)

hemolysis (p. 287)

infant botulism (p. 285)

isolate (p. 279)

lichen (p. 302)

lithotrophy (p. 296)

nitrogen fixation (p. 297)

polymicrobial (p. 278)

Proteobacteria (p. 280)

shelter species (p. 281)

spirillum (p. 301)

Spirochaetes (p. 280)

spirochete (p. 301)

strain (p. 279)

zoonosis (p. 280)

Objectives Assessment Questions

10.1 Bacterial Diversity: A Quick Glimpse

1. Which group of organisms provides oxygen for us to breathe?

   a. Cyanobacteria
   b. Gram-negative bacteria
   c. Spirochetes
   d. Chlamydias

2. Which bacteria have tightly coiled cells with internal flagella?

   a. Cyanobacteria
   b. Gram-negative bacteria
   c. Spirochetes
   d. Chlamydias

3. View Figure 10.1. Which two genera are evolutionarily most closely related?

   a. *Escherichia* and *Pseudomonas*
   b. *Helicobacter* and *Nitrospira*
   c. *Helicobacter* and *Rickettsia*
   d. *Bacillus* and *Synechococcus*

10.2 Gram-Positive Phyla: Firmicutes and Actinobacteria

4. Which member of the Firmicutes <u>lacks</u> a thick cell wall that excludes many antibiotics?

   a. *Clostridium botulinum*, the cause of botulism
   b. *Clostridioides difficile*, an enteric organism that thrives when normal enteric biota have been eliminated by use of antibiotics
   c. *Mycobacterium tuberculosis*, the cause of tuberculosis
   d. *Mycoplasma pneumoniae*, a cause of pneumonia

5. Which Gram-positive bacterium/bacteria endanger human health instead of providing benefits?

   a. *Listeria* species, which grow in milk and cheese, even in the refrigerator
   b. *Lactobacillus* species, which grow in milk during production of yogurt and cheese
   c. *Streptomyces* species, which produce antibiotics such as streptomycin
   d. *Staphylococcus epidermidis*, which is found in the human skin

10.3 Gram-Negative Phyla: Proteobacteria and Bacteroidetes

6. Which bacterium/bacteria oxidize minerals instead of organic food?

   a. *Pseudomonas aeruginosa*, a common wound contaminant
   b. *Nitrobacter* species, growing in soil
   c. *Neisseria meningitidis*, a cause of meningitis
   d. *Bradyrhizobium*, a genus of nitrogen-fixing bacteria

7. Which human-associated Gram-negative bacterium/bacteria grow <u>only</u> with extremely low oxygen?

   a. *Helicobacter pylori*, the cause of gastritis
   b. *Yersinia pestis*, the cause of plague
   c. *Salmonella enterica*, a cause of food poisoning
   d. *Bacteroides*, enteric bacteria that help digest our food but cause abscesses if they escape the intestine and enter surrounding tissues

8. Which of the following is unique to Gram-negative bacteria?

   a. Peptidoglycan
   b. Lipopolysaccharide
   c. Teichoic acid
   d. Flagella

10.4 Phylum Spirochaetes: Twisted-Cell Pathogens and Symbionts

9. Which structural finding in a microscope would rule out the organism that causes Lyme disease?

   a. Round, coccoid cells
   b. Flexible, tightly coiled cells
   c. Under electron microscopy, intracellular flagella
   d. Cell motility

10. Which organisms possess mutualistic spirochetes?

    a. Humans
    b. Lichens
    c. Termites
    d. Amebas

11. The agent of Lyme disease is transmitted by which vector?
   a. Ticks
   b. Termites
   c. Bats
   d. Amebas

10.5 Phyla Cyanobacteria and Chlamydiae

12. Which trait do cyanobacteria lack?
   a. Production of oxygen through photosynthesis
   b. Nitrogen fixation
   c. Growth in globular clusters
   d. Intracellular flagella

13. Which organelle of eukaryotes descended from a common ancestor of cyanobacteria?
   a. Mitochondrion
   b. Chloroplast
   c. Golgi body
   d. Nucleus

14. Which bacterial structure(s) do chlamydiae lack?
   a. Ribosomes
   b. Outer membrane
   c. Thick peptidoglycan cell wall
   d. Inner membrane

15. Which life stage is absent from chlamydiae?
   a. Endospore
   b. Reticulate body
   c. Stage with netlike uncondensed DNA
   d. Elementary body

16. Which two organisms are <u>least</u> likely to grow in the same environment?
   a. *Chlamydia* and *Nitrobacter*
   b. *Streptomyces coelicolor* and *Escherichia coli*
   c. *Mycobacterium tuberculosis* and *Mycoplasma pneumoniae*
   d. *Clostridium botulinum* and *Bacillus subtilis*
   e. *Lactococcus* and *Lactobacillus*

Review Questions

1. Which major phylum of bacteria shows a great variety of lifestyles, including disease, even though all members have a similar, distinctive cell structure?

2. In which major phylum of bacteria do all members show the same kind of metabolism yet vary widely in form and habitat?

3. Discuss major pathogens in each of the seven bacterial phyla highlighted in this chapter.

4. Explain a key or unique feature for each of the seven bacterial phyla highlighted in this chapter.

Clinical Correlation Questions

1. If a patient has pneumonia, how might you discover whether the cause is a mycoplasma, a bacterium such as *Streptococcus*, or a virus?

2. An *E. coli* strain isolated from an outbreak of diarrheal disease displays an unusual combination of virulence factors not seen in any other single strain of *E. coli*. Describe how the strain may have evolved.

3. A patient observes a round, red lesion that goes away after several weeks. If the lesion is caused by a spirochete, which of three diseases could it be? For each disease, what subsequent symptoms may occur?

4. The variety of species in gut microbiomes was identified for two individuals. The microbiome from patient A included a very diverse number of species, compared to patient B, whose microbiome was much less diverse. Which of the two patients is more likely to suffer from chronic intestinal distress? Why?

5. Review Case History 10.1, about the man (Michael) with secondary tuberculosis. Why was Michael placed in isolation?

 For more assessment, visit digital.wwnorton.com/michum2

11

Eukaryotic Microbes and Invertebrate Infectious Agents

A Desert Sand Fly Parasite

SCENARIO Mario, a 29-year-old pizza delivery worker in Indianapolis, had an open sore on his arm that would not heal. The sore developed a red depression in the middle, surrounded by a raised border. It was not painful, and Mario postponed seeking treatment because he was busy working two jobs and managing his family of three children.

SIGNS AND SYMPTOMS The sore began to heal, but after 2 months additional sores formed on Mario's skin and within his mouth. He now experienced fever, swollen lymph nodes, and weight loss.

TESTING AND DIAGNOSIS Mario finally went to a clinic and reported his symptoms to a physician assistant. The physician assistant asked whether he had traveled out of the country. Mario reported that 6 months before onset of symptoms, he had returned from a tour in Iraq as a sergeant in the US Army. The physician assistant observed that the appearance of the sores was typical of cutaneous leishmaniasis. Microscopy and PCR confirmed the presence of the protozoan *Leishmania donovani*, a parasite commonly acquired by soldiers in Iraq, where it is transmitted by sand flies.

Sand Fly, the Vector that Transmits *Leishmania* Parasites

Skin Lesions Typical of Leishmaniasis

TREATMENT The physician assistant prescribed sodium stibogluconate (Pentostam), a mineral containing antimony, which kills the parasite. The sores healed, but Mario was told that he could never be sure the parasite was completely eradicated; symptoms might recur. For this reason, all US military personnel who have been deployed in Iraq and Afghanistan are restricted from blood donations for one year.

LM   10 μm

Leishmania, **Single-Celled Parasite**
Giemsa stain reveals the nucleus and whiplike flagellum.

The infectious agent that Mario acquired in the opening case history, *Leishmania donovani*, is a protozoan, a single-celled eukaryotic parasite. *Leishmania* is commonly found in tropical and subtropical regions of South America and Asia. The protozoan causes the disease leishmaniasis, which shows three different forms: (1) cutaneous leishmaniasis, or skin infection; (2) mucocutaneous leishmaniasis, in which infection invades the mucous membranes (mouth and nostrils), as well as the skin; and (3) visceral leishmaniasis, or kala-azar, infection of internal organs, including the liver and spleen. The mucocutaneous and visceral forms are serious and can be fatal.

As discussed in Chapter 5, eukaryotic cells are very different from bacteria. The eukaryotic cell of *Leishmania* possesses a nucleus (see photo in the chapter-opening case history) and a whiplike flagellum powered by ATP, unlike bacterial rotary flagella. Because the cell structure of eukaryotic microbes is so similar to that of human cells, far fewer antimicrobials are available against eukaryotes than against bacteria. In addition, the treatment agents, such as antimony-containing compounds, tend to have worse side effects on the host tissues.

Parasitic protozoa such as *Leishmania* often have complex life cycles, sometimes involving multiple hosts (**Figure 11.1**). In *Leishmania*, the promastigote, or flagellated form of the protozoan, is carried by a sand fly (Figure 11.1, step 1), which injects it into the skin of the host. In the bloodstream, the promastigotes are phagocytosed by macrophages (white blood cells), which attempt to destroy them (step 2). But the promastigotes survive within the macrophage and develop into amastigotes (unflagellated forms) that multiply intracellularly (step 3). The macrophage bursts, releasing amastigotes that multiply within various host tissues, causing sores (step 4). When another sand fly takes a blood meal (step 5), it ingests infected macrophages full of amastigotes. The amastigotes develop into promastigotes within the sand fly's midgut (step 6). The promastigotes multiply and migrate to the proboscis, where they can be transmitted to the next host that the sand fly bites.

Besides *Leishmania*, other protozoa cause diseases such as malaria and trypanosomiasis (sleeping sickness). Protozoan diseases cause tremendous mortality and morbidity among humans

and animals. And fungi, another group of eukaryotes, cause opportunistic infections among people with weakened immune systems. At the same time, microbial eukaryotes—including protozoa, fungi, and algae—play key roles in all ecosystems. For example, protozoa and fungi participate in microbial catabolism during wastewater treatment, helping to break down organic wastes to carbon dioxide.

In this chapter we will explore the diversity of microbial eukaryotes, including fungi, algae, and protozoa. We will describe the branching of multicellular animals and plants from the eukaryotic family tree, and we will explore the form and function of eukaryotes as pathogens and as partners in our ecosystem. In addition, we will introduce certain differentiated animals (helminths and arthropods) that act as infectious agents.

11.1 Eukaryotes: An Overview

What is the structure of eukaryotic cells? As presented in Chapter 5, all eukaryotic cells, whether human or microbial, possess a nucleus and other key membranous organelles, and nearly all possess mitochondria. Membranous organelles enable some microbial eukaryotes to grow to a million times larger than a bacterial cell. Yet others, particularly marine algae, have been downsized by evolution to less than 2 μm, comparable in size to *Escherichia coli*. Even so, such tiny eukaryotes, such as the alga *Ostreococcus tauri* (**Figure 11.2**), possess all the key organelles of a eukaryote (presented in Section 5.6). These organelles include the nucleus, containing linear chromosomes; mitochondria, the powerhouses of respiration; the endomembrane

Figure 11.1 Stages in the Life Cycle of *Leishmania*

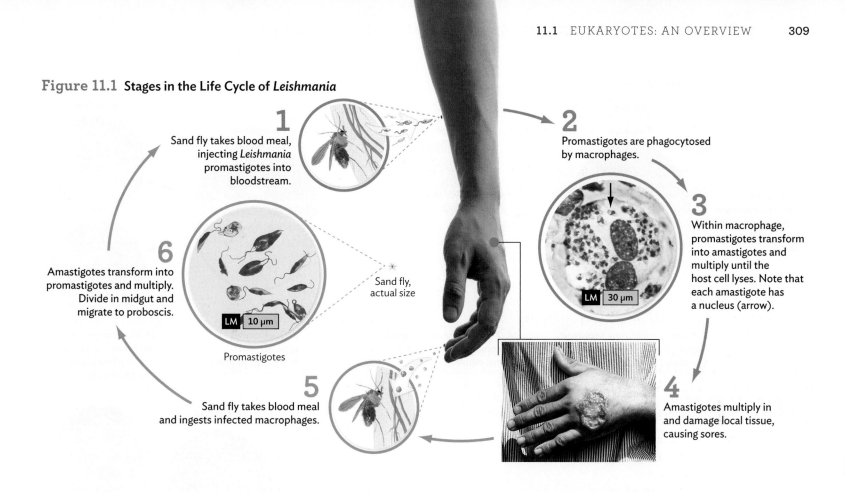

1 Sand fly takes blood meal, injecting *Leishmania* promastigotes into bloodstream.

2 Promastigotes are phagocytosed by macrophages.

3 Within macrophage, promastigotes transform into amastigotes and multiply until the host cell lyses. Note that each amastigote has a nucleus (arrow).

LM 30 µm

4 Amastigotes multiply in and damage local tissue, causing sores.

5 Sand fly takes blood meal and ingests infected macrophages.

6 Amastigotes transform into promastigotes and multiply. Divide in midgut and migrate to proboscis.

Promastigotes

LM 10 µm

Sand fly, actual size

---

system and Golgi, for intracellular transport; and a chloroplast, for photosynthesis.

Reproduction of Eukaryotic Cells

All eukaryotes have linear chromosomes that must divide by mitosis. The "ends" of linear chromosomes require special means of replication not required for most bacteria with their circular chromosomes. DNA replication occurs during a portion of interphase, the active state of the cell. Once DNA replication is complete, the cell halts most of its enzymatic activities and prepares for mitosis (**Figure 11.3A**). **Mitosis** is the process of segregating the two copies of all chromosomes evenly into the daughter cells. The process ensures that each daughter cell receives a full set of daughter chromosomes.

In most cases, mitosis involves four phases:

- **Prophase.** Paired chromosomes (postreplication) condense into short rods. In most species the nuclear membrane dissolves.

- **Metaphase.** The chromosomes arrange themselves along a plane across the equator of the cell.

- **Anaphase.** The pairs separate, pulled by spindle fibers composed of microtubules that contract. The separated chromosomes are pulled toward opposite poles of the cell.

- **Telophase.** The chromosomes decondense (become long and thin again), and a nuclear membrane forms around each set. The cell completes division of its cytoplasm, forming two daughter cells.

Figure 11.2 Even a Tiny Eukaryote Has Organelles

Ribosomes

TEM 0.50 µm

Nucleus

Mitochondrion

Chloroplast

Golgi

Granules

Ostreococcus tauri, visualized by cryo-EM (**A**) and in 3D by tomography, colorized (**B**).

Figure 11.3 Mitosis and Meiosis

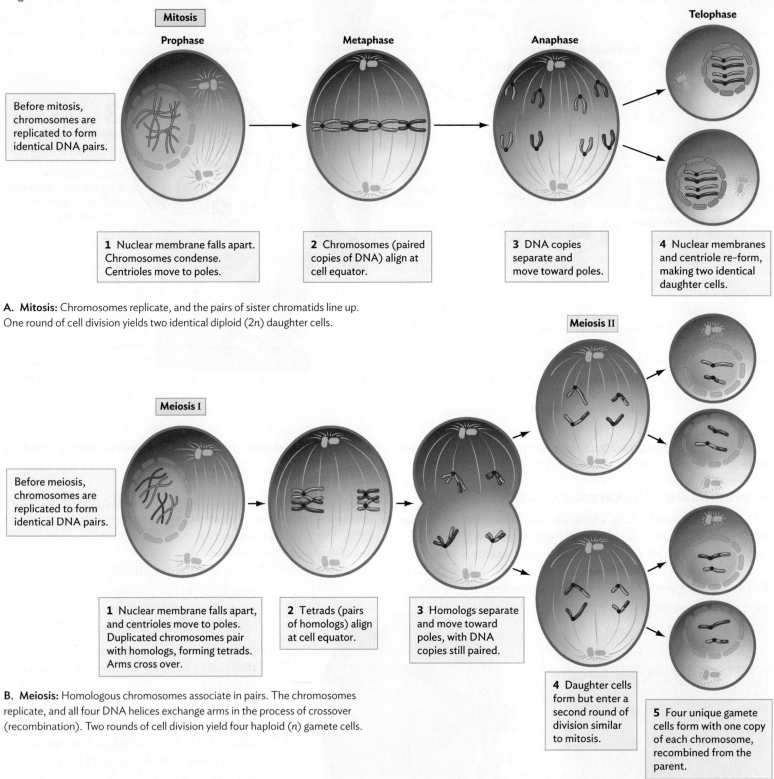

A. Mitosis: Chromosomes replicate, and the pairs of sister chromatids line up. One round of cell division yields two identical diploid (2*n*) daughter cells.

B. Meiosis: Homologous chromosomes associate in pairs. The chromosomes replicate, and all four DNA helices exchange arms in the process of crossover (recombination). Two rounds of cell division yield four haploid (*n*) gamete cells.

Many eukaryotic microbes, such as yeasts, can proliferate indefinitely by mitosis, a process called **asexual reproduction** or **vegetative reproduction**. But most eukaryotes, single-celled or multicellular, also have the option of **sexual reproduction**. Sexual reproduction requires the reassortment of genetic material from different chromosomes. A sexual life cycle alternates between cells that are **diploid** (2*n*, containing two copies of each chromosome) and sex cells that are **haploid** (*n*, containing a single copy of each chromosome). Two of the haploid sex cells (called gametes) can join each other by fertilization to regenerate a diploid cell (called a **zygote**). The diploid thus possesses two homologs of each chromosome—that is, two versions of the same chromosome from two different parents.

The process of gamete formation requires a special modification of mitotic cell division called **meiosis** (Figure 11.3B). Meiosis includes two cell divisions: meiosis I and meiosis II. Like mitosis, meiosis I must be preceded by replication of all chromosomes (during interphase). So, a diploid ($2n$) cell temporarily has four copies of each DNA homolog. In meiosis, unlike mitosis, prophase I requires that each replicated pair line up with its homolog, the homologous chromosome inherited from the other parent. Now the aligned homologs exchange portions of their DNA. This genetic exchange reassorts the gene versions from the two parents and increases genetic diversity in the next generation.

In meiosis I, the paired homologs separate during metaphase I and anaphase I. A short telophase occurs, in which each daughter cell now has $2n$ DNA helices, but as n pairs containing a chromosome from each parent. The chromosome pairs then separate during meiosis II, which includes prophase II, metaphase II, anaphase II, and telophase II. The result is four haploid (n) cells, as seen in Figure 11.3B. For many microbes, the haploid forms may undergo asexual (vegetative) reproduction. At some point, the haploid forms may develop into specialized cells that reunite through fertilization, restoring a diploid form (the zygote).

What is the consequence of alternation between haploid (n) and diploid ($2n$) forms? Consider a protozoan parasite that grows within a human host. The haploid form of the parasite requires fewer resources and generates fewer varieties of coat proteins that might activate the host's immune system. But the diploid form generates novel combinations of genes that may provide an advantage when the environment changes or when the parasite enters a new host. Thus, most eukaryotic microbes maintain the option of a sexual cycle that alternates between haploid and diploid. In some cases, one form exists only briefly; for example, the malarial parasite proliferates mainly as a haploid but undergoes a brief cycle of fertilization and meiosis within the insect vector.

Diversity of Microbial Eukaryotes

Microbial eukaryotes are classified traditionally as fungi, protozoa, and algae. For most of human history, life was understood in terms of multicellular eukaryotes. There were animals (creatures that move to obtain food) and plants (rooted organisms that grow in sunlight). **Fungi** (singular, **fungus**) lack photosynthesis yet were considered a form of plant because they grow on the soil or other substrate. As a result, **mycology**, the study of fungi, was often included with botany, the study of plants. Later, microscopists came to recognize microscopic forms of fungi such as hyphae (filaments of cells) and yeasts (single cells); thus, fungi were studied also by microbiologists. Surprisingly, however, fungi and microsporidians show close genetic relatedness to multicellular animals.

Other microscopic life forms, such as amebas and paramecia, are motile and appear more like microscopic animals. These animal-like organisms were called **protozoa** (singular, **protozoan**), meaning "first animals." Microscopic life forms containing green chloroplasts were called **algae** (singular, **alga**) and were thought of as primitive plants. But some protozoa turned out to have chloroplasts, and some algae turned out to be motile with flagella. Some authors refer to algae and protozoa collectively as **protists**.

The microscopic forms of fungi, protozoa, and algae are all considered "microbial eukaryotes." So, how do these microbes relate to multicellular animals and plants? Plants show close genetic relationship with the "primary algae," green algae that evolved from a common chloroplast-bearing ancestor. Animals, however, are most closely related to fungi. The relatedness of animals to fungi came as a surprise because animals were thought to be more similar to protists, which seem to resemble animals in their motility. DNA sequence analysis shows that several different groups of protists are more different from each other—and from animals—than animals are from fungi.

Major kinds of microbial eukaryotes are summarized in Table 11.1. Although all these groups include microbial forms, they also include species that can be observed by the unaided eye (such as giant amebas) as well as multicellular forms (such as mushrooms and kelps). Thus, the term "microbe" serves as a working description but does not define a strict category of life.

Major groups of microbial eukaryotes include the following:

- **Fungi** grow in chains called hyphae or as single-celled yeasts. They have absorptive nutrition, digesting many complex plant polymers. Most are nonmotile. **Microsporidians** are spore-forming organisms closely related to fungi.

- **Amebas** are single-celled protists of highly variable shape that form **pseudopods**, locomotor extensions of cytoplasm enclosed by the cell membrane.

- **Ciliates** are single-celled protists with short, whiplike motility organelles called cilia. Most are free-living, but some are human parasites.

- **Trypanosomes** and **metamonads** are flagellated protists with complex parasitic life cycles involving developmental stages within hosts.

- **Algae** are protists containing chloroplasts that conduct photosynthesis, related to those of plants. Some algae are single-celled, whereas others grow as filaments or sheets. **Dinoflagellates** possess chloroplasts obtained by their ancestral cells via endosymbiosis of a green alga.

These groups are considered "true" microbes, although they include large macroscopic forms, such as mushrooms (fungal fruiting bodies) and sheets of kelp (algae). Certain other multicellular parasites are <u>not</u> considered microbes, although they may be too small to see, such as mites and worms. These invertebrate animals

Table 11.1
Eukaryotic Microbes and Multicellular Infectious Agents

Major Category	Example	Major Category	Example

Fungi and microsporidians
Filaments or yeasts that usually lack motility; tough cell walls made of polysaccharides; heterotrophs by absorptive nutrition; include decomposers and parasites

Aspergillus

Amebas (amoebas)
Heterotrophic protists with amorphous shape; motile with pseudopods

Chaos carolinensis

Ciliates
Protists with cilia to capture prey; most are free-living, but some are parasites

Balantidium coli

Apicomplexans
Intracellular parasites with complex life cycles; apical complex enables attachment and penetration of host cell

Plasmodium falciparum

Trypanosomes and metamonads
Extracellular parasites with complex life cycles

Trypanosoma cruzi

Algae
Chloroplast-containing protists that conduct photosynthesis; dinoflagellates produce neurotoxins

Flagella

Chlamydomonas

Helminths
Multicellular worms, including nematodes (roundworms), cestodes (tapeworms), and trematodes (flukes); include parasites, as well as free-living members of soil ecosystems

Taenia solium
(scolex of tapeworm)

Arthropods
Multicellular insects and related microscopic organisms; include parasites and free-living members of soil ecosystems

Body louse

have fully differentiated organ systems, sometimes comparable in complexity to those of vertebrates. The dynamics of transmission and infection of invertebrate parasites parallel those of microbial pathogens. Therefore, these animal parasites are often covered by health professionals under the category of "eukaryotic microbiology." In particular:

- **Helminths** are multicellular worms, including the nematodes (roundworms), cestodes (tapeworms), and trematodes (flukes).
- **Arthropods** include insects such as fleas and lice, as well as noninsects such as mites.

Worms and arthropods are a major source of morbidity and mortality worldwide. A quarter of the world's population is parasitized by worms, including at least 10% of people in the United States. The presence of worms causes nutritional impairment and developmental delays. The horror of infection by worms and arthropods inspired the classic science fiction film *Alien*, which depicts an intelligent extraterrestrial parasite.

SECTION SUMMARY

- **Microbial eukaryotes** may undergo asexual (vegetative) reproduction. Alternatively, they may undergo a sexual cycle involving meiosis and fertilization.
- **Fungi** grow by absorptive nutrition, as single cells or by extending multicellular hyphae. **Microsporidians** are related to fungi.
- **Amebas** are single-celled protists with pseudopod motility.
- **Ciliates** are single-celled protists covered with hairlike microtubular organelles.
- **Trypanosomes** and **metamonads** are flagellated single-celled protists with complex parasitic cycles.
- **Algae** are protists that contain chloroplasts and conduct photosynthesis. **Dinoflagellates** are flagellated protozoa with chloroplasts.
- **Invertebrate parasites** include helminths and arthropods.

11.2 Fungi and Microsporidians

SECTION OBJECTIVES

- Describe the diseases caused by yeasts.
- Describe the structure of the fungal mycelium.
- Explain how fungi impact the environment.
- Describe how different types of fungi and microsporidians cause disease.

How do fungi contribute to our ecosystems? Fungi recycle wood and leaves, including substances, such as lignin, that animals cannot digest. Many are saprophytes (live on dead organisms). They decompose animal bodies; all of us someday will feed the fungi.

Fungi produce antibiotics such as penicillin and contribute to food products such as beer and cheese. Underground, fungal filaments called **mycorrhizae** (singular, **mycorrhiza**) extend the root

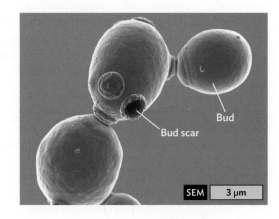

Figure 11.4
***Candida albicans*, a Unicellular Fungus (Yeast)**

Bud
Bud scar
SEM   3 μm

systems of trees, forming a nutritional "internet" that interconnects plant communities. Mycorrhizae could have inspired the fictional underground tree network shown in the 2009 film *Avatar*.

At the same time, some fungi are pathogens that threaten the lives of humans, animals, and plants. The yeast *Candida albicans* (Figure 11.4) is normally carried by most people, but it can kill a susceptible patient. The related yeast *Candida auris* is a deadly pathogen spreading in hospitals. *C. auris* is resistant to nearly all antifungal agents and has a mortality rate of 35%.

Yeasts Are Single-Celled Fungi

Single-celled fungi are known as **yeasts**. Yeasts have the advantage of rapid growth and dispersal in aqueous environments. Many different single-celled yeasts evolved independently in different fungal taxa. Baker's yeast, *Saccharomyces cerevisiae*, is used to leaven bread and to brew wine and beer. Other yeasts, such as *Candida albicans*, are important members of human vaginal biota. But *C. albicans* is also an opportunistic pathogen, causing infections such as thrush in the mouth of patients with AIDS. Other yeasts that are opportunistic pathogens, commonly infecting AIDS patients, include *Pneumocystis jirovecii* (formerly *Pneumocystis carinii*), a yeast-form ascomycete, and *Cryptococcus neoformans*, a yeast-form basidiomycete.

Some yeasts are asexual, whereas others, such as *S. cerevisiae*, alternate between diploid (2*n*) and haploid (*n*) forms, a life cycle known as alternation of generations (Figure 11.5). For *S. cerevisiae*, both haploid and diploid forms are single cells. The haploid (*n*) cells undergo several generations of mitosis and budding. This asexual reproduction is referred to as vegetative growth.

At some point, however, various stress conditions, such as nutrient limitation, may lead the haploid cells to develop into gametes. The gametes are of two mating types. Cells of the two mating types may fertilize each other by fusing together and combining their nuclei to make a diploid (2*n*) zygote. The diploid cell—like the haploid cell—may undergo vegetative budding and reproduction. Under stress, however, a 2*n* cell may undergo meiosis, regenerating haploid cells. The haploid progeny possess chromosomes reassorted and recombined from those of their "parental" gametes.

Figure 11.5 Yeast Alternation of Generations

The life cycle of *Saccharomyces cerevisiae*. Haploid (*n*) cells reproduce by budding (asexually). For sexual reproduction, a pair of opposite mating types fuse, forming a diploid (2*n*) cell. The 2*n* cell reproduces by budding (asexual) until it undergoes meiosis (sexual). Meiosis generates an ascus full of ascospores, which germinate as haploid cells.

MATING
(cell fusion and nuclear fusion)

2*n*
Diploid

Haploid mating
cells in contact

Mated haploids
(zygote) budding

n

Budding cells

2*n*

Vegetative cell

Vegetative cell

Ascus with
ascospores

n   2*n*

MEIOSIS
(haploid cells)

Bud scar

SEM   5 µm

Bud

Saccharomyces cerevisiae
Bud and bud scars left by
earlier buds.

Baker's yeast
Saccharomyces cerevisiae
is used to leaven bread
and to brew wine and beer.

Some fungi, such as *C. albicans*, can grow either as yeast or as filamentous fungi, depending on the environment; these are known as "dimorphic" fungi. Another dimorphic fungal pathogen is *Blastomyces dermatitidis*, the cause of blastomycosis, a type of pneumonia. *B. dermatitidis* grows as a yeast within the infected lung, but in culture and in soil environments it forms multicellular filaments. Filamentous fungi are described next.

Filamentous Fungi

CASE HISTORY 11.1

Fungus Growing in the Lung

Frank, a 45-year-old man, visited the hospital with respiratory distress and a high fever. Frank had diabetes mellitus and had smoked for 25 years. His white blood cell (WBC) count was 24,000/µl, and he had a C-reactive protein concentration of 24 mg/ml—both signs that his immune system was fighting an infection. His chest X-ray showed an abnormal shadow, and his chest computed tomography (CT) scan showed thickened bronchial walls. He was coughing up black sputum. The physician prescribed an antibiotic, cefotiam, but Frank's condition worsened. By the tenth day he was producing 100 ml of sputum a day, and the CT scan of his lungs showed multiple lesions. Compare the normal lung scan (**Figure 11.6A**) with the scan of lungs infiltrated by an infectious agent (**Figure 11.6B**). Since antibacterial antibiotics had failed, Frank was tested for fungi. A culture of his sputum and PCR amplification of fungal DNA revealed the fungus *Aspergillus fumigatus* (**Figure 11.6C**). Frank was treated with the antifungal agents itraconazole, meropenem, and amphotericin B, but he died on the twenty-fifth day.

Aspergillus is a filamentous fungus of the ascomycete group (see Table 11.2). Filaments of *Aspergillus* typically grow on soil or indoors as basement mold or on air conditioner coils. It forms spores that can reach significant concentrations in the air, where they are taken into the lungs and may cause allergic reactions. In patients with impaired immune function, *Aspergillus* spores may germinate and grow "fungus balls" that require surgical removal. The fungus is a common source of hospital-acquired (nosocomial) infection. In Frank's case, however, the infection was community acquired. Frank's predisposing conditions (depressed immune function due to diabetes and smoking) had allowed the fungal infection to take hold.

Figure 11.6 Lungs Infected by *Aspergillus*

A. Chest CT, healthy lungs.

B. Lungs infiltrated by *Aspergillus* mycelium.

C. *Aspergillus* mycelium invading the lungs.

Figure 11.7 Fungi Grow Hyphae with Cell Walls of Chitin

A. A mycelium: *Penicillium* fungal hyphae form the skin of Brie cheese.

B. Chitin consists of polymers of *N*-acetylglucosamine.

C. Fungal hyphae extend and form branches, generating a mycelium.

Many fungi grow as multicellular branched filaments called **hyphae** (singular, **hypha**; **Figure 11.7A**). The hyphae undergo a life cycle that produces spores for dissemination. Inhalation of fungal spores can lead to aspergillosis and histoplasmosis, conditions particularly dangerous for immunocompromised patients. Other fungi cause skin infections such as ringworm and athlete's foot (medical term, *tinea pedis*). The fungi most commonly found in humans or in our local environment are presented in **Table 11.2**.

Hyphae have tough cell walls that are composed of **chitin**, an acetylated amino-polysaccharide stronger than steel (**Figure 11.7B**). Its strength derives from multiple hydrogen bonds between fibers. Hyphae grow by extending multinucleated cells (**Figure 11.7C**). As a fungal hypha extends, its nuclei divide mitotically without cell division. The hyphae form branching tufts called a **mycelium** (plural, **mycelia**). The mycelium may grow large enough to be seen as a fuzzy colony or puffball. As the hypha expands, its chitinous cell wall enables it to penetrate softer materials, such as plant or animal cells.

The biosynthesis of chitin provides a target for antifungal agents such as nikkomycin Z. Another drug target is the membrane lipid

impact

Mold after Hurricane Katrina

In the wake of Hurricane Katrina in 2005, much of the city of New Orleans was flooded. Entire neighborhoods were submerged for days or weeks. The flooding of homes full of drywall and organic materials made ideal conditions for the growth of "mold," which generally consists of airborne ascomycete fungi. Mold grew not only on the materials submerged but also on the surfaces above, exposed to water-saturated air. Figure 1 shows how mold grew up to 3 feet above the flood line (the highest level submerged) within the kitchen of a flooded home. Unfortunately, most homeowner insurance policies covered only damage "up to the flood line."

Mold growing on damp surfaces releases spores into the air. Inhaled mold spores cause respiratory problems, even at relatively low levels (such as 2,000 spores per cubic meter). In New Orleans after the waters receded, the mold counts in the air inside homes reached as high as 650,000 spores per cubic meter. Such spore levels cause asthma attacks and hypersensitivity pneumonitis, a condition in which fever rises and the lungs fail. All remediation workers were advised to wear an N95 respirator mask. The Natural Resources Defense Council (NRDC) found that even after all contaminated items—including furniture, carpets, and drywall down to the studs—had been removed from flooded homes, airborne spore counts still reached 70,000.

The most common types of mold found by the NRDC researchers were ascomycetes such as *Cladosporium* (Figure 2). *Cladosporium* species form black or dark green colonies of conidia that readily break off and become airborne. Other ascomycetes found at high levels included *Aspergillus* and *Penicillium* species. These organisms occur at low levels even in clean air. Their spores cause allergy attacks and infect immunocompromised individuals. Some ascomycete molds produce mycotoxins, fungal toxins that cause long-term health problems. In addition, some flooded homes showed significant levels of *Stachybotrys* species, known as "black mold," a possible cause of neurological problems and immunosuppression. Once such molds establish growth, their eradication is a major challenge.

Figure 1 Mold Growing above the Flood Line

An undergraduate student volunteer indicates mold colonies growing above the flood line on a kitchen wall in a New Orleans home, 6 months after flooding.

DISCUSSION QUESTION

What can homeowners and communities do to decrease the incidence and prevalence of harmful airborne fungi?

Figure 2 *Cladosporium* Mold

A. Conidiophores (branched hyphae) producing conidia (asexual spores) of *Cladosporium cladosporioides* (left).

B. *Cladosporium* mold growing on media in a Petri dish (right).

LM 5 µm

Table 11.2
Fungi and Microsporidians

Phylum and Genus	Habitat or Disease
Ascomycetes	Form asci, pods of eight spores
Aspergillus	Plant saprophyte; produces aflatoxin; infects lungs of immunocompromised patients
Blastomyces	Blastomycosis; opportunistic infections of the skin
Candida	Yeast infections; thrush; opportunistic
Geomyces	White-nose syndrome in bats
Histoplasma	Histoplasmosis, lung disease
Magnaporthe	Rice blast, the most important disease of rice farming
Penicillium	Grows on bread and fruit; original source of penicillin
Saccharomyces	Ferments beer and bread dough
Stachybotrys	Black mold; contaminates dwellings
Trichophyton	Ringworm and athlete's foot skin infections
Basidiomycetes	Mushrooms and mycorrhizae
Amanita	Extremely poisonous; produces alpha-amanitin toxin
Cryptococcus	Opportunistic infection of patients with AIDS
Lycoperdon	Edible mushrooms
Chytridiomycetes	Motile zoospores with single flagellum
Batrachochytrium	Frog epizootic in South America
Neocallimastix	Digests plant material within bovine rumen
Zygomycetes	Saprophytes, insect pathogens, and mycorrhizae
Mucor	Soil mold; opportunistic infections
Rhizopus	Bread mold; food production (tempeh)
Microsporidia	Related to fungi
Encephalitozoon	Opportunistic infection of patients with AIDS

Filamentous fungi can expand at great length by asexual division (mitosis). Eventually, however, they run out of nutrients, and the cells undergo meiosis to form gametes. The gametes develop into spores for dispersal. An example of a spore-forming life cycle is that of ascomycete fungi such as *Aspergillus* or *Morchella* (**Figure 11.8**). In an ascomycete, the tips of the diploid mycelium undergo meiosis and develop into haploid **ascospores** within pods called **asci** (singular, **ascus**). The ascospores disperse until they germinate in a region of fresh resources. Each ascus undergoes cell divisions to form a haploid mycelium, which is male or female. The male and female structures fuse, followed by migration of all the male nuclei into the female structure. The paired nuclei then undergo several rounds of mitotic division while migrating into the growing mycelium. In the mycelial tips, the paired nuclei finally fuse (becoming 2*n*) and re-form a diploid mycelium. Overall, the diploid-haploid life cycle enables a fungus to respond genetically to environmental change by reassorting its genes through meiosis and fertilization. Gene reassortment provides new genotypes, some of which may increase survival in the changed environment.

Some fungi pack their asci (or other spore-producing forms) into a larger structure called a **fruiting body**. Ascomycetes such as *Aspergillus* and *Penicillium* mold form relatively small fruiting bodies, called conidiophores, that disperse airborne spores. Conidiospore-forming ascomycetes are the major type of mold associated with dampness in human dwellings; they caused massive damage to homes flooded in the wake of Hurricane Katrina (see **IMPACT**).

The fruiting bodies of other ascomycete fungi can be impressively large and edible as prized foods. Morels (*Morchella hortensis*; see Figure 11.8) and truffles (*Tuber aestivum*) form large, mushroom-like fruiting bodies whose ascospores are dispersed by animals attracted by their delicious flavor. The famous truffles grow underground, where human collectors traditionally use muzzled pigs to detect and unearth them.

SECTION SUMMARY

- **Single-celled fungi are called yeasts.** The yeast *Saccharomyces cerevisiae* is used to make fermented drinks and to leaven bread. *Candida albicans* is a part of normal human biota, but certain strains can cause opportunistic infections. *Candida auris* is an emerging drug-resistant pathogen.
- **Filamentous fungi grow as multinucleated hyphae.** Hyphae are exceptionally strong filaments that can penetrate many materials. A branching mass of hyphae is called a mycelium.
- **Fungi have cell walls of chitin.** Chitin biosynthesis is a target for antifungal agents.
- **Fungi are heterotrophs that absorb nutrients.** Most fungi in nature absorb nutrients from decaying organisms, although some are pathogens that infect animal or plant hosts.
- **Ascomycete fungi form asci that disperse spores.** Spores of fungi such as *Aspergillus* can spread disease or cause allergies.

ergosterol, an analog of cholesterol not found in animals or plants. Ergosterol biosynthesis is inhibited by antifungal agents called triazoles. Most fungal forms are nonmotile (with the exception of chytridiomycete zoospores). Fungi cannot ingest particulate food, as do protists. This is because their cell walls cannot part and re-form like the flexible pellicle of amebas and ciliates does (discussed next, in Section 11.3). Instead, fungi secrete digestive enzymes and then absorb the broken-down molecules from their environment.

Figure 11.8 **The Life Cycle of an Ascomycete**

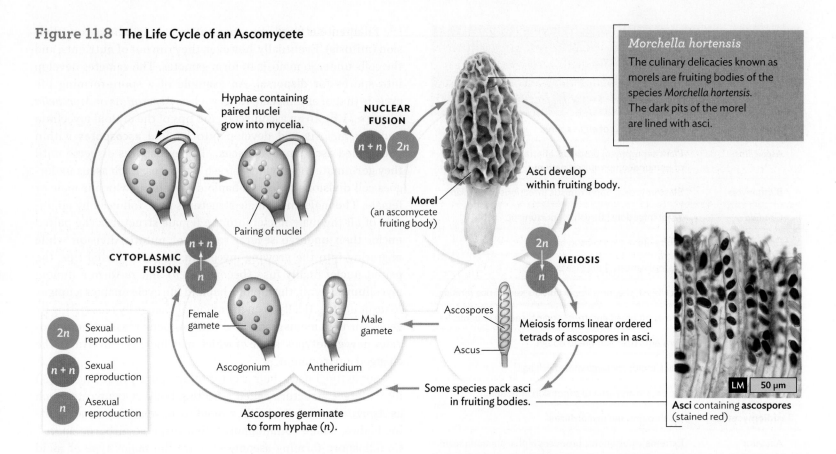

Hyphae containing paired nuclei grow into mycelia.

NUCLEAR FUSION

n + n   *2n*

CYTOPLASMIC FUSION

n + n
n

Pairing of nuclei

Morchella hortensis
The culinary delicacies known as morels are fruiting bodies of the species *Morchella hortensis*. The dark pits of the morel are lined with asci.

Morel (an ascomycete fruiting body)

Asci develop within fruiting body.

2n
MEIOSIS
n

Female gamete

Male gamete

Ascogonium   Antheridium

Ascospores

Ascus

Meiosis forms linear ordered tetrads of ascospores in asci.

Some species pack asci in fruiting bodies.

Ascospores germinate to form hyphae (*n*).

2n Sexual reproduction

n + n Sexual reproduction

n Asexual reproduction

LM   50 µm

Asci containing **ascospores** (stained red)

Thought Question 11.1   For a fungus, what are the advantages of growing a mycelium? What are the advantages of growing as a single-celled yeast?

11.3 Protozoa: From Amebas to Trypanosomes

SECTION OBJECTIVES

- Describe the form and motility of amebas.
- Describe two diseases caused by amebas.
- Describe the distinctive traits of ciliates and apicomplexan parasites.
- Outline the complex life cycle of *Plasmodium falciparum* and *Plasmodium vivax*, the parasites that cause malaria.
- Describe the diseases caused by trypanosomes, giardia, and trichomonads.

Amebas

Amebas (also spelled "amoebas") are voracious predators, consuming other protists and bacteria. Some amebas can reach several millimeters in size—a single cell that is large enough to eat small invertebrates. There are two major groups of amebas: the lobed amebas, with large, bulky pseudopods (**Figure 11.9A**);

Figure 11.9 **Amebas**

A. *Amoeba proteus* extends its lobed pseudopods (colorized LM).

Pseudopod

LM   200 µm

B. A foraminiferan extends needlelike pseudopods through its shell of calcium carbonate.

LM   500 µm

and the filamentous amebas, with thin, needlelike pseudopods (**Figure 11.9B**). Some amebas with needlelike pseudopods, such as the **foraminiferans** (forams), form inorganic shells made of calcium carbonate (limestone). Fossil foraminiferan shells are common in rock formations derived from ancient seas. Foraminiferan shells formed the white cliffs of Dover in Britain and the stone used to build the Egyptian pyramids.

The lobed amebas (the more familiar kind) have pseudopods that extend lobes of cytoplasm through cytoplasmic streaming. Cytoplasmic streaming is the flow of nutrients and other components through the cytoplasm, mediated by actin proteins. Most lobed amebas are free-living in soil or water, where they prey on bacteria and small arthropods. They play an important role in the food web of soil and aquatic ecosystems. Some amebas are slime molds, in which individual amebas converge to form a fruiting body. Certain ameba species, however, cause diseases such as dysentery and meningitis.

INTESTINAL AMEBIASIS

CASE HISTORY 11.2
Spring Break Surprise

Brianna, a 22-year-old college student, had returned recently from spring break at the beach in Acapulco, Mexico. She presented at the college clinic complaining of severe abdominal cramps and fever. She also had nausea and bloody diarrhea with mucus. The nurse practitioner collected stool samples and sent them to the laboratory for analysis. Cultures were negative for bacterial pathogens. A trichrome stain was performed, revealing amebas of the **trophozoite** (active feeding) stage (**Figure 11.10A**). The stained amebas contained ingested red blood cells (RBCs). The appearance of the amebas was characteristic of *Entamoeba*

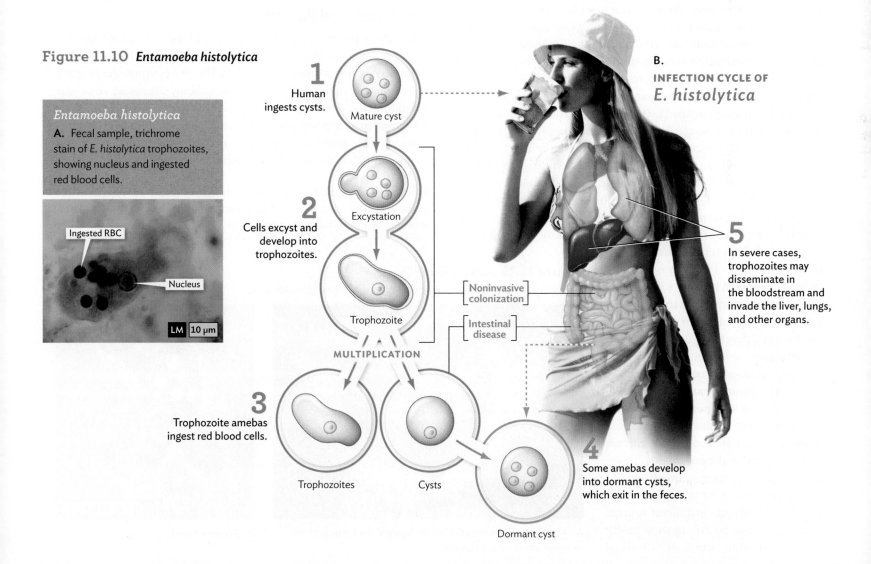

Figure 11.10 *Entamoeba histolytica*

Entamoeba histolytica

A. Fecal sample, trichrome stain of *E. histolytica* trophozoites, showing nucleus and ingested red blood cells.

Ingested RBC

Nucleus

LM 10 μm

1 Human ingests cysts.

Mature cyst

Excystation

2 Cells excyst and develop into trophozoites.

Trophozoite

MULTIPLICATION

3 Trophozoite amebas ingest red blood cells.

Trophozoites

Cysts

Dormant cyst

4 Some amebas develop into dormant cysts, which exit in the feces.

Noninvasive colonization

Intestinal disease

B.
INFECTION CYCLE OF
E. histolytica

5 In severe cases, trophozoites may disseminate in the bloodstream and invade the liver, lungs, and other organs.

histolytica, the cause of amebic dysentery (bloody diarrhea), or intestinal amebiasis. Brianna was treated with paromomycin (an aminoglycoside that inhibits protein synthesis) for 7 days, and her symptoms resolved.

Entamoeba histolytica is a serious cause of intestinal illness in much of the developing world, where it can lead to malnutrition and death. The amebas are usually ingested in the form of **cysts**, dormant cells encased in a tough coating (**Figure 11.10B**). Cysts can remain viable for many weeks in contaminated water supplies, such as that from which Brianna apparently drank. The ingested cysts reach the intestine, where they "excyst" and develop into trophozoites, amebas with pseudopods that phagocytose prey. In the intestinal lumen and in the bloodstream, the amebas ingest red blood cells (RBCs). As the amebas run short of food, they develop into dormant cysts, which exit in the feces—a "fecal-oral" route of transmission. In severe cases, however, the trophozoites may invade the liver, lungs, and other organs.

The symptoms of dysentery can be caused by a number of different viruses and bacterial pathogens, which require different kinds of treatment; thus, the pathogen must be identified correctly. Common bacterial pathogens such as *Shigella* may be detected by immunological tests. But to diagnose the presence of amebas and other intestinal parasites requires microscopy because their cell size and shape are distinctive. A procedure called Wheatley's trichrome stain reveals the parasites among human cells and fecal material (see Figure 11.10A). The trichrome stain involves treatment with iodine-alcohol solution, followed by a triple-stain solution. The cytoplasm of protozoa stains green or blue, whereas the nuclear membrane and ingested RBCs stain bright red. Cyst forms of the ameba stain bluish purple. Nevertheless, it must be noted that the closely related non-pathogen *Entamoeba dispar* is indistinguishable from *E. histolytica* under the microscope and is found ten times more often, so it is possible to miss the real pathogen.

Amebic dysentery is best avoided by careful choice of fluid consumption. When traveling in regions with limited sanitation, it is important to drink only filtered or bottled water. Avoid fresh produce that may be washed in contaminated water.

In most cases, an otherwise healthy host can eliminate amebic infection within 2 weeks. Repeated infections, however, may cause malnutrition. Sometimes the infection leads to chronic carrier status, and occasionally, infection becomes systemic, involving the liver, lungs, and other organs. Systemic infection by amebas is difficult to treat and can be fatal. So, where good health care is available, cases of intestinal

amebiasis are treated early with anti-amebic agents. The drug paromomycin specifically targets the small-subunit ribosomal RNA of microbial eukaryotes. An alternative drug, metronidazole, targets anaerobic protozoa and bacteria.

OTHER DISEASES INVOLVING AMEBAS Several other diseases can involve different species of amebas. For example, primary amebic meningoencephalitis, an infection of the brain, can be caused by the amebas *Naegleria fowleri* and *Balamuthia mandrillaris*. The amebas may enter the body through the nose during swimming in a contaminated pool. Free-living soil amebas such as *Acanthamoeba* may contaminate contact lens cleaning solutions, causing keratitis, an infection of the cornea (**Figure 11.11**). Wearers of reusable contact lenses have a keratitis infection rate of 2 per 1,000. Finally, free-living amebas may play host to bacterial pathogens such as *Legionella pneumophila,* the cause of legionellosis, an often fatal form of pneumonia (discussed in Section 20.3). Amebas carrying intracellular *L. pneumophila* often contaminate water supplies, air ducts, and even dental irrigation solutions. The host amebas enable the pathogen's persistence and transmission to human hosts.

Predators and Parasites

What other protozoa have interesting ways to catch prey? Ciliates are covered with **cilia** (singular, **cilium**), short, hairlike organelles composed of microtubules (**Figure 11.12**). The beating cilia propel a stream of prey to the oral groove (mouth). Many ciliates are voracious predators that swallow prey nearly their own size (**Table 11.3**). Other predatory protozoa are called flagellates, possessing pairs of flagella—long, whiplike organelles. Like cilia, protozoan flagella are composed of microtubules. Their whip-like motion is powered by ATP.

⊙ **Note:** Eukaryotic flagella and cilia are very different from bacterial flagella, which are helical protein filaments. Bacterial helical flagella rotate, powered by the proton motive force across the cell membrane.

Figure 11.11 Amebic Keratitis
Amebic keratitis commonly results from improperly cleaned contact lenses.

A. *Acanthamoeba* infects the keratin layers of the cornea, which becomes inflamed and clouded.

B. *Acanthamoeba.*

Figure 11.12 The Cortex of a Ciliate

Beneath the pellicle (outermost cell layer) of a ciliated protist lie flattened sacs of fluid called alveoli. The cilia, each composed of microtubules, are rooted in a network of lateral microtubules. Parasomal sacs take up nutrients and form endocytic vesicles.

Ciliates

Ciliates are cilia-covered, heterotrophic protists that consume algae and smaller protists and are in turn consumed by amebas. A medically important ciliate is *Balantidium coli*, a cause of dysentery. Powered by ATP hydrolysis, the cilia beat in coordinated waves that maximize the efficiency of motility.

Table 11.3
Ciliates and Apicomplexans

Phylum and Genus	Habitat or Disease
Ciliates (ciliated protists)	
Paramecium	Pond predator of algae and bacteria
Stentor	Stalked, ciliated predator; cilia draw prey into mouth; wastewater treatment community
Apicomplexans (parasites with apical complex; lack flagella or cilia; undergo complex life cycles)	
Babesia	Babesiosis, blood infection spread by ticks
Cryptosporidium	Intestinal infection, opportunistic; waterborne
Plasmodium	Malaria; carried by mosquito vector
Toxoplasma	Toxoplasmosis; transmitted to humans by cats

Cilia serve two functions:

- **Cell propulsion.** Coordinated waves of beating cilia, usually covering the cell surface, propel the cell forward.
- **Food acquisition.** By generating water currents into the mouth of the cell, a ring of cilia around the mouth brings food into the cell.

One of the best-studied ciliates is the paramecium (**Figure 11.13A**). Paramecia feed on bacteria by trapping them in the oral groove and then consuming them through enzymes within digestive vacuoles. Paramecia can also take up smaller particles through endocytosis by specialized pores in their cortex called parasomal sacs. Paramecia in turn are consumed by larger ciliates, such as *Didinium* (**Figure 11.13B**).

Ciliates have a complex cell form called "alveolate," referring to the flattened vacuoles called alveoli (singular, alveolus) within

Figure 11.13 Ciliate Predators

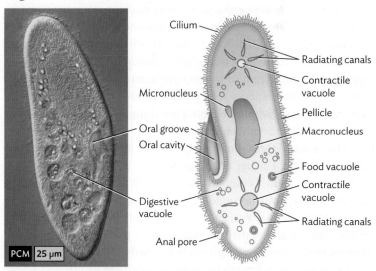

A. A paramecium has digestive vacuoles and an oral groove to ingest prey (colorized phase contrast).

B. *Didinium* consuming *Paramecium*. Both are ciliates.

the outer cortex (Figure 11.12). Some alveoli contain plates of stiff material, such as protein, polysaccharide, or mineral. Most alveolate protists possess other kinds of cortical organelles in addition to alveoli. These include extrusomes (organelles that extrude defensive enzymes or toxins), bands of microtubules for reinforcement, and whiplike cilia or flagella.

Without a rigid cell wall, how do ciliates maintain osmotic balance? Ciliates generally grow in aqueous solutions with salt and organic concentrations that are lower than those of their cytoplasm. Thus, water concentration outside is higher, and water tends to run into the cell. The excess water is expelled by a contractile vacuole, which withdraws water from the cytoplasm or contracts to expel it. Contractile vacuoles are widespread among protists and algae, but their mode of action has been studied mostly in paramecia. As water enters the cell from a low-solute environment, the contractile vacuole takes up the water through an elaborate network of intracellular channels and then contracts to expel the water through a pore, thus preventing osmotic shock and lysis (see Section 5.3).

Ciliates, like other eukaryotes, have nuclei containing chromosomes that undergo division by mitosis and generate gametes by meiosis. Unique to ciliates, however, the cell maintains two types of nuclei: the micronucleus, whose genes maintain genetic fidelity for sexual reproduction; and the macronucleus, containing multiple gene copies that express gene products throughout the large cell. During asexual reproduction, both micronuclear and macronuclear chromosomes undergo replication.

For sexual reproduction, a ciliate undergoes **conjugation** (**Figure 11.14**). During conjugation, two paramecia of opposite mating types form a cytoplasmic bridge. Within each conjugating cell, the 2n micronucleus undergoes meiosis to form four n gametic nuclei. (Meiosis was described in Section 11.1.) Three out of four of the gametic nuclei disintegrate, as does the macronucleus. The haploid micronuclei undergo mitosis, forming two daughter micronuclei. Daughter micronuclei from each cell are exchanged across the cytoplasmic bridge and then fuse with their respective counterparts, restoring a 2n micronucleus to each cell. The bridged cells come apart, and each micronucleus generates a new macronucleus. The new macronucleus now expresses gene products.

Note: Distinguish the conjugation of ciliates (with exchange of micronuclei) from the very different process of bacterial conjugation (one-way transfer of DNA from a donor bacterium to a recipient).

Apicomplexans

Apicomplexans form a major group of parasites of humans and other animals (Table 11.3). Their name derives from the "apical complex," a specialized structure that facilitates entry of the parasite into a host cell. Like the ciliates, apicomplexans are alveolates possessing an elaborate cortex composed of alveoli, pores, and microtubules. But as parasites, apicomplexans have undergone extensive reductive evolution, losing their flagella and cilia.

Figure 11.14 Conjugation

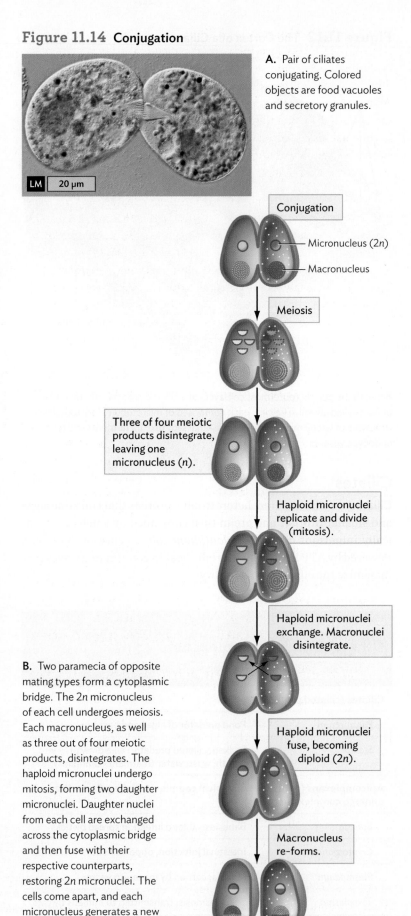

A. Pair of ciliates conjugating. Colored objects are food vacuoles and secretory granules.

LM   20 µm

Conjugation

Micronucleus (2n)

Macronucleus

Meiosis

Three of four meiotic products disintegrate, leaving one micronucleus (n).

Haploid micronuclei replicate and divide (mitosis).

Haploid micronuclei exchange. Macronuclei disintegrate.

Haploid micronuclei fuse, becoming diploid (2n).

Macronucleus re-forms.

B. Two paramecia of opposite mating types form a cytoplasmic bridge. The 2n micronucleus of each cell undergoes meiosis. Each macronucleus, as well as three out of four meiotic products, disintegrates. The haploid micronuclei undergo mitosis, forming two daughter micronuclei. Daughter nuclei from each cell are exchanged across the cytoplasmic bridge and then fuse with their respective counterparts, restoring 2n micronuclei. The cells come apart, and each micronucleus generates a new macronucleus.

Parasitic protists may have life cycles involving multiple hosts. For example, the malaria parasite (*Plasmodium*) alternates between the human host and the *Anopheles* mosquito. Generally, one type of animal constitutes the **definitive host**, the host where the parasite matures and may reproduce sexually; in the case of malaria, the definitive host is the mosquito. The alternate host (the human in this case) is called the **intermediate host**. The intermediate host supports asexual proliferation of the parasite and is required for transmission to the next insect host.

MALARIA The best-known apicomplexan parasites are *Plasmodium falciparum* and *Plasmodium vivax*, the main causes of malaria, the most important parasitic disease of humans worldwide (**Figure 11.15A**). *P. falciparum* and *P. vivax* are carried by mosquitoes, which transmit the parasite to humans when the insect's proboscis (mouthparts) penetrates the skin. The disease is endemic in areas inhabited by 40% of the world's population; each year malaria infects hundreds of millions of people and kills more than a million African children. As people travel, malaria has increasingly showed up in the United States as well.

The full life cycle of the malaria parasite includes developmental stages in the mosquito, followed by transmission to the human host, where the parasites undergo stages in the liver and in the blood (see **Figure 21.21**). In the liver, the transmitted parasites develop into the "merozoite" form, which invades RBCs. The merozoite first contacts a red blood cell through interaction between its apical complex (the structure at the penetrating tip of the cell) and receptor proteins on the RBC (**Figure 11.15B**). The apical complex contains a pair of secretory organelles that are capped by a ring of microtubules and that inject enzymes that aid entry of the parasite. The tip of the apical complex penetrates the host cell, enabling secretion of lipids and enzymes that facilitate invasion. Eventually, the entire merozoite enters the host cell, leaving no traces of the parasite on the host cell surface. Thus, the internalized parasite (ring stage) becomes invisible to the immune system until its progeny burst out. Some progeny are then picked up by mosquitoes, where they undergo a sexual life cycle and are transmitted to new hosts.

The malarial parasites acquire resistance rapidly and no longer respond to drugs, such as quinine, that nearly eliminated the disease half a century ago. The life cycle and molecular properties of *P. falciparum* have been studied extensively for clues to the development of new antibiotics and vaccines. The life cycle of *P. falciparum* and the etiology of malaria are discussed in Chapter 21.

TOXOPLASMOSIS An apicomplexan parasite common in the United States is *Toxoplasma gondii*, a parasite carried by cats and transmissible to humans. Approximately 30% of the world's population is infected with *T. gondii*. Fortunately, the infection is usually asymptomatic in humans because of our immune system, but the parasite can cross the placental barrier and harm a developing fetus. In rare cases, *T. gondii* can enter the brain, where it can alter host behavior. Some epidemiological evidence suggests a link between *T. gondii* infection and some cases of schizophrenia. The life cycle of *T. gondii* and the etiology of toxoplasmosis are discussed in Chapter 21.

BABESIOSIS An emerging human pathogen in the United States is the blood parasite *Babesia microti*. *B. microti* is spread by deer ticks, the same agents that spread Lyme disease—yet another unfortunate risk of a walk in the northeastern woods. *B. microti*, however, may also be spread by blood contact and transfusions. As in malaria, the parasite multiplies within RBCs. The symptoms of *B. microti* infection can mimic those of Lyme disease (caused by the tick-borne bacterium *Borrelia burgdorferi*) and are often misdiagnosed. Babesiosis is discussed further in **Section 21.5**.

WATERBORNE APICOMPLEXANS Several apicomplexans are emerging waterborne parasites. *Cryptosporidium parvum* infects the intestine of immunocompromised patients, causing watery diarrhea. An obligate intracellular pathogen, the organism generates a cyst form transmitted by the fecal-oral route. *Cryptosporidium* resists chlorination; it can cause major outbreaks of intestinal

Figure 11.15 Malaria

A. Red blood cells (RBCs) infected with *Plasmodium falciparum*, an apicomplexan parasite that causes malaria (stained purple with a dye that interacts with DNA).

B. The merozoite form of *P. falciparum* invades a red blood cell. The apical complex facilitates invasion and then dissolves as the merozoite transforms into an intracellular form.

disease, such as the one in 1993 when more than 400,000 people were sickened in Milwaukee, Wisconsin. Another cause of diarrhea is *Cyclospora*, also a danger for immunocompromised patients. *Cyclospora* causes occasional outbreaks from fecally contaminated produce.

Trypanosomes and Metamonads

What other motile parasites cause human disease? Euglenids are a group of protists that have whiplike flagella and chloroplasts arising from secondary endosymbiosis, similar to the dinoflagellates (see Section 11.4) but more distantly related. Some of the euglenids, however, lost their chloroplasts and evolved into obligate parasites called **trypanosomes**. Trypanosomes cause devastating diseases in humans and animals, such as sleeping sickness (*Trypanosoma brucei*), Chagas disease (*Trypanosoma cruzi*), and leishmaniasis (*Leishmania donovani*; see the chapter-opening case history). Another group of motile parasites is the **metamonads**. A common metamonad is *Giardia intestinalis*, the cause of intestinal giardiasis.

TRYPANOSOMES A trypanosome consists of an elongated cell with a single flagellum. The trypanosome has a unique organelle called the "kinetoplast," which evolved as a specialized mitochondrion that provides energy for the flagellum.

Trypanosomiasis, also known as African sleeping sickness, is a major killer of humans and livestock; in some sub-Saharan countries, trypanosomiasis is the second-biggest killer after AIDS. The parasite, *Trypanosoma brucei* (**Figure 11.16A**), is carried by the tsetse fly. *T. brucei* multiplies in the bloodstream of the host animal, causing repeated cycles of proliferation and fever that cause increasing drowsiness, ultimately leading to death, if untreated. This trypanosome is known for its extraordinary degree of **antigenic variation**, the ability to generate numerous different versions of its coat protein and thereby evade antibodies of the host immune system. The trypanosome's genome includes 200 different active versions of its variant surface glycoprotein (VSG), the antigen that induces the immune response—as well as 1,600 different "silent" versions that can recombine with "active" VSG to make additional variations. In effect, the trypanosome overwhelms the host immune system by continually generating new antigenic forms until the host repertoire of antibodies is exhausted.

Diagnosis of trypanosomiasis requires microscopic detection of the parasite in centrifuged blood. The parasite invades many body systems, causing cardiac and kidney dysfunction. In a late stage, the trypanosomes cross the blood-brain barrier, invading the central nervous system. Neurological impairment leads to disruption of the sleep cycle—hence the term "sleeping sickness"—ultimately leading to death.

To infect its human host, the trypanosome needs to interconvert among several different forms of its life cycle. Several of these forms are shown in Figure 11.16. For example, the "slender form" (Figure 11.16A) proliferates in the bloodstream of a human or other mammal (definitive host). But the "stumpy form" does not proliferate in the mammal; it needs to be ingested by the tsetse fly (intermediate host), where it lives as the asexual stage. Only the stumpy form can survive the tsetse midgut to develop into the forms that can grow there (procyclic and epimastigote).

Figure 11.16
Trypanosomes Cause Sleeping Sickness

Trypanosoma brucei
A. Cells of the slender form.

Slender form

PCM   13 μm

B.
LIFE CYCLE OF THE Trypanosome

INITIATION OF DIFFERENTIATION
A few trypanosomes differentiate into the stumpy form.

Slender (*proliferative*) → Stumpy (*nondividing*)

Upon ingestion via a blood meal, the stumpy form differentiates to the procyclic form, which multiplies in the insect midgut.

TRANSMISSION COMPETENCE
The slender form infects the animal via an insect bite.

Procyclic (*proliferative*)

Epimastigote (*proliferative*)

The epimastigote form multiplies in the insect's salivary gland.

Figure 11.17 *Giardia intestinalis*, a Metamonad Intestinal Parasite

LM 2 µm

SEM 5 µm

A. *Giardia intestinalis* is commonly ingested from mountain streams or transmitted in day-care centers.

B. Multiple *Giardia* cells attached to a gerbil intestinal lining.

The salivary gland can transmit the parasite to the next host, where the slender form develops again. Trypanosomiasis is discussed further in Section 21.5.

The molecular basis of conversion among the different developmental forms of a parasite offers targets for new antiparasitic agents. Discovering new drugs against trypanosomes and malarial parasites is urgently needed to decrease the global impact of these devastating diseases.

METAMONADS The metamonads are a group of flagellated parasites distantly related to other protozoa. A common example is *Giardia intestinalis*, the cause of intestinal giardiasis (Figure 11.17). In the United States, *Giardia* is a frequent nemesis of infant day-care centers, where a significant proportion of caregivers are carriers. *Giardia* is also endemic in wildlife and contaminates freshwater streams crossed by bears and other animals. The parasite can contaminate community water supplies and is endemic in some Russian cities. *Giardia* and other metamonads are noted for their anaerobic metabolism and their complete absence of mitochondria, lost through reductive evolution; their metabolism requires the anaerobic intestinal environment of their hosts.

Another common metamonad is *Trichomonas vaginalis*, the cause of trichomoniasis, one of the most common sexually transmitted infections (discussed in Chapter 23). In North America, the asymptomatic infection rate among young adults is estimated at 50%. *T. vaginalis* grows in the urethra and in the vagina. Men rarely report symptoms, but women may experience pain on urination. Untreated infections lead to complications during pregnancy. The infection is treated with the antiparasitic drug metronidazole.

SECTION SUMMARY

- **Amebas move by using pseudopods.** Lobed amebas have large, lobular pseudopods, whereas filamentous amebas have needle-shaped pseudopods.

- **Free-living amebas are predators.** Predatory amebas play an important role in the food web of soil and aquatic communities. They may also harbor pathogenic bacteria.

- ***Entamoeba histolytica* causes intestinal amebiasis.** The amebas develop into cysts that are expelled in feces and remain dormant for long periods in water.

- ***Naegleria fowleri* and *Balamuthia mandrillaris* cause meningoencephalitis.**

- **Ciliates are predatory protozoa with cilia.** Cilia provide motility and generate currents that draw prey into the oral groove.

- **Apicomplexans are obligate parasites lacking cilia or flagella.** Important parasites cause malaria (*Plasmodium falciparum*) and toxoplasmosis (*Toxoplasma gondii*). Waterborne opportunistic pathogens include *Cryptosporidium parvum* and *Cyclospora*.

- **Trypanosomes are obligate parasites that have a single flagellum.** Trypanosomes undergo complex life cycles. They cause sleeping sickness (*Trypanosoma brucei*), Chagas disease (*Trypanosoma cruzi*), and leishmaniasis (*Leishmania donovani*).

- **Metamonads are parasites or endosymbionts with highly degenerate cells.** The intestinal parasite *Giardia intestinalis* causes giardiasis. The urogenital parasite *Trichomonas vaginalis* causes trichomoniasis.

Thought Question 11.2 How do you think amebas evolved to carry symbiotic bacteria? How do these ameba-carried bacteria endanger humans?

Thought Question 11.3 What are the relative advantages and limitations of flagellar motility, ciliate motility, and pseudopodial motility of amebas? Which kind of natural habitat or body environment might favor one kind of motility over another?

Thought Question 11.4 Do you think most eukaryotic parasites have genomes that are larger or smaller than those of free-living microbial eukaryotes? Why might parasite genomes be larger—and why might they be smaller?

11.4 Algae and Dinoflagellates

SECTION OBJECTIVES

- Explain the environmental significance of algae and their relatedness to plants.
- Describe the environmental role and health hazards of dinoflagellates.

How do algae affect our health and environment? Eukaryotic microbes flourish in soil and water habitats, where they enrich the food web as producers and consumers. Major photosynthetic producers are the algae. In freshwater and marine ecology, the algae, together with photosynthetic bacteria (cyanobacteria), are known as phytoplankton (discussed in Chapter 27). Algae and photosynthetic bacteria form the fundamental base of Earth's food chain for all living things. They fix most of Earth's carbon and produce most of the oxygen we breathe.

Algae are protists that conduct photosynthesis by using chloroplasts. Chloroplasts evolved from ingested cyanobacteria through the process of endosymbiosis. An ancient proto-alga engulfed a photosynthetic cyanobacterium, perhaps with the aim of digesting it. But instead of being digested, the cyanobacterium provided photosynthetic carbon and received nutrients and protection in return from its host cell. Eventually, the cyanobacteria evolved into algal chloroplasts, organelles with genomes and structure greatly diminished compared with the original cyanobacteria. Some ancient algae then evolved into multicellular plants. In soil and water ecosystems, algae are consumed by many heterotrophic protists, such as amebas (discussed in the previous section).

Note: Cyanobacteria that cause toxic blooms on lakes are sometimes incorrectly called "algae." The so-called algal blooms on lakes usually consist of cyanobacteria.

Green and Red Algae

Green algae make up the group Chlorophyta. Like plants, they are autotrophs, growing entirely through photosynthesis and absorption of minerals. Many green algae are unicellular, such as *Chlamydomonas reinhardtii*, a unicellular chlorophyte that serves as an important model system for genetics. *C. reinhardtii* has a symmetrical pair of flagella—a common pattern for green and red algae. The alga swims forward by bending its flagella back toward the cell, as if doing the breaststroke. *C. reinhardtii* cells are mostly haploid (n), reproducing by asexual cell division. But when conditions change, the haploid algae mate to undergo meiosis and form gametes. The gametes swim away and fuse with others to form $2n$ zygotes, whose reassorted genes may improve fitness in the new environment. Thus, *C. reinhardtii* undergoes environment-triggered alternation between diploid ($2n$) and haploid (n) forms. This alternation of diploid and haploid forms is similar to that of some fungal yeasts (see Section 11.2).

Other chlorophytes develop more complex multicellular forms. *Volvox* forms hollow spherical colonies, like a geodesic dome (**Figure 11.18A**), composed of cells similar to those of *Chlamydomonas*. *Spirogyra* is a filamentous pond dweller known for its spiral chloroplasts. In marine water, *Cymopolia* forms relatively large, calcified stalks of cells (**Figure 11.18B**). Still other marine algae grow in undulating sheets, such as *Ulva*, the "sea lettuce" familiar to bathers. Sheets of *Ulva* can extend over many square meters, although they are only two cells thick.

The second group of "true algae," also derived from primary endosymbiosis, is the red algae, or Rhodophyta. Rhodophytes are colored red by the pigment phycoerythrin, which absorbs efficiently in the blue and green range that chlorophytes fail to

Figure 11.18 Green (Primary) Algae

A. *Volvox* forms colonies of flagellated cells; the colony may grow to a size of 1–3 mm. Progeny colonies form within the larger sphere until the sphere bursts, releasing them.

B. *Cymopolia* forms calcified stalks of cells.

absorb. Because the red algae absorb wavelengths missed by the green algae, they can colonize deeper marine habitats, below the green algal populations. The most famous red alga is *Porphyra*, which forms large sheets harvested for food. Originally devised in Japan, sheets of *Porphyra* are used as nori to wrap sushi.

Algae Derived from Secondary Endosymbiosis

The original endosymbiosis event that gave rise to green and red algae was the uptake of a cyanobacterium that evolved into the chloroplast (discussed in Section 11.1). This past evolutionary event—in a common ancestor of all plants—is known as **primary endosymbiosis**. Later kinds of algae evolved through **secondary endosymbiosis**, in which one of the "true algae" was engulfed and incorporated again, retaining the chloroplast. Different events of engulfment by different kinds of protists led to the evolution of different types of algae, such as dinoflagellates, kelps, and diatoms. Algae derived from secondary endosymbiosis show "mixotrophic" nutrition, involving both photosynthesis and heterotrophy. For example, dinoflagellates are predators of smaller protists.

DINOFLAGELLATES The **dinoflagellates** are a major group of marine algae, essential to marine food chains (**Figure 11.19A**). Marine water may glow at night from *Noctiluca*, a dinoflagellate that exhibits bioluminescence. Other dinoflagellates make carotenoid pigments that are bright red. Blooms of red dinoflagellates such as *Karenia brevis* cause the famous red tide, which may have inspired the biblical story of the plague in which water turns to "blood" (**Figure 11.19B**). *K. brevis* releases neurotoxins that can be absorbed by shellfish, poisoning human consumers months or years later. In aquatic and coastal regions, such as the Chesapeake Bay, *Pfiesteria* dinoflagellates cause algal blooms and poison fish.

The armor-plated appearance of a dinoflagellate results from its stiff alveolar plates, composed of proteins or calcified polysaccharides. The complex outer cortex includes various toxin-emitting extrusomes and endocytic pores, as well as a species-specific

Figure 11.19 Dinoflagellates

A. *Karenia* dinoflagellates. Each cell has one flagellum wrapped around it, whereas the other flagellum provides whiplike motion (colorized SEM).

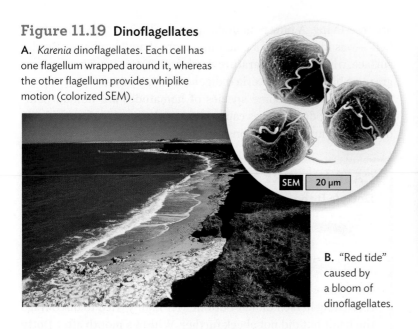

SEM   20 μm

B. "Red tide" caused by a bloom of dinoflagellates.

Figure 11.20 Kelp Forest

Sargassum natans is the basis of the Sargasso Sea. The brown alga forms stalks with leaflike blades and round gas bladders to keep the alga afloat.

pattern of alveolar plates. Like ciliates, dinoflagellates are highly motile, but instead of numerous short cilia, dinoflagellates possess just two long flagella, one of which wraps along a crevice encircling the cell.

Dinoflagellates are called **secondary endosymbiont algae** because their evolution includes two events of endosymbiosis: the original algal endosymbiosis, in which a protist engulfed a cyanobacterium, the precursor of the chloroplast; and a second endosymbiosis, in which another protist engulfed the chloroplast-bearing alga. Dinoflagellates retain the predatory ability of the second engulfing protist, as well as the chloroplast derived from the ancestral alga.

Certain dinoflagellates inhabit other organisms as mutualists, providing sugars from photosynthesis in exchange for a protected habitat. Their hosts include sponges, sea anemones, and, most important, reef-building corals. Coral endosymbionts, known as zooxanthellae, are vital to reef growth. The zooxanthellae are temperature sensitive, and their health is endangered by global warming. Rising temperatures in the ocean lead to coral bleaching (the expulsion of zooxanthellae), after which the coral dies.

DIATOMS AND KELPS Diatoms derived from a different secondary endosymbiotic event. They grow intricate cell walls of silicate. Also known as "brown algae," diatoms can accumulate in aquarium tanks. Other kinds of secondary endosymbiont algae are structured more like plants, including seaweeds such as kelps and *Sargassum*. *Sargassum* (**Figure 11.20**) forms the basis of the Sargasso Sea, a region of the Atlantic Ocean bounded by currents that maintain a relatively intact mass of algae. The *Sargassum* grows stalks with leaflike blades and round gas bladders that keep the alga afloat, supporting a unique community of fish and invertebrates at the surface of the open ocean. While *Sargassum* supports an important ecosystem, climate change has led to its overgrowth and disruption of beaches in the Caribbean.

SECTION SUMMARY

- **Algae are protists with chloroplasts that conduct photosynthesis.** They are important producers of biomass in all ecosystems. Green algae (Chlorophyta) and red algae (Rhodophyta) are autotrophic photosynthesizers, closely related to green plants.

- **Dinoflagellates have a stiff, plated cortex and a pair of flagella.** Dinoflagellates contain chloroplasts derived from secondary endosymbiosis, but they also act as predators.

- **Brown algae include diatoms, as well as seaweeds such as kelps and *Sargassum*.**

11.5 Invertebrate Parasites: Helminths and Arthropods

SECTION OBJECTIVES

- Describe the parasitic cycles of worms, including roundworms, flukes, and tapeworms.

- Describe the parasitic cycles of arthropods, including ticks, mites, and bloodsucking insects.

What kinds of parasites are actually multicellular animals? Infectious parasites include many invertebrate animals, primarily worms and arthropods. Although they are not, strictly speaking, microbes, some require a microscope to visualize. Their infection cycles resemble those of microbial pathogens.

Invertebrate parasites, like the protozoan parasites discussed earlier, commonly cycle through more than one host, such as human and insect or human and domestic animal. Typically there is a definitive host (where sexual reproduction occurs) and an intermediate host (asexual reproduction only). In some cases, the human or animal may be an **incidental host**—that is, a host in which the parasite may grow and cause serious morbidity but cannot be transmitted to other hosts.

Helminths

Parasitic worms are called **helminths**. Helminths are wormlike invertebrates that do not share a unique clade (taxonomic group sharing common ancestor). While most worms are nonparasitic and provide important ecosystem functions (such as that of earthworms), the parasitic worms cause a tremendous global burden of morbidity and mortality.

Worms are multicellular animals that possess fully differentiated organs. Although they are invertebrates, their body plans have surprising complexity, in some cases comparable to that of vertebrates. The free-living soil nematode *Caenorhabditis elegans* (**Figure 11.21**) serves as a major model system for the study of human development. The entire genome of *C. elegans* has been mapped, and many of the worm's genes have homologs in the human genome that are involved in vital processes related to neural development, aging, and cancer. Helminths include three categories:

- **Nematodes** (**roundworms**) are cylindrical (hence the term "roundworm"), with a digestive tube that ends in an anus. Their reproductive forms are mostly hermaphrodites (possessing female and male reproductive organs) although a few are males.

- **Trematodes** (**flukes**) are oval-shaped flatworms, with a digestive tube that ends in a cecum (no outlet). They are hermaphrodites.

- **Cestodes** (**tapeworms**) are parasitic flatworms, absorbing nutrients through their skin. The sucker head structure grows a long "tail" of segments containing male and female reproductive structures.

NEMATODES Nematodes are diverse animals, with species adapted to all kinds of ecosystems; about half the known species are free-living. A new nematode species—dubbed "the worm from hell"—was identified in a gold mine more than 3 km below Earth's surface, where the temperature reaches 48°C (118°F). These worms graze on biofilms of bacteria indigenous to the mine.

Unfortunately, some species of nematodes are human parasites. Nematode parasites cause some of the most widespread and debilitating human infections worldwide.

CASE HISTORY 11.3
Dotty's Itch

Dotty was the 2-year-old daughter of a Texas petroleum executive. The family included four children and three dogs. Dotty often played outdoors with the other children and ranch animals. She developed a habit of scratching herself on her bottom. Her nanny tried to discourage the habit but did not check further. Within a month after Dotty started doing this, Dotty's mother, an attorney for the petroleum firm, noticed that other members of the family had developed a similar habit, which they ascribed to discomfort with the humid summer. One evening, Dotty's mother changed her daughter's diaper and noticed white, threadlike worms a few millimeters long. A trip to the pediatrician revealed pinworms, as well as pinworm eggs (**Figure 11.22**). The eggs were observed on a "tape test," in which a piece of tape was pressed to Dotty's anus and then placed on a slide for examination by light microscopy. The physician prescribed mebendazole (a microtubule inhibitor) for the entire family, including the nanny and the maid who served dinner. Dotty's mother asked about the dogs; she was told that dogs do not carry pinworms, although they do carry other worms transmissible to humans. Because the drug mebendazole affects only mature worms, further treatment 2 weeks later was required to attack the eggs and immature worms that had grown in that time. Ultimately, it took 6 months of treatments to eliminate the infection from the household.

Pinworms are small, white nematodes of the species *Enterobius vermicularis*. They infect only humans and are the most prevalent helminth infection in the United States and Europe, possibly infecting 25%–50% of the population. Children commonly ingest pinworms by sucking dirty fingers. Eggs can also be ingested from airborne dust. In the small intestine, the larvae hatch and grow, absorbing nutrition from the host. Usually, larval growth is unnoticed unless there is an exceptionally large number of worms. The female worms migrate to the anus, where they release their eggs. The perianal (around the anus) movement of worms causes irritation, leading to scratching of the anus; in the process, the host's fingernails pick up eggs for transfer elsewhere.

When left untreated, pinworms can lead to complications, such as malabsorption of nutrients and anorexia related to the host's immune response. In addition, secondary bacterial infection

Figure 11.21 *Caenorhabditis elegans*

LM   50 µm

This free-living nematode is a famous model system for studying human development, cancer, and aging. Nematodes commonly grow in soil.

Figure 11.22 Pinworms

Enterobius vermicularis

A. A pinworm (3–10 mm long).

LM 1 mm

B.
LIFE CYCLE OF
E. vermicularis

1
Human ingests
pinworm eggs.

LM 50 µm

Pinworm eggs,
observed in a tape test.

2
In the small intestine,
the larvae hatch and grow.

3
Female worms migrate to the anus,
where they release their eggs.

4
Scratching of irritated
anus promotes egg
transmission to new host.

may occur at the perianal site of scratching or at the vulva or other places where the worms may migrate by mistake. The eggs persist for several weeks outside and are transmitted easily by contact with hands or with contaminated bedding. They are also transmitted by sexual contact.

Because transmission occurs readily within a household, all family members must be treated. The drug mebendazole, a selective inhibitor of worm microtubule biosynthesis, is effective against a broad spectrum of nematodes, including roundworm (*Ascaris*), whipworm, threadworm, and hookworm.

Pinworms are exclusive to humans, but other nematodes infect domestic animals and can be transmitted to humans. The roundworm *Toxocara* infects 14% of people in the United States and is contracted mainly from dogs. Toxocariasis commonly lacks symptoms, but complications can occur, such as blindness. To avoid parasites, all dogs and cats should receive monthly medication.

Worms called **hookworms** (*Ancylostoma duodenale* and *Necator americanus*) are common in the southern United States and in tropical regions around the world. Their cycle differs from that of pinworms. Unlike pinworms, hookworm larvae grow in the soil, feeding on soil microbes until they penetrate the host skin, often through the soles of bare feet. The hookworms travel through the blood until they penetrate the alveoli in the lungs and then are coughed up to the throat and ingested. Within the small intestine, the hookworms latch onto the intestinal lining and suck blood.

Their offspring then exit in the feces. Large numbers of hookworms can cause anemia and protein insufficiency.

Another parasitic nematode is *Trichinella spiralis*, the cause of trichinosis. *T. spiralis* is named for the spiral-form cysts made by worm larvae embedded in muscle. The parasite may be transmitted to humans who eat raw or undercooked pork; this is why cooking directions traditionally specify that pork should be "well done." Bear meat and horse meat can also carry *T. spiralis*. Once ingested, the larvae mature in the intestinal mucosa, mate, and produce progeny larvae. The new larvae now encyst in the host muscles, where they may cause pain and eventually death. In recent years, trichinosis has become rare in industrial countries, but it remains a problem in developing countries.

In tropical regions, severely debilitating worm infections are caused by various kinds of tiny, **filarial worms** that grow within the blood, lymph, or subcutaneous fatty tissues. Most species are transmitted by biting flies, although some are transmitted by aquatic arthropods. Once introduced by an insect, the filariae (singular, filaria) reproduce and multiply to enormous numbers throughout the host's surface tissues. This large number of worms is necessary to ensure a high population within the tiny amount of blood meal taken by an insect, enabling transmission to the next host. For some species, such as *Brugia malayi*, the large numbers of worms in the lymphatic system may block the lymphatic ducts, leading to swelling of the body parts (a condition called

Figure 11.23
Roundworm (*Ascaris lumbricoides*) Infection of the Intestine

More than 400 roundworms were removed surgically from the intestine of a 5-year-old child in India.

Figure 11.24 Liver Fluke (*Clonorchis sinensis*)

elephantiasis). More will be said about elephantiasis in Chapter 21 (see Figure 21.28). Other filarial diseases include river blindness (*Onchocerca volvulus*), in which worm-induced corneal inflammation leads to blindness; and guinea worm disease (*Dracunculus medinensis*), in which worms form painful blisters on the skin until they are released outside into water.

Roundworms such as *Ascaris lumbricoides* can grow to relatively large sizes within the human digestive tract, and their effects may be lethal (**Figure 11.23**). *Ascaris* worms cause infection by ingestion of eggs from soil. The worm eggs then develop into worms within the digestive tract, where they increase in size indefinitely and can obstruct the intestine. In some cases, worms may invade the lungs or other organs, causing life-threatening damage.

Within the intestine, the worms produce eggs that exit the anus. Defecation outdoors in soil may lead to transmission to other people who unknowingly pick up the eggs on hands or feet. Roundworms infect a billion people worldwide, primarily in tropical areas, as well as the southern United States. Antihelminthic medications treat roundworms successfully when caught early, but the removal of larger worms may require surgical intervention.

TREMATODES Trematodes (flukes) infect many kinds of animals, though most species require a mollusk as their primary host. Trematodes are flatworms (phylum Platyhelminthes). Like roundworms, they have an internalized mouth, "throat" (pharynx), and digestive tube, but the tube ends in one or more pouches called ceca (singular, cecum). Because the worm lacks an anus, it must expel its wastes back out the mouth. A fluke usually has two suckers—one near its mouth and one on the ventral side of the body. The mouth sucker ingests nutrients from the host.

Different fluke species preferentially attach to different internal organs, such as the liver, the lung, or the intestine. The phenomenon of a liver fluke was popularized on TV by an episode of *The X-Files* called "The Host." A real-life example of a liver fluke would be the liver fluke, *Clonorchis sinensis* (**Figure 11.24**). The fluke arises from eggs ingested by a snail, where the larvae develop into a free-swimming form. The swimming larvae then penetrate the skin

and flesh of a fish, where they develop into a secondary form. If the fish is raw or undercooked when ingested by a person, the parasite may attach to the person's bile duct, where it grows and matures, producing eggs that exit through the bowel and anus. This fluke can grow to 2.5 cm; others can grow as long as several meters. Prolonged growth of the fluke is associated with secondary infections and predisposition to cancer.

CESTODES Tapeworms are transmitted through larvae embedded in uncooked meat. Different species are found in pork (*Taenia solium*), beef (*Taenia saginata*), and fish (*Diphyllobothrium*). Infection with fish tapeworms has historically been very common in regions of Asia and South America, where raw fish is routinely consumed. In the United States, the growing popularity of raw-fish dishes such as sushi and ceviche has led to an increase in the incidence of fish tapeworm infections.

Tapeworms are flatworms (Platyhelminthes), and their form is very different from that of both flukes and roundworms. The tapeworm is composed of a head and successive segments, which form out of the head and then are gradually pushed down in a tail that may grow 2–15 meters in length (**Figure 11.25**). The head of the tapeworm, called the "scolex," has a special sucker device. Each segment of the tapeworm contains both male and female reproductive organs, which may fertilize each other or cross-fertilize with another segment. Tapeworm segments packed with offspring pass out through the anus of the host, where the segments appear like moving bits of rice. The segments fall to the ground, where some other host may ingest them. A tapeworm may grow for many years inside the digestive tract of a host without notice, until some of its offspring migrate to the muscle, brain, or liver. In these internal organs, the worms may form cysts, leading to muscle pain and neurological symptoms such as seizures.

From a global standpoint, parasitic worms cause substantial health problems, particularly for children, in whom malnutrition leads to stunted mental and physical development. At the same time, because worms are so prevalent, a hypothesis has been proposed that the human immune system evolved for optimal function in the presence of a modest number of helminths. This hypothesis is supported by the negative correlations observed between the prevalence of helminths and prevalence of autoimmune disorders such as asthma, eczema, and Crohn's disease. Such diseases are common in urban areas with good hygiene but are rare in rural areas with poor hygiene and high worm prevalence. The hypothesis suggests that in

Figure 11.25 Tapeworm

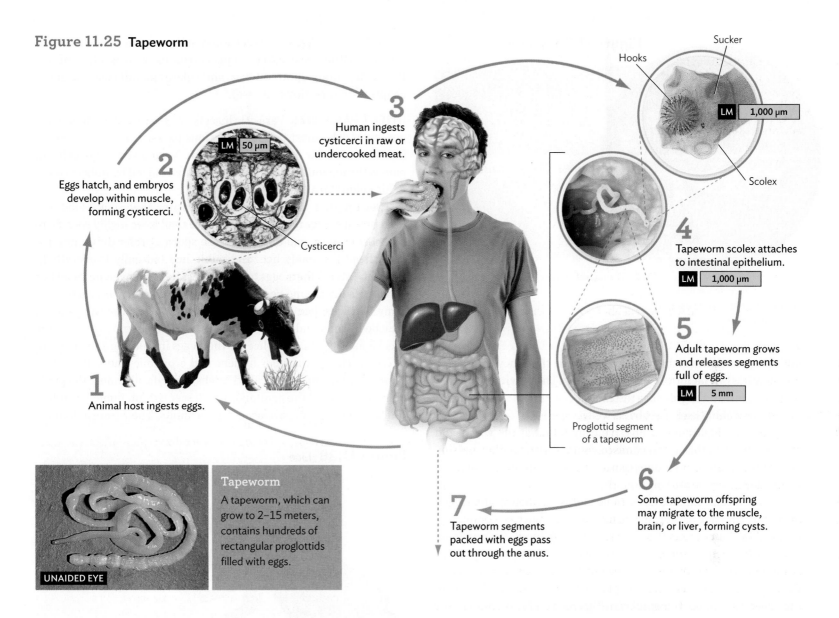

3 Human ingests cysticerci in raw or undercooked meat.

2 Eggs hatch, and embryos develop within muscle, forming cysticerci.

LM 50 µm

Cysticerci

Hooks   Sucker

LM 1,000 µm

Scolex

4 Tapeworm scolex attaches to intestinal epithelium.

LM 1,000 µm

1 Animal host ingests eggs.

5 Adult tapeworm grows and releases segments full of eggs.

LM 5 mm

Proglottid segment of a tapeworm

6 Some tapeworm offspring may migrate to the muscle, brain, or liver, forming cysts.

7 Tapeworm segments packed with eggs pass out through the anus.

Tapeworm
A tapeworm, which can grow to 2–15 meters, contains hundreds of rectangular proglottids filled with eggs.

UNAIDED EYE

---

the absence of helminths, components of the immune system that evolved to fight worms attack the body's own tissues instead. The concept has led researchers to test the possibility that modest exposure to helminths may help prevent autoimmune disorders (also discussed in Chapter 16). Although the hypothesis is unproven, it is being tested in clinical trials in which Crohn's patients are exposed to pig parasites that stimulate the human immune system but cannot complete an infection cycle in humans.

Arthropods

Arthropods are invertebrate animals with an exoskeleton and jointed appendages; the group includes insects and arachnids (spiders and mites). Many arthropods are free-living, but others are parasites with complex life cycles. Some arthropods are **ectoparasites**, attaching to the surface of a human or other vertebrate. Others are parasites that burrow into the flesh. Some arthropod parasites suck blood and then fall off, whereas others

inject their eggs to develop for many weeks or months inside the host. As we saw for worms, arthropod parasites of humans may be acquired from the environment, or they may be highly infectious, acquired from other infected humans.

ARACHNID PARASITES Members of the arachnid group Acari are eight-legged **mites** and **ticks**. They include many species that are parasites of animals and plants throughout terrestrial habitats. Most people have *Dermodex* mites living in their hair follicles and facial pores.

Certain types of mites cause mange (in animals) and scabies (in animals or humans), a condition also known as acariasis (**Figure 11.26**). Scabies in humans is caused most often by the "human itch mite," *Sarcoptes scabiei*. The mites attach to the skin by suckers and then burrow under the skin by using special mouthparts and cutting surfaces on their forelegs. As the mites burrow into the skin, they lay eggs that hatch into larvae. The larvae come out of the skin to attach to a hair follicle, where they feed and molt

Figure 11.26　Scabies

A. *Sarcoptes scabiei* var. *hominis*, the mite that causes scabies in humans.

B. The rash of scabies.

until they reach the eight-legged adult stage; then they drop off to find new hosts. The movement of mites and larvae within the skin causes intense itching, as does the allergic reaction to the eggs. Itching and scratching lead to scabs, as well as hair loss.

Mites are highly infectious, typically found at the wrists and other body parts that often come in contact with another person. Mites spread rapidly in groups of people in close quarters, such as a school dormitory or nursing home.

Ticks (order Ixodida) resemble mites in their eight-legged form, but they are larger and usually do not burrow completely in the skin. Ticks are ectoparasites that suck blood for nutrients to produce eggs and then fall off to disperse their progeny. Ticks often go unnoticed by their hosts, but they are vectors for many serious disease agents, including bacteria, protozoa, and viruses. The deer tick, *Ixodes scapularis* (**Figure 11.27**), is famous for carrying the spirochete of Lyme disease: *Borrelia burgdorferi*. Once the tick attaches to the host, the spirochete slowly migrates

Figure 11.27　Deer Tick, Obtaining Blood Meal

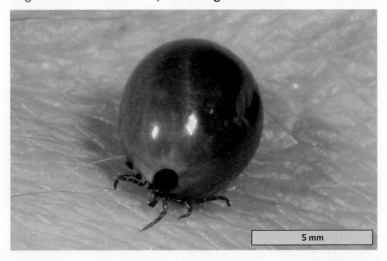

from the tick's digestive tract to its salivary gland and then into the host. Other ticks carry anaplasmosis, babesiosis, ehrlichiosis, Rocky Mountain spotted fever, and tularemia (all covered in the chapters on infectious disease).

INSECT PARASITES Parasitic insects are six-legged, with or without wings. The most famous infectious parasitic insects are sucking lice (suborder Anoplura). Lice have been given many colorful names throughout history and were immortalized by Robert Burns in his poem "To a Louse."

Sucking lice are wingless ectoparasites that suck blood and then produce eggs (**Figure 11.28**). Lice tend to be highly specific to one host species, and different louse species prefer different body sites: head lice, body lice, and pubic lice (sexually transmitted). Head lice attach their eggs to a hair. Despite good hygiene, head lice spread readily among people in close quarters, such as schools and army barracks; lice are difficult to eliminate because the eggs are attached firmly and are not killed by normal shampoos, requiring special medication. Body lice lay their eggs in clothing and are more often associated with poor hygiene and economic status. Pubic lice, *Pthirus pubis*, commonly called "crabs," specifically infect the pubic hair (hair around the genitals). To a lesser extent, pubic lice infect the hair of eyebrows, eyelashes, and armpits. Their transmission is

Figure 11.28　Lice

A. Head louse.

B. Head lice, laying eggs in hair.

usually through sexual contact (although the dismayed sexual partner is sometimes assured that it "could have been the eyebrows").

Another human parasite common throughout history is the flea; as poet Ogden Nash wrote, "Adam had 'em." Fleas (order Siphonaptera) are wingless insect parasites in which the adults jump instead of crawl, so they spend less time than lice directly on their host, although they can jump on and off. After sucking blood, fleas produce larvae that spin a cocoon and can remain dormant many months, until they sense the presence of a host. Fleas often alternate among different animal hosts, such as cats or dogs, although a given flea species usually has preference for a particular animal species. Fleas carry various kinds of pathogens, most famously the plague bacterium *Yersinia pestis* (see Section 10.3).

Other insects, such as mosquitoes, alight briefly on their host to suck blood and then depart to produce eggs. Mosquito eggs require stagnant water to develop into larvae that pupate. Mosquitoes themselves are technically not parasites, but they carry and transmit many kinds of pathogens, such as the malaria parasite and the yellow fever virus.

A different kind of reproductive strategy is that of the botfly (such as *Dermatobia hominis*, which seeks a human host). The adult botfly does not feed at all, but it attaches its eggs to a smaller fly, seemingly innocuous, which then deposits the botfly eggs on a host animal or human. Each egg burrows under the skin, where it develops as a larva. The parasitic larvae do not kill the host, but their presence can be painful. The larva ultimately crawls out from the host and falls off to the ground in order to pupate.

A growing problem worldwide is the bedbug (Figure 11.29). Human bedbugs (*Cimex lectularius*) find their hosts at night while their human targets are asleep; the bugs detect the host mainly from the exhaled carbon dioxide. Unnoticed, the bedbugs suck the host's blood and depart to lay eggs. Their bites lead to allergic reactions and psychological trauma. Bedbugs were nearly eradicated from the United States in the mid-twentieth century by the use of DDT

and other pesticides. Today, bedbugs resist the older pesticides and can be killed only by newer chemicals that are restricted because of their side effects. Elimination is possible but costly, and the persistent bugs spread especially fast through the homes of senior citizens who cannot afford home treatment. Hotels, movie theaters, and homes of all economic levels are now affected. Some bedbugs reportedly carry dangerous pathogens, such as MRSA, although their ability to transmit disease remains unproven.

SECTION SUMMARY

- **Nematodes are roundworms with a complete digestive tube.** Many nematodes are free-living, but others are parasites of vertebrates, such as the roundworm *Ascaris*, pinworms, and hookworms.

- **Trematodes (flukes) are flatworms with suckers and a digestive tube but no anus.** Most flukes are parasites requiring multiple hosts, the primary host being a shellfish.

- **Cestodes (tapeworms) are segmented flatworms.** The scolex attaches within the intestine, growing a tail of segments, each of which contains reproductive organs. Larvae may encyst within muscle or brain tissue.

- **Parasitic worms can be avoided by practicing good hygiene (washing hands), wearing shoes, and thoroughly cooking meat or fish.**

- **Arthropod parasites include arachnids (eight-legged) and insects (six-legged).** Many arthropods are ectoparasites, attaching to the exterior of their host, although others burrow into their host.

- **Ticks suck blood and then fall off to lay eggs.** Ticks transmit numerous kinds of pathogens.

- **Human itch mites cause scabies.** Scabies mites burrow under the skin and lay eggs, which develop into larvae and feed on the skin.

- **Bloodsucking insects include lice, fleas, and bedbugs.**

Thought Question 11.5 The *X-Files* episode "The Host" depicts a human-sized worm as a "liver fluke" that is claimed to have a "scolex" and to regenerate from a cut piece. Which two kinds of worms are confused in this description?

Thought Question 11.6 Overall, why do eukaryotic parasites show such a wide range of size? What selective forces favor large size, and what favors small size?

Figure 11.29 Bedbugs

1 mm

A. Bedbugs were collected from an infested home.

B. Bedbug on skin.

Perspective

Although eukaryotic cells are generally larger and more complex than bacteria, there exists an extraordinary diversity of microbial eukaryotes. These organisms, including fungi, algae, and protists, serve many key roles in our ecosystems. They take part in food webs, especially in the ocean, that provide essential food for animals large and small. We will discuss these ecosystems further in Chapter 27.

At the same time, many fungi and protists parasitize humans, as do worms and arthropods, causing some of the most debilitating conditions worldwide. Diseases such as malaria and trypanosomiasis affect the habitability of major regions of the globe. And everywhere, people from all socioeconomic groups need to be aware of eukaryotic parasites as a medical concern.

Key Terms

alga (pp. 311, 326)
ameba (pp. 311, 318)
antigenic variation (p. 324)
apicomplexan (p. 322)
arthropod (pp. 313, 331)
ascospore (p. 317)
ascus (p. 317)
asexual reproduction (p. 310)
cestode (p. 328)
chitin (p. 315)
ciliate (pp. 311, 321)
cilium (p. 320)
conjugation (p. 322)
cyst (p. 320)

definitive host (p. 323)
dinoflagellate (pp. 311, 326)
diploid (p. 310)
ectoparasite (p. 331)
filarial worm (p. 329)
fluke (p. 328)
foraminiferan (p. 319)
fruiting body (p. 317)
fungus (pp. 311)
haploid (p. 310)
helminth (pp. 313, 328)
hookworm (p. 329)
hypha (p. 315)
incidental host (p. 327)

intermediate host (p. 323)
meiosis (p. 311)
metamonad (pp. 311, 324)
microsporidian (p. 311)
mite (p. 331)
mitosis (p. 309)
mycelium (p. 315)
mycology (p. 311)
mycorrhiza (p. 313)
nematode (p. 328)
pinworm (p. 328)
primary endosymbiosis (p. 326)
protist (p. 311)
protozoan (p. 311)

pseudopod (p. 311)
roundworm (p. 328)
secondary endosymbiont alga
  (p. 327)
secondary endosymbiosis (p. 326)
sexual reproduction (p. 310)
tapeworm (p. 328)
tick (p. 331)
trematode (p. 328)
trophozoite (p. 319)
trypanosome (pp. 311, 324)
vegetative reproduction (p. 310)
yeast (p. 313)
zygote (p. 311)

Objectives Assessment Questions

11.1 Eukaryotes: An Overview

1. Which eukaryotic microbes usually lack motility?
   a. Fungi
   b. Algae
   c. Amebas
   d. Trypanosomes

2. Which infectious parasites are actually invertebrate animals, not true microbes?
   a. Fungi
   b. Trypanosomes
   c. Microsporidians
   d. Helminths

3. Which eukaryotic microbes have long, whiplike organelles?
   a. Dinoflagellates
   b. Ciliated protists such as paramecia
   c. Fungi
   d. Amebas

11.2 Fungi and Microsporidians

4. Most fungi are multicellular, but which of these fungi is single-celled (a yeast)?
   a. *Aspergillus*
   b. *Candida albicans*
   c. *Penicillium roqueforti*
   d. *Morchella hortensis*

5. Which fungal process is inhibited by triazoles?
   a. Ergosterol biosynthesis
   b. Chitin biosynthesis
   c. Fruiting-body formation
   d. Ascus development

6. Which fungus, known as "black mold," commonly endangers health?
   a. Mycorrhiza
   b. *Aspergillus*
   c. *Stachybotrys*
   d. *Candida*

11.3 Protozoa: From Amebas to Trypanosomes

7. Which kind of ameba does/do <u>not</u> commonly cause human disease?
   a. *Entamoeba histolytica*
   b. *Naegleria fowleri*
   c. Foraminiferans
   d. *Acanthamoeba* species

8. Which organism(s) swim by beating hairlike cilia?
   a. *Balantidium coli*
   b. *Giardia intestinalis*
   c. *Pfiesteria* species
   d. *Plasmodium falciparum*

9. Which apicomplexan pathogen is commonly harbored by cats?
   a. *Babesia microti*
   b. *Toxoplasma gondii*
   c. *Plasmodium falciparum*
   d. *Cryptosporidium parvum*

10. Which apicomplexan parasite invades blood cells?
    a. *Plasmodium vivax*
    b. *Trypanosoma cruzi*
    c. *Cryptosporidium*
    d. *Cyclospora*

11. Which is <u>not</u> a trait of the life cycle of *Trypanosoma brucei*?
    a. Motility via a single flagellum powered by the kinetoplast
    b. Antigenic variation, producing 200 different forms of its variant surface glycoprotein
    c. Development within the tsetse fly
    d. Transmission via the fecal-oral route

12. *Giardia intestinalis* is commonly transmitted via _____.
   a. infant day-care centers
   b. coughing and sneezing
   c. mosquitoes
   d. lactation

13. Which parasite is transmitted by sand flies?
   a. *Trypanosoma cruzi*
   b. *Giardia intestinalis*
   c. *Leishmania donovani*
   d. *Toxoplasma*

11.4 Algae and Dinoflagellates

14. Which trait do many algae possess, but not plants?
   a. Chloroplasts
   b. Cell wall of cellulose or other polysaccharide
   c. Nucleus
   d. Flagella for motility

15. Which photosynthetic protist(s) produce toxins and cause red tide?
   a. *Chlamydomonas*
   b. Dinoflagellates
   c. Kelps
   d. *Sargassum*

11.5 Invertebrate Parasites: Helminths and Arthropods

16. The highly prevalent infectious agent *Enterobius vermicularis* is what type of organism?
   a. Fluke
   b. Nematode
   c. Tapeworm
   d. Hookworm

17. Which insect ectoparasites are transmitted primarily through sexual contact?
   a. Ticks
   b. Fleas
   c. Bedbugs
   d. Pubic lice

18. Which parasites preferentially attach to the liver?
   a. Tapeworms
   b. Nematodes
   c. Flukes
   d. Hookworms

Review Questions

1. Summarize the key traits of fungi. What do fungi have in common with protists, and how do they differ?

2. Explain the alternation of diploid and haploid forms in yeasts and in algae. Why do eukaryotic microbes undergo a sexual cycle instead of just vegetative reproduction?

3. Explain the roles of fungi and protists in ecosystems. Why are they important in wastewater treatment? In animal digestion?

4. Suppose a patient has symptoms of dysentery, and a laboratory technician finds amebas in a microscope slide. The patient is treated with anti-amebic drugs, but the symptoms remain. What explanations are possible?

5. A physician diagnoses Lyme disease in a patient and then advises testing as well for babesiosis. Why?

6. Outline the three major kinds of helminths. Explain what these parasites have in common and how they differ in terms of structure and life cycle.

7. Explain why some parasites need more than one host in their life cycle.

8. What is an ectoparasite? Which arthropods are considered ectoparasites, and which are not?

Clinical Correlation Questions

1. Two weeks after a beach vacation, a 25-year-old store clerk develops intestinal symptoms, including nausea and bloody diarrhea. Antibacterial antibiotics fail to resolve the symptoms, and stool studies rule out a bacterial diarrhea. A trichrome stain of the patient's blood reveals a parasite showing ingested RBCs. What parasite is the likely cause of infection? What is its mode of transmission? Which antibiotic will most likely resolve the symptoms?

2. A 65-year-old woman has been hospitalized for 3 weeks for chronic obstructive pulmonary disease (COPD), when she develops lung pain and exacerbation of her shortness of breath. A CT scan shows a clouded region, indicating a mass in the left lung. Broad-spectrum antibacterial agents fail to resolve the infection. Which organism is a probable cause? What is the treatment?

12

Viruses

Virus from a Needlestick

SCENARIO In 1990, Christine, age 25, was a critical care nurse in a hospital in Phoenix, Arizona. She sustained a needlestick injury while attempting to draw a blood sample from a patient chronically infected with hepatitis C virus (HCV).

Hepatitis C
Healthy liver (top) compared to liver with cirrhosis.

RECURRENCE Ten years later, in 2000, Christine experienced an unexplained fever. Given her history, she was tested for liver function. The liver enzymes aspartate aminotransferase and alanine aminotransferase were elevated (indicating liver damage). A PCR test of a blood sample revealed hepatitis C viral RNA. Christine was again treated with interferon and ribavirin, but the side effects were severe and the drugs were less effective than before. Over the next 15 years, her liver function declined, and the organ showed cirrhosis (liver deterioration, with healthy tissue replaced by fibrosis). In 2015, Christine's physician treated her with a newly devised combination of two antiviral drugs: sofosbuvir (inhibits RNA-dependent RNA polymerase) and simeprevir (inhibits protease for maturation of viral proteins). This combined therapy has a cure rate of 90% and can stabilize liver function (but not reverse damage) until a donor becomes available for liver transplant. But the drugs cost $94,500 for a 12-week treatment, and Christine had to seek assistance to pay for it.

SIGNS AND SYMPTOMS Six weeks after the needle injury, Christine experienced fever and other flu-like symptoms. Because of the needlestick, hepatitis C infection of the liver was suspected.

TESTING Christine's liver function test (LFT) levels fluctuated. A hepatitis C rapid antibody test was positive.

DIAGNOSIS AND TREATMENT Christine was put on a course of ribavirin (an RNA synthesis inhibitor) and pegylated interferon-alpha (interferon protein with polyethylene glycol chains that stabilize interferon in the blood). She experienced a number of side effects, including hemolytic anemia (due to ribavirin buildup in the red blood cells) and fatigue, hair thinning, and depression (due to interferon injection). Her virus titer decreased to very low levels. The therapy was discontinued because the side effects were greater than the effects of the virus.

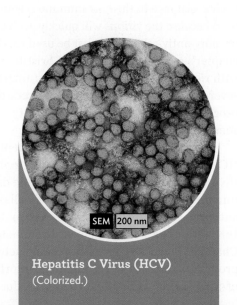

SEM 200 nm

Hepatitis C Virus (HCV)
(Colorized.)

Hepatitis C virus (HCV) causes chronic liver disease in about 1% of the world's population. In the United States, HCV is the most prevalent blood-borne disease. Most people living with HCV have few or no symptoms; but the high prevalence of the virus means that the death rate due to HCV (in those patients with severe disease) will soon overtake the death rate of AIDS.

HCV is a virus. A **virus** is a noncellular particle with a genome contained by a **capsid** (protein coat). The virus takes over a cell to manufacture progeny. Viruses require a host cell to reproduce. The term "virus" refers to the particles and their replication cycle, whereas the term **virion** refers specifically to the structure of the virus particle. Unlike cells, some viruses (such as HCV) have a genome of RNA. Other viruses (such as herpesvirus) have DNA genomes.

Without treatment, the course of disease with HCV varies greatly. About 15% of those infected clear the virus completely, whereas the rest may live for several decades before becoming aware of liver damage. However, 5% eventually experience liver cancer. Liver failure may necessitate a transplant; HCV is now the leading cause of liver transplant in the United States, but the waiting list is long. Unless the virus is first cured, the new liver becomes infected with the virus still present in the blood.

RNA viruses such as HCV have high error rates during replication and therefore high rates of mutation. Within a patient, mutant strains rise and fall, depending on their virulence and on their resistance to the host immune system. No single drug can be used because the virions will quickly evolve resistance; instead, two or more antiviral agents must be used together. That is why, in the chapter's opening case history, Christine received dual treatment with ribavirin and interferon. Unfortunately, antiviral agents have severe side effects because they must retard the body's own cellular processes in order to attack viral activity; so, healthy cells suffer to some extent along with infected ones.

Ribavirin is an effective antiviral agent for many patients with HCV. Ribavirin is an RNA analog that inhibits replication of viral RNA (Section 13.7). Unfortunately, the drug can also inhibit the function of erythrocytes (red blood cells). Interferon is another antiviral agent, often administered in tandem with ribavirin. Interferon is a host protein of the innate immune system (see Chapter 15) whose gene is cloned in bacteria for pharmaceutical production. Cloned or recombinant interferon is a good example of a lifesaving recombinant DNA product of the biotechnology industry (discussed in Chapters 8 and 27). Interferon has many effects on viral proliferation, as well as on normal cell growth. Although ribavirin and interferon did not help Christine, these drugs do control symptoms for many patients.

Greater success rates are achieved with new drugs developed through research, such as the combined treatment with sofosbuvir and simeprevir used by Christine's physician. In some cases triple therapy is prescribed, such as a combination of simeprevir with ribavirin and interferon-alpha.

Hepatitis C is one of many important diseases caused by viruses. Viruses have enormous impact on human lives and populations and on life throughout all ecosystems. For instance, in the open ocean, viruses are the main source of predation that dissipates massive algal blooms. Every kind of cellular organism known, even the smallest of bacteria, is infected by viruses.

Thought Question 12.1   Why are relatively few antiviral agents available, compared with antibiotics? And why are the side effects of antivirals more harmful?

12.1 Virus Infection and Viral Genomes

Viruses are ubiquitous, infecting every taxonomic group of organisms, including bacteria, eukaryotes, and archaea. In marine ecosystems, viruses act as major predators and sequester significant amounts of nutrients in their structure. For humans, viruses cause many forms of illness, whose influence on our history and culture would be hard to overstate. More people died in the influenza pandemic of 1918 than in the battles of World War I. In the past 30 years, the AIDS pandemic caused by HIV has killed 25 million people worldwide. In 2019, a new coronavirus, SARS-CoV-2, caused the pandemic of COVID-19 (coronavirus disease of 2019), a respiratory illness that killed nearly

1,000,000 people by October 2020 and caused the most severe economic disruption since the Great Depression.

Viruses are part of our daily lives. The most frequent infections of college students are due to respiratory pathogens such as rhinovirus (the common cold) and Epstein-Barr virus (infectious mononucleosis), as well as sexually transmitted viruses such as herpes simplex virus (HSV) and papillomavirus (cutaneous warts, genital warts, and genital cancers). Viruses also impact human industry; for example, bacteriophages (literally, "bacteria eaters") infect industrial cultures of *Lactococcus* during the production of yogurt and cheese. Plant pathogens such as cauliflower mosaic virus and rice dwarf virus cause major losses in agriculture.

And yet, viruses have made astonishing contributions to medical research and biotechnology. Bacteriophages are used routinely as **cloning vectors**, small genomes into which foreign genes can be inserted and cloned for gene technology. Some of the most lethal human viruses, such as HIV, have been used to make harmless vectors for human gene therapy.

In natural environments, viruses fill important niches in all ecosystems. Viruses limit the population density of hosts, such as marine algae. Algal blooms grow so large that they show up in satellite photos from space; only viruses can dissipate such blooms. Because viruses dissipate dense populations, they also select for host diversity. Each viral species has a limited host range and requires a critical population density to sustain the chain of infection. This population density is, however, lower than that sustainable by the nutrient resources in the absence of virus. Thus, a given algal species can increase only so much before its bloom dissipates. The resources then support other species resistant to the given virus (but susceptible to other viruses). Thus, overall, viruses prevent the dominance of any one host species. From bacteriophages in our colon to cold-tolerant viruses of Antarctica, viruses promote ecosystem diversity.

History of the Virus Concept

Our concept of a virus has developed over the past century. In the late eighteenth century, microbiologists puzzled over infectious agents that passed through a filter too small to permit bacterial cells. The Dutch microbiologist Martinus Beijerinck (1851–1931) proposed that a "virus" was some kind of infectious "fluid." But the "fluid" turned out to consist of small infectious particles that we now refer to as virions. Some types of virions are geometrically symmetrical particles that can actually be crystallized.

When the first viruses were discovered, they were defined as nonliving. The viruses known at that time lacked metabolism to use energy or conduct biosynthesis, and certain viruses, such as tobacco mosaic virus, could be crystallized like inert chemicals. But some microbiologists have always questioned this position. Bacteria such as *Clostridium* and *Bacillus* species form endospores that can remain viable for thousands of years. The endospores exist as inert particles, like a virus, but they can germinate outside any host cell. Intracellular bacteria such as *Rickettsia* and *Chlamydia*

Figure 12.1 Mimivirus Within an Ameba
About 300 nm across, the mimivirus is larger than some bacteria.

TEM | 100 nm

species metabolize only during growth in a host cell. These obligate intracellular pathogens develop inert spore-like forms that survive outside the host, comparable to virus particles, except that they possess ribosomes. The status of "viral life" remains contested, but possession of ribosomes is generally accepted as a key requirement for existence as a cell.

In 2003, Didier Raoult and colleagues at Aix-Marseille University, France, described a new kind of virus that is larger than some bacteria and whose genome is larger than some bacterial genomes. The virus was named mimivirus because it "mimics" a cell (**Figure 12.1**). Amebas phagocytose the virion as they do with bacteria they eat. But once ingested, the virus takes over the ameba's cytoplasm, forming intracellular "virus factories" that resemble a contained living organism. Mimivirus may be capable of infecting humans, as suggested by the finding of antibodies to this virus in patients with chronic obstructive pulmonary disease (COPD).

How Did Viruses Originate?

The ancient origin of viruses remains a mystery, but the form of modern viruses suggests three intriguing models with implications for human health.

- **Virus-first model.** This model states that ancient viruses were cellular life forms that predated cellular organisms. The argument for this model is that viruses contain many genes found in no living cells. An objection to this model, however, is that no viruses are ever found to replicate outside host cells. Therefore, viral existence always requires the existence of a host cell.

- **Reduction model.** Viruses might have evolved from parasitic cells, by genome reduction. Parasitic cells such as those of chlamydias show loss of functions that host cells provide. Perhaps such parasites could undergo further reductive evolution, ending up as acellular viruses.

- **Escape model.** Some viruses, particularly smaller ones, might have arisen from cell parts that "escaped" and evolved the ability to infect other cells. The reverse transcriptase of retroviruses such as HIV resembles telomerase, the RNA-containing enzyme that regenerates chromosome ends. Such viruses might have evolved from a telomerase that escaped from a cell.

Another startling discovery about viral genomes—both large and small—is that many have been incorporated into the human genome, contributing a substantial part of our own DNA sequence. Viral evolution tells us much about human evolution.

Viruses Replicate in Host Cells

The effects of viral disease arise from the replication of viruses within host cells, which either destroys or debilitates the cells. Viruses also induce host responses that debilitate the host, such as overreaction by the immune system (see Chapters 15 and 16). Some viruses alter host cell genomes to cause cancer, such as leukemia caused by human T-lymphotrophic virus (HTLV) and the head, neck, and genital cancers caused by human papillomavirus (HPV; see Section 12.4).

The human gut teems with **bacteriophages** (phages), viruses that infect bacteria. An example is phage T2, which infects *Escherichia coli* (**Figure 12.2A**). Most bacteriophages insert their genome into the host cell, leaving the empty capsid outside. Within the bacterial cytoplasm, the phage genome directs the replication of progeny **virions** (virus particles). The virions are released when the host cell lyses. Bacteriophages are being studied for their use in "phage therapy," the practice of applying a bacteriophage to infect and kill antibiotic-resistant bacteria. For example, in 2017 a patient suffering from *Mycobacterium abscessus* infection was successfully treated with engineered "mycophages" that infect the pathogen. In the future, phages may be engineered for personalized treatment of various bacterial infections.

In the lab, phages can be observed in a Petri dish as a **plaque**, a clear spot against a lawn of bacterial cells (**Figure 12.2B**). The bacteria grow to completely cover the surface of the agar medium, rendering it opaque—except where bacteria are lysed by phages. Each plaque arises from a single virion or bacteriophage. The bacteriophage lyses a host cell and spreads progeny to infect adjacent cells. Eventually, enough bacterial cells are lysed, making the plaque visible to the naked eye. Plaques can be counted as representing individual infective virions from a phage suspension.

An example of a virus that infects humans is measles virus (**Figure 12.2C**). The measles virus has an envelope that is derived from the host cell plasma membrane when the virus exits the host cell. When the virus enters a new host cell, the envelope fuses with the host cell plasma membrane, releasing the viral contents into the cell's cytoplasm. After replicating within the infected cell, newly formed measles virions become enveloped by host cell membrane as they bud out of the host cell. The spreading virus generates a

Figure 12.2 Virus Infections

A. Bacteriophage T2 particles form a semicrystalline array within an *E. coli* cell.

B. Bacteriophage infection forms plaques of lysed cells on a lawn of bacteria.

C. Measles virions bud out of human cells in tissue culture.

D. A child infected with measles shows a rash of red spots.

E. A tobacco leaf section is packed with tobacco mosaic virus particles.

F. A tomato leaf infected by tobacco mosaic virus shows a mottled appearance.

rash of red spots on the skin of infected patients (**Figure 12.2D**). Untreated cases of measles can be fatal (1 in 500 cases).

Tobacco mosaic virus (TMV) infects a wide range of plant species, its virions accumulating to high numbers within plant cells (**Figure 12.2E**). The progeny virions then travel through interconnections to neighboring cells. Infection by TMV results in mottled leaves and stunted growth (**Figure 12.2F**). Plant viruses cause major economic losses in agriculture worldwide.

As new virions are produced, these virus particles need to find new cells to infect—and ultimately a new host. The process of reaching and infecting a new host is called **transmission**. Different viruses have different mechanisms and efficiencies of transmission. For example, the measles virus is transmitted by droplets of respiratory fluid. Measles transmission is highly efficient. By contrast, HIV is transmitted via blood or sexual contact. Transmission of HIV is relatively inefficient, especially via sexual contact; multiple contacts may occur before HIV infection is established in a new host. But still, most HIV infections occur by sexual contact.

Each species of virus infects a particular group of host species, known as the **host range**. Some viruses can infect only a single species; for example, HIV infects only humans. Close relatives of humans, such as the chimpanzee, are not infected, although they are susceptible to a virus that shares ancestry with HIV, simian immunodeficiency virus (SIV). In contrast, West Nile virus, transmitted by mosquitoes, has a much broader host range, including many species of birds and mammals.

The term **tissue tropism** refers to the range of tissue types a virus can infect. Some viruses have broad tropism; for example, Ebola virus infects many different tissues throughout the body. Other viruses have a narrow tropism; for example, rabies virus specifically infects nervous tissues, whereas influenza virus specifically infects cells of the respiratory epithelium. Both tropism and host range depend on various host factors, most importantly the surface receptor molecules—typically, the specific proteins on a cell surface that the virus particle can bind to. For example, avian strains of influenza virus (also known as the avian flu) bind effectively to the host cell receptors in birds, but those viruses bind less effectively to receptors in the upper respiratory tissues of humans.

In all cases, viruses use a relatively small number of virus-encoded proteins to commandeer the metabolism of their hosts. This situation has profound consequences for medical therapy:

- **Antiviral agents are hard to discover.** Because most viral pathogens have relatively few specific parts (compared with bacterial pathogens), there are relatively few targets for antibiotic design.

- **Antiviral agents have severe side effects.** Because viral replication involves so many host cell processes, in combination with the few viral components, agents that disrupt viral infection usually harm the host as well.

- **Viral genomes mutate fast, even faster than bacteria.** The small genomes of viruses enable them to mutate rapidly within the host. So, no one antiviral agent will work for long. New agents with new targets must be found.

Fortunately, many viruses, such as those that cause colds and influenza, are soon eliminated by a healthy immune system. But others, such as hepatitis C virus and HIV, may lead to a lifetime infection. That is because these viruses are capable of hiding within host cells in a latent state, where they are inaccessible to the host immune system. Virology supports the old adage that prevention is better than cure. But as long as people are getting infected, we need new antiviral agents. And the only way to find them is by studying virus structure and function.

ⓞ **Note:** Virology has two uses of the word "latent." A virus may persist in the cell in a "latent state" ("latency")—that is, present but inactive until a molecular signal reactivates viral replication. But after infection, "latent period" refers to the time during which a virus is multiplying within a host but cannot yet transmit infection to a new host.

Viral Genomes

The key determinant of viral infection is its genome (**Table 12.1**). Viruses of the same genome type (such as double-stranded DNA) are more likely to share ancestry with each other than with viruses of a different type of genome (such as RNA). The classification model based on genome type was devised by David Baltimore, who, with Renato Dulbecco and Howard Temin, was awarded the Nobel Prize in Physiology or Medicine in 1975 for discovering how tumor viruses cause cancer.

Viral genomes are more diverse in composition than the genomes of cells, which are always made up of double-stranded DNA. A viral genome is composed of either DNA or RNA, depending on the viral species; and in different species, the genome may be double-stranded or single-stranded (see Figure 12.3). As mentioned earlier, each viral genome (DNA or RNA) is enclosed within a protein coat, the capsid. The capsid keeps the viral genome intact outside the host. In some species, the capsid is further encased by an **envelope** formed out of host membranes with embedded viral envelope proteins. The capsid (and envelope, if present) must provide a mechanism of infection of the next host cell. The infection mechanism must be one of two types: (1) injection of viral genome into the host cell or (2) uptake into the host cell, followed by disassembly, or **uncoating**. In either case, the original virion loses its own structure and its own identity as such. Loss of individual identity is necessary to generate viral progeny.

The **Baltimore model** classification of viral groups is based on whether the genome is composed of RNA or DNA; whether it is single- or double-stranded; and, if single-stranded, whether the strand directly encodes proteins or requires synthesis of a complement that encodes proteins (**Figure 12.3**). The type of genome determines how the virus will manufacture progeny virions. For example, an RNA virus such as coronavirus or influenza virus lacks DNA and therefore cannot use a cellular DNA polymerase to replicate its genes; the virus must encode an **RNA-dependent RNA polymerase**. Assuming a common host, the different means of mRNA production generate distinct groups of viruses with shared ancestry.

Table 12.1
Baltimore Groups of Viruses

Virus Example	Taxonomic Group with Examples	Virus Example	Taxonomic Group with Examples

Group I.
Double-stranded DNA viruses
± DNA

Bacteriophage lambda infects *Escherichia coli*.
Herpesviruses cause chickenpox, genital infections, and birth defects.
Human papillomavirus strains cause warts and tumors.
Mimivirus, one of the largest known viruses, infects amebas.

TEM | 30 nm
Phage lambda

Group II.
Single-stranded DNA viruses
+ DNA

Bacteriophage M13 infects *E. coli*.
Parvoviruses cause disease in cats, dogs, and other animals.
Geminiviruses infect tomatoes and other plants.

TEM | 100 nm
Geminivirus

Group III.
Double-stranded RNA viruses
± RNA

Birnaviruses infect fish.
Reoviruses such as rotavirus cause severe diarrhea in infants. Other reoviruses are in clinical trials to fight tumors (by oncolysis).
Varicosaviruses infect plants.

TEM | 80 nm
Rotavirus

Group IV.
(+) sense single-stranded RNA viruses
+ RNA

Tobacco mosaic virus infects plants.
Coronaviruses cause colds, as well as severe respiratory disease (COVID-19).
Flaviviruses cause hepatitis C, West Nile disease, yellow fever, and dengue fever.
Poliovirus infects human intestinal epithelium and nerves.

Model | 2 nm
Coronavirus

Group V.
(–) sense single-stranded RNA viruses
– RNA

Orthomyxoviruses cause influenza.
Filoviruses such as Ebola virus cause severe hemorrhagic disease.
Paramyxoviruses cause measles and mumps.
Rhabdovirus causes rabies.

TEM | 30 nm
Rabies virus

Group VI.
Retroviruses (RNA reverse-transcribing viruses)
+ RNA → DNA

Feline leukemia virus (FeLV), **Rous sarcoma virus (RSV)**, and **avian leukosis virus (ALV)** cause cancer.
Lentiviruses include **human immuno-deficiency virus (HIV)**, the cause of AIDS, and **simian immunodeficiency virus (SIV)**, the cause of simian AIDS.

TEM | 30 nm
Human immunodeficiency virus

Group VII.
Pararetroviruses (DNA reverse-transcribing viruses)
DNA ⇄ + RNA

Caulimoviruses (such as cauliflower mosaic virus, or CaMV) infect many kinds of vegetables. CaMV provides the best vector tools for plant biotechnology.
Hepadnaviruses such as hepatitis B virus infect the human liver.

TEM | 200 nm
Caulimovirus

Figure 12.3 Baltimore Classification of Viral Genomes

There are seven categories of viral genomes, which consist of RNA or DNA and are either single-stranded (ss) or double-stranded (ds), and either (+) or (−) sense. Each kind of genome requires a different route to mRNA and a different replication mechanism.

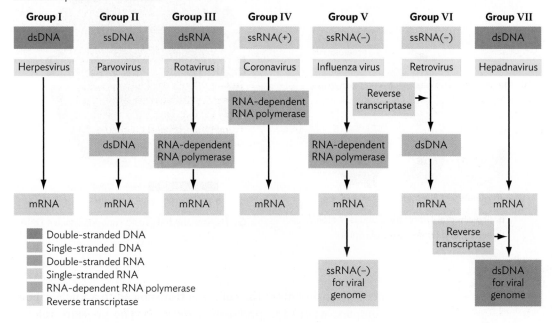

The seven Baltimore groups of viruses, based on genome type and mRNA generation, are outlined in Figure 12.3. Examples of each group are presented in Table 12.1. Note, however, that within each genome type, very different viruses have evolved in hosts as different as humans and bacteria. It is likely that a given form of genome (such as single-stranded DNA) evolved independently in different viral lineages.

The type of genome (DNA versus RNA, single- versus double-stranded) determines how the virus infects the cell and can influence the course of disease in the patient. Double-stranded DNA viruses (group I), such as herpesviruses and poxviruses (poxviruses cause smallpox and cowpox), make their own DNA polymerase or use that of the host for genome replication. Their genes can be transcribed directly by a host RNA polymerase. An exception is the poxviruses, which encode their own RNA polymerase to make mRNA.

By contrast, single-stranded DNA viruses (group II), such as canine parvovirus, require the host DNA polymerase to generate the complementary DNA strand. The double-stranded DNA is then transcribed by host RNA polymerase to make mRNA.

Double-stranded RNA viruses (group III) require a viral RNA-dependent RNA polymerase to generate mRNA by transcribing directly from the RNA genome. An example is rotavirus, which causes diarrhea in children. Such viruses must package a viral RNA polymerase with their genome before exiting the host cell.

Single-stranded RNA viruses, such as hepatitis C virus and SARS coronavirus (group IV), have a positive-sense (+) strand (the coding strand) that can serve directly as mRNA to be translated to

viral proteins. They do need to make an RNA-dependent RNA polymerase to synthesize a template (−) strand complementary to the (+) strand. The (+) strand for progeny virions is then replicated from the (−) template.

Other RNA viruses, such as influenza virus, actually package the complementary sequence of RNA, the (−) strand, instead of the coding strand. With their (−) RNA, these viruses (group V) must also package a viral RNA-dependent RNA polymerase for transcribing (−) RNA to (+) mRNA, which is translated by ribosomes. Later, the RNA-dependent RNA polymerase uses the (+) RNA to synthesize genomic (−) RNA. The (−) RNA genomes are then packaged into progeny virions.

A special case is the **retroviruses**, or **RNA reverse-transcribing viruses** (group VI). The retroviruses, such as HIV and feline leukemia virus, have genomes that consist of (+) strand RNA. They package a **reverse transcriptase**, which transcribes the RNA into a double-stranded DNA (see Section 12.6).

⊙ Note: In nomenclature, families of viruses are designated by Latin names with the suffix "-viridae"—for example, "Papillomaviridae." Common forms of such family names are also used—for example, "papillomaviruses." Within a family, a virus species is simply capitalized, as in "Papillomavirus."

SECTION SUMMARY

- **Viruses are noncellular particles (virions)** containing a genome that replicates within a host cell.
- **Viruses may have evolved from ancestral cells** or from parts of cells.
- **A given virus has a host range** that may be narrow or broad.
- **A given type of virus may have a genome that consists of DNA or RNA,** single-stranded or double-stranded.

12.2 Virus Structure and Diversity

SECTION OBJECTIVES

- Describe the different types of viral structure, and give an example of each.
- Explain the significance of viral genome size.
- Describe the nature of viroids and prions.
- Explain why viruses evolve so fast.

Figure 12.4 Herpesvirus: Icosahedral Capsid Symmetry

Depth into virion 0 10 30 50 65 nm

A. Icosahedral capsid of herpes simplex 1 (HSV-1), envelope removed. Imaging of the capsid structure is based on computational analysis of cryo-electron microscopy (cryo-EM). Images of 146 virus particles were combined digitally to obtain this model of the capsid at 2-nm resolution.

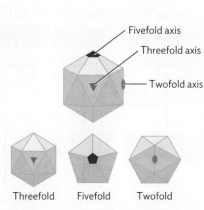

Fivefold axis
Threefold axis
Twofold axis

Threefold   Fivefold   Twofold

B. Icosahedral symmetry includes fivefold, threefold, and twofold axes of rotation.

Capsid proteins
DNA

Depth into virion 0 10 30 50 65 nm

C. The icosahedral capsid contains spooled DNA.

What different kinds of structure can a virus have? An important basis for classification of viruses is their **capsid**, the protein package that encloses their genome. Viral species typically show one of four structural types of capsids: icosahedral, filamentous, complex tailed, or amorphous (lacking symmetry).

Virus Structure: Capsid and Envelope

The International Committee on Taxonomy of Viruses (ICTV) devised a classification scheme that includes factors such as capsid form (icosahedral or filamentous), envelope (present or absent), and host range.

ICOSAHEDRAL CAPSIDS Many viral capsids show radial symmetry, like a crystalline object. The advantage of symmetry is that it forms a package out of repeating protein units generated by a small number of genes. The smaller the viral genome, the more genome copies can be synthesized from the host cell's limited supply of nucleotides.

The form of a radially symmetrical virion is based on an icosahedron, a polyhedron with 20 identical triangular faces. An example of an icosahedral capsid is that of herpes simplex virus (**Figure 12.4A**). Each triangular face of the capsid is determined by the same genes encoding the same protein subunits. No matter what the pattern of subunits in the triangle is, the structure overall exhibits rotational symmetry characteristic of an icosahedron (**Figure 12.4B**). Within the icosahedral capsid, the herpesvirus genome (composed of double-stranded DNA) is spooled tightly (**Figure 12.4C**), the DNA packed under high pressure. When the

herpes capsid enters the cell, it is transported to a nuclear pore complex, where the pressure is released. The pressure release drives viral DNA into the host nucleus.

ENVELOPE Some icosahedral viruses, such as poliovirus and papillomavirus, have only a protein capsid. Other viruses, such as herpesviruses, also have an **envelope** derived from the host nuclear or endoplasmic reticulum membrane (**Figure 12.5**). The envelope bristles with virus-encoded **spike proteins** that plug the membrane onto the capsid. The spike proteins enable the virus to attach and infect the next host cell.

Between the envelope and the capsid may be additional proteins called **tegument** (see Figure 12.5). Tegument proteins are

Figure 12.5 Envelope and Tegument Surround the Herpes Capsid

Envelope
Tegument
Capsid

A. Section showing envelope and tegument proteins surrounding the capsid (cryo-TEM).

B. Cutaway reconstruction: spike envelope glycoproteins (yellow), envelope membrane (dark blue), tegument proteins (orange), and capsid (light blue; cryo-EM tomography).

Figure 12.6 Filamentous Viruses

A. The helical filament of tobacco mosaic virus (TMV) contains a single-stranded RNA genome coiled inside. Image reconstruction is based on X-ray crystallography.

B. Ebola virions.

expressed during infection of a host cell and then packaged in the virion during envelope formation. Both viral and host proteins may be packaged. As the virus enters a new host cell, the viral envelope fuses with the host cell membrane in a way that opens the viral contents to the cytoplasm. The tegument proteins are released in the cytoplasm, where they help viral replication.

FILAMENTOUS VIRUSES Filamentous viruses include tobacco mosaic virus (Figure 12.6A) as well as animal viruses such as Ebola virus, which causes a swiftly fatal disease of humans and related primates (Figure 12.6B). Many bacteriophages are filamentous; for example, the filamentous phage CTXφ integrates its sequence into the genome of *Vibrio cholerae*, where it carries the genes for the deadly cholera toxin (discussed in Chapter 4).

Filamentous viruses show helical symmetry. The pattern of capsid monomers forms a helical tube around the genome, which is wound helically within the tube (see Figure 12.6A). The length of the helical capsid may extend up to 50 times its width, generating a flexible filament. Unlike the icosahedral capsid, which has a fixed size, the helical capsid can vary in length to accommodate different lengths of nucleic acid. This variable length is convenient for genetic engineering vectors (viruses with genomes artificially recombined to carry genes into cells).

TAILED BACTERIOPHAGES Some bacteriophages have complex multipart structures. In a tailed phage, the icosahedral protein package, called the "head," is attached to an elaborate delivery device. For example, bacteriophage T4 (Figure 12.7) has an icosahedral head containing the pressure-packed DNA, attached to a helical "neck" that channels the nucleic acid into the host cell. The neck is surrounded by the tail sheath, with six jointed tail fibers that attach the bacteriophage to the host cell surface. Within the tail, an "injector" penetrates the host cell envelope, and the release of pressure within the head propels the DNA into the host cytoplasm.

Figure 12.7 Bacteriophage T4 Capsid

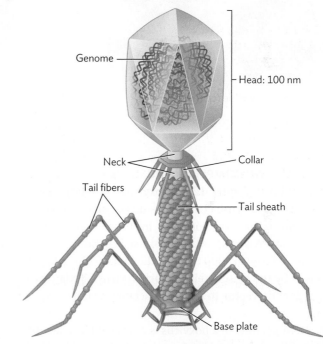

A. Phage T4 particle with protein capsid containing a packaged double-stranded DNA genome. The head is attached to a neck surrounded by a tail sheath, with tail fibers that attach to the host cell's surface. After attachment, the sheath contracts and the core penetrates the cell surface, injecting the phage's genome.

B. *E. coli* infected by phage T4 (colorized blue).

Figure 12.8 Vaccinia Virus

A. The vaccinia virus particle (colorized TEM).

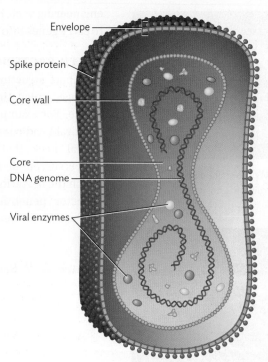

B. The pox virion consists of a core containing DNA and enzymes, surrounded by an envelope with embedded spike proteins.

AMORPHOUS OR COMPLEX VIRUSES Some viruses have no symmetrical form (**Figure 12.8**). Poxviruses are large asymmetrical viruses with no symmetrical capsid. The size of vaccinia virus (360 nm) approaches that of some bacteria. Vaccinia virus, also known as "cowpox virus," is the virus used by Edward Jenner and others around the year 1800 to make the first vaccine against smallpox (caused by the closely related variola virus). Today, vaccinia is used as a genetic vector to develop vaccines against many animal diseases.

The DNA genome of a poxvirus is contained by a flexible "core wall" that also encloses a large number of enzymes and accessory proteins, similar to a cell's cytoplasm. The double-stranded DNA genome resembles that of a cell, except that it is stabilized by covalent connection of its two strands at each end. The core is enclosed loosely by a viral envelope studded with spike proteins.

Thought Question 12.2   Which viruses have a narrow host range, and which have a broad host range?

Viral Genome Size

How do viruses arrange their genetic information? The genomes of viruses, especially RNA viruses, can be surprisingly small; and the ability of such small entities to take over cells is remarkable. Viruses evolve rapidly, enabling them to evade host defenses and even to "jump" into new host species.

The genome of an RNA virus can be as small as three genes. **Figure 12.9A** depicts the genome of avian leukosis virus (ALV), a well-studied **retrovirus** (reverse-transcribing virus) that causes lymphoma in chickens. The ALV genome has three protein-encoding genes: *gag*, *pol*, and *env*. Retroviruses have small genomes whose synthesis consumes minimal resources, thus maximizing the number of virions that can be made from an infected cell.

Other kinds of viral genomes are large, approaching the size of cellular genomes. Double-stranded DNA viruses have especially large genomes, encoding numerous enzymes and regulatory proteins similar to those of cells. For example, the genome of herpes simplex virus 1, a cause of cold sores and genital herpes, spans 152 kilobases and encodes more than 70 gene products (**Figure 12.9B**). These herpesvirus products include capsid and envelope proteins, DNA replication proteins, and accessory proteins that manage viral replication or interact with host immune cells. Other herpes genes encode latency proteins that can maintain the virus in a latent state within the host cell. Overall, herpes genomes have evolved to be highly regulated, with numerous extra products that have small but important effects on viral replication and host interaction.

Viroids and Prions

Some infectious agents are so limited in content that they cannot even be considered viruses. **Viroids** are virus-like infectious agents in which an RNA genome is itself the entire infectious particle; there is no protective capsid. Most viroids infect plants, including many kinds of fruits and vegetables. For example, citrus viroids cause economic losses in the citrus industry. The viroid typically consists of a circular, single-stranded molecule of RNA that doubles back on itself to form base pairs interrupted by short, unpaired loops. This unusual circularized form avoids breakdown by host RNase enzymes. The RNA folds up into a globular structure that interacts with host cell proteins. Some viroids have catalytic ability, comparable to enzymes made of protein. An RNA molecule capable of catalyzing a reaction is called a ribozyme. The small hepatitis delta (hepatitis D) viral RNA acts

Figure 12.9 Simple Viral Genomes

Genome of ALV (RNA)   7.2 kb (RNA bases)

gag   *pol*   *env*

LTR   LTR

Core   Enzymes   Envelope

In this book, drawings of single-stranded sequences have flat edges, whereas drawings of double-stranded, helical DNA are cylindrical.

Genome of HSV-1 (DNA)   152 kb

IR$_L$ IR$_S$

Regulation   Capsid assembly   Envelope proteins   DNA replication

A. The single-stranded RNA genome of avian leukosis virus, a retrovirus. Three genes (*gag*, *pol*, and *env*) encode polypeptides that are eventually cleaved to form a total of nine functional products. LTR = long terminal repeat; blue sections indicate noncoding RNA.

B. The double-stranded DNA genome of herpes simplex virus (HSV) spans 152,000 base pairs encoding more than 70 gene products. The types of products encoded are shown by color: regulation (green), capsid assembly (yellow), envelope proteins (orange), and DNA replication (brown). IR$_L$ and IR$_S$ are, respectively, long and short inverted-repeat sequences.

like a viroid ribozyme (**Figure 12.10**). Hepatitis delta virus coinfects human liver cells along with hepatitis B virus, increasing the mortality rate of disease to 20%.

Yet another class of infectious agents has no nucleic acid at all. These agents, known as **prions**, are aberrant (wrongly folded) proteins arising out of a preexisting cell. Prions gained notoriety when they were implicated in "mad cow disease," a brain condition in cattle that, after being transmitted to humans who consumed beef from diseased cows, caused human brain disease. The bovine prions resembled a human protein closely enough to cause misfolding of native human proteins and thus damage the brain cells. The pathology resembled that of the inherited brain condition

Creutzfeldt-Jakob disease. Other diseases caused by prion transmission include scrapie, a brain disease of sheep; and kuru, a degenerative brain disease formerly endemic among the Fore people of New Guinea. In the early twentieth century, the Fore people customarily consumed the brains of deceased relatives. When this practice stopped, the disease kuru disappeared.

A prion is an abnormal form of a normally occurring brain cell protein (**Figure 12.11**). The prion form of the protein (called PrP^C) acts by binding to normally folded proteins of the same class and altering their conformation to that of the prion, PrP^{Sc}. The misfolded protein then alters the conformation of other normal subunits, forming harmful aggregates and ultimately killing the cell. In

Figure 12.10 Hepatitis Delta Ribozyme

The portion of the RNA of hepatitis delta virus that catalyzes cleavage of itself. The X indicates the remainder of the RNA molecule that gets cleaved off.

Figure 12.11 Prions Cause Disease

A. Normal conformation   **B.** Abnormal conformation

The conformation (folding pattern) of a normal, noninfectious, brain PrP^C protein (**A**) compared with that of an abnormal, infectious prion (**B**). The abnormal form "recruits" normally folded PrP^C proteins and changes their conformation into the abnormal (PrP^{Sc}) form.

the brain, prion-induced cell death leads to tissue deterioration and dementia (discussed in Chapter 24).

Prion diseases can be initiated by infection with an aberrant protein. More rarely, the cascade of protein misfolding can start with the spontaneous misfolding of an endogenous host protein. The chance of spontaneous unfolding is greatly increased in individuals who inherit certain gene variants that encode the protein; thus, spontaneous prion diseases can be inherited genetically. Overall, prion diseases are unique in that they can be transmitted by an infective protein instead of by DNA or RNA, and they propagate conformational change of existing molecules without actually synthesizing new infective molecules.

Viral Diversity and Evolution

Like cellular organisms, viruses evolve through genome change and natural selection (see Section 9.3). Their small genome size and small number of parts enable viruses to mutate tenfold or a hundredfold faster than their host cells. Rapid mutation and evolution of a virus leads to **antigenic drift**, the accumulation of random mutations that lead to viruses whose mutant proteins are no longer recognized by host antibodies. Antigenic drift generates new strains of viruses that cause serious disease; for example, antigenic drift continually generates new influenza strains requiring repeated immunization.

Over extended time, viral evolution generates new kinds of viruses that cause different diseases in different hosts. **Figure 12.12** shows a phylogeny for herpesviruses, an ancient group of viruses infecting many animals. We see that over time, different diseases evolve, infecting the same or different species. Members of the herpesvirus family cause several important human diseases, including chickenpox (varicella-zoster virus), cold sores and genital lesions (herpes simplex 1 and 2), birth defects (cytomegalovirus), infectious mononucleosis (Epstein-Barr virus), and Kaposi's sarcoma (human herpesvirus 8, not shown).

In general, viruses evolve at different levels, all of which are significant for medicine and epidemiology:

- **Within a host community.** Viruses evolve to preferentially infect different species—for example, equine herpesvirus (in horses) versus murine herpesvirus (in mice).

- **Within a viral species population.** Different viral strains arise that vary in infectivity and virulence. Closely related viruses may cause diseases that are similar (herpes simplex 1 and 2) or different (varicella-zoster versus cytomegalovirus).

- **Within an individual organism.** Viruses evolve variants that resist therapeutic agents. Extremely mutable viruses, such as hepatitis C virus and HIV, evolve a "quasispecies" of diverse strains that infect different tissues within a host.

Within natural environments, viruses evolve in response to host diversity. The most diverse communities tend to generate diverse viruses. A virus emerging from the environment into a

Figure 12.12 Phylogeny of Herpesvirus Genomes

Phylogeny of human and animal herpesviruses, based on whole-genome sequence analysis comparing groups of genes with similar function. Numbers measure genetic divergence.

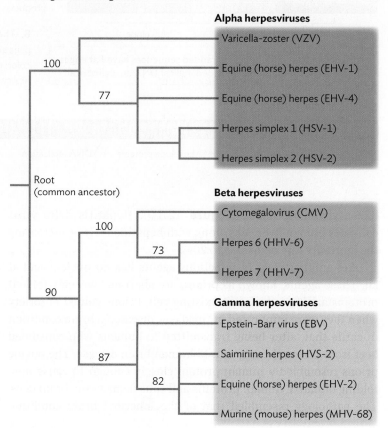

human population, such as the new coronavirus in 2019, may cause a serious epidemic among hosts whose immune systems have no prior exposure to the new pathogen. See **IMPACT** for a particularly compelling case of human viral pathogens emerging among African primate hosts.

Now that we have introduced the major principles of virology, we are ready to examine the culture and replication cycles of a bacteriophage and of three different human-infecting viruses: papillomavirus (dsDNA), influenza virus [(−) sense RNA], and HIV (retrovirus).

SECTION SUMMARY

- **The virus capsid consists of repeated protein subunits**—a structure that maximizes the structural capacity while minimizing the number of genes needed for construction. Capsids may be icosahedral, filamentous, tailed, or amorphous.

- **Some types of virus have an envelope** derived from the host membrane. The envelope contains embedded spike proteins.

- **Viral genomes can be large or small.** RNA genomes usually have few genes, whereas double-stranded DNA viral genomes may approach the size of cellular genomes.

- **Viroids and prions are infectious particles that are not viruses.** Viroids consist of an RNA circle that self-hybridizes. Prions consist of aberrantly folded cell proteins.

- **Viruses evolve into strains that cause diverse diseases in diverse hosts and resist therapeutic drugs.**

Thought Question 12.3   What are the advantages to the virus of having a small genome? What are the advantages of a large genome?

12.3 Virus Replication and Culture

SECTION OBJECTIVES
- Explain how viral genomes enter cells.
- Describe the lytic and lysogenic cycles of bacteriophage infection.
- Explain how viruses are cultured using host cells.

How do viruses find their host cell, subvert its defenses, and take over its metabolism? Viruses show surprisingly diverse mechanisms for infection and replication. Each molecular step of infection offers a potential target for antiviral agents. In this section we examine bacteriophages as a model; subsequent sections present examples of human viruses with DNA and RNA genomes.

One way or another, all viral replication cycles must achieve the following tasks:

- **Host recognition and attachment.** Viruses must contact and adhere to a host cell that can support their particular reproductive mechanism.

- **Genome entry.** The viral genome (either isolated or within an endocytosed virion) must enter the host cell and gain access to the cell's machinery for gene expression.

- **Assembly of progeny virions.** Viral components must be expressed and assembled. Components usually "self-assemble"; that is, the joining of their parts is favored by the energy release of spontaneous assembly.

- **Exit and transmission.** Progeny virions must exit the host cell and then reach new host cells to infect. In the case of multicellular organisms, the virus must eventually reach other multicellular hosts (epidemiology is discussed in Chapter 26).

A virus needs to contact and attach to the surface of an appropriate host cell, one that its own genetics has evolved to take over. A virus binds specifically to a **surface receptor**, a protein on the host cell surface that is specific to the host species and that binds to a specific viral part. An example is the tail fibers of bacteriophage T4, which bind to the bacterial cell-surface lipopolysaccharide (LPS) (**Figure 12.13**).

Why would a host cell evolve to make a surface protein that attaches to a virus? Actually, the receptor for a virus is a protein with an important function for the host cell, but the virus has evolved to

Figure 12.13 **Bacteriophage Genome Insertion**

Phage T4 attaches to bacterium.

Sheath contracts, and viral DNA enters bacterium.

A bacteriophage attaches to the cell surface by its tail fibers and then contracts to inject its DNA. The empty capsid remains outside as a "ghost."

take advantage of the protein. For example, the CD4 surface protein of T cells (see Chapter 16) is often called an "HIV receptor," although it evolved as a signaling protein of the immune system.

Bacteriophage Replication

Bacteriophages interact with bacteria in all ecosystems, from soil and water to the human gut. Phages decrease population density, increase host diversity, and transfer genes among diverse bacteria. Phages are relatively easy to study in the laboratory, where their host bacteria can be grown in tube and plate cultures (methods are presented in Chapter 6). Thus, phage-host systems provide important models for human virus infection and for virus culture.

We consider here the extensively studied bacteriophage lambda, which infects the common enteric bacterium *Escherichia coli*. Enteric bacteriophages are an important part of the ecology of the human intestine, where they transfer genes between bacteria by transduction (Section 9.1).

How does bacteriophage lambda know which bacteria to infect within the mixed population of our colon? The phage virion binds specifically to the maltose porin in the outer membrane of *E. coli*. Although the maltose porin protein is often called "lambda receptor protein," it actually evolved in *E. coli* as a way to obtain the sugar maltose for catabolism. Thus, natural selection maintains the maltose porin in *E. coli*, despite the danger of phage infection.

So, how does the inert phage particle penetrate the castle-like fortress of the bacterial cell? The phage injects its genome through the cell envelope into the cytoplasm, thus avoiding the need for the capsid to penetrate the tough cell wall. The sheath of the phage neck tube contracts, bringing the head near the cell surface to insert its DNA (see Figure 12.13). The pressure of the spooled DNA—as high as 50 atmospheres—is released, expelling the DNA into the cell. After inserting its genome, the phage capsid remains outside, attached to the cell surface. The empty capsid is termed a "ghost" because of its pale appearance in an electron micrograph. (In contrast, animal viruses enter their hosts by binding to receptors that

impact

Monkey Business and Deadly Viruses

In 2014, an outbreak of Ebola virus devastated the countries of Liberia, Sierra Leone, and Guinea, and some cases reached countries outside of Africa. Ebola virus, HIV, and monkeypox virus—where did they come from? These very different viruses all came from apes and monkeys, our nearest relatives in Africa, the birthplace of humans and other primate species. As primates (including humans) evolved, our viruses evolved with us—and continue to do so. A virus needs to find new hosts. It may jump from one species to a closely related one. Thus, a region such as Africa, with rich diversity of our closest animal relatives, is also a rich source of emerging viral pathogens. For example, Ebola virus causes outbreaks in gorillas and chimpanzees, killing large numbers of animals. The dead animals are then found by hunters, who bring them home to feed their families. This consumption of "bushmeat" can then lead to a human outbreak of Ebola.

Nathan Wolfe is a professor at Stanford University who founded the Global Viral Forecasting Institute (now the Laboratory for Research in Complex Systems), an organization that studies emerging viruses (Figure 1). Wolfe's aim is to prevent future pandemics by actually predicting viral emergence before a pandemic occurs. For example, Wolfe's team reported an outbreak of monkeypox among humans in the Democratic Republic of the Congo. He showed that the outbreak correlated with the decline in smallpox vaccination following the extinction of smallpox; the vaccination had previously cross-protected humans against monkeypox transmitted from monkeys. A slight mutation of monkeypox could lead to a global outbreak of smallpox-like disease. In fact, Wolfe argues, humanity has been extremely "lucky" with recent pandemics, such as AIDS: if HIV could actually spread as easily as small-pox, think how many people would have died. In the COVID-19 pandemic, a mutant coronavirus caused widespread mortality.

In his work, Wolfe travels among African villages, testing monkeys for new viruses. In one study, he found monkeys harboring new kinds of human T-lymphotrophic viruses (HTLVs), retroviruses related to HIV. He reported two strains, HTLV-3 and HTLV-4, never before seen, their effects on humans unknown. Many more such strains must exist in the wild. To avoid introducing such strains into humans, Wolfe

Figure 1 Nathan Wolfe, Director of the Laboratory for Research in Complex Systems

Wolfe hunts for emerging viruses in African monkeys, with the aim of preventing future pandemics.

trigger endocytosis or fusion with the host cell membrane, to be discussed shortly.)

When a virulent phage injects its genome into a cell, it immediately begins a replication cycle to form as many progeny phage particles as possible (Figure 12.14A). The process involves replicating the phage genome and then expressing phage mRNA to make enzymes and capsid proteins. Ultimately, the phage particles assemble, and **lysis** (rupture of the host cell) follows, releasing progeny phages (**lytic infection**). Some phages, such as phage T4, digest the host DNA to increase the efficiency of urges the local hunters to avoid eating monkeys, especially dead monkeys they find in the forest—monkeys that may have died of viral disease. But the local people need these dead monkeys to feed their families; for them, the alternative may be starvation.

Although the circumstantial evidence is compelling, can we actually prove the host source of a given virus? At the University of Alabama at Birmingham, professor Beatrice Hahn used laboratory methods to do just that, tracing the precise origin of HIV. The cause of AIDS, HIV has one of the highest mutation rates of any virus, and it evolves quickly. Hahn performed statistical comparisons of various HIV strains with strains of simian immunodeficiency virus (SIV), viruses closely related to HIV that cause immunodeficiency in monkeys. She sequenced the genomes of 50 new SIV strains from chimpanzees in southern Cameroon. Some of the SIV strains she found were even more closely related to HIV than any SIV strain found before. Her work established the statistical likelihood that the major HIV strain affecting North America and Europe (group M) diverged specifically from SIV strains in chimpanzees of Cameroon (Figure 2).

The emergence of HIV group M in humans occurred well before 1960. But in 2009, researchers reported a much more recent emergence of a novel HIV strain in a Cameroonian woman—a strain never before seen in humans. This strain showed closest relatedness to an SIV strain in gorillas. Thus, HIV was "caught in the act" of emerging from our primate relatives.

Although primates are a source of deadly viruses, they also provide our key source of information on how to combat these pathogens. For instance, SIV infects certain primate species without causing any symptoms. A monkey called the sooty mangabey shows no symptoms, despite high SIV titers, whereas the rhesus macaque develops simian AIDS. By studying the host response of monkeys that resist the virus, we may derive clues suggesting therapies for human AIDS.

DISCUSSION QUESTION

In regions where monkeypox is endemic, should smallpox vaccination be resumed?

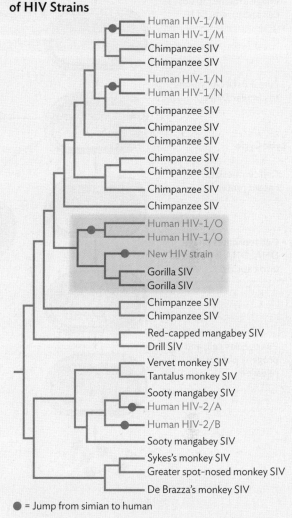

Figure 2 Beatrice Hahn's Phylogeny of HIV Strains

● = Jump from simian to human

The phylogeny of HIV strains shows that groups M and N diverged from SIV in chimpanzees, whereas a novel HIV strain arose from gorilla SIV (shaded orange). Strains of HIV-2 arose from SIV in sooty mangabeys.

phage production; others, such as phage lambda, retain the host DNA for potential **lysogeny**. Lysogeny is a process in which the phage genome becomes integrated in the genome of the host cell (**Figure 12.14B**). Lysogenic phages may carry toxin genes, such as the gene that expresses Shiga toxin in *E. coli* O157:H7.

We now examine in more detail the alternative "fates" of lytic infection (lysis) and lysogeny. Although we focus on bacteriophages, similar alternative strategies for replication—for example, rapid virus replication versus hiding in a host cell—are exploited by human viruses such as HIV and herpesviruses.

Figure 12.14 Lysis and Lysogeny

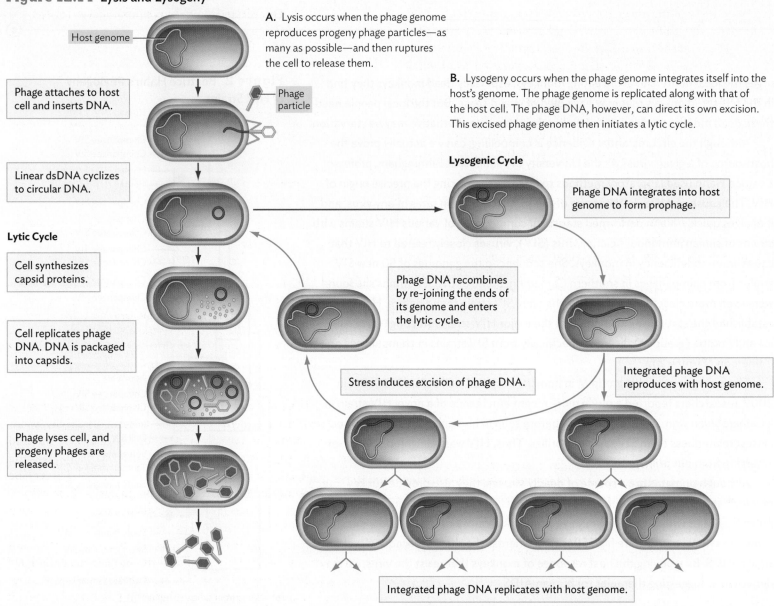

Host genome

A. Lysis occurs when the phage genome reproduces progeny phage particles—as many as possible—and then ruptures the cell to release them.

Phage attaches to host cell and inserts DNA.

Phage particle

B. Lysogeny occurs when the phage genome integrates itself into the host's genome. The phage genome is replicated along with that of the host cell. The phage DNA, however, can direct its own excision. This excised phage genome then initiates a lytic cycle.

Linear dsDNA cyclizes to circular DNA.

Lysogenic Cycle

Phage DNA integrates into host genome to form prophage.

Lytic Cycle

Cell synthesizes capsid proteins.

Phage DNA recombines by re-joining the ends of its genome and enters the lytic cycle.

Cell replicates phage DNA. DNA is packaged into capsids.

Integrated phage DNA reproduces with host genome.

Stress induces excision of phage DNA.

Phage lyses cell, and progeny phages are released.

Integrated phage DNA replicates with host genome.

LYTIC INFECTION After the phage inserts its DNA into the cell, the phage genes are expressed by the host cell RNA polymerase and ribosomes. "Early genes" are those expressed early in the lytic cycle—those needed for transcribing DNA and other early steps to produce progeny phages. Other phage-expressed proteins then work together with the host's cellular enzymes and ribosomes to replicate the phage genome and produce phage capsid proteins. The capsid proteins self-assemble into capsids and package the phage genomes—a process that takes place in defined stages, like a factory assembly line. Finally, a "late gene" from the phage genome expresses an enzyme that lyses the host cell wall, releasing the mature virions. Lysis is also referred to as a burst, and the number of virus particles released is called the **burst size**.

Some phages (such as phage T4) reproduce entirely by the lytic cycle and thus are called **virulent phages**. Phages that can undergo lysogeny, such as phage lambda, are called **temperate phages**.

A temperate phage such as phage lambda can infect and lyse cells in the same manner as a virulent phage. A temperate phage can also exploit the alternative pathway, lysogeny, which involves integrating its genome into that of the host cell.

LYSOGENY Phage lambda has a linear genome of double-stranded DNA, which circularizes upon entry into the cell. To lysogenize, the circularized genome recombines into that of the host by site-specific recombination of DNA. In site-specific recombination, an enzyme aligns the phage genome with the host DNA and exchanges the sugar-phosphate backbone links with those of the host genome. The DNA exchange process integrates the phage genome into that of the host. The integrated phage genome is called a **prophage**. A similar integration event occurs in human viruses, such as when the DNA copy of an HIV genome integrates into the genome of a host T cell (see Section 12.6).

Integration of the phage lambda genome as a prophage results in lysogeny, a condition in which the phage genome is replicated along with that of the host cell as the host reproduces. However, the prophage retains the ability to spontaneously generate a lytic burst of phage. For lysis to occur, the prophage (integrated phage genome) directs its own excision from the host genome. The two ends of the phage genome are cut, come apart from the host DNA molecule, and then are re-joined so that the phage DNA is once again circularized. The phage DNA then initiates a lytic cycle, destroying the host cell and releasing phage particles.

For a temperate virus (a virus capable of lysogeny), the "decision" between lysogeny and lysis is determined by proteins that bind DNA and repress the transcription of genes for virus replication. Exit from lysogeny into lysis can occur at random, or it can be triggered by environmental stress such as UV light, which damages the cell's DNA. The regulatory switch of lysogeny responds to environmental cues indicating the likelihood that the host cell will survive and continue to propagate the phage genome. When a cell grows under optimal conditions, it is more likely that the phage DNA will remain inactive, whereas events that threaten host survival will trigger a lytic burst. An analogous phenomenon occurs in human viral infections such as herpes, in which environmental stress triggers reactivation of a virus that was dormant within cells (a latent infection). Reactivation of a latent herpes infection results in painful outbreaks of skin lesions.

During the exit from bacteriophage lysogeny (or during the latent growth of animal viruses), a virus can acquire host genes and pass them on to other host cells. In bacteria, this process of viral transfer of host genes is known as **transduction** (discussed in Section 9.1). A transducing bacteriophage can pick up a bit of host genome and transfer it to a new host cell. In some forms of transduction, the entire phage genome is replaced by host DNA packaged in the phage capsid, resulting in a virus particle that transfers only host DNA. In other cases, only a small part of the host genome is transferred, attached to the viral genome.

The integration and excision of viral genomes make extraordinary contributions to the evolution of host genomes, including pathogens. For example, the CTXφ prophage encodes cholera toxin in *Vibrio cholerae*, and prophages encode toxins for *Corynebacterium diphtheriae* (diphtheria) and *Clostridium botulinum* (botulism). In the laboratory, the ability of phages to transfer genes is exploited in developing vectors (carrier genomes) for biotechnology.

Culturing Viruses

To study any living microorganism, we must culture it in the laboratory. A complication of virus culture is the need to grow the virus within a host cell. So, any virus culture system must be a double culture of host cells plus viruses.

Bacteriophages may be produced in **batch culture**, a culture in an enclosed vessel of liquid medium. Batch culture enables

Figure 12.15 One-Step Growth Curve for a Bacteriophage

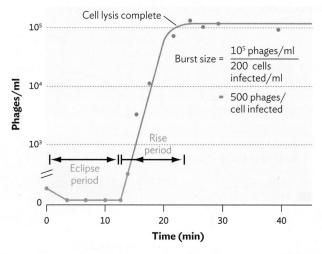

After initial infection of a liquid culture of host cells, the titer of virus drops near zero as all virions attach to the host. During the eclipse period, progeny phages are being assembled within the cell. As cells lyse (the rise period), phages are released until they reach the final plateau. The replication cycle is typically complete within less than an hour.

growth of a large population of viruses for study. A phage sample is inoculated into a growing culture of bacteria, usually in a culture tube or a flask. The culture fluid is then sampled over time and assayed for phage particles. The growth pattern usually takes the form of a **one-step growth curve** (Figure 12.15). To observe one cycle of phage reproduction, we must add phages to host cells at a multiplicity of infection (MOI), the ratio of phage to cells, such that every host cell is infected. The phage particles immediately adsorb to surface receptors of host cells and inject their DNA. As a result, phages are virtually undetectable in the growth medium for a short period after infection called the **eclipse period**. The progeny phages within the bacterial cell before it lyses can be seen by transmission electron microcopy (see Figure 12.2A).

As cells begin to lyse and liberate progeny phages, the culture enters the rise period, in which phage particles begin appearing in the growth medium. The rise period ends when all the progeny phages have been liberated from their host cells. The number of phages produced per infected host cell is called the **burst size**. We can estimate the burst size by dividing the concentration of progeny phages by the original concentration of phages added to the host cells, assuming that every phage particle infected a cell.

An important culture technique is the growth of individual colonies on a solid substrate that prevents dispersal throughout the medium, as described in Chapters 1 and 6. Plate culture of colonies enables us to isolate a population of microbes descended from a common progenitor. But how can we isolate viruses as "colonies"? Although viruses can be obtained at incredibly high concentrations, they disperse in suspension. Even on a solid medium, viruses never form a solid visible mass comparable to the mass of cells that constitutes a cellular colony.

Figure 12.16 Pour-Plate Culture of Bacteriophages on Host Bacterial Lawn

1 Add phages to bacteria.

E. coli in rich broth culture

Phage stock

2 Add phage-infected bacteria to molten top agar.

Molten top agar

3 Pour immediately onto the agar plate.

50°C H₂O bath

4 Rotate to spread evenly. Agar solidifies.

Bottom agar plate

5 Incubate at 37°C overnight.

Bacterial lawn

Soft agar (0.75%)
Bottom agar (1.5%)

Each plaque contains about 10⁶ phages from one parent.

A. A suspension of bacteria in rich broth culture is inoculated with a low proportion of phage particles to host cells. Therefore, only a few of the bacteria become infected immediately, while the rest continue to grow. Each plaque arises from a single infected bacterium that bursts, its phage particles diffusing to infect neighboring cells.

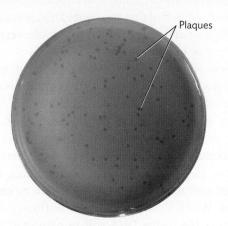

Plaques

B. Phage lambda plaques on a lawn of *Escherichia coli* K-12.

serial dilution in the same way we would measure a suspension of bacteria.

CULTURING ANIMAL VIRUSES For animal viruses, the plaque assay has to be modified for host cells in tissue culture (**Figure 12.17A**). The host cells are usually an immortalized cell line—that is, a line of cells from a cancer in which genetic control of growth is eliminated, so the cells can double in culture indefinitely.

Tissue culture involves the growth of human or animal cells in a monolayer on the surface of a dish containing fluid medium. But the fluid medium would quickly disperse any viruses released by lysed cells. To culture viruses, the tissue culture procedure is modified as follows. First, the tissue culture is inoculated with a virus suspension (Figure 12.17A, step 1). Then, after sufficient time to allow for viral attachment to cells, the culture fluid is removed, including any unattached virions (step 2). The fluid is then replaced by a gelatin medium (step 3). The gel retards the dispersal of viruses from infected cells, and as the host cells die, plaques are observed (step 4). **Figure 12.17B** shows the appearance of plaques of human coronavirus (HCoV NL63) cultured on colon carcinoma cells. This coronavirus causes mild respiratory symptoms; it is related to SARS-CoV-2, the cause of COVID-19.

The remaining three sections of this chapter describe the replication and pathology of three viruses that infect humans: human papillomavirus (HPV), influenza virus, and human immunodeficiency virus (HIV). Each example reveals a different kind of replication cycle with very different implications for the patient.

In bacteriophage plate culture (**Figure 12.16**), bacteriophages from a single progenitor lyse their surrounding host cells, forming a clear area called a **plaque**. To perform a plaque assay of bacteriophages, we mix a diluted suspension of bacteriophages with bacterial cells in soft agar (agar at low concentration) and then pour the mixture over an agar plate. Phage particles diffuse through the soft agar until they reach a host bacterium. Where no bacteriophages are present, the bacteria grow homogeneously as a "lawn," an opaque sheet over the surface (confluent growth). Where there is a bacteriophage, it infects a cell, replicates, and spreads progeny phages to adjacent cells, killing them as well. The loss of cells results in a round, clear area seemingly cut out of the bacterial lawn; the tiny clear areas are the plaques. Plaques can be counted and used to calculate the concentration of phage particles, or **plaque-forming units (PFUs)**, in a given suspension of liquid culture. The phage concentration can be measured by

Figure 12.17 Culture of Human Coronavirus

A. Modified plaque assay using cultured host cells. The addition of gelatin medium (step 3) retards the dispersal of progeny virions from infected cells, restricting new infections to neighboring cells. The result is a visible clearing of cells (a plaque) in the monolayer.

1 Infect monolayer with virus.

2 Remove liquid medium.

3 Add gelatin medium.

4 Virus reproduces. Host cells lyse, forming plaques.

Plaque

B. Plaque assay in which human coronavirus HCoV NL63 suspension was plated on a monolayer of colon carcinoma cells.

SECTION SUMMARY

- **Virions attach to host cell surface receptors.** The receptors, proteins with important host function, determine which host cells the virus can infect.

- **Lytic replication lyses the host cell.** A virulent bacteriophage injects its DNA into a host cell, where it uses host gene expression machinery to produce progeny virions.

- **Lysogeny maintains the phage genome within that of the host.** A phage that can undergo lysogeny is called a **temperate phage**. In lysogeny, the viral genome is integrated into that of the host cell, which then replicates the phage genome along with its own. Under stress, a temperate phage can initiate a lytic cycle.

- **Batch culture of viruses generates a step curve.** For a period of time, there are no new virions; then, as infected cells lyse, the viral titer rises rapidly.

- **Viruses require culture within host cells.** Bacteriophages may be cultured either in batch culture or as isolated plaques on a bacterial lawn. Animal viruses are grown in tissue culture.

Thought Question 12.4 Given the mechanism by which viruses infect cells, how might an organism evolve to become resistant to a viral infection?

12.4 Papillomavirus: DNA Genome

SECTION OBJECTIVES

- Describe the genome of papillomavirus.

- Explain the replication cycle of papillomavirus, and explain how papillomavirus causes cancer.

CASE HISTORY 12.1

Genital Warts from a Virus

Sean was a 19-year-old college sophomore at a large midwestern university. Sean visited the campus health center, where he told the nurse practitioner that the shaft of his penis had a raised spot. The spot had caused no discomfort yet, but Sean knew it was abnormal. The practitioner told Sean that he probably had human papillomavirus (HPV), a cause of genital warts (**Figure 12.18A**). Although some HPV strains cause warts on external skin, genital HPV is transmitted only by sexual intercourse. The practitioner conducted a physical exam and discovered additional raised spots on Sean's anus. Sean told the practitioner that he had vaginal sex with women and had also experimented sexually with his fraternity brothers. The practitioner told Sean that the warts would probably go away on their own if he maintained good sleep and health habits, keeping his immune system strong, but some HPV strains can persist for years without symptoms. The warts could be removed by freezing or chemical treatment, but the virus would persist. Meanwhile, the presence of HPV suggested that Sean was likely to contract related strains causing cancer of the penis or anus. The practitioner recommended use of condoms, although HPV may infect areas that condoms do not protect. She also suggested that Sean recommend the Gardasil HPV vaccine to his parents for his younger brother and sister before they became sexually active.

How prevalent is papillomavirus? In the United States, the genital strains of human papillomavirus (HPV), including those that cause cancers of the cervix, penis, and anus, infect 80% of adults. The discovery that cervical cancer is caused by HPV earned German researcher Harald zur Hausen the 2008 Nobel Prize in Physiology

Figure 12.18 Human Papillomavirus Infection

A. Certain strains of human papillomavirus (HPV) cause warts on the genitals or anus.

B. HPV infects basal epithelial cells, where the DNA uncoats but remains dormant. As cells differentiate, new virions are synthesized and released by shedding cells. Some HPV proteins transform host cells into cancer cells.

or Medicine (he shared the prize with Françoise Barré-Sinoussi and Luc Montagnier, the researchers who discovered that HIV causes AIDS; see Section 12.6). The genital strains of HPV are highly infectious through sexual contact, including oral contact leading to throat cancers. Although condoms offer partial protection, greater protection is afforded by vaccines such as Gardasil, but the vaccine is effective only before exposure to the virus. Gardasil is now recommended by the US Centers for Disease Control and Prevention for all girls and boys by age 11–12. The vaccine protects against four of the most prevalent strains, including HPV-16, the leading cause of genital cancer and throat cancer. The only sure way to avoid genital HPV, however, is to abstain from sexual contact or to have sex only in an exclusive relationship.

Like many viruses, HPV shows tropism, the ability to infect only specific organs or tissues. HPV has a narrow tropism: strains infect specifically the skin (cutaneous epithelium; **Figure 12.18B**) or the mucous membranes. Tropism is usually determined by specific cell-surface protein receptors (just as bacteriophage attachment requires specific host protein receptors). For HPV, the receptors for some strains are heparan sulfate proteoglycans (HSPGs), cell-surface proteins involved in wound healing. The cell needs these proteins for important functions, and HPV evolved to take advantage of them.

In the affected tissues, the virus must gain access to the actively dividing cells of the basal layer, usually through a tiny wound in the tissue. Virions are endocytosed by the basal cells. Their replication, however, is inhibited until the basal cells start to differentiate into keratinocytes (mature epithelial cells). As the epithelial layers reach the surface and slough off, progeny virions are shed. In other infected cells, however, the HPV virions become latent, persisting in the cell for months or years. These latent viral genomes may

induce the host cells to form abnormal growths, such as warts or cancers. Different HPV strains cause skin warts, genital warts, and genital cancers; that is why, in Case History 12.1, Sean was told that the penile warts would not cause cancer but similarly transmitted HPV strains could.

Structurally, the papillomaviruses are small icosahedral viruses, about twice the size of a ribosome. The virion has no envelope (**Figure 12.19A**). Its genome consists of a circular, double-stranded DNA (Baltimore Group I; see Table 12.1 and Figure 12.3). The papillomavirus genome is relatively small, compared with the genomes of herpesvirus or mimivirus; for example, HPV-16 has fewer than 8,000 base pairs, encoding only eight gene products (**Figure 12.19B**). The genes are expressed in overlapping **reading frames**. Reading frames represent the three different ways that an RNA sequence can be read to define triplet codons (presented in Chapter 8). Using three reading frames allows partial overlap of the genes and maximizes the efficiency of the information content in the shortest possible genome. The shorter the viral genome, the greater the number of virions that can be made with the DNA nucleotides available within a host cell.

Because HPV replication requires the developmental transition of basal cells into cells of the epidermis (outer skin layers; see Figure 12.18B), the virus is difficult to grow in culture for study. HPV is often eliminated by the host immune system, but if infection persists, no effective drug therapies are known. To discover treatments for HPV, researchers must identify drug targets by studying the virus's replication cycle (**Figure 12.20**).

To infect a host cell and initiate replication, the HPV virion binds to cell-surface receptors (such as the heparan wound-healing proteins). The bound virion then becomes endocytosed by the host cell (Figure 12.20, step 1). The virion then travels through

Figure 12.19 Human Papillomavirus HPV-16, a Cause of Genital and Anal Cancers

A. HPV virion; diameter, 55 nm.

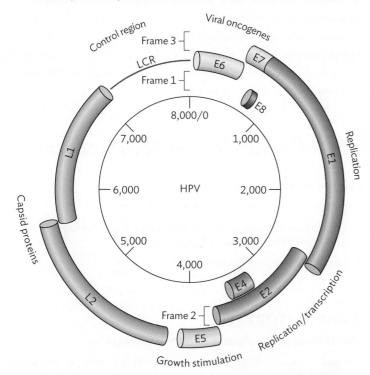

B. Genome of HPV-16. The DNA genome is circular; the "starting" base position (base 1) is collocated with base 7,906. The genome encodes eight gene products, some of which overlap as shown. "E" proteins come from early genes; "L" proteins, from late genes. LCR = long control region.

Figure 12.20 Papillomavirus Replication Cycle

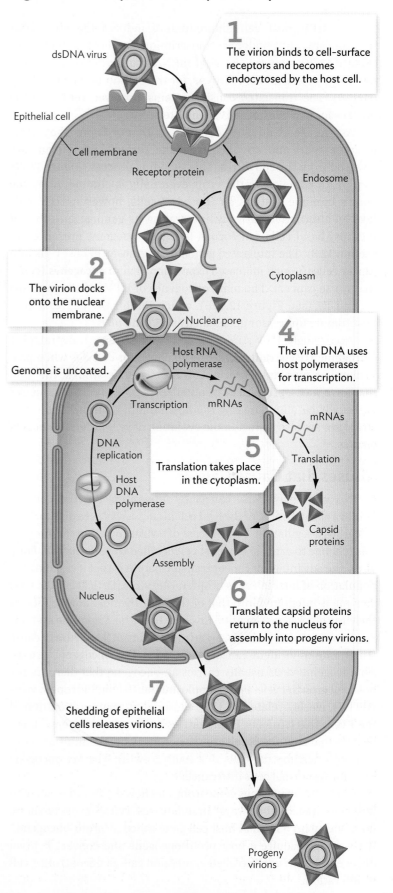

1 The virion binds to cell-surface receptors and becomes endocytosed by the host cell.

2 The virion docks onto the nuclear membrane.

3 Genome is uncoated.

4 The viral DNA uses host polymerases for transcription.

5 Translation takes place in the cytoplasm.

6 Translated capsid proteins return to the nucleus for assembly into progeny virions.

7 Shedding of epithelial cells releases virions.

the endomembrane system (described in Chapter 4). To dock at a nuclear pore complex (step 2), the virus uncoats; that is, the capsid releases its genome (step 3). For HPV, the uncoated genome must enter the nucleus for all replication events. Nuclear replication is typical of many DNA viruses (such as herpes), but other DNA viruses, such as smallpox and mimiviruses, replicate entirely in the cytoplasm.

For HPV, host cell differentiation induces the viral DNA to replicate and undergo transcription by host polymerases (Figure 12.20, step 4). The viral mRNA transcripts then exit the nuclear pores, as do host mRNAs, for translation in the cytoplasm (step 5). The translated capsid proteins, however, return to the nucleus for assembly of the virions (step 6).

As the keratinocytes complete differentiation, they start to come apart and are shed from the surface. The virions are released from the cell during this shedding process (Figure 12.20, step 7). In some cases the virus eventually is eliminated by the body's immune system. In the basal cells, however, HPV may take an alternative pathway, integrating its genome into that of the host basal cells (analogous to phage lysogeny, presented in Section 12.3). The integrated genome can transform host cells into cancer cells through increased expression of viral **oncogenes** (genes that cause cancer). The main oncogenes of HPV encode proteins E6 and E7 (see Figure 12.19B), which inhibit the expression of host tumor suppressor genes such as *p53* and *pRB*.

Accordingly, HPV strains exhibit two different viral strategies. Oncogenic HPV strains persist in the body for decades while producing low numbers of virions. Non-oncogenic strains, the kind that form warts, produce large numbers of progeny virions but can eventually be eliminated by the host immune system. These two strategies are also shown in the evolution of very different kinds of viruses.

Viruses that Cause Cancer

A DNA virus may cause cancer by integrating its viral genome within a chromosome of the host cell in such a way as to disrupt host cell growth regulation. The process of a virus inducing carcinogenesis (change to cancer) in a host cell is called **transformation**. For a virus, the advantage of cancer transformation is that it expands the population of infected cells proliferating virus particles. If the viral genome is integrated into that of the host, the host now replicates the viral genome in a location invisible to the host immune system.

Some retroviruses (reverse-transcribing RNA viruses) cause a high rate of virulent cancer; these are called oncogenic viruses. An oncogenic virus usually carries an oncogene that disrupts the host cell growth cycle. For example, human T-lymphotrophic virus (HTLV), closely related to HIV, causes abnormal proliferation of the T cells that it infects. HTLV carries the *tax* oncogene, encoding a protein that stimulates cell growth despite contact inhibition (contact with neighboring cells that halts growth). The *tax* oncogene transforms normal cells into cancer cells.

Where do oncogenes come from, and how do they interact with host cells in such subtle ways? In some cases, a viral oncogene shows homology with a normal host cell gene called a "proto-oncogene." A virus may acquire a host proto-oncogene and convert it to an oncogene whose uncontrolled expression causes uncontrolled cell proliferation.

Note: Distinguish between two uses of the term "transformation." A cell can be transformed with DNA, meaning that the cell takes up exogenous DNA into its own genome. In a different sense, a cell can be transformed by a virus, converting to a cancer cell (oncogenesis).

SECTION SUMMARY

- **Papillomaviruses have a small genome of double-stranded DNA.** The virion is icosahedral and has no envelope.

- **Human papillomavirus (HPV) shows tropism for the skin or for the genital mucosa.** Different strains cause warts or cancers of the skin or genitals. HPV-16 is the main cause of cervical cancer. Most sexually active people acquire HPV strains.

- **HPV virions bind specific cell-surface proteins of the basal epithelial layer.** The virion is then endocytosed, and the capsid is uncoated.

- **The HPV genome replicates within the nucleus of differentiating cells.** Progeny virions are transported to the cytoplasm, where they escape the host cell as it is sloughed off from the epithelium.

- **Some HPV genomes integrate into the host genome.** Integrated HPV genomes can persist for many years and transform the cell to cancer.

- **Viral oncogenes transform a host cell to cause cancer.**

Thought Question 12.5   What are the advantages and disadvantages to the virus of replication by the host polymerase, compared with using a polymerase encoded by its own genome?

12.5 Influenza Virus: RNA Genome

SECTION OBJECTIVES

- Describe the structure of the influenza virion.

- Explain how the form of the influenza genome enables rapid evolution of novel strains that cause pandemics.

- Describe the replication cycle of influenza virus.

CASE HISTORY 12.2

An Influenza Pandemic Reaches College

Eighteen-year-old Aisha was a first-year student at a small private college in New England. In the fall of 2009, the campus implemented management procedures for the expected H1N1 influenza pandemic. Students with flu symptoms were told to "self-isolate" and to report their status on a college web form. By the end of October, 50 students were reporting each week.

One night, Aisha felt hot, flushed, and unsteady on her feet. Her thermometer showed a temperature of 39°C (102°F). She started coughing, and she felt extremely tired. She managed to find the web form on her computer, although her arm was shaking and made false starts. Finally, she typed her information and confirmed her status report. A health worker arrived at her room wearing a face mask. The health worker gave Aisha a face mask and told her to collect her essential personal items and books; Aisha did the best she could in her fevered state. The health worker drove her to a separate residence reserved for suspected H1N1 cases. There, Aisha lay in a bed and tried to sleep. In the morning a nurse's aide left a tray of food and medication outside Aisha's isolation room, on the hallway floor. The medications included oral oseltamivir (Tamiflu) and inhaled zanamivir (Relenza). Aisha had to drag herself out of bed to open the door and get the tray. She was required to stay in the isolation room until 2 days after her temperature returned to normal. She and most other infected students made a full recovery, although one student was hospitalized with life-threatening complications.

Influenza A virus, an orthomyxovirus (in Baltimore Group V; see Table 12.1 and Figure 12.3), is one of the most common life-threatening viruses in the United States. The virus infects cells of the upper respiratory mucosa, causing fever, sore throat, headache, and other symptoms. Each year, influenza A infects approximately 10% of the US population, causing about 36,000 deaths annually. The elderly are most susceptible, but in an epidemic year mortality rises among young people.

Influenza strains show antigenic drift in that their envelope proteins continually mutate, evading the host immune system; as a result, new strains emerge annually—the "seasonal flu." In addition, beyond the annual variant strains, at wider intervals extremely virulent strains of influenza emerge that cause pandemic mortality. For example, the famous influenza pandemic of 1918 infected 20% of the world's population and killed more people than World War I did. The 1918 strain arose as a mutant form of an influenza strain infecting birds. Such extremely virulent strains usually arise through **antigenic shift**, in which genes reassort from two or more different influenza viruses. Antigenic shift represents a larger change than antigenic drift, the smaller mutations observed in seasonal influenza strains. Today, other "avian influenza" strains continue to emerge by viral reassortment in animal hosts. As of this writing, however, the known avian strains have limited tropism for the human respiratory tract, and none have mutated to a form readily transmitted between humans (discussed further in Chapter 20).

Antigenic shifts and drifts enable viruses to evade the host immune system. Influenza type A is the most capable of undergoing genetic change, so it more commonly leads to seasonal disease. Type A influenza is classified into subtypes based on the kind of hemagglutinin (H) and neuraminidase (N) expressed on the viral surface, giving rise to some very well-known strains, such as H1N1 (swine flu) and H5N1 (avian flu).

In 2009, a highly transmissible strain related to swine influenzas ("swine flu") spread rapidly around the world. Public health officials feared that hospitals would overflow and that the severity of illness could resemble that of the 1918 pandemic. That is why colleges took measures for isolation, as experienced by Aisha in Case History 12.2. In fact, the 2009 flu did infect large numbers of people and strained our health systems, especially at institutions such as colleges. Fortunately, the symptoms turned out to be mild; the virus had limited virulence, compared with the strains in 1918 and 2006. A future strain, however, might emerge combining high transmission (as in 2009) with high human mortality (as for avian flu in 2006).

What causes the sudden appearance of new pandemic strains of influenza virus? The cause can be understood from the structure of the influenza virion and its mode of replication.

Influenza Virion Structure

The influenza virion is asymmetrical; it has no fixed capsid (**Figure 12.21**). Its genome consists of (−) sense RNA, in contrast to hepatitis C virus, which has (+) sense RNA.

Figure 12.21 Structure of Influenza Virus

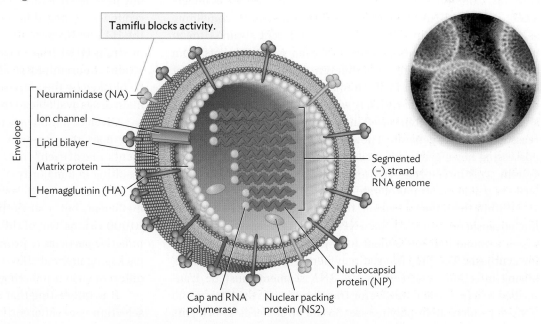

Diagram of influenza virion structure, showing the envelope (brown), envelope proteins (HA and NA), matrix protein (yellow), RNA segments with attached polymerase, and nuclear packing protein NS2. **Inset:** An influenza A virion. The brush-like border coating the envelope consists of glycoproteins, hemagglutinin (HA), and neuraminidase (NA). Envelope proteins involved in receptor binding (HA and NA) are also called "spike" proteins.

Figure 12.22 **Reassortment between Human and Avian Strains Generates Exceptionally Virulent Influenza Strains**

A. The reconstructed strain of the 1918 pandemic influenza virus is studied by Dr. Terrence Tumpey, microbiologist at the Centers for Disease Control and Prevention, Atlanta. The 1918 strain may hold clues to combating pandemic influenza today.

B. Exceptionally virulent strains arise from reassortment of human-infecting influenza viruses by antigenic shift. Antigenic shift occurs when a human-infecting influenza virus acquires a chromosomal segment from a strain that evolved within ducks or swine, agricultural animals that live in proximity to humans.

A bizarre feature of the influenza genome is that its genomic RNA is divided among eight chromosome segments. Each segment contains a different essential gene. When influenza virus infects a cell, the eight RNA segments can **reassort** as virions combining segments from a coinfecting virus of a different strain. Human influenza RNA segments can reassort even with segments from a strain that normally infects a different host species and cannot complete its replication in the human cell (**Figure 12.22**). Such reassortment is particularly likely with influenza strains from birds and pigs, which are often raised in close proximity to humans. The reassortment may generate a new strain able to infect humans but possessing novel gene sequences unfamiliar to the host immune system. This interspecies reassortment is what leads to the most virulent pandemic strains of influenza.

Within the influenza virion, the RNA segments are coated by nucleocapsid protein (NP). Each NP-coated RNA segment also possesses a bound RNA-dependent RNA polymerase complex, poised for synthesis. The prepackaged polymerase is necessary because, during infection, each (−) strand RNA segment must be transcribed to a (+) strand mRNA for translation—but the host has no RNA-dependent RNA polymerase. So the virus needs to bring its own, a copy of which is all set to go on each segment.

The virion's RNA segments are loosely contained by a shell of matrix proteins. The envelope originates from the phospholipid membrane of the host cell. The membrane incorporates

viral proteins, such as hemagglutinin (HA) and neuraminidase (NA), that peg the membrane to the matrix, maintaining an intact but flexible structure. The HA and NA proteins differ slightly in sequence among different strains. The sequence differences of the HA and NA proteins give rise to the H and N numbers, such as in strain H1N1 (the so-called swine flu) and H5N1 (an "avian flu" strain). Neuraminidase also acts as an enzyme whose activity can be blocked by the antiviral agent oseltamivir (Tamiflu), one of the main drugs available to treat influenza.

Given that influenza infection requires all eight RNA segments, how does the assembly mechanism package exactly eight segments, one of each? In fact, the segments are packaged imperfectly, resulting in a majority of defective particles. As a result, less than 1% of influenza particles are capable of infection. This may sound inefficient, but a single infected cell can produce an astonishing 10,000 virions. Out of this number, there will be more than enough infective particles to propagate the virus. Furthermore, the inexact packaging method allows great flexibility for reassortment among different virions infecting the same host cell.

It is interesting that influenza virus has responded to natural selection very differently than HPV. HPV has evolved the tiniest possible genome, generating as many virions as possible out of limited resources, while persisting in the host. Influenza virus, on the other hand, has evolved a genome that maximizes gene reassortment in order to evade the immune response, while still generating large quantities of virions to infect the next host. So, evolution leads to different results, given different conditions of host cells and transmission.

Influenza Replication Cycle

What determines how influenza spreads—and how can we stop it? As discussed in Chapter 20, surprisingly small details of the host receptor structure may determine whether a strain of influenza virus will spread directly between humans. Accordingly, avian influenza strain H5N1 is rarely transmitted between humans. However, a mutation in the HA gene could allow the envelope protein to bind human cells in the upper respiratory tract, which would increase transmission and possibly lead to an influenza pandemic. Much research addresses hemagglutinin (HA) and other viral proteins as targets of possible drugs.

The influenza virus replication cycle is shown in **Figure 12.23**. An influenza virion attaches to a cell when its HA envelope protein binds to a host cell receptor, a glycoprotein (Figure 12.23, step 1). After the influenza virion undergoes endocytosis, the endocytic vesicle fuses with a lysosome (step 2). The low pH of the lysosome contents causes hydrogen ions to leak through the virion's ion channel, an important drug target. The hydrogen ions (low pH) cause the virion to disassemble and fuse with the endocytic membrane. As the membranes fuse, the virion contents are released into the cytoplasm, including all the viral (−) RNA segments with their prepackaged polymerases (step 3). The RNA segments pass through nuclear pore complexes into the nucleus. Within the nucleus, the viral RNA-dependent RNA polymerases that are attached to each (−) RNA synthesize (+) RNA for mRNA (step 4). Viral transcription is primed by capped host RNA—a process that is the target of the antiviral drug baloxavir marboxil. The mRNA then returns to the cytoplasm for translation by host cell ribosomes (step 5).

Some of the mRNA molecules encode envelope proteins, which are made by ribosomes attached to the endoplasmic reticulum (ER; Figure 12.23, step 6). The ER transports the envelope proteins to the Golgi and ultimately to the cell membrane, where they will coat progeny virions. The NP proteins needed to package the viral RNA (packaging proteins) must return to the nucleus (step 7). In the nucleus, the (+) RNA strands now serve as templates for RNA-dependent RNA polymerase to make complementary (−) RNA segments for the progeny viral genomes (step 8). The (−) RNA viral genome segments become coated with proteins (step 9). At last, the protein-coated (−) RNA segments exit the nucleus and are transported to the cell membrane for packaging into a loose capsid of M1 matrix proteins (step 10). As they assemble, the capsids acquire envelope membrane from the host, incorporating viral envelope proteins. The mature virions now bud out (step 11) in a massive release of virions that destroys the host cell.

The final budding step requires the action of neuraminidase (NA), an enzyme embedded in the envelope of the progeny virions; the enzyme cleaves host surface sugar molecules that bind exiting virions to the cell surface. NA is the target of oseltamivir (Tamiflu)

and zanamivir (Relenza)—at present the most effective drugs for decreasing the symptoms of influenza.

Compared with the papillomavirus cycle, the replication cycle of influenza virus is very rapid, causing swift destruction of mucosal cells. But the host immune response is also swift. If successful, the host immune response completely eliminates the flu virus within 1 or 2 weeks. Unlike HPV, there is no long-term persistence in the host cells. So, influenza virus is committed to rapid infection and highly efficient host-to-host transmission—as demonstrated during the 2009 H1N1 outbreak.

In the next section we examine one more virus, with another very different infection strategy: the retrovirus HIV, the cause of AIDS. Like influenza virus, HIV has an RNA genome with a high mutation rate. But the HIV virion generates a DNA copy of its genome that, like human papillomavirus, can integrate into the host genome.

SECTION SUMMARY

- **Influenza virus causes periodic pandemics of respiratory disease.** New virulent strains arise through reassortment of viral genome segments from human, avian, and swine strains.

- **Antigenic shift** is the reassortment of genome segments from a viral strain that infects one species with segments from a strain that infects a different species. Antigenic shift can lead to highly virulent strains of a virus that may cause widespread outbreaks of infection.

- **Antigenic drift** is the occurrence of small mutations that continually generate slightly different forms of a particular virus.

- **Influenza virus consists of segmented (−) strand RNA.** Each segment is packaged with nucleocapsid proteins. Segments from different strains reassort by coinfection of a host cell.

- **Nucleocapsid and matrix proteins enclose the RNA segments of influenza virus.** The matrix is enclosed by an envelope containing spike envelope proteins HA and NA. The HA proteins mediate virion attachment. NA is an enzyme that enhances the release of progeny virions from the host cell.

- **The influenza virion is internalized by endocytosis, followed by lysosome fusion.** Low pH triggers viral envelope fusion with the endosome membrane, releasing the viral contents into the cytoplasm.

- **Viral (−) RNA segments attached to RNA-dependent RNA polymerase enter the nucleus.** The RNA-dependent RNA polymerase makes new (+) sense and (−) sense strands. The (+) sense viral mRNAs return to the cytoplasm for translation, while (−) sense progeny genome segments are packaged into virions.

- **Influenza virions are assembled at the cell membrane.** The capsid, matrix, and (−) strand RNA components are packaged and enclosed by host cell membrane that contains viral envelope proteins. Completed virions bud out and destroy the host cell.

Thought Question 12.6 Researchers in Hong Kong are conducting a regular survey of the genomes of influenza strains emerging in a particular swine-processing facility, where swine are shipped from many different regions. What might the researchers learn that would be important for human public health?

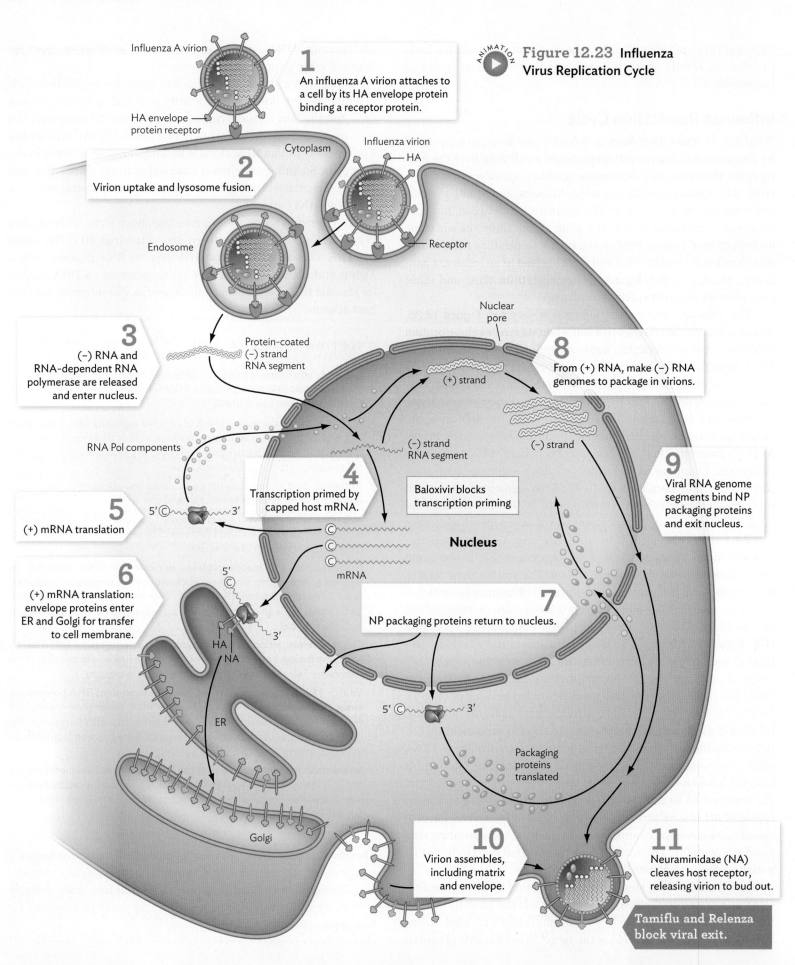

Influenza A virion

HA envelope protein receptor

1 An influenza A virion attaches to a cell by its HA envelope protein binding a receptor protein.

ANIMATION

Figure 12.23 Influenza Virus Replication Cycle

Cytoplasm

Influenza virion

HA

2 Virion uptake and lysosome fusion.

Endosome

Receptor

3 (−) RNA and RNA-dependent RNA polymerase are released and enter nucleus.

Protein-coated (−) strand RNA segment

Nuclear pore

(+) strand

8 From (+) RNA, make (−) RNA genomes to package in virions.

RNA Pol components

(−) strand RNA segment

(−) strand

4 Transcription primed by capped host mRNA.

Baloxivir blocks transcription priming

9 Viral RNA genome segments bind NP packaging proteins and exit nucleus.

5 (+) mRNA translation

5′ Ⓒ   3′

Ⓒ
Ⓒ
Ⓒ

mRNA

Nucleus

6 (+) mRNA translation: envelope proteins enter ER and Golgi for transfer to cell membrane.

5′ Ⓒ

HA
NA

3′

7 NP packaging proteins return to nucleus.

ER

5′ Ⓒ   3′

Packaging proteins translated

Golgi

10 Virion assembles, including matrix and envelope.

11 Neuraminidase (NA) cleaves host receptor, releasing virion to bud out.

Tamiflu and Relenza block viral exit.

12.6 Human Immunodeficiency Virus: Reverse Transcription

SECTION OBJECTIVES

- Describe the structure of the HIV virion.
- Explain the replication cycle and epidemiology of HIV.
- Describe the positive roles of retroviruses in human health.

CASE HISTORY 12.3

Lifelong Infection by a Retrovirus

 At 24 years of age, Ralph had just been promoted at the car dealership where he worked in sales. He was ready to propose to his girlfriend, a bank clerk. Several years had passed since Ralph had engaged in risky behaviors, but he and his girlfriend both decided to get tested for STIs. That is how Ralph learned he was HIV-positive. The diagnosis came as a surprise because Ralph had no symptoms of disease. His T-cell count, however, had dipped to 500 cells/μl (normal is about 1,200 cells/μl). Now he faced the dilemma of how to avoid infecting his future wife and how to conceive healthy children.

The doctor informed Ralph of his treatment options. The current recommendation of the Centers for Disease Control and Prevention (CDC) is to commence antiretroviral therapy as soon as an individual tests positive for the virus. A single daily pill was prescribed, containing tenofovir and emtricitabine (reverse transcriptase inhibitors, together marketed as Truvada). The advantage of early treatment with Truvada is the early halt in decline of T cells and the decrease in transmission risk. Ralph was also told of options for conceiving children without infection, such as the "sperm washing" procedure that eliminates HIV from sperm for artificial insemination. He had a lot to discuss with his girlfriend.

What is human immunodeficiency virus (HIV), and what is its global impact today? HIV is the causative agent of acquired immunodeficiency syndrome (AIDS) and is a member of the lentivirus family (in Baltimore Group VI; see Table 12.1 and Figure 12.3). A **lentivirus** is a retrovirus that causes infections that progress slowly over many years. According to the United Nations, approximately 33 million people globally are estimated to be living with HIV, and 3 million people die of AIDS annually. In the United States, the first cases of AIDS were reported in 1981; three decades later, more than a million Americans are infected with HIV, and more than half a million have died of AIDS. Worldwide, HIV infects 1 in every 100 adults, equally among women and men. Countries in southern Africa have already experienced population decline; in South Africa, approximately 20% of adults test positive for HIV.

Figure 12.24 HIV Titer Increase and T-Cell Decline over Time

An initial rise in virus titer is fought off by the immune system, while the T-cell count declines and rebounds. Over years, the virus titer remains low while T-cell count steadily declines. At a T-cell level of about 350 cells/μl, the constitutional symptoms of AIDS (for example, weight loss and fatigue) appear, followed by opportunistic infections and death.

The experience of Ralph in Case History 12.3 is typical of HIV infection in developed countries. At some point, Ralph must have acquired HIV through high-risk behaviors such as sexual contact (genital or anal) or sharing of needles. The initial infection often goes unnoticed, as flu-like symptoms subside, or there may be no symptoms at all. During this early period (**Figure 12.24**), the number of CD4+ T lymphocytes declines while the number of virions in the blood rises. T lymphocytes (or T cells) are white blood cells of the active immune system; they are needed to activate the B cells that generate antibodies (explained in detail in Chapter 16). During HIV infection, an immune response is mounted, leading to a rebound of T cells and a drop in virion numbers. But the immune response is unable to prevent the next phase of HIV infection: a slow, steady loss of T cells (the period called "clinical latency"), until a level is reached at which the immune system fails (about 350 cells/μl). At this critical level, without treatment the patient usually experiences the constitutional symptoms of AIDS, such as weight loss, fatigue, fever, and swollen lymph nodes. Opportunistic infections soon follow because the immune system can no longer respond (discussed in Chapter 21).

HIV Structure

HIV and its causative role in AIDS were discovered by French virologists Luc Montagnier and Françoise Barré-Sinoussi, at the Pasteur Institute, building on Robert Gallo's studies of retroviruses at the US National Institutes of Health (**Figure 12.25**). The virion has an unusual conical-icosahedral capsid enclosed in an envelope with spike proteins (**Figure 12.26**). After three decades of research,

Figure 12.25 HIV Discoverers

A. Françoise Barré-Sinoussi, at the Pasteur Institute, worked with Luc Montagnier to discover the virus that causes AIDS.

B. Luc Montagnier and Robert Gallo agree to collaborate on development of an AIDS vaccine, 2002.

Table 12.2
Antiretroviral Drugs Active against HIV

Drug	Mode of Action
Azidothymidine (AZT), also called zidovudine (ZDV)	Nucleoside reverse transcriptase inhibitor (base analog that halts elongation)
Emtricitabine (Truvada component)	Cytidine analog, reverse transcriptase inhibitor
Enfuvirtide	Fusion inhibitor (prevents viral envelope fusion with host cell membrane)
Maraviroc	Entry inhibitor (blocks spike protein binding to coreceptor CCR5)
Nevirapine	Nonnucleoside reverse transcriptase inhibitor (binds reverse transcriptase and inhibits activity)
Raltegravir	Integrase strand transfer inhibitor (blocks integration of DNA genome copy into host cell genome)
Ritonavir	Protease inhibitor (inhibits the protease PR, which cleaves initial protein to make final proteins)
Tenofovir (Truvada component)	Nucleoside reverse transcriptase inhibitor (base analog that halts elongation)

there is no cure for infection and no vaccine. But study of the HIV replication cycle has revealed a surprising array of effective therapeutic agents that enable a normal life and prevent transmission (Table 12.2). In developed countries, treatment effectively prolongs the life of people with AIDS. Unfortunately, the cost leaves treatment out of reach for the majority of those infected worldwide.

The key polymerase of all retroviruses, including HIV, is **reverse transcriptase**. Reverse transcriptase is an enzyme that uses RNA as a template to synthesize DNA and then uses the new DNA as a template for the complementary DNA strand. Reverse transcriptase has the highest error rate of any known polymerase, possibly as high as one base in ten. For this reason, fewer than 1 in 1,000 progeny virions released in the blood are infectious. But the prodigious output of virions may reach 10 billion daily. So, a large number will be infectious, and many of these will have mutations that evade the immune system and antiretroviral drugs.

To delay the loss of effectiveness due to viral mutation, HIV drugs are always taken in combinations of two or three, such as the Truvada combination of tenofovir and emtricitabine recommended for Ralph. Consequently, a mutant virion would need to have two or three lucky mutations to survive the drugs. But resistance inevitably

Figure 12.26
HIV-1 Particles

HIV virions from infected human lymphoid tissue culture (thin section).

TEM   300 nm

occurs, and as people live for longer periods with HIV, new drugs are needed to replace the old ones.

The structure of the HIV virion is depicted in **Figure 12.27**. The conical core of capsid subunits encloses two copies of the RNA genome, together with reverse transcriptase and other enzymes. The presence of two genome copies is a unique feature of retroviruses. Each RNA genome copy is coated with nucleocapsid proteins similar in function to those of influenza virus. For priming DNA synthesis (reverse transcription), each RNA is complexed with a tRNA derived from the previously infected host cell. The capsid also contains about 50 copies of reverse transcriptase (RT) and protease, as well as a DNA integration factor—all of which are potential targets for drugs (see Table 12.2). The capsid is surrounded by matrix proteins, which reinforce the host-derived phospholipid viral envelope. The envelope is pegged by spike proteins composed of the subunits TM and SU. As in influenza virus, the spike proteins play crucial roles in host attachment and entry.

The HIV genome includes three main genes that are found in all retroviruses: *gag*, *pol*, and *env*. The *gag* sequence encodes capsid, nucleocapsid, and matrix proteins; *pol* encodes reverse transcriptase (RT), integrase, and protease; and *env* encodes envelope proteins. The *gag* and *pol* sequences overlap, but (as in papillomavirus) they are translated in different reading frames. Surprisingly, each gene can express two or three different products. The gene's different products arise by different start sites of transcription and by cleavage of multiproduct proteins. So, the retroviral genome

Figure 12.27 HIV-1 Structure and Genome

Complex retrovirus (HIV-1)

A. Internal structure of the HIV-1 virion, color-coded to match the genome in panel B. In the genome sequence, the staggered levels indicate three different reading frames. Each virion contains two copies of the RNA genome plus multiple copies of reverse transcriptase (RT) and protease enclosed within a conical capsid (capsid subunits plus subunits of a host protein, cyclophilin A). The capsid is surrounded by matrix subunits, which reinforce the host-derived phospholipid membrane, pegged by spike proteins (SU, TM).

HIV-1 genome

B. The genome encodes the virion proteins, as well as six accessory proteins that are expressed within the infected host cell and regulate the replication cycle. LTR = long terminal repeat.

evolves for extreme economy of RNA content, maximizing the information potential of its tiny genome.

HIV Replication Cycle

The first cells infected by HIV are T lymphocytes (T cells) that possess cell-surface proteins called CD4 (the "receptor") and CCR5 (the "coreceptor"). As we have seen for other virus-binding host proteins, these receptors have important host functions. CD4 helps the T cell activate the immune response (discussed in Chapter 16). CCR5 is a receptor for a chemokine, an immune

system signal (also discussed in Chapter 16). The CD4 function is essential for the host, but CCR5 is dispensable. Thus, people who carry a genetic defect for CCR5 have a functional immune system but resist infection by HIV-1, the main strain of HIV causing AIDS. The drug maraviroc was designed to prevent HIV infection by blocking CCR5.

As an HIV virion binds the CD4 and CCR5 receptors (Figure 12.28, step 1), its envelope fuses with the cell membrane of the T cell (step 2). This fusion step can be blocked by drugs such as enfuvirtide. Unlike influenza virus, the HIV core is released directly into the cytoplasm, without endocytosis. The two RNA genome copies are then reverse-transcribed by reverse transcriptase enzyme (step 3). Reverse transcription actually forms double-stranded DNA—a process that involves cleaving away the original RNA template and replacing it with DNA. Reverse transcription can be inhibited by azidothymidine (AZT) and emtricitabine, which are base analogs (molecules with structure similar to a nucleobase). These base analogs are chain terminators; they cannot attach the next nucleotide.

The double-stranded DNA copy of the HIV genome enters the nucleus through a nuclear pore (Figure 12.28, step 4), where it integrates as a **provirus** (or episome) at a random position in a host chromosome (step 5). Integration is a critical step; it ensures permanent infection of the host. The integrase enzyme can be blocked by inhibitors such as raltegravir. The entire length of the integrated HIV genome is transcribed to RNA by the host RNA polymerase II (step 6). Some of the RNA copies exit the nucleus to serve as mRNA for translation to proteins (step 7). Proteins are synthesized in alternative versions, such as Gag-Pol. The formation of alternative versions involves cleavage by a viral protease. The viral protease breaks up the initial proteins translated from *gag* and *pol* into smaller, active proteins. Viral protease can be blocked by protease inhibitors, another major category of antiviral drugs (see Table 12.2).

Some full-length RNA transcripts exit the nucleus to be packaged as genomes for progeny virions (Figure 12.28, step 8). Meanwhile, Env (envelope) proteins are made within the endoplasmic reticulum (ER; step 9). They pass through the Golgi for glycosylation (adding sugar chains) and packaging and are exported to the cell membrane (step 10). At the membrane, Env proteins interact with the core particle as it forms from the RNA dimers plus Gag-Pol peptides (step 11). The virions then bud out of the host cell—a process that requires cleavage of a host attachment protein called tetherin (step 12).

An alternative to viral budding is **cell fusion**, mediated by the binding of Env in the membrane to CD4 receptors on a neighboring cell. Cell fusion occurs because, unlike influenza virus, which rapidly destroys the host cell, HIV can persist in a host cell for long periods. The HIV-infected cell buds out virions while continuing to display Env proteins on the plasma membrane. Thus, an infected cell can fuse with an uninfected cell, and HIV core particles can enter the new cell through their fused cytoplasm. The fusion of

Figure 12.28 HIV Replication Cycle

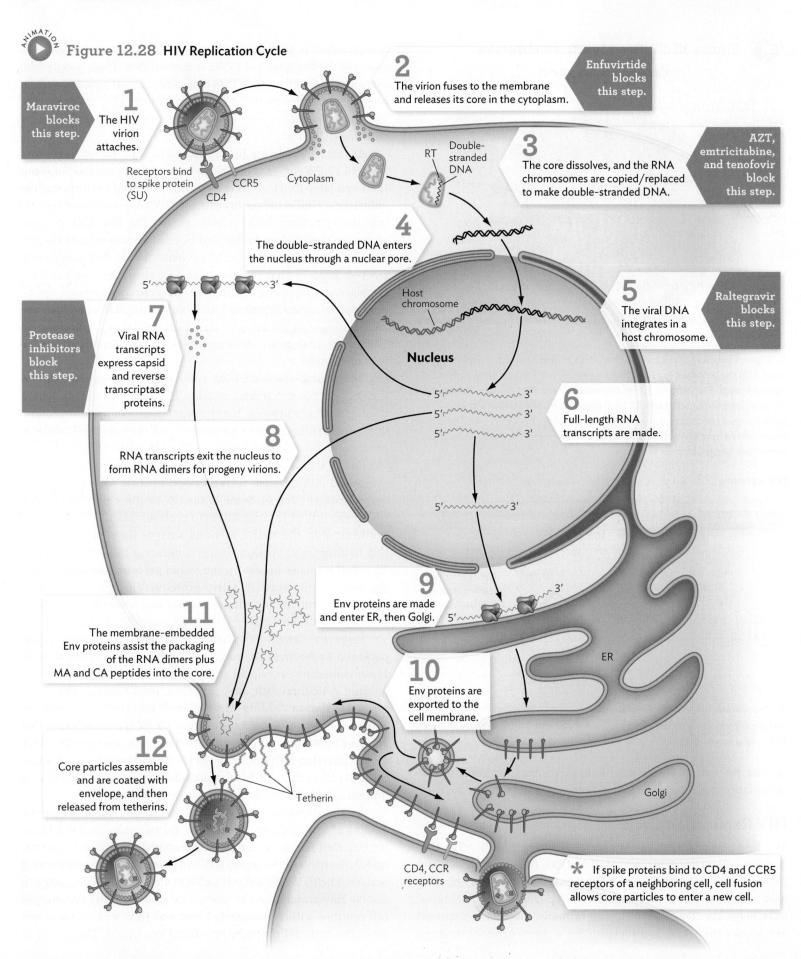

Maraviroc blocks this step.

1 The HIV virion attaches.

Receptors bind to spike protein (SU)
CD4   CCR5   Cytoplasm

2 The virion fuses to the membrane and releases its core in the cytoplasm.

Enfuvirtide blocks this step.

RT   Double-stranded DNA

3 The core dissolves, and the RNA chromosomes are copied/replaced to make double-stranded DNA.

AZT, emtricitabine, and tenofovir block this step.

4 The double-stranded DNA enters the nucleus through a nuclear pore.

Host chromosome

Nucleus

5 The viral DNA integrates in a host chromosome.

Raltegravir blocks this step.

Protease inhibitors block this step.

7 Viral RNA transcripts express capsid and reverse transcriptase proteins.

6 Full-length RNA transcripts are made.

8 RNA transcripts exit the nucleus to form RNA dimers for progeny virions.

9 Env proteins are made and enter ER, then Golgi.

11 The membrane-embedded Env proteins assist the packaging of the RNA dimers plus MA and CA peptides into the core.

10 Env proteins are exported to the cell membrane.

ER

12 Core particles assemble and are coated with envelope, and then released from tetherins.

Tetherin

Golgi

CD4, CCR receptors

✱ If spike proteins bind to CD4 and CCR5 receptors of a neighboring cell, cell fusion allows core particles to enter a new cell.

Figure 12.29 HIV Transfer through a Nanotube

A T cell infected with HIV can transfer virions to an uninfected cell through a nanotubular connection. The two T cells are tagged by different fluorescent labels (red and green).

many cells can form a giant multinucleated cell called a "syncytium" (plural, syncytia). Cell fusion enables HIV to infect neighboring cells without exposure to the immune system. Even more insidious, HIV-infected cells can form long "nanotubes" through which virions can transfer to uninfected T cells (**Figure 12.29**).

Retroviruses and the Human Genome

Suppose an integrated HIV genome mutated and lost the ability to produce progeny. What would happen to its genome? The integrated genome would be "trapped," replicating only as part of the host genome. Over many generations in its host, the viral sequence would inevitably accumulate more mutations. In fact, the human genome is riddled with remains of decaying retroviral genomes, collectively known as **retroelements**. Some retroelements contain all the genomic elements of a retrovirus, including *gag*, *env*, and *pol* genes, yet never generate virions. Presumably, the integrated retroviral genome lost this ability by mutation. Other elements, known as "retrotransposons," retain only partial retroviral elements but may maintain a reverse transcriptase to copy themselves into other genome locations. Amazingly, about half the sequence of the human genome appears to have originated from viruses and retroelements.

Could a virus as deadly as HIV be used to improve human health? In fact, the exceptional ability of lentiviruses to deliver their genomes into a human genome makes them the most promising source of vectors for gene therapy. To be used for gene therapy, the integration sequences are maintained while the virulence genes are removed and replaced by human genes to be delivered by the vector. In 2009, a lentiviral vector was used to successfully treat two children with an inherited neurodegenerative disease, X-linked adrenoleukodystrophy (ALD). The vector transferred a corrected version of the gene into the children's bone marrow stem cells, which then migrated into the brain and halted disease progression. In 2013, the first child was cured of leukemia using an HIV-derived vector to carry genes into her T cells. The transferred genes caused her T cells to fight the cancer. Looking to the future, we can expect many health-transforming applications from these deadly viruses.

SECTION SUMMARY

- **Human immunodeficiency virus (HIV) has caused an ongoing pandemic of acquired immunodeficiency syndrome (AIDS).** Although no cure exists, the molecular biology of HIV replication has led to drugs that prevent transmission and enable infected individuals to lead a normal life.

- **Reverse transcriptase uses the RNA genome of HIV as an initial template to synthesize double-stranded DNA, which integrates into the host cell's DNA.** The HIV core particle contains two copies of its RNA genome, each bound to a primer (host tRNA) and reverse transcriptase (RT). The core is surrounded by an envelope containing spike proteins.

- **HIV envelope proteins bind the CD4 receptor** of T lymphocytes together with the chemokine receptor CCR5. Following virion-receptor binding and envelope membrane fusion, HIV virions are released into the cytoplasm.

- **Double-stranded DNA is synthesized from the HIV RNA by reverse transcriptase.** The retroviral DNA integrates into the host genome. This genome integration results in permanent infection of the host.

- **HIV mRNAs are expressed and exported to the cytoplasm for translation.** Envelope proteins are translated at the endoplasmic reticulum and exported to the cell membrane.

- **HIV virions are assembled at the cell membrane**, where virions are released slowly, without lysis. Alternatively, virions can infect other T cells through intercellular nanotubes or through cell fusion to form syncytia.

- **Ancient retroviral sequences persist within the human genome.** Retroviruses can be used to develop vectors for gene therapy, correcting the symptoms of inherited genetic defects.

Thought Question 12.7 How do attachment and entry of HIV resemble attachment and entry of influenza virus? How do attachment and entry differ between these two viruses?

Perspective

Viral infections show important commonalities, as well as intriguing diversity. Virus particles (and their genomes) range in size from a few components to assemblages approaching the complexity of cells. The advantage of simplicity is the minimal requirement for resources, whereas the advantage of complexity is the fine-tuning of mechanisms for evading host defenses. Viral genomes are remarkably diverse; they can be RNA or DNA, single- or double-stranded, linear or circular. Different viruses evolve to take advantage of different host cell proteins in order to recognize and gain entry into host cells and then replicate within the host cells. Finally, each virus needs a means of transmission to the next host.

Research in virology offers hope for new drugs and cures for humanity's worst plagues, as well as remedies for the devastating diseases of agricultural plants and animals. At the same time, it's sobering to note that despite the enormous volumes we now know about viruses such as HIV and influenza virus, the AIDS pandemic continues, and we face the emergence of new pandemics such as COVID-19. The pathology and epidemiology of viral diseases are discussed further in the chapters of Part V.

Key Terms

antigenic drift (p. 348)
antigenic shift (p. 359)
bacteriophage (p. 340)
Baltimore model (p. 341)
batch culture (p. 353)
burst size (pp. 352, 353)
capsid (pp. 338, 344)
cell fusion (p. 365)
cloning vector (p. 339)
eclipse period (p. 353)
envelope (pp. 341, 344)
filamentous virus (p. 345)

host range (p. 341)
lentivirus (p. 363)
lysis (p. 350)
lysogeny (p. 351)
lytic infection (p. 350)
oncogene (p. 358)
one-step growth curve (p. 353)
plaque (pp. 340, 354)
plaque-forming unit (PFU) (p. 354)
prion (p. 347)
prophage (p. 352)

provirus (p. 365)
reading frame (p. 356)
reassort (p. 360)
retroelement (p. 367)
retrovirus (pp. 343, 346)
reverse transcriptase (pp. 343, 364)
RNA-dependent RNA polymerase (p. 341)
RNA reverse-transcribing virus (p. 343)
spike protein (p. 344)

surface receptor (p. 349)
tegument (p. 344)
temperate phage (p. 352)
tissue tropism (p. 341)
transduction (p. 353)
transformation (p. 358)
transmission (p. 341)
uncoating (p. 341)
virion (pp. 338, 340)
viroid (p. 346)
virulent phage (p. 352)
virus (p. 338)

Objectives Assessment Questions

12.1 Virus Infection and Viral Genomes

1. Which of the following traits does not apply to viruses?
   a. Most virus particles pass through a filter.
   b. Isolated particles conduct their own metabolism.
   c. Viruses have a genome of nucleic acid contained by a protein capsid.
   d. Some classes of virus particles are enclosed by a phospholipid envelope.

2. Which of these traits of mimivirus differs from a cell?
   a. The mimivirus genome contains many cellular genes.
   b. The mimivirus particle is larger than some bacteria.
   c. Within a host ameba, mimivirus forms "viral factories" that resemble a contained living organism.
   d. Mimivirus has no ribosomes of its own; it must use the host ribosomes.

3. Tissue tropism is _____.
   a. the range of tissues a virus can infect
   b. the mechanism of viral transmission to a tissue
   c. the means of viral entry into cells
   d. the conversion of tissue to cancer

4. The genome of a virus may consist of any of these except _____.
   a. phospholipid
   b. single-stranded RNA
   c. single-stranded DNA
   d. double-stranded DNA

12.2 Virus Structure and Diversity

5. Each of these viruses has an icosahedral capsid except _____.
   a. poliovirus
   b. herpesvirus
   c. Ebola virus
   d. papillomavirus

6. An example of an asymmetrical virus particle is _____.
   a. poliovirus
   b. papillomavirus
   c. smallpox virus
   d. herpesvirus

12.3 Virus Replication and Culture

7. The property of viruses that provides access to a host cell is _____.
   a. binding of a virion to a specific host receptor protein
   b. viral genome encoding of DNA polymerase
   c. self-assembly of virions within the cytoplasm
   d. viral use of an enzyme that lyses the host cell membrane

8. Which of these steps is not part of a bacteriophage replication cycle?
   a. The virion binds to a host cell surface protein that provides an important host function.
   b. The virion injects its genome into the host cell.
   c. The capsid of the virion enters the host cytoplasm.
   d. The viral genome directs host ribosomes to synthesize viral components.

9. The function of bacteriophage tail fibers is to _____.
   a. propel the virion toward a host cell
   b. attach the virion to a host cell
   c. uncoat the virion within a cell
   d. help the progeny virions exit a host cell

12.4 Papillomavirus: DNA Genome

10. Which of the following steps of the papillomavirus replication cycle leads to cancer transformation?
    a. The HPV virion becomes endocytosed by the host cell.
    b. The virus uncoats, and its genome enters the nucleus for replication.
    c. The viral genome integrates into the host genome and causes increased expression of oncogenes.
    d. Progeny virions are released during host cell shedding.

11. Which enzyme does HPV use to replicate its genome?
    a. Reverse transcriptase
    b. RNA-dependent RNA polymerase
    c. Host DNA polymerase
    d. Viral DNA polymerase

12. New HPV virions are released from the body by _____.
    a. exocytosis
    b. endocytosis
    c. becoming enveloped in cell membrane
    d. shedding of host cells

12.5 Influenza Virus: RNA Genome

13. The form of the capsid of influenza virus is _____.
    a. icosahedral
    b. filamentous
    c. complex tailed
    d. asymmetrical

14. Which step of the influenza virus replication cycle leads to evolution of virulent hybrid strains?

    a. The influenza virion is endocytosed by the cell and enters the cytoplasm.

    b. The viral genome is composed of multiple segments that get transcribed by RNA-dependent RNA polymerase and reassort with segments from a coinfecting virus.

    c. Viral proteins are translated by ribosomes in the cytoplasm and are then transported by the ER to form progeny virions.

    d. Progeny virions bud out of the infected host cell with the help of neuraminidase (NA), which cleaves host surface molecules that capture the virions.

15. Which pathology is <u>not</u> caused by influenza virus?

    a. Cancer   **c.** Shortness of breath

    b. High fever   **d.** Coughing

12.6 Human Immunodeficiency Virus: Reverse Transcription

16. Which structural feature of HIV determines the cell type that virions can infect?

    a. Virion envelope proteins specifically interact with the CD4 and CCR5 cell-surface proteins of a T lymphocyte.

    b. The capsid dissolves within the T-cell cytoplasm, releasing packaged reverse transcriptase to generate a double-stranded DNA copy of the single-stranded RNA genome.

    c. The double-stranded DNA copy of the HIV genome enters the nucleus and integrates as a provirus.

    d. The integrated HIV genome is transcribed by host RNA transcriptase, expressing viral proteins.

17. Which drug is <u>not</u> used for HIV?

    a. Azidothymidine, which prevents reverse transcriptase from elongating DNA

    b. Raltegravir, which blocks integrase from integrating a viral DNA copy into the host genome

    c. Oseltamivir, which binds neuraminidase, preventing release of virions as they exit the host cell

    d. Protease inhibitors, which block the viral protease that cleaves Gag-Pol to form proteins with different functions

18. Following initial infection with HIV, the disease progresses as follows:

    a. HIV virions gradually increase in titer throughout the host lifetime.

    b. An initial burst of virions is followed by a long latent period, during which T cells decline.

    c. The infected T cells decline and then increase and remain steady.

    d. T cells rapidly decline, and virions rapidly reach a high titer, which is sustained.

Review Questions

1. Compare and contrast the forms of icosahedral and filamentous (helical) viruses, citing specific examples.

2. How do viral genomes gain entry into cells in bacteria and animals?

3. Explain the structure and function of the seven Baltimore groups of viral genomes.

4. How do viral genomes interact with host genomes, and what are the consequences for host evolution?

5. Compare the lytic and lysogenic replication cycles of bacteriophages. What are the strengths and limitations of each?

6. Compare the replication cycles of RNA viruses and DNA viruses in human hosts. What are the strengths and limitations of each?

7. Explain the plate titer procedure for enumerating viable bacteriophages. How must this procedure be modified to titer animal viruses?

8. Explain how a pure isolate of a virus can be obtained. How do the procedures differ from those used for isolating bacteria?

9. Explain the generation of the step curve of virus proliferation. Why is virus proliferation generally observed as a single step, or generation, in contrast to the life cycles of cellular microbes, outlined in Chapter 6?

10. How do influenza virions gain access to the host cytoplasm?

11. How does influenza virus manage the replication and packaging of its segmented genome? What is the consequence of genome segmentation for virus evolution?

12. How does HIV provide ready-made components for replication of its RNA genome? How and why does reverse transcriptase make double-stranded DNA?

13. What is the role of protease in HIV replication? What is the significance of protease for AIDS therapy?

14. How did ancient retroviruses participate in the evolution of the human genome? What role do retroviruses play in gene therapy?

Clinical Correlation Questions

1. A 10-year-old child comes down with a fever and presents with large, raised sores typical of a pox disease. The child has recently acquired a pet rabbit from a store that also sells prairie dogs. The prairie dogs were obtained from an exotic pet dealer that also sells Gambian rats from Africa. What is the infectious agent?

2. During a routine physical checkup, a 60-year-old woman is advised to be tested for a virus with high prevalence in her age-group. To her surprise, the test comes back positive. The virus, which damages the liver, is known to be transmitted by sexual contact, needlestick, and tattoos applied under unsanitary conditions. The patient can recall none of these behaviors, although it is hard to be certain of events from decades before. What is the pathogen?

For more assessment, visit digital.wwnorton.com/michum2

13

Sterilization, Disinfection, and Antibiotic Therapy

[A Needless Death]

SCENARIO Bill, a new nurse practitioner, was part of a hospital team caring for 9-year-old Jill, a girl seriously ill from a methicillin-resistant *Staphylococcus aureus* (MRSA) infection. On this particular day, Bill was changing the girl's bandages and suddenly remembered he needed to ask Brenda, the nurse covering the desk on the floor, a quick question about when rounds would begin. He ducked out of the patient's room and walked to the nurse's desk to ask. Brenda turned to find the answer while Bill absentmindedly tapped his still-gloved fingers on the desk. Once Bill got his answer, he thanked Brenda and returned to the patient's room. Meanwhile, Anna, a physician assistant, stopped by the desk that Bill had touched to read the chart of 15-year-old Penny, a patient recovering from an appendectomy. Penny was recovering well, but 1 week later she was dead from a MRSA infection.

DIAGNOSIS MRSA causes serious life-threatening infections that are easily transmitted in a hospital environment.

FOLLOW-UP Everyone who enters an MRSA patient's room must wash their hands with antiseptic soap, put on sterile gloves, and don a paper gown to cover their clothes. The procedure must be reversed when leaving the room.

SEM · 750 nm

"The Superbug"
Methicillin-resistant *Staphylococcus aureus* (MRSA) is a highly pathogenic strain of *S. aureus* that causes nosocomial and community-acquired infections. The photomicrograph depicts MRSA cells (shown in red) destroying a human neutrophil (yellow) trying to kill bacteria.

CHAPTER OBJECTIVES

After reading this chapter, you will be able to:

- Present the modes of action, microbial spectrum, and clinical use of disinfectants and other antimicrobial agents used to treat infections.

- Discuss how a given antibiotic is deemed clinically effective.

- Describe the roles of empirical therapy, targeted drug therapy, and synergism in treating infections.

- Explain how selective toxicity and the mechanisms of antibiotic resistance influence the development of new antibiotics.

Antiseptics, disinfectants, and antibiotics are powerful tools in the fight to prevent disease. So are personal protection devices, such as gloves and gowns. However, these tools must all be used conscientiously and properly to be effective. The case just described illustrates a major failure in hospital infection control.

Bill should have removed his gloves and washed his hands with antiseptic soap before leaving the patient's room—even for a quick question. By not doing so, he ran the risk of spreading MRSA to others in the hospital. Brenda, having noticed Bill tapping his gloved hand on the desk, should have reminded him of procedure and then wiped the desk down with disinfectant before Anna had a chance to pick up the MRSA contaminating the desk. Anna then unwittingly transmitted the microbe to the appendectomy patient in the other room, with tragic results.

Infectious diseases have threatened human existence for more than 200,000 years. In fact, the short life spans of our ancestors were due mostly to deadly infections, not to the venom of a snake or the bite of a lion. Even in the eighteenth century, life expectancy was only 35–40 years, in large measure because of unstoppable infections. Fortunately, the discoveries of disinfectants more than 200 years ago and of antibiotics almost 100 years ago have played a major role in increasing average life expectancy. Helped by disinfectants, life expectancy in 1900 was 45–50 years in the United States. Antibiotics made an even larger impact. We can now expect to live, on average, 75–80 years. Of course, the increase is not due solely to antibiotics. Other factors, such as improved nutrition, more widely available health care, and fewer encounters with snakes and lions, also are important.

In this chapter we will describe how we control and (when necessary) kill microorganisms. Heat-based sterilization techniques, chemical disinfection, and antimicrobial chemotherapies are now part of our everyday routine. However, success in curbing infections is now threatened by the possibility that today's antibiotics will become useless. Decades-long abuse in how antibiotics were used has led bacteria to become antibiotic resistant. The surge of antibiotic-resistant pathogens has produced a crisis in which the human race is now vulnerable to many infectious diseases once thought conquered. One example is tuberculosis, whose dramatic resurgence over recent years has been fueled by new multidrug-resistant strains of *Mycobacterium tuberculosis*.

13.1 Concepts of Sterilization and Disinfection

SECTION OBJECTIVES

- Distinguish between sterilization, disinfection, antisepsis, and sanitation.

- Differentiate bacteriostatic agents from bactericidal agents.

- Discuss decimal reduction time and its use in sterilizing material.

Have sterilization and disinfection improved our lives? Without a doubt. What today would be a simple infected cut posed a serious risk of death in years past. A trip to the surgeon, who worked with unsterilized scalpels, was tantamount to playing Russian roulette. Improvements in sanitation, sterilization, disinfection, and antisepsis have greatly curtailed the incidence of infectious diseases and helped increase our life expectancy since 1900 by more than 30 years.

Terms Used to Describe Antimicrobial Measures

- **Sterilization** is the process by which <u>all</u> living cells, spores, and viruses on an object are destroyed.

- **Disinfection** is the killing, or removal, of <u>disease-producing</u> organisms from <u>inanimate</u> surfaces; it does not necessarily result in sterilization. Pathogens are killed, but other microbes may survive. A chemical liquid used to kill bacteria on inanimate objects is called a **disinfectant**.

- **Antisepsis** (also known as degermation) is similar to disinfection, but it applies to removing pathogens from the surface of <u>living</u> tissues, such as the skin. Antiseptic chemicals are usually not as toxic as disinfectants, which frequently damage living tissues.

- **Sanitation**, closely related to disinfection, consists of reducing the microbial population to safe levels and usually involves cleaning an object, as well as disinfection.

Antimicrobial chemicals fall into two broad classes: those that kill microbes (cidal agents) and those that inhibit or control growth (static agents). Agents that kill microbes can be subcategorized as bactericidal, algicidal, fungicidal, or virucidal, depending on what type of microbe is killed. If the agent merely inhibits growth, then it is called bacteriostatic, algistatic, fungistatic, or virustatic. For example, the antibiotic gentamicin is **bactericidal** in action, whereas chloramphenicol is **bacteriostatic**. You may have heard the term "germicidal." How does this term differ from what we just described? A **germ** is a microorganism, but especially one that causes disease. Consequently, the terms **germicide** and **germicidal** pertain to any antiseptic or disinfectant chemical that kills germs (bacterial, viral, or fungal). Note that germicidal agents do not necessarily kill spores.

Although the emphasis is placed on antimicrobials acting on pathogens, it is important to note that these agents can also kill or prevent the growth of nonpathogens. In fact, many public health standards are based on the <u>total</u> number of microorganisms found on an object, regardless of their pathogenic potential. For instance, to gain public health certification, the restaurants we frequent must demonstrate low numbers of total bacteria in their food preparation areas.

Sterilizing and Cidal Agents Kill Exponentially

How do we know when sterilization has been achieved? The challenge of sterilization is to be certain when every last microbe in a liquid or on a solid has been killed. Microbes in a culture exposed to a lethal chemical or condition (such as heat) do not all die instantaneously. Microbes die at an exponential rate, much the way they grow (considered in Chapter 6). Exponential death means that viable cell numbers decrease in equal fractions at constant intervals.

Why don't all cells die instantly when exposed to a lethal chemical or physical agent? Cells contain thousands of different proteins (and other types of targets) and thousands of molecules of each. Not all proteins in the cytoplasm, lipids in a membrane, or genes in a chromosome are damaged by an agent simultaneously. Damage accumulates over time. Only when enough molecules of an <u>essential</u> protein or gene are damaged will the cell die. Several factors influence the speed at which lethal damage accumulates:

- Initial population size (the larger the population, the longer it takes to decrease it to a specific number)

- Population composition (are spores involved?)

- Agent concentration, or dose of radiation

- Duration of exposure

- Presence of organic material (such as blood or feces) that can inhibit disinfectant action; also called "organic load"

Although the effect of agent concentration seems intuitively obvious, only over a narrow range will an increase in concentration produce an increase in death rate. Increases above a certain level might not accelerate killing at all. For example, a 70% ethyl alcohol solution is actually better than pure ethyl alcohol at killing organisms because some water is needed to help the ethanol penetrate cells, where the ethanol can dehydrate proteins.

The next question is: How do we compare the efficacy of different disinfecting or sterilizing agents? One way is to measure their decimal reduction times. **Decimal reduction time (D-value)** is the length of time it takes an agent (or condition) to kill 90% of the population, leaving 10% of the original population alive. For example, imagine that a culture of 100 million bacterial cells is heated to 100°C (**Figure 13.1**). After 1 minute, 90 million cells are killed and the number of viable cells remaining is now 10 million (10% of the original viable count). The D-value, therefore, is 1 minute. An agent (chemical or physical) with a D-value of 5 minutes takes longer to kill than an agent with a D-value of 1 minute.

You might then ask: Why, if 90% of a population is killed in 1 minute, aren't the remaining 10% killed in the next minute?

Figure 13.1 The Microbial Death Curve and the Determination of D-Values

Bacteria were exposed to a temperature of 100°C, and survivors were measured by viable count. The D-value is the time required to kill 90% of cells (that is, for the viable cell count to drop by one $\log_{10}$ unit). In this example, the D-value at 100°C (D_{100}) is approximately 1 minute. The bacteria in the test tube are colored green if they are viable and red if they are nonviable (dead). Realize that the dead bacteria remain targets for the agent, which is why the live cells in the 90%-killed tube do not die faster than those in the 100%-living tube.

The reason is that the agent maintains the same D-value (1 minute) no matter how many cells are killed. Remember, the dead cells are still there in solution; their proteins, DNA, and so on, still absorb "hits" of the chemical agent. So, although there are fewer viable cells after 1 minute of treatment, each viable cell has the same random chance of suffering a lethal hit as when the treatment began. Death rate, therefore, is an exponential function.

In practice, a sterilization procedure is designed so that the length of treatment is several D-values <u>longer</u> than the time needed, theoretically, to decrease the population to one microbe. This over-kill precaution ensures that few, if any, microbes survive.

SECTION SUMMARY

- **Sterilization** kills all living organisms, **disinfection** kills pathogens on inanimate objects, **antisepsis** is the removal of potential pathogens from the surfaces of living tissues, and **sanitation** combines disinfection with the cleaning of an object.

- **Antimicrobial compounds** can have a static (inhibitory) effect on growth or a cidal (killing) effect on viability. With respect to bacteria, agents can be bacteriostatic or bactericidal.

- **The D-value** is the time (or dose, in the case of irradiation) it takes an antimicrobial treatment to reduce the number of organisms to 10% of the original value.

Thought Question 13.1   A hand-sanitizing solution inoculated with 2×10^6 *Staphylococcus aureus* cells exhibits a D-value of 5 minutes. What will the D-value be if ten times more *S. aureus* is added?

13.2 Physical Agents That Kill Microbes

SECTION OBJECTIVES

- Compare autoclave sterilization with pasteurization.
- Explain the difference between refrigeration and lyophilization.
- Describe the uses of filtration and laminar flow biological safety cabinets.
- List conditions in which irradiation can be used to sterilize.

Can microbes be controlled without chemicals? Physical agents are often used to kill microbes or control their growth. Commonly used physical control measures are temperature extremes, pressure (usually combined with temperature), filtration, and irradiation.

High Temperature and Pressure

Even though microbes were discovered less than 400 years ago, heating of food products has been a practice for more than 5,000 years. Moist heat kills better than dry heat, thanks to the ability of water to penetrate cells. Many bacteria, for instance, easily withstand 100°C (212°F) in dry heat but not 100°C in boiling water. We humans are not so different, finding it easier to endure a temperature of 32°C (90°F) in dry Arizona than in humid Louisiana.

Boiling water (100°C) can kill most vegetative (actively growing) organisms, but spores are built to withstand this abuse, and hyperthermophiles actually prefer it (see Table 6.3). Killing spores and thermophiles usually requires combining high pressure and temperature. At high pressure, the boiling point of water rises to a temperature rarely experienced by microbes living at sea level. Even endospores quickly die under these conditions. This combination of pressure and temperature is the principle behind sterilization using the steam autoclave (Figure 13.2). Standard conditions for steam-sterilizing endospores are 121°C (250°F) at 15 psi (pounds per square inch) for 20 minutes. Time can vary, however, depending on the volume of material to be sterilized. These are also the conditions produced in pressure cookers used for home canning of vegetables.

Failure to adhere to the parameters of steam sterilization can have deadly consequences, even in your own home. Take, for instance, *Clostridium botulinum*, a spore-forming soil microbe that commonly contaminates fruits and vegetables preserved by home canning. The improper use of a pressure cooker while canning will allow clostridial spores to survive. Once the can or jar is cool, the spores will germinate and begin producing their deadly toxin while the can sits on a shelf waiting to be opened. Once ingested, the toxin makes its way to the nervous system and paralyzes the victim. Several incidents of this disease, called botulism, occur each year in the United States. (For more on botulism, see Chapters 18 and 24.)

The food industry uses several parameters to evaluate the efficiency of heat-killing. The D-value has already been described. Another measure the industry uses is called 12D, the amount of time required to kill 10^{12} spores (or reduce a population by 12 logs). Because finding 10^{12} spores in a food is highly unlikely, a 12D treatment (12 × D-value time) will produce food that is highly safe. This measurement is extremely important to the canning industry, which must ensure that canned goods do not contain spores of *C. botulinum*.

Pasteurization

Originally devised by Louis Pasteur to save products of the French wine industry from devastating bacterial spoilage, **pasteurization** today involves heating a particular food (such as milk) to a moderately high temperature long enough to kill *Coxiella burnetii*, the causative agent of Q fever, and the most heat-resistant, non-spore-forming pathogen known. The goal of pasteurization is not to sterilize milk (1 ml of pasteurized milk may still contain 20,000 bacteria); its aim is to kill pathogens without affecting the texture, color, or taste of the product. Too much heat can make milk or other foods inedible.

Many different time and temperature combinations can be used for pasteurization. The low-temperature long-time (LTLT) process involves bringing milk to a temperature of 63°C (145°F) for 30 minutes. In contrast, the high-temperature short-time (HTST) method, also called flash pasteurization, brings the milk to a temperature of 72°C (162°F) for only 15 seconds. Both processes accomplish the same thing: the destruction of *C. burnetii*, as well as *Mycobacterium tuberculosis*, *Salmonella*, and other potential pathogens.

A process known as ultra-high-temperature (UHT) treatment—134°C, or 273°F, for 1–2 seconds—can actually produce sterilized milk with an unrefrigerated shelf life of up to 6 months. This is important, especially in developing countries, where refrigeration is not always available. In the United States, milk labeled "organic" is sterilized by UHT.

Cold

Low temperatures have two basic purposes in microbiology: to temper growth and to preserve strains. Bacteria not only grow more slowly in cold but also die more slowly. Refrigeration temperatures (4°C–8°C, or 39°F–46°F) are used for food preservation because most pathogens are mesophilic: they grow best between 20°C (68°F) and 40°C (104°F) and grow poorly, if at all, at cold temperatures. One exception is the Gram-positive bacillus *Listeria monocytogenes*, which can grow reasonably well in the cold and causes disease when ingested (see Chapter 22).

Long-term storage usually requires placing bacteria in a glycerol solution at very low temperatures (–70°C, or –94°F). Glycerol prevents the production of razor-sharp ice crystals that can pierce cells from without or within, and it keeps cells from dying. Another technique, called **lyophilization**, freeze-dries microbial cultures for long-term storage. In this technique, cultures are quickly frozen at very low temperatures to prevent ice crystal formation and placed under vacuum. The vacuum causes frozen water to transition directly from solid to gas (sublimation), leaving just the cells in the form of a powder. Freeze-dried organisms stay viable for years. Viruses and mammalian cells must be kept at extremely low temperatures (–196°C, or –321°F), submerged in liquid nitrogen. Cells freeze so quickly that ice crystals do not have time to form.

Filtration

Before a liquid pharmaceutical drug is administered intravenously or orally to a patient in the hospital, the solution must be sterile. However, many drugs are sensitive to heat or chemical sterilization. These solutions can be sterilized by being passed through sterile filters having tiny pore sizes. The filters "sift" the microbes out of the fluid, much like coffee filters remove coffee grounds from your morning coffee. Filtration through micropore filters with pore sizes of 0.2 μm can remove microbial cells, but not viruses, from solutions. To remove viruses, pore sizes of 0.02 μm are necessary. Filter sterilization is also useful in standard microbiology labs (**Figure 13.3**).

Figure 13.2 The Steam Autoclave

A. Steam supplied from an external boiler is stored in the autoclave's steam jacket. The opening of a valve allows steam to enter the steam chamber and expel air. When pure steam hits a valve in the air ejector line, the valve closes and chamber pressure and temperature build to the proper levels. After the appropriate period of time, another valve opens to allow steam to escape the chamber.

B. Two autoclaves in a microbiology laboratory.

Figure 13.3 Membrane Filtration

A. Syringe apparatus and bottle top filter.

B. *Staphylococcus epidermidis* on 0.2 μm pore size filters (magnification 8,000×).

Air can also be sterilized by filtration. Infectious agents can become airborne in laboratory or hospital environments. A variety of personal protective devices—from simple surgical masks to sophisticated air purifiers—depend on filtration to remove organisms from air before inhalation. **Laminar flow biological safety cabinets** are elaborate and effective ventilated workbenches. Inside these cabinets, air is forced through high-efficiency particulate air (HEPA) filters to remove more than 99.9% of airborne particulate material 0.3 μm or larger. Biosafety cabinets are critical to protect individuals working with highly pathogenic material (see Figure 2.17A and B).

Irradiation

Public health authorities worldwide are increasingly concerned about food contaminated with pathogenic microorganisms such as *Salmonella* species, *Escherichia coli* O157:H7, *Listeria monocytogenes*, and *Yersinia enterocolitica*. Irradiation, in which food is bombarded with high-energy electromagnetic radiation, has long been a potent strategy for sterilizing food after harvesting. The food consumed by NASA astronauts, for example, has for some time been sterilized by irradiation as a safeguard against food-borne illness in space. Foods do not become radioactive when irradiated, and

any reactive molecules produced when high-energy particles are absorbed by food dissipate almost immediately (these reactive molecules are also produced by cooking, by the way).

Ultraviolet (UV) light, which causes the dimerization (linking) of pyrimidine nucleotides in DNA, is one form of sterilizing radiation. However, UV light does not penetrate liquids or solids well; it is useful only for surface sterilization. Three sources of irradiation that are more effective are gamma rays (cobalt-60), electron beams, and X-rays. X-rays and gamma rays can break the DNA backbone. The dosage of any of these treatments is usually measured in a unit called the gray (Gy), which is the amount of energy transferred to food, microbe, or other substance being irradiated. A single chest X-ray delivers roughly half a milligray (1 mGy = 0.001 Gy). To kill *Salmonella*, freshly slaughtered chicken must be irradiated at up to 4.5 kilograys (kGy), about 7 million times the energy of a single chest X-ray. Table 13.1 lists examples of foods approved for irradiation in the United States.

When microbes in food are irradiated, intracellular water and other molecules absorb the energy and generate extremely short-lived but highly reactive chemicals (oxygen radicals) that damage DNA. Unless the organism repairs this damage, it will die while trying to replicate. A microbe's sensitivity to irradiation depends on the size of its genome, the rate at which damaged DNA can be

Table 13.1

Examples of Foods Approved for Irradiation in the United States

Approval Year	Food	Dose	Purpose
1963	Wheat flour	0.2–0.5 kGy	Control mold
1964	White potatoes	0.05–0.15 kGy	Inhibit sprouting
1986	Pork	0.3–1.0 kGy	Kill trichina (*Trichinella*) parasites
1986	Fruits and vegetables	1.0 kGy	Control insects; increase shelf life
1986	Herbs and spices	30 kGy	Sterilize
1992	Poultry	1.5–3.0 kGy	Reduce number of bacterial pathogens
1997	Meat	4.5 kGy	Reduce number of bacterial pathogens
2006	Shellfish	5.5 kGy	Reduce number of *Vibrio* species and other pathogens
2010	Lettuce and spinach	4.0 kGy	Reduce number of bacterial pathogens

repaired, and other factors. Whether the irradiated food is frozen or fresh also matters because it takes a higher dose of radiation to kill microbes in frozen foods.

The size of the DNA "target" is a major factor in radiation efficacy. Parasites and insect pests, which have large amounts of DNA, are rapidly killed by extremely low doses of radiation. It takes more radiation to kill bacteria because they have less DNA per cell unit (less target per cell). It takes even more radiation to kill a bacterial spore because they contain little water, the source of most ionizing damage to DNA. Viral pathogens have the smallest amount of nucleic acid, making them resistant to irradiation doses approved for foods. Infectious prion particles associated with bovine spongiform encephalopathy (BSE, also known as mad cow disease; Section 24.6) contain no nucleic acid and are inactivated by irradiation only at extremely high doses. So, irradiation of food is effective in eliminating parasites and bacteria but is woefully inadequate for eliminating viruses or prions.

Note: Electromagnetic radiation emitted by microwave ovens does not directly kill bacteria. However, the heat generated when electromagnetic radiation excites water molecules in an organism will kill the organism if the temperature attained is high enough.

SECTION SUMMARY

- **The autoclave** uses high pressure to achieve temperatures that will sterilize objects or solutions.
- **Pasteurization** is a heating process designed to kill specific pathogens in milk and other food products.
- **Refrigeration** is used to prevent microbial growth in foods. Extreme cold (freezing) is used to preserve bacteria.
- **Lyophilization** is a method of freeze-drying that permits living cultures to remain viable for years.
- **Filtration** can remove cells from a solution and from the air, but it cannot remove the smallest viruses.
- **Irradiation** can kill pathogens in foods without damaging the food itself.

Thought Question 13.2 How would you test the killing efficacy of an autoclave?

13.3 Chemical Agents of Disinfection

SECTION OBJECTIVES

- List factors that can affect the efficacy of a disinfectant.
- Explain how disinfectants are compared.
- List the basic forms of commercial disinfectants and their various uses.
- Describe the emergence of microbial resistance to disinfectants.

How can we kill microbes on items that would be damaged by physical agents? A cruise ship sailing around Cuba in February 2019 had an emergency. Thirty-six passengers suddenly developed severe vomiting and diarrhea. These symptoms were determined later to be caused by the highly contagious norovirus. Affected passengers were quarantined to their cabins, and hand sanitizer was liberally distributed. If you follow the news, you know that cruise ships are no strangers to norovirus outbreaks. But how do you go about decontaminating an entire ship? You certainly can't use extreme heat.

Physical agents such as heat are very effective, but their use is not always practical (kitchen countertops and cruise ship surfaces, for instance, won't fit in an autoclave) or possible (skin). In these instances, chemical agents are the best approach. In the case of the cruise ship, crews cleaned the entire vessel with disinfectant three times. But what makes a good disinfectant? A number of factors influence the efficacy of a given chemical agent. These include:

- **The presence of organic matter.** A chemical placed on a dirty surface (as in a blood spill) will bind to the organic material present, lowering the agent's effectiveness against microbes. The presence of organic material must be factored into any estimate of how long to disinfect a surface or object.
- **The kinds of organisms present.** Ideally, the agent should be effective against a broad range of pathogens. Table 13.2 presents a classification of disinfectants by their spectrum of activity.
- **Corrosiveness.** The disinfectant should not corrode the surface or, in the case of an antiseptic, damage skin.
- **Stability, odor, and surface tension.** The chemical should be stable upon storage, possess a neutral or pleasant odor, and have a low surface tension so that it can penetrate cracks and crevices.

But how do you know whether you're using the most effective disinfectant? Two ways to compare the efficacy of disinfectants are the phenol coefficient (no longer used) and the use-dilution test. A third way, the disk diffusion test, will be described in Section 13.4 when we discuss antibiotics.

Measuring the Efficacy of Disinfectants

Phenol, first introduced by Joseph Lister in 1867 to reduce the incidence of surgical infections, is no longer used as a disinfectant because of its toxic effects. However, its derivatives, such as cresols and *ortho*-phenylphenol, are still in use. Some versions of the household product Lysol contain a mixture of phenolics. Phenolics are useful disinfectants because they destroy protein structure, work well in the presence of organic material, and remain active on surfaces long after application.

The **phenol coefficient test**, now of little more than historical significance, was the method once used to compare disinfectants. In this test, parallel dilutions of phenol and the agent to be tested were

Table 13.2
CDC Levels of Disinfection Based on Range of Microbes Affected

CDC Disinfection Level[a]	Vegetative Bacterial Cells	Mycobacteria	Spores	Fungi	Viruses	Examples
High	+	+	+	+	+	Ethylene oxide, glutaraldehyde, formaldehyde
Intermediate	+	+	−	+	+	Phenolics, halogens
Low	+	−	−	+	+/−	Alcohols, quaternary ammonium compounds

[a]High-level disinfection kills all organisms, except <u>high</u> levels of bacterial spores or prions. Intermediate-level disinfection kills mycobacteria, most viruses, and bacteria. Low-level disinfection kills some viruses and bacteria with a chemical germicide registered as a hospital disinfectant by the EPA.

inoculated with a fixed number of bacteria. The highest dilution (lowest concentration) of a test disinfectant that kills all the bacteria after 10 minutes but not 5 minutes (called the maximum effective dilution) was compared with the highest dilution (lowest concentration) of phenol that did the same. If the test disinfectant killed bacteria at lower concentrations than phenol could, the agent tested was considered the better disinfectant.

Today, elaborate methods outlined by the Environmental Protection Agency (EPA) evaluate the effectiveness of a disinfectant. Some methods examine efficacy on specific organisms, such as *Clostridioides difficile*. Other methods test effectiveness in different situations, like biofilms on porous or nonporous surfaces. One such test is called the use-dilution test.

The **use-dilution test** determines how well a disinfectant kills microorganisms dried onto stainless steel, a typical nonporous clinical surface, or dried onto ceramic tile, a porous surface. The test organism is air-dried onto a surface and exposed to different dilutions of a disinfectant for a fixed time (for example, 10 minutes). The entire surface is then placed in broth containing chemicals that neutralize the disinfectant. Successful disinfection results in no bacterial growth. The most effective disinfectant is the one that works at the lowest dilution.

Commercial Disinfectants

Many different types of disinfectants are available (Table 13.3). Ethanol, iodine, chlorine, and surfactants (such as detergents) are all used to reduce or eliminate microbial content from commercial products (see Appendix 3, Figure A3.1, for structures of various disinfectants). Ethanol, iodine, and chlorine are highly reactive compounds that damage proteins, lipids, and DNA. Iodine complexed with an organic carrier forms an iodophor, a compound that is water-soluble, stable, nonstaining, and capable of releasing iodine slowly to avoid skin irritation. Wescodyne and Betadine (trade names) are iodophors used, respectively, for the surgical preparation of skin and for wounds. Chlorine is a disinfectant recommended for general laboratory and hospital disinfection. Chlorine compounds are also typically used to decontaminate cruise ships.

Detergents can also be antimicrobial agents. The hydrophobic (water-hating, or fat-loving) and hydrophilic (water-loving) ends of detergent molecules will emulsify fat into water. Cationic (positively charged) detergents are useful as disinfectants because they contain positive charges that can penetrate the negatively charged bacterial cell and disrupt membranes. Anionic detergents, which are negatively charged, are not antimicrobial but do help in the mechanical removal of bacteria from surfaces.

Low-molecular-weight aldehydes (formula R-CHO) such as formaldehyde are highly reactive, combining with and inactivating proteins and nucleic acids. This trait makes them useful disinfectants too.

GAS STERILIZATION   Disposable plasticware such as Petri dishes, syringes, sutures, and catheters are not amenable to heat sterilization or liquid disinfection. These materials are best sterilized with antimicrobial gases (or with gamma irradiation). Ethylene oxide gas (EtO) is very effective. It destroys cell proteins by alkylation, is microbicidal and sporicidal, and rapidly penetrates packing materials, including plastic wraps. To carry out this form of sterilization, objects to be sterilized are placed in an ethylene oxide chamber, which resembles an autoclave (Figure 13.4), and

Figure 13.4
Ethylene Oxide Chamber

Once the gas sterilization process is complete (usually in 1–6 hours), the used EtO gas is vented from the chamber (before the door is opened) and destroyed via chemical "scrubbing" by passage through sulfuric acid solution or by passage over a catalytic oxidizer.

Table 13.3
Some Categories of Disinfectants

Disinfectant	Mode of Action	Uses	Toxicity	Comments	CDC Level of Activity[a]
Phenolics	Denatures proteins	Standard for comparison	Can cause dermatitis; vapor affects lungs	Works in the presence of organics; rarely used except as a standard for comparison	Low to intermediate
Cresol *Ortho*-phenylphenol		Disinfection			
Bisphenols (two linked phenol groups) Lysol disinfectant (o-benzyl-p-chlorophenol)	Disrupt membranes and proteins	Disinfection of solid surfaces	Very toxic if swallowed	Bacteria are developing resistance	Intermediate
Triclosan (a chlorinated bisphenol)	Inhibits fatty acid synthesis at low concentrations	Some toothpastes and skin products; FDA banned use in soaps			
Halogens Chlorine gas	Corrode tissues; destroy proteins	Disinfection of drinking water; sanitation; disinfection of pools, hospital equipment; wound antisepsis	Gas highly toxic	Sporicidal; inactivated by organics; corrosive to metals	Intermediate
Hypochlorite (bleach)	Yields toxic reaction products	Water purification; disinfect contaminated surfaces; infectious body fluids	Solution damages skin, eyes, mucous membranes		Intermediate
Iodine tinctures (mixed with alcohol or water)	Damage membranes	Topical antiseptic	Can irritate tissue; toxic if ingested	Sporicidal; corrosive to metals	Intermediate
Iodophors (iodine mixed with solubilizing agent— e.g., povidone-iodine)		Antiseptic prep for surgery; hand scrubbing; disinfection of equipment			Intermediate
Chlorhexidine (complex of chlorine and two phenol rings)	Disrupts membranes; denatures proteins	Hand scrubbing; preparation of skin for surgical incision; dental hygiene; contact lens solutions; mouthwash; cleaning of skin	Low toxicity	Inactivated by anionic compounds; avoid food or drink for 1 hour after oral use	Intermediate
Alcohols 70% ethanol Isopropanol	Cause dehydration; denature proteins; dissolve membrane lipids	Cleaning of instruments; cleaning of skin	Dries skin; toxic if ingested	Flammable; fast acting	Intermediate
Hydrogen peroxide	Hydroxyl radicals attack membrane lipids, DNA, protein	Wound cleaning; sterilization of contact lenses	Will whiten skin	Sporicidal; works in presence of organic matter	High
Quaternary ammonium compounds (cationic agents)	Disrupts membrane	Sanitation of floors, walls, blood pressure cuffs	Toxic if ingested; irritates mucous membranes	Not sporicidal	Low
Glutaraldehyde	Alkylation damages protein, DNA, and RNA • Binds membranes, inhibits transport • Cross-links proteins and DNA	Disinfection of medical equipment (e.g., endoscopes and respiratory therapy equipment)	Irritates skin, mucous membranes	Sporicidal; works in presence of organics	High
Formaldehyde	Reacts with protein, DNA, and RNA; cross-links proteins and DNA	Embalming; preservation of anatomical specimens	Pungent odor, irritating fumes; carcinogen	Sporicidal; slow action	High
Ethylene oxide gas	Alkylation modifies proteins and DNA	Sterilization of disposable plasticware	Carcinogen; damages eyes and lungs	Sporicidal; explosive	High

[a]High-level disinfection kills all organisms, except <u>high</u> levels of bacterial spores and prions. Intermediate-level disinfection kills mycobacteria, most viruses, and bacteria. Low-level disinfection kills some viruses and bacteria with a chemical germicide registered as a hospital disinfectant by the EPA.

are exposed to EtO at 700 mg/l. After 4 hours at 54°C (129°F), the object will be sterilized. Unfortunately, EtO can explode! A less hazardous gas sterilant is beta-propiolactone. It does not penetrate as well as EtO, but it decomposes after a few hours, which makes it easier to dispose of than EtO.

ANTIMICROBIAL TOUCH SURFACES AND TEXTILES   Despite widespread use of disinfectants and antibiotics by medical personnel, hospital-acquired infections remain a major concern. A promising technology that helps reduce infections in hospitals, businesses, and homes embeds copper (Cu) in the surfaces that people touch. Metallic copper releases toxic copper ions (Cu^+) that trigger the lysis of bacterial membranes within minutes, although the mechanism involved is unclear. Companies are now incorporating metallic copper into objects such as computer keyboards, handrails, door releases, hospital bed rails, and the arms of visitor's chairs.

Several antimicrobials are also being incorporated into textile fibers for wearing. The goal is to interrupt the transmission of infectious agents. Quaternary ammonium compounds, certain metals such as silver, and some antimicrobial compounds extracted from herbs and plants have been used or are being tested.

Bacterial Resistance to Disinfectants

You probably know that bacteria are becoming increasingly resistant to antibiotics used to treat infections. But can bacteria also develop resistance to chemical disinfectants that prevent infections? The answer is yes—and no. It is difficult for a bacterium to develop resistance to chemical agents that easily diffuse into a cell and have multiple targets. Iodine is one example. However, disinfectants that have multiple targets at high concentrations may have only a single target at lower concentrations—a situation that can foster the development of resistance. For instance, triclosan (a halogenated bisphenol compound previously used in many soaps and deodorants) is bactericidal at high concentrations because it targets several cell constituents. At low concentrations, however, triclosan only inhibits fatty acid synthesis and is merely bacteriostatic. Some organisms have evolved resistance to triclosan at low concentrations through mutations that altered the low-dose target protein, a fatty acid synthesis enzyme. Consequently, in 2016 the FDA banned triclosan in consumer antiseptic washes, although it is still deemed safe for some toothpastes as a way to prevent gingivitis.

Low-level resistance to disinfectants can also be achieved through membrane-spanning, multidrug efflux pumps (see Section 13.6). For instance, a multidrug efflux system of *Pseudomonas aeruginosa*, a Gram-negative bacterium that causes infections in burn and cystic fibrosis patients, can pump several different biocides, detergents, and organic solvents out of the cell, thereby reducing their efficacy. This finding has led many clinicians to advocate caution in the widespread use of certain chemical disinfectants.

Finally, biofilms can impart resistance against disinfectants. The extracellular matrix proteins and polysaccharides that bind together cells in a biofilm can also bind disinfectants, slowing their penetration to deeper recesses of the biofilm. Cells deep in the biofilm also have a stress-induced physiology better suited to resist disinfectant action. Finally, metabolic collaboration between multiple species in a mixed biofilm can be protective. For instance, enzymes from one species can shield a nonproducing species from a chemical insult, much like a big brother protecting a little brother.

SECTION SUMMARY

- **Factors that influence the efficacy of disinfectants** include the presence of organic matter; the variety of microbes present; and the agent's corrosiveness, stability, odor, and surface tension.
- **Phenolics, halogens (chlorine and iodine), alcohols, hydrogen peroxide, cationic agents, aldehydes, and gases** are general types of chemical disinfectants.
- **The phenol coefficient and use-dilution tests** are used to compare the efficacy of chemical disinfectants.
- **Ethylene oxide or beta-propiolactone** gases can be used to sterilize disposable plasticware.
- **Bacterial resistance** to some disinfectants is emerging.

13.4 Concepts of Antibiotic Therapy

SECTION OBJECTIVES

- Explain the concept of selective toxicity and its importance for clinically useful antibiotics.
- Explain what is meant by spectrum of activity, and discuss its clinical importance.
- Discuss how the effectiveness of an antibiotic is tested, and differentiate minimum inhibitory concentration (MIC) from minimum bactericidal concentration (MBC).
- Describe the chemotherapeutic index.

How do we eliminate pathogens from the body? We can't use disinfectants, which are often toxic to humans. Disinfectants lack specificity in their action and can affect animal as well as microbial cells. Instead, we use antibiotics. **Antibiotics** (as made in nature) are chemical compounds synthesized by one microbe to selectively kill other microbes. When purified and administered to patients suffering from an infectious disease, antibiotics can produce seemingly miraculous recoveries (**Figure 13.5**). Although many clinically useful antibiotics are made biologically, others are synthesized chemically. The generic term for these natural and synthetic antibiotics is **chemotherapeutic agents**. The following case history illustrates several important principles of antimicrobial therapy.

CASE HISTORY 13.1

Molly's Meningitis

Normal CSF   Infected CSF

Three-year-old Molly was brought to the emergency department crying. She had a stiff neck and high fever (40°C, or 104°F). A Gram stain of her cerebrospinal fluid (CSF), which was cloudy, revealed Gram-negative rods. When asked, Molly's parents said that their daughter had not received any of the standard child-hood vaccinations. The attending physician diagnosed Molly with meningitis and, suspecting that the Gram-negative rod *Haemophilus influenzae* was the cause, immediately prescribed intravenous ampicillin. Within hours Molly began to improve, and she was released 2 days later. Just before she left, a report from the clinical microbiology labora-tory confirmed that the organism was *H. influenzae* type b and that it was susceptible to ampicillin.

The antibiotic used to cure Molly in this case history, ampicillin, is a modified form of penicillin, the drug accidentally discovered by Alexander Fleming in 1929 (Section 1.3). Fleming's discovery marked the start of the golden age of antibiotics. However, as early as 1904 the German physician Paul Ehrlich (1854–1915) predicted that any successful antimicrobial compound would have to be a "magic bullet" that <u>selectively</u> kills or inhibits the pathogen but not the host. Ehrlich's "magic bullet" concept is now known as **selective toxicity**.

Antibiotics Exhibit Selective Toxicity

Selective toxicity is possible because key aspects of a bacterial pathogen's physiology are strikingly different from those of eukaryotes. For example, bacterial antibiotic targets include pep-tidoglycan, which eukaryotic cells lack, and ribosomes, which are structurally distinct between these two domains of life. So, chem-icals such as penicillin or ampicillin (which prevent peptidoglycan synthesis) and tetracycline (which binds to bacterial 30S ribosomal subunits) inhibit bacterial growth but are essentially invisible to host cells because they do not interact with them at low doses. In contrast to bacteria, the ribosomes and cell membranes of <u>eukary-otic</u> pathogens are very similar to those of their hosts. As a result, far fewer antibiotics can selectively target eukaryotic pathogens.

Even with a high degree of selective toxicity, some antibiotics have unintended side effects that harm the patient. Antibiotic tox-icities are unrelated to their antimicrobial modes of action. For example, chloramphenicol, a drug that targets bacterial 50S ribo-somal subunits, can interfere with the development of blood cells in bone marrow—a phenomenon that may result in aplastic anemia (failure to produce red blood cells). The toxicity of an antibiotic can also depend on the age of the patient. Ciprofloxacin, for instance,

Figure 13.5

The Power of Antibiotics

Pictures taken in 1942, shortly after the introduction of penicillin, show the improvement in a child suffering from an infection at different stages after treatment. Panels 5 and 6 show her fully recovered.

Day 4

Day 9

Recovered

can cause defects in human bone growth plates and should not be administered to pregnant women or children. Metabolism by the body can also be a problem. For instance, the liver metabolizes and detoxifies foreign chemicals, including antibiotics, and can become damaged by the toxic metabolic products. Likewise, the kidney, which excretes drugs and their metabolites, can be damaged by some antibiotics.

Problems can arise even if the drug does not directly impact mammalian physiology. For example, many people develop an extreme allergic sensitivity toward penicillin—a situation in which the treatment of an infection may end up being worse than the infec-tion itself. Physicians must be aware of these allergies and use alter-native antibiotics to avoid harming their patients. As a student of microbiology, you must distinguish between drug <u>susceptibility</u> and drug <u>sensitivity</u>. A microbe is susceptible to the drug's action, but a human can develop an allergic sensitivity to the drug.

Antimicrobial Spectrum of Activity

Because no single chemotherapeutic agent affects all microbes, antimicrobials are classified by the types of organisms they do affect. This classification defines the drug's **spectrum of activity**.

Thus, we have antifungal, antibacterial, antiprotozoal, and antiviral agents. The term "antibiotic" is usually reserved for chemotherapeutic agents that affect bacteria. Even within a group, one agent may have a very narrow spectrum of activity, meaning it affects only a few species, whereas another antibiotic may inhibit many species. For instance, penicillin has a relatively narrow spectrum of activity, killing primarily Gram-positive bacteria. However, ampicillin, the drug used in Case History 13.1, is like penicillin but with an added amino group that helps the drug penetrate the Gram-negative outer membrane. As a result of this chemical modification, ampicillin kills Gram-positive and Gram-negative organisms and is described as having a broader spectrum of activity than penicillin. There are also antimicrobials with an extremely narrow spectrum of activity. One example is isoniazid, which is clinically useful only against *Mycobacterium tuberculosis*, the microbe that causes tuberculosis.

Bacteriostatic and Bactericidal Antibiotics

Patients typically believe that all antibiotics kill their intended targets. This is a misconception. Many drugs simply prevent the growth of the organism and let the body's immune system dispatch the intruding microbe. Consequently, antimicrobials are also classified on the basis of whether they kill the microbe. As defined earlier, an antibiotic is bactericidal if it kills the target microbe; it is bacteriostatic if it merely prevents bacterial growth. The ampicillin used in Molly's case is bactericidal toward susceptible organisms.

Minimum Inhibitory Concentration

How do we compare antibiotics? The in vitro effectiveness of an antibiotic is determined by measuring how little of it is needed to stop growth—in other words, its **minimum inhibitory concentration (MIC)**, defined as the lowest concentration of the drug that will prevent the growth of an organism. But the MIC for any one drug will differ among different bacterial species. For example, the MIC of ampicillin needed to stop the growth of *Staphylococcus aureus* will be different from that needed to inhibit *Shigella dysenteriae*. The reasons for the differences include the ease with which the drug penetrates each cell and the affinity of the drug for its molecular target.

So, how do we measure MIC? As shown in **Figure 13.6**, an antibiotic is serially diluted along a row of test tubes containing nutrient broth. After dilution, the organism to be tested is inoculated at identical low population densities into each tube, and the tubes are usually incubated overnight. Growth of the organism is seen as turbidity. Note that in Figure 13.6, the tubes with the highest concentration of drug were clear, indicating no growth. The tube containing the MIC is the tube with the <u>lowest</u> concentration of drug that shows no growth. In this case, it was the 1.0-μg/ml tube. However, the MIC does not tell whether a drug is bacteriostatic or bactericidal.

MIC determinations are great for estimating a single drug's effectiveness against a bacterial pathogen isolated from a patient, but tube dilutions are not very practical when 20 or more different drugs need to be screened. Dilutions take time—time that the

Figure 13.6 Determining Minimum Inhibitory Concentration (MIC)

μg/ml

8.0   4.0   2.0   1.0   0.5   0.25   0.125   0.06

Tetracycline serially diluted

In this series of tubes, tetracycline was twofold diluted serially, starting at 8 μg/ml (tube at far left). To perform this serial twofold dilution, 1 ml of fluid from the 8-μg/ml drug–containing tube was mixed with 1 ml of media in the next tube (making 4 μg/ml); next, 1 ml of fluid from the 4-μg/ml tube was added to 1 ml in the next tube (making 2 μg/ml); and so forth. Each tube was then inoculated with an equal number of bacteria. Turbidity in a tube indicates that the antibiotic concentration was not sufficient to inhibit growth. The MIC in this example is 1.0 μg/ml.

technician, not to mention the patient, may not have. The technician can reduce processing time by using a strip test (like the Etest shown in **Figure 13.7**), which avoids the need for dilutions. The strip, containing a gradient of antibiotic, is placed over a fresh lawn of bacteria spread on an agar plate. While the bacteria are trying to grow, the drug diffuses out of the strip and into the medium. Drug diffusing from the more concentrated areas of the strip will achieve and maintain MIC farther away from the strip than will drug diffusing from the less concentrated areas of the strip. The result is an elliptical **zone of inhibition** where the antibiotic has stopped bacterial growth. The MIC is the point at which the elliptical zone intersects with the strip.

KIRBY-BAUER DISK SUSCEPTIBILITY TEST Clinical labs may process 100 or more bacterial isolates in one day, so individual MIC determinations as described earlier are not practical. Although the strip test eliminates the time and effort needed to make dilutions, it would take many plates to test 20 different antibiotics against just one bacterial isolate. However, a simplified agar diffusion test that can handle 12 antibiotics on one plate makes evaluating antibiotic susceptibility a manageable task.

Named for its inventors, the **Kirby-Bauer assay** uses a series of round filter-paper disks impregnated with different antibiotics

(Figure 13.8). A dispenser delivers up to 12 disks simultaneously to the surface of an agar plate covered by a bacterial lawn. The standardized medium used is called Mueller-Hinton agar. Each disk is marked to indicate the drug used. During incubation, the drugs diffuse away from the disks into the surrounding agar and inhibit growth of the lawn to different distances. The zones of inhibition vary in width depending on the antibiotic used, the concentration of drug in the disk, and the susceptibility of the organism to the drug. The diameter of the zone around an antibiotic disk correlates to the MIC of that antibiotic against the organism tested. The clinical laboratory in Case History 13.1 used the Kirby-Bauer test to determine the susceptibility of Molly's pathogen to various antibiotics.

The Kirby-Bauer test has been standardized to ensure reproducibility so that results from a lab in California will match those in Alabama, Ohio, or any other state. The following parameters have been standardized:

- Size of the agar plate
- Depth of the medium
- Composition of the medium
- Number of organisms spread on the agar plate
- Size of the disks
- Concentrations of antibiotics in the disks
- Incubation temperature

Figure 13.7 An MIC Strip Test

Numbers reflect the relative concentrations of antibiotic present at various points within the zone of inhibition. The concentrations along the periphery of the clear zone are equal and reflect the MIC, which in this case is 0.032 μg/ml.

The Etest (developed by AB Biodisk) is a commercially prepared strip that produces a gradient of antibiotic concentration (μg/ml) when placed on an agar plate. The MIC corresponds to the point where bacterial growth crosses the numbered strip. C/T = ceftolozane/tazobactam.

Figure 13.8 The Kirby-Bauer Disk Susceptibility Test

A. Antibiotic disk dispenser used to deliver up to 12 disks to the surface of a Mueller-Hinton plate. The device is placed over the plate, and the plunger is depressed to deliver the disks.

B. Disks impregnated with different antibiotics are placed on a freshly laid lawn of bacteria and incubated overnight. The clear zones around specific disks indicate growth inhibition. Shown are the results with normal methicillin-sensitive *Staphylococcus aureus* (MSSA).

C. Results with methicillin-resistant *S. aureus* (MRSA). Note the lack of inhibition around several of the disks, including the oxacillin disk (arrow). This strain is resistant to both methicillin and oxacillin.

C = chloramphenicol; CC = clindamycin; CZ = cefazolin; E = erythromycin; NOR = norfloxacin; OX = oxacillin; P = penicillin; RA = rifampin; SAM = sulbactam-ampicillin; SXT = sulfamethoxazole-trimethoprim; TE = tetracycline; VA = vancomycin.

Changing any one of these parameters will alter the size of the zone of inhibition and give an incorrect assessment of susceptibility.

Neither the Kirby-Bauer method nor the tube dilution techniques can tell you whether a drug is bactericidal because you don't know whether the cells within the zone of inhibition (Kirby-Bauer) or in the tubes with no growth (MIC) are dead or remain viable but just stopped growing. To determine the **minimum bactericidal concentration (MBC)** of an antibiotic, a tube dilution test is performed and then the antibiotic is removed. To remove antibiotic, any cells remaining in the no-growth tubes

Figure 13.9 Automated MIC Determination

Each column in the microtiter plate contains a concentration gradient of one antibiotic. An oxidation-reduction indicator in each well will turn red when bacteria in the well grow. The top row contains no antibiotic and serves as a growth control. The MIC of each antibiotic corresponds to the first well in each column (reading from top to bottom) that shows no growth (no red pellet). Concentration range for each drug from second well to last well: Amp (ampicillin), 2–128 µg/ml; Tet (tetracycline), 1–64 µg/ml; Chl (chloramphenicol), 2–128 µg/ml; Nal (nalidixic acid), 2–128 µg/ml; Ceph (cephalothin), 2–128 µg/ml; Str (streptomycin), 2–128 µg/ml; Gent (gentamicin), 1–64 µg/ml. Note that in this example, the MIC results for tetracycline and nalidixic acid varied between duplicates.

are first forced to the bottom of the tube by centrifugation. The antibiotic-containing medium is removed, and cells at the bottom of the tube are resuspended in fresh medium. If cells grow, they were not killed. The <u>lowest</u> concentration of drug that does <u>not</u> show growth represents the MBC.

Almost any antibiotic can be bactericidal at a high concentration (MBC) and bacteriostatic at a lower concentration (MIC). However, a drug is deemed to have a bactericidal mode of action if its MBC is no more than four times its MIC.

AUTOMATED ANTIBIOTIC SUSCEPTIBILITY INSTRUMENTS

Recent advances in computer technology have made it possible to automate MIC determinations. Multiple antibiotics can be diluted in a 96-well microtiter plate and the plate optically scanned to monitor increases in cell number (**Figure 13.9**). Results can be seen within a few hours instead of overnight—precious time when dealing with life-threatening infections.

Determining Whether an Antibiotic Is Clinically Useful

The results of an MIC determination or a disk diffusion test must be evaluated to determine whether a drug is clinically useful. A high concentration of drug may stop growth in the laboratory but not work in the patient. For an antibiotic to stop bacterial growth in the patient, the drug's concentration in tissue must remain higher than the MIC at all times.

How long a drug stays in tissue (called the drug's half-life) depends on how quickly the antibiotic is removed from the body via secretion by the kidney or destruction in the liver. Clinical usefulness of the drug also depends on whether side effects appear at concentrations needed to affect the pathogen. As shown in **Figure 13.10**, as long as the concentration of the drug in the tissue or blood remains higher than the MIC (without side effects appearing), the drug will be effective. The clinician can keep the drug's concentration at sufficient levels either by administering a higher dose, which runs the risk of side effects, or by prescribing a second dose before tissue levels from the first dose drop below the MIC. This is why patients are told to take doses of some antibiotics three times a day and other antibiotics only once a day.

Data from a disk diffusion test are compared with a table that specifies whether a zone size is wide enough (meaning the MIC is low enough) to be clinically useful. **Table 13.4** shows susceptibility data for *Staphylococcus aureus* against a half dozen antibiotics. The zone size chosen to indicate whether a pathogen is susceptible to the drug is based on knowing the average tissue level for each antibiotic that can be safely attained (average attainable tissue concentration).

Figure 13.10 Correlation between MIC and Serum or Tissue Level of an Antibiotic

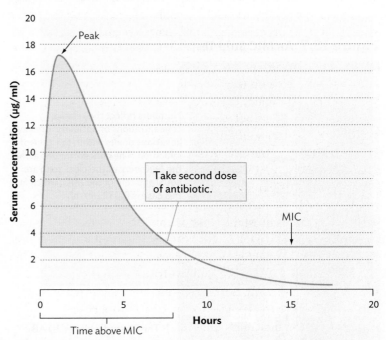

This graph illustrates the serum level of ampicillin over time. Notice how long the serum level of the antibiotic remains higher than the MIC (about 8 hours). Once the concentration falls below the MIC, owing to destruction in the liver or clearance through the kidneys and secretion, the drug fails to control the infectious agent. To maintain a serum level higher than the MIC, a second dose would be taken. The shaded area of the curve represents time above MIC and explains why the doctor asks you to take three pills of ampicillin a day, one every 8 hours.

THE CHEMOTHERAPEUTIC INDEX

One of the most important decisions a clinician must make when treating an infection is which antibiotic to prescribe. In addition to knowing whether the organism is susceptible to the antibiotic, the clinician must know whether the attainable tissue level of the antibiotic is higher than the MIC and must understand the relationship between the therapeutic dose and the toxic dose of the drug. **Therapeutic dose** is the <u>minimum</u> dose per kilogram of body weight that stops pathogen growth. **Toxic dose** is the <u>maximum</u> dose the patient can tolerate. The ratio of toxic dose to therapeutic dose is called the **chemotherapeutic index**. Ideally, the therapeutic dose is much <u>lower</u> than the toxic dose. The higher the chemotherapeutic index, the safer the drug. For example, drug A, with a toxic dose of 1 milligram per kilogram of body weight and a therapeutic dose of 0.1 milligram per kilogram of body weight, has a good chemotherapeutic index: 1 mg per kg/0.1 mg per kg = 10. But drug B, with toxic and therapeutic doses equal to 0.5 milligram per kilogram of body weight, has an unacceptable chemotherapeutic index: 1.

You should also be aware that combinations of antibiotics can be **synergistic**, meaning each drug alone may work poorly but can work well if combined (their combined effect is greater than their additive effect). A common synergistic pairing used for serious enterococcus infections of the blood (sepsis) and of the heart valve (endocarditis) is an aminoglycoside combined with vancomycin. Neither drug alone is very effective, but when combined, vancomycin (a cell wall synthesis inhibitor) weakens the enterococcus cell wall, thereby facilitating entry of the aminoglycoside into the bacterial cell. Once inside the cell, the aminoglycoside can inhibit protein synthesis.

Other antibiotic combinations can be **antagonistic**, meaning their mechanisms of action interfere with each other. Antagonism can occur when a bacteriostatic agent is combined with a bactericidal agent. For example, penicillin is bactericidal; it kills growing cells that are actively making peptidoglycan but has little effect on nongrowing cells. Macrolides, in contrast, are bacteriostatic and merely prevent growth. So, when a macrolide and penicillin are combined, the macrolide's mechanism of action—stopping growth—is antagonistic to penicillin's activity, which relies on growing cells.

SECTION SUMMARY

- **Antimicrobial agents** are produced naturally by microorganisms and synthetically in laboratories.
- **Selective toxicity** refers to the ability of an antibiotic to attack a unique component of microbial physiology that is missing or distinctly different from eukaryotic physiology.

Table 13.4
Susceptibility Results for *Staphylococcus aureus*

Antibiotic	Quantity in Disk (µg)	Zone of Inhibition Diameter (mm)[a]		
		Resistant	Intermediate	Susceptible
Ampicillin	10	<12	12–13	>13
Chloramphenicol	30	<13	13–17	>17
Erythromycin	15	<14	14–17	>17
Gentamicin	10	≤12.5	—	>12.5
Streptomycin	10	<12	12–14	>14
Tetracycline	30	<15	15–18	>18

[a]The zone of inhibition is the diameter of no growth around an antibiotic disk on an agar plate.

- **Antibiotic side effects** on mammalian physiology can limit the clinical usefulness of an agent.
- **Antibiotic spectrum of activity** refers to the range of microbes that a given drug affects.
- **Bactericidal** antibiotics kill microbes; **bacteriostatic** antibiotics inhibit their growth.
- **Minimum inhibitory concentration (MIC)** of a drug is the minimum concentration that will inhibit growth. **Minimum bactericidal concentration (MBC)** is the minimum concentration that will kill the target cell. When correlated with average attainable tissue levels of the antibiotic, the MIC can predict the effectiveness of an antibiotic in treating disease.
- **MIC is measured** by serial dilution techniques but can be approximated with the Kirby-Bauer disk diffusion technique.
- **Chemotherapeutic index** is the ratio of the toxic dose of an antibiotic to its therapeutic dose.

Thought Question 13.3 During a laboratory exercise you must determine the MIC of the protein synthesis inhibitor tobramycin against *Listeria monocytogenes*, a food-borne pathogen. The initial concentration of tobramycin in a tube of broth is 1,000 µg/ml. This tube is labeled "tube 1." You make serial twofold dilutions from this tube into nine other broth tubes. All ten tubes are inoculated, and 24 hours later turbidity is observed in tubes 6–10. What is the MIC?

Thought Question 13.4 You are testing whether a new antibiotic will be a good treatment choice for a patient with a staph infection. The Kirby-Bauer test using the organism from the patient shows a 15-mm zone of inhibition around the disk containing this drug. Clearly, the organism is susceptible. But you conclude from other studies that the drug will not be effective in the patient. Provide a potential explanation for this conclusion.

13.5 Antibacterial Agents

SECTION OBJECTIVES

- Classify different antibiotics by their molecular targets.
- Describe the process of cell wall synthesis, and indicate the targets for various cell wall antibiotics.
- Distinguish transcription inhibitors from translation inhibitors.

How do antibiotics work? As noted earlier, antibiotics exhibit selective toxicity because they disturb enzymes or structures unique to the bacterial target cell. Most antibiotics affect one of the following aspects of a microbe's physiology:

- Cell wall synthesis
- Cell membrane integrity
- DNA synthesis
- RNA synthesis
- Protein synthesis
- Metabolism

Chapters 5, 7, and 8 describe the cell structures and processes targeted by antibiotics. The following sections describe the salient features of these drugs and their modes of action. Summaries are provided in Table 13.5 and Figure 13.11.

Antibiotics That Target the Cell Wall

Bacterial cell walls are an obvious candidate for selective toxicity because their main component, peptidoglycan, does not exist in mammalian cells; thus, antibiotics that target peptidoglycan synthesis should selectively kill bacteria. The antibiotic used in Case History 13.1, amoxicillin, killed the bacterial pathogen by targeting cell wall (peptidoglycan) synthesis. Figure 13.12A outlines peptidoglycan synthesis. Initially, sugar molecules called *N*-acetylglucosamine (NAG) and *N*-acetylmuramic acid (NAM) are made by the cell and linked together into long peptidoglycan chains assembled at the cell wall (see Section 5.3). Disaccharide units of NAG and NAM with a short peptide side chain are made in the cytoplasm and then shuttled across the membrane by a lipid carrier molecule (Figure 13.12A, steps 1–4). The disaccharide units are

Table 13.5

Targets of Antibacterial Agents

Target	Antibiotic Example(s)
Cell wall synthesis	Penicillins, cephalosporins, bacitracin, vancomycin
Cell membrane integrity	Polymyxins, gramicidin (antibacterial) Amphotericin, imidazoles (antifungal)
Mycolic acid synthesis	Isoniazid (*Mycobacterium tuberculosis*)
DNA replication and integrity	Quinolones, metronidazole, nitrofurantoin, some antiviral compounds
Transcription	Rifampin, fidaxomicin (*Clostridioides difficile*)
Protein synthesis	Chloramphenicol, tetracyclines, aminoglycosides, macrolides, lincosamides
Intermediary metabolism	Sulfonamides, trimethoprim (inhibit folic acid synthesis) Bedaquiline (inhibits ATP synthase in *M. tuberculosis*) Platensimycin (inhibits bacterial fatty acid synthesis)

Figure 13.11 Bacterial Targets of Antimicrobial Drugs

Depicted here are six broad categories of bacterial physiology that enable selective toxicity. The general classes of antibiotics that affect these physiological processes are identified.

Metabolic Inhibitors
Sulfonamides
Trimethoprim
Metronidazole

DNA Replication Inhibitors
Quinolones

Substrate
Active site
Enzyme
DNA
New DNA
RNA polymerase
DNA polymerase
Unwinding
DNA Gyrase
RNA
New DNA
Ribosome

RNA Polymerase Inhibitors
Rifampin
Myxopyronins

Protein Synthesis Inhibitors
50S ribosome subunit
Chloramphenicol
Macrolides (erythromycin)
Clindamycin
Oxazolidinones (linezolid)
Streptogramins
(quinupristin-dalfopristin)

30S ribosome subunit
Aminoglycosides (gentamicin)
Tetracyclines (doxycycline)

Cell Membrane Damage
Polymyxins

Peptidoglycan Synthesis

Cell Wall Inhibitors
Penicillins
Cephalosporins
Vancomycin
Bacitracin
Monobactams

Figure 13.12 Bacterial Cell Wall Synthesis Is a Target of Certain Antibiotics

A. Cell wall (peptidoglycan) synthesis. NAG and NAM are made and joined as units in the cytoplasm (steps 1–3). Each unit is then transported across the cell membrane by a specialized lipid carrier molecule called bactoprenol (step 4). Enzymes then link the new unit to a preexisting chain of peptidoglycan (step 5) and catalyze a cross-link between peptide chains sticking out from NAM molecules on parallel chains (step 6). Penicillin, vancomycin, and cephalosporins inhibit the activity of enzymes that cross-link the peptide side chains.

B. The cross-linking between strands shown in panel A provides the cagelike structure shown here. Antibiotics that inhibit cross-linking and chain elongation compromise the integrity of the cage.

assembled outside the cytoplasmic membrane to form long strands of peptidoglycan (step 5). Adjacent strands are then "snapped" together by cross-linking the short peptide side chains that spring from individual NAM molecules (step 6). Cross-linking produces a rigid, cagelike macromolecular structure essential for maintaining cell shape (**Figure 13.12B**). Figure 13.12A also indicates where several antibiotics target various stages of this assembly process.

BETA-LACTAM ANTIBIOTICS The enzymes that attach the disaccharide units to preexisting peptidoglycan and produce the peptide cross-links are collectively called **penicillin-binding proteins (PBPs)** because penicillin and related antibiotics bind to them. Penicillin is a naturally produced antibiotic that contains a critical beta-lactam ring structure that chemically resembles a piece of the peptidoglycan peptide side chain (**Figure 13.13A and B**). The structure of penicillin G is shown in **Figure 13.13C**. Because penicillin looks like the peptidoglycan side chain, the drug will bind to enzymes that mediate peptide crosslinking and inhibit their activity. The lack of stabilizing peptide crosslinks is a disaster for bacteria that are "trying" to grow larger before they divide (**Figure 13.14**). An intact cell wall acts as a rigid "exoskeleton" that restrains the hydrostatic pressure from within the cell. Penicillin's action compromises the cell wall, so a growing cell will eventually burst from internal pressure, like a child's balloon with too much air. Penicillin, then, is a bactericidal drug.

Penicillin, though, has only a narrow spectrum of activity. It kills Gram-positive organisms better than Gram-negative bacteria because the Gram-negative outer membrane blocks penicillin's entry. Pharmacologists have overcome this limitation by adding chemical groups to the basic ring structure of penicillin that can change the antimicrobial spectrum, stability, and/or intestinal absorption of the derivative penicillin (see Appendix 3, Figure A3.2). For instance, ampicillin, the drug used to treat Molly in Case History 13.1, is a modified version of penicillin that more easily penetrates the outer membrane. Ampicillin, then, works on both groups of bacteria, giving it a broader spectrum of activity than penicillin has.

Figure 13.13 The Structure of Penicillins

D-Alanine-D-alanine

A. The D-alanine-D-alanine end of the normal peptidoglycan peptide side chain (shown here) resembles the beta-lactam ring of all penicillin drugs.

Basic structure

B. The beta-lactam core structure of all penicillins.

Penicillin G

C. Penicillin G includes the basic structure shown in B with the R group shown here; see Appendix 3, Figure A3.2.

Figure 13.14 Effect of a Penicillin-like Antibiotic on *E. coli*

◀— Cell length
◀— Bulge depth

Time 71 min | Time 72 min 25 s | Time 72 min 26 s

Bulge forms

Time 0 | Time 70 min | Time 70 min

Lysis

LM   2 μm

Time-lapse light microscopy follows the effect of the antibiotic cephalexin over 73 minutes of growth. Arrows indicate directions of cell growth (red) and bulge expansion (blue). Cephalexin weakens the cell wall of growing cells and causes a bulge to form. The bulge enlarges until the cell lyses.

Cephalosporins, such as the cephalexin used in Figure 13.14, are another type of beta-lactam antibiotic originally discovered in nature but modified in the laboratory (a type of **semisynthetic antimicrobial**) to fight microbes that are naturally resistant to penicillins (especially *Pseudomonas aeruginosa*). Over the years, pharmaceutical chemists have modified the basic structure of cephalosporin in ways that improve the drug's effectiveness against penicillin-resistant pathogens. Each modification is increasingly complex and produces what is referred to as a new "generation" of cephalosporins. There are currently five generations of this semisynthetic antibiotic (**Figure 13.15**). Ceftriaxone, for example, is a widely used third-generation cephalosporin.

OTHER ANTIBIOTICS THAT TARGET PEPTIDOGLYCAN SYNTHESIS Bacitracin, produced by *Bacillus subtilis* and *Bacillus licheniformis*, is a large polypeptide molecule that binds to the bactoprenol lipid carrier molecule that normally transports the disaccharide units of peptidoglycan across the cell membrane to the growing chain (see Figure 13.12, steps 4 and 7). Resistance to bacitracin can develop if the organism has an efficient drug export system that pumps bacitracin out of the cell (discussed in Section 13.6). Normally, bacitracin is used in combination with

Figure 13.15 Cephalosporin Generations

Representative examples of cephalosporins. With each successive generation, the side groups become more complex. Yellow-highlighted areas indicate the core structure of each of the cephalosporins, with beta-lactam rings.

Cephalexin

(1st generation)

A. First generation: cephalexin (Keflex).

Cefoxitin

(2nd generation)

B. Second generation: cefoxitin.

Ceftriaxone

(3rd generation)

C. Third generation: ceftriaxone.

Cefepime

(4th generation)

D. Fourth generation: cefepime.

Ceftaroline

(5th generation)

E. Fifth generation: ceftaroline.

polymyxin and neomycin in the over-the-counter cream Neosporin. Neosporin is applied only topically because, if ingested, bacitracin can have serious side effects, including kidney damage.

Cycloserine, made by *Streptomyces garyphalus*, is one of several antimicrobials used to treat tuberculosis (see Appendix 3, Figure A3.3A). Cycloserine inhibits two enzymes that make

Figure 13.16 Penicillin Resistance

The enzyme beta-lactamase, produced by bacteria carrying the beta-lactamase gene, will destroy the critical beta-lactam ring and confer penicillin resistance. Some derivatives of penicillin (for example, oxacillin) block beta-lactamase action and will kill the bacteria.

β-Lactam ring

β-Lactamase

β-Lactamase breaks a bond in the β-lactam ring of penicillin to disable the molecule. Bacteria with this enzyme can resist the effects of penicillin and some other β-lactam antibiotics.

Penicillin

Penicilloic acid

part of the peptide side chain on NAM (see Figure 13.12, step 1). As a result, the incomplete pentapeptide side chains prevent cross-linking and compromise peptidoglycan integrity.

Vancomycin is a very large and complex glycopeptide produced by *Amycolatopsis orientalis* (see Appendix 3, Figure A3.3B). The drug binds to part of the peptide end of a newly exported NAM-NAG disaccharide and blocks the cross-linking of the peptide side chains of peptidoglycan (see Figure 13.12, step 6). The microbial mechanisms of resistance to vancomycin and penicillin are very different, so vancomycin is particularly useful against penicillin-resistant bacteria. Potential side effects of vancomycin include kidney damage (nephrotoxicity) and hearing loss (ototoxicity), although these are very rare.

It is important to note that antibiotics targeting cell wall biosynthesis generally kill only growing cells. These drugs do not affect static, or stationary-phase, cells because in this state the cell has no need for new peptidoglycan.

MICROBIAL RESISTANCE TO CELL WALL-INHIBITING ANTIBIOTICS Antibiotics were heralded as the "magic bullets" that Paul Ehrlich predicted in 1904. Unfortunately, we have learned repeatedly over the past 80 years that the microbial world does not sit idly by and politely die. Microbes constantly evolve to become resistant to new antibiotics. Bacterial resistance to penicillin was first recognized in the 1940s as the cause of occasional treatment failures. Today, resistance to many different antibiotics is a growing concern throughout the world.

Bacteria develop resistance to penicillin in two basic ways. The first is by inheriting a gene encoding one of the beta-lactamase enzymes, which cleave the critical beta-lactam ring structure of this class of antibiotics (**Figure 13.16**). The beta-lactamase enzyme is transported out of the cell and into the surrounding medium (for Gram-positives) or the periplasm (for Gram-negatives), where it will destroy penicillin before the drug even reaches the cell. Bacteria that produce beta-lactamase are still susceptible to certain modified penicillins and cephalosporins that are no longer susceptible to the enzyme. Methicillin, for example, works well against most

beta-lactamase-producing microbes. However, a new enzyme, called the New Delhi metallo-beta-lactamase, found predominantly in *Escherichia coli* and *Klebsiella pneumoniae*, destroys the beta-lactam ring of all the penicillins, carbapenems, and cephalosporins.

Note: Because of stability issues, methicillin is no longer used to treat infections and has been replaced by the more stable oxacillin. Organisms resistant to methicillin are also resistant to oxacillin. Nevertheless, the term "methicillin-resistant *Staphylococcus aureus*" (MRSA) is still used.

Bacteria develop the second form of penicillin resistance when a key penicillin-binding protein is altered by mutation. The altered protein can still synthesize cell wall peptidoglycan but will no longer bind to the antibiotic. Most methicillin-resistant bacteria use this strategy. As described in the opening case of this chapter, hospitals take a special interest in methicillin-resistant *S. aureus* (MRSA) because very few drugs can kill it. One of the few remaining antibiotics effective against MRSA is vancomycin. Unfortunately, resistance to this drug is also developing.

Drugs That Target the Bacterial Membrane

Poking holes in a bacterial cytoplasmic membrane is another effective way to kill bacteria. A group of peptide antibiotics uses this strategy. **Gramicidin**, produced by *Brevibacillus brevis*, is a cyclic peptide composed of 15 alternating D- and L-amino acids. It inserts into the membrane as a dimer, forming a cation channel (a kind of hole) that enables positive ions to freely cross the membrane and disrupt membrane polarity—similar to short-circuiting a battery (**Figure 13.17**).

Polymyxin B (from *Paenibacillus polymyxa*)—a drug belonging to another group of polypeptide antibiotics, the **polymyxins**—has a positively charged polypeptide ring that binds to the outer

Figure 13.17 Gramicidin Is a Peptide Antibiotic That Affects Membrane Integrity

Gramicidin forms a cation channel across cell membranes through which hydrogen, sodium, and potassium ions can freely pass.

Hydrophilic pore

Outside of cell

Hydrophobic side chains

K⁺

Cytosol

and inner membranes of bacteria, both of which are negatively charged. Polymyxin B dissolves the inner membrane by disrupting phospholipid interactions, much like a detergent. This antibiotic is used only topically to treat or prevent infection. It is never ingested because it also damages human cell membranes. However, polymyxin B has been fused to some bandage materials used to treat burn patients, who are particularly susceptible to Gram-negative infections (such as *Pseudomonas aeruginosa*).

Besides destroying membrane integrity, do any antibiotics <u>prevent</u> membrane synthesis? A new class of antibiotic does just that. The drug platensimycin, made by *Streptomyces platensis* isolated from South African soil, binds to a bacterial enzyme that synthesizes fatty acids, a component of cell membrane. The drug exhibits broad-spectrum bacteriostatic activity, acting on Gram-positive and Gram-negative bacteria. It is one of only four entirely new antibiotics developed in the last four decades. Unfortunately, its rapid clearance from the body has prevented it from being used clinically.

Drugs That Affect DNA Replication and Integrity

Bacteria generally make and maintain their DNA by using enzymes that closely resemble those of mammals. As a result, you might think it impossible to selectively target bacterial DNA synthesis. Surprisingly, there are a few unique features of bacterial DNA synthesis that make it a target for antibiotics.

CASE HISTORY 13.2
When Time Is Critical

Lauren, a 23-year-old woman from Dallas, Texas, arrived at the emergency department by ambulance with fever, chills, and severe muscle aches. She developed a nonproductive cough, had difficulty breathing, felt pleuritic chest pain (sharp pain when inhaling), and became hypotensive (low blood pressure). An X-ray showed a lower-lobe infiltrate in the lungs. She was initially treated with ceftriaxone, but her condition worsened. Two days later, the clinical laboratory reported the presence of the Gram-negative anaerobe *Fusobacterium necrophorum* in blood cultures. The patient was diagnosed with pneumonia and treated with metronidazole, a DNA-damaging agent specific for anaerobes. She recovered fully.

This case illustrates a common approach to infectious diseases by clinicians. A critically ill patient clearly suffers from an infectious disease and needs immediate treatment. But because the identity of the organism is not yet known, the clinician will use an antimicrobial that has a broad spectrum of activity, hoping it's broad enough to "cover" the usual pathogens that could cause the disease (in Lauren's case, pneumonia coupled with sepsis). The approach is called **empirical therapy**. However, in Lauren's case

Figure 13.18 **Activation of Metronidazole**

A. Metronidazole enters the anaerobe and is reduced (receives single-electron transfers) by anaerobic metabolism. The resulting free-radical form can attack DNA in the organism.

B. *Trichomonas vaginalis*, a protozoan that causes vaginal infections. Metronidazole is very effective at treating *Trichomonas* infections.

the empirical treatment, ceftriaxone, which is very effective against Gram-positive and Gram-negative bacteria, did not cover the anaerobe, so metronidazole was added to the treatment regimen.

METRONIDAZOLE Also known by the trade name Flagyl, metronidazole is an example of a prodrug, a drug that is harmless until activated by reduction. Metronidazole is activated after being metabolized by microbial protein cofactors (ferredoxin) found in anaerobic or microaerophilic bacteria such as *Bacteroides* and *Fusobacterium* (**Figure 13.18A**). Aerobic microbes are resistant because they lack the electron transport proteins to reduce metronidazole. Once activated in anaerobes, the compound begins nicking DNA at random, thus killing the cell. Because the etiologic agent in Case History 13.2 was an anaerobe, metronidazole was an effective therapy. Metronidazole is also effective against protozoa such as *Giardia*, *Entamoeba*, and *Trichomonas*, where it is activated in special organelles (**Figure 13.18B**). Human cells also lack the anaerobiosis-associated cofactors and are thus unharmed by the drug.

QUINOLONES Recall that during replication, the unwinding of DNA at a replication fork introduces positive supercoils ahead of the moving fork (Section 8.2). If these positive supercoils are not removed, the torsional stress of too many twists builds, and DNA polymerase cannot move forward. Replication halts, and the cell dies. To reverse the positive supercoils, the enzyme DNA gyrase introduces negative supercoils ahead of the fork (see Figure 8.11). However, a group of antibiotics called quinolones can inactivate DNA gyrase. Without DNA gyrase, positive supercoils accumulate and the pathogen cannot multiply. Because bacterial DNA gyrases are structurally distinct from their mammalian counterparts, quinolone antibiotics will not affect mammalian DNA replication. Various chemical modifications, such as the addition of fluorine and amine groups, have increased the antimicrobial spectrum and bloodstream half-life of these drugs. Ciprofloxacin and levofloxacin are quinolones commonly used to treat a variety of infections (Table 13.6).

Table 13.6
Clinical Uses, Possible Side Effects, and Modes of Action for Antimicrobials

Antimicrobial	Clinical Use	Possible Side Effects	Mode of Action
Aminoglycosides • Gentamicin • Kanamycin • Tobramycin • Amikacin • Streptomycin	Broad spectrum of activity but used mainly for infections caused by Gram-negative bacteria, such as *Escherichia coli* and *Klebsiella*, and particularly *Pseudomonas aeruginosa*; effective against aerobic bacteria (not obligate/facultative anaerobes) and tularemia	Hearing loss, vertigo, kidney damage	Bactericidal; binds to the bacterial 30S ribosomal subunit, inhibits the translocation of peptidyl-tRNA, and causes misreading of mRNA
Carbapenems • Imipenem • Meropenem	Broad spectrum; bactericidal for both Gram-positive and Gram-negative organisms and therefore useful for empirical broad-spectrum antibacterial coverage (*Note:* MRSA resistance to this class)	Gastrointestinal upset, diarrhea, nausea, seizures, headache, rash and allergic reactions	Bactericidal; inhibits cell wall synthesis
Cephalosporins, 2nd generation • Cefoxitin	Extended spectrum; good Gram-positive coverage	Gastrointestinal upset, diarrhea, nausea (if alcohol taken concurrently), allergic reactions	Bactericidal; inhibits cell wall synthesis
Cephalosporins, 3rd generation • Cefixime • Ceftazidime • Ceftriaxone	Extended spectrum; improved Gram-negative coverage		
Cephalosporins, 4th generation • Cefepime	Extended spectrum; covers *Pseudomonas*		
Cephalosporins, 5th generation • Ceftobiprole	Extended spectrum; used for MRSA		
Chloramphenicol	Broad spectrum; *Staphylococcus, Neisseria meningitidis, Streptococcus pneumoniae, Haemophilus influenzae*	Aplastic anemia (decreases in red blood cells, white blood cells, and platelets)	Bacteriostatic; binds to ribosome and inhibits protein synthesis
Glycopeptides • Vancomycin • Teicoplanin	Narrow spectrum; Gram-positive coverage: *Staphylococcus* (MRSA), *Streptococcus, Enterococcus, Clostridium*	Kidney damage, tinnitus (ear ringing), diarrhea, nausea	Bactericidal; inhibits cell wall synthesis
Glycylcyclines • Tigecycline	Broad spectrum; intravenous injection for complicated infections; *Enterococcus faecalis*, MRSA, others	Similar to tetracycline; precautions for pregnant women and children	Bacteriostatic; protein synthesis inhibitor
Lincosamides • Clindamycin • Lincomycin	Broad spectrum; used for serious staphylococcal, pneumococcal, and streptococcal infections in penicillin-allergic patients; also used for anaerobic infections; used topically for acne	Possible pseudomembranous enterocolitis	Bacteriostatic; inhibits protein synthesis

(continued on next page)

Table 13.6

Clinical Uses, Possible Side Effects, and Modes of Action for Antimicrobials (*continued*)

Antimicrobial	Clinical Use	Possible Side Effects	Mode of Action
Lipopeptides • Daptomycin	Narrow spectrum; Gram-positive microbes	Muscle weakness, eosinophilic pneumonia, diarrhea	Bactericidal; binds to membranes, causes leaks and loss of membrane potential (energy drain)
Macrolides • Azithromycin • Clarithromycin • Erythromycin	Narrow spectrum; streptococcal infections, syphilis, respiratory tract infections, mycobacterial infections, Lyme disease, pertussis, chlamydia	Nausea, vomiting, diarrhea, jaundice	Bacteriostatic; binds to ribosome, inhibits protein synthesis
Monobactams • Aztreonam	Narrow spectrum; intravenous; Gram-negative bacteria, including *Pseudomonas*	Rash, diarrhea, vomiting	Bactericidal; inhibits cell wall synthesis
Oxazolidinones • Linezolid	Narrow spectrum; Gram-positive bacteria, including vancomycin-resistant enterococci, methicillin-resistant *Staphylococcus aureus*	Long-term use promotes bone marrow suppression and low platelet counts	Bacteriostatic; inhibits protein synthesis
Penicillins • Penicillin G • Amoxicillin • Cloxacillin • Oxacillin	Varies from narrow spectrum to broad; *Streptococcus*, *Treponema* (syphilis), *Borrelia* (Lyme disease)	Gastrointestinal upset, allergy	Bactericidal; inhibits cell wall synthesis
Polypeptides • Bacitracin • Polymyxin B • Gramicidin	Topical infections; bacitracin (Gram-positive), polymyxin B (Gram-negative), gramicidin (Gram-positive and Gram-negative)	Kidney damage, nerve damage	Bactericidal; bacitracin (cell wall); polymyxin B and gramicidin (membrane integrity)
Quinolones • Ciprofloxacin • Levofloxacin	Broad spectrum; Gram-positive and Gram-negative coverage; used for urinary tract, sexually transmitted, gastrointestinal, respiratory tract, and bone infections	Inhibition of bone growth in children, tendonitis in older patients, seizures (rare)	Bactericidal; inhibits topoisomerases (DNA gyrase and Topo IV)
Rifampin	Broad spectrum; Gram-positives, mycobacteria (tuberculosis)	Reddish orange body fluids	Bactericidal; binds RNA polymerase, prevents transcription
Streptogramins • Quinupristin-dalfopristin	Narrow spectrum; vancomycin-resistant *Staphylococcus aureus* (VRSA) and *Enterococcus* (VRE)	Nausea, diarrhea, headache	In combination, bactericidal; binds ribosome, inhibits protein synthesis
Sulfonamides • Trimethoprim-sulfamethoxazole	Narrow spectrum; mainly Gram-negatives; urinary tract infections	Nausea, allergy, kidney failure, sensitivity to light	Bacteriostatic; inhibits folic acid synthesis
Tetracyclines • Doxycycline • Minocycline	Broad spectrum; syphilis, chlamydia, Lyme disease, mycoplasmas, rickettsias	Gastrointestinal upset, sensitivity to light, depression of bone growth, staining of teeth	Bacteriostatic; binds ribosome, inhibits protein synthesis

RNA Synthesis Inhibitors

To grow and cause disease, bacterial pathogens must transcribe and translate their genes. If we can stop either of those processes, bacteria will not grow and the host's immune system can kill them. Many clinically useful antibiotics can stop protein synthesis, but only a few can interfere with transcription. One example is rifampin, the best-known member of the rifamycin family of antibiotics (see Appendix 3, Figure A3.4). Rifampin (also called rifampicin) selectively binds to bacterial RNA polymerase and prevents newly made mRNA from exiting the enzyme (see Figure 8.13).

Figure 13.19
Rifamycin in Urine

Rifampin is excreted in body fluids, turning them reddish orange. Shown is a urine bag from a hospital patient undergoing rifampin treatment for a cerebral abscess.

Consequently, the polymerase stops in its tracks. Rifampin is often used to treat tuberculosis or meningococcal meningitis. Curiously, because rifampin is reddish orange, it turns bodily secretions, including breast milk, orange (**Figure 13.19**). The astute physician will warn the patient of this highly visible but harmless side effect to avoid unnecessary anxiety when the patient's urine changes color.

Protein Synthesis Inhibitors

Protein synthesis was initially described in Chapter 8. The molecular complexity of translation provides numerous sites that can be attacked by antibiotics. Fortunately, the structural differences between prokaryotic and eukaryotic ribosomes enable antibiotics to be selectively toxic, affecting only bacterial protein synthesis.

Inhibitors of protein synthesis can be classified into several groups according to their structures (see Appendix 3, Figure A3.5) and mechanisms of action (**Figure 13.20**). Note that most protein synthesis inhibitors are, by and large, bacteriostatic. These antibiotics generally work by binding to and interfering with the function of bacterial rRNA, which differs from eukaryotic rRNA. Features of these antibiotics are discussed in the next section.

CASE HISTORY 13.3
Treatment Meets Allergy

Sixteen-year-old Jamal arrived at the emergency department after 2 days of fever, malaise, chills, and neck stiffness. His most notable symptom was a painful, red, rapidly spreading rash covering the right side of his face. Throat cultures taken on admission revealed group A *Streptococcus pyogenes*, indicating that the rash was a case of erysipelas (described in Chapter 19). Ordinarily penicillin would be the drug of choice because *S. pyogenes* is very susceptible to it and no penicillin-resistant *S. pyogenes* has ever been seen clinically. However, Jamal was known to be allergic to this antibiotic. So instead, he was given azithromycin.

When a patient is known to be allergic to the usual drug of choice, a structurally distinct, but still effective, drug is used. Often, that drug will be one that inhibits protein synthesis—in this case, the macrolide azithromycin. Drugs that inhibit protein synthesis can be subdivided into different groups according to their structures and the part of the translation machine that is targeted. Ribosomes have two major subunits (see Section 8.3). One class of antibiotics interferes with the 30S small subunit, and the other targets the 50S large subunit.

AMINOGLYCOSIDES   Aminoglycosides include streptomycin, gentamicin (see Appendix 3, Figure A3.5A), and tobramycin. These broad-spectrum antibiotics are effective against Gram-positive and Gram-negative bacteria. The aminoglycosides bind to the 30S ribosomal subunit to cause misreading of the mRNA and perhaps inhibit the movement of peptidyl-tRNA from the A site to the P site. Unlike most protein synthesis inhibitors, aminoglycosides are bactericidal, but how they kill is not clear (misreading of mRNA could jumble critical proteins or generate reactive oxygen species). The potential toxic effects of these antibiotics include damage to the auditory nerve (hearing loss, ototoxicity) and

Figure 13.20 Modes of Action of Common, Clinically Useful Antibiotics That Inhibit Bacterial Protein Synthesis

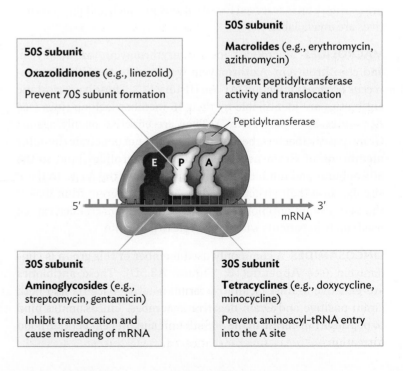

50S subunit

Oxazolidinones (e.g., linezolid)

Prevent 70S subunit formation

50S subunit

Macrolides (e.g., erythromycin, azithromycin)

Prevent peptidyltransferase activity and translocation

Peptidyltransferase

E   P   A

5′ —— 3′
mRNA

30S subunit

Aminoglycosides (e.g., streptomycin, gentamicin)

Inhibit translocation and cause misreading of mRNA

30S subunit

Tetracyclines (e.g., doxycycline, minocycline)

Prevent aminoacyl-tRNA entry into the A site

damage to kidney function. Approximately 0.5%–3% of patients treated with gentamicin suffer from ototoxicity.

TETRACYCLINES   This group includes the commonly used drug doxycycline. Tetracyclines are broad-spectrum, bacteriostatic antibiotics that feature four fused cyclic rings—hence the name (see Appendix 3, Figure A3.5B). These antibiotics bind to the 30S subunit and prevent tRNAs from entering the A site. Doxycycline is used to treat early stages of Lyme disease (*Borrelia burgdorferi*), acne (*Cutibacterium acnes*; formerly *Propionibacterium acnes*), and other infections. Tetracyclines are not recommended for children under 8 years old because the drugs can be incorporated into developing teeth, giving them a yellowish brown discoloration. Tetracyclines are also phototoxic, causing an increased tendency for sunburn.

GLYCYLCYCLINES   Tigecycline is a commonly used member of this class. It is a broad-spectrum, bacteriostatic drug that looks and acts like tetracycline. It prevents protein synthesis by binding to the 30S ribosomal subunit and inhibits the entry of aminoacyl-tRNA into the A site. Its major advantage over tetracycline is that it can function in tetracycline-resistant cells. Tetracycline resistance in pathogens is due mostly to the presence of specific efflux pumps in the bacterial membrane that remove the drug from the cell (see Section 13.6). Tigecycline, however, is <u>not</u> readily pumped out of the cell by those pumps. A disadvantage of tigecycline is that it cannot be taken orally and must be slowly administered intravenously.

CHLORAMPHENICOL   This antibiotic prevents peptide bond formation by inhibiting peptidyltransferase in the 50S subunit (see Appendix 3, Figure A3.5E). It is bacteriostatic and broad spectrum but can have some toxic side effects. For instance, chloramphenicol can inhibit bone marrow development and lead to decreased production of red blood cells (called aplastic anemia). As a result, the use of this drug is reserved for situations in which suitable alternatives are unavailable.

MACROLIDES   Macrolides include erythromycin, azithromycin, and clarithromycin. Azithromycin was used to treat the *Streptococcus pyogenes* infection in Case History 13.3. These bacteriostatic antibiotics are identifiable by their 16-member lactone rings (see Appendix 3, Figure A3.5C). They are effective mainly against Gram-positive bacteria because the drugs do not penetrate the outer membrane of Gram-negative bacteria. Macrolides bind to the 50S subunit and inhibit tRNA movement from the A site to the P site. Because their structure is dramatically different from that of the beta-lactam antibiotics, such as penicillin, macrolides can be used safely in patients who are allergic to penicillin.

LINCOSAMIDES   A commonly used member of this group is clindamycin (see Appendix 3, Figure A3.5D). These antibiotics are effective against facultative Gram-positive bacteria and both Gram-positive and Gram-negative anaerobes. Lincosamides bind to peptidyltransferase in the 50S subunit and prevent peptide bond formation.

OXAZOLIDINONES   These wholly synthetic antibiotics (see Appendix 3, Figure A3.5F, linezolid) have an activity unique among protein synthesis inhibitors. Oxazolidinones bind to the 50S subunit and prevent <u>assembly</u> of the 70S ribosome. The drugs are effective against Gram-positive organisms, but their use is restricted, serving as a last-resort antibiotic for MRSA infections. Gram-negative bacteria are resistant to this class of drugs because the outer membrane limits their entry, and multidrug efflux pumps hasten the exit of whatever amount manages to enter.

STREPTOGRAMINS   This group includes the cyclic peptide drugs quinupristin and dalfopristin. The two are combined in an antibiotic (trade name, Synercid) that has a Gram-positive spectrum of activity. They bind to different sites on the 50S subunit and act synergistically. Dalfopristin blocks tRNA entry into the A site, while quinupristin blocks exit of a growing peptide from the ribosome. Individually, the drugs are bacteriostatic, but in combination they exhibit bactericidal activity. This drug is a last line of defense against life-threatening infections with vancomycin-resistant *Enterococcus faecium* and MRSA. To delay the development of resistance, quinupristin-dalfopristin is held in reserve for these infections.

Drugs That Affect Intermediary Metabolism

SULFONAMIDES   The sulfonamide (sulfa) drugs are known as antimetabolites because they interfere with the synthesis of metabolic intermediates. Ultimately, the sulfa drugs act to inhibit the synthesis of nucleic acids. Sulfa drugs such as sulfanilamide or sulfamethoxazole prevent the synthesis of folic acid, an important cofactor in the synthesis of nucleic acid precursors (**Figure 13.21**).

All organisms use folic acid to synthesize nucleic acids, so why are the sulfa drugs selectively toxic to bacteria? The selectivity occurs because mammals do not synthesize folic acid. Higher mammals generally rely on bacteria and green leafy vegetables as sources of folic acid. Bacteria make folic acid from the combination of PABA, glutamic acid, and pteridine (Figure 13.21B). Sulfanilamide, a structural analog of PABA, competes for one of the enzymes in the bacterial folic acid pathway and inhibits folic acid production (see Figure 13.21C). Because humans lack that pathway, sulfa drugs are selectively toxic to bacteria. Sulfamethoxazole is typically used in combination with trimethoprim, a drug that inhibits a later step where folic acid (dihydrofolic acid) is converted to tetrahydrofolic acid. The combination, called Bactrim, is very effective, especially in treating urinary tract infections.

SECTION SUMMARY

- **Selective toxicity** is achieved by targeting a process that occurs only in bacteria, not host cells.
- **Antibiotic targets** include cell wall synthesis, cell membrane integrity, membrane synthesis, DNA synthesis, RNA synthesis, protein synthesis, and metabolism.

Figure 13.21 Mode of Action of Sulfonamides

H_2N—⬡—COOH
Para-aminobenzoic acid (PABA)

H_2N—⬡—SO_2NH_2
Sulfanilamide (SFA)

A. The structures of PABA and sulfanilamide are very similar.

Pteridine (P) — N—⬡—C—N — Glutamic acid (G)
PABA

Folic acid

B. PABA, pteridine, and glutamic acid combine to make the vitamin folic acid.

Normal folic acid formation

P PABA G

Enzyme

P PABA G → Synthesis → Folic acid

Folic acid formation blocked by sulfanilamide (SFA)

P SFA G

Enzyme

P SFA G → No synthesis → P SFA G

C. Normal synthesis of folic acid requires that all three components engage the active site of the biosynthetic enzyme. The sulfa drugs replace PABA at the active site. The sulfur group, however, will not form a peptide bond with glutamic acid, and the size of sulfanilamide sterically hinders binding of pteridine, so folic acid cannot be made.

- **Antibiotics targeting the cell wall** bind to the enzymes and lipid carrier proteins involved with peptidoglycan synthesis and cross-linking.
- **Antibiotics interfering with DNA** include the antimetabolite sulfa drugs that inhibit nucleotide synthesis; quinolones that inhibit DNA topoisomerases; and the drug metronidazole, which, when activated, randomly nicks the phosphodiester backbone.
- **The RNA synthesis inhibitor** rifampin binds RNA polymerase and stops transcription.
- **Aminoglycosides** cause misreading of mRNA. **Tetracyclines** and **glycylcyclines** prevent tRNA entry into the A site. All bind to the 30S subunit of the bacterial ribosome.
- **Macrolides** inhibit tRNA movement from the A site to the P site, **chloramphenicol** and **lincosamides** prevent peptide bond formation, **oxazolidinones** prevent formation of the 70S initiation complex, and **streptogramins** block tRNA entry into the A site and peptide exit from the ribosome. All bind to the 50S subunit of the bacterial ribosome.

Thought Question 13.5 Besides fostering antibiotic resistance and potential tissue toxicity, what might be another detrimental aspect of antibiotic therapy on human health?

Thought Question 13.6 Why would a bacteriostatic antibiotic that only <u>inhibits</u> growth of a pathogenic bacterium be useful for treating an infection? (*Hint:* Is the human body a quiet bystander during an infection?)

13.6 The Challenges of Antibiotic Resistance

SECTION OBJECTIVES
- Describe the six basic mechanisms of antibiotic resistance.
- Explain how drug resistance develops.
- Outline strategies to counteract drug resistance.

Can pathogens defend themselves from antibiotic attack? In this section we discuss the various strategies microbes use to become resistant to antibiotics.

CASE HISTORY 13.4

Boy Meets Drug-Resistant Pathogen

Brian, a 14-year-old boy with fever (39°C, or 102.2°F), chills, and left-sided pleuritic chest pain (pain when inhaling) was referred to a hospital emergency department by his general practitioner. A chest X-ray showed left-lower-lobe pneumonia. The boy reported that he was allergic to amoxicillin and cephalosporins (as a child he had developed a rash to these agents) and had been taking daily doxycycline (tetracycline) for the previous 3 months to treat mild acne. He was admitted to the hospital and treated with intravenous erythromycin because of his reported beta-lactam allergies, but he continued to feel sick. The day after admission, both sputum and blood cultures grew *Streptococcus pneumoniae*. After 48 hours, antibiotic susceptibility results indicated that the microbe was resistant to penicillin, erythromycin, and tetracycline. Armed with this information, the clinician immediately changed antibiotic treatment to vancomycin, a drug still effective in the face of these resistance mechanisms. The boy's fever resolved over the next 12 hours, and he made a slow but full recovery over the next week.

Unfortunately, the scenario presented in this case is far too common and has become an extremely serious concern. The Centers for Disease Control and Prevention estimates that 23,000 people in the United States die every year from infections that cannot be cured because of antibiotic resistance. **Figure 13.22** illustrates the rapid rise of penicillin resistance in *S. pneumoniae* in the United States.

Figure 13.22 Penicillin-Resistant *Streptococcus pneumoniae* (United States, 1985–2006)

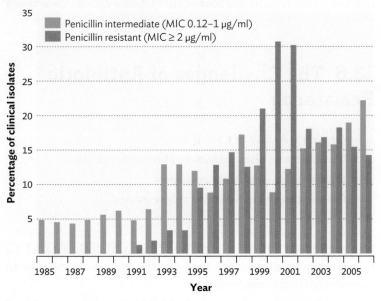

Legend:
- Penicillin intermediate (MIC 0.12–1 µg/ml)
- Penicillin resistant (MIC ≥ 2 µg/ml)

Y-axis: Percentage of clinical isolates
X-axis: Year (1985, 1987, 1989, 1991, 1993, 1995, 1997, 1999, 2001, 2003, 2005)

Penicillin was introduced in 1942. The first report of penicillin-resistant *S. pneumoniae* was in Spain in 1979. This graph tracks the increase in resistant strains in the United States. Strains resistant to intermediate levels of the drug have fewer altered penicillin-binding proteins than do strains resistant to higher levels of the drug. Intermediate resistance means that the achievable tissue level of the drug may not be sufficient to inhibit growth of the strain in question. The numbers of penicillin-resistant isolates steadily increased until 2001 and then dropped with the introduction of the pneumococcal vaccine for children.

There are three basic ways a microbe can become resistant to an antibiotic (**Figure 13.23**). The resistant organism can (1) prevent intracellular accumulation of the antibiotic, (2) prevent the antibiotic from binding to its intracellular target, or (3) dislodge the antibiotic from its target. These three strategies involve six basic mechanisms.

Mechanisms That Prevent Intracellular Accumulation

— **Destroy the antibiotic before it enters the cell.** One example is the enzyme beta-lactamase (or penicillinase), which is made exclusively to destroy penicillins (discussed in Section 13.5).

— **Pump the antibiotic out of the cell by using specific transport proteins (for example, tetracycline export) or nonspecific transport proteins.** The objective is to pump drugs out of the cell faster than the drugs can get in. Protein pumps vary in structure. Single-component pumps reside in the cytoplasmic membrane of Gram-negative and Gram-positive bacteria. Multicomponent drug efflux pumps are present in Gram-negative bacteria only (discussed next). Tetracycline resistance, for instance, is typically due to an efflux pump.

— **Decrease membrane permeability across the outer membrane.** Many antibiotics such as beta-lactams, tetracyclines, and fluoroquinolones have to pass through outer membrane porins to gain access to the target cell. Gram-negative bacteria, however, can express alternative porins with pores too narrow to allow penetration.

Mechanisms That Prevent Binding

— **Modify the target so that it no longer binds the antibiotic.** For example, a modified penicillin-binding protein enables methicillin resistance; an altered DNA gyrase produces resistance to quinolones; and an altered ribosomal protein can cause resistance to some aminoglycosides (discussed in Section 13.5).

Figure 13.23 Alternative Mechanisms of Antibiotic Resistance

Antibiotic resistance genes can be plasmid-borne, or they can be part of the chromosome. A specific antibiotic resistance gene will carry out only one of the six mechanisms shown.

Strategy 1: Prevent entry.
- Destroy antibiotic. Beta-lactamase
- Pump antibiotic out. Efflux pump
- Reduce membrane permeability (porin).

Strategy 2: Prevent binding.
- Alter the target. Mutated component / Antibiotic / Ribosome
- Modify the antibiotic. Modification enzyme

Strategy 3: Dislodge antibiotic.
- Protect the ribosome. Ribosome protection protein / Ribosome / Macrolide

Figure 13.24 Development and Selection of an Antibiotic-Resistant Pathogen in a Patient

Antibiotic-susceptible pathogen

Replication

Patient's lung is infected.

Spontaneous antibiotic-resistant mutant

Patient is sick with pneumonia.

Antibiotic administered

Patient starts to recover.

Antibiotic still present, but antibiotic-resistant mutant replicates

Patient relapses.

An antibiotic-susceptible bacterial pathogen (blue) infects the lung and multiplies to cause pneumonia. A spontaneous mutant (pink) that is resistant to an antibiotic arises as the cells replicate. When the patient is treated with the antibiotic, the susceptible bacteria die but the antibiotic-resistant cells survive and multiply. The patient relapses and risks transmitting the antibiotic-resistant strain to a new person.

— **Add modifying groups that inactivate the antibiotic.** As one example, three different classes of enzymes modify and inactivate aminoglycoside antibiotics.

Mechanism That Dislodges the Antibiotic

— **Ribosome protection (or rescue).** Gram-positive organisms can produce proteins (for example, MsrE or TetO) that bind to ribosomes and dislodge antibiotics bound near the peptidyltransferase site.

A particularly dangerous type of drug resistance that falls under the category "preventing accumulation" involves **multidrug resistance (MDR) efflux pumps** present in the cell membrane. One pump can export many different kinds of antibiotics with little regard for structure. Many important pathogens express MDR pumps. Strains of the pathogen *Mycobacterium tuberculosis*, for instance, have developed resistance to multiple drugs in part because of MDR pumps. Approximately 2 million people die from tuberculosis annually, mostly in developing nations. Chemists typically try to tweak the structure of an antibiotic to overcome a specific type of resistance mechanism, but the MDR pumps act on an exceptionally wide range of antibiotics, almost without regard to structure.

How Does Drug Resistance Develop?

Antibiotic resistance can arise spontaneously in a patient through mutation (discussed in Section 8.5) or through gene duplication followed by random mutations that "repurpose" the duplicated gene or genes. For instance, the MDR pumps evolved from genes encoding other transport mechanisms. The process leading to a drug-resistant bacterial cell, however, is a rare event and doesn't require that the antibiotic be present. As illustrated in **Figure 13.24**, an antibiotic-susceptible pathogen that infects the lung of a patient can multiply to become hundreds of millions of siblings before a single antibiotic-resistant cell spontaneously emerges. At this point the patient, now symptomatic, is treated with that antibiotic. All of the bacteria except the one resistant cell and its descendants are killed. By killing the sensitive cells, the antibiotic has selected for survival of the drug-resistant strain. The patient may even start to feel better. However, the resistant cells grow and repopulate the lung, causing the patient to relapse and possibly transmit the drug-resistant strain to someone else.

Once an antibiotic resistance gene is established in an organism, horizontal gene transfer mechanisms such as conjugation, described in Chapter 9, can move the gene from one organism to another and from one species to another. In fact, organisms that make antibiotics also have antibiotic resistance genes to protect themselves. Several of those genes have ended up in bacterial pathogens via horizontal gene transfer mechanisms. Transfer of a drug resistance gene is particularly easy if the gene has been incorporated into a plasmid.

But why is antibiotic resistance so rampant? Consider the following case: A young mother takes her 4-year-old child to the physician. The child is screaming because he has an extremely painful sore throat. Simply looking at the throat is not diagnostic. The raw tissue could mean that the child is suffering from a bacterial infection, in which case antibiotics are needed. Alternatively, a virus could be the culprit, a situation where antibiotics do nothing but pacify the parent. The clinician will often prescribe an antibiotic without ever knowing the cause of disease. The problem is this: the more often an antibiotic is unnessesarily used, the more

impact

Antibiotic Treatment Failures: It's Not Only Resistance

This chapter explores ways we control bacterial growth and how random gene transfers or mutations can lead to heritable drug resistance. But might bacteria have a way to tolerate antibiotics without developing heritable resistance? We ask this question to try to understand chronic bacterial infections.

Chronic infections present a formidable therapeutic challenge to clinicians. An infection can be treated with antibiotics and appear to resolve; then, once the drug is stopped, the infection roars back. Curiously, though, the resurgent pathogen is not usually resistant to antibiotics in vitro. This type of infection becomes almost impossible to eradicate. The most vexing example is that of the incurable *Pseudomonas aeruginosa* infection of cystic fibrosis patients (Figure 1).

We now think that these resurging infections are explained by a long-known but unsolved puzzle of microbiology—antibiotic-tolerant **persister cells**—cells that neither grow nor die in the presence of bactericidal agents. Joseph Bigger, in 1944, noticed that penicillin would lyse a growing culture of *Staphylococcus aureus*, but a few persister cells always survived. These persisters were not mutants made permanently resistant through mutation. They acted as though dormant.

In any biofilm or population of late-exponential-phase cells, persister cells that tolerate antibiotic treatment can be found. Figure 2 shows in vitro and in vivo biofilms of *P. aeruginosa*. It is now thought that persisters help explain antibiotic treatment failures and even latent bacterial infections, such as recrudescent typhus or latent tuberculosis. Recrudescent, or recurrent, typhus can develop in someone

Figure 1 Chest X-Ray of Adult Cystic Fibrosis Patient with Chronic *Pseudomonas aeruginosa* Infection

Note that the left lung (right side of the image) is cloudy, indicating it is filled with fluid. In medical terminology, this is called left-sided consolidation and pleural effusion.

Figure 2 Biofilms of *Pseudomonas aeruginosa*

A. *P. aeruginosa* grown in vitro (green fluorescent protein–tagged bacteria studied by confocal laser scanning microscopy; top view and two side views that indicate height of the biofilm).

B. *P. aeruginosa* grown in vivo in the lungs of a cystic fibrosis patient who died of chronic lung infection (hematoxylin-eosin stained). The arrows indicate the biofilms in both panels.

Figure 3 Persister Cells of *Pseudomonas aeruginosa*

Dose-dependent killing of a clinical strain isolated from an 8-month-old patient. The antibiotic ofloxacin was added to stationary-phase cultures of this *P. aeruginosa* isolate. Concentrations were 0, 5, 25, or 100 µg/ml. After an 8-hour incubation, surviving persister cells were plated for colony count. The survivors of antibiotic treatment were not antibiotic resistant by standard tests.

who was cured of the disease typhus (*Rickettsia prowazekii*) with antibiotics, only to develop the disease again years later. Persister cells of this pathogen are thought to remain undetectable in the patient until something triggers them to grow again. **Figure 3** illustrates the presence of persister cells in a clinical isolate of *P. aeruginosa*.

How might dormancy explain antibiotic tolerance? If persisters are dormant and have little or no cell wall synthesis, no translation, and no topoisomerase activity, then even if antibiotics can bind to proteins or structures that carry out those activities, their functions cannot be corrupted. Therefore, antibiotics are tolerated at the price of not growing. An antibiotic can kill all susceptible bacteria in an infection, but the remaining persister cells serve as a source of regrowth and infection once the antibiotic is removed. You might say that persister cells "play dead" the way some animals do when threatened by a predator. The molecular mechanisms of persistence include ATP depletion and self-imposed slowing of translation.

DISCUSSION QUESTION
How would you go about identifying new antimicrobials that specifically target persister cells?

----

opportunities there are to select for an antibiotic-resistant organism. The presence of the drug does not <u>cause</u> resistance but will kill off or inhibit the growth of competing bacteria that are sensitive, thereby allowing a spontaneously resistant organism to grow to large numbers.

Antibiotic Stewardship

Clinicians, pharmacists, and laboratory personnel are now being trained in something called antibiotic stewardship—a set of guidelines designed to slow the development of antibiotic-resistant microbes. One goal is to know when antimicrobials should be administered. For instance, when a patient presents with simple cold-like symptoms, which are usually caused by viruses, antibiotics should not be given until a bacterial etiology is confirmed. Abscesses, however, are always caused by bacteria and can quickly develop into more serious disseminated infections. As a result, empirical antibiotic therapy is often employed without the clinician ever knowing the organism's name.

Generally speaking, antibiotics should be empirically prescribed in life-threatening situations such as sepsis, severe pneumonia, or meningitis, where time is of the essence. Then, antibiotic usage should be de-escalated once an infectious agent and its resistance pattern have been identified. For instance, a patient with sepsis being treated empirically with broad-spectrum antibiotics (sometimes involving two or more drugs) should be switched to a suitable narrower-spectrum antibiotic once lab results are known, and unnecessary antibiotics should be discontinued. Following these guidelines will limit unnecessary antibiotic use and lower the chance that antibiotic resistance can develop.

Another proposed source of antibiotic resistance was the widespread practice of adding antibiotics to animal feed. No one quite knows why, but giving animals antibiotics along with their food makes them larger and therefore more profitable. Some estimates suggest that 70% of all antibiotics used in the United States are fed to healthy livestock. The consequence is that the animals may serve as incubators for the development of antibiotic resistance. Even if the resistance develops in nonpathogens, the antibiotic resistance genes produced can be transferred to pathogens by horizontal gene transfer. Finally, as of January 2017, the US Department of Agriculture banned the use of medically important antibiotics in animal feed.

Be aware that treatment failures for infectious diseases are not due only to genetically heritable antibiotic resistance mechanisms; there is a nongenetic cause called persistence in which infecting bacteria "play dead" when an antibiotic is present, only to grow again when the antibiotic is gone (see **IMPACT**).

Fighting Drug Resistance

Several strategies are being used to stay one step ahead of drug-resistant pathogens. In some instances, dummy target compounds that inactivate resistance enzymes have been developed.

Clavulanic acid, for example, is a compound sometimes used in combination with penicillins such as amoxicillin. Clavulanic acid, a beta-lactam compound with no antimicrobial effect, competitively binds and ties up beta-lactamases secreted from penicillin-resistant bacteria. Amoxicillin, then, escapes destruction, enters the cell, and kills the bacterium. A different strategy to combat drug resistance is to alter the structure of the antibiotic in a way that sterically hinders (blocks) the access of modifying enzymes.

Many new compounds with antibacterial activity are currently being tested for their efficacy in humans. One of the most intriguing is cannabidiol, a nonpsychoactive chemical compound extracted from cannabis (marijuana) and hemp plants. Australian scientists in 2019 found that cannabidiol will kill a wide range of Gram-positive pathogens—even those that are resistant to other antibiotics. The drug, already approved by the FDA for treating epilepsy, is also under investigation for treating anxiety, pain, and inflammation.

Finally, there are a variety of actions individuals can take to limit infections and thus the use of antibiotics. These include washing hands frequently, getting vaccinated, avoiding the use of antibiotics for viral infections, and refusing to use leftover antibiotics from friends. Finally, when you do need to use an antibiotic, be sure to complete the full course. Stopping halfway through because you feel better will leave bacterial survivors that can regrow and develop resistance.

SECTION SUMMARY

- **Antibiotic resistance** is a growing problem worldwide.
- **Mechanisms of antibiotic resistance** include modifying the antibiotic, destroying the antibiotic, altering the target to reduce affinity, and pumping the antibiotic out of the cell.
- **Multidrug resistance pumps** use promiscuous binding sites to bind antibiotics of diverse structure.
- **Antibiotic resistance can arise spontaneously** through mutation, can be inherited by gene exchange mechanisms, or can develop through gene duplication and mutational reengineering.
- **Indiscriminate use of antibiotics** has significantly contributed to the rise in antibiotic resistance.
- **Measures to counter antibiotic resistance** include chemically altering the antibiotic, using combination antibiotic therapy, and adding a chemical decoy.

Thought Question 13.7 How might bacteria such as *Streptomyces* and *Actinomyces* species avoid committing suicide when they make their antibiotics?

Thought Question 13.8 Jill goes up the hill with Jack, whom she has just met at a local bar, and engages in unprotected sex. The next morning, Jill becomes worried that Jack, someone she knows nothing about, might have given her a sexually transmitted infection (STI). She remembers that she still has some ciprofloxacin pills left over from an earlier respiratory tract infection (the original infection resolved after a few days, and she forgot to take the rest of the treatment). Should she take the rest of the cipro prophylactically to prevent a possible STI? Explain why or why not.

13.7 Antiviral Agents

SECTION OBJECTIVES

- Describe the basic types of antiviral agents and their targets.
- Discuss selective toxicity for antiviral drugs.
- Explain the spectrum of activity for antiviral drugs.

Why aren't there more antiviral agents? A father pleading with a physician to give his child antibiotics when the infant is suffering with a cold is heart-wrenching. Unfortunately, the common cold is caused by a virus, so antibiotics designed for bacteria can't touch it. The physician, who can do nothing, feels terrible but wonders why there aren't more antiviral drugs. After all, more infections are caused by viruses than by bacteria. The reason we don't have more antiviral drugs is that selective toxicity is much harder to achieve for viruses than for bacteria. Viruses by their very nature use host cell functions to make copies of themselves. So, a drug that hurts the virus is likely to also harm the patient. Nevertheless, there are several useful antiviral agents for which selective viral targets have been found and exploited. Some of these agents are listed in **Table 13.7**. Select examples are discussed in this section.

Agents That Prevent Virus Uncoating or Release

Membrane-coated viruses are vulnerable at two stages: when the virus is invading the host cell and after viral propagation, when the progeny viruses release from the host cell. The flu virus presents a good example of both.

CASE HISTORY 13.5
Baby Phil Fights the Flu

Philip, a 9-month-old infant, suffered from a series of chronic problems after birth, including respiratory syncytial virus bronchiolitis (infection and inflammation of the bronchioles) and neonatal group B streptococcal sepsis. (Neonatal sepsis happens soon after birth and is often caused by Lancefield group B *Streptococcus agalactiae*, passed to the newborn from an infected mother as described in Chapter 23.) Because of his weakened state,

Baby Phil was fitted with a gastrostomy feeding tube so that his parents could feed him. Within a week of being sent home, however, Philip was rushed to the Johns Hopkins Hospital with an acute onset of fever, cough, regurgitation from his gastrostomy feeding tube, and dehydration. Physical exam confirmed that the boy had a severe cough resulting in respiratory distress, a rapid heart rate, and moderate dehydration. A nasopharyngeal aspirate (the nasopharynx is the nasal part of the pharynx) was positive for influenza A antigen. Philip was diagnosed with influenza and treated with the antiviral drug oseltamivir. He gradually improved and was discharged 4 days after admission.

Influenza can be a particularly dangerous disease. An unusually severe form of influenza A (H3N2) spread across the United States in 2017. Over 30,000 people were hospitalized (58% were over 65 years old). The flu vaccine that year was only about 40% effective, but the availability of antiviral agents shortened the disease course. There are now three selective antiviral targets: viral M2 protein, required for virus entry; neuraminidase, needed for viral exit; and a viral endonuclease, the newest target, which is critical for viral replication.

VIRAL M2 PROTEIN   Influenza M2 protein is crucial for the flu virus to enter a cell. As described in Section 12.5, viral entry starts when hemagglutinin in the envelope of influenza virus binds to host membrane sialic acid receptors. The virus then enters the cell enveloped in an endosome. Normal acidification of the endosome causes M2 proteins to form channels in the viral envelope, which enables the space between the virus and the viral envelope to acidify. Acidification causes the viral membrane to fuse with the endosome membrane, releasing the virus into the cytoplasm. Amantadine

Table 13.7
Examples of Antiviral Agents

Virus	Agent	Mechanism of Action	Result
Influenza virus	Amantadine	Inhibits viral M2 protein	Prevents viral uncoating
	Zanamivir, oseltamivir (Tamiflu)	Neuraminidase inhibitor	Prevents viral release
Herpes simplex virus and varicella-zoster virus (shingles)	Acyclovir	Guanosine analog	Halts DNA synthesis
	Famciclovir, valacyclovir	Prodrugs of penciclovir and acyclovir, respectively; guanosine analogs	Halts DNA synthesis
Hepatitis B, chronic	Lamivudine	Cytosine nucleoside analog	Halts DNA synthesis
	Entecavir	Guanosine nucleoside analog	Halts DNA synthesis
Hepatitis C, chronic	Sofosbuvir	Nucleotide analog	Halts replication
	Simeprevir	Protease inhibitor	Prevents processing of viral polyprotein
Cytomegalovirus	Ganciclovir	Similar to acyclovir	Halts DNA synthesis
	Foscarnet	Analog of inorganic phosphate	Binds and inhibits virus-specific DNA polymerase
Respiratory syncytial virus and chronic hepatitis C	Ribavirin	RNA virus mutagen	Causes catastrophic replication errors
HIV	Zidovudine (ZDV), also called azidothymidine (AZT)	Nucleoside analog, resembles thymine	Inhibits reverse transcriptase
		Nonnucleoside; binds to allosteric site	Inhibits reverse transcriptase
	Nevirapine	Nucleotide analog, resembles adenosine monophosphate (AMP)	Inhibits reverse transcriptases of HIV and hepatitis B virus
	Tenofovir		
		Protease inhibitor	Prevents viral maturation
	Nelfinavir, indinavir, lopinavir, darunavir	Integrase inhibitor	Prevents integration into host genome
	Raltegravir	CCR5 entry inhibitor	Prevents viral entry into host cells
	Maraviroc		

Figure 13.25 Inhibitors of Influenza Proteins

Amantadine

A. Amantadine inhibits the M2 protein.

Zanamivir

B. Zanamivir (Relenza) inhibits neuraminidase.

TEM 25 nm

C. Influenza virus, showing the viral envelope (orange) containing hemagglutinin and neuraminidase (proteins are too small to see).

(Figure 13.25A) is a specific inhibitor of the influenza M2 protein. Amantadine is useful only against influenza type A—not against type B strains. Unfortunately, many amantadine-resistant strains have developed, in part because of the widespread use of amantadine by Chinese poultry farmers. As a result, the drug is no longer recommended as a treatment for influenza.

NEURAMINIDASE   The viral envelope protein neuraminidase is the second target. Antiflu drugs zanamivir (Relenza) and oseltamivir (Tamiflu) are **neuraminidase inhibitors** that act against types A and B influenza strains (Figure 13.25B). Neuraminidase on the viral envelope allows virus particles to leave the cell in which they were made. Neuraminidase inhibitors prevent this release and cause the virus particles to aggregate at the cell surface, reducing the number of virus particles released.

VIRAL ENDONUCLEASE   The third, and newest target, is a viral endonuclease that steals a piece of host mRNA and attaches it to viral RNA. This stolen component is required to replicate and translate influenza RNA. The new drug, called baloxavir marboxil (sold as Xofluza) inhibits the activity of the influenza endonuclease in type A and B strains and stops the replication of both. Xofluza is effective against the growing number of flu viruses that are resistant to M2 and neuraminidase inhibitors.

The neuraminidase and endonuclease inhibitors, when used within 48 hours of disease onset, decrease viral shedding and reduce the duration of influenza symptoms by approximately 1 day. Given that flu symptoms generally last only 3–10 days, this treatment may not seem greatly beneficial. But for elderly or immunocompromised patients, as in Baby Phil's case, shortening the flu's course can minimize damage to the lungs. The healthier a lung is, the less susceptible it will be to secondary bacterial infections such as bronchitis or a life-threatening pneumonia.

Antiviral DNA and RNA Synthesis Inhibitors

Many antiviral agents work by inhibiting viral DNA synthesis. These drugs chemically resemble normal DNA nucleosides, but once one is phosphorylated and added to a replicating chain of DNA, replication stops. It's like adding a defective link at the end of a metal chain so that additional links cannot be added. The drugs are composed of normal adenine, guanine, cytosine, or thymine molecules attached to an analog of deoxyribose. Zidovudine and acyclovir (Figure 13.26) are two examples. Viral enzymes will add phosphate to the defective deoxyribose group to form chain-terminating nucleotide analogs. A DNA polymerase can then add one of the nucleotide analog molecules onto the end of a replicating viral DNA strand in place of a normal nucleotide. Once incorporated, however, the analog cannot accept any new nucleotides needed for chain elongation.

These DNA chain–terminating analogs are selectively toxic because viral polymerases more readily incorporate nucleotide analogs into their nucleic acid than do the more selective host cell polymerases. In addition, host DNA polymerase has an effective proofreading activity (missing in the viral polymerase) that removes the analog from host DNA.

The antiviral DNA synthesis inhibitors work on DNA viruses such as varicella-zoster virus (chickenpox) and herpes simplex virus or on RNA retroviruses such as HIV, but not on RNA viruses like influenza. Ribavirin, however, is a drug that, when metabolized, produces a compound resembling purine RNA nucleosides. Ribavirin will inhibit RNA metabolism by RNA viruses such as influenza and hepatitis C.

Figure 13.26 Antiviral Agents That Prevent DNA Synthesis

Zidovudine (ZDV) and acyclovir are structural analogs of DNA bases. Because the analogs have no 3′ OH (yellow highlight) to which another nucleotide can add, DNA chain elongation ceases.

Zidovudine

Deoxyribonucleotide containing thymine

A. Zidovudine (ZDV) is an analog of thymine.

Acyclovir

Deoxyguanosine

B. Acyclovir is an analog of guanine.

HIV: When Life Gets Real

Bob and Carol, a recently married, young couple from Tacoma, Washington, came to the community clinic for prenatal care. Carol was 19 and reportedly 2 months pregnant with her first child. She denied intravenous (IV) drug use or a history of other sexual partners and had no history of sexually transmitted infection; however, a routine prenatal HIV antibody screen was reported as positive for HIV-1. Careful questioning revealed that Bob had a history of IV drug use. An HIV antibody screen for him was also positive. The lab results indicated that Carol had a low viral load—fewer than 1,000 copies per milliliter of blood—and a normal CD4 T-cell lymphocyte count. Bob had an HIV viral load of 10,000 copies per milliliter and a lower CD4 count: 150 cells/µl. Antiretroviral therapy was initiated for both husband and wife (even though her CD4 count was normal). Because she was pregnant, Carol's regimen avoided drugs that could potentially harm the fetus. Both his regimen and hers included two nucleotide reverse transcriptase inhibitors plus a protease inhibitor. After 7 months of treatment, Bob's viral load had decreased and his CD4 T-cell count was approaching normal, and Carol delivered an HIV-free baby girl.

Human immunodeficiency virus (HIV) is an RNA retrovirus that uses its reverse transcriptase to convert its HIV RNA genome into DNA. HIV DNA then integrates into host nuclear DNA to form a provirus (discussed in Section 12.6). The virus targets specific white blood cells (T cells; discussed in Chapters 15 and 16) and slowly destroys parts of the immune system, leaving its victims helpless against secondary infections. HIV has killed hundreds of thousands of people in the United States since the HIV pandemic began in 1981, but being diagnosed with HIV is no longer a death sentence. Advances in antiretroviral therapy (ART) over the past 10 years have transformed HIV into a manageable chronic condition. At least half, if not most, of the HIV-infected people in the United States now live long enough to die from diseases of aging, such as heart attacks or strokes. A number of drugs are the heroes of this new paradigm.

Nucleoside and Nonnucleoside Reverse Transcriptase Inhibitors

The antiretroviral nucleoside inhibitor zidovudine (ZDV), also known as azidothymidine (AZT), was described earlier. It is a thymidine analog recognized by reverse transcriptase. Once incorporated into a replicating HIV DNA molecule, the DNA chain–terminating property of AZT prevents further DNA synthesis. Because HIV can be transmitted from mother to neonate at delivery or during breast-feeding, treatment of the mother and the child are important steps to prevent transmission.

Tenofovir is a nucleotide analog similar to adenosine monophosphate that is recognized by HIV reverse transcriptase and also by the reverse transcriptase of hepatitis B virus. It, too, functions as a chain terminator. Tenofovir is often given after needle stick accidents to prevent transmission of HIV or HepB.

Besides these nucleoside inhibitors, there are nonnucleoside reverse transcriptase inhibitors. For example, the drug delavirdine binds directly to reverse transcriptase and inactivates the enzyme.

Viral Protease Inhibitors

To make optimum use of its short provirus DNA sequence, HIV makes long, nonfunctional polypeptide chains that are cleaved by an HIV protease into the functional proteins and enzymes needed to replicate and produce new virions. For example, one HIV mRNA produces a polyprotein called Gag-Pol. Once made, Gag-Pol is cleaved by HIV protease to make reverse transcriptase and integrase. Protease inhibitors, such as nelfinavir (Viracept) and lopinavir (Kaletra), are a powerful class of drugs that block the HIV protease. When the protease is inactivated, even though new virus particles are made the polyprotein remains uncleaved and the virus cannot mature to infect other cells; progress of the disease stalls. It is important to note that protease inhibitors do not cure AIDS; they can only decrease the number of infectious copies of HIV.

Viral Entry Inhibitors

Another way to stop HIV is to prevent the virus from infecting cells in the first place. Drugs called entry inhibitors do just that, stop entry. There are two types of entry inhibitors. CCR5 inhibitors such as maraviroc block viral envelope protein gp120 (also known as SU) from binding to host surface protein CCR5. CCR5 is a coreceptor that, together with CD4, is needed for HIV to bind to certain T cells (see Section 12.6 and Figure 12.28 for details). As a result of CCR5 inhibition, HIV never attaches. Fusion inhibitors such as enfuvirtide, in contrast, do not prevent initial binding but prevent HIV membranes from fusing with T-cell membranes and thereby halt viral entry. Imagine trying to enter a room through a door. The action of CCR5 inhibitors is like removing the doorknob so there is nothing to grab onto. The action of fusion inhibitors, however, is like gluing the door shut. You can grab the knob but still cannot open the door.

HIV Treatment Regimens and HIV Controllers

Because HIV can mutate rapidly and become resistant to single-drug therapies, HIV treatments today involve combinations of three or more antiretroviral drugs. This therapeutic strategy was called **highly active antiretroviral therapy (HAART)**, but the name has been changed to simply **antiretroviral therapy (ART)**. Current ART regimens include three drugs: usually two nucleoside reverse transcriptase inhibitors plus a protease inhibitor, a nonnucleoside reverse transcriptase inhibitor, or an integrase inhibitor. Integrase inhibitors block the enzyme needed to insert viral DNA into the host genome. You can learn more about HIV and treatment regimens in Section 23.4.

Only two people have been functionally cured of HIV. Their cures came after receiving a bone marrow transplant from a donor who lacked CCR5, the coreceptor for HIV. The transplant was initiated because both patients had cancer, but the CCR5-deficient donor was chosen on purpose. As hoped, the transplants repopulated the HIV-positive cancer patients' blood systems with HIV-resistant T cells. No further antiretroviral therapy was needed.

Besides these two cures, a few patients given early and aggressive ART regimens maintain very low viral loads (<400 per milliliter of blood) even after stopping ART. These people are called posttreatment controllers. The viral loads in all other patients that stop ART will rebound to large numbers. The posttreatment controllers are not cured; they still have very low viral loads (<400 per milliliter of blood).

SECTION SUMMARY

- **There are fewer antiviral agents** than antibacterial agents because it is harder to identify viral targets that provide selective toxicity. The spectrum of activity of an antiviral agent is linked to its molecular target: for instance, RNA virus inhibitors, DNA virus inhibitors, or retrovirus inhibitors.

- **Preventing viral attachment to, or release from, host cells** characterizes mechanisms of action for antiviral agents such as amantadine and zanamivir, used to treat influenza virus.

- **Inhibiting DNA synthesis** is the mode of action for many antiviral agents, although they work only for DNA viruses and retroviruses.

- **HIV treatments** include reverse transcriptase inhibitors, which prevent synthesis of DNA; protease inhibitors, which prevent the maturation of viral polyproteins into active forms; and entry inhibitors, which prevent HIV from infecting cells.

13.8 Antifungal and Antiprotozoal Agents

SECTION OBJECTIVES

- Describe the major targets of antifungal and antiprotozoal agents.
- Discuss selective toxicity for both types of antimicrobial agents.
- Identify agents associated with serious side effects.

Can antimicrobial drugs be used to treat disease-causing eukaryotic microbes such as fungi and protozoa? The basic structure and physiology of eukaryotic microbes are so similar to those of human cells that you might think anything that could kill one of these microbes would surely kill our own cells. Fortunately, if you do think that, you are wrong.

Antifungal Agents

Fungal infections are much more difficult to treat than bacterial infections—in part because fungal physiology is more similar to that of humans than is bacterial physiology. Fungi also have an efficient drug detoxification system that modifies and inactivates many antibiotics. Nevertheless, it is possible to treat fungal infections if repeated applications of antifungal agents are given to keep the level of unmodified drug above MIC levels.

CASE HISTORY 13.7
Blastomycosis

Enrique, a 37-year-old male, presented to a Florida hospital's emergency department with persistent fever, malaise, and a painful mass in his right arm. He denied trauma to the arm. His white blood cell count was elevated, at 27,000 cells/μl, and chest X-ray revealed a left-lung infiltrate. Bronchoscopy revealed granulomatous (walled-off) inflammation containing a single yeastlike mass. Incision and drainage were performed on the arm mass, and cultures were obtained. Serum cryptococcal antigen tests were negative, as were tests for *Bartonella henselae* and *Toxoplasma*. Cultures from the right arm grew a fungal form similar to that identified from the bronchoscopy specimens. A tentative diagnosis of *Blastomyces dermatitidis* was confirmed using PCR. The patient was placed on intravenous amphotericin B, and his fever and leukocytosis (elevated white cell count) subsided within days. His medication was then changed to oral fluconazole for 6 months.

Fungal infections are described as being either superficial mycoses, such as athlete's foot, or systemic mycoses, such as blastomycosis (as in this case history). Because fungal and human biochemical pathways are very similar, finding selectively toxic antifungal agents has been difficult. Fortunately, there are some differences between a few fungal and human pathways that we can exploit. Currently, there are six categories of available antifungal agents:

Table 13.8
Major Antifungal Agents and Their Common Uses

Class of antifungal	Drug	Clinical Application[a]								
		Systemic Mycoses				Opportunistic Mycoses				
		Coccidioido-mycosis	Histo-plasmosis	Blasto-mycosis	Paracoccidioido-mycosis	Aspergil-losis	Candidi-asis[b]	Crypto-coccosis	Dermato-phytosis	Other
Polyenes (Fungicidal)	Amphotericin B (intravenous)	+	+	+	+	+	+	+	−	
	Nystatin (oral; topical)	−	−	−	−	−	+ mc	−	−	
	Natamycin (topical; eyedrops)	−	−	−	−	+ Superficial	+ Superficial	−	−	Mycotic keratitis
Azoles (Fungistatic), Imidazoles—Topical	Clotrimazole	−	−	−	−	−	+ mc	−	+	
	Miconazole (Monistat)	−	−	−	−	−	+ mc	−	+	
	Ketoconazole	−	−	+	−	−	+	−	+	Dandruff shampoo
Azoles (Fungistatic), Triazoles—Oral	Itraconazole	+	+	+	+	+	+ mc	+	+	Sporotrichosis
	Fluconazole	+	+	+	?[c]	−	+	+	+	Sporotrichosis
Allylamines—Topical or Oral (Fungicidal)	Terbinafine (Lamisil)	−	−	−	−	−	−	−	+	
Echinocandins—Intravenous	Caspofungin	−	−	−	−	+	+	−	−	
Griseofulvin—Oral (Fungistatic)	Griseofulvin	−	−	−	−	−	−	−	+	
Antimetabolites (Fungistatic or Fungicidal)	5-Fluorocytosine[d] (flucytosine)	−	−	−	−	+	+	+	−	Phaeohypho-mycosis

[a] + indicates that the drug inhibits growth of the disease-causing agent; − indicates that the drug is not useful for the disease.
[b] mc = mucocutaneous but not systemic candidiasis.
[c] Insufficient data.
[d] Used only in combination with amphotericin B.

- **Polyenes** (nystatin, amphotericin B), which disrupt membrane integrity
- **Azoles** (imidazoles and triazoles), which interfere with the synthesis of ergosterol, a cholesterol specific to fungi
- **Allylamines** (such as terbinafine, Lamisil), which also inhibit ergosterol synthesis
- **Echinocandins** (caspofungin), which block fungal cell wall synthesis
- **Griseofulvin**, which blocks cell division
- **Flucytosine**, which inhibits DNA synthesis
- **Orotomides** (a new class, not yet FDA approved), which stop pyrimidine biosynthesis

Superficial mycoses and systemic mycoses require very different treatments. Table 13.8 lists a number of clinically used antifungal agents and their use for different mycotic infections.

Superficial mycoses include infections of the skin, hair, and nails, as well as *Candida* infections of moist skin and mucous membranes (for example, vaginal yeast infections). Imidazole-containing drugs (clotrimazole and miconazole, with a fungistatic mode of action) are often used topically in creams for superficial mycoses (see Appendix 3, Figure A3.6A). These antifungal agents disrupt the fungal membrane by selectively inhibiting fungal enzymes of sterol synthesis but not inhibiting those of humans.

Triazoles such as itraconazole are administered orally, are fungistatic, and are less susceptible to metabolic degradation than are imidazoles. Lamisil (a terbinafine compound, fungicidal) is a different type of agent that selectively inhibits ergosterol synthesis in fungi. It is currently a popular antifungal agent used to treat superficial mycoses. More chronic dermatophytic infections typically require another antifungal agent, called griseofulvin, produced by a *Penicillium* species (see Appendix 3, Figure A3.6B). **Griseofulvin** (a fungistatic) disrupts the mitotic spindle in fungi (but not in humans) needed for cell division. This disruption stops growth but does not kill the fungus. As the infected hair, skin, or nails grow and are replaced, the fungus is shed. Griseofulvin is typically taken orally.

Vaginal yeast infections caused by *Candida* are often treated with nystatin, a polyene fungicidal agent synthesized by *Streptomyces* that forms membrane pores (see Appendix 3, Figure A3.6C). The name "nystatin" came about because two of the people who discovered it worked for the New York State Public Health Department. Nystatin may be given orally or as a cream but is not available in injectable form.

The serious, sometimes fatal, consequences of systemic mycoses require more aggressive therapy. The drugs used in these instances include intravenous **amphotericin B** (produced by *Streptomyces*; see Appendix 3, Figure A3.6D) and oral fluconazole, a triazole compound. This was the regimen used in Case History 13.7. Amphotericin B binds to the ergosterol present in fungal but not mammalian membranes and destroys membrane integrity. Fluconazole, on the other hand, makes membranes defective by inhibiting the synthesis of ergosterol, like the imidazoles do.

Typically, curing systemic fungal infections requires long-term treatment to prevent disease relapse. Extended treatment is required because fungi grow slowly. But even with long-term therapy, a fair number of patients do not respond. Fortunately, a new class of antifungal agents called echinocandins (caspofungin) has helped 70% of patients who do not tolerate other treatments.

Unfortunately, long-term treatment also increases the chances that the patient will exhibit toxic side effects of these drugs. For example, amphotericin B can damage kidneys, griseofulvin can trigger severe skin reactions, and caspofungin can damage the liver.

Antiprotozoal Agents

The major protozoal parasites include members of *Babesia*, *Entamoeba*, *Giardia*, *Leishmania*, *Plasmodium*, *Trypanosoma*, and *Trichomonas*. The diseases they cause will be described throughout Chapters 19–24. The good news about protozoal infections is that their incidence in developed countries is very low. The bad news is that these same infections are much too common in areas of the world that are impoverished and lack adequate sanitation, potable water, and mosquito control. For instance, there were 219 million cases of malaria worldwide in 2017, caused by various species of *Plasmodium*—mostly in Central America, South America, and Asia. Fortunately, several potent antiprotozoal agents are available.

Several agents have antimetabolic functions. Pyrimethamine and sulfamethoxazole impede synthesis of folic acid (effective against *Toxoplasma*). Miltefosine interferes with *Leishmania* cytochrome *c* oxidase in mitochondria and obstructs Ca^{2+} homeostasis. Aminoquinolines such as chloroquine are antiplasmodial drugs that concentrate in the acid vesicles of malarial parasites. The increase in vesicle pH blocks the ability of the parasite to use hemoglobin and rid itself of toxic by-products. The agent iodoquinol chelates (binds) ferrous iron, which prevents the iron from being used. The drug is used primarily to treat cases of amebic dysentery caused by the ameba *Entamoeba histolytica*. Finally, eflornithine acts by inhibiting the synthesis of putrescine, a polyamine that *Trypanosoma* species need to proliferate.

Two protein synthesis inhibitors are commonly used to treat infections by protozoa. Paromomycin is an aminoglycoside used to treat dysentery caused by *Entamoeba histolytica*. The antibiotic binds to 70S ribosomes in mitochondria and cytosolic 80S ribosomes. But how paromomycin actually inhibits protein synthesis in protozoa is unclear. Azithromycin, a macrolide, is often used to treat infections by *Babesia*.

A few antiprotozoal agents generate free radicals that damage macromolecules. Nitroimidazoles such as metronidazole (Figure 13.18) and benznidazole are reduced in vivo, under low-oxygen conditions, to make nitro radicals that can damage cell structures, including DNA. Mammalian cells are unaffected because their aerobic metabolism doesn't allow reduction of the nitro group. Nitroimidazoles have been used to treat infections by *Giardia* and the trypanosomes that cause Chagas disease. Another free radical–producing drug is artemisinin, which was originally extracted from the *Artemisia annua* shrub from Africa. In red blood cells, artemisinin releases free radicals that eradicate parasites. Because *Plasmodium* species have become resistant to many aminoquinolines, artemisinin has become the drug of choice for treating malaria.

- **Fungal infections** are difficult to treat because of similarities in human and fungal physiologies.

- **Azole-containing** antifungal agents (imidazoles and triazoles) and **allylamines** (Lamisil) inhibit ergosterol synthesis.

- **Griseofulvin** inhibits mitotic spindle formation.

- **Nystatin** and **amphotericin B** produce membrane pores and destroy membrane integrity.

- **Flucytosine** blocks DNA synthesis.

- **Antiprotozoal drugs** include antimetabolites, protein synthesis inhibitors, and free radical generators.

Perspective

Microbiology as a science was founded on the need to understand and control microbial growth in the hope of conquering human disease. We have largely succeeded. Infectious diseases dreaded for centuries (such as the plague and tuberculosis) were successfully defeated, or so we thought. Today, a crisis of antibiotic resistance looms because of the irresponsible use of antimicrobial agents and because very few new, clinically useful antimicrobials have been discovered over the past 30 years. It is now possible to imagine a world in which effective antibiotics do not exist. However, the impending crisis of antibiotic resistance has given scientists and pharmaceutical companies an incentive to redouble efforts to find new drugs and impose measures that ensure the responsible use of existing antibiotics. The goal is nothing short of survival for our species.

Key Terms

allylamine (p. 405)

amphotericin B (p. 406)

antagonistic (p. 385)

antibiotic (p. 380)

antiretroviral therapy (ART) (p. 404)

antisepsis (p. 372)

azole (p. 405)

bacitracin (p. 388)

bactericidal (p. 373)

bacteriostatic (p. 373)

cephalosporin (p. 388)

chemotherapeutic agent (p. 380)

chemotherapeutic index (p. 385)

cycloserine (p. 388)

decimal reduction time (D-value) (p. 373)

disinfectant (p. 372)

disinfection (p. 372)

echinocandin (p. 405)

empirical therapy (p. 390)

flucytosine (p. 405)

germ (p. 373)

germicidal (p. 373)

germicide (p. 373)

gramicidin (p. 389)

griseofulvin (pp. 405, 406)

highly active antiretroviral therapy (HAART) (p. 404)

Kirby-Bauer assay (p. 382)

laminar flow biological safety cabinet (p. 376)

lyophilization (p. 375)

minimum bactericidal concentration (MBC) (p. 383)

minimum inhibitory concentration (MIC) (p. 382)

multidrug resistance (MDR) efflux pump (p. 397)

neuraminidase inhibitor (p. 402)

orotomide (p. 405)

pasteurization (p. 374)

penicillin-binding protein (PBP) (p. 387)

persister cell (p. 398)

phenol coefficient test (p. 377)

polyene (p. 405)

polymyxin (p. 389)

sanitation (p. 373)

selective toxicity (p. 381)

semisynthetic antimicrobial (p. 388)

spectrum of activity (p. 381)

sterilization (p. 372)

synergistic (p. 385)

therapeutic dose (p. 385)

toxic dose (p. 385)

use-dilution test (p. 378)

vancomycin (p. 389)

zone of inhibition (p. 382)

Objectives Assessment Questions

13.1 Concepts of Sterilization and Disinfection

1. _____ is the process by which a chemical kills disease-producing microbes on inanimate surfaces.
   a. Sanitation
   b. Disinfection
   c. Sterilization
   d. Antisepsis

2. The effectiveness of a chemical agent is determined by _____.
   a. the Kirby-Bauer test
   b. decimal reduction time
   c. complete death reduction time
   d. spore death time

3. Agent A acting on *Staphylococcus aureus* has a D-value of 5 minutes. If the population of microbes starts at a cell density of 5×10^8 CFUs/ml, how many viable cells will remain after 15 minutes?
   a. 5×10^7 CFUs/ml
   b. 1×10^7 CFUs/ml
   c. 5×10^6 CFUs/ml
   d. 5×10^5 CFUs/ml
   e. 1×10^6 CFUs/ml

13.2 Physical Agents That Kill Microbes

4. Standard autoclave conditions are _____.
   a. 101°C, 10 pounds of pressure, 30 minutes
   b. 121°C, 15 pounds of pressure, 20 minutes
   c. 72°C, atmospheric pressure, 15 seconds
   d. 121°C, atmospheric pressure, 15 seconds
   e. 121°C, 20 pounds of pressure, 15 minutes

5. An injectable drug solution filtered through a 0.2-μm filter should be considered sterile. True or false?

6. Pasteurization conditions are designed to kill which of the following microbes from milk?
   a. *Coxiella burnetii*
   b. *Staphylococcus aureus*
   c. *Mycobacterium tuberculosis*
   d. *Brucella abortus*
   e. *Clostridium botulinum*

7. You are developing a liquid food supplement for populations of developing nations after natural disasters. It must have a long shelf life without refrigeration. Which microbial control technique would best accomplish your goal?
   a. Filtration
   b. Pasteurization
   c. Gamma irradiation
   d. Steam sterilization
   e. Disinfection

13.3 Chemical Agents of Disinfection

8. In a use-dilution test against *Staphylococcus aureus*, agent A at 0.01 M killed all the bacteria with 10 minutes of exposure, while agent B did the same at 0.005 M. Which one would make the better antiseptic?
   a. Agent A
   b. Agent B
   c. Both are equally useful.
   d. More data are needed.

9. A chemical agent that can kill mycobacteria and spores but not prions is considered a _____.
   a. low-level disinfectant
   b. intermediate-level disinfectant
   c. high-level disinfectant
   d. high-level antiseptic
   e. c and d

10. Which of the following method(s) is/are effective for sterilizing plasticware such as Petri plates?
    a. Steam sterilization
    b. Chemical disinfection
    c. Ethylene oxide gas
    d. Irradiation
    e. c and d
    f. b and c

11. Which of the following is useful as a disinfectant but not as an antiseptic?
   a. An iodophor
   b. Triclosan
   c. A cationic agent
   d. Glutaraldehyde
   e. Chlorhexidine

12. A patient just treated in the emergency department left a considerable amount of blood on the metal examination table. Which disinfectant is best for decontaminating the table?
   a. Glutaraldehyde
   b. Ethylene oxide
   c. Formaldehyde
   d. Hypochlorite
   e. Chlorhexidine
   f. a and d
   g. c and d

13.4 Concepts of Antibiotic Therapy

13. A series of culture tubes containing dilutions of an antibiotic are inoculated with a bacterial pathogen. Tube 1 has the highest concentration of drug, and tube 6 has the lowest. Growth occurs in tubes 5 and 6; no growth is seen in tubes 1–4.
   a. Which tube represents the MIC?
   b. Which tube represents the MBC?

14. The figure shows the effectiveness of two antibiotics on a bacterial pathogen isolated from spinal fluid. Which of the following could explain why a clinician might choose to treat the patient with drug A?
   a. Drug B might not penetrate into the brain.
   b. Drug B might be known to have a toxic effect on the patient (patient allergy).
   c. The level of drug A in tissue might be higher and maintained longer than that of drug B.
   d. All of the above
   e. a and b only

Drug A

Drug B

13.5 Antibacterial Agents

15. Which of the following antibiotic classes and host targets are matched correctly?
   a. Penicillin / membrane synthesis
   b. Aminoglycoside (e.g., gentamicin) / 30S ribosome
   c. Quinolone (e.g., ciprofloxacin) / folic acid synthesis
   d. Sulfonamide / DNA gyrase
   e. Macrolide (e.g., azithromycin) / 30S ribosome

16. A bacterial pathogen that produces the enzyme beta-lactamase will be resistant to which of the following antibiotics?
   a. Oxacillin
   b. Azithromycin
   c. Ceftriaxone
   d. Penicillin
   e. Oxazolidinone

17. In the figure, identify which of the antibiotics vancomycin, bacitracin, cephalosporins, or cycloserine affect cell wall synthesis at each of points A, B, and C.

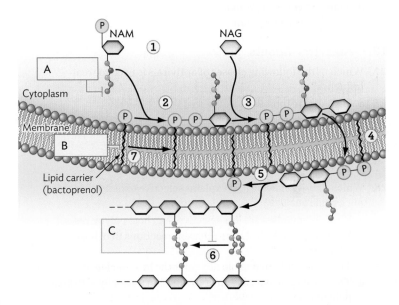

13.6 The Challenges of Antibiotic Resistance

18. _____ is a single bacterial system that can provide resistance to multiple classes of antibiotics simultaneously.
   a. A multidrug efflux pump
   b. A beta-lactamase
   c. An enzyme that modifies antibiotic structure
   d. An enzyme that blocks a drug's access to its target

19. Clavulanic acid _____.
   a. inhibits protein synthesis
   b. acts as a decoy for beta-lactamase
   c. disrupts membrane integrity
   d. acts as a decoy for enzymes that modify and inactivate aminoglycosides
   e. inhibits folic acid synthesis

Match the antibiotic to the resistance mechanism that bacteria use against it.

Antibiotic	Resistance Mechanism
20. Penicillin	a. Alter target
21. Methicillin	b. Degrade antibiotic
22. Quinolones	c. Modify antibiotic
23. Aminoglycosides	d. Pump antibiotic out of cell
24. Tetracyclines	

25. Which of the following antibiotics is/are bactericidal?

  a. Chloramphenicol
  b. Gentamicin
  c. Oxacillin
  d. Ceftriaxone
  e. Doxycycline

26. Bacterial resistance to ciprofloxacin is due to _____.

  a. an altered ribosomal subunit
  b. an altered topoisomerase
  c. production of a degradation enzyme
  d. a missing cell wall protein
  e. a dedicated efflux pump

13.7 Antiviral Agents

27. Which of the following classes of antivirals is used to treat influenza?

  a. DNA gyrase inhibitors
  b. Reverse transcriptase inhibitors
  c. DNA polymerase inhibitors
  d. Neuraminidase inhibitors
  e. Protease inhibitors

28. Which of the following best explains selective toxicity of antiviral nucleoside analogs?

  a. Viral polymerases are less selective than host polymerases.
  b. Viral polymerases catalyze different reactions than host polymerases.
  c. The nucleoside analogs do not enter the host nucleus.
  d. The antiviral nucleoside analogs function normally with host polymerases.
  e. There is a host enzyme that converts the analog incorporated in DNA to the proper nucleotide.

Match the antiviral drug to its target or mode of action.

Drug	Target or Mode of Action
29. Zanamivir	a. CCR5
30. Viracept	b. Fusion inhibitor
31. Enfuvirtide	c. Neuraminidase inhibitor
32. Ribavirin	d. Viral DNA replication
33. Acyclovir	e. Viral RNA synthesis
	f. Protease inhibitor

13.8 Antifungal and Antiprotozoal Agents

34. Among the antifungal agents, _____ interferes with ergosterol synthesis, _____ inhibits DNA synthesis, and _____ disrupts the mitotic spindle.

  a. flucytosine
  b. a triazole or allylamine
  c. amphotericin B
  d. griseofulvin
  e. linezolid

35. A male patient complains of an itchy irritation in his genital region. Your diagnosis is a superficial fungal infection. Which of the following drugs is the usual treatment for this disease?

  a. Itraconazole   d. Amphotericin B
  b. Miconazole   e. Nystatin
  c. Griseofulvin

36. A female patient enters the hospital in severe respiratory distress. The lab report indicates that she has a lung infection caused by the fungus *Histoplasma*. Which of the following drugs would be most appropriate to give the patient?

  a. Amoxicillin   d. Amphotericin B
  b. Nystatin   e. Terbinafine (Lamisil)
  c. Acyclovir

37. A 45-year-old man returned to the United States from a trip to Nigeria. Within a few days, he began experiencing bouts of high fever followed by shaking chills. The doctor in the emergency department diagnosed malaria. Which of the following drugs should be prescribed?

  a. Azithromycin   c. Artemisinin
  b. Metronidazole   d. Chloroquine
  e. Eflornithine

Review Questions

1. List and briefly explain the various means by which humans control microbial growth. What is a D-value? What is a phenol coefficient?

2. How are pasteurization and sterilization different?

3. What is the difference between a disinfectant and an antiseptic?

4. How is the effectiveness of a disinfectant measured?

5. How do microbes prevent the growth of other microbes?

6. What is selective toxicity? Provide examples.

7. Explain the difference between antibiotic susceptibility and antibiotic sensitivity.

8. What is meant by "spectrum of antibiotic activity"?

9. Provide examples of bacteriostatic and bactericidal antibiotics.

10. What is the Kirby-Bauer test? Does it tell whether a drug is bacteriostatic or bactericidal?

11. Give examples of drugs that target each of the following: bacterial cell wall synthesis, RNA synthesis, protein synthesis, and DNA replication.

12. How do antibiotic-producing microorganisms prevent suicide?

13. Why is antibiotic resistance a growing problem?

14. What are the four basic mechanisms of antibiotic resistance?

15. What are some mechanisms used to combat the development of drug resistance?

16. Why are there few antiviral agents available to treat disease?

17. How does amantadine inhibit influenza?

18. Discuss the general modes of action of antifungal and antiprotozoal agents.

Clinical Correlation Questions

1. As part of your senior project, you have been isolating strains of *Escherichia coli* from your stool at 3-week intervals and have checked antibiotic resistance patterns on each isolate. The strains are sensitive to all antibiotics tested except ciprofloxacin (a quinolone antibiotic). Halfway through your project, you develop a bloody diarrhea caused by *Shigella sonnei*, another Gram-negative rod. The physician prescribes amoxicillin for 7 days, which doesn't work, so he switches you to ciprofloxacin. Two weeks after stopping the treatment, you resume isolating *E. coli* from your stool. Now you find that several of the most recently isolated *E. coli* strains are resistant to ampicillin and ciprofloxacin. Explain how this could happen.

2. A patient is admitted to the hospital with a systemic fungal infection. You want to give intravenous amphotericin B. Blood work shows elevated blood urea nitrogen (BUN) and creatinine levels. Explain why you should or should not administer amphotericin B. Use the Internet to search for the significance of elevated BUN and creatinine.

3. A 70-year-old patient with signs of sepsis is rushed to the emergency department. The physician quickly orders empirical antibiotic therapy with imipenem and sends a blood sample for culture and antibiotic susceptibility testing. Imipenem is a carbapenem with a very broad spectrum of activity that includes Gram-positives, Gram-negatives, *Pseudomonas*, and anaerobes. Within 12 hours the patient has significantly improved. The next day the lab reports having isolated *Streptococcus pneumoniae* from the blood. The pathogen was found to be sensitive to imipenem, vancomycin, amoxicillin, tetracycline, chloramphenicol, and trimethoprim-sulfamethoxazole but was resistant to levofloxacin. Do these results change the physician's treatment strategy? Why and how?

4. Refer to Case History 13.1. Why do you think the physician treating Molly wanted to know Molly's immunization status?

5. Review Case History 13.3. Because Jamal was allergic to penicillin (the usual drug of choice for fighting off *Streptococcus pyogenes* infection), he was given azithromycin. Why wasn't he given oxacillin instead, which is structurally different from penicillin?

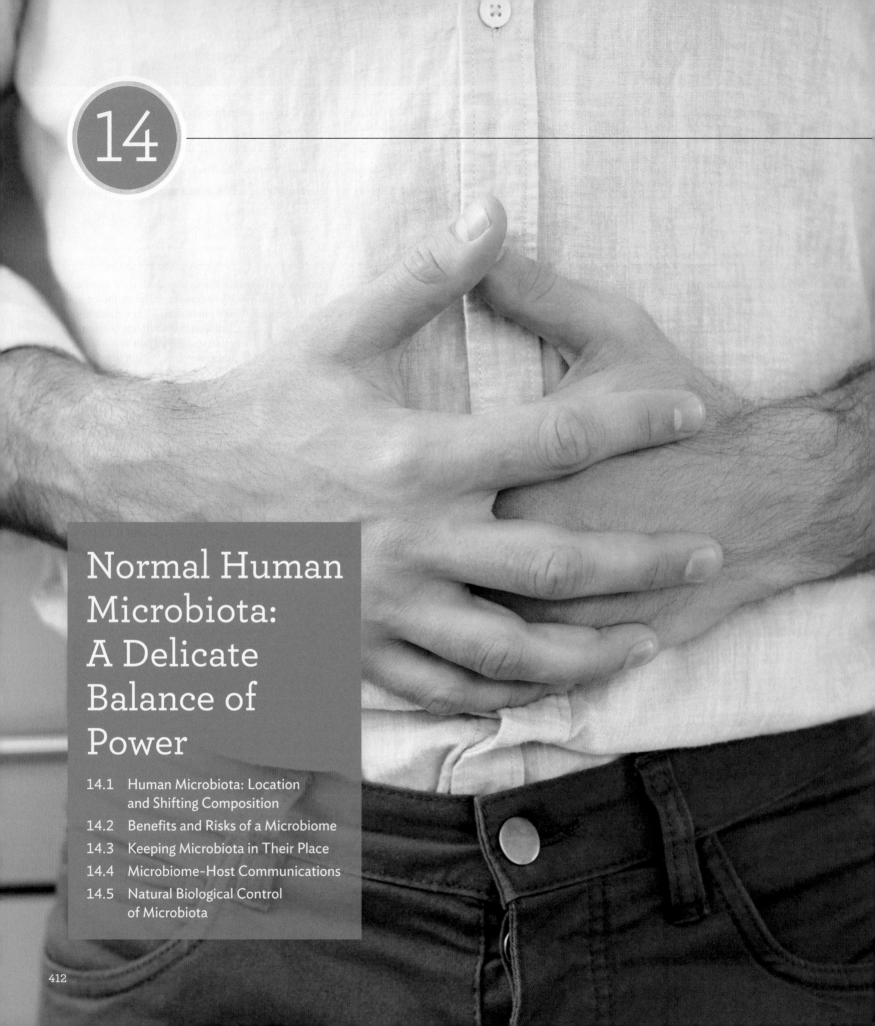

14

Normal Human Microbiota: A Delicate Balance of Power

[Case of the Unsatisfactory Stool]

SCENARIO A few months ago, Jason, a 38-year-old software salesman, visited his physician to complain of "unsatisfactory stools." The frequency of his stools was about two to three times a day but fluctuated from week to week. Sometimes his stools were semisolid and covered with thick mucus; other times they were characterized by small, hard, elongated pellets. He also complained of having weekly episodes of severe cramping pain all over his abdomen that would come and go during the course of a day. He said he belched frequently and would sometimes experience an ineffectual urge to pass stools, occasionally passing only gas.

Colonoscopy
In this procedure, a fiber-optic endoscope is inserted from rectum to cecum. The endoscope can also take tissue samples. Although IBS is painful, the colons of patients, as viewed by colonoscopy, have a normal appearance. On the left is a normal colon, as viewed by colonoscopy. On the right is the colon of a person with inflammatory bowel disease, which is more serious than IBS. The mucosa is denuded, and active bleeding is seen.

DIAGNOSIS Jason was ultimately diagnosed with a nebulous disease called irritable bowel syndrome (IBS). IBS affects 10%–15% of the US population. Although microbes are involved, IBS is not an infection. Jason's distress was most likely caused by a shift in the microbial composition of his intestinal contents. A change in the mix of bacteria and archaea residing in the intestine can lead to inflammation in the bowel and the symptoms Jason experienced. A research study Jason enrolled in showed that his gut microbiome contained lower than typical numbers of *Lactobacillus*, *Bifidobacterium*, and *Faecalibacterium prausnitzii*.

OUTCOME The clinician recommended that Jason switch to a FODMAP diet low in fermentable carbohydrates and ingest probiotics containing microorganisms that normally inhabit the intestine, such as *Bifidobacterium* and *Lactobacillus* species. The hope was that these bacteria could help restore the microbial composition of his intestine to a healthier balance and thus reduce inflammation. Jason was also evaluated for depression, which can worsen symptoms. In this case, the probiotic treatment worked (it doesn't always), and within a week, Jason's gastrointestinal symptoms disappeared, much to his relief.

SEM | 5 μm

Intestinal Microbiome
Intestinal bacteria in the small intestine are essential to our survival. Among other things, they help train the immune system, digest carbohydrates that we cannot, and provide vitamins our bodies need.

Note: Do not confuse IBS with inflammatory bowel disease (IBD). IBD is a more serious and complicated illness. Two common forms of IBD are ulcerative colitis, whose symptoms include small ulcers and abscesses in the colon and rectum, and Crohn's disease, in which any part of the gastrointestinal tract can become inflamed and ulcerated. Although less severe, IBS is nonetheless very disruptive and painful. IBS and IBD are both associated with altered microbiota, but IBD also appears to involve a more severe dysfunction of the immune system than does IBS.

Like it or not, your body teems with microbial hitchhikers. You might even say that we are more microbe than human because the human body contains about twice as many bacterial cells as it does nucleated human cells. All the bacteria, archaea, fungi, and protozoa that inhabit our bodies are called the human **microbiome** or **microbiota**. These microbes play critically important roles in our health and development. Jason discovered that the balance between members of the gut microbiome is very important. We will discuss how probiotics and, in severe cases, fecal microbiota transplants (FMTs) can correct imbalances and alleviate some gastrointestinal ailments.

Our microbiome has many curious features. For instance, despite the importance of the microbiome to human health, the majority of contributing species cannot be grown easily in the laboratory. Their growth in a host depends on factors yet to be determined, which are provided by the human microbiome itself.

You might also be surprised to learn that microbes and their human hosts chemically "talk" to each other. Chemical signals, including hormones, are sent back and forth, altering gene expression and physiology of all three domains that make a human: Eukarya, Archaea, and Bacteria. Today, considerable research is directed toward understanding the composition of a healthy microbiome and its relationship to human health.

This chapter will address several issues of the host-microbe relationship. For instance:

- What organisms make up our microbiota?
- Does the composition of an individual's microbiota change over time?
- Why do humans maintain normal microbiota if microbes also serve as sources of infection?
- How does our body keep members of our microbiota in their place?
- What cost-benefit analysis might explain the evolution of the host-microbe relationship?

14.1 Human Microbiota: Location and Shifting Composition

How and when do we become colonized by microbes? We once thought that a fetus growing in the amniotic sac was sterile and that colonization by microbes began only after birth. That view is now challenged by finding bacteria in the first stool (meconium) passed by a newborn <u>before</u> its first meal. To be present in meconium, presumably these organisms must have colonized the fetal intestine in the womb. Another surprise was that the placenta, also once thought to be sterile, may itself have a microbiome. Be aware that these results are very controversial (other studies failed to find a placental microbiome) but, if proven true, it would mean that we are constantly exposed to microbes, from the womb to the tomb.

What is not controversial is that once the baby breaks out of the embryonic membrane, it is exposed to a whole array of microbes residing in the birth canal and in the awaiting world. As a result, humans—and all other higher life forms—are heavily populated with bacteria. Colonization typically occurs where our tissues come in contact with the external environment (for example, mouth, skin, and parts of the genitourinary tract). The more inaccessible internal body sites are typically sterile. For example, most internal organs, as well as the blood and cerebrospinal fluid, should not harbor any bacteria. The presence of pathogenic microbes at these sites indicates that an infection is under way.

Current research tells us that our microbiome is critically important to our health—even if the average person doesn't realize it's there. We now understand that the bacteria colonizing our bodies may collectively be as important to us as a kidney or liver. Some scientists even call the microbiome a "virtual organ system." For instance, our body's cells are constantly bathed in metabolites produced by gut bacteria. It is thought that these microbial metabolites, which circulate in our blood, significantly and favorably influence human health and development. Members of our microbiome also play an important role in preventing pathogens from colonizing us, a phenomenon called microbial antagonism. Antagonism is a consequence of normal microbiota occupying body sites that pathogens "covet" and their ability to create an environment that is inhospitable to, or inhibits, pathogen growth.

Figure 14.1 Examples of Normal Microbiota Present at Various Colonizing Sites

All these species can be found as part of the microbiota without causing disease. Species highlighted in yellow are notable for their predominance at the site. Species in red font are potential pathogens. *Staphylococcus* species are differentiated by whether they can produce coagulase, an enzyme that coagulates serum. *Staphylococcus aureus* is generally coagulase-positive (there are some coagulase-negative strains), whereas other species of staphylococci normally found on skin are coagulase-negative. Some organisms—*Streptococcus pneumoniae*, for example—are pathogens but can be found as normal microbiota in some individuals. (Note that viridans streptococci is not an actual species but represents a collection of streptococci that can turn blood agar green.)

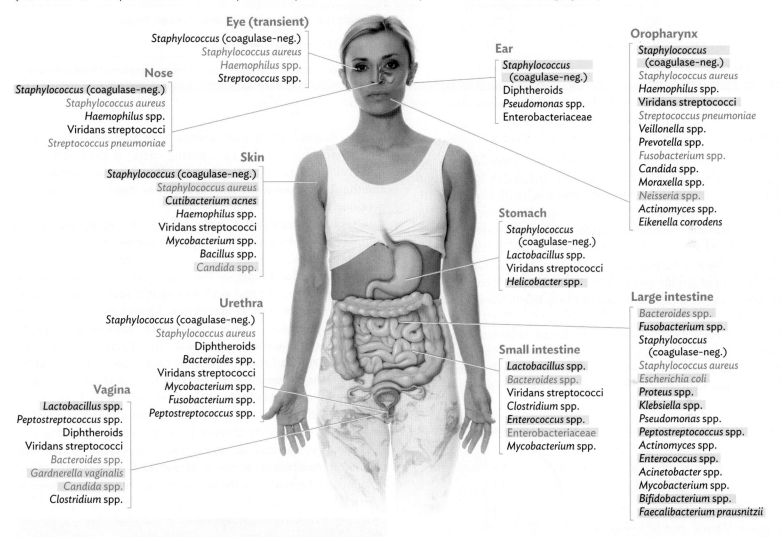

Eye (transient)
Staphylococcus (coagulase-neg.)
Staphylococcus aureus
Haemophilus spp.
Streptococcus spp.

Nose
Staphylococcus (coagulase-neg.)
Staphylococcus aureus
Haemophilus spp.
Viridans streptococci
Streptococcus pneumoniae

Skin
Staphylococcus (coagulase-neg.)
Staphylococcus aureus
Cutibacterium acnes
Haemophilus spp.
Viridans streptococci
Mycobacterium spp.
Bacillus spp.
Candida spp.

Urethra
Staphylococcus (coagulase-neg.)
Staphylococcus aureus
Diphtheroids
Bacteroides spp.
Viridans streptococci
Mycobacterium spp.
Fusobacterium spp.
Peptostreptococcus spp.

Vagina
Lactobacillus spp.
Peptostreptococcus spp.
Diphtheroids
Viridans streptococci
Bacteroides spp.
Gardnerella vaginalis
Candida spp.
Clostridium spp.

Ear
Staphylococcus (coagulase-neg.)
Diphtheroids
Pseudomonas spp.
Enterobacteriaceae

Stomach
Staphylococcus (coagulase-neg.)
Lactobacillus spp.
Viridans streptococci
Helicobacter spp.

Small intestine
Lactobacillus spp.
Bacteroides spp.
Viridans streptococci
Clostridium spp.
Enterococcus spp.
Enterobacteriaceae
Mycobacterium spp.

Oropharynx
Staphylococcus (coagulase-neg.)
Staphylococcus aureus
Haemophilus spp.
Viridans streptococci
Streptococcus pneumoniae
Veillonella spp.
Prevotella spp.
Fusobacterium spp.
Candida spp.
Moraxella spp.
Neisseria spp.
Actinomyces spp.
Eikenella corrodens

Large intestine
Bacteroides spp.
Fusobacterium spp.
Staphylococcus (coagulase-neg.)
Staphylococcus aureus
Escherichia coli
Proteus spp.
Klebsiella spp.
Pseudomonas spp.
Peptostreptococcus spp.
Actinomyces spp.
Enterococcus spp.
Acinetobacter spp.
Mycobacterium spp.
Bifidobacterium spp.
Faecalibacterium prausnitzii

The majority of species that make up our microbiome are unknown and have never been grown in the lab. However, as part of the Human Microbiome Project, new DNA sequencing technologies have enabled scientists to sequence the genomes, or many pieces of them, from all the microbes at a particular body site. The process, called metagenomic DNA sequencing, has already provided valuable insights about our microbiome and about host-microbe relationships. For instance, researchers who have compared the intestinal microbiomes of lean and obese humans discovered vast differences in gut microbiota. It is now thought that colonization by different species of bacteria influences the efficiency of digestion and absorption of nutrients.

This metagenomic cataloging of microbes in a person's microbiome has revealed that each person's microbiome is unique, like fingerprints. The composition of a person's microbiome is not set in stone, however. An individual's microbiome can evolve over time with a change in diet or after making new human (or animal) friends.

In this section, we describe some of the organisms that make up the human microbiome and discuss why they populate different sites on the human body. **Figure 14.1** illustrates various human body sites colonized by microbiota and provides examples of the microbial species found there. **Table 14.1** lists the four prominent ecosystems of humans along with the number of microorganisms that typically inhabit each (called the **bioburden**). The table also includes the ratio of aerobes to anaerobes (note that the body has anaerobic parts, such as the intestine) and the sources of colonizing species. Viruses are also part of the human microbiome (called the **virome**), probably equal to or greater in number than the bacteria they prey upon. But because little is known about how they impact human health, we will not dwell on them here.

Table 14.1
The Prominent Bacterial Ecosystems of Humans

Microbial Reservoir	Total Bioburden or Colony-Forming Units	Aerobe-to-Anaerobe Ratio[a]	How Acquired over a Lifetime
Skin	10^{4-6}/sweat gland	1:10	Initially birth canal; then oral and external environments
Mouth	10^{6-8}	1:10	Initially birth canal, caregiver; then food, water, fingers
Genitourinary tract	10^{8-9}/vagina/urethra	1:100	Surrounding external environment
Intestine (colon)	10^{11}/cm^3	1:1,000	Possibly as a fetus; then baby formula, mother, ingestion of food and water

[a]Note that the more anaerobic the body site, the more anaerobes are present. Also note that even aerobic sites contain anaerobes.

Skin

The human adult, on average, is covered with 2 square meters (more than 21 square feet) of skin (**epidermis**) populated by 10^{12} microorganisms. Skin, as do all colonized body sites, harbors persistently colonizing (normal) and transiently colonizing microbiota. But even a persistently colonizing microbe exhibits diversity in that different strains colonize at different times. Thus, our microbiota do not comprise a static population but represent a vibrantly changing mixture of strains and species. Of course, some combinations of species may be unique and stable on an individual (see **IMPACT**).

Several features of the epidermis make it difficult to colonize. Large expanses of skin are subject to drying, although some areas harbor enough moisture to support microbial growth; moist areas include the scalp, ear, armpit, and genital and anal regions. The skin also has an acidic pH (pH 4–6) due to the secretion of organic acids by oil and sweat glands. As noted in Section 6.4, organic acids inhibit microbial growth by lowering bacterial cytoplasmic pH. Epidermal secretions are also high in salt and low in water activity, and they contain enzymes such as lysozyme that degrade bacterial peptidoglycan.

Despite these hurdles, many species of bacteria manage to colonize the epidermis (1,000 species are estimated). Most of these organisms are Gram-positive because they tend to be more resistant to salt and dryness. *Staphylococcus epidermidis*, various *Bacillus* species, and yeast such as *Candida* species are common examples of normal skin microbiota. Though normal, they are not always innocuous. One member of the skin's resident microbiota, the Gram-positive anaerobic rod *Cutibacterium acnes* (formerly *Propionibacterium acnes*), causes acne, a very visible plague of adolescence. Curiously, this species is closely related to a species that makes Swiss cheese (see Section 27.3). Increased hormonal activity during the teen years stimulates oil production by the sebaceous glands (**Figure 14.2**). *C. acnes* readily degrades the triglycerides in this oil, turning them into free fatty acids that promote inflammation of the oil glands. One consequence of the inflammatory response is the formation of a blackhead (known as

Figure 14.2 Anatomy of the Skin and the Development of Acne

Epidermis
Dermis
Subcutaneous tissue

Hair follicle   Sebaceous gland   Blood vessels   Sweat gland

A. Location of sebaceous glands in the skin.

B. Acne on a patient's back.

Figure 14.3 Examples of Anaerobic Oral Microbiota and Periodontal Disease

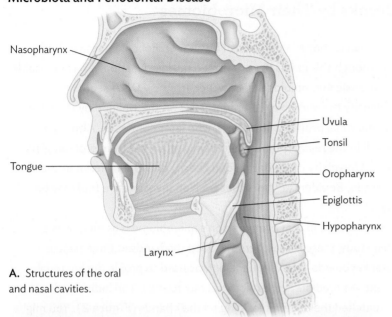

A. Structures of the oral and nasal cavities.

B. *Prevotella* (colorized).

C. *Fusobacterium*.

D. Symptoms of periodontal disease include red, swollen gums; bleeding gums; gum shrinkage; and teeth drifting apart.

a comedo), a plug of fluid and keratin that forms in the duct of the sebaceous gland (the oils turn black from oxidation). The result is the typical skin eruptions of acne (see Chapter 19). Because of its bacterial basis, treatments for acne include tetracycline to kill the bacteria.

The Eye

Because the eye is exposed to the outside environment, we might expect it to be heavily colonized. However, colonization is inhibited by the presence of antimicrobial factors such as lysozyme in the tears that continually rinse the eye surface (conjunctiva). Despite this, a few, probably transient, colonizing bacteria can be found on the conjunctiva and ocular surfaces. Members of five phyla have been found: mostly Proteobacteria and Firmicutes but also Actinobacteria, Bacteroidetes, and Deinococcus-Thermus. Skin microbiota such as *Staphylococcus epidermidis* and diphtheroids (Gram-positive rods that look like clubs), as well as some Gram-negative rods such as *Pseudomonas*, *Rhizobium/Bradyrhizobium*, and *Acinetobacter* manage to make the eye their home without causing damage. Occasionally, bacteria such as *Streptococcus pneumoniae* and *Haemophilus influenzae*, as well as some viruses, can cause an ocular disease known as pink eye (Section 19.6), a condition in which the inflamed conjunctiva becomes pink or red and emits a watery discharge.

Oral and Nasal Cavities

Within hours of birth, a human infant's mouth becomes colonized with nonpathogenic *Neisseria* species (Gram-negative cocci); *Streptococcus*, *Actinomyces*, and *Lactobacillus* species (all Gram-positive); and some yeasts. These organisms come from the environment surrounding the newborn, such as the mother's skin and garments. As teeth emerge in the newborn, the anaerobic nature of the space between teeth and gums supports the growth of anaerobes such as *Prevotella* and *Fusobacterium* (Figure 14.3). Whatever the organism, colonizers of the oral cavity must be able to adhere to surfaces such as teeth and gums (gingiva) to avoid mechanical removal and flushing into the acidic stomach. Organisms from 14 different phyla can be found in the mouth, with most species coming from Firmicutes, Bacteriodetes, and Proteobacteria. All told, the teeth and gingival crevices are colonized by more than 500 species of bacteria.

Organisms such as *Streptococcus mutans*, which attaches to tooth enamel, and *Streptococcus salivarius*, which binds gingival surfaces, secrete a polysaccharide or peptide polymer called a glycocalyx that enables the bacteria to firmly adhere to oral surfaces and to each other. They are but two of the microbes found in dental plaque—a sticky, yellowish biofilm that builds up on teeth. The acidic fermentation products of these organisms demineralize and dissolve tooth enamel, which leads to dental caries (tooth decay).

Important microbial habitats of the throat include the **nasopharynx**, which is the area leading from the nose to the oral cavity, and the **oropharynx**, which lies between the soft palate and the upper edge of the epiglottis (see Figure 14.3A). Organisms such as *Staphylococcus aureus* and *Staphylococcus epidermidis* populate these sites. The nasopharynx and oropharynx can also harbor relatively harmless streptococci such as *S. salivarius* and *S. oralis*, as well as *S. mutans*. Other oropharyngeal organisms include a large number of diphtheroids and small Gram-negative rods like *Eikenella corrodens* and *Moraxella catarrhalis*. Within the tonsillar crypts (small pits along the tonsil surface) lie anaerobic species of *Prevotella*, *Porphyromonas*, and *Fusobacterium*.

Even though oral microbiota are normally harmless, they do hold potential for disease. Dental procedures, for instance,

impact

Catching Crooks by Their Microbiomes

Imagine this: A man is shot in a dark alley (Figure 1). Police detectives find a gun nearby. Although this gun is the confirmed murder weapon, it has no usable fingerprints. Forensic microbiologists, however, use new DNA sequencing technologies to identify the microbial community left on the weapon by the killer and matched the microbial profile to a suspect, who later confesses. Although that approach is still in the realm of science fiction, a study led by microbiologists Rob Knight (now at University of California San Diego) and Noah Fierer, of the University of Colorado, Boulder, suggests that such an investigative tool may be available one day.

Knight and Fierer reasoned that every person's microbiome is as unique as a fingerprint. In their study, these researchers swabbed the hands of three people and their computer keyboards. DNA sequencing was used to profile the bacterial populations present. As predicted, the microbial mix found on an individual's keyboard closely matched the bacteria residing on their hands (Figure 2). You might ask why not just extract human DNA from a smudged fingerprint and identify the killer that way? A fingerprint contains very few human cells but many bacterial cells, so obtaining human DNA from a fingerprint would be challenging.

The team then compared the bacterial DNA signatures of the computer mice and user's palms with a database of microbiomes taken from the hands of 270 other people as part of the Human Skin Microbiome project. In all nine cases, the bacterial profile on a computer mouse was more similar to that of the owner's palm than to any other hand in the database.

A more rigorous approach was taken by scientists at the University of North Texas. They used rapid DNA sequencing methods to examine 286 bacterial genetic markers in 22 microbial taxa found on skin to profile 51 people. The method not only identified the variety of species present in a sample but also pinned down individual strains of each species by examining sequence diversity within those markers. The approach is similar to using a tire tread pattern to identify the make of a suspect's car and then analyzing slight irregularities in the tread pattern to find the *specific* car and its owner. This approach proved 100% accurate in identifying individuals.

Can this technique work in real-world applications? One problem is the possibility of cross-contamination by two or more people touching the same object. The mixture of bacteria might resemble that of a third person. Despite these important questions, microbiome forensics is an intriguing prospect.

Figure 1

Figure 2 UV Light Showing Germs on a Keyboard

DISCUSSION QUESTION

How might the analysis of a necrobiome—the community of organisms found on or around a decomposing body—aid in forensics?

Figure 14.4 Gross Pathology of Subacute Bacterial Endocarditis Involving the Mitral Valve

Vegetations

The left ventricle of the heart has been opened to show mitral valve fibrin vegetations due to infection. These growths are not present in a normal heart.

can cause these organisms to enter the bloodstream, transiently producing what is called **bacteremia** (presence of bacteria in the bloodstream; Section 21.3). Normal immune mechanisms clear these transient bacteremias quite easily, but in patients who have a heart mitral valve prolapse (heart murmur), microbes can occasionally become trapped in the defective valve and form bacterial vegetations. Vegetations are biofilms formed by a large number of bacterial cells encased within a mix of secreted glycocalyx and fibrin produced by clotting blood. The disease state is called subacute bacterial endocarditis; "subacute" because it develops slowly (**Figure 14.4**). Once buried deep within a vegetation, the microbes are extremely difficult to kill with antibiotics. Recall, too, that oral bacteria played prominent roles in the bite wound suffered by Cody after his fistfight in the opening case of Chapter 10.

The Respiratory Tract

The respiratory tract has a surface area exposed to air of about 750 square feet. Originally thought to be sterile, the trachea and even deep areas of the lung are now known to contain low numbers of normal microbiota. Numbers range from approximately 10 to 100 bacteria per 1,000 human cells. The predominant genera are from the Bacteroidetes and Firmicutes phyla, including *Prevotella, Veillonella,* and *Streptococcus.*

The upper respiratory tract, of course, contains numerous microbes, but many organisms entering the nasopharynx are trapped in the nose by cilia that beat toward the pharynx. The ultimate destination of these microbial interlopers is the acidic stomach—and death. Microorganisms that make it into the trachea are trapped by mucus produced by ciliated epithelial cells that line the airways leading to the lungs. The ciliated mucous lining of the trachea, bronchi, and bronchioles makes up the **mucociliary escalator** (**Figure 14.5**). The mucociliary escalator constantly sweeps foreign particles up and out of the lungs. This is extremely important for preventing respiratory infections. When

the mucociliary escalator fails or is overwhelmed by inhaling too many infectious microbes, infections can result. Such infections include the common cold, caused by rhinoviruses, or pneumonia, caused by *Streptococcus pneumoniae.*

As with the gut microbiome, the lung microbiome can contribute to the development of lung mucosal immunity. However, the lung microbiome can also add to the severity of chronic airway diseases such as chronic obstructive pulmonary disease (COPD). Lungs damaged by disease (for example, by cystic fibrosis or emphysema) are populated by a number of different bacterial species that can exacerbate the disease state. Many of the organisms found are potentially pathogenic, such as *S. pneumoniae, Pseudomonas aeruginosa, Haemophilus influenzae,* and *Klebsiella pneumoniae.* Others do not typically cause respiratory infections, including *Staphylococcus epidermidis,* some species of *Streptococcus* and *Corynebacterium,* and some fungi, such as *Candida* species. Nevertheless, these organisms can form biofilm communities in the bronchioles of damaged lungs and interfere with lung function.

The Stomach

We have known for more than a century that the stomach contents are acidic and that gastric acidity can kill bacteria. Just how important that acidity is for protection against microbes is illustrated by the infection caused by *Vibrio cholerae,* the causative agent of cholera. Cholera is a severe diarrheal disease endemic to many of the poorer countries of the world. The toxin produced by this organism acts on the intestinal lining to cause voluminous diarrhea—as much as 10 liters a day (as described in the case history that opens Chapter 4). Death can occur rapidly as a result of dehydration and shock. Although cholera actually infects the intestine—not the

Figure 14.5 Lung Structure and Mucociliary Escalator

A. Basic lung anatomy.

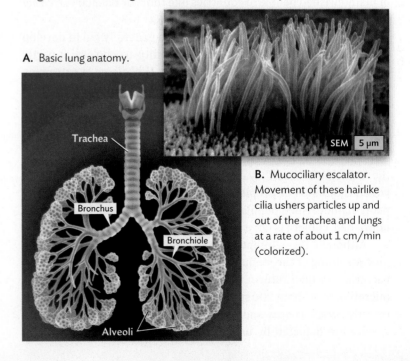

Trachea

Bronchus

Bronchiole

Alveoli

SEM 5 µm

B. Mucociliary escalator. Movement of these hairlike cilia ushers particles up and out of the trachea and lungs at a rate of about 1 cm/min (colorized).

Figure 14.6 *Helicobacter pylori* Defies Stomach Acidity

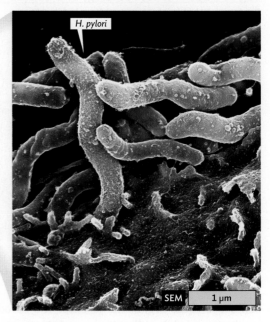

This figure shows the stomach at increasing levels of magnification to reveal *H. pylori* growing in the mucus lining the stomach epithelium. *H. pylori* bacteria attach to gastric epithelial cells and induce specific changes in cellular function, such as an increase in the expression of laminin receptor 1, a protein associated with the development of stomach cancer.

stomach—the bacteria must survive passage through the stomach to reach the intestine.

The importance of stomach acidity in preventing infections is probably best illustrated by a US Army study conducted in the mid-twentieth century. The study was supposed to test potential cholera vaccines for their efficacy. But after scientists administered huge numbers of virulent *V. cholerae* organisms to healthy volunteers (as a control group), very few of the volunteers contracted cholera. This finding was confusing because hundreds of thousands of malnourished people in countries like India or Bangladesh easily contract the disease.

Why did volunteers given live cholera generally fail to develop the disease? It turns out that susceptibility to cholera is partly related to differences in stomach acidity. *V. cholerae* is extremely sensitive to low pH; even a pH of 4 will readily kill it. Because the pH in the stomachs of healthy volunteers was pH 4 or below, *V. cholerae* was easily killed. Malnourished people, however, suffer from achlorhydria (loss of stomach acid). This permissive environment enables ingested cholera microbes to survive and enter the intestine, where they can cause disease. Cholera epidemics in which thousands of people die, even today, typically occur among poor, malnourished populations. Because the US populace is better nourished, cholera is not a major problem in the United States.

Although the contents of the stomach are very acidic, the mucous lining of the stomach is much less so. It is there that some bacteria can take refuge. In fact, the average stomach harbors a microbiome of about 200 species, mainly from the phyla Firmicutes (mostly *Streptococcus* and *Veillonella*) and Proteobacteria. These species are detected by using cultural and molecular techniques

such as the polymerase chain reaction (PCR) to detect and sequence 16S rRNA genes (see Section 10.1). A classic stomach pathogen is *Helicobacter pylori* (see Section 10.3). This organism has a remarkable ability to resist acidic pH. It can survive at pH 1 in stomach contents by using the enzyme urease to generate ammonia and raise the local pH. However, it will not grow at pH 1. It can, however, grow and divide in the stomach's mucous lining, where the pH is closer to 5 or 6 (Figure 14.6). The US Centers for Disease Control and Prevention (CDC) estimates that *Helicobacter* colonizes the stomachs of half the world's population. Most of the time, *Helicobacter* does not cause any apparent problem, but on occasion, the organism can produce gastric ulcers and even cancer.

The Intestine

CASE HISTORY 14.1

Case of the Greasy Stool

Renita, a 67-year-old woman, complained to her physician of having persistent diarrhea with weight loss, bloating, and excess flatulence (passing gas). Renita also said her stools looked greasy, which the physician knew was a sign that ingested fat was not being absorbed. A blood test revealed that the woman's vitamin B_{12} level was abnormally low. Cultures of fecal specimens failed to find any typical bacterial pathogens that could cause diarrhea. However, the decreased vitamin B_{12} level and excess fat in the stool suggested the woman had a malabsorption disease.

The physician, initially perplexed, suddenly realized that the cause could be small intestine bacterial overgrowth disease, a diagnosis confirmed by aspirating fluid from the jejunum (small intestine). The laboratory discovered abnormally high numbers of facultative and anaerobic bacteria in the fluid, including *Streptococcus*, *Escherichia coli*, *Lactobacillus*, and *Bacteroides*. Fermentation by these organisms produced the bloating and flatulence. Antibiotic treatment corrected the syndrome.

This case illustrates the remarkable balance that must be maintained between a host and its microbiota. Bacteria considered healthy members of one part of the intestine, the colon, can become irksome if they begin to populate another area, the jejunum. How does the healthy body avoid this?

The gastrointestinal tract beyond the stomach consists of an extremely long tube made of several sections, each of which provides a unique environment that supports the growth of different bacterial species. Pancreatic secretions, including bile (at pH 10), enter the intestine at a point just past the stomach and raise the intestinal pH to about pH 8. This increase in pH immediately relieves the acid stress placed on microbes. The relatively high pH and bile content of the duodenum and jejunum (the section of small intestine just past the duodenum; **Figure 14.7**) allow colonization by only a few residents, mostly Gram-positive Firmicutes (enterococci and lactobacilli) and Actinobacteria (diphtheroids). Gram-positive organisms are, in general, very sensitive to bile salts as a result of lacking

Figure 14.8 Human Feces Showing Food Particles and Resident Bacteria (Colorized)

an outer membrane. However, the specific Gram-positive species that live in the small intestine possess a bile salt hydrolase that can destroy bile salts if they enter the cell.

There are only an estimated 10^7 bacteria per milliliter of distal jejunum aspirate. The problem in Case History 14.1 was that the bile content was decreased in the patient, allowing more bacteria to grow. Because there were so many bacteria, nutrients in the small intestine were consumed faster by bacteria than by the host. This caused the malabsorption symptoms. For instance, not enough vitamin B_{12} remained to be absorbed, hence the low blood level of B_{12}. Because bile also solubilizes ingested fat so it can be absorbed, the decreased bile content led to increased fat content in Renita's stools and the greasy appearance.

In contrast to the jejunum, the distal parts of the human intestine (ileum and colon) have a slightly acidic pH (pH 5–7) and lower levels of bile salts. These conditions support a more diverse and vibrant ecosystem. The intestine actually contains roughly 10^{11}–10^{13} bacteria per gram of feces (**Figure 14.8**). This generally anaerobic environment is populated by both anaerobic and facultative microbes in a ratio of about 1,000 anaerobes to 1 facultative organism. Why is the intestine anaerobic? The small amount of oxygen that diffuses from the intestinal wall into the lumen is immediately consumed by the facultative microbes, rendering the environment anaerobic.

WHICH BACTERIA COLONIZE OUR INTESTINE? Most estimates place the number of unique bacterial species living in the gut to be about 500 or more! Why so diverse? Intestinal bacteria have special talents that allow them to bind and colonize the intestinal wall. One reason why such a diverse mix of species is supported in the intestine is that different species of bacteria attach to different host cell receptors. Also, the many different food sources within the intestine support metabolically diverse groups of microbes.

Figure 14.7 The Human Gastrointestinal Tract

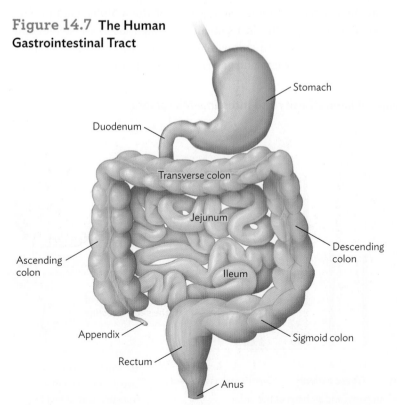

In addition to the structures shown here, the liver and gallbladder are considered parts of the gastrointestinal system.

But how did members of the gut microbiome get there? The fetal intestine was thought to be sterile, but we now find that contents of the fetal intestine may contain bacteria, possibly donated from the placenta. Microbial colonization of the new child continues during birth and is influenced by the mode of delivery (birth canal versus cesarean) and type of feeding (formula versus breast milk). Formula-fed babies are initially colonized by a mixture of Enterobacteriaceae, streptococci, and staphylococci. The microbiota of breast-fed babies, however, consists primarily of *Bifidobacterium* species that easily digest human milk oligosaccharides (lactose with long side chains). One reason some people promote breast-feeding over bottle-feeding is that the metabolic products of *Bifidobacterium* have been shown to protect against infection by some intestinal pathogens. But there is much we do not know about colonization of the newborn human.

Only recently have we really started to learn what microbes inhabit our intestine, mainly because we cannot culture most of them. The use of culture-independent techniques such as PCR—a method to amplify DNA present in small quantities for further study—and new DNA sequencing methods can identify the nonculturable species through differences in their 16S rRNA genes. The results reveal the identity, diversity, and variation among individual members of the human microbiome. In one example, 395 unique microbial "phylotypes" were identified in humans, but 60% were unknown at the species level, and 80% have never been cultured. Furthermore, scientists observed huge differences between individuals in what phylotypes were present. (Phylotypes are organisms that can be assigned to different phyla; see Section 10.1.)

All resident organisms normally present in the intestine are considered normal microbiota and are usually harmless if they stay where they belong. But if they accidentally escape into nearby tissues, they can cause serious disease (for instance, appendicitis).

Organisms that inhabit the intestine are mostly from the Bacteroidetes and Firmicutes phyla, but many other phyla are represented. For instance, the most well-known gut microbe, *E. coli*, is from the phylum Proteobacteria, but it is a minor member of the gut. One of the most abundant bacteria in the gut is *Faecalibacterium prausnitzii*, a member of the Firmicutes that produces a large amount of the fermentation by-product butyrate. Butyrate serves as a major energy source for epithelial cells in the colon and can inhibit inflammation. Other well-known gut microbes include anaerobes such as *Bacteroides* species (Gram-negative rods, from phylum Bacteroidetes) and two firmicute genera, *Peptostreptococcus* and the spore-forming *Clostridium*. Besides bacteria, other common inhabitants are the yeast *Candida albicans* and protozoa such as *Trichomonas hominis* and *Entamoeba hartmanni*.

Bacteroides thetaiotaomicron is a particularly important gut microbe. Much of its genome is dedicated to taking the complex carbohydrates we eat and breaking them down into products we can use. We humans actually absorb 15%–20% of our daily caloric intake in this way. As a whole, gut microbiota perform a number of useful functions, such as fermenting unused energy substrates, training the immune system, preventing growth of pathogenic bacteria, regulating the development of the gut, producing vitamins for the host (such as biotin and vitamin K), and producing hormones that direct the host to store fats. Additional bacterial inhabitants of the gut with important roles in human health will be mentioned later.

This broad diversity of species that make up the gut microbiome is important for human health, as we will describe later. Maintaining that diversity, however, is not trivial. Despite the buffet-like assortment of foods available in the intestine, many species still end up vying for the same nutrient, so competition for food is fierce. Perturbing the balance of power with antibiotics, diet, or stress can reduce gut diversity and lead to disease.

IS THERE A CONNECTION BETWEEN OUR MICROBIOTA AND OBESITY? Obesity is a growing public health concern influenced by the increased consumption of high-energy foods and decreased physical activity. Recent evidence from many scientists suggests that gut microbiota may influence obesity by increasing energy-harvesting efficiency. One of the simpler studies supporting this hypothesis used high-throughput DNA sequencing and metagenomics to compare the composition of microbiota in normal-weight, obese, and gastric bypass patients.

Figure 14.9 Relationship between Obesity and Intestinal Microbiota

A. Relative numbers of bacteria, archaea, and archaea of the order Methanobacteriales in different normal-weight, obese, and gastric bypass patients, measured as ribosomal RNA genes quantified by real-time PCR.

The small intestine is connected to the stomach pouch.

Unused portion of the small intestine.

B. Illustration of a gastric bypass, in which the intestine is attached to a small, surgically engineered stomach pouch.

All three groups in the study had highly diverse species of bacteria present, but the archaea present were mainly H_2-oxidizing methanogens (order Methanobacteriales) that produce the greenhouse gas methane. The surprise was that obese individuals had much higher numbers of the H_2-using, methane-producing methanogenic archaea than did normal-weight or gastric bypass individuals (**Figure 14.9**).

How might a methanogen influence weight gain? **Figure 14.10** illustrates that H_2 buildup from bacterial fermentation normally slows fermentation, possibly by inhibiting NADH dehydrogenase. NADH dehydrogenase regenerates NAD^+ needed for fermentation (NAD metabolism is discussed in Chapter 7). Oxidation of H_2 by methanogens would remove that block and allow more efficient regeneration of NAD^+. More efficient regeneration of NAD^+ would enable bacteria to ferment <u>more</u> plant polysaccharides and produce <u>more</u> fermentation end products such as acetate and butyrate, which human cells readily use. The result is weight gain. Consequently, the weight loss following gastric bypass may reflect the double impact of altering the gut microbiota as well as decreasing food ingestion. Be aware that many other intriguing studies suggest an even more complex relationship between the intestinal microbiome and obesity, as well as diabetes.

Genitourinary Tract

Much of the genitourinary tract is normally free from microbes (**Figure 14.11**). These areas include the kidneys, which remove waste products from the blood; and the ureters, which remove the urine from the kidneys. The urinary bladder, which holds urine until it is excreted, was thought to be sterile but is now known to harbor small numbers of microbes (mainly anaerobes). The distal urethra, because of its proximity to the outside world, normally contains *Staphylococcus epidermidis*, *Enterococcus* species, and a few members of Enterobacteriaceae. Some of these organisms can cause urinary tract infections (UTIs), such as cystitis, if they can make their way into the bladder. Urinary catheterization, for instance, can introduce these bacteria into the bladder.

The large surface area and associated secretions of the female genital tract make it a rich environment for microbes. Composition of the vaginal microbiome changes with personal hygiene, sexual intercourse, the menstrual cycle, changing nutrients and

Figure 14.10 **Model of Metabolic Collaboration**

The presence of H_2-oxidizing methanogens in the human intestine removes H_2 produced by bacterial fermentation. The reduction of H_2 levels is thought to relieve H_2 end product inhibition of bacterial fermentation (possibly via NADH dehydrogenase). Enhanced fermentation produces excess short-chain fatty acids that are absorbed into the host's intestinal epithelial cells and metabolized as food.

1 H_2 buildup from bacterial fermentation normally feedback-inhibits fermentation.

2 H_2 oxidation by methanogens, if present, counteracts H_2 buildup, allowing fermentation to continue.

Inhibition

Polysaccharide

CO_2 CH_4

Bacteria H_2 Methanogen

Short-chain fatty acid

3 Fermentation end-products such as acetate and butyrate are readily used by human cells, resulting in weight gain.

Intestinal epithelial cells

Figure 14.11 **The Urogenital Tract in Women**

Right adrenal gland
Right kidney
Right renal artery
Right renal vein
Renal pelvis
Vena cava
Right ureter
Right fallopian tube
Right ovary
Bladder
Vagina

Left adrenal gland
Left kidney
Left renal artery
Left renal vein
Abdominal aorta
Left ureter
Uterus
Left fallopian tube
Left ovary
Urethra

Blood in renal arteries carries waste products into the kidney. The waste products are removed by filtration and passed down the ureters as urine into the bladder. Urine is expelled through the urethra. The figure also shows the location of the adrenal glands, fallopian tubes, and uterus. Areas colonized by microbiota are the urinary bladder, vagina, and urethra (labels are highlighted). All other organs represented are sterile.

pH, and the onset of menopause. The mildly acidic nature of vaginal secretions (approximately pH 4.5) discourages the growth of many microbes. As a result, the acid-tolerant *Lactobacillus crispatus* is among the most populous vaginal species. Healthy women appear to fall into two broad categories: 70% have lactobacilli as the primary type of vaginal microbiota, whereas 30% have mixed species with fewer lactobacilli. Women in the latter group appear to be more susceptible to sexually transmitted diseases.

The balance between competing species of vaginal microbiota is crucial to preventing infection. Antibiotic therapy to treat a bacterial infection anywhere in the body can also affect the vaginal microbiota. The resulting imbalance can result in a yeast infection caused by *Candida albicans*, another member of the vaginal microbiome. *C. albicans*, of course, is not susceptible to the antibiotics designed to kill bacteria.

SECTION SUMMARY

- **The normal microbiome** present on body surfaces constantly changes.
- **Normal microbiota can cause disease** if microbes breach the surfaces they colonize and gain access to deeper tissues or the circulation.
- **Colonization of skin** by microbes is difficult owing to surface dryness, acidic pH, high salinity, and the presence of degradative enzymes.
- **Skin microbiota** consist primarily of Gram-positive microbes, including *Cutibacterium acnes*, which can cause acne.
- **Microbe-free areas of the body** are cerebrospinal fluid, the urinary tract (excluding the bladder and urethra), and other internal organs that are not open to the outside environment. The eyes are generally microbe-free, although some microbes can be present for short periods.
- **Oral and nasal surfaces** are colonized by aerobic and anaerobic microbes.
- **The ileum and colon of the intestine** are populated by 10^{11}–10^{13} microbes per gram of feces in a ratio of approximately 1,000 anaerobes to 1 facultative organism. Composition is diverse but consists mainly of Bacteroidetes and Firmicutes.
- **Collaboration between bacterial and archaeal species** in the intestine can form new, more efficient pathways that degrade food.
- **The urinary bladder contains a small microbiome,** mostly anaerobic, whereas the **vagina has a diverse microbiome** that antagonizes colonization by pathogens.

Thought Question 14.1   How can an anaerobic microorganism grow on skin or in the mouth, both of which are exposed to air?

Thought Question 14.2   What should an immunocompromised person with mitral valve prolapse do to prevent formation of subacute bacterial endocarditis when visiting the dentist?

Thought Question 14.3   Two men, both 55, ate raw oysters at a popular oyster bar. Two days later, one man is hospitalized with a life-threatening blood infection (septicemia) caused by the Gram-negative curved rod *Vibrio vulnificus*. The other man suffers no ill effects. The only difference between the two men is that the one infected was taking an over-the-counter proton pump inhibitor to reduce his stomach acidity and treat his persistent indigestion. What does this tell you about the infecting organism?

14.2 Benefits and Risks of a Microbiome

SECTION OBJECTIVES

- Describe ways that microbiota benefit the health of their host.
- Explain the risks of having a microbiome, especially to immunocompromised hosts.
- Discuss how host-microbiome relationships are studied in the laboratory.
- Describe how the compositional balance of our microbiome can be compromised and the potential effects on the individual.

Should we view colonizing bacteria as hostile armies encamped at our body's gates (also called portals of entry) just waiting for the opportunity to invade? No. Bacteria do not possess hostile intent, only the need to find food. Because the needs of our normal microbiota are met at the colonizing site, these microbes, for the most part, have not evolved tools to pass beyond their resident niche. In fact, members of our microbiota are of tremendous benefit to us, but as with any relationship, there is risk.

Benefits of a Microbiome

Members of our microbiota are often beneficial (mutualistic). Normal microbiota help us digest food and synthesize compounds needed by the body (for example, vitamin B_{12}). Our microbiota also interfere with the colonization of pathogens (called microbial antagonism) by competing for attachment receptors on host cells, competing for food sources, and synthesizing antimicrobial compounds such as fermentation end products. As noted earlier, the acidic fermentation products produced by lactobacilli in the female genital tract help maintain a low pH that dissuades colonization by various pathogenic microbes.

Our microbiota also shape and enhance our immune system. The constant sampling of microbiota by mucosal cells of the immune system (as is done, for example, in Peyer's patches in the intestine) keeps the immune system sharp. Peyer's patches will be discussed in Section 14.3 and Section 15.2. In addition, bacterial proteins such as catalase can act as immunomodulins. **Immunomodulins** made by normal microbiota growing on mucosal surfaces modify the secretion of host proteins such as cytokines and tumor necrosis factor that influence the immune response. **Cytokines** are small secreted host proteins that bind to various immune system cells and regulate their function. (Cytokines will be discussed in Section 15.3 and in Chapter 16.)

Some toxic bacterial products are "double-edged swords," having both positive and negative effects. **Enterotoxins**, for example, are proteins secreted by some Gram-negative pathogens that damage the small intestine of the host and cause diarrhea. But the enterotoxins may also protect against colorectal cancer by activating membrane calcium channels in intestinal epithelial cells

(cells lining the intestine). Increasing calcium transport through the membrane turns on an antiproliferative pathway that dampens host cell division, which provides resistance to colon cancer.

⊙ **Note:** Do not confuse the terms "exotoxin," "enterotoxin," and "endotoxin." An exotoxin refers to any protein with toxic activity that is secreted from the bacterium. An enterotoxin is an exotoxin that specifically affects the intestinal mucosa. Endotoxin is not a protein at all but is a part of the microbial cell envelope (for example, lipopolysaccharide that makes up the Gram-negative outer membrane; see Section 5.3).

The microbiome might also be engineered to serve as a novel vaccine delivery system that can be developed quickly and deployed rapidly. A species selected from the microbiome can be genetically engineered to produce proteins from pathogenic species and display them on their cell surfaces. Colonization of a host by the modified strain will elicit an immune response to the cloned protein and provide immunity to the pathogen. Organisms being tested as potential vaccine delivery systems include *Streptococcus salivarius* (part of the oral and nasopharyngeal microbiome), *Lactococcus* species (residents of the intestinal tract), and *Lactobacillus* species (present in the urogenital tract and rectal area).

The benefit that microbiota offer to their hosts is especially apparent in gnotobiotic animals. A **gnotobiotic animal** is an animal that is germ-free or one in which all the microbial species present are known. Developing a gnotobiotic colony of animals involves delivering offspring by cesarean section under aseptic conditions in an isolator (**Figure 14.12**). The newborn is moved to a separate isolator where all entering air, water, and food are sterilized. Once gnotobiotic animals are established, the colony is maintained by normal mating between the members. Germ-free animals, however, often have poorly developed immune systems, lower cardiac output, and thin intestinal walls, and they are more susceptible to infection by pathogens. The lesson from these animals is that the presence of normal microbiota continually challenges the immune system and keeps it active. Note, too, that the ability to deliver germ-free animals argues against the existence of a fetal microbiome.

Risks of a Microbiome

Despite the enormous benefits, maintaining resident microbiota is still a risky business. Even though microbes of our normal microbiota, unlike true pathogens, have not evolved mechanisms to specifically breach human defenses, accidental penetration beyond these sites can cause serious infections. A cancerous lesion in the colon, for example, can provide a passageway for microbes to enter deeper tissues, where they can begin to grow unabated. *E. coli*, a facultative anaerobe that is harmless in the gut, can invade other tissues through surgical wounds, causing abscesses that persist and even lead to gangrene after abdominal surgery (**Figure 14.13**).

By and large, host defense systems work well to prevent microbiota from breaching mucosal barriers and will kill them if they do breach. However, these defenses can fail in people whose immune

Figure 14.12 Gnotobiotic Unit at North Carolina State University

The room contains flexible film isolators for housing mice in axenic, or gnotobiotic, conditions. Another room is used to sterilize supplies needed in the flexible film isolators.

systems are compromised by medical treatments like anticancer drugs or by diseases such as serum complement factor deficiencies. Serum complement is a host defense system that attacks bacterial membranes and attracts phagocytes to an infection (see Section 15.7). Any person with a weakened immune system is

Figure 14.13 Abscess of Colon Caused by *Escherichia coli*

This infection, shown by computed tomography (CT) scan, was an unfortunate result of bowel surgery. Organisms escaped from the intestine and initiated infection in the abdominal wall (arrow). CT scans use X-ray images taken from many different angles to produce detailed images of internal organs.

called an **immunocompromised host** and is vulnerable to infection by members of the microbiome called opportunistic pathogens. **Opportunistic pathogens** rarely cause disease in people with healthy immune systems.

Our intestine, with its diverse microbiome, can also be considered an incubator for antibiotic resistance. The nearly indiscriminate use of antibiotics—as when antibiotics are used to treat viral diseases—selects for resident bacteria that spontaneously developed resistance. Antibiotic treatment can completely change the intestinal mix of microbes. Bacteria that develop resistance to an antibiotic can overgrow in the presence of the antibiotic and become a more dominant presence once the antibiotic treatment has stopped. Antibiotic resistance genes from these bacteria can then be passed to others, including pathogens, by transformation (see Figure 9.2), by conjugation (if the resistance genes are on a plasmid; see Figure 9.4), or by resident viruses via transduction (see Figure 9.6).

Microbiome Dysbiosis

Human health depends on maintaining a balance among the various members of the gut microbiome. A variety of situations, however, can alter that balance to produce a condition called **dysbiosis**. Infections, antibiotics, new foods, emotional stress: all of them can produce dysbiosis. Dysbiosis is now thought to influence behaviors or even cause some human diseases. For example, changes in the gut microbiome can produce inflammatory responses, which manifest as irritable bowel syndrome or more severe inflammatory bowel diseases like Crohn's (mentioned in the opening case history), or deadly infections such as pseudomembranous enterocolitis (**Figure 14.14**).

Are specific species within one's gut microbiome important? Some bacteria have been implicated as contributing to intestinal health, such as *Faecalibacterium*, *Bacteroides*, *Akkermansia*, and many others. Some research suggests that specific members of the microbiome per se are not as important as the presence of a core set of genes distributed among the microbiome. An individual's microbiome is in a constant state of flux, like a hotel with guests constantly checking in and checking out. It is the hotel's staff, with their individual core skill sets, that keep the enterprise going. If the plumber quits the hotel, it is only a matter of time before the plumbing fails. Similarly, if a shift in the microbiome leads to the loss of some core bacterial genes, a human metabolic dysfunction can result. Thus, if our microbiome can be considered a "virtual organ system," it is a potentially fickle one, apt to change with

Figure 14.14 Pseudomembranous Enterocolitis Caused by *Clostridioides difficile*

On the left is a healthy colon as viewed by colonoscopy. On the right is the colon of a person with pseudomembranous enterocolitis. *C. difficile* is often a part of the normal intestinal microbiota, but antibiotic treatment, especially following surgery, can kill off competing microbes, leaving *C. difficile* to grow unabated. A toxin produced by the organism damages and kills host epithelial cells, causing exudative plaques to form on the intestinal wall. The small plaques eventually coalesce to form a large pseudomembrane that can slough off into the intestinal contents.

Exudative plaques

little provocation. Later, we will discuss how dysbiosis within the gut microbiome can lead to human disease and altered behavior.

How can balance be restored? Some people favor restoring the natural microbial balance by eating plant fibers that humans cannot digest but that can be digested by beneficial members of the microbiome. These plant substances are called **prebiotics**. Prebiotic fiber is present in many natural foods, such as fruits, vegetables, and whole grains. Other people try to restore balance by ingesting living microbes such as the lactobacilli present in yogurt, which are called **probiotics** (from the Greek meaning "for life"). Taking these supplements is thought to restore balance to the microbial community and return the host to good health. The most commonly used probiotic bacteria are *Lactobacillus* and *Bifidobacterium* species (**Figure 14.15**).

Figure 14.15 Commonly Used Probiotic Microorganisms

A. *Bifidobacterium* (colorized); note the Y shape of some cells.

B. *Lactobacillus acidophilus* (colorized).

An extremely exciting development in treating extreme dysbiosis is the fecal transplant, more properly called **fecal microbiota transplant (FMT)**, in which the intestinal microbiome of a healthy person is transferred to a person with severe intestinal disease such as colitis. The fecal transplant process is discussed in Section 14.5.

SECTION SUMMARY

- **Benefits of microbiota** include interference with pathogen colonization, maintenance of our immune system, stimulation of immunomodulatory protein production, and potential use as vaccine delivery vehicles.

- **Gnotobiotic animals** are germ-free or colonized with a known set of microbes.

- **Risks of normal microbiota** include accidental infection, especially in immunocompromised hosts; the potential for developing new antibiotic resistance mechanisms; and a susceptibility to some metabolic diseases when the balance of microbial species is disturbed.

- **Dysbiosis is an imbalance in the composition of a microbiome**, primarily referred to in terms of the intestine. Intestinal dysbiosis can be brought on by stress or antibiotic therapy, the effects of which can affect many aspects of human health.

Thought Question 14.4 People who take antibiotics for an upper respiratory tract infection often develop antibiotic-associated diarrhea. However, pathogenic microbes are not the usual cause of this diarrhea. Using what you have read here combined with Internet searches, formulate a general hypothesis that might explain this connection. (*Hint:* Look up osmotic diarrhea.)

14.3 Keeping Microbiota in Their Place

SECTION OBJECTIVES

- List and compare the various barriers that separate the microbiome from its host.

- Explain the function of tight junctions in preventing infection by microbiota.

- Discuss the role of innate immunity in preventing infection by microbiota.

How are members of our microbiota kept from slipping past our own cells and infecting us? In the previous sections, we mentioned several ways in which normal microbiota benefit human health. However, members must be kept in their place because those that escape can cause disease. Several physical barriers act to prevent escape.

Epithelial Barriers

The major barrier separating us from our microbiota is a layer of glycoprotein-covered epithelial cells lining the mucosa of our digestive, genitourinary, and respiratory tracts. The epithelium of

Figure 14.16 Tight Junctions

A. Epithelial cell tight junctions imaged in a culture of human colon cells. Proteins (occludin) that reside specifically at tight junctions are stained with antibodies conjugated to a fluorescent dye.

FM   30 μm

Lipid bilayer

Tight transmembrane junctional protein

Cytoplasm

Intercellular space

Solute molecule

B. Schematic representation of two adjoining epithelial cell membranes fused to each other by tight junctions. Transmembrane junctional proteins in the membrane of two adjacent cells are clustered along the points of contact.

most mucosal surfaces consists of interconnected, polarized epithelial cells whose basal membrane separates them from underlying connective tissue. The epithelial layers lining the mucosa are held together by so-called tight junctions formed between cells (**Figure 14.16**). The molecular "glue" of tight junctions is a series of molecules connected to the cytoskeleton of epithelial cells. Tight junctions act as a strictly regulated port of entry that opens and closes in response to various signals. These signals include small secreted host peptides called cytokines that originate in the epithelium and lamina propria (a layer of connective tissue that lies beneath mucosal epithelium). Tight junctions are key to controlling intestinal diffusion but also effectively prevent our microbiota from gaining access to deeper tissues.

So how do bacteria get through the seemingly impervious epithelial barrier? Sometimes microtraumas, such as small lesions that damage the epithelial layer, will provide a temporary gateway to deeper tissues. Cancerous lesions, for example, can perforate this layer in the intestine; gunshot and knife wounds provide pathways for infection as well. But cancer and trauma are not the only means by which bacteria breach the epithelial barrier. The mucosa is also speckled with small patches of lymphoid tissue associated with the immune response. These areas, such as Peyer's patches in the intestine (see Section 15.2), sample normal bacteria to help maintain a low-level immune response. Normal microbiota may accidentally penetrate these areas. Many pathogens, on the other hand, actively invade these regions. We will discuss how the host copes with these breaches in Chapters 15 and 16.

CASE HISTORY 14.2

The Deadly "Hangover"

N. meningitidis

Hiroshi, a 19-year-old college student, arrived at the emergency department complaining of a fever and lethargy that started the previous evening. He originally suspected that the symptoms were due to a hangover from a fraternity party held 2 days earlier. However, Hiroshi became alarmed when turning his head became painful and when tiny red blotches suddenly appeared on his skin. The physician recognized this as a **petechial rash** (red spots on the skin that do not blanch, or turn white, when pressed). Blood cultures were obtained, and Hiroshi underwent a lumbar puncture to obtain cerebrospinal fluid (CSF) for culture. He was immediately admitted to the hospital and treated with intravenous antibiotics. The next day, blood and CSF cultures grew a deadly Gram-negative diplococcus— *Neisseria meningitidis*. Hiroshi had bacterial meningitis.

Neisseria meningitidis is a bacterium that can reside in the nasopharynx of some individuals, called carriers, without causing disease. Carriers are immune but can transmit the organism to a susceptible person. Transmission is more frequent among people housed in close quarters, such as on a college campus. *N. meningitidis* will initially colonize the nasopharynx of a newly infected person and cause a sore throat. From there the organism can gain access to the bloodstream. Usually, this is where the infection stops, as various immune mechanisms engage. However, sometimes the bacterium, while coursing through the bloodstream, can penetrate into the central nervous system (CNS) through the nearly impervious blood-brain barrier, a brain lining that bars the passage of most microbes and molecules. Once this barrier is breached, *N. meningitidis* can cause potentially deadly meningitis. The mechanisms used by pathogens to penetrate the blood-brain barrier will be revealed in Chapter 24.

The Blood-Brain Barrier

"I think, therefore I am." This famous assertion by the fifteenth-century philosopher René Descartes underscores the importance of the human brain. Nature has taken note of the brain's importance by evolving extraordinary measures of protection. The skull, an obvious example, protects the brain from flying fists and falling branches. But there is an equally important, if less apparent, protective structure inside the brain called the **blood-brain barrier**. The blood-brain barrier is composed of extremely tight junctions between the endothelial cells of capillaries feeding the brain. The section of the blood-brain barrier along the choroid plexus (called the blood–cerebrospinal fluid barrier) occurs between epithelial cells that surround capillaries (**Figure 14.17A**) and controls the passage of substances from the blood into the CSF that bathes the brain and spinal cord. Cerebrospinal fluid, made primarily by the choroid plexus, has the composition of blood plasma but normally contains no blood cells—red or white.

The blood-brain barrier is a highly selective permeability barrier that lets essential metabolites, such as oxygen and glucose, pass from the blood to the brain but blocks most molecules with a molecular mass greater than 500 daltons. This means that everything from hormones and neurotransmitters to viruses and bacteria are refused access to brain tissue (unless there is a specific transport system; see **Table 14.2**). It also means that many antibiotics, which would otherwise be capable of treating infections, cannot pass through the blood-brain barrier.

The blood-brain barrier is composed of thin, flat endothelial cells that form the walls of blood capillaries. In most parts of the body, endothelial cells overlap and form tight junctions. Though considered tight, cell-cell junctions in most of the body are leaky enough to let some materials slip between adjacent endothelial cells and thus through the wall of the blood vessel. Most endothelial cells also have active vesicle transport systems that can carry compounds from one side of the cell to the other. This is how hormones

Figure 14.17 The Blood-Brain Barrier

A. The choroid plexus in the brain is composed of a rich capillary bed and forms the blood–cerebrospinal fluid barrier.

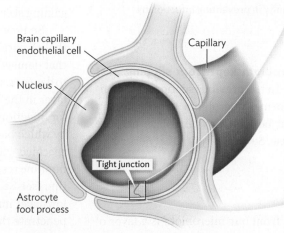

Brain capillary endothelial cell

Nucleus

Capillary

Tight junction

Astrocyte foot process

B. Diagram of a brain capillary in cross section showing an endothelial tight junction.

TEM 200 nm

C. Electron micrograph of boxed area in (B), showing the appearance of a tight junction between neighboring endothelial cells (arrow).

Table 14.2 **Some Materials That Cross the Blood-Brain Barrier**	
Freely Diffusible	**Requiring Carrier Systems in the Cells**
Oxygen	Glucose
Carbon dioxide	Metabolic by-products
Alcohol	Tryptophan and tyrosine (precursors of neurohormones such as serotonin, which does not cross the barrier)
	L-dopa (a precursor of dopamine, which does not cross the barrier)

and nutrient molecules get to tissues. Unfortunately, potentially harmful agents, such as toxins, viruses, and the occasional bacterium, can slip through as well.

In brain capillaries, endothelial cells form extremely tight junctions with no breaks for leakage (**Figure 14.17B** and **C**). These junctions, which block the passage of most things except for very small molecules, are a crucial component of the blood-brain barrier. Brain endothelial cells also contain few intracellular vesicles, the lack of which adds to the barrier. The blood-brain barrier makes it very difficult for most pathogens to penetrate into the brain and keeps any microbiome organism that accidentally enters the bloodstream from gaining access to the central nervous system. However, some pathogenic bacteria (such as *N. meningitidis*) and viruses (such as eastern equine encephalitis virus) have developed ways to slip though the blood-brain barrier—usually at the choroid plexus—and cause disease, as we saw in Case History 14.2.

The Maternofetal (Placental) Barrier

CASE HISTORY 14.3
Born Infected

L. monocytogenes

A 2.2-pound (1 kg) female newborn (neonate) was born 2 months premature. The tiny baby had extreme difficulty breathing and had to be intubated (a breathing tube was inserted). The mother, at the time of admission, had complained of mild diarrhea and abnormal abdominal pain unrelated to her pregnancy. The infectious disease doctor who was called in to consult on the case immediately recognized the likely problem and ordered blood cultures performed on the infant. The infant was also started on intravenous antibiotics. Two days later, the lab reported finding a Gram-positive bacillus—*Listeria monocytogenes*—in the infant's blood. This same organism was the cause of the mother's diarrhea. The mother had unwittingly ingested some unpasteurized cheese contaminated with this pathogen and developed listeriosis. The organism entered the mother's bloodstream and was transmitted across the placenta to the fetus prior to birth.

The human fetus is primarily germ-free (the fetal intestine may be an exception), even though areas of the pregnant mother teem with microorganisms. The placenta plays a major role in keeping a fetus pathogen-free. In mammals, the placenta separates the fetal and maternal circulations and keeps stray bacteria that find their way into the mother's bloodstream away from the fetus. Capillaries of the human placenta are the sites of maternofetal exchange (the area where substances are exchanged between the mother's blood and that of the fetus). The placental membrane (or "placental barrier") contains branches of fetal capillaries and is in direct contact with maternal blood (**Figure 14.18**). The placenta enables the

Figure 14.18 The Maternofetal Barrier
The placental barrier separates the maternal and fetal circulations.

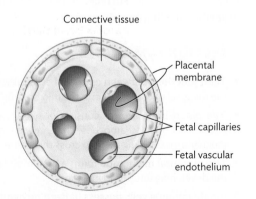

A. Fetus attached to the placenta. The circle indicates the location of the maternofetal barrier.

B. Overview of the maternal and fetal circulations at the placenta. Maternal blood enters the intervillous spaces that surround the villi containing the fetal capillaries. Fetal capillaries merge to form the umbilical veins and arteries that transport oxygen and nutrients to the fetus and carbon dioxide and wastes away from the fetus.

C. Cross section of the villus, showing the placental membrane and fetal capillaries. The cells in maternal and fetal blood never mix.

transfer of oxygen, carbon dioxide, nutrients, and waste products between the mother and the fetus without maternal and fetal blood ever mixing. The endothelial cells that form fetal capillaries are, as in the case for other capillaries, connected by tight junctions. Very few species of bacteria that find their way into the mother's circulation can pass through the placental membrane and endothelial cells to infect the developing fetus. *Listeria monocytogenes*, however, is one of the few that can. *Listeria* crosses the placenta first by invading placental cells and then by using a unique form of motility (actin tails) to pass from cell to cell (discussed in Chapter 18).

Innate Immune Mechanisms

Mucosa, epithelial cells, and tight junctions are effective barriers that keep our microbiota in their proper place. However, breaches in these barriers do occur, and when they do, our immune system is poised to keep the bacterial threats at bay. As the next two chapters will discuss, there are two basic types of immunity. The first is nonspecific, innate immunity, in which host cells that form mucosal surfaces or circulate in the blood can recognize the presence of a microorganism and respond immediately. The second form is adaptive immunity, which takes longer to develop and includes the production of antibodies and specialized immune cells. Innate mechanisms actually sample the microbiome and then modulate the host's immune response to limit the numbers and diversity of microbes at the colonization sites.

How do cells of the innate immune system recognize microbiota? These cells possess sensitive membrane and cytoplasmic receptors, called pattern recognition receptors (PRRs). These PRRs bind various microbe-associated molecular patterns (MAMPs)—molecules such as peptidoglycan or lipopolysaccharide (**Figure 14.19**). When PRRs on host cells lining the intestine bind to bacteria trying to breach the epithelial layer, the host cells will synthesize small proteins called cytokines. The cytokines attract phagocytic white cells to the area and stimulate the production of special antibodies that enter the intestine. The cytokine-stimulated immune response then suppresses the growth of microbiota at the intestinal mucosal surface. Aspects of innate immunity will be described in Chapter 15.

SECTION SUMMARY

- **Tight junctions between epithelial cells** prevent bacteria from breaching the mucosal barriers that form the microbiome-human interface.

- **The blood-brain barrier** consists of extremely tight junctions between endothelial cells. Pathogens that cause meningitis possess special mechanisms enabling them to cross this barrier.

- **The placental membrane** forms a maternofetal barrier that helps protect the developing fetus from infection.

- **Innate immune cells possess pattern recognition receptors (PRRs)** that help control the numbers and composition of our microbiota.

Thought Question 14.5 Why do many Gram-positive microbes that grow on the skin, such as *Staphylococcus epidermidis*, grow poorly or not at all in the gut? (*Hint:* Think about differences between skin and intestinal contents.)

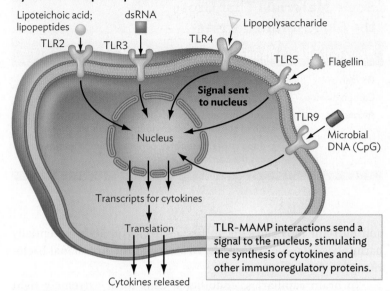

Figure 14.19 Host Pattern Recognition Receptors and Cytokines Help Keep Normal Microbiota in Balance

Different pieces of microbes, such as peptidoglycan, lipopeptides, lipopolysaccharides, flagellin, DNA (from bacteria), and double-stranded RNA (dsRNA; from viruses), are called microbe-associated molecular patterns (MAMPs). MAMPs bind to different host cell pattern recognition receptors (PRRs), such as membrane-associated Toll-like receptors (TLRs). A PRR-MAMP interaction sends a signal to synthesize cytokines that can either promote or inhibit inflammation (depending on the cytokine) or can stimulate antibody production. These processes control microbial proliferation at colonized body sites (the intestine, for instance) and help shape the diversity of our microbiota.

Thought Question 14.6 What caused the petechial rash in Case History 14.2?

Thought Question 14.7 How might normal microbiota escape the intestine and cause disease at <u>other</u> body sites?

14.4 Microbiome-Host Communications

SECTION OBJECTIVES

- Describe mechanisms by which our microbiota communicate with our cells and tissues.

- Explain ways that our cells and tissues communicate with our microbiota.

- Provide examples of the benefits resulting from microbiota-host interactions.

- Argue the case for declaring our microbiome a "virtual organ system" essential to our survival as a species.

If the microbiome is so important to human health, do the parties involved communicate with each other? In sheer numbers, our microbiota outnumber our own nucleated cells about 2-fold. With all of the metabolic products produced by both sides

Figure 14.20 Interspecies and Interdomain Influences on Gene Expression

Communication at a colonized site can be viewed as a three-way "discussion." First, bacterial species communicate with each other (quorum sensing) to stimulate the production of chemicals or proteins that enhance survival. Second, some of those microbial products influence host gene expression and function. Finally, the host cells make compounds that affect the growth of microbiota. The result could be regarded as a three-way chemical conversation.

of this partnership, surely they sense each other's existence. This section will discuss the chemical and neurological "conversations" that take place between you and your microbiome. We start with the chemical conversations each side uses to influence the other. Figure 14.20 summarizes these complex interspecies and interdomain "discussions."

Does the Microbiome Talk to Itself?

The first aspect of interspecies chemical discussions involves microbe-to-microbe communications that take place among different members of the microbiome. Bacteria, in a process called quorum sensing (see Section 6.6), can communicate with each other by using a variety of autoinducer molecules that they themselves secrete. The secreted autoinducers accumulate in the environment and enter nearby bacteria of the same or different species. Sensing these chemicals, the receiving species alters its expression of specific genes whose products can inhibit or stimulate growth (Section 9.1). In this way, microbe-to-microbe quorum sensing shapes the composition of the microbiome and can influence the formation of composite biochemical pathways in which gene products from different bacterial species work together to form new metabolic systems.

How Does the Microbiome Influence Its Host?

Studies that look for small molecules present in cells or body fluids (the science of metabolomics) find that mammalians are awash in metabolic products made by host microbiota. For example,

a study with mice has shown that many end products of bacterial metabolism are present in blood (though the blood itself is sterile). Another study with humans discovered several microbial end products in urine. What are these microbial products doing there? Are they influencing gene expression in host cells? The answer appears to be yes.

A study of gnotobiotic mice whose intestines were colonized with only two members of the normal human microbiome, *Eubacterium rectale* (a firmicute bacterium) and *Bacteroides thetaiotaomicron*, resulted in a remarkable discovery. Colonization by these two bacteria changed the expression of 508 host genes. Of the 508 host genes, 112 were involved with host cell growth and proliferation and 130 were associated with cell death. Many of the genes identified are critically important in controlling self-renewal of the colonic epithelium. The data suggest that our microbial tenants actively participate in our colonic health and well-being.

Other examples of bacteria influencing host expression are mounting. For example, one report suggests that the microbial product indole (produced when *E. coli* degrades tryptophan) can "tell" host intestinal epithelial cells to tighten tight junctions and increase the production of anti-inflammatory products such as the cytokine interleukin 10 (IL-10). As noted earlier, cytokines are small, secreted host peptides that can be either pro-inflammatory or anti-inflammatory. Another report found that polysaccharide A secreted by *Bacteroides fragilis* also prevents intestinal inflammation by modulating the production of certain host cytokines (Figure 14.21).

How Does the Host Influence the Microbiome?

Animal studies suggest that perturbations (or agitations) of behavior, such as stress, can change the composition of the microbiome; these changes may foster increased vulnerability to inflammatory

Figure 14.21 Polysaccharide A (PSA) from *Bacteroides fragilis*

A *Bacteroides* capsule composed of polysaccharide A is shown.

stimuli in the gastrointestinal tract. Evidence also suggests that hormones such as epinephrine and norepinephrine, which are released following stress, can influence the composition of the intestinal microbiota. For example, these neurotransmitters have been shown to stimulate the growth of pathogenic and non-pathogenic *E. coli* in vitro and to influence their adherence to the mucosa. More than 50 different types of bacteria, occupying major niches of the human body, have been shown to respond to neuroendocrine hormones (**Figure 14.22**). **Figure 14.23** illustrates the dramatic effect of norepinephrine on biofilm formation by *Staphylococcus epidermidis*.

Microbial endocrinology, as the field is called, has focused mainly on how pathogens respond to target host's hormones. We are just beginning to question how hormones and other host products might influence the local ecology of natural microbiota.

The Microbiome-Gut-Brain Axis

Nerves that connect the human gastrointestinal tract to the CNS form something called the gut-brain axis (**Figure 14.24**). The axis is composed of the central nervous system, neuroendocrine and autonomic nervous systems, the enteric nervous system, and gut microbiota. The vagus nerve is an important part of the gut-brain axis. Neurochemicals made by the brain initiate efferent vagus nerve impulses that travel from the brain to the gut, whereas gut bacteria make neurotransmitter-like chemicals that can fire afferent signals back along the axis to the brain. Consequently, brain activity will alter gut functions, but the microbiome can also alter brain function. Exciting new research suggests that the gut microbiome can influence many aspects of health through its communications with the brain via the so-called microbiome-gut-brain axis.

Gut microbes, for instance, may affect your mood. It is clear that some members of the microbiome can make compounds that mimic feel-good neurotransmitters such as dopamine. Other microbes produce chemicals associated with fear recognition (5-hydroxytryptamine) or anger (4-aminobutanoic acid, commonly referred to as GABA because of its former name, "gamma-aminobutyric acid"). Some microbial products can even influence the production of host receptors associated with depression. Recent **metagenomic studies** with humans have identified three genera of Firmicutes in the gut that <u>correlate</u> with depression. *Faecalibacterium* and *Coprococcus* were more prevalent in individuals considered happy, while people identified as depressed were <u>depleted</u> for *Coprococcus* and *Dialister*. (However, recall that a correlation does not prove causation.)

Figure 14.22 The Spectrum of Known Neuroendocrine-Responsive Bacteria

Respiratory pathogens
Bordetella bronchiseptica
B. pertussis
Pseudomonas aeruginosa
Klebsiella pneumoniae

Cardiac pathogens
Enterococcus faecalis

Enteric microbiota and pathogens
Escherichia coli
Campylobacter jejuni
Proteus mirabilis
Salmonella enterica
Shigella sonnei
S. flexneri
Listeria monocytogenes
Vibrio parahaemolyticus
V. mimicus
V. vulnificus
Yersinia enterocolitica

Oral/peridontal pathogens
Actinomyces
Campylobacter
Capnocytophaga
Eikenella
Eubacterium
Fusobacterium
Leptotrichia
Neisseria
Peptostreptococcus
Prevotella
Streptococcus

Skin microbiota and pathogens
Staphylococcus epidermidis
S. capitis
S. saprophyticus
S. haemolyticus
S. hominis
S. aureus

Figure 14.23 Effect of Norepinephrine on *Staphylococcus epidermidis* Biofilm Formation
Bacteria were added to slides and grown for 24 hours.

A. No norepinephrine; the control.

B. 0.1 mM norepinephrine. The slide prepared in the presence of hormone showed extensive biofilm formation and considerably more exopolysaccharide coat than the slide prepared without hormone.

Figure 14.24 The Gut-Brain Axis

Signals through the gut-brain axis of nerves, such as the vagus nerve, pass back and forth from the central nervous system (CNS) to the gut.

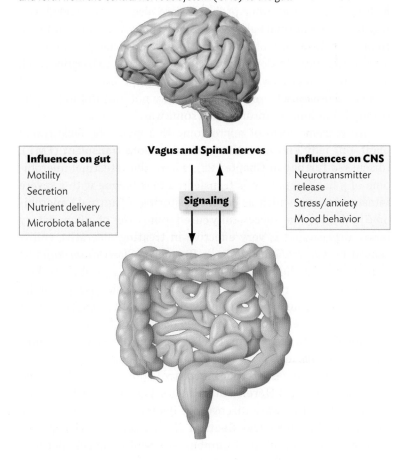

Vagus and Spinal nerves

Influences on gut
Motility
Secretion
Nutrient delivery
Microbiota balance

Signaling

Influences on CNS
Neurotransmitter release
Stress/anxiety
Mood behavior

The gut microbiome may also influence behaviors such as the craving for certain foods . . . even for chocolate! Remember that stomach bypass experiment? After a gastric bypass, the gut microbiome undergoes a dramatic shift and the patient experiences a major change in cravings. Most lose interest in treats and no longer enjoy fatty foods. Precisely why is unclear. It's proposed that the shift in microbiome composition changes what microbial neurosignaling compounds are made and, consequently, how the gut-brain axis is stimulated. The result is stimulation of brain areas that influence cravings.

And then there's the pregnant woman who famously craves pickles and ice cream at 3:00 a.m. Hormones released by the mother and the fetus during pregnancy can also alter a woman's gut microbiome and, consequently, tweak the gut signals sent along the vagus nerve. Taste bud receptors can be changed, for instance. While these studies are exciting, it's important to realize that definitive proof for gut bacteria affecting mood and behavior is lacking.

Intriguing research has also suggested a link between gut dysbiosis and the development of certain neurological diseases: Parkinson's disease, for instance. Parkinson's is a neurodegenerative disease characterized by the accumulation of specific intracellular protein inclusion bodies (Lewy bodies) that destroy neurons in the CNS. The hypothesis is that intestinal dysbiosis can trigger a form of inflammation that initiates Lewy body formations in the intestinal section of the vagus nerve. The protein aggregates then travel up the nerve to eventually reach the brain stem. Similar potential links have also been made between dysbiosis of the gut microbiome and Alzheimer's disease.

Microbiome as Organ System?

Our new understanding of the human microbiome and its importance to human health raises an interesting question: Should the microbiome be considered an organ system of the human body? An organ system can be defined as a relatively independent part of the body that performs one or more specific functions. Does your microbiome fit this description?

To start this line of reasoning, remember that your body has about twice the number of microbial cells as human nucleated cells and 100 times more microbial genes than human genes. So, in terms of sheer numbers of cells and genes, your body is more microbial than animal. Because microbiomes are also contained in defined areas, they can be considered relatively independent parts of the body. As far as performing specific functions, microbes in the intestine, for example, contribute to our digestion, stimulate a healthy immune system, and promote human tissue development.

We can also measure the importance of an organ by noting the pathologies that develop when it is defective. Research shows that defects in microbiome composition have dire effects on human health, as already mentioned. For example, inflammatory bowel disease and obesity can result from microbiome imbalances.

Now let's consider inheritance of the microbiome. Genes for your traditional organs were vertically inherited from your parents. Your microbiome was inherited, at least in part, from your mother.

These arguments have led many scientists and health care professionals to now consider the microbiome a "virtual organ system" that can no longer be ignored when considering the physiology of the human body. Treatments, such as fecal transplants, that target microbiota will become increasingly important tools in the medical toolbox.

SECTION SUMMARY

- **Products of the microbiome** can influence host gene expression.
- **Hormones** made by the host can alter the microbiome.
- **Autoinducers** made by different microbial species can influence cross-species gene expression and overall microbiome function via quorum-sensing mechanisms.
- **Many now consider the microbiome to be a virtual organ system.**
- **The gut-brain axis** is a bidirectional communication pathway that connects the central nervous system to the intestinal tract. Intestinal dysbiosis affects the gut-brain axis, especially via the vagus nerve, to impact host mood and behavior; dysbiosis may even cause certain neurological disorders.

14.5 Natural Biological Control of Microbiota

Are the members of our microbiota always one big, happy extended microbial family? Not exactly. As in any large community, conflicts will break out. In fact, pitting microbe against microbe is an effective way to prevent disease in humans and animals. One of the hallmarks of a healthy ecosystem is the presence of a diversity of organisms. This is true not only for tropical rain forests and coral reefs but also for the complex ecosystems of human skin and the intestinal tract. In these environments, the presence of harmless microbial species can retard the growth of undesired pathogens. Naturally occurring staphylococci on human skin, for example, produce short-chain fatty acids that inhibit the growth of pathogenic strains. *Lactobacillus acidophilus* colonizes the vaginal epithelium during childbearing years and establishes the low pH that inhibits the growth of pathogens. Another illustration is the human intestine, which is populated by as many as 500 microbial species. Most of these species are nonpathogenic organisms that exist in symbiosis with their human host. However, the native microbiome inhibits the growth of numerous pathogens through vigorous competition for food and the production of inhibitory weak acids by fermentation.

Microbiome Therapy: Probiotics, Fecal Transplants, and Phage

As important as our microbiome appears to be for human health, can we manipulate it? One way to gain a healthier microbiome is to ingest microbes thought to be beneficial. Microbial competition has been widely exploited to improve human health in a process called probiotics, described in Section 14.2. A probiotic is a food or supplement that contains live microorganisms thought to improve intestinal microbial balance. In an agricultural example, newborn baby chicks are fed a microbial cocktail of normal members of the microbiome designed to quickly colonize the intestinal tract and prevent colonization by *Salmonella*, a frequent contaminant of factory-farmed chicken. In humans, *Lactobacillus* and Y-shaped *Bifidobacterium* (Figure 14.15) have been used to prevent and treat diarrhea in children.

Several probiotics are commercially available in food (yogurt) or pill form that can correct noninfectious diarrhea in adults, including irritable bowel syndrome mentioned in the opening case history. Yogurt and other probiotics are also recommended as a way to restore normal balance to gut microbes following antibiotic treatment. However, other research indicates that probiotic use after antibiotics can slow the microbiome's return to baseline level, although it's unclear whether that is important. Daily ingestion of an over-the-counter probiotic is typically not harmful unless the person is immunocompromised in some way.

An extreme form of microbiome therapy is the fecal transplant, officially known as the **fecal microbiota transplant (FMT)**, (discussed further in Chapter 22), where the intestinal microbiome of a healthy person is transferred to someone with a severe intestinal disease such as colitis. Restoring a "normal" microbiome in this way has successfully cured some cases of inflammatory bowel disease and is very effective in treating ulcerative colitis caused by *Clostridioides difficile* (formerly *Clostridium difficile*) in patients who did not respond to antibiotics. A fecal transplant can be administered to patients rectally by colonoscopy, orally via endoscopy, and more recently with ingestible capsules that selectively deliver a specific, purified cocktail of healthy microbiome microbes to the patient. FMT, once considered wacky, is now mainstream therapy.

Phage therapy is another biocontrol method, first described in 1907 by Félix d'Hérelle at France's Pasteur Institute, long before antibiotics were discovered. Bacteriophages are viruses that prey on bacteria (see Section 12.1). Each bacterial species is susceptible to a limited number of specific phages. Because phage infection often culminates in bacterial lysis, treating infectious diseases with a phage targeted to the pathogen was considered feasible. At one time, doctors used phages as medical treatment for illnesses including cholera and typhoid fever. In some cases, a liquid containing the phage was poured into an open wound. In other cases, phages were given orally, introduced via aerosol, or injected.

Sometimes the phage treatments worked; sometimes they did not. When antibiotics came into the mainstream, phage therapy largely faded. Today, because antibiotic-resistant pathogens are a growing threat, the idea of phage therapy has enjoyed renewed interest. But you might wonder whether bacteria can also develop resistance to phages. Won't the same resistance problem develop? Although bacteria can develop resistance to the phages that infect them, realize that both bacteria and phages coevolve. Because phages evolve and antibiotics do not, it is much easier to identify a new phage able to infect phage-resistant bacteria than it is to find a new antibiotic able to inhibit the growth of antibiotic-resistant bacteria. Another way to limit the development of phage-resistant bacteria is to simultaneously use three different phages during treatment. The pathogen is unlikely to develop resistance against all three phages at once. The potential of phage therapy has spurred several biotechnology companies to develop effective bacteriophage-based treatments. Several are in various stages of clinical trials.

Impact of a Disappearing Microbiome

As human health has improved in developing countries, unexplained new diseases have arisen. One hypothesis for the appearance of new diseases has been called the "disappearing microbiota hypothesis," otherwise known as the "hygiene hypothesis." Beginning in the nineteenth century and accelerating until today, dramatic changes in human behavior have influenced the makeup of our microbiota (Table 14.3). For example, the increased use of clean drinking water has interrupted the natural transmission of microorganisms to and between humans. Antibiotic use clears the body of many of its normal (as well as pathogenic) microbial species, leaving ecological voids that may be filled by species not as beneficial as those displaced.

Some scientists postulate that an important factor driving modern allergic and metabolic diseases may be the loss of our "ancestral" microbiota. Decreased transmission of normal microbiome species due to increased hygiene and antibiotics must mean that some microbes that once colonized humans long ago no longer do. The ancient human microbiome most certainly looked different from that of the present day. The idea is that our disappearing microbiome may have altered human physiology and immune development in ways that encourage today's allergies and obesity.

To give one example, the Gram-negative bacterium *Helicobacter pylori* at one time colonized 80% of humans in the United States and Europe. Within two or three generations, *H. pylori* prevalence dropped below 10%. The diminished presence of *H. pylori* has altered gastric secretory, hormonal, and immune physiology. What if a lesser-known ancestral indigenous organism disappeared from the colon, mouth, or vagina? Could that change have contributed to some of the diseases (for example, diabetes, asthma, and obesity) that are becoming more prevalent? The hope is that the Human Microbiome Project will begin to answer these and other questions surrounding the relationship between humans and their microbes.

SECTION SUMMARY

- **Biocontrol** is the use of one microbe to control the growth of another.
- **Probiotics** contain certain microbes that, when ingested, help restore balance to intestinal microbiota.
- **Fecal microbiota transplants** have proven useful in curing recurrent *C. difficile* colitis.
- **Phage therapy** offers a possible alternative to antibiotics in the face of rising antibiotic resistance.
- **The increase in diseases such as allergies** over the past century may be caused by changes in the human microbiome.

Thought Question 14.8 Your dorm mate, Teresa, has confided to you that she developed diarrhea after taking antibiotics for a few days to treat an upper respiratory tract infection. She still has a week to go on the antibiotics. To try to cure her diarrhea, she started eating active-culture yogurt, which in the past has helped her "cure" occasional episodes of diarrhea. Tell her why she has diarrhea and why the yogurt is not working this time.

Table 14.3
Changes in Human Behavior That Can Alter the Composition of Microbiota

Change	Consequence
Clean water	Reduces fecal transmission of bacteria
Cesarean deliveries	Reduces vaginal transmission
Use of preterm antibiotics	Reduces vaginal transmission
Reduced breast-feeding	Reduces cutaneous transmission and altered immunological environment
Smaller family size	Reduces early-life transmission
Widespread antibiotic use	Accelerates change of microbiome composition
Increased bathing/showering and use of antibacterial soaps	Accelerates change of skin microbiome composition
Use of mercury amalgam dental fillings	Accelerates change of oral cavity microbiome composition

Perspective

As we learn more and more about the human microbiome and its positive impact on human health and development, many of us have come to think of it as a new kind of organ system, albeit a constantly changing one. It has even been proposed that our adaptive immune system evolved, in part, to accommodate our microbiome. As a result, we reap the benefits of these microbes while keeping them from invading and killing us. One of the puzzles we will address in later chapters is how the intestine simultaneously accommodates the indigenous microbiome, maintains an intact mucosal barrier, and can tell the difference between symbiotic or mutualistic species from dangerous pathogens.

Key Terms

bacteremia (p. 419)
bioburden (p. 415)
blood-brain barrier (p. 428)
cytokine (p. 424)
dysbiosis (p. 426)
enterotoxin (p. 424)

epidermis (p. 416)
**fecal microbiota transplant
(FMT)** (p. 427, 434)
gnotobiotic animal (p. 425)
immunocompromised host (p. 426)
immunomodulin (p. 424)

metagenomic study (p. 432)
microbiome (p. 414)
microbiota (p. 414)
mucociliary escalator (p. 419)
nasopharynx (p. 417)
opportunistic pathogen (p. 426)

oropharynx (p. 417)
petechial rash (p. 428)
prebiotic (p. 426)
probiotic (p. 426)
virome (p. 415)

Objectives Assessment Questions

14.1 Human Microbiota: Location and Shifting Composition

1. Which of the following body sites is considered sterile?
    a. Urethra
    b. Kidney
    c. Small intestine
    d. Stomach
    e. Tonsils

2. *Bifidobacterium* is a common inhabitant of the _____.
    a. skin
    b. oral cavity
    c. lung
    d. intestine
    e. urinary tract

3. The collection of microbes associated with a human body site is referred to as _____.

4. Which of the following factors will influence the types and relative numbers of microbes inhabiting the intestine?
    a. Antibiotic usage
    b. Types of ingested foods
    c. Housemates
    d. All of the above
    e. None of the above

14.2 Benefits and Risks of a Microbiome

5. The benefits of having a microbiome include which of the following?
    a. The production of cytokines by intestinal microbes
    b. Interfering with pathogen colonization
    c. Stimulation of neuron development
    d. All of the above
    e. b and c only
    f. None of the above

6. An organism that is part of normal microbiota but can cause infection of an immunocompromised host is called a/an _____.

7. A germ-free animal is also called a _____ animal.

8. A germ-free animal is characterized by _____.
    a. increased numbers of T cells
    b. diminished brain activity
    c. increased cardiac output
    d. thin intestinal walls
    e. neutrophils that exhibit enhanced phagocytic activity

14.3 Keeping Microbiota in Their Place

9. Endothelial cells are a major part of which barrier?
    a. Blood-brain
    b. Maternofetal
    c. Peyer's patches
    d. All of the above
    e. a and b only
    f. None of the above

10. Which of the following is a host structure that recognizes microbial structures?
    a. Tight junction
    b. Toll-like receptor
    c. Lipopolysaccharide
    d. Petechia
    e. Firmicute
    f. All of the above
    g. None of the above

11. Which of the following structures is designed to sample microbiota to maintain a low immune response to microbiota?
    a. Maternofetal barrier
    b. Blood-brain barrier
    c. Peyer's patch
    d. Petechia
    e. Pattern recognition receptor

14.4 Microbiome–Host Communications

12. Normal microbiota can influence host cell gene expression. True or false?

13. Hormones released from host cells can change bacterial gene expression. True or false?

14. You perform an experiment using germ-free mice to study host-microbe interactions. Germ-free mice have an abnormally low level of T cells (a type of white blood cell) in their spleens. You find that transferring feces from a normal mouse to the intestines of germ-free mice restores normal T-cell levels over time. Which of the following is the most likely conclusion?
    a. Host cells secrete a factor that permits growth of the fecal transplant organisms.
    b. Donor T cells present in the transplant repopulated the recipient's spleen.

    c. Factors produced by the recipient's original microbiota inhibited production of T cells.

    d. Factors produced by donor microbiota stimulated production of T cells.

15. Our microbiome could be considered an organ system for which of the following reasons?

    a. Microbiota contribute to our caloric intake.

    b. The microbiome is located in specific parts of the body.

    c. Microbiota and human cells communicate with each other.

    d. A person's microbiome can be inherited.

    e. All of the above

    f. None of the above

14.5 Natural Biological Control of Microbiota

16. Which of the following organisms is Y-shaped and is considered a probiotic bacterial organism?

    a. *Salmonella enterica*

    b. *Bifidobacterium*

    c. *Staphylococcus aureus*

    d. *Lactobacillus*

    e. *Staphylococcus epidermidis*

17. Phage therapy is considered a viable alternative to antibiotics because _____.

    a. antibiotic-resistant bacteria are more susceptible to phage

    b. phages can evolve to infect phage-resistant bacteria

    c. antibiotics will not affect phage-resistant bacteria

    d. All of the above

    e. None of the above

18. Microbiome transplants are performed only between relatives. True or false?

19. The improvement in hygiene practices over the past century has been associated with _____.

    a. increased cases of intestinal infections

    b. increased cases of allergies

    c. decreased life span

    d. an increase in antibiotic-resistant organisms

    e. increased microbiota

Review Questions

1. Name some sterile body sites.

2. What body sites are colonized by a microbiome?

3. Under what circumstances can members of the normal microbiome cause disease?

4. Why is the microbiome beneficial to the host?

5. What are probiotics? How do they help maintain health?

6. How does the lung avoid being colonized?

7. What is a gnotobiotic animal?

8. Describe the various barriers that separate the microbiome from host tissues.

Clinical Correlation Questions

1. A 50-year-old man, after returning to the United States from a trip to India, develops excessive watery diarrhea. The diagnosis is cholera caused by *Vibrio cholerae*, a very acid-sensitive, Gram-negative rod. Healthy individuals do not ordinarily develop cholera unless they ingest very large numbers of organisms. The patient's history reveals that he did not drink tap water or swim in lakes, bays, or seas during his visit. Previously he was quite healthy except for recurrent bouts of indigestion that he self-medicated with Prilosec (a medication that reduces production of stomach acid). Explain how this individual developed cholera.

2. You accidentally ingest food contaminated with a bacterial species harboring an antibiotic resistance plasmid. Fortunately, the organism does not survive in the intestine. Nevertheless, how might this organism influence your intestinal microbiota?

3. How would you explain the following set of experimental results? (1) Germ-free mice have an altered ratio of certain immune system white blood cells. (2) A fecal transplant from a healthy mouse to a germ-free mouse restores the normal ratio of white cells. (3) Boiling the fecal transplant material for 10 minutes eliminates all culturable microbes, but transplanting this material still reverses the white cell defect. (4) Autoclaving the fecal material destroys its ability to restore the immune system.

4. You are participating in a study designed to explore how the microbiome can influence clinical outcomes in patients with bipolar disorder, formerly called manic depression. The condition is associated with extreme mood swings. Your research has found that individuals with bipolar disorder eat significantly <u>fewer</u> polyunsaturated fatty acids; these are the healthy fats that come mostly from vegetable oil and fatty fish. You suspect that this diet has an effect on microbiome composition. Discuss how you would explore the hypothesis that bipolar disorder could be treated by diet and perhaps FMT.

15

The Immune System: Inflammation and Innate Immunity

[A Tale of Two Tourists]

SCENARIO Two vacationing tourists, unknown to each other, visited the Alabama gulf coast over the same weekend in August and decided to take a swim in the Gulf of Mexico. One, a 12-year-old girl named Alexis, had a cut on her arm caused by a fall onto the sharp edge of a scooter. The other vacationer, Brad, a man 66 years of age, had recently cut his leg on a nail protruding from a wooden railing at his beach condo. Within moments after Brad and Alexis entered the Gulf of Mexico to swim, several small Gram-negative microbes invaded both of their bodies through those cuts. Four days later, the young girl was heading back to Birmingham in the back of the family car, oblivious to the battle recently waged in her bloodstream. At the same time, Brad lay dead of an aggressive blood infection.

Cellulitis Caused by *Vibrio vulnificus*
V. vulnificus can rapidly produce hemorrhagic bullae (the fluid- and blood-filled vesicle on the left is a bulla) and purpuric macules (the large purple rashes on the right). The vesicle fluid and bloodstream will contain *V. vulnificus* bacteria.

SEM | 1 μm

Vibrio vulnificus
Vibrio vulnificus is a curved, Gram-negative rod that grows in saline environments. It can enter the body via wounds or ingestion of contaminated seafood such as oysters. The organism causes mild gastroenteritis in otherwise healthy people, but it can cause life-threatening blood and tissue infections in people with an immune system compromised by disease or drug therapy.

BRAD'S CLINICAL PROGRESSION Within a day after his swim, Brad's leg began to swell, was warm to the touch, and started to hurt. He put off going to the emergency department until day 2, when he realized his leg was now extremely painful and had started to darken, and he felt feverish. At the hospital, he had a fever of 39°C (102.2°F) and his blood pressure was abnormally low. Samples of wound material containing pus and a blood culture were collected and sent to the diagnostic laboratory to determine the identity of the pathogen. Results of the tests, however, would not be known for at least 48 hours. Meanwhile, Brad was admitted to the hospital and placed on empirical antibiotic therapy to try to stop the infection. By day 3, areas of his leg were turning black and becoming gangrenous, and the swelling had spread up and down his leg. His fever remained high and he became unresponsive. The antibiotics were not working. Brad died the next day. The lab report, now completed but too late to matter, indicated that both the wound and blood cultures contained *Vibrio vulnificus*.

CHAPTER OBJECTIVES

After reading this chapter, you will be able to:

- Describe the two major systems of immunity and explain their separate yet interdependent roles.
- Discuss the relationship between microbial structure and innate immune mechanisms.
- Identify features of the innate immune system that affect acute versus chronic inflammation.
- Explain how activating innate immunity influences disease symptoms.

Both swimmers in the opening case history were infected by the same pathogen, *Vibrio vulnificus*, a Gram-negative rod that lives in salt water. The organism can grow rapidly in blood (it has a 10-minute generation time) and cause massive blood and tissue infections. The infections can be so severe that parts of the body seem to dissolve. So, why did Alexis live and Brad die? As is the case for most healthy people, a variety of nonspecific, innate immune factors present in Alexis's body killed the invading pathogen before it could multiply. Brad, on the other hand, was an alcoholic with liver disease. Because the liver supplies critically important nonspecific defense proteins like complement (Section 15.7), Brad's diseased liver left him vulnerable to infection and disease.

In contrast to slow adaptive immune mechanisms, which must "see" a specific microbe before making immune cells and antibodies to fight it, innate immune mechanisms provide immediate and almost blind protection against many different microbes. These innate immune mechanisms often mean the difference between life and death, as they did for Alexis and do in our own daily lives.

◉ **Note:** Antibodies are small, Y-shaped proteins in blood that bind and tag invading pathogens. Different antibodies bind different bacteria and viruses. Innate immune mechanisms on their own can kill pathogens but are more aggressive if antibodies also coat the pathogen.

In Chapter 14 we discussed innate physical and chemical barriers, such as skin and stomach acid, that keep our microbiota at bay and help protect us from pathogens. This chapter provides an overview of the cellular makeup of our immune system, focusing on the nonadaptive, innate mechanisms that first engage pathogens after they breach the body's physical barriers.

15.1 Overview of the Immune System

SECTION OBJECTIVES

- Distinguish between innate immunity and adaptive immunity.
- Describe the various cells and organs that constitute the immune system.
- Discuss the white blood cell count differential and its role in diagnosing infectious disease.

What happens in our body after a bacterium, virus, fungus, or parasite enters? Microorganisms that slip through our body's physical barriers are met with a more aggressive series of defenses collectively called the immune system. The **immune system** is a collection of organs, tissues, cells, and cell products that work together to differentiate self (host cells) from nonself (invaders) and rid the body of nonself substances (like potentially pathogenic organisms) that do not belong. The immune system is remarkable in its capacity to respond to nearly any foreign molecular structure.

The Two Branches of Immunity: Innate and Adaptive

There are two broad branches of mammalian immunity: **innate immunity**, often called nonadaptive immunity, and **adaptive immunity**. Innate immune responses are immediate and not very specific about what they target. In contrast, adaptive responses take a week or so to develop but are highly specific to the given pathogen. Once primed, however, adaptive immunity responds swiftly if the same pathogen reenters the body at a later date. The protection these two systems afford the human body might be compared to the defenses used to protect medieval castles (**Figure 15.1**). For instance, the innate immune system is analogous to an alligator-stocked moat surrounding the castle. The alligators can indiscriminately attack any intruder, whether friend or foe (it's a myth, though, that alligators were placed in medieval moats). The adaptive immune system, on the other hand, is more like the king's archers. The archers, upon hearing a commotion outside the castle wall, climb to the rampart atop the castle, where they identify and selectively target hostile intruders with arrows (think antibodies).

The nonadaptive cellular responses (the alligators) of innate immunity are present even before birth. These innate mechanisms attack intruding microbes as soon as they enter the body (our first line of defense), and in a kind of molecular "call for backup" they send chemical signals to the more targeted adaptive immune system (the archers), alerting it that an infection is under way. Thus, innate immune mechanisms buy precious time for the more specific second line of defense to engage. Recall the stages of disease illustrated in Figure 2.8. The innate defense system operates continuously from the time of infection through the incubation period and until the infection ends. Adaptive immunity becomes evident

Figure 15.1 The Immune System as Castle Defense

In this analogy, the castle plays the role of our body under attack. The castle walls form a protective barrier, like skin. The barrels of boiling oil used to douse any who attempt to climb the castle walls can be compared to chemical defenses of innate immunity. The alligators swimming in the moat represent white blood cells, such as macrophages or neutrophils, that indiscriminately drown or attack anyone who tries to enter. The archers represent our adaptive immune system (B cells and plasma cells), which must first mount the towers, see the invaders, and then specifically target them with their arrows (antibodies). Swordsmen battling invaders inside the castle can be compared to cytotoxic T cells of the adaptive immune system (discussed in the next chapter), which specifically target infected host cells.

only toward the end of the prodromal phase (the period when the first signs of disease are evident) and is usually maintained for long periods after the infection clears.

Although this chapter focuses on innate immunity, it is important to recognize that innate and adaptive immune mechanisms intertwine in ways that support and enhance each other's effectiveness. Adaptive immunity reacts to very specific structures called antigens. An **antigen** is any chemical, compound, or structure foreign to the body that will elicit an adaptive immune response such as the production of antibodies. As its name implies, the adaptive immune response to a specific antigen does not begin until the body "sees" that antigen. Adaptive immune mechanisms can recognize billions of different antigenic structures and launch a directed attack against each one. Once such an attack is under way, the immune system stores a "memory" of the exposure in the form of specific memory cells. Any later encounter with the microbe will reactivate the appropriate memory cells, which will then rapidly multiply, mobilize, engage, and destroy the intruder. Antibodies and adaptive immunity are described more completely in Chapter 16.

The two branches of immunity, innate and adaptive, can be illustrated by the immune response to infection by *Neisseria gonorrhoeae*, the bacterial pathogen that causes the sexually transmitted infection gonorrhea. A rapidly acting part of the innate immune response is **complement**, composed of numerous soluble protein factors constantly present in the blood. Within moments of an initial bloodstream infection, complement "attacks" by binding to and making holes in the bacterial membrane, thereby killing the microbe (see Section 15.7). The second wave of the immune response, adaptive immunity, will then generate specific antibodies that bind any *N. gonorrhoeae* cells that escaped the innate mechanisms. These antibodies, however, are not made until well after the organism infects a person. Together, the forces of innate and adaptive immunity can ward off disease caused by *N. gonorrhoeae*. In fact, anti-*Neisseria* antibodies bound to *Neisseria* cells will actually stimulate killing by complement.

Although innate and adaptive immunity are often treated as separate entities, certain kinds of cells and organs participate in both. We begin our discussion of immunity by describing the various cells and organs that make up the whole of the immune system.

Cells of the Innate and Adaptive Immune Systems

Blood is composed of red blood cells (erythrocytes), white blood cells (leukocytes), and platelets (**Figure 15.2**). The different types of white blood cells develop from a precursor stem cell produced in bone marrow throughout life (**Figure 15.3**). The process of making blood cells, called hematopoiesis, is complex, involving a number of differentiation stages. The precursor stem cell can differentiate into myeloid stem cells and lymphoid stem cells, which give rise to white blood cells (WBCs); and into erythroid stem cells, which give rise to red blood cells and platelets. Myeloid stem cells differentiate into the WBCs of innate immunity; lymphoid stem cells generate the WBCs of adaptive immunity. The WBCs of innate immunity include:

- Polymorphonuclear leukocytes (PMNs)
- Monocytes
- Macrophages
- Dendritic cells
- Mast cells
- Natural killer cells (lymphoid lineage but part of innate immunity)

Figure 15.2 Red and White Blood Cells

This micrograph illustrates the relative sizes and shapes of red blood cells, white blood cells (leukocytes), and platelets. Leukocytes shown are macrophage (green), neutrophil (large white cell), and lymphocyte (pink).

Figure 15.3 Development of White Blood Cells of the Immune System

Pluripotent stem cells in bone marrow divide to form three lineages. One lineage consists of the myeloid stem cells, which develop into polymorphonuclear leukocytes (PMNs) and monocytes, which function primarily as part of innate immunity. The second lineage, erythroid stem cells, produce red blood cells and platelets. The third lineage, lymphoid stem cells, ultimately form natural killer cells, B cells, and T cells. B and T cells are involved in adaptive immunity. The B cells mature in bone marrow ("B" is for bone), and T cells mature in the thymus. Colors identify groups of differentiated cells that arise from the same progenitor.

POLYMORPHONUCLEAR LEUKOCYTES Polymorphonuclear leukocytes (PMNs), also called granulocytes, have multilobed nuclei, differentiate from an intermediate cell called the myeloblast, and contain enzyme-rich lysosome organelles (**Figure 15.4A**). PMNs are of several types named for their different staining characteristics. Each cell type has a different function. **Neutrophils**, making up nearly all PMNs in the blood, can engulf microbes by a process called **phagocytosis**. For phagocytosis, the directed movement of cytoplasm generates armlike membrane protrusions called pseudopods that reach out from the cell toward a target microbe. Pseudopods have surface receptors that attach directly to the pathogen or attach indirectly by binding to complement or to antibodies already bound to the pathogen. The pseudopods then engulf the pathogen, which ends up in a **phagosome** vacuole (**Figure 15.5**). The phagocyte then kills microbes by fusing enzyme-engorged lysosomes with phagosomes. Enzymes spilling from the lysosome

into the phagosome will destroy various components of the microbe and, ultimately, the microbe itself.

Two other types of PMNs are **basophils**, which stain with basic dyes, and **eosinophils** (**Figure 15.4B**), which stain with the acidic dye eosin. These cell types inefficiently phagocytose (engulf) microbes, but they do release products, like major basic protein, that are toxic to the microbe. These WBCs also release chemical mediators, called vasoactive agents, that have effects on the diameter and permeability of blood vessels—effects important to inflammation (discussed in Section 15.3).

MAST CELLS Mast cells contain many granules rich in histamine and heparin. They are similar to basophils in structure but differentiate in a lineage separate from PMNs. Unlike PMNs, mast cells are residents of connective tissues and mucosa (mucus-secreting tissue) and do not circulate in the bloodstream. Basophils and mast cells not only have roles in inflammation and wound healing but also contain high-affinity receptors for a class of antibody (immunoglobulin E, or IgE) associated with allergic responses. We will describe the role of these receptors in allergy in Chapter 16.

Figure 15.4 Selected Types of White Blood Cells

A. Neutrophils (PMNs). Note the multilobed nuclei.

B. Eosinophils stain with eosin, an acidic dye.

C. Monocyte, with a single nucleus.

D. Lymphocyte (B cell or T cell), with its large single nucleus.

Figure 15.5 Basic Process of Phagocytosis (Details Found in Section 15.4)

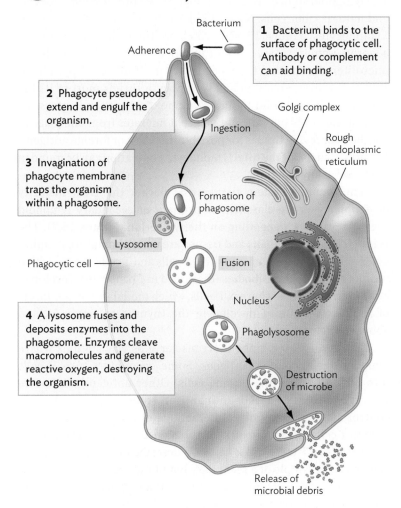

Bacterium

Adherence

1 Bacterium binds to the surface of phagocytic cell. Antibody or complement can aid binding.

2 Phagocyte pseudopods extend and engulf the organism.

Ingestion

Golgi complex

3 Invagination of phagocyte membrane traps the organism within a phagosome.

Rough endoplasmic reticulum

Formation of phagosome

Lysosome

Phagocytic cell

Fusion

Nucleus

4 A lysosome fuses and deposits enzymes into the phagosome. Enzymes cleave macromolecules and generate reactive oxygen, destroying the organism.

Phagolysosome

Destruction of microbe

Release of microbial debris

Note: Because neutrophils make up nearly all PMNs, the terms "neutrophil" and "PMN" are often used interchangeably.

MONOCYTES, MACROPHAGES, AND DENDRITIC CELLS Monocytes (Figure 15.4C) are WBCs with a single nucleus (not multilobed as in a PMN); they engulf (phagocytose) foreign material and are the precursors of two other cell types: macrophages and dendritic cells (see Figure 15.3). Monocytes circulating in the blood migrate out of blood vessels into various tissues, where they then differentiate into macrophages and dendritic cells. **Macrophages** (Figure 15.6A) are phagocytic and form a major part of an amorphous **mononuclear phagocyte system (MPS)**, previously called the reticuloendothelial system. The MPS is widely distributed throughout the body and is composed of monocytes that circulate in the blood, as well as macrophages that are dispersed in tissues. In addition to macrophages, the system is composed of specialized endothelial cells, called Kupffer cells, that line the sinusoids (special capillaries) of the liver, spleen, and bone marrow.

Macrophages are present in most tissues of the body and are the cells most likely to make first contact with invading pathogens. Macrophages are like sentries with two functions. As part of innate immunity, they kill invaders directly. Protoplasmic protrusions from the macrophage surface extend and clasp nearby bacteria, pulling them into the cell (Figure 15.5, steps 1 and 2, and **Figure 15.6B**). Once ingested by the macrophage, the bacteria are destroyed within a phagolysosome, as illustrated in Figure 15.5. Then, in the first step of adaptive immunity, called antigen processing, the remnants are degraded into smaller peptide antigens and are presented on the macrophage cell surface. Lymphocytes known as **T cells** can bind to antigens displayed on the macrophage and initiate an adaptive immune response. Different T cells are specific for different antigens. Cells, such as macrophages, that introduce or "present" bacterial antigens to cells of the adaptive immune system are called

Figure 15.6 Macrophages and Dendritic Cells

A. A macrophage (20 μm long) at the site of a skin wound (colorized SEM).

B. Membrane protrusions from a macrophage detecting and engulfing 1.5-μm-long *E. coli* (pink; colorized SEM).

C. Dendritic cell (colorized SEM).

antigen-presenting cells (APCs). In Chapter 16 we will discuss how the specific immune response develops.

In addition to macrophages, **dendritic cells** (**Figure 15.6C**) can also take up, process, and present small antigens on their cell surface to T cells. Dendritic cells, not to be confused with dendrites of the nervous system, are located in the spleen, lymph nodes, and mucous membranes. Dendritic cells of one type, called Langerhans cells, are also present in skin. Dendritic cells are different from macrophages in structure and in the fact that they take up small soluble antigens from the surroundings in addition to phagocytosing whole bacteria. Dendritic cells are very important to the adaptive immune process.

WHITE BLOOD CELL DIFFERENTIALS IN HEALTHY AND INFECTED PATIENTS Today's clinical laboratory can take a blood sample and automatically count the total number of WBCs and identify their various cell types—neutrophils, eosinophils, basophils, lymphocytes, and monocytes (**Table 15.1**)—providing what is called a **white blood cell (WBC) differential**. The results can offer the clinician important clues about the cause of a patient's illness. For example, for a patient presenting with a high fever and headache, abnormally high total WBC numbers could indicate infection or allergy. Elevated neutrophils or their precursors, called band cells, would suggest a bacterial infection, whereas elevated lymphocytes would suggest a viral infection. Alternatively, an increase in eosinophils can indicate an allergy or a parasitic infection. Remember, there can be other explanations for a particular WBC profile (see the discussion of lymphoid cancers in Chapter 17), so it is important to view the WBC differential results in the context of the patient's symptoms. Did our sample patient, for instance, exhibit hallmark symptoms of viral or bacterial disease? If not, there could be a noninfectious cause for the abnormal profile.

Lymphoid Organs

The lymphoid stem cells noted in Figure 15.3 produce lymphocytes and natural killer cells. Natural killer cells will be described later. Lymphocytes (**Figure 15.4D**), which are the main participants in adaptive immunity, are present in blood at about 2,500 cells per microliter, accounting for more than one-third of all peripheral WBCs. However, an individual lymphocyte spends most of its life within specialized solid tissues (lymphoid organs) and enters the bloodstream only periodically, where it migrates from one place to another, surveying tissues for possible infection or foreign antigens. No more than 1% of the total lymphocyte population can be found in the blood at any one time.

The immune system tissues that harbor the great majority of lymphocytes are classified as primary or secondary lymphoid organs or tissues, depending on their function (**Figure 15.7**). The primary lymphoid organs and tissues are where immature lymphocytes mature into antigen-sensitive B cells and T cells. **B cells**, which ultimately produce antibodies (discussed in Chapter 16), develop in bone marrow tissue, whereas T cells, which modulate various facets of adaptive immunity, develop in the thymus, an organ located above the heart.

The secondary lymphoid organs serve as way stations where lymphocytes encounter antigens. Some encounters turn B cells into antibody-secreting plasma cells. Other antigen-lymphocyte interactions change T cells into antigen-specific helper cells that control adaptive immunity. Interactions of a third kind convert certain T cells into cytotoxic killers that destroy infected body cells. The details of T cells and B cells are described in Chapter 16. The spleen is an example of a secondary lymphoid organ. It directly filters blood and detects microorganisms. Macrophages in the spleen engulf and destroy these organisms and then migrate to other

Table 15.1
Guidelines for Interpreting WBC Differentials[a]

	Normal	Acute Bacterial Infection	Viral Infection	Parasitic Infection	Allergy
Total WBC Count	4,500–11,000/µl	Elevated: 12,000–30,000	Elevated	Elevated	Elevated
Differential					
Neutrophils	54%–62%	Increased numbers and more immature forms (band cells)	Can be reduced		
Eosinophils	1%–3%			Increased	Increased
Basophils	0%–0.75%				Increased
Lymphocytes	25%–33%		Increased		
Monocytes	3%–7%	Typically increase with chronic infections (e.g., tuberculosis, brucellosis, subacute bacterial endocarditis, rickettsiosis, many protozoan infections)			

[a]This table provides general guidelines. Individual organisms or certain noninfectious medical conditions or immunological defects can alter the findings. A blank space indicates little change in the parameter.

Figure 15.7 Relative Locations of Lymphoid Organs

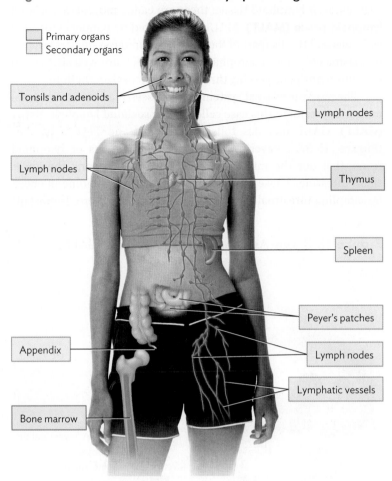

Primary organs
Secondary organs

Tonsils and adenoids

Lymph nodes

Lymph nodes

Thymus

Spleen

Peyer's patches

Appendix

Lymph nodes

Lymphatic vessels

Bone marrow

- **Myeloid bone marrow stem cells** differentiate to form cells of the innate immune system—phagocytic PMNs, monocytes, macrophages, antigen-presenting dendritic cells, and mast cells.
- **Lymphoid stem cells** differentiate into natural killer cells (part of the innate immune system) and lymphocytes (cells of the adaptive immune system). Lymphocytes are classified as B cells (which ultimately produce antibodies) and T cells (which regulate adaptive immunity or kill infected cells).

Thought Question 15.1 Invertebrates such as clams possess a strong innate immune system but do not possess adaptive immune mechanisms. Why do vertebrates, humans in particular, need an adaptive immune system in addition to the innate one?

15.2 Innate Host Defenses: Keeping the Microbial Hordes at Bay

SECTION OBJECTIVES

- Describe the physical barriers to infection.
- Explain the basic roles of SALT, GALT, and M cells in innate immunity.
- Describe chemical barriers employed by cells of the innate immune system.

We have broadly described the various cells and organs of the immune system, but what combination of cells, structures, and processes constitutes innate immunity? To continue our earlier analogy (see Figure 15.1), the first lines of castle defense include physical barriers (the watery moat and castle wall), chemical barriers (boiling oil tossed onto invaders trying to scale the wall), and "cellular" barriers (hand-to-hand combat) once the wall is breached. Similarly, the body's initial, innate defenses against infectious disease are composed of physical, chemical, and cellular barriers designed to prevent a pathogen's access to host tissues. Although generally described as nonspecific, some of our innate defense systems are more specific than others in recognizing invaders, as you will see.

Physical Barriers to Infection

As described in Chapter 14, the first line of defense against any potential microbial invader (either a member of the microbiome or a pathogen) is found wherever the body interfaces with the environment. These interfaces (skin, lung, gastrointestinal tract, genitourinary tract, and oral cavities) have similar defense strategies, although each has unique characteristics.

SKIN Skin is the protective shield separating you from the sea of microbes that inhabit your surroundings. Few microorganisms can penetrate skin because of the thick keratin armor produced by closely packed cells called keratinocytes. Keratin protein is a hard

secondary lymphoid organs to present pieces (the antigens) of the microbe to the B and T cells, which then become activated.

Lymph nodes, another kind of secondary lymphoid organ, are strategically positioned around the body where lymphatic vessels converge (for example, under the armpits). Lymphatic vessels are similar to blood vessels, but instead of carrying blood they drain **lymph**, made of interstitial fluid and WBCs (including macrophages) that bathe nearby tissues. Along with lymph, lymph nodes trap microorganisms that drain from infected tissues. Lymphoid tissues are also found in the mucosal regions of the gut and respiratory tract—for example, Peyer's patches and gut-associated lymphoid tissue, or GALT, discussed in the next section. Other secondary lymphoid organs are the tonsils, adenoids, and appendix.

SECTION SUMMARY

- **The immune system** consists of both innate and adaptive mechanisms that recognize and then contain or kill pathogens.
- **Innate immunity** includes physical and chemical barriers and some cellular responses to various microbial structures.
- **Adaptive immunity** is a cellular response to specific structures (antigens), which generates a memory of the first exposure. This memory enables a rapid, massive response upon later reexposure to the same antigen.

substance (hair and fingernails are made of it) that is not degraded by known microbial enzymes. An oily substance (sebum) produced by the sebaceous glands also covers and protects the skin. Its slightly acidic pH inhibits bacterial growth. Competition among species also limits colonization by pathogens, and microorganisms that manage to adhere are continually removed by the constant shedding of outer epithelial skin layers.

A consortium of specialized cells called **skin-associated lymphoid tissue (SALT)**, located just under the skin, recognizes microbes that may slip through the physical barrier. **Langerhans cells**, specialized dendritic cells that can also phagocytose microbes, make up a significant portion of SALT. Once a Langerhans cell has ingested a microbe, the cell migrates by ameboid movement to nearby lymph nodes. Once it reaches the lymph nodes, it "presents" antigenic parts of the microbe to the immune system to activate adaptive immunity.

⊙ **Note:** Do not confuse phagocytic Langerhans cells with the pancreatic "islets of Langerhans," which secrete insulin.

MUCOUS MEMBRANES Mucosal surfaces form the largest interface between the human host and the environment: 2,000–3,000 square feet, roughly the area of a basketball court. The intestine alone is about 25 feet long. This large surface presents a huge containment problem. Mucous membranes must be selectively permeable to nutrients and waste products but form a staunch barrier against invading pathogens. Mucosal membranes are covered with special tight-knit epithelial layers that support this barrier function (see Figure 14.16). Multiple layers of box-shaped, stratified squamous epithelial cells secrete mucus that coats the mucosal surfaces and traps microbes. Compounds within mucus can serve as a food source for some members of the microbiome, but other secreted compounds—like the enzymes **lysozyme** (which cleaves cell wall peptidoglycan) and lactoperoxidase (which produces toxic superoxide radicals)—can kill an organism trapped in the mucus.

In addition to these nonspecific defense mechanisms, there are also semispecific innate immune mechanisms associated with mucosal surfaces. These mechanisms are like using touch in a dark room to distinguish a soft pillow from a dangerous knife: you don't know what kind of knife it is, but you know it's bad. Host cells, even mucosal epithelial cells, have evolved mechanisms to distinguish harmless compounds from dangerous microbes. Certain conserved structures on microbes, called **microbe-associated molecular patterns (MAMPs)**, are recognized by host cell surface receptors collectively called pattern recognition receptors (PRRs). The Toll-like receptors discussed later, in Section 15.5, are examples of PRRs. MAMPs include structures that are unique to bacteria, such as peptidoglycan (the cagelike, cell wall macromolecule) and lipopolysaccharide (LPS; a component of Gram-negative outer membranes). Once a MAMP is recognized by a body cell, the cell sends out chemicals that can activate innate and adaptive immune system cells.

As mentioned earlier, mucosal surfaces also contain scattered aggregates of lymphoid tissue; these are called **mucosa-associated lymphoid tissue (MALT)**. MALT is populated by lymphocytes such as T cells and B cells (part of the adaptive immune system), as well as by plasma cells and macrophages, all of which are well situated to encounter antigens passing through the mucosal epithelium.

The gastrointestinal system also possesses an innate mucosal immune system, in this case called **gut-associated lymphoid tissue (GALT)**. GALT includes tonsils, adenoids, and Peyer's patches (Figure 15.8A). **Peyer's patches** are aggregations of lymphoid tissue that dot the intestinal surface, primarily along the lower small intestine. These tissues help keep our microbiome in check by sampling intestinal microbes to "see" what is there. Important

Figure 15.8 Gut-Associated Lymphoid Tissue (GALT)

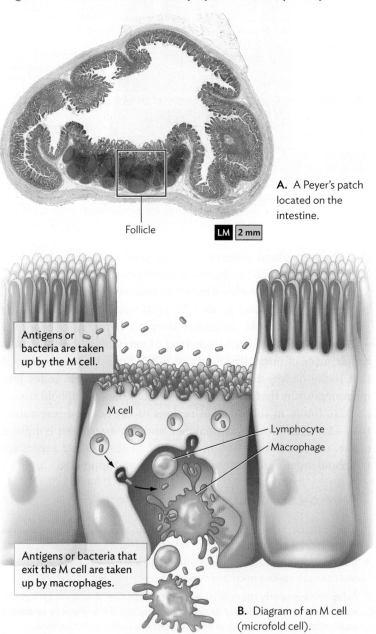

A. A Peyer's patch located on the intestine.

Follicle

LM 2 mm

Antigens or bacteria are taken up by the M cell.

M cell

Lymphocyte

Macrophage

Antigens or bacteria that exit the M cell are taken up by macrophages.

B. Diagram of an M cell (microfold cell).

components of Peyer's patches are the specialized **M cells** wedged between epithelial cells. "M" stands for "microfold," which describes their appearance (**Figure 15.8B**). M cells take up microbes from the intestine and release them, or pieces of them, into a pocket formed on the opposite (basolateral) side of the cell. Other innate immune system cells, such as macrophages, gather here and collect the organisms that emerge. The macrophages engulf and try to kill the organism, and they place small, degraded components (antigens) of the organism on their cell surface in a process called antigen presentation. Other immune system cells can then scan the macrophage surface to recognize those antigens. The process of antigen presentation will be discussed in Chapter 16.

M cells, then, are extremely important for the development of mucosal immunity to pathogens. However, M cells can also serve as a gateway for some pathogens, like *Salmonella*, to gain entry to the body and cause disease.

LUNGS The lung also has a formidable defense. In addition to the respiratory mucociliary escalator discussed in Section 14.1, microorganisms larger than 100 μm become trapped by hairs and cilia lining the nasal cavity and trigger the forceful expulsion of air from the lungs (a sneeze). The sneeze is designed to clear the organism from the respiratory tract. Organisms that make it to the alveoli (see Figure 14.5A), however, are not easily expelled by a sneeze but are met by phagocytic cells called alveolar macrophages. These cells can ingest and kill most bacteria. Alveolar macrophages also send out chemical signals that attract other cells of the innate and adaptive immune systems to the infection site.

Chemical Barriers to Infection

Chemical barriers to infection exist at numerous body sites. For instance, the acid pH of the stomach is lethal to most bacteria, and lysozyme in tears will degrade the cell walls of Gram-positive bacteria (see Chapter 14). Potentially pathogenic bacteria can also be killed by superoxide radicals generated by host enzymes such as lactoperoxidase, present in a variety of cell types. Another set of antimicrobial chemicals consists of small (containing 29–47 amino acids), cationic (positively charged) peptides called **defensins**. Defensins function as important components of innate immunity against microbial infection and are made by a variety of human cells. These antimicrobial peptides kill by destroying the microbial cytoplasmic membrane. They are effective against Gram-positive and Gram-negative bacteria, fungi, and even some enveloped viruses—viruses, like HIV, that are wrapped with eukaryotic membranes donned when the virus buds from a host cell. Defensins are produced by many human cells, including cells of the skin, lung, genitourinary tract, and gastrointestinal tract.

SECTION SUMMARY

- **Skin defenses** against invading microbes include closely packed keratinocytes and a SALT lymphoid system made up largely of phagocytic Langerhans cells.

- **Mucous-membrane defenses** involve secreted enzymes, cytokines, and GALT components, such as Peyer's patches, that contain phagocytic M cells.
- **M cells** in gut-associated lymphoid tissues sample bacterial cells at their surface and release pieces of them to immune system cells.
- **Phagocytic alveolar macrophages** inhabit lung tissues, contributing to nonspecific defense.
- **Chemical barriers against disease** include cationic defensins, acid pH in the stomach, and superoxide produced by certain cells.
- **Microbe-associated molecular patterns (MAMPs)** are recognized by pattern recognition receptors (PRRs) found on many host cells, such as macrophages. Binding triggers the release of chemical signaling molecules that activate innate and adaptive immune mechanisms.

Thought Question 15.2 Why do defensins have to be so small? Do defensins kill normal microbiota?

15.3 The Inflammatory Response

SECTION OBJECTIVES
- List the five cardinal signs of inflammation.
- Outline the process of inflammation.
- Describe the roles of vasoactive factors and cytokines.
- Differentiate acute from chronic inflammation.

What is inflammation, and how does it develop? The boil shown in **Figure 15.9** is an acute, or quickly developing, inflammatory response triggered by infection with the organism *Staphylococcus aureus*. Inflammation is a critical innate defense in the war between microbial invaders and their hosts. Through inflammation, phagocytic cells (such as neutrophils) normally confined to the bloodstream can enter infected areas within tissues. Movement of these

Figure 15.9 Inflammation Caused by Infection

A. Boil resulting from infection of a hair follicle by *Staphylococcus aureus*.

SEM 2 μm

B. SEM of the organism.

neutrophils out of blood vessels is called **extravasation** (discussed shortly). Once at the infection site, the neutrophils begin engulfing microbes. The white pus associated with an infection is teeming with these WBCs. Inflammation is marked by five cardinal signs that were first described more than 2,000 years ago: <u>h</u>eat (warmth at the site), <u>e</u>dema (swelling from fluid accumulation), <u>r</u>edness, <u>p</u>ain, and <u>a</u>ltered function or movement at the affected site (use the acronym HERPA to remember).

CASE HISTORY 15.1

A Case of Nonstick Neutrophils and Recurring Infections

Constance, a 7-month-old female, has suffered from recurrent infections since the age of 2 weeks. She has been frequently admitted to the hospital for *Staphylococcus aureus* sepsis (blood infection), hepatosplenomegaly (enlarged liver and spleen), oral thrush caused by *Candida albicans*, and poor weight gain. Repeated treatments with different antibiotics have only temporarily resolved her infections. Two days ago, her alarmed mother and father once again brought Constance to the emergency department at the University of Chicago Medical Center. Constance was lethargic, didn't move much, and had a high fever. She also struggled for breath. Her blood work revealed high WBC counts (59,000/μl; normal is 4,000–11,000), with a higher-than-normal level of neutrophils in her blood (neutrophilia). A chest X-ray showed inflammation in both lungs. (**Figure 15.10** compares a healthy, clear lung with one showing an inflammatory infiltrate caused by infection.) It was no surprise, then, when the lab technician reported finding *S. aureus* in Constance's blood. Her doctor ordered intravenous antibiotic therapy and was determined to find the cause of these repeated infections. He suspected a serious defect in Constance's immune system. Laboratory analyses found that her lymphocyte numbers, overall serum immunoglobulin (antibody) levels, as well as liver and kidney functions, were all normal. No help there. Then, flow cytometry analysis of surface proteins present on Constance's peripheral blood neutrophils revealed a complete absence of a critically important membrane protein. The doctor's heart sank. It was not good news.

Constance was repeatedly infected with *Staphylococcus aureus* and other microbes because she had a rare but serious inherited defect in her neutrophils, one of the first responders to infection. The defect, called leukocyte adhesion deficiency (LAD), was not in the number of neutrophils she had; in fact, the excessive number of neutrophils in her blood was a clue to the problem. Her neutrophils lacked the "sticky" cell-surface **integrin** proteins needed to get these white cells out of the circulation and to a site of infection—a critical part of the inflammatory response. Unfortunately, patients with LAD usually do not survive beyond 2 years of age.

⊙ **Note:** Flow cytometry was initially described in Chapter 6 as a way to count live versus dead cells. It can also be used to identify the presence of specific surface proteins if the cells are first treated with a fluorescent antibody that binds those proteins. Then, as the cells move past the laser in the flow cytometer, the instrument records fluorescence to determine how many cells with those proteins are present.

Extravasation and the Symptoms of Acute Inflammation

Although many things can trigger inflammation, we will focus on how microbes cause the response. The process begins with the infection itself. **Figure 15.11** presents a simplistic view of inflammation; **Figure 15.12** provides clarifying details. Microorganisms introduced into the body—for example, on a wood splinter—will begin to grow and produce compounds that damage host cells (Figures 15.11A and 15.12, step 1). Resident macrophages (the sentries of the immune system) that wander into the infected area engulf these organisms and then release inflammatory mediators, some of which (chemoattractants) "call" for more help (Figures 15.11B and 15.12, step 2).

The mediators also include **vasoactive factors** such as leukotrienes, platelet-activating factor, and prostaglandins. These vasoactive factors increase vascular permeability by loosening tight junctions between endothelial cells, and they widen blood vessels by relaxing the smooth muscles that surround them (Figures 15.11C and 15.12,

Figure 15.10 X-Ray Images of Lung

A. This image of a healthy, clear lung is purportedly that of the famous actress Marilyn Monroe, circa 1950.

B. Bilateral inflammation (cloudiness) caused by *Mycobacterium tuberculosis*. The nodular cloudiness is due to the infiltration of white cells into affected areas.

ANIMATION ▶ **Figure 15.11 Basic Inflammatory Response**

Neutrophils (a type of phagocyte) circulate freely through blood vessels and can squeeze between cells in the walls of a capillary (extravasation) to the infection site. The neutrophils then engulf and destroy any pathogens they encounter.

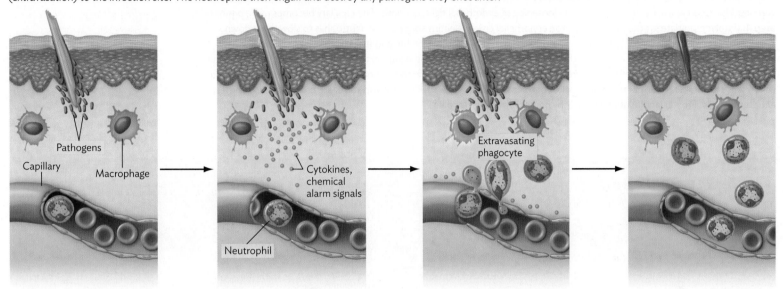

A. The host becomes infected via a splinter.

B. Resident macrophages engulf pathogens and release cytokines.

C. Vasoactive factors and cytokines help deliver additional phagocytes.

D. Some cytokines initiate healing as pathogens are destroyed.

step 3). The later process, called **vasodilation**, slows blood flow and, as a result, increases blood volume in the affected area. The now more permeable vessel allows blood plasma to escape into tissues, causing edema, and enables neutrophils to leave the capillaries—a process called **extravasation** (Figures 15.11C and D, and 15.12, step 4). The neutrophils then swarm the infection site. The events cause localized swelling, redness, and heat.

How does neutrophil extravasation happen? In addition to vasoactive factors, macrophages secrete small protein molecules called cytokines (Figure 15.12, step 3). Cytokines are signaling peptides that bind to specific membrane receptors on target cells and have powerful effects on the functions of those cells. Many kinds of cytokines exist, all of which are involved in regulating the immune response. Cytokines such as **interleukin 1 (IL-1)** and **tumor necrosis factor alpha (TNF-α)** released by macrophages stimulate the production of adhesion molecules (**selectins**) on the inner lining of the capillaries (step 4). The selectins snag neutrophils zooming by in the bloodstream, slow them down, and cause them to roll along the endothelium.

The rolling neutrophils are activated by other inflammatory mediators to produce, and display, **integrin** adhesion molecules on their surface. Integrins were the molecules missing from Constance's neutrophils in Case History 15.1. The integrins enable neutrophils to lock onto endothelial adhesion molecules ICAM-1 (intracellular adhesion molecule 1) and VCAM-1 (vascular cell adhesion molecule 1), which stop the neutrophils from rolling (Figure 15.12, steps 4 and 5). The neutrophils then flatten and squeeze through spaces between endothelial cells (extravasation)

to enter the tissues, where they can help macrophages attack the invading microbes.

While all of this is taking place, additional mechanisms further increase vascular permeability. Damaged tissue cells in the area of inflammation release **bradykinin**, a nine-amino-acid polypeptide that can directly loosen tight junctions between endothelial cells (see Figure 15.12, steps 3 and 6). Bradykinin molecules also bind to mast cells in the area, causing the mast cells to degranulate. Degranulation releases antibacterial compounds that can kill microbes in the area, and histamine that further loosens endothelial cell junctions. The result is that even more fluid and neutrophils enter tissues.

Where does the pain come from during acute inflammation? Along with its effect on mast cells, bradykinin will attach to endothelial cells and induce synthesis of prostaglandins that, in turn, stimulate local nerve endings to cause pain (Figure 15.12, step 7). This pain alerts the patient to the location of the infection. Popular pain medications target a key enzyme in prostaglandin synthesis: cyclooxygenase (COX). Aspirin, as well as the prescribed anti-inflammatory agent celecoxib (Celebrex), are COX inhibitors that prevent the synthesis of prostaglandins and thus reduce inflammatory pain. These drugs are also called nonsteroidal anti-inflammatory drugs, or NSAIDs.

Once neutrophils have passed through the vascular wall, chemotactic factors are needed to lure them to the site of infection. Macrophages, damaged host cells, and even the bacteria themselves make these chemotactic factors. For example, macrophages and damaged cells release the chemokine peptides CXCL8

Figure 15.12 Steps of the Acute Inflammatory Response

Step 1: Traumatic introduction of microbes through skin. Damaged host cells release factors that help initiate inflammation.
Step 2: Local macrophages engulf microbes and release a variety of cytokines and chemokines. **Step 3:** Cytokines and other compounds cause capillary vasodilation and the loosening of endothelial tight junctions. The capillary becomes more permeable.
Step 4: Cytokines initiate synthesis of selectins on endothelial cells. Selectins snag passing neutrophils to slow them down. Other cytokines induce neutrophils to synthesize integrins that more tightly bind the endothelial adhesion molecules ICAM-1 and VCAM-1. **Step 5:** Neutrophils squeeze through the more permeable endothelium (extravasation). **Step 6:** Bradykinin made by damaged host cells contributes to the increase in vascular permeability and activates nearby mast cells to degranulate. Mast cells release histamine to further increase capillary permeability. **Step 7:** Bradykinin will also stimulate prostaglandin synthesis in endothelial cells. Prostaglandins stimulate nearby nerve cells to register pain.

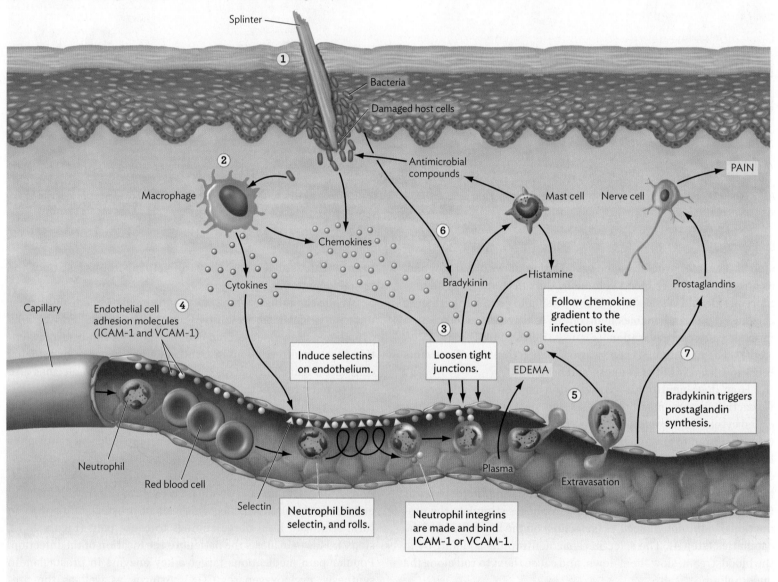

(formerly called IL-8) and CCL2 (formerly MCP-1). **Chemokines** are cytokines that have chemotactic properties. As the factors diffuse away from the infection site, they form a concentration gradient that neutrophils can follow, like bread crumbs, to the site of infection. The factors bind to neutrophil surface receptors and stimulate pseudopod projections aimed toward the microbes (Figure 15.12, step 5).

Let's return to Constance, from Case History 15.1. She had plenty of neutrophils in her bloodstream, which should be a good thing. However, the number of neutrophils in her blood (59,000/µl) was <u>exceptionally</u> high. The reason is that her blood neutrophils lacked integrin and could not leave the vessels to enter the infected tissues. As a result, her neutrophils accumulated in the circulation, locked out from doing battle.

Acute inflammation is your body's rapid response to cell damage, including damage caused by infection. But the symptoms of inflammation—heat, edema, redness, and pain—are not directly caused by substances produced by the microbe. Instead, these symptoms are the result of the body's reaction to the microbes' presence. Once the infection is resolved, the neutrophils, cytokines, and chemotactic factors are removed and the tissue heals. Chronic inflammation, however, is another story.

Chronic Inflammation

Chronic inflammation results from the <u>persistent</u> presence of a foreign body. This type of inflammation inevitably causes permanent tissue damage, even though the body attempts repair. Chronic inflammation has many causes. For example, infectious organisms such as *Mycobacterium tuberculosis, Actinomyces bovis,* and various pathogenic protozoa can avoid or resist host defenses (**Figure 15.13**). As a result, these pathogens persist at the site and continually stimulate the basic inflammatory response. Nonliving, irritant material such as wood splinters, inhaled asbestos particles, and surgical implants can also lead to chronic inflammatory responses.

During a chronic infection, the body attempts to wall off the site of inflammation by forming a **granuloma**. A granuloma begins as an aggregation of mononuclear inflammatory cells surrounded by a rim of lymphocytes. The body then deposits fibroconnective tissue around the lesion, causing tissue hardening known as fibrosis.

M. tuberculosis uses granuloma formation to survive for years, even decades, in an infected person. To form the granuloma, cells of *M. tuberculosis* are initially engulfed by lung macrophages and reside in phagosomes. The organisms then block phagosome-lysosome fusion, thereby ensuring that they can survive intracellularly. However, intracellular *M. tuberculosis* secretes a toxin that promotes necroptosis, a host cell death program that results in more robust replication of mycobacteria, strong inflammation, and disease progression. As more white cells are recruited to the area, they too are killed. Over time, a characteristic granuloma is formed (Figure 15.13C) that shields the microbe from further attacks by the immune system.

SECTION SUMMARY

- **Inflammation** begins with a mechanism (extravasation) that moves neutrophils from the bloodstream into infected tissues.

- **Macrophages** in tissues engulf microorganisms and then release vasoactive factors that increase vascular permeability and cytokines that stimulate production of selectin receptors on vascular endothelial cells.

- **Chemoattractant molecules** (for example, chemokines released from damaged cells) summon neutrophils.

- **Extravasation** is the process by which neutrophils pass between cells of the endothelial wall. Once out of the circulation, the neutrophils travel to the infection site.

Figure 15.13 Chronic Inflammation

FM   5 µm

A. The fluorescent green organisms shown here are *Mycobacterium tuberculosis* within macrophages (red) in a tuberculosis abscess. A thick, waxy cell wall, along with the ability to inhibit phagocyte killing mechanisms, helps mycobacteria survive inside macrophages for prolonged periods. The continual stimulation of an inflammatory response leads to chronic inflammation.

B. Fish tank granuloma. *Mycobacterium marinum* can cause tuberculosis-like infection in fish. The organism can accidentally enter the human body through an open wound or abrasion during cleaning of an aquarium containing infected fish. The infection is first noted as a lesion that heals very slowly on the hand or forearm and often forms a granuloma at the site that contains live organisms.

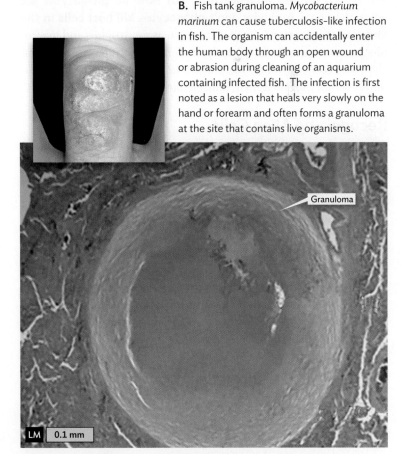

Granuloma

LM   0.1 mm

C. Cross section of liver showing a necrotizing granuloma caused by *M. tuberculosis* that spread through the bloodstream to this site after escaping the lung.

- **Bradykinin** helps loosen the tight junctions between endothelial cells to promote extravasation and produces pain in the affected area.
- **Chronic inflammation** results from the persistent presence of a foreign object.

Thought Question 15.3   How does increasing the diameter of a blood vessel affect blood flow and blood pressure?

Thought Question 15.4   You have discovered a drug that can decrease the production of selectin molecules. What would a decrease in selectin do to innate immunity, and for what noninfectious medical conditions might this drug be therapeutically useful?

15.4 Phagocytosis: A Closer Look

SECTION OBJECTIVES

- Describe how phagocytes recognize foreign cells.
- Explain the mechanisms phagocytes use to kill engulfed cells.
- List some ways that pathogens avoid the consequences of phagocytosis.
- Discuss how our body ends the inflammatory process.

We have explored phagocytosis, but how do phagocytes kill what they eat? And why don't phagocytes kill host cells in the absence of infection?   Phagocytes such as neutrophils and macrophages use a variety of oxygen-dependent and oxygen-independent chemical and enzymatic techniques to kill what they eat. Fortunately for us, they do not kill indiscriminately.

Recognizing Alien Cells

Phagocytosis is an effective nonspecific immune response. However, it would be disastrous for WBCs to indiscriminately phagocytose both host and nonhost structures. Fail-safe controls built into our immune defenses prevent this from happening. Without restraints, the mammalian immune system would constantly attack itself, causing more problems than it solves. When one of these fail-safe mechanisms fails, autoimmune diseases can arise.

For phagocytosis to proceed, macrophages and neutrophils must first "see" the surface of a particle as being foreign; note the macrophage pseudopods extending toward the monocyte (self) and *E. coli* (foreign) in Figure 15.14. When a phagocyte surface meets with the surface of another body cell, the phagocyte becomes temporarily paralyzed (inept at pseudopod formation). This paralysis is triggered by the binding of WBC membrane glycoproteins to inhibitory glycoproteins present on other host cell membranes. Paralysis allows the phagocyte to evaluate whether the other cell is friend or foe, self or nonself. The inhibitory glycoprotein on human cells is called CD47. Because invading bacteria lack the CD47 inhibitory surface molecules, they can readily be engulfed.

Figure 15.14 Phagocytosis

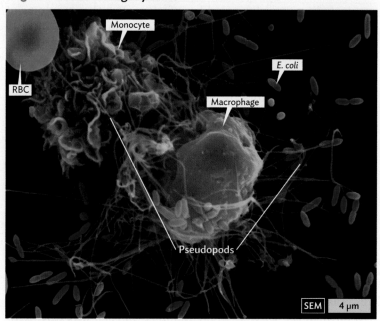

A macrophage and a monocyte in lung pleural cavity (colorized SEM). The smooth, reddish-pink cell with extended pseudopods is a macrophage engulfing *E. coli* (via phagocytosis). The ruffled purple cell is a monocyte. Monocytes are essentially circulating macrophages that can also rush to the scene of an infection. Notice the macrophage pseudopods reaching toward the monocyte. Once contact is made between a macrophage pseudopod and the monocyte, further movement of the pseudopod stops to avoid attacking self.

Although many bacteria, such as *Mycobacterium* or *Listeria* species, are easily recognized and engulfed by phagocytosis, others, like *Streptococcus pneumoniae* (a cause of pneumonia), possess polysaccharide capsules that are too slippery for pseudopods to grab (Figure 15.15A). This is where innate immunity and adaptive immunity join forces in a process known as **opsonization** (Figure 15.15B). Adaptive immunity produces Y-shaped molecules called **antibodies** (see Section 16.1). The two arms of an antibody bind to structures called **antigens**. For instance, the two arms of a capsule antibody bind to polysaccharide antigens that make up the capsule. Thousands of these antibodies, then, will coat the capsular surface of the pneumococcus. Because the two antigen-binding arms of each antibody are attached to the capsule, the tail of each antibody, called the Fc region, points outward and away from the bacterium. Like two strips of Velcro, the Fc regions of these antibodies bind to specific Fc receptors on phagocyte cell surfaces. As a result, the phagocyte can better grab and engulf the invader.

Killing Mechanisms of Phagocytes

During phagocytosis, the cytoplasmic membrane of the phagocyte flows around, and then engulfs, the bacterium, producing an intracellular phagosome (described in Figure 15.5). Subsequent phagosome-lysosome fusion (producing a phagolysosome) permits

Figure 15.15 Opsonization

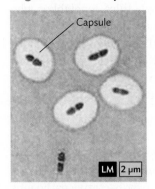

A. *Streptococcus pneumoniae* cells surrounded by capsules (India ink preparation). The slippery nature of the polysaccharide capsule makes phagocytosis more difficult.

Opsonization: Antibodies bind to bacteria, and the Fc portion of the antibodies binds to receptors on the macrophage surface.

Extracellular bacteria approach macrophage.

Extracellular bacteria

Antigen-binding sites on antibody

Fc receptor

Antibody

Bacterial antigen

Fc region

Fc receptor

Ingestion by macrophage: Antibodies link bacteria and macrophage, aiding phagocytosis.

Digestion in phagolysosome

B. Opsonization is a process that facilitates phagocytosis. Here, macrophage receptors bind to the Fc region of antibodies already bound to bacteria.

both oxygen-independent (anaerobic) and oxygen-dependent (aerobic) killing pathways and other reactive intermediate pathways. These mechanisms are like the boiling oil tossed over the castle wall in Figure 15.1. Oxygen-independent mechanisms include enzymes like lysozyme that destroy the cell wall, compounds such as lactoferrin that bind iron and keep it away from the microbe, and defensins (small, cationic antimicrobial peptides; see Section 15.2).

Oxygen-dependent mechanisms kill by making various highly reactive oxygen radicals that bind and damage lipids, proteins, DNA, and RNA. Cellular enzymes in the phagosome membrane generate superoxide radicals ($\cdot O_2^-$), hydrogen peroxide (H_2O_2), and hydroxyl radicals ($\cdot OH$). Myeloperoxidase, present only in neutrophils, converts hydrogen peroxide and chloride ions to the highly reactive compound hypochlorous acid (HOCl).

Macrophages, mast cells, and neutrophils also generate reactive nitrogen intermediates that serve as potent cytotoxic agents. Nitric oxide (NO) is synthesized by NO synthetase. Further oxidation of NO by oxygen yields nitrite ions (NO_2^-) and nitrate ions (NO_3^-). These mechanisms cause the large increase in oxygen consumption noted during phagocytosis, called the **oxidative burst**. The reactive chemical species formed during the oxidative burst do little to harm the phagocyte because the burst is limited to the phagolysosome and because the various reactive oxygen species, such as superoxide, are very short-lived.

Surviving Phagocytosis

Phagocytosis is very effective at killing microbes. However, some pathogens have evolved ingenious ways to avoid being killed by phagocytosis. Some bacterial intracellular pathogens, such as *Coxiella,* have evolved to live within the toxic phagolysosome. Others, such as *Shigella* and *Listeria,* escape from the phagosome

before phagosome-lysosome fusion can take place. These organisms then live in the cytoplasm. Members of another group (*Salmonella* and *Mycobacterium tuberculosis*) remain in the phagosome but secrete proteins that prevent fusion with lysosomes. Some of these pathogens, like *Shigella*, can also kill the host cell (macrophages for *Shigella*) by triggering a host process called **apoptosis**. Death of the macrophage saves other *Shigella* cells from phagocytosis. Apoptosis is actually a normal host process in which a host cell kills itself, but in a way that limits inflammatory damage to surrounding cells. Apoptosis is important during embryonic development. In a developing fetus, for example, cells that form webs between fingers and toes will undergo apoptosis before birth to release free digits. Apoptosis also plays a role in ending inflammation, which we discuss next.

Ending Inflammation

During an inflammatory response, many neutrophils converge on an infected area (commonly seen as pus). But what happens once the infection is cleared? Neutrophils are like tiny time bombs loaded with destructive enzymes and signaling molecules. Because the life of a neutrophil is quite short (7 days) and there are no chemoattractants to draw neutrophils <u>away</u> from the former site of infection, cleanup crews of long-lived phagocytes, such as macrophages, are called on to quickly remove neutrophils from the area. Once a neutrophil has finished its job, the cell undergoes apoptosis, which conserves membrane integrity (preventing the uncontrolled release of toxic compounds) and allows phagocytosis by macrophages (**Figure 15.16**). Once engulfed by the macrophage, the dead neutrophil can no longer release its toxic chemical weaponry into recovering tissues. Recovery from inflammation (drainage through the lymphatic system and wound healing) can begin.

Figure 15.16 A Macrophage Ingesting an Apoptotic Neutrophil

A macrophage contacts and prepares to ingest an apoptotic neutrophil during a lung inflammatory response. Apoptotic debris of previously ingested neutrophils accumulates in phagosomes within the macrophage.

SECTION SUMMARY

- **Phagocytosis** is selective for particles recognized as foreign to the body.

- **Oxygen-independent and oxygen-dependent mechanisms of killing** are initiated by fusion between a lysosome and a bacteria-containing phagosome.

- **The oxidative burst**, a large increase in oxygen consumption during phagocytosis, results in the production of superoxide radicals, nitric oxide, and other reactive oxygen species.

- **Intracellular pathogens have evolved ways to escape** from the phagosome, prevent phagosome-lysosome fusion, or survive in a toxic phagolysosome.

- **Inflammation subsides** as neutrophils at the site of infection undergo apoptosis and macrophages clean up cellular debris via phagocytosis.

15.5 Interferon, Natural Killer Cells, and Toll-Like Receptors

SECTION OBJECTIVES

- Describe the host-specific, antimicrobial function of interferons.

- Explain how natural killer cells first recognize and then kill infected host cells.

- Discuss Toll-like and NOD-like receptors and their contribution to innate immunity.

Can human cells under attack warn others of danger? When a residential community is threatened by a thief, a neighbor who has been robbed alerts others to take precautions. And if the thief is discovered, the authorities are called to arrest him (or her). Similarly, interferon peptides secreted by infected cells warn healthy host cells of a nearby infection, and natural killer cells seek out and then destroy cells already infected.

Interferons

Interferons are low-molecular-mass cytokines (15–20 kDa) produced by many eukaryotic cells in response to intracellular infection—that is, infection by viruses or by bacterial pathogens that can grow intracellularly (*Listeria*, for instance). Infected cells will release interferon that can bind to nearby, uninfected host cells and trigger an increase in the uninfected cell's general antiviral defenses. Interferons can protect cells against <u>many</u> different pathogens, but they work only on cells of the same host species. For instance, human interferon protects only human cells against viral infections and fails to protect mouse cells against any viruses.

There are two general types of interferons, which differ in the receptors they bind and the responses they generate (Table 15.2). Type I interferons have high antiviral potency; they consist of IFN-alpha (IFN-α), IFN-beta (IFN-β), and IFN-omega (IFN-ω). Type II interferon, IFN-gamma (IFN-γ), has more of an immunomodulatory function that will be discussed in the next chapter.

Type I interferons bind to specific receptors on uninfected host cells and induce the synthesis of two classes of proteins: double-stranded-RNA-activated endonucleases, which can cleave viral RNA, and protein kinases that phosphorylate and inactivate eukaryotic initiation factor eIF2, which is required to translate viral RNA. These mechanisms affect both RNA and DNA viruses because protein synthesis is required for the propagation of all viruses.

Type II interferon functions by activating macrophages, natural killer cells, and T cells to increase the number of **major histocompatibility complex (MHC)** antigens on their surfaces. MHC proteins are important for recognizing self and for presenting

| Table 15.2 **Classes of Interferons** | | | |
Type	Examples	Function	Mechanism of Action[a]
I	IFN-alpha IFN-beta IFN-omega	Antiviral	Induce dsRNA endonuclease, eIF2 kinase (prevents translation of viral RNA)
II	IFN-gamma	Immuno-modulatory	Triggers signal cascades in macrophages, NK cells, and T cells; increases MHC II on cell surfaces

[a]dsRNA = double-stranded RNA; eIF2 = eukaryotic initiation factor 2; NK = natural killer; MHC = major histocompatibility complex.

Figure 15.17 Natural Killer Cells

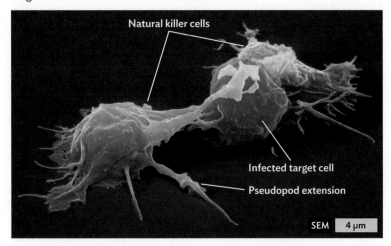

Natural killer (NK) cells attack eukaryotic cells infected by microbes, not the microbes themselves. Perforin produced by the NK cell punctures the membrane of target cells, causing them to burst.

foreign antigens to the adaptive immune system. They will be discussed more fully in Chapter 16.

Natural Killer Cells

Infected or cancerous body cells can be a major problem for the host. Infected cells can hide a pathogenic microbe from the immune system, whereas cancer cells can take over and kill the host. A class of lymphoid cells called **natural killer (NK) cells** can identify and handle these situations. Instead of directly killing microbes, the mission of NK cells is to seek out and destroy infected host cells and cells that have been transformed into cancer cells

(Figure 15.17). NK cells recognize changes in the cell-surface proteins of infected or cancerous cells and release chemicals that kill only those cells.

Natural killer cells recognize their targets in two basic ways. One involves MHC class I molecules, and the other uses antibody Fc receptors. Body cells can display two classes of MHC molecules on the outside of the cell membrane. MHC I, present on all host cells, is an indicator of "self." (MHC II molecules, present only on certain immune cell types, will be discussed in Chapter 16.) NK cells have specific receptors that bind to self MHC I molecules on the surfaces of other cells in the body. An NK cell that "touches" a host cell with a compatible self MHC I protein will not attack that cell. However, if a host cell lacks MHC class I molecules, as can happen during infection or as a result of cancer, NK cells will perceive the host cell as foreign and a potential threat. The NK cell will insert a pore-forming protein called **perforin** into the membrane of the target cell, through which cytotoxic enzymes are delivered (Figure 15.18).

Natural killer cells also contain antibody-binding Fc receptors on their cell surface. The second killing mechanism, called **antibody-dependent cell-mediated cytotoxicity (ADCC)**, occurs when the Fc receptor on the NK cell links to an antibody-coated host cell. Why would host cells be coated with antibodies? During their replication, many viruses place viral proteins in the membrane of the infected cell. Because the viral proteins protrude from the host membrane, antibodies to those proteins will coat the compromised cell, tagging it for ADCC. Once the compromised cell is targeted, the NK cell kills it by inserting a perforin molecule. The ADCC killing mechanism is another example of cooperation between innate immunity (NK cells) and adaptive immunity (antibody-producing lymphocytes).

Figure 15.18 Natural Killer Cell Killing a Target Cell

A. Natural killer cells recognize cells that have lost expression of MHC class I molecules. NK cells store cytotoxic molecules such as perforin in secretory lysosome granules. Recognition of an infected cell triggers the release of these cytotoxic molecules (yellow fluorescence) into target cell membranes which kills the target cell. Cell cytoplasm is green.

B. Perforin forms a pore in the target cell membrane and destroys membrane integrity. Cytotoxic proteases, called granzymes, are also released from the NK cell, pass through the perforin pore, and enter the target cell. The granzymes trigger apoptosis of the target cell.

Toll-Like Receptors, NOD Proteins, and MAMPs

CASE HISTORY 15.2

An Immune Defect That Takes a Toll

Bobby, an 11-year-old boy from Nebraska, suffered his entire life from recurring life-threatening infections caused mostly by the Gram-positive bacteria *Staphylococcus aureus* and *Streptococcus pneumoniae*. His infections included meningitis, osteomyelitis (bone infection), and arthritis caused by *S. pneumoniae*, as well as episodes of septicemia, osteomyelitis, and recurrent boils—all caused by *S. aureus*. Tests to find a reason for these recurring infections were initially disappointing. The results of all standard immunological tests were normal, including T-cell lymphocyte responses, blood antibody levels, and antibody responses to injected proteins and polysaccharides. This means his adaptive immune system was functioning. The numbers of monocytes/macrophages in his blood were also normal. However, pro-inflammatory cytokine levels (indicators of innate immunity) measured during the latest infection were considerably lower than expected. Further tests proved that Bobby had inherited an innate immune defect in something called Toll-like receptor signaling. Currently, there is no cure for this immunodeficiency. To stem the tide of infections, Bobby was placed on long-term, preventive antibiotic treatment.

Toll-like receptors (TLRs) on host cells enable the innate immune system to respond rapidly to infection. Obviously, the faster the body can detect pathogens, the more quickly it can begin to cure itself. However, the adaptive immune system takes several days to make antibodies capable of targeting a microbe (discussed in Chapter 16). Without being rapidly confronted by an innate immune system, pathogens would be free to grow and cause serious disease. Fortunately, bacteria and viruses have telltale structural patterns that immediately label them as being foreign. Microbe-associated molecular patterns (MAMPs) such as peptidoglycan, flagellin, single- or double-stranded RNA and lipoteichoic acids are not found in tissues unless bacteria or viruses are present. The MAMPs are recognized by host cell TLRs (Table 15.3). TLRs, along with NLRs (described shortly), are called **pattern recognition receptors (PRRs)**. Once bound to a

Table 15.3
Examples of Toll-Like Receptors and NOD-Like Receptors

Receptor	MAMPs Recognized[a]	Source	Host Cells	Location
TLR1	Lipopeptides	Bacteria	Monocytes/macrophages, dendritic cells, B cells	Cell surface
TLR2	Glycolipids, lipoteichoic acids	Bacteria	Monocytes/macrophages, dendritic cells, mast cells	Cell surface
TLR3	Double-stranded RNA	Viruses	Dendritic cells, B cells	Cell compartment
TLR4	Lipopolysaccharide, heat-shock proteins	Bacteria	Monocytes/macrophages, dendritic cells, mast cells, intestinal epithelium	Cell surface
TLR5	Flagellin	Bacteria	Monocytes/macrophages, dendritic cells, intestinal epithelium	Cell surface
TLR6	Diacyl lipopeptides	*Mycoplasma*	Monocytes/macrophages, mast cells, B cells	Cell surface
TLR7	Single-stranded RNA in endosomes	Viruses	Monocytes/macrophages, B cells, dendritic cells	Cell compartment
TLR9	Unmethylated CpG residues in DNA	Bacteria	Monocytes/macrophages	Cell compartment
NOD1	Component of Gram-negative peptidoglycan	Gram-negative bacteria	Many cell types	Inflammasome
NOD2	Peptidoglycan component	Bacteria	Macrophages, dendritic cells, epithelia of lung and GI tract	Inflammasome
NLRP-3	Peptidoglycan	Bacteria	Many cell types	Inflammasome
NLRP-4	Flagellin, CpG, ATP, dsRNA	Bacteria	Many cell types	Inflammasome

[a]CpG = cytosine-phosphate-guanine sequence in DNA.

Figure 15.19

Cell-Surface TLRs and Cytoplasmic NLRs Recognize MAMPs

Toll-like receptors (TLRs) bind extracellular microbe-associated molecular patterns (MAMPs) and transmit a signal through a cascade of other proteins to the nucleus. The signal cascade activates transcription of specific cytokine genes. (*Note:* TLR3, TLR7, and TLR9, not shown in this figure, are located in endosomal membranes.) Cytoplasmic NOD-like receptors (NLRs) bind MAMPs generated primarily by intracellular pathogens and trigger a signal cascade different from that used by TLRs. NLRs also trigger inflammasome assemblies that convert precursor cytokines to functional cytokines.

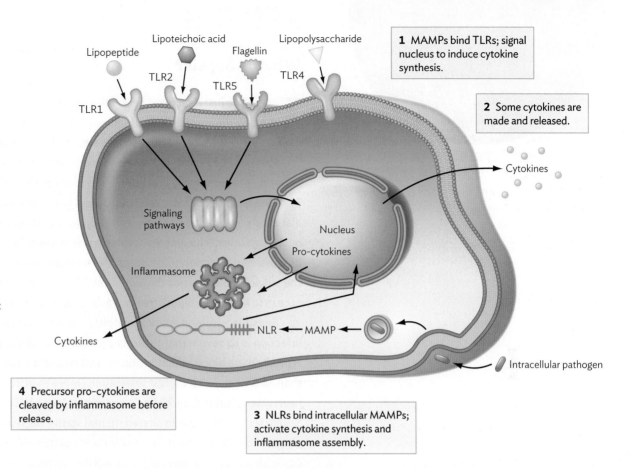

1 MAMPs bind TLRs; signal nucleus to induce cytokine synthesis.

2 Some cytokines are made and released.

4 Precursor pro-cytokines are cleaved by inflammasome before release.

3 NLRs bind intracellular MAMPs; activate cytokine synthesis and inflammasome assembly.

MAMP, the TLR sends a signal to the interior of the cell to start making interferon and pro-inflammatory cytokines that will speed up the inflammatory response (**Figure 15.19**).

First discovered in insects, Toll-like receptors are evolutionarily conserved glycoproteins present on the cell membranes of many eukaryotic genera. (Note, however, that some TLRs, such as TLR7, are located only in the membranes of intracellular vacuoles.) Humans have numerous Toll-like receptors, each of which recognizes different MAMPs present on pathogenic microorganisms, making them an innate defense mechanism with some degree of specificity (see Table 15.3). For example, TLR2 binds to a glycolipid on mycobacteria (lipoarabinomannan), a protein-carbohydrate complex on yeast (zymosan), and lipopolysaccharide (LPS) from spirochetes. TLR4, on the other hand, binds LPS from Gram-negative bacteria. Note that these receptors bind fragments of structures released from the microbe. They don't interact with the whole organism.

The cytokines secreted after a MAMP-TLR interaction float away to bind receptors on various cells of the immune system and direct those cells to engage the invading microbe (inflammation; see Figure 15.12). Bobby, in Case History 15.2, had a defect in one of the intracellular proteins that receives the signal from TLRs. Consequently, battle-ready levels of cytokines could not be made, and the acute inflammatory process failed. Cytokines can influence the function of cells that are part of innate immunity, adaptive immunity, or both.

Although TLRs are restricted to membranes (some surface, some vacuolar), certain cytoplasmic proteins can also recognize MAMPs. **NOD-like receptor (NLR)** proteins, so named because of their similarity to a family of plant proteins (NOD factors), are important intracellular sensors of MAMPs. When bound to a MAMP, NLRs trigger a signaling pathway different from that used by TLRs. The result, though, is the same: activation of cytokine genes in the nucleus. NLR activation also initiates assembly of cytoplasmic, multiprotein complexes called inflammasomes that proteolytically cleave precursor pro-cytokines into active cytokines. Once made, the cytokines are secreted and bind to target immune system cells to initiate the inflammation and adaptive immune mechanisms needed to respond to intracellular pathogens.

The severity of Bobby's condition illustrates how important TLRs and innate immunity are for preventing serious disease—even when the adaptive immune system is functional. If Bobby lives long enough, however, his adaptive immunity (production of antibodies, for example) against these pathogens will build to a point where preventive antibiotic therapy can probably be stopped. TLRs and NLRs also play important roles in the medical condition called sepsis (see **IMPACT**).

Besides infection-driven inflammation triggered by MAMP interactions with TLRs and NLRs, there is sterile inflammation, in which human cells damaged by trauma or by decreased blood flow release damage-associated molecular pattern (DAMP) molecules, such as ATP, hyaluronic acid (a large polysaccharide found

impact

What Is Sepsis?

We just described how inflammation develops in tissues. But many of the mechanisms of tissue inflammation also play roles in bloodstream infections. The results, however, can be catastrophic.

Vanessa, a 22-year-old fashion model, entered the hospital for what was to be a routine plastic surgery procedure. The procedure went well and she was released from the hospital, but she returned 3 days later extremely ill. Her blood pressure was alarmingly low, her heart rate was rapid, she was disoriented, and she had a dangerously high fever. The emergency department physician diagnosed her with sepsis, a bloodstream infection that causes life-threatening illness. It appears that in this case the surgical field was not thoroughly sterilized, and contaminating bacteria gained access to Vanessa's bloodstream. Vanessa was near death.

Sepsis is a serious medical condition in which a pro-inflammatory response to an infection is so severe that the inflammation can kill the patient. A patient with sepsis typically has a blood infection and at least two of the following symptoms: fever (greater than 38.5°C, or 101.3°F), fast respiration (more than 20 breaths per minute), quickened heart rate (greater than 90 beats per minute), or a WBC count greater than 12,000/μl. Severe sepsis can result in altered mental status, decreased urine output, and decreased platelet counts. A patient who also exhibits extremely low blood pressure is said to be in septic shock; 30% of these patients will die. Surprisingly, what causes these symptoms is the host <u>response</u> to the bacterial pathogen, not the pathogen itself.

Bacterial MAMPs trigger sepsis by interacting with TLRs and NLRs in host cells. For Gram-negative organisms, lipopolysaccharide (LPS, or endotoxin) in the outer membrane stimulates innate immune cells via Toll-like receptor 4 (TLR4). For Gram-positive organisms like *Staphylococcus aureus*, sepsis starts when cell wall components such as lipoteichoic acid and peptidoglycan bind to the PRRs in these cells. Bacterial toxins called superantigens can also produce septic shock (discussed in Section 18.4). Regardless of what initiates sepsis, the cells activated by MAMPs or by superantigens release a "storm" of pro-inflammatory cytokines such as TNF-alpha and IL-1. These pro-inflammatory cytokines affect the body's thermostat (see Section 15.6) and damage endothelial cells lining blood vessels.

Figure 1 illustrates the basic mechanisms (or pathophysiology) of sepsis. Because blood vessels are damaged, proteins that are normally hidden from blood by endothelial cells are exposed. One such protein is tissue factor (TF), the primary cellular initiator of blood coagulation. Once exposed, TF interacts with coagulation factors in blood to initiate coagulation, and small blood clots (microclots) begin to form. The microclots block capillaries, causing diminished blood flow and poor oxygen delivery to affected organs. Clots form in many locations throughout the body, producing what is called disseminated intravascular coagulation (DIC). All of this coagulation causes a drop in blood platelet levels, which are important for clotting. The irony is that now, because platelet numbers are low, it takes <u>longer</u> for blood to coagulate.

Figure 1   The Triggering of Sepsis by LPS and Lipoteichoic Acid

The MAMPs released by bacteria, such as lipoteichoic acid (Gram-positive bacteria) or LPS (Gram-negative bacteria), will bind TLRs on macrophages and other cell types. The activated cells release TNF-alpha and IL-1 pro-inflammatory cytokines. Damage to endothelial cells exposes tissue factor, which triggers the coagulation cascade. Fibrin and platelets form clots in the microvasculature and block the flow of oxygenated blood to various organs. TNF-alpha and IL-1 have other effects that can lead to direct tissue injury and elevated body temperature (fever).

A consequence of damaged blood vessels and decreased clotting is that blood begins seeping out of the damaged vessels and into tissues. The result is a rash that begins as a series of pinpoint red splotches on the skin called petechiae (singular, petechia). The petechiae can enlarge and coalesce into large purple purpuric rashes (Figure 2). Massive leakage of blood into organs decreases the effective blood volume, which produces a severe drop in blood pressure. The patient goes into shock and often dies.

So how is sepsis treated? Intravenous fluids are initially used to try to raise blood pressure, but if that fails, meaning that the patient is in septic shock, vasopressor drugs are administered to constrict smooth muscles around blood vessels. The narrowing of vessels will increase blood pressure. Antibiotics are the only way to kill or inhibit growth of the organism. However, even dead organisms will release endotoxin or peptidoglycan, which can make the sepsis even worse before it gets better. The best hope for recovery is to recognize sepsis early, before it becomes too severe and treatment becomes futile. In Vanessa's case, proper treatment and antibiotics saved her life.

DISCUSSION QUESTION

Is sepsis contagious?

Figure 2   Disseminated Intravascular Coagulation (DIC)

A 1-year-old boy has massive purpuras as a result of DIC caused by *Neisseria meningitidis* septicemia.

in tissues), and histones, all of which also interact with certain PRRs. Like MAMPs, DAMPS trigger synthesis of pro-inflammatory cytokines that produce inflammation. The activation of innate immunity by DAMPs alerts the host to damage and leads to a repair response.

- **Interferons are host-specific** molecules that can nonspecifically interfere with viral replication (type I) and modulate the immune system (type II).
- **Natural killer (NK) cells** are a class of white blood cell that destroy cancer cells or cells harboring microorganisms.
- **Natural killer cells target host cells** that either have lost MHC class I receptors as a result of infection or cancer or are coated with antibody (antibody-dependent cell-mediated cytotoxicity, ADCC).
- **Natural killer cells kill** by inserting perforin pores into the membranes of target cells.
- **Membrane-embedded Toll-like receptors (TLRs) and intracytoplasmic NOD-like receptors (NLRs)** recognize different microbe-associated molecular patterns (MAMPs) and activate synthesis and secretion of cytokines.

Thought Question 15.5 Both NK cells and neutrophils have Fc receptors on their cell membranes. NK cells successfully use these receptors to carry out ADCC on virus-infected cells that are coated with antibody bound to the viral proteins inserted in the host cell's membrane. Why don't neutrophils phagocytose antibody-coated bacterial cells?

Thought Question 15.6 *Shigella flexneri* is a Gram-negative rod that causes bacillary dysentery (bloody diarrhea). The organism is ingested and enters M cells in the intestine. After exiting the M cell, *S. flexneri* forcibly enters nearby macrophages and escapes from the phagocytic vacuole into the cytoplasm. *Shigella* triggers the activation of NLRs and cytokine release, but at a level much lower than you would expect. Propose a mechanism that *Shigella* might use to limit cytokine production. (*Hint:* *Shigella* has a protein secretion system that can inject virulence proteins into the host cell.)

Thought Question 15.7 Why haven't microbes evolved to avoid being recognized by Toll-like receptors?

Thought Question 15.8 Can you explain why the infections in Bobby's case (Case History 15.2) were caused by Gram-positive cocci and not Gram-negative bacteria (such as *Salmonella*) or viruses? (*Hint:* *Staphylococcus aureus* and *Streptococcus pneumoniae* are extracellular pathogens, whereas viruses and *Salmonella* are intracellular pathogens.)

15.6 Fever

What is fever, and why is it a good thing? To understand fever (elevated body temperature), we first need to know how the body controls its temperature, a process called thermoregulation. Heat in a human body is produced as a consequence of metabolic reactions. The liver and muscles are major generators of heat and will warm blood that passes through them. In a healthy person, body temperature is kept constant within a very small range, despite large differences in surrounding temperature and physical activity. The normal healthy adult oral temperature is 36°C–38°C (96.8°F–100.4°F), with the average being 37°C (98.6°F). Rectal temperature is the body's <u>core</u> temperature and is about 1°C (1.8°F) higher than oral temperature. Fever is an oral (or ear) temperature above 38°C (100.4°F).

So, how do we normally maintain a 37°C (98.6°F) body temperature while standing in a winter snowstorm or enduring a summer heat wave? Thermoregulation starts with heat sensors in skin, in large organs, and along the spinal cord. The sensors constantly send information about your body's temperature to your thermostat, the hypothalamus, which is a small region of the brain located near the brain stem. The hypothalamus controls how much warm blood flows through the skin and subcutaneous areas. The goal is to maintain your body's core temperature at the expense of peripheral temperature. When we are cold, vasoconstriction narrows peripheral blood vessels so that less blood flows through those vessels. The result is that less heat is released from the periphery and core heat is preserved. Shivering begins as a way to generate more heat. When we are hot, vasodilation <u>increases</u> blood flow to the periphery to quickly release heat. Sweat is made to further cool the body through evaporation.

Fever is a natural reaction to infection and is usually accompanied by general symptoms, such as sweating, shivering, and feeling cold. Substances that cause fever are known as **pyrogens**. External pyrogens (for example, certain bacterial toxins) originate outside the body, whereas internal pyrogens (such as interferon, tumor necrosis factor, and IL-6) are made by the body itself. External pyrogens generally cause fever by inducing the release of internal pyrogens. The pyrogenic cytokines bind to receptors on neurons near the thermoregulatory center in the hypothalamus, thus stimulating production of the enzyme phospholipase A2. Phospholipase A2 is required to make prostaglandins, which then change the responsiveness of the thermosensitive neurons that make up the thermoregulatory center. In other words, pyrogens turn up the thermostat. The body thinks it is cold, so the hypothalamus sends signals via the autonomic nervous system (which acts below the level of consciousness) to constrict peripheral blood vessels (vasoconstriction). Heat is not released and builds up to cause fever. Involuntary muscle contractions (shivering and chills) also generate heat.

What are the advantages and disadvantages of fever? The obvious disadvantage of fever is the discomfort the patient feels and, if the fever rises too high (above 42°C, or 107.6°F), irreversible brain damage can occur. However, because the ideal growth temperature for many microbes is 37°C (98.6°F), elevated temperature can place the organism outside its "comfort zone" of growth. There is also evidence that fever reduces iron availability to bacteria. Slower growth of the pathogen gives the body's immune system time to subdue the infection before it is too late. Elevated body temperature also stimulates various aspects of the immune response, including the production and function of neutrophils and the trafficking of immune cells to lymph nodes. Consequently, interventions that reduce a moderate fever caused by infection may be counterproductive to a speedy recovery.

SECTION SUMMARY

- **The hypothalamus** acts as the body's thermostat.
- **External and internal pyrogens** elevate body temperature by stimulating production of prostaglandins.
- **Prostaglandins** change the responsiveness of thermosensitive neurons in the hypothalamus.
- **Fever** raises body temperature above the optimal level for bacterial and viral growth and may reduce iron availability for bacterial pathogens.

Thought Question 15.9 If fever limits bacterial growth, why would bacteria make pyrogenic toxins that increase body temperature?

15.7 Complement

SECTION OBJECTIVES

- Define complement.
- Distinguish between the classical, alternative, and lectin pathways of complement activation.
- Discuss the role of complement in host defense.

White blood cells engulf and kill pathogens, but can simple serum proteins also kill microbes? Yes; a series of 20 serum proteins called complement factors also attack bacterial invaders. Complement is extremely critical to preventing infections. But sometimes something goes wrong.

CASE HISTORY 15.3

A Little Complement Goes a Long Way

 Derek, a 16-year-old high school chess prodigy, was rushed by his parents to the St. Michael's Hospital emergency department in Toronto, Canada. The staff found Derek very confused and disoriented. His parents said that Derek had complained of a sore throat a few days earlier but over the previous 24 hours had developed a very high fever and a severe headache. When examined, Derek had a stiff neck and many reddish

spots, called petechiae, over his body. The attending physician performed a lumbar puncture, in which a needle is inserted into the spinal column between two lower back vertebrae to remove cerebrospinal fluid (CSF). Normally clear, Derek's CSF was cloudy, looking like dilute milk. The physician diagnosed meningitis. A few hours later, the hospital laboratory reported that the CSF contained Gram-negative diplococci, and a rapid test indicated *Neisseria meningitidis*. The doctor was not surprised because this was Derek's third bout with meningitis caused by this organism. The other infections had occurred when Derek was 9 and 13 years old. After this last episode, the physician suspected that Derek might have a genetic deficiency in one of his blood complement components. The laboratory confirmed that Derek's blood had a deficiency in complement factor C8. That was the problem!

We will soon explain why the C8 deficiency led to recurrent meningitis. But first, understand that meningitis is a serious disease that can rapidly turn fatal. *Neisseria meningitidis*, only one of the possible causes of this disease, first infects the nasopharynx to cause a sore throat. Sometimes the bacteria penetrate into the bloodstream and are carried throughout the body. From the bloodstream, *N. meningitidis* can cross the blood-brain barrier and enter the brain and spinal column. The inflammation that results causes increased cranial pressure, which produces the headache, stiff neck, and disorientation. Over their lifetimes, most patients contend with only a single bout of meningitis caused by *N. meningitidis*. Recurring bouts of *Neisseria* meningitis are quite uncommon unless the patient, like Derek, has a defect in one or more of the complement cascade proteins.

Any localized tissue infection can become deadly once an organism gains access to the bloodstream to cause septicemia. During septicemia, the organism can spread (disseminate) to far-flung destinations throughout the body and cause secondary infections called metastatic infections. Even a trip to the dentist can result in oral bacteria gaining entry to the blood. Why doesn't septicemia occur more often than it does? It turns out that a series of 20 proteins circulating in the blood but made in the liver plays a huge role in preventing blood infections. These proteins are collectively called complement. Complement was first discovered as a heat-labile component of blood that enhances, or complements, the killing effect of antibodies on bacteria. Several complement components are proteases that sequentially cleave each other in a cascade. Triggering the complement cascade produces a number of outcomes. Pores called membrane attack complexes (MACs), composed of four complement components, are inserted into membranes and cause cytoplasmic leaks. In addition, pieces of the complement system can attract WBCs and facilitate phagocytosis (opsonization).

Recall Brad from the opening case of this chapter, the 66-year-old alcoholic man with liver disease who died from an infection with *Vibrio vulnificus*. Because complement factors originate in the liver, Brad's liver disease left him complement deficient and more susceptible to blood infection.

Pathways of Complement Activation

There are three complement-activating pathways, officially dubbed the classical pathway, the alternative pathway, and the lectin pathway. The <u>classical complement pathway</u> depends on antibody, so it links the innate immunity of complement with the adaptive immune system. However, because the adaptive immune system can take days to make antibody, this system does not immediately provide protection. The <u>alternative pathway</u> does not require antibody for activation and can therefore attack invading microbes long before a specific immune response can be launched. In terms of response time, the lectin pathway falls between the other two. The <u>lectin pathway</u> requires synthesis of mannose-binding lectin by the liver in response to macrophage cytokines released after a macrophage engages the pathogen. This lectin is released to the bloodstream and will coat the surfaces of invading microbes. Microbes coated with lectin will activate complement that bumps into it. A major goal of the complement cascade is to insert pores into target microbial membranes. The pores destroy membrane integrity and kill the pathogen.

All complement pathways converge at the complement C3 protein (**Figure 15.20**). The steps are complex but simplified here for discussion.

1. **The alternative pathway.** The alternative pathway starts with the spontaneous cleavage of factor C3 into two parts: C3a and C3b. The C3b fragment combines with factor B to form a C3 convertase that quickens C3 cleavage. The C3b fragment also binds to carbohydrates on cell surfaces, such as LPS on Gram-negative outer membranes, and becomes part of a complex (called C5 convertase) that cleaves a factor called C5 into C5a and C5b fragments. (You can think of C3b as a sticky "grenade" tossed at microbes in the bloodstream. Once it hits, an "explosion" of sequential reactions happens.) The C5a fragment made as a result of C5 convertase chemoattracts PMNs. Meanwhile, the C5b fragment enters the bacterial membrane and directs the assembly of proteins C6, C7, C8, and C9 into a pore, the so-called membrane attack complex, which kills the bacterial cell. The cascade we have just described, from C5 convertase to pore formation, is <u>identical</u> in all complement activation pathways. Now, remember Derek's C8 factor deficiency from Case History 15.3? Because Derek lacked the C8 protein, his complement cascade was defective and could not kill *Neisseria meningitidis* in his bloodstream. The result was recurrent meningitis.

2. **The classical pathway.** The classical pathway starts with antibody bound to a bacterial cell. The Fc tail portion of the antibody binds and activates the protease of a large, spindly-looking complement factor called C1. C1 then cleaves two other factors: C4 and C2. Parts of C2 and C4 unite to form

Figure 15.20 Complement Cascades

Three pathways can activate complement. The three pathways begin differently but converge at C3. **A.** The alternative pathway starts with the spontaneous cleavage of C3. The C3b fragment (called C5 convertase) binds to bacterial surfaces (like LPS) and cleaves C5. The C5b fragment enters the membrane and directs the assembly of the C6-C7-C8-C9 pore, which kills the bacterial cell. The cascade from C3b to pore formation is common to all pathways. **B.** The classical pathway starts with antibody bound to a bacterial cell. The Fc tail of the antibody binds and activates the protease of C1. C1 mediates the cleavage of C4 and, indirectly, C2. Parts of C2 and C4 combine to form a C3 convertase, which speeds the cleavage of C3. **C.** In the lectin pathway, lectin, made by the liver, circulates in the blood and binds mannan (a polymer of mannose) on bacterial surfaces. Lectin mediates cleavage of C4 and C2. As a result, C3 convertase forms, and the cascade continues as before. C3a and C5a can act as chemoattractants and anaphylatoxins, whereas C3b can function as either an opsonin or a C5 convertase.

another C3 convertase (different from the one in the alternative pathway), which speeds the cleavage of C3 (again forming C3b), thereby propelling the rest of the cascade.

3. **The lectin pathway.** In the lectin pathway, lectin, made by the liver, circulates in the blood and binds carbohydrate structures (mannose residues) on bacterial surfaces. Lectin

mediates cleavage of C4—and, indirectly, C2—the same factors used in the classical pathway. As a result, C3 convertase once again forms, the C3b "grenade" is made, and the cascade continues as before.

When complement interacts with Gram-negative cells, lysozyme (present in serum) enters through outer membrane pores and cleaves peptidoglycan, making the cell membrane more susceptible to the membrane attack complex. Gram-positive bacteria are resistant to complement because they lack an outer membrane (no LPS to efficiently start the cascade) and have a thick peptidoglycan layer that hinders access of complement components. However, even in the absence of LPS, antibody-coated Gram-positive bacteria can activate complement via the classical pathway, and several complement factors can intensify inflammation. For instance, C3a and C5a fragments act as chemoattractants to call in neutrophils and work as anaphylatoxins, compounds that trigger the release of vasoactive factors like histamine from endothelial cells, mast cells, or phagocytes. In addition, C3b can function as an opsonin to facilitate phagocytosis. Because phagocytes contain C3b receptors on their surfaces, it is easier for them to grab and engulf cells coated with C3b.

Regulating Complement Activation

Membrane attack complexes are not selective about the membranes they attack. So, how do normal body cells prevent self-destruction following complement activation? Regulatory factors can intervene at several places. One such factor is the host cell surface protein CD59 (called protectin). This protein prevents C9 from polymerizing on host membranes. Another regulatory mechanism hinges on a serum protein called **factor H**. Factor H normally binds to the surfaces of host cells, not pathogens, and protects the host cell from complement attack by inactivating C3b. Some bacteria, such as certain strains of *Neisseria gonorrhoeae*, have evolved to protect themselves from complement by binding factor H—cloaking themselves in the host's own protective factor.

C-Reactive Protein and Complement

As noted earlier, inflammation is associated with the production of various cytokines by macrophages. Within hours of an infection, some of these cytokines (IL-1, TNF-α, and IL-6) travel to the liver. There, they stimulate the synthesis of several so-called acute-phase reactant proteins, including **C-reactive protein (CRP)**. Named for its ability to activate complement, CRP will bind to bacterial cell surfaces but not to host cell membranes. Once stuck to the bacterial cell surface, CRP binds C1 of the classical complement pathway and initiates the complement cascade. C-reactive protein essentially takes the place of antibody in the classical pathway.

⊙ **Note:** The term "acute" in "acute-phase reactants" refers to the rapid, or acute, development of an infection. Despite their name, acute-phase reactants accompany both acute and chronic inflammatory states. They can also appear in association with trauma, infarction, inflammatory arthritis, and various neoplasms.

SECTION SUMMARY

- **Complement** is a series of 20 proteins naturally present in serum.
- **Activation of the complement cascade** results in a pore being introduced into target membranes.
- **The three pathways for activation** are the classical pathway, the alternative pathway, and the lectin pathway. They differ in the portion of the pathway that leads to C3 cleavage to C3b.
- **The cascade of protein factors** C3 to C3b to C5 to C5b and the membrane assembly of C5b, C6, C7, C8, and C9 produce the membrane attack complex in target membranes.
- **C-reactive protein** in serum is activated when bound to microbial structures and will convert C3 to C3b, which can also start the complement cascade.

Thought Question 15.10 Why does complement have to be activated? Why is it not always active and ready to kill invading microbes?

Perspective

The various "hardwired" innate immune mechanisms described in this chapter keep our normal microbiota at bay and provide an effective first line of defense against potential pathogens. The next chapter addresses what happens when microbes breach these innate defenses. Unlike the general protective mechanisms discussed in this chapter, the system of acquired immunity generates a molecular defense specifically tailored to a given pathogen. Keep in mind that the innate and adaptive immune systems "talk" to each other and collaborate to present the most effective response to a given pathogen, whether bacterial, viral, or parasitic.

Key Terms

adaptive immunity (p. 440)

antibody (p. 452)

antibody-dependent cell-mediated
    cytotoxicity (ADCC) (p. 455)

antigen (pp. 441, 452)

antigen-presenting cell (APC)
    (p. 444)

apoptosis (p. 453)

basophil (p. 442)

B cell (p. 444)

bradykinin (p. 449)

chemokine (p. 450)

complement (p. 441)

C-reactive protein (CRP) (p. 463)

defensin (p. 447)

dendritic cell (p. 444)

eosinophil (p. 442)

extravasation (pp. 448, 449)

factor H (p. 463)

granuloma (p. 451)

gut-associated lymphoid tissue
    (GALT) (p. 446)

immune system (p. 440)

innate immunity (p. 440)

integrin (pp. 448, 449)

interferon (p. 454)

interleukin 1 (IL-1) (p. 449)

Langerhans cell (p. 446)

lymph (p. 445)

lymph node (p. 445)

lysozyme (p. 446)

macrophage (p. 443)

major histocompatibility complex
    (MHC) (p. 454)

mast cell (p. 442)

M cell (p. 447)

microbe-associated molecular
    pattern (MAMP) (p. 446)

monocyte (p. 443)

mononuclear phagocyte system
    (MPS) (p. 443)

mucosa-associated lymphoid
    tissue (MALT) (p. 446)

natural killer (NK) cell (p. 455)

neutrophil (p. 442)

NOD-like receptor (NLR) (p. 457)

opsonization (p. 452)

oxidative burst (p. 453)

pattern recognition receptor
    (PRR) (p. 456)

perforin (p. 455)

Peyer's patch (p. 446)

phagocytosis (p. 442)

phagosome (p. 442)

polymorphonuclear leukocyte
    (PMN) (p. 442)

pyrogen (p. 460)

selectin (p. 449)

sepsis (p. 458)

skin-associated lymphoid tissue
    (SALT) (p. 446)

T cell (p. 443)

Toll-like receptor (TLR) (p. 456)

tumor necrosis factor alpha
    (TNF-α) (p. 449)

vasoactive factor (p. 448)

vasodilation (p. 449)

white blood cell (WBC)
    differential (p. 444)

Objectives Assessment Questions

15.1 Overview of the Immune System

1. Which of the following immune system components is considered part of the innate system?

   a. Antigen
   b. Complement
   c. Antibody
   d. B-cell lymphocytes
   e. T-cell lymphocytes

2. Which one of the following is considered a primary lymphoid organ?

   a. Peyer's patch
   b. Thymus
   c. Lymph node in the armpit
   d. Spleen
   e. Liver

3. A 35-year-old man complains of a fever and fatigue. His temperature is 38.3°C (101°F). You suspect a bacterial infection and order a CBC (complete blood count) with differential. Which of the following is consistent with a serious bacterial infection?

   a. Leukocyte count of 14,000/μl; 60% neutrophils; 10% eosinophils
   b. Leukocyte count of 11,000/μl; 55% neutrophils; 30% lymphocytes
   c. Leukocyte count of 15,000/μl; 60% neutrophils; 5% basophils
   d. Leukocyte count of 19,000/μl; 80% neutrophils; 20% lymphocytes
   e. Leukocyte count of 9,000/μl; 60% neutrophils; 25% lymphocytes

15.2 Innate Host Defenses: Keeping the Microbial Hordes at Bay

4. Which of the following cell types is part of SALT?

   a. B cell
   b. Macrophage
   c. Langerhans cell
   d. M cell
   e. Peyer's cell

5. M cells _____.

   a. sample bacteria and other antigens from the intestine
   b. are a major producer of defensins
   c. contribute to the acidic intestinal environment
   d. are equivalent to alveolar macrophages
   e. are equivalent to Langerhans cells

6. Microbial structures recognized by host receptors are called _____.

15.3 The Inflammatory Response
Match the following:

7. Selectin   a. Receptor on endothelial cells

8. Chemokines   b. Attract macrophages

9. Bradykinin   c. Edema

10. ICAM-1   d. Fever

11. A 1-year-old patient suffers from repeated infections caused by Gram-positive bacteria. Laboratory studies find 49,000 neutrophils per microliter of blood, but all antibody types were at normal levels for healthy individuals. Which of the following most likely explains this result and the patient's symptoms?
    a. Excess integrin
    b. Lack of integrin
    c. Low levels of bradykinin
    d. Excess selectin
    e. Low monocyte level

12. List the five features of inflammation.

13. A granuloma is a sign of ____ infection.

15.4 Phagocytosis: A Closer Look

14. The following process or protein prevents phagocytes from phagocytosing host cells.
    a. Apoptosis
    b. CD47
    c. Integrin
    d. Fc receptors
    e. VCAM-1

15. Which of the following processes ends inflammation at a healed site of infection?
    a. Anti-neutrophil antibodies cause apoptosis.
    b. Dendritic cells engulf the dead neutrophil.
    c. Chemoattractants draw neutrophils away.
    d. Dead neutrophils release chemicals.
    e. Macrophages engulf dead neutrophils.

16. A 5-year-old boy develops an infection with *Streptococcus pneumoniae*, an organism encased with a thick, slippery carbohydrate capsule. Which of the following helps phagocytes engulf these organisms?
    a. Defensin release
    b. NADH oxidase
    c. Fc receptors
    d. Integrin
    e. CD47

17. *Shigella dysenteriae*, a Gram-negative bacillus that causes bloody diarrhea, survives in a phagocyte by ____.
    a. preventing the oxidative burst
    b. growing in the phagolysosome
    c. preventing fusion between phagosomes and lysosomes
    d. triggering apoptosis of the phagocyte
    e. blocking integrin

15.5 Interferon, Natural Killer Cells, and Toll–Like Receptors

18. Interferon is a host-specific but pathogen-nonspecific defense mechanism. True or false?

19. Type I interferons can prevent viral infections by ____.
    a. activating natural killer cells
    b. triggering apoptosis of an infected host cell
    c. inducing production of antiviral antibodies
    d. stimulating phagocytosis
    e. inducing production of double-stranded RNA endonucleases

20. Which of the following mechanisms is shared by natural killer cells and phagocytes?
    a. Fc receptors
    b. Perforin
    c. Phagocytosis
    d. Selectins
    e. Antibody-dependent cell cytotoxicity

21. Toll-like receptors bind to which of the following?
    a. Interferon
    b. Cytokines
    c. Surface MHC I molecules
    d. Microbial structures
    e. CD47

15.6 Fever

22. Vasodilation will cause a decrease in body temperature. True or false?

23. Which of the following organs or tissues serves as the body's thermostat?
    a. Liver
    b. Hypothalamus
    c. Thymus
    d. Bone marrow
    e. Pituitary

15.7 Complement

24. The classical, alternative, and lectin pathways of complement share which one of the following?
    a. C1
    b. Antibodies
    c. C5
    d. C2
    e. C4

25. Some bacterial pathogens can prevent complement attack by coating themselves with which of these proteins?
    a. CD59
    b. Factor H
    c. Factor C5
    d. Perforin
    e. C-reactive protein

26. Which of the following statements is true about complement?
    a. Factor C3a acts as an opsonin.
    b. Factor C3b directs assembly of the pore.
    c. Factor C5b acts as an opsonin.
    d. Factor C5b directs assembly of the pore.
    e. Factor C1 is part of the alternative pathway.

27. An increase in C-reactive protein in the blood is an indication of inflammation, especially during an infection. Which of the following scenarios represents the most likely path to C-reactive protein increase during an infection?
    a. MAMPs bind TLRs in macrophages, and cytokines are released, triggering CRP production in the liver.
    b. MAMPs bind to liver cells directly and trigger C-reactive protein production.
    c. Bacterial infection triggers complement activation, and factor C3a travels to the liver and activates C-reactive protein production.

Review Questions

1. Name and describe various types of innate immunity.

2. What are Toll-like and NOD-like receptors, and how are they different?

3. How does the lung avoid being colonized?

4. Describe GALT and SALT.

5. Describe some chemical barriers to infection.

6. Discuss the different types of white blood cells.

7. What is a lymphoid organ?

8. Outline the process of inflammation.

9. How are neutrophils removed from a body site recovering from an infection?

10. Explain why phagocytes do not indiscriminately phagocytose body cells.

11. What is interferon?

12. Describe antibody-dependent cell-mediated cytotoxicity.

13. How is the complement cascade triggered?

14. How does complement kill bacteria?

15. Discuss the differences between the alternative and classical pathways of complement activation.

16. How does a host prevent membrane attack complexes from being formed in host cells?

17. Why might fever be helpful in fighting infection?

Clinical Correlation Questions

1. Danny, a 2-year-old boy with recurrent *Staphylococcus aureus* infections and oral thrush, was brought to the emergency department with a high fever and rapid breathing. A complete blood count showed 58,000 leukocytes/μl (normal is 4,000–11,000) and neutrophilia. Bacterial cultures of his blood were positive for *S. aureus*. Laboratory tests showed normal antibody and lymphocyte levels, but Danny's white cells were deficient in integrin. Explain why, with such a high level of white cells in his blood, Danny could not fight off infections.

2. Patients with paroxysmal nocturnal hemoglobinuria develop dark urine at night and anemia caused by lysis of red blood cells. The defect in these patients is a lack of CD59 protein. Explain why this defect leads to lysis of red blood cells and what might be an effective treatment.

3. Laboratory results from a 15-year-old boy infected with *Salmonella* showed that he had produced a large amount of some cytokines.

He returned a year later infected again, but with a different strain of *Salmonella*. Testing this time showed that the boy had produced low amounts of those cytokines. Research efforts revealed that the difference was due to differences in LPS from the two strains. Explain how LPS from two different strains of the same organism would cause different responses in the same boy. (*Hint:* It has something to do with pattern recognition receptors.)

4. Many viruses, such as cytomegalovirus (CMV), can cause host cells to lose MHC I surface proteins during the course of a disease. Natural killer (NK) cells can recognize this loss and attack the infected host cells in an attempt to kill the infecting virus. CMV, however, can deploy countermeasures for the loss of MHC I proteins. What might one of these countermeasures be?

16

The Immune System: Adaptive Immunity

SEVERE COMBINED IMMUNODEFICIENCY DISEASE

David Phillip Vetter was born in 1971 without an adaptive immune system. Better known as the "bubble boy," he was the only human known to live in a plastic, germ-free bubble for his entire 12 years of life (1971–1984). He suffered from a genetic disease known as severe combined immunodeficiency (SCID), in which the patient forms no T-cell lymphocytes and has dysfunctional (or sometimes no) B-cell lymphocytes. Severe SCID patients cannot launch a meaningful adaptive immune defense against any invading microbe—whether pathogen or normal microbiota.

PROTECTIVE ISOLATION
Because his older brother had died of SCID (his sister was healthy), physicians knew David might have the disorder too. Consequently, David was transferred to a sterile environment within seconds after birth to await a bone marrow transplant. Water, air, food, diapers, and clothes all were disinfected with special cleaning agents before entering his sterile plastic bubble. He was handled only through special plastic gloves attached to the wall. He lived for 12 years physically isolated in this plastic, sterile environment. His only ventures out of the bubble were in a NASA-designed sterile space suit.

DESPERATE TREATMENT
Fearing that David's health could no longer endure, doctors desperately attempted a bone marrow transplant from his sister in 1984. What they didn't realize was that the transplanted tissue contained an undetectable amount of Epstein-Barr virus (a cause of lymphoma). Because David's defective immune system left him defenseless against the virus, he became extremely ill and developed multiple tumors. There was no choice but to remove him from his protective bubble for treatment. Finally, his mother was able to touch her son's skin, but it was to be the first and last time. He slipped into a coma and died—just before his thirteenth birthday.

Life in a Bubble
For 12 years David lived inside this sterile plastic bubble, unable to touch the skin of another human being.

SEM | 4 μm

T Cell Meets Antigen
A T cell (blue) interacts with antigens presented on an antigen-presenting cell (yellow). Lacking T cells, David's immune system was helpless against all manner of invading cells.

CHAPTER OBJECTIVES

After reading this chapter, you will be able to:

- Discuss the development of primary and secondary antibody responses and how they correlate to the stages of a disease.
- Describe ways that an antibody can participate in adaptive and innate immunity.
- Demonstrate how T cells control the immune system and play an active role in immune function.
- Explain how the immune system avoids reacting against self.

The case of David Vetter illustrates how much we depend on our adaptive immune system for survival. The innate immune system that we described in Chapter 15 is certainly important, but it cannot by itself protect us from the microorganisms that surround us. The key cells of adaptive immunity are lymphocytes called B cells and T cells. B cells ("B" because they develop in bone marrow) produce antibodies. T cells ("T" because they develop in the thymus) are divided into two general groups. One group of T cells, called helper T cells, orchestrates the overall adaptive immune response. The second group, called cytotoxic T cells, attacks infected host cells. It is no wonder that SCID patients, who lack these lymphocytes, are at the mercy of microbes.

Remarkably, all it takes to subvert the entire adaptive immune process is a small defect in a single gene. There are several genetic forms of SCID, but two are the most common. In one form, linked to chromosome 20, patients lack an enzyme called adenosine deaminase and accumulate amounts of deoxyadenosine triphosphate that are toxic to lymphocytes. In the second form, linked to the X chromosome, the patient cannot produce an important cytokine receptor that both guides T-cell development and is used to communicate with B cells. David Vetter had this second form. (Cytokines were introduced in Section 15.3.) Fortunately, even though a small genetic change can destroy adaptive immunity, only one in every million people develops SCID.

In this chapter we begin with a general description of the two systems of adaptive immunity and the factors that influence how well foreign proteins, lipids, and microbes elicit adaptive immune responses. We will then explore how B cells ultimately differentiate into plasma cells and make antibodies (defining what is called humoral, or circulating, immunity) and how certain T cells develop to directly kill infected host cells in what is called cell-mediated or cellular immunity. You will also learn about a type of T cell that helps control the balance between humoral and cell-mediated responses to a given infection. Ultimately, you will appreciate why adaptive immunity is a major reason the human race still exists.

16.1 Basics of Adaptive Immunity

SECTION OBJECTIVES

- Explain the relationships between antigens, epitopes, immunogens, and haptens.
- Define antigen presentation and antigen-presenting cells.
- Differentiate humoral immunity from cellular immunity, as well as the basic roles of B cells versus T cells, in those systems.
- Describe what makes one antigen more immunogenic than another.
- Discuss antigenic (immunological) specificity and its application to vaccine development.

Why do we call this form of immunity "adaptive"? As explained in Chapter 15, innate immune mechanisms are present from birth. Adaptive immunity, in contrast, develops as the need arises. For instance, adaptive immunity against malaria develops only after the individual has encountered the plasmodial parasite that causes the disease; that is, the immune response <u>adapts</u> to the presence of the new organism. Recall Figure 2.8, which illustrates how an infecting microbe first increases in number before the adaptive system gears up. As a whole, the adaptive immune response is a complex, interconnected, and cross-regulated defense network. And it has a memory.

⊙ **Note:** The terms "adaptive immune response" and "immune response" are often used interchangeably.

The Big Picture

What triggers an immune response, and how long does it take to develop? After being exposed to an invading microbe, your adaptive immunity takes at least 3 or 4 days to develop. However, the immune system does not recognize the <u>whole</u> microbe all at once; it recognizes innumerable <u>pieces</u> of it called antigens (**Figure 16.1**). An

Figure 16.1 Breakdown of Antigens, Epitopes, and Immunogens

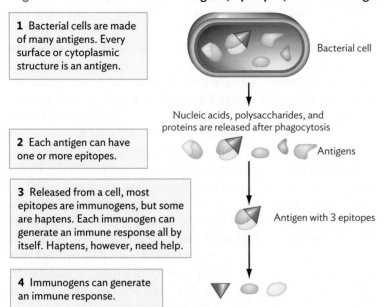

1 Bacterial cells are made of many antigens. Every surface or cytoplasmic structure is an antigen.

Bacterial cell

Nucleic acids, polysaccharides, and proteins are released after phagocytosis

2 Each antigen can have one or more epitopes.

Antigens

3 Released from a cell, most epitopes are immunogens, but some are haptens. Each immunogen can generate an immune response all by itself. Haptens, however, need help.

Antigen with 3 epitopes

4 Immunogens can generate an immune response.

Every protein, nucleic acid, lipid, and carbohydrate in (or on) the cell can be recognized by the immune system, so they are all antigens. Antigens can have one or many different epitopes (a large protein, for example, will have many epitopes). Some epitopes are immunogenic; other epitopes act as haptens. So, all immunogens are antigens, but not all antigens are immunogens (some are antigenic haptens). Each color and shape represents a different epitope.

antigen is any molecule, such as a protein, that will elicit an immune response when introduced into a person. When the antigen by itself can elicit antibody production, it is termed an **immunogen**. Many antigens have even smaller segments, called **epitopes** or **antigenic determinants**, that elicit an immune response. Besides proteins, antigenic structures in the cell include complex polysaccharides, nucleic acids, and lipids. From birth, the naive immune system is capable of recognizing and responding to billions of possible foreign (but not self) antigens.

The adaptive immune response begins when cells of the innate immune system (macrophages and dendritic cells) engulf a microbe and break it down into component antigens (**Figure 16.2**, step 1). These phagocytic macrophages and dendritic cells are also called **antigen-presenting cells (APCs)** because they place antigens and epitopes on their cell surface in order to "present" them to T lymphocytes. Imagine a stolen automobile entering a "chop shop," where the car is taken apart and individual parts are sold (presented) to someone waiting outside. For presentation, APCs place the antigens on cell-surface proteins called major histocompatibility complex (MHC) proteins.

Figure 16.2 Overview of the Adaptive Immune System

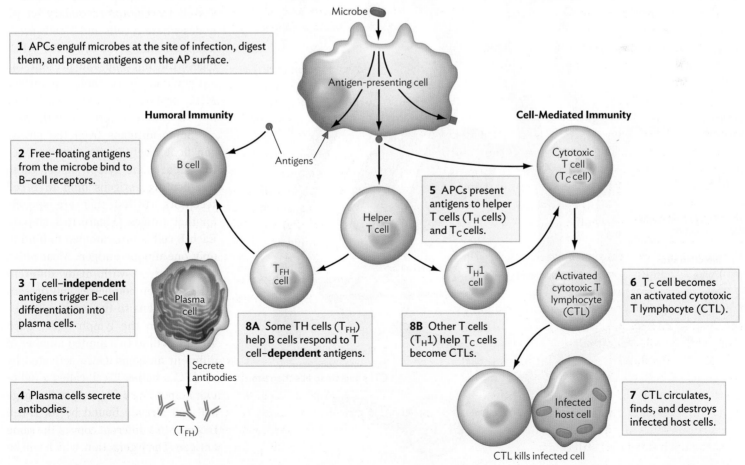

Microbe

1 APCs engulf microbes at the site of infection, digest them, and present antigens on the AP surface.

Antigen-presenting cell

Humoral Immunity

Cell-Mediated Immunity

B cell

Antigens

Cytotoxic T cell (T$_C$ cell)

2 Free-floating antigens from the microbe bind to B–cell receptors.

5 APCs present antigens to helper T cells (T$_H$ cells) and T$_C$ cells.

Helper T cell

3 T cell–**independent** antigens trigger B-cell differentiation into plasma cells.

Plasma cell

T$_{FH}$ cell

T$_H$1 cell

Activated cytotoxic T lymphocyte (CTL)

6 T$_C$ cell becomes an activated cytotoxic T lymphocyte (CTL).

8A Some TH cells (T$_{FH}$) help B cells respond to T cell-**dependent** antigens.

8B Other T cells (T$_H$1) help T$_C$ cells become CTLs.

4 Plasma cells secrete antibodies.

Secrete antibodies

(T$_{FH}$)

Infected host cell

7 CTL circulates, finds, and destroys infected host cells.

CTL kills infected cell

Adaptive immunity comes in two forms: humoral and cell-mediated. In **humoral immunity**, B cells bind to free-floating microbial antigens released from phagocytic cells or the microbe itself. However, each B cell binds to only one antigenic structure. When an antigen binds to a matched B cell, the B cell divides to make many copies (Figure 16.2, steps 2 and 3). Most of the copies become **plasma cells** that secrete large amounts of an antibody specific to the triggering antigen (step 4). An **antibody** is a Y-shaped protein that circulates in the bloodstream, waiting to bind to a matched antigen on an infecting pathogen. In fact, the system's name "humoral" comes from a Latin word meaning "related to body fluids."

The other arm of adaptive immunity, called **cell-mediated immunity**, begins when an APC presents an antigen to an inactive **cytotoxic T cell**, or **T_c cell** (Figure 16.2, step 5). That interaction converts a T_C cell into an <u>activated</u> cytotoxic T lymphocyte (CTL), which will seek and destroy host cells infected with microbes that express the same antigen (steps 6 and 7).

In truth, the humoral and cellular immune responses are not separate but intertwined, each relying on some facet of the other to work efficiently. The cells that link the two forms of adaptive immunity are called **helper T cells (T_H cells)**. T_H cells determine whether antibody- or cell-mediated mechanisms will predominate in response to an antigen. APCs activate T_H cells by presenting foreign antigens to a T_H cell whose receptor can bind to the same antigen (Figure 16.2, step 5). One class of helper T cell helps B cells become plasma cells (step 8A). A second class of helper T cell (T_H1) secretes cytokines that activate cytotoxic T cells (step 8B). So, the immune response to a microbe is really a cluster of responses made by thousands of different B cells, cytotoxic T cells, and helper T cells toward many different epitopes that came from a microbe digested by an APC. We will present clarifying details of these steps later.

Where in the body does all of this happen? Antigen-presenting cells such as macrophages and dendritic cells encounter microbes at the site of an infection (**Figure 16.3**, step 1). After engulfing the pathogen, the APCs migrate along lymphatic vessels to regional secondary lymph nodes where B cells and T cells await (step 2). Along the way, the APCs digest the microbe and present the antigenic pieces on their cell-surface MHC proteins. We will reveal later how APCs actually present antigens.

Some antigens from the pathogen float freely in the lymph. Once they reach the lymph node, these free-floating antigens bind to B-cell receptors (BCRs) that are specific for that antigen (Figure 16.3, step 3). Each B cell is programmed to bind to only one unique antigen. Meanwhile, APCs adorned with many different MHC-antigen complexes will present those antigens to the various T cells stationed in the lymph node. Each T cell can bind to a <u>limited</u> number of different antigens. Once activated by an APC, a helper T cell called a follicular helper T cell (T_{FH} cell) will engage a B cell already bound by its B-cell receptor to a different copy of the same antigen. The interaction, which will be

Figure 16.3 Staging Areas of the Immune Response

3 In the lymph node, free antigens interact with B cells; APCs engage T cells; activated B cells differentiate into plasma cells and memory cells; CTL cells are activated.

4 Plasma cells, memory cells, and CTLs leave the lymph node and enter the circulation.

2 APCs present antigens on their surface and move into regional lymph nodes.

1 APCs engulf pathogens at sites of infection.

5 Plasma cells and memory cells enter bone marrow. CTLs migrate to infection site.

Plasma cell   Memory cell   CTL

Lung

Regional lymph node

B cell   T cell

APC

Oxygenated blood

Lymphatic vessel

Infection site

APC

Heart

Deoxygenated blood

Plasma cell

Memory cell

Bone

CTL

described in Section 16.4, causes the B cell to proliferate and differentiate into plasma cells or memory cells.

Once plasma cells and memory cells are made, they migrate into the bloodstream and then to bone marrow, where they remain, although some of these cells periodically recirculate through blood. Plasma cells release antibodies into blood, while memory cells await a second encounter with the antigen at some later date (Figure 16.3, steps 4 and 5). Meanwhile, the activated cytotoxic T lymphocytes (CTLs) also leave the lymph node, but they migrate through the bloodstream to tissues where they attack any infected host cells. We will share details of these processes later. For a more detailed picture of the circulatory system, see Section 21.1.

Note that antibodies, as well as B- and T-cell receptors, demonstrate varying levels of **immunological specificity**, also referred to as **antigenic specificity**. Immunological specificity is the degree to which an antibody, or receptor, distinguishes between similar-looking antigens. An antibody that can bind any one of several similar-looking but different antigens is said to have low immunological specificity. In contrast, an antibody that binds only one of those antigenic structures has high immunological specificity.

Antigenicity

Are some antigens better than others at generating an immune response? What about our own antigens? Do we react to them? Even though our immune system is capable of reacting to any molecular structure, the strength of its reaction to different antigens, especially our own, is limited.

Antigenicity, or **immunogenicity**, measures how well an antigen elicits an immune response (for instance, how much specific antibody is produced). One antigen can be more immunogenic than another. For example, proteins are the strongest antigens, but carbohydrates can also elicit immune reactions. Nucleic acids and lipids are usually weaker antigens because they are made of relatively uniform repeating units and are very flexible. As such, they do not easily interact with antibodies. Proteins are more effective antigens for three reasons: they form a variety of shapes, they maintain their three-dimensional shape, and they are made of many different amino acids that can be assembled in many different combinations. These features enable proteins to interact strongly with antibodies in the bloodstream and with B- and T-cell receptors.

Antigen presentation by antigen-presenting cells also plays a role in how immunogenic an antigen is. Antigen presentation is a necessary step before most immune responses can occur. For example, the larger the antigen, the more likely it is that antigen-presenting phagocytic cells such as macrophages and dendritic cells will "see" and engulf it. The larger the antigen is, the more easily the APC can degrade it and present the epitopes on the APC's cell-surface MHC molecules.

A **hapten**, however, is a special form of antigen. Haptens are very small molecules, generally less than 1,000 daltons (a measure of mass). Haptens are so small, in fact, that they cannot elicit an immune response unless they are attached to a larger carrier protein. An example of a hapten is the antibiotic penicillin, a serious cause of allergic reactions in some individuals (see Chapter 17). The tiny molecule penicillin must bind to another protein before the immune system can recognize it. The reason haptens by themselves cannot <u>elicit</u> an immune response is that haptens are too small to bind to MHC molecules on APCs, so they cannot be presented to T cells. Thus, there are two types of antigens: immunogens that elicit an immune response by themselves and haptens that must be <u>attached</u> to an immunogen in order to generate an immune response.

As you might expect, the body must regulate the immune system carefully so that a response is not leveled against itself. In effect, the immune system must become "blind" to its own antigens; as a result, the host will often be blind to foreign antigens that resemble epitopes of its own cells. The more complex a foreign protein is, the more likely it will possess unique antigenic determinants that a lymphocyte can recognize as nonself. The further an antigen structure is from "self," the greater its immunogenicity will be.

Immunological Specificity

Imagine a row of doors in a hotel corridor. Each door has a lock that is different from all the other locks. Each guest is given a key that unlocks only one door. You could say that each guest's key is highly specific. This analogy is similar to immunological specificity, in which an adaptive immune response to one antigen is generally not effective against a different antigen (that is, one immune response "key" fits only one antigen "door lock"). Stated in terms of infectious disease, an immune response to smallpox virus will not protect someone against the plague bacillus (*Yersinia pestis*), which is antigenically different from the smallpox virus.

Although immunological specificity is important, it is not absolute. In the late eighteenth century, long before viruses were discovered, an English country physician named Edward Jenner (see Figure 1.15) learned how to protect townsfolk from smallpox, an extremely deadly disease, by inoculating them with scrapings from lesions produced by a tamer disease, cowpox. In so doing, Jenner unwittingly transferred the vaccinia virus that causes cowpox (see **Figure 16.4**) to villagers susceptible to the more dangerous variola virus, which causes smallpox. The resulting immune reaction to vaccinia produced an effective cross-protection against variola and thus prevented smallpox. This example illustrates that an immune reaction against one organism or virus may be sufficient to protect against an antigenically related, if not identical, organism.

In honor of Jenner's discovery, the term **vaccination** describes the technique of exposing individuals to "tame" microbes to protect the person against the more dangerous, pathogenic strains. Vaccination has been used to protect humans against many microbial and viral pathogens. Today, most vaccines—the material used to vaccinate someone—come in one of three forms: killed pathogens, live but crippled (**attenuated**) strains of the pathogen, or inactivated microbial toxins (for example, diphtheria toxin). In Chapter 17 we discuss vaccines more deeply.

Cross-protection will not occur, however, if the structures of two proteins differ significantly. A good example is the common

Figure 16.4 Antigenic Specificity Is the Basis of Vaccination

Edward Jenner recognized that disease symptoms were similar between the deadly smallpox disease and less severe cowpox disease, and he used cowpox scrapings to vaccinate humans against smallpox.

A. White pox pustules covering a smallpox patient.

TEM 150 nm

B. The smallpox virus, variola major. The dumbbell-shaped, membrane-enclosed nucleic acid core is visible.

TEM 100 nm

C. The vaccinia virus that causes cowpox.

Figure 16.5 Neutralizing Antibodies Prevent Rhinovirus Attachment to Cell Receptors

Rhinovirus

Antibody-coated rhinovirus

Cell receptor (ICAM-1)

Cell

Antibody to virus receptor protein

A. This real-world molecular model illustrates the complexity of the rhinovirus capsid and shows the attachment of the virus to the cell-surface molecule ICAM-1 (colored reddish brown).

B. Rhinovirus coated with protective (neutralizing) antibodies (green) that block the ICAM-1 receptors on the virus, making it impossible for the virus to attach to and infect the host cell.

cold, caused by hundreds of closely related rhinovirus strains (rhinitis, a runny nose, is one of the symptoms of this viral disease). Rhinoviruses use a viral capsid protein to bind the intercellular adhesion molecule ICAM-1 on host cell membranes. Once bound, the membrane invaginates to pull the virus into the cytoplasm. The cell is infected.

Unfortunately, infection with one strain of rhinovirus will not immunize the victim against a second strain because virus proteins that attach to the host's ICAM-1 surface protein differ dramatically in structure between different strains of rhinovirus (**Figure 16.5A**). As a result, so-called neutralizing antibodies that bind to the attachment protein on one strain of rhinovirus and block infection (**Figure 16.5B**) fail to bind an attachment protein from a different strain. In our hotel analogy, a key that is specific to one lock will not work on a different lock. However, if scientists can devise an antibody with broad specificity—that is, one that acts like a master key in the hotel—then a single vaccine might protect us from colds caused by numerous strains of rhinovirus.

SECTION SUMMARY

- **An antigen** can elicit an antibody response. An antigen is usually made of many different epitopes (antigenic determinants), each of which binds to a different, specific antibody.

- **Antigen-presenting cells**, such as macrophages and dendritic cells, degrade microbial pathogens and present distinct pieces on their cell surface.

- **Humoral immunity** against infection is the result of a specific B cell binding to a specific antigen. The B cell differentiates into a plasma cell that secretes antibodies directed against the antigen.

  - **Cellular immunity** involves cytotoxic T lymphocytes that directly kill infected host cells.

  - **Helper T cells (T_H cells) and cytotoxic T cells (T_C cells) are activated** only by binding to APC-presented antigens.

  - **Activated follicular helper T cells (T_{FH} cells)** stimulate B-cell differentiation into plasma cells. Activated T_H1 helper cells assist T_C cells in becoming activated cytotoxic T lymphocytes (CTLs).

  - **Proteins are better immunogens** than nucleic acids and lipids because proteins have more diverse chemical structures.

  - **Antigenic or immunological specificity** means that antibody made to bind one epitope will not bind to different epitopes (although some weak cross-binding can occur).

Thought Question 16.1 Your body makes B cells and T cells capable of reacting to your own self antigens. The consequence of letting such reaction against self happen, however, would be a disaster. In what simple way might your body prevent this disaster?

Thought Question 16.2 How does an antibody that recognizes a viral coat protein prevent infection by the associated virus?

16.2 Antibody Structure, Diversity, and Response

What are antibodies, and why are they so important? Antibodies are the keys to immunological specificity. They are glycoproteins made by the body in response to antigens. Antibodies belong to the superfamily of proteins called **immunoglobulins**. This superfamily also includes the major histocompatibility complex (MHC) proteins and the B-cell and T-cell receptors, described later.

Like a miniature "smart bomb," an antibody circulates through blood and lymph, ignoring all antigens except for its specific target antigen. When an antibody finds its antigenic match, it binds to the antigen and initiates several events that can destroy the target (see the discussion of complement, Section 15.7). While some antibodies float freely, others are strategically situated on the surface of B cells as part of the B-cell receptor (described later). These surface antibodies enable B lymphocytes to recognize specific antigens.

Basic Antibody Structure

An antibody consists of four polypeptide chains: two long **heavy chains** and two short **light chains**, colored green, in **Figure 16.6**. The four polypeptides combine to form a Y-shaped, tetrameric (four-subunit) structure held together by disulfide (–S–S–) bonds. Two disulfide bonds connect the two identical heavy chains to each other, and each light chain uses a single disulfide bond to attach to one of the heavy chains.

The antigen-binding sites of an antibody (marked in yellow in Figure 16.6) form at the amino-terminal (NH_3) ends of the light and heavy chains. One antibody molecule possesses two identical antigen-binding sites, one on each "arm" of the Y-shaped molecule. The two binding sites per molecule enable antibodies to cross-link identical antigens—a feature critical to antibody function.

You might think that all antibodies are alike except for their antigen-binding sites. The human body actually produces five classes of antibodies, defined by five different types of heavy chains called alpha (α), mu (μ), gamma (γ), delta (δ), and epsilon (ε). Antibody classes, or **isotypes**, are named according to their heavy chains. So, antibodies containing the alpha heavy chain are called IgA, while those with mu, gamma, delta, and epsilon heavy chains are called, respectively, IgM, IgG, IgD, and IgE.

Heavy-chain classes differ one from another at a region in their amino acid sequence called the **constant region**, denoted C_H (Figure 16.6). For example, all alpha chain (IgA) antibodies have the same amino acid sequence in their constant regions, but the constant-region sequence in all gamma chain (IgG) antibodies is different from that of IgA. The different heavy chains impart distinct purposes to each antibody class. For instance, IgA antibodies

Figure 16.6 Basic Antibody Structure

The Fc region is the part of an antibody that reacts with complement or binds to the surface of cells such as macrophages.

Each antibody contains two long heavy chains (long green proteins) and two shorter light chains (shorter green proteins) held together by disulfide bonds. The Y-shaped structure contains two antigen-binding sites, one at each arm of the molecule. Each antigen-binding site is formed by the variable amino acid sequences (yellow) at the amino-terminal ends of a heavy chain–light chain pair. The functional parts of the antibody are the $F(ab)_2$ region and the Fc region. The $F(ab)_2$ region binds antigens, and the Fc region can bind to complement or to Fc receptors on phagocytic cell membranes. Antibody isotype is defined by unchanging amino acid sequences in the heavy-chain constant regions (C_H1, C_H2, C_H3). For instance, all IgG molecules in all humans have the same amino acid sequences in these regions. Allotype is defined by constant-region amino acid sequences within an isotype antibody class (for example, IgG) that are identical within a single person but that differ among individuals. Idiotype differences within a single person are defined by the variable sequences (V_L, V_H) in the antigen-binding sites that bind to different antigens.

are secreted onto mucosal surfaces to intercept pathogens trying to sneak into our body. Other antibody classes circulate only in blood.

In contrast to the five heavy chains, there are only two classes of light chains—kappa (κ) and lambda (λ)—defined by their own constant regions (C_L). Every antibody contains either two kappa chains or two lambda chains, but never one of each.

The antigen-binding part of an antibody, regardless of the heavy-chain class, is formed by highly variable amino acid sequences at the amino-terminal ends (yellow sections) of the light and heavy chains. By "variable" we mean that the amino acid sequence of an antigen-binding site that binds, say, peptidoglycan, is very different from the amino acid sequence of an antigen-binding site that binds to a different antigen, like flagellar proteins. These sequence differences alter the shape of the antigen-binding site, allowing each antibody to bind to a different antigen. The **variable regions** are referred to as the V_L (light-chain) regions and the V_H (heavy-chain) regions (Figure 16.6). Note that antibodies bind to antigens by noncovalent interactions (for instance, ionic, or hydrophobic, bonds), so antigen-antibody binding is reversible.

In addition to the two "arms" that bind antigen, called the **F(ab)$_2$ region**, every antibody contains a "tail," called the **Fc region** (Figure 16.6). The Fc tail region of an antibody is not involved in antigen recognition but is very important for anchoring antibodies to the surface of certain host cells and for binding components of the complement system.

Antibody Isotypes, Allotypes, and Idiotypes

We often discuss antibody diversity in terms of the amino acid differences that distinguish antibodies between mammalian species (**isotype**), between different individuals within a species (**allotype**), and between different antibodies within an individual (**idiotype**). Understanding the differences between antibody isotype, allotype, and idiotype is important because these terms reflect different levels of antibody diversity. Put most simply, a single person has five antibody isotypes: IgG, IgA, IgD, IgE, and IgM. But each antibody isotype (IgG, for instance) in an individual has allotypic amino acid sequence differences compared with the same isotype (IgG, in this case) in another individual. That is, all IgG molecules in an individual are of the same allotype. Finally, the millions of molecules of the single IgG antibody allotype in a person will contain millions of idiotypic differences that reflect the antigen specificity of each molecule. The synthesis and function of these different levels of antibody diversity will be described in the next several sections.

Antibody Isotype Functions and "Superstructures"

All antibody isotypes have the same basic structure. However, each isotype has a unique "superstructure." Some isotypes function as monomers (IgG), whereas others are dimers (IgA), and each isotype is designed to carry out a different task. Some key properties of the five different immunoglobulin classes are listed in Table 16.1.

IgG is the simplest, most abundant antibody in blood and tissue fluids, and it provides the majority of antibody-based immunity against invading microbes. Although it is a monomer (an example of which is illustrated in Figure 16.6), it carries out several missions for the immune system. First, IgG can **opsonize** microbes; that is, it can bind to antigens on an organism's surface to make the organism more susceptible to phagocytes. Like a gymnast doing a handstand, the antigen-binding sites of an antibody will attach to a microbe, leaving the antibody's Fc region extending outward (like the gymnast's feet), pointing away from the cell. Phagocytes possess surface Fc receptors that can then attach to the protruding Fc regions (the feet) to gain a firmer grip on the microbe, facilitating phagocytosis (discussed in Section 15.3). Second, IgG can

Table 16.1
Properties of Human Immunoglobulins

Property	IgG[a]	IgA	IgM	IgD	IgE
Serum half-life (days)	7–21	6	10	3	2
% Total serum Ig	70%	15%–20%	5%–10%	0.2%	0.002%
Antigen-binding sites	2	2–4	2–10	2	2
Produced by fetus	Poor, if at all	Poor, if at all	Yes	?	Poor, if at all
Transmitted across placenta	Yes	No	No	No	No
Binds complement	Yes	No	Yes	No	No
Opsonizing	Yes	No	No	No	No
Binds mast cells	No	No	No	No	Yes

[a]IgG has four subtypes: IgG1 through IgG4.

directly neutralize viruses by binding to virus attachment sites, blocking the virus from invading host cells. Other important roles for IgG include being one of only two antibody types that can activate complement by the classical pathway (described in Section 15.7) and being the <u>only</u> antibody capable of crossing the placenta to provide passive immunity to the fetus.

IgA, containing the alpha heavy chain, is called a secretory antibody because it is secreted across mucosal surfaces. In secretions, IgA is a dimer whose monomeric parts are held together at their Fc regions by a protein called the J chain (**Figure 16.7A**). Plasma cells within mucosal membranes synthesize the IgA dimer. Each dimer then passes through a mucosal cell, where a sixth protein, the secretory piece, is wrapped around the IgA dimer during the final secretion process. The secreted molecule, now called secretory IgA (sIgA), is found in tears, breast milk, and saliva, as well as on other mucosal surfaces. The molecule sIgA is important for immunity against pathogens that use mucosal membranes as portals of entry. Most pathogens enter the body through mucosal surfaces.

Circulating **IgM**, containing the mu heavy chain, is a huge, Ferris wheel–shaped molecule formed from five monomeric immunoglobulins tethered together by the J-chain protein (**Figure 16.7B**). IgM can also be found as a monomer attached by its Fc region to the surface of B cells, where it forms part of the B-cell antigen receptor (described later). During the early stages of an immune response, IgM is the <u>first</u> antibody isotype detected in blood. Unlike the smaller IgG immunoglobulins, IgM is so large that it cannot cross the placenta or easily enter tissues (see Table 16.1).

IgD, containing the delta heavy chain, is a monomer present at very low levels in the blood but abundant on the surface of B cells. Attached to the cell surface by its Fc region, IgD, along with monomeric IgM, can bind antigen and signal B cells to differentiate and make antibody. IgD neither binds complement nor crosses the placenta.

IgE, defined by the epsilon heavy chain, is found mostly on the surface of mast cells and basophils. Mast cells and basophils contain granules loaded with inflammatory mediators. The primary role of IgE is to amplify the body's response to invaders. Once secreted into the blood, IgE attaches to mast cells (**Figure 16.8A and B**), again by way of its Fc region. Then, like a Venus flytrap, the antibody sits and waits until one or both of its antigen-binding sites binds to a specific antigen. When an antigen cross-links two surface IgE molecules, the mast cell granules empty their contents into the surroundings (called degranulation). Histamine and other pharmacological mediators are released to help orchestrate acute inflammation, described in Chapter 15. The system can also produce severe allergic hypersensitivities, such as anaphylaxis, and milder forms, such as hay fever (**Figure 16.8C and D**). We discuss allergic hypersensitivity in Section 17.3.

Primary and Secondary Antibody Responses

When you are first infected with a microorganism or have been given a vaccine, what happens before antibodies begin to appear in your blood is called the **primary antibody response**. As outlined in Section 16.1, antigens from the pathogen bind to B cells that are able to make antibodies against only that antigen. The activated B cells then proliferate and differentiate into high-output antibody-secreting

Figure 16.7 Structures of IgA and IgM

The antibodies are made as multimers of two (IgA) or five (IgM) immunoglobulin molecules.

A. IgA is secreted as a dimer held together by the secretory piece and the J chain. IgA can bind four identical antigens.

B. IgM forms a pentamer held together by disulfide bonds and the J chain. IgM can bind ten antigens.

Figure 16.8 Mast Cells

A. Scanning EM of a mast cell.

B. Transmission EM of a mast cell in cross section, showing granules (arrow).

C. As a result of degranulation, IgE-coated mast cells release histamine and other pharmacological mediators that cause hay fever.

D. When released in the upper respiratory tract, the chemical mediators produced by mast cells cause the symptoms of runny nose and sneezing.

plasma cells (**Figure 16.9**) and memory B cells. Memory B cells do not secrete antibodies themselves but are primed to quickly become antibody-secreting plasma cells if the host is <u>reinfected</u> with the same pathogen at a later date—months or years after the initial infection (or given a second booster dose of vaccine). Consequently, secondary exposure to the antigen triggers an almost instantaneous **secondary antibody response** (**Figure 16.10**).

The net result of the primary antibody response is the early synthesis and secretion of pentameric IgM molecules specifically directed against the antigen, or immunogen. Later in the primary response, a process known as **isotype switching** (also called **class switching**) occurs in activated

Figure 16.9 B Cell and Activated Plasma Cell

After binding antigen, a B cell differentiates into an antibody-secreting plasma cell. Notice the increased size of the plasma cell compared with the B cell. The plasma cell cytoplasm is stuffed with endoplasmic reticulum needed to support the incredible increase in protein synthesis associated with antibody production.

A. Resting B cell.

B. Active plasma cell.

Figure 16.10 Primary versus Secondary Antibody Response

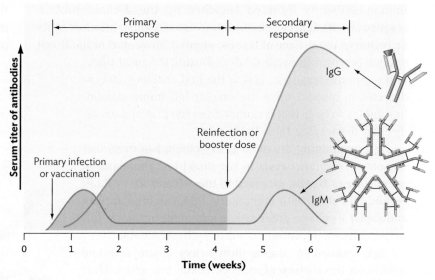

Primary vaccination or infection leads to the early synthesis of IgM, followed by a class switch to IgG. Reinfection or a second, booster dose of a vaccine results in a more rapid antibody response consisting mainly of IgG because of memory B cells formed during the primary response. Be aware that the time course and level of antibody made vary with the immunogen and the host.

B cells. The predominant antibody type produced switches from IgM to IgG. Antibodies made during the primary response, though specific for the immunogen, are actually not of the highest affinity. Affinity is a measure of the "tightness" by which an antibody binds to its antigen. As you will see, mechanisms to increase antibody affinity occur later in the immune response. Eventually, as the pathogen is cleared from the body, the levels of both IgG and IgM decline because the plasma cells that produced them die. Most plasma cells die within 3 months.

Memory B cells made during the primary response are maintained in the body because, unlike plasma cells, they continue to divide. Memory B cells make up approximately 40% of the circulating B-cell population. During the secondary response, most memory B cells switch from IgM production to IgG production as they become plasma cells. Huge amounts of IgG antibody are secreted from these plasma cells. These new and improved IgG antibodies have a higher specificity and affinity for the antigen than the IgM antibodies produced during the primary response (see "Memory B Cells Make Better Antibodies" later in this section). Unfortunately, infection by some viral pathogens, most notably measles virus, can destroy the body's immune memory of encounters with other pathogens (see **IMPACT**).

The speedy production of antibody during the secondary response is why immunization works to protect you against infectious diseases. During a natural infection, an aggressive pathogen can do considerable harm while the body is gearing up for the primary response. Vaccination prior to infection can prevent this harm. In vaccination, an innocuous version of a pathogen (or a harmless piece of it) can be injected into a person to trigger the primary response without producing disease (or, at worst, producing only a mild form of the illness). Immunization primes the immune system to respond efficiently and without delay upon encountering the real pathogen. You will find more on vaccines in Chapter 17.

B-Cell Receptors, Antigens, and Clonal Selection

Your body produces many <u>billions</u> of naive and memory B cells specific for different antigens. A naive B cell is one that has not previously encountered antigen. To start a primary or secondary antibody response, a B cell (or memory B cell) must first bind to its matched antigen. But how does a B cell find the right antigen? The answer is the **B-cell receptor (BCR)**, the key component of which is an antibody specific for the antigen.

It turns out that a naive B cell can produce antibody, but only monomeric IgM and IgD, which have <u>identical</u> antigen specificities. These two antibody classes are not secreted but are displayed like tiny satellite dishes on the B-cell surface, anchored by their Fc regions as part of the B-cell receptor (**Figure 16.11**). Each B cell may have upwards of 50,000 B-cell receptors, all of which can bind identical antigens.

When two adjacent B-cell receptors on one B cell bind <u>identical epitopes</u> on the same microbe (think about the repetitive structure of a virus capsid), the surface B-cell receptors begin to cluster in a process called capping. **Capping** sends a signal into the nucleus (**Figure 16.12**) that activates the B cell, causing it to proliferate into many clones (or copies) of itself. The process is called **clonal selection** (**Figure 16.13**). The antigen therefore selects which B cell among many will ultimately differentiate into antibody-secreting plasma cells and memory B cells.

Figure 16.11 A B-Cell Receptor

The B-cell receptor is formed as a complex between a monomeric IgM and proteins Igα and Igβ in the membrane. Igα and Igβ are not immunoglobulins themselves, but associate with the antibody immunoglobulin part of the receptor. The inset shows how the BCR will be represented in other figures.

Figure 16.12
Capping and Activation of the B cell

Two B-cell receptors can bind two identical epitopes on a pathogen. The resulting capping process initiates a signal cascade that activates differentiation and proliferation of the B cell independent of T-cell help. Antigens with repeating epitopes, such as polysaccharides, can directly cross-link B-cell receptors.

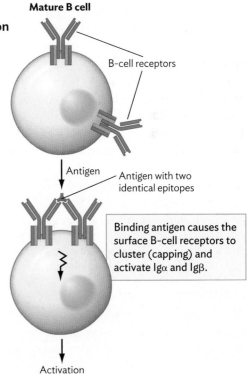

How Are Antibodies Made to Nonrepeating Antigens?

We just described how antigens with repeating epitopes activate B cells by cross-linking B-cell receptors. These antigens are called T cell–independent antigens. Such cross-linking is great for macromolecules like viral capsids, nucleic acids, polysaccharides, and

impact

Measles Virus: The Memory-Cell Assassin

You have just learned some things about immune memory, including that it is composed of memory B cells and T cells (yes, there are memory T cells). Immune memory is what enables vaccines to protect you and your child from serious infections that you haven't yet had. However, anti-vaxxers (people who think vaccination is bad for children) argue that <u>natural</u> infection by pathogens provides excellent protection against future infection and that serious complications are rare for childhood diseases. Anti-vaxxers also believe that there is a connection between vaccines and autism. So, who needs vaccines?

<u>Nearly everyone</u> needs vaccines! While it is true that a natural infection provides protection against subsequent infection by the same pathogen, an unvaccinated child will suffer serious symptoms leveled by the natural infection. It is also true that serious complications from a natural infection are uncommon, but when they do occur, those complications can be life-altering and even deadly. In addition, pathogens from a naturally infected child can be transmitted to other children, some of whom may be immunodeficient and thus more likely to develop those deadly complications. Finally, the linchpin of the anti-vaxxers' argument—a connection between vaccination and autism—has been proved wrong time and time again. Vaccinations do <u>not</u> cause autism.

There is another good reason to vaccinate your child, especially for measles. Natural infection with measles virus is known to severely diminish immune memory. Suppose a child has been naturally infected with mumps, rubella (German measles), and pneumonia bacteria (remember the child was not vaccinated) and then contracts measles. The measles virus will infect the child's memory B cells and T cells, including those made against mumps, rubella, and the pneumonia bacteria. The virus doesn't kill those cells, but the cells will place measles virus antigens on their surface MHC molecules. The memory cells are now targets for activated cytotoxic T lymphocytes (CTLs) that recognize measles antigens. The CTLs will seek out and destroy immune memories of the earlier infections. The child has effectively developed "immunological amnesia," once again becoming vulnerable to those pathogens.

So, how can you prevent immunological amnesia? Vaccinate for measles before your child contracts it.

DISCUSSION QUESTION

Should vaccinations for preventable diseases be mandated for children entering school? If not, why not? If so, why? And would any exemptions be necessary?

------

lipopolysaccharides, all of which have repeating structures. But most proteins do not contain multiple repeating units. A typical protein is a mosaic of many different small, discrete, single epitopes, making it difficult to cross-link two B-cell receptors on the same cell. B-cell responses to these types of antigens require assistance from **helper T cells (T$_H$ cells)**. As a result, B-cell activation can take place by one of two routes: one that does not involve the use of T cells (the T cell–independent route) and one that does (the T cell–dependent route). How helper T cells foster B-cell activation will be discussed in Section 16.4.

Generating Antibody Diversity

Before we discuss how T cells influence antibody production, let's see how antibody diversity happens. Each of us can synthesize an estimated 10^{11} different antibodies. Because each B cell displays antibodies to only one antigenic determinant, it follows that the immune system can potentially produce 10^{11} different B cells displaying 10^{11} different B-cell antigen receptors—just not all at one time. Each person possesses only about 1,000 genes or gene segments involved in antibody formation. So, how are 10^{11} different antibodies made from only 10^3 genes? It turns out that antibody genes can rearrange within the genome of a differentiating B cell (and only a B cell) to make millions of possible combinations. Three steps are involved: the rearrangement of antibody gene segments (or cassettes), the random introduction of mutations, and the generation of different codons during gene rearrangements. In humans, the process of generating antibody diversity takes place constantly over a lifetime.

Put simply, antibody diversity begins in bone marrow, where precursor B cells randomly rearrange a set of immunoglobulin genes to make an antigen-binding site (the V$_L$ and V$_H$ segments of the antibody mentioned earlier). The process is similar to randomly picking 3 blocks from a box of 200 differently shaped Lego pieces and hooking them together. Millions of 3-block combinations can be made from the 200 pieces. Think of one such combination as a single antigen-binding site, the VDJ region shown in **Figure 16.14A**. Each B cell randomly selects one such combination and sticks with it. Once the DNA segment encoding the antigen-binding site is made, <u>all</u> of the offspring from that B cell will make antibodies with that same antigen-binding site.

Once assembled, DNA for the antigen-binding site will sit in front (meaning the 5′ end) of a string of genes encoding the five different heavy chains: IgM, IgD, IgG, IgE, and IgA, in that order. Initially, only IgM and IgD antibodies are made. These antibodies

Figure 16.13 Clonal Selection Theory

The B-cell population is composed of individual cells that have specificity for different antigens. When a B-cell surface antibody contacts its matched antigen (the antigen to which the B-cell antibody receptors bind), an intracellular signal is generated and the B cell is activated, leading to proliferation of that clone (clonal expansion) and differentiation into plasma cells and memory B cells.

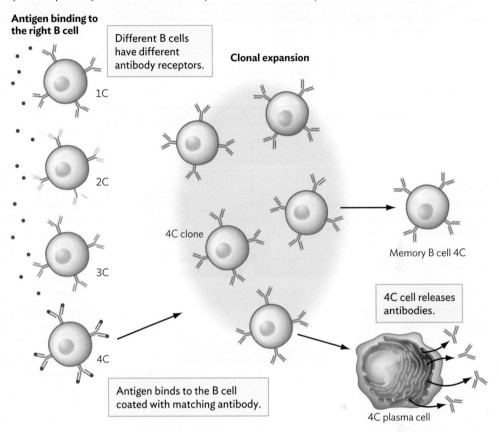

are not secreted but are placed on the surface of the B cell to make B-cell receptors. At this stage, the B cell will move from bone marrow to a secondary lymphoid tissue.

A B cell that never encounters its matched antigen eventually dies. But the lucky B cell that binds to its predetermined antigen will proliferate and become antibody-secreting plasma cells or memory cells. As mentioned earlier, most of these plasma cells secrete IgM. But eventually, some of the B-cell clones will undergo another series of gene rearrangements that will move a <u>different</u> heavy-chain gene (IgG, IgE, or IgA) next to the antigen-binding-site DNA (**Figure 16.14B**). Which heavy chain is moved depends on the cytokine cocktail present in the area. The B cell has now undergone an isotype switch (class switch), so the B cell makes a new antibody isotype: IgG, IgA, or IgE. Only one isotype is made per B-cell clone. But whichever antibody is made, it will have the same antigen-binding site.

Memory B Cells Make Better Antibodies

Memory B cells are B cells that have already undergone class switching (that is, they are committed to making IgG, IgA, or IgE), and they are very long-lived. The memory-B-cell line can replicate

Figure 16.14 Genetics of Antigen-Binding-Site Formation and the Isotype Switch

Formation of antigen-binding site

Germ-line DNA

Heavy-chain antigen-binding-site genes   Heavy-chain isotype genes

V₁ V₂ V₃ V₄ D₁ D₂ D₃ D₄ J₁ J₂ J₃ J₄ μ δ γ ε α

Rearrangement of genes to form VDJ antigen-binding site

Naive B-cell DNA   V₂ D₄ J₃ μ δ γ ε α

IgM heavy chain made

Isotype switch   Deleted

Recombination to switch heavy-chain isotype

V₂ D₄ J₃ α

IgA heavy chain made

B. After binding to its matched antigen in a lymph node, the B cell proliferates to make many copies. Some of these copies undergo an isotype switch that changes the heavy chain, in this case to IgA. The same VDJ antibody-binding sequence is linked to the new heavy chain.

A. A series of 132 V, D, and J genes is rearranged in the germ-line cell to form a segment of DNA that encodes part of the antibody's antigen-binding site. The result is a naive B cell that has not yet seen its matched antigen. Note that only a small subset of the V, D, and J genes is actually shown in this model. In these cells, only IgM and a small number of IgD heavy chains are made and used to make the B-cell receptor.

- **The Fc "tail" portion** of an antibody can bind to specific receptors on host cells. This binding is antigen <u>in</u>dependent.

- **Antibody "isotype," "allotype," and "idiotype"** refer to amino acid sequence differences found at the different levels of antibody diversity. Antibody **isotype** is defined by sequences that are unique to a species. **Allotype** is defined by isotype sequences that are consistent within one individual but different from those of another individual of the same species. **Idiotype** is defined by sequence differences found within one person's individual antibodies, usually at antigen-binding sites.

- **Antibody diversity** is generated during B-cell development by recombining (mixing and splicing) about 1,000 DNA gene cassettes and introducing mutations within DNA segments corresponding to the antigen-binding site.

- **The primary antibody response** to an antigen begins when B cells differentiate into antibody-producing plasma cells and memory B cells. IgM antibodies are generally the first class of antibodies secreted during the primary response. **Isotype switching** occurs late during the primary response when a subclass of B cells switches from making IgM to making other antibody isotypes (for instance, IgG).

- **The secondary antibody response** occurs during subsequent exposures to an antigen and arises because memory B cells are activated. IgG is the predominant antibody made.

- **Clonal selection** is the rapid proliferation of a subset of B cells during the primary or secondary antibody response.

- **A B-cell receptor (BCR)** consists of a protein complex in the B-cell membrane that includes an antibody. Binding of antigen to the antibody in a B-cell receptor triggers B-cell proliferation and differentiation.

- **T cell–independent immunity** is possible when an antigen possesses multiple repeating epitopes that bind multiple B-cell receptors in the process called capping.

Thought Question 16.3 The mother of a newborn is diagnosed with rubella, a viral disease. Infection of the fetus can lead to serious consequences for the newborn. How can you use antibody production to determine whether the newborn was infected in utero?

Thought Question 16.4 Why do immunizations lose their effectiveness over time?

Thought Question 16.5 IgM is the first antibody produced during a primary immune response. Of the five different types of antibodies, why would the body want IgM to be the first produced?

---

and live for many years, guarding us against microbial assault. Plasma cells, however, usually die within months.

During its lifetime, a memory B cell, like other B cells, will mutate the DNA that encodes the antigen-binding region of its antibody genes (the VDJ regions described earlier). These mutations can increase (or decrease) the affinity of antibody produced during the secondary response. However, memory B cells that produce higher-affinity antibodies are maintained in the body. How? As an immune response clears an infection, antigen becomes scarce, so only B cells with the highest-affinity B-cell receptor can bind to the vanishing antigen and become activated. This overall process is known as **affinity maturation**, meaning that with repeated exposures to the same antigen, a host will produce better and better memory B cells that make antibodies with better and better affinities for the antigen.

SECTION SUMMARY

- **Antibodies are Y-shaped** molecules that contain two heavy chains and two light chains. **There are five classes (isotypes) of heavy chains.** Each antibody isotype is defined by the sequence of the heavy chain.

- **Each antibody molecule contains two "arms" that have identical antigen-binding sites** formed by the hypervariable ends of a heavy chain–light chain pair.

16.3 T Cells: The Link between Humoral and Cell-Mediated Immunity

SECTION OBJECTIVES

- Explain the three major types of T cells.
- Discuss the role of major histocompatibility complex proteins in the adaptive immune response.
- Outline the two routes of antigen processing and presentation.
- Describe a T-cell receptor (TCR), and explain its role in T-cell activation.
- Describe processes that minimize autoimmunity.
- Describe the role of cytokines in directing adaptive immunity toward humoral or cytotoxic responses.

Do the two branches of the immune system (antibody-mediated immunity and cell-mediated immunity) work together? During an infection, the two types of immunity will communicate with each other to tilt the response toward whichever branch is best for eliminating the specific infection—much the way the army and air force communicate to determine who will take the lead in assaulting an enemy position. As mentioned earlier, T cells are the key to that balance. T cells develop in the thymus (rather than in the bone marrow, where B cells develop) and contain key surface proteins different from those of B cells. In general, T cells are divided into two broad groups by proteins on their surface called cell differentiation (CD) proteins. The important CD proteins are CD4 and CD8 (Table 16.2). Helper T cells display CD4 on their surface, whereas cytotoxic T cells (T_C cells) display CD8. Helper T cells (CD4) can stimulate either antibody- or cell-mediated immunity. CD8 T cells

carry out cell-mediated immunity. How the balance is achieved will become apparent later.

Helper T cells come in several models. The first model is the precursor T_H0 cell, which can be stimulated by APCs to become the other T-cell types—for example, follicular helper T cells (T_{FH}) that stimulate B cells to differentiate into plasma cells (antibody-mediated immunity) or T_H1 cells that activate cytotoxic T cells (cell-mediated immunity). The ratio of different T_H cells produced during an infection is influenced by the relative mix of cytokines in the area. The result tilts the immune response more toward antibody-mediated immunity (required to deal with extracellular pathogens) or cell-mediated immunity (used to handle intracellular pathogens). To be of any use, however, T cells must be activated only by foreign antigens and trained to avoid recognizing self antigens.

⊙ **Note:** T_H2 cells mainly stimulate B cells to switch classes and make IgE antibody useful against parasites such as helminths (worms). T_H17 cells mainly help B cells switch to IgG subtypes but also produce a pro-inflammatory cytokine (IL-17) that stimulates the recruitment of monocytes and macrophages to the site of an infection or inflammation. Treg cells (regulatory T cells) suppress the responses of other immune cells.

How T Cells Recognize Antigens

ANTIGEN-PRESENTING CELLS PLACE ANTIGENS ON MHC COMPLEXES To be clear, T cells do not bind free-floating antigen. T cells bind only to antigens that are presented on the surface of APCs. As illustrated in **Figure 16.15**, the APC surface proteins that present antigens to T cells are known as **major histocompatibility complex (MHC)** proteins. MHC proteins differ between species and among individuals within a species. Those proteins help determine

Table 16.2
Major Classes of T Cells[a]

T-Cell Type	CD Coreceptor	MHC Restriction	Cytokines Produced	Major Function
T_H0 cell	CD4	Class II	Wide range	Differentiate into functional T_H cells (below)
Helper T cell types				
T_{FH}	CD4	Class II	IL-12, IL-4	Activate B cells
T_H1	CD4	Class II	IFN-γ, IL-2, TNF-β	Activate cytotoxic T cells
T_H2	CD4	Class II	IL-4, IL-5, IL-6	B-cell helper; IgE class switch
T_H17	CD4	Class II	IL-17, IL-23, TNF-α	Pro-inflammatory; IgG subtype class switch
Treg	CD4	Class II	IL-10, TGF-β[b]	Anti-inflammatory
Cytotoxic T cell (T_C)	CD8	Class I	IFN-γ, TNF	Kill virus-infected and cancer cells

[a]CD4 and CD8 proteins are located on the surfaces of different T cells. MHC I proteins are present on all nucleated cells.
Antigen-presenting cell surfaces have MHC I proteins but also have MHC II proteins.

[b]TGF = transforming growth factor.

Figure 16.15 Major Histocompatibility Complexes

Major histocompatibility proteins

A. MHC class I molecules are composed of a large alpha chain and a small peptide called β₂-microglobulin. MHC class II molecules contain an alpha chain and a beta chain. The peptide-binding regions of both classes show variability in amino acid sequence that yields different shapes and grooves. Peptide antigens nestle in the grooves and are held there, awaiting interaction with T-cell receptors. CD8 T cells recognize antigen peptides associated with class I molecules; CD4 T cells recognize peptides bound to class II molecules.

MHC-antigen binding

B. Top view of a peptide antigen (red) nestled in the MHC peptide-binding site.

C. Cartoon versions of MHC I–antigen and MHC II–antigen complexes used in subsequent figures.

whether a given antigen is recognized as coming from the host (a self antigen) or from another source (a foreign antigen) in a phenomenon called histocompatibility (hence the name "major histocompatibility complex").

Two classes of MHC molecules are found on cell surfaces. **Class I MHC molecules** are found on all nucleated cells (not red blood cells). **Class II MHC molecules**, in contrast, are expressed only by dendritic cells, macrophages, and B cells—the "professional" APCs. Thus, every nucleated host cell can present antigens from an invading pathogen (a virus, for example) on its MHC I proteins. But professional APCs can present a pathogen's antigens on <u>either</u> MHC I or MHC II proteins.

Note that each MHC can bind a broad range of antigens and that each antigen-presenting cell expresses several different MHC proteins. Multiple MHC molecules enable a single APC to present many different antigens. Unlike antibodies, DNA rearrangements are not involved in the making of MHC molecules.

Why are there two MHC complexes? The two complexes tell the immune response which system—antibodies or cytotoxic T cells—is better for fighting an infection. MHC I antigen presentation activates the cytotoxic-T-cell pathways needed to combat intracellular pathogens, whereas MHC II antigen presentation means that extracellular pathogens are present, so antibody production would be the way to go. But how are these antigens trafficked to the right MHC protein?

Endogenous antigens made by viruses and intracellular bacteria that grow within the cytoplasm of a host cell become attached

to class I MHC molecules stationed on the endoplasmic reticulum's inner surface (**Figure 16.16A**, steps 1 and 2). Vesicles then taxi the MHC I–antigen complexes to the cell surface (step 3). The vesicle membranes fuse with the cell membrane (step 4), placing the MHC I–antigen complexes on the outer surface of the cell, where CD8 T cells can find them.

In contrast to antigens synthesized <u>inside</u> the cell, antigens produced outside the APC (called exogenous antigens) enter professional APCs via phagocytosis. The antigens then attach to class II MHC molecules affixed to the membranes of an acidic phagosome or lysosome (**Figure 16.16B**). The MHC class II–peptide complex is then carried to the cell surface where CD4 T cells can find them. Like MHC I, MHC II molecules are initially assembled in the endoplasmic reticulum but include a protein that blocks the antigen-binding groove. The blocking protein is degraded before an MHC II–carrying vesicle fuses with a phagosome that contains microbial antigens.

⊙ **Note:** MHC molecules present only peptide antigens. Carbohydrate antigens cannot attach to the binding clefts of MHCs but will trigger antibody production independent of T cells.

T-CELL RECEPTORS RECOGNIZE MHC-ANTIGEN COMPLEXES

Once an MHC-antigen complex is presented on a professional APC surface, T cells can interact with that cell using one of its T-cell receptors. Professional antigen-presenting cells and naive T cells interact within the lymph nodes, the spleen, or Peyer's patches in the gut.

Figure 16.16 Antigen Presentation on Class I and Class II MHC Proteins

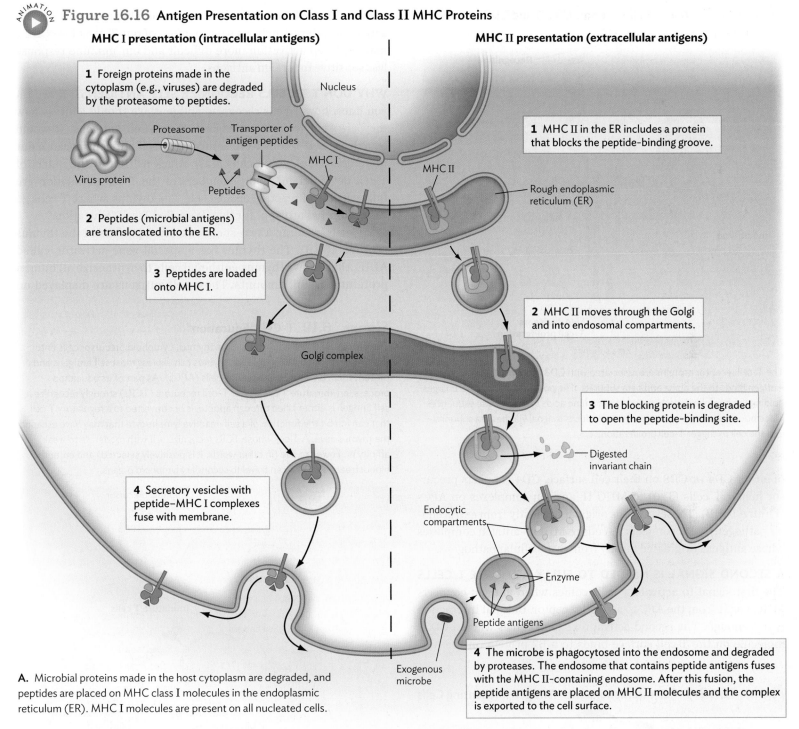

MHC I presentation (intracellular antigens)

1 Foreign proteins made in the cytoplasm (e.g., viruses) are degraded by the proteasome to peptides.

Nucleus

Proteasome

Transporter of antigen peptides

MHC I

Virus protein

Peptides

2 Peptides (microbial antigens) are translocated into the ER.

3 Peptides are loaded onto MHC I.

Golgi complex

4 Secretory vesicles with peptide–MHC I complexes fuse with membrane.

MHC II presentation (extracellular antigens)

1 MHC II in the ER includes a protein that blocks the peptide-binding groove.

MHC II

Rough endoplasmic reticulum (ER)

2 MHC II moves through the Golgi and into endosomal compartments.

3 The blocking protein is degraded to open the peptide-binding site.

Digested invariant chain

Endocytic compartments

Enzyme

Peptide antigens

Exogenous microbe

4 The microbe is phagocytosed into the endosome and degraded by proteases. The endosome that contains peptide antigens fuses with the MHC II-containing endosome. After this fusion, the peptide antigens are placed on MHC II molecules and the complex is exported to the cell surface.

A. Microbial proteins made in the host cytoplasm are degraded, and peptides are placed on MHC class I molecules in the endoplasmic reticulum (ER). MHC I molecules are present on all nucleated cells.

B. Microbial proteins made outside the cell are endocytosed, degraded in the endosome, and placed on MHC class II molecules. MHC II molecules are present only on macrophages, dendritic cells, and B cells, the so-called professional antigen-presenting cells.

The **T-cell receptor (TCR)** is composed of several transmembrane proteins (**Figure 16.17**). The part used to recognize antigen is composed of two molecules: alpha and beta. Much like antibodies, the alpha and beta proteins of the TCR are formed from gene clusters that undergo gene rearrangements analogous to, but different from, antibody genes.

The TCR alpha and beta proteins are found in a complex with four other peptides called the CD3 complex. When the TCR binds to an MHC-antigen complex, the CD3 proteins send a signal to the nucleus that triggers proliferation of the T cell. Thus, T cells, like B cells, are clonally selected to multiply.

CD4 AND CD8 SURFACE PROTEINS HELP T CELLS DISCRIMINATE BETWEEN MHC-ANTIGEN COMPLEXES Recall that the two main subsets of T cells are distinguished by the presence

Figure 16.17   The T-Cell Receptor (TCR) and CD3 Complex

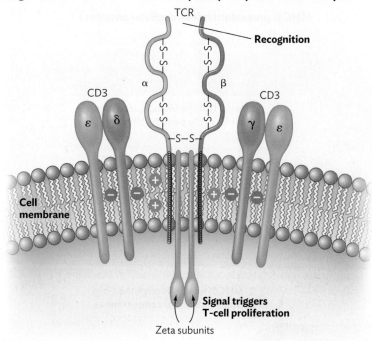

The T-cell receptor proteins are associated with CD3 proteins at the cell surface. Antigen binds to the alpha and beta subunits. The positive and negative charges holding the complex together come from amino acids in the peptide sequences. Once bound to antigen, the complex transduces a signal via the zeta subunits into the cell to trigger T-cell proliferation.

of either CD4 or CD8 on their cell surface. CD4 connects precursor helper T cells (T_H0) to MHC II–antigen complexes on APCs (**Figure 16.18**). Those antigens originated mostly from extracellular pathogens. CD8 connects T_C cells to MHC I–antigen complexes whose antigens originated mostly from intracellular pathogens.

A SECOND SIGNAL IS NEEDED TO FULLY ACTIVATE T CELLS
The first signal to activate T cells comes when antigen connects MHC I or II from the APC to a T-cell receptor. But that interaction is not enough. The second activation signal comes when a CD28 molecule on the T-cell surface binds to a B7 protein on the APC

Figure 16.18   T-Cell Interactions with Antigen-Presenting Cells

Cytotoxic T cells (CD8 T cells) use CD8 surface structures to bind MHC I–antigen complexes. Helper T cells use CD4 surface structures to bind MHC II–antigen complexes. Professional APCs have MHC I and MHC II complexes. Other body cells have only MHC I.

surface (Figure 16.18). Once that interaction happens, the T cell is activated. Interestingly, as we age, more of our T cells lose CD28, making T-cell activation more difficult and our immune response less sensitive to foreign antigens.

WHY DON'T T CELLS RECOGNIZE SELF ANTIGENS? Now that you know how T cells recognize antigens, we should address how they are kept from recognizing our own self antigens. It would be a disaster if T cells were activated by self antigens because our immune system would constantly attack us! But at the outset of T-cell development, some members of the T-cell population are fully capable of reacting against self. The thymus, where T cells go to develop, has an "app" for that. It's called T-cell "education."

T cells undergo a two-stage education process in the thymus (**Figure 16.19**). The thymus has a special gene activator, called AIRE, that allows thymus epithelial cells to synthesize all human proteins in small amounts. Those self antigens are displayed on

Figure 16.19   T-Cell "Education"

The process of T-cell education in the thymus. Lymphoid precursor cells enter the thymus to become T cells. The thymus can express most self antigens and present them on antigen-presenting cells (APCs). As part of its education process, an immature T cell whose T-cell receptors (TCRs) strongly recognize a self antigen is either killed through apoptosis or converted to a regulatory T cell that can inhibit the function of a self-reactive lymphocyte that may have escaped the thymus alive. A T cell whose TCRs recognize self with moderate to weak affinity is allowed to live (in other words, it is positively selected) and enter the bloodstream, where it can travel to secondary lymphoid organs.

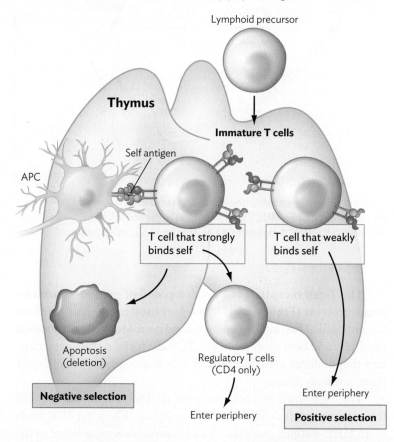

thymus cell MHC molecules (thymus epithelial cells have MHC I and II molecules). The first education process is to <u>delete</u> (kill) T cells with dysfunctional TCRs that bind too weakly, or not at all, to the MHC-antigen complexes. T cells bearing functional T-cell receptors are allowed to live. This process is called **positive selection**. These T cells are allowed to live because T-cell receptors have to bind MHC proteins in order to recognize any MHC-antigen complex.

The second phase of T-cell education deletes T cells that recognize self MHC–self peptide complexes <u>too strongly</u>. This process is called **negative selection**. Negative selection is important because T cells that bind self MHC complexes <u>too</u> tightly would constantly react to our own MHC molecules, regardless of which antigen was attached. The surviving T cells—those whose T-cell receptors have intermediate affinity for MHC-antigen complexes—leave the thymus to seed secondary lymphoid organs like the spleen and lymph nodes. Almost 95% of T cells entering the thymus die during the negative selection processes. Note, however, that some self-reactive T cells are allowed to survive and are converted to **regulatory T cells (Tregs)**. Regulatory T cells can block the activation of harmful, self-reactive (autoimmune) lymphocytes that escape deletion and enter the circulation. Chapter 17 will discuss situations where T cell education fails and someone develops autoimmunity toward a self-antigen; for example, the autoimmune disease called Graves disease.

You might wonder how someone who has had their thymus removed (a treatment for myasthenia gravis caused by a tumor in the thymus) can live if the organ is critical for T-cell maturation and for deleting self-reactive T cells. Actually, within a few years after birth, the thymus begins to regress, and eventually, little function remains in adults. Fortunately, a large amount of T-cell education occurs during fetal development. Once a T cell is educated, reserve pools of these T cells are maintained throughout life outside of the thymus. So, adults without a thymus can live relatively normal lives. Babies born without a thymus, however, have a severe, life-threatening T-cell deficit.

Superantigens and Cytokine Storms

You now know that normal antigens must be processed and presented by antigen-presenting cells to T cells in order to activate our immune system. When an antigen links a T cell to an antigen-presenting cell, both cell types secrete certain cytokines that help move the immune process along. Because the interaction is antigen-specific, only a few T cells in our body will have the proper T-cell receptor able to recognize that antigen—in the range of 1–100 cells in a million. The proliferation of the T cells through antigen-dependent activation is what increases T-cell number and focuses the immune response to that specific antigen.

What's scary is that some proteins, called **superantigens**, do not require antigen processing and presentation to stimulate the synthesis of <u>massive</u> amounts of cytokines (called a **cytokine storm**). In fact, recognition as an antigen is not even involved. As illustrated in **Figure 16.20**, superantigens, such as staphylococcal toxic shock syndrome toxin (TSST), bind to the <u>outside</u> of T-cell receptors on T cells and to the MHC molecules on antigen-presenting cells (for example, macrophages). Because <u>antigenic specificity is not involved</u>, superantigens can activate large numbers of T cells—many more T cells than in a typical, antigen-specific immune reaction (up to 20% of T cells in the body can be activated). Once stimulated by superantigens, these T cells and APCs release massive amounts of inflammatory cytokines. The resultant cytokine storm can overwhelm the host immune system and cause severe damage to tissues and organs (see Chapter 15 IMPACT).

Throughout Chapters 15 and 16 we have mentioned many different cytokines secreted by various immune, infected, or damaged cells. These cytokines have a profound influence over the direction and intensity of the immune response. Table 16.3 lists a fraction of the many different cytokines produced by various cell types and describes their influence.

Some cytokines (called chemokines) attract innate immune system cells to the infection site. Other cytokines influence whether humoral or cell-mediated immune mechanisms dominate. Cytokines such as IL-3 can affect the innate immune system by stimulating the production of macrophages and neutrophils needed to clear an infection. The tumor necrosis family of cytokines can trigger fever and various inflammatory processes that can damage host cells (see Section 15.6).

Figure 16.20 Superantigens

Antigen presentation of normal antigens by APCs (left) requires the antigen to bind within the binding pockets of MHC and TCR molecules. Superantigens, however, do not require processing (right). They can bind directly to outer aspects of the TCR and MHC proteins, linking and activating the two cell types. Because antigen specificity is not involved, many T cells will be activated. CHO indicates carbohydrate additions to the protein.

Table 16.3
Select Cytokines That Modulate the Immune Response

Cytokine[a]	Sample Sources	General Functions
IL-1	Many cell types, including endothelial cells, fibroblasts, neuronal cells, epithelial cells, macrophages	Affects differentiation and activity of cells in inflammatory response; influences central nervous system functions; acts as endogenous pyrogen
IL-2	T cells	Stimulates T-cell and B-cell proliferation
IL-3	T cells	Stimulates production of macrophages, neutrophils, mast cells, others
IL-4	T_H2 cells, mast cells	Promotes differentiation of CD4 T cells into T_H2 helper T cells; promotes proliferation of B cells
IL-5	T_H2 cells	Chemoattracts eosinophils; activates B cells and eosinophils
IL-6	T cells, macrophages, fibroblasts, endothelial cells, hepatocytes, neuronal cells	Stimulates T-cell and B-cell growth; stimulates production of acute-phase proteins
IL-8 (CXCL8)	Monocytes, endothelial cells, T cells, keratinocytes, neutrophils	Chemoattracts PMNs; promotes migration of PMNs through endothelium
IL-10	T_H2 cells, B cells, macrophages, keratinocytes	Inhibits production of IFN-γ, IL-1, TNF-α, IL-6 by macrophages
IL-17	T_H17 cells	Increases chemokine production to recruit monocytes and neutrophils
IFN-α/β	T cells, B cells, macrophages, fibroblasts	Promotes antiviral activity
IFN-γ	T_H1 cells, cytotoxic T cells, NK cells	Activates T cells, NK cells, macrophages
TNF-α	T cells, macrophages, NK cells	Exerts wide variety of immunomodulatory effects
TNF-β	T cells, B cells	Exerts wide variety of immunomodulatory effects

[a]IL = interleukin; IFN = interferon; TNF = tumor necrosis factor.

SECTION SUMMARY

- **T_H0 precursor cells**, which have surface CD4 molecules, can make T_H1, T_H2, and T_H17 cells, which support B-cell differentiation into antibody-producing plasma cells. T_H1 cells also drive cell-mediated immunity by CD8 T cells.

- **T cells recognize antigen** presented only on self MHC–containing APCs.

- **The major histocompatibility complex (MHC)** consists of membrane proteins with variable regions that can bind antigens. Class I MHC molecules are on all nucleated cells, whereas professional antigen-presenting cells contain both class I and class II MHC molecules.

- **Professional antigen-presenting cells (APCs)**, such as dendritic cells, present antigens synthesized during an intracellular infection on their surface class I MHC molecules, but they place antigens from engulfed microbes or allergens on their class II MHC molecules. T cells recognize antigens bound to MHC.

- **T-cell receptors (TCRs)** are antigen-specific T-cell surface proteins that recognize MHC-antigen complexes on APCs.

- **CD4 on helper T cells** helps TCRs bind to MHC II–antigen complexes. CD8 on cytotoxic T cells helps TCRs bind to MHC I–antigen complexes.

- **T-cell receptors must be able to weakly bind self MHC–antigen complexes.** T cells whose TCRs bind too strongly to self MHC–antigen complexes are killed to prevent reacting against self.

- **Superantigens** stimulate T cells by nonspecifically linking the T-cell receptor on a T cell with MHC II on an APC. Antigen processing and presentation is not required. Both cell types release massive amounts of cytokines that cause immune system chaos that severely damages tissues.

Thought Question 16.6   How do viruses stimulate antibody production if they are intracellular pathogens and their intracellularly synthesized and processed epitopes are presented only by MHC I surface receptors? Wouldn't presentation by MHC I trigger only T_C activation and cell-mediated immunity? (See Figure 16.16.)

16.4 Completing the Immune Response

SECTION OBJECTIVES

- Explain how a TCR participates in B-cell activation.
- Discuss CD40 and CD40 ligand interaction.
- Explain why memory B cells do not always need T-cell help for activation.
- Explain how cytotoxic T cells become activated.
- Discuss how activated T_C lymphocytes (CTLs) recognize their target host cells.

How do all of these processes work together to resolve an infection? Up to this point, we have provided an overview of the immune system and have examined some required interactions between T cells and antigen-presenting cells. **Figure 16.21** reviews

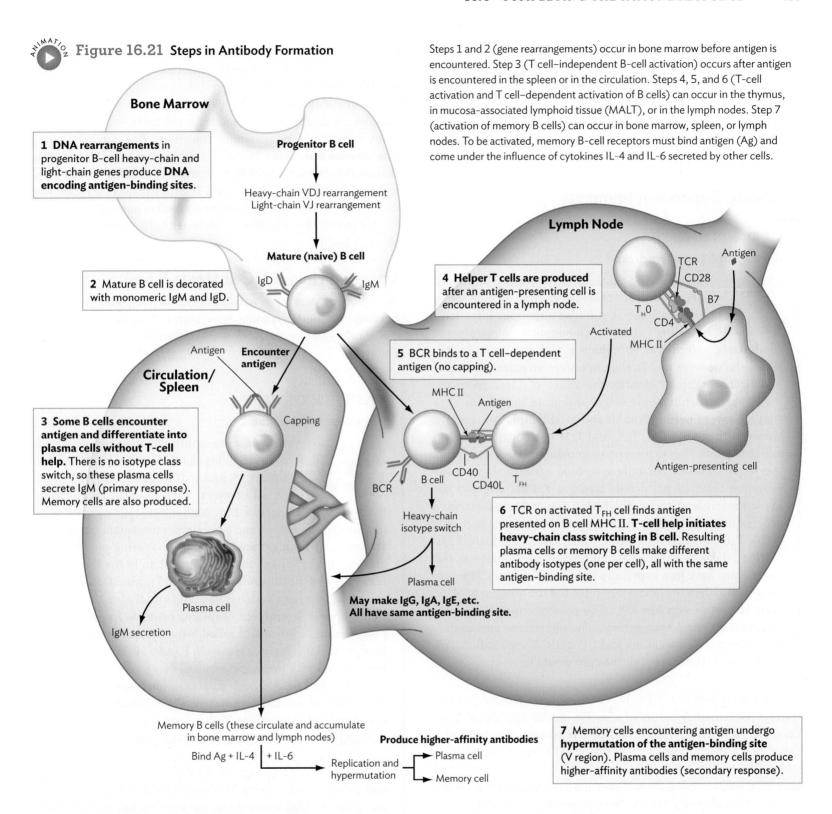

Figure 16.21 Steps in Antibody Formation

Steps 1 and 2 (gene rearrangements) occur in bone marrow before antigen is encountered. Step 3 (T cell–independent B-cell activation) occurs after antigen is encountered in the spleen or in the circulation. Steps 4, 5, and 6 (T-cell activation and T cell–dependent activation of B cells) can occur in the thymus, in mucosa-associated lymphoid tissue (MALT), or in the lymph nodes. Step 7 (activation of memory B cells) can occur in bone marrow, spleen, or lymph nodes. To be activated, memory B-cell receptors must bind antigen (Ag) and come under the influence of cytokines IL-4 and IL-6 secreted by other cells.

Bone Marrow

1 DNA rearrangements in progenitor B-cell heavy-chain and light-chain genes produce **DNA encoding antigen-binding sites**.

Progenitor B cell

Heavy-chain VDJ rearrangement
Light-chain VJ rearrangement

Mature (naive) B cell

2 Mature B cell is decorated with monomeric IgM and IgD.

IgD   IgM

Lymph Node

TCR
CD28
B7
T_H0
CD4
MHC II
Antigen

4 Helper T cells are produced after an antigen-presenting cell is encountered in a lymph node.

5 BCR binds to a T cell–dependent antigen (no capping).

Activated

MHC II   Antigen

Antigen   **Encounter antigen**

Circulation/ Spleen   Capping

3 Some B cells encounter antigen and differentiate into plasma cells without T-cell help. There is no isotype class switch, so these plasma cells secrete IgM (primary response). Memory cells are also produced.

BCR   B cell   CD40   T_FH
CD40L

Heavy-chain isotype switch

Antigen-presenting cell

6 TCR on activated T_FH cell finds antigen presented on B cell MHC II. **T-cell help initiates heavy-chain class switching in B cell.** Resulting plasma cells or memory B cells make different antibody isotypes (one per cell), all with the same antigen-binding site.

Plasma cell

Plasma cell

Plasma cell

IgM secretion

May make IgG, IgA, IgE, etc. All have same antigen-binding site.

Memory B cells (these circulate and accumulate in bone marrow and lymph nodes)

Produce higher-affinity antibodies

Bind Ag + IL-4   + IL-6

Replication and hypermutation → Plasma cell
→ Memory cell

7 Memory cells encountering antigen undergo **hypermutation of the antigen-binding site** (V region). Plasma cells and memory cells produce higher-affinity antibodies (secondary response).

the steps of antibody production starting from B-cell development in the bone marrow (steps 1–2); T cell–independent antibody production in the circulation or spleen (step 3); T-cell activation in the lymph node (step 4); and finally, T cell–dependent activation of B cells (steps 5 and 6). We will refer to Figure 16.21 often during this discussion.

In the previous section we explained how T cells recognize antigens and how T_H0 cells become activated. Realize that as

T_H0 cells become activated, the mix of cytokines produced during an infection will influence whether a T_H0 cell converts to a T_{FH}, T_H1, T_H2, or T_H17 cell. The ratios of different subsets can change the direction and severity of the immune response. For instance, predominance of T_{FH} cells will stimulate B cells to become plasma cells and secrete antibody. However, the more T_H1 cells are made, the more cytotoxic T cells can be stimulated, favoring cell-mediated immunity. On the other hand, if T_H17 cells are made in significant numbers, they will promote inflammation by secreting cytokine IL-17. IL-17 stimulates nearby host cells to synthesize pro-inflammatory cytokines that, among other things, attract neutrophils to a site of infection.

Let's now complete the story of the immune response by explaining what happens after a T cell has encountered antigen. Along the way, we will use case histories to explore what can go wrong.

Antibody-Dependent Immunity

CASE HISTORY 16.1

Bare Lymphocyte Syndrome

Fatima is the third child of healthy Moroccan parents who are themselves first cousins. From birth, Fatima has suffered numerous infections, including varicella pneumonia, disseminated candidiasis, enterovirus encephalitis, and *E. coli* meningitis. When she was 3 years old, she underwent an extensive immunological workup to identify the cause of these recurring illnesses. Her white blood cell counts were in the reference range, as were her neutrophil and total lymphocyte counts. However, her total IgG, IgA, and IgM antibody levels were extremely low. This extreme immunoglobulin deficiency was the likely cause of her failure to fend off infections. However, the team of physicians treating Fatima still did not know the cause of the antibody deficiency. More tests were ordered to examine key protein surface markers on B cells (IgD), T_H0 cells (CD4), T_C cells (CD8), and APCs (MHC I and MHC II classes). Fatima's B-cell numbers were normal, but her T cells were low, especially the CD4 T-cell set. The surprise came when the laboratory reported the absence of MHC class II proteins on her APC. The presence of MHC class I proteins, however, was normal. The physicians now understood the problem—Fatima had a genetic disease called bare lymphocyte syndrome—but the cure would be risky.

How do B cells lose MHC II receptors? This case nicely illustrates a fundamental feature of adaptive immunity. Fatima's lymphocytes (B cells) contained no MHC II receptors because of a defect in a transcriptional activator needed to express the MHC II genes. The lack of MHC II on these cells is why they are called bare lymphocytes, even though they have other surface proteins. Actually, the deficiency extends to <u>all</u> professional APCs (dendritic cells, macrophages, B cells). Fatima's plight underscores the importance of MHC receptors to immunity.

A Failure to Communicate

The absence of MHC II on Fatima's APCs disrupted communications with T_H0 cells. Without MHC II, APCs could not present antigen to T_H0 cells. Without antigen presentation, T_H0 cells could not proliferate (note that Fatima had a low CD4 T-cell count) and could

not differentiate into T_{FH}, T_H1, T_H2, or T_H17 cells. Without activated helper T cells to "talk" to, B cells cannot initiate heavy-chain class switching, and most cannot even differentiate into plasma cells. This is why blood levels of <u>all</u> Fatima's antibody classes were so low. The result was that Fatima could not make enough antibody of any isotype to fend off simple infections. Her one hope was a **bone marrow transplant**. In this procedure, her bone marrow was destroyed and replaced with bone marrow containing healthy stem cells able to make MHC II proteins. The transplant in this case was successful, and her antibody levels returned to normal.

The question now is, How do activated helper T cells communicate with B cells?

Activated Helper T Cell Meets B Cell

The following case of a patient overproducing IgM provides a backdrop for understanding how helper T cells communicate with B cells—the next step in B-cell differentiation and antibody production.

CASE HISTORY 16.2

Hyper IgM Syndrome

Nathan is a 10-month-old male who has suffered much in his short life. Since birth, he's had recurrent serious extracellular bacterial infections, all of which were resolved by antibiotic treatment. On Christmas Day, Nathan's mother discovered he had another high fever and a nonproductive cough (no mucus). He was also having trouble breathing. Fearing another serious infection, Nathan's mother rushed him to the hospital. An X-ray showed the lung infiltrate seen in **Figure 16.22A**. A tube was placed through Nathan's nose and into his lung, and a solution was used to wash out deep-lung contents for microscopic examination (a procedure called bronchoalveolar lavage). The fluid yielded the organisms seen in **Figure 16.22B**. Nathan was hospitalized with pneumonia caused by *Pneumocystis jirovecii*, a yeastlike fungus. Although most of us have been infected with this organism, our immune system almost always prevents disease. Finding the disease in Nathan suggested a serious immune dysfunction. An immunological workup revealed mild neutropenia (lower-than-normal numbers of neutrophils) but normal numbers of T and B cells, normal levels of complement, and normal complement activity. However, Nathan's blood exhibited exceptionally high levels of IgM but very low IgG and IgA levels. The physician suspected a specific defect in Nathan's T cells.

Disease caused by *Pneumocystis jirovecii* is rare in healthy individuals but common in patients with defects in their T cells. For instance, the organism is a frequent cause of death for patients infected with human immunodeficiency virus (HIV) who develop acquired immunodeficiency syndrome (AIDS). HIV targets T cells and severely reduces their numbers in blood. Because Nathan had

Figure 16.22 *Pneumocystis jirovecii* **Lung Infection**

A. X-ray image of a patient with *P. jirovecii* fungal infection. The lung infiltrate has the appearance of ground glass.

B. Cysts of the organism obtained in bronchoalveolar lavage (fluid is used to wash out some deep-lung contents).

a normal T-cell blood count, HIV was not the culprit. Nathan's recurrent infections were due to a disease called hyper IgM syndrome, which is caused by another communication breakdown between immune cells.

As mentioned earlier, T cell–<u>independent</u> antigens like polysaccharides have repetitive structures that can cross-link B-cell receptors in the process called capping (see Figure 16.12 and Figure 16.21, step 3). The second activation signal for T cell–independent antigens occurs when the B cell binds to complement factor C3 deposited on the surface of the pathogen (see Section 15.7). Alternatively, the second activation signal can involve an interaction between a microbe-associated molecular pattern (MAMP) and a Toll-like receptor on the B cell (see Section 15.5). This activation does not result in heavy-chain class switching; only IgM is secreted by the resulting plasma cell.

B cells responding to a T cell–<u>dependent</u> antigen also need two signals to become activated. The first signal requires the antigen to bind to a BCR (step 5, no cross-linking involved). Next, the antigen-BCR complex is internalized by endocytosis and processed, and antigenic epitopes are presented on the B cell's MHC II proteins (Figure 16.21, also step 5). The B cell now requires a helper T cell—usually a T_{FH} cell (shown being made in step 4; details found in Figure 16.18 and Section 16.3). That T_{FH} cell will have a T-cell receptor able to bind to the antigen presented by the B cell–MHC II complex (step 6). This cell-cell contact enables the second signal (also step 6): an interaction between a B-cell surface protein called CD40 and a T-cell protein called CD40 ligand (CD40L, or CD154). Once activated by the helper T cell, a B cell can undergo heavy-chain class switching to make IgG, IgA, or other antibodies and differentiate into plasma cells.

This second signal was the problem with Nathan in Case History 16.2. Nathan had a mutation in the gene encoding CD40L, so his T_H cells lacked CD40 ligand. His T-cell and B-cell numbers remained normal, but his B cells would not undergo heavy-chain class switching. Therefore, any plasma cells made could make <u>only</u> IgM. The result in such cases is hyper IgM syndrome. Note that most of the IgM produced is <u>not</u> antigen-specific because it does not result from clonal selection in response to antigen. The nonspecific IgM explains why Nathan suffered recurrent infections.

T Cells and the Secondary Antibody Response

During the primary response, B cells generally need T-cell help to become plasma cells. But what about the secondary response? Do memory B cells also need T-cell help? During the secondary response, memory B cells still need help to become plasma cells but do not need direct contact with helper T cells. Memory B cells with antigen bound to their B-cell receptors can respond to the soluble IL-4 and IL-6 cytokines secreted by activated helper T cells without having direct contact with the T_H cell. IL-4 stimulates B-cell proliferation, while IL-6 directs differentiation into antibody-secreting plasma cells (see Figure 16.21, step 7).

Cell-Mediated Immunity

As introduced in Section 16.1 and Figure 16.2, the other branch of the adaptive immune system, cell-mediated immunity, relies on cytotoxic T cells (T_C cells) and activated cytotoxic T lymphocytes (CTLs) to directly attack intracellularly infected host cells, a place that antibodies cannot reach. Recall, too, that both cell types have the surface protein CD8, which enables TCRs on inactive T_C cells to engage MHC I–antigen complexes on antigen-presenting cells (Section 16.3 and Figure 16.18). In this section we explain how T_C cells are activated to become killer cytotoxic T lymphocytes and how CTLs kill infected host cells. Cell-mediated immunity is crucial for clearing infections caused by intracellular pathogens, such as viruses. However, engaging this branch of immunity can come at a price, as the following case history demonstrates.

CASE HISTORY 16.3
The Run-Down Mechanic

Luke, a 24-year-old mechanic from Egypt Beach, Massachusetts, was feeling run-down. Over the past month, he hadn't been eating much and was always tired, and he had recently noticed pain just under his right rib cage. Also recently, his urine had turned the color of tea, and his wife noticed that the whites of his eyes were taking on a yellowish tinge. Luke finally went to the doctor, who suspected Luke had cirrhosis, a liver disease. This preliminary diagnosis was based on Luke's symptoms. His abdominal pain was located over the liver, and the color of his urine and the jaundice evident in his eyes all pointed toward liver disease. Luke did not drink alcohol and had never left the country, but when questioned, he admitted to taking intravenous drugs. The doctor suspected hepatitis and ordered a series of blood tests to check for the presence of liver enzymes, anti-hepatitis antibodies, and hepatitis viral DNA or RNA in Luke's blood. The tests revealed a high level of hepatitis B viral DNA. Luke had hepatitis B disease caused by hepatitis B virus (HBV).

There are several types of hepatitis viruses (types A–G). Hepatitis B virus is an enveloped DNA virus (discussed in Chapters 12 and 22) that can be transmitted by sexual intercourse, blood transfusion, or the use of contaminated needles. It seems likely that the virus was transmitted to Luke by a contaminated needle, given his intravenous drug use. The main target of hepatitis B virus is the liver, where the virus replicates in hepatocytes (liver cells). However, the damage to the liver is not caused by the virus per se. It is caused by activated cytotoxic T lymphocytes called in to fight the infection. This section will describe how cytotoxic T cells are activated and how they target infected cells.

CYTOTOXIC T LYMPHOCYTES As with other lymphocytes, cytotoxic T cells (T_C cells) circulate in the blood and lymph. Along the way, T_C cells pause in lymph nodes to survey APCs for recognizable antigens. To become activated, T_C cells require <u>three</u> signals. First, a T-cell receptor on a T_C cell must bind to a matched antigen presented on an APC's MHC I complex (**Figure 16.23**, step 1). Remember that the CD8 protein on T_C cells recognizes MHC I surface proteins but not MHC II. The second signal for activation is the interaction between CD28 on the T cell and the B7 molecule on the APC, the same as for helper T cells.

T_C cells also need a little help (signal 3) in the form of a cytokine called IL-2, produced by the T_H1 class of helper T cells (Figure 16.23, step 2). T_H1 helper cells are formed in a manner similar to T_{FH} in Figure 16.21, except the cytokine cocktail present during differentiation is different. Upon binding IL-2, the T_C cell begins to proliferate (clonal expansion), gains cytotoxic activity, and becomes an activated cytotoxic T lymphocyte (CTL). The CTL leaves the lymph node, travels to the site of infection, and kills any cell bearing the same peptide–MHC class I complex that triggered T_C differentiation (Figure 16.23, step 3B). In Case History 16.3, those would be liver cells infected with hepatitis B.

Cytotoxic T lymphocytes kill infected target cells by releasing the contents of their granules: granzyme and perforin. Perforin becomes a pore in the target cell membrane. Granzymes will then enter the target cell through the perforin pore. Once in the cytoplasm of the infected cell, granzymes cleave and activate certain host proteins that trigger apoptosis and cell death (**Figure 16.24**).

Why do CTLs more effectively clear viral infections than B cells and antibodies do? A major reason is that intracellular pathogens (for example, viruses) hide inside host cells, where they are protected from antibody. Consequently, these pathogens are best killed when the harboring host cell is also sacrificed by CTLs. Because all nucleated cells have class I MHC molecules, all of Luke's HBV-infected

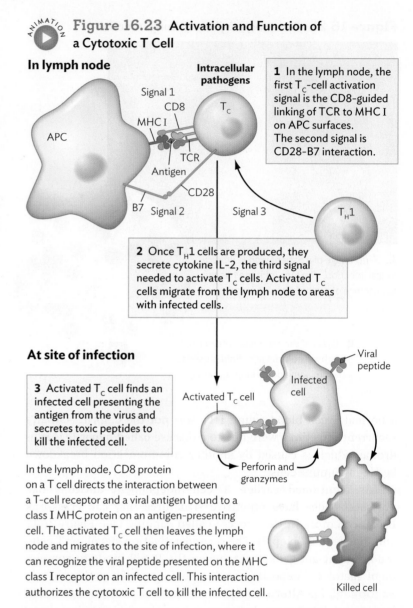

Figure 16.23 **Activation and Function of a Cytotoxic T Cell**

In lymph node

1 In the lymph node, the first T_C-cell activation signal is the CD8-guided linking of TCR to MHC I on APC surfaces. The second signal is CD28-B7 interaction.

2 Once T_H1 cells are produced, they secrete cytokine IL-2, the third signal needed to activate T_C cells. Activated T_C cells migrate from the lymph node to areas with infected cells.

At site of infection

3 Activated T_C cell finds an infected cell presenting the antigen from the virus and secretes toxic peptides to kill the infected cell.

In the lymph node, CD8 protein on a T cell directs the interaction between a T-cell receptor and a viral antigen bound to a class I MHC protein on an antigen-presenting cell. The activated T_C cell then leaves the lymph node and migrates to the site of infection, where it can recognize the viral peptide presented on the MHC class I receptor on an infected cell. This interaction authorizes the cytotoxic T cell to kill the infected cell.

cells will have HBV antigens presented on MHC I complexes. The MHC I–HBV antigen complexes will be recognized by the TCR of an appropriate CTL. The CTL will then kill the infected cell and stop the pathogen from spreading to other host cells.

While this self-sacrifice is important for survival, cellular immunity is a double-edged sword. During Luke's hepatitis B virus infection (see Case History 16.3), activated cytotoxic T lymphocytes caused nearly all of the liver injury by killing the infected liver cells. Nevertheless, by killing infected cells, cytotoxic T cells also eliminate the virus. However, HBV is almost never completely eliminated, even with interferon-alpha therapy (which was done in Luke's case to activate intracellular antiviral mechanisms). Still, the therapy did reduce Luke's viral load, and his liver did recover some degree of health.

So, there it is: the story of our immune system. As illustrated by the tragic tale of David Phillip Vetter in the chapter's opening case history, we cannot survive without it.

Figure 16.24 Death by Cytotoxic T Lymphocyte

An activated cytotoxic T lymphocyte recognizes the target cell through TCR antigen–MHC I interaction. The CTL introduces a perforin pore into the target cell membrane, which granzymes use as a portal. Granzymes start a cascade of events leading to mitochondrial damage that triggers apoptosis and DNA fragmentation.

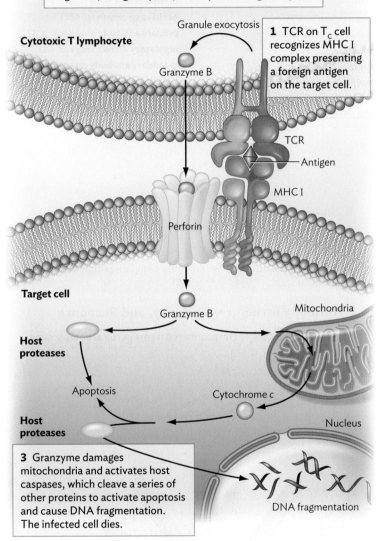

2 T$_C$ cell secretes perforin, which makes a pore in the target cell, and granzymes, which pass through the pore.

Granule exocytosis

Cytotoxic T lymphocyte

Granzyme B

1 TCR on T$_C$ cell recognizes MHC I complex presenting a foreign antigen on the target cell.

TCR

Antigen

MHC I

Perforin

Target cell

Granzyme B

Mitochondria

Host proteases

Apoptosis

Cytochrome c

Host proteases

Nucleus

3 Granzyme damages mitochondria and activates host caspases, which cleave a series of other proteins to activate apoptosis and cause DNA fragmentation. The infected cell dies.

DNA fragmentation

SECTION SUMMARY

- **Activation of a T$_H$0 cell** requires two signals: binding of T-cell receptor CD4 to an MHC II–antigen complex on an antigen-presenting cell and CD28-B7 interaction.

- **T$_H$0 cell differentiation** to T$_{FH}$, T$_H$1, T$_H$2, or T$_H$17 cells is influenced by different cytokine "cocktails" secreted by macrophages and natural killer cells during an infection.

- **Activation of a B cell** into an antibody-producing plasma cell usually requires two signals. The first is antigen binding to a B-cell receptor, followed

by endocytosis, presentation on MHC II, and binding of the MHC-antigen complex to a TCR of a T$_H$ cell activated by the same antigen. As part of this second signal, CD40 ligand on the T cell binds to CD40 on the B cell. Both interactions are required.

- **Activation of memory cells** requires antigen binding to a BCR and cytokines made by T$_H$ cells. Direct B cell–T cell interaction is not involved.

- **Activation of cytotoxic T cells** into activated cytotoxic T lymphocytes requires three signals. Signal 1 involves T-cell receptor–CD8 molecules recognizing MHC I–antigen complexes on APCs. Signal 2 is a CD28-B7 interaction. Signal 3 is stimulation by cytokine IL-2 made by activated T$_H$1 helper cells. Once activated, cytotoxic T lymphocytes (CTLs) destroy infected host cells.

Thought Question 16.7   IL-10 is considered an anti-inflammatory cytokine because it down-regulates production of MHC II receptors. How would down-regulating MHC II on macrophages affect the immune response?

Thought Question 16.8   Why does attaching a hapten to a carrier protein enable production of antihapten antibodies?

Thought Question 16.9   Transplant rejection is a major concern in transplanting most tissues because host T$_C$ cells can recognize allotypic MHC on donor cells. So, why are corneas easily transplanted from a donor to just about any other person? (*Hint:* Do corneas have blood vessels?)

Thought Question 16.10   Cytotoxic T cells (T$_C$ cells) lyse the membranes of host cells carrying viruses or bacteria. Why does this lysis help clear the infection rather than facilitate the spread of those organisms?

Thought Question 16.11   Why does the immune system work so well against infection by some pathogens (for example, poliovirus) but not others (for example, *Staphylococcus aureus*)? Immunity raised by the poliovirus vaccine will prevent subsequent infection by poliovirus, but someone with an *S. aureus* infection (and the attendant immune response) can be reinfected many times by *S. aureus*.

Perspective

Immunity is a remarkable feat of nature. The realization that any one person's immune system can recognize and respond to virtually any molecular structure and yet remain selectively "blind" to the body's own antigens is hard to comprehend. But even more remarkable, some pathogenic microbes have found ways to outmaneuver the interconnected redundancies and safeguards of the immune system. In this chapter we outlined how a normal immune system functions. In the process we provided a few examples of immune dysfunction, but there is more to discuss. The following chapter will tell more of what happens when the immune system fails (immunodeficiencies) or attacks itself to produce autoimmune diseases. In addition, Chapter 17 will explain some of the technological applications of immunology.

Key Terms

affinity maturation (p. 482)
allotype (p. 476)
antibody (p. 472)
antigen (p. 471)
antigenic determinant (p. 471)
antigenicity (p. 473)
antigenic specificity (p. 473)
antigen-presenting cell (APC) (p. 471)
attenuated (p. 474)
B-cell receptor (BCR) (p. 479)
bone marrow transplant (p. 490)
capping (p. 479)
cell-mediated immunity (p. 472)

class I MHC molecule (p. 484)
class II MHC molecule (p. 484)
class switching (p. 478)
clonal selection (p. 479)
constant region (p. 475)
cytokine storm (p. 487)
cytotoxic T cell (T$_C$ cell) (p. 472)
epitope (p. 471)
F(ab)$_2$ region (p. 476)
Fc region (p. 476)
hapten (p. 473)
heavy chain (p. 475)
helper T cell (T$_H$ cell) (pp. 472, 481)
humoral immunity (p. 472)

idiotype (p. 476)
IgA (p. 477)
IgD (p. 477)
IgE (p. 477)
IgG (p. 476)
IgM (p. 477)
immunogen (p. 471)
immunogenicity (p. 473)
immunoglobulin (p. 475)
immunological specificity (p. 473)
isotype (pp. 475, 476)
isotype switching (p. 478)
light chain (p. 475)

major histocompatibility complex (MHC) (p. 483)
memory B cell (p. 481)
negative selection (p. 487)
opsonize (p. 476)
plasma cell (p. 472)
positive selection (p. 487)
primary antibody response (p. 477)
regulatory T cell (Treg) (p. 487)
secondary antibody response (p. 478)
superantigen (p. 487)
T-cell receptor (TCR) (p. 485)
vaccination (p. 473)
variable region (p. 476)

Objectives Assessment Questions

16.1 Basics of Adaptive Immunity

1. A small segment of any antigen that can elicit an immune response is called a(n) _____.
   a. Fc factor
   b. epitope
   c. hapten
   d. cytokine
   e. antibody

2. Which of the following secrete antibodies during an immune response?
   a. Helper T cells
   b. Macrophages
   c. B cells
   d. Plasma cells
   e. Langerhans cells

3. Which cell type is most directly involved with coordinating humoral and cellular immune responses to an infection?
   a. Neutrophil
   b. T cell
   c. Macrophage
   d. B cell
   e. Plasma cell

4. All "professional" antigen-presenting cells _____.
   a. make antibody
   b. secrete perforin
   c. inject processed antigen into B cells
   d. present antigen on MHC surface receptors
   e. make complement

5. Which of the following are the most effective immunogens?
   a. Lipids
   b. Polysaccharides

   c. Nucleic acids
   d. Small chemicals such as penicillin
   e. Proteins

6. The ability of an antibody to bind only a specific, single structure is an example of _____.

16.2 Antibody Structure, Diversity, and Response

7. Match the following to the numbered items in the accompanying structure.
   a. Heavy chain
   b. Light chain
   c. Antigen-binding site
   d. Region that attaches to cell membranes
   e. Region that binds complement factor C1

8. Idiotypic sequences are found in which numbered part of the structure shown in the previous question?
   a. Part 1
   b. Part 2
   c. Part 3
   d. Part 4
   e. Part 5

9. Two antibody class molecules, IgG and IgM, that recognize the same epitope and are from the same person differ from each other in their _____.
   a. light-chain constant regions
   b. heavy-chain constant regions
   c. antigen-binding sites
   d. variable regions

10. Which class of antibody has a form secreted across mucosal tissues of the intestine or lung?
    a. IgG
    b. IgM
    c. IgD
    d. IgA
    e. IgE

11. All immunoglobulin family proteins are antibodies. True or false?

12. Which of the following antibody isotypes is predominantly made following a secondary exposure to an antigen (a pathogen, for instance)?
    a. IgA
    b. IgD
    c. IgE
    d. IgG
    e. IgM

13. B-cell receptors have which of the following as their primary component?
    a. IgM or IgD
    b. Major histocompatibility proteins
    c. IgM only
    d. IgG or IgD
    e. IgE only

14. Clonal selection occurs when _____.
    a. B cells proliferate to produce new B cells with different B-cell receptor surface proteins
    b. B cells proliferate to produce new B cells with identical B-cell receptor surface proteins
    c. isotype switching occurs
    d. capping of B-cell receptor molecules takes place
    e. antigenic diversity is generated

15. Memory B cells undergo which of the following processes?
    a. Class switching of antibody isotype
    b. Affinity maturation of the antigen-binding site
    c. Antibody secretion
    d. Synthesis of IgG B-cell receptor surface proteins
    e. Switch from making kappa light chains to lambda light chains

16.3 T Cells: The Link between Humoral and Cell-Mediated Immunity

16. The following cells participate most directly in adaptive cell-mediated immunity.
    a. Dendritic cells
    b. Natural killer cells
    c. Activated CD4 T_H2 cells
    d. Neutrophils
    e. Activated CD8 T cells

17. A patient whose antigen-presenting cells lack the protein called the transporter of antigen peptides will be defective in the _____.
    a. stimulation of cytotoxic T cells
    b. generation of T_H1 cells
    c. generation of T_H2 cells
    d. stimulation of B cells
    e. deletion of self-reacting T cells

18. To function properly, all T cells must _____.
    a. bind strongly to some self antigen
    b. not bind to any self antigen
    c. express the AIRE protein
    d. bind weakly to some self antigen
    e. bind CD8 on antigen-presenting cells

19. MHC II complexes on antigen-presenting cells are loaded with _____ produced peptides.
    a. extracellularly
    b. intracellularly
    c. intracellularly or extracellularly

20. In addition to T-cell receptor complexes, T cells must express surface marker _____ in order to selectively recognize an MHC class II–antigen complex on antigen-presenting cells.

21. Which cell type is needed to activate CD8 T cells?
    a. T_H2 cell
    b. Dendritic cell
    c. Plasma cell
    d. Epithelial cell
    e. Natural killer cell

22. Cytotoxic T cells target infected cells that present antigen on ____.
    a. MHC I receptors
    b. MHC II receptors
    c. MHC III receptors
    d. Fas ligand
    e. CD8

23. To activate T cells, superantigens must ____.
    a. bind to the T-cell receptors and B-cell receptors in their antigen-binding sites
    b. bind to the antigen-binding region of MHC and outside of the antigen-binding region of the T-cell receptors
    c. bind to T-cell receptors only
    d. bind to MHC only
    e. bind to T-cell receptors and MHC outside of their antigen-binding regions

16.4 Completing the Immune Response

24. How does a lack of CD40 ligand on T cells affect a person's immune response?
    a. Inability to make activated cytotoxic T cells
    b. Increased synthesis of IgG antibody
    c. Inability to synthesize antibody
    d. Inability to undergo isotype switching
    e. Hyperproliferation of activated B cells

Match the immune response events with the location at which they occur.

Event	Location
25. Gene rearrangement to form the antigen-binding site (VDJ)	a. Bone marrow b. Thymus
26. T-cell education	c. Bloodstream d. Lymph node
27. Isotype switching	
28. B-cell activation by APCs	
29. T-cell help for B cells	

30. A lung infection with *Pneumocystis jirovecii* is an indication of which condition?
    a. Deficiency in B-cell numbers
    b. Lack of functional antigen-presenting cells
    c. Excess CD40L on T cells
    d. B cells lacking B-cell receptors
    e. Deficiency in T-cell numbers or function

31. Hepatitis B virus infection produces liver damage because ____.
    a. activated CD4+ T lymphocytes attack infected cells
    b. infected cells coated with anti-HBV antibodies are subject to antibody-mediated complement fixation
    c. replication of HBV triggers liver cell apoptosis
    d. activated CD8+ T lymphocytes attack infected cells

Review Questions

1. Define antigen, epitope, hapten, and antigenic determinant.

2. What is the basic difference between humoral immunity and cellular immunity?

3. Why are proteins better immunogens than nucleic acids?

4. What makes IgA antibody different from IgG?

5. Explain isotypic, allotypic, and idiotypic differences in antibodies.

6. What is antigen processing?

7. Discuss differences in the primary and secondary antibody responses.

8. Outline the basic steps that turn a B cell into a plasma cell.

9. What is isotype switching?

10. What signals are needed to activate helper T cells?

11. What signals activate cytotoxic T cells?

12. What influences T_H0 conversion to T_H1 or T_H2 cells?

13. How do superantigens activate T cells?

Clinical Correlation Questions

1. A 5-year-old patient presents to his pediatrician with another of what has been a series of infections. Flow cytometry of the boy's blood reveals that the patient is deficient in CD4 helper T cells. His CD8 T cells and B cells are at normal levels. What do you expect to see if you measure his serum IgM, IgG, IgA, and IgE levels? How might the results explain the repeated infections?

2. A patient arrives from Africa with cycling fevers and chills. She is diagnosed with malaria. The malarial parasite grows inside red blood cells (RBCs). Explain whether or not cytotoxic T cells will kill RBCs infected with the malarial parasite.

3. Tumor cells are cells that have lost the ability to control their replication. These cells also produce a number of abnormal cytoplasmic proteins that are not recognized as self by the immune system. So, most often they are killed by the immune system. But sometimes they are not killed and a tumor forms. Hypothesize about how tumor cells can escape CD8+ T-cell surveillance to form tumors.

4. In Case 16.1, Fatima had bare lymphocyte syndrome, in which her antigen-presenting cells lacked MHC II, leading to antibody deficiencies. How do you think her cell-mediated response fared?

Vaccines, Diagnostics, and Immune Disorders

[Kindergarten Contagion]

SCENARIO Maureen, a 56-year-old woman undergoing cancer treatment, awoke in her home 5 days ago with a sore throat, fever, and runny nose. Two days later, she developed a violent cough and had trouble breathing. Her alarmed husband rushed her to the hospital.

Viral pneumontitis
Chest X-ray of pneumonia showing infiltrate mostly in the left lung (arrows). Dark areas of the film indicate air space and white areas indicate bone or an infiltration of fluid and pus.

Typical measles rash

PATIENT HISTORY Maureen is a beloved kindergarten teacher in Maine who has been receiving rituximab chemotherapy for the past 6 months for a B-cell lymphoma. Rituximab is a monoclonal antibody that binds to a protein on B cells. Natural killer cells recognize the antibody and destroy the cancerous B cells, which are more susceptible than normal B cells. Until this episode she was handling her treatment well. Maureen visits her physician regularly and has received all childhood and adult vaccinations.

SURPRISING LAB RESULTS A blood count revealed that Maureen was neutropenic (low white cell count), and a chest X-ray indicated that she had pneumonitis (a lung infiltrate). She was admitted to the hospital and empirically treated with cefepime (a 4th-generation cephalosporin antibiotic). However, tests for bacterial infection were negative. The next day she developed a maculopapular rash on her face, raising the suspicion of measles. Real-time PCR (see Section 8.4) from a cheek swab was positive for measles virus, but surprisingly she had no IgG or IgM anti-measles antibodies. Maureen had measles pneumonia.

TEM 100 nm

Measles virions
A host-derived membrane envelope (red) surrounds each virus nucleocapsid with its single-strand, negative-sense RNA (green).

OUTCOME Maureen slowly recovered after being treated with an antiviral drug (ribavirin) and IV immunoglobulins. Physicians later learned that one of her 4-year-old students had been in the incubation phase of measles before Maureen's illness. That student, along with several others, had not been vaccinated before attending school.

CHAPTER OBJECTIVES

After reading this chapter, you will be able to:

- Explain the key concepts of vaccination.
- Describe common immunological tools and their use.
- Explain the types of hypersensitivities, their treatments, and prevention.
- Describe the various malignancies associated with the immune system.
- Categorize immunodeficiencies of the innate and adaptive immune systems.
- Discuss the positive and negative aspects of autoimmunity.

Maureen was lucky she recovered. The chemotherapy left her immunocompromised with low B-cell counts and, therefore, highly susceptible to the childhood diseases she had been vaccinated against. Her low B-cell count also explains why she didn't develop IgG or IgM antibodies to the measles virus. Maureen's plight highlights why vaccination is so important in a population. So, how did she contract the disease?

Measles is highly contagious. Estimates are that 90% of unvaccinated people who encounter someone with measles will contract the disease. You can add immunocompromised patients like Maureen to that group. Because her 4-year-old student was in the incubation stage of measles, he appeared fine but was shedding the virus for several days <u>before</u> his symptoms appeared. Unwittingly, the 4-year-old infected his beloved, but immunocompromised, schoolteacher.

Had all the students in the teacher's class been vaccinated with the measles, mumps, and rubella vaccine, Maureen might have been spared. As mentioned in Chapter 16 IMPACT, many parents in the United States are choosing not to vaccinate their children because of unsubstantiated fears that vaccination can lead to autism. Dozens of studies have shown that there is no difference in autism rates between vaccinated and unvaccinated children. But despite this proof, some parents are still reluctant to vaccinate their children. Consequently, the number of measles cases has skyrocketed in recent years; more on that later.

We begin this chapter by discussing how vaccines work and explaining the tools used to examine a patient's immune status. We then examine immune system–related diseases such as hypersensitivities and immunodeficiencies, as well as cancers and autoimmune diseases. Remember, anything that perturbs a patient's immune system will increase susceptibility to infectious disease.

17.1 Vaccines and Immunization

SECTION OBJECTIVES

- Distinguish active from passive immunization.
- List key features of an effective vaccine.
- Explain the different types of vaccines.
- Discuss the concept of herd immunity.

What are vaccines, and how do they work?   The history of vaccination was chronicled in Chapter 1. Its origin can be traced back more than 1,000 years to China and India (Section 1.3). The early attempts at vaccination were crude and fraught with peril. Healthy people were intentionally infected with material from a diseased person—for example, material from a smallpox pustule containing pathogenic variola virus—in an attempt to protect them from later disease. The process was called "variolation." Unfortunately, about 1 of every 100 deliberately infected patients died from smallpox. Things improved 700 years later, when the less virulent but closely related vaccinia virus (cowpox) was used. The process was thereafter called vaccination or, now, also immunization. The material used to vaccinate someone is called a **vaccine**.

⊙ **Note:** There is a subtle but important difference between vaccination and immunization. Vaccination is the process of introducing a vaccine into the body, in the hope of provoking an immune response. For a variety of reasons, vaccination may occasionally fail to provide immunity. If immunity does develop, you can say the person has been immunized, and the process is rightly called immunization.

The type of immunization just described is called **active immunization**. An antigen such as a weakened or killed pathogen injected into a person stimulates the immune system to actively make antibody and antigen-specific T cells. The resulting cellular and humoral responses then retain a memory of the antigen. Another type of immunization, termed **passive immunization**, is really a form of immunotherapy. Examples of passive immunization include the injection of immunoglobulin or hyperimmune globulin to treat the infections of immunodeficient patients. Maureen in the opening case history was treated with **intravenous immunoglobulin (IVIG)** prepared from 1,000 or more donors, some of whom had anti-measles antibodies.

Antitoxins (antibodies that inactivate microbial exotoxins) are another form of passive immunization. Microbial exotoxins are proteins that pathogenic bacteria secrete to damage eukaryotic host cells. Antitoxin sera are usually obtained from animals, such as horses. Preparations of horse sera effective against diphtheria, tetanus, and botulism toxins are available. However, injecting horse sera into humans can cause serious allergic reactions, such as serum sickness or anaphylaxis (Section 17.3). Another form of passive immunization is breast-feeding because a mother's milk contains IgA antibodies that can protect her child from ingested pathogens.

What Makes an Effective Vaccine?

The best vaccine in terms of long-term protection is a <u>natural</u> infection—if you survive, that is. In a natural infection, the microbe engages all aspects of the immune system. In contrast to natural infection, an effective <u>commercial</u> vaccine should not harm the person being vaccinated. It should stimulate B-cell (antibody) and T-cell (cell-mediated) responses and produce long-term memory (that is, it should have a lasting effect). Vaccination should not require many boosters to achieve protection, and it must, of course, protect against the natural pathogen. The reason for booster doses is that secondary exposure to an antigen provides a more robust and long-lasting immunity. The most useful vaccines should be inexpensive and easy to administer, and have a long shelf life. These last points are especially true for vaccines used in developing countries, where storage and access to the population are serious challenges.

What form should a vaccine take? The answer depends on the pathogen and which of its antigens generate a protective immune response. Even though every antigen of a pathogen can produce an immune response, most do not produce a <u>protective</u> immune response that will prevent disease. Vaccines come in four basic forms:

1. Killed whole cells or inactivated viruses
2. Live, attenuated bacteria or viruses
3. Purified components (antigenic subunits) of an infectious agent (acellular or toxoid)
4. DNA or RNA vaccines (not yet commercially available)

Table 17.1 lists many vaccines that are currently used. It notes what form they come in, how they are delivered (by injection, orally, or by other means), and when they are given. **Table 17.2** shows the schedule recommended by the Centers for Disease Control and Prevention for administering vaccinations, including boosters, to children.

Killed whole-cell or inactivated-virus vaccines (cholera and rabies, for example) use microbes inactivated by formalin or phenol. For these vaccines to work, the organism must be dead or inactivated, but its ability to stimulate an immune response (its antigenicity) to the protective antigens must remain intact.

As effective as killed vaccines are, vaccines made from living microbes, called **attenuated** ("crippled") vaccines, are usually better. Live, attenuated versions of a pathogen can arise when the organism is repeatedly grown on artificial growth media for long periods. The attenuated strains arise because mutations develop in pathogenicity genes required for growth in the host only (the "use it or lose it" principle). Attenuated strains make better vaccines than killed microbes because the act of replicating in a host, even a little, leads to a more robust immune response. The response is better because replication generates increased antigen load and stimulates immune responses targeted to the natural pathogen (for example, pathogens that infect mucosal membranes stimulate the synthesis

Table 17.1
Approved Vaccines

Vaccine Type/Disease	Route of Delivery[a]	Routine or Targeted Use	Comments on Use[b]
Whole-Cell, Killed Bacteria			
Cholera	SubQ	Travel to endemic areas	Short-term protection
Plague	SubQ		No longer available
Live, Attenuated Bacteria			
Tuberculosis (BCG)[c]	Intradermal	Only high-risk groups in US; routine in many other countries	
Typhoid fever	Oral	Travel to endemic areas	Low effectiveness
Acellular Vaccines, Bacteria			
Anthrax (protective antigen, PA)	IM	Only high-risk groups	
Meningococcal meningitis (capsule)	SubQ	Routine in children	
Haemophilus influenzae type b meningitis (capsule)	IM	Routine in children	
Pneumococcal pneumonia (capsule)	IM or SubQ	Routine in young and elderly	
Pertussis (aP; inactive pertussis toxin, filamentous hemagglutinin, pertactin)	IM	Routine in children	Given in a mixture with tetanus and diphtheria toxoids (DTaP vaccine)

(continued on next page)

Table 17.1
Approved Vaccines (*continued*)

Vaccine Type/Disease	Route of Delivery[a]	Routine or Targeted Use	Comments on Use[b]
Toxoids			
Diphtheria	IM	Routine in children	All three given in combination with *Bordetella pertussis* pertactin and filamentous hemagglutinin
Tetanus	IM	Routine in children	
Pertussis	IM	Routine in children	
Botulism toxin	IM	Only highly exposed individuals	
Viruses, Inactivated			
Hepatitis A	IM	Routine in children	
Influenza	IM	Routine, seasonal	Recombinant vaccines are under development
Japanese encephalitis	SubQ	Travel to endemic areas	
Polio (Salk)	IM	Routine in children	
Rabies	IM	Victim of animal bite	Killed virus and recombinant vaccine are available
Viruses, Attenuated			
Adenovirus	Oral	Military	
Chickenpox/shingles (varicella/herpes zoster)	SubQ	Routine in children and adults	In children, given at the same time as MMRV; adults receive different dose
Influenza	Intranasal	Routine in children	
Measles (rubeola)	SubQ	Routine in children	Given as part of MMRV
Mumps	SubQ	Routine in children	Given as part of MMRV
Polio (Sabin)	Oral		Not currently recommended in US
Rotavirus	Oral	Routine in children	
Rubella	SubQ	Routine in children	Given as part of MMRV
Smallpox (vaccinia)	Puncture	Voluntary for high-risk groups; some military	
Yellow fever	Inhaled	Travel to endemic area	
Virus Subunit			
Hepatitis B (HepB surface antigen, HBsAg)	IM	Routine at birth	
Influenza			Under development
Human papillomavirus (Gardasil)	IM	Indicated for males and females aged 9–26	Capsid proteins from types 6, 11, 16, 18

[a]IM = intramuscular; SubQ = subcutaneous.
[b]DTaP = diphtheria, tetanus, and acellular pertussis multivalent vaccine; MMRV = measles, mumps, and rubella multivalent vaccine.
[c]BCG = Bacillus Calmette-Guérin vaccine.

Table 17.2

Recommended Immunization Schedule for Children and Adolescents[a]

Vaccine	Birth	1 Month	2 Months	4 Months	6 Months	9 Months	12 Months	15–18 Months	24 Months	4–6 Years	11–12 Years
Hepatitis B[b]	HepB	HepB			HepB					HepB series	
Rotavirus			Rota	Rota	Rota						
Diphtheria, tetanus, acellular pertussis[c]			DTaP	DTaP	DTaP			DTaP		DTaP	Tdap
Haemophilus influenzae type b[d]			Hib	Hib	Hib		Hib				
Inactivated poliovirus			IPV	IPV			IPV			IPV	
Measles, mumps, rubella							MMR			MMR	MMR
Varicella							Varicella			Varicella	
Meningococcal (ACWY)[e]							Administer to high-risk children				MCV4
Meningococcal (B)											MenB (10–12 yr)
Pneumococcal[f]			PCV13	PCV13	PCV13		PCV13		PPSV23		
Influenza[g]					Influenza (yearly)						
Hepatitis A[h]							HepA (2-dose series)			Hep A	
Human papillomavirus[i]											HPV

[a]This schedule indicates the recommended ages for routine administration of currently licensed childhood vaccines, as of January 1, 2020.

▬ Range of recommended ages for each vaccine dose. ▬ Catch-up immunization. ▬ Assessment at age 11–12 years.

[b]Hepatitis B vaccine (HepB). All newborns should receive monovalent HepB, administered soon after birth and before hospital discharge.

[c]Diphtheria, tetanus, and acellular pertussis (DTaP) vaccine. Tdap is a modified vaccine with lower doses of diphtheria and tetanus toxoids.

[d]*Haemophilus influenzae* type b (Hib) conjugate vaccine.

[e]Meningococcal conjugate vaccine (MCV4). MCV4 should be administered to all children at age 11–12 years, as well as to unvaccinated adolescents at high school entry (age 15 years). The vaccine contains four types of capsules: A, C, W, and Y.

[f]Pneumococcal vaccine. The 13-valent pneumococcal conjugate vaccine (PCV13) is recommended for all children aged 2–23 months and for certain children aged 24–59 months. The final dose in the series should be administered at age ≥12 months. Pneumococcal polysaccharide vaccine (PPSV23) is a 23-valent vaccine recommended in addition to PCV13 for certain high-risk groups. PPSV23 is also recommended for people over 65.

[g]Inactivated influenza vaccine should be administered annually starting at age 6 months. Live, attenuated influenza vaccine should not be given until 2 years and not to immunocompromised individuals.

[h]Hepatitis A vaccine (HepA). HepA is recommended for all children at age 1 year (12–23 months).

[i]Human papillomavirus (HPV) vaccine: HPV9 (Gardasil). A 2-dose series of HPV vaccine should be administered on a schedule of 0 and 6 months to all adolescents aged 9–15 years. *Source:* Adapted from the Centers for Disease Control and Prevention (http://www.cdc.gov).

of IgA whereas pathogens infecting the circulatory system predominantly stimulate IgG synthesis).

In **toxoid vaccines**, the toxic activity of a microbial toxin is inactivated without its antigenicity being compromised. Toxoid vaccines stimulate the production of antitoxin antibodies in the person vaccinated. If the vaccinated person is infected with the pathogen later, the antitoxin antibodies will neutralize the toxin and enable host defenses to destroy the pathogen. As with most vaccines, toxoid vaccines usually require several booster shots to achieve a protective level of immunity, which eventually wanes. For example, the tetanus and diphtheria vaccines given during childhood should be readministered every 10 years to maintain effective immunity. Because many older adults have not received boosters, they are prone to these diseases.

Subunit vaccines are made from fragments of a microorganism or toxin. The antigen (for instance, a toxin subunit) can be purified directly from the pathogen, or the genes encoding these subunits can be introduced into another microbe that then produces large amounts of the desired antigen. The vaccine resulting from the latter method is called a **recombinant subunit vaccine**. For example, the viral capsid antigens that make up Gardasil, one of the vaccines given to prevent cervical cancer by human papillomavirus, are synthesized by a genetically modified yeast.

Conjugated vaccines physically link a highly immunogenic protein (such as diphtheria toxoid) with a poorly immunogenic capsule polysaccharide, as from *Haemophilus influenzae* type b, to boost the immune reaction to the polysaccharide. Why is this linkage necessary? Polysaccharides are T cell–independent antigens that directly activate B cells by cross-linking B-cell receptors. But without T-cell help, the antibody response is only partially activated. Conjugated vaccines are made to solicit T-cell help. A B cell whose B-cell receptor binds to the polysaccharide part of a conjugated vaccine will endocytose the polysaccharide with the attached diphtheria protein. Antigenic small pieces (epitopes) of diphtheria protein are then processed and presented on the B-cell MHC II receptors. Helper T cells that recognize the diphtheria epitope can then offer help and fully activate the B cell, even though the B cell is programmed to make anti-capsule antibody and not anti-diphtheria toxin antibody. Conjugated vaccines are particularly important for producing effective immunity to capsular antigens in children.

DNA vaccines and RNA vaccines are a new, emerging technology. Instead of injecting an antigen into a person, we will eventually inject DNA or RNA encoding that antigen directly into muscle (**Figure 17.1**). In this technique, which is currently limited to animal testing, the muscle cell makes the microbial protein from the foreign nucleic acid, and a healthy immune response ensues. Why is this approach better than simply injecting the protein directly? Having a host cell synthesize a pathogen's protein is the best way to generate a potent immune response because a small amount of any foreign protein synthesized in a host cell will be treated as endogenous antigen, degraded, and the pieces presented on the surface

Table 17.3
Immune Responses to Vaccines

Vaccine Type	Response		
	Innate	Cell-Mediated	Humoral
Live, attenuated	+	+	+
DNA		+	+
Dead, whole cell			+
Subunit (e.g., toxoid)			+

of that cell's MHC I molecules for presentation to CD8 cytotoxic T cells. What's more, some of the protein can be picked up by professional antigen-presenting cells that place the antigen's epitopes onto MHC II molecules for presentation to helper T cells. This process is akin to what occurs during natural infection (discussed in Chapter 16). By contrast, proteins simply injected into a person are not processed the same way and may not generate a response as complete or long-lasting. We should stress that nucleic acid vaccines are still experimental, although their future looks promising.

Adjuvants are not themselves vaccines but are substances that enhance the effectiveness of a vaccine. Think of an adjuvant as a sports team's coach. The coach doesn't play the sport but helps the players play better. Starting in the 1930s, aluminum compounds (called alums) have been used as adjuvants for killed, subunit, and toxoid vaccines, and they are used even today. Alums recruit various subsets of leukocytes, such as neutrophils, macrophages, and dendritic cells, to the site of vaccine injection. What follows is increased expression of chemokines and cytokines that stimulate the immune response. Alums also boost MHC II production, thereby improving antigen presentation to helper T cells and driving antibody responses. Some vaccines use other adjuvants, such as oil and lipid preparations, as well as the nucleotide CpG (cytosine-phosphate-guanine), which mimics bacterial and viral DNA. The latter engages Toll-like receptors.

Administering Vaccines

Vaccines can be administered by intramuscular injection, subcutaneous injection, oral ingestion, inhalation, or gene gun. Choosing the proper route of administration is important for getting a response appropriate for the pathogen in question. For instance, do you mostly need a circulating antibody response or a mucosal response? Oral vaccines will generate a greater gut mucosa immune response than a circulatory antibody response, for example. **Table 17.3** summarizes the immune responses generated by different categories of vaccines. Clearly, live, attenuated vaccines generate the broadest immune response.

Figure 17.1
DNA Vaccination

This gene gun, manufactured by Bio-Rad Laboratories, uses a blast of helium to fire tiny gold pellets carrying harmless snippets of a pathogen's DNA into the skin. The DNA infiltrates the nucleus of the skin and muscle cells. The DNA forces the cell to produce pathogen proteins, triggering the immune system. Although this is a promising technology, no commercial DNA vaccines are yet available.

Multiple vaccines can be given simultaneously and safely. The majority of vaccines are administered during childhood, when the diseases can be the most devastating (Table 17.2). Almost all vaccines are given in multiple doses, called booster doses, to provide longer-lasting immunity. An exception is the influenza vaccine, which is given in a single dose but changes every year. Most vaccines are not administered until patients are at least 2 months of age because maternal antibody crossing through the placenta to the fetus (or to a newborn through breast milk) will persist for a short time in the newborn, temporarily protecting the baby from disease and possibly dampening the response to a vaccine antigen.

Herd Immunity

Must everyone in a community be vaccinated to be protected? Not really. In practice it is impossible to immunize all humans against any given disease, even in developed countries such as the United States. But generally, the infection rate for a given infection after an aggressive vaccination program is much lower than you would expect from the number of people vaccinated. For instance, early in the twentieth century, measles (caused by the rubeola virus) was a common and potentially deadly disease. In 1954, before measles vaccination, about 4,000 cases of measles occurred for every 1 million unprotected people. Today, with greater than 90% of the US population being vaccinated, we still have 31 million <u>unvaccinated</u>, and thus vulnerable, people. So, you might still expect to see 4,000 cases of measles for every 1 million <u>unprotected</u> individuals. Surprisingly, the rate is less than one case per million unvaccinated people, and most cases originate with someone returning from a country where the disease is endemic. So, how are the millions of unvaccinated people protected?

The answer is herd immunity. **Herd immunity** occurs when a large portion of a population is immunized (**Figure 17.2**). Because the vaccinated individuals in the "herd," or community, will not become infected, they cannot spread the disease to others. At the very least, the presence of these individuals slows progression of the disease throughout the populace. It is estimated that immunizing three-fourths or more of a population can greatly reduce the incidence of a disease by cutting off transmission. Be aware, though, that the level of coverage needed for a given pathogen varies with the disease involved (the herd protection threshold for measles is greater than 93%). Herd immunity works only for diseases that are contagious. Herd immunity is not possible for infections, like tetanus, that are not contagious.

Opposition to Vaccines

The data are indisputable. Vaccines have literally saved millions of people from certain death. Study after study shows that the incidence of illness and death by a given pathogen dramatically declines once an effective vaccination program is under way. Take measles, discussed in the previous section. **Figure 17.3** illustrates how sharply the number of people infected with measles dropped after

Figure 17.2 Herd Immunity

A. Two infected and contagious people can rapidly spread disease throughout an uninfected population.

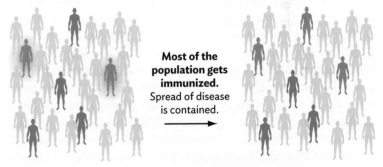

B. Immunizing only a few healthy people in a population does not slow the spread of disease among the unvaccinated (greater than 90% in the example still become ill), but those who are vaccinated will be protected.

C. Vaccinating a large fraction of the population will prevent most of those unvaccinated (75% in the example) from encountering sick and contagious individuals.

the advent of the measles vaccine in the early 1960s. Widespread use of measles vaccine has led to a greater than 99.99% reduction in measles cases in the United States compared with the prevaccine era. In 2012, only 55 cases of measles were reported in the United States. The refusal of some people to be vaccinated, however, is starting to reverse that trend.

Nearly all people who receive vaccines suffer no, or only mild, reactions, such as fever or soreness at the injection site. Very rarely

Figure 17.3 Effect of Measles Vaccine on Number of Reported Measles Cases

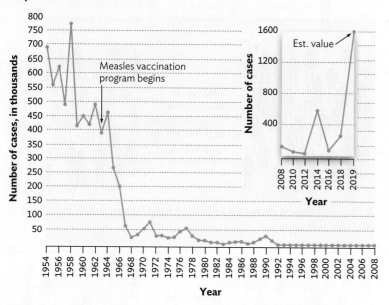

Widespread measles vaccination began in 1963. The result was a greater-than-99% reduction in measles cases in the United States compared with the prevaccine era.

do more serious side effects, like allergic reactions, occur. Vaccines are <u>extremely</u> safe. However, unsettling, erroneous information circulating on the Internet purports a link between vaccinations and other diseases, such as diabetes or autism. No well-controlled scientific study supports these claims. In fact, one study examining 1.25 million children failed to link autism to vaccinations with the measles, mumps, and rubella vaccine. Clearly, the risks associated with contracting preventable infectious diseases far outweigh the minimal risk associated with being vaccinated against them. As proof of this point, the undervaccination of children in the United States starting in the 1980s and '90s has led to an increase in cases of measles. Although measles was declared eradicated from the United States in 2000, the number of cases estimated for 2019 is more than 1,600 (the first 9 months of 2019 produced 1,244 confirmed cases). That level of measles has not been seen in over 25 years (Figure 17.3 inset).

This example shows that failing to vaccinate makes a human population highly susceptible to these preventable infections and forecasts an inevitable resurgence of serious diseases with potentially deadly outcomes.

Note: Live, attenuated vaccines, like the one for measles, should <u>not</u> be given to immunocompromised patients or to people who live with an immunocompromised patient. A competent immune system is needed to stop growth of the live, albeit crippled, microbe. Even a crippled virus can cause disease in patients who have an immunodeficiency. Nevertheless, many immunocompromised patients can receive and respond to subunit vaccines, depending on the nature of the immunodeficiency.

SECTION SUMMARY

- **Active immunization** is the process of introducing an antigen into a person to activate an immune response. If the antigen is a pathogen, or part of a pathogen, the individual will be protected from future infection.
- **Passive immunization, or immunotherapy,** involves injecting protective antibodies into an individual. **Antitoxins** are a form of passive immunization.
- **Vaccine types include** killed or inactivated pathogens; live, attenuated pathogens; inactivated toxins (toxoids); fragments of a pathogen (subunit vaccines); conjugated vaccines; and DNA vaccines.
- **Vaccines can be administered** orally, intranasally, by injection, or by DNA gun.
- **Herd immunity** to a communicable disease arises in a population when the number of people immunized is sufficient to make transmission of the infectious agent to an unimmunized person less likely.

Thought Question 17.1 Explain why the immune response to a ribosomal protein of *Streptococcus pneumoniae* is not protective, but an immune response to its capsule <u>is</u> protective. That is, why isn't *S. pneumoniae* ribosomal protein a good vaccine?

17.2 Diagnostic Tools of Immunology

SECTION OBJECTIVES

- Describe how immunoprecipitation assays are used to measure immunoglobulin levels.
- Explain agglutination and its use in blood typing and pathogen detection.
- Distinguish direct, indirect, and sandwich forms of enzyme-linked immunosorbent assays (ELISAs).
- Discuss how monoclonal antibodies are made and are used in fluorescent antibody techniques (immunofluorescence microscopy, western blots, and flow cytometry).

How do we know when someone has an IgA deficiency or a deficit of T cells? Many of the case histories we present rely heavily on quantifying certain antibodies or the types of white blood cells (WBCs) present in blood. Here we introduce the diagnostic tools used to make those determinations. These methods are crucial to diagnosing infectious diseases and immune dysfunctions.

Serology is the branch of immunology that reveals the contents of serum, primarily the concentrations of antibodies and antigens. But before interpreting any serological test, it is important to know the specificity and sensitivity of the test. **Specificity** measures how <u>selective</u> an antibody is toward an antigen. An antibody that binds very dissimilar antigens has a very low specificity. **Sensitivity**, on the other hand, reflects how many antigen or antibody molecules must be present for a test to turn positive (described further in Section 26.4). A highly sensitive test detects very small amounts of antigen or antibody. Thus, the most effective serological test is both highly specific and highly sensitive.

Figure 17.4 Basis of Immunoprecipitation

Only when the numbers of epitopes and antigen-binding sites are roughly equivalent will a large complex form and fall out of solution.

Antigen

Antibody

No complex is possible because there are more antigens than antigen-binding sites.

A. Antigen excess.

Complexes are possible when antigen numbers equal the number of antigen-binding sites.

B. Equivalence.

No complex is possible because there are more antigen-binding sites than antigens.

C. Antibody excess.

Note: Do not confuse blood plasma with serum. Plasma is the fluid portion of blood minus the cells. This material contains antibodies and fibrinogen, and it can clot. Technicians prepare plasma by spinning a fresh tube of blood containing an anticoagulant to prevent clotting. Serum is the clear fluid left after blood clots. It will contain antibodies.

Immunoprecipitation and Agglutination

Immunoprecipitation and agglutination are two phenomena that enable us to measure antigen concentrations. Recall that a typical antibody molecule has two antigen-binding sites, each of which binds identical antigens. Because each antibody can bind to two antigen molecules, antibodies can cross-link antigens having two or more epitopes in solution. The sequential cross-linking of antigens builds antigen-antibody complexes too large to remain in solution. The complexes fall out of solution (that is, precipitate; **Figure 17.4**). The phenomenon, called **immunoprecipitation**, is observed only in vitro, where the concentration of antigen and antibody can be manipulated experimentally. Immunoprecipitation takes place only when there are approximately equal amounts of antigen and antibody. The presence of too many antigen molecules (antigen excess; Figure 17.4A) or too few antigen molecules (antibody excess; Figure 17.4C) results in complexes that are too small to immunoprecipitate. Large complexes are formed only at an appropriate antigen/antibody ratio called equivalence (Figure 17.4B), the point where the number of antigenic sites roughly equals the number of antigen-binding sites.

The **radial immunodiffusion assay (RIA)** is an immunoprecipitation test that can be used to measure the concentration of an antigen. The antigen being measured could even be another antibody (**Figure 17.5**). In the RIA technique, a solution containing an antigen is placed into a well cut into an agarose gel (a substance similar to Jell-O) that is impregnated with antibody to that antigen. Antigen in the well diffuses outward until reaching a zone of equivalence with the embedded antibody. At that point, antigen-antibody complexes precipitate to form a ring around the well. Remember that antigen concentration progressively <u>decreases</u> as the antigen diffuses farther away from the well. Consequently, the higher the concentration of antigen in the well, the farther the antigen will have to diffuse before the zone of equivalence is reached and a ring of precipitation forms. Lab techs determine the concentration of antigen in a solution by comparing the radius of immunoprecipitation generated by the solution against a standard curve that plots known concentrations of that antigen.

The agarose gel in the plate shown in Figure 17.5 contains antibodies (anti-IgA antibodies) specific for an antigen—here the antigen is IgA. Different concentrations of the antigen are placed in each well. The antigen diffuses from each well into the agarose. A ring of precipitation forms where the concentration of the diffusing antigen reaches a zone of equivalence with the antibody in the plate. The more antigen placed in the well, the larger the diameter of the precipitation ring. Well A was loaded with more IgA antigen than well B.

The RIA technique can be used to measure the amount of IgA, or any other antibody class, in a patient's serum. Remember, antibodies are proteins and, as such, are themselves antigens. Thus,

Figure 17.5 Radial Immunodiffusion

The agarose gel contains antibodies specific for a specific antigen. A known or unknown amount of the antigen IgA is added to each well. The diameter of the ring reflects how much antigen there was in the well.

More IgA antigen in well A

Less IgA antigen in well B

Figure 17.6 Slide Agglutination

A. Red blood cells cross-linked with IgM against a surface antigen. Sizes of cells and antibodies are not proportional.

B. Blood typing in which a drop of blood is mixed with anti-B (left) or anti-A (right) antibodies. The blood type in this instance is A.

C and D. Agglutination used to identify group B beta-hemolytic streptococci (GBS). Cells from suspected GBS colonies (C) are mixed with antiserum against group B streptococci (D). Cells on the left in panel D are agglutination-positive, meaning they are group B streptococci. Cells on the right are agglutination-negative and not group B streptococci.

a rabbit injected with human IgA will make antibodies directed against human IgA (rabbit anti–human IgA). The rabbit anti–human IgA antibodies can be embedded into the agarose of an immunodiffusion plate, and sera from various humans can be placed into different wells. The serum with the most IgA will produce the widest ring of precipitation. Serum from a patient with an IgA deficiency has very little IgA and will produce a very narrow ring of precipitation.

Agglutination, like precipitation, is based on antibody-antigen cross-linking. The difference is that precipitation measures small soluble antigens, whereas agglutination measures large insoluble antigens like whole cells (red blood cells [RBCs] or bacteria; Figure 17.6A) or latex beads coated with an antigen. Because cells are large, agglutination reactions usually require pentameric IgM antibody for cross-linking. Agglutination reactions are used to determine the A, B, AB, and O blood groups of RBCs. In this test, two drops of a patient's blood are placed on a slide. Antibody to the A antigen is mixed with one blood drop, and anti-B antibody is mixed with the second drop. Interaction between antigen and antibody causes the blood cells to clump, a positive test indicating the presence of antigen (Figure 17.6B).

Agglutination can also be used to identify bacterial pathogens. Figure 17.6C shows colonies of group B beta-hemolytic streptococci that cause serious neonatal infections. In nature, many strains of beta-hemolytic streptococci are not pathogenic. Consequently, it is of utmost importance to quickly determine whether a beta-hemolytic streptococcus isolated from an expectant mother is group B. Cells from one of those colonies will agglutinate when mixed with antibody directed against group B streptococci (Figure 17.6D).

Enzyme-Linked Immunosorbent Assay (ELISA)

Visualizing immunoprecipitation requires large amounts of antigen and antibody in just the right balance. Subtler antigen-antibody interactions can be detected by an **enzyme-linked antibody** test in which an enzyme is attached (conjugated) to a specific detection antibody. The enzymes typically used are horseradish peroxidase or alkaline phosphatase, which can convert a colorless substrate to a colorful product. The technique that uses this approach is called **enzyme-linked immunosorbent assay (ELISA)**, sometimes referred to as an **enzyme immunoassay (EIA)**. ELISA/EIA tests can detect antibody or antigen and do it quantitatively.

As shown in **Figure 17.7**, ELISA comes in three basic forms: direct, indirect, and sandwich. **Direct ELISA** (Figure 17.7A) starts by binding antigen from a patient's serum or tissue to the well of a plastic microtiter dish. We can detect the specific antigen of interest (a virus protein, for instance) by adding an enzyme-conjugated, antigen-specific antibody to the well. After unbound antibody is washed off, substrate for the conjugated enzyme is added. The intensity of color change reflects how much antigen initially bound to the well and thus was present in the patient sample.

Indirect ELISA (Figure 17.7B) is best used to detect how much anti-pathogen antibody is present in a patient's serum. Here an excess of known antigen, like hepatitis antigen, is bound to microtiter plate wells and the patient's serum is added. Any anti-hepatitis antibody (called the primary antibody) present in the patient's serum will be captured by the hepatitis antigen fixed to the well. Unrelated antibodies are washed away. Next, enzyme-conjugated secondary antibody that can bind to the captured primary antibody is added (for instance, anti–human IgG antibody made in rabbits). After substrate for the conjugated enzyme is added, the amount of color change produced reflects how much of the patient's anti–hepatitis IgG antibody was captured by the hepatitis antigen bound to the well. The method is called indirect ELISA because the secondary detection antibody does not directly bind to the pathogen's antigen. One advantage of this and the following method (sandwich ELISA) is that you need only one enzyme-conjugated detection antibody (for instance, rabbit-made anti–human IgG) to detect any primary human IgG antibody specific for any antigen.

Sandwich ELISA (Figure 17.7C) is an indirect assay that begins when a capture antibody is attached directly to the microtiter well.

Figure 17.7 Basic Forms of Enzyme Immunoassay

Direct ELISA is used to detect the presence of an antigen (Ag) in a patient's blood. Indirect ELISA is used to detect specific antibody (Ab) in a patient's serum. Sandwich ELISA can detect either Ag or Ab in a patient sample.

Direct assay

Indirect assay

Capture assay "sandwich"

A. An antigen in the patient's blood is immobilized by direct adsorption to the assay plate and detected by addition of enzyme-conjugated antibody.

B. A known antigen is directly adsorbed to the assay plate. Serum Ab to the antigen is captured by the plate antigen. Secondary enzyme-conjugated Ab is then used to detect the captured Ab.

C. An antigen-specific Ab is directly adsorbed to the assay plate. Antigen from the patient sample is captured by the plate antibody. Unlabeled secondary Ab to the Ag binds to the captured antigen. Enzyme-conjugated Ab is then used to detect the secondary Ab.

The capture antibody is specific for a given antigen, which could be a pathogen protein or a patient's antibody. Patient sample is added, and the antigen of interest, if present, is captured. Next, unlabeled, antigen-specific secondary antibody—usually IgG—is added, followed by enzyme-conjugated antibody to the secondary antibody.

TITERS Now that you know what techniques are used to detect antigens, how does the laboratory quantitate an antigen—for example, the level of an antibody or the amount of a virus present in a patient's serum? The measurement used to express the antibody or antigen concentration in a solution is called a **titer**. An antibody titer is a measurement of how much antibody in serum will recognize a particular epitope and how tightly the antibodies bind to that epitope. A titer is expressed as the greatest <u>dilution</u> of serum that still gives a positive result in an assay. To illustrate, imagine two bottles of water containing different intensities of blue dye. Bottle A is dark blue, but bottle B is much less blue and clearly contains less dye. You can estimate how much less dye by serially diluting the water from each bottle into a series of tubes containing clear water. With each dilution the water becomes less and less blue until, eventually, you do not see any blue. Bottle A, which started out the most blue, can be diluted more than bottle B before you no longer see color. If the dilutions were carried out as serial twofold dilutions (1:2, 1:4, 1:8, and so on) and the last tube showing color was a 1:128 dilution for bottle A and 1:16 for bottle B, the titers would be the <u>reciprocal</u> values, 128 and 16. Bottle A, therefore, contained 8 times more dye than bottle B (128/16).

The same basic procedure can be done to determine an antibody titer. The patient's serum is serially diluted twofold (1:2, 1:4;

1:8; 1:16, 1:32, 1:64, 1:128, and so on), and the dilutions are analyzed by ELISA. How much antibody is present in serum is reflected by how far you can dilute the serum and still see a color change. The reciprocal of the highest dilution that still shows a color change is called the antibody titer.

Figure 17.8A illustrates how sandwich ELISA can detect the serum quantity (titer) of an IgM antibody that binds to a specific antigen—in this case dengue virus antigen. The procedure is performed in a plastic microtiter dish—a dish with many wells (**Figure 17.8B**). Capture antibodies against human IgM are first attached to the bottom of each well. The patient's serum is added to the first well and then serially diluted into other wells. The serum IgM in each well is allowed to bind to the anti-IgM stuck to the plastic. Everything not bound is washed off. Next, purified dengue virus antigen is added to the well. Dengue antigen will stick only to the captured anti-dengue IgM from the patient. An enzyme-linked, anti-dengue antibody is then added. It binds to the IgM-captured dengue antigen. After unbound enzyme-linked antibody is washed away, the enzyme substrate is added. A visible color is produced when the enzyme reacts with the substrate. The amount of anti-dengue IgM present in the patient's serum dictates how many wells down the dilution series will exhibit color change.

Sandwich ELISA can also detect an <u>antigen</u> in a patient's blood—for instance, hepatitis antigen (**Figure 17.8C**). Here the wells of a microtiter plate are coated with anti-hepatitis antibody (made from a rabbit, for example). The patient's serum is added, and the antibodies attached to the well will capture any hepatitis antigen present. A second anti-hepatitis antibody (IgG made from a mouse, say) is added and binds to the captured hepatitis antigen

ANIMATION **Figure 17.8** Enzyme-Linked Immunosorbent Assay (ELISA)

A. In an IgM-capture ELISA, a plate is coated with antibody directed against the constant region of human IgM. Twofold dilutions of the patient's serum are added to the microtiter plate wells. All the IgM in serum is captured. Known antigen—such as a dengue virus antigen, which will bind only to captured patient IgM specific to dengue antigen—is then added. The next addition is an anti–dengue antigen antibody conjugated to a detection enzyme, followed by a chromogenic (color-producing) substrate for the enzyme (not shown). A reaction with the substrate will cause a visible color change.

B. The antibody titer is determined by the serial twofold dilution of patient serum from top to bottom of the dish (dilutions are marked). The titers from patient samples increase from left to right, with column four being the highest.

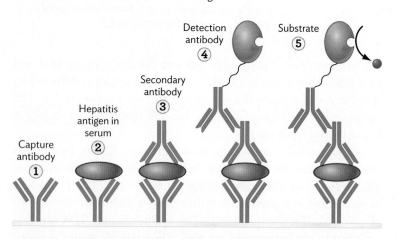

C. Detecting hepatitis antigen in blood. **Step 1:** Plate is coated with an anti-hepatitis capture antibody. **Step 2:** Serum sample is added. Any hepatitis antigen present binds to capture antibody. **Step 3:** Secondary antibody is added (mouse anti-hepatitis IgG) and binds to antigen. **Step 4:** Enzyme-linked detection antibody (rabbit anti–mouse IgG) is added and binds to secondary antibody. **Step 5:** Substrate is added; enzyme converts substrate to colored form. Capture and secondary antibodies must be from different animals (rabbit and mouse) so that the detection antibody (anti–mouse) will bind only to secondary antibody.

(making a "sandwich"). To see how much hepatitis antigen was captured, enzyme-linked anti–mouse IgG is added, followed by the enzyme substrate. The more hepatitis antigen captured, the greater the color change.

One way to diagnose viral and some bacterial infections is to document a large increase in antibody levels (or titer) as a patient progresses from the acute phase of a disease to the convalescent phase. ELISA is a convenient way to do this. A fourfold or greater increase in titer from the acute to convalescent stage of a disease is diagnostic for a given microbial cause (for example, antibody titer against West Nile virus rises from 32 in acute stage to 128 in convalescence).

Monoclonal Antibodies

Many different techniques in diagnostic microbiology require epitope-specific antibodies of a single isotype (for instance, IgG or IgM), known as **monoclonal antibodies**. Monoclonal antibodies originate from a single antibody-producing B-cell clone. **Polyclonal antibody** is antiserum that contains all antibody isotypes with specificities directed against many different epitopes of an antigen. A polyclonal mix of antibodies reduces the specificity and sensitivity needed for many immunological tests. So, how do we generate the more specific monoclonal antibodies?

The process is illustrated in **Figure 17.9**. Mice or rabbits are injected with a primary dose of antigen and then, some days or weeks later, a booster dose. B cells are then removed from the animal's spleen, sorted into separate wells of microtiter plates (one B cell per well), and stimulated to make antibody. Antibodies from each well can be screened by ELISA for specificity against the desired epitope—say, a hepatitis capsid protein. That's the easy part. Keeping the clone alive to make monoclonal antibody over many years is hard. Scientists solved the life expectancy problem by fusing an antibody-producing B cell with myeloma cancer cells that divide indefinitely. The myeloma cells do not themselves make antibody. The resulting immortalized hybrid clone is called a **hybridoma**. Hybridoma cells divide indefinitely and make single-epitope, single-isotype, monoclonal antibodies.

Monoclonal antibodies have several uses. In diagnostic tests they are used to detect antibodies or antigens (see the discussion of ELISA in the previous section), to treat viral diseases previously deemed untreatable, to classify strains of a bacterial species (called typing), and to identify cells or molecules in an organism.

Immunofluorescence Microscopy

Fluorescence microscopy was first described in Section 3.5. The technique can be used to directly find pathogenic microorganisms in tissue (direct immunofluorescence) or to reveal pathogen-specific antibodies in serum (indirect immunofluorescence). In the **direct immunofluorescence test** (Figure 17.10A), a tissue or fluid suspected of harboring the pathogen is placed on a slide, and specific, fluorescently labeled, anti-pathogen antibody is added. If the pathogen is in the tissue, it will fluoresce when viewed through a fluorescence microscope, indicating a positive test.

In the **indirect immunofluorescence test** (Figure 17.10B), the technician affixes a known pathogen to a glass slide and floods the slide with patient serum. If the serum contains pathogen-specific antibodies—the result of infection—those antibodies will bind to the microbe stuck to the slide. Then, a fluorescently tagged anti-human antibody is applied. If the patient serum has antibodies to the pathogen, the pathogen will now fluoresce. The etiologic agent for syphilis (*Treponema pallidum*) is often detected with the direct test, whereas the indirect test can identify patients with later stages of the disease, when the organism may be difficult to find but antibodies are still present (Section 23.4).

Immunoblotting (Western Blot)

Another important research technique involving antibodies is the **western blot**, which is used to detect the presence of a specific protein in cell extracts (Figure 17.11A). Proteins from an extract are separated by size, using sodium dodecyl sulfate–polyacrylamide gel electrophoresis (SDS-PAGE). The proteins are transferred from the gel onto a polyvinylidene difluoride (PVDF) membrane by an electrical current. Next the membrane is flooded with a primary antibody that will bind to a specific protein. The membrane is then probed with a secondary, enzyme-linked antibody that specifically binds to the primary antibody (for example, anti-IgG antibody). Adding enzyme substrate sets off either a color change or a luminescent reaction wherever an antibody sandwich has assembled. Only protein bands to which the primary antibody has bound will be detected. The technique is used to determine whether, and at what levels, specific proteins are present in a cell extract. Figure 17.11B shows a western blot of patient sera to find HIV proteins.

Flow Cytometry versus FACS

Several case histories in preceding chapters report that a patient had low levels of B cells or T cells and sometimes even deficits in a specific class of T cells. For instance, Fatima, the child with bare lymphocyte syndrome in Case History 16.1, lacked T cells. How does the lab quantify the different types of lymphocytes?

Figure 17.9 Production of Monoclonal Antibodies

Immunization of animals with a selected antigen stimulates antibody-forming B cells to produce a range of antibodies with various specificities. Collections of B cells are fused with tumor (myeloma) cells to produce immortalized hybridoma cells.

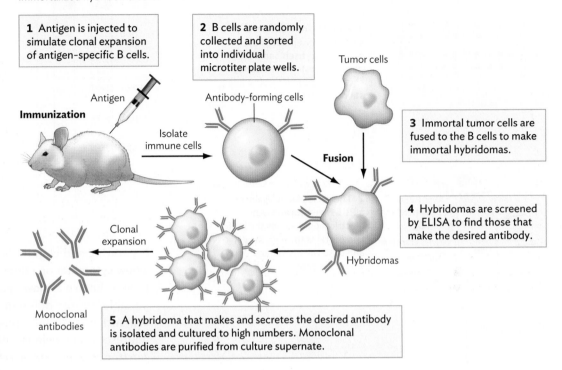

1 Antigen is injected to simulate clonal expansion of antigen-specific B cells.

2 B cells are randomly collected and sorted into individual microtiter plate wells.

Tumor cells

Antigen

Immunization

Antibody-forming cells

Isolate immune cells

Fusion

3 Immortal tumor cells are fused to the B cells to make immortal hybridomas.

4 Hybridomas are screened by ELISA to find those that make the desired antibody.

Clonal expansion

Hybridomas

Monoclonal antibodies

5 A hybridoma that makes and secretes the desired antibody is isolated and cultured to high numbers. Monoclonal antibodies are purified from culture supernate.

Figure 17.10 Immunofluorescence Microscopy

Direct immunofluorescence test

Fluorescently tagged anti-treponeme Ab

Patient tissue

T. pallidum

Indirect immunofluorescence test

Fluorescently tagged anti-human Ab

Anti-treponeme Ab (patient)

T. pallidum

FM 10 μm

A. The direct test uses fluorescently labeled anti-pathogen antibodies (Ab) to directly detect the presence of a pathogen in host tissue (here, *Treponema pallidum*).

B. The indirect test uses known pathogens placed on a slide to capture anti-pathogen antibodies in a patient's serum. Then, fluorescently tagged anti-human antibodies are added to detect the captured antibodies, if any. **Inset:** An indirect fluorescence assay detecting antibodies to *Helicobacter pylori*, a cause of stomach ulcers.

Figure 17.11 Western Blot Analysis

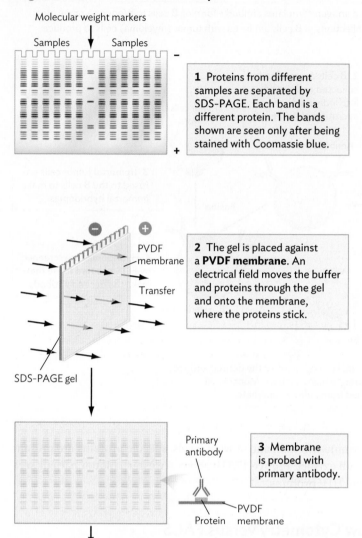

Molecular weight markers

Samples   Samples

1 Proteins from different samples are separated by SDS-PAGE. Each band is a different protein. The bands shown are seen only after being stained with Coomassie blue.

PVDF membrane

Transfer

SDS-PAGE gel

2 The gel is placed against a **PVDF membrane**. An electrical field moves the buffer and proteins through the gel and onto the membrane, where the proteins stick.

Primary antibody

3 Membrane is probed with primary antibody.

PVDF membrane

Protein

Secondary antibody

4 Visualize with secondary antibody.

Horseradish peroxidase

A. Western blot can identify specific proteins within a mixture of many proteins. Proteins in different sample extracts are applied to separate wells of a polyacrylamide gel (top of the gel). The proteins are separated by size after an electrical field is applied through the gel (smaller proteins travel farther down the gel). Separated proteins are transferred (blotted) onto a polyvinylidene difluoride (PVDF) membrane, which is then probed with a primary antibody against the protein of interest. An enzyme-linked secondary antibody against the primary antibody is used to find where on the gel the primary antibody bound. Adding enzyme substrate causes a color change in those bands. The protein detected in this example has different sizes in some samples and is absent in other samples.

1   2   3

gp160
gp120

p66
p55
p51

gp41

p31

p24

p17
p15

B. Western blot to detect HIV proteins in patient sera. In this example, several anti-HIV antibodies are combined to detect multiple proteins. Lane 1 is a positive control showing many HIV proteins; lane 2 is an intermediate positive showing the presence of some HIV proteins; lane 3 is a negative control.

Separating and counting cells in a patient's blood requires either a **flow cytometer** or a **fluorescence-activated cell sorter (FACS)** in a process known as **flow cytometry** (**Figure 17.12**). The technique is possible because different types of cells are decorated with unique cell-surface antigens. Thus, you can "tag" a specific cell type with a fluorescent monoclonal antibody that binds only to the unique antigen (for example, anti-CD4 for helper T cells).

The fluorescent antibody is added to blood, unbound antibody is removed, and the blood cells are injected into the flow cytometer. The blood passes through a tiny orifice that delivers the cells single file past a laser (**Figure 17.13A**). The laser activates the fluorescent chemical (fluorophore) linked to the antibody attached to the cell, and a detector measures the intensity of the emitted light. Multiple detectors can be used to measure fluorescence from different

Figure 17.12 Fluorescence-Activated Cell Sorter

Dr. Robert Barrington (left) and Bryant Hanks at the University of South Alabama use a FACS machine to research systemic lupus erythematosus. The arrow shows where samples are injected into the machine.

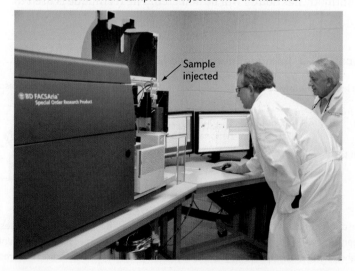

Sample injected

Figure 17.13 Using Flow Cytometry to Count and Sort CD4 and CD8 T Cells

Cell tagging

1 Fluorescently tagged antibodies to CD4 and CD8 are added to a blood sample.

Flow cytometry

2 Cells are passed single file past a laser, which excites the fluorophore.

3 Separate detectors read the level of fluorescence from each tag as each cell passes.

Cell counting

4 Sorter counts each cell as CD4+ or CD8+ as it passes.

A. Anti-CD4 and anti-CD8 antibodies tagged with different fluorescent dyes are added to a mixture of cells. The dyes in this example are PerCP (red emission) and fluorescein isothiocyanate (FITC; green emission). Each dye will fluoresce a different color. The mixture of cells is passed, single file, past the cell sorter's laser. Separate detectors read the level of fluorescence from the two dyes as each cell passes. In doing so, the machine counts the CD4 and CD8 T cells.

A negative sign (−) indicates the absence of CD protein; a positive sign (+) indicates the presence of a CD protein.

B. Results of CD4 counts (x-axis) relative to CD8 counts (y-axis) from two patients. The left-hand plot shows a normal situation. Lower-right-quadrant cells express CD4 but not CD8 (the helper-T-cell set), whereas upper-left-quadrant cells express CD8 but not CD4 (cytotoxic T cells). Lower-left-quadrant cells express neither antigen. The right-hand plot shows the presence of cells that express both the CD4 and the CD8 antigen (arrow, upper right quadrant), which is highly irregular. Cells with abnormal antigen expression may indicate a T-cell malignancy.

C. Schematic of a FACS apparatus (bidirectional sorting). A tiny positive or negative charge is added to each cell as it passes the detector. Deflector panels will then move (sort) the charged cells into separate collection vessels for further study.

fluorophores that emit different colors. The results are graphically displayed in several ways. **Figure 17.13B** shows one approach and illustrates how the technique can detect aberrant T cells resulting from cancer. A FACS machine can not only count the different cell types but also sort them into separate tubes for later analysis (**Figure 17.13C**).

To separate cells, the FACS machine imparts a positive or negative charge on a passing cell (like a hand stamp placed on someone entering a bar). Deflecting plates move the positively and negatively charged cells in opposite directions as they fall through the instrument. A cell that fluoresces green can be given a positive charge, and a different cell, one that emits red light, can be given a negative charge. As the cells continue to move through the machine, the deflector plates pull the charged cells in opposite directions, and the cells are routed to separate collection vessels. The sorted cells can then be used for other tests.

SECTION SUMMARY

- **Immunological tools** can detect antigens or antibodies (which are themselves antigens) in a patient's serum or tissues.
- **Radial immunodiffusion** measures the concentration of an antibody or soluble antigen in a serum sample. Molecules from a serum sample radially diffuse outward from a well

cut into agarose embedded, respectively, with either antigen or antibody. Molecule concentration in a well is proportional to the diameter of the immunoprecipitation ring that forms around the well.

- **Agglutination** is an antigen-antibody reaction that detects insoluble antigens on whole cells. Antibody binding will link cells, causing them to aggregate.

- **Enzyme-linked immunosorbent assays (ELISAs)** use enzyme-linked antibodies to detect specific antibodies or antigens in a patient's serum. The serum target molecule is first captured by matched antigen or antibody attached to wells of a microtiter dish. Next, enzyme-conjugated secondary antibody is added to the well, followed by substrate for the enzyme.

- **Monoclonal antibodies** are secreted by immortalized hybridomas made by the fusion of a B cell with a myeloma cell. Hybridomas produce single-isotype antibodies with single-epitope specificity. Monoclonal antibodies are used for ELISA, immunofluorescence microscopy, and immunoblotting.

- **Immunofluorescence microscopy** uses fluorescently tagged monoclonal antibodies to identify antigens in blood or tissue samples.

- **Immunoblotting** uses monoclonal or polyclonal antibodies to detect specific proteins in a sample after polyacrylamide gel electrophoresis.

- **Flow cytometry** uses fluorescently tagged monoclonal antibodies to count specific cell types in a mixed sample. A fluorescence-activated cell sorter can then separate and collect the different cell types.

17.3 Hypersensitivity

SECTION OBJECTIVES

- Explain what an allergen is.
- Describe the development, pharmacology, and treatment of type I hypersensitivity.
- Differentiate type II and type III hypersensitivities.
- Discuss type IV hypersensitivity in terms of which cells are involved and how it can be used as a diagnostic tool.

Have you ever overreacted to a small provocation? Your immune system can do the same thing. Sometimes the immune system overreacts to a foreign antigen and causes more damage than the antigen alone might cause. For example, your immune system's reaction to a certain type of pollen can be more irritating (hay fever) than the effect of the pollen itself. In other cases, a foreign antigen may possess structures <u>similar</u> in shape to those of host structures. These antigens can trick the immune system into reacting against self. Both types of immune miscues are examples of allergic hypersensitivity reactions, and the antigen causing the reaction is called an **allergen**. There are four types of hypersensitivity reactions, as described in Table 17.4. The first three types are antibody-mediated; the fourth is cell-mediated.

Stages of Type I Hypersensitivity

CASE HISTORY 17.1

Sting of Fear: Type I Hypersensitivity

During spring break, 9-year-old Brian was walking with his mother at the zoo when he was stung by a bee. Within minutes, Brian began sweating and itching. His chest started to tighten, and he had tremendous difficulty breathing. Terrified, he looked to his equally frightened mother for help.

This is a classic example of severe **type I (immediate) hypersensitivity**. The case involved a severe reaction, sometimes called anaphylaxis. Type I hypersensitivity reactions can also be mild and localized, as in hay fever, also called allergic rhinitis. Symptoms of type I hypersensitivities arise when an allergen—an antigen that elicits an allergic reaction—enters a susceptible person's body and triggers the release of pharmacologically active chemicals. For

Table 17.4
Summary of Hypersensitivity Reactions

Type	Description	Time of Onset	Mechanism[a]	Manifestations
I	IgE-mediated hypersensitivity	2–30 min	Ag induces cross-linking of IgE bound to mast cells with release of vasoactive mediators	Systemic anaphylaxis, local reactions, hay fever, asthma, eczema
II	Antibody-mediated cytotoxic hypersensitivity	5–8 h	Ab directed against cell-surface antigens mediates cell destruction via ADCC[b] or complement	Blood transfusion reactions, hemolytic disease of the newborn, autoimmune hemolytic anemia, rheumatic fever
III	Immune complex–mediated hypersensitivity	2–8 days	Ag-Ab complexes deposited at various sites induce mast cell degranulation via Fc receptor; neutrophil degranulation damages tissue (localized reaction)	Systemic reactions, disseminated rash, arthritis, glomerulonephritis
IV	Cell-mediated hypersensitivity	24–72 h	Memory T_H1 cells release cytokines that recruit and activate macrophages	Contact dermatitis, tubercular lesions

[a]Ab = antibody; Ag = antigen.
[b]ADCC = antibody-dependent cell-mediated cytotoxicity.

ANIMATION ▶ **Figure 17.14** **Type I Hypersensitivity Reactions**

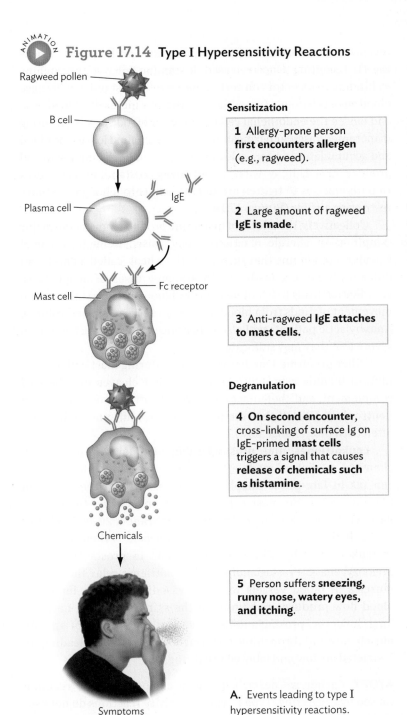

Sensitization

1 Allergy-prone person **first encounters allergen** (e.g., ragweed).

2 Large amount of ragweed **IgE is made**.

3 Anti-ragweed **IgE attaches to mast cells**.

Degranulation

4 **On second encounter**, cross-linking of surface Ig on IgE-primed **mast cells** triggers a signal that causes **release of chemicals such as histamine**.

5 Person suffers **sneezing, runny nose, watery eyes, and itching**.

A. Events leading to type I hypersensitivity reactions.

B. Peanut allergy. This case involved hives, but symptoms can include diarrhea, nausea, and life-threatening anaphylaxis.

C. Latex allergy, a problem for some health care professionals. Symptoms range from itching and redness to breathing difficulties. Nonlatex, hypoallergenic gloves are available for these individuals.

Normal bronchiole   Asthmatic bronchiole

D. Bronchoconstriction. During an asthma attack, smooth muscles in the bronchioles of the lung constrict and decrease airflow in the airways. Inflammation or excess mucus secretion can further decrease airflow.

instance, inhaling pollen causes the release of these chemicals in the lungs of hay fever sufferers and produces symptoms that include rhinitis, watery and itchy eyes, and sneezing. By contrast, a bee sting on a person allergic to bee venom causes systemic release of these chemicals as the venom circulates throughout the body. The result is **anaphylaxis**, whose symptoms include swelling from fluid accumulation in tissues (edema), low blood pressure, cardiovascular collapse causing heart arrhythmias, and severe breathing difficulties that sometimes end with suffocation. Anaphylaxis kills more than 1,500 people per year in the United States.

Type I hypersensitivity develops in two stages. The first stage is **sensitization**, which occurs when the patient first encounters the allergen. On <u>initial</u> exposure, an allergen elicits the production of IgE antibodies specific to the allergen (bee venom in Case History 17.1). The Fc portions of these antibodies bind to Fc receptors on the surfaces of mast cells (**Figure 17.14A**, steps 1–3). IgE-coated mast cells are then said to be <u>sensitized</u>. Recall that IgE primarily amplifies the body's response to invaders by causing mast cells to release inflammatory mediators (see Section 16.2). Allergens elicit higher levels of IgE in certain people; why some people are more affected than others is unclear, but a genetic predisposition appears to exist.

The second stage of type I hypersensitivity occurs within minutes of a <u>second</u> exposure to an allergen, when the allergen reacts with the IgE-coated mast cells. Most allergen molecules contain several identical antigenic sites. Thus, during a second exposure, a single allergen particle with multiple epitopes can bind to adjacent IgE molecules protruding like hands from the mast cell surface, forming a bridge between the two IgE molecules (Figure 17.14A, step 4). This cross-linked complex launches a complex signaling cascade that causes the mast cell granules to quickly migrate to the cell surface and release their contents in a process called **degranulation**.

PHARMACOLOGY OF TYPE I HYPERSENSITIVITY Mast cell degranulation releases chemicals with potent pharmacological activities. In humans, the most important are histamine, prostaglandins, leukotrienes, and chemotactic factors. The severity of an allergic reaction—whether life-threatening anaphylaxis as for the bee sting, or a less serious, more localized reaction such as contact dermatitis or hay fever—depends on several factors: the allergen, the route of allergen entry (injection, skin contact, inhalation), and the number of sensitized mast cells present. For example, an injected allergen like bee venom or an ingested allergen like penicillin or peanuts (**Figure 17.14B**) can circulate via the blood and trigger the degranulation of larger numbers of sensitized mast cells than can an allergen that touches only skin, such as a latex glove (**Figure 17.14C**).

Note: Although an allergy to latex can be caused by type I hypersensitivity reactions, latex allergies are more commonly the result of type IV hypersensitivity reactions (contact dermatitis), described later.

Histamine is a chemical synthesized and stored in mast cell granules before an encounter with allergen. Histamine causes the symptoms of allergy by binding to several different histamine receptors present on different body cells. Antihistamines, a common treatment for less severe forms of type I hypersensitivity, are structurally similar to histamine and work by blocking histamine

receptors (for instance, diphenhydramine, or Benadryl, blocks the H1 receptor). Once released, histamine will produce itching by binding to receptors on certain nerve cells. Histamine will cause blood vessels to dilate (vasodilation), thereby increasing blood flow, and loosen the endothelial cell junctions of blood vessels, resulting in increased vascular permeability. As a result, fluid leaves the blood and accumulates in tissues. Fluid that accumulates under skin will produce raised, itchy welts called **hives**. Histamine also causes smooth muscles to tighten around bronchi (bronchoconstriction), restricting airflow (**Figure 17.14D**).

Collectively, the effects of histamine produce some of the major symptoms of allergic reactions: itchy, raised welts (also called wheals); a reddening that surrounds the wheal (called a flare); and difficulty breathing (asthma). In systemic cases of anaphylaxis, vasodilation leads to hypotension (low blood pressure) because the movement of fluids from blood into tissue decreases blood volume. Anaphylactic patients will also experience bronchoconstriction so severe that they may suffocate.

Other preformed mediators released during mast cell degranulation include a protease (tryptase) that cleaves members of complement, and chemotactic factors that attract eosinophils and neutrophils. (Recall that the terms "neutrophil" and "PMN" are often used interchangeably.)

Leukotrienes and **prostaglandins** are membrane-derived chemical mediators made <u>after</u> the mast cell binds to an allergen (as in late-phase anaphylaxis, discussed shortly). Allergen binding to IgE on mast cells activates an enzyme (phospholipase A) that releases from the cell membrane a 20-carbon fatty acid called arachidonic acid. Arachidonic acid is then converted to leukotrienes by 5-lipoxygenase A and to prostaglandins by cyclooxygenase. Prostaglandins and leukotrienes relax the smooth muscles of blood vessels to cause vasodilation. The increased blood flow produces reddening of the skin. If released near the lung, however, prostaglandins cause the contraction of smooth muscle around bronchioles (bronchoconstriction), leading to restricted airflow and labored breathing.

ATOPY Earlier we described a severe type I allergic reaction caused by a bee sting. Type I hypersensitivity reactions do not usually involve the whole body. Most type I reactions are more localized and cause what is called atopic ("out of place") disease. Hay fever, or allergic rhinitis, is a common atopic disease of the respiratory tract that can be triggered by dust mite feces (**Figure 17.15**), animal skin or hair (dander), or certain types of grass or weed pollens. Hay fever affects the eyes, nose, and upper respiratory tract. Atopic asthma, another form of type I hypersensitivity resulting from inhaled allergens, affects the lower respiratory tract and is characterized by wheezing and difficulty breathing.

TREATING TYPE I HYPERSENSITIVITY Antihistamines effectively treat allergic rhinitis, but the more important chemical mediators of asthma are the leukotrienes produced during **late-phase anaphylaxis**. In late-phase anaphylaxis, mast cells release

Figure 17.15 Example of an Inhaled Allergen

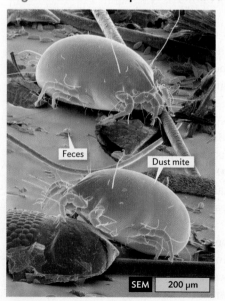

Dust mites (200–600 μm) and dust mite feces (colorized SEM). Dust mites roam every home, but not everyone is allergic to dust mite feces. Vacuuming regularly (using a HEPA filter) and using dustproof or allergen-blocking covers on mattresses and pillows can control dust mite allergies. Washing bedding weekly in hot water (54.5°C) and freezing nonwashable items can kill dust mites.

Feces

Dust mite

SEM   200 μm

chemotactic factors that call in eosinophils. Eosinophils entering the affected area produce high levels of leukotrienes that, like histamine, cause vasoconstriction and inflammation. At this point, antihistamines have little impact. Effective treatment includes inhaled steroids to minimize inflammation as well as bronchodilators (for example, albuterol) to widen bronchioles and facilitate breathing. A more recently developed oral medication, montelukast (Singulair), blocks leukotriene receptors in the lung. During severe asthma attacks, injected epinephrine is critical. Epinephrine will open airways to ease breathing through direct hormonal action.

A relatively new treatment for allergic asthma, omalizumab (Xolair), can prevent sensitization. Omalizumab is a monoclonal antibody administered by injection every 2–4 weeks that selectively binds IgE. It prevents IgE from binding to Fc receptors on mast cells, so it prevents sensitization. The drug is used to treat patients who do not respond adequately to inhaled steroids. Long-term treatment of an allergy requires identifying the cause, which is often not known. A clinician can identify the offending allergen by individually injecting small amounts of various allergens into a patient's skin (Figure 17.16). The tested skin is then examined for reactions. Spots showing wheals indicate that the patient is sensitive to the corresponding allergen.

People sensitized to allergens are not necessarily doomed to suffer with the allergy their entire lives. A clinical treatment called **desensitization**, or allergy vaccination, can sometimes prevent anaphylaxis. Desensitization is the process of injecting small but increasing doses of allergen over a period of months. The process is thought to produce IgG molecules that circulate, bind, and neutralize allergens before they contact sensitized mast cells. In addition, regulatory T cells (Treg cells) are generated to prevent isotype switching to IgE production. Desensitization has been useful in cases of asthma and bee sting allergies (but not food allergies), although results vary and are by no means guaranteed.

Why don't all antigens generate type I hypersensitivity? Most antigens do not elicit high levels of IgE antibodies. What makes certain antigens (but not others) capable of eliciting high IgE levels, or what makes different individuals prone to different allergies, is unclear, but the differences may be related to the presence or absence of antigen-specific regulatory T cells (see IMPACT). The next two types of hypersensitivity are IgG-dependent and do not involve mast cells.

Stages of Type II Hypersensitivity

CASE HISTORY 17.2

Transfusion Confusion: Type II Hypersensitivity

Richard, who has blood type B, was undergoing surgery after a car accident and accidentally received a transfusion with type A blood. Within hours, Richard experienced chills and lowered blood pressure. His urine turned red with blood. All these symptoms indicate an ABO blood group incompatibility.

This case history describes a form of **type II hypersensitivity**. Type II hypersensitivity starts with specific IgG antibody binding to cell-surface antigens—in this case, donor RBCs. The major characteristic that differentiates type II from type III hypersensitivity (considered next) is that type II hypersensitivity involves antibody binding to cell-surface antigens, whereas type III hypersensitivity involves antibody binding to <u>soluble</u> antigens. The classic type II hypersensitivity reaction is illustrated by blood group incompatibilities arising from transfusions.

Human blood types fall into four major groups—A, B, AB, and O—in a classification system called the **ABO blood group system**. (Rh blood groups will be described next.) Blood group incompatibility occurs when specific antibodies in the serum of an individual of one blood type (A or B) bind to antigens on RBCs of a different blood type. Type B individuals carry specific anti-A antibodies in their bloodstream, whereas type A individuals carry specific anti-B antibodies (Table 17.5). These antibodies explain why a type A person cannot receive blood from a type B individual and vice versa. Donor RBCs that contain the B antigen are destroyed (lysed) by anti-B antibodies when transfused into a type A person. Lysis occurs as a result of classical complement activation (see Section 15.7) or antibody-dependent, cell-mediated cytotoxicity (Section 15.5).

Figure 17.16 Allergy Skin Testing

Skin testing is generally the test of choice to evaluate allergies because it is a quick, accurate, nearly painless procedure. It can easily be performed in a clinic and provides results in less than 20 minutes.

A. In a skin test, drops of allergen extracts are placed on the forearms or the back and then pricked with a plastic device or a small needle.

B. Allergies reveal themselves as red, itchy, mosquito bite–like bumps called wheals. These itchy spots typically go away on their own in less than an hour.

Major Human Blood Types and Donor Compatibility

| Blood Type Antigen | Serum Antibodies | Recipient | Donor |
		Acceptable Donor Blood Types	Unacceptable Donor Blood Types
A	Anti-B	A, O	B, AB
B	Anti-A	B, O	A, AB
AB[a]	Neither A nor B	A, B, AB, or O	
O	Anti-A and Anti-B	O[b]	A, B, AB

[a]Universal recipients (blood group AB and Rh-positive—that is, AB-positive) can receive blood from donors of any blood type.
[b]Universal donors (blood type O and Rh-negative—that is, O-negative) can donate blood to recipients of any blood type.

Type AB people carry neither antibody, and type O individuals, whose RBCs contain neither A nor B antigen, carry both anti-A and anti-B antibodies. Consequently, type AB people, having neither anti-A nor -B antibodies, are called universal recipients because they can receive RBCs of any blood type. In contrast, type O people are called universal donors because their RBCs, devoid of A and B antigens, can be donated to type A, B, AB, or O individuals. Note, however, that true universal recipients must also possess the Rh antigen on red blood cells (thus they are called AB-positive), and universal donors must lack the Rh antigen (they are called O-negative).

Note: Researchers think that anti–group A and anti–group B antibodies arise early in life, after encounters with environmental antigens whose structures mimic blood group antigens. For instance, anti–group A antibodies in a type B person are thought to arise after an encounter with influenza virus, which possesses epitopes similar to those of the A blood group. For a similar reason, anti–group B antibodies are believed to develop in type A individuals after encounters with Gram-negative bacteria in their intestine.

Another form of type II hypersensitivity is **Rh incompatibility disease** of the fetus and newborn, also known as erythroblastosis fetalis (hemolytic disease of the newborn; **Figure 17.17A**). Similar to ABO antigens, Rh antigens are chemical groups on RBCs. A mother who is Rh-negative can react to a fetus that inherited the Rh$^+$ gene from the father. Incompatibility between mother (Rh$^-$) and fetus (Rh$^+$) can lead to a hypersensitivity reaction in the fetus, seen mostly in a second pregnancy (**Figure 17.17B**).

When an Rh$^-$ mother gives birth to an Rh$^+$ firstborn child, fetal Rh$^+$ RBCs can incidentally enter the mother's bloodstream during delivery. The Rh$^-$ mother's immune system will make anti-Rh$^+$ IgG antibodies. The presence of anti-Rh$^+$ antibodies (IgG) is a problem during a subsequent pregnancy because they can cross the placenta to attack an Rh$^+$ fetus. Fortunately, giving the mother anti-Rh$^+$ immunoglobulin (called RhoGAM) by intramuscular injection at 30 weeks gestation and within a day after the birth of each Rh$^+$ child can prevent Rh incompatibility disease. This antibody binds the fetal Rh$^+$ antigen, preventing the mother's immune system from making Rh$^+$ antibody and the attendant memory B cells. Passive immunity lasts only 4–6 weeks—long enough to clear the fetal antigens from the mother's bloodstream.

Why doesn't hemolytic disease of the newborn happen when a fetus having type AB blood is born to a type A mother? Doesn't the mother have anti-B antibodies? She does, but by and large they are IgM antibodies. IgM antibodies do not cross the placenta and cannot reach the bloodstream of the fetus. Anti-Rh$^+$ antibodies are IgG and can cross the placenta.

Stages of Type III Hypersensitivity

CASE HISTORY 17.3

The Hives Have It: Type III Hypersensitivity

A 16-year-old girl, Elaine, is treated for acne with minocycline (a tetracycline derivative). Two weeks later she develops a fever and an urticarial rash (**Figure 17.18**). Urticaria—a.k.a. hives— is an itchy rash caused by tiny amounts of fluid that leak from blood vessels just under the skin surface. Elaine also complains of muscle and joint pains. Her symptoms resolve after 5 days of treatment with anti-inflammatory steroids.

Elaine's case is an example of **type III hypersensitivity**, also called **immune complex disease**. In contrast to type II hypersensitivity, type III hypersensitivity is initiated when IgG antibody binds to soluble antigen, which in this case is minocycline. Normally, macrophages readily engulf and clear antigen-antibody complexes, but excessive amounts of immune complexes may overwhelm the system and circulate in the blood (**Figure 17.19**). The complexes can embed in vessel walls (particularly in the kidneys and joints) and bind complement factor C1 of the classical pathway (Section 15.7). The binding of C1 triggers a cascade in which certain complement fragments recruit and activate polymorphonuclear leukocytes (PMNs). The complement-activated PMNs release lysosomal proteases and reactive oxygen species that damage host cells in the area. The resulting inflammation can cause skin eruptions (called urticaria; Figure 17.18), damage the glomeruli of the kidneys (sites of blood filtration), and cause joint pain.

Serum sickness is a form of type III hypersensitivity reaction caused when a host is injected with antiserum or other proteins derived from animals, such as antivenoms or hormones. These individuals developed antibodies to the animal proteins from previous exposures. Note that a common hypersensitivity to a different

Figure 17.17 Type II Hypersensitivity

Rh incompatibility disease of the newborn.

A. This infant presented with jaundice 8 weeks after birth. The cause was hemolytic disease of the newborn owing to Rh incompatibility. Severe cases can lead to intrauterine death of the fetus.

antibiotic, penicillin, is usually a type I anaphylaxis whose symptoms appear within minutes of antigen exposure. But penicillin can also cause a longer-term type III hypersensitivity reaction.

Stages of Type IV Hypersensitivity

CASE HISTORY 17.4

TB or Not TB? Type IV Hypersensitivity

The patient, Asuncion, is an 8-year-old girl from Argentina. She received the BCG vaccine for tuberculosis about 1 year before coming to the United States. Upon entering school in the United States, she has to take a skin test for tuberculosis (TB). Asuncion tries to refuse but is told it is a requirement. She allows the nurse to apply the test to her arm. Three days later, the test site has a large red lesion and the skin is starting to slough (peel off). Does she have TB?

Type IV hypersensitivity, also known as **delayed-type hypersensitivity (DTH)**, is the only class of hypersensitivity triggered by antigen-specific T cells. Because T cells need time to react and proliferate to cause a response, a reaction generally takes 24–72 hours to appear. This delay distinguishes type IV hypersensitivity from the more rapid antibody-mediated allergic reactions (types I–III). In the case study, Asuncion was initially sensitized by vaccination with BCG

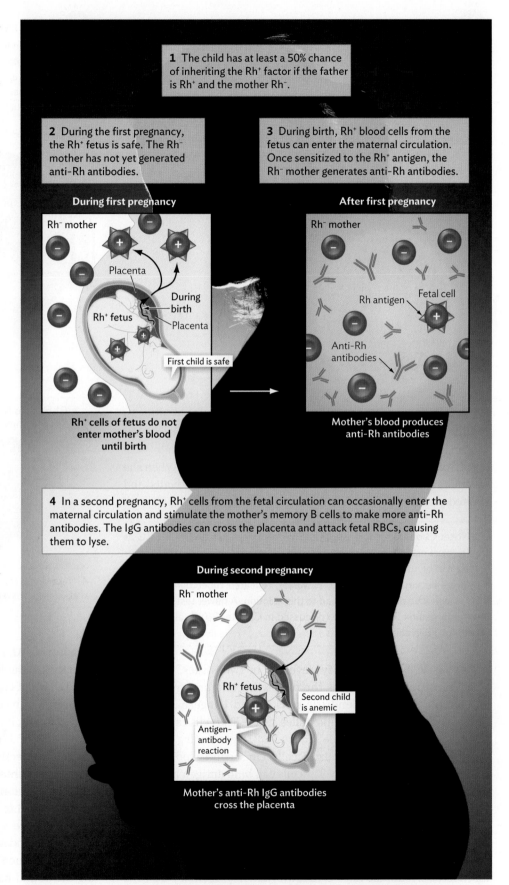

1 The child has at least a 50% chance of inheriting the Rh⁺ factor if the father is Rh⁺ and the mother Rh⁻.

2 During the first pregnancy, the Rh⁺ fetus is safe. The Rh⁻ mother has not yet generated anti-Rh antibodies.

3 During birth, Rh⁺ blood cells from the fetus can enter the maternal circulation. Once sensitized to the Rh⁺ antigen, the Rh⁻ mother generates anti-Rh antibodies.

During first pregnancy

Rh⁻ mother

Placenta

During birth

Rh⁺ fetus

Placenta

First child is safe

Rh⁺ cells of fetus do not enter mother's blood until birth

After first pregnancy

Rh⁻ mother

Rh antigen   Fetal cell

Anti-Rh antibodies

Mother's blood produces anti-Rh antibodies

4 In a second pregnancy, Rh⁺ cells from the fetal circulation can occasionally enter the maternal circulation and stimulate the mother's memory B cells to make more anti-Rh antibodies. The IgG antibodies can cross the placenta and attack fetal RBCs, causing them to lyse.

During second pregnancy

Rh⁻ mother

Rh⁺ fetus

Second child is anemic

Antigen–antibody reaction

Mother's anti-Rh IgG antibodies cross the placenta

B. Development of anti-Rh antibodies during pregnancy and Rh incompatibility during a second pregnancy.

Figure 17.18 Type III Hypersensitivity

Serum sickness can cause an urticarial rash, a form of type III hypersensitivity that can follow the administration of serum. The rash is similar to that formed after the administration of nonprotein drugs (for example, minocycline).

Figure 17.19 Mechanism of Type III Hypersensitivity

Step 1: Small complexes of antibody (Ab) bound to antigen (Ag) deposit on vascular walls. **Steps 2 and 3:** Complement factor C1 binds to the Fc regions of the antibodies and triggers a complement cascade. **Step 4:** Products of the cascade attract neutrophils that deposit lysosomal components that damage the vessel wall endothelial cells.

1 Soluble Ag and Ab form complexes that embed in vessel wall.

Immune complex formed

2 C1 binds to Fc region of complex.

3 C1 triggers complement cascade that attracts neutrophils.

4 Neutrophils release lysosomal contents that damage vessel and tissues.

(an attenuated strain of *Mycobacterium bovis*, a close relative of *M. tuberculosis*). Because the organism grows intracellularly, the vaccination produced a cell-mediated immunity, complete with preactivated memory T cells (similar to memory B cells). When she was reinoculated by the skin test, the memory T cells activated and elicited a localized reaction at the site of injection.

⊙ **Note:** Current evidence suggests that people vaccinated with BCG as children will not typically have a positive reaction as adults. The tuberculin test can be used to screen adults for tuberculosis regardless of whether they were vaccinated with BCG. However, repeated tuberculin testing in BCG-vaccinated people could lead to this reaction. A newer test, called the interferon-gamma release assay (discussed in Chapter 20), is also used to screen for TB in children or adults.

Type IV hypersensitivity develops in two stages. The first stage, sensitization, involves the processing and presentation of antigen on cutaneous dendritic cells (called Langerhans cells). These APCs travel to the lymph nodes, where T_H0 cells can react to them as described earlier (see Figure 16.21), generating activated T cells and a subset of memory T cells. On second exposure, two routes lead to DTH. Memory T_H1 cells can bind antigen presented on class II MHC receptors on APCs at the infection site. T_H1 cells release the inflammatory cytokines IFN-gamma, TNF-beta, and IL-2 (Figure 17.20A), which recruit macrophages and PMNs to the site and then activate macrophages and natural killer (NK) cells to release inflammatory mediators that damage innocent, uninfected bystander host cells. Alternatively, memory T_C cells (cytotoxic T cells) can bind antigen presented on class I MHC

receptors of *M. tuberculosis*–infected cells, become activated, and directly kill the host cell presenting the antigen (Figure 17.20B). A hallmark of type IV hypersensitivity is that transferring WBCs, not serum, can confer the sensitivity to a naive animal. The reason is that T cells, not serum antibodies, bring about the reaction.

Forms of DTH reactions include allergic contact dermatitis (such as a reaction to poison ivy or latex) and allograft rejection after transplantation surgery. An allograft is a tissue graft from a donor who belongs to the same species (human) but is not genetically identical to the recipient. The antigens involved with contact dermatitis are usually small haptens that are not immunogenic until they bind and modify normal host proteins (Figure 17.21). The hapten-modified protein is processed, and fragments containing the bound hapten are presented on MHC I complexes on the surface of antigen-presenting cells.

ANIMATION ▶ **Figure 17.20 Mechanisms of Damage in Delayed-Type Hypersensitivity (Type IV)**

1 T_H1 cells at the site of infection **react to antigen** on dendritic cells.

3 PMN cells attack the tissues in the area of the antigen and cause damage.

Antigen

Dendritic cell

Cellular infiltrate

Swelling

PMN

Blood vessel

T_H1

T cell

Site of infection

Macrophage

Recruitment

Cytokine secretion

2 T cells secrete cytokines that "call in" macrophages and PMNs from the circulation.

A. T_H1 cells react to antigen on dendritic cells.

Cytotoxic T cells recognize allergen-altered epithelial cells.

T_C T_C

MHC I   T-cell receptor

Antigen-altered epithelial cells

Granzymes and cytokines trigger apoptosis.

Cell lysis/ necrosis, apoptosis

B. Antigen-sensitized cytotoxic T cells recognize allergen haptens attached to proteins of other cells. This recognition triggers granzyme release and cytokine production that triggers apoptosis of the target cell.

Allergies of all types have been on the increase for many years. The reason why, however, has remained an unsolved mystery. Recent research now suggests that the answer may be our obsession with cleanliness, as described in **IMPACT**.

Figure 17.21 Contact Dermatitis

Pentadecylcatechols are chemicals present on the surface of poison ivy leaves. Pentadecylcatechols are haptens that bind to proteins in dermal cells, where the chemicals can activate T cells. The result of the ensuing delayed-type hypersensitivity is contact dermatitis, whose hallmarks are insistent itching and fluid-filled blisters.

An unsaturated pentadecylcatechol

SECTION SUMMARY

- **Allergens** cause the host immune system to overrespond or react against self.
- **Type I hypersensitivity** involves IgE antibodies bound to mast cells by the antibody Fc region. Binding of antigen to the bound IgE causes mast cell degranulation. Symptoms occur within minutes of exposure.
- **Type II hypersensitivity** (for example, Rh incompatibility) involves IgG antibodies binding to cell-surface antigens. This process can trigger cell-mediated cytotoxicity or activate the complement cascade. Symptoms occur within hours of exposure.
- **Type III hypersensitivity** is an immune complex disease involving IgG antibodies forming complexes with small, soluble antigens. The complexes can activate complement or bind to mast cells. The result is the recruitment of PMNs that damage host cells in the area. Symptoms may take weeks to form.
- **Type IV hypersensitivity** (delayed-type hypersensitivity) involves antigen-specific T cells. First, T_H1 cells release cytokines that activate macrophages and NK cells. Macrophages and NK cells cause tissue damage. Second, activated T_C cells can directly kill any host cells that present the antigen. Reaction (for instance, skin rash) is seen within a few days of exposure.

Thought Question 17.2 Why do individuals with type A blood have anti-B and not anti-A antibodies?

Thought Question 17.3 Does a peanut allergy rely more on T_H1 cells or on T_H2 cells?

Thought Question 17.4 Why would you consider blood-type-O people "universal donors" but not universal acceptors of RBCs from type O, A, B, or AB individuals? Who might be considered a universal recipient?

Thought Question 17.5 Patients with primary immunodeficiencies can receive multiple IV immunoglobulin treatments with only minor side effects. Given that IgG antibodies from different individuals have allotypic antigenic differences, why aren't serious hypersensitivity reactions to IVIG treatment more common?

impact

Let Them Eat Dirt!? The Revised Hygiene Hypothesis and Allergies

The incidence of asthma and other hypersensitivity diseases in children has skyrocketed since 1980. Today, an average of 1 of every 10 children in the United States has asthma. In 1980, it was closer to 1 in 30. What might account for this disturbing trend? How many of your own friends have asthma?

An early clue came after the reunification of Germany in the late 1990s when the pediatrician Erika von Mutius compared the allergy and asthma rates of East and West Germany. She expected that children growing up in the poorer, dirtier, less healthful cities of East Germany would have more allergies than youngsters in West Germany, with its cleaner and more modern environment. She found the opposite. Children in polluted areas of East Germany had less intense allergic reactions and fewer cases of asthma than children in West Germany. What does this mean?

Many studies now support what has been called the "hygiene hypothesis." The early version of the hypothesis tied childhood infections to the proper "training" of the immune system. Clearly, the advent of vaccines and improved hygiene have reduced the incidence of these natural infections in the developed world. The hypothesis suggested that without these infections, the immune system was improperly trained to ignore natural irritants like pollen and would overreact to them, producing allergic reactions. It turns out that this hypothesis was too simple. The increased incidence of allergies seems to be less about the decline in infections and more about decreased diversity in the microbiome. (Chapter 14 discusses the microbiome.)

The revised hygiene hypothesis currently states that children who play with many other children or animals early in life are exposed to more microbes in nature (Figure 1). Many of these innocuous microbes can colonize intestines and produce an expanded, more diverse microbiome that trains young immune systems to tolerate irritants that might cause allergies. The hypothesis further suggests that the extremely clean household environments often found in the developed world derail the critical postnatal period of immune system training. In other words, the young child's environment can be "too clean" to effectively challenge a maturing immune system. One piece of evidence supporting the revised hygiene hypothesis is that killing neonatal gut flora with antibiotics has been associated with a greater risk of developing asthma.

Why does the immune system need to be educated in selective tolerance? Before birth, the fetal immune system is suppressed to prevent it from rejecting maternal tissue. Such a low default setting is necessary before birth—when the

Figure 1 Child Undergoing Early Antigen Exposure

mother is giving the fetus her antibodies. But immediately after birth, the child's own immune system must take over and learn how to fend for itself.

So, how might more bacterial diversity help protect a young child from allergies? **Figure 2** presents one model for protection. The antigenic diversity that comes with an expansive microbiome is thought to help drive and maintain selection of a diverse set of foreign antigen–specific regulatory T cells in the gut and lungs (Section 16.3). Foreign antigens like those associated with food, gut microbiota, and ingested or inhaled allergens are presented on dendritic cell MHC II molecules. The MHC II–presented antigens then interact with the T-cell receptors of regulatory T cells (Treg cells) to cause Treg clonal expansion (Section 16.3).

The more diverse the microbiome is, the more diverse the Treg cell population will be. An immune system lacking these Treg cells will be set on a "hair trigger," ready to react strongly to any innocuous foreign antigens for which a Treg cell doesn't exist (Figure 2A). Think of Treg cells as a series of traffic cops, each assigned to control the flow of traffic through a different intersection. If one traffic cop fails to show up, chaos ensues at that intersection. By analogy, if a Treg cell that reacts to ragweed is not generated early in life, the child can develop hay fever.

This army of Treg cells can down-regulate inappropriate immune responses to foreign antigens by secreting anti-inflammatory cytokines like IL-10 (Figure 2B). Collectively, these cytokines can inhibit mast cell degranulation and suppress T_H1, T_H2, and T_H17 functions that are associated, respectively, with the activation of cytotoxic T cells, the IgE class switch, and inflammation. All of these are important components of various allergic responses. It also appears that changing the microbiome in one mucosal compartment (the intestine) can affect the immune system in another compartment (the lungs). These cross-organ effects show communication between different mucosal immune compartments. Thus, it appears that the mucosal immune system is a system-wide organ in which stimulation in one compartment (lung) can lead to changes in distal areas.

Figure 2 How Microbes Might Prevent Allergies

A. A limited microbiome does not generate a diverse set of regulatory T cells that can recognize and respond to certain foreign antigens, like allergens.

B. A diverse microbiome will generate and maintain a diverse set of regulatory T cells (Treg cells) that can limit inflammatory T_H1, T_H2, and T_H17 responses to allergens and inhibit mast cell degranulation.

The take-home message from all of these studies is that, as with a muscle, exercising your immune system by engaging microbes strengthens your immunity.

DISCUSSION QUESTION

If exposure to more environmental microbes helps train a healthy immune system, should we stop washing our hands?

17.4 Immunodeficiencies and Lymphoid Cancers

SECTION OBJECTIVES

- Distinguish primary and secondary immunodeficiencies.
- Describe the common forms of severe combined immunodeficiencies.
- Compare the underlying defects and outcomes of bare lymphocyte syndrome, X-linked agammaglobulinemia, hyper IgM syndrome, selective IgA deficiency, and complement deficiencies.
- Discuss the relationship between our immune system and cancer.
- Distinguish lymphoma, leukemia, plasmocytoma, and multiple myeloma.
- Explain the defects leading to B-cell neoplasms.

CASE HISTORY 17.5

The Missing Thymus

Erica's pregnancy went smoothly, even though she knew from a prenatal sonogram that her baby, Silas, would be born with a cleft lip and palate. But those were two easily correctable malformations (**Figure 17.22A**). After Silas was born, however, blood tests revealed he had a low WBC count, and flow cytometry discovered a lack of T cells, which led doctors to check his thymus. They were startled when a computed tomography (CT) scan of his chest found <u>no</u> evidence of a thymus (**Figure 17.22B**). Without a thymus, Silas could not make T cells. A subsequent DNA test, called fluorescence in situ hybridization (FISH), revealed a deletion in chromosome 22 (**Figure 17.22C**). The deletion, called 22q11.2, is known to cause **DiGeorge syndrome**, an immunodeficiency disease that can produce many congenital defects, including cleft palate and thymus aplasia (no thymus). Unless Silas received a thymus transplant, he would probably die from an infection before he was 2 years old. Luckily, Erica's baby received a transplant from a matched donor and is alive today.

Silas's case reveals an immune system that is both remarkably resilient and frighteningly fragile. Its layers of innate and adaptive mechanisms, when fully functioning, provide a nearly impenetrable defense against infection. With one missing gene, however, the system breaks down, its owner now vulnerable to the microbial world. Immunodeficiency diseases are disorders in which part of a person's immune system is missing or diminished. Defects can occur in cell-mediated (T cell) and/or antibody-mediated (B cell) pathways.

Figure 17.22 DiGeorge Syndrome

A. Silas was born with a rare congenital disease called DiGeorge syndrome that left him with a cleft palate and the inability to fend off infections. He is shown here at the age of 1 year, before (left) and after (right) palate surgery.

B. Thymus of an infant, over the heart. This organ is important for T-cell development. A patient with DiGeorge syndrome is left with little to no thymus.

C. The fluorescence in situ hybridization (FISH) assay can detect deletions of the 22q11 region. Fluorescent DNA probes bind in the area of the deletion. Image shows one normal chromosome 22 with one orange signal and one green signal, and a del (22) chromosome with only one green signal.

Either way, the afflicted person has trouble clearing infections, usually the first sign of an immunodeficiency (**Table 17.6**).

For Silas, the lack of a thymus caused a severe depletion of his T cells. Individuals with T-cell defects are generally susceptible to infection by viruses, fungi, and some protozoa. In contrast, patients defective in antibody formation (B cells) are chiefly prone to multiple infections with encapsulated bacterial pathogens such as streptococci, *Haemophilus*, and the meningococcus *Neisseria meningitidis*; and with the *Giardia* protozoa. Recurrent infections with encapsulated bacteria happen because the patient cannot make the antibodies needed to coat the pathogen's capsule and aid phagocytosis.

Table 17.6
Characteristic Infections of Primary Immunodeficiencies

Immune Component	Primary Pathogen	Primary Site(s)	Clinical Example
T cells	Intracellular bacteria, viruses, protozoa, fungi	Nonspecific	Severe combined immunodeficiency (SCID), DiGeorge syndrome
B cells	*Streptococcus pneumoniae, Streptococcus pyogenes, Haemophilus influenzae*	Lung, skin, central nervous system (CNS)	Deficiency in immunoglobulin G (IgG), IgM
	Enteric bacteria and viruses	Gastrointestinal tract, nose, eye	Deficiency in IgA
Phagocytes	*Staphylococcus* spp., *Klebsiella* spp., *Pseudomonas* spp.	Lung, skin, regional lymph node	Chronic granulomatous disease
Complement	*Neisseria* spp., *Haemophilus* spp., *S. pneumoniae*, other *Streptococcus* spp.	CNS, lung, skin	C3, factors I and H, late C components

How are primary immunodeficiencies different from secondary immunodeficiencies? **Primary immunodeficiencies** have a genetic basis and usually manifest in early childhood, like Silas's did. **Secondary immunodeficiencies**, however, can be acquired at any age as a consequence of certain infections (HIV, for instance), the use of immunosuppressive drugs (as with the kindergarten teacher in the chapter-opening case history), or radiation therapy. In practice, an immunodeficiency should be suspected in any person experiencing recurrent, serious infections (Table 17.6).

How common are immunodeficiencies? Estimates are that 1 out of every 2,000 babies is born with a primary (heritable) immunodeficiency. More than 400 primary immunodeficiency diseases exist. B-cell defects causing antibody deficiencies are the most common, accounting for 50%–60% of primary immunodeficiencies. T-cell disorders account for about 5%–10% of primary immunodeficiencies.

Chapter 16 described several immune system deficiencies. Examples included severe combined immunodeficiency, or SCID (covered in the opening case history); bare lymphocyte syndrome (Case History 16.1); and hyper IgM syndrome (Case History 16.2). In this discussion we will expand on those diseases and

introduce some other common immunodeficiencies. **Table 17.7** lists key features of these dysfunctions and indicates applicable case histories in the text. The immune system schematic in **Figure 17.23** shows where various primary immunodeficiencies make their presence known.

Figure 17.23 Correlation between Primary Immunodeficiencies and Defects in the Immune System

See the text for details of specific diseases. Each X marks the defect leading to an immunodeficiency. The quantity of cells or serum antibody levels are low downstream of an X.

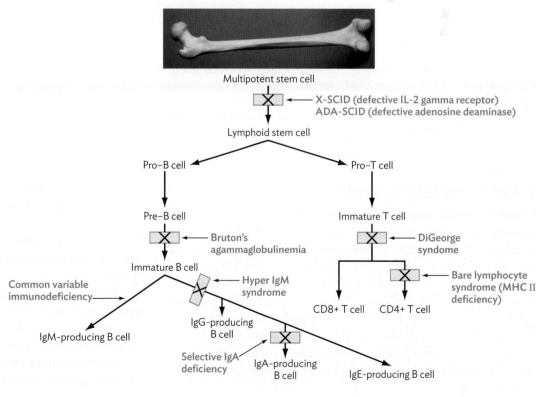

Table 17.7

Examples of Primary Immunodeficiencies[a]

Cells Involved	Disease	Cause[b]	Case History	T-Cell Level	B-Cell Level	Immunoglobulin Levels
T cells	X-SCID	Defective gamma chain in receptors for interleukin 2 (IL-2); X-linked	Chap. 16 opening case	Absent	Normal but not functioning	All low
	ADA-SCID	Adenosine deaminase deficiency		Absent	Absent	All low
	DiGeorge syndrome	Defective development of thymus and parathyroid glands	17.5	Absent	Variable	Variable
	Bare lymphocyte syndrome	Major histocompatibility complex (MHC) II deficiency; defective gamma-chain IL-2 receptor	16.1	Low CD4 T cells	Normal	All low
B cells	X-linked agammaglobulinemia (Bruton's disease)	Defective tyrosine kinase; X-linked		Normal	Absent or low	All low
	Common variable immunodeficiency (CVID)	Defective B-cell development into plasma cells; cause unclear		Normal	Normal or low	IgG and IgA low; IgM sometimes low
	Hyper IgM syndrome	T-cell CD40 ligand defect; X-linked	16.2	Normal	Normal	Elevated IgM; others low
	Selective IgA deficiency	Unknown; failure to switch classes; most common immunodeficiency disorder		Normal	Normal	IgA low
Phagocytes	Leukocyte adhesion deficiency	Lack of integrins; failure of extravasation	15.1	Normal	Normal	Normal
	IRAK deficiency[c]	Toll-like receptor signaling, low cytokine secretion	15.2	Normal	Normal	Normal
	Chronic granulomatous disease	NADPH oxidase deficiency; no respiratory burst; X-linked		Normal	Normal	Normal
	Complement deficiency	Early components; late components	15.3	Normal	Normal	Normal

[a]Abnormal lab results are in color.

[b]X-linked means that the mutation involved is on the X chromosome, so the disease is more prevalent in males.

[c]IRAK = IL-1 receptor–associated kinase.

T-Cell Immunodeficiencies

Severe combined immunodeficiencies (SCIDs), introduced in the opening case history of Chapter 16, are characterized by a lack of T cells. But B-cell and NK cell numbers can also be low, high, or even normal, depending on the variety of SCID syndrome. Mutations in any one of 12 genes can lead to a SCID disease. The most common form involves a defective gene on the X chromosome, producing a defective interleukin 2 (IL-2) receptor (the gamma chain) on T cells. Because other cytokine receptors share the gamma chain, this defect prevents many cytokines from signaling T cells to proliferate. Note that because this is an X-linked disorder, it is most commonly seen in males.

The second-most-common form of SCID develops from a deficiency in adenosine deaminase (ADA), an enzyme that breaks down and recycles adenosine. Diminished recycling of adenosine prevents DNA synthesis and causes apoptosis of T-cell, B-cell, and NK cell precursors. These patients lack all three subsets of lymphocytes.

By 6 months of age, most SCID infants develop infections caused by yeasts, viruses, or *Pneumocystis jirovecii*. Treatments can involve antibiotics, bone marrow stem cell transplantation to generate new T and B cells, ADA enzyme replacement by injection, or **intravenous immunoglobulin (IVIG)**. The protective effect of IVIG against infection is only short-lived, lasting between

2 weeks and 3 months. Long-lasting cure requires the introduction of a functional ADA gene into the patient's genome through gene therapy.

DiGeorge syndrome, suffered by Silas in Case History 17.5, results from mutations in one of several genes, such as those on chromosome 22. The mutations affect specific aspects of embryogenesis. Infants with DiGeorge syndrome can have low-set ears, a midline facial cleft, a small receding mandible (jaw), and a congenital heart disorder. Affected babies also have incomplete development or absence of thymus and parathyroid glands, causing T-cell deficiency and hypoparathyroidism, respectively. Recurrent infections begin soon after birth, but the degree of immunodeficiency varies considerably. Complete DiGeorge syndrome (no thymus) is fatal unless the patient receives a transplant of cultured thymus tissue.

MHC II deficiency (**bare lymphocyte syndrome**; discussed in Case History 16.1) is a rare disease caused by a failure to express MHC II molecules on antigen-presenting cells. The mutations involved are <u>not</u> in the MHC II genes themselves but occur in genes encoding <u>regulators</u> of the MHC II genes. The deficiency produces severe defects in cellular and humoral immunity. Without MHC II, antigens cannot be presented to T_H0 cells (Section 16.3); T_H0 cells, then, will not form T_{FH}, T_H1, or other types of helper T cells. Without T_H cells to "talk" to, B cells cannot initiate heavy-chain class switching or differentiate into plasma cells, explaining why no antibodies are made. Cell-mediated immunity is defective because without T_H1 helper cells, cytotoxic T cells cannot be activated. As you might guess, patients with this disease are extremely vulnerable to infections.

B-Cell Immunodeficiencies

Patients with **X-linked agammaglobulinemia (Bruton's disease)** have mutations in the Bruton tyrosine kinase gene (*BTK*) on the X chromosome and produce very little immunoglobulin (agammaglobulinemia). The BTK enzyme is essential for B-cell development and maturation; without it, no B cells—and hence no antibodies—can form. As a result, male infants have small tonsils and do not develop lymph nodes. They suffer from recurrent pyogenic (pus-producing) lung, sinus, and skin infections caused by encapsulated bacteria (for example, *Streptococcus pneumoniae*, *Haemophilus influenzae*). Treatment of infections requires IVIG and antibiotics.

Common variable immunodeficiency (CVID), or adult-onset hypogammaglobulinemia ("hypo" meaning low levels), is characterized by low immunoglobulin (Ig) levels. B cells look normal but do not develop into Ig-producing plasma cells. However, the degree and type of serum immunoglobulin deficiency vary from patient to patient. Some patients' B cells secrete IgM but do not undergo a class switch to IgG or IgA. B cells in other patients may produce almost no antibodies. In most patients the molecular defect is

Figure 17.24 Defect That Causes Hyper IgM Syndrome

A. T-cell receptors (TCRs) on a helper T cell (T_H2) recognize which B cell to help by binding to antigen (Ag) presented on B-cell MHC II. Binding triggers differentiation into an antibody-secreting plasma cell. Interaction between T-cell CD40 ligand (CD40L) and CD40 on the B cell authorizes the B cell to switch antibody isotype.

B. T cells that lack CD40L cannot authorize a class switch. As a result, the plasma cells produce IgM only.

unknown. Infections tend to develop in adulthood and most commonly affect the upper and/or lower respiratory tracts. Treatment requires IVIG.

Patients with **hyper IgM syndrome** (Case History 16.2) have normal or elevated serum IgM levels but low levels (or complete absence) of other antibodies. Most cases are caused by mutations in an X-linked gene that encodes a membrane protein called CD40 ligand (CD154) on surfaces of activated helper T cells (Figure 17.24). Normally, CD40 ligand on T_H cells interacts with B cells (and signals the switch from IgM to IgA, IgG, or IgE production). B cells will not switch isotypes without CD40L binding. Hyper IgM syndrome is characterized by recurrent pyogenic bacterial infections during the first 2 years of life (similar to X-linked agammaglobulinemia). Diagnosis comes from detecting normal or elevated serum levels of IgM and low levels, or an absence, of other immunoglobulins. Treatment is monthly IVIG. Bone marrow transplantation can be curative for the X-linked form if an MHC-identical sibling donor is available.

Selective IgA deficiency is the most common primary immunodeficiency. Patients lack secretory IgA, but the exact cause is largely unknown. The disease is called "selective" because the levels of other serum immunoglobulins, such as IgM and IgG, are normal. B cells can switch isotype classes—just not to IgA. These patients have normal T-cell, neutrophil, and complement system function that can ward off many types of infections but still suffer recurrent mucosal infections of the ear, lung, or GI tract. This syndrome has no specific treatment other than antibiotics to cure infections.

The incidence of autoantibodies (antibodies to the body's own tissue) in IgA-deficient patients is high. As a result, autoimmune diseases are common and include rheumatoid arthritis and systemic lupus erythematosus (symptoms include sore and swollen joints of the hands or knees, a facial rash, and anemia).

Complement Deficiencies

Complement in serum is important for opsonizing and killing bacteria. **Complement deficiencies** are inherited as autosomal traits (that is, not linked to the X chromosome) and can occur in the early complement components (C1–C4) or late complement components (C5–C9; we describe complement in Section 15.7). Early complement components bind bacteria and facilitate phagocytosis (opsonization) of encapsulated bacteria, so patients with these defects are more susceptible to infections with *Streptococcus pneumoniae*, *Streptococcus pyogenes*, and *Haemophilus influenzae*. The late components form the bacteriolytic membrane attack complex (MAC). Defects in MAC formation will render a person more susceptible to Gram-negative infections, most notably by *Neisseria meningitidis*. There is no specific cure for complement deficiencies. Infected patients are treated with antibiotics and can receive plasma that contains complement. Patients with complement deficiencies display normal humoral and T-cell responses and should receive normal immunizations.

Immune Surveillance

Can defects in the immune system lead to cancer? **Cancer**, or a **neoplasm**, is defined as new growth of abnormal cells. **Tumors** are masses of abnormal cells that form within a tissue. A tumor can be **benign** (meaning it is a slow-growing, self-contained mass that does not infiltrate other tissues) or **malignant** (abnormal cells with rapid, uncontrolled growth that can spread to form tumors in other tissues). Spontaneous genetic damage can transform a normal cell into a malignant, cancerous one.

All people, including you, have probably had a cancerous cell develop in their body. Fortunately, an important function of our immune system is **immune surveillance**, a process that detects and eliminates cancer cells as they arise. This search-and-destroy function is possible because cancer cells are marked with surface antigens not present on normal cells. Our immune system can recognize and target those antigens.

Without immune surveillance, tumors would arise often. Consequently, dysregulation (that is, impaired regulation) of the immune system can result in a variety of cancers, particularly of lymphoid cells. Patients with primary immunodeficiencies, AIDS patients, and immunosuppressed transplant patients are particularly prone to cancer. The malignancies most commonly manifest as aggressive B-cell lymphomas and are often associated with Epstein-Barr virus (EBV) infection (Chapter 21). EBV, the cause of infectious mononucleosis, is a DNA virus that infects B cells and becomes latent by forming the equivalent of a circular plasmid. Expression of some EBV genes can cause a cell to become malignant.

Cancers of Lymphoid Tissue

Lymphoid cancers break down into four groups known as lymphomas, leukemias, plasmocytomas, and multiple myelomas. A **lymphoma** is a solid mass in a lymphoid organ (lymph node, spleen, thymus). **Leukemia** refers to malignant lymphoid cells found primarily in the circulation or bone marrow. Plasma cells can also become malignant. Cancerous growth of a plasma cell at a single site is called a **plasmocytoma**. If cancerous plasma cells appear at multiple sites, mainly throughout bone, the condition is called **multiple myeloma**. The following section describes some lymphoid tissue cancers.

B-CELL NEOPLASMS Cancers arising in B cells often come about when genes whose products normally control cell division accidentally move near the powerful promoter of an immunoglobulin gene. The division control gene will overexpress and either stimulate cell division or prevent cell death (apoptosis). Either way, the affected cell continues to replicate. For instance, **Burkitt's lymphoma** (which can also manifest as a leukemia) develops when the Epstein-Barr virus infects B cells and causes a cell proliferation gene, c-*MYC*, to move (translocate) next to highly expressed antibody heavy-chain or light-chain genes (Section 16.2). The result is uncontrolled cell division.

Another example is **follicular lymphoma** (Figure 17.25A). Follicular lymphoma develops when a gene that normally prevents apoptosis in a B cell, *BCL2*, moves and comes under the control of immunoglobulin gene promoters. The resulting overexpression

Figure 17.25 Lymphoid Cancers

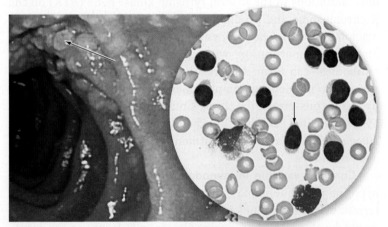

A. Follicular lymphoma of the duodenum. The follicles (arrow) consist of monoclonal IgA. A *BCL2* gene translocation caused extensive expression of BCL2 protein in the follicle.

B. Chronic lymphocytic leukemia (CLL). Blood smear showing abnormally high numbers of lymphocytes (arrow). These neoplastic lymphocytes show round to slightly irregular nuclei and scant cytoplasm. CLL is the most common leukemia in adults.

Figure 17.26 Reed-Sternberg Cell of Hodgkin's Lymphoma

The classic neoplastic Reed-Sternberg cell is 20–50 µm, with a large bilobed or multilobed nucleus. A large, round, eosinophilic central nucleolus is always present.

Figure 17.27 Serum Electrophoresis of Serum Immunoglobulins

A. Normal serum electrophoresis.   **B.** Multiple myeloma spike.

Fraction	Relative %	mg/ml
Albumin	54.2	44
Alpha 1	3.3	2.7
Alpha 2	10.3	8.4
Beta	9.8	8
Gamma	22.4	18.4+

Electrophoresis of serum from a healthy patient (A) and a patient with multiple myeloma (an immunoglobulin-secreting tumor; B). Electrophoresis gels are shown in the lower panels. Densitometry tracings of the protein bands in the gels are shown in the upper panels. The tracing in B shows an abnormally narrow gamma-globulin band, indicating the presence of a monoclonal (and therefore abnormal) immunoglobulin. Here the concentration of this protein is about 18 mg/ml, and the "+" on the printout indicates that this value is higher than normal for gamma globulin (reference range = 6–17 mg/ml). In other cases, such a monoclonal protein may be present at much higher concentrations.

of *BCL2* prevents cell death, enabling continued replication and tumor development. These lymph node cancers proliferate slowly but continuously, resulting in a long clinical course.

The most common leukemia in North America and western Europe is **chronic lymphocytic leukemia** (Figure 17.25B). Most cases involve a mutation that eliminates a short RNA molecule that normally inhibits expression of the *BCL2* gene. As noted already, overexpression of *BCL2* stops cell death, and a B-cell malignancy results. To battle this malignancy, physicians inject the patient with antibodies against the B cell–specific surface antigen CD20. The antibody-CD20 complex initiates an immune response that kills B-cell lymphomas.

Hodgkin's lymphoma is a highly survivable B-cell cancer characterized by the spread of painless tumors from one lymph node group to another. Symptoms include fever, night sweats, and weight loss. The tumors contain rapidly dividing, strikingly multinucleated Reed-Sternberg cells derived from B cells (Figure 17.26). Any lymphomas lacking Reed-Sternberg cells are called **non-Hodgkin's lymphomas**. Patients with Hodgkin's lymphoma often have a history of infectious mononucleosis caused by Epstein-Barr virus (EBV). EBV stimulates synthesis of the regulatory factor NF-κB (nuclear factor kappa B) that inhibits apoptosis while stimulating cell proliferation. The resulting Reed-Sternberg cells release cytokines such as tumor necrosis factor (TNF) that contribute to the fever and chronic inflammation evident in these patients. Treatment of early-stage Hodgkin's involves a combination of chemotherapy and targeted radiation treatment. Survival rates approach 90%.

PLASMA CELL NEOPLASMS (MULTIPLE MYELOMA) Multiple myeloma accounts for about 1% of all cancers and typically occurs in older patients (more than 66 years old). Because these tumors form in bone marrow, patients can experience excruciating bone pain. Cancerous plasma cells multiply to large numbers and secrete

massive amounts of antibody. Secreted antibody light chains can form clumps called amyloid that deposit in, and damage, various organs, particularly the kidney. Free light chains secreted from multiple myeloma cells can even be secreted in urine and are referred to as Bence-Jones proteins.

A single malignant plasma cell generates a group of offspring that make antibody reactive toward a single epitope. The dramatic overproduction of a monospecific antibody (also called a monoclonal antibody) can be seen when blood serum proteins are separated by electrophoresis (Figure 17.27). Excess immunoglobulin registers as a spike in the electrophoresis readout. The levels of all other antibodies are severely decreased, causing an immunosuppression that makes these patients more susceptible to infection.

SECTION SUMMARY

- **Immunodeficiency patients** usually present with recurrent infections.
- **T-cell deficiencies** account for 5%–10% of all immunodeficiencies, including **severe combined immunodeficiencies**, **DiGeorge syndrome**, and **bare lymphocyte syndrome**.
- **B-cell deficiencies** are the most common immunodeficiencies. They include **X-linked agammaglobulinemia (Bruton's disease or Bruton's agammaglobulinemia)**, **common variable immunodeficiency**, **hyper IgM syndrome**, and **selective IgA deficiency**.

- **Intravenous immunoglobulin (IVIG)** treatment is useful for treating infections in immunodeficiency patients.

- **Complement deficiencies** can result in recurrent infections because of poor opsonization (C1–C4 defects) or a failure to kill through MAC formation (C5–C9 defects).

- **Immune surveillance** by the immune system normally eliminates cancer cells as they arise.

- **Immunodeficient patients** are more prone to cancers.

- **Lymphoid cancers** include lymphomas (tumors in a lymphoid organ), **leukemia** (cancerous lymphoid cells in circulation or bone marrow), **plasmocytoma** (cancerous plasma cell growth at a single location), and **multiple myeloma** (multiple plasmocytomas, usually in bone).

- **Lymphoid cancers** arise when certain natural cell cycle regulatory genes (c-*MYC* or *BCL2*) are overexpressed after moving to and coming under the control of a heavy- or light-chain gene. **Epstein-Barr virus** can cause these rearrangements.

- **Multiple myeloma cells** secrete high levels of a specific antibody.

Thought Question 17.6   Which cell type of the immune system would be missing in a patient lacking surface MHC I molecules? *Hint:* Which lymphocyte requires antigen presentation by MHC I?

Thought Question 17.7   The class of immunoglobulin used to treat or prevent infections in immunodeficient patients is IgG pooled from 1,000–2,000 random individuals. The donors were not chosen because of documented prior exposure to the infectious agent being treated. So, why would this IgG protect the patient?

17.5 Autoimmunity and Organ Transplants

SECTION OBJECTIVES

- Recall how the immune system differentiates self from nonself (foreign) antigens.

- Discuss how autoreactive antibodies can be formed.

- Describe the key features of the autoimmune diseases lupus, Graves' disease, diabetes, rheumatoid arthritis, and myasthenia gravis.

- Explain the basis of transplant rejection and graft-versus-host reactions.

If my immune system is "blind" to my own antigens, how can I develop autoimmune disease?   The ability to distinguish between self antigens and foreign antigens is crucial to our survival; without this self-tolerance, our immune system would constantly attack us from within. Occasionally, however, an individual's immune system loses tolerance against some self antigens and mounts an abnormal immune response against the person's own tissues. The attack can involve antibodies or T cells and is called an **autoimmune response**. Recall that tolerance to self (Section 16.3) comes from deleting T cells in the thymus that bind too strongly to self antigens. Similarly, a self-reacting B cell residing in bone marrow can be deleted through apoptosis if the B cell is exposed to high doses of its BCR-matched self antigen .

Autoimmune responses may or may not produce disease. Almost 30% of the population, including you, will have an autoimmune antibody by age 65, but many will not exhibit disease.

The mechanisms leading to autoimmune disease are essentially hypersensitivity reactions. An autoimmune response results in disease when an autoantibody or autoimmune T cell damages tissue components. Tissue damage can develop when antibody to a self antigen activates antibody-dependent cell-mediated cytotoxicity (ADCC) mechanisms (see Section 15.5). In this scenario, autoantibodies attach to host cell self antigen, the Fc regions of those antibodies bind to Fc receptors on NK cells, and the NK cells kill the tissue cell.

How might autoimmune antibodies form? One proposed mechanism starts with the occasional escape of a self-reacting B cell from the negative selection process. Self-reacting B cells that escape this negative selection are usually not a problem because the specific helper T cells needed to activate them were successfully deleted from the T-cell population. Sometimes, however, a self-reacting B cell—a B cell that makes antibodies against a self antigen—can be activated without <u>specific</u> T-cell assistance. A self-reactive B cell can take up, process, and present a self antigen or a foreign antigen that mimics a self antigen. If the antigen <u>also</u> contains a "nonself" epitope (possibly because of a mutation; Figure 17.28), this nonself epitope will also be presented on the B cell's MHC II molecules. When this happens, T cells reacting to the nonself epitope activate that self-reactive B cell. The self-reactive B cell then pumps out autoantibody to the <u>self</u> antigen, not the <u>nonself</u> antigen. The autoantibodies set about attacking the host.

Some microbial antigens can also trigger autoimmune reactions. A good example is the M protein, a cell-surface protein that covers *Streptococcus pyogenes*, a microbe that causes strep throat and the autoimmune disease rheumatic fever. M protein, which covers the cell surface like fuzz on a tennis ball, contains an epitope that resembles cardiac antigen (such resemblance is known as molecular mimicry). But the heart-like epitope is flanked by streptococcal epitopes that are not related to the human host. The heart-like epitope can bind surface antibody on a self-reacting B cell and be taken up. Because the nonself epitope and the heart-like epitope are contained within the same protein, the nonself (noncardiac) epitope will "piggyback" its way into the B cell and, as just described, can be used by helper T cells to activate the autoreactive B cells. Antibodies to the cardiac antigen are made, bind to cardiac tissue, and trigger autoimmune disease. With *S. pyogenes*, cardiac tissue is damaged and rheumatic fever results. The following section presents other examples of well-known autoimmune diseases.

Figure 17.28 Formation of Autoimmune Antibodies

The model starts with a B cell expressing surface antibody (a B-cell receptor, or BCR) to self epitope A.

A. No T-cell help; no autoantibody production. An antigen from self can contain two epitopes: a normal self epitope A linked to a normal self epitope B. A B cell expressing antibody to one of these self epitopes can present both of the epitopes on MHC molecules. But helper T cells are not available to either self epitope, so no T cell can help stimulate antibody production.

B. T$_{FH}$ cell available that can bind an altered self epitope. If the antigen from self contains a normal self epitope A and an abnormal (mutated) self epitope B, autoantibodies (antibodies directed against self) may be produced. The BCR binds epitope A and processes the linked epitopes, presenting both on the B-cell MHC II receptors. If T$_{FH}$ cells with a T-cell receptor (TCR) that can bind the altered epitope B are present, they can help activate the B cell, which differentiates into a plasma cell that makes autoantibodies to epitope A. Alternatively, the pentagon-shaped antigen can be a foreign epitope (for instance, microbial) that mimics a host self epitope, and the square-shaped antigen can be another foreign epitope that does not mimic host.

Autoimmune Diseases

Many different human autoimmune diseases exist. You may be surprised to learn that some well-known diseases, such as type 1 diabetes, are actually caused by autoimmune mechanisms. As you explore these examples, think about how the immune response was "tricked" into reacting against its own self-interest.

SYSTEMIC LUPUS ERYTHEMATOSUS (SLE)

CASE HISTORY 17.6
Rash of a "Butterfly"

Tamil, a 26-year-old woman, had recently developed joint pain in her wrists, fingers, and ankles. She had also noticed a strange rash over her nose and cheeks that reminded her of a butterfly (**Figure 17.29A**). She told her physician that the redness grew worse after she was in the sun for only an hour. A blood test taken during her visit showed a normal blood cell differential count. However, her serum antinuclear antibody test was highly positive, with a 1:256 dilution; normal levels are negative at 1:8 dilution (we discuss antibody titers in Section 17.2). Further tests revealed that her serum complement levels were much lower than normal. On the basis of the clinical and laboratory findings, the physician told Tamil she had developed systemic lupus erythematosus (SLE). SLE is an autoimmune disease in which complexes form between autoantibodies and the patient's own proteins, such as nuclear material (DNA and histones, for example). The antigen-antibody complexes deposit in various tissues and trigger inflammation.

Figure 17.29 Systemic Lupus Erythematosus

A. Classic butterfly rash of lupus.

B. Granular deposits of antigen-antibody complexes in the glomerulus of the kidney, viewed by immunofluorescence microscopy using fluorescently labeled antibody. The green areas would be absent in a healthy kidney.

The butterfly rash and antinuclear antibodies are hallmarks of SLE disease. The illness is really a type III hypersensitivity reaction in which autoantibodies mistakenly made to many different host antigens bind those antigens and form circulating antigen-antibody complexes. The complexes deposit in the basement membranes of various organs, especially kidneys (**Figure 17.29B**), and activate the classical complement pathway (Chapter 15), causing inflammation and depleting serum complement levels. Inflammation of the kidney leads to leakage of protein and RBCs into the urine.

What causes lupus remains a mystery. Viral infections leading to loss of regulatory T cells that help limit an inflammatory process, as well as hormonal changes during pregnancy, are suspected triggers. Although lupus has no cure, a variety of treatments can reduce symptoms, limit damage to vital organs, and reduce the risk of recurrence.

GRAVES' DISEASE, HASHIMOTO'S DISEASE, AND MYASTHENIA GRAVIS Other autoimmune diseases involve the production of autoantibodies that bind and block certain cell receptors. **Graves' disease** (hyperthyroidism), for example, occurs when an autoantibody is made that binds to the thyroid-stimulating hormone (TSH) receptor present on thyroid gland cells. Autoantibody binding mimics TSH binding and abnormally stimulates production of the thyroid hormones triiodothyronine (T3) and thyroxine (T4). Abnormally high levels of T3 and T4 increase metabolic rate, making the patient hyperenergetic. The thyroid is not destroyed but instead enlarges to form a goiter (**Figure 17.30**). By contrast, **Hashimoto's disease** is a form of <u>hypo</u>thyroidism in which autoantibodies and T cells destroy thyroid follicle cells and thyroglobulin (the protein that produces T3 and T4) instead of stimulating thyroglobulin production. Consequently, patients with Hashimoto's are always tired or fatigued. The inflammation enlarges the thyroid to form a goiter that eventually atrophies.

Myasthenia gravis is a disease whose most striking symptom is profound muscle weakness that eventually affects all skeletal muscles. **Figure 17.31** shows a patient with a drooping eyelid resulting from this disease. The disease is caused by autoantibodies that bind to and block muscle receptors for acetylcholine. Thus, myasthenia gravis is a type II hypersensitivity disease. Acetylcholine is a chemical released from nerves that is needed to transmit nerve impulses across a synaptic junction—here, to a muscle.

DIABETES Type 1 **diabetes** is an autoimmune disease that usually occurs during childhood when T cells attack and destroy the insulin-producing islet cells in the pancreas. The autoimmune reaction involves autoreactive CD4 T cells and cytotoxic T cells (CD8+). As such, it is a form of type IV hypersensitivity. (Unlike patients with type 1 diabetes, those with type 2 continue to produce insulin, but their cells no longer respond to it.) The lack of insulin in type 1 diabetics leads to excessive levels of blood glucose, increased hunger, frequent urination, and a great thirst.

RHEUMATOID ARTHRITIS Rheumatoid arthritis is a chronic autoimmune disease characterized by fatigue and chronically inflamed synovium (soft tissue that lines the joints). One out of

Figure 17.30 Graves' Disease

In this Graves' disease patient, autoantibodies to TSH receptors in the thyroid gland constantly stimulate the gland, causing it to enlarge into a goiter and continually produce hormones T3 and T4. Hallmarks of the condition are an enlarged thyroid, bulging eyes (exophthalmos), heat intolerance, increased energy, difficulty sleeping, diarrhea, and anxiety.

Figure 17.31 Myasthenia Gravis

Autoantibodies against acetylcholine receptors on muscle cells inhibit the transmission of nerve impulses to muscle. The typical symptom is weakness of muscles that gets worse with activity and improves with rest. The muscles around the eyes are usually affected first. This patient's right eyelid muscles do not move properly.

Figure 17.32 Rheumatoid Arthritis

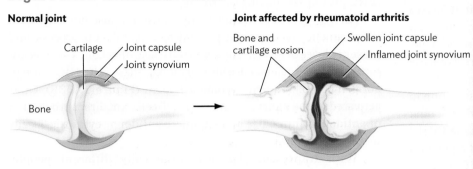

Normal joint

Cartilage
Joint capsule
Joint synovium
Bone

Joint affected by rheumatoid arthritis

Bone and
cartilage erosion
Swollen joint capsule
Inflamed joint synovium

A. Inflammation of the synovium at a joint leads to erosion of bone and cartilage and swelling of the joint capsule.

B. Deformation of hands caused by rheumatoid arthritis.

every 100 people, mostly women, is afflicted. The inflammatory reaction erodes cartilage and bone around the joint, usually in the wrist, ankle, or cervical spine (**Figure 17.32**). The inflamed joint becomes painful, disfigured, and densely packed with lymphocytes, dendritic cells, NK cells, and plasma cells. Inflammation begins with autoantibodies (rheumatoid factors) and immune complex formation in joints that trigger type III hypersensitivity reactions.

Eventually, macrophages and autoreactive CD4 T cells produce pro-inflammatory cytokines, such as TNF-alpha, that worsen the damage. Nonsteroidal anti-inflammatory drugs (NSAIDs) are typically used for early treatment, but more aggressive treatments for more severe disease involve TNF inhibitors methotrexate and etanercept (Enbrel). **Table 17.8** lists other important autoimmune diseases.

Table 17.8
Examples of Autoimmune Diseases

Disease	Autoantigen	Pathology
Type II Hypersensitivity—Mediated by Antibody to Cell-Surface Antigens		
Acute rheumatic fever	Streptococcal M protein, cardiomyocytes	Myocarditis; scarring of heart valves
Autoimmune hemolytic anemia	Rh blood group	Destruction of red blood cells by complement phagocytosis
Goodpasture syndrome	Basement membrane collagen	Pulmonary hemorrhage; glomerulonephritis
Graves' disease	Thyroid-stimulating hormone receptor	Antibody-stimulated T3, T4 production; hyperthyroidism
Myasthenia gravis	Acetylcholine receptor	Electrical transmission interrupted by antibody; progressive muscular weakness
Type III Hypersensitivity—Mediated by Antibody Complexes with Small Soluble Antigens (Immune Complex)		
Systemic lupus erythematosus	DNA, histones, ribosomes	Arthritis; vasculitis; glomerulonephritis
Type IV Hypersensitivity—Mediated by Antigen-Specific T Cells		
Type 1 diabetes	Pancreatic beta cell antigen	Beta cell destruction
Multiple sclerosis	Myelin protein	Demyelination of axons

Organ Donation and Transplant Rejection

One problem with having an immune system is that it makes transplanting an organ from one person to another, called an **allograft**, a very dangerous proposition. Transplants are possible, however, because we have learned how to minimize the risk that the recipient's immune system will reject the donated tissue.

Why are transplants dangerous? Recall that after selection in the thymus, T cells have T-cell receptors (TCRs) that will not bind to self MHC proteins on other cells of the body <u>unless</u> the self MHC has bound to a foreign antigen. However, about 5% of a person's T cells can recognize and directly bind to <u>foreign</u> MHC proteins on transplanted donor cells. In this <u>direct route</u> of transplant rejection, the recipient's T cells see the foreign MHC complexes on donor cells as equal to a complex of self MHC bound to a foreign antigen (**Figure 17.33A**). Once activated by the direct route,

cytotoxic T cells will destroy all donor cells containing the foreign MHC proteins, rejecting the tissue.

An <u>indirect route</u> to transplant rejection is also possible when the recipient's own antigen-presenting cells take up, process, and present <u>pieces</u> of foreign antigens from the transplant. T-cell receptors on recipient helper T cells can bind to these self MHC–foreign antigen complexes and become activated (**Figure 17.33B**). Once activated by the indirect route, helper T cells will direct production of antidonor antibodies and stimulate antidonor cytotoxic T cells to destroy the foreign tissue.

With transplants between genetically different people, another form of medical complication is possible. The complication is called **graft-versus-host disease (GVHD)**, in which immune cells carried within donor tissue attack the recipient host's tissues. The disease is seen most often following a bone marrow transplant. Because immune cells from the donor's bone marrow can disseminate, many different host organ systems can be affected.

Clearly, MHC type matching between donor and recipient is important for a successful transplant; the closer the match, the lower the likelihood of transplant rejection. However, blood transfusions do not require MHC matching. Type A blood from one person can be transfused to a second type A person regardless of MHC compatibility. The reason is that human RBCs are not nucleated, and because only nucleated cells express MHC, human RBCs do not contain MHC proteins. Thus, MHC incompatibility does not cause RBCs to be rejected.

Figure 17.33 Transplant Rejection

Direct route

Foreign MHC
Foreign antigen
Recipient T cell
TCR
APC graft cell
CD8
Donor tissue rejection

A. In the direct route to tissue rejection, T cells from the recipient recognize the foreign MHC–foreign peptide complexes on donor (graft) cells and become activated. Activated cytotoxic T cells will subsequently kill any donor cells that express the foreign antigen.

Indirect route

Self MHC
Foreign antigen
Recipient T cell
TCR
APC recipient
CD4
Donor tissue rejection

B. In the indirect route, pieces of donor antigens are presented on the surface of recipient APCs. The recipient's T cells recognize the self MHC–foreign peptide complexes and become activated. Activated helper T cells will stimulate antidonor antibody production and enhance cytotoxic-T-cell activity.

SECTION SUMMARY

- **Autoimmune disease** is caused by the presence of lymphocytes that react to self.

- **Autoreactive B cells** (B cells that make antibody directed against self epitopes) are activated when they present a coprocessed, nonself epitope on their surface MHC. T cells that recognize the nonself epitope activate the B cell, which then secretes antibody against the self epitope.

- **Cytotoxic T cells can produce autoimmune disease** by killing host cells that make a self protein that closely resembles a foreign antigen.

- **Transplant rejection** involves cytotoxic T cells that can recognize non–self MHC proteins on transplanted tissues.

- **Recipient T cells can be activated** directly by binding to non–self MHC on graft cells or indirectly when proteins from the donor graft are presented on the host's self APCs.

- **Red blood cells are not subject to transplant rejection** because they do not possess MHC proteins.

Thought Question 17.8 Do bone marrow transplants in a patient with severe combined immunodeficiency require immunosuppressive chemotherapy?

Perspective

As complex and as highly regulated as our immune system is, it should be no surprise that things can go wrong. Hypersensitivities, immunodeficiencies, cancer, autoimmune attacks—all these dysfunctions arise as a consequence of our immune system having so many moving parts. Yet despite all the complexity and unintended consequences of the immune system, our very existence depends on its remaining vigilant 24 hours a day, 7 days a week, for 80–100 years. When it fails, we die. What other machine can stand that level of pressure?

This chapter concludes our unit on the immune system, but it does not end our discussion of immunity. Upcoming chapters describe the many infectious agents that challenge our immune system and, often, turn it against us to inflict damage (immunopathology). As we describe the conflicts between host and pathogen, continue to think about which parts of the immune system are engaged and when. Recall, too, the immunological tools that can measure which of us—human or microbe—is winning.

Key Terms

ABO blood group system (p. 517)

active immunization (p. 500)

adjuvant (p. 504)

agglutination (p. 508)

allergen (p. 514)

allograft (p. 534)

anaphylaxis (p. 515)

antitoxin (p. 500)

attenuated (p. 501)

autoimmune response (p. 530)

bare lymphocyte syndrome (p. 527)

benign (p. 528)

Bruton's disease (p. 527)

Burkitt's lymphoma (p. 528)

cancer (p. 528)

chronic lymphocytic leukemia (p. 529)

common variable immunodeficiency (CVID) (p. 527)

complement deficiency (p. 528)

conjugated vaccine (p. 504)

degranulation (p. 516)

delayed-type hypersensitivity (DTH) (p. 519)

desensitization (p. 517)

diabetes (p. 532)

DiGeorge syndrome (pp. 524, 527)

direct ELISA (p. 508)

direct immunofluorescence test (p. 510)

DNA vaccine (p. 504)

enzyme immunoassay (EIA) (p. 508)

enzyme-linked antibody (p. 508)

enzyme-linked immunosorbent assay (ELISA) (p. 508)

flow cytometer (p. 512)

flow cytometry (p. 512)

fluorescence-activated cell sorter (FACS) (p. 512)

follicular lymphoma (p. 528)

graft-versus-host disease (GVHD) (p. 534)

Graves' disease (p. 532)

Hashimoto's disease (p. 532)

herd immunity (p. 505)

histamine (p. 516)

hives (p. 516)

Hodgkin's lymphoma (p. 529)

hybridoma (p. 510)

hyper IgM syndrome (p. 527)

immediate hypersensitivity (p. 514)

immune complex disease (p. 518)

immune surveillance (p. 528)

immunoprecipitation (p. 507)

indirect ELISA (p. 508)

indirect immunofluorescence test (p. 511)

intravenous immunoglobulin (IVIG) (pp. 500, 526)

late-phase anaphylaxis (p. 516)

leukemia (p. 528)

leukotriene (p. 516)

lymphoma (p. 528)

malignant (p. 528)

MHC II deficiency (p. 527)

monoclonal antibody (p. 510)

multiple myeloma (p. 528)

myasthenia gravis (p. 532)

neoplasm (p. 528)

non-Hodgkin's lymphoma (p. 529)

passive immunization (p. 500)

plasmocytoma (p. 528)

polyclonal antibody (p. 510)

primary immunodeficiency (p. 525)

prostaglandin (p. 516)

radial immunodiffusion assay (RIA) (p. 507)

recombinant subunit vaccine (p. 504)

rheumatoid arthritis (p. 532)

Rh incompatibility disease (p. 518)

sandwich ELISA (p. 508)

secondary immunodeficiency (p. 525)

selective IgA deficiency (p. 527)

sensitivity (p. 506)

sensitization (p. 515)

serology (p. 506)

serum sickness (p. 518)

severe combined immunodeficiency (SCID) (p. 526)

specificity (p. 506)

subunit vaccine (p. 504)

titer (p. 509)

toxoid vaccine (p. 503)

tumor (p. 528)

type I hypersensitivity (p. 514)

type II hypersensitivity (p. 517)

type III hypersensitivity (p. 518)

type IV hypersensitivity (p. 519)

vaccine (p. 500)

western blot (p. 511)

X-linked agammaglobulinemia (p. 527)

Objectives Assessment Questions

17.1 Vaccines and Immunization

1. Which of the following is an example of passive immunization?
   a. Inactivated virus
   b. Natural infection
   c. Live, attenuated vaccine
   d. DNA vaccine
   e. Breast milk

2. The best protection against a second infection by a pathogen is which of the following?
   a. Inactivated-virus vaccine
   b. IV immunoglobulin
   c. Subunit vaccine
   d. Live, attenuated vaccine
   e. Natural infection

3. The concept whereby unvaccinated people are protected from infection by those in the community who are vaccinated is called ____.

17.2 Diagnostic Tools of Immunology

4. An ELISA was performed on the sera of three patients to detect antibody against hepatitis B virus. Given the titer results shown here, which patient's serum had the most anti–hepatitis B antibody?
   a. Patient 1: dilution, 1:30
   b. Patient 2: dilution, 1:1,350
   c. Patient 3: dilution, 1:640
   d. There's not enough information to answer the question.

5. Which of the following is the best method for differentiating a patient's various T-cell populations and counting the cells in each?
   a. Western blot
   b. Radial immunoprecipitation
   c. Immunofluorescence microscopy
   d. Flow cytometry
   e. ELISA

The radial immunoprecipitation gels shown here are impregnated with different anti-immunoglobulin antibodies as indicated. Sera from a control healthy individual and three patients were placed in the wells. Match the immunoprecipitation pattern for each patient with the corresponding immunodeficiency disease.

Patient

	Control	A	B	C
IgM	●	●	●	●
IgG	●	●	●	●
IgA	●	●	·	⦿
IgE	●	⦿	·	●

Patient	Disease
6. Patient A	**a.** Complement deficiency
7. Patient B	**b.** Hyper IgM syndrome
8. Patient C	**c.** X-linked agammaglobulinemia
	d. Selective IgA deficiency

17.3 Hypersensitivity

9. Which of the following hypersensitivity reactions is antibody-independent?

   a. Type I
   b. Type II
   c. Type III
   d. Type IV

10. Omalizumab prevents anaphylaxis by which of the following means?

   a. Increasing epinephrine production to stop mast cell degranulation
   b. Killing sensitized mast cells
   c. Blocking IgE from binding to mast cells
   d. Contracting smooth muscle around bronchioles
   e. a and c
   f. b and c
   g. c and d

This figure illustrates the initiation of an allergic reaction to a bee sting. Label the numbered items.

11. ____

12. ____

13. ____

14. ____

15. In Case History 17.2, Richard, blood type B, needed a blood transfusion. Assume that the blood bank ran out of type B RBCs. What blood type could be substituted?

Match each of the following scenarios with the appropriate responses. Each question has <u>two</u> correct answers.

Scenario	Response
16. Patient is injected with antiserum made from horse blood; 12 days later, patient develops fever, urticarial rash, and joint pains.	**a.** Type I hypersensitivity **b.** Type II hypersensitivity **c.** Type III hypersensitivity **d.** Type IV hypersensitivity
17. Mouse A has an allergy to a chemical placed on its skin; mouse B does not. However, when mouse B is transfused with WBCs from mouse A, mouse B becomes allergic to the same chemical.	**e.** T_H1 cells **f.** Immune complexes **g.** IgE **h.** Cell lysis
18. A 5-year-old boy with type A blood was in a serious car accident and required a transfusion. Mistakenly, he received type AB blood. Within hours he was experiencing chills, his blood pressure dropped, and his urine was tinged red.	

17.4 Immunodeficiencies and Lymphoid Cancers

19. Human immunodeficiency virus causes ____.

   a. a primary immunodeficiency
   b. a secondary immunodeficiency
   c. a tertiary immunodeficiency
   d. agammaglobulinemia
   e. IgA deficiency

20. The most common form of severe combined immunodeficiency is due to ____.

   a. adenosine deaminase deficiency
   b. diminished production of IL-2
   c. decreased surface CD154 (CD40 ligand)
   d. a lack of IL-2 receptor
   e. tyrosine kinase deficiency

21. A 1-month-old child with a low WBC count, low-set ears, a cleft palate, and an underdeveloped thymus most likely has which one of the following?

   a. IgA deficiency
   b. X-linked agammaglobulinemia
   c. DiGeorge syndrome
   d. Bare lymphocyte syndrome
   e. Hyper IgM syndrome

22. Treatment of patients who have X-linked agammaglobulinemia includes which one of the following?

   a. IV administration of immunoglobulin
   b. Thymus transplant
   c. T-cell transfusion
   d. B-cell transfusion
   e. Complement transfusion

23. A solid mass in a lymphoid organ is called a ＿＿＿.
    a. multiple myeloma
    b. plasmocytoma
    c. leukemia
    d. lymphoma
    e. All of the above
    f. None of the above

24. The cell shown here is diagnostic for ＿＿＿.
    a. follicular lymphoma
    b. Burkitt's lymphoma
    c. chronic lymphocytic leukemia
    d. plasmocytoma
    e. Hodgkin's lymphoma

25. Infection with Epstein-Barr virus is linked to ＿＿＿.
    a. follicular lymphoma
    b. non-Hodgkin's lymphoma
    c. chronic lymphocytic leukemia
    d. plasmocytoma
    e. Burkitt's lymphoma

26. The process by which lymphocytes scan for cancerous cell growth is called ＿＿＿.

17.5 Autoimmunity and Organ Transplants

27. Autoantibodies are formed when ＿＿＿.
    a. molecular mimicry enables a mismatched T_H cell to activate a B cell that escaped deletion
    b. T and B cells that are autoreactive to the same antigen escape negative selection
    c. a B cell undergoes isotype switching
    d. a superantigen activates a T cell and an APC

28. The pathophysiology of Graves' disease involves ＿＿＿.
    a. antibody that targets acetylcholine receptor
    b. cytotoxic T cells that destroy pancreas beta islet cells
    c. cytotoxic T cells that destroy thyroid follicle cells
    d. antibody that targets myelin protein
    e. antibody that targets thyroid-stimulating hormone receptor

29. Why can type A RBCs be transplanted (transfused) to type A individuals with mismatched MHC complexes?
    a. MHC type II complexes are absent on RBCs.
    b. MHC type I complexes are absent on RBCs.
    c. A nucleus is absent in RBCs.
    d. Type A antibodies in the recipient do not recognize the donor's red blood cells.
    e. The recipient's T cells recognize the donor's blood cells as self.

Review Questions

1. What are the features of an effective vaccine?

2. Describe the different types of vaccines.

3. Explain the role of adjuvants in certain vaccines.

4. What is herd immunity, and what type of vaccine will not result in herd immunity?

5. Why must many vaccines be administered multiple times?

6. What are immunoprecipitation and radial immunodiffusion?

7. How is an antibody titer determined?

8. How can immunofluorescence help a clinician diagnose an infectious disease?

9. What are monoclonal antibodies?

10. How can the number of CD4 and CD8 T cells be counted in blood?

11. Name some B-cell and T-cell immunodeficiencies, and list key features of each.

12. What are the differences between a lymphoma, leukemia, plasmocytoma, and multiple myeloma?

13. How are the genes c-*MYC* and *BCL2* involved in lymphocytic cancers?

14. What are Reed-Sternberg cells?

15. What is immune surveillance?

16. Describe the differences between hypersensitivity types I, II, III, and IV.

17. How does a B cell programmed to make an antibody against self become activated in the absence of specific T-cell help?

18. How does atopy differ from anaphylaxis?

19. How is IgE involved in allergic hypersensitivity?

20. Describe autoimmune diseases that are based on type II, type III, and type IV hypersensitivities.

21. Discuss the differences between transplant rejections of nucleated cells and rejection of blood cells.

Clinical Correlation Questions

1. An Rh-negative mother is pregnant with an Rh-positive fetus. This is her second Rh-positive child. The fetus, however, is showing signs of distress because its red blood cells are lysing. You want to transfuse the fetus through its umbilical cord with new RBCs to replace those lost. What cells should you use (Rh-positive or Rh-negative), and why?

2. A 6-month-old girl has suffered recurrent bacterial infections almost since birth. Her white blood cell count when she was healthy was 85,000/μl (normal is 5,000–10,800/μl), with 70% neutrophils. Flow cytometry analysis was performed on her peripheral blood leukocytes. Cell markers Fcγ (Fc receptor, a neutrophil surface marker) and CD18 (an integrin) were measured. Results from a healthy control and the patient are shown here:

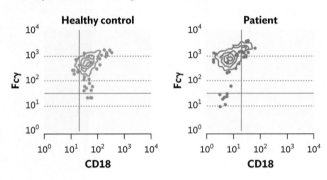

Which immunodeficiency disease does the patient have: hyper IgM syndrome, bare lymphocyte syndrome, leukocyte adhesion deficiency, or chronic granulomatous disease?

3. For the past few days, a 21-year-old woman has been feeling extremely tired, and she has a sore throat and enlarged regional lymph nodes. Her clinician suspects she has mononucleosis caused by the Epstein-Barr virus. The clinician orders an antibody blood test to confirm her suspicion. From the techniques discussed in this chapter, design an antibody test to determine whether the woman was recently infected with EBV.

Microbial Pathogenesis

[Panic, September 2001]

SCENARIO Robert Stevens, a 63-year-old photo editor for a Florida tabloid, was looking forward to a nature trip to North Carolina later in the week. Before he left, someone asked him to examine a strange love letter to Jennifer Lopez that the paper's offices received. Co-workers later said the letter was sprinkled with an unknown white powder. The letter was ultimately deemed a hoax and discarded. Eight days later, while in North Carolina, Robert started feeling ill with muscle aches, nausea, and fever. The symptoms waxed and waned over the 3-day trip. Then on October 2, the day after he arrived home, he awoke from sleep feverish, vomiting, and mentally confused. His wife rushed him to the emergency department.

2001 Bioterror Attack
Robert Stevens, a photo editor from Florida, was the first victim of the anthrax attack of 2001.

SIGNS AND SYMPTOMS His temperature in the ER was 39°C (102.2°F). Cerebrospinal fluid (CSF) and blood cultures were collected for microbiological examination, but the clinician who performed the spinal tap immediately knew something was terribly wrong. The CSF in the tube was cloudy when it should have been clear.

TEST RESULTS The lab later reported the CSF had a protein level of 666 milligrams per deciliter (mg/dl; too high), a glucose level of 57 mg/dl (within expected limits), and a white blood cell (WBC) count of 4,700/microliter (μl), mostly polymorphonuclear leukocytes (there shouldn't be any). An X-ray of Robert's lungs also showed an infiltrate. Both the CSF and blood cultures grew Gram-positive rods.

OUTCOME Within hours after admission, Robert had a generalized grand mal seizure and was intubated to help him breathe. On October 5, his third day in the hospital, Robert developed unresponsive hypotension (low blood pressure), underwent cardiac arrest, and died despite desperate attempts at resuscitation. The 63-year-old photo editor had become the first victim of the anthrax bioterrorism attack of 2001.

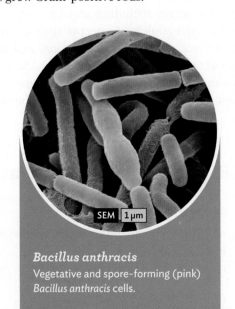

SEM | 1 μm

Bacillus anthracis
Vegetative and spore-forming (pink) *Bacillus anthracis* cells.

The 2001 anthrax attack, which came on the heels of the 9/11 airliner attacks on New York and Washington, DC, led to widespread panic across the country. Someone had spread weaponized spores (which can float in air longer than natural spores) of *Bacillus anthracis* via the US mail system. Who was responsible? To date many people have been investigated, but no one has been brought to trial.

Robert's inhalational anthrax developed after endospores of *B. anthracis*, presumably packed with the letter, deposited deep into the alveolar spaces of his lung. (Someone must inhale many spores—more than 10,000—to get sick.) In the lung, macrophages engulf dormant endospores and transport them to regional lymph nodes. There the spores germinate, becoming a growing form of the bacterium that causes toxemia (toxin release in the bloodstream) and massive septicemia.

The genome (chromosome) of *B. anthracis* is very similar to that of another common soil organism, *Bacillus cereus*. These two Gram-positive rods, identical in appearance, are both found in soil. Why is one a killer and the other not? For *B. anthracis* and *B. cereus*, the differences between friend and foe primarily involve two plasmids present in *B. anthracis*—pXO1 and pXO2—that encode the anthrax toxin and capsule biosynthesis genes, respectively. We will describe how the toxin works in Section 18.4.

Recall that Chapter 2 discussed the basic concepts of infectious disease, including disease stages, infection cycles, and basic epidemiology. Concepts such as virulence, pathogenicity, reservoirs, modes of transmission, and endemic, epidemic, and pandemic diseases were covered there. The current chapter assumes you have mastered those concepts. Now we will discuss how pathogens come to be and describe the clever strategies they use to commandeer a host.

18.1 The Making of a Pathogen

What distinguishes a pathogen from a nonpathogen? As outlined in Chapter 2, the process by which a pathogen causes disease is called **pathogenesis**, which includes the following steps:

1. Entry into host
2. Attachment and colonization of tissue(s)
3. Avoidance of host immunity
4. Host damage
5. Exit from host

REPEAT

Pathogens can be distinguished from nonpathogens by the presence of **virulence factors**, which help establish the pathogen in the host, disturb host functions, and enable transmission to new hosts. These will be described later. Many virulence genes reside in **pathogenicity islands**, segments of a pathogen's chromosome acquired by horizontal gene transfers such as transformation, transduction, and conjugation (Chapter 9). Genes for other virulence factors reside in plasmids. For example, the genes for the diarrhea-producing labile toxin of certain *Escherichia coli* strains are borne on plasmids. Virulence genes can also be part of phage genomes that lysogenize the pathogen. One example is the diphtheria toxin gene of *Corynebacterium diphtheriae* carried by the lysogenic bacteriophage called beta phage. All these factors contribute to disease, but keep in mind that the degree of harm caused by an infection depends both on the pathogen's virulence mechanisms and on immunopathology—how aggressively the immune response tries to destroy the pathogen.

What do pathogenicity, or virulence, genes do? Some encode molecular "grappling hooks," such as pili that attach to host cells. Once attached, some pathogens can secrete toxins that injure the host cell. Other virulence genes encode enzymes that exploit host physiology. For example, the coagulase of *Staphylococcus aureus* encases the microbe in nets of host fibrin that protect *S. aureus* from host inflammatory responses. Virulence gene products made by some bacterial pathogens can even "reprogram" host cells. These organisms inject proteins directly into the host cell to disrupt normal signaling pathways. This reprogramming causes the target cell to engulf and provide a safe haven for the bacterium, undergo a programmed cell death (apoptosis), or facilitate an even more intimate host-pathogen attachment. All these factors enhance the pathogen's disease-producing capability, or pathogenicity. So, where did these virulence genes come from?

Pathogen Evolution through Horizontal Gene Transfer

Chapter 9 explained that bacteria can share genetic content through transformation, conjugation, and transduction. These horizontal gene transfer (HGT) processes play major roles in the evolution of bacterial pathogens. New genes acquired by a non-pathogen through HGT can provide a growth advantage in natural environments but can sometimes enable a nonpathogen to enter and survive in eukaryotic hosts, transforming the nonpathogen into a pathogen. Likewise, viruses that infect eukaryotic cells can evolve to become more aggressive pathogens by "stealing" new genes from the host or from other viruses that happen to coinfect the same cell. If the new genes provide an advantage to the pathogen, the genes will be retained and enhanced through random mutations.

One example of HGT's affecting the evolution of a pathogen is the Gram-positive organism *Streptococcus agalactiae*, also called group B streptococcus for its capsular antigen type. *S. agalactiae* is the leading pathogen affecting neonates in the developed world. The microbe is part of the normal vaginal microbiota of some women and can be transferred to a baby passing through the birth canal. The organism is a major cause of newborn septicemia, which can lead to long-term illness and even death.

At least five clusters of *S. agalactiae* strains are recognized, which vary in the severity of disease they can cause. Scientists at the Pasteur Institute discovered how these different strain clusters may have evolved. The scientists showed that different *S. agalactiae* strains exchange large portions (20–330 kilobases [kb]) of their chromosome via conjugative elements, similar to F factor, located at different positions around the chromosome. The distribution of these genes among different groups of *S. agalactiae* has produced strains with varied degrees of virulence. This work demonstrated that large conjugal exchanges can contribute significantly to genome evolution and pathogenicity.

The *S. agalactiae* example illustrates that the infection process is like a chess match, with each side, human and microbe, trying to evolutionarily outmaneuver the other. The result is that host and pathogen coevolve.

How to Spot a Pathogenicity Island

What is a pathogenicity island and what does it look like? Horizontal gene transfers can move whole blocks of DNA (more than 10 kb) from one organism to another, placing the blocks of genes directly in the chromosome in what is called a **genomic island**. If the island increases the "fitness" of a microorganism (pathogen) during interaction with a host, it is called a pathogenicity island. New, emerging pathogens continue to evolve this way. Genomic islands in general have several signature features:

- The relative numbers of GC and AT base pairs in a DNA molecule are expressed as the GC/AT ratio. The chromosome of *E. coli*, for instance, has a GC/AT ratio of about 50/50, or a GC content of 50%. The GC/AT ratio of a genomic island, however, is typically <u>different</u> from that of the rest of the chromosome. In **Figure 18.1**, for example, the GC content of the host chromosome averages 50%. But somewhere along the genome, a 50-kb region sticks out on the graph, showing a GC content of 40%. This area probably reflects the GC content of the microbe that long ago donated the island—a donor microbe that likely no longer exists.

- Genomic islands are typically flanked by genes with homology to phage or plasmid genes. These genes were probably once associated with a phage or plasmid that moved the island from one organism to another.

Figure 18.1 Model Pathogenicity Island

A. The guanine + cytosine (GC) content of the island (about 40%) is different from that of the core *E. coli* genome (50%).

B. Schematic model of a pathogenicity island. The DNA block is linked to a transfer RNA (tRNA) gene and is flanked by direct repeats (DR) that may be "footprints" of a transposon or a virus-mediated transfer. The island also typically includes the integrase gene (*int*) and insertion sequences (IS), which may be remnants of transposition (Chapter 9).

- Genomic islands will encode clusters of genes that contribute to the organism's fitness while in a host, such as protein export systems that secrete toxins (for example, type III secretion systems that inject toxic proteins directly into target host cells).

Whether embedded in a chromosomal pathogenicity island, carried by a phage, or hosted by a plasmid, a pathogen's virulence genes function together to carry out the five steps of pathogenesis described earlier. Remarkably, the expression of these genes can even be coordinated to the stage of pathogenesis in which they are used; for instance, cell attachment genes are expressed before cell entry genes, followed by genes needed to grow and then exit the host. The next three sections discuss virulence factors that carry out the five steps of bacterial pathogenesis.

SECTION SUMMARY

- **Pathogens evolve** from nonpathogens via horizontal gene transfers, followed by random mutations and selective pressures imposed by growth in a host.
- **Genomic islands** are DNA sequences within a species that are acquired by horizontal gene transfer (conjugation, transduction, or transformation) from a different species. Genomic islands contain features, such as altered GC content and the remnant genes of phages or plasmids, that mark them as different from the rest of the genome.
- **Pathogenicity islands** are genomic islands that encode virulence factors.
- **Virulence genes** encode proteins that enhance the organism's disease-causing ability. Many virulence genes can be found within pathogenicity islands; some virulence genes are located outside an obvious genomic island or reside in plasmids.

Thought Question 18.1 What is the significance of the CSF results in the case that opens this chapter?

18.2 Pathogen Entry and Colonization

SECTION OBJECTIVES
- Explain why pathogens need to attach to host cells.
- Describe various microbial attachment techniques.
- Differentiate the structure and function of type I and type IV pili.
- Discuss the role of biofilms in pathogenicity.

Why is it so difficult for an infected body to rid itself of pathogens? As described in Chapter 2, different pathogens enter the body via different routes. Inhalation, ingestion, and sex, as well as injection by wound or insect vector, are common routes of entry. The entry route that enables a given pathogen to produce an infection is the one offering the least resistance and the easiest access to

tissues compatible with growth of that pathogen. One of the most important features of compatibility is an attachment site to which the pathogen can moor itself and stay put.

Attachment

The human body has many ways to exclude pathogens. The lungs use a mucociliary escalator (see Figure 14.5) to rid themselves of foreign bodies, the intestine uses peristaltic action to ensure its contents are constantly flowing, and the bladder uses contraction to propel urine through the urethra with tremendous force. How do bacteria ever manage to stay around long enough to cause problems? Like a person grasping a telephone pole during a hurricane, successful pathogens moving through the body manage to grab host cells and tenaciously hold on. Thus, the first step toward infection is attachment, also called adhesion. An **adhesin** is the general term for any microbial factor that promotes attachment.

We described in Chapter 12 how viruses use capsid or envelope proteins to attach to specific host cell receptors. Bacteria have similar strategies to bind a host. As Section 5.5 describes, some bacteria can sprout hairlike appendages called **pili** (singular, pilus; also called **fimbriae**), whose tips contain receptors for mammalian cell surface structures. Or bacteria can use various adherence proteins or other molecules on the bacterial surface. Sometimes both pili and nonpilus adhesins are used. Table 18.1 summarizes bacterial attachment strategies.

Pilus Assembly

How bacteria assemble pili on their cell surfaces is an engineering marvel. Regardless of pilus type, the process is similar. The shafts of pili are cylindrical structures composed of identical pilin protein subunits. Adorning the tip of the pilus is the adhesin protein that binds to host receptors (Figure 18.2). In addition to these structural components, many other proteins work together as a machine

Figure 18.2 Attachment Pili

TEM   16 nm

High-resolution micrograph showing a type I pilus, whose proteins are encoded by the *fim* operon. The FimH adhesin at the tip (arrow) is the protein that binds to the cell receptor.

Table 18.1
Examples of Bacterial Adhesins

Bacterium	Adhesin	Host Receptor	Attachment Site	Disease(s)
Streptococcus pyogenes	Protein F	Amino terminus of fibronectin	Pharyngeal epithelium	Sore throat
Streptococcus mutans	Glucan	Salivary glycoprotein	Pellicle of tooth	Dental caries
Streptococcus salivarius	Lipoteichoic acid	Unknown	Buccal epithelium of tongue	None
Streptococcus pneumoniae	Cell-bound protein	*N*-acetylhexosamine galactose disaccharide	Mucosal epithelium	Pneumonia
Staphylococcus aureus	Cell-bound protein	Amino terminus of fibronectin	Mucosal epithelium	Various
Neisseria gonorrhoeae	*N*-methylphenylalanine pili	Glucosamine galactose carbohydrate	Urethral/cervical epithelium	Gonorrhea, pelvic inflammatory disease
Enterotoxigenic *E. coli*	Type I fimbriae (pili)	Species-specific carbohydrate(s)	Intestinal epithelium	Diarrhea
Uropathogenic *E. coli*	Type I fimbriae (pili)	Complex carbohydrate	Urethral epithelium	Urethritis
Uropathogenic *E. coli*	P pili (pyelonephritis-associated pili)	P blood group	Upper urinary tract	Pyelonephritis
Bordetella pertussis	Pili ("filamentous hemagglutinin")	Galactose on sulfated glycolipids	Respiratory epithelium	Whooping cough
Vibrio cholerae	*N*-methylphenylalanine pili	Fucose and mannose carbohydrate	Intestinal epithelium	Cholera
Treponema pallidum	Peptide in outer membrane	Surface protein (fibronectin)	Mucosal epithelium	Syphilis
Mycoplasma	Membrane protein	Sialic acid	Respiratory epithelium	Pneumonia
Chlamydia	Lipooligosaccharide, OmcB surface protein	Sulfonated glycosaminoglycans	Conjunctival or urethral epithelium	Conjunctivitis, urethritis, pelvic inflammatory disease
Corynebacterium diphtheriae	Pili	Unknown	Pharyngeal epithelium	Diphtheria

to assemble the structure. As shown in **Figure 18.3**, assembly of several types of pili starts with the tip protein, which will ultimately bind to carbohydrates on host membranes after the pilus is complete. Once the tip is set at the bacterial surface, identical pilin subunits are strung together to form the shaft, forcing the tip protein farther from the surface.

Figure 18.4A shows *E. coli* attached to microvilli lining the intestine. In the inset, you can see actual pili linking bacteria to epithelial cells. But pili are more than simple microbial "stickpins" affixing a bacterium to a host cell membrane. Pathogens and normal microbiota are often subjected to strong shear forces from fast-moving bodily fluids like urine in the urethra, mucus during a sneeze or cough, or intestinal contents moved by peristalsis. To counter these shear forces, some pili tighten their "grip" on target cells (acting more like Velcro than a pin; **Figure 18.4B**), making it extremely difficult to pry the organism from its target host cell. In this way the organism resists being flushed away.

Types of Pili

The classification of pili is based largely on protein sequence information (deduced from DNA sequence) and details of their assembly. We will consider only two basic classes, type I and type IV pili. **Type I pili** are static, hairlike appendages with specialized tip proteins that can bind to specific substrates. For example, some type I pili on *E. coli* (members of our normal microbiota) bind to mannose residues present on many host cells. Other type I pili, such as the pyelonephritis adhesion pili (Pap) of uropathogenic *E. coli*, bind to a sugar group called the P blood group antigen (a digalactoside; unrelated to ABO blood groups) present on urinary tract host cell surfaces.

Whereas type I pili are static structures "patiently" waiting to bind their receptor, **type IV pili** are more dynamic. Type IV pili are thin and flexible, and they repeatedly extend and retract, giving their bacterial owner a spastic movement called twitching motility. Twitching motility occurs when the pilus elongates, attaches to a surface, and then disassembles from the base, which shortens the

Figure 18.3   Assembly of Type I Pili

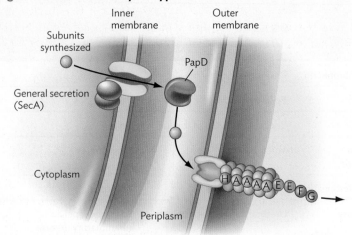

The figure illustrates pyelonephritis-associated pilus (Pap) assembly in *E. coli* but also represents other type I pili such as Fim in Figure 18.2. The Sec system secretes proteins to the periplasm, where PapD chaperones them to the site of assembly. Assembly starts with the tip protein, PapG, marked at the far right, which ultimately binds to carbohydrates on host membranes. The subunits fit together like puzzle pieces at the base of a growing pilus. The arrow at the head of the elongating pilus indicates the direction of pilus growth.

pilus and pulls the cell forward (**Figure 18.5A**). This mechanism is akin to using a grappling hook to scale a building. All type IV pili use similar secretion and assembly machinery involving at least a dozen proteins.

Type IV pili have an important role in pathogenesis and are found in a broad spectrum of Gram-negative bacteria. Pathogens with type IV pili include *Vibrio cholerae*, *Pseudomonas aeruginosa*, certain pathogenic strains of *E. coli*, *Neisseria meningitidis*, and

N. gonorrhoeae. The type IV pili of *N. meningitidis*, shown in **Figure 18.5B**, enable this organism to cross the blood-brain barrier and cause bacterial meningitis.

Nonpilus Adhesins

Bacteria also carry nonpilus adhesin proteins that stick to host tissues by way of host proteins such as integrin, which normally promotes interactions between host cells, and fibronectin, a host protein that mediates cell-cell adhesion in tissues (**Figure 18.6**). Examples of nonpilus adhesins include *Bordetella* surface pertactin (which binds to host integrin), *Streptococcus* protein F (binds to fibronectin), *Streptococcus* M protein (binds to fibronectin and the complement regulatory factor H), and intimin of enteropathogenic *E. coli* (binds to Tir, a receptor for intimin that *E. coli* itself injects into the host cell membrane).

Fimbriae (pili) and nonpilus attachment proteins often collaborate to secure a pathogen to a host cell. Pili will initiate binding between bacterium and host, after which a nonpilus attachment protein provides a more intimate attachment. For *Neisseria gonorrhoeae*, once the type IV pilus has attached to the mucosal epithelial cell, the filamentous pilus retracts, pulling the bacterium down onto the host cell membrane. Tight secondary binding is then mediated by the neisserial Opa membrane proteins—another example of a nonpilus adhesin (Opa is so named because of the <u>opa</u>city it adds to colony appearance).

Type IV pili also help *Neisseria meningitidis* cells stick to each other in the throat, making a biofilm. After a while, some pili are modified and detach from the surface. Detachment gives individual diplococci the freedom to disseminate through the bloodstream and cause sepsis or meningitis (see **IMPACT**). Inhibiting

Figure 18.4
Bacterial Attachment by Pili

A. *E. coli* attached to the surface of the intestine. Inset shows uropathogenic *E. coli* binding to epithelial cells by pili.

B. Elementary student hangs from a Velcro wall. Much as the stress of gravity strengthens the individual Velcro connections, shear stresses strengthen the connection between pilus tip adhesion proteins and their host receptor molecules.

Figure 18.5 Type IV Pili

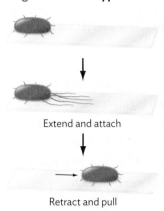

Extend and attach

Retract and pull

Type IV pilus

SEM   1 μm

A. Type IV pili produce twitching motility by extending one or more pili, attaching to a surface, and retracting the pilus, which pulls the bacterium to the point of attachment. Repeating that cycle gives the organism the appearance of twitching forward.

B. Type IV pili (green) are essential for *Neisseria meningitidis* (red) to interact with brain endothelial cells. The diameter of each cell in this diplococcal pair is approximately 0.8 μm.

adhesion can make a pathogenic organism nonvirulent. Therefore, making anti-adhesin antibodies after an infection (or vaccination) is an important part of protective immunity.

Colonization and Biofilm Development

Once a bacterial pathogen attaches to a host cell, it must be able to gain nutrients and grow. As discussed in Section 6.6, bacteria in most environments form biofilms: organized, high-density communities of cells that embed themselves in self-made exopolymer matrices. Biofilms allow microorganisms to adhere to any surface, living or nonliving, and help the bacteria survive in hostile environments. Within a single biofilm, you can find stratified differences in the expression of surface molecules, antibiotic resistance, nutrient use, and virulence factors. Bacteria in biofilms also coordinate their behavior through cell-cell communication by using secreted chemical signals (quorum sensing).

Biofilms are important features in chronic infections of oral, lung, and urogenital (bladder) tissues. For example, *Pseudomonas aeruginosa* causes a life-threatening, chronic lung infection in individuals with cystic fibrosis (CF). This microbe grows as biofilm aggregates within mucus from CF patients and causes inflammation. Insufficient mucociliary clearance of the CF lung contributes to *P. aeruginosa* biofilm formation. Biofilms are also important in periodontitis (gum disease), indwelling catheter infections, infections of artificial heart valves, chronic urinary tract infections, recurrent tonsillitis, rhinosinusitis (inflammation of nasal sinuses), chronic otitis media (middle ear infection), chronic wound infections, and osteomyelitis (bone infection).

Infections involving biofilms may linger for months, years, or even a lifetime. **Figure 18.7** shows a chronic infection of skin (a pressure ulcer). The chronic presence of these organisms in biofilms continually stimulates innate immune mechanisms through interactions with Toll-like receptors (TLRs). Recall that TLRs on host cells bind to microbe-associated molecular patterns (MAMPs), such as lipopolysaccharide or flagellar proteins, and initiate the secretion of cytokines that stimulate the immune system. Continual stimulation of the immune system causes chronic inflammation. If

Figure 18.6 Nonpilus Adhesins

M-protein fibrils

TEM   0.5 μm

A. M-protein surface fibrils on *Streptococcus pyogenes*.

Cilia

Bacteria

Nonciliated cells

SEM   2 μm

B. *Bordetella pertussis* colonizing tracheal epithelial cells. This organism uses a surface protein called pertactin, as well as a pilus called filamentous hemagglutinin, to bind bronchial cells.

Figure 18.7 A Bacterial Biofilm Infection

Bacteria

SEM   7 μm

A pressure ulcer showing bacteria of different morphotypes (rods and cocci) colonizing the wound.

impact

Neisseria meningitidis: Training Single-Cell Infiltrators

For many pathogens, very few organisms need to be transmitted to cause disease in a newly infected person—sometimes serious disease. How can such small numbers of bacteria manage to outwit our highly trained immune system? A possible answer can be found in our own human experience. In any war, an invading army first establishes a base of operations (a beachhead), from which small infiltration teams are sent deeper into enemy locations. It appears that many pathogens, including the Gram-negative bacterium *Neisseria meningitidis*, carry out a strikingly similar strategy to overwhelm their host.

N. meningitidis is a leading worldwide cause of fatal sepsis and meningitis. After infecting the nasopharynx via respiratory aerosols, the pathogen uses type IV pili (described in this chapter) to adhere to throat epithelial cells (**Figure 1A**). Once attached, the bacterium multiplies and forms larger biofilm aggregates, also held together by type IV pili, but this time the contacts are between bacteria (**Figure 1B and C**). An *N. meningitidis* biofilm causes few problems while in the throat. But occasionally, a few bacterial cells detach from the aggregate and pass through the throat into blood and then from the blood into the brain—essentially acting as infiltration teams to establish serious infection elsewhere in the body. But why do individual cells only <u>occasionally</u> leave the aggregate? Guillaume Duménil and colleagues at the Université Paris Descartes think they have part of the answer—and it has to do with those pili.

N. meningitidis has an enzyme (PptB) that adds phosphoglycerol to certain residues in pilin proteins, the stuff of which pili are made. This modification stops type IV pili from attaching to other bacteria. As a consequence, these bacteria detach from the aggregate and disseminate. These lone cell scouts can better migrate across the epithelium and enter the bloodstream. More intriguing is that the modification is a <u>regulated</u> process that does not happen all the time. Transcription of the *pptB* gene increases only <u>after</u> bacteria attach to and grow on epithelial cells. Consequently, the pathogen's base of operations (the biofilm) is established in the throat before the infiltration teams are released to cause serious disease. More on the pathogenesis of *N. meningitidis* is found in Section 24.4.

DISCUSSION QUESTION

Would a compound that inhibits PptB function be an effective antimicrobial? Why or why not?

Figure 1 Building a Biofilm on Host Cells

A. Initial adhesion occurs with individual or small groups of bacteria (arrow points to a diplococcus).

B. After initial adhesion, bacteria proliferate and form large bacterial aggregates known as microcolonies.

C. A cluster of *Neisseria* (pseudocolored in orange) attached to a human epithelial cell. Networks of type IV pili cover the bacteria. When the fibers retract, they pull with tremendous force on the surface of a human cell, changing the cell surface. Type IV fiber retraction actually stimulates gene expression in microvilli beneath the bacteria.

Figure 18.8 Biofilm Forming inside a Needleless Connector

Staphylococci forming a biofilm secrete a polysaccharide matrix that holds cells together. Needleless connectors connect syringes or bags of medication to an indwelling catheter tube inserted into a patient's vein.

Needleless connector

Staphylococcus cell

Secreted polysaccharide matrix

EM   5 µm

the biofilm is associated with an implanted medical device such as an artificial knee or heart valve, the device may have to be replaced. Antibiotics alone won't work. Today, catheter-related bloodstream infections caused by biofilms that form on indwelling catheters are the most serious and costly health care–associated infections (Figure 18.8).

The structure and group physiology of biofilm infections are important clinically. Bacteria in biofilms can tolerate antimicrobial compounds and persist despite sustained host defenses. White blood cells, for instance, cannot easily penetrate biofilms, so biofilm infections are hard to cure. Tolerance to antibiotics may be caused by poor nutrient penetration into the deeper regions of the biofilm, leading to a stationary phase–like dormancy. Bacterial factors important to biofilm formation include type IV pili, quorum-sensing structural genes and regulators, and extracellular matrix synthesis. Interfering with cell-cell signaling can prevent or limit biofilm formation and may offer a target for new antimicrobial therapies.

SECTION SUMMARY

- **Bacteria use pili and nonpilus adhesins** to attach to host cells.
- **Type I pili** produce a static attachment to the host cell, whereas **type IV pili** repeatedly assemble and disassemble.
- **Nonpilus adhesins** are bacterial surface proteins or other molecules that can tighten interactions between bacteria and target cells.
- **Biofilms** play an important role in chronic infections by enabling persistent adherence to host cells and resistance to bacterial host defenses and antimicrobial agents.

18.3 Surviving within a Host

SECTION OBJECTIVES

- Explain why and how pathogens distinguish between intracellular and extracellular existence.
- Describe various strategies pathogens use to avoid the immune system.
- Distinguish extracellular pathogens from facultative and obligate intracellular pathogens.
- Discuss the three main ways intracellular pathogens avoid intracellular destruction.

If the immune system is so good at killing microbes, how do we ever get sick? Can pathogens somehow hide from the immune system? Once inside a human host, a successful pathogen must avoid detection and destruction for as long as possible. Many products of a pathogen's virulence arsenal help the microbe escape or resist innate immune mechanisms. Other products are dedicated to stealth—that is, hiding from the immune system. But how do pathogens even know that they are in a host?

A pathogen uses the same regulatory mechanisms that sense environmental conditions in a pond to learn that it's in a host. The microbe monitors conditions such as temperature, as well as iron, magnesium, and hydrogen ion (pH) concentrations unique to the host environment. Some pathogens can sense the presence of host hormones, such as epinephrine, and can communicate with members of the microbiome through chemical signaling. Combinations of these conditions help inform the pathogen of its whereabouts and signal when virulence genes should be engaged.

However, the main thing a pathogen must do—before, during, and after attachment to tissues—is survive host assaults. The tools pathogens use to survive depend, in part, on whether the microbe is an extracellular pathogen, an intracellular pathogen, or a little bit of both.

Extracellular Pathogens

Extracellular pathogens such as *Streptococcus pyogenes* are always exposed to the host immune system. To survive, these pathogens have evolved several immune avoidance, or cloaking, strategies (Figure 18.9). Many bacteria, such as *Streptococcus pneumoniae* and *Neisseria meningitidis*, produce a thick polysaccharide capsule that envelops the cell. The capsule coats bacterial cell wall components that phagocytes normally use for attachment (see Section 15.4). The slippery nature of capsules makes the bacterial cell difficult for phagocytes to "grab." Fortunately, immune defense mechanisms can eventually circumvent this avoidance strategy by producing opsonizing antibodies (immunoglobulin G) against the capsule itself (see Section 15.4). The Fc regions of antibodies that bind to the capsule stick out away from the bacterium. The Fc receptors on a phagocyte's pseudopods can latch on to antibody Fc regions and phagocytize the pathogen.

Figure 18.9 **Strategies Used by Extracellular Pathogens to Avoid or Alter Innate and Adaptive Immunity**

Some pathogens use cell-surface proteins to sequester antibodies and avoid phagocytosis. *Staphylococcus aureus* has a cell wall protein called **protein A** that binds to the Fc region of antibodies, hiding the bacteria from phagocytes. This works in two ways. Protein A can bind to the Fc region of an antibody and cause the antigen-binding sites to point <u>away</u> from the microbe. However, even if the antigen-binding site binds to one bacterium, protein A from a second bacterium can bind to the Fc region and block phagocyte recognition.

Many pathogens are masters of illusion when it comes to the immune system. Some pathogens can alter their antigenic structure to temporarily become invisible to the immune system. These pathogens have multiple genes encoding the same protein, such as flagellin or pilin, but the antigenic structure of each version is different. By sequentially expressing these alternative proteins, the pathogen will confuse the immune system. For example, an immune system that recognizes cells expressing flagellin A will not recognize cells making flagellin B. Therefore, *Salmonella* that switched from making flagellin A to flagellin B will become invisible to immune response, at least for a time.

Pathogenic bacteria, and viruses, can also use imitation (mimicry) to confuse the immune system. Some pathogens, for instance, make proteins that look like cytokines. These fake factors can manipulate the balance of helper T cells and send immunity down the wrong path for combating the microbe. Other virulence factors can manipulate the expression of host cytokines. *Yersinia enterocolitica* (one cause of gastroenteritis) can inhibit production of pro-inflammatory cytokines such as tumor necrosis factor (TNF) and interleukins 1 and 8 (IL-1 and IL-8). Mycobacteria that cause tuberculosis and leprosy induce the production of anti-inflammatory cytokines. Both strategies will decrease the inflammation that would otherwise limit bacterial growth.

Other microbes can trigger apoptosis (programmed cell death) in target host cells. The pathogen makes proteins that enter the host cell and trigger this "suicide" program. If the macrophage self-destructs, it cannot destroy the microbe.

Cell-cell communication via quorum sensing can also help pathogens evade the immune system. Recall that in quorum sensing the concentration of a secreted chemical **autoinducer** molecule will rise as the number of bacteria in an infected space increases (described in Sections 6.6, 9.1, and 14.4). Once the number of bacteria and, hence, the concentration of an autoinducer molecule reaches a certain level, the pathogen detects the autoinducer and turns on virulence gene expression.

Why would a pathogen employ quorum sensing to regulate virulence factors? One reason may be to prevent alerting the host that it is under attack before enough microbes can accumulate. Tripping the host's alarms too early would make eliminating infection easy. However, releasing toxins and proteases only <u>after</u> large numbers of bacteria have amassed could overwhelm the host. In fact, all of the aforementioned strategies work toward that end. Fortunately, in most cases the human immune system catches on to these tricks and uses redundant humoral and cellular mechanisms to resolve these infections.

Intracellular Pathogens

In contrast to extracellular pathogens, **intracellular pathogens** avoid innate and humoral immune mechanisms by invading and living inside host cells as unwanted guests. Hiding intracellularly offers the pathogen temporary safe harbor because antibodies and phagocytic cells will not penetrate live host cells (although cytotoxic T cells and NK cells can kill infected cells). Some bacteria dedicate their entire lifestyle to intracellular parasitism. *Rickettsia*, for example, for reasons still unknown, will not grow outside a living eukaryotic cell. Other microbes, such as *Salmonella* and *Shigella*, are considered **facultative intracellular pathogens** because they can live either inside host cells or free from them. We will discuss how intracellular pathogens get into host cells later, but once they are inside, how do they withstand intracellular attempts to kill them?

CASE HISTORY 18.1

A Threatened Fetus

Shakia was in her fourth month of pregnancy and excited to become a mother. A 33-year-old African-American from Indianapolis, Shakia had been in good health until about a week ago, when she started to experience abdominal pain, fever, muscle aches, and diarrhea. She visited her obstetrician to make sure it was nothing serious. When the obstetrician asked whether she had eaten anything unusual in the past few weeks, Shakia recalled eating some soft Mexican cheese about 3 weeks earlier during a visit to Oregon with her husband. The physician sent a sample of Shakia's blood and stool to the local diagnostic laboratory and prescribed the antibiotic ampicillin as treatment. The lab reported finding a Gram-positive, non-spore-forming bacillus called *Listeria monocytogenes* in both the fecal and blood samples. Shakia had mild gastroenteritis and septicemia. The septicemia placed her unborn child at serious risk of infection because this bacterium can cross the placental barrier. If not treated quickly, an infected fetus has a 70% chance of dying from meningitis and multiple organ system failure. The doctor hoped she treated Shakia in time.

Fortunately, Shakia was treated in time and delivered a healthy baby girl. *L. monocytogenes* is a facultative intracellular bacterial pathogen that causes a mild gastrointestinal illness in healthy adults but can infect a fetus in the womb or a newborn during parturition (birth). The organism, which grows at both body and refrigerator temperatures, is usually ingested in contaminated soft cheeses or luncheon meats and avoids the immune system by growing inside intestinal epithelial cells and macrophages. Using similar tactics, the pathogen can penetrate and cross the placental barrier to infect a fetus. More details about *Listeria* infections are found in Chapter 22. Here we use Shakia's case to introduce how some pathogens adopt an intracellular lifestyle to avoid a host's immune system.

Bacterial pathogens that seek refuge in a host cell initially enter by induced phagocytosis and end up in a phagosome. Once inside the phagosome, intracellular pathogens have three options to avoid being killed by a phagolysosome (**Figure 18.10**). They can survive in a phagolysosome (fate 1), prevent phagosome-lysosome fusion (fate 2), or escape the phagosome (fate 3).

FATE 1: THRIVING UNDER STRESS In what could be called the "grin and bear it" strategy, some intracellular pathogens are adapted to grow in the harsh environment of the phagolysosome. *Coxiella burnetii*, for example, grows well in the acidic phagolysosome environment (Figure 18.10, fate 1, and **Figure 18.11A**). This obligate intracellular organism causes a flu-like illness called Q fever (query fever), with symptoms including sore throat, muscle aches, headache, and high fever. It has a mortality rate of about 1%, so most people recover. The organism allows phagosome-lysosome fusion because the resulting acidic environment enables *Coxiella* to survive and grow.

FATE 2: INHIBITING PHAGOSOME-LYSOSOME FUSION Some intracellular pathogens avoid the hazard of lysosomal enzymes by preventing the lysosome from fusing with the phagosome. *Salmonella*, *Mycobacterium*, *Legionella*, and *Chlamydia* are good examples. For example, *Legionella pneumophila* grows inside alveolar macrophage phagosomes and produces the potentially fatal Legionnaires' disease, so named for the veterans' group that suffered the first recognized outbreak in 1976 (Section 20.3). The organism, once inside a phagosome, secretes proteins through the vesicle membrane and into the cytoplasm. These bacterial proteins interfere with the cell-signaling pathways that cause phagosome-lysosome fusion. The result is that *L. pneumophila* can grow in a friendlier, less toxic vesicle. The organism survives in the macrophage, which then shuttles (disseminates) the organism to the lymph nodes and circulation (Figure 18.10, fate 2).

FATE 3: ESCAPE FROM THE PHAGOSOME The Gram-negative bacillus *Shigella dysenteriae* and the Gram-positive bacillus *Listeria monocytogenes* (Case History 18.1) both cause food-borne gastrointestinal disease and, in the process, engineer their entry into host cells (Section 22.5). Neither pathogen stays inside a phagosome. They both use hemolysins to break out of the phagosome vacuole before it fuses with a lysosome that contains lethal enzymes. Once free in the cytoplasm, these facultative intracellular bacilli are thought to enjoy unrestricted growth, having found ways to redirect host cell function to their own ends.

A fascinating aspect of escaping the phagosome involves motility. In vitro, *Shigella* and *Listeria* are both nonmotile (no flagella) at 37°C, but they manage to move around inside the host cell. How do they move? These species are equipped with a special device at one end of the cell that mediates host cell actin polymerization behind the bacterium. The result is called an actin tail that propels the organism forward through the cell until it reaches a membrane (**Figure 18.11B**). The membrane is then pushed into an adjacent cell, where the organism once again ends up in a vacuole. This strategy allows the microbe to spread from cell to cell without ever encountering the extracellular environment, where it would be vulnerable to attack.

Another bacterial pathogen that breaks out of the phagosome is the obligate intracellular, Gram-negative rod *Rickettsia prowazekii*, the cause of epidemic typhus. Unlike *Shigella* or *Listeria*, *R. prowazekii* appears to be an "energy parasite" that can transport adenosine triphosphate (ATP) from the host cytoplasm and exchange it for spent adenosine diphosphate (ADP) in the bacterium's cytoplasm. However, this ability does not explain its obligate intracellular status because giving *Rickettsia* ATP outside a host does not allow the bacterium to grow. Other factors that prevent *Rickettsia* from growing outside a eukaryotic cell remain to be discovered.

Along with surviving in the host, bacterial pathogens can fight back using toxins that can kill host cells directly or indirectly; these strategies we cover in the next section.

SECTION SUMMARY

- **Pathogens can evade the immune system** by hiding in capsules, changing their surface proteins, triggering apoptosis, or using molecular mimicry.
- **Cell-to-cell regulatory systems** regulate virulence gene expression in response to the environment.
- **Quorum sensing** will prevent pathogens from synthesizing virulence proteins too early during infection.
- **Intracellular bacterial pathogens** try to avoid the immune system by growing inside host cells. Pathogens use different mechanisms to avoid intracellular death.
- **Survival in a phagolysosome** requires that the intracellular pathogen have a specialized physiology that can counter the chemical attacks levied against it.
- **Some pathogens can inhibit host phagosome-lysosome fusion** and multiply inside host phagosomes.
- **Certain pathogens use hemolysins** to escape from the phagosome and grow in the host cytoplasm.
- **Some microbes use actin tails** to move within and between host cells.

ANIMATION

Figure 18.10 **Alternative Fates of Intracellular Pathogens**

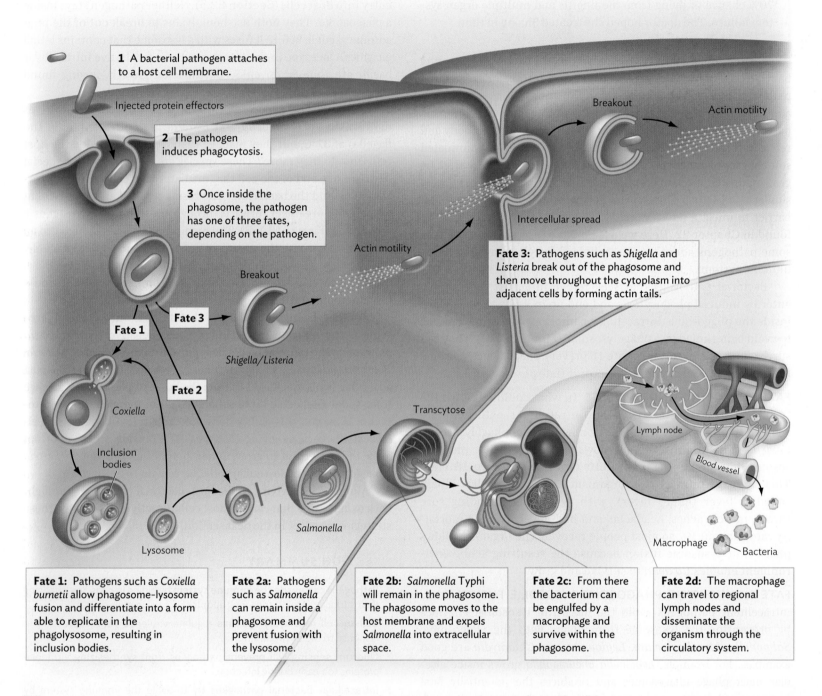

1 A bacterial pathogen attaches to a host cell membrane.

Injected protein effectors

2 The pathogen induces phagocytosis.

3 Once inside the phagosome, the pathogen has one of three fates, depending on the pathogen.

Breakout

Actin motility

Breakout

Actin motility

Intercellular spread

Fate 3: Pathogens such as *Shigella* and *Listeria* break out of the phagosome and then move throughout the cytoplasm into adjacent cells by forming actin tails.

Fate 3

Breakout

Fate 1

Shigella/Listeria

Fate 2

Coxiella

Inclusion bodies

Transcytose

Lymph node

Blood vessel

Lysosome

Salmonella

Macrophage

Bacteria

Fate 1: Pathogens such as *Coxiella burnetii* allow phagosome-lysosome fusion and differentiate into a form able to replicate in the phagolysosome, resulting in inclusion bodies.

Fate 2a: Pathogens such as *Salmonella* can remain inside a phagosome and prevent fusion with the lysosome.

Fate 2b: *Salmonella* Typhi will remain in the phagosome. The phagosome moves to the host membrane and expels *Salmonella* into extracellular space.

Fate 2c: From there the bacterium can be engulfed by a macrophage and survive within the phagosome.

Fate 2d: The macrophage can travel to regional lymph nodes and disseminate the organism through the circulatory system.

Different pathogens have different strategies to survive in a host cell. Some pathogens tolerate phagolysosome fusion (*Coxiella*), others prevent phagolysosome fusion (*Salmonella* and *Legionella*), and still others escape the phagosome to replicate in the cytoplasm (*Shigella* and *Listeria*). *Salmonella enterica* serovar Typhi, the cause of typhoid fever, invades intestinal cells, prevents phagolysosome fusion, exits the intestinal cell, and invades awaiting macrophages. The bacterium can survive inside the macrophage, which then shuttles the pathogen to lymph nodes and to the circulation.

Figure 18.11 Intracellular Pathogens

Coxiella burnetii

A. A typical vacuole in mouse macrophage cells infected with *Coxiella burnetii* at 2 hours (left) and 6 hours (right) after infection. The organism, which prefers to live in the acidified vacuole, undergoes a form of differentiation that changes its shape and alters its interactions with the host cell. *Coxiella* is an obligate intracellular pathogen.

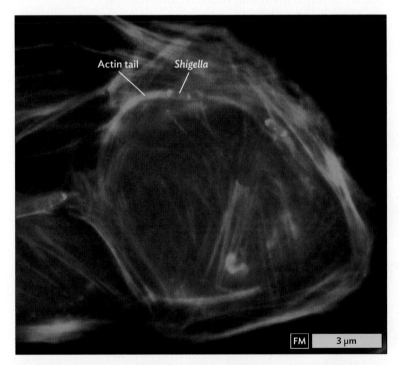

B. *Shigella flexneri* (stained red) is a facultative intracellular pathogen that escapes from the phagosome and polymerizes host cytoplasmic actin to make actin tails (fluorescently stained green). The actin tails propel *Shigella* through the cytoplasm.

Thought Question 18.2 How can one determine whether a bacterium is an intracellular parasite?

Thought Question 18.3 Why might killing a host be a bad strategy for a pathogen?

18.4 Host Damage and Manipulation

SECTION OBJECTIVES

- Describe the nine basic cellular targets for bacterial toxins.
- Explain the modes of action for staphylococcal alpha toxin, cholera toxin, diphtheria toxin, Shiga toxin, and anthrax toxin.
- Differentiate endotoxin from exotoxin.
- Describe secretion systems pathogens use to export exotoxins/effector proteins.
- Discuss the evolutionary relationships between secretion systems and other microbial structures.

CASE HISTORY 18.2

The Telltale Cough

 Will, a formerly rambunctious 5-year-old from San Francisco, was taken to the emergency department after having a severe cough for 2 weeks. His illness started as a runny nose, dry cough, and low-grade fever. But within days the coughing would come in violent fits lasting up to 1 minute (called a paroxysm). The nurse practitioner attending Will observed one of these fits. Unable to breathe between coughs, the boy became cyanotic (turned blue). When the coughing spell finally ended, Will gasped, desperate for breath. The air rushing to fill his lungs made an ear-piercing whooping sound. This telltale "whoop" led the nurse practitioner to suspect whooping cough, a disease that is highly contagious in its early stages. In the later stages (>4 weeks), the patient is no longer contagious because the bacterial agent, *Bordetella pertussis*, has succumbed to the immune response. However, the telltale cough will persist for weeks while the lungs repair the damage caused by the pertussis toxins. The nurse practitioner, now wearing a mask, took a nasopharyngeal swab sample (**Figure 18.12**) for bacteriological culture and prescribed the antibiotic azithromycin (a member of the macrolide class of antibiotics, which inhibit protein synthesis; Section 13.5). Because the organism takes so long to grow in the laboratory (5–12 days) and because the organism is hard to find late in the disease, a blood sample was also drawn to test for anti–*B. pertussis* immunoglobulin A antibody. By the next day, the serum test was positive for *B. pertussis* antibodies. The pathogen eventually also grew in the culture. Will, indeed, had whooping cough.

Can pathogens damage host tissues directly, or is all damage the result of immunopathology? Immunopathology is important, of course. An aggressive immune response to an infection and the inflammation that comes with it can inflict serious damage on a

Figure 18.12 Nasopharyngeal Swab

host's organs and tissues. Massive amounts of cytokines produced and released at the site of infection can call in and activate neutrophils and macrophages, and a variety of T cells, all of which can damage the host along with the pathogen.

But after attaching to the host, many pathogens, including *B. pertussis* in the above case, can secrete protein toxins (called **exotoxins**) that alter host cell function, disrupt the immune system, or outright kill the host cell to unlock their nutrients (because dead host cells ultimately lyse). Bacterial pathogens have developed an impressive array of toxins that take advantage of different key host proteins or structures. Gram-negative bacteria also possess a nonprotein yet toxic compound called **endotoxin** that can hyperactivate host immune systems to harmful levels. Endotoxin is part of the outer membrane and will be described later.

Exotoxin Modes of Action

Microbial exotoxins are categorized on the basis of their cellular targets and mechanisms of action, from lysing cell membranes to disrupting neurotransmitter secretion. The basic modes of action are summarized in **Figure 18.13** and **Table 18.2**. Notice how exotoxins can target many different host mechanisms.

Figure 18.13 Actions of Bacterial Exotoxins
The figure summarizes the modes of action for many bacterial exotoxins.

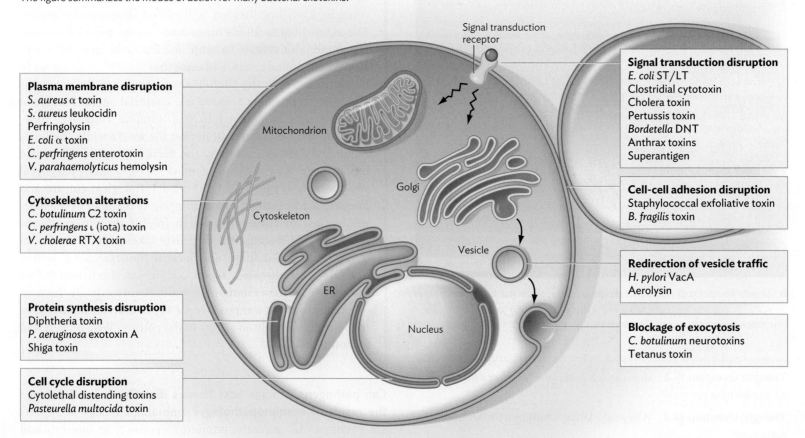

- **Plasma membrane disruption.** Toxins in this group, exemplified by the alpha (α) toxin of *Staphylococcus aureus*, form pores in host cell membranes that cause cell contents to leak.

- **Cytoskeleton alterations.** Toxins in this group can cause host cell actin to polymerize or depolymerize. Actin is a protein in the cytoplasmic microfilaments that maintain cell shape and act like "train tracks" along which vesicles filled with protein cargo move. *Vibrio cholerae* RTX toxin, for example, depolymerizes actin and alters cell shape. Other cytoskeleton-altering exotoxins enable pathogens to break into the host cell. These toxins force host cell membranes to wrap around the bacterium, pulling the invader into the cytoplasm.

- **Protein synthesis disruption.** Toxins in this group, exemplified by diphtheria and Shiga toxins, target eukaryotic ribosomes and prevent protein synthesis.

- **Cell cycle disruption.** These toxins either stop (*E. coli* cytolethal distending toxin) or stimulate (*Pasteurella multocida* toxin) host cell division.

- **Signal transduction disruption.** This type of toxin subverts host cell second messenger pathways by increasing or decreasing the synthesis of critical signaling molecules such as cyclic adenosine monophosphate or cAMP (both cholera and pertussis toxins increase cAMP synthesis).

- **Cell-cell adhesion disruption.** Some protease exotoxins cleave proteins that bind host cells together. One such toxin, exfoliative toxin of *Staphylococcus aureus*, breaks the connection between dermis and epidermis, giving victims a gruesome scalded skin appearance.

- **Redirection of vesicle traffic.** The major toxin in this class (VacA of *Helicobacter pylori*) has several modes of action depending on the host cell. The most visually striking effect is its ability to cause vacuolization, the fusion of many intracellular vesicles.

- **Blockage of exocytosis.** This group includes two protease exotoxins that interrupt the movement of host cytoplasmic vesicles to membranes (known as exocytosis). One example is tetanus toxin, which prevents nerve cells from releasing the inhibitory neurotransmitter GABA (4-aminobutanoic acid). The frightening result is paralyzing muscle spasms.

- **Excessive activation of the immune response (superantigens).** These exotoxins activate the immune system without being processed by antigen-presenting cells (APCs) and are potent triggers of fever. The pyrogenic (fever-producing) toxins of *Staphylococcus aureus* (such as toxic shock syndrome toxin) and *Streptococcus pyogenes* are examples of superantigen exotoxins.

Figure 18.14 AB Toxins

A. A typical AB toxin consists of an A subunit joined noncovalently to a pentameric B subunit.

B. Many AB toxins are ADP-ribosyltransferase enzymes that modify protein structure and function.

GENERAL CONCEPT: TWO-SUBUNIT AB TOXINS Many, but not all, bacterial toxins have two subunits, usually called A and B. These two-subunit toxins are called **AB toxins**. The A subunit is toxic; the B subunit binds host cell receptors. Thus, the B subunit for each toxin delivers the A subunit to the host cell. Many AB toxins have five B monomers arranged as a ring, in the center of which is nestled a single A subunit (**Figure 18.14A**). Toxins with this configuration are called AB5 toxins.

The A subunits of several AB toxins posses an **ADP-ribosyltransferase** enzymatic activity that changes specific host cell functions. The ADP-ribosyltransferase activity removes the ADP-ribose group from a nicotinamide adenine dinucleotide (NAD) molecule and transfers it to a target host protein (**Figure 18.14B**). Sometimes function of the host protein is destroyed (for example, protein synthesis is disrupted when diphtheria toxin ADP-ribosylates a eukaryotic ribosomal protein called elongation factor 2); other times, the modified target protein is locked into an active form that is insensitive to regulatory feedback control (for example, cAMP synthesis continues unchecked in the presence of cholera toxin).

The following sections describe five mechanisms for iconic exotoxins. Later chapters will cover the diseases they influence.

TOXINS THAT LYSE CELLS: *STAPHYLOCOCCUS* ALPHA TOXIN
Staphylococcus aureus, an organism that causes boils and blood infections, produces a hemolysin called **alpha toxin**. Alpha toxin forms a seven-membered transmembrane pore in target cell membranes. The resulting leakage of cell constituents and influx of

Table 18.2

Characteristics of Bacterial Exotoxins[a]

Toxin(s)	Organism(s)	Mode(s) of Action	Host Target(s)/ Target Tissue(s)	Disease(s)	Toxin Implicated in Disease[b]
Damage Plasma Membranes					
Aerolysin	*Aeromonas hydrophila*	Pore former	RBCs, leukocytes	Diarrhea	(Yes)
Perfringolysin O	*Clostridium perfringens*	Pore former	RBCs, leukocytes	Gas gangrene[c]	Unknown
Hemolysin[d]	*Escherichia coli*	Pore former	RBCs, leukocytes	UTIs	(Yes)
Listeriolysin O	*Listeria monocytogenes*	Pore former	RBCs, leukocytes	Food-borne systemic illness, meningitis	Yes
Alpha toxin	*Staphylococcus aureus*	Pore former	RBCs, leukocytes	Abscesses[c]	(Yes)
Panton-Valentine leukocidin	*Staphylococcus aureus*	Pore former	RBCs, leukocytes	Abscesses, necrotizing pneumonia	(Yes)
Pneumolysin	*Streptococcus pneumoniae*	Pore former	RBCs, leukocytes	Pneumonia[c]	(Yes)
Streptolysin O	*Streptococcus pyogenes*	Pore former	RBCs, leukocytes	Strep throat, scarlet fever	Unknown
Disrupt Cytoskeleton					
Vibrio cholerae RTX toxin	*Vibrio cholerae*	Actin depolymerization	Cross-links actin fibers	Cholera	Unknown
C2 toxin	*Clostridium botulinum*	ADP-ribosyltransferase	Monomeric G-actin	Botulism	Unknown
Iota toxin	*Clostridium perfringens*	ADP-ribosyltransferase	Actin	Gas gangrene[c]	(Yes)
Inhibit Protein Synthesis					
Diphtheria toxin	*Corynebacterium diphtheriae*	ADP-ribosyltransferase	Elongation factor 2	Diphtheria	Yes
Shiga toxins	*Escherichia coli, Shigella dysenteriae*	N-glycosidase	28S rRNA	HC and HUS	Yes
Exotoxin A	*Pseudomonas aeruginosa*	ADP-ribosyltransferase	Elongation factor 2	Pneumonia[c]	(Yes)
Disrupt Cell Cycle					
Cytolethal distending toxin	*Escherichia coli, Campylobacter, Haemophilus ducreyi*, others	DNase	DNA damage, triggers G_2 cell cycle arrest	Diarrhea, chancroid, others	Unknown
P. multocida toxin	*Pasteurella multocida*	Mitogen (also activates Rho GTPases)	Nucleus (encourages cell division)	Wound infection	(Yes)
Activate Second Messenger Pathways					
Cytotoxic necrotizing factor	*Escherichia coli*	Deamidase	Rho G proteins	UTIs	Unknown
Cholera toxin (CT), labile toxin (LT)	*Vibrio cholerae* (CT), *Escherichia coli* (LT)	ADP-ribosyltransferase	G proteins	Diarrhea	Yes
Heat-stable toxin (ST)[d]	*Escherichia coli*	Stimulates guanylate cyclase	Guanylate cyclase receptor	Diarrhea	Yes
Enteroaggregative *E. coli* heat-stable toxin	*Escherichia coli*	ST-like?	Unknown	Diarrhea	Unknown

(continued on next page)

Table 18.2
Characteristics of Bacterial Exotoxins[a] (continued)

Toxin(s)	Organism(s)	Mode(s) of Action	Host Target(s)/ Target Tissue(s)	Disease(s)	Toxin Implicated in Disease[b]
Activate Second Messenger Pathways (continued)					
Edema factor	Bacillus anthracis	Adenylate cyclase	ATP	Anthrax	Yes
Dermonecrotic toxin	Bordetella pertussis	Deamidase	Rho G proteins	Rhinitis	(Yes)
Pertussis toxin	Bordetella pertussis	ADP-ribosyltransferase	G protein(s)	Pertussis (whooping cough)	Yes
C3 toxin	Clostridium botulinum	ADP-ribosyltransferase	Rho G protein	Botulism	Unknown
Toxin A	Clostridioides difficile	Glucosyltransferase	Rho G protein(s)	Diarrhea/PC	(Yes)
Toxin B	Clostridioides difficile	Glucosyltransferase	Rho G protein(s)	Diarrhea/PC	Unknown
Cholera toxin	Vibrio cholerae	ADP-ribosyltransferase	G protein(s)	Cholera	Yes
Lethal factor	Bacillus anthracis	Metalloprotease	MAPKK1/MAPKK2	Anthrax	Yes
Disrupt Cell-Cell Adhesion					
Exfoliative toxins	Staphylococcus aureus	Serine protease, superantigen	Desmoglein; TCR and MHC II	Scalded skin syndrome[c]	Yes
B. fragilis toxin	Enterotoxigenic Bacteroides fragilis	Metalloprotease	E-cadherin	Diarrhea, inflammatory bowel disease	Yes
Redirect Vesicle Traffic					
VacA	Helicobacter pylori	Large vacuole formation, apoptosis	Receptor-like protein tyrosine phosphatase, sphingomyelin	Gastric ulcers, gastric cancer	(Yes)
Interrupt Exocytosis					
Neurotoxins A–G	Clostridium botulinum	Zinc metalloprotease	VAMP/synaptobrevin, SNAP-25 syntaxin	Botulism	Yes
Tetanus toxin	Clostridium tetani	Zinc metalloprotease	VAMP/synaptobrevin	Tetanus	Yes
Activate Immune Response (Superantigens)					
Enterotoxins	Staphylococcus aureus	Superantigen	TCR and MHC II; medullary emetic center (vomit center)	Food poisoning[c]	Yes
Exfoliative toxins	Staphylococcus aureus	Serine protease, superantigen	Desmoglein; TCR and MHC II	Scalded skin syndrome[c]	Yes
Toxic shock syndrome toxin	Staphylococcus aureus	Superantigen	TCR and MHC II	Toxic shock syndrome[c]	Yes
Pyrogenic exotoxins	Streptococcus pyogenes	Superantigens	TCR and MHC II	Toxic shock syndrome, scarlet fever	Yes, Yes
Lethal factor	Bacillus anthracis	Metalloprotease	MAPKK1/MAPKK2	Anthrax	Yes

[a]ADP, adenosine diphosphate; ATP, adenosine triphosphate; CT, cholera toxin; DNase, deoxyribonuclease; GTPase, guanosine triphosphatase; HC, hemorrhagic colitis; HUS, hemolytic uremic syndrome; LT, heat-labile toxin; MAPKK, mitogen-activated protein kinase kinase; MHC II, major histocompatibility complex class II; PC, antibiotic-associated pseudomembranous colitis; RBCs, red blood cells; rRNA, ribosomal RNA; SNAP-25, synaptosomal-associated protein, 25 kilodaltons; ST, heat-stable toxin; TCR, T-cell receptor; UTI, urinary tract infection; VacA, vacuolating toxin; VAMP, vesicle-associated membrane protein.

[b]Yes = strong causal relationship between toxin and disease; (Yes) = animal model or cell culture shows role in pathogenesis.

[c]Other diseases are also associated with the organism.

[d]Toxin is also produced by other genera of bacteria.

Figure 18.15 Alpha Hemolysin of *Staphylococcus aureus*

Alpha hemolysin

Cross section of alpha hemolysin

Beta hemolysis by *S. aureus*

A. Three-dimensional structure of the pore complex comprising seven monomeric proteins.

B. Cross section showing the channel.

C. A blood agar plate inoculated with *S. aureus*. The organism secretes the alpha toxin, which diffuses away from the producing colony. The toxin forms pores in the red blood cells embedded in the agar, causing them to lyse, which is visible as a clear area surrounding each colony. The clear form of hemolysis is called beta, or complete, hemolysis even though the toxin is called alpha toxin. See Section 25.2 for more information.

Figure 18.16 Healthy Intestinal Function (Absorption and Secretion)

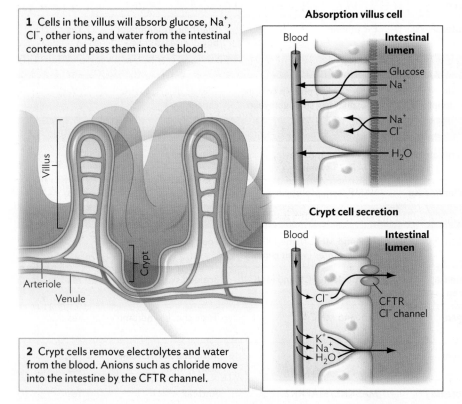

1 Cells in the villus will absorb glucose, Na⁺, Cl⁻, other ions, and water from the intestinal contents and pass them into the blood.

Absorption villus cell

Blood — **Intestinal lumen**

Glucose
Na⁺
Na⁺
Cl⁻
H_2O

Crypt cell secretion

Blood — **Intestinal lumen**

Cl⁻
CFTR Cl⁻ channel
K⁺
Na⁺
H_2O

Villus
Crypt
Arteriole
Venule

2 Crypt cells remove electrolytes and water from the blood. Anions such as chloride move into the intestine by the CFTR channel.

Electrolytes and water are absorbed from the intestinal lumen by intestinal villus cells and secreted from the body into the lumen by crypt cells. This figure shows some of the mechanisms used. The key to toxin-induced diarrhea is the chloride channel known as cystic fibrosis transmembrane conductance regulator (CFTR). Both cholera and labile toxins increase cAMP levels in crypt cells, stimulating the CFTR. CFTR increases chloride secretion into the lumen, causing a charge imbalance across the membrane (negative outside) that draws positively charged potassium and sodium ions out. The resultant salt imbalance across the membrane pulls water into the intestinal lumen, producing diarrhea (see Figure 4.1).

fluid cause the target cell to burst. **Figure 18.15A and B** shows a complete pore and a cutaway view that exposes the channel. Diagnostic microbiology laboratories visualize hemolysins (proteins that lyse red blood cells [RBCs]) such as alpha toxin by inoculating bacteria onto agar plates containing sheep RBCs (**Figure 18.15C**). The clear, yellow zones of lysed RBCs around the *S. aureus* colonies indicate that the microbe secretes a hemolysin.

TOXINS THAT DISRUPT SIGNAL TRANSDUCTION: *VIBRIO CHOLERAE, ESCHERICHIA COLI,* AND *BORDETELLA* TOXINS *Vibrio cholerae* produces a severe diarrheal disease called cholera. This disease typically afflicts malnourished people of poor countries such as Pakistan (as the Chapter 4 opening case describes) or countries in which war or natural disasters (such as the aftermath of the 2010 earthquake in Haiti) disrupt access to clean water. *V. cholerae* produces an enterotoxin nearly identical to one produced by some strains of *E. coli* associated with "traveler's diarrhea" (these strains are called enterotoxigenic *E. coli,* or ETEC). Enterotoxins specifically affect the intestine. The *E. coli* enterotoxin is called **labile toxin (LT)** because heat easily destroys it. Cholera toxin (CT) and LT are both AB5 toxins with identical modes of action: they increase the level of cAMP made inside the host cell. The mode of action for these toxins was first discussed in Chapter 4 and illustrated in Figure 4.1.

Recall that normal intestinal transport mechanisms absorb NaCl, other ions (electrolytes), and water from material flowing through the intestine (**Figure 18.16**). This absorption produces well-formed feces with little

Figure 18.17 Pathogenesis of Enterotoxigenic *Escherichia coli*

Enterotoxigenic *E. coli* (ETEC) adheres to the intestinal epithelium. ETEC binds to the fingerlike villi (colored green) on the apical surface. Like *Vibrio cholerae*, ETEC does not invade the host cell.

water and salt content. CT and LT, however, reverse this process by altering host signal transduction.

After *V. cholerae* or ETEC attaches to the cells lining the intestinal villi (**Figure 18.17**), the organisms secrete their AB toxins (**Figure 4.1B**). The B subunits stick to a carbohydrate surface structure (ganglioside GM1) on eukaryotic cell membranes and deliver the A subunit within an endocytic vacuole into the target cell. The A subunit possesses the toxic part of the molecule, an ADP-ribosyltransferase.

As **Figure 4.1B** outlined, the toxic ADP-ribosyltransferase part of the A subunit is eventually exported into the cytoplasm, where it modifies (ADP-ribosylates) a membrane-bound regulatory protein called a G protein. The G protein controls the cAMP-producing enzyme adenylyl cyclase. Once modified, the G factor overstimulates adenylyl cyclase to make excessive amounts of cAMP.

A cascade of subsequent events produces diarrhea. Excessive cAMP <u>activates</u> secretion of Cl⁻ via the cystic fibrosis transmembrane conductance regulator (CFTR) present in crypt cells. The Cl⁻ transporter is called CFTR because a defect in this protein causes the lung disease cystic fibrosis. Activated CFTR transports negatively charged chloride ions <u>out</u> of the cell and <u>into</u> the intestinal lumen. The electrical imbalance that develops across the cell membrane then draws positively charged sodium and other ions into the intestinal lumen. This massive exodus of ions produces an osmotic imbalance between the inside (fewer ions) and outside (more ions) of the cells. In an attempt to balance osmolarity, a large amount of water also leaves the cells and enters the intestinal lumen. The result is watery stools (diarrhea).

How does diarrhea benefit the pathogen? The more diarrhea that is produced and expelled, the more bacteria are disseminated throughout the environment. Consequently, infectious diarrhea increases the chance that another host will ingest the organism to ensure survival of the species. Diarrhea can also benefit a pathogen by <u>decreasing</u> competition with other organisms as they are flushed away. In fact, nearly all organisms in the diarrheal fluids of cholera patients are *V. cholerae* bacteria.

Different pathogens have evolved to use other ways to alter host cAMP levels. The Gram-negative pathogen that causes whooping cough, *Bordetella pertussis* (discussed in Case History 18.2), also secretes an ADP-ribosylating toxin called pertussis toxin that enters host lung cells and activates adenylyl cyclase activity. Runaway cAMP synthesis also causes fluid accumulation, or edema, this time in the lung.

In addition to pertussis toxin, *B. pertussis* secretes a "stealth" adenylyl cyclase (CyaA) that remains inactive until it enters host cells. The inactive CyaA toxin becomes active when it binds the host calcium-binding protein calmodulin. The resulting increase in cAMP levels inappropriately triggers certain host cell signaling pathways that disrupt tight junctions and damage lung cells.

TOXINS THAT INHIBIT PROTEIN SYNTHESIS: DIPHTHERIA TOXIN When an army tries to conquer a new land, an early goal is to destroy the industrial base of the foe. If the people under siege cannot make the materials and weapons needed to survive, the territory will fall. In much the same way, pathogens deploy bacterial exotoxins to destroy a host's synthetic base: its ability to synthesize proteins. *Corynebacterium diphtheriae*, for example (**Figure 18.18**), produces a potent toxin that inhibits protein synthesis in human cells and causes the disease diphtheria (described in the case history that opens Chapter 9 and in Section 20.3). The pathogen remains in the pharynx while the toxin spreads throughout the body in the circulatory system and causes many symptoms, including heart damage. It is the toxin, not the organism's growth, that produces these symptoms. The gene encoding the toxin actually resides in the genome of a bacteriophage that lysogenizes *C. diphtheriae*. The phage, called beta phage, remains functional but is a prime example of how a phage genome may, over many generations, degenerate to become a pathogenicity island, losing its former identity as a phage genome.

Figure 18.18
Corynebacterium diphtheriae

Methylene blue stain of *C. diphtheriae*. Note the club-like appearance of the organisms.

Diphtheria toxin is also an ADP-ribosyltransferase and another example of an AB toxin. Diphtheria toxin is synthesized as a single polypeptide that is later cleaved into a B subunit that binds to the target cell receptor on the host membrane and an A subunit containing the ADP-ribosylase activity. The two peptides cling to each other by a disulfide bond.

The AB toxin enters the cell by endocytosis. When the inside of the endosome vesicle acidifies (via H⁺-pumping ATPases), the disulfide bonds holding the A and B subunits together are reduced (broken), allowing subunit A to pass through the vesicle membrane. Once in the cytoplasm, the A subunit modifies an important protein synthesis factor called eukaryotic elongation factor 2. This act halts protein synthesis and kills the cell. Today, thanks to intense vaccination efforts using inactivated toxin (Section 17.1), diphtheria is rare in the United States.

Why is diphtheria toxin made? Iron availability appears to be a key factor. The body holds its iron very tightly in proteins such as lactoferrin and ferritin. So, to an invading microbe, the body seems to be an iron-poor environment. Limiting iron in this way makes it hard for the pathogen to grow. However, production of diphtheria toxin is activated under low iron conditions (Section 8.4), which offers the organism a way to rob the host's iron stores by killing the cell. The pathogen then uses transport mechanisms to "vacuum" up the released iron.

SHIGA TOXIN Shiga toxin also destroys host protein synthesis, but it does so by a mechanism different from that of diphtheria toxin. *Shigella flexneri* and *E. coli* O157:H7 (also known as enterohemorrhagic *E. coli*) cause food-borne diseases whose symptoms include bloody diarrhea. These organisms produce an important AB5-type toxin called **Shiga toxin** (or Shiga-like toxin). The gene (*stx*) encoding Shiga toxin is part of a phage genome integrated into the bacterial chromosome. The A subunit, upon entering the host, destroys protein synthesis by cleaving 28S rRNA in eukaryotic ribosomes (ricin toxin has a similar mode of action). This rRNA is needed to identify the beginning of an mRNA that has entered a ribosome. Strains that produce high levels of Shiga toxin are associated with acute kidney failure known as hemolytic uremic syndrome. Like diphtheria toxin, iron limitation is also a key factor for inducing the expression of *stx*.

THREE-COMPONENT EXOTOXIN: ANTHRAX A century ago, anthrax (caused by *Bacillus anthracis*, pictured in the chapter opening) was mainly a disease of cattle and sheep. Humans acquired the disease only by accident. Today we fear the deliberate dispersion of *B. anthracis* spores from the air ducts of heavily populated buildings or by its shipment through the mail (see the opening case of this chapter). What makes this Gram-positive, spore-forming microbe so dangerous is the secretion of a three-component toxin (encoded in a plasmid). Anthrax toxin is composed of a membrane-binding protein (B subunit) and two different enzyme components (A subunits). The B subunit of the toxin is called **protective antigen (PA)** because antibodies to this protein protect

Figure 18.19 *Bacillus anthracis* **Anthrax Toxin Mode of Entry**
Anthrax toxin from *B. anthracis* enters the host cell (EF = edema factor; LF = lethal factor; PA = protective antigen).

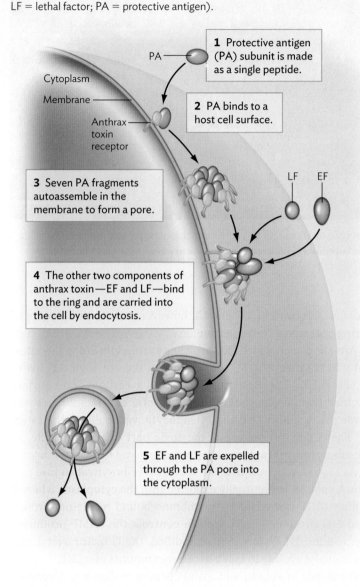

1 Protective antigen (PA) subunit is made as a single peptide.

2 PA binds to a host cell surface.

3 Seven PA fragments autoassemble in the membrane to form a pore.

4 The other two components of anthrax toxin—EF and LF—bind to the ring and are carried into the cell by endocytosis.

5 EF and LF are expelled through the PA pore into the cytoplasm.

Cytoplasm
Membrane
Anthrax toxin receptor
PA
LF   EF

hosts from disease. The pathogen secretes PA, which binds to host cell surface receptors and autoassembles in the membrane as a seven-membered pore (**Figure 18.19**, steps 1–3). The toxic components (A subunits) of anthrax toxin, called **edema factor (EF)** and **lethal factor (LF)**, bind to the PA pores, and endocytosis carries them into the cell (Figure 18.19, steps 4 and 5). The two proteins then pass through the vacuole's pore into the host cytoplasm.

EF and LF are enzymes that attack the signaling functions of the host cell. Edema factor is an adenylate cyclase that remains inactive until it enters the cytoplasm, where it binds host calmodulin (much like the exotoxin from *Bordetella pertussis* discussed above). This binding inactivates calmodulin but activates adenylyl cyclase, resulting in the overproduction of cAMP. The result is the secretion of electrolytes and water that accumulate outside cells in affected tissues (edema). Edema in the lung can impair breathing.

The third component of anthrax toxin, LF, is a protease that cleaves components of critical regulatory cascades that affect cell growth and proliferation. One consequence of subverting these regulatory cascades is a failure to produce chemokines that call immune cells to the affected area to fight the infection. Lacking chemokines, the affected cell is essentially "gagged" and cannot call for help.

We have examined only a few of the many exotoxins that pathogens employ. The chapters on infectious disease will describe some of the others, including tetanus and botulism toxins. What our brief sampling reveals, however, is the evolutionary "ingenuity" pathogens use to inflict damage to their host.

Only Gram-Negative Bacteria Make Endotoxin (LPS)

Endotoxin is an important virulence factor common to all Gram-negative bacteria. Not to be confused with secreted <u>exo</u>toxins, endotoxin is part of the **lipopolysaccharide (LPS)** that forms the outer leaflet of the Gram-negative outer membrane (Section 5.3). LPS is composed of modules of lipid A (the endotoxic part of LPS), core oligosaccharide, and a repeating polysaccharide chain (Figure 18.20). When Gram-negative bacteria die, endotoxin is released. Endotoxin can then bind toll-like receptors on macrophages or B cells to trigger the release of proinflammatory cytokines such as TNF-α, interferon, and IL-1 (Sections 15.5

and 16.3). The massive release of these active agents, known as a cytokine storm, causes a variety of symptoms, such as:

- Fever
- Activation of clotting factors leading to disseminated intravascular coagulation
- Activation of the alternative complement pathway
- Vasodilation, leading to hypotension (low blood pressure)
- Shock due to hypotension
- Death when other symptoms are severe

⊙ **Note:** Lipopolysaccharides are also the outer-membrane structures called O antigens used to classify different strains of *E. coli* as well as other Gram-negative organisms (for example, *E. coli* O157 versus *E. coli* O111).

You can see the role of endotoxins in disease in infections with the Gram-negative diplococcus *Neisseria meningitidis* (Figure 18.21A), a major cause of bacterial meningitis. *N. meningitidis* has, as part of its pathogenesis, a septicemic phase where the organism can replicate to large numbers in the bloodstream. The large amount of endotoxin present massively depletes clotting factors. This depletion leads to internal bleeding, most prominently displayed to a physician as small pinpoint hemorrhages called **petechiae** on the patient's hands and feet (Figure 18.21B; see also Chapter 15 IMPACT). Capillary bleeding near the surface of the skin causes the petechiae. Gram-negative sepsis is always treated with antibiotics; however, this strategy comes with a calculated risk. During treatment, the enormous release of endotoxin from dead bacteria could kill the patient. Untreated Gram-negative sepsis is, however, almost always fatal, so treating it, albeit risky, is imperative.

Secretion of Virulence Proteins

How do exotoxins get out of the bacterial cell? The previous sections stressed that many bacterial pathogens secrete proteins that destroy, cripple, or subvert host target cells. The bacterial toxins

Figure 18.20 Endotoxin

LPS membrane

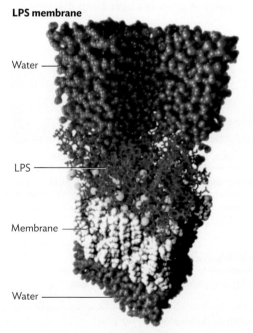

Water

LPS

Membrane

Water

Gram-negative bacterial endotoxin (LPS)

Lipid A

O-specific oligosaccharide subunit

(outer)   (inner)

Core oligosaccharide

O-specific polysaccharide chain   **Core glycolipid**

A. Model of a lipopolysaccharide (LPS) membrane of *Pseudomonas aeruginosa*, consisting of 16 LPS molecules (red) and 48 ethylamine phospholipid molecules (white).

B. Basic structure of endotoxin, showing the repeating O-antigen side chain (farthest from the microbe) and the membrane-proximal core oligosaccharide and lipid A (contains endotoxic activity).

Figure 18.21
Effect of *Neisseria meningitidis* Endotoxin

A. *N. meningitidis.*

B. Petechial rash caused by *N. meningitidis.*

Here we describe some secretion systems crucial to delivering toxins or effector proteins to the external environment or to target host cells. A particularly interesting aspect of these secretory systems is that they structurally resemble other molecular machines of bacteria. Secretion systems evolved when genes encoding these other systems duplicated and then diverged via mutation to have different functions (paralogs; see Section 9.4). Secretion systems that are evolutionarily related to other molecular machines include:

- Type II protein secretion, paralogous to type IV pili
- Type III protein secretion, paralogous to flagellar synthesis machinery
- Type IV protein secretion, paralogous to conjugation machinery

Table 18.3 gives examples of virulence proteins associated with these export systems. All these systems are associated with Gram-negative bacterial pathogens.

Before describing specific virulence protein export systems, we should briefly discuss the **general secretion pathway**. Present in both Gram-negative and Gram-positive bacteria, the general secretion pathway (called SecA dependent) moves many proteins from the cytoplasm across the cytoplasmic membrane and into the external environment (for Gram-positive bacteria) or into the periplasm (for Gram-negative bacteria, as shown in Figure 18.3). If the protein is a virulence protein (exotoxin or effector protein), special secretion systems can capture and secrete those proteins across the outer membrane. Some secretion systems can snatch virulence proteins directly from the cytoplasm, move them across the two bacterial membranes (bypassing the general secretion system), and expel them from the cell. To date, nine systems able to secrete virulence proteins have been discovered. We will describe three.

described earlier are all secreted into the surrounding environment, where they float until chance intervenes and they hit a membrane-binding site. However, other pathogens attach themselves to a tissue and <u>inject</u> bacterial proteins called effector proteins directly into the host cell cytoplasm. These effector proteins may not kill the cell but can redirect host pathways to benefit the microbe.

Table 18.3
Secretion Systems for Bacterial Toxins

Secretion Type	Features	Examples of Proteins Secreted
I	SecA dependent[a]; delivers one effector per system	*E. coli* alpha hemolysin, *Bordetella pertussis* adenylate cyclase
II	SecA dependent; derived from type IV pili genes	*Pseudomonas aeruginosa* exotoxin A, elastase, cholera toxin
III	SecA independent; syringe-like appearance; derived from flagella genes; secrete multiple effectors from the bacterial cytoplasm directly into the target host cell cytoplasm	*Yersinia* Yop proteins, *Salmonella* Sip proteins, enteropathogenic *E. coli* (EPEC) EspA proteins, TirA
IV	Derived from conjugational DNA transfer genes; secrete multiple effectors	*B. pertussis* toxin, *Helicobacter* CagA
V	Autotransporter; transport to the periplasm is SecA dependent; secretes itself through the outer membrane; delivers one effector per system	Gonococcal and *Haemophilus influenzae* IgA proteases
VI	Derived from phage tail genes; delivers a single effector	*Burkholderia* and *Vibrio cholerae* VgrG

[a]The general secretory system, which requires SecA protein, transports some virulence proteins to the periplasm of Gram-negative bacteria. These toxins are then transported across the outer membrane by a separate transporter.

Figure 18.22 Type II Secretion

A general secretion system, called Sec dependent, transports a type II secretion protein from the cytoplasm to the periplasm of a Gram-negative bacterium. The protein is then loaded into the type II secretion mechanism, where polymerization of the pilus-like ramrod propels the protein across the outer membrane and into the surrounding medium. After secretion, the ramrod depolymerizes from the base.

TYPE II SECRETION SYSTEMS ARE PISTONS Type II secretion offers a clear example of how nature has modified the blueprints of one system to do a different task. DNA sequence analysis shows that the genes used to make a type IV pilus (see Section 18.2) were duplicated at some point during evolution and evolved through mutation to serve as a protein secretion mechanism. Recall that type IV pili have the unusual ability to extend and retract from the outer membrane, producing the twitching motility of *Neisseria* and *Pseudomonas*. Type II protein secretion mechanisms mirror this property. Virulence proteins to be secreted via type II systems first make their way to the periplasm via the general secretion pathway. Type II secretion systems then use a pilus-like structure as a piston to ram the folded proteins through an outer membrane

pore structure and into the surrounding void (**Figure 18.22**). Piston action occurs via cyclic assembly and disassembly of pilus-like proteins. Cholera toxin is a well-known example of a protein expelled by a type II secretion mechanism.

TYPE III SECRETION IS A SYRINGE The etiological agents of Black Death and various forms of diarrhea (caused by species of *Yersinia*, *Salmonella*, and *Shigella*) can take bacterial virulence proteins made in the cytoplasm and <u>inject</u> them directly into the eukaryotic cell cytoplasm without the protein ever getting into the extracellular environment. Direct delivery is a good idea because it eliminates the dilution that happens when a toxin is secreted into the extracellular medium. This strategy also avoids the need to tailor the toxin to fit an existing host receptor. That is, the same export system can deliver several different proteins to the same cell.

How does direct delivery happen? Some microbes use tiny molecular syringes called **type III secretion** systems (T3SS) embedded in their membranes to inject proteins directly into the host cytoplasm. **Figure 18.23** shows T3SS needles and a model of the complex. Cell-to-cell contact between host and bacterium triggers T3SS secretion of effector proteins into the target cell. Type III systems are evolutionarily related to flagellar genes, whose products export flagellin proteins through the center of a growing flagellum.

Virulence proteins secreted by type III systems hijack normal host cell signaling pathways. Some secreted proteins can cause dramatic, actin-driven rearrangements of host membranes that ultimately engulf the microbe (**Figure 18.24**). Engulfment is not, however, the only virulence strategy associated with type III systems. Some pathogenic strains (also called pathovars) of *E. coli* use type III secretion to <u>avoid</u> entering the host cell. These systems deliver effector proteins that alter host actin polymerization and cause the host cell to erect an elevated membrane pedestal on which the *E. coli* perch (**Figure 18.25**).

Figure 18.23 Needle Complex of the *Salmonella* Type III Secretion System

Unlike other secretion systems, the type III mechanism injects proteins directly from the bacterial cytoplasm into the host cytoplasm. The proteins in these systems are related to flagellar assembly proteins.

A. Purified type III needle complexes from *Salmonella*.

B. Schematic of the *S.* Typhimurium needle complex penetrating a host cell membrane.

Figure 18.24 *Shigella* Invades a Host Cell Ruffle Produced as a Result of Type III Secretion

Shigella flexneri is shown entering a human cell. The bacterium interacts with the host cell surface and injects (via its type III secretion apparatus) invasin proteins, which choreograph a local actin-rich membrane ruffle at the host cell. The ruffle engulfs the bacterium and eventually disassembles, internalizing the bacterium within an endosome vacuole.

Figure 18.25 *E. coli* Engineers Membrane Pedestals

Some pathovars of *E. coli* use type III secretion systems to inject effector proteins into host cells that redirect host actin polymerization to build a membrane pedestal. The pedestal helps prevent engulfment of this extracellular pathogen.

and outer membranes, but not necessarily across the host cell membrane.

Salmonella: A Model of Bacterial Pathogenesis

Earlier we explained what happens to the facultative intracellular pathogen *Salmonella enterica* after it enters a host cell. Now that we have discussed type III secretion, we can explain how the organism uses one type III secretion system to invade the intestinal cells and a second to make a safe space, called a *Salmonella*-containing vacuole, devoid of harsh lysosomal proteins. After someone ingests food or water contaminated with *Salmonella*, the bacterium attaches to and invades M cells interspersed along the intestinal wall within Peyer's patches. Recall that M cells are specialized intestinal epithelial cells (see Figure 15.8B) that sample microbiota for the immune system. *Salmonella*, after invading M cells, causes an inflammatory response that leads to diarrhea.

To break into the M cell, *Salmonella* uses a type III protein secretion system encoded by pathogenicity island 1 (SPI-1) to deliver a cocktail of at least 13 different protein toxins (called effector proteins) directly into the cytosol of the M cell (Figure 18.26). These effector proteins interfere with signal transduction cascades and modulate the host response. One mission of these effectors is to rearrange the cytoskeleton, ruffling the eukaryotic membrane

TYPE IV SECRETION RESEMBLES CONJUGATION SYSTEMS As Section 9.1 describes, many bacteria can transfer DNA from donor to recipient cells via a cell-cell contact system known as conjugation. In some pathogens, evolution has modified the conjugation systems into new systems that export proteins. For example, the bacterium that causes whooping cough in humans, *Bordetella pertussis*, uses a **type IV secretion** system to export pertussis toxin to the medium. *Helicobacter pylori* uses a type IV system to inject CagA exotoxin directly into gastric epithelial cells. Type IV systems move proteins directly out of the bacterial cytoplasm and across the cytoplasmic

Figure 18.26 Schematic Overview of *Salmonella* Pathogenesis

Salmonella injects effector proteins into a host M cell to affect the activity of host proteins that trigger cell death (apoptosis), influence gene expression (via host proteins Cdc42 and JNK), influence electrolyte movements, and rearrange actin. Actin rearrangement ruffles the host membrane to engulf the organism.

around the extracellular microbe (Figure 18.24), which starts the process of engulfment. *Salmonella* induces this response to avoid the normal endocytic process that could destroy it.

At the end of this process, *Salmonella* finds itself in a vacuole (phagosome). Normally, an enzyme-packed lysosome would then fuse with the phagosome and release its contents in an effort to kill the invader. *Salmonella*, however, possesses a second pathogenicity island called SPI-2 that subverts this host response. SPI-2 uses a different type III secretion system to inject proteins that alter vesicle trafficking (fate 2, as described in Figure 18.10), thereby reducing phagosome-lysosome fusion and sparing the intracellular bacteria.

SECTION SUMMARY

- **Nine categories of protein exotoxins exist,** based on their host cell target. These include toxins that disrupt membranes, inhibit protein synthesis, act as superantigens, restructure the cytoskeleton, cleave cell-cell adhesion proteins, alter host cell signaling, redirect vacuolar traffic, prevent exocytosis, and disrupt the cell cycle.

- *Staphylococcus aureus* **alpha toxin** forms pores in host cell membranes.

- **AB subunit toxins** are common among bacterial toxins. The B subunit always promotes penetration through host cell membranes, and the A subunit has toxic activity. Some AB toxins add ADP-ribose to specific host proteins. These include **cholera toxin**, *E. coli* **labile toxin**, and **pertussis toxin**, which alter host cAMP production, and **diphtheria toxin**, which inactivates eukaryotic protein synthesis.

- **Shiga toxin** is an AB5 exotoxin that cleaves host cell 28S rRNA in host cell ribosomes.

- **Anthrax toxin** is a three-part AB toxin with one B subunit (protective antigen) and two different A subunits that affect cAMP levels (edema factor) and cleave host signaling proteins (lethal factor).

- **Lipopolysaccharide,** also known as endotoxin, is an integral component of Gram-negative outer membranes and an important virulence factor that triggers massive release of cytokines from host cells. The indiscriminate release of cytokines can cause fever, shock, and death.

- **Many pathogens use specific protein secretion pathways** to deliver exotoxins. Type II, III, and IV secretion systems discussed in this chapter are restricted to Gram-negative rods.

- **Type II secretion** systems use a pilus-like extension/retraction mechanism to push proteins from the periplasm out of the cell. These systems evolved from type IV pilus genes.

- **Type III secretion** uses a molecular syringe to inject proteins from the bacterial cytoplasm into the host cytoplasm. These systems evolved from flagellar genes.

- **Type IV secretion** systems use conjugation-like machinery to export proteins from the cytoplasm. These systems evolved from conjugation genes.

Thought Question 18.4   Antibodies to which subunit of cholera toxin will best protect a person from the toxin's effects?

Thought Question 18.5   Would patients with iron overload (excess free iron in the blood) be more susceptible to infection?

18.5 Viral Pathogenesis

SECTION OBJECTIVES

- Discuss antigenic variation in rhinovirus and influenza viruses.
- Explain how human immunodeficiency virus (HIV) targets T cells and describe the roles of Nef and Tat proteins in HIV pathogenesis.
- Outline how human papillomavirus (HPV) causes unregulated cell division.
- Explain latency of herpesvirus and the role of microRNAs in helping the virus reactivate.

Are virulence mechanisms of viruses different from those of bacteria? The pathogenic strategies of viruses are similar to and yet distinctly different from those of bacteria. On their own outside a host cell, viruses are inert objects, lacking motility and unable to replicate. Like tiny molecular water mines, they float in the environment until they bump into a suitable target cell. Once in a host cell, viruses have no need to transport food, expel waste, or generate energy because the host does all that. But each virus must still attach to a host cell and subvert its biochemistry, directing it to make more virus particles.

Chapter 12 already discussed basic virus replication strategies. However, pathogenesis involves more than replication. The progression of disease depends on where in the body the virus replicates, how it interacts with the immune system, and how the virus subverts host cell functions. This section will describe some of these concepts by using a few select viruses as examples. Chapters 19–24 will then build on these concepts to describe viral diseases of various organ systems. **Figure 18.27** presents a summary of the viral strategies of pathogenesis that we will describe.

Rhinovirus versus Influenza Virus

You have probably wondered why you repeatedly get colds if the immune system is so effective. Shouldn't you be protected from recurring infections after you have one cold? The problem is that the common cold is caused by several different types of viruses, including more than 100 known serotypes of **rhinovirus**, a small (30 nm), nonenveloped, single-stranded, positive-sense RNA virus. Serotypes of a virus (or bacterial species) differ from one another in the antigenic structure of a specific protein.

The host receptor for a subgroup of these viruses is intracellular adhesion molecule 1 (ICAM-1; see Figure 16.5), which is also important for the extravasation (movement) of leukocytes through blood vessels (see Section 15.3). Many differences among this subgroup of rhinoviruses can be traced to the structure of one of the rhinovirus capsid proteins, VP1, which binds to host ICAM-1. Because of differences in VP1 between strains, antibodies that neutralize one strain of rhinovirus will not neutralize a different strain. So, we repeatedly contract colds because we are infected with different strains.

Figure 18.27 Strategies Viruses Use to Survive in Host Cells

No single virus uses all these strategies.

Unlike many viruses that prefer to grow at 37°C (98.6 °F), rhinoviruses replicate most efficiently at 33°C (91.4 °F). This preference in growth temperature explains why rhinoviruses grow only in the cooler regions of the respiratory tract, such as nasal passages. These viruses cause a runny nose because virus-infected cells release bradykinin and histamine, both of which cause fluid loss from local blood vessels (discussed in Section 15.3).

Certain strains of coronavirus also cause common cold symptoms. However, recent variants of coronavirus cause the newer, deadlier diseases known as severe acute respiratory syndrome (SARS-CoV), Middle East respiratory syndrome coronavirus (MERS-CoV), and SARS-CoV-2, the cause of the recent COVID-19 pandemic, which will be discussed in Chapter 26.

Influenza infection proceeds differently from rhinovirus infections (Chapter 20), even though the two are often confused by the public. Influenza virus was introduced in Section 12.5 as a single-stranded, negative-sense RNA virus with a segmented genome. Influenza virus can be extremely dangerous. After being inhaled, the virus establishes a local upper respiratory tract infection, where it targets and kills mucus-secreting, ciliated epithelial cells, destroying this primary defense. If the virus spreads to the lower respiratory tract, it can cause the shedding of bronchial and alveolar epithelium down to the basement membrane (a thin sheet of fibers lying below mucosal epithelial cells). Shedding these lung cells will compromise oxygen and CO_2 exchange, making breathing difficult.

Flu virus MAMPs interact with Toll-like and NOD-like receptors on host cells. Once activated, the receptors trigger the release of interferons that produce systemic symptoms, such as headaches and muscle aches. Highly virulent strains can produce such severe lung inflammation that the patient succumbs to respiratory failure. Another consequence of influenza infection, even for moderate infections, is diminished natural defenses. The patient becomes more susceptible to secondary infections with other lung pathogens, such as the bacteria *Haemophilus influenzae* or *Streptococcus pneumoniae*. These secondary infections can also lead to death.

What makes influenza so frightening is that every 10–15 years the virus undergoes a major genetic alteration, called an **antigenic shift**. An antigenic shift can produce a pandemic strain to which almost no one on Earth is immune (actual pandemics typically occur in 30-year cycles). Such a shift occurs when two different strains of influenza that usually infect different animals (for instance, bats, birds, or pigs) accidentally infect the same cell in one animal, usually a pig or chicken. Recall that influenza has a segmented RNA genome (Section 12.5). Thus, as two influenza viruses disgorge (eject) their eight nucleic acid segments into the same cell, the segments mix and reassort to generate a chimeric virus that may pair a hemagglutinin segment from one strain with the neuraminidase segment from the second strain. Recall that hemagglutinin is required for virus entry into cells and neuraminidase is required for virus release. The new influenza virus may then "jump" from the infected animal to humans. A pandemic can develop if the versions of hemagglutinin or neuraminidase in the new virus have not previously been seen in a human influenza strain. Antibodies that recognize the old hemagglutinin or neuraminidase structures will not recognize and neutralize the new ones.

Another, more frequent but smaller type of antigenic change called **antigenic drift** also takes place in the flu virus and other RNA viruses. Antigenic drift happens when error-prone viral

RNA polymerase introduces random mutations in hemagglutinin. Antigenic shifts and drifts are why a new vaccine (the flu shot) is required each year. These shots are highly recommended for the elderly, who are more susceptible to the disease, and are a good idea for everyone else.

Human Immunodeficiency Virus

Acquired immunodeficiency syndrome (AIDS) became an important disease in the late twentieth century and continues to grip us today. Some 30 million to 40 million cases exist worldwide. Infection by the retrovirus HIV (human immunodeficiency virus) causes the patient to become severely immunocompromised and susceptible to a wide variety of infectious diseases. Death in patients with AIDS is usually the result of these secondary infections. Section 12.6 discusses the molecular biology of HIV replication, and Section 23.4 will describe the disease; here we will deal with mechanisms of HIV pathogenesis.

In contrast to rhinovirus, which binds to a single cell receptor (ICAM-1, for instance), HIV simultaneously binds multiple cell receptors via its spike protein, glycoprotein 120 (gp120) (Section 12.6; see Figure 12.28). HIV spike protein binds host cell CD4 protein (meaning T helper cells are the usual targets), and at the same time it binds a nearby chemokine receptor called CCR5 (see Figure 12.28). By binding to the chemokine receptor, HIV can block chemokine binding and disconnect communication between the target cell and the immune system.

The virus also makes a protein (negative factor, Nef) that prevents T-cell apoptosis, which would destroy the virus. Apoptosis is a process triggered in distressed human cells, as would occur during an infection. Apoptosis is a protective measure that kills the infected host cell and any viruses attempting to replicate. You could say the apoptotic cell "takes one for the team." HIV, however, has evolved to keep its host cell alive, at least for a while. As a result of Nef function, T cells live longer so HIV virus can replicate longer (Figure 18.28A).

Infected lymphocytes eventually do die from attack by CD8 cytotoxic T cells and NK cells despite the presence of Nef.

Figure 18.28 HIV and AIDS

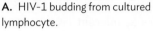

A. HIV-1 budding from cultured lymphocyte.

B. Kaposi's sarcoma (oval spots).

Consequently, the levels of these T cells decline dramatically over time and cause an immunodeficiency in an infected patient. Monitoring T-cell numbers, therefore, is a crucial diagnostic tool because simply having a detectable HIV load does not mean someone is overtly ill. Only when T-cell numbers fall to around $300/\mu l$ of blood (the healthy range is $500–1,500/\mu l$) will the signs of early symptomatic HIV infection appear, such as repeated secondary infections (see Section 23.4).

Another important cause of AIDS symptoms is the HIV Tat protein (trans-activator of the HIV promoter). A trans-activator is a protein that can diffuse through the cytoplasm to bind and activate transcription of a target gene sequence. The virus needs Tat protein in order to replicate, but infected monocytes also <u>release</u> Tat. The extracellular Tat protein enters other cells, where it alters host regulatory cascades and gene expression. Tat decreases synthesis of major histocompatibility complex I (MHC I), up-regulates and down-regulates various cytokines, and stimulates apoptosis, all of which help disassemble the immune system.

Because Tat protein plays a central role in viral replication and subduing the immune system, some believe that a vaccine using inactivated Tat will help prevent AIDS. Table 18.4 presents other HIV proteins important to the pathogenesis of HIV and AIDS.

Table 18.4
HIV Proteins That Affect Pathogenesis

Protein	Role(s) in Pathogenesis
Tat (transcriptional trans-activator)	Accelerates HIV transcription by host polymerase; is also secreted from host cells and, upon entering an uninfected cell, can trigger apoptosis (programmed cell death, called bystander apoptosis)
Vpr (viral protein R)	Regulates the import of HIV into the nucleus before the viral cDNA integrates into the host chromosome
Nef (negative factor)	Down-regulates CD4 and MHC class I proteins, protecting the infected cell from killing by CD8 cytotoxic T cells; slows apoptosis in infected T cells; secreted Nef activates apoptosis in uninfected lymphocytes
Vif (virion infectivity factor)	Counteracts a host cellular protein that naturally inhibits HIV replication
Vpu (viral protein U)	Counteracts a host cellular protein that inhibits virus assembly and export

Note that an immune system weakened by HIV renders the victim more susceptible to infections by other pathogens such as human herpesvirus 8 (HHV-8) that can cause Kaposi's sarcoma (**Figure 18.28B**). Kaposi's sarcoma is a malignancy of endothelial or lymphatic cells and can arise anywhere—gastrointestinal tract, mouth, lungs, skin, or brain.

Human Papillomavirus

CASE HISTORY 18.3
Sex, Warts, and Cancer

 Jeneen is a 45-year-old migrant crop worker from Belize. Over the past 2 months she has noticed abnormal vaginal bleeding accompanied by a foul-smelling yellowish vaginal discharge. She was also experiencing increasing pain in her pelvis. She visited a free clinic near the California farm where she worked, hoping a simple medicine would help. The clinic doctor performed a pelvic exam and used a magnifying colposcope to view her cervix. He saw an advanced stage of cervical cancer (**Figure 18.29**). He took a sample for DNA testing and a Pap smear to confirm the neoplasm. Jeneen admitted having six sexual partners over the years. She also recalled seeing a wartlike growth in her genital area about 20 years ago. At the time, she was too embarrassed to go to the doctor—but the warts disappeared after a few months, so she thought nothing of it. A week after her visit to the free clinic, the Pap smear results confirmed his diagnosis: late-stage cervical cancer. The DNA test revealed the presence of human papillomavirus DNA. Tragically, Jeneen would never return to Belize.

Human papillomavirus (HPV) is a common DNA virus that infects skin cells, causing them to divide excessively (**Figure 18.30A**). The result is warts (**Figure 18.30B**). HPV can produce warts on the feet, hands, vocal cords, mouth, and genital organs. More than 60 types of HPV have been identified so far. Each type infects a different part of the body, with specific strains of HPV associated with cervical and penile cancer. Genital HPV

Figure 18.29 **HPV-Associated Cervical Cancer**

Healthy   **Cancerous**

Colposcopic views of a healthy cervix and a cancerous cervix. A colposcope is an instrument used to examine the cervix and vagina.

Figure 18.30
Human Papillomavirus

A. Human papillomavirus.

TEM 50 nm

B. Hand warts.

Nondisease state: Normally, pRb neutralizes E2F, a protein needed to make essential cell division proteins. E2F stimulates cell division.

Disease state: HPV viral protein E7 chaperones human protein pRb toward degradation.

C. Part of the strategy HPV uses to stimulate cell division in infected cells. Protein pRb normally helps limit cell division by inactivating E2F. Viral E7 protein, however, inactivates pRb. E2F, therefore, keeps stimulating cell division, so the cell divides uncontrollably and becomes cancerous.

is, in fact, one of the most common sexually transmitted diseases among college students. Chapters 19 and 23 further discuss HPV diseases.

Why does HPV cause excessive cell division? The answer is self-preservation. Because HPV needs host replication machinery to make progeny virus, the more often HPV makes host cells divide, the more often virus particles can be made. But how does HPV stimulate host cell division? Several mechanisms are involved. In one, HPV synthesizes a viral protein called E7 that inactivates a host protein (pRb) that normally <u>limits</u> host cell division. The host pRb protein prevents the transcription of genes needed for mitosis (c-*myc*) and for cell cycle progression toward mitosis (cyclin E, *ccnE*). When present, pRb will stop cells from dividing (**Figure 18.30C**).

HPV protein E7, however, causes excessive cell division by sending pRb to the cell's **proteasome**. Like a wood chipper that shreds unwanted tree limbs, a cell's proteasome degrades damaged or unneeded cellular proteins. The damaged proteins are marked for destruction by a protein modification called **ubiquitination**. The HPV E7 protein stimulates ubiquitination of <u>normal</u> pRb and sends it to its doom. With host pRb protein removed, cell division control is reduced, so more host cells and viruses are made. Skin warts are the result, or in the worst-case scenario, cervical or penile cancer develops (usually 15 years or more after the initial lesion). Even after the initial infection resolves (90% resolve within 2 years), the virus may remain latent in tissues and can reactivate years later if the immune system becomes impaired. Fortunately, researchers have developed an effective HPV vaccine that will protect against most cervical cancer if given before exposure (Section 23.4 will discuss this vaccine).

Latent Herpesvirus

Herpesviruses are double-stranded DNA viruses that cause a variety of diseases, including cold sores (herpes simplex virus type 1, HSV-1), genital herpes (herpes simplex virus type 2, HSV-2), mononucleosis, and some cancers (Burkitt's lymphoma, caused by Epstein-Barr virus). Later chapters will discuss many of these further. Key features of HSV infections are rapid replication in epithelial cells and the development of a latent state after infecting sensory nerve cells. Like a bear waking from hibernation, active virus can emerge from latency years after the primary infection. Reactivation of latent herpes simplex virus is most commonly noticed as recurrent cold sores around the mouth or as genital herpes lesions.

HSV DNA, normally linear in a capsid, circularizes after entering the neuronal cell nucleus and becomes latent by shutting down viral replication genes. The mechanism of shutdown is not yet clear. The proviral DNA condenses but remains separate from host DNA in the nucleus (the HSV genome rarely integrates into the host DNA). The latent virus DNA can reactivate by unknown mechanisms after exposure to UV light, emotional stress, fever, or immune suppression. The results are cold sores or recurrent genital lesions. However, viral replication can trigger apoptosis in which the infected cell kills itself to destroy the virus and spare nearby host cells. Herpesvirus, however, avoids triggering host cell apoptosis (and its own demise) during virus reactivation. How?

Latent herpesvirus DNA produces several small RNA molecules called microRNAs (miRNAs, 20–25 nucleotides long) that interfere with the host cell apoptosis program. This process is especially important when a latent virus reactivates. The miRNAs inhibit the two known pathways that trigger apoptosis, although the inhibitory mechanism remains unclear. In a clever turn, the latent virus DNA makes only miRNA and no viral proteins. Because no foreign proteins are present to trigger an immune response, the host cannot detect the infected cell as long as the virus remains latent. And by virtue of inhibiting apoptotic pathways, once the virus reactivates, the infected cell will live long enough to make more virus.

Viral Attacks on Adaptive Immunity

Efficient as our immune system is, many viral pathogens have developed ways to evade the adaptive immune response. We already discussed HIV's ability to infect and destroy T cells, but other viruses use different subversive mechanisms. For example, many viruses produce proteins that down-regulate production of class I MHC proteins on infected cell surfaces. This down-regulation limits antigen presentation because MHC I is needed for that process.

On the other hand, losing MHC I should expose the infected host cell to natural killer cells because natural killer cells attack body cells that lack MHC I (discussed in Section 15.5). To surmount this obstacle, human cytomegalovirus places a decoy MHC I–like molecule on the surface of infected cells—an example of molecular mimicry. These decoys bind inhibitory receptors on natural killer cell surfaces and block cytotoxicity, leaving the virus and infected cell unscathed.

Viruses such as rotavirus, a major cause of childhood diarrhea, can also direct the host to degrade proteins that activate the expression of important inflammatory cytokines. The result is less inflammation, which enables the pathogen to survive. Regardless of whether you think viruses are living or nonliving (a question posed in Section 12.1), they certainly possess an impressive array of pathogenicity tools.

SECTION SUMMARY

- **Multiple reinfection** of the same person by a virus can occur when different strains of the virus have minor sequence differences in attachment proteins (for example, capsid or envelope). Because of these differences, neutralizing antibodies made during a previous infection are ineffective against the new strain.

- **Viral infection** can diminish the immune response by lowering MHC I levels on host cells or by altering cytokine production (HIV, for example). The result can increase a patient's susceptibility to other, less virulent microbes.

- **Antigenic shifts** arise from the reassortment of viral antigen genes (as in the hemagglutinin and neuraminidase antigens of influenza) and can produce a new pandemic strain of the virus. **Antigenic drifts** are less dramatic changes in a viral antigen that can lead to epidemics.

• **Animals coinfected with two viruses** (either similar strains or even different viruses) can serve as incubators for antigenic shifts (influenza) or the evolution of new viruses.

• **Some pathogenic viruses can become latent** after infection. Herpesvirus, for example, becomes latent, does not express viral proteins, and escapes immune surveillance mechanisms. Reactivated herpesviruses produce microRNA molecules that prevent the host cell from undergoing apoptosis.

Thought Question 18.6   Why do rhinovirus infections fail to progress beyond the nasopharynx?

Thought Question 18.7   How do you think an infection with one type of pathogen could influence coinfection with a second pathogen?

Thought Question 18.8   Why don't viruses such as herpesvirus or measles virus undergo antigenic shifts?

18.6 Protozoan and Fungal Pathogenesis

SECTION OBJECTIVES

• Describe different ways protozoa invade host cells.

• Correlate the immune response to protozoan-induced damage to host tissues.

• Explain several mechanisms protozoa use to avoid the immune system.

• Discuss fungal pathogens with respect to their transmission, the types of diseases they cause, the role of cell differentiation, and host immune competence.

We have seen how bacteria and viruses cause human disease, but what about eukaryotic microbes? Sure, the sizes of eukaryotic pathogens are vastly different from those of bacteria and viruses, but are the mechanisms they use to cause disease also different? This section explores how protozoa and fungi cause infection, pointing out what is similar to bacteria and viruses and what is unique.

Protozoan Pathogenesis

Protozoa cause many different human diseases, including malaria (*Plasmodium*), sleeping sickness (*Trypanosoma*; **Figure 18.31A**), amebic dysentery (*Entamoeba*), and trichomoniasis, a sexually transmitted vaginitis (*Trichomonas*). Protozoan infections will produce tissue damage, but the pathogens, in general, do not produce toxins as potent as those of bacteria. Exceptions include certain flagellated protists (dinoflagellates) that form the so-called red tides floating along the Gulf of Mexico. The red tide dinoflagellates are not infectious, but inhaling sea spray containing the toxins can

Figure 18.31 African Sleeping Sickness

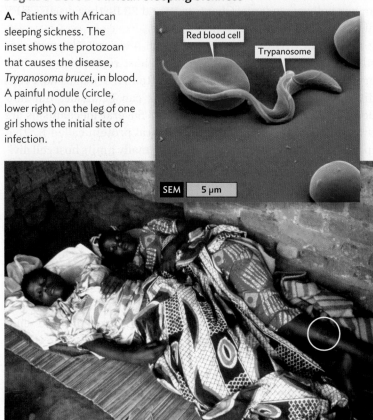

A. Patients with African sleeping sickness. The inset shows the protozoan that causes the disease, *Trypanosoma brucei*, in blood. A painful nodule (circle, lower right) on the leg of one girl shows the initial site of infection.

Antibodies to VSG develop

| Trypanosome coated with VSG | One cell switched to "blue" VSG. The "green" VSG protozoa are killed (ghost white cells). | "Blue" VSG cells repopulate blood. |

B. Antigenic variation as an immune-avoidance mechanism. The figure illustrates *T. brucei* coated with one type of variant surface glycoprotein (VSG) antigen (green). Eventually, antibodies build up that can attack the "green" form of VSG and kill the cells. However, a few protozoa will begin expressing a different VSG (blue) that the antibody does not recognize. These variants survive and repopulate the blood. This cycle continues because the *T. brucei* genome contains hundreds of silent *vsg* genes that can become activated—one at a time.

produce respiratory symptoms, and skin contact will cause a painful rash.

With few exceptions (for example, *Entamoeba* species that penetrate the intestine's mucosal barriers), damage from protozoan infections is due primarily to the persistence of the organism in tissues and the resulting immunopathology. Damaging immune reactions can be triggered by protozoan antigens or by host antigens released as a result of pathogen growth. However, pathogenic protozoa employ clever strategies that help them avoid or escape the immune system, leading to persistent protozoan growth. These immune-avoidance strategies include antigenic masking, antigenic variation, intracellular location, and immunosuppression.

Antigenic masking. Some protozoa escape immune detection by covering themselves with host antigens. Malarial parasites, for instance, can bind to the Fc region of any antibody, a property similar to protein A of *Staphylococcus aureus*. Binding these nonspecific antibodies masks the parasite's antigens from binding parasite-specific antibodies.

Antigenic variation. Another strategy protozoan parasites use to hide from the immune system is to tinker with their surface antigens, changing them during an infection. Parasites covered with the new antigens will escape the immune response made against the original antigens. Trypanosomes (the cause of African sleeping sickness and Chagas disease), for example, do not have an intracellular stage, making them an easy target for the immune system. Indeed, macrophages clear many trypanosomes from blood. Trypanosomes that survive, however, do so because they have swapped to a different antigenic form of variant surface glycoprotein (VSG; **Figure 18.31B**). Each trypanosome contains a large repertoire of VSG genes with different sequences, but only one VSG gene is expressed at a time. Switching to a new VSG allows the pathogen to avoid the antibody response and produce successive surges of parasitemia (parasites in the blood). Antigenic variation and recurrent bouts of parasitemia cause progressively more serious symptoms such as heart abnormalities, making the development of a vaccine extremely difficult.

Intracellular location. The intracellular lifestyle of some protozoan parasites helps delay their recognition by the immune system. Even once they are discovered, being inside a host cell can still protect the parasite from the direct effects of the host's immune response. How do protozoa enter host cells? Different pathogens use different mechanisms. For example, *Leishmania*, a protozoan transmitted by tiny sand flies, engages normal phagocytic receptors on macrophages to engineer its entry. By contrast, *Toxoplasma* species, which probably infect 10%–20% of the US population, use an actin-myosin motor of their own to forcibly drive themselves into a host cell (**Figure 18.32**). Although *Toxoplasma* infects at least an estimated 50% of the world's population, the disease toxoplasmosis is rare unless the patient is immunocompromised (for instance, an AIDS patient or the fetus of a pregnant mother).

Fortunately, an intracellular lifestyle offers protozoa only temporary refuge from the immune system. Some *Plasmodium* parasites that cause malaria, for instance, first grow in hepatocytes and then inside RBCs, temporarily hiding from the

Figure 18.32 *Toxoplasma* Entry into Host Cells

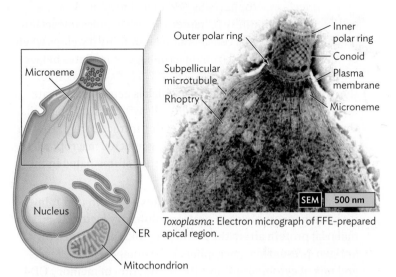

Toxoplasma: Electron micrograph of FFE-prepared apical region.

A. View of *Toxoplasma*. The apical end of *Toxoplasma* (see inset) sports a myosin motor attached to the parasite's actin filaments. This motor is used to carry out gliding motility. Microneme organelles secrete several microneme (MIC) proteins that facilitate attachment. ER = endoplasmic reticulum; FFE = freeze-fracture-etch.

B. *Toxoplasma* approaching a host cell membrane uses MIC proteins (red) to attach (1 and 2). The myosin motor propels the organism through the membrane without the cell forming a phagocytic vacuole (3). A protease (green) located at the parasite's posterior cleaves the adhesin and allows internalization (4 and 5).

immune system. Eventually, however, the pathogen must place special proteins on the host cell surface that are needed for the parasite's survival and other proteins that cause the infected RBC to stick to blood vessel walls. The cell-mediated immune system will now detect and mount a response to these surface antigens.

Immunosuppression. Another way protozoa survive in a host is to suppress the host's immune system, which most parasitic protozoa do to some degree. This suppression delays the detection of the parasite and reduces the ability of the immune system to kill it or hinder its growth. One good example is *Leishmania donovani*, which, as stated above, infects macrophages as part of its life cycle. As a rule, cell-mediated immunity (Chapter 16) best handles intracellular pathogens. Normal cell-mediated immunity develops when cytokine IL-12 secreted from macrophages stimulates helper T cells to differentiate into T_H1 cells. T_H1 cells then secrete the cytokine interferon-gamma, which is needed to activate cytotoxic CD8 T cells. Cytotoxic T cells will then kill host cells that present antigens from an infecting intracellular pathogen. *L. donovani* avoids this fate by keeping infected macrophages from producing IL-12, undermining the cell-mediated immunity mechanism.

Plasmodium, on the other hand, makes a protein that mimics human macrophage migration inhibitory factor. This malarial protein alters the blend of cytokines secreted by antigen-presenting macrophages. The inappropriate levels and mix of cytokines inhibit the production of memory CD4+ T cells, so the recovering patient does not develop protective immunity and becomes susceptible to subsequent plasmodial infections.

In summary, protozoan pathogens have infection strategies—like immune avoidance—that are similar to those of bacteria and viruses. But the mechanisms used to carry out these strategies are very different.

Fungal Pathogenesis

Medically important fungi include species that produce superficial mycoses of skin or hair (dermatomycoses caused by *Trichosporon* and *Malassezia*) and species that cause systemic infections (*Cryptococcus*, *Histoplasma*, *Blastomyces*, *Coccidioides*, and *Paracoccidioides*). *Trichosporon* species can infect hair, skin, and nails. *Malassezia* species infect skin to produce hyperpigmented patches (a disease called tinea versicolor). Systemic fungal pathogens, however, cause serious respiratory, central nervous system, GI tract, and tissue diseases.

Section 11.2 explained that there are two types of fungi: molds that grow as filaments, called hyphae or mycelia, and single-celled yeast, which bud to replicate. Fungi can reproduce sexually and asexually, but most medically important fungi replicate asexually in nature by forming tiny spores called conidia. As with all diseases covered in this chapter, discussions about specific fungal infections (mycoses) take place in the disease chapters (Chapters 19 through 24). Here we will describe some general strategies and tools of fungal pathogenesis.

Growth. Many medically important fungi are thermally dimorphic, meaning they can grow as yeast or in hyphal forms that produce conidia. Most grow as hyphae at 25°C, but when the temperature rises to 37°C (body temperature) they grow as yeast, the most pathogenic form. Thermally dimorphic fungi include *Histoplasma*, *Blastomyces*, *Coccidioides*, *Paracoccidioides*, *Cryptococcus*, and *Sporothrix*, among

others. *Candida albicans*, a cause of vaginal yeast infections and a member of the human microbiome, is also dimorphic, but the transition is not triggered by temperature. *Candida* grows as yeast when in soil, but after colonizing the body it can replicate in yeast or hyphal forms. The hyphae facilitate tissue penetration during infection. Some nondimorphic fungal pathogens also exist. These include *Aspergillus fumigatus*, *Pneumocystis jirovecii*, and the dermatophytes.

Entering the host. Fungal infections are usually acquired from the external environment rather than by person-to-person (or animal-to-person) transmission. For instance, inhalation of infectious conidia generated by molds growing in soil is the usual infection route (respiratory). Some of these molds are ubiquitous around the world, such as *Aspergillus*. Others are endemic to specific geographical areas, such as the Ohio and Mississippi river valleys (*Histoplasma* and *Blastomyces*) or the southwestern United States (*Coccidioides*). Other fungi produce disease only after an injury introduces them past the skin or mucosal barriers. For example, *Sporothrix schenckii* infection, called sporotrichosis or rose handler's disease, can develop when a thorn covered with *Sporothrix* spores pricks the skin.

Immunocompromised patients are at highest risk for fungal infections, making most fungal pathogens opportunistic pathogens. A few, like meningitis-causing *Cryptococcus neoformans*, can cause disease in healthy people. Some potentially pathogenic fungi, such as the yeast *Candida albicans*, are even members of the human microbiome but can cause opportunistic infections if they breach normal innate defense mechanisms.

Injuring the host. How do fungal pathogens cause injury? Many fungi produce **mycotoxins** in the environment that can affect human health, but only when ingested. Unlike bacterial toxins, mycotoxins are not involved with the pathogenesis of infectious fungi. Fungi also lack components analogous to the endotoxin of Gram-negative bacteria. Host injury during most fungal infections is due primarily to collateral damage produced by the immune response fighting the infection.

Healthy people wield an innate immunity effective against most fungal infections. Important innate host receptors that recognize fungal elements include a lectin-like structure on phagocytes (dectin-1), which binds glucan on fungal walls, and Toll-like receptors (TLR2, TLR4). In most instances, neutrophils and alveolar macrophages can kill the conidia of fungi. However, fungal pathogens have ways to avoid immune recognition and impede immune-mediated clearance. For instance, the polysaccharide capsule that envelops *Cryptococcus neoformans*, an important cause of meningitis, hides surface epitopes from host pattern recognition receptors and prevents the binding of complement proteins. Secreted capsule material can also inhibit immune cell function.

Candida albicans can bind complement factors in ways that interfere with phagocytosis. And when thermally dimorphic fungi convert into the pathogenic yeastlike state, the resulting changes in surface structures are poorly recognized by TLRs.

Some pathogenic fungi are also able to survive after being engulfed by immune cells. For instance, the intracellular dimorphic transition from yeast to hyphae by *C. albicans* favors escape from phagocytic immune cells. Some fungi produce substances, such as melanin, that interfere with oxidative killing by phagocytes. The yeast forms of *Histoplasma capsulatum* and *C. neoformans* can actually adapt to live and multiply within macrophages by interfering with lysosomal killing mechanisms.

Despite the pathogenicity features just described, clinical and experimental evidence indicate that the most important factor in fighting off fungal infections is cellular immunity. Fungi, themselves, do not have effective mechanisms that counteract cellular immunity. Consequently, most patients with invasive mycoses have defects in cellular immunity, such as neutropenia; defects in neutrophil function; or insufficient interferon-gamma production by T_H1 cells needed to activate macrophages. These deficiencies can develop from steroid treatment, leukemia, lymphoma, transplantation, and AIDS.

SECTION SUMMARY

- **Protozoan parasites cause tissue damage** that is primarily the result of immunopathology.
- **Antigenic masking** occurs when the parasite coats itself with host proteins to hide from the immune system.
- **Growing inside host cells** offers some parasites temporary sanctuary from immune recognition.
- **Periodically changing cell-surface antigens** allows a parasite to evade the immune response to the previous antigen.
- **Altering the pattern of host cytokine production** is one way parasites can actively suppress the immune response to their presence.
- **Fungal pathogens are typically opportunistic pathogens**, unlikely to cause disease in immunocompetent individuals.

- **Most fungal pathogens are dimorphic**, able to grow in yeast forms (usually pathogenic) or as filamentous hyphae. **Thermally dimorphic fungi** typically grow as yeast when causing disease (37°C) but form hyphae that produce infectious conidia in soils (25°C).
- **Fungi do not produce toxins** that contribute to disease, but they can block immune recognition by producing capsules and through changing surface antigens when morphing from conidia to yeast.

Thought Question 18.9 Person-to-person transmission of a thermally dimorphic lung pathogen such as *Histoplasma* is very difficult. Disease most often initiates after inhaling organisms from the external environment. What is it about thermally dimorphic fungi that can explain why person-to-person transmission is difficult?

Perspective

By now, it should be clear that attachment, immune avoidance, and subversion of host signaling pathways are common goals of most successful pathogens, be they bacterial, viral, or parasitic. But even with all we know, many aspects of infectious disease remain a mystery. Chapters 19–24 describe a variety of infectious diseases. Understanding the basic strategies of pathogenesis and the types of molecular weapons pathogens can wield will help you appreciate how infectious diseases are diagnosed, treated, and—sometimes—prevented.

You will learn, too, in the coming chapters, that new pathogens are constantly emerging. Over the last few decades, we have seen the development of HIV, SARS-CoV and SARS-CoV-2, avian flu, hantavirus, West Nile virus, and *E. coli* O157:H7, as well as the reemergence of "flesh-eating" streptococci, to name but a few. Can we ever stop pathogens from emerging? Probably not. For every countermeasure we develop, nature designs an effective counter-countermeasure. Our hope is that continuing research into the molecular basis of pathogenesis and antimicrobial pharmacology will keep us one step ahead.

Key Terms

AB toxin (p. 555)
adhesin (p. 544)
ADP-ribosyltransferase (p. 555)
alpha toxin (p. 555)
antigenic drift (p. 566)
antigenic shift (p. 566)
autoinducer (p. 550)
edema factor (EF) (p. 560)
endotoxin (pp. 554, 561)
exotoxin (p. 554)

facultative intracellular pathogen
(p. 550)
fimbria (p. 544)
general secretion pathway (p. 562)
genomic island (p. 543)
influenza (p. 566)
intracellular pathogen (p. 550)
labile toxin (LT) (p. 558)
lethal factor (LF) (p. 560)
lipopolysaccharide (p. 561)

mycotoxin (p. 572)
pathogenesis (p. 542)
pathogenicity island (p. 542)
petechia (p. 561)
pilus (p. 544)
proteasome (p. 569)
protective antigen (PA) (p. 560)
protein A (p. 550)
rhinovirus (p. 565)
Shiga toxin (p. 560)

type I pilus (p. 545)
type II secretion (p. 563)
type III secretion (p. 563)
type IV pilus (p. 545)
type IV secretion (p. 564)
ubiquitination (p. 569)
virulence factor (p. 542)

Objectives Assessment Questions

18.1 The Making of a Pathogen

1. The accompanying graph plots guanine plus cytosine content across the genome of a bacterial pathogen whose overall GC content is 55%. Which of the following chromosome segments could represent a pathogenicity island?

   a. Segment A
   b. Segment B
   c. Segment C
   d. Segment D
   e. Segments A and C
   f. Segments B and D

2. Which of the following mechanisms contributes directly to pathogen evolution?

   a. Conjugation
   b. Type III secretion
   c. Exocytosis
   d. ADP-ribosyltransferase
   e. All of the above

18.2 Pathogen Entry and Colonization

3. Bacteria lacking pili cannot attach to host cells. True or false?

4. Twitching motility is carried out by _____.

5. Chronic infections are usually associated with _____.

   a. antibiotic-resistant bacteria
   b. biofilm-producing bacteria
   c. capsule-forming bacteria
   d. Gram-negative bacteria
   e. exotoxin-producing bacteria

6. _____ pili repeatedly assemble and disassemble.

7. *Bordetella* pertactin is an example of a(n) _____.

   a. pilus
   b. capsule
   c. nonpilus adhesin
   d. secreted effector protein
   e. endotoxin

8. Pilus contact between a pathogen and a host cell _____.

   a. delivers exotoxins into cells via the pilus
   b. enables the pathogen to avoid the immune system
   c. extracts nutrients from the host cell
   d. prevents evacuation of the pathogen
   e. All of the above
   f. None of the above

18.3 Surviving within a Host

9. *Salmonella enterica* serovar Typhimurium, while causing an infection, _____.

  a. moves intracellularly via actin tails
  b. remains in the host cell phagosome
  c. prefers growth within a phagolysosome
  d. uses a polysaccharide capsule to avoid phagocytosis
  e. makes protein A that binds antibody Fc regions

10. Which one of the following bacteria is an obligate intracellular pathogen?

  a. *Salmonella*   **d.** *Shigella*
  b. *Staphylococcus*   **e.** *Rickettsia*
  c. *Listeria*

11. *Listeria monocytogenes* _____.

  a. moves intracellularly via actin tails
  b. remains in the host cell phagosome
  c. prefers growth within a phagolysosome
  d. uses a polysaccharide capsule to avoid phagocytosis
  e. makes protein A that binds antibody Fc regions

12. Which of the following strategies helps a bacterial pathogen avoid the immune system?

  a. Delay pilus formation until physical contact with a target cell
  b. Immediately synthesize virulence factors upon entering a host
  c. Delay synthesis of virulence factors until a certain bacterial cell number is reached
  d. Turn off the secretion of all bacterial proteins
  e. Maintain an extracellular presence in the bloodstream

18.4 Host Damage and Manipulation

13. Which of the following toxins does calmodulin activate?

  a. Diphtheria toxin   **d.** Shiga toxin
  b. Cholera toxin   **e.** Anthrax EF toxin
  c. Staphylococcal alpha toxin

14. Which of the following toxins ADP-ribosylates a G-factor?

  a. Diphtheria toxin
  b. *Escherichia coli* LT
  c. Staphylococcal alpha toxin
  d. Shiga toxin
  e. Anthrax EF toxin

15. Which of the following toxins stimulates cytokine production through a superantigen mechanism?

  a. Diphtheria toxin
  b. Shiga toxin
  c. Toxic shock syndrome toxin
  d. Tetanus toxin
  e. Staphylococcal alpha toxin

16. Endotoxin is associated with _____.

  a. *Escherichia coli*
  b. *Staphylococcus aureus*
  c. *Bacillus anthracis*
  d. *Streptococcus pyogenes*
  e. influenza virus

17. Type III secretion systems are paralogous to _____.

  a. chromosome replication systems
  b. conjugation systems
  c. transduction systems
  d. flagellum synthesis systems
  e. type IV pili

18. Type II secretion systems _____.

  a. export DNA and protein
  b. inject proteins directly into the host cytoplasm
  c. secrete proteins into the surrounding environment
  d. import secreted host proteins into the bacterium
  e. are homologous to conjugation systems

18.5 Viral Pathogenesis

19. Which of the following best explains why one person can develop multiple colds over a lifetime?

  a. Ineffective B-cell memory for the rhinovirus
  b. Multiple rhinovirus strains that possess different VP1 structures
  c. A constantly changing ICAM-1 structure in hosts
  d. Ineffective antibody response to the rhinovirus
  e. Different strains of rhinovirus attach to different host cell receptors

Match the following proteins or RNAs to their actions:

Protein or RNA	Action
20. HIV glycoprotein 120	**a.** Accelerating viral transcription
21. HIV Tat protein	**b.** Preventing MHC I processing
22. HPV E7 protein	**c.** CD4 binding
23. Herpesvirus microRNAs	**d.** Preventing apoptosis
	e. Stimulating host cell division

24. Cytomegalovirus, which can cause a mononucleosis-type disease, prevents the immune system from destroying infected cells by _____.

  a. down-regulating MHC II on infected cells
  b. inhibiting production of anti-inflammatory cytokines
  c. placing a molecular mimic of MHC I on an infected cell's surface
  d. placing a mimic of an Fc receptor on an infected cell's surface
  e. preventing phagosome-lysosome fusion

18.6 Protozoan and Fungal Pathogenesis

Match the following organisms to the method of entry into host cells:

Organism	Method of entry
25. *Toxoplasma*	a. Antibody-mediated entry
26. *Leishmania*	b. Actin-myosin motor
	c. Phagocytic cell receptors

27. Half the world's population suffers from the disease toxoplasmosis. True or false?

28. Trypanosomes avoid the immune system by varying their surface antigens. True or false?

29. By which of the following actions do *Leishmania* protozoa avoid the immune system?
    a. Up-regulate TNF-α in infected macrophages
    b. Vary their surface antigens
    c. Inhibit plasma cell formation
    d. Invade red blood cells
    e. Down-regulate IL-12 production in infected macrophages
    f. None of the above

30. When fungi infect an immune-competent host, which of the following actions can they use to avoid the immune system? You may choose multiple answers.
    a. Secrete effector proteins that ubiquitylate host proteins
    b. Make a polysaccharide capsule
    c. Alter surface proteins during infection
    d. Invade and kill host immune cells
    e. Alter host MHC I levels

31. Which of the following is/are nondimorphic pathogenic fungi? You may choose multiple answers.
    a. *Aspergillus fumigatus*
    b. *Blastomyces dermatitidis*
    c. *Trichophyton*
    d. *Histoplasma capsulatum*
    e. *Cryptococcus neoformans*

Review Questions

1. Describe the basic features of a pathogenicity island.

2. Explain various ways bacteria can attach to host cell surfaces.

3. Describe the basic steps by which pili are assembled on the bacterial cell surface. How do type I and type IV pili differ?

4. Explain the broad categories of toxin modes of action.

5. What is ADP-ribosylation and how does it contribute to pathogenesis?

6. Explain the differences between exotoxins and endotoxins.

7. Explain the mechanisms of secretion that type II and type III protein secretion systems carry out. What are the paralogous origins of these systems?

8. Describe the key features of *Salmonella* pathogenesis.

9. What different mechanisms do intracellular bacterial pathogens use to survive within the infected host cell?

10. Describe different molecular strategies that microbes use to avoid the immune system.

11. How do bacteria determine whether they are in a host environment?

12. What is antigenic shift?

13. How is attachment of influenza virus different from HIV attachment? Why does HIV produce an immune deficiency, whereas influenza does not?

14. Explain in basic terms how human papillomavirus can trigger cancer.

15. Describe two basic mechanisms protozoan pathogens use to enter host cells.

16. What mechanisms do protozoa use to avoid the immune system?

17. Which protozoan phyla include human pathogens? (You may need to consult Chapter 11.)

18. Discuss dimorphic fungi and name important factors in establishing a fungal infection.

Clinical Correlation Questions

1. You have just learned that your cousin's newborn boy, only 2 days old, has listeriosis caused by *Listeria monocytogenes*. What can you tell her family about how the child most likely acquired the disease?

2. In the above case, which of the following test results would best indicate the child contracted the disease in utero? (You should review Table 16.1.) Positive indicates an elevated level of the indicated antibody class.

	Anti-*Listeria* Antibodies			
	Child		Mother	
Scenario	IgM	IgG	IgM	IgG
1	Positive	Negative	Negative	Positive
2	Negative	Positive	Positive	Positive
3	Negative	Negative	Negative	Positive
4	Positive	Positive	Negative	Positive

3. In 2011, a new form of pathogenic *E. coli* emerged in Europe that caused bloody diarrhea and a kidney disease called hemolytic uremic syndrome in many victims. The organism (O104:H4) was unusual in that it carried a gene for a distinct form of enteroaggregative pilus, produced high levels of Shiga toxin, and carried a distinct set of other virulence genes never seen assembled in the same strain. What is the most likely explanation for the appearance of this new organism?

4. Cytomegalovirus can cause a life-threatening disease of newborns. During infection, the virus tags MHC I molecules with ubiquitin in infected cells. Ubiquitination causes MHC I molecules to degrade before they can be transported to the cell surface. What is the most likely immunological consequence of this action?

5. In the chapter-opening case, the newsman was infected with *Bacillus anthracis*. There is a vaccine suitable for preventing this disease. However, the vaccine is not available to the general public because of risk-versus-benefit considerations. What component of the pathogen would be best to include in this vaccine? Use your knowledge of anthrax and vaccines (from Chapter 17) to answer this question.

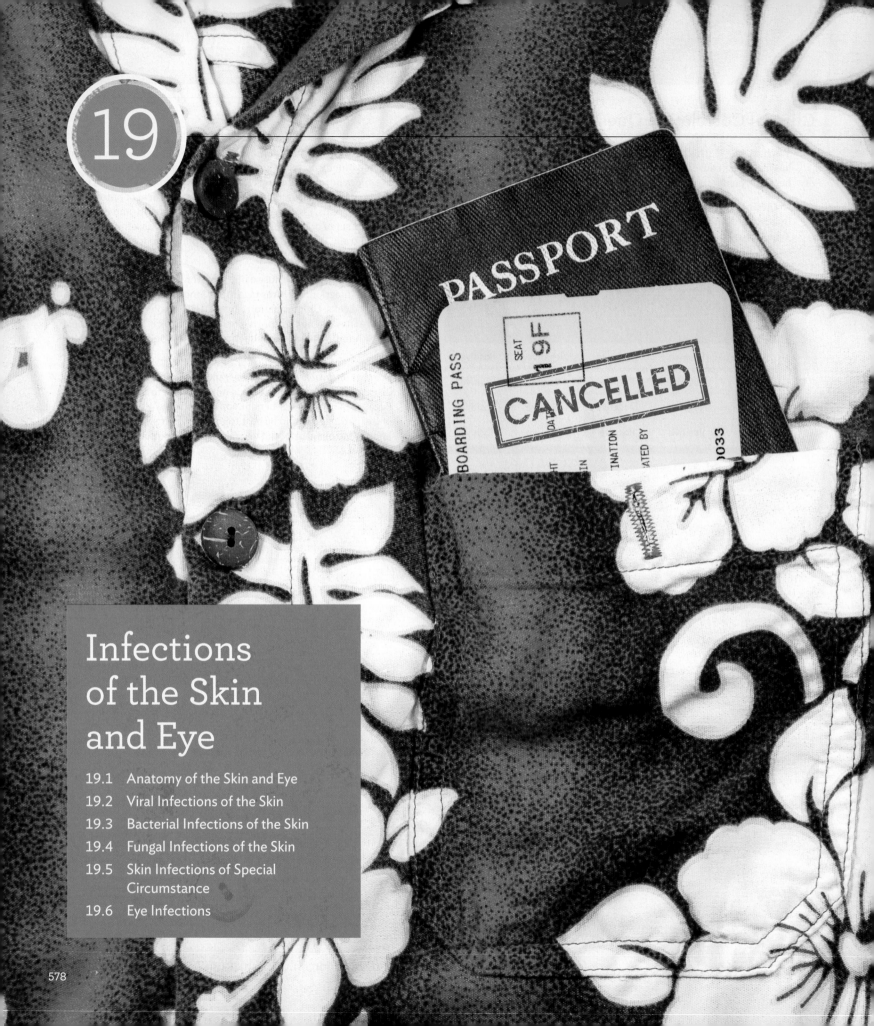

19

Infections of the Skin and Eye

[Rash Ruins Vacation]

SCENARIO Nate, a 69-year-old type 2 diabetic, awoke in his Naples, Italy, hotel room with a tingling sensation that extended from his spine around to the middle of his rib cage, like a thin, invisible ribbon.

SIGNS AND SYMPTOMS The discomfort increased until the next morning, when it became severe pain. Small blisters revealed themselves along an irritated red swath of skin. The hotel concierge referred Nate to the local doctor.

DIAGNOSIS The Italian doctor diagnosed Nate with St. Anthony's fire, a disease, the doctor said, caused by the same virus that causes chickenpox. Nate was surprised to learn that the chickenpox virus he had as a small boy had long ago nestled at the root of one of his spinal nerves. It had stayed there until this trip, when it began replicating, causing the painful skin eruptions along the route of that nerve (dermatome).

TREATMENT The doctor prescribed antiviral and pain relief medications. He also told Nate to get plenty of rest and advised him to make sure that his blood sugar levels were well controlled. He told Nate to follow up with his regular physician when back in the States.

Shingles
Latent VZV emerges from spinal nerve to cause skin lesions along a dermatome.

Chickenpox
Varicella-zoster virus emerges from skin cells to cause chickenpox vesicles. These are very infectious.

FOLLOW-UP Nate cut short his trip to Italy and returned home to the United States, where his doctor confirmed the diagnosis and treatment plan. The doctor told Nate that in Italy, St. Anthony's fire is another name for shingles, a disease caused by the same virus that causes chickenpox.

RESOLUTION Nate eventually recovered from shingles with the help of his antiviral medication and rest, and he was finally able to return to Italy months later for a trip that turned out very nicely this time.

VZV capsid and DNA

Viral envelope

TEM | 200 nm

ANIMATION ▶

Varicella-Zoster Virus
VZV is a DNA virus that gains a membrane envelope when it buds from an infected cell. The viral envelope fuses with the membrane of a new target cell.

Taken for granted as just the covering for other body tissues, human skin is a barrier deflecting continuous assaults on our body, and it acts as a mirror that can reflect our health status. Many infectious diseases manifest themselves as skin rashes or eruptions. Varicella-zoster virus (VZV), the virus that interrupted Nate's Italian vacation, causes the itchy, crusty, vesicular lesions of childhood chickenpox and the painful rash of shingles (also called herpes zoster) experienced by older adults. Although the virus erupts from the skin, the most common portal of entry is the respiratory tract. VZV, however, does not typically damage other body tissues but travels through blood to the dermis, where the virus actively replicates.

Varicella-zoster is not, of course, the only microbe that can cause skin infections. Measles, smallpox, warts, and cold sores, among other skin diseases, are all viral in origin. Various bacteria can cause boils, burn infections, and so-called flesh-eating disease. Fungi, too, can cause skin infections, including athlete's foot and cutaneous candidiasis. A few of these diseases and their causative agents can be diagnosed solely on the basis of signs and symptoms. Other skin infections require laboratory testing. Some skin diseases, such as shingles, can be prevented through vaccination, some can be treated by antimicrobials, and some must simply run their course. This chapter focuses not only on skin but also on infections of the eye, because of the eye's close proximity to the skin.

19.1 Anatomy of the Skin and Eye

Why is the skin considered an organ? Our skin is made of a combination of cells that work together to provide a physical barrier that protects us from pathogenic microbes and helps us regulate our body temperature and avoid dehydration. The **skin** is the largest human organ, covering, on average, 16–22 square feet of surface. It serves as an effective barrier for blocking microbial access to deeper tissues (see Section 15.2), even though the skin itself serves as host to 1 trillion microbes.

Structure of the Skin

The skin, with its superficial layer (**epidermis**) and deeper layer (**dermis**), covers the human body in varying thicknesses ranging from approximately 0.05 to 5 mm. However, the microscopic structure of the skin is the same regardless of its thickness.

The epidermis consists of five layers of epithelial cells. The epithelial cells produce and secrete substances that act as waterproofing agents to reduce water loss from the body surface (**Figure 19.1**). The outermost layer—stratum corneum ("horn layer")—consists of dead keratinocytes that are shed roughly every 2 weeks. Keratinocytes are the cells that make the tough, fibrous, structural protein keratin. The deepest layer of epidermis—stratum basale (also called stratum germinativum)—consists mostly of stem cells (called basal cells), as well as melanocytes (cells that produce melanin). Melanin is the pigment that gives skin its color.

Basal cells continually divide and produce keratinocytes and more basal cells. The newly made keratinocytes push the older keratinocytes upward through the other epidermal layers, where these keratinocytes mature and produce increasing amounts of keratin. The mature keratinocytes eventually reach the surface, where they maintain the superficial layer of the skin by replacing dead keratinocytes that are shed. Langerhans cells are star-shaped, macrophage-like dendritic cells that reside in the epidermis above the basal layer. The job of Langerhans cells is to survey the epidermis and remove foreign antigens (for example, from an infection), and carry those foreign antigens to a nearby lymph node for presentation to T lymphocytes (see Section 16.1).

The dermis is the layer under the epidermis and consists mostly of connective tissue. Here, you can find the types of cells in

Figure 19.1 Structure of the Skin

The skin is composed of two main layers: epidermis and dermis. The epidermis contains five layers (strata) featuring progressively maturing keratinocytes.

Viral infections of the skin
Measles (rubeola)
German measles (rubella)
Fifth disease (erythema infectiosum)
Roseola
Chickenpox
Shingles
Cold sores
Warts
Hand, foot and mouth disease
Smallpox

Bacterial infections of the skin
Folliculitis, furuncle, carbuncle, scalded skin syndrome
Necrotizing fasciitis
Erysipelas
Cellulitis
Acne

Fungal infections of the skin
Dermatophytosis
Tinea versicolor
Candidiasis

Skin infections of special circumstance
Burns
Wounds
Gas gangrene
Cutaneous anthrax

Stratum lucidum
Langerhans cell
Keratinocytes
Stratum corneum
Stratum granulosum
Stratum spinosum
Stratum basale (or germinativum)
Fibroblast

Vein   Artery   Hair follicle   Sweat gland

Epidermis
Stratum corneum
Stratum lucidum
Stratum granulosum
Stratum spinosum
Stratum basale (or germinativum)

Dermis
Fibroblast
Vein
Artery
Sweat gland
Hair follicle

Hypodermis
Also called superficial fascia or the subcutaneous layers. It is located directly under the dermis and consists mostly of adipose and connective tissue.

Stratum corneum
Stratum lucidum
Stratum granulosum
Stratum spinosum
Stratum basale

connective tissue, such as white blood cells and fibroblasts, as well as mast cells and macrophages. The dermis hosts blood vessels, nerves, hair follicles, and sweat glands. The hypodermis, also called superficial fascia or the subcutaneous layer, is located directly under the dermis and consists mostly of adipose and connective tissues.

What Is a Rash?

A change in color and texture of the skin is usually referred to as a rash. **Exanthem** is the name given to any widespread skin rash accompanied by systemic symptoms such as fever, malaise, and headache. **Enanthems** are rashes on mucous membranes. Exanthems and enanthems are usually caused by an infectious agent, such as a virus, and represent a reaction to a toxin produced by the organism, damage to the skin by the organism, or an immune response. Aside from infections, a variety of other situations can affect the skin and produce a rash. Allergies to drugs can produce noninfectious rashes (see Section 17.3). A number of factors, such as the form and type of the lesion, its location, whether it is spreading, and the pattern of spread, as well as other associated symptoms, have to be considered in order to identify the cause as infectious or otherwise.

Figure 19.2 Skin Rashes

A skin rash consists of lesions that may be localized to one part of the body or may cover the entire surface (generalized). The lesions that make up a rash are classified by their shape and contents.

A. A macule is a flat lesion that cannot be palpated, like a freckle.

B. A vesicle forms from accumulation of fluid under the epidermis.

C. A papule is a raised lesion resulting from accumulation of material, infectious or otherwise, in the dermis.

D. A pus-filled raised lesion on the skin is called a pustule. It is usually the result of a buildup of the cellular debris of inflammatory cells, with or without microorganisms, under the epidermis.

There are different types of rashes. A macular rash is a flat, often red segment of skin, called a **macule**, that is less than 1 cm in diameter. A papular rash is a small, circumscribed, solid, elevated lesion called a **papule**. When a papule fills with pus, the lesion is called a **pustule**. A **maculopapular rash** is essentially a reddened papule. Sometimes, as in herpes infections of the skin, the rash consists of **vesicles** (small blisters) and is called a vesicular rash. **Figure 19.2** illustrates various types of lesions.

Mucous Membranes

Mucous membranes (called **mucosae**; singular, **mucosa**), which are epithelial in origin, line the inside of the body and are continuous with the skin in several places. The insides of your eyelids, gastrointestinal tract, and urogenital tract, for example, are covered by mucous membranes. Not all mucous membranes produce mucus (for example, those in the eye or mouth), but they all serve as a barrier to protect the structures they cover. We can think of mucous membranes as "the skin for inside the body." In this chapter we focus on the infections of the skin and the external parts of the eye. Infections of the mucosae will be discussed in Chapters 22 (gastrointestinal diseases) and 23 (urogenital diseases).

Structure of the Eye

Our eyes are our organs of sight; they help us see the world, and what we see shapes our actions and reactions. Each eye has external and internal parts (**Figure 19.3**). The external parts, from the outside in, consist of the eyelid and outer part of the globe (the eyeball).

The part of the eyeball that is outside the skull includes the cornea, the lens, the iris, the pupil, and the sclera (the white part of the eye). The outside parts of the eye, except the cornea, are covered with the conjunctiva, a moist, transparent mucous membrane that is continuous with the conjunctiva that covers the inside of the eyelids. The conjunctiva becomes red and feels uncomfortable when we have a cold or a seasonal allergy attack—a condition called **conjunctivitis** (Figure 19.3D).

The inside of the eye is lined by neuronal tissue that originates from the optic nerve and is called the retina. The retina is orange in color. The macula is in the center of the retina; this is where the best vision (central vision) occurs. A viscous and transparent fluid called vitreous humor fills the inside of the globe, mostly to maintain the eye's shape. Eye infections are always considered very serious because they have the potential to cause blindness.

SECTION SUMMARY

- **Skin** is the largest organ of the body.
- **Epidermis** consists of five layers of epithelial cells. Keratinocytes made in the deepest layer move upward through the other layers and replace dead cells shed from the outermost layer.
- **Langerhans cells** in the epidermis sample antigens for the immune system.
- **Dermis** is made of connective tissue and contains blood vessels, hair follicles, nerves, mast cells, and macrophages.

Figure 19.3 Structure of the Eye

The eye is a ball-shaped structure called a globe situated inside the eye socket of the skull. Muscles connected to the exterior of the globe help move it in different directions as we desire.

A. The external eye consists of the eyelids, the cornea, the sclera, the iris, and the iris opening, called the pupil.

B. The inside of the eye is lined by the retina, which consists of different types of neurons and is orange in color. The macula is the central part of the retina.

C. Conjunctiva.

D. Conjunctivitis, inflammation of the conjunctiva.

Viral infections of the eye
Conjunctivitis
Keratitis

Bacterial infections of the eye
Conjunctivitis
Trachoma
Ophthalmia neonatorum

Fungal and parasitic infections of the eye
Fungal keratitis
Parasitic keratitis

- **Enanthems and exanthems** are rashes associated with systemic symptoms. The forms of rashes include macules (flat, red), papules (solid, elevated), vesicles (fluid filled), and pustules (pus-filled papules).

- **Eye anatomy** includes external structures (conjunctiva and cornea) and internal structures: **retina**, orange-colored neuronal tissue lining the inside of the eye; and **macula**, the center of the retina, where the best vision occurs.

19.2 Viral Infections of the Skin

SECTION OBJECTIVES

- Classify viruses that cause skin rashes on the basis of their genome structure.
- Explain the pathogenesis of viral skin infection sequelae.
- Outline how to diagnose the causative agent of a viral skin infection according to the patient's signs and symptoms.
- Recommend treatment and prevention options for different viral skin infections.

Why do we get viral skin infections? Many viruses prefer replicating in our skin cells, and we see the evidence of their visit as a rash. Some viruses, such as varicella, come for repeat visits, each time producing a different kind of rash (for example, chickenpox

or shingles). Some produce small, infected blisters, called vesicles; others produce growths such as warts or tumors. The majority of rash-causing viruses access our body first through our respiratory tract; examples are varicella-zoster virus (VZV) and rubella virus (German measles). But others, such as papillomavirus (warts), directly attack the epithelial cells of the skin. **Table 19.1** lists several examples of viruses that affect the skin and eye.

Viruses That Produce Macular Rashes

CASE HISTORY 19.1

Moyo's Macular Rash

It was a September day in 2009 when Assaggi, a young mother from Mudzengerere village, Zimbabwe, carried her little boy, Moyo, 2 miles to the Mosomo clinic. When they arrived, Moyo appeared confused and nonresponsive. He had a high fever and even seized once during the examination. The clinic doctor diagnosed probable viral encephalitis. The physician, noticing that Moyo had a macular rash over much of his body, asked Assaggi whether her son had ever been vaccinated for measles. She was evasive at first but finally admitted that her son had not been

Table 19.1

Viral Diseases of the Skin and Eye

Disease	Symptoms	Complications	Virus	Transmission	Diagnosis	Treatment	Prevention[a]
Measles (rubeola)	High fever, conjunctivitis, Koplik's spots, maculopapular rash	Myocarditis, pericarditis, ADEM, SSPE[b]	Measles virus	Respiratory droplets, direct contact	Clinical presentation; serology	Treat symptoms; antiviral ribavirin is used for the immuno-compromised	Vaccine available (MMR routine childhood vaccination)
German measles (rubella)	Low-grade fever, maculopapular rash	Congenital rubella syndrome	Rubella virus	Respiratory droplets, direct contact	Clinical presentation; viral culture and qRT-PCR[c]	Treat symptoms	Vaccine available (MMR routine childhood vaccination)
Erythema infectiosum (fifth disease)	Slapped-cheek rash	Sickle cell disease patients: severe anemia; pregnant women: miscarriage, spontaneous abortion, or fetal defect	Parvovirus B19	Respiratory droplets; vertical and transfusion transmission also possible but not common	Clinical presentation; PCR	Treat symptoms; immunoglobulin is used for the immuno-compromised	No vaccine available
Roseola infantum (exanthem subitum, sixth disease)	High fever, macular or maculopapular red rash on body (not face), erythematous tympanic membrane	Possible neurological symptoms; rarely, encephalitis	Human herpesviruses 1 and 2 (HHV-1, HHV-2)	Respiratory droplets	Clinical presentation	Treat symptoms	No vaccine available
Chickenpox	Generalized pruritic vesicular rash	Shingles	Varicella-zoster virus	Respiratory droplets	Clinical presentation; FAMA, ELISA, and PCR[d]	Treat symptoms, avoid ibuprofen	Vaccine available (routine childhood vaccination)
Shingles	Painful vesicular rash with a dermatomal distribution	Keratitis	Varicella-zoster virus	The virus can be transmitted from a person who has shingles to a person who has not had chickenpox and has not been vaccinated; such transmission can cause chickenpox (not shingles) in the recipient	Clinical presentation; Tzanck smear; viral culture; PCR; direct fluorescent antibody staining	Antiviral treatment: valacyclovir, famciclovir, acyclovir	Vaccine available
Cold sores	Blisters, usually around mouth (or genitals)	Herpetic whitlow, herpes gladiatorum, herpetic keratitis, encephalitis	Herpes simplex virus 1 (HSV-1), sometimes HSV-2	Contact	Clinical presentation; Tzanck smear; viral culture; serological assays; PCR	HSV-1: acyclovir, famciclovir, valacyclovir; HSV-2: valacyclovir	No vaccine available
Common warts (verruca vulgaris)	Small, hard growths on the skin; some subtypes cause genital warts	Cosmetic complications	Human papillomavirus (HPV)	Contact	Clinical presentation	Removal by freezing, burning, or surgery	Vaccine now available for some strains of HPV that cause genital warts (see Section 23.4)

(continued on next page)

Table 19.1
Viral Diseases of the Skin and Eye (*continued*)

Disease	Symptoms	Complications	Virus	Transmission	Diagnosis	Treatment	Prevention[a]
Hand, foot and mouth disease	Fever, painful oral lesions, rash on palms of hands and soles of feet	Myocarditis, meningitis, pneumonia	Coxsackievirus A16, enterovirus 71	Contact with respiratory or oral fluids, blisters, or feces of infected people	Clinical presentation; qRT-PCR of throat or stool samples	Supportive therapy	No vaccine available in US
Smallpox	Enanthem on oral mucosa followed by a pustular, red, and crusted exanthem on skin	Generalized sepsis, pneumonia, encephalitis, blindness, death from multi-organ failure, severe scarring in survivors	Variola major	Respiratory droplets	Eradicated and considered a biological weapon; suspected cases must be reported immediately, and diagnosis and treatment is per CDC protocols	Per CDC protocols	Vaccine available

[a]Childhood vaccination schedule is found in Table 17.2; MMR = measles, mumps, and rubella.
[b]ADEM = acute disseminated encephalomyelitis; SSPE = subacute sclerosing panencephalitis.
[c]qRT-PCR = quantitative reverse-transcription polymerase chain reaction.
[d]ELISA = enzyme-linked immunosorbent assay; FAMA = fluorescent antibody to membrane antigen.

vaccinated. The clinic doctor had been seeing many cases of measles recently, and a significant number of the affected children had died. He heard that the death toll from the measles epidemic of 2009 in Zimbabwe stood at 570—5% of all suspected cases. He started Moyo on the antiviral medication ribavirin, but the doctor was not optimistic. A week later, Moyo was dead.

Mr. Fibion Rupiya, the village elder, sought assistance from the police and health officials after realizing that several children who recently had died had shown symptoms of measles. The children belonged to a religious sect that disdained immunization. As a result of Mr. Rupiya's alert, more than 100 children in Mashonaland Central Province were forcibly vaccinated against measles. Some children were locked up in bedrooms by their parents, and others were hidden in the bush, but health officials managed to find the children and vaccinate them.

MEASLES (RUBEOLA) Measles is one of the six exanthem diseases of childhood and is sometimes called first disease (Table 19.2). While we generally think of it as non-life-threatening, measles caused an estimated 142,300 fatalities worldwide among children younger than 15 years in 2018, even though an effective vaccine has been available since 1963. The measles virus is a member of the Paramyxoviridae family (negative-sense, single-stranded RNA viruses) that replicates inside the host cytoplasm (see Section 12.1) and causes a very contagious disease. The portal of entry for measles virus is either the respiratory system or the conjunctiva. The virus replicates in the lungs, moves to the regional lymph nodes, and produces a viremia (presence of virus in the blood) that spreads the virus throughout the body. An 8- to 10-day incubation period culminates in the prodromal phase (see Section 2.2), with symptoms resembling those of the common cold or flu. The patient runs a fever as high as 40°C (104°F), has no appetite, and feels tired and run down (malaise). Conjunctivitis is a common problem and can be severe in some cases.

The maculopapular rash of measles is mediated by the patient's immune response. The immune system's cytotoxic T cells target the infected endothelial cells of small blood vessels in the dermis. In an effort to kill infected cells, the cytotoxic T cells damage the vessels, causing the rash. This immune reaction, however, is critical

Table 19.2
Exanthems of Childhood

Disease	Old Name	Section
Measles	First disease	19.2
Scarlet fever	Second disease	20.3
Rubella	Third disease	19.2
Filatow-Dukes' disease, believed to be scalded skin syndrome	Fourth disease	19.3
Erythema infectiosum	Fifth disease	19.2
Roseola infantum	Sixth disease	19.2

to controlling the virus. T cell–deficient individuals do not have the rash but do have uncontrolled disease and usually die.

Measles usually starts with an enanthem, called Koplik's spots, which are white spots on the buccal mucosa (inner lining of the cheeks; Figure 19.4A). Koplik's spots are **pathognomonic**—a sign characteristic of a particular condition or disease—for measles. The enanthem is then followed by a descending rash. This rash—an exanthem—appears first at the hairline and then spreads to the head, neck, body, arms, and legs. Sometimes, although rarely, it even spreads to the palms of the hands and soles of the feet (Figure 19.4B). The fever peaks over the next 2–3 days, and the patient experiences respiratory symptoms (including sore throat and cough) and conjunctivitis. The rash starts to fade about 4 days after it appears and is gone by day 7. The fever also subsides 4 days after the exanthem appears, but the cough can persist for another 2 weeks.

Figure 19.4 The Rash of Measles

A. Koplik's spots, a white enanthem in the mouth.

B. The exanthem of measles, a maculopapular rash, on the back.

Immunocompromised patients infected with measles virus can develop extremely serious complications, including viral encephalitis (brain inflammation) that can cause death, blindness due to ulceration of the cornea, myocarditis (inflammation of heart muscle) or pericarditis (inflammation of the pericardial membrane covering the heart), and a variety of gastrointestinal maladies.

Serious complications of measles in immunocompetent patients include **acute disseminated encephalomyelitis (ADEM)**, a disease that strips off the myelin covering of affected nerves. ADEM is an autoimmune disease that presents during the recovery phase. A delayed complication of measles called **subacute sclerosing panencephalitis (SSPE)** is a progressive degenerative disease of the central nervous system that occurs 7–10 years after natural measles infection. SSPE is caused by a defective variant strain of measles virus that produces a persistent infection. ADEM and SSPE are <u>not</u> caused by attenuated vaccine strain.

Diagnosis of measles is based first on clinical presentation. Confirmation is provided by serological tests that measure the presence of anti–measles virus antibodies in the patient's blood. Diagnosis is more challenging in immunocompromised patients because they present atypically (possibly no rash) and have a less-than-optimal immune response (low antibody level). In this situation, viral cultures and the histological study of affected tissues are required.

Treatment of measles is generally supportive. Supportive therapy (or comfort care) increases patient comfort instead of producing a cure. The virus itself is eliminated by the patient's immune system. Fortunately, the disease can easily be prevented through the use of the multivalent MMR vaccine containing attenuated measles, mumps, and rubella viruses (see Table 17.2).

Vaccination had nearly eliminated measles from the United States by the year 2000. Only 50 cases a year were seen, mostly in people returning from overseas. However, the irrational fear some parents have about vaccinating their children has caused a resurgence in measles cases in the United States. In 2019, 1,276 confirmed cases of measles had been reported to the CDC from 31 states—the largest number of cases in the United States since 1992. In two instances in Los Angeles, infected individuals exposed a large number of people at UCLA and California State University. As a result, California officials had to put more than 650 students and staff from the two campuses on quarantine. The question being debated now, nationwide, is whether vaccination, for measles and other diseases, should be required for admission to and attendance at schools. It is clear, however, that outbreaks will not stop until vaccination rates increase.

GERMAN MEASLES (RUBELLA) German measles, or rubella (also called third disease or three-day measles), was initially publicized in German literature, earning rubella its common name. German measles produces a rash similar to measles (rubeola) and was thought to be a type of measles when first described in the 1750s. However, the disease is actually caused by a completely different virus—rubella, a member of the Togaviridae family, which is a single-stranded RNA (positive sense) enveloped virus. Rubella was pronounced eliminated from the United States in 2004 but is still endemic in many parts of the world, including Africa, the Middle East, and South and Southeast Asia.

Little is known about the pathogenesis of the rubella virus. We know that infection is acquired by inhalation of aerosolized respiratory particles. Initial replication takes place in the cytoplasm of cells lining the nasopharynx and in the neighboring lymph nodes. A viremia ensues during the first week of the 12- to 23-day incubation period. The pinpoint maculopapular pink rash (smaller than in rubeola) erupts suddenly on the head and spreads to the body and extremities, much like measles. Like the rash of measles, this rash is thought to be immune related and does not darken and scab over, but it spreads much faster and usually has a short duration, 1–3 days (leading to the colloquial name "three-day measles"). A low-grade fever and enlargement of the lymph nodes of the head and neck usually accompany the appearance of the rash.

German measles is very mild, usually subclinical (lacking clinical symptoms) and self-limiting (resolving on its own) in children. In adults, however, German measles has a more difficult course, often involving joint pain, particularly in the knees, wrists, and fingers. Bacterial superinfections and other complications are extremely rare. Rubella virus can pass through the placenta and cause a variety of defects in a developing fetus, a complication known as congenital rubella syndrome. Infants born with this syndrome usually suffer from hearing loss, but intellectual disabilities or eye or cardiovascular defects can also occur. The risk of fetal infection, along with severity of congenital problems and number of fetal organs involved, is highest when the fetus is exposed during the first trimester of pregnancy.

Because of the mild nature of the disease, diagnostic tests are not indicated unless there is a suspicion of congenital rubella syndrome. Enzyme-linked immunosorbent assays (ELISAs) or enzyme immunocapture techniques (see Section 17.2) can detect rubella-specific antibodies in blood. Quantitative reverse-transcription PCR (qRT-PCR; see Section 25.3) can detect the virus RNA in cord blood or placenta samples to confirm congenital rubella infections. Positive viral cultures obtained from maternal nasopharyngeal secretions are also diagnostic. Supportive and symptomatic treatments are the only available options for patients suffering from rubella infections. The disease and its potential complications are easily prevented through vaccination with the MMR multivalent vaccine.

FIFTH DISEASE (ERYTHEMA INFECTIOSUM) **Erythema infectiosum**, also known as fifth disease, is caused by human parvovirus B19. The disease was first reported by Anton Tschamer in 1889, but its cause was discovered, accidentally, in 1975 in sample 19 of panel B of a routine screening of blood for hepatitis B. Parvovirus B19 is a small, nonenveloped, single-stranded DNA virus of the Parvoviridae family. Exposure to the virus via the respiratory tract is followed by viral attachment to and replication in erythrocyte progenitor cells (cells that grow and change into red blood cells). The infection results in a viremia within 7–10 days. The virus binds the P antigen (globoside), which is expressed abundantly on the membrane of red blood cells. Another unknown receptor, however, is probably required to infect respiratory cells. The body responds to the viremia by producing antiviral antibodies and mounting an inflammatory response that produces tumor necrosis factor alpha (TNF-α), interferon-gamma (IFN-γ), and interleukins 2 and 6 (IL-2 and IL-6).

Prodromal symptoms of mild fever, flu-like symptoms, and sometimes arthralgia (joint pain) coincide with viremia. The initial rash is pathognomonic for erythema infectiosum and is often described as a "slapped-cheek rash" because it is a malar (on the cheek), erythematous (reddened), warm, and nonpruritic (nonitchy) rash that appears as the viremia is resolving (**Figure 19.5A**). The rash is immune-mediated. The slapped-cheek rash is followed by the appearance of a red (or gray) papular enanthem on the palate and throat. The third stage is a maculopapular rash that forms on the body and limbs, with areas of clearing that give it the appearance of lace (reticular rash; **Figure 19.5B**). The rash may also present in young adults as a papular purpuric rash (flat, dark purple spots) in a "glove" and "sock" distribution on the hands and feet (**Figure 19.5C**). The disease is benign in most individuals but can

Figure 19.5 Erythema Infectiosum (Fifth Disease)

A. The initial rash appears on the cheek and has a slapped-cheek appearance.

B. Reticular rash (lacelike appearance).

C. The rash of fifth disease on the hand of a young adult infected with parvovirus B19.

become life-threatening in some patients—for example, in people with chronic anemias or who are immunosuppressed.

The slapped-cheek rash is pathognomonic, but detecting immunoglobulin M (IgM) to parvovirus B19 or using quantitative PCR to identify viral DNA is definitive. Treatment is supportive except in immunocompromised patients, in which case intravenous (IV, inside the vein) immunoglobulin is administered. There is no vaccine.

⊙ **Note:** To produce immunoglobulin, technicians pool blood donated from at least 1,000 people who have been tested to be free of blood-borne diseases such as HIV or hepatitis. The antibody proteins are then separated out of the whole blood. Immunoglobulin can be used to treat a number of diseases but runs the risk of hypersensitivity reactions.

ROSEOLA INFANTUM Roseola infantum—also called roseola, exanthem subitum (meaning sudden rash), pseudorubella, sixth disease, and the three-day fever—is a viral disease of very young children, usually under 3 years of age. It is transmitted by respiratory secretions or saliva. The disease is described as 3–5 days of very high fever (over 40°C, or 104°F), prompting many an emergency department visit. Resolution of the fever is followed by a sudden macular or maculopapular red rash all over the body that blanches (turns white) when touched. The rash is probably caused by antigen-antibody complexes. Human herpesvirus 6 (HHV-6) and, less frequently, human herpesvirus 7 (HHV-7) are usually the cause of the illness. Both HHV-6 and HHV-7 are double-stranded DNA viruses related to the herpesvirus that causes cold sores. Herpesviruses are described in Section 12.1. Roseola cases are sporadic. The disease is mild and is contagious only during the febrile ("with fever") stage. Once the fever has dissipated, the child does not need to be isolated from the day-care population. The virus is actually present in most people, remaining **latent** (hidden and not manifesting disease) within myeloid and bone marrow progenitor cells. HHV-6 and HHV-7, like most herpesviruses, can reactivate later in life, particularly in immunocompromised people.

Diagnostic tests are not done for roseola unless there is an unusual presentation of the disease (when, for example, the rash and fever happen simultaneously). Complications are rare. There is no vaccine.

Viruses That Produce Papular and Pustular Rashes

Some viral infections make a patient's skin erupt into a rash that resembles small pimples filled with clear liquid or pus. Some of these rashes leave a permanent round scar on the skin called a "pockmark," whereas others resolve without scarring their victims. Some viral diseases can start with a simple raised rash but end up producing life-threatening systemic symptoms affecting more than one organ system. One of these, the dreaded smallpox, has fortunately been eradicated as a disease. However, the smallpox virus is still preserved in secure American and Russian freezers. Other viral diseases have not been eradicated but can be prevented by vaccination (chickenpox) or treated more effectively than in the past (cold sores).

CHICKENPOX Chickenpox is caused by a very contagious member of the Herpesviridae family, varicella-zoster virus (VZV; see the opening case of this chapter). The VZV capsid contains a linear double-stranded DNA genome and is surrounded by a lipid envelope acquired when the virus buds from an infected cell. Initial exposure to the virus causes chickenpox—usually a mild childhood disease that resolves by itself. The disease resolves, but the virus does not. Instead, viral DNA takes refuge in clusters of nerve cells (ganglia) and their fibers arising from the section of the spinal cord called the dorsal root ganglia. In 20% of patients, the virus can emerge later in life, travel back down the nerve fibers to the skin, and cause the painful disease known as shingles (herpes zoster).

Varicella-zoster virus is contracted by inhaling infectious particles from skin lesions. Once inhaled, the virus is presumed to replicate in the nasopharynx and then infects the regional lymph nodes, leading to viremia. A second round of viral replication takes place in the liver and the spleen, followed by a secondary viremia 14–16 days postinfection. VZV can then invade capillary endothelial cells and the deepest layer of epidermis. VZV infection of epidermal cells produces fluid accumulation that results in the characteristic vesicles (see Figure 19.2B). Children do not usually have prodromal symptoms. The disease first announces itself as an itchy rash on the child's face, back, chest, and belly and includes maculopapules, vesicles, pustules, and scabs. The clear liquid inside the vesicles contains the viral particles that can be transferred either by contact or inhalation when aerosolized.

The course of the disease is usually mild in children but can become life-threatening in immunocompromised victims. Because the immune system in these patients cannot subdue the virus, a persistent viremia develops that can result in pneumonia (lung infection), encephalopathy (brain disorder), and dissemination of the virus—a condition called "progressive varicella." However, even immunocompetent patients can develop a complication as they age. During the acute phase of the childhood disease, viruses in the vesicles infect the nerve endings of the skin and travel along these nerves to the ganglia, where they hide in a dormant state. Dormancy, or latency, is established when the linear viral DNA circularizes and becomes an episome (Chapter 18) upon entering the nucleus of the host cell, and it can last for decades. Under certain ill-defined situations (including stress), the neurotropic (nerve-loving) virus reactivates and begins replicating.

After reactivation, the replicating virus particles travel along the sensory nerves of the skin to produce a localized, painful, dermatomal rash known as shingles (illustrated in the chapter opening). A **dermatome** is the area of the skin served by a specific nerve. The rash usually runs in a strip from the back around to the chest (the term "shingles" is derived from the Latin word for "girdle"). Whereas chickenpox is usually a mild disease of childhood, the agony of shingles is typically reserved for older adults. Varicella-zoster-specific cell-mediated immunity decreases as the person gets older, which may explain why the incidence of shingles progressively increases with age.

A diagnosis of chickenpox or shingles is usually made clinically, but a variety of other techniques may be used, including culture of the virus from the vesicular fluid, fluorescent antibody to membrane antigen (FAMA) to detect virus proteins on host membranes, enzyme-linked immunosorbent assay (ELISA) to quantify antiviral antibodies (see Section 17.2), and the polymerase chain reaction (PCR) tests to detect viral DNA (see Section 8.6). These tests are used to investigate a difficult case or to establish previous exposure or immunity.

Chickenpox is ordinarily treated with antihistamines, oatmeal baths, and calamine lotion to reduce and control the intense itching, and with acetaminophen to reduce fever and pain. Aspirin is not used to treat viral infections because it is associated with Reye's syndrome, involving sudden encephalopathy and liver damage. Inhibitors of viral replication, such as acyclovir (a guanosine analog), are commonly used to treat shingles, but for chickenpox they are used only in severe or complicated cases.

Vaccination against varicella is part of the routine childhood immunization schedule (see Table 17.2). The vaccine is a live, attenuated form of VZV. Protection from this vaccine (or infection) wanes, so the CDC recommends that healthy people over 50 years of age should be vaccinated with the newer recombinant zoster vaccine to prevent shingles, regardless of whether they remember having had chickenpox in the past. Zoster vaccine live (ZVL), however, can still be used for healthy adults who are 60 years or older.

COLD SORES AND GENITAL HERPES **Cold sores** and **genital herpes** infections are caused by herpes simplex viruses 1 and 2 (HSV-1 and HSV-2), distant relatives of VZV. HSV-1 and HSV-2 are enveloped, linear, double-stranded DNA viruses. The virus replicates in the host nucleus, then the viral core exits the nucleus and picks up its envelope from the Golgi apparatus of the host cell. Best known for infecting the skin and mucous membranes, HSVs are also able to infect the central nervous system and occasionally the visceral (internal) organs. These viruses are transmitted by direct contact. They replicate in mucosal surfaces or epidermis and eventually enter the local sensory or autonomic nerve endings that serve as a path to the virus's final destination: the neuronal cell in the ganglia, where the virus becomes latent. The primary infections may be subclinical or symptomatic, but reactivation of the latent virus always results in symptoms as the virus moves along the nerve toward the skin or mucous membrane. Severity and recurrence of infection with either virus is dictated by the immune status of the person who has acquired it. HSV infections are serious and difficult to manage in immunocompromised transplant patients or in people infected with human immunodeficiency virus (HIV).

Cold sores (also called fever blisters or herpes labialis) are caused primarily by HSV-1 (Figure 19.6A). HSV-1 is very contagious and is present in either active or latent form in 60%–90% of older adults. Currently, there is no vaccine against HSV-1, and the virus is easily spread from person to person by contact. The virus remains dormant in the trigeminal nerve ganglion that links the central nervous system and the face. The major symptom of initial or reactivated infections is the appearance of several painful vesicles on **erythematous** (red and inflamed) skin.

Figure 19.6 Herpesvirus Lesions

A. Cold sores on the upper lip.

B. Herpetic whitlow. Herpes simplex infection around the fingernails is caused in children who suck their thumbs when they have a cold sore and in health care workers who fail to use latex gloves.

C. Erythema multiforme target lesions in a patient infected with HSV-1.

D. Herpes corneal infection stained with a fluorescent dye. The blood vessels penetrate the infected cornea to deliver neutrophils to the area, which is normally devoid of blood vessels.

Infection site

Note: A canker sore is a painful, flat, round or oval open sore in the mouth. Canker sores do not contain virus, are not the same as cold sores, and are not contagious. The etiology of canker sores is unknown, but they can be caused by emotional stress, hormonal changes, vitamin deficiencies, or, indirectly, viral infections.

Primary infections with HSV-1 are usually associated with fever, sore throat, muscle aches, and cervical lymphadenopathy (inflammation of lymph nodes of the neck), along with the vesicular lesions, lasting up to 2 weeks. Reactivation of the infection is not associated with systemic symptoms. Some patients sense the reactivation of the disease because they have prodromal symptoms, including pain, tingling, or a burning sensation before the characteristic vesicles appear. The vesicles most commonly form at the outer border of the lips, but the virus present in these lesions can also infect other parts of the body where the skin is disrupted. In children and some health care professionals (dentists in particular), abrasions of the skin of the fingers result in an HSV-1 infection referred to as **herpetic whitlow** (Figure 19.6B). Rugby players and wrestlers suffer outbreaks of HSV-1 infections on their neck, face, or arm that are aptly named **herpes gladiatorum**. HSV-1 DNA has also been detected in lesions of erythema multiforme, an immune-mediated condition characterized by target-shaped lesions (Figure 19.6C), suggesting an association between the two. In immunocompetent people, cold sores heal by crusting over within a few days to a week.

HSV-1 can set up a chronic infection that may involve the deeper parts of the skin in immunocompromised patients. The virus is also the leading infectious cause of blindness in the United States. In **herpetic keratitis**, virus reactivated in the trigeminal ganglion travels the ophthalmic branch of the trigeminal nerve and infects the cornea. HSV-1 replication then triggers a severe inflammatory response that can destroy the cornea (Figure 19.6D).

Definitive diagnosis of HSV-1 disease is typically made when the virus is cultured from a vesicle, but now real-time PCR assays provide sensitive and rapid identification. A variety of serological tests, immunofluorescent staining (see Section 17.2), and the Tzanck smear test can also be used to verify the presence of the virus. HSV-1 and VZV alter the shape of the cells they infect. In the Tzanck smear test, the base of a lesion is scraped, and the cells are smeared on a microscope slide, stained, and examined microscopically for telltale giant multinucleated cells, syncytia (Figure 19.7).

Antiviral medications, such as acyclovir, valacyclovir, and famciclovir, are used during primary infections to reduce the pain and duration of lesions, shorten the fever period, and decrease

Figure 19.7 Tzanck Smear

A Giemsa stain of cells scraped from the base of a herpes simplex vesicle shows a giant multinucleated cell in the center of the picture. Varicella-zoster virus can also give rise to giant multinucleated cells.

viral shedding. Reactivation infections can be treated with topical antiviral drugs.

The other medically important herpesvirus, HSV-2, causes primarily genital herpes, a sexually transmitted infection that will be discussed in Chapter 23. It is important to note that the tissue specificity of HSV-1 and HSV-2 is not absolute. It is possible for HSV-1 to infect the genital skin and mucosa and for HSV-2 to infect the oral tissue.

WARTS Warts (or **papillomas**) have long appeared on children's fingers (Figure 19.8). How to remove them has been the subject of some colorful discussions. Mark Twain gives an account of Tom Sawyer and Huck Finn discussing the validity of touching a dead cat or plunging their hand in spunk water (rain that pools in a dead tree stump) to get rid of warts! Myths about the cause

Figure 19.8 Warts

A. A common presentation of hand warts.

B. Extreme hand warts.

of warts rival those about their cure; both frogs and toads have often been designated as the supposed culprits. We now know that warts are caused by some subtypes of human papillomavirus (HPV). The Papillomaviridae family contains several hundred subtypes of small, nonenveloped viruses with a double-stranded circular DNA genome that infect mammals, birds, and even tortoises and turtles.

HPV is transmitted to skin by contact, but host- and tissue-specific strains also exist, so the subtype that causes warts on the soles of the feet is different from the one that causes warts on the palms of the hands. HPV probably enters the cell via an endosome by a receptor-mediated mechanism. Once inside, HPV DNA leaves the endosome and enters the nucleus. Virus-synthesized proteins then interfere with cell proliferation controls (discussed in Sections 12.4 and 18.5), and the infected epithelial cell replicates uncontrollably, producing those unsightly (but noncancerous) warts. Treatments for cutaneous warts include freezing, burning (electrically or chemically), and surgical removal—all more effective than "spunk water" but none as imaginative!

Other subtypes of the virus, particularly subtypes 6, 11, 16, and 18, are sexually transmitted (see Section 23.4 and Case History 18.3). HPV-6 and HPV-11 infect the mucous membranes of the anal and genital regions and cause 90% of genital warts (**condylomata acuminata**; singular **condyloma acuminatum**). Human papillomavirus subtypes 16 and 18 have more serious consequences; they are linked to 70% of human cervical cancers. Vaccines prepared from hollow virus capsids are now available. The vaccine, however, must be administered before the recipient (male or female) is sexually active to ensure effectiveness.

HAND, FOOT AND MOUTH DISEASE Hand, foot and mouth disease (HFMD) is a contagious disease caused by positive-sense, single-stranded RNA enteroviruses of the Picornaviridae family. The majority of HFMD cases are caused by coxsackievirus A16 and enterovirus 71 (EV-71), although other coxsackieviruses may rarely cause the disease as well. HFMD is a mild disease that usually affects very young children (under 5 years of age). The infected child has a fever and vague symptoms (such as not feeling well and poor appetite). A painful rash that starts as red spots begins appearing in the back of the mouth, on the palms of the hands, and on the soles of the feet 2 days after the fever starts. The red spots become blister-like and painful, and the child does not eat enough food or drink enough water. The major complication of the disease, therefore, is dehydration; in some cases, it is severe enough to require giving the child intravenous fluids. Life-threatening complications such as severe neurological or heart disease may rarely occur. Neurological complications are usually related to EV-71 infection. Heart-related complications are seen when the disease is caused by coxsackievirus B (see Section 21.2).

Diagnosis of HFMD is usually based on patient history and the presenting clinical signs and symptoms, but isolation of the virus from throat or stool samples may also be required. Treatment is

Figure 19.9 Smallpox Virus

The image shows variola major, the etiologic agent of smallpox, in color. The virus is unusual among DNA viruses because it replicates in the cytoplasm of host cells rather than in the nucleus.

limited to reducing fever and pain with over-the-counter medications, except aspirin, which is associated with Reye's syndrome. Handwashing, keeping surfaces clean, and avoiding contact with an infected person are the only means of prevention because there are no vaccines available for this disease.

⊙ **Note:** HFMD is different from the foot-and-mouth disease of livestock (also called hoof-and-mouth disease), which is caused by the genus Aphthovirus.

SMALLPOX Smallpox is the dreaded disease caused by the large, enveloped, DNA-containing, "dumbbell"-shaped variola virus (**Figure 19.9**). Its brutal history includes killing almost 10% of the world's population at one time. Humans are the only known reservoir of variola, which was eradicated from the human population in 1980 thanks to an effective vaccine and aggressive vaccination campaigns.

Variola is a member of the Poxviridae family. The virus is divided into two variants, variola major and variola minor, based on the severity of the symptoms they cause. The two variants are genetically very similar to each other and to the vaccinia virus. Vaccinia virus causes cowpox and is the virus used to make a vaccine against smallpox (see Figure 16.4C). Smallpox produces a characteristic vesicular rash on the skin (**Figure 19.10**), from which dried particles can aerosolize and transmit the disease.

The smallpox virus was transmitted from person to person by direct or indirect contact. The infective particles would drop onto the mucosal membranes of the upper airways or alveoli. This was the direct, and most common, mode of disease transmission. Indirect transmission of the disease was also possible, but less common, when a person inhaled small aerosolized particles or handled fomites containing the virus. The virus would replicate in the lymph

Figure 19.10 The Rash of Smallpox

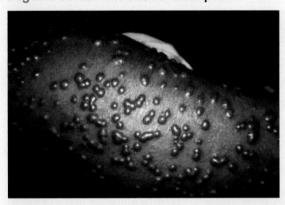

The rash of smallpox starts as vesicular and progresses to pustular.

nodes and lymphoid organs. Then, through two sequential episodes of viremia, it would infect internal organs and bone marrow before showing up as small spots on the oral mucosa (enanthem). Finally, the pox pustule would appear on the skin (exanthem). The virus could be isolated from the kidneys, liver, and bone marrow, as well as from the vesicular rash that finally appeared on the skin (see Figure 19.10).

There are no FDA-approved treatments for smallpox, and the virus is considered a possible bioterrorism threat. We no longer vaccinate the general public against smallpox. Government officials, however, have stockpiled the vaccine in strategic centers all across the United States. Definitive diagnosis of smallpox can be made on the basis of viral cultures and serology. The CDC has developed specific algorithms for diagnosis of variola and necessary responses if a suspected case presents. Closely related monkeypox remains endemic in Africa, infecting animals and occasionally people. The disease produces smallpox-like symptoms in humans but is not as lethal.

SECTION SUMMARY

- **Viruses causing macular rashes** include measles (rubeola), rubella (German measles), parvovirus B19 (erythema infectiosum), and human herpesvirus 6 (roseola).
- **Viruses causing vesicular/pustular** rashes include varicella-zoster (chickenpox, shingles), herpes simplex 1 (cold sores), and variola (smallpox).
- **The respiratory portal of entry** is used by the varicella-zoster, measles, rubella, and variola viruses. The viruses are transmitted by inhalation of respiratory droplets or aerosols; they replicate in lymph nodes and lymphoid tissue and pass via lymphatics and blood to the skin.
- **Skin is the portal of entry** for HSV-1 and papillomavirus, which are transmitted by direct skin contact.
- **Diagnosis of viral skin diseases** is usually based on clinical presentation, but laboratory tests can be used in certain instances, especially in cases involving severe disease in immunocompromised patients.
- **Treatment of viral skin diseases** usually involves supportive therapy. However, antivirals can be used for treating shingles and complications of most of the other viruses. Vaccination is recommended for measles, rubella, and VZV.

Thought Question 19.1 A child brought to a US clinic exhibits the symptoms of German measles, yet the disease was declared eliminated within the United States in 2004. How might the child have contracted the disease?

Thought Question 19.2 Explain why cytotoxic T cells do not attack nerve cells containing latent viruses such as VZV or herpesviruses.

Thought Question 19.3 Is smallpox vaccination required for anyone? Why or why not?

19.3 Bacterial Infections of the Skin

SECTION OBJECTIVES

- Name the bacterial pathogens that cause common skin infections.
- Categorize skin infections by the physiology of the bacteria causing them.
- List clinical and laboratory means of diagnosing bacterial skin infections.
- Develop a prevention and treatment plan for bacterial skin infections.

How do bacteria infect our skin if the skin is supposed to be a protective barrier? Minor insults to the skin, such as an abrasion (superficial skin damage) or even a paper cut, can give bacteria a way in. Healthy individuals develop infections of the skin only rarely. People with underlying immunosuppressive diseases such as diabetes, however, are at much higher risk. Skin infections caused by bacteria range from simple boils to severe, complicated "flesh-eating" diseases (Table 19.3).

A number of different bacterial species can cause infections of the skin and mucous membranes, but Gram-positive *Staphylococcus* and *Streptococcus* species are the major contributors. Staphylococci resemble microscopic clusters of grapes, whereas streptococci appear as strings of pearls (Figure 19.11). The pathogens from each are capable of invading and causing disease in different tissues. *Bacillus anthracis* is another Gram-positive organism that can infect our skin to cause cutaneous anthrax, a zoonotic disease (transmitted from animals to humans). Ingestion of anthrax spores may cause gastrointestinal disease in rare cases. Dissemination of *B. anthracis* to the lungs causes a deadly disease: pulmonary anthrax (see Section 20.3).

Staphylococcal Skin Infections

Staphylococci are Gram-positive bacteria associated with skin infections. Among these bacteria are *Staphylococcus epidermidis*, a normal member of skin microbiota (see Chapter 14), and *Staphylococcus aureus*, which can cause serious skin infections,

Figure 19.11 *Staphylococcus* **and** *Streptococcus*

SEM | 3 μm

A. Staphylococci grow in clusters.

SEM | 1 μm

B. Streptococci grow in chains.

as well as other infections. *S. aureus,* often a normal inhabitant of the nares (nostrils), can infect a cut or gain access to the dermis via a hair follicle. Once *S. aureus* infects the skin, its growth triggers an inflammatory response that attracts macrophages and neutrophils. Pus forming at infected sites is composed of bacteria and white cells, most of which are dead. *S. aureus* possesses a number of enzymes that contribute to disease, including coagulase, which helps coat the bacterium with fibrin and walls off the infection from the immune system and antibiotics. The result is an **abscess**. *S. aureus* produces coagulase and *S. epidermidis* does not, making the former a primary pathogen and the latter an opportunistic pathogen (see Section 2.1). As a result of coagulase activity, "staph" infections generally require surgical drainage as well as antibiotic therapy. Note that coagulase production correlates with the organism's ability to produce exotoxins that cause host tissue damage and weaken host defenses.

S. aureus infection of the hair follicles can be superficial or deep. A superficial infection is referred to as **folliculitis** (Figure 19.12A). A **furuncle**, or **boil**, is a painful abscess of a hair

follicle (Figure 19.12B). **Carbuncles** are lesions that result when several boils join together (Figure 19.12C). Diagnosis of these infections is usually made clinically and is based on their appearance. Folliculitis and most furuncles normally do not require antibiotic treatment and can resolve on their own. Carbuncles and some furuncles have to be drained and treated with antibiotics to resolve the infection.

Other staphylococcal diseases are caused by toxin-producing strains in which the organism remains localized, perhaps in an abscess, but the toxin disseminates (spreads) via the bloodstream. Some strains of *S. aureus* produce toxic shock syndrome toxin (TSST), a superantigen that can lead to serious disease (staphylococcal toxins are discussed further in Section 18.4). Superantigens, unlike other antigens, are able to indiscriminately stimulate a large number of T cells because antigen recognition is not involved (see Section 16.3). Some other strains of *S. aureus* produce a toxin called exfoliative toxin that causes a blistering disease in children called staphylococcal scalded skin syndrome (SSSS), or just **scalded skin syndrome** (Figure 19.12D). Exfoliative toxin, like TSST, is a superantigen, but it also cleaves a skin cell adhesion molecule. Without adhesion, the epidermis will separate from the underlying skin layer to produce blisters in which the skin appears to be scalded by hot water. Remember, the exfoliated skin is not infected. The toxin disseminates from a staphylococcal infection located elsewhere in the body.

A particularly dangerous strain of *S. aureus* called **methicillin-resistant** *Staphylococcus aureus* **(MRSA)** has emerged over the past several decades. *S. aureus* infections are commonly treated with penicillin-like drugs, such as methicillin, that prevent cell wall synthesis. (Virtually all *S. aureus* strains are now

Figure 19.12 **Staphylococcal Infections of the Skin**

A. Folliculitis.

B. A boil, or furuncle.

C. A carbuncle.

D. Scalded skin syndrome.

Table 19.3

Bacterial Diseases of the Skin and Eye

Disease	Symptoms	Complications	Bacteria	Transmission	Diagnosis	Treatment	Prevention
Folliculitis	Infected hair follicles	Skin abscess	*Staphylococcus aureus*	Disease not transmitted from person to person[a]	Clinical presentation; Gram stain (usually not necessary); specimen culture (usually not necessary)	Warm compress for folliculitis and uncomplicated furuncles; topical antibiotics for uncomplicated nonbullous impetigo; oral or systemic antibiotics as warranted, usually inhibitors of cell wall synthesis such as oxacillin, or vancomycin for MRSA; hospitalization and IV antibiotics for complicated cases	No vaccine available
Furuncle (boil)	Red, swollen, painful skin mass	Scarring, sepsis, necrotizing fasciitis	Usually polymicrobial, *Staphylococcus aureus*	Disease not transmitted from person to person[a]	Clinical presentation; Gram stain (usually not necessary); specimen culture (usually not necessary)	Warm, moist compress; incision and drainage; systemic antibiotic sometimes indicated	No vaccine available
Carbuncle	Red, painful, draining mass (coalescence of boils), fever, malaise	Scarring, sepsis, necrotizing fasciitis	Usually polymicrobial, *Staphylococcus aureus*	Disease not transmitted from person to person[a]	Clinical presentation; Gram stain (usually not necessary); isolation of bacteria from specimen in culture (usually not necessary)	Warm, moist compress; incision and drainage; systemic antibiotic sometimes indicated	No vaccine available
Impetigo	Small papular lesions on an erythematous base that often weep (ooze) and crust	Complications are rare; include scarring, cellulitis, acute post-streptococcal glomerulonephritis	*Staphylococcus aureus* and *Streptococcus pyogenes*, Gram-positive cocci	Disease transmitted from person to person	Clinical presentation; Gram stain if confirmation needed; culture (rarely done)	Topical or systemic antibiotics, depending on patient; oral antibiotic in complicated cases: dicloxacillin or amoxicillin-clavulanate; if high prevalence of community MRSA (>10%), then clindamycin, trimethoprim-sulfamethoxazole, and others are considered	No vaccine available
Bullous impetigo	Large blister-like lesions full of a clear yellowish liquid	Cellulitis, staphylococcal scalded skin syndrome	*Staphylococcus aureus*	Disease not transmitted from person to person[a]	Clinical presentation; Gram stain for confirmation; culture (rarely done)		No vaccine available
Staphylococcal scalded skin syndrome (SSSS)	Generalized peeling of skin, usually in infants (generalized form of bullous impetigo), fever, chills, weakness	Dehydration, secondary bacterial infections of skin, septicemia, cellulitis, shock	Exfoliative toxin–producing strains of *Staphylococcus aureus*	Disease not transmitted from person to person[a]	Clinical presentation; histology showing skin damage; Gram stain; culture	IV antibiotics with penicillinase-resistant penicillins such as cloxacillin	No vaccine available
Erysipelas	Fiery red, raised, well-demarcated rash with an uneven surface (orange peel)	Systemic symptoms, necrotizing fasciitis, meningitis	*Streptococcus pyogenes*	Disease not transmitted from person to person[a]	Clinical presentation; culture of needle aspirates of skin lesions; blood culture	Oral amoxicillin for very mild cases; IV or IM (intramuscular) penicillin V or G	No vaccine available

(continued on next page)

Table 19.3

Bacterial Diseases of the Skin and Eye (*continued*)

Disease	Symptoms	Complications	Bacteria	Transmission	Diagnosis	Treatment	Prevention
Cellulitis	Red, hot swollen area of the skin that is tender to palpation	Skin abscess, fever, inflammation of lymphatic vessels (lymphangitis), sepsis	*Staphylococcus aureus* and *Streptococcus pyogenes* most common	Autoinfection; disease not transmitted from person to person[a]	Clinical presentation; blood culture if systemic symptoms (like fever) are present	Oral antibiotics; dicloxacillin or cephalexin usually used	No vaccine available
Necrotizing fasciitis	Rapidly progressing cellulitis; lesion spreads along the fascia (deep to the skin); systemic symptoms	Sepsis, scarring and disfigurement, loss of limb, death	*Streptococcus pyogenes*, majority of cases; others: *Staphylococcus aureus*, *Clostridium perfringens*, *Vibrio vulnificus*	Transmission differs, depending on the organism	Clinical presentation; skin biopsy and culture; blood tests; CT scan	Surgical debridement and leaving the wound open to heal from inside out; antibiotics per culture and sensitivity	No vaccine available
Burn infections	Pus, possibly with green hue with *Pseudomonas*	Delayed, incomplete, or lack of wound healing; sepsis; death	*Pseudomonas aeruginosa* (Gram-negative); various Gram-positive cocci and Gram-negative bacilli	Transmitted from environmental sources or by autoinfection	Clinical presentation; culture and sensitivity	IV antibiotics and hospitalization; surgical debridement as indicated	No vaccine available
Gas gangrene	Necrotic tissue including muscle, pus, foul-smelling odor, much gas and pain	Disfigurement, amputation, liver damage, kidney failure, sepsis, shock, coma	*Clostridium perfringens*	Environmental sources; autoinfection	Clinical presentation; culture and sensitivity	Surgical debridement and IV antibiotics, clindamycin	No vaccine available
Cutaneous anthrax	Ulceration with blackened center	Sepsis, pneumonia, death	*Bacillus anthracis*	Environmental sources; direct contact with infected animals or people	Isolation of bacteria from clinical specimens per CDC; serology	Doxycycline, penicillin, or ciprofloxacin	Avoid contact with contaminated animals; no vaccine available
Pink eye, bacterial conjunctivitis	Red, watery, irritated eye with purulent discharge, sticky eyelids	Conjunctival scarring and obstruction of tear ducts	*Haemophilus influenzae*, *Streptococcus pneumoniae*	Person-to-person contact	Clinical presentation; Gram stain of the purulent discharge (rarely done)	Stop contact lens use; topical ophthalmic antibiotics: erythromycin, fluoroquinolones, azithromycin, tobramycin, or gentamicin; exclusion from school	No vaccine available[b]
Trachoma	Eye pain and redness, itchy eye, turned-in eyelashes	Corneal opacification and blindness	*Chlamydia trachomatis*	Person-to-person contact	Clinical presentation using WHO simple grading system; tissue culture of *C. trachomatis*; PCR	Oral azithromycin; topical tetracycline or topical plus oral sulfonamides are used as alternatives if azithromycin cannot be used	No vaccine available
Ophthalmia neonatorum	Red eye with ocular discharge	Corneal ulceration and blindness	*Chlamydia trachomatis*, *Neisseria gonorrhoeae*; other bacteria possible	Transmitted from infected mother during birth	Culture of ocular specimen; Gram stain of smear of an ocular sample; direct fluorescence antibody test; PCR; rapid enzyme immunoassay	Oral erythromycin for *C. trachomatis*; IV or IM ceftriaxone for *N. gonorrhoeae*; other antibiotics as indicated for other bacteria	No vaccine available

[a]Organisms causing the disease can be transferred from person to person, but the disease is generally not. Once an organism is transferred, it can colonize the new host.
[b]Vaccines are available to protect against *H. influenzae* and *S. pneumoniae* respiratory tract infections but not conjunctivitis.

resistant to penicillin.) Most *S. aureus* strains (about 60% of clinical isolates) have now evolved to resist methicillin. MRSA has developed resistance to methicillin and many other penicillin-like drugs through a mutation that alters one of the proteins (called a penicillin-binding protein, PBP) involved in cell wall synthesis. Because methicillin is normally used as a first line of defense against staphylococcal infections, treatment failure of MRSA can be life-threatening, and alternative drugs, such as vancomycin (a cell wall synthesis inhibitor unrelated to penicillin), must be used.

MRSA first appeared as nosocomial infections—infections that occur after a patient enters a health care setting (Section 26.2). These infections are also known as hospital-acquired infections or health care–associated infections. Today, these antibiotic-resistant organisms are no longer confined to hospitals. Individuals who have not been in a hospital are being infected with MRSA in what are called community-acquired infections. These infections are occurring at an epidemic rate in the United States (an incidence of approximately 15 per 100,000 bloodstream infections in 2017). MRSA now causes most staphylococcal skin infections. A health care provider can no longer assume that a patient walking into the office with a staphylococcal infection will respond to methicillin (oxacillin). Vancomycin and linezolid are the antibiotics typically used to treat MRSA infections.

- -

Note: Methicillin is no longer used in the United States. Oxacillin, which is from the same class of antibiotics, is used instead. The strains tested for resistance, therefore, are tested against oxacillin, but the name MRSA has not been changed. Strains that are resistant to oxacillin are also resistant to methicillin.

- -

CASE HISTORY 19.2
Leili's Crusted Lip

Leili, a 3-year-old girl, was brought to the pediatrician by her mother, who was worried about a rash on Leili's lip. It had started 4 days earlier as a little bump above her lip that spread to the corner of her mouth (Figure 19.13A). The base of the rash was red and covered by pustules. The little girl said that her rash hurt, and the clinician noticed a honey-colored crust on the ruptured pustules. Leili's mother said that her daughter had not been trying to scratch the rash, nor was the area around it hot to the touch. The doctor gently swabbed a sample from one of Leili's sores and sent it to the lab, where the causative agent would be identified and its antibiotic susceptibility determined. Once the lab results came back, the doctor informed Leili's mother that her little girl had impetigo, a skin infection often caused by bacteria called staphylococci, although streptococci can also cause it. The doctor explained that impetigo is a very contagious disease, and she gave Leili's mother a topical antibiotic to use. Treating topically rather than orally would accomplish two goals: it would reduce the chance that the bacteria would develop drug resistance, and because Leili had a very mild case of impetigo, the antibiotic in the gel should easily penetrate the pustules and kill the organism. Leili and her mother left but returned a week later for a follow-up exam. The rash was gone, replaced by Leili's smile. The pediatrician explained that the cause was, in fact, methicillin-sensitive *Staphylococcus aureus* (MSSA), not MRSA. Leili's mother, in turn, reported that another child at Leili's day care had impetigo before Leili, and that is where Leili probably acquired the infection.

Figure 19.13 Impetigo

Impetigo is usually caused by *Staphylococcus aureus*, although *Streptococcus pyogenes* can also be an agent.

A. Nonbullous impetigo presents with honey-colored, crusted lesions on a violaceous (purplish red) base with no fluid content.

B. Bullous impetigo is uncommon and has fluid-filled, blister-like lesions.

C. Ecthyma is a rare form of impetigo that is invasive.

Impetigo is a contagious skin infection that affects mostly young children. The majority of cases are caused by *Staphylococcus aureus*, but some cases are caused by group A beta-hemolytic streptococci (discussed later).

There are two main types of impetigo: **nonbullous impetigo** (see Figure 19.13A), the most common form (the kind that Leili had), seen as honey-crusted lesions (this type is often called simply impetigo); and **bullous impetigo** (Figure 19.13B), fluid-filled and blister-like. The nonbullous form starts as superficial "bumps" that become papules and vesicles before they form the characteristic pustules on erythematous skin. The fluid from the pustules leaks out as they break open ("weep") and forms a honey-colored crust. In bullous impetigo, the vesicles grow larger and become **bullae** (blisters; singular, **bulla**) that are full of clear yellow fluid. Strains of *Staphylococcus aureus* that produce exfoliative toxin A can cause bullous impetigo. Scalded skin syndrome is a generalized form of bullous impetigo seen mainly in infants (see Figure 19.12D and Section 18.4).

A minority of cases of impetigo result from streptococci, mostly *Streptococcus pyogenes*, a microbe normally associated with sore throat infections (pharyngitis). Among the many strains of *Streptococcus pyogenes* that cause infections, some will put the patient at higher risk of complications after the infection is resolved. These complications are referred to as post-streptococcal sequelae (conditions developing after the infection is over) and include post-streptococcal acute glomerulonephritis (inflammation of the kidney caused by antigen-antibody complexes) and rheumatic fever caused by self-reactive antibodies (discussed in Section 20.3). The risk for acute glomerulonephritis appears to be higher for patients infected with impetigo-producing strains of *Streptococcus pyogenes* than for patients infected with pharyngitis-associated strains. It is unclear whether there is strain specificity for rheumatic fever.

Diagnosis of impetigo is based on clinical signs and symptoms and can be confirmed by isolation of the bacteria from the lesions. It is important to treat impetigo to avoid complications such as scalded skin syndrome. In Case History 19.2, Leili was treated with a topical antibiotic because her case was mild. Oral antibiotics are better suited for more severe cases. The antibiotic used should be active against Gram-positive cocci.

Streptococcal Skin Infections

CASE HISTORY 19.3
The Voracious Bacteria

 One weekend in June, Cassi was camping with her three children. She suffered a minor cut on her finger, which she bandaged properly. She also injured the left side of her body while playing sports with her kids. Not thinking much of either minor injury, she went home and went to bed. Two days later, Cassi was extremely ill. Her symptoms included vomiting, diarrhea, and fever. She was also in severe pain where she had injured her side, and the area had begun to bruise (the skin did not look broken). By the next day, she could barely get out of bed, and by the end of the night, she was breathing with difficulty and could not see. Her side began to leak fluid and blood. Cassi was admitted to the hospital in septic shock, with no detectable blood pressure. She was given vasopressors (medications that constrict the blood vessels) to raise her blood pressure to the normal level. An infectious disease specialist looking at the wound (which was similar to the one pictured in Figure 19.14) diagnosed the problem as necrotizing fasciitis, and Cassi was rushed into surgery. In an effort to save her life, about 7% of her body surface was removed. Because the large wound infection in her side would need to resolve before a skin graft could be performed to repair it, the hole in Cassi's body was left wide open. Meanwhile, the continued use of vasopressors caused gangrene in her fingers and lower extremities. After nearly 3 months and several operations, including amputation, Cassi recovered.

What kind of organism can cause this horrific disease? **Necrotizing fasciitis**, also known as flesh-eating disease, is rare; the CDC estimates an incidence of 700 cases annually. The disease is often caused by the Gram-positive bacterium *Streptococcus pyogenes*, which may be the reason it used to be called streptococcal gangrene. The disease is caused when microbes penetrate the skin through a microscopic cut or abrasion. The bacteria spread in the subcutaneous tissue, causing death and destruction of fat and fascia without harming the skin itself, so patients often do not seek help until it is too late. Fascia is the sheath of thin, fibrous tissue

Figure 19.14 Necrotizing Fasciitis Caused by *Streptococcus pyogenes*
Early-stage wound of necrotizing fasciitis.

that covers muscles and organs. There are two different types of the disease: 1 and 2. Generally speaking, type 1 is polymicrobial, whereas type 2 is due to one microorganism, usually *Streptococcus pyogenes* and sometimes *Staphylococcus aureus*. Other organisms that can cause necrotizing fasciitis include *Clostridium perfringens* (a Gram-positive, anaerobic rod found in soil) and, in immunocompromised hosts (for example, alcoholics or diabetics), *Vibrio vulnificus* (a Gram-negative marine bacillus).

Treatment of necrotizing fasciitis caused by any bacterial species requires rapid and aggressive surgical removal of the affected tissue (debridement), coupled with aggressive IV antibiotic therapy. Antibiotic treatment alone is almost always ineffective because of insufficient blood supply to the affected tissues. Therapy includes antibiotics such as clindamycin and metronidazole, which act against anaerobes and Gram-positive cocci, and gentamicin, a drug particularly effective against Gram-negative microbes.

The incidence of necrotizing fasciitis may have risen owing to the increased use of nonsteroidal anti-inflammatory drugs, which accentuate a person's susceptibility to infection by *Streptococcus pyogenes*. In Case History 19.3, Cassi probably had *S. pyogenes* on her skin when the injury to her side occurred. The injured area probably suffered an invisible microabrasion (microscopic scrape), providing a good growth environment for the organism and enabling the cocci to secrete potent toxins that trigger inflammation and mediate death of the surrounding tissues. Why some patients infected with *S. pyogenes* develop necrotizing fasciitis while others do not may be related to the immunological status of the patient.

Different strains of *Streptococcus pyogenes* have different virulence factors, giving these microbes much versatility. Virulence factors include:

- A capsule that, when thick, will help the organism avoid phagocytosis by macrophages
- The pilus-like M protein (more than 80 different serotypes), which binds a complement regulatory protein (factor H) and protects the microbe from opsonization by complement (see Section 15.7)
- A cell wall containing lipoteichoic acid, thought to facilitate adherence to pharyngeal epithelial cells
- Enzymes that lyse blood cells (streptolysins)
- Enzymes that degrade DNA (deoxyribonucleases), fibrin (streptokinase), and connective tissue (hyaluronidase), making pus less viscous and allowing the organism to spread rapidly
- Peptidoglycan, capable of activating the alternative complement pathway and a microbe-associated molecular pattern (MAMP) that binds NOD-like receptors, causing inflammation (see Section 15.5)

S. pyogenes is also capable of producing several exotoxins called **streptococcal pyogenic exotoxins (SPEs)**. SPEs are superantigens. The massive amounts of cytokines released in response to SPEs can produce impressive levels of inflammation and lead to shock (cardiovascular collapse, extremely low blood pressure). Formerly called erythrogenic toxins, SPEs are associated with scarlet fever (see Section 20.3), streptococcal toxic shock syndrome, and necrotizing fasciitis.

The natural reservoir for *S. pyogenes* is the human nasopharynx and parts of the skin. The innate and adaptive immune systems of carriers normally protect the host from disease. Of course, not everyone harbors the pathogenic strains. The virulent strains, however, can be spread between carriers and non-carriers.

Why are these streptococci classified as group A beta-hemolytic? A large number of streptococci produce and secrete a hemolysin (an enzyme capable of lysing red blood cells). When plated onto agar containing red blood cells, hemolysin-secreting strains lyse red blood cells around the colony to produce a clear zone called beta hemolysis. These streptococci are subclassified in a clinical laboratory into groups A–O according to their cell wall antigens (see Section 25.2). *S. pyogenes* is the main pathogen among the group A beta-hemolytic streptococci (also called GAS) and is often used synonymously with the term group A strep, although this is not completely accurate. It is true that all *S. pyogenes* bacteria are group A streptococci, but not all group A streptococci are *S. pyogenes*.

Our final point about GAS involves the role of the pilus-like M protein in post-streptococcal sequelae. As noted earlier, rheumatic fever can develop after resolution of a primary GAS infection. This sequela is thought to be the result of immunological cross-reactivity between specific GAS M-protein antigens and host antigens. In a case of mistaken identity, the autoreactive B cells that make antibodies against cardiac antigen appear to be activated by the bacterial M-protein antigen. Once made, the autoreactive antibodies begin to attack the self antigen of the heart. The ensuing antigen-antibody complexes trigger an inflammatory reaction that damages those tissues and produces rheumatic fever (see Section 17.5).

ERYSIPELAS Erysipelas is another disease instigated by *Streptococcus pyogenes*. The disease involves the upper layer of dermis and then spreads to the superficial lymphatics. The "rash" of erysipelas may appear on the face but is most commonly seen on the lower extremities (Figure 19.15A). Erysipelas is an acute infection that begins typically as a small erythematous patch and progresses to a fiery-red, shiny plaque. Erysipelas lesions on the face cover the cheeks and the bridge of the nose, sometimes giving them a "butterfly" appearance (Figure 19.15B). The patient has swollen lymph nodes (lymphadenopathy) and often has fever and other systemic symptoms. Streptococci are usually the cause, but the bacteria cannot be isolated from the surface of the affected area because the infection is in the deeper dermis. Patients must be treated with antibiotics because septicemia can occur quickly.

⊙ Note: In England and the United States, "St. Anthony's fire" refers to erysipelas, whereas in Italy and Malta, the colorful name is used for herpes zoster (shingles).

Figure 19.15 Erysipelas

Streptococcus pyogenes, and rarely other bacteria, can spread to the deeper layers of skin, causing a dangerous infection.

A. Infection of the leg. Note the clear demarcation between infected and noninfected areas.

B. Classic rash on the face.

CELLULITIS Cellulitis is a term commonly used to indicate an uncomplicated non-necrotizing inflammation of the dermis related to acute infection. Cellulitis does not involve the fascia or muscles but is characterized by localized pain, swelling, tenderness, erythema, and warmth (**Figure 19.16**). Cellulitis is a relatively common infection. It develops more slowly than erysipelas and involves the deeper levels of dermis, including the subcutaneous fat and connective tissue. The infection is often a complication of a wound

Figure 19.16 Cellulitis

Note the swelling of the affected hand and lack of epidermal involvement.

infection. Although *Streptococcus pyogenes* is the most frequent cause of cellulitis in immunocompetent adults, a number of bacterial species, including *Staphylococcus aureus*, Gram-negative bacilli, and anaerobes, can cause this skin infection. Diagnosis is based on clinical symptoms, history, and bacterial cultures of the original infection. Cellulitis is treated by aggressive antibiotic therapy administered by mouth, although intravenous administration of the antibiotic may be required, depending on the severity of the case.

ACNE Acne vulgaris is America's most common disease, affecting 60%–70% of Americans at some time during their lives. Twenty percent will have severe acne, resulting in permanent physical scarring. Acne usually affects the face, but any area of skin with a dense population of follicles, such as the upper part of the chest or the back, can also be affected. Not all acne is the same. **Comedone acne** results from blocked hair follicles or pores called comedones, which can be open (blackheads) or closed (whiteheads). Inflammatory acne arises from inflamed macules, papules, pustules, and nodules (aggregations of cells resembling papules but larger than 0.5 cm). The most severe type of acne is cystic or nodular acne, consisting of painful, fluid-filled cysts or nodules, respectively, under the skin surface (**Figure 19.17**).

Several factors contribute to acne development. The first is a genetic predisposition in which epithelial cells in a follicle replicate more often than normal (hyperproliferation), contributing to

Figure 19.17 Cystic Acne

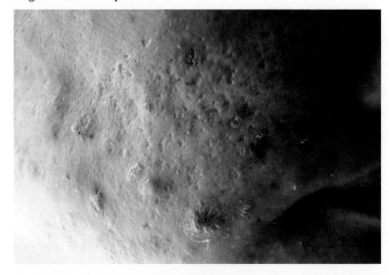

plugging of the follicle. A child has a much greater chance of developing acne if both parents had acne than if neither parent had the condition. High levels of certain hormones can cause hyperproliferation and excessive sebum production (nodular acne). Sebum, a waxy, oily substance secreted by sebaceous glands, can plug the follicle (see Figure 19.1). Finally, the Gram-positive, anaerobic bacterium *Cutibacterium acnes* (formerly *Propionibacterium acnes*) plays an important role. A normal member of skin microbiota, *C. acnes* uses the triglycerides in sebum as a nutrient. The growing numbers of this organism promote inflammation by binding to Toll-like receptor 2 on macrophages and neutrophils, which then release pro-inflammatory cytokines such as TNF-α (Section 15.5 and Table 15.3).

Topical retinoids and topical antibiotics are the first line of treatment for acne. Retinoids have some anti-inflammatory effects, can make the skin less oily, and decrease the number of *C. acnes* so that they can help unplug clogged skin pores. Oral retinoids, such as isotretinoin, and antibiotics are sometimes used for severe acne. Oral retinoids, however, present a risk of severe side effects, like birth defects if used during pregnancy, and inflammation of the pancreas (pancreatitis).

A cautionary word about bacterial skin infections: Skin lesions are a rich source of bacteria. Without medical intervention, the bacteria in these lesions can enter the bloodstream and cause bacteremia. Once in the blood, the bacteria can disseminate to other body sites and cause focal or metastatic lesions or produce life-threatening sepsis (see Chapter 15 IMPACT).

Note: Proper cleaning of wounds and cuts will reduce the number of microorganisms in the site, which will, in turn, enhance healing.

SECTION SUMMARY

- *Streptococcus pyogenes* (also known as group A streptococci, or GAS) and *Staphylococcus aureus* are common bacterial causes of skin infections.
- **Folliculitis, scalded skin syndrome, and impetigo** are usually caused by *Staphylococcus aureus*.
- *Staphylococcus aureus* virulence factors involved with skin infections include coagulase, toxic shock syndrome toxin (TSST), and exfoliative toxin.
- **Methicillin-resistant *Staphylococcus aureus* (MRSA)** has become an important cause of community-acquired staphylococcal infections.
- **Necrotizing fasciitis** is usually caused by *Streptococcus pyogenes*, but it can also be caused by other pathogens.
- *Streptococcus pyogenes* virulence factors involved with skin infections include capsule, M protein, and streptococcal pyogenic exotoxins (SPEs).
- **Cellulitis** is a skin infection caused by a variety of bacterial species.
- **Erysipelas** is a skin infection caused by *Streptococcus pyogenes*.
- **Acne** is influenced by genetics, hormone levels, and the bacterium *Cutibacterium acnes*.
- **Infections of the skin** can disseminate via the bloodstream to other sites in the body.

Thought Question 19.4 Why do viral skin infections often lead to secondary bacterial infections, whereas bacterial skin infections do not lead to secondary viral infections?

19.4 Fungal Infections of the Skin

SECTION OBJECTIVES

- Outline common fungal skin infections according to their etiology.
- Compare dermatophytes with other fungi that commonly infect the skin.
- Explain how common fungal infections are diagnosed clinically.

What are fungi? Fungi, which include molds and yeasts, are eukaryotic microbes. Fungi grow as either filamentous or single cells that bud to replicate (as in yeasts), as presented in Section 11.2. **Dermatophytes**, from the Greek for "skin plants," are fungi that "love" human skin. Dermatophytes find cool, moist, keratinized tissues (skin, hair follicles, nails) a perfect growth environment. Dermatophytic fungi invade the epidermis and elicit an inflammatory response. *Epidermophyton, Trichophyton,* and *Microsporum* cause the majority of dermatophytic infections (Table 19.4). Diseases caused by these fungi are named after the location of the infection and not the fungus itself. Medieval doctors, ignorant of microscopic fungi but well versed in Latin, called the infections "tinea" (Latin for "gnawing worm"). So, if the lesion sits on the scalp, it is called **tinea capitis**; on the body, **tinea corporis**; in the fold of the groin, **tinea cruris** (jock itch); and on the foot, **tinea pedis** (athlete's foot). Collectively, these infections are also called **ringworm**, even though worms are not involved. Fungal infection of the nails is called **tinea unguium** or onychomycosis.

Tinea capitis is a fungal infection of the scalp (primarily in small children) that has two different varieties: black dot tinea capitis (BDTC) and gray patch tinea capitis (GPTC). The black dot form of tinea capitis, caused by *Trichophyton tonsurans*, is the most common form in the United States. The name comes from the fact that infected hair may break off at the scalp, leaving black dots. Gray patch tinea capitis, caused by *Microsporum canis*, is transmitted to humans primarily from cats and dogs. Again, the hair breaks off above the scalp, but the scalp frosts over (Figure 19.18A).

Tinea corporis refers to ringworm on the body surface, excluding the groin (Figure 19.18B). The lesion is round or annular (ring-shaped) and pruritic (itchy). The ring expands as the fungus grows outward from the initial site of infection. Jock itch, or tinea cruris, is a form of tinea corporis that affects the crural fold (the groin). Jock itch is more prevalent in men. The main cause of jock itch in the United States is *Trichophyton rubrum*. Athlete's foot (or tinea pedis) is caused by either *Trichophyton mentagrophytes* (acute athlete's foot) or *Trichophyton rubrum* (see Figure 2.2).

Table 19.4

Fungal and Parasitic Diseases of the Skin and Eye

Disease	Symptoms	Complications	Fungus or Parasite	Transmission	Diagnosis	Treatment	Prevention
Dermatomycosis	Dry, scaly lesions	Scarring, secondary bacterial infections, cellulitis	Dermatophytes: *Epidermophyton, Trichophyton, Microsporum*	Environmental sources; person-to-person contact	KOH preparation[a]	Topical azole compounds (such as clotrimazole); systemic azole compounds if indicated (fluconazole)	No vaccine available
Tinea versicolor	Small, flat, discolored lesions of skin, variable colors; does not itch	Cosmetic complications	*Malassezia* (genus of yeast)	Environmental sources	Clinical presentation; KOH preparation[a]	Oral or topical azole antifungals	No vaccine available
Candidal intertrigo	Patchy, smooth-looking lesions	Can become invasive in the immuno-compromised	*Candida albicans* (major cause), other *Candida* spp.	Autoinfection in the immuno-compromised	Clinical presentation; microscopic evaluation of clinical specimen; culture	Treatment depends on patient age, immune status, and disease presentation; clotrimazole and nystatin as indicated; maintain oral hygiene	No vaccine available
Fungal keratitis	Eye pain and redness, sensitivity to light, blurred vision, eye discharge	Corneal scarring, blindness	*Fusarium, Aspergillus, Candida* spp.	Environmental sources	Clinical presentation; isolation of fungus in culture from corneal scrapings for definitive diagnosis; microscopic examination of corneal sample; PCR	Antifungal medications: topical natamycin for superficial infections (*Aspergillus* and *Fusarium*); systemic antifungals for deep infections: amphotericin B, fluconazole, voriconazole	No vaccine available
Parasitic keratitis	Corneal lesions	Encephalitis	*Acanthamoeba* spp.	Environmental sources	Direct visualization of corneal ulcer under slit lamp microscopy; isolation of *Acanthamoeba* in culture; PCR	6–12 months of antiseptic treatment of corneal lesions; pain management	No vaccine available

[a]KOH = potassium hydroxide.

Spores of these fungi are shed from the skin of an infected person and stick to the skin of uninfected individuals who may be using the same gym.

Tinea versicolor is a long-term (chronic) infection of the skin, prevalent mostly in warm, moist climates. Its cause is a genus of yeast (*Malassezia*), not a dermatophyte. This yeast is actually part of the normal skin biota and is not contagious. *Malassezia* is a dimorphic yeast, meaning it can have both yeast and mycelial (hyphal) forms. Pathogenesis begins when the round yeast form converts to hyphal form and invades the stratum corneum. The lesions of tinea versicolor are usually small, with a sharp border, and hypopigmented (**Figure 19.18C**). Hypopigmentation is thought to occur when azelaic acid produced by the infecting yeast damages the melanocytes of the skin.

Candida species are dimorphic yeasts that form part of the normal biota of the gastrointestinal tract, vaginal tract, oral cavity, and even skin. Best known for causing vaginal infections, *Candida* (usually *C. albicans*) can also infect the skin (in its mycelial form), as well as mucous membranes and body organs. In healthy individuals, *Candida* can overgrow and cause disease in the oral cavity or vagina as a result of changes in the local microbiota. The fungus can cause serious problems in patients who are immunocompromised. On the skin, *Candida* infects areas where the skin touches and rubs together, such as between the fingers, under the arm, or in the groin.

Figure 19.18 Various Forms of Ringworm

A. Gray patch tinea capitis.

B. Tinea corporis. Multiple lesions can come together to form a flower petal shape.

C. The hypopigmented lesions of tinea versicolor caused by the yeast *Malassezia*.

The most common of these skin infections is referred to as **candidal intertrigo** (Figure 19.19). Normally kept at bay by macrophages, neutrophils, and T-cell immunity, *C. albicans* can take advantage of defects in these cell types (as in neutropenia, abnormally low numbers of neutrophils) to penetrate skin microabrasions and gain a foothold.

Diagnosis of fungal skin diseases is based on their clinical appearance and on microscopic examination of potassium hydroxide (KOH) preparations (see Section 25.1) of skin flakes or hair. The KOH destroys the skin cells but not the more resilient walls of mycelia or spores, which can be seen under a light microscope. These fungi can also be cultured on a special selective medium called Sabouraud agar (an acidic, pH 5.6 medium containing antibiotics to thwart the growth of bacterial contaminants and culture fungi). A variety of antifungal medications are used to treat these diseases. The most common, the imidazole compounds (such as clotrimazole), can be bought over the counter. The antimicrobials either directly interrupt fungal membrane integrity or interfere with the synthesis of sterols required for fungal membrane synthesis. (Antifungal agents are discussed further in Section 13.8.)

Figure 19.19 Candidal Intertrigo

Candida infection between fingers.

SECTION SUMMARY

- **Dermatophytes** are fungi that cause skin infections.
- **Fungal skin diseases are named by location**, not by fungal species—for example, tinea capitis (scalp), tinea corporis (body), tinea cruris (groin), tinea pedis (feet).
- **Ringworm** refers to the circular appearance of lesions as the fungus extends in all directions.
- *Malassezia* **and** *Candida* **are dimorphic yeasts** that can cause skin infections.
- **KOH preparations** are used to diagnose fungal diseases of the skin.

Thought Question 19.5 Why do some people develop fungal infections but others do not?

19.5 Skin Infections of Special Circumstance

SECTION OBJECTIVES

- Discuss why *Pseudomonas aeruginosa* is a common cause of wound infections.
- Categorize organisms that are commonly isolated from diabetic foot wounds on the basis of their Gram stain status.
- Explain the roles of anaerobic and aerobic bacteria in the pathogenesis of gangrene.

Why does physical injury, even when it is planned, as with surgery, increase the risk of skin infections? Some microorganisms infect human skin only when its integrity is compromised. This section discusses skin infections that involve unusual precipitating factors (such as burns or traumatic injuries) or a chronic disease (such as diabetes; see IMPACT).

Burn Wound Infections

Each year in the United States, about 40,000 people are hospitalized with burn infections; about 4,000 die. Roughly 500 of those deaths are due to infection. The skin is a primary barrier against infection, so when this barrier is damaged, as in a burn, pathogens have a direct route to infiltrate the body and cause disease. Victims with burns covering more than 10% of their body are very likely to develop serious infections (Figure 19.20).

Burns obviously alter the innate immune features of the skin, but they also subvert other arms of the immune system. T-cell activity declines as a result of an increase in the number of regulatory T cells and a decrease in the number of helper T cells. The levels of inflammatory cytokines and complement also decrease.

Figure 19.20 Burn Wound Infected with *Aspergillus*

Burn wound of a 7-year-old boy who developed an infection with the fungus *Aspergillus* 3 weeks after the burn. Sadly, the boy died, despite aggressive medical treatment.

In addition, burns decrease neutrophil infiltration and phagocytosis, diminishing their bactericidal activity. All of these factors make the exposed tissues of the victim more susceptible to infection, while destruction of local blood and lymphatic vessels prevents immune cells and systemically administered antibiotics from being delivered to the infection site.

Immediately after a severe burn, the surface of the wound is essentially heat sterilized. So why does covering a burn immediately with sterile bandages often fail to prevent infection? For one thing, deep structures that survive the initial burn injury, such as sweat glands and hair follicles, often contain staphylococci, which will quickly colonize the wound surface. Then, 5–7 days after the burn, other Gram-negative and Gram-positive bacteria colonize the wound. These microbes typically come from the patient's gastrointestinal tract or upper respiratory tract, or from the hospital environment, and are inadvertently transferred to the wound by health care workers. Serious burn wound infections can produce cellulitis, necrotizing fasciitis, and even sepsis. MRSA is perhaps the most common cause of burn wound infections. Gram-negative bacteria and fungi such as *Candida* and *Aspergillus* species can also cause dangerous infections.

Pseudomonas aeruginosa (a Gram-negative, aerobic rod) is a major cause of serious wound infections owing to its wealth of virulence factors and its high level of antibiotic resistance. *P. aeruginosa* is found nearly everywhere in the environment. It is not a primary pathogen but rather an opportunistic pathogen able to infect tissues when one of the body's defenses is down. Most clinical specimens produce a green pigment called pyocyanin, which gives *Pseudomonas* colonies, and the infected pus, a greenish hue. The organism's toxin repertoire includes exotoxin A and exotoxin S, which are ADP-ribosyltransferases (see Section 18.4). These toxins, respectively, inactivate host protein synthesis and interfere with intracellular host signaling pathways. Both can cause cell death. The organism also secretes phospholipases and proteases that contribute to tissue invasion and disease. Many of the genes encoding the virulence factors are controlled by quorum-sensing mechanisms, discussed in Chapter 6 (see Section 6.6). *P. aeruginosa* is resistant to a great many antibiotics, making treatment of burn wounds challenging.

Gangrene

Gangrene is localized necrosis (tissue death). There are two types of gangrene: dry gangrene, which is caused when the blood supply to a tissue is cut off, and wet gangrene, which is caused by an infection. Dry gangrene usually affects fingers and toes and is seen in cases of severe frostbite and sometimes in individuals who have diabetes or certain autoimmune diseases. Wet gangrene is a necrosis due to an infection and can affect an internal or external part of the body. Anaerobic infections affecting the genital area (Fournier's gangrene) and infections caused by *Clostridium* species that produce large amounts of gas (gas gangrene) are also classified as wet gangrenes.

Figure 1 *Candida albicans*

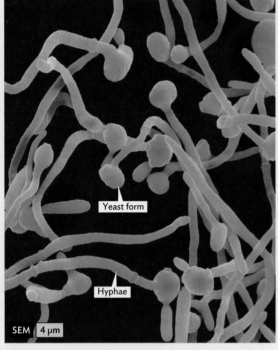

Yeast form

Hyphae

SEM 4 μm

Yeast and hyphal forms of *Candida albicans*.

Figure 2 **Diabetic Foot Ulcer**

Diabetic foot ulcers often result from minor trauma that goes unnoticed. Patients suffering from a neuropathic ulcer often do not feel the pain and can use the affected foot without any difficulty. Note the pus in the wound and the redness in the periphery, signifying cellulitis.

impact

Of Skin and Wounds and Diabetes

Most of us have suffered an occasional blister from a tight-fitting shoe or a minute cut when we accidentally stepped on a small object while walking barefoot in the yard or along the sidewalk. We simply protect the small injury, and soon it is healed and forgotten. Remarkably, the same blister or cut may cost a diabetic patient the amputation of a foot. Wounds take longer to heal in diabetics and are more prone to infection than in nondiabetics. But why is this so?

An increase in plasma glucose above normal physiological levels (such as after a meal) is normally met with the production of additional insulin by beta cells of the pancreas. The insulin binds to its receptors on the surfaces of other body cells (for example, muscle cells) and helps those cells take up the additional glucose, which is then used as fuel. Type 1 diabetes is a disease in which the production of insulin is stopped as a result of damage to the pancreatic beta cells. In type 2 diabetes, insulin is made but the insulin receptors on target cells become less responsive to the hormone. The result in either case is widely fluctuating levels of blood glucose (because the glucose is not transported and burned efficiently), even when the disease is treated and well controlled. Elevated blood glucose levels are the reason for diverse complications that affect all body systems, including the skin.

Diabetic patients are more susceptible to fungal skin infections caused by *Candida* (**Figure 1**) and *Rhizopus* species. Because tissues in diabetic patients cannot easily transport glucose, serum glucose levels remain high after a meal. Despite being bathed in glucose, cells are actually starving and must resort to fatty acid degradation for energy production. Fatty acid degradation leads to ketone production. *Rhizopus* species possess a ketone reductase that allows them to grow in high levels of ketone and cause infections. Why diabetics experience higher rates of *Candida albicans* infection is not known, but the suppressed immune status of these patients may be the culprit.

Neuropathy (nerve damage) is a common complication of diabetes and can affect a variety of nerves. Neuropathy may result when elevated glucose levels cause microvascular damage to vessels that supply nerves. The most commonly affected are the sensory nerves of the feet. A patient with sensory neuropathy in the foot may lose the ability to feel pain in that foot. As a result, a minor foot trauma can go unnoticed and quickly become a serious infected foot wound (**Figure 2**). In fact, the majority of diabetic foot wounds result from simple traumas (such as abrasion from a tight-fitting shoe) that may become seriously infected, making diabetes the number one cause of nontraumatic lower-extremity amputation in the United States.

Diabetic foot wounds are also more susceptible to infection because the patient is immunocompromised. Although we do not know exactly how the immune system is damaged by diabetes, we know that essential neutrophil functions, such as chemotaxis, adhesion, and phagocytosis, are affected. In addition, C3-mediated phagocytosis of bacteria is defective (see Section 15.7).

Infection also inhibits wound healing. Some diabetic foot ulcers become chronic and heal very slowly. Chronic wounds can be infected repeatedly, making prolonged antibiotic therapy a necessity. A chronic wound can extend to the bone if untreated (or if the treatment is unsuccessful) and cause a bone infection (osteomyelitis). Gram-positive cocci (*Staphylococcus*, *Streptococcus*, *Enterococcus*), as well as Gram-negative *Pseudomonas* and anaerobic species, including *Bacteroides* and *Clostridium* species, are commonly isolated from these infected wounds.

Treatment of diabetic foot wounds includes debriding dead tissue, providing antibiotic therapy, and reducing pressure on the wound (called off-loading) by using specially designed shoes and shoe inserts. Health care providers are attempting to reduce the incidence of diabetes-related amputations by aggressively promoting the prevention and treatment of diabetic foot wounds. Prevention is promoted by neuropathy screening and patient education. Among the treatment techniques are custom-made off-loading devices, artificial skin, and even sterile maggot therapy (Figure 3).

Infection is one reason diabetes is such a serious health concern. It is predicted that one-third of all Americans could suffer from diabetes by 2050. A majority of these will be type 2 cases that could be prevented through diet and exercise.

DISCUSSION QUESTION

Infected diabetic foot wounds are treated with antibiotics, but the therapy is usually not very successful. Why do you think that is?

Figure 3 Maggot Treatment of a Diabetic Foot Infection

Maggots can rid a diabetic patient of an infection caused by methicillin-resistant *Staphylococcus aureus* (MRSA). Note that this person has already had several toes amputated.

CASE HISTORY 19.4

Shot in the Black Hills of South Dakota

Jennifer remembered the day she was shot but not the day they had to remove her leg. She was camping in the middle of the Black Hills of South Dakota when an unseen hunter tried to shoot a deer. He missed his prey, but the errant bullet traveled a great distance and tore through Jennifer's thigh. In tremendous pain, Jennifer fashioned a tourniquet to stop the bleeding and then struggled for 2 days to hike the 10 miles back to civilization. By the time she arrived at a hospital, the wound and underlying muscle were clearly infected. The area around the wound was reddened and contained fluid-filled blisters, and parts were starting to turn black. The drainage coming from the wound contained pus and was foul-smelling—all signs of gas gangrene. The surgical team worked to debride the dead tissue and placed her on intravenous antibiotic therapy (clindamycin). The microbiology laboratory reported finding *Clostridium perfringens* in the samples sent. Despite treatment, including putting the affected limb in a high-concentration oxygen chamber, the infection continued spreading farther up Jennifer's leg. The surgeons finally decided that amputation was the only way to save her life.

Gas gangrene, also called clostridial myonecrosis because muscle tissue is affected, is an extremely serious, life-threatening disease caused mainly by *Clostridium perfringens*, a Gram-positive, spore-forming, anaerobic bacillus (Figure 19.21). *C. perfringens* is a true saprophyte, growing only on dead tissue. The organism is widely distributed throughout the environment and is also found

Figure 19.21 Gas Gangrene Caused by *Clostridium perfringens*

A. A patient with massive leg involvement. Note the blackened areas and the bullae.

SEM 2 μm

B. *Clostridium perfringens*.

in the intestine. Being ubiquitous in soil, *C. perfringens* commonly infects traumatic wounds (knife or gunshot wounds, for example), but infections can occur spontaneously, with no obvious precipitating event. Infections usually start out mixed with aerobic organisms that consume oxygen and provide an anaerobic environment for the clostridia. *C. perfringens* is highly fermentative and produces considerable amounts of foul-smelling gases that can be trapped by skin and produce bullae. Pockets of gas in muscle can choke off blood supply (ischemia) and contribute to necrosis, which produces more food for the pathogen.

C. perfringens produces a number of toxins (up to 17) that promote the spread of the organism through the host's tissues. The secreted toxins kill host tissue ahead of the advancing pathogen. The many strains of *C. perfringens* are classified into types A–E, based not on their antigenic properties but on the combination of toxins they produce. The toxins include lecithinases (alpha toxin), collagenase, hyaluronidase, fibrinolysin, and leukocidins that destroy neutrophils. Some strains also produce *C. perfringens* enterotoxin (CPE), which can lead to a gastrointestinal disease (see Section 22.5).

Treatment requires debridement and excision of surrounding tissue. Amputation is necessary in many cases. Antibiotics alone are not sufficient because they do not penetrate ischemic muscles (muscles with poor blood supply) enough to be effective. Penicillin or clindamycin is given along with surgical treatment to prevent spreading of the microbe. In addition to surgery and antibiotics, hyperbaric oxygen therapy, in which the affected limb is placed in a high-pressure oxygen chamber, is used to inhibit the growth of and kill the anaerobic *C. perfringens*.

C. perfringens infections are often so severe that by the time a diagnosis is made, curing the infection with antibiotics is impossible. Furthermore, the extent of injury caused by the infection may leave muscle tissues so damaged that the body cannot replace the lost structures. Note that gas gangrene is not associated only with limbs. Any body part with necrotic muscle can be affected (for example, the abdominal wall after surgery).

SECTION SUMMARY

- **Burns affect several aspects of the immune system.** The physical barrier of skin is destroyed, and T-cell and neutrophil responses to infection are undermined.

- **Microbes that cause burn infections** include staphylococci and streptococci residing in hair follicles, as well as Gram-negative bacteria originating from the gastrointestinal tract (for example, *E. coli*, *Klebsiella*) or from the environment (such as *Pseudomonas aeruginosa*).

- ***Pseudomonas aeruginosa* is an opportunistic pathogen** that expresses many virulence factors, including exotoxins A and S (which kill host cells), and enzymes that promote tissue invasion.

- **Wound injuries** can host life-threatening infections by organisms such as *Clostridium perfringens*, a Gram-positive, anaerobic, spore-forming rod commonly found in soil and the gastrointestinal tract.

- **Gas gangrene** (clostridial myonecrosis) is associated with necrotic tissue, including muscle. *Clostridium perfringens* is the usual cause.

Thought Question 19.6 Why do skin wounds tend to be polymicrobial, meaning you usually can find more than one species of bacteria in the wound?

19.6 Eye Infections

SECTION OBJECTIVES

- Identify viral, bacterial, and fungal causes of eye infections.
- Describe the symptoms and etiologies of pink eye.
- Explain how the pathogenesis of infectious keratitis differs from that of endophthalmitis.
- Outline the relationship between corneal epithelium and fungal keratitis.

Are all eye infections serious? Actually, some infectious diseases of the eye are self-limiting, but others can cause blindness.

Infections of the eye are quite common. Infections of the external parts of the eye include conjunctivitis and keratitis. "Pink eye" is a general, nonspecific term for conjunctivitis (inflammation of the conjunctiva), which can be due to an infection, trauma, or an allergic reaction (Figure 19.22). **Keratitis** is inflammation of the cornea, which is a sight-threatening process. A hallmark of bacterial keratitis is its rapid progression; corneal destruction may be complete in 24–48 hours with some of the more virulent bacteria. Infections of the inner structures of the eye (**endophthalmitis**) are extremely uncommon and almost always result either from direct spread of a superficial eye infection or from seeding of the internal eye structures with bacteria carried by blood from another infected site (hematogenous extension), such as a liver abscess.

Figure 19.22 Pink Eye, the Common Name for Conjunctivitis

An inflamed conjunctiva is red, giving rise to the term "pink eye." The arrow shows one of the blood vessels that have become prominent. When the eye shows such prominent vessels, it is said to be "injected." Pink eye is commonly caused by adenoviruses or allergies.

Viral Eye Infections

Viral conjunctivitis is seen as a diffuse pinkness of the conjunctiva, often called pink eye. Conjunctivitis is usually associated with a respiratory tract infection (typically an adenoviral cold or sore throat). The eyelid may be swollen, and the eye itself is red and produces a lot of tears and sometimes a mucoid (mucus-like) discharge. Contact with a person who has viral conjunctivitis can easily spread the virus, but how readily a subsequent infection forms depends on the virus. Viral conjunctivitis is a self-limiting condition that has no treatment. To prevent spreading the infection to others, or reinfecting themselves, patients are advised to wash hands often and change towels, not to share towels, to avoid touching their eyes with their hands, and to stay home as long their eye (or eyes) have a discharge.

Viral keratitis is a medical emergency that can lead to blindness. Herpesviruses are a major cause of keratitis and keratoconjunctivitis in the United States, with herpes simplex 1 causing the majority cases. The primary lesion may not cause much damage, but recurrence of the lesion due to reactivation of the virus can cause corneal scarring and blindness (Figure 19.23). The antiviral medications acyclovir and trifluridine are used to treat herpes keratitis. A 14-day treatment course is reported to improve healing of the ulcers, and long-term use of these antiviral medications decreases the rate of recurrence of the lesions.

The trigeminal nerve is a cranial nerve (a nerve originating from the brain) with several divisions. One of the divisions, the ophthalmic, extends into the eye and to the tip of the nose to gather sensory information and send it back to the brain. A shingles outbreak along the ophthalmic division of the trigeminal nerve (Figure 19.24) will result in the eruption of vesicular lesions on the forehead, eyelids, and nose. This eruption may even spread to the eye itself, causing corneal inflammation, eye pain, and sensitivity—a condition called **herpes zoster ophthalmicus**. The virus can spread deep into the inner

Figure 19.23 Scarring Caused by Viral Keratitis

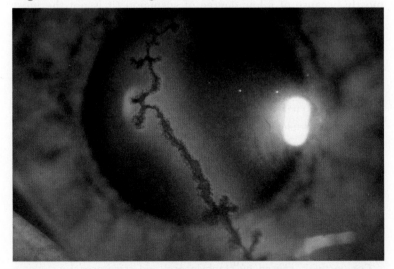

Yellow fluorescein dye added to the affected eye reveals a branching (dendritic) scarring ulcer on the cornea.

Figure 19.24 Herpes Zoster Ophthalmicus

A. An outbreak of shingles along the ophthalmic division of the trigeminal nerve causes vesicular lesions to erupt on the forehead, eyelids, and nose.

B. Corneal inflammation and scarring in herpes zoster ophthalmicus.

structures of the eye and cause necrosis (death) of the retina (acute retinal necrosis) in some immunocompromised patients. VZV infection of the eye threatens a patient's sight and is considered a medical emergency. Oral antiviral medications such as acyclovir, valacyclovir, and famciclovir are used to treat herpes zoster infection.

Bacterial Eye Infections

Bacterial conjunctivitis usually affects one eye and is acute, painful, and purulent. *Haemophilus influenzae* eye infections, however, initially look like viral or allergic conjunctivitis. Conjunctivitis due to the common pyogenic (pus-producing) bacteria, such as staph and strep, causes marked irritation and a stringy, opaque, grayish or yellowish mucopurulent discharge that may cause the lids to stick together, especially after sleep (Figure 19.25). *Chlamydia trachomatis* (see Figure 23.17) and *Neisseria gonorrhoeae* (see Figure 23.15)—two species that cause serious infections in the reproductive system (see Section 23.4)—also cause the majority of acute bacterial conjunctivitis. How this happens will be discussed shortly. Potential complications of bacterial conjunctivitis include spreading of infection to the cornea and blindness. With proper antibiotic treatment, however, these infections usually resolve before they spread to the cornea.

Chlamydia trachomatis, an obligate intracellular bacterial pathogen, is an important cause of conjunctivitis. Inclusion conjunctivitis is the *C. trachomatis* infection of a newborn's conjunctiva that is contracted by newborns passing through an infected vagina during birth (these particular strains cause an important sexually transmitted infection). Fortunately, inclusion conjunctivitis is usually self-limiting. The disease is prevented by prophylactically treating all newborns with antibiotic eye drops (erythromycin) as soon as they are born.

Figure 19.25 Purulent Bacterial Conjunctivitis

Bacterial infection of the eye can result in severe conjunctivitis. The purulent material on the cornea is extending to the corner of the eye. The eyelashes are covered in pus as well and appear wet and sticky.

A second, more serious disease caused by some strains of *C. trachomatis* is **trachoma**. Trachoma is the most frequent cause of infectious blindness in the world. The organism is spread by direct contact with eye, nose, and throat secretions. The patient initially notices a mild conjunctivitis, but soon the cornea becomes vascularized, and leukocytes infiltrate the affected area. The conjunctiva of the inner eyelid becomes pebbled and causes a turning in of the eyelashes (trichiasis). As the patient blinks, the eyelashes rake across the cornea. The cornea can become ulcerated and scarred, causing blindness. If diagnosed early, trachoma can be cured by oral (azithromycin) and topical (tetracycline) antibiotic treatment.

Neisseria gonorrhoeae can cause a serious eye disease of infants called **ophthalmia neonatorum** or neonatal conjunctivitis. As with inclusion conjunctivitis, the organism is transmitted from infected mother to newborn during passage through the birth canal. Prophylactic treatment with erythromycin eye drops will also prevent this disease.

This year, many millions of people worldwide will develop eye infections. Trachoma alone affects 6 million people. Almost all of these infections are treatable. Unfortunately, because of the cost, people who are poor will fail to get treated, and many will go blind.

Fungal and Parasitic Eye Infections

Fungi and parasites usually have a very hard time infecting our eyes because they cannot penetrate the intact corneal epithelium. But even a very small injury to the eye, such as a minor abrasion, can provide these organisms with an opportunity to cause ocular infections. The resulting corneal infection can lead to inflammation, necrosis, and even blindness. Wearing a contact lens or getting a foreign body (an eyelash, a grain of sand, a small piece of dirt) in your eye may cause an abrasion sufficient for fungi to gain entry. A large portion of ocular fungal and parasitic infections are seen among people who practice poor contact lens hygiene.

Fungal keratitis is most often caused by species of *Fusarium*, *Aspergillus* (both found in the environment), and *Candida* (part of the human microbiota; **Figure 19.26**). The incidence of fungal keratitis is extremely variable in different regions, but it seems to be higher in warm, moist climates.

Definitive diagnosis of fungal keratitis is made by culturing fungi from a corneal scraping specimen. PCR amplification and confocal microscopy, if available, are also useful. How fungal eye infections are treated depends on whether the infection is superficial or invasive. Invasive lesions require systemic therapy (oral or intravenous delivery of antifungal agents) and often surgery. Natamycin 5% solution (for filamentous fungi) and amphotericin B (for yeast) are the current drugs of choice for superficial lesions (see Section 13.8).

Parasitic keratitis is due mostly to *Acanthamoeba* species (see Section 11.3), protozoa ubiquitous in nature (air, water, and soil); in heating, ventilation, and air-conditioning equipment; and even in sewer systems. The parasite can exist as a cyst for years when conditions are unfavorable. Ocular *Acanthamoeba* infection makes the eye red, painful, and teary. The patient will avoid light (photophobia) and suffer visual loss. If untreated, the infection may cause permanent visual impairment (**Figure 19.27A**) or spread to the brain (**Figure 19.27B**).

Acanthamoeba is increasingly recognized as an important cause of keratitis in non–contact lens wearers, but wearing contact lenses is currently the main risk factor for this disease. Contaminated lens solution and lack of proper contact lens hygiene are the main contributors to the risk in patients who wear contact lenses.

Diagnosis of *Acanthamoeba* ocular infections can be difficult, mostly because of a low degree of suspicion (it is not the most likely cause of keratitis), but chronic inflammation of the cornea and resistance to other antimicrobial treatments is an indication. Definitive diagnosis requires culturing the organisms from a corneal specimen. PCR and confocal microscopy, in conjunction with evaluation, are also very useful. Prolonged treatment (6–12 months) with antiseptic agents and ocular and systemic pain medications is used to help eradicate the parasite from the cornea, but surgery may be required to restore vision.

SECTION SUMMARY

- **Eye infections** usually start as a conjunctivitis and, if unresolved, develop into keratitis.
- **Pink eye** is usually caused by an adenovirus or by *Haemophilus influenzae*.
- **Conjunctivitis caused by staphylococci and streptococci** includes a mucopurulent discharge that can produce crusting of the affected eye.

Figure 19.26 Fungi That Cause Ocular Infections

Fungi can infect an injured cornea when they are accidentally introduced into the eye. *Fusarium* (**A**) and *Aspergillus* (**B**) are molds naturally found on plants, in water, and in soil. *Candida* (**C**) is part of the human microbiota.

Conidiophore

Conidia

LM 10 μm

A.

Cross section of conidial head

Conidia

Conidiophore

LM 10 μm

B.

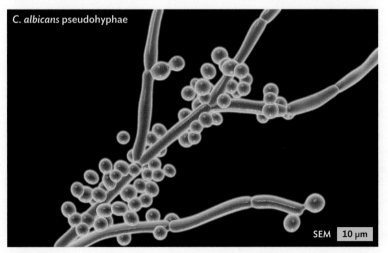

C. albicans pseudohyphae

SEM 10 μm

C.

Figure 19.27 *Acanthamoeba* Keratitis

Acanthamoeba species are free-living protozoa that are found in air, water, and soil and can cause corneal infection and inflammation.

Ring infiltrate

LM 15 μm

A. *Acanthamoeba* keratitis.

B. *Acanthamoeba* cyst (arrow) in brain tissue.

- **The leading cause of bacterial infectious blindness is *Chlamydia trachomatis***, an obligate intracellular pathogen. Blindness results from trichiasis of the cornea.

- ***Chlamydia trachomatis* and *Neisseria gonorrhoeae* can cause blindness in newborns.** The organisms are transmitted during passage through an infected birth canal, but disease is prevented by the application of erythromycin eye drops at birth.

- ***Acanthamoeba*** species, protozoa found ubiquitously in nature, can cause parasitic keratitis.

- **Fungal keratitis** is most often caused by species of *Fusarium*, *Aspergillus* (both found in the environment), and *Candida* (part of human microbiota).

Thought Question 19.7 Why is eye infection by *Acanthamoeba* often associated with the use of contact lenses?

Perspective

Skin truly is a mirror that reflects the status of our health. The color, tone, and texture of skin can change in response to nutrition, health, hormonal status, and the state of our immune system. Skin appearance also changes as a consequence of many microbial diseases. The infections discussed in this chapter manifest themselves as blatant skin rashes or lesions. However, keep in mind that infections elsewhere in the body, such as streptococcal sore throat caused by GAS or typhoid fever caused by *Salmonella* Typhi, also produce striking skin manifestations (scarlet fever and rose spots), even though the skin itself is not infected. These diseases are described in other chapters. Other infectious diseases, such as anthrax, tularemia, and plague, include infected skin lesions but were not discussed here because the etiologic agent is more typically associated with disseminated disease. These, too, will be covered later. Clearly, skin rashes and lesions, unlike beauty, often reflect something more than skin deep.

Figure 19.28 Summary: Microbes That Infect the Skin and Eyes

Viruses	Genome[a]	Disease
Varicella-zoster	DNA, ds	Chickenpox/shingles
HSV-1 (sometimes HSV-2)	DNA, ds	Cold sores/genital lesions
HHV-6 and HHV-7	DNA, ds	Roseola infantum
Papillomavirus	DNA, ds	Warts
Variola	DNA, ds	Smallpox
Adenoviruses	DNA, ds	Pink eye
Parvovirus B19	DNA, ss	Erythema infectiosum
Measles virus	RNA, ss, negative sense	Measles (rubeola)
Rubella virus	RNA, ss, positive sense	Rubella

Bacteria	Morphology	Disease
Bacillus anthracis	Rods, Gram-positive, facultative	Cutaneous anthrax
Enterococcus spp.	Cocci, Gram-positive, facultative	Wound infections
Staphylococcus aureus	Cocci, Gram-positive, facultative	Impetigo, folliculitis, furuncles, carbuncles, scalded skin syndrome, cellulitis
Streptococcus pyogenes	Cocci, Gram-positive, facultative	Erysipelas, necrotizing fasciitis, cellulitis, ecthyma
Haemophilus influenzae	Coccobacillus, Gram-negative, aerobic	Pink eye
Chlamydia trachomatis	Pleomorphic,[b] obligate intracellular parasite	Eye infections
Pseudomonas aeruginosa	Rods, Gram-negative, aerobic	Wound infections
Bacteroides spp.	Rods, Gram-negative, anaerobic	Wound infections
Vibrio vulnificus	Rods, Gram-positive, aerobic	Necrotizing fasciitis
Clostridium perfringens	Rods, Gram-positive, anaerobic	Gas gangrene, necrotizing fasciitis
Cutibacterium acnes	Rods, Gram-positive, anaerobic	Acne vulgaris

Fungi	Type	Disease
Epidermophyton spp.	Dermatophytes	Tinea capitis; tinea corporis, including tinea cruris (jock itch); tinea pedis (athlete's foot)
Trichophyton spp.	Dermatophytes	
Microsporum spp.	Dermatophytes	
Rhizopus spp.	Mold	Diabetes-related wound infections
Malassezia spp.	Yeast	Tinea versicolor
Candida spp.	Yeast, dimorphic; can form true hyphae and pseudohyphae	Diabetes-related wound infections

[a]ds = double-stranded; ss = single-stranded.
[b]Chlamydias have a structure resembling the Gram-negative cell wall that is devoid of peptidoglycan, despite having the genes for peptidoglycan synthesis. These bacteria do not stain well, and if they do stain, they stain Gram-negative-like.

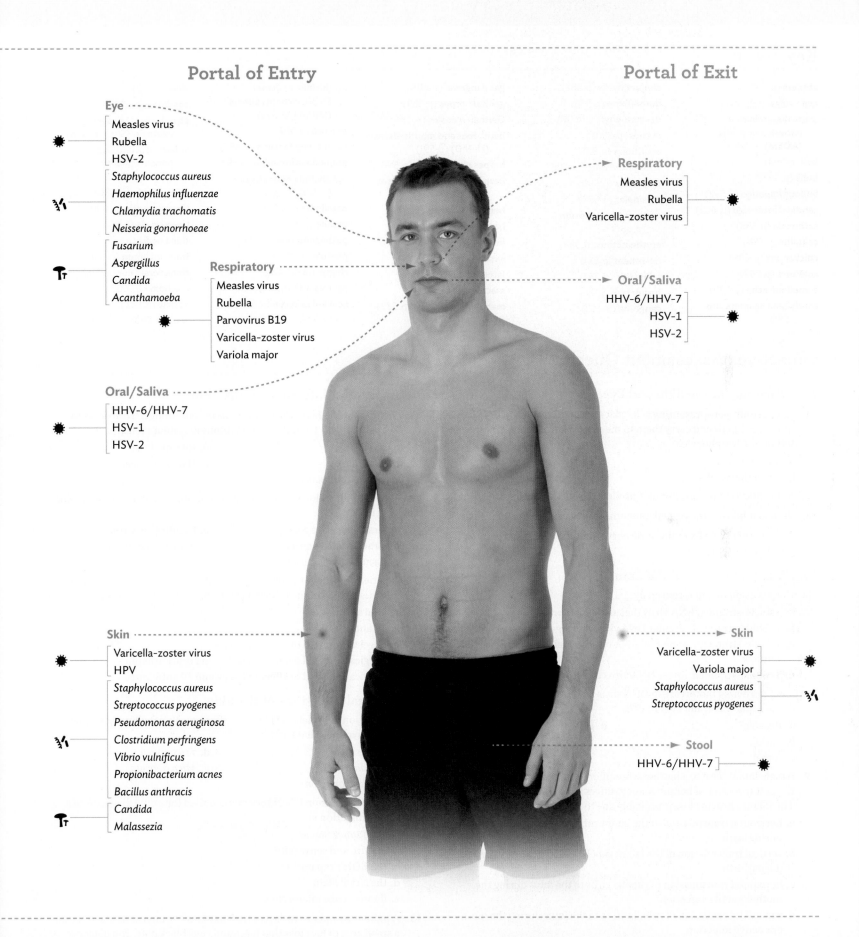

Portal of Entry

Eye
- Measles virus
- Rubella
- HSV-2

- *Staphylococcus aureus*
- *Haemophilus influenzae*
- *Chlamydia trachomatis*
- *Neisseria gonorrhoeae*

- *Fusarium*
- *Aspergillus*
- *Candida*
- *Acanthamoeba*

Respiratory
- Measles virus
- Rubella
- Parvovirus B19
- Varicella-zoster virus
- Variola major

Oral/Saliva
- HHV-6/HHV-7
- HSV-1
- HSV-2

Skin
- Varicella-zoster virus
- HPV

- *Staphylococcus aureus*
- *Streptococcus pyogenes*
- *Pseudomonas aeruginosa*
- *Clostridium perfringens*
- *Vibrio vulnificus*
- *Propionibacterium acnes*
- *Bacillus anthracis*

- *Candida*
- *Malassezia*

Portal of Exit

Respiratory
- Measles virus
- Rubella
- Varicella-zoster virus

Oral/Saliva
- HHV-6/HHV-7
- HSV-1
- HSV-2

Skin
- Varicella-zoster virus
- Variola major
- *Staphylococcus aureus*
- *Streptococcus pyogenes*

Stool
- HHV-6/HHV-7

Key Terms

abscess (p. 593)
acne vulgaris (p. 599)
acute disseminated encephalomyelitis (ADEM) (p. 586)
boil (p. 593)
bulla (p. 597)
bullous impetigo (p. 597)
candidal intertrigo (p. 602)
carbuncle (p. 593)
cellulitis (p. 599)
chickenpox (p. 588)
cold sore (p. 589)
comedone acne (p. 599)
condyloma acuminatum (p. 591)

conjunctivitis (p. 582)
dermatome (p. 588)
dermatophyte (p. 600)
dermis (p. 580)
enanthem (p. 581)
endophthalmitis (p. 606)
epidermis (p. 580)
erysipelas (p. 598)
erythema infectiosum (p. 587)
erythematous (p. 589)
exanthem (p. 581)
folliculitis (p. 593)
furuncle (p. 593)
gangrene (p. 603)

gas gangrene (p. 605)
genital herpes (p. 589)
German measles (p. 586)
hand, foot and mouth disease (HFMD) (p. 591)
herpes gladiatorum (p. 590)
herpes zoster ophthalmicus (p. 607)
herpetic keratitis (p. 590)
herpetic whitlow (p. 590)
impetigo (p. 597)
keratitis (p. 606)
latent (p. 588)
macule (p. 582)
maculopapular rash (p. 582)
measles (p. 585)

methicillin-resistant *Staphylococcus aureus* (MRSA) (p. 593)
mucosa (p. 582)
necrotizing fasciitis (p. 597)
nonbullous impetigo (p. 597)
ophthalmia neonatorum (p. 608)
papilloma (p. 590)
papule (p. 582)
pathognomonic (p. 586)
pustule (p. 582)
ringworm (p. 600)
roseola infantum (p. 588)
scalded skin syndrome (p. 593)

skin (p. 580)
smallpox (p. 591)
streptococcal pyogenic exotoxin (SPE) (p. 598)
subacute sclerosing panencephalitis (SSPE) (p. 586)
tinea capitis (p. 600)
tinea corporis (p. 600)
tinea cruris (p. 600)
tinea pedis (p. 600)
tinea unguium (p. 600)
tinea versicolor (p. 601)
trachoma (p. 608)
vesicle (p. 582)
wart (p. 590)

Objectives Assessment Questions

19.1 Anatomy of the Skin and Eye

1. _____ are antigen-presenting cells that survey the dermis to remove foreign antigens and carry them to the nearby lymph node for presentation to T lymphocytes.
   a. Keratinocytes
   b. Langerhans cells
   c. Melanocytes
   d. Basal cells

2. Keratinocytes are continuously produced from _____ cells.

3. Vesicular lesions are formed under the layer of skin called the _____.

4. Tissue originating from the optic nerve lines the outside structures of the eye. True or false?

19.2 Viral Infections of the Skin

5. Herpes gladiatorum is caused by _____.
   a. a single-stranded RNA virus that also causes warts
   b. a double-stranded, circular DNA virus that also causes warts
   c. a single-stranded, segmental RNA virus that also causes herpetic whitlow
   d. a double-stranded, linear DNA virus that also causes cold sores

6. _____ is a feared bioterrorism DNA virus and the etiologic agent of a disease that was eradicated in 1979.
   a. Roseola
   b. Variola
   c. Rubella
   d. HSV-1
   e. HPV

7. A male infant born to a mother who suffered rubella infection during the first trimester of her pregnancy suffers from severe hearing loss. The infant's hearing loss is probably due to _____.
   a. horizontal transmission of the latent rubella virus to the newborn during birth
   b. vertical transmission of the latent rubella virus to the newborn during birth
   c. horizontal transmission of rubella virus to the fetus during the mother's active infection
   d. vertical transmission of rubella virus to the fetus during the mother's active infection

19.3 Bacterial Infections of the Skin

8. Which enzyme helps *Staphylococcus aureus* produce an abscess to keep itself away from the human immune system's reach?
   a. Gyrase
   b. Coagulase
   c. Deoxyribonuclease
   d. Ribonuclease
   e. Hyaluronidase

9. Which of the following are the most common type of bacteria causing skin infections?
   a. Gram-negative rods
   b. Gram-positive cocci
   c. Facultative anaerobes
   d. Facultative aerobes
   e. Acid-fast bacteria

10. A patient is brought to the hospital suffering from severe necrotizing fasciitis. Which of the following is the best treatment option?
    a. Oral antibiotics
    b. IV antibiotics
    c. Surgical removal of affected tissue
    d. Surgical removal of affected tissue and oral antibiotics
    e. Surgical removal of affected tissue and IV antibiotics

19.4 Fungal Infections of the Skin

11. Production of azelaic acid by _____ damages skin melanocytes and causes hypopigmentation.
    a. *Malassezia*
    b. *Candida*
    c. *Epidermophyton*
    d. *Trichophyton*
    e. *Microsporum*

12. The most common laboratory method used for initial diagnosis of a fungal infection is _____.
    a. the Tzanck smear
    b. culture and sensitivity
    c. the KOH preparation
    d. the Gram stain
    e. fluorescence microscopy

13. A patient complains of "a patch of hair loss." Your examination shows a small area of hair loss that has numerous black dots. You diagnose

black dot tinea capitis and suspect that this patient has a fungal infection he might have acquired by _____.

   a. wearing an old pair of athletic shoes

   b. playing with a household dog or cat

   c. wearing tight underwear and pants

   d. swimming in a gym's crowded pool

19.5 Skin Infections of Special Circumstance

14. Which of the following contributes to an increased risk of wound infection in burn patients?

   a. Decreased humoral response

   b. Increased inflammatory response

   c. Decreased cell-mediated immunity

   d. Increased superficial vascularity

15. *Pseudomonas aeruginosa* _____.

   a. is a pathogen found only in hospitals

   b. is a ubiquitous opportunistic pathogen

   c. infects burn wounds only

   d. is a Gram-positive anaerobe

16. What types of microorganisms are commonly isolated from diabetic foot wounds?

   a. Fungi   c. Bacteria

   b. Protozoa   d. Viruses

17. Which of the following is true regarding gas gangrene?

   a. It may be caused by aerobes.

   b. It results from infection of superficial wounds by any anaerobe.

   c. It occurs only in trauma (knife or gunshot) wounds.

   d. It often is due to polymicrobial wound infection with aerobes and *Clostridium perfringens*.

19.6 Eye Infections

18. A 7-year-old child is suffering from red, watery eyes and a runny nose. He feels very well otherwise and has no other symptoms. His mother reports that "he had no symptoms last night and woke up this morning with both eyes looking like this." Which of the following best describes the condition of his eyes?

   a. Acute allergy attack, causing conjunctivitis   c. Acute viral eye infection

                                 d. Acute viral keratitis

   b. Acute bacterial eye infection   e. Acute fungal keratitis

19. Inflammation of _____ can have a variety of etiologies, including infections, and is often referred to as "pink eye."

20. A patient suffering from inflammation of the inner structure of the eye after a traumatic eye injury (penetrating foreign body or surgery) is said to have _____.

   a. conjunctivitis   d. keratitis

   b. scleritis   e. keratoconjunctivitis

   c. endophthalmitis

Review Questions

1. What are keratinocytes and Langerhans cells?

2. Describe the basic types of skin rashes and lesions. Give an example of each.

3. What is the pathophysiology of most skin rashes?

4. Discuss the pathogenesis of congenital rubella syndrome.

5. Discuss complications of varicella and rubeola diseases.

6. What is fifth disease?

7. What causes warts? Can warts be prevented?

8. What microbe can cause cold sores and an ocular disease? How does it cause both diseases?

9. Explain the pathogenesis of the organism that causes furuncles.

10. Name the causes of impetigo. What virulence factors do these organisms produce?

11. Why do some strains of *Streptococcus pyogenes* cause necrotizing fasciitis, whereas others do not?

12. What is the difference between erysipelas and impetigo?

13. What is gas gangrene?

14. Describe tinea versicolor and name the causative agent.

15. List important etiologic agents of burn infections. What is their source?

16. Why is *Pseudomonas aeruginosa* such an important pathogen?

17. What is the difference between conjunctivitis and keratitis?

Clinical Correlation Questions

1. A 7-year-old child has a nonpruritic malar skin rash and now has white papular enanthem on her palate. Name the probable causative agent and best preventative strategy.

2. A 5-year-old girl has a rash on her face that includes a few small pustules. Microscopic and biochemical examination of the bacteria cultured from the exudate shows catalase-negative (Section 6.5; Section 25.3), Gram-positive cocci that look like a string of pearls. Which pathogen—*S. aureus, S. epidermidis, S. pyogenes, C. perfringens,* or *P. aeruginosa*—causes the small pustules on the face, and why?

3. A patient complains of eye pain. Examination shows a dendritic lesion on the cornea of the affected eye. The doctor suspects a virus caused the lesion and obtains corneal scrapings for microscopic examination. What is the doctor likely to see?

4. A child is in the hospital diagnosed with staphylococcal scalded skin syndrome (SSSS). Microbiological culture of the blister fluid does not grow any organism. Why were no organisms found?

 For more assessment, visit digital.wwnorton.com/michum2

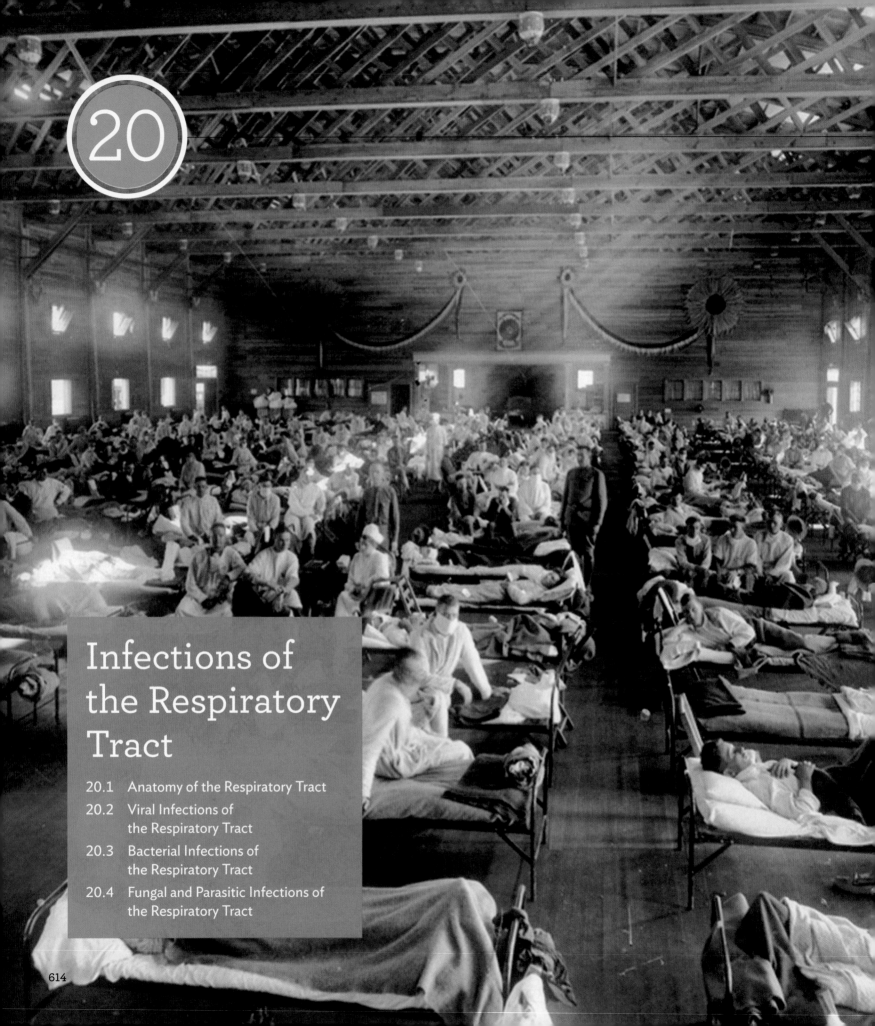

Infections of the Respiratory Tract

[The Ominous Cough]

SCENARIO It was the summer of 1918, and 20-year-old Floyd, a US Army private stationed in Kansas, began to feel ill. He visited the camp doctor complaining of a cough, a headache, and some chills. Within a day, Floyd's fever reached 40°C (104°F). He became incoherent and lost consciousness. When Floyd awoke, he found himself in a large barracks lying on a cot among a long line of cots filled with sick soldiers.

SIGNS AND SYMPTOMS His cough was severe. Floyd saw dark, purplish brown spots appear on the face of the soldier lying on the cot next to his. The soldier began to cough blood, and eventually his face turned blue from lack of oxygen—a sign that death was imminent. Doctors could do little to help.

INFLUENZA
FREQUENTLY COMPLICATED WITH
PNEUMONIA
IS PREVALENT AT THIS TIME THROUGHOUT AMERICA.

THIS THEATRE IS CO-OPERATING WITH THE DEPARTMENT OF HEALTH.

YOU MUST DO THE SAME
IF YOU HAVE A COLD AND ARE COUGHING AND SNEEZING. DO NOT ENTER THIS THEATRE

GO HOME AND GO TO BED UNTIL YOU ARE WELL

Coughing, Sneezing or Spitting Will Not Be Permitted In The Theatre. In case you must cough or Sneeze, do so in your own handkerchief, and if the Coughing or Sneezing Persists Leave The Theatre At Once.

This Theatre has agreed to co-operate with the Department Of Health in disseminating the truth about Influenza, and thus serve a great educational purpose.

HELP US TO KEEP CHICAGO THE HEALTHIEST CITY IN THE WORLD

JOHN DILL ROBERTSON
COMMISSIONER OF HEALTH

1918 Warning Poster
A public health poster in 1918 warning that flu can develop into what we now know as secondary bacterial pneumonia. No effective treatment was known.

Influenza virus

TMV

TEM 1 µm

First Electron Micrograph of Influenza Virus (1941)
This sample was from someone who died of flu. The tiny gray dots are the dreaded virus. The tubelike tobacco mosaic virus (TMV; Figure 12.5) was added to the sample for comparison.

RESOLUTION Fortunately, after about 2 weeks, Floyd started to improve. Many others were not so lucky and died. No one knew what caused the disease raging through the camp or how to stop it.

What started with a sudden cough and chills for nearly 500 million people—one-third of the world's population at the time—ended with pneumonia and death for more than 50 million worldwide. They all suffered from the viral influenza pandemic that came to be known as the "Spanish flu." The pandemic of 1918 is thought to have begun with 18 cases in early March of 1918 in Haskell County, Kansas, and spread to European civilians and soldiers who were fighting in the Great War (World War I) by May of 1918. The outbreak in Spain, which led to the pandemic's name, actually happened later. At the time, the United States and other Allied countries were reluctant to report the illness among the soldiers in an effort to maintain morale. Consequently, the first news of the epidemic wasn't reported until newspapers in Spain broke the story, and the epidemic of 1918 came to be known as the Spanish flu. Later decades saw the pandemic flus of 1957 and of 1968. Later pandemics were less deadly, thanks in part to a better understanding of the virus, the infection, and its treatment. But no one knows what would happen if a strain similar to the 1918 flu were to strike again.

Viral infections of the respiratory tract continue to challenge the medical community, owing to a lack of definitive treatments and the occurrence of secondary bacterial infections. Today, preventive measures such as vaccination, much improved treatment offered in intensive care units (ICUs), and the availability of antiviral medications have lessened the mortality of influenza virus. But although antibiotics to treat dangerous secondary bacterial infections give flu patients a fighting chance, overuse of those drugs has encouraged the emergence of new antibiotic-resistant bacteria, threatening the effectiveness of current medications and technology.

Lung and upper respiratory tract infections are among the most common human diseases. Many different viruses, bacteria, fungi, and parasites can cause pneumonia. Successful lung pathogens come equipped with attachment mechanisms and countermeasures against various lung defenses (such as alveolar macrophages). The most prevalent respiratory diseases are of viral origin, and most viral infections of the respiratory tract (such as the common cold) rarely spread beyond the lung. Fortunately, most viral diseases of the respiratory tract are self-limiting and typically resolve within 2 weeks. Lung damage caused by a primary viral infection, however, can lead to secondary infections by bacteria, endangering the weakened patient.

20.1 Anatomy of the Respiratory Tract

Are all respiratory tract infections the same? No. Respiratory tract infections are often classified by whether they affect the upper or lower respiratory tract. The **upper respiratory tract** includes the upper airways, consisting of the nasal passages, oral cavity, and pharynx (**Figure 20.1A**). Other structures in the head that have direct or indirect communication with the upper airways include the passages to the sinuses (small air-filled cavities in the skull), lacrimal (tear) ducts, and the middle ear. The lowest structure in the upper respiratory tract is the larynx (the voice box). Infections in the upper airways can easily spread to these associated structures, and vice versa.

The **lower respiratory tract** (**Figure 20.1B**) consists of a series of diverging tubes that eventually terminate in alveoli, small sacs where oxygen from the inhaled air is exchanged for carbon dioxide in the blood in a process called respiration. The tubes leading to the alveoli include the trachea, right and left main bronchi, bronchioles, terminal bronchioles, and alveolar ducts. Note that some people assign the trachea as being part of the upper respiratory tract. All the respiratory tubes are warm and moist and remove impurities from inhaled air as it travels to the alveoli. The cells that

Figure 20.1 **Anatomy of the Human Respiratory Tract**

A. Upper respiratory tract. Oral and nasal cavities open into the pharynx, forming the upper part of the human respiratory tract. The pharynx is the hollow space shared by the nasal and oral cavities. Both the esophagus, the tube that carries food from the mouth to the stomach, and the trachea, or windpipe (part of the lower respiratory tract), are continuous with the pharynx. Different portions of the pharynx are named for the structure closest to them. The nasopharynx, for example, is the part of the pharynx attached to and adjacent to the nose. Inflammation of the pharynx is a common respiratory problem called pharyngitis, or sore throat.

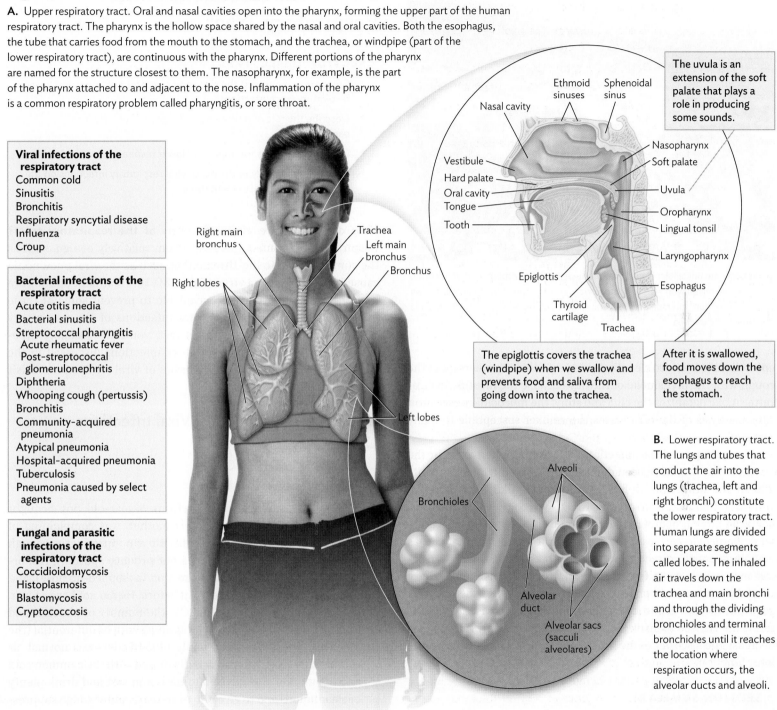

Viral infections of the respiratory tract
Common cold
Sinusitis
Bronchitis
Respiratory syncytial disease
Influenza
Croup

Bacterial infections of the respiratory tract
Acute otitis media
Bacterial sinusitis
Streptococcal pharyngitis
  Acute rheumatic fever
  Post-streptococcal glomerulonephritis
Diphtheria
Whooping cough (pertussis)
Bronchitis
Community-acquired pneumonia
Atypical pneumonia
Hospital-acquired pneumonia
Tuberculosis
Pneumonia caused by select agents

Fungal and parasitic infections of the respiratory tract
Coccidioidomycosis
Histoplasmosis
Blastomycosis
Cryptococcosis

The uvula is an extension of the soft palate that plays a role in producing some sounds.

The epiglottis covers the trachea (windpipe) when we swallow and prevents food and saliva from going down into the trachea.

After it is swallowed, food moves down the esophagus to reach the stomach.

B. Lower respiratory tract. The lungs and tubes that conduct the air into the lungs (trachea, left and right bronchi) constitute the lower respiratory tract. Human lungs are divided into separate segments called lobes. The inhaled air travels down the trachea and main bronchi and through the dividing bronchioles and terminal bronchioles until it reaches the location where respiration occurs, the alveolar ducts and alveoli.

line the interior of the respiratory tubes (the lumen) are epithelial cells, which have moving appendages called cilia that are extensions of the cellular membrane. Cilia remove foreign particles, including microorganisms trapped in the sticky mucus. Mucus is formed when mucin, which is secreted by goblet cells interspersed among the epithelial cells, combines with water (**Figure 20.2**). The cilia "beat" like a whip upward, moving the mucus and its trapped particles up toward the oral and nasal cavities for clearance and expulsion. The ciliated epithelium, referred to as the **mucociliary escalator** (or mucociliary elevator), works to keep the lungs devoid of microorganisms. Any impairment of the mucociliary escalator results in the accumulation of inhaled particles in the airways and a reflexive cough. Heavy smokers, for example, have a tendency to cough when they wake up in the morning because their ciliated

Figure 20.2 Tracheal Lining

Cilia

Ciliated cells

Goblet cells

LM 30 μm

Goblet cells are unicellular glands that produce mucin, a compound that becomes mucus when mixed with water. This colorized micrograph of tracheal cells shows goblet cells (green) interspersed between ciliated columnar epithelial cells that line the lumen of a trachea.

epithelium is trying to clear their lungs of debris. Smokers quiet the cough with the first smoke of the day because the nicotine in cigarette smoke paralyzes the cilia and stops the mucociliary escalator. This paralysis of the cilia leaves the smoker susceptible to many diseases.

Secondary bacterial infections of the lung often arise in the airways after they have been damaged by a viral infection. Patients who do not drink enough fluids will become mildly dehydrated, a condition that leads to increased mucus viscosity. Thick mucus or bacterial growth hampers the movement of the mucociliary escalator. *Bordetella pertussis,* for example, the bacterium that causes whooping cough, directly interacts with the mucociliary escalator and slows it down (see Figure 18.6B). A sluggish mucociliary escalator makes it difficult to expel the microbe, so the patient's susceptibility to secondary bacterial infection increases. Cold sufferers are advised to drink plenty of fluids because hydration helps decrease the viscosity of the mucus and improve mucociliary escalator function.

SECTION SUMMARY

- **Respiratory infections** affect either the upper or lower respiratory tract.
- **The upper respiratory tract** consists of nasal and oral cavities, the pharynx, and associated structures (ears, sinuses, larynx).
- **The lower respiratory tract** consists of the trachea, bronchi, bronchioles, terminal bronchioles, alveolar ducts, and alveoli.
- **Alveolar ducts** lead to the alveoli, where the process of respiration takes place.
- **Microorganisms in the lung** are trapped by mucus produced by goblet cells, swept out by the cilia of the mucociliary escalator, or killed by alveolar macrophages. These mechanisms are part of the innate immune system.

20.2 Viral Infections of the Respiratory Tract

SECTION OBJECTIVES

- Compare and contrast the signs and symptoms of upper and lower respiratory tract infections.
- Describe the role of mutations in the genesis of influenza pandemics.
- List the risk factors for upper and lower respiratory tract infections.
- Identify the etiology of the major viral respiratory infections on the basis of their clinical presentation.

How dangerous are viral infections of the respiratory tract? Respiratory tract infections range from relatively benign, such as the common cold, to life-threatening, such as the viral pneumonia caused by the influenza virus (Table 20.1). Treatment options are limited, and few vaccines are available to prevent some of these infections. The result is that viral infections of the respiratory tract are annoyingly common. In fact, you have probably experienced most of the viral respiratory infections discussed in the following section. To review the basics of viral replication, revisit Section 12.3.

Upper Respiratory Tract Viral Infections

CASE HISTORY 20.1

Jacob's Runny Nose

Jacob, a rambunctious 5-year-old boy brought to the clinic by his mother, was complaining of a scratchy sore throat, a nonproductive cough (a cough that does not produce sputum), rhinorrhea (a runny nose), nasal congestion (a stopped-up nose), and a headache that started the night before. He did not have a fever. His complete blood count (CBC)—the number of platelets and red and white blood cells in blood, along with its differential (the different types of red and white blood cells)—was normal, as was his chest X-ray. Jacob was diagnosed with the common cold and his mother was advised to have him rest and drink plenty of clear liquids and to give him over-the-counter cough suppressants and either ibuprofen or acetaminophen for pain.

Why do we keep catching colds? The **common cold** is the most frequent viral infection of the upper respiratory tract. We remain susceptible to "catching" a cold because the symptoms are caused by more than 200 viral subtypes, and immunity to one subtype does not protect us against any other subtypes (see Section 16.1). The average person suffers a few colds a year—more episodes for those of us who are younger than for those of us who are older. The annual economic impact of simple colds is enormous. Children can catch as

Table 20.1

Viral Diseases of the Respiratory System

Disease	Symptoms	Complications	Virus	Transmission	Diagnosis	Treatment(s)	Prevention
Common cold	Runny nose, congested nose, sore throat	Rare; infection of sinuses, ear, throat, or lung	Rhinoviruses	Person-to-person contact	Clinical symptoms	Supportive treatment	Hand washing; no vaccine available
Sinusitis	Sinus congestion, facial pain, headache	Bacterial sinusitis that has purulent, sometimes malodorous, nasal discharge and fever	Viruses that cause common cold	Typically not spread from one person to another	Clinical presentation	Pain relievers; nasal corticosteroids; antibiotics upon progression to bacterial infection	No vaccine available
Croup	Rare in adults; seallike bark, hoarseness, cough, fever, respiratory distress	Bacterial respiratory infections	Parainfluenza (80%), other viruses (20%)	Direct contact, respiratory droplets	Clinical presentation; chest X-ray	Corticosteroids; bronchodilators; humidified air; oxygen	Hand washing; no vaccine available
Influenza	Fever, chills, cough, chest pain, sore throat, muscle pain	Secondary bacterial pneumonia, asthma flare-ups	Influenza A and B viruses	Respiratory droplets or contact	Clinical presentation; viral culture; rapid antigen testing; serology; qRT-PCR[a]	Zanamivir (Relenza), oseltamivir (Tamiflu), peramivir (Rapivab); less commonly: amantadine, rimantadine; newest: baloxavir marboxil (Xofluza)	Seasonal vaccines available
Respiratory syncytial disease	Fever, chills, cough, chest pain, respiratory wheezing	In very young infants and children, heart failure, low blood pressure, and seizures	Respiratory syncytial virus (RSV)	Respiratory droplets	Clinical presentation; rapid viral testing (rapid immunoassay)	For infants and children: hydration; oxygen; bronchodilators (to relieve contraction and constriction of bronchi); palivizumab (monoclonal antibody to F surface glycoprotein of RSV) for premature infants, per criteria. For adults: supportive therapy; ribavirin as indicated	No vaccine available
COVID-19	Flu-like symptoms, dry cough, fever, shortness of breath	Respiratory distress, respiratory failure, death	SARS-CoV-2	Respiratory droplets, fecal-oral route	qRT-PCR, antibody testing, point of care antigen test	Supportive therapy	Strict personal hygiene; vaccine candidates in clinical trials

[a]qRT-PCR, quantitative reverse transcription–polymerase chain reaction.

many as eight and adults as many as five colds every year. It is very difficult to accurately calculate the cost because no one is actually keeping records, but the estimates, based on small-population studies of several thousand US households in 2003, are in the billions of dollars. Our economy spends approximately $17 billion in direct medical costs (buying over-the-counter medications, for example) and $22.5 billion in indirect costs (workdays lost) every year!

Rhinoviruses are the most common cause of colds, with more than 100 subtypes, but other prominent players include some coronaviruses and influenza, parainfluenza, and respiratory syncytial viruses. Colds in adults are milder and shorter than in children. In adults, a scratchy or sore throat and a runny, congested nose are usually the only symptoms, and these clear up within a week. In children, the course of a cold can take as long as 2 weeks. The good news is that the common cold is almost always self-limiting and does not cause any serious illness. Complications of a cold are uncommon, although after a cold the very young, the immunocompromised, or patients with lung disease may suffer inflammation or infection of the sinuses, middle ear, throat, or lungs.

Note: Contrary to common belief, the color of nasal discharge does not indicate the presence or absence of bacteria. Yellow or white nasal discharge may reflect an increase in the number of polymorphonuclear leukocytes (PMNs). A green discharge is thought to result from an enzymatic activity of these cells.

The viruses that cause a cold can invade through upper respiratory tract openings or through ocular mucosa, which means that cold viruses can easily spread from person to person by contact or by inhalation of large or small aerosols. Given these modes of transmission, the only effective preventive measures against spreading a cold are frequent handwashing and using the lower arm (rather than the hand) to cover the nose or mouth when sneezing or coughing.

Many time-honored remedies are aimed at alleviating the symptoms of a cold, such as drinking plenty of liquids, but there are no medications approved by the Food and Drug Administration (FDA) for treating the common cold, and there is no vaccine. A variety of over-the-counter medications that promise prevention or treatment are on the market, but their efficacy, contraindications, and side effects are unknown.

Symptoms of the common cold are mimicked by a number of other viral pathogens at the onset of their infective cycle. These include respiratory syncytial virus (RSV) and influenza (which will be discussed shortly), as well as early stages of rubella and rubeola viral infections (see Section 19.2).

Another common viral infection of the upper respiratory tract is **viral sinusitis**. Viral infections of the upper airways often spread to the sinuses and cause inflammation and congestion of the sinuses (known as sinusitis). The sinus cavities are lined with mucous membranes, which become inflamed when infected. The inflammation and congestion of the sinuses, however, can narrow passages that connect them to the nasal cavity, impeding the flow of mucus. Sinus congestion commonly causes the headache that we feel when our nose is "stopped up" from a bad cold. Viral sinusitis usually resolves on its own, as does a cold. The congestion and poor drainage of mucus in the sinuses, however, provide a breeding ground for bacteria and may lead to bacterial sinusitis or middle ear infections, as we will discuss in Section 20.3.

Lower Respiratory Tract Viral Infections

Infections of the lower respiratory tract, regardless of whether their etiology is viral, bacterial, or fungal, are called **bronchitis** when the bronchial tubes are involved and **pneumonia** when the lungs are affected. Viral bronchitis is a self-limiting disease. The most common viral causes of bronchitis are influenza A and B, parainfluenza, coronavirus (types 1–3), rhinovirus, respiratory syncytial virus, and human metapneumovirus. A cough (dry or sputum-producing) that lasts more than 5 days is the main symptom of the disease, which is diagnosed clinically and treated with cough suppressants and occasionally an inhaled medication (usually a steroid) to reduce inflammation of the bronchial tubes.

Clinical presentation of pneumonia is somewhat variable, depending on the causative agent, but all types include shortness of breath and cough. Pneumonia can result from many viruses, but respiratory syncytial virus (RSV) and influenza viruses cause the majority of cases.

RESPIRATORY SYNCYTIAL VIRUS DISEASE Respiratory syncytial disease is an important viral lung infection caused by **respiratory syncytial virus (RSV)**, an enveloped, negative-sense, single-stranded RNA virus of the Paramyxoviridae family (Figure 20.3).

Figure 20.3 Respiratory Syncytial Virus (RSV)

Five virions (one of which is shown by the red arrow) can be seen budding from an infected cell. The black arrow points to a viral nucleic acid core.

RSV derives its name from the fusion of adjacent infected cells into a **syncytium** (plural, **syncytia**), a giant cell containing many nuclei. Cell fusion is caused by viral proteins that travel to the infected cell's membrane, causing fusion of membranes between neighboring cells.

RSV is transmitted from person to person. Infections occur when the virus comes in contact with mucous membranes of the eyes, mouth, or nose or through the inhalation of droplets generated by a sneeze or cough. The virus causes disease mostly in the very young. By age 2, all children are thought to have been infected by it. The course of infection in healthy children is relatively benign.

RSV illness usually begins with fever, runny nose, cough, and sometimes wheezing. The virus moves down the respiratory tract to cause more severe disease in some children. RSV is the most common cause of **bronchiolitis** (inflammation of the bronchioles) and pneumonia among infants and children younger than 1 year. Infected infants 6 months of age or younger often require hospitalization.

RSV attachment to respiratory epithelial cells triggers an immune response that includes production of chemokines (see Section 15.3) and recruitment of inflammatory cells but fails to confer immunity against reinfection. The extent of the inflammatory response impacts the severity of presenting symptoms and depends on the patient's age, overall health, and genetic background. In general, healthy adults will have only cold-like symptoms, mainly rhinitis (a stuffy nose); but some adults experience dangerous lower respiratory tract disease. Risk of a more severe illness is greater among the elderly or people with compromised cardiac, pulmonary, or immune systems. As yet, there is no vaccine.

Presumptive diagnosis of RSV is usually made on the basis of clinical presentation. A number of other viruses, including adenovirus and influenza and parainfluenza viruses, can cause signs and symptoms similar to RSV. A list of conditions with similar signs and symptoms is called a differential diagnosis. A definitive diagnosis confirms the exact etiology of a disease (in this case RSV) among those listed in a differential diagnosis. Definitive diagnosis of RSV requires isolating the virus from the patient's respiratory secretions, which can be collected from a variety of sources including the nose, throat, or even the bronchi and lungs. Real-time quantitative reverse transcription–polymerase chain reaction (real-time qRT-PCR; see Section 8.6) and rapid antigen testing are used to make a diagnosis. Rapid immunoassays (see Section 17.2) and serology tests (see Section 17.2) are also available. Serological testing is not a good choice in infants because the presence of maternal antibodies is likely to result in a false positive.

Treatment options for RSV infections are currently limited to relieving the symptoms and using the one medication available. A nucleoside analog (ribavirin; Section 13.7) that inhibits RSV replication has been approved by the FDA to treat severe cases of RSV. Animal studies have shown an association between birth defects and ribavirin use during pregnancy.

Ribavirin, therefore, is contraindicated in pregnant women and their partners.

INFLUENZA Influenza is caused by single-stranded, negative-sense enveloped RNA viruses of the Orthomyxoviridae family. Although some influenza strains reach pandemic status (such as the one that caused the Spanish flu described at the beginning of this chapter), most strains cause seasonal outbreaks in limited regions. Seasonal influenza can infect people of all ages. Influenza viruses are divided into types A, B, and C on the basis of their antigenic determinants, which are small segments of a microbe capable of eliciting an immune response (see Section 16.1). Type A and B influenza viruses can both result in "the flu," but type A causes the majority of cases. Type A viruses infect a variety of animals as well as people. This virus is constantly changing and is the cause of large flu epidemics and occasional pandemics. Type B viruses are found primarily in humans and usually result in a milder form of illness than type A. Type B viruses do not cause pandemics. Type C causes only a mild respiratory disease and no epidemics or pandemics, so it is not classified into subtypes and there is no vaccine to protect against it.

The pathogenicity of influenza viruses is largely dependent on three of their proteins: hemagglutinin (HA), neuraminidase (NA), and M2 (see Sections 12.5 and 18.5). Hemagglutinin is a transmembrane protein that forms the spikes on the viral surface (Figure 20.4A). The HA spikes grab the *N*-acetylneuraminic acid (sialic acid) residues present on the outer surface of the respiratory epithelial cells to gain entry (Figure 20.4B). The host cell tries to defend itself by engulfing the bound virus via endocytosis. Captivity in the vacuole, however, is short-lived.

The low pH of the endoplasmic vacuole helps the influenza virus remove proteins covering its genome (uncoating), making

Figure 20.4 Influenza

A. Hemagglutinin (HA) spikes on the viral surface.

B. Influenza virion (circled) attached to cilia of an epithelial cell.

its RNA accessible and ready for replication. Acidification of the vacuole also changes the three-dimensional shape (conformation) of the viral protein M2. The conformational change leads to formation of a pore in the membrane of the vacuole and allows the viral genetic material to leave the endosomal compartment and enter the host cell cytoplasm. When inside the cytoplasm, the viral RNA is replicated, producing progeny that will be transported to the cell surface, once again bound to sialic acid residues of the host cell.

As the influenza virion buds out of the host cell membrane, viral neuraminidase cleaves the sialic acid residues that anchor the budding virus particles to the infected cell surface while also preventing the viruses from clumping together. The progeny are now free to find and bind an uninfected cell of their own.

Influenza viruses are designated as avian flu strains or swine flu strains. The difference in targeting an animal host is related to the ability of different flu viruses to bind to different sialic acid residues in the lungs of these animals. Two forms of sialic acid residues are associated with cell-surface glycoproteins. The two forms differ in their linkages to galactose. One form is called alpha-2,3 and the second form is called alpha-2,6. The human respiratory tract contains receptors for both forms. Avian flu strains are generally those that bind to alpha-2,3 sialic acid residues. Birds, such as ducks, principally have the alpha-2,3 sialic acid receptors in their guts and lack the alpha-2,6 receptors. Pigs, on the other hand, have both types of sialic acid receptors. Flu viruses that bind to the alpha-2,6 form are generally called swine flu strains, although both types of viruses can infect pigs. Human lungs also contain alpha-2,3 and alpha-2,6 receptors, so both avian and swine flu strains can infect humans.

Why do some influenza strains cause more severe disease in humans than other strains? The reason is not entirely clear, but recent studies indicate several features of the virus affect virulence, such as the type of hemagglutinin (see Section 18.5). An earlier hypothesis, which correlated virulence of the virus strain with the distribution of different host sialic acid receptors in the human respiratory tract, is likely incorrect.

Influenza A viruses can infect humans, pigs, and birds. Multiple strains of influenza A can sometimes infect the same cell at the same time (for example, in a duck landing on a pig farm in Thailand). When this happens, the eight RNA segments of each segmented influenza genome (see Section 12.5) will uncoat, mix in the cytoplasm, and re-sort to form a new influenza virus containing a new form of the hemagglutinin gene. This process of RNA segment mixing is known as an **antigenic shift** (see Figure 12.22). Antigenic shifts typically happen every 10–15 years. In between these large antigenic shifts, smaller so-called **antigenic drifts** in HA and NA structures can occur because the influenza RNA polymerase is very error prone, introducing numerous mutations that can ultimately alter the HA protein's structure. Antigenic shifts can result in pandemics and antigenic drifts are usually responsible for local outbreaks.

Antigenic shifts and drifts allow viruses to evade the host immune system. Influenza type A is most capable of undergoing genetic change, so it more commonly leads to seasonal disease. Type A influenza is further classified into subtypes based on the kind of hemagglutinin (H) and neuraminidase (N) expressed on the viral surface, giving rise to some very well-known strains, such as H1N1 (swine flu) or H5N1 (avian flu).

Every year, the Centers for Disease Control and Prevention (CDC) and the World Health Organization (WHO) identify influenza strains that cause the most cases of flu during that year. Flu outbreaks are monitored all over the world, including areas where living conditions foster close contact between humans and the animals that serve as incubators for new influenza strains. Close contact (for example, a farmer that shares living quarters with his ducks) can facilitate the virus "jumping" from the animal to humans and the start of a possible new pandemic. The identified strains are then used to generate a multivalent vaccine for the coming year. Research is currently under way to develop a single universal vaccine effective against seasonal flu from year to year (see IMPACT).

Influenza viruses used for inactivated (killed-virus) vaccines are usually grown in fertilized chicken eggs (Figure 20.5). Two different preparations of the flu vaccine are made: an inactivated vaccine that can be injected intramuscularly and an intranasal live, attenuated vaccine (vaccine with weakened virus) that is administered as a nasal spray. The attenuated intranasal vaccine offers more effective protection because the attenuated virus actually replicates in the host. The intranasal vaccine is used for healthy people who are 2-49 years old or those with egg allergies and a healthy immune system. The injectable vaccine is used for the immunocompromised and those who care for them. Flu vaccines must be administered annually in advance of (usually October) and during the flu season. The effectiveness of a flu vaccine varies from 50% to 80%, depending on how well the vaccine matches the

Figure 20.5 Growing Viruses to Make the Flu Vaccine

Factory workers in China examine the eggs used to grow the influenza virus for production of the inactivated flu vaccine. The live, attenuated vaccine is prepared in mammalian cells.

infecting strain. The length of the flu season varies from year to year but commonly starts in November and ends in May, with the majority of cases occurring in February.

The newest agent to treat influenza, baloxavir marboxil (trade name, Xofluza) was approved by the FDA in 2018. Baloxavir marboxil is an inhibitor of the influenza virus's endonuclease activity (discussed in Section 13.7). Other long-standing antiviral agents able to effectively treat the flu include those that inhibit the M2 surface protein (amantadine and ramantadine) and those that inhibit neuraminidase activity. Amantadine and ramantadine are rarely used anymore because most flu viruses have become resistant to them and because they cause central nervous system side effects, such as hallucinations and dizziness. Antiviral agents now used preferentially are the neuraminidase inhibitors peramivir (Rapivab), zanamivir (Relenza), and oseltamivir (Tamiflu; Section 13.7).

COVID-19 COVID-19 is a highly transmissible disease caused by severe acute respiratory syndrome coronavirus 2 (SARS-CoV-2), a single-stranded, positive-sense, enveloped RNA virus of the Coronaviridae family. Coronaviruses, in general, produce diseases in mammals and birds. Until 2003, the only coronavirus disease humans suffered from was the common cold. In that year, a member of the coronavirus family produced a dangerous disease called severe acute respiratory syndrome (SARS) that killed 10% of its victims. The virus itself was called SARS-associated coronavirus. Fortunately, fears that SARS would develop into a worldwide pandemic were not realized. A second dangerous coronavirus emerged in 2012 to cause Middle East respiratory syndrome (MERS), another contagious, deadly disease that frightened the world with the prospect of a pandemic. That pandemic, too, failed to develop. The coronavirus pandemic finally arrived in 2019 as the respiratory disease called COVID-19. COVID-19 emerged in December of 2019 in Wuhan, the capital city of Hubei province of China. The outbreak was declared a pandemic by the World Health Organization on March 11, 2020, when 118,000 cases of the disease were detected in 110 countries and territories around the world, including the United States. (Events leading to the pandemic are described in Chapter 26.)

COVID-19 patients initially present with flu-like symptoms, primarily cough, fever, and shortness of breath. Some infected individuals respond to the virus by overproducing a number of inflammatory cytokines in a phenomenon called a cytokine storm (Section 16.3). The disease in these patients can escalate to trouble breathing, new-onset confusion, and bluish lips. Patients can progress quickly to bilateral (in both lungs) pneumonia and needing a ventilator to breathe. Multiple organ failure and death result in severe cases. Individuals most at risk for severe disease include adults over 60 years old, immunosuppressed patients, people with diabetes, and people with underlying diseases of the heart and lungs. The average fatality rate of symptomatic patients is 2%–5%. However, many SARS-CoV-2 infected individuals are asymptomatic, so the actual mortality rate will be accurately determined post-pandemic. Diagnosis of COVID-19 requires qRT-PCR amplification of viral RNA from nasal or oropharyngeal specimens. Although corticosteroids and a new antiviral drug, remdesivir, are now used to reduce the severity of the disease, the mainstay of treatment remains supportive care. Remdesivir is a nucleoside analog that was given an emergency authorization by the FDA for use with COVID-19. Considerable efforts are underway to identify pharmaceuticals that stop SARS-CoV-2 replication and several vaccine candidates are in clinical trials (see Chapter 26).

CROUP Some viruses can infect the larynx and then spread to the trachea and even the bronchi, a condition called **croup**, or **laryngotracheobronchitis (LTB)**. Young children (6 months to 5 years of age) are at highest risk for contracting the disease. Typically, the croup patient has had a fever and a runny nose for a few days before coming to the clinic with a "barking" cough. The barking cough (or the seallike bark) is caused by swelling of the larynx. The most common causes of croup are the parainfluenza viruses, especially types 1 and 2. Note that viruses that cause bronchitis in adults can also cause croup in children. Treatment of croup depends on the severity of the major signs and symptoms of the disease. Children who have only mild disease (no barking cough at rest) are treated at home with humidified air and often steroids to reduce airway inflammation. Those with severe croup are usually hospitalized and treated with oxygen. The majority of croup cases are mild and resolve within a week, but severe cases may lead to respiratory failure.

SECTION SUMMARY

- **Viruses cause the majority of upper respiratory tract infections.**
- **The common cold is the most common viral infection** of the upper respiratory tract. A majority of the 200 viral subtypes that cause the common cold are rhinoviruses.
- **Influenza viruses and respiratory syncytial virus (RSV)** can cause both upper and lower respiratory infections.
- **Pneumonia** is a lung infection. Influenza virus is a major cause of viral pneumonia.
- **Bronchiolitis,** inflammation of the bronchioles, is the most frequent presentation of a lower respiratory RSV infection in infants and children younger than 1 year.
- **RSV-infected cells fuse to form syncytia,** giant cells containing many nuclei.
- **Influenza viruses affect people of all ages,** and their pathogenicity is dependent on three of their proteins: hemagglutinin (HA), neuraminidase (NA), and the M2 protein.
- **Influenza virus subtypes** are based on their HA and NA proteins.
- **Antigenic shifts and drifts** in the HA gene are caused by frequent mutations and result in different influenza strains. Shifting and drifting HA antigens help the virus evade the host's immune system.
- **Effectiveness of the annual multivalent flu vaccines** depends on how well vaccine antigens match those of the infecting strain.

Thought Question 20.1   For which populations would the intranasal flu vaccine not be recommended, and why?

---

impact

The Search for a Universal Flu Shot

In this chapter, we're learning how we suffer some illnesses because different microbes find different parts of our respiratory tract desirable for their growth and colonization. Among the most famous of these microbes are the influenza viruses. We have to get vaccinated every year because influenza virus changes just enough every year to avoid our body's defenses.

Wouldn't it be nice to have a flu vaccine that lasts for as long as we live—or at least much longer than a year? A universal flu shot would free us from fear of the flu and reduce the overall cost of flu vaccine production and administration, not to mention decreasing flu-associated deaths and health care costs. It would also make the scientist who discovers it the next Edward Jenner.

The problem is the flu virus and its ever-changing outer protein, hemagglutinin, which looks like a series of lollipops with ball-shaped tips anchored to the viral coat by their stalks. Current vaccines are directed against the most prominently displayed, but highly mutable, part of the virus—the head of hemagglutinin. The stalk of the HA protein is a lot less mutable than the tip and can generate an immune response. This genetic stability should mean that the stalk is a much better vaccine candidate than the seasonally changing HA tip (Figure 1). But the stalk is mostly hidden from our immune system by the ball-shaped head of the protein, which presents difficulties in using it as a vaccine target. Scientists at the National Institute of Allergy and Infectious Diseases (NIAID), part of the National Institutes of Health (NIH), are focusing on development of a universal flu vaccine by using different novel approaches. One is a vaccine made from microscopic pieces (nanoparticles) of HA stem that seems promising in animal testing. Another is using virus-like particles that cause an immune response but cannot replicate. A third approach uses a DNA-based vaccine. A fourth tactic uses antigenic peptide sequences from many flu viruses to generate a potential universal flu vaccine. In May of 2018, NIAID launched phase II clinical trials of the latter. As this research continues, we remain hopeful that one of these approaches can give us a universal flu vaccine. It is also feasible that antibodies made in the lab by one of these methods could be an effective way to treat flu victims.

Proteins inside the viral envelope are in principle much better candidates for a universal vaccine because the genes producing them mutate much less frequently than do the genes for surface proteins when the virus replicates. The viral envelope, however, sequesters these proteins (such as the HA stalk) from the reach of antibodies made against them by our immune system. But flu vaccine designers may have identified a vulnerability of the influenza virus: one of the internal viral proteins, influenza nucleoprotein, is stripped off from viral nucleic acid and displayed on infected cell surfaces. Antibodies made to target this protein should recognize the surface nucleoprotein and mark the infected cell for destruction by cytotoxic T cells. Elimination of infected cells would also make virus replication within them impossible and therefore reduce the severity of the disease.

Figure 1 Potential Targets for a Universal Vaccine against Influenza Virus

The genes for the hemagglutinin stalk and M2 proteins of the influenza virus mutate much less frequently than the gene for hemagglutinin, making their protein products good targets for generating a universal flu vaccine.

Another vaccine candidate is the protein M2. M2 is mostly hidden inside the virion envelope, but a portion of M2 is exposed outside the virus. Because this exposed part is immunogenic and genetically stable, it might be considered a candidate for a universal vaccine. The M2 protein of animal influenza viruses, however, is slightly different from the M2 protein of human strains. Consequently, a hybrid virus, such as swine flu (H1N1), may escape antibodies raised against the human M2.

The answer to the dilemma of how to make a universal flu vaccine may be to design a multivalent version that stimulates production of antibodies to the HA stalk of hemagglutinin, the viral nucleoprotein, and the different M2 proteins. Research on this is under way and promising, but for now roll up your sleeves once a year if you wish to be protected.

DISCUSSION QUESTION

Could a universal vaccine for flu eradicate the influenza virus? Why or why not?

20.3 Bacterial Infections of the Respiratory Tract

SECTION OBJECTIVES

- List the bacterial etiologies of respiratory tract diseases and the disease's anatomical location.
- Describe the relationships between host and bacteria that contribute to the pathogenesis of respiratory tract infections.
- Compare and contrast clinical presentations of bacterial infections of the respiratory tract.
- Develop a prevention plan for different bacterial respiratory pathogens.

Is there a relationship between viral and bacterial infections of the respiratory tract? Viral infections often lead to congestion, dehydration, and weakening of the immune system. The damaged tissue is more susceptible to bacterial colonization and disease. A variety of bacteria cause respiratory infections (Table 20.2), with impressive economic and personal consequences. For example, US children 0–4 years of age have an estimated 13.7 million cases of acute otitis media (middle ear infection) per year with an estimated economic burden of $3.8 billion. Lung infections cause approximately 1 million hospitalizations in the United States annually. Development of vaccines against some bacterial pathogens, especially *Streptococcus pneumoniae* and *Haemophilus influenzae*, has reduced the US incidence of lung infections and epiglottitis (inflammation of the epiglottis, the cartilage covering the opening of the windpipe when we swallow). People who come down with these infections still face high mortality rates—currently about 10% for streptococcal pneumonia—despite antibiotic therapy.

Upper Respiratory Tract Bacterial Infections

CASE HISTORY 20.2
A Big Pain in a Little Ear

Chloe, a generally healthy 2-year-old child, was brought to the clinic by her father. The child had a fever since the previous night, was irritable, and kept pulling on her left ear. Chloe had also had a runny nose for the past 2 weeks. Her temperature was 38.5°C (101.3°F) at the clinic, and the physical exam showed a red and swollen ear, an erythematous (reddened) ear canal, and a bulging tympanic membrane. Chloe was diagnosed with acute otitis media and prescribed an antibiotic that she had to take for 7 days.

What is the biggest risk factor for getting a bacterial upper respiratory infection (URI)? Viruses that infect and damage the upper respiratory tract create conditions suitable for a secondary

Table 20.2

Bacterial Diseases of the Respiratory System

Disease	Symptoms	Complications	Bacteria	Transmission	Diagnosis	Treatment(s)	Prevention[a]
Acute otitis media (AOM)	Earache, ear discharge, fever, irritability, sense of blockage in the ear, decreased hearing	Rare; hearing loss, meningitis	*Streptococcus pneumoniae, Haemophilus influenzae, Moraxella catarrhalis*	Spread of bacteria from throat to middle ear	Clinical presentation; definitive diagnosis by culture	Pain relievers; oral amoxicillin or amoxicillin-clavulanate, intramuscular ceftriaxone; antibiotic eardrops; surgical placement of tube in the ear to enhance drainage	Vaccine available[b]
Bacterial sinusitis	Sinus congestion, feeling pressure behind the eye, facial pain, foul-smelling breath and/or nasal discharge	Meningitis, brain abscess, orbital cellulitis (inflammation and infection of the eye socket)	Same as acute otitis media	Spread of bacteria from adjacent infected structures to sinuses	Clinical presentation	Pain relievers; antipyretics (medications to bring down fever); amoxicillin or amoxicillin-clavulanate	No vaccine available[a]
Pharyngitis	Sore throat, fever, pain swallowing, tender lymph nodes, headache, nausea, and vomiting	AOM in children, abscess behind tonsils, scarlet fever (some strains), acute rheumatic fever (some strains), glomerulonephritis (some strains)	*Streptococcus pyogenes*	Person-to-person contact	Clinical diagnosis based on Centor score[c]; rapid antigen detection test; definitive diagnosis by culture	Beta-lactams (penicillin V or amoxicillin) preferred; macrolides when beta-lactams are not an option (e.g., penicillin allergies); pain relievers and antipyretics	No vaccine available
Diphtheria	Sore throat, low-grade fever, formation of a membrane over tonsils and soft palate, toxin-related symptoms within 6–10 days: irregular heartbeat, stupor, coma, death	Airway obstruction, septicemia, endocarditis (nontoxigenic strains)	*Corynebacterium diphtheriae*	Respiratory droplets, person-to-person contact	Culture; PCR; amplification of *dtx* gene from clinical specimen	Equine antitoxin (available through CDC); erythromycin or penicillin G	Vaccine available
Whooping cough (pertussis)	Prolonged (>2 weeks) paroxysmal cough (sudden and rapidly repeated); "whoop" noise when breathing in after the cough paroxysm, followed by vomiting	Pneumonia, seizure, severe respiratory failure	*Bordetella pertussis*	Respiratory droplets	Clinical presentation; PCR; culture	Azithromycin, clarithromycin; trimethoprim-sulfamethoxazole used as alternative but contraindicated for pregnant or nursing women and infants <2 months	Vaccine available

(continued on next page)

Table 20.2
Bacterial Diseases of the Respiratory System (continued)

Disease	Symptoms	Complications	Bacteria	Transmission	Diagnosis	Treatment(s)	Prevention[a]
Bacterial bronchitis	Cough, shortness of breath on exertion, wheezing, may have fever or sore throat	Acute disease associated with asthma; long-term respiratory decline if chronic	*Mycoplasma pneumoniae, Chlamydophila pneumoniae, Bordetella pertussis, Haemophilus influenzae, Streptococcus pneumoniae, Moraxella catarrhalis*	Respiratory droplets	Clinical presentation; culture (rarely done)	Inhaled medications to dilate and reduce inflammation of bronchial wall; antibiotics, depending on patient and microbial agent	Vaccine available only for *B. pertussis, H. influenzae,* and *S. pneumoniae*[a,b]
Lobar (pneumococcal) pneumonia	Cough, shortness of breath, fever, chills, chest pain	Meningitis, kidney failure, heart failure	*Streptococcus pneumoniae*	Person-to-person contact, respiratory droplets	Clinical presentation; chest X-ray	Antibiotics; macrolides, fluoroquinolones; beta-lactams, cephalosporin as alternative antibiotics	Vaccine available[b]
Walking (atypical) pneumonia	Fever, sore throat, nonproductive (dry) cough, sweating, chest pain	Ear infections, skin rashes, anemia	*Mycoplasma pneumoniae*	Person-to-person contact	Clinical presentation	Macrolides; fluoroquinolones, tetracyclines	No vaccine available
Legionellosis (Legionnaires' disease)	Fever, chest pain, coughing up blood-tinged sputum, muscle pain, abdominal pain, vomiting and diarrhea	Respiratory failure, shock (sudden drop in blood pressure), kidney failure	*Legionella pneumophila*	Inhalation of aerosols, mists, sprays, or droplets of contaminated water sources	Culture; detection of *L. pneumophila* antigen in urine	Macrolides or quinolones[d]	No vaccine available
Pseudomonas pneumonia	Cough, fever, shortness of breath, chest pain	Declining lung function, death	*Pseudomonas aeruginosa*	Contaminated water droplets and instruments	Culture; clinical history and presentation	A two-drug combination is recommended, such as quinolones and aminoglycosides	No vaccine available
Tuberculosis	Productive cough progressing to coughing up blood; fever, loss of appetite, night sweats, weight loss, fatigue	Lung abscess, spread of TB to other organs	*Mycobacterium tuberculosis*	Inhalation of droplets	Definitive diagnosis by culture; tuberculin (Mantoux) skin test; interferon-gamma release assay	Numerous treatment protocols, depending on patient specifics; initial simultaneous four-drug therapy with rifampin, isoniazid, ethambutol, and pyrazinamide	Vaccine available
Inhalation anthrax (woolsorter's disease)	Common cold symptoms progressing to fever, muscle aches, weakness, shortness of breath	Shock, death	*Bacillus anthracis*	Inhalation of spores	Culture; serology	Ciprofloxacin, doxycycline	Vaccine available

[a]The childhood vaccine schedule is shown in Table 17.2.

[b]Pneumococcal vaccines protect against a minority of the more than 90 strains of *S. pneumoniae* (23 in PPSV23 for adults and 13 in PCV13 for children). Vaccine against *H. influenzae* type b (Hib) is also available, but it does not offer protection against nontypeable *H. influenzae.*

[c]W. J. McIsaac et al. 1998. *Can. Assoc. Med. J.* 158(1):79.

[d]Per Infectious Diseases Society of America (IDSA) guidelines.

Figure 20.6 **The Changing Angle of the Eustachian Tube**

The Eustachian tube drains from the middle ear into the nasal cavity. In infants and young toddlers, the tube is shorter and almost horizontal, which makes drainage less optimal. As a result, bacteria can get trapped in the middle ear and cause an infection. As we get older, the Eustachian tube lengthens and its path becomes angled, making drainage of the middle ear much easier and ear infections much less likely.

bacterial infection. Some URIs occur more often in certain age groups. For example, infections of the middle ear are most common in infants and toddlers like Chloe, thanks to the angle of the Eustachian tube, which connects the middle ear to the nasal cavity (**Figure 20.6**). We discuss bacterial URIs, including the risk factors for acquiring them, in this section.

OTITIS MEDIA The most common diagnosis made during a sick-child visit to the pediatrician's office is middle ear infection, or **acute otitis media (AOM)** (**Figure 20.7**). Most children (80%–90%) will have at least one episode of AOM by the time they are 3 years old. The number of episodes is higher for those who attend a day-care center or who live with a sibling who has a history of frequent AOM.

Signs and symptoms of AOM are wide-ranging, and diagnosis is generally made on the basis of clinical presentation and physical exam. Diagnosis through culturing middle ear fluid is difficult to interpret and rarely performed. Because the nasopharynx is near the middle ear, microbes that naturally colonize the nasopharynx can easily reach the middle ear, but those microbes do not always cause AOM.

Infections of the middle ear are caused by viruses as well as bacteria, both Gram-positive and Gram-negative. Only 20% of AOM cases are purely viral, but viral upper respiratory tract infections are the most common antecedent event to bacterial AOM. Bacteria migrate from the nasopharynx to the middle ear, usually following a common cold, and cause what is generally a self-limiting disease. Immunocompromised patients, however, can have a complicated

Figure 20.7 **Acute Otitis Media**

A. Viewing the tympanic membrane with an otoscope.

B. Otoscopic view of a healthy tympanic membrane.

C. Otoscopic view of an infected tympanic membrane.

course leading to the more invasive disease process of **epiglottitis** (inflammation of the epiglottis), pneumonia, or meningitis.

Three bacterial species, *Streptococcus pneumoniae*, *Haemophilus influenzae*, and *Moraxella catarrhalis*, are the most frequent causes of AOM in otherwise healthy children. All three species are capable of colonizing a child's nasopharynx early in life, but *S. pneumoniae* is the number one culprit. *S. pneumoniae* is a facultatively anaerobic, Gram-positive, alpha-hemolytic coccus that possesses a thick polysaccharide capsule (**Figure 20.8**). The makeup of the

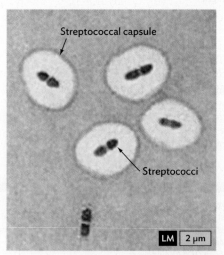

Figure 20.8
Streptococcus pneumoniae
Photomicrograph of *S. pneumoniae*, one of the most common causes of acute middle ear infections, pneumonia, and sinusitis.

polysaccharide capsule differs from strain to strain and is thought to determine the extent of the bacterium's invasiveness (the propensity to cause bacteremia and more severe disease). More than 90 different serotypes of *S. pneumoniae* have been identified on the basis of capsule structure; all 90 serotypes are capable of causing infections. *S. pneumoniae* strains causing AOM are generally of the noninvasive serotypes but seem to be associated with an increased risk of childhood asthma. A multivalent polysaccharide vaccine that protects against more invasive serotypes of *S. pneumoniae* now appears to reduce annual cases of AOM. The important role of *S. pneumoniae* as a cause of pneumonia will be described later.

The second most common cause of AOM is *Haemophilus influenzae*, a Gram-negative bacillus once thought to be the cause of influenza. The organism is fastidious but can be grown on chocolate agar, which contains growth factors X (hemin) and V (NAD), in a CO_2 incubator (see the case history at the beginning of Chapter 6). Some strains have an outer capsule and some do not. Six serotypes of encapsulated strains (serotypes a–f) can be detected by antisera that react with the capsule's polysaccharide. *H. influenzae* type b (Hib) is the most invasive and can cause severe diseases such as meningitis, bacteremia, pneumonia, and epiglottitis in infants and very young children. These infections are now rare thanks to a Hib polysaccharide vaccine included routinely in childhood vaccinations.

The nonencapsulated varieties of *H. influenzae* are said to be nontypeable because they do not react to any capsule antisera. The genomes of nontypeable *H. influenzae* strains differ greatly from each other. Some strains, for example, have the genes for the capsule but do not express them. Other strains do not have the capsule genes at all. Traditionally, the nontypeable strains have been classified, or "biotyped," on the basis of their ability to produce indole, urease, and ornithine decarboxylase. Eight biotypes have been identified, some of which are linked to clinical syndromes such as Brazilian purpuric fever (biotype 3) and neonatal sepsis (biotype 4). The majority of *H. influenzae* infections of the upper and lower respiratory tracts are now caused by the nontypeable strains; no particular biotype has been identified as the main culprit.

Because the Hib vaccine does not confer any immunity toward nonencapsulated strains, nontypeable *H. influenzae* has become the second most common cause of AOM and conjunctivitis (inflammation of the conjunctiva) in children. These bacteria are capable of invading and surviving inside respiratory tract epithelial cells. Nontypeable strains can also form biofilms in the middle ear. As described in Chapter 6, bacteria growing in biofilms are more resistant to antibiotics than the same bacteria growing individually, which may explain why some cases of AOM are refractory to antibiotic treatment.

Transmission of *H. influenzae* from person to person occurs through contact with infected respiratory secretions or by inhaling aerosolized droplets. A number of different classes of antibiotics are effective against *H. influenzae*, but the beta-lactam antibiotics, particularly amoxicillin, remain the treatment of choice. Some strains

of *H. influenzae*, however, are resistant to beta-lactam antibiotics because they produce an enzyme (beta-lactamase; Section 13.5) that inactivates the antibiotic.

The third most common cause of AOM in children, *Moraxella catarrhalis*, is a Gram-negative diplococcus, but it causes far less disease than *S. pneumoniae* and *H. influenzae*. Other bacteria, such as *Chlamydia trachomatis*, have also been implicated in AOM. Treatment of AOM usually involves systemic or local antibiotics and depends on the severity of infection, type of bacteria, and patient's age and health status.

BACTERIAL SINUSITIS The same pathogens that cause bacterial middle ear infections and pneumonia (as will be discussed shortly) can infect the sinuses. This is why **bacterial sinusitis** often accompanies middle ear infection in children or pneumonia in adults. Inflammation and congestion of the sinuses obstruct the sinuses and the mucus flow. Accumulation of mucus in a warm and relatively airy cavity supplies bacterial pathogens with a nice breeding ground. The pressures behind the eyes, pain in the face, and foul-smelling nasal discharge or breath are common symptoms of bacterial sinusitis. Diagnosis is usually made on the basis of clinical presentation, and the patient is treated with antibiotics.

PHARYNGITIS

CASE HISTORY 20.3

The Contagious Football Player with GAS

Dallin, a 16-year-old high school athlete, came to the clinic complaining of sore throat. He did not have a cough or any cold or flu symptoms. His cervical nodes (lymph nodes in the neck) were enlarged and tender to palpation, there was an exudate visible on his tonsils and pharynx (pharyngotonsillar region), and his temperature was 39.8°C (103.6°F). The physician assistant (PA) told Dallin that he probably had a strep throat, but she could not perform a rapid strep test to confirm this because they had run out of the strips. The PA then swabbed the back of Dallin's throat and sent the sample to the lab for culture and antibiotic sensitivity testing. Dallin was given an antibiotic empirically (based on clinical experience and in the absence of definitive diagnosis). He was told that he was contagious and should rest and avoid contact with others until better. He followed the recommendations, recovered, and was back playing football in a few days. The lab isolated *Streptococcus pyogenes*, also called group A streptococcus (GAS), from Dallin's pharyngeal swab, confirming the diagnosis.

A patient complaining of a sore throat is usually suffering from **pharyngitis** (inflammation of the pharynx), although the cause may instead be **tonsillitis** (inflammation of the tonsils), **laryngitis** (inflammation of the larynx), or **peritonsillar abscess**

Table 20.3
Etiology of Infectious Pharyngitis

Viruses	
Respiratory viruses	• Rhinoviruses • Coronavirus • Respiratory syncytial virus (RSV) • Parainfluenza virus • Influenza A and B viruses • Adenovirus
Other viruses	• Herpes simplex virus types 1 and 2 (HSV-1 and HSV-2) • Enteroviruses (coxsackie A) • Epstein-Barr virus (infectious mononucleosis) • Cytomegalovirus (CMV) • Human immunodeficiency virus (HIV)

Bacteria	
Gram-positive	• *Streptococcus*, group A (most common bacterial cause) • *Streptococcus*, groups C and G • *Corynebacterium diphtheriae* • *Arcanobacterium haemolyticum*
Gram-negative	• *Neisseria gonorrhoeae* • *Neisseria meningitidis* • *Francisella tularensis* (rare)
Other	• *Mycoplasma pneumoniae*

Fungi	
	• *Coccidioides immitis* (retropharyngeal abscess) • *Candida albicans* (thrush)

Figure 20.9 "Strep" Throat

Note the reddened area on the pharynx and the enlarged tonsils.

(an abscess in the pharynx). Differential diagnosis for a patient complaining of sore throat includes a variety of viral and bacterial etiologies (Table 20.3), including infectious mononucleosis (see Section 21.2) and diphtheria (as will be discussed shortly). A sore throat may also be due to a noninfectious cause, such as gastroesophageal reflux disease (GERD; frequent heartburn) or thyroid dysfunction, but the associated signs and symptoms of those diseases easily distinguish them from pharyngitis caused by infectious agents.

Pharyngitis is caused primarily by viruses, particularly those of the common cold and flu. The disease usually resolves without intervention, regardless of viral or bacterial origin. Bacteria, such as the *S. pyogenes* infecting Dallin in Case History 20.3, cause less than 20% of the pharyngitis cases, but those cases are the most severe and have the highest potential for detrimental outcomes if untreated, like rheumatic fever (which will be discussed shortly). Bacterial infections of the throat can be caused by *S. pyogenes*, *Neisseria gonorrhoeae*, *Chlamydophila pneumoniae*, and *Chlamydophila psittaci*. Streptococcal species other than *S. pneumoniae* are divided into groups A–U on the basis of antigenic composition of their carbohydrate peptidoglycans (discussed further in Section 25.2). Streptococcal groups A, B, and C are the most common streptococcal causes of pharyngitis. Group A streptococci (GAS)—*S. pyogenes* in particular—are the bacteria often equated to sore throat in the public's mind. The notoriety is well deserved because group A streptococci cause the more severe cases of the disease.

Streptococcal pharyngitis, or strep throat, is contagious and is spread through person-to-person contact or by indirect contact with fomites (contaminated objects), like sharing a can of soda. The hallmarks of streptococcal pharyngitis (Figure 20.9) are sudden onset of high fever and sore throat; tender, enlarged cervical lymph nodes; exudate on the tonsils; and absence of a cough. These symptoms will usually resolve within 5 days even when untreated. Treatment with antibiotics, however, will quickly stop replication of the pathogen. Patients should complete the full course of treatment to ensure the pathogen is eradicated and to prevent development of antibiotic resistance (see Section 13.6). A set of criteria based on hallmark signs and symptoms (known as Centor criteria) may be used to identify the likelihood of a bacterial infection in patients complaining of a sore throat.

Some strains of *S. pyogenes* produce exotoxins called streptococcal pyogenic exotoxins (SPEs) that can cause fever and a red rash called scarlet fever, or scarletina (described in Section 19.3). The rash usually starts on the head and neck and eventually spreads to the trunk of the body and then to the arms and legs (Figure 20.10). Some patients with scarlet fever may also have a very distinctive red "bumpy" tongue referred to as a "strawberry tongue" (Figure 20.11).

Definitive diagnosis of streptococcal pharyngitis is made from culturing the organism from a patient's pharyngeal **exudates** (see Section 25.1). An exudate is the protein- and cell-rich fluid that oozes out of tissues because of inflammation. A rapid strep test that employs immunoassay technology is available and widely used (Figure 20.12). The test is highly specific for *S. pyogenes*. Patients with *S. pyogenes* pharyngitis are considered noncontagious 24 hours after initiation of antibiotic treatment.

Figure 20.10
The Rash of Scarlet Fever
The rash starts on the trunk and spreads to the extremities but spares the palms of the hands and soles of the feet.

Figure 20.11
The Strawberry Tongue of Scarlet Fever

Figure 20.12 A Rapid Strep Test

The rapid strep test is an in-office immunochromatography test done by a clinician to determine the presence of group A (Gr A) beta-hemolytic streptococci (*Streptococcus pyogenes*). The test begins when streptococcal cell wall antigens extracted from a throat swab are added to a sample well containing red-tagged anti-strep-A antibody. The Gr A antigen binds to the antibody, and the complex moves along a membrane to an area embedded with a different anti-strep-A antibody (larger rectangular area). Any red-tagged antigen-antibody complexes present will be captured by the membrane-fixed antibody and generate a vertical red line (a positive test). The horizontal blue control line indicates that the reagents were added properly. Another control is the small red line (smaller rectangular area), indicating the presence of test kit group C (Gr C) antigen also added to the sample well. Note that a positive group A strep test is not an absolute indication of a strep infection because some people are strep carriers, in which case the strep bacteria are growing in their throats without causing any symptoms.

Antibiotics that inhibit cell wall synthesis, the beta-lactams (such as penicillin), are commonly the first line of therapy. Macrolide antibiotics (for example, erythromycin) are used for patients with a penicillin allergy. The relative ease by which streptococcal infections can be cured has reduced complications of streptococcal pharyngitis to a remarkable degree.

STREPTOCOCCAL SEQUELAE Sequelae are pathological conditions that can result after a primary disease has run its course. Sequelae from streptococcal infections are caused by the immune response to the bacteria. *S. pyogenes* uses M proteins to attach to host cells. M proteins are structures that cover the *S. pyogenes* surface like fuzz on a tennis ball. Some strains of *S. pyogenes* possess M proteins with epitopes similar to those of some human cells, especially cardiac or kidney cells; this phenomenon is known as molecular mimicry. When our body produces antibodies against these M proteins, those antibodies can bind (cross-react) with our own cells or tissue components in an autoimmune reaction (in which our antibodies act against us), resulting in serious sequelae (see Section 17.5).

Post-streptococcal sequelae include rheumatic fever (cardiac disease) and glomerulonephritis (kidney disease), which develop long after the causative organism has been cleared from the body. Whether an individual develops rheumatic fever or glomerulonephritis depends on the antigenic type of M protein made by the infecting GAS strain.

Patients with **acute rheumatic fever (ARF)** are generally young children, 4–9 years of age, with a recent history of sore throat approximately 3 weeks prior to presentation. The illness starts with a high fever and one or more symptoms resulting from damage to the patient's heart, joints, skin, or nervous system. The initial presenting symptom of rheumatic fever is pain in a large joint (arthritis), which typically starts in the knees and spreads to the ankles, elbows, and wrists successively. The joints are not affected all at the same time. Instead, improvement of symptoms in the affected joints coincides with the onset of symptoms in other joints (for example, the ankles)—a phenomenon called migrating, or migratory,

Figure 20.13 Heart Valve Damaged by Rheumatic Fever

Inflammation caused by heart cross-reacting anti-M-protein antibodies damages the heart valve, making it unable to close as needed.

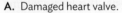

A. Damaged heart valve.   **B.** Healthy heart valve.

arthritis. The X-ray of the affected joints may show some effusion (fluid accumulation), but the inflammation is normally minimal compared with the pain, and the synovial fluid is aseptic (without microbes).

Acute rheumatic fever affects heart tissue through complex interactions between T cells, cross-reacting anti-heart antibodies (triggered by streptococcal M protein), and complement. The result is inflammation of the different layers of tissue covering the heart, including the innermost layer, which also forms the heart valves (see Section 21.1). Pancarditis, simultaneous inflammation of different heart tissues, is almost always accompanied by

valvulitis (inflammation of a valve) and onset of a heart murmur (an abnormal sound) or changes in an existing one (**Figure 20.13**). Patients suffering from rheumatic heart disease will complain of shortness of breath and fatigue as well as varying degrees of chest discomfort or pain.

Skin manifestations associated with ARF include a rash that usually appears early in the disease course. This rash is referred to as the **erythema marginatum** (*marginatum*, "having a distinct margin") or erythema annulare (*annulare*, "ring-shaped"). The circular edge of the lesion is erythematous (red), and the center is paler by comparison (**Figure 20.14**). Skin manifestations may also include subcutaneous nodules (**Figure 20.15**). The nodules are firm and painless and usually resolve within a month. The skin overlying the nodules does not have a rash or ulceration.

Streptococcal pharyngitis may also result in an acute glomerulonephritis (inflammation of renal tissue), which can be asymptomatic with microscopic hematuria (blood in urine). Alternatively, the disease may manifest as acute renal failure, hypertension, edema, and proteinuria (excessive protein in the urine) accompanied by reddish brown urine (Coca-Cola colored)—collectively referred to as the nephritic syndrome. Strains causing symptoms of **post-streptococcal glomerulonephritis** are said to be nephritogenic (causing inflammation of kidneys) and are thought to cause immune complex–mediated glomerulonephritis.

Reemerging Upper Respiratory Tract Infections

Many diseases over the last century were close to being eliminated but are now experiencing an unwelcome resurgence, even in developed countries such as the United States. The anti-vaccination

Figure 20.14 Erythema Marginatum

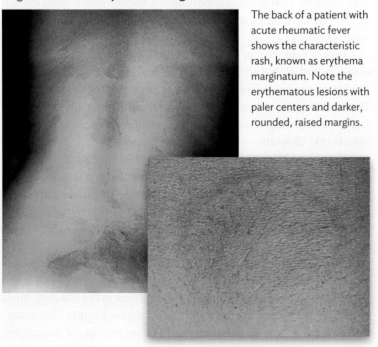

The back of a patient with acute rheumatic fever shows the characteristic rash, known as erythema marginatum. Note the erythematous lesions with paler centers and darker, rounded, raised margins.

Figure 20.15 Rheumatic Subcutaneous Nodules

These nodules are usually located near tendons or over a bony surface or prominence. The overlying skin is not inflamed and typically can be moved over the nodules. The diameter varies from a few millimeters to 1–2 centimeters.

movement has significantly contributed to the resurgence of such diseases as whooping cough. Fear of vaccination is fueled, in part, by the unfounded notion that vaccination can lead to autism. Unfortunately, as more and more families deny their children vaccines, we will continue to see these deadly diseases reemerge.

CASE HISTORY 20.4
The "Bull Neck"

Ramona, a 23-year-old Hispanic mother living in New Mexico, brought her 3-year-old daughter into the emergency department. The child was lethargic, had a fever of 40°C (104°F), and was having difficulty breathing. The mother explained that the family arrived in the United States from El Salvador the previous week. The attending physician noted an extreme swelling of the child's cervical lymph nodes, which were giving the girl a thick, "bull neck" appearance (Figure 20.16). The doctor also noticed the beginnings of a membranous growth at the back of the child's throat that was beginning to obstruct the trachea. It was grayish in color and bled when scraped. The distraught mother admitted that the child had not received any vaccinations before arriving in New Mexico. Suspecting the nature of the child's illness, the physician immediately admitted the child to the hospital and ordered administration of penicillin and a specific antitoxin. The culture results of a throat swab sent to the microbiology lab confirmed the physician's suspicion. The root of the child's disease was *Corynebacterium diphtheriae*, which causes diphtheria. Ramona recovered within 10 days and was later referred to a local pediatrician for evaluation and updating of routine vaccinations for childhood diseases.

DIPHTHERIA Among the many bacterial causes of sore throat, *Corynebacterium diphtheriae* is one that was almost eliminated from the developing world, thanks to routine immunization practices. However, this disease remains a threat from global travel to regions lacking effective immunization. **Diphtheria** is a reportable disease caused by *C. diphtheriae*, a Gram-positive rod that exists in two major forms, lysogenized and nonlysogenized. The toxigenic form is lysogenized and produces a toxin (diphtheria toxin, an inhibitor of host protein synthesis) whose gene, *dtx*, is encoded by the beta phage genome embedded in the *C. diphtheriae* chromosome (see Section 18.4). The nontoxigenic form of the bacterium is not lysogenized with the phage. Both forms of this species can infect the upper respiratory tract and the skin. Toxigenic strains infecting the throat produce toxin that can cause cardiac and nervous system dysfunction. The bacteria remain localized in the pharynx, but the toxin is free to disseminate via the circulation. Cutaneous infections with toxigenic diphtheria, however, rarely result in systemic manifestations.

Figure 20.16
Bull Neck
Swollen lymph nodes give the characteristic "bull neck" appearance in a patient suffering from diphtheria.

Upper respiratory tract infection with *C. diphtheriae* results in a sore throat, cervical lymphadenopathy, a low-grade fever, and a "growing" thick, grayish membrane (made of dead cells, debris, and fibrin) at the back of the throat. This **pseudomembrane** gives the disease its name, diphtheria, which is derived from the Greek word for "leather" (Figure 20.17A). Infections of the skin are seen as lesions that do not heal (Figure 20.17B) and are sometimes covered by a gray membrane. Cardiac manifestations often start as the local symptoms begin to improve and can vary from shortness of breath and irregular heartbeat to heart blocks (problems with the electrical system of the heart) and cardiac dilation (enlargement of heart chambers due to muscle thinning and weakness). The severity of neurological symptoms depends on how quickly the disease is treated with antitoxin. Symptoms can include weakness, paralysis of the local tissue (particularly the soft palate), paralysis of cranial nerve III (the oculomotor nerve, causing double vision or an inability of the pupils to react to light), or total paralysis.

Figure 20.17 Diphtheria

A. Pseudomembrane produced by *Corynebacterium diphtheriae* is thick and leatherlike.

B. Cutaneous lesion caused by *C. diphtheriae*.

Transmission of *C. diphtheriae* is person-to-person, with asymptomatic carriers serving as the only reservoir. Presumptive diagnosis is made when an erythematous throat develops exudates that merge to form a pseudomembrane. The pseudomembrane adheres tightly to the underlying tissue and bleeds when scraped to remove a sample. Culturing the bacteria from patient's throat or skin lesion and verifying toxin production will make the definitive diagnosis. Toxin production can be verified by PCR amplification (see Section 8.6) of the bacterial *dtx* gene or by laying a filter paper strip impregnated with diphtheria antitoxin over a streak of the suspect bacterium. If toxin is made, the antibody against the toxin (the antitoxin) will react with its antigen (the toxin) as it diffuses away from the streak. The reaction produces a line of precipitation called an immunoprecipitate (see Section 17.2).

Antibiotics and antitoxin are used simultaneously to treat the disease and should be administered early for best outcomes. The antitoxin is a hyperimmune serum—plasma with a high concentration of a certain antibody—that is produced in horses and carries a 10% risk of hypersensitivity and serum sickness (a type III hypersensitivity reaction; see Section 17.3). Of course, the best medicine is prevention by vaccination. The vaccine contains inactivated diphtheria toxin (called a toxoid), whose toxic activity has been destroyed while leaving its antigenic properties intact. Diphtheria toxoid is part of the **DTaP series** of vaccinations, which also includes tetanus toxoid and an acellular pertussis vaccine (see Table 17.2).

WHOOPING COUGH Whooping cough, or **pertussis**, is another disease reemerging in developed countries. It is a highly contagious illness caused by the Gram-negative bacillus *Bordetella pertussis*. The infection is acquired when a person inhales aerosolized bacteria-containing droplets spread by the cough of someone with the illness. After an incubation period of 7–21 days, the disease progresses through catarrhal, paroxysmal, and convalescent phases. Symptoms of the catarrhal phase (runny nose and sinus congestion) are indistinguishable from a viral upper respiratory tract infection. The paroxysmal phase is marked by a violent paroxysmal (rapidly repeated) cough. The cough is followed sometimes by vomiting but always by a struggling deep breath that makes a whooping noise, giving the disease its name. The paroxysms of coughing gradually disappear during the next 2–3 weeks as the patient recovers in the convalescent phase.

The pathogenesis of *B. pertussis* is still the subject of much research. Its entire genome has been sequenced, and a number of biologically active molecules and virulence factors have been identified. Among them are adhesion factors such as filamentous hemagglutinin (FHA) and tracheal colonization factor—two surface proteins that latch the bacterium onto cilia in the respiratory tract (Figure 20.18). The pathogen also produces a number of toxins, including an adenylyl cyclase (CyaA) activated by contact with calmodulin in host cells, a tracheal cytotoxin that is a component of

Figure 20.18 *Bordetella* Binding to Ciliated Epithelial Cells

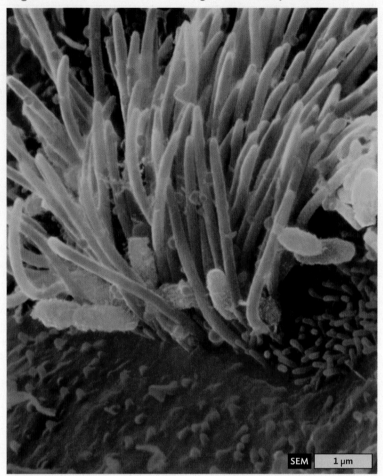

Bordetella pertussis colonizes tracheal epithelial cells and adheres to their cilia. This adherence and the resulting damage impedes the mucociliary escalator that normally removes bacteria from the lung. Bacteria have been pseudocolored green.

Bordetella's peptidoglycan, and the two-component pertussis toxin that activates host adenylyl cyclase (toxins, including CyaA, are described in Section 18.4). The functions of some of these toxins are known, whereas the functions of others are unclear. Increased production of cyclic adenosine monophosphate (cAMP), for example, causes the accumulation of fluid in the lungs in much the same way that cholera toxin causes diarrhea.

The incidence of whooping cough had been declining for decades, thanks to effective vaccines. The original vaccine was made from killed cells but was replaced by an acellular vaccine composed only of purified pertussis toxin and FHA to minimize side effects. Unfortunately, the incidence of disease has started to increase (Figure 20.19). The increase may have been due, in part, to parents' refusal to vaccinate their children as well as reduced effectiveness of the new vaccine. Waning immunity among adults and better diagnostic and surveillance practices (particularly diagnosis by PCR)

Figure 20.19 Incidence of Whooping Cough in the United States

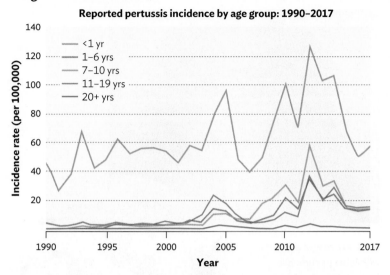

Reported pertussis incidence by age group: 1990–2017

In the early 1990s, parents started to refuse to vaccinate their children because of some studies that seemed to link vaccination to autism. The link has been scientifically rejected. This graph shows the reported pertussis incidence (per 100,000 individuals) by age group in the United States from 1990 to 2017.

may account for some of the increase. Since 2012, the incidence has decreased, possibly due to a change in the acellular pertussis vaccine to include pertactin, an attachment protein of *B. pertussis*.

Early treatment is very important for individuals infected with pertussis, especially infants. Patients who are treated within the first 2 weeks of the disease may avoid paroxysmal coughing and severe symptoms. Macrolides such as azithromycin and clarithromycin are the antibiotics of choice for treatment, but trimethoprim-sulfamethoxazole can be used under special conditions.

Lower Respiratory Tract Bacterial Infections

Bacteria account for the majority of lower respiratory tract infections. Aerobic bacteria find the human lung particularly desirable to assault. The bacteria that attack our lungs scare us for a very good reason: they cause severe diseases with high mortality rates.

BRONCHITIS Inflammation of the bronchi (known as bronchitis) is usually an acute and self-limiting disease with multiple viral and bacterial etiologies. A productive cough is generally the only presenting symptom. At the beginning, an upper respiratory tract infection is indistinguishable from acute bronchitis. Acute bronchitis is usually suspected when a patient's cough persists for more than 5 days. Viruses are the usual cause of acute bronchitis, but some bacteria that cause pneumonia, such as *Mycoplasma*, *Chlamydophila*, and *Bordetella* species, can produce acute bronchitis as well.

Chronic bronchitis affects more than 40% of smokers and is associated with increased production of mucus and a decline in pulmonary function. Smoking is the major but not the only risk factor for chronic bronchitis. Long-term exposure to cigarette smoke, environmental exposure to air pollutants, and repeated lower respiratory tract infections during childhood can also increase the risk of chronic bronchitis. The viscous (sticky) mucus that is not removed can then serve as a breeding ground for bacterial pathogens. Bacteria associated with chronic bronchitis are the same as those causing other upper and lower respiratory tract infections. Patients who suffer from chronic bronchitis are at increased risk for chronic obstructive pulmonary disease (COPD) and death. Around 70%–80% of COPD patients develop respiratory tract infections, nearly half of which are due to bacteria.

COMMUNITY-ACQUIRED PNEUMONIA The term "pneumonia" refers to an infection that causes inflammation in the lung alveoli but does not imply a specific infectious agent. Most cases of pneumonia are attributed to a few bacteria (see Table 20.2), and most cases, regardless of etiology, occur in the winter months because the infectious agents are more easily transmitted from person to person when people are confined indoors. Cases of pneumonia acquired outside a hospital or health care setting are said to be community acquired.

Pneumonias and their bacterial causes are often classified as "typical" or "atypical." The term *atypical* was coined by Hobart Reimann in 1938 to distinguish the first diagnosed human cases of pneumonia caused by *Mycoplasma pneumoniae* from cases caused by *Streptococcus pneumoniae*. Other bacteria, such as *Legionella* and *Chlamydophila*, also produce atypical pneumonias. What makes an atypical pneumonia atypical? Generally, atypical pneumonias affect multiple organ systems, usually produce a normal white blood cell (WBC) count, and produce symptoms that appear gradually, often mimicking upper respiratory infections at the onset. The terms typical and atypical, however, do not have a solid clinical basis, despite their common use. Organisms that produce typical pneumonias can also present atypically.

Lung infections can be contracted during a hospital stay (hospital-acquired, or nosocomial, pneumonias) or outside a hospital setting (community-acquired pneumonias). Community-acquired pneumonias result from a different set of bacteria than hospital-acquired pneumonias. How a particular pneumonia presents depends on the etiologic agent, the length and severity of infection, and the patient's overall health status. However they were acquired, lung infections almost always present with fever and cough. Some patients with pneumonia are able to continue normal activities for quite some time without seeking help, giving rise to the term "walking pneumonia."

Access of microorganisms to the lungs is restricted. Several defense mechanisms contribute to lung innate immunity, including the mucociliary escalator, respiratory epithelium, and alveolar macrophages. The mucociliary escalator continually traps and removes offending agents. The respiratory epithelium produces a number of cytokines, chemokines, and antimicrobial peptides, such as lysozymes and defensins, to defend against the invading pathogens (see Section 15.2). Alveolar macrophages provide further help deeper in the lungs and send for neutrophils (a type of PMN) to come and help with the fight. Organisms that have developed the tools to survive these defenses are able to gain access to the lungs and set up infection (see Chapter 18).

CASE HISTORY 20.5

Night Sweats in a Nursing Home

 James, an 80-year-old resident of a New Jersey nursing home, had a fever accompanied by a productive cough with brown sputum (mucous secretions of the lung that can be coughed up). James reported to the attending physician that he was short of breath, had pain on the right side of his chest, and suffered from night sweats. Blood tests revealed that his white blood cell (WBC) count was 14,000/μl (reference range, 4,500–10,000), composed of 77% neutrophils (reference range, 40%–60%) and 20% band cells (immature neutrophils; reference range, 0%–3%). The chest X-ray revealed a right upper lobe infiltrate (Figure 20.20A). From this information, the clinician made a diagnosis of pneumonia. Microscopic examination of the patient's sputum revealed Gram-positive cocci surrounded by a capsule, occurring in pairs and short chains (Figure 20.20B). Bacteriological culture of the patient's sputum and blood yielded *Streptococcus pneumoniae*.

S. pneumoniae (also called pneumococcus) accounts for about 25% of community-acquired cases of pneumonia. **Pneumococcal pneumonia** occurs mostly among the elderly (like James in our case history) as well as in smokers and immunocompromised individuals, including diabetics and alcoholics. A breakdown of pneumonia cases by causative organism is shown in Figure 20.20C. *S. pneumoniae* is also able to cause upper respiratory tract infections such as otitis media, sinusitis, and meningitis, as discussed earlier.

Our upper airways are often colonized by one of the 91 serotypes of *S. pneumoniae*. Colonization is usually a benign process that leads to a carrier state. The noses and throats of 30%–70% of a given population can contain *S. pneumoniae*. The pneumococcus is highly host adapted, making humans its only natural reservoir. Initial colonization is facilitated by a slippery polysaccharide capsule that allows the organism to avoid engulfment by phagocytic cells and removal by the mucociliary escalator; this helps *S. pneumoniae* gain access to the respiratory epithelium. Once these bacteria reach the epithelium, the thick capsule becomes a hindrance to attachment and changes to a thin, transparent one.

Figure 20.20 Pneumonia Caused by *Streptococcus pneumoniae*

A. X-ray view of a patient with lobar pneumonia. Infiltration in the right upper lobe (RUL) of the lung is caused by *S. pneumoniae*. The sharp lower border represents the upper boundary of the right middle lobe (RML) fissure (arrow).

B. Micrograph of *S. pneumoniae* from a sputum sample, showing numerous polymorphonuclear leukocytes (PMNs) and extracellular diplococci in pairs and short chains.

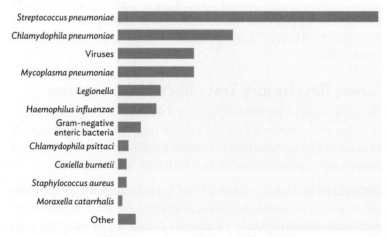

C. Relative incidence of pneumonia caused by various microorganisms.

A large number of virulence factors will then contribute to invasion of the host and onset of the disease.

The microbe reaches the lung, usually after being aspirated, and will grow in the nutrient-rich edema fluid of the alveolar spaces. Neutrophils and alveolar macrophages then arrive to try to stop the infection. They are called into the area from the circulation by chemoattractant chemokines released by damaged alveolar cells. The polysaccharide capsule of the pneumococcus, however, makes phagocytosis very difficult.

In an otherwise healthy adult, pneumococcal pneumonia usually involves one lobe of the lungs; thus it is sometimes called lobar pneumonia. The infiltration of PMNs and fluid leads to the characteristic diffuse cloudy areas seen on X-rays (see Figure 20.20A). In contrast, infants, young children, and the elderly more commonly develop an infection in other parts of the lungs, such as around the air vessels (bronchi), causing bronchopneumonia.

The white cell count in James's case history is telling. He had an elevated WBC count (reference range is 4,500–10,000/µl) and elevated band cells (reference range is 0%–3%)—increases indicative of a bacterial, rather than viral, infection. The number of neutrophils (PMNs), the front-line combatants against bacterial infection, rises in response to the presence of microbes, as reflected in the elevated WBC count. The sudden need for a larger quantity of neutrophils causes an increase in release of immature neutrophils from the bone marrow. The immature neutrophils have band-shaped nuclei and are called "band cells" or "bands." Historically, these cells were first depicted on the left side of a linear diagrammatic classification of white blood cells. Ever since, the increase in percentage of bands in the peripheral blood has been noted as a "left shift." Elevated numbers of WBCs and a left shift (Figure 20.21) indicate a bacterial infection.

Pneumococcal pneumonia among residents of nursing homes is projected to have a 10%–30% mortality rate. There are two different pneumococcal vaccines: the pneumococcal polysaccharide vaccine (PPSV23) and the pneumococcal conjugate vaccine (PCV13). There are more than 90 antigenic types of pneumococcal capsular polysaccharide; PPSV23 contains the 23 polysaccharide types that most often cause disease and

is recommended for individuals over 65 as well as for those who are immunocompromised. In Case History 20.5, James had not received the vaccine.

Young children are also vaccinated against pneumococcus, but not with PPSV23 because a child's immune system is not developed enough to recognize and respond to the free polysaccharide antigens in PPSV23. The polysaccharide antigens must be joined (conjugated) to a protein so they can be recognized by the child's immune system (see Section 17.1). So far, polysaccharides from 13 of the more dangerous 23 *S. pneumoniae* strains have been conjugated to proteins. The result is a 13-valent vaccine called pneumococcal conjugate vaccine, or PCV13. PCV13 is used to immunize children 2–59 months of age or children 5–18 years of age if they have certain medical conditions (such as sickle cell disease, a compromised immune system, or HIV infection) and have not received the vaccine before.

In addition to causing pneumonia, *S. pneumoniae* can invade the bloodstream and the membrane covering of the brain (meningitis). Individuals with health problems such as liver disease, AIDS, and organ transplants are even more likely to die from these diseases because of compromised immune systems.

An emerging infectious disease problem throughout the United States and the world is the increasing resistance of *S. pneumoniae* to antibiotics. At least 30% of the strains isolated are already resistant to penicillin, the former drug of choice for treating the disease. Why antibiotic resistance is on the rise for *S. pneumoniae* and other microbes is discussed in Chapter 13.

ATYPICAL PNEUMONIA *Mycoplasma pneumoniae* is perhaps the best-known cause of "atypical" bacterial pneumonia, but other bacteria, such as *Legionella* and *Chlamydophila*, also produce atypical pneumonia. *Mycoplasma* species lack a rigid cell wall (no peptidoglycan), usually exist in a filamentous form, and are the smallest known bacterial cells capable of independent growth. *M. pneumoniae* can be grown on medium supplemented with serum in the presence or absence of oxygen, but the organism is fastidious and takes 3 weeks to grow in culture. Transmission is person-to-person, involving the inhalation of aerosolized infected droplets. *M. pneumoniae* uses a specialized tip containing adherence proteins to attach itself to the epithelium of the host cell (Figure 20.22). From this perch it can acquire nutrients from the host cell.

Symptoms of *M. pneumoniae* infection are insidious, starting with headache and malaise and progressing to a cough that is usually nonproductive. A low-grade fever may be present, and chills are common. Pneumonia resulting from an infection with *Mycoplasma* is more common in school-age children and young adults attending college and relatively uncommon in infants and young children, in sharp contrast to the disease caused by *Streptococcus pneumoniae*, which predominantly affects the very young and very old. *Mycoplasma* infections activate the immune system, but the

Figure 20.21 Left Shift

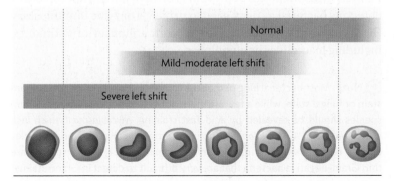

Increasing neutrophil maturity

Infections increase the body's demand for neutrophils. A faster neutrophil production rate leads to the appearance of immature neutrophils and is referred to as a left shift.

Figure 20.22 *Mycoplasma pneumoniae*

Hamster trachea ring infected with *M. pneumoniae*. Note the orientation of the mycoplasmas through their specialized tip-like organelles, which permit close association with the respiratory epithelium. The area within the red rectangle shows a *M. pneumoniae* structure, called the terminal organelle, that helps the bacterium attach to the host cell.

response does not protect against reinfection. The exact mechanism of pathogenesis is not known, but the host's immune reaction to the organism is thought to be the cause of pathology, rather than damage directly inflicted by the bacteria.

Patients infected with *M. pneumoniae* may have subclinical anemia with a positive Coombs test or may develop a positive cold agglutinin test 2 weeks postinfection. A Coombs test detects immunoglobulin G (IgG) antibodies in the patient's blood that can react with antigens on the same patient's red blood cell membrane and destroy it. Some diseases (including *M. pneumoniae* infections) can result in production of such antibodies. **Cold agglutinins** are antibodies that can agglutinate red blood cells at 4°C (39.2°F) but not at 37°C (98.6°F); those antibodies are produced only in response to certain diseases. White blood cell count is typically normal but can be slightly elevated. Diagnosis usually involves the use of enzyme-linked immunosorbent assay (ELISA) to detect a >4-fold increase in anti-mycoplasma IgM and IgG titers in acute sera (from patients with active disease) compared with convalescent sera (from patients who are recovering). qPCR testing to detect presence of the organism is also available, but the sensitivity varies from 40% to 90%, depending on the timing of the sample collection. Treatment, usually macrolide antibiotics, is based on clinical presentation and serology.

Although we think of respiratory infections as being typically spread from person to person through aerosolization of respiratory secretions, that is not always the case. The natural environment, for example, can serve as a reservoir for some lung pathogens. Knowing where disease agents may reside is important both in prevention and in diagnosis.

LEGIONELLOSIS (LEGIONNAIRES' DISEASE) *Legionella pneumophila* is an aerobic Gram-negative bacterium that causes an atypical pneumonia. The first recognized outbreak developed after a meeting of the American Legion in Philadelphia, Pennsylvania, in 1976, which is how the disease came to be known as Legionnaires' disease, or **legionellosis**. The infected legionnaires presented with fever, various systemic symptoms, shortness of breath, and an initial productive cough that often progressed to production of a blood-tinged sputum. Tests subsequently showed that a less virulent form of the organism caused an earlier outbreak in Pontiac, Michigan, and was dubbed Pontiac fever. Today, we know that legionellosis is a community-acquired illness that is not spread person-to-person.

L. pneumophila contaminates various water sources, ranging from lakes to the hot-water or air-conditioning distribution systems of hospitals, hotels, and other large buildings (the sources of infection in the Pennsylvania and Michigan outbreaks). Transmission is through the inhalation of contaminated water droplets. The organism is an intracellular pathogen capable of growing inside alveolar macrophages and free-living amebas. After being phagocytosed, the bacterium prevents fusion of the phagosome with the lysosome (see Section 18.3).

Pathogenesis is facilitated by a number of virulence factors, including flagella, exotoxins, endotoxin, and proteins that help the organism prevent fusion of the phagosome and lysosome—particularly the products of the genes *dot* (defective organelle trafficking) and *icm* (intracellular multiplication). Diagnosis of legionellosis is complicated because these bacteria are difficult to grow in the lab and not all laboratories have the capability to culture *L. pneumophila*. As a result, the number of cases reported is thought to be an underestimate. Diagnosis requires either isolating the organism on buffered charcoal yeast extract (BCYE) agar supplemented with antibiotics to prevent growth of normal microbiota or detecting bacterial antigen shed in urine via enzyme immunoassay (EIA). Legionellosis can be treated with a number of antibiotics, including macrolides or quinolones.

ⓞ **Note:** Most bacterial lung pathogens can be observed in sputum by Gram stain or fungal stain, while a few are more difficult to discern. *Mycobacterium* species should be revealed by acid-fast staining. *Mycoplasma* (which lack a cell wall) and *Chlamydophila* (obligate intracellular pathogens) species cannot be seen with standard staining techniques. *Legionella* species are thin Gram-negative bacteria that are very difficult to detect in Gram-stained sputum.

HOSPITAL-ACQUIRED PNEUMONIA The CDC defines **health care–associated infections (HAIs)**, formerly known as nosocomial infections, as "infections that patients acquire during the course of receiving health care treatment for other conditions." A variety of bacteria, both pathogenic and opportunistic, cause HAIs in almost every body system. *Pseudomonas aeruginosa*, described in Chapter 19 as an important cause of burn infections, is also the most common cause of hospital-acquired pneumonia.

P. aeruginosa is an opportunistic Gram-negative pathogen with a large arsenal of virulence factors that can be used once the bacterium breaches primary host defenses. The bacterium is a rare agent of community-acquired pneumonia but a major contributor to pneumonia and mortality in patients with cystic fibrosis (CF)—an inherited disease that affects secretory glands producing mucus and sweat—and in people who are hospitalized. Hospital-acquired pneumonia caused by *P. aeruginosa* is an acute disease with a high mortality rate, and yet the organism is able to colonize the lungs of CF patients for long periods of time, causing a chronic infection. *P. aeruginosa* achieves this long-term colonization by forming a biofilm in the lungs of CF patients (Section 6.6). Biofilm formation is facilitated by decreased sialylation (addition of sialic acid to cell-surface structures) of CF lung cells, which exposes more bacterial receptors. The bacteria, in turn, shorten their "O" lipopolysaccharide side chain (see Sections 10.3 and 18.4) and down-regulate the production of their toxins to further facilitate biofilm formation. *Pseudomonas* strains associated with CF also produce a thick, gelatinous slime layer, made of alginate, that severely retards cilial movement. The alginate layer also gives colonies of these strains a very mucoid appearance. Colonization as a biofilm results in chronic inflammation, reduction of lung function, and finally death due to respiratory failure.

Other Gram-negative bacilli that cause hospital-acquired pneumonias include *Escherichia coli*, *Klebsiella pneumoniae*, *Enterobacter* species, and *Acinetobacter* species. Gram-positive cocci associated with hospital-acquired pneumonias are *Staphylococcus aureus*, including methicillin-resistant *S. aureus* (MRSA), and *Streptococcus* species. Treatment of hospital-acquired lung infections is difficult because most of the organisms exhibit resistance to multiple antibiotics.

TUBERCULOSIS Until recently, **tuberculosis (TB)**, caused by the acid-fast bacillus *Mycobacterium tuberculosis*, was considered of passing historical significance to physicians practicing in the developed world. In 1985, however, owing primarily to the newly recognized HIV pandemic and a growing indigent population, TB resurfaced in developed countries, especially in inner-city hospitals. In 1991, highly virulent **multidrug-resistant (MDR)** tuberculosis strains were reported. These strains not only produce fulminant (rapid-onset) and fatal disease among patients infected with HIV (the time between TB exposure and death is 2–7 months) but also have proved highly infectious to healthy people without HIV. The **tuberculin skin test (TST)**, which will be discussed shortly, indicates exposure to tuberculosis bacilli when positive. Conversion rates (changing from a negative to a positive test) of up to 50% are reported in health care workers exposed to MDR

strains. An immune system exposed to *M. tuberculosis* generates a T cell–mediated response to *M. tuberculosis*, providing the basis for the skin test. This is true regardless of whether the exposed individual later develops disease or not.

M. tuberculosis causes primarily a respiratory infection but can disseminate through the bloodstream to produce abscesses in many different organ systems. Disseminated disease is called "miliary" TB because the size of the infected nodules, called **tubercles**, approximates the size of millet seeds in bird feed. *M. tuberculosis* bacteria are spread from person to person (no animal reservoir) through aerosolization of respiratory secretions. *Mycobacterium bovis*, which normally causes TB disease in cattle, can also infect humans and cause TB, especially in children, but this accounts for a minority of TB cases. Once either species enters the lung, the bacteria are phagocytosed by macrophages and survive sheltered within modified phagolysosomes. A delayed-type hypersensitivity response results, and small, hard tubercles (granulomas) form around the site of infection. Over time, the tubercles develop into caseous lesions that have a cheese-like consistency and can calcify into the hardened Ghon complexes seen on typical X-rays (**Figure 20.23**). The trapped bacilli are usually killed by the immune system. If not, they are contained by the immune system, often for extended periods of time, a condition called **latent tuberculosis infection (LTBI)**. A person with LTBI is not infectious (**Figure 20.24**).

In some LTBI cases, however, *M. tuberculosis* overcomes the imposed confinement of the immune system, multiplies, and after a period of 4–12 weeks causes primary disease. People with a weakened immune system and the very young are often among those who manifest primary disease, exhibiting a productive cough that generates sputum and experiencing fever, night sweats, and weight loss. These patients are said to have **active tuberculosis disease**, and they are contagious. They will also test positive in the tuberculin skin test, owing to delayed-type hypersensitivity.

Figure 20.23 Calcified Ghon Complex of Tuberculosis

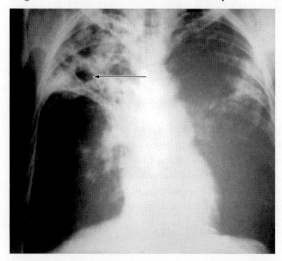

The arrow points to a Ghon complex in the right upper lobe of a patient's lung. Note the difference in appearance compared with Figure 20.20A.

Figure 20.24 Disease Progression of Tuberculosis

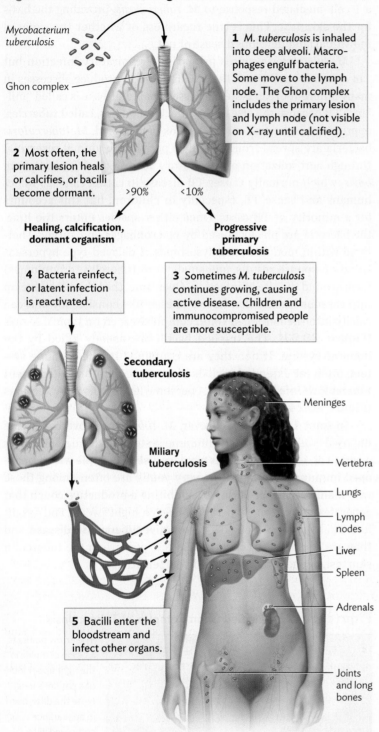

Mycobacterium tuberculosis

Ghon complex

1 *M. tuberculosis* is inhaled into deep alveoli. Macrophages engulf bacteria. Some move to the lymph node. The Ghon complex includes the primary lesion and lymph node (not visible on X-ray until calcified).

2 Most often, the primary lesion heals or calcifies, or bacilli become dormant.

>90%   <10%

Healing, calcification, dormant organism   **Progressive primary tuberculosis**

4 Bacteria reinfect, or latent infection is reactivated.

3 Sometimes *M. tuberculosis* continues growing, causing active disease. Children and immunocompromised people are more susceptible.

Secondary tuberculosis

Miliary tuberculosis

Meninges

Vertebra

Lungs

Lymph nodes

Liver

Spleen

Adrenals

5 Bacilli enter the bloodstream and infect other organs.

Joints and long bones

(1) The disease begins as primary tuberculosis. (2) Most patients never develop disease, or the organism becomes latent within granulomas. (3) A few patients develop progressive primary tuberculosis because the immune system failed to control the infection, allowing it to spread to other lung regions. (4) Secondary tuberculosis develops upon reinfection or reactivation of latent bacilli. (5) More damage occurs in the lung: bacilli disseminate to other organs (including reseeding of the lung) to produce miliary TB.

Figure 20.25 Cavity Formation in Tuberculosis

A. A cavity is seen (inside the rectangle) on a chest X-ray.

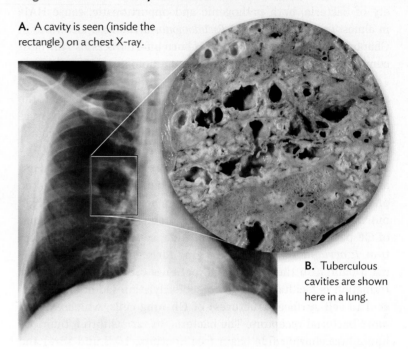

B. Tuberculous cavities are shown here in a lung.

Latent TB can be reactivated months or years after initial infection, a condition called **secondary tuberculosis** or **reactivation tuberculosis**. Secondary TB commonly occurs in people who become immunocompromised. The tubercles described earlier form during secondary TB. The symptoms of secondary TB are more serious than those of primary TB, mainly due to uninhibited replication of the bacteria. Those symptoms include severe coughing, greenish or bloody sputum, low-grade fever, night sweats, and weight loss. The gradual wasting of the body led to the older name for tuberculosis—consumption. Among patients with secondary TB, 40% show cavity formation on a chest X-ray (**Figure 20.25**). Some new and extremely powerful anti-inflammatory medications used to treat other conditions may weaken the patient's immune system just enough to cause reactivation of TB.

TB bacilli usually cause pulmonary disease, but they can disseminate and gain access to other parts of the body by invading the lymph nodes of the lungs. From there they spread into the blood. The bacilli are then carried throughout the body, giving them a chance to infect other organs, such as the adrenal glands or bones. When an organ other than the lungs is infected with *M. tuberculosis*, the disease is said to be extrapulmonary. Patients who have only extrapulmonary TB are usually not contagious. TB bacilli can cause a disseminated infection, often in people who are immunocompromised.

About 10% of people who have a healthy immune system and have been exposed to TB may develop TB. The infection risk depends on the frequency and duration of exposure to people with active TB disease. Health care workers, immunocompromised patients, people who live or work in high-risk congregate settings (such as group homes or intermediate-care facilities), and children and adolescents who live with a patient with active disease are at higher risk of contracting the infection. The elderly and patients who take immunosuppressive drugs, on the other hand, are at higher risk of developing secondary (reactivation) TB.

Screening for LTBI or TB disease involves chest X-ray, the Mantoux tuberculin skin test, interferon-gamma release assay (IGRA), nucleic acid amplification testing (NAAT), and examination of sputum by microscopy and culture. The CDC has developed extensive guidelines for screening and interpretation of test results on the basis of exposure risk, general health, and immune status of the individual tested. A chest X-ray is helpful in a patient with signs and symptoms of disease but is very difficult to interpret as a screening tool. A positive **Mantoux tuberculin skin test** indicates a delayed-type hypersensitivity reaction to *M. tuberculosis* proteins (called purified protein derivative, or PPD) injected under the skin. The result of the test is read 48–72 hours after placement of the antigen. An induration (area of hard, raised skin) larger than 5–15 mm at the injection site is interpreted as a positive test, depending on the health and risk factors of the tested patient. Some people born outside the United States and some US health care or military personnel may have been inoculated against TB with a BCG vaccine. The BCG vaccine is a live culture of bacillus of Calmette-Guérin strain of *M. bovis*. Individuals who have received BCG vaccination will have a positive TB test.

◉ **Note:** A positive tuberculin reaction does not mean that the patient has tuberculosis but does mean that the patient was at one time exposed to tubercle bacilli. In most cases, the immune response will kill the organism when it infects the lung. But memory T cells generated during the response will persist and can respond years later to the injection of PPD, producing a positive tuberculin test reaction.

Interferon-gamma release assay (IGRA) is a blood test that measures the release of interferon-gamma from white blood cells upon exposure to mycobacteria. BCG vaccination will not result in a false-positive IGRA blood test. IGRA is the preferred method of TB screening for people who may not return for TST reading or those who have received BCG vaccination.

The tuberculin skin test and IGRA do not distinguish between latent infection (LTBI) and active disease. Amplification of the mycobacterial DNA from patient blood and/or visualization of *M. tuberculosis* bacilli in patient sputum under the microscope is used to detect disease, and the diagnosis is confirmed by culturing TB bacilli from the patient's sputum or tissue. Culturing is also used to monitor treatment progress. Negative cultures are a required criterion to declare the patient free of TB disease.

Figure 20.26 Ziehl-Neelsen Staining of Mycobacteria

Mycobacteria are able to resist decoloration when they are washed with acidified ethanol, and under the microscope they appear as red bacilli in tissue specimens or culture samples counterstained with methylene blue. Ziehl-Neelsen staining is also referred to as acid-fast staining.

Mycobacterium species have a Gram-positive cell wall but cannot be Gram stained because about 60% of their cell wall is lipid (mycolic acid), called a waxy coat; see Section 5.3. The Ziehl-Neelsen stain (also called acid-fast stain) is used instead of the Gram stain when a sample is suspected to contain mycobacteria. The bacteria are fixed to a glass slide, flooded by carbolfuchsin dye, and then washed with acidified ethanol. The acidified ethanol washes away the dye from other bacteria but not from mycobacteria, which retain the dye and appear as red bacilli under the microscope (**Figure 20.26**), which is why mycobacterial species are called **acid-fast bacilli (AFB)**. AFB must be isolated in culture to confirm diagnosis. Other laboratory techniques used to grow or identify *Mycobacterium* are discussed in Sections 25.2 and 25.3.

Ten drugs are currently approved by the FDA for tuberculosis treatment. Initial treatment of active disease is aggressive and involves a regimen of four drugs called first-line drugs (or drugs of choice): isoniazid, rifampin, pyrazinamide, and ethambutol, given over several months. MDR strains are defined as being resistant to two or more first-line drugs. MDR strains are treated with a nine-drug regimen. **Extensively drug-resistant tuberculosis (XDR-TB)** strains are resistant to three or more second-line drugs (drugs used when drugs of choice fail) as well as two (isoniazid and rifampin) or more first-line drugs. These strains are almost untreatable.

Pneumonia Caused by Select Agents

Bioterrorism is defined as the intentional release of biological agents to bring about disease or death to humans, livestock, or crops. The fear that some infectious agents may be used for such nefarious purposes has led the CDC to develop a list of select agents (see Table 2.4) whose use and distribution among researchers are highly restricted. One such agent is *Bacillus anthracis*.

B. anthracis is a spore-forming, Gram-positive rod that primarily causes infections of animals (for example, cattle). The spores can survive for decades in soil and under proper conditions can even germinate and multiply there. Infection of humans usually results from accidental exposure to infected animals or animal products contaminated with spores (for instance, animal hides). *B. anthracis* can infect a number of tissues, among them skin (wound infections), lungs, and gastrointestinal tract (rare). Spores can be inhaled if aerosolized and produce inhalation anthrax. Weaponized spores are specially treated to increase the time they can remain airborne. Despite the fact that pulmonary anthrax is contracted by inhalation, transmission does not usually occur from person to person (too few organisms or spores are aerosolized to support person-to-person transmission). The lethality of the bacterium is due to its three-subunit toxin (see Section 18.4). The core subunit, called protective antigen (PA), helps the trivalent toxin enter the host cell. The core subunit is called protective antigen because antibodies to this protein will protect hosts from disease. The other two subunits, the edema factor (EF) and lethal factor (LF), attack internal host metabolic components to disrupt host cell function and growth. The severity of disease produced, as well as the ability to store and aerosolize spores, makes *B. anthracis* a possible agent for bioterrorism (as described in the case that opens Chapter 18).

Inhalation anthrax has a prodromal phase that lasts 5 days, during which the patient suffers from mostly nonspecific symptoms resembling those of influenza. Some patients may have additional symptoms of hemoptysis (coughing up blood), dyspnea (shortness of breath), chest pain, and nausea. Once there is a clinical suspicion of inhalation anthrax, rapid diagnosis and antibiotic treatment is critical because death can occur within 48 hours. Local and state health authorities must be informed and must follow prescribed containment protocols to limit spread of the disease. Treatment is most effective during this time, but the rapid disease course makes diagnosis before death very difficult. A vaccine is available but administered only to high-risk groups such as military or laboratory personnel.

Many microbes designated as select agents cause deadly pneumonias and can be transmitted by aerosolization (though not always by person-to-person contact), including *Coxiella burnetii* (the most infectious organism known), *Yersinia pestis* (the cause of plague), *Francisella tularensis*, *Coccidioides immitis* (fungal infection), Lassa fever virus, and Ebola virus.

SECTION SUMMARY

- **Three bacterial species,** *Streptococcus pneumoniae*, *Haemophilus influenzae*, and *Moraxella catarrhalis*, are the most frequent causes of acute otitis media (AOM) in otherwise healthy children.

- ***S. pneumoniae*, the most frequent etiology of AOM,** is identified on the basis of its antiphagocytic polysaccharide capsule structure. All identified serotypes can cause infection, but the relative invasiveness of *S. pneumoniae* strains depends on capsule structure.

- ***H. influenzae* strains may be encapsulated (typeable) or lack a capsule (nontypeable).** Encapsulated *H. influenzae* subtype b (Hib) is the most invasive and can cause epiglottitis and meningitis in addition to AOM.

- **Unencapsulated, nontypeable *H. influenzae*** cause the majority of *H. influenzae*–related infections. The Hib vaccine does not protect against infections with nontypeable strains.

- ***Streptococcus pyogenes*** is the most common bacterial cause of pharyngitis.

- **Rheumatic fever and post-streptococcal glomerulonephritis** are sequelae of infections by some serotypes of *S. pyogenes*.

- **Scarlet fever is a consequence of pharyngitis** caused by certain serotypes of *S. pyogenes* that produce streptococcal pyogenic exotoxin (SPE).

- ***Streptococcus pneumoniae* is the most common cause of bacterial pneumonia.** The pneumococcus organism normally infects one of the lobes of the lungs, which has given rise to the term "lobar pneumonia."

- ***Mycoplasma pneumoniae*** lacks a cell wall, usually exists in a filamentous form, and results in pneumonia with an atypical presentation.

- ***Corynebacterium diphtheriae*** is a Gram-positive rod that can be lysogenized by the phage carrying the *dtx* gene that encodes the diphtheria toxin. ***C. diphtheriae* can cause upper respiratory tract and skin infections.** Circulating toxin can cause systemic disease.

- **Whooping cough** is caused by the Gram-negative bacillus *Bordetella pertussis*, which produces several toxins, including the pertussis toxin.

- ***Pseudomonas aeruginosa*** is an opportunistic Gram-negative bacillus that is the most common cause of infection in burn units and the leading cause of hospital-acquired pneumonias.

- ***Legionella pneumophila*** is an aerobic Gram-negative bacillus that contaminates various environmental and industrial water sources.

- ***Mycobacterium tuberculosis*** is an acid-fast bacillus that causes TB, a chronic form of pneumonia that is reemerging as a result of an increased number of immunosuppressed patients.

- **Highly virulent multidrug-resistant (MDR) strains of** *M. tuberculosis* can produce a fulminant and fatal disease among HIV-infected individuals and others who are severely immunosuppressed.

- ***Bacillus anthracis*** is a spore-forming Gram-positive bacterium that normally causes disease in animals and can occasionally cause skin lesions in humans who come in contact with an infected animal.

- **Inhalation anthrax** is a highly uncommon, often lethal form of pneumonia caused when spores of *B. anthracis* are inhaled.

Thought Question 20.2 Why are smokers at higher risk for bacterial bronchitis?

Thought Question 20.3 The TB skin test is usually negative in AIDS patients. Why?

Thought Question 20.4 Which have the higher risk of a bad outcome, hospital- or community-acquired pneumonias, and why?

20.4 Fungal and Parasitic Infections of the Respiratory Tract

SECTION OBJECTIVES

- Identify the most common etiologies of fungal lower respiratory tract infections.
- Describe the differences between fungal and bacterial lower respiratory tract infections.
- Name the risk factors associated with fungal respiratory tract infections.

CASE HISTORY 20.6

A Boxer's Fight to Survive

Tyrrell, a 35-year-old male boxer who installs home insulation for a living, was recently admitted to a Maryland hospital when he presented at the emergency department complaining of difficulty walking, fever, chills, night sweats, and a recent 10-kg (22-lb) weight loss. Tyrrell admitted that he does not wear a protective mask, as he should, while working. He denied having prior pneumonia, sinus infection, arthritis, hematuria (blood in the urine), numbness, or muscle weakness. Tyrrell also denied any history of intravenous drug use and reported he had been in a monogamous relationship for the past 4 years. There was no history of travel outside the area for the past year. A review of Tyrrell's past medical history showed that he had visited the emergency department 6 months earlier with flu-like symptoms, a chronic cough that produced blood-tinged white sputum, shortness of breath, loss of appetite, and weight loss. One month prior to current admission, he developed some painless subcutaneous nodules and became so short of breath that he could no longer continue boxing. At that time, an X-ray taken in the emergency department showed infiltrates in both lungs, indicating pneumonia (Figure 20.27A). A tuberculosis skin test was negative. He was given a prescription for the antibiotic azithromycin (an antibiotic commonly used to treat bacterial infections of the respiratory tract) and discharged. The cough and weight loss continued. Physical exam at the time of current admission revealed several painful subcutaneous nodules on Tyrrell's left thigh, one filled with pus, and a tender tibia that prevented him from walking (Figure 20.27B). A CBC obtained at the time of current admission showed normal counts and differential.

What are the main symptomatic differences between respiratory infections caused by fungi and those caused by viruses or bacteria? Respiratory infections caused by viruses and bacteria have clearly defined clinical presentations. In contrast, fungal infections of the respiratory tract present with a constellation of vague and sometimes seemingly unrelated symptoms. In Tyrrell's case, the expression of frankly purulent material (pus) from the left leg nodule suggests an infectious process. His recent history of chronic cough suggests that the infection in this patient probably started in the lung, after which the organism disseminated throughout the body via the bloodstream. Fungus is a probable cause, given the chronic nature of the patient's symptoms. Tuberculosis, caused by the bacterium *Mycobacterium tuberculosis*, also causes a chronic lung infection and might have been suspected except for the negative TB skin test.

Figure 20.27 Pneumonia and Metastatic Disease Caused by *Blastomyces dermatitidis*

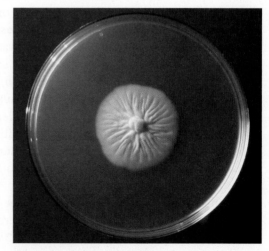

A. Infiltrates in both right and left lungs (arrows).

B. Metastatic leg lesion at the tibia.

C. Fungal colony of *B. dermatitidis*.

Figure 20.28 *Histoplasma capsulatum*

This clinical sample was visualized with methenamine silver stain.

Histoplasma capsulatum

LM   4 µm

The most likely fungal causes of infection in this case history are the endemic mycoses, such as histoplasmosis, blastomycosis, and coccidioidomycosis. **Coccidioidomycosis**, also called valley fever, is caused by a variety of *Coccidioides* species, including *Coccidioides immitis*. The fungus is endemic in the southwestern United States. Tyrrell had never traveled to the western United States, ruling out exposure to *Coccidioides*. **Histoplasmosis** is caused by *Histoplasma capsulatum* (**Figure 20.28**), a dimorphic fungus, meaning it exists in two different forms depending on growth temperature. *H. capsulatum* lives as a mold with a filamentous form in soil (about 25°C) and changes its form to a small budding yeast once it enters the human lung (37°C). In the United States, *H. capsulatum* is found mostly in the Ohio and Mississippi river valleys. Histoplasmosis most commonly presents as a flu-like pulmonary illness, with erythema nodosum (tender red bumps on the skin) and arthritis (swollen joints) or arthralgia (joint pain), none of which the patient had. **Blastomycosis** can disseminate to the lung, skin, bone, and genitourinary tract, which is consistent with the pattern of organ involvement seen in this patient. Blastomycosis is endemic to the eastern US. *Cryptococcus neoformans*, an encapsulated yeast that causes **cryptococcosis**, could also be considered. The most prevalent clinical form of cryptococcosis, however, is meningoencephalitis in AIDS patients, although disease can also involve the skin, lungs, prostate gland, urinary tract, eyes, myocardium, bones, and joints. Cryptococcosis, therefore, was considered but rejected.

Amphotericin B, an antifungal agent (see Section 13.8), was ultimately given to Tyrrell. His fever lowered almost immediately, and the skin nodules diminished. After 2 weeks, a fungus was found in the cultures of the nodule biopsy, bronchoalveolar lavage

(washes), and urine (**Figure 20.27C**). This fungus was identified by a DNA probe as *Blastomyces dermatitidis*, confirming the diagnosis of blastomycosis.

B. dermatitidis is a dimorphic fungus, which means that, like *H. capsulatum*, it can exist in two different forms: a hyphal mold at 25°C that converts to a large budding yeast at 37°C (see Section 18.6). Like *H. capsulatum*, this fungus also resides in the soil of the Ohio and Mississippi river valleys and the southeastern United States. The portal of entry is the respiratory tract, and infection is usually associated with occupational exposure, demolition projects, and recreational activities in wooded areas along waterways, such as camping or caving, where there is moist soil with a high content of organic matter and spores. The incubation period ranges from 21 to 106 days. This patient most likely inhaled conidia (fungal spores) from the soil while crawling underneath houses installing insulation. Fungal respiratory tract infections are acquired from the environment and not transmitted person-to-person by respiratory aerosols. The physician learned that Tyrrell had used only a T-shirt to cover his mouth and nose, an ineffective method of keeping spores from entering the respiratory tract. He should have worn a respirator.

Several critical features of this case help differentiate it from the preceding case of pneumococcal pneumonia. First, the initial macrolide antibiotic, azithromycin, should have killed most bacterial sources of infection. Second, the X-ray finding of diffuse infiltrate is more indicative of fungal lung infection than pneumococcal infection. The blood count was also a clue. Bacterial infections usually cause an increase in WBCs, especially in the percentage of band cells (immature neutrophils), but Tyrrell's blood count and differential (the percentage of each type of white blood cell in the sample) were within expected limits. Finally, the metastatic lesions (infectious lesions that develop at a secondary site away from the initial site of infection) on the leg were in no way consistent with *Streptococcus pneumoniae*. Many infectious diseases start out as a localized infection but end up disseminating throughout the body to cause metastatic lesions.

In addition to viral, bacterial, and fungal pneumonias, parasites such as *Strongyloides stercoralis* and *Toxoplasma gondii* can (but rarely do) cause pneumonia. Parasitic pneumonias are highly uncommon and present almost exclusively in HIV patients or result from dissemination from another infected site of the body. *Dirofilaria*, *Paragonimus*, and *Entamoeba histolytica* infections of the lung are also seen in patients with compromised immune systems but are even rarer.

With the exception of *Paragonimus*, a flatworm that can cause a lung infection if ingested in undercooked crayfish or crab, worms do not usually infect the lungs. Hookworms and *Ascaris* larvae can, however, migrate from the intestines to the lungs and trigger the infiltration of eosinophils, causing an eosinophilic pneumonia (see Section 22.6).

SECTION SUMMARY

- **Fungal infections of the lungs** can result from inhalation of fungi or their dissemination from a primary infection site to the lung.
- **The most common causes of fungal lung infections** are *Histoplasma capsulatum, Cryptococcus neoformans, Blastomyces dermatitidis,* and *Coccidioides immitis.*
- **Immunosuppression and environmental exposures** are the two most important risk factors associated with acquiring a fungal lung infection.

Thought Question 20.5 What is an indication that a chronic cough may be due to a fungal infection?

Perspective

It is hard to overstate the significance of respiratory tract infections today and, indeed, throughout history. Respiratory diseases have resulted in major changes in our behavior and major shifts in population dynamics. The possibility that an infectious agent could spread quickly and globally was demonstrated by the SARS virus, first reported as the cause of a severe respiratory illness in Asia in February 2003. The virus spread rapidly to Europe and North and South America, causing 8,098 cases of disease and 774 deaths worldwide according to the WHO. Although global interventions by health officials prevented a pandemic, the outbreak dramatically illustrated the potential for air travel and globalization to spread deadly infections. Outbreaks of MERS, the virus that causes Middle East respiratory syndrome, and Ebola virus followed SARS, providing the world with repeated, ominous warnings that a pandemic was inevitable. Then, in 2019, the SARS-CoV-2 virus emerged in China and quickly spread around the world, producing the now historic COVID-19 pandemic. COVID-19 has painfully demonstrated that a coordinated global response to disease outbreaks is needed not only to prevent the resurgence of diseases once considered eliminated from the developed world but also to contain emerging new diseases and the costs imposed by infectious epidemics.

Figure 20.29 **Summary: Microbes That Infect the Respiratory Tract**

Viruses	Genome	Disease
Adenoviruses	DNA	Common cold
Respiratory syncytial virus	RNA, ss,[a] negative sense	RSV disease, sometimes common cold
Metapneumovirus	RNA, ss, negative sense	Common cold
Influenza virus	RNA, ss, negative sense	Influenza, sometimes common cold
Parainfluenza viruses	RNA, ss, negative sense	Croup, common cold
Rhinoviruses	RNA, ss, positive sense	Common cold
Enteroviruses	RNA, ss, positive sense	Common cold
Coronaviruses	RNA, ss, positive sense	Common cold, SARS, MERS, COVID-19

Bacteria	Morphology	Disease
Mycobacterium tuberculosis	Acid-fast bacilli	Tuberculosis
Streptococcus pneumoniae	Gram-positive cocci	Otitis media, sinusitis, pneumonia, meningitis
Bacillus anthracis	Gram-positive rod	Inhalation anthrax
Corynebacterium diphtheriae	Gram-positive rod	Diphtheria
Bordetella pertussis	Gram-negative coccobacilli	Whooping cough
Pseudomonas aeruginosa	Gram-negative rod	Pneumonia
Legionella pneumophila	Gram-negative rod	Legionellosis (Legionnaires' disease)
Chlamydophila pneumoniae	N/A	Pneumonia
Chlamydophila psittaci	N/A	Pneumonia
Mycoplasma pneumoniae	N/A	Pneumonia

Fungi	Type	Disease
Histoplasma capsulatum	Dimorphic yeast	Pneumonia
Coccidioides immitis	Dimorphic yeast	Pneumonia
Cryptococcus species	Dimorphic yeast	Pneumonia
Blastomyces dermatitidis	Dimorphic yeast	Pneumonia

[a]ss = single-stranded.

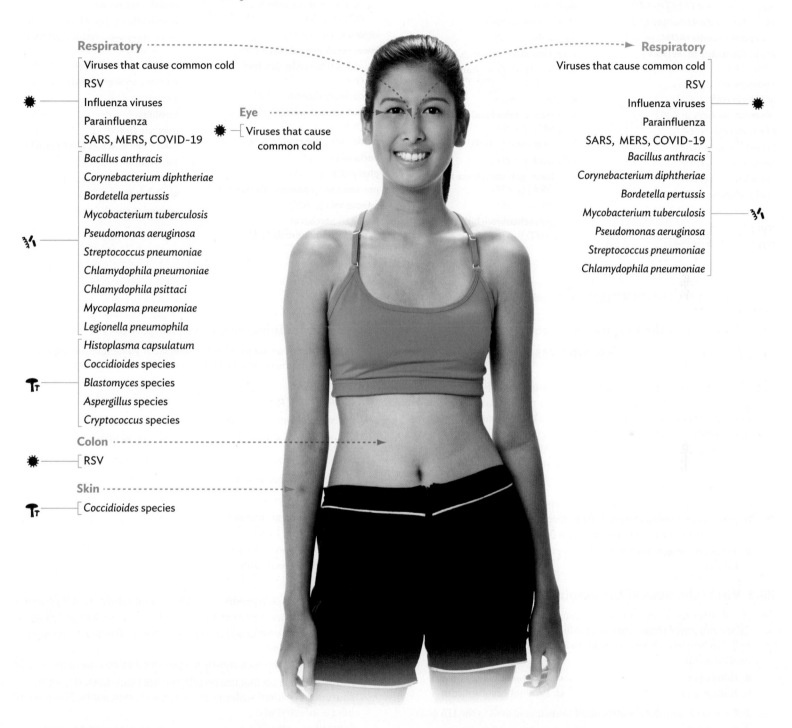

Portal of Entry

Respiratory
Viruses that cause common cold
RSV
Influenza viruses
Parainfluenza
SARS, MERS, COVID-19
Bacillus anthracis
Corynebacterium diphtheriae
Bordetella pertussis
Mycobacterium tuberculosis
Pseudomonas aeruginosa
Streptococcus pneumoniae
Chlamydophila pneumoniae
Chlamydophila psittaci
Mycoplasma pneumoniae
Legionella pneumophila
Histoplasma capsulatum
Coccidioides species
Blastomyces species
Aspergillus species
Cryptococcus species

Eye
Viruses that cause common cold

Colon
RSV

Skin
Coccidioides species

Portal of Exit

Respiratory
Viruses that cause common cold
RSV
Influenza viruses
Parainfluenza
SARS, MERS, COVID-19
Bacillus anthracis
Corynebacterium diphtheriae
Bordetella pertussis
Mycobacterium tuberculosis
Pseudomonas aeruginosa
Streptococcus pneumoniae
Chlamydophila pneumoniae

Key Terms

acid-fast bacillus (AFB) (p. 641)
active tuberculosis disease (p. 639)
acute otitis media (AOM) (p. 628)
acute rheumatic fever (ARF)
   (p. 631)
antigenic drift (p. 622)
antigenic shift (p. 622)
bacterial sinusitis (p. 629)
blastomycosis (p. 644)
bronchiolitis (p. 621)
bronchitis (p. 620)
coccidioidomycosis (p. 644)
cold agglutinin (p. 638)
common cold (p. 618)
croup (p. 623)
cryptococcosis (p. 644)

diphtheria (p. 633)
DTaP series (p. 634)
epiglottitis (p. 628)
erythema marginatum (p. 632)
extensively drug-resistant
   tuberculosis (XDR-TB) (p. 641)
exudate (p. 630)
health care–associated infection
   (HAI) (p. 638)
histoplasmosis (p. 644)
influenza (p. 621)
interferon-gamma release assay
   (IGRA) (p. 641)
laryngitis (p. 629)
laryngotracheobronchitis (LTB)
   (p. 623)

latent tuberculosis infection
   (LTBI) (p. 639)
legionellosis (p. 638)
lower respiratory tract (p. 616)
Mantoux tuberculin skin test
   (p. 641)
mucociliary escalator (p. 617)
multidrug-resistant (MDR)
   (p. 639)
peritonsillar abscess (p. 629)
pertussis (p. 634)
pharyngitis (p. 629)
pneumococcal pneumonia (p. 636)
pneumonia (p. 620)
post-streptococcal
   glomerulonephritis (p. 632)

pseudomembrane (p. 633)
reactivation tuberculosis (p. 640)
respiratory syncytial virus (RSV)
   (p. 620)
secondary tuberculosis (p. 640)
streptococcal pharyngitis (p. 630)
syncytium (p. 621)
tonsillitis (p. 629)
tubercle (p. 639)
tuberculin skin test (TST) (p. 639)
tuberculosis (TB) (p. 639)
upper respiratory tract (p. 616)
viral sinusitis (p. 620)
whooping cough (p. 634)

Objectives Assessment Questions

20.1 Anatomy of the Respiratory Tract

1. The larynx is a part of the upper respiratory tract that helps us generate sounds. True or false?

2. A child swallowed a large coin that became lodged in his pharynx. He was barely able to breathe. The paramedics made a hole in the child's neck to insert a tube and help him breathe. Which structure was punctured by the paramedics?
   a. Bronchial tube
   b. Trachea
   c. Alveolar duct
   d. Bronchiole

3. The mucociliary escalator acts by _____.
   a. enhancing movement of air in and out of the bronchi
   b. preventing foreign materials from reaching the lungs
   c. moving CO_2 out of the lungs so it can be exhaled
   d. producing mucus only when we have a respiratory tract infection

20.2 Viral Infections of the Respiratory Tract

4. Nasal congestion, sore throat, and cough can be present in both upper and lower respiratory tract infections. Which of the following will differentiate a lower respiratory tract infection in an otherwise healthy adult?
   a. Itchy eyes
   b. Runny nose
   c. Shortness of breath
   d. Production of sputum

5. Influenza pandemics are attributed to antigenic drift in the HA gene. True or false?

6. The error-prone replication of genomic material in influenza virus A necessitates production of a new flu vaccine annually. True or false?

7. _____ are at highest risk for developing RSV pneumonia.
   a. The elderly
   b. Adolescents
   c. Adults
   d. Infants

20.3 Bacterial Infections of the Respiratory Tract

Match the infectious agent with the disease it causes, choosing from the following list. (Some questions may have more than one answer.)

Infectious agent
8. *Streptococcus pyogenes*
9. *Mycobacterium tuberculosis*
10. *Haemophilus influenzae*
11. *Pseudomonas aeruginosa*
12. *Streptococcus pneumoniae*

Disease
a. Acute otitis media
b. Pharyngitis
c. Bronchitis
d. Pneumonia

13. _____ is the most effective prevention for whooping cough.
   a. Childhood vaccination
   b. Respiratory isolation
   c. Antibiotic prophylaxis
   d. Wearing a respirator

14. Scarlet fever is _____.
   a. a sequela of pharyngitis caused by certain serotypes of *S. pyogenes*
   b. a consequence of pharyngitis caused by toxin-producing *S. pyogenes*
   c. a skin rash accompanying most streptococcal upper respiratory tract infections
   d. a common manifestation of streptococcal infections in the elderly

15. A small bacterium that has no cell wall has been isolated from the sputum of a patient suffering from atypical pneumonia. This patient's lung is infected with _____.
   a. *Bacillus anthracis*
   b. *Streptococcus pneumoniae*
   c. *Legionella pneumophila*
   d. *Mycoplasma pneumoniae*
   e. *Mycobacterium tuberculosis*

16. Patients who have TB are treated with a combination of four different first-line anti-tuberculosis drugs. True or false?

17. *Mycobacterium bovis* infects cattle and cannot cause tuberculosis in humans. True or false?

20.4 Fungal and Parasitic Infections of the Respiratory Tract

18. Blood-tinged sputum can be a presenting symptom of TB as well as fungal pneumonias. True or false?

19. Which of the following laboratory test results indicates a fungal etiology in a patient suffering from pneumonia?

a. A left shift

b. Normal values

c. Increased neutrophils

d. Decreased neutrophils

20. A patient with a _____ lung infection is probably infected with HIV.

a. *Histoplasma*

b. *Blastomyces*

c. *Cryptococcus*

d. *Coccidioides*

21. Coccidioidomycosis is endemic in the _____.

a. northern United States

b. southern United States

c. eastern United States

d. western United States

Review Questions

1. What anatomical structures form the respiratory tract?

2. Describe the mucociliary escalator and its function.

3. What are the differences between primary and secondary bacterial infections of the respiratory tract?

4. List the symptoms and etiologic agents of the common cold.

5. What is the relationship of RSV to bronchiolitis in infants?

6. How are appropriate strains of influenza selected each year for production of flu vaccine?

7. Why does croup cause a "barking cough"?

8. Name the top three causes of acute otitis media.

9. What distinguishes bacterial sinusitis from viral sinusitis?

10. What respiratory infections are caused by *Streptococcus pyogenes*?

11. Describe the sequelae related to infections with some *S. pyogenes* serotypes.

12. How are streptococcal species (other than *Streptococcus pneumoniae*) divided into different groups?

13. What types of vaccines are available against *S. pneumoniae*? Who is vaccinated with which vaccine?

14. What is atypical pneumonia, and what organisms cause it?

15. What are cold agglutinins, and what is their relationship to walking pneumonia?

16. Who is susceptible to pneumonia resulting from *Pseudomonas aeruginosa*, and how does the organism colonize some patients' lungs?

17. Describe the differences between latent, active, primary, secondary, multidrug-resistant, and extensively drug-resistant TB.

18. Describe the Mantoux skin test and IGRA test.

19. What is an acid-fast bacillus?

20. What organism causes inhalation anthrax?

Clinical Correlation Questions

1. A 65-year-old female is in your office complaining of a chronic cough. She says she started coughing a few months ago. The cough was dry and did not bother her much at first, but now she is coughing a lot and brings up blood-tinged sputum. She takes only one medication—the new arthritis pills you prescribed a few months back. She denies any shortness of breath or any travel in the past year and says she has been mostly home taking care of her grandchildren. She also denies any other complaints. She is overweight but has no history of any other medical problems. Her temperature, heart rate, and respiratory rate are within expected limits, and there are no lesions on her skin. What organism do you suspect is causing the problem, and why?

2. The chest X-ray of a patient suffering from cough, high fever, and shortness of breath shows infiltrates in only the middle lobe of her right lung. Complete blood count with differential shows markedly elevated white blood cells and a decided left shift. What is the most likely diagnosis: bacterial, viral, fungal, or parasitic pneumonia?

3. Charles is an 8-year-old boy who is brought to the clinic complaining of a persistent dry cough for the past 2 months. His mother reports that Charles had what appeared to be a cold with a sore throat and a low-grade fever 2 months ago. Charles says he recovered from his cold and sore throat after about a week or so but has not stopped coughing. Charles has one of his paroxysmal cough episodes while in the office.

You witness him coughing hard and hear a loud sound when he finally is able to stop coughing and inhales. Charles has received no childhood immunization—his parents refused. Charles's temperature, respiration, and heart rate are within expected limits; his chest X-ray shows clear, healthy lungs; and he does not appear to be in any distress. What organism is most likely the cause of Charles's symptoms?

4. Yolanda, a 45-year-old nurse, is brought to the emergency room by her husband. She is barely able to walk on her own because she becomes severely short of breath. She has a fever (38°C) and a blood oxygen saturation of 89% (expected is 95%–99%). Yolanda's husband tells the doctor that they came back from a hiking trip just a week ago. They had stayed in a beautiful cabin, taken long hikes during the day, and soaked in the cabin's hot tub when they got back. Yolanda had started having a dry cough shortly after returning home. The cough became productive (produced sputum) and she had muscle aches and intermittent diarrhea in addition to being very tired and short of breath. Yolanda is admitted to the hospital and put on oxygen and wide-spectrum antibiotics after blood, urine, and sputum samples are obtained. Sputum is obtained for staining and culture. Chest X-ray shows patchy infiltrates in her lower lungs. Gram stain of her sputum shows no bacteria. Acid-fast stain, TB skin test, and IGRA are negative. What is the most likely cause of Yolanda's infection, and how is it confirmed?

 For more assessment, visit digital.wwnorton.com/michum2

21

Systemic Infections

[Too Tired for Soccer]

SCENARIO Mira, a 16-year-old soccer player, complained of a sore throat and woke up in a cold sweat 2 days in a row. She felt very tired but wanted to go to soccer practice anyway. Her mother was reluctant to give permission unless Mira went to her pediatrician first.

LABS Mira was not feeling well when they got to the clinic. The nurse practitioner (NP) examined her and found that Mira had a mild fever, pharyngitis, swollen tonsils, and some swollen lymph nodes in her neck. She swabbed Mira's throat to check for infectious mononucleosis (mono) and strep throat.

Infectious Mononucleosis
Enlarged tonsils with overlying exudates are one of the signs of infectious mononucleosis, caused by Epstein-Barr virus. In young children, the tonsils can become large enough to touch each other and obstruct air flow.

LAB RESULTS The mono test was positive. It was at this point that Mira recalled her boyfriend complaining of a sore throat a couple of weeks back.

TREATMENT Mira was instructed to rest, take ibuprofen for pain, and call the office if any of her symptoms got worse. She was told not to take aspirin because it could cause a severe reaction. The NP emphasized that Mira should not play soccer or engage in any other strenuous physical activity for 4 weeks because her spleen could be enlarged as a result of infection. An enlarged spleen could burst (rupture) if she were hit or fell onto the left side of her body. The NP further explained that a ruptured spleen causes internal bleeding, a life-threatening condition.

Atypical Lymphocyte
Epstein-Barr virus (EBV) infection of B cells produces the large atypical lymphocytes found in blood smears of patients with infectious mononucleosis.

FOLLOW-UP By week 4, Mira called the clinic to see whether she could start playing soccer again. She was asked to come to the clinic for a follow-up visit first. The NP examined Mira and told her she could return to playing soccer because her spleen was not enlarged. Mira was happy to go back to her soccer practice that afternoon and resume her normal life.

CHAPTER OBJECTIVES

After reading this chapter, you will be able to:

• Correlate the anatomy and physiology of the circulatory system with the infectious processes affecting them.

• Describe symptoms that can help differentiate circulatory system infections caused by viruses from those caused by bacteria and parasites.

• Relate prevention strategies based on transmission routes for microbes that cause systemic infections.

• Choose appropriate treatment and prevention modalities for microbes that cause systemic infections.

Mira's infectious mononucleosis (IM or mono for short) is a good example of a **systemic infection**, an infection that starts in one part of the body but eventually affects many other sites. Microbes that cause systemic infections travel from an initial infection site throughout the body, using blood or lymph as a highway. The signs and symptoms of systemic disease can arise from dissemination of an infection or from the generalized response of the immune system to a localized infection. The virus that causes mono, Epstein-Barr virus, originally infected the epithelium of Mira's oropharynx, causing a sore throat. Then the virus spread through her body and caused her systemic problems: enlarged lymph nodes, mild fever, fatigue, and night sweats.

Many systemic infections are bacterial diseases, such as plague (*Yersinia pestis*), tularemia (*Francisella tularensis*), and septicemia (caused by many species of bacteria). But viruses can cause systemic infections that affect many organ systems as well. Systemic viral diseases include mononucleosis, as in the case just described, but also more frightening diseases, such as the hemorrhagic fevers (Ebola, Marburg), and even some childhood diseases discussed in other chapters (chickenpox, measles). Recall that, in this book, diseases are introduced in the context of the organ system that displays the disease's most prominent sign or symptom (in the context of the skin for chickenpox and measles, for example). This chapter focuses on diseases caused by organisms that primarily infect cells of the circulation or travel in the bloodstream to other locations in the body to infect multiple organs simultaneously.

21.1 Anatomy of the Cardiovascular and Lymphatic Systems

SECTION OBJECTIVES

• List the structural components of the cardiovascular and lymphatic systems and their relationship to one another.

• Compare and contrast the systemic and pulmonary circulations.

• Explain the relationship between the circulatory systems and systemic infections.

What is the most important role of the cardiovascular system? The cardiovascular system provides nutrients and oxygen to all organs of the body, but its most important role is to deliver oxygen to the brain. A quote from Rainer Maria Rilke reads, "All the soarings of my mind begin in my blood." In fact, no one's brain can function, let alone soar, without an ample supply of oxygenated blood. Without a functioning brain, all other organs will fail to function. Fortunately, our body is equipped with a highly efficient circulatory system of blood dispersal, retrieval, and reoxygenation. Fluids and white blood cells that leave the circulation to nourish tissues or fight infections are collected by another network called the **lymphatic system**. To understand how pathogens disseminate via the circulatory highways and byways, we must first review the basic anatomy of these systems.

Anatomy of the Heart

The organ that delivers oxygenated blood is the heart, a pump with four chambers attached to "pipes" (blood vessels). These blood vessels form two circuits: one between the heart and the lungs, called **pulmonary circulation**, and one between the heart and all of the body (including the lungs), called **systemic circulation** (**Figure 21.1**). Pulmonary circulation takes oxygen-depleted blood coming into the heart from all locations of the body and moves that blood to the lungs. In the lungs, blood is replenished with oxygen and then returned to the heart. Systemic circulation then delivers oxygen-replete blood to the entire body and collects oxygen-depleted blood and carries it back to the heart.

The four chambers of the human heart are the right and left atria, which receive the blood coming into the heart, and the right and left ventricles, which pump the blood out of the heart. Each atrium is separated from its ventricle by a one-way valve that allows the blood to flow only from atrium to ventricle. Large vessels bring the blood into or out of the heart. Vessels that move the blood away from the heart are called arteries, and those that bring the blood back to the heart are called veins. Blood flowing in the systemic arteries is oxygenated, but the blood in the systemic veins is oxygen depleted and has a high concentration of CO_2.

Figure 21.1 The Cardiovascular System

Systemic viral infections
Infectious mononucleosis
Burkitt's lymphoma
Cytomegalovirus infections
Dengue fever

Viral infections of the heart
Cardiomyopathy
Pericarditis
Myocarditis

Systemic bacterial infections
Systemic inflammatory
  response syndrome (SIRS)
Bacteremia
Sepsis
Septicemia
Septic shock
Plague
Lyme disease
Typhoid fever
Rocky Mountain spotted fever
Tularemia
Cat scratch fever

**Bacterial infections of
  the heart**
Subacute bacterial
  endocarditis
Infectious endocarditis
Acute bacterial endocarditis
Prosthetic valve endocarditis

Systemic parasitic infections
Malaria
Babesiosis
Chagas disease
Toxoplasmosis
Leishmaniasis
Filariasis

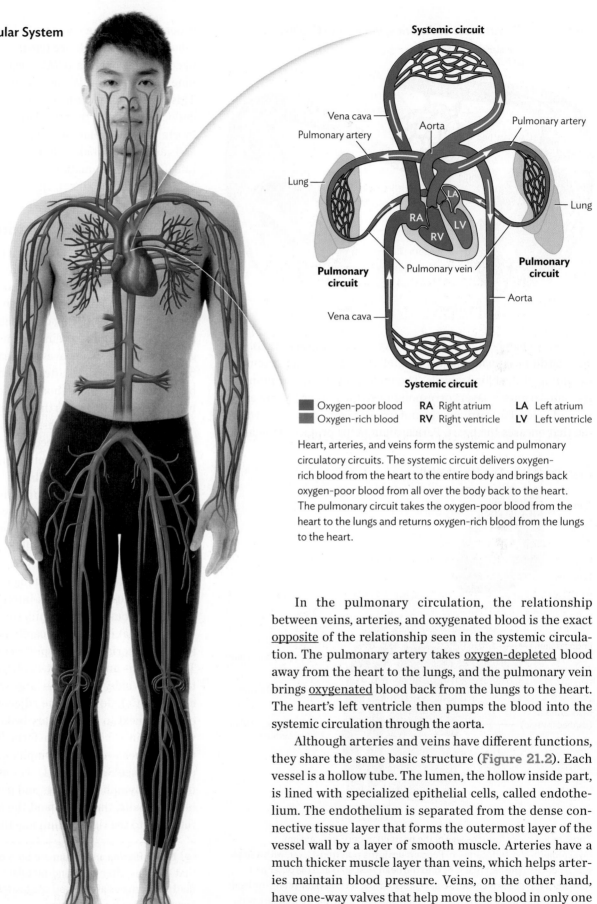

▮ Oxygen-poor blood	**RA** Right atrium	**LA** Left atrium
▮ Oxygen-rich blood	**RV** Right ventricle	**LV** Left ventricle

Heart, arteries, and veins form the systemic and pulmonary circulatory circuits. The systemic circuit delivers oxygen-rich blood from the heart to the entire body and brings back oxygen-poor blood from all over the body back to the heart. The pulmonary circuit takes the oxygen-poor blood from the heart to the lungs and returns oxygen-rich blood from the lungs to the heart.

In the pulmonary circulation, the relationship between veins, arteries, and oxygenated blood is the exact <u>opposite</u> of the relationship seen in the systemic circulation. The pulmonary artery takes <u>oxygen-depleted</u> blood away from the heart to the lungs, and the pulmonary vein brings <u>oxygenated</u> blood back from the lungs to the heart. The heart's left ventricle then pumps the blood into the systemic circulation through the aorta.

Although arteries and veins have different functions, they share the same basic structure (**Figure 21.2**). Each vessel is a hollow tube. The lumen, the hollow inside part, is lined with specialized epithelial cells, called endothelium. The endothelium is separated from the dense connective tissue layer that forms the outermost layer of the vessel wall by a layer of smooth muscle. Arteries have a much thicker muscle layer than veins, which helps arteries maintain blood pressure. Veins, on the other hand, have one-way valves that help move the blood in only one direction.

Figure 21.2 Structure of Arteries, Veins, and Capillaries

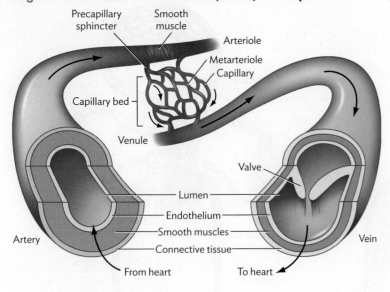

Heart chambers have three layers as well (**Figure 21.3**). The endocardium is made mainly of endothelial cells and resembles the lining of the blood vessels. The middle layer, myocardium, is a specialized muscle layer corresponding to the muscle layer of arteries and veins. Pericardium, the outermost layer, is a double-walled sac that encases the heart. Pericardium helps protect the heart as

Figure 21.3 The Heart Wall

Three tissue layers form the wall of the heart chambers: pericardium, myocardium, and endocardium. Pericardium is the outermost layer, with a small space between the part that is next to the heart (visceral) and the outer part (parietal); this space is called the pericardial cavity. Infections affecting the heart can cause inflammation and accumulation of fluids inside the pericardial cavity, which puts pressure on the heart muscle and impedes proper contraction.

it beats against the chest wall. Normally, a small amount of fluid (about 25 ml) fills the sac (cavity). However, some viral infections, like cytomegalovirus (CMV), cause an increase in the fluid volume in the pericardial sac, a condition called pericardial effusion. This condition leads to chest pain and shortness of breath. The endocardium is further specialized at the origin of the great vessels (aorta and pulmonary artery) and the junctions between the atria and ventricles to make one-way valves. Serious diseases result from infection or inflammation of different layers of the heart wall or any of its valves.

Anatomy of the Circulation

The aorta, the body's largest artery, originates from the left ventricle and delivers oxygen- and nutrient-rich blood to the entire body. The pulmonary artery originates from the right ventricle and takes the deoxygenated blood from the heart to the lungs to become replete with oxygen; this oxygenated blood is returned to the left atrium via pulmonary veins. A one-way valve at the origin of each great vessel (where it leaves the heart) prevents blood from coming back into the ventricles after it has been pumped out. Deoxygenated blood collected from all body tissues returns to the right atrium by two large veins, known as the superior and inferior vena cava.

As the large blood vessels branch out, they become successively smaller until they become tissue capillaries consisting of just one layer of endothelial cells. Small branches of arteries called arterioles connect to capillaries through smaller blood vessels named metarterioles. The pressure in the arterial part of capillaries is still high enough to move plasma out of the vessel and into the tissue so the tissue can get food and oxygen. The pressure at the venous part of the capillaries is lower than that at the arterial side. The low pressure on the venous side allows oxygen- and nutrient-depleted plasma, along with some waste products generated by tissues, to flow back into the venous capillaries. The arterial and venous sides of capillaries are attached to each other and make up what is called the capillary bed. Not all the fluid that flows from the arterial side of capillaries flows back into the venous side. The oxygen- and nutrient-depleted plasma that did not get into the venous side, along with large waste molecules and white blood cells (WBCs), flow into the adjacent small lymphatic vessels that originate next to the capillary beds (**Figure 21.4**). Small lymphatic vessels join each other to form larger ones. The lymph travels through these vessels to lymph nodes and becomes richer in lymphocytes (see Section 15.1). Eventually, the lymph flows through the larger lymphatic vessels and pours into the large veins near the heart. Finally, the lymph and the oxygen-depleted blood are then returned to the right atrium together.

⊙ **Note:** Plasma and serum are both blood derivatives. Plasma is the fluid that remains after cells and platelets have been removed. Serum is the fluid that remains after blood coagulates, so serum lacks cells, platelets, and factors needed for coagulation.

Figure 21.4 The Lymphatic System

Tonsils

Lymph nodes of the neck

Thoracic duct

Thymus gland

Lymph vessels of the chest

Lymph nodes of the axillary region

Spleen

Lymph vessel of the abdomen

Lymph nodes of the groin

Appendix

Lymph nodes behind the knee

A. The lymphatic system is a network of organs, lymph nodes, lymph ducts, and lymph vessels; this network collects and transports lymph from tissues to the bloodstream.

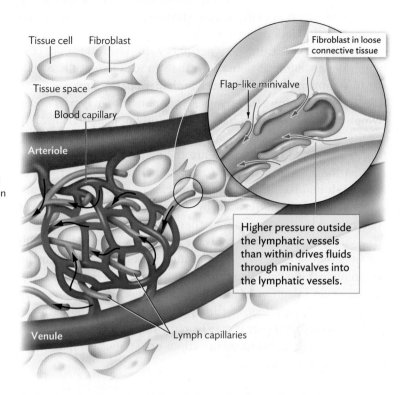

Tissue cell   Fibroblast

Tissue space

Blood capillary

Arteriole

Fibroblast in loose connective tissue

Flap-like minivalve

Higher pressure outside the lymphatic vessels than within drives fluids through minivalves into the lymphatic vessels.

Venule   Lymph capillaries

B. The lymphatic capillary beds are adjacent to the capillary beds. Pressure in the interstitial space drives fluid through minivalves into the lymphatic vessels.

The constant flow and high pressure of the blood makes it exceedingly difficult for microbes to attach themselves to the heart's structures, but cardiac infections do occur. Viral and bacterial infections of the cardiac muscle often inflict irreparable damage. The infected muscle can lose its ability to contract properly (a condition called **cardiomyopathy**) and can even fail.

General Aspects of Systemic Infections

How common are direct infections of cardiovascular and lymphatic tissues? Cardiovascular tissue infections are uncommon; when they occur, they usually develop as an existing localized infection spreads. Microorganisms or their toxins can affect any of the heart's three layers, cause inflammation, and inflict damage. Inflammation of endocardium (**endocarditis**), myocardium (**myocarditis**), or pericardium (**pericarditis**) is often seen as a complication of infectious diseases. On rare occasions, all three tissue layers of the heart are affected at the same time—a condition called **pancarditis**.

More commonly, infectious agents that are able to hitch a ride in the blood or lymphatic flow will find themselves free to access different body tissues and organs and cause a variety of systemic symptoms. General signs and symptoms of a systemic infection usually include fever, malaise, and muscle aches, along with more organ-specific manifestations, depending on which organs are affected.

Lymph flows from the periphery to (and through) the lymph nodes before it pours into the venous circulation. Swelling of lymph nodes is termed **lymphadenopathy** and can have infectious or noninfectious causes. Inflammation of lymph nodes (**lymphadenitis**) or lymphatic vessels (**lymphangitis**) usually results from an infection—viral, bacterial, fungal, or parasitic (Figure 21.5). Skin infections are the major cause of lymphangitis as the infection travels from the primary infected site to the lymph nodes. The most common causes of lymphangitis are acute skin infections caused by streptococci or staphylococci (bacterial skin infections are further discussed in Section 19.3).

Lymph nodes serve as traps for organisms carried away from the infection site by lymph. The numerous lymphocytes in the lymph nodes help fight these organisms (described in Chapter 16). Spread of an infection to a lymph node causes an inflammatory reaction: the lymph node swells (lymphadenopathy) and becomes tender, and the skin over it may be red (lymphadenitis). Cellulitis, abscess formation, and blood infections are serious complications of infectious lymphadenitis.

SECTION SUMMARY

- **The circulatory system** consists of the heart and blood vessels and is divided into pulmonary and systemic circuits.

- **Endocarditis, myocarditis, and pericarditis** are terms describing inflammation of various layers of the heart.

- **Capillaries** are the site where nutrients and oxygen are delivered from blood to the tissues and waste products generated by tissues are returned to the blood.

- **Lymphatic vessels collect waste products, tissue fluids, and white blood cells,** channeling them through lymph nodes and ultimately back into the venous system prior to reaching the heart wall.

- **Lymphadenopathy** describes swollen lymph nodes. **Lymphadenitis** describes swollen, inflamed, and painful lymph nodes that are infected. **Lymphangitis** is an inflamed lymph vessel seen as a red streak extending from a localized skin infection.

Figure 21.5 Infections of the Lymphatic System

A. Lymphangitis.

B. Lymphadenitis.

Thought Question 21.1 Why do the red streaks of lymphangitis always move toward the heart?

21.2 Systemic Viral Infections

SECTION OBJECTIVES

- Explain the pathogenesis of each systemic viral infection discussed.
- Relate the diagnoses of systemic viral infections to the pathogenesis of the specific virus causing the disease.
- Relate the signs and symptoms of each systemic viral infection to the effect of the respective virus on different organ systems.
- List the treatment and prevention options for different systemic viral infections.

How dangerous are systemic viral infections? Well, it depends on the health of your immune system. Systemic viral infections are uncommon in the general healthy population in North America, but when they occur, they are caused mostly by one of three rather famous members of the herpes family of viruses: Epstein-Barr virus (EBV), the cause of infectious mononucleosis (mono); herpes simplex virus (HSV), the cause of cold sores and genital herpes; and cytomegalovirus (CMV), the cause of an infection similar to mono. All three viruses establish latency (meaning they remain present without producing any symptoms) after infection, regardless of a person's immune status, and all three cause severe disease and complications in immunocompromised patients. Immunocompetent individuals, on the other hand, tend to be asymptomatic or to have a milder presentation of these viral diseases. We limit our discussion of the herpes family of viruses in this chapter to EBV and CMV. Herpes simplex viruses 1 and 2 cause primary infections, discussed in Sections 19.2 and 23.5, respectively. Systemic dissemination of HSV occurs in immunocompromised individuals and results in viral meningitis (see Section 24.3).

Other viruses, some old and some emerging, can also cause systemic disease (Table 21.1). Most are rarely encountered in North America, but climate change, the ease of travel, and the wide-reaching routes of international commerce are open doors that may allow any of these illnesses to reach our shores. Knowing about them is the single most effective preventive tool.

Infectious Mononucleosis: The "Kissing Disease"

What is the "kissing disease"? The virus that causes **infectious mononucleosis**, the **Epstein-Barr virus (EBV)**, is shed in the saliva of the person harboring it.

Table 21.1
Systemic Viral Diseases

Disease	Symptoms	Complications	Virus	Transmission	Diagnosis	Treatment	Prevention
Infectious mononucleosis	Pharyngitis, fever, lymphadenopathy, fatigue, splenomegaly	Bacterial throat infections, spleen rupture, anemia, nervous system problems, death	Epstein-Barr virus (EBV)	Saliva	Testing for antibodies to EBV antigens	Treatment of symptoms; activity restriction	No vaccine available
Cytomegalovirus mononucleosis-like syndrome[a]	Similar to those of infectious mononucleosis	Transplacental transmission causing CID; retinal disease and a variety of other complications possible in newborns and immunocompromised patients	Cytomegalovirus	Saliva	"Owl's eye"; inclusion bodies; serology; viral culture; DNA studies		No vaccine available
Cytomegalic inclusion disease (CID)	Premature birth; small head size; seizures; lung, liver, or spleen problems at birth	Deafness, mental retardation, seizures, jaundice, purpura	Cytomegalovirus	Transmitted vertically from mother to fetus			
Dengue fever, dengue hemorrhagic fever (DHF)	Fever, nausea, and vomiting; rash; aches and pains — Additional symptoms in DHF: severe hemorrhage, respiratory distress, organ failure, and shock	Progression to severe dengue with intense severe abdominal pain, liver inflammation, persistent vomiting, hemorrhage, abnormally low blood pressure, death	Dengue virus	*Aedes* species mosquitoes; vertical transmission possible but rare; exposure to infected blood or transplantation	Virus isolation; PCR; serology; positive tourniquet test[b]	Hospitalization; supportive care, including hydration, pain management, and blood transfusion in DHF cases	Vaccine available but used only under certain conditions
Yellow fever	High fever, chills, joint and muscle pain, nausea, vomiting	Progression to toxic form of the disease; vomiting blood; jaundice; petechiae or purpura; bleeding from mouth, nose, eyes; multiorgan failure and death	Yellow fever virus	*Aedes aegypti* mosquito	Virus isolation; serology	Hospitalization to control bleeding, prevent shock, and provide hydration and supportive care	Live, attenuated vaccine for travelers, residents, and laboratory staff
Lassa fever	Fever, retrosternal and generalized pain, vomiting, diarrhea, encephalitis, neurological symptoms, mucosal bleeding	Various degrees of hearing loss that becomes permanent in 1/3 of cases; spontaneous abortion	Lassa virus	Contact with *Mastomys* rodent urine or droppings or with blood or body fluids of an infected person	Virus isolation, detection of anti–Lassa virus antibodies, Lassa antigen, qRT-PCR[c]	Hospitalization, supportive care, antiviral therapy with ribavirin	No vaccine available
Chikungunya fever (CHIK fever, knuckle fever)	High fever, rigors, severe joint pain, rash, nausea, vomiting	Persistent joint pain for several months or years in some patients; occasional neurological, heart, or GI problems	Chikungunya virus	*Aedes aegypti* or *Aedes albopictus* (Asian tiger mosquito)	Virus isolation; serology	Treatment of symptoms	Experimental vaccine in clinical trials

(continued on next page)

Table 21.1
Systemic Viral Diseases (*continued*)

Disease	Symptoms	Complications	Virus	Transmission	Diagnosis	Treatment	Prevention
Crimean Congo hemorrhagic fever (CCHF)	High fever; back, joint, and stomach pain; vomiting, sore throat, petechiae on the palate, sore eyes; bleeding and jaundice in severe cases	Persistent joint pain for months to years	Nairovirus, also known as Asian Ebola virus and Congo virus	Ixodid (hard) ticks; contact with blood, body fluids, and excretions of infected animals or humans	ELISA,[d] qRT-PCR; isolation of virus in culture	Supportive treatment; investigational drugs available but not approved by FDA	No worldwide vaccine available
Rift Valley fever	Mild form: flu-like symptoms; some mild cases with retinal lesions Severe form: meningoencephalitis, heart failure, jaundice, purpura, hemorrhage	Inflammation of the retina, leading to visual loss in 1%–10% of cases	Rift Valley fever virus	Direct or indirect contact with blood or organs of infected animals; *Aedes* and *Culex* mosquitoes; consuming unpasteurized or uncooked animal products from infected animals	Virus isolation; ELISA for presence of IgM antibodies specific to virus; qRT-PCR	Supportive treatment	No vaccine currently available
Marburg hemorrhagic fever	Acute high fever with severe headache, severe malaise, severe hemorrhage from several sites, shock	Prolonged recovery, inflammation of the testis, spinal cord, eye, or parotid gland; prolonged hepatitis, death	Marburg virus	Contact with unknown reservoir in nature or close contact with a person with Marburg virus infection or with their body fluids	Antigen detection by ELISA; qRT-PCR; serology in patients who are recovering; skin biopsy for postmortem diagnosis	Supportive treatment	No vaccine available
Ebola hemorrhagic fever	Acute high fever with severe headache, severe malaise, nausea, vomiting; lymphadenopathy; altered mental status; severe hemorrhage from several sites; shock	Death; prolonged recovery for the rare survivors, with possible sensory problems, headaches, hepatitis, and testicular inflammation	Ebola virus	Contact with unknown reservoir in nature or close contact with a person with Ebola infection or with their body fluids	Clinical presentation; ELISA; antigen detection; serology; isolation of virus in culture	Supportive treatment	Vaccine available

[a]Also called heterophile-negative mononucleosis.
[b]Blood pressure cuff is left inflated at halfway between systolic and diastolic pressure for 5 minutes. Appearance of 20 or more petechiae on a 2.5-cm² patch of skin on forearm is interpreted as a positive test.
[c]qRT-PCR = quantitative reverse transcription–polymerase chain reaction.
[d]ELISA = enzyme-linked immunosorbent assay.

Exchange of saliva, as often happens with kissing, passes the virus from one person to another, earning this infection the title "kissing disease." Once Mira, the patient in the opening case history, realized she had mononucleosis, it didn't take her long to recall a certain kiss with a certain boy who was not feeling well. The disease was first described as an infectious disease by Nil Filatov in 1887 (and independently again by Emil Pfeiffer in 1889). Pfeiffer coined the name "glandular fever" because of its hallmark symptoms of fever, sore throat, and enlarged lymph nodes (**Figure 21.6A**). In 1920, Thomas Sprunt and Frank Evans described seeing an elevated number of monocytes in the blood of patients suffering from glandular fever, and the disease became known as infectious mononucleosis in the United States. Scientists later discovered that the "monocytes" were in fact misshapen B cells infected with the virus. EBV was identified as the etiologic agent of infectious mononucleosis in 1967 when a technician working on a cell line harboring EBV developed

Figure 21.6 Infectious Mononucleosis and Epstein-Barr Virus

A. Symptoms of mononucleosis include swollen cervical lymph nodes (often bilateral), sore throat, fever, and fatigue.

B. A normal lymphocyte in peripheral blood.

C. Large, irregular, atypical lymphocytes (resembling monocytes) seen in the peripheral blood of a patient with infectious mononucleosis.

infectious mononucleosis and subsequently developed antibodies to the virus.

Epstein-Barr virus is a member of the herpes family of viruses, also known as human herpesvirus 4 (HHV-4). More than 95% of adults worldwide are infected with EBV. In fact, nearly all children in developing countries are exposed to EBV by age 4. The majority of those infected (more than 70%) will have no symptoms thanks to an effective cellular immune response. On the other hand, all who do become infected with EBV will shed the virus in their saliva for many weeks until the infection becomes latent. The latent infection will last the rest of the victims' lives, making them a reservoir for the virus. Humans are the only known reservoir.

EBV initially infects and replicates in the oropharyngeal epithelium, tonsils, and salivary glands. B cells become infected when they come in contact with the epithelium of the oropharynx or crypts of the tonsils. The virus is not found in the blood, but the infected B cells spread the infection throughout the body. Infected B cells in blood smears have an atypical appearance that resembles that of monocytes. This mistake led to the disease name, mononucleosis (Figure 21.6B and C). Normally, the immune system controls EBV infection by using mainly cytotoxic T cells. The proliferation of B and T cells results in the enlargement of lymph nodes (lymphadenitis) and of organs such as the liver and spleen that remove old or abnormal cells from the peripheral circulation.

What kind of laboratory test might distinguish mononucleosis from other lymphadenopathies? A rapid blood test called the "monospot test" may be used. EBV nonspecifically stimulates all infected B cells to proliferate and differentiate into plasma cells and long-lived memory B cells (see Section 16.4). Latently infected memory B cells are the reason EBV produces a lifelong infection. The plasma cells produced from EBV-infected B cells will secrete whatever antibody their B-cell progenitors were programmed to make. Since a wide variety of B cells are infected, a variety of different antibodies will be produced. These include EBV-specific antibodies as well as an assortment of nonspecific antibodies. Among the nonspecific antibodies are those that react against horse and sheep red blood cells (RBCs), collectively known as **heterophile antibodies**. The presence of heterophile antibodies is the basis of the monospot test used to screen for mono, as was the case for Mira. The monospot test is a latex agglutination reaction (see Section 17.2) in which patient serum is added to latex beads coated with horse RBC antigens. The monospot test can produce both false-positive and false-negative results and is now used as a screening tool in patients with clear signs and symptoms of mono. Tests for EBV-specific immunoglobulin M and G (IgM and IgG) antibodies are available and are used to confirm diagnosis.

Anyone can be infected with EBV at any age. The age of initial infection seems to dictate the disease's outcome. Children infected with EBV are usually asymptomatic or have fewer symptoms. Adults infected with EBV, on the other hand, are at higher risk of neurological complications such as encephalitis (brain inflammation).

In the United States, the typical EBV infection occurs mainly in young adults and produces a sore throat, fever, and generalized lymphadenopathy that often includes the lymph nodes behind the ear, on the neck, under the arm, or in the groin. Disease symptoms appear about 5 weeks after infection. Most patients with infectious mononucleosis also suffer from extreme fatigue and difficulty concentrating. Examination of the throat usually shows enlarged tonsils with exudates (see the chapter-opening case history), which can lead to a mistaken diagnosis of strep throat.

Most cases of infectious mononucleosis are mild and resolve by themselves after 2–4 weeks. The fatigue and difficulty concentrating associated with most cases of the disease, however, can linger months after recovery from the acute illness.

Figure 21.7 Burkitt's Lymphoma

A. Front view of the face of a child showing a large tumor in the jaw due to Burkitt's lymphoma.

B. The surface of the skin has begun to ulcerate over the growth. It is thought that malaria in childhood alters the body's response to the virus, causing cells to become cancerous.

The dissemination of EBV can unfortunately cause a variety of other complications, including liver enlargement (**hepatomegaly**) and jaundice (especially in patients over 40 years of age) or enlargement of the spleen (**splenomegaly**), which is seen in 50%–60% of patients. The possibility of splenic rupture explains why the nurse practitioner warned Mira about playing sports. In addition to affecting the liver and spleen, EBV can cause neurological problems, such as meningitis, encephalitis, or Guillain-Barré syndrome (see Chapters 22 and 24), or even kidney failure. About 5% of patients may develop a rash, usually on their arm and body. The neurological complications of EBV infection are more common in adults and rare in children.

Symptoms of mono are very similar to those of bacterial pharyngitis (sore throat). A patient with pharyngitis may be empirically treated with ampicillin because the clinician initially suspects strep throat. Patients with mono who are mistakenly treated with ampicillin almost always develop a rash, but this does not mean that they will have a reaction to penicillin in the future. Once the antibiotic is stopped, the rash subsides. The exact cause of the rash is unknown. Mononucleosis is typically a self-limiting disease requiring only supportive care, such as activity restriction and pain management. Death due to infectious mononucleosis is rare but can happen as a result of splenic rupture or bacterial superinfections.

Cancers Associated with Epstein-Barr Virus

Epstein-Barr virus is associated with a number of cancers of lymphoid origin, including Hodgkin's disease, nasopharyngeal carcinoma, and perhaps tonsillar carcinoma, but the strongest association is with endemic **Burkitt's lymphoma** (Figure 21.7). A lymphoma is a cancer of lymphocytes, and Burkitt's lymphoma is a cancer of B lymphocytes in which tumors form primarily in the upper and lower jaws and orbit of the eye. EBV, as a result of its integration into human DNA, generates DNA rearrangements that can move a cell proliferation gene next to a highly expressed antibody promoter (see Section 17.4).

Burkitt's lymphoma was first described by Denis Parsons Burkitt (Figure 21.8A), a surgeon who in 1961 presented the results of his research in Uganda in a talk titled "The Commonest Children's Cancer in Tropical Africa—A Hitherto Unrecognized Syndrome." The "endemic pediatric" cancer he described eventually became known as Burkitt's lymphoma. Among attendees of the lecture was a pathologist and electron microscopist, Michael Anthony Epstein. In 1964, Michael Epstein and Yvonne Barr (Figure 21.8B), along with Bert Achong, reported finding a new strain of virus in cultured lymphoblasts from Burkitt's lymphoma. This new virus is what we now know as the Epstein-Barr virus.

Figure 21.8 Making the Connection between a Cancer and a Virus

A. Denis Parsons Burkitt first described the lymphoma that bears his name.

B. Michael Epstein (left) and Yvonne Barr (right), along with Bert Achong, discovered the viral cause of Burkitt's lymphoma. The link to mononucleosis was discovered by sheer chance, when a technician handling a Burkitt's lymphoma cell line developed mono.

There are three different categories of Burkitt's lymphoma based on disease presentation: endemic (found exclusively in Africa), sporadic (found worldwide), and immunodeficiency associated. The EBV virus is found in 95% of the endemic variety of Burkitt's lymphoma, compared to 10%–20% of the sporadic cases, and in 40%–50% of the immunodeficiency-related cases (individuals infected with HIV). The endemic variety of Burkitt's lymphoma (eBL) is the most common type of childhood cancer in equatorial Africa, where malaria is also prevalent. We know that malaria increases the risk of endemic Burkitt's lymphoma, but we do not know why. The immunodeficiency-related variety of Burkitt's lymphoma is seen mostly in HIV-positive patients, even those with high CD4+ T-cell counts (see Section 23.4). This finding suggests that immune suppression by HIV (low CD4+ count) is not the only mechanism by which HIV-positive patients develop these tumors.

Burkitt's lymphoma grows rapidly and should be aggressively treated with a variety of anticancer chemotherapies. Treatment for EBV itself is not necessary because once the cancer is detected, the chromosomal rearrangement has already occurred and the virus plays no further role.

Cytomegalovirus Infections

How are cytomegalovirus infections different from other herpesvirus infections? **Cytomegalovirus (CMV)** is a double-stranded DNA virus, also called human herpesvirus 5 (HHV-5). CMV initially replicates in epithelial cells (typically in the salivary glands) but eventually infects circulating lymphocytes and monocytes, the kidneys, and different glands. CMV, like most herpesviruses, can become latent in host cell nuclei as episomes (Chapter 18). The difference between CMV and many other herpesvirus infections lies in the severity of disease. CMV infections are usually asymptomatic or mild, except when they affect fetuses or someone who is immunocompromised.

CMV infections are incredibly common worldwide. In the United States, 50%–80% of adults are infected with the virus by the time they are 40 years old. We know this because the virus leaves "footprints" in the form of antibodies. CMV infection usually happens in childhood. People infected with CMV are usually asymptomatic but will shed the virus in different body fluids for a long time (months to years in urine and saliva). The viral shedding period is generally longer for children than for adults. The main mode of transmission is contact with contaminated saliva, particularly in day-care centers and among family members. Transmission by sexual contact or from transplanted tissue is also possible. Primary CMV infection is mostly asymptomatic in immunocompetent people, but a minority of healthy people may develop an illness resembling mononucleosis—known as CMV mononucleosis-like syndrome or heterophile-negative mononucleosis—with similar nonspecific symptoms.

THE IMMUNE RESPONSE TO CYTOMEGALOVIRUS INFECTION
Cell-mediated immunity is the main defense against primary CMV infection following initial contact, and it stops secondary infections when latent CMV virus is reactivated. Consequently, suppressing cellular immunity can lead to serious primary as well as secondary infections. Reactivation of latent CMV viral DNA can happen anytime in life. Secondary CMV infection can also develop when a person harboring a latent CMV becomes infected with a new strain of virus, a process called **reinfection**.

Anything that suppresses the immune system is a major risk factor for CMV reactivation, so patients on immunosuppressive therapy, such as those on certain medications after an organ transplant or with medical conditions such as AIDS, are at higher risk of reactivation and symptomatic infection. Symptomatic CMV disease in immunocompromised individuals can affect almost every organ of the body, resulting in a fever of unknown origin, pneumonia, hepatitis, encephalitis, myelitis, colitis, uveitis (inflammation of the eye's interior), retinitis, and neuropathy.

CONGENITAL CYTOMEGALOVIRUS INFECTION Unlike most viral pathogens, CMV can pass transplacentally from infected mother to fetus. The reported vertical transmission rate is nearly 75%, but the great majority of infected newborn infants (80%) are asymptomatic. Symptomatic infected newborns are said to have **congenital cytomegalovirus infection** or **cytomegalic inclusion disease (CID)**. Generalized infection can occur in the infant, causing complications such as low birth weight, microcephaly (abnormally small head), seizures, a petechial rash similar to the rash of congenital rubella syndrome, and moderate hepatosplenomegaly (enlarged liver and spleen) with jaundice. Though severe cases can be fatal, with supportive treatment most infants with CMV disease will survive. However, from 80% to 90% of severe cases will have complications within the first few years of life. The most common abnormality associated with CID is hearing loss, but the disease also remains one of the leading causes of mental retardation in United States.

Note: CMV is a member of the so-called TORCH microbes that can cross the placenta and infect a fetus to cause severe congenital defects. The **TORCH complex** includes *Toxoplasma*, rubella, cytomegalovirus, and herpes simplex 2. The O stands for others, such as coxsackievirus, syphilis, varicella-zoster virus, HIV, parvovirus B19, and *Listeria*.

CID was initially named in 1905 when researchers noticed that cells obtained from infants with the disease looked distinctly different from other cells under the microscope. The infected cells were much larger than normal (sometimes called **giant cells**) and had inclusion bodies that gave them the appearance of an owl's eyes (**Figure 21.9**). The owl's-eye appearance remains an indication of CMV infection even today. Congenital CMV infection results in permanent problems in nearly 20% of people affected. Epstein-Barr virus and cytomegalovirus infections can be distinguished from each other by specific diagnostic approaches.

Figure 21.9
Cytomegalic Inclusion Disease

A hematoxylin- and eosin-stained tissue sample from a lung infected with CMV shows the resulting enlargement of an infected cell and two large inclusion bodies that have the appearance of an owl's eyes.

LM   20 µm

Other Systemic Viral Diseases

Though uncommon in the United States, several serious systemic viral diseases are seen in other parts of the world. Some of the viral agents—such as the four closely related viruses that cause dengue fever (DENV-1–4) from the Flaviviridae family—impose a tremendous burden on health and resources. Dengue fever is endemic throughout the tropics and subtropics. Victims of dengue often have contortions due to intense joint and muscle pain. For this reason, dengue fever is also known as "breakbone fever." Dengue viruses are called arboviruses because they are transmitted by an arthropod (an insect vector such as a mosquito). The World Health Organization (WHO) and Centers for Disease Control and Prevention (CDC) estimate that, of almost 400 million people worldwide who are infected with dengue each year, 100 million develop symptoms and 22,000 die from severe dengue, known as **dengue hemorrhagic fever (DHF)**. There were 2.35 million cases of dengue fever in the Americas in 2015 (mostly in South America and the Caribbean). Among reported cases of dengue fever, 10,200 involved the most severe variant of the disease, DHF. The CDC reported 408 cases of dengue virus infection in the United States during the first 8 months of 2019.

Reemergence of certain systemic viral diseases is especially troubling. The Chikungunya virus, or CHIKV, of the Togaviridae family, for example, is reemerging. CHIKV causes millions of cases, which are characterized by debilitating polyarthralgia (pain in several joints) and fever, in areas around the Indian Ocean, and the virus has now spread to Europe. Ebola virus—a lipid-enveloped, twisted, threadlike RNA virus (see Figure 2.5A) that is endemic to Africa—has a frightening reputation for the hemorrhagic fever it causes. Disease outbreaks of Ebola are characterized by acute (rapid) onset of fever, severe muscle pains, and horrible bleeding from multiple orifices (nose, mouth, anus, and vagina; Figure 2.19) and often result in death (50%–60% mortality). The virus is easily transmitted person-to-person by contact with contaminated body fluids. In June 2019, the CDC activated emergency protocols as a result of a new outbreak of Ebola in the Democratic Republic of Congo. The largest Ebola outbreak included 28,600 cases and claimed 11,325 lives in West Africa during a 2.5-year period that ended in 2016 after a 2-year intense effort. During outbreaks, the disease can easily spread worldwide when a few foreign volunteers who treat the victims unknowingly become infected with the virus and bring it back to their home countries, including the United States. Fortunately, major containment efforts have been successful in preventing the spread of the disease so far. An effective Ebola vaccine, Ervebo, received approval from the US Food and Drug Administration (FDA) in December 2019 and is indicated for individuals 18 years of age or older.

A large number of other systemic disease–producing viruses are transmitted by either mosquitoes or rodents and often cause high fevers and hemorrhage (see Table 21.1). Simple, practical measures, such as mosquito and rodent control, have tremendous potential to improve the lives of millions around the world, save billions of dollars, and help keep us healthy.

Viral Infections of the Heart

As noted earlier, pericarditis, myocarditis, and endocarditis can result when the various layers of the heart are infected by microbes that have entered the circulation. Viral pericarditis is more common than bacterial, which will be discussed later. Viral infection of the pericardium leads to an inflammation and a collection of fluid in the pericardial cavity that increases intrapericardial pressure, resulting in chest pain and low fever. Among the many viral causes of pericarditis and myocarditis are coxsackieviruses (particularly coxsackievirus B; see Section 19.2), echovirus, Epstein-Barr virus, and HIV (HIV is covered further in Section 23.4). Infections of the myocardium can make heart muscle lose its ability to contract properly (Figure 21.10). Treatment of viral heart infections is usually supportive and may require draining the pericardial sac for pericarditis. Antiviral therapy has not proven effective and is not typically recommended. Most patients, however, recover.

SECTION SUMMARY

- **Infectious mononucleosis** is caused by the Epstein-Barr virus. Symptoms are extreme tiredness and swollen cervical lymph nodes. Transmission is usually via saliva.

- **Epstein-Barr virus also causes Burkitt's lymphoma,** a B-cell cancer. EBV can integrate into the DNA of a host cell and cause a rearrangement that dramatically stimulates proliferation.

- **Cytomegalovirus** is a herpesvirus that infects epithelial cells, leukocytes, monocytes, and lymphocytes, as well as the kidneys and various glands. CMV disease is typically mild unless the patient is immunocompromised.

- **CMV can be transmitted vertically to a fetus,** causing a generalized infection.

- **Dengue virus, Ebola virus, and Chikungunya virus** also cause systemic disease.

- **Cardiomyopathies** can be caused by echovirus, Epstein-Barr virus, and HIV.

Thought Question 21.2   What is the significance of a patient having a positive EBV IgM test but a negative EBV IgG test?

Figure 21.10 Myocarditis

A. An echocardiogram showing normal-sized heart chambers (LA = left atrium; LV = left ventricle; RA = right atrium; RV = right ventricle).

B. An echocardiogram showing a small left ventricle with thickened walls (arrows). The inflammatory process caused by a viral infection can result in left ventricular wall thickening. Thickening of myocardium reduces the size of the left ventricle, which means the heart cannot pump enough blood into the circulation or even to its own muscle tissue. Eventually, the diminished blood supply severely damages the heart muscle.

21.3 Systemic Bacterial Infections

SECTION OBJECTIVES

- Compare the systemic signs and symptoms of different bacterial circulatory infections.
- Generate a list of organ malfunctions according to the pathogenesis of each bacterial pathogen discussed.
- Relate the pathogenesis of the systemic bacterial infections discussed with the symptoms and progression of each disease.
- Summarize the treatment and prevention plans for the various systemic bacterial infections discussed.

How do bacteria gain access to our circulatory systems? Bacterial infections of the circulation fall into two broad categories: infections that primarily grow in the bloodstream to cause **sepsis** and those that disseminate through the blood to infect a variety of other organs. The latter group often produces lymphadenitis along the way, although not all organisms that produce lymphadenitis produce systemic disease.

Sepsis and Septic Shock

The majority of systemic bacterial infections start innocuously enough when a few microorganisms enter the bloodstream. This happens more often than you think. For example, when a dental

hygienist cleans your teeth, bacteria can enter the bloodstream. The result of this breach is called **bacteremia**. Fortunately, your immune system will quickly dispatch these invaders before they can begin to grow and cause disease.

Septicemia is said to occur when a pathogen that has entered the bloodstream overcomes our innate defensive mechanisms, such as complement, and begins to replicate to high numbers. Pathogens that cause septicemia can enter the bloodstream in a number of ways. Microbes can escape from a local infection (an abscess, for example) situated anywhere in the body, or gain access via a bleeding wound, or enter by transfusion (see **IMPACT**) with contaminated blood or blood products. Gram-positive, Gram-negative, aerobic, and anaerobic bacteria can all produce septicemia under the right conditions. Any septicemia will lead to serious medical emergencies.

One such emergency is **systemic inflammatory response syndrome (SIRS)**, a set of symptoms produced by a variety of infectious and noninfectious causes (**Figure 21.11**). A patient with SIRS has either too high or too low a body temperature, rapid heart and breathing rates, and an abnormal WBC count. SIRS caused by an infection is referred to as **sepsis** (commonly known as blood

Figure 21.11 Overlap of Infection, Sepsis, and Systemic Inflammatory Response Syndrome (SIRS)

Not all infections result in sepsis, and not all SIRS patients have sepsis.

Blood-borne infection

impact

Transfusion-Associated Infections

One of the greatest gifts you can give is blood. Donated blood has saved countless lives worldwide. If you lose excessive blood from a car accident or during surgery, blood is life. But as you can see from this chapter and from others to follow, infections by many microorganisms arise from dissemination in blood. How can we be sure that the blood we receive in an emergency is safe? Consider the case of Travis.

Travis, a 12-year-old African-American male with sickle-cell disease, was brought to the clinic by his older sister for a checkup. He appeared healthy, but lab results of his blood revealed that Travis had a hemoglobin value of 6 (expected range is 13–17) and a hematocrit (ratio of cell volume in the blood to the plasma volume) of 18% (expected is 41%–53%). On the basis of these findings, the clinic doctor decided a transfusion was required. Travis had had transfusions before and knew what to expect (Figure 1). He laid back and started texting his friend about weekend plans. Thirty minutes after starting the transfusion, Travis began feeling ill. He was hot, progressively light-headed, and had chills. The nurse took his temperature and found he had a fever of 39.9°C (103.8°F). When the doctor examined Travis, he noted that Travis was diaphoretic (profusely sweating), his blood pressure was 60/30, and he was pale, agitated, and tachycardic (heart beating at a faster-than-normal rate). What was happening?

Travis was not having a transfusion reaction, which is what you might expect. Travis was suffering from transfusion-related sepsis by a Gram-negative rod, *Yersinia enterocolitica*. How could this happen? Don't hospitals test donated blood for infectious diseases? Well, they do test for some diseases but not all. In this case, testing the blood immediately after donation would probably have missed the culprit pathogen. *Y. enterocolitica* is a food-borne pathogen that can penetrate the intestinal wall, cause lymphadenitis, and disseminate. Most healthy people to whom this happens are asymptomatic because the number of bacteria in the blood is typically very low and the immune system quickly kills the organism. However, if this person donates blood before the bacteremia is cleared, the donated blood could contain low numbers of *Y. enterocolitica*. But a low number of bacteria in blood will not trigger sepsis within 30 minutes. So what happened?

Y. enterocolitica is one of a few pathogens that can actually grow at refrigeration temperature (*Listeria* is another). As the bag of donated blood hung in the refrigerator awaiting use, bacteria grew to high levels, and the amount of LPS present in the blood transfused into Travis's arm was enough to quickly trigger septic shock. Travis was treated with antimicrobial drugs and recovered.

Laboratory testing of donated blood is intended to ensure that transfusion recipients receive the safest possible blood products. As of 2019, such testing consists of determining the ABO blood group and Rh blood type of the donated unit and testing for red cell antibodies to avoid transfusion reactions. In addition, screening for the

Figure 1 Blood Transfusion

A boy with sickle-cell disease being transfused with blood.

infectious disease agents shown in Table 1 is also performed. Nucleic acid testing (polymerase chain reaction, or PCR) and enzyme immunoassay (EIA) are the methods used to screen for these agents. Notice that not all pathogens are screened, and some are screened only under select circumstances. For example, blood is routinely screened for the malarial parasite when donated in an endemic country but not in nonendemic countries such as the United States. This is due, in part, to cost-benefit calculations. Transmission of many diseases by transfusion is rare, and even when they are present in blood, the low numbers of some agents preclude detection. In a case like Travis's, testing is done after a problem arises. The clinic staff immediately disconnected Travis's transfusion bag and sent it to the laboratory for bacteriological examination, which revealed the presence of high numbers of *Y. enterocolitica*.

Table 1
Infectious Agents Screened in Donated Blood

Routinely Screened		Discussed in
HIV-1 and HIV-2	Human immunodeficiency viruses	Chapter 23
HTLV-1 and HTLV-2	Human T-lymphotropic viruses[a]	
HBV	Hepatitis B virus	Chapter 22
HCV	Hepatitis C virus	Chapter 22
WNV	West Nile virus	Chapter 24
Zika	Zika virus	Chapter 24
Treponema pallidum	Agent of syphilis	Chapter 23
Screened in Endemic Countries		
Trypanosoma cruzi	Chagas disease	Chapter 21
Plasmodium	Malaria	Chapter 21
Screened in Blood Intended for Immunosuppressed Patients, Neonates, and Pregnant Women[b]		
CMV	Cytomegalovirus	Chapter 21

[a]Human T-lymphotropic viruses are retroviruses known to cause T-cell leukemia and lymphoma.
[b]CMV screening of blood is not required for immunocompetent donors, but donated blood intended for immunocompromised recipients should be screened prior to release for clinical use.

In this age of rapid transportation and mobility, a US citizen might travel to a country endemic for a disease such as malaria, be infected, and return to donate blood before symptoms arise. But if you have ever given blood, you know that donors must fill out a fairly lengthy questionnaire packed with queries about a donor's travel, sexual habits, drug use, and so on—all kept extremely confidential. A person who just returned from the Amazon rain forest, where malaria is endemic, will declare his travel and then learn that a donation cannot be made for at least 6 months. When disease is acquired through blood transfusion, as in Travis's case, the blood bank will check the donor's records, and all units of that donor's blood will be destroyed.

The CDC is one of the federal agencies responsible for ensuring the safety of the US blood supply through investigations and surveillance. The second agency, the FDA, regulates how blood donations are collected and how blood is transfused.

DISCUSSION QUESTION
Zika virus was recently added to the list of infectious agents screened by blood bank laboratories. What warrants adding a pathogen to that list?

----

poisoning; see Chapter 15 IMPACT). When sepsis progresses to involve the kidneys or the heart, these organs begin to fail, and the patient is said to be in severe sepsis. Severe sepsis can progress to septic shock.

"Shock" is a general term used to describe cardiovascular collapse, meaning the heart fails and blood pressure is not high enough to perfuse the tissues (a condition called **hypotension**, or low blood pressure). **Septic shock**, the catastrophic fall of blood pressure due to severe sepsis, is a complex process. During septic shock, host macrophages or T cells interact with circulating microbial exotoxins called superantigens (see Section 16.3) or microbe-associated molecular patterns (MAMPs; Section 15.2), such as lipopolysaccharide (LPS) in Gram-negative organisms or teichoic acid in Gram-positive ones. The activated host cells then release massive amounts of cytokines that promote inflammation, cause vasodilation, and trigger coagulation.

Vasodilation during sepsis allows fluid to leave the blood vessels, contributing to hypotension. Excessive coagulation that causes clotting in the small blood vessels throughout the body is known as **disseminated intravascular coagulation (DIC)**; this process damages organs and can lead to death. The excessive clotting ultimately uses up a large portion of platelets. As a result, observing a low platelet count is an indicator—along with fever and low blood pressure—of septic shock and DIC. Note that the reference range for platelet count is 130,000–400,000 per microliter of blood. Ironically, the low platelet count results in slower clotting times for these patients. Blood, unable to clot, can leak from vessels below the skin to produce a petechial rash (Figure 21.12).

Organisms commonly associated with septicemia and septic shock include both Gram-negative bacteria (*E. coli*, *Klebsiella*, *Yersinia enterocolitica*, *Salmonella* Typhi, *Francisella tularensis*, and *Neisseria meningitidis*) and Gram-positive bacteria (*Staphylococcus aureus*, *Streptococcus pneumoniae*, enterococci, and *Bacillus anthracis*). Even anaerobes that are normal inhabitants of the intestine (for example, *Bacteroides fragilis*) can be lethal if they escape the intestine and enter the blood, as might happen following surgery. This is one reason surgical patients are given massive doses of antibiotics immediately before and after surgery.

How do we diagnose and treat septicemia? The obvious way to identify the cause of septicemia is to identify the presence of a bacterium in the blood. Blood cultures involve taking samples of a patient's blood from two different locations (such as two different arms; see Section 25.1). Growth of the same organism in cultures taken from two body sites rules out inadvertent contamination with skin flora, which would probably yield growth in only one culture. The patient's blood is added to liquid culture medium and incubated at 37°C (98.6°F). Incubations should be done aerobically

Figure 21.12 Petechial and Purpuric Rashes

A. A petechial rash on a patient's lower leg.

B. A larger, more extensive purpuric rash on the leg of a 9-month-old patient.

and anaerobically. Treatment of septicemia depends on the organism identified. A patient with suspected septicemia is empirically treated, usually with a broad-spectrum cephalosporin antibiotic, until the causative organism is identified.

Systemic Infections Involving Multiple Organs

Several organisms, some of which can also produce sepsis, use the bloodstream as a transit system to infect other organs. We describe only selected multiple-organ diseases here; others are listed in Table 21.2.

CASE HISTORY 21.1

Prairie Dogs Are Not Good Companions

A 25-year-old New Mexico rancher was admitted to an El Paso hospital because of a 2-day history of headache, chills, and fever (40°C, 104°F). The day before admission, he began vomiting. The day of admission, an orange-sized, painful swelling in the right groin area was noted (Figure 21.13A). A lymph node aspirate and a smear of peripheral blood were reported to contain Gram-negative rods that exhibited bipolar staining (the ends of the bacilli stained more densely than the middle; Figure 21.13B). The patient's WBC count was 24,700/μl (reference range is 4,300–10,800/μl), and platelet count was 72,000/μl (expected is 130,000–400,000/μl). In the 2 weeks prior to becoming ill, the patient had trapped, killed, and skinned two prairie dogs, four coyotes, and one bobcat. The patient had cut his left hand shortly before skinning a prairie dog. PCR and typical biochemical testing of a Gram-negative rod isolated from blood cultures identified the organism as *Yersinia pestis*, the organism that causes plague. The patient received an antibiotic cocktail of gentamicin and tetracycline. He eventually recovered after 6 weeks in intensive care.

Table 21.2
Major Systemic Bacterial Diseases

Disease	Symptoms	Complications	Bacteria	Transmission	Diagnosis	Treatment	Prevention
Lyme disease	<u>Stage 1</u>: rash <u>Stage 2</u>: chills, headache, malaise, systemic involvement <u>Stage 3</u>: neurological changes	<u>Untreated patients</u>: neurological problems, arthritis, heart rhythm abnormalities <u>Treated patients</u>: posttreatment Lyme disease syndrome[a]	*Borrelia burgdorferi*	Deer tick	<u>In endemic areas</u>: erythema migrans ("bull's-eye" rash) <u>In nonendemic areas</u>: erythema migrans and positive antibodies against *B. burgdorferi*; positive antibodies against *B. burgdorferi* plus involvement of at least one organ	Amoxicillin; doxycycline	No vaccine available
Brucellosis (undulant fever, Mediterranean fever, Malta fever)	Fever, weakness, sweats	Splenomegaly, osteomyelitis, endocarditis, others	*Brucella abortus*	Contact with animal products or consumption of unpasteurized milk	Positive culture; serology; PCR[b]	Tetracycline plus rifampin is treatment of choice	No vaccine available
Leptospirosis	Fever, photophobia, headache, abdominal pain, skin rash	Liver involvement and jaundice	*Leptospira interrogans*	Contact with urine of infected animals	IgM antibodies against *Leptospira* detected by EIA; MAT; DFM[c]	Erythromycin, penicillin	Vaccine for animals available
Epidemic typhus	Chills, fever, cough, headache, muscle pain, maculopapular rash	Splenomegaly, coma	*Rickettsia prowazekii*	Human louse, flying squirrel, flea	Serology; culture	Tetracycline, chloramphenicol	No vaccine available
Rocky Mountain spotted fever (RMSF)	Fever, headache, rash starting on wrists and ankles spreading to trunk of body	Hearing loss, mental disability, limb or digit amputation	*Rickettsia rickettsii*	Ticks	Serology	Doxycycline	No vaccine available
Tularemia (rabbit fever, deer fly fever)	Fever, chills, headache, muscle pain, abdominal pain, pneumonia, death	No long-term complications	*Francisella tularensis*	Rabbits, rodents, deer flies, ticks, mosquitoes	Serology; PCR; histology of tissue samples	Gentamicin, streptomycin	Vaccine available but not currently in the US
Typhoid fever	Septicemia, chills, fever, hypotension, rash (rose spots)	Intestinal holes (perforation) and/or intestinal bleeding in third week of illness	*Salmonella enterica* serovar Typhi	Fecal-oral route	Blood culture; serology (Widal test)	Fluoroquinolones, ceftriaxone	Vaccines available
Salmonellosis (typhoid-like fever)	Septicemia, chills, fever, hypotension	Bacteremia, arthritis, aneurysm	*Salmonella enterica* serovar Choleraesuis	Fecal-oral route	Blood culture	Fluoroquinolones, ceftriaxone	No vaccine available
Vibriosis	Serious in immunocompromised patients; fever, chills, multiorgan damage, death	No long-term complications	*Vibrio vulnificus*	Contact with seawater, raw oysters	Blood culture	Tetracycline plus aminoglycoside	No vaccine available
Bubonic plague	Buboes (swollen lymph glands), high fever, chills, headache, chest pain, shortness of breath, cough	Pneumonia, septicemia	*Yersinia pestis*	Rodents, rodent fleas, human respiratory aerosol; potential bioterrorism agent	Culture	Levofloxacin; streptomycin or tetracycline	Vaccine available but not currently in US

[a]Fatigue and muscle and joint pain that can last for more than 6 months; sometimes referred to as chronic Lyme disease.
[b]PCR = polymerase chain reaction.
[c]DFM = dark-field microscopy; EIA = enzyme immunoassay; MAT = microscopic agglutination test.

Figure 21.13 Plague

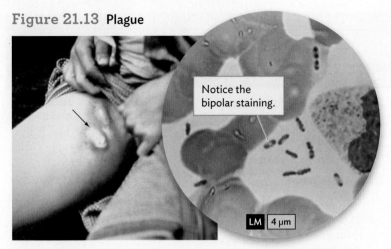

A. Classic bubo (swollen lymph node) of bubonic plague.

B. *Yersinia pestis*, showing bipolar staining using Wright stain.

Notice the bipolar staining.

LM 4 μm

PLAGUE Plague is caused by the bacterium *Yersinia pestis*, an organism that can infect both humans and animals. During the Middle Ages, the disease, known as the Black Death, killed more than a third of Europe's population. Such was the horror it evoked that invading armies would actually catapult dead plague victims into embattled fortresses—perhaps the first use of biowarfare.

Y. pestis is present in the United States and is endemic in 17 western states. It is normally transmitted from animal to animal, typically rodents such as rats and prairie dogs, by the bite of infected fleas. **Figure 21.14** illustrates the infection cycles of the plague bacillus: the sylvatic cycle (transmission in the wild) and the urban cycle (transmission in urban settings). Humans are not usually part of the natural infection cycle. However, in the absence of an animal host, the flea can take a blood meal from humans and thereby transmit the disease to them (an example of a zoonotic disease). During the Middle Ages, urban rats venturing back and forth to the countryside became infected by the fleas of wild rodents that served as a reservoir. Upon returning to the city, the rat fleas passed the organism on to other rats, which then died in large numbers.

Figure 21.14 The Cycles of Plague

The sylvatic cycle (cycle 1) occurs in the wild, where fleas transmit the organism between rodents. An accidental interaction with urban rats can trigger a similar urban cycle (cycle 2). Humans can be infected through contact with infected fleas coming from either cycle. Flea bite transmission initiates bubonic plague symptoms that can progress to pneumonic plague. Pneumonic plague is highly infectious, which can cause epidemic spread of the disease.

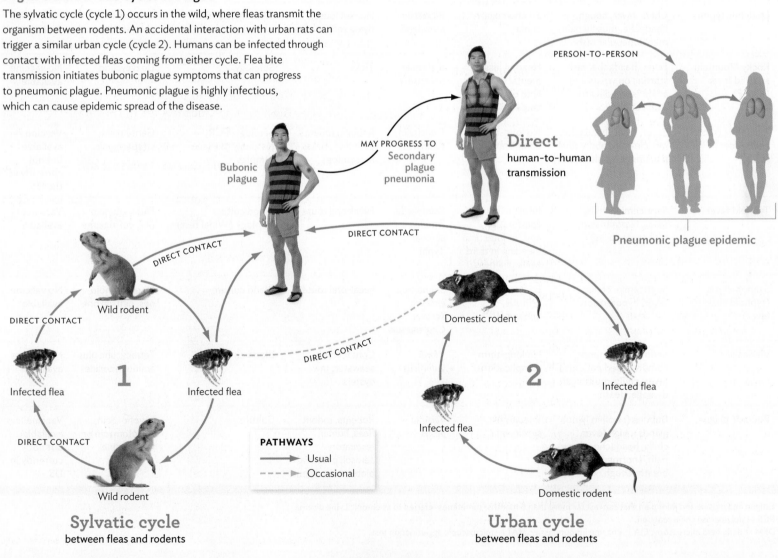

The rat fleas, deprived of their normal meal, were forced to feed on city dwellers, passing the disease on to them.

Individuals bitten by an infected flea or infected through a cut while skinning an infected animal first exhibit the symptoms (like those of the young rancher in Case History 21.1) of **bubonic plague**. Bubonic plague emerges as the organism moves from the site of infection to the lymph nodes, producing characteristically enlarged nodes called buboes (Figure 21.13A). From the lymph nodes, the pathogen can enter the bloodstream, causing **septicemic plague**. In this phase, the patient can go into shock from the massive amount of endotoxin in the bloodstream. Neither bubonic nor septicemic plague is passed from person to person. As the organism courses through the bloodstream, however, it will invade the lungs and produce **pneumonic plague**, which one person can easily transmit to another person through aerosol droplets generated by coughing. Pneumonic plague is the most dangerous form of the disease because it can kill quickly and spread rapidly through a population. Pneumonic plague is so virulent that an untreated patient can die within 24–48 hours. For this reason, the organism is usually identified postmortem.

Y. pestis has numerous virulence factors, including so-called V and W cell-surface lipoprotein antigens that inhibit phagocytosis. Another factor, the F1 protein, a capsule surface antigen, helps block phagocytosis in mammalian hosts. Certain biofilms (see Section 6.6) formed by *Y. pestis* are also important. An extracellular matrix synthesized by *Y. pestis* produces an adherent biofilm in the flea midgut that contributes to flea-to-mammal transmission. The biofilm blocks flea digestion, making the flea feel "starved" even after a blood meal. Therefore, the flea jumps from host to host in a futile effort to feel full. As the flea tries to take a blood meal, the blockage causes the insect to regurgitate bacteria into the wound. This curious effect of *Y. pestis* on the insect vector is another unique reason for how plague spreads so quickly.

Y. pestis uses type III secretion systems (see Section 18.4) to inject virulence proteins (YopB and YopD) into host cell membranes. Injection of the Yop proteins disrupts the actin cytoskeleton and helps the organism evade phagocytosis. By evading phagocytosis, the organism avoids triggering an inflammatory response and produces massive tissue colonization.

Plague has disappeared from Europe; the last major outbreak occurred in 1772. The reason for its disappearance is not known but was probably the result of multiple factors, including human intervention. Although it wasn't until the nineteenth century that doctors understood how germs could cause disease, Europeans recognized by the sixteenth century that plague was contagious and could be carried from one area to another. Beginning in the late seventeenth century, governments set up a medical boundary, or *cordon sanitaire*, between Europe and the areas to the east from which epidemics came. Ships traveling west from the Ottoman Empire were forced to wait in quarantine before passengers and cargo could be unloaded. Those who attempted to evade medical quarantine were shot.

LYME DISEASE

CASE HISTORY 21.2
Bull's-Eye Rash

 Brad, an otherwise healthy 9-year-old from Connecticut, developed a fever and a large (8 cm) reddish rash with a clear center (**erythema migrans**) on his arm (**Figure 21.15A**). He also had some left facial nerve palsy (paralysis). A week before, Brad had returned from a Boy Scout camping trip to the local woods, where he did a lot of hiking. When asked by his physician, Brad admitted finding a tick on his stomach while in the woods but thought little of it. The doctor ordered serological tests for *Borrelia burgdorferi* (the organism that causes Lyme disease), *Rickettsia rickettsii* (which produces Rocky Mountain spotted fever), and *Ehrlichia chaffeensis* (which causes ehrlichiosis). The ELISA for *B. burgdorferi* came back positive, confirming a diagnosis of Lyme disease. The boy was given a 3-week regimen of doxycycline (a tetracycline derivative), which led to resolution of the rash and palsy.

Lyme arthritis was first reported in Lyme, Connecticut, in the 1970s, but the causative organism, *Borrelia burgdorferi*, was not identified until 1982. Since then, **Lyme disease** (also known as **Lyme borreliosis**) has become the most common vector-borne illness in the United States and is considered an emerging infectious disease. The main endemic areas are the northeastern coastal area from Maine to Maryland; Wisconsin and Minnesota; and northern California and Oregon. Lyme disease is also common to parts of Europe and can be found in Sweden, Germany, Austria, Switzerland, and Russia.

B. burgdorferi is a spirochete (**Figure 21.15B**) transmitted to humans by ixodid ticks (deer ticks or hard ticks; **Figure 21.16**). In

Figure 21.15 Lyme Disease

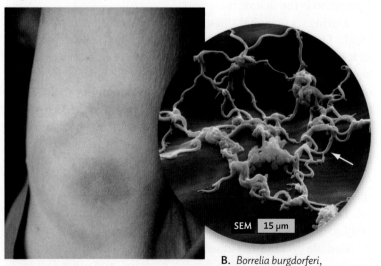

A. Erythema migrans rash.

B. *Borrelia burgdorferi,* the agent of Lyme disease (arrow).

SEM   15 μm

Figure 21.16
Ixodes scapularis **Vector and Its Host Associations**

B. *I. scapularis* (deer tick, also known as blacklegged tick).

Tick stages are drawn actual size. Ticks engorged with blood will be larger.

A. Life cycle of *I. scapularis*. The life cycle from eggs to adult takes 2 years to complete. As the ticks develop, they are attracted to the barberry bush (shown). Removing the bush from wooded areas associated with Lyme disease can dramatically lower the incidence of disease.

the northeastern and central United States, where most cases occur, the deer tick *Ixodes scapularis* transmits the spirochete, usually during the summer months. In the western United States, *I. pacificus* is the tick vector.

During its nymphal stage (the stage after taking its first blood meal), *I. scapularis* is the size of a poppy seed. Its bite is painless, so it is easily overlooked. Infection takes place when the tick feeds because the spirochete is regurgitated into the host. However, the organism grows in the tick's digestive tract and takes about 2 days to make its way to the tick's salivary gland, so if the tick is removed before that time, the patient will likely not be infected. Once transferred to the human, the microbe can travel rapidly via the bloodstream to any area in the body, but it prefers to grow in skin, nerve tissue, synovium (joint lining), and the conduction system of the heart.

How *Borrelia* infects the human host and spreads through the body so quickly is not well understood, largely because of its complicated life cycle and its unusual genome. We know that the organism uses surface proteins, adhesion proteins, and other factors to establish infection and to neutralize the complement system of the host. *Borrelia* has enzymes that may contribute to the clinical symptoms of Lyme disease. One enzyme cleaves the proteoglycans found in host connective tissues like tendons, ligaments, and cartilage found in joints. The damage to the connective tissue can cause inflammation and pain.

The *B. burgdorferi* genome, which consists of a large linear chromosome and 21 linear and circular plasmids, has been sequenced. The organism's genomic sequence has helped identify two plasmids important in *Borrelia* infections. The complete array of *Borrelia* virulence factors and pathogenic mechanisms, however, has yet to be determined.

Lyme disease has three general stages. The first two stages are part of early Lyme disease. Stage 1 is localized dissemination of *B. burgdorferi*, which occurs 3–30 days after the initial tick bite. Approximately 75% of patients experience a "bull's-eye" or "target" rash called erythema migrans, usually at the site of the tick bite, that varies in appearance (Figure 21.15). The lesions contain *B. burgdorferi* and are infiltrated with lymphocytes. As the bacteria move away (migrate) from the site of the bite, the redness of the rash moves outward, leaving a pale center. This stage is often associated with constitutional symptoms, such as fever, myalgias (muscle pain), arthralgias (joint pain), and headache.

Stage 2 occurs weeks to months after the initial infection, after *B. burgdorferi* spreads from the blood to other organs (hematogenous spread). In this stage, the patient can be quite ill with malaise, myalgias, arthritis (joint inflammation), and arthralgias, as well as neurological or cardiac involvement. The inflammation causing the joint pain of Lyme disease is thought to be an autoimmune reaction (our antibodies reacting against us). Lyme arthritis affects mostly large joints, especially the knees. Interestingly, only one knee is affected at a time, but the pain can alternate between knees. Common neurological manifestations include **Bell's palsy** (facial paralysis, described in Chapter 24 IMPACT), inflammation of spinal nerve roots, and chronic meningitis. The most common cardiac manifestation is an irregular heart rhythm.

Late Lyme disease (stage 3) occurs months to years later and can involve the synovium (the membrane that lines joints), nervous system, and skin, though skin involvement is more common in Europe than in North America. Arthritis occurs in the majority of patients with borreliosis and lasts from weeks to months in any given joint. Analysis of synovial fluid typically shows a much higher than normal WBC count. Late neurological involvement may include peripheral neuropathy and encephalopathy, manifested as memory, mood, and sleep disturbances.

Treatment with antibiotics is recommended for all stages of Lyme disease but is most effective in the early stages. Treatment for early Lyme disease (stage 1) is a course of doxycycline for 14–28 days. Lyme arthritis is usually slow to respond to antibiotic therapy. Despite antimicrobial drug treatment, patients with persistent active arthritis and persistently positive PCR tests may benefit from repeated treatment with injected antibiotics.

Curiously, 25% of people infected with *B. burgdorferi* never develop erythema migrans, and many infected individuals do not recall having a tick bite. Patients with Lyme disease may present with arthritis as their first complaint. Therefore, knowing that the patient lives in or near, or has recently traveled to, an endemic area can provide a critical clue. Recent studies also show that ticks are commonly coinfected with other bacterial pathogens such as *Ehrlichia* or *Anaplasma* species. Patients with Lyme disease who are coinfected with these organisms often have symptoms that can complicate diagnosis and may contribute to chronic Lyme disease.

Systemic Infections Caused by Intracellular Pathogens

Plague and borreliosis are diseases caused by extracellular pathogens—bacteria that grow outside host cells. Some bacteria that produce systemic infections are intracellular pathogens, with various modes of transmission and disease courses.

Salmonella enterica serovar Typhi, for example, is a Gram-negative rod acquired through ingestion. As described in Section 18.4, *S.* Typhi (and other, nontyphoidal salmonellae) are facultative intracellular pathogens that invade intestinal M cells (growing inside phagosomes) and exit into the lamina propria, where they invade tissue macrophages. The macrophages head for the circulation and taxi *S.* Typhi through the bloodstream. Along the way, the organism can infect a variety of organ systems (lymph nodes, liver, spleen, kidney, bone marrow). The result is **typhoid fever**, a potentially fatal disease. Lymphadenopathy, a fever of 39.4°C–40°C (103°F–104°F), confusion, and a distended abdomen, as well as a short-lived but characteristic maculopapular rash on the skin (called rose spots) that contains bacteria (**Figure 21.17**), are trademarks of typhoid fever. From the bloodstream, the organisms can sometimes infect the gallbladder, turning a person who survived infection into a chronic carrier (for example, look up Typhoid Mary online). The gallbladder of a chronic carrier will constantly secrete

Figure 21.17 Rose Spots of Typhoid Fever

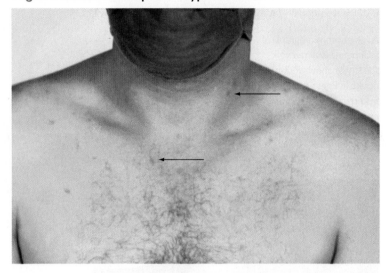

The rash is limited to the trunk of the body.

bile containing *S.* Typhi into the intestines, which is how the disease is transmitted. Food preparers who chronically shed *S.* Typhi in their feces and fail to wash their hands after bathroom use can introduce the organism into food and infect their patrons. Humans are the only reservoirs for this pathogenic variety of *Salmonella*. Typhoid fever can be treated with antibiotics (Table 21.2).

Rocky Mountain spotted fever (RMSF) is caused by *Rickettsia rickettsii* (**Figure 21.18A**), and epidemic typhus is caused by *Rickettsia prowazekii*. Both species are obligate intracellular bacteria, meaning they replicate only inside a host cell's cytoplasm. RMSF and epidemic typhus are two of several diseases collectively called **rickettsioses**. *R. rickettsii* and *R. prowazekii* are transmitted from reservoir animals to humans by the bite of arthropod vectors; ticks are the vector for RMSF, and human body lice are the vector for epidemic typhus (see Table 21.2). While biting, these ticks and lice also defecate, depositing the rickettsia at the bite site. Subsequent scratching by the host forces the organism into the bite. Rickettsia then attach to the surface of the endothelial cells and induce their own phagocytosis. After phagocytosis, the organisms escape the phagosome and replicate to large numbers in the cell's cytoplasm. Eventually, the infected cells burst and the organism disseminates to infect additional endothelial cells and organs. Symptoms for both diseases include chills, confusion, fever, headache, muscle aches, and sometimes photophobia. A major sign differentiating the two diseases is the rash that develops. The rash for RMSF usually begins on the wrists and ankles and then spreads toward the center of the body (**Figure 21.18B**). In contrast, epidemic typhus rashes begin on the trunk. Both diseases are treated with antibiotics, usually doxycycline.

Other pathogens that can cause systemic disease include facultative intracellular pathogens such as the Gram-negative rods *Francisella tularensis*, three *Brucella* species, and *Bartonella henselae*. *F. tularensis* is transmitted by ingestion, wound infection,

Figure 21.18 Rocky Mountain Spotted Fever

RMSF is a potentially fatal infection transmitted to humans by ticks.

A. A photomicrograph of a Gimenez-stained yolk sac smear shows *Rickettsia rickettsii*. Eggs are used to grow rickettsia in the laboratory because the organism is an obligate intracellular parasite.

B. The classic rash of RMSF often begins on the wrists and ankles and spreads rapidly toward the center of the body. The rash may also be present on the palms and soles.

or ocular contamination from infected small wild animals and causes **tularemia** (or **rabbit fever**). Three *Brucella* species, transmitted by ingesting milk or meat from infected cattle (*Brucella abortus*), pigs (*Brucella suis*), or goats (*Brucella melitensis*), are agents of brucellosis. *B. henselae*, a pathogen of cats that can be transmitted to humans by cat scratch or flea bite (it's not clear which), causes **cat scratch disease**. *Ehrlichia* and *Anaplasma* are two obligate intracellular pathogens that produce flu-like illnesses that are transmitted from deer to humans by tick bite.

SECTION SUMMARY

- **Septicemia** is caused by many Gram-positive and Gram-negative bacterial pathogens and can lead to disseminated, systemic disease. **Septic shock** is sepsis in which the patient becomes hypotensive.
- **Blood cultures** are useful in diagnosing septicemia and endocarditis.
- **Plague** has sylvatic and urban infection cycles involving transmission between fleas and rats. The bites of *Yersinia pestis*–infected fleas lead to bubonic plague. Aerosolized respiratory secretions will directly spread *Y. pestis* pneumonic plague from person to person (no insect vector).
- **Lyme disease** is caused by the spirochete *Borrelia burgdorferi*, which is transmitted from animal reservoirs to humans by the bite of *Ixodes* ticks. The initial presentation of the disease is a "bull's-eye" rash, also called erythema migrans.
- **Typhoid fever,** caused by *Salmonella* Typhi and transmitted by ingestion, causes high fever and is marked by rose spots forming on the patient's trunk. The organism is a facultative intracellular pathogen.

- **Rocky Mountain spotted fever** (*Rickettsia rickettsii*) and **epidemic typhus** (*Rickettsia prowazekii*) are caused by obligate intracellular pathogens that grow in the cytoplasm of endothelial cells. The rashes produced by these organisms either start on the wrists and ankles and spread to the trunk (RMSF) or start on the trunk and move outward (epidemic typhus).

Thought Question 21.3 A patient diagnosed with SIRS has blood samples taken from different body sites, including percutaneous (through the skin) and two different IV sites. The laboratory reports the isolation of *Staphylococcus epidermidis* from blood samples drawn from one of the IV sites. The remaining cultures, including the second IV site, show no growth. The patient is noted to have improved since being admitted 2 days ago. The attending physician does not change the diagnosis of the patient from SIRS to sepsis but orders continued daily blood cultures from percutaneous and IV sites. What may be the reason for this decision?

Thought Question 21.4 LPS from Gram-negative bacteria is a potent cause of septic shock. Consequently, is it a good idea to treat septic patients with a bactericidal antibiotic?

Thought Question 21.5 From what you know about the transmission of RMSF and epidemic typhus, which one is associated with war-torn countries?

21.4 Bacterial Infections of the Heart

SECTION OBJECTIVES

- Name and define the different manifestations of bacterial infections of the heart.
- Compare and contrast different types of endocarditis.
- Discuss the possible links between bacterial infections and atherosclerosis.

Are heart infections considered systemic infections? No, but they result from the circulatory spread of microbes that originate from other body sites. In addition, organisms infecting the heart can break free to circulate and infect other organs. Bacterial infections of the heart include pericarditis (infection of the sac surrounding the heart), myocarditis (infection of heart muscle), and endocarditis (inflammation of the inner layer of the heart) (see Figure 21.3). The inflammation of the heart known as rheumatic fever (discussed in Chapter 19) is not an infection per se but an autoimmune reaction that follows streptococcal pharyngitis caused by certain strains of *Streptococcus pyogenes* (see Chapters 17 and 19).

Pericarditis, when it occurs, usually develops after a respiratory tract infection. The advent of antibiotics has made pericarditis rare. Bacterial myocarditis is also rare in immunocompetent patients and is usually seen as a consequence of overwhelming sepsis. Endocarditis is by far the most common form of bacterial cardiac infection and is the focus of the following section. Table 21.3 lists bacteria that typically cause heart infections.

Bacterial Endocarditis

CASE HISTORY 21.3
Dental Procedure Leads to "Heartache"

 Elizabeth was 58 years old and had a history of mitral valve prolapse (a common congenital condition in which a heart valve does not close properly). She was on immunosuppressive therapy following a kidney transplant. She had recently been admitted to the hospital complaining of fatigue, intermittent fevers for 5 weeks, and headaches for 3 weeks—symptoms the physician recognized as possible indications of endocarditis. Elizabeth reported having a dental procedure a few weeks prior to the onset of symptoms, but she forgot to take the recommended prophylactic antibiotic beforehand. A sample of her blood placed in a liquid bacteriological medium grew Gram-positive cocci, which turned out to be *Streptococcus mutans*, a bacterial species associated with dental caries. With the finding of bacteria in the bloodstream, the diagnosis of bacterial endocarditis was confirmed. Elizabeth began a 1-month course of intravenous penicillin G and gentamicin therapy and eventually recovered to normal health.

Infectious endocarditis can be viral, fungal, or bacterial in origin. The viral and fungal forms of the disease are very rare and limited mostly to patients with transplanted heart valves. Endocarditis caused by bacteria, on the other hand, can be a consequence of many bacterial diseases, such as brucellosis, gonorrhea, psittacosis, staphylococcal and streptococcal infections, and Q fever caused by *Coxiella burnetii*. The most common causes are found in Table 21.3.

Endocarditis is traditionally classified as acute or subacute, based on the pathogenic organism involved and the speed of clinical presentation. **Subacute bacterial endocarditis (SBE)** has a slow onset with vague symptoms. It is usually caused by a bacterial infection of a heart valve. Subacute bacterial endocarditis infections are usually (but not always) caused by a viridans streptococcus from the oral microbiota (for example, *Streptococcus mutans*, a common cause of dental caries). **Viridans streptococci** is a general term for normal microbiota streptococci whose colonies produce green alpha hemolysis on blood agar (viridans, from the Greek *viridis*, "to be green"). Most patients who develop infective endocarditis have mitral valve prolapse (90%), although this is often not the case when patients are intravenous drug abusers or have hospital-acquired (nosocomial) infections.

As in Case History 21.3, bacterial endocarditis can begin at the dentist's office, although very rarely. Following a dental procedure (such as tooth restoration) or in someone with gingivitis, oral microbiota can transiently enter the bloodstream and circulate. *S. mutans*, which is not normally a serious health problem, can become lodged onto damaged heart valves, grow as a biofilm, and secrete a thick layer of polysaccharides (called glycocalyx) that

Table 21.3
Organisms That Cause Infections of the Heart

Pericarditis (rare)	Myocarditis (rare)	Endocarditis
Haemophilus influenzae	*Streptococcus pyogenes*	*Streptococcus mutans* (oral source)
Neisseria meningitidis	*Staphylococcus aureus*	*Streptococcus gallolyticus* (formerly *S. bovis*; intestinal source)
Streptococcus pneumoniae	*Corynebacterium diphtheriae* (disseminated toxin)	*Enterococcus* (intestinal source)
Streptococcus pyogenes	*Borrelia burgdorferi*	*Staphylococcus* spp.
Staphylococcus aureus	*Leptospira interrogans*	**HACEK** organisms[a]
	Rickettsia spp.	

[a]HACEK organisms include *Haemophilus parainfluenzae*; *Aggregatibacter aphrophilus* and *Aggregatibacter actinomycetemcomitans* (formerly *Actinobacillus actinomycetemcomitans*); *Cardiobacterium hominis*; *Eikenella corrodens*; and *Kingella kingae*.

encases the microbes and forms a vegetation on the valve, damaging it further (**Figure 21.19A and B**). It is also possible for a piece of the vegetation to break off and be carried by blood to the brain, where the piece can become lodged in one of the blood vessels of the brain and cause a stroke. If untreated, endocarditis can be fatal within 6 weeks to a year. Other bacteria that are important include intestinal bacteria such as *Streptococcus gallolyticus* or *Enterococcus* species. These organisms can escape the intestine of patients with colorectal cancer and enter the bloodstream.

When virulent organisms, such as *Staphylococcus aureus*, gain access to cardiac tissue, a rapidly progressive and highly destructive infection ensues: **acute bacterial endocarditis (ABE)**. Symptoms of acute endocarditis include fever, pronounced valvular regurgitation (backflow of blood through the valve), and abscess formation.

Most patients with subacute bacterial endocarditis present with a fever that lasts several weeks. They also complain of nonspecific symptoms, such as cough, shortness of breath, joint pain, diarrhea, and abdominal or flank pain. Endocarditis is suspected in any patient who has a heart murmur and an unexplained fever for at least 1 week. Sharing needles or using a contaminated needle can introduce bacteria directly into the bloodstream, so subacute bacterial endocarditis should be considered in any intravenous drug user who has a fever, even in the absence of a heart murmur. Definitive diagnosis requires blood cultures that grow bacteria, even with strong clinical suspicion for bacterial endocarditis.

Curing endocarditis is difficult because the microbes are usually sealed away in a nearly impenetrable glycocalyx. Consequently, eradicating microorganisms from the vegetation almost always requires hospitalization, where high doses of intravenous antibiotic therapy can be administered and monitored. Antibiotic therapy

Figure 21.19 View of Bacterial Endocarditis

C. An infected prosthetic heart valve showing bacterial endocarditis (granulated tissue at center, shown by arrow). When infection occurs early (within 2 months after surgery), it is likely that organisms gained entry during the operative period.

A. An open normal mitral valve shows the cords that tether it to the heart wall.

B. Close-up of native mitral valve endocarditis, showing vegetation, a microbial biofilm encased in a thick glycocalyx coating.

usually continues for at least a month, and in extreme cases, surgery may be necessary to repair or replace the damaged heart valve.

As a preventive measure, patients with certain heart conditions (an artificial heart valve, for instance) or on immunosuppressive therapy (such as Elizabeth in Case History 21.3) are prophylactically treated with penicillin or azithromycin 1 hour before a dental procedure. The idea is that the antibiotic will kill any oral bacteria that enter the bloodstream during the procedure and prevent the development of endocarditis. Although prophylactic therapy was previously recommended for patients with heart valve prolapse (a common condition), recent data suggest that doing so offers no significant preventive benefit.

Prosthetic Valve Endocarditis

For patients with a severely defective heart valve, replacing it with a prosthetic valve is critical. However, the surgical process can result in the artificial valve becoming infected, a condition called **prosthetic valve endocarditis (PVE)**. *Staphylococcus epidermidis* and *S. aureus* are the organisms that most commonly cause prosthetic valve endocarditis (Figure 21.19C). Many of these strains are methicillin resistant and extremely hard to treat. As with native valve endocarditis, the symptoms of PVE are vague. The most common symptoms are fever and chills accompanied by fatigue, anorexia, back pain, and weight loss. Prolonged intravenous antibiotics coupled with possible surgical removal and replacement of the valve are important treatment options. As noted earlier, patients with prosthetic heart valves should be prophylactically treated with penicillin or azithromycin 1 hour before certain dental procedures.

Atherosclerosis and Coronary Artery Disease

Microbes may also be involved in the process of **atherosclerosis** (hardening of the arteries). Atherosclerosis is an inflammatory reaction in response to the deposition of "bad" cholesterol (low-density

lipoprotein, or LDL) inside the arterial wall. Atherosclerosis leads to narrowing or blockage of arteries and can be life-threatening. Coronary artery disease, for instance, is the result of atherosclerotic deposits in the arteries of the heart, which can produce a heart attack if blood flow is occluded.

The causes of atherosclerosis have been studied and debated for many years. It now appears that microbes may have an important role in the development of this disease. *Chlamydophila pneumoniae* (see Section 20.3), for example, is suspected of involvement. Infection of alveolar macrophages by *C. pneumoniae* may lead to circulating "activated" macrophages. Activated macrophages generate inflammatory cytokines and factors that result in lowered high-density cholesterol (high-density lipoprotein or HDL, the "good" cholesterol) and increased clotting—both of which can damage endothelial cells and form plaques. Some studies reported evidence of periodontal disease pathogens such as *Porphyromonas gingivalis* within atherosclerotic plaques.

SECTION SUMMARY

- **Pericarditis** is an infection of the outer covering of the heart that sometimes follows a respiratory infection.
- **Myocarditis** is an infection of the heart muscle.
- **Endocarditis** is an infection of the inner heart lining that typically involves heart valves.
- **Bacterial endocarditis** can have acute or subacute onsets.
- **Subacute bacterial endocarditis** is usually an endogenous infection caused by *S. mutans*.
- ***Chlamydophila pneumoniae* infection** may contribute to atherosclerosis.

Thought Question 21.6 A blood culture was taken from a 63-year-old previously healthy patient with suspected subacute bacterial endocarditis. The laboratory reports *Streptococcus gallolyticus* present in three out of three bottles. Seeing this result, the physician suspects that the patient has another underlying condition. What is it? (*Hint:* Where is the organism normally found in the body?)

Thought Question 21.7 Why is stroke a complication of bacterial endocarditis, especially endocarditis of the mitral valve?

21.5 Systemic Parasitic Infections

SECTION OBJECTIVES

- Construct a list of systemic parasitic infections according to the type and mode of transmission of each parasite.
- Relate the different stages of systemic parasitic diseases to the growth cycle of the infecting agent.
- List the organ systems affected by each systemic parasitic infection discussed.
- Devise a plan for preventing infection by each systemic parasite discussed.

What are the major risk factors for acquiring a systemic parasitic infection? Poverty, a compromised immune system, and travel to endemic areas are among the most common risk factors. Although systemic infections caused by protozoa and helminths (see Section 11.1) are uncommon in developed countries, these diseases are among the most prevalent and devastating worldwide. Be mindful that, within a matter of hours, someone infected with one of these diseases can fly into your city's airport. Table 21.4 summarizes important features and treatments of systemic parasitic infections.

Malaria and Mosquitoes

Why is malaria the most devastating infectious disease known? In 2017, 219 million people developed malaria worldwide and 435,000 of these people, mostly children, died. Fortunately, the disease is relatively rare in the United States; only about 1,700 cases occurred in 2017. Almost all reported cases of malaria in the United States are acquired as a result of international travel to endemic areas (Figure 21.20). Once again, ease of travel illustrates the diagnostic importance of knowing a patient's history.

Malaria is caused by four species of *Plasmodium*: *P. falciparum* (the most deadly), *P. malariae*, *P. vivax*, and *P. ovale*. The life cycle of *Plasmodium* is complex and involves two cycles—an asexual erythrocytic cycle in humans (the intermediate host) and a sexual cycle

Figure 21.20 Malaria, a Major Disease Worldwide
Areas of the world where malaria transmission occurs.

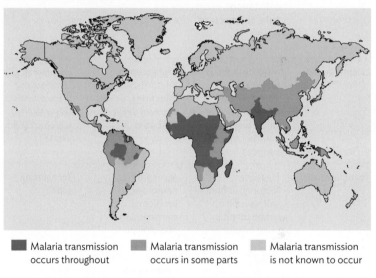

■ Malaria transmission occurs throughout   ■ Malaria transmission occurs in some parts   ■ Malaria transmission is not known to occur

Table 21.4
Selected Systemic Diseases Caused by Protozoa and Helminths

Disease	Symptoms	Complications	Parasite	Transmission	Diagnosis	Treatment	Prevention
Protozoa							
Malaria	Fever (may be every 48 or 72 hours but usually is random), chills, headache, vomiting	Inflammation of the brain, difficulty breathing, low blood glucose, anemia, death	*Plasmodium* spp.: *P. falciparum* is most common	*Anopheles* mosquitoes; via blood transfusion; can be transmitted from mother to child	Microscopy, thin and thick blood smears of patient; RDT to detect malaria antigens, PCR[a] to determine parasite species, serology	Therapy depends on type of infection; ACT[b] recommended by WHO for uncomplicated cases; chloroquine, mefloquine, quinine, and quinidine; doxycycline, tetracycline, and clindamycin in combination with quinine	No vaccine available; experimental vaccine in clinical trials since 2009, chemoprophylaxis offered to travelers to endemic areas

(continued on next page)

Table 21.4

Selected Systemic Diseases Caused by Protozoa and Helminths (*continued*)

Disease	Symptoms	Complications	Parasite	Transmission	Diagnosis	Treatment	Prevention
Babesiosis	Flu-like symptoms, then fever, shaking chills, sweats, muscle and joint pain, shortness of breath, hemolytic anemia, hemoglobinuria	Low and unstable blood glucose, low platelet count, DIC[c] leading to blood clots and hemorrhage, multi-organ failure, death	*Babesia* spp.: *B. microti* in northeastern US, *B. duncani* in Washington and California, *B. divergens* in Europe	Hard-bodied ticks: *Ixodes scapularis*; *I. ricinus* in Europe	Microscopy, thin and thick blood smears of patient, molecular and serological methods available at CDC	Antimicrobial choice depends on severity of disease and age of patient; atovaquone along with azithromycin, clindamycin, and quinine are used	No vaccine available
Chagas disease (American try-panosomiasis)	Fever, malaise, nausea, vomiting, and diarrhea; if progresses to cardiac form, then palpitations, chest pain, shortness of breath	Heart failure, enlargement of esophagus, enlargement of colon	*Trypanosoma cruzi*	Reduviid bug,[d] blood transfusion, vertical transmission, organ transplant, food-borne or laboratory exposure (rare)	Microscopy; serology; PCR for acute phase; culture for acute phase	Benznidazole and nifurtimox are effective but not approved for use in US; available from CDC	No vaccine available
Leishmaniasis	Mucocutaneous leishmaniasis: skin ulcer can disseminate and cause ulceration on other mucosal surfaces	Disfigurement, hemorrhage, infections, death, preterm delivery or stillbirths	*Leishmania* spp.: *L. mexicana*, *L. braziliensis* (New World), *L. tropica* (Old World)	Sand fly bite (cannot penetrate clothing)	Detection of parasite or its DNA in biopsied tissue samples (CDC helps with diagnosis)	IV pentavalent antimonies (sodium stibogluconate or meglumine antimoniate), amphotericin B	No vaccine available
Toxoplasmosis	Headache, seizures, abnormal gait, jaundice, lymphadenitis, fever in adults	Myocarditis, neurological problems; transmission of organism to fetus may cause mental retardation, neurological abnormalities, microcephaly, hydrocephalus, anemia in newborns[e]	*Toxoplasma gondii*	Inhaling airborne spores, contact with contaminated cat feces, or ingestion of contaminated food; transfusion, organ transplant, vertical transmission to fetus	Serology (positive in 60% of cases), direct observation of parasite in body fluids, molecular techniques, isolation of parasite from body fluids	Different complex regimens; drug of choice is pyrimethamine plus sulfadiazine for 3–4 weeks (or clindamycin if patient has sulfa allergies)	No vaccine available
Helminths							
Schistosomiasis (bilharziasis)	Abdominal pain, fever, chills, diarrhea, hepatitis, dysentery, pneumonitis	Dermatitis, cellulitis, lymphangitis	*Schistosoma* spp.: *S. haematobium*, *S. mansoni*, and *S. japonicum*	Contact with contaminated water containing free-living infectious organisms	Biopsy sample showing eggs; eggs in feces; serology	Praziquantel	No vaccine available
Lymphatic filariasis	Fever, inguinal or axillary lymphadenopathy (painful), limb or genital swelling		*Wuchereria bancrofti*	Mosquito-borne (*Aedes*; *Anopheles*)	Microfilariae in peripheral blood smear, biopsy, serology in patients with active infection	Diethylcarba-mazine and ivermectin	No vaccine available

[a]PCR = polymerase chain reaction; RDT = rapid antigen diagnostic testing.
[b]ACT = artemisinin-based combination therapy.
[c]DIC = disseminated intravascular coagulation.
[d]Also called triatomine bugs.
[e]Gestational stage of development at the time of fetal infection determines extent of complications.

Figure 21.21 Malaria: Life Cycle of *Plasmodium falciparum* between Mosquito and Human Host

7 Zygotes produce sporozoites that are released and delivered to a new host.

Mosquito salivary gland

IN THE MOSQUITO

Mosquito midgut

Female gametes

2n

n

Male gametes

6 Gametes fuse in the insect midgut to make zygotes (sexual cycle).

Fertilization

5 Some merozoites develop into male and female gametocytes. Mosquitoes taking a blood meal ingest the gametes.

1 Infected mosquito bites human. Mosquito salivary gland delivers sporozoites to human bloodstream.

2 Sporozoites travel to liver.

Liver cells

IN THE LIVER

3 Sporozoites divide to form merozoites, which are released into the bloodstream.

IN THE BLOODSTREAM

n

4 Merozoites repeatedly infect and lyse red blood cells. Cell fragments produce cyclic fevers (asexual erythrocytic cycle).

Schizont bursts, releasing merozoites

Ring stage

Schizont

TEM   10 μm

Plasmodium falciparum
Schizont after completion of division. A residual body of the organism (yellow-green) is left over after division. The erythrocyte has lysed and only a ghost cell remains; no cytoplasm is seen surrounding the merozoites just being released. Free merozoites are seen outside the membrane.

in mosquitoes (the definitive host). **Figure 21.21** summarizes the key stages in the life cycle and transmission of *Plasmodium*.

In the erythrocytic cycle, the organisms enter the bloodstream through the bite of an infected female *Anopheles* mosquito (the mosquito injects a small amount of saliva containing *Plasmodium*). The haploid sporozoites travel immediately to the liver, where they undergo asexual fission to produce merozoites. Released from the liver, the merozoites attach to and penetrate RBCs (see **Section 11.4**), where *Plasmodium* consumes hemoglobin and

Figure 21.22
Plasmodium falciparum within Red Blood Cells

Note the multiple trophozoite ring forms (shown by arrows) of the protozoan within red blood cells. The dark dots are nuclei, and the "empty" space is the food vacuole (Wright stain).

enlarges into a trophozoite. A trophozoite is the so-called ring form of the parasite seen within a red blood cell in **Figure 21.22**. The protist nucleus divides, so that the cell, now called a schizont, contains up to 20 or so nuclei. The schizont then divides to make the smaller, haploid merozoites (see Figure 21.21, step 4). The glutted red blood cell eventually lyses, releasing merozoites that can infect new RBCs.

Sudden, synchronized release of the merozoites and red cell debris triggers the telltale symptoms of malaria—violent, shaking chills followed by high fever and sweating. The erythrocytic cycle (and thus the symptoms) repeats every 48–72 hours. After several cycles, the patient goes into a remission lasting several weeks to months, after which a relapse occurs.

Much of today's research focuses on why malarial relapse happens. Why does the immune system fail to eliminate the parasite after the first episode? When *Plasmodium* invades the RBCs, it lines the outer surface of the blood cells' membranes with a sticky protein, PfEMP1, that causes the blood cells to stick to the sides of blood vessels. That process removes the parasite from circulation, but the protein cannot protect the parasite from patrolling macrophages, which eventually detect the invader and recruit other immune cells to fight it. So, during a malarial infection, a small percentage of each generation of parasites switches to a different version of PfEMP1 that the body has never seen before. In its new disguise, *P. falciparum* can invade more RBCs and cause another wave of fever, headaches, nausea, and chills. The body now has to repeat the recognition and attack responses all over again.

Diagnosis of malaria involves seeing the protist within erythrocytes (the Wright stain; see Figure 21.22 and Figure 11.18A) or using serology to identify antimalarial antibodies. Treatment regimens include chloroquine or mefloquine, which kill *Plasmodium* in erythrocytic asexual stages, and primaquine, effective in the exoerythrocytic stages. The chloroquine family of drugs acts by interfering with the detoxification of heme generated from hemoglobin digestion. Malaria parasites accumulate the hemoglobin released from RBCs in plasmodial lysosomes, where digestion occurs. The parasites use the amino acids from hemoglobin to grow but find free heme toxic. To prevent overeating themselves to death from accumulating too much heme, the organism detoxifies heme via polymerization, which produces a black pigment. Many antimalarial drugs, such as chloroquine, prevent polymerization by binding to the heme. As a result, the increased iron level (from heme) kills the parasite. The antiparasitic drugs actually used to treat active disease are presented in Table 21.4.

Chloroquine is given prophylactically to persons traveling to endemic areas. Unfortunately, *Plasmodium* has been developing resistance to these drugs, forcing development of new ones such as artemisinin, which acts by generating free radicals in the parasite that damage numerous proteins. The antigenic shape-shifting carried out by this parasite has so far stymied development of an effective vaccine. Prevention involves eliminating the vector, avoiding their bites, and screening the blood supply.

Babesiosis and Ticks

Babesiosis, caused by *Babesia microti*, is an emerging protozoan disease with similarities to malaria, except *Babesia* is transmitted to humans by ticks that have fed on infected white-footed mice and other small mammals. As with Lyme disease, most cases in the United States occur in the Northeast (20% of babesiosis patients are coinfected with *Borrelia burgdorferi*). The organism grows inside RBCs and produces a characteristic cell arrangement called the Maltese cross formation (**Figure 21.23**). Immunocompetent patients are typically asymptomatic or have a mild flu-like illness (generalized weakness, fatigue, fever, anorexia). But the disease can produce prolonged illness in some patient subpopulations (for example, the elderly) and can be fatal in patients who have had a splenectomy, have received an organ transplant, or are immunosuppressed.

Chagas Disease and the Reduviid Bug

Chagas disease is a zoonosis endemic to the Americas (it is also called American trypanosomiasis). The illness is caused by the hemoflagellate protozoan *Trypanosoma cruzi* (**Figure 21.24A**), which was discovered in 1910 by Carlos Chagas, a Brazilian microbiologist. The WHO estimates that, of almost 8 million people infected with *T. cruzi* worldwide (mainly in 21 Latin American countries), nearly 10,000 die each year from the disease. *T. cruzi* is

Figure 21.23 *Babesia*-Infected Erythrocytes
Wright-Giemsa–stained smears are shown.

A. Ringlike trophozoite.   **B.** Two merozoites.   **C.** Maltese cross (tetrad).

Figure 21.24 Chagas Disease

A. *Trypanosoma cruzi* (arrow), the cause of Chagas disease.

B. Reduviid bug, the vector.

C. Romaña sign.

found in both wild and domestic animals at scattered locales from southern Argentina to the southern United States.

T. cruzi lives and grows in the gut of a bloodsucking insect (a hematophagous triatomine) called the reduviid bug (genus *Reduvius*). The reduviid likes to make its home near a source of blood—for example, in the walls of mud or stone huts (**Figure 21.24B**). The insect's meal is blood, and its nasty habit of defecating while eating means that it leaves its feces at the site of the bite (usually on the face). The person or animal, in response to the sting of the insect, inadvertently rubs the feces along with its trypanosome content into the broken skin of the bite site. Because the reduviid often bites people around the mouth, it has earned the nickname "the kissing bug." The sclera of the eye can also serve as a portal of entry when rubbed with contaminated fingers. Doing so leads to a painless edema of the tissue around the eye, called the Romaña sign (**Figure 21.24C**) and may be followed by facial edema.

Within a week after transmission, most patients develop a mild acute form of Chagas disease with symptoms of fever and lymphadenopathy. A minority of infected people will have a more violent form of acute disease with systemic symptoms that include enlargement of the liver and spleen, neurological symptoms, and even myocarditis. These symptoms resolve spontaneously in almost all cases of acute disease, but the parasite remains in most victims for an indefinite period of time. The acute phase of trypanosomiasis (Chagas disease) is treated with nifurtimox or benznidazole, whose modes of action remain unclear.

After 4–6 weeks, patients who have not been cured develop the chronic form of the disease. Chronic disease presents itself in some 20%–30% of patients. The organism attacks different organs, including the heart, the gastrointestinal tract, and the nervous system, long after the initial infection (sometimes more than 20 years later). Death from Chagas disease is usually due to heart failure

caused when myocarditis destroys the heart muscle. The management of chronic Chagas disease is primarily supportive, although some studies suggest that benznidazole may be beneficial.

T. cruzi can be transmitted vertically from mother to fetus during gestation but cannot pass from person to person. Transfusion of contaminated blood, organ transplant, and consumption of contaminated uncooked food can also transmit the disease. Laboratory exposure is a rare cause of Chagas disease. Preventing congenital, transfusion-related, and organ transplant–related transmission of the disease is the main focus of control efforts in regions, such as the United States, where the disease is found but is not widespread.

Toxoplasmosis and Cats

Do you own a cat? If so, pay close attention to this section. **Toxoplasmosis** is a disease caused by *Toxoplasma gondii* (**Figure 21.25** and Figure 18.32), a single-celled, spore-forming obligate intracellular protozoan (coccidian) with a complicated life cycle (**Figure 21.26**).

Figure 21.25 *Toxoplasma gondii* Cysts

A. Microscopic cysts containing *T. gondii* develop in the tissues of many vertebrates. Here, in mouse brain tissue, thousands of resting parasites (stained red) are enveloped by a thin parasite cyst wall.

B. A tissue cyst with its many bradyzoites.

Figure 21.26 Life Cycle of *Toxoplasma gondii*

The only known definitive hosts for *T. gondii* are domestic cats and their relatives. Infected cats can release up to 100 million gametes (unsporulated oocytes) every day for 7–21 days. The oocytes will sporulate in the environment and become infective, which will complete their growth cycle in the intermediate host.

4 Cats (definitive hosts) become infected by eating material containing toxoplasma tissue cysts.

DEFINITIVE HOST (cat)

Bradyzoite

Tachyzoite

Cyst

LM 20 μm

3 The ingested sporulated oocytes transform into tachyzoites and move into brain and muscle tissue, where they become cyst bradyzoites.

The oocyte contains two sporocysts

1 Infected cats shed unsporulated oocytes.

Cat feces

Fecal oocyte

LM 20 μm

2 Oocytes sporulate, become infective, and are ingested by the intermediate hosts.

INTERMEDIATE HOSTS

Tissue cyst

5 Fecal oocytes contaminate kitty litter, food, and water.

6 Humans can become infected from handling contaminated materials via hand-to-mouth transfer, . . .

ingesting fecal-contaminated fruits, vegetables, or water, . . .

7 ingesting meat from infected animals, . . .

8 by blood transfusion, . . .

9 or by transplacental transmission to a fetus.

The parasite starts its life cycle in its definitive host, a cat (the feline stage), and completes it in an intermediate host, a bird or a mammal (the nonfeline stage). Cats do not become sick when they ingest spores of *T. gondii*, but their intestinal tract serves as a breeding ground. Here, spores undergo the changes necessary to produce gametes that fuse together to form zygotes. The zygotes then form a rigid wall around themselves and are ready to be released.

When defecating, infected cats can release up to 100 million of these walled-off gametes each day for 7–21 days. Ambient temperature and exposure to air induce sporulation and maturation of the spores. Cats, other mammals, and birds ingest the spores (sporozoites) when eating contaminated food or drinking contaminated water. Stomach acidity of the intermediate host dissolves the wall of the cyst containing the spores and releases them into the small intestine. The spores enter the small intestine's epithelial cells and rapidly divide until the population of parasites is so high that the cell ruptures and releases them. The rapidly growing protozoan cells are called **tachyzoites** because of their rapid ("tachy") division. Tachyzoites can invade all <u>nucleated</u> cells, so they spread rapidly through blood (without infecting RBCs, which lack nuclei) and all organs. The tachyzoites then change into a slow-dividing version of themselves (**bradyzoites**) and become encased in a cyst as the host immune system fights back and eliminates the tachyzoites. The major sites of cyst formation are the host's muscles and central nervous system (brain and spinal cord). The cysts represent a chronic stage of the infection.

If the intermediate host is an immunocompetent human, the humoral and cell-mediated immune responses of that host will eliminate most of the tachyzoites (but not the cysts). Infected individuals suffer no or very mild symptoms. However, suppression of a host's immune system can cause spontaneous release of parasites from these cysts. The released parasites then transform into the invasive tachyzoite form and cause more serious disease, including brain infection.

Congenital disease is another serious consequence of toxoplasmosis. *T. gondii* can cross the placenta from a parasitemic mother (one with parasites in her blood) to her developing fetus. The severity of fetal disease depends on the gestational age at transmission. Generally, infections early in pregnancy are more severe and result in a classic triad of chorioretinitis (inflammation of the retina and vascular layer of the eye), hydrocephalus (abnormal accumulation of cerebrospinal fluid in the brain), and intracranial calcifications. A large percentage of infected newborns are asymptomatic but can develop symptoms months after birth.

Congenital toxoplasmosis is a preventable disease. Expectant mothers should be screened for antibodies to *T. gondii* and treated if the test is positive. The currently recommended drugs for *T. gondii* infection act primarily against the tachyzoite form but do not eradicate the encysted form (bradyzoite). Pyrimethamine is the most effective agent and is included in most drug regimens.

Do *Toxoplasma* cysts in the brain have any effect on a host's behavior? Oxford University zoologists showed that *T. gondii* appears to alter the brain chemistry of rats, making them more likely to seek out cats. Infection thus makes a rat more likely to be killed and the parasite more likely to end up in a cat—the only host in which it can complete the reproductive step of its life cycle. In humans, several studies have shown a correlation between exposure to *Toxoplasma* and schizophrenia, although no causal link has been proven.

Leishmaniasis and the Sand Fly

Another group of protozoan diseases, collectively called **leishmaniasis**, come in two basic forms: a cutaneous form, produced by *Leishmania tropica* and *L. braziliensis*, and a visceral form, caused by *L. donovani*. The visceral form can grow at core body temperatures (98.6°F), whereas the cutaneous form requires the cooler temperature of skin (91°F). The protozoa causing both clinical syndromes are transmitted by the female sand fly (**Figure 21.27A**). Smaller than a mosquito, the sand fly is found in tropical areas,

Figure 21.27 Visceral Leishmaniasis

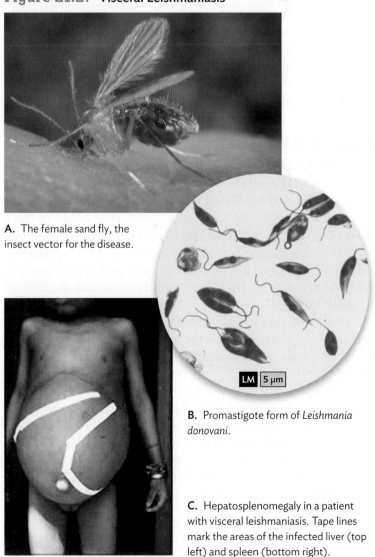

A. The female sand fly, the insect vector for the disease.

B. Promastigote form of *Leishmania donovani*.

LM 5 μm

C. Hepatosplenomegaly in a patient with visceral leishmaniasis. Tape lines mark the areas of the infected liver (top left) and spleen (bottom right).

around the Mediterranean, and in the Middle East. Its tiny size allows this vector to penetrate the mesh of standard mosquito netting.

Leishmania spends part of its life cycle in the gut of the sand fly, but the protozoan's life cycle is completed in a vertebrate host. The sand fly carries the motile form of the protozoan (called a pro-mastigote; **Figure 21.27B**). Multiplication of the pathogen blocks the insect's esophagus. When taking a blood meal, the infected sand fly regurgitates the organism with saliva into the bite. After being introduced into the skin by a bite, the promastigote loses its flagellum and becomes an amastigote that multiplies inside tissue macrophages. In visceral leishmaniasis, macrophages carry the organism throughout the body.

The symptoms of visceral leishmaniasis include fever, hepato-splenomegaly (enlarged spleen and liver) resulting from acceler-ated production of phagocytic blood cells (**Figure 21.27C**), wasting and weakness, lymphadenopathy (often), and a characteristic darkening of the skin (from which the Indian name for the disease, kala-azar, or black fever, originated).

In visceral leishmaniasis, patients may die within a matter of months from hemorrhage, severe anemia, secondary bacterial infections of mucous membranes, bacterial pneumonia, septicemia, tuberculosis, dysentery, or measles. Antiparasitic drugs containing the chemical element antimony are commonly used for treatment, although an oral drug, miltefosine, is used in some countries. Milte-fosine is thought to trigger cell death programs in leishmania.

Lymphatic Filariasis and Filarial Worms

Filariasis describes a group of diseases caused by nematodes in the family Filarioidea. More than 120 million people world-wide are infected by filarial worms, primarily in Africa, Asia, the eastern Pacific, and Indonesia. The disease comes in three forms that involve different filarial species: a cutaneous group (*Loa loa*, *Onchocerca volvulus*, and *Mansonella streptocerca*), a lymphatic group (*Wuchereria bancrofti* and *Brugia malayi*), and a body cavity group (*Mansonella perstans* and *Mansonella ozzardi*).

Our focus here is on lymphatic filariasis because of its sys-temic nature. Understanding the life cycles of filarial nematodes is important for understanding transmission (**Figure 21.28**). An infection starts when larvae are transmitted from mosquitoes or flies to humans (the definitive host) and develop into adult worms that can reproduce. For the lymphatic group of filarial worms, this occurs in the lymphatic system. The adults mate in the lymphatic system and produce thousands of microfilariae per day that enter the bloodstream. Adult worms can live and produce microfilar-iae for at least 5 years. As they course through the bloodstream, microfilariae can be ingested by a second mosquito (intermediate

host) taking a blood meal. Microfilariae in the mosquito take 10–14 days to develop into stage 3 larvae, which can be transmitted to the human host by insect bite.

The presence of adult worms in the lymphatic system stim-ulates an impressive inflammatory response that can produce lymphatic obstruction. Because the lymphatic circulation moves lymph (white cells and recovered tissue fluid) from the extremities toward the heart, a lymphatic obstruction means that lymph will accumulate behind the blockage. The result can be a grotesque swelling of a limb (known as **elephantiasis**) or, in males, the scrotum (a condition called **hydrocele**); see Figure 21.28.

Diagnosis of lymphatic filariasis is based on finding microfi-lariae in blood smears or detecting filarial antigen in blood, with or without seeing microfilariae. Antihelminthic treatments may include ivermectin (paralyzes the parasite), diethylcarbamazine (action unclear), and albendazole (decreases ATP production), among others. Surgical intervention can correct hydrocele but has been less successful with limb elephantiasis.

Many other pathogenic helminths infect multiple organ sys-tems after migrating in the blood. We have chosen to discuss these worms in whatever system they primarily infect or that serves as the portal of entry and exit (for example, schistosomiasis is covered with gastrointestinal infections).

SECTION SUMMARY

- **Malaria** is caused by *Plasmodium* species. These are transmitted from mosquitoes (the definitive host) to humans (intermediate host), where the organism invades and replicates within red blood cells.

- **Babesiosis**, caused by *Babesia microti*, is transmitted by ticks to humans. The infection is potentially lethal in those who are immunodeficient.

- **Chagas disease** develops following infection by the protozoan *Trypanosoma cruzi*. The tiny reduviid bug transmits the parasite from animal to humans, usually during a bite to the face.

- **Toxoplasmosis** is caused by *Toxoplasma gondii*. *T. gondii* is usually transmitted from cats (definitive host) to humans (intermediate host) who accidentally ingest sporozoites present in infected cat feces.

- **Congenital toxoplasmosis** results from transplacental transmission of *T. gondii* from a parasitemic mother to a developing fetus.

- **Leishmaniasis** is caused by several species of *Leishmania* protozoa transmitted to humans via the sand fly.

- **Lymphatic filariasis** is a disease caused by a variety of nematode species, including *Wuchereria bancrofti*, and is potentially disfiguring. Disease starts when a mosquito (the intermediate host) transmits larvae of the worm to the human (the definitive host). Larvae enter the lymphatic system and develop into adults that block lymphatic circulation.

Thought Question 21.8 Sickle-cell disease occurs in people who have two copies of a defective gene for hemoglobin production, *HbS*.

Figure 21.28 Lymphatic Filariasis

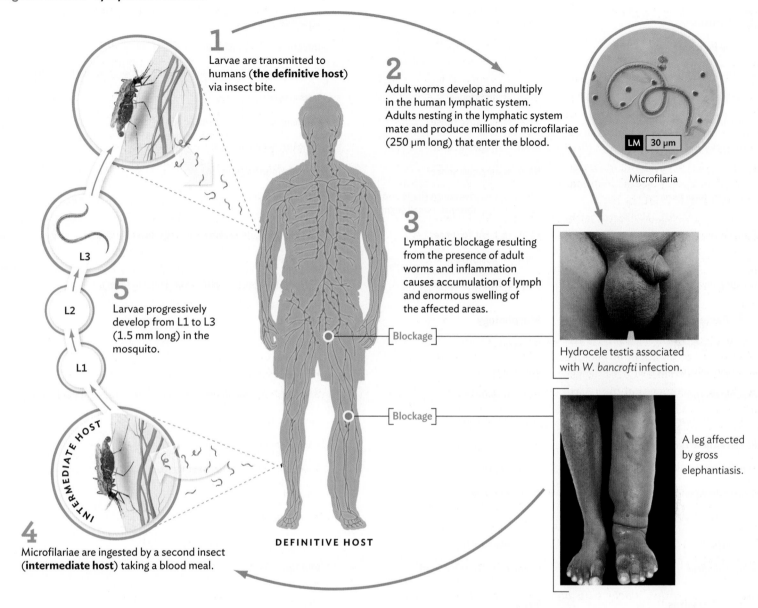

1 Larvae are transmitted to humans (**the definitive host**) via insect bite.

2 Adult worms develop and multiply in the human lymphatic system. Adults nesting in the lymphatic system mate and produce millions of microfilariae (250 μm long) that enter the blood.

Microfilaria

LM 30 μm

3 Lymphatic blockage resulting from the presence of adult worms and inflammation causes accumulation of lymph and enormous swelling of the affected areas.

[Blockage]

[Blockage]

Hydrocele testis associated with *W. bancrofti* infection.

A leg affected by gross elephantiasis.

5 Larvae progressively develop from L1 to L3 (1.5 mm long) in the mosquito.

L3

L2

L1

INTERMEDIATE HOST

DEFINITIVE HOST

4 Microfilariae are ingested by a second insect (**intermediate host**) taking a blood meal.

This defective hemoglobin will polymerize into long polymers under low-oxygen conditions, causing RBCs to morph from round cells to cells shaped like sickles. Sickle-cell <u>trait</u> occurs in people who have one normal and one mutant copy of the hemoglobin gene. The RBCs of people with sickle-cell trait do not change shape under low oxygen because 50% of their hemoglobin is normal and the rest abnormal. People with sickle-cell trait are resistant to malaria. What could explain this phenomenon? (*Hint: Plasmodium* parasites consume a lot of oxygen during growth.)

Perspective

Many of the bacteria, viruses, fungi, protozoa, and worms we discuss with specific organ systems also disseminate systemically when given the chance. In fact, many microbial pathogens could be considered "explorers" in which "scouts" from an initial infection are sent through the circulatory system to seek new sources of food (organ systems) and possible shelter from an immune system bent on their destruction. For example, *Neisseria meningitidis* causes meningitis but requires sepsis to do so. Keep dissemination in mind as you progress further through these infectious disease chapters. Take note of which organisms can move between organ systems.

Figure 21.29 Summary: Microbes That Infect the Circulatory System

Viruses	Genome	Disease
Epstein-Barr virus	DNA, ds	Infectious mononucleosis
Cytomegalovirus	DNA, ds	Cytomegalic inclusion disease, CMV mononucleosis-like syndrome[a]
Lassa virus	RNA, ss, ambisense[b]	Lassa fever
Nairovirus[c]	RNA, ss, negative sense	Crimean-Congo hemorrhagic fever (CCHF)
Marburg virus	RNA, ss, negative sense	Marburg hemorrhagic fever
Ebola virus	RNA, ss, negative sense	Ebola hemorrhagic fever
Rift Valley fever virus	RNA, ss, negative sense (large and medium segments), ambisense (small segment)	Rift Valley fever
Dengue virus	RNA, ss, positive sense	Dengue fever, dengue hemorrhagic fever (DHF)
Yellow fever virus	RNA, ss, positive sense	Yellow fever
Chikungunya virus	RNA, ss, positive sense	Chikungunya fever (CHIK fever, knuckle fever)

Bacteria	Morphology	Disease
Borrelia burgdorferi	Spirochete	Lyme disease
Leptospira interrogans	Spirochete	Leptospirosis
Brucella abortus	Gram-negative rod	Brucellosis (undulant fever, Mediterranean fever, Malta fever)
Rickettsia prowazekii	Gram-negative rod	Epidemic typhus
Rickettsia rickettsii	Gram-negative rod	Rocky Mountain spotted fever
Francisella tularensis	Gram-negative rod	Tularemia (rabbit fever, deer fly fever)
Salmonella enterica serovar Typhi	Gram-negative rod	Typhoid fever
Salmonella enterica serovar Choleraesuis	Gram-negative rod	Typhoid fever
Yersinia pestis	Gram-negative rod	Plague (bubonic, pneumonic)
Vibrio vulnificus	Gram-negative curved rod	Vibriosis

Parasites	Type	Disease
Plasmodium spp.[d]	Protozoan	Malaria
Toxoplasma gondii	Protozoan	Toxoplasmosis
Babesia spp.[e]	Protozoan	Babesiosis
Trypanosoma cruzi	Protozoan	Chagas disease
Leishmania spp.[f]	Protozoan	Leishmaniasis
Schistosoma spp.[g]	Helminth	Schistosomiasis (bilharziasis)
Wuchereria bancrofti	Helminth	Lymphatic filariasis

[a]Also known as heterophile-negative mononucleosis.
[b]Ambisense = an RNA strand that has partially positive polarity (positive sense) and partially negative polarity (negative sense), ds = double-stranded; ss = single-stranded.
[c]Also known as Asian Ebola virus and Congo virus.
[d]*P. falciparum* is the most common cause.
[e]*B. microti* in northeastern US; *B. duncani* in Washington and California; *B. divergens* in Europe.
[f]*L. mexicana, L. braziliensis* (New World), *L. tropica* (Old World).
[g]*S. haematobium, S. mansoni, S. japonicum.*

Portal of Entry

Respiratory
- *Yersinia pestis*
- *Toxoplasma gondii*

Oral/Saliva
- EBV
- CMV
- *Brucella* spp.
- *Leptospira interrogans*
- *Salmonella enterica* serovar Typhi
- *Salmonella enterica* serovar Choleraesuis
- *Trypanosoma cruzi*

Skin
- Dengue virus
- Yellow fever virus
- Lassa virus
- Chikungunya virus
- Rift Valley fever virus
- Marburg virus
- Ebola virus
- Nairovirus
- *Borrelia burgdorferi*
- *Rickettsia*
- *Francisella tularensis*
- *Vibrio vulnificus*
- *Plasmodium* spp.
- *Babesia* spp.
- *Leishmania*
- *Schistosoma mansoni*
- *Wuchereria bancrofti*

Blood
- HIV

Portal of Exit

Respiratory
- *Yersinia pestis*

Oral/Saliva
- EBV
- CMV

Stool
- *Schistosoma mansoni*
- *Salmonella enterica* serovar Typhi
- *Salmonella enterica* serovar Choleraesuis

Blood
- Dengue virus
- Yellow fever virus
- Lassa virus
- Chikungunya virus
- Rift Valley fever virus
- Marburg virus
- Ebola virus
- Nairovirus

Key Terms

acute bacterial endocarditis (ABE) (p. 673)

atherosclerosis (p. 674)

babesiosis (p. 678)

bacteremia (p. 663)

Bell's palsy (p. 670)

bradyzoite (p. 681)

bubonic plague (p. 669)

Burkitt's lymphoma (p. 660)

cardiomyopathy (p. 655)

cat scratch disease (p. 672)

Chagas disease (p. 678)

congenital cytomegalovirus infection (p. 661)

cytomegalic inclusion disease (CID) (p. 661)

cytomegalovirus (CMV) (p. 661)

dengue hemorrhagic fever (DHF) (p. 662)

disseminated intravascular coagulation (DIC) (p. 666)

elephantiasis (p. 682)

endocarditis (p. 655)

Epstein-Barr virus (EBV) (p. 656)

erythema migrans (p. 669)

filariasis (p. 682)

giant cell (p. 661)

HACEK (p. 673)

hepatomegaly (p. 660)

heterophile antibody (p. 659)

hydrocele (p. 682)

hypotension (p. 666)

infectious mononucleosis (p. 656)

leishmaniasis (p. 681)

Lyme borreliosis (p. 669)

Lyme disease (p. 669)

lymphadenitis (p. 656)

lymphadenopathy (p. 656)

lymphangitis (p. 656)

lymphatic system (p. 652)

malaria (p. 675)

myocarditis (p. 655)

pancarditis (p. 655)

pericarditis (p. 655)

plague (p. 668)

pneumonic plague (p. 669)

prosthetic valve endocarditis (PVE) (p. 674)

pulmonary circulation (p. 652)

rabbit fever (p. 672)

reinfection (p. 661)

rickettsiosis (p. 671)

Rocky Mountain spotted fever (RMSF) (p. 671)

sepsis (p. 663)

septicemia (p. 663)

septicemic plague (p. 669)

septic shock (p. 666)

splenomegaly (p. 660)

subacute bacterial endocarditis (SBE) (p. 673)

systemic circulation (p. 652)

systemic infection (p. 652)

systemic inflammatory response syndrome (SIRS) (p. 663)

tachyzoite (p. 681)

TORCH complex (p. 661)

toxoplasmosis (p. 679)

tularemia (p. 672)

typhoid fever (p. 671)

viridans streptococcus (p. 673)

Objectives Assessment Questions

21.1 Anatomy of the Cardiovascular and Lymphatic Systems

1. The _____ delivers oxygen and nutrients to the entire body.
   a. bronchial artery
   b. aorta
   c. pulmonary artery
   d. superior vena cava

2. The main function of the pulmonary circulation is to provide lungs with oxygenated blood. True or false?

3. A patient is in your office with an inflamed lymph node under her arm. The swollen node is warm and painful to the touch. Which of the following best explains this condition?
   a. Simple noninfectious lymphadenopathy
   b. Lymphangitis
   c. Lymphadenitis
   d. A systemic infection
   e. Pulmonary circulatory insufficiency

21.2 Systemic Viral Infections

4. _____ infected with human herpesvirus 4 (HHV-4) have an atypical appearance and resemble monocytes.
   a. Oropharyngeal epithelial cells
   b. Tonsillar epithelial cells
   c. B cells
   d. Salivary gland cells

5. Which of the following is the basis for the monospot test?
   a. IgM to viral capsid antigen
   b. IgG to viral capsid antigen
   c. IgM to early antigen
   d. Antibody to nuclear antigen
   e. Heterophile antibodies

6. What is the most dangerous form of transmission of CMV from a mother to a child?
   a. Breast-feeding
   b. Transplacental
   c. Contact
   d. None of the above

Match the following:

7. Dengue fever virus   a. Owl's eyes

8. Kissing disease   b. Splenomegaly

9. Congenital CMV   c. Burkitt's lymphoma

10. EBV and malaria   d. Hemorrhagic fever

21.3 Systemic Bacterial Infections

Match the following:

Condition	Characteristic
11. SIRS	a. Systemic inflammatory response syndrome caused by an infectious agent
12. Sepsis	b. Sepsis with one or more signs of organ dysfunction
13. Severe sepsis	c. May have a noninfectious etiology
14. Septic shock	d. Need vasopressors to maintain systolic blood pressure of 90 mm Hg or mean arterial pressure of 70 mm Hg
15. Bacteremia	e. Presence of bacteria in the blood

16. Bell's palsy is more likely to be a manifestation of _____ than of any other stage of Lyme disease.

    a. stage 1
    b. stage 2
    c. stage 3
    d. stage 4

17. _____ is a contagious disease caused by Gram-negative intracellular bacteria that can colonize the gallbladder of an infected person, making that person a carrier.

    a. Rocky Mountain spotted fever
    b. Tularemia
    c. Lyme disease
    d. Infective endocarditis
    e. Typhoid fever

21.4 Bacterial Infections of the Heart

18. Which of the following is the more common etiology of prosthetic valve endocarditis?

    a. *Streptococcus mutans*
    b. *Staphylococcus aureus*
    c. *Enterococcus* species
    d. *Streptococcus gallolyticus*

19. Inflammation of the three structural layers of the heart is referred to as _____.

    a. pancarditis
    b. endocarditis
    c. myocarditis
    d. epicarditis

20. Activation of alveolar macrophages by *Chlamydophila pneumoniae* is linked to decreased plasma HDL levels. True or false?

21.5 Systemic Parasitic Infections

21. A man bitten by an *Anopheles* mosquito has cycles of fever because the protozoa introduced into his system _____.

    a. survive in his liver indefinitely
    b. undergo antigenic variation
    c. are too small to be recognized by macrophages
    d. resemble host red blood cell precursors
    e. paralyze the host immune system

22. Which of the following is a symptom of a disease caused by a nematode?

    a. Darkening of the skin (kala-azar, or black fever)
    b. Chorioretinitis (inflammation of the membrane covering the retina)
    c. Hydrocele (grotesque swelling of the scrotum)
    d. Romaña sign (painless edema of the tissue around the eye)
    e. Waves of fever and chills repeating every 48–72 hours

Review Questions

1. What is systemic infection?

2. What is the function of the lymphatic system?

3. Compare and contrast systemic and pulmonary circulation.

4. Relate the structure of the heart and the layers of the heart wall to organisms that can affect them.

5. What are the most common causes of systemic viral infections?

6. What organs are affected in infectious mononucleosis?

7. What are the signs and symptoms of infectious mononucleosis?

8. Describe the relationships between EBV, Burkitt's lymphoma, and malaria.

9. What is TORCH, and why is it important?

10. Describe cytomegalic inclusion disease (CID).

11. What is a hemorrhagic fever, and what viruses can cause it?

12. Differentiate between bacteremia, SIRS, sepsis, and septic shock.

13. What is the mechanism of disseminated intravascular coagulation (DIC)?

14. Describe how *Yersinia pestis* uses its type III secretion system differently than *Salmonella*.

15. Which is more contagious, bubonic plague or pneumonic plague? Why?

16. Describe the different stages of Lyme disease.

17. Name intracellular pathogens and parasites that cause systemic infections and the diseases they cause.

18. Describe why symptoms of malaria follow a cycle of 48–72 hours.

19. Name the mode of transmission for each parasite discussed in this chapter.

20. Which of the systemic diseases described in this chapter can be prevented by vaccination?

Clinical Correlation Questions

1. A patient presents with a solitary enlarged lymph node in his groin. The bacteria isolated from lymph node aspirates of this patient are reported as Gram-negative bacilli with bipolar staining. Is this patient infected with *Yersinia pestis*, *Borrelia burgdorferi*, *Salmonella* Typhi, or *Rickettsia rickettsii*? Explain.

2. A patient is diagnosed with sepsis. You have ordered blood cultures, but you have to give antibiotics until the blood culture results come back. You order IV antibiotics. The antibiotics you order should be effective against what class or classes of microorganisms? Explain.

3. A 22-year-old otherwise healthy marathon runner presents to the clinic complaining of palpitations along with pain and swelling in his right ankle for the past 3 weeks. He reports that he has had similar episodes on and off for the past 6 months. His electrocardiogram shows an abnormal heart rhythm. Examination of his feet reveals neuropathy in the right foot. Synovial fluid aspirate shows 25,000 WBCs, but the culture is negative. His medical history is benign and his vaccinations are up to date. He regularly travels to Wisconsin on business and recalls a circular rash on his left arm following a morning run when he was there about a year ago. What is the most likely bacterial cause of this young man's symptoms, and how is it transmitted?

4. A 34-year-old woman presented to an Oklahoma clinic complaining of having palpitations and being short of breath and overly tired. An electrocardiogram revealed a cardiac arrhythmia that could explain her symptoms. She was a legal immigrant from Brazil who arrived in the United States 10 years ago. She said that her family lived in a remote village in Brazil in a mud hut because they were very poor. Her medical history was unremarkable, but she recalled waking up one morning when she was about 15 years old with a badly swollen but painless area around her eye. The swelling went down after a few weeks, but she remembered developing a very high fever afterward and that her stomach got very large. Eventually, those symptoms disappeared. What infectious disease could cause these symptoms?

Infections of the Digestive System

Dehydration: A Toddler's Plight

SCENARIO Dao Ming is a 2-year-old who woke up lethargic, with watery, nonbloody diarrhea. His mother kept him home from the day-care center he normally attends in downtown Portland, Maine. Ming's diarrhea became more pronounced as the day wore on, he seemed to be urinating less, and he started to develop a fever. By the next morning, Ming's eyes started to appear sunken and his diarrhea persisted. His parents rushed him to the emergency department (ED) of a nearby hospital. When asked, Ming's parents said their son had not received any vaccinations.

Intravenous (IV) Rehydration
Excessive diarrhea can produce dangerous levels of dehydration, especially in children. Rehydration by IV is often used to replace the fluids and electrolytes lost.

0.9% Sodium Chloride Injection USP
500 mL

SIGNS AND SYMPTOMS Ming had a fever of 39.5°C (103.1°F), an elevated heart rate, and rapid breathing. The ED physician determined that Ming was dehydrated. The doctor ordered a complete blood count (CBC).

LABS The lab reported that Ming's white cell counts were normal, so the doctor began to suspect Ming's diarrhea had a viral origin. An enzyme immunoassay of the boy's stool sample identified the presence of rotavirus, a virus that can be prevented by vaccination.

RESOLUTION The boy was admitted to the hospital, where he received intravenous fluids to help rehydrate him and oxygen to help his breathing. With the supportive therapy, Ming's diarrhea disappeared within a day, and by week's end he was back playing in day care.

TEM 50 μm

Rotavirus
Rotavirus is the most common cause of diarrheal disease in children around the globe.

CHAPTER OBJECTIVES

After reading this chapter, you will be able to:

- Correlate the anatomy and physiology of the digestive system with the infectious diseases affecting them.
- Distinguish gastrointestinal infections by their presenting signs and symptoms.
- Relate the virulence mechanisms of gastrointestinal microbes to the presentation of diseases they cause.
- Discuss effective prevention and treatment for different infectious diseases of the gastrointestinal system.

Raise your hand if you have never had diarrhea! Nearly everyone has experienced diarrhea, a condition characterized by frequent loose bowel movements accompanied by abdominal cramps. Hundreds of millions of cases occur each year in the United States. The loose stools usually result from inflammation of the gastrointestinal tract (a condition called gastroenteritis) due to viral or bacterial growth or toxin production. These infections or intoxications cause large amounts of water and electrolytes to leave the intestinal cells and enter the intestinal lumen. Consequently, the patient not only suffers diarrhea but also can become dangerously dehydrated; the risk of dehydration is particularly high for the very young, like Ming, and the very old. Most diarrheal disease is viral in origin, with rotavirus and norovirus being the primary culprits. Among bacteria, *Salmonella* (a Gram-negative bacillus) and *Campylobacter* (a Gram-negative helical bacillus) vie for being the most frequent cause. Immunoassay (see Section 17.2 and Figure 17.7) revealed that rotavirus caused Ming's symptoms.

Rotavirus is the most common cause of severe diarrhea among children worldwide, resulting in the hospitalization of 55,000–70,000 children each year in the United States and more than 200,000 emergency department visits annually among children under 5 years of age. Because it is commonly spread in day-care centers, it's estimated that every child will have been infected with rotavirus by the age of 5. Fortunately, most children have a mild case, and in countries with easily accessible health care, few children die of rotavirus infections. In developing countries, on the other hand, rotavirus infections are a major cause of death among children younger than 5 years of age every year. Availability of a vaccine, however, has reduced the number of rotavirus-related deaths in young children (less than 5 years old) worldwide, from 527,000 in 2004 to an estimated 215,000 in 2013. In the United States, approximately 20–60 children younger than 5 die each year as a result of rotavirus infection.

Diarrhea is the ailment that comes to mind when we think of digestive tract diseases, but not all diseases of the digestive system involve diarrhea. For example, mumps is an infection of the parotid gland (one of the salivary glands), and hepatitis arises from infections of the liver, which is considered part of the digestive system. We will start our discussion by reviewing the anatomy of the digestive system, including the oral cavity and gastrointestinal tract, and by describing general features of relevant infectious syndromes.

22.1 Anatomy of the Digestive System

SECTION OBJECTIVES

- List the structural components of the digestive system and describe the relationships between them.
- Name the functions of different structures of the digestive system.
- Explain the roles of specialized cells of the intestinal tract (chief cells, M cells, parietal cells, and mucous neck cells).

Which structures of the human digestive system do microbial pathogens encounter? The human digestive system includes the oral cavity; the gastrointestinal, or GI, tract (also called the alimentary canal); and its associated structures: salivary glands, liver, gallbladder, and pancreas (Figure 22.1). The gastrointestinal tract is a long tube extending from mouth to anus, divided into structural and functional sections. Different sections of the GI tract work together to deny ingested bacteria access to the rest of the body, to process food and absorb nutrients, and, ultimately, to expel the unwanted digested material as feces (also called stool). Realize that wherever food goes, microbes go. Pathogens that gain access to the GI tract in food can colonize anywhere from the mouth to the anus; some can penetrate certain tissues to gain access to the circulation.

The mouth opens to the oral cavity, whose basic anatomy was introduced in Section 20.1 (see Figure 20.1). Figure 22.1 shows the anatomy of teeth, an important infection site. Teeth protrude

Figure 22.1 The Gastrointestinal Tract

Viral diseases of the digestive system
Viral gastroenteritis
  Rotaviruses
  Norovirus
  Other pathogenic
    gastrointestinal viruses
Hepatitis A, B, C, D, E
Mumps

Bacterial diseases of the digestive system
Peptic ulcer disease
Intestinal diseases caused by
 Gram-negative bacteria
  Gastroenteritis, *E. coli*
  Shigellosis
  Salmonelloses
  Yersiniosis
  Gastroenteritis,
    Campylobacter
  Cholera
  Noncholera vibriosis
Intestinal diseases caused by
 Gram-positive bacteria
  Listeriosis
  Antibiotic-associated colitis
  Food poisoning

Parasitic diseases of the digestive system
Protozoan infections of the
 digestive system
  Giardiasis
  Amebic dysentery
  Cyclosporiasis
Helminthic infections of the
 digestive system
  Whipworm
  Pinworm
  Tapeworm
  Ascariasis
  Hookworm
  Roundworm
  Trichinosis
  Schistosomiasis

Diseases of the mouth
Tooth decay
Gingivitis and periodontal
  disease
Thrush

Cariogenic bacteria can dissolve the tooth enamel and cause tooth decay. The root is encased in cementum (also a calcium substance). Underneath the hard enamel and cementum shells lies a porous dentin layer, and under that, pulp (made of connective tissue and cells). Nerves and blood vessels penetrate the pulp through the canal inside the tooth's root.

Once food enters the mouth (the oral cavity), it is mixed with saliva produced by the salivary glands and chopped into small pieces through mastication (chewing). Saliva helps lubricate food to protect teeth and soft tissue while salivary amylase and lipase

through a mucosal tissue called gingiva, or gums, which are vulnerable to infection. The part of a tooth protruding outward through the gum is called the crown; the part below the gum is called the root. The crown is covered in an incredibly strong material called enamel (made of hydroxyapatite, a calcium phosphate polymer).

Figure 22.2 Salivary Glands

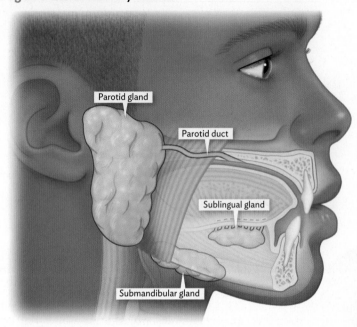

Parotid gland

Parotid duct

Sublingual gland

Submandibular gland

break down starches and some fats. There are several salivary glands; the major ones are the parotid, submandibular, and sublingual glands (**Figure 22.2**). The parotids are the largest of the salivary glands and sit just forward (anterior) of the ears. Their microbiological significance comes from being the infection site of mumps virus.

Once it has left the mouth, masticated food passes through the pharynx to enter the esophagus. The esophagus is a muscular

tube that extends from the pharynx through an opening in the diaphragm and connects to the stomach. The diaphragm is a muscular structure that separates the contents of the abdominal cavity from the contents of the chest. The job of the esophagus is to propel the food toward the stomach. The esophagus is equipped with two sphincters that prevent reflux of digested material from unexpectedly passing up from the stomach and possibly into the airways. The upper esophageal sphincter, made of skeletal muscle, sits near the oropharynx. Farther down, just before the stomach, is the lower esophageal sphincter, made of smooth muscle.

After passing through the esophagus, food enters the stomach, a muscular sac shaped like a J. The serious business of digestion begins here with digestive enzymes and highly acidic (pH 1.5–2) stomach secretions (gastric juice) that prevent growth of most (but not all) microbial pathogens. *Shigella*, for example, has mechanisms that enable it to survive stomach acidity and cause severe gastrointestinal disease (discussed in Section 22.5). The interior stomach lining is made of epithelial cells and is often called the **gastric mucosa** (**Figure 22.3A**). The gastric mucosa is not smooth but is dotted with innumerable tiny holes that are the openings of **gastric pits**. Gastric pits are formed by invaginations of the gastric mucosa and are lined with several specialized cell types, shown in **Figure 22.3B**. Different cell types secrete different digestive enzymes: chief cells secrete lipase and pepsinogen (pepsinogen is converted to pepsin when it comes in contact with stomach acid), parietal cells secrete hydrochloric acid, and mucous neck cells secrete mucus and bicarbonate. The mucus and bicarbonate form a thick mucus layer (the gastric mucosal barrier). This layer coats the gastric mucosa's epithelial cells and protects them from the harsh acidity of gastric juice. *Helicobacter pylori* bacteria can infect the human stomach's mucosa and cause a majority of gastric ulcers (**Figure 22.3C**).

Figure 22.3 The Gastric Mucosa and Gastric Pits

Stomach lumen

Substance secreted

Cell type

Mucous neck cell
- Mucus (protects lining)
- Bicarbonate

Parietal cell
- Gastric acid (HCl)
- Intrinsic factor (B_{12} absorption)

Enterochromaffin-like cell
- Histamine (stimulates acid)

Chief cell
- Pepsinogen/pepsin
- Gastric lipase

D cell
- Somatostatin (inhibits acid)

G cell
- Gastrin (stimulates acid secretion)

Stomach pit

A. The epithelial cells that form the gastric mucosa. The holes are the openings of gastric pits.

B. Diagram of a gastric pit and the various cell types that are involved in secreting mucus and bicarbonate (mucous neck cells), acid (parietal cells), digestive enzymes (chief cells), and cells that control acid secretion (G cells).

C. Arrows show *Helicobacter pylori* bacteria in a stomach pit.

Food in the stomach is broken down physically by stomach muscle contractions and chemically by gastric juice to form "chyme." Chyme eventually moves through another ring-shaped orifice, named the pyloric sphincter, and enters the duodenum, the first part of the human small intestine. The 22 feet of small intestine is coiled serpentlike through the abdomen. The small intestine is divided into three sections: the duodenum, closest to the stomach and shortest in length; the jejunum, the middle section; and the ileum, the longest section, which joins to the large intestine, or colon. Viral and bacterial infections of the intestines usually lead to diarrheal diseases, some of which are just a nuisance, whereas others can be life-threatening. *Salmonella*, for example, is an important GI pathogen that enters the small intestine and attaches to specialized cells (microfold cells, or M cells) present in aggregates of lymphoid tissue called Peyer's patches (Section 15.2). M cells in Peyer's patches normally sample antigens (including bacteria) from the gut lumen and deliver them to underlying immune cells. *Salmonella* uses this route to gain access to the deeper tissues of the gut.

The liver, gallbladder, and pancreas connect to the duodenum, where they deposit bile salts, acids, and pancreatic digestive enzymes that further break down chyme, especially the fatty components. A wavelike motion produced by intestinal muscles (**peristalsis**) propels the chyme through the intestine. Most of the digestion and absorption of nutrients takes place in the small intestine, so when the chyme enters the large intestine, it's almost devoid of nutrients.

The colon (approximately 5 feet long) is also divided into three sections: the ascending colon, transverse colon, and descending colon, which ends in the rectum and anus. The appendix is located at the beginning of the colon, where it is connected to the small intestine. The job of the large intestine is to absorb whatever water and nutrients remain in chyme—a process by which chyme and its bacterial contents are concentrated into a semisolid waste called **feces**. Feces are finally expelled into the rectum and out of the body through the anal sphincter. Expulsion of feces provides a key transmission route for enteric pathogens. For this reason, sanitary disposal of feces is a major public health concern.

SECTION SUMMARY

- **Teeth consist of a crown covered in enamel above the gum and a root below the gum**, encased in cementum. The tooth's interior contains dentin and, at the center, pulp. Nerves enter through a canal inside the root.
- **Food entering the mouth is masticated** by teeth and mixed with saliva made by salivary glands.
- **The esophagus propels food to the stomach**, where acidic secretions released from the gastric mucosa begin breaking food down to chyme.
- Chyme moves through the pyloric sphincter into the duodenum.
- **The gallbladder, liver, and pancreas deliver bile and various enzymes** to chyme in the duodenum. The bile and enzymes help break down the chyme to more digestible particles as the chyme moves through the jejunum and ileum of the small intestine.

- **The large intestine (colon) removes any remaining nutrients and water**, concentrating the chyme into feces, which exit through the anal sphincter.

22.2 Infections of the Oral Cavity

SECTION OBJECTIVES

- List microbes found in the normal microbiota of the oral cavity.
- Analyze how biofilms contribute to infections of the oral cavity.
- Compare the signs and symptoms of common oral cavity infections.
- Explain how oral cavity infections impact diseases elsewhere in the body.

Does our oral cavity have any natural defense against infections? Yes. Human saliva contains proteins, such as lysozyme, that help inactivate the microscopic invaders. The mucosa of the digestive system, including that of the oral cavity, produces immunoglobulin A (IgA), which is thought to have some antimicrobial activity as well. The major defense against microbial infections, however, is left to the extremes of pH in the stomach and small intestine. In this section we will discuss common infectious diseases of the oral cavity caused by bacteria and fungi. (The major viral infection of the oral cavity—cold sores, caused by herpes simplex virus—was discussed in Section 19.2.)

Tooth Decay

How do microbes infect the gum and teeth? Tooth and gum disease is associated with the formation of "plaque," the stuff dental hygienists scrape off your teeth (**Figure 22.4A**). Plaque starts as a biofilm formed by microbiota of the oral cavity at the margin of the teeth and gum. The normal microbiota in plaque include *Streptococcus*, *Lactobacillus*, *Peptostreptococcus*, *Veillonella*, and diphtheroid species (nonpathogenic corynebacteria). Bacteria in the biofilm produce a substance that covers and protects the biofilm from antibiotics and other chemicals. As the biofilm thickens, it is referred to as **dental plaque**.

If not removed, plaque can build up, and the bacteria living there will ferment any sugars left on the teeth and make acidic end products. Acid is the one thing that can dissolve enamel. We develop **dental caries**, or tooth decay (also called a cavity), when acid eats away at the enamel, producing crevices for more bacteria to grow and make more acid that dissolves more enamel (**Figure 22.4B**). Eventually, the decay penetrates to the nerve and we feel pain. Two major types of bacteria, *Streptococcus mutans* and *Lactobacillus* species, cause most dental caries. *Lactobacillus*, however, is also an important probiotic. So please, before giving up yogurt, reread the Chapter 14 opening case.

Dentists treat dental caries by removing decay (drilling) and filling the cavity with resins. Eating less sugar, brushing, and flossing regularly all help prevent the development of caries.

Figure 22.4 Dental Caries and Gingivitis

A. Dental hygienists use sharp tools to remove plaque from your teeth.

B. Bacteria in plaque produce acids that dissolve enamel to form a cavity (dental caries).

C. Heavy plaque buildup (called tartar) can trigger inflammation of the gingiva around your teeth, the beginning of gingivitis.

A dentist can also seal your teeth with a resin coating to prevent the formation of surface pits produced by grinding teeth. Surface pits serve as a niche where plaque can form. Finally, drinking fluoride-treated water (provided by most municipalities) helps harden tooth enamel.

Gingivitis and Periodontal Disease

Plaque left on the teeth eventually absorbs minerals and thickens into dental caries. As tartar forms below the gumline, it causes inflammation of the gums, called **gingivitis** (**Figure 22.4C**), which can lead to more serious **periodontal disease**, or periodontitis (**Figure 22.5**). In periodontitis, the gum bleeds easily and recedes from the tooth. As the disease progresses, the density of bone holding the tooth in place also decreases. Periodontitis is most often caused by *Aggregatibacter* (formerly *Actinobacillus*) *actinomycetemcomitans* and *Porphyromonas gingivalis,* but other species, particularly *Bacteroides forsythus,* are also involved.

Dental caries and periodontal disease can cause tooth loss or abscess formation when the infection spreads to the pulp. Treatment of these infections depends on the type of infection and almost always includes systemic antibiotic therapy. As you may recall from

Figure 22.5 Periodontal Disease

Gingivitis can lead to periodontitis (periodontal disease), in which the gums withdraw from the teeth (arrow).

the Chapter 8 opening case, the symptoms of periodontal disease include painful, swollen gums that easily bleed when touched.

In its most serious form, gingivitis can become a necrotizing infection (necrotizing ulcerative gingivitis) commonly called **trench mouth** or **Vincent's angina**. Trench mouth is easy to recognize. Symptoms include painful, bleeding, bright red gums covered with a gray film. Craterlike canker sores form between the teeth and gums, and the patient has bad breath, a foul taste in the mouth, and fever. Swollen lymph nodes can appear around the head, neck, or jaw. Eating and swallowing are painful. Treatment of trench mouth includes the removal of dead gum tissue, followed by hydrogen peroxide mouthwash and antibiotics.

Periodontal infections involve complex microbiomes that can include 200–500 species of bacteria, including pathogens such as *P. gingivalis.* The proximity of these organisms to the circulation can facilitate bacteremia and the systemic spread of bacterial products able to produce systemic inflammation. Evidence suggests that bacteria associated with periodontitis may contribute to adverse pregnancy outcomes (miscarriages, for instance). These bacteria are also thought to be linked to diseases such as cancer, atherosclerosis, and others. Some oral pathogens, like *Fusobacterium nucleatum,* can cause pneumonia if they are aspirated into the lung. It is likely that effective management of periodontal disease and proper oral care may lessen the incidence or severity of some systemic diseases.

Thrush

Thrush, a disease common to neonates and infants, is first noticed as a white coating in the child's mouth resembling cottage cheese. The disease is caused by a species of yeast, *Candida albicans.* Thrush lesions often start as tiny areas that enlarge to white patches on oral mucosa (**Figure 22.6**). *C. albicans* can colonize the adult mouth, vagina, and other areas, some of which can be sources

Figure 22.6 Thrush

Lesions caused by *Candida albicans* often start as tiny areas that enlarge to white patches on oral mucosa. In adults, thrush is a sign of a compromised immune system.

of infection during birth or after. Topical treatment (oral spray or a swab) with a nystatin solution is curative. Adults with severely compromised immune systems or those on lengthy antibiotic therapy are likely to develop thrush.

SECTION SUMMARY

- **Plaque forms as a biofilm** at the border of teeth and gums.
- **Cavities (caries) form as bacteria in plaque ferment carbohydrates** and make acid that dissolves enamel. *Streptococcus mutans* and *Lactobacillus* species cause most acid production.
- **Gingivitis forms when plaque builds to tartar**, which starts an inflammation at the gumline.
- Periodontal disease eventually develops if gingivitis is left untreated. Gums recede and bone density declines.
- Trench mouth is a serious necrotizing infection that develops from neglected gingivitis.
- **Thrush is an infection of oral mucosa** caused by the yeast *Candida albicans*.

22.3 Gastrointestinal Syndromes

SECTION OBJECTIVES

- List the most common signs and symptoms associated with gastrointestinal infections.
- Correlate microbial pathogenic mechanisms with the presenting signs and symptoms of gastrointestinal infections.
- Outline the general mechanisms of diarrhea.

Why does the gastrointestinal tract get so many infections? The oral cavity and the GI tract are portals between the inner us and the outer world; thus, numerous microbes pass through them. Some of these microbes overstay their welcome to erode teeth and

destroy gums, as described in the previous section, or they infect GI tissues and impair their function. Each gastrointestinal pathogen has what you might call its own tissue proclivities or specificities (tropisms), so different microbes infect different parts of the digestive tract. For example, M cells (Section 15.2) are used by *Shigella* or *Salmonella* as a conduit to cause infection. Before we delve into the variety of viruses, bacteria, and parasites involved, a general description of the attendant diseases will set the stage.

Types of Diarrhea

What is diarrhea? **Diarrhea** is defined by the World Health Organization as having three or more loose (not formed) stools per day. Normally, intestinal mucosal cells absorb water (8–10 liters per day) <u>from</u> the intestinal contents, essentially drying out stool. Diarrhea occurs when water is not absorbed or when it actually <u>leaves</u> intestinal cells and <u>enters</u> the intestinal lumen. The excess water loosens stool and <u>voilà</u>!—diarrhea. Several situations can cause water to be drawn from the intestinal lining:

- **Osmotic diarrhea** follows the intake of nonabsorbable substrates, such as lactulose, a synthetic sugar, which is used to treat constipation. Because lactulose is not absorbed, it remains within the intestinal lumen and increases osmolarity (concentration of solute particles per liter). When the osmolarity of intestinal contents is greater than the internal osmolarity of mucosal cells, water leaves the cells and enters the lumen in an attempt to equalize osmotic pressure. The result is osmotic diarrhea. Infectious organisms (such as rotavirus) that prevent nutrient absorption can cause osmotic diarrhea. Osmosis is discussed in Section 4.5.

- **Secretory diarrhea** develops when microbes cause mucosal cells to increase ion secretion, as seen with cholera toxin. The secretion (exit) of electrolytes produces an electrolyte imbalance across the cell membrane (more electrolytes outside than inside the cell). Water then leaves the cells to equilibrate the imbalance.

- **Inflammatory diarrhea** occurs when the mucosal lining of the intestine is inflamed (red and swollen), as through an infection. In response to infection, the release of inflammatory cytokines attracts polymorphonuclear leukocytes (PMNs) that inadvertently damage the epithelial layers. Damage to the intestinal wall diminishes the intestine's ability to absorb nutrients and water. If the damage is severe enough, red and white blood cells can enter the stool (a condition called bloody diarrhea, or dysentery). *Shigella* and *Salmonella* species, among others, can cause infections leading to inflammatory diarrhea.

- **Motility-related diarrhea** occurs when food moves so quickly through the intestine (hypermotility) that there isn't sufficient time to absorb water or nutrients. Among various causes of intestinal hypermotility are enterotoxins (toxins that are specific for intestinal mucosa) produced by some microbes, such as rotavirus (described later).

Stress, food allergies, and a number of microbial etiologies cause inflammation of different parts of the GI tract, which often manifests as diarrhea. The following general terms are used to describe the inflammation of different parts of the GI tract:

- **Gastritis** refers to inflammation of the stomach lining.
- **Gastroenteritis** has evolved into a nonspecific term for any inflammation along the GI tract.
- **Enteritis** is defined as inflammation mainly of the small intestine.
- **Enterocolitis** is inflammation of the colon and small intestine.
- **Colitis** refers to inflammation of the large intestine (colon).

You might wonder why some microbes prefer the chaos of diarrhea to the relative stability of a nice commensal relationship. After all, isn't finding a niche and sticking to it the goal of every microbe? The simple answer is that diarrhea enables the dissemination of microbes that might otherwise kill their host or be killed themselves after too aggressively provoking the host's immune system. Dissemination helps the microbe reach new hosts and proliferate. A diarrhea-causing microbe is like a serial bank robber fleeing from city to city to avoid capture and find more banks to rob.

Signs and Symptoms of Hepatitis

Hepatitis is an inflammation of the liver (Figure 22.7) in response to infection, overconsumption of alcohol, exposure to toxic chemicals (such as the dry-cleaning solvent carbon tetrachloride), some medications (including over-the-counter medications such as acetaminophen, or Tylenol), or autoimmune reactions. Several viruses from different virus families can infect liver cells, as can some amebas and worms. The inflammatory response to these infections damages the liver and ultimately causes hepatitis. Symptoms (see Figure 22.7) include an enlarged, painful liver; jaundice (yellowing of eyes or skin); dark urine; pale or clay-colored stool; low-grade fever; nausea and vomiting; and weight loss.

Because the liver is damaged, bilirubin (an orange-yellow breakdown product of heme that is normally cleared by the liver) spills over into the bloodstream (hyperbilirubinemia) and is secreted in urine. Hyperbilirubinemia manifests in the patient as jaundice and dark urine. The liver also uses bilirubin to make bile, which is deposited into the small intestine and produces the brown color of feces. A damaged liver fails to deliver adequate amounts of bile, resulting in clay-colored stool. A key laboratory indicator of liver damage is the elevation of liver enzymes, such as aspartate aminotransferase (AST) and alanine aminotransferase (ALT), in serum. The various causes of viral hepatitis will be described in the next section.

SECTION SUMMARY

- **Infectious diarrhea can be classified** as osmotic, secretory, inflammatory, or motility related.
- **Inflammation at different intestinal locations** produces gastritis (stomach), enteritis (small intestine), colitis (colon), enterocolitis (distal

Figure 22.7 Hepatitis

A. Healthy liver.

B. Diseased liver resulting from chronic hepatitis.

C. Yellowing of the conjunctiva of the eye in jaundice.

D. Hyperbilirubinemia is a symptom of hepatitis in which urine turns brown (bilirubinuria) from excess bilirubin secreted from a damaged liver.

ileum and proximal colon), or gastroenteritis (stomach and small intestine. "Gastroenteritis" is also a general term for gastrointestinal symptoms).

- **Hepatitis is inflammation of the liver.** Symptoms include jaundice, dark urine, and clay-colored stool. The elevation of serum liver enzymes AST and ALT is an indicator of liver damage.

22.4 Viral Infections of the Gastrointestinal Tract

SECTION OBJECTIVES

- Match gastrointestinal symptoms to the viral infections that cause them.
- List routes of transmission of specific viral diseases of the gastrointestinal tract.
- Discuss viral pathogenic mechanisms that lead to specific disease signs and symptoms in the gastrointestinal tract.

How do viruses affect your GI tract? A wide range of viruses can affect your digestive system. Their effects range from swollen salivary glands to diarrhea and hepatitis. Some viral infections of the GI tract can even spread to the brain. In fact, viral infections cause the majority of diarrheal disease worldwide, as well as deaths

Table 22.1

Viral Diseases of the Digestive System

Disease	Symptoms	Complications	Virus	Transmission	Diagnosis	Treatment	Prevention[a]
Acute gastroenteritis (mostly children under 3)	Watery, dark green diarrhea	Dehydration	Rotavirus	Fecal-oral route	Rapid detection of viral antigen in stool, qRT-PCR[b] for RNA to identify strains	Rehydration, treatment of symptoms	Vaccines available
Acute gastroenteritis, also called stomach flu (older children and adults)	Watery diarrhea, vomiting, headache, fever, chills, myalgia	Dehydration	Norovirus	Fecal-oral route	qRT-PCR for RNA	Rehydration, treatment of symptoms	No vaccine available
Acute gastroenteritis (mostly adults)	Watery diarrhea, vomiting, headache, fever, chills, myalgia	Dehydration	Sapoviruses	Fecal-oral route	Electron microscopy, qRT-PCR	Rehydration, treatment of symptoms	No vaccine available
Acute gastroenteritis (people with weakened immune system)	Diarrhea, vomiting, abdominal pain, fever	Dehydration, severe disease	Adenovirus, types 40 and 41	Fecal-oral route	EIA,[c] PCR for RNA	Self-limiting; rehydration	No vaccine available
Hepatitis	Variable degree of nausea, vomiting, abdominal pain, jaundice, dark urine, clay-colored stool	Vary depending on causative viral agent; see Table 22.2	Hepatitis A–E viruses (from different families)	Varies depending on the virus; see Table 22.2	Varies depending on the causative viral agent; see Table 22.2	Varies depending on the causative viral agent	Vaccines available against hepatitis A and B viruses; vaccine against hepatitis E available in China only
Mumps (parotitis)	Prodromal chills, fever, headache, swollen parotid glands, pain with chewing and swallowing	Meningitis, encephalitis, orchitis	Mumps virus	Aerosolization, fomites	Viral culture, qRT-PCR, serum IgM or rise in IgG titer	Supportive treatment	Vaccine available (childhood immunizations)

[a]The childhood vaccine schedule is shown in Table 17.2.
[b]qRT-PCR = quantitative reverse transcription–PCR.
[c]EIA = enzyme immunoassay to detect viral protein.

from diarrheal disease. Some viruses that affect the digestive system are spread by aerosols, others rely on fecal-oral transmission (Section 2.4), and some hepatitis viruses are sexually transmitted. Table 22.1 presents features of common viral diseases that affect the digestive system.

Viral Gastroenteritis

Many people wrongly think that most cases of diarrhea are caused by a bacterial agent. Actually, viruses cause more intestinal disease than do any bacterial species. Rotavirus has been the main cause of infectious diarrhea, but its incidence is declining now that we have

an effective vaccine. Lesser-known viruses that cause gastroenteritis include members of the positive-sense, single-stranded RNA viruses of the Caliciviridae family (similar to norovirus); adenovirus strains 40 and 41 (double-stranded DNA), which can cause diarrhea that lasts 10–14 days; astroviruses (single-stranded RNA); and toroviruses (positive-sense, single-stranded RNA), which affect primarily infants under 1 year of age.

ROTAVIRUS GASTROENTERITIS Rotavirus is a nonenveloped, segmented, double-stranded RNA virus spread by the fecal-oral route. The virus is highly infectious, is endemic around the globe, and affects all age groups, although children between 6 and

24 months are the most severely affected. The Centers for Disease Control and Prevention (CDC) estimates that by age 5, all children have had a rotavirus infection.

The rotavirus incubation period is approximately 2 days, after which the patient suffers frequent episodes of watery, dark green, explosive diarrhea. The dark green color of the stool is caused by increased bile excretion. The patient can also feel nauseated, vomit, and have abdominal cramping. Diarrhea is caused in part by the virus destroying the intestinal epithelial cells, thus disrupting the absorptive function that these cells normally perform (and causing osmotic diarrhea). In addition, a virus-encoded **enterotoxin** (a toxin that affects the intestines) stimulates intestinal nerves to overactivate intestinal motility to produce hyperperistalsis (increased intestinal motility). This enterotoxin also stimulates the vagus nerve, a long cranial nerve (cranial nerve X, the tenth cranial nerve) connecting the brain to many organs, including the stomach. Overstimulation of the vagus nerve causes stomach muscles to involuntarily contract, making the patient vomit.

Severe dehydration and electrolyte loss due to the diarrhea will cause death unless supportive measures, such as fluid and electrolyte replacement, are undertaken. There is no curative treatment, but most patients recover if rehydrated properly, as did Dao Ming in the opening case history of this chapter. Few deaths from rotavirus occur in the United States. The global impact of this disease is significant. For instance, more than 200,000 children worldwide died from this viral diarrhea in 2016. The mortality and incidence of this disease has declined since 2006 in the United States because of the availability of rotavirus vaccines. Two different rotavirus vaccines are currently licensed for use by US health care providers. Both vaccines are administered as oral drops. The first dose of both should be administered before the infant is 15 weeks of age. The routine childhood immunization schedule is shown in Table 17.2. Protective antibodies conferred by vaccination or infection are primarily IgA (Chapter 16) and can be transferred in mother's milk.

NOROVIRUS GASTROENTERITIS With the advent of the rotavirus vaccine, **norovirus** (formerly Norwalk virus) is becoming the most common nonbacterial cause of gastroenteritis worldwide. Norovirus is a nonenveloped, positive-sense, single-stranded RNA virus transmitted by the fecal-oral route (**Figure 22.8A**). The classic scenario for norovirus outbreaks is the cruise ship in which tens of people get violently ill simultaneously, as a result of contaminated food and direct person-to-person contact (**Figure 22.8B**). Norovirus outbreaks are actually more common wherever many people are in a small area. The virus can spread quickly in nursing homes, dormitories, hotels, and so on.

About 24 hours after infection, patients experience a sudden onset of vomiting, abdominal cramps, and watery diarrhea. As with rotavirus infections, diarrhea results when norovirus destroys the absorptive capacity of the cells that line the intestinal lumen. Norovirus infections are sometimes called "stomach flu" because, in addition to diarrhea, patients suffer from headache, fever, chills, and muscle aches (myalgia)—symptoms similar to those of

Figure 22.8 Norovirus

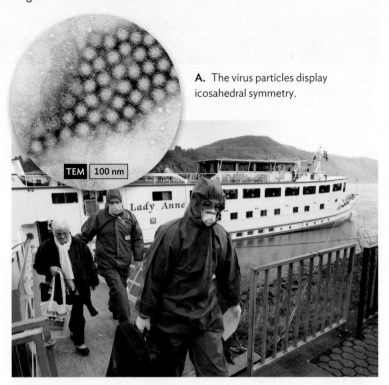

A. The virus particles display icosahedral symmetry.

TEM | 100 nm

B. An outbreak of norovirus diarrhea hit the cruise ship *Lady Anne* on the Rhine River in Germany in November 2008. Here, an elderly woman is leaving the cruise ship with two Red Cross paramedics in full protective gear.

influenza. The symptoms mercifully last only about 12–24 hours. All age groups are affected, although the principal victims are adults. Treatment is supportive, and no vaccine is available.

Hepatitis

CASE HISTORY 22.1
"Hurling" in Hawaii

Three weeks ago, Katherine, a 17-year-old from Waikiki, Hawaii, developed nausea, fever, and aches. She initially thought it was the flu. She felt extremely tired, dozed through school auditorium events, and was unable to participate in golf tournaments or the school debating team. About 2 days ago, Katherine could not keep food down and was taken to the emergency department (ED), where she was given an antinausea shot. The shot didn't work; she continued to vomit and made a second visit to the ED. She noted that her urine seemed dark, and she told the physician that she had not been hospitalized before this event and had not traveled outside the United States. She denied using drugs or having any sexual partners. The practitioner noticed a yellow tinge to the white part of her eyes. A blood test showed that she was positive for anti–hepatitis A virus (HAV) antibodies.

Table 22.2
Characteristics of Hepatitis Viruses

Feature	Hepatitis A Virus (HAV)	Hepatitis B Virus (HBV)	Hepatitis C Virus (HCV)	Hepatitis D Virus (HDV)	Hepatitis E Virus (HEV)
Common Name	Infectious hepatitis virus	Serum hepatitis virus	Posttransfusion non-A, non-B hepatitis virus	Delta agent	Enteric non-A, non-B hepatitis virus
Virus Family and Structure	Picornavirus; nonenveloped; positive-sense ssRNA	Hepadnavirus; enveloped; partially ds, circular DNA	Flavivirus; enveloped; positive-sense ssRNA	Viroid-like; enveloped; circular ssRNA	Orthohepevirus: calicivirus-like capsid; nonenveloped; positive-sense ssRNA
Transmission	Fecal-oral	Parenteral, sexual (blood/body fluid transmission)	Parenteral, sexual (blood/body fluid transmission)	Parenteral, sexual (blood/body fluid transmission)	Fecal-oral
Onset	Abrupt	Insidious	Insidious	Abrupt	Abrupt
Incubation Period (Days)	15–50	60–150	14–180	15–64	15–60
Severity	Mild	Occasionally severe	Usually subclinical; 70%–80% chronicity (become chronic)	Usually mild when it coinfects with HBV; often severe when it superinfects an HBV carrier	Nonpregnant patients, mild; pregnant women, severe
Diagnosis	Clinical presentation and history; anti-hepatitis A virus IgM test	Clinical presentation; serum levels of HBsAg, HBeAg, and anti-HBc IgM[a]	Serology; molecular assay for HCV RNA	Detection of HDV RNA by qRT-PCR; detection of serum IgG and IgM antibodies against hepatitis D	Clinical presentation and history; IgM anti-HEV in the serum by ELISA
Treatment	Supportive care	Interferon-α; antiviral drugs; vaccine against hepatitis A to prevent secondary HAV infection	Treatment plans are individualized and may include interferon-α and/or ribavirin; vaccine against hepatitis A and B	Interferon-α; antiviral lamivudine has been tried, but efficacy is questionable; liver transplant	Usually self-limiting; nucleoside reverse transcriptase inhibitors in severe cases
Chronic Status (Carrier State)	No (only acute)	Yes (acute or chronic)	Yes (acute or chronic)	Yes (acute or chronic)	No (acute or chronic)
Associated with:	No other diseases	Primary hepatocellular carcinoma, cirrhosis	Primary hepatocellular carcinoma, cirrhosis	Cirrhosis, fulminant hepatitis	No other diseases

[a]HBsAg is an HBV surface antigen; HBeAg is an HBV protein present during active infection, and HBc is an HBV capsid protein and another indicator of active infection.

Katherine's symptoms are typical of acute viral hepatitis caused by one of the hepatitis viruses: A, B, C, or E. (There is also a hepatitis D virus, but HDV is defective and produces disease only if it coinfects a patient with hepatitis B virus.) The initial disease symptoms of fever, nausea, vomiting, dark urine, clay-colored stool, and often jaundice can be seen in all patients suffering from hepatitis, regardless of the viral cause. Despite producing similar symptoms, the different viruses that cause hepatitis do not even belong to the same virus family (Table 22.2). They have different morphologies and different genome structures, and ultimately produce different clinical pictures. The good news about hepatitis is the ongoing decline in disease incidence. The not-so-good news is that hepatitis diseases are still dangerous and often incurable.

HEPATITIS A Hepatitis A virus (HAV) is a nonenveloped, positive-sense, single-stranded RNA virus transmitted by the fecal-oral route, so overcrowding and lack of personal hygiene can increase the risk of getting this highly contagious disease. Eating undercooked seafood collected from contaminated waters and traveling to developing countries are also risk factors for HAV infection. Once inside the body, the virus replicates in the liver during an incubation period of 2–7 weeks. By the time jaundice appears, the number of virus particles in the blood and feces has started to drop, making the virus difficult to find. Fortunately, our immune system eventually comes to our rescue, and hepatitis A infections almost always go away without medical intervention (they are self-limiting) in healthy people. Hepatitis A has no asymptomatic carrier state or links to future liver disease.

Health care professionals use serological tests to determine whether a person is currently infected with hepatitis A virus or has previously been infected. Acutely infected individuals have anti-HAV antibodies of the IgM type (Section 16.2), as in Katherine's case, whereas those who were previously infected with the virus (or vaccinated) will have anti-HAV antibodies of the IgG type (and probably little or no IgM). People infected with this virus acquire long-term immunity. Children are now routinely immunized with HAV-inactivated viral vaccines.

HEPATITIS B Hepatitis B virus (HBV), an enveloped DNA virus, is transmitted through the exchange of blood or other body fluids. For instance, blood is exchanged during sexual intercourse or when drug users share needles. HBV does not have a fecal-oral transmission. It is the only hepatitis virus that has a DNA genome, which is small but codes for four viral proteins. Health care professionals can identify the extent and severity of HBV disease by immunologically detecting either the proteins (for example, HBsAg) or the antibody response to them. HBsAg is an HBV surface antigen made in such large amounts during acute infection that it accumulates as tubular structures in the blood (Figure 22.9).

Once a person becomes infected, there is a relatively long incubation period of 8–21 weeks before typical hepatitis symptoms slowly (insidiously) appear. Only 30%–50% of infected immunocompetent individuals older than 5 years of age become symptomatic with this acute disease, and only 1% of them develop

acute liver failure. Most people who are infected develop long-term chronic infections lasting years. Chronically infected but otherwise asymptomatic individuals essentially serve as the reservoir for HBV. Immunosuppressed patients and children younger than 5 years of age rarely develop acute disease because they lack an adequate inflammatory response. They are asymptomatic but remain chronically infected.

HBsAg, as well as IgM to HBsAg, can be found in the blood as soon as symptoms appear. In acute infections, levels of both decrease as symptoms decrease. If HBsAg persists in blood samples for longer than 6 months, the patient has chronic HBV. IgM antibody will decrease even in chronically infected patients, but IgG antibodies to hepatitis will remain high.

During chronic infection, HBV DNA and antigens are found in organs and tissues other than the liver and may be the reservoirs for chronic infection. Two-thirds of people with chronic HBV show no symptoms, but the other third have a chronic active infection in which the liver is slowly and continuously damaged over time. The accumulated damage leads to cirrhosis (scarring of the liver) and liver failure. Chronic active infection also increases the risk of liver cancer. Note that patients chronically infected with HBV can develop a rapidly fatal disease if infected with hepatitis D virus.

The acute infection is treated only with supportive measures. Treatment of chronic hepatitis B infections includes long-term antiviral medications to suppress viral replication, but the length of treatment depends on the patient. The best weapon against becoming infected with HBV is prevention. Vaccination against this virus is now part of routine childhood immunizations. The vaccine consists of purified HBsAg prepared from recombinant yeast that were engineered to produce this protein in quantity.

Hepatitis D virus (HDV) is a positive-sense, single-stranded RNA virus that has an HDV core and an HBsAg coat derived from a coinfecting (simultaneously infecting) hepatitis B virus. HDV cannot replicate unless it coinfects with HBV or infects a chronic carrier of HBV. The result is a rapidly progressive disease that destroys the liver. Fortunately, HDV infections are uncommon in the United States. There are no vaccines available for hepatitis D.

HEPATITIS C Hepatitis C virus (HCV) is an enveloped, positive-sense, single-stranded RNA virus transmitted by blood and body fluids (as described in the case history that opens Chapter 12). HCV causes an acute infection within the first 6 months of exposure. The acute infection is mild and self-limiting, so the patient often does not show any symptoms. Liver function (and serum enzyme levels) in some patients can even return to normal after the initial acute disease. Patients who do not clear the virus after 180 days are diagnosed as having chronic disease. Most people infected with hepatitis C virus (70%–80%) develop a chronic infection, even with normal liver function.

In the United States, an estimated 2.4 million individuals have hepatitis C. In 2017, there were 17,253 reported HepC-related deaths. Most people with chronic hepatitis C infection (60%–70%)

Figure 22.9 Hepatitis B

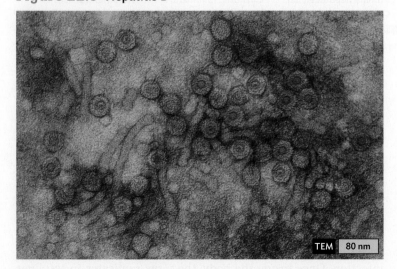

The dark brown circles are infectious hepatitis B virions. The long tubular structures are made from HepB surface antigen, HBsAg (colorized TEM).

have active liver disease, including fatty liver (accumulation of fat in the liver) leading to liver failure, and are at risk of developing liver cancer (hepatocellular carcinoma). The less intense inflammation process associated with HCV means that most chronic infections take a long time to cause liver failure. The stronger the immune response, the more rapidly liver disease develops.

Diagnosis for both acute and chronic disease is made by serology to detect antibody to hepatitis C and molecular assays that detect or quantify HCV viral DNA. The CDC recommends that people at risk be screened for this infection and that a specialist be consulted about the treatment of those who have the disease. HCV treatment is based on the guidelines set by the Infectious Diseases Society of America (IDSA) and the American Association for the Study of Liver Diseases (AASLD). Acute HCV infections are effectively treated with interferon (IFN) and the antiviral drug ribavirin, which reduces the synthesis of viral DNA. A number of effective FDA-approved antiviral medications are available, and some result in cure rates over 90%. The antiviral or combination of antivirals used depends on the HCV strain infecting the patient and on its genome.

It is important to remember that a person infected with any strain of HepC virus who has been treated and cleared of that strain can become infected with a different strain of HepC. So, patients cured of HepC should continue avoiding risk factors, such as using IV drugs and having sex with HepC-infected individuals, to prevent reinfection. No vaccines are available for hepatitis C.

HEPATITIS E Hepatitis E virus (HEV) is a positive-sense, single-stranded RNA virus that is rare in the United States but common in many other parts of the world. The virus is transmitted through the fecal-oral route, usually when a person drinks contaminated water. After an incubation period of 2–8 weeks, often a mild and self-limiting disease develops. The patient experiences muscle aches, fatigue, mild fever, nausea, loss of appetite, and mild icteric (jaundice) symptoms. Fulminant disease—one that develops rapidly and is serious—is more likely in people who are pregnant, already have liver disease, or are malnourished. There is some evidence of vertical transmission (from mother to fetus) when the mother is infected during the third trimester of pregnancy. Chronic HEV infection is increasingly seen in solid-organ transplant recipients and other immunosuppressed patients in developed countries.

Hepatitis E infection can be diagnosed by serology. Detection of the virus in bile, stool, or liver cells by qRT-PCR or immuno-electron microscopy is also possible; however, commercial tests are not currently available, and samples can be processed only by specialized labs. A vaccine has been developed but is not approved for use in the United States. Treatment is supportive.

Mumps

Mumps is a highly infectious, self-limiting, viral infection of the parotid gland caused by a **paramyxovirus**, a single-stranded RNA virus of the Paramyxoviridae family (a different paramyxovirus

Figure 22.10 Mumps

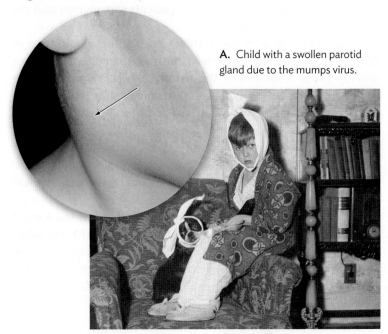

A. Child with a swollen parotid gland due to the mumps virus.

B. Before the advent of the mumps vaccine in 1967, children with mumps were generally quarantined at home as a way to contain an epidemic. The head wrap was used to limit jaw movement that caused pain.

causes measles). The virus is shed in saliva and can be spread through sneezing, coughing, and other direct contact with an infected person's saliva. The virus initially replicates in the upper respiratory tract and then hitchhikes through the bloodstream to the parotid gland. Prodromal symptoms become noticeable 2 or 3 weeks after infection and include chills, headache, loss of appetite, and a lack of energy. Swelling of the salivary glands in the face (**parotitis**) generally occurs within 12–24 hours of the prodromal symptoms and usually disappears by the seventh day (**Figure 22.10**). Patients often complain of pain when chewing or swallowing and can have a fever as high as 40°C (104°F). Fortunately, after contracting mumps people become immune to the disease, regardless of how mild or severe their symptoms were.

Routine childhood vaccination has made mumps an uncommon occurrence. Unvaccinated children and adults, however, are still at risk. Although most cases of mumps are uncomplicated, complications can occur. Surprisingly, adults who contract mumps are more prone to disease sequelae. In one complication, seen in 15% of adult cases, the covering of the brain and spinal cord becomes inflamed—a condition known as meningitis. Symptoms of viral meningitis include a stiff neck, headache, vomiting, lack of energy, and pain with bending the head or flexing the neck. Mumps meningitis usually resolves within 7 days and has no lingering effects. However, the mumps virus can spread into the brain itself, causing inflammation (encephalitis). Symptoms of mumps encephalitis include seizures, an inability to feel pain, and high fever. Patients typically recover completely within 14 days, although some are left with seizure disorders. About 1 in 100 patients with mumps encephalitis die from the complication.

Swelling of the testicles (**orchitis**) is a painful complication in 25% of postpubescent males who contract mumps. Orchitis can develop about 7 days after the parotitis stage. Painful swelling can occur in one or both testicles, accompanied by fever, nausea, and headache. Symptoms usually subside after 5–7 days, although the testicles can remain tender for weeks. Mumps orchitis sometimes causes long-term complications of testicular atrophy and infertility. Girls occasionally suffer an inflammation of the ovaries, or **oophoritis**, as a complication of mumps, but this condition is less painful than orchitis in boys.

The first vaccine against mumps was licensed in the United States in 1967 and is now a component of the MMRV (measles, mumps, rubella, and varicella multivalent vaccine). The vaccine is given in two doses, at 1 and 4 years of age. Because of widespread vaccination, the disease rate in the United States has fallen by 99%.

SECTION SUMMARY

- **Rotavirus is the most common cause of diarrhea**, especially among children. Transmitted by the fecal-oral route, the virus causes osmotic (antiabsorptive) diarrhea and enterotoxin-induced hyperperistalsis. A vaccine is available and recommended for children.

- **Norovirus is a major cause of diarrhea among adults** (antiabsorptive mechanism). Fecal-oral and person-to-person transmission can cause outbreaks among people in close quarters. No vaccine is available, but disease resolves quickly.

- **Hepatitis is caused by five different viruses.** Hepatitis A virus (HAV) is transmitted by the fecal-oral route (often via contaminated seafood). A vaccine is available and recommended for children.

- **Hepatitis B virus (HBV) can cause acute or chronic hepatitis.** It is transmitted by blood or other body fluids. A vaccine containing HBsAg is recommended for older children. Antiviral treatments are used for active disease. **Hepatitis D virus (HDV)** is defective (viroid-like) and infects only with HBV. HDV infection increases the severity of disease in patients with chronic HepB.

- **Hepatitis C virus (HCV) can cause a mild, acute infection that can resolve quickly but often becomes chronic** and will slowly, over years, produce liver damage. There is no vaccine.

- **Hepatitis E virus (HEV) is usually a mild disease** transmitted through the fecal-oral route. A vaccine has been developed but is not approved for use in the United States.

- **Mumps is a viral infection of the parotid gland (parotitis)** that spreads through aerosolization and disseminates via blood to the parotid gland. Mumps vaccine is recommended for children.

Thought Question 22.1 If hepatitis viruses all replicate in the liver and are released into the intestine in bile, why is it that only HAV and HEV can be transmitted by the fecal-oral route?

Thought Question 22.2 Why are people immunized against HBV also protected from HDV?

22.5 Bacterial Infections of the Gastrointestinal Tract

SECTION OBJECTIVES

- Categorize GI infections by bacterial pathogenic mechanisms.
- Correlate bacterial virulence mechanisms to GI disease symptoms.
- Relate the treatment of GI infections to virulence of GI pathogens.
- Evaluate treatment and prevention methods used for bacterial infections of the GI tract.

How do bacterial GI infections compare with infections caused by viruses? Acute bacterial infections of the gastrointestinal tract tend to be more severe than viral infections. A large part of this difference is due to the toxins bacteria produce and the inflammatory response they provoke. On the other hand, bacterial infections can usually be treated with antibiotics, whereas viral infections cannot. Table 22.3 summarizes the bacterial diseases described in this section.

CASE HISTORY 22.2

It's Not What You Eat

Gary is a 34-year-old accountant who immigrated to Nebraska from Poland 7 years ago. Since his teenage years he has been bothered by episodes of epigastric pain (pain around the stomach), nausea, and heartburn after eating certain foods. Gary was told to avoid those foods and take antacids as needed. Antacids usually alleviated the symptoms. Over the years, he received several courses of treatment with cimetidine (Tagamet) or famotidine (Pepcid) to reduce acid secretion and provide relief. Recently, Gary's physician ordered an upper-GI endoscopy, in which a long, thin tube tipped with a camera and light source was inserted into Gary's mouth and down into his stomach (Figure 22.11). The view through the endoscope showed some reddened areas in the antrum (bottom part) of the stomach. The endoscope was also equipped with a small clawlike structure that was used to obtain a small tissue sample (biopsy) from the lining of Gary's stomach. A urease test performed on the antral biopsy turned positive in 20 minutes. Histological examination of the biopsy sample confirmed moderate chronic active gastritis (inflammation of the stomach lining) and revealed the presence of numerous spiral-shaped organisms. Cultures of the antral biopsy sample were positive for the bacterium *Helicobacter pylori*.

Peptic Ulcers

Painful and sometimes life-threatening gastric and duodenal ulcers, collectively known as **peptic ulcers** (Figure 22.12), were for many years blamed on spicy foods and stress. These factors

Table 22.3
Bacterial Diseases of the Digestive System

Disease	Symptoms	Complications	Bacteria	Transmission	Diagnosis	Treatment	Prevention
Gram-Negative Bacteria							
Cholera	Acute large-volume (10–14 liters) watery diarrhea ("rice water" stool), thirst, altered mental status	Electrolyte imbalance leading to cardiac and renal failure and death	*Vibrio cholerae*	Fecal-oral route	Isolation from stool	Replacement of fluids and electrolytes; antibiotics in cases of severe dehydration; antibiotic choice and dosage depend on age (adult vs. children of different ages): doxycycline, azithromycin, ciprofloxacin	Vaccine available but not in the US
Colitis, *E. coli* (bacillary dysentery)	Bloody diarrhea, fever	Dehydration, chronic diarrhea, septicemia, pneumonia	Enteroinvasive *E. coli*[a]	Fecal-oral route	Isolation from stool	Possibly fluoroquinolones; rehydration	No vaccine available
Enterocolitis, *Campylobacter*	Prodrome of fever and malaise leading to watery and bloody diarrhea, periumbilical[b] abdominal pain	Guillain–Barré syndrome (*C. jejuni*)	*Campylobacter* spp.	Fecal-oral route	Isolation from stool; dark-field microscopy; Gram stain with carbolfuchsin counterstain	Rehydration; antibiotics may be required in severe cases and cases with high fever	No vaccine available
Enterocolitis, *E. coli*	Acute (often bloody) diarrhea, abdominal cramps, painful defecation	HUS, TTP[c]	Entero-hemorrhagic *E. coli* (*E. coli* O157:H7)	Fecal-oral route	Isolation from stool	Rehydration; treatment of symptoms (antidiarrheal medications may increase risk of HUS in children; narcotics and NSAIDs not recommended)	No vaccine available
Gastroenteritis, *E. coli* (traveler's diarrhea)	Watery diarrhea, low-grade fever or no fever, abdominal cramps, nausea	Dehydration	Enterotoxigenic *E. coli*	Fecal-oral route	Clinical presentation; isolation from stool	Treatment of symptoms; fluid and electrolyte replacement; antibiotics (1–3 days with fluoroquinolones or azithromycin)	No vaccine available
Peptic ulcer disease	Pain associated with eating	GERD,[d] increased risk of esophageal and gastric cancer	*Helicobacter pylori*	Fecal-oral route	Urea breath test; detection of *H. pylori* antibody in blood; detection of *H. pylori* antigen in stool; detection of urease in biopsy samples	Depends on patient and type of ulcer; reduction of stomach acid production (proton pump inhibitors); two or three antibiotics given together	No vaccine available
Salmonellosis (*Salmonella* ileocolitis)	Diarrhea with some blood and white cells, fever, abdominal pain and cramps, nausea, sometimes vomiting	Sepsis, osteomyelitis	*Salmonella* spp. (nontyphoidal)	Fecal-oral route	Isolation from stool	Treatment of symptoms; antibiotics not indicated because they increase risk of carrier state	No vaccine available

(continued on next page)

Table 22.3

Bacterial Diseases of the Digestive System (*continued*)

Disease	Symptoms	Complications	Bacteria	Transmission	Diagnosis	Treatment	Prevention
Shigellosis (bacillary dysentery)	Bloody diarrhea, abdominal pain and cramps, tenesmus,[e] fever	May produce HUS, TTP[c]	*Shigella* spp., especially *S. dysenteriae*	Fecal-oral route	Isolation from stool	Oral rehydration; antibiotics may be indicated in some cases	Vaccine being developed but not available
Vibriosis	Acute watery diarrhea, sometimes bloody, abdominal cramps and pain, sometimes fever	Soft-tissue infections, organ failure, shock	*Vibrio parahaemolyticus*	Fecal-oral route (particularly contaminated seafood, especially raw oysters)	Isolation from stool	Usually self-limiting; replacement of fluids; tetracycline and ciprofloxacin	No vaccine available
Yersiniosis	Severe right-lower-quadrant abdominal pain, fever, diarrhea, possible blood in stool	Liver and spleen abscesses, serious GI problems such as perforation	*Yersinia enterocolitica*	Fecal-oral route	Isolation from stool; isolation of *Y. enterocolitica* DNA from stool by PCR	Rehydration; antibiotics may be required in cases of very high fever	No vaccine available
Gram-Positive Bacteria							
Antibiotic-associated diarrhea and pseudomem-branous colitis	Watery or mucoid diarrhea, fever, abdominal pain and/or cramps	Bowel perforation, SIRS,[f] sepsis, others	*Clostridioides difficile*	Fecal-oral route	Clinical presentation; PCR to identify *tox* gene in stool specimen; EIA to detect toxins in stool	Depends on severity of symptoms; removal of offending antibiotic in all cases; supportive treatment may be sufficient for mild cases; vancomycin and metronidazole as indicated for severe cases	No vaccine available
Gastroenteritis, *Clostridium perfringens* (food poisoning)	Antibiotic-associated diarrhea, watery	Segmental necrosis of jejunum (enteritis necroticans or pigbel), which is life-threatening	*Clostridium perfringens*	Fecal-oral route	Detection of toxin in stool samples	Removal of offending antibiotic; oral rehydration	No vaccine available
Gastroenteritis, *Staphylococcus aureus* (food poisoning)	Acute vomiting and diarrhea, abdominal pain and cramps, sometimes fever	Dehydration	*Staphylococcus aureus*	Ingestion of food contaminated with toxin	Clinical presentation; examination of food for presence of toxin	Self-limiting; rehydration	No vaccine available
Listeriosis	Watery diarrhea	Meningitis in immuno-compromised; systemic disease in neonates	*Listeria monocytogenes*	Fecal-oral route (refrigerated pâté and meat spreads, soft cheeses, hot dogs and lunch meats, unpasteurized milk)	Isolation from stool, blood, or CNS[g]	Oral rehydration; IV antibiotics for septicemia or meningitis	No vaccine available

[a]Disease presentation for enteroinvasive *E. coli* is indistinguishable from that of shigellosis.

[b]Periumbilical = around the umbilicus (belly button).

[c]HUS = hemolytic uremic syndrome; TTP = thrombotic thrombocytopenic purpura.

[d]GERD = gastroesophageal reflux disease.

[e]Tenesmus = feeling as though you're going to have a bowel movement but being unable to have one.

[f]SIRS = systemic inflammatory response syndrome.

[g]CNS = central nervous system.

Figure 22.11
Endoscopy
Endoscopy is used to view the stomach lining.

Upper endoscopy

Endoscope

Figure 22.12 Peptic Ulcers

Erosion of the lining of the stomach or duodenum is called an ulcer. Continued erosion of an ulcer can lead to bleeding. Patients with peptic ulcers often experience pain on the upper-left side of their abdomen.

A. The white arrow indicates a gastric ulcer.

were believed to cause increased production of acid that ate away at the stomach lining, even though the gastric mucosa was known to be well protected from stomach acid, which can have a pH as low as 1.5. Then, in the 1980s, there was a seismic shift in our understanding of ulcers. Australians Robin Warren and Barry Marshall discovered unusual Gram-negative, helical bacteria present in the biopsy samples of gastric ulcers and proposed that bacteria, not pepperoni, caused ulcers. The hypothesis was viewed with skepticism and declared as heresy by the established medical community. Faced with disbelief bordering on ridicule, Marshall drank a vial of the helical organisms and waited. A week later, he began vomiting and suffered other painful symptoms of gastritis. The young intern could not have been happier. He had proved his point. We now know that this curly-shaped microbe, *Helicobacter pylori*, causes the vast majority of stomach and duodenal ulcers (**Figure 22.13**). In fact, *H. pylori* is the only known cause of infectious gastritis.

The discovery of *Helicobacter pylori* (a Gram-negative member of Helicobacteraceae) and its association with gastric ulcer disease led to an upheaval in gastroenterology. Prior to this discovery, treatment focused on suppressing acid production, which did not provide long-term relief. Therapy now includes antimicrobial treatment to kill the bacteria and acid suppression therapy to prevent further inflammation while the ulcer heals. Warren and Marshall (who recovered from his gastritis) received the 2005 Nobel Prize in Physiology or Medicine for their groundbreaking work.

We now know that *H. pylori* causes one of the most common chronic infections in adults. The risk of infection is 10% in young adults and increases to more than 50% for those who are 60 or older. *H. pylori* can inhabit the human stomach because of

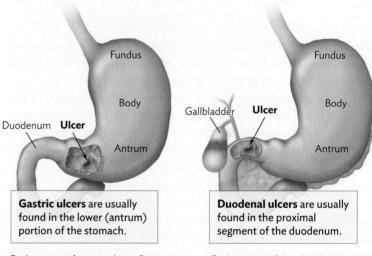

Gastric ulcers are usually found in the lower (antrum) portion of the stomach.

Duodenal ulcers are usually found in the proximal segment of the duodenum.

B. Location of gastric ulcers. Pain usually occurs when the stomach is empty and may temporarily be relieved after eating, which dilutes stomach acidity.

C. Location of duodenal ulcers. Pain typically starts about 2 hours after eating as the acidic contents of the stomach and bile acids empty into the duodenum.

a bacterial enzyme, called urease, that converts urea to ammonia. The alkaline ammonia neutralizes gastric acidity in and around *H. pylori*, enabling it to survive (but not grow) in the pH 1.5–2 stomach environment. This temporary protection from stomach acidity gives *Helicobacter* time to swim (using a tuft of polar flagella) to the relative safety of the mucus layer covering the stomach lining. In contrast to the stomach itself, the mucus layer has a hospitable pH of about 6. The bacteria then colonize the stomach by attaching to receptors on gastric epithelial cells. The presence of urease, lipopolysaccharides, and some other antigenic

Figure 22.13 A Bacterial Cause of Gastric Ulcers

A. *Helicobacter pylori* in stomach crypts (colorized SEM).

B. Australian physician Barry Marshall was so sure he was right about the cause of stomach ulcers that he swallowed bacteria to prove his point.

macromolecules produced by *H. pylori* initiates an immune response, which in turn leads to inflammation (gastritis), which in some people can progress to ulcers.

The signs and symptoms of gastritis vary from person to person but often include pain in the upper abdomen and bloating. Victims also complain of nausea and/or vomiting and a lack of appetite. Of those infected, 15%–20% develop gastric or duodenal ulcers. Individuals with gastric ulcer disease can suffer with dyspepsia (indigestion, upset stomach) and a hunger-like pain in the upper abdomen that temporarily abates after eating or using antacids. Patients with duodenal ulcer disease develop upper abdominal pain and fullness, bloating, belching, nausea, and sometimes even vomiting <u>after</u> they eat. The pain results from ulcerated tissue coming in contact with the bile acids injected into the duodenum by the gallbladder.

Ulcers can become severe enough to "eat" through the lining of the stomach and cause bleeding, and they may completely perforate the wall of the stomach to form a hole. Perforated ulcers are serious because blood and the contents of the stomach will leak into the abdominal cavity. The patient with a perforated ulcer will have a sudden sharp pain in the epigastric region of the abdomen that radiates to the back. The patient's condition worsens rapidly within 12 hours of onset and, without surgical intervention, will end in a catastrophic loss of blood pressure and collapse. Patients who have a bleeding ulcer may also have dark (tar-colored) stool and in some cases low RBC counts. In addition to ulcers, *H. pylori* infection is a risk factor for stomach cancer and may be involved in the development of gastroesophageal reflux disease, or GERD (repeated partial regurgitation of gastric acid that damages the

esophagus). Long-term *H. pylori* infection causes chronic inflammation of the stomach lining, leading to precancerous changes that may increase the risk of gastric malignancy.

Helicobacter pylori infection can be diagnosed in several ways. In one, a small sample of stomach tissue is removed by endoscopy (as with Gary in Case History 22.2) and analyzed for the presence of bacteria, urease, or abnormal changes in the stomach tissue. Other common procedures include testing the stool for the presence of *H. pylori* antigen and using assays (enzyme-linked immunosorbent assay, or ELISA) to detect serum antibodies against *Helicobacter* CagA protein (a cytotoxin). ELISA methodology is described in Section 17.2.

Patients with *H. pylori* infection are treated with drugs that reduce stomach acidity, such as a proton pump inhibitor and a combination of high doses of at least two antibiotics (for example, metronidazole, amoxicillin, and clarithromycin) for 14 days. The most serious concern about *H. pylori* infection is the strong link between the infection and an increased risk of developing stomach cancer or MALT (mucosa-associated lymphoid tissue) lymphomas, a type of cancer.

Intestinal Diseases Caused by Gram-Negative Bacteria

A number of bacteria and bacterial toxins cause intestinal diseases marked by secretory or inflammatory diarrhea. **Figure 22.14** illustrates how different Gram-negative and Gram-positive pathogens colonize and affect different parts of the intestines. Gram-negative bacteria cause the majority of intestinal infections. Enterotoxigenic *E. coli* and *Vibrio cholerae* produce a secretory diarrhea with large volumes of watery stool, whereas enterohemorrhagic *E. coli* causes a bloody diarrhea, despite being

Figure 22.14 Intestinal Locations of Bacterial Infections

Different bacteria show a preference for different parts of the human small and large intestines to cause severe morbidity and mortality. EHEC = enterohemorrhagic *E. coli*; EIEC = enteroinvasive *E. coli*; ETEC = enterotoxigenic *E. coli*.

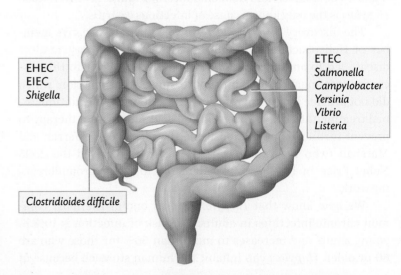

EHEC
EIEC
Shigella

ETEC
Salmonella
Campylobacter
Yersinia
Vibrio
Listeria

Clostridioides difficile

noninvasive. *Salmonella, Shigella, Campylobacter*, and entero-invasive *E. coli*, on the other hand, invade the intestinal lumen, resulting in a diarrhea that includes blood and fecal leukocytes (white blood cells in the stool). In this section we describe the pathogenesis and clinical presentations of diarrheal diseases caused by Gram-negative bacteria.

CASE HISTORY 22.3
Field Trip Diarrhea

Tammy, a 6-year-old girl from Montgomery County, Pennsylvania, arrived at a hospital's emergency department with bloody diarrhea, a temperature of 39°C (102.2°F), abdominal cramping, and vomiting. She was admitted to the hospital 5 days after a kindergarten field trip to the local dairy farm. The child's health history was otherwise unremarkable. At the time of hospital admission, her parents were asked about Tammy's activities during the trip. The parents confirmed that Tammy had bought a snack while at the farm. The lab reported that her fecal smear was positive for leukocytes and contained Gram-negative rods that produced Shiga toxins 1 and 2. Further testing of the isolate's serotype revealed *E. coli* O157:H7. By this time, Tammy had developed additional problems. Her face and hands had become puffy, she had a decreased urine output despite being given IV fluids (suggesting kidney damage), and she was beginning to develop some neurological abnormalities. Laboratory analyses of blood samples revealed thrombocytopenia (reduced blood platelet count) and confirmed hemolytic uremic syndrome, or HUS (progressive renal failure). Tammy was treated with intravenous fluid and electrolyte replacement. Antibiotics were not administered, and she eventually recovered.

Pathogenic bacteria that cause diarrhea (called **enteric pathogens**) are often more aggressive than viral pathogens, so they can cause more extensive epithelial damage. Some bacteria can invade and destroy the intestinal epithelial cells, whereas others remain stuck to the surface of the host cell and secrete toxins that damage the underlying cells. In either case, cytokines produced by the damaged cells can call in neutrophils and trigger inflammation. If the damage and inflammation are severe enough, as in Tammy's case, white blood cells or blood mixed with stool can be seen by the patient or discovered in the lab. WBCs in stool are rarely seen when the case is viral.

You might wonder why antibiotics were not administered to Tammy. Although antibiotic treatment of infectious gastroenteritis seems intuitive, it's rarely used (there are exceptions). Most GI infections are viral, so antibiotics are ineffective. Gastroenteritis caused by bacteria usually resolves spontaneously, so antibiotics are unnecessary. Of course, severe systemic disease stemming from

gastroenteritis (for example, *Salmonella* infection; see Section 21.3) can develop, often in the young or elderly. In addition, diseases such as typhoid fever (caused by *Salmonella enterica* serovar Typhi) or bacillary dysentery (caused by *Shigella sonnei*) will respond well to antibiotics. The most important treatment for diarrheal disease is the replacement of fluid and electrolytes either intravenously or through oral rehydration. We begin with *Escherichia coli*, the cause of Tammy's distress, and will describe what caused Tammy's puffy face, renal failure, and neurological signs.

***ESCHERICHIA COLI* GASTROENTERITIS** Although numerous varieties of *E. coli* are part of our normal microbiota, there are at least six different classes (**pathovars**) of diarrheagenic (diarrhea-causing) *E. coli* strains. A pathovar is actually a group of different *E. coli* strains that have different O and H antigens but possess the same combination of virulence genes whose products produce a similar pathology. Thus, a single species (for instance, *E. coli*) can have many pathovars, each of which causes a different disease.

E. coli pathovars can be differentiated by their telltale O and H cell-surface antigens, which may be identified by agglutination tests. The **O antigen** is part of the bacterium's LPS, and the **H antigen** is a flagellar protein (Section 18.4). So, "O157:H7" denotes the specific version of LPS (O157) and flagellar protein (H7) found on *E. coli* O157:H7. Three notable categories of diarrheagenic *E. coli* are described here.

- **Enterotoxigenic *E. coli* (ETEC)** strains, as the name implies, secrete enterotoxins. One such toxin, called **labile toxin** (heat-labile toxin), or **LT**, has a mode of action identical to that of cholera toxin, which was initially discussed in Sections 4.1 (see Figure 4.1) and 18.4. ETEC causes a secretory, watery diarrhea that can quickly dehydrate a patient. The organism has a fecal-oral route of transmission and no animal reservoirs.

- **Enteroinvasive *E. coli* (EIEC)** is the only pathovar of *E. coli* that invades epithelial host cells. EIEC is genetically very similar to *Shigella* species and even possesses the same plasmid required for invasion. As with ETEC strains, EIEC does not have an animal reservoir and is transmitted by the fecal-oral route. The invasive nature of EIEC means that the patient will present with a bloody diarrhea, much like that caused by *Shigella*, which we will discuss later.

- **Enterohemorrhagic *E. coli* (EHEC)** O157:H7 and other serotypes of this pathovar are recently recognized pathogens that cause serious disease in humans. An important difference from other diarrheagenic *E. coli* is that EHEC strains have animal reservoirs; that is, the bacteria reside in the intestines of cattle and other animals. The animal reservoirs increase the chances of human contact. Transmission is fecal-oral, as you will see.

Case History 22.3 did not involve Tammy only. Numerous other children visiting the farm that day also developed diarrhea with similar symptoms. Using the epidemiological investigative tools described in Chapter 26, the CDC identified the source of the disease outbreak by interviewing 51 infected patients and 92 controls (children who visited the farm but did not become ill). Infected patients were more likely than controls to have had physical contact with cattle. All 216 cattle on the farm were sampled by rectal swab, and 13% yielded *E. coli* O157:H7 with a DNA restriction pattern exactly the same as the one isolated from the patients. This finding indicated that the cattle were the source of infection. Activities that promoted hand-mouth contact, such as biting nails and purchasing food from an outdoor concession, were more common among the children who contracted disease (indicating a fecal-oral route of infection). Furthermore, areas for eating were not separated from areas for interacting with farm animals. Visitors could touch cattle, calves, sheep, goats, llamas, chickens, and a pig while eating and drinking. Handwashing facilities were unsupervised and lacked soap, and disposable hand towels were out of the children's reach. All these conditions provided ample opportunity for infection.

E. coli O157:H7 rarely affects the health of the reservoir animal. But when an infected steer is slaughtered, fecal matter can contaminate the carcass, despite the manufacturer's considerable efforts to prevent such contamination. Grinding the tainted meat into hamburger distributes the microbe throughout. As a result, cooking burgers to 71.1°C (160°F) is essential to kill any existing EHEC. Also be aware that cross-contamination between foods is possible. Using the same cutting board to prepare meat and salad is highly likely to contaminate the salad, which will not be cooked.

Despite its common association with hamburger, vegetarians are not safe from this organism. During heavy rains, waste from a cattle farm can easily wash into nearby vegetable fields. If the cattle waste contains *E. coli* O157:H7, the crops will be contaminated. One such outbreak occurred in 2006, when spinach from certain areas of California became contaminated with this pathogen, prompting a nationwide recall of bagged spinach and a month without spinach salad.

E. coli O157:H7 and, indeed, most EHEC strains have a remarkably low oral infectious dose because these pathogens are particularly capable of living in the acidic maelstrom of the stomach. *E. coli* survives stomach acidity by using an impressive level of acid resistance that rivals that of the gastric pathogen *Helicobacter pylori*. These acid resistance mechanisms permit *E. coli* to survive in the acidic stomach and enable a mere 10–100 individual organisms to cause disease.

Once the organism has entered the intestine, the attachment of all pathovars of *E. coli* begins with specialized pili acting as grappling hooks that tether the bacteria to the host cell. Some pathovars, such as EHEC and enteropathogenic *E. coli* (EPEC), go further. These strains inject bacterial proteins directly into the host cytoplasm by using syringe-like type III secretion systems (discussed in Section 18.4), establishing one-way communication. Some of these proteins instruct the host to form a membrane **pedestal** that hoists the organism up and away from the cell while maintaining intimate

Figure 22.15 Attaching and Effacing Lesion Produced by *E. coli*

This lesion was produced by an EPEC strain. Identical effacing lesions are produced by EHEC.

contact with it (see Figure 18.25). The pedestal prevents engulfment of the organism, keeping it away from intracellular defenses.

The host, for its part, emits a "cry for help" by way of cytokine secretion that initiates an inflammatory response. The resulting loss of microvilli produces what is called an **attaching and effacing lesion** (Figure 22.15). An antiabsorptive inflammatory diarrhea results, complete with blood and white cells entering the stool.

HEMOLYTIC UREMIC SYNDROME (HUS) AND THROMBOTIC THROMBOCYTOPENIC PURPURA (TTP) In Case History 22.3, Tammy's bloody diarrhea was accompanied by decreased urine output, a sign of kidney damage. The kidney damage was caused by two toxins (Shiga toxins 1 and 2, described in Section 18.4) that are encoded by bacteriophage genes embedded in the bacterial chromosome. These toxins are identical to the Shiga toxins produced by *Shigella dysenteriae* (discussed shortly). The toxins cleave host ribosomal RNA, which halts protein synthesis and kills the host cell. Endothelial cells in the kidney and brain are the most vulnerable. Endothelial damage triggers the formation of tiny platelet-fibrin clots that block blood vessels, leading to two major syndromes—**hemolytic uremic syndrome (HUS)** and **thrombotic thrombocytopenic purpura (TTP)**. HUS occurs when the microclots are limited to the kidney. The microclots clog the tiny blood vessels in this organ and cause decreased urine output, which can lead to kidney failure and death. Decreased urine output means that fluid is being retained in the body. Fluid retention explains Tammy's puffy face and hands.

In TTP, the clots occur throughout the circulatory system, causing reddish skin hemorrhages called petechiae and purpura (see Figure 21.12). Petechiae (small, pinpoint, purplish red spots) and purpura (purple patches) result from hemorrhages under the skin. Neurological symptoms (including confusion, severe headaches, and possibly coma) then arise from microhemorrhages in the brain. The hemorrhaging occurs because platelets needed for normal clotting are removed from circulation as they form the microclots. The decreased number of platelets is called **thrombocytopenia**.

The development of HUS, as in Tammy's case, is a common consequence of *E. coli* O157:H7 infection, especially in children. Unfortunately, HUS can be treated only with supportive care, such as blood transfusions and dialysis throughout the critical period until kidney function resumes. It's important to note that antibiotic treatment can actually increase the release of Shiga toxins from these organisms and trigger HUS. As a result, antimicrobial therapy is not recommended.

Note that many Shiga toxin–producing strains of *E. coli* are not EHEC strains; that is, they fail to cause hemorrhagic disease. So, **Shiga toxin–producing E. coli (STEC)** is a blanket term used for any *E. coli* that carries a Shiga toxin gene.

In this section we have focused on three diarrheagenic pathovars of *E. coli*, but there are others. These include **enteroaggregative E. coli (EAEC)**, which possesses unique bundle-forming pili, and **enteropathogenic E. coli (EPEC)**, similar to EHEC but without Shiga toxin. Each of these pathovars causes diarrhea, but they are isolated less frequently than EHEC.

SHIGELLOSIS Pathogenic species of *Shigella* (a Gram-negative bacillus, member of the family Enterobacteriaceae) cause **shigellosis**, a bloody diarrhea also called **bacillary dysentery**. The organisms are genetically very similar to EIEC (taxonomists, in fact, argue that *Shigella* is really a strain of *Escherichia*). *Shigella* lacks flagella (so no H antigen) and therefore is nonmotile. The structures of LPS O antigens divide *Shigella* into 45 antigenic serotypes, separated into four groups, or species (A–D). Two species cause the majority of shigellosis cases. *Shigella sonnei* (group D) is the most common pathovar in the United States, whereas *Shigella dysenteriae* (group A) is predominant elsewhere in the world.

The symptoms of shigellosis begin after a short incubation period (1–4 days) with severe abdominal cramping, high-grade fever, vomiting, loss of appetite (anorexia), and large-volume watery diarrhea. Later the patient experiences an uncontrollable urge to defecate (tenesmus) and fecal incontinence and defecates a small volume of mucoid diarrhea with visible red blood present. *Shigella dysenteriae*—but none of the other *Shigella* species—produces Shiga toxins that can lead to HUS and TTP. Treatment includes supportive oral rehydration therapy, and in contrast to EHEC, antibiotic treatment (usually a quinolone) is recommended to prevent a transient carrier state. Without antibiotic treatment, *Shigella* can continue to shed in stool for weeks after symptoms disappear, and the recovered patient will serve as a source of infection.

The pathogenesis of *Shigella*, which is similar to that of EIEC, is outlined in **Figure 22.16**. The organisms are very acid resistant, which means they easily survive stomach acidity (infectious dose is 10–100 ingested organisms). Once in the intestine, the bacteria are carried into the colon and attach to microfold (M) cells in Peyer's patches (see Figure 15.8). The microbes then engineer their entry (invasion) into the M cell (Figure 22.16, step 1). M cells are a common conduit for bacterial pathogens that invade the intestinal mucosa. *Shigella* (as well as EIEC) uses a virulence plasmid–encoded type III secretion system to inject bacterial effector proteins into the host cell. These effector proteins alter host actin polymerization and cause a membrane ruffle to form around the organism, bringing *Shigella* into the host cell (shown in Figure 18.24). The bacillus, now within a phagosome, uses an enzyme to lyse the phagosome membrane and escape to the cytoplasm (step 2).

A remarkable thing then happens in the cytoplasm. These bacteria, lacking flagella, nevertheless begin to move. As described in Section 18.3, proteins at one end of the bacillus begin to polymerize actin tails that move the organism through the cytoplasm. Formation of actin tails propels the microbe directly from one host cell to another. Because *Shigella* species can grow outside or inside a eukaryotic host cell, they are considered facultative intracellular pathogens.

The organism can also exit the basolateral side of the M cell, where it's met by a macrophage (Figure 22.16, step 3). *Shigella* continues to grow in the macrophage and triggers apoptosis. The dying macrophage sends out the pro-inflammatory cytokine IL-1 beta, and the bacteria are released into the lamina propria (connective tissue underneath intestinal epithelium). *Shigella* then enters the basolateral side of epithelial cells, escapes the vacuole, and uses actin motility (actin tails) to move into adjacent cells, where the phagosome escape process repeats itself (step 4). Stressed by the invader, host epithelia release another pro-inflammatory cytokine, IL-8.

The infected epithelial cells begin to die, destroying the integrity of the mucosal surface (Figure 22.16, step 5). At the same time, PMNs—called to the scene by the cytokines—arrive. The PMNs engulf and destroy the bacteria, thereby preventing dissemination (the infection remains localized). These white cells add to the intestinal damage and pass through compromised epithelial cell-cell junctions into the intestinal lumen (step 6). The rampant destruction also allows blood to enter the stool, resulting in the bloody diarrhea.

Shigella is transmitted by the fecal-oral route from person to person or via contaminated food or water. However, *Shigella* does not have an animal reservoir (other than apes and monkeys). Because of its ability to resist gastric acid, few organisms are needed to cause disease. This low infectious dose means that *Shigella* is easily transferred in day-care facilities between children, who, by most accounts, have limited hygiene skills.

SALMONELLOSIS The genus *Salmonella* (Gram-negative bacillus, member of the family Enterobacteriaceae) consists of only two species: *S. bongori* (found in cold-blooded animals, or ectotherms) and the more clinically significant *S. enterica*. Any gastrointestinal disease caused by a *Salmonella* species is called **salmonellosis**. *S. enterica* includes more than two thousand

Figure 22.16 Pathogenesis of *Shigella* Infection

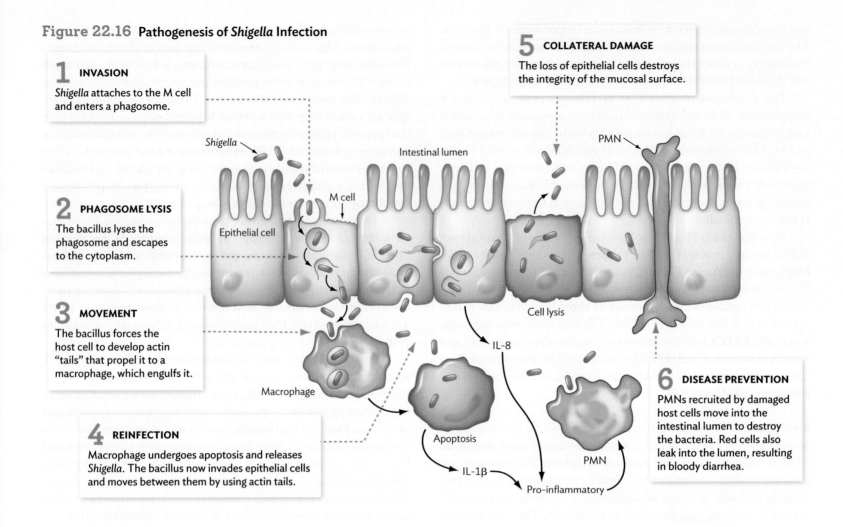

1 INVASION
Shigella attaches to the M cell and enters a phagosome.

2 PHAGOSOME LYSIS
The bacillus lyses the phagosome and escapes to the cytoplasm.

3 MOVEMENT
The bacillus forces the host cell to develop actin "tails" that propel it to a macrophage, which engulfs it.

4 REINFECTION
Macrophage undergoes apoptosis and releases *Shigella*. The bacillus now invades epithelial cells and moves between them by using actin tails.

5 COLLATERAL DAMAGE
The loss of epithelial cells destroys the integrity of the mucosal surface.

6 DISEASE PREVENTION
PMNs recruited by damaged host cells move into the intestinal lumen to destroy the bacteria. Red cells also leak into the lumen, resulting in bloody diarrhea.

Shigella

Intestinal lumen

PMN

M cell

Epithelial cell

Cell lysis

IL-8

Macrophage

Apoptosis

PMN

IL-1β

Pro-inflammatory

---

infectious serovars (equivalent to serotypes, based on O and H antigens). Fortunately, all these serovars cause only two basic clinical diseases: **typhoid fever** (or **enteric fever**) and enterocolitis. The nomenclature for these organisms has undergone extensive revision over the years. Serovars were previously given species status, such as *typhimurium*, *heidelberg*, and *dublin*; note that these strains were often named after the city in which they were first isolated. Today the strains are referred to as follows: "*S. enterica* serovar Typhimurium" (no italics for the serovar) or simply "*S.* Typhimurium" for convenience.

⊙ **Note:** Typhoid fever is an ancient disease whose name comes from the Greek *typhos*, meaning an ethereal smoke or cloud believed to cause disease and madness. In the advanced stages of typhoid fever, the patient's mental state is "clouded."

Enteric fevers caused by *S.* Typhi, *S.* Paratyphi, and *S.* Choleraesuis are actually systemic infections that sometimes include abdominal symptoms (Section 21.3). Regardless of GI involvement, the disease starts with the ingestion of bacteria in contaminated food or water (fecal-oral transmission), which is why it's included in this

chapter. *S.* Typhi and *S.* Paratyphi do not have animal reservoirs. Transmission therefore requires both symptomatic and asymptomatic human carriers to contaminate food or water with fecal material containing these organisms.

The symptoms of enteric fever occur between 1 and 3 weeks after ingestion. The classic presentation begins with a stepwise increase in fever to over 40°C (104°F; Figure 22.17) and bacteremia during the first week, followed by abdominal pain, continued fever, and chills. In some individuals, at fever's peak a salmon-colored rash (called rose spots) containing *Salmonella* can appear on the patient's trunk and abdomen (see Figure 21.17). Diarrhea can also develop, especially in children, although some patients report constipation. The spleen and liver can become enlarged (**hepatosplenomegaly**), and the patient may develop septic shock (see Chapter 15 IMPACT), although the disease usually resolves after about 3–4 weeks. Treatment includes antibiotic therapy: fluoroquinolones for susceptible strains or ceftriaxone for multidrug-resistant strains. Disease caused by *S.* Typhi can be prevented with vaccines—a live, oral attenuated vaccine or a purified capsule polysaccharide vaccine (Vi antigen, discussed shortly). However, because only 200–300 cases

Figure 22.17 Typhoid Fever

Stepwise increase in fever over the first week. A patient's temperature was monitored twice daily (closed circles) and graphed. The graph shows daily rises and falls of the patient's fever. The fever crescendos to its highest by day 7 and then gradually recedes to normal range by day 21.

of typhoid fever are reported each year in the United States, the vaccines are reserved for high-risk populations, such as military personnel and travelers to endemic areas.

Gastrointestinal disease caused by other serovars of *Salmonella enterica*, such as Typhimurium, Heidelberg, or Dublin (known as nontyphoidal strains), does not typically develop into septicemia

or septic shock unless the host is immunocompromised. (Case History 6.1 presented a patient with salmonellosis.) These organisms do have prominent animal reservoirs and compete with *Campylobacter jejuni* for the title of "most common bacterial cause of diarrhea" in the United States. Poultry, including chicken, chicken eggs, and turkey, is the most common source of infection; the presence of these pathogens in reptiles such as pet store turtles sometimes causes direct transmission to pet owners.

Symptoms of enteritis are indistinguishable from those caused by other enteric pathogens. Nausea, vomiting, and diarrhea occur 6–48 hours after ingestion. Stools are normally loose and nonbloody, although dysentery may develop in some patients. Fever usually resolves within 3 days; diarrhea, within a week. Uncomplicated enteritis caused by *Salmonella* should not be treated with antibiotics because the infection is usually short-lived and self-limiting. However, septicemia and invasive disease caused by some *Salmonella* serovars must be treated with antibiotics (often fluoroquinolones). *S.* Choleraesuis, for example, causes little, if any, diarrhea but rapidly penetrates the intestinal mucosa and invades the bloodstream (septicemia) to cause disseminated disease.

Salmonella is not as acid resistant as *E. coli* or *Shigella* and, as a result, has a higher infectious dose (10,000–100,000 organisms ingested). Once in the small intestine, *Salmonella* binds to M cells present in Peyer's patches in the ileum (Figure 22.18, step 1). In Section 18.4 we described *Salmonella*'s use of type III

Figure 22.18 Pathogenesis of *Salmonella* Infection

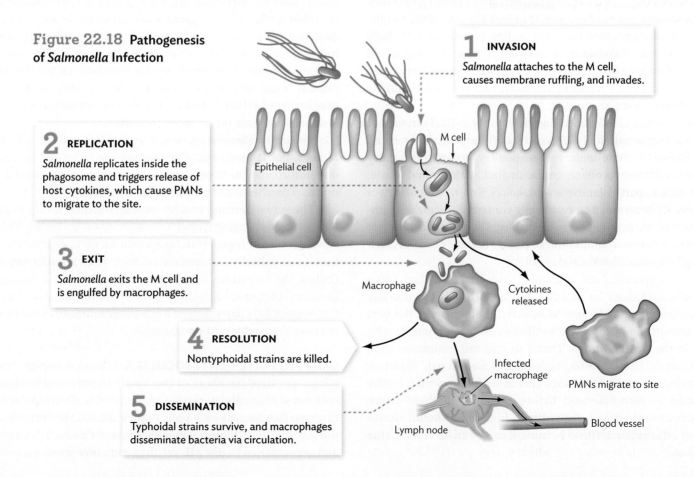

1 INVASION
Salmonella attaches to the M cell, causes membrane ruffling, and invades.

2 REPLICATION
Salmonella replicates inside the phagosome and triggers release of host cytokines, which cause PMNs to migrate to the site.

3 EXIT
Salmonella exits the M cell and is engulfed by macrophages.

4 RESOLUTION
Nontyphoidal strains are killed.

5 DISSEMINATION
Typhoidal strains survive, and macrophages disseminate bacteria via circulation.

M cell

Epithelial cell

Macrophage

Cytokines released

Infected macrophage

PMNs migrate to site

Lymph node

Blood vessel

secretion systems to engineer its entry into M cells and limit phagosome-lysosome fusion. Unlike *Shigella, Salmonella* remains in the phagosome and replicates there. As the organism replicates, the phagosome migrates to the basolateral side of the M cell (step 2). Cytokines released by the infected cells start the inflammatory process (described in Section 15.3), which includes a call to attract PMNs. *Salmonella* bacteria are ultimately released from the basolateral side of the M cell (step 3), where they are met by macrophages and PMNs.

The nontyphoidal strains of *S. enterica* are killed after they are engulfed by these phagocytic cells (Figure 22.18, step 4) and do not survive the inflammatory process. But inflammation severely compromises the integrity of the intestinal mucosa and causes the loss of fluid (diarrhea) as well as some blood and white cells (inflammatory diarrhea), although the stools usually are not obviously bloody. In contrast, the typhoidal strains of *Salmonella* can survive within macrophages, which then serve as cellular "Trojan horses" carrying the organism to the regional lymph nodes and eventually the bloodstream (step 5). At this stage the patient's fever begins to rise, and sepsis becomes a risk.

During its travels, *S.* Typhi is protected from the immune system by a capsule called **Vi Ag** (virulence antigen) unique to *S.* Typhi. As a result, the organism can ride the bloodstream unassailed and disseminate around the body. The pathogen eventually takes refuge in the gallbladder and periodically reenters the colon to be shed in feces. The colonized patient, after recovery, becomes an asymptomatic carrier who can transmit the organism to others through contaminated food. Up to 10% of untreated typhoid patients can shed *Salmonella* in feces for as long 3 months, and 1%–4% can become chronic carriers. Treatments to end the carrier state involve at least 4 weeks of antibiotic therapy (amoxicillin or trimethoprim-sulfamethoxazole).

The incidence of infection caused by nontyphoidal serovars of *Salmonella* is approximately 15 per 100,000 individuals. As noted earlier, sources include undercooked poultry, ground beef, dairy products, and fresh produce contaminated with animal waste. Chicken eggs in particular are a problem. *S.* Enteritidis can infect the ovaries of hens, and as a result, egg contents become contaminated before an eggshell forms. Washing the eggshell does not remove the inside contamination, so adding raw eggs to food (such as eggnog) comes with some risk.

Cases of enteric fever are much less frequent than cases of enteritis and must be reported to the CDC for monitoring. Quickly identifying and treating infected individuals is the best way to prevent an outbreak. The disease is not commonly contracted in the United States. Travel to endemic countries, such as Bangladesh, China, India, Indonesia, Laos, Nepal, Pakistan, or Vietnam, comes with a higher risk of infection, which is why vaccination is recommended before traveling to these areas. Practitioners evaluating a patient with high fever should always ask about international travel because it can be an invaluable clue to the cause.

YERSINIOSIS Any disease caused by a *Yersinia* species (Gram-negative bacilli, members of the family Enterobacteriaceae) is called **yersiniosis**. The infamous *Y. pestis* is transmitted by flea bite, not ingestion, and produces a systemic yersiniosis called **plague** or, more ominously, the Black Death. Because plague is a systemic disease and does not typically include GI involvement, it was discussed in the previous chapter (Section 21.3). Two *Yersinia* species that do cause gastrointestinal yersiniosis are *Y. enterocolitica* and *Y. pseudotuberculosis*.

These organisms are typically found in wild and domestic animals and are transmitted to humans via contaminated food (milk, meat, or vegetables). Following ingestion, *Y. enterocolitica* infects M cells in the ileum and colon. Here, after a few days, it causes symptoms of enteritis or enterocolitis involving low-grade fever, abdominal cramps, and, in some cases, nausea and vomiting. Diarrhea, which may contain blood and white cells, can linger for up to 2 weeks before ending. As with most uncomplicated diarrheas, antibiotic therapy is not recommended.

Yersinia enterocolitica infections can cause problems outside the intestine if the bacteria make it to the mesenteric lymph nodes. These nodes are embedded in the membrane (mesentery) that suspends the jejunum and ileum from the posterior wall of the abdomen. The result is mesenteric adenitis (**adenitis** is inflammation of a gland or lymph node). Pain associated with mesenteric adenitis is easily confused with appendicitis. For instance, the patient exhibits rebound tenderness (pain after pressure to the lower-right quadrant of the abdomen is applied and then released), a symptom usually associated with an inflamed appendix. In some cases the infection can disseminate via the bloodstream to other organs, causing focal infections. Curing complications such as these requires antibiotic treatment (usually an aminoglycoside, quinolone, or third-generation cephalosporin).

The other *Yersinia* species, *Y. pseudotuberculosis*, also infects M cells, but it more easily penetrates the intestinal mucosa and disseminates to cause septicemia or metastatic infections in other organs. Diarrhea is not a common manifestation.

These two *Yersinia* species have an unusual ability to grow at temperatures ranging from 1°C to 45°C (33.8°F–113°F). This ability to grow at refrigeration temperatures (4°C, or 39.2°F) means that food initially contaminated with few *Yersinia* organisms (below the infectious dose) can, when stored in a refrigerator, become laden with huge numbers of these bacteria over time. Consequently, what was once a relatively safe food to eat (unlikely to cause disease) is now dangerous.

***CAMPYLOBACTER* ENTEROCOLITIS** *Campylobacter* species (Gram-negative members of the family Campylobacteraceae) are corkscrew-shaped bacteria that are motile thanks to bipolar flagella (a single flagellum at each pole; Figure 22.19). The organisms are microaerophilic (they prefer low oxygen conditions, 5%, for growth) and are sensitive to low pH, yet they somehow survive a variety of

Figure 22.19 *Campylobacter*

B. Bipolar flagella enable *Campylobacter* to move.

A. The bacterium *Campylobacter* has a helical shape.

inhospitable environments to cause infection. The infectious dose is about 500 bacteria.

Campylobacter species (mostly *C. jejuni* and *C. coli*) represent one of the most common causes of diarrhea worldwide. These bacteria are part of the normal intestinal microbiota of poultry and cattle and are transmitted to humans through consumption of contaminated and undercooked meat, contaminated water, or unpasteurized milk. *Campylobacter* harbors a large plasmid, called pVir, and can cause bloody diarrhea. The organism uses flagella and adhesins to colonize the epithelium in the distal small intestine and then invades the mucosa through M cells to produce invasive disease.

The incubation period following ingestion is about 2–5 days, followed by a brief (24–hour) prodromal phase consisting of high fever, generalized pain, and sometimes dizziness. Severe abdominal pain, sometimes localized to the area around the belly button (periumbilical), and diarrhea with up to ten bowel movements a day begin suddenly. Nausea is common, and vomiting may also be present. These symptoms, as with enterocolitis caused by *Yersinia enterocolitica*, mimic those of appendicitis and can be mistaken for it. Blood and mucus can appear in the stool of infected patients. The signs and symptoms of infection in children vary depending on the age of the child. Severe complications such as meningitis and seizures or convulsion can be seen but are not common. Neonates can become infected during birth if their mother is a carrier, or they may acquire the organism as a hospital-acquired infection while in the hospital nursery.

As with most GI pathogens, our immune system is quite effective at confining *Campylobacter* to the GI tract, and the disease usually ends within a week. Oral rehydration therapy is thus usually treatment enough, and antibiotics are unnecessary.

However, antibiotic therapy can be used for patients suffering with prolonged symptoms (diarrhea more than a week) and severe abdominal pain with bloody stools.

Although normally self-limiting, *Campylobacter*-associated disease in immunocompromised patients can take a more complicated course. These patients not only experience a more severe GI disease but also can develop disseminated infections resulting in **pancreatitis** (inflammation of the pancreas), **cholecystitis** (gallbladder infection), or **peritonitis** (infection of the peritoneum), as well as **myocarditis** (infection of heart muscle) or **pericarditis** (infection of the heart sac, or pericardium). These infections <u>must</u> be treated with antibiotics.

Campylobacter species are naturally resistant to a number of antibiotics, including penicillins, trimethoprim, and most cephalosporins. The drug of choice for treating severe gastroenteritis is erythromycin (a macrolide), but the organisms are also sensitive to fluoroquinolones, clindamycin, tetracyclines, aminoglycosides, and chloramphenicol. Unfortunately, pathogenic strains resistant to macrolides and fluoroquinolones have already emerged and are prevalent in some parts of the world.

Following *Campylobacter* infection, a small number of patients develop an ascending muscle weakness known as **Guillain-Barré syndrome (GBS)**, an autoimmune disease that appears to be triggered by cross-reactivity between *Campylobacter* lipopolysaccharide structures and gangliosides (lipid-carbohydrate molecules) on nerve fibers. Antibodies to the *Campylobacter* LPS bind to nerve gangliosides and begin an immune complex disease process (Chapter 17) in which cytotoxic lymphocytes and macrophages infiltrate the nerves and remove the protective myelin sheath (**Figure 22.20**). Most GBS patients make a good recovery but may require hospitalization. About 10% die. Although other infectious agents, mostly viral, have been associated with Guillain-Barré

Figure 22.20 Guillain-Barré Syndrome

Infiltration of cytotoxic T lymphocytes (T_C cells) and macrophages demyelinates the nerve, causing nerve impulses to dissipate before they can reach a muscle.

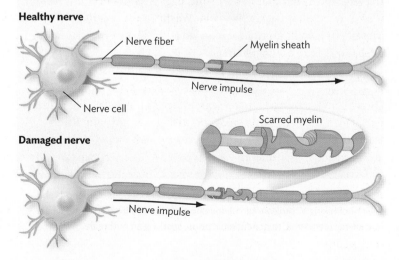

syndrome, *Campylobacter* is linked to 40% of cases and is the number one antecedent cause of this disease.

CHOLERA AND OTHER VIBRIOSES The case history that opens Chapter 4 describes the predicament of Akal, a 10-year-old Pakistani boy suffering from a disease caused by *Vibrio cholerae*. Any disease caused by a *Vibrio* species is called **vibriosis**. Members of the genus *Vibrio* (a Gram-negative curved rod belonging to the family Vibrionaceae) are primarily aquatic organisms that grow well in brackish waters found in estuaries (where ocean and freshwater mix, such as bays). These halophilic ("salt-loving") bacteria are motile via a single polar flagellum (**Figure 22.21A**).

Vibrio species that cause disease are divided into cholera and noncholera vibrios. **Cholera**, caused by *V. cholerae*, is a severe, watery diarrhea in which the patient can lose 10–15 liters of fluid a day. Serotyped by their LPS O antigens, strains of *V. cholerae* that cause epidemic cholera are designated serotype O1 (classical and El Tor biotypes) and O139. Non-O1/O139 strains are occasionally associated with outbreaks of cholera, but they more commonly cause wound infections or sepsis in immunocompromised hosts.

After the 2010 earthquake in Haiti, an outbreak of cholera infected more than 640,000 people and left 8,000 dead. The disease is thought to have been brought to the island by Nepalese soldiers who joined the UN peacekeeping force deployed to Haiti

Figure 22.21 *Vibrio cholerae*

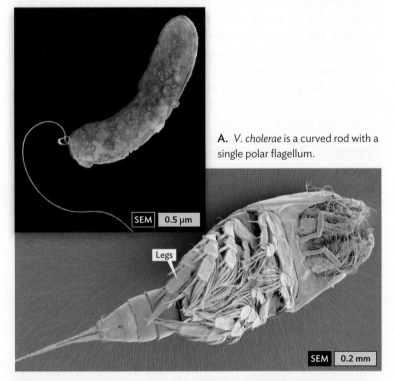

A. *V. cholerae* is a curved rod with a single polar flagellum.

SEM 0.5 μm

Legs

SEM 0.2 mm

B. Cholera organisms can be carried on the legs of copepods. Copepods and their hitchhiking *V. cholerae* can be transported via ships from seaports where *V. cholerae* is endemic to nonendemic ports, sparking an outbreak.

after the quake. Although cholera is typically found in developing countries with poor hygiene facilities, occasional outbreaks caused by the O1 serotype have appeared along the gulf coast of the United States.

Members of the genus are extremely acid sensitive, so millions of *V. cholerae* bacteria must be ingested to cause cholera. The infectious dose can be dramatically lower in patients taking antacids or proton pump inhibitors, such as omeprazole (Prilosec), that decrease stomach acidity. Survivors that make their way to the small intestine attach to epithelia there and secrete a toxin called **cholera toxin**, which causes a secretory diarrhea. Cholera toxin (Section 18.4) stimulates the secretion of electrolytes from intestinal epithelial cells, followed by water. The result is a watery diarrhea that has the appearance of rice water (the water drained from boiled rice). Because *V. cholerae* is not invasive, no fever occurs, and fecal blood and white cells are absent.

Treatment requires rehydration via an oral or intravenous route. Because large volumes of water and electrolytes are lost, blood can become extremely viscous, and organs will not function properly. Without rehydration therapy, death comes to 50% of patients. **Figure 22.22** shows the approach used during large cholera outbreaks in endemic areas. Patients lie in "cholera cots" containing a hole under which a bucket is placed. Because the victim is too weak to move, diarrheal fluid is collected in the bucket, its volume is measured, and an equal volume of replacement fluid is given intravenously.

When treated with fluid replacement only, cholera is generally self-limiting, and mortality drops to less than 1%. But antibiotic treatment (doxycycline) will shorten the duration and volume of fluid loss and hasten clearance of the organism from stool. Two different oral vaccines are available outside the United States. The CDC, however, does not recommend cholera vaccination for US travelers because their risk of getting cholera is low and the vaccines offer only partial protection (up to 50%). Prevention strategies in poor, endemic areas can be rather crude but effective. For example, *V. cholerae* tends to cling to the legs of copepods that live in brackish water (**Figure 22.21B**). The simple act of filtering water collected from these areas through the finely woven sari cloth worn by Indian women removes the copepods and with them the majority of the vibrios.

Several species of **noncholera *Vibrio*** cause noncholera diseases. The most important species are *V. parahaemolyticus* and *V. vulnificus*. Others, described earlier, are classified as non-O1 *V. cholerae*. *Vibrio parahaemolyticus* is widespread in aquatic environments. Outbreaks are associated with eating undercooked seafood, especially raw oysters. The pathogenic mechanism is unclear, but the organism can cause a watery or, less commonly, bloody diarrhea.

V. vulnificus infections include sepsis (which occurs in patients with liver disease, such as cirrhosis) and wound disease that can affect people even with no underlying disorder (as

Figure 22.22 Mass Treatment of Cholera Victims

A. Cholera cots used to collect the copious watery diarrhea from incapacitated patients.

B. A child being treated on a cholera cot at a clinic in Bangladesh. Note the IV lines used to replace fluids and electrolytes.

described in the case history that opens Chapter 15). As with all waterborne, pathogenic vibrios, infections with *V. vulnificus* typically occur in coastal areas in the warm summer months, when the organism grows best. Wounds can become infected from swimming or fishing in endemic coastal areas. Alternatively, individuals with decreased stomach acid can contract disease through ingestion (of raw oysters, for instance). The species grows extremely fast in blood (10-minute generation time); as a result, full-blown, life-threatening sepsis can develop quickly. Wound infections can begin to swell within hours of infection and become an extremely

Figure 22.23 Cellulitis Caused by *Vibrio vulnificus* Infection

The bubbles under the skin, called bullae, are actually large blisters filled with fluid. The fluid is a result of the inflammatory response to the infection and contains blood and *V. vulnificus*.

painful cellulitis that can develop into necrotizing disease (Figure 22.23). Left untreated, the mortality rate is an incredible 50%. Treatment must include doxycycline.

Intestinal Diseases Caused by Gram-Positive Bacteria

Only a few species of Gram-positive bacteria that cause GI diseases are worthy of mention. These include *Listeria monocytogenes*, *Clostridioides difficile*, *Clostridium perfringens*, and *Staphylococcus aureus*. We first discussed *C. perfringens* in Section 19.5 because it's the etiologic agent of gas gangrene. Some species of *C. perfringens*, however, produce an enterotoxin called *C. perfringens* enterotoxin (CPE), which causes abdominal cramps and diarrhea that usually resolve in 24 hours. A major symptom caused by intestinal pathogens is diarrhea, and most diarrheal diseases caused by Gram-positive pathogens are related to the toxins they produce. Definitive diagnosis of this pathogen is made when a technician cultures the offending agent or, in some cases, detects the toxin produced.

LISTERIOSIS

CASE HISTORY 22.4

Premature Arrival

A 2.1-kg (4.7-pound) female neonate was born at 32 weeks. A normal neonate averages 3.17 kg, or 7 pounds, and is born at 40 weeks. The baby was clearly in distress and was intubated at birth because of poor respiratory effort. A chest X-ray showed an infiltrate in the baby's right lung. Her mother, Linda, presented at the time of delivery with lower-abdominal pain, a temperature of 39°C (102.2°F), and a white cell count of 25,000/μl (normal is 4,000–10,000). Blood cultures from Linda and her child grew Gram-positive coccobacilli (Figure 22.24). Mother and baby fully recovered after antibiotic treatment. The physician later learned that Linda had recently visited Mexico, where she had eaten an unpasteurized soft cheese called *queso blanco*.

Linda initially experienced mild gastrointestinal distress caused by *Listeria monocytogenes* (a Gram-positive, non-spore-forming rod and member of the family Listeriaceae), an organism found in the gastrointestinal tracts of many animals and in

Figure 22.24 **Listeriosis and *Listeria monocytogenes***

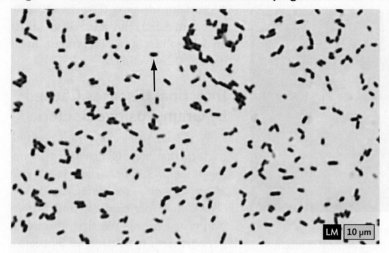

A sample from a patient with listeriosis shows characteristic Gram-positive rods of *Listeria* (arrow). Listeriosis in adults is much more common among patients infected with HIV than among members of the general population.

natural environments. At some point the organism entered her bloodstream (as indicated by her fever and elevated WBC count) and crossed the placental barrier into her fetus (vertical transmission). Her premature baby was born with septicemia and pneumonia and could have developed meningitis.

In healthy adults, *L. monocytogenes* usually causes a mild GI illness called **listeriosis**. The pathogen, however, can cause life-threatening septicemia, meningitis, and/or pneumonia in newborns, the elderly, or immunocompromised people such as HIV patients. Approximately 2,500 cases of listeriosis occur each year in the United States, leading to about 500 deaths. Children and adults become infected from eating contaminated foods. Person-to-person passage is rare (fecal-oral transmission does not occur). However, infected mothers can vertically transmit the pathogen to their fetus (as in Linda's case). *L. monocytogenes* grows not only at normal body temperature but also in the cold. As a result, the organism is found in processed and unprocessed foods, especially soft cheeses, hot dogs, lunch meats, milk, and salads stored in the refrigerator. These foods can be contaminated if made with unpasteurized raw milk or when processed foods come in contact with the organism in processing plants. Because *Listeria* multiplies in the cold, the risk of infection increases with prolonged refrigeration of these foods.

Listeria engineers its entry into epithelial cells of the small intestine, breaks out of the phagosome, and zooms around the host cell by using actin-based motility, just as *Shigella* does. Actin motility enables the microbe to spread from cell to cell without having to encounter the immune system. Actin-based motility also explains *Listeria*'s ability to cross the blood-brain barrier to cause

meningitis, as well as the maternal-fetal barrier to cause neonatal septicemia. The organism can survive in macrophages, which can, like taxicabs, shuttle the pathogen through the bloodstream. The immune system effectively halts most infections at this stage, but dissemination will produce serious disease in immunocompromised patients.

Ampicillin is the treatment of choice for *Listeria* infections. There are no vaccines, so the best prevention strategy is to fully cook meats, wash vegetables, and avoid unpasteurized dairy products.

ANTIBIOTIC-ASSOCIATED COLITIS Although we normally expect antibiotics to cure infections, sometimes antibiotic treatment can actually trigger GI disease. High doses of almost any antibiotic (including quinolones, penicillins, cephalosporins, and clindamycin) may cause diarrhea or sometimes a serious disease called antibiotic-associated colitis (inflammation of the colon). This condition is also called pseudomembranous colitis (or pseudomembranous enterocolitis if the small intestine is involved). Taking oral antibiotics will kill most normal intestinal microbiota, but the naturally resistant Gram-positive, spore-forming anaerobe *Clostridioides difficile* is an exception (see **IMPACT**).

Unrestrained by microbial competition, *C. difficile* will grow in the intestine and produce specific toxins that can damage intestinal cells. The organism's growth can lead to inflammation of the intestinal mucosa and the formation of plaques along the intestinal wall (see Figure 14.14). The plaques are yellowish white and made of sloughed-off intestinal epithelial cells, inflammatory cell debris, fibrin, and mucin. The plaques coalesce to form a pseudomembrane. Because they block the intestinal mucosa, pseudomembranes cause malabsorption of nutrients and water, resulting in diarrhea. As the pseudomembrane enlarges, it begins to slough off and pass into the stool.

C. difficile produces three different toxins: A, B, and CDT. CDT is related to *Clostridium perfringens*, and the toxin's role in the pathogenesis of *C. difficile* is not clear. Toxins A and B, which are closely related to each other (63% amino acid homology), can glucosylate (add a glucose molecule to) host cell proteins that make up the cytoskeleton. Altering the cytoskeleton changes the cell's shape and weakens tight junctions, leading to fluid leakage and hemorrhage. The toxins also cause an increase in production of tumor necrosis factor alpha (TNF-α), which in turn causes a large inflammatory response. The genes for *C. difficile* toxins and their regulators are all located on a large, horizontally transferred DNA sequence called a pathogenicity island (Section 9.4).

Why some people suffer only from diarrhea while others have the life-threatening pseudomembranous colitis is not clear. Diagnosis of this disease involves PCR identification of the organism or immunological identification of the toxin in fecal samples. Metronidazole or vancomycin is appropriate treatment for this organism, although nonpharmacological therapies, such as probiotics, are also used (see Section 14.1).

STAPHYLOCOCCUS AUREUS FOOD POISONING Like a ventriloquist who can make her voice sound as though it's coming from another room, *Staphylococcus aureus* can cause intestinal disease without really being in the intestine. We've all heard of the local church or school picnic where scores of people become violently ill within hours of eating unrefrigerated potato salad. *S. aureus* is the usual culprit in these disasters. Not all strains of *S. aureus* cause food poisoning, but those that secrete enterotoxins into tainted foods, such as pies, turkey dressing, or potato salad, can cause serious GI distress. Heat-stable *S. aureus* enterotoxins are a cause of **food poisoning**, which is defined as an illness caused by eating food contaminated with viruses, bacteria, and bacterial and other toxins.

After ingestion, the *S. aureus* enterotoxin travels to the intestine, where it enters the bloodstream and stimulates nerves leading to the vomit center in the brain. Because the toxin is preformed, symptoms occur quickly after ingestion. Within 2–6 hours, the poisoned patient begins vomiting and experiences severe stomach cramps and possibly diarrhea. The disease, though violent, is not life-threatening and usually resolves spontaneously within 24–48 hours. In contrast, diarrheas caused by infectious agents, such as *Salmonella enterica*, that must first grow in the victim, do not occur until 12–24 hours after ingestion, sometimes longer. A clinician noting quick onset of symptoms in a patient will immediately suspect staphylococcal food poisoning. Antibiotic treatment is not needed for staph food poisoning.

SECTION SUMMARY

- *Helicobacter pylori* **causes gastric and duodenal ulcers.** The organism uses urease to survive gastric acidity before attaching to the gastric mucosa, where it is protected by the gastric mucus layer.
- **Key pathovars of** *Escherichia coli* **include enterotoxigenic** *E. coli*, **enteroinvasive** *E. coli*, **and enterohemorrhagic** *E. coli*.
- **Invasive** *Shigella* **species produce colitis.** *Shigella dysenteriae* produces Shiga toxins, moves in and between cells by actin-mediated motility, and does not disseminate beyond the intestine.
- *Salmonella enterica* **includes thousands of serovars** that invade M cells in the small intestine and remain in vacuoles, limiting phagosome-lysosome fusion. *S.* Typhi enters macrophages and disseminates to produce typhoid fever with little GI involvement. Other serovars elicit inflammation, cause enteritis with bloody diarrhea, and usually do not disseminate.
- *Yersinia enterocolitica* **and** *Y. pseudotuberculosis* **invade M cells**, produce enteritis, sometimes disseminate to produce adenitis, and grow at refrigeration temperatures.
- *Campylobacter* **species invade M cells**, cause inflammatory diarrhea, and can trigger autoimmune **Guillain-Barré syndrome**.
- *Vibrio* **species are marine organisms** that include *V. cholerae* (cause of cholera), *V. parahaemolyticus* (found in undercooked seafood), and *V. vulnificus* (may cause necrotizing wound infections and sepsis in immunodeficient patients).

- *Listeria monocytogenes* **causes mild gastroenteritis and more serious diseases** by invading M cells in the small intestine and moving from cell to cell, crossing placental and blood-brain barriers; the organism can cause meningitis in the elderly and immunocompromised.
- *Clostridioides difficile* **causes antibiotic-associated diarrhea and pseudomembranous colitis.** The disease follows antibiotic treatment and causes antiabsorptive diarrhea. Antibiotic treatment is generally not recommended for uncomplicated bacterial diarrhea; oral rehydration therapy is critical to treat bacterial diarrhea.
- **Staphylococcal food poisoning is a toxigenic disease** caused by a staphylococcal enterotoxin; vomiting and diarrhea begin 2–6 hours after ingestion.

Thought Question 22.3 Why is it important to reduce stomach acidity in a patient with a stomach ulcer?

Thought Question 22.4 How might the ability of *Yersinia enterocolitica* to grow in a cold environment be used to more effectively isolate the organism from a patient's stool?

Thought Question 22.5 A college friend tells you she has been suffering with watery diarrhea for 2 days and wonders what could be causing it. You find out she has also experienced abdominal pain. What viruses or bacteria would you place in your list of differential diagnoses (list of possible causes)? What if she tells you there is blood in her stool?

Thought Question 22.6 A stool culture of a patient afflicted with *Staphylococcus aureus* food poisoning reveals no *S. aureus*. Where might you find the *S. aureus* that caused the food poisoning?

22.6 Parasitic Infections of the Gastrointestinal Tract

SECTION OBJECTIVES

- Distinguish GI infections caused by protozoa from those caused by helminths.
- Describe the role of parasitic life cycles in the transmission and pathogenesis of parasitic GI diseases.
- Generate a list of the diagnostic, prevention, and treatment options available for the parasitic diseases discussed.

What infections are caused by eukaryotic pathogens? As we learned in Chapter 11, many microbial pathogens are eukaryotes. In the developed nations of the world, we tend to associate GI disease with viruses and bacteria, but in fact, waterborne parasites such as *Giardia*, *Cryptosporidium*, and roundworms cause numerous infections in the United States. The developing world contends with millions upon millions of cases of parasitic diseases caused by protozoa and helminths (worms). In this section we will

impact

Fecal Transplant (Bacteriotherapy)

So, you could not believe your eyes when you read the title of this IMPACT, and you read it again. Indeed, fecal transplant does sound strange, to say the least, but it's a real approach to help cure some very dangerous GI ailments, such as recurrent *Clostridioides difficile* infections. And it makes sense, given the close relationship between nutrition, normal gut microbiota, and gastrointestinal health. We are born with minimal microbiota in our GI tract, but the GI tract rapidly becomes colonized by desirable microbes that help us use food and stave off some diseases. As discussed in Chapter 14, the number of bacteria residing in the human intestine is smallest in the duodenum and grows larger until it reaches a remarkable 10^{11}–10^{13} colony-forming units per milliliter (CFUs/ml) in the distal colon. This microbial menagerie includes a diverse array of known and unknown microbes. Changes in the population of bacteria colonizing our intestines can lead to disease—for example, pseudomembranous colitis caused by *Clostridioides difficile* after normal gut microbiota have been killed by antibiotics. The fecal matter of patients suffering from pseudomembranous colitis has significantly fewer Firmicutes and Bacteroidetes, for example, than do samples from healthy individuals.

Patients are at risk of developing hospital-acquired, or nosocomial, infections during a hospital stay. Among the more dangerous hospital-acquired infections are those involving antibiotic-resistant strains of *C. difficile*. Patients who develop a *C. difficile* infection in the hospital are at higher risk for serious complications and frequent relapse than are nonhospitalized patients. In some instances, the only way to save the patient's life is to remove the patient's colon (colectomy), a surgical procedure that carries its own risks, such as injury to other nearby organs (including the bladder and small intestine) and blood clot formation.

If the precipitating cause of *C. difficile* infection is an imbalance in the intestinal microbiota, then repopulating the intestines of patients with normal microbiota may be the better answer. Normal microbiota from a healthy individual can be "transplanted" into the patient through a variety of methods, including a nasogastric tube (inserted into the stomach through the nose) or a colonoscope, depending on the patient's condition (Figure 1). Another approach is to place selected members of a healthy gut microbiome in a capsule taken orally. Fecal transplantation is not a new procedure. Hieronymus Fabricius ab Aquapendente (1537–1619), a surgeon in one of Europe's most prestigious institutes of learning, the University of Padua, Italy, performed the first transplantation of enteric bacteria, although he didn't know what bacteria were at the time (Figure 2). The procedure fell out of favor, for obvious reasons, until 300 years later, in 1958, when fecal transplants successfully cured four patients suffering from pseudomembranous colitis. *C. difficile* had yet to be identified as the cause of this disease in 1958.

The technique has been used sporadically until recent years, when the emergence of drug-resistant and virulent strains of *C. difficile* gave rise to more and more treatment failure of pseudomembranous colitis and popularized bacteriotherapy. How effective is this

Figure 1 Transplanting Microbiota to Cure *Clostridioides difficile* Infection

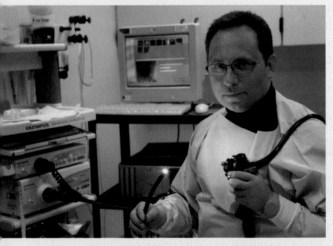

Dr. Alexander Khoruts, a gastroenterologist at the University of Minnesota, is shown here holding a colonoscope. Dr. Khoruts used the gut bacteria of a patient's husband to cure her chronic and debilitating *C. difficile* infection.

Figure 2 Hieronymus Fabricius ab Aquapendente

method? According to a January 2013 report in the *New England Journal of Medicine* online, fecal transplant cured 81% of patients suffering from pseudomembranous colitis, compared with 31% of those who received antibiotics.

A number of other serious GI diseases, including colon cancer and inflammatory bowel diseases, are also linked to changes in nutrition and the composition of microbes that colonize the intestine. Replenishment of normal gut microbiota has shown impressive success rates in patients who suffer from Crohn's disease and ulcerative colitis, two chronic and very painful inflammatory bowel diseases. This is very good news. The better news is the mounting evidence that proper nutrition, which supports a healthy microbiome, protects against the onset of some of these GI diseases and even against colon cancer. A healthy gut means a healthy you. So if you eat your fruits and vegetables to keep the microbes in your bowel happy and active, they will protect you.

DISCUSSION QUESTION

What are some options to promote a healthy gut microbiome?

Figure 22.25 *Giardia intestinalis*

This protist is a major cause of diarrhea in the world.

A. Cysts present in fecal matter. **B.** Trophozoite form.

discuss several diseases caused by eukaryotic microbes (summarized in Table 22.4).

Protozoan Infections of the Digestive System

Protozoa such as paramecia and amebas are familiar to most students of biology, but many are surprised to learn that some species of protists cause serious human disease. For instance, *Entamoeba histolytica* and *Cryptosporidium parvum* cause the diarrheal diseases amebic dysentery and cryptosporidiosis, respectively. Some diseases have great importance to public health. "Notifiable diseases" are those that must be reported to the local health department or to the CDC when diagnosed by a health care provider or a laboratory (further discussed in Section 26.3). Cryptosporidiosis, for example, is a notifiable disease.

GIARDIASIS The flagellated, anaerobic protozoan *Giardia intestinalis* (also known as *Giardia lamblia*) is a major cause of diarrhea throughout the world. In the United States alone, *G. lamblia* causes an estimated 1.2 million cases of **giardiasis** (diarrhea caused by *Giardia*). *G. intestinalis* is very infectious and can be found in various rodents, deer, cattle, and even household pets in addition to humans. *Giardia intestinalis* enters a human or other host as a cyst present in drinking water contaminated by feces (**Figure 22.25A**). Ingestion of as few as 25 cysts can lead to disease. Following ingestion, the hard, outer coating of the cyst is dissolved by the action of digestive juices to produce a **trophozoite** (**Figure 22.25B**), which attaches itself to the wall of the small intestine and reproduces. Offspring quickly encyst and are excreted from the host's body.

Asymptomatic carriers of *G. intestinalis* are common. It has been estimated that anywhere from 1% to 30% of children in US day-care centers are carriers. Disease usually manifests as greasy stools alternating between a greasy or watery diarrhea, loose stools, and constipation. However, some patients experience explosive diarrhea. Diagnosis is made after a technician observes the cysts or trophozoite forms of the protozoan in feces. Metronidazole is a drug often used to cure the disease. As for prevention, proper treatment of community water supplies is essential.

Table 22.4

Parasitic Diseases of the Digestive System

Disease	Symptoms	Complications	Parasite	Transmission	Diagnosis	Treatment	Prevention
Protozoa							
Amebic dysentery (amebiasis)	Bloody diarrhea, fever, abdominal pain	Small-bowel obstruction, intestinal perforation, liver abscess	*Entamoeba histolytica*	Fecal-oral route	Microscopic detection of trophozoites or cysts; isolation in culture; isolation of *E. histolytica* antigen and DNA in stool (PCR, qRT-PCR); clinical presentation with travel history	Amebicides (paromomycin, diloxanide, iodoquinol) given intraluminally (into the bowel lumen)	No vaccine available
Cryptosporidiosis	Watery diarrhea, fever, stomach cramps and pain	Malnutrition, weight loss	*Cryptosporidium* spp.	Fecal-oral route	Microscopic examination of stool samples (different techniques); enzyme immunoassay to detect antigens	Antiparasitic medications	No vaccine available
Cyclosporiasis	Watery diarrhea, flu-like symptoms, stomach cramping	Malabsorption of nutrients, cholecystitis (inflammation of gallbladder)	*Cyclospora cayetanensis*	Fecal-oral route	Identification of oocytes in stool (multiple stool specimens on different days needed)	Trimethoprim-sulfamethoxazole	No vaccine available
Giardiasis (beaver fever, traveler's diarrhea)	Diarrhea, flatulence, bloating, weight loss, abdominal pain, greasy stools	Malabsorption of nutrients	*Giardia intestinalis*	Fecal-oral route (contaminated water)	Visualization of cysts by microscopy; antigen detection in stool or intestinal samples by EIA; *Giardia*-specific DNA (qRT-PCR)	Asymptomatic cases may not be treated; antiparasitic medications in all symptomatic cases; metronidazole, tinidazole, others	No vaccine available
Helminths							
Blood fluke (schistosomiasis)	Dysentery, abdominal pain, fever, chills, GI bleeding, pneumonitis, enlarged liver and spleen	Damage to small intestine, causing death; encephalitis	*Schistosoma mansoni*	Penetration of skin; eggs eliminated in feces and urine	Visualization of eggs in biopsy specimen; antibody detection by ELISA, followed by immunoblot for confirmation	Antiparasitic medications: praziquantel	No vaccine available
Hookworm	Recurrent epigastric pain, nausea, flatulence, itching and rash at site of entry, sometimes anemia	Impaired growth and cognitive development in children, increased maternal mortality in pregnant women, pneumonitis (inflammation of the lung)	*Necator americanus*	Penetration of skin; eggs released in stool	Visualization of eggs in stool	Albendazole, mebendazole, pyrantel pamoate	No vaccine available
Liver fluke (fascioliasis, the common liver fluke)	Can be asymptomatic; acute and chronic phases of infection; nausea, vomiting, fever, abdominal pain, enlarged liver, anemia, malaise	Inflammation and obstruction of bile duct, jaundice	*Fasciola hepatica*	Fecal-oral route	Visualization of eggs in stool, more than one specimen needed; EIA with excretory-secretory antigens combined with confirmation of positives by immunoblot	Triclabendazole is drug of choice	No vaccine available

(continued on next page)

Table 22.4
Parasitic Diseases of the Digestive System (*continued*)

Disease	Symptoms	Complications	Parasite	Transmission	Diagnosis	Treatment	Prevention
Liver fluke (Oriental or Chinese liver fluke)	Mostly asymptomatic; enlarged liver, upper abdominal pain, anorexia	Liver inflammation, cancer of bile ducts	*Opisthorchis* and *Clonorchis*	Fecal-oral route	Visualization of eggs in stool; liver biopsy	Praziquantel, albendazole are drugs of choice	No vaccine available
Pinworm	Itchy anus, sometimes diarrhea	Secondary bacterial infections, sleep disturbance from itching	*Enterobius vermicularis*	Fecal-oral route	Eggs on rectal swab; worms on cellophane, Scotch tape applied to anus in morning before bathing	Pyrantel pamoate, mebendazole, albendazole	No vaccine available
Roundworm (ascariasis)	Eosinophilic pneumonia, lung irritation, cough; small-bowel obstruction, abdominal pain, vomiting, diarrhea or constipation, peritonitis	Malabsorption of nutrients, small-bowel obstruction, hepatic and biliary complications	*Ascaris lumbricoides*	Fecal-oral route	Visualization of eggs in stool (may also be found in vomitus or sputum)	Albendazole, mebendazole, ivermectin	No vaccine available
Tapeworm (fish)	Abdominal pain or discomfort, diarrhea, enlarged liver, weakness, weight loss	Anemia, intestinal blockage, vitamin B_{12} deficiency	*Diphyllobothrium latum*	Fecal-oral route	Visualization of eggs or parasite in stool	Antihelminthic medication: praziquantel, niclosamide are used most often	No vaccine available
Tapeworm (pork)	Diarrhea, nausea, vomiting, visual disturbances	Meningitis, increased risk of epilepsy	*Taenia solium*	Fecal-oral route	Antigen detection in stool sample by ELISA; visualization of eggs in stool	Antihelminthic medications with steroids, depending on type of infection	No vaccine available
Threadworm (strongyloidiasis)	Asymptomatic or burning pain in abdomen, nausea and vomiting, alternating watery diarrhea and constipation, itchy skin rash	Nonspecific symptoms, abdominal pain, bloating, intermittent diarrhea and constipation; sepsis and pneumonitis in immuno-compromised, with high mortality (60%–85%)	*Strongyloides stercoralis*	Penetration of skin; eggs can hatch in the intestine, eggs and larvae released in stool	Serial stool examination for larvae (gold standard); serology; marked eosinophilia (increased eosinophils)	Ivermectin, albendazole, thiabendazole	No vaccine available
Trichinosis	Nausea, vomiting, diarrhea, abdominal pain, fever, muscle weakness	Rare except in severe cases; arthritis, CNS and cardiac involvement, pneumonia, death due to heart or respiratory paralysis	*Trichinella* spp.	Ingestion of contaminated raw pork	Serology; muscle biopsy; elevated muscle enzymes and eosinophilia	Steroids; mebendazole, albendazole	No vaccine available
Whipworm (trichuriasis)	Painful passage of watery stool that contains mucus and blood, nausea, fatigue, abdominal pain, loss of appetite	Anemia, growth retardation, and cognitive impairment in children	*Trichuris trichiura*	Fecal-oral route	Visualization of eggs in stool	Albendazole, mebendazole, pyrantel pamoate	No vaccine available

AMEBIC DYSENTERY (AMEBIASIS) Although you rarely hear about this organism, 10% of the world's population is infected with *Entamoeba histolytica. E. histolytica* is an ameba that alternates between a motile trophozoite form and an infectious cyst form that survives for weeks in the environment. These forms can be differentiated under the microscope by the number of nuclei they contain. The cysts contain four nuclei, whereas trophozoites contain only one (see Figure 11.12).

Following ingestion of cysts, trophozoites are released in the small intestine, and in most individuals they remain in the large intestine as harmless commensals. The trophozoites replicate, and some become infectious cysts that are shed in stool. Trophozoites are sensitive to acid and are not infectious. In some unfortunate individuals, 2–6 weeks after they ingest the cysts the trophozoites emerge, and the protozoan penetrates the intestinal mucosa, causing colitis (**amebic dysentery**, or **amebiasis**), or invades the bloodstream to produce abscesses in the liver, lungs, or brain. Amebic dysentery is marked by diarrhea, bloody stools, fever, abdominal pain, and weight loss. Liver abscesses will result in hepatitis symptoms.

Diagnosis of amebic dysentery is tricky because visually similar nonpathogenic *Entamoeba* species are among the normal intestinal microbiota. The laboratory must identify trophozoites that have ingested red blood cells in stool or colonic abscesses to make the diagnosis. Prevention of disease relies on adequate water sanitation. Treatment of active disease includes iodoquinol, which will kill intestinal cysts and trophozoites by an unknown mechanism but will not penetrate the intestinal mucosa. Extraintestinal infections are treated with metronidazole, which will penetrate to deeper tissues and kill disseminated organisms.

CRYPTOSPORIDIOSIS *Cryptosporidium* species (*Cryptosporidium hominis* and *C. parvum*) are widely distributed in the world and cause self-limiting diarrheal disease (**cryptosporidiosis**) in immunocompetent hosts. Following ingestion, oocysts dissolve (excyst) and release sporozoites that then enter intestinal epithelial cells. There the parasites undergo asexual and then sexual cycles to produce infective oocysts. These oocysts are excreted in stool (Figure 22.26). The oocysts are quite hardy and resist killing even in chlorinated water. So, both drinking water and recreational water, which can be contaminated by animal waste containing *Cryptosporidium*, are sources of infection.

About a week after ingestion, the patient develops a watery diarrhea that lasts 1–2 weeks. The mechanism underlying this secretory diarrhea is unknown. Symptoms can include abdominal pain, nausea, anorexia, fever, and weight loss. In contrast to this self-limiting disease, immunocompromised AIDS patients with CD4 counts of less than 100/µl will develop a chronic, remarkably profuse, life-threatening diarrhea (1–25 liters per day) and suffer severe fluid and electrolyte depletion.

Because oocytes of *Cryptosporidium* are tiny (4–6 µm in diameter), their discovery in stool relies on a variety of specific tests

Figure 22.26 *Cryptosporidium* **Oocysts**

Rounded oocysts stained with safranin (pink, 4–6 µm in diameter) are seen in feces and contain the infective sporozoite stage.

(the lab should be informed that you're looking for this parasite). A modified acid-fast stain, direct immunofluorescence, or immunoassay can be used. Treatment of immunocompetent patients includes oral rehydration and the drug nitazoxanide, which kills the protozoan by interfering with its anaerobic metabolism. Unfortunately, the drug is ineffective in AIDS patients. Antiretroviral treatments have resulted in a decreased incidence of cryptosporidiosis in developing countries. In the United States, for example, the incidence of the disease in patients with HIV is less than 1 in 1,000. In developing countries, however, cryptosporidiosis is found in the stool of 74% of AIDS patients suffering from chronic diarrhea.

CYCLOSPORIASIS *Cyclospora cayetanensis* causes illness (**cyclosporiasis**) ranging from asymptomatic to diarrhea, as well as flu-like symptoms, flatulence, belching, and fatigue. The disease is generally self-limiting but can be treated with trimethoprim-sulfamethoxazole if needed. The organism makes oocysts in the intestine (Figure 22.27), but the oocysts do not become infective until 1–2 weeks after being passed in the stool. So, direct fecal-oral transmission cannot occur (this inability differentiates *Cyclospora* from *Cryptosporidium*, whose oocytes show fecal-oral transmission). The *Cyclospora* oocytes need a few days or weeks in the environment to sporulate and produce sporocysts so that they can become infectious and cause disease. A sporulated oocyst contains two sporocysts, each harboring an elongated sporozoite. Fresh produce and water contaminated with oocytes can then serve as vehicles for transmission, when they are ingested. The oocyst shell dissolves in the GI tract, freeing the sporozoites to invade the small intestine's epithelial cells. Once inside the intestinal epithelial cells, the sporozoites undergo asexual multiplication, followed by sexual development, to mature into spherical oocysts that can be detected in the stool of infected patients.

Figure 22.27 Stages in the Life Cycle of *Cyclospora cayetanensis*

Oocysts are excreted in feces unsporulated (noninfective). After 1–2 weeks outside the host, they sporulate and become infective.

A. Undifferentiated oocyst.

B. Sporulating oocyst containing two maturing sporocysts. An ingested sporulated oocyst can lead to disease.

C. Ruptured oocyst releasing one of its two immature sporocysts.

D. One free sporocyst is shown with two free sporozoites.

Helminthic Infections of the Digestive System

CASE HISTORY 22.5

Worming Through

Ileana, a high-ranking US government official, traveled to the Near East on a diplomatic mission. During a formal reception, she ate some delicious steak tartare (raw beef), a traditional dish in that region. Three months later, she noticed thin white rectangular segments in her stool (approximately 1 × 2 × 0.2 cm). She experienced nausea, in part from the parasite and in part from the sight of worms in her stool. Laboratory studies revealed that the segments were proglottids (body segments) of *Taenia saginata*, a tapeworm (**Figure 22.28A**). Ileana's stool also contained eggs of this worm (**Figure 22.28B**). She was reassured by her physician, who told her that this infection is unlikely to have clinical consequences in a healthy person. The physician prescribed niclosamide, which eliminated the rest of the tapeworm.

Helminths (worms) were first described in Section 11.5. The parasitic helminths, ranging in size from incredibly tiny (0.3 mm) to ridiculously huge (25 meters, or 82 feet), fall into three groups: **roundworms (nematodes)**, **flukes (trematodes)**, and **tapeworms (cestodes)**. Trematodes and cestodes are classified together as **flatworms** (**platyhelminths**). Symptoms of helminthic infections are caused primarily by the immune response provoked by these worms rather than by pathogenicity factors such as toxins. A telltale sign that practitioners look for when diagnosing worm infections is an increase in the number of eosinophils present in peripheral blood counts (**eosinophilia**). The role of eosinophils in clearing helminthic infections remains controversial, but these WBCs are clearly drawn to sites of worm infections and are the cause of a considerable amount of the inflammatory pathology.

HELMINTH LIFE CYCLES AND TRANSMISSION TO HUMANS So, how does one manage to contract worm infections? The individual worm's life cycle shows that helminths infect humans in

Figure 22.28 Tapeworm

A. *Taenia saginata* can grow to several meters in length.

B. *T. saginata* egg found in stool

Figure 22.29 Life Cycles and Transmission of Helminthic Parasites

The four different paths that worms take to develop from eggs to larvae to adults back to eggs.

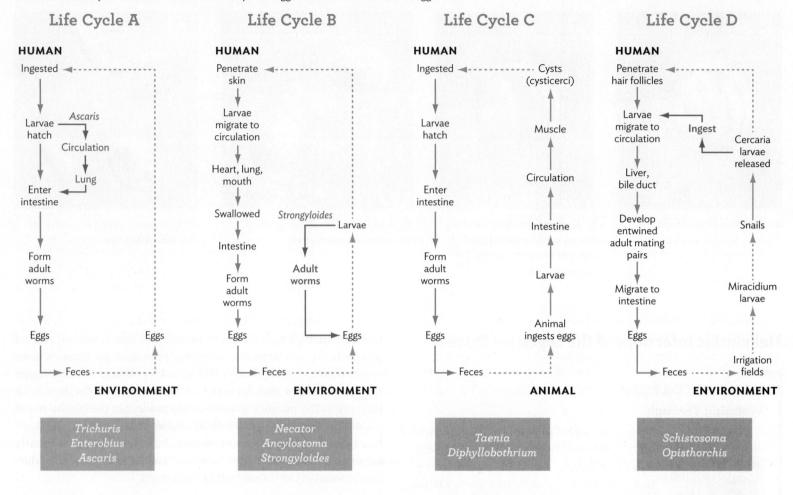

four general ways, in this book classified as life cycles A–D. Life cycle A involves the fecal-oral route of transmission; life cycle B is a fecal-environment-skin route; life cycle C involves the fecal–intermediate host–oral route; and life cycle D is a combination of life cycles A and B. Be aware that some helminths require an **intermediate host** to complete their life cycle. A **definitive host** is one in which the adult worms develop and lay eggs. All worms have one or more definitive hosts, but only a few worms have intermediate hosts. **Figure 22.29** provides a summary of the four primary ways in which helminths infect humans.

Some helminths use life cycle A. Eggs from certain roundworm species (such as *Trichuris*, *Enterobius*, and *Ascaris*) can contaminate food or water and be ingested. The larvae hatch in the stomach, attach to the intestine, and enter intestinal tissue to become male and female worms. Worm sex then produces more eggs, usually in huge quantities. The eggs are then excreted in feces and await ingestion by a new host. The *Ascaris* life cycle is a little more complicated. Before *Ascaris* larvae can develop into adults, they migrate through the circulation, enter the lungs, break into

the alveoli, and ascend to the oral cavity, where they are again swallowed. Only then can adults develop and lay eggs.

Life cycle B is preferred by other species. After being excreted in feces, eggs from hookworms such as *Necator americanus* and *Ancylostoma duodenale* hatch in the environment and develop into larvae. Infection results when the larvae actively penetrate human skin (now you know one of the reasons your mother told you not to play outside in bare feet). Once in the lymphatic and blood circulations, larvae travel to the heart and lungs and ultimately the throat. Larvae in the oral cavity are then swallowed with saliva (as with *Ascaris*) and travel to the small intestine, where they anchor (attach) and feed on blood. The larvae develop into adults that produce eggs, which are then excreted in feces to begin the next cycle of transmission. Eggs will not hatch in the intestine.

A third group of helminths prefers to go through an intermediate host and use life cycle C. Tapeworms such as *Taenia*, *Diphyllobothrium*, and *Echinococcus* mature in the human intestine (definitive host), releasing eggs in feces. Grazing animals (intermediate hosts)

Figure 22.30 Whipworm (*Trichuris trichiura*)

A. The narrow end of the worm attaches to the intestine, while the fatter end hangs in the intestinal lumen to sample nutrients.

B. A colon heavily infested with whipworms.

C. Whipworm eggs, which measure 50–55 μm in length, are easily identified by their plugged ends.

ingest the eggs, and stomach acid releases the larvae. Liberated larvae travel to the animal's GI system, enter the circulatory system, invade muscle, and become encased in cysts (called **cysticerci**; singular, **cysticercus**). Humans who eat meat containing cysticerci become infected. Cooking potentially contaminated foods such as pork and fish will break this cycle.

Finally, certain helminths use life cycle D. The flatworms *Schistosoma* and *Opisthorchis* have fork-tailed larvae called **cercariae** (singular, **cercaria**), which are present in contaminated water (often irrigated fields or ponds) and can directly burrow through the hair follicles of a wading human or be ingested. The larvae migrate through the circulation to the liver and bile ducts, where they develop into male and female flatworms that become permanently entwined as mating pairs. The writhing pairs migrate to blood vessels in the small intestine, where they lodge, feed on blood, and lay eggs. The eggs are then released into irrigation fields through fertilization, defecation, or urination. Once in the water, the eggs release ciliated larvae, called **miracidia** (singular, **miracidium**), that seek out nearby snails (intermediate hosts) and burrow into them. In the snail, the larvae morph into the large, fork-tailed cercaria that started this cycle. Cercaria are released from snails by the thousands into the surrounding water, where they await new, unsuspecting human hosts.

Clinical manifestations of helminthic diseases can be divided into four basic categories: intestinal distress alone, intestinal distress associated with

migratory symptoms (caused by migrating larvae), muscle involvement, and liver disease.

HELMINTHS THAT CAUSE INTESTINAL DISTRESS The **whipworm** (*Trichuris trichiura*) is so named because it resembles a whip (**Figure 22.30A**). Infecting humans via life cycle A, adult worms form in the intestine. When these worms burrow into the intestinal wall, the wall hemorrhages, making gateways for secondary bacterial infection (**Figure 22.30B**). A heavily infected individual can experience dysentery, loss of muscle tone, and even a prolapsed (protruding) rectum.

Pinworm caused by *Enterobius vermicularis* is the most common worm disease of children; it was described in Case History 11.3. The worms are very small and use life cycle A for transmission (**Figure 22.31A**). Females tend to migrate from the anus at night and lay their eggs in the perianal region. A simple

Figure 22.31 Pinworm (*Enterobius vermicularis*)

A. *Enterobius* up close.

B. Pinworm eggs.

Figure 22.32 Tapeworm

Notice the head with hooklets and suckers, and the segmented proglottids.

Proglottid

SEM   10 mm

piece of Scotch tape applied to the anus for a few seconds at night will catch eggs that the lab can microscopically identify (Figure 22.31B). Symptoms of pinworm infection include pronounced anal itching and sometimes diarrhea. Itching can interrupt the child's sleep and promote self-reinfection. As the child scratches, eggs stick to the child's fingers, and when those fingers are placed back in the child's mouth, the eggs are reingested.

Tapeworms include *Taenia saginata* from beef (discussed in Case History 22.5), *Taenia solium* from pig, and *Diphyllobothrium latum* from fish. We become infected with these parasites when we eat contaminated raw or undercooked beef, pork, or fish. Tapeworms are visually impressive segmented worms, whose

head (**scolex**; plural, **scolices**) contains a central ring of hooklets flanked at four corners by suckers that attach to the intestine (Figure 22.32). The body segments, called **proglottids**, form eggs and break off from the main worm (quite disconcerting to the person confronted with 2–3 feet of worm parts exiting the anus). Adult worms can release up to 1 million eggs per day this way. For such a large parasite, tapeworms cause relatively few symptoms, which include nausea, vague complaints of abdominal pain, and the occasional discovery of proglottids in feces (as Ileana discovered in Case History 22.5). The nutrient demands of the large worm can also cause the host to lose weight.

HELMINTHS THAT CAUSE INTESTINAL DISTRESS WITH LARVAL MIGRATION SYMPTOMS Ascariasis is caused by the largest of the roundworms—*Ascaris lumbricoides*, 40 cm long (Figure 22.33A)—and is the most common nematode infection in the world. Symptoms of larval migration to the lung include an irritating nonproductive cough and a burning sensation. Intestinal infection is usually asymptomatic, unless a large clump of entangled worms causes a small-bowel obstruction, in which case symptoms can include abdominal pain, vomiting, constipation or diarrhea, appendicitis, and peritonitis (Figure 22.33B). An estimated 1.2–1.5 billion people in tropical and subtropical regions have *Ascaris* in their GI tract. The number of cases in the United States is estimated at 4 million. Diagnosis involves finding appropriately shaped eggs in stool (Figure 22.33C).

Hookworms, such as *Necator americanus* (Figure 22.34A) and *Ancylostoma duodenale*, are nematodes that get their common name both from the vicious-looking oral cutting plates they use to anchor themselves to the intestinal wall to feed on patient's blood and from the hook on their tail (Figure 22.34B). Following the course of life cycle B, larvae from these species hatch in the environment

Figure 22.33 *Ascaris lumbricoides*

SEM   1 mm

A. The head of *Ascaris lumbricoides*.

2 cm

B. Cluster of *Ascaris* worms in a resected bowel (part of the bowel that has been removed).

LM   25 μm

C. *Ascaris* eggs. Note the thick, bumpy coat.

Figure 22.34 Hookworm

A. *Necator americanus*. Note the hook on the tail (arrow).

B. Hookworm sucker.

C. Larvae migrating under the skin.

D. Hookworm egg.

and penetrate the skin, usually the feet, and then follow a visible serpentine path to the lungs and mouth (**Figure 22.34C**). Most infections are asymptomatic, but the larvae can cause itchy dermatitis (ground itch) at the site of penetration and a mild, transient pneumonia as they migrate into the lungs. Once the larvae are in the intestine (after being swallowed), there may be gastrointestinal pain and inflammatory diarrhea. The main symptom is iron deficiency.

Strongyloides stercoralis (**Figure 22.35**) is a nematode (roundworm) that has a dramatically modified B-type life cycle. The added wrinkle is that adult worms can form in the human intestine <u>or</u> in the soil and lay eggs in both environments. To illustrate, let's start a cycle with larvae shed in feces from an infected human. The larvae in soil can do one of two things: stay as they are and develop into adults that mate and lay eggs (they maintain a soil cycle of replication) or change into infective **filariform** larvae that are able to penetrate human skin to start a new infection. Eggs produced in soil can also hatch filariform larvae capable of penetrating human skin. Larvae that end up penetrating the skin, again usually the feet, migrate through the bloodstream to the lungs and mouth, are swallowed, and make their way to the intestine. There they develop into females that can reproduce without males (via **parthenogenesis**). These adult worms release eggs that hatch in the intestinal mucosa, and the larvae are again shed in feces. This mode of egg hatching contrasts with that of hookworms, whose eggs must leave the intestine to hatch.

Disease caused by *S. stercoralis* is called **strongyloidiasis**. It starts with an intensely itchy, red serpentine skin rash at the initial infection site, made by larvae as they migrate. The threadlike eruption can move at an alarming pace (to the patient), 10 cm per hour. Most patients, however, have few other symptoms. Sometimes as the worms burrow into the duodenum or jejunum, the patient experiences gastric pain. A heavy infection can lead to small-bowel obstruction, but host immunity usually keeps the disease contained, even with autoreinfection. Patients already on immunosuppressive agents (corticosteroids) can suffer hyperinfection and the

production of enormous numbers of larvae. The larvae can disseminate and cause symptoms of pneumonia, as well as liver, kidney, and central nervous system involvement.

HELMINTHIC DISEASE WITH MUSCLE INVOLVEMENT Trichinosis, caused by *Trichinella* species such as *Trichinella spiralis*, is transmitted by the ingestion of pork or other meat in which cysts of this parasite are embedded. The *Trichinella* life cycle is different from the others we've described in that it all takes place in a mammalian host (no need for an intermediate host). After ingestion, the envelopes of ingested cysts dissolve in the stomach, freeing the larvae, which then burrow into intestinal mucosa. There they develop into adults, mate, and produce eggs that hatch, and the resulting larvae enter the circulatory systems (blood and lymph). All tissues are subject to larval invasion, but

Figure 22.35
Strongyloidiasis Caused by *Strongyloides stercoralis*

An adult free-living female *S. stercoralis* (about 3 mm) alongside a smaller rhabditoid larva (noninfectious, produced in patient). Notice the developing eggs in the adult female (arrow).

Figure 22.36 Encysted Larvae of *Trichinella spiralis* in Muscle
The *T. spiralis* cyst (arrow) is about 1 mm in length.

final development happens when coiled larvae enter muscle and become encased as cysts (**Figure 22.36**). Transmission takes place when a new host eats the infected muscle of a dead host. Because humans generally do not engage in cannibalism, we contract disease after eating contaminated meat from another animal, such as pig or wild boar.

Disease severity ranges from unnoticeable to life-threatening, depending on how many larvae were ingested. Symptoms, when present, start with nausea, abdominal pain, diarrhea, fever, and sweating. Once larvae migrate into muscles, the patient experiences intense muscle pain. Heart and brain involvement can be lethal. In most cases, symptoms slowly subside, but once the larvae have encysted, there is no cure. The best prevention is to thoroughly cook pork and wild meats.

HELMINTHIC DISEASE WITH LIVER INVOLVEMENT Schistosomes (species of the genus *Schistosoma*; **Figure 22.37A–C**) and the liver flukes *Opisthorchis* (**Figure 22.37D**) and *Fasciola* are trematodes that replicate via life cycle D, as noted earlier, and use snails as their intermediate hosts. The patient with **schistosomiasis** will first notice itchiness at the site where the worm entered the body. Next come fever, chills, diarrhea, and a cough. Because schistosomes coat their bodies with host blood proteins, the parasite can avoid the immune system for years. However, with this chronic infection comes liver enlargement.

Unlike *Schistosoma*, *Opisthorchis* does not enter a human host by penetrating the skin but enters when the person ingests cercarial larvae present in poorly cooked or raw fish (intermediate host). In **opisthorchiasis**, larvae enter the intestine and migrate to the bile duct, where they mature and shed eggs into the intestine. Unless heavily infected, most victims are asymptomatic. Moderately to heavily infected patients complain of vague upper-right-quadrant abdominal pain. Symptoms come from toxic irritation of the bile duct and granulomas in the liver.

DIAGNOSING AND TREATING HELMINTHIC DISEASES Most of the worm infections that we just discussed are relatively rare in countries, such as the United States, that have extensive sanitation systems and effective hygiene practices. In developing countries that are still building the infrastructure and health standards needed for their prevention, however, these diseases are common. Easy travel between countries means that travel history in a patient with diarrheal disease can suggest possible worm infestations. Diagnosis always includes a finding of eosinophilia in peripheral

Figure 22.37 Liver Flukes

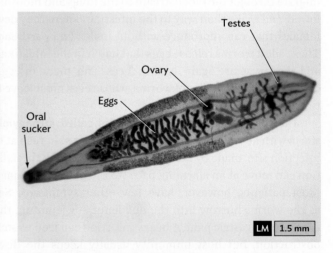

A. A mated pair of *Schistosoma mansoni*, the human blood fluke. The female fits inside a groove on the larger male's body.

B. Egg of *S. mansoni*. Note the lateral spine, a hallmark of this organism recognized by clinical parasitologists.

C. Fork-tailed cercaria of *S. mansoni*.

D. *Opisthorchis sinensis*. Shown is the hermaphroditic larva seen in a mammalian host. The tail of *Opisthorchis* is not forked.

blood and the discovery of appropriately shaped eggs in stool samples. Treatment varies with the organism. Antihelminthic drugs are available (as noted in Table 22.4), but in general they have poor systemic absorption and tend to either paralyze the worm or starve it. As a result, it's difficult to ensure complete elimination of worms.

SECTION SUMMARY

- **Protozoa causing diarrhea** include *Giardia intestinalis*, *Entamoeba histolytica*, *Cryptosporidium*, and *Cyclospora*.

- **Protozoan disease typically starts with the ingestion of oocysts**, which dissolve and release trophozoites that attach and invade the intestinal wall. From this locale, more cysts are made that pass in feces.

- **Helminthic diseases are caused by three families of worms:** nematodes, trematodes, and cestodes.

- **A helminth life cycle has three major stages:** egg, larva, and adult. Transmission to humans can be fecal-oral, transdermal, or vector-borne, or it can occur through a life cycle using both an intermediate and a definitive host.

- **Worms that cause intestinal distress** include *Trichuris*, *Enterobius*, and *Taenia*.

- **Worms that cause intestinal distress with migratory symptoms** include *Ascaris*, *Necator*, *Ancylostoma*, and *Strongyloides*.

- ***Trichinella* causes disease with muscle involvement.**

- **Worms that cause liver disease** include *Schistosoma*, *Opisthorchis*, and *Fasciola*.

- **Eosinophilia** is a key marker for helminthic disease.

- **Stool exams** looking for eggs and parasites are important for diagnosing protozoan and worm diseases.

Perspective

Because of its open access to the outside world, its rich nutrients, and its abundant nooks and crannies, the GI tract is like a four-star hotel to microbes. Their lengths of stay vary, depending on what we eat, which attachment tools are available, and which attachment sites are open (any room at the inn?). How are these and other factors that influence the composition of our normal GI microbiota affected by the presence of pathogens? Although competition with abundant normal microbiota and effective mucosal immunity pose daunting obstacles that pathogens must overcome, the GI tract plays host to more infectious microbes than any other organ system. One reason infections are so common in the GI tract is the open-access aspect of the system. Another reason involves the weapons that pathogens wield to "soften the battlefield." Toxins that increase intestinal motility flush less tenacious organisms from the intestine. And the ability of some pathogens to invade host cells removes competition altogether.

Be aware, too, that a pathogen's ability to perturb normal microbiota and intestinal function has far-reaching consequences for the host. For example, cholera toxin's ability to pump electrolytes out of the host will affect muscle contraction (heart), nerve impulses, and other aspects of host physiology that you wouldn't intuitively connect to diarrhea. Finally, note that because of inadequate sanitation, diarrhea is one of the greatest causes of death, especially of children, throughout the world.

Figure 22.38 Summary: Microbes That Infect the Digestive System

⚫ Viruses	Genome	Disease
Hepatitis B virus	DNA, ds (partially), circular	Hepatitis
Adenoviruses, types 40 and 41	DNA, ds, linear	Acute gastroenteritis primarily in children
Rotavirus	RNA, ds, linear, segmented	Acute diarrhea in children under 3 years of age
Hepatitis D virus[a]	RNA, ss, negative sense	Hepatitis
Mumps virus	RNA, ss, negative sense	Mumps
Hepatitis A virus	RNA, ss, positive sense	Hepatitis
Hepatitis C virus	RNA, ss, positive sense	Hepatitis
Hepatitis E virus	RNA, ss, positive sense	Hepatitis
Norovirus	RNA, ss, positive sense	Acute gastroenteritis, also called stomach flu
Sapoviruses	RNA, ss, positive sense	Acute gastroenteritis in adults

〰 Bacteria	Morphology	Disease
Campylobacter spp.	Gram-negative	Enterocolitis
E. coli (EHEC—e.g., *E. coli* O157:H7)	Gram-negative	Diarrhea, may progress to bloody
E. coli (EIEC)	Gram-negative	Bacillary dysentery
E. coli (ETEC)	Gram-negative	Traveler's diarrhea
Helicobacter pylori	Gram-negative	Peptic ulcer
Salmonella spp.	Gram-negative	Salmonellosis (enterocolitis)
Shigella spp.	Gram-negative	Shigellosis (bacillary dysentery)
Vibrio cholerae	Gram-negative	Cholera
Vibrio parahaemolyticus	Gram-negative	Vibriosis
Yersinia enterocolitica	Gram-negative	Yersiniosis
Clostridioides difficile	Gram-positive	Antibiotic-associated diarrhea (pseudomembranous colitis)
Clostridium perfringens	Gram-positive	Gastroenteritis (due to intoxication)
Listeria monocytogenes	Gram-positive	Listeriosis
Staphylococcus aureus	Gram-positive	Gastroenteritis (food poisoning due to intoxication)

〰 Parasites		Disease

Protozoa

Entamoeba histolytica	No flagella/NA[b]	Amebic dysentery
Giardia intestinalis	Flagellated/NA	Giardiasis

↻ Helminths	Type/Life Cycle (A–D, as defined in Figure 22.29)	
Ascaris lumbricoides	Nematode/A	Ascariasis
Enterobius vermicularis	Nematode/A	Pinworm
Trichuris trichiura	Nematode/A	Whipworm (trichuriasis)
Necator americanus	Nematode/B	Hookworm
Strongyloides stercoralis	Nematode/modified B	Strongyloidiasis
Trichinella spp.	Nematode/wholly within mammalian host	Trichinosis
Diphyllobothrium latum	Cestode/C	Tapeworm (fish)
Taenia solium	Cestode/C	Tapeworm (pork)
Fasciola hepatica	Trematode/D	Liver fluke
Opisthorchis and *Clonorchis*	Trematode/D	Liver fluke
Schistosoma mansoni	Trematode/D	Blood fluke, schistosomiasis

[a]Able to replicate and cause disease only when coinfecting with hepatitis B virus.
[b]NA = not applicable.

Portal of Entry

Portal of Exit

Blood
- Hepatitis B, C, and D viruses

Blood
- Hepatitis B, C, and D viruses

Respiratory
- Mumps virus
- *Enterobius vermicularis*

Respiratory
- Mumps virus

Oral
- Hepatitis A and E viruses
- Rotavirus
- Norovirus
- Sapovirus
- Adenoviruses 40 and 41
- *Helicobacter pylori*
- *Escherichia coli*
- *Shigella*
- *Salmonella* spp.
- *Yersinia enterocolitica*
- *Campylobacter* spp.
- *Vibrio cholerae*
- *Vibrio parahaemolyticus*
- *Listeria monocytogenes*
- *Clostridium perfringens*
- Toxins
  (e.g., *Staphylococcus aureus*)
- *Entamoeba histolytica*
- *Giardia intestinalis*
- *Ascaris lumbricoides*
- *Enterobius vermicularis*
- *Trichuris trichiura*
- *Taenia solium*
- *Diphyllobothrium*
- *Opisthorchis* and *Clonorchis*
- *Fasciola hepatica*

Stool
- Hepatitis A and E viruses
- Rotavirus
- Norovirus
- Sapoviruses
- Adenoviruses 40 and 41
- *Helicobacter pylori*
- *Escherichia coli*
- *Shigella*
- *Salmonella* spp.
- *Yersinia enterocolitica*
- *Campylobacter* spp.
- *Vibrio cholerae*
- *Vibrio parahaemolyticus*
- *Listeria monocytogenes*
- *Clostridium perfringens*
- Toxins (e.g., *Staphylococcus aureus*)
- *Entamoeba histolytica*
- *Giardia intestinalis*
- *Ascaris lumbricoides*
- *Enterobius vermicularis*
- *Trichuris trichiura*
- *Taenia solium*
- *Diphyllobothrium*
- *Opisthorchis* and *Clonorchis*
- *Fasciola hepatica*
- *Necator americanus*
- *Strongyloides*
- *Schistosoma mansoni*

Skin
- Hepatitis B, C, and D viruses
- *Necator americanus*
- *Strongyloides*
- *Schistosoma mansoni*

Key Terms

adenitis (p. 714)

amebiasis (p. 724)

amebic dysentery (p. 724)

ascariasis (p. 728)

attaching and effacing lesion (p. 710)

bacillary dysentery (p. 711)

cercaria (p. 727)

cestode (p. 725)

cholecystitis (p. 715)

cholera (p. 716)

cholera toxin (p. 716)

colitis (p. 698)

cryptosporidiosis (p. 724)

cyclosporiasis (p. 724)

cysticercus (p. 727)

definitive host (p. 726)

dental caries (p. 695)

dental plaque (p. 695)

diarrhea (p. 697)

enteric fever (p. 712)

enteric pathogen (p. 709)

enteritis (p. 698)

enteroaggregative E. coli (EAEC) (p. 711)

enterocolitis (p. 698)

enterohemorrhagic E. coli (EHEC) (p. 709)

enteroinvasive E. coli (EIEC) (p. 709)

enteropathogenic E. coli (EPEC) (p. 711)

enterotoxigenic E. coli (ETEC) (p. 709)

enterotoxin (p. 700)

eosinophilia (p. 725)

feces (p. 695)

filariform (p. 729)

flatworm (p. 725)

fluke (p. 725)

food poisoning (p. 719)

gastric mucosa (p. 694)

gastric pit (p. 694)

gastritis (p. 698)

gastroenteritis (p. 698)

giardiasis (p. 721)

gingivitis (p. 696)

Guillain-Barré syndrome (GBS) (p. 715)

H antigen (p. 709)

hemolytic uremic syndrome (HUS) (p. 710)

hepatitis (p. 698)

hepatitis A virus (HAV) (p. 701)

hepatitis B virus (HBV) (p. 702)

hepatitis C virus (HCV) (p. 702)

hepatitis D virus (HDV) (p. 702)

hepatitis E virus (HEV) (p. 703)

hepatosplenomegaly (p. 712)

hookworm (p. 728)

inflammatory diarrhea (p. 697)

intermediate host (p. 726)

labile toxin (LT) (p. 709)

listeriosis (p. 718)

miracidium (p. 727)

motility-related diarrhea (p. 697)

mumps (p. 703)

myocarditis (p. 715)

nematode (p. 725)

noncholera Vibrio (p. 716)

norovirus (p. 700)

O antigen (p. 709)

oophoritis (p. 704)

opisthorchiasis (p. 730)

orchitis (p. 704)

osmotic diarrhea (p. 697)

pancreatitis (p. 715)

paramyxovirus (p. 703)

parotitis (p. 703)

parthenogenesis (p. 729)

pathovar (p. 709)

pedestal (p. 710)

peptic ulcer (p. 704)

pericarditis (p. 715)

periodontal disease (p. 696)

peristalsis (p. 695)

peritonitis (p. 715)

pinworm (p. 727)

plague (p. 714)

platyhelminth (p. 725)

proglottid (p. 728)

rotavirus (p. 699)

roundworm (p. 725)

salmonellosis (p. 711)

schistosomiasis (p. 730)

scolex (p. 728)

secretory diarrhea (p. 697)

Shiga toxin–producing E. coli (STEC) (p. 711)

shigellosis (p. 711)

strongyloidiasis (p. 729)

tapeworm (pp. 725, 728)

thrombocytopenia (p. 711)

thrombotic thrombocytopenic purpura (TTP) (p. 710)

thrush (p. 696)

trematode (p. 725)

trench mouth (p. 696)

trichinosis (p. 729)

trophozoite (p. 721)

typhoid fever (p. 712)

Vi Ag (p. 714)

vibriosis (p. 716)

Vincent's angina (p. 696)

whipworm (p. 727)

yersiniosis (p. 714)

Objectives Assessment Questions

22.1 Anatomy of the Digestive System

1. The _____ is physically attached to the stomach.
   a. esophagus
   b. duodenum
   c. ileum
   d. jejunum
   e. a and b only

2. A patient is diagnosed with appendicitis. Which of the following parts of his digestive system is affected?
   a. Esophagus
   b. Stomach
   c. Small intestine
   d. Large intestine

3. The lining of the human stomach is protected by a layer of mucus and bicarbonate produced by the parietal cells of the stomach. True or false?

4. Removing a large section in the middle of the small intestine will result in _____.
   a. inadequate absorption of nutrients
   b. increased absorption of water and electrolytes
   c. loss of bile and pancreatic juices
   d. increased gastroesophageal reflux

22.2 Infections of the Oral Cavity

5. Which of the following bacterial genera is part of the normal microbiota of the oral cavity?
   a. *Streptococcus*
   b. *Lactobacillus*
   c. *Peptostreptococcus*
   d. All of the above
   e. None of the above

6. Which of the following statements is true?
   a. Biofilm buildup results in the formation of dental plaque, which, if not removed, can result in tooth decay.
   b. Bacteria in the biofilm produce a substance that covers and protects the biofilm from antibiotics and other chemicals.
   c. Both a and b

Match the following:

7. Bleeding; red gums
8. Tooth decay
9. Thrush

   a. *Candida albicans*
   b. Trench mouth
   c. Plaque formation

10. Which of the following diseases is linked to gingivitis?
    a. Pneumonitis
    b. Vincent's angina
    c. Gastroenteritis
    d. Colitis

22.3 Gastrointestinal Syndromes

11. _____ refers to inflammation of the small intestine.
    a. Gastritis
    b. Enteritis
    c. Colitis
    d. Hepatitis
    e. Esophagitis

12. A patient with cholera will have _____.
    a. inflammatory diarrhea
    b. increased absorption of water and electrolytes
    c. secretory diarrhea
    d. intestinal hypermotility

13. Bacterial infections of the GI tract are the only cause of inflammation in the digestive system. True or false?

22.4 Viral Infections of the Gastrointestinal Tract

Match the following:

14. Hepatitis B
15. Hepatitis C virus
16. Mumps virus
17. Norovirus

    a. Diarrhea
    b. ss positive-sense RNA genome
    c. Orchitis
    d. DNA genome

18. Inflammation of the _____ may result in hyperbilirubinemia and clay-colored stool.
    a. liver
    b. intestinal mucosa
    c. salivary glands
    d. gastric mucosa

19. Hepatitis D can replicate only inside a cell that is coinfected with _____.
    a. hepatitis E
    b. hepatitis B
    c. hepatitis C
    d. hepatitis G

22.5 Bacterial Infections of the Gastrointestinal Tract

20. An enterotoxin secreted by _____ growing in food is an important cause of food poisoning.
    a. *Salmonella enterica* serovar Typhimurium
    b. *Vibrio cholerae*
    c. *Staphylococcus aureus*
    d. *Streptococcus pyogenes*
    e. Enteroinvasive *E. coli*

21. A patient who has bloody diarrhea is diagnosed with shigellosis. *Shigella* avoided this patient's immune system by producing _____.
    a. attaching and effacing lesions
    b. Shiga toxin 1
    c. Shiga toxin 2
    d. pedestals
    e. actin tails

22. Patients suffering from _____–induced diarrhea are treated with antibiotics to reduce the possibility of a carrier state.
    a. *Vibrio cholerae*
    b. *Salmonella* Typhimurium
    c. *Vibrio vulnificus*
    d. *Salmonella* Typhi
    e. *Escherichia coli*

23. The most effective preventive measure against cholera is _____.
    a. good sanitation
    b. vaccination
    c. prophylactic antibiotics
    d. adequate intake of fluids

24. *Salmonella* Typhi has an animal reservoir. True or false?

22.6 Parasitic Infections of the Gastrointestinal Tract

25. _____ is caused by protozoa.
    a. Cryptosporidiosis
    b. Strongyloidiasis
    c. Trichinosis
    d. Schistosomiasis
    e. Opisthorchiasis

26. A patient complains of a nonproductive cough accompanied by a burning sensation and is suffering from abdominal pain and vomiting. She is diagnosed with small-bowel obstruction. Examination of her stool reveals roundworms and worm eggs. The reason for the patient's cough is _____.
    a. migration of worms to the lungs
    b. migration of larvae to the lungs
    c. a systemic reaction to the infection
    d. an associated pneumonia

27. *Strongyloides stercoralis* can cause infection following _____.
    a. ingestion of larvae
    b. larval penetration of the skin
    c. ingestion of the small roundworms
    d. a and c only
    e. a and b only

28. A 25-year-old day-care teacher is diagnosed with parasitic diarrhea after a stool examination shows flagellated protozoa and trophozoites. What is the treatment for this infection?
    a. Proper hydration
    b. Electrolyte replacement
    c. Metronidazole
    d. Vancomycin
    e. Penicillin

Review Questions

1. What are the different parts of the digestive system?

2. What is the importance of the acidic pH of the stomach?

3. What is the relationship between biofilm formation in the mouth and dental plaques?

4. What are the differences between gingivitis, periodontitis, dental caries, and trench mouth?

5. Which bacteria often cause periodontitis?

6. Describe thrush and its causative agent.

7. How do gastritis, gastroenteritis, colitis, and enterocolitis differ from one another?

8. What are the types of diarrhea?

9. List the signs and symptoms of hepatitis.

10. How do the viruses that cause hepatitis differ from one another?

11. Describe how rotavirus increases intestinal motility.

12. What part of the digestive system is infected with mumps virus?

13. Name viral infections of the digestive system that can be prevented by vaccination.

14. Why is hepatitis D virus found only in coinfection with hepatitis B virus?

15. How does *Helicobacter pylori* cause an infection in the stomach?

16. How is *H. pylori* infection diagnosed?

17. What is hemolytic uremic syndrome (HUS)?

18. Which enteric pathogens can cause a bloody diarrhea?

19. List enteric pathogens that produce an enterotoxin.

20. How do different bacterial toxins produce GI disease?

21. How are various pathovars of *E. coli* different from one another?

22. Describe how *Shigella* causes shigellosis.

23. How do various salmonelloses differ from one another, and how are they similar?

24. What are the complications of *Campylobacter jejuni* infections?

25. Name risk factors associated with diseases caused by members of different *Vibrio* species.

26. How are the treatments for *Vibrio cholerae* and *Vibrio vulnificus* different from each other, and why?

27. What is listeriosis, and how is it diagnosed and treated?

28. What are the complications of listeriosis?

29. How is *Listeria*'s ability to grow in the cold related to its pathogenesis?

30. How does *Staphylococcus aureus* cause food poisoning?

31. What is giardiasis, and who is at risk of acquiring it?

32. What organism causes amebic dysentery, and how?

33. Define cryptosporidiosis, and describe its diagnosis and treatment.

34. List the gastrointestinal helminthic infections, their modes of transmission, and how they are treated.

35. Describe the four life cycles used by helminths.

36. List the signs and symptoms for different helminthic diseases.

37. Describe the best preventive methods used against viral, bacterial, and parasitic infections of the digestive system.

Clinical Correlation Questions

1. A patient who has just returned from Mexico presents to the emergency department complaining of voluminous watery diarrhea. Vitals are normal, and a stool sample contains no blood, leukocytes, or cysts. The physician assistant on the case tells the patient that he most likely has traveler's diarrhea. What is the causative organism, and how does it cause diarrhea?

2. A 24-year-old male presents to the ER with diarrhea, high fever (40°C, or 104°F) and chills, malaise, and small red spots on his trunk. The patient reports that he has had a fever for days and that the fever was initially low grade but has been getting higher and higher. Blood culture reveals a facultative intracellular organism. What is the most likely diagnosis? Explain.

3. A patient with a 2-week history of watery diarrhea presents to the ER complaining of right-lower-quadrant pain and a low-grade fever. He is diagnosed with appendicitis and taken to surgery. The patient's appendix is found to be normal during the surgery, but he is found to have mesenteric lymphadenitis and terminal ileitis. What bacterial pathogen is most likely to cause this patient's symptoms? Explain.

4. A 20-year-old female presents to the emergency department complaining of nausea, recurrent stomach pain, shortness of breath, and a burning sensation in the lungs. Physical exam shows that the patient does not have a fever, but blood work indicates anemia. What is the most likely cause?

23

Infections of the Urinary and Reproductive Tracts

[Frequent Urination]

SCENARIO Lois woke up with a headache one morning, feeling a bit disoriented and unwell. She is 87 years old and lives by herself. She had not slept well the night before because she'd had to get up several times to urinate. Lois is diabetic, but her blood sugar the night before had been 120 mg/dl (normal is 100) and it measured 180 mg/dl in the morning. She went about her day, taking her insulin shot and eating her breakfast. The breakfast did not stay with her long, and after several bouts of vomiting she called her son.

Urine Specimen Cup
Patients can use these sterile cups to collect their own urine. Lois, however, was poorly responsive and needed to have a catheter inserted into her bladder.

SIGNS AND SYMPTOMS Lois was almost nonresponsive by the time her son and daughter-in-law arrived. She had no fever—not unusual for the elderly during an infection. Her family members were familiar with the situation; it had happened before, but never so severely. They took Lois to the ER, where her temperature measured below normal; her heart rate was faster than normal and her blood pressure dangerously low. Lois was immediately given fluids, and blood and urine samples were taken for laboratory analysis.

LABORATORY RESULTS Gram stain of the urine sample showed Gram-negative rods. Cultures of urine and blood showed bacteria and confirmed everyone's suspicions: she had a severe urinary tract infection (UTI), and the bacteria had spread to her bloodstream, causing sepsis (blood infection; see Chapter 21). Lois was diagnosed with urosepsis, a term often used to describe a UTI complicated by sepsis. The organism in this case was found to be a uropathogenic strain of *Escherichia coli*.

TREATMENT Lois was transferred to the intensive care unit (ICU), where she was treated with IV antibiotics, fluids, and her routine medications. Lois fully recovered and was sent home with extensive patient education materials given to her, her son, and her daughter-in-law.

SEM 4 μm

Bacterial Biofilm in the Urinary Bladder
Uropathogenic *Escherichia coli* (green) covers the surface epithelium of a mouse bladder (purple) while a leukocyte (blue) attacks the organism. Uropathogenic *E. coli* can form a biofilm in the bladder, making it very difficult to rid the bladder of infection.

CHAPTER OBJECTIVES

After reading this chapter, you will be able to:

- Correlate the anatomy and physiology of the urinary and reproductive tracts with the infectious diseases affecting them.

- Identify different genitourinary infections by their presenting signs and symptoms.

- Describe similarities and differences between sexually transmitted and non–sexually transmitted infections of the genitourinary system.

- Explain effective prevention and treatment methods for infectious diseases affecting the urinary and reproductive tracts.

ois's problem in the opening case history started with a bladder infection with no clear symptoms. The bladder infection progressed rapidly, and by the time she got to the ER, she had **urosepsis** and was in danger of losing her life. How could this happen?

Urine made in the kidneys is normally considered sterile, so the presence of bacteria in urine (called **bacteriuria**) is considered abnormal. Bacteriuria might or might not produce symptoms such as a burning sensation while urinating. An infection anywhere from the kidney to the bladder is called a **urinary tract infection (UTI)**. UTIs are the most common type of bacterial infection in humans, ranking alongside respiratory infections, such as bronchitis and pneumonia, and GI tract infections. UTIs have a large economic impact, resulting in innumerable office visits, more than 100,000 hospital admissions, and $1.6 billion in medical expenses each year.

Longtime diabetics like Lois often do not have the burning pain usually associated with a UTI because of the nerve damage (neuropathy) caused by elevated blood glucose levels. The process of aging itself also seems to reduce the severity of UTI symptoms. For example, up to 50% of women over 65 years of age report having no symptoms even when they have confirmed bacteriuria.

To better understand infections of the urinary tract, we will first review the structures of the urinary system and their functions. The main job of the urinary tract is to rid the body of metabolic waste products, whereas the primary role of the reproductive tract is to propagate the human species. Infections in either system can undermine vital functions and have serious consequences. The formation and excretion of urine, for instance, appears so mundane, yet its interruption can cause extreme pain and even threaten life.

23.1 Anatomy of the Urinary Tract

SECTION OBJECTIVES

- List the function of each structural component of the male and female urinary tracts.

- Explain the relationship between the upper and lower urinary tract anatomy and physiology.

- Discuss the significance of structural differences between the male and female urinary tract as they relate to infections.

What is the job of our urinary tract? Normal metabolism generates a lot of harmful waste products. The urinary tract, which includes the kidneys, ureters, urinary bladder, and urethra, removes the fluid waste and any excess water and electrolytes from the bloodstream and then expels the product: urine (**Figure 23.1**). Large waste molecules (for example, unneeded proteins or lipids) are eliminated mostly by the liver or gastrointestinal tract. Kidneys act as a filter to eliminate unwanted small molecules from the blood while maintaining plasma electrolytes and blood pH in the normal physiological range.

Renal arteries bring unfiltered blood from the heart to each kidney via the aorta, and renal veins return filtered blood from each kidney to the heart via the inferior vena cava. A ureter originates from each kidney and ends in the urinary bladder. Urine collects in the bladder until it is full; the urine is expelled through the urethra.

The urinary system is divided into upper and lower tracts. Kidneys and ureters are considered the upper tract; the urinary bladder and urethra form the lower tract.

Figure 23.1 **The Urinary Tract**

Renal arteries bring unfiltered blood from the heart to each kidney via the aorta. Renal veins return renal blood from each kidney to the heart via the inferior vena cava. A ureter originates from each kidney and ends in the urinary bladder. Urine collects in the bladder until it is full; the urine is expelled through the urethra.

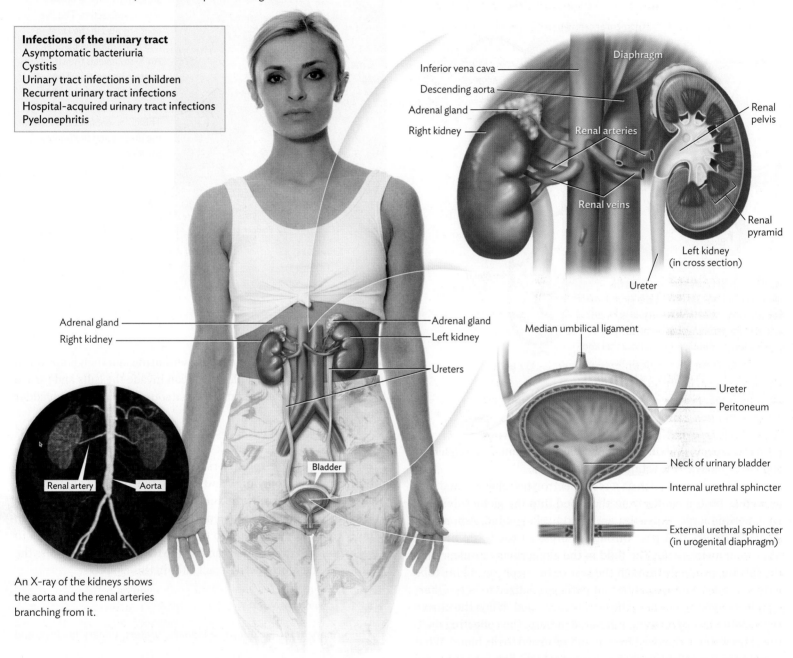

Infections of the urinary tract
Asymptomatic bacteriuria
Cystitis
Urinary tract infections in children
Recurrent urinary tract infections
Hospital-acquired urinary tract infections
Pyelonephritis

Diaphragm

Inferior vena cava

Descending aorta

Adrenal gland

Right kidney

Renal arteries

Renal veins

Renal pelvis

Renal pyramid

Left kidney (in cross section)

Ureter

Adrenal gland

Right kidney

Adrenal gland

Left kidney

Ureters

Bladder

Median umbilical ligament

Ureter

Peritoneum

Neck of urinary bladder

Internal urethral sphincter

External urethral sphincter (in urogenital diaphragm)

Renal artery

Aorta

An X-ray of the kidneys shows the aorta and the renal arteries branching from it.

The kidneys are two bean-shaped organs that sit in the abdominal cavity on both sides of the descending aorta with an adrenal gland perched on top of each. The renal artery and vein enter at the concave part of each kidney, called the renal hilum. The renal pelvis, the beginning part of the ureter, exits the kidney at the renal hilum (see upper right inset in Figure 23.1).

Each kidney is made of many tubular structures called nephrons—roughly a million of them—parts of which make the renal pyramids. Each nephron has a horseshoe-shaped structure called a glomerular capsule (Bowman's capsule) that surrounds a ball of blood capillaries known as a glomerulus (plural, glomeruli; **Figure 23.2**). It is the glomerulus that becomes damaged in acute

Figure 23.2 Nephron Structure

A. Nephrons are the filtering units of the kidneys. Each nephron starts at the glomerular capsule (Bowman's capsule) and continues as a long tube that ends in the collecting duct.

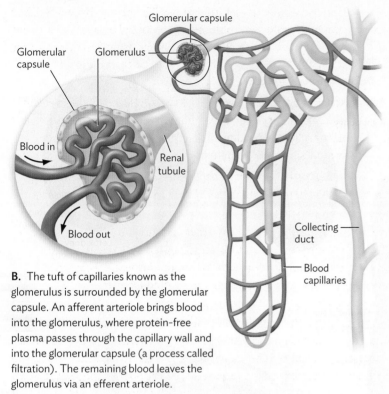

B. The tuft of capillaries known as the glomerulus is surrounded by the glomerular capsule. An afferent arteriole brings blood into the glomerulus, where protein-free plasma passes through the capillary wall and into the glomerular capsule (a process called filtration). The remaining blood leaves the glomerulus via an efferent arteriole.

Figure 23.3 Ureters, and the Urinary Bladder

Contrast dye is given intravenously when structural abnormality in the urinary tract is suspected. The dye concentrates in the kidneys and flows into the bladder. An X-ray reveals the structures containing the dye. The resulting image is called an intravenous pyelogram (IVP). This color-enhanced IVP shows the ureters and the urinary bladder.

post-streptococcal glomerulonephritis, an autoimmune sequela of some streptococcal infections (Section 20.3).

Small molecules such as water, electrolytes, glucose, and CO_2 as well as protons move from the blood into the glomerular capsule, leaving large molecules and blood cells behind. Among the small molecules that pass through is urea, a toxic waste product of protein metabolism. The fluid in the glomerular capsule, called the filtrate, proceeds through the rest of the nephron, a long tube with several structurally different parts specialized to keep different electrolytes in or out of the filtrate as needed. When the filtrate reaches the last part of the tubular structure, the collecting duct, the extra water is removed from it and returned to the blood. What remains in the collecting duct is urine that will flow into the renal pelvis, the beginning of the ureter. Each ureter is about 25–30 cm (10–12 inches) long, starting at a kidney and ending in the bladder. Ureters have muscular walls that continuously contract and relax to propel urine down to the urinary bladder.

The human urinary bladder is a triangle-shaped muscular organ that can collect and store 400–600 ml of urine (Figure 23.3). Inside, the bladder is lined with specialized epithelium (called transitional epithelium) that consists of several layers of cells put together in a way that can easily accommodate the changes in bladder volume as the bladder fills with urine and then empties. In some people, uropathogenic *E. coli* can invade the cells and form a biofilm inside these cells and serve as a source for recurrent bladder infections.

Two small circular muscles, called the internal and external urethral sphincters, mark the beginning and end of the urethra. Relaxing these sphincters allows urine to flow through the urethral meatus (opening) and out of the body (urination). In males, the urethra forms the central tubular structure of the penis and spans the length of the penile shaft. The seminal fluid, which usually contains sperm, has to travel through the male urethra to be delivered into the vaginal canal of the female. We will discuss the anatomy of the reproductive tract later in this chapter.

SECTION SUMMARY

- **The urinary tract** includes the kidneys, ureters, urinary bladder, and urethra.
- The **kidneys filter waste products** from blood to make urine, the **ureters deliver urine** to the bladder, the **bladder stores** the urine, and the **urethra passes urine** outside the body.
- **Nephrons** include the **glomerular capsule (Bowman's capsule)**, which surrounds capillaries of the **glomerulus**, where filtration occurs.
- **Urine leaves the nephron** through the collecting duct and passes through the renal pelvis and the ureters to reach the bladder.
- **Urethral sphincters** control urine flow from the bladder.
- The opening of the urethra to the outside of the body is **called the urethral meatus**.

23.2 Urinary Tract Infections

SECTION OBJECTIVES

- Discuss the role of physical and hormonal factors in the development of urinary tract infections in women.
- Explain the relationships between lower and upper urinary tract infections.
- Compare and contrast the signs and symptoms of common urinary tract infections.
- List treatments for the most common etiologies of urinary tract infections.

How big of a problem are urinary tract infections? A huge one! Everyone is at risk of getting a UTI, but women suffer most cases. In fact, women have a greater than 50% chance of getting a urinary tract infection sometime during their lifetime, and 20%–40% of those infected will have recurrent infections. A UTI is one of the most common hospital-acquired (nosocomial) infections and the fifteenth-most-common diagnosis made by family physicians during office visits. There are two types of urinary tract infections: bladder infections (**cystitis**) and kidney infections (**pyelonephritis**).

Why do women get urinary tract infections more often than men? The perineum (the distance between anus and vagina or between anus and base of the penis) is roughly twice as long in males as in females. The short female perineum places the female urethra adjacent to the vaginal and rectal openings, which are colonized by some potential pathogens. The combination of a short urethra and proximity to pathogens is thought to be why women are at increased risk of UTIs. The longer male urethra makes it more difficult for bacteria to reach the urinary bladder.

The risk of a bladder infection in young women is also related to sexual activity and is proportional to the frequency of intercourse. In postmenopausal women, bladder incontinence, due partially to lower estrogen (female hormone) levels, increases the risk for UTIs. Hormonal changes of pregnancy and menopause also reduce the normal muscle tone in the urethral wall, causing it to become wider and shorter, further increasing the chances of getting a lower UTI. For men, instrumentation, such as the insertion of a catheter through the urethra to retrieve urine from the bladder, is the main risk factor for getting a UTI.

Diabetes increases the risk of UTIs for two reasons. First, diabetic patients usually have an impaired immune system. Second, in diabetic patients, unlike nondiabetics, urine contains glucose, making it a fertile environment for bacteria to grow. A common consequence of long-term diabetes is nerve damage (diabetic neuropathy). Damage to nerves that serve the bladder can diminish a patient's ability to feel the symptoms of the UTI, increasing the chances of a dangerous outcome. The patient in the chapter-opening case history, Lois, most likely had bladder nerve damage that destroyed her ability to feel the symptoms of her UTI and led to progression of the infection and sepsis.

Pathogenesis of Urinary Tract Infections

How do we get urinary tract infections? The newly discovered, but mostly unculturable, microbiome of the bladder is not the cause of cystitis (Section 14.1). Pathogenic microorganisms from other sources, however, can "find" their way into the urinary tract and establish an infection in one of three basic ways:

- **Infection from the urethra to the bladder.** Bacteria residing along the superficial urogenital membranes of the urethra can ascend to the bladder. This route of infection is more common in women than in men.

- **Ascending infection to the kidney.** Organisms from an established infection in the bladder ascend along the ureter to infect the kidney.

- **Descending infection from the kidney.** Bacteria can travel via the ureters from an infected kidney into the bladder. Kidney infections may arise when microorganisms are deposited in the kidneys from the bloodstream.

Several organisms can cause a UTI. In Lois's case, the laboratory found the organism to be a Gram-negative bacillus that ferments lactose, suggesting *E. coli* as the likely culprit. Given that the normal habitat of *E. coli* is the gastrointestinal tract, the initial UTI suffered by the patient was probably the result of an inadvertent introduction of the microbe into the urethra (which then made its way up the urethra and into the bladder). Patients may suffer recurring UTIs as a result of organisms entering the bladder by a descending route or organisms emerging from a pod-like biofilm inside the bladder (see **IMPACT**).

Urine is bacteriostatic to most of the Gram-positive microbiota inhabiting the perineum and the vagina, such as lactobacilli, corynebacteria, and *Staphylococcus epidermidis*. In contrast, Gram-negative rods not only thrive in urine but also may be adapted to cause UTIs through specialized pili. As a result, most UTIs are caused by facultative Gram-negative rods from the GI tract. Uropathogenic strains of *Escherichia coli*, for example, typically have P-type pili, with a terminal receptor for the P antigen (a rather appropriate name for a bladder-specific virulence factor). The P antigen—expressed by approximately 75% of the population—is a blood group marker found on cells lining the perineum and urinary tract. Individuals expressing P antigen are particularly susceptible to UTIs. The most common etiologic agents of UTIs are the uropathogenic *E. coli* (75% of all UTIs), *Klebsiella*, *Proteus*, *Pseudomonas*, *Enterobacter*, *Staphylococcus aureus*, *Enterococcus*, *Chlamydia*, fungi, and *Staphylococcus saprophyticus*.

Although some UTIs resolve spontaneously, others can progress to destroy the kidney or, via Gram-negative septicemia, the host. As a result, antibiotic therapy is recommended. Infections of the lower urinary tract are usually afebrile (without fever) except

impact

Recurrent UTIs: The Hide-and-Seek Strategy of Uropathogens

Doctors used to think that women suffering from recurrent bladder infections were repeatedly infected as a result of sexual activity or poor hygiene—a presumption many women with chronic infections found frustrating and offensive. Evidence now indicates that bacteria may hide inside bladder cells, forming a reservoir for reinfection.

Figure 1 **Bladder Pods of Uropathogenic *E. coli***

A. Increasing magnifications of large intracellular communities of uropathogenic *E. coli* (UTI89) inside pods on the surface of a mouse bladder infected for 24 hours.

B. Hematoxylin- and eosin-stained sections of *E. coli*–infected mouse bladders show a bacterial factory 6 hours after inoculation (top panel) and a pod 24 hours after inoculation (bottom panel). At 24 hours, bacteria in the pod were densely packed and shorter, and filled the host cell.

Specialized colonies of *E. coli* have been found living in the surface layer of cells lining the bladders of infected mice (**Figure 1A**). Though the mouse served as a model system, the bacteria used were originally isolated from humans. These strains of *E. coli* cause about 80% of all urinary tract infections. The bacteria use pili to latch on to proteins coating the bladder cells. Those proteins form a protective substance, known as uroplakin, that strengthens the epithelial cells and shields them from toxins that may build up in the urine. The bacteria can slip under the protective barrier and penetrate superficial cells lining the bladder. Once inside the cell, the microbes begin to multiply, and the epithelial cell fills with bacteria also coated with uroplakin, taking on the appearance of a pod (**Figure 1B**).

The podlike *E. coli* biofilms we describe here are notoriously resistant to antibiotic treatment and attacks by the immune system. Living inside host cells and surrounded by uroplakin, the *E. coli* in the bladder pods have devised a clever mechanism for surviving. Bacteria living on the edges of the biofilm may break free, leading to subsequent infection. Understanding the cycle of infection and pod formation may help researchers design drugs to prevent new UTIs or stop established ones.

Although this work was done with mice, intracellular bacterial communities of *E. coli* have also been identified in some women with recurrent cystitis. Elsewhere in the body, biofilms hiding inside cells may act as reservoirs for other hard-to-treat chronic or recurring infections, such as ear infections.

DISCUSSION QUESTION

How might intracellular pods of *E. coli* in the bladder contribute to the antibiotic resistance of uropathogenic *E. coli*?

in young children. In older patients, like the elderly Lois, UTIs frequently show atypical symptoms, including delirium, which disappears when the UTI is treated.

Identification of culturable microorganisms in the urine makes the definitive diagnosis of a UTI. A urine sample must be collected and sent for urinalysis and culture. Urinalysis has three parts: determining the color and clarity of the sample, microscopic examination, and biochemical studies performed with a dipstick, also called a dry reagent strip (Figure 23.4). A urine dipstick is used to check for the presence of leukocyte esterase and nitrites. Leukocyte esterase is an enzyme of white blood cells (WBCs) and is present in the urine only when WBCs are present, even if they are not intact. Presence of nitrite in the urine points to the presence of bacteria, which can form nitrite from the nitrate normally present in the urine. A noncontaminated urine sample from a healthy individual should not contain any WBCs. Clinicians often make a presumptive diagnosis of a

Figure 23.4 Urinalysis

Urine is collected in a sterile cup by the clean-catch technique. A dry reagent strip (a dipstick) is then inserted into the patient's urine, and color changes on the strip indicate the diagnosis. Normal results are on the far left. Progressively abnormal results are shown toward the right.

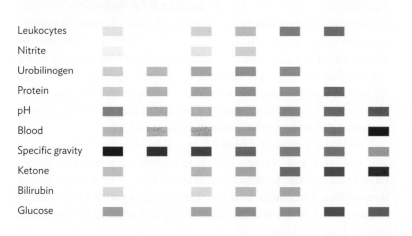

UTI on the basis of the patient's signs and symptoms and start empiric therapy to reduce the risk of serious complications.

In most patient populations the treatment of asymptomatic bacteriuria is not necessary, and consequently, screening is not recommended. An important exception is pregnant women, for whom asymptomatic infection carries significant risks and treatment provides obvious benefits.

Symptomatic bacterial UTIs are treated with antibiotics such as trimethoprim-sulfamethoxazole. Choosing the appropriate antibiotic depends on the infecting bacteria and their drug resistance profile, as well as the age, sex, and health status of the patient. Most clinicians agree that increased intake of liquids to "flush" the system is helpful, and cranberry juice is a favorite among some because it contains proanthocyanidins (tannins) that have antimicrobial qualities. The enthusiasm for cranberry juice as a cure is not universal, but it's often recommended nevertheless.

Bacterial Infections of the Urinary Tract

As mentioned earlier, two types of UTIs exist: cystitis (bladder infection) and pyelonephritis (kidney infection). Several organisms can cause either type of infection with similar signs and symptoms (Table 23.1). Inflammation of the urethra, or **urethritis**, is signified by pain or burning upon urination and urethral discharge. For two reasons, we will defer discussion of urethritis until we review reproductive diseases (Section 23.4). First, women and men with urethritis (gonococcal and nongonococcal) typically contract the disease as a result of sexual intercourse. Second, in men, the urethra is shared by the urinary and reproductive tracts.

ASYMPTOMATIC BACTERIURIA Interestingly, the presence of bacteria in the urine (bacteriuria) often does not produce any signs or symptoms. In these cases the person is said to have asymptomatic bacteriuria. How can patients with bacteriuria be asymptomatic? That's a good question with no definitive answer. The microbiology of asymptomatic bacteriuria is the same as for other uncomplicated UTIs. In some cases the lack of symptoms may be due to changes in the pathogens, making them less virulent; or the immune system may be less responsive; or a combination of the two.

Although asymptomatic bacteriuria is typically not considered problematic, it can be a serious concern for pregnant women and the elderly. Asymptomatic bacteriuria in pregnant women is associated with low birth weight of the newborn, preterm labor, and perinatal death (fetal death just prior to birth). Production of massive amounts of female hormones during pregnancy loosens the walls of the ureters (tonic relaxation of ureters), allowing bacteria in the bladder easier access to the kidney and increasing the risk of kidney infection (pyelonephritis) in the mother. Left untreated, 20%–30% of pregnant mothers with asymptomatic bacteriuria progress to pyelonephritis. Asymptomatic bacteriuria in pregnant women is treated aggressively because of its serious consequences.

Table 23.1
Infections of the Urinary Tract

Disease	Symptoms	Complications	Organisms	Transmission	Diagnosis	Treatment	Prevention
Cystitis (lower urinary tract infection)	Dysuria (burning and/or pain) upon urination, frequency (increased urination frequency), and urgency (sudden urge to urinate)	Acute pyelonephritis, bacteremia, sepsis	*E. coli* (75%–90% of cases), *Staphylococcus saprophyticus* (5%–15% of cases), other Gram-negative rods; rarely: group B streptococci, *Mycobacterium tuberculosis*, *Mycoplasma, Candida*, viruses, *Chlamydia, Schistosoma haematobium*	Self-inoculation or hospital acquired	Clinical presentation; urinalysis; urine dipstick for nitrite and leukocyte esterase; urine culture	Oral antibiotics (nitrofurantoin, trimethoprim-sulfamethoxazole, fluoroquinolones); special regimens recommended for men and pregnant women	No vaccine available
Pyelonephritis	Flank (costovertebral angle) pain, midback pain and tenderness, nausea, fever, chills, symptoms of lower urinary tract infection, cloudy and sometimes foul-smelling urine	Kidney necrosis or abscess, sepsis, preterm labor in pregnant women	*Escherichia coli, Staphylococcus saprophyticus*; less common: *Klebsiella, Proteus, Enterobacter, Clostridium, Candida* in diabetics	Ascending infection from lower urinary tract or hematogenous spread from a distant focus of infection	Urine culture; clinical symptoms; imaging (rarely)	Oral or IV antibiotic therapy, depending on disease presentation; choice of antibiotic, depending on age, sex, and pregnancy status of patient	No vaccine available

Women, the elderly, and patients with structural or functional abnormalities of the genitourinary tract (also called the urogenital tract) are most at risk. Patients who are undergoing a urological procedure or will be having hip arthroplasty (hip replacement surgery) are also screened and treated to reduce specific postsurgical infections. The prevalence of asymptomatic bacteriuria increases with age and depends on the patient population.

CYSTITIS

CASE HISTORY 23.1
When It Hurts to Pee

Lashandra is 24 years old and had been experiencing increased frequency of urination and dysuria (pain or burning with urination) over the previous 3 days. This was the first time Lashandra had ever suffered from persisting dysuria. She consulted her general practitioner, who requested a midstream specimen of urine. Upon microscopic examination, the urine was found to contain more than 50 leukocytes/μl (normal is fewer than 5) and 35 red blood cells/μl (normal is 3–20). Epithelial cells from the urogenital area contaminate a midstream specimen if the specimen is not collected properly. No squamous epithelial cells (skin cells) were seen, indicating a well-collected midstream catch. The urine culture plated on agar medium yielded more than 10^5 colonies/ml (meaning more than 10^5 organisms/ml in the urine) of a facultative, anaerobic, Gram-negative bacillus that can ferment lactose.

The first question to ask in this case is whether the patient had a significant UTI. The answer is yes. The purpose of the midstream urine collection is to provide laboratory data to make this determination. Urine becomes contaminated with normal microbiota that adhere to the urethral wall. In a midstream collection, the patient urinates briefly, stops to position a sterile collection jar, then resumes urinating in the jar (see Section 25.1). The initial brief period of urination minimizes the number of contaminating microbes in the sample by washing away organisms clinging to the urethral opening before the sample is collected. Nevertheless, the collected sample will contain low numbers of organisms representing normal microbiota. A diagnosis of UTI is made when the number of bacteria in the sample becomes greater than 10^5/ml. (However, that number can be as low as 1,000/ml in some symptomatic women.) In the case history, Lashandra was symptomatic and had enough bacteria in her urine to indicate a UTI.

Do you know what a bladder infection feels like? Chances are you answered yes because by young adulthood most of us either have known someone who has suffered a UTI or have experienced it ourselves. Uncomplicated (no fever) acute cystitis starts with a sudden onset of **dysuria** (painful or difficult urination), increased

Table 23.2

Signs and Symptoms of Urinary Tract Infections in Different Populations

Patient Population	Dysuria	Frequency	Urgency	Nocturia	Fever	Other
Newborn	+	N/A	–	–	+	Failure to thrive, irritability, sepsis, jaundice, vomiting
Infants–preschoolers	+ (preschoolers)	–	+	–	+ (mostly infants)	Failure to thrive, diarrhea, vomiting, abdominal or flank pain, urgency and new-onset urinary incontinence (preschoolers)
School-age children	+	+	+	–	Possible	New-onset incontinence, vomiting, abdominal or flank pain
Adults (cystitis)	+	+	+	+	–	Hematuria and suprapubic pain; systemic symptoms may be present in complicated cases
Adults (pyelonephritis)	–	–	–	–	+	Costovertebral tenderness, flank pain, chills, oliguria, hematuria, nausea, vomiting
Elderly	In only 12%	In only 10%	In only 2%	+ (may be chronic due to other causes)	–	Nonspecific symptoms in most, new or worsening confusion, new or worsening incontinence, functional decline, falls

frequency of urination, urgency, burning upon urination, and nocturia (waking up several times at night to urinate). Some patients may have fever and lower abdominal pain, but these are uncommon symptoms. Most elderly patients (77%), like Lois, whom you met in the chapter-opening case, have nonspecific symptoms. They feel confused, their incontinence gets worse, they may suffer a fall, or they may be unable to function at their usual levels. Children's symptoms are different, depending on their age, and often include a fever when they are very young (Table 23.2).

URINARY TRACT INFECTIONS IN CHILDREN The main risk factor for developing a UTI in children is incomplete emptying of the bladder. Some children do not empty their bladder properly because of congenital abnormalities of the urinary tract. Others (the majority) may postpone emptying their bladder ("hold it") for a variety of reasons; they do not want to stop playing or they do not want to use a public toilet, for example. In male infants less than 6 months of age, lack of circumcision increases the risk of UTIs. Colonization of the foreskin during the early stages of life may increase the chance of bacteria accessing the urethra. The risk profile, however, seems to equalize between circumcised and uncircumcised infant populations at about 6 months of age.

Childhood UTIs can lead to serious damage and scarring of the kidney if not treated promptly, which can cause kidney-related disease in adulthood. Fortunately, UTIs are uncommon in children except in the very young (less than 6 months old). Incidence (number of cases in a population) of UTIs in children decreases as children grow older; for example, the overall prevalence of febrile (with fever) UTIs in female children decreases from 7.5% in infants 0–3 months old to 2.1% in girls older than 12 years of age.

RECURRENT URINARY TRACT INFECTIONS Some women suffer from recurrent urinary tract infections. In fact, recurrent UTIs (defined as either two or more UTIs in 6 months or three or more UTIs in 12 months) are common among young and postmenopausal women. Incomplete emptying of the bladder is thought to be the culprit after menopause, but sexual activity and use of spermicides are risk factors for recurrent UTIs in young women. Prophylactic (preventive) continuous antibiotic therapy is used in both groups to prevent the recurring infections. Postcoital (after intercourse) antibiotic prophylaxis is an alternative to continuous therapy for treating recurring infections in young women. Other factors are also thought to be involved (see IMPACT).

HOSPITAL-ACQUIRED URINARY TRACT INFECTIONS Urinary tract infections account for 20%–30% of hospital-acquired (nosocomial) infections, making them the most common infection developed during a hospital stay. Most of these infections are associated with urinary catheterization, which is performed on nearly 25% of hospitalized patients (Figure 23.5A and B). The organisms isolated from catheter-associated infections (CAIs) are more likely to have some antibiotic resistance. *Escherichia coli* causes most hospital-acquired UTIs, but several other bacteria are also implicated, including *Proteus mirabilis, Pseudomonas*

Figure 23.5 Urinary Catheterization

A. A urinary catheter.

C. Biofilm formed by *Pseudomonas aeruginosa* on the surface of a urinary catheter.

B. Urinary catheters are inserted through the urethra into the bladder. The balloon at the end of the catheter is inflated to keep it anchored inside the bladder. The urine flows through the catheter and is collected inside a sterile bag.

aeruginosa, Enterococcus faecalis, and *Staphylococcus epidermidis*. **Figure 23.5C** shows a biofilm of bacteria forming on the inner surface of a recently used urinary catheter. The biofilm protects the bacteria from the mechanical flow of urine, host defenses, and antibiotics, making bacterial elimination difficult. While in the patient, the catheter can shed organisms into the bladder and produce an infection.

PYELONEPHRITIS Pyelonephritis (kidney infection) is not a common disease, but like cystitis, it's more prevalent in women than in men. Patients suffering from pyelonephritis typically have a high fever (greater than 38°C, or 100.4°F), chills, **oliguria** (production of a less-than-normal volume of urine), flank pain, tenderness at the point between the last rib and the lumbar vertebrae (**costovertebral tenderness; Figure 23.6**), and nausea.

Figure 23.6 Costovertebral Angle and Costovertebral Tenderness
The point between the last rib and the lumbar vertebrae is called the costovertebral angle. Tenderness at this point is a sign of pyelonephritis.

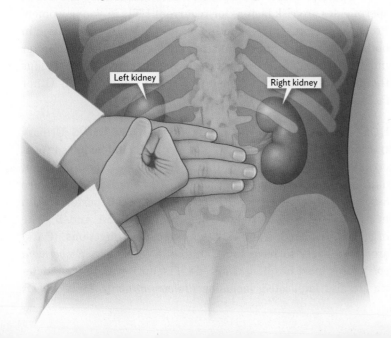

The most common way bacteria reach the kidneys is to ascend from the lower urinary tract along the ureters. Bacterial flagella and various pili—such as the type IV pili that help Gram-negative bacteria move by twitching motility (Section 18.2)—enable bacteria to swim or climb up the ureters. Septicemic spread from infections at other body sites is also possible, but rare.

The presentation of pyelonephritis varies from mild to severe. Mild infections can be treated with an antibiotic on an outpatient basis. Severe pyelonephritis, on the other hand, can cause collapse of the patient's cardiovascular system (shock), threatening the patient's life. Hospitalization in an intensive care unit (ICU) and intravenous administration of antibiotics are necessary to resolve the infection.

SECTION SUMMARY

• **Urinary tract infections** are extremely common, especially in women. **Diagnosis of a UTI** includes finding a significant number of bacteria in urine (typically greater than 10^5 organisms/ml of urine). Common causes include uropathogenic *Escherichia coli*, *Klebsiella*, *Staphylococcus aureus*, *Enterococcus*, and *Staphylococcus saprophyticus*.

- **Bladder infections** can be caused by perianal skin microbes contaminating a urethra or by bacteria descending from an infected kidney. **Cystitis** is a simple bladder infection whose symptoms include dysuria, increased frequency of urination, urgency, and burning upon urination.

- **UTIs in early life** are more prevalent in children less than 6 months old and uncommon in children over 12 years of age.

- **Catheter-associated UTIs** are the most common hospital-acquired infection.

- **Asymptomatic UTIs** are characterized by significant numbers of organisms in the urine (greater than 10^5 organisms/ml of urine) but a lack of symptoms.

- **Kidney infections (pyelonephritis)** can arise from bladder infection by microbes that ascend the ureter or occasionally from septicemic blood. Patients with pyelonephritis usually have high fever, oliguria, costovertebral tenderness, and nausea.

Thought Question 23.1 Why are urinary tract infections among the most commonly acquired nosocomial (hospital-acquired) infections?

Thought Question 23.2 Why would someone with pyelonephritis have a lower urine output?

23.3 Anatomy of the Reproductive Tract

SECTION OBJECTIVES

- List the structural components of the male and female reproductive tracts.
- Explain the function of each component of the male and female reproductive tracts.
- Discuss the significance of structural differences between the male and female reproductive tracts as they relate to infections.

Note: Gender identity is an important issue because not everyone identifies as strictly "male" or "female." The fluidity of gender identity can make discussions of sexually transmitted infections difficult. To simplify our discussion of reproductive tract infections, we will use the terms "male" and "female" solely to indicate basic reproductive anatomy, not gender.

Do reproductive organs serve any function other than sexual intercourse? Both male and female reproductive systems include sex-specific reproductive organs. These include testes (male) and ovaries (female) that have endocrine and exocrine functions (**Figure 23.7**). Endocrine glands secrete hormones into the

---

Figure 23.7 The Male and Female Reproductive Systems

Viral infections of the reproductive tract
Genital warts (HPV)
Genital herpes
Neonatal herpes
HIV
Hepatitis B and C infections

Bacterial and protozoan infections of the reproductive tract
Syphilis
Gonorrhea
Chlamydial infection
Lymphogranuloma venereum
Chancroid
Trichomoniasis
Pelvic inflammatory disease (PID)

Non-sexually transmitted infections of the reproductive tract
Bacterial vaginosis
Fournier's gangrene
Vulvovaginal candidiasis

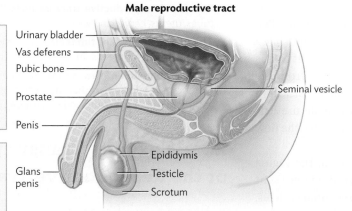

Male reproductive tract

Urinary bladder
Vas deferens
Pubic bone
Prostate
Penis
Glans penis
Epididymis
Testicle
Scrotum
Seminal vesicle

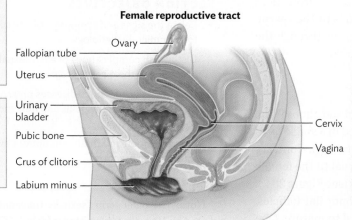

Female reproductive tract

Ovary
Fallopian tube
Uterus
Urinary bladder
Pubic bone
Crus of clitoris
Labium minus
Cervix
Vagina

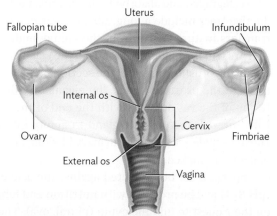

Uterus
Fallopian tube
Infundibulum
Internal os
Ovary
Cervix
Fimbriae
External os
Vagina

A. Ovaries are located on each side of the uterus but not attached to it.

Fallopian tube
Ovary

B. A normal uterine (fallopian) tube is shown draping over a normal ovary.

bloodstream and affect multiple organ systems. Exocrine glands secrete enzymes and some chemical hormones into the immediate external vicinity, where their effects are localized. The different morphologies of the male and female reproductive tracts notwithstanding, the ultimate physiological function of the two systems is the same: to produce a human offspring, which occurs through heterosexual vaginal intercourse, unless alternative fertility methods are used. Unfortunately, the warm and moist environment of the genital mucosa, so necessary for procreation, also provides an ideal environment for viruses, bacteria, fungi, and parasites to grow, wreaking havoc with the reproductive system. Once an organism establishes an infection in one person's reproductive tract, it can usually be transferred to others via sexual contact. Sexually transmitted infections of the human reproductive system are all communicable, and most have to be reported to public health officials upon diagnosis.

Male Reproductive System

The main job of the male reproductive system is to produce sperm and deliver it to the female reproductive system. The male reproductive tract has both external and internal parts (see Figure 23.7); the external parts include the penis, scrotum, testes (singular, testis) housed in the scrotum, and epididymides (singular, epididymis).

Spermatozoa (sperm) are produced in the testes and mature as they wind their way through the epididymides, where they are stored until delivered to the female reproductive tract via the penis or are degraded and absorbed. The internal parts of the male reproductive tract include a long tube, the vas deferens (or "vas"), that transfers mature sperm from the epididymides to the penis. The sperm's trip through the vas ends in the ejaculatory ducts that open inside the prostatic urethra, the section of male urethra that passes through the prostate gland. The ejaculatory duct forms from the fusion of the vas deferens with the end of the seminal vesicle and runs through the prostate gland to empty into the male urethra. During a male erection, the urinary bladder, and therefore the flow of urine, is blocked.

Sperm must be protected against the acidic pH of the vagina (pH 3–4) and be provided with nutrition and lubrication once sent on their quest to find the ovum (plural, ova). The seminal vesicles, prostate, and bulbourethral glands provide the needed protection, nutrition, and lubrication by adding different fluids to the sperm. The result of this process is the ejaculate that travels through the urethra upon orgasm and is delivered to its destination.

Female Reproductive System

The female reproductive system is able to "discern and choose" the best sperm candidate from the millions introduced in a single ejaculate to fertilize the usually one existing ovum (the egg). Contrary to its male counterpart, the entire system is internal to the female body and consists of the vagina, uterus, and ovaries (see Figure 23.7). Each ovary contains approximately 200,000 primordial follicles. Every month, some of these follicles are recruited to mature. One

of the recruited follicles eventually becomes dominant and releases an oocyte into the peritoneal cavity; the others are degraded and absorbed.

Many oocytes are lost in the peritoneal cavity of the pelvis because there is no direct contact between the uterine tubes (also called the fallopian tubes) and the ovaries (see Figure 23.7). As soon as the oocyte is released, the funnel-shaped end of the uterine tube (the infundibulum) bends over the ovary, and the fingerlike projections on its edge (called fimbriae) perform a sweeping movement to capture the released ovum and move it into the tube. The fallopian tube is the site of fertilization of the ovum.

The fertilized egg will eventually move through the long fallopian tube and enter the uterus, a muscular organ whose chamber is lined with a specialized layer of tissue called endometrium. The uterus ends at the cervix (neck of the womb), which is also highly muscular and attached to the top of the vagina, or birth canal. The vagina is a muscular and distensible tubular structure, 8–10 cm (3–4 inches) long and lined with a mucous membrane. Several glands located around the vaginal entrance (vaginal introitus) secrete lubricating material necessary to facilitate vaginal intercourse and delivery of a baby.

SECTION SUMMARY

- **The male reproductive tract** includes the penis, testes, epididymides, vas deferens, seminal vesicles, and prostate.
- **Testes** are housed in the scrotum.
- **The female reproductive tract** includes the ovaries, fallopian (uterine) tubes, uterus, cervix, and vagina.
- **The uterine (fallopian) tubes** are not physically connected to the ovaries, and many oocytes are lost in the peritoneal cavity of the pelvis.
- **The infundibulum of the uterine tube** bends over the ovary as soon as the oocyte is released, so that the sweeping movement of its fimbriae can capture the released ovum and move it into the tube.

23.4 Sexually Transmitted Infections of the Reproductive Tract

SECTION OBJECTIVES

- Relate the signs and symptoms of sexually transmitted infections to their microbial etiology.
- Distinguish complications of each sexually transmitted infection according to the disease progression.
- Describe how infectious diseases present differently in biologically male and female individuals and how those differences impact communicability.
- List prevention and treatment options available for sexually transmitted infections.

◉ **Note:** The older term "sexually transmitted disease (STD)" has been replaced by "sexually transmitted infection (STI)" because not all infections transmitted by sexual contact result in disease. A particularly important example is human immunodeficiency virus (HIV). A person infected with HIV usually does not develop the disease called acquired immunodeficiency syndrome (AIDS) for many years.

Is sexual intercourse the only way to get a sexually transmitted infection? No. **Sexually transmitted infections (STIs)** are defined as infections transmitted primarily through sexual activity, which may include genital, oral-genital, oral-anal, or anal-genital contact. Some infections of the reproductive tract, however, are not exclusively sexually transmitted and will be discussed separately, in Section 23.5. Some STIs cause overt symptoms, and some are transiently asymptomatic. For example, HIV is a virus that can be harbored for long periods before symptoms appear.

Organisms or viruses that cause STIs are generally very susceptible to drying and require direct physical contact with mucous membranes for transmission. Because sex can take many forms besides intercourse, these microbes can initiate disease in the urogenital tract, rectum, or oral cavities. Microbes that infect the human reproductive tract include viruses, bacteria, fungi, and protozoa. Generally speaking, risk-taking behaviors, such as unprotected sex, having multiple sexual partners, and drug abuse (including alcohol), are major contributors to acquiring these infections.

The best way to prevent sexually transmitted infections is to abstain from sexual activity—not a popular option for most people. Staying in a long-term, mutually monogamous relationship with someone who is uninfected is the next best thing. Correct and consistent use of latex male condoms is very effective in reducing the risk of transmitting STIs transferred by genital fluids, such as HIV, gonorrhea, chlamydia, and trichomoniasis. However, condom use reduces the risk of ulcer-producing STIs (syphilis, herpes, genital warts) only if the condom covers the infected site or the ulcer.

A patient who has symptoms of one STI is very likely to have other STIs as well. For example, patients who have gonorrhea are highly likely to be infected with another type of bacterium, *Chlamydia*.

Viral Infections of the Reproductive System

Most viral sexually transmitted infections are caused by one of three virus species, as shown in Table 23.3: human papillomavirus (HPV), herpes simplex virus (HSV), and human immunodeficiency virus (HIV). The initial infection with these viruses varies from asymptomatic (often the case) to severe. The infected person, even when asymptomatic, can pass the virus to a healthy person through sexual contact.

GENITAL WARTS

CASE HISTORY 23.2

Freeze What?

 Jamie, a 20-year-old college student from New Mexico, was in the clinic complaining of a "growth" around her "private area." She said she had no other symptoms and denied any itching or discomfort in the affected area. Her medical history was benign. She said she liked to drink a lot and admitted to having unprotected sex with multiple partners. Physical examination showed warts near the vaginal introitus (vaginal opening) but no edema (swelling) or erythema (reddening). The doctor diagnosed genital warts and discussed different treatment options, including cryotherapy, which freezes the warts. Jamie was shocked but picked cryotherapy anyway, even though the procedure had to be repeated every 1–2 weeks. Her warts disappeared after the third session, but she was told to come back to the clinic every 3 months for 1 year to check for recurrence. She followed up as directed and was lucky enough to be free of any genital warts 1 year later.

Genital warts, or **condylomata acuminata** (singular, **condyloma acuminatum**), are the most common sexually transmitted viral infection, caused by a small, nonenveloped DNA virus called human papillomavirus (HPV). Specific subtypes of HPV produce cauliflower-shaped, soft, flesh-colored lesions on infected patients' genitals or rectum (Figure 23.8). Genital warts resemble common warts found elsewhere on the skin (see Section 19.2). Most HPV infections occur in women, but the risk of infection is also significant for men who have sex with other men, or MSM (homosexual or bisexual males). For some patients, like Jamie, there are no uncomfortable symptoms. Other patients, particularly those with many lesions, can experience itching, pain, or tenderness.

Note: Epidemiologists in the 1990s invented the term "men who have sex with other men" (MSM) because some males choose not to accept a homosexual or bisexual identity.

Figure 23.8
Anal Condylomata Acuminata (Genital Warts) Caused by Human Papillomavirus (HPV)

Table 23.3

Viral Sexually Transmitted Infections of the Reproductive Tract

Disease	Symptoms	Complications	Virus	Transmission	Diagnosis	Treatment	Prevention
Genital warts	Soft-tissue growths (wartlike), recurrent	Laryngeal or respiratory infections in infants born vaginally to infected mothers	Human papillomavirus (HPV)	Sexual contact	Clinical presentation of the lesion; biopsy	Removal of warts by surgery, cryotherapy, medications (podofilox), other methods	Recombinant quadriva-lent human papillomavirus (HPV) vaccine (Gardasil), condoms
Genital herpes	Clusters of erythematous papules and vesicles, burning, pain, itching, and sometimes fever, headache, and muscle pain	Genital scarring, infection in infants born vaginally to infected mothers with active lesions or mothers shedding the virus (neonatal herpes)	Herpes simplex virus (HSV-1 and HSV-2)	Sexual contact	Classic clinical presentation; viral culture, PCR, serology	Antiviral medications: acyclovir, famciclovir, valacyclovir; different treatment regimens for first episode, recurrent episodes, and suppression; sexual partners should be screened and treated	No vaccine available; condoms
HIV/AIDS	Primary infection: majority will have no symptoms after acquiring the virus; some people will have acute illness with fever, muscle and joint pain, lymphadenopathy, and rash on the trunk of the body; symptoms vary, depending on progression of disease and the patient	Opportunistic infections, malignancies, other morbidities defined by CDC as AIDS-defining conditions	Human immunodeficiency virus (HIV)	Blood or body fluids, transplacental, sexual contact	Extensive guidelines for diagnosis recommended by CDC, WHO, and Infectious Diseases Society of America; criteria different for patients <18 months of age; ELISA, immuno-fluorescence assay, nucleic acid (RNA and DNA) tests, HIV p24 antigen test, viral cultures	Antiretroviral therapy based primarily on CD4+ T-cell count and viral load; prophylactic treatment to prevent opportunistic infections; other treatments as needed	No vaccine available; condoms

HPV subtypes that cause genital warts are part of a large virus family whose members produce warts on the upper and lower extremities (see Section 19.2). Infection with some subtypes of HPV is also linked to cervical, penile, anal, and throat cancers. The subtypes associated with cervical cancer are divided into low-risk and high-risk groups, depending on the strength of their association with malignancy. Subtypes 16 and 18 of the high-risk group are the two most commonly isolated from cervical cancer tissue.

Clinicians diagnose HPV usually by inspecting the affected area, as was the case for Jamie, but biopsy may be required in rare complicated cases. Several treatment options are available for the infected person, including topical medications (for example, interferon-alpha), cryotherapy, and surgical methods. Treatments are designed to remove the warts but do not cure the viral infection. The success rate of the treatments varies, but cryotherapy and electrosurgery (burning the tissue by delivering high-frequency electrical current to it) are the most effective methods.

The recombinant HPV vaccine, Gardasil, offers protection against serotypes 6, 11, 16, and 18; vaccination is recommended for girls and boys at age 11 or 12 years. The vaccine is only effective if administered before any sexual activity begins. A 2019 report showed a remarkable 50% decrease in the rate of cervical cancer in vaccinated women.

Figure 23.9 **Genital Herpes**

HSV-2 lesions on the genitals of an infected woman.

GENITAL HERPES Infection of mucosal surfaces and skin of the anogenital area by herpes simplex viruses 1 and 2 (Section 19.2) is called **genital herpes** or herpes genitalis (**Figure 23.9**). HSV-2 has a higher affinity for the genital skin and mucosa than does HSV-1. But the tissue specificity of HSV-1 and HSV-2 is not absolute. It is possible for HSV-1 to infect genital skin and mucosa and for HSV-2 to infect oral tissue. HSV infection is a worldwide problem, with some 50 million people thought to have it.

A person without antibodies to HSV-1 or HSV-2 (meaning no prior HSV-1 or HSV-2 infection) who has a positive PCR test for herpesvirus DNA is said to have a **primary infection**. Two to twelve days after initial exposure, the hallmark lesions of primary infection appear. The lesions, usually painful, start as erythematous papules (red, slightly raised areas of skin) that progress to vesicles and finally pustules (see Figure 23.9). Nearly half of all primary genital herpes cases are caused by HSV-1, but recurrent infections are caused primarily by HSV-2.

A person with a primary infection may be asymptomatic (also described as "subclinical") or severely ill, or may exhibit any gradation between those two extremes. The signs and symptoms of a primary infection, in addition to the characteristic lesions, can include burning, pain, itching, dysuria, and even systemic symptoms such as headache or muscle ache and fever. A person with a severe primary genital herpes infection suffers from almost all of these signs and symptoms.

After the initial outbreak, the lesion resolves, and the virus moves to a bundle of nerves at the base of the spine. There the virus remains inactive for a time called the **latent period**, during which there are no symptoms (see Sections 18.5 and 19.2). The latent period sets the stage for recurrent episodes of the disease.

A recurrent episode of genital herpes occurs when the virus becomes activated (sometimes triggered by stress) and travels through nerves to the skin's surface, causing an outbreak of ulcers. These recurrent episodes tend to be milder than the initial outbreak but are still infectious. Recurrent disease can also involve the brain—a condition known as herpes encephalitis (see Section 24.3).

Clinical diagnosis of HSV is made when the characteristic lesions are observed. Samples from the vesicles or pustules are then collected and tested for presence of virus to confirm the clinical diagnosis of herpes (Section 19.2). Confirmation of diagnosis is particularly important in HIV-positive patients because genital herpes presents differently in these patients. There is no cure for genital herpes and no vaccine to help prevent it. Antiviral medications (for instance, acyclovir) are used to help reduce both the severity of the symptoms and the frequency of recurring episodes.

NEONATAL HERPES Pregnant mothers with genital herpes can vertically transmit the virus to their baby before (transplacental transmission), during, or immediately after delivery; a baby so infected is said to have **neonatal herpes**. Some mothers may not even be aware that their disease is active. They can shed the virus but have no lesions. The incidence of neonatal disease is approximately 1 of every 3,200 births. Neonatal herpes may appear only as a skin infection, with small, fluid-filled vesicles that can rupture, crust over, and finally heal (**Figure 23.10**). However, the virus can also spread (disseminate) through the bloodstream and affect many parts of the infant's body, including the brain (herpes encephalitis), liver, lungs, and kidneys. These neonates are extremely ill. They can appear blue (cyanosis), have rapid breathing (tachypnea), appear jaundiced, have a low body temperature (hypothermia), and bleed easily.

Figure 23.10 **Neonatal Herpes**

Pregnant mothers who have genital herpes may transmit the virus to their baby as the newborn is passing through the birth canal.

Guidelines for preventing vertical HSV transmission include the following:

- In the absence of active lesions or prodromal symptoms, allow vaginal delivery.

- At the time of delivery, consider obtaining a herpes culture from the mother's cervix and from the site of recurrence, as well as from the newborn baby.

- If there is an active herpetic lesion, perform cesarean delivery, preferably within 4–6 hours of membrane rupture.

- Treat the mother with suppressive antiviral agents, such as acyclovir or famciclovir, prior to delivery.

Untreated newborns diagnosed with disseminated HSV have a mortality rate of 60%. Treatment with antiviral agents reduces mortality to 35%.

HIV INFECTION **Human immunodeficiency virus (HIV)** is thought to have originated in the 1920s when the related simian immunodeficiency virus (SIV) jumped via contaminated blood from chimpanzees to the humans who hunted them. The first human case of HIV was identified in 1959 in the Belgian Congo (today's Democratic Republic of the Congo). HIV has since spread across the globe as a result of human travel, first entering the United States in 1966–1972 from Haiti.

HIV caused the greatest pandemic of the late twentieth century and remains a serious problem today, especially in Africa. The World Health Organization (WHO) estimates that 28.5 million people in sub-Saharan Africa live with HIV, representing nearly 71.2% of all global cases. The Centers for Disease Control and Prevention (CDC) estimates that 1.2 million people in the United States are infected with the virus. HIV has claimed the lives of more than 33 million people worldwide, over half a million in the United States alone. The molecular biology and virulence of HIV, as well as its targeting of CD4+ T cells, are discussed in Chapters 12 and 18. In this section we focus on the disease that HIV causes: **acquired immunodeficiency syndrome (AIDS)**, or end-stage HIV disease.

HIV is a lentivirus (a genus of retroviruses; Section 12.6) with an RNA genome, a member of the Retroviridae family, and is a prominent example of viruses that can be transmitted either sexually (vaginally, orally, anally) or through direct contact with body fluids, such as blood transfusion or the sharing of hypodermic needles by intravenous drug users. HIV is not transmitted by kissing, tears, or mosquito bites. It can, however, be transferred from mother to fetus through the placenta (transplacental transfer). We discuss HIV infection here because sexual contact remains the major route of transmission.

The 2018 data available from the CDC for the United States show that gay, bisexual, and other men who have sex with men (MSM) constitute the largest group of newly diagnosed patients (70%), followed by heterosexuals (24%) and IV drug abusers (10%). Women, overall, accounted for 19% of cases in 2017. The global incidence of HIV infection, however, has generally been trending downward, in part because of multinational funding for test-and-treat policies to provide education, screening, and treatment.

HIV infection can be diagnosed shortly after it is acquired, by either nucleic acid amplifications tests (NAAT) within 10–33 days or by an antigen antibody test within 18–45 days post-exposure. Having entered the bloodstream, HIV infects CD4+ T cells (Section 18.5) and replicates very rapidly, producing a billion particles a day. The number of CD4+ cells in the blood (CD4+ count) is determined by flow cytometry (Section 17.2). The concentration of virus in the blood, expressed as the number of viral RNA copies per milliliter of blood, is called the **viral load**. Viral load is determined by quantitative PCR (see Section 8.6) to detect HIV-specific genes such as *gag*, *pol*, or *nef*. The viral load and CD4+ count are important prognostic factors that the CDC uses to monitor the progression of HIV infection for clinical treatment and epidemiological purposes.

Clinical progression of HIV infection is now divided into three stages by the CDC: The first is **acute HIV infection** leading to seroconversion (finding antibodies to the virus in patient's blood). The second stage is **clinical latency**, an asymptomatic stage involving a slow, steady loss of T cells and few, if any, symptoms. The third stage is **AIDS**, the end-stage HIV infection. The body normally replaces the entire CD4 cell population every 2 weeks but cannot keep up with the pace of destruction caused by rapid viral replication. The number of CD4+ cells will eventually start to decline, leading to this final stage.

During the acute HIV stage most patients are asymptomatic, but nearly half have nonspecific symptoms defined as "mononucleosis-like," including headache, low-grade fever, lymphadenopathy, and a maculopapular rash (see Chapter 19) on the trunk. The acute symptoms usually last 1–2 weeks, during which time anti-HIV antibodies will develop. Most patients change from HIV-negative to HIV-positive (seroconvert) within 4–10 weeks of exposure. More than 95% of exposed individuals have seroconverted 6 months after exposure.

The disease next enters the clinical latency stage, when the virus still replicates but at a very low level. The number of CD4+ T cells slowly declines during this stage. Clinical latency can last as long as 10 years without treatment and for decades with proper antiviral therapy. Although most patients do not exhibit symptoms during clinical latency, some patients have swollen lymph nodes (lymphadenopathy). Patients with two or more swollen lymph nodes outside the groin region for 3 months or more during the latent period are said to have persistent generalized lymphadenopathy, or PGL. Specialized cells in the lymph node "trap" the virus, resulting in enlargement of the nodes; the number of virus particles in the peripheral blood (the viral load) is low at this point, and there is a high concentration of extracellular HIV in these nodes. The enlarged lymph nodes inevitably are destroyed by the ever-increasing concentration of virus, and they release their viral content into the peripheral blood.

The decrease in CD4+ T-cell numbers during the latent stage eventually leads to the symptomatic HIV infection stage, AIDS.

Figure 23.11 AIDS Indicator Opportunistic Infections

LM | 8 μm

A. Oral candidiasis (thrush) is normally not seen in adults unless they are immunosuppressed. The white patches are caused by secondary infection by the yeast *Candida albicans*.

B. *Pneumocystis jirovecii* infection of the lung will garner an AIDS diagnosis. Note the crushed-spheres appearance of the fungus. These organisms are 2–6 μm in diameter.

Diagnosis of AIDS is based on CD4+ T-cell counts, determined by flow cytometry (Section 17.2) or by defined specific clinical conditions. These conditions include unusual infections, high fever, diarrhea that persists for over a month, some cancers, and peripheral neuropathy. The debilitated immune system leaves the victim susceptible to secondary infections such as thrush (**Figure 23.11A**), persistent vaginal candidiasis, listeriosis, or shingles in more than one dermatome (an area of skin supplied mainly by a single spinal nerve).

The CDC guidelines define AIDS as any of the following:

- A CD4+ cell count of less than 200 cells/μl, regardless of the presence or absence of any other symptoms;

- A percentage of CD4+ T lymphocytes relative to total lymphocytes that falls below 14%; and/or

- The presence of an AIDS indicator condition (Table 23.4).

For example, once the CD4+ T-cell population falls below 500 cells/μl, opportunistic AIDS indicator infections such as pneumonia caused by *Pneumocystis jirovecii* (**Figure 23.11B**) or candidiasis of the esophagus, trachea, bronchi, or lungs can develop. Various unusual cancers are also on the list of AIDS indicator conditions. For instance, **Kaposi's sarcoma**, which is a common cancer seen in AIDS patients (see Figure 18.28B) and is caused by human herpesvirus type 8 (HHV-8), is a devastating condition associated with AIDS.

Diagnosing an HIV infection requires detecting anti-HIV antibodies, determining the CD4+ T-cell count, and quantifying the viral load of a patient. Quantitative assay to determine viral load is done via quantitative PCR to detect HIV-specific genes such as *gag, pol,* or *nef.* The gene *nef* encodes a protein that interacts with host cell signal transduction components. (Because HIV is an RNA virus, quantitative PCR starts with conversion of HIV RNA to DNA

Table 23.4

AIDS-Defining Conditions in Adults and Adolescents

- Bacterial infections, multiple or recurrent[a]
- Candidiasis of bronchi, trachea, or lungs
- Candidiasis of esophagus[b]
- Cervical cancer, invasive[c]
- Coccidioidomycosis, disseminated or extrapulmonary
- Cryptococcosis, extrapulmonary
- Cryptosporidiosis, chronic intestinal (>1 month's duration)
- Cytomegalovirus disease (other than liver, spleen, or nodes), onset at age >1 month
- Cytomegalovirus retinitis (with loss of vision)[b]
- Encephalopathy, HIV related
- Herpes simplex: chronic ulcers (>1 month's duration) or bronchitis, pneumonitis, or esophagitis (onset at age >1 month)
- Histoplasmosis, disseminated or extrapulmonary
- Isosporiasis, chronic intestinal (>1 month's duration)
- Kaposi's sarcoma[b]
- Lymphoid interstitial pneumonia or pulmonary lymphoid hyperplasia complex[a,b]
- Lymphoma, Burkitt's (or equivalent term)
- Lymphoma, immunoblastic (or equivalent term)
- Lymphoma, primary, of brain
- *Mycobacterium avium* complex or *Mycobacterium kansasii*, disseminated or extrapulmonary[b]
- *Mycobacterium tuberculosis* of any site, pulmonary,[b,c] disseminated,[b] or extrapulmonary[b]
- *Mycobacterium*, other species or unidentified species, disseminated[b] or extrapulmonary[b]
- *Pneumocystis jirovecii* pneumonia[b]
- Pneumonia, recurrent[b,c]
- Progressive multifocal leukoencephalopathy
- *Salmonella* septicemia, recurrent
- Toxoplasmosis of brain, onset at age >1 month[b]
- Wasting syndrome attributed to HIV

[a]Only among children less than 13 years of age.
[b]Condition that might be diagnosed presumptively.
[c]Only among adults and adolescents at least 13 years of age.

by a reverse transcriptase, followed by amplification by a thermostable DNA polymerase; see Sections 8.6 and 25.3.) Diagnosis of AIDS is also based on CD4+ count and/or the presence of AIDS indicator conditions, as mentioned already.

New treatment regimens called antiretroviral therapy (ART) have made HIV a treatable chronic disease in the United States. Combinations of drugs from different pharmacological classes are used as recommended by CDC guidelines that take individual patient history, viral load, and CD4+ counts into consideration. Several classes of pharmaceuticals with different modes of action are now available for use in ART. These pharmaceuticals include HIV reverse transcriptase inhibitors, HIV protease inhibitors,

HIV cDNA (complementary DNA) integrase inhibitors, and CCR5 antagonists that inhibit the entry of virus into the cell. More information on these and other HIV drugs can be found in Section 13.7. The goal of ART is to keep the viral load in peripheral blood to fewer than 50 copies per milliliter and the CD4+ count to greater than 500 cells per microliter.

AIDS can be a devastating disease. Can treating at-risk populations with antivirals before exposure be effective at preventing infection? It can! This strategy is known as preexposure prophylaxis (PrEP). In fact, the FDA has approved the daily use of an HIV medicine, Truvada (emtricitabine-tenofovir), by healthy but high-risk people hoping to lower their risk of infection by a sexual partner. The two drugs are nucleoside analogs of cytosine and adenosine, respectively, that inhibit HIV reverse transcriptase.

HIV-infected individuals are routinely screened and treated for other infections because AIDS-associated death is usually caused by the opportunistic pathogens noted earlier. Prophylactic antibiotic treatment, for example, is routinely used to prevent some recurrent infections in affected patients.

Remarkably, some HIV-infected individuals can actually control the initial viremia of the infection. These patients remain asymptomatic for years with no antiviral therapy and are called **long-term nonprogressors**. An estimated 4%–7% of HIV-infected individuals qualify for long-term nonprogressor status. A low viral load in these patients (1,000–10,000 viral copies/ml) is thought to prevent their progression to more severe disease. The lack of progression may be due to an effective cytotoxic-T-cell response. A very small percentage of HIV-seropositive patients, called **elite controllers**, maintain a high CD4+ cell count and show negligible or no viremia (fewer than 50–75 viral copies/ml). These infected individuals also have highly effective cytotoxic T cells.

There have been two reports of remission (or functional cure) of HIV infection (no detectable HIV in peripheral blood). Both of the HIV-positive patients were suffering from leukemia. Their treatment regimens included chemotherapy and having their own bone marrow destroyed and replaced by healthy bone marrow from suitable donors. The healthy donors' cells in both cases happened to lack a coreceptor for HIV called CCR5. The WBCs produced by the transplanted marrow in these two patients could not be infected with HIV because the cells did not have CCR5 coreceptor. So the virus was eliminated from the peripheral blood. Treating HIV-positive patients with chemotherapy and bone marrow transplant procedures to address their infection is obviously not a feasible option. These two cases, however, demonstrate CCR5 as an important treatment target and show that HIV remission is possible. Remember, a person who is HIV-positive does not necessarily have AIDS; the disease may take years to develop. A vaccine is not yet available to prevent AIDS, in part because the envelope proteins of the virus (see Figure 12.27) typically change their antigenic shape.

HEPATITIS B AND C INFECTIONS Hepatitis viruses and the diseases they cause are described in connection with GI tract infections because the primary target of the infection is the liver. However, two of the viruses, HBV and HCV, can be sexually transmitted. Chapter 22 presents details of these viruses and their treatments.

Bacterial and Protozoan Infections of the Reproductive Tract

The latest available estimates of the global incidence of STIs indicate that every year, 376 million individuals acquire a curable STI. Most of these STI cases (about 90%) occur in developing countries, which house 75% of the world's population. Reduced access to health care (a direct result of the extreme poverty in these regions), lack of reproductive health services, war, and population growth are all to blame. In the United States, the annual number of acquired STIs rose steadily from 2014 to 2019. The incidence of STIs reported for 2018 was 2.5 million people. The CDC estimates that at least half of all new cases of STIs occur in sexually active young men and women. Undoubtedly, having unprotected sex plays a prominent role here, as well as in the developing world.

SYPHILIS

CASE HISTORY 23.3

The Great Imitator

 Donna, a pregnant 18-year-old woman, came to the Nassau County, New York, urgent-care clinic with a low-grade fever, malaise, and headache. She was sent home with a diagnosis of influenza. She again sought treatment 7 days later, after she discovered a macular rash (flat, red) developing on her trunk, arms, palms of her hands, and soles of her feet. She thought maybe she was allergic to the new soap she was using, but the rash was not itchy or painful. Further questioning of the patient revealed that 1 year earlier, she had a painless ulcer on her vagina that had healed spontaneously. She was diagnosed with secondary syphilis, a stage of the disease that can mimic many other conditions. She was given a single intramuscular injection of penicillin and told that her sexual partners had to be treated as well.

The vaginal ulcer, the long latent period, and secondary development of rash on the hands and feet described in the case history are classic symptoms of **syphilis** (Table 23.5). Syphilis was recognized as a disease as early as the sixteenth century, but the organism that causes it, the spirochete *Treponema pallidum*, was not discovered until 1905 (Figure 23.12A). The illness has several stages. The disease has often been called the "great imitator" because its symptoms in the second stage, as exhibited in the case history, can mimic those of many other diseases. The incubation stage can last from 2 to 6 weeks after transmission, during which time the organism multiplies and spreads throughout the

Figure 23.12 Syphilis

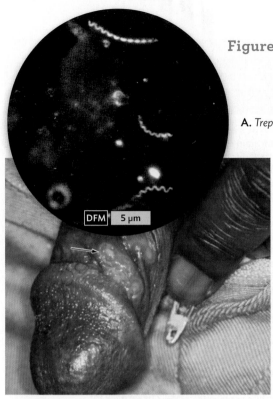

A. *Treponema pallidum.*

B. Chancre of primary syphilis.

C. Macular rash of secondary syphilis.

D. Gumma on the face of a patient with tertiary syphilis.

Table 23.5
Bacterial and Protozoan Sexually Transmitted Infections of the Reproductive Tract

Disease	Symptoms	Complications	Organism(s)	Transmission	Diagnosis	Treatment	Prevention
Syphilis	*Primary:* chancre (painless ulcer) *Secondary:* generalized rash, including palms of hands and soles of feet, generalized nontender lymphadenopathy, fatigue, fever *Tertiary:* Loss of coordination of movement (tabes dorsalis), progressive dementia, stroke and other CNS abnormalities; abnormal pupillary reaction to light; gummas (soft, noncancerous tumors); thoracic aortic aneurysm	*Complications of primary syphilis:* secondary disease *Complications of secondary syphilis:* latent disease leading to tertiary stage *Complications of tertiary syphilis:* disability and death	*Treponema pallidum*	Sexual contact	Definitive diagnosis based on dark-field microscopy and detection of *T. pallidum* in lesion or tissue samples; presumptive diagnosis based on positive results of two types of serology tests; clinical presentation	Alternative treatments for pregnant women and children; sexual partners should be screened and treated; benzathine penicillin G (BPG) is the drug of choice for males and non-pregnant females *Primary and secondary syphilis:* BPG 2.4 million units IM (intramuscular) single dose; azithromycin 2 mg orally is used as well *Tertiary syphilis:* BPG 2.4 million units IM once a week for 3 weeks if no neurosyphilis; complicated regimens for tertiary stage with neurosyphilis	No vaccine available

(continued on next page)

Table 23.5

Bacterial and Protozoan Sexually Transmitted Infections of the Reproductive Tract (continued)

Disease	Symptoms	Complications	Organism(s)	Transmission	Diagnosis	Treatment	Prevention
Gonorrhea	*In females, gonococcal cervicitis:* usually no symptoms; if symptoms are present, they include vaginal discharge, dyspareunia (pain when having vaginal intercourse), vaginal bleeding after sex. *In males, gonococcal urethritis:* mucopurulent or purulent discharge, dysuria, urethral itching	*Complications of gonococcal cervicitis in females:* pelvic inflammatory disease (PID). *Complications of gonococcal urethritis in males:* epididymitis, disseminated infection	*Neisseria gonorrhoeae* (also called gonococcus, has >100 serotypes)	Sexual contact	>5 WBCs in Gram stain of urethral discharge, WBCs containing Gram-negative diplococci; clinical presentation; positive leukocyte esterase test on first-void urine; culture; nucleic acid amplification test from urethral swab or urine	Ceftriaxone 250 mg IM single dose plus either azithromycin 1 g orally single dose or doxycycline 100 mg orally twice a day for 7 days; alternative treatments available if needed; sexual partners should be screened and treated	No vaccine available
Nongonococcal urethritis	Dysuria, urethral discharge, itching	*In females:* PID. *In males:* prostatitis and epididymitis	*Chlamydia trachomatis* is the most common cause (20% of cases); *Mycoplasma genitalium* (9%); adenovirus (4%); herpes simplex virus (2%)	Sexual contact	>5 WBCs in Gram stain of urethral discharge; absence of WBCs containing Gram-negative diplococci; nucleic acid amplification test from urethral swab or urine; positive leukocyte esterase test on first-void urine; clinical presentation	Azithromycin 1 g orally single dose or doxycycline 100 mg orally twice daily for 7 days; alternative regimens if needed; sexual partners should be screened and treated	No vaccine available
Lymphogranuloma venereum (LGV)	Unilateral tender inflammation of groin lymph nodes, anal pain and discharge if anal exposure, constipation	*In females:* inflammation of cervix and fallopian tubes, neuropathy. *In males:* chronic disease, lymphatic scarring leading to genital (scrotal) elephantiasis, penile ulcers, neuropathy	*Chlamydia trachomatis*	Sexual contact	Culture of swab of the lesion or aspirate of lymph node; direct immunofluorescence test; nucleic acid tests	Doxycycline 100 mg orally twice daily for 21 days (recommended treatment); other regimens; prolonged treatment may be needed for HIV-positive patients	No vaccine available
Chancroid (also called soft chancre)	Acute and extremely painful multiple genital ulcers that start as papules or pustules	Secondary infections and fistulas (hollow passages between two epithelium-lined body organs); high risk of having HIV or other STI	*Haemophilus ducreyi*	Sexual contact (highly infectious)	Isolation of *H. ducreyi* in culture from genital ulcer samples; clinical presentation	Four different antibiotic regimens recommended: azithromycin 1 g orally single dose, ceftriaxone 250 mg IM single dose, ciprofloxacin orally multiple doses (contraindicated during pregnancy and lactation), erythromycin multiple doses; sexual partners should be examined and treated within 10 days	No vaccine available

(continued on next page)

Table 23.5

Bacterial and Protozoan Sexually Transmitted Infections of the Reproductive Tract (*continued*)

Disease	Symptoms	Complications	Organism(s)	Transmission	Diagnosis	Treatment	Prevention
Trichomoniasis vaginitis	*In females:* copious malodorous vaginal discharge, vulvar itching and burning *In males:* either asymptomatic or only urethritis symptoms	Adverse pregnancy outcome, including premature labor and low birth weight in newborns	*Trichomonas vaginalis*	Sexual contact	Nucleic acid amplification tests; culture for *T. vaginalis* from vaginal discharge or from urethral swab, urine, or semen; motile *T. vaginalis* on wet mount; PCR	Metronidazole 2 g orally single dose or tinidazole 2 g orally single dose; alternative treatments available	No vaccine available

body. **Primary syphilis** is an inflammatory reaction at the site of infection called a **chancre** (**Figure 23.12B**). About a centimeter in diameter, the chancre is painless and hard, and it contains spirochetes. Patients are usually too embarrassed to seek medical attention, and because it is painless, they hope it will just go away.

The initial lesion does go away after several weeks, and without scarring. The disease has now entered the early latent stage. During this stage, which can be as long as 5 years, symptoms may be absent, but at any time, as described in Case History 23.3, the infected person can develop the rash typical of **secondary syphilis** (**Figure 23.12C**). The rash can be similar to that produced by many diseases. The patient remains contagious in this stage. The symptoms eventually resolve, and the patient enters the late latent stage of syphilis. Some patients eventually progress over years to **tertiary syphilis** and develop many cardiovascular or nervous system–associated symptoms. Soft growths resembling tumors, called **gummas**, are often found during later stages of tertiary syphilis. A gumma has necrotic (dead) tissue in the center surrounded by inflammatory and fibrotic cells (**Figure 23.12D**). Gummas are usually found in the liver, but they can form in the mouth, bone, brain, heart, and skin as well. Gummas in long bones are painful. The patient may develop dementia and eventually dies from the disease.

Neurological symptoms resulting from syphilis can develop at any stage of the disease and are referred to as **neurosyphilis**. Syphilis patients whose cerebrospinal fluid (CSF) shows an elevated WBC count (more than 20 cells/μl; normal is 0–5 cells/μl) or who have a reactive CSF **Venereal Disease Research Laboratory (VDRL) test** have developed neurosyphilis. The VDRL test and the **rapid plasma reagin (RPR) test**, a modification of the VDRL test, are nonspecific tests (described later) that detect antibodies against a nontreponemal heart antigen called cardiolipin. A positive VDRL or RPR test is followed by a specific test, such as direct or indirect fluorescent antibody staining (fluorescent treponemal antibody test [FTA-ABS]; see Section 17.2, Figure 17.10).

The symptoms of neurosyphilis vary widely. Early neurosyphilis develops during the primary and secondary phases of the disease and may be asymptomatic. Alternatively, the meninges (membranes enveloping the brain) and structures of the eyes or ears may be affected. The results can be meningitis, ocular complications, or tinnitus and hearing loss. Late neurosyphilis involves damage to the central nervous system, particularly the brain. Neurological manifestations during tertiary syphilis are mostly **general paresis** or **tabes dorsalis**. General paresis of late neurosyphilis refers to problems of mental function associated with late syphilis. The neurological problems that arise include defects in p̲ersonality, a̲ffect, r̲eflexes, e̲ye, s̲ensorium (illusions, delusions, or hallucinations), i̲ntellect, and s̲peech. The first letters of these traits spell the word "paresis." "Tabes dorsalis" refers to demyelination of axons in certain parts of the spinal cord (the dorsal root and dorsal root ganglia), which causes abnormal gait and degeneration of the weight-bearing joints (Charcot joint).

According to the CDC, a positive result from the RPR or VDRL serology tests is sufficient for a presumptive diagnosis of syphilis during any stage of disease. A positive PCR (looking for bacterial genes) or dark-field examination of the material from the lesion (looking for highly motile spirochetes) will make the definitive diagnosis during the early and secondary phases of the disease.

A diagnosis of tertiary syphilis is based solely on clinical presentation and serology. The diagnosis of latent syphilis, however, begins with serological testing but has several additional criteria. One criterion, for example, is detecting the organism in tissues by using fluorescent antibody.

Syphilis is particularly dangerous in pregnant women. The treponeme can cross the placental barrier and infect the fetus, causing **congenital syphilis**. At birth, infected newborns may have perforated palates and other congenital defects, including notched teeth that become visible only when permanent teeth emerge (**Figure 23.13**). Women should be screened for syphilis as part of their prenatal testing to prevent these congenital infections.

Figure 23.13 Notched Teeth in an Adult Caused by Congenital Syphilis

The treponeme can cross the placenta and infect the fetus. Notched teeth (also called Hutchinson's teeth) indicate congenital syphilis.

Figure 23.14 Examples of Serological Tests for Syphilis

A. The RPR (rapid plasma reagin) test screens for nonspecific antibodies that may indicate infection with *Treponema pallidum*. A positive test (+) shows agglutination of cardiolipin-coated charcoal particles.

T. pallidum

HIV

B. A positive enzyme immunoassay for a specific *T. pallidum* antigen is compared with one done for HIV. The red line indicates a positive reaction.

Common serological tests for syphilis include nontreponemal tests used for screening and treponemal-specific tests used for definitive diagnosis. Examples of a nontreponemal test are the VDRL and RPR tests (**Figure 23.14A**). Reagin is an antibody to the lecithin-cholesterol antigen called cardiolipin, which is not a treponemal antigen. Infection with *T. pallidum* elicits reagin antibodies that react against cardiolipin. In the RPR test, a mixture of charcoal particles coated with cardiolipin is added to a patient's serum. If anti-cardiolipin antibodies (reagin) are present, the charcoal beads clump, representing a positive test (reactive circles in Figure 23.14A). The RPR test is very sensitive, but false positives are common, which is why a more specific serological test should be performed if an RPR test is positive.

Treponemal-specific tests more definitively determine whether a patient is infected with *T. pallidum*. Some are direct immunofluorescence tests to detect treponemal antigen in patient tissue; others are indirect immunofluorescence (Figure 17.10B) or immunochromatography (**Figure 23.14B**) tests that examine patient serum for the presence of anti-treponemal antibodies (see Section 17.2).

Benzathine penicillin G is the antibiotic of choice for syphilis. It is given at 2.4 million units in a single intramuscular injection for primary and secondary disease and at 7.2 million units once weekly for 3 weeks in tertiary disease without evidence of neurosyphilis. Other antibiotics are used in patients with penicillin allergies. Treatment regimens for patients with <u>neuro</u>syphilis always include high doses of antibiotics, usually penicillin or ceftriaxone given intravenously or by daily injections. Patients who are allergic to penicillin can be desensitized to penicillin first (Section 17.3) or given other antibiotics. Antibiotics are useful for eradicating the organism, but there is no vaccine, and cure does not confer immunity.

A serious consequence of antibiotic treatment is the **Jarisch-Herxheimer reaction**. After treatment begins, the dying treponemes release inflammatory molecules that trigger a cytokine cascade leading to symptoms that include myalgias, fever, headache, and tachycardia. Sometimes the Jarisch-Herxheimer reaction exacerbates whatever current syphilitic lesions are manifested (for example, rash or chancre). The reaction is common, develops within several hours after the start of antibiotic treatment, and usually clears within 24 hours after onset.

Columbus and the New World theory of syphilis. An outbreak of syphilis that spread throughout Europe soon after Christopher Columbus and his crew returned from the Americas (1493) led to the theory that Columbus brought the treponeme to Europe from the New World. Whether Columbus's crew carried syphilis to or from the New World has been debated for 500 years. A recent genetic study may have resolved the issue. By sequencing 26 geographically distributed strains of pathogenic syphilis, scientists discovered that the sexually transmitted syphilis strains most closely resemble the non–sexually transmitted, yaws-causing strains of *Treponema pallidum* from South America. Thus, the crew of the Columbus voyages may have brought smallpox and measles to America but returned to Europe with syphilis.

Figure 23.15 *Neisseria gonorrhoeae*

A. Within pus-filled exudates, the Gram-negative diplococci are often found inside polymorphonuclear cells (PMNs). The intracellular bacteria in this case are no longer viable, having been killed by the antimicrobial mechanisms of the white cell.

B. Colonies of *N. gonorrhoeae* growing on chocolate agar (agar plates containing heat-lysed red blood cells that turn the medium chocolate-brown).

C. *N. gonorrhoeae* binding to epithelial cells (colorized SEM).

The Tuskegee experiment. Unfortunately, much of what we know about syphilis is the result of the infamous Tuskegee experiment conducted in the 1930s in Alabama. The study was titled "Untreated Syphilis in the Negro Male." Through dubious means and deception, a group of African-American males was enlisted in a study that promised treatment but whose real purpose was to observe how the disease progressed without treatment. Today, such experiments are barred, thanks to strict institutional review board (IRB) oversight that requires human subjects to sign informed-consent forms. An interesting treatise on the Tuskegee experiment can be found on the website of the National Center for Case Study Teaching in Science (search term "bad blood").

GONORRHEA

CASE HISTORY 23.4

Symptom: Urethral Discharge

Rick, a 22-year-old mechanic from Honolulu, Hawaii, saw his family doctor for treatment of painful urination and urethral discharge. His past medical history was unremarkable. Rick was sexually active, with three regular and several "one time, good time" partners. Physical examination was benign except for a prevalent urethral discharge. The discharge was Gram-stained and sent for culture. The Gram stain revealed many pus cells, some of which contained a large number of phagocytosed Gram-negative diplococci (**Figure 23.15A**)—a finding consistent with *Neisseria gonorrhoeae* infection. Blood was drawn for syphilis serology, which proved negative. Rick was given an intramuscular injection of ceftriaxone (250 mg) to treat the gonococcus, and oral doxycycline (100 mg, two times a day)

was prescribed for 7 days to empirically treat any possible coinfection with chlamydia. The bacteriology lab recovered the bacteria seen in the Gram-stained smear of the urethral discharge. The organism produced characteristic colonies on chocolate agar and was identified as *Neisseria gonorrhoeae*. Chocolate agar plates contain heat-lysed red blood cells that turn the medium chocolate-brown (**Figure 23.15B**). The case was then reported to the state public health department so they could contact Rick's prior and current sexual partners and treat them. Upon his return visit, Rick's symptoms had resolved, and a repeat culture was negative.

The disease here is classic **gonorrhea** (colloquially known as "the clap") caused by *Neisseria gonorrhoeae*, a Gram-negative diplococcus. A characteristic that distinguishes *Neisseria* infections from *Chlamydia* infections (described later) is that bacterial cells are seen in gonorrheal discharges but not in chlamydial discharges. Gonorrhea has been a problem for centuries and remains epidemic in the United States today. Symptoms generally occur 2–7 days after infection but can take as long as 30 days to develop. Most infected men (85%–90%) exhibit symptoms such as painful urination, yellowish white discharge from the penis, and sometimes swelling of the testicles and penis. The Greek physician Galen (129–ca. 199 CE) originally mistook the discharge for semen. This error led to the name gonorrhea, which means "flow of seed."

In contrast to men, most infected women (80%) do not exhibit symptoms, and they constitute the major reservoir of the organism (**Figure 23.16**). If they are asymptomatic, they have no reason to seek treatment; thus they can spread the disease. When symptoms are present, they are usually mild. A symptomatic woman will experience a painful burning sensation when urinating and will notice vaginal discharge that is yellow or occasionally bloody. She may also

Figure 23.16 Gonococcal Infection

Gonococcal infection in women can be asymptomatic even when cervicitis is present.

A. Normal cervix.

B. Gonococcal cervicitis. Notice the reddening (arrow).

complain of cramps or pain in her lower abdomen, sometimes with fever or nausea. As the infection spreads throughout the reproductive organs (uterus and fallopian tubes), pelvic inflammatory disease (PID) occurs.

N. gonorrhoeae is generally serum sensitive (does not survive in blood) owing to the organism's sensitivity to complement. Complement is a series of 20 host serum proteins that can attract white blood cells, facilitate phagocytosis, and insert holes (pores) in the bacterial membrane, causing cytoplasmic leaks (Section 15.7). However, certain serum-resistant strains of *N. gonorrhoeae* can make their way to the bloodstream and cause disseminated infections throughout the body. As a result, both sexes can develop purulent arthritis (joint fluid containing pus), endocarditis, or even meningitis. An infected mother can also infect her newborn during parturition (birth), leading to a serious eye infection called **ophthalmia neonatorum**. Because of this risk and because most infected women are asymptomatic, all newborns receive antimicrobial eyedrops at birth. Because adults engage in a variety of sexual practices, *N. gonorrhoeae* can also infect the anus or the pharynx, where it can develop into a mild sore throat. These infections generally remain unrecognized until a sexual partner presents with a more typical form of genitourinary gonorrhea. The organism can periodically change the shapes (immunogenicity) of its various surface antigens. It can also bind to CD4+ T cells, similar to how it binds to epithelial cells. *N. gonorrhoeae* binding to T cells inhibits their activation and proliferation to become memory T cells (Figure 23.15C). So, an infection with *N. gonorrhoeae* does not cause lasting immunity in the host, and reinfection is possible.

Treatment for gonorrhea is complicated because of widespread resistance to penicillin, tetracycline, and quinolones. *N. gonorrhoeae* infections are now typically treated with a single intramuscular (IM) injection of ceftriaxone plus either oral azithromycin or oral doxycycline to empirically treat possible chlamydia infection. No serological test or vaccine for gonorrhea exists because the organism frequently changes the structure of its surface antigens.

Gonorrhea is epidemic in the United States and is a notifiable disease (a disease that must be reported to federal, state, or local officials when diagnosed). Syphilis and chlamydial infections are also notifiable diseases. Patients with STIs are questioned about their sexual contacts, enabling public health officials to track, and slow, the spread of STIs by treating potentially infected people.

CHLAMYDIAL INFECTION *Chlamydia trachomatis* is a Gram-negative-like obligate intracellular pathogen with several serotypes; genital serotypes infect the reproductive tract whereas ocular serotypes infect the eye (trachoma; Section 19.6). According to the CDC, **chlamydia** is the most frequently reported sexually transmitted infection in the United States, with almost 1.8 million cases reported in 2018 alone. But many of the infected people are completely unaware of their infection. Three-fourths of infected women, for instance, have no symptoms.

Chlamydias are unusual Gram-negative-like organisms with a unique developmental cycle. They are obligate intracellular pathogens that start as small, nonreplicating infectious elementary bodies that enter target eukaryotic cells. Once inside vacuoles, the pathogens begin to enlarge into replicating reticulate bodies (Figure 23.17). As the vacuole fills, now called an inclusion, the reticulate bodies divide to become new nonreplicating elementary bodies. *C. trachomatis* can cause STIs, pneumonia, and trachoma.

People most at risk of developing genitourinary tract infections caused by *Chlamydia* are young, sexually active men (15–24) and women (14–24); anybody who has recently changed sexual partners; and anybody who has recently had another sexually transmitted infection. *Chlamydia* is the major cause of nongonococcal urethritis in the United States. The spread of *Chlamydia* has also been linked to its ability to bind to sperm in a process called "hitchhiking." This interaction is thought to help the organism spread to females. Left untreated, *Chlamydia*, like *N. gonorrhoeae*, can cause pelvic inflammatory disease in women. Men left untreated can suffer urethral and testicular infections and a serious form of arthritis.

Nongonococcal urethritis is diagnosed when the Gram stain of a patient's urethral mucopurulent or purulent discharge (defined as 5 or more white blood cells per oil immersion field) reveals no organisms (*Chlamydia* does not stain). Differential diagnosis of nongonococcal urethritis includes several etiologies, both infectious and noninfectious. When a chlamydial infection is suspected, samples of the patient's urethral discharge are sent for nucleic acid amplification tests such as PCR to confirm a diagnosis. *Chlamydia* infections can be treated with either azithromycin (single 1-gram dose orally) or doxycycline (100 mg twice a day orally for 7 days). Doxycycline cannot be used during pregnancy because tetracycline derivatives interfere with bone development, so for pregnant women the choice is either the single-dose azithromycin or amoxicillin (500 mg three times a day orally for 7 days).

Figure 23.17
Replication Cycle of *Chlamydia*

1 Elementary bodies (EBs) bind and enter eukaryotic cell by phagocytosis.

2 Elementary body differentiates into the larger reticulate body (RB).

3 Reticulate bodies replicate.

4 Reticulate bodies differentiate to elementary bodies and form inclusions.

5 Elementary bodies are released.

EBs

RBs

Attachment

Chlamydia trachomatis

C. trachomatis–containing vacuole in an infected cell, showing an elementary body, a reticulate body, and an intermediate form.

Intermediate form

Reticulate body

Elementary body

TEM | 400 nm

LYMPHOGRANULOMA VENEREUM Lymphogranuloma venereum (LGV) is a chlamydial disease involving only the L1, L2, and L3 serotypes of *Chlamydia trachomatis*. These serotypes are more virulent and invasive than other chlamydial serotypes. The organism gains entrance through breaks in mucosal surfaces and

Figure 23.18 **Lymphogranuloma Venereum**

Painful lymphadenopathy caused by *Chlamydia trachomatis*.

travels via the lymphatic vessels to multiply within mononuclear phagocytes (macrophages and monocytes) in regional lymph nodes. LGV is initially characterized by self-limited genital papules or ulcers, followed in several weeks by painful inguinal and/or femoral lymphadenopathy, which may be the only clinical manifestation at presentation (**Figure 23.18**). Rectal lesions may also be seen, especially in LGV patients who engage in receptive anal intercourse. In these cases, rectal pain, discharge, and bleeding may be confused with other gastrointestinal conditions, such as colitis.

CHANCROID *Haemophilus ducreyi* is a Gram-negative, fastidious (having complex nutritional requirements for growth) coccobacillus that causes painful genital ulcers and inflammation of the regional lymph nodes (**Figure 23.19**). The painful lesions are usually on the external genitals and are called **chancroid**, or soft chancre. Chancroid is common in the tropics and in developing countries but uncommon in the United States. Only three cases were reported to the CDC in 2018. Patients with chancroid are likely to have syphilis simultaneously. Chancroid and syphilis increase the risk of acquiring an HIV infection.

TRICHOMONIASIS *Trichomonas vaginalis* is a flagellated protozoan (**Figure 23.20**) that causes an unpleasant sexually transmitted vaginal infection called **trichomoniasis**. Approximately 2–3 million infections occur each year in the United States. *Trichomonas* can infect both men and women, but men are usually asymptomatic. Even among infected women, 25%–50% are considered asymptomatic carriers.

There is no cyst in the life cycle of *T. vaginalis*, and transmission is via the trophozoite stage only, a stage of protozoan life cycle involving active, feeding cells (**Section 11.3**). Growth of this protozoan can cause vaginal pH to rise above pH 4.5 (normal vaginal pH varies from 4.2 to 5.0, depending on the individual's ethnicity). The female

Figure 23.19 Chancroid

Infection with *Haemophilus ducreyi* causes a painful lesion called chancroid, in contrast to the painless lesion of primary syphilis, which is called a chancre.

Figure 23.21 Strawberry Cervix

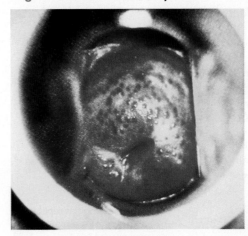

Infection by *Trichomonas vaginalis* causes a purulent discharge and lesions on the cervix referred to as "strawberry cervix." The discharge is usually thin but can appear white, yellow, or green. Compare with normal cervix in Figure 23.16A.

patient with trichomoniasis may complain of vaginal itching and/or burning, pain during intercourse, and a musty vaginal odor. An abnormal yellowish green vaginal discharge (often frothy) may also be present. In 2%–5% of patients, physical exam shows punctate (spotty or mottled), hemorrhagic lesions on the vaginal wall and cervix, referred to as "strawberry cervix" (**Figure 23.21**). Males

Figure 23.20
Trichomonas vaginalis

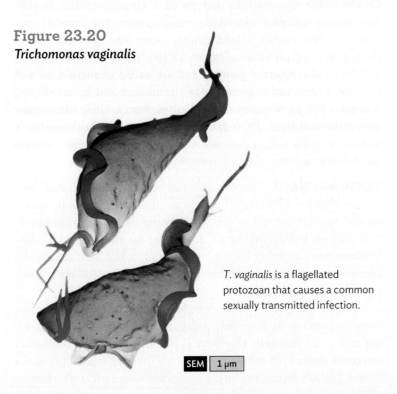

T. vaginalis is a flagellated protozoan that causes a common sexually transmitted infection.

SEM 1 μm

may complain of painful urination (dysuria), urethral or testicular pain, and lower abdominal pain.

Complications of trichomoniasis include an increased risk of infection with HIV in both men and women. *T. vaginalis* infections may lead to prostatitis, epididymitis, and infertility in men. In women, the infection can progress to pelvic inflammatory disease and a type of cervical cancer. Infected pregnant women have an increased risk of premature rupture of the amniotic sac, preterm labor, and low-birth-weight infants. Pregnant women coinfected with HIV have a higher risk of transmitting the virus to their infant, in addition to the other possible complications associated with trichomoniasis during pregnancy.

Diagnosis is usually made from clinical presentation, but organisms in clinical specimens can also be detected by a PCR test that amplifies ribosomal RNA (the Aptima *Trichomonas vaginalis* assay) or by microscopic observation of the flagellated protozoan in secretions. The nucleic acid test, however, is more sensitive than the microscopic examination of secretions. Polymorphonuclear cells (PMNs), the primary host defense against the organism, can also be seen microscopically in the vaginal secretion of an infected patient. Metronidazole is the drug of choice for treating trichomoniasis. But pregnant patients treated with metronidazole have a higher risk of preterm labor.

PELVIC INFLAMMATORY DISEASE Pelvic inflammatory disease (PID) is a damaging infection of the uterus and fallopian tubes caused by several microbial species, including *Neisseria gonorrhoeae* and *Chlamydia*, when these organisms directly spread from the vagina to these other structures. PID is often subclinical, but symptomatic patients can present with severe low-abdominal pain (constant, dull, aching pain or cramping pain), fever, nausea, and vomiting. Pelvic exam of these patients shows an inflamed cervix that is extremely tender and painful when moved (see Figure 23.16). A vaginal discharge containing white blood cells is common. The damage produced by PID can lead to tubal pregnancies, infertility, and chronic pelvic pain.

SECTION SUMMARY

- **Sexually transmitted infections (STIs)** are transmitted through genital, oral-genital, or anal-genital contact and must be reported to public health officials.

- **Microbes that cause STIs are susceptible to drying** and require direct contact with mucous membranes for transmission.

- **Patients with one STI** are likely to be infected with multiple sexually transmitted microbes.

- **Genital warts** (condylomata acuminata), the most common sexually transmitted viral infection, are caused by specific strains of HPV (usually subtypes 6, 11, 16, and 18). A multivalent vaccine is available for those subtypes.

- **Genital herpes** is marked by anogenital ulcerations caused by herpesvirus types 1 and 2 (HSV-1 and HSV-2). From an initial infection, the virus can become latent in spinal nerves and reemerge to cause recurrent episodes.

- **Babies can acquire neonatal herpes** by vertical transmission from pregnant mothers infected with HSV before, during, or after birth.

- **Human immunodeficiency virus (HIV) infection** proceeds through three stages: acute HIV infection (coincident with seroconversion), clinical latency, and then acquired immunodeficiency syndrome (AIDS). The number of circulating CD4+ T cells in a patient and the viral load are important prognostic indicators of the disease.

- **A CD4+ T-cell count of less than 200 cells/µl or the presence of an AIDS indicator condition is defined as AIDS.** HIV depletion of CD4+ T cells to below 500 cells/µl results in lethal secondary infections and cancers.

- **Syphilis, gonorrhea, and chlamydial diseases** are the most common sexually transmitted bacterial infections.

- **Syphilis has three stages:** primary (localized, painless chancres), secondary (widespread lesions over the body that can mimic other skin rashes), and tertiary (cardiovascular and nervous system lesions). Penicillin is the treatment of choice.

- **Gonorrhea is marked** by a purulent urethral discharge in men and a vaginal discharge in women, although women are often asymptomatic. Cephalosporins are the antibiotic treatment of choice.

- ***Chlamydia* is the major cause of nongonococcal urethritis.** *Chlamydia* species are Gram-negative, obligate intracellular bacteria. Azithromycin or doxycycline is used for treatment.

- ***Trichomonas vaginalis*** is a flagellated protozoan that causes trichomoniasis, a sexually transmitted vaginal disease. The reservoirs for this organism are the male urethra and the female vagina.

Thought Question 23.3 Why would HIV disease lead to shingles in more than one dermatome?

Thought Question 23.4 What aspect of the immune system is affected by diminished numbers of CD4+ T cells?

Thought Question 23.5 *Neisseria meningitidis* readily causes meningitis, yet *N. gonorrhoeae* does not. Why?

Thought Question 23.6 How would you treat pelvic inflammatory disease?

Thought Question 23.7 A patient in your office complains of a high fever and diarrhea of 3 weeks' duration. Upon physical exam, you notice he has swollen lymph nodes. You also discover that he is an IV drug user. Blood work reveals a negative RPR test and a CD4+ count of 500 cells/µl (normal is 1,000–1,400 cells/µl). What might you conclude from these results, and what other test should be run?

23.5 Non–Sexually Transmitted Infections of the Reproductive Tract

SECTION OBJECTIVES

- Relate signs and symptoms of non–sexually transmitted infections to their microbial etiology.

- Explain the role of the vaginal microbiome in the prevention and/or development of non–sexually transmitted infections of the reproductive tract.

- List treatment options available for non–sexually transmitted infections of the reproductive tract.

Are all infections of the reproductive tract only sexually transmitted? No, some genital tract infections can develop in women when microbes present on the perineum are introduced into the vagina or when certain organisms in the vaginal microbiome normally present in low numbers grow to unusually large numbers. The resulting vaginal infection (**vaginosis**; plural, vaginoses) and inflammation (**vaginitis**) are commonly seen in women of childbearing years. Hormonal changes that take place over the menstrual cycle or during pregnancy, as well as sexual activity, can change the composition of resident microbiota. Antibiotic treatment can also alter the balanced microbial population of the vagina, leading to a takeover by resident yeasts or anaerobes. A common example is a *Candida* yeast infection suffered by women when their vaginal microbiota is inadvertently killed by treatment of another infection with a short course of antibiotics. *Candida* then finds itself unopposed for growth (Table 23.6).

In males, infection of the epididymides (**epididymitis**), testes (**orchitis**), or prostate gland (**prostatitis**) can develop in several ways. Surgery or instrumentation (for example, the insertion of a urinary catheter) can obstruct urine flow that normally flushes organisms from the system. Systemic diseases or immunosuppression can provide an opportunity for organisms to spread from nearby areas. The main cause of bacterial prostatitis (inflammation and infection of the prostate) is *E. coli* (87.5% of cases), which either invades the gland from the adjacent rectum or extends from a UTI. Chronic prostatitis is often nonbacterial, but *E. coli* is the most common cause when the condition is secondary to an infection.

Table 23.6

Bacterial and Fungal Non–Sexually Transmitted Infections of the Reproductive Tract

Disease	Symptoms	Complications	Organism(s)	Transmission	Diagnosis	Treatment	Prevention
Bacterial vaginosis (BV)	Foul-smelling discharge	Premature delivery, postpartum endometritis (infection and inflammation of the endometrium)	*Gardnerella vaginalis* (major cause); several other minor etiologies, including *Mycoplasma hominis* and *Peptostreptococcus* spp.	Self-inoculation or overgrowth of existing *G. vaginalis*	Clinical diagnosis (Amsel criteria) requires three of the following: • Homogeneous thin, white or gray discharge coating vaginal walls • >20% clue cells present on wet mount microscopy of vaginal saline preparation • Vaginal pH >4.5 • Fishy odor of vaginal discharge before or after addition of 10% potassium hydroxide (KOH), a positive "whiff" test Gram stain also used in diagnosis	Topical or oral metronidazole or clindamycin; treatment regimen is different for pregnant women	No vaccine available
Fournier's gangrene	Prodromal fever 2–7 days prior to other symptoms; severe pain; tenderness in the genital area can be accompanied by itching and edema of the overlying skin, progressing to wounds with drainage and gangrene	Various systemic effects, including septic shock	Polymicrobial necrotizing fasciitis of genitals (particularly penis and scrotum), including anaerobic organisms, Enterobacteriaceae, streptococcal and staphylococcal species, and fungi	Surgical or other trauma to the penis, scrotum, perineum, or anal areas	Culture of the wound; blood and urine cultures; clinical presentation	Surgery to remove necrotic tissue; antibiotic therapy; tetanus prophylaxis	No vaccine available
Vulvovaginitis	Acute vaginal discharge, vulvar itching and pain that can result in dysuria	Chronic candidal vulvovaginitis	*Candida albicans* (80%–92%) and other *Candida* species; overgrowth of personal microbiota after immunosuppression, hormonal changes, trauma, or antibiotic therapy	Self-inoculation or overgrowth of existing *C. albicans*	Clinical presentation; wet prep or Gram stain showing yeasts, hyphae, or pseudohyphae; culture	Topical antifungals (butoconazole, clotrimazole, miconazole, nystatin, tioconazole, terconazole) or oral antifungals (fluconazole); treatment depends on the type of infection: uncomplicated, complicated, or recurrent	No vaccine available

Bacterial Vaginosis

The large surface area and associated secretions of the female reproductive tract make it a rich environment for microbes. Composition of the vaginal microbiota changes with the menstrual cycle, owing to changing nutrients and pH. The mildly acidic nature of vaginal secretions (approximately pH 4.5) discourages the growth of many microbes. As a result, the acid-tolerant *Lactobacillus acidophilus* is among the most populous vaginal species. As with all microbial ecosystems associated with the human body, the balance between competing species is crucial to preventing infection. Antibiotic therapy to treat an infection can also cause the loss of normal microbiota. In the vagina, this disruption allows infection by *Candida albicans* (yeast infection), which is not susceptible to antibiotics designed to kill bacteria.

A decrease in the number of lactobacilli among the vaginal microbiota is thought to be the major culprit in the pathogenesis of bacterial vaginosis, a common infection in women of childbearing years. *Lactobacillus* species produce lactic acid (a strong microbicide that lowers the pH of the vagina to an acidic 3.4–4.5) and antimicrobial compounds such as hydrogen peroxide and bacteriocins. An acidic pH makes the vaginal canal an inhospitable environment for many pathogens. Lactic acid is an effective microbicide against both HIV and *Neisseria gonorrhoeae*. Bacteriocins secreted by lactobacilli are narrow-spectrum bactericidal proteins that increase the permeability of the target microbe's cell membrane.

Not all women have lactobacilli as the major species of their vaginal microbiota. Recent research using culture-independent methods shows that several microbial species other than lactobacilli can be the dominant members of normal vaginal microbiota in some women. The percentage of women colonized by non-lactobacilli-dominated vaginal microbiota appears to vary with ethnic background. Genetics, socioeconomic and behavioral backgrounds, and environmental factors are hypothesized to cause these variations.

Bacterial vaginosis (BV) is not sexually transmitted, but sexual activity is a risk factor. Women are at higher risk for BV if they have multiple sexual partners or a new sexual partner, douche (rinse out their vagina with water or a cleansing agent), or lack lactobacilli. Clinicians should use their own judgment in determining whether patients at high risk of BV should be screened. Low-risk patients, however, should not be screened.

Bacterial vaginosis can be polymicrobial or result from overgrowth of several indigenous anaerobic bacteria. The vaginal tissue of the patient affected with BV shows no sign of inflammation (this is why it's called vaginosis and not vaginitis), and in most cases the patient has no symptom other than a thin discharge. In some, particularly those cases caused by overgrowth of *Gardnerella vaginalis*, the discharge is grayish and foul smelling (musty or fishy) and the patient complains of dysuria and pruritis (itchiness). Pregnant women with BV may develop serious complications, such as premature delivery,

Figure 23.22 Clue Cells

Samples obtained from the grayish discharge of bacterial vaginosis show "clue cells," vaginal epithelial cells with bacteria (usually *Gardnerella vaginalis*) stuck to them, similar to the one shown here.

spontaneous abortion, or postpartum (after delivery) endometritis (inflammation of the endometrium).

Diagnosis of BV is based on clinical presentation, a vaginal pH greater than 4.5, presence of a film of thin, grayish white discharge coating the vaginal wall, and more than 20% clue cells. Clue cells in a vaginal sample are vaginal epithelial cells covered with bacteria, giving them a stippled look (**Figure 23.22**). Oral and topical antibiotics (usually metronidazole) are used to treat BV in both pregnant and nonpregnant women. Antibiotic therapy of pregnant women with BV reduces the risk of preterm delivery and spontaneous abortion. Oral or vaginal probiotics containing *Lactobacillus* species are also used to treat bacterial vaginosis.

Fournier's Gangrene

Fournier's gangrene is a form of necrotizing fasciitis infection that most often involves the soft tissues of the male genitals, although women can also contract the disease. The infection involves a mixed population of aerobic and anaerobic microorganisms—typically streptococci, clostridia, staphylococci, *Klebsiella*, and other enteric bacteria. The most common sources of these agents are skin and urogenital areas. Patients with Fournier's gangrene, such as the one pictured in **Figure 23.23**, usually have an underlying immune dysfunction. Risk factors for Fournier's gangrene include diabetes, IV drug use, trauma, recent surgery, immunosuppression (for example, to treat cirrhosis or malignancy), peripheral vascular disease, and morbid obesity. Treatment of this disease includes broad-spectrum antibiotics (imipenem or meropenem) and surgical debridement. The mortality rate for Fournier's gangrene, a urological emergency, varies from 3% to 45%, depending on how far the disease progressed before diagnosis.

Figure 23.23 Fournier's Gangrene

This patient presented to the emergency department with exquisite (intense) scrotal pain. A foul-smelling, purulent fluid emerged when the scrotum was surgically opened, exposing necrotic tissue inside.

Vulvovaginal Candidiasis

Candida is part of the normal vaginal microbiota of most women. **Vulvovaginal candidiasis** (vulvovaginitis), usually called a yeast infection, is neither an opportunistic nor a sexually transmitted infection; rather, it is a *Candida* overgrowth problem. Diabetes, hormone replacement therapy, taking birth control pills, taking corticosteroids, or short-term antibiotic treatment will increase the risk of vulvovaginitis caused by *Candida* because these conditions alter the balance of the normal vaginal microbiota.

Yeast infections are very common during childbearing years. Most premenopausal women (75%) have suffered at least one episode of yeast infection and recall the intense vulvar itching and irritation that causes external dysuria (painful urination without urethritis). Different *Candida* species can cause the infection, but *Candida albicans* is the culprit in most cases (80%–92%). The infection may or may not be accompanied by a curd-like (or cottage cheese–like) discharge, but there is no foul odor (Figure 23.24).

About half of all patients who get a yeast infection will have a recurrence. If a woman gets four or more *Candida* infections, she is said to have recurrent yeast infections; these are thought to be caused by species other than *Candida albicans* (for instance, *Candida glabrata*). HIV-infected women have a higher incidence of vulvovaginal candidiasis and suffer a longer duration of symptoms, but the severity of symptoms is not any different from that in non-HIV-infected women. Vaginal pH remains unchanged (about 4.5), in contrast to what is seen in *Trichomonas* infection.

To make a diagnosis, 10% potassium hydroxide is added to a wet mount of the discharge (or material scraped from the infected area) and examined under the microscope. Budding yeast will be seen in positive cases.

Azole drugs, described in Chapter 13, are believed to block the biosynthesis of ergosterol present in yeast membranes. Over-the-counter topical antimycotic drugs (vaginal suppositories) can achieve cure rates in excess of 80%. The only <u>oral</u> azole agent approved by the US Food and Drug Administration (FDA) for treating vulvovaginitis is fluconazole, which also achieves a high rate of cure. Therapeutic concentrations of this drug are maintained in vaginal secretions for at least 72 hours after the ingestion of a single 150-mg tablet.

Figure 23.24 Vulvovaginal Candidiasis

A. Vulvovaginal candidiasis as seen through a speculum. Compare with normal cervix in Figure 23.16A.

B. Budding *Candida albicans* cells are seen in a Gram stain of vaginal secretions.

SECTION SUMMARY

- **Vaginosis** is a bacterial infection of the vagina (often caused by the anaerobe *Gardnerella vaginalis*) resulting from an imbalance in the vaginal microbiota.

- **Epididymitis, orchitis, and prostatitis** are caused by local bacteria when urine flow is obstructed or when the immune system is suppressed.

- **Fournier's gangrene** is a form of necrotizing fasciitis that most often involves the soft tissues of the male genitals, although women can also contract the disease. A mixed population of aerobic and anaerobic microorganisms, typically enteric bacteria, commonly causes this infection.

- **Candidiasis**, commonly called a yeast infection, is neither an opportunistic nor a sexually transmitted infection; it results from overgrowth of *Candida* species.

Perspective

Cystitis, pyelonephritis, and other infections of the urinary tract are common in medical practice. Clinicians realize that UTI presentations can vary, depending on the ages of patients and their comorbidities. Remember, too, that infections starting in the urinary tract can spread to other organ systems (urosepsis), and infections focused elsewhere in the body can also end up causing UTIs. The toll on the lives of patients, especially women, and on the nation's economy makes UTIs an important, though somewhat embarrassing, burden on society.

The societal burden of STIs is also great. The CDC reports a 14% increase in diagnosed syphilis cases from 1991 to 2019. The reported increase in gonorrhea cases is even more striking: 18.6% during the 2016–2017 period. In contrast, the number of new HIV infections after 2010 has mostly decreased or remained stable in the United States, depending on age and socioeconomic groups. The success is due to aggressive and excellent public education. STIs have an added social dimension. The stigma of having an STI makes it difficult for some people to visit a clinic to be checked and treated. Others are evaluated and start appropriate treatment but do not complete the treatment. People who are infected but do not know, refuse to find out, or ignore treatment are the ones who pass the infection to their partners. Public education efforts that aim to increase testing and teach the benefits of proper treatment will go a long way toward addressing the problem of STIs.

Of course, the impact of STIs on society is not new. Humans have been afflicted with these diseases for thousands of years. So, why have we been so ineffective in preventing these infections? Our failure is due in part to the ingenuity of microbes in evading our immune system and their ability to mutate and become resistant to the medications used against them. We have also helped them survive and flourish by our desire to have sex and our propensity to ignore effective preventive methods, simple as they are, such as the use of condoms.

Figure 23.25 Summary: Microbes That Infect the Genitourinary System

Viruses	Genome[a]	Disease
HSV-1	DNA, ds	Genital herpes
HSV-2	DNA, ds	Genital herpes
Human papillomavirus	DNA, ds	Genital warts
Hepatitis B	DNA, ds (partially)	Hepatitis
Hepatitis C	RNA, ss, positive sense	Hepatitis
Human immunodeficiency virus	RNA, ss, positive sense (two copies of the genome in a core capsid)	HIV/AIDS

Bacteria	Morphology	Disease(s)
Enterobacter spp.	Gram-negative, bacilli	Cystitis, pyelonephritis
Escherichia coli	Gram-negative, bacilli	Epididymitis, cystitis, prostatitis, pyelonephritis
Klebsiella pneumoniae	Gram-negative, bacilli	Cystitis, pyelonephritis
Proteus mirabilis	Gram-negative, bacilli	Cystitis, pyelonephritis
Pseudomonas aeruginosa	Gram-negative, bacilli	Cystitis, pyelonephritis
Neisseria gonorrhoeae	Gram-negative, cocci	Gonorrhea, also called gonococcal urethritis (STI)
Haemophilus ducreyi	Gram-negative, coccobacilli	Chancroid
Chlamydia trachomatis	Gram-negative-like,[b] coccoid or rod-shaped	Chlamydial diseases; nongonococcal urethritis and lymphogranuloma venereum (STI)
Enterococcus spp.	Gram-positive, cocci[c]	Cystitis, pyelonephritis
Staphylococcus aureus	Gram-positive, cocci	Cystitis, pyelonephritis
Staphylococcus saprophyticus	Gram-positive, cocci	Cystitis, pyelonephritis
Gardnerella vaginalis	Gram-variable, coccobacilli	Bacterial vaginosis
Treponema pallidum	Spirochete	Syphilis (STI)

Fungi	Type	Disease
Candida albicans	Yeast	Vulvovaginitis, cystitis in diabetic patients

Parasites	Type	Disease
Trichomonas vaginalis	Protozoan	Trichomoniasis

[a]ds = double-stranded; ss = single-stranded.
[b]Chlamydia trachomatis is an obligate intracellular aerobic organism, very difficult to stain but closely related to Gram-negative bacteria.
[c]Enterococci usually occur in pairs and sometimes in short chains.

Portal of Entry

Portal of Exit

Mouth, Oral Cavity
[HSV-1 and HSV-2

Mouth, Oral Cavity
HSV-1 and HSV-2]

Blood
[Hepatitis B and hepatitis C
HIV

Blood
Hepatitis B and hepatitis C
HIV]

**Anus, Perineum,
Vagina, Penis, or
Self-Inoculation**

**Anus, Perineum,
Vagina, or Penis**

[HBV
HCV
HIV
HPV
HSV-1 and HSV-2

HBV
HCV
HIV
HPV
HSV-1 and HSV-2]

Chlamydia trachomatis
Enterobacter spp.
Enterococcus spp.
Escherichia coli
Gardnerella vaginalis
Haemophilus ducreyi
Klebsiella pneumoniae
Neisseria gonorrhoeae
Proteus mirabilis
Pseudomonas aeruginosa
Staphylococcus aureus
Staphylococcus saprophyticus
Treponema pallidum

Chlamydia trachomatis
Enterobacter spp.
Enterococcus spp.
Escherichia coli
Gardnerella vaginalis
Haemophilus ducreyi
Klebsiella pneumoniae
Neisseria gonorrhoeae
Proteus mirabilis
Pseudomonas aeruginosa
Staphylococcus aureus
Staphylococcus saprophyticus
Treponema pallidum

[*Candida albicans*

Candida albicans]

[*Trichomonas vaginalis*

Trichomonas vaginalis]

Key Terms

acquired immunodeficiency syndrome (AIDS) (p. 754)

acute HIV infection (p. 754)

bacteriuria (p. 740)

chancre (p. 759)

chancroid (p. 763)

chlamydia (p. 762)

clinical latency (p. 754)

condyloma acuminatum (p. 751)

congenital syphilis (p. 759)

costovertebral tenderness (p. 748)

cystitis (p. 743)

dysuria (p. 746)

elite controller (p. 756)

epididymitis (p. 765)

Fournier's gangrene (p. 767)

general paresis (p. 759)

genital herpes (p. 753)

gonorrhea (p. 761)

gumma (p. 759)

human immunodeficiency virus (HIV) (p. 754)

Jarisch-Herxheimer reaction (p. 760)

Kaposi's sarcoma (p. 755)

latent period (p. 753)

long-term nonprogressor (p. 756)

lymphogranuloma venereum (LGV) (p. 763)

neonatal herpes (p. 753)

neurosyphilis (p. 759)

oliguria (p. 748)

ophthalmia neonatorum (p. 762)

orchitis (p. 765)

pelvic inflammatory disease (PID) (p. 764)

primary infection (p. 753)

primary syphilis (p. 759)

prostatitis (p. 765)

pyelonephritis (p. 743)

rapid plasma reagin (RPR) test (p. 759)

secondary syphilis (p. 759)

sexually transmitted infection (STI) (p. 751)

syphilis (p. 756)

tabes dorsalis (p. 759)

tertiary syphilis (p. 759)

trichomoniasis (p. 763)

urethritis (p. 745)

urinary tract infection (UTI) (p. 740)

urosepsis (p. 740)

vaginitis (p. 765)

vaginosis (p. 765)

Venereal Disease Research Laboratory (VDRL) test (p. 759)

viral load (p. 754)

vulvovaginal candidiasis (p. 768)

Objectives Assessment Questions

23.1 Anatomy of the Urinary Tract

1. Which structure(s) of the urinary tract act as a filter to remove unwanted material from blood?
   a. Kidneys
   b. Ureters
   c. Urinary bladder
   d. Urethra
   e. Urethral meatus

2. The urine is transported to the urinary bladder via _____.
   a. nephrons
   b. the urethra
   c. ureters
   d. kidneys

3. Which of the following is a major difference between the male and female urinary tracts?
   a. Number of ureters
   b. Size of the perineum
   c. Structure of the urethral meatus
   d. Length of the urethra
   e. Bladder lining

23.2 Urinary Tract Infections

4. Which of the following is a risk factor for developing bladder infections in young, nonpregnant women?
   a. Sexual activity
   b. Low plasma estrogen concentration
   c. Bladder incontinence
   d. Elevated plasma estrogen concentration

5. A male hospitalized patient who had a urinary catheter inserted to enable urination is now running a high fever and feels a lot of pain when tapped at the costovertebral angle. The patient's blood pressure is within normal limits, and his heart rate is 110 (normal is 60–100). Urine culture shows *Proteus mirabilis*; blood cultures are negative. Which of the following statements is probably true?
   a. This patient has pyelonephritis resulting from an ascending bladder infection.
   b. This patient has a bladder infection resulting from a descending kidney infection.
   c. This patient has a systemic infection that has spread to his kidneys.
   d. Continuous flow of urine from kidneys to the bladder prevents the infection from spreading between these organs.
   e. Urinary tract infections never cause a fever.

6. The majority of UTIs are caused by Gram-negative bacteria because urine is bacteriostatic to Gram-positive bacteria. True or false?

7. The majority of cystitis cases in children are due to _____.
   a. congenital structural abnormalities that prevent complete emptying of the bladder
   b. children postponing emptying their bladder
   c. the shorter length of urethra in younger children
   d. continuous changes in the normal microbiota of children's perineum

8. Which of the following UTIs is a risk factor for developing cardiovascular shock?

    a. Cystitis

    b. Urethritis

    c. Pyelonephritis

    d. Acute glomerulonephritis

23.3 Anatomy of the Reproductive Tract

9. The _____ is/are external to the human body.

    a. ovaries

    b. vagina

    c. testes

    d. seminal vesicles

10. Which of the following statements about the female reproductive tract is correct?

    a. The female reproductive tract has internal and external components.

    b. Every month only one follicle is recruited to mature into an ovum.

    c. There is no direct contact between the ovaries and the fallopian tubes.

    d. The cervix forms the end of the birth canal.

11. Sexually transmitted infections are passed from one person to another only through intercourse. True or false?

23.4 Sexually Transmitted Infections of the Reproductive Tract

12. Infection with certain serotypes of _____ increases the risk of cervical cancer.

    a. human papillomavirus

    b. human immunodeficiency virus

    c. herpes simplex 1

    d. herpes simplex 2

Match the following:

13. Vesicular lesions	a. Mononucleosis-like symptoms
14. Acute HIV infection	b. CD4+
15. HPV vaccine	c. AIDS
16. Kaposi's sarcoma	d. Gardasil
17. Host protein that HIV binds to	e. HSV

18. Women serve as the natural reservoir for _____ because they are usually asymptomatic when infected with this microbe.

    a. *Chlamydia trachomatis*

    b. *Treponema pallidum*

    c. *Haemophilus ducreyi*

    d. *Neisseria gonorrhoeae*

    e. *Trichomonas vaginalis*

19. A grayish yellow discharge with a fishy odor, along with punctate (spotty or mottled) hemorrhagic lesions on the cervix (strawberry cervix), is indicative of infection by _____.

    a. *Chlamydia trachomatis*

    b. *Treponema pallidum*

    c. *Haemophilus ducreyi*

    d. *Neisseria gonorrhoeae*

    e. *Trichomonas vaginalis*

23.5 Non–Sexually Transmitted Infections of the Reproductive Tract

20. A young woman visiting her clinician complains of intense vulvar itching and burning upon urination. Physical exam reveals a whitish, odorless, thick discharge and inflamed, erythematous external genital tissue. Her cervix is not tender. The most likely diagnosis is _____.

    a. candidal vulvovaginitis

    b. bacterial vaginosis

    c. gonococcal urethritis

    d. nongonococcal urethritis

    e. trichomoniasis

21. A 45-year-old truck driver with a 10-year history of diabetes presents to the clinic, complaining of a wound in his groin that extends into his scrotum. He denies using illicit drugs or having any trauma in the area. You send a sample from the wound to the lab for identification and diagnose him with Fournier's gangrene. Which of the following is the lab most likely to report?

    a. Gram-negative rods normally found in the GI tract

    b. Gram-positive rods normally found on the skin

    c. A mixed population of aerobic and anaerobic organisms

    d. An obligate intracellular parasite

    e. A flagellated anaerobic protozoan

22. Which of the following is used to treat bacterial vaginosis?

    a. Metronidazole

    b. Probiotics with lactobacillus

    c. Topical azole compounds

    d. a and b

    e. a and c

Review Questions

1. Describe areas of the urinary and reproductive tracts associated with infectious disease. Explain how microbes gain access to these structures.

2. Describe the difference between cystitis, pyelonephritis, vaginosis, and vulvovaginitis. What organisms are the most common causes of these diseases?

3. When can asymptomatic bacteriuria be a dangerous condition?

4. Discuss how the presentation of UTIs can differ between children, adults, and the elderly.

5. Define recurrent UTIs. What are some possible causes of recurrent UTIs?

6. Explain how UTIs and pyelonephritis can be differentiated in a patient.

7. What are the most common viral diseases of the reproductive tract? Describe key features of these viruses (structure and target cells).

8. Explain how patient presentation would help you differentiate the diseases you cited in question 7.

9. Which sexually transmitted infections have a latent period sandwiched between symptomatic periods?

10. Discuss neonatal disease involving sexually transmitted agents. How can you distinguish a neonate infected with one or another of these agents?

11. Describe the stages of HIV infection and the parameters used to characterize those stages. What is the significance of a CD4+ count of less than 500 cells/μl? Less than 200 cells/μl?

12. How do the lesions of syphilis, chancroid, and LGV differ?

13. Discuss how the diagnosis and treatment of gonorrhea differ from the diagnosis and treatment of syphilis.

14. What is pelvic inflammatory disease, and what unique dangers does it pose?

15. Which pathogen causes the most common sexually transmitted infection? What is unique about the growth of this pathogen?

16. What is a characteristic feature of lymphogranuloma venereum?

17. For which STIs are vaccines available?

18. Describe the features of Fournier's gangrene. Who is at risk?

19. Explain how changes in the vaginal microbiome can be a risk factor for bacterial vaginosis.

Clinical Correlation Questions

1. A young woman is complaining of a fishy-smelling vaginal discharge. Physical exam does not reveal any inflammation. The cervix is normal and nontender. Microscopic examination of the discharge reveals clue cells. What organism is the most likely etiology of this woman's infection? How did this patient acquire this infection?

2. A 25-year-old pregnant female is in the clinic, complaining of a maculopapular rash that extends into the palms of her hands and soles of her feet. Physical exam shows an abnormal gait. You explain to the patient that when her infant is born, it may have several diseases and abnormalities. What physical abnormalities is the newborn likely to have?

3. In Case History 23.4, Rick from Honolulu developed gonorrhea and was successfully treated. He was told that his sexual partners also had to be identified and treated. What are the possible difficulties and complications of this action?

24

Infections of the Central Nervous System

[Diagnosis Comes Too Late]

SCENARIO Miguel, a 19-year-old young man from Michoacán, Mexico, worked at a sugarcane plantation in Louisiana. One day Miguel felt extremely tired, his left shoulder hurt, and his left hand was numb. Over the next 3 days, his condition worsened, and he was taken to a New Orleans hospital.

DISEASE PROGRESSION AND ADDITIONAL TESTS Miguel developed a fever of 38.4°C (101.1°F) and had trouble breathing, which prompted intubation (insertion of a tube into the trachea). During the next several days, Miguel became unresponsive to external stimuli, developed fixed and dilated pupils, and had episodes of bradycardia (slow heart rate) and hypothermia (low core body temperature). His CSF now had 87 WBCs/μl with 97% lymphocytes and an elevated protein level of 233 milligrams per deciliter (normal level, 15–45 mg/dl). An electroencephalogram (EEG) was consistent with encephalitis. Bacterial, viral, and fungal cultures of blood and CSF were negative, as were serological tests for HIV, syphilis, herpes simplex virus, arboviruses, Lyme disease, and autoimmune neuropathies.

PHYSICAL EXAMINATION AND LAB WORK Physical examination at the hospital revealed numbness, tingling, pinprick sensation (paresthesia) of his left shoulder, weakness of his left hand, generalized areflexia (absence of reflexes), and drooping of his left upper eyelid. Cerebrospinal fluid (CSF) obtained from a lumbar puncture (see Figure 24.4) had a mildly elevated white blood cell (WBC) count of 8 cells/microliter (normal level, 0–5 cells/μl), a normal glucose level, and no organisms upon staining. Miguel was admitted to the intensive-care unit with suspected viral encephalitis or meningitis.

DIAGNOSIS AND CONCLUSION
A diagnosis of rabies was suspected on the basis of the clinical history, even though animal exposure was unknown. On day 18, rabies was confirmed by finding rabies virus–specific IgG and IgM in Miguel's CSF and serum. After the doctor discussed Miguel's prognosis with the family and a subsequent EEG showed severe cortical impairment, Miguel was extubated and died shortly thereafter. Rabies virus antigen was detected in postmortem brain tissues, and antigenic typing determined it was a vampire bat rabies virus variant, which was confirmed by nucleic acid amplification and sequencing.

Adapted from CDC, 2011, *Morbidity and Mortality Weekly Report* 60(31):1050–1052.

TEM 50 nm

Rabies Virus
Rabies virus is an enveloped, bullet-shaped, negative-sense RNA virus (Rhabdoviridae family) transmitted to humans by the bite of a rabid animal. The virus infects the peripheral nervous system, moves to the central nervous system, and causes encephalitis (brain inflammation), which ultimately results in death.

CHAPTER OBJECTIVES

After reading this chapter, you will be able to:

- Correlate the anatomy and physiology of the nervous system with the infectious processes affecting them.
- Describe symptoms that can help differentiate nervous system infections caused by viruses from those caused by bacteria and parasites.
- Relate pathogenic mechanisms to disease prevention strategies for microbes that infect the nervous system.
- Correlate the physiology of each pathogen with its antimicrobial treatment.

Rabies, caused by the rabies virus, is one of the deadliest of all infectious diseases known to humans; fortunately, it is rarely seen in the United States. Rabies virus is endemic in different mammalian animal populations, including foxes, dogs, cats, and bats. The infected animals usually transfer the virus to us by depositing saliva with rabies virus in our flesh when they bite us or after casual contact with our injured skin. Once within the human body, rabies virus attacks the central nervous system and almost always causes death. Anyone who has had a dog or a cat knows that pets must be vaccinated against rabies. It is because of pet vaccination programs that the number of rabies cases has drastically declined in the United States. Other microorganisms can also attack our nervous system, but thanks to a vigilant immune system and our knowledge of infectious disease, we can fight these other microbes more effectively than we can rabies. That is not to say that other infections of the nervous system are harmless; on the contrary, most have the potential to cause immense damage and even death.

24.1 Anatomy of the Central Nervous System

SECTION OBJECTIVES

- Describe the functional organization of the central nervous system.
- Explain how meninges and cerebrospinal fluid protect the central nervous system.
- Describe the differences between two major types of neurons.
- Correlate the anatomy of the central nervous system to the symptoms when it is infected.

How is the nervous system organized? The central nervous system is the control center of the body. It houses our cognition, makes sense of our environment, formulates and helps dictate responses to all stimuli, enacts reflexes, and regulates every voluntary and involuntary action, from walking and talking body temperature and heart rate, and it does it all seamlessly. Our nervous system has two main parts (**Figure 24.1**): the **central nervous system (CNS)** and the **peripheral nervous system (PNS)**. The CNS controls and coordinates everything, while the PNS collects data from the periphery (parts of the body outside the brain) and from the outside environment and communicates that information to the CNS. You might envision the CNS as a computer's hard drive with different kinds of memories and operating systems to classify, store, and analyze data received from the PNS. The PNS provides the needed circuitry and instruments to collect and input data from the body and the environment and then conveys orders from the CNS to the periphery. Damage to either system can have dire consequences.

Peripheral nerve damage (neuropathy) is a complication of systemic diseases such as diabetes and several infectious diseases, such as leprosy (see Case History 5.1), syphilis, and Lyme disease. In this chapter, however, we will focus on diseases of the CNS.

The nervous system has, of course, defensive measures in place. The CNS is protected from most harm by several mechanisms. The hard bony structures that house the CNS, the skull and vertebrae, are physical barriers to possible trauma from the outside world, such as a blow to the head. Three thin membranes known as the **meninges** (singular, **meninx**) separate the brain and spinal cord from the bony structures surrounding them. The meninges include an inner layer next to the brain called pia mater, a middle layer named arachnoid mater, and an outer layer next to the skull called dura mater (**Figure 24.2**). Specialized projections of pia mater, called the choroid plexus, secrete a clear fluid derived from blood called **cerebrospinal fluid (CSF)** into the subarachnoid space.

We produce approximately 500–700 ml of CSF every day. The optimal CSF volume in an average adult is around 140–270 ml. The extra CSF produced by a person's choroid plexus is taken up by specialized structures called arachnoid villi, which will discard it into the venous circulation. The pulsation of local arteries helps circulate the CSF around the CNS. This means that CSF has a normal pressure associated with it. Infections in the CNS can result in severe inflammation of the meninges and occlusion of the arachnoid villi, which in turn can cause an increase in CSF pressure. The bony skull and vertebrae are not able to expand, so increased CSF volume will lead to increased CSF pressure on the brain itself. Elevation of CSF pressure is a very serious condition that can cause damage to the brain or spinal cord.

Figure 24.1 The Nervous System

Central nervous system
Peripheral nervous system

12 cranial nerves

Cerebrum

Cerebellum

8 cervical nerves

Spinal cord

Spinal nerves

12 thoracic nerves

Chain of vertebral ganglia, part of the nerve network of the autonomic system.

5 lumbar nerves

5 sacral nerves

Neurons are specialized cells of the nervous system. Information is transferred between axons and dendrites through the cell body. Some axons are covered in myelin, a fatty molecule produced by Schwann cells that protects the axon and accelerates information transfer.

FM 100 µm

Axon terminal

Myelin sheath

Schwann cell

Axon

Cell body

Nucleus

Dendrite

Major diseases of the nervous system

Viral diseases
Viral meningitis
Viral encephalitis
Poliomyelitis

Bacterial diseases
Bacterial meningitis
Meningococcal meningitis
Listeriosis
Neonatal meningitis

Bacterial neurotoxins
Botulism
Tetanus

Fungal diseases
Candida meningitis
Coccidioidal meningitis
Cryptococcal meningitis

Parasitic and prion diseases
Protozoal diseases
　Toxoplasmosis
　African sleeping sickness
　Naegleria meningoencephalitis
　Acanthamoeba
Prion diseases
　Bovine spongiform
　　encephalopathies
　Creutzfeldt-Jakob disease

Finally, a firewall of sorts, called the **blood-brain barrier**, prevents most microorganisms and large molecules from gaining access to the CNS. The blood-brain barrier exists because the endothelial cells that make up the CNS capillaries are attached to one another by tight junctions (see Section 14.3). These specialized structures prevent bacteria, large molecules, and hydrophilic molecules from moving out of the circulating blood and into the CSF. Unfortunately, the permeability of the barrier can be compromised by inflammation and other impairments, allowing bacteria, fungi, parasites, or even large molecules to reach the CNS and cause infection. Moreover, the blood-brain barrier has the unintended consequence of hindering the delivery of antimicrobial agents and medicines that are not lipid-soluble.

The CNS and PNS consist mainly of specialized cells called neurons. Neurons come in different shapes and sizes but fall into two general categories: **sensory neurons** send information from the sensors in the periphery (skin, gastrointestinal tract, and so on) to the CNS, and **motor neurons** send information from the CNS to the periphery.

Each neuron has a cell body and specialized thin extensions called axons and dendrites (see Figure 24.1 inset). Dendrites bring information to the cell body, whereas axons send information from the cell body. Some axons are covered by a whitish-colored sheath composed of a fatty molecule called myelin. The myelin sheath insulates and protects the axons it covers. Information travels faster along myelinated axons than it does along those

Figure 24.2 The Meninges

Skull   Dura mater   Arachnoid mater   Subarachnoid space   Pia mater

Brain tissue

The three membranous layers—dura mater, arachnoid mater, and pia mater—form the meninges that cover the human brain and spinal cord. The space deep between the pia mater and arachnoid mater is called the subarachnoid space. The cerebrospinal fluid flows through the subarachnoid space.

that are unmyelinated. When myelin is destroyed, transmission of nerve impulses is impaired. Loss of the myelin sheath contributes to certain nervous system disorders, including multiple sclerosis and Guillain-Barré syndrome (see Figure 22.20).

The human brain consists of myelinated (white matter) and unmyelinated (gray matter) neurons and is divided into functional and structural parts. The brain stem, located at the base of the brain, controls vital involuntary bodily functions, such as our heartbeat and breathing. The spinal cord extends from the brain stem into the vertebral canal, also called the spinal cord cavity. We

have 33 vertebrae stacked from our neck to our tailbone (coccyx). The vertebral canal is the inner space made by the stacked vertebrae. A pair of spinal nerves exit each vertebra, one from each side, and innervate (provide nerve supply to) the structures around that vertebra (Figure 24.3). Skin at the level of each vertebra is innervated by the corresponding spinal nerve and is referred to as a dermatome (see Section 19.2).

The spinal cord extends almost the entire length of the vertebral canal until it stops around lumbar (lower torso) vertebrae 1 and 2, depending on the person. When a person is suspected of having a CNS infection, a sample of CSF is removed by inserting a needle into the subarachnoid space between the third and fourth lumbar vertebrae, where there is no chance of injuring the spinal cord (Figure 24.4). This procedure is called a **lumbar puncture** or spinal tap. Analysis of the CSF will give clues that help identify the cause of CNS infection or inflammation.

SECTION SUMMARY

- **The CNS** includes the brain, encased in the skull, and the spinal cord, encased in the vertebral column.
- **The PNS collects information and transmits it via sensory neurons to the CNS,** which processes that information and returns signals through motor neurons of the PNS, directing a response.
- **The meninges** are three membranes (dura mater, arachnoid mater, and pia mater) that cover the brain and spinal cord. CSF fills the space between the arachnoid and pia mater.
- **The choroid plexus** comprises projections of pia mater that secrete CSF.
- **Neurons** are cells of the nervous system with short extensions, called dendrites, that bring information to the cell body and long extensions, called axons, that send information from the cell body. Some axons are myelinated.

Figure 24.3 Vertebrae

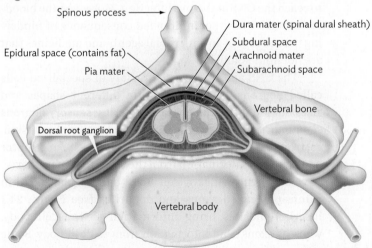

Spinous process

Dura mater (spinal dural sheath)
Subdural space
Arachnoid mater
Subarachnoid space

Epidural space (contains fat)

Pia mater

Vertebral bone

Dorsal root ganglion

Vertebral body

A. Each vertebra has a body, a spinous process, and a central opening. The spinal cord is separated from the bony structure of the vertebrae by the spinal meninges, which are continuous with the brain meninges. The space between the outermost layer of the meninges (dura mater) and the spinous bone is called the subdural space.

Epidural space

Spinal cord
Pia mater
Arachnoid mater
Dura mater

Spinal nerve

Vertebra

Spinous process

Intervertebral disk

Back

Internal organs

B. Intervertebral disks cushion the contact between the vertebrae. Damage or rupture of these disks can cause severe back pain in many people.

Figure 24.4 Lumbar Puncture

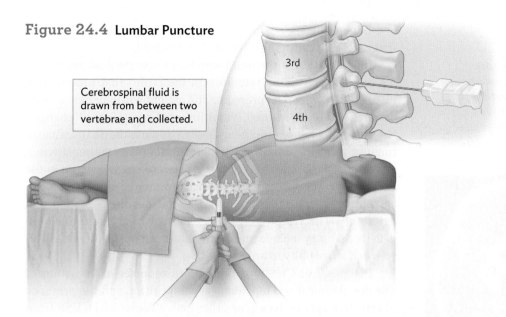

Cerebrospinal fluid is drawn from between two vertebrae and collected.

3rd

4th

A. A patient lies on their side (knees up, head down). A spinal tap needle is inserted between the third and fourth lumbar vertebrae and into the subarachnoid space using aseptic technique. The procedure is normally performed under local anesthesia.

B. CSF is allowed to drip from the needle into a collection vial.

24.2 Overview of Infectious Diseases of the Central Nervous System

SECTION OBJECTIVES

- Categorize CNS infections by the anatomical structures affected.
- Use key laboratory information to distinguish types of CNS infections.
- Compare and contrast the roles of meninges and cerebrospinal fluid in the pathogenesis of CNS infections.

How serious are brain infections? A brain infection must be diagnosed quickly and properly to avoid severe disease and possible death. A timely diagnosis can make the difference between life and death—or at least better quality of life versus worse. Yet CNS infections are easily misdiagnosed because of vague initial presenting signs and symptoms. Most CNS infections start with flu-like symptoms and nonspecific neurological manifestations such as a headache, dizziness, or fatigue. In this section, we will discuss the general concepts of CNS infections and lay the groundwork for the rest of the chapter.

How do microorganisms reach the CNS? The path to the CNS varies, depending on the invading agent. Most microbes first infect distal sites in the body and then travel to the CNS. Other pathogens reach the CNS via an infection in an adjacent bone or one of the structures closely associated with the CNS—such as eyes, sinuses, pharynx, or ears. Toxins made by organisms embedded elsewhere in the body can travel along and up the peripheral nerves to reach the CNS. A few organisms target the CNS directly; for example, some mosquito- and tick-borne viruses enter the circulation as a result of an insect bite and travel directly to the CNS.

Damage to the CNS as a result of infection can range from mild to severe. Usually, as with Miguel in the opening case history, symptoms become gradually more serious. The rabies patient initially presented with relatively mild and vague symptoms (generalized fatigue, left shoulder pain, and left-hand numbness), but as the infection progressed, his symptoms worsened and he lost some normal reflexes, such as the reactive narrowing of his pupils to light. Eventually, he lost the ability to breathe on his own.

The signs and symptoms of infectious diseases of the CNS do not usually point to a specific offending agent (bacterium versus virus, for example). Health care providers must identify compositional changes in blood or cerebrospinal fluid to make a correct diagnosis. Ironically, the major obstacle in treating CNS infections is the natural blood-brain barrier that prevents harmful substances and microorganisms from entering the CNS; this same barrier also blocks entry of many antimicrobial agents.

Three major types of diseases are associated with CNS infections: meningitis, encephalitis (plural, encephalitides), and myelitis. **Meningitis**, or inflammation of meninges, develops when the meninges are infected by a microbe or irritated by a chemical or biochemical. Symptoms include severe headache (due to increased intracranial pressure), fever, and confusion. **Encephalitis**, or inflammation of the brain, can also result from an infection or a reaction to a chemical or biochemical agent. Encephalitis can occur as a primary or secondary disease. **Primary encephalitis** is the result of a direct attack on the brain, as in rabies. **Secondary encephalitis** ensues when the infection or inflammation spreads from the meninges to the brain. Bacterial and fungal encephalitides are relatively rare and usually occur when an infection from another site, usually the meninges, extends from that location into the brain itself.

Figure 24.5 Computed Tomography Can Reveal Brain Damage from Encephalitis

In computed tomography (CT), computers are used to generate a three-dimensional image of an object from a large series of two-dimensional X-rays taken around a single axis of rotation. A contrast agent (or contrast dye) is material that can be swallowed by the patient or injected into their vein that will enhance the images obtained. The dye cannot pass through an intact blood-brain barrier. When the blood-brain barrier is damaged, as in encephalitis, the dye will leak across the damaged capillaries and accumulate in the affected areas.

Normal brain

Encephalitis

A. CT scan of a normal brain.

B. CT scan showing an accumulation of contrast dye around areas damaged by encephalitis.

Clinically, the distinction between the presenting signs and symptoms of meningitis and encephalitis is that encephalitis will cause disruption and deficiencies in the patient's cognitive, motor, and sensory functions. The nature of the deficiency depends on the area of the brain that has been damaged. Imaging studies of a patient's brain are usually performed to exclude other causes of brain dysfunction that may have similar symptoms (for example, a brain abscess or tumor). **Figure 24.5** illustrates the difference between a healthy brain and one damaged by encephalitis.

Viruses can also cause **myelitis** (inflammation of the spinal cord) and **radiculitis** (inflammation of the spinal nerve root, where the spinal nerve exits the vertebra). A combination of terms is used when the disease affects more than one site. For example, **meningoencephalitis** denotes simultaneous inflammation of the meninges and brain, whether by infection or chemical insult.

CNS infections are generally rare. When they do occur, however, viruses are the most common cause, especially of the primary encephalitides. Most of these viral infections, however, are mild. We become infected with these viruses when we cross paths with their reservoirs, such as ticks, birds, or other humans. Transmission of viral encephalitides is often through an insect bite. Depending on the virus, however, transmission can also happen after inhaling contaminated respiratory droplets, eating or drinking contaminated food or liquids, or through direct contact. Typically, an incubation period of variable length is followed by symptomatic disease, which may be mild or severe. Even when presentation is mild, encephalitis

is a potentially serious disease with neurological manifestations that can persist after the infection has resolved.

Inflammation of the brain may lead to cerebral edema (swelling and excessive fluid retention in the brain) that produces an increase in intracranial pressure. The patient will experience fever, headache, fatigue, nausea, vomiting, and neurological changes. Neurological changes can be as benign as **paresthesia** (numbness, tingling, pinprick sensation), **hyperesthesia** (extreme sensitivity), and **areflexia** (absence of reflexes), as suffered by the victim in the chapter-opening rabies case. Changes in mental status, memory, and temper are also common, as are confusion, disorientation, and irritability. The symptoms of encephalitis can progress to severe deficits that may include muscle weakness, an unsteady gait, paralysis, and an inability to breathe.

Diagnosis of CNS infections often requires examination of CSF for chemical and biochemical composition and the presence of infective agents, blood, or other cells (**Table 24.1**). CSF from a healthy person is usually clear, containing no cells. **Pleocytosis** (increased number of white blood cells [WBCs] in CSF) is a common feature of almost any meningitis case. The type of WBCs found may offer a clue as to the cause. Finding polymorphonuclear leukocytes (PMNs) in the CSF suggests a bacterial infection, whereas the presence of lymphocytes typically reflects a viral infection. CSF is best obtained <u>before</u> initiating antibacterial treatment so as to avoid killing bacterial pathogens before the laboratory can grow them. As noted earlier (see Figure 24.4), a spinal tap (lumbar puncture) involves inserting a needle between the third and fourth lumbar vertebrae into the subarachnoid space and collecting samples of the cerebrospinal fluid. Usually four sequential 1-ml sterile samples are obtained and sent to the lab to determine glucose and protein concentration, white and red blood cell count with differential (identifying different types of cells), Gram stain, and cultures as indicated. (We discuss specimen collection and processing in Section 25.1.) The decreased CSF glucose levels in bacterial

Table 24.1
Typical Cerebrospinal Fluid Specimen Values

CSF Specimen	Number of WBCs/μl	Glucose, mg/dl	Protein, mg/dl
Normal[a]	0–5	40–70	15–50
Viral infection	<100, mostly lymphocytes	Usually normal	Usually normal or mildly elevated
Bacterial infection	100 to >1,000, mostly PMNs	Usually decreased	Mildly or markedly elevated
Fungal infection	Varies	Low	Usually mildly elevated

[a]Normal CSF is colorless and clear, with ≤5 white blood cells (WBCs)/μl, primarily lymphocytes, some monocytes, and no neutrophils.

meningitis may be due to blocked glucose transport. Increased CSF protein concentration during a CNS infection is attributed to damage to the blood-brain barrier that results in leakage of protein from blood into the CSF.

Treatment of CNS viral infections is largely supportive and is aimed at reducing inflammation of the brain and intracranial pressure. Some antiviral agents, like ribavirin, are used to treat selected viral encephalitides on an investigative basis, but no medications are currently approved for treatment of viral cases. In contrast, bacterial, fungal, and parasitic infections of the CNS are aggressively treated with antimicrobial agents in combination with supportive measures. Permanent damage or death remains a significant threat with these latter organisms, even with quick diagnosis and immediate treatment. As always, an ounce of prevention is worth a pound of cure.

Vaccination is the best means of prevention. Unfortunately, vaccines are available for only a minority of viruses and bacteria that affect the CNS, and no vaccines exist for parasitic or fungal culprits (as you will see in Tables 24.2 and 24.3). Therefore, preventive measures largely aim to avoid contact between us and potential infectious agents. For example, reducing insect populations by fumigation, venturing outdoors only at certain times to minimize insect or arthropod exposures, and reducing contact with infected people and animals help reduce the incidence of infection. State and local health departments, in cooperation with the Centers for Disease Control and Prevention (CDC), help monitor and publicize outbreaks. This joint effort makes an enormous difference in reducing rates of infection and ensuring that the public knows when to seek medical treatment.

Controlling mosquito populations by means other than fumigation is also being researched. On November 3, 2017, the US Environmental Protection Agency (EPA) approved the release of mosquitoes infected by the bacterium *Wolbachia pipientis* as a means to combat the Asian "tiger mosquito" (*Aedes albopictus*). The Asian tiger mosquito can spread several viruses that can cause CNS infections, such as West Nile virus, dengue virus, and Saint Louis encephalitis virus, among others. *Wolbachia pipientis* is not part of the natural *Aedes albopictus* microbiome, but when introduced into the mosquito it prevents replication of these viruses and shortens the mosquito's life span.

SECTION SUMMARY

- **CNS infections** generally start with flu-like symptoms, including headache, dizziness, and fatigue.
- **Microbial routes to the CNS** can be direct (for example, following insect vector bites) or indirect (via dissemination from a distal infected site).
- **Analysis of the composition of CSF** can aid in diagnosing whether a disease is of bacterial, viral, or other origin. Pleocytosis (increased number of WBCs in CSF) is a sign of meningitis.
- **The three types of CNS infection** are meningitis (inflammation of meninges), encephalitis (inflammation of the brain), and myelitis (inflammation of the spinal cord).

- **Viruses cause most CNS infections,** especially encephalitides, but generally produce mild disease that goes undiagnosed.
- **Treatment of viral CSF infections** is largely supportive. Bacterial CSF infections can be treated with antibiotics.
- **Vaccines** are available for only a limited number of organisms that cause CNS infections.

Thought Question 24.1   What will cause a CSF sample to contain a large number of red blood cells (RBCs)?

24.3 Viral Infections of the Central Nervous System

SECTION OBJECTIVES
- Classify viral CNS infections by their mode of transmission.
- Explain the role of reservoirs in pathogenesis and prevention of viral diseases of the CNS.
- Outline clinical diagnoses of viral CNS infections according to the patient's history, signs, and symptoms.
- Recommend treatment and prevention options for different viral CNS infections.

What are the different kinds of viral CNS infections? Meningitis and encephalitis are the two major categories of viral CNS infections. CNS infections caused by viruses can be extremely dangerous. Fortunately, serious viral infections of the CNS are uncommon and typically resolve favorably in patients with a healthy immune system. Most viruses that attack the CNS tend to infect the meninges and cause what is called **aseptic meningitis** because no bacteria are present. In fact, 30% of patients suffering from herpes simplex virus type 2 (HSV-2) infections initially present with meningitis. But some viruses that attack CNS prefer the brain tissue instead of meninges and produce symptoms of encephalitis. Herpes simplex virus type 1 (HSV-1), for example, causes most fatal cases of viral encephalitis. Herpesviruses are also associated with a strange neurological manifestation called Bell's palsy (see **IMPACT**).

Although viral infections of the meninges and brain tissue are more common, viruses can also attack the myelin sheath of myelinated nerves. Damage to the myelin sheath is irreversible. Table 24.2 summarizes key features of many of the viruses that affect the CNS; some of the more common viral CNS diseases are discussed here. You have already heard about the most famous one: rabies.

Viral Encephalitis

Viral encephalitis is categorized as either primary or secondary. Among the causes of secondary encephalitis are viral meningitis caused by HSV-1, reactivation of latent human herpesviruses (except HHV-8, which causes Kaposi's sarcoma), autoimmune reactions following viral infection, and childhood diseases such as

impact

Why Does Only Half of My Forehead Wrinkle?

How strange must it feel to suddenly start losing control of half of your face? You try to smile, but only half of your face will show it. You are now frowning because you are mad at your face for not taking orders from your brain, but you notice that only half of your forehead will wrinkle. Is this a bad dream? No, worse: it's reality and it's called Bell's palsy, a condition that affects nearly 40,000 people every year in the United States.

What is Bell's palsy? Bell's palsy is weakness or paralysis of the muscles of one side of the face (Figure 1). Some viral infections can result in inflammation of cranial nerve VII (CN VII), the facial nerve. The inflammation damages the nerve so that it no longer can relay messages from the brain to the muscles of the face. So when the affected person wants to smile, only one side of the face (the side with a functioning CN VII) will smile; the other side remains expressionless.

The cause of Bell's palsy is often not known, and the disorder does not disproportionately affect any race or sex. Viral infections are not the only cause of facial nerve palsy. Other causes include autoimmune diseases, trauma, and congenital abnormalities. Among viruses, the one most strongly associated with Bell's palsy is herpes simplex virus type 1 (HSV-1), the same virus that causes cold sores (see Figure 12.4 and Section 19.2). Other members of the herpesvirus family, such as herpes zoster virus, cytomegalovirus, and Epstein-Barr virus, have also been implicated.

Patients feel like their face is pulled to one side and that they may drool. Those affected may be sensitive to sound, lose their sense of taste, have difficulty drinking or swallowing, have food fall out of their mouth, and have difficulty closing one of their eyes. The onset of symptoms is sudden and usually takes no more than 2–3 days to manifest. Although recovery is usually rapid, it can take weeks or even months.

No specific diagnostic tests exist for Bell's palsy, although tests are often run to exclude other causes of facial paralysis (for example, tumors, multiple sclerosis, other brain lesions, or Lyme disease). No specific treatments for this disorder are available, but eye patches and/or eye lubricants (eyedrops or eye ointments) are used to prevent eye dryness, corneal lesions, and corneal scarring.

So what should you do if half of your face suddenly starts feeling numb? You should immediately seek medical care to make sure it's not something a lot more serious, and a lot less exotic, than Bell's palsy.

DISCUSSION QUESTION

Would you expect a powerful anti-inflammatory drug, an antiviral drug, or a combination of the two to be effective therapy for Bell's palsy? Why?

Figure 1 Bell's Palsy

The muscles on the left side of this patient's face do not receive the commands sent from his brain—a result of Bell's palsy.

A. The patient cannot raise his left eyebrow.

B. The patient cannot raise the left corner of his mouth.

C. The patient cannot close his left eye, and the left corner of his mouth is drooping.

Table 24.2

Viral Infections of the Central Nervous System

Virus	Incubation Period	Meningitis[a]	Encephalitis[a]	Transmission	Diagnosis	Treatment	Prevention
Rabies virus	2 weeks–12 months	Rare	+	Mammalian bite or contact	Isolation of virus from clinical samples; serology	Therapy using post-exposure prophylactic rabies vaccine	Avoiding infected mammals; available vaccine is used for post-exposure treatment only
Western equine encephalitis virus	5–10 days	+	+	Mosquitoes	IgM-based serology including ELISA; demonstration of virus or viral antigen[b]	Supportive care; management of symptoms	Avoiding mosquitoes; vaccine available for wild animals; no vaccine for human use
Eastern equine encephalitis virus	5–10 days	Rare	+	Mosquitoes	Clinical presentation; serology; viral culture; postmortem detection of viral antigen in brain	Supportive care; management of symptoms	Avoiding mosquitoes; no vaccine available
Venezuelan equine encephalitis virus	1–5 days		+	Mosquitoes	Clinical presentation; isolation of virus, from blood or throat swab, in culture	Supportive care; management of symptoms	Avoiding mosquitoes; human and horse vaccines available but human vaccine not available in US
West Nile virus	2–14 days	Uncommon	+	Mosquitoes	Clinical presentation; serology; nucleic acid amplification; tests; viral isolation from blood or CSF	Supportive care; IFN-α 2b and IVIG[c]	Avoiding mosquitoes; no vaccine available
Saint Louis encephalitis virus	5–15 days	+	+	Mosquitoes	Clinical presentation of difficulty urinating (dysuria) or tremors in the elderly; leukocytosis with left shift; serology	Supportive care	Avoiding mosquitoes; no vaccine available
Japanese encephalitis virus	5–15 days	+	+	Mosquitoes	Detection of viral-specific antibody in serum or CSF by ELISA; viral isolation	Supportive care	Avoiding mosquitoes; vaccine available for travelers to Asia
Murray Valley encephalitis virus	1–4 weeks		+	Mosquitoes	Isolation of virus or viral RNA in clinical specimen; serology	Supportive care	Avoiding mosquitoes; no vaccine available
Dengue virus	2–7 days		+	Mosquitoes	Serology; qRT-PCR; viral isolation during acute phase of hemorrhagic fever[d]	Supportive care; blood transfusion as needed	Avoiding mosquitoes; vaccine available in some countries (limited effectiveness)
Powassan virus	7–30 days	+	+	Ticks	Clinical presentation; serology; CSF analyses	Supportive care	Avoiding ticks; no vaccine available

(continued on next page)

Table 24.2

Viral Infections of the Central Nervous System (*continued*)

Virus	Incubation Period	Meningitis[a]	Encephalitis[a]	Transmission	Diagnosis	Treatment	Prevention
LaCrosse agent of California encephalitis[e]	5–15 days	Rare	+	Mosquitoes	IgM-capture assays of serum and CSF; CBC with differential shows WBC count of ≥20,000 with a left shift; CSF analyses[f]	Supportive care	Avoiding mosquitoes; no vaccine available
Colorado tick fever virus	7–10 days		+	Ticks	Serology; isolation of virus from blood or CSF	Supportive care	Avoiding ticks; no vaccine available
Coxsackievirus, echovirus	3–10 days	+	Rare	Person to person	Clinical presentation of conjunctivitis and rash; viral isolation from blood, CSF, stool, or respiratory secretions; PCR; serology	Supportive care	No vaccine available
Human immuno-deficiency virus	5–70 days for those with acute HIV infection	+	Meningoen-cephalitis	Person to person	Extensive guidelines for diagnosis[g]; testing criteria different for patients <18 months of age; ELISA, immunofluorescence assay, nucleic acid tests, HIV p24 antigen test, viral cultures	Antiretroviral therapy based primarily on CD4+ cell count and monitoring viral load; prophylactic treatment to prevent opportunistic infections; other treatments as needed	No vaccine available
Herpes simplex virus type 1	Variable, subsequent to primary lesion	Rare	+	Not applicable	HSV lesions on skin or mucous membranes; detection of HSV-1 DNA by PCR	Supportive care in ICU; IV acyclovir[h]	No vaccine available
Herpes simplex virus type 2	2–4 weeks after delivery	+	Rare	From mother to fetus or from mother to newborn during birth	Clinical presentation; PCR of CSF sample; PCR of blood or skin lesions as supportive evidence	Supportive care in ICU; IV acyclovir	No vaccine available
Lymphocytic choriomeningitis virus	1–3 weeks	+		Exposure to rodent secretions or excretions	Viral cultures; serology; ELISA available through CDC; PCR amplification of viral RNA (not commercially available)	Supportive care; anti-inflammatory medications	Rodent control; no vaccine available
Poliovirus	3–35 days	+	+	Fecal-oral; direct person-to-person contact; contact with infected phlegm or mucus	Isolation of virus from throat washings, spinal fluid, or cultures; PCR; serology	Supportive care	Vaccine available

[a]A + sign designates that the virus causes the specified disease.
[b]ELISA = enzyme-linked immunosorbent assay; IgM = immunoglobulin M.
[c]IFN-α = interferon-alpha; IVIG = intravenous immunoglobulin.
[d]qRT-PCR = quantitative reverse transcription–polymerase chain reaction.
[e]California group of viruses.
[f]CBC = complete blood count.
[g]Recommended by Centers for Disease Control and Prevention (CDC), World Health Organization (WHO), and Infectious Diseases Society of America (IDSA).
[h]ICU = intensive care unit; IV = intravenous.

measles and rubella (which are now rare, thanks to vaccination). The deadliest of all infectious diseases—rabies—is a primary viral encephalitis.

RABIES Rabies is a zoonotic disease causing some 30,000–70,000 human fatalities annually worldwide. The agent is an enveloped, single-stranded, negative-sense RNA virus and a member of the genus Lyssavirus in the Rhabdoviridae family (see Chapter 12). Viruses in this family have an elongated bullet shape (**Figure 24.6**) and include two famous members, rabies virus and vesicular stomatitis virus (VSV). VSV is a major pathogen of cattle that can be transferred to humans. VSV infection in humans produces flu-like symptoms and sometimes oral lesions that resemble cold sores, but the infection usually resolves by itself—a stark contrast to rabies.

Rabies virus can infect all mammals, but the endemic reservoir species varies with geographic locale. The highest incidence of human rabies infections is found in rural areas of Asia and Africa, mostly among children. Dogs are the main reservoir on these continents, with dog bites the main route of transmission. In contrast, canine infections in the United States are almost nonexistent because of vigilant animal control programs, vaccination of raccoons and coyotes with oral rabies vaccine for wildlife (Raboral V-RG), and routine vaccination of household pets. The oral vaccine for wildlife is a recombinant vaccine packaged in a fish meal–based bait and distributed into the wildlife habitat. Despite those efforts, 23 confirmed cases of rabies have occurred in the United States between 2008 and 2017. Eight of those were linked to travel outside the United States and its territories. Because rabies is endemic in certain animal populations outside the United States, rabies should be included in the differential diagnosis of any patient who presents with symptoms of encephalitis and who has recently traveled abroad.

Rabies virus is a **neurotropic virus**, meaning it prefers to replicate in nerves. The virus spreads through the body of an infected animal or person and then into the salivary glands in the later stages of the disease. When the infected animal bites someone, the virus is transferred with saliva into the bite wound. The virus replicates in the wound for several weeks before it infects the local nerve; eventually, it travels up that nerve to the spinal cord and ultimately to the brain. Although usually transmitted by animal bite, humans can also acquire the infection through small cuts or abrasions or through mucous membranes that come in contact with an infected animal or their saliva. Contact or even bites from infected animals can be so minor that they don't concern the person. The wound heals and the incident is often forgotten, as happened with the

Figure 24.6 Rabies

A. Rabies virus has an elongated shape frequently referred to as "bullet-shaped."

B. The virus (arrow), shown here in great numbers in hamster brain, initially replicates in the bite wound and subsequently moves along the nerve to the brain. Accumulations of large numbers of the virus in intracellular inclusions are called Negri bodies.

migrant worker in the opening case history. Appearance of symptoms finally drives the patient to seek medical attention.

Initial symptoms of the disease during the prodromal stage are very nonspecific. The area near the wound or contact site is usually (50%–80% of the time) pruritic (itchy), painful, and hypersensitive to any stimulus (hyperesthesia). The patient will feel tired and have a fever and a headache—all typical signs and symptoms of a viral infection. Some patients may have nausea and vomiting, and some may feel and act anxious, even agitated. The wound will heal. After an incubation period that varies, depending on the viral variant (generally 2–3 months), neurological symptoms will ensue and force the patient to seek medical care.

Rabies appears in three basic forms: encephalitic, paralytic, and an atypical form usually associated with bat bites. Most rabies cases (80%) are encephalitic, or "furious"—the form typically portrayed in motion pictures, in which the affected animal or person fears drinking water (hydrophobia), is aggressive, and may bite. In paralytic rabies, the patient exhibits a weakness in all four limbs (quadriparesis) that can be misdiagnosed as Guillain-Barré syndrome (see Section 22.5).

A few rabies victims present atypically with neuropathic pain (pain due to nerve damage, like the shoulder pain in the opening case history), sensory or motor nerve manifestations (our patient had numbness in his left hand), abnormal movements of the bitten limb, and eventual signs of brain stem damage, such as difficulty breathing. Regardless of initial presentation, rabies will progress to coma and finally death.

The rabies virus has several variants that can be identified (as in our opening case history) by using antigenic typing or nucleic acid techniques. Identifying the variant is particularly important because that knowledge allows public health workers to identify the infected animal population. With this strategy, most US cases

Figure 24.7 Silver-Haired Bats

A. The silver color of the fur on their backs has earned these animals the alias "silver-haired bats." Silver-haired bats are small mammals and the main transmitters of rabies virus to humans in the United States.

B. Negri bodies, cytoplasmic particles that contain rabies DNA, can be seen postmortem in brain neurons.

have been linked to contact with silver-haired bats (**Figure 24.7A**). Bites from these bats are also a problem in Mexico and Canada. Foxes were once a major reservoir in Canada, but the distribution of rabies vaccine in bait significantly reduced transmission in that wild population.

Diagnosis of rabies prior to death requires several tests including quantitative reverse transcription–polymerase chain reaction (qRT-PCR) identification of virus in patient's saliva, serum, spinal fluid, skin, and hair follicles. This is followed by testing of the serum and spinal fluid for antibodies to the rabies virus. Diagnosis is confirmed postmortem when Negri bodies are seen in the brain neurons. Negri bodies are cytoplasmic particles (inclusions) that contain rabies RNA (**Figure 24.7B**). Rabies is often misdiagnosed, probably owing to the rarity of cases. The misdiagnosis in organ donors can result in rabies in organ recipients.

Treatment of rabies includes administering rabies immunoglobulin and rabies vaccine. Unfortunately, rabies treatment is successful only if given soon after the exposure and before appearance of the neurological symptoms. Inquiring about animal bites or contacts is therefore very important in patients who complain about sudden onset of symptoms that indicate CNS infection. Health care providers use an algorithm to determine whether a bite victim is at risk of rabies and needs treatment.

MOSQUITO-BORNE ENCEPHALITIS Many of the viruses that cause encephalitis are transmitted to humans by arthropod

vectors—that is, mosquitoes and ticks. We discuss two widely recognized viruses here and summarize others in Table 24.2.

CASE HISTORY 24.1

Death on Golden Pond

In August, Mr. C brought his 21-year-old son, Rich, to a New Jersey emergency department. Rich appeared dazed and had trouble responding to simple commands. When questioned about his son's activities over the past few months, Mr. C told the physician that Rich planned to enter veterinary school in the fall and had spent the month of July relaxing and sunning himself on New Jersey beaches. Aside from the shore, his favorite locale was a pond in a wooded area near a horse farm. On the afternoon before admission, Rich became lethargic and tired. He returned home and went to bed. That evening, his father woke him for supper, but Rich was confused and had no appetite. By 11 p.m., Rich had a fever of 40.7°C (105.3°F) and could not respond to questions. A few hours later, when his father had trouble rousing him, he brought Rich to the ER. Over the next week, Rich's condition worsened to the point where his limbs were paralyzed. Two weeks later, he died.

Serum samples taken upon entering the hospital and a few days before he died showed a sixfold rise in antibody titer to eastern equine encephalitis (EEE) virus. Brain autopsy showed many small foci of necrosis in both the gray and white matter. The serology and autopsy results confirmed the diagnosis of EEE virus infection as the cause of death.

Eastern equine encephalitis (EEE) virus, a member of the Togaviridae family (positive-sense, single-stranded RNA genome), is transmitted from bird to bird by mosquitoes. Horses and humans contract the disease the same way (**Figure 24.8**). Viruses transmitted by insect bites are generally referred to as **arboviruses**. Rich contracted it inadvertently while visiting the pond. The disease, an encephalitis, is often fatal (35% mortality) but fortunately rare. One reason human EEE disease is rare is that the species of mosquito that usually transmits the virus between marsh birds does not prey on humans. But sometimes a human-specific mosquito bites an infected bird and then transmits EEE virus to humans. Another reason human disease is rare is that the virus generally does not fare well in the human body. People infected with EEE virus often have no apparent illness because the immune system thwarts viral replication. However, as already noted, mortality is high in those individuals who do develop disease.

An interesting point in the case history is that diagnosis relied on detecting an increase in antibody titer to the virus. The titer of a specific antibody in serum is determined by how far serum can be diluted and still produce a visible antigen-antibody reaction (see Section 17.2). Serological tests such as enzyme-linked immunosorbent assay (ELISA), discussed in Section 25.4, are used to determine titer. Typically, the body has not had time to generate

Figure 24.8 Eastern Equine Encephalitis Virus

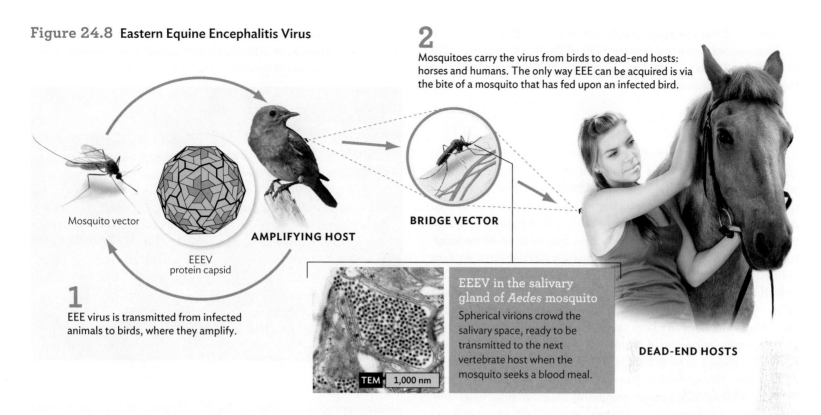

2
Mosquitoes carry the virus from birds to dead-end hosts: horses and humans. The only way EEE can be acquired is via the bite of a mosquito that has fed upon an infected bird.

Mosquito vector

AMPLIFYING HOST

EEEV protein capsid

BRIDGE VECTOR

1
EEE virus is transmitted from infected animals to birds, where they amplify.

EEEV in the salivary gland of *Aedes* mosquito
Spherical virions crowd the salivary space, ready to be transmitted to the next vertebrate host when the mosquito seeks a blood meal.

TEM | 1,000 nm

DEAD-END HOSTS

large amounts of specific antibodies when disease symptoms first appear. After a week or so, often when the patient is nearing recovery (convalescence), antibody titers have risen manyfold. The general rule is that a greater than fourfold rise in antibody titer between acute disease and convalescence (or in this case death) indicates that the patient has had the disease. Although this knowledge could not help the patient in this case history, it is valuable for public health and prevention strategies.

Eastern equine encephalitis virus is one of several viruses transmitted by mosquitoes. Other members of the Togaviridae family and of two other arbovirus families (flaviviruses and bunyaviruses) also cause mosquito-borne primary encephalitis (see Table 24.2). West Nile virus holds an equal place of prominence and causes as much concern as EEE.

WEST NILE ENCEPHALITIS Five major diseases are caused by flaviviruses (positive-sense, single-stranded RNA), all transmitted by mosquitoes. **West Nile virus (WNV)** and three other flaviviruses cause encephalitis or meningitis. The fifth virus, dengue virus, has different serotypes that cause dengue fever, dengue hemorrhagic fever (DHF), or dengue shock syndrome (DSS), as described in Section 21.2. Exposure to any one serotype fails to confer immunity to the others.

West Nile virus (Figure 24.9) is endemic in bird populations and transmitted by mosquitoes. WNV was first discovered in the West Nile district of Uganda in 1937 and remained mainly in eastern Europe, Asia, West Africa, and the Middle East. A European outbreak of West Nile encephalitis in 1962 was followed nearly 40 years later by its arrival and outbreak in New York. The virus then spread through the country. Now only a few states can boast of

being West Nile–free. The strain of West Nile that caused the New York epidemic is almost identical to one found in a goose farm in Israel in 1968, making travel a likely mode of introduction of the virus into the United States.

Most people infected with the virus do not develop disease, thanks to their immune system. Most of those who do develop disease have a flu-like illness, but some develop encephalitis. Rigorous surveillance and insect control programs have reduced cases of WNV infections.

A virus related to WNV, Zika virus, is transmitted from monkeys (the Zika reservoir) to humans by an *Aedes aegypti* mosquito. Zika usually produces mild disease in humans unless it is

Figure 24.9 Molecular Model of West Nile Virus

The genetic material of WNV (RNA) is surrounded by a nucleocapsid (smooth red area), which is enclosed in an outer protein envelope containing attachment proteins (multicolored).

transmitted from a pregnant mother to her fetus. Zika infection of the fetus causes abnormal development of the brain (microcephaly, or small head). There have been no reports of Zika transmission in the United States since 2017.

Viral Meningitis

Viral meningitis is normally not as severe as bacterial meningitis and often tends to be a mild, self-limiting disease in adults with healthy immune systems. The patient with viral meningitis has a headache and a fever, which may be accompanied by nausea and vomiting, photophobia, muscle ache, and malaise. The symptoms of viral meningitis are largely the same as those of meningitis caused by other etiologies and generally persist for 7–10 days before they resolve. Most cases of viral meningitis (80%–85%) are caused by enteroviruses (positive-sense, ssRNA viruses, members of the Picornaviridae family). Poliovirus is a prominent enterovirus that infects the CNS to cause paralysis and meningitis. Non-polio enteroviruses, particularly echoviruses and coxsackieviruses, are the major causes of viral meningitis in both children and adults. Arboviruses (insect-borne viruses) that cause encephalitis, such as EEE and West Nile viruses, commonly cause meningitis as well. Many other viruses, such as influenza and HSV-2, can also cause meningitis. Childhood vaccinations have all but eliminated meningitis cases that were once caused by mumps and measles viruses.

A combination of patient history, epidemiology, and diagnostic information may point to a viral cause as opposed to a bacterial (or other) cause. However, because bacterial meningitis is a severe disease with a bad prognosis, any suspected case of meningitis is initially treated as a bacterial case, and antimicrobial agents are administered empirically.

Meningitis is classified as aseptic if CSF and blood cultures are negative for bacteria. As with encephalitis, most cases of meningitis are viral. "Aseptic meningitis" is an old term used to denote any meningitis that is not bacterial. Viral infections that start as meningitis and then spread to the brain are classified as meningoencephalitis. The specific viral etiologic agent can be determined by subjecting CSF, blood, or sometimes tissues to viral culture, antibody detection, or nucleic acid–based methods.

Treatment for viral meningitis is limited mostly to providing the patient with symptomatic relief, such as reducing pain and fever, and offering supportive care, such as rest in a quiet room with dim lighting and intravenous fluids for patients with prolonged emesis (vomiting). Several antiviral medications are available, but they are not indicated to treat viral meningitis. Acyclovir is the exception. Acyclovir is used empirically in some cases involving neonates and immunocompromised individuals suspected of having viral meningitis, particularly HSV meningitis, but its use is limited to very special cases by highly specialized physicians.

Lymphocytic choriomeningitis virus (LCMV) is an enveloped, single-stranded, negative-sense RNA virus of the Arenaviridae family that can cause aseptic meningitis, encephalitis, or

Figure 24.10 Lymphocytic Choriomeningitis Virus

LCMV is a member of the Arenaviridae family. The glycoprotein protuberances on the enveloped surface of the virus are used to attach to and enter target cells.

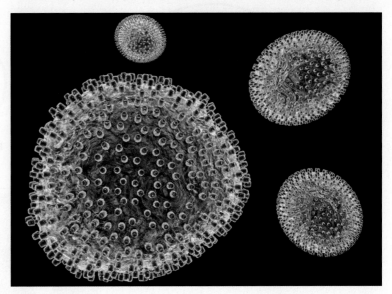

meningoencephalitis (**Figure 24.10**). Household mice are natural reservoirs for the virus, but the virus causes them no harm. Other rodents are not natural reservoirs for this virus but can acquire it from mice. Pet store rodents have also been identified as a source of LCMV. LCMV is shed in the saliva, urine, and feces of the infected rodent for life. The virus is transmitted to humans through inhalation of aerosolized secretions or through direct contact with them.

LCMV meningitis is distinguished from other forms of aseptic meningitis by the unusually elevated number of lymphocytes present in the CSF of the infected person (>1,000/μl). Some infected people will remain well, whereas others develop a flu-like disease with fever, muscle aches, headache, nausea, and vomiting; the disease resolves by itself. A few days after seeming to recover, the patient will suffer a second, more severe phase of the disease that includes fever, a bad headache, and a stiff neck. Acute encephalitis may also be part of the secondary phase of the disease. LCMV patients with acute encephalitis present with sensory or motor deficits, including paralysis, myelitis, and hydrocephalus (increased fluid on the brain). The virus is not transmitted from person to person but can be vertically transmitted from mother to fetus. Vertical transmission in humans, an increased incidence in urban areas, and the severe second-phase symptoms have gained this virus prominence. Patients affected by LCMV are given supportive therapy, mainly analgesics, because currently no effective antiviral treatments are available.

Poliomyelitis

Poliovirus is a very contagious enterovirus of the Picornaviridae family that easily spreads either by the fecal-oral route or directly from person to person via saliva exchange. Once inside a new host, the virus multiplies in the throat or gastrointestinal (GI) tract and soon is shed in the saliva or stool of the infected person. Most infected people remain asymptomatic but still shed the virus and transmit it to others. In a few cases, some of the virions produced in the intestine get into the circulation, eventually reach the CNS, and cause one of three types of **poliomyelitis (polio)**, whose symptoms range in severity. The patient can exhibit mild disease with nonspecific symptoms such as headache and malaise; a nonparalytic form of the disease; or the more serious paralytic polio. Less than 1% of clinical cases of polio are paralytic. The incubation period for any of these diseases can be as short as 3 days or as long as 35 days, but symptoms usually appear after 1–3 weeks.

Patients with mild polio syndrome will have the symptoms typical of most viral infections—fever, malaise, headache, and muscle ache. Nonparalytic disease starts with a prodromal phase with symptoms resembling those of the mild form of disease, which then progresses to aseptic meningitis. Nonparalytic disease lasts up to 10 days before it resolves.

Patients with the paralytic form of polio suffer from unilateral muscle weakness and diminished muscle tone (flaccid paralysis) that reaches its full presentation in about 3 days. Paralytic polio can affect almost any part of the CNS, including the cranial nerves, and cause severe disability or even death. Among the critical nerves that can be affected is the phrenic nerve, which controls the movements of the diaphragm—the muscle that separates the chest from the abdomen and plays a major role in ventilation. Damage to the phrenic nerve will interfere with breathing. During the peak years of polio infections (1940s and 1950s), a negative-pressure ventilator was used to help patients breathe. The machine was dubbed the "iron lung." Three iron lungs are still in use in the United States, helping victims of polio (**Figure 24.11**).

Thirty to forty years after recovery, victims of paralytic polio may experience new weakness or sudden worsening of existing weakness, along with increased muscle pain. These symptoms define a condition called post-polio syndrome (PPS). PPS is not an infection process, and people experiencing the syndrome do not shed poliovirus. The weakness can occasionally spread to muscle groups that were not originally affected. The appearance of PPS is insidious: its progression is slow, and the prognosis depends on severity of the original disease. Patients who originally had a mild case of polio will probably have a mild case of PPS. Patients who suffered from a severe case of polio are at risk of a severe case of PPS. The only available treatment for PPS is physical therapy and management of symptoms.

Diagnosis of polio includes isolation of poliovirus, usually from the stool or less commonly from throat or CSF. However, finding the virus in the stool of a patient exhibiting muscle weakness or paralysis does not prove the patient has polio. Poliovirus can exist in the gut, even in a patient who has been vaccinated against polio. So recovery of the virus from the stool alone is not enough. Other causes of muscle weakness and acute paralysis, such as non-polio enteroviruses, Japanese encephalitis virus, and coxsackievirus A24, have

Figure 24.11 The Iron Lung

Martha Mason, of Lattimore, North Carolina, became paralyzed after contracting polio at age 11. She passed away in 2009 after living in an iron lung for 60 years. Her friend's grandson is shown reading aloud to her during a visit.

to be ruled out. Observing virus-specific changes to infected cell cultures (cytopathic effect) is also very helpful. Treatment for polio patients and those with PPS includes rehabilitation and exercises to help prevent further atrophy of affected muscles. Pain is relieved by use of different pain relievers.

Fortunately, polio has been eradicated from the Western Hemisphere, thanks to vaccination and the efforts of the Pan American Health Organization (vaccination schedule is shown in Table 17.2). Two types of polio vaccine exist. Jonas Edward Salk developed the first polio vaccine, which contains inactivated forms of all three serotypes of poliovirus. The inactivated vaccine does not contain active virus. The **Salk vaccine** must be injected for immunity to develop. The second vaccine, developed by Albert Bruce Sabin, is a live, **attenuated** (diminished virulence) vaccine that contains mutated forms of all three serotypes of the poliovirus. The **Sabin vaccine** is given orally, which affords these mutated viruses a chance to briefly replicate in intestinal cells and generate a more natural immunity. Unfortunately, attenuated viruses can, at times, revert (change back through a mutation) to their active, virulent form. For the Sabin vaccine, the result is **vaccine-associated paralytic poliomyelitis (VAPP)**. Because of VAPP, the Advisory Committee on Immunization Practices, in the year 2000, recommended exclusive use of the inactivated (Salk) polio vaccine in the United States. The last case of VAPP in the United States was seen in 1999. Oral vaccines, however, are still used in some developing countries because they are cheaper than the injectable alternative.

Global eradication of wild poliovirus type 2 was certified in 2015. All polio vaccinations were switched from oral trivalent covering types 1, 2, and 3 polioviruses to bivalent vaccine covering polioviruses types 1 and 3. Wild poliovirus type 3 was certified as globally eradicated in 2019. Attenuated polio virus vaccine strains (Sabin vaccine) can, on rare occasion, revert to virulence and cause polio in the recipient. The affected person can then transmit the virulent strain to others and cause outbreaks in communities with low vaccination rates. Outbreaks like this resulted in a total of 161 cases in Africa and China between January 2018 and June 2019. A vaccination rate of 100% is needed to prevent the outbreaks and maintain the eradication of poliovirus.

Is polio still a significant disease? Visit the HealthMap .org site and do an advanced search for cases of polio worldwide over the past year. The new cases shown are vaccine associated. Vaccination, however, has dramatically reduced the annual numbers of diagnosed cases around the world from hundreds of thousands to around 161. This feat offers hope that one day we will eradicate even the vaccine-derived poliomyelitis disease.

SECTION SUMMARY

- **Viral encephalitis** can occur as a primary infection (rabies) or as a secondary consequence to an infection elsewhere (HSV-1).

- **Rabies virus** has a variety of animal reservoirs. Pathogenesis starts with an animal bite or transfer of saliva from the infected animal to human skin bearing a small cut. The virus then infects nerves in the area and moves along those nerves to the spinal cord and to the brain. In later stages, the virus travels to the salivary glands, where it can be transmitted with saliva by a bite.

- **Symptoms of rabies** are initially vague (headache, fever, nausea). Then 2–3 months later, neurological symptoms begin (hydrophobia, the fear of water; aerophobia, fear of fresh air; aggression; confusion) and may include quadriparesis (weakness of all four limbs). The disease progresses to coma and death. Disease can be prevented if rabies immunoglobulin and rabies vaccine are administered before symptoms begin.

- **Eastern equine encephalitis virus** is an arbovirus endemic to birds. It is transmitted between birds by mosquito vectors, which can also transmit the virus to humans. Disease in humans (encephalitis) is rare but serious.

- **West Nile virus** is also endemic to birds and transmitted by mosquito bite. The disease is similar to EEE but typically less severe.

- **Viral meningitis** (also called aseptic meningitis) is less severe than bacterial meningitis. Major causes are enteroviruses, arboviruses, HSV-2, and lymphocytic choriomeningitis virus. Treatment is supportive.

- **Poliomyelitis** is a very contagious disease caused by poliovirus, an enterovirus of the Picornaviridae family. The disease can manifest in either nonparalytic or paralytic forms; paralytic polio can affect almost any part of the CNS.

Thought Question 24.2 What steps could you take to limit your exposure to arboviruses?

24.4 Bacterial Infections of the Central Nervous System

SECTION OBJECTIVES

- Identify risk factors for each of the bacterial CNS diseases.
- Describe how different bacteria overcome host defenses to cause CNS disease.
- List patient history information and clinical signs and symptoms specific to each bacterial pathogen of the CNS.
- Relate pathogenesis of bacterial CNS diseases to prevention and treatment modalities used.

How do bacteria infect the central nervous system? CNS infections involving bacteria affect the meninges primarily, but from the meninges, bacteria can spread to the brain to cause meningoencephalitis and brain abscesses (**Figure 24.12**). The initial symptoms of bacterial meningitis, headache and fever, are nonspecific, but the disease progresses quickly into a dangerous, life-threatening situation. Infection of the meninges and the resulting inflammation, for example, will interfere with the natural flow of CSF, leading to increased intracranial pressure and brain dysfunction (**encephalopathy**), despite the lack of a brain infection. The very young and the very old are particularly at risk of contracting bacterial meningitis. Effective treatments are available in most cases, but the diagnosis must be made rapidly. Unfortunately, the mortality rate remains significant even with timely diagnosis and treatment. **Table 24.3** lists various bacteria that can cause meningitis.

◉ **Note:** Encephalopathy (disease of the brain tissue) is not the same as encephalitis (inflammation of the brain).

Figure 24.12 Magnetic Resonance Imaging Can Reveal a Brain Abscess

A. MRI of normal brain tissue.   **B.** MRI showing a brain abscess (arrow).

Table 24.3
Bacterial Infections of the Central Nervous System

Bacterial Species	Group at Risk	Transmission	Diagnosis	Treatment[a]	Prevention
Neisseria meningitidis	All ages[b]	From nasopharynx via bacteremia	Clinical presentation; CSF and/or blood culture; Gram stain of CSF; PCR of clinical samples is possible	Penicillin G or ampicillin if penicillin sensitive; ceftriaxone or cefotaxime if penicillin resistant; other alternative antibiotics[c]	Vaccine available
Streptococcus pneumoniae	All ages	Extension from other foci in respiratory tract	Clinical presentation; CSF analyses; bacterial culture, Gram stain; PCR; CBC with differential	Vancomycin + ceftriaxone or cefotaxime; most strains resistant to penicillin	Vaccine available
Streptococcus agalactiae (group B streptococci)	Neonates (infants <3 months of age)	Vertical transmission from mother; infection of mucous membranes causing sepsis and spreading to CNS	Clinical presentation; CSF analyses; blood culture	Antibiotic treatment different for neonates and premature infants[d]	No vaccine available
Listeria monocytogenes	Neonates and elderly people	Vertical transmission through placenta; fecal-oral route	Clinical presentation; CSF analyses; CSF, blood, or amniotic fluid culture; Gram stain	Ampicillin or penicillin G[c]	No vaccine available
Neonatal meningitis *Escherichia coli*	Neonates and elderly people	Various	Clinical presentation; CSF analyses; CSF and/or blood culture; Gram stain	Third-generation cephalosporins; alternatively aztreonam, fluoroquinolone, meropenem, trimethoprim-sulfamethoxazole, or ampicillin	No vaccine available
Pseudomonas aeruginosa	Hospitalized patients; immunocompromised people	Bacteremia	Clinical presentation; CSF analyses; CSF and/or blood culture; Gram stain	Cefepime or ceftazidime, aztreonam, ciprofloxacin, meropenem, an aminoglycoside antibiotic	No vaccine available
Coagulase-negative *Staphylococcus*	All ages	Bacteremia originating from a foreign body or wound	Clinical presentation; CSF analyses; CSF and/or blood culture; Gram stain	Nafcillin or oxacillin if methicillin susceptible; vancomycin or linezolid if methicillin resistant	No vaccine available
Staphylococcus aureus	All age groups	Direct and indirect contact and hematogenous spread	Clinical presentation; CSF and/or blood culture; Gram stain	Oxicillin or naficillin; vancomycin if methicillin resistant; other antibiotics[e]	No vaccine available
Haemophilus influenzae	Unvaccinated adults and children	Extension from upper respiratory tract foci	Clinical presentation; CSF analyses; CSF and/or blood culture; Gram stain	Ampicillin, ceftriaxone, cefotaxime; third-generation cephalosporins if beta-lactamase-positive	Vaccine available
Mycobacterium tuberculosis	HIV-infected people	Disseminated tuberculosis	Clinical presentation; CSF analyses; CSF and/or blood culture; chest X-ray	Variable[f]	No vaccine available
Treponema pallidum (syphilitic aseptic meningitis)	Syphilis patients	Vertical transmission causing congenital syphilis; complication of syphilis	Clinical presentation; CSF analyses; cerebral angiography; VDRL or RPR blood test; FTA-ABS test to confirm syphilis[g]	Variable[f]	No vaccine available

(continued on next page)

Table 24.3

Bacterial Infections of the Central Nervous System (continued)

Bacterial Species	Group at Risk	Transmission	Diagnosis	Treatment[a]	Prevention
Borrelia burgdorferi	All ages	Complication of Lyme disease[h]	Clinical presentation; CSF analyses	IV antibiotic[i]; ceftriaxone, cefotaxime, and penicillin G usually used	No vaccine available
Bacteroides fragilis	Immunocompromised people	Bacteremia	Clinical presentation; CSF analyses	Metronidazole	No vaccine available
Fusobacterium spp.	Immunocompromised people	Bacteremia	Clinical presentation; CSF analyses	Metronidazole	No vaccine available

[a]Suspected cases of meningitis are treated empirically until etiology is determined. Empirical antibiotic therapy can include a corticosteroid, combination of a ceftriaxone and cefotaxime + vancomycin + doxycycline (for bacterial tick-borne etiology during tick season) and acyclovir.

[b]Exposure to infected individuals requires prophylactic antibiotic treatment.

[c]Per Infectious Diseases Society of America (IDSA) guidelines.

[d]Per extensive guidelines from the American College of Obstetricians and Gynecologists and American Academy of Pediatrics.

[e]Antibiotic choice is based on the resistance profile of the infecting strain.

[f]Per Centers for Disease Control and Prevention (CDC) guidelines.

[g]Angiography = a procedure in which a dye is injected into the blood to study blood and/or lymph vessels by X-ray; FTA-ABS = fluorescent treponemal antibody test; RPR = rapid plasma reagin; VDRL = Venereal Disease Research Laboratory.

[h]Lyme disease is an autoimmune disease, not an actual infection of the CNS.

[i]Antibiotic choice and dosage depend on patient's age.

Bacterial Meningitis

Meningitis is considered a medical emergency. Hallmark symptoms include sudden onset of severe headache accompanied by neck pain, fever, and **nuchal rigidity** (stiffness of the neck resulting in an inability to touch chin to chest). Meningitis can be caused by bacteria, viruses, fungi, and parasites, but symptoms alone cannot determine the cause. Patient history can provide clues (for instance, was the patient exposed to a certain agent?), but laboratory results are needed for a definitive diagnosis.

Because bacteria cause the most serious forms of meningitis, any person presenting with the aforementioned symptoms is presumed to have bacterial meningitis until proven otherwise and is immediately treated empirically with broad-spectrum antibiotics (Chapter 13). If it is determined that the offending agent is not bacterial, antibiotic treatment can be stopped. The severity of bacterial meningitis is due in part to the fact that bacteria can cause other systemic, life-threatening problems in addition to inflammation of the meninges. *Neisseria meningitidis,* for example, can cause sepsis and disseminated intravascular coagulation. Complications are a major reason antimicrobial treatment of meningitis too often fails.

Three bacteria—*Streptococcus pneumoniae, Neisseria meningitidis,* and *Haemophilus influenzae* type b—have historically received top billing as major etiologies of bacterial meningitis. *S. pneumoniae* is the number one cause of meningitis in infants, children, and the elderly, whereas *N. meningitidis* most commonly causes the disease in adolescents and young adults. *H. influenzae* type b may cause meningitis in young children. Aggressive vaccination programs, however, have significantly reduced the number of meningitis cases caused by these organisms (see Section 20.3). *Listeria monocytogenes,* which has no vaccine, is a relatively new arrival to the scene. *Listeria* infections are more common in pregnant women and their newborns and in elderly and immunocompromised patients.

As noted earlier, patient history—particularly travel and exposure history—as well as findings from the physical exam can help suggest causes of bacterial meningitis, but a definitive diagnosis requires laboratory analysis of CSF samples. CSF from cases of bacterial meningitis will contain increased numbers of white cells (PMNs), an increase in protein content, and a decrease in glucose concentration (all of which argue against viral causes). Definitive diagnosis is based on identifying the bacterial agent from CSF and/or blood. Because CSF is drawn from a sterile body site, a quick, direct Gram stain of CSF material can be extremely informative. For instance, finding a Gram-positive diplococcus in CSF strongly suggests infection by *S. pneumoniae.* Gram-negative diplococci or Gram-negative rods would suggest *N. meningitidis* or *H. influenzae,* respectively. Even though the results are quick and strongly indicate cause, Gram staining must be followed by culture and identification (see Section 25.2).

CASE HISTORY 24.2

The Purple Rash

In April 2001, Abdul, a 4-month-old infant from Saudi Arabia, was hospitalized with fever, tender neck, and purplish, nonblanching spots (petechiae and purpuric spots) on his trunk (**Figure 24.13A**). Suspecting meningitis, the clinician took a CSF sample and examined it by Gram stain. The smear revealed Gram-negative diplococci inside PMNs. The CSF looked cloudy with 900 leukocytes (white blood cells)/µl. The presence of *Neisseria meningitidis* in CSF was confirmed by culture. The child was treated with cefotaxime and made a full recovery. His father, the person who brought Abdul in, was clinically well. However, *N. meningitidis* was isolated from his throat (oropharynx), as well as from the throat of the patient's 2-year-old brother. Surprisingly, the father had previously been vaccinated against four serotypes of *N. meningitidis* (A, C, Y, and W135). All three specimens were sent to the WHO (World Health Organization), which confirmed meningococcus serotype W135. DNA analysis of the *N. meningitidis* in the three specimens found them to be indistinguishable, meaning that the father and his children were infected with the same strain of *N. meningitidis*.

MENINGOCOCCAL MENINGITIS The meningococcus *Neisseria meningitidis* (shown in **Figure 24.13B**) is a Gram-negative diplococcus that can colonize the human oropharynx (throat), where it causes mild, if any, disease. At any given time, 10%–20% of the healthy population can be colonized and asymptomatic, like Abdul's 2-year-old brother. The organism spreads directly by person-to-person contact or indirectly via fomites or droplet nuclei from sneezing. In the throat, *N. meningitidis* uses type IV pili to adhere to and then invade epithelial cells. The organism moves from one side of the host cell to the other (**transcytosis**), exits the epithelial cell, and enters the bloodstream, where the real problem arises.

Meningococci have a modified form of outer membrane lipopolysaccharide, called lipooligosaccharide (LOS), that lacks the O antigen side chains of typical Gram-negative LPS but retains endotoxic activity (Section 18.4). As the bacterium grows, it releases fragments of its outer membrane, called blebs, into the host circulation. Because blebs contain LOS endotoxin, they trigger the release of cytokines that increase capillary permeability. LOS also causes the depletion of clotting factors. The combination of massive release of cytokines and depletion of clotting factors causes blood to leak from capillaries under the skin and produces pinpoint red spots called **petechiae**. Petechiae may enlarge and merge to form a purplish patch called a **purpura;** both were seen on Abdul's skin. This type of rash does not lose color when pressed (it is nonblanching).

Figure 24.13 Bacterial Meningitis

Meningococcal disease is dreaded by parents and medical practitioners alike for its rapid onset and the difficulty of obtaining a timely and accurate diagnosis.

A. Purpuric spots produced by local intravascular coagulation due to *Neisseria meningitidis* endotoxin. The rash in meningitis typically has petechial (small) and purpuric (large) components.

B. *N. meningitidis.*

C. Normal brain.

D. Autopsy specimen of a brain affected by meningitis due to *Streptococcus pneumoniae*. Note the greening of the brain, compared with the pink normal brain.

Unlike the gonococcus *Neisseria gonorrhoeae* (see Section 23.4), *N. meningitidis* is highly resistant to complement, owing to its production of a polysaccharide capsule. Complement, part of our innate immune system, puts a hole in the bacterial membrane in order to destroy it (see Section 15.7). The meningococcus capsule prevents complement from accessing the bacterial membrane, which allows the microbe to produce a transient blood infection (bacteremia) and reach the blood-brain barrier.

Sinus and ear infections with *N. meningitidis* can extend directly to the meninges, whereas septicemic spread requires passage through the blood-brain barrier. Once confronted with the blood-brain barrier, *N. meningitidis* again uses type IV pili to adhere to endothelial cells but does not enter the host cell. Instead, the interaction loosens the tight junctions between cells and disrupts the blood-brain barrier. The organism then squeezes through these spaces (like a child squeezing through the hole in a fence) and encounters the meninges. Once in the CSF, microbes can multiply almost at will. **Figure 24.13C and D** shows the remarkable damage that *N. meningitidis* and other microbes, such as *Streptococcus pneumoniae*, cause in the brain.

Several antigenic capsule types, called type-specific capsules, are produced by different strains of pathogenic *N. meningitidis*: types A, B, C, W135, and Y. Types A, C, Y, and W135 are usually associated with epidemic infections seen among people kept in close quarters, such as college students or military personnel. Type B meningococcus is typically involved in sporadic infections. However, large outbreaks of type B in the United States occurred at Princeton University and the University of California, Santa Barbara, in December 2013.

Antibodies to capsular antigens are used to type the organisms causing an outbreak. Knowing the capsular type of organism involved in each case helps determine whether the disease cases are related and where the infection may have started. In Abdul's case, a polyvalent reagent containing antibodies to four capsular types was used to confirm *N. meningitidis*. The test could not distinguish between the four serotypes tested. Further testing was necessary to see whether the strains from the father and his children were the same.

Meningococcal meningitis is highly communicable by the respiratory route. Transmission is most common among people in closed communities, such as families, college dormitories, and military barracks. As a result, close contacts, such as the parents or siblings of any patient with meningococcal disease, should receive antimicrobial prophylactic treatment within 24 hours of diagnosis; a single dose of ciprofloxacin (a quinolone antibiotic, discussed in Section 13.5) can be given to adults, and 2 days of rifampicin treatment can be given to children.

A tetravalent vaccine containing the A, C, Y, and W135 capsular structures is available to prevent disease by these strains. Because of growing concern about the transmission of meningitis among teenagers, the tetravalent vaccine is now recommended for children between ages 11 and 12 and is required by many colleges. Vaccination does not prevent colonization (as with Abdul's father in Case History 24.2) or prevent spread of the bacteria from the colonized person to others.

Vaccine to group B capsule is available and is recommended for those at risk, including those without a spleen, people with complement component deficiency, or patients taking complement inhibitor drugs. The CDC also recommends the vaccine for microbiologists who work with *N. meningitidis*, people exposed during an outbreak of *N. meningitidis* type B in their community, and young adults 16–23 years of age. Two meningococcal type B vaccines are available. These vaccines contain outer-membrane vesicles and proteins from *N. meningitidis* type B but not its capsular antigen. Many colleges now include vaccination against *N. meningitidis* type B as an admissions requirement.

Although the capsule can protect the meningococcus against serum complement, once antibody against the capsule is made, the organism is easily killed via the classical complement cascade (see Section 15.7). This is important because patients with complement deficiencies are susceptible to repeated *N. meningitidis* (or *S. pneumoniae*) infections. As a result, children as young as 2 years old with diagnosed complement deficiencies are given the meningitis vaccine to boost other protective immune mechanisms.

LISTERIOSIS *Listeria monocytogenes* seems like just another food-borne pathogen that causes a few days of diarrhea and fever. That is true if you have a competent immune system. However, newborns and elderly and immunocompromised people who encounter *L. monocytogenes* are at increased risk of a much more invasive disease course that includes meningitis, rhombencephalitis (encephalitis of the brain stem), and brain abscess.

L. monocytogenes is a Gram-positive, facultative anaerobic rod and a facultative intracellular pathogen that is a major cause of zoonotic disease. Disease caused by *L. monocytogenes* is called **listeriosis** (see Sections 18.3 and 22.5 and Case Histories 18.1 and 22.4). *Listeria* is able to survive and grow in cold temperatures (Section 22.5); outbreaks of listeriosis are often linked to foods that were kept in the refrigerator. Eating contaminated foods, especially soft foods such as some unpasteurized soft cheeses, can introduce the organism into our GI tract. In the intestine, the organism invades epithelial cells, breaks out of the engulfing phagosome, and moves in and between cells by using host cell actin tail formation, much like *Shigella* (see Figure 18.24). *Listeria* invades macrophages and disseminates in the bloodstream, where it can encounter the blood-brain barrier. *L. monocytogenes* causes up to 10% of all community-acquired bacterial meningitis and has a high mortality rate (up to 26%).

Listeriosis during pregnancy is associated with a very high percentage of fetal infection (up to 90%) and fetal death (stillbirth or spontaneous abortion). The death rate of infected, live-born infants is much lower than that of their fetal counterparts but still is about 20%. Pregnant women who have listeriosis will have fever and often a backache and do not feel well. *L. monocytogenes* infection must be investigated by taking a detailed dietary history from any pregnant woman who has a fever and flu-like symptoms. Diagnosis of meningitis or neonatal infection by *Listeria* can be made by isolating the organism from blood or CSF cultures. No vaccine is available, but antibiotic treatment (ampicillin) is effective.

NEONATAL MENINGITIS Pathogens are usually transferred from mother to newborn through contact with contaminated vaginal secretions during birth and sometimes by directly crossing the maternofetal barrier before birth (*Listeria*, for instance, can infect the newborn by either of these routes). Neonatal meningitis can also be nosocomial (hospital-acquired) or community-acquired after birth.

Meningitis in patients less than a week old is often attributable to transmission of bacteria during birth. *Neisseria meningitidis* is rarely the cause because the microbe is not found in vaginal secretions, nor does it cross the maternofetal barrier. Rather, neonatal meningitis commonly involves infection by *Escherichia coli* (NMEC) and group B streptococcus (*Streptococcus agalactiae*). The American College of Obstetricians and Gynecologists and American College of Nurse-Midwives recommend that pregnant women be tested for group B streptococcus (GBS) when they are 35–37 weeks pregnant. Women who test positive for GBS are then treated with an antibiotic during labor to prevent GBS neonatal meningitis. Other Gram-negative organisms (such as *Klebsiella*) and some Gram-positive cocci (such as *Enterococcus*, *Staphylococcus aureus*, and *Listeria monocytogenes*) can cause meningitis in the newborn as well.

The infected infant will have neurological problems such as lethargy, irritability, poor muscle tone, and (occasionally) febrile seizures due to an abnormally high body temperature. Infants with neonatal meningitis feed poorly and may have nausea and vomiting, diarrhea, and even difficulty breathing. Definitive diagnosis is made on the basis of Gram stain and culture of a CSF sample, but infants suspected of having meningitis are empirically treated with ampicillin and an aminoglycoside antibiotic until the offending agent is identified. Lumbar puncture is repeated at 24–48 hours after treatment begins to make sure that sterility of CSF is restored.

The incidence of neonatal meningitis has remained relatively unchanged for the past 40 years, but the mortality rate has dropped significantly. The prognosis seems to depend mainly on the type of organism involved, whether the infant is preterm or full term, and the age at diagnosis. Neurological sequelae such as seizures and developmental delays continue to characterize a high percentage of long-term outcomes.

Bacterial Neurotoxins That Cause Paralysis

Imagine a disease that causes complete loss of muscle function. Using secreted exotoxins, two microbes cause lethal paralytic diseases. In one instance, the victim suffers a flaccid paralysis in which the muscles, including muscles of respiration, go limp, making breathing difficult. Voluntary muscles fail to respond to the mind's will because the botulinum toxin interferes with neural transmission. This disease, called **botulism**, is typically food-borne and is caused by an obligate anaerobe, the Gram-positive, spore-forming bacillus named *Clostridium botulinum*.

CASE HISTORY 24.3
Botulism—The Vicious Green Beans

Mitch, a 47-year-old resident of Oklahoma, was admitted to the hospital with rapid onset of progressive dizziness, blurred vision, slurred speech, difficulty swallowing, and nausea. Findings on examination included drooping eyelids, facial paralysis, impaired gag reflex, and no fever. He developed breathing difficulties and required mechanical ventilation. The patient reported that during the 24 hours before onset of symptoms, he had eaten home-canned green beans and a stew containing roast beef and potatoes. Analysis of the patient's stool detected botulinum type A toxin, but no *Clostridium botulinum* organisms were found. The patient was hospitalized for 49 days, including 42 days on mechanical ventilation, before being discharged.

Paralysis is among the most feared symptoms of CNS damage. Two bacteria of the same genus produce two very potent neurotoxins, botulinum and tetanus toxins, that cause paralysis and often lead to death. These toxins are produced by *Clostridium botulinum* and *Clostridium tetani*, respectively. The two toxins share 30%–40% identity. They have similar structures and nearly identical modes of action. Each toxin is initially made as a single peptide that is cleaved after secretion to form two fragments, heavy and light chains (analogous to the AB toxins; Section 18.4). These heavy and light chains remain tethered to each other by a disulfide bond (Figure 24.14). Each heavy chain has a domain that binds to a receptor molecule (ganglioside) on the nerve cell membrane and a translocation domain that makes a pore in the nerve cell. The toxin (the light chain) then passes through the pore and into the cell. The light chains are proteases that disrupt the intracellular movement of exocytic vesicles containing neurotransmitters needed for muscles to contract or relax.

Figure 24.14 Basic Structure of Tetanus and Botulinum Toxins

1 The binding domain binds to receptor molecules (gangliosides) of the nerve cell.

2 The translocation domain makes a pore for passage of the toxin.

3 The protease toxin disrupts release of neurotransmitter.

COO⁻
Ganglioside-binding domain
Heavy chain
Translocation domain
Light chain
Zn   Catalytic domain
NH₂

Schematic diagram of the domains of tetanus and botulinum toxins after cleavage.

Both botulinum and tetanus toxin enter at peripheral nerve ends where nerve meets muscle (the neuromuscular junction). The botulinum toxin binds to target membranes and is brought into the cell by **endocytosis**. The low pH that forms in the endosome cleaves the disulfide bond holding the two parts of the toxin together. Next, a channel is assembled through the membrane and the light chain (toxin) is released into the cytoplasm. The toxic subunit then cleaves key host proteins involved in the **exocytosis** of vesicles containing neurotransmitters (**Figure 24.15**), disrupting release of the neurotransmitters and causing disease.

With such a high degree of mechanistic similarity, why do tetanus and botulinum toxins have such drastically different effects? The answer, as you will see, is that they act on two different types of neurons. Botulism is an intoxication resulting from the ingestion of preformed toxin in poorly prepared, contaminated foods (whether the organism is also ingested is immaterial), as in Case History 24.3. Once botulinum toxin enters a peripheral nerve, it will immediately cleave peptides that help with exocytic release of the neurotransmitter acetylcholine (see Figure 24.15D). Without acetylcholine to activate nerve transmission, muscles will not contract and the patient is paralyzed. Botulism is a disease of the peripheral nervous system.

Occasionally, germination of *C. botulinum* in an infected wound or in the anaerobic environment of the GI tract will allow these organisms to produce exotoxin and cause disease. In food, the spores of *C. botulinum* germinate, and the cells that emerge produce toxin. Because the organism is an anaerobe, an anaerobic environment must be present. Home-canning processes are designed to remove oxygen in the canned food as well as sterilize the food. Sometimes the sterilization is not complete but all the oxygen is removed—a perfect environment for the surviving spores to germinate and for *C. botulinum* cells to produce exotoxin. After ingestion, the toxin is absorbed from the intestine. As in Case History 24.3, the organism is usually absent from stool samples, which is why serological identification of the toxin is required.

A rare form of botulism, called infant botulism or "floppy head syndrome," can arise when infants accidentally ingest soil containing *C. botulinum* spores or are fed honey that may harbor the spores. Ingested spores can potentially germinate in the GI tract of infants younger than 12 months of age. The vegetative cells that grow in the intestine and secrete toxin will cause "floppy head syndrome." Incomplete colonization of the GI tract with microbiota and a less-than-optimal immune system may explain why only very young infants are susceptible to this condition.

Although botulism is now rare, the botulinum toxin has gained renewed interest because it is considered a select biological toxic agent of potential use to bioterrorists. Botulinum toxin is considered one of the deadliest poisons in the world. The toxin can be absorbed through the eyes, mucous membranes, respiratory tract, or cuts in the skin. Commercial interest, however, is even higher. Botulinum toxin can safely relax muscles in a localized area if injected in small doses. Botulism toxin administered in this

Figure 24.15 Mechanism of Action of Botulinum Toxin

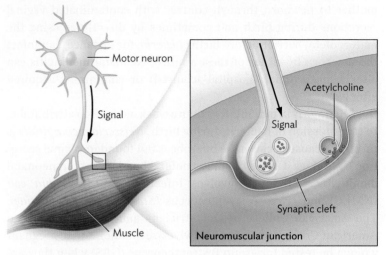

A. The neuromuscular junction. Inset shows vesicles filled with neurotransmitters.

B. Toxin binds via the heavy chain.

C. Toxin is endocytosed into the nerve terminal.

D. The small subunit of the toxin enters the cytoplasm, cleaves its target, and makes the nerve terminal unable to release acetylcholine.

1 Disulfide bond is reduced; heavy chain binds to membrane.

2 Light chain enters cytoplasm.

3 Light chain cleaves proteins, blocks exocytosis.

manner is known commercially as Botox; it is used cosmetically by plastic surgeons to reduce facial wrinkles and to treat migraine headaches.

In striking contrast to botulism is **tetanus**, a very painful disease in which muscles continuously and involuntarily contract (called

Figure 24.16 Tetanus and Tetany

A. Photomicrograph of *Clostridium tetani*.

B. Painting from Charles Bell that depicts the spasms associated with tetanus.

Figure 24.17 Retrograde Movement of Tetanus Toxin to an Inhibitory Neuron

Tetanus toxin enters the nervous system at the neuromuscular junction and travels up the axon to an inhibitory neuron of the central nervous system. There the toxin cleaves proteins associated with the exocytosis of vesicles containing inhibitory neurotransmitters. In the absence of inhibitory signals (gamma-aminobutyric acid [4-aminobutanoic acid], GABA), excitatory signals from acetycholine (ACh) prevail, stimulating muscle contraction.

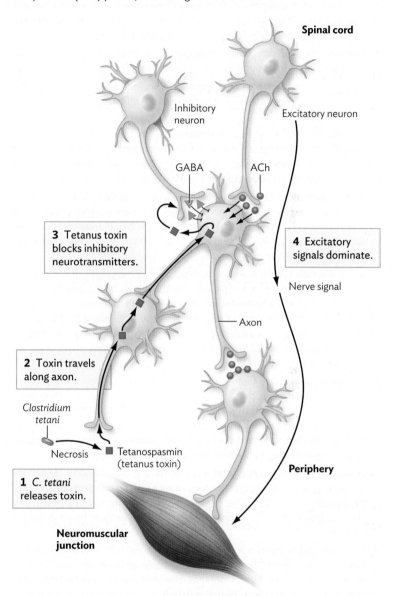

tetany or spastic paralysis). Tetanus is caused by **tetanospasmin**, a potent exotoxin made by another species of *Clostridium, C. tetani* (**Figure 24.16A**). Tetanospasmin interferes with neural transmission, but in contrast to botulinum toxin, it causes <u>excessive</u> nerve signaling to muscles, forcing the victim's muscles to contract. The jaw muscles (masseters) are the first to be affected, giving the disease tetanus its common name, "lockjaw." Eventually, other muscles are affected; the back will arch grotesquely while the arms flex and the legs extend (**Figure 24.16B**). The patient remains locked this way until death. Spasms can be strong enough to fracture the patient's vertebrae. In both botulism and tetanus, death can result from asphyxiation.

Tetanus, unlike botulism, is not a food-borne disease. *C. tetani* spores are introduced into the body by trauma (such as by stepping on a dirty nail). Necrotic tissue then provides the anaerobic environment required for germination. Growing, vegetative cells release tetanospasmin, which enters the peripheral nerve cells at the site of injury. The toxin travels up axons toward the spinal column, where it becomes fixed on presynaptic inhibitory motor neurons (presynaptic neurons communicate with other neurons and not with the muscle directly). The function of vesicles in these neurons is to release <u>inhibitory</u> neurotransmitters that dampen nerve impulses. Tetanus toxin blocks release of the inhibitory neurotransmitters GABA (gamma-aminobutyric acid [4-aminobutanoic acid]) and glycine into the synaptic cleft, which leaves excitatory nerve impulses unchecked. As a result, excitatory impulses come too frequently and produce the generalized muscle spasms characteristic of tetanus (**Figure 24.17**).

Botulism and tetanus are preventable diseases. No vaccine exists for botulism; therefore, proper cooking and canning of food is essential to prevent botulism. Tetanus vaccine (a toxoid) is part of routine childhood vaccinations. The vaccine does not confer lifelong immunity, and a booster is needed every 10 years. People who may have been exposed to the bacterium—for example, by cutting themselves on a sharp object—have their wounds debrided and are given a booster shot.

Antibiotics are of no use in treating either disease because bacteria themselves are not the cause. Injection of antitoxins and supportive care to the patient constitute the mainstay treatment for both diseases.

SECTION SUMMARY

- **Hallmarks of meningitis** include severe headache accompanied by neck pain, nuchal rigidity, and fever followed by lethargy and confusion. Empirical antibiotic treatment is imperative when meningitis is suspected.

- **The neurological deficits of encephalitis,** such as uneven gait, do not occur unless the infection has spread to the brain (meningoencephalitis).

- **Common agents of bacterial meningitis** include *Neisseria meningitidis, Streptococcus pneumoniae, Haemophilus influenzae,* and *Listeria monocytogenes.*

- **CSF results indicating bacterial meningitis** include pleocytosis (usually neutrophils), increased protein concentration, decreased glucose concentrations, and a Gram stain showing bacteria. Definitive diagnosis, however, requires isolation of the organism from CSF.

- *N. meningitidis,* a Gram-negative diplococcus, spreads from person to person by aerosolization, colonizes the oropharynx, and occasionally penetrates the epithelium to gain entrance to the bloodstream. Vaccination with *N. meningitidis* A, C, Y, and W135 capsular antigens is recommended for children between ages 11 and 12.

- *N. meningitidis* **type B** is an uncommon cause of meningitis that can affect people without a spleen, people with complement component deficiency, or patients taking complement inhibitor drugs. The CDC also recommends the vaccine for microbiologists who work with *N. meningitidis,* people exposed during an outbreak of *N. meningitidis* type B in their community, and young adults 16–23 years of age. Two different types of vaccines are available for this organism.

- *L. monocytogenes,* a Gram-positive rod, is a food-borne, facultative intracellular pathogen that can survive and grow in cold temperatures. *L. monocytogenes* penetrates the intestinal mucosa, invades macrophages, and disseminates through the bloodstream. The organism typically causes mild diarrhea but can cross the blood-brain barrier and infect meninges in very young, very old, and immunocompromised patients.

- **Common agents of neonatal meningitis** are *Streptococcus agalactiae* (group B streptococcus, transmitted during birth), *Escherichia coli* (NMEC, transmitted after birth), and *L. monocytogenes* (transmitted through the placenta before birth, by vaginal secretions during birth, or by other means after birth).

- **Botulism and tetanus** are toxigenic diseases caused by toxins from the Gram-positive, spore-forming bacteria *Clostridium botulinum* and *C. tetani,* respectively. These toxins prevent the release of neurotransmitters from nerves. In botulism, flaccid paralysis arises from lack of excitatory neurotransmitters (acetylcholine); in tetanus, spastic paralysis arises from lack of inhibitory neurotransmitters (GABA).

Thought Question 24.3 How do the actions of tetanus toxin and botulinum toxin help the bacteria colonize or obtain nutrients?

Thought Question 24.4 If *Clostridium botulinum* is an anaerobe, how might botulinum toxin get into foods?

24.5 Fungal Infections of the Central Nervous System

SECTION OBJECTIVES

- Name fungi that cause CNS disease.
- Describe the role of host immunity in fungal infections of the CNS.
- Define fungal CNS infections in the context of environmental exposure risks.
- Identify diagnostic methods and treatment options for different fungal CNS infections.

Fungal infections (**mycoses**; singular, **mycosis**) of the CNS are rare and are caused mostly by dissemination of a primary infection elsewhere in the body. People most commonly affected have a compromised immune system (as in AIDS patients or in the elderly) or have undergone a neurosurgical procedure to implant a shunt, a device used to reduce intracranial pressure. The shunt can serve as a conduit into the brain for bacterial and fungal pathogens. Patients who suffer from intractable back pain and are not candidates for surgery are often given a spinal steroid shot to relieve their pain. In 2012, a contaminated batch of steroid medications was injected into the spinal columns of some of these patients, which resulted in a multistate outbreak of fungal meningitis. As with neurological procedures, spinal injection of contaminated pharmaceuticals bypassed the hosts' innate immune defenses and caused meningitis.

The most common fungus to invade the CNS is *Candida*, particularly *Candida albicans*. CNS infections caused by *Candida* include meningitis and brain abscesses. **Candida meningitis** is often seen in premature neonates, whereas brain abscesses occur predominantly in immunocompromised patients of any age. Diagnosis is made by isolating *Candida* from the CSF. Treatment depends on the individual patient's circumstances and includes use of antifungal agents such as amphotericin B, fluconazole, and echinocandins.

Coccidioides immitis and *C. posadasii* are fungi that grow in dry desert soil. *Coccidioides* species are endemic in the southwestern United States. The dry climate makes their mycelia easily breakable. Slight air movements break the dry mycelia into single-celled spores called **arthroconidia** (singular, **arthroconidium**) that can float in air for extended periods. Inhaling the spores causes a primary fungal infection, a pneumonia called valley fever, 1–3 weeks after exposure (see Section 20.4). Fatigue and joint pain lasting for extended periods (months) may be the only symptoms of the infection, giving it the nickname "desert rheumatism." The fungus can disseminate and cause meningitis after a delay lasting weeks or months.

The symptoms of **coccidioidal meningitis** are very nonspecific, with a persistent headache being the most prominent. *Coccidioides* should be in the differential diagnosis when a patient with severe, progressing headache, blurred vision, and altered mental status has a history of travel to arid areas. The travel history does not need to be extensive. For instance, a short stopover (hours to days) in Phoenix, Arizona, may qualify. Coccidioidal meningitis is extremely lethal unless it is diagnosed and treated quickly; about 95% of patients who are not treated die within 2 years. The definitive diagnosis is made by isolating the fungus from the CSF. Patients with confirmed coccidioidal CNS infection are treated with fluconazole or itraconazole for life.

Cryptococcus neoformans var. *neoformans* (abbreviated as *C. neoformans*) is a fungus often present in pigeon and chicken droppings that finds its way into the soil. *C. neoformans*, notable for its thick, protective capsule, can spread to the CNS through the bloodstream after initially infecting the respiratory tract when inhaled. Extension of the disseminated fungus through the bloodstream often causes meningoencephalitis, especially in immunocompromised patients. More than 80% of cryptococcus meningitis cases are associated with AIDS, making **cryptococcal meningoencephalitis** an AIDS-defining infection. The incidence of the disease has decreased in HIV patients who have access to antiviral medication (because their immune systems recover), but worldwide nearly a million cases are diagnosed annually, with a 60% mortality rate.

Another variety of cryptococcus, *Cryptococcus neoformans* var. *gattii* (abbreviated as *C. gattii*), is an emerging cause of meningitis in the northwestern United States, particularly Oregon. This fungus also infects birds; it is spread by bird droppings and causes pulmonary disease and meningitis when inhaled, as does *C. neoformans*. But in striking contrast to *C. neoformans*, *C. gattii* readily causes disease in immunocompetent individuals. The organism infects a variety of trees around the world, including the river red gum trees (such as eucalyptus) in Australia, where most cases of *C. gattii* infection occur.

India ink, a stain, is added to CSF to visualize *Cryptococcus*. The ink will stain only the background and does not affect the extracellular polysaccharide capsule of the organism, making the capsules appear as halos around the cryptococci. Identification of *Cryptococcus*, with its prominent capsule, in the CSF sample makes the definitive diagnosis (**Figure 24.18**). Patients suffering from cryptococcal meningoencephalitis are currently treated with amphotericin B, fluconazole, and flucytosine, in a complicated triphasic regimen that ends with maintenance therapy.

(◉) **Note:** The taxonomic classification of *Cryptococcus* species is changing. For example, *C. neoformans* var. *neoformans* has been renamed *C. deneoformans*, and *C. neoformans* var. *gattii* is now divided into five different species. We retained the abbreviations *C. neoformans* and *C. gattii* in the text for simplicity and because the medical field still uses the original names, at least for now.

Figure 24.18 *Cryptococcus neoformans*

An India ink stain of cerebrospinal fluid shows the thick capsule of *C. neoformans*.

SECTION SUMMARY

- **Fungal infections of the CNS generally result from dissemination of the agent from a primary infection elsewhere,** typically in immunocompromised patients.
- ***Candida albicans* is the most common fungal cause of meningitis.**
- ***Coccidioides* is found in arid environments and is endemic in the southwestern United States.** Infection usually begins as pneumonia (valley fever) but can progress to meningitis. Lifetime treatment with antifungal agents is recommended.
- **Infection by *Cryptococcus* begins as a pneumonia** that can spread via the bloodstream to the brain (meningoencephalitis). Diagnosis is made upon seeing yeast with a prominent capsule in CSF preparations.

24.6 Parasitic Infections and Prion Diseases of the Central Nervous System

SECTION OBJECTIVES

- Compare and contrast parasites and prions.
- Evaluate acquisition of parasitic CNS infections.
- Explain the role of host immune responses in parasitic infections and prion diseases of the CNS.
- Name prevention and treatment options for parasitic infections and prion diseases of the CNS.

What is the risk of acquiring a CNS disease caused by a parasite or by prions? The risk of a parasitic infection of the CNS is very small in healthy people with a competent immune system. We commonly come in contact with protozoa (for example, *Toxoplasma*), and our body fights them off effectively. Under certain conditions, however, the protozoa invade and cause serious disease. (Protozoa are eukaryotic parasites whose biochemistry resembles our own; they are introduced in Chapter 11.) By contrast, prions are infectious proteins made by our own cells. In this section, we discuss some of the more common protozoa and briefly discuss prions.

Protozoal Diseases

Toxoplasma gondii is a parasitic protozoan that can infect most mammals, including humans. It is one of the world's most common parasites. Its primary host is the cat, which can be infected by eating contaminated meat or the feces of infected animals. The parasite can also be transmitted vertically from mother to offspring. Contact with cats, especially with cat feces, is the primary mode of transmission to humans. Over half of the world's population is thought to be infected with this parasite. Most infections are asymptomatic, but the disease **toxoplasmosis** typically begins as a respiratory infection that resolves (see Section 21.5 and Figure 21.26). The organism can, however, form cysts in muscles and nervous tissues, including the brain, and remain latent for years unbeknownst to the host. If the patient becomes immunocompromised (for example, the patient has to be on immunosuppressive therapy), the organism can begin to grow and damage the brain. Symptoms include confusion, fever, headache, retinal inflammation that causes blurred vision, and seizures. Antiparasitic treatment is available, as is a vaccine for cats.

Pregnant women are usually asymptomatic when they contract toxoplasmosis. *Toxoplasma*, however, can be vertically transmitted from an infected mother to her fetus. Transmission during the first trimester (first 3 months) can cause stillbirth or serious sequelae in the newborn. The infected fetus may not have any symptoms at birth (in 70%–90% of cases) but can later develop complications such as mental disability and blindness.

Deep in rural sub-Saharan Africa, the dreaded tsetse fly roams and carries with it the protozoa that cause **African sleeping sickness**, *Trypanosoma brucei rhodesiense* and *T. brucei gambiense* (**Figure 24.19**). Reservoirs for *T. brucei rhodesiense* include wild animals (lions, pigs, and hyenas), domestic animals (cows and goats), and humans. Humans are the main reservoir for *T. brucei gambiense*. The tsetse fly becomes infected after taking a blood meal from an infected person. The trypanosome replicates in the fly's gut and migrates to the salivary glands, where the microbe develops into its infective stage. When the fly bites a new host, trypanosome-laden saliva is deposited in the wound, where the microbe then multiplies and produces a painful lesion. Eventually, the parasite migrates to the lymphatic and blood circulations, triggering fever and lymphadenopathy (swollen lymph nodes). The later stage of trypanosomiasis involves CNS invasion, at which point the patient develops daytime sleepiness but nighttime insomnia, a listless gaze, tremors, halting speech, and a shuffling gait. In the final phase, the progressive neurological damage ends in coma and death.

Figure 24.19 African Sleeping Sickness

A. *Trypanosoma brucei gambiense* in a peripheral smear of blood.

B. The tsetse fly transmits the trypanosomes from infected people or animals to healthy people.

Trypanosoma species such as *T. brucei*, the cause of African sleeping sickness, and *T. cruzi*, the cause of Chagas disease (see Section 21.5), are successful parasites that manage to escape the host's immune response by a complex mechanism of antigen switching described in Section 18.6. The knowledge of antigen switching has led us to the first steps in developing an anti-trypanosome vaccine.

Eflornithine is a highly effective anti-trypanosome drug used to treat African sleeping sickness, although others are available. The disease is not seen elsewhere in the world (unless an infected tourist has returned from Africa). But 10,000 cases are reported each year in Africa, although more are suspected.

Naegleria fowleri is an ameba that produces an acute, and usually lethal, CNS infection (**Figure 24.20**). The organism is found in freshwater but also in geothermal pools, heated pools, aquaria, and sewage. The *Naegleria* developmental cycle includes cysts, trophozoites (the amebic form that infects humans), and flagellated forms. Trophozoites infect humans or animals by entering the olfactory nerve, following the path of the nerve directly to the brain. The organism then literally eats through the blood-brain barrier and causes ***Naegleria* meningoencephalitis**.

Symptoms of an infection with *N. fowleri* start with loss of taste and smell, after which headaches, nausea, and stiff neck develop. The patient experiences hallucinations and death, usually within 14 days of exposure. Diagnosis, made by finding *N. fowleri* trophozoites in CSF, is usually a death sentence, in part because the disease is rare and its progress rapid. There are no clinical trials or standard guidelines for treatment. Amphotericin B and miltefosine are the two most agreed-upon drugs. Early diagnosis and treatment while cooling the body below normal temperature is thought to be key to survival and favorable disease outcome. The CDC provides 24/7 specimen collection and shipping, diagnosis, and treatment guidance. Fortunately, contact with *N. fowleri* does not always result in

Figure 24.20 *Naegleria fowleri*

A. *N. fowleri* in infected brain tissue.

White blood cell
Red blood cell
Naegleria fowleri
Blood vessel

TEM 15 µm

Figure 24.21
Acanthamoeba

This unusual-looking ameba can infect a corneal epithelium damaged with microscopic abrasions and travel along the optic nerve to cause CNS disease.

PCM 15 µm

Cyst

Trophozoite

Flagellated

PCM 5 µm

B. *N. fowleri* in various stages. The scale bar is accurate for all panels.

infection. Our immune system fends off *Naegleria* infections quite well. Only 30 or so cases have been reported since 2000 despite common exposure to this protozoan. *Naegleria* was once a pathogen of warmer climates, but global warming appears to be causing this pathogen to migrate to lakes farther and farther north. Minnesota saw its first two cases in 2010 and 2012.

Acanthamoeba (**Figure 24.21**) is another ameba that can infect the brain. The organism is found worldwide in water and soil and can be transmitted into the eye by contaminated contact lenses, or it can enter the body through inhalation or breaks in the skin. Most everyone is exposed to this organism at some time in life, but few people contract disease. The most common symptom is keratitis (inflammation of the cornea). In immunocompromised individuals, the organism can move up the optic nerve into the brain to cause encephalitis. Although few in number, most cases of **Acanthamoeba encephalitis** are fatal.

Onset of encephalitis is slow (weeks to months) and the symptoms are similar to those of *Naegleria* encephalitis. Ketoconazole, miconazole, and trimethoprim-sulfamethoxazole can be effective against this organism, but diagnosis is usually made too late.

Prion Diseases

Prions are not microorganisms. Instead, they are unusual nonliving infectious agents that have been implicated as the cause of a series of relatively rare but invariably fatal brain diseases (**Table 24.4**). Prions are infectious agents that do not have a nucleic acid genome and do not carry out any metabolic processes (**Section 12.2**). The

Table 24.4
Prion Diseases

Disease	Susceptible Animal	Incubation Period	Disease Characteristics
Creutzfeldt-Jakob (sporadic, familial, new variant [vCJD])	Human	Months (vCJD) to years	Spongiform encephalopathy (degenerative brain disease)
Kuru	Human	Months to years	Spongiform encephalopathy
Gerstmann-Straussler-Scheinker syndrome	Human	Months to years	Genetic neurodegenerative disease
Fatal familial insomnia	Human	Months to years	Genetic neurodegenerative disease with untreatable insomnia
Mad cow disease	Cattle	5 years	Spongiform encephalopathy
Scrapie	Sheep	2–5 years	Spongiform encephalopathy
Chronic wasting disease	Deer	Months to years	Spongiform encephalopathy

Figure 24.22 Transmissible Spongiform Encephalopathies

Improperly folded proteins stack to form aggregates called plaques. Plaques appear as vacuoles in the brain tissue.

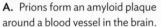

A. Prions form an amyloid plaque around a blood vessel in the brain.

B. Normal brain section.

C. Section of brain taken from a victim of Creutzfeldt-Jakob disease. Note the "Swiss cheese" appearance, indicating brain damage. Arrows point to two of the larger vacuoles.

idea that a protein alone could mediate an infection was thought to be impossible. Yet the **prion** is now recognized as an infectious, misfolded brain protein that resists inactivation by procedures that normally destroy proteins. The discovery that proteins alone can transmit an infectious disease surprised the scientific community. Prion-mediated diseases are especially worrisome because prions resist destruction by many chemical agents and remain active after heating at extremely high temperatures. In some documented cases, sterilized surgical instruments, originally used on a prion-infected person, still held infectious prions and transmitted the disease to a later surgery patient.

How can a nonliving entity without nucleic acid be an infectious agent? Prions associated with human brain disease are thought to be aberrantly folded forms of a normal brain protein called PrP (see Section 12.2). The thought is that when a prion is introduced into the body and manages to enter the brain, it interacts with normally folded forms of the protein, causing them to refold incorrectly. The improperly folded proteins fit together like Lego blocks to produce damaging aggregated structures, known as **amyloid** plaques, within the brain (Figure 24.22).

Some investigators suggest that living infectious agents, such as *Spiroplasma* (a spiral-shaped genus of bacteria whose members lack cell walls), can somehow change the shape of normal brain protein, converting it to an infectious prion protein. This proposal is not, however, universally accepted and has yet to be proved.

Prion diseases are often called **transmissible spongiform encephalopathies (TSEs)** because the postmortem appearance of the brain includes large, spongy vacuoles in the cortex and cerebellum. These plaques are visible in a brain sample from a victim of one of these diseases, **Creutzfeldt-Jakob disease (CJD)** (Figure 24.22C). Typically, onset of CJD symptoms occurs about age 60, and about 90% of individuals die within 1 year. In the early stages of disease, people may have failing memory, behavioral changes, lack of coordination, and visual disturbances. As the illness progresses, mental deterioration becomes

pronounced, and involuntary movements, blindness, weakness of extremities, and coma may occur.

Most mammalian species appear to develop prion diseases. Since 1996, mounting evidence has pointed to a causal relationship between outbreaks in Europe of a disease in cattle called **bovine spongiform encephalopathy (BSE)**, or "mad cow disease," and a disease in humans called **variant Creutzfeldt-Jakob disease (vCJD)**. Both disorders are invariably fatal brain disorders. They have unusually long incubation periods (measured in years) and are caused by an unconventional transmissible agent.

Prion diseases are extremely rare. For example, a total of 26 cases of BSE have been reported in North America as of August 2018: six in the United States and 20 in Canada. Thanks to aggressive surveillance efforts in the United States and Canada it is unlikely, but not impossible, that BSE will be a major food-borne hazard to humans in this country. The CDC monitors the trends and current incidence of typical and variant CJD in the United States.

SECTION SUMMARY

- ***Toxoplasma gondii*** is a protozoal parasite typically transmitted from cats to humans. Infection is common, but disease is rare and usually confined to immunocompromised patients. Latent brain cysts of infected individuals can break open (excyst) and the organisms can grow, destroying brain tissue.

- ***Trypanosoma brucei gambiense*** is the protozoan that causes African sleeping sickness. The organism is transmitted by the bite of the tsetse fly in sub-Saharan Africa.

- ***Naegleria fowleri*** is a common environmental ameba that causes occasional cases of meningoencephalitis.

- ***Acanthamoeba*** is a parasitic ameba that can cause keratitis, but in an immunocompromised patient, the ameba can move into the brain to cause encephalitis.

- **Transmissible spongiform encephalopathies,** such as Creutzfeldt-Jakob disease, are believed to be caused by nonliving, proteinaceous agents called prions.

Perspective

Any infectious disease comes with some degree of angst. The common cold, for instance, is usually suffered with only a mild sense of annoyance, whereas a bout of bloody diarrhea causes anxiety. Gonorrhea, too, produces anxiety, but also embarrassment and anger. However, few infections are met with the level of fear brought by a CNS infection. The severity of bacterial meningitis and the slow mental and physical decline caused by Creutzfeldt-Jakob disease help drive this fear. Most people don't realize that the CNS is the most heavily fortified and protected system in the body. Our immune system and the blood-brain barrier make the CNS almost impenetrable to microbes. As a result, CNS infections are infrequent and serious disease is extremely rare. For example, there were 350 confirmed cases of meningococcal meningitis in the United States in 2017 and only three cases of neuroinvasive eastern equine encephalitis that same year—out of 319 million people. Nevertheless, adhering to simple precautions such as hand washing, wearing insect repellent outdoors, and vaccinating pets can lower the risk even more.

Figure 24.23 Summary: Microbes That Infect the Central Nervous System

Viruses	Genome[a]	Disease
HSV-1	DNA, ds	Encephalitis, meningitis (rare)
HSV-2	DNA, ds	Meningitis
Colorado tick fever virus	RNA, ds, segmented	Encephalitis
LaCrosse agent of California encephalitis	RNA, segmented, negative sense	Encephalitis
Lymphocytic choriomeningitis virus	RNA, segmented	Meningitis
Rabies virus	RNA, ss, negative sense	Encephalitis, meningitis (rare)
Western equine encephalitis virus	RNA, ss, positive sense	Encephalitis, meningitis
Eastern equine encephalitis virus	RNA, ss, positive sense	Encephalitis, meningitis (rare)
Venezuelan equine encephalitis virus	RNA, ss, positive sense	Encephalitis
West Nile virus	RNA, ss, positive sense	Encephalitis, meningitis (rare)
Saint Louis encephalitis virus	RNA, ss, positive sense	Encephalitis, meningitis
Japanese encephalitis virus	RNA, ss, positive sense	Encephalitis, meningitis (rare)
Murray Valley encephalitis virus	RNA, ss, positive sense	Encephalitis
Dengue virus	RNA, ss, positive sense	Encephalitis
Powassan virus	RNA, ss, positive sense	Encephalitis
Poliovirus	RNA, ss, positive sense	Encephalitis, meningitis
Coxsackievirus, echovirus	RNA, ss, positive sense	Encephalitis (rare), meningitis
Human immunodeficiency virus	RNA, ss, positive sense	Meningoencephalitis, meningitis

Bacteria	Morphology	Disease
Mycobacterium tuberculosis	Acid-fast, bacilli	Meningitis
Escherichia coli	Gram-negative, bacilli	Meningitis
Haemophilus influenzae	Gram-negative, bacilli	Meningitis
Bacteroides fragilis	Gram-negative, bacilli	Meningitis
Pseudomonas aeruginosa	Gram-negative, bacilli	Meningitis
Listeria monocytogenes	Gram-negative, bacilli	Meningitis
Neisseria meningitidis	Gram-negative, bacilli	Meningitis
Streptococcus pneumoniae	Gram-positive, cocci	Meningitis
Streptococcus agalactiae, group B streptococci	Gram-positive, cocci	Meningitis
Coagulase-negative Staphylococcus	Gram-positive, cocci	Meningitis
Staphylococcus aureus	Gram-positive, cocci	Meningitis
Treponema pallidum	Spirochete	Meningitis
Borrelia burgdorferi	Spirochete	Meningitis

Fungi	Type	Disease
Candida albicans	Yeast	Meningitis, brain abscesses
Coccidioides species	Spherules (yeast form)	Valley fever, meningitis
Cryptococcus species	Yeast	Meningitis

Parasites	Type	Disease
Toxoplasma gondii	Protozoan	Encephalopathy
Trypanosoma brucei gambiense	Protozoan	African sleeping sickness
Naegleria fowleri	Ameba	Meningoencephalitis
Acanthamoeba	Ameba	Keratitis, encephalitis

[a]ds = double-stranded; ss = single-stranded.

Portal of Entry

Eye
Acanthamoeba

Respiratory
Lymphocytic choriomeningitis virus
Neisseria meningitidis
Streptococcus pneumoniae
Haemophilus influenzae
Naegleria
Acanthamoeba
Coccidioides species
Cryptococcus species

Oral/GI
Poliovirus
Coxsackievirus, echovirus
Botulinum toxin
Listeria monocytogenes

Blood
Candida albicans

Skin
Rabies
West Nile virus
Eastern equine encephalitis
HSV-1 and HSV-2
Tetanus toxin

Urinary Tract/Sepsis
Escherichia coli

Portal of Exit

Respiratory
Neisseria meningitidis
Streptococcus pneumoniae
Haemophilus influenzae

Oral
Rabies virus
Poliovirus
Coxsackievirus, echovirus
Tetanus and botulinum toxins
Listeria monocytogenes

Skin
HSV-1 and HSV-2

Stool
Poliovirus
Coxsackievirus, echovirus
Escherichia coli
Listeria monocytogenes

Key Terms

Acanthamoeba encephalitis (p. 803)
African sleeping sickness (p. 802)
amyloid (p. 804)
arbovirus (p. 788)
areflexia (p. 782)
arthroconidium (p. 800)
aseptic meningitis (p. 783)
attenuated (p. 791)
blood-brain barrier (p. 779)
botulism (p. 797)
bovine spongiform encephalopathy (BSE) (p. 804)
Candida meningitis (p. 800)
central nervous system (CNS) (p. 778)
cerebrospinal fluid (CSF) (p. 778)
coccidioidal meningitis (p. 800)
Creutzfeldt-Jakob disease (CJD) (p. 804)

cryptococcal meningoencephalitis (p. 801)
eastern equine encephalitis (EEE) virus (p. 788)
encephalitis (p. 781)
encephalopathy (p. 792)
endocytosis (p. 798)
exocytosis (p. 798)
hyperesthesia (p. 782)
listeriosis (p. 796)
lumbar puncture (p. 780)
lymphocytic choriomeningitis virus (LCMV) (p. 790)
meningitis (p. 781)
meningoencephalitis (p. 782)
meninx (p. 778)
motor neuron (p. 779)
mycosis (p. 800)

myelitis (p. 782)
Naegleria meningoencephalitis (p. 802)
neurotropic virus (p. 787)
nuchal rigidity (p. 794)
paresthesia (p. 782)
peripheral nervous system (PNS) (p. 778)
petechia (p. 795)
pleocytosis (p. 782)
poliomyelitis (polio) (p. 791)
primary encephalitis (p. 781)
prion (p. 804)
purpura (p. 795)
rabies (p. 778)
radiculitis (p. 782)
Sabin vaccine (p. 791)
Salk vaccine (p. 791)

secondary encephalitis (p. 781)
sensory neuron (p. 779)
tetanospasmin (p. 799)
tetanus (p. 798)
toxoplasmosis (p. 802)
transcytosis (p. 795)
transmissible spongiform encephalopathy (TSE) (p. 804)
vaccine-associated paralytic poliomyelitis (VAPP) (p. 791)
variant Creutzfeldt-Jakob disease (vCJD) (p. 804)
West Nile virus (WNV) (p. 789)

Objectives Assessment Questions

24.1 Anatomy of the Central Nervous System

1. A patient suffering from encephalitis has diminished cognitive function. Which structure(s) are affected?
   a. Brain
   b. Spinal cord
   c. Peripheral sensory neurons
   d. Peripheral motor neurons

2. CNS infections are relatively rare, mostly because of _____.
   a. meninges
   b. cerebrospinal fluid
   c. the blood-brain barrier
   d. the skull and bony structures

3. To obtain CSF for analysis, the needle must enter the space between _____.
   a. brain and spinal cord
   b. vertebral bone and dura mater
   c. dura mater and arachnoid mater
   d. arachnoid mater and pia mater
   e. arachnoid mater and subarachnoid mater

4. Severe meningitis can cause damage to the brain because of increased CSF pressure resulting from injury to _____.
   a. the choroid plexus
   b. arachnoid villi
   c. the blood-brain barrier
   d. bony structures

24.2 Overview of Infectious Diseases of the Central Nervous System

5. A patient admitted to the hospital with meningitis becomes unable to answer questions properly, in addition to his initial symptoms. You identify this new symptom as a deficit in the patient's cognitive function. Which of the following terms best describes this patient's condition?
   a. Complicated meningitis
   b. Primary encephalitis
   c. Meningoencephalitis
   d. Radiculitis
   e. Myelitis

6. A CSF sample from a patient suffering from viral meningitis will most likely show an increased number of _____.
   a. red blood cells
   b. polymorphonuclear cells
   c. lymphocytes
   d. macrophages
   e. endothelial cells

7. A 20-year-old is in the hospital with a high fever, severe headache, neck pain, nausea, vomiting, and confusion. Examination of a sample of his CSF shows many PMNs. What is the most likely cause of his headache?
   a. Inflammation of his meninges
   b. Disruption of his blood-brain barrier
   c. Inflammation of his spinal cord
   d. Demyelination of his sensory neurons
   e. Demyelination of his motor neurons

24.3 Viral Infections of the Central Nervous System

8. Which of the following CNS infections is transmitted via aerosolization?
   a. Rabies
   b. Lymphocytic choriomeningitis
   c. Eastern equine encephalitis
   d. West Nile encephalitis
   e. HSV-1 meningitis

9. Poliomyelitis is transmitted from person to person through the fecal-oral route. True or false?

10. Eastern equine encephalitis (EEE) is a relatively rare disease in humans because _____.
    a. close human and horse contacts are not common
    b. mosquitoes that transfer the virus from horses to humans are rare
    c. ticks that transfer the virus do not usually prey on humans
    d. mosquitoes that transfer the virus between birds do not usually prey on humans

24.4 Bacterial Infections of the Central Nervous System

11. *Neisseria meningitidis* can attach to the endothelial cells of the blood-brain barrier. The organism can penetrate the barrier later because attachment of the organism _____.
    a. loosens the tight junctions between endothelial cells
    b. leads to its endocytosis by the endothelial cells
    c. results in its phagocytosis
    d. reduces the local complement activity

12. An 8-month-old infant is brought to the emergency department by her parents, who report that she started having constipation and poor feeding, which progressed to lethargy and now difficulty breathing. The parents report that symptoms progressed over the past 24 hours. History shows that the child's last meal included a few bites of a natural peanut butter and honey. Spores from which of the following organisms might have contributed to this disease process?
    a. *Neisseria meningitidis*
    b. *Streptococcus pneumoniae*
    c. *Clostridium botulinum*
    d. *Escherichia coli*
    e. *Clostridium tetani*

13. A woman in her fourth month of pregnancy is in the clinic complaining of a fever, flu-like symptoms, and not feeling well. Physical exam of the patient and laboratory analysis of her urine rule out any respiratory or urinary tract infections. She admits that her diet has consisted of only cheeses and crackers; everything else makes her nauseated. You are concerned about an infection by which of the following organisms?
    a. *Clostridium botulinum*
    b. *Escherichia coli*
    c. *Staphylococcus aureus*
    d. *Listeria monocytogenes*
    e. *Streptococcus pneumoniae*

24.5 Fungal Infections of the Central Nervous System

14. A major risk factor for acquiring fungal CNS infections is _____.
    a. environmental exposure
    b. being immunocompromised
    c. proximity to arthropod vectors
    d. a and b
    e. a and c

15. A patient develops a fungal pneumonia a week after his vacation in Arizona to take pictures of the desert areas. He is now suffering from severe headache, blurred vision, and altered mental status. This patient is likely to be infected with _____.
    a. *Coccidioides immitis*
    b. *Cryptococcus neoformans*
    c. *Candida albicans*
    d. *Toxoplasma gondii*
    e. *Naegleria fowleri*

24.6 Parasitic Infections and Prion Diseases of the Central Nervous System

16. A patient diagnosed with a parasitic keratitis is now suffering from meningitis after the parasite spread from his eye to his brain via the optic nerve. Which of the following is the most likely parasite?
    a. *Toxoplasma gondii*
    b. *Trypanosoma brucei*
    c. *Acanthamoeba* species
    d. *Cryptococcus neoformans*
    e. *Coccidioides posadasii*

17. _____ is the drug of choice to treat *Naegleria* meningoencephalitis.

18. The etiologic agent of transmissible spongiform encephalitis (TSE) is _____.
    a. a protozoan
    b. a protein
    c. a small ribonucleic acid
    d. an ameba
    e. a nucleic acid

Review Questions

1. What is the basic anatomy of the CNS, and how does it relate to the PNS?

2. What is the source of cerebrospinal fluid?

3. What is the blood-brain barrier?

4. What is a lumbar puncture? How is it done?

5. What are three paths microbes can take to gain access to the CNS?

6. What are the three basic types of CNS infections?

7. How do the symptoms of encephalitis and meningitis differ?

8. What findings in an analysis of CSF would suggest viral versus bacterial meningitis?

9. What is the difference between primary and secondary encephalitis?

10. Discuss the pathogenesis, symptoms, treatment, and prevention of rabies, eastern equine encephalitis, and West Nile virus encephalitis.

11. Compare the presenting symptoms of viral meningitis with those of bacterial meningitis.

12. List the major causes of viral meningitis.

13. For which patient populations are viral CNS infections most dangerous?

14. List the major causes of bacterial meningitis and neonatal meningitis. When should antibiotics be administered?

15. Travel history can yield important clues as to the identity of which organisms that cause CNS infections?

16. For *Neisseria meningitidis* and *Listeria monocytogenes*, discuss the pathogenesis, symptoms, treatment, and prevention of disease.

17. Discuss the pathogenesis of botulism and tetanus. How can these diseases be prevented?

18. Discuss key features of major fungal infections of the CNS.

19. Discuss the nature of prions and how they are thought to cause disease.

Clinical Correlation Questions

1. A 1-week-old infant presents with a temperature of 38.9°C (102°F). The child is listless but cries when attempts are made to move her head. Her cerebrospinal fluid contains 180 white blood cells/μl (normal is ≤5), most of which are neutrophils. The CSF glucose concentration is lower than normal, whereas the protein concentration is markedly above normal. From this information, what do you suspect is the causative agent? What treatment would you recommend?

2. A 63-year-old female presents to the emergency department of a local hospital, complaining of having a crooked smile since receiving an injection at her doctor's office earlier in the morning. When asked to smile, the patient complies, but only half of her mouth opens to a proper smile, giving her smile a crooked appearance. Physical exam shows droopy eyelids and cockeyed eyebrows. The patient denies having any similar experience in the past or any trauma to her face. Which medical specialist did she see earlier, and which injection did she receive?

3. A person from Maine developed symptoms of rabies but reports no contact with dogs, bats, or other known reservoirs of rabies the virus. Patient history reveals he enjoys boating, is a vegetarian, and received a corneal transplant about 1 year ago. Explain how he might have contracted rabies.

25

Diagnostic Clinical Microbiology

[Dazed and Confused in New Jersey]

PRESENTATION   Jean, a 63-year-old woman from Plainfield, New Jersey, was brought by ambulance to the Muhlenberg Regional Medical Center, accompanied by her husband. On arrival she was disoriented and nonresponsive. Her husband said that Jean is normally quite active, but over the past 24 hours she had become increasingly tired and eventually could not answer simple questions. A stroke was suspected.

PHYSICAL EXAM AND PRELIMINARY LAB RESULTS   Jean showed no obvious signs of headache or unilateral muscle weakness (weakness on one side), which argued against a stroke. Blood was drawn and, because she could not walk to the bathroom, she was catheterized to collect urine. The lab reported that her white blood cell (WBC) count was normal, but the nurse attending to Jean noticed the urine in her collection bag was cloudy. A fresh urine specimen was obtained directly from her catheter and subjected to a rapid dipstick test. In just minutes, the test revealed high levels of nitrites, indicating bacterial growth, and the enzyme leukocyte esterase, signifying the presence of WBCs. Jean had a urinary tract infection (UTI).

CONFIRMATORY LABORATORY DIAGNOSIS The urine specimen contained more than 100,000 bacteria/milliliter (ml). Pink colonies growing on MacConkey agar indicated the organism was a Gram-negative rod that ferments lactose and was probably *E. coli*, a common cause of UTIs. On day 3, tests showed that the biochemical capabilities of the isolate matched those of *E. coli*, confirming the presumptive diagnosis. However, Jean was discharged long before these results were known.

TREATMENT AND OUTCOME   The physician, knowing Jean had a UTI, started her on an intravenous quinolone antibiotic a mere 4 hours after admission. Within 24 hours, Jean had regained full mental ability and was able to walk out of the hospital.

Urine Collection Bag
A nurse empties a urine collection bag. Because it may have been collecting for hours, this urine is unsuitable for bacteriological culture. Suitable urine is taken straight from the catheter or by suprapubic puncture.

Urinalysis
A urine dipstick used for rapid analysis. The plastic strip is dipped into fresh urine. Patches that monitor pH, nitrate, leukocyte esterase, protein, blood, glucose, bilirubin, and ketones change color. The presence of nitrite, leukocyte esterase, or both in the urine indicates urinary tract infection.

Escherichia coli
Escherichia coli growing on a MacConkey lactose plate. The colonies turn red from acid produced by lactose fermentation. Acidic pH enables colonies to absorb dye from the medium.

The symptoms in this case may appear strange for a urinary tract infection, but they are quite common among older patients (Chapter 23). Older patients with UTIs often do not notice the usual symptoms of flank pain or burning when urinating. Their altered mental status is due to the toxicity of the infection, which may spread into the bloodstream (sometimes called urosepsis; see the opening case history for Chapter 23). Without quick action, the patient could have died. The nurse's keen awareness and the rapid urine dipstick result certainly avoided a tragic outcome.

Many of the case histories sprinkled throughout this textbook name a pathogenic culprit without describing how it was identified. In this chapter we focus on identifying pathogens, a major objective of clinical microbiology. For instance, you may wonder how clinical microbiologists know which organisms to suspect in a given case. And you might be curious about what tests unequivocally identify the right pathogen. Another puzzle is how the real pathogen is found among thousands of normal microbiota. The task would seem similar to finding the proverbial needle in a haystack.

Our coverage of clinical microbiology is, by necessity, selective. Not all techniques, nor all diseases, can be described. Our goal is simply to demonstrate the general principles and problem-solving approaches used to identify pathogens. The modern tools of clinical microbiology and the technologists who brandish them are indispensable to clinicians.

25.1 The Importance of Clinical Microbiology and Specimen Collection

What is the first step in investigating any infectious disease? As in any good detective mystery, the first step is to identify the most likely suspects. This can be accomplished, in part, from observing the disease symptoms in a patient and knowing what organisms typically produce those symptoms. It also helps if the physician is aware of a similar disease outbreak under way in the community. Beyond these clues, the etiological agent must be identified by using biochemical, molecular, serological, or antigen detection strategies.

Why take the time to identify an infectious agent? This is the first question many students ask when contemplating the effort and expense required to identify the genus and species of an infection-causing organism. Why not simply treat the patient with an antibiotic and be done with it? Although this approach sounds appealing, identifying an infectious agent is important for several compelling reasons:

• **Many bacteria are resistant to certain antibiotics.** As discussed in Chapter 13, antibiotic resistance is an increasingly serious global problem. Understanding which antibiotics can treat an infectious agent helps the clinician avoid prescribing the wrong drug. Nevertheless, life-threatening emergencies such as meningitis force clinicians to make their best guess as to which antibiotic(s) will target (or "cover") the likely pathogens before knowing the pathogen's identity. This approach is called **empirical therapy**. But sometimes the clinician's selection is wrong. The clinical laboratory must still identify the causative organism and its antibiotic resistance pattern so the clinician can be certain that the prescribed treatment was correct.

• **Antibiotic-resistant pathogens are spreading across the world.** A major task of national and international health organizations, such as the US Centers for Disease Control and Prevention (CDC), is to track the global spread of antibiotic-resistant pathogens. To minimize treatment failures, we must know what pathogens are causing disease across the globe and characterize their antibiotic resistance. A case in point is *Neisseria gonorrhoeae*. Before 1970, most clinical isolates were susceptible to penicillin, a drug widely used to treat gonorrhea. Today, most clinical isolates are penicillin-resistant.

Figure 25.1 Outbreak of Diarrhea in Babies

Infants attending the same day-care center can spread shigellosis to each other via the fecal-oral route. Parents, however, may take each child to a different doctor. Diagnostic laboratories will identify *Shigella* in each case and report this to local public health officials, but because the affected children are scattered around the city, the source may not be evident. By questioning the parents, public health officials will discover that all the affected children attend the same day-care center. Dots in this city map represent children under 5 years of age in the area. Arrows indicate children who attend the same day-care center. Red arrows and dots represent sick children, whereas blue arrows indicate healthy children.

- **Specific pathogens are associated with secondary disease complications.** Clinicians must identify a disease-causing pathogen to predict possible secondary complications known as disease sequelae. For example, children whose sore throats are caused by certain strains of *Streptococcus pyogenes* can develop rheumatic fever long after the infection has resolved. Rheumatic fever is a serious disease that can damage the heart and other organs (Section 20.3). Knowing early on that *S. pyogenes* has caused a child's sore throat allows the physician to prescribe antibiotics that will quickly eradicate the infection and prevent dangerous sequelae from developing.

- **Tracking the spread of a disease can lead to its source.** Consider a situation in which eight infants scattered throughout a city develop bloody diarrhea (Figure 25.1). The clinical lab identifies the same strain of *Shigella sonnei*, a Gram-negative bacillus, as the cause in each case. Finding the same strain in all cases suggests that the pathogens probably originated from the same source. Shigellosis is transferred person-to-person by the fecal-oral route (Sections 2.4 and 22.5). Upon learning that the same strain of *Shigella* infected several patients, public health officials will question parents and might learn that all the children attend the same

day-care center. By testing the other children and workers in that center, officials can stop the infection from spreading further. Chapter 26 covers this investigative process, called epidemiology, more completely.

Specimen Collection

How do microbiologists "catch" pathogens hiding in clinical samples? To identify disease-causing microbes, clinicians and nurses must collect, and the laboratory must process, a wide variety of clinical specimens. Specimens can include simple cotton swabs of sore throats, in which the swab is placed into a liquid transport medium before being sent to the lab, as well as urine and stool samples that are transported directly. Table 25.1 lists some common pathogens and diseases, the general procedures labs use to detect these pathogens, and the clinical samples involved. Each procedure requires a hypothesis as to the possible agent, and the savvy technologist has to know how the likely organisms grow and survive (see Chapter 6). We will explain these procedures later in this chapter. Here we discuss how the clinical specimens are collected and processed.

CASE HISTORY 25.1

A Literal Pain in the . . . Rectum

A 4-year-old boy, Evan, was admitted to the hospital to evaluate and treat persistent pain in the rectal area. His problem began about 1 week earlier with ill-defined pain in the same region. He had a WBC count of 24,900 per microliter (μl) with 87% granulocytes (neutrophils). An abdominal computed tomography scan revealed an abscess (walled-off region of infectious bacteria) next to his rectum. A needle aspiration of the abscess drained 20 ml of yellowish, foul-smelling fluid. The specimen was plated on blood agar under aerobic conditions, but the culture was negative despite evidence of bacteria in the abscess drainage. (Blood agar is a rich medium containing 5% sheep red blood cells [RBCs].) Why didn't the infectious agent grow?

This problem is related to specimen collection and processing. Internal abscesses located near the gastrointestinal tract are usually anaerobic infections, in this case caused by the Gram-negative rod *Bacteroides fragilis*, a strict anaerobe (Figure 25.2A). Intestinal microbes, most of which are anaerobic, can sometimes escape the intestine if the organ is damaged. If the body's immune system cannot eliminate the infection, the tissues form a fibrous layer that walls off the infected region (an abscess). The specimen here should have been collected under anaerobic conditions by aspiration into a nitrogen-filled tube before transport to the clinical lab. Alternatively, a swab of the abscess material can be inserted into a special transport tube that has a built-in system to eliminate

Table 25.1
Identification Procedures for Selected Diseases [Reference Table]

Agent	Disease	Examples of Diagnostic Tests[a]
Bacteria		
Corynebacterium diphtheriae	Diphtheria	Culture of material from nose and throat on special medium; in vivo or in vitro tests for toxin
Bordetella pertussis	Pertussis (whooping cough)	Staining smears of nasopharyngeal secretions with fluorescent antibody; ELISA for toxin in respiratory secretions; culture on special medium; PCR
Legionella pneumophila	Legionellosis; Legionnaires' disease; Pontiac fever	Culture on special medium; enzyme immunoassay to detect antigen in urine; PCR and nucleic acid probes to identify *Legionella* nucleic acid
Campylobacter spp.	Campylobacteriosis	Isolation of bacteria from stool on selective medium (CAMPY blood agar plates) incubated at 42°C in a microaerophilic atmosphere; PCR of stool samples
Leptospira interrogans	Leptospirosis	Serological tests early in the illness and after 2–3 weeks to detect rise in antibody titer; PCR
Listeria monocytogenes	Listeriosis	Culture of blood and spinal fluid; selective and enrichment cultures performed on food samples to grow potential pathogens; DNA probe to rapidly identify colonies
Chlamydia trachomatis	Chlamydial genital infections	Monoclonal antibody to identify *C. trachomatis* antigen in urine or pus; nucleic acid probes
Treponema pallidum	Syphilis	Direct fluorescent antibody staining from tissue samples; serological tests
Francisella tularensis	Tularemia	Culture of blood on cysteine-containing medium; fluorescent antibody stain of pus; detection of rise in antibody titer (specialized lab)
Yersinia pestis	Plague	Fluorescent antibody or ELISA to identify capsular antigen; PCR; IgM serology (specialized lab)
Viruses		
Rhinovirus	Common cold	Specific antibodies to identify strain from nasal secretions (specialized lab)
Influenza virus	Flu	Rapid antigen detection or viral culture of nasal secretions; serology to compare influenza antibody levels in blood samples taken during acute and convalescent stages of illness; qRT-PCR
Hantavirus	Hantavirus pulmonary syndrome	Electron microscopy or monoclonal antibody to detect antigen in tissues; ELISA and Western blot tests for IgG and IgM antibodies in patient's blood (specialized lab)
Herpes simplex virus	Different strains cause cold sores, ocular lesions, and genital lesions	Viral culture; fluorescent antibody or DNA probes to identify the viral antigen in clinical material (specialized lab)
Mumps virus	Mumps	Rise in serum antibody titer or presence of IgM antibody to mumps virus in patient's blood; viral culture (specialized lab)
Rotavirus	Diarrhea	ELISA of diarrheal stool for virus; qRT-PCR
Human immunodeficiency virus	AIDS	Detection of antibody to HIV antigen in patient's blood; Western blot; PCR
Coronavirus SARS-CoV-2	COVID-19	qRT-PCR targeting the nucleocapsid gene; nucleocapsid antibody test; antigen detection test
Fungi		
Coccidioides immitis	Coccidioidomycosis; infection in lung can disseminate to almost any tissue	Observation of large, thick-walled, round spherules from clinical specimens; PCR identification
Histoplasma capsulatum	Histoplasmosis; infection of lung; sometimes disseminates	Examination of stained material from pus, sputum, tissue, etc., and finding intracellular *H. capsulatum* yeast phase; blood tests for antibody to the organism; antigen detection in urine or blood

[a]ELISA = enzyme-linked immunosorbent assay; Ig = immunoglobulin; PCR = polymerase chain reaction; qRT-PCR = quantitative reverse transcription–polymerase chain reaction.

Figure 25.2 Anaerobic Infection

A. Gram stain of *Bacteroides fragilis* (1.5–4 μm long).

LM 8 μm

1 Remove tube from package.

2 Remove plunger with attached swab. Collect sample.

3 Reinsert swab and press plunger through stopper so that inner tube drops to bottom of outer tube.

4 Mix by swirling. Transport to laboratory.

- Plastic foil package
- Plunger
- Stopper
- Swab
- Inner tube
- Palladium catalyst
- Anaerobic indicator

B. Vacutainer anaerobic specimen collector. Plunging the inner tube to the bottom will activate a built-in oxygen elimination system. The anaerobic indicator changes color when anaerobic conditions have been achieved.

oxygen (**Figure 25.2B**). However, because the specimen was collected, transported, and handled in air, the oxygen probably killed many of these anaerobes.

B. fragilis, however, has a stress response system that permits the organism to survive for 1 or 2 days in oxygen, so some bacteria may have survived transport. The laboratory still had a chance to find the pathogen. This raises the second problem in the case study. Once the lab received the specimen, they cultured it only under aerobic conditions. The lab should have also incubated a series of plates anaerobically (see **Figure 6.23**). This contrived case shows the importance of both proper specimen collection and proper processing.

Two broad categories of specimens exist: specimens taken from body sites that are normally sterile and specimens taken from sites teeming with normal microbiota. Collecting specimens from each type of site comes with its own special considerations.

Samples from Sterile Body Sites

Some body sites should be sterile (contain no microorganisms) when collected from a healthy individual. These sites include blood, cerebrospinal fluid, pleural fluid from the space that lines the outside of the lung, synovial fluid from joints, peritoneal fluid from the abdominal cavity, and any tissues from internal organs. Because these sites are sterile, specimens can be plated onto nonselective agar media (blood or chocolate agar, for example) as well as selective media. Nonselective media can be used because any organism found in these specimens is considered significant. Bladder urine and lung tissues were once thought to be sterile but are now known to have a normal microbiome. Collection from those sites will be described in the next section.

BLOOD CULTURES Blood, of course, should be sterile. Blood cultures are used to evaluate possible blood infections (septicemia) or sepsis (see Chapter 15 IMPACT and Section 21.3). Sepsis must be considered in any patient who presents with a very high fever, is experiencing chills, and appears disoriented. How, then, is a blood sample taken and processed for blood culture?

In an adult, 5–10 ml of blood is collected by **venipuncture** (less blood is taken from pediatric patients). The blood is transferred into blood culture bottles that will be incubated aerobically and anaerobically (**Figure 25.3A**). These bottles all contain a pH-sensitive fluorescent sensor. If bacteria grow in the bottle, they will release CO_2, which lowers the pH of the medium and triggers fluorescence. The bottles are placed in an automated incubator that detects the fluorescence and sounds an alarm when growth is detected (**Figure 25.3B**). Samples from a positive bottle are then plated onto bacteriological agar (such as blood agar or MacConkey agar) to begin the identification process.

An inherent problem with collecting blood for a blood culture is contamination by skin microbiota. Although the area around the needle stick is swabbed with an antiseptic, sometimes skin bacteria survive and contaminate the needle as it is inserted. The organism is then transferred with blood into the culture bottle. Thus *Staphylococcus epidermidis* and *Cutibacterium acnes* (formerly *Propionibacterium acnes*), two prominent members of the skin microbiome, commonly contaminate blood cultures. One way to screen for contaminants is to take blood from <u>two</u> body sites (left arm and right arm, for instance). If the same organism grows in both cultures, it's most likely the agent. However, contamination is the likely conclusion if an organism is found in one culture but not the other or if each culture yields a different organism.

CEREBROSPINAL FLUID CULTURES Meningitis is a rapidly progressive disease that can kill quickly (Chapter 24). So, identifying the microbial cause is crucial and time-sensitive. The lumbar

Figure 25.3 Blood Cultures

A. Typical blood culture bottles.

B. A blood culture incubator that monitors growth.

Figure 25.4 *Neisseria meningitidis* in a CSF Sample

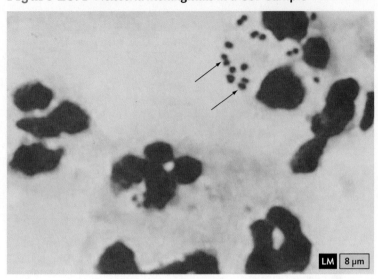

Gram stain of CSF from a meningitis patient infected with *N. meningitidis*. Arrows indicate bacteria found within infiltrating polymorphonuclear leukocytes.

>60% of serum glucose). Higher-than-normal levels of protein and lower-than-normal glucose is a sign of bacterial infection. The CSF is then cultured onto blood agar and chocolate agar media, and/or rapid molecular tests such as polymerase chain reaction (PCR) are performed to identify the agent.

PLEURAL, SYNOVIAL, AND PERITONEAL FLUID CULTURES

Infection or inflammation of the pleural cavity around the lungs, the joints, or the peritoneal cavity can cause excessive fluid accumulation at these body sites. If a clinician suspects an infection at these sites, samples of the fluids are aspirated using a sterile syringe and then cultured for the presence of bacteria or fungi. Aspirated fluids can be viewed microscopically for inflammatory cells (polymorphonuclear leukocytes, macrophages) but are typically cultured on agar plates incubated both aerobically and anaerobically.

Samples from Sites with Normal Microbiota

Identifying pathogens at sites that contain normal microbiota is harder. A stool (fecal) sample, for instance, is normally teeming with microbial species. These specimens are typically plated onto selective media (for example, MacConkey, Hektoen, or CNA agar) to eliminate or decrease the number of normal microbiota that might contaminate the specimen. The following four sections describe techniques used to collect specimens from nonsterile body sites.

SKIN, ABSCESS, THROAT, AND NASOPHARYNGEAL SPECIMENS

Skin samples are taken when a mycotic or viral infection (papillomavirus, for example) is suspected. With the help of a sterile scalpel, skin is scraped from the active border areas of lesions. If a fungal infection is suspected, scrapings are placed in a drop of 10% KOH (potassium hydroxide) solution to dissolve tissue. The preparation is then examined under a bright-field microscope to

puncture, commonly called a spinal tap (seen in Figure 24.4), is the most common method for collecting cerebrospinal fluid (CSF). As Section 24.1 first described, the patient lies on their side, with knees pulled up toward the chest and chin tucked down (Figure 24.4A). After disinfecting and anesthetizing the skin, a health care worker inserts a spinal tap needle, usually between the third and fourth lumbar vertebrae (located in the lower spine safely below the end of the spinal cord). Once the needle is properly positioned in the subarachnoid space (the space between the spinal cord and its covering, the meninges), CSF pressures can be measured and fluid can be collected for testing. Spinal fluid slowly drips out of the needle and is collected in a sterile tube.

A portion of CSF is spun in a centrifuge at low speed to concentrate any cells that may be present, and the sediment is Gram stained. No cells should be present unless an infection is under way (Figure 25.4). Any bacteria or WBCs seen should be reported immediately. Other important tests performed on CSF include determining the amount of protein (normally low; <45 milligrams [mg]/100 ml) and glucose (normally high; >60 mg/100 ml or

Figure 25.5 Throat and Nasopharyngeal Samples

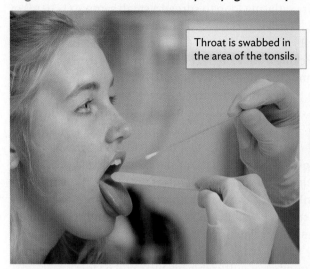

Throat is swabbed in the area of the tonsils.

Swab

Tonsil

Tongue depressor

Nasopharynx

A. For a throat sample, a sterile cotton-tipped applicator is used to swab the back of the throat around the tonsils and uvula. A sterile tongue depressor allows unobstructed access to the throat.

B. For a nasopharyngeal sample, sterile cotton on a flexible wire is inserted through the nose and back to the nasopharynx.

detect fungal elements (hyphae, conidia, ascospores; Section 11.2). If a viral lesion is suspected, samples are transferred to 5 ml of phosphate-buffered glycerin virus transport medium at pH 7.6 and rushed to the lab. There the samples are analyzed for viral antigens by a variety of immunological tests.

Samples from draining abscesses and inflamed throat tissues are collected using sterile cotton swabs (Figure 25.5A). Swabs should be placed in specialized liquid nutrient medium for transport to the lab. Upon arrival, the sample will be used to inoculate blood and chocolate agars.

Deep wound abscesses are collected by needle aspiration when an anaerobic infection is suspected (as in Case History 25.1). The needle of a syringe is inserted into the abscess and a portion of pus is withdrawn (aspirated) into the syringe. The aspirate should be transferred quickly to a transport vial containing no oxygen.

Special nasopharyngeal swabs are used to collect secretions from the nasopharynx (Figure 25.5B). Some viruses and bacterial pathogens, including *Bordetella pertussis*, specifically infect this area. The swab is a cotton-tipped flexible wire that is passed through the nose and into the nasopharynx. Again, the swab should be placed in specialized liquid medium for transport to the laboratory.

SPECIMENS OF LUNG CONTENTS Sputum comes from deep lung secretions and is the result of inflammation. It is not saliva. A lung infection results in a considerable amount of sputum, which can be expectorated for oral collection. As a result, sputum is an appropriate fluid to sample from patients suspected of having tuberculosis or pneumonia. Collection is usually performed in the morning, with the patient sometimes sequestered in a

sputum collection booth to prevent aerosolization (Figure 25.6A). Generally, a 2-ml volume is adequate. The sample should contain material recently discharged from the bronchial tree with minimal saliva content. If the patient has difficulty producing a sputum specimen (Figure 25.6B), sputum production may be induced by inhaling a warm aerosol of sterile 5%–10% sodium chloride in water produced by a nebulizer. Figure 25.6C shows an acid-fast stain of sputum containing *Mycobacterium tuberculosis* (described further in Section 25.2).

For patients having trouble producing sputum, **bronchoalveolar lavage** (BAL) is used to collect lung material, including bacterial pathogens. A bronchoscope is first passed through the nose or mouth and into the lungs of a sedated patient. Sterile fluid is squirted into the lungs and then collected by suction for culture or examination. BAL is useful for diagnosing fungal infections like aspergillosis and bacterial infections like tuberculosis. Note that although the lung was previously thought to be sterile, it is now known to contain small numbers of normal microbiota.

STOOL SPECIMENS Fecal stool samples are collected from patients complaining of diarrhea. These samples can be either cup samples (a teaspoon amount of feces) or a rectal swab. The laboratory will plate stool samples on a variety of selective and differential media (MacConkey, Hektoen; Section 6.2) or on more specialized media designed for *Campylobacter* or *Vibrio* species. The stool may also be examined for parasite eggs (ova) and the parasites themselves.

URINE SPECIMENS Urine in the bladder of a healthy individual was once considered sterile. We now know that the bladder has a

Figure 25.6 Collecting Sputum

A. A patient with tuberculosis has coughed up sputum and is spitting it into a sterile container. (The patient is sitting in a special sputum collection booth that prevents the spread of tubercle bacilli. The booth is decontaminated between uses.)

B. Sputum sample.

C. Acid-fast stain of *Mycobacterium tuberculosis* in sputum. Note the acid-fast bacilli (reddish, shown by arrow) among the normal microbiota and white blood cells in the sputum, which are not acid-fast and appear blue.

normal microbiome, although its role in human health is unclear. Why didn't we see it before? Urinary bladder microbiota are primarily anaerobic and not culturable under typical aerobic conditions. So for all practical purposes, clinical microbiologists consider urine in the bladder to be sterile, or nearly so. Symptoms of urinary tract infections appear only when significant numbers of aerobically culturable bacteria appear in urine.

To determine bacterial numbers in urine, the technician uses a calibrated inoculating loop to spread a small sample (1–10 µl) of urine over the entire surface of a blood agar plate and incubates the plate aerobically for 24 hours (**Figure 25.7A and B**). Each colony observed came from a single cell in the urine. If 10 µl was spread on the plate and 10 colonies grew, then 10 bacterial cells were present in the 10 µl of urine. Results, however, are reported as "cells per milliliter of urine," which here would be 1,000 colony-forming units (CFUs)/ml of urine (10 cells/0.01 ml = 1,000 cells/ml).

Urine is commonly collected through a **midstream clean-catch technique**, which the patient performs. In this procedure, the patient cleans the external genitals with a sterile wipe containing an antiseptic. The patient partially urinates, to wash as many organisms as possible out of the urethra, and then collects 5–15 ml of the midstream urine in a sterile cup. This urine sample will probably not be truly sterile because of urethral contamination, but the number of culturable bacteria will be low.

In situations where a patient is immobilized in a hospital or at home, urinary catheterization is used to assist urination. This method also offers a convenient way to collect urine for bacteriological examination. When collected from a catheterized patient, urine should contain few, if any, aerobically culturable bacteria unless an infection is present. Catheterization involves passing a thin, sterile tube (a Foley balloon catheter) through the urethra and directly into the bladder (**Figure 25.7C**). A balloon at the end of the tube is inflated to keep the catheter in place. Unfortunately, inserting the catheter through the nonsterile urethra can <u>introduce</u> culturable organisms into the bladder and cause an infection (Section 23.2). Also, once a patient is catheterized, urine samples should be collected only from the catheter, <u>never</u> from the collection bag. Urine may sit for hours in the collection bag, so organisms initially present at small, insignificant numbers have time to replicate to large numbers even if the patient does not have a UTI.

The clinical laboratory determines how many organisms per milliliter are present in urine and informs the physician whether an infection is present. Finding more than 1,000 organisms of a single colony type per milliliter of midstream clean-catch urine indicates a UTI in a patient with symptoms (burning pain upon urination, increased frequency of urination). These patients should be treated. However, 10,000 or fewer bacteria in a clean-catch specimen is usually considered normal in an <u>asymptomatic</u> patient. In fact, asymptomatic patients with 100,000 or more bacteria per milliliter of urine (called asymptomatic bacteriuria) are usually not treated unless they are pregnant or will be undergoing an invasive urinary tract procedure.

Urine collection in infants is a special problem. The infant cannot perform a clean catch and won't leave a Foley balloon catheter in place. Short-term catheterization without inflating a balloon is tried first, but in rare instances the situation may call for a **suprapubic puncture**. In this procedure, a sterile needle is inserted through the abdomen into the bladder and urine is removed by aspiration (**Figure 25.7D**). Any bacterial number greater than 1,000 per ml (of a single species) suggests a UTI. Affixing a sterile bag to the infant's genitals with a sticky strip is no longer recommended because of very high false-positive results.

Handling Biological Specimens

CASE HISTORY 25.2
Microbe Hunter Infects Self

On July 15, an Alabama microbiologist was taken to the emergency department with acute onset of generalized malaise, fever, and diffuse myalgias (muscle aches). She received a prescription for oral antibiotics and was released. On July 16, she became tachycardic (high heart rate) and hypotensive (low blood pressure) and returned to the hospital. She died 3 hours later. Blood cultures were positive for *Neisseria meningitidis* serogroup C. Three days before symptoms appeared, the patient had prepared a Gram stain from the blood culture of a patient later shown to have meningococcal disease; the microbiologist also handled agar plates containing CSF cultures from the same patient. Co-workers reported that, in the laboratory, aspirating fluids from blood culture bottles was typically performed at the open lab bench—without using biosafety cabinets, eye protection, or masks. Testing at CDC indicated that the isolates from both patients were indistinguishable. This hospital lab infrequently processed isolates of *N. meningitidis* and had not processed another meningococcal isolate during the previous 4 years.

Section 2.4 described the universal precautions that research and clinical lab personnel must take to avoid infecting themselves with the pathogens they handle. Recall that pathogens are grouped by their level of risk from low to high (risk levels 1–4). Risk level, in turn, dictates the biological safety level (BSL 1–4; Table 2.4) needed to protect the scientist and to prevent the pathogen from escaping the lab. However, clinical laboratory workers do not know which pathogen, if any, lurks in the samples they process. Consequently, lab workers must expect the worst. Then again, typical community hospitals rarely, if ever, encounter risk level 3 or 4 pathogens such as *Brucella* species or Ebola virus, respectively. Routinely requiring level 4 containment procedures, including airlocks and positive-pressure suits, to process urine, blood, and stool would be overreacting.

In general, most clinical specimens are handled using level 2 biosafety procedures (including the use of a biosafety cabinet) if aerosolization is unlikely (streaking an agar plate, for example). When risk level 3 pathogens are suspected, such as *Mycobacterium tuberculosis*, culture manipulations must be performed under level 3 biosafety conditions, including negative-pressure airflow into the room. Biosafety level 4 conditions are used only when the severity of a patient's symptoms suggest infection by a risk level 4 agent such as hantavirus or Ebola virus. As of this

Figure 25.7 Collecting a Urine Specimen

A. Calibrated loops used to streak urine samples onto agar media.

B. Result of plated urine sample. Each colony arose from a single organism present in the urine at collection.

C. Urinary catheter showing placement in the urethra. A balloon attached to the end of the catheter (not shown) is inflated in the bladder to keep the catheter in place.

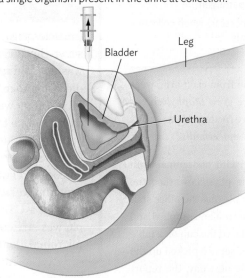

D. Suprapubic puncture. A sterile needle is inserted through the skin above the pubic arch and into the bladder.

writing (September 2020), the guidelines for handling the recent COVID-19 pandemic virus, SARS-CoV-2, are to conduct qRT-PCR testing of patient samples under biosafety level 2 conditions, but biosafety level 3 conditions are required to culture the virus.

What about biosafety procedures for nurses or technologists collecting patient samples? The nurse or phlebotomist (the person who collects blood) <u>must</u> thoroughly wash their hands and wear latex gloves when collecting or handling a specimen. This measure not only protects the hospital employee from contracting a disease but also will prevent transmission of potential pathogens from the provider to the patient. More strident protective measures must be taken when collecting risk level 3 and 4 pathogens. Chapter 26 talks more about hospital-acquired infections.

In addition to gloves, <u>sterile</u> collection devices such as syringes, catheters, collection cups, and cotton swabs must be used. A swab or collection cup accidentally contaminated by touch or sneeze can introduce microorganisms into the sample and complicate laboratory identification. Conversely, contaminated syringes or catheters used to collect samples can introduce new pathogens into the patient, causing septicemia or urinary tract disease.

SECTION SUMMARY

- **Identifying a pathogen** enables us to:
  - —**Use appropriate antibiotics**, if necessary.
  - —**Anticipate possible sequelae.**
  - —**Track and limit the spread** of the disease.
- **Special collection and biosafety precautions** are necessary when collecting specimens from abscesses of suspected anaerobic etiology.
- **Common specimens** include sputum, throat, stool, urine from the bladder, and skin, all of which are contaminated with normal microbiota, as well as specimens from sterile sites such as blood and CSF.
- **Specimens** from sites containing normal microbiota must be processed on selective media to find the potential bacterial pathogen.
- **Appropriate biosafety protective measures** are necessary when collecting or handling potentially infectious biological materials.

Thought Question 25.1 Two blood cultures, one from each arm, were taken from a patient with high fever. One culture grew *Staphylococcus epidermidis*, but the other blood culture was negative (no organisms grew). Is the patient suffering from septicemia caused by *S. epidermidis*?

Thought Question 25.2 A 30-year-old woman with abdominal pain went to her physician. After the examination, the physician asked the patient to collect a midstream clean-catch urine sample that the office staff would send to the lab across town for analysis. The woman complied and handed the collection cup to the nurse. The nurse placed the cup on a table at the nurses' station. Three hours later, the courier service picked up the specimen and transported it to the laboratory. The next day, the report came back "greater than 200,000 CFUs/ml; multiple colony types; sample unsuitable for analysis." Why was this determination made?

25.2 Pathogen Identification Using Biochemical and Protein Profiles

SECTION OBJECTIVES

- Explain how biochemical profiles and result algorithms can be used to identify microbial pathogens.
- Discuss the major biochemical features that broadly differentiate Gram-negative and Gram-positive pathogens.
- Describe the Lancefield classification and its importance to clinical microbiology.
- Explain the difference between selective and differential media and their uses in clinical microbiology.
- Discuss differences between the acid-fast stain and Gram stain.
- Explain how generating a whole-cell protein profile with mass spectrometry can be used to rapidly identify bacterial pathogens.

Because thousands of bacterial species and viruses can cause disease, how can a laboratory quickly, sometimes within hours, identify which one is causing an infection? The solution, in part, comes from knowing the biochemical and enzymatic features of these bacteria. Because no two species have the same biochemical "signature," the clinical lab can look for reactions, or combinations of reactions, unique to a given species. This section discusses how to identify bacterial pathogens from their biochemistry.

CASE HISTORY 25.3

Medical Detective Work

 Cindy, a 38-year-old woman with no significant previous medical history, came to the emergency department complaining of a mild sore throat persisting for 3 days. Her symptoms included arthralgia (joint pain), myalgia (muscle pain), and low-grade fever. The day before, she had a severe headache with neck stiffness, nausea, and vomiting. She was not taking medications, had no known drug allergies, and did not smoke. Cindy lived with her husband and two children, all of whom were well. Cerebrospinal fluid (CSF) was collected from a spinal tap and sent to the diagnostic laboratory. The CSF appeared cloudy (it should be clear) and contained 871 WBCs (mainly neutrophils) per microliter (normal is 0–10/μl); the glucose level was 1 mg/dl (normal is 50–80 mg/dl); and the total protein level was 417 mg/dl (normal is under 45 mg/dl). Gram stain of a CSF smear revealed Gram-negative rods. The sample was streaked onto chocolate, blood, and Hektoen agars for microbial identification.

As discussed in Sections 24.4 and 25.1, low glucose and elevated protein levels in CSF indicate <u>bacterial</u> infection, not viral. The increase in WBCs revealed that the woman's immune system was trying to fight the disease. The presence of Gram-negative rods in the CSF smear confirmed a diagnosis of bacterial meningitis because CSF should be sterile. Now it was up to the clinical lab to determine the etiological agent.

Biochemical Algorithms to Identify Bacteria

How does a clinical microbiology laboratory handle incoming specimens such as the one considered in Case History 25.3? What are the first steps toward identifying the etiological agent of a disease? Over the years, clinical microbiologists have developed algorithms (step-by-step problem-solving procedures) that expose the most likely cause of a given infectious disease. For instance, only a limited number of microbes cause meningitis. The microbiologist poses a series of binary yes-or-no questions about the clinical specimen in the form of biochemical or serological tests. Typical questions in this case might include the following:

- Is an organism seen in the CSF of a patient with symptoms of meningitis?

- Is the organism Gram-positive or Gram-negative?

- Does it stain acid-fast?

Answers to a first round of questions will dictate the next series of tests to be used.

Because speed is essential in deciding how to treat the patient, a series of tests is not always carried out sequentially. To save time, a slew of tests are carried out simultaneously, but the results are <u>interpreted</u> sequentially, based on the algorithm. In Case History 25.3, for instance, consider the most common causes of bacterial meningitis (discussed in Chapter 24): *Neisseria meningitidis*, *Streptococcus pneumoniae*, *Haemophilus influenzae*, and *Escherichia coli*. A microbiologist Gram-stained the CSF sample and simultaneously plated it onto three media— chocolate agar, blood agar, and Hektoen agar. Chocolate agar is an extremely nutrient-rich medium that looks brown owing to the presence of heat-lysed RBCs (**Figure 25.8A and B**). Because it is so nutrient-rich, all four organisms will grow on chocolate agar. However, nutritionally fastidious organisms such as

Figure 25.8 **Chocolate Agar and Hektoen Agar: Two Widely Used Clinical Media**

A. Uninoculated chocolate agar. Its color is due to gently lysed red blood cells (RBCs) that provide a rich source of nutrients for fastidious bacteria.

B. Chocolate agar inoculated with *Neisseria gonorrhoeae*. This organism will not grow well on typical blood agar because important nutrients remain locked within intact RBCs, which *N. gonorrhoeae* cannot lyse.

C. Uninoculated Hektoen agar, which contains lactose, peptone, bile salts, thiosulfate, an iron salt, and the pH indicators bromothymol blue and acid fuchsin—the bile salts prevent growth of Gram-positive microbes.

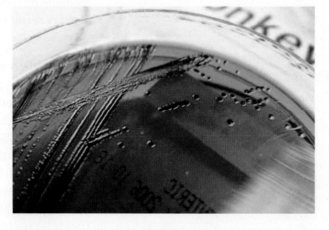

D. Hektoen agar inoculated with *Escherichia coli*. This organism ferments lactose to produce acidic fermentation products that give the medium an orange color owing to the pH indicators acid fuchsin and bromophenol blue.

E. Hektoen agar inoculated with *Salmonella enterica*. This organism does not ferment lactose but instead grows on the peptone amino acids. The resulting amines are alkaline and produce a more intense blue color with bromothymol blue. *Salmonella* species also produce hydrogen sulfide gas from the thiosulfate. Hydrogen sulfide reacts with the medium's iron salt to produce an insoluble, black iron sulfide precipitate visible in the center of the colonies.

N. meningitidis and *H. influenzae* will not grow well, if at all, on ordinary blood agar because these bacteria cannot lyse RBCs and release required nutrients. Less fastidious organisms, such as *S. pneumoniae* and *E. coli*, will grow on blood agar, but only *E. coli* can grow on Hektoen agar (**Figure 25.8C–E**), which is a selective and differential medium for enteric Gram-negative rods. Section 6.1 describes differential and selective media.

In the case history, the Gram stain of the CSF revealed Gram-negative rods, which ruled out *Neisseria* (a Gram-negative diplococcus) and *S. pneumoniae* (a Gram-positive diplococcus). The organism in CSF did grow on blood agar, which eliminated *H. influenzae* (a Gram-negative, nonenteric rod) as a candidate. The organism also grew on Hektoen agar, where it produced orange, lactose-fermenting colonies. Therefore, the organism was a Gram-negative, enteric rod—probably *E. coli*. Additional biochemical tests confirming the identity of the organism had to be carried out, but this simple example shows how simultaneous tests can be interpreted.

IDENTIFYING ENTERIC GRAM-NEGATIVE BACTERIA The Gram-negative bacterium afflicting Cindy in Case History 25.3 was subjected to a battery of 36 biochemical tests in an automated microbial identification instrument described below. A simplified example of the identification process is shown in **Figure 25.9**. The analytical profile index (API 20E) strip can test 20 metabolic processes. Such a test battery is called a **probabilistic indicator**. The results of all the individual tests are compared with those from a database of pathogen tests, which lists probabilities of a positive or negative result for each test and for each organism. The strip requires overnight incubation and tests whether the organism can ferment a series of carbohydrates or breakdown certain amino acids. It also provides evidence of specific products of fermentation. The results appear as colored reactions in each chamber and are scored as positive or negative, depending on the color (**Table 25.2**). For example, in the indole chamber (ninth well from the left in Figure 25.9B), a red reaction is positive and indicates that the organism can produce indole from tryptophan. A colorless chamber would be a negative result.

The results of the API chambers can be interpreted as a dichotomous key, with lab personnel making a stepwise interpretation starting with a key reaction, such as lactose fermentation (tagged ONPG in Table 25.2). The reaction is read as positive or negative on the basis of the color. Then, following a printed flowchart, the technologist would go to the next key reaction—say, indole production—and read it as positive or negative. If the organism is a lactose fermenter and the indole test is positive, the choices have been narrowed to *E. coli* or *Klebsiella* species. Another reaction is read to distinguish between the remaining choices. The process continues until a single species is identified.

A simplified example is shown in **Figure 25.10** for a limited number of enteric Gram-negative species. In reality, making a definitive identification takes many more reactions than the 11 shown. This is due to species differences with respect to a single reaction. For example, Figure 25.10 shows that *Klebsiella pneumoniae* and *K. oxytoca* exhibit opposite indole reactions, even though they are of the same genus. Even within a given species, only a certain percentage of strains might be positive for a given reaction. The inherent danger in using a dichotomous key is that one anomalous result can lead to an incorrect identification.

Most clinical laboratories in the United States and Europe now use more sophisticated automated identification systems. These include VITEK by bioMérieux, MicroScan WalkAway plus by Beckman Coulter, and BD Phoenix by Becton, Dickinson and Company (**Figure 25.11A**). All use cards or plates with 30 or more biochemical or enzymatic reactions (**Figure 25.11B and C**). As with the API chambers, the results appear as colored reactions readable by computer. The BD Phoenix system, for example, will convert results into a ten-digit code used to identify genus and species.

Figure 25.9 API 20E Strip Technology to Biochemically Identify Enterobacteriaceae

A. Uninoculated API strip. Each well contains a different medium that tests for a specific biochemical capability. The numbers correspond to those in Table 25.2. A color change in a well after 24-hour incubation indicates a positive or negative reaction (see Table 25.2).

B. API results for *E. coli*. Plus (+) and minus (–) indicate positive and negative reactions, respectively.

C. API results for *Proteus mirabilis*.

IDENTIFYING NONENTERIC GRAM-NEGATIVE BACTERIA The procedures we have outlined will accurately identify members of Enterobacteriaceae, but pathogenic Gram-negative bacilli can also be found in other phylogenetic families. One possibility in Case History 25.3 is that the Gram-negative bacillus seen in CSF smears would <u>not</u> grow on blood agar or on the other selective media. Instead, it would grow as small, glistening colonies on chocolate agar. This result would implicate the Gram-negative rod *Haemophilus influenzae*.

Table 25.2
Reading the API 20E [Reference Table]

Test	Substrate	Reaction Tested	Negative Results	Positive Results
1 ONPG	ONPG[a]	Beta-galactosidase	Colorless	Yellow
2 ADH	Arginine	Arginine dihydrolase	Yellow	Red/orange
3 LDC	Lysine	Lysine decarboxylase	Yellow	Red/orange
4 ODC	Ornithine	Ornithine decarboxylase	Yellow	Red/orange
5 CIT	Citrate	Citrate utilization	Pale green/yellow	Blue-green/blue
6 H$_2$S	Sodium thiosulfate	H$_2$S production	Colorless/gray	Black deposit
7 URE	Urea	Urea hydrolysis	Yellow	Red/orange
8 TDA	Tryptophan	Deaminase	Yellow	Brown-red
9 IND	Tryptophan	Indole production	Yellow	Red (2 min)
10 VP	Sodium pyruvate	Acetoin production	Colorless	Pink/red (10 min)
11 GEL	Charcoal gelatin	Gelatinase	No diffusion of black	Black diffuse
12 GLU	Glucose	Fermentation/oxidation	Blue/blue-green	Yellow
13 MAN	Mannitol	Fermentation/oxidation	Blue/blue-green	Yellow
14 INO	Inositol	Fermentation/oxidation	Blue/blue-green	Yellow
15 SOR	Sorbitol	Fermentation/oxidation	Blue/blue-green	Yellow
16 RHA	Rhamnose	Fermentation/oxidation	Blue/blue-green	Yellow
17 SAC	Sucrose	Fermentation/oxidation	Blue/blue-green	Yellow
18 MEL	Melibiose	Fermentation/oxidation	Blue/blue-green	Yellow
19 AMY	Amygdalin	Fermentation/oxidation	Blue/blue-green	Yellow
20 ARA	Arabinose	Fermentation/oxidation	Blue/blue-green	Yellow

[a]ONPG = o-nitrophenyl beta-D-galactoside.

Growth of *H. influenzae* requires hemin (X factor) and NAD (V factor). Therefore, confirming the identity as *H. influenzae* involves growing the organism on agar medium containing hemin and NAD. This can be done by placing small filter paper disks containing these compounds on a nutrient agar surface (not blood agar) covered with the organism. *H. influenzae* will grow only around a strip containing both X and V factors. Alternatively, X and V factors can be incorporated into Mueller-Hinton agar, as shown in **Figure 25.12A**. The organism grows on chocolate medium because the lysed RBCs release these factors. Although XV growth phenotype is still used for identification, fluorescent antibody staining (discussed shortly) is more specific.

The lab would have used a completely different identification scheme in the preceding meningitis case had workers found a Gram-negative diplococcus (rather than a Gram-negative rod). A Gram-negative diplococcus would suggest *Neisseria meningitidis*. *N. gonorrhoeae* is also possible, although less likely because the gonococcus lacks the protective capsule that meningococci use to survive in the bloodstream. Without this capsule, *N. gonorrhoeae* is not as resistant to serum complement and is unlikely to disseminate to the meninges.

The first test to determine whether the organism is a species of *Neisseria* is the **cytochrome oxidase test** (**Figure 25.12B**). In this test, a few drops of the colorless reagent N,N,N',N'-tetramethyl-p-phenylenediamine dihydrochloride are applied to the suspect colonies. The reaction, which takes place only if the organism possesses both cytochrome oxidase and cytochrome c, turns the p-phenylenediamine reagent (and the colony) a deep purple/

Figure 25.10
Simplified Biochemical Algorithm to Identify Gram-Negative Rods

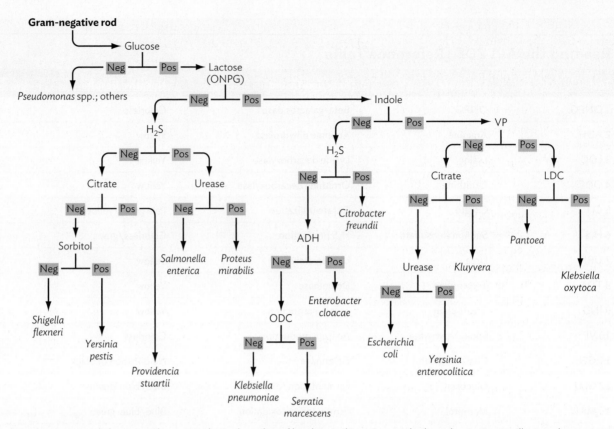

The diagram presents a dichotomous key using a limited number of biochemical reactions and selected organisms to illustrate how biochemistry is used to identify species. For example, if our specimen is positive for glucose and lactose fermentations, makes indole and acetoin (VP), and degrades lysine (LDC), it is identified as *Klebsiella oxytoca*. Abbreviations and reactions: ADH (arginine dihydrolase) = stepwise degradation of arginine to citrulline and ornithine; citrate = use of citrate as a carbon source; H_2S = production of hydrogen sulfide gas; indole = production of indole from tryptophan; lactose (ONPG) = fermentation of lactose produce acid (*o*-nitrophenyl beta-D-galactoside, cleaved by beta-galactosidase); LDC (lysine decarboxylase) = cleavage of lysine to produce CO_2 and cadaverine; ODC (ornithine decarboxylase) = cleavage of ornithine to make CO_2 and putrescine; sorbitol = fermentation of sorbitol to produce acid; urease = production of CO_2 and ammonia from urea; VP (Voges-Proskauer test) = production of acetoin or 2,3-butanediol. (*Note:* "Neg" for *Pseudomonas* in terms of glucose indicates an inability to ferment glucose. *Pseudomonas* can still use glucose as a carbon source.)

Figure 25.11 Automated Microbiology Identification System

A. The BD Phoenix 100 system uses plates with many reaction wells and a computerized plate reader to automatically identify pathogenic bacteria. Such an instrument automatically generates and evaluates the combination of reactions that develop as the bacteria grow.

B. Microbial identification plate.

C. Loading a multiwell ID plate with a multichannel pipettor. All wells are simultaneously loaded with the same volume and number of bacteria.

black. Although many bacteria possess cytochrome oxidase, *Neisseria* is one of only a few genera whose members also contain cytochrome *c* in their membranes, which is why this organism tests oxidase positive. Other oxidase-positive bacteria include *Pseudomonas, Haemophilus, Bordetella, Brucella,* and *Campylobacter* species—all Gram-negative rods. All Enterobacteriaceae, however, are oxidase-negative because they lack cytochrome *c*.

Finding an oxidase-positive, Gram-negative diplococcus in Cindy's case would strongly indicate that the pathogen is a member of the genus *Neisseria*. Differentiating between species of *Neisseria* then involves testing their ability to grow on certain carbohydrates. Once the organism is identified as *N. meningitidis,* its capsule type can be determined by immunofluorescent antibody staining.

Figure 25.12 *Haemophilus influenzae* **Growth Factors and** *Neisseria meningitidis* **Oxidase Reaction**

A. *H. influenzae* will grow on an agar plate (here, Mueller-Hinton agar) only when the medium has been fortified with both X factor (hemin) and V factor (NAD) but not with either one alone.

B. Oxidase-positive reaction for *N. meningitidis*. Oxidase reagent (which is colorless) was dropped onto colonies of *N. meningitidis* grown on chocolate agar. The test is called the cytochrome oxidase test, but it really tests for cytochrome *c*.

IDENTIFYING GRAM-POSITIVE PYOGENIC COCCI Recall the woman with necrotizing fasciitis (see Case History 19.3). How did the laboratory determine that the etiological agent was *Streptococcus pyogenes*? A sample algorithm, or flowchart (Figure 25.13), shows how. The physician sends a cotton swab containing a sample from a lesion to the clinical laboratory. The laboratory technologist streaks the material onto several media: (1) blood agar (on which both Gram-positive and Gram-negative organisms will grow), (2) blood agar containing the inhibitors colistin and nalidixic acid (called a **CNA plate**; only Gram-positive organisms will grow on this agar), and (3) MacConkey agar (on which only Gram-negative bacilli will grow) (see Section 6.1). The suspect organism in this case grows on the CNA and blood plates. Because the organisms grow in the presence of the compounds in CNA that inhibit growth of Gram-negative organisms, one would immediately suspect the organism to be Gram-positive. This is an assumption borne out by the actual Gram stain.

⊙ **Note:** Though skin is normally populated by many different microorganisms that could potentially contaminate a wound (like the necrotizing fasciitis in Case History 19.3), clinical samples from an infected lesion are populated primarily by the etiological agent. This occurs because the pathogenic microbe outgrows normal microbiota. Using selective media to isolate the infectious agent will further simplify diagnosis by reducing growth of any normal microbiome members still present.

The algorithm tells the lab technologist that because the organism is a Gram-positive coccus, the next step is to test for **catalase** production. Catalase, which converts hydrogen peroxide

(H_2O_2) to O_2 and H_2O, clearly distinguishes staphylococci from streptococci (remember that the morphology, or shape, of Gram-stained cells alone is not reliable enough to make that distinction). The **catalase test** is performed by mixing a colony with a drop of H_2O_2 on a glass slide. Effusive bubbling due to the release of oxygen indicates catalase activity (see Figure 25.13). Staphylococci are catalase-positive, whereas streptococci are catalase-negative. Many other organisms possess catalase activity, including the Gram-negative rod *E. coli*. However, on the basis of the algorithm, *E. coli* would not be considered because it does not grow on CNA agar and is not Gram-positive.

⊙ **Note:** When performing a catalase test from colonies grown on blood agar, be sure not to transfer any of the agar because red blood cells also contain catalase.

Having established that the organism is catalase-negative, the technologist examines the blood agar plate for evidence of **hemolysis**—the microbe's ability to directly lyse red RBCs. Three types of colonies are possible: nonhemolytic, alpha-hemolytic, and beta-hemolytic.

1. **Nonhemolytic** streptococci grown on blood agar do not produce a lytic zone around their colonies.

2. **Alpha-hemolytic** strains produce large amounts of hydrogen peroxide that oxidize the heme iron within intact RBCs to produce a green product. These streptococci produce a green zone around their colonies as a result of **alpha hemolysis**—even though the RBCs remain intact. (For example, *Streptococcus*

Figure 25.13 Algorithm to Identify Gram-Positive Pathogenic Cocci

The red arrows follow the identification of *Streptococcus pyogenes*. The bacitracin and optochin results are designated "positive" if the organism is susceptible (note the zone of no growth around the disk) and "negative" if the organism is resistant to the agent.

Streptococcus pneumoniae

Streptococcus pyogenes

mutans, a cause of dental caries and subacute bacterial endocarditis, is alpha-hemolytic.)

3. **Beta-hemolytic** streptococci produce a completely clear zone of true hemolysis surrounding their colonies. Complete or **beta hemolysis** is caused by secreted bacterial enzymes called hemolysins that lyse the membranes of RBCs. The flowchart in Figure 25.13 indicates that the organism from Case History 19.3 was beta-hemolytic.

The final relevant test in this flowchart involves susceptibility to the antibiotic bacitracin, which identifies the most pathogenic

group of beta-hemolytic streptococci. The beta-hemolytic streptococci are subdivided into many groups, known as **Lancefield groups** (or **Lancefield classification**), based on differences in the composition of their cell wall carbohydrate antigens. These cell wall differences are distinguished from each other immunologically and divide the streptococci into Lancefield groups A–U. This classification scheme is named after Rebecca Lancefield, who was the first to use immunoprecipitation to group the streptococci. Group A beta-hemolytic streptococci (also called GAS), defined as the species *S. pyogenes*, cause most streptococcal diseases.

Unfortunately, the Lancefield classification procedure is somewhat time-consuming and not readily amenable as a rapid identification method. However, group A beta-hemolytic streptococci are uniformly susceptible to the antibiotic bacitracin. Therefore, a simple bacitracin disk susceptibility test can be used to indicate group A beta-hemolytic streptococci (that is, *S. pyogenes*; Figure 25.13). But beware: many bacteria are bacitracin sensitive.

As with the catalase test, the bacitracin test must be used as part of an algorithm to be useful for identification. The technologist must follow the appropriate algorithm before assigning importance to this or any other test result. For instance, whether an alpha-hemolytic organism is bacitracin susceptible is irrelevant. Some such organisms may exist, but they are not associated with disease. The organism in our case of necrotizing fasciitis, however, was beta-hemolytic, so bacitracin susceptibility indicated that the organism was *S. pyogenes*.

The other tests named in Figure 25.13 are equally important for identifying Gram-positive infectious agents. For example, *Streptococcus pneumoniae* is an important cause of pneumonia. Like *S. pyogenes*, *S. pneumoniae* is a catalase-negative, Gram-positive coccus; unlike *S. pyogenes*, it is alpha-hemolytic. Optochin susceptibility is a property closely associated with *S. pneumoniae*, whereas other alpha-hemolytic strains of streptococci are resistant to this compound. So, an **optochin susceptibility disk test** is a useful tool to identify *S. pneumoniae*.

Coagulase is a key reaction used to distinguish the pathogen *Staphylococcus aureus*—which causes boils and bone infections—from other staphylococci, such as the normal skin species *S. epidermidis*. To conduct a coagulase test, a tube of plasma is inoculated with the suspect organism. If the organism is *S. aureus*, it will secrete the enzyme coagulase. Coagulase will convert fibrinogen to fibrin and produce a clotted, or coagulated, tube of plasma. (More recently, diagnostic labs are using an antibody assay method to more quickly detect coagulase [immunochromatographic enzyme immunoassay; see Section 25.5].) Coagulase-negative staphylococci can still be medically significant, however. *S. saprophyticus*, for instance, is an important cause of urinary tract infections. It can be distinguished from *S. epidermidis* on the basis of resistance to the antibiotic novobiocin.

IDENTIFYING ACID-FAST BACTERIA A few pathogenic bacteria cannot be Gram stained, most notably *Mycobacterium tuberculosis*, the dreaded agent of tuberculosis (Section 20.3). Mycobacteria do not Gram stain well because the organisms sport a waxy outer coat composed of **mycolic acid** that resists penetration by most dyes. However, a harsher procedure called the **acid-fast stain**, a technique described by German scientists Franz Ziehl and Friedrich Neelsen in 1882, can stain the organism. The original Ziehl-Neelsen acid-fast stain used phenol and heat to drive carbolfuchsin (a red dye) into mycobacterial cells on glass slides. Destaining with an acid alcohol solution removes the stain from all bacterial pathogens <u>except</u> mycobacteria. After rinsing, the slide is counterstained with methylene blue. Mycobacteria will be seen as curved, red rods (called acid-fast bacilli), whereas everything else looks blue, including other bacteria (Figure 25.6C).

A more modern version of the acid-fast stain uses the fluorescent dye called auramine O to stain the mycolic acid. This fluorescent dye also resists removal by an acid alcohol wash, and

Figure 25.14 **Fluorescent Acid-Fast Stain and Growth of *Mycobacterium tuberculosis***

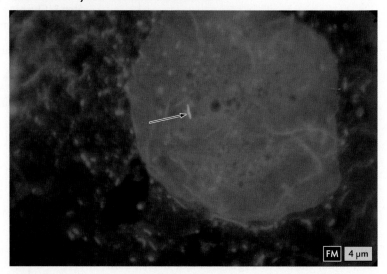

A. Auramine O acid-fast stain of *M. tuberculosis*. Figure 25.6C shows a standard acid-fast stain.

B. Löwenstein-Jensen medium enables growth of mycobacterial species, some of which grow extremely slowly. The colonies look like bread crumbs.

when observed under a fluorescence microscope, the dye causes mycobacteria to fluoresce bright yellow (**Figure 25.14A**). A presumptive diagnosis of tuberculosis is made after finding acid-fast bacilli in a patient's sputum.

Because *M. tuberculosis* grows slowly, taking 4 weeks to form colonies on agar media, other fast-growing organisms present in sputum can easily overgrow tubercle bacilli on ordinary media. Löwenstein-Jensen (LJ) medium is a selective medium designed to grow *M. tuberculosis*. LJ medium contains antibiotics to inhibit the growth of other bacteria and uses coagulated eggs and potato starch as nutrients. The medium is typically used as slanted agar in a test tube rather than a petri plate (**Figure 25.14B**). Biochemical tests can identify *M. tuberculosis* from among nontuberculosis mycobacteria, but newer nucleic acid–based methods (described in Section 25.3) can do this much faster than traditional culture-based techniques.

Pathogen Identification by Protein Signature

Traditional biochemical identification can be a frustrating, time-consuming process that takes at least 2–3 days to complete. However, a new, faster technology that also probes bacterial biochemistry has been introduced into many clinical labs. It is called **mass spectrometry (MS)**. More specifically, the technique is called matrix-assisted laser desorption ionization–time of flight mass spectrometry or **MALDI-TOF MS**. MALDI-TOF MS can provide results in minutes compared to traditional biochemical profile methods. The technique, shown in **Figure 25.15**, is based on the fact that each pathogen has a specific protein composition that defines its own unique protein signature or fingerprint.

Briefly explained, bacterial cells from a colony are fixed in a matrix (Figure 25.15, step 1) and irradiated by a laser beam (step 2). The laser releases proteins from the fixed cells and places a charge on them (ionizes them). The intact, but positively charged, proteins enter a vacuum where they are accelerated by an electrostatic field (step 3). The charged proteins then race toward a detector at different speeds that depend on each ion's charge and mass. Large (heavy) proteins with a high mass-to-charge ratio take longer to reach the detector (called <u>time of flight</u>) than do smaller (lighter) proteins having a low mass-to-charge ratio. The resulting pattern of peaks, called the protein signature (step 4), is compared against a growing database derived from different species of microorganisms previously subjected to the procedure. The procedure can be used on positive blood cultures (sepsis) or directly on urine samples (cystitis) and cerebrospinal fluid (meningitis). Many large and small laboratories are rapidly moving toward use of this new technology.

Figure 25.15 MALDI-TOF MS Identification of Bacterial Pathogens

Step 1: Bacterial colony is spotted into a well of the sample plate. **Step 2:** Laser ionizes proteins. **Step 3:** Charged proteins of different sizes travel at different speeds to the detector. **Step 4:** Mass/charge peaks form a signature used to identify the organism. Each peak represents a different protein.

SECTION SUMMARY

- **Selective media** inhibit growth of one group of organisms while permitting the growth of others (such as Gram-positive versus Gram-negative bacteria). This technique is often used to prevent growth of normal microbiota while permitting the growth of pathogens.

- **Differential media** exploit the unique biochemical properties of a pathogen to distinguish it from similar-looking nonpathogens.

- **Identification through biochemical tests** uses a dichotomous key in which each test eliminates possible pathogens until only one remains. **Glucose and lactose fermentation,** for instance, are often the first tests used to separate species of Gram-negative bacilli.

- **Automated systems** can simultaneously evaluate 30 or more biochemical tests and determine the probability that the biochemical profile of a pathogen recovered from a patient matches that of a given species.

- **The catalase test** distinguishes the **Gram-positive cocci** recovered from a patient as *Staphylococcus* (catalase positive) or *Streptococcus* (catalase negative). The **coagulase test** differentiates *Staphylococcus aureus* (coagulase positive) from *Staphylococcus epidermidis* (coagulase negative).

- **Hemolysis** differentiates catalase-negative, Gram-positive cocci (*Streptococcus*) into nonhemolytic, alpha-hemolytic, and beta-hemolytic groups. **Bacitracin susceptibility** of a beta-hemolytic streptococcus indicates the strain is *S. pyogenes*.

- **The Lancefield classification** differentiates beta-hemolytic streptococci into serological groups. Group A streptococci (*S. pyogenes*) and group B streptococci (*S. agalactiae*) are the most common pathogens.

- **The acid-fast stain** differentiates *Mycobacterium* and some *Actinomyces* species from all other bacteria.

- **Mass spectrometry** is now used to identify pathogens by determining the molecular weight of their proteins. The result is a protein signature that is unique to each pathogen.

Thought Question 25.3 Use Figure 25.13 to identify the organism in the following case. A sample was taken from a boil located on the arm of a 5-year-old boy. Bacteriological examination revealed the presence of Gram-positive cocci that were also catalase-positive, coagulase-positive, and novobiocin-resistant.

Thought Question 25.4 Staphylococci from blood cultures were isolated from two body sites, but the laboratory says the organisms are probably not the cause of the patient's disease. Why might the laboratory technologist reach that conclusion?

25.3 Pathogen Identification Using DNA Methods

SECTION OBJECTIVES

- Discuss how the polymerase chain reaction can be used to identify pathogens in a clinical sample and the role of real-time PCR in clinical diagnosis.

- List the advantages and disadvantages of PCR, restriction fragment length polymorphisms, and biochemical strategies of pathogen identification.

What is the quickest way to identify a pathogen from a clinical specimen? From the time the clinical lab receives a specimen, it can take days, or sometimes weeks, to determine the biochemical profile of the likely agent. This can be a problem if the clinician has chosen the wrong empirical antimicrobial therapy to treat the patient. Mass spectrometry is fast, but it requires extremely expensive instrumentation.

Identifications made from clinical specimens by nucleic acid tests are very fast. Because we now know the genome sequences of many pathogens, clinical labs can use rapid, nucleic acid–dependent techniques to identify a particular pathogen. Detecting DNA using PCR, for example, takes only a few hours (Section 8.6). DNA/RNA

detection methods are especially useful for viruses, which otherwise require tissue culture techniques to grow the virus or serology to detect an increase of antiviral antibodies. Recall that significant quantities of antibodies may not appear in serum until the patient is recovering. Also, molecular detection techniques may provide the <u>only</u> means to identify newly recognized, or emerging, pathogens.

Polymerase Chain Reaction

The polymerase chain reaction (PCR) is the most widely used molecular method in the clinical laboratory's diagnostic tool kit. DNA primers that bind to unique genes in a pathogen's genome can be used to specifically amplify DNA or RNA present in a clinical specimen. Successful amplification can be seen as an appropriately sized fragment in agarose gels after electrophoresis or, more typically, by real-time techniques, which we will discuss later.

Why do we need PCR to detect these nucleic acids? Without the PCR amplification steps, clinical samples usually provide too little nucleic acid from infecting microorganisms to be detected. For example, *Mycobacterium tuberculosis*, the cause of tuberculosis, can take weeks to grow on standard bacteriological media. Detecting its DNA in the specimen, however, yields a rapid diagnosis. Unfortunately, clinical samples have only minute amounts of pathogen DNA to detect. PCR, however, amplifies *M. tuberculosis* nucleic acid, turning one copy of DNA into billions of copies.

Preparing a clinical specimen for PCR by extracting the DNA or RNA usually takes less than an hour. PCR itself is completed in 2–3 hours. What might take 2–3 days (or sometimes weeks) using biochemical algorithms may take less than a day (sometimes less than 10 minutes) with molecular strategies. Most clinical laboratories are equipped with the **thermocyclers** needed to carry out PCR identification of known pathogens. A thermocycler is an instrument that precisely and rapidly cycles the temperatures required for the melting, annealing, and polymerizing steps of PCR.

Table 25.3 lists instances in which DNA detection tests are useful. An example in **Figure 25.16** shows using PCR to type different strains of the anaerobic pathogen *Clostridium botulinum*, the cause of food-borne botulism (Section 24.4). Unlike *Streptococcus pneumoniae*, these organisms are not typed serologically. Rather, *C. botulinum* is divided into types based on the neurotoxin genes they possess. In this example, **multiplex PCR** was used to simultaneously search for these toxin genes. Multiplex PCR uses multiple sets of primers, one pair for each gene, combined in one tube with a specimen. One must ensure that the primers chosen yield different-sized products, do not interfere with each other, and do not produce false PCR products that can confound interpretation. Multiplex PCR can help identify sets of specific genes present in a single species or can screen for multiple pathogens in a clinical sample. In the latter case, primer sets are designed to amplify genes unique to each pathogen.

Table 25.3
Examples of Nucleic Acid–Based Tests Used to Detect Pathogens [Reference Table]

Test	Intended Use	Specimen for PCR	Transport Conditions (Temperature and Maximum Time for Arrival at Lab)
Detection of HIV proviral DNA	To diagnose HIV infection in newborns; resolve indeterminate serological results	>2 ml of whole blood in EDTA tube[a]	Room temperature, within 24 hours after collection
Detection of *Bordetella pertussis*	To diagnose whooping cough (pertussis)	Nasopharyngeal swab	Room temperature, within 24 hours after collection
Detection of *Borrelia burgdorferi*	To diagnose/monitor Lyme disease	Tick, skin biopsy, >1 ml of spinal fluid, urine, blood	Wrap tick in moist tissue in small plastic bag; 4°C (cold packs), within 24 hours after collection
Detection of *Rickettsia rickettsii*	To diagnose/monitor Rocky Mountain spotted fever	Tick, >2 ml of whole blood in EDTA tube, skin biopsy	Biopsied skin lesion kept at 4°C; wrap tick in moist tissue in small plastic bag; biopsied skin lesion should be tested within 24 hours after collection
Detection of *Ehrlichia chaffeensis*	To diagnose/monitor human monocytic ehrlichiosis	Tick, >2 ml of whole blood in EDTA tube	Wrap tick in moist tissue in small plastic bag; room temperature, within 24 hours after collection
Detection of norovirus and other viruses	To diagnose viral gastroenteritis; to investigate food-borne and waterborne outbreaks	>1 ml of diarrheal stool in sterile container	4°C (cold packs), within 72 hours after collection
Typing of *Mycobacterium tuberculosis*	To determine relatedness of *M. tuberculosis* isolates; to investigate suspected outbreaks	Sputum, urine	Room temperature
Typing of various Gram-negative and Gram-positive bacteria	To determine relatedness of isolates; to investigate suspected nosocomial or food-borne outbreaks	Food, feces	4°C (cold packs), within 24 hours after collection (if *Campylobacter* is suspected, transport at room temperature)

[a]EDTA = ethylenediaminetetraacetic acid; this compound prevents coagulation by chelating (binding) ions.

Figure 25.16 Using Multiplex PCR to Identify *Clostridium botulinum*

C. botulinum cells are typed by which toxin genes a strain possesses, not on the basis of surface antigens. Isolates can be typed in a single round of PCR that includes primer pairs for each of the four major toxin genes: types A, B, E, and F. Because multiple products are sought in a single reaction, this is called multiplex PCR. Each lane in the agarose gel was loaded with multiplex products from different isolates of *C. botulinum* and subjected to electrophoresis. Recall from Section 8.6 that DNA, which is negatively charged because of its phosphate backbone, is loaded into the top of the gel and moves toward a positive electrode at the bottom. The slower-moving fragments (toward top of gel) are larger than those that have moved farther down the gel toward the positive pole. Lane 1, DNA size markers; lanes 2–5, individual strains of *C. botulinum*; lanes 6–8, mixtures of strains as indicated. bp = base pairs.

Quantitative Reverse Transcription–Polymerase Chain Reaction

CASE HISTORY 25.4

A Stiff Neck and a Stealthy Mosquito

A 55-year-old man complaining of headache, high fever, and neck stiffness was admitted to a local hospital. The man appeared confused and disoriented. He also complained of muscle weakness. History indicated that he had received several mosquito bites approximately 2 weeks before. CSF and blood specimens were sent to the lab. The report the next day indicated no bacterial infection. However, quantitative reverse transcription–PCR tests for various possible viral causes revealed that the patient was suffering from West Nile virus.

West Nile virus (WNV) is primarily an infection of birds and culicine mosquitoes (a group of mosquitoes that can transmit human diseases). Humans and horses serve as incidental, dead-end hosts for WNV. An incidental host is not part of the organism's normal life cycle. Section 24.3 describes other features of WNV disease. Here we focus on diagnosis.

Many laboratories are not equipped to carry out the special tissue culture techniques needed to grow viruses. Consequently, viral infections, including human WNV infections, are usually diagnosed by measuring the patient's antibody response. For instance, the presence of West Nile virus–specific immunoglobulin M (IgM) in cerebrospinal fluid is a good indicator of current WNV infection, but it's indirect and not conclusive. Quantitative reverse transcription–PCR is a molecular test that can quickly reveal the presence of the virus's RNA in blood or tissue samples.

Quantitative reverse transcription–PCR (qRT-PCR) is used routinely for the high-throughput diagnosis of viral pathogens such as WNV. Because the genome of WNV is made of single-stranded RNA, its genome must first be converted to complementary DNA (cDNA) by using an enzyme called reverse transcriptase before PCR can be attempted. The quantitative advantage of qRT-PCR is that you can estimate the number of virus particles present in the sample from the number of viral RNA molecules there are. Another advantage is that qRT-PCR does not require gel electrophoresis.

The basic technique is as follows: Viral RNA is extracted from the sample and converted to cDNA by using a single, virus-specific primer and the enzyme reverse transcriptase (**Figure 25.17**). The more virus particles in the sample, the more cDNA product is made. The cDNA is then amplified by PCR using two specific primers and a heat-stable DNA polymerase such as Taq polymerase (Figure 8.26 describes how PCR exponentially increases copies of DNA). The trick with qRT-PCR is in the method used to detect and quantify amplification.

In one method, a third, fluorescent DNA oligonucleotide (called a **molecular beacon probe**) is added to the PCR tube (**Figure 25.18**).

Figure 25.17 Reverse Transcriptase Synthesis of DNA

Conversion of mRNA to cDNA by reverse transcription

1 A virus-specific DNA primer is added to an RNA prep from a clinical sample.

2 Reverse transcriptase (RT) copies first cDNA strand.

3 RT digests and displaces mRNA and copies second strand of cDNA.

4 The result is double-stranded cDNA.

Reverse transcriptase (RT) uses mRNA as a template to make DNA. The DNA can then be amplified by using standard PCR methods. A specific oligonucleotide primer added to the reaction tube will anneal to the 3′ end of the viral RNA, and RT will synthesize the first strand of a complementary DNA (cDNA). Many RT enzymes also have an RNase activity that will degrade the original RNA template as the enzyme synthesizes the second strand of cDNA. Subsequent amplification by PCR requires adding a thermostable DNA polymerase (Taq polymerase) that can withstand the denaturing and DNA synthesis temperatures required to amplify the cDNA.

Figure 25.18 Real-Time Polymerase Chain Reaction

1 The reporter probe contains a fluorescent dye and a quenching dye, so no fluorescence is emitted.

2 Target DNA is denatured at 95°C (only ssDNA shown).

3 Temperature is lowered to 55°C. The reporter probe anneals downstream of a DNA primer. Still no fluorescence.

4 Temperature is raised to 72°C. Taq polymerase extends upstream DNA and degrades reporter. The release of the fluorescent dye from the vicinity of the quencher allows fluorescence, which is measured.

Taq polymerase extends an upstream primer and reaches the downstream reporter probe, where a 5′-to-3′ exonuclease activity degrades the probe. This releases the fluorescent dye from the vicinity of the quencher, so it can now fluoresce.

impact

Next-Generation Pathogen ID: Saved by Metagenomics

This chapter has described many strategies clinical microbiologists use to identify pathogens. But sometimes the standard methods fail. What then? This story describes what can be done and illustrates the future of diagnostic microbiology. In 2014, a 14-year-old boy with severe combined immunodeficiency (SCID) was in a medically induced coma at a Wisconsin hospital. His brain had swollen with fluid as a result of encephalitis, but the cause of the disease was a mystery. The usual diagnostic tests and antibiotic treatments did not provide answers or relief. In a last-ditch attempt to find the infection's cause, the boy's physicians called Joseph DeRisi and Charles Chiu at the University of California, San Francisco. Drs. DeRisi and Chiu are renowned for their use of next-generation DNA sequencing to identify pathogens.

Next-generation DNA sequencing is incredibly fast. The process, simplified here, begins by cutting a genome into millions of small, but overlapping, bits. Each single-stranded DNA molecule is randomly attached at one end to a small slide. When attachment is complete, the DNA fragments resemble a field of wheat waving in the wind (**Figure 1A**). All the strands are then sequenced together, one base at a time. The sequencer adds fluorescent DNA nucleotides labeled with one of four differently colored fluorescent tags (**Figure 1B**, first round). Nucleotides that bind to the first base of the thousands of different strands will fluoresce in their assigned colors and the instrument takes a photographic record. The sequencer then adds another round of nucleotides to label the second base of each template and removes the fluorescent tag from the previous base. This process is repeated until all the bases of all the templates have been sequenced. Photos taken at the end of each round are stacked by the computer, and the strings of colors for each strand are translated into DNA sequences. The computer program then compares

Figure 1. Next-Generation Sequencing

Strands of fragmented genomic DNA for sequencing

A. Small fragments (approximately 200 bp) of single-stranded DNA from a genome are affixed to a slide. There can be many thousands of fragments on a single slide. Each strand is amplified to produce a tuft of identical strands at each site (not shown).

Mix of fluorescently tagged deoxyribonucleotides added

1st round

Strand 1   Strand 2

Template sequence

T C

Deoxyribonucleotide key

A G T C

— Untagged

— Fluorescently tagged

Primers

2nd round   **6th round**

Template sequence   Template

A C   T C
T C   T G
              T A
              G C
              A C
              T C

B. Simplified steps of the sequencing process. Shown is a small portion of a sequencing slide with two different strands of genomic DNA. A DNA primer that binds to all fragments is added along with fluorescently tagged deoxyribonucleotides (see deoxyribonucleotide key). Pairing occurs at the first base of each strand in round 1. Unused fluorescent nucleotides are removed and a photo is taken to show the color of the fluorescent tag on the first base. A new batch of tagged deoxynucleotides is added to label the second base of the sequence and removes fluorescence from the previously added base (indicated here by dulled colors). The process is repeated until the sequence is finished. The sequencer reads the photos after each round to determine the sequence.

Figure 2. *Leptospira* Species

SEM 5 µm

all the sequences, finds those that overlap, and assembles the mass of information into a genome. An entire genome can now be sequenced within a couple of hours.

Using this technology, DeRisi and Chiu performed an unbiased metagenomic analysis of <u>all</u> the DNA contained in the boy's cerebrospinal fluid. No culturing was needed! After 3 million DNA fragments were sequenced, the answer was apparent. There, staring out from the sequences of the boy's genome, were DNA sequences of the spirochete *Leptospira santarosai*, an easily treatable organism. After being given intravenous high-dose penicillin for 7 days, the once-doomed boy eventually recovered.

Leptospirosis is a zoonotic disease. The organism (Figure 2) is excreted in the urine of infected animals, usually into nearby bodies of freshwater. The boy probably contracted the disease during a trip to Puerto Rico, where he swam in freshwater. This metagenomic strategy to detect pathogens is not widely used yet because of its complexity and cost, but someday it could be used to diagnose any infection.

DISCUSSION QUESTION

Discuss why this technique would be useful to identify a pathogen in CSF but problematic for identifying a pathogen in stool.

The beacon probe contains a fluorescent dye at its 5′ end and a chemical-quenching dye at the 3′ end that absorbs the energy emitted from the fluorescent dye. Like a pillow covering a flashlight, no light is emitted so long as the intact probe tethers the two chemicals close to each other. During amplification, the beacon probe binds to the target cDNA between the binding sites of the two other primers. In a successful amplification, Taq or another heat-stable polymerase will synthesize DNA inward from the two outside primers and degrade the centrally placed beacon probe oligonucleotide as it passes through that area. This cleavage separates the dye from the quencher, and the fluorescent dye begins to fluoresce. The more cDNA present to begin with, the fewer amplification cycles needed to register a fluorescence increase over background (**Figure 25.19**). Because fluorescence is measured during the amplification (time-consuming gel electrophoresis is not involved), the process is also called real-time PCR.

In addition to the standard nucleic acid–based methods just described, a new DNA technology is being used to identify hard-to-find infectious agents. The method involves rapid, next-generation DNA sequencing of patient samples. The **IMPACT** feature shows how this method, though not universally available, successfully saved a boy's life.

CASE HISTORY 25.5

Tracking a Slow-Growing Pathogen

From August through December, 45 patients in the pediatric surgery unit of a New Delhi hospital developed postoperative wound infections. Of these patients, 42 were outpatients, whereas three were inpatients who had undergone major surgery. The diseases ranged in severity from chronic ear infections to bacteremia from using hemodialysis equipment. Thirty-two clinical samples of pus and wound exudates stained positive for acid-fast bacilli (Figure 25.14A) and were cultured on Löwenstein-Jensen slants for several weeks (Figure 25.14B). Biochemical tests indicated that the organisms from all patients were mycobacteria. PCR of the 16S rRNA gene identified the slow-growing organism as *Mycobacterium abscessus*. Using DNA fingerprint analysis (restriction fragment length polymorphisms) of different *M. abscessus* strains isolated from patients and from different locations around of the hospital, infection control workers traced the source of the outbreak to the tap water in the operating room and to a defective autoclaving process. (The defective autoclave process resulted from a leaky vacuum pump and faulty pressure gauge in the autoclave).

As noted in Section 25.2, one can use the acid-fast stain to make a presumptive diagnosis of *Mycobacterium tuberculosis*. Although

Figure 25.19 Results of Real-Time PCR

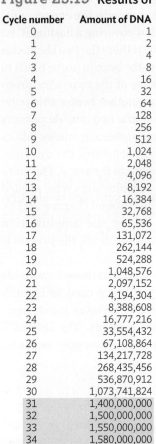

Cycle number	Amount of DNA
0	1
1	2
2	4
3	8
4	16
5	32
6	64
7	128
8	256
9	512
10	1,024
11	2,048
12	4,096
13	8,192
14	16,384
15	32,768
16	65,536
17	131,072
18	262,144
19	524,288
20	1,048,576
21	2,097,152
22	4,194,304
23	8,388,608
24	16,777,216
25	33,554,432
26	67,108,864
27	134,217,728
28	268,435,456
29	536,870,912
30	1,073,741,824
31	1,400,000,000
32	1,500,000,000
33	1,550,000,000
34	1,580,000,000

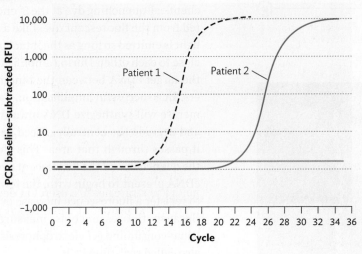

A. The exponential increase in PCR products after each cycle of hybridization and polymerization is shown. The switch to yellow indicates the point where the increase in product plateaus because the primers have been exhausted.

B. An increase in relative fluorescence units (RFUs) occurs as the fluorescent dye is released from the dual-labeled probe during real-time PCR. In patient 1, the dashed black curve shows that fluorescent PCR product remained flat for the first 10 cycles because the amount of DNA made, and therefore the level of fluorescent dye released, remained below background level (horizontal blue line). After cycle 10, fluorescence increased logarithmically. For patient 2, the solid red curve representing PCR product remained below detection for 22 cycles and then increased. The more DNA present at the start, the sooner the RFU values will increase over background (that is, the fewer heating cycles will be needed to see the increase over background). The slope eventually decreases because the fluorescent probe has become limiting.

the acid-fast stain is very useful, sometimes organisms cannot be found in a sputum sample. And, even if they are found, a definitive diagnosis requires confirmatory tests. This case history points out an underlying problem with some biochemical identifications. Growing enough organisms to perform these biochemical tests took weeks. An alternative PCR test that probed for *Mycobacterium*-specific genes encoding 16S rRNA could have been performed directly on the clinical specimen. This alternative test would have confirmed the diagnosis immediately, much faster than biochemical methods. Instead, in the case presented, the PCR test was not performed until after the cells were grown in the lab.

Restriction Fragment Length Polymorphism Analysis

But how did the lab trace the source of the pathogen? The DNA-based "fingerprint" method used in this case was **restriction fragment length polymorphism (RFLP)** analysis. Genome sequences from different strains of a single species are slightly different. Because of these small sequence differences, the number and location of restriction enzyme sites in the chromosome will differ. As a result, restriction enzyme digestion of genomic DNA

from different species will produce uniquely sized DNA fragments, like fingerprints. The clinical lab compared the restriction patterns produced from mycobacteria isolated from different areas throughout the New Delhi hospital in the case history. The results allowed the infection control staff to trace the source of all the infections to the operating room's tap water.

One method of RFLP analysis is shown in **Figure 25.20**. DNA samples extracted from different strains of *Mycobacterium tuberculosis* are cut with the same restriction enzyme, and the digested fragments are separated by gel electrophoresis (**Figure 25.20B**). The gel is then probed with a radioactively labeled fragment of insertion sequence IS*6110* (a fluorescent probe can also be used). IS*6110* is a repetitive DNA sequence present at multiple sites in the chromosome. The radioactive probe can hybridize only to DNA fragments that contain IS*6110* sequences. Strains of *M. tuberculosis* obtained from different outbreaks would display different hybridization patterns, similar to a bar code. Strains isolated from the same outbreak would look similar, if not identical (see Figure 25.20B).

SECTION SUMMARY

- **DNA-based technologies** can identify bacterial pathogens faster than traditional culture methods.
- **The polymerase chain reaction (PCR)** can identify a pathogen in a specimen within 1–2 hours.
- **Multiplex PCR** can, in a single reaction, detect two or more pathogens in a sample.
- **Real-time qRT-PCR (quantitative reverse-transcription PCR)** can detect the presence and quantity of RNA viruses in a sample.

Figure 25.20 DNA Fingerprinting of *Mycobacterium tuberculosis*

IS*6110*, a repetitive DNA sequence

1 Restriction endonucleases cleave chromosomal DNA at restriction sites.

Restriction sites

2 Some DNA fragments contain IS*6110* (yellow).

3 Gel electrophoresis separates fragments by size.

A. Genetic fingerprinting of *M. tuberculosis* isolates. All DNA fragment sizes produced by restriction enzyme cleavage are shown. Fragments containing insertion sequence IS*6110* are marked yellow.

4 Fragments containing IS*6110* hybridize to a specific radioactive probe.

1 1 2 2 3 3 4 4 5 5 6 6 7 7 8 8 9 9 9 101010

B. Each lane in the gel contains DNA extracted from different isolates. Fragments containing IS*6110* hybridize to a specific radioactive probe and become visible. A characteristic banding pattern (fingerprint) appears for each isolate. Isolates with similar banding patterns are assigned the same number in this example.

- **Restriction fragment length polymorphism (RFLP) analysis** can be used to identify different strains of a pathogen and trace an outbreak of a disease to its source.

Thought Question 25.5 How would you design a multiplex PCR test for bacteria using real-time quantitative PCR? A good example would be for *Chlamydia trachomatis* and *Neisseria gonorrhoeae,* two common causes of sexually transmitted disease.

25.4 Pathogen Identification Using Immunological Methods

SECTION OBJECTIVES

- Describe how enzyme-linked immunosorbent assays can identify microbial antigens present in clinical samples or antimicrobial antibodies present in serum.
- Discuss the fluorescent antibody staining technique.

In addition to biochemical and nucleic acid–based tests, immunological strategies are useful in detecting a pathogen or identifying the "footprints" of a pathogen's past or recent presence. When <u>direct</u> detection of a pathogen is not possible, detection of specific IgM antibodies or microbial antigens may also help diagnose an acute infection.

Enzyme-Linked Immunosorbent Assay

CASE HISTORY 25.6
Bloody Death in West Africa

During March 2014, about 60 people in the southeastern forest regions of Guinea, Africa, became ill with high fever, diarrhea, headache, vomiting, and gastrointestinal bleeding from the rectum. At least 23 people died. The presumptive diagnosis was Ebola hemorrhagic fever. Laboratory confirmation included viral antigen and antibody detection by enzyme-linked immunosorbent assay (ELISA). Lab-confirmed Ebola patients were defined as patients who were positive for either Ebola virus antigen or Ebola IgG antibody. Once the virus was identified, rigorous quarantine mechanisms were implemented in the hope of preventing outbreaks in other villages and countries. Despite these efforts, by August 2014 the Ebola epidemic had spread into the neighboring countries of Sierra Leone, Liberia, and (by air travel) Nigeria. By then the outbreak involved approximately 1,300 presumed or confirmed cases, more than half of whom died. As of January 2015, a mere 9 months after it started, the Ebola crisis in West Africa had claimed 7,000 lives and showed little sign of stopping.

Ebola is a horrific disease with a death rate of more than 50% and no antiviral agent able to cure the infection. However, a new vaccine has been approved by the World Health Organization (Section 21.2). Nevertheless, early lab diagnosis using ELISA can lead to quick implementation of supportive measures (fluid replacement, blood pressure maintenance) that can improve chances of survival. As first described in Section 17.2, an ELISA

Figure 25.21 Antibody-Capture Enzyme-Linked Immunosorbent Assay

A. Antibody-capture ELISA is used to detect anti-Ebola antibodies circulating in patient serum. The 96-well plate can be used to make dilutions of a single patient's serum, to more precisely determine the amount of anti-Ebola antibody, or it can be used to test samples from multiple patients.

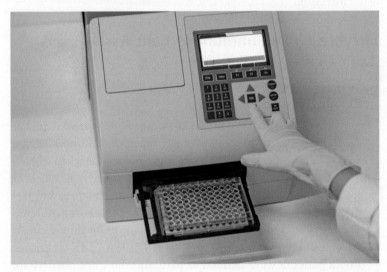

B. An ELISA plate reader.

can detect antibodies or antigens. The **antibody capture** ELISA detects serum antibodies. It is carried out in a 96-well microtiter plate, which allows simultaneous testing of multiple patient serum samples. An antigen from the virus (here, Ebola) is attached (or adsorbed) to the plastic of the wells (**Figure 25.21A**). Patient serum is then added. If Ebola-specific antibodies are present in the serum, they will bind to the antigen attached to the microtiter plate. The antigen-antibody complex is then reacted with a detection antibody, rabbit anti–human immunoglobulin G (IgG), to which an enzyme (for example, horseradish peroxidase) has been attached (or conjugated). The resulting antigen-antibody-antibody-enzyme chain links the enzyme to the well. After unbound conjugated antibody is washed out, a chromogenic substrate for the enzyme is added (for example, tetramethylbenzidine). If enzyme-conjugated antibody has bound to any

human IgG affixed to the well, the enzyme will convert the substrate to a colored product (blue for tetramethylbenzidine). The amount of colored product formed, measured as absorbance by an ELISA plate reader (**Figure 25.21B**), correlates to the amount of anti-Ebola antibody present in the patient's serum sample (Figure 25.21A). Quantitative measurements of patient antibodies can also be made by diluting patient sera in the ELISA plate wells and comparing acute versus convalescent sera (**Section 17.2**).

Antigen capture is another ELISA technique. In this instance, reference anti-Ebola antibody, rather than viral antigen, is adsorbed to the wells of a microtiter plate (**Figure 25.22**). Patient serum is then added to the wells. If the patient's serum contains Ebola antigen, the antibody tethered to the well will capture it directly. Then

Figure 25.22 Antigen-Capture Enzyme-Linked Immunosorbent Assay

Antigen-capture, or direct, ELISA captures Ebola antigens circulating in patient serum. The steps of this procedure are similar to those shown in Figure 25.21.

a second, enzyme-conjugated detection antibody against the Ebola antigen is added. The more antigen present in the serum, and thus captured in the well, the more enzyme-linked antibody will bind to the well. Adding the appropriate chromogenic substrate will yield a measurable colored product.

Antibody against Ebola may be easier to detect than Ebola antigen at later stages of the disease because antibodies will be present at higher levels than the viral antigen itself. But because a delay occurs between the time virus is first present in serum and when the body manages to make antibodies, directly detecting viral antigen offers a speedier diagnosis.

A previous case history also shows the utility of ELISA in differentiating between acute and chronic infections. During a discussion of cytotoxic T cells, Case History 16.3 introduced Luke, a 24-year-old man who was jaundiced, had pain in his upper right abdomen (liver), and was extremely tired. The case mentioned that serological tests were used to prove he had an acute infection with hepatitis B virus. Those tests were ELISA procedures. A direct ELISA identified the presence of HBsAg (hepatitis B surface antigen, a soluble viral coat protein) in the patient's serum, and an indirect ELISA discovered IgM anti-HBsAg antibody, both of which develop soon after infection. The presence of both markers is a sign of acute hepatitis B disease (Section 22.4). When the disease becomes chronic, however, the IgM antibody disappears; IgG anti-HBsAg antibody and the HBsAg protein (the latter indicating active infection) remain elevated.

Fluorescent Antibody Staining

In Case History 20.5, an 80-year-old nursing home resident contracted pneumonia caused by *Streptococcus pneumoniae*. The lab diagnosis was probably made by using the biochemical algorithm described earlier. However, more than 80 serological types of *S. pneumoniae* exist, each containing a different capsular antigen. How can the lab identify which antigenic type has caused the infection? One way is to stain the organism with antibodies.

Figure 25.23A shows the result of staining a smear of isolated streptococcus with fluorescently tagged antibodies directed against a specific antigenic type of capsule. This approach is called **fluorescent antibody staining**. Viewed under a fluorescence microscope, the organism "glows" green when the right antibody binds to the capsule (called direct immunofluorescence microscopy). For pneumonia, this knowledge probably will not help in treating the individual patient, but its broader value is in determining whether a single type of organism caused an outbreak of pneumonia, which in turn is of epidemiological value for identifying the source of the bacterium.

On the other hand, fluorescent antibody staining techniques are crucial for rapidly identifying organisms that are difficult to grow. Infected tissues can be subjected to direct fluorescent antibody staining. **Figure 25.23B**, for example, shows a direct

Figure 25.23 Fluorescent Antibody Staining

A. *Streptococcus pneumoniae* capsule. The capsule is the green halo. The center of the halo is the cell.

B. *Legionella pneumophila* (approximately 1 μm) from a respiratory tract specimen.

fluorescent antibody stain of pleural fluid from a patient with Legionnaires' disease.

Diagnostic Overview: Which Tools for Which Pathogens?

Space does not permit listing every method used to identify all infectious agents, but Table 25.1 shows some additional examples. In general, you will notice that easily cultured bacteria are currently grown in the lab, after which biochemical tests are performed. Mass spectrometry is starting to replace these biochemical and agar plate phenotype approaches. Difficult-to-grow bacterial, viral, and fungal species are typically identified through immunological or nucleic acid–based techniques. Immunological techniques can directly identify microbes in infected tissues or indirectly measure a rise in antibody titer (Chapter 17). Nucleic acid-based detection methods include PCR, qRT-PCR, and RFLP analysis. Often these tests can be performed directly on patient samples, with no need to grow the pathogen first. Eukaryotic microbial parasites such as *Plasmodium* species (the cause of malaria), *Giardia intestinalis* (which causes giardiasis, a diarrheal disease), and *Entamoeba histolytica* (the cause of amebic dysentery) can be identified through their telltale morphologies under the microscope, making biochemical tests unnecessary.

The last section of this chapter discusses diagnostic tests that can be performed without an expensive laboratory. These are called point-of-care tests.

SECTION SUMMARY

- **The presence of serum antibodies made in response to a specific pathogen** can be revealed and quantified by using indirect ELISA techniques.
- **The presence of virus antigens in serum** can be detected by using direct ELISA or real-time quantitative PCR.
- **Fluorescent antibody staining** can rapidly identify organisms or antigens present in tissues.

Thought Question 25.6   Why does finding IgM to West Nile virus indicate current infection? Why wouldn't finding IgG do the same?

Thought Question 25.7   Specific antibodies against an infectious agent can persist for years in the bloodstream. So how can antibody titers be used to diagnose diseases such as infectious mononucleosis? Couldn't the antibody be from an old infection?

25.5 Point-of-Care Rapid Diagnostics

SECTION OBJECTIVES

- Differentiate between assay specificity and sensitivity.
- Explain the basic immunochromatographic assay used in point-of-care diagnostic kits.
- Discuss the advantages and disadvantages of point-of-care diagnostics.

Is there a quick, inexpensive way to screen sick patients for infectious agents? Conventional diagnosis of an infection often requires sending a clinical specimen to a far-away laboratory, followed by a considerable delay in obtaining results. Some patients inevitably lose patience and fail to attend follow-up appointments. **Point-of-care (POC) laboratory tests** address these problems because they are designed to be used directly at the site of patient care, such as physicians' offices, outpatient clinics, intensive-care units, emergency departments, hospital laboratories, and even patients' homes. Most patients are happy to wait 40–50 minutes for a rapid POC test result in order to receive immediate treatment or reassurance. But how do we know a POC test is reliable?

Sensitivity and Specificity of Diagnostic Assays

Before we describe types of POC assays, you must first understand the difference between an assay's **sensitivity** and **specificity**. Let's start with an analogy. Take a common bathroom scale and a laboratory scale, and place an apple on each. Most bathroom scales will not detect the weight of an apple. However, a laboratory scale will easily register the apple's weight. Thus, the laboratory scale is <u>more sensitive</u> than the bathroom scale. Now place an apple and then an orange on the same laboratory scale. The scale that was very sensitive to the apple's weight cannot distinguish an apple from an orange. You could say that the laboratory scale is very sensitive but poorly specific.

With respect to infectious disease, sensitivity and specificity address two questions you must ask when analyzing a new diagnostic test:

1. If a patient <u>has</u> a disease, how often will the test be positive (a measure of <u>sensitivity</u>)?
2. If a patient does <u>not</u> have a disease, how often will the test be negative (a measure of <u>specificity</u>)?

An assay with high sensitivity and high specificity is a valuable diagnostic tool, whereas an assay with low sensitivity or low specificity is useless.

A test's <u>sensitivity</u> reflects how <u>small</u> a concentration of a microbe, antibody, or antigen the test can detect. A highly sensitive assay can detect the presence of minute quantities of hepatitis B antigen, for example. A less sensitive test may require larger amounts of that antigen before the test turns positive. Any sample containing less than that amount of antigen would yield a **false-negative reaction**, meaning the antigen was present—that is, the patient has the disease—but the test could not detect it. Sensitivity is measured by testing a new assay on a group of people known to <u>have</u> the disease. A test sensitivity of 0.95 indicates that the test will be positive for 95 of 100 people with the disease, but it will be negative for five patients with the disease (false negatives).

In contrast to sensitivity, <u>specificity</u> measures how well an assay can distinguish between two closely related antigens (think apples and oranges in our analogy). A test highly specific for a *Mycobacterium tuberculosis* antigen, for example, is positive <u>only</u> for the *M. tuberculosis* antigen but negative for a similar antigen from a different *Mycobacterium* species. A less specific test might turn positive when <u>either</u> antigen is tested. The antigen from the other species could produce a so-called **false-positive reaction**. Someone with a false-positive reaction does <u>not</u> have the disease, but the new assay suggests the person does. Specificity is measured by testing a new assay on a group of people who are known to <u>not have</u> the disease. A test whose specificity is 0.95 means that the test will be <u>negative</u> for 95 of 100 patients without disease, but five disease-free patients will falsely test positive. The higher the sensitivity and specificity values, the better the test.

Commercial Point-of-Care Tests

Commercial POC tests are widely available to diagnose bacterial and viral infections and parasitic diseases, including malaria (Table 25.4). However, as convenient as these tests are, sensitivity may be compromised in the quest for a speedy result.

Table 25.4

Some Point-of-Care Rapid Diagnosis Test Kits for Infectious Diseases[a] [Reference Table]

Pathogen or Disease	Test Type	Sample	Indication	Performance Sensitivity (%)	Specificity (%)	Notes
Bacterial						
Chlamydia	ICT	Vaginal swab, urine	Screening, suspicion of PID	98.7	99.4	Detects antigen
Neisseria gonorrhoeae	PCR	Urine	Urethral exudate	98	99	Detects antigen
Syphilis	ICT	Blood	Screening	90–95	90–95	Detects antibody
Group A streptococci	EIA	Pharyngeal swab	Sore throat	92	100	Confirm negative swabs
Legionella	ICT	Urine	Severe pneumonia, risk factors for legionellosis	77.7	99	Detects only serotype 1 antigen reliably
Pneumococcal pneumonia	ICT	Urine (pleural fluid, CSF)	Severe pneumonia (empyema,[b] meningitis)	86	94	Detects capsule antigen
Clostridioides difficile	ICT	Stool	Antibiotic-associated diarrhea	92.8	92.6	Detects toxin; notably less sensitive than cultures or PCR
Streptococcus agalactiae	PCR	Vaginal swab	Peripartum detection of colonization	92	96	Detects antigen
Protozoal						
Malaria	ICT	Blood	Fever in returning traveler	99.7	94.2	Sensitivity better for *Plasmodium falciparum*
Trichomonas	ICT	Vaginal swab	Symptoms of vaginitis	92	100	
Viral						
Influenza	ICT	Nasopharyngeal swab	Flu-like symptoms	95	95	False negatives are more likely during outbreaks
Respiratory syncytial virus	ICT	Nasopharyngeal swab	Viral symptoms, especially during the winter season	93	93	Detects antigen
Human immunodeficiency virus	ICT	Blood or oral fluid	Screening, preventing vertical transmission	99.6	99.3	Detects anti-HIV antibodies
Dengue fever	ICT	Blood	Screening in endemic regions	90	100	Detects IgG-IgM antibodies
Epstein-Barr virus/ mononucleosis	ICT	Blood	Symptoms (swollen lymph nodes, sore throat)	90	100	Detects IgM heterophile antibodies
Rotavirus	ICT	Stool	Diarrhea in child	88	99	Detects antigen
Hepatitis B	ICT	Blood	Jaundice, pain over liver	95	100	Detects HBsAg
Rubella	ICT	Blood	Pregnancy	99	99	Detects IgG antibody

[a]CSF = cerebrospinal fluid; EIA = enzyme immunoassay; HBsAg = hepatitis B surface antigen; HIV = human immunodeficiency virus; ICT = immunochromatographic test; IgG = immunoglobulin G; IgM = immunoglobulin M; PID = pelvic inflammatory disease.
[b]Empyema is a collection of pus in the pleural space.

Figure 25.24 Principle of Immunochromatographic Rapid Diagnostic Tests

- ◆ C-ps derived from *S. pneumoniae*
- ▲ ● Other components
- ⵤ Anti–pneumococcal C-ps colloidal gold–labeled rabbit polyclonal antibody
- ⵤⵤ Anti–pneumococcal C-ps solid-phase rabbit polyclonal antibody
- ⊥⊥ Solid-phase goat anti–rabbit IgG antibody

A. The example is a test for the presence of *Streptococcus pneumoniae* in sputum. An extract of sputum is placed onto one end of the strip where *S. pneumoniae* capsular polysaccharide antigen (C-ps), if present, will bind anti–pneumococcal C-ps polyclonal antibodies tagged with colloidal gold. The resulting immunoconjugates move along the strip by capillary action and are captured by different anti C-ps antibodies affixed to the strip. The resulting sandwich conjugate forms a colored test line. The presence of both a test and a control line indicates a positive result, whereas the appearance of only a single control line indicates a negative result.

B. Immunochromatographic test for *Legionella* antigen in urine (Thermo Scientific™ Xpect™ Legionella Test). Urine placed on left strip came from a patient with legionellosis.

Infectious-disease specialists and clinical microbiologists should be aware of the indications and limitations of each rapid test, so that they can use them appropriately and correctly interpret their results. Some tests aren't sensitive enough and should therefore be coupled with confirmatory tests when the results are negative (false negatives, such as the *Streptococcus pyogenes* rapid antigen detection test), whereas the results of other POC tests need to be confirmed when positive (false positives, such as malaria).

The typical POC test involves an **immunochromatographic assay**. The test for *Streptococcus pneumoniae* capsular polysaccharide antigen (C-ps), shown in **Figure 25.24A**, is one example. This particular immunochromatographic test, called a "red colloidal gold" test, involves extracting the relevant antigen (here, C-ps) from a clinical specimen and placing a few drops of the extract on a test strip containing rabbit antibodies to the antigen. The antibodies have red colloidal gold particles attached to them. The antigen-antibody complexes that form move by capillary action to the upper level of the strip, where they are captured by a line of more anti-antigen antibodies embedded in the strip, forming a sandwich. The colloidal gold particles accumulate and eventually produce a red test line, indicating the presence of the antigen. By contrast, rabbit antibodies not

bound to the antigen pass through the test line but then are captured by goat anti–rabbit IgG antibodies on a second line, once again forming a red line, as a control showing that the strip components are working. **Figure 25.24B** shows the results of a Thermo Scientific Xpect test for *Legionella* antigens in urine.

The Future of Point-of-Care Diagnostics: Lab-on-a-Chip

Imagine a day when all the agar plates, test tubes, culture flasks, and incubators used in a clinical microbiology laboratory are shrunk to the size of a single credit card (**Figure 25.25**). This **lab-on-a-chip (LOC)** could be used in doctors' offices or bring diagnostic capability to remote locations of the world. It may sound like science fiction, but that day is not too far off.

An LOC is possible because of microfluidics, whereby liquid flows through tiny channels etched into a small silicon chip. Strategically placed electrodes on the chip move the sample along a predefined route. In one LOC used to detect pathogens, part of the slide is filled with nanoparticles (10–1,000 nanometers in diameter) that adhere to pathogens in a sample drop of water. The sample moves to a chamber containing antibodies that bind the pathogen and cause the attached particles to agglutinate (see Chapter 17). An optical sensor detects agglutination as scattered light. Similar LOC technologies attach fluorescent dyes to the antibodies.

Another chip-like technology released in 2020 by Abbott Laboratories uses a PCR-type technique that does not rely on thermocycling. The instrument has an extraction chamber where nucleic acids are extracted from samples, a reaction chamber that uses fluorescent probes (the beacon technology described above) to amplify a specific gene or RNA released from the pathogen, and a detector that measures the light emitted as the fluorescent dye is released from the beacon probe. This technology was used during the 2019–2020 COVID-19 pandemic to detect RNA from the SARS-CoV-2 coronavirus in as little as 5 minutes.

In summary, POC rapid tests have several advantages and some disadvantages. Advantages of POC tests are as follows:

- Results are fast, easy to interpret, and require less training to perform, and culturing is not required.
- The clinician can immediately initiate specific antibiotic therapy.
- They help avoid unnecessary antibiotic use in the case of a viral infection.
- They enable the clinician to quickly recognize a chain of infection (Section 26.1) among patients with similar symptoms.
- They enable rapid notification of patients who may be difficult to reach later.

Figure 25.25 Model of a PCR-Based Lab-on-a-Chip Used to Detect Pathogens

New technologies offer hope that the power of an entire laboratory can be miniaturized to a series of chambers on a small chip.

Disadvantages of POC tests include the following:

- They provide no data about pathogen antibiotic sensitivity.
- There is an increased risk of the technologist becoming infected.
- Double or multiple infections are more likely to be overlooked than in culture.

The newer molecular tests (described earlier) exhibit better sensitivity and specificity than most immunochromatographic assays, but they require more expensive instrumentation and training. In the coming years, further evolution of POC tests may lead to new diagnostic approaches. These new approaches could include panel testing that targets all possible pathogens suspected in a specific clinical setting. The development of next-generation multiplex tests based on serology or molecular techniques will certainly facilitate quicker diagnosis and improved patient care.

SECTION SUMMARY

- **Point-of-care diagnostic tests** can rapidly identify the cause of an infectious disease.
- **Sensitivity** measures the likelihood that an assay will be positive for a patient who has the disease.
- **Specificity** reflects the likelihood that an assay will be negative for a patient free from the disease.
- **Immunochromatography** is the primary platform for POC testing.
- **An advantage of point-of-care tests** is that specific therapy can be initiated quickly after a positive result.
- **A drawback to point-of-care tests** is that they may miss simultaneous, multiple infections.

Perspective

In 1880, Robert Koch used the surface of a potato slice to culture and isolate the agent of anthrax. Ever since, clinical microbiologists have tried to design faster and more accurate ways to collect, detect, and identify pathogenic microorganisms. Those efforts shaped the strategies now used to diagnose infectious diseases. The clinical microbiology laboratory today is undergoing another new and exciting technological revolution in which culture-based methods are slowly giving way to molecular techniques such as PCR, DNA sequencing, and mass spectrometry. Someday, clinicians in remote areas of the world may be able to use a single chip to identify every possible pathogen in any clinical specimen.

As this chapter ends, reflect on earlier case histories and try to envision which techniques were used to identify the pathogen. Then think ahead about how these methods might help epidemiologists predict local outbreaks and the spread of pandemics (Chapter 26). Finally, know that whatever your eventual role in health care—as a clinician, technician, nurse, parent, or patient—you will benefit from understanding specimen collection and the basic diagnostic strategies used to reveal a pathogen's identity.

Key Terms

acid-fast stain (p. 829)

alpha hemolysis (p. 827)

alpha-hemolytic (p. 827)

antibody capture (p. 838)

antigen capture (p. 838)

beta hemolysis (p. 828)

beta-hemolytic (p. 828)

bronchoalveolar lavage (BAL) (p. 819)

catalase (p. 827)

catalase test (p. 827)

CNA plate (p. 827)

cytochrome oxidase test (p. 825)

empirical therapy (p. 814)

false-negative reaction (p. 840)

false-positive reaction (p. 840)

fluorescent antibody staining (p. 839)

hemolysis (p. 827)

immunochromatographic assay (p. 842)

lab-on-a-chip (LOC) (p. 842)

Lancefield classification (p. 828)

Lancefield groups (p. 828)

MALDI-TOF MS (p. 830)

mass spectrometry (MS) (p. 830)

midstream clean-catch technique (p. 820)

molecular beacon probe (p. 833)

multiplex PCR (p. 831)

mycolic acid (p. 829)

nonhemolytic (p. 827)

optochin susceptibility disk test (p. 829)

point-of-care (POC) laboratory test (p. 840)

probabilistic indicator (p. 824)

quantitative reverse transcription–PCR (qRT-PCR) (p. 833)

restriction fragment length polymorphism (RFLP) (p. 836)

sensitivity (p. 840)

specificity (p. 840)

suprapubic puncture (p. 820)

thermocycler (p. 831)

venipuncture (p. 817)

Objectives Assessment Questions

25.1 The Importance of Clinical Microbiology and Specimen Collection

1. Of the following, which is the best clinical reason to identify the etiologic agent causing an infection?

   a. Accurately predicts disease severity

   b. Alerts the clinician to possible disease sequelae

   c. Pinpoints the source of the infection

   d. Indicates whether gloves should be worn when examining the patient

2. Twenty-five students from two schools were stricken with bacterial diarrhea. All the patients went to a countywide fair 3 days ago that 400 students attended. Which of the following would best confirm that the county picnic was the source of infection?

   a. Different bacterial species were isolated from the sick students but all had the same antibiotic resistance pattern.

   b. The same bacterial species was found in each of the 400 students and all specimens had the same antibiotic resistance pattern.

   c. The same bacterial species with the same antibiotic resistance pattern was found in all the sick students and in a vendor at the picnic.

   d. The same bacterial species but with different antibiotic resistance patterns was found in all the sick students and in a vendor at the picnic.

   e. c and d

3. Which of the following specimens, when collected properly, will contain normal microbiota in a healthy patient?

   a. Cerebrospinal fluid

   b. Pleural fluid

   c. Throat swab

   d. Blood culture

   e. Tissue biopsy sample

4. _____ is the procedure used to collect cerebrospinal fluid.

5. A 13-year-old boy has a high fever and is disoriented, but his blood pressure is normal. Blood cultures are obtained and are incubated aerobically and anaerobically. After overnight incubation, growth is observed in all blood culture bottles. Which of the following is the most defensible conclusion?

   a. The sample was contaminated because an aerobe and an anaerobe are unlikely to both be present in the patient's blood.

   b. The patient has septicemia caused by an organism that is facultative with respect to oxygen.

   c. The patient has septicemia and the organism is aerobic.

   d. The patient has septicemia caused by a Gram-negative rod.

   e. The antibiotic resistance patterns of the organisms isolated from all blood culture bottles will be identical.

6. Which of the following specimens should undergo centrifugation and Gram staining of the sediment?

   a. Blood

   b. Stool

   c. Urine

   d. CSF

   e. Throat

25.2 Pathogen Identification Using Biochemical and Protein Profiles

7. Which of the following organisms will grow on chocolate agar but not blood agar?

   a. *Neisseria meningitidis*

   b. *Escherichia coli*

   c. *Salmonella* Typhimurium

   d. *Staphylococcus aureus*

   e. *Treponema pallidum*

8. A urine specimen from a patient who experienced burning pain upon urination contained Gram-negative, lactose-positive, indole-positive rods that were urease negative. Which of the following bacterial species has these characteristics?

a. *Staphylococcus saprophiticus*
b. *Escherichia coli*
c. *Proteus mirabilis*
d. *Staphylococcus aureus*
e. *Streptococcus pneumoniae*

9. A fecal sample from a patient with diarrhea contained a bacterium that produced white colonies on MacConkey medium, produced colonies with black centers on Hektoen agar, and is catalase positive but did not grow on CNA blood agar. Which of the following bacterial species has this biochemical profile?

a. *Streptococcus gallolyticus*
b. *Shigella flexneri*
c. *Escherichia coli*
d. *Staphylococcus aureus*
e. *Salmonella enterica*

10. An organism isolated from the bloodstream of a person with necrotizing fasciitis grows on CNA blood agar, is beta-hemolytic on blood agar, does not grow on MacConkey agar, is catalase negative, and is serologically classified as group A. The most likely organism is _____ and the serologic test used is called _____.

11. _____ prevents the Gram staining of *Mycobacterium tuberculosis*.

Diagnostic microbiology matching: Match the description with the organism.

Description	Organism
12. Oxidase-positive coccus	a. *Mycobacterium tuberculosis*
13. Lactose-fermenting bacillus	b. *Streptococcus pneumoniae*
14. Hydrogen sulfide–producing bacillus	c. *Staphylococcus aureus*
15. Coccus with serologically typed capsule	d. *Neisseria gonorrhoeae*
	e. *Escherichia coli*
16. Catalase-positive, coagulase-positive coccus	f. *Salmonella enterica*
	g. *Yersinia pestis*
17. Acid-fast bacillus	

For each of the six diseases listed, choose the pathogen that is the most likely cause.

Disease	Pathogen
18. Abscess	a. *Mycobacterium tuberculosis*
19. Diarrhea	b. *Streptococcus pneumoniae*
20. Vaginitis	c. *Staphylococcus aureus*
21. Cystitis	d. *Neisseria gonorrhoeae*
22. Pneumonia	e. *Escherichia coli*
23. Reactivation pneumonia	f. *Salmonella enterica*
	g. *Yersinia pestis*

25.3 Pathogen Identification Using DNA Methods

24. Bacterial pathogens can be typed in three ways: _____, _____, and/or _____.

25. Three patients in the same hospital developed pneumonia caused by *Streptococcus pneumoniae* during their stay. The infection control specialist suspected that the same strain of *S. pneumoniae* infected all three patients and that a common source existed. PCR assays used to examine the isolates from each patient produced DNA fragment products of identical size, but after further study, the infection control expert concluded the three patients were not infected with the same strain. Which of the following would best support that conclusion?

a. No common source of *S. pneumoniae* was identified.
b. The DNA sequences of the PCR fragments were different.
c. The isolates formed different-sized colonies on blood agar.
d. The patients were present in the hospital on different days.

25.4 Pathogen Identification Using Immunological Methods

26. Draw the scheme for an indirect fluorescent antibody staining technique.

27. ELISA was used to measure hepatitis A antibody levels in five patients (viewed in a microtiter plate). From the results shown, which patient(s) probably have the disease? Each patient's serum underwent serial twofold dilutions. (In the diagram, a = serum collected during the acute phase of the disease; c = serum collected during the convalescent phase of the disease.)

a. Patient 3
b. Patients 3 and 4
c. Patients 3, 4, and 5
d. Patients 1, 3, 4, and 5
e. Patient 2

25.5 Point-of-Care Rapid Diagnostics

28. Which of the following statements can you make once you know that a detection assay has a specificity of 0.95?

a. A patient with disease has a 95% chance of testing positive with the assay.

b. A patient with disease has a 5% chance of testing negative with the assay.

c. A patient without disease has a 95% chance of testing negative with the assay.

d. A patient without disease has a 95% chance of testing positive with the assay.

29. A point-of-care test will provide useful information about the antibiotic susceptibility of a pathogen. True or false?

30. Evaluate the immunochromatographic results shown. Which option represents a proper positive result for the test antigen?

31. You are given two tubes containing different amounts of the same antigen and a third control tube containing no antigen. You have designed two different ELISAs to test for the presence of that antigen. You find the following:

- Both tests are negative when the control tube is used.
- Assay A turns positive when both antigen-containing tubes are analyzed.
- Assay B turns positive only with sample from antigen-containing tube 1.

Assay B is _____ than Assay A.
a. more sensitive
b. more specific
c. less sensitive
d. less specific

Review Questions

1. Why is identifying the genus and species of a pathogen important?

2. What is an API strip, and how is it used in clinical microbiology?

3. Describe three examples of selective media.

4. If a colony on a nutrient agar plate is catalase positive, does this mean it is made up of Gram-positive microorganisms? Why or why not?

5. Describe the types of hemolysis visualized on blood agar.

6. What is the clinical significance of a group A, beta-hemolytic streptococcus?

7. Describe the appropriate samples one should collect in the following cases:
   a. Severe headache, stiff neck, high fever
   b. High fever, altered mental status, but flexible neck
   c. Abdominal pain in lower-right quadrant, nausea

8. How does one distinguish *Staphylococcus aureus* from *S. epidermidis*?

9. Why are PCR identification tests preferable to biochemical approaches?

10. How is qRT-PCR performed?

11. Describe an ELISA.

12. Name some sterile and nonsterile body sites.

13. List seven common types of clinical specimens collected for bacteriological examination.

14. How can genomics help identify nonculturable pathogens?

15. What is a point-of-care test? What are the limitations of point-of-care tests?

Clinical Correlation Questions

1. In the case history that opens Chapter 2, Brandon had a small, round, painless lesion on his penis. The lesion exuded a clear fluid. The physician sent a sample of the fluid to the clinical laboratory, which found highly motile, corkscrew-shaped bacteria. How would the lab definitively diagnose *Treponema pallidum*?

2. Construct a short case history in which the causative agent is respiratory syncytial virus. Describe the patient, the signs and symptoms, the clinical specimens you would collect, what tests you would run, and the results needed to prove the patient had RSV. (This exercise may be done in groups or individually.)

3. Describe how you can use DNA-based methods to quickly determine whether a slow-growing bacterial pathogen, such as *Mycobacterium tuberculosis* isolated from a patient, carries an antibiotic resistance gene.

4. A 40-year-old male arrives at the emergency department with a high fever and low blood pressure. He screams when you, the examining clinician, try to turn his head. He also has small reddish spots on his arms and legs. Answer the following questions. What is your preliminary diagnosis? What samples should be taken for bacteriologic examination? What tests should be run on those samples?

Epidemiology: Tracking Infectious Diseases

[Mystery of the Microcephalic Newborns]

SCENARIO One morning in 2015, a young woman from Recife, Brazil, was in the fourth month of her pregnancy when she awoke with a fever, a fiery red rash, and pain in her joints. Her symptoms went away within a few days, so she went about her life, oblivious to what was to come. Five months later, in September, she delivered her baby and immediately knew from the look on the obstetrician's face that something was terribly wrong. Her son had an abnormally small head and incomplete brain development—something called microcephaly. The young mother was devastated.

A child born with microcephaly in Brazil
Stunted brain development results in an unusually small cranium at birth.

Computed tomography image of brain calcifications (arrow) caused by congenital Zika virus infection

STATUS Fortunately, the number of Zika infections overall has dropped significantly because of mosquito control measures and personal protection strategies such as using mosquito repellent. The World Health Organization (WHO) and the US Centers for Disease Control and Prevention (CDC) continue to carefully monitor the spread of Zika virus throughout the Americas by intensively screening mosquitoes.

INITIAL CLUES In August 2015, Dr. Vanessa van der Linden, a neurologist in Recife, started noticing an increasing number of newborns with microcephaly. Normally, months would pass without her seeing a single case. Now she saw three or four on any given day. Serological tests ruled out the usual suspects, measles and toxoplasma, but imaging found unusual patterns of calcification in affected brains—a possible sign of viral necrosis. Then Vanessa's mother, Ana, called. Ana was a neuropediatrician in another hospital. Ana had just seen seven microcephalic babies in <u>one day</u>. Many of the mothers reported having a rash early during pregnancy. Was a new infectious agent involved?

EPIDEMIOLOGY Vanessa alerted the state health secretary about the spike in microcephaly cases. This was the first step in the long epidemiological journey linking Zika virus, a positive-sense RNA virus that normally causes benign, mosquito-transmitted disease in adults, to the development of microcephaly in Zika virus–infected fetal brains. Between 2015 and 2018, at least 4,000 babies in Brazil (116 in the USA) had been born with microcephaly and other birth defects linked to Zika virus. Estimates are that about 1 in 7 babies exposed to Zika virus in the womb will develop at least one developmental defect within a year.

Zika virus

TEM 100 nm

Zika Virus
This enveloped, single-stranded RNA virus (Flavivirus family) is usually transmitted by mosquito bite and typically causes a mild rash with fever. However, the virus can cross the placental barrier and cause devastating fetal disease.

CHAPTER OBJECTIVES

After reading this chapter, you will be able to:

- Explain the epidemiologic basis for endemic disease, outbreaks, epidemics, and pandemics.
- Describe how epidemiologists track health care–associated infections and community outbreaks of a disease.
- Discuss the basic methods used to trace epidemics and pandemics to a source.
- Explain the impact of technology and climate on the epidemiology of infectious disease.
- Outline factors that influence the emergence of new pathogens.

ere is a startling fact. Throughout history, infectious diseases have killed more people than all of Earth's wars . . . combined. The opening case of this chapter highlights the important roles that clinical microbiology and epidemiology play in preventing infectious illnesses and in identifying new emerging diseases like microcephaly caused by Zika virus. Our present success in controlling the spread of disease is due largely to worldwide surveillance agencies such as the World Health Organization (WHO) and Centers for Disease Control and Prevention (CDC) that can detect outbreaks quickly, before major epidemics develop. These agencies rely on data submitted by smaller clinical microbiology laboratories, scattered throughout the world, that help local clinicians diagnose infectious diseases.

This chapter will explain the basic concepts of epidemiology and discuss how epidemiologists manage to track infectious diseases, identify "patient zero," and recognize and contain emerging diseases. Finally, we'll consider bioterrorism. Clinical scientists and epidemiologists are trained to detect bioterrorist attacks by using the same principles they employ to detect naturally occurring infectious diseases.

26.1 Principles and Tools of Epidemiology

SECTION OBJECTIVES

- Explain the four major links in a chain of infection.
- Differentiate between descriptive and analytical epidemiology.
- Define endemic, epidemic, and pandemic, and explain the difference between disease prevalence and incidence.
- Explain the three steps of surveillance and describe how molecular approaches can contribute to surveillance.

What is epidemiology, and why is it important? The word **epidemiology** is derived from Greek, loosely translated as "the study of that which befalls man" (*epi*, "on or upon"; *demos*, "people";

and *logos*, "the study of"). Infectious disease epidemiology examines how far a known disease has spread, determines the source of a disease outbreak, and identifies the factors that influence how many individuals will contract the disease. Epidemiological principles are also used to determine the effectiveness of treatments and to identify emerging disease syndromes, such as SARS (severe acute respiratory syndrome), MERS (Middle East respiratory syndrome), microcephaly caused by Zika virus infection, and now the respiratory disease COVID-19 caused by the virus SARS-CoV-2 that, as of May 2020, has rapidly spread around the world. **We already covered some important concepts of epidemiology in Chapter 2** when we discussed the basics of infectious disease, including routes of transmission. Now we will explore how those principles are used to track disease.

To an epidemiologist, data are clues: clues that include locations of outbreaks, where infected persons traveled, what they ate, and whom they met. The infectious disease epidemiologist scrutinizes these clues, looking for trends that will pinpoint the source of an infectious agent and then predict where it will spread. With this knowledge comes the power to stop or limit a disease's march through a community, a country, or even the world. Remember our discussions about transmission, virulence, reservoir, portal of entry and exit, and immunocompetence in Chapter 2? All of these factors are important in what is called the **chain of infection** (introduced in Chapter 1). A chain of infection has four major links (**Figure 26.1**): the organism (its virulence and invasiveness), the agent's reservoir (animal, human, or environment), its mode of

Figure 26.1 Links in the Chain of Infection

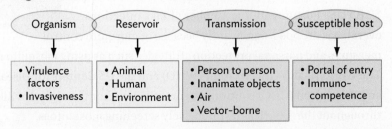

transmission (person-to-person or via air, vectors, or inanimate objects, including food), and its host (portal of entry, immunocompetence). Influencing any one link can break the chain of infection.

John Snow (1813–1858), Father of Epidemiology

The first known case in which the source of a disease outbreak was methodically investigated took place in the mid-nineteenth century. During a serious outbreak of cholera in London in 1854, John Snow (Figure 26.2A) used a map to plot the locations of all the diarrheal cases he learned about. The source of the infection was unknown ("experts" once believed it was transmitted through a contagious power in the mist, or miasma), but Snow thought that if the cases clustered geographically, he might gain a clue as to the actual source. Water in that part of London was pumped from separate wells located in various neighborhoods. Snow's map revealed a close association between the density of cholera cases and a single well located on Broad Street (Figure 26.2B). Simply removing the pump handle of the Broad Street well ended the epidemic, proving that the well water was the source of infection (removing the pump handle broke the transmission link in the chain of infection). This approach succeeded brilliantly, even though the infectious agent that causes cholera, *Vibrio cholerae*, was not recognized until 1905, more than 50 years later. How did the cholera organism get into the water? It seems the well was inadvertently contaminated by sewage. Today, epidemiologists still identify clusters of patients afflicted with a given infectious disease and follow the trail back to potential sources

of the infection. The science of medical statistics, founded by the British nurse Florence Nightingale, is also critical to this effort (Section 1.3). Her techniques proved that soldiers fighting in the Crimean war died primarily from infectious disease and not from bullets or poor nutrition.

Types of Epidemiology

Epidemiologists are detectives, but they are reporters, too. And as any reporter will tell you, a good story describes the what, who, where, when, and why (or how) of an event. For the epidemiologist, the five Ws are case definition (what), person (who), place (where), time (when), and causes, risk factors, and modes of transmission (why and how).

DESCRIPTIVE EPIDEMIOLOGY Every infectious disease has patterns. Sometimes the disease is most frequent during certain times of the year, like influenza in the winter and West Nile virus (WNV) in summer. Some diseases are concentrated in certain geographic locations; examples are Lyme disease in the northeastern United States and dengue fever in Central and South America. Other infections most often affect certain types of people. Respiratory syncytial virus predominantly infects children; Creutzfeldt-Jakob disease mainly afflicts older adults; and *Pneumocystis* pneumonia is seen only among immunocompromised individuals. By collecting data relating to person, place, and time (called **descriptive epidemiology**), an investigator can identify geographic areas or population groups that have high rates, or elevated risks, for a particular disease.

ANALYTICAL EPIDEMIOLOGY An epidemiologist will take the person, place, and time data of descriptive epidemiology and formulate hypotheses about the cause of the disease and the possible risk factors involved. **Analytical epidemiology** tests those hypotheses. Testing can involve either experimental or observational studies. In an **experimental study**, for instance, you might use two groups of patients to compare a new vaccine, such as the new malaria vaccine (RTS/S), against a **placebo** (an inactive substance with no therapeutic or prophylactic effect, used as a control). In this type of experimental study, you would ask whether the vaccinated group contracted the disease less often than the placebo group.

An **observational study** can also test hypotheses. Here an epidemiologist will observe the exposure and

Figure 26.2 Early Epidemiology

A. John Snow.

B. Map of London that Snow used to pinpoint the source of the cholera outbreak in 1854. The Broad Street well is found within the red circle. Each black bar represents a death from cholera.

disease status of each study participant. John Snow's study of cholera in London was observational. Two main types of observational studies exist. In a **cohort study**, an investigator observes whether each participant is exposed to an agent and then notes who develops disease. The exposed group (or cohort) is compared with the unexposed cohort. In a **case-control study**, investigators enroll a group of people with a disease and a control group of people without disease. Researchers then compare previous exposures between the groups. For instance, during an outbreak of hepatitis A, you might ask whether the case patients had exposure to a type of food more often than did the control group. If so, that food could be the source of the infection.

Endemic, Epidemic, or Pandemic?

The terms endemic, epidemic, and pandemic reflect how many cases of a disease are observed over time in a given geographic locale (county, state, country, or world). As noted in Chapter 2, certain diseases are considered **endemic**, or native to certain areas (**Figure 26.3**). An **endemic disease** exhibits a steady (usually low) number of cases in a location. For example, *Campylobacter jejuni* is endemic in much of the United States, as are chickenpox and *Shigella*. Other diseases, such as that caused by *Borrelia burgdorferi* (Lyme disease), are endemic to smaller areas (the northeastern United States for Lyme disease). A disease endemic to an area usually has a reservoir there (for instance, the white-footed mouse is the reservoir for *B. burgdorferi*). Recall that a reservoir is an animal, bird, or arthropod (such as mosquitoes or ticks) that harbors the infectious agent and is indigenous to a geographic area. Humans become infected when they come in contact with the reservoir or with an arthropod vector that feeds on the animal reservoir. If you are planning a trip overseas and wonder what endemic diseases

lurk at your destination, consult the *CDC Yellow Book* for advice on endemic disease in various locales.

A nonendemic area is free of a disease because the disease has no reservoir there. However, even a locale considered nonendemic may still develop sporadic cases of the disease. This can happen when a person living in a nonendemic area receives a contaminated package containing imported foods sent from an endemic area or when people travel to an endemic area where they become infected, only to return home and develop disease. For instance, even though measles has <u>not</u> been considered endemic in the United States since 2000, sporadic cases of measles (50–100 per year) have occurred when someone lacking immunity returns from a country where measles is endemic, such as the United Kingdom. Outbreaks in the United Kingdom are thought to be caused by decreased MMR vaccination of children, a consequence of baseless speculation that vaccines are linked to autism and allergies. In 2015, a similar situation developed in the United States after someone with measles visited Disneyland. The virus infected several tourists, who returned home and spread the disease among 176 mostly unvaccinated children in 17 states. Endemic disease is important, but when the number of cases of a disease rises <u>above</u> the endemic level, we have an **outbreak**. Local outbreaks can develop into more serious **epidemics** (Figure 26.3) when the case numbers rise rapidly and the affected geographic area widens. The difference between an outbreak and an epidemic varies by disease. For influenza, an epidemic is declared if the percent mortality rises above a certain percentage (about 7%). An epidemic for meningococcal meningitis, however, is declared when the attack rate rises above 15 cases per 100,000 people.

⊙ **Note:** Recall from Section 2.2 that morbidity refers to the incidence of a disease state and that mortality measures how many patients died from the disease.

Epidemics develop, in part, because of rapid and direct human-to-human transmission. However, an endemic disease can become epidemic if the population of the reservoir increases, which allows for more frequent human contact, or if the infectious agent evolves to spread directly from person to person, bypassing the need for a reservoir. The latter is the concern with the H5N1 avian flu virus, which is endemic in animals and birds in Asia. **Figure 26.4** illustrates how humans living near poultry in some countries provides ample opportunity for the virus to jump between species. Although H5N1 does not easily spread between humans now, it could evolve to do so.

Out-of-control epidemics that spread across continents are called **pandemics** (AIDS, influenza, and COVID-19, for example). Pandemics may be long-lived—such as the bubonic plague pandemic in the fourteenth century and the AIDS pandemic in the late twentieth and early twenty-first centuries—or they may be short-lived, as with the 1918 flu pandemic. Table 2.3 lists some pandemics that struck the world. The fear of a pandemic can lead to extreme public health measures. When the cholera pandemic hit Chicago

Figure 26.3 Difference between Endemic and Epidemic Disease

An endemic disease is continually present at a low frequency in a population of constant size and composition. A sudden rise in disease frequency constitutes an outbreak or epidemic (if the upswing is not due to many infected patients moving into the area).

Figure 26.4 The Avian Flu Virus H5N1 Is Endemic in Asian Birds and Animals

Live poultry for sale in Vietnam.

in the 1860s, for instance, it led to quarantine of the entire city and spawned new laws governing drinking water and sewage treatment. We will discuss the COVID-19 situation later.

⊙ **Note:** "Outbreak" is also a generic term applied to everything from small local increases in case numbers all the way to large case increases across the globe.

Disease Prevalence versus Incidence

How do epidemiologists determine if efforts to limit an epidemic are succeeding? When discussing a disease, epidemiologists distinguish between prevalence and incidence. **Prevalence** describes the total number of active cases of a disease in a given location regardless of when the case first developed (think of it as the disease's burden on the society). **Incidence**, however, refers to the number of <u>new</u> cases of a disease in that location over a specified time, usually a year. Incidence is the risk of acquiring a disease. Take a city of 100,000 people. For chronic diseases, such as tuberculosis, incidence rates can provide better insight into whether efforts to limit disease are succeeding. In our imaginary city, prevalence might remain at 0.1% (100 cases per 100,000) over each of 2 years even though new cases develop. The steady prevalence rate happens because some patients who contracted the disease in year 1 might have been cured (or died) before year 2 but were replaced by newly developed cases.

However, if the incidence of <u>new</u> cases per year (per 100,000 previously uninfected individuals) rises from 20 cases/100,000 in year 1 to 60 cases/100,000 in year 2, then efforts to stem the disease are failing. By contrast, if the incidence drops from 20 new cases to 3 per 100,000, then control measures are working. An endemic disease will maintain a relatively constant prevalence and incidence rate. Outbreaks are marked by an increase in both incidence and prevalence. This is what happened during the 2014 and 2018 Ebola epidemics in West Africa and in 2020 with COVID-19.

Incidence Statistics

Incidence statistics can be broken down by sex, race, age, occupation, and many other categories. The data can then be presented to reflect geographic or yearly trends. For example, **Figure 26.5** presents the US geographic distribution of Lyme disease over the course of 15 years and the distribution of syphilis in 2017. Note the geographic clusters of disease. Lyme disease occurs mostly in the Northeast but has spread over time (Figure 26.5A). The distribution of cases corresponds to the geographic migration of the *Ixodes* tick vector. In contrast to Lyme disease, syphilis is more prominent in the South (Figure 26.5B).

Plotting case incidence over time can yield other insights. **Figure 26.6A**, for instance, tracks the incidence of US measles

Figure 26.5 Geographic Representation of Disease Incidence

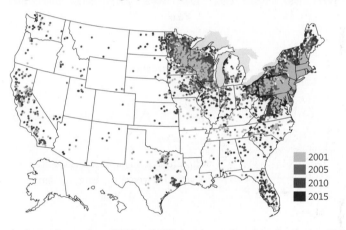

A. Lyme disease from 2001 to 2015. Incidence (number of cases per 100,000 population) is highest in the northeastern United States. All data were obtained from the CDC. Each dot represents one case. The map illustrates how Lyme disease is spreading across the United States.

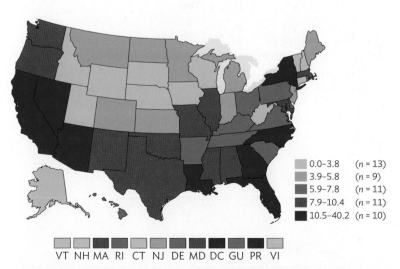

B. Syphilis incidence in 2017. Data are reported as number of cases per 100,000 population.

Figure 26.6 Disease Incidence of Measles, Mumps, and Pertussis in the United States

Data used were obtained from the CDC.

A. Measles incidence

B. Mumps incidence

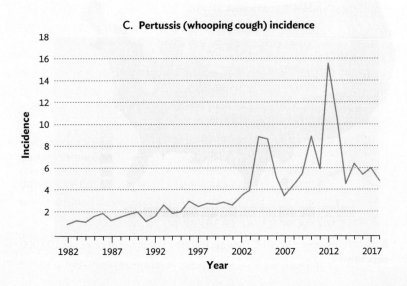

C. Pertussis (whooping cough) incidence

A. Measles incidence (per 100,000 population) from 1977 to 2018. Inset magnifies the 2000–2018 region of the larger graph and presents incidence on a log scale. Measles vaccine was introduced in 1963. Measles is no longer considered endemic in the United States.

B. Mumps incidence (per 100,000 population) from 1987 to 2018. Inset magnifies the 2002–2018 region of the larger graph and presents incidence on a log scale. Mumps vaccine was introduced in 1967. The institution of a second dose of combined MMR vaccine in 1989 led to historically low morbidity until 2006. A 2006 outbreak of more than 6,000 cases, affecting primarily college students aged 18–24 years in the Midwest, caused the CDC's Advisory Committee on Immunization Practices to update its vaccination recommendations for mumps.

C. Pertussis incidence (per 100,000 population) from 1982 to 2018. The recent spikes in cases appear to be related to diminished coverage by vaccination and the waning of protection among older adults. The acellular pertussis vaccine was introduced in 1996 as part of the DTaP series, replacing the older, killed, whole-cell vaccine used since 1948.

cases from 1977 to 2018. Measles vaccine was introduced in 1963, after which the annual incidence of new cases per year fell from 26 per 100,000 to only 2–3 per 100,000. An unexpected spike of cases in 1990 (resulting from a decrease in vaccinations; 75% of victims were unvaccinated) prompted a change in US vaccination practices from administering a single dose at 1 year of age to two doses, given at 1 year and then at 4–6 years old. By the year 2000, measles was declared eradicated from the United States. Only 86 cases were confirmed that year, and most were linked to traveler-imported cases from other countries. However, by 2019 the number of US measles cases had increased to over 1,200, meaning measles could once again become endemic in the United States.

Figure 26.6B tracks the incidence of mumps, which decreased after the introduction of mumps vaccine in 1967. The institution of a second dose of combined MMR (measles, mumps, and rubella) vaccine in 1989, although primarily aimed at measles, led to historically low morbidity for mumps until 2006. The 2006 outbreak of more than 6,000 cases affected primarily college students aged 18–24 years in the Midwest. The outbreak was surprising, given the high level of vaccine coverage. Health officials concluded that the immune response for mumps faded more rapidly than was thought and that some people did not receive the second dose of vaccine. (Immunity wanes when memory B and T cells are not periodically stimulated by exposure to antigen.) This situation left many immunized people at risk. As a result, the CDC's Advisory Committee on Immunization Practices updated its vaccination recommendations in 2006 to ensure that children received two doses of mumps vaccine and that health care workers were immune or revaccinated.

The increasing incidence of pertussis, or whooping cough, shown in **Figure 26.6C**, is not entirely clear. Factors thought to be involved include decreased rates of vaccination, a possibly less effective vaccine, and improved detection. The acellular pertussis vaccine was introduced in 1996 as part of the DTaP series, replacing the older, killed, whole-cell vaccine used since 1948. It is now recommended that older people, whose immunity to pertussis is waning, receive a booster vaccination with Tdap. Tdap is a multivalent vaccine containing full-strength tetanus toxoid combined with diphtheria toxoid and acellular pertussis antigens in doses lower than those of the standard DTaP vaccine. These examples illustrate how epidemiologists, by monitoring incidence and prevalence trends, can anticipate epidemics and contain them.

Thought Question 26.1 In 2011, it was reported that 77 people in British Columbia, Canada, contracted mumps. Seventy percent of those cases were aged 18–35 years. How would you have linked these cases, and what do you think led to the outbreak?

Tools of Epidemiology

The principles just outlined—combined with knowledge of disease transmission, the presence of effective diagnostic laboratories, and vigilant surveillance programs—are all vital for tracking diseases on local and global scales. An interactive network linking health care professionals to local, state, and national agencies forms the early-warning system we rely on to keep us one step ahead of pathogenic microbes. This section describes some of the tools they employ.

CLINICAL MICROBIOLOGY LABORATORIES Infectious disease epidemiology depends on dedicated clinical microbiology laboratories to identify infectious agents. These laboratories are located in or are associated with thousands of hospitals around the world. In addition, most (if not all) hospitals send samples of difficult cases or unusual organisms to regional reference labs equipped to run technically challenging tests and handle extremely dangerous or hard-to-grow pathogens (for instance, *Francisella*). Though important for individual patient care, this information also yields a wealth of data useful for epidemiological studies. Processing this information is part of surveilling an infectious disease.

SURVEILLANCE Surveillance is a three-stage process. The first stage is to see whether a patient's disease fits the CDC case definition of a specific infection. Validating an infection requires an infection control practitioner (ICP) to review a patient's chart and see whether the patient had the signs and symptom of an infection. Surveillance definitions cover infections of the bloodstream, urinary tract, lung (pneumonia), and surgical sites. The second stage of surveillance is to identify the case's geographic location and the patient's demographic profile (for example, age, sex, and race). The third stage is to use this information to identify trends.

Surveillance traditionally required laborious manual data entry and assessment. Now, integrated software solutions assess incoming data from microbiology laboratories and other online sources. By reducing the need for data entry, this software significantly reduces the data workload of ICPs, freeing them to concentrate on clinical surveillance.

Epidemiological early-warning systems require an extensive organization that coordinates information from many sources. In the United States, that duty falls to the CDC. On the world stage, it is the WHO. Any disease considered highly dangerous or infectious is first reported to local public health centers, usually within 48 hours of diagnosis. The local centers forward that information to their state agencies, which then report to the CDC in Atlanta. This is how authorities in 2013 recognized that an unusual *Neisseria meningitidis* group B outbreak occurred at the University of California, Santa Barbara campus.

MOLECULAR SURVEILLANCE Molecular approaches to infectious disease surveillance can also prevent or limit the international spread of disease. A pandemic of pulmonary tuberculosis currently affects more than 2 billion people. Many *Mycobacterium tuberculosis* infections are caused by multidrug-resistant (MDR) strains that are difficult, if not impossible, to kill with existing antibiotics (see Section 20.3). MDR tuberculosis is especially serious among refugee populations attempting to flee war-torn countries. Measures to ensure that refugees are screened for tuberculosis as they enter neighboring countries are vital to contain the pandemic. Although chest X-rays are often mandatory, a positive image will be obtained only if the disease is at a relatively advanced stage. X-rays will not identify actively infected individuals who have not developed the characteristic lung tubercles (Section 20.3).

Unfortunately, the acid-fast staining of sputum samples (Section 25.2) also fails to detect individuals at an early stage of infection. However, polymerase chain reaction (PCR) amplification techniques are much more sensitive for identifying these individuals. Increased sensitivity means that PCR amplification can detect DNA from very small numbers of mycobacterial cells, compared to the higher numbers of cells required for a positive acid-fast stain. As time progresses, PCR surveillance strategies will be used more often to track the worldwide ebb and flow of microbial diseases. Epidemiologists already use PCR and other DNA strategies for epidemiological purposes to type (that is, determine the relatedness of) different microbial isolates by generating a complex DNA profile specific for a particular strain.

SECTION SUMMARY

- **Epidemiology** is a field that examines factors that determine the distribution and source of disease.

- **Descriptive epidemiology** identifies geographic areas or groups of people that have a high incidence of a disease.

- **Analytical epidemiology** uses experimental or observational studies (cohort or case-control) to test hypotheses about the cause of a disease or the risk factors associated with that disease.

- **Endemic, epidemic, and pandemic** refer, respectively, to a disease that is always present in a locale, an increase of disease cases (outbreak) in a geographic area, and the worldwide outbreak of a disease.

- **Incidence** measures the number of new cases of a disease that develop in a geographic area within a specified time frame. **Prevalence** reflects the total number of **active cases** (old or new) of a disease in a specified area.

- **Surveillance techniques** help epidemiologists detect new outbreaks of a disease and predict when and where a disease may spread. International, national, and local health organizations and their laboratories work together to develop case descriptions and collect incidence data needed to track an infectious disease.

Thought Question 26.2 A recent outbreak of *E. coli* O157:H7 disease has been detected. How long will it take from the time the first victim eats contaminated food before a laboratory confirmation of this organism is established and a public health lab determines that strains from several cases are identical? As you work through this timeline, consider every event that must occur and how long each takes.

26.2 Tracking Health Care–Associated Infections

SECTION OBJECTIVES

- Discuss the nature of health care–associated infections and list the two most common forms.
- List key questions that can help identify the source of a hospital-acquired infection.
- Discuss procedures used to prevent hospital-acquired infections.

CASE HISTORY 26.1

Death by Negligence

Stephanie was a bright 7-year-old girl who came to her rural Manitoba hospital for a tonsillectomy. She was a little scared, but her mother was with her and she liked the surgeon, Dr. Ryan, and his funny bear ties. Sometimes she would even grab his tie while Dr. Ryan was examining her (**Figure 26.7**). However, she didn't much like the IV catheter Dr. Ryan had to insert in her arm. It hurt a little, but she was brave and didn't cry much. The surgery went well and Stephanie was recovering in her hospital room a few hours later. However, during the night the nurse noted her temperature had risen to 102°F. By the next day, Stephanie's heart was racing and her breathing was labored. She was rushed to the pediatric intensive care unit, and empirical antibiotic therapy was initiated. Despite this treatment, her blood pressure began to fall. Stephanie was diagnosed with postsurgical septic shock. To everyone's horror, Stephanie died 2 days later of methicillin-resistant *Staphylococcus aureus* (MRSA) sepsis. The organism recovered from her blood was a MRSA strain identical to one causing a wound infection in a 57-year-old man two floors down.

How could this happen? Infections contracted in a hospital setting are called **hospital-acquired infections**, formerly known as

Figure 26.7 Opportunity for a Health Care–Associated Infection

As the doctor examines this child, he could accidentally transmit an infectious agent if his lab coat, tie, or stethoscope is contaminated with a pathogen from another patient.

nosocomial infections. But hospitals are not the only medical care facilities that can transmit infections to their patients. Outpatient clinics, nursing homes, rehabilitation centers, physicians' offices, and home health care providers are other sources of infectious agents that can be transmitted to patients. Thus, the more general term **health care–associated infections (HAIs)** is used to identify any infection that involves a health care provider or facility as the source. The classic example is a patient initially admitted for medical treatment, here a tonsillectomy, who then develops a serious infection. The source of the microbe is somewhere in the hospital. But where? You will later learn the unfortunate circumstance that led to Stephanie's deadly MRSA infection.

Health care–associated infections are a major problem for hospitals, both nationally and internationally. The gravity of the situation is evident in the following statistics: 5%–10% of hospital patients develop health care–associated infections. Of 2 million US patients who develop HAIs every year, about 90,000 will die. Many of those infections are caused by antibiotic-resistant bacteria. It is sobering that approximately one-third of all health care–associated infections could be prevented if appropriate surveillance and preventive actions were taken by hospital staff. Table 26.1 lists major types of health care–associated infections and risk factors. Of course, even a single patient contracting an HAI is tragic, but even more troubling is when several patients become infected with the same organism from the same source.

Ignaz Semmelweis, "Savior of Mothers"

The best-known and perhaps earliest case of hospital epidemiology was initiated by Ignaz Semmelweis, a Hungarian physician practicing in a hospital in Vienna, Austria, in 1847

Table 26.1

Typical Health Care–Associated Infections

Infection Type	Percentage of Total HAIs	Category of Infection	Risk Factors	Common Etiologic Agents
Urinary tract	20–30	Catheter-associated	Catheters, other instrumentation	*Escherichia coli, Candida* spp.
Pneumonia	15–20	Ventilator-associated	Aspiration of endogenous microbiota, ventilators	*Streptococcus pneumoniae, Haemophilus, Pseudomonas, Staphylococcus aureus, Klebsiella pneumoniae, Acinetobacter*
Surgical wounds	20–30	Surgical site–associated	Surgeon's technical skill, patient's underlying diseases (diabetes, obesity), age, presence of surgical drains	*Staphylococcus aureus*, coagulase-negative staphylococci, enteric bacteria, anaerobes
Septicemia	5–10	Central line–associated	Intravascular devices (catheters)	Derived from microbiota at the site: *Staphylococcus aureus*, enterococci, Gram-negative bacilli, *Candida*
Diarrhea	15–20	Antibiotic-associated	Antibiotic treatment	*Clostridioides difficile*

(**Figure 26.8**). All hospital obstetric wards of the day struggled with a deadly disease called puerperal sepsis, or childbed fever, a lethal blood infection that many women contracted after giving birth in the hospital (10%–35% mortality). Semmelweis noticed that the incidence of the disease, now known to be caused by *Streptococcus pyogenes*, increased during certain times of the year. He then realized that these times coincided with an infusion of medical students into the clinic. Tracking the movements of these students, he found they came straight from the autopsy room to the maternity ward to assist in deliveries. Semmelweis made the connection that medical student and physician hands were probably transmitting "cadaverous particles" (what we now call germs) from the dead bodies in the autopsy room to the women in the maternity ward.

The value of handwashing was not recognized in the mid-1800s. Germ theory was still in its infancy and highly controversial. Physicians scoffed at handwashing unless they were covered in blood and guts. Nevertheless, Semmelweis insisted that all people entering the clinic wash their hands with an antiseptic (a chlorinated lime solution) before touching any patient. The audacity of this physician, a Hungarian no less, caused an uproar in the prejudiced Viennese medical community.

Although Semmelweis's edicts led to a sharp decline in the incidence of puerperal sepsis, his success went unrewarded. It is not clear whether rejection of his work contributed to his slow mental decline, but this "savior of mothers"—as he was called by those he saved—eventually developed dementia and ended up confined to a darkened cell in a mental institution, where he died in 1865 from, ironically, a blood infection. But in the long run, his advocacy of antisepsis became a key precept of hospital care.

Identifying the Source of Hospital-Acquired Infections

So how do infection control personnel identify and contain sources of infection in hospitals? Infectious disease epidemiologists are essentially detectives who must ask probing questions in the midst of an outbreak:

- **Is each patient infected with an identical isolate of the pathogenic organism?** If the isolates of MRSA are different, for example, there may be multiple sources. Immunological or DNA approaches are typically used to determine relatedness. In Case History 26.1, both MRSA isolates were identical by DNA analyses.

Figure 26.8 Ignaz Semmelweis (1818–1865)

Figure 26.9 Important Infection Control Measures

A. Foam hand sanitizer outside a patient's room.

B. Disposable gown and gloves.

- **Has any member of the hospital staff made contact with all the patients?** If a hospital employee harbors the organism as part of their normal microbiota, that employee could be the common source. In the case history, someone attending to the 57-year-old MRSA patient may have become contaminated with MRSA and accidentally transferred it to Stephanie.

- **Did all patients come in contact with the same medical instrument, the same lot of medical supplies (such as IV fluid, indwelling catheters, or intubation tubes), or the same food or water?** Any of these might be contaminated.

- **Were the affected patients in contact with each other at any time, enabling person-to-person transmission?**

- **If the infectious agent is airborne, could air ducts linking the rooms have problems?**

Figure 26.10 Peripherally Inserted Central Catheter (PICC Line)

PICC

Superior
vena cava

Heart

PICC line
enters body

This is a form of long-term, intravenous access used for several purposes, including nutrition or extended antibiotic therapy. A PICC line is inserted into a peripheral vein (for instance, in the arm) and then advanced to the heart, where it rests in the superior vena cava.

- **Have hospital staff, sinks, surfaces, and carts been tested for presence of the organism?** Water is the most common source of hospital-acquired infections.

- **Have standard infection control measures been maintained?**

Answers to these questions will often, but not always, reveal the source of the infections. Once identified, the problem can be rectified and recurrence can be prevented by using effective infection control measures.

Infection Control Measures

So how can hospital-acquired infections be prevented? Hand sanitation, using soap or alcohol-based or alcohol-free hand sanitizers, is one important factor. Most hospitals have posted dispensers of foam hand sanitizer outside and inside each patient's room (**Figure 26.9A**). Everyone who enters or exits the room must use these dispensers.

In addition to handwashing, disposable paper gowns and latex gloves (**Figure 26.9B**) are worn by anyone entering the rooms of immunocompromised patients or of patients with extremely dangerous infectious diseases (such as MRSA). These measures protect the susceptible patient and, for "gowning," also protect the visitor (health care worker or relative). Disposing of gowns and gloves before leaving the room will also prevent spread of the organism to others in the hospital.

Aseptic technique is also crucial to good patient care. Asepsis must be maintained whenever a procedure such as inserting a catheter into the bladder or a vein is performed. The site must be prepared with an antiseptic, and sterile surgical drapes and gloves should be used when inserting intravascular lines such as a peripherally inserted central catheter (**Figure 26.10**). Equipment used for procedures should be sterilized by heat, if possible, or should come from undamaged sterile packaging. Surfaces of tables, doorknobs, and the like should be disinfected and linens should be cleaned to prevent accidental transfer of pathogens to other people.

So how did Stephanie end up with sepsis in Case History 26.1, and how could it have been prevented? Before Stephanie's surgery, Dr. Ryan was examining the 57-year-old MRSA patient's wound when his tie accidentally grazed the wound area (he did not don a protective gown before entering the room). *Staphylococcus aureus* was transferred to the tie. Later, while Stephanie was playing with Dr. Ryan's tie, the organisms moved from the tie to Stephanie's arm. Inserting the catheter in her arm inadvertently introduced the pathogen into her bloodstream. Two days later Stephanie was dead. Hospital surveillance procedures later identified five more of Dr. Ryan's patients who were contaminated with the same strain of MRSA. This tragedy could have been prevented if Dr. Ryan had simply not worn a tie or kept his tie tucked behind a lab coat or under a surgical gown while examining patients. This and other scenarios raised awareness that clothing is an important potential vector for transmitting disease in hospital settings.

"Daisy Chain" Nosocomial Infections

Patients, especially elderly, long-term-care patients, are often moved between hospitals and nursing home facilities. Transferred patients take with them any bacteria they harbor, which may initiate a **daisy-chain nosocomial infection**: an insidious series of linked infections. For example, in Florida, a severely wounded trauma victim developed a lung infection with a multidrug-resistant strain of *Salmonella enterica* serovar Senftenberg. As diagrammed in **Figure 26.11**, 19 other cases of *S. enterica* Senftenberg developed in that hospital. One of those patients was transferred to a

Figure 26.11 "Daisy Chain" Movement of an Antibiotic-Resistant Strain of *Salmonella enterica* Serovar Senftenberg between Hospitals

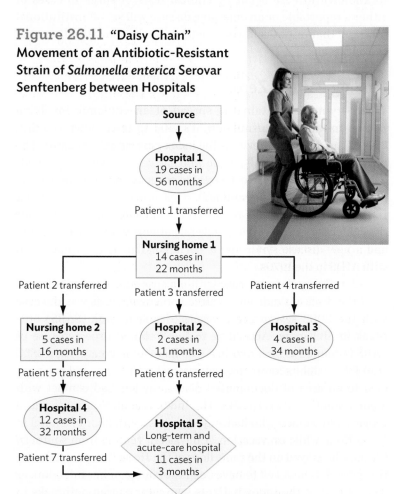

The diagram portrays how a single infected patient transferred between two health care facilities can initiate a string of infections in multiple facilities as later infected patients are transferred to other facilities. Each arrow indicates that a different infected patient was transferred. A breakdown in aseptic procedures most likely led to the secondary infections.

nursing home facility, after which 14 new cases developed. Some of those new patients were transferred to additional facilities, leading to yet more cases. Repeated breakdown of standard infection control measures by hospital personnel apparently caused the daisy-chain transmission of the organism to patients in different hospitals. The case also illustrates how organisms, such as MRSA, that were previously restricted to hospital environments can escape to cause community-acquired infections.

Where do we stand today with hospital-acquired infections? Have things improved? Available statistics through 2018 for US hospitals indicate that the number of infections actually observed, compared with what was predicted from prior years, has fallen an average of 10% since 2015 and about 30% since 2008. So the situation is improving, but much work remains.

SECTION SUMMARY

- **Health care–associated infections (HAIs)** are a major concern for hospital staff and other health care professionals.
- **The source of single or multiple HAIs in a single facility can often be identified** by answering questions such as:
  —Are all strains of the organism identical?
  —Did any hospital staff have access to all the patients?
  —Did all the patients use a particular medical device?
  —Were standard infection control measures enforced?
- **Infection control measures** can prevent most HAIs. These measures include frequent handwashing, using gowns and sterile gloves to interact with infected or immunocompromised patients, and using aseptic technique when inserting intravascular lines or urinary catheters.
- **Transferring an infected patient** between health care facilities can spread HAIs to the new facility.

Thought Question 26.3 An outbreak of MRSA occurred in a community hospital. The infection control team identified a nurse harboring the same strain of MRSA as part of his normal microbiota. The nurse, however, refused treatment because he wasn't sick. Should the hospital administrators fire the nurse? Should all hospital personnel be screened for MRSA as a condition of employment?

Thought Question 26.4 Imagine a large metropolitan hospital in which eight serious infections with MRSA have occurred, and you must determine the source of infection so it can be removed. Using common bacteriological and molecular techniques, how would you accomplish this task?

26.3 Tracking Global Disease Trends

SECTION OBJECTIVES

- Discuss how global disease trends are recognized.
- Explain the importance of finding the index case of an outbreak.
- Employ online sources to track worldwide disease trends.

CASE HISTORY 26.2

The Case of the Killer Sprouts

Between May and June 2011, six residents of the United States developed striking cases of bloody diarrhea. Five of them developed life-threatening hemolytic uremic syndrome and one eventually died. All the cases were caused by the same serotype of *E. coli*, O104:H4, an enteroaggregative pathovar of *E. coli* that normally doesn't produce Shiga toxin. However, this one did! Another oddity was that the six cases did not cluster in a single location; they were widely dispersed among five different states. During this same time, many people in Europe started to die from hemolytic uremic syndrome. By June, 3,228 cases and 35 deaths were reported in Germany and France. The organism isolated in each instance was the same odd pathovar of *E. coli* O104:H4 that produced Shiga toxin. Public health officials on both continents realized that all the cases in the three different countries had to be linked, but how?

How do we track global outbreaks of infectious diseases and even predict when one may be coming our way? Microbiological characterizations of the pathogens in each case were initially carried out by local clinical labs. These labs then shared the patient strains of *E. coli* with the CDC in the United States and the WHO in Europe. More detailed genetic analyses conducted by the scientists at both health organizations revealed that the same unique strain of *E. coli* was involved, now called EAEC-STEC O104:H4 because it has features of both enteroaggregative *E. coli* (EAEC) and Shiga toxin–producing *E. coli* (STEC). Great! But how were the Americans infected? Epidemiologists at the CDC discovered that five of the six sickened Americans had traveled to Germany recently, where they were infected just before returning to the United States. The sixth victim was a close relative of one of those travelers.

Meanwhile, once the common pathogen was identified, the hunt for the source began. Restaurant cohort studies were conducted to identify which foods all the victims had eaten. Traceback investigations were then used to pinpoint where contamination may have occurred along the food production and transportation chains. The results tied the source to mixed raw fenugreek bean sprouts grown on a farm in Uelzen, Germany. How the organisms contaminated the sprouts in the first place remains a mystery.

Procedures to track global outbreaks of disease are similar to those used at local and national levels but are more complex. Stopping the intercontinental outbreak of O104:H4 disease in this case began with data collection at the local level, but communication and cooperation between the health agencies of several nations was crucial.

Local Physician Notification of Health Organizations

As noted earlier, epidemiologists first recognize that an epidemic is under way because physicians report instances of what are known as reportable diseases to state health agencies. A disease is classified as reportable, or notifiable, because of its severity and transmissibility (Table 26.2). Local agencies pass these reports to central health organizations such as the CDC and WHO. These organizations use the data to track disease incidence around the world, looking for sudden upsurges. In addition to reportable diseases, diseases of <u>unknown</u> etiology are also reported to health authorities because a new emerging disease, not on the list of notifiable diseases, can be detected as a cluster of patients with unusual symptoms or combinations of symptoms. This approach is known as **syndromic surveillance** or **symptomatic surveillance**. A new disease manifests with common symptoms (for example, the cough and fever of SARS, described later) but cannot be linked to a known disease agent by clinical tests. A surge in cases of either a reportable or an emerging disease will set off institutional "alarms" that initiate efforts to determine the source and cause of the outbreak.

Finding Patient Zero

When trying to contain the spread of an epidemic involving human-to-human transmission, it's vital to track down the first case of the disease (known as the **index case** or **patient zero**). The next step is to identify all people who had contact with the individual so that each contact can be treated or separated from the general population (that is, **quarantined**). This process is called contact tracing. When a new disease arises, the epidemiological search for the index case starts only after several patients have been diagnosed and a new disease syndrome is declared. This is what happened with AIDS in the 1970s.

Identifying an index case within a specific community is easier if the disease syndrome is already recognized, as was the case with the 2003 severe acute respiratory syndrome (SARS) outbreak in Singapore. According to the WHO, a suspected case of SARS is defined as an individual who has a fever greater than 38°C (100.4°F), exhibits lower respiratory tract symptoms, and has traveled to an area of documented disease or has had contact with a person afflicted with SARS. The index case in Singapore was a 23-year-old woman who had stayed on the ninth floor of a hotel in Hong Kong while on vacation. A physician visiting from southern China, who stayed on the same floor of the Hong Kong hotel during this period, is believed to have been the source of her infection, as well as that of the index patients who later caused outbreaks in Vietnam and Canada.

During the last week of February, the woman, who had returned to Singapore, developed fever, headache, and a dry cough. She was admitted to Tan Tock Seng Hospital, Singapore, on March 1 with a low white blood cell count and patchy consolidation in

Table 26.2
Notifiable Infectious Diseases (Centers for Disease Control and Prevention)

Bacterial		
Anthrax	Hansen's disease (leprosy)	Salmonellosis (non–typhoid fever types)
Botulism	Legionellosis	Shigellosis
Brucellosis	Leptospirosis	Staphylococcal enterotoxin
Campylobacter infection	Listeriosis	Streptococcal invasive disease
Chlamydia infection	Lyme disease	*Streptococcus pneumoniae* invasive disease
Cholera	Meningitis, infectious	Syphilis
Diphtheria	Pertussis	Tuberculosis
Ehrlichiosis	Plague	Tularemia
E. coli O157:H7 infection	Psittacosis	Typhoid fever
Gonorrhea	Q fever	Vancomycin-resistant *Staphylococcus aureus*
Haemophilus influenzae invasive disease	Rocky Mountain spotted fever	(VRSA)

Viral		
Dengue fever	Measles	Smallpox
Ebola	Mumps	Varicella (chickenpox), fatal cases only
Hantavirus infection	Poliomyelitis	(all types)
Hepatitis, viral	Rabies	Yellow fever
HIV infection	Rubella and congenital rubella syndrome	

Fungal		
Coccidioidomycosis		

Parasitic		
Amebiasis	Giardiasis	
Cryptosporidiosis	Malaria	
Cyclosporiasis	Microsporidiosis	

the lobes of the right lung. Tests for the usual microbial suspects (*Legionella, Chlamydia, Mycoplasma*) were negative. Electron microscopy of nasopharyngeal aspirates showed virus particles with widely spaced clublike projections that resemble a crown (**Figure 26.12A**). The pathogen was a new type of coronavirus, a member of a large group of enveloped, positive-sense RNA viruses, some of which cause common colds. At the time of the patient's admission to the hospital, the clinical features and highly infectious nature of SARS were not known. So, for the first 6 days of hospitalization, she was in a general ward, without barrier infection control measures. During this period, the index patient infected at least 20 other individuals, including hospital staff, nearby patients, and visitors. **Figure 26.13** presents a network diagram showing how the disease spread between patients.

Within weeks, the WHO named the disease in China severe acute respiratory syndrome (SARS) and issued travel alerts (discussed further in Section 26.5). These alerts allowed Singapore health officials to rapidly identify the index patient and her contacts, limiting spread of the illness. Ultimately, SARS killed fewer than 1,000 victims worldwide, although thousands more became ill and recovered. Implementation of infection control procedures that included the wearing of surgical masks by the public

Figure 26.12 Severe Acute Respiratory Syndrome

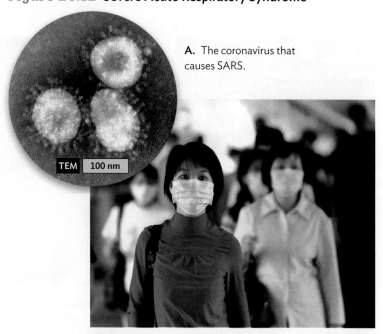

A. The coronavirus that causes SARS.

TEM 100 nm

B. Citizens of China, including the military, donned surgical masks in 2003 to slow the spread of SARS.

Figure 26.13 Network Diagram of SARS in Singapore

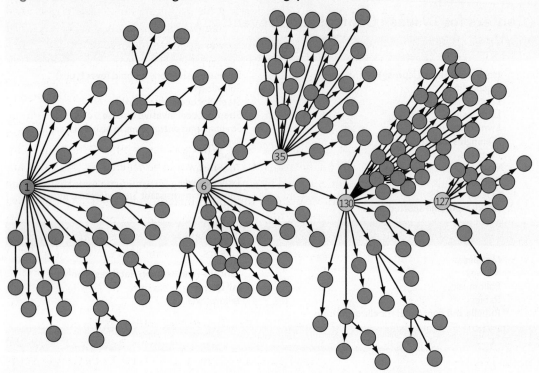

The diagram maps the connections between SARS patients during the initial outbreak. Patient 1 is "patient zero," or the index case, in this instance (red circle). The numbered cases were considered "superspreaders" because they infected more than 10 contacts (yellow circles). For example, patient 6 was a nurse who attended to patient zero. Patient 35, a diabetic patient with heart disease, initially stayed in the same room as patient 6.

helped prevent a potential pandemic (**Figure 26.12B**). After 2004, SARS was not considered much of a threat... until 2019, when a new, more dangerous SARS coronavirus called SARS-CoV-2 emerged in China.

Obstacles to Global Surveillance

In a perfect world, every province in every country on every continent would have modern clinical laboratories, efficient case reporting, and effective political systems needed to quickly detect and limit outbreaks of infectious disease. Unfortunately, the resources of developing countries are too limited to reach public health surveillance goals. The CDC and the WHO often mobilize teams of scientists and medical personnel, sending them to remote parts of the world to help, especially when global health is threatened. Teams often encounter desperate conditions in which local medical personnel struggle against overwhelming odds (**Figure 26.14**).

Figure 26.14 CDC Mission to Monitor Monkeypox in the Congo Basin
Monkeypox is a close relative of smallpox and is endemic to central Africa.

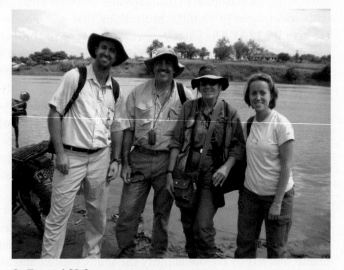

A. A local market in Boende, at the equator in Democratic Republic of the Congo, sells typical "bush meat." Electric power in this area is inconsistent, so no refrigeration exists.

B. A regional clinic/hospital pharmacy that supports the local area of up to 5,000 people. They have about four syringes and needles that are used over and over again.

C. Team of CDC scientists sent to Boende in 2011 to survey monkeypox.

SECTION SUMMARY

- **Tracking a global outbreak** begins with physicians and clinical laboratories notifying local health agencies, the CDC, and the WHO of reportable diseases and diseases of unknown cause.
- **Finding patient zero** is important to contain the spread of disease.
- **Rapid response to and containment of a new disease** can prevent an outbreak from becoming a pandemic.

Thought Question 26.5 Discuss the reasons some diseases spread quickly through a population, whereas others take a long time.

26.4 Detecting Emerging Microbial Diseases

SECTION OBJECTIVES

- Describe the relationship between animal health and human disease.
- Explain how molecular techniques can identify previously unrecognized and unculturable pathogens.
- Discuss how technology contributes to the evolution of emerging pathogens and the spread of infectious disease.
- Correlate the rise in reemerging diseases with the AIDS pandemic.
- Explain how ecology and climate change influence the spread of infectious diseases.

How do scientists predict and stop the emergence of a new infectious disease? A group of factors have simultaneously converged over recent decades to create the perfect incubator for new emerging and ancient reemerging infectious diseases. The most important of these factors include genetic adaptability of microbes, increased global travel and transportation, altered host susceptibility, climate change, aggressive land use, and human demographics and behavior. Also, both public and animal health infrastructures have broken down, poverty has increased, and social inequality persists. Humans are responsible for most of these factors. As a result, the struggle between humans and microbes has intensified. Microbes now have more opportunities to inhabit new niches, cross species boundaries, and travel worldwide to establish infections in new populations of people and animals. Public health initiatives and epidemiology are crucial to keep the struggle balanced in our favor.

The COVID-19 Pandemic: The Perfect Storm

CASE HISTORY 26.3

It Began in Wuhan (but Didn't Stay There)

Peng Yinhua was a 29-year-old respiratory and critical illness doctor at First People's Hospital in the Jiangxia District of the city of Wuhan, China.

Near the end of December 2019, Peng found himself heading to work, daydreaming about his upcoming wedding; a long-awaited ceremony that would take place in just a matter of days. But he couldn't know what was about to happen. Later that morning the patients started arriving. There were only a few at first, but the few grew to nearly 20 that day. Each of them had a high fever and a persistent cough and struggled to breathe. Radiographs showed that most had bilateral lung infiltrates characteristic of pneumonia. The next day more patients arrived, and even more the day after that. What was happening? Peng found himself on the front lines of what would be called the COVID-19 pandemic, caused by a new emerging coronavirus eventually called SARS-CoV-2 (see thumbnail image). Peng knew he would have to postpone his wedding because he was needed here. Sadly, despite taking medical precautions, Peng himself was infected while treating his patients. By January 25, 2020, nearly 700 patients with COVID-19 disease had been admitted to the hospitals of Wuhan. One of those admitted was Peng. Five days later, this selfless doctor died from respiratory failure. His wedding invitations, never mailed, were still in his desk.

The global pandemic known as COVID-19 disease (coronavirus disease 2019) continues to unfold as we write (May 2020). Starting in China, the SARS-CoV-2 virus quickly spread to every country in the world, including the United States. **Figure 26.15** shows an isolate of the virus recovered from the first US patient. By October 2020, the world had suffered over 32 million cases of COVID-19 and nearly one million deaths. That's about 3,000 times the 11,000 cases (1,700 deaths) produced during the SARS (2004) and MERS (2012) epidemics combined.

At this moment during the pandemic, no one knows when or if COVID-19 will be effectively contained. You, the reader, will know

Figure 26.15 Extracellular SARS-CoV-2 Viral Isolate from the First US Case of COVID-19

Black dots show the viral genome in this cross-sectional electron micrograph.

that answer. Nevertheless, we can offer in these pages a sense of our hope, inspired by how quickly scientists have learned so much about this virus, but also reveal our panic as the disease moves across the globe. For instance, the pandemic has curtailed global trade, causing stock market prices around the world to crash. The Dow Jones Industrial Average lost about 35% of its value by mid-March. But there is hope. The following section describes how we got here and where we stand.

Containment efforts. In an attempt to stop the disease from spreading beyond Wuhan, Chinese authorities took the drastic step of placing Wuhan, a city of 11 million people, under quarantine—no one in, no one out, and no cars on the roads. The city appeared as a ghost town. Markets were stripped of food as fast as they could be restocked (**Figure 26.16A**). These steps were the equivalent of what the French called *"le cordon sanitaire"* used by the Duke of Richelieu in 1821 along the Pyrenees Mountains in an attempt to stop yellow fever (and liberal ideas) from Spain spreading into France. But as drastic and as well-meaning as the Chinese effort was, people still managed to leave the city. As a result, the virus made its way to other parts of China, Hong Kong, South Korea, and Japan. In Japan, a cruise ship carrying over 2,000 passengers was quarantined in Yokohama for 2 weeks after several passengers became ill. By the end of those 2 weeks, 690 passengers tested positive for the new virus. As the outbreaks continued to expand to more and more countries, all large public events, such as the 2020 Tokyo Olympics, were canceled or postponed to curb transmission.

The first case of COVID-19 in the United States was identified in Washington State on January 19, 2020. Ultimately, New York became the epicenter of the disease, seeded mostly by infected people arriving from Europe. Starting from the first case identified on March 1 until April 8, 2020, New York State handled 150,000 COVID-19 cases resulting in over 6,000 deaths: an onslaught of disease that overwhelmed the capacities of many hospitals. Protective masks, gloves, gowns, and ventilators (needed to treat the dying) were running low

and hospital staff themselves were falling ill. By mid-April, the United States had suffered a total of 420,000 cases, resulting in over 14,000 deaths. Those numbers increased to over 7 million cases and 205,000 deaths by October. In an effort to break the link of transmission, nearly everyone in the United States spent months sheltering at home, venturing out wearing face masks, but only to the supermarket—no restaurants, no malls, no movie theaters, no family gatherings, no school, and no work (except from home). We've avoided anything involving person-to-person social contact. As a result, we are starting to see the numbers of new cases leveling off, but our economy is in crisis. Funny note: as people rushed to supermarkets to prepare for the arrival of SARS-CoV-2, the first item to run out was, of all things, toilet paper (**Figure 26.16B**).

Identifying the cause. Efforts to understand and control this emerging disease have been rapid. On December 31, 2019, the WHO was first informed of a cluster of pneumonia cases of unknown cause detected in Wuhan. Within three weeks (January 20, 2020), scientists had used well-known nucleic acid and antibody tests to rule out SARS-CoV, MERS-CoV, influenza, avian influenza, and other known respiratory viruses. Genome sequence analysis of samples taken from the first COVID-19 cases revealed the cause was a novel coronavirus similar to the SARS and MERS viruses. On February 11, 2020, the disease was officially named COVID-19 and the virus was dubbed SARS-CoV-2. Note that having separate names for a virus and its disease is not new: think of HIV and the disease AIDS.

The sequence data quickly led to a diagnostic PCR test specific for the viral RNA genome. This and similar tests are used around the world to confirm cases of COVID-19 disease. Rapid antibody tests that can identify everyone who is or was previously infected have been developed. The specificity and sensitivity of the tests were initially poor, but have improved (as of September, 2020). An accurate antibody test is essential to conduct the contact tracing needed to stop the spread of this virus. Most recently, an antigen detection test has been developed that will, within 15 minutes, register the presence of SARS-CoV-2 nucleocapsid antigen in nose or throat secretions. Scientists in the United Kingdom and the United States are even trying to train medical detection dogs that can sniff out COVID-19 individuals, much as they can detect people with cancer or Parkinson's disease.

By scrutinizing the viral genome sequences, scientists also identified a potential vaccine candidate. The coat spike protein of SARS-CoV-2, which binds to host receptors, was found to

Figure 26.16 Pandemic Panic Strains Food Supplies from China to Texas

A. A supermarket in Wuhan, China, on January 23, 2020.

B. A supermarket in Austin, Texas, on March 13, 2020.

be unique among coronaviruses (shown in the image for Case History 26.3). Scientists realized that an mRNA vaccine encoding those sequences could be made and injected into a host. Host ribosomes would then translate the mRNA into spike protein antigens that would stimulate the immune system. In May 2020, one such potential mRNA vaccine had been shown to stimulate the production of antibodies able to inactive the virus. Meanwhile, about 70 other vaccine candidates are also undergoing clinical trials. Fortunately, the SARS-CoV-2 RNA virus is equipped with a proofreading RNA polymerase that limits mutability. Consequently, once a vaccine is developed, yearly immunizations should not be necessary.

Where did the virus come from? Genome analysis also suggested that the virus originated in bats, a common incubator for emerging viral diseases. Scientists think the virus passed from bats through an unknown intermediate animal host before it underwent an animal-to-human spillover event linked to live-animal Asian markets.

The connection between animals and humans in the evolution of new infectious organisms cannot be overstated. Bacteria and viruses that infect only animals can evolve to "jump" from animal to human hosts and, in so doing, emerge as a new human infectious disease. Areas of the world such as China and Southeast Asia, where a diversity of wild and domestic animals live near humans (sometimes in the same room), are often where these new diseases first erupt. Crowded Asian animal markets, for instance, are opportune places for the transmission of pathogens from animals to humans (**Figure 26.4**).

Epidemiology. Epidemiologists, using the techniques described earlier, have learned much about the disease, its transmission, and potential victims. The primary symptoms are fever, cough, and shortness of breath that can evolve into pneumonia. Many patients first noticed a loss of taste and smell before the fever started. In severe cases, patients can also experience diarrhea and vomiting. The patients most affected include older people and people with compromised immune systems. Most, but not all, patients who have died from this infection are in these groups. Children seem less susceptible to the virus and typically develop only mild symptoms. Why children are generally spared is not known. However, by May 2020, some children infected with the virus developed a serious, but rare, inflammatory disease called pediatric multisystem inflammatory syndrome (PMIS). Symptoms include high fever, rash, abdominal pain, and skin peeling from the hands and feet.

The COVID-19 pandemic has spread at a frighteningly fast pace. Transmission is primarily person-to-person via respiratory droplets shed by symptomatic patients but also by infected but asymptomatic people, a feature that complicates contact tracing efforts. It is estimated that one infected person can pass the virus to 5–6 others, a number referred to as the **R number**. To stop spread of the disease, the R number must fall below 1. As for mortality, approximately 2%–5% of people known to be infected have died, compared with a 0.1% mortality rate for seasonal flu and 10% mortality for the original SARS. But we don't yet know how many asymptomatically infected individuals there are, so the actual mortality rate for COVID-19 might be lower than 2%, and the R number might be higher.

The WHO declared COVID-19 a pandemic on March 11, 2020, having met the three criteria needed for that declaration; the virus spreads person-to-person, it kills, and it traveled worldwide. Out of 195 countries in the world, 97 had developed cases of this disease by early March. By April, every country was infected. A graph comparing the increase in cases from various countries reveals that the rate of increase was greatest in the United States (**Figure 26.17**). The good news is that the curves in some places are starting to flatten, which means that fewer new cases per day are being reported. What's important now, mid-pandemic, is to keep focused on containment. In addition to asking everyone to wear face coverings to prevent transmission (unfortunately not everyone does), efforts are directed toward identifying and quarantining infected individuals, tracing their contacts, and canceling large and small group events. Most every school was closed during the spring, switching to online teaching to keep students learning. By fall, some schools cautiously reopened with a variety of new safety procedures in place. Meanwhile, we all hold our breaths and wash our hands (repeatedly),

Figure 26.17 Progression of the COVID-19 Pandemic in Countries around the World

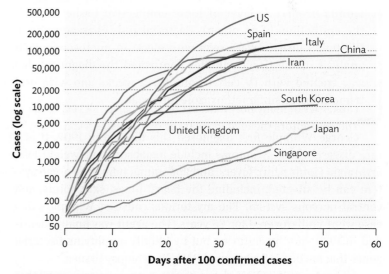

March 30, 2020. To normalize the stages of disease outbreak, each country's curve starts the day after the first 100 cases were reported. The number of reported cases in the United States has risen faster than in other countries at similar stages. Within 6 months (September 23, 2020) the number of cases (and deaths) in these countries had risen to: US, 6.97 million (202,000); Spain, 704,00 (31,118); Italy, 304,000 (35,781); China, 85,314 (4,634); Iran, 436,000 (25,015); South Korea, 23,341 (393); United Kingdom, 416,000 (41,902); Japan, 80,041 (1,520); and Singapore, 57,654 (27).

wondering when the virus will find us. Some estimates predict that 40%–70% of the world's adult population could be infected within a year. You, sitting in the future, will know how bad it was.

Genomic Strategies to Identify Unculturable Pathogens

Microbiologists have developed many strategies to identify the causes of infectious diseases. Robert Koch in the late nineteenth century devised a set of postulates that, when followed, can identify the agent of a new disease (Section 1.3). One important feature of Koch's postulates calls for growing the suspected organism in pure culture. However, for some bacterial diseases the agent cannot be cultured. How might such pathogens be identified?

CASE HISTORY 26.4

Curious Case of the 6-Year Diarrhea

 Joe, a 50-year-old man from Topeka, Kansas, had an abrupt onset of watery diarrhea with an impressive stool frequency of up to ten times every 24 hours. This episode was the latest in a 6-year history of illness that included recurring fevers, flu-like symptoms, profuse night sweats, painful joint swelling, and gripping lower-abdominal pain, especially after meals. Blood and stool cultures tested negative for known infectious agents. A flexible sigmoidoscopy showed only diffuse mild erythema in the bowel, but its appearance was consistent with that of malabsorption of nutrients. A duodenal biopsy to look for the organisms of disease showed large macrophages in the lamina propria (a layer of loose connective tissue beneath the epithelium of an organ). Bacterial rods characteristic of Whipple's disease were seen with electron microscopy. PCR analysis of the tissue samples confirmed the diagnosis. Joe was treated with IV ceftriaxone for 2 weeks and with oral trimethoprim-sulfamethoxazole for 1 year.

George Whipple first diagnosed this disease in 1907. It is a rare disease whose symptoms include malabsorption, weight loss, arthralgia (joint pain), fevers, and abdominal pain. Any organ system can be affected, including the heart, lungs, skin, joints, and central nervous system. The mysterious cause of Whipple's disease went undiscovered for 85 years. The agent was finally identified in 1992—not by culturing but by blindly amplifying bacterial genes that encode 16S ribosomal RNA from biopsy tissues.

All bacterial 16S rRNA genes have some sequence regions that are highly homologous across different species and other sequences that are unique to a species. Tissues from patients diagnosed with Whipple's disease were subjected to PCR analysis using DNA primers that amplified regions of unique 16S rDNA sequences nestled between nonspecific sequences. If a bacterial agent was present, this PCR approach would amplify a DNA fragment corresponding

Figure 26.18 Whipple's Disease

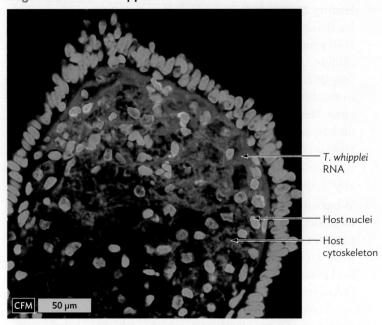

Fluorescent in situ hybridization of a small intestinal biopsy in a case of Whipple's disease is shown. In this test, a fluorescently tagged DNA probe (blue) that hybridizes specifically to *Tropheryma whipplei* RNA is added to the tissue. Other fluorescent probes are used to visualize host nuclei (green) and cytoskeleton (red). Magnification approximately 200×.

to the agent's unique 16S rRNA gene sequence. All tissues produced such a fragment whose sequence was similar to actinomycetes but unlike any of the known species. The organism is a Gram-positive, soil-dwelling actinomycete named *Tropheryma whipplei* in honor of the physician who first recognized the disease. **Figure 26.18** shows the result of an in situ fluorescent DNA probe used to reveal *T. whipplei* RNA in a tissue biopsy sample.

This story reveals that Koch's postulates (see Figure 1.13) must sometimes be modified when identifying the cause of a new disease. Here, the organism could not be cultured in pure form in the lab but was found, via molecular techniques, in all instances of the disease. More recently, clinical samples taken from normally sterile sites (such as blood, bone, and cerebrospinal fluid) were subjected to this 16S rDNA approach. From this experiment, 111 novel genera of uncultured bacteria were identified. Their roles in the infectious process have yet to be determined.

Technology Influences the Emergence and Spread of Infectious Agents

In the 1970s, medical science was claiming victory over infectious disease. Yet today, despite all we know of microbes—despite the many ways we have to combat microbial diseases—our species still lives at the mercy of the microbe. *Borrelia burgdorferi*, MRSA, SARS-CoV-2, Ebola, *E. coli* O157:H7, HIV, "flesh eating" strepto-cocci, hantavirus—all these and many other new infectious agents

Figure 26.19 Visualizing Human Mobility Networks

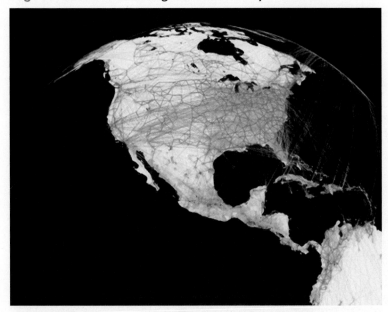

Lines show short-range transportation and long-range airline connections capable of spreading disease. Commuting patterns strongly correlate with how a pandemic evolves.

have emerged over the last 40 years. Worse yet, forgotten scourges, such as tuberculosis, have reappeared. What happened?

Part of the answer is progress itself. Travel by jet, the use of blood banks, and suburban sprawl have all opened new avenues of infection called human mobility networks (**Figure 26.19**). People unwittingly infected by a new disease in Asia or Africa, traveling by jet, can bring the pathogen to any other country in the world within hours. The person may not even show symptoms until days or weeks after the trip. This means that diseases can spread faster and farther than ever. Rapid intercontinental travel is why SARS-CoV-2 spread so quickly around the world. Newly emerged blood-borne pathogens can also spread by blood transfusion—a major problem with HIV before an accurate blood test was developed to screen all donated blood.

Although human encroachments into the tropical rain forests have often been blamed for the emergence of new pathogens, one need go no farther than the Connecticut woodlands to find such developments. *Borrelia burgdorferi*, the spirochete that causes Lyme disease, lives on deer and white-footed mice and is passed between these hosts by the deer tick (**Figure 21.16**). Humans have crossed paths with these animals for years. Why, then, have we suddenly become susceptible to disease? The answer appears to be suburban development. In the wild, foxes and bobcats hunt the mice that carry the Lyme agent. As developers clear land and build roads and houses, the numbers of predators shrink, leaving infected mice and ticks to proliferate. Humans in these developed areas are more likely to be bitten by an infected tick and contract the disease than in prior decades.

Many other examples show the unintended consequences of technology and development in breeding disease. Here are just a few:

- **Mad cow disease.** Modern farming practices (North America and Europe) of feeding livestock the remains of other animals helped spread transmissible spongiform encephalopathies similar to Creutzfeldt-Jakob disease associated with prions (Section 24.6). Because the prion is infectious, the brain matter from one case of mad cow disease could end up infecting hundreds of other cattle, which in turn increases the chance that the disease could spread to humans.

- **Babesiosis.** Often mistaken for Lyme disease, babesiosis is caused by *Babesia microti*. This protist is also carried by the deer tick, which transmits the agent from mice.

- **Hepatitis C.** Transfusions and transplants spread this blood-borne viral disease.

- **Influenza.** Live poultry markets in Asia serve as breeding grounds for avian flu viruses, which can jump to humans.

- **Enterohemorrhagic *E. coli* (for example, *E. coli* O157:H7).** Modern meat-processing plants can accidentally grind trace amounts of these acid-resistant, fecal organisms into beef while making hamburgers.

Natural environmental events can also trigger surges in the incidence of unusual diseases. A good example involves the hantavirus (also known as Sin Nombre virus). An unprecedented outbreak of hantavirus pulmonary disease occurred in the Four Corners area of Arizona, New Mexico, Colorado, and Utah in 1993 and then again in Yosemite National Park in 2012. Heavy rains in each case led to greater-than-normal growth of plants that serve as food for deer mice. The resulting increases in deer mice, which can carry the virus and shed it in their urine, made it more likely that humans and infected mice would come in contact.

You might wonder whether all zoonotic pathogens, which leap from animals to humans, can undergo human-to-human transmission. Luckily for us, many pathogens such as rabies virus, West Nile virus, and *B. burgdorferi* find humans to be dead-end hosts, incapable of spreading the disease to other humans. Zoonotic diseases that cannot transmit from one person to another fail for one or more of the following reasons:

- No opportunity to transmit

- No convenient portal of exit

- An inability to reach a concentration in human blood or secretions sufficient for an infectious dose

For example, for *B. burgdorferi* to transfer from one person to another, a deer tick would have to bite an infected victim and then hop to a new human host. This scenario is highly unlikely because tick transfer between two people is rare. Likewise, people with rabies seldom have an opportunity to bite another person—and even if they did, the concentration of virus in human

Figure 26.20 Locations of Some Emerging and Reemerging Infectious Diseases
Examples shown represent extreme increases in reported cases. Many of these diseases, such as HIV and cholera, are widespread but show alarming increases in the areas indicated.

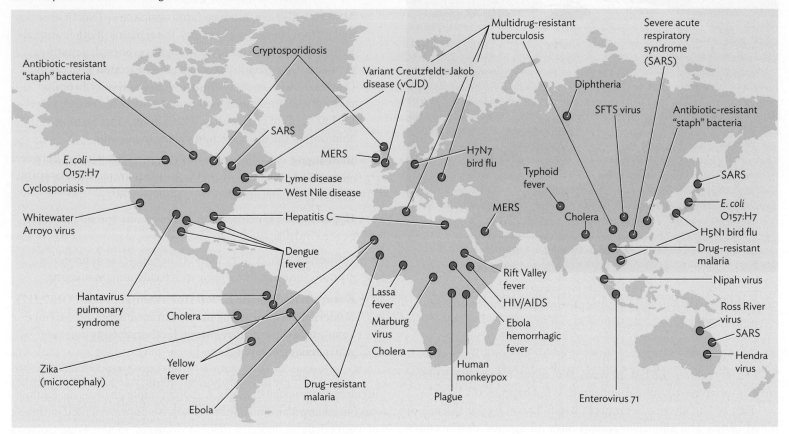

saliva is thought to be too low. West Nile virus is not transmitted between people because the virus does not reach a concentration in a victim's blood high enough for a mosquito bite to transfer an infectious dose. However, epidemiologists do worry about zoonotic pathogens evolving to transfer from person to person (see the earlier discussion of SARS-CoV-2).

Reemerging Pathogens

Figure 26.20 is a world map showing the general locations of emerging and reemerging diseases. A reemerging disease is one that was thought to be under control but whose incidence has risen. For example, the incidence rate of tuberculosis (TB) in the United States was dropping sharply until about 1979 and then rose dramatically until the 1990s. The trigger for its reemergence was the AIDS pandemic that began in 1980. Immunocompromised AIDS patients are highly susceptible to infection by many organisms, including *Mycobacterium tuberculosis*.

M. tuberculosis reemerged among non-AIDS patients owing to the development of drug-resistant strains. Drug resistance in *M. tuberculosis* developed largely because patients failed to

complete their full courses of antibiotic treatment. Treatment usually involves three or more antibiotics to reduce the risk that resistance to any one drug will develop. Many patients failed to take all three drugs simultaneously, allowing the organism to develop resistance to one drug at a time until it became resistant to all of them. These multidrug-resistant (MDR) and extensively drug-resistant (XDR) strains are almost impossible to kill. The link between noncompliance and the development of drug resistance is the primary reason why many TB patients must now take their drugs in the presence of a physician or another member of the medical staff. Fortunately, the incidence of TB since 2000 has fallen once again, partly because of early detection and more effective treatments for HIV. However, drug-resistant TB strains cause an increasing percentage of total cases.

As described in Section 26.3, highly infectious diseases are identified by using aggressive epidemiological surveillance. Communications between local, national, and world health organizations are crucial and helped expose the rise in TB. To see how emerging diseases are monitored, search the Internet for a website called ProMED-mail. There you will find daily reports from around the world that describe new outbreaks of infectious diseases. For example, on April 16, 2014, the site reported on an Ebola outbreak in Guinea, Africa, that involved 197 people and 122 deaths. Twenty-four health care workers fell ill, 13 of whom died. The source of the virus was not known, but it was a new strain of Ebola. In hindsight, we now know that this early report marked the leading edge of a major Ebola epidemic in West Africa.

A ProMED post from January 3, 2020, reported an outbreak of a mysterious lung infection in Wuhan, China, involving 44 people. That report foretold the COVID-19 pandemic currently sweeping the globe. Tracking ProMED reports will allow you, the fledgling microbiologist, to monitor disease trends and spot potential pandemics almost in real time.

Infectious Disease, Climate, and the One Health Initiative

Weather has considerable influence over when and where diseases happen. For example, many diseases appear seasonally. Influenza and pneumonia are more prevalent in winter because more people are huddled together in close quarters passing around pathogens; plus the influenza virus appears to survive better in cold/dry conditions than in hot/humid ones. In contrast, Lyme disease from *Borrelia burgdorferi* and diarrhea caused by *Vibrio parahaemolyticus* occur more often in summer. Lyme disease happens in summer because people venture outdoors more, which increases the chance of contact with tick vectors. Similarly, *V. parahaemolyticus* outbreaks flare in summer because warm summer gulf waters promote bacterial growth, which increases seafood contamination.

But there is another, more serious, problem. The world's climate has been changing slowly but significantly over the last 50 years—largely as a result of humans pumping enormous amounts of greenhouse gases, such as carbon dioxide and methane, into the atmosphere. Greenhouse gases trap heat in the atmosphere, which has caused the world's average yearly temperature to rise by about 1.5°F since 1970. Effects of this warming on climate are evident as disappearing glaciers, rising sea levels, mounting ferocity of hurricanes and tornadoes, extended droughts, increased rainfall (floods), dust storms, and heat waves. These climate changes are bad enough, but they can also affect the epidemiology of infectious disease.

Climate change brings more severe and extreme weather to many parts of the world. The new normal is more rain, heavier snowstorms, and more and greater floods, among other changes. Decreased sanitation in the wake of flooding will propagate gastrointestinal infections. Increasing temperatures and flooding will also influence animal migrations—and the migration of the microbes they carry. Thus, as temperate and even cold regions of the world warm, diseases previously restricted to tropical climates have started migrating into new geographic areas.

CASE HISTORY 26.5
What Happened in Kathmandu?

 Lilawati Awasthi, a 50-year-old woman, lived in a remote mountainous area of Nepal, a landlocked country situated along the Himalayan Mountains next to India in South Asia. In September 2019, Lilawati developed a fever that came with a searing headache and eye pain. Her joints hurt and she tired easily. When the disease failed to go away, she went to the hospital. An antibody test revealed that she had dengue fever, a disease that before 2004 was unheard of in the high-altitude cities of Nepal, such as Kathmandu. Between August and October 2019, 9,000 people in Nepal were hospitalized with the disease. *Aedes aegypti*, the mosquito vector that carries the dengue virus, never used to survive at such high altitudes because it was too cold. What happened?

Climate change happened! Rising air temperatures have extended the habitat of mosquito and tick vectors to higher mountain elevations and wider latitudes. The result is that dengue virus, Zika virus, and *Plasmodium* (the protozoan cause of malaria), among other infectious agents, have increased their geographic range. It's predicted, for instance, that the global range of the *Anopheles* and *Aedes* mosquito vectors that can carry these agents will increase between 20% and 40% by 2030. Extending a vector's geographic range can also provide new hosts for pathogens and spark new transmission cycles that promote pathogen evolution.

As should now be apparent, the natural environment harbors known pathogens and fosters the emergence of new ones. Knowledge that pathogen reservoirs exist in animals, arthropods, and plants has given rise to a collaborative effort among clinicians, scientists, veterinarians, and ecologists called the **One Health Initiative**. The goal of the One Health Initiative is to control human health through animal health and vice versa. For example, using a vaccine-loaded bait to vaccinate wild rodents can decrease Lyme disease in humans. The idea has been extended to plant pathology because some bacteria are pathogens of both plants and humans. For example, *Pantoea agglomerans*, formerly called *Enterobacter agglomerans*, can cause leaf blight on rice and other plants and can also cause sepsis in immunocompromised humans. Some enteric bacteria can even live within the vascular system of some edible plants. Lettuce vasculature, for instance, can harbor *E. coli*, *Klebsiella*, and *Salmonella*.

Only by integrating our knowledge of the environment and ecology with epidemiology can we completely understand how disease trends will change. Anticipating these changes will be crucial to devise and implement effective disease prevention strategies. A multi-state outbreak of *E. coli* O157:H7 in 2006 involving 200 people illustrates the inextricable link between animal health, ecology, and human health. The O157:H7 outbreak was traced to certain batches of organic bagged spinach sold in grocery stores. A collaborative One Health investigation uncovered a series of unfortunate events that led to the contamination. It began on one California farm with a herd of cows that were infected with O157:H7. The cows, of course, deposited their contaminated feces on the ground. Wild hogs traipsed through the contaminated cow dung and raced through nearby muddy spinach fields, dropping the pathogen as they went. The pathogen contaminated the leaves and is thought to have been drawn into the vasculature of the leafy vegetable with irrigation water. After farmers harvested the contaminated spinach, it was bagged and sold throughout the US. Nearly 300 people who ate the contaminated spinach developed bloody diarrhea. Three of the victims died. Clearly, we need a holistic approach involving experts in animal health, ecology, and human health to understand, protect, and promote the health of all.

SECTION SUMMARY

- **Cooperation between governmental health agencies** can lead to rapid recognition of an emerging disease.

- **Molecular methods, such as PCR, can help identify new pathogens** that cannot be cultured.

- **Emerging diseases** can spread quickly around the world as a result of air travel.

- **Modern technology and urban growth** have provided opportunities for new diseases to emerge.

- **The increasing number of HIV-associated immunocompromised patients** has played an important role in reemerging infections.

- **The effect of climate change** on the expansion of tropical and temperate temperature zones and the warming of cold climates has also broadened the range of animal and arthropod vectors of infectious diseases.

- **Multidisciplinary collaboration** among ecologists, veterinarians, clinicians, and other scientists is necessary to devise appropriate strategies aimed at disease intervention and epidemic prevention.

Thought Question 26.6. Why would transmission of the SARS-CoV-2 virus by asymptomatic people be problematic?

26.5 Bioterrorism and Biodefense

SECTION OBJECTIVES
- Define the term select agent.
- List some potential bioweapons.
- Discuss the properties of an effective bioweapon.

"A wide-scale bioterrorism attack would create mass panic and overwhelm most existing state and local systems within a few days," said Michael T. Osterholm, director of the Center for Infectious Disease Research and Policy at the University of Minnesota. "We know this from simulation exercises." These words were spoken on October 9, 2001.

Less than a month after the September 11, 2001, attack on the World Trade Center, some unknown person(s) sent weapons-grade anthrax spores through the US mail (see the Chapter 18 opening case). Fortunately, only five people died and a mere 25 became ill. Despite the clumsy nature of the attack, the impact was enormous. More than 10,000 people took a 2-month course of antibiotics after possible exposure, and mail deliveries throughout the country were affected. The simple act of opening an envelope suddenly became risky business. The epidemiological principles described in this chapter were used to trace the origin of the anthrax strain and identify who had contact with affected individuals.

After this attack and other events, the CDC and the National Institutes of Health (NIH) assembled a list of select agents (marked in blue in Table 2.4) that could potentially be used as bioweapons. A **select agent** is an infectious agent (or toxin) that has a high virulence or mortality rate, is highly contagious, lacks a protective vaccine, or cannot be effectively treated with antimicrobial drugs or other pharmaceuticals. Microorganisms considered bioweapons can be used to conduct **biowarfare**, which aims to inflict massive casualties, or bioterrorism, which may have few casualties but causes widespread psychological trauma.

Although the list of select agents is recent, biowarfare is not new. In the Middle Ages, catapults were used to fling dead victims of the Black Death (plague caused by *Yersinia pestis*) over castle walls. Later, during the French and Indian War in the eighteenth century, a British officer, Colonel Henry Bouquet, appears to have distributed two smallpox-infested blankets and a similarly infested handkerchief to Native Americans. Another incident took place during World War II. The Imperial Japanese Army experimented with infectious disease weapons, using Chinese prisoners as guinea pigs. Even the United States has participated through developing weapons-grade anthrax spores. The strain of *Bacillus anthracis* used in the 2001 anthrax attack purportedly had the same genetic signature as a strain originally researched at Fort Detrick, a US Army base where research into possible bioweapons was conducted in the 1950s and 1960s. The base now hosts elements of the US biological defense program.

The first documented act of bioterrorism in the United States occurred in 1984, when followers of the cult leader Bhagwan Shree Rajneesh tried to control a local election in the city of The Dalles, Oregon, by infecting restaurant salad bars with *Salmonella* (Figure 26.21). The goal was to keep non-Rajneesh voters away from the voting booths. More than 700 people became ill. Eventually, vials of *Salmonella* were found in a clinic laboratory on the cult's compound. Dr. Michael Skeels, director of the Oregon State Public Health Laboratory, discovered the vials during a surprise raid by local, state, and federal agencies. The CDC in Atlanta confirmed that the strain found matched the strain that sickened the

Figure 26.21 Bioterror Attack in the United States, 1984

A. Rajneeshee bioterrorists spread *Salmonella* onto salad bars in 11 different restaurants operating in The Dalles, Oregon, in 1984.

B. Mug shot of Rajneesh in 1985 taken by the Oregon Department of Corrections, Multnomah County, Oregon.

use of anthrax as a weapon. Once spores hit the ground, the threat of infection is limited. Weapons-grade spores are very finely ground so that they stay airborne longer. But as we saw, the letter-borne dispersal system did not generate many victims. Nevertheless, the potential threat of weapons-grade anthrax on the battlefield led the US military in 2006 to resume vaccinating all soldiers serving in Iraq, Afghanistan, and South Korea.

An effective bioweapon would be one that capitalizes on person-to-person transmission. In an easily transmitted disease, one infected person could disseminate disease to scores of others within 1 or 2 days. So for generating massive numbers of deaths, anthrax was a poor choice. But the goal of most terrorists is not to kill many people but rather to terrorize them. In that regard, the anthrax attack succeeded.

restaurant patrons. Rajneesh was never charged with the crime (although his subordinates were), but he pleaded guilty to immigration violations and received a 10-year suspended sentence and a fine of $400,000. Deported and barred from reentering the United States for 5 years, he died in India in 1990.

How effective are bioweapons? The method by which a biological agent is dispersed plays a large role in its effectiveness as a weapon. Only a few people became ill during the 2001 anthrax attack (**Figure 26.22A**), in part because of effective epidemiological surveillance but mainly because anthrax is difficult to disperse and cannot easily be transmitted from one person to another. A person must inhale thousands of spores to contract disease, which means that effective dispersal of the spores is critical for the

The most effective bioweapon in terms of inflicting death (biowarfare) would have a low infectious dose, be easily transmitted between people, and be one to which a large percentage of the population is susceptible. Smallpox (caused by variola major virus) fits these criteria and would be the bioweapon of choice (**Figure 26.22B**). Fortunately, however, smallpox has been eradicated (almost) from the face of the earth and is not easily obtained. Only two WHO-sanctioned laboratories still harbor the virus—one in the United States and one in Russia. It is believed that the virus has been destroyed in all other laboratories (although in 2014 some long-forgotten, unsecured, sealed vials of smallpox virus were found stored in a closet at a Food and Drug Administration lab in Washington, DC). Because smallpox is a serious biowarfare agent, the last two smallpox repositories must remain secure.

Figure 26.22 Dealing with Bioterrorism

Although the good news is that smallpox disease has been eradicated, the bad news is that no civilian born after 1970 has been vaccinated (except some military and laboratory workers). As a result, most anyone under 45 years of age is susceptible to smallpox. Even people who received the smallpox vaccination more than 45 years ago are at risk because their protective antibody titers have diminished. A terrorist attack with smallpox would cause terrible numbers of deaths. The vaccine (vaccinia virus) does, however, exist. Were a smallpox attack to be launched, the vaccine would be rapidly administered to limit the

A. Members of a hazardous materials team near Capitol Hill during the anthrax attacks in 2001.

B. An Illinois man suffering from smallpox, 1912.

impact

What's Blowing in the Wind? Quick Pathogen Detection Systems Guard against Bioterrorism

In April 1979, workers at a secret Soviet biological weapons facility neglected to replace a filter in a laboratory ventilation system, and a cloud of highly weaponized *Bacillus anthracis* spores quickly spewed into the air outside. The deadly plume drifted downwind, infecting humans and cattle across a wide area. The resulting outbreak ultimately killed more than 64 people. Had a rapid pathogen detection system been available and deployed at the facility, all those lives could have been saved. Unfortunately, 40 years ago such technology was only the stuff of science fiction. Today, that technology exists in small, portable forms.

Current surveillance systems include the BioWatch program run by the CDC and the Biohazard Detection System (BDS) employed by the US Postal Service. The BioWatch system places air samplers in undisclosed cities to collect particulates. The filters are then processed in designated BioWatch laboratories that use PCR technology to test for several biological pathogens. The BDS system collects air samples from mail-canceling equipment and uses an automated system to screen for *B. anthracis*. In addition to these safeguards, small, portable devices that can identify the presence of multiple pathogens within minutes are needed.

Indeed, several multipathogen molecular detection platforms have been developed. Some involve machines that detect antigen-antibody interactions; other systems use automated PCR to amplify specific pathogen genes and detect the products through hybridization to immobilized oligonucleotides. As described in Section 25.5, the wet chemistries for all these detection systems are carried out in a roughly 2-inch square called a lab-on-a-chip (LOC).

Figure 1 The BioFlash-E Biological Identifier

Figure 2 Principle of the Bioelectronic Sensor Used in the BioFlash-E Biological Identifier

1 B cells are exposed to bioagents in test sample.

Antibodies

Antigen cross-linked antibodies

2 B-cell antibodies bind to the agent.

DAG = diacylglycerol; IP_3 = 1,4,5-inositol trisphosphate; PIP_3 = phosphatidylinositol 3,4,5-trisphosphate.

Tyrosine kinase

3 Cross-linking triggers a phosphorylation cascade.

Phospholipase C

PIP_3

$IP_3 + DAG$

B cell with added aequorin gene

4 Calcium influx activates aequorin.

Ca^{2+}

Open Ca^{2+} channels

Aequorin

Light emission

Ca^{2+}

Endoplasmic reticulum

5 Photons released hit the photon detector.

An interesting addition to the field of quick pathogen detection is the BioFlash-E Biological Identifier (previously known as PANTHER, for Pathogen Notification for Threatening Environmental Releases). This device, developed in 2008, can detect and identify a pathogen in as little as 3 minutes—significantly faster than traditional methods requiring isolation and growth of the pathogen (which can take 48 hours or longer). The BioFlash-E Biological Identifier (Figure 1) uses a cell-based technology that can detect as few as a dozen particles of a pathogen per liter of air. Currently, it can detect 21 pathogens, including the potential bioterrorism agents of anthrax, plague, smallpox, and tularemia.

The device uses an array of B cells, each displaying antibodies to a particular bacterium or virus. The cells are engineered to emit photons (particles of light) when they detect their target pathogen. How do they do this? The B cells have been bioengineered to express the gene that encodes aequorin, a calcium-sensitive bioluminescent protein. When a pathogen surface antigen cross-links the appropriate B-cell surface antibodies, a signal transduction cascade produces a rapid influx of calcium (Figure 2). Aequorin in the B cell will luminesce within seconds of calcium influx, and a photon detector detects the light emitted. The device then lists any pathogens found.

Quick pathogen detection technologies are constantly being improved in the effort to defend against bioterrorism. Small, quick pathogen detection devices could be used in buildings, subways, and other public areas. Eventually, such technology will supplant the classical clinical microbiology practices that must grow the microbe to identify it. There is hope, for example, that such devices can be used on farms or in food-processing plants to test for contamination by *E. coli*, *Salmonella*, or other food-borne pathogens. Another potential application is in medical diagnostics, where the technology could test patient samples, giving rapid results without having to send samples to a lab. Wherever it's used, successful quick pathogen detection should enable local and global public health centers to quickly detect potential bioterrorism threats and rapidly mobilize response teams.

DISCUSSION QUESTION

Discuss which types of genes you might test to identify pathogens by using a rapid PCR detection technology.

----

spread of disease. Nevertheless, the economic and psychological impact of a smallpox epidemic would be devastating.

Research with organisms considered select agents is tightly regulated. Because *Y. pestis*, for instance, is a select agent, lab personnel working with it must now possess security clearance with the Department of Justice. The laboratory must also register with the CDC to legally possess this pathogen. Furthermore, access to the lab and the organism must be tightly controlled.

Recall from Section 2.4 that biosafety containment has four levels. Agents designated level 1 require the lowest level of containment, whereas level 4 agents are highly virulent and require the strictest containment procedures.

Much has improved since Michael Osterholm offered his dire assessment of a wide-scale bioterrorism attack. Education and surveillance procedures have been bolstered, and new detection technologies are being developed (IMPACT). We will probably never be fully protected from attack, biological or otherwise, but recent efforts, including the Strategic National Stockpile of medicines and medical supplies, have improved the situation.

SECTION SUMMARY

- **Select agents** are bacterial and viral pathogens that the CDC considers potential bioweapons.
- **Bioweapons,** when they have been used, typically kill few people but create great fear.
- **Effective bioweapons** would be transmitted person to person, have a low infectious dose, be easily transmissible, and have a large fraction of the target population susceptible.
- **Research with select agents** is tightly regulated.

Thought Question 26.7. Would lack of an effective treatment be an important feature of an ideal biowarfare agent? Why?

Perspective

This chapter has examined the basic epidemiological principles used to detect infectious disease outbreaks, whether they occurred naturally, were associated with health care, or resulted from malicious intent. As you can see, controlling disease spread is a daunting and ever-changing task. This challenge is upon us for several reasons, not the least of which is because pathogens evolve. Centuries-old pathogens evolve to elude the immune system and to avoid eradication by antibiotics. Meanwhile, new pathogens such as SARS-CoV-2 keep emerging from the shrinking jungles and impoverished areas of the world. To stop the parade of new pathogens that threaten us, the world must take a more considered approach toward land use, confront the issue of climate change, and move more aggressively to improve sanitation around the globe.

Key Terms

analytical epidemiology (p. 851)
biowarfare (p. 870)
case-control study (p. 852)
chain of infection (p. 850)
cohort study (p. 852)
daisy-chain nosocomial infection (p. 859)
descriptive epidemiology (p. 851)
endemic disease (p. 852)
epidemic (p. 852)

epidemiology (p. 850)
experimental study (p. 851)
health care–associated infection (HAI) (p. 856)
hospital-acquired infection (p. 856)
incidence (p. 853)
index case (p. 860)
observational study (p. 851)
One Health Initiative (p. 869)
outbreak (p. 852)

pandemic (p. 852)
patient zero (p. 860)
placebo (p. 851)
prevalence (p. 853)
quarantined (p. 860)
R number (p. 865)
select agent (p. 870)
symptomatic surveillance (p. 860)
syndromic surveillance (p. 860)

Objectives Assessment Questions

26.1 Principles and Tools of Epidemiology

1. The four links in the chain of infection are ____, ____, ____, and ____.

Which link in the chain of infection will the following items interrupt?

2. Handwashing: ____

3. Antibiotics: ____

4. Insect eradication: ____

5. Vaccine: ____

6. An increase in disease incidence above a local county's normal endemic level is called a(n) ____.
    a. epidemic
    b. outbreak
    c. pandemic
    d. acute infection

7. Patients with Lyme disease were asked how often they walk through woods, use insect repellent, and wear shorts or pants. A second set of patients without Lyme disease from the same practice were asked the same questions, and the responses from the two groups were compared. This is an example of ____.
    a. experimental epidemiology
    b. a cohort study

c. a case-control study
d. descriptive epidemiology

8. The number of new cases of a disease arising in a population over a given period is called ____.

26.2 Tracking Health Care–Associated Infections

9. Which of the following is the most common hospital-acquired infection in the United States?
    a. Sepsis
    b. Pneumonia
    c. Meningitis
    d. Urinary tract infections
    e. Diarrhea

10. Twelve cases of MRSA occurred over the course of a year in the hospital. What data will most definitively tell you whether a common source exists?
    a. One doctor and three nurses attended to all the affected patients.
    b. All affected patients ate food from the hospital cafeteria.
    c. The genetic analysis found that all the MRSA strains were identical.
    d. All the affected patients were treated by the same nurse.
    e. All the affected patients had contact with each other.

11. You are about to enter the hospital room of a patient who had an appendectomy. If you planned to walk to the bed but not touch the patient, which of the following should you do?

    a. Because you will not touch the patient, no special precautions are needed.

    b. Don a protective gown.

    c. Put on sterile gloves.

    d. Sanitize hands before entering the room.

    e. c and d

    f. b, c, and d

26.3 Tracking Global Disease Trends

12. What is the usual first step in identifying a pandemic?

    a. Local hospitals and physicians notify city or state health agencies about patients with a specific disease.

    b. State agencies report cases to the national health agencies (such as the CDC).

    c. National health agencies report cases to the WHO.

    d. Epidemiologists use data sent to WHO to identify the global locations and the numbers of cases of a disease.

13. In any epidemic or pandemic, why is identifying the index case important?

    a. To treat the patient

    b. To identify where the index case traveled and whom that person contacted

    c. To quarantine that person

    d. To identify the food that person has eaten

    e. a and b

    f. b and c

26.4 Detecting Emerging Microbial Diseases

14. Climate change can widen the spread of an infectious disease. True or false?

15. A man who contracted dengue fever in Brazil and traveled by jet to Canada would be the index case for an outbreak of dengue fever in Canada. True or false?

True or false? The following are recognized as <u>directly</u> contributing to the <u>evolution</u> of an emerging pathogen:

16. Antibiotic resistance

17. Climate change

18. Acquired immunodeficiency syndrome

19. Air travel

20. Deforestation

21. Vaccination

22. Fifteen people developed symptoms of muscle aches, high fever, and vomiting. You suspect a blood-borne pathogen. But evidence of typical agents known to cause these symptoms was not found, and cultures of blood for bacterial, viral, and fungal agents failed to grow anything. Which of the following strategies would most likely identify a new agent?

    a. Screen each patient for elevated antibody titers.

    b. Screen blood by PCR for rRNA genes and sequence the products.

    c. Determine whether the patients crossed paths.

    d. Test whether a blood transfusion will transfer the disease to another human.

    e. All of the above.

    f. None of the above.

26.5 Bioterrorism and Biodefense

23. NIH designates an agent that might be used for biowarfare as a(n) _____.

24. *Salmonella enterica* has been used as a bioweapon but is not considered a select agent because _____.

    a. infections can be treated with antibiotics

    b. a vaccine exists

    c. diagnostic laboratories can easily identify it

    d. it has no animal reservoirs

    e. it is difficult to grow

25. Which of the following characteristics contributed to the decision to add *Yersinia pestis* to the list of select agents?

    a. Infections cannot be treated with antibiotics.

    b. It is easily transmitted from person to person.

    c. Diagnostic laboratories can easily identify it.

    d. It has no animal reservoirs.

Review Questions

1. Who is John Snow?

2. How is a pandemic different from an epidemic?

3. Explain the difference between prevalence and incidence of disease.

4. How can an endemic disease become an epidemic?

5. Discuss three ways to graphically represent epidemiological data and what information can be gleaned from each.

6. What are the three stages of disease surveillance?

7. What is a hospital-acquired infection? How can it be contracted?

8. Discuss five questions that infectious disease personnel should ask in response to a hospital-acquired infection.

9. Describe some common infection control measures.

10. How are global outbreaks of disease identified?

11. Briefly describe four emerging diseases.

12. How can genomics help identify nonculturable pathogens?

13. What factors contribute to the development and spread of emerging infectious diseases?

14. Why are diseases reemerging that were long thought to be on the decline?

15. What is a select agent? Name four select agents and the diseases they cause.

16. Discuss whether a bioweapon needs to be efficient to be effective.

Clinical Correlation Questions

1. A local hospital received nine patients with symptoms of diarrhea, cramps, and low-grade fever. *Listeria monocytogenes* was isolated from the feces of all victims. All victims had attended an outdoor community block party on April 19. Five people who attended the party but were not ill were also surveyed. From the accompanying table, list trends you find in the data and indicate what is the likely source of the infection. As a member of the local health department, how would you contain the outbreak knowing the food source?

Patient	Age	Sex	Cohabitation	Illness	Onset Date and Time	Foods Ingested				
						Baked Ham	Potato Salad	Soft Cheese	Milk	Ice Cream
1	55	M		N		Y	N	Y	N	Y
2	49	M		Y	4/21, 9 a.m.	Y	N	Y	Y	Y
3	85	F		Y	4/22, 12 p.m.	N	Y	Y	N	N
4	62	M	Lives with patient 5	Y	4/24, 11 a.m.	Y	Y	N	Y	Y
5	61	F	Lives with patient 4	Y	4/21, 6 a.m.	N	N	Y	N	Y
6	56	M		Y	4/22, 12 a.m.	N	Y	Y	N	N
7	69	M		Y	4/22, 1 p.m.	Y	N	Y	N	N
8	35	F	Lives with patient 9	N		N	N	N	Y	Y
9	45	F	Lives with patient 8	Y	4/24, 3 p.m.	N	Y	Y	Y	Y
10	39	F		Y	4/21, 4 a.m.	Y	N	Y	Y	Y
11	25	M		N		N	Y	N	Y	N
12	48	M		Y	4/20, 11 p.m.	Y	N	Y	N	Y
13	24	F		N		N	Y	N	Y	Y
14	55	M		N		Y	Y	N	N	N

2. Graphically represent an epidemic with respect to time and cases.

3. On the ProMED-mail website (promedmail.org), look at "Latest Posts" to view recent outbreaks recorded by the WHO. What outbreaks happened throughout the world during the current year? You can also view the world map on this site to search for specific diseases.

4. In September 2009, a 39-year-old woman from Henan province in China entered a hospital with a high fever, accompanied by vomiting and diarrhea. Blood work revealed a low platelet count (thrombocytopenia) and low white cell count (leukopenia). Tests for known pathogens, bacterial and viral, came up negative. Over the next 2 years, 171 patients were diagnosed with what was called severe fever with thrombocytopenia syndrome (SFTS). The mortality rate was about 30%. Epidemiologists discovered that 98% of the patients were farmers who lived and worked in wooded, hilly areas. Ticks in wooded areas near the farms were thought to be involved. How would you investigate this disease to identify the agent and prove ticks were the vector? What would make the SFTS virus a global concern?

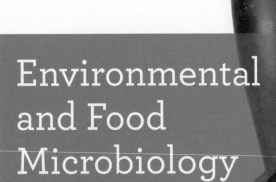

Environmental and Food Microbiology

SCENARIO In Huntington, West Virginia, the US Army Corps - - - - - - - - - - - - - - - - - - ▶ of Engineers had a problem. For decades, the army had produced the compound 2,4,6-trinitrotoluene (TNT), an explosive used in the manufacture of munitions. Inevitably, some of the TNT entered the soil.

120-mm Mortar Bomb
Contains explosive TNT.

Windrow Composting
A machine turns the soil at regular intervals to balance anaerobic and aerobic microbial digestion.

CONTAMINATION In 1988, after the ordnance works was shut down, the Environmental Protection Agency (EPA) determined that the manufacturing plant had leaked unsafe levels of TNT into the surrounding environment. TNT is toxic to humans and wildlife, causing anemia and liver disease. For many years, Huntington's water and soil had to be contained without effective treatment.

WEST VIRGINIA

● **Huntington**

TREATMENT In 2003, the army chose to conduct bioremediation, a process ◀ - - - - - in which living organisms decompose contaminating material. The bioremediation process used is called windrow composting. Windrows are elongated hills of soil that are turned by a machine to regulate the availability of oxygen for compost. In composting, the soil is combined with carbohydrate-rich material such as vegetable waste or wood chips. Bacteria in the soil catabolize these materials, along with the TNT. The TNT is catabolized to carbon dioxide.

OUTCOMES At the Huntington base, windrow composting removed more than 99% of the contaminating TNT. Thus, bacterial catabolism helped restore the local environment. It saved the army more than half a million dollars on containment measures that would not have removed the contamination.

2,4,6-Trinitrotoluene (TNT)
Explosive from bomb manufacture contaminates soil.

CHAPTER OBJECTIVES

After reading this chapter, you will be able to:

- Describe the essential roles of microbes in ecosystems that support all life, including humans.
- Explain how we use microbes to produce food and beverages.
- Describe how we protect food from microbial spoilage.
- Explain how microbes cycle essential materials, including human wastes, throughout the biosphere.

In this chapter's opening case history, the "patient" is not an individual person but a sickened habitat—a polluted environment in Huntington, West Virginia. A Cornell University study estimates that, worldwide, environmental pollution causes 40% of human deaths. Our health depends on the quality of our environment, which provides the water we drink and the soil in which we grow our food. The contributions of an environment to human health are called **ecosystem services**. When an environment is polluted, its ecosystem services are lost or degraded.

How can a contaminated environment be restored? Restoration can involve the metabolism of living microbes—a technology called **bioremediation**. Bioremediation is a waste management technique that uses organisms to remove or neutralize pollutants from a contaminated site. During bioremediation, surprisingly, microbes can decompose even highly toxic compounds such as TNT. In fact, the very property that makes TNT useful as an explosive—the huge release of energy as it breaks down—enables certain bacteria to use it as food.

First, under anaerobic conditions (lack of oxygen within the compost pile), some species of bacteria use the TNT molecule to accept electrons from the carbohydrate-rich compost. Next, the compost material is aerated by the rotation machine (see the photo in the opening case history). This process enables aerobic respiration, a key cell metabolism you learned about in Chapter 7. Aerobic respiration breaks down the composted TNT via the tricarboxylic acid (TCA) cycle, releasing CO_2 and H_2O and yielding energy for microbes to build biomass. So, TNT composting joins the environment's endless cycle of food webs in which organisms form living bodies and decompose them. And microbes are the key players.

Microbial communities are the foundation of Earth's biosphere, shaping the environments inhabited by plants and animals. Of all organisms, microbes grow in the widest range of habitats, from Antarctic lakes to the acid drainage of iron mines, as well as the air, water, and soil that surround us. In the ocean, vast quantities of microbes produce the biomass that ultimately feeds fish and humans. In forests and fields, microbes decompose plant material, generating fertile soil. And in all environments, microbes associate with animals and plants. Our skin microbes protect us from pathogens, while our intestinal microbes digest complex

plant fibers (discussed in Chapter 14). Food production generates artificial microbial ecosystems. On a global scale, microbes cycle the elements, consuming CO_2 and fixing nitrogen (incorporating into biomass) for all living things. In this chapter we explore how microbes interact with each other and with their many diverse habitats on Earth in ways that affect our lives and health.

27.1 Microbes in Ecosystems

SECTION OBJECTIVES

- Describe the various roles of microbes in natural ecosystems and the kinds of ecosystem services that microbes provide.
- Explain how physical factors determine which microbes grow in a given habitat.
- Outline how microbes participate in the food chains of ocean, soil, and wetland.
- Explain the phenomenon of biochemical oxygen demand and the cause of dead zones.

Can any organism exist by itself? "No man is an island"—nor is any microbe. All organisms exist within an **ecosystem** consisting of populations of species plus their habitat or environment. A **population** is a group of individuals of one species living in a common location. Within an ecosystem, each population of organisms fills a specific **niche**. The niche is a set of traits that defines the conditions for an organism to live and the organism's relations with other members of the ecosystem. For example, the niche of *Staphylococcus epidermidis* is that of a human symbiont, a heterotroph that catabolizes hydrocarbons from human sweat. The habitat of *S. epidermidis* is the human epidermis, and the bacterium's metabolism releases organic acids that suppress pathogen growth.

Microbes Fill Unique Niches in Ecosystems

In every ecosystem, certain key roles are filled only by microbes. Only bacteria and archaea fix nitrogen gas (N_2) into ammonia, which plants incorporate into protein and nucleic acids. Without bacterial nitrogen fixation, life on Earth would grind to a halt, starved of nitrogen. Legume plants such as soybeans actually harbor symbiotic bacteria called rhizobia that fix nitrogen within the plant cells forming root nodules (**Figure 27.1**). In a different example, only bacteria and fungi degrade lignin (wood biomass) and other complex plant polymers, releasing their carbon for uptake into food chains (**Figure 27.2**). Without decomposition, plant biomass would pile up (think autumn leaves), smothering new life.

In addition, microbial metabolism evolves so fast and in so many diverse ways that microbial genomes can respond to new challenges in our environment— even the alarming rise of waste plastics in our oceans. Marine bacteria are evolving to eat plastic, though not fast enough to consume all the waste. If an energy-yielding reaction exists, some microbe must have evolved—or will evolve—to use it. For this reason, we can search contaminated environments (such as Huntington's TNT-contaminated soil) for species of microbes that we could use for bioremediation.

What determines the environmental contribution of a microbe? What a microbe gives to and takes from its ecosystem depends on two things:

- **The microbe's genome.** Each genome encodes a set of enzymes that can interconvert molecules (discussed in Chapters 7 and 8). For example, the genomes of cyanobacteria encode the enzymes and electron transport proteins for photosynthesis. Microbial genomes share content through mobile genes (acquired by transfer mechanisms, discussed in Chapter 9).

- **The microbe's environment.** The environment's physical and chemical factors, such as temperature, nutrient supplies, oxygen availability, and pH, determine the range of metabolic options. In a wetland, for example, if hydrogen gas is available from decomposing biomass, *Rhodopseudomonas palustris* will oxidize hydrogen to water. On the other hand, if light is available without oxygen, as in a sewage-polluted lake, *R. palustris* will use anaerobic photosynthesis to split hydrogen sulfide. Either way, *R. palustris* forms biomass for consumers in the food web.

All organisms depend, directly or indirectly, on the presence of other organisms. Some populations cooperate by reciprocally supplying each other's needs. Cooperation by organisms may be incidental, as when hydrogen-oxidizing bacteria use H_2 from fermenting bacteria. Or cooperation may involve mutualism, a highly developed partnership in which two species have coevolved to support each other (discussed in Section 27.2).

Figure 27.1 Soybeans Carry Bacterial Partners That Fix Nitrogen

A. Soybean farm.

B. Cross section through a soy root cell that contains nitrogen-fixing bacteroids of *Bradyrhizobium japonicum*.

Carbon Assimilation and Dissimilation: The Food Web

The interactions between microbes and their ecosystems include two common roles of metabolic input and output, often referred to as assimilation and dissimilation, respectively. The processes of metabolism were discussed in Chapter 7, but ecology sees them from a different point of view.

Assimilation refers to processes by which organisms acquire an element, such as carbon from CO_2, to build into cell parts. Organisms that produce biomass from inorganic carbon (usually CO_2 or bicarbonate ions) are called **primary producers**. Producers are a key source of productivity for other members of the ecosystem, particularly the **consumers** that feed on producers.

Figure 27.2 White Rot Fungi Decompose a Tree Stump

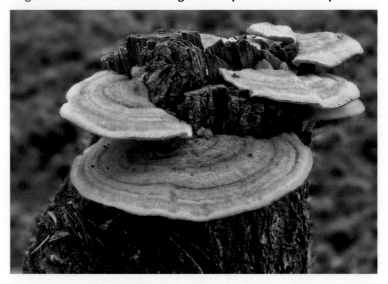

Fungi catabolize lignin from a dead tree stump.

Dissimilation is the process of breaking down organic nutrients to inorganic minerals such as CO_2 and NO_2^-, usually through oxidation. These minerals are then released to the environment. An example of a nutritive process causing dissimilation is aerobic respiration. Microbial dissimilation releases minerals for uptake by plants, and it provides the basis of wastewater treatment (see Section 27.6). Waste removal is key for every life form, and wastewater treatment is fundamental to modern human societies.

The major interactions among organisms in the biosphere are dominated by the production and transformation of **biomass**—the bodies of living organisms. To obtain energy and materials for biomass, all organisms participate in a **food web** (Figure 27.3). A food web describes the ways in which various organisms consume each other, as well as each other's products. Levels of consumption are called **trophic levels**. Each trophic level represents the consumption of biomass of organisms from the previous level, lower in the food web. At each trophic level, the fraction of biomass retained by the consumer is small; most is dissipated as CO_2 to provide energy.

Every food web depends on primary producers for two things:

- **Absorbing energy from outside the ecosystem.** In most cases, the ultimate source of energy is the Sun, and the producers are phototrophs.

- **Assimilating minerals into biomass.** The biomass of producers is then passed on to subsequent trophic levels.

The majority of carbon in Earth's biosphere is assimilated by oxygen-producing phototrophs such as cyanobacteria, algae, and plants. Certain important ecosystems are founded on carbon fixation by lithotrophs (microbes that oxidize minerals). For example, in hydrothermal vent communities, bacteria oxidize hydrogen sulfide to fix CO_2, capturing both gases as they well up from Earth's crust. The vent communities, however, require oxygen generated by phototrophs from above.

In addition to producers, all ecosystems include consumers that acquire nutrients from producers and ultimately dissimilate biomass by respiration, returning carbon to the atmosphere (see Figure 27.3).

Consumers constitute several trophic levels based on their distance from the primary producers. The first-level consumers, generally called **grazers**, directly feed on producers. Grazers include many kinds of life, from paramecia consuming algae to cows eating grass. Grazers overall (including protists, insects, and vertebrate herbivores) convert 90% of the carbon back to CO_2 through respiratory metabolism, yielding energy. Consumers occupying the next level, often called **predators**, feed on the grazers. Predators convert another 90% of remaining biomass back to atmospheric CO_2. In microbial ecosystems, the trophic relationships are often highly complex because a given microbe may act as both producer and consumer.

At each trophic level, some organisms die, and their bodies are consumed by **decomposers**. Decomposition returns carbon and minerals to the environment for use by producers. Decomposers are invariably microbes (fungi or bacteria). They have particularly versatile digestive enzymes capable of breaking down complex molecules such as lignin. Without decomposers, carbon and minerals needed by phototrophs would be locked away by ever-increasing mounds of dead biomass.

Instead, all biomass is recycled, and all the energy acquired by ecosystems is eventually converted to heat. Energy conversion to heat results in increase of entropy (disorder) as defined by the second law of thermodynamics (introduced in Chapter 4).

Environmental Limiting Factors

In all environments, key physical factors such as oxygen and temperature limit the ability of microbes to grow. Physical factors have different effects on different species; thus, physical factors determine which microbial species inhabit a given environment.

OXYGEN The availability of oxygen determines whether microbes use molecular oxygen as an electron acceptor during cellular respiration on organic compounds produced by other organisms (discussed in Chapter 7). Aerobic respiration is favored in well-tilled soil, where maximal surface area of soil particles is exposed to oxygen. Aerobic respiration enables microbes to maximize the breakdown of food substrates to CO_2.

In anaerobic environments, such as waterlogged soil and deep sediment, microbes use minerals such as iron (Fe^{3+}) and nitrite (NO_2^-) to oxidize organic compounds supplied by other organisms (anaerobic respiration). But these anaerobic electron acceptors are soon used up, so breakdown to CO_2 takes much longer without oxygen. Anaerobic environments allow much slower rates of assimilation and dissimilation than occurs in the presence of oxygen. The total volume of anaerobic microbial communities, however, includes vast regions underground that exceed the volume of our aboveground oxygenated biosphere. Underground communities of anaerobic bacteria and archaea filter water for our wells—an ecosystem service crucial for human health.

SALINITY High salinity (salt concentration) limits the growth of microbes adapted to freshwater conditions. Salt-ridden land is also

Figure 27.3 Food Web: All Levels Include Microbes

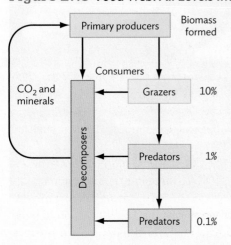

Primary producers fix CO_2 into biomass; grazers (consumers) feed on primary producers, and predators consume the grazers and smaller predators. Decomposers consume dead bodies of organisms from every level, recirculating their minerals to the producers. Percentages indicate the fraction of original CO_2 converted to biomass at each trophic level.

excluded from agriculture, influencing human health. On the other hand, many microbial species have adapted to high salinity (they are called halophiles). The haloarchaea, for instance, bloom in population as a body of water shrinks and becomes hypersaline. Haloarchaea grow in salt ponds that process salt for human consumption.

PH Acidity (low pH) is important geologically because a high concentration of hydrogen ions accelerates the release of reduced minerals from exposed rock. Extreme acidity is often produced by lithotrophic acidophiles, such as those at iron mines. The increasing acidity enhances further release of minerals to be oxidized while excluding acid-sensitive competitors. Acid mine drainage leaks into streams and lakes, killing fish and endangering human health.

Microbes in Freshwater

Freshwater habitats include lakes, rivers, and wetlands. Freshwater is usually less than 0.1% salt. The water is often shallow, so phototrophs have access to the sediment. Thus, lakes tend to contain organic nutrients that support a moderate density of microbial life (**Figure 27.4**). That microbial growth is the reason lake water more often appears green or brown rather than clear.

In the upper layers of water, the concentration of heterotrophic microorganisms determines the **biochemical oxygen demand** (also called biological oxygen demand), or **BOD**, the amount of oxygen removed from the water by aerobic respiration. The upper layers of a lake may have such a low concentration of organisms that the BOD is low and thus the dissolved oxygen content is high. The BOD rises, however, when excess sewage is introduced. Human and industrial sewage contains organic compounds rich in nitrogen as well as carbon. Microbes such as Proteobacteria and Actinobacteria readily oxidize these organic compounds by aerobic respiration. Sewage microbes also release ammonia, which nitrifying microbes oxidize to nitrates. Oxygen use by microbial respiration deprives fish of oxygen, leading to a "zone of hypoxia" where fish cannot grow.

Figure 27.4 Microbes in a Drop of Pond Water

Filamentous green algae

Diatom

Figure 27.5 Algal Bloom

Fertilizer runoff caused this algal bloom in Toledo, Ohio, in 2014.

A lake that receives excessive nutrients, such as runoff from agricultural fertilizer or septic systems, shows increased BOD and may become **eutrophic**. In a eutrophic lake, the nutrients support growth of algae and bacteria to high densities, causing a bloom, commonly called an **algal bloom** (**Figure 27.5**). Often, the "algal bloom" actually consists of cyanobacteria called *Microcystis*. When the blooming cells die, their carbon is consumed by heterotrophic bacteria, whose respiration removes all the oxygen. But all fish and invertebrates need to breathe oxygen, so in a eutrophic lake, fish die off owing to lack of oxygen.

Blooms of algae or cyanobacteria increasingly threaten our water supplies. For example, in 2014, fertilizer runoff in Lake Erie caused a bloom of *Microcystis* that released a toxin called **microcystin**. If swallowed, microcystin causes intestinal illness, skin rashes, and liver damage. The toxin contaminated the water supply of Toledo, a major city in Ohio. For several days, drinking water was banned for half a million people.

Blooms increase at higher temperatures, where cyanobacteria grow faster than other microbial phototrophs. Thus, cyanobacterial blooms associated with climate change increasingly threaten water systems worldwide.

Microbes of the Oceans

The oceans cover more than two-thirds of Earth's surface, reaching depths of several kilometers and forming an immense habitat. The salt concentration is much higher than in freshwater, typically 3.5%. This high salt level prevents growth of freshwater bacteria and pathogens, although some salt-tolerant pathogens, such as *Vibrio cholerae*, grow well.

In the oceans, most of the CO_2 fixation and biomass production are performed by the smallest inhabitants, tiny cyanobacteria such as *Prochlorococcus* (**Figure 27.6**). Marine photosynthetic bacteria and algae fix a large portion of the world's carbon dioxide into biomass, and they produce much of the oxygen we breathe

Figure 27.6 Microbes within Food Webs

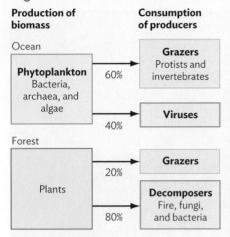

A. In marine ecosystems, the main primary producers are cyanobacteria and algae. Viruses break down both producer and consumer microbes. In a forest, the major producers are trees, much of whose biomass is poorly digested by grazers; 80% of biomass is decomposed by fungi and bacteria.

500 nm

B. *Prochlorococcus*, a marine cyanobacterium that produces much of the world's oxygen.

(**Figure 27.7A**). Many are free-floating, single-celled microbes, whereas others grow as tiny biofilms on bits of shell and detritus called "marine snow" (**Figure 27.7B**). Still other "microbes" may form long sheets, such as kelps (eukaryotic algae, discussed in Chapter 11). Bacteria and algae support vast marine ecosystems with many trophic levels, ultimately including whales, the largest animals on Earth. Fish depend on a vast invisible microbial food web, and thus their numbers remain limited, despite the seemingly huge volume of ocean. Thus, marine fish populations are especially vulnerable to overharvesting and to any environmental change that disrupts the food chain.

An important determinant of marine habitat is the **thermocline**, a depth at which temperature decreases steeply and water density increases. At the thermocline, a population of heterotrophs will peak, feeding on organic matter that settles from above.

Coastal regions show the highest concentration of nutrients and living organisms and the least light penetration. The open ocean is largely oligotrophic (having an extremely low concentration of nutrients and organisms).

The ocean floor experiences extreme pressure beneath several kilometers of water. Nevertheless, the deep ocean supports vast, unimagined realms of life including unique bacteria and archaea. Organisms that live there are barophiles, adapted to high pressure (discussed in Chapter 6). The reduced minerals from thermal vents support huge ecosystems of lithotrophic bacteria and archaea. Some are mutualists of giant clams and tube worms. The clams and tube worms evolved in symbiosis with bacteria that oxidize hydrogen sulfide emanating from the vent. That means these animals have blood full of hydrogen sulfide—a poison that would kill most animals. The ocean floor provides so many reduced inorganic minerals that bacterial redox reactions with oxygen generate electricity. As a result, the entire interface between floor and water acts as a charged battery.

Dead Zones

In the upper layers of the ocean, as we saw for freshwater lakes, the concentration of heterotrophic bacteria and protozoa determines the BOD. When sewage is introduced, oxygen-consuming

Figure 27.7 Marine Microbes

A. Bioluminescent dinoflagellates at the ocean shore, Vaadhoo Island, Maldives.

FM   50 µm

B. Marine snow, particles of organic and inorganic matter colonized by bacteria.

Figure 27.8 The Dead Zone in the Gulf of Mexico

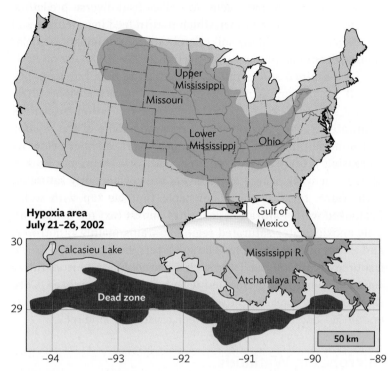

A. Map of the Mississippi River drainage, which empties into the Gulf of Mexico. Every summer, drainage high in organic carbon and nitrogen causes algal blooms, leading to hypoxia and the death of fish.

B. Aerial view of the oil leaked from the *Deepwater Horizon* oil well in the Gulf of Mexico, 2010.

bacteria multiply, and BOD rises. Until recently, BOD was considered a local issue, affecting the health of individual lakes and rivers. Today, however, large regions of ocean have become **dead zones**, or zones of hypoxia, devoid of most fish and invertebrates.

A well-known dead zone is a region in the Gulf of Mexico off the coast of Louisiana where the Mississippi River releases about 40% of the US drainage to the sea (**Figure 27.8A**). Over its long, meandering course, which includes inputs from the Ohio and Missouri Rivers, the Mississippi builds up high levels of organic wastes, as well as nitrates from agricultural fertilizer. When these nitrogen-rich substances flow rapidly out to the gulf in the spring, they lift the nitrogen limitation on algal growth and feed massive blooms of cyanobacteria such as *Trichodesmium*. The algal bloom population then crashes, often by viral infection; but the released carbon is consumed by heterotrophic bacteria. The heterotrophs respire, thus using up the available oxygen and raising the BOD. The resulting hypoxia (lack of oxygen) kills off the fish, shellfish, and crustaceans over a region equivalent in size to the state of New Jersey. A dead zone can be expanded by a petroleum spill such as that caused by the *Deepwater Horizon* oil well blowout in 2010 (**Figure 27.8B**).

Another effect of dead zones is the anaerobic metabolism of nitrogen. Nitrogen-rich effluents (liquid waste) from fertilizer runoff are oxidized by lithotrophs to nitrate. (Lithotrophy is presented in Chapter 7.) Then, in the dead zone, where oxygen is low, some bacteria use nitrate as an alternative electron acceptor for anaerobic respiration. These anaerobic respiring bacteria now release some of the reduced nitrogen in the form of a gas, nitrous oxide (N_2O). Nitrous oxide is a highly potent greenhouse gas; it also reacts catalytically with ozone, depleting the ozone layer.

What can be done about dead zones and about the polluted drainages that lead to them? In the long run, prevention is far better than cleanup. To prevent dead zones, we must avoid the release of industrial pollutants such as petroleum. In addition, all the communities throughout the river drainage area must treat their wastes to eliminate nitrogenous materials before disposal.

Microbes in Soil

In contrast to the ocean, where the base of producers is almost entirely microbial, on land the major producers and fixers of CO_2 are macroscopic plants. Plants generate detritus—discarded biomass such as leaves and stems—that requires decomposition by fungi and bacteria. Detritus decomposition helps to form soil, one of the densest sources of microbes on Earth. Soil is a complex mixture of decaying organic and mineral matter that feeds vast communities of microbes, many of which grow as biofilms on soil particles (**Figure 27.9**). And soil-based agriculture is the major source of food for our planet's human inhabitants. The qualities of a given soil—oxygenated or water saturated, acidic or alkaline, salty or fresh, nutrient-rich or nutrient-poor—define what food can be grown and whether the human community will eat or starve.

Soil's surface layer consists of dark, organic detritus, such as shreds of leaves fallen from plants. The detritus is in the

Figure 27.9 A Soil Particle Colonized by Microbes

Actinobacteria decompose plant material.

Bacterial microcolonies (*Pseudomonas*, *Bacillus*, *Streptomyces*)

Protozoa prey on bacteria.

Fungi decompose dead plant material.

100 μm

plant cells. Some bacteria live off the rhizosphere nutrients and protect plants from pathogens. Microbes feed diverse predators, such as nematodes and fungi, which in turn feed invertebrate and vertebrate animals. Many soil microbes have mutualistic associations with animals or plants (see Section 27.2).

Well-drained soil is full of oxygen, as well as nutrients liberated by decomposing microbes and used by plants. The deeper soil experiences periods of water saturation from rain. Rainwater leaches (dissolves and removes) some of the organic and mineral nutrients from the upper layers. Farther below lie increasing proportions of minerals and rock fragments broken off from bedrock. These lower layers are permanently saturated with water, forming the water table that we tap with wells. Prolonged water saturation generates anoxia (oxygen depletion). This anoxic, water-saturated region contains mainly lithotrophs and anaerobic heterotrophs. The soil layers finally end at bedrock, a source of mineral nutrients such as carbonates and iron. Interestingly, bedrock is permeated with bacteria and archaea. Core samples show that crustal rock as deep as 3 km down contains endoliths—bacteria growing between crystals of solid rock.

earliest stages of decomposition by microbes (**Figure 27.10**), primarily fungi such as *Mycena* species, as well as actinomycete bacteria (filamentous bacteria that produce antibiotics; discussed in Chapter 10) such as *Streptomyces*. *Streptomyces* generates chemicals whose odors give soil its characteristic smell.

The top layers of soil feature a food web of extraordinary complexity (**Figure 27.11**). The major producers are green plants, whose root systems feed predators, scavengers, and mutualists. Another source of organic matter from plants is the **rhizosphere**, the region of soil surrounding plant roots. The rhizosphere contains proteins and sugars released by the roots, as well as sloughed-off

Microbes in Wetlands

A **wetland** is defined as a region of land that undergoes seasonal fluctuations in water level. Sometimes the land is dry and oxygenated, while at other times it is water saturated and anoxic (lacking oxygen; **Figure 27.12**). Anoxic soil supports fermenting bacteria, as well as bacteria that conduct anaerobic respiration using nitrate and other alternative electron acceptors. Wetlands give us many ecosystem services; for example, the Everglades filter much of the water supply for Florida communities. Because wetlands have access to rich minerals, oxygen, and sunlight all at once, they produce biomass at the highest rate of all ecosystems.

Figure 27.10 Soil Microbes

C. A bacteria-feeding nematode has special mouthparts to consume bacteria. Other species of nematodes graze on plant roots or fungi.

A. Hyphae of actinomycete bacteria, *Streptomyces griseus*. *Streptomyces* bacteria give the soil its characteristic odor.

B. Bacteria grow on the surfaces of fungi.

Figure 27.11 The Soil Food Web

On land, plants are the major producers (fixing CO_2 into biomass). Detritus from plants is decomposed by fungi and bacteria, which feed protists and small invertebrates such as nematodes. Protists and small invertebrates are consumed by larger invertebrates and vertebrate animals.

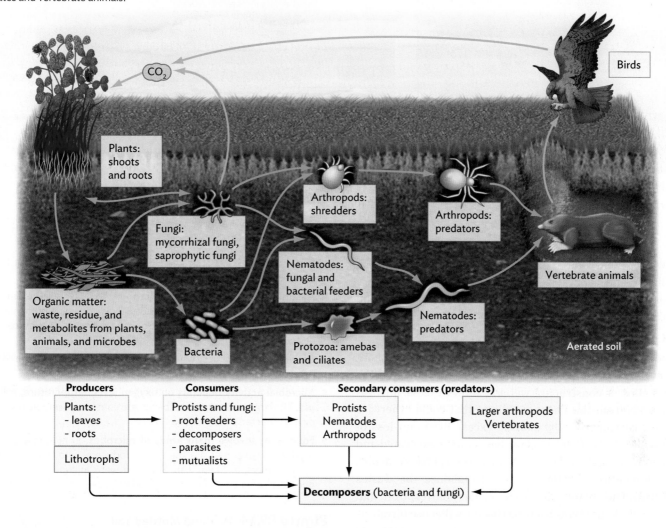

Figure 27.12 Wetland Soil

A true wetland experiences periods of water saturation alternating with dry soil. Soil that is water saturated becomes anoxic because O_2 diffuses slowly through water.

CASE HISTORY 27.1

Desert Kidneys: A Constructed Wetland

In 1994, the city of Phoenix, Arizona, faced challenges to its water supply. In a desert, water is scarce, and quality is hard to maintain. The city's water department needed to decrease the organic nitrogen content of its wastewater, which threatened the river downstream. A study proposed the construction of a wetland environment to receive the treated effluent from the Phoenix sewage plant. Over the next 10 years, the city constructed a series of experimental wetlands, building the large-scale Tres Rios Constructed Wetlands Project (**Figure 27.13A**). By 2010, the constructed wetland encompassed 800 acres and filtered 170 million gallons per day. Besides controlling nitrogen and organic content, the wetland restores

Figure 27.13 Constructed Wetlands Purify Wastewater

A. The Tres Rios constructed wetland in Arizona purifies municipal wastewater and enriches the river ecosystem.

B. Wetland plants and associated microbes purify wastewater as it flows through.

groundwater, offers a carbon offset (net CO_2 removed from the atmosphere), and generates a wildlife-rich recreational area. The Tres Rios wetland now attracts beavers, bald eagles, and bobcats, as well as many human visitors.

Wetlands are called the "kidneys" of our environment because they filter organic wastes and recycle nutrients. In the case of Tres Rios, a constructed wetland filters excess nitrogenous compounds from the Phoenix effluent that could otherwise harm the downstream river by causing excess BOD and loss of fish. To construct the wetland, shallow basins were excavated and pipes were installed that allow wastewater to disperse gradually, giving environmental microbes time to metabolize the wastes (**Figure 27.13B**). The organic carbon feeds anaerobic respirers, and the nitrogen feeds denitrifying bacteria (bacteria that use nitrate as an electron acceptor). Soil bacteria and archaea release nutrients that encourage growth of wetland plants.

How does a wetland manage change in water flow? Wetland soil undergoes periods of anoxic water saturation, leading to what is known as hydric soil. Hydric soil is characterized by "mottles," patterns of color and paleness (**Figure 27.14**). The reddish brown portions (for example, surrounding a plant root) result from oxidized iron (Fe^{3+}). The gray portions indicate loss of iron in its water-soluble reduced form (Fe^{2+}) generated by anaerobic respiration. Anaerobic denitrifiers (bacteria using nitrate to oxidize organic food) remove nitrate from water before it enters the water table—one of the many ways that wetlands protect our water supply.

SECTION SUMMARY

- **Bioremediation is a waste management technique** that uses living organisms to remove or neutralize pollutants from a contaminated site.
- **Microbial populations fill unique niches in ecosystems.** Every environment on Earth, including those below ground, contains microbes.

In principle, every chemical reaction that yields energy might be used by some kind of microbe.

- **Assimilation occurs when microbes fix essential elements into biomass,** which recycles within ecosystems, including human agriculture. Some key elements, such as nitrogen, are fixed solely by bacteria and archaea.
- **Dissimilation occurs when consumers catabolize the bodies of producers,** generating CO_2 and releasing heat energy.
- **Microbial activity depends on oxygen level, temperature, salinity, and pH.** The largest volume of our biosphere contains anaerobic (living without oxygen) bacteria and archaea.
- **Food webs involve many kinds of microbes.** In marine and freshwater food webs, microbes are the main producers, with many trophic levels of consumers, including fish. Soil food webs include a complex array of microbial producers, consumers, predators, decomposers, and mutualists.

Figure 27.14 Wetland Mottled Soil

Alternating periods of water saturation and dryness give the soil a mottled color. The reddish brown portions around root holes result from oxidized iron (Fe^{3+}), whereas the gray portions indicate Fe^{2+} generated by anaerobic respiration and leached out by water.

- **Wetland soils alternate aerated (dry) with anoxic (water-saturated) conditions.** Wetland soils are among the most productive ecosystems.
- **Dead zones** are caused by massive runoff of organic effluents, rich in carbon and nitrogen, into the rivers and ultimately the ocean. High BOD accelerates heterotrophic respiration and depletes oxygen needed by fish.

27.2 Microbial Symbiosis with Animals

SECTION OBJECTIVES

- Explain the different kinds of microbe-animal symbioses, including mutualism, synergism, commensalism, amensalism, and parasitism.
- Describe the digestive communities of animals.
- Explain how microbial communities contribute to human health.

Why do some microbes live within animals, including our own body? Outside our body, environmental microbes provide benefits at a distance. For example, aquatic cyanobacteria produce oxygen that is breathed by organisms around the globe. But other kinds of interactions require intimate association between microbes or between a microbial population and animals or plants (Table 27.1). Intimate association between organisms of different species is called **symbiosis** (plural, **symbioses**). Symbiotic associations include a full range of positive through negative relationships. Multicellular animals and plants provide unique habitats for symbiotic microbes. All animals coevolve with their specific associated microbes, some of which can grow nowhere else. Many are beneficial to the human body. Others, of course, can be highly adapted pathogens, as presented in Chapters 18–24.

Classes of Symbiosis

The most highly developed forms of symbiosis involve partner species that require each other for survival—a phenomenon termed **mutualism**. A famous example of mutualism is that of termites, which harbor bacteria and protists that digest the cellulose from wood. Termites absolutely require their digestive microbes, and the microbes cannot live outside the termite. This mutualism between termite and gut microbe is an obligate association for both partners.

The term "mutualism" also applies to a positive association in which both partners benefit from their interaction, although they may exist without each other, such as the microbial community of the bovine digestive tract (discussed shortly). A less intimate cooperation may also be called **synergism**, in which both species benefit but can grow independently and show fewer specific cellular interactions. For example, on human skin, *Staphylococcus epidermidis* and related species produce short-chain fatty acids whose uptake inhibits growth of yeasts and pathogens. The staphylococci receive nutrients from human skin secretions, and the bacteria outcompete inhibited microbes for scarce water and nutrients. Nevertheless, staphylococci and human beings can live without each other.

An interaction that benefits one partner only is called **commensalism**. Commensalism is difficult to define in practice because "commensal" microbes often provide a hidden benefit to their host. For example, gut bacteria such as *Bacteroides* species were considered commensals until it was discovered that their metabolism aids human digestion. *Bacteroides* species break down complex plant polysaccharides to sugars that humans can digest—much as termite microbes digest wood and bovine rumen microbes digest grass (discussed in the next section).

A relationship in which one member benefits at the expense of another is **parasitism**. Generally, the parasite benefits while harming

Table 27.1
Associations Involving Microorganisms

Type of interaction	Effects of interaction	Example
Mutualism	Organisms grow in an intimate relationship between two or more specific partners. Both partner species benefit and may fail to grow independently.	Lichens consist of fungi and algae (in some cases, cyanobacteria) growing together in a complex layered structure. Each species requires the presence of the other.
Synergism	Both species benefit through growth, but the partners are easily separated and either partner can grow independently of the other.	Human colonic bacteria ferment, releasing H_2 and CO_2, which methanogens convert to methane. The methanogens gain energy, and the bacteria benefit energetically from the removal of their fermentation products.
Commensalism	One species benefits, while the partner species neither benefits nor is harmed.	In wetlands, *Beggiatoa* bacteria oxidize H_2S for energy. Removal of H_2S enables growth of other microbes for whom H_2S is toxic. The other microbes are not known to benefit *Beggiatoa*.
Parasitism	One species (the parasite) benefits at the expense of the other, a specific host. The relationship is usually obligatory for the parasite.	*Legionella pneumophila*, the cause of legionellosis, parasitizes amebas in natural aquatic habitats. Within the human lung, *L. pneumophila* parasitizes macrophages.
Amensalism	One species harms others, without receiving benefit. The relationship is nonspecific.	*Lactobacillus* species produce lactic acid, decreasing the pH to levels that inhibit the growth of other bacteria.

a specific host. Parasitic bacteria may be intracellular, such as the rickettsias, which cause diseases such as Rocky Mountain spotted fever. Other parasites such as *Giardia* are extracellular, although they grow within a digestive tract.

More general relationships in which one partner is harmed are called **amensalism**. In amensalism, organisms are harmed, for instance, by the toxic waste products of another organism that does not receive specific benefit. For example, lactic acid bacteria such as *Lactobacillus* species produce acid that inhibits the growth of other kinds of bacteria.

The distinction between mutualist and parasite is often subtle. Lichens consist of a mutualistic association between fungi and algae, but environmental changes can convert the fungus to a parasite. On the other hand, parasitic microbes may coevolve with a host to the point that each depends on the other for optimal health. For example, the incidence of certain allergies and immune disorders appears to correlate with lack of exposure to pathogens and parasites.

Animals Host Microbial Digestive Communities

As discussed in Chapter 14, many kinds of microbes inhabit parts of the human body, as well as other animal bodies. The human skin, digestive tract, and mucous membranes harbor complex communities that have food webs with competitive and cooperative interactions, much like the free-living communities in soil or water.

Our understanding of human-associated microbial communities leads to surprising new treatments for human disease. An example is fecal transplant, or bacteriotherapy (discussed in Chapter 22 IMPACT). Fecal bacteriotherapy produces "miracle cures" for many patients with recurrent *Clostridioides difficile* infection and may help others with irritable bowel syndrome (see Section 14.5). In these cases, we don't even know the identity of the bacteria involved, but a balanced community of microbes from

a healthy person can colonize a diseased colon and return it to normal function. The finding is particularly remarkable given that individuals differ substantially in their microbial populations; there must be many possible combinations compatible with health. Yet a given population from a given healthy person can restore health to another person—even with a colon severely distressed by inflammation and bleeding. Physicians now argue that the colonic microbial community constitutes an "organ" of the human body, as essential as our other human organs.

Human microbiota also interact positively with our immune system and generate defenses against pathogens. For example, *Bifidobacterium* down-regulates human secretion of cytokines (molecules that cause inflammation). Other colonic bacteria produce bacteriocins (molecules that prevent pathogens from adhering to the gut lining).

Many vertebrate and invertebrate animals host digestive communities of microbes, including thousands of species of bacteria, protists, and archaea. The anaerobic (low-oxygen) contents of the human colon contains many fermenters and methanogens.

Some of these bacteria and archaea feed on intestinal mucus, whereas others digest complex plant fibers, thus providing up to 15% of our caloric intake. Our gut microbiota enable us to acquire nutrition from complex plant fibers that we could not otherwise digest. The most extensively studied digestive communities are those of ruminants, such as cattle and sheep. Throughout most of human civilization, ruminants have provided protein-rich food, textile fibers, and mechanical work. A historical reference is the biblical injunction to consume an animal that "is cleft-footed and chews the cud"—that is, "ruminates," or redigests its food in the fermentation chamber known as the rumen (**Figure 27.15**).

The bovine gut system has four chambers (Figure 27.15A). The reticulum receives feed from the esophagus, sending large pieces to the rumen for initial digestion. After initial digestion, feed is regurgitated for rechewing and then returned to the rumen. In the rumen, feed is broken down to small particles and fermented slowly by thousands of species of microbes. The partially digested feed passes to the omasum, which absorbs water and short-chain acids produced by fermentation. The abomasum then decreases pH and secretes enzymes to digest proteins before sending its contents to the colon for further nutrient absorption and waste excretion.

Different microbes fill different niches in ruminal metabolism. Cattle grown on relatively poor forage (that is, forage high in complex plant content) show a high proportion of ruminal fungi. By contrast, cattle fed a

Figure 27.15 The Bovine Rumen

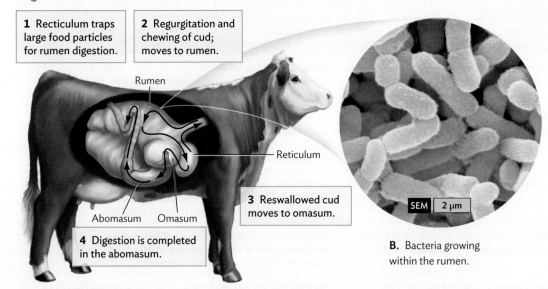

1 Recticulum traps large food particles for rumen digestion.

2 Regurgitation and chewing of cud; moves to rumen.

3 Reswallowed cud moves to omasum.

4 Digestion is completed in the abomasum.

Rumen

Reticulum

Abomasum   Omasum

A. The rumen is the largest of four chambers in the bovine stomach.

SEM   2 µm

B. Bacteria growing within the rumen.

high-cellulose diet, such as hay, grow faster and show cellulose-fermenting bacteria such as *Ruminococcus* and *Fibrobacter*. Today, commercial agriculture has largely shifted from hay to grain, whose starch content leads to even more rapid digestion and faster growth of the animal. Unfortunately, rapid starch digestion favors sugar fermenters such as *Streptococcus bovis* and *E. coli*, which generate higher acid levels and gases, leading to "starch bloat." Furthermore, rumen acidity selects for acid-resistant pathogens such as the *E. coli* strain O157:H7.

One problem with ruminal fermentation is the bacterial production of H_2 and CO_2 through mixed-acid fermentation. The H_2 and CO_2 from fermentation are converted by methanogens to methane. A cow can produce 250–500 liters of methane per day—so much that a cannula inserted into the rumen releases enough of the gas to light a flame. Methane wastes valuable carbon from feed and adds a large amount of greenhouse gas to global warming.

Diseases of Animals and Humans

Like humans, animals are subject to many microbial pathogens. Animal pathogens affect agriculture; for example, in 2007 in the United Kingdom, an outbreak of hoof-and-mouth disease (caused by aphthovirus) occurred in cattle. In hoof-and-mouth disease (also called foot-and-mouth disease), the virus causes blisters on the mucous membranes and hooves, impairing the animal's ability to feed and stand. The outbreak required farmers to slaughter thousands of dairy cattle to contain the spread of disease.

Hoof-and-mouth disease does not infect humans, but other pathogens, such as *E. coli* O157:H7, may be carried by animals and then transferred to humans, where they cause disease. An animal population that carries pathogens transmissible to humans is called a reservoir of infection (discussed in Chapter 2). Many kinds of animal pathogens can be picked up by humans, even if humans are not the preferred host (Table 27.2). An important example is *Toxoplasma*, a cat parasite that can infect humans and cause congenital defects in a developing fetus (the replication cycle is presented in Figure 21.26). Emerging diseases can arise from "exotic" pets whose microbiota are unfamiliar to physicians. For example, in 2003 the United States experienced an outbreak of monkeypox transmitted by pet prairie dogs. The prairie dogs contracted the virus from a Gambian giant rat imported by the same pet distributor. Monkeypox virus is closely related to smallpox virus and could someday mutate into a similar virulent strain.

A disease acquired by humans from animals is called a **zoonosis** (plural, **zoonoses**). Typically, a zoonotic strain, such as "avian flu" (avian-adapted strains of influenza), fails to transmit effectively between humans. But the danger is that once established in humans, an animal pathogen may mutate into a strain exceptionally deadly to humans—and exceptionally virulent—because the humans lack immunity. The H1N1 influenza strain of 2009 is an example of a newly emerged, virulent strain. Fortunately, H1N1 turned out to have a relatively low mortality rate, but had it mutated to a deadly

Table 27.2
Animal Diseases Transmitted to Humans

Disease	Pathogen	Transmission to Humans
Anthrax	*Bacillus anthracis*	Cattle or sheep transmit through contact with hide or wool
Ascariasis	Roundworm (*Ascaris*)	Ingestion of contaminated soil, fruits, or vegetables; transmitted via feces, often of dogs or cats
Brucellosis	*Brucella* spp.	Cattle, goats, pigs, and dogs can transmit pathogen through infected meat or milk
Bubonic plague	*Yersinia pestis*	Fleas from prairie dogs or rats transmit the bacteria to humans
Cat scratch disease	*Bartonella* spp.	Cats transmit pathogen through bite or scratch, usually to children
Monkeypox	Orthopoxvirus	African monkeys and rodents, including giant rats sold in the United States as pets, transmit a milder smallpox-like disease to humans
Rabies	Lyssavirus	Dogs, raccoons, bats, and other mammals transmit pathogen through bite to humans
Salmonellosis	*Salmonella enterica*	Pet turtles and other reptiles spread pathogen through contact, usually to children; also spread by contact with raw chicken and eggs
Toxoplasmosis	*Toxoplasma gondii*	Cat feces and undercooked contaminated pork transmit pathogen to humans
Tularemia	*Francisella tularensis*	Rodents transmit pathogen to humans via fly or tick bite; more rarely transmitted through consumption of contaminated food or water
West Nile encephalitis	Flavivirus	Horses and birds spread pathogen via mosquitoes to humans

form, the result could have been a pandemic similar to the flu pandemic of 1918.

Human diseases can also be transmitted to animals. For example, diseases such as polio and measles have been transmitted to endangered apes under study by researchers. Some researchers are attempting to vaccinate endangered apes against human viruses, as well as endemic ape viruses such as Ebola.

The incidence of vector-borne and zoonotic diseases is challenging to predict. However, the Centers for Disease Control and Prevention (CDC) warns that environmental change can increase the range of animals and pathogen vectors such as ticks and mosquitoes. Rising global temperatures are already increasing the spread of human-animal pathogens. An important example of such a pathogen is West Nile virus (WNV), the cause of West Nile encephalitis, previously uncommon in the United States. West Nile virus is now spreading among birds and horses, transmitted by a tropical mosquito whose range is expanding northward, thanks to global warming. These mosquitoes increasingly transmit the virus to humans in the Northern Hemisphere.

Besides their direct interactions with other forms of life, microbes have enormous impact on the planet as a whole. Their metabolism generates much of our planet's atmosphere and geology, through biogeochemical cycles (see the next section).

SECTION SUMMARY

- **Symbiosis is an intimate association** between organisms of different species. The association may cause benefits or harm.

- **Mutualism is a form of symbiosis in which each partner species benefits from the other.** The relationship is usually obligatory for growth of one or both partners.

- **Synergism is a relationship in which both species benefit.** The partners are easily separated, and either partner can grow independently of the other.

- **Commensalism is an association in which one partner benefits and the other is unharmed.** It may, however, be difficult to prove that the second partner receives zero benefit.

- **Parasitism is a form of symbiosis in which one species grows at the expense of another, host organism.**

- **The rumen of ruminant animals is a complex microbial digestive chamber.** Rumen microbes digest complex plant materials. Gut fermenters contribute to digestion in many vertebrate animals, including humans.

- **Environmental conditions facilitate disease transmission between humans and animals.** An animal population that can transmit disease to humans is called a reservoir of infection.

Thought Question 27.1 Explain what mutualism and parasitism have in common and how they differ. Describe an example of a relationship that combines aspects of both.

Thought Question 27.2 Why do the microbes of digestive communities leave food value for the animal host? How can the animal obtain nourishment from waste products that the microbes could not use?

27.3 Biogeochemical Cycles

SECTION OBJECTIVES

- Describe how microbes participate in the global cycling of carbon by converting carbon between its reduced and oxidized forms.
- Explain the implications of global carbon cycling for human health.
- Outline the unique roles of microbes in cycling different forms of nitrogen.
- Explain how human technology perturbs the global nitrogen cycle.

How do microbes help cycle elements of life throughout Earth's biosphere? Even microbes have limits: no microbe, or any other living creature, can convert one element to another. Living organisms conduct chemical reactions, not nuclear reactions. Because organisms cannot generate their own elements, they need to get them from their environment. So, organisms acquire their elements either from nonliving components of their environment, such as by fixing atmospheric CO_2, or from other organisms, by grazing, predation, or decomposition. Furthermore, all organisms recycle their components to the biosphere. The partners in this recycling include abiotic entities such as air, water, and minerals, as well as biotic entities such as predators and decomposers. Collectively, the metabolic interactions of microbial communities with the biotic and abiotic components of their ecosystems are known as biogeochemistry or geomicrobiology.

Cycles of Elements for Life

Which elements need to be recycled and made available for life? In most organisms, six elements predominate: carbon, oxygen, nitrogen, hydrogen, phosphorus, and sulfur (discussed in Chapter 6). All these elements flow in biogeochemical cycles of nutrients throughout the biosphere's biotic and abiotic components. The environmental levels of key elements can limit biological productivity. For example, the concentration of iron limits the marine populations of phytoplankton, which depend on iron supplied by wind currents.

Biogeochemical cycles include both biological components (such as phototrophs that consume CO_2 and heterotrophs that release CO_2) and geological components (such as volcanoes that release CO_2 and oceans that absorb CO_2). The major part of the biosphere containing significant amounts of an element needed for life is called a **reservoir** of that element. Each reservoir acts both as a source of that element for living organisms and as a sink to which the element returns. For example, the ocean is an important reservoir for carbon (CO_2 in equilibrium with HCO_3^-, bicarbonate ion).

How do we study microbial cycling on a global scale? How do we figure out whether ecosystems are net sources or sinks of CO_2? Figure 27.16A shows a tower used for sampling atmospheric CO_2 as part of a global effort to monitor carbon flux (change in atmospheric concentration). The CO_2 is measured by infrared absorption spectroscopy, a method in which the pattern of infrared radiation

Figure 27.16
Global Atmospheric CO_2 Is Rising

A. At Tonzi Ranch, a FLUXNET tower is used for atmospheric CO_2 sampling as part of a global effort to monitor carbon flux.

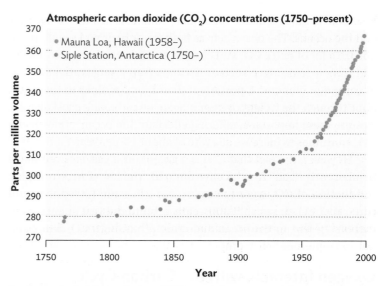

Atmospheric carbon dioxide (CO_2) concentrations (1750–present)

- Mauna Loa, Hawaii (1958–)
- Siple Station, Antarctica (1750–)

B. Atmospheric CO_2 levels since 1750 reveal the increase of this greenhouse gas accompanying the rise in industrial consumption of fossil fuels. Microbial CO_2 fixation helps limit CO_2 increase in the atmosphere.

absorbed by the gas molecules reveals which gases are present. The results of such readings at stations around the globe have led to a consensus among scientists that CO_2 levels are rising rapidly, faster than ever before on Earth (**Figure 27.16B**). The burning of fossil fuels is producing more CO_2 than the biosphere can absorb. The CO_2 traps solar radiation as heat—a process known as the **greenhouse effect**. The greenhouse effect of CO_2 and other gases is raising Earth's temperature at an unprecedented rate. This observation has enormous political and economic implications.

Will the human-induced global warming cause mass extinctions of Earth's species? Or can we use our knowledge of microbial ecology to channel microbial activities into recovering the balance—for example, by increasing microbial CO_2 fixation? Throughout most of this book, we have focused on the growth of individual organisms. Now we examine how the collective metabolic activities of microbial populations contribute to global cycles of elements throughout Earth's biosphere. And we consider the ways that we humans enlist microbes to manage our environment.

The Carbon Cycle

All food webs involve influx and efflux of carbon. From a global standpoint, where is carbon found on Earth? The major source of carbon is carbonate rock in Earth's crust. But Earth's crust is the source least accessible to the biosphere above ground. Crustal rock provides carbon only to organisms at the surface and to subsurface microbes that grow extremely slowly. So, subsurface carbon turnover is very slow.

The carbon reservoir that cycles most rapidly through living bodies is that of the atmosphere, a source of CO_2 for photosynthesis and lithotrophy (**Figure 27.17**). The atmosphere also acts as a **sink** (a place that accepts an element) for CO_2 produced by heterotrophy and by geological outgassing from volcanoes. But the atmospheric

Figure 27.17 **The Carbon Cycle: Aerobic and Anaerobic**

Aerobic and anaerobic conversions of carbon. In an aerobic environment, photosynthesis generates molecular oxygen (O_2), which enables the most efficient metabolism by heterotrophs and lithotrophs. In an anaerobic environment, photosynthesis generates oxidized minerals but no oxygen gas. Fermentation generates organic carbon products, as well as CO_2 and H_2. In the absence of oxygen, methanogens convert carbon dioxide (CO_2) and molecular hydrogen (H_2) to methane (CH_4), one of the most potent greenhouse gases. Blue = reduction of carbon; red = oxidation of carbon; black = fermentation; [CH_2O] = organic biomass; yellow highlight = greenhouse gas.

reservoir is much smaller than other sources, such as crustal rock and the oceans. The ocean acts as both a reservoir and a sink; it can absorb a lot of extra CO_2 and convert the gas to carbonate ions. In addition, marine algae such as diatoms trap a substantial amount of carbon in biomass. A portion of their biomass sinks to the ocean floor through the weight of their silicate or carbonate exoskeletons. Despite these important buffering effects of the ocean, atmospheric CO_2 continues to increase at a rate of about 1% per year.

However, the fact that CO_2 is a greenhouse gas does not make it inherently "bad" for the environment. Indeed, if heterotrophic production of CO_2 were to cease altogether, phototrophs would run out of CO_2 in roughly 300 years, despite the vast quantities of carbon present in the ocean and crust. Thus, both CO_2 fixers and heterotrophs need each other.

Oxygen Interacts with the Carbon Cycle

The global cycle of carbon in the biosphere is closely linked to the cycles of oxygen and hydrogen, elements to which most carbon is bonded. Overall, carbon cycles between CO_2 and various reduced forms of carbon, including biomass (living material). But the results of carbon cycling depend on the presence of molecular oxygen (O_2). Oxygen-rich ecosystems include the photic (light-receiving) zone of oceans and the surface of terrestrial habitats, such as soil. In an aerated (or oxic) habitat, such as the open ocean, the ecosystem absorbs enough light for the rate of photosynthesis to exceed the rate of heterotrophy. Microbial and plant photosynthesis fixes CO_2 into biomass. Biomass includes much carbohydrate material with approximately twice the number of hydrogens as carbon, so it is designated by the shorthand $[CH_2O]$. Phototrophs include bacteria and protists, as well as plants. Aerobic CO_2 fixation by bacteria and plants releases O_2. The O_2 is then used by heterotrophs (such as bacteria, protists, and animals) to convert $[CH_2O]$ back to CO_2. In the presence of light, a net excess of O_2 is released.

Compared with oxygen-rich environments, anoxic environments support lower rates of biomass production. Anaerobic conversion of CO_2 to biomass is done mainly by bacteria and archaea. Vast, permanently anaerobic habitats extend several kilometers below Earth's surface, encompassing greater volume than the rest of the biosphere put together. In these habitats, endolithic bacteria inhabit the interstices of rock crystals. In soil and water, anaerobic metabolism includes fermentation of organic carbon sources, as well as respiration and lithotrophy with alternative electron acceptors such as nitrate, ferric iron (Fe^{3+}), and sulfate.

Anaerobic decomposition by microbes is one stage in the formation of fossil fuels such as oil and natural gas (primarily methane). Recall that methanogens—methanogenic archaea—are a form of life very divergent genetically from bacteria (discussed in Chapter 10). Methanogens obtain energy by forming methane from H_2, CO_2, and other fermentation products of anaerobic bacteria.

Anoxic environments near the surface also favor methanogens producing methane. In the Arctic, increasing temperatures are accelerating methane release. Methane is a greenhouse gas that in turn amplifies global warming. Another source of concern is the methane hydrates accumulating in deep marine sediments, generated by huge deep-ocean communities of methanogens. Warming the methane hydrate releases methane gas, which quickly rises to the atmosphere. Geological evidence suggests that rapid methane release accompanied the retreat of the glaciers during ice age transitions. A rapid methane release today could cause irreversible climate change.

The Nitrogen Cycle

Another major element that cycles largely by microbial conversion is nitrogen (**Figure 27.18**). Nitrogen is found in more different oxidation states than any other major biological element.

Figure 27.18 The Global Nitrogen Cycle

Prokaryotic conversions of nitrogen occur throughout the biosphere. Blue = reduction; red = oxidation; black = redox-neutral; yellow highlight = greenhouse gas.

Conversion between these oxidation states requires several metabolic processes that occur only in bacteria and archaea. These processes include nitrogen fixation, ammonia conversion to nitrate, and the reduction of gases such as nitrogen dioxide (NO_2), nitrogen monoxide (NO), and nitrous oxide (N_2O). Without bacteria and archaea, the global nitrogen cycle would not exist.

Where is nitrogen found on Earth? Much nitrogen is found in Earth's crust in the form of ammonium salts in rock, but this form is largely inaccessible to microbes. The major accessible source of nitrogen is the atmosphere (79% nitrogen gas, N_2). However, N_2 is a highly stable molecule that requires enormous input of reducing energy before assimilation is possible. So, for many natural ecosystems and most forms of agriculture, nitrogen is the limiting nutrient for primary productivity (production of biomass by photosynthesis).

Until recently in Earth's history, N_2 entered ecosystems only via nitrogen-fixing bacteria and archaea. In the twentieth century, however, the Haber process was invented for artificial nitrogen fixation to generate fertilizers for agriculture. In the Haber process, N_2 is hydrogenated to ammonia (NH_3) by methane under extreme heat and pressure. Today, the Haber process for producing fertilizers accounts for approximately 30%–50% of all nitrogen fixed on Earth. Other human activities, such as burning of fuels and use of nitrogenous fertilizers, contribute to oxidized nitrogen pollutants such as nitrous oxide (N_2O), a potent greenhouse gas. The nitrogen cycle is now the most perturbed of the major biogeochemical cycles.

As discussed in Chapter 7, bacteria fix nitrogen by the overall reduction of N_2 into two molecules of ammonium ion (NH_4^+). Plants and bacteria can combine ammonium ion with carbon skeletons (such as TCA cycle intermediates) to form amino acids. But given the energy expense and the need to exclude oxygen when fixing N_2, many species of bacteria, as well as all plants and animals, have lost the nitrogen fixation pathway by reductive evolution (the loss of a trait in the absence of need). The need to fix nitrogen and make ammonia disappeared in these life forms because other bacteria and archaea that are present in every ecosystem still fix nitrogen. Nitrogen-fixing bacteria in the soil include anaerobes such as *Clostridium* species and Gram-negative species of *Azotobacter* and *Geobacter*, as well as obligate respirers such as *Pseudomonas*. In ocean and freshwater, major nitrogen fixers are cyanobacteria such as *Synechococcus* species.

Nitrogen is also fixed by bacterial symbionts of plants, such as *Rhizobium* and *Frankia*. Within an ecosystem, nitrogen fixers ultimately make the reduced nitrogen available for assimilation by nonfixing microbes and plants, either directly through symbiotic association (such as that between rhizobia and legumes) or indirectly through predation (marine cyanobacteria) and decomposition (soil bacteria). Symbiotic nitrogen fixation by rhizobial bacteria is important for agriculture because it increases the yield of crops such as soybeans.

The fixed forms of nitrogen (ammonium ion and ammonia) in soil or water are quickly oxidized for energy by **nitrifiers**, bacteria that can oxidize ammonia to nitrite (NO_2^-), or nitrite to nitrate (NO_3^-). The oxidation of ammonia to nitrite and nitrate is called **nitrification**. Nitrification is a form of lithotrophy, an energy generation pathway involving oxidation of minerals. Production of both nitrite and nitrate generates acid, which can acidify the soil. Nitrate produced in the soil is assimilated by plants and bacteria nearly as quickly as ammonium ion. But in agriculture, intensive fertilization generates a large excess of ammonia and thus a buildup of nitrites and nitrates. These ions are highly soluble in water, and they readily diffuse into rivers and lakes. Nitrogen buildup in water adds nutrients that contribute to algal blooms and subsequent anoxic dead zones.

Aquatic nitrate reacts with organic compounds to form toxic nitrosamines. Nitrate influx also relieves the nitrogen limit on algae, causing algal blooms and raising BOD. Chronic nitrate influx causes eutrophication and die-off of fish. Human consumption of nitrate in drinking water can lead to **methemoglobinemia**, a blood disorder in which hemoglobin is inactivated. Methemoglobinemia occurs in infants whose stomachs are not yet acidic enough to inhibit the growth of bacteria that convert nitrate to nitrite (Chapter 7). Nitrite oxidizes the iron in hemoglobin, decreasing the protein's ability to release oxygen. The failure to carry oxygen leads to a bluish appearance, one cause of "blue baby syndrome." Nitrite-induced blue baby syndrome is a problem in intensively cultivated agricultural regions, such as Kansas and Nebraska (**Figure 27.19**).

The global cycles we have described for carbon and nitrogen have counterparts in the cycling of sulfur, phosphorus, iron, and other essential elements. Sudden losses and influxes of any of these elements can perturb an ecosystem. For instance, marine algae are often limited by scarcity of iron. Thus, when a sandstorm off the Sahara desert blows iron into the ocean, it may cause an algal bloom.

Figure 27.19 Nitrate Runoff Contaminates Groundwater

Nitrate in drinking water is especially prevalent in agricultural regions of the United States. Bacterial ammonification of fertilizer, followed by nitrification, generates nitrate and nitrite. Nitrate and nitrite runoff from oxidized nitrogenous fertilizers pollutes streams and groundwater.

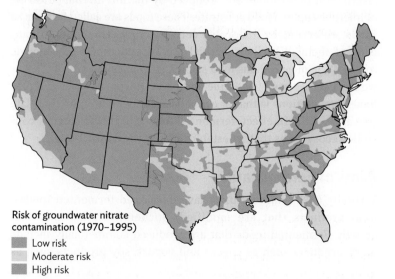

Risk of groundwater nitrate contamination (1970–1995)

▨ Low risk
▨ Moderate risk
▨ High risk

SECTION SUMMARY

- **The most accessible reservoir for carbon is the atmosphere (CO_2).** Atmospheric carbon is severely perturbed by the burning of fossil fuels.

- **Photosynthetic bacteria and algae cycle much of the CO_2 in the biosphere.** Oxygen released by phototrophs is used by aerobic heterotrophs and lithotrophs.

- **Anaerobic environments cycle carbon through bacteria and archaea.** Bacteria conduct fermentation and anaerobic respiration. Methanogens release methane, much of which is removed by methane-oxidizing bacteria and archaea.

- **Microbial decomposition returns CO_2 to the atmosphere.**

- **Nitrogen gas is fixed into ammonium ion (NH_4^+) only by bacteria and archaea.** Other organisms can incorporate ammonium ion into amino acids and additional biological molecules.

- **Industrial fixation of nitrogen to make fertilizers has perturbed the global nitrogen cycle.** Excessive nitrogen fixation destabilizes freshwater and marine ecosystems, leading to algal blooms and dead zones.

27.4 Fermented Foods and Beverages

SECTION OBJECTIVES

- Describe the roles of bacteria and fungi in the production of cheese and fermented soy foods.
- Outline the process of ethanolic fermentation in the production of beer and wine.
- Explain why fermented foods and beverages retain high food value.

If "you are what you eat," are you a microbe? You consume microbes every day as part of the food you put in your mouth. Some foods, such as mushrooms and seaweed (sushi covering), are considered microbial products. Others—including plant foods, from salad to fruit—contain microbes growing within the plant's vasculature. Still other foods are produced by the intentional culturing of microbes upon a food substrate. These foods are called **fermented foods**, referring to anaerobic fermentation, the predominant class of metabolism in microbial food production. They include some of our most famous food staples, such as bread and cheese, as well as beverages such as wine and beer. One of the most complex food fermentations is that of cocoa bean fermentation for chocolate (see **IMPACT**). In effect, fermented foods are the product of controlled ecosystems that we establish for our own nutrition.

Aims of Fermentation

Virtually all human cultures have developed fermented foods—food products that are modified biochemically by microbial growth. Fermented foods that are produced commercially include dairy products such as cheese and yogurt, soy products such as miso (from Japan) and tempeh (from Indonesia), vegetable products such as sauerkraut (Europe) and kimchi (Korea), and various forms of cured meats and sausages. Alcoholic beverages are made from grapes and other fruits (wine), grains (beer and liquor), and agave (tequila).

Why do we eat foods that have been altered by microbial growth? Food fermentation bestows certain benefits:

- **It preserves food.** Certain bacteria, particularly *Lactobacillus* and *Lactococcus* species, metabolize only a narrow range of nutrients (such as sugars) before their waste products (acids and alcohols) build up and inhibit further growth. Thus, acid buildup keeps the product stable for much longer than the original food substrate would be. That is why yogurt and cheese last much longer than fresh milk. Other fermentation products that preserve food include ethanol (alcoholic beverages) and ammonia (fermented soybean products).

- **It improves digestibility.** Microbial action breaks down fibrous macromolecules and tenderizes the product, making it easier for humans to digest. Meat and vegetable products are tenderized by fermentation.

- **It adds nutrients and flavors.** Microbial metabolism generates vitamins, particularly vitamin B_{12}, as well as flavor molecules, such as esters and sulfur compounds.

The nature of fermented foods depends on the quality of the fermented substrate, as well as on the microbial species and the type of biochemistry they perform. Traditional fermented foods depend on indigenous microbiota—that is, microbes found naturally in association with the food substrate—or on starter cultures derived from a previous fermentation, as in yogurt or sourdough fermentation. Commercial fermented foods use highly engineered microbial strains to inoculate their cultures, although in some cases indigenous microbiota still participate. For example, cheeses aged in the same caves for centuries include fermenting organisms that persist in the air and the containers used.

The most common metabolism in food fermentation is anaerobic fermentation of glucose (discussed in Chapter 7). Glucose is fermented to lactic acid (lactic acid fermentation) in cheeses and sausages, primarily by lactic acid bacteria such as *Lactobacillus*. A second-stage fermentation of lactic acid to propionic acid (propionic acid fermentation) by *Propionibacterium* generates the special flavor of Swiss and related cheeses. Some kinds of vegetable fermentation, as in sauerkraut, involve production of lactic acid, ethanol, and carbon dioxide (heterolactic fermentation) by *Leuconostoc*. Fermentation to ethanol plus carbon dioxide without lactic acid (ethanolic fermentation) is conducted by yeast during bread leavening and production of alcoholic beverages. In some food products, particularly those fermented by *Bacillus* species, proteolysis and amino acid catabolism generate ammonia in amounts that raise the pH (alkaline fermentation). For example, alkaline fermentation forms the soybean product natto.

Other products require the growth of mold, such as the mold-spiked Roquefort cheese. Mold growth requires some oxygen for aerobic respiration. Respiration must be limited, however, to avoid excessive decomposition of the food substrate and loss of food value.

Fermented Dairy Products

Microbial fermentation of milk converts a liquid suspension of proteins and carbohydrates to a semisolid or solid product, such as yogurt or cheese. The practice of milk fermentation probably arose among ancient herders who collected the milk of their pack animals but had no way to prevent the rapid growth of bacteria. The milk was stored in a container such as the stomach of a slaughtered animal. After hours of travel, the combined action of stomach enzymes (rennet) and lactic acid–producing bacteria caused the milk proteins to denature (misfold) and coagulate into a semisolid substance.

In the first stage of cheese production, the coagulated protein is called **curd**. The curd naturally separates from the liquid portion, called whey. Both curds and whey can be eaten, as in the nursery rhyme "Little Miss Muffet." The curds, however, are particularly valuable for their concentrated protein content.

A cheese is any milk product from a mammal (usually cow, sheep, or goat) in which the milk protein coagulates to form a semisolid curd. How does milk coagulate? The major organic components of cow's milk are milk fat (about 4% unless skimmed), protein (3.3%), and the sugar lactose (4.7%). Milk starts out at about pH 6.6, slightly acidic. At this pH, the milk proteins are completely soluble in water; otherwise, they would clog the animal's udder as the milk came out. Fermentation generally begins with bacteria such as *Lactobacillus* and *Streptococcus* (Figure 27.20). As bacteria ferment milk sugars to lactic acid, the pH starts to decline. Because no aerobic respiration occurs, only lactic acid is made in

Figure 27.20 **Bacterial Community within Emmentaler Cheese**

Bacterial species include *Lactobacillus helveticus* and *Streptococcus thermophilus*.

Figure 27.21 **Cheese Varieties**

A. Cottage cheese, an unripened, perishable cheese.

B. Emmentaler Swiss cheese, with eyes produced by carbon dioxide fermentation.

C. Feta cheese, a soft cheese from goat's milk, preserved in brine.

D. Roquefort, a medium-hard cheese ripened by spiking with *Penicillium roqueforti*.

large quantities that acidify the milk until the bacteria can no longer grow. Halting bacterial growth minimizes the oxidation of amino acids, thereby maintaining food quality.

Milk contains suspended droplets of hydrophobic proteins called **caseins**. As the pH of milk declines below pH 5, the casein molecules misfold (or "denature"). The tangling of denatured caseins generates a gel-like network throughout the milk, trapping other substances, such as droplets of milk fat. This protein network generates the semisolid texture of **yogurt**, a simple product of milk acidified by lactic acid bacteria.

In soft cheeses such as cottage cheese (Figure 27.21A), bacterial lactic acid acidification is the only cause of coagulation. Other kinds of cheese, such as Swiss cheese (Figure 27.21B), include an additional step of casein coagulation by proteases such as chymosin, one of several enzymes found in rennet. Rennet is a mixture of enzymes traditionally obtained from a calf abomasum (fourth stomach). Today, cheese makers also use microbial sources of chymosin.

The water-soluble portion of milk proteins, about one-third of the total casein, enters the whey and is lost from the curd.

impact

Chocolate: The Mystery Fermentation

Chocolate, the product of the cocoa bean, *Theobroma cacao*, or "food of the gods," requires one of the most complex fermentations of any food. For all the sales of chocolate, totaling 2.5 billion kilograms per year worldwide, no "starter culture" has yet been standardized to ferment the cocoa bean. Instead, cocoa beans harvested in Africa or South America are heaped in mounds upon plantain leaves for fermentation by indigenous microorganisms—essentially the same way cocoa has been processed for thousands of years (Figure 1). The beans must ferment immediately where they are harvested; they cannot be exported and fermented later.

The fermentation occurs outside the cocoa bean, within the pulp that clings to the beans after they are removed from the cocoa fruit. The pulp contains 1% pectin (a complex polysaccharide found in fruits) plus amino acids and minerals. These nutrients support growth of many kinds of microbes. Brazilian microbiologist Rosane Schwan, of the Federal University of Lavras, Brazil, analyzed the microbial community (Figure 2). Schwan defined three stages of succession:

1. **Yeast fermentation (anaerobic).** The pulp initially is full of citric acid (pH 3.6), which favors growth of yeasts such as *Candida* and *Saccharomyces*. The yeasts consume citric acid, increasing pH. They also degrade pectin into glucose and fructose, allowing the pulp to liquefy. As the liquefied pulp drains, its remaining sugars are fermented to ethanol, CO_2, and acetate. These products eventually rise to levels that inhibit yeast growth. The ethanol and acetate penetrate the bean embryo, preventing germination and disrupting cell membranes. The disrupted cells release enzymes that generate key molecules of chocolate. For example, theobromine is a stimulant similar to caffeine, and 2-phenylethylamine is a neurotransmitter associated with the pleasure response.

2. **Lactic acid fermenters (anaerobic).** The consumption of citric acid by yeast increases the pH to 4.2, enabling growth of lactic acid bacteria. *Lactobacillus* species ferment sugars and citric acid to acetate, CO_2, and lactic acid. When fermentable substrates are used up, *Lactobacillus* stops growing.

3. **Acetic acid bacteria (aerobic).** The beans are turned over and mixed periodically to permit access to oxygen. With oxygen, aerobic bacteria such as *Acetobacter* respire, converting the ethanol and acids to CO_2. The consumption of acids neutralizes undesirable acidity. The bean is also penetrated by oxygen, which oxidizes key cocoa components, such as polyphenols. Polyphenol oxidation generates the brown color of cocoa and contributes flavor.

Oxidative respiration within the mound of beans generates heat faster than it can dissipate, raising the temperature as high as 50°C. The high temperature halts bacterial fermentation. After the pulp drains, the beans are dried and roasted. The

Figure 1 Cocoa Beans

A. Cocoa fruit, showing beans encased in mucilage (white coating).

B. Fruit pulp ferments, liquefies, and drains away, while the beans acidify and turn brown.

Figure 2 Microbial Succession during Cocoa Pulp Fermentation

A. Rosane Schwan harvests cocoa beans for her research on cocoa fermentation.

1 Yeasts degrade pectin; ferment sugars and citrate to ethanol and acetate.

2 *Lactobacillus* ferments sugars and citrate to lactate and acetate.

3 *Acetobacter* oxidizes ethanol.

Turn beans, adding O_2.

B. Yeasts generate ethanol, which lactic acid bacteria convert to lactate and acetic acid bacteria oxidize to CO_2. CFUs = colony-forming units.

roasting process completes the transformation of cocoa. Without the preceding fermentation process, no flavor would develop. Cocoa liquor and cocoa butter are extracted from the fermented beans and then recombined with sugar and other components to make "cocoa mass" (Figure 3). The cocoa mass is stirred for several days to achieve a smooth texture; then it is molded into the decorative forms known as chocolate.

The quality of chocolate rests largely on the original process of pulp fermentation, whose details remain poorly understood. Schwan is working to develop a defined starter culture, a community of microbes to produce a predictable high quality of flavor. At present, however, one of the world's most refined and highly prized commercial food products still depends on the indigenous microbial fermentation of cocoa pulp.

Figure 3 Chocolate Manufacture

Cocoa mass contains cocoa butter and liquor extracted from the cocoa beans, mixed with sugar and other ingredients.

DISCUSSION QUESTION

Cocoa bean fermentation works best when conducted in the same location where the cocoa trees grow. Why might this be? What problems might occur if the beans are transported to a distant location for fermentation?

To retain the whey protein, the cheese is treated at high temperature, which denatures even the whey protein, so it's retained in the curd. Aging—that is, incubation under controlled humidity for several months—further develops flavor. Some cheeses, such as feta, are brined (treated with concentrated salt; Figure 27.21C) to limit bacterial growth and develop flavor. Other cheeses are "ripened" by inoculation with mold, either to form a surface crust, as in Brie and Camembert, or to spike deep into the cheese, as in blue cheese or Roquefort (Figure 27.21D).

Soybean Fermentation

Soybeans offer one of the best sources of vegetable protein and are indispensable for the diet of millions of people, particularly in Southeast Asia. But soybeans also contain substances that decrease their nutritive value, such as phytate, a sugar phosphate that inhibits iron absorption by the intestine. Soybeans also contain lectins, proteins that bind to cell-surface glycoproteins within the human body. At high concentration, lectins may upset digestion and induce autoimmune diseases. Protease inhibitors interfere with chymotrypsin and trypsin, thus decreasing the amount of protein that can be obtained from soy-based food. But all these biochemical drawbacks of soybeans are diminished by microbial fermentation, while the protein content remains comparable to that of the unfermented bean (40%).

A major fermented soy product is tempeh, a staple food of Indonesia and other countries in Southeast Asia (Figure 27.22). Tempeh consists of soybeans fermented by *Rhizopus oligosporus*, a common bread mold. The mold grows as a white mycelium that permeates the beans, joining them into a solid cake. The final product has a mushroom-like taste and is served fried or grilled like a hamburger. Besides decreasing the negative factors of soy, the mold growth breaks down proteins into more digestible peptides and amino acids. During World War II, tempeh was fed to American prisoners of war held by the Japanese. Tempeh was later credited with saving the lives of prisoners whose dysentery and malnutrition had impaired their ability to absorb intact proteins.

Alcoholic Beverages

Ethanolic fermentation of grain or fruit was important to early civilizations because it yielded a source of water free of pathogens. Traditional forms of beer also provided essential vitamins supplied by the unfiltered yeasts. Ethanol provides a significant source of caloric intake, but it is also a toxin that impairs mental function. A modest level of ethanol enters the human circulation naturally from intestinal microbiota, equivalent to a fraction of a drink per day. For protection, the human liver produces the enzyme alcohol dehydrogenase, which detoxifies ethanol. In a healthy liver, alcohol dehydrogenase can metabolize small amounts of alcohol without harm. However, excess alcohol consumption can overload the liver's capacity for detoxification and permanently damage the liver and brain.

BEER: ALCOHOLIC FERMENTATION OF GRAIN Beer production is one of the most ancient fermentation practices and is depicted in the statuary of ancient Egyptian tombs dated to 5,000 years ago (Figure 27.23A). The earliest Sumerian beers were made from bread soaked in water and fermented. Today, most beer is produced commercially in giant vats that ferment barley (Figure 27.23B). Production of high-quality beer is a complex process with many steps. The barley grains must germinate (sprout as a seedling) in order for enzymes from the grain to break down starch (long-chain carbohydrates) into maltose (disaccharide composed of two molecules of glucose). The fermenting grains are mashed in water and cooked down. Flowers of the hops plant are added for flavor and for their antibacterial properties, which increase the beer's shelf life. The glucose then is fermented by the yeast, producing ethanol and CO_2. Finally, the beer is filtered to remove yeast and other solids.

The yeast in the beer mash ferment most of the maltose to ethanol (Figure 27.24). In addition, side products in lower concentration (such as long-chain alcohols and esters) contribute flavors. But they also make unpleasant off-flavors if present in too great

Figure 27.22 Tempeh, a Mold-Fermented Soy Product

A. Fried tempeh.

B. *Rhizopus oligosporus* mold, used to make tempeh.

Figure 27.23 Beer Production: Ancient and Modern

A. Making beer in ancient Egypt, circa 3000 BCE. The mash is stirred in earthen jars.

B. Fermentors in a modern brewery.

an amount. The off-flavors result from the presence of oxygen. Oxidative respiration diverts some pyruvate into acetaldehyde and diacetyl (shaded green in Figure 27.24), two compounds that cause off-flavors. Although beer fermentation is mostly anaerobic, the yeast requires a small amount of oxygen to synthesize some of its cell components.

WINE: ALCOHOLIC FERMENTATION OF FRUIT The fermentation of fruit gives rise to wine, another class of alcoholic products of enormous historical and cultural significance. Grapes (*Vitis vinifera*) produce the best-known wines, but wines and distilled liquors are also made from apples, plums, and other fruits. The key difference between fermentation of fruits and fermentation of grains is the exceptionally high monosaccharide content in fruits. Grape juice, for example, can contain concentrations of glucose and fructose as high as 15%. The availability of simple sugars enables yeast to begin fermenting immediately, without preliminary breakdown of long-chain carbohydrates, as in the fermentation of grain for beer.

Figure 27.24 Alcoholic Fermentation in Beer and Wine

Yeast fermentation generates ethanol in substantial quantities. The biosynthesis of amino acids generates by-products that contribute both desirable flavors (long-chain alcohols) and off-flavors (acetaldehyde and diacetyl).

SECTION SUMMARY

- **Fermented foods preserve food value** because the anaerobic fermentation products (acids or alcohol) limit microbial growth and leave most of the protein intact.

- **Milk curd** forms by lactic acid fermentation and rennet proteolysis. The cleaved casein peptides coagulate to form a semisolid curd. The main fermentative organisms are lactic acid bacteria.

- **Cheese varieties** include unripened cheese, as well as ripened cheeses involving subsequent steps, such as brining or mold growth. Cheese flavors are generated by minor side products of fermentation.

- **Soy fermentation** to tempeh and other products improves digestibility and decreases undesirable soy components.

- **Beer derives from alcoholic fermentation of grain.** Barley grains are germinated, allowing enzymes to break down the starch to maltose for yeast fermentation. Secondary products of grain fermentation, such as long-chain alcohols and esters, generate the special flavors of beer.

- **Wine derives from alcoholic fermentation of fruit**, most commonly grapes. The grape sugar (glucose) is fermented by yeast to alcohol.

- -

Thought Question 27.3 In food production, why does oxygen allow excessive breakdown of a food substrate, compared with anaerobic fermentation?

Thought Question 27.4 Compare and contrast the role of fermenting organisms in the production of cheese and alcoholic beverages.

- -

27.5 Food Spoilage and Preservation

SECTION OBJECTIVES

- Explain the difference between food spoilage and food contamination.

- Outline the different processes of spoilage of dairy products, meat, seafood, and plant foods. Explain why different types of food spoil differently.

- Describe the various physical and chemical approaches to food preservation.

If microbes help us make fine foods, then why does food spoil? Excess microbial growth can decrease nutrients and form toxins. We humans have always competed with microbes for our nutrients. When early humans killed an animal, microbes immediately began to consume its flesh. Because meat perished so fast, it made economic sense to share the kill immediately and consume all of it as soon as possible. Vegetables might last longer, but eventually they succumbed to mold and rot. Yet how does one define "rot" versus "food production"? To a surprising degree, the definition of spoilage depends on cultural practice. What is sour milk to one person may be buttermilk to another, and meat that one society considers spoiled, another might consider merely aged.

From our earliest history, societies developed practices that control and minimize microbial degradation of food. Practices such as drying, smoking, and canning enabled humans to survive winters and dry seasons on stored food. Such practices are called **food preservation**. Modern food preservation includes antimicrobial agents, chemical substances that either kill microbes or slow their growth, and physical treatment with heat and pressure.

Food Spoilage and Food Contamination

After food is harvested, several kinds of chemical changes occur. Some begin instantly, whereas others take several days to develop. Some changes, such as meat tenderizing, may be considered desirable; others, such as putrefaction (protein decomposition), render food unfit for consumption. The major classes of food change include:

- **Enzymatic processes.** After an animal dies, its flesh undergoes proteolysis by its own enzymes. Limited proteolysis tenderizes meat. Plants after harvest undergo other changes; in harvested corn, for example, the sugar rapidly converts to starch. That is why vegetables taste sweetest immediately after harvest, before the sugar is lost.

- **Chemical reactions with the environment.** The most common abiotic chemical reactions involve oxidation by air—for example, lipid autooxidation, which generates rancid odors.

- **Microbiological processes.** Microbes from the food's surface begin to consume it—some immediately, others later in succession—generating a wide range of chemical products. In meat, the digestive tract's microbial community is an important source of decay.

Microbial activity can aid food production, but it can also have various undesirable effects. Two classes of microbial effects are distinguished: food spoilage and food contamination with pathogens.

Food spoilage refers to microbial changes that render a product obviously unfit or unpalatable for consumption. For example, rancid milk and putrefied meats are unpalatable and contain metabolic products that may be harmful to human health, such as oxidized fatty acids or organic amines. Food that is capable of spoilage is referred to as perishable food. Signs of food spoilage include a change in appearance from that of the normal form, such as a color change, a change in texture, an unpleasant odor, or an undesirable taste.

Different pathways of microbial metabolism lead to different kinds of spoilage. Sour flavors result from acidic fermentation products, as in sour milk. Alkaline products generate bitter flavor. Oxidation, particularly of fats, causes rancidity, whereas general decomposition of proteins and amino acids leads to putrefaction. The particularly noxious odors of putrefaction derive from amino acid breakdown products that often have apt names, such as the amines cadaverine and putrescine and the aromatic product skatole.

Food contamination refers to the presence of microbial pathogens that cause human disease, usually gastrointestinal illness (called **food poisoning**). For example, food contaminated with rotavirus is the major cause of food poisoning in young children. Some kinds of food poisoning are caused by a toxin produced by microbes in the food, even if the microbes do not colonize the consumer. An important example is botulism, caused by botulinum toxin produced by *Clostridium botulinum*. Pathogens usually go unnoticed as food is consumed because their numbers are low. Even the freshest-appearing food may cause serious illness if it has been contaminated with a small number of pathogens.

How Food Spoils

Different foods spoil in different ways, depending on their nutrient content, the microbial species, and environmental factors such as temperature.

DAIRY PRODUCTS Milk and other dairy products contain carbon sources, such as lactose, protein, and fat. In fresh milk, the nutrient most available for microbial catabolism is lactose, which commonly supports anaerobic fermentation to sour milk. Fermentation by the right mix of microbes, however, leads to yogurt and cheese production, as previously described. Under certain conditions, bitter off-flavors may be produced by bacterial degradation of proteins. The release of amines causes a rise in pH.

Dairy products can be spoiled or contaminated by **psychrotrophs**, which grow at lower environmental temperatures including temperatures typical of refrigeration (2°C–4°C). Psychrotrophs may also grow at human body temperature and include pathogens such as *Pseudomonas* and *Listeria* species. *Listeria monocytogenes* cause serious illness, such as the multistate listeriosis outbreak from ice cream in 2015.

Hard cheeses are less susceptible than milk to general spoilage because of their solid structure and lowered water activity (water available to microbes). However, cheeses can grow mold on their surface. Historically, the surface growth of *Penicillium* strains led to the invention of new kinds of cheeses. But other kinds of mold, such as *Aspergillus*, produce toxins and undesirable flavors.

MEAT AND POULTRY Meat in the slaughterhouse is easily contaminated with bacteria from hide, hooves, and intestinal contents. Muscle tissue offers high water content, which supports microbial growth, as well as rich nutrients, including glycogen, peptides, and amino acids. The breakdown of peptides and amino acids produces the undesirable odorants that define spoilage, such as cadaverine and putrescine.

Meat also contains fat, or adipose tissue, but the lipids are largely unavailable to microbial action because they consist of insoluble fat (triacylglycerides). Instead, meat lipids commonly spoil abiotically by autooxidation (reaction with oxygen) of unsaturated fatty acids, independent of microbial activity. Thus, when meats are exposed to air during storage, they turn rancid—particularly meats such as pork, which contains highly unsaturated lipids. Autooxidation can be prevented by anaerobic storage, such as vacuum packing, which also prevents growth of aerobic microorganisms.

In industrialized societies, the most significant determinant of microbial populations in meat spoilage is the practice of refrigeration. Refrigeration prolongs the shelf life of meat because contaminating microbes from a thermoregulated animal are inhibited at low temperature (see Chapter 6). Nevertheless, refrigerated meat products such as lunch meats can be contaminated by psychrotrophs such as *Listeria*.

SEAFOOD Fish and other seafood contain substantial amounts of protein and lipids, as well as amines such as trimethylamine oxide (TMAO). Fish spoils more rapidly than meat and poultry for several reasons. First, fish do not thermoregulate, and they inhabit relatively low-temperature environments. Because fish grow in low-temperature environments, their surface microorganisms tend to be psychrotrophic and thus grow during refrigeration. In addition, marine fish contain high levels of the osmoprotectant TMAO, which bacteria reduce to trimethylamine as an alternative to oxygen—a form of anaerobic respiration. Trimethylamine is a volatile amine that gives seafood its "fishy" smell. Finally, the rapid microbial breakdown of proteins and amino acids leads to foul-smelling amines and sulfur compounds.

PLANT FOODS Fruits, vegetables, and grains spoil differently from animal foods because of their high carbohydrate content and their relatively low water content. The low water content of plant foods usually translates into considerably longer shelf life than for animal-based foods. Carbohydrates favor microbial fermentation to acids or alcohols that limit further decomposition, and this microbial action can be managed to produce fermented foods, as described in Section 27.4.

Plant pathogens rarely infect humans but may destroy the plant before harvest. Most plant pathogens are fungi or fungus-like pathogens, such as the oomycete that caused the Irish potato famine. Plant pathogens continue to devastate local economies and cause shortages worldwide. For example, the witches'-broom fungus *Moniliophthora perniciosa* causes a fungal disease of cocoa trees that has drastically cut Latin American cocoa production.

After harvest, various molds and bacteria can soften and wilt plant foods by producing enzymes that degrade the pectins and celluloses that give plants their structure. In general, the more processed the food, the greater the opportunities for spoilage. For example, citrus fruits generally last for several weeks, but peeled oranges are susceptible to spoilage by Gram-negative bacteria.

Pathogens Contaminate Food

Intestinal pathogens spread readily because many pathogens can be transmitted through food without any outward sign that the food is spoiled. The CDC estimates there are 76 million cases of gastrointestinal (GI) illness a year in this country, usually spread through water or food. Thus, one in four Americans experiences GI illness in a given year. **Table 27.3** summarizes the top pathogens that contaminate our food.

CASE HISTORY 27.2
Peanut Butter Cracker Surprise

 In the fall of 2008, several individuals, many of them children, were sickened by the common food pathogen *Salmonella enterica*. Most infected individuals developed diarrhea, fever, and abdominal cramps 12–72 hours after infection, and symptoms lasted 4–7 days. Over the next 6 months, cases were reported from nearly all US states, ultimately totaling 700 people (**Figure 27.25A**). The curve of the outbreak (cases rising and then falling) followed the profile of a single-source epidemic, in which all infections are ultimately traced to one source. All cases of illness showed a common strain of the pathogen *S. enterica* serovar Typhimurium (**Figure 27.25B**).

CDC researchers investigated the outbreak by comparing the food intake histories of ill persons with those of matched controls. The researchers found a statistical association between illness and intake of peanut butter, eventually narrowed to a specific brand of peanut butter sold to institutions. As the epidemic grew, cases emerged in which the contaminated food product was crackers filled with peanut butter cream. Ultimately, the peanut butter and cream were traced back to peanuts from a single factory in Georgia.

This case of *Salmonella* contamination is a typical example of a **point-source epidemic**, in which food is the common source of a pathogen (discussed in Chapter 26). But this case illustrates how modern food industries pose new challenges to finding the source. The peanuts from the Georgia factory ended up in many secondary products, from peanut butter to crackers filled with peanut butter cream. The cream contained peanuts from a secondary supplier to the company that had made the crackers.

At the original food plant, the source of *Salmonella* contamination could not be identified precisely because too much time had passed since the peanuts had shipped—but the plant records showed that product samples had tested positive for *Salmonella*. Instead of discarding the product, the plant had retested the samples until they "tested negative." Many health violations were cited, including gaps in the walls and dirt buildup throughout the plant.

An important question is: How could the CDC know which *Salmonella* cases reported from around the country were part of this one outbreak? DNA analysis was used to show that all cases shared a common strain of the pathogen *S. enterica* serovar Typhimurium. (A serovar is a strain whose surface proteins elicit a

Table 27.3
Food-Borne Pathogens in the United States[a]

Pathogen	Incidence and Transmission	Course of Illness
Norovirus (Norwalk and Norwalk-like viruses)	Most common cause of diarrhea; also called "stomach flu" (no connection with influenza). An estimated 180,000 cases occur per year. Transmitted mainly by virus-contaminated food and water. Infection rates are highest under conditions of crowding in close quarters, such as inside a ship or a nursing home.	Disease lasts 1 or 2 days. Includes vomiting, diarrhea, and abdominal pain; headache and low-grade fever may occur.
Salmonella	Most common food-borne cause of death; more than 1 million cases per year; estimated 600 deaths per year. Transmission nearly always through food—raw, undercooked, or recontaminated after cooking, especially eggs, poultry, and meat; also contaminates dairy products, seafood, fruits, and vegetables.	Causes GI disease that includes diarrhea, fever, and abdominal cramps lasting 4–7 days. Fatal cases are most common in immunocompromised patients.
Campylobacter	More than 1 million cases of campylobacteriosis per year; estimated 100 deaths per year. Grows in poultry without causing symptoms. Transmission is mainly through raw and undercooked poultry; contaminates half of poultry sold. Occurs less often in dairy products or in foods contaminated after cooking.	In humans, usually causes severe bloody diarrhea, fever, and abdominal cramps lasting 7 days. Fatal cases are most common in immunocompromised patients. May trigger Guillain-Barré Syndrome.
Escherichia coli O157:H7	An emerging pathogen, first recognized in a hamburger outbreak in 1982; now known to infect 73,000 people yearly, including 60 deaths per year. Grows in cattle without causing symptoms. Transmitted through ground beef; also through unpasteurized cider and from produce, where it grows as an endophyte.	In humans, usually causes severe bloody diarrhea and abdominal cramps lasting 5–10 days. About 5% of patients, especially children and elderly, develop hemolytic uremic syndrome, in which red blood cells are destroyed and the kidneys fail.
Clostridium botulinum	Causes about 100 cases per year of botulism, with a 50% fatality rate if untreated. Grows in improperly home-canned foods, more rarely in commercially canned low-acid foods and improperly stored leftovers such as baked potatoes. Spores occur in honey, endangering infants under 2 years of age.	Botulinum toxin from growing bacteria causes progressive paralysis, with blurred vision, drooping eyelids, slurred speech, difficulty swallowing, and muscle weakness. Infant botulism causes lethargy and impaired muscle tone, leading to paralysis.
Listeria monocytogenes	Grows in animals without causing symptoms. Animal feces may contaminate water, which is then used to wash vegetables. Transmission occurs mainly through vegetables washed in contaminated water and through soft cheeses. Listeria is psychrotrophic, growing at refrigeration temperatures.	Listeriosis involves fever, muscle aches, and sometimes GI symptoms. In pregnant women, symptoms may be mild but lead to serious complications for the unborn child.
Shigella	Infects about 18,000 people a year in the United States; in developing countries, Shigella infections are endemic in most communities. Transmission occurs through fecal-oral contact or from foods washed in contaminated water.	Shigellosis involves GI symptoms such as diarrhea, fever, and stomach cramps, usually lasting 7–10 days. Complications are rare.
Staphylococcus aureus	Best known as the cause of skin infections transmitted through open wounds. However, can also be transmitted through high-protein foods such as ham, dairy products, and cream pastries.	S. aureus causes toxic shock syndrome. Can also cause food poisoning via preformed toxins.
Toxoplasma gondii	A parasite believed to infect 60,000 people annually, most with no symptoms. In a few cases, serious disease results. Transmitted through contact with feces of infected animals, particularly cats, or through contaminated foods such as pork.	Toxoplasmosis causes mild flu-like symptoms, but in pregnant women its transmission to the unborn child can lead to severe neurological defects, including death. Neurological complications also occur in immunocompromised patients.
Vibrio vulnificus	A free-living marine organism that contaminates seafood or open wounds. About 200 cases per year are reported.	V. vulnificus can infect the bloodstream, causing septic shock. Threatens mainly people with preexisting conditions such as liver disease.

[a]Ten major food-borne pathogens highlighted by the US Public Health Service, in order of importance.

Figure 27.25 *Salmonella enterica* **Outbreak from Contaminated Peanut Butter**

(*n* = 696 for whom information was reported as of April 20, 2009)

A. Infected individuals reported to the CDC from September 1, 2008, through April 20, 2009.

Humans with illness

Jar 1 of peanut butter

Jar 2 of peanut butter

B. Electrophoretic separation of restriction-digest DNA fragments from a bacterial strain isolated from humans with illness. The peanut butter in jar 1 contained the same strain of *Salmonella enterica*; the peanut butter in jar 2 contained a different strain.

distinctive immune response.) The strain was identified by analysis of its genomic DNA cleaved by restriction endonucleases.

After cleavage, fragments of the *Salmonella* DNA samples from infected patients were separated by pulsed-field gel electrophoresis (see Figure 27.25B). The pulsed field optimizes separation of the largest fragments. The distance each fragment moves is visualized as a band in the gel. The band pattern, or "fingerprint," of *Salmonella* DNA from infected patients showed fragment lengths matching those of *Salmonella* DNA from the peanut butter sample (labeled "Jar 1" in Figure 25.27B). The band pattern from peanut butter sample 1 was different from that of another contaminated peanut butter sample (labeled "Jar 2"); the bacterium in sample 2 proved to be unrelated to this *Salmonella* outbreak.

This case illustrates several troubling features of food contamination in modern society. It shows the consequence of a food production plant's failure to follow regulations and the failure of health inspectors to enforce them. The contaminated product shipped out to a diverse array of institutions, such as schools, and to secondary producers, such as cookie manufacturers, who incorporated the peanut butter ingredient. The bacteria then remained viable in contaminated food products for many months, sickening people long after the contamination event had occurred.

Food contamination events, such as the peanut butter *Salmonella* contamination, commonly lead to point-source epidemics, in which each infection is traced directly to a common source. By contrast, a **person-to-person epidemic** occurs when one infected individual infects one or more others (person-to-person transmission). Factors that affect the spread of disease (the organism, reservoir, mode of transmission, and host) are collectively known as the **chain of infection** (discussed in Chapter 26). An example of a person-to-person epidemic would be a norovirus outbreak, in which contaminated food can sicken a few people, who then spread the infection to others.

Food Safety

In our everyday lives, what can people do to avoid illness from food-borne pathogens? We can practice **food safety** in many ways. The most important food safety practice is washing our hands and utensils before handling food. In addition, we should always avoid eating undercooked animal foods, and check the standard temperatures required to kill pathogens during cooking. Such standards are devised through research by public health agencies, most notably the Centers for Disease Control and Prevention (CDC). The CDC has published guidelines for food safety that we all can practice (Table 27.4).

Food Preservation

Cultural practices and cuisines have long evolved so as to limit food spoilage. Such practices include cooking (heat treatment), addition of spices (chemical preservation), and fermentation (partial microbial digestion). In modern commercial food production, methods to prevent spoilage and contamination are based on fundamental principles of physics and biochemistry that limit microbial growth (discussed in Chapter 6).

PHYSICAL MEANS OF PRESERVATION Specific food preservation processes based on temperature, pressure, or other physical factors include:

- **Dehydration and freeze-drying.** Removal of water prevents microbial growth. Water is removed either by application of heat or by freezing under vacuum (known as freeze-drying or lyophilization). Drying is especially effective for vegetables and pasta. The disadvantage of drying is that some nutrients are broken down.

Table 27.4
Ten Dangerous Food Safety Mistakes—from the CDC

You can protect your family by avoiding these common food safety mistakes

Mistake #1: Eating risky foods if you are more susceptible to food poisoning

Why It's a Mistake: Anyone can get food poisoning. But some people are more likely to get sick and to have a more serious illness. People at greater risk include:

- Adults aged 65 and older
- Children younger than 5 years
- People who have health problems or who take medicines that suppress the body's immune system
- Pregnant women

Solution: People who are more likely to get food poisoning should not eat the following:

- Undercooked or raw animal products (such as meat, chicken, turkey, eggs, or seafood)
- Raw or lightly cooked sprouts
- Unpasteurized (raw) milk and juices
- Soft cheese (such as queso fresco), unless it is labeled as made with pasteurized milk

Mistake #2: Not washing your hands

Why It's a Mistake: Pathogens on your hands can get on food and make it unsafe.

Solution: Wash hands the right way—for 20 seconds with soap and running water. Wash hands before, during, and after preparing food; before eating; and after using the toilet or changing a child's diaper.

Mistake #3: Washing meat, chicken, or turkey

Why It's a Mistake: Washing raw meat, chicken, turkey, or eggs can spread pathogens. Pathogens can hide on your sink, countertops, and other surfaces in your kitchen. Those pathogens can be transferred to other foods, such as salads or fruit.

Solution: Don't wash meat, chicken, turkey, or eggs. Cooking them thoroughly will kill harmful germs.

Mistake #4: Peeling fruits and vegetables without washing them first

Why It's a Mistake: Fruits and vegetables may have pathogens on their peel or skin. It's easy to transfer those pathogens to the inside of fruits and vegetables when you cut or peel them.

Solution: Wash all fruits and vegetables under running water even if you're going to peel them. Use a clean vegetable brush to scrub firm fruits and vegetables like melons, avocados, and cucumbers.

Mistake #5: Putting cooked meat back on a plate that held raw meat

Why It's a Mistake: Pathogens from the raw meat can spread to the cooked meat.

Solution: Always use separate plates for raw meat and cooked meat. The same rule applies to chicken, turkey, and seafood.

Mistake #6: Not cooking meat, chicken, turkey, seafood, or eggs thoroughly

Why It's a Mistake: Cooked food is safe only after it's been cooked to a high enough temperature to kill bacteria.

Solution: Use a food thermometer to make sure you cook food to a safe internal temperature.

- 145°F for whole cuts of beef, pork, veal, and lamb (then allow the meat to rest for 3 minutes before carving or eating)
- 160°F for ground meats, such as beef and pork
- 165°F for all poultry, including ground chicken and turkey
- 165°F for leftovers and casseroles
- 145°F for fresh ham (raw)
- 145°F for seafood, or cook until flesh is opaque

If you won't be serving hot food right away, keep it hot (at 140°F or above) until serving.

Mistake #7: Eating raw batter or dough, including cookie dough, and other foods with uncooked eggs or uncooked flour

Why It's a Mistake: Flour and uncooked eggs may contain *Escherichia coli*, *Salmonella*, or other harmful bacteria.

Solution: Cook or bake flour and eggs thoroughly. Don't eat foods that contain raw or undercooked eggs, such as runny eggs, or homemade mayonnaise, hollandaise sauce, and eggnog. Don't eat raw (uncooked) dough or batter that contains either flour or eggs. Wash hands, work surfaces, and utensils thoroughly after contact with flour and raw dough.

Mistake #8: Tasting or smelling food to see if it's still good

Why It's a Mistake: You can't taste, smell, or see the germs that cause food poisoning. Tasting only a tiny amount can make you very sick.

Solution: Check a storage-time chart to see how long you can store food safely. When the time is up, throw it out.

(continued on next page)

Table 27.4

Ten Dangerous Food Safety Mistakes—from the CDC (*continued*)

Mistake #9: Thawing or marinating food on the counter

Why It's a Mistake: Pathogens can multiply very quickly at room temperature.

Solution: **Thaw food safely.** You can thaw it in the refrigerator, in cold water, or in the microwave. Always marinate food in the refrigerator, no matter what kind of marinade you're using.

Mistake #10: Leaving food out too long before putting it in the fridge

Why It's a Mistake: Harmful germs can grow in perishable foods (such as meat, chicken, turkey, seafood, eggs, cut fruit, cooked rice, and leftovers) if you leave them out of the refrigerator 2 hours or longer.

Solution: **Put perishable foods in the refrigerator within 2 hours** or within 1 hour if the food is exposed to a temperature over 32°C (90°F), such as in a hot car. Divide roasts and large portions of food into smaller containers so they will chill quickly.

Source: Modified from: https://www.cdc.gov/foodsafety/ten-dangerous-mistakes.html.

- **Refrigeration and freezing.** Refrigeration temperature (typically 2°C–4°C, or 35°F–39°F) slows microbial growth, as shown in an experiment comparing bacterial growth in ground beef at different temperatures (Figure 27.26). However, refrigeration also selects for psychrotrophs, such as *Listeria*. Freezing halts the growth of most microbes, but preexisting contaminant strains often survive to grow again when the food is thawed. This is why deep-frozen turkeys, for example, can still cause *Salmonella* poisoning, especially if the interior is not fully thawed before roasting.

- **Controlled or modified atmosphere.** Food can be packed under vacuum or stored under atmospheres with decreased oxygen or increased CO_2. Controlled atmospheres limit abiotic oxidation, as well as microbial growth. For example, CO_2

Figure 27.26 Bacterial Growth in Ground Beef

The growth rate of total aerobic bacteria in ground beef declines at lower storage temperatures. CFUs = colony-forming units.

storage is particularly effective for extending the shelf life of apples.

- **Pasteurization.** Invented by Louis Pasteur, pasteurization is a short-term heat treatment designed to decrease microbial contamination with minimal effect on food value and texture. For example, milk is commonly pasteurized at 72°C (162°F) for 15 seconds followed by quick cooling to 4°C (39°F). Pasteurization is most effective for extending the shelf life of liquid foods with consistent, well-understood microbiota, such as milk and fruit juices.

- **Canning.** In canning, the most widespread and effective means of long-term food storage, food is cooked under pressure to attain a temperature high enough to destroy endospores (typically 121°C, or 250°F). Commercial canning effectively eliminates microbial contaminants, except in very rare cases. The main drawback of canning is that it incurs some loss of food value, particularly that of nutrients that decay fast, such as vitamins, as well as loss of desirable food texture and taste.

- **Ionizing radiation.** Exposure to ionizing radiation, known as food irradiation, effectively sterilizes many kinds of food for long-term storage. The main concerns about food irradiation are its potential for unknown effects on food chemistry and the hazards of the irradiation process itself for personnel involved in food processing. Nevertheless, irradiation is highly effective at eliminating pathogens that would otherwise cause serious illness—for example, pathogenic *E. coli* strains in ground beef.

CHEMICAL MEANS OF PRESERVATION Many kinds of chemicals are used to preserve foods. Major classes of chemical preservatives include:

- **Salt (NaCl).** Salting is an ancient means of preserving meat and vegetables. Salt decreases water activity, causing microbes to lose water by osmosis. So, high salt inhibits microbial growth.

- **Sugar.** High sugar content also decreases water activity and inhibits microbial growth. Sugar is commonly added to fruit as part of the canning process.

- **Acids.** While microbial fermentation can preserve foods by acidification, an alternative approach is to add acids directly. Organic acids commonly used to preserve food include benzoic acid, sorbic acid, and propionic acid. The acids are generally added as salts: sodium benzoate, potassium sorbate, sodium propionate. These acids act by crossing the cell membrane in the protonated form (at low external pH) and then releasing their protons at the higher intracellular pH. For this reason, they work best in foods that already have moderate acidity (pH 5–6), such as dried fruits and processed cheeses.

- **Other organic compounds.** Numerous organic compounds, both traditional and synthetic, have antimicrobial properties. For example, cinnamon and cloves contain the benzene derivative eugenol, a potent antimicrobial agent.

- **Inorganic compounds.** Besides salt (NaCl), inorganic food preservatives include phosphates, nitrites, and sulfites. Nitrites are added to meat products such as salamis and hot dogs, whereas sulfites are added to dried fruits and wine. These minerals inhibit aerobic respiration of bacteria, and their effectiveness is enhanced at low pH. However, they may have harmful effects on humans; nitrites can be converted to toxic nitrosamines, and sulfites cause allergic reactions in some people.

- **Food spoilage** refers to chemical changes that render food unfit for consumption. Food spoils through degradation by enzymes within the food, through spontaneous chemical reactions, and through microbial metabolism.

- **Different foods spoil in different ways.** Dairy products can be soured by excessive fermentation or made bitter by bacterial proteolysis. Meat and poultry are putrefied by decomposition of amino acids. Fish and other seafood harbor psychrotrophic bacteria that grow under refrigeration. Vegetables spoil by excess growth of bacteria and molds.

- **Food contamination, the presence in food of microbial pathogens, leads to food poisoning.** Food poisoning is human disease caused by foodborne pathogens or the presence of toxins produced by microbial growth. Food harvesting, processing, and shared consumption are all activities that spread pathogens.

- **Food safety** refers to everyday practices that people can undertake to avoid food poisoning, such as washing hands and avoiding undercooked animal products.

- **Food preservation** includes physical treatments such as freezing and canning, as well as the addition of chemical preservatives such as benzoates and nitrites.

Thought Question 27.5 If you were a food safety regulator, which pathogen in Table 27.3 would you consider your top priority? Defend your answer by discussing factors such as numerical incidence of infection, severity of disease, and economic losses due to illness.

27.6 Wastewater Treatment

What is the fate of all the food we eat—and where do our wastes go? Human society generates tremendous amounts of organic wastes, from human sewage as well as industrial effluents. These wastes are rich in organic compounds containing reduced carbon and nitrogen, such as short-chain fatty acids and urea. Introduction of these compounds into an ecosystem disrupts the ecological balance, leading to eutrophication and dead zones, as discussed in Section 27.1.

The Water Cycle

Untreated human and industrial wastes eventually enter the water cycle (or hydrologic cycle; **Figure 27.27**). Water evaporates from lakes and is carried up from underground by trees, which release the water to the air (transpiration). Atmospheric water condenses in clouds and then precipitates as rain or snow. The precipitated water flows down to lakes and the oceans, carrying wastes from human activity.

In water, bacteria quickly consume the organic molecules as food, combining them with oxygen. But oxygen diffuses slowly through water. When the waste is highly concentrated, the bacteria oxidize it faster than the oxygen is replaced. Microbial respiration then competes with respiration by fish, invertebrates, and amphibians for the limited supply of oxygen dissolved in water, raising the BOD. High BOD can cause blooms of algae or cyanobacteria. Algal blooms lead to die-off of fish and other aquatic animals and form dead zones, as discussed in Section 27.1. To prevent such disruption, human and industrial wastes must undergo treatment before being released into the ecosystem.

Water Treatment

Two common approaches to community wastewater treatment are **wastewater treatment plants** and **wetland filtration**. Both approaches depend on microbes to remove organic carbon and nitrogen from water before it returns to aquatic systems and

Figure 27.27 The Hydrologic Cycle Interacts with the Carbon Cycle

B. Bottled water samples are measured for dissolved oxygen over time; the rate of decrease of dissolved oxygen indicates biochemical oxygen demand (BOD).

A. The hydrologic cycle carries bacteria and organic carbon into groundwater and aquatic systems.

ultimately the oceans. However, more specialized polluters require additional, specialized treatments. For example, hospital wastes include large amounts of pathogens and dangerous objects such as needles. Particularly dangerous wastes may require more drastic methods, such as incineration.

In industrialized nations, all municipalities use some form of wastewater treatment plant that includes microbial metabolism (**Figure 27.28**). The purpose of wastewater treatment is to

decrease the BOD and the level of human pathogens before water is returned to local rivers. The wastewater treatment plant is the final destination for all household and industrial liquid wastes passing through the municipal sewage system. It may come as a surprise to learn that a modern wastewater plant can convert sewage to water that exceeds all government standards for humans to drink—and in many cities treated sewage is, in fact, returned to the source of drinking water.

A typical plant includes the following treatment stages:

- **Preliminary treatment** consists of screens that remove solid debris, such as sticks, dead animals, and hygiene items.
- **Primary treatment** includes fine screens and sedimentation tanks that remove insoluble particles. The particles eventually are recombined with the solid products of wastewater

Figure 27.28 Wastewater Treatment Plant

A. In a typical municipal treatment plant, wastewater undergoes primary treatment (filtering and settling), secondary treatment (microbial decomposition), and tertiary treatment (chlorination or other chemical treatments).

B. Aeration basin for secondary treatment.

Figure 27.29 Microbes in Wastewater Treatment

Stalked ciliate

A. *Nocardia* species, filamentous bacteria, from flocs formed during secondary treatment.

B. Flocs with a stalked ciliate, which preys on bacteria.

treatment (known as sludge). The sludge ultimately is used for fertilizer or landfill or is incinerated.

- **Secondary treatment** consists primarily of microbial ecosystems that decompose the soluble organic content of wastewater. Two phases of secondary treatment are aerobic and anaerobic digestion. In aerobic digestion, the microbes form particulate flocs (aggregate particles) of biofilm.

- **Tertiary treatment** (advanced wastewater treatment) includes chemical applications such as chlorination or ozone or UV treatment to eliminate pathogens.

The microbial ecosystems of aerobic digestion require continual aeration to maximize the microbes' aerobic respiration, which breaks down pollutants to carbon dioxide and nitrates. The microbes typically include bacilli such as *Pseudomonas*, as well as filamentous species such as *Nocardia* (**Figure 27.29A**). Optimal treatment depends on the ratio of filamentous to single-celled bacteria: enough filaments to hold together the flocs for sedimentation, but not so many that they trap air and cause flocs to float and foam, preventing sedimentation. The flocs are sedimented as sludge, also known as **activated sludge**, referring to their microbial activity.

Besides bacteria, the ecosystem of activated sludge includes filamentous methanogens, which metabolize short-chain molecules such as acetate within the anaerobic interior of flocs. In addition, the bacteria are preyed upon by protists such as stalked ciliates (**Figure 27.29B**), swimming ciliates, and amebas, as well as invertebrates such as rotifers and nematodes. These predators serve a valuable function of limiting the free-floating population of bacteria, enabling the bulk of the biomass to be removed by sedimentation.

An important phase of secondary treatment is anaerobic digestion. Anaerobic digestion occurs without oxygen; it involves anaerobic respiration and fermentation. Fermentation produces hydrogen and carbon dioxide gases, which methanogens convert to large quantities of methane. In some cases, the methane can be harvested as natural gas for fuel.

Wastewater treatment plants are remarkably effective at converting human wastes to ecologically safe water and ultimately human drinking water. The plants are, however, impractical for purifying the runoff from large agricultural operations. For large-scale alternatives to treatment plants, communities and agricultural operations are looking to **wetland restoration**. Much of our current water supply is filtered and purified by natural wetlands, such as the Florida Everglades (**Figure 27.30A**). And agricultural operations replace treatment plants with constructed wetlands. **Figure 27.30B** shows a hog farm where a series of terraced wetlands was built to drain liquefied manure. The wetlands were found to produce fewer odors and to remove organics more efficiently and at a lower cost than do traditional filtration plants.

SECTION SUMMARY

- **In the water cycle**, water precipitates from the atmosphere into rivers, lakes, and oceans, where it enters living organisms. Organisms expel water containing wastes. The water ultimately evaporates, thus returning to the atmosphere.

- **Wastewater treatment** involves primary treatment (filtration and sedimentation), secondary treatment (microbial decomposition of organic matter), and tertiary treatment (chemical or UV treatment of microbes).

- **Secondary treatment includes aerobic and anaerobic digestion.** Aerobic digestion requires aeration and converts organic matter to carbon dioxide. Anaerobic digestion occurs without aeration and generates methane, as well as carbon dioxide.

- **Wastewater treatment cuts down BOD.** Microbial communities decompose the soluble organic content that causes algal blooms and dead zones.

- **Wetlands filter water naturally.** Wetland filtration helps purify groundwater entering aquifers.

Thought Question 27.6 What would happen if wastewater treatment lacked microbial predators? Why would the result be harmful?

Figure 27.30 Water Filtration by Wetlands

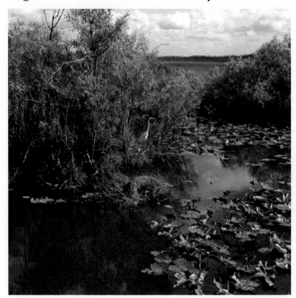

A. The marshes of the Everglades act as natural filters for the aquifers of southern Florida.

B. A filtering system installed by Steve Kerns on his hog farm in Taylor County, Iowa. A series of hillside terraces forms wetlands containing bacteria that purify hog manure and wastewater.

Perspective

The function of our entire biosphere depends on our microbial partners to cycle key elements and acquire energy to drive the food web. Microbes colonize a vast array of habitats, from the deserts and oceans to the roots of plants and the digestive tracts of animals. Wherever found, microbes both respond to and modify the environment that surrounds them. They provide key ecosystem services, such as biomass production and waste filtration. More directly, we use bacteria and fungi to make fermented foods, from cheese to champagne. On a global scale, we depend on microbial participation in biogeochemical cycles. Yet today, for the first time in history, human technology rivals the ability of microbes to perturb those cycles. In effect, our planet is a patient in need of better health management (**Figure 27.31**). To manage and moderate our alterations—for our own survival and that of the biosphere—our fate still depends on the microbes.

Figure 27.31 Planet Earth Isn't Well

Key Terms

activated sludge (p. 910)
algal bloom (p. 883)
amensalism (p. 890)
assimilation (p. 881)
biochemical oxygen demand (BOD) (p. 883)
biomass (p. 882)
bioremediation (p. 880)
casein (p. 897)
chain of infection (p. 905)
commensalism (p. 889)
consumer (p. 881)
curd (p. 897)
dead zone (p. 885)
decomposer (p. 882)

dissimilation (p. 882)
ecosystem (p. 880)
ecosystem service (p. 880)
eutrophic (p. 883)
fermented food (p. 896)
food contamination (p. 902)
food poisoning (p. 902)
food preservation (p. 902)
food safety (p. 905)
food spoilage (p. 902)
food web (p. 882)
grazer (p. 882)
greenhouse effect (p. 893)
methemoglobinemia (p. 895)
microcystin (p. 883)

mutualism (p. 889)
niche (p. 880)
nitrification (p. 895)
nitrifier (p. 895)
parasitism (p. 889)
person-to-person epidemic (p. 905)
point-source epidemic (p. 903)
population (p. 880)
predator (p. 882)
preliminary treatment (p. 909)
primary producer (p. 881)
primary treatment (p. 909)
psychrotroph (p. 902)
reservoir (p. 892)

rhizosphere (p. 886)
secondary treatment (p. 910)
sink (p. 893)
symbiosis (p. 889)
synergism (p. 889)
tertiary treatment (p. 910)
thermocline (p. 884)
trophic level (p. 882)
wastewater treatment plant (p. 908)
wetland (p. 886)
wetland filtration (p. 908)
wetland restoration (p. 910)
yogurt (p. 897)
zoonosis (p. 891)

Objectives Assessment Questions

27.1 Microbes in Ecosystems

1. A habitat in which the major phototrophs consist of the smallest unicellular bacteria is _____.
   a. forest
   b. ocean
   c. desert
   d. wetland

2. Which of the following is not a function of wetland microbes?
   a. Decomposing and recycling dead animal and plant bodies
   b. Filtering nitrogenous organic wastes from human wastewater
   c. Providing mineral nutrients for vascular plants
   d. Production of oxygen gas by *Prochlorococcus*

3. Biochemical oxygen demand is increased by _____.
   a. excessive bacterial nitrogen fixation
   b. excessive organic substrates for microbial aerobic respiration
   c. sulfur pollution involving oxidation
   d. aquatic nitrates and nitrites

27.2 Microbial Symbiosis with Animals

4. A relationship in which one partner benefits while the other partner receives neither benefit nor harm is called _____.
   a. mutualism
   b. synergism
   c. commensalism
   d. parasitism

5. Which statement is not true of the bovine rumen?
   a. The rumen supports many species of bacteria, archaea, and protozoa.
   b. The rumen is well supplied with oxygen from the breathing animal.
   c. Rumen bacteria catabolize plant polysaccharides that the bovine enzymes cannot otherwise digest.
   d. Methanogens convert CO_2 and H_2 (products of bacterial fermentation) to methane.

6. Which of the following bacteria does/do not contribute positively to human health?
   a. *Shigella flexneri*
   b. *Bacteroides* species
   c. *Staphylococcus epidermidis*
   d. *Bifidobacterium* species

27.3 Biogeochemical Cycles

7. Which part of the global carbon cycle is performed only by microbes?
   a. Respiration of organic compounds, releasing CO_2
   b. CO_2 fixation into biomass
   c. Fermentation, releasing organic acids
   d. Methanogenesis, reducing CO_2 to methane

8. Which part of the global nitrogen cycle is performed by many plants and animals in addition to microbes?
   a. Nitrogen gas fixation into ammonium ion
   b. Ammonium ion incorporation into amino acids
   c. Oxidation of ammonia to form nitrite and nitrate (nitrification)
   d. Reduction of nitrate to nitrous oxide by anaerobic respiration (denitrification)

9. Which form of nitrogen is a potent greenhouse gas?
   a. Nitrogen gas (N_2)
   b. Nitrous oxide (N_2O)
   c. Nitrate (NO_3^-)
   d. Nitrite (NO_2^-)

27.4 Fermented Foods and Beverages

10. A soft cheese in which milk proteins are coagulated solely by bacterial lactic acid fermentation is _____.
    a. cottage cheese
    b. Gouda cheese
    c. Swiss cheese
    d. Brie cheese

11. Which of the following is not a step of beer production?
    a. Germination of barley grains
    b. Mashing the grains in water
    c. Removal of toxic ethanol
    d. Filtration to remove the yeast

12. Cheese provides exceptional food value because _____.
    a. it is fermented by nitrogen-fixing bacteria
    b. bacterial fermentation breaks down all the organic molecules
    c. bacterial fermentation breaks down mainly sugars, leaving most protein intact
    d. fermentation adds valuable minerals to the food product

27.5 Food Spoilage and Preservation

13. Which process is especially common in seafood spoilage?

   a. Carbohydrates are fermented to acids or alcohols by indigenous microbes.

   b. Lipids spoil by oxidation of unsaturated fatty acids, causing rancidity.

   c. Sugars rapidly ferment to lactic acid, turning the food sour.

   d. Anaerobic respiration by bacteria converts trimethylamine oxide to trimethylamine, causing a distinctive unpleasant odor.

14. Which of these food preservation methods requires subsequent cold storage?

   a. Ionizing radiation

   b. Canning

   c. Pasteurization

   d. Packaging under CO_2 atmosphere

15. Which microbial pathogen can grow in food under refrigeration?

   a. *Salmonella enterica*   **c.** *Campylobacter jejuni*

   b. Norovirus   **d.** *Listeria monocytogenes*

27.6 Wastewater Treatment

16. Which process of the water cycle leads to increased BOD in a freshwater lake?

   a. Water condenses in clouds.

   b. Water precipitates as rain or snow.

   c. Agricultural runoff enters streams.

   d. From the lake, water evaporates in the atmosphere.

17. Which stage of water treatment does <u>not</u> involve microbes?

   a. Primary treatment by screens and sedimentation

   b. Secondary treatment by aerobic respiration

   c. Secondary treatment by <u>an</u>aerobic respiration

   d. Floc sedimentation as sludge

18. Which microbial process decreases the BOD of wastewater?

   a. Nitrogen fixation

   b. Aerobic or anaerobic respiration

   c. Lithotrophy

   d. Photosynthesis

Review Questions

1. What unique functions do microbes perform in ecosystems?

2. What kinds of microbial metabolism are favored in aerated environments? In anoxic environments?

3. Can a microbe be both a producer and a consumer? Explain.

4. Explain the microbial relationships in various forms of symbiosis, including mutualism, commensalism, and parasitism. Outline an example of each, detailing the contributions of each partner.

5. Compare and contrast the marine food web with the soil food web. What kinds of organisms are the producers and consumers? How many trophic levels are typically found?

6. Explain how the carbon cycle differs in oxygenated and anoxic environments.

7. Describe the various roles of microbes in the nitrogen cycle. Explain which roles are unique to microbes.

8. What are the advantages of fermentation for a food product? How does acidic fermentation contribute to the formation of different cheeses? What is the role of different kinds of metabolism performed by different microbial species?

9. Compare and contrast acidic and ethanolic fermentation processes. What different kinds of foods are produced by each type of fermentation?

10. Explain the difference between food spoilage and food poisoning.

11. What are the most important food-borne pathogens, based on infection rates? Based on mortality rates?

12. What are the major means of preserving food? Compare and contrast their strengths and limitations.

13. Explain the role of biochemical oxygen demand (BOD) in water quality and how it may be perturbed by human pollution.

14. Outline the function of a wastewater treatment plant. Include the phases of primary, secondary, and tertiary treatment. Explain the roles of microbes in these phases of water treatment.

Clinical Correlation Questions

1. A community is cited by the EPA for discharging excess nitrogen into the local river. The community decides to construct a wetland system to filter its wastewater. Over a period of years, the wetland grows a dense array of plants and attracts wildlife. The water that flows out of the wetland into the stream shows low organic carbon and nitrogen content—well below the acceptable EPA limits. Explain why this system enhances the health of the human community.

2. Shortly after takeoff, six passengers in a plane fall ill with diarrhea and vomiting. These passengers are all members of a tour group that has shared dining experiences. The passengers vomit into air sickness bags and have to travel through the aisle to the restroom with their air sickness bags and diarrhea. The source of their illness is confirmed to be norovirus. Subsequently, over the next 12 days after the flight, seven passengers who were not members of the tour group report illness that points to norovirus. Does this case represent a point-source epidemic, a chain of infection, or both? Explain.

Figure A1.1 The Periodic Table of the Elements

Main-group elements

Transitional elements

Main-group elements

Legend: Metals · Nonmetals · Metalloids · Noble gases

Callouts: Atomic number (number of protons) · Chemical symbol · Atomic mass (average of all isotopes)

Group	1 / 1A	2 / 2A	3 / 3B	4 / 4B	5 / 5B	6 / 6B	7 / 7B	8	9 (8B)	10	11 / 1B	12 / 2B	13 / 3A	14 / 4A	15 / 5A	16 / 6A	17 / 7A	18 / 8A
1	1 H 1.00794																	2 He 4.00260
2	3 Li 6.941	4 Be 9.01218											5 B 10.811	6 C 12.011	7 N 14.0067	8 O 15.9994	9 F 18.9984	10 Ne 20.1797
3	11 Na 22.9898	12 Mg 24.3050											13 Al 26.9815	14 Si 28.0855	15 P 30.9738	16 S 32.066	17 Cl 35.4527	18 Ar 39.948
4	19 K 39.0983	20 Ca 40.078	21 Sc 44.9559	22 Ti 47.88	23 V 50.9415	24 Cr 51.9961	25 Mn 54.9381	26 Fe 55.847	27 Co 58.9332	28 Ni 58.693	29 Cu 63.546	30 Zn 65.39	31 Ga 69.723	32 Ge 72.61	33 As 74.9216	34 Se 78.96	35 Br 79.904	36 Kr 83.80
5	37 Rb 85.4678	38 Sr 87.62	39 Y 88.9059	40 Zr 91.224	41 Nb 92.9064	42 Mo 95.94	43 Tc (98)	44 Ru 101.07	45 Rh 102.906	46 Pd 106.42	47 Ag 107.868	48 Cd 112.411	49 In 114.818	50 Sn 118.710	51 Sb 121.76	52 Te 127.60	53 I 126.904	54 Xe 131.29
6	55 Cs 132.905	56 Ba 137.327	57 La 138.906	72 Hf 178.49	73 Ta 180.948	74 W 183.84	75 Re 186.207	76 Os 190.23	77 Ir 192.22	78 Pt 195.08	79 Au 196.967	80 Hg 200.59	81 Tl 204.383	82 Pb 207.2	83 Bi 208.980	84 Po (209)	85 At (210)	86 Rn (222)
7	87 Fr (223)	88 Ra 226.025	89 Ac 227.028	104 Rf (261)	105 Db (262)	106 Sg (263)	107 Bh (262)	108 Hs (265)	109 Mt (266)	110 Ds (281)	111 Rg (280)	112 Cn (285)	113 Nh (286)	114 Fl (289)	115 Mc (289)	116 Lv (293)	117 Ts (294)	118 Og (294)

Lanthanide series

58 Ce 140.115	59 Pr 140.908	60 Nd 144.24	61 Pm (145)	62 Sm 150.36	63 Eu 151.965	64 Gd 157.25	65 Tb 158.925	66 Dy 162.50	67 Ho 164.930	68 Er 167.26	69 Tm 168.934	70 Yb 173.04	71 Lu 174.967

Actinide series

90 Th 232.038	91 Pa 231.036	92 U 238.029	93 Np 237.048	94 Pu (244)	95 Am (243)	96 Cm (247)	97 Bk (247)	98 Cf (251)	99 Es (252)	100 Fm (257)	101 Md (258)	102 No (259)	103 Lr (260)

Symbol	Name	Symbol	Name	Symbol	Name	Symbol	Name	Symbol	Name
Ac	Actinium	Cu	Copper	Fe	Iron	Os	Osmium	Ag	Silver
Al	Aluminum	Cm	Curium	Kr	Krypton	O	Oxygen	Na	Sodium
Am	Americium	Ds	Darmstadtium	La	Lanthanum	Pd	Palladium	Sr	Strontium
Sb	Antimony	Db	Dubnium	Lr	Lawrencium	P	Phosphorus	S	Sulfur
Ar	Argon	Dy	Dysprosium	Pb	Lead	Pt	Platinum	Ta	Tantalum
As	Arsenic	Es	Einsteinium	Li	Lithium	Pu	Plutonium	Tc	Technetium
At	Astatine	Er	Erbium	Lv	Livermorium	Po	Polonium	Te	Tellurium
Ba	Barium	Eu	Europium	Lu	Lutetium	K	Potassium	Tn	Tennessine
Bk	Berkelium	Fm	Fermium	Mg	Magnesium	Pr	Praseodymium	Tb	Terbium
Be	Beryllium	Fl	Flerovium	Mn	Manganese	Pm	Promethium	Tl	Thallium
Bi	Bismuth	F	Fluorine	Mt	Meitnerium	Pa	Protactinium	Th	Thorium
Bh	Bohrium	Fr	Francium	Md	Mendelevium	Ra	Radium	Tm	Thulium
B	Boron	Gd	Gadolinium	Hg	Mercury	Rn	Radon	Sn	Tin
Br	Bromine	Ga	Gallium	Mo	Molybdenum	Re	Rhenium	Ti	Titanium
Cd	Cadmium	Ge	Germanium	Mc	Moscovium	Rh	Rhodium	W	Tungsten
Ca	Calcium	Au	Gold	Nd	Neodymium	Rg	Roentgenium	U	Uranium
Cf	Californium	Hf	Hafnium	Ne	Neon	Rb	Rubidium	V	Vanadium
C	Carbon	Hs	Hassium	Np	Neptunium	Ru	Ruthenium	Xe	Xenon
Ce	Cerium	He	Helium	Ni	Nickel	Rf	Rutherfordium	Yb	Ytterbium
Cs	Cesium	Ho	Holmium	Nh	Nihonium	Sm	Samarium	Y	Yttrium
Cl	Chlorine	H	Hydrogen	Nb	Niobium	Sc	Scandium	Zn	Zinc
Cr	Chromium	In	Indium	N	Nitrogen	Sg	Seaborgium	Zr	Zirconium
Co	Cobalt	I	Iodine	No	Nobelium	Se	Selenium		
Cn	Copernicium	Ir	Iridium	Og	Oganesson	Si	Silicon		

Figure A2.1 Embden-Meyerhof-Parnas Pathway (Glycolysis)

EMP Stage 1: Glucose → 2 G3P

α-Glucose

1 ATP phosphorylates glucose to glucose 6-phosphate.

hexokinase

ATP
ADP

α-Glucose 6-phosphate

2 Glucose 6-phosphate isomerizes to fructose 6-phosphate.

phosphoglucose isomerase

α-Fructose 6-phosphate

3 ATP phosphorylates fructose 6-phosphate to fructose 1,6-bisphosphate.

phosphofructokinase

ATP
ADP

β-Fructose 1,6-bisphosphate

4–5 Cleavage produces G3P and DHAP, which isomerizes to another G3P.

aldolase

triose phosphate isomerase

Dihydroxyacetone phosphate (DHAP)

Glyceraldehyde 3-phosphate (G3P)

EMP Stage 2: 2 G3P → 2 Pyruvate

Glyceraldehyde 3-phosphate

glyceraldehyde 3-phosphate dehydrogenase

NAD⁺, Pᵢ
NADH + H⁺

6 Each G3P loses 2 e⁻ and is phosphorylated to 1,3-bisphosphoglycerate.

1,3-Bisphosphoglycerate

phosphoglycerate kinase

ADP
ATP

7 Phosphoryl group is lost to ATP, forming 3-phosphoglycerate.

3-Phosphoglycerate + H⁺

phosphoglycerate mutase

8 3-Phosphoglycerate isomerizes to 2-phosphoglycerate.

2-Phosphoglycerate + H⁺

enolase

H_2O

9 H_2O is lost, forming phosphoenolpyruvate (PEP).

Phosphoenolpyruvate (PEP) + H⁺

pyruvate kinase

ADP
ATP

10 Phosphoryl group is lost to ATP, forming pyruvate.

Pyruvate + H⁺

Figure A2.2 Entner-Doudoroff Pathway of Glucose Catabolism

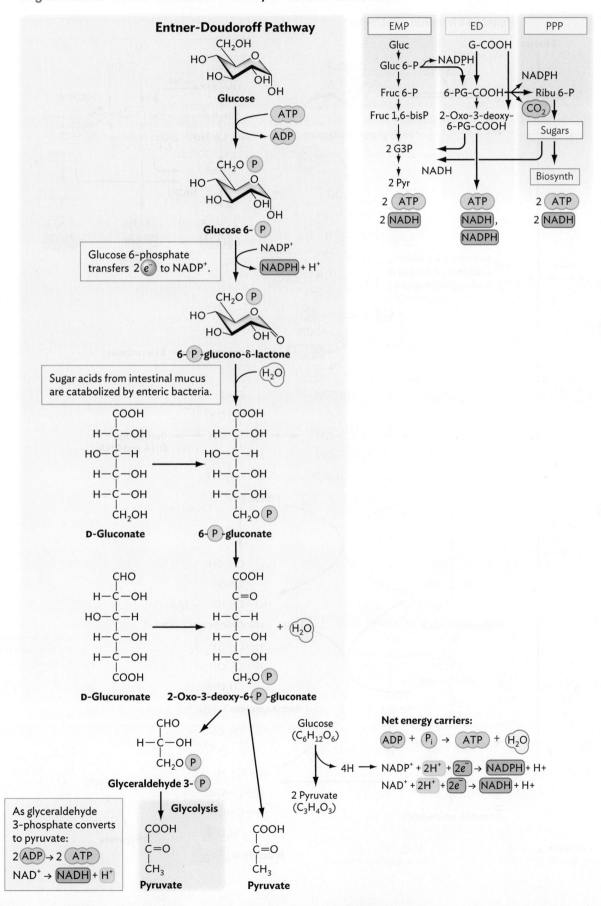

Figure A2.3 Pentose Phosphate Pathway of Glucose Catabolism

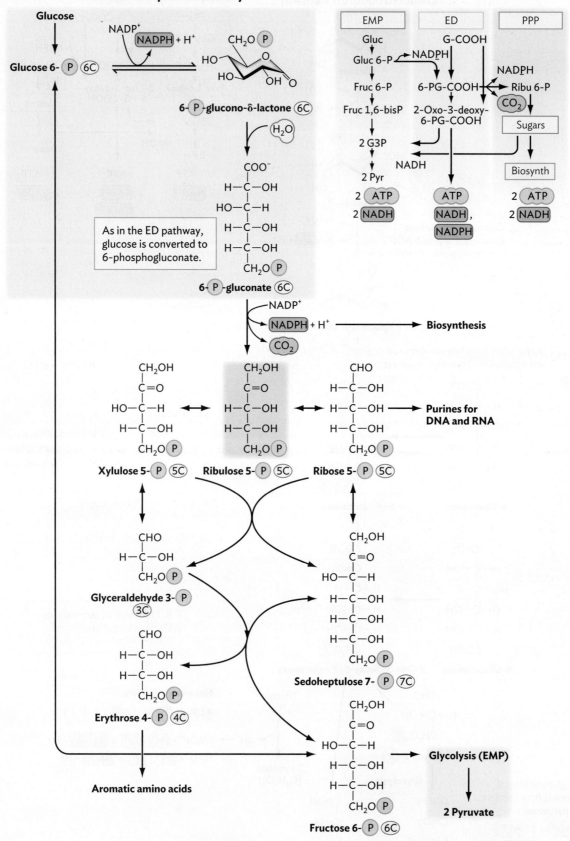

Pentose Phosphate Pathway

Figure A2.4 The Tricarboxylic Acid (TCA) Cycle

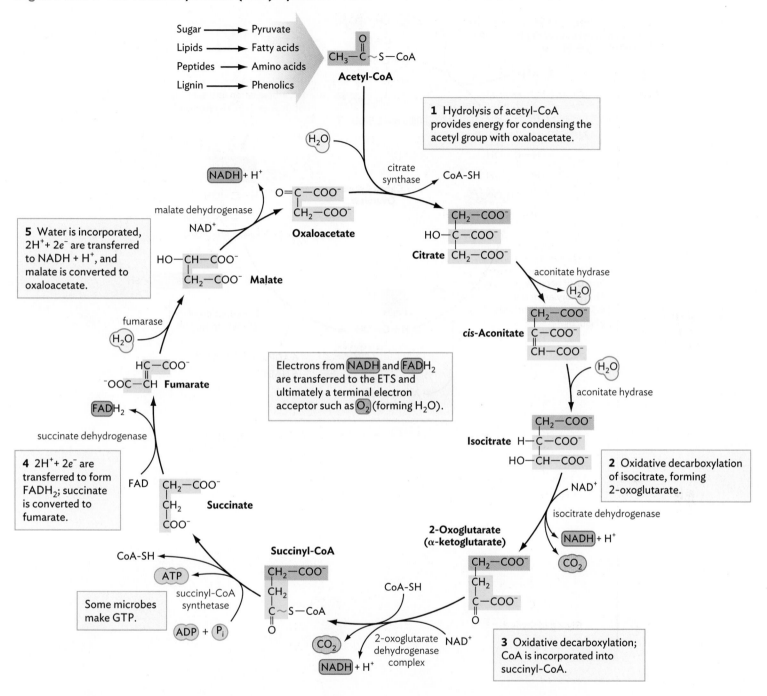

Sugar ⟶ Pyruvate
Lipids ⟶ Fatty acids
Peptides ⟶ Amino acids
Lignin ⟶ Phenolics

Acetyl-CoA

1 Hydrolysis of acetyl-CoA provides energy for condensing the acetyl group with oxaloacetate.

citrate synthase

CoA-SH

Oxaloacetate

Citrate

malate dehydrogenase

NAD⁺

NADH + H⁺

5 Water is incorporated, 2H⁺ + 2e⁻ are transferred to NADH + H⁺, and malate is converted to oxaloacetate.

Malate

aconitate hydrase

cis-Aconitate

aconitate hydrase

fumarase

Fumarate

Isocitrate

Electrons from NADH and FADH₂ are transferred to the ETS and ultimately a terminal electron acceptor such as O₂ (forming H₂O).

2 Oxidative decarboxylation of isocitrate, forming 2-oxoglutarate.

isocitrate dehydrogenase

NAD⁺

NADH + H⁺

CO₂

succinate dehydrogenase

FADH₂

FAD

4 2H⁺ + 2e⁻ are transferred to form FADH₂; succinate is converted to fumarate.

Succinate

2-Oxoglutarate (α-ketoglutarate)

CoA-SH

ATP

succinyl-CoA synthetase

Some microbes make GTP.

ADP + Pᵢ

Succinyl-CoA

CO₂

NADH + H⁺

2-oxoglutarate dehydrogenase complex

CoA-SH

NAD⁺

3 Oxidative decarboxylation; CoA is incorporated into succinyl-CoA.

Figure A2.5 Calvin-Benson Cycle of Carbon Dioxide Fixation

3 Phosphorylation of ribulose 5-(P) to ribulose 1,5-bis-(P)

1 Carboxylation and splitting of (6C) → 2 (3C) by Rubisco

2 Reduction of RCOO⁻ to RCOH

Calvin Cycle: Overview

Ribulose 1,5-bis-(P)

Ribulose 5-(P)

(Sugar-intermediates)

Five G3P become phosphorylated; one G3P enters biosynthesis of glucose.

3-(P)-glycerate (PGA)

Glyceraldehyde 3-(P) (6 G3P)

Amino acids

Glucose

Figure A2.6   The Mitochondrial Electron Transport System (ETS)

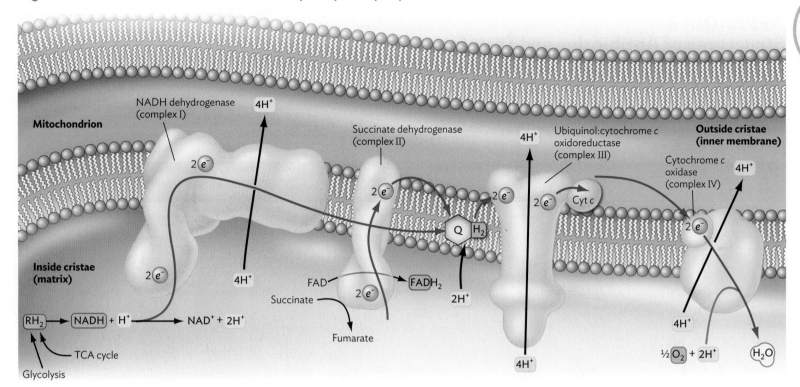

APPENDIX 3
Structures of Antibiotics and Disinfectants

Figure A3.1 Structures of Some Common Disinfectants, Antiseptics, and Surfactants

Refer to Section 13.3.

Phenolics	Alcohols	Aldehydes	Quaternary Ammonium Compounds	Gases
Phenol*	Ethanol	Formaldehyde*	Cetylpyridinium chloride	Ethylene oxide*
Hexachlorophene	Isopropanol (rubbing alcohol)	Glutaraldehyde*	Benzalkonium chloride (mixture)	Betapropiolactone*

*= Disinfectant; not to be used on skin or other living tissues.

Figure A3.2 The Structures of Penicillins

Basic structure   **D-Alanine-D-alanine**

A. The beta-lactam ring of all penicillin drugs resembles a part of the peptidoglycan structure (D-alanine-D-alanine of the side chain).

6-Aminopenicillanic acid

B. 6-Aminopenicillanic acid is the core penicillin molecule from which derivatives are made.

C. Derivatives of penicillin that vary in the R group, as highlighted by a yellow box.

Penicillin G   **Ampicillin**

Carbenicillin   **Methicillin**   **Oxacillin**

Figure A3.3 Structure of Cycloserine and Vancomycin

A. Cycloserine, an analog of D-alanine, is one of several drugs used to treat tuberculosis.

B. Vancomycin is a cyclic glycopeptide made by *Amycolatopsis orientalis*. This antibiotic is synthesized by an exceedingly complex biochemical pathway. Me = methyl.

Figure A3.4 Rifamycin

Rifampin (a member of the rifamycin group) binds to the RNA polymerase beta subunit to halt transcription.

Figure A3.5
Protein Synthesis Inhibitors

Gentamicin

A. The aminoglycoside gentamicin.

Doxycycline

B. The tetracycline doxycycline.

Erythromycin D

C. The macrolide antibiotic erythromycin D.

Clindamycin

D. The lincosamide antibiotic clindamycin.

Chloramphenicol

E. Chloramphenicol.

Linezolid

F. The oxazolidinone linezolid.

Figure A3.6
Examples of Antifungal Agents

Clotrimazole

A. Clotrimazole belongs to the group of imidazole antifungals, so named because they all contain an imidazole ring.

Griseofulvin

B. Griseofulvin is produced by *Penicillium griseofulvum*.

Nystatin

C. Nystatin is a polyene macrolide produced by *Streptomyces noursei*.

Amphotericin B

D. Amphotericin B is a polyene produced by *S. nodosus*.

[Answers]

Chapter 1

THOUGHT QUESTIONS

1.1 Human eyes cannot see individual microbes without magnification. Furthermore, microbes are so prevalent in the environment that even if they had been seen, it would have been hard to tell which ones cause disease and which do not. By the late nineteenth century, microbiologists were discovering ways to distinguish microbial traits, such as the Gram-negative and Gram-positive cell wall structures. The sequencing of microbial DNA has vastly improved our ability to define the precise causative agents of disease.

1.2 To demonstrate that the virus causes influenza, it is necessary to culture the virus. However, the virus must be cultured together with cells to infect, such as human cells in tissue culture. Moreover, aspects of the influenza replication cycle and transmission must be studied in nonhuman animals such as guinea pigs. Using guinea pigs, it is possible to obtain influenza virus from the sick animals only (postulate 1), to replicate the virus in tissue culture (postulate 2), and then to show that the reisolated virus can infect guinea pigs again (postulate 3), producing influenza virus again (postulate 4).

1.3 We do not fully understand why some pathogens generate an effective, long-term immune response whereas others do not. Pathogens that readily evade the immune system usually have genomes that can change rapidly, enabling the microbe to display different surface proteins that the immune system does not yet recognize. Pathogens whose genomes mutate the most rapidly (such as influenza virus and HIV) are the most challenging to develop effective vaccines against.

1.4 Microbes from deep in Earth's crust would have evolved and adapted to conditions very different from those of the human body, such as the absence of oxygen and the scarcity of organic molecules. It is unlikely that such microbes could grow in a human body. Microbes that cause disease generally have evolved for many generations in intimate association with humans or with animals closely related to humans.

OBJECTIVES ASSESSMENT QUESTIONS

1.1 From Germ to Genome: What Is a Microbe?

1. **a.** A nucleus with a nuclear membrane
2. **d.** They have never been shown to cause disease.
3. **a.** A genome that undergoes replication

1.2 Microbes Shape Human History

4. **c.** Catherine of Siena
5. **d.** Louis Pasteur
6. **c.** AIDS

1.3 Medical Microbiology and Immunology

7. **b.** Both charts represent the rise and fall of an infectious disease.
8. **c.** When the microbe is introduced into a healthy, susceptible host, the same disease occurs.
9. **a.** Antibiotics reproduce themselves to form more antibiotics.

1.4 Microbes in Our Environment

10. **b.** Lithotrophs
11. **c.** Cyanobacteria
12. **b.** Nitrogen fixation

1.5 The DNA Revolution

13. **d.** Rosalind Franklin
14. **c.** Claire Fraser
15. **c.** Polymerase chain reaction (PCR) amplification of DNA

CLINICAL CORRELATION QUESTIONS

1. The tests must have revealed no sign of bacteria, and the symptoms are typical of viral pneumonia. Antibiotics do not affect viruses, but only bacterial pathogens. Therefore, antibiotics would not be prescribed; instead, the patient is offered supportive therapy that minimizes the damage caused by the virus and helps the body recover.

2. Koch's first postulate states that if a microbe is the causative agent of disease, it is not found in healthy hosts. Many types of *E. coli* bacteria, however, grow normally in the human intestine. If the particular strain is a known pathogen (such as *E. coli* O157:H7) or an emerging strain not commonly found, it may turn out to be the cause of the patient's intestinal illness.

Chapter 2

THOUGHT QUESTIONS

2.1 Because it takes fewer cells to cause disease, the microbe with the smaller LD_{50} (5×10^4) is the more virulent one.

2.2 *S.* Typhi is host adapted to infect only humans. It will not infect other hosts, such as mice, because the bacterium either cannot attach to host cells in other animals (the right receptors are missing) or is unable to counter their immune mechanisms. *S.* Typhimurium has a broader host range but lacks some of the tools that *S.* Typhi uses to cause severe disease in humans.

2.3 The goal of any microbe is to maintain its species. If a microbe does not have an opportunity to easily spread to a new host, killing its host is tantamount to suicide.

2.4 The key in this case is that you know blood should not harbor any microorganisms. So, the question is not whether you always find *E. coli* in someone with septicemia, but rather, do you ever find a case of septicemia when there is no bacteria in the patient's blood? The answer is no. Yet the blood of anyone with clear signs of septicemia will contain some species of bacteria. The species of pathogen can be different for different cases.

2.5 Mosquitoes that bite humans do not transfer blood from one person into another; they transmit only mosquito saliva. HIV does not replicate in mosquito cells and is quickly inactivated. In contrast, yellow fever virus can replicate in mosquito cells and can survive for long periods. Another important factor is that once mosquitoes are fed, they do not immediately bite another person. HIV will be inactivated long before another meal is needed. Yellow fever virus, however, will replicate in the insect, so it can be found in saliva and can be transmitted.

2.6 Various factors can account for the fact that only 20% of your contacts became ill. The main factors involve the underlying health and immune status of each person. Most of your contacts will have been "exposed," but some individuals' immune systems clear the pathogen more effectively than others. In addition, some of your contacts might use hand sanitizers liberally.

OBJECTIVES ASSESSMENT QUESTIONS

2.1 Normal Microbiota versus Pathogens

1. **c.** Virulence
2. **b.** Immune avoidance
3. **b.** The infectious dose 50% is 1,000 organisms.

2.2 Basic Concepts of Disease

4. nosocomial infection
5. **a.** sign
6. A secondary infection is an infection that follows a primary infection. Secondary infections are typically caused by an organism different from the one that caused the primary infection.
7. Mortality
8. **c.** the illness phase

2.3 Infection Cycles and Disease Transmission

9. **c.** Animal reservoir
10. transplacental transmission or vertical transmission

11. **b.** The immune system

12. endemic

2.4 Disease Portals and Biosafety

13. **e.** Parenteral

14. **d.** Level 4

15. **d.** Negative-pressure room ventilation

2.5 Host Factors in Disease

16. **e.** All of the above

17. **a.** Respiratory

18. immunopathogenesis

19. **g.** All of the above

2.6 Global Change and Emerging Infectious Diseases

20. **d.** Leptospirosis

21. **e.** a and c only

CLINICAL CORRELATION QUESTIONS

1. Answers will vary.

2. Knowing the portal of <u>entry</u> for a particular microbe enables individuals to protect themselves against contracting disease from someone else. Knowing the portal of <u>exit</u> enables a patient with disease to take measures to prevent transmitting the disease to a healthy person. For instance, knowing that the portal of exit for influenza virus is the respiratory tract could lead the infected patient to wear a sterile mask when venturing out.

3. The hide and tissues of an animal that died of anthrax can be riddled with anthrax cells and spores. The hide can be contaminated when anthrax cells or spores become aerosolized from a lung infection or when contaminated soil is disturbed. The farmer's hands probably became contaminated with anthrax when he moved the cattle for burial. The portal of entry for his infection was likely a small, unnoticed cut on his cheek. The farmer must have transferred the spores or cells from his hands to his cheek when he touched his face. The spores germinated and grew, and the immune system walled off the infection, forming the eschar.

4. Many diseases can cause acute and chronic infections, including hepatitis B and C and tuberculosis. Often the disease starts as an acute infection, but as the immune system starts to fight, the pathogen can find ways to limit the immune system's effectiveness and remain viable in some cells (e.g., hepatocytes in the case of hepatitis).

Chapter 3

THOUGHT QUESTIONS

3.1 The part of the wavefront that emerges first travels faster than the portion still in the glass, causing the wavefront to bend toward the surface of the glass. Ultimately, the wave travels in the same direction and with the same speed as it did before entering the glass. The path of the emerging light ray is parallel to the path of the light ray entering the glass and is shifted over by an amount dependent on the thickness of the glass.

This refraction will alter the path of the light beam and decrease the amount of light reaching the microscope's lens.

3.2 In theory, an angle of 90° for the light cone would produce the highest resolution. However, a 90° angle generates a cone of 180°—that is, a cone collapsed flat. Such a "cone" would require the object to sit in the same position as the objective lens—in other words, to have a focal distance of zero. In practice, the cone of light needs to be somewhat less than 180° to allow room for the object and to avoid substantial aberrations (light-distorting properties) in the lens material.

3.3 You can tell whether the cell is out of focus or actually hollow by rotating the fine-focus knob to move the objective up and down while observing the specimen carefully. If the hollow shape appears to be the sharpest image possible, it is probably a hollow cell. If the hollow shape turns momentarily into a sharp, dark cell, it was probably a full cell out of focus before.

3.4 Both dark-field and fluorescence microscopy enable detection (but not resolution) of objects whose dimensions are below the wavelength of light. Dark-field technique is based on light scattering, which detects all small objects without discrimination. The technique is relatively simple and inexpensive. Fluorescence requires more complex optics to separate light at different wavelengths for excitation and emission. The fluorescence technique, however, provides a way to label specific parts of cells, such as cell membrane or DNA, or particular species of microbes by use of fluorescent antibody tags.

3.5 Light microscopy could answer questions such as: What is the overall shape of this cell? Does it form individual cells or chains? Is the organism motile? Only light microscopy can visualize an organism alive. Electron microscopy can answer questions about internal and external subcellular structures. For example, does a bacterial cell possess external structures such as flagella or pili? If the dimensions of the unknown microbe are smaller than the lower limits of a light microscope's resolution, EM may be the only way to observe the organism. Viruses are often characterized by shape, and this shape is observed by electron microscopy.

OBJECTIVES ASSESSMENT QUESTIONS

3.1 Observing Microbes

1. **d.** Bacteria are smaller than the resolution distance of the human retina.

2. **a.** Bright-field light microscopy

3.2 Optics and Properties of Light

3. **b.** The object must be fixed to a slide.

4. **c.** Light bends, or "refracts," as it enters a substance that slows its speed.

3.3 Bright-Field Microscopy

5. **a.** the condenser

6. **b.** the objective lens

3.4 Stained Samples

7. **b.** the ring form and gametocyte stage of the malaria parasite *Plasmodium falciparum*

8. **d.** It requires an antibody stain.

3.5 Advanced Microscopy

9. **a.** fluorescence microscopy, with a fluorophore attached to a cell-surface–specific antibody

10. **d.** electron microscopy

11. **a.** SEM. **b.** Dark-field microscopy. The cell bodies appear "overexposed" by scattered light, but the technique reveals flagella, whose width would be too narrow for resolution. **c.** Gram stain, bright-field microscopy. **d.** TEM of thin section.

CLINICAL CORRELATION QUESTIONS

1. *Salmonella enterica* is likely. Other possibilities would include *Listeria monocytogenes* or *Escherichia coli*.

2. *Mycobacterium tuberculosis*

Chapter 4

THOUGHT QUESTIONS

4.1 NaCl dissolves in the form of Na^+ and Cl^-. The charges on these ions readily interact with the partial charges on water molecules, which are polar. By contrast, vegetable oil consists of a hydrocarbon with no charge and no polarization. Hydrophobic forces cause hydrocarbons to associate with each other but not with water.

4.2 The surface of a membrane-embedded protein has hydrophobic amino acid residues such as leucine and phenylalanine, which can interact favorably with the hydrophobic hydrocarbon chains of phospholipids. Within the middle of the protein, charged and polar side chains of amino acid residues will bind to each other, thus helping hold the protein together.

4.3 The bacteria in your intestine receive the food you eat, which is composed of a huge range of different kinds of materials, especially the complex macromolecules of plants. These diverse materials require enzymes to break down. Each type of enzyme can "fit" only one specific type of substrate, such as starch, cellulose, or pectin (fruit polymer), among thousands of others. To make these enzymes, digestive bacteria evolve genomes with many genes that encode different enzymes with binding sites for different substrates. By contrast, pathogens in your blood encounter a limited range of substrates like simple sugars and peptides. Consequently, the genomes of pathogens usually encode fewer types of digestive enzymes.

4.4 According to the periodic table, the atomic weight of Na is 23 and that of Cl is 35.5. So, 1 mole of NaCl weighs 23 grams + 35.5 grams = 58.5 grams. The weight of 1 mole (58.5 grams) times 0.154 mole per liter gives 9 grams of NaCl to weigh out for 1 liter of saline.

4.5 The hydrolysis of ATP has a high energy of activation; that is, the reaction requires a large input of energy to reach the transition state between ATP and ADP. This activation energy can be decreased by an enzyme that binds ATP in its reaction site. Within a cell, such ATP-hydrolyzing enzymes are regulated to avoid spending all the ATP. Examples of ATP-hydrolyzing enzymes are discussed in Chapter 7.

4.6 When the bacterium *M. tuberculosis* is taken up by phagocytosis, the phagocytic vesicle fuses with a lysosome, which secretes enzymes of host defense. The lysosomal enzymes digest bacterial proteins and oxidize various cell components. *M. tuberculosis* uses several strategies to avoid destruction in the fused phagolysosome. One way is through an exceptionally thick, complex cell wall and envelope. The envelope contains molecules that the lysosomal enzymes cannot digest or oxidize.

OBJECTIVES ASSESSMENT QUESTIONS

4.1 Elements, Bonding, and Water

1. **e.** Helium
2. **a.** Butane (CH_3–CH_2–CH_2–CH_3)
3. **a.** NaCl

4.2 Lipids and Sugars

4. **b.** Hydrocarbon chain
5. **c.** an amino group
6. **d.** Glycoprotein

4.3 Nucleic Acids and Proteins

7. **a.** a thiol (–SH)
8. **b.** Secondary structure
9. **a.** Uracil

4.4 Biochemical Reactions

10. **a.** DNA synthesis
11. **d.** Increase in temperature from 25°C to 35°C
12. **c.** Breaking down a protein

4.5 Membranes and Transport

13. **b.** Hydrocarbon chain
14. **c.** Alanine
15. **c.** A cell placed in distilled water takes up water and expands.

4.6 Endocytosis and Phagocytosis

16. **d.** *Vibrio cholerae*
17. **b.** Phospholipids
18. **d.** Ameba

CLINICAL CORRELATION QUESTIONS

1. A patient with cholera often has multiple biochemical needs because of the effects of cholera or of preexisting malnutrition. Vitamins and sugar may well be needed but will not address the urgent need to restore electrolyte balance. Antibiotics could kill the bacteria, but in this instance the patient's physiological recovery is a higher priority. The top priority is to provide NaCl to replace the ions lost via the transport proteins that were inappropriately activated by the cholera toxin.

2. When *S. enterica* bacteria infect a patient, the bacteria are endocytosed by the cells of the intestinal lining. Within the cell, the bacteria form intracellular membrane vesicles that protect the bacteria from host defenses. By contrast,

V. cholerae bacteria do not enter human cells. Within the intestinal lumen, the bacteria release a toxin that causes excess flux of water from the human epithelium. The water carries most bacteria out of the body into the environment, where they can infect the next host.

3. The molecule, called an "RNA analog," can bind to the active site of RNA-dependent RNA polymerase. The analog prevents RNA synthesis, thus halting replication of influenza virus in the patient.

Chapter 5

THOUGHT QUESTIONS

5.1 In a prokaryotic cell, the molecules that occur in the greatest number (besides water) are inorganic ions (250 million per cell). They are also the smallest molecules. Peptidoglycan and DNA are found in the lowest number, although these macromolecules are large and complex. A prokaryotic cell contains 100 times as many lipid molecules as strands of RNA because lipids are small structural molecules, highly packed. RNA molecules are long macromolecules that are either packed into complexes (such as ribosomal RNA) or found as temporary information carriers (messenger RNA), present only as needed to make proteins.

5.2 The cell envelopes differ greatly for Gram-positive and Gram-negative bacteria. Gram-positive bacteria have thick cell walls, which protect them from some antibiotics and toxins, but they lack the outer membrane of Gram-negative bacteria. The outer membrane contains porins that exclude small molecules, including antibiotics. Both Gram-positive and Gram-negative bacterial envelopes contain drug efflux pumps that exclude antibiotics, but they are different pumps that exclude different types of antibiotics. In addition, Gram-positive and Gram-negative bacteria have diverged deeply through evolution, so their fundamental structures differ in amino acid sequence even where the function is the same. For example, the ATP synthase of *Mycobacterium tuberculosis* is sufficiently distinct to be targeted by drugs that do not affect Gram-negative bacteria.

5.3 The antibiotic killed the Gram-negative bacteria, causing lysis of their cells. The lysed cells released LPS into the blood. LPS is an endotoxin that causes shock.

5.4 Degeneration of protective traits is a common problem for research focusing on microbes that can produce 30 generations overnight. Their rapid reproductive rate gives ample opportunity for spontaneous mutations to accumulate over an experimental timescale. For the S-layer, in a laboratory test tube free of predators or viruses, mutant bacteria that fail to produce the thick protein layer would save energy compared with S-layer synthesizers and would therefore grow faster. Such mutants would quickly take over a rapidly growing population. For flagella, the motility apparatus requires 50 different genes generating different protein parts. Cells that acquire mutations eliminating expression of the motility apparatus gain an energy advantage over

cells that continue to invest energy in motors. In a natural environment, the nonmotile cells lose out in competition for nutrients, despite their energy advantage. In a lab, however, cells are cultured in uniform environments, such as a shaking test tube, where motility confers no advantage. These culture conditions lead to evolutionary degeneration of motility.

5.5 The disadvantage of a contractile vacuole is that it requires a continuous input of energy to bail out the water. On the other hand, the contractile vacuole permits the existence of a cell that is flexible enough to engulf other cells as prey. But a cell wall does not have to spend energy (other than the initial energy to synthesize it). A cell wall is inflexible and does not allow another cell to be engulfed as prey.

5.6 Many human tissue types have cilia—most notably the lining of the respiratory tract. The respiratory cilia contract in waves to push mucus and foreign particles up and out of the airways. In the auditory canal of the ear, very specialized cells called "hair cells" possess cilia that flex in response to sound waves. The one type of human cell that has a flagellum is the spermatozoon (sperm cell). Sperm use the flagellum to swim up the cervix and reach the egg for fertilization.

OBJECTIVES ASSESSMENT QUESTIONS

5.1 The Bacterial Cell: An Overview

1. **d.** Lipopolysaccharide (LPS)
2. **c.** Water molecules
3. **b.** Cell membrane (inner membrane)

5.2 Bacterial Membranes and Transport

4. **a.** Protein synthesis in the cytoplasm
5. **b.** Facilitated diffusion of oxygen across a membrane
6. **a.** TolC efflux complex

5.3 The Bacterial Cell Wall and Outer Layers

7. **a.** Cell wall
8. **d.** Mycolic acids
9. **d.** Outer membrane

5.4 The Nucleoid and Bacterial Cell Division

10. **c.** DNA is contained within a nuclear membrane.
11. **a.** Completion of the septum to separate two daughter cells
12. **c.** A circular molecule with loops bound by DNA-binding proteins

5.5 Specialized Structures of Bacteria

13. **b.** Stalk
14. **c.** Flagella generate a whiplike motion powered by ATP.
15. **a.** Chemotaxis

5.6 The Eukaryotic Cell

16. **d.** Mitochondrion
17. **b.** Peptidoglycan cell wall
18. **b.** Chloroplast

CLINICAL CORRELATION QUESTIONS

1. A skin biopsy could reveal *Staphylococcus* species as a cause of infection, but the cells

would be small, round cocci, without flagella. *Proteus mirabilis* is unlikely to be found in skin. *Mycobacterium leprae* might be found in skin, but the bacteria consist of small, nonmotile rods. *Borrelia burgdorferi*, however, is a narrow spiral-shaped cell called a spirochete. Its internally embedded flagella cause the cell to twist, generating motility. The patient may have borreliosis, also known as Lyme disease.

2. The skin biopsy reveals bacteria that stain acid-fast. Acid-fast rods are usually mycobacteria, either *Mycobacterium tuberculosis* or *Mycobacterium leprae*. The skin lesions with loss of sensation suggest the disease leprosy. The bacterium *M. leprae* is one of the slowest-growing pathogens known; it can take several weeks to grow colonies in culture.

Chapter 6

THOUGHT QUESTIONS

6.1 Blood agar can be considered differential because different species growing on blood have different abilities to lyse the red blood cells in the agar. Some species do not lyse, others completely lyse (secreted hemolysin produces complete clearing around a colony), and still others only partially lyse the red blood cells (secreted hemolysin produces a greening around the colony). As a result, the blood agar will differentiate between hemolytic and nonhemolytic bacteria. Because the medium is very rich and does not prevent the growth of any organism, it is not considered selective.

6.2 The answer is asynchronous growth. Not all cells replicate at the same instant. Even a culture started with one cell will, over time, produce millions of offspring that begin division at slightly different times. The result is a smooth curve.

6.3 After 2 hours, the minimal medium will contain $1,000 \times 2^2 = 4,000$ bacteria, whereas the rich medium will contain only $10 \times 2^{(2h)(3\,div/h)} = 640$ bacteria. After 4 hours, the minimal medium will contain 1.6×10^5 bacteria, whereas the rich medium will have surpassed this count, reaching nearly 4.1×10^5 bacteria.

6.4 On its own, the bacterium cannot break RBC membranes, so X and V factors remain inaccessible. However, during the preparation of chocolate agar, heat treatment is used to lyse RBCs and release their contents. As a result, *H. influenzae* can grow on chocolate agar but not on blood agar.

6.5 At first the growth of the heterotroph might outpace the growth of the autotroph, using the carbon sources faster than the autotroph can make them. As the organic carbon sources diminish through consumption, growth of the heterotroph decreases. However, the CO_2 formed by the heterotroph will allow the autotroph to grow and make more organic carbon. Ultimately, the ecosystem comes into balance.

6.6 Food preservation traditionally includes water exclusion by salt, as seen in hams, back bacon, and salted fish, or by high concentration of sugar, as in canned fruit and jellies. The lower a_w prevents microbial growth. Dehydrating foods with an

evaporating dehydrator will also prevent microbial growth.

6.7 Anaerobes incorporate oxygen from their carbon sources (for example, CO_2 and carbohydrates such as glucose), all of which contain oxygen. This form of oxygen will not damage the cells.

6.8 A synergistic relationship occurs between facultative organisms and anaerobes within a tooth biofilm. The facultative microbes consume oxygen within the biofilm microenvironment, which allows anaerobes to grow underneath them.

6.9 Home canning involves placing fresh fruit or vegetable preparations into a jar and heating the jar in a pressure cooker. The high pressure prevents water from boiling until it reaches 120°C (248°F), a temperature high enough to kill spores. The heat also drives oxygen from the jar. Because *C. botulinum* is found in soil, so are its spores. If these heat-resistant spores contaminate the vegetables being canned and the canning process is incomplete, spores will survive. Once the canned food cools, the spores germinate in the anaerobic environment and the viable bacteria begin secreting toxin. If the food containing toxin is ingested, the unwitting victim will contract botulism.

OBJECTIVES ASSESSMENT QUESTIONS

6.1 Culturing and Counting Bacteria

1. **e.** Plates B and C
2. **c.** Differential medium
3. **c.** 3.5×10^4
4. **b.** Viable count

6.2 The Growth Cycle

5. **b.** 2
6. **f.** 7×10^{11}
7. **a.** Decrease growth rate

6.3 Microbial Nutrition

8. defined minimal medium
9. **c.** autotroph
10. Clockwise from top: N_2, nitrogen fixers, NH_4^+, nitrifiers, NO_3^-, denitrifiers

6.4 Environmental Limits on Microbial Growth

11. barophiles or piezophiles
12. **c.** mesophiles
13. **d.** The growth rate in exponential phase will be slower.
14. **e.** Membranes become too flexible.

6.5 Living with Oxygen

15. **c.** Anaerobes lack enzymes that detoxify oxygen by-products.
16. False
17. False
18. **d.** Intestine
19. **e.** Anaerobes and facultative organisms
20. **d.** Strict anaerobe

6.6 Microbial Communities and Cell Differentiation

21. quorum sensing
22. extracellular polymeric substances (EPS)

23. **b.** Chronic
24. **f.** a, b, and c only

CLINICAL CORRELATION QUESTIONS

1. Brian ate the salad when it was fresh and few live *Salmonella* organisms were present. If the salad was not refrigerated properly or if it was left out for hours at room temperature, the pathogen could have grown to numbers high enough to cause disease in Brianna. (Recall infectious dose 50% from Chapter 2.)

2. The proton pump inhibitor that Padman takes decreases his stomach acidity. *V. cholerae*, which is very sensitive to acid, can survive better in his stomach than in Gagandeep's. The number of organisms that reached Padman's intestine surpassed the infectious dose, so he became ill.

3. The urease produced by *H. pylori* converted enough urea in the medium to ammonia to change the pH of the medium from pH 2 to over pH 5, a pH in which this organism can grow.

4. The pathogen affecting the prosthetic knee (probably *Staphylococcus aureus*) most likely formed a biofilm on the metal. Because of the community nature of the biofilm, some organisms within the biofilm developed transient resistance to the antibiotic, due either to molecular mechanisms or to poor penetration of the antibiotic into the biofilm. The prosthetic may need to be replaced.

Chapter 7

THOUGHT QUESTIONS

7.1 In habitats where free glucose monomers are scarce, there are still many glucose units linked in polymers such as starch and cellulose. If bacteria have enzymes such as amylase to break starch down into monomers, then the glucose can be catabolized by glycolysis or other pathways. Even if glucose is unavailable, enzymes may convert other sugar monomers, such as fructose or galactose, to glucose or other intermediates of glucose catabolism. Finally, it's always possible that glucose (for instance, the blood sugar of an animal host) will suddenly become available to a microbe.

7.2 Many different kinds of sugars can be converted to glucose, which is then fermented to acids. The acids of fermentation provide a simple, inexpensive test by causing a color change in a pH-sensitive indicator dye. The different kinds of sugars are useful because different species of bacteria possess different combinations of enzymes to convert the sugars; many different combinations are possible, enabling us to distinguish many species of bacteria.

7.3 In healthy lung alveoli, *P. aeruginosa* would have ample access to oxygen as a terminal electron acceptor; thus, the pathogen would conduct aerobic respiration. As it grows into a biofilm, oxygen would be used up, and the bacteria could start respiring with nitrate. In the soil, the bacteria would quickly use up any oxygen in competition with neighboring microbes. With oxygen gone, *P. aeruginosa* would soon start expressing oxidoreductases that can transfer electrons to alternative electron acceptors—such as nitrate, if available. Similarly, in the lungs of cystic fibrosis patients, infection by *P. aeruginosa* depletes oxygen, so the pathogen expresses its oxidoreductase to respire using nitrate.

7.4 Pathogens within the lumen of the digestive tract are most likely to synthesize their own amino acids because they compete intensively with other closely related species. As available amino acids are used up, the ability to make one's own amino acids might enable one species to outcompete its neighbors. By contrast, obligate intracellular pathogens are most likely to undergo reductive (degenerative) evolution, losing genes that encode enzymes of amino acid biosynthesis. Intracellular pathogens have an ample supply of amino acids from their host, and they have no competitors. However, there is selective pressure to lose unneeded genes because the supply of DNA monomers from the host is limited.

7.5 Without synthesizing its own fatty acids, *M. genitalium* must grow its membranes by using preformed fatty acids obtained from host cells within the blood. For mycoplasmas cultured in the laboratory, the composition of the membrane will vary, depending on which fatty acids are provided in the growth medium.

OBJECTIVES ASSESSMENT QUESTIONS

7.1 Energy for Life

1. **b.** fermentation
2. **d.** carbon dioxide
3. **a.** Combustion

7.2 Catabolism: The Microbial Buffet

4. **a.** Cellulose
5. **c.** It is an electron donor formed during catabolism.
6. **c.** The pentose phosphate pathway

7.3 Glucose Catabolism, Fermentation, and the TCA Cycle

7. **c.** Glycolysis has a cycle of intermediates that incorporate a carbon skeleton and regenerate the original intermediate.
8. **a.** Acetyl-CoA

7.4 Respiration and Lithotrophy

9. **c.** The oxidoreductase directly phosphorylates ADP to make ATP.
10. **b.** *Geobacter* catabolizes small organic molecules and donates electrons to an electrode in a fuel cell.
11. **d.** Anaerobic respiration reducing nitrate to nitrite

7.5 Photosynthesis and Carbon Fixation

12. **a.** Cyanobacteria
13. **d.** CO_2 fixation via the Calvin cycle
14. **a.** Dinoflagellates in red tide

7.6 Biosynthesis and Nitrogen Fixation

15. **d.** all of the above
16. **b.** glutamate

CLINICAL CORRELATION QUESTIONS

1. The enteric pathogen formed orange colonies on Hektoen agar. It was thus identified as *Escherichia coli*.

2. The physician should test immediately for the presence of saxitoxin, a toxic substance produced by red tide algae.

Chapter 8

THOUGHT QUESTIONS

8.1 Bacteria that are growing rapidly need to replicate their chromosomes in time for the cell to divide. If DNA replication is blocked, then the cells may try to divide while the two copies of the chromosome are not yet complete. If they do, the cells will lose their DNA as they divide and will die. On the other hand, when cells form a biofilm, most of them enter a state in which they divide slowly or not at all. The thick layers of cells and the formation of extracellular polysaccharides (EPS) may restrict entry of the antibiotic. Furthermore, cells that are not dividing may be unharmed by an antibiotic that blocks DNA replication.

8.2 Multiple promoters enable a pathogen to respond to different environmental conditions. In particular, a pathogen needs to "know" when it finds itself within a host versus the outside environment. A vector-borne pathogen, such as the plague bacillus *Yersinia pestis*, needs to have promoters that receive signals of at least three environments: the human host, the flea host, and the outside (soil and air). Each environment requires the pathogen to express different proteins. For example, within the human host, signals such as elevated temperature and blood iron cause *Y. pestis* to secrete toxins that would be useless outside.

8.3 The *dtxR* gene specifies a transcriptional repressor. If the *dtxR* gene is deleted, the expression of diphtheria toxin may become constitutive (turned on continuously). Constitutive expression of the toxin may decrease fitness of the bacterium under conditions outside the host body.

8.4 One way to become resistant to an antibiotic is for the genes encoding the target structure, such as the ribosome, to acquire a mutation resulting in an altered structure. The altered target structure now fails to bind the antibiotic. Another way to become resistant is for a cell to acquire a plasmid encoding an enzyme that cleaves the antibiotic. A third possibility is that a plasmid can encode a protein complex that exports the antibiotic out of the cell.

8.5 One possibility is that a large segment of DNA could be inverted within a bacterial chromosome. Within the inverted region, most of the genes would be intact and could still be expressed. Although the map of the bacterial genome would be profoundly altered, the bacterium's phenotype (observed traits) might show little change.

OBJECTIVES ASSESSMENT QUESTIONS

8.1 Bacterial Genomes

1. **a.** The genome is composed of double-helical DNA.
2. **c.** When a plasmid is transferred into a new cell, it usually changes the species of the host organism.
3. **b.** Intron

8.2 DNA Replication

4. **a.** An RNA primer is synthesized.
5. **c.** to generate negative supercoils in DNA
6. **b.** to connect Okazaki fragments

8.3 Gene Expression: From DNA to RNA and Protein

7. **c.** Messenger RNA (mRNA)
8. **b.** The incoming aminoacyl-tRNA binds the A site of the ribosome.
9. **a.** The first formylmethionyl-tRNA binds to the mRNA.

8.4 Regulation of Gene Expression

10. **d.** A translation initiation sequence binds a translational repressor.
11. **b.** DNA phase variation occurs.
12. **d.** A two-component regulator transmits information from an external signal to the DNA inside the cell.

8.5 Mutation and Repair

13. **b.** Silent mutation
14. **b.** SOS repair
15. **a.** deletion of a base pair

8.6 Biotechnology

16. **a.** Gel electrophoresis
17. **d.** PCR amplification of a bacterial gene
18. **d.** fluorophores (fluorescent molecules)

CLINICAL CORRELATION QUESTIONS

1. Increasingly, clinical isolates of *Enterococcus faecium* show resistance to multiple antibiotics. For this reason *E. faecium* is designated an ESKAPE pathogen. The main tests to perform are MIC (minimum inhibitory concentration) for several antibiotics. Unfortunately, in the United States most clinical isolates of *E. faecium* are now resistant to the former drugs of choice: ampicillin and vancomycin. The antibiotics linezolid and daptomycin are effective for some patients. In some cases, infection may be resolved by synergistic (combination) therapy with daptomycin and tigecycline.

2. Tests such as palpation of the uterus and Gram stain of the discharge could suggest the presence of a sexually transmitted infection. The presence of Gram-negative diplococci suggests *Neisseria gonorrhoeae*. To confirm this identification, the clinician could perform a standard DNA PCR test called the nucleic acid amplification test (NAAT).

Chapter 9

THOUGHT QUESTIONS

9.1 A *recA* mutant can be transformed and transduced, but whether the incoming DNA is maintained depends on the DNA. For example, transformation of plasmids can happen in a *recA* mutant because plasmid replication does not require plasmids to recombine with the chromosome. Transformation of chromosomal DNA cannot take place, however, because the donor DNA cannot be inherited without integrating into the recipient chromosome (requires RecA). The same is true for generalized transduction of chromosomal genes: no horizontal transfer is involved. Specialized transduction, however, can occur because the phage genome includes its own recombinase. Similarly, transposition can occur in a *recA* mutant because the transposon carries its own transposase gene; RecA is not needed.

9.2 The last piece of DNA to transfer from an Hfr cell into an F⁻ cell is the F factor and *oriT*. Because it can take 100 minutes or more to transfer a whole chromosome, only rarely will an entire chromosome transfer from one cell to another before the conjugation bridge breaks. Consequently, most Hfr transfers do not result in transfer of *oriT* and thus cannot initiate conjugation.

9.3 Phage DNA may survive because sometimes the modification enzyme reaches and modifies the foreign DNA restriction site before the restriction enzyme cuts the site. Once methylated, the phage DNA is shielded from restriction, and the methylated molecule replicates unchallenged. In addition, not all phage DNAs have the DNA sequence targeted by a host's particular set of restriction enzymes.

9.4 There are several possible routes. A bacteriophage can package host DNA containing the transposon. Once the DNA enters the recipient via transduction, the transposase can be expressed and mediate the "hop" from donor to recipient DNA. If the donor cell contains a conjugative plasmid, the transposon can hop into the plasmid, and the plasmid can move by conjugation into the recipient. The transposon can then hop into that cell's chromosome. Transformation is also possible. Finally, a few transposons are capable of conjugation themselves.

9.5 Parasites undergo reductive evolution for traits whose functions are provided by the host, such as metabolism of complex food molecules and biosynthesis of amino acids available from the host. The loss of these traits, and the DNA that encodes them, saves energy and thus confers a competitive advantage compared with other members of the parasite species. At the same time, parasites evolve specialized traits that enhance their survival within the host. For example, the malaria parasite has evolved an apparatus that enables it to attach to and penetrate a red blood cell. *Giardia* has evolved a strong flagellum that enables dissemination within the host intestine, as well as molecular defenses against the host immune system.

9.6 There are three bases per codon, so the ORF can encode 400 amino acids. The average molecular weight of an amino acid is 110. Therefore, the hypothetical protein will be approximately 44,000 daltons (44 kDa).

9.7 The bacterium would key out as *Beggiatoa* species. If the cell width were measured as 0.9 μm, however, the identification would proceed down a completely different, wrong track at the first step of the key. This potential for misidentification is one disadvantage of the dichotomous key.

OBJECTIVES ASSESSMENT QUESTIONS

9.1 Bacterial Gene Transfers

1. **b.** transduction

2. **a.** transfer of a multidrug resistance plasmid

3. **b.** to prevent replication of an invading bacteriophage

9.2 Origins and Evolution

4. **d.** Genetically distinct organisms emerge from environments, and they prevail over preexisting organisms.

5. **b.** The gene must acquire mutations at a steady rate over time in all organisms compared.

6. **d.** *Vibrio*

9.3 Natural Selection

7. **c.** The number of bacteria containing the resistance plasmid will decline because replicating the plasmid costs energy.

8. **b.** *T. pallidum* has evolved virulence genes that enable it to take advantage of host cells.

9. **c.** the random mutation that conferred resistance also conferred deleterious effects on growth

9.4 Bioinformatics

10. **c.** *S. pneumoniae* isolates were selected for genome analysis, which revealed the gene functions that were shared and the ones that were distinct to different species.

11. **a.** Two genes within an organism's genome show very similar DNA sequence, but they encode enzymes that act on different substrates.

9.5 Microbial Taxonomy

12. **d.** Two genomes show very different proportions of GC versus AT content.

13. **a.** a combination of DNA-based tests and phenotypic traits

CLINICAL CORRELATION QUESTIONS

1. Pathogens are exposed to changing stressful environments owing to host defenses and antibiotic treatments—conditions expected to favor mutators. Therefore, mutator phenotypes might accelerate the evolution of pathogenic strains by increasing variation of surface antigens to avoid the immune system, as well as by facilitating acquisition of pathogenic determinants and antibiotic resistance.

2. The short answer is that *S.* Typhi underwent reductive evolution to limit its host range. You could generally hypothesize that an ancestor of *S.* Typhi lost one or more genes encoding proteins needed to attach to, or replicate in, nonhuman cells. For instance, some research indicates that the pili of *S.* Typhi include mutations that reduce binding to nonhuman cell receptors. A more recent study also found that a protein injected by many serovars of *Salmonella* into nonhuman macrophage cells is missing from *S.* Typhi. Macrophages are white blood cells that wander through tissues and kill bacteria. To cause disease, pathogenic *Salmonella* must survive and grow within macrophages by injecting proteins that alter macrophage function. *S.* Typhi, however, lacks a protein needed to grow in nonhuman macrophages but is fully able to survive in human macrophages because that protein is not needed.

3. The short answer is to isolate any DNA fragments present in the boy's spinal fluid and sequence them using rapid, next-generation DNA sequencing techniques. A computer then compares the sequence data with the vast database of known, sequenced pathogens. The agent in this case was *Leptospira interrogans*, a spirochete that usually causes kidney infections but can spread to the brain to cause encephalitis. Infected animals secrete the organism in the urine, and humans who drink contaminated water or who have open wounds can become infected. The team of scientists sequenced 3 million fragments of DNA over 2 days. Software developed by Dr. Charles Chiu set aside any sequences matching human DNA and focused on what was left. Within about 90 minutes, the computer determined the answer. DNA sequences from *Leptospira* were found. The antibiotic penicillin was administered, and Joshua's brain swelling almost immediately began to subside.

Chapter 10

THOUGHT QUESTIONS

10.1 Climate change certainly influences bacterial diversity. As some environments warm and others cool, the microbes suited to those conditions will shift as well. As new opportunities arise for gene exchange, new competitions will dictate a need to alter biochemical pathways, and some species will disappear while others will emerge. The exact impact of climate change on pathogen evolution is unknown, but the natural engineering and reengineering of genes in environmental species responding to climate change could produce the raw material needed to evolve new pathogens.

10.2 *M. tuberculosis* has porins in its envelope that enable nutrients to enter, albeit slowly. The bacterium grows slowly in a place where it has no competitors—namely, intracellularly within the macrophage.

10.3 Being restricted to an intracellular lifestyle would severely limit opportunities for cross-species gene exchange. The likelihood of two bacterial species infecting the same host cell is small. However, the human pathogen *Legionella pneumophila* infects amebas that harbor several other symbiotic bacterial species that have donated DNA sequences to this organism. Whether a species such as *R. prowazekii* can incorporate DNA from bacteria engulfed and killed by its eukaryotic host is unknown.

10.4 External flagella, such as those of *E. coli*, propel the cell effectively in a dilute watery environment such as a pond. In water, the cell can freely swim and tumble to find a new direction. By contrast, in a viscous solution such as mucus, the cell cannot tumble. The twisting motion of the entire spirochete enables penetration of mucus and other viscous regions of a host organism.

OBJECTIVES ASSESSMENT QUESTIONS

10.1 Bacterial Diversity: A Quick Glimpse

1. **a.** Cyanobacteria

2. **c.** Spirochetes

3. **a.** *Escherichia* and *Pseudomonas*

10.2 Gram-Positive Phyla: Firmicutes and Actinobacteria

4. **d.** *Mycoplasma pneumoniae*, a cause of pneumonia

5. **a.** *Listeria* species, which grow in milk and cheese, even in the refrigerator

10.3 Gram-Negative Phyla: Proteobacteria and Bacteroidetes

6. **b.** *Nitrobacter* species, growing in soil

7. **d.** *Bacteroides*, enteric bacteria that help digest our food but cause abscesses if they escape the intestine and enter surrounding tissues

8. **b.** Lipopolysaccharide

10.4 Phylum Spirochaetes: Twisted-Cell Pathogens and Symbionts

9. **a.** Round, coccoid cells

10. **c.** Termites

11. **a.** Ticks

10.5 Phyla Cyanobacteria and Chlamydiae

12. **d.** Intracellular flagella

13. **b.** Chloroplast

14. **c.** Thick peptidoglycan cell wall

15. **a.** Endospore

16. **a.** *Chlamydia* and *Nitrobacter*

CLINICAL CORRELATION QUESTIONS

1. A Gram stain for the Gram-positive cell wall will reveal whether a patient's sputum contains bacteria such as *Streptococcus pneumoniae*. The mycoplasma, however, has no cell wall, and the virus is not a cell. Mycoplasma or virus would have to be detected by a test for serum antibodies or by a PCR test for the pathogen's DNA.

2. One possible scenario is that an ancestral strain of the new pathogenic strain of *E. coli* received one or more pathogenicity genes via horizontal gene transfer(s) from one or more other strains of *E. coli*. The result would be a new pathogenic strain of *E. coli* with an unusual combination of virulence factors.

3. The disease could be syphilis (caused by *Treponema pallidum* subspecies *pallidum*), Lyme disease (caused by *Borrelia burgdorferi*), or yaws (caused by *Treponema pallidum* subspecies *pertenue*). The lesion of syphilis (acquired by sexual contact) disappears, but later a macular rash of multiple spots appears on other parts of the body; the pathogen can also cross the placenta and infect a fetus. Yaws is caused by a different subspecies of *T. pallidum*; it is transmitted by casual contact but not across the placenta. The original lesion can be followed by multiple lesions distributed over the body. If the disease is Lyme disease, the pathogenic spirochete *Borrelia burgdorferi* can later spread to the joints and the nervous system.

4. Patient B, with a less diverse microbiome, is more likely to experience chronic intestinal distress. The more diverse microbiome of patient A can better prevent colonization by unwanted pathogens. Furthermore, patient B's immune system is not properly trained to ignore changes in the microbiome

composition that might occur over time. The immune system could launch a low-grade inflammatory response to the new members. A microbiome-trained immune system is discussed further in Chapter 14.

5. Tuberculosis is a contagious disease transmitted by aerosolization due to coughing. Patients are isolated in a private room with limited visitor access to decrease the chance that the organism will be transmitted to other people. Visitors and staff who enter the room must wear protective masks, disposable gowns, and gloves.

Chapter 11

THOUGHT QUESTIONS

11.1 Mycelial growth enables coordinated expansion into a food resource, such as a host tissue. The large size of the mycelium enables it to avoid being engulfed by amebas or host macrophages. Single-celled yeasts, however, can disperse readily to new habitats, finding new resources to exploit.

11.2 Amebas engulf bacteria as food. But some bacteria have molecular mechanisms to avoid digestion within the ameba. If the bacteria grow inside the ameba, their respiration may provide nutrients such as vitamins that are useful to the ameba. At the same time, the ameba provides food molecules, and an endosymbiosis may evolve. However, some bacteria, such as *Legionella*, evolve a partnership with amebas but act as infectious agents within host cells such as macrophages. Inhaled water droplets may contain amebas with intracellular *Legionella* bacteria, which can cause legionellosis.

11.3 Flagellar motility is the fastest, enabling rapid propulsion of a protist through a dilute watery habitat. However, if the flagellum gets caught or damaged, the cell cannot move. Cilia provide robust movement through a wide range of environments, including places where there are large particles to "dodge." Cilia also provide a way to generate currents that direct food into the cell's oral groove. However, cilia may be less effective in a highly viscous medium. Pseudopods enable an ameba to crawl through crevices and highly viscous material. Pseudopodial motility is slow, however, and ineffective in open water.

11.4 Natural selection tends to shrink the genome of a parasite because the smaller the genome, the more progeny can be made from a given host, a limited resource. In addition, the host can provide many complex materials that the parasite need not make, such as amino acids and vitamins. At the same time, however, the parasite needs special mechanisms to attach and manipulate the host and evade the host's defenses—mechanisms that require extra genes to encode. Consequently, apicomplexans have evolved the apical complex, which penetrates the host cell, and trypanosomes have evolved a complex series of coat proteins that evade the host's immune system. Genes encoding these coat proteins expand the size of the parasite's genome.

11.5 The *X-Files* episode referred to the worm as a "fluke" that attached to the bile duct of a liver, but the parasite's head was depicted as a "scolex." The scolex is a specialized form with hooks and suckers found on a tapeworm, not a fluke. In addition, the ability of a cutoff piece (a segment) to reproduce is characteristic of a tapeworm, not a fluke.

11.6 The advantage of large size, for a parasite, is that larger organisms can consume more food to produce large numbers of progeny. On the other hand, the larger parasite is more likely to shorten the life span of the host. The smaller the parasite, the greater the number of individuals that can be made with finite resources. So, many kinds of parasites alternate between large and small developmental stages. An extreme example is tapeworms, which may grow to many meters in length, with a large number of independently reproductive segments. The progeny, however, are microscopic, so large numbers can be made.

OBJECTIVES ASSESSMENT QUESTIONS

11.1 Eukaryotes: An Overview

1. **a.** Fungi

2. **d.** Helminths

3. **a.** Dinoflagellates

11.2 Fungi and Microsporidians

4. **b.** *Candida albicans*

5. **a.** Ergosterol biosynthesis

6. **c.** *Stachybotrys*

11.3 Protozoa: From Amebas to Trypanosomes

7. **c.** Foraminiferans

8. **a.** *Balantidium coli*

9. **b.** *Toxoplasma gondii*

10. **a.** *Plasmodium vivax*

11. **d.** Transmission via the fecal-oral route

12. **a.** infant day-care centers

13. **c.** *Leishmania donovani*

11.4 Algae and Dinoflagellates

14. **d.** Flagella for motility

15. **b.** Dinoflagellates

11.5 Invertebrate Parasites: Helminths and Arthropods

16. **b.** Nematode

17. **d.** Pubic lice

18. **c.** Flukes

CLINICAL CORRELATION QUESTIONS

1. The parasite is an ameba, *Entamoeba histolytica*, which engulfs red blood cells by phagocytosis. The amebas are transmitted through contaminated water. The antiparasitic drug paromomycin will eliminate the infection.

2. *Aspergillus* species is the most probable cause. *Aspergillus* may grow as a fungal ball localized in the lung, which may become invasive. Antifungal agents such as itraconazole may be attempted, but often surgical removal is required.

Chapter 12

THOUGHT QUESTIONS

12.1 Antibiotics are molecules that specifically attack bacteria, not human cells. Antiviral agents, however, must inhibit virus production within an infected host cell. Because virus production involves host components, antiviral agents may harm host cells, including cells uninfected by virus. It's hard to find an antiviral agent that successfully blocks viruses without harming the host.

12.2 A virus with a narrow host range is human immunodeficiency virus (HIV). HIV causes AIDS only in humans, not even in chimpanzees, our nearest genetic relative. Poliovirus infects humans and chimpanzees but not nonprimate animals. A virus with a potentially broad host range is influenza virus. Influenza strains can infect different species of animals, but most strains cause disease in only one type of animal (humans, pigs, or birds). On the other hand, rabies virus has a broad host range, infecting many kinds of mammals.

12.3 For a virus, the advantage of a small genome is that the smaller the genome, the greater the number of progeny virions that can be formed from the nucleotides available. The host cell contains a limited supply of nucleotides, which may determine the number of virions made. A smaller genome saves space in a small capsid, and smaller capsids require less energy and resources than larger ones. On the other hand, a large viral genome enables the virus to encode products with many extra functions that help control the host cell. For example, herpesviruses encode products that modulate the host immune system, preventing it from stopping the virus.

12.4 For viruses of animal cells, the crucial first step is for the virus to bind to a host cell surface. Surface binding requires a viral coat protein or spike protein to specifically recognize a receptor protein on the host cell membrane. But the host population may acquire a mutation in which the receptor protein is absent or has a changed peptide structure that no longer binds the viral spike protein. Mutation in the receptor gene can prevent infection by the virus.

12.5 The advantage of genome replication by the host polymerase is that the virus does not need to encode a polymerase in its own genome, so its genome can be smaller. The advantage of the virus encoding its own polymerase is that the viral polymerase can be optimized for the needs of the virus; for example, the viral polymerase can be error prone, generating mutant progeny that escape the immune system. In addition, RNA viruses must encode their own polymerase because the host cells lack any polymerase that synthesizes RNA from an RNA template.

12.6 Swine are known to be an important source of reassorted strains of influenza virus that can cause human pandemics. For example, an avian strain and a human strain can coinfect swine and generate a new strain that infects humans. The infectivity of influenza strains depends on the sequence and binding properties of their envelope proteins. Researchers can study the envelope proteins of influenza strains appearing in swine and predict whether these strains may "escape" into human populations. If a dangerous strain is found, it might be possible to devise a vaccine ahead of an outbreak.

12.7 Both HIV and influenza virus attach to their host cell by their envelope spike proteins binding to host receptors. For HIV-1, the Env protein binds receptors CD4 and CCR5; for influenza, the HA envelope protein binds to host glycoproteins. However, the two viruses enter their host cells by different routes. Influenza virus induces formation of an endocytic vesicle, whose acidification triggers membrane fusion and release of the core contents into the host cytoplasm. HIV virions, however, do not induce endocytosis and do not require acidification. The HIV envelope fuses with the cell membrane to release the core particle into the cytoplasm. In addition, HIV virions can travel between cells via thin tubes of cytoplasm (nanotubules) that connect an infected cell with an uninfected cell or through the merging of two cells to form a syncytium.

OBJECTIVES ASSESSMENT QUESTIONS

12.1 Virus Infection and Viral Genomes

1. **b.** Isolated particles conduct their own metabolism.

2. **d.** Mimivirus has no ribosomes of its own; it must use the host ribosomes.

3. **a.** the range of tissues a virus can infect

4. **a.** phospholipid

12.2 Virus Structure and Diversity

5. **c.** Ebola virus

6. **c.** smallpox virus

12.3 Virus Replication and Culture

7. **a.** binding of a virion to a specific host receptor protein

8. **c.** The capsid of the virion enters the host cytoplasm.

9. **b.** attach the virion to a host cell

12.4 Papillomavirus: DNA Genome

10. **c.** The viral genome integrates into the host genome and causes increased expression of oncogenes.

11. **c.** Host DNA polymerase

12. **d.** shedding of host cells

12.5 Influenza Virus: RNA Genome

13. **d.** asymmetrical

14. **b.** The viral genome is composed of multiple segments that get transcribed by RNA-dependent RNA polymerase and reassort with segments from a coinfecting virus.

15. **a.** Cancer

12.6 Human Immunodeficiency Virus: Reverse Transcription

16. **a.** Virion envelope proteins specifically interact with the CD4 and CCR5 cell-surface proteins of a T lymphocyte.

17. **c.** Oseltamivir, which binds neuraminidase, preventing release of virions as they exit the host cell

18. **b.** An initial burst of virions is followed by a long latent period, during which T cells decline.

CLINICAL CORRELATION QUESTIONS

1. The infectious agent is monkeypox virus, which is endemic in Africa among monkeys and rodents, and is also frequently transmitted to humans. The infection is prevented by smallpox vaccination, but because smallpox has been eliminated in the wild, most people no longer receive vaccination and therefore remain vulnerable to other contagious poxviruses.

2. The pathogen is hepatitis C virus, which shows elevated prevalence in individuals born between 1945 and 1964. Most infected individuals show no symptoms, but hidden damage may occur to the liver, ultimately requiring liver transplant.

Chapter 13

THOUGHT QUESTIONS

13.1 The D-value, the time required to decrease a microbial population to 10% of the original, will still be 5 minutes. D-value does not change with cell number. It will take 5 minutes longer to kill <u>all</u> the organisms, however.

13.2 Construct a death curve by measuring the survival of a known quantity of spores (for example, of *Bacillus stearothermophilus*) after autoclaving for various lengths of time. Spores should be used because they are more resistant to heat than is any vegetative cell. Typically, autoclaves are regularly checked with spore strips that change color once the endospores are no longer viable.

13.3 The MIC = 62.5 μg/ml, the concentration in tube 5, the last tube with no growth. (Tube 5 is a 16-fold dilution of tube 1: 1,000/500/250/125/62.5.)

13.4 One possibility is that the drug's MIC is 100 μg/ml using a tube dilution test but the average attainable concentration of the drug in tissues or blood is only 10 μg/ml. Anything higher leads to toxic side effects in the patient. So, by the Kirby-Bauer test, the drug will inhibit growth of the bacterium, but only at a concentration beyond what the patient can tolerate. However, if the average attainable tissue level of the drug were greater than the MIC, the drug would be effective.

13.5 Members of the human microbiome are also susceptible to antibiotics. Because the microbiome has numerous positive effects on human health, such as honing the immune system and keeping potential pathogens at bay, altering the composition of the microbiome can have serious consequences. One prominent example is the development of a severe *Clostridioides difficile* gastrointestinal infection called pseudomembranous enterocolitis.

13.6 Inhibiting the growth of the pathogen buys time for the host's immune system to generate antibody and produce cytotoxic T cells that can clear the organism. Chapters 15 and 16 discuss the immune response.

13.7 Along with possessing genes whose products synthesize the antibiotic, these organisms have genes whose products either flush the drug from the cell or protect their own sensitive cell targets. Some of these genes have been horizontally transferred to other bacteria and are the source of spreading antibiotic resistance.

13.8 Self-treatment using leftover or shared antibiotics does not ensure adequate treatment for an STI or any other infection. Different STIs may require different antibiotics to effect a cure. Self-treatment with inappropriate antibiotics can also lead to resistant STI microbes or medical complications requiring extended medical care. A physician, however, can prescribe prophylactic antibiotics in certain situations. For instance, rifampin is prescribed for family members of a patient with meningitis.

OBJECTIVES ASSESSMENT QUESTIONS

13.1 Concepts of Sterilization and Disinfection

1. **b.** Disinfection
2. **b.** decimal reduction time
3. **d.** 5×10^5 CFUs/ml

13.2 Physical Agents That Kill Microbes

4. **b.** 121°C, 15 pounds of pressure, 20 minutes
5. False
6. **a.** *Coxiella burnetii*
7. **c.** Gamma irradiation

13.3 Chemical Agents of Disinfection

8. **d.** More data are needed.
9. **c.** high-level disinfectant
10. **e.** c and d
11. **d.** Glutaraldehyde
12. **f.** a and d

13.4 Concepts of Antibiotic Therapy

13a. Tube 4
13b. Not enough information is available to know.
14. **d.** All of the above

13.5 Antibacterial Agents

15. **b.** Aminoglycoside (e.g., gentamicin) / 30S ribosome
16. **d.** Penicillin
17. A: cycloserine; B: bacitracin; C: cephalosporins, vancomycin

13.6 The Challenges of Antibiotic Resistance

18. **a.** A multidrug efflux pump
19. **b.** acts as a decoy for beta-lactamase
20. **b.** Degrade antibiotic
21. **a.** Alter target
22. **a.** Alter target
23. **c.** Modify antibiotic
24. **d.** Pump antibiotic out of cell
25. **b.** Gentamicin, **c.** Oxacillin, and **d.** Ceftriaxone
26. **b.** an altered topoisomerase

13.7 Antiviral Agents

27. **d.** Neuraminidase inhibitors
28. **a.** Viral polymerases are less selective than host polymerases.
29. **c.** Neuraminidase inhibitor
30. **f.** Protease inhibitor
31. **b.** Fusion inhibitor
32. **e.** Viral RNA synthesis
33. **d.** Viral DNA replication

13.8 Antifungal and Antiprotozoal Agents

34. **b.** a triazole or allylamine; **a.** flucytosine; **d.** griseofulvin
35. **b.** miconazole
36. **d.** Amphotericin B
37. **c.** Artemisinin

CLINICAL CORRELATION QUESTIONS

1. The most likely scenario is that a plasmid expressing an ampicillin resistance gene was transferred from *Shigella* to some of your resident *E. coli*. Those strains of *E. coli*, already expressing ciprofloxacin resistance, survived when you were being treated with ampicillin.

2. Increases in BUN and creatinine levels in blood are signs of kidney damage. A damaged kidney does not efficiently filter or excrete urea and creatinine, causing these compounds to accumulate in blood. Amphotericin B is known to exhibit nephrotoxicity, so administering this drug to someone who already has a damaged kidney could lead to even greater damage.

3. The report included antibiotic susceptibility results. Good antibiotic stewardship would be to immediately switch treatment from the very broad-spectrum antibiotic, imipenem, to a narrower-spectrum antibiotic, amoxicillin, to which this organism is susceptible. The reason is to limit the chance that the pathogen or members of the microbiome develop resistance to the broad-spectrum drug. The saying "Why use a bazooka to kill a fly when a flyswatter will do?" also applies.

4. Had Molly received the standard childhood vaccinations, which include *Haemophilus influenzae* type b capsular antigen, the likelihood that her meningitis was caused by *H. influenzae* would be low. The clinician would have placed other Gram-negative rods higher on the differential.

5. Jamal's allergy to penicillin extended to other antibiotics whose structures are quite similar to penicillin. Oxacillin is structurally different enough from penicillin to treat simple penicillin-resistant bacteria but not different enough to fool the immune system. However, the structure of azithromycin, a macrolide antibiotic, is vastly different from penicillin, making it safe for penicillin-sensitive patients to use. For similar reasons, clinicians also avoid using first-generation cephalosporins for these patients. Cephalosporins of third or later generations are also usually safe to use because

their structures diverge far enough from penicillin.

Chapter 14

THOUGHT QUESTIONS

14.1 Facultative organisms living in proximity to the anaerobes will deplete the oxygen in the environment, especially around nooks and crannies (for example, between teeth and gums or in gingival pockets) that would ordinarily prevent anaerobes from growing. These small spaces have limited access to oxygen.

14.2 Immunocompromised individuals with mitral valve prolapse should premedicate with a high dose of amoxicillin or a cephalosporin before even simple dental procedures, such as tooth extractions or cleaning. An <u>immunocompetent</u> patient with mitral valve prolapse, however, need not premedicate for simple dental procedures because premedication has no effect on whether that patient will develop endocarditis. But the immunocompetent patient with mitral valve prolapse should prememdicate if they are about to undergo <u>surgery</u> at mucosal surfaces, like a tonsillectomy.

14.3 *Vibrio vulnificus*, a normal inhabitant of brackish water where oysters are farmed, is a common contaminant of these bivalves. However, it is extremely acid-sensitive. Normal stomach acidity readily kills this pathogen, even if large numbers of the bacteria are ingested. Taking drugs that decrease stomach acidity can allow *V. vulnificus* to survive long enough to reach the intestine, where it can penetrate the epithelium and gain entrance to the circulatory system.

14.4 Part of the answer is that antibiotic treatment lowers the numbers of intestinal microbes that digest nutrients such as complex carbohydrates. Left in the bowel, these carbohydrates set up an osmotic imbalance that draws water out of the intestinal cells. Excess water in the bowel produces loose stools and diarrhea. Bile acids that escape absorption in the small intestine are another factor. These bile acids are usually metabolized by bacteria in the large intestine. When bacterial composition and numbers are altered by antibiotic treatment, unmetabolized dihydroxy bile acids, which are potent secretory agents, contribute to the development of secretory diarrhea in the colon. Finally, though not proven, altering the composition of the microbiome may interfere with interdomain chemical signals that contribute to proper water absorption.

14.5 Bile salts present in the intestine (not on the skin) easily gain access to and destroy cytoplasmic membranes of Gram-positive organisms (unless the organism possesses bile hydrolases). Gram-negative microbes have extra protection in the form of an outer membrane and so can survive better in the intestine.

14.6 The rash was caused by lipopolysaccharide (LPS; also called endotoxin) present in the outer membrane of the Gram-negative bacteria.

The LPS is released from the organism when it dies and can trigger the coagulation system, which produces tiny clots that can block capillaries. When blockage occurs in skin capillaries, the tiny red blotches are seen.

14.7 Normal microbiota can escape through intestinal perforations resulting from gunshot or knife wounds, surgery, or cancer.

14.8 The antibiotic Teresa is taking probably altered her intestinal microbiota, which are now out of balance. This imbalance can lead to diarrhea, although why is not always clear (see Thought Question 14.4 regarding osmotic diarrhea). The bacteria in yogurt can help restore balance and resolve diarrhea (it can also reverse constipation). Live and active yogurt products are made with *Lactobacillus, Bifidobacterium,* and *Streptococcus* species, organisms that can have a probiotic effect on the intestine. Unfortunately, because they are live bacteria, the antibiotics Teresa is taking may also kill these organisms, thereby negating any probiotic effect.

OBJECTIVES ASSESSMENT QUESTIONS

14.1 Human Microbiota: Location and Shifting Composition

1. **b.** Kidney
2. **d.** intestine
3. microbiota or microbiome
4. **d.** All of the above

14.2 Benefits and Risks of a Microbiome

5. **b.** Interfering with pathogen colonization
6. opportunistic pathogen
7. gnotobiotic
8. **d.** thin intestinal walls

14.3 Keeping Microbiota in Their Place

9. **e.** a and b only
10. **b.** Toll-like receptor
11. **c.** Peyer's patch

14.4 Microbiota-Host Communications

12. True
13. True
14. **d.** Factors produced by donor microbiota stimulated production of T cells.
15. **e.** All of the above

14.5 Natural Biological Control of Microbiota

16. **b.** *Bifidobacterium*
17. **b.** phages can evolve to infect phage-resistant bacteria
18. False
19. **b.** increased cases of allergies

CLINICAL CORRELATION QUESTIONS

1. The man avoided drinking or swimming in water that could potentially be contaminated with large numbers of *V. cholerae*. Hence, he most likely ingested only small numbers of this pathogen from other sources. Small numbers of ingested vibrios are easily killed by the acidic stomach. However, the man was treating his indigestion with Prilosec, a proton pump inhibitor. The medication slows stomach acidification, which means that the stomach will have a less acidic pH that *V. cholerae* can survive. As a result, he was at increased risk of developing cholera, even if only small numbers of this pathogen were ingested.

2. The antibiotic resistance plasmid could be transferred by transformation or transduction to a bacterial member of your normal microbiota—one capable of maintaining that plasmid. If you are later treated with the corresponding antibiotic, the drug-resistant bacterial species will survive and grow while other members are killed. When the antibiotic is removed, the antibiotic-resistant strain, now present in large numbers, could limit the growth of returning microbiota and change the balance of competing species.

3. There is a spore-forming bacterium present in mouse feces that cannot be grown in the laboratory. It will survive boiling but not autoclave temperatures. As part of the mouse microbiome, this organism (or set of organisms) stimulates the innate immune system, which balances the ratio of white blood cells. This is a true scenario. Search the Internet for "segmented filamentous bacteria and the immune system."

4. The process is complex, but put simply, you might start by evaluating whether bipolar patients on a diet high in polyunsaturated fatty acids reported improved psychiatric outcomes (mood stabilization). If so, you could use metagenomics to characterize the members of the gut microbiome in bipolar patients versus control patients and correlate those findings with their diets. You would look for species of bacteria that are consistently (more or less) abundant in bipolar patients on high versus low polyunsaturated fatty acid diets. You might then consider isolating the species most abundant in people placed on a diet high in polyunsaturated fatty acids and transplanting that species to a new group of bipolar patients placed on high or low polyunsaturated fatty acid diets. Patients having the transplanted microbes and who eat a high polyunsaturated fatty acid diet might experience the best outcomes.

Chapter 15

THOUGHT QUESTIONS

15.1 One hypothesis is that adaptive immunity developed because of a need to protect offspring. Most invertebrates produce many offspring, so chances that the species could be wiped out by disease are small. They have no need for a memory-based immunity to maintain the species. Vertebrates, in contrast, generally have few offspring and would be highly susceptible to pathogens. Evolution of adaptive immunity would be a potent system of defense for otherwise vulnerable offspring. An alternative hypothesis is that vertebrates possess highly diverse microbiota, whereas invertebrates do not. Vertebrates derive great benefit from harboring thousands of resident microbes. Consequently, vertebrate evolution has probably dampened the innate immune system compared with the lowly clam to allow a microbiome to exist. But that deficit has been balanced with an adaptive immune system that copes with members of its microbiome (as well as pathogens) that periodically breach the innate system.

15.2 Defensins need to be small so that they can penetrate the outer membrane of Gram-negative organisms and the thick peptidoglycan maze of Gram-positive organisms. Defensins do kill normal microbiota and, in fact, are part of what keeps levels of the normal intestinal microbiota in check.

15.3 The heart pumps at a constant force. Vessels can widen (dilate) and narrow (constrict) to change how much blood they can hold (capacity). When vessels constrict, their capacity to hold blood is reduced, allowing more blood to return faster to the heart, from which it is pumped. One consequence of constriction is that blood pressure increases. Conversely, when vessels dilate, their capacity to hold blood is increased, allowing less blood to return to the heart and thus slowing the flow of blood through those vessels. As a result, blood pressure decreases.

15.4 Selectin molecules on the surface of endothelial cells catch neutrophils passing in the bloodstream. Production of selectins increases in response to inflammatory cytokines secreted from a nearby infection site or as a result of coronary artery disease (atherosclerotic vessels) or autoimmune diseases (for instance, lupus). Higher numbers of selectin molecules attract higher numbers of neutrophils to the site of inflammation, which can cause more damage. So, decreasing selectin expression could help limit autoimmune or other inflammatory diseases. Selectin inhibitors are called cell migration inhibitors and have been approved for the treatment of asthma, psoriasis, and COPD (chronic obstructive pulmonary disease).

15.5 One reason is that CD47 present on all host cell surfaces prevents neutrophils from phagocytosing host cells, whether or not the host cell is coated by antibody. CD47 does not inhibit NK cell action. MHC I surface molecules will inhibit NK cell activity, but infected cells often have lower amounts of MHC I on their cell surface, so NK cells can carry out ADCC.

15.6 While in the macrophage, *Shigella* secretes a series of bacterial proteins that interfere with the regulatory cascades needed to activate cytokine gene expression in the host cell. The decrease in cytokine release dampens the inflammatory response.

15.7 The MAMPs recognized by Toll-like receptors are essential parts of the microbes. They cannot be altered sufficiently by mutation to remain functional and yet not be recognized by a TLR.

15.8 The problem in Bobby's case was a defect in TLR signaling. NLR signaling, however, was normal. Activation of either receptor type will activate cytokine production and secretion. Finding low cytokine levels after Bobby's infections suggests that intracellular NLRs were not activated. MAMPS of Gram-positive organisms like *Staphylococcus aureus* and *Streptococcus pneumoniae* will bind mostly surface TLRs rather than intracellular NLRs. Viruses and *Salmonella* are intracellular pathogens whose MAMPs can interact with NLRs even if the TLR pathways are defective.

15.9 The pyrogenic toxins have other effects that compromise and damage the host. The toxins can induce cytokines that damage local host cells or confuse the immune system. These effects can provide the pathogen with nutrients and help it hide from the immune system longer.

15.10 If C3b were constantly made at significant levels, the complement cascade would target host cell membranes. The alternative pathway is constantly active but at very low levels, plus the functional life of C3b is very short (milliseconds) if it does not bind carbohydrates on membranes. The problem is that C3b can bind carbohydrates on both host and bacterial membranes. So, if C3a were constantly made at high levels, it would overwhelm the regulatory mechanisms that prevent low levels of C3b from starting a cascade on eukaryotic membranes.

OBJECTIVES ASSESSMENT QUESTIONS

15.1 Overview of the Immune System

1. **b.** Complement
2. **b.** Thymus
3. **d.** Leukocyte count of 19,000/μl; 80% neutrophils; 25% lymphocytes

15.2 Innate Host Defenses: Keeping the Microbial Hordes at Bay

4. **c.** Langerhans cell
5. **a.** sample bacteria and other antigens from the intestine
6. microbe-associated molecular patterns (MAMPs)

15.3 The Inflammatory Response

7. **a.** Receptor on endothelial cells
8. **b.** Attract macrophages
9. **c.** Edema
10. **a.** Receptor on endothelial cells
11. **b.** Lack of integrin
12. Heat, edema, redness, pain, altered function
13. chronic

15.4 Phagocytosis: A Closer Look

14. **b.** CD47
15. **e.** Macrophages engulf dead neutrophils.
16. **c.** Fc receptors
17. **d.** triggering apoptosis of the phagocyte

15.5 Interferon, Natural Killer Cells, and Toll-Like Receptors

18. True
19. **e.** inducing production of double-stranded RNA endonucleases
20. **a.** Fc receptors
21. **d.** Microbial structures

15.6 Fever

22. True
23. **b.** Hypothalamus

15.7 Complement

24. **c.** C5
25. **b.** Factor H
26. **d.** Factor C5b directs assembly of the pore.
27. **a.** MAMPs bind TLRs in macrophages, and cytokines are released, triggering CRP production in the liver.

CLINICAL CORRELATION QUESTIONS

1. The level of white cells in Danny's blood was high because their production in his bone marrow had increased in response to cytokines produced at the site of infection and because, once in the circulation, these white cells, lacking integrin, could not extravasate out of capillaries to the site of infection. Unable to reach the infection site, these cells could not fight the infection.

2. CD59 is a cell-surface protein that binds to complement factor C5a and prevents formation of the MAC pore complex on host cells. Red blood cells left unprotected by CD59 will lyse as a result of MAC formation. The current best way to treat this deficiency is to administer antibody to C5a. Removing C5a in this way will prevent MAC formation in red blood cells. Eculizumab is a commercially available anti-C5a monoclonal antibody used to treat these patients. Monoclonal antibodies are described in Chapter 17.

3. The two different strains of *Salmonella* added different structural modifications to their LPS molecules. The different modifications altered the ability of the two LPS molecules to interact with the boy's TLR4 pattern recognition receptors. As a result, LPS from the second strain did not trigger as strong a TLR signal, so fewer cytokines were made. The number of cytokines produced in response to an infection is important because it influences the degree of inflammation that results. The degree of inflammation, in turn, impacts the duration and severity of symptoms.

4. CMV makes decoy molecules that look like MHC I proteins and places the decoys on the host cell surface to fool NK cells into bypassing the infected cell.

Chapter 16

THOUGHT QUESTIONS

16.1 Mechanisms during the development of lymphocytes delete (kill) most of the T cells and B cells that can recognize self antigens. Those lymphocytes that escape can cause autoimmune diseases later in life.

16.2 Neutralizing antibodies usually bind attachment proteins on the virus and physically block them from binding to host cell receptors (see Figure 16.5). Some antibodies to enveloped viruses might trigger the complement cascade, destroying the membrane and infectivity.

16.3 Because maternal IgM antibodies cannot cross the placenta, finding IgM antibodies to rubella antigens in the newborn's circulation would indicate that the fetus was infected and initiated its own immune response.

16.4 Because memory B cells eventually die. Without some exposure to antigen (for example, a vaccine booster shot), those memory cells will not be replaced.

16.5 IgM is a pentamer possessing ten antigen-binding sites. The affinity of the antigen-binding sites, however, is not as strong as that made after a heavy-chain switch, which includes the generation and selection of mutations in the variable regions that more tenaciously bind antigen. IgM makes up for its poor individual binding site affinity by having 10 antigen binding sites that can simultaneously bind multiple copies of an antigen. As a result, IgM remains attached to the pathogen and interacts with other components of the immune system that rapidly clear the pathogen from the bloodstream.

16.6 There are probably several ways this can happen. One way is if "dead" virus particles enter the cell by endocytosis rather than infection. The resulting processed epitopes can be placed on MHC II surface receptors, where they can promote activation of T_H0 cells. The resulting activated T_{FH} cells can stimulate B-cell differentiation into plasma cells that will produce antibodies.

16.7 The macrophage, also an antigen-presenting cell, relies on surface MHC II to present epitopes to T_H0 cells. Consequently, down-regulating MHC II on macrophages would limit the stimulation of T_H0 cells and diminish the production of helper T cells. The lack of helper T cells would translate into limited humoral and cell-mediated responses.

16.8 B cells with a B-cell receptor that can bind a hapten can take up hapten but cannot present the hapten to a helper T cell. The same B cell can also take up the hapten bound to a carrier molecule, and because the carrier molecule is larger than the hapten, the B cell will present the carrier epitope to the helper T cell. The helper T cell stimulates the B cell, which was already programmed to make antihapten antibody, to differentiate into plasma and memory B cells.

16.9 The cornea is not normally vascularized (it lacks blood vessels). So even though corneal cells express MHC proteins, circulating host T cells do not have an opportunity to interact with them. The cornea will not be rejected. Such nonvascularized tissues are called immune-privileged sites.

16.10 For viruses, T_C-cell destruction of the target host cell destroys the machinery that viruses use to replicate. So, the death of host cells prevents the synthesis of more virus. In addition, some bacterial pathogens and all viruses use host cells to hide from the immune system. Destroying the infected cell exposes pathogens to the rest of the immune system.

16.11 (1) The immune system might fail to prevent reinfection if a pathogen has many strains that differ in their surface antigen structure. Memory B or T cells from the last infection cannot recognize the new antigen. (2) A pathogen could also produce different antigenic types of toxins (e.g., botulism toxin). (3) A pathogen may produce substances that interfere with a full-blown immune response so that effective or long-lasting immune memory is not established. (4) A pathogen (e.g., *S. aureus*) that infects skin and then walls itself off in an abscess can limit its interaction with the immune system. Once again, the body may not develop long-lasting immunity.

OBJECTIVES ASSESSMENT QUESTIONS

16.1 Basics of Adaptive Immunity

1. **b.** epitope
2. **d.** Plasma cells
3. **b.** T cell
4. **d.** present antigen on MHC surface receptors
5. **e.** Proteins
6. antigenic (or immunological) specificity

16.2 Antibody Structure, Diversity, and Response

7. **a.** 3; **b.** 2; **c.** 1; **d.** 4; **e.** 5
8. **a.** Part 1
9. **b.** heavy-chain constant regions
10. **d.** IgA
11. False
12. **d.** IgG
13. **a.** IgM or IgD
14. **b.** B cells proliferate to produce new B cells with identical B-cell receptor surface proteins
15. **b.** Affinity maturation of the antigen-binding site

16.3 T Cells: The Link between Humoral and Cell-Mediated Immunity

16. **e.** Activated CD8 T cells
17. **a.** stimulation of cytotoxic T cells
18. **d.** bind weakly to some self antigen
19. **a.** extracellularly
20. CD4
21. **b.** Dendritic cell
22. **a.** MHC I receptors
23. **e.** bind to T-cell receptors and MHC outside their antigen-binding regions

16.4 Completing the Immune Response

24. **d.** Inability to undergo isotype switching
25. **a.** Bone marrow
26. **b.** Thymus

27. **d.** Lymph node
28. **d.** Lymph node
29. **d.** Lymph node
30. **e.** Deficiency in T-cell numbers or function
31. **d.** activated CD8+ T lymphocytes attack infected cells

CLINICAL CORRELATION QUESTIONS

1. His blood will contain normal or elevated levels of IgM but very low levels of the other classes of antibodies. IgM is present because some B cells can become plasma cells and make some IgM antibody even without T-cell help. T-cell help, however, is required for isotype switching to other antibody types. For this reason, IgM is made, but a switch to IgG or other antibody types is rare. Repeated infections happen because IgM is not a high-affinity antibody (T-cell help is needed to trigger affinity maturation of the antigen-binding sites). In addition, IgM does not contain Fc regions that can bind Fc receptors on phagocytes. The lack of IgG, which has these features, means that bacterial pathogens can overwhelm the minimal protection afforded by IgM. In addition, without CD4 T_H1 cells, CD8 cytotoxic T cells are not properly activated.

2. Cytotoxic T cells will not kill infected red blood cells. Red blood cells, which are not nucleated, will not have surface MHC I receptors that can present antigen to CD8 cytotoxic T cells.

3. The main way tumor cells can avoid CD8+ cytotoxic T lymphocytes is to reduce MHC I production in the tumor cell through mutation or regulation. The tumor cell, then, will not effectively present the abnormal protein antigens on its cell surface for the CD8+ cytotoxic T cells to recognize.

4. The CTL response would also be deficient. Without T_H cell development via MHC II antigen presentation, Fatima would not have the T_H1 cells required to complete T_C-cell activation. Recall that T_H1 cells provide signal 3 for T_C activation: cytokine IL-2.

Chapter 17

THOUGHT QUESTIONS

17.1 Antigen-presenting cells can present pieces of the bacterial ribosomal proteins on their surface. B cells and T cells can recognize those antigens to produce humoral and cell-mediated responses. However, the resulting antibodies and cytotoxic T cells will not "see" that ribosomal antigen in a living *S. pneumoniae* cell because the antigen is internal. Antibodies will be able to attack capsular antigen because it's a surface structure. (Cytotoxic T cells are not as important in this case because *S. pneumoniae* is not an intracellular pathogen.)

17.2 Because the B-cell population that would react to type A antigen was deleted during B-cell maturation.

17.3 T_H2 cells, which are needed to activate B cells to make the IgE antibodies.

17.4 Blood-group-O red blood cells do not contain A or B antigens. They can be given to group A or B patients because the cells will not be attacked by anti-B or anti-A antibodies, respectively, present in those patients. Group-O blood can be universally given to A, B, AB, or O patients. (Some minor blood groups could cause problems.) Blood-group-O people cannot receive blood from any other group because their serum contains antibodies against A and B antigens. If those patients received blood of the wrong type, they would experience a type II hypersensitivity reaction. Universal recipients are people with blood group AB. Because they have neither anti-A nor anti-B antibodies, universal recipients can receive red blood cells from donors with type A, B, or O.

17.5 First, severely immunodeficient patients will not develop an antibody response to foreign proteins. Second, each lot of immunoglobulin will come from different individuals, so even if the patient develops an immune response to antigens in one lot, those antigens will be different in a second lot.

17.6 CD8 cytotoxic T cells are low in number when MHC I is missing because antigen-presenting cells lacking MHC I cannot activate cytotoxic T cells to proliferate. The disease is called bare lymphocyte syndrome type 1. In addition to having a low number of CD8 T cells, these patients have deficient NK cells that cannot effectively kill infected host cells missing MHC I molecules (see Section 15.5). The APCs in these patients lack surface MHC I but actually do contain <u>intracellular</u> MHC I. The mutation for this form of bare lymphocyte syndrome is in the *TAP* gene, whose product helps transport antigens into the endoplasmic reticulum for loading on MHC I. Without loading, the MHC I molecule is not transported to the cell surface. Patients with this disorder typically develop chronic lung infections in late childhood.

17.7 The use of many donors helps to ensure that the preparation contains a broad spectrum of protective antibodies and is relatively uniform. Most people are naturally infected by the same pathogens that commonly infect immunodeficient patients. However, the donor immune systems effectively fought these infections. Therefore, the donated serum will contain the appropriate antibodies. The content of specific antibodies in pooled immunoglobulin is routinely performed only for antibodies to measles, poliovirus, and hepatitis B surface antigen.

In contrast to immunoglobulin, hyperimmune globulin (HIG) comes from individuals whose sera contain high titers (levels) of antibodies to specific pathogens. HIG preparations are available for the hepatitis B (HBIG), cytomegalovirus (CMVIG), varicella-zoster (VariZIG), rabies, and vaccinia viruses, as well as the tetanus and botulinum toxins.

17.8 To prevent tissue rejection, no. Because the patient has no T cells to recognize foreign antigens, the transplant is not rejected. However, there is the possibility of graft-versus-host disease. Methods to prevent GVHD include removing T cells from the donor graft or treating the patient to suppress T cells in the graft.

OBJECTIVES ASSESSMENT QUESTIONS

17.1 Vaccines and Immunization

1. **e.** Breast milk
2. **e.** Natural infection
3. herd immunity

17.2 Diagnostic Tools of Immunology

4. **b.** Patient 2: dilution, 1:1,350
5. **d.** Flow cytometry
6. **b.** Hyper IgM syndrome
7. **c.** X-linked agammaglobulinemia
8. **d.** Selective IgA deficiency

17.3 Hypersensitivity

9. **d.** Type IV
10. **c.** Blocking IgE from binding to mast cells
11. Allergen
12. IgE
13. Mast cells or basophil
14. Histamine, prostaglandins, and leukotrienes
15. Blood type O could be substituted.
16. **c.** Type III hypersensitivity, and **f.** Immune complexes
17. **d.** Type IV hypersensitivity, and **e.** T_H1 cells
18. **b.** Type II hypersensitivity, and **h.** Cell lysis

17.4 Immunodeficiencies and Lymphoid Cancers

19. **b.** a secondary immunodeficiency
20. **d.** a lack of IL-2 receptor
21. **c.** DiGeorge syndrome
22. **a.** IV administration of immunoglobulin
23. **d.** lymphoma
24. **e.** Hodgkin's lymphoma
25. **e.** Burkitt's lymphoma
26. immune surveillance

17.5 Autoimmunity and Organ Transplants

27. **a.** molecular mimicry enables a mismatched T_H cell to activate a B cell that escaped deletion
28. **e.** antibody that targets thyroid-stimulating hormone receptor
29. **b.** MHC type I complexes are absent on RBCs.

CLINICAL CORRELATION QUESTIONS

1. Rh-negative cells, because the mother's anti-Rh⁺ antibodies have crossed the placenta and will continue to attack any newly transfused Rh⁺ red blood cells in the fetus. An Rh-positive fetus does not make antibodies to Rh-negative blood cells (no foreign antigen is present on Rh-negative cells).

2. Leukocyte adhesion deficiency. The leftward shift of cells in the first graph (patient) relative to the second panel (control) indicate that the patient's leukocytes (PMNs and lymphocytes) have a CD18 integrin deficiency (Y-axis). In contrast, both sets of leukocytes have plenty of Fc receptors (X-axis). Presence of the Fc receptor suggests that the leukocytes are plentiful. The loss of CD18 means those leukocytes cannot extravasate from the bloodstream into tissues when called. Therefore, many leukocytes accumulate in blood. So, although the elevated WBC count indicates that the patient is responding to an infection, leukocytes cannot reach the site of infection. The patient, therefore, is more susceptible to recurrent infections.

3. IgM and IgG sandwich ELISA tests (Figure 17.8A) can be used to determine whether the patient has a recent EBV infection. Microtiter plate wells are coated with either anti–human IgM antibody or anti-IgG antibody. Serial twofold dilutions of patient sera are added to the series of wells. Patient IgG or patient IgM will be captured in the appropriate wells. After washing, EBV antigen is added to all the wells. Any patient anti-EBV IgG or IgM captured in the wells will bind to the added EBV antigen. Finally, a known anti-EBV antibody conjugated to a reporter enzyme or fluorescently tagged is added to the wells. This reporter antibody will tag EBV antigen that has bound to captured patient IgM or IgG. From the dilutions, a titer of anti-EBV antibody can be determined. If the patient was recently infected with EBV, her anti-EBV IgM antibody titer will be high, whereas the anti-EBV IgG titer will be low. High anti-EBV IgM titer but low IgG titer indicates that she is undergoing a primary response to EBV infection.

Chapter 18

THOUGHT QUESTIONS

18.1 The results clearly indicate bacterial meningitis. Cerebrospinal fluid in a healthy person is sterile. It contains few, if any, WBCs and almost no protein (<40 mg/dl), but it will contain a fair amount of glucose (40–70 mg/dl). In bacterial meningitis, WBCs (mostly neutrophils) enter the CSF to attack the bacteria, raising WBC levels in CSF. The polymorphonuclear leukocytes (PMNs) and the many bacteria present all contain protein, which dramatically increases the CSF protein level. Bacteria and PMNs will consume glucose; if the blood-brain barrier is disrupted, glucose transport is faulty. CSF glucose levels will thus decrease. Although an obvious decrease in CSF glucose did not happen in this case, the other indications were clear.

18.2 Microscopic examination to see whether bacteria are found within cultured mammalian cells is usually not satisfactory. The difficulty lies in determining whether the organism is inside the host cell or just bound to its surface—or, if inside, whether the bacterium is alive or dead. One common approach is to add to infected cell monolayers an antibiotic that can kill the microbe but will not penetrate the mammalian cells—such as the protein synthesis inhibitor gentamicin (we did not discuss this previously). A bacterium that invades a host cell will gain sanctuary from gentamicin and grow intracellularly. Extracellular bacteria and bacteria attached to the host cell are killed. Intracellular bacteria are then released by lysing the mammalian cell membranes with a gentle detergent and plating the lysate on agar media. Each colony that forms represents a single intracellular bacterium. Of course, this approach will work only if the microorganism is not an obligate intracellular parasite.

18.3 The goal of any microbe is to maintain its species. If a microbe does not have an opportunity to easily spread to a new host, killing its host would cause its own death.

18.4 Antibodies to the B subunit will be more protective. Inactivating the B subunit will prevent the toxin from binding to cell membranes. The A-subunit active site is typically sequestered in these toxins and inaccessible to antibody. Furthermore, once the A subunit has entered a host cell, antibodies cannot enter and neutralize it.

18.5 With a few exceptions, withholding iron from potential pathogens is a host defense strategy because when iron is plentiful, the microbe does not have to expend energy to get iron and so can readily grow. On the other hand, low iron levels can also be a signal to express various virulence genes, so for some organisms, a high iron level might hinder infection.

18.6 For one thing, rhinoviruses are susceptible to acid pH (pH 3.0; not mentioned in the text), so they cannot replicate in the gastrointestinal tract. They also grow best at 33°C, which may help explain their predilection for the cooler environs of the nasal mucosa.

18.7 It will depend on the combination of pathogens. A pathogen that shares an important antigenic feature with a second pathogen could elicit an immune response that will dispatch the second pathogen. However, a pathogen that suppresses the immune response could make the patient more susceptible to secondary infection by other pathogens. For example, HIV depletion of T cells will make the patient more susceptible to bacterial, fungal, protozoan, and viral pathogens.

18.8 Because their genomes are not segmented, these viruses do not undergo the reassortment necessary to quickly alter the antigenic forms of key viral proteins.

18.9 The infectious forms of most fungal pathogens are the spores or conidia, which easily become airborne and can be inhaled. Spores in thermally dimorphic fungi are produced when hyphae are made at 25°C during growth outside the body; they are not made at 37°C, the temperature at which yeast forms are produced. While these large yeast forms can be present in respiratory aerosols, not enough of them could enter the lung or travel deep enough to cause disease.

OBJECTIVES ASSESSMENT QUESTIONS

18.1 The Making of a Pathogen

1. **e.** Segments A and C
2. **a.** Conjugation

18.2 Pathogen Entry and Colonization

3. False
4. type IV pili
5. **b.** biofilm-producing bacteria
6. Type IV

7. **c.** nonpilus adhesin

8. **d.** prevents evacuation of the pathogen

18.3 Surviving within the Host

9. **b.** remains in the host cell phagosome

10. **e.** *Rickettsia*

11. **a.** moves intracellularly via actin tails

12. **c.** Delay synthesis of virulence factors until a certain bacterial cell number is reached

18.4 Host Damage and Manipulation

13. **e.** Anthrax EF toxin

14. **b.** *Escherichia coli* LT

15. **c.** Toxic shock syndrome toxin

16. **a.** *Escherichia coli*

17. **d.** flagellum synthesis systems

18. **c.** secrete proteins into the surrounding environment

18.5 Viral Pathogenesis

19. **b.** Multiple rhinovirus strains that possess different VP1 structures

20. **c.** CD4 binding

21. **a.** Accelerating viral transcription

22. **e.** Stimulating host cell division

23. **d.** Preventing apoptosis

24. **c.** placing a molecular mimic of MHC I on an infected cell's surface

18.6 Protozoan and Fungal Pathogenesis

25. **b.** Actin-myosin motor

26. **c.** Phagocytic cell receptors

27. False

28. True

29. **e.** Down-regulate IL-12 production in infected macrophages

30. **b.** Make a polysaccharide capsule, **c.** Alter surface proteins during infection

31. **a.** *Aspergillus fumigatus*, **c.** *Trichophyton*

CLINICAL CORRELATION QUESTIONS

1. The pathogen infected the pregnant mother and passed through the placenta to the fetus.

2. Scenario 4. An infected fetus will produce IgM. IgM cannot cross from mother to fetus, so finding IgM in the fetus means the fetus was infected. This rules out scenarios 2 and 3. Since the mother was infected, she would produce IgG that can cross the placenta. Thus, you would expect mother's IgG antibodies to be in the fetus, so scenario 1 is not likely.

3. Horizontal transfer of genes from other strains of *E. coli*.

4. Decreased killing of infected cells by cytotoxic T cells.

5. Anthrax produces a potent exotoxin composed of three subunits: PA, EF, and LF. The PA component binds to host membranes and mediates the entry of the toxic subunits. The major component of an anthrax vaccine, therefore, should be the PA component because antibodies against it will prevent entry of the toxic subunits into target cells.

Chapter 19

THOUGHT QUESTIONS

19.1 A patient history would reveal whether the child was born in an endemic area outside the United States, had recently traveled to an endemic area, or had contact with someone else who recently returned from an endemic area. Individuals born in the United States who have not been vaccinated against German measles will be at high risk of contracting German measles if they come in contact with an infected person or travel to endemic areas.

19.2 The viral DNA does not express any proteins to present on the host membrane. Thus, cytotoxic T cells are blind to the presence of these latently infected cells.

19.3 The general population is not vaccinated against smallpox because the disease was eradicated. Laboratory workers who work with the smallpox virus should be vaccinated because of the possibility of accidental exposure. Some members of the military are vaccinated because smallpox is a possible weapon of biowarfare.

19.4 The integrity of the skin is a major deterrent to bacterial infections. Viral skin lesions produce physical breaks in that defense. The breaks allow potentially pathogenic bacteria that may be colonizing the outermost layer of epidermis to gain access to living skin tissues. The reverse doesn't generally happen because viruses capable of causing skin infections are not normal biota of the skin.

19.5 As with any infection, several factors are involved. Differences in various aspects of innate and adaptive immunity are important, as are the number of breaks in the skin through which the fungus can enter (opportunity). Immunocompromised patients generally have a higher risk of infection, fungal or otherwise.

19.6 Wounds are surrounded by healthy tissue colonized by normal microbiota. The injured tissue may then be infected by multiple species of bacteria originating from the surrounding microbiota.

19.7 The main source of *Acanthamoeba*, a free-living protozoan, is contaminated water. No one knows why the risk of infection with this organism is higher in contact lens wearers, but using contaminated water for washing hands prior to putting in contact lenses or contaminating the lens solution may be the reason.

OBJECTIVES ASSESSMENT QUESTIONS

19.1 Anatomy of the Skin and Eye

1. **b.** Langerhans cells

2. basal

3. epidermis

4. False

19.2 Viral Infections of the Skin

5. **d.** a double-stranded, linear DNA virus that also causes cold sores

6. **b.** Variola

7. **d.** vertical transmission of rubella virus to the fetus during the mother's active infection

19.3 Bacterial Infections of the Skin

8. **b.** Coagulase

9. **b.** Gram-positive cocci

10. **e.** Surgical removal of affected tissue and IV antibiotics

19.4 Fungal Infections of the Skin

11. **a.** *Malassezia*

12. **c.** the KOH preparation

13. **b.** playing with a household dog or cat

19.5 Skin Infections of Special Circumstance

14. **c.** Decreased cell-mediated immunity

15. **b.** is a ubiquitous opportunistic pathogen

16. **c.** Bacteria

17. **d.** It often is due to polymicrobial wound infection with aerobes and *Clostridium perfringens*.

19.6 Eye Infections

18. **a.** Acute allergy attack, causing conjunctivitis

19. the conjunctiva

20. **c.** endophthalmitis

CLINICAL CORRELATION QUESTIONS

1. The most likely cause of infection is the measles virus. The white papular enanthem that appears on the palate and throat is referred to as Koplik's spots. Vaccination with attenuated virus is the means by which this disease is prevented (see Table 17.2).

2. *S. pyogenes* causes the pustules. *C. perfringens* (a Gram-positive anaerobe) and *P. aeruginosa* (a Gram-negative aerobe) both are rod-shaped bacteria (bacilli). *S. aureus* and *S. epidermidis* are catalase-positive. *S. pyogenes* is catalase-negative. Staphylococci appear as clusters, much like a cluster of grapes under the microscope, but *Streptococcus pyogenes* looks like pearls on a string. The only bacterial species consistent with the lab results is *S. pyogenes*.

3. The ophthalmologist is likely to see multi-nucleated giant cells (Tzanck cells). Herpes simplex viruses can infect the cornea and cause a dendritic lesion. Scrapings from the lesion spread on a glass slide and viewed under the microscope (Tzanck smear) reveal multinucleated giant cells called Tzanck cells.

4. The initial diagnosis is correct. The blisters of SSSS are caused by the toxic shock syndrome toxin (TSST) produced by *Staphylococcus aureus*. The toxin is secreted by staph that has caused an infection elsewhere in the body and disseminates hematogenously to the skin from that site, so the organisms are not found in the blisters.

Chapter 20

THOUGHT QUESTIONS

20.1 The intranasal vaccine contains live, attenuated virus capable of replication. People who are immunosuppressed cannot fight the attenuated virus and may contract influenza as a result.

20.2 Smoking paralyzes the mucociliary escalator, so the bacteria trapped in the mucus are not moved out of the bronchi, giving them ample opportunity to cause infection.

20.3 The TB skin test is based on a T cell–mediated immune response. AIDS patients suffer from immunodeficiency as a result of a markedly reduced number of helper T cells.

20.4 Hospital-acquired pneumonias are potentially worse for two reasons. First, hospitalized patients are not healthy, so their immune system is less able to fight an infection. Second, bacteria in a hospital setting are more likely to be resistant to antibiotics because they are continually exposed to different antibiotics. The blend of a compromised immune system and antibiotic-resistant bacteria is a dangerous combination that makes hospital-acquired pneumonias a dangerous proposition.

20.5 Failure of antibiotic therapy to resolve a pneumonia, together with atypical laboratory findings (negative bacterial cultures, negative tuberculin test, negative immunologic tests for common viral causes of infection, relatively normal WBC counts), may indicate a fungal pneumonia.

OBJECTIVES ASSESSMENT QUESTIONS

20.1 Anatomy of the Respiratory Tract
1. True
2. **b.** Trachea
3. **b.** preventing foreign materials from reaching the lungs

20.2 Viral Infections of the Respiratory Tract
4. **c.** Shortness of breath
5. False
6. True
7. **d.** Infants

20.3 Bacterial Infections of the Respiratory Tract
8. **b.** Pharyngitis
9. **d.** Pneumonia
10. **a.** Acute otitis media, **c.** Bronchitis, **d.** Pneumonia
11. **d.** Pneumonia
12. **a.** Acute otitis media, **c.** Bronchitis, **d.** Pneumonia
13. **a.** Childhood vaccination
14. **b.** a consequence of pharyngitis caused by toxin-producing *S. pyogenes*
15. **d.** *Mycoplasma pneumoniae*
16. True
17. False

20.4 Fungal and Parasitic Infections of the Respiratory Tract
18. True
19. **b.** Normal values
20. **c.** *Cryptococcus*
21. **d.** western United States

CLINICAL CORRELATION QUESTIONS

1. The most likely organism causing the clinical picture given is *Mycobacterium tuberculosis*. There is no history of travel or immune suppression, so fungal causes are not the prime suspect. *Mycoplasma pneumoniae* and *Bordetella pertussis* can also cause a chronic cough but will not progress to coughing up a blood-tinged sputum.

2. The decided left shift (increased number of neutrophils) indicates a bacterial pneumonia. *Streptococcus pneumoniae* is the most likely cause, based on the clinical presentation and the typical involvement of one lobe of the lung (lobar pneumonia).

3. *Bordetella pertussis*. The distinct sound of inspiration made after a paroxysm of cough has given the disease its name: whooping cough. Because the paroxysmal cough of whooping cough continues for quite some time after the bacteria are gone, the disease was called the 100-day cough.

4. *Legionella pneumophila* is the most likely cause. Urine test for *Legionella pneumophila* antigen is positive and Yolanda is treated accordingly. The sputum cultures grew *L. pneumophila*, a thin, Gram-negative, flagellated bacterium. The likely source of infection was contaminated hot tub water at the vacation rental cabin.

Chapter 21

THOUGHT QUESTIONS

21.1 Streaks move toward the heart because the lymph in lymph vessels moves toward the heart, where it can reenter the venous system. Lymph never flows away from the heart.

21.2 Because IgM is the first antibody that appears after an infection, the patient probably has an acute infection with EBV. Had the pattern been reversed, the patient would probably be recovering from the disease or had already recovered.

21.3 Diagnosis of sepsis requires positive blood cultures. In this case, only one blood sample exhibits growth, which usually indicates contamination with skin microbiota during specimen handling, not sepsis. The continuation of daily blood cultures is used to monitor the patient closely, making sure SIRS does not change to sepsis.

21.4 Keep in mind that a bactericidal antibiotic could cause the release of more LPS from killed bacteria. Treatment with any antibiotic is not a good choice, but it is usually the only choice.

21.5 Epidemic typhus. Body lice are associated with decreased sanitary conditions and require close contact between victims in order to migrate from one host to another. Both of these unsettling requirements are features of war.

21.6 *Streptococcus gallolyticus* and enterococci are typical members of the intestinal microbiome. The physician would suspect intestinal cancer or some other breach in the integrity of the intestine that would allow *S. gallolyticus* access to the bloodstream.

21.7 Bacteria infecting the heart valves (especially the mitral valve) form a mass, called vegetation. A piece of the vegetation can break off and be transported by the bloodstream to the brain's blood vessels and occlude them. The occlusion prevents blood perfusion at the site and causes an ischemic (inadequate blood supply) stroke.

21.8 The answer is not entirely clear. One hypothesis is that plasmodia infecting RBCs in heterozygous individuals consume so much oxygen that these cells sickle and are marked for destruction by phagocytes, which will also destroy the parasite. Another hypothesis is that sickle-cell trait RBCs produce higher levels of oxygen radicals that can directly kill the parasite.

OBJECTIVES ASSESSMENT QUESTIONS

21.1 Anatomy of the Cardiovascular and Lymphatic Systems
1. **b.** aorta
2. False
3. **c.** Lymphadenitis

21.2 Systemic Viral Infections
4. **c.** B cells
5. **e.** Heterophile antibodies
6. **b.** Transplacental
7. **d.** Hemorrhagic fever
8. **b.** Splenomegaly
9. **a.** Owl's eyes
10. **c.** Burkitt's lymphoma

21.3 Systemic Bacterial Infections
11. **c.** May have a noninfectious etiology
12. **a.** Systemic inflammatory response syndrome caused by an infectious agent
13. **b.** Sepsis with one or more signs of organ dysfunction
14. **d.** Need vasopressors to maintain systolic blood pressure of 90 mm Hg or mean arterial pressure of 70 mm Hg
15. **e.** Presence of bacteria in the blood
16. **b.** stage 2
17. **e.** Typhoid fever

21.4 Bacterial Infections of the Heart
18. **b.** *Staphylococcus aureus*
19. **a.** pancarditis
20. True

21.5 Systemic Parasitic Infections
21. **b.** undergo antigenic variation
22. **c.** Hydrocele (grotesque swelling of the scrotum)

CLINICAL CORRELATION QUESTIONS

1. *Yersinia pestis*. *Borrelia burgdorferi* is a spirochete, not a bacillus. *Salmonella* Typhi and *Rickettsia rickettsii* cause generalized lymphadenopathy and do not exhibit bipolar staining.

2. Antibiotics effective against Gram-positive, Gram-negative, and anaerobic bacteria. Viruses and parasites do not cause septicemia.

3. *Borrelia burgdorferi*, which is transmitted by tick bite.

4. American trypanosomiasis or Chagas disease. She was probably in the chronic stage of Chagas disease when the cardiac symptoms and fatigue brought her to the clinic.

Chapter 22

THOUGHT QUESTIONS

22.1 HAV and HEV are nonenveloped viruses. Nonenveloped viruses are more resistant than enveloped viruses to the effects of intestinal bile and stomach acid.

22.2 Because HDV requires HBV to replicate. Someone immunized against HBV can be infected with HDV, but no disease can develop.

22.3 Reducing stomach acidity helps protect the lining of the stomach, which in turn promotes healing of the existing ulcer and alleviates stomach pain and discomfort.

22.4 The many normal microbiota present in infected stool could obscure the presence of low numbers of *Yersinia* causing disease. Placing the stool in a refrigerator for several days (called cold enrichment) can increase the number of *Yersinia* organisms present (because they grow) while inhibiting growth of normal microbiota. Cold enrichment combined with various selective media makes it easier for the laboratory to identify the presence of *Yersinia* in stool samples.

22.5 For watery diarrhea, norovirus is a prime candidate (because of the patient's age), but rotavirus, ETEC, *Vibrio cholerae* (if she visited an endemic area), and *Salmonella* (it may be hard to see blood in stool without a laboratory test) should be considered as well. Absence of vomiting and the duration of diarrhea place staphylococcal enterotoxin low on the differential list; absence of blood in stool reduces the likelihood of *Shigella*. For bloody diarrhea, the following organisms should be considered: *Shigella*, *Salmonella* (though not always), EHEC, *Yersinia enterocolitica*, *Campylobacter* (though not always), *Entamoeba*, and some of the helminths (although unlikely if she has not traveled outside the country).

22.6 *Staphylococcus aureus* food poisoning is caused by the ingestion of preformed toxin present in food contaminated with *S. aureus*. You could find the organism in the offending food (if any remains). Remember, the enterotoxin is heat stable. If the food was stored and then heated, the toxin will remain active but the organism may be killed, so you may not find the organism in the food.

OBJECTIVES ASSESSMENT QUESTIONS

22.1 Anatomy of the Digestive System

1. **e.** a and b only
2. **d.** Large intestine
3. False
4. **a.** inadequate absorption of nutrients

22.2 Infections of the Oral Cavity

5. **d.** All of the above
6. **c.** Both a and b

7. **b.** Trench mouth
8. **c.** Plaque formation
9. **a.** *Candida albicans*
10. **b.** Vincent's angina

22.3 Gastrointestinal Syndromes

11. **b.** Enteritis
12. **c.** secretory diarrhea
13. False

22.4 Viral Infections of the Gastrointestinal Tract

14. **d.** DNA genome
15. **b.** ss positive-sense RNA genome
16. **c.** Orchitis
17. **a.** Diarrhea
18. **a.** liver
19. **b.** hepatitis B

22.5 Bacterial Infections of the Gastrointestinal Tract

20. **c.** *Staphylococcus aureus*
21. **e.** actin tails
22. **d.** *Salmonella* Typhi
23. **a.** good sanitation
24. False

22.6 Parasitic Infections of the Gastrointestinal Tract

25. **a.** Cryptosporidiosis
26. **b.** migration of larvae to the lungs
27. **b.** larval penetration of the skin
28. **c.** Metronidazole

CLINICAL CORRELATION QUESTIONS

1. The causative agent is enterotoxigenic *E. coli*. The organism produces a heat-labile toxin that increases cAMP production in the host, resulting in the secretion of electrolytes followed by water into the lumen of the intestine.

2. Typhoid fever. Typhoid fever is caused by *Salmonella enterica* serovar Typhi, a facultative intracellular organism. The originally low-grade fever that keeps going up every day (crescendos) and the red spots on the trunk (likely to be rose spots) indicate typhoid fever.

3. *Yersinia enterocolitica*. The differential diagnosis of this patient's symptoms includes several different bacteria. *Y. enterocolitica*, however, is the only etiology that would cause mesenteric lymphadenitis and terminal ileitis and symptomatically mimic appendicitis.

4. The patient is infected with *Necator americanus*. The burning sensation in the lungs is due to the larvae that traveled to the lungs to mature. Lack of fever indicates a nonbacterial cause. Anemia is due to the organism's feeding on the patient's blood.

Chapter 23

THOUGHT QUESTIONS

23.1 Several factors contribute. The organisms can piggyback on a catheter and into the bladder

during insertion. Presence of a catheter in the urethra keeps the urethral meatus open, allowing microbes better access to the bladder. Patients who are in the hospital and catheterized, such as postsurgical patients or patients in the intensive care unit, have a compromised immune system and are more susceptible to infections.

23.2 Infection of the kidney can damage the nephrons and thereby decrease filtration through the kidney. Less water from blood enters the nephrons, so less urine is produced.

23.3 Shingles (described in Chapter 19) involves reemergence of HSV from a spinal nerve. Each spinal nerve serves a different section of skin (dermatome). In a typical patient, the immune system keeps outbreaks in check, but an outbreak of shingles limited to a single dermatome or adjacent dermatomes is possible. Immunosuppression in early AIDS leads to activation of latent HSV that may be present in multiple spinal nerves.

23.4 Both antibody and cell mediated immune responses will be affected because CD4+ helper T cells are involved in both responses. CD4+ T_{FH} cells are required for adaptive antibody responses and CD4+ T_H1 cells are needed for effective cell-mediated adaptive immunity. HIV can infect and kill both types of CD4+ T cells.

23.5 *N. meningitidis* possesses a polysaccharide capsule that protects the organism from complement and phagocytosis in serum. The organism is "serum resistant" and can disseminate through the blood to the brain. *N. gonorrhoeae* does not possess this capsule and is usually sensitive to serum complement.

23.6 Broad-spectrum antibiotics should be used for all potential causes of PID, but antibiotics that treat *Neisseria* and *Chlamydia* should always be included. Surgical intervention may be necessary.

23.7 The patient is likely to have an HIV infection, and the gastrointestinal symptoms and diarrhea may be due to a secondary infection. This patient should be tested for HIV to determine the appropriate course of treatment.

OBJECTIVES ASSESSMENT QUESTIONS

23.1 Anatomy of the Urinary Tract

1. **a.** Kidneys
2. **c.** ureters
3. **d.** The length of the urethra

23.2 Urinary Tract Infections

4. **a.** Sexual activity
5. **a.** This patient has pyelonephritis resulting from an ascending bladder infection.
6. True
7. **b.** children postponing emptying their bladder
8. **c.** Pyelonephritis

23.3 Anatomy of the Reproductive Tract

9. **c.** testes
10. **c.** There is no direct contact between the ovaries and the fallopian tubes.
11. False

23.4 Sexually Transmitted Infections of the Reproductive Tract

12. **a.** human papillomavirus
13. **e.** HSV
14. **a.** Mononucleosis-like symptoms
15. **d.** Gardasil
16. **c.** AIDS
17. **b.** CD4+
18. **d.** *Neisseria gonorrhoeae*
19. **e.** *Trichomonas vaginalis*

23.5 Non–Sexually Transmitted Infections of the Reproductive Tract

20. **a.** candidal vulvovaginitis
21. **c.** A mixed population of aerobic and anaerobic organisms
22. **d.** a and b

CLINICAL CORRELATION QUESTIONS

1. *Gardnerella vaginalis* is the most likely cause. A decrease in the number of lactobacilli of the normal vaginal microbiota leads to an increase in the vaginal pH. *G. vaginalis*, another member of the normal vaginal microbiota, grows better in slightly alkaline pH. Overgrowth of *G. vaginalis* causes bacterial vaginosis and the symptoms described in the case.

2. Notched teeth, a perforated palate that will need surgery to repair, and other defects.

3. The difficulties are that he may be reluctant to identify his partners, he may not know the contact information of all his sexual partners, or he may not be forthcoming about some of them. In addition, the contacts themselves may not want to be approached and may not want to identify their current partners.

Chapter 24

THOUGHT QUESTIONS

24.1 A traumatic lumbar puncture can cause overt bleeding at the site and contaminate the sample. The presence of many RBCs in a CSF sample obtained without any trauma often indicates malignancy.

24.2 Avoid the outdoors in endemic areas if possible; but when outdoors, use insect repellent and wear long-sleeved shirts and long pants when hiking in areas where vectors are endemic.

24.3 This is a difficult question to answer. Few scientists have speculated. Recall that the toxins are encoded by genes in resident bacteriophages that became part of the clostridial genome through horizontal transfer from some other source. Because the organisms (vegetative as well as spores) normally reside in soil, the function of these toxins may have something to do with survival in that habitat. The toxin's effect on humans may simply be an unfortunate accident. However, with tetanus toxin, the organism may offer a benefit because muscle spasms could limit oxygen delivery to infected tissues, enabling a more anaerobic environment for growth. Cell death may also release iron or other nutrients useful to *Clostridium tetani*.

24.4 The most common way botulinum toxin gets into food today is via home canning. The canning process involves heating food in jars to very high temperatures. The heat destroys microorganisms and drives out oxygen; both processes help preserve foods. If the jars are not heated enough to ensure sterilization, spores of *C. botulinum* will survive. When the jars are cooled for storage at room temperature, the spores germinate; the organism then grows in the anaerobic medium and releases the toxin. When the food is eaten, the toxin is eaten, too.

OBJECTIVES ASSESSMENT QUESTIONS

24.1 Anatomy of the Central Nervous System

1. **a.** Brain
2. **c.** the blood-brain barrier
3. **d.** arachnoid mater and pia mater
4. **b.** arachnoid villi

24.2 Overview of Infectious Diseases of the Central Nervous System

5. **c.** Meningoencephalitis
6. **c.** lymphocytes
7. **a.** Inflammation of his meninges

24.3 Viral Infections of the Central Nervous System

8. **b.** Lymphocytic choriomeningitis
9. True
10. **d.** mosquitoes that transfer the virus between birds do not usually prey on humans

24.4 Bacterial Infections of the Central Nervous System

11. **a.** loosens the tight junctions between endothelial cells
12. **c.** *Clostridium botulinum*
13. **d.** *Listeria monocytogenes*

24.5 Fungal Infections of the Central Nervous System

14. **b.** being immunocompromised
15. **a.** *Coccidioides immitis*

24.6 Parasitic and Prion Infections of the Central Nervous System

16. **c.** *Acanthamoeba* species
17. Amphotericin B
18. **b.** a protein

CLINICAL CORRELATION QUESTIONS

1. This is neonatal meningitis. Gram-positive bacteria, especially *Streptococcus agalactiae*, and Gram-negative bacteria, often *Escherichia coli*, are the most likely etiologies because of the infant's age. The infant should be treated with antibiotic(s) to cover both Gram-positive and Gram-negative bacteria until the results of culture and sensitivities are obtained from the clinical laboratory.

2. A dermatologist who gave the patient a Botox (botulinum toxin) injection in the face. The botulinum toxin relaxes the facial muscles, which in turn reduces the appearance of wrinkles. Several factors, including administration of a larger-than-needed dose or injury to a nearby nerve, could have contributed to the facial paralysis of this patient.

3. The source of infection is most likely the corneal transplant. Although it is extremely rare, a case like this was reported by the CDC in 2013. The source of a transplant was a 20-year-old male whose cause of death was originally recorded as encephalitis, an often lethal condition resulting from viral infections. Donated organs are routinely screened for infectious agents, but rabies virus is not among those screened. Following the death of the corneal transplant patient, the remaining tissues from the donor also tested positive for rabies virus. The RNA sequence of the virus isolated from the transplanted cornea and from the other tissues of the donor were identical.

Chapter 25

THOUGHT QUESTIONS

25.1 Probably not. *S. epidermidis* is a common inhabitant of the skin and could easily have contaminated the needle when blood was taken from the patient. The fact that only one of the two cultures grew this organism supports this conclusion. If the patient had really been infected with *S. epidermidis*, both blood cultures would have grown this organism.

25.2 Although the CFU number is high enough to consider relevant, UTIs are typically caused by a single organism. The fact that the lab found many different colony types suggests a problem with specimen collection. In this case, not refrigerating the sample allowed the small number of urethral contaminants to overgrow the specimen.

25.3 *Staphylococcus aureus.* The novobiocin test is irrelevant in this situation.

25.4 To be a valid cause of septicemia, two strains isolated from blood samples taken from two sites should be identical. For the lab to shed suspicion on these strains of staphylococci, the two strains might have had different antibiotic susceptibility patterns or polymorphisms in DNA restriction patterns, suggesting they are not identical.

25.5 The sample would be processed to extract DNA. RNA extraction is not necessary for detecting bacteria but is required to detect RNA viruses. Two primer sets would be designed to amplify specific genes from *C. trachomatis* and *N. gonorrhoeae*, respectively. Two beacon reporter probes that bind between each primer set are made, each with a different fluorescent dye attached to one end and a quencher attached to the other. Real-time PCR is performed on the sample, using the two primer sets and two reporter probes. Different color emissions will result, depending on which organism is present.

25.6 Upon infection with any organism, IgM antibodies are the first to rise (discussed in Chapter 15). After a short time, the levels of IgM decline as IgG levels rise. IgG, however, can remain in serum for years (produced by memory B cells), making it a poor indicator of current infection.

25.7 During the course of a disease, the body's immune system increases the amount of antibody made specifically against the infectious agent. Thus, one compares the antibody titer in a blood sample taken from a patient in the active, or acute, phase of disease with the antibody titer several weeks later, when the patient is in the recovery, or convalescent, stage. Seeing a greater-than-fourfold rise in a specific antibody titer (for example, in mononucleosis) indicates that the patient's immune system was responding to the specific agent. Remember, simply finding IgG against an organism or virus in serum indicates only that the patient was exposed to that microbe at some time in the past. One could look for a specific increase in a specific IgM (as in Thought Question 25.6), but those reagents are not available for all infectious diseases.

OBJECTIVES ASSESSMENT QUESTIONS

25.1 The Importance of Clinical Microbiology and Specimen Collection

1. **b.** Alerts the clinician to possible disease sequelae

2. **c.** The same bacterial species with the same antibiotic resistance pattern was found in all the sick students and in a vendor at the picnic.

3. **c.** Throat swab

4. Lumbar puncture or spinal tap

5. **b.** The patient has septicemia caused by an organism that is facultative with respect to oxygen.

6. **d.** CSF

25.2 Pathogen Identification Using Biochemical and Protein Profiles

7. **a.** *Neisseria meningitidis*

8. **b.** *Escherichia coli*

9. **e.** *Salmonella enterica*

10. *Streptococcus pyogenes*; the Lancefield classification

11. A waxy coat or mycolic acid

12. **d**

13. **e**

14. **f**

15. **b**

16. **c**

17. **a**

18. **c**

19. **f**

20. **d**

21. **e**

22. **b**

23. **a**

25.3 Pathogen Identification Using DNA Methods

24. biochemically, antigenically (or serologically), genetically

25. **b.** The DNA sequences of the PCR fragments were different.

25.4 Pathogen Identification Using Immunological Methods

26.

1 Microbial antigen is dried on a glass slide and treated with a chemical fixative.

2 Dilutions of patient serum are incubated with the antigen on the slide and then rinsed. Specific antibodies in the serum bind to the microbial antigen.

3 An enzyme-linked secondary antibody (conjugate) is added.

Anti–human IgG

27. **c.** Patients 3, 4, and 5 (all are ≥ 4-fold increases)

25.5 Point-of-Care Rapid Diagnostics

28. **c.** A patient *without* disease has a 95% chance of testing negative with the assay.

29. False

30. Option C

31. **c.** less sensitive

CLINICAL CORRELATION QUESTIONS

1. The laboratory could use dark-field microscopy to see live, motile, corkscrew-shaped, spirochete bacteria, but this would have to be done soon after collection because these anaerobic organisms die quickly in air. Serological tests would be used to detect anti-treponemal antibody in the patient's blood (RPR, VDRL; Section 23.4). The laboratory could also use fluorescent anti-treponemal antibody and immunofluorescence microscopy to "stain" the organisms present in tissue scraped from the chancre (Section 17.2). This test would provide the most definitive diagnosis. Culture is not possible because, as mentioned in Chapter 10, *T. pallidum* has not yet been cultured continuously in laboratory medium.

2. Answers will vary.

3. Use PCR to amplify the suspected antibiotic resistance gene and monitor its production by using real-time PCR. For example, if the gene is *rpoD*, which encodes RNA polymerase, you can sequence the PCR product and look for mutations known to produce a rifampin-resistant protein.

4. Preliminary diagnosis is bacterial sepsis complicated by possible meningitis, although viral meningitis or encephalitis is possible. Blood samples for culture should be taken from both arms and incubated both aerobically and anaerobically. A CSF sample should also be collected, Gram stained for bacteria, and cultured on blood agar and chocolate agar. CSF should also be tested for viral causes: PCR testing for herpes simplex virus and qRT-PCR testing for HIV, dengue (if travel to endemic areas), enteroviruses, West Nile virus, and any other arboviruses suggested by travel to endemic countries. The red rash argues against a viral cause. A bacterial cause is most likely and should be immediately treated empirically, with antibiotics, until a cause is determined.

Chapter 26

THOUGHT QUESTIONS

26.1 Each case must match an official case description of mumps (fever, headache, muscle aches, tiredness, and loss of appetite, followed by swelling of salivary glands). Each patient would be tested for IgM antibody against mumps virus (why IgM? See Section 16.2), and epidemiologists would then look for whether the patients had been in contact with each other. The epidemiologists would also examine vaccination records for compliance. The outbreak in 2011 primarily affected people who had not been vaccinated or had received only one dose of the MMR vaccine.

26.2 Once a patient eats contaminated food, it can take 2–3 days to become ill and 1–5 days to seek treatment for that illness. At that point a stool sample will be collected and sent to the hospital laboratory. It will take the organism overnight to grow on agar media, and lab personnel will take another day to perform the biochemical tests needed to identify the organism as *E. coli* O157:H7. The organism is then shipped to a public health lab (0–2 days in shipping), and then 1–4 days will be needed for the public health lab to perform DNA "fingerprinting" to determine the relatedness between strains received from different cases. So it takes at least 1 week from the time of ingestion for public health agencies to declare an outbreak with a common source. As molecular detection techniques improve, that time will shorten.

26.3 These are clinical dilemma questions. The hospital cannot allow the nurse to continue treating patients without confirming that he was rid of the organism. The hospital might offer the nurse a position without patient contact, but he could still transmit the organism to another staff member who does have contact with patients. Screening all hospital staff for MRSA is controversial and not standard practice, although some states screen all ICU patients for the organism.

26.4 Samples from affected patients and hospital staff would be screened for MRSA. DNA from each strain would be analyzed by multilocus sequence typing (MLST) or DNA sequencing of conserved genes such as *coa* (coagulase). Strains from patients will have identical patterns or sequences if they came from the same source. The source would harbor MRSA with the same pattern. The source may be a staff member or an inanimate object, such as surgical equipment or a ventilation apparatus. A connection between patients and the colonized staff member or instruments must also be established.

26.5 One factor is mode of transmission: airborne diseases can spread more quickly than food-borne diseases, for instance. Sexually transmitted diseases spread more slowly still. Another factor is whether the disease produces a large number asymptomatic carriers who can transmit the pathogen before showing symptoms. These people can contribute to a rapid but "silent" spread of the disease. In contrast, a disease that does not produce asymptomatic carriers is more easily detected and contained. Herd immunity is another factor. Herd immunity is based on the number of individuals within a population who are resistant to a disease. Someone immune to the disease cannot pass it on. The more immune people there are in a population (or herd), the more slowly the disease will spread to susceptible people.

26.6 Transmission by infected but asymptomatic people is problematic because health officials won't know where these "healthy" shedders are. Officials don't know who the shedders are because people are not typically tested for a disease unless they exhibit symptoms. Asymptomatic disease carriers can move freely through screening systems designed to identify symptoms such as fever. With most infections, people who were infected but never show any sign of disease have killed the pathogen and cannot transmit the agent. However, look up "Typhoid Mary" online.

26.7 Not having an effective treatment for a potential biowarfare agent might sound like a good idea to the terrorist, but it would actually be a really terrible idea. The people wanting to use the agent could accidentally incapacitate or kill themselves before they could deploy the pathogen.

OBJECTIVES ASSESSMENT QUESTIONS

26.1 Principles and Tools of Epidemiology

1. organism, reservoir, mode of transmission, and host
2. Mode of transmission
3. Organism
4. Reservoir
5. Host
6. **b.** outbreak
7. **c.** a case-control study
8. incidence

26.2 Tracking Health Care–Associated Infections

9. **d.** Urinary tract infections
10. **c.** The genetic analysis found that all the MRSA strains were identical.
11. **d.** Sanitize hands before entering the room.

26.3 Tracking Global Disease Trends

12. **a.** Local hospitals and physicians notify city or state health agencies about patients with a specific disease.
13. **e.** a and b

26.4 Detecting Emerging Microbial Diseases

14. True
15. True
16. False
17. False
18. False
19. False
20. False
21. False
22. **b.** Screen blood by PCR for rRNA genes and sequence the products.

26.5 Bioterrorism and Biodefense

23. select agent
24. **a.** infections can be treated with antibiotics
25. **b.** It is easily transmitted from person to person.

CLINICAL CORRELATION QUESTIONS

1. Everyone who became ill either ate soft cheese or lived with someone who ate soft cheese at the block party. Older people were more likely to get the disease. To contain the outbreak, you would identify the brand of soft cheese involved and notify the Food and Drug Administration. The FDA would test different lots of soft cheese from the company to determine the extent of the contamination and decide whether or not to order a recall.

2. Answers will vary.

3. Answers will vary.

4. Epidemiologists would need to find a potential infectious agent that is present in all symptomatic patients but not in healthy people in the same geographic area. The genome of the pathogen should be sequenced and PCR approaches could be used to determine if all patients were infected by the same pathogen. Once an immunoassay for a pathogen antigen is developed, patients can be screened for antibody levels. Patient homes should also be screened for the presence of ticks, and the ticks would be analyzed for the presence of the same pathogen. In the case of SFTS, these criteria were all met. The pathogen was a new RNA virus in the Bunyaviridae family of hemorrhagic fever-causing viruses that includes hantavirus. What would make SFTS a major global concern is demonstrating that the virus was also able to spread via person-to-person transmission. Person-to-person transmission would mean the disease could be easily and quickly spread through populations. Combined with easy air travel, the disease would disseminate around the globe. Fortunately, person-to-person spread has not been observed with SFTS.

Chapter 27

THOUGHT QUESTIONS

27.1 In mutualism, both partner species benefit from an intimate relationship. In parasitism, one species grows at the expense of a host. A relationship that combines both would be that of a lichen. Lichens consist of a mutualism between algae and fungi, in which the algae provide organic carbon for the fungus, whose structure protects the algae. But depending on environmental conditions, the fungus may actually consume the algae faster than they grow. Another example is the role of *Bacteroides* bacteria in the human digestive tract. Some species of *Bacteroides* enhance human digestion, but when they escape the colon and invade surrounding tissues, they may cause abscesses.

27.2 The digestive community is usually anaerobic. Anaerobic catabolism lacks sufficient electron acceptors (such as O_2 and nitrate) to accept all the electrons from organic food. Therefore, organic molecules such as acetate and sugars may remain to be absorbed by the intestinal lining and enter the bloodstream. The blood contains oxygen, which enables complete respiration to carbon dioxide. Thus, the host animal obtains energy from reactions unavailable to the digestive community bacteria.

27.3 Oxygen oxidizes food molecules such as sugars and amino acids. Oxidation decreases the energy yield available for the consumer of the food and generates toxic products such as nitrosamines. The oxidation of amino acids decreases the usable protein content of food. By contrast, bacterial anaerobic fermentation breaks down sugars but leaves most of the protein intact.

27.4 In cheese production, the fermenting bacteria generate lactic acid, which lowers the pH of the food product. The acidity limits microbial growth and preserves the cheese. In the production of alcoholic beverages, the fermentation products of yeast are ethanol and carbon dioxide. The ethanol is retained as a major part of the product. In some beverages, the carbon dioxide gas remains dissolved under pressure but forms bubbles once the bottle is opened.

27.5 This question has no one correct answer. A top priority might be *Campylobacter* because it infects a million people annually, causing substantial morbidity and economic loss. On the other hand, *Clostridium botulinum* contamination is rare, but botulism has a particularly high death rate. Accordingly, the regulated standards for canning foods are very high. Still another case to consider is norovirus. While norovirus infections are rarely fatal, the person-to-person transmission rate is high, leading to large outbreaks associated with institutions such as cruise lines and nursing homes. So, the social and economic consequences of norovirus are severe.

27.6 Within the ecosystem of an aerobic digester in secondary treatment, bacteria are consumed by microbial predators such as ciliated protists. The protists convert most of the bacterial biomass to CO_2, which dissipates in the atmosphere. Without microbial predators, the bacterial cellular biomass could build up to unacceptable levels, trapping carbon and clogging the treatment system.

OBJECTIVES ASSESSMENT QUESTIONS

27.1 Microbes in Ecosystems

1. **b.** ocean
2. **d.** Production of oxygen gas by *Prochlorococcus*
3. **b.** excessive organic substrates for microbial aerobic respiration

27.2 Microbial Symbiosis with Animals

4. **c.** commensalism
5. **b.** The rumen is well supplied with oxygen from the breathing animal.
6. **a.** *Shigella flexneri*

27.3 Biogeochemical Cycles

7. **d.** Methanogenesis, reducing CO_2 to methane
8. **b.** Ammonium ion incorporation into amino acids
9. **b.** Nitrous oxide (N_2O)

27.4 Fermented Foods and Beverages

10. **a.** cottage cheese
11. **c.** Removal of toxic ethanol
12. **c.** bacterial fermentation breaks down mainly sugars, leaving most protein intact

27.5 Food Spoilage and Preservation

13. **d.** Anaerobic respiration by bacteria converts trimethylamine oxide to trimethylamine, causing a distinctive unpleasant odor.
14. **c.** Pasteurization
15. **d.** *Listeria monocytogenes*

27.6 Wastewater Treatment

16. **c.** Agricultural runoff enters streams.
17. **a.** Primary treatment by screens and sedimentation
18. **b.** Aerobic or anaerobic respiration

CLINICAL CORRELATION QUESTIONS

1. When organic nitrogen-rich molecules enter a river, they can lead to blooms of algae that produce toxins harmful to people who drink the water. In addition, the organic nitrogen can be released as ammonia, which nitrifiers convert to nitrites and nitrates; these ions are toxic, especially to infants. Instead, in the case reported here, wetland bacteria break down organic molecules from wastewater and facilitate incorporation of carbon and nitrogen into the biomass of living organisms in the wetland ecosystem. The water that comes out is purer, enhancing the health of the river and contributing to healthy water supplies.

2. The initial six passengers probably all became sick following ingestion of a shared food source. But the subsequent illness of other passengers over the next 2 weeks is the result of a chain of infection transmitted by the first ill passengers. Norovirus is often acquired from contaminated food, but it also transmits readily from person to person.

[Glossary]

ABC transporter An ATP-powered transport system that contains an ATP-binding cassette and a transmembrane channel. (5)

ABO blood group system A system that classifies blood on the basis of the presence of antigenic carbohydrates A and B on the surface of red blood cells. (17)

abscess A localized region of pus surrounded by inflamed tissue. (19)

absorption In optics, the capacity of a material to absorb light. (3)

AB toxin An exotoxin composed of two subunits. The toxic A subunit is delivered to the host cell by the B subunit. (18)

Acanthamoeba encephalitis A rare but often fatal inflammation of the brain caused by the ameba *Acanthamoeba*. (24)

acid-fast bacillus (AFB) A descriptive name for mycobacteria; with their mycolic acid–rich cell walls, they stain positive with the Ziehl-Neelsen stain (or acid-fast stain). (20)

acid-fast stain A diagnostic stain for mycobacteria and actinomycetes, which retain the dye carbol-fuchsin because of mycolic acids in the cell wall. (3, 25)

acidophile An organism with optimal growth in acidic environments (generally defined as below pH 5). Compare with *alkaliphile* and *neutralophile*. (6)

acme The point during the illness phase of an infectious disease when symptoms are most severe. (2)

acne vulgaris *or* acne A common chronic disease of the skin resulting from inflammation and/or blockage of hair follicles and their associated glands; involves mainly the face, chest, and back and is characterized by comedones, inflamed papular and pustular lesions, and inflamed cysts. (19)

acquired immunodeficiency syndrome (AIDS) A disease caused by human immunodeficiency virus (HIV) that leads to the destruction of T cells and the inability to fight off opportunistic infections. (23)

actin filament See **microfilament**. (5)

Actinobacteria A phylum of Gram-positive bacteria that have thick, tough cell walls and grow as branched, mycelium-like forms. Some members are commonly referred to as actinomycetes. Examples include *Streptomyces* and *Mycobacterium*. (10)

activated cytotoxic T lymphocyte (CTL) See **cytotoxic T cell**. (16)

activated sludge Sedimented flocs consisting of wastewater and a combination of filamentous and single-celled microbes and invertebrates designed to reduce the biochemical oxygen demand and pathogen load in wastewater. (27)

activation energy The energy needed for reactants to reach the transition state between reactants and products. (4)

activator A regulatory protein that can bind to a specific DNA sequence and increase the transcription of genes. (8)

active immunization Injection of a weakened or killed pathogen or of an antigenically active part of a pathogen with the intent of stimulating the adaptive immune system to produce specific antibodies and antigen-specific T cells. Compare with *passive immunization*. (17)

active transport The energy-requiring movement of molecules across a membrane against their electrochemical gradient. Compare with *passive transport*. (5)

active tuberculosis disease In tuberculosis patients, the stage when they are contagious and exhibit primary disease symptoms, including coughing sputum, fever, night sweats, and weight loss. (20)

acute bacterial endocarditis (ABE) Bacterial infection of the heart by a virulent microbe that causes severe inflammation of the heart lining (endocardium) and valves. Compare with *subacute bacterial endocarditis*. (21)

acute disseminated encephalomyelitis (ADEM) An autoimmune disease that follows an acute viral infection (such as measles), causing inflammation of the spinal cord and brain. (19)

acute HIV infection The first stage of HIV infection, in which a patient first contains detectable levels of antibodies against HIV antigens. (23)

acute infection A disease in which symptoms develop rapidly and resolve quickly (within a few days or weeks). Compare with *chronic infection*. (2)

acute otitis media (AOM) Bacterial infection of the middle ear, typically found in children following an upper respiratory viral infection; usually caused by *Streptococcus pneumoniae*, *Haemophilus influenzae*, or *Moraxella catarrhalis*. (20)

acute rheumatic fever (ARF) An autoimmune sequela of streptococcal infection; results from cross-reactivity of antibodies produced against certain streptococcal M protein epitopes with epitopes of the heart, joints, and nervous system. (20)

adaptive immunity Immune responses activated by a specific antigen and mediated by B cells and T cells. Compare with *innate immunity*. (15)

adenitis Inflammation of a lymph node or gland. (22)

adenosine triphosphate (ATP) A ribonucleotide with three phosphates and the base adenine. It has many functions in the cell, including as a precursor for RNA synthesis and as an energy carrier. (7)

adherence The ability to attach to a substrate. (5)

adhesin Any cell-surface factor that promotes attachment of an organism to a substrate. (2, 18)

adjuvant A substance that enhances the body's immune response to a vaccine antigen. (17)

ADP-ribosyltransferase An activity of some bacterial toxins that enzymatically transfers the ADP-ribose group from NAD^+ to target proteins, altering the target protein's structure and function. (18)

adult-onset hypogammaglobulinemia See **common variable immunodeficiency**. (17)

advanced wastewater treatment See **tertiary treatment (of wastewater)**. (27)

aerobic respiration The use of oxygen as the final electron acceptor of an electron transport system (electron transport chain), where a proton gradient is generated and used to drive ATP synthesis. Compare with *anaerobic respiration*. (6, 7)

aerosol A suspension of droplets in air formed by sneezing or coughing; the droplets can contain bacterial and viral pathogens. (2)

aerotolerant Not using oxygen for metabolism but able to grow in the presence of oxygen. (6)

affinity maturation The process by which the antigen-binding site of an antibody population gains increased affinity for its target antigen or epitope during an adaptive immune response. (16)

African sleeping sickness A protozoan disease caused by *Trypanosoma brucei* and transmitted via the bite of the tsetse fly. The disease begins with a localized lesion, followed by lymphadenopathy, fever, and damage to the central nervous system leading to coma and death. (24)

agar A polymer of galactose that is used as a gelling agent. (1)

agglutination A technique for measuring insoluble antigens on whole cells. For example, technicians type ABO blood groups by using antibodies against the A and B antigens and looking for clumping of cells. (17)

AIDS See **acquired immunodeficiency syndrome**. (23)

airborne transmission In disease, the transfer of a pathogen from one host to another through air via a respiratory route. (2)

alga (*pl.* algae) A microbial eukaryote that contains chloroplasts and conducts photosynthesis. (1, 11)

algal bloom A high density of algae or cyanobacteria resulting from acceleration of growth caused by an overabundance of nutrients. (27)

alkaliphile An organism with optimal growth in alkaline environments (above pH 9). Compare with *acidophile* and *neutralophile*. (6)

allergen An antigen that causes an allergic hypersensitivity reaction. (17)

allograft Transplanted tissue from a donor with MHC genes different from those of the recipient. Unless the recipient is immunosuppressed, the transplant is rejected. (17)

allotype An amino acid difference in the antibody constant region that distinguishes individuals within a species. Compare with *idiotype* and *isotype*. (16)

allylamine Any of a class of antifungal agents that inhibit ergosterol synthesis. Terbinafine (Lamisil) is one example. (13)

alpha helix A protein secondary structure in which a portion of the polypeptide chain forms a helix held in place by hydrogen bonding between an amino hydrogen and the carbonyl of the fourth next aminoacyl group along the peptide backbone. Compare with *beta sheet*. (4)

alpha hemolysis The ability to oxidize heme iron within intact red blood cells in a blood agar plate, generating green color; used as a diagnostic test to identify species and strains of bacteria. Compare with *beta hemolysis*. (25)

alpha-hemolytic Referring to a bacterial species that produces a green color of partial hemolysis around colonies growing on blood agar, such as *Streptococcus mutans* or *S. pneumoniae*. Compare with *beta-hemolytic* and *nonhemolytic*. (25)

alpha toxin A toxin produced by *Staphylococcus aureus* that lyses red blood cell; a hemolysin. (18)

alveolar macrophage A type of macrophage, located in the lung alveoli, that phagocytoses foreign material. (15)

ameba *or* amoeba A protist that moves via pseudopods. (11)

amebic dysentery *or* amebiasis (*pl.* amebiases) A gastrointestinal disease caused by invasion of the intestinal mucosa or the bloodstream by *Entamoeba histolytica*. Symptoms include colitis, diarrhea, fever, bloody stool, abdominal pain, and sometimes hepatitis. (22)

amensalism An interaction between species that harms one partner but not the other. The interaction is less intimate than symbiosis. (27)

amino acid The monomer unit of proteins. Each amino acid contains a central carbon covalently bonded to a hydrogen, an amino group, a carboxyl group, and a side chain. An exception is proline, in which the side chain is cyclized with the central carbon. (4)

amoeba See **ameba**. (11)

amphotericin B An antifungal agent that binds the fungus-specific sterol ergosterol and destroys membrane integrity. (13)

amyloid Clumps of proteins, generally associated with pathology, that are deposited in tissue. In the brain, beta-amyloid deposits are associated with transmissible spongiform encephalopathy (prions) and Alzheimer's disease. In the kidney and other organs, amyloid accumulation refers to clumps of secreted antibody light chains, generated by neoplastic plasma cells. (17, 24)

anabolism See **biosynthesis**. (7)

anaerobic respiration The use of a molecule other than oxygen as the final electron acceptor of an electron transport chain. Compare with *aerobic respiration*. (6)

analytical epidemiology The testing of hypotheses by using experimental or observational methods to establish the cause of disease and associated risk factors based on descriptive epidemiological data. (26)

anaphylaxis A type I hypersensitivity reaction that is caused by the introduction of an allergen into the bloodstream, causing sensitized mast cells to release large amounts of histamine and other immune mediators. Responses can include a rapid drop in blood pressure, constriction of the airways, and swelling of the epiglottis. (17)

anion A negatively charged ion. Compare with *cation*. (4)

annotation Interpreting the information in genome sequences, including identifying genes and predicting gene function. (9)

antagonistic Referring to antibiotics used in combinations whose mechanisms of action interfere with one another and hence diminish their effectiveness. Compare with *synergistic*. (13)

antibiotic A molecule that can kill or inhibit the growth of selected microorganisms. (1, 13)

antibody A host defense protein (a type of immunoglobulin), produced by differentiated B cells, that binds to a specific antigen. (15, 16)

antibody capture An immunoassay that detects whether a patient's serum contains antibodies to a specific antigen. In this procedure, reference antigen coats the wells of the ELISA plate, and patient serum is added to look for antibodies in the serum that will bind to the reference antigen. (25)

antibody-dependent cell-mediated cytotoxicity (ADCC) The process by which natural killer cells destroy viral protein–expressing, antibody-coated host cells. (15)

antibody stain An imaging procedure that uses an antibody to visualize cell components recognized by the antibody with high specificity. The stain is an antibody either linked to an enzyme that converts a colorless reactant to a colored product or linked to a fluorescent molecule that can be seen by fluorescence microscopy. (3)

antigen A compound, recognized as foreign by the cell, that elicits an adaptive immune response. (15, 16)

antigen capture An immunoassay that detects an antigen directly in a patient's serum. In this procedure, a reference antibody coating the wells of an ELISA plate will capture a specific viral or bacterial antigen that may be present in patient serum. Secondary, enzyme-linked antibody that can also bind the pathogen's antigen is used to quantify the amount of antigen captured. (25)

antigenic determinant See **epitope**. (16)

antigenic drift Random mutations in a viral genome that cause minor changes in the structure of viral surface antigens, such as influenza hemagglutinin

and neuraminidase. The consequence is a new viral strain that might better evade the host immune system. (12, 18, 20)

antigenicity Also called *immunogenicity*. A measure of the effectiveness of an antigen in eliciting an immune response. (16)

antigenic shift A major change in a viral antigen that occurs when two (or more) strains of a virus with a segmented genome infect the same host cell and reassort their genome segments. The resulting chimeric virus can express a new combination of surface antigens (such as influenza hemagglutinin and neuraminidase) that can alter virulence or enable the infection of new hosts. (12, 18, 20)

antigenic specificity See **immunological specificity**. (16)

antigenic variation The ability of some protozoan parasites to express multiple variants of their surface protein antigens, thereby overwhelming the host's ability to produce antibodies against the parasite and enabling it to evade the host immune response. (11)

antigen-presenting cell (APC) An immune cell that can process antigens into epitopes and display those epitopes on the cell surface for recognition by naive T cells. (15, 16)

antiport Coupled transport in which the molecules being transported move in opposite directions across the membrane. Compare with *symport*. (5)

antiretroviral therapy (ART) A three-drug cocktail that targets retroviruses and is highly effective at inhibiting the replication of HIV in patients; formerly known as *highly active antiretroviral therapy (HAART)*. (13)

antisepsis Also called *degermation*. The removal of pathogens from living tissues. Compare with *disinfection*. (13)

antiseptic A chemical liquid used to kill microbes on living tissues. (1, 13)

antitoxin An antibody that binds to and inactivates microbial exotoxins. (17)

apicomplexan Any of the eukaryotic phylum Apicomplexa, which contains parasitic alveolates that possess an apical complex used for entry into a host cell. (11)

apoptosis (*pl.* apoptoses) Programmed cell death of eukaryotic cells. (2, 15)

arabinogalactan A polymer of galactose and arabinose; an important structural component of the mycobacterial cell wall. (5)

arbovirus Any of a group of viruses transmitted by arthropod vectors, especially insects. (24)

archaeon (*pl.* archaea) A prokaryotic organism that is a member of the domain Archaea, distinct from bacteria and eukaryotes. Compare with *bacterium*. (1)

areflexia Absence of reflexes. (24)

arthroconidium (*pl.* arthroconidia) A fungal spore produced from the fragmentation of specialized hyphae. It is the infectious agent of *Coccidioides immitis* and *C. posadasii*, which can cause coccidioidomycosis. (24)

arthropod   Any member of the phylum Arthropoda, characterized by an exoskeleton, segmented body, and appendages with joints. Many are parasites. (11)

artifact   A structure that is incorrectly interpreted when viewed through a microscope. (3)

ascariasis (*pl.* ascariases)   An infection caused by *Ascaris lumbricoides*, a parasitic roundworm. After ingestion, the larvae migrate first to the lungs and then to the intestines. (22)

ascospore   The spore produced by an ascomycete fungus. (11)

ascus (*pl.* asci)   The case that contains the spores of an ascomycete fungus. (11)

aseptic   Free of microbes. (1)

aseptic meningitis   Inflammation of meninges due to nonbacterial causes, including viruses. (24)

asexual reproduction   Also called *vegetative reproduction*. Reproduction of a cell by fission or mitosis to form identical daughter cells. Compare with *sexual reproduction*. (11)

assimilation   An organism's acquisition of an element, such as carbon from CO_2, to build into body parts. (27)

atherosclerosis   Hardening of the arteries caused by an inflammatory response to the accumulation of low-density lipoprotein inside the arterial walls, leading to narrowing or blockage of arterial lumen. (21)

atom   The fundamental particle of an element. It is made of a positively charged nucleus, composed of protons and neutrons, and of negatively charged electrons that encircle the nucleus. (4)

atomic number   The number of protons in an atom; it is unique for each element. (4)

atomic weight   The average mass number of the isotopes of an atom. (4)

attaching and effacing lesion   A localized loss of intestinal microvilli as a result of an inflammatory response to the presence of enteropathogenic or enterohemorrhagic *Escherichia coli*. (22)

attenuated   Referring to live but crippled pathogens. Attenuated vaccines contain crippled viruses or bacteria that are injected into a host to stimulate an adaptive immune response. (16, 17, 24)

autoclave   A device that uses pressurized steam to sterilize materials by raising the temperature above the boiling point of water at standard pressure. (1)

autoimmune response   A pathology caused by lymphocytes that can react to self antigens. (17)

autoinducer   A secreted molecule that induces quorum-sensing behavior in bacteria. (6, 18)

autotroph   An organism that can reduce CO_2 to produce organic carbon compounds for biosynthesis. Compare with *heterotroph*. (6, 7)

azole   Any of a class of antifungal agents that interfere with the synthesis of ergosterol. Imidazole is one example. (13)

babesiosis   An emerging disease caused by the protozoan *Babesia microti* and transmitted by the bite of an infected tick. (21)

bacillary dysentery   See **shigellosis**. (22)

bacillus (*pl.* bacilli)   A bacterium or archaeon with a linear shape. (3)

bacitracin   A topical antibiotic that affects cell wall synthesis. (13)

bacteremia   Presence of bacteria in the blood. (14, 21)

bacterial sinusitis   Inflammation and congestion of the paranasal sinuses caused by bacterial infection. (20)

bactericidal   Having the ability to kill bacterial cells. Compare with *bacteriostatic*. (13)

bacteriophage   Also called *phage*. A virus that infects bacteria. (9, 12)

bacteriostatic   Having the ability to inhibit the growth of bacterial cells. Compare with *bactericidal*. (13)

bacterium (*pl.* bacteria)   A prokaryotic organism that is a member of the domain Bacteria, distinct from archaea and eukaryotes. Compare with *archaeon*. (1)

bacteriuria   The presence of bacteria in the urine. (23)

Bacteroidetes   A phylum of Gram-negative bacilli, most of which are obligate anaerobes, such as *Bacteroides*. (10)

Baltimore model   A classification scheme, devised by David Baltimore, that organizes viruses by genome type and method of replication. (12)

bare lymphocyte syndrome   See **MHC II deficiency**. (17)

barophile   Also called *piezophile*. An organism that requires high pressure to grow. (6)

base excision repair   A DNA repair mechanism that cleaves damaged bases off the sugar-phosphate backbone. After endonuclease cleaves the phosphodiester backbone at the damaged site, a new, correct DNA strand is synthesized complementary to the undamaged strand. (8)

base-pair   See **hybridize**. (8)

base substitution   See **point mutation**. (8)

basophil   A white blood cell that stains with basic dyes and secretes compounds that aid innate immunity. (15)

batch culture   The growth of bacteria in a closed system without additional input of nutrients; oxygen may be provided. (6, 12)

B cell   An adaptive immune cell that develops in the bone marrow and gives rise to antibody-producing plasma cells. Along with T cells, they are antigen-specific cells that mediate or regulate the adaptive immune response. (15)

B-cell receptor (BCR)   A B-cell membrane protein complex containing an antibody in association with the proteins Igα and Igβ. (16)

Bell's palsy   Paralysis on one side of the face. (21)

benign   Referring to tumors that are slow-growing and self-contained; they do not spread to infiltrate other tissues. (17)

beta hemolysis   The ability to completely lyse red blood cells in a blood agar plate; used as a diagnostic test to identify species and strains of bacteria. Compare with *alpha hemolysis*. (25)

beta-hemolytic   Referring to a bacterial species that can lyse red blood cells completely, such as *Streptococcus pyogenes*. Compare with *alpha-hemolytic* and *nonhemolytic*. (25)

beta sheet   A protein secondary structure that arises from hydrogen bonding between amino and carbonyl groups along neighboring stretches of the polypeptide backbone to form a sheet. Compare with *alpha helix*. (4)

binary fission   The process of replication in which one cell divides to form two genetically equivalent daughter cells of equal size. (6)

bioburden   The number of microorganisms found in a pharmaceutical product or in a body part. (14)

biochemical oxygen demand (BOD)   Also called *biological oxygen demand*. The amount of oxygen removed from an environment (usually water or wastewater) by aerobic respiration. (27)

biofilm   A community of microbes growing on a solid surface. (6)

biogeochemical cycling   The global interconversion of various inorganic and organic forms of elements. (27)

bioinformatics   A discipline at the intersection of biology and computing that analyzes gene and protein sequence data. (9)

biological oxygen demand   See **biochemical oxygen demand**. (27)

biomass   The dry weight of organic matter found in the bodies of living organisms. (27)

bioremediation   The use of living organisms to detoxify environmental contaminants. (27)

biosynthesis   Also called *anabolism*. The synthesis of complex biomolecules from smaller precursors. Compare with *catabolism*. (7)

biotechnology   An applied field of biology in which DNA and genomes are analyzed and artificial molecules of DNA are constructed for various purposes, such as forensics, agriculture, food safety, and medical research. (8)

biowarfare   The use of pathogenic microbes to inflict mass casualties or cause widespread panic and fear. (26)

Black Death   See **plague**. (22)

blastomycosis   Primary fungal infection of the lungs caused by inhalation of *Blastomyces dermatitidis* spores, which can be followed by disseminated disease affecting the skin, bone, and genitourinary organs. (20)

blood-brain barrier   A selectively permeable membrane made up of tightly packed capillaries that supply blood to the brain and spinal cord. Large molecules and most pathogens cannot permeate the narrow spaces. Fat-soluble (lipophilic) molecules and oxygen can dissolve through the capillary cell membranes and are absorbed into the brain. (14, 24)

blood poisoning   See **sepsis**. (21)

boil   See **furuncle**. (19)

bone marrow transplant   A medical procedure in which a person's bone marrow is destroyed and then replaced with bone marrow that contains healthy stem cells for the proper production of lymphocytes and other blood cells. (16)

botulinum *or* **botulism toxin** *or* **Botox**   The protein toxin produced by the bacterium *Clostridium botulinum*; causes botulism and is used to relax muscle spasms and reduce wrinkles. (10)

botulism   A deadly food-borne disease involving flaccid muscle paralysis; caused by *Clostridium botulinum* toxin. (24)

bovine spongiform encephalopathy (BSE)   Also called *mad cow disease*. A disease caused by prions that infect cattle, characterized by a slow deterioration of brain functions and invariably fatal. (24)

bradykinin   A cell signaling molecule that promotes extravasation, activates mast cells, and stimulates pain perception. (15)

bradyzoite   The slowly dividing cyst-enclosed form of *Toxoplasma gondii*. Compare with *tachyzoite*. (21)

bright-field microscopy   A type of light microscopy in which the specimen absorbs light and appears dark against a light background. (3)

bronchiolitis   Inflammation of the bronchioles. (20)

bronchitis   Inflammation of the bronchi caused by a viral or bacterial infection. (20)

bronchoalveolar lavage (BAL)   A procedure that uses a bronchoscope to introduce sterile saline solution into the lung and then collect the fluid for further analysis. (25)

Bruton's disease *or* **Bruton's agammaglobulinemia**   See **X-linked agammaglobulinemia**. (17)

bubonic plague   A disease caused by *Yersinia pestis*; characterized by swollen lymph nodes that often turn black. (21)

budding   A form of reproduction in which mitosis of the mother cell generates daughter cells of unequal size. (11)

bulla (*pl.* bullae)   A fluid-filled sac, or a blister, such as those associated with bullous impetigo. (19)

bullous impetigo   A contagious skin infection caused by *Staphylococcus aureus* strains that produce exfoliative toxin A; characterized by fluid-filled blisters. Compare with *impetigo*. (19)

Burkitt's lymphoma   A B-cell cancer associated with Epstein-Barr virus infection. (17, 21)

burst size   The number of virus particles released from a lysed host cell. (12)

Calvin cycle   The metabolic pathway of carbon dioxide fixation in which the CO_2-condensing step is catalyzed by Rubisco. It is found in chloroplasts and many bacteria. (7)

cancer   Also called *neoplasm*. Uncontrolled division of abnormal cells. (17)

candidal intertrigo   A skin infection caused by *Candida* species where skin rubs together, such as between toes or under the arms. (19)

***Candida* meningitis**   A fungal infection of the meninges caused by *Candida* species, often seen in premature neonates. (24)

capping   The clustering of B-cell receptor molecules at one pole of a B cell after receptors have been cross-linked by antigens or epitopes. (16)

capsid   The protein shell that surrounds a virion's nucleic acid. Within an enveloped virus, such as HIV, the capsid may be called a *core particle*. (12)

capsule   A slippery outer layer, composed of polysaccharides, that surrounds the cell envelope of some bacteria. (5)

carbohydrate   A class of biomolecules made of the atoms carbon, hydrogen, and oxygen; composed of simple sugars and sugars covalently bound together to form larger structures, including disaccharides, oligosaccharides, and polysaccharides. (4)

carbon fixation *or* **carbon dioxide (CO_2) fixation**   The enzymatic covalent incorporation of inorganic carbon dioxide (CO_2) into an organic compound. (7)

carbuncle   A painful lesion resulting from the joining of several furuncles. (19)

cardiomyopathy   A decrease in the ability of the heart muscle to contract. (21)

carrier   A person who harbors a disease agent but does not have symptoms of the disease. (2)

case-control study   An observational epidemiological study comparing a disease (or affected) group with a control group that is disease-free (or unaffected). The investigator assesses each individual for previous exposure to a possible causal agent. (26)

casein   Any of a family of hydrophobic milk proteins found as suspended droplets in milk. (27)

catabolism   The cellular breakdown of large molecules into smaller molecules, releasing energy. Compare with *biosynthesis*. (7)

catabolite   A molecule, formed by the enzymatic digestion of food molecules, that is fed into metabolic energy pathways. (7)

catalase   An enzyme that catalyzes the conversion of hydrogen peroxide to molecular oxygen and water. (25)

catalase test   A rapid diagnostic test that looks for bubbles due to the release of O_2 when hydrogen peroxide is added to a bacterial colony. (25)

catalysis   The process of accelerating the rate of a chemical reaction; in cells, it is always accomplished with the assistance of enzymes. (4)

catalyst   A molecule or substance that increases the rate of a chemical reaction without being consumed in the reaction. (4)

cation   A positively charged ion. Compare with *anion*. (4)

cat scratch disease   An infection caused by *Bartonella henselae* following a bite or scratch from an infected cat. (21)

cavity   See **dental caries**. (22)

cell   The smallest unit of life, composed of a membrane-enclosed compartment of water solution containing molecules that carry out metabolism. (1)

cell envelope   The cell membrane, cell wall, and (for Gram-negative species) outer membrane of a bacterium. (5)

cell fusion   A method of viral spreading in which an infected host cell fuses with an uninfected cell, allowing the virus particles to enter the uninfected cell through their common cytoplasm. (12)

cell-mediated immunity   Also called *cellular immunity*. A type of adaptive immunity employing mainly T-cell lymphocytes. (16)

cell membrane   Also called *cytoplasmic membrane* or *plasma membrane*. The phospholipid bilayer that encloses the cytoplasm. (4, 5)

cellular immunity   See **cell-mediated immunity**. (16)

cellulitis   A spreading superficial infection (inflammation) of the skin characterized by localized pain, swelling, tenderness, erythema, and warmth. (19)

cell wall   A rigid structure outside the cell membrane. The molecular composition depends on the organism; in bacteria, it is composed of peptidoglycan. (5)

central nervous system (CNS)   The brain and spinal cord. Compare with *peripheral nervous system*. (24)

cephalosporin   Any of a class of semisynthetic beta-lactam antibiotics that are used to fight penicillin-resistant microbes. (13)

cercaria (*pl.* cercariae)   The fork-tailed larva of the parasitic flatworms *Schistosoma* and *Opisthorchis*. (22)

cerebrospinal fluid (CSF)   A clear fluid, derived from blood, that bathes the spinal cord and brain. (24)

cestode   See **tapeworm**. (11, 22)

Chagas disease   A zoonotic illness caused by the protozoan *Trypanosoma cruzi* and transmitted by the bite of the bloodsucking reduviid bug. (21)

chain of infection   Factors that affect the spread of disease, which include the organism, reservoir, mode of transmission, and host. (1, 26, 27)

chancre   A painless, hard lesion due to an inflammatory reaction at the site of infection with *Treponema pallidum*, which causes syphilis. (23)

chancroid   A sexually transmitted infection, caused by *Haemophilus ducreyi*, that is characterized by painful ulcers on the genitals. (23)

chemically defined minimal medium   See **minimal medium**. (6)

chemoautotroph   Also called *lithotroph*. An organism that oxidizes inorganic compounds to yield energy and reduce carbon dioxide. (6)

chemokine   Any of a group of proteins, produced by damaged or infected tissues, that stimulate the migration of white blood cells toward the site of infection. (15)

chemostat   A continuous culture system in which the introduced medium contains a limiting nutrient. (6)

chemotaxis   The ability of an organism to move toward or away from specific chemicals. (5)

chemotherapeutic agent   An antibiotic made by microbes or modified or synthesized in the laboratory. (13)

chemotherapeutic index   A measure of an antibiotic's safety; the ratio of the toxic dose to the therapeutic dose. (13)

chemotrophy   Metabolism that yields energy from oxidation-reduction reactions without using light energy. Compare with *phototrophy*. (6, 7)

chickenpox   A childhood disease, caused by the varicella-zoster virus, that presents as an itchy rash over much of the upper body. (19)

chitin   A polymer of *N*-acetylglucosamine that is a major structural component of fungal cell walls. (11)

chlamydia   1. An obligate, intracellular parasitic bacterium of the phylum Chlamydiae. (10) 2. The most frequently reported sexually transmitted infection in the United States. Symptoms range from none, to a burning sensation upon urination, to sterility. (23)

Chlamydiae   A phylum of bacteria that are obligate intracellular pathogens causing diseases such as pneumonia, trachoma, and chlamydia. (10)

chlorophyll   A magnesium-containing porphyrin pigment that captures light energy at the start of photosynthesis. (7)

cholecystitis   Inflammation of the gallbladder. (22)

cholera   A serious gastrointestinal illness caused by *Vibrio cholerae* and characterized by excessive watery diarrhea. (22)

cholera toxin   The *Vibrio cholerae* protein that enters host intestinal cells, alters cAMP production, and causes the symptoms of cholera. The protein is an example of an AB toxin. (22)

chromosome   A molecule of DNA (for a cell) or DNA or RNA (for a virus); it makes up part or all of an organism's genome. (8)

chronic infection   A disease in which symptoms develop gradually (over weeks or months) and are slow to resolve. Compare with *acute infection*. (2)

chronic lymphocytic leukemia   The most common type of leukemia; it is a B-cell cancer often associated with an abnormality in specific micro-RNA production. (17)

ciliate   A motile protozoan, typically identified by having many cilia on the cell surface and two different-sized nuclei. (11)

cilium (*pl.* cilia)   A short, hairlike structure of eukaryotes composed of microtubules; structurally similar to the flagellum. Cilia beat in waves to propel the cell. (11)

citric acid cycle   See **tricarboxylic acid cycle**. (7)

clade   A group of organisms that includes an ancestral species and all its descendants. (9)

clap   See **gonorrhea**. (23)

class I MHC molecule   Any of a class of surface proteins, found on all nucleated cells of the human body (absent from red blood cells and platelets), that present cytoplasmically synthesized foreign antigen epitopes (e.g., virus epitopes) to cytotoxic T cells. (16)

class II MHC molecule   Any of a class of surface proteins, found on antigen-presenting cells (dendritic cells, macrophages, and B cells), that present phagocytosed foreign antigen epitopes to helper T cells. (16)

class switching   See **isotype switching**. (16)

clinical latency   The second stage of HIV infection, in which T cells are slowly and steadily lost but few, if any, symptoms appear. (23)

clonal selection   The rapid proliferation of a subset of B cells during the primary or secondary antibody response. (16)

cloning vector   A small genome, often a plasmid or a virus, into which foreign DNA can be inserted for cloning. (12)

clostridial myonecrosis   See **gas gangrene**. (19)

CNA plate   Blood agar medium containing the inhibitors colistin and nalidixic acid; a selective medium that permits the growth of Gram-positive organisms. (25)

coccidioidal meningitis   A fungal infection of the meninges caused by *Coccidioides immitis* or *C. posadasii*. (24)

coccidioidomycosis   Also called *valley fever*. A fungal infection of the lungs caused by inhalation of the spores of a variety of *Coccidioides* species, including *Coccidioides immitis*. (20)

coccus (*pl.* cocci)   A bacterium or archaeon with a spherical shape. (3)

coding strand   Also called *sense strand*. The strand of DNA containing the sequence of nucleotides that is equivalent to the mRNA sequence synthesized from the template strand of DNA. (8)

codon   A set of three nucleotides within mRNA that encodes a particular amino acid. (8)

cohort study   An observational epidemiological study comparing an exposed group with a control group that is not exposed to a particular disease-causing agent. The investigator assesses whether each individual in the study contracts the disease. (26)

cold agglutinin   An antibody that can agglutinate red blood cells at 4°C but not at 37°C. (20)

cold sore   Also called *fever blister* or *herpes labialis*. A small blister occurring on the lips that is caused by herpes simplex virus 1. (19)

colitis   Inflammation of the large intestine. (22)

colonization   The ability of a microorganism to affix itself to a host surface or an environmental habitat, where it can then replicate. (2)

colony   A visible cluster of microbes on a plate, all derived from a single founding microbe. (1, 6)

colony-forming unit (CFU)   A measure of the number of viable bacteria or fungi present in a sample that can grow to form colonies on an agar plate. (6)

colorization   The addition of color to a microscopic image to provide contrast and/or highlight components of the image. (3)

comedone acne   A type of acne resulting from blocked hair follicles or pores. (19)

commensalism   A symbiotic relationship that benefits only one partner. (27)

common cold   The most frequent viral infection of the upper respiratory tract; it is a self-limiting infection, commonly caused by a rhinovirus, but can also be caused by a coronavirus or respiratory syncytial virus (RSV). (20)

common variable immunodeficiency (CVID)   Also called *adult-onset hypogammaglobulinemia*. A disorder characterized by various types of serum immunoglobulin deficiencies, including low levels of immunoglobulin isotypes IgG, IgA, and IgE, and an impaired ability to produce specific antibodies after antigen exposure. (17)

competence factor   A species-specific secreted bacterial protein that induces competence for transformation. (9)

complement   Innate immunity proteins in the blood that form holes in bacterial membranes, killing the bacteria. Some components attract phagocytes; others can coat bacteria and promote phagocytosis. (15)

complement deficiency   An immunodeficiency characterized by autosomally inherited defects in the complement pathway, resulting in increased susceptibility to infections with *Neisseria* and other pyogenic bacteria. (17)

complex   A macromolecule made of multiple protein subunits that work in a coordinated manner. (4)

complex medium   Also called *rich medium*. A nutrient-rich growth solution including undefined chemical components such as beef broth. Compare with *minimal medium* and *synthetic medium*. (6)

compound microscope   A microscope with multiple lenses to compensate for lens aberration and increase magnification. (3)

condensation reaction   In biochemistry, the formation of a covalent bond between two molecules that in the process releases a water molecule. (4)

condenser   In a microscope, a lens that focuses parallel light rays from the light source onto a small area of the specimen to improve the resolution of the objective lens. (3)

condyloma acuminatum (*pl.* condylomata acuminata)   A genital wart caused by human papillomavirus. (19, 23)

confluent   Describing a mode of growth that results in a lawn of organisms covering a surface. (6)

congenital cytomegalovirus infection   Also called *cytomegalic inclusion disease*. A condition caused by cytomegalovirus contracted in utero. (21)

congenital syphilis   Syphilis contracted in utero. (23)

conjugated vaccine   A vaccine that combines a highly immunogenic protein with a poorly immunogenic bacterial capsule polysaccharide to boost the immune reaction to the polysaccharide. (17)

conjugation   Horizontal gene transmission involving cell-to-cell contact. In bacteria, pili draw together the donor and recipient cell envelopes, and a protein complex transmits DNA across. In ciliated eukaryotes, a conjugation bridge forms between two cells, connecting their cytoplasm and allowing the exchange of micronuclei. (9, 11)

conjunctivitis   Inflammation of the conjunctiva of the eye. (19)

constant region   The region of an antibody that defines the class of a heavy chain (isotype) or light chain. (16)

consumer   An organism that acquires nutrients from producers, either directly or indirectly. (27)

continuous culture   A culture system in which new medium is continuously added to replace old medium. (6)

contractile vacuole   An organelle in eukaryotic microbes that pumps water out of the cell. (5)

contrast   Differential absorption or reflection of electromagnetic radiation between an object and its background that allows the object to be distinguished from the background. (3)

convalescence   The stage of an infectious disease after symptoms have disappeared and the patient is returning to health. (2)

core particle   See **capsid**. (12)

core polysaccharide   A sugar chain that attaches to the glucosamine of lipopolysaccharides and extends outside the cell. (5)

costovertebral tenderness   Tenderness between the last rib and the lumbar vertebrae; an indicator of pyelonephritis. (23)

counterstain   A secondary stain used to visualize cells that do not retain the first stain. (3)

coupled transport   The movement of a substance against its electrochemical gradient (from lower to higher concentration or from opposite charge to like charge), using the energy provided by the simultaneous movement of a different chemical down its electrochemical gradient. (5)

covalent bond   A bond between two atoms that share a pair of electrons. (4)

C-reactive protein (CRP)   A protein that stimulates the complement cascade; it is induced in the liver by cytokines. Elevated levels in the blood are associated with inflammation of any kind, including infection. (15)

Creutzfeldt-Jakob disease (CJD)   A disease caused by prions that infect humans, characterized by a slow deterioration of brain functions and invariably fatal. (24)

croup   Also called *laryngotracheobronchitis*. Viral infection of the larynx, trachea, and bronchi that results in a "barking" cough; usually caused by parainfluenza viruses. (20)

cryptococcal meningoencephalitis   A fungal infection of the central nervous system caused by certain strains of *Cryptococcus*. It is a serious concern among immunocompromised patients. (24)

cryptococcosis (*pl.* cryptococcoses)   A serious, potentially fatal fungal infection of the lungs or meninges caused by *Cryptococcus neoformans*. Cryptococcal meningitis is usually a hallmark infection for AIDS. (20)

cryptosporidiosis (*pl.* cryptosporidioses)   An intestinal infection caused by the parasite *Cryptosporidium*. (22)

curd   Coagulated milk proteins produced by the combined action of lactic acid–producing bacteria and stomach enzymes of certain mammals, such as cattle. (27)

Cyanobacteria   A phylum of bacteria that are phototrophs, the only bacteria that produce oxygen. (10)

cycloserine   An antibiotic that inhibits peptidoglycan synthesis. (13)

cyclosporiasis   A gastrointestinal disease caused by infection with *Cyclospora cayetanensis* sporocysts; presentation varies from asymptomatic to diarrhea, as well as flu-like symptoms. (22)

cyst   1. A tough, protective capsule in tissue that encloses the larva of a parasitic worm within an infected animal. 2. An infectious form of many protozoan parasites, usually passed in feces. (11)

cysticercus (*pl.* cysticerci)   The cystic phase of certain tapeworms, such as *Taenia*. The cysts are found in the muscle of infected cattle, pigs, and goats. (22)

cystitis   Inflammation of the urinary bladder often caused by infection. (23)

cytochrome oxidase test   A diagnostic test for the presence of *Neisseria*. The colorimetric test is positive only if the bacteria contain both cytochrome oxidase and cytochrome *c*. (25)

cytokine   Any of a group of small signaling proteins secreted by human cells that bind to receptors on endothelial and immune system cells, altering the behavior of those cells. (14)

cytokine storm   A massive production of cytokines leading to excessive tissue damage, inflammation, and shock. (16)

cytomegalic inclusion disease (CID)   See **congenital cytomegalovirus infection**. (21)

cytomegalovirus (CMV)   Also called *human herpesvirus 5*. A double-stranded DNA virus that causes a mononucleosis-like disease in adults and cytomegalic inclusion disease in neonates. (21)

cytoplasm   Also called *cytosol*. The aqueous solution contained by the cell membrane in all cells and outside the nucleus (in eukaryotes). (4)

cytoplasmic membrane   See **cell membrane**. (5)

cytoskeleton   A collection of filamentous proteins that impart structure to cells, aid movement of cells, and facilitate transport of components within cells. In a eukaryote, components of the cytoskeleton include microfilaments, intermediate filaments, and microtubules. (5)

cytosol   See **cytoplasm**. (4)

cytotoxic T cell (T$_C$ cell)   A T cell that expresses CD8 on its cell surface, binds MHC I–presented epitopes, and can secrete toxic proteins such as perforin and granzymes. Once activated, T$_C$ cells are known as *activated cytotoxic T lymphocytes*, or *CTLs*. (16)

daisy-chain nosocomial infection   An infection that begins within one hospital or medical facility and is then spread to other facilities by infected patients who are moved from place to place. (26)

dark-field microscopy   A type of light microscopy in which the specimen scatters light rays and appears light against a dark background. (3)

daughter strand   One of the two growing strands formed during DNA replication; it is complementary to the *parental strand*, which serves as its template. (8)

dead zone   Also called *zone of hypoxia*. An anoxic region of an ocean, devoid of most fish and invertebrates. (27)

death phase   A period of cell culture, following stationary phase, during which bacteria die faster than they replicate. (6)

death rate   The rate of halving of population density under a deleterious condition. (6)

decimal reduction time (D-value)   The length of time it takes for a treatment to kill 90% of a microbial population, and hence a measure of the efficacy of the treatment. (13)

decline phase   The stage of an infectious disease in which symptoms begin to subside. (2)

decomposer   An organism that consumes dead biomass. (27)

defensin   A type of small, positively charged peptide, produced by animal tissues, that destroys the cell membranes of invading microbes. (15)

definitive host   The host organism in which a parasitic worm develops and undergoes sexual reproduction. (11, 22)

degenerative evolution   See **reductive evolution**. (5, 9)

degermation   See **antisepsis**. (13)

degranulation   A process in which intracellular vesicles (granules) within mast cells, neutrophils, natural killer cells, and cytotoxic T cells release their toxic molecules to the external environment. (17)

delayed-type hypersensitivity (DTH)   See **type IV hypersensitivity**. (17)

deletion   The loss of one or more nucleotides from a DNA sequence. Compare with *insertion*. (8)

denatured   Describing loss of protein structure (and hence function) as a result of extreme temperature or a chemical agent. (4)

dendritic cell   An antigen-presenting white blood cell that primarily takes up small soluble antigens from its surroundings. (15)

dengue hemorrhagic fever (DHF)   The most severe form of dengue fever, with symptoms that include hemorrhaging, respiratory distress, extreme bone pain, and organ failure. (21)

denitrification   The process by which some bacteria and archaea reduce nitrate (NO_3^-) to nitrite (NO_2^-), diatomic nitrogen (N_2), and in some cases ammonia (NH_3) to yield energy. Compare with *nitrification* and *nitrogen fixation*. (6)

dental caries   Also called *cavity*. Tooth decay, which is caused by acidic products of bacterial fermentation that destroy tooth enamel. (22)

dental plaque   A thickening biofilm of microorganisms that forms at the margin of the teeth and gums. (22)

depth of field   In a microscope, a region of the optical column over which a specimen appears in reasonable focus. (3)

dermatome   An area of skin served by a specific spinal nerve. (19)

dermatophyte   A fungal pathogen that colonizes human skin. (19)

dermis   The deeper layer of skin located below the epidermis and consisting mostly of connective tissue. (19)

descriptive epidemiology   The collection of data concerning the people, places, and times of events related to a disease. (26)

desensitization   A clinical treatment to decrease allergic reactions by exposing patients to small doses of an allergen. (17)

diabetes   A chronic systemic disease caused by abnormal plasma glucose levels due to faulty glucose metabolism. (17)

diaphragm   A device in a microscope that varies the diameter of the light column, changing the amount of light admitted. (3)

diarrhea   A malady of the gastrointestinal tract characterized by having three or more loose stools a day. (22)

dichotomous key   A tool for identifying organisms, in which a series of yes/no decisions successively narrows down the possible categories of species. (9)

differential   A list of possible causes of an infection or disease. (2)

differential medium   A growth medium that can distinguish between various bacteria on the basis of metabolic differences. Compare with *selective medium*. (6)

differential stain   A stain that differentiates among objects by staining only particular types of cells or specific subcellular structures. (3)

diffusion   The energy-independent net movement of a substance from a region of higher concentration to a region of lower concentration. (4)

DiGeorge syndrome   A recessive genetic immuno-deficiency disorder causing underdevelopment or complete lack of the thymus, underdevelopment of some facial features, and twitching of extremities. (17)

dilution streaking   See **isolation streaking**. (6)

dinoflagellate   Any of a group of secondary endo-symbiont algae that conduct photosynthesis and have two flagella, one of which is wrapped distinctively around the cell equator. (11)

diphtheria   A systemic disease caused by the dissemination of a toxin produced by *Corynebacterium diphtheriae*. (20)

diploid   Describing an organism (or cell) that contains two copies of each chromosome in its cells. Compare with *haploid*. (11)

direct contact   In disease, the spread of a pathogen either through intimate contact or by aerosolization. (2)

direct ELISA   An immunoassay that detects a specific antigen bound to a microtiter plate with enzyme-conjugated antibodies against that specific antigen and a colorless substrate that the enzyme converts to a colored product. (17)

direct immunofluorescence test   A diagnostic method for detecting microbial pathogens in a patient sample by adding fluorescently labeled anti-pathogen antibodies to a sample of tissue or fluid (blood, saliva, cerebrospinal fluid) and then observing under fluorescence light microscopy to see the now-fluorescent pathogen in the sample. (17)

disaccharide   A carbohydrate composed of two simple sugars joined by a covalent bond. (4)

disease   The disruption of the normal structure or function of any body part, organ, or system that is recognized by a characteristic set of symptoms. (2)

disinfectant   A chemical liquid used to kill bacteria on inanimate objects. Not to be used on living tissues. (13)

disinfection   The removal of pathogens from inanimate surfaces. Compare with *antisepsis*. (13)

disseminated intravascular coagulation (DIC)   A consequence of septic shock in which excessive coagulation occurs, causing clots to form in organs throughout the body. (21)

dissimilation   An organism's catabolism or oxidation of nutrients to inorganic minerals that are released into the environment. (27)

divergence   The accumulation of genetic differences between two lineages after evolutionary separation; the differences arise from mutations, gene mobility, and natural selection. It can lead to the formation of new species. (9)

DNA ligase   An enzyme that cells use to form a covalent bond at a nick in the phosphodiester backbone; also used in molecular biology laboratories to join pieces of DNA. (8)

DNA polymerase   Any of a group of enzymes that catalyze DNA replication. (8)

DNA sequencing   A technique to determine the order of bases in a DNA sample. (1)

DNA vaccine   An experimental vaccine that contains the DNA for the antigen rather than the fully formed antigen. (17)

domain   1. In protein structure, the portion of a protein that possesses a defined function, such as binding DNA. (4) 2. In taxonomy, one of three major subdivisions of life: Archaea, Bacteria, and Eukarya. (9)

doubling time   See **generation time**. (6)

down-regulation   See **repression**. (8)

DTaP series   A vaccine against diphtheria toxin, tetanus toxin, and pertussis antigens. (20)

dysbiosis   An imbalance in the composition of a microbiome that can be brought on by stress or antibiotic therapy. It may cause some human diseases or even influence behaviors. (14)

dysuria   Painful or difficult urination. (23)

eastern equine encephalitis (EEE) virus   A positive-sense, single-stranded RNA virus transmitted to birds, humans, and horses via a mosquito vector. (24)

echinocandin   Any of a class of antifungal agents that block cell wall synthesis. (13)

eclipse period   The time during the replication cycle of a virus after viral genome injection into a host cell but before complete virions are formed. (12)

ecosystem   A community of species plus their environment (habitat). (27)

ecosystem service   The benefits for human society conferred by a natural ecosystem. (27)

ectoparasite   A harmful organism that colonizes the surface of a host. (2, 11)

edema factor (EF)   A component of anthrax toxin with adenylate cyclase activity. (18)

efflux transporter   A type of membrane-embedded transporter that expels waste products from cells. (5)

electrolyte   See **ion**. (4)

electron microscopy (EM)   A form of microscopy in which a beam of electrons accelerated through a voltage potential is focused by magnetic lenses onto a specimen. (3)

electron transport system (ETS) *or* **electron transport chain**   A series of membrane-embedded proteins that convert the energy of redox reactions into a proton potential. (6, 7)

electroporation   A laboratory technique that temporarily makes the cell membrane more leaky to allow the uptake of DNA. (9)

element   A pure chemical substance composed of one type of atom; it is given a unique identifier, its atomic number, which equals the number of protons in its nucleus. (4)

elephantiasis (*pl.* elephantiases)   A symptom of several diseases of the lymphatic system, including infection by filarial worms, characterized by blockage of lymphatic vessels that leads to grotesque swelling of the limbs. (21)

elite controller   A rare HIV-positive individual who maintains a high CD4$^+$ T-cell number and shows little or no viremia. (23)

elongation   The middle phase of DNA synthesis, in which DNA polymerase adds dNTPs to the growing daughter strands; also the middle-phase events in transcription and in translation. (8)

Embden-Meyerhof-Parnas pathway (EMP pathway)   See **glycolysis**. (7)

emerging disease   A disease that has recently appeared in a population. (2)

empirical therapy   The administration of antibiotics according to a patient's symptoms and a clinician's understanding of infectious disease before the pathogen has been identified; used in life-threatening situations, such as meningitis. (13, 25)

empty magnification   Magnification without an increase in resolution. (3)

enanthem   A rash on a mucous membrane. Compare with *exanthem*. (19)

encephalitis (*pl.* encephalitides)   Inflammation of the brain. (24)

encephalopathy   Diseased or damaged brain tissue, such as that resulting from an infection. (24)

endemic disease   A disease that is always present in a population, although the prevalence of infection may be low. (2, 26)

endergonic   Describing a reaction that requires an input of energy to proceed. Compare with *exergonic*. (4)

endocarditis   Inflammation of the endocardium (heart's inner lining), usually involving the heart valves. (21)

endocytosis   The invagination of the cell membrane to form a vesicle that contains extracellular material. (4, 24)

endomembrane system   A series of membranous organelles that organize uptake, transport, digestion, and expulsion of particles through a eukaryotic cell. It includes endosomes, lysosomes, the endoplasmic reticulum, the nuclear envelope, and the Golgi complex. (5)

endoparasite   A harmful organism that lives inside a host. (2)

endophthalmitis   An infection of the inner eye. (19)

endoplasmic reticulum (ER)   A complex membranous organelle of eukaryotic cells where the synthesis of proteins and lipids occurs. (5)

endosome   A membrane-enclosed compartment inside a eukaryotic cell; a component of the endomembrane system, originating from the Golgi. (4)

endospore   A durable, inert, heat-resistant bacterial spore that can remain viable for thousands of years. (10)

endospore stain A differential stain that is specific for the endospore coat of various bacteria, typically a firmicute species. (3)

endosymbiont An organism that lives as a symbiont inside another organism. (1, 9)

endosymbiosis An intimate association between different species in which one partner grows within the body or cell of another organism. (5)

endotoxin A lipopolysaccharide in the outer membrane of Gram-negative bacteria that becomes toxic to the host after the bacterial cell has lysed. (5, 18)

energy The ability to do work. (4, 7)

enriched medium A growth solution for fastidious bacteria, consisting of complex medium plus additional components. (6)

enrichment culture The use of selective growth media and specific incubation conditions to allow only certain microbes to grow. (1)

enteric fever See **typhoid fever**. (22)

enteric pathogen A pathogenic bacterium that causes diarrhea. (22)

enteritis Inflammation of the small intestine. (22)

enteroaggregative E. coli (EAEC) Any of a group of diarrheagenic *Escherichia coli* strains that have unique bundle-forming pili to adhere to and colonize the intestinal mucosa. (22)

enterocolitis Inflammation of the colon and small intestine. (22)

enterohemorrhagic E. coli (EHEC) Any of a group of diarrheagenic *Escherichia coli* strains that secrete Shiga toxin and induce intestinal cells on which they attach to form pedestals; the bacterial strains are found in animal reservoirs and cause hemorrhagic disease (bloody stool). (22)

enteroinvasive E. coli (EIEC) Any of a group of diarrheagenic *Escherichia coli* strains that invade epithelial cells. (22)

enteropathogenic E. coli (EPEC) Any of a group of diarrheagenic *Escherichia coli* strains that form pedestals similar to those that occur in enterohemorrhagic *E. coli* but typically lack toxins. (22)

enterotoxigenic E. coli (ETEC) Any of a group of diarrheagenic *Escherichia coli* strains that secrete enterotoxins. (22)

enterotoxin A toxic protein, made by microbes, that damages the intestine of the host and causes diarrhea and/or vomiting; produced by some Gram-negative and Gram-positive pathogens. (14, 22)

entropy A measure of the disorder in a system. (4, 7)

envelope For a virus, a membrane enclosing the capsid, or core particle. (12)

enzyme A biological catalyst; a protein or RNA molecule that can speed up the rate of a reaction without itself being changed. (4)

enzyme-linked antibody An antibody conjugated to an enzyme that converts a colorless substrate to a colored product. (17)

enzyme-linked immunosorbent assay (ELISA) or enzyme immunoassay (EIA) A set of diagnostic techniques that use enzyme-linked antibodies to detect specific antigens, pathogens, or antibodies. (17)

eosinophil A white blood cell that stains with the acidic dye eosin and secretes compounds that facilitate innate immunity. (15)

eosinophilia An abnormally large number of eosinophils in the blood. (22)

epidemic A disease outbreak in which many individuals in a population become infected over a short time. Compare with *pandemic*. (2, 26)

epidemiology The study of disease in a population, including how far a known disease has spread, the source of an outbreak, and the factors that influence how many individuals will contract it. (26)

epidermis The outer protective cell layer in most multicellular animals. (14, 19)

epididymitis Inflammation of an epididymis, one of two coiled tubes that carry semen from the testicles. (23)

epiglottitis Inflammation of the epiglottis. (20)

epitope Also called *antigenic determinant*. A small segment of an antigen that can elicit an immune response. An antigen can have many epitopes. (16)

Epstein-Barr virus Also called *human herpesvirus 4*. A double-stranded DNA virus that commonly causes infectious mononucleosis and is strongly associated with Burkitt's lymphoma. (21)

erysipelas An acute infection of the upper layer of dermis that spreads to the superficial lymphatic vessels and is caused by streptococcal bacteria, typically *Streptococcus pyogenes*, and that is characterized by a fiery-red rash, lymphadenopathy, and fever. (19)

erythema infectiosum Also called *fifth disease*. A disease that is caused by parvovirus B19 and causes a nonpruritic malar rash, "slapped-cheek" rash, along with mild fever and flu-like symptoms. (19)

erythema marginatum A skin rash with distinct margins, usually on the trunk or extremities, that occurs in some children who have contracted acute rheumatic fever. (20)

erythema migrans A bull's-eye rash characteristic of borreliosis (Lyme disease). (21)

erythematous Reddened and inflamed (describing skin). (19)

erythroblastosis fetalis See **Rh incompatibility disease**. (17)

ESKAPE pathogen Any of the six pathogens designated by U.S. medical authorities as the leading causes of life-threatening hospital-acquired infections: *Enterococcus faecium*, *Staphylococcus aureus*, *Klebsiella pneumoniae*, *Acinetobacter baumannii*, *Pseudomonas aeruginosa*, and *Enterobacter* species. (8)

essential nutrient A compound that an organism cannot synthesize and must acquire from the environment in order to survive. (6)

etiology The organism, virus, or toxin that causes a disease. (2)

eukaryote An organism whose cells contain a nucleus. All eukaryotes are members of the domain Eukarya. Compare with *prokaryote*. (1, 5)

eutrophic Referring to a freshwater or marine environment in which oversupply of electron donors (usually organic matter) leads to oxygen depletion and hence loss of animal life. (27)

exanthem Any widespread rash on the skin that is accompanied by fever, malaise, or headache. Compare with *enanthem*. (19)

exanthem subitum See **roseola infantum**. (19)

exergonic Describing a spontaneous reaction that releases free energy. Compare with *endergonic*. (4)

exocytosis The fusion of vesicles with the cell membrane to release vesicle contents into the extracellular milieu. (5, 24)

exotoxin Any of a group of protein toxins, secreted by bacteria, that kill or damage host cells. (18)

experimental study A controlled analysis designed to test a question or hypothesis. (26)

exponential growth Growth in which the population size doubles at a fixed rate. (6)

exponential phase Also called *logarithmic (log) phase*. A period of cell culture, occurring after lag phase, during which bacteria grow exponentially at their maximum possible rate based on the growth conditions. (6)

extensively drug-resistant tuberculosis (XDR-TB) Disease caused by strains of *Mycobacterium tuberculosis* that are resistant to at least two first-line tuberculosis drugs and at least three second-line drugs. (20)

extracellular polymeric substances (EPS) Polysaccharides and entrapped materials that form a thick extracellular matrix around microbes in a biofilm. (6)

extravasation The movement of white blood cells out of blood vessels and into surrounding infected tissue. (15)

extremophile An organism that grows only in an extreme environment—that is, an environment including one or more conditions that are "extreme" in relation to the conditions for human life, such as very high or very low pH. (6)

exudate The protein- and cell-rich fluid that oozes out of tissues because of inflammation. (20)

eyepiece See **ocular lens**. (3)

F(ab)$_2$ region The two linked antigen-binding arms of an antibody molecule. (16)

facilitated diffusion Passive transport across a membrane that is facilitated by transport proteins. (5)

factor H A normal serum protein that prevents the inadvertent activation of complement in the absence of infection. (15)

facultative Having two alternative growth states or conditions, such as aerobic and anaerobic. (6)

facultative anaerobe An organism that can grow under either aerobic or anaerobic conditions (in the presence or absence of oxygen). Compare with *strict aerobe* and *strict anaerobe*. (6)

facultative intracellular pathogen A pathogen that can live either inside host cells or outside host cells. (18)

false-negative reaction Negative result from a diagnostic test of a patient who has the disease or disease-causing pathogen, but the test is not sensitive enough to detect its presence. See also *sensitivity*. (25)

false-positive reaction   Positive result from a diagnostic test of a patient who is free of the disease or disease-causing pathogen, but the test is not specific enough to detect only the relevant parameter. See also *specificity*. (25)

F⁻ cell   A cell lacking F factor that can receive the F factor from an F⁺ cell during conjugation. (9)

F⁺ cell   The DNA donor cell that transmits the fertility factor F to an F⁻ cell during conjugation. (9)

Fc region   The region of an antibody that binds to specific receptors on host cells in an antigen-independent manner. It is found in the carboxy-terminal "tail" region of the antibody. (16)

fecal microbiota transplant (FMT) *or* fecal transplant   Transfer of the intestinal microbiome of a healthy donor to a patient suffering from a severe intestinal disease. (14)

fecal-oral route   A route of transmission of pathogenic microbes in which organisms are ingested in contaminated food or water and then defecated, so they again contaminate the food or water supply. (2)

feces   A mixture of chyme and intestinal bacteria concentrated into a semisolid waste material in the large intestine. (22)

fermentation   1. The production of ATP via substrate-level phosphorylation, using organic compounds as both electron donors and electron acceptors. 2. Industrial fermentation is the production of microbial products that are made by microbes grown in fermentation vessels; it may include respiratory metabolism to maximize microbial growth. (1, 6, 7)

fermented food   A food product that is biochemically modified by microbial growth. (27)

fertility factor (F factor)   A specific plasmid (transferred by an F⁺ donor cell) that contains the genes needed for pilus formation and DNA export. (9)

fever blister   See **cold sore**. (19)

fifth disease   See **erythema infectiosum**. (19)

filamentous virus   A virus consisting of a helical capsid surrounding a single-stranded nucleic acid. (12)

filarial worm   A parasitic nematode that grows within blood, lymph, or subcutaneous fatty tissue. (11)

filariasis (*pl.* filariases)   Any of a group of diseases caused by nematodes (roundworms) in the family Filarioidea. (21)

filariform   Referring to the long, slender larval form of hookworms and parasitic nematodes, which can penetrate human skin, migrate to the intestines, and establish a new infection. (22)

fimbria (*pl.* fimbriae)   See **pilus**. (5, 18)

Firmicutes   A phylum of Gram-positive bacteria (bacilli or cocci) that have thick, tough cell walls. Some species produce endospores. Examples include *Bacillus*, *Clostridium*, *Clostridioides*, *Streptococcus*, and *Staphylococcus*. (10)

first disease   See **measles**. (19)

first law of thermodynamics   Energy can be transferred from place to place, and it can be transformed, but energy cannot be created or destroyed. The first law is a statement about the conservation of energy. (4)

fixation   The process of adhering cells to a slide by a chemical or heat treatment. (3)

flagellum (*pl.* flagella)   A filamentous structure for motility. In prokaryotes, a helical protein filament attached to a rotary motor; in eukaryotes, an undulating cell membrane–enclosed complex of microtubules and ATP-driven motor proteins. (5)

flatworm   Also called *platyhelminth*. Any member of the phylum Platyhelminthes, which comprises morphologically simple invertebrate animals, including the parasitic trematodes and cestodes. (22)

flesh-eating disease   See **necrotizing fasciitis**. (19)

flow cytometer   A device that can count cells. (6, 17)

flow cytometry   A diagnostic and analytical method for detecting, counting, and distinguishing types of cells by suspending them in a stream of liquid that forces cells to move single file between a laser and a detector. (17)

flucytosine   An antifungal agent that inhibits DNA synthesis. (13)

fluke   Also called *trematode*. Any of a family of parasitic flatworms that cause infections of the blood, lungs, liver, and intestines. (11, 22)

fluorescence   The emission of light from a molecule that absorbs light of a specific wavelength and emits light of a longer wavelength. (3)

fluorescence-activated cell sorter (FACS)   A device that uses differences in fluorescence to count cells (by flow cytometry) and sort them into fractions. (6, 17)

fluorescence microscopy (FM)   Microscopy in which a fluorescent specimen is illuminated with light in the excitation wavelength range and observed in the emission wavelength range. (3)

fluorescent antibody staining   A diagnostic and analytical method of identifying microorganisms by visualizing them under a fluorescence microscope after they have been exposed to antibodies conjugated to a fluorophore. (25)

fluorophore   A fluorescent molecule used to stain specimens for fluorescence microscopy. (3)

focal plane   A plane that contains the focal point for a given lens. (3)

focal point   The position at which light rays intersect after passing through a lens. (3)

follicular lymphoma   A non-Hodgkin's lymphoma in which the *BCL2* gene translocates near an immunoglobulin heavy-chain promoter, causing overexpression of *BCL2*, which prevents cell death. (17)

folliculitis   Inflammation of hair follicles, often caused by a *Staphylococcus aureus* infection. (19)

fomite   An inanimate object on which pathogens can be transmitted from one host to another. (2)

food contamination   The presence in food of microbial pathogens that cause human disease. (27)

food poisoning   A disease caused by the ingestion of food contaminated with microbial pathogens or toxins. (22, 27)

food preservation   Methods for keeping food from spoiling owing to microbial contamination; includes drying, smoking, canning, acidification (pickling), addition of antimicrobial agents, and killing of microbes with heat or pressure. (27)

food safety   Practices for handling food that help people avoid illness from food-borne pathogens. (27)

food spoilage   Microbial changes that render a food unfit or unpalatable for consumption. (27)

food web   A network of interactions in which organisms obtain nutrients from or provide nutrients for each other—for example, by predation or by mutualism. (27)

foraminiferan *or* foram   An ameba with a calcium carbonate shell and a helical arrangement of chambers. (11)

Fournier's gangrene   A necrotizing fasciitis infection of the perineum and genitals. (23)

F-prime (F′) plasmid *or* F-prime (F′) factor   A fertility factor plasmid that contains some chromosomal (host) DNA. (9)

frameshift mutation   A gene mutation in which the insertion or deletion of one or more nucleotides causes a shift in the codon reading frame. (8)

free energy change (ΔG)   In a chemical reaction, a measure of how much energy that is available to do work is released (negative free energy change) or required (positive free energy change) as the reaction proceeds. (4)

freeze-drying   See **lyophilization**. (13)

fruiting body   A multicellular fungal or bacterial reproductive structure. (11)

functional genomics   A field of molecular biology that uses data from genomic research to describe the relationships between genes, proteins, and protein functions. (9)

functional group   A cluster of covalently bonded atoms that behaves with specific properties and functions as a unit. (4)

fungus (*pl.* fungi)   A heterotrophic eukaryote with chitinous cell walls; nutrition is by absorption. Fungi include Eumycota but traditionally may refer to fungus-like protists such as the oomycetes. (1, 11)

furuncle   Also called *boil*. An abscess of a hair follicle. (19)

gangrene   A condition in which a substantial portion of tissue (often in the extremities) becomes necrotic. (19)

gas gangrene   Also called *clostridial myonecrosis*. A form of gangrene caused by bacterial infection, usually by *Clostridium perfringens*. (19)

gastric mucosa   The layer of epithelial cells that line the stomach. (22)

gastric pit   Any of the many invaginations of the gastric mucosa lined with secretory cells involved in digestion. (22)

gastritis   Inflammation of the stomach lining. (22)

gastroenteritis   Inflammation of the gastrointestinal tract. (22)

gel electrophoresis   An analytical method for separating a mixture of nucleic acids or proteins on the basis of their size and charge, by subjecting them to an electrical field as the molecules migrate through a sieving gel medium. (8)

gene   A distinct series of nucleotides within DNA that has a distinct function (regulatory) or whose product has a distinct function (protein). The functional unit of heredity. (8)

generalized recombination   Also called *homologous recombination*. Recombination between two DNA molecules that share long regions of homology. Compare with *site-specific recombination*. (9)

generalized transduction   Transduction in which any bacterial donor gene can be transferred to a recipient cell. Compare with *specialized transduction*. (9)

general paresis   A condition of progressive dementia and partial paralysis in the later stages of syphilis. (23)

general secretion pathway   A mechanism in bacteria for moving proteins from the cytoplasm across the cell membrane (or the inner membrane of Gram-negative bacteria). (18)

generation time   Also called *doubling time*. The species-specific period for doubling of a population (e.g., by bacterial cell division) in a given environment, assuming no depletion of resources. (6)

genetic drift   The process by which two species diverge as a result of random mutations. (9)

genital herpes   Herpes genitalis, an infection of the external genitals caused by herpes simplex viruses 1 and 2. (19, 23)

genome   The complete genetic content of an organism; the sequence of all the nucleotides in a haploid set of chromosomes. (1, 4, 8)

genomic island   A region of DNA sequence whose properties indicate that it has been transferred from another genome. Usually comprises a set of genes with shared function, such as pathogenicity or symbiosis support. (9, 18)

genotype   All or part of the genetic constitution of an individual organism or cell. (8)

germ   A microorganism that causes disease. (13)

German measles   Also called *rubella*, *third disease*, or *three-day measles*. A childhood disease, caused by the rubella virus, that produces a rash, fever, and swollen lymph nodes. (19)

germicidal   Able to kill cells but not spores. (13)

germicide   A chemical that kills germs. (13)

germination   The activation of a dormant spore to generate a vegetative cell. (6)

germ theory of disease   The theory that specific diseases are caused by specific microbes. (1)

giant cell   An abnormally large cell formed by fusion of several cells, as seen in cytomegalovirus infection. (21)

giardiasis (*pl.* giardiases)   A diarrheal disease caused by the parasitic protozoan *Giardia intestinalis* (formerly *G. lamblia*). (22)

gingivitis   Inflammation of the gums. (22)

global regulator   Any protein that controls the expression of multiple genes. (8)

glycan   See **polysaccharide**. (4, 5)

glycolysis   Also called *Embden-Meyerhof-Parnas pathway*. The catabolic pathway of glucose oxidation to pyruvate, in which glucose 6-phosphate isomerizes to fructose 6-phosphate, ultimately yielding 2 pyruvate, 2 ATP, and 2 NADH. (7)

gnotobiotic animal   An animal that is germ-free or colonized by a known set of microbes. (14)

Golgi *or* Golgi complex *or* Golgi apparatus   A membranous organelle that modifies proteins and helps sort them to the correct eukaryotic cell compartment. (5)

gonorrhea   Also called *clap*. A sexually transmitted infection caused by *Neisseria gonorrhoeae*. (23)

graft-versus-host disease (GVHD)   A medical complication following a transplant between genetically different people, in which immune cells carried within donor tissue attack the recipient host's tissues. (17)

gramicidin   A peptide antibiotic that disrupts membrane integrity. (13)

Gram-negative   Describing cells that do not retain the crystal violet Gram stain. (3, 5)

Gram-positive   Describing cells that retain the crystal violet Gram stain and appear dark purple after staining. (3, 5)

Gram stain   A differential stain that distinguishes cells that possess a thick cell wall and retain a positively charged stain (Gram-positive) from cells that have a thin cell wall and outer membrane and fail to retain the stain (Gram-negative). (3)

granulocyte   See **polymorphonuclear leukocyte**. (15)

granuloma   A thick lesion formed around a site of infection. (15)

Graves' disease   An autoimmune disorder in which an excess of thyroid hormone is produced, causing hyperthyroidism. (17)

grazer   A first-level consumer, feeding directly on producers. (27)

greenhouse effect   The rise in Earth's temperature that is caused when CO_2 and other greenhouse gases trap solar heat as radiation. (27)

griseofulvin   An antifungal agent that inhibits cell division. (13)

growth factor   A compound needed for the growth of only certain cells. (6)

Guillain-Barré syndrome (GBS)   An autoimmune disease with ascending muscle weakness due to nerve damage, often resulting from a prior *Campylobacter* infection. (22)

gumma   A soft, tumorlike growth resulting from the effects of tertiary syphilis. (23)

gut-associated lymphoid tissue (GALT)   Lymphoid tissues, such as tonsils and adenoids, that are found in conjunction with the gastrointestinal tract and contain immune cells. (15)

gyrase   A topoisomerase that introduces negative supercoils in double-stranded DNA; enables DNA to untwist during DNA replication. (8)

HACEK   A collection of fastidious Gram-negative species that can cause endocarditis. HACEK organisms include <u>H</u>aemophilus parainfluenzae, <u>A</u>ggregatibacter aphrophilus and A. actinomycetemcomitans, <u>C</u>ardiobacterium hominis, <u>E</u>ikenella corrodens, and <u>K</u>ingella kingae. (21)

half-life   The amount of time it takes for one-half of a radioactive sample to decay. (4)

halophile   An organism that requires a high extracellular salt (NaCl) concentration for optimal growth. (6)

hand, foot and mouth disease (HFMD)   A relatively mild childhood disease characterized by fever and rash of the hands, feet, and mouth that is usually caused by coxsackievirus A16 or enterovirus 71. (19)

H antigen   The flagellin proteins that make up flagella and are recognized by the adaptive immune system. (22)

haploid   Describing an organism (or cell) that contains one copy of each chromosome in its cells. Compare with *diploid*. (11)

hapten   A small compound that must be conjugated to a larger carrier antigen in order to elicit an antibody response. (16)

Hashimoto's disease   A type of hypothyroid disease caused by an autoimmune disorder in which antibodies and T cells destroy self thyroid cells. (17)

health care–associated infection (HAI)   Any infection contracted by a patient receiving treatment for other conditions within a health care setting. A *hospital-acquired infection* is one type of HAI. (20, 26)

heavy chain   The larger of the two protein types that make up an antibody. Each antibody contains two heavy chains and two light chains. Compare with *light chain*. (16)

helminth   A multicellular worm that is a parasite. (11)

helper T cell (T_H cell)   A T cell that expresses CD4 on its cell surface and secretes cytokines that modulate B-cell isotype, or class, switching. (16)

hemolysis   The lysing of red blood cells; used diagnostically to identify species of streptococci and staphylococci. (10, 25)

hemolytic uremic syndrome (HUS)   A life-threatening sequela of infections with Shiga toxin–producing pathogens, such as *Escherichia coli* O157:H7, in which microclots in the renal blood supply decrease kidney function and reduce urine output. (22)

hepatitis   Inflammation of the liver, caused by infection or by exposure to a toxic substance. (22)

hepatitis A virus (HAV)   A picornavirus, transmitted by the fecal-oral route, that causes a self-limiting infectious disease. (22)

hepatitis B virus (HBV)   A hepadnavirus, transmitted by blood and other body fluids, that causes acute and chronic infections of the liver. (22)

hepatitis C virus (HCV)   A flavivirus, transmitted primarily by blood but also by sexual contact, that causes chronic liver infections; it can cross the placenta. (22)

hepatitis D virus (HDV)   A delta virus that exacerbates HBV infections; HDV is an incomplete virus requiring HBV to replicate. (22)

hepatitis E virus (HEV)   An orthohepevirus, transmitted by the fecal-oral route, most often through contaminated water, that causes a self-limiting infection. (22)

hepatomegaly   Enlargement of the liver. (21)

hepatosplenomegaly   Enlargement of the liver and spleen. (22)

herd immunity   The concept that an unvaccinated individual is protected from person-to-person transmission of a pathogen because most members of the community were vaccinated against that pathogen, lessening the likelihood that the unvaccinated person will come in contact with an infected person. (17)

herpes genitalis   See **genital herpes**. (23)

herpes gladiatorum   A skin infection of the neck, face, or arm caused by herpes simplex virus 1. (19)

herpes labialis   See **cold sore**. (19)

herpes zoster ophthalmicus   A disease caused by the reactivation of dormant herpes zoster virus specifically along the first division of the trigeminal nerve, causing vesicles to erupt over the forehead, eyelids, nose, and possibly eye. (19)

herpetic keratitis   Inflammation of the cornea caused by herpes simplex infection of the eye. (19)

herpetic whitlow   A skin infection of the fingers caused by herpes simplex virus 1. (19)

heterophile antibody   A random collection of antibodies produced in response to Epstein-Barr virus infection of B cells; they react against diverse antigens from different species. Anti–horse red blood cell antibodies are the basis of the monospot test for infectious mononucleosis. (21)

heterotroph   An organism that uses external sources of organic carbon compounds for biosynthesis. Compare with *autotroph*. (6, 7)

Hfr cell   A bacterial cell capable of high-frequency recombination caused by the presence of a chromosomally integrated F factor. (9)

highly active antiretroviral therapy (HAART)   See **antiretroviral therapy**. (13)

histamine   A cell mediator molecule released by mast cells that dilates blood vessels and causes some of the symptoms associated with type I hypersensitivity. (17)

histoplasmosis   A fungal infection of the lungs caused by *Histoplasma capsulatum*. (20)

HIV   See **human immunodeficiency virus**. (23)

hives   Raised, itchy, fluid-filled welts on the skin. (17)

Hodgkin's lymphoma   A cancer of the lymphocytes sometimes associated with prior infectious mononucleosis caused by Epstein-Barr virus and characterized by the presence of Reed-Sternberg cells. Compare with *non-Hodgkin's lymphoma*. (17)

holdfast   An adhesion factor, secreted by the tip of a stalk, that firmly attaches an organism to a substrate. (5)

homolog   One of a group of genes or proteins that share ancestry and have similar DNA sequences and amino acid sequences. (8)

homologous   Similar, referring to DNA sequences derived from a common ancestral gene. (9)

homologous recombination   See **generalized recombination**. (9)

hookworm   Any of a group of soil-dwelling parasitic nematodes that penetrate the host skin and migrate to the intestines, where they multiply. (11, 22)

horizontal gene transfer   The passage of genes from one genome into another, nonprogeny genome. Compare with *vertical gene transfer*. (9)

horizontal transmission   In disease, the transfer of a pathogen from one organism into another, nonprogeny organism. Compare with *vertical transmission*. (2)

hospital-acquired infection   A health care–associated infection that is contracted in a hospital setting; formerly known as *nosocomial infection*. (20, 26)

host range   All the species that can be infected by a given pathogen. (2, 12)

housekeeping gene   A gene that encodes an essential function found in all cells (e.g., a gene for ribosomal RNA). (9)

human herpesvirus 4 (HHV-4)   See **Epstein-Barr virus**. (21)

human herpesvirus 5 (HHV-5)   See **cytomegalovirus**. (21)

human immunodeficiency virus (HIV)   A lentivirus, transmitted by blood, body fluids, and sexual contact, that targets CD4$^+$ helper T cells and causes a chronic infection that can lead to end-stage HIV disease, or AIDS; it can cross the placenta. (12, 18, 23)

humoral immunity   A type of adaptive immunity mediated by antibodies. (16)

hybridize   Also called *base-pair*. For a nucleic acid strand, to form base pairs with a complementary strand from a different source. (8)

hybridoma   An immortal cell line created by the fusion of a monoclonal antibody–producing B cell and a myeloma cell. (17)

hydrocele   A buildup of fluid in the scrotum that causes it to swell. (21)

hydrogen bond   An electrostatic attraction between a hydrogen that is bonded to an oxygen or nitrogen and a second, nearby oxygen or nitrogen. (4)

hydrolysis   The cleaving of a bond by the addition of a water molecule. (7)

hydrophilic   Soluble in water; either ionic or polar. (4)

hydrophobic   Insoluble in water; nonpolar. (4)

hyperesthesia   Extreme sensitivity to stimuli of one or more senses, especially skin. (24)

hyper IgM syndrome   Any of a group of genetic diseases characterized by an overproduction of IgM antibodies and a lack of other antibody classes due to defects in class-switching proteins. (17)

hyperosmotic   See **hypertonic**. (4)

hyperthermophile   An organism adapted for optimal growth at extremely high temperatures, generally above 80°C (176°F) and as high as 121°C (250°F). (6)

hypertonic   Also called *hyperosmotic*. Having more solutes than another environment separated by a semipermeable membrane. Water will tend to flow toward the hypertonic solution. (4)

hypha (*pl.* hyphae)   The threadlike filament that forms the mycelium of a fungus. (11)

hypoosmotic   See **hypotonic**. (4)

hypotension   Low blood pressure. (21)

hypotonic   Also called *hypoosmotic*. Having fewer solutes than another environment separated by a semipermeable membrane. Water will tend to flow away from the hypotonic solution. (4)

idiotype   A difference in amino acid sequence, usually in the antigen-binding site of an antibody, that distinguishes different antibodies within any isotype class of antibodies in an individual. Compare with *allotype* and *isotype*. (16)

IgA   Immunoglobulin A, an antibody isotype that contains the alpha heavy chain. In monomeric form it can be found in blood. It can be secreted in dimeric form and is found in tears, saliva, breast milk, and so on. (16)

IgD   Immunoglobulin D, an antibody isotype that contains the delta heavy chain. It is found on B-cell membranes. (16)

IgE   Immunoglobulin E, an antibody isotype that contains the epsilon heavy chain. It is involved in degranulation of mast cells. (16)

IgG   Immunoglobulin G, an antibody isotype that contains the gamma heavy chain. It is found in serum. (16)

IgM   Immunoglobulin M, the first antibody isotype detected during the early stages of an immune response. It contains the mu heavy chain and is found as a pentamer in serum. (16)

illness phase   The stage of an infectious disease when signs and symptoms are evident and the immune system is combating the infection. (2)

immediate hypersensitivity   See **type I hypersensitivity**. (17)

immersion oil   An oil with a refractive index similar to that of glass, which minimizes light ray loss at wide angles, thereby minimizing wavefront interference and maximizing resolution. (3)

immune complex disease   See **type III hypersensitivity**. (17)

immune surveillance   Identification and elimination of tumor cells by the immune system before they become clinically detectable. (17)

immune system   An organism's cellular defense system against pathogens. (1, 15)

immunity   Resistance to a specific disease. (1)

immunization   See **vaccination**. (1, 16)

immunochromatographic assay   A simple rapid assay to detect specific antigens or antibodies in a patient sample (blood, urine, or cerebrospinal fluid). (25)

immunocompromised host   An animal with a weakened immune system. (14)

immunogen   An antigen that, by itself, can elicit antibody production. (16)

immunogenicity   See **antigenicity**. (16)

immunoglobulin   Any of a family of proteins that includes antibodies and B-cell receptors. (16)

immunological specificity   Also called *antigenic specificity*. The degree to which an antibody, or receptor, distinguishes between similar-looking antigens. (16)

immunomodulin   A protein made by normal microbiota that influences the host immune response by modifying the secretion of host proteins, such as cytokines. (14)

immunopathology   Damage caused to tissues as the result of an immune response. (2)

immunoprecipitation   The antibody-mediated cross-linking of antigens to form large, insoluble complexes. Immunoprecipitation is used in research labs and is normally seen only in vitro. (17)

impetigo   Also called *nonbullous impetigo*. A contagious skin infection of children caused by *Staphylococcus aureus* and by group A beta-hemolytic streptococci; characterized by red sores around the nose, mouth, and neck that break open and form a honey-colored crust. Compare with *bullous impetigo*. (19)

incidence   In referring to a disease, the number of new cases in a given sample or location over a specified period. Compare with *prevalence*. (2, 26)

incidental host   A host in which a pathogen or parasite can grow and cause disease but not be transmitted to other hosts. (11)

incubation phase   The time between initial infection of a host by a microbe and the manifestation of signs and symptoms of disease. (2)

index case   Also called *patient zero*. The first case of an infectious disease in a location and an important piece of data for helping contain the spread of disease. (26)

indirect ELISA   An immunoassay in which an antigen fixed to a plastic surface is used to capture specific antibody from a patient's serum; a secondary antibody able to bind the captured antibody is added next; the secondary antibody is conjugated either to a fluorescent molecule or to an enzyme that can catalyze a reaction that turns a colorless substrate to a colored product. (17)

indirect immunofluorescence test   A diagnostic method for detecting specific anti-pathogen antibodies in a patient by attaching the pathogen to a slide, adding patient serum followed by fluorescently labeled anti-human antibodies, and looking under fluorescence light microscopy for fluorescence, which indicates that the serum contains antibodies against the pathogen. (17)

indirect transmission   In disease, the transfer of a pathogen from host to host through an intermediary such as an inanimate object, animal vector, or contaminated food and water. (2)

induction   Also called *up-regulation*. Increased transcription of a target gene due to an inducer binding a repressor; in the absence of inducer, the repressor prevents transcription. (8)

infant botulism   Infant paralysis by botulism toxin following the ingestion and germination of *Clostridium botulinum* spores. (10)

infection   The growth of a pathogen or parasite in or on a host. (2)

infection cycle   The route of transmission of an infectious disease. (2)

infectious disease   A disease that can be transferred from host to host. Infectious diseases are caused by pathogenic bacteria, viruses, or fungi or by parasitic helminths or protists. (2)

infectious dose 50% (ID_{50})   The number of bacteria or virions required to cause disease symptoms in 50% of an experimental group of hosts. Compare with *lethal dose 50%*. (2)

infectious mononucleosis   Highly contagious disease with flu-like symptoms, caused by the very common Epstein-Barr virus. (21)

inflammatory diarrhea   Diarrhea caused by inflammation and inflammation-induced damage of the intestinal mucosa that diminishes the intestines' ability to absorb nutrients and water. (22)

influenza   A respiratory tract infection caused by strains of the influenza virus, a member of the orthomyxoviruses. (18, 20)

initiation   The first phase of DNA synthesis, in which the DNA helix unwinds, RNA primers attach, and DNA polymerase and accessory proteins assemble on the promoter; also the initial events in transcription and in translation. (8)

innate immunity   Also called *nonadaptive immunity*. The system of nonspecific mechanisms that the body uses for protecting against pathogens. Compare with *adaptive immunity*. (15)

inner membrane *or* inner cell membrane   In Gram-negative bacteria, the membrane in contact with the cytoplasm, equivalent to the cell membrane. Compare with *outer membrane*. (5)

insertion   The addition of one or more nucleotides to the middle of a DNA sequence. Compare with *deletion*. (8)

insertion sequence (IS)   A simple transposable element consisting of a transposase gene flanked by short, inverted-repeat sequences that are the target of transposase. (9)

integrin   Any of a family of host cell membrane proteins that mediate adhesion of cells to one another and to the extracellular matrix. (15)

interferon   Any of a family of cytokines that inhibit viral replication. (15)

interferon-gamma release assay (IGRA)   A blood test that measures release of interferon gamma by leukocytes upon exposure to mycobacteria; a tuberculosis screening method. (20)

interleukin 1 (IL-1)   A cytokine released by macrophages. (15)

intermediate filament   A eukaryotic cytoskeletal protein filament composed of various proteins, depending on the cell type. (5)

intermediate host   A transient host in which a parasite undergoes asexual reproduction before being transmitted to a definitive host. (11, 22)

intracellular pathogen   A pathogen that lives within a host cell. (18)

intravenous immunoglobulin (IVIG)   Pooled IgG extracted from the plasma of many blood donors. The antibodies are delivered intravenously to patients who cannot make their own antibodies. (17)

invasion   The entry of a pathogen into a living human or nonhuman animal cell, where it then lives. (2)

invasiveness   The ability of a pathogen to spread rapidly through host tissues. (2)

inversion   A reversal of the orientation of a DNA fragment within a chromosome. It may activate or repress the transcription of a particular gene. (8)

ion   Also called *electrolyte*. An atom or molecule containing a charge—that is, a number of electrons greater or smaller than the number of protons. (4)

ionic bond   A bond between ions of positive and negative charge. (4)

isolate   A microbe that has been obtained from a specific location and grown in pure culture. (10)

isolation streaking   Also called *dilution streaking*. A method for separating a mixture of cells in a solution so that they make isolated, individual colonies on agar plates. (6)

isomer   In chemistry, one of two or more compounds or ions that contain the same number of atoms but differ in structural arrangement and properties. (4)

isosmotic   See **isotonic**. (4)

isotonic   Also called *isosmotic*. Being in osmotic balance; having equal concentrations of solutes on both sides of a semipermeable membrane. A cell in an isotonic environment will neither gain nor lose water. (4)

isotope   A form of an element that differs by the number of neutrons in the nucleus. For example, the most abundant isotope of carbon is carbon-12 (six protons plus six neutrons), but the isotopes carbon-13 (seven neutrons) and carbon-14 (eight neutrons) also exist. (4)

isotype   A species-specific antibody class, defined by the structure of the constant region of the heavy chain. IgG, IgA, IgM, IgD, and IgE are the five isotypes. Compare with *allotype* and *idiotype*. (16)

isotype switching   Also called *class switching*. A change in the predominant antibody isotype produced by a B cell. (16)

Jarisch-Herxheimer reaction   An acute reaction that occurs within hours after the start of antibiotic treatment for syphilis; an inflammatory response triggered by dying treponemes, resulting in fever, headaches, tachycardia, and myalgias. (23)

Kaposi's sarcoma   A cancer caused by human herpesvirus 8; a hallmark of AIDS. (23)

keratitis   Inflammation of the cornea. (19)

Kirby-Bauer assay   A method for determining antibiotic susceptibility. Antibiotic-impregnated disks are placed on an agar plate whose surface has been confluently inoculated with a test organism. The antibiotic diffuses away from the disk and inhibits growth of susceptible bacteria. The width of the inhibitory zone is proportional to the susceptibility of the organism. (13)

Koch's postulates   Four criteria, developed by Robert Koch, that should be met for a microbe to be designated the causative agent of an infectious disease. (1)

Krebs cycle   See **tricarboxylic acid cycle**. (7)

labile toxin (LT)   An *Escherichia coli* enterotoxin, destroyed by heat, that increases host cellular cAMP concentrations. (18, 22)

lab-on-a-chip (LOC)   A diagnostic or analytical instrument that integrates multiple laboratory functions into a tiny silicon chip with incorporated microfluidics. (25)

lagging strand   During DNA replication, the daughter strand of DNA that is synthesized as a series of short fragments that are then ligated. Compare with *leading strand*. (8)

lag phase   A period of cell culture, occurring right after inoculation into new medium, during which bacteria grow slowly or not at all. (6)

laminar flow biological safety cabinet   An air filtration appliance that removes pathogenic microbes from within the cabinet. (13)

Lancefield groups *or* Lancefield classification   A scheme to assign beta-hemolytic streptococci to groups A–U, based on differences in their cell wall carbohydrate antigens. (25)

Langerhans cell   A specialized, phagocytic dendritic cell that is the predominant cell type in skin-associated lymphoid tissue. (15)

laryngitis   Inflammation of the larynx. (20)

laryngotracheobronchitis (LTB)   See **croup**. (20)

late log phase   A period of cell culture, occurring toward the end of log phase, during which the exponential growth rate of bacteria slows. (6)

latent   Referring to a virus or infection that is hidden and not manifesting symptoms of disease. (19)

latent period   The time in the viral life cycle when progeny virions have formed but are still within the host cell. (23)

latent state   A period of the infection process during which a pathogenic agent is dormant in the host and cannot be cultured. (2)

latent tuberculosis infection (LTBI)   A condition in which a person is infected with *Mycobacterium tuberculosis* but is asymptomatic and is not infectious to others. (20)

late-phase anaphylaxis   A stage of anaphylaxis in which mast cells in the affected area release leukotrienes that cause inflammation and vasoconstriction. (17)

leading strand   During DNA replication, the daughter strand of DNA that is formed by continuous elongation in the 5′-to-3′ direction. Compare with *lagging strand*. (8)

legionellosis *or* Legionnaires' disease   An acute upper respiratory tract infection caused by certain strains of *Legionella* bacteria. (20)

leishmaniasis (*pl.* leishmaniases)   A disease caused by infection with *Leishmania* species. (21)

lentivirus   Any of a family of retroviruses with a long incubation period. An example is HIV. (12)

lethal dose 50% (LD$_{50}$)   A measure of virulence; the number of bacteria or virions required to kill 50% of an experimental group of hosts. Compare with *infection dose 50%*. (2)

lethal factor (LF)   A component of anthrax toxin that cleaves host protein kinases. (18)

leukemia   A disease in which malignant lymphoid cells circulate in the blood or bone marrow. (17)

leukotriene   Any of a group of inflammation-mediating lipids that are derived from arachidonic acid secreted by mast cells, macrophages, and other cells; leukotrienes cause smooth-muscle contraction, mucus secretion, and activation of leukocytes, leading to inflammation and allergic-reaction symptoms. (17)

lichen   A symbiotic organism composed of a fungal partner and an algal or cyanobacterial partner. (10)

light chain   The smaller of the two protein types that make up an antibody. Each antibody contains two heavy chains and two light chains. Compare with *heavy chain*. (16)

light microscopy (LM)   Observation of a microscopic object based on light absorption and transmission. (3)

lipid   Any of a class of organic molecules that is partially or completely hydrophobic and often contains long hydrocarbon chains. (4)

lipid A   The portion of a lipopolysaccharide that is anchored in the outer membrane of the Gram-negative bacterial envelope; also functions as an endotoxin. (5)

lipopolysaccharide (LPS)   Any of a class of structurally unique phospholipids found in the outer leaflet of the outer membrane in Gram-negative bacteria. Many are endotoxins. (5, 18)

listeriosis (*pl.* listerioses)   A food-borne disease caused by *Listeria monocytogenes*. (22, 24)

lithotroph   See **chemoautotroph**. (1, 6)

lithotrophy   Metabolic oxidation of inorganic compounds to yield energy used to reduce carbon dioxide. Compare with *organotrophy*. (6, 7, 10)

lockjaw   See **tetanus**. (24)

logarithmic (log) phase   See **exponential phase**. (6)

long-term nonprogressor   An HIV-positive person who has remained asymptomatic for years without antiviral therapy. (23)

lower respiratory tract   Lower airways of the body, which include the trachea, bronchi, and lungs. (20)

lumbar puncture   Also called *spinal tap*. A medical procedure used to withdraw cerebrospinal fluid from a patient. (24)

lumen   1. The interior of an intracellular membrane-enclosed compartment. (5) 2. The cavity of a tubular organ (intestine) or part (blood vessel). (22)

Lyme borreliosis *or* Lyme disease   A tick-borne disease caused by *Borrelia burgdorferi* that may involve skin lesions and arthritis. (21)

lymph   Interstitial fluid and white blood cells (including macrophages) that bathe nearby tissues. (15)

lymphadenitis   Inflammation of lymph nodes. (21)

lymphadenopathy   Swelling of lymph nodes. (21)

lymphangitis   Inflammation of a lymphatic vessel, often seen as a red streak extending from a localized bacterial skin infection. (21)

lymphatic system   Network of lymph-carrying vessels and peripheral lymphoid tissues through which extracellular fluid passes before it is collected and returned to the venous circulation. (21)

lymph node   A secondary lymphoid organ, formed by the convergence of lymphatic vessels, where antigen-presenting cells and lymphocytes interact. (15)

lymphocytic choriomeningitis virus (LCMV)   An arenavirus transmitted from rodents to humans that causes meningitis, encephalitis, and meningo-encephalitis. (24)

lymphogranuloma venereum (LGV)   A sexually transmitted infection caused by especially virulent serotypes of *Chlamydia trachomatis*. (23)

lymphoma   A solid tumor found in a lymphoid organ. (17)

lyophilization   Also called *freeze-drying*. The removal of water from food, by freezing under vacuum, to limit microbial growth. (13)

lysis   Rupture of a cell by a break in the cell wall and membrane. (4, 12)

lysogeny   A viral replication cycle in which the viral genome integrates into and replicates with the host genome but retains the ability to initiate host cell lysis. (12)

lysosome   An acidic eukaryotic organelle that aids digestion of molecules. Not found in plant cells. (4, 5)

lysozyme   A hydrolytic enzyme, secreted by eukaryotic cells, that degrades bacterial cell wall peptidoglycan. (15)

lytic infection   A viral replication cycle in which progeny virions are released from the host cell by virus-induced rupturing of the cell membrane. (12)

MacConkey medium   A selective and differential medium that selects for Gram-negative bacteria and can differentiate between fermenters and nonfermenters of a specific sugar, such as lactose (lactose MacConkey medium). (6)

macromolecule   A large organic molecule, such as a protein or nucleic acid, typically containing thousands of carbon-carbon bonds. (4)

macrophage   A mononuclear, phagocytic, antigen-presenting cell of the immune system that is found in tissues. (15)

macule   A flat, red skin rash less than 1 cm across. (19)

maculopapular rash   A reddened papule. (19)

mad cow disease   See **bovine spongiform encephalopathy**. (24)

magnetosome   An organelle that contains the mineral magnetite and thereby enables microbes to sense a magnetic field. (5)

magnification   An increase in the apparent size of a viewed object as an optical image. (3)

major histocompatibility complex (MHC)   Transmembrane cell proteins important for recognizing self and for presenting foreign antigens to the adaptive immune system. (15, 16)

malaise   Sensations of fatigue and mild discomfort in various locations of the body. (2)

malaria   A disease caused by the protozoan *Plasmodium falciparum* and transmitted by mosquitoes. (21)

MALDI-TOF MS   Matrix-assisted laser desorption time of flight mass spectrometry; a technique in which macromolecules such as proteins are fragmented and ionized, and their time of flight provides information on the molecular mass. (25)

malignant   Referring to tumors whose cells are undergoing rapid, unchecked growth. (17)

Mantoux tuberculin skin test   See **tuberculin skin test**. (20)

mass number   The mass of an atom's nucleus, which is equal to the sum of the number of protons and neutrons. (4)

mass spectrometry (MS)   A technique in which defined fragments of a macromolecule are analyzed by particle mass. (25)

mast cell   A white blood cell that secretes proteins that aid innate immunity. Mast cells reside in connective tissues and mucosa and do not circulate in the bloodstream. (15)

M cell   A specialized epithelial cell (microfold cell) within Peyer's patches lining the intestine; M cells transport antigens and microbes from the intestines across the epithelial barrier to macrophages. (15)

measles   Also called *rubeola* or *first disease*. A childhood disease caused by a paramyxovirus; characterized by flu-like symptoms and a maculopapular rash caused by cytotoxic-T-cell damage to small blood vessels. (19)

mechanical vector   An animal, usually a fly, that can pick up infectious agents on its body and transmit them through physical contact. (2)

medical history   See **patient history**. (2)

meiosis   A form of cell division by which a diploid eukaryotic cell generates haploid sex cells that contain recombinant chromosomes. (11)

membrane-permeant weak acid   An acid that exists in charged and uncharged forms, such as acetic acid. The uncharged form can penetrate the cell membrane. (4)

membrane-permeant weak base   A base that exists in charged and uncharged forms, such as methylamine. The uncharged form can penetrate the cell membrane. (4)

memory B cell   A long-lived type of lymphocyte preprogrammed to produce a specific antibody. After encountering their activating antigen, memory B cells differentiate into antibody-producing plasma cells. (16)

meninges (*sing.* meninx)   Three membranes that envelop the brain and spinal cord; from innermost to outermost, they are the pia mater, arachnoid mater, and dura mater. (24)

meningitis   Inflammation of the meninges. (24)

meningoencephalitis   Inflammation of both the meninges and the brain. (24)

mesophile   An organism adapted for optimal growth between 20°C (68°F) and 40°C (104°F). (6)

messenger RNA (mRNA)   An RNA molecule that encodes a protein. (8)

metagenomic study   An analytical method that uses DNA sequencing to identify all the organisms directly from an environmental sample. (14)

metamonad   Any of a phylum of flagellated protozoa that have no mitochondria. Several metamonads, such as *Giardia intestinalis*, parasitize humans. (11)

methanogenesis   An energy-yielding metabolic process that produces methane, commonly from hydrogen gas and oxidized one- or two-carbon compounds. It is unique to archaea. (7)

methemoglobinemia   A disorder in which red blood cells contain excess methemoglobin, a form of hemoglobin that transports oxygen to tissues less efficiently, causing tissues to starve (that is, become hypoxic). (27)

methicillin-resistant *Staphylococcus aureus* (MRSA)   Any strain of *S. aureus* that is resistant to the first-line antibiotics methicillin and oxacillin. (19)

methyl mismatch repair   A DNA repair system that fixes misincorporation of a nucleotide after DNA synthesis. The unmethylated daughter strand is corrected to complement the methylated parental strand. (8)

MHC II deficiency   Also called *bare lymphocyte syndrome*. A rare genetic disease in which MHC II molecules are absent on antigen-presenting cells, resulting in greatly increased susceptibility to many kinds of infection. (17)

microaerophilic   Requiring oxygen at a concentration lower than that of the atmosphere but unable to grow in high-oxygen environments. (6)

microbe   An organism or virus too small to be seen with the unaided human eye. (1)

microbe-associated molecular pattern (MAMP)   Molecules associated with groups of microbes, both pathogenic and nonpathogenic, that are recognized by cells of the innate immune system. (15)

microbiota *or* microbiome   The total community of microbes associated with an organism (such as the human body) or with a defined habitat (such as soil or plants). (1, 2, 14)

microcystin   A toxin, produced by the cyanobacterium *Microcystis*, that, if swallowed, causes intestinal illness, skin rashes, and liver damage. (27)

microfilament   Also called *actin filament*. A eukaryotic cytoskeletal protein filament composed of polymerized actin. (5)

microfossil   A microscopic fossil in which calcium carbonate deposits have filled in the form of ancient microbial cells. (9)

microscope   A tool that increases the magnification of specimens to enable viewing at higher resolution. (3)

microsporidian   Any of a large phylum of obligate intracellular fungal parasites that produce spores. (11)

microtubule   A eukaryotic cytoskeletal protein filament composed of polymerized tubulin. (5)

midstream clean-catch technique   A sample collection method that yields urine with a reduced number of contaminating bacteria from the urethra. (25)

minimal medium   Also called *chemically defined minimal medium*. A synthetic medium in which the components are limited to only what a particular organism (or group of organisms) needs to grow. Compare with *complex medium* and *synthetic medium*. (6)

minimum bactericidal concentration (MBC)   The lowest concentration of an antibiotic that kills the test cells. Compare with *minimum inhibitory concentration*. (13)

minimum inhibitory concentration (MIC)   The lowest concentration of a drug that will prevent the growth of an organism. Compare with *minimum bactericidal concentration*. (13)

miracidium (*pl.* miracidia)   A ciliated larva of pathogenic flatworms. (22)

missense mutation   A point mutation that alters the sequence of a single codon, leading to a single amino acid substitution in a protein. (8)

mite   A small arachnid. Many mites are parasites that burrow into the skin and lay eggs. When inhaled, some may cause allergic reactions. (11)

mitosis   The orderly replication and segregation of eukaryotic chromosomes, usually prior to cell division. (11)

mixotroph   An organism capable of both photosynthetic and heterotrophic metabolism. (7)

molarity   A unit of concentration, measured as the number of moles of solute per liter of solution and expressed as molar (abbreviated M). (4)

mole   The amount of a chemical substance that contains 6.02×10^{23} molecules of that substance; Avogadro's number equals 6.02×10^{23}. (4)

molecular beacon probe   An oligonucleotide with a fluorescent dye on one end and a quencher on the other that fluoresces only when the dye and quencher molecules are separated by degradation; used as a molecular biology reagent to quantify the results of qRT-PCR. (25)

molecular clock   The use of DNA or RNA sequence information to measure the time of divergence between different species. (9)

molecular formula   A notation indicating the number and type of atoms in a molecule. For example, H_2O is the molecular formula for water. (4)

molecule   A substance composed of two or more atoms that are sharing electrons covalently. (4)

monoclonal antibody   An epitope-specific antibody of a single isotype. It originates from a single antibody-producing B-cell clone (17)

monocyte   A white blood cell with a single nucleus that can differentiate into a macrophage or a dendritic cell. (15)

mononuclear phagocyte system (MPS)   A collection of cells that can phagocytose and sequester extracellular material. (15)

monosaccharide   Also called *simple sugar*. The monomer unit of carbohydrates. Monosaccharides have the molecular formula $(CH_2O)_n$. (4)

morbidity   The number of people who are sick from a disease in a given population. Compare with *mortality*. (2)

mortality   The number of people who have died in a population in a given area or over a given period, often measured with respect to a specific disease. Compare with *morbidity*. (2)

motility   The ability to generate self-directed movement. (5)

motility-related diarrhea   Diarrhea caused by hypermotility of the intestinal tract, causing food to move faster than normal in the intestines, so there isn't enough time to absorb water and nutrients. (22)

motor neuron   A nerve cell that sends signals from the central nervous system to muscles and glands to produce movement. Compare with *sensory neuron*. (24)

mucociliary escalator *or* mucociliary elevator   The ciliated mucous lining of the trachea, bronchi, and bronchioles that sweeps foreign particles up and away from the lungs. (14, 20)

mucosa (*pl.* mucosae)   A mucous membrane. (19)

mucosa-associated lymphoid tissue (MALT)   A system of concentrated sites of lymphoid tissue closely associated with mucosal surfaces, whose function includes surveying the antigens and pathogens that pass through the mucosae. (15)

multidrug resistance (MDR) efflux pump   A transmembrane protein pump that can export many kinds of antibiotics largely independent of antibiotic structure. (13)

multidrug-resistant (MDR)   Referring to a strain of pathogen that is resistant to at least two first-line drugs. (20)

multilocus sequence typing (MLST)   A diagnostic DNA sequencing method that types multiple genetic loci from patient-isolated bacterial cultures in order to identify strain variability. (9)

multiple myeloma   A cancerous growth of plasma cells at multiple sites in the body, especially in the bone marrow. Compare with *plasmocytoma*. (17)

multiplex PCR   A polymerase chain reaction that uses multiple pairs of oligonucleotide primers to amplify several different DNA sequences simultaneously. (25)

mumps   A childhood disease, caused by a paramyxovirus, that infects the major salivary glands, causing them to swell as a result of inflammation. (22)

murein   See **peptidoglycan**. (5)

murein lipoprotein   The major lipoprotein that connects the outer membrane of Gram-negative bacteria to the peptidoglycan cell wall. (5)

mutation   A heritable change in a DNA sequence. (8)

mutator   Any strain of a microorganism that has an abnormally high frequency of mutations; results from mutation of a gene encoding a DNA proofreading enzyme. (8)

mutualism   A symbiotic relationship in which both partners benefit. (1, 2, 27)

myasthenia gravis   An autoimmune disease characterized by severe muscle weakness; it is caused by autoantibodies against acetylcholine receptors at neuromuscular junctions. (17)

mycelium (*pl.* mycelia)   A mass of branched fungal hyphae that projects into the air (aerial mycelium) or into the growth substrate (substrate mycelium). (11)

mycolic acid   Any of a diverse class of sugar-linked fatty acids found in the cell envelopes of mycobacteria such as *Mycobacterium tuberculosis*. (5, 25)

mycology   The study of fungi. (11)

mycorrhiza (*pl.* mycorrhizae)   A fungus involved in an intimate mutualism with plant roots, in which nutrients are exchanged. (11)

mycosis (*pl.* mycoses)   A fungal infection. (24)

mycotoxin   A substance in the environment, produced by certain fungi, that can affect human health when ingested. (18)

myelitis   Inflammation of the spinal cord. (24)

myocarditis   Inflammation of the heart myocardium. (21, 22)

***Naegleria* meningoencephalitis**   A life-threatening infection of the central nervous system caused by the pathogenic ameba *Naegleria fowleri*. (24)

nasopharynx   The passage leading from the nose to the oral cavity. (14)

native conformation   The fully folded, functional form of a protein. (4)

natural killer (NK) cell   A lymphocyte that kills some tumor cells and cells infected with a virus or bacteria; an important component of innate immunity. (15)

natural selection   A change in the frequency of genes in a population under environmental conditions that favor some genes over others. (9)

necrotizing fasciitis   Also called *flesh-eating disease*. A severe skin infection usually caused by the Gram-positive coccus *Streptococcus pyogenes*. (19)

necrotizing ulcerative gingivitis   See **trench mouth**. (22)

negative selection   In immunology, the destruction of T cells bearing T-cell receptors that bind strongly to self MHC proteins displayed on thymus epithelial cells. Compare with *positive selection*. (16)

negative stain   A stain that colors the background and leaves the specimen unstained. (3)

nematode   See **roundworm**. (11, 22)

neonatal conjunctivitis   See **ophthalmia neonatorum**. (19)

neonatal herpes   Infection of a newborn by herpesvirus transmitted vertically from its mother, who has genital herpes. (23)

neoplasm   See **cancer**. (17)

neuraminidase inhibitor   Any of a class of anti-influenza drugs that target neuraminidase on the viral envelope and decrease the number of virus particles produced. (13)

neurosyphilis   Neurological symptoms resulting from syphilis at any stage of the disease, caused by infection of the central nervous system by *Treponema pallidum*. (23)

neurotropic virus   A virus that preferentially infects neurons. (24)

neutralophile   An organism with optimal growth in environments between pH 5 and pH 8. Compare with *acidophile* and *alkaliphile*. (6)

neutrophil   A white blood cell of the innate immune system that can phagocytose and kill microbes. (15)

niche   An organism's environmental requirements for existence and its relations with other members of the ecosystem. (27)

nicotinamide adenine dinucleotide (NADH)   The reduced form of the energy carrier NAD. In the cell, NADH can donate electrons, and the oxidized form NAD^+ can accept electrons. (7)

nitrification   The process by which some bacteria and archaea oxidize reduced nitrogen compounds to nitrite or nitrate. Compare with *denitrification* and *nitrogen fixation*. (6, 27)

nitrifier   An organism that converts reduced nitrogen compounds to nitrite or nitrate. (27)

nitrogen cycle   The biogeochemical transformations of nitrogen compounds in the biosphere, including nitrification, denitrification, and nitrogen fixation. (6)

nitrogen fixation   The process by which some bacteria and archaea reduce diatomic nitrogen gas (N_2) to two molecules of ammonium ion (NH_4^+). Compare with *denitrification* and *nitrification*. (1, 6, 10)

nitrogenous base   See **nucleobase**. (4)

NOD-like receptor (NLR)   A eukaryotic cytoplasmic protein that recognizes particular microbe-associated molecular patterns present on microorganisms. (15)

nonadaptive immunity   See **innate immunity**. (15)

nonbullous impetigo   See **impetigo**. (19)

noncholera *Vibrio*   Any pathogenic *Vibrio* species other than *V. cholerae*. (22)

nonhemolytic   Referring to a bacterial species that is unable to lyse red blood cells. Compare with *alpha-hemolytic* and *beta-hemolytic*. (25)

non-Hodgkin's lymphoma   Any lymphoma that does not contain Reed-Sternberg cells and thus is not a *Hodgkin's lymphoma*. (17)

nonpolar covalent bond   A covalent bond in which the electrons are distributed equally between two atoms. (4)

nonsense mutation   A mutation that changes an amino acid codon into a premature stop codon. (8)

norovirus   A nonenveloped single-stranded RNA virus, transmitted by the fecal-oral route, that is a major cause of gastroenteritis; formerly known as *Norwalk virus*. (22)

Norwalk virus   See **norovirus**. (22)

nosocomial infection   See **hospital-acquired infection**. (20, 26)

nuchal rigidity   An inability to bend the neck forward and touch chin to chest; the symptom is a sign that the meninges are irritated and is a hallmark of meningitis. (24)

nuclear pore complex   A large complex of proteins that traverse the nuclear membranes, providing selective movement of molecules between the cytoplasm and the nucleus within a cell. (5)

nucleobase   Also called *nitrogenous base*. A planar, heteroaromatic nitrogen-containing base that is one component of nucleotides in nucleic acids; nucleobases determine the information content of DNA and RNA. (4, 8)

nucleoid   The region within a prokaryotic cell that contains the compacted loops of the cell's chromosome. (5)

nucleolus (*pl.* nucleoli)   A region inside the nucleus where ribosome assembly begins. (5)

nucleotide   The monomer unit of nucleic acids, consisting of a five-carbon sugar, a phosphate, and a nitrogenous base. (4, 8)

numerical aperture   The product of the refractive index of the medium and sin θ. As numerical aperture increases, the magnification increases. (3)

O antigen   See **O polysaccharide**. (5, 22)

objective lens   In a compound microscope, the lens that is closest to the specimen and generates magnification. (3)

observational study   Epidemiological approach in which the exposure and disease status of each study participant is monitored and recorded. (26)

ocular lens   Also called *eyepiece*. In a compound microscope, the lens situated closest to the observer's eye. (3)

oliguria   The production of abnormally low amounts of urine. (23)

oncogene   A gene that, through mutation or inappropriate expression, can lead to cancer. (12)

One Health Initiative   A collaboration dedicated to improving the lives of all species by integrating human medicine, veterinary medicine, and environmental science. (26)

one-step growth curve   Graphical representation of a lytic replication life cycle—from attachment through infection, formation of progeny, and cell lysis—in a system with a fixed number of host cells. (12)

onychomycosis   See **tinea unguium**. (19)

oophoritis   Inflammation of the ovaries. (22)

open reading frame (ORF)   A DNA sequence predicted to encode a protein. (9)

operon   A collection of genes that are in tandem on a chromosome and are transcribed into a single RNA. (8)

ophthalmia neonatorum   Neonatal conjunctivitis, a serious eye infection of a newborn caused by *Neisseria gonorrhoeae* or *Chlamydia trachomatis* transferred from the mother to the newborn during parturition. (19, 23)

opisthorchiasis (*pl.* opisthorchiases)   Any infection that is caused by liver flukes, *Opisthorchis* species, following ingestion of larvae; infected hosts are often asymptomatic but can develop diarrhea, abdominal pain, and mild fever. (22)

O polysaccharide   Also called *O antigen*. A sugar chain in Gram-negative bacteria that connects to the core polysaccharide of lipopolysaccharides. (5, 22)

opportunistic pathogen   A microbe that normally is not pathogenic but can cause infection or disease in an immunocompromised host organism. (2, 14)

opsonization   The process of coating pathogens with IgG antibodies or complement to facilitate phagocytosis. (15, 16)

optical density   A measure of how many particles are suspended in a solution, based on light scattering by the suspended particles. (6)

optochin susceptibility disk test   A diagnostic method for identifying *Streptococcus pneumoniae* by its susceptibility to the antibiotic optochin. (25)

orbital   A region of space occupied by electrons around an atomic nucleus. (4)

orchitis   Inflammation of the testicles. (22, 23)

organelle   A membrane-enclosed compartment within eukaryotic cells that serves a specific function. (4)

organic molecule   A molecule that contains a carbon-carbon bond. (4)

organotrophy   Metabolic oxidation of organic compounds to yield energy without absorption of light. Compare with *lithotrophy*. (6, 7)

origin (*oriC*)   The region of a bacterial chromosome where DNA replication initiates. (8)

origin of replication (*ori*)   A DNA sequence where DNA replication initiates. In a bacterial chromosome, this site is also attached to the cell envelope. (5)

oropharynx   The area between the soft palate and the upper edge of the epiglottis. (14)

orotomide   Any of a class of antifungal agents that stop pyrimidine synthesis. (13)

ortholog   A gene (or protein) present in more than one species that is derived from a common ancestral gene and encodes the same function in each of the species. Compare with *paralog*. (9)

osmolarity   A measure of the number of solute molecules in a solution. (6)

osmosis   The diffusion of water from regions of high water concentration (low solute) to regions of low water concentration (high solute) across a semipermeable membrane. (4)

osmotic diarrhea   Diarrhea caused by the inability of the intestine to absorb nutrients from chyme, resulting in an intestinal lumen that is hypertonic compared with the mucosal cells of the intestines. (22)

osmotic shock   A sudden increase in osmotic pressure within a cell that is moved rapidly into a hypotonic environment. (4)

outbreak   A situation in which the number of cases of a disease exceeds the endemic level of that disease. (26)

outer membrane   In Gram-negative bacteria, a membrane outside the cell wall. Compare with *inner membrane*. (5)

oxidation   Loss of an electron. Compare with *reduction*. (4)

oxidative burst   A large increase in the oxygen consumption of immune cells during phagocytosis of pathogens as the immune cells produce oxygen radicals to kill the pathogen. (15)

oxidative phosphorylation   The process of ATP formation by respiration, via an electron transport system that uses diatomic oxygen as a final electron acceptor and generates a proton gradient across a membrane for the production of ATP via ATP synthase. (7)

oxidized   Describing a molecule that has lost an electron to an oxidizing agent during a redox reaction. Compare with *reduced*. (7)

oxidizing agent   A molecule that accepts electrons during a redox reaction. Compare with *reducing agent*. (4)

oxidoreductase   An electron transport system protein that accepts electrons from one molecule (oxidizing that molecule) and donates electrons to a second molecule (reducing the second molecule). (7)

oxygenic photosynthesis   A form of phototrophy in which the excited electrons reduce CO_2 to form sugars and release O_2. (7)

pancarditis   Inflammation of all three layers of the heart. (21)

pancreatitis   Inflammation of the pancreas. (22)

pandemic   An epidemic that occurs over a wide geographic area. Compare with *epidemic*. (2, 26)

papilloma   See **wart**. (19)

papule   A small, solid, elevated skin lesion. (19)

paralog   A gene that arises by gene duplication within a species and evolves to carry out a different function from that of the original gene; also the protein expressed by that gene. Compare with *ortholog*. (9)

paramyxovirus   Any of the single-stranded RNA viruses in the family Paramyxoviridae, which includes the mumps and measles viruses. (22)

parasite   Any bacterium, virus, fungus, protozoan (protist), or helminth that colonizes and harms its host; the term commonly refers to protozoa and to invertebrates. (1, 2)

parasitism   A symbiotic relationship in which one member benefits and the other is harmed. (27)

parental strand   One of the two strands of a DNA double helix; the strand used as a template for the synthesis of a *daughter strand* during DNA replication. (8)

parenteral route   The introduction of materials into the body other than through the gastrointestinal system—typically via intravenous or intramuscular injection. (2)

paresthesia   Sensation of tingling, prickling, burning, or numbness on the skin due to pressure on or damage to peripheral nerves. (24)

parfocal   In a microscope with multiple objective lenses, setting of the objective lenses at different heights so that focus is maintained when switching between lenses. (3)

parotitis   Inflammation of the major (parotid) salivary glands. (22)

parthenogenesis   Reproduction that results from the development of unfertilized (usually female) gametes. (22)

passive immunization   Injection of antibodies or antiserum into a naive patient to provide immediate, but temporary, protection against a toxin or pathogen or to help a patient coping with certain autoimmune diseases. Compare with *active immunization*. (17)

passive transport   The net movement of molecules across a membrane without energy expenditure by the cell. Compare with *active transport*. (5)

pasteurization   The heating of food at a temperature and time combination that will kill pathogens such as *Mycobacterium tuberculosis*. (13)

pathogen   A bacterial, viral, fungal, protozoan, or helminthic agent of disease; among health professionals, pathogens typically are limited to bacteria, viruses, and fungi. (1, 2)

pathogenesis   The process microbes use to cause disease in a host. (18)

pathogenicity   The ability of a microorganism to cause disease. (2)

pathogenicity island   A genomic island that contains virulence factors and may have been transferred from another genome. (9, 18)

pathognomonic   Referring to a sign or symptom that is characteristic of a particular disease. (19)

pathovar   A group of strains within a bacterial species that have the same virulence factors and induce a similar pathology but that differ in traits such as antigen specificity. (22)

patient history   Also called *medical history*. Written summary of a conversation between a health care professional and a patient that aids the health care professional in developing a diagnosis. (2)

patient zero   See **index case**. (26)

pattern recognition receptor (PRR)   A receptor on a cell of the innate immune system that recognizes common molecular patterns on microbial surfaces. (15)

pedestal   An actin-based, raised surface structure produced by a host epithelial cell in response to binding certain bacterial pathogens. (22)

pellicle   A protective cover around some protists; it is made of flexible, membranous layers reinforced by microtubules. (5)

pelvic inflammatory disease (PID)   Inflammation of the uterus and fallopian tubes, usually due to infection by either *Neisseria gonorrhoeae* or *Chlamydia trachomatis*. (23)

penicillin-binding protein (PBP)   Any of a group of bacterial proteins involved in cell wall synthesis that are the target of penicillin or other beta-lactam antibiotics. (13)

pentose phosphate pathway (PPP)   An alternative glycolytic pathway in which glucose 6-phosphate is first oxidized and then decarboxylated to ribulose 5-phosphate, ultimately generating 1 ATP and 2 NADPH. (7)

peptic ulcer   A break in the lining of the stomach or duodenum resulting from inflammation associated with *Helicobacter pylori* infection of the gastric mucosa. (22)

peptide bond   The covalent bond that links two amino acid monomers. (8)

peptidoglycan   Also called *murein*. A polymer of peptide-linked chains of amino sugars; a major component of the bacterial cell wall. (5)

perforin   A cytotoxic protein, secreted by cytotoxic T cells and natural killer cells, that forms pores in target cell membranes. (15)

pericarditis   Inflammation of the heart pericardium. (21, 22)

periodontal disease *or* periodontitis   A progressive bacterial disease of the gums; can include inflammation and bleeding of the gums, gum recession around the roots of the teeth, and erosion of the teeth. (22)

peripheral membrane protein   A protein that is associated with a membrane but does not span it. (4)

peripheral nervous system (PNS)   Network of nerves that delivers sensory information to the brain and spinal cord and returns motor commands to muscles and other parts of the body. Compare with *central nervous system*. (24)

periplasm   The region between the outer and inner membranes of Gram-negative bacteria; it contains water, solutes, and a variety of transport proteins. (5)

peristalsis   Waves of contraction passing along the walls of the intestines that move chyme through the intestines. (22)

peritonitis   Inflammation of the peritoneum, a life-threatening condition if not treated promptly. (22)

peritonsillar abscess   An abscess in the pharynx. (20)

permease   A substrate-specific carrier protein in the membrane. (5)

persister cell   A pathogenic cell within a host that neither grows nor dies when exposed to appropriate antimicrobials; this tolerance is due not to genetic mutation but to the cell's existence in a transient, dormant state. (13)

person-to-person epidemic   An epidemic that occurs when one infected individual infects one or more others. (27)

pertussis   See **whooping cough**. (20)

petechia (*pl.* petechiae)   A small red spot on the skin due to blood leaking from capillaries under the skin, resulting from depletion of clotting factors. (18, 24)

petechial rash   Tiny red spots on the skin that do not blanch when pressed. (14)

Petri dish *or* Petri plate   A round dish with vertical walls covered by an inverted dish of slightly larger diameter. The smaller dish can be filled with a substrate for growing microbes. (1)

Peyer's patch   An aggregation of lymphoid tissue found in the lower small intestine. (15)

phage   See **bacteriophage**. (9, 12)

phagocytosis   A form of endocytosis in which a large extracellular particle is brought into the cell. (4, 15)

phagosome   A large intracellular vesicle that forms as a result of phagocytosis. (4, 15)

pharyngitis   Also called a *sore throat*. Inflammation of the pharynx. (20)

phase-contrast microscopy   Microscopy in which contrast is generated by variable values of refractive index within the specimen, leading to additive and subtractive interference patterns between the light that penetrates the specimen and the light transmitted outside the specimen. (3)

phenol coefficient test   A test of the ability of a disinfectant to kill bacteria; the higher the coefficient, the more effective the disinfectant. (13)

phenotype   The outward expression of a genotype. (8)

phospholipid   The major component of membranes. A typical phospholipid is composed of a core of glycerol to which two fatty acids and a modified phosphate group are attached. (4)

phospholipid bilayer   A structure comprising two layers of phospholipids; the hydrocarbon fatty acid tails face the interior of the bilayer, and the charged phosphate groups face the cytoplasm and the extracellular environment. The cell membrane is a phospholipid bilayer. (4)

photolysis   The first energy-yielding phase of photosynthesis; the light-driven separation of an electron from a molecule, coupled to an electron transport system. (7)

photosynthesis   The metabolic ability to absorb solar energy and convert it to chemical energy for biosynthesis. Autotrophic photosynthesis, or photoautotrophy, includes CO_2 fixation. (7)

photosystem I   A protein complex that harvests light from a chlorophyll, donates an electron to an electron transport system and receives an electron from a small molecule such as H_2S or H_2O, and stores energy in the form of NADPH. (7)

photosystem II   A protein complex that harvests light from a chlorophyll, donates an electron to an electron transport system, and stores energy in the form of a proton potential. (7)

phototrophy   Metabolism that yields energy from chemical reactions powered by the absorption of light. Compare with *chemotrophy*. (6, 7)

phylogenetic tree   A diagram that depicts the relationships between genetically related organisms by estimating the relative amounts of evolutionary divergence among species. (9)

phylogeny   The historical relationships between genetically related organisms; a measurement of genetic relatedness. (9)

piezophile   See **barophile**. (6)

pilin   The protein monomer that polymerizes to form a pilus. (5)

pilus (*pl.* pili)   Also called *fimbria*. A straight protein filament composed of a tube of pilin protein monomers that extend from the bacterial cell envelope. (5, 18)

pinworm   A small, white nematode, of the species *Enterobius vermicularis*, that infects humans. (11, 22)

placebo   An inactive substance used as a control in an experiment to determine the effectiveness of a medicine or drug. (26)

plague   Also called *Black Death*. A life-threatening disease caused by *Yersinia pestis*. (21, 22)

plaque   A cell-free zone on a lawn of bacterial cells caused by viral lysis. (12)

plaque-forming unit (PFU)   A measure of the concentration of phage particles in liquid culture. (12)

plasma cell   A fully differentiated antibody-producing cell. (16)

plasma membrane   See **cell membrane**. (4, 5)

plasmid   An extrachromosomal genetic element that may be present in some cells. (8)

plasmocytoma   A plasma cell tumor found at a single site in the body, typically in the bone or soft tissue; can develop into *multiple myeloma*. (17)

platyhelminth   See **flatworm**. (22)

pleocytosis   Elevated number of white blood cells in cerebrospinal fluid. (24)

pneumococcal pneumonia   Pneumonia caused by *Streptococcus pneumoniae*. (20)

pneumonia Infection of the lungs that can be caused by a variety of bacteria, viruses, fungi, and parasites. (20)

pneumonic plague A highly virulent and contagious *Yersinia pestis* lung infection. (21)

point mutation Also called *base substitution*. A change in a single nucleotide within a nucleic acid sequence. (8)

point-of-care (POC) laboratory test A rapid diagnostic test that is run at the site of patient care. (25)

point-source epidemic An epidemic in which multiple cases of a disease appear over a short period because of exposure to a common source of contaminated food or water. (27)

polar covalent bond A covalent bond in which the electrons are distributed unequally between two atoms. (4)

polio *or* **poliomyelitis** A group of neurological diseases caused by migration of poliovirus into the cerebrospinal fluid. (24)

polyamine A molecule that contains multiple amine groups and is positively charged at neutral pH. (5)

polyclonal antibody A complex mixture of epitope-specific antibodies directed against the many epitopes of an antigen; it is produced after the antigen is injected into a laboratory animal (e.g., rabbit). Each epitope-specific antibody is made by a different B cell–derived plasma cell. (17)

polyene Any of a class of antifungal agents that disrupt the integrity of the cell membrane. (13)

polymerase chain reaction (PCR) A method to amplify DNA in vitro by using many cycles of DNA denaturation, primer annealing, and DNA polymerization with a heat-stable polymerase. (8)

polymicrobial Referring to an infection involving several or many types of pathogens. (10)

polymorphonuclear leukocyte (PMN) Also called *granulocyte*. A white blood cell with a multilobed nucleus and cytoplasmic granules. PMNs are classified as neutrophils, basophils, and eosinophils. (15)

polymyxin Any of a class of antibiotics that bind to and dissolve the inner membrane of bacteria. (13)

polysaccharide Also called *glycan*. A carbohydrate polymer composed of simple sugars. (4)

population Members of one species that live in a common location. (27)

porin A transmembrane protein complex that allows movement of specific molecules across the cell membrane, and in Gram-negative bacteria, across the outer membrane. (5)

positive selection In immunology, the survival of T cells bearing T-cell receptors that do not recognize self MHC proteins displayed on thymus epithelial cells. Compare with *negative selection*. (16)

post-streptococcal glomerulonephritis Sequela of a *Streptococcus pyogenes* infection that can cause symptoms from inflammation of renal tissue to kidney failure. (20)

power See **total magnification**. (3)

prebiotic A food or nutritional supplement that is ingested to improve health by providing fibers that can be digested by beneficial bacteria. (14)

predator A consumer that feeds on grazers or on other lower-level predators. (27)

preliminary treatment (of wastewater) The first stage of wastewater treatment, which uses screens and traps to remove solid debris from the sewage. (27)

prevalence In referring to a disease, the total number of active cases in a given sample or location. Compare with *incidence*. (26)

primary antibody response The production of antibodies upon first exposure to a particular antigen. B cells become activated and differentiate into plasma cells and memory B cells. Compare with *secondary antibody response*. (16)

primary encephalitis Inflammation of the brain due to a virus or other pathogen that directly attacks the brain. (24)

primary endosymbiosis An ancient endosymbiotic event between a cyanobacterium and a eukaryotic cell, which led eventually to the evolution of green and red algae after the cyanobacterium evolved into a chloroplast. (11)

primary immunodeficiency A disease or disorder of the immune system that is the result of a genetic or developmental error. (17)

primary infection First-time infection of a particular individual by any given pathogen. (23)

primary pathogen A disease-causing microbe that can breach the defenses of a healthy host. (2)

primary producer An organism that produces biomass (reduced carbon) from inorganic carbon sources such as CO_2. (27)

primary structure (of proteins) The first level of protein organization, consisting of the linear sequence of amino acid residues. Compare with *secondary structure* and *tertiary structure*. (4)

primary syphilis The initial inflammatory reaction (chancre) at the site of infection with *Treponema pallidum*. (23)

primary treatment (of wastewater) An early stage of wastewater treatment designed to mechanically remove large insoluble particles from sewage. (27)

prion An infectious agent that causes the propagation of misfolded host proteins; it consists of a defective version of the host protein. (12, 24)

probabilistic indicator A means of quickly identifying microbes in the clinical setting on the basis of a battery of biochemical tests performed simultaneously on an isolated strain. (9, 25)

probiotic A food or nutritional supplement that contains live microorganisms and is ingested to improve health by promoting beneficial bacteria. (14)

prodromal phase *or* **prodrome** The stage of a disease characterized by vague, generalized symptoms, such as headache, fever, and malaise, that precede disease-specific symptoms. (2)

proglottid A segment of a tapeworm that contains both male and female reproductive organs. (22)

prokaryote An organism whose cell or cells lack a nucleus. Prokaryotes include both bacteria and archaea. Compare with *eukaryote*. (1, 5)

promoter A noncoding DNA regulatory region immediately upstream of a structural gene; needed for transcription initiation. (8)

prophage A phage genome that has integrated into a bacterial host genome. (9, 12)

prostaglandin Any of a group of inflammation-mediating lipid products of arachidonic acid metabolism that are secreted by mast cells activated by an allergen; prostaglandins cause smooth-muscle contraction, mucus secretion, and activation of leukocytes, leading to inflammation and allergic-reaction symptoms. (17)

prostatitis Inflammation of the prostate gland. (23)

prosthetic valve endocarditis (PVE) A life-threatening bacterial infection of a prosthetic heart valve following heart valve surgery. (21)

proteasome An intracellular protein complex where unneeded or damaged proteins are degraded. (18)

protective antigen (PA) The core subunit of anthrax toxin, so called because immunity to this protein protects against disease. (18)

protein A A *Staphylococcus aureus* cell wall protein that binds to the Fc region of antibodies, hiding the *S. aureus* cells from phagocytes. (18)

Proteobacteria A very diverse phylum of Gram-negative bacteria that includes *Escherichia coli* and *Salmonella enterica*. Recent genome sequence data suggest that Proteobacteria should be elevated to superphylum status. (10)

proteome All the proteins expressed in a cell at a given time. The complete proteome includes all the proteins that the cell can express under any condition. The expressed proteome is the set of proteins made under a given set of conditions. (5)

protist A single-celled eukaryotic microbe, usually heterotrophic and motile; not a fungus. (1, 11)

proton motive force (PMF) The potential energy of the concentration gradient of protons (hydrogen ions, H^+) plus the charge difference across a membrane. (5, 7)

protozoan (*pl.* protozoa) A heterotrophic eukaryotic microbe, usually motile, that is not a fungus. (1, 11)

provirus A viral genome that is integrated into the host cell genome. (12)

pseudomembrane A membrane-like material, such as the one made of dead cells, bacteria, and fibrin that forms at the back of the throat in a patient with diphtheria. (20)

pseudopod A flexible locomotor extension of cytoplasm bounded by the cell membrane. (4, 11)

pseudorubella See **roseola infantum**. (19)

psychrophile An organism adapted for optimal growth at temperatures below 20°C (68°F). (6)

psychrotroph An organism, such as *Listeria*, that grows well at temperatures below 20°C, but equally well or faster at higher temperatures. (27)

public health A profession in which the health and disease of large populations are monitored and assessed. (1)

pulmonary circulation The system of vessels that moves blood between the heart and the lungs. (21)

pure culture   1. A culture containing only a single strain or species of microorganism. (6) 2. A large number of microorganisms that all descended from a single cell. (1)

purpura   A patch of purplish discoloration on the skin due to hemorrhaging of capillaries. (24)

pustule   A papule that has filled with pus. (19)

pyelonephritis   Infection and inflammation of the kidney. (23)

pyrimidine dimer   A form of DNA damage caused by exposure to ultraviolet light, which results in the covalent bonding of two adjacent pyrimidines on a DNA strand. If not repaired, the dimer will be read incorrectly during DNA replication, resulting in a mutation to the DNA. (8)

pyrogen   Any substance that induces fever. (15)

quantitative reverse transcription–PCR (qRT-PCR)   A technique that uses an oligonucleotide probe, such as a molecular beacon, which is cleaved during each PCR cycle, providing a signal (such as fluorescence) that can be measured and correlated with the amount of product generated after each PCR cycle. (25)

quarantined   Separated from the general population to limit the spread of infection. (26)

quorum sensing   The ability of bacteria to sense the presence of other bacteria via secreted chemical signals called autoinducers. (6)

rabbit fever   See **tularemia**. (21)

rabies   A deadly infectious disease caused by the rabies virus, which invades the central nervous system, including the brain. (24)

radial immunodiffusion assay (RIA)   A technique in which a ring of precipitation is visualized in an agarose gel impregnated with antibody. Antigen placed within a well diffuses outward until reaching a zone of equivalence where antigen-antibody complexes precipitate and form a ring. (17)

radiculitis   Inflammation of a spinal nerve root. (24)

rapid plasma reagin (RPR) test   A nonspecific test for syphilis that looks for the presence of anti-cardiolipin antibodies, which form as a result of infection by the syphilis spirochete. (23)

reactivation tuberculosis   See **secondary tuberculosis**. (20)

reading frame   A sequence of nucleotides in DNA or RNA whose nonoverlapping triplets can potentially be translated into a polypeptide. (8, 12)

reassortment   The process by which genome segments come together from different influenza strains infecting the same host cell. (12)

receptor   A molecule on the surface of a host cell that can be bound by a microbial adhesin protein. (2)

recombinant DNA   A DNA molecule, prepared in a laboratory, that contains sequences from two or more source organisms (often different species). (8)

recombinant subunit vaccine   A vaccine containing synthetic antigen made in the laboratory by the use of recombinant DNA technology. (17)

recombination   The enzyme-mediated process by which two DNA molecules exchange sections of DNA by cutting and splicing their helix backbones. (9)

reduced   Describing a molecule that has gained an electron from a reducing agent during a redox reaction. Compare with *oxidized*. (7)

reducing agent   A molecule that donates electrons during a redox reaction. Compare with *oxidizing agent*. (4)

reduction   Gain of an electron. Compare with *oxidation*. (4)

reductive evolution   Also called *degenerative evolution*. The loss or mutation of DNA encoding traits that are not under any selection pressure. (5, 9)

reflection   Deflection of an incident light ray by an object at an angle equal to the incident angle. (3)

refraction   The bending and slowing of light as it passes through a substance. (3)

refractive index   The ratio of the speed of light in a vacuum divided by the speed of light through a medium, such as water or glass. (3)

regulatory T cell (Treg)   A T cell that regulates the activity of another T cell, usually by suppressing its activity. (16)

reinfection   A later infection by the same microorganism, such as infection with a new strain of cytomegalovirus in a person who is CMV-seropositive. (21)

rennet   A mixture of stomach enzymes from ruminants that is used to curdle milk in the production of yogurt and cheese. (27)

replication fork   During DNA synthesis, the region of the chromosome that is being unwound. (8)

repression   Also called *down-regulation*. Decreased transcription of a target gene due to a repressor protein binding the operator-promoter sequence upstream of the coding gene. (8)

repressor   A regulatory protein that can bind to a specific DNA sequence and decrease the transcription of genes. (8)

reservoir   1. An organism that maintains a virus or bacterial pathogen in an area by serving as a host. (2) 2. The major part of the biosphere that contains a significant amount of an element needed for life. (27)

resolution   The smallest distance between two objects at which they can still be distinguished as separate objects. (3)

respiration   The oxidation of reduced organic electron donors through a series of membrane-embedded electron carriers to a final electron acceptor. The energy derived from the redox reactions is stored as an electrochemical gradient across the membrane and may be harnessed to produce ATP. (7)

respiratory route   Inhalation of airborne pathogens that then enter the respiratory system of the host. (2)

respiratory syncytial virus (RSV)   A virus that causes bronchitis, bronchiolitis, and pneumonia. (20)

restriction endonuclease *or* **restriction enzyme**   Any of a group of bacterial enzymes that cleave double-stranded DNA within a specific short sequence, usually a palindrome. (8)

restriction enzyme site   A DNA sequence recognized and cleaved by a restriction endonuclease. (8)

restriction fragment length polymorphism (RFLP)   A difference in DNA sequence between strains or individuals that is identified as a difference in the length of DNA fragments cut by restriction endonucleases. (25)

retroelement   A mobile genetic element in the genome of an organism. (12)

retrovirus   Also called *RNA reverse-transcribing virus*. A single-stranded RNA virus that uses reverse transcriptase to generate a double-stranded DNA copy of its genome. (12)

reverse transcriptase   An enzyme that produces a double-stranded DNA molecule from a single-stranded RNA template. (12)

rheumatoid arthritis   A chronic autoimmune disease causing inflammation and pain in many joints, especially those of the hands and feet. (17)

Rh incompatibility disease   Also known as *erythroblastosis fetalis*. A form of type II hypersensitivity caused when an Rh⁻ mother produces antibodies that destroy her fetus's Rh⁺ red blood cells. (17)

rhinovirus   A small (30 nm), nonenveloped, single-stranded, positive-sense RNA virus with more than 100 known serotypes; the most frequent cause of the common cold. (18)

rhizobium (pl. rhizobia)   A bacterial species of the order Rhizobiales that forms highly specific mutualistic associations with plants in which the bacteria form intracellular bacteroids that fix nitrogen for the plant. (7)

rhizosphere   The soil environment surrounding plant roots. (27)

rich medium   See **complex medium**. (6)

rickettsiosis (pl. rickettsioses)   Any of a collection of diseases, each caused by a species of *Rickettsia*, that present with similar symptoms. (21)

ringworm   Any of various forms of dermatophytosis. (19)

RNA-dependent RNA polymerase   An enzyme that produces an RNA complementary to a template RNA strand. (12)

RNA polymerase   Any of a group of enzymes that catalyze synthesis of an RNA copy from a DNA template during transcription. (8)

RNA primer   A short segment of RNA, complementary to a specific region of a chromosome, that is required for DNA polymerase to carry out DNA replication. (8)

RNA reverse-transcribing virus   See **retrovirus**. (12)

R number   The estimated number of people to whom an infected person can pass a pathogen. To stop spread of the disease, the R number must fall below 1. (26)

Rocky Mountain spotted fever (RMSF)   Tick-borne disease caused by *Rickettsia rickettsii* that can be fatal if not treated promptly. (21)

roseola infantum *or* **roseola**   Also called *exanthem subitum, pseudorubella, sixth disease*, or *three-day fever*. A disease of young children in which a high fever is followed by a rash, caused by human herpesvirus 6 or, less frequently, human herpesvirus 7. (19)

rotavirus   A nonenveloped, double-stranded RNA virus that causes severe diarrhea in children. (22)

roundworm   Also called *nematode*. Any of a large family of tubular worms that includes parasitic pinworms, hookworms, and whipworms. (11, 22)

rubella   See **German measles**. (19)

rubeola   See **measles**. (19)

Rubisco   Ribulose 1,5-bisphosphate carboxylase oxygenase, the enzyme that catalyzes carbon fixation in the Calvin cycle. (7)

Sabin vaccine   Oral polio vaccine made with live, attenuated virus. (24)

Salk vaccine   Injected polio vaccine made with inactivated virus. (24)

salmonellosis (*pl.* salmonelloses)   Any infection caused by *Salmonella*. (22)

sandwich ELISA   An immunoassay designed to capture the antigen of interest between an immobilized anti-target capture antibody and the same or another anti-target antibody in solution. (17)

sanitation   The killing or removal of pathogens through the safe disposal of waste material and through disinfection and proper cleaning. (13)

saturated   Describing a fatty acid that contains all single bonds between adjacent carbons, so each carbon is bonded to the maximum number of hydrogen atoms. Compare with *unsaturated*. (4)

scalded skin syndrome   Also called *staphylococcal scalded skin syndrome*. The destructive consequence of *Staphylococcus aureus* exfoliative exotoxin, which destroys adhesion molecules that hold cells in the host epidermis to the underlying dermis, leaving the skin severely blistered. (19)

scale bar   A micrograph annotation whose length corresponds to the actual size magnified, such as 5 μm. Like a ruler, a scale bar applies to length measurements throughout a micrograph. (3)

scanning electron microscopy (SEM)   Electron microscopy in which the electron beams scan across the specimen's surface to reveal the three-dimensional topology of the specimen. (3)

scattering   Interaction of light with an object that results in propagation of spherical light waves at relatively low intensity. (3)

schistosomiasis (*pl.* schistosomiases)   A serious helminthic infection, caused by *Schistosoma* species of trematodes (flukes), that can damage a wide variety of organs, including intestines, bladder, liver, lungs, and spleen. (22)

scolex (*pl.* scolices)   The head of a tapeworm. (22)

secondary antibody response   A memory B cell–mediated rapid increase in the production of antibodies in response to a repeat exposure to a particular antigen. Compare with *primary antibody response*. (16)

secondary encephalitis   Inflammation of the brain caused by a spreading infection or inflammation of the meninges. (24)

secondary endosymbiont alga   An algal species that evolved from a secondary endosymbiotic event. (11)

secondary endosymbiosis   An ancient endosymbiotic event between a descendant of primary endosymbiosis and another eukaryotic cell. (11)

secondary immunodeficiency   A disease or disorder of the immune system acquired as a result of infection, radiation treatment, or use of immunosuppressive drugs. (17)

secondary structure (of proteins)   The second level of protein organization, consisting of repeating patterns, especially the alpha helix and the beta sheet. Compare with *primary structure* and *tertiary structure*. (4)

secondary syphilis   The second stage of syphilis, signified by a rash that may appear at some point after the primary latent stage of syphilis. (23)

secondary treatment (of wastewater)   An intermediate stage of wastewater treatment that uses microbiological processes to decompose soluble organic matter in the wastewater and thus reduce its biochemical oxygen demand. (27)

secondary tuberculosis   Also called *reactivation tuberculosis*. A new round of serious disease that is caused by *Mycobacterium tuberculosis* in immunocompromised patients with latent tuberculosis. (20)

second law of thermodynamics   In every transfer or transformation of energy, some energy is lost, and as a result, the entropy of the system will increase. (4)

secreted   Describing molecules that are transported out of a cell for a specific purpose, such as to signal another cell. (8)

secretory diarrhea   Diarrhea caused by microbial action (usually a toxin) that induces mucosal cells to increase secretion of ions, making the intestinal lumen hypertonic. (22)

select agent   A pathogen or toxin from a pathogen that could serve as a bioweapon because it meets some or all of the following criteria: it is highly virulent, it causes a high rate of mortality, it is very contagious, no vaccine exists against it, and treatment with antimicrobial drugs or other pharmaceuticals is largely ineffective. (26)

selectin   Any of a family of cell adhesion molecules. (15)

selective IgA deficiency   An immunodeficiency disease in which the patient fails to produce IgA but produces relatively normal levels of IgG and IgM. (17)

selective medium   A medium that allows the growth of certain species or strains of organisms but not others. Compare with *differential medium*. (6)

selective toxicity   The ability of a drug, at a given dose, to harm a pathogen without harming its host. (13)

semisynthetic antimicrobial   A drug made when a precursor antimicrobial that was isolated from a natural source is chemically modified. (13)

sense strand   See **coding strand**. (8)

sensitivity   For a diagnostic test, a parameter that describes how <u>small</u> a concentration of a microbe, antibody, or antigen a test can detect if it is present in the sample. Compare with *specificity*. (17, 25)

sensitization   The first stage of type I hypersensitivity, in which a naive patient encounters the allergen, which elicits production of anti-allergen IgE that binds to mast cells, making them reactive to the presence of allergen. (17)

sensory neuron   A nerve cell that transmits sensory information about the environment and state of the organism to the central nervous system. Compare with *motor neuron*. (24)

sepsis   Also called *blood poisoning*. A systemic inflammatory response that is triggered by an infection and is so extreme that it can kill the patient. (15, 21)

septation   The formation of a septum, a new section of cell wall and envelope that separates two prokaryotic daughter cells. (5)

septicemia   A condition where a pathogen that has entered the bloodstream overcomes our innate defensive mechanisms and begins to replicate to large numbers. (21)

septicemic plague   Infection of the bloodstream by *Yersinia pestis*. (21)

septic shock   Catastrophic fall in blood pressure owing to severe sepsis. (21)

septum (*pl.* septa)   A plate of cell wall and envelope that forms to separate two daughter cells. (5)

sequela (*pl.* sequelae)   A serious, harmful immunological consequence of disease that occurs after the infection itself is over; an example is rheumatic fever. (2)

serology   The branch of immunology that analyzes the contents of serum. (17)

serum   The noncell, liquid component of blood that remains after blood coagulation. (16)

serum sickness   A type III hypersensitivity reaction following an injection of animal-derived protein, such as antiserum or antivenom. (17)

severe combined immunodeficiency (SCID)   An immunodeficiency in which T-cell responses are absent. There are multiple genetic causes, but SCID is invariably fatal unless T-cell immunity can be restored by transplantation or gene therapy. (17)

sex pilus   A pilus specialized to attach a DNA-donating bacterium to a recipient bacterium, enabling DNA transfer through a conjugation apparatus. (5)

sexually transmitted infection (STI)   An infection transmitted primarily through sexual activity; formerly known as *sexually transmitted disease (STD)*. (23)

sexual reproduction   Reproduction involving the joining of gametes generated by meiosis. Compare with *asexual reproduction*. (11)

shelter species   An organism, often an invertebrate or protist, that harbors microbes that exchange genetic information and undergo evolution within the host, occasionally leading to changes that allow a microbe to cross species barriers and infect a new host species, such as a human. (10)

Shiga toxin   An exotoxin produced by *Shigella dysenteriae* and *Escherichia coli* O157:H7 (and some other *E. coli* pathovars) that disables host cell protein synthesis. (18)

Shiga toxin–producing *E. coli* (STEC)   Any of a group of *Escherichia coli* strains that can produce Shiga toxin. (22)

shigellosis (*pl.* shigelloses)   Also called *bacillary dysentery*. Bloody diarrhea caused by pathogenic *Shigella* species. (22)

sigma factor   A protein needed to bind RNA polymerase to a DNA promoter for the initiation of transcription in bacteria. (8)

sign   A consequence of a disease in a patient that another person can observe, such as a runny nose or fever. Compare with *symptom*. (2)

silent mutation   A mutation that does not change the amino acid sequence encoded by an open reading frame. The changed codon encodes the same amino acid as the original codon. (8)

simple sugar   See **monosaccharide**. (4)

sink   A part of the biosphere that can receive or assimilate significant quantities of an element; it may be biotic (as in plants fixing carbon) or abiotic (as in the ocean absorbing carbon dioxide). (27)

site-specific recombination   Recombination between two DNA molecules that do not share long regions of homology but do contain short regions of homology specifically recognized by the recombination enzyme. Compare with *generalized recombination*. (9)

sixth disease   See **roseola infantum**. (19)

skin   The largest organ in the human body, providing a physical barrier to infection, protecting underlying organs, aiding in thermoregulation and water loss, synthesizing vitamin D, and sensing various aspects of the outside world, such as heat, cold, and pressure. (19)

skin-associated lymphoid tissue (SALT)   Immune cells, such as dendritic cells, located under the skin that help eliminate bacteria that have breached the skin's surface. (15)

S-layer   A crystalline protein surface layer replacing or outside the cell wall in many species of archaea and bacteria. (5)

smallpox   A deadly human disease caused by the variola virus. (19)

sore throat   See **pharyngitis**. (20)

specialized transduction   Transduction in which the phage can transfer only a specific, limited number of bacterial donor genes to the recipient cell. Compare with *generalized transduction*. (9)

species   A single, specific type of organism, designated by a genus and species name. (1)

specificity   For a diagnostic test, a parameter that describes how well it can distinguish among closely related targets. Compare with *sensitivity*. (17, 25)

spectrophotometer   An instrument for measuring the amount of light absorbed by a solution when different wavelengths of light are passed through that solution; used to calculate the concentration of substances in solution. (6)

spectrum of activity   The group of pathogens for which an antimicrobial agent is effective. (13)

spike protein   A viral glycoprotein that connects the membrane to the capsid or the matrix and may be involved in viral binding to host cell receptors. (12)

spinal tap   See **lumbar puncture**. (24)

spirillum (*pl.* spirilla)   A Gram-negative bacterium that has a helical body shape. (10)

Spirochaetes   A phylum of bacteria that are tightly coiled cells with periplasmic flagella that double back along a cell body enclosed by a sheath. *Borrelia burgdorferi* is one example. (10)

spirochete   A bacterium that has a tight, flexible spiral shape; a member of the phylum Spirochaetes. (3, 10)

splenomegaly   Enlargement of the spleen. (21)

spontaneous generation   The theory, much debated in the nineteenth century, that under current Earth conditions life can arise spontaneously from nonliving matter. (1)

spread plate   A method to grow separate bacterial colonies by plating serial dilutions of a liquid culture. (6)

staining   The process of treating microscopic specimens with a stain to enhance their detection or to visualize specific cell components. (3)

stalk   An extension of the cytoplasm and envelope that attaches a microbe to a substrate. (5)

staphylococcal scalded skin syndrome (SSSS)   See **scalded skin syndrome**. (19)

staphylococcus   Any member of the Gram-positive bacterial genus *Staphylococcus*, which forms a hexagonal arrangement of round cells resulting from septation in random orientations. (5)

stationary phase   A period of cell culture, following exponential phase, during which bacteria appear to stop growing. (6)

STD   See **sexually transmitted infection**. (23)

sterilization   The destruction of all cells, spores, and viruses on an object. (13)

strain   A genetic variant of a microbial species. (10)

streptococcal pharyngitis   Strep throat; inflammation of the pharynx caused by a *Streptococcus pyogenes* infection. (20)

streptococcal pyogenic exotoxin (SPE)   An exotoxin produced by some strains of *Streptococcus pyogenes*; multiple forms; all are superantigens; associated with streptococcal toxic shock syndrome and scarlet fever. (19)

strict aerobe   An organism that performs aerobic respiration and can grow only under aerobic conditions (in the presence of oxygen). Compare with *facultative anaerobe* and *strict anaerobe*. (6)

strict anaerobe   An organism that performs anaerobic respiration and can grow only under anaerobic conditions (in the absence of oxygen). Compare with *facultative anaerobe* and *strict aerobe*. (6)

strongyloidiasis   Any infection caused by parasitic nematodes of the genus *Strongyloides*. (22)

structural formula   A representation of molecular structure in which each covalent bond is shown as a line between atoms. (4)

subacute bacterial endocarditis (SBE)   Bacterial infection of a heart valve characterized by slow onset and vague symptoms; usually caused by a viridans streptococcus from the normal oral flora. Compare with *acute bacterial endocarditis*. (21)

subacute sclerosing panencephalitis (SSPE)   A sequela of measles, possibly due to a latent measles infection, that manifests as a slow, progressive degenerative disease of the central nervous system. (19)

substrate-binding protein   A membrane protein that binds specific substrates and delivers them to their cognate ABC transporter. (5)

subunit vaccine   A vaccine containing only isolated antigens of a pathogen or fragments of a pathogen containing the antigens. (17)

superantigen   A molecule that directly stimulates T cells without undergoing antigen-presenting-cell processing and surface presentation. (16)

supercoil   An extra twist or turn found in DNA, either positive (increasing DNA winding) or negative (decreasing DNA winding). (8)

suprapubic puncture   A method for collecting sterile urine by inserting a needle through the abdomen and directly into the bladder. (25)

surface receptor   A host cell membrane protein that is recognized and bound by a microbe having a complementary binding molecule on its surface. (12)

symbiont   An organism that lives in close association with another organism. (6)

symbiosis (*pl.* symbioses)   The intimate association of two species. (1, 27)

symport   Coupled transport in which the molecules being transported move in the same direction across the membrane. Compare with *antiport*. (5)

symptom   A sensation related to disease that only a patient can feel, such as a headache, numbness, or fatigue. Compare with *sign*. (2)

symptomatic surveillance   See **syndromic surveillance**. (26)

syncytium (*pl.* syncytia)   A large multinucleated cell formed by the fusion of many uninucleate cells. (20)

syndrome   A collection of signs and symptoms that occur together and signify a particular disease. (2)

syndromic surveillance   Also called *symptomatic surveillance*. Methods to identify clusters of patients with an unusual combination of symptoms that may represent a new emerging infectious disease. (26)

synergism   Cooperation between species in which both species benefit but can grow independently. The cooperation is less intimate than symbiosis. (27)

synergistic   Referring to antibiotics used in combinations whose mechanisms of action complement one another and hence intensify their effectiveness. Compare with *antagonistic*. (13)

synthetic medium   A bacterial growth solution that contains a defined set of known components. Compare with *complex medium* and *minimal medium*. (6)

syphilis   A sexually transmitted infection caused by the spirochete *Treponema pallidum*. (23)

systemic circulation   The system of vessels that moves blood between the heart and the rest of the body. (21)

systemic infection   An infection that spreads from its site of origin to many other parts of the body. (21)

systemic inflammatory response syndrome (SIRS)   A life-threatening condition whose symptoms include abnormal body temperature, rapid heart and breathing rates, and abnormal white cell counts. Patients with SIRS and a known infection are classified as having sepsis. (21)

tabes dorsalis   Loss of coordinated movement, especially as a result of syphilis. (23)

tachyzoite   The rapidly dividing motile form of *Toxoplasma gondii*; these cells spread throughout the host. Compare with *bradyzoite*. (21)

tapeworm   Also called *cestode*. Any of a family of parasitic flatworms that infect many animal species via the fecal-oral route of transmission. (11, 22)

taxonomy   The description of distinct life forms and their organization into categories. (9)

T cell   Also called *T lymphocyte*. An adaptive immune cell that develops in the thymus and can give rise to antigen-specific helper T cells and cytotoxic T cells. (12, 15)

T-cell receptor (TCR)   A T-cell membrane protein complex that recognizes and binds to MHC-antigen complexes on the surface of antigen-presenting cells. The interaction triggers differentiation of the T cell. (16)

tegument   The contents of a virion between the capsid and the envelope. (12)

teichoic acid   A chain of phosphodiester-linked glycerol or ribitol that threads through and reinforces the cell wall in Gram-positive bacteria. (5)

temperate phage   A bacteriophage capable of lysogeny. Compare with *virulent phage*. (12)

template strand   A DNA strand (or an RNA strand in some viruses) that is used as a template for the synthesis of mRNA. (8)

termination   The final phase of DNA synthesis, in which DNA replication stops because DNA polymerase lands on a termination site; also the final events in transcription and in translation. (8)

tertiary structure (of proteins)   The third level of protein organization, consisting of the unique three-dimensional shape of a protein. Compare with *primary structure* and *secondary structure*. (4)

tertiary syphilis   A final stage of syphilis, manifested by cardiovascular and nervous system symptoms. (23)

tertiary treatment (of wastewater)   Also called *advanced wastewater treatment*. A late stage of wastewater treatment designed to remove remaining contaminants such as heavy metals, phosphorus, and pathogens. (27)

tetanospasmin   The tetanus-causing potent exotoxin produced by *Clostridium tetani*. (24)

tetanus   Also called *lockjaw*. A deadly disease caused by the *Clostridium tetani* exotoxin tetanospasmin, which interferes with nerve transmission and forces uncontrollable muscle contraction. (24)

therapeutic dose   The minimum dose of an antibiotic or antimicrobial per kilogram of body weight that stops the growth of a pathogen in a patient. Compare with *toxic dose*. (13)

thermocline   A region of the ocean where temperature decreases steeply with depth, and water density increases. (27)

thermocycler   Laboratory instrument that rapidly and precisely cycles the temperature during polymerase chain reaction. (25)

thermophile   An organism adapted for optimal growth at high temperatures, usually 55°C (131°F) or higher. (6)

third disease   See **German measles**. (19)

three-day fever   See **roseola infantum**. (19)

three-day measles   See **German measles**. (19)

thrombocytopenia   Low platelet count. (22)

thrombotic thrombocytopenic purpura (TTP)   A life-threatening sequela of Shiga toxin–producing pathogens, such as *Escherichia coli* O157:H7, in which microclots occur throughout the circulation owing to a drop in platelets. (22)

thrush   A disease of neonates, infants, and immunocompromised adults, caused by *Candida albicans* and characterized by a white coating in the mouth. (22)

tick   An arachnid ectoparasite that sucks host blood and, in the process, can deliver or receive a variety of microbial pathogens. (11)

tinea capitis   Dermatophytosis (ringworm) of the scalp. (19)

tinea corporis   Dermatophytosis (ringworm) of the body, primarily the arms and legs. (19)

tinea cruris   Dermatophytosis (ringworm) of the groin. (19)

tinea pedis   Dermatophytosis (ringworm) of the feet. (19)

tinea unguium   Also called *onychomycosis*. Fungal infection of the nails. (19)

tinea versicolor   A chronic infection of the skin caused by the yeast *Malassezia*. (19)

tissue tropism   The range of host tissue that a pathogen can infect. (12)

titer   A measure of concentration used to express the effective amount of antibody or antigen; also used to measure viral load. (17)

T lymphocyte   See **T cell**. (12)

Toll-like receptor (TLR)   Any of a family of eukaryotic transmembrane glycoproteins that recognize a particular microbe-associated molecular pattern present on pathogenic microorganisms. (15)

tonsillitis   Inflammation of the tonsils. (20)

TORCH complex   Collection of microbes able to cross the placenta and infect the fetus; they include *Ṯoxoplasma*, o̱thers, ṟubella, c̱ytomegalovirus, and ẖerpes simplex 2 virus. (21)

total magnification   Also called *power*. The magnification of the ocular lens multiplied by the magnification of the objective lens. (3)

toxic dose   The maximum dose of an antibiotic or antimicrobial that is tolerated by a patient. Compare with *therapeutic dose*. (13)

toxoid vaccine   A vaccine composed of an inactivated microbial toxin. (17)

toxoplasmosis (*pl.* toxoplasmoses)   A protozoan disease, caused by *Toxoplasma gondii*, that is typically asymptomatic but can be severe in immunocompromised patients and in fetuses. (21, 24)

trachoma   A serious disease of the eye caused by *Chlamydia trachomatis*; damage to the cornea can lead to irreversible blindness. (19)

transcription   The synthesis of RNA complementary to a DNA template. (8)

transcytosis   Transport of vesicle-bound macromolecules across a cell; common in epithelial cells. The process is co-opted by some pathogens to aid in invasion of host tissue. (24)

transduction   The transfer of host genes between bacterial cells via a bacteriophage. (9, 12)

transfer RNA (tRNA)   A noncoding RNA molecule that carries an amino acid to the acceptor site of a ribosome for addition onto a growing peptide chain. (8)

transformasome   A bacterial cell membrane protein complex that imports external DNA during transformation. (9)

transformation   1. The internalization of free DNA from the environment into bacterial cells. (8, 9) 2. The viral conversion of a normal cell to a cancer cell. (12)

translation   The synthesis of proteins by a ribosome by decoding the sequence of mRNA. (8)

transmembrane protein   A protein with one or more membrane-spanning regions. (4)

transmissible spongiform encephalopathy (TSE)   Any of a group of prion diseases characterized by the formation of large, spongy vacuoles in the brain. (24)

transmission   In disease, the transfer of a pathogen from one host to another. (12)

transmission electron microscopy (TEM)   Electron microscopy in which electron beams are transmitted through a thin specimen to reveal internal structure. (3)

transovarial transmission   The transfer of a pathogen from parent to offspring by infection of the egg cell; typically seen in insects. It is a type of *vertical transmission*. (2)

transplacental transmission   In disease, the transfer of a pathogen from an infected woman to her fetus across the placenta. It is a type of *vertical transmission*. (2)

transposable element   Also called *transposon*. A segment of DNA that can move from one DNA region to another. (9)

transposase   An enzyme that catalyzes the transfer of a transposable DNA sequence from one DNA location to another; often encoded by the transposable element. (9)

transposition   The process of moving a transposable element from one DNA location to another. (9)

transposon   See **transposable element**. (9)

trematode   See **fluke**. (11, 22)

trench mouth   Also called *Vincent's angina* or *necrotizing ulcerative gingivitis*. A severe form of gingivitis. (22)

tricarboxylic acid cycle (TCA cycle)   Also called *citric acid cycle* or *Krebs cycle*. A metabolic cycle that catabolizes the acetyl group from acetyl-CoA into two CO_2 molecules and produces NADH, $FADH_2$, and ATP. (7)

trichiasis   A condition in which eyelashes grow back toward the eye and touch the cornea or conjunctiva, as in trachoma. (19)

trichinosis (*pl.* trichinoses)   A helminthic infection, caused by the ingestion of *Trichinella* cysts embedded in pork or other meat, that affects muscles. (22)

trichomoniasis   A sexually transmitted infection of the urogenital tract that is caused by the protozoan *Trichomonas vaginalis*; often causes vaginitis. (23)

trophic level   A level of the food web representing the consumption of biomass of organisms from a previous level. (27)

trophozoite   The active feeding stage of certain protozoa, such as amebas and apicomplexans. (11, 22)

trypanosome   A parasitic protozoan that has a cortical skeleton of microtubules culminating in a long flagellum. Trypanosomes are the cause of diseases such as Chagas disease (*Trypanosoma cruzi*) and African trypanosomiasis (*T. brucei*). (11)

tubercle   A nodule seen in various organs that is characteristic of tuberculosis. (20)

tuberculin skin test (TST)   Also called *Mantoux tuberculin skin test*. Diagnostic test for assessing whether a person is or has been infected with *Mycobacterium tuberculosis*. Positive readout is a delayed-type hypersensitivity reaction to tuberculin purified protein derivative. (20)

tuberculosis (TB)   An infectious disease of the lungs caused by *Mycobacterium tuberculosis*. (20)

tularemia   Also called *rabbit fever*. Disease caused by infection with the bacterium *Francisella tularensis*; symptoms and severity of the disease depend on the virulence of the infecting strain and how the bacteria enter the host. (21)

tumor   A mass of abnormal cells that forms within a tissue. (17)

tumor necrosis factor alpha (TNF-α)   A cytokine involved in systemic inflammation. (15)

turbidity   The cloudiness or opacity of a solution due to suspended solids, such as microbes. (6)

type I hypersensitivity   Also called *immediate hypersensitivity*. An IgE-mediated allergic reaction that causes degranulation of mast cells within minutes of exposure to the antigen. The severe reaction known as anaphylaxis is triggered by type I hypersensitivity. (17)

type I pilus   A pilus that adheres to mannose residues on host cell surfaces. (18)

type II hypersensitivity   An immune response in which antibodies bind to the patient's own cell-surface antigens or to foreign antigens adsorbed onto the patient's cells. Antibody binding triggers cell-mediated cytotoxicity or activation of the complement cascade. (17)

type II secretion   A bacterial protein secretion system that uses a type IV pilus–like extension/retraction mechanism to push proteins out of the cell. (18)

type III hypersensitivity   Also called *immune complex disease*. An immune response triggered when IgG antibody binds to an excess of soluble foreign antigen in the blood. The immune complexes deposit in small blood vessels, where they interact with complement to initiate an inflammatory response. (17)

type III secretion   A bacterial protein secretion system that uses a molecular syringe to inject bacterial proteins into the host cytoplasm. (18)

type IV hypersensitivity   Also called *delayed-type hypersensitivity*. An immune response that develops 24–72 hours after exposure to an antigen that the immune system recognizes as foreign. The response is triggered by antigen-specific T cells. It is delayed because the T cells need time to proliferate after being activated by the allergen. (17)

type IV pilus   A dynamic pilus that can repeatedly assemble and disassemble; it mediates twitching motility. (18)

type IV secretion   A bacterial protein secretion system that is homologous to the bacterial conjugation apparatus. (18)

typhoid fever   Also called *enteric fever*. A life-threatening systemic disease of humans, transmitted by the fecal-oral route, that is caused by *Salmonella* Typhi. (21, 22)

ubiquitination   Posttranslational attachment of the protein ubiquitin to another protein, which alters protein function or location or signals that the protein is to be degraded. (18)

uncoating   The release of a viral genome from its capsid after entry of the virion into a host cell. (12)

unsaturated   Describing a fatty acid that contains one or more double bonds between adjacent carbons. Compare with *saturated*. (4)

upper respiratory tract   Upper airways of the body, which include the nasal passages, oral cavity, pharynx, and larynx. (20)

up-regulation   See **induction**. (8)

urethritis   Inflammation of the urethra. (23)

urinary tract infection (UTI)   Any infection of the urinary tract; includes cystitis and pyelonephritis. (23)

urogenital route   Entry of a pathogen via the urethra or vagina. (2)

urosepsis   Life-threatening infection of the blood after a urinary tract infection. (23)

use-dilution test   A method for measuring the efficacy of disinfectants against microbes dried onto a surface. (13)

vaccination   Also called *immunization*. Exposure of an individual to a weakened version of a microbe or a microbial antigen, in the hope it will provoke immunity that will prevent future disease by the virulent microbe. In technical terms, "immunization" refers to the development of immunity after a person has been vaccinated or exposed to a pathogen. (16)

vaccine   The material used to vaccinate someone. (17)

vaccine-associated paralytic poliomyelitis (VAPP)   Polio acquired as the result of vaccination with a Sabin vaccine whose attenuated virus reverted to a virulent form. (24)

vaginitis   Inflammation of the vagina. (23)

vaginosis (*pl.* vaginoses)   Infection of the vagina. (23)

valley fever   See **coccidioidomycosis**. (20)

vancomycin   A glycopeptide antibiotic that inhibits bacterial cell wall synthesis by binding the D-alanine–D-alanine dipeptide—a mechanism that is distinct from penicillin's inhibitory mechanism. (13)

variable region   The amino-terminal portion of antibody light and heavy chains that confers specificity to antigen binding and defines the antibody idiotype. (16)

variant Creutzfeldt-Jakob disease (vCJD)   A human form of mad cow disease (bovine spongiform encephalopathy), characterized by a slow deterioration of brain functions and invariably fatal. (24)

vasoactive factor   A cell signaling molecule that increases capillary permeability. (15)

vasodilation   Widening of blood vessels. (15)

vector   1. An organism (e.g., insect) that can carry infectious agents from one animal to another. (2) 2. In molecular biology, a molecule of DNA into which exogenous DNA can be inserted to be cloned. (8)

vegetative reproduction   See **asexual reproduction**. (11)

vehicle transmission   In disease, the movement of infectious agents between hosts via contaminated fomites, food, water, or air. (2)

Venereal Disease Research Laboratory (VDRL) test   A nonspecific test for syphilis that looks for the presence of anti-cardiolipin antibodies in blood or cerebrospinal fluid. (23)

venipuncture   The puncture of a vein with a needle to either withdraw blood or administer an intravenous injection. (25)

verruca vulgaris   See **wart**. (19)

vertical gene transfer   The passage of genes from parent to offspring through reproduction. Compare with *horizontal gene transfer*. (9)

vertical transmission   In disease, the transfer of a pathogen from parent to offspring. *Transovarial transmission* and *transplacental transmission* are types of vertical transmission. Compare with *horizontal transmission*. (2)

vesicle   1. A small, membrane-enclosed sphere found within a cell. (4) 2. A small blister on the skin. (19)

viable   Capable of replicating—for instance, by forming a colony on an agar plate. (6)

Vi Ag   Virulence antigen, a polysaccharide capsule on the surface of *Salmonella* Typhi that renders the bacteria resistant to phagocytosis and serum complement. (22)

vibriosis (*pl.* vibrioses)   Any infection caused by a *Vibrio* species. (22)

Vincent's angina   See **trench mouth**. (22)

viral load   The concentration of virus in the blood. (23)

viral sinusitis   Inflammation and congestion of the sinuses due to a viral infection of the upper respiratory tract. (20)

viridans streptococcus   Streptococci that are alpha-hemolytic on blood agar; most of the members are colonizing microbiota of the oral cavity. (21)

virion   A virus particle. (12)

viroid   An infectious, naked nucleic acid. (12)

virome   All the viruses that inhabit a particular organism or environment. (14)

virulence   A measure of the severity of a disease caused by a pathogenic agent. (2)

virulence factor   A trait of a pathogen that enhances the pathogen's disease-producing capability. (18)

virulent phage   A bacteriophage that reproduces entirely by a lytic cycle. Compare with *temperate phage*. (12)

virus   A noncellular particle containing a genome that can replicate only inside a host cell. (1, 12)

vulvovaginal candidiasis *or* vulvovaginitis   Also called *yeast infection*. Overgrowth of the vaginal commensal fungus *Candida*. (23)

wart   Also called *papilloma*. A small growth on the skin caused by human papillomavirus. The occurrence of common warts is a condition called verruca vulgaris. (19)

wastewater treatment plant   A facility for treating wastewater that uses a series of wastewater transformations designed to lower biochemical oxygen demand and eliminate human pathogens before water is returned to local rivers. (27)

water activity   A measure of water that is not bound to solutes and is available for use by organisms. (6)

wavelength   The distance between two adjacent peaks in the electromagnetic spectrum. (3)

western blot   A technique to detect specific proteins. Proteins are subjected to gel electrophoresis, transferred to a blot, and probed with enzyme-linked or fluorescently tagged antibodies that specifically bind the protein of interest. (17)

West Nile virus (WNV)   An RNA virus that is endemic to birds and is transmitted by mosquitoes to the dead-end hosts humans and horses, where it causes West Nile encephalitis. (24)

wetland   A region of land that undergoes seasonal fluctuations in water level and aeration. (27)

wetland filtration   A wastewater treatment method that uses wetland microbial ecosystems to remove organic carbon and nitrogen from the wastewater. (27)

wetland restoration   The repair and reconstruction of wetlands and their ecosystems that were previously degraded by pollutants and urban or suburban development. (27)

wet mount   A technique to view living microbes under a microscope by placing the microbes in water on a slide under a coverslip. (3)

whipworm   Any of the soil-dwelling helminthic parasites of the genus *Trichuris*, which burrow into the wall of the intestines, causing frequent bloody stools and rectal prolapse. (22)

white blood cell (WBC) differential   A laboratory test that counts the types of white blood cells in a patient's blood. (15)

whooping cough   Also called *pertussis*. A highly contagious respiratory infection, caused by *Bordetella pertussis*, whose defining symptom is a paroxysmal cough that lasts for weeks to months. (20)

X-linked agammaglobulinemia   Also called *Bruton's disease* or *Bruton's agammaglobulinemia*. A disease in which the body fails to make B cells, owing to mutation in the X-linked *BTK* gene; the gene product, BTK, is a protein tyrosine kinase essential for normal B-cell development. (17)

X-ray crystallography (X-ray)   A technique to determine the positions of atoms (atomic coordinates) within a molecule or molecular complex, based on the diffraction of X-rays by the molecule. (3)

yeast   A unicellular fungus. (11)

yeast infection   See **vulvovaginal candidiasis**. (23)

yersiniosis (*pl.* yersinioses)   Any infection caused by a *Yersinia* species (22)

yogurt   A semisolid food produced through acidification of milk by lactic acid–producing bacteria. (27)

zone of hypoxia   See **dead zone**. (27)

zone of inhibition   A region of no bacterial growth on an agar plate owing to the diffusion of a test antibiotic; correlates with the minimum inhibitory concentration. (13)

zoonotic disease *or* zoonosis (*pl.* zoonoses)   An infection that normally affects animals but can be transmitted to humans. (2, 10, 27)

zygote   The diploid cell produced by fusion of a male gamete and a female gamete. (11)

Credits

Front Matter

p. iii (all): Shutterstock; p. v (left): Sattva78/Shutterstock; p. v (right): GlobalStock/Getty Images; p. vi (top left): DigitalVision Vectors/Getty Images; p. vi (bottom left): DRB Images/Getty Images; p. vi (right): Mammuth/Getty Images; p. vii (top): Image Source/Corbis via Getty Images; p. vii (bottom): Shutterstock; p. viii (left): Fotolia; p. viii (top right): Sdart/iStock; p. viii (bottom right): 4x6/iStock; p. ix (left): Alexandr Zagibalov/Shutterstock; p. ix (right): Shutterstock; p. x (top): Studioloco/Shutterstock; p. x (bottom): Lightwavemedia/Shutterstock; p. xi (left): 4x6/Getty Images; p. xi (right): Getty Images; p. xii (top): Luis Molinero/Shutterstock; p. xii (bottom): Alamy Stock Photo; p. xiii (top left): age fotostock; p. xiii (bottom left): iStockphoto; p. xiii (right): 4x6/iStockphoto; p. xiv (left): FrankyLiu/Shutterstock; p. xiv (top right): Getty Images; p. xiv (bottom right): Majivecka/Shutterstock; p. xv (left): Ambrophoto/Shutterstock; p. xv (right): Michaeljung/Shutterstock; p. xvi: Getty Images; p. xvii: John W. Foster; p. xvii: Charity M. Mason; p. xvii: Joan Slonczewski; p. xxii (wooden blocks): Vivid Pixels/Shutterstock; p. xxii (insect icons): Alexander Ryabintsev/Shutterstock; p. xxvii (top left): AP Photo/Felipe Dana; p. xxvii (bottom left): Courtesy of Lara Brandão M.D., Rio De Janeiro, Brazil.; p. xxvii (right): AMI IMAGES/Science Source; p. xxvii (center): Michaeljung/Shutterstock; p. xxvii (coronavirus icon); p. xxiv: GlobalStock/Getty Images; p. xxvi (cat feces): Jonnysek/canstockphoto; p. xxvi (Tachyzoite): CDC/Dr. L.L. Moore, Jr.; p. xxvi (pregnant woman): Kupicoo/Getty Images; p. xxvi (pig): Hidesy/iStockphoto; p. xxvi (blood bag): DieterMeyrel/iStockphoto; p. xxvi (bird with seed): Chas53/iStockphoto; p. xxvi (meat on fork): © Food Collection/age footstock; p. xxvi (sheep): Eric Isselee/Shutterstock; p. xxvi (mouse): Shutterstock; p. xxvi (strawberry on fork): Pixomar/Shutterstock; p. xxvi (cat litter box): Axel Bueckert/Shutterstock; p. xxvi (cat): Rosa Jay/Shutterstock; p. xxvi (glass of water): Luis Molinero/Shutterstock; p. xxvi (tissue cyst): Jitender P. Dubey/USDA/Science Source; p. xxvi (fetus): Medi-Mation/Science Source; p. xxvi (cat with mouse): Getty Images; p. xxvi (fecal oocyte): TGPhoto/Alamy Stock Photo; p. xxvii (left): Medi-Mation/Science Source; p. xxvii (center): Kupicoo/Getty Images; p. xxvii (right): CDC/Dr. L.L. Moore, Jr.

Chapter 1

p. 2: Courtesy of Anne DeMarinis; p. 2 (inset): BSIP SA/Alamy Stock Photo; p. 3 (left): Courtesy of Dr. Charles L. Daley/National Jewish Health; p. 3 (center): Sattva78/Shutterstock; p. 3 (right): CDC/Dr. Ray Butler; p. 4: Sattva78/Shutterstock; p. 5 (left): Dr. Ralf Wagner/http://www.dr-ralf-wagner.de; p. 5 (center left): Scimat/Science Source; p. 5 (center right): Electron Microscope Laboratory, University of California, Berkeley; p. 5 (right): Dr. Linda M. Stannard, University of Cape Town/Science Source; p. 6 (left): Image-BROKER/Alamy Stock Photo; p. 6 (right): Hans L. Bonnevier, Johner/Johner Images/Getty Images; p. 10 (top left): Bridgeman Images; p. 10 (top right): Vanessa Vick/Getty Images; p. 10 (bottom): © Arte & Immagini srl/CORBIS/Corbis via Getty Images; p. 11 (left): SCIENCE SOURCE/Science Source/Science Source; p. 11 (top right): Bettmann/Getty Images; p. 11 (bottom): Brian J. Ford; p. 12: The Print Collector/Alamy Stock Photo; p. 13 (left): © CORBIS/Corbis via Getty Images; p. 13 (right): Smith Collection/Gado/Getty Images; p. 14: Edith Czech/Shutterstock; p. 15 (top left): CDC; p. 15 (top right): Dennis Kunkel Microscopy/Science Source; p. 15 (bottom left): Robert Koch-Institut, Berlin; p. 15 (bottom center): Robert Koch-Institut, Berlin; p. 15 (bottom right): Sueddeutsche Zeitung Photo/Alamy Stock Photo; p. 16 (top): Dennis Kunkel/Science Source; p. 16 (center): SPL/Science Source; p. 16 (bottom): ImageBROKER/Alamy Stock Photo; p. 17 (left): Courtesy of the U.S. National Library of Medicine; p. 17 (right): Dr. Michael Gabridge/Visuals Unlimited, Inc.; p. 18: Cynthia Lee/Alamy Stock Photo; p. 19 (left): Hulton Archive/Getty Images; p. 19 (center): Bettmann/Getty Images; p. 19 (right): © CORBIS/Corbis via Getty Images; p. 20 (top): Institut Pasteur; p. 20 (bottom left): Bettmann/Getty Images; p. 20 (bottom right): Mediscan/Visuals Unlimited, Inc.; p. 21: From Grunewald et al. Three-Dimensional Structure of Herpes Simplex Virus from Cryo-Electron Tomography, Science 21 November 2003: Vol. 302 no. 5649 pp. 1396–1398. Reprinted with permission from AAAS.; p. 22: Joan Slonczewski; p. 23 (top): "Colonization of Mucin by Human Intestinal Bacteria and Establishment of Biofilm Communities in a Two-Stage Continuous Culture System" by S. Macfarlane et al., *Applied and Environmental Microbiology*, 2005, Volume 71, Issue 11, pp. 7483–7492, DOI: 10.1128/aem.71.11.7483–7492.2005. Reproduced with permission from American Society for Microbiology; p. 23 (bottom): Nancy R. Schiff/Getty Images; p. 24 (left): Tony Craddock/Science Source; p. 24 (right): AP Photo; p. 25 (left): National Library of Medicine/Science Source; p. 25 (center): Omikron/Science Source; p. 25 (right): A. Barrington Brown/Science Source; p. 26 (left): Bettmann/Getty Images; p. 26 (right): Institute for Genome Sciences, University of Maryland School of Medicine; p. 27: Joan Slonczewski, Kenyon College.

Chapter 2

p. 30: Pando Hall/Digital Vision/Getty Images; p. 31 (center): GlobalStock/Getty Images; p. 31 (left): SPL/Science Source; p. 31 (right): Scott Camazine/Kallista Images/Superstock; p. 32: GlobalStock/Getty Images; p. 33 (top): Reprinted from *Current Biology*, Vol 22 /Issue 7, D.A. Shifrin et al., "Enterocyte Microvillus-Derived Vesicles Detoxify Bacterial Products and Regulate Epithelial-Microbial Interactions", pp. 627–631, 2012, with permission from Elsevier; p. 33 (bottom left): Jane Shemilt/Science Source; p. 33 (bottom right): CDC; p. 33 (bottom right inset): CDC/Libero Ajello, Ph.D.; p. 34 (top left): BSIP SA/Alamy Stock Photo; p. 34 (top right): CDC Photo; DPDx; p. 34 (bottom left): CDC/Lois Norman; p. 34 (bottom right): CDC/Hermann; p. 35 (left): CDC/Frederick A. Murphy; p. 35 (right): Tommy Trenchard/Alamy Stock Photo; p. 36 (left): Santos, R.L., Tsolis, R.M., Bäumler, A.J., & Adams, L.G.. (2003). Pathogenesis of Salmonella-induced enteritis. *Brazilian Journal of Medical and Biological Research*, 36(1), 03–12. https://doi.org/10.1590/S0100-879X2003000100002; p. 36 (right): Matthew Welch, University of California Berkeley; p. 38: Jamie Wilson/Shutterstock; p. 39 (top): Melinda Fawver/Shutterstock; p. 39 (bottom): CDC; p. 40 (bottom): Blend Images/Shutterstock; p. 40 (top): NIAID/Science Source; p. 41 (top): Courtesy of New Zealand Medical Journal. Karabay, Ogu, et al. "Medical Image. Tularaemic Cervical Lymphadenopathy." *The New Zealand Medical Journal*, vol. 120, no. 1248, 2007, pp. U2403.; p. 41 (bottom): NIAID/Science Source; p. 42 (top): Shutterstock; p. 42 (bottom left): Shutterstock; p. 42 (center left): iStock; p. 42 (center right): iStock; p. 42 (right): Shutterstock; p. 44 (top): CDC/Brian Judd; p. 44 (bottom left): Science History Images/Alamy Stock Photo; p. 44 (bottom right): CDC; p. 44 (bottom right inset): Science Source; p. 45 (left): GK Hart/Vikki Hart/Photodisc/Getty Images; p. 45 (right): FLHC 85/Alamy Stock Photo; p. 47: FogStock/Vico Images/Alin Dragulin/Getty Images; p. 48 (top): Time Life Pictures/National Archives/The LIFE Picture Collection via Getty Images; p. 48 (bottom): Duda Vasilii/Shutterstock; p. 49: PongMoji/Shutterstock; p. 51 (top left): Will & Deni McIntyre/Science Source; p. 51 (bottom): Image courtesy of J. Craig Venter Institute; p. 51 (top right): Kevin Karem/CDC; p. 53: Shinichi Murata/Science Source; p. 54 (top): NIAID/Science Source; p. 54 (center): Alain Pol, ISM/Science Source; p. 54 (bottom left): NASA Landstat/University of Maryland Global Land Cover Facility; p. 54 (bottom right): NASA image by Jeff Schmaltz, MODIS Rapid Responses Team at NASA GSFC; p. 55: © George Kourounis; Fig. 2.20 (p. 56): Figure from "Amazon deforestation drives malaria transmission, and malaria burden reduces forest clearing," by Andrew J. MacDonald, Erin A. Mordecai. *Proceedings of the National Academy of Sciences,* Oct. 2019, 116 (44) 22212–22218; DOI: 10.1073/pnas.1905315116.

Chapter 3

p. 60 (center): Stable/Shutterstock; p. 60 (top left): Louella938/Shutterstock; p. 60 (bottom): Ruslan Ivantsov/Shutterstock; p. 60 (top right): Eye of Science/Science Source; p. 61 (right): CDC/Dr. George

Source; p. 300 (bottom left): Martin M. Rotker/Science Source; p. 300 (bottom right): CDC/Dr. Peter Perine; p. 300 (bottom center): James Cavallini/Science Source; p. 302 (top): National Geographic Image Collection/Alamy Stock Photo; p. 302 (center left): Dr. Robert Calentine/Visuals Unlimited, Inc.; p. 302 (center right): Dr. Peter Siver/Visuals Unlimited, Inc.; p. 302 (bottom): Ed Reschke/Getty Images; p. 303: CNRI/Science Source; p. 295 (prairie dog 1): Life on White/Photodics/Getty Images; p. 295 (prairie dog 2): Eric Isselee/Shutterstock.

Chapter 11

p. 306 (dog tags): Getty Images; p. 306 (parasites): SPL/Science Source; p. 307 (center): Alexandr Zagibalov/Shutterstock; p. 307 (left): CDC/ Frank Collins; p. 307 (top right): BSIP SA/Alamy Stock Photo; p. 307 (bottom right): Michael Abbey/Science Source; p. 308: Alexandr Zagibalov/Shutterstock; p. 309 (promastigotes): Michael Abbey/Science Source; p. 309 (promastigotes inside micrphage): Sinclair Stammers/Science Source; p. 309 (arm): DreamBig/Shutterstock; p. 309 (hand with sores): CDC/Dr. D.S. Martin; p. 309 (bottom): Gregory P. Henderson, Lu Gan, and Grant J. Jensen. 3-D Ultrastructure of O. tauri. *PLoS ONE.* 2007; 2(8): e749. 2007 August 15.; p. 312 (Aspergillus): Dennis Kunkel/Science Source; p. 312 (Chlamydomonas): Dr. Peter Siver/Visuals Unlimited, Inc.; p. 312 (Chaos carolinensis): Michael Abbey/Visuals Unlimited; p. 312: Dennis Kunkel Microscopy/Science Source; p. 312 (Trypanosoma cruzi): CDC/Blaine Mathison; p. 312 (Taenia solium): SPL/Science Source; p. 312 (Balantidium coli): Science History Images/Alamy Stock Photo; p. 312 (body louse): Gary Retherford/Science Source; p. 312 (Plasmodium falciparum): CDC/Dr. Greene; Steven Glenn, Laboratory & Consultation Division; p. 313: Nano Creative/Science Source; p. 314 (top): Eye of Science/Science Source; p. 314 (center): Alexander Dashewsky/Shutterstock; p. 314 (bottom): Dennis Kunkel/Science Source; p. 315 (top left): © Suatcanozan/Dreamstime.com; p. 315 (center left): Steven Needell/Science Source; p. 315 (bottom left): Luo, B., Zhang, L., Hu, C. et al. "Clinical analysis of 68 patients with pulmonary mycosis in China." *Multidiscip Respir Med* 6, 278 (2011). https://doi.org/10.1186/2049-6958-6-5-278; p. 315 (right): Andrew Syred/Science Source; p. 316 (top): Scimat/Science Source; p. 316 (center): Joan Slonczewski; p. 316 (bottom left): Science Photo Library/Alamy Stock Photo; p. 316 (bottom right): Shay Levy/Alamy Stock Photo; p. 318 (top left): Stephen B. Goodwin/Shutterstock; p. 318 (top right): Getty Images; p. 318 (center): M. I. Walker/Science Source; p. 318 (bottom): NHPA/Superstock; p. 319: CDC/Dr. Mae Melvin; Dr. Greene; p. 319 (bottom left): CDC/Dr. Mae Melvin; Dr. Greene; p. 319 (bottom right): Sinclair Lea/age footstock; p. 319 (glass): age footstock; p. 320 (left): ISM/BARRAQUER, Barcelona, ISM/Science Photo Library; p. 320 (right): An update on Acanthamoeba keratitis: diagnosis, pathogenesis and treatment. Jacob Lorenzo-Morales, Naveed A. Khan and Julia Walochnik. Parasite, 22 (2015) 10.DOI: https://doi.org/10.1051/parasite/2015010; p. 321 (top): M. I. Walker/Science Source; p. 321 (bottom): Greg Antipa/Science Source; p. 322: Gerd Guenther/Science Source; p. 323: age fotostock/Alamy Stock Photo; p. 324 (left): CDC/Dr. Mae Melvin; p. 324 (cow): Robertharding/Alamy Stock Photo; p. 324 (insect overhead): The

Natural History Museum/Alamy Stock Photo; p. 324 (insect sideview): Pascal Goetgheluck/Science Source; p. 325 (left): Dennis Kunkel/Science Source; p. 325 (right): CDC/Dr. Stan Erlandsen; p. 326 (left): Science History Images/Alamy Stock Photo; p. 326 (right): Premaphotos/Alamy Stock Photo; p. 327 (center): Dr. David Phillips/Visuals Unlimited, Inc.; p. 327 (left): Bill Bachman/Science Source; p. 327 (right): Masa Ushioda/Alamy Stock Photo; p. 328 (bottom): Sinclair Stammers/Science Source; p. 328 (top): Dr. Robert Calentine/Visuals Unlimited, Inc.; p. 329 (left): CDC; p. 329 (center right): Corbis; p. 329 (center left): Loisjoy Thurstun/Alamy Stock Photo; p. 329 (right): CDC; p. 330 (left): Sinclair Stammers/Science Source; p. 330 (right): Dr. Arthur Siegelman/Visuals Unlimited, Inc.; p. 331 (man): Trendsetter Images/Shutterstock; p. 331 (cow): Iakov Filimonov/Shutterstock; p. 331 (LM of eggs): Power and Syred/Science Source; p. 331 (tapeworm mouth): Biophoto Associates/Science Source; p. 331 (tapeworm scolex): Juan Gaertner/Shutterstock; p. 331 (tapeworm segment): M. I. Walker/Science Source; p. 331 (full tapeworm): Science Source; p. 332 (top left): The Natural History Museum/Alamy Stock Photo; p. 332 (top right): Dr. P. Marazzi/Science Source; p. 332 (bottom left): Scott Camazine/Science Source; p. 332 (center): Dr. Keith Wheeler/Science Source; p. 332 (bottom right): Dr. Martin Oeggerli/Visuals Unlimited, Inc.; p. 333 (left): AP Photo/Terry Gilliam; p. 333 (right): Piotr Naskrecki/Centers for Disease Control and Prevention/Science Source.

Chapter 12

p. 336 (glove): iStockphoto; p. 336 (syringe): Shutterstock; p. 337 (center): Shutterstock; p. 337 (top left): Martin M. Rotker/Science Source; p. 337 (bottom left): Martin M. Rotker/Science Source; p. 337 (right): Dennis Kunkel Microscopy/Science Source; p. 338: Shutterstock; p. 339: Dr. Raoult/Science Source; p. 340 (top left): Lee D. Simon/Stammers/Science Source; p. 340 (top right): Dr. Edward Chan/Visuals Unlimited, Inc.; p. 340 (center left): NIBSC/Science Source; p. 340 (center right): Lowell Georgia/Science Source; p. 340 (bottom left): Thomas A. Shalla; ASSEMBLY AND AGGREGATION OF TOBACCO MOSAIC VIRUS IN TOMATO LEAFLETS . *J Cell Biol* 1 May 1964; 21 (2): 253–264. DOI: 10.1083/jcb.21.2.253; p. 340 (bottom right): Norm Thomas/Science Source; p. 342 (Phage Lambda): Biophoto Associates/Science Source; p. 342 (Rabies virus): Eye of Science/Science Source; p. 342 (Geminivirus): Cindy L. McKenzie/USDA; p. 342 (HIV): James Cavallini/Science Source; p. 342 (Caulimovirus): CENTRE FOR BIOIMAGING, ROTHAMSTED RESEARCH/Science Source; p. 342 (Coronavirus): National Infection Service/Science Source; p. 342 (Rotavirus): Mary Martin/Science Source; p. 344 (top left): Matthew L. Baker et al. 2003. J. Mol. Biol. 331(2):447-56.; p. 344 (top right): "Visualization of Tegument-Capsid Interactions and DNA in Intact Herpes Simplex Virus Type 1 Virions" by Z. Hong Zhou et al., *Journal of Virology* Apr 1999, 73 (4) 3210–3218; DOI: 10.1128/JVI.73.4.3210-3218.1999; p. 344 (bottom left): National Institute for Arthritis, Musculoskeletal and Skin Diseases, NIH; p. 344 (bottom right): Alasdair Steven, NIH; p. 345 (top): R.W. Horne/Biophoto Associates/Science Source; p. 345 (center): Everett Collection Historical/ Alamy Stock Photo; p. 345 (bottom): George Chapman/Visuals Unlimited, Inc.;

p. 346: Dr. Klaus Boller/Science Source; p. 350 (top): Lightpoet/Shutterstock; p. 350 (bottom): Jonathan Torgovnik/Getty Images; p. 354: Lynn Thomason; p. 355 (left): Herzog et al. *Virology Journal* 2008 5:138 doi:10.1186/1743-422X-5-138; p. 355 (right): BSIP/Universal Images Group via Getty Images; p. 356: Dr. Ken Greer/Visuals Unlimited, Inc.; p. 357: NCI/Science Source; p. 358: Johan Larson/Shutterstock; p. 359: BSIP SA/Alamy Stock Photo; p. 360 (left): CDC/Douglas Jordan; p. 360 (center left): Shutterstock; p. 360 (center right): Shutterstock; p. 360 (right): Stelios Filippou/Adobe Stock; p. 363: BSIP/UIG Via Getty Images; p. 364 (top left): Bob Strong/REUTERS/Newscom; p. 364 (top right): John Mottem/AFP via Getty Images; p. 364 (bottom): Dr. Klaus Boller/Science Source; p. 367: Reprinted from *Cellular Immunology*, Vol. 254/Issue 2, E.A. Eugenin et al., "Tunneling nanotubes (TNT) are induced by HIV-infection of macrophages: A potential mechanism for intercellular HIV trafficking" pp. 142–148, 2009, with permission from Elsevier.

Chapter 13

p. 370 (top): Anton Prado Photo/Shutterstock; p. 370 (center): James Marvin Phelps/Shutterstock; p. 370 (bottom): Wk1003mike/Shutterstock; p. 371 (left): B-D-S Piotr Marcinski/Shutterstock; p. 371 (center): Studioloco/Shutterstock; p. 371 (right): BSIP SA/Alamy Stock Photo; p. 372: Studioloco/Shutterstock; p. 375: John W. Foster; p. 376: Courtesy of Douglas F. Bray; p. 378: Kindly provided by Jonathan Lintott of Andersen Caledonia; p. 381 (left): Biophoto Associates/Science Source; p. 381 (right): Photographs used with permission from Mayo Clin Proc. 1,2. 1. Mayo Clinic Proceedings legacy – May 2019: The research and practical use of antibiotics [published online only May 1, 2014]. https://els-jbs-prod-cdn.literatumonline.com/pb/assets/raw/Health%20Advance/journals/jmcp/jmcp_la_89_5_1_2-1398374782743.pdf. Accessed November 8, 2019. 2. Herrell WE, Smith HL. Further observations on the clinical use of penicillin. Proc Staff Meet Mayo Clin. 1943;18 (5)65-76.13; p. 382: Courtesy of Dr. John W. Foster; p. 383 (left): Courtesy of bioMérieux; p. 383 (right, all): Courtesy of Dr. John W. Foster; p. 384: "Evaluation of a Scanner-Assisted Colorimetric MIC Method for Susceptibility Testing of Gram-Negative Fermentative Bacteria" by M. Rahman et al., *Applied and Environmental Microbiology* Apr 2004, 70 (4) 2398–2403; DOI: 10.1128/AEM.70.4.2398-2403.2004; p. 388: Yao Z, Kahne D, Kishony R. Distinct single-cell morphological dynamics under beta-lactam antibiotics. *Mol Cell.* 2012 Dec 14;48(5):705-12. DOI: 10.1016/j.molcel.2012.09.016. Epub 2012 Oct 25. PMID: 23103254; PMCID: PMC3525771.; p. 390 (left): CDC/Dr. V. R. Dowell, Jr.; p. 390 (right): Dr. David Phillips/Visuals Unlimited, Inc.; p. 393 (top): Courtesy of LITFL.com; p. 393 (bottom): CDC/Melissa Brower; p. 395: © Francis James Grogan/Dreamstime.com; p. 398 (top left): Kwangshin Kim/Science Source; p. 398: Mohan, K. et al., "Empyema due to a highly transmissible Pseudomonas aeruginosa strain in an adult cystic fibrosis patient" *Journal of Medical Microbiology*, May 2010 vol. 59 no. 5 614–616 p. 398 (right, both): *Future Microbiology*, Vol. 5, No. 11, Pages 1663–1674, DOI: 10.2217/fmb.10.125, (doi:10.2217/fmb.10.125); p. 400: James Cavallini/Science Source; p. 402: CDC/Erskine. L. Palmer, Ph.D.; M. L. Martin; p. 403: Eye of Science/

Chapter 18

p. 540 (envelope): Shutterstock; p. 540 (smudge): Shutterstock; p. 540 (wood): Shutterstock; p. 541 (center): Alamy Stock Photo; p. 541 (left): Getty Images; p. 541 (right): © Dennis Kunkel Microscopy, Inc./ Visuals Unlimited/Corbis; p. 542: Alamy Stock Photo; p. 544: Figure 1B from: Commensal and pathogenic Escherichia coli use a common pilus adherence factor for epithelial cell colonization. María A. Rendón et al., *Proceedings of the National Academy of Sciences* Jun 2007, 104 (25) 10637–10642; DOI: 10.1073/ pnas.0704104104. Copyright (2007) National Academy of Sciences, U.S.A.; p. 546 (left): "Interaction of Enteropathogenic Escherichia coli with Human Intestinal Mucosa: Role of Effector Proteins in Brush Border Remodeling and Formation of Attaching and Effacing Lesions" by Robert K Shaw et al., *Infection and Immunity* Jan 2005, 73 (2) 1243–1251; DOI: 10.1128/ IAI.73.2.1243-1251.2005; p. 546 (left inset): Figure 1B from: Commensal and pathogenic Escherichia coli use a common pilus adherence factor for epithelial cell colonization. María A. Rendón et al., *Proceedings of the National Academy of Sciences* Jun 2007, 104 (25) 10637–10642; DOI: 10.1073/pnas.0704104104. Copyright (2007) National Academy of Sciences, U.S.A.; p. 546 (right): © ZUMA Press, Inc./Alamy Stock Photo; p. 547 (top): The Type 4 Pili of Enterohemorrhagic Escherichia coli O157:H7 Are Multipurpose Structures with Pathogenic Attributes. Juan Xicohtencatl-Cortes et al., *Journal of Bacteriology* Dec 2008, 191 (1) 411–421; DOI: 10.1128/JB.01306-08; p. 547 (bottom left): Maria Fazio and Vincent A. Fischetti, Ph.D. with permission. The laboratory of Bacterial Pathogenesis and Immunology, Rockefeller University.; p. 547 (bottom center): NIBSC/Science Source; p. 547 (bottom right): James, G. A. et al. (2008), Biofilms in chronic wounds. Wound Repair and Regeneration, 16: 37–44.; p. 548 (top): Dennis Kunkel Microscopy/Science Source; p. 548 (center): "Posttranslational Modification of Pili upon Cell Contact Triggers N. meningitidis Dissemination" by Chamot-Rooke, J. et al. *Science* 11 February 2011: Vol. 331 no. 6018 pp. 778–782. Reprinted with permission from AAAS; p. 548 (bottom): Courtesy of D.L. Higashi and M. So; p. 549: CDC/Rodney M. Donlan, Ph.D.; Janice Carr; p. 549 (inset): StockPassion/Shutterstock; p. 550: age fotostock/Alamy Stock Photo; p. 553 (top left): "Actin-Based Motility of Intracellular Microbial Pathogens" by M.B. Goldberg, *Microbiology and Molecular Biology Reviews*, 2001, Volume 65, Issue 4, pp. 595–626, DOI: 10.1128/MMBR.65.4.595-626.2001. Reproduced with permission from American Society for Microbiology; p. 553 (bottom left): "Coxiella burnetii Exhibits Morphological Change and Delays Phagolysosomal Fusion after Internalization by J774A.1 Cells" by D. Howe and L.P. Mallavia, *Infection and Immunity*, Volume 68, Issue 7, 2000, pp. 3815–3821, DOI: 10.1128/IAI.68.7.3815-3821.2000. Reproduced with permission from American Society for Microbiology; p. 553 (right): VSstockstudio/Shutterstock; p. 554: Anton Zabielskyi/Shutterstock; p. 558: Banos/ Alamy Stock Photo; p. 559 (top): Stephanie Schuller/ Science Source; p. 559 (bottom): CDC/Dr. P.B. Smith; p. 561: Reprinted from *Biophysical Journal*, Vol. 81/ Issue 2, R. Lins and T.P. Straatsma, "Computer Simulation of the Rough Lipopolysaccharide Membrane of *Pseudomonas aeruginosa*" pp. 1037–1046, 2001, DOI: 10.1016/S0006-3495(01)75761-X, with permission from Elsevier; p. 562 (top): Dennis Kunkel Microscopy/ Science Source; p. 562 (bottom): Mediscan/Visuals Unlimited, Inc.; p. 563: From "Type III Secretion Machines: Bacterial Devices for Protein Delivery into Host Cells" by J.E. Galàn and A. Collmer, *Science*, 1999, Vol. 284, Issue 5418, pp. 1322–1328. Reprinted with permission from AAAS; p. 564 (top left): From "Bacterial Invasion: The Paradigms of Enteroinvasive Pathogens" by P. Cossart and P. Sansonetti, *Science*, 2004, Volume 304, Issue 5668, pp. 242–248, DOI: 10.1126/science.1090124. Reprinted with permission from AAAS; p. 564 (top right): SPL/Science Source; p.564 (bottom): FogStock/Vico Images/Alin Dragulin/ Getty Images; p. 567 (left): CDC/C. Goldsmith, P. Feorino, E. L. Palmer, W. R. McManus; p. 567 (right): SPL/Science Source; p. 568 (top left): Pidjoe/Getty Images; p. 568 (bottom left): SPL/Science Source; p. 568 (bottom right): Yu,B. et al.Emerging Optical Techniques for Detection of Oral, Cervical and Anal Cancer in Low-Resource Settings. Austin J Biomed Eng. 2014;1(2): 1007. (Photo courtesy of Daron Ferris, MD).; p. 568 (top right): NCI/Science Source; p. 568 (bottom right): Clinical Photography, Central Manchester University Hospitals NHS Foundation Trust, UK/Science Source; p. 570 (inset): Eye of Science/Science Source; p. 570: Andy Crump, TDR, World Health Organization/Science Source; p. 571: Republished by permission of Co. of Biologists, Ltd. Morrissette et al. *Cell Sci* 110, 35–42, Jan. 1, 1997.

Chapter 19

p. 578 (Hawaiian shirt): Dstephens/Getty Images; p. 578 (boarding pass): Photka/Shutterstock; p. 578 (cancelled stamp): Aquir/Shutterstock; p. 579 (top left): Dr P. Marazzi/Science Source; p. 579 (bottom left): age fotostock/SuperStock; p. 579 (center): age footstock; p. 579 (right): Science Photo Library-Heather Davies/Getty Images; p. 580: age footstock; p. 581 (left): Obak/Shutterstock; p. 581 (right): Jose Luis Calvo/Shutterstock; p. 582 (left): Dr P. Marazzi/Science Source; p. 582 (center left): Dr P. Marazzi/Science Source; p. 582 (center right): James Cavallini/Science Source; p. 582 (right): Scott Camazine/Science Source; p. 583 (top): Macia/iStock; p. 583 (center): Tim McClean/iStock; p. 583 (bottom): AP Photo/Ben Curtis; p. 586 (top): Dr P. Marazzi/ Science Source; p. 586 (bottom): Biophoto Associates/ Science Source; p. 587 (left): Dr P. Marazzi/Science Source; p. 587 (center): Courtesy of Joshua E. Lane, M.D. Reprinted with the permission of American Family Physicians. 2006 Nov 15;74(10):1765-1766.; p. 587 (right): Harel et al. Papular Purpuric Rash Due to Parvovirus B19 with Distribution on the Distal Extremities and the Face, *Clin Infect Dis.* (2002) 35 (12): 1558–1561 by permission of Oxford University Press.; p. 589 (top left): Sergii Chepulskyi/Shutterstock; p. 589 (top right): SPL/Science Source; p. 589 (bottom left): Dr. Ken Greer/Visuals Unlimited, Inc.; p. 589 (bottom right): Dr. P. Marazzi/Science Source; p. 590 (top): CDC; p. 590 (bottom left): Marcel Jancovic/Shutterstock; p. 590 (bottom right): CNRI/ Science Source; p. 591: Eye of Science/Science Source; p. 592: CDC/Dr. John Noble, Jr.; p. 593 (top left): Dennis Kunkel Microscopy/Science Source; p. 593 (top right): CDC/Sarah Bailey Cutchin; p. 593 (bottom left): Dr. Chris Hale/Science Source; p. 593 (bottom center left): SPL/Science Source; p. 593 (bottom center right): Dr. Ken Greer/Visuals Unlimited, Inc.; p. 593 (bottom right): Dr. Ken Greer/Visuals Unlimited, Inc.; p. 596 (top): Banos/Alamy Stock Photo; p. 596 (bottom left): Dr P. Marazzi/Science Source; p. 596 (bottom center): SPL/Science Source; p. 596 (bottom right): Courtesy of Loretta Davis, MD/Dermatology, Medical College of Georgia; p. 597 (left): Zoonar GmbH/Alamy Stock Photo; p. 597 (right): AndaPhoto/Shutterstock; p. 599 (top left): TisforThan/Shutterstock; p. 599 (top right): Dr P. Marazzi/Science Source; p. 599 (bottom left): Dr P. Marazzi/Science Source; p. 599 (bottom right): TLaoPhotography/Shutterstock; p. 602 (top left): Dr. Ken Greer/Visuals Unlimited, Inc.; p. 602 (top center): Dr. Harout Tanielian/Science Source; p. 602 (top right): Dr P. Marazzi/Science Source; p. 602 (bottom): CNRI/Science Source; p. 603: Photo provided by Jennifer Bartholomew from www.aspergillus.org. uk; p. 604 (top): Dennis Kunkel Microscopy/Science Source; p. 604 (center): Dennis Kunkel Microscopy/ Science Source; p. 604 (bottom): Dr. P. Marazzi/ Science Source; p. 605 (top): Phovoir/Shutterstock; p. 605 (bottom left): Volker Steger/Science Source; p. 605 (bottom center): Schröpfer, E., Rauthe, S. & Meyer, T. Diagnosis and misdiagnosis of necrotizing soft tissue infections: three case reports. Cases Journal 1, 252 (2008). https://doi.org/10.1186/1757-1626-1-252 © 2008 Schröpfer et al; licensee BioMed Central Ltd, Cases Journal 2008,1:252; p. 605 (bottom right): Dr. David Phillips/Visuals Unlimited, Inc.; p. 606: Paul Whitten/Science Source; p. 607 (top left): Reprinted from: Sampathkumar, P, Drage, LA, Martin, DP. Herpes zoster (shingles) and Postherpetic Neuralgia. Mayo Clin Proc 2009; 84:274-80. © 2009, with permission from Elsevier.; p. 607 (top right): © Dr. Charles J. Ball/ CORBIS/Corbis via Getty Images; p.607 (bottom): Argentum/Science Source; p. 608: Courtesy of Joseph Sowka; p. 609 (top left): CDC/Dr. Libero Ajello; p. 609 (center left): M. I. Walker/Science Source; p. 609 (bottom left): Gary Carlson/Science Source; p. 609 (top center): CDC/DPDx; George Healy, Ph.D.; p. 609 (top right): Photo courtesy of Dan B. Jones, M.D.; p. 611: Obak/Shutterstock.

Chapter 20

p. 614: U.S. Army/Science Source; p. 615 (left): Everett Collection/Newscom; p. 615 (center): iStockphoto; p. 615 (right): Bettmann/Getty Images; p. 616: iStockphoto; p. 617: Phil Date/Shutterstock; p. 618 (top): Science Photo Library/Alamy Stock Photo; p. 618 (bottom): Tom Gowanlock/Dreamstime.com; p. 620: J.L. Carson; p. 621 (left): BSIP SA/Alamy Stock Photo; p. 621 (right): The University of North Carolina at Chapel Hill/J.L. Carson; p. 622: AP Photo/Greg Baker; p. 624: Scott Camazine/Visuals Unlimited, Inc.; p. 625: John Wollwerth/Shutterstock; p. 628 (top): Professor Tony Wright, Institute of Laryngology & Otology/ Science Source; p. 628 (bottom): CDC; p. 630: CDC/ Heinz F. Eichenwald, MD; p. 631 (top left): Biophoto Associates/Science Source; p. 631 (top right): Science VU/Visuals Unlimited, Inc.; p. 631: John W. Foster; p. 632 (top left): CNRI/Science Source; p. 632 (top right): Biophoto Associates/Science Source; p. 632 (bottom left): Binotto, MA, Guilherme, L, Tanaka, AC. Rheumatic fever. Images Paediatr Cardiol 2002; 11:12. Copyright © 2002 Images in Paediatric Cardiology.; p. 632 (bottom right): CDC/Emory Univ.; Dr. Sellers; p. 633 (top right): CDC; p. 633 (top left): © iantfoto/

Chapter 21

Chapter 22

Chapter 23

Chapter 24

Chapter 25

Chapter 26

Index

Page numbers followed by f or t denote figures or tables, respectively. Page numbers set in **bold** type refer to glossary terms.

A

ABC transporters (ATP-binding cassette), **125–26**
aberrations, defined, 71, 72
ABO blood group system, 508, **517–18**, 664
abscess, 425, **593**
absorption, defined, **69**
AB toxins, **555**, 558, 559, 560, 718
AB5 toxins, 88, 89f, 105, 555, 558, 560
Acanthamoeba, 320, 601t, 608, 609f, 803
Acanthamoeba encephalitis, **803**
acariasis, 331
acetic acid, 12, 109, 189, 195, 199
Acetobacter, 898, 899f
acetylcholine, 532
acetyl-CoA
  from amino acids, 189, 190f
  fatty acid synthesis, 207, 208f, 209
  from lipid catabolism, 189, 196
  from petroleum catabolism, 189, 196, 198
  from pyruvate, 190f, 196
  structure, 189f
  in TCA cycle, 196, 198
achlorhydria, 52, 420
Achong, Bert, 660
acid–base chemistry. *See also* pH
  enzyme reaction rates, 104–5
  membrane-permeant weak acids and bases, 108–9, 170
  microbial responses to changes, 166t, 169–71
  organic acids and bases, ionization, 105
  water, 105
acid-fast bacillus (AFB), **641**
acid-fast bacteria, biochemical profiles of, 829–30
acid-fast stains, **78**, 133, 287, 289, 290f, 641, **829**, 835–36
Acidithiobacillus, 297
acidophiles, 166t, **170**, 278, 883
acidosis, 105, 197
Acinetobacter, hospital-acquired pneumonia, 639
Acinetobacter baumannii, 216
acme, defined, 38
acne (acne vulgaris)
  blackhead formation, 416–17, 599
  comedone acne, 599
  Cutibacterium acnes, 189, 289, 394, 416, 600
  cystic or nodular acne, 599, 600
  doxycycline for, 394, 395
  inflammatory acne, 599
  overview, 416f, **599–600**
  tetracycline for, 394, 417
acquired immunodeficiency syndrome (AIDS). *See* AIDS
actin, 36f, 144, 555, 563, 564f
Actinobacteria. *See also specific types*
  acid-fast stain, 287

hierarchy of taxa, 271t
high GC content, 271t, 282, 287
overview, **280**, 287–91
phylogeny, 279f
selected species, 283t
Actinomyces, 417
Actinomyces bovis, 451
actinomycetes. *See also specific types*
  acid-fast stain, 133, 287
  Amycolatopsis orientalis, 128f, 129
  antibiotics from, 128f, 129, 133, 176, 288, 886
  arthrospores, 288
  cell differentiation, 176
  complex cell walls, 133
  filaments, 128, 287–88
  Gardnerella vaginalis, 288
  Nocardia, 288
  Streptomyces coelicolor, 271, 287–88
  Streptomyces griseus, 886f
  streptomycetes, antibiotic production, 4, 287f, 288
  streptomycin from, 133, 271
  vancomycin from, 128f, 129, 133
actin tails, 36f, 144, 285, 286f, 430, 551, 553f
activated sludge, **910**
activation energy (E_a), **104**, 188
activators, **228**
active immunization, **500**
active transport, **124–26**
active tuberculosis disease, **639–40**
acute bacterial endocarditis (ABE), **673**
acute disseminated encephalomyelitis (ADEM), **586**
acute glomerulonephritis. *See* post-streptococcal glomerulonephritis
acute HIV infection, **754**. *See also* HIV (human immunodeficiency virus)
acute infections, **34**, 37t
acute inflammatory response
  chemical signals leading to, 448–49
  extravasation, 448, 449, 450f, 565
  HERPA, 448, 451
  overview, 447–48, 449f, 450f
acute otitis media (AOM)
  Haemophilus influenzae, 232f, 628, 629
  incidence and economic impact, 625
  Moraxella catarrhalis, 628, 629
  overview, 625, 626t, **628–29**
  Streptococcus pneumoniae, 628
  symptoms and diagnosis, 625, 628
acute rheumatic fever (ARF), 38, 530, 597, 598, **631–32**, 672, 815
acyclovir
  genital herpes, 752t, 753, 754
  guanosine nucleobase analog, 401t, 402, 589
  herpes keratitis, 607
  neonatal herpes, 754
  primary herpes infections, 584t, 590
  shingles, 584t, 589, 607

structure, 402f
viral meningitis, 790
adaptive immunity. *See also* cell-mediated immunity; humoral immunity; lymphocytes
  cells of adaptive immune system, 441, 442f, 471–73
  comparison of innate and adaptive immunity, 440–41, 470
  downregulation by viruses, 569
  immunological specificity, 473–74, 475, 476, 479, 481f
  overview, **440**, 470–74
  reaction to antigens, 441
  white blood cells from lymphoid stem cells, 441, 442f
adenine, 12, 97
adenitis, **714**
adenosine, 97f
adenosine deaminase, 470, 526
adenosine triphosphate (ATP)
  conversion to cAMP, 88, 89f, 102–3
  as energy carrier in catabolism, 12, 190–91, 202
  energy metabolism, 97, **190**
  from glucose respiration, 199–200, 202
  hydrolysis by ABC transporters, 126
  phosphorylation of ADP, 191, 200
  structure, 97f, 191f
adenylate cyclase, 88, 89f, 103, 560
adherence, defined, **137**
adhesins. *See also* pili
  binding to receptors, 36
  defined, **32**, 36, **544**
  examples of bacterial adhesins, 545t
  nonpilus adhesins, 546–47
  type I pili, 544f
adjuvant, **504**
ADP-ribosyltransferases, **555**, 559, 560, 603
adrenoleukodystrophy (ALD), 367
Aedes aegypti, 44f, 869
aequorin, 873
aerobes
  anaerobes *vs.*, 173–75
  examples of aerobic and anaerobic microbes, 174t
  microaerophilic organisms, 174–75, 390
  oxygen-related growth zones in a test tube, 173, 174f
  strict aerobes, 173
aerobic lithotrophy, 188
aerobic organotrophy, 187–88
aerobic respiration, defined, **173**, **188**, 191–92
aerosols, 41, **43**, 44f, 45, 48, 50
aerotolerant anaerobes, **174**
affinity maturation, 478–79, **482**
African sleeping sickness (trypanosomiasis), 64, 308, 324–25, 570f, 571, **802**
agar, **16–17**, 153
agarose gel, 236, 237f, 507–8
age and disease susceptibility, 52
agglutination, **508**
Aggregatibacter actinomycetemcomitans, 213, 696

hepatitis C virus (HCV)
    antiviral agents, 401t, 702
    electron microscopy, 337f
    evolution, 348
    high mutation rate, 338
    lifetime infection, 337, 338, 341, 702–3
    liver cancer caused by, 338
    liver transplants for, 337, 338
    overview, 337–38, 701t, **702–3**
    polymerase chain reaction viral RNA test, 337
    RNA genome, 337, 338, 343, 359, 701t, 702
    transmission, 48, 337, 702, 756, 867
    treatment, 703
hepatitis D virus (HDV, delta agent), 346–47, 701t, **702**
hepatitis E virus (HEV), 701t, **703**
hepatomegaly, **660**
hepatosplenomegaly, 448, 661, 681f, 682, **712**
herd immunity, **505**
HERPA (heat, edema, redness, pain, altered function
    or movement), 448, 451
herpes gladiatorum, **590**
herpes simplex viruses (HSV). *See also* herpesviruses
    Bell's palsy, 783, 784
    cold sores, 34, 144, 569, 584t, 589–90
    erythema multiforme, 589f, 590
    evolution, 348
    genital herpes, 569, 589, 590, 752t, 753
    herpes encephalitis, 753, 783, 786t
    herpes gladiatorum, 590
    herpes meningitis, 783, 786t, 790
    herpes simplex virus type 2 (HSV-2), 348, 589, 590,
        753, 783, 786t, 790
    herpetic keratitis, 589f, 590, 607
    herpetic whitlow, 589f, 590
    HIV patients, 753
    HSV-1 genome, 346, 347f
    HSV-1 structure, 21f, 344f
    icosahedral capsid, 344
    latent period, 569, 656, 753
    neonatal herpes, 753–54
    primary infections, 590, 753
    reactivated herpesvirus, 34, 353, 569, 589, 607, 753
    sexual transmission, 339
herpesviruses. *See also* cytomegalovirus; herpes sim-
    plex viruses; varicella-zoster virus
    capsid proteins, 21
    DNA genome, 21, 338, 342t, 343, 344, 346, 569
    DNA polymerase, 343
    equine herpes (EHV-1, EHV-2), 348
    evolution, 346, 348
    icosahedral capsid, 21, 344
    Kaposi's sarcoma (human herpesvirus 8), 348, 567f,
        568, 755, 783
    latent infection reactivation, 34f, 353, 569, 588, 589,
        607, 753
    latent state, 353, 569, 588
    microRNAs (miRNA), 569
    murine herpes (MHV-68), 348
    in nucleus, 142, 144
    pathogenesis, 569
    phylogeny, 348f
    prevention of apoptosis, 569
    roseola (human herpesvirus 6 or 7), 584t, 588
    size, 64f
    spike proteins, 344
    structure, 6, 21f
    systemic infections, 656

    tegument, 344f
    travel through microtubules, 144
    viral envelope, 344
herpes zoster. *See* shingles; varicella-zoster virus
herpes zoster ophthalmicus, **607**
herpetic keratitis, 589f, **590**, 607
herpetic whitlow, 589f, **590**
Hesse, Angelina, 16–17
Hesse, Walther, 16–17
heterophile antibodies, **659**
heterotrophs, **163**, 164f, 164t, **187**
hexoses, 95
Hfr cells, **251**, 252f, 254
highly active antiretroviral therapy (HAART), **404**
Hippocrates, 290
histamine, overview, **516**
Histoplasma capsulatum, 573, 644
histoplasmosis, 315, 572, **644**
HIV (human immunodeficiency virus). *See also* AIDS;
    antiretroviral drugs
    acute HIV infection, 754
    animal origins, 47
    antiretroviral therapy (ART), 404, 755–56
    binding to cell receptors, 107, 567
    budding from host cell, 82, 364f, 365, 567f
    cause of AIDS, 10, 18, 20, 47, 356, 363, 754–56
    cell fusion, 365, 366f, 367
    in cell nucleus, 142
    clinical latency, 363, 754
    conical-icosahedral capsid, 363
    constitutional symptoms of AIDS, 363
    discovery, 20, 356, 363
    early symptomatic HIV infection stage, 567, 754–55
    electron microscopy, 82, 342f
    elite controllers, 756
    endemic disease in Africa, 46, 363
    evolution, 348
    "functional cure" or remission of HIV infection, 756
    glycoprotein 120 (gp120), 403, 567
    herpes simplex virus infections, 753
    high mutation rates, 234, 351, 364, 403
    host range, 341
    infection, overview, 363, 752t, 754–56
    influence on reemerging tuberculosis, 868
    intercellular transmission, 285
    Kaposi's sarcoma (human herpesvirus 8), 348, 567f,
        568, 755
    latent stage, 751, 754
    lifetime infection, 341, 363
    long-term nonprogressors, 756
    meningoencephalitis, 786t, 801
    nanotube formation, 367
    negative factor (Nef), 567
    nonpathogenic strain used against leukemia, 6,
        26, 367
    nucleocapsid protein (NP), 364
    pandemic, 45, 46, 338, 403, 754, 852
    pathogenesis, 567–68
    persistent generalized lymphadenopathy
        (PGL), 754
    phylogeny of HIV strains, 351f
    polymerase chain reaction (PCR) tests, 18
    posttreatment controllers, 404
    prevention of T-cell apoptosis, 567
    prophylactic antibiotic therapy, 756
    proteins that affect pathogenesis, 567t
    provirus formation, 365, 403

    replication cycle, 365, 366f, 367
    reverse transcriptase, 340, 343, 364–65, 366f, 403
    RNA genome, 82, 96, 343, 364–65, 754
    seroconversion, 754
    sexual transmission, 46, 754
    from simian immunodeficiency virus (SIV),
        351, 754
    "sperm washing" procedure, 363
    spike proteins, 363, 364, 365f, 567
    structure and genome, 363–65
    syncytium formation, 367
    Tat protein (trans-activator of the HIV
        promoter), 567
    T-cell CCR5 coreceptors and, 365, 403, 404,
        567, 756
    T-cell CD4 receptors and, 107, 349, 365, 403,
        567, 754–55
    T cell decline, 363, 491, 567, 754
    transmission mechanism, 46, 48, 341, 754
    transplacental transmission, 754
    use as vector for gene therapy, 6, 26, 339, 367
    viral envelope, 363, 364, 365, 756
    viral load, 403, 754, 755–56
    virus, overview, 363, 754
    virus titer, 363f
hives (urticarial rash), 515f, **516**, 518, 520f
Hodgkin, Dorothy, 8t, 25
Hodgkin's lymphoma, **529**, 660
holdfasts, **137**
homologous DNA sequences, 235, **249**, 251, 253,
    254, 267
homologous recombination, 235, **254**
homologs (homologous genes), **217–18**, 267
hoof-and-mouth disease, 591, 891
Hooke, Robert, 10, 11f
hookworms
    Ancylostoma duodenale, 329, 726, 728
    eosinophilic pneumonia, 644
    life cycle, 329, 726, 728–29
    Necator americanus, 329, 726, 728, 729f
    overview, 5, **329**, 722t, **728–29**
hopanoids, 108
horizontal gene transfer. *See also* conjugation; recom-
    bination; transformation
    antibiotic resistance genes, 250, 254, 397, 399
    bacteriophages, 248
    bioinformatics and, 266–69
    DNA restriction and modification, 253–54
    between domains, 261
    dtx gene in diphtheria, 248
    GC content, 268, 543
    generalized transduction, 252–53
    genomic islands, 268, 543–44
    natural selection and, 261–65
    overview, **248**
    pathogen evolution, 281f, 543
    pathogenicity islands, 542, 543–44
    prophages, 253
    specialized transduction, 253
    Streptococcus pneumoniae, 266–67, 268
    transduction, 251–53, 353
    virulence genes, 248, 258
horizontal transmission, **42**, 43–44
hospital-acquired infections. *See also* health care–
    associated infections; nosocomial infections
    Aspergillus, 314
    Campylobacter, 715

Taxonomy of Major Pathogens

The theme of this textbook is to present infectious diseases as a health professional sees them—by organ system. But it can also be useful to consider a single organism and summarize the different diseases it can cause. This table, arranged taxonomically, does this and directs you to pages covering each disease. The table also lists key features of each pathogen and key aspects of each disease. The table will help you see microbiology from two converging perspectives: those of the organ system and the pathogen. Some bacterial names are highlighted in red or blue. This approach provides a means to quickly identify Gram-negative and Gram-positive organisms, respectively. Pronunciations of microbe names are best learned using an online search engine. A good site is atsu.edu/faculty/chamberlain/website/studio.htm.

● = Gram-negative   ● = Gram-positive

NAME MICROBE FEATURES	DISEASE	PAGES	KEY FEATURES OF PATHOGENESIS / VACCINE
BACTERIA			

PHYLUM PROTEOBACTERIA (all members of this phylum are Gram-negative)
CLASS ALPHAPROTEOBACTERIA

NAME MICROBE FEATURES	DISEASE	PAGES	KEY FEATURES OF PATHOGENESIS / VACCINE
● *Anaplasma phagocytophilum*			
Amorphous; obligate intracellular pathogen	Anaplasmosis	671–72	Tick-borne; constitutional symptoms; zoonosis
● *Bartonella henselae*			
Bacillus; facultative intracellular pathogen	Cat scratch disease	671	Contracted from fleas, body lice, and ticks; fever and headache
● *Brucella* spp.			
Bacillus; facultative intracellular pathogen	Brucellosis	667, 671–72	Food-borne (e.g., unpasteurized milk); undulant fever and septicemia; zoonosis
● *Ehrlichia* spp.			
Bacillus; obligate intracellular pathogen	Ehrlichiosis	671–72	Tick-borne; constitutional symptoms
● *Rickettsia prowazekii*			
Bacillus; obligate intracellular pathogen	Epidemic typhus	551, 669, 671	Louse-borne; constitutional symptoms and rash; potential bioterror agent
● *Rickettsia rickettsii*			
Bacillus; obligate intracellular pathogen	Rocky Mountain spotted fever	669, 671–72	Tick-borne; constitutional symptoms and rash
● *Rickettsia typhi*			
Bacillus; obligate intracellular pathogen	Endemic murine typhus	34, 667, 671	Tick- or flea-borne; constitutional symptoms and rash

CLASS BETAPROTEOBACTERIA

NAME MICROBE FEATURES	DISEASE	PAGES	KEY FEATURES OF PATHOGENESIS / VACCINE
● *Bordetella pertussis*			
Bacillus; facultative intracellular pathogen	Whooping cough	618, 626, 634–35	Upper respiratory tract; calmodulin-activated adenylate cyclase toxin
● *Burkholderia pseudomallei*			
Bacillus; facultative intracellular pathogen	Melioidosis		Inhalation; constitutional symptoms and pneumonia
● *Neisseria gonorrhoeae*			
Diplococcus; facultative intracellular pathogen	Gonorrhea	671, 761-62, 764	Sexual transmission; purulent discharge
	Pelvic inflammatory disease	762, 764	Upper female reproductive tract; severe pelvic pain
	Ophthalmia neonatorum	608, 762	Transmitted during birth; newborn eye disease
● *Neisseria meningitidis*			
Diplococcus; facultative intracellular pathogen	Meningitis	561, 608, 794–96	Person-to-person transmission; high fever, headache, and petechial rash; type IV pili aid in invasion of epithelial cells; resistant to complement, owing to polysaccharide capsule; vaccine is available
	Sepsis	459, 546, 548, 794	Fever, petechial rash; endotoxin is responsible for septic shock symptoms
● *Spirillum minus*			
Spiral	Rat-bite fever		Zoonosis; rash, fever, arthritis

CLASS GAMMAPROTEOBACTERIA

NAME MICROBE FEATURES	DISEASE	PAGES	KEY FEATURES OF PATHOGENESIS / VACCINE
● *Coxiella burnetii*			
Obligate intracellular pathogen	Q fever	551, 552	Zoonosis; inhalation of contaminated material after animal birth; pneumonia

Taxonomy of Major Pathogens

NAME / MICROBE FEATURES	DISEASE	PAGES	KEY FEATURES OF PATHOGENESIS / VACCINE
Escherichia coli			
Bacillus		130, 250, 254	O and H antigens are immunogenic cell-surface components that are used to serotype *E. coli*
	Cystitis	739, 742–44, 813	Infects the bladder
	Gastroenteritis	546, 708–11, 860	Multiple forms
	Neonatal meningitis	797	Infects the meninges; K1 capsule protects against immune response
	Pyelonephritis	743, 748	Infects the kidneys
	Sepsis/urosepsis	663	High fever, low blood pressure
Francisella tularensis			
Bacillus; facultative intracellular pathogen	Tularemia	48–49, 296, 642, 671–72	Zoonosis; transmitted via inhalation, wound, or ingestion; constitutional symptoms and lymphadenopathy
Haemophilus ducreyi			
Bacillus	Chancroid	758, 763–64	Sexual transmission; painful genital ulcers
Haemophilus influenzae			
Bacillus	Conjunctivitis	607, 629	Its capsule is an important virulence factor; X and V factors are necessary for growth
	Epiglottitis	625, 628	
	Meningitis	151–52, 793–94	
	Otitis media	628–29	
	Pneumonia	296	
Klebsiella pneumoniae			
Bacillus	Pneumonia	419, 639, 824	Its capsule is an important virulence factor
Legionella pneumophila			
Bacillus; facultative intracellular pathogen	Legionellosis	80–81, 638	Pneumonia symptoms
Moraxella catarrhalis			
Diplococcus	Otitis media	628–29	Middle ear infection; pain
Pasteurella multocida			
Coccobacillus	Wound infections	555	Transmitted via animal bites
Pseudomonas aeruginosa			
Bacillus	Conjunctivitis		Opportunistic pathogen; multiple toxins
	Wound	603	
	Otitis externa		
	Pneumonia	627, 639	
Pseudomonas fluorescens			
Bacillus	Septicemia		High fever, low blood pressure
Salmonella enterica nontyphoidal serovars			
Bacillus; facultative intracellular pathogen	Salmonellosis	713–14	Food-borne; diarrhea
Salmonella enterica serovar Typhi			
Bacillus; facultative intracellular pathogen	Typhoid fever	666, 671, 711–14	Food-borne; high fever, low blood pressure; Vi capsule; vaccine is available
Shigella spp.			
Bacillus; facultative intracellular pathogen	Bacillary dysentery	551, 560, 710, 711	Food-borne and person-to-person transmission; low infectious dose; bloody diarrhea
Vibrio cholerae			
Curved bacillus	Cholera	555, 558, 716, 717	Watery diarrhea; ADP-ribosylating toxin affects adenylyl cyclase
Vibrio parahaemolyticus			
Curved bacillus	Gastroenteritis	716	Diarrhea
Vibrio vulnificus			
Curved bacillus	Gastroenteritis	716–17	Transmitted by ingestion of contaminated oysters; found in brackish water
	Wound infections		
	Sepsis		

Taxonomy of Major Pathogens

NAME MICROBE FEATURES	DISEASE	PAGES	KEY FEATURES OF PATHOGENESIS / VACCINE
Yersinia enterocolitica			
Bacillus	Gastroenteritis	664–65, 714–15	Diarrhea, mesenteric lymphadenitis
Yersinia pestis			
Bacillus	Plague	666, 668–69	High fever, pneumonia, and buboes; adenylyl cyclase and plasminogen activator toxins, as well as *Yersinia* outer proteins, are key to pathogenesis; potential bioterror agent
Yersinia pseudotuberculosis			
Bacillus	Gastroenteritis	294–95, 714	Diarrhea, mesenteric lymphadenitis
CLASS EPSILONPROTEOBACTERIA			
Campylobacter jejuni			
Curved bacillus; facultative intracellular pathogen	Gastroenteritis	714–16	Ingestion; major source is poultry
Helicobacter pylori			
Helical bacillus	Gastritis Peptic ulcers Gastric cancer	704, 707–08	Epigastric pain; bacterial urease and the *cag* pathogenicity island are key virulence factors
PHYLUM FIRMICUTES (all pathogenic members of this phylum are Gram-positive) **SPORE FORMERS**			
Bacillus anthracis			
Bacillus; facultative aerobe	Anthrax, three forms: • Cutaneous • Pulmonary • Intestinal (rare) Any form can cause sepsis and pneumonia	541, 560–61, 592, 642	Tripartite toxin is key virulence factor; bioterror agent
Clostridioides difficile			
Bacillus; anaerobe	Pseudomembranous colitis	718–19	Malabsorption and diarrhea; three toxins that alter cytoskeleton
Clostridium botulinum			
Bacillus; anaerobe	Botulism	797–800	Ingestion of toxin; inhibits neurotransmitter release; flaccid paralysis
Clostridium tetani			
Bacillus; anaerobe	Tetanus	555, 797–800	Wound infection; toxin inhibits neurotransmitter release; spastic paralysis
NON-SPORE FORMERS			
Enterococcus spp.			
Coccus	Sepsis Urinary tract infection Endocarditis Intra-abdominal abscesses	280–81, 748	Resistant to multiple antibiotics
Listeria monocytogenes			
Bacillus; facultative intracellular pathogen	Listeriosis	550–51, 718, 794, 796	Ingestion, transplacental transmission; pathogen polymerizes host actin to transfer between cells; mild diarrhea, meningitis, sepsis; more serious disease in neonates, the elderly, and the immunocompromised
Mycoplasma pneumoniae			
Lack cell wall (no meaningful Gram reaction); adherence tip	"Walking" pneumonia	635, 637–38	Respiratory tract; cold agglutinins
Staphylococcus aureus			
Coccus, grapelike clusters; coagulase-positive	Folliculitis	593	Hair follicle
	Impetigo	576	Crusty lesions
	Food poisoning	719	Toxin ingestion
	Scalded skin syndrome	555, 593, 797	Toxin dissemination
	Toxic shock syndrome	286, 555, 557, 593	Toxin dissemination
	Endocarditis	673	Heart valve
	MRSA	371, 856	Antibiotic resistant; can cause any of these diseases
Staphylococcus saprophyticus			
Coccus, clusters; coagulase-negative	Cystitis	743, 746	Painful urination

Taxonomy of Major Pathogens

NAME MICROBE FEATURES	DISEASE	PAGES	KEY FEATURES OF PATHOGENESIS / VACCINE
Streptococcus agalactiae			
Coccus, chains; group B strep	Neonatal sepsis and meningitis	793, 797	Transmitted during birth
Streptococcus mutans			
Coccus, chains; viridans strep	Dental caries and endocarditis	673, 695, 827–28	Normal oral microbiota
Streptococcus pneumoniae			
Diplococcus, capsule		637	Its capsule is a key pathogenicity factor; vaccine is available
	Meningitis	793, 796	Typically contracted by children
	Otitis media	625, 628–29	Middle ear
	Pneumonia	636–37	Elderly
	Sepsis	666	Any age
Streptococcus pyogenes			
	Acute glomerulonephritis	631–32, 742	Immune complex; affects the kidneys
	Impetigo	597	Crusty skin lesion
	Erysipelas	393, 598	Reddened skin
Coccus chains; group A, β-hemolysis, M protein	Strep throat (pharyngitis)	629	Inflamed throat and tonsils
	Necrotizing fasciitis	597–98	SPE toxins, invasive
	Puerperal sepsis	856–57	Uterine infection
	Rheumatic fever	631–32	Autoimmune, heart
	Scarlet fever	630	Rash; SPE toxin
	Toxic shock syndrome	593, 598	Toxin dissemination

PHYLUM   ACTINOBACTERIA

NAME MICROBE FEATURES	DISEASE	PAGES	KEY FEATURES OF PATHOGENESIS / VACCINE
Actinomyces israelii			
Gram-positive bacillus, branched	Actinomycosis		Head and neck abscesses; pus contains "sulfur" granules of clumped rods
Corynebacterium diphtheriae			
Gram-positive bacillus	Diphtheria	559–60, 633	Upper respiratory tract, pseudomembrane in throat; ADP-ribosylating toxin; vaccine is available
Cutibacterium acnes			
Gram-positive bacillus	Acne	419, 600	Lives in hair follicles and pores, digests sebum
Gardnerella vaginalis			
Gram-variable coccobacillus	Vaginosis	288, 767	Opportunistic; normal vaginal microbiota
Mycobacterium abscessus			
Does not Gram-stain; acid-fast bacillus; mycolic acid waxy coat; rapidly growing mycobacterium	Abscess Chronic lung disease	340, 835	Immunodeficient patients at risk
Mycobacterium avium and *Mycobacterium intracellulare*			
Does not Gram-stain; acid-fast bacillus; mycolic acid waxy coat; rapidly growing mycobacterium	Primarily a lung pathogen in immunocompromised patients	291	Ingestion, respiratory; associated with AIDS
Mycobacterium bovis			
Does not Gram-stain; acid-fast bacillus; mycolic acid waxy coat; rapidly growing mycobacterium	Tuberculosis	291, 520	Zoonosis; ingestion, direct contact; primarily GI but can progress to lung; fever and night sweats
Mycobacterium leprae			
Does not Gram-stain; acid-fast bacillus; mycolic acid waxy coat	Leprosy, two types: • Lepromatous • Tuberculoid	132–33, 289	Severity varies with immune status
Mycobacterium tuberculosis			
Does not Gram-stain; acid-fast bacillus; mycolic acid waxy coat; slow-growing mycobacterium; facultative intracellular pathogen	Tuberculosis	290, 639–41	Inhalation; primarily lung, but also kidney, spine, brain; fever and night sweats; BCG vaccine is available

Taxonomy of Major Pathogens

NAME MICROBE FEATURES	DISEASE	PAGES	KEY FEATURES OF PATHOGENESIS / VACCINE
Mycobacterium ulcerans			
Does not Gram-stain; acid-fast bacillus; mycolic acid waxy coat	Buruli ulcer		Necrotic ulcer; major virulence factor is a cytotoxin
Nocardia asteroides			
Gram-positive; partial acid-fast bacillus; filamentous	Nocardiosis: cutaneous, pulmonary, or disseminated forms	288, 910	Necrosis and abscess formation

PHYLUM CHLAMYDIAE (obligate intracellular)

NAME MICROBE FEATURES	DISEASE	PAGES	KEY FEATURES OF PATHOGENESIS / VACCINE
Chlamydia trachomatis			
	Inclusion conjunctivitis	607	
	Trachoma	608, 762	
Bacillus or coccoid (no meaningful Gram reaction); lacks peptidoglycan	Pelvic inflammatory disease	762, 792	Direct contact or sexual transmission; human reservoir; elementary body is infectious; reticulate body is noninfectious; forms cell inclusions
	Lymphogranuloma venereum	763	
	Urethritis	762	
Chlamydophila pneumoniae			
Bacillus or coccoid (no meaningful Gram reaction); lacks peptidoglycan	Pneumonia	630, 635, 674	Human reservoir; other features are similar to those of *C. trachomatis*
Chlamydophila psittaci			
Bacillus or coccoid (no meaningful Gram reaction); lacks peptidoglycan	Psittacosis	630	Transmission is by inhalation; lung infection; birds are reservoir; other features are similar to those of *C. trachomatis*

PHYLUM SPIROCHAETES (spirochetes)

NAME MICROBE FEATURES	DISEASE	PAGES	KEY FEATURES OF PATHOGENESIS / VACCINE
Borrelia spp.			
Flagella within periplasmic space	Relapsing fever	299	Animal reservoir; tick- or louse-borne; repeated 2-day fever cycles
Borrelia burgdorferi			
Flagella within periplasmic space	Lyme disease	117–18, 669–71, 851–52, 869	Animal reservoir; tick-borne; erythema migrans (bull's-eye rash)
Leptospira interrogans			
Cell ends are hooked; flagella within periplasmic space	Leptospirosis	56, 66, 667	Transmission by contact with animal urine, water; high fever, chills, and jaundice
Treponema pallidum			
Flagella within periplasmic space	Syphilis	300, 759–60	Sexual transmission; painless genital ulcers

PHYLUM BACTEROIDETES

NAME MICROBE FEATURES	DISEASE	PAGES	KEY FEATURES OF PATHOGENESIS / VACCINE
Bacteroides fragilis			
Gram-negative; anaerobe	All body sites	173, 815, 817	Opportunistic; normal intestinal microbiota
Capnocytophaga canimorsus			
Gram-negative; anaerobe	Infected bite wounds; septic shock		Canine and feline mouth microbiota
Porphyromonas spp.			
Gram-negative; anaerobe	Periodontal disease	213, 696	Oral microbiota; can invade cells
Prevotella intermedia			
Gram-negative; anaerobe	Acute necrotizing gingivitis	417	Opportunistic; normal oral microbiota

PHYLUM FUSOBACTERIA

NAME MICROBE FEATURES	DISEASE	PAGES	KEY FEATURES OF PATHOGENESIS / VACCINE
Streptobacillus moniliformis			
	Rat-bite fever		Fever, chills, myalgia, polyarthritis

Taxonomy of Major Pathogens

NAME MICROBE FEATURES	DISEASE	PAGES	KEY FEATURES OF PATHOGENESIS / VACCINE

🍄 FUNGI

DIVISION ZYGOMYCETES

Mucor spp.

Not dimorphic; no septa	Opportunistic infections; mucormycosis		Diabetes major risk factor; infections of facial sinuses, lung, gastrointestinal tract

Rhizopus spp.

Not dimorphic; no septa	Opportunistic infections; mucormycosis	604	Diabetes major risk factor; infections of facial sinuses, lung, gastrointestinal tract

PHYLUM MICROSPORIDIA

Encephalitozoon intestinalis

Obligate intracellular pathogen	Opportunistic infections		Germinating spore infects host cells

DIVISION ASCOMYCOTA (ASCOMYCETES)

Aspergillus fumigatus

Tuft of spores at ends of aerial hyphae; not dimorphic; hyphae grow in body	Aspergillosis	314–17, 572, 603, 608	Inhaled spores; lung infection; immunocompromised are susceptible

Blastomyces dermatitidis

Dimorphic; yeast form grows in body	Blastomycosis	404, 644, 572, 643–44	Inhaled spores; lung infection (flu-like), disseminated abscesses; immunocompromised are susceptible

Candida albicans

	Candidiasis	572, 601–02	General term
	Vulvovaginitis	768	Infects immunocompetent
Dimorphic; yeast and hyphal forms grow in body	Sepsis		Infects immunocompromised
	Pneumonia	419	Infects immunocompromised
	Meningitis	800	Infects immunocompromised
	Oral thrush	697–98	Infects immunocompromised

Coccidioides immitis

Dimorphic; spherule form grows in host	Coccidioidomycosis	642, 644, 800	Endemic in San Joaquin Valley; found in soil; respiratory infection

Histoplasma capsulatum

Dimorphic; yeast form grows in body	Histoplasmosis	572, 644	Inhaled spores; endemic to Ohio, Missouri, and Mississippi River valleys; found in soil; pneumonia

Microsporum spp., *Trichophyton* spp.

Not dimorphic; septate hyphae	Ringworm, athlete's foot	600	Transmitted between people and animals or by fomites

Pneumocystis jirovecii

Looks like crushed Ping-Pong balls	Pneumonia; opportunistic pathogen	526, 755	Sign of HIV; does not respond to antifungal agents; trimethoprim-sulfamethoxazole works

Sporothrix schenckii

Dimorphic; yeast form grows in host	Sporotrichosis	572	Associated with thorns of roses; transmitted via skin pricks; suppurating nodules along lymphatic vessels

DIVISION BASIDIOMYCOTA (BASIDIOMYCETES)

Cryptococcus spp.

Yeast; not dimorphic; thick capsule	Cryptococcosis; meningitis from a pulmonary infection	313, 644, 801	Its capsule is its key virulence factor

Malassezia furfur

Yeast; not dimorphic	Tinea versicolor; dandruff; folliculitis	572, 601–02	

Taxonomy of Major Pathogens

NAME MICROBE FEATURES	DISEASE	PAGES	KEY FEATURES OF PATHOGENESIS / VACCINE
PROTISTA (PROTOZOA)			
PHYLUM DIPLOMONADIDA			
Giardia intestinalis (formerly *G. lamblia*)			
Flagellated	Amebic dysentery	319–20, 721–22, 724	Bloody diarrhea; endemic in countries with poor sanitation
PHYLUM PARABASALIA			
Trichomonas vaginalis			
Flagellated	Trichomoniasis resulting in vaginitis, cervicitis, and pelvic inflammatory disease	570, 763–64	Sexually transmitted
PHYLUM EUGLENOZOA			
Leishmania spp.		307, 572, 681–82	Transmitted by sand fly; endemic in Africa, Asia, Middle East, Mediterranean, India, and Central and South America
Intracellular protozoan	Cutaneous leishmaniasis		Cutaneous ulcers
	Visceral leishmaniasis		Lethal systemic disease
Trypanosoma brucei			
Flagellated	African trypanosomiasis, also called sleeping sickness	324, 571–72, 802	Transmitted by tsetse fly bite; endemic to tropical Africa; fever, rashes, headaches, excessive sleepiness, and anorexia
Trypanosoma cruzi			
Flagellated	Chagas disease	678–79, 802	Transmitted by reduviid bug; endemic in Mexico, Central and South America; fever, malaise, nausea; can progress to a cardiomyopathy; Romaña sign
PHYLUM PERCOLOZOA			
Naegleria fowleri			
Ameba	Meningoencephalitis	802	Travels up the nose to the brain
PHYLUM APICOMPLEXA			
Babesia microti			
Trophozoite ring form and Maltese cross forms seen in red blood cells	Babesiosis	323, 678, 867	Infects red blood cells; transmitted by ticks
Cryptosporidium spp.			
Oocytes pass through water filter pores larger than 1 μm	Cryptosporidiosis	724	Waterborne; diarrhea; immunocompromised are at risk; associated with HIV patients
Cyclospora cayetanensis			
Oocytes do not pass through water filter pores smaller than 10 μm	Cyclosporiasis	724	Endemic in areas of Mexico, Central and South America, East Asia, and Middle East; cyclic diarrhea
Plasmodium spp.			
Ring-form trophozoites seen in red blood cells	Malaria	323, 677–78	Transmitted by *Anopheles* mosquito; infects red blood cells; repeated cycles of fever, chills, and sweats
Toxoplasma gondii			
Obligate intracellular protozoan; oocytes passed in animal feces	Toxoplasmosis	323, 675, 677–78	Usually asymptomatic except for immunocompromised; can cause encephalitis; congenital toxoplasmosis results when parasite crosses the placenta
PHYLUM AMOEBOZOA (AMEBAS)			
Acanthamoeba spp.		608, 803	
Cysts formed, passed in stool	Encephalitis		Found in lake water and soil; serious, often fatal, for immunocompromised
	Keratitis		Contaminated contact lens solution
Entamoeba histolytica			
Cysts and trophozoites passed in stool	Amebiasis	319–320, 724	Bloody diarrhea; amebic dysentery; chronic diarrhea; liver abscess

Taxonomy of Major Pathogens

NAME MICROBE FEATURES	DISEASE	PAGES	KEY FEATURES OF PATHOGENESIS / VACCINE
PHYLUM CILIOPHORA (CILIATES)			
Balantidium coli			
Cysts formed, passed in feces	Balantidiasis; dysentery	321	Pig reservoir; transmitted by ingestion of contaminated food or water

⌇ HELMINTHS

PHYLUM PLATYHELMINTHES			
Echinococcus granulosus			
Tapeworm in definitive animal hosts; in humans, larvae migrate to organs and form cysts	Hydatid disease	726	Zoonosis; eggs ingested; liver and other organ abscesses
Paragonimus spp.			
Flatworm	Lung fluke, rare in US	641	Ingestion of contaminated, undercooked crab or crayfish
Schistosoma spp. *S. haematobium, S. japonicum, S. mansoni*			
S. mansoni eggs in feces have a lateral spine	Schistosomiasis (not in US)	727, 730	Waterborne; snails release larval forms that penetrate skin and infect the intestines, urinary tract, and sometimes the brain; disease caused by immune reaction to eggs in tissues; endemic in Africa, South America, Caribbean, Middle East, and Asia
Schistosoma spp.			
Different from schistosomiasis species	Swimmer's itch (cercarial dermatidis)	727, 730	Freshwater contamination by these worms
Taenia spp.			
Worms have a head (scolex) with attachment suckers	Tapeworm infections	725, 728	Transmitted by ingestion of poorly cooked beef or pork; GI symptoms, passage of proglottids in stool
	Cysticercosis		Extraintestinal encysted larvae in muscles, eye, or brain

PHYLUM NEMATODA (NEMATODES)			
Ascaris lumbricoides (intestinal roundworm)			
Can grow 15–35 cm in length	Ascariasis	726, 728	Ingestion of eggs in soil or food; larvae migrate from intestine to lungs; respiratory or GI symptoms
Brugia malayi			
Adult females 6 cm long; males 2 cm long	Lymphatic filariasis	329–30, 682	Transmission by mosquito bite; microfilariae enter the lymphatic system, and adults block circulation of lymph
Enterobius vermicularis			
Can grow 2–13 cm in length	Pinworm infection	328–29, 726–28	Usually pediatric; exits the anus at night
Necator americanus, Ancylostoma duodenale			
Eggs passed in stool	Hookworm infection	726, 728–29	Larvae enter skin from soil, migrate to lung, expectorated (coughed up) into esophagus, and to intestine
Toxocara spp.			
Eggs shed in animal feces	Toxocariasis	329	Dogs and cats usual hosts; humans accidentally ingest contaminated soil or undercooked hosts (rabbits); larvae penetrate intestinal wall, enter bloodstream, disseminate to variety of tissues
Trichinella spiralis			
Transmitted exclusively by predation; no environmental forms	Trichinosis	729–30	Ingestion of larvae in undercooked meat; larvae migrate through capillaries to muscle and form painful intramuscular cysts
Trichuris trichiura			
Eggs in feces are barrel-shaped and have bipolar protuberances	Whipworm infection	726–27	Fecal-oral transmission; loose stools, dysentery, and rectal prolapse

✹ VIRUSES

DNA VIRUSES (most replicate in nucleus)			
Cytomegalovirus			
Enveloped, herpesvirus family	Mononucleosis-like syndrome	656, 661–62	Dangerous to immunocompromised people; characteristic inclusion bodies visible by staining as "owl's eye" inclusion bodies
	Encephalitis		

Taxonomy of Major Pathogens

NAME MICROBE FEATURES	DISEASE	PAGES	KEY FEATURES OF PATHOGENESIS / VACCINE
Epstein-Barr virus			
Enveloped, herpesvirus family, (human herpesvirus 4)	Infectious mononucleosis	528–29, 658–661	Infects B cells, becomes latent; cervical lymphadenitis
	Burkitt's lymphoma		B-cell lymphoma
Hepatitis B virus			
Enveloped; HBs antigen; hepadnavirus	Hepatitis	491–92, 698, 702	Dark urine, fever, nausea, yellowing skin and eyes, weakness; HBs antigen is assayed to detect presence and degree of viral infection; vaccine is available
Herpes simplex virus			
	Cold sores	589–90	
	Keratitis	589–90, 607	
Enveloped; herpesvirus family	Genital herpes	589–90, 753	Latent forms in neural ganglia
	Encephalitis	683, 783, 786	
	Neonatal herpes	653–54	
Human herpesvirus 6			
Enveloped; herpesvirus family	Roseola	588	High fever, sometimes febrile seizure, maculopapular rash
Human papillomavirus			
Infects keratinocytes	Warts	568–69, 590–91	Stimulates hyperplasia of infected host cells; vaccine is available
	Genital warts	751–52	
Monkeypox virus			
Enveloped; poxvirus	Monkeypox	592, 862, 891	Zoonosis; similar but less infectious and less deadly than smallpox
Parvovirus B19			
Parvovirus	Fifth disease (slapped-cheek rash)	343, 577–78	Lacelike rash
Varicella-zoster virus			
Enveloped; herpesvirus	Chickenpox	588–89	Primary infection; vesicular lesions; vaccine is available
	Shingles	579–80, 607	Reactivation infection arising from latent forms of the virus; vaccine is available
Variola major/minor			
Enveloped; poxvirus; eradicated from Earth except for two secure labs	Smallpox	473, 591–92	Replicates in cytoplasm; pustular skin lesions; highly lethal; potential bioterror agent; vaccine is available

RNA VIRUSES

NAME MICROBE FEATURES	DISEASE	PAGES	KEY FEATURES OF PATHOGENESIS / VACCINE
Coronavirus			Virus is spread by respiratory droplets
Single-stranded; (+) sense; enveloped; coronaviruses; SARS-CoV, SARS-CoV-2, MERS-CoV	Cold	620	Rhinorrhea (runny nose)
	SARS	860–62	Respiratory failure
	MERS	566, 850, 860	Respiratory failure
	COVID-19	338–39, 566, 623, 863–66	Multi-organ dysfunction; vaccine in development
Coxsackievirus			
Single-stranded; (+) sense; picornavirus	Hand, foot and mouth disease	591, 662, 790–91	Complications include myocarditis
	Herpangina		
Dengue virus			
Single-stranded; (+) sense; enveloped; flavivirus	Dengue fever (breakbone fever)	869	Transmitted by *Aedes* mosquito; endemic in Caribbean, South America, and Asia; high fever; vaccine available (given only to people with prior dengue infection)
Eastern equine encephalitis virus			
Single-stranded; (+) sense; enveloped; togavirus	Encephalitis	788–89	Transmission by *Aedes* or *Culex* mosquitoes; systemic illness with headache, fever, chills, muscle and joint pain; or neurological disease that can include coma or death
Ebola virus			
Single-stranded; (−) sense; enveloped; filovirus	Ebola	662, 658, 837	Diarrhea, vomiting, excessive hemorrhaging, death; virus is spread by contact with body fluids; vaccine approved by WHO
Hantavirus			
Single-stranded; (−) sense; enveloped; bunyavirus	Hantavirus pulmonary syndrome	821, 867	Transmitted by inhalation of dried rodent urine or feces

Taxonomy of Major Pathogens

NAME MICROBE FEATURES	DISEASE	PAGES	KEY FEATURES OF PATHOGENESIS / VACCINE
Hepatitis A virus			
Single-stranded; (+) sense; picornavirus	Hepatitis	700–702	Dark urine, fever, nausea, yellowing skin and eyes, weakness; vaccine is available
Hepatitis C virus			
Single-stranded; (+) sense; enveloped; flavivirus	Hepatitis	289–90, 337–38, 702–03	Dark urine, fever, nausea, yellowing skin and eyes, weakness
Hepatitis D virus			
Single-stranded; (−) sense; circular genome; enveloped; delta virus	Hepatitis	346–47, 702	Dark urine, fever, nausea, yellowing skin and eyes, weakness; needs hepatitis B virus as helper
Hepatitis E virus			
Single-stranded; (+) sense; Hepeviridae family	Hepatitis	703	Dark urine, fever, nausea, yellowing skin and eyes, weakness
Human immunodeficiency virus (HIV)			
Single-stranded; (−) sense; Retroviridae family	Acquired immunodeficiency syndrome (AIDS)	363–65, 567, 754–56	Transmitted by blood or sexual contact; virus targets T cells; mononucleosis-like symptoms to increased opportunistic infections and cancers
Influenza virus			
Single-stranded; segmented genome; enveloped; orthomyxovirus	Influenza (flu)	358–59, 400–01, 615, 891–92	Cough, fever, sore throat, headache, tachycardia, weakness; antigenic shift and drift aid its pathogenesis; vaccine is available
Measles virus (rubeola)			
Single-stranded; (−) sense; enveloped; paramyxovirus	Measles	583, 585–86, 620, 852	Cough, fever, runny nose, maculopapular rash, Koplik spots; vaccine is available
Mumps virus			
Single-stranded; (−) sense; enveloped; paramyxovirus	Mumps	703–04, 854	Swollen, painful salivary glands; vaccine is available
Norovirus			
Single-stranded; (+) sense; calcivirus	Gastroenteritis	692, 700	Outbreaks associated with cruise ships; watery diarrhea
Poliovirus			
Single-stranded; (+) sense; picornavirus	Poliomyelitis	790–92	Vaccine is available
Rabies virus			
Single-stranded; (−) sense; enveloped; rhabdovirus	Encephalitis	777–78, 782, 787–88	Transmitted to humans from bite of infected animal; multisystem illness ranging from fever and fatigue to symptoms of encephalitis (neck rigidity, paralysis, convulsions, sensitivity to light) and death
Respiratory syncytial virus			
Single-stranded, (−) sense; enveloped; paramyxovirus	RSV disease	620–21	Lower respiratory tract; infects children; low-grade fever, cough, difficulty breathing, wheezing
Rhinovirus			
Single-stranded; (+) sense; picornavirus	Common cold	565–66, 618, 620	Rhinorrhea (runny nose)
Rotavirus			
Double-stranded; segmented genome	Gastroenteritis	691–92, 699–700	Usually infects children; watery diarrhea; vaccine is available
Rubella virus			
Single-stranded; (+) sense; enveloped; togavirus	Rubella (German measles)	586–87, 587, 620, 661, 704	Eye pain, conjunctivitis, sore throat, headache; vaccine is available
West Nile virus			
Single-stranded; (+) sense; enveloped; flavivirus	Encephalitis	789, 851, 883, 892–93	Transmitted to humans from animal reservoirs by *Culex* mosquitoes; systemic illness with headache, fever, chills, muscle and joint pain; or neurological disease that can include paralysis or death
Yellow fever virus			
Single-stranded; (+) sense; enveloped; flavivirus	Yellow fever	44, 657	*Aedes* mosquito–borne; endemic in parts of South America and Africa; jaundice; vaccine is available
Zika virus			
Single-stranded; (+) sense; enveloped; flavivirus	Zika fever	789–90, 849	*Aedes* mosquito–borne; endemic in parts of Africa, South America, and Southeast Asia; usually mild but associated with fetal microcephaly